CHEMISTRY

Principles, Patterns, and Applications

Bruce Averill

Patricia Eldredge

With contributions by

C. Alton Hassell

Baylor University

Daniel J. Stasko

University of Southern Maine

PEARSON

Benjamin Cummings

San Francisco Boston New York

Cape Town Hong Kong London Madrid Mexico City

Montreal Munich Paris Singapore Sydney Tokyo Toronto

Editor-in-Chief	Adam Black, Ph.D.
Publisher	Jim Smith
Director of Development	Kay Ueno
Art Developmental Editors	Sonia DiVittorio, Hilair Chism, Blake Kim
Text Developmental Editors	Moira Lerner, Ruth Steyn
Developmental Editor, Glossary and Appendices	John Murdzek
Project Editors	Katie Conley, Lisa Leung, Kate Brayton
Assistant Editor	Cinnamon Hearst
Editorial Assistant	Kristin Rose Field
Executive Producer, Media	Claire Masson
Executive Managing Editor	Erin Gregg
Managing Editor	Corinne Benson
Senior Marketing Manager	Scott Dustan
Market Development Managers	Josh Frost, Susan Winslow
Photo Editor	Travis Amos
Photo Researcher	Brian Donnelly
Art Editor	Kelly Murphy
Director, Image Resource Center	Melinda Patelli
Rights and Permissions Manager	Zina Arabia
Project Manager	Crystal Clifton, Progressive Publishing Alternatives
Composition	Progressive Information Technologies
Illustrators	Imagineering Media Services, Inc.
Graph Designers	New York Charts and Diagrams
Manufacturing Manager	Pam Augspurger
Text Designer	Mark Ong
Cover Designer	Mark Ong
Text and Cover Printer	Quebecor World Dubuque
Cover Photo	The Belovsov-Zhabotinsky reaction. (See Chapter 14, chapter opening image, for a complete description.) © Fritz Goro.

Library of Congress Cataloging-in-Publication Data

Averill, Bruce.
Chemistry: principles, patterns, and applications / Bruce Averill, Patricia Eldredge; with contributions by C. Alton Hassell, Daniel J. Stasko.—1st ed.
v. cm.
Includes index.
ISBN 0-8053-3803-9
1. Chemistry—Textbooks. 2. Chemistry, Physical and theoretical—Textbooks. 3. Environmental chemistry—Textbooks. 4. Chemistry, Technical—Textbooks. I. Eldredge, Patricia. II. Title.
QD31.3.A94 2006
540—dc22

2005022980

ISBN 0-8053-3799-7 with MasteringGeneralChemistry (Student Edition)
ISBN 0-8053-3803-9 without MasteringGeneralChemistry (Student Edition)
ISBN 0-8053-4886-7 with MasteringGeneralChemistry (Instructor Edition)

PEARSON
Benjamin Cummings

www.aw-bc.com

1 2 3 4 5 6 7 8 9 10 — QWD—09 08 07 06

To Harvey, who opened the door

About the Authors

Bruce Averill

After growing up in New England, Bruce Averill received his B.S. with high honors in chemistry at Michigan State University in 1969, and his Ph.D. in inorganic chemistry at MIT in 1973. After three years as an NIH and NSF Postdoctoral Fellow at Brandeis University and the University of Wisconsin, he began his independent academic career at Michigan State University in 1976. He moved to the University of Virginia in 1982, and was promoted to Professor in 1988. In 1994, Dr. Averill moved to the University of Amsterdam in the Netherlands as Professor of Biochemistry. In 2001, he returned to the United States where he is a Distinguished University Professor at the University of Toledo. Dr. Averill's research focuses on the role of metal ions in biology.

While in Europe, Dr. Averill headed an EU-funded network of seven research groups from seven different European countries. In addition, he lead a 22-member team investigating biocatalysis within the E. C. Slater Institute of the University of Amsterdam.

From 2004 to 2005, Dr. Averill was a Jefferson Science Fellow at the U.S. State Department, where he acted as a senior science and technology adviser in the Bureau of Western Hemisphere Affairs and focussed on energy issues in Latin America. He has been asked to return to the State Department for 2006 and 2007 as a William C. Foster Fellow in the Bureau of Political-Military Affairs to work on R&D issues in the area of critical infrastructure protection and cybersecurity.

Dr. Averill's published papers are frequently cited by other researchers, and he has been invited to give more than 100 presentations at educational and research institutions and at national and international scientific meetings. Dr. Averill has been an Honorary Woodrow Wilson Fellow, an NSF Predoctoral Fellow, an NIH and NSF Postdoctoral Fellow, and an Alfred P. Sloan Foundation Fellow; he has also received an NSF Special Creativity Award.

Dr. Averill has published more than 140 articles on chemical, physical, and biological subjects in refereed journals, 15 chapters in books, and more than 80 abstracts from national and international meetings. In addition, he has co-edited a graduate text on catalysis and taught courses at all levels, including general chemistry, biochemistry, advanced inorganic, and physical methods.

Aside from his research program, Dr. Averill is an enthusiastic sailor and an avid reader. He also enjoys traveling with his family, and at some point in the future he would like to sail around the world in a classic wooden boat.

Patricia Eldredge

Having been raised in the U.S. diplomatic service, Patricia Eldredge has traveled and lived around the world. After receiving a B.A. in Spanish language and literature from Ohio State University, Dr. Eldredge developed an interest in chemistry while studying general chemistry at Kent State University. She obtained a B.S. in chemistry from the University of Central Florida. Following several years as an analytical research chemist in industry, she began her graduate studies at the University of Virginia and obtained her Ph.D. in inorganic chemistry from the University of North Carolina at Chapel Hill. In 1989, Dr. Eldredge was named the Science Policy Fellow for the American Chemical Society. While in Washington, D.C., she examined the impact of changes in federal funding priorities on academic research funding. She was awarded a Postdoctoral Research Fellowship with Oak Ridge Associated Universities, working with the U.S. Department of Energy on heterogeneous catalysis and coal liquefaction. Subsequently, she returned to the University of Virginia as a Research Scientist and a member of the General Faculty. In 1992, Dr. Eldredge moved to Europe for several years. While there, she studied advanced Maritime Engineering, Materials, and Oceanography at the University of Southampton in England, arising from her keen interest in naval architecture. Since her return to the United States in 2002, she has been a Visiting Assistant Professor and a Senior Research Scientist at the University of Toledo. Her current research interests include the use of protein scaffolds to synthesize biologically relevant clusters.

Dr. Eldredge has published more than a dozen articles dealing with synthetic inorganic chemistry and catalysis, including several seminal studies describing new synthetic approaches to metal–sulfur clusters. She has also been awarded a patent for her work on catalytic coal liquefaction. Her diverse teaching experience includes courses on chemistry for the life sciences, introductory chemistry, general, organic, and analytical chemistry.

When not writing scientific papers or textbooks, Dr. Eldredge enjoys traveling, reading political biographies, sailing high-performance vessels under rigorous conditions, and caring for her fourth child, her pet Havanese.

Preface to the Instructor

In this new millenium, as the world faces new and extreme challenges, the importance of acquiring a solid foundation in chemical principles has become increasingly central to understanding the challenges that lie ahead. Moreover, as the world becomes more integrated and interdependent, so too do the scientific disciplines. The divisions between fields such as chemistry, physics, biology, environmental sciences, geology, and materials science, among others, have become less clearly defined. This text addresses the closer relationships developing among various disciplines and shows the relevance of chemistry to contemporary issues in a friendly and approachable manner.

Because of the enthusiasm of the majority of first-year chemistry students for biologically and medically relevant topics, this text uses an integrated approach that includes explicit discussions of biological and environmental applications of chemistry. Topics relevant to materials science are also introduced in order to meet the more specific needs of engineering students. To integrate this material, simple organic structures, nomenclature, and reactions are introduced very early in the text, and both organic and inorganic examples are used wherever possible. This approach emphasizes the distinctions between ionic and covalent bonding, thus enhancing the students' chance of success in the organic chemistry course that traditionally follows general chemistry.

Our overall goal is to produce a text that introduces the students to the relevance and excitement of chemistry. Although much of first-year chemistry is taught as a service course, there is no reason that the intrinsic excitement and potential of chemistry cannot be the focal point of the text and the course. We emphasize the positive aspects of chemistry and its relationship to students' lives; this approach requires bringing in applications early and often. Unfortunately, we cannot assume that students in these courses today are highly motivated to study chemistry for its own sake. The explicit discussion of biological, environmental, and materials applications from a chemical perspective is intended to motivate the students and help them appreciate the relevance of chemistry to their lives. Material that has traditionally been relegated to boxes, and perhaps perceived as peripheral by the students, has been incorporated into the text to serve as a learning tool.

To begin the discussion of chemistry rapidly, the traditional first chapter introducing units, significant figures, conversion factors, dimensional analysis, and so on has been reorganized in this book. The material has been placed in the chapters where the relevant concepts are first introduced, thus providing three advantages: it eliminates the tedium of the traditional approach, which introduces mathematical operations at the outset, and thus avoids the perception that chemistry is a mathematics course; it avoids the early introduction of operations such as logarithms and exponents, which are typically not encountered again for several chapters and may easily be forgotten when they are needed; and it provides a review for those students who have already had relatively sophisticated high school chemistry and math courses, although the sections are designed primarily for students unfamiliar with the topics.

Our specific objectives include the following:

- To write the text at a level suitable for science majors, but using a less formal writing style that will appeal to modern students.
- To produce a *truly* integrated text that gives the student who takes only a single year of chemistry an overview of the most important subdisciplines of chemistry,

including organic, inorganic, biological, materials, environmental, and nuclear chemistry, thus emphasizing unifying concepts.

- To introduce fundamental concepts in the first two-thirds of the chapter and then applications relevant to the health sciences or engineers, thus providing a flexible text that can be tailored to the specific needs and interests of the audience.
- To ensure the accuracy of the material presented, which is enhanced by the authors' breadth of professional and research experience.
- To produce a spare, clean, uncluttered text that is not distracting to the student, one in which each piece of art serves as a pedagogical device.
- To introduce the distinction between ionic and covalent bonding and reactions early in the text, and to continue to build on this foundation in the subsequent discussions while emphasizing the relationship between structure and reactivity.
- To use established pedagogical devices to maximize students' ability to learn directly from the text. Copious worked examples in the text, problem-solving strategies, and similar unworked exercises with solutions are included. End-of-chapter problems are designed to ensure that students have grasped major concepts in addition to testing their ability to solve numerical problems. Problems emphasizing applications are drawn from many disciplines.
- To emphasize an intuitive and predictive approach to problem solving that relies on a thorough understanding of key concepts and recognition of important patterns rather than on memorization. Many patterns are indicated throughout the text by a "Note the Pattern" feature in the margin.

The text is organized by units that discuss introductory concepts, atomic and molecular structure, the states of matter, kinetics and equilibria, and descriptive inorganic chemistry. The text divides the traditional chapter on liquids and solids into two chapters in order to expand the coverage of important topics such as semiconductors and superconductors, polymers, and engineering materials. Part V, included in the complete edition of the text, is a systematic summary of the descriptive chemistry of the elements organized by position in the periodic table; it is designed to bring together the key concepts introduced in the preceding chapters: chemical bonding, molecular structure, kinetics, and equilibrium. A great deal of descriptive chemistry will have been introduced prior to this point, but only in ways that are germane to particular points of interest.

In summary, our hope is that this text represents a step in the evolution of the general chemistry textbook toward one that reflects the increasing overlap between chemistry and other disciplines. Most important, the text discusses exciting and relevant aspects of biological, environmental, and materials science that are usually relegated to the last few chapters, and it provides a format that allows the instructor to tailor the emphasis to the needs of the class. By the end of Part I (Chapter 5), the student will have been introduced to environmental topics such as acid rain, the ozone layer, and periodic extinctions, and to biological topics such as antibiotics and the caloric content of foods. Nonetheless, the new material is presented in a way that minimally perturbs the traditional sequence of topics in a first-year course, making the adaptation easier for instructors.

Supplements

A full set of print and media supplements is available for the student and the instructor, which are designed to enhance in-class presentations, to engage students in classroom discussion, and to assist students outside the classroom.

For the instructor:

Instructor Solutions Manual: Complete solutions to all of the end-of-chapter problems in the textbook.

Instructor Guide: Includes chapter overviews and outlines, suggestions for lecture demonstrations, teaching tips, and a guide to the print and media resources available for each chapter.

Test Bank (print and electronic formats): More than 1000 questions to use in creating tests, quizzes, and homework assignments.

Transparency Acetates: 300 full-color illustrations and tables to enhance classroom lectures.

Instructor Resource CD-ROM: All images and tables from the book in jpeg and PowerPoint format, a PowerPoint lecture outline for each chapter, Clicker Questions, interactive graphs, interactive 3-D molecular structures, and links to student resources online.

Clicker Questions (for use with Classroom Response Systems): Five questions per chapter written to inspire in-class discussion and test student understanding of material.

PowerPoint Lecture Outline: Hundreds of editable slides for in-class presentations.

MasteringGeneralChemistry™: The most advanced online homework and tutorial system available provides thousands of problems and tutorials with automatic grading, immediate wrong-answer specific feedback, and simpler questions upon request. Problems include randomized numerical and algebraic answers and dimensional analysis. Instructors can compare individual student and class results against data collected through pre-testing of students nationwide.

For the student:

Student Solutions Manual: Detailed, worked-out solutions to selected end-of-chapter problems in the textbook.

Chemistry Place for General Chemistry: Features chapter quizzes, interactive 3-D molecular structures, interactive graphs, and InterAct Math for General Chemistry.

Study Guide: Study and learning objectives, chapter overviews, problem-solving tips, and practice tests.

MasteringGeneralChemistry: The first adaptive-learning tutorial system that grades students' homework automatically and provides Socratic tutorials with feedback specific to errors, hints and simpler subproblems upon demand, and motivation with partial credit. Pre-tested on students nationally, the system is uniquely able to respond to students' needs, effectively tutoring and motivating their learning of the concepts and strengthening their problem-solving skills.

Acknowledgments

Although putting a text together is always a team effort, there are several individuals at Benjamin Cummings whom we would like to particularly acknowledge for their tireless efforts and commitment to this project. We would especially like to thank the following individuals: Jim Smith for believing so strongly in this project; Kay Ueno for her superhuman efforts in the face of excruciating deadlines (we still have to get out for a sail!); Linda Davis, who, despite all the twists and turns, made sure that it happened; Sonia DiVittorio, whose enormous dedication and attention to detail produced a stellar art program, while keeping track of where everything was at any given time; and Moira Nelson, a truly impressive master of the American English language, who taught us that Los Angeles and Washington, D.C., aren't so different after all. Special thanks, too, are due to Neil Weinstein, who helped enormously in scrubbing the page proofs of errors and inconsistencies in data, and to Mike Helmstadter, whose expertise with Excel was crucial in generating many of the plots used in the figures. We would also like to thank the rest of the team at Benjamin Cummings for creating such a cordial and supportive environment despite the pressures of production.

Finally, to Tonya McCarley, thanks.

Reviewers

Several drafts of the manuscript were informed by the meticulous and considered comments of the instructors listed here. Their review of our work was invaluable to us as we polished the book. We acknowledge and thank them for their generous efforts on our behalf.

Dawood Afzal
Truman State University

Thomas E. Albrecht-Schmitt
Auburn University

Jeffrey Appling
Clemson University

David Atwood
University of Kentucky

Jim D. Atwood
State University of New York at Buffalo

Debbie Beard
Mississippi State University

Kevin Bennett
Hood College

Richard Biagioni
Southern Missouri State University

Robert S. Boikess
Rutgers University, The State University of New Jersey

Steven Boone
Central Missouri State University

Allen Clabo
Frances Marion University

Carl David
University of Connecticut, Storrs

Sonja Davison
Tarrant County College, Northeast

Nordulf Debye
Towson University

Dru DeLaet
Southern Utah University

Patrick Desrochers
University of Central Arkansas

Jane DeWitt
San Francisco State University

Michael Doherty
East Stroudsburg University

W. Travis Dungan
Trinity Valley Community College

Delbert J. Eatough
Brigham Young University

Marly Eidsness
University of Georgia

Paul Farnsworth
Brigham Young University

Michael Freitas
Ohio State University

Roy Garvey
North Dakota State University

Jim Geiger
Michigan State University

Brian Gilbert
Linfield College

Alexander Golger
Boston University

Lara Gossage
Hutchinson Community College

John M. Halpin
New York University

Greg Hartland
University of Notre Dame

Dale Hawley
Kansas State University

Gregory Kent Haynes
Morgan State University

James W. Hershberger
Miami University

Carl Hoeger
University of California, San Diego

Don Hood
Saint Louis University

John B. Hopkins
Louisiana State University

Jayanthi Jacob
Indiana University–Purdue University

David Johnson
University of Dayton

Lori Jones
Guelph University

Jeff Keaffaber
University of Florida

Philip C. Keller
University of Arizona

Michael E. Ketterer
Northern Arizona University

Raj Khanna
University of Maryland

Paul Kiprof
University of Minnesota, Duluth

Patrick Kolniak
Louisiana State University

Jeremy Kua
University of San Diego

Jothi V. Kumar
North Carolina A&T State University

Donald Land
University of California, Davis

Alan Levine
University of Louisiana, Lafayette

Scott B. Lewis
James Madison University

Robley J. Light
Florida State University

Da-hong Lu
Fitchburg State College

Joel T. Mague
Tulane University

Pshemak Maslak
Penn State University

Hitoshi Masui
Kent State University

Elmo Mawk
Texas A&M University

C. Michael McCallum
University of the Pacific

Robert McIntyre
East Carolina University

Abdul K. Mohammed
North Carolina A&T State University

Kathy Nabona
Austin Community College, Northridge

Cheuk-Yiu Ng
University of California, Davis

Frazier W. Nyasulu
University of Washington

Gay L. Olivier-Lilley
Point Loma Nazarene University

Joseph Pavelites
Purdue University, North Central

James Pazun
Pfeiffer University

Earl Pearson
Middle Tennessee State University

Lee Pedersen
University of North Carolina at Chapel Hill

Cathrine E. Reck
Indiana University

James H. Reho
East Carolina University

Jeff Roberts
University of Minnesota

John Selegue
University of Kentucky

Robert Sharp
University of Michigan

Jerald Simon
Frostburg State University

J. T. (Dotie) Sipowska
University of Michigan

Sheila Smith
University of Michigan, Dearborn

Michael Sommer
University of Wyoming

James Stickler
Allegany College of Maryland

Michael Stone
Vanderbilt University

Shane Street
University of Alabama, Tuscaloosa

Donald Thompson
University of Missouri–Columbia

Douglas Tobias
University of California, Irvine

Bilin P. Tsai
University of Minnesota, Duluth

Tom Tullius
Boston University

John B. Vincent
The University of Alabama

Yan Waguespack
University of Maryland, Eastern Shore

Thomas Webb
Auburn University

Mark Whitener
Montclair State University

Marcy Whitney
University of Alabama

Kathryn Williams
University of Florida

Kurt Winkelmann
Florida Institute of Technology

Troy Wood
State University of New York, Buffalo

Catherine Woytowicz
George Washington University

David Young
Ohio University, Athens

Class Testers

In addition, we are grateful to these professors who class tested portions of the early manuscript and student class testers who provided insights into what students need to effectively learn chemistry. We wish to acknowledge these instructors and their students:

Jamie Adcock
University of Tennessee, Knoxville

Adegboye Adeyemo
Savannah State University

Lisa Arnold
South Georgia College

Dale Arrington
South Dakota School of Mines

Karen Atkinson
Bunker Hill Community College

John Barry
Houston Community College, Town and Country

Krishna Bhat
Philadelphia University

Christine Bilicki
Pasadena City College

Rose Boll
University of Tennessee, Knoxville

Bill Brescoe
Tulsa Community College

Steve Burns
St. Thomas Aquinas College

Timothy Champion
Johnson C. Smith University

Thomas Chasteen
Sam Houston State University

Walter Cleland
University of Mississippi

Zee Ding
Queensland University of Technology

Judy Dirbas
Grossmont College

Marly Eidsness
University of Georgia

Cristina Fermin-Ennis
Gordon College

Richard Frazee
Rowan University

Neal Gray
University of Texas, Tyler

William Griffin
Bunker Hill Community College

Greg Hale
University of Texas, Arlington

Jessica Harper
Antelope Valley College

Alton Hassell
Baylor University

Michael Hauser
St. Louis Community College

Scott Hendrix
University of Tampa

Daniel Huchital
Florida Atlantic University

Steven Hughes
Carl Albert State College

Lawrence Kennard
Walters State Community College

Jeffrey Kovac
University of Tennessee, Knoxville

James Lankford
Saint Andrew's Presbyterian College

Debra Leedy
Arizona State University

Larry Manno
Triton College

Carol Martinez
TVI Community College, Albuquerque

Lydia Martinez-Rivera
University of Texas, San Antonio

Graeme Matthews
Florida Community College, Jacksonville

Keith McCleary
Adrian College

Larry McRae
Berry College

Gary Mercer
Boise State University

Matt Merril
Florida State University

David Nachman
Mesa Community College

Bill Newman
Mohawk Valley Community College

Jason Overby
College of Charleston

Linda Pallack
Washington and Jefferson College

James Pazun
Pfeiffer University

John Penrose
Jefferson Community College

Rafaelle Perez
University of South Florida

Joanna Petridou-Fischer
Spokane Falls Community College

Dale Powers
Elmira College

David Prentice
Coastal Carolina University

Victoria Prevatt
Tulsa Community College

Lisa Price
Bennett College for Women

Laura Pytlewski
Triton Community College

Gerald Ramelow
McNeese State University

Mitch Rhea
Chattanooga State

Lyle Roelofs
Haverford College

Steven Rowley
Middlesex County College

J. B. Schlenoff
Florida State University

Raymond Scott
Mary Washington College

Lisa Seagraves
Haywood Community College

Shirish Shah
Towson University

George Smith
Herkimer County College

Zihan Song
Savannah State University

Paris Svoronos
Queensborough Community College

Richard Terry
Brigham Young University

John Vincent
University of Alabama

Kjirsten Wayman
Humboldt State University

John Weide
Mesa Community College

Neil Weinstein
Santa Fe Community College

Barry West
Trident Technical College

Drew Wolfe
Hillsborough Community College

Servet Yatin
Quincy College

Lynne Zeman
Kirkwood Community College

Xueli Zou
California State University, Chico

Forum Participants

Beneficial to our crafting of this text have been the numerous Chemistry Forum participants, who took time to read and remark thoughtfully on our manuscript as it was in progress and share their ideas about the future of chemical education. Participants at the Chemistry Forums held in cities across the country include the following people:

Michael Abraham
University of Oklahoma

William Adeniyi
North Carolina A&T State University

Ramesh Arasasingham
University of California, Irvine

Yiyan Bai
HCC, Central

Monica Baloga
Florida Institute of Technology

Mufeed Basti
North Carolina A&T State University

Rich Bauer
Arizona State University

Debbie Beard
Mississippi State University

Jo A. Beran
Texas A&M, Kingsville

Wolfgang Bertsch
University of Alabama

Christine Bilicki
Pasadena City College

Bob Blake
Texas Tech University

Rose Boll
University of Tennessee, Knoxville

Philip Brucat
University of Florida

Kenneth Busch
Baylor University

Marianna Busch
Baylor University

Donnie Byers
Johnson County Community College

Brandon Cruikshank
Northern Arizona University

Mapi Cuevas
Santa Fe Community College

Nordulf Debye
Towson University

Michael Denniston
Georgia Perimeter College, Clarkston

Patrick Desrochers
University of Central Arkansas

Deanna Dunlavy
New Mexico State University

Marly Eidsness
University of Georgia

Thomas Engel
University of Washington

Deborah Exton
University of Oregon

Steven Foster
Mississippi State University

Mark Freilich
University of Memphis

Elizabeth Gardner
University of Texas, El Paso

Roy Garvey
North Dakota State University

John Gelder
Oklahoma State University

Eric Goll
Brookdale Community College

John Goodwin
Coastal Carolina

Thomas Greenbowe
Iowa State University

Asif Habib
University of Wisconsin, Waukesha

Jerry Haky
Florida Atlantic University

Greg Hale
University of Texas, Arlington

C. Alton Hassell
Baylor University

Gregory Haynes
Morgan State University

Claudia Hein
Diablo Valley College

Louise Hellwig
Morgan State University

David Hobbs
Montana Tech at University of Montana

Jim Holler
University of Kentucky

John Hopkins
Louisiana State University

Susan Hornbuckle
Clayton College and State University

James Hovick
University of North Carolina, Charlotte

Thomas Huang
East Tennessee State University

Denley Jacobson
North Dakota State University

Phillip Keller
University of Arizona

Debbie Koeck
Texas State University

Jeffrey Kovack
University of Tennessee, Knoxville

Jothi Kumar
North Carolina A&T State University

Charles Kutal
University of Georgia

Robley Light
Florida State University

Larry Manno
Triton College

Pam Marks
Arizona State University

Carol Martinez
Albuquerque Technical Vocational College

Selah Massoud
Louisiana University, Lafayette

Graeme Matthews
Florida Community College, Jacksonville

Joe Elmo Mawk
Texas A&M University

Maryann McDermott Jones
University of Maryland

Robert A. McIntyre
East Carolina University

Abdul Mohammed
North Carolina A&T State University

John Nelson
University of Nevada, Reno

Frazier Nyasulu
University of Washington

Greg Oswald
North Dakota State University

Jason Overby
College of Charleston

Gholam Pahlavan
Houston Community College

Colleen Partigianoni
Ferris State University

Cindy Phelps
California State University, Chico

Shawn Phillips
Vanderbilt University

Louis Pignolet
University of Minnesota

Laura Pytlewski
Triton Community College

William Quintana
New Mexico State University

Catherine Reck
Indiana University

John Richardson
University of Louisville

Jill Robinson
Indiana University

Peter Roessle
Georgia Perimeter College, Decatur

Jimmy Rogers
University of Texas, Arlington

Svein Saebo
Mississippi State University

James Schlegel
Rutgers University, Newark

Pat Schroeder
Johnson County Community College

Jack Selegue
University of Kentucky

John Sheridan
Rutgers University

Don Siegel
Rutgers University

Brett Simpson
Coastal Carolina

Jerry Skelton
St. Johns River Community College

Sheila Smith
University of Michigan, Dearborn

Sherril A. Soman
Grand Valley State University

Mark Sulkes
Tulane University

Ann Sullivan
John Sergeant Reynolds Community College

Susan Swope
Plymouth University

Will Tappen
San Diego Mesa College

James Terner
Virginia Commonwealth University

Donald Thompson
Oklahoma State University

Michael Topp
University of Pennsylvania

John Turner
University of Tennessee

Julian Tyson
University of Massachusetts, Amherst

Chris Uzomba
Austin Community College

Robert Vergenz
University of North Florida

Ed Walters
University of New Mexico

Philip Watson
Oregon State University

Thomas Webb
Auburn University

Neil Weinstein
Santa Fe Community College

Steven Weitstock
Indiana University

M. Stanley Whittingham
State University of New York, Binghamton

Alex Williamson
North Carolina A&T State University

Kim Woodrum
University of Kentucky

Vaneica Young
University of Florida

Lin Zhu
University of Georgia

Problem Advisory Board Members

All of the end-of-chapter problems were examined for consistency, clarity, and relevance. We warmly thank the following instructors for their contributions. Their efforts helped us considerably.

Jamie Adcock
University of Tennessee

William Adeniyi
North Carolina A&T State University

Mufeed Basti
North Carolina A&T State University

Salah M. Blaih
Kent State University, Trumbull

Mary Joan Bojoan
Pennsylvania State University

Jerry Burns
Pellissippi State Technical Community College

Frank Carey
Wharton County Junior College

Karen Eichstadt
Ohio University

Marly Eidsness
University of Georgia

Gregory Kent Haynes
Morgan State University

Melissa Hines
Cornell University

Denley Jacobson
North Dakota State University

Neil Kestner
Louisiana State University

Charles Kirkpatrick
St. Louis University

John Kovacs
University of Tennessee

Mark Kubinec
University of California, Berkeley

Barbara Lewis
Clemson University

Lauren McMillis
Ohio University

Matthew J. Mio
University of Detroit Mercy

Daniel Moriarty
Siena College

Gregory Oswald
North Dakota State University

Dale Powers
Elmira College

Lydia J. Martinez Rivera
University of Texas at San Antonio

Theodore Sakano
Rockland Community College

Fred Safarowic
Passaic County Community College

Pat Schroeder
Johnson County Community College

Dotie Sipowska
University of Minnesota, Duluth

Sheila Smith
University of Michigan, Dearborn

Bilin Tsai
University of Minnesota, Duluth

Jon Turner
University of Tennessee

M. Stanley Whittingham
State University of New York at Binghamton

Kurt Winkleman
Florida Institute of Technology

We welcome feedback from colleagues and students who use this text. Please send your comments to the Chemistry Editor, Benjamin Cummings, 1301 Sansome Street, San Francisco, California 94111.

Bruce Averill
Patricia Eldredge

Brief Contents

Detailed Contents

RELEVANT, INNOVATIVE, INFORMATIVE

Chemistry: Principles, Patterns, and Applications is rich in applications that make chemistry relevant to students' lives and provides an innovative approach to learning principles, recognizing patterns, and solving problems.

What the student needs to know

Chapter openers provide **Learning Objectives** and **Chapter Outlines**.

1 Introduction to Chemistry

LEARNING OBJECTIVES

- To recognize the breadth, depth, and scope of chemistry
- To understand what is meant by the scientific method
- To be able to classify matter
- To understand the development of the atomic model
- To know the meaning of isotopes and atomic masses
- To become familiar with the periodic table and to be able to use it as a predictive tool

CHAPTER OUTLINE

Selected chapters include **Essential Skill** sets, which help students build the requisite skills for understanding the equations and working the problems featured in the chapter.

ESSENTIAL SKILLS 1

Units of Measurement
Scientific Notation
Significant Figures
Accuracy and Precision

As you begin your study of college chemistry, those of you who do not intend to become professional chemists may well wonder why you need to study chemistry. In fact, as you will soon discover, a basic understanding of chemistry is useful in a wide range of disciplines and career paths. You will also discover that an understanding of chemistry helps you make informed decisions about many issues that affect you, your community, and your world. A major goal of this text is to demonstrate the importance of chemistry in your daily life and in our collective understanding of both the physical world we occupy and the biological realm of which we are a part. The objective of this chapter is to introduce the breadth, the importance, and some of the challenges of modern chemistry, and to present some of the fundamental concepts and definitions you will need to understand how chemists think and work.

An atomic corral for electrons. A "corral" of 48 Fe atoms (yellow-orange) on a smooth Cu surface (cyan-purple) confine the electrons on the surface of the Cu, producing a pattern of "ripples" in the distribution of the electrons. Scientists assembled the 713-pm-diameter corral by individually positioning Fe atoms with the tip of a scanning tunneling microscope.

Contemporary applications embody modern chemistry ...

14.5 ○ Half-Lives and Radioactive Decay Kinetics

Half-Lives

MGC Mastering Gen Chem

Another approach to describing reaction rates is based on the time required for the concentration of a reactant to decrease to one-half its initial value. This period of time is called the **half-life** of the reaction, written as $t_{1/2}$. Thus, the half-life of a reaction is the time required for the reactant concentration to decrease from $[A]_0$ to $[A]_0/2$. If two reactions have the same order, the faster reaction will have a shorter half-life and the slower reaction will have a longer half-life.

The half-life of a first-order reaction under a given set of reaction conditions is a constant. This is not true for zeroth- or second-order reactions. Most important, the half-life of a first-order reaction is *independent of the concentration of the reactants.* This becomes evident when we rearrange the integrated rate law for a first-order reaction (Equation 14.23) to produce the following equation:

$$\ln\frac{[A]_0}{[A]} = kt \qquad (14.31)$$

TABLE 14.7 Half-lives and applications of some radioactive isotopes

Radioactive Isotope	Half-Life	Typical Uses
Hydrogen-3 (tritium)	12.26 yr	Biochemical tracer
Carbon-11	20.39 min	PET scans (biomedical imaging)
Carbon-14	5730 yr	Dating of artifacts
Sodium-24	14.659 min	Cardiovascular system tracer
Phosphorus-32	14.3 days	Biochemical tracer
		Dating of rocks

EXAMPLE 14.10

The anticancer drug cisplatin hydrolyzes in water with a rate constant of 1.5×10^{-3} min^{-1} at pH 7.0 and 25°C. Calculate the half-life for the hydrolysis reaction under these conditions. If a freshly prepared solution of cisplatin has a concentration of 0.053 *M*, what will be the concentration of cisplatin after five half-lives? After 10 half-lives? What is the percent completion of the reaction after five half-lives? After 10 half-lives?

Given Rate constant, initial concentration, and number of half-lives

Asked for Half life, final concentrations, and percent completion

Contemporary applications in text discussions, illustrations, worked examples, and problem sets ...

Figure 8.1 **A plot of potential energy vs. internuclear distance for the interaction between a gaseous Na^+ ion and a gaseous Cl^- ion.** Note that the energy of the system reaches a minimum at a particular distance, r_0, where the attractive and repulsive interactions are balanced. Interactive Graph

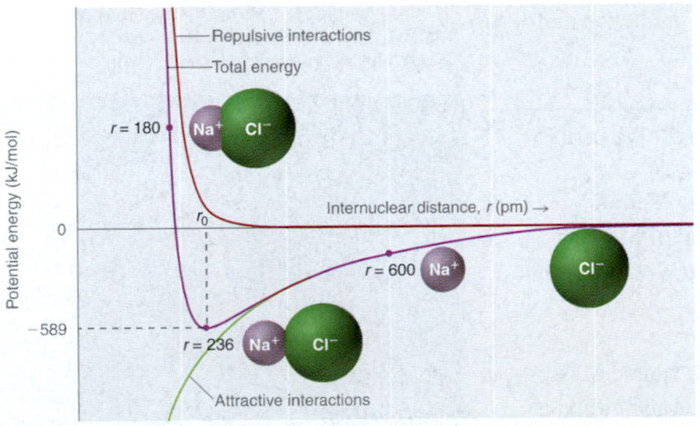

Key graphs are provided in an interactive format online at The Chemistry Place to enhance understanding.

The Ozone Layer and Chlorofluorocarbons

Each year since the mid-1970s, scientists have noted a disappearance of approximately 70% of the ozone (O_3) layer above Antarctica during the Antarctic spring, creating what is commonly known as the "ozone hole." In 2003, the Antarctic ozone hole measured 11.1 million square miles, slightly larger than the size of North America but smaller than the largest ever recorded, in the year 2000, when the hole covered 11.5 million square miles and for the first time extended over a populated area—the city of Punta Arenas, Chile (population 120,000).

September 1979 September 198

... showing students how and why chemistry is relevant

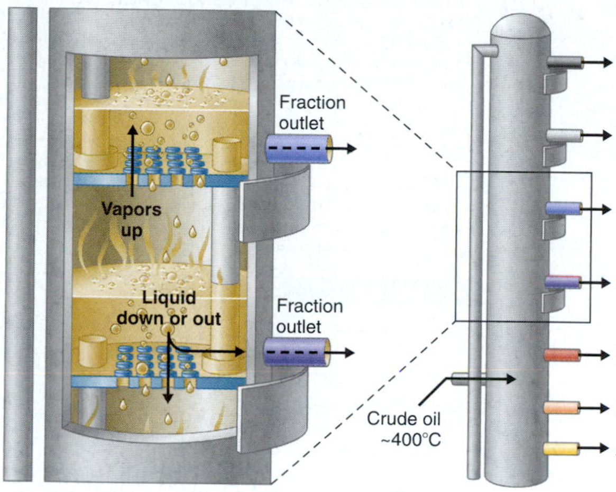

	Number of carbons	Boiling point range	Uses
Gases	1–4	0–30°C	Bottled and natural gas
Naphthas	5–10	30–180°C	Gasoline
Kerosenes	10–16	180–260°C	Kerosene for home heaters, jet fuel
Gas oils	16–60	260–350°C	Diesel fuel, feedstock for cracking
Lubricants	>60	350–575°C	Motor oil, feedstock for cracking
Fuel oil	>70	>490°C	Candles, fuel oil for ships and power stations
Asphalt	>80	>580°C	Roofing tar, road tar

(a) Petroleum distillation tower (b) Petroleum fractions

Figure 2.20 **Distillation of petroleum.**

... provide students with an introduction to the relevance of modern chemistry in their everyday lives.

Figure 4.20 **Rust formation.** The corrosion process involves an oxidation-reduction reaction in which metallic iron is converted to $Fe(OH)_3$, a reddish-brown solid.

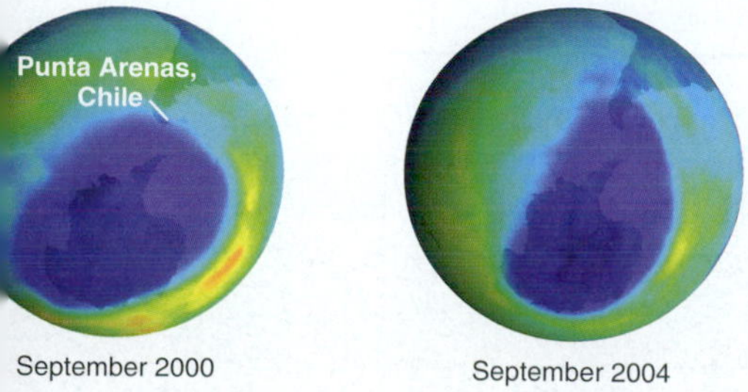

Figure 3.14 **Satellite photos of Earth reveal the increased size of the Antarctic ozone hole with time.** Dark blue colors correspond to the thinnest ozone, while light blue, green, yellow, orange, and red indicate progressively thicker ozone.

Build problem-solving and pattern recognition skills

Note the Pattern
Students learn to recognize important, recurring patterns encountered both in text discussions and in problems.

Worked examples avoid providing students with algorithmic solutions, and instead suggest strategies specific to each problem, thus promoting development of the analytical skills necessary for successful problem solving.

MasteringGeneralChemistry
Key concepts and problem-solving skills are reinforced through assignable interactive tutorials online at **www.masteringgenchem.com.**

Practice exercises reinforce problem-solving skills by enabling students to check their understanding of the strategy employed in the previous example.

Note the pattern
The lattice energy is usually the most important energy factor in determining the stability of an ionic compound.

If the formation of ionic lattices that contain multiply charged ions is so energetically favorable, why does CsF contain Cs^+ and F^- ions rather than Cs^{2+} and F^{2-} ions? If we assume that U for a $Cs^{2+}F^{2-}$ salt would be approximately the same as U for BaO, the formation of a lattice containing Cs^{2+} and F^{2-} ions would release 2291 kJ/mol (3048 kJ/mol − 756.9 kJ/mol) more energy than one containing Cs^+ and F^- ions. To form the Cs^{2+} ion from Cs^+, however, would require removing a 5p electron from a filled inner shell, which calls for a great deal of energy: I_2 = 2234.4 kJ/mol for Cs. Furthermore, forming an F^{2-} ion is expected to be even more energetically unfavorable than forming an O^{2-} ion. Not only is an electron being added to an already negatively charged ion, but because the F^- ion has a filled 2p subshell, the added electron would have to occupy an empty high-energy 3s orbital. Cesium fluoride, therefore, is not $Cs^{2+}F^{2-}$ because the energy cost of forming the doubly charged ions would be higher than the additional lattice energy that would be gained.

EXAMPLE 8.3

Use data from Figure 7.13, Tables 7.5, 8.2, and 8.3, and Appendix A to calculate the lattice energy of MgH_2.

Given Chemical compound, data from figures and tables

Asked for Lattice energy

MGC Introduction to Lewis Structures

Strategy

A Write a series of stepwise reactions for forming MgH_2 from the elements via the gaseous ions.

B Use Hess's law and data from the specified figures and tables to calculate the lattice energy.

Solution

A Hess's law allows us to use a thermochemical cycle (a Born–Haber cycle) to calculate the lattice energy for a given compound. We begin by writing reactions in which we form the component ions from the elements in a stepwise manner and then assemble the ionic solid:

(1)	$Mg(s) \longrightarrow Mg(g)$	$\Delta H_1 = \Delta H_{sub}(Mg)$
(2)	$Mg(g) \longrightarrow Mg^{2+}(g) + 2e^-$	$\Delta H_2 = I_1(Mg) + I_2(Mg)$
(3)	$H_2(g) \longrightarrow 2H(g)$	$\Delta H_3 = D(H_2)$
(4)	$2H(g) + 2e^- \longrightarrow 2H^-(g)$	$\Delta H_4 = 2EA(H)$
(5)	$Mg^{2+}(g) + 2H^-(g) \longrightarrow MgH_2(s)$	$\Delta H_5 = -U$
	$Mg(s) + H_2(g) \longrightarrow MgH_2(s)$	$\Delta H = \Delta H_f$

B Table 7.5 lists first and second ionization energies for the Period-3 elements [$I_1(Mg)$ = 737.7 kJ/mol, $I_2(Mg)$ = 1450.7 kJ/mol]. First electron affinities for all elements are given in Figure 7.13 [$EA(H)$ = −72.8 kJ/mol]. Table 8.2 lists selected enthalpies of sublimation [$\Delta H_{sub}(Mg)$ = 147.1 kJ/mol]. Table 8.3 lists selected bond dissociation energies [$D(H_2)$ = 436.0 kJ/mol]. Enthalpies of formation (ΔH_f = −75.3 kJ/mol for MgH_2) are listed in Appendix A. From Hess's law, ΔH_f is equal to the sum of the enthalpy changes for Reactions 1–5:

$$\begin{aligned}\Delta H_f &= \Delta H_1 + \Delta H_2 + \Delta H_3 + \Delta H_4 + \Delta H_5 \\ &= \Delta H_{sub}(Mg) + [I_1(Mg) + I_2(Mg)] + D(H_2) + 2EA(H) - U \\ -75.3\text{ kJ/mol} &= 147.1\text{ kJ/mol} + (737.7\text{ kJ/mol} + 1450.7\text{ kJ/mol}) + 436.0\text{ kJ/mol} \\ &\quad + 2(-72.8\text{ kJ/mol}) - U \\ U &= 2701.2\text{ kJ/mol}\end{aligned}$$

EXERCISE 14.9

1,3-Butadiene (CH_2=CH–CH=CH_2) is a volatile and reactive organic molecule used in the production of rubber. Above room temperature it reacts slowly to form products. Concentrations of 1,3-butadiene (C_4H_6) as a function of time at 326°C are listed in the table along with ln[C_4H_6] and the reciprocal concentrations. Graph the data as concentration versus t. Then determine the order of the reaction in C_4H_6, the rate law, and the rate constant for the reaction.

Time (s)	[C_4H_6] (M)	ln[C_4H_6]	1/[C_4H_6] (M^{-1})
0	1.72×10^{-2}	−4.063	58.1
900	1.43×10^{-2}	−4.247	69.9
1800	1.23×10^{-2}	−4.398	81.3
3600	9.52×10^{-3}	−4.654	105
6000	7.30×10^{-3}	−4.920	137

Answer

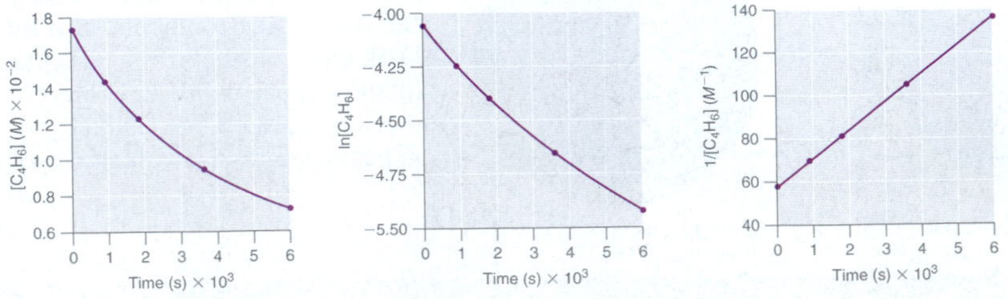

Second order in C_4H_6; rate = $k[C_4H_6]^2$; $k = 2.05 \times 10^3\ M^{-1} \cdot s^{-1}$

Reinforce learning with end-of-chapter materials

Review: Key Concepts, Key Terms, Key Equations

Practice: Conceptual, Numerical, and Applied Problems

SUMMARY AND KEY TERMS

14.0 Introduction (p. 1)

Chemical kinetics is the study of **reaction rates,** changes in the concentrations of reactants and products with time.

14.1 Factors That Affect Reaction Rates (p. 2)

Factors that influence the rates of chemical reactions include concentration of reactants, temperature, physical state of reactants and their dispersion, solvent, and presence of a catalyst.

14.2 Reaction Rates and Rate Laws (p. 4)

Reaction rates are reported either as the **average rate** over a period of time or as the **instantaneous rate** at a single time.

The **rate law** for a reaction is a mathematical relationship between the reaction rate and the concentrations of species in solution. Rate laws can be expressed either as a **differential rate law,** describing the change in reactant or product concentrations as a function of time, or as an **integrated rate law,** describing the actual concentrations of reactants or products as a function of time. The **rate constant,** k, of a rate law is a constant of proportionality between reaction rate and reactant concentration. The power to which a concentration is raised in a rate...

first-order reaction is a constant that is related to the rate constant for the reaction: $t_{1/2} = 0.693/k$.

Radioactive decay reactions are first-order reactions. The **rate of decay,** or **activity,** of a sample of a radioactive substance is the decrease in the number of radioactive nuclei per unit time.

14.6 Reaction Rates — A Microscopic View (p. 34)

A **reaction mechanism** is the microscopic path by which reactants are transformed into products. Each step is an **elementary reaction.** Species that are formed in one step and consumed in another are **intermediates.** Each elementary step can be described in terms of its **molecularity,** the number of molecules that collide in that step. The slowest step in a reaction mechanism is the **rate-determining step. Chain reactions** consist of three kinds of reaction: initiation, propagation, and termination. Intermediates in chain reactions are often called **radicals,** species that have an unpaired valence electron.

14.7 The Collision Model of Chemical Kinetics (p. 39)

... collision ... potential ... n energy ... arrange- ... **ted com-** ... ature, the ... tions that

KEY EQUATIONS

General definition of rate	$\text{rate} = \frac{\Delta[B]}{\Delta t} = -\frac{\Delta[A]}{\Delta t}$	(14.4)
General form of rate law	$\text{rate} = k[A]^m[B]^n$	(14.10)
Zeroth-order reaction	$\text{rate} = -\frac{\Delta[A]}{\Delta t} = k$	(14.16)
	$[A] = [A]_0 - kt$	(14.17)
First-order reaction	$\text{rate} = -\frac{\Delta[A]}{\Delta t} = k[A]$	(14.21)
	$[A] = [A]_0 e^{-kt}$	(14.22)
	$\ln[A] = \ln[A]_0 - kt$	(14.23)
Second-order reaction	$\text{rate} = -\frac{\Delta[A]}{\Delta t} = k[A]^2$	(14.25)
	$\frac{1}{[A]} = \frac{1}{[A]_0} + kt$	(14.26)
Half-life of first-order reaction	$t_{1/2} = \frac{0.693}{k}$	(14.32)
Radioactive decay	$A = kN$	(14.34)
Arrhenius equation	$k = Ae^{-E_a/RT}$	(14.43)

Approximately 100 problems covering conceptual, numerical, and applied material—building from simple to advanced—are found at the end of every chapter.

QUESTIONS AND PROBLEMS

Answers to odd-numbered problems are given in the Appendix. Complete solutions to even-numbered problems are provided in the Student Solutions Manual.

CONCEPTUAL

14.1 Factors That Affect Reaction Rates

1. What information can you obtain by studying the chemical kinetics of a reaction? Does a balanced chemical equation provide the same information? Why or why not?
2. If you were given the task of determining whether to proceed with a particular reaction in an industrial facility, why would studying the chemical kinetics of the reaction be important to you?
3. What is the ... of each of these factors with the rate of ...

reaction rates significantly relative to the same reaction run in the presence of a heterogeneous catalyst. What is the reason for anticipating that the relative rate will increase?

8. Water has a dielectric constant more than two times greater than that of methanol (dielectric constant = 80.1 for H_2O and 32.7 for CH_3OH). Which would be your solvent of choice for a substitution reaction between an ionic compound and a polar reagent, both of which are soluble in either methanol or water? Why?

14.2 Reaction Rates and Rate Laws

9. Explain why the rate of a reaction is generally fastest at early time intervals. For the reaction $A + B \longrightarrow C$, what would a plot of concentration versus time look like?
10. Explain the differences between a differential rate law and an integrated rate law. What ...

Paired problems reinforce key problem-solving skills.

Most can be assigned for online homework or tests with automatic grading at www.masteringgenchem.com.

NUMERICAL

This section includes "paired problems" (marked by brackets) that require similar problem-solving skills.

14.6 Reaction Rates — A Microscopic View

63. Cyclopropane, a mild anesthetic, rearranges to propylene via a collision that produces and destroys an energized species. Here are the important steps in this rearrangement.

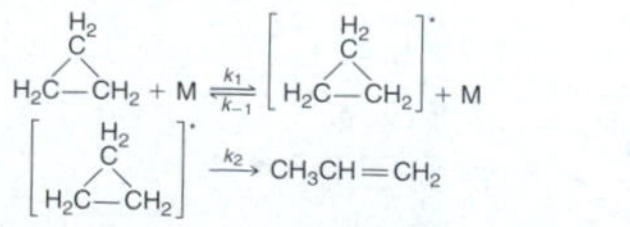

where M is any molecule, including cyclopropane. Only those cyclopropane molecules with sufficient energy (denoted with an asterisk) can rearrange to propylene. Which step determines the rate constant of the overall reaction?

64. Above approximately 500 K, the reaction between NO_2 and CO to produce CO_2 and NO follows the second-order rate law $\Delta[CO_2]/\Delta t = k[NO_2][CO]$. At lower temperatures, however, the rate law is $\Delta[CO_2]/\Delta t = k'[NO_2]^2$, for which it is known

Problems based on relevant applications from modern chemistry mirror the text and reinforce both the theoretical and the practical nature of chemistry.

APPLICATIONS

77. Atmospheric chemistry in the region below the clouds of Venus appears to be dominated by reactions of sulfur and carbon-containing compounds. Included in representative elementary reaction steps are the following:

$$SO_2 + CO \longrightarrow SO + CO_2$$
$$SO + CO \longrightarrow S + CO_2$$
$$SO + SO_2 \longrightarrow S + SO_3$$

For each elementary reaction, write an expression for the reaction rate in terms of changes in the concentrations of each reactant and product with time.

78. In acid, nitriles hydrolyze to produce a carboxylic acid and ammonium ion. For example, acetonitrile, a substance used to extract fatty acids from fish liver oils, is hydrolyzed to acetic acid via the reaction

$$CH_3C{\equiv}N + 2H_2O + H^+ \longrightarrow CH_3\overset{\overset{\large O}{\|}}{C}OH + NH_4^+$$

Express the rate of the reaction in terms of changes in the concentrations of each reactant and each product with time.

Teaching with art

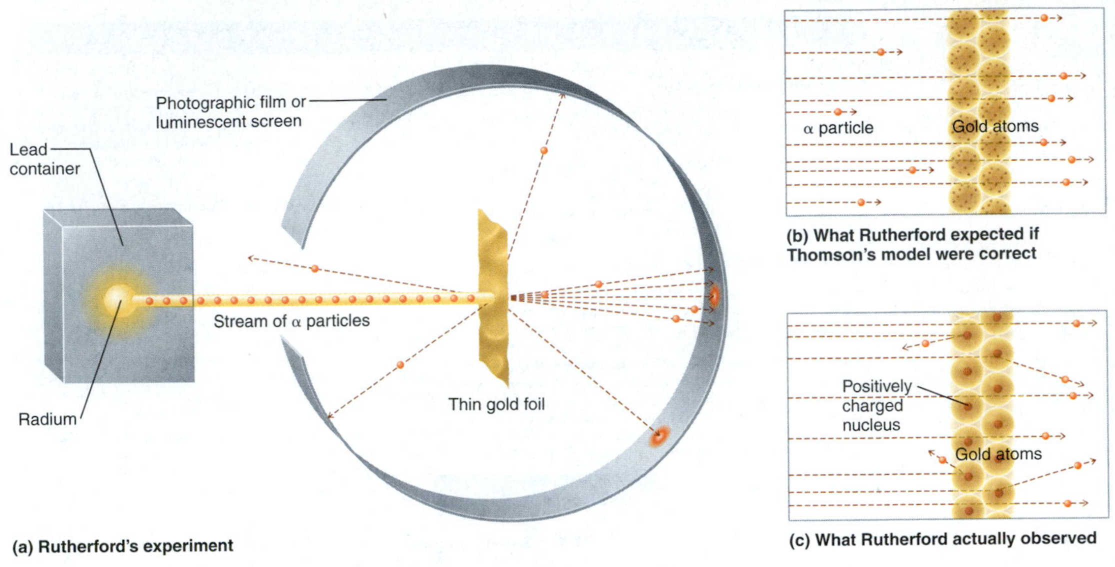

Figure 1.21 **A summary of Rutherford's experiments.**

An historical experiment illustrated and analyzed. Part (a) tells the experimental story, while parts (b) and (c) compare hypothesis versus observation at an atomic level, prompting students to think like an experimental scientist.

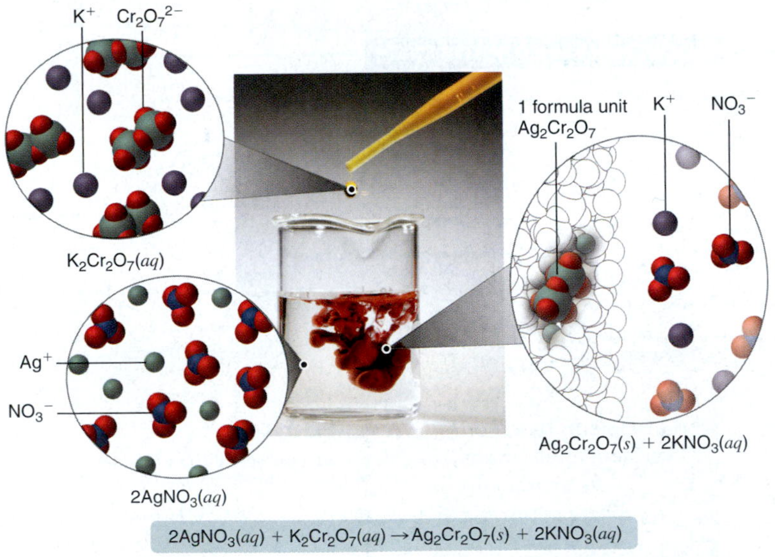

Figure 4.11 **Illustration of what happens at the molecular level when solutions of $AgNO_3$ and $K_2Cr_2O_7$ are mixed.** The Ag^+ and $Cr_2O_7^{2-}$ ions form a red precipitate of solid $Ag_2Cr_2O_7$, while the K^+ and NO_3^- ions remain in solution.

Molecular blowups help students visualize a precipitation reaction explored in the text. Highlighting of formula units and ions in the art makes the stoichiometrically related compositions of reactants and products very tangible.

A clear, consistent, and accurate art program helps students to ...

Teaching with art

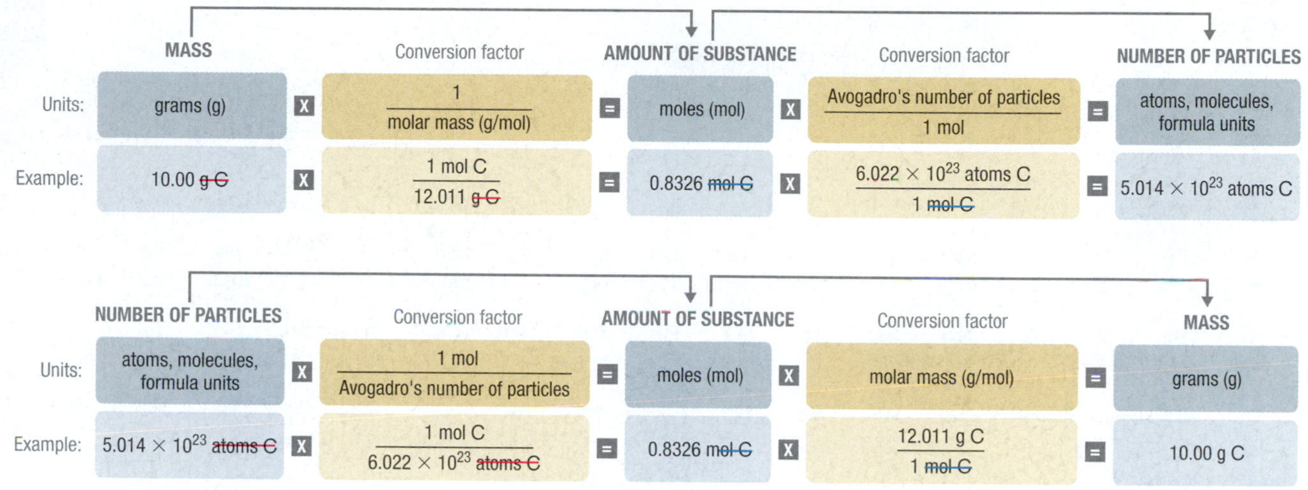

Figure 3.2 **A flowchart for converting between mass, moles, and numbers of atoms, molecules, or formula units.**

Mass/mole conversions are an essential skill. Flowcharts in Chapters 3 and 4 therefore guide students through every step. Paired examples allow students to confirm for themselves that the general procedure really works if properly implemented.

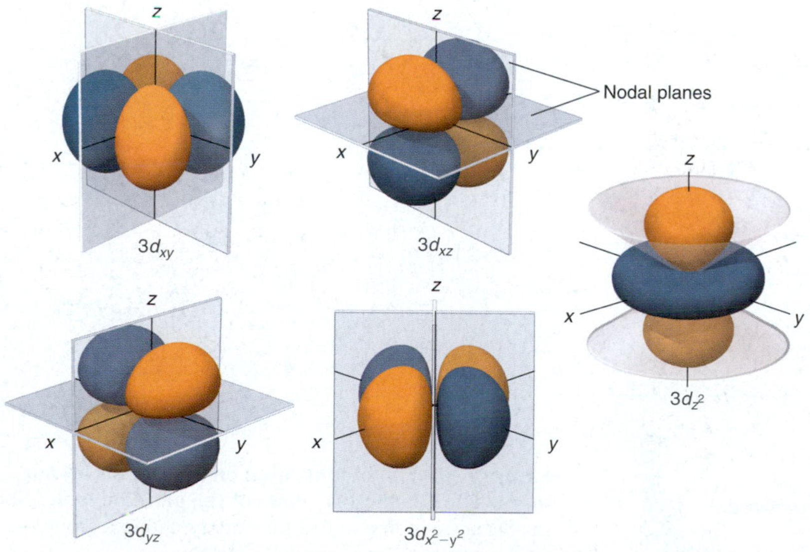

Figure 6.26 **The five equivalent 3*d* orbitals of the hydrogen atom.**

... visualize and more effectively learn key concepts and principles.

Orbitals in Chapter 6 are rendered based on Schrödinger's Equation to ensure that shapes and sizes of lobes are mathematically accurate—unlike the "elongated dumb-bells" still shown in some textbooks. Orange and blue are used to distinguish positive and negative phases of wave function, a distinction that will become important in later discussions of bonding and molecular orbital theory.

Teaching with art

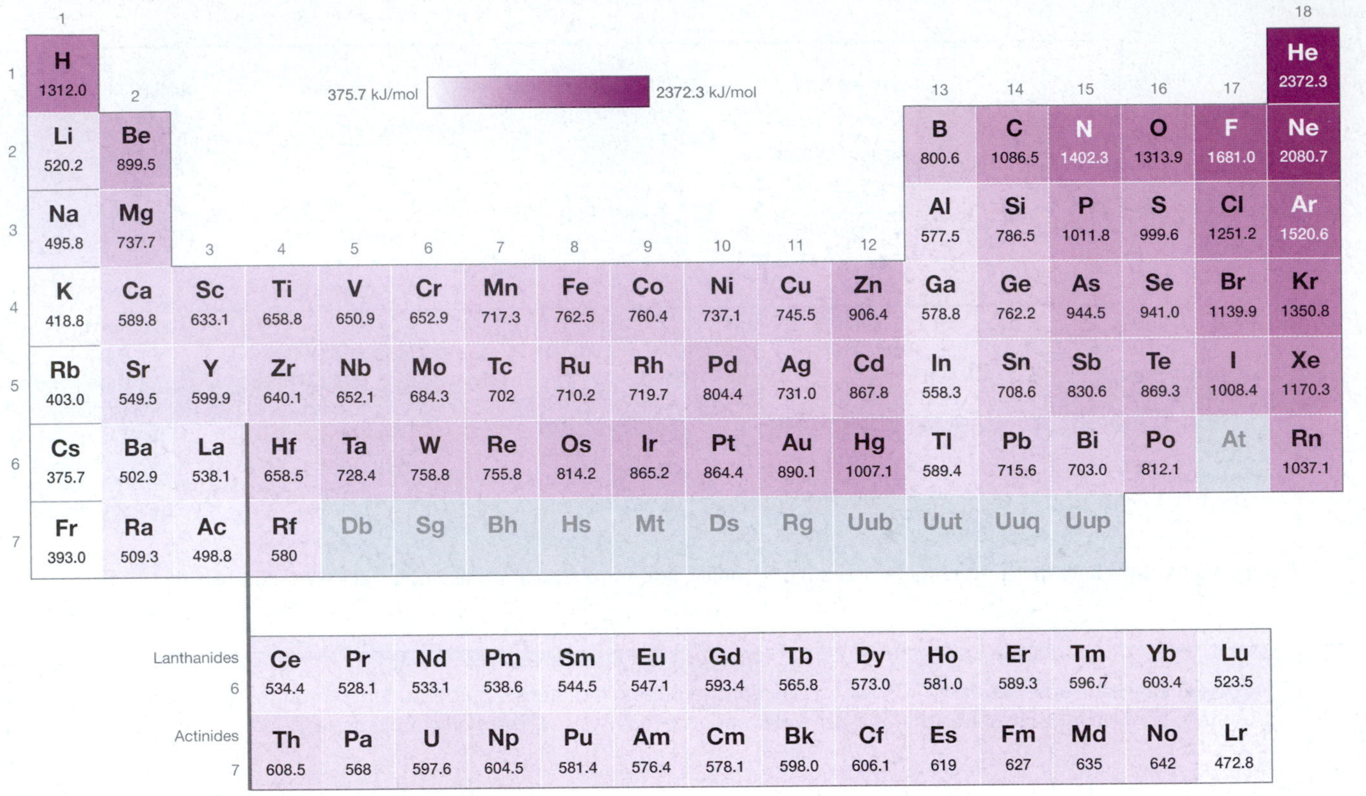

Figure 7.11 First ionization energies of the elements. The relative darkness of boxes illustrate the relative energies of the atoms.

All periodic trends are plotted from real data and consistently styled and color coded. Consistent presentations make it possible for students to easily compare the different trends covered in Chapter 7. Multiple styles of presentation accommodate differing visualization skills among students (some think most comfortably in 2D, others more comfortably in 3D) and are mutually reinforcing.

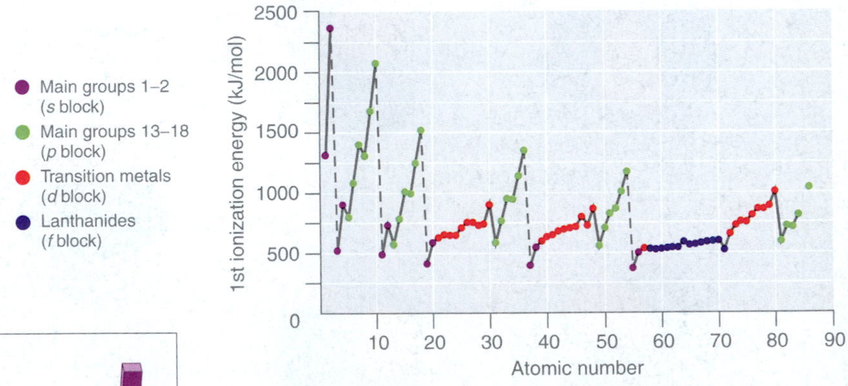

Figure 7.10 Plot of ionization energy versus atomic number for the first six rows of the periodic table. Note the decrease in first ionization energy within a group (most easily seen here for Groups 1 and 18).

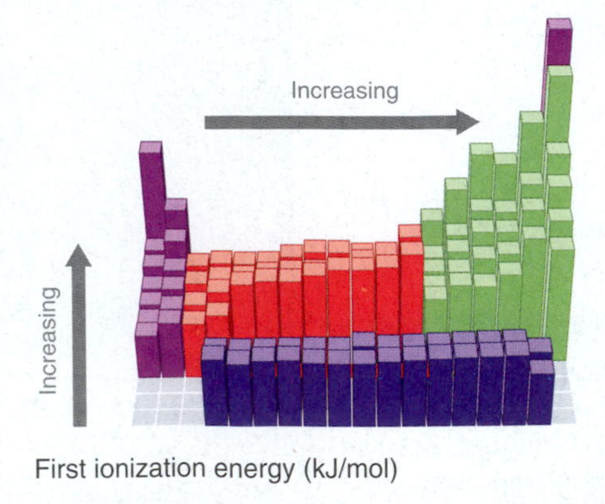

MasteringGeneralChemistry™ ... the most advanced and educationally effective online homework and tutorial system available

Students turn to **MasteringGeneral-Chemistry**™ as a personalized tutor available 24/7 to help them improve problem-solving skills and prepare for exams. Instructors assign homework problems and tutorials and can use data to pinpoint areas of difficulty for individual students or for the class as a whole.

Every tutorial and homework problem has been pre-tested on thousands of students to ensure that they are accurate and effective. Additional benefits to this student-data based approach:

- **Create rigorous and instructive homework assignments** that combine testing with tutoring. Multistep tutorials provide wrong-answer specific feedback, simpler problems upon request, and a variety of answer types.
- **Assign problems based on data drawn from pre-testing,** such as estimated time to complete each problem, level of difficulty, and answer type (for example, molecule drawing, dimensional analysis, orbital diagrams, algebraic solutions, and randomized numbers).
- **Adjust the focus and the pace of your course** based on data collected from homework problems. For example, data lets you see which problems stump students, as well as which steps taken to solve problems cause the most confusion.
- **Check the work of an individual student** in unprecedented detail, including time spent on each step of a problem, wrong answers submitted at every step, how much help was asked for, and how many practice problems a student worked.
- **Compare your results against the "national average"** problem by problem, step by step, class by class, and year by year.

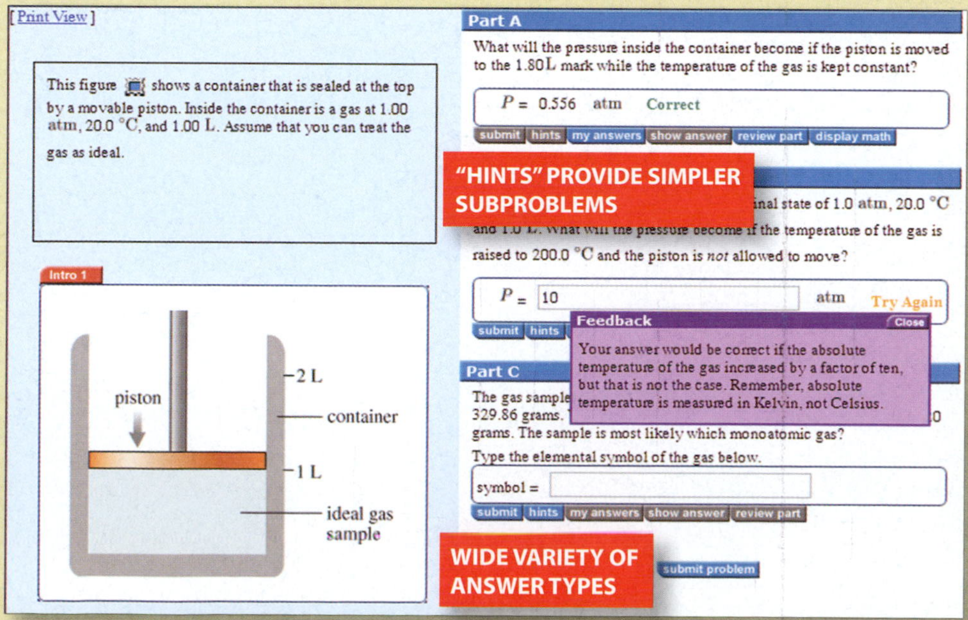

www.masteringgenchem.com

1 Introduction to Chemistry

LEARNING OBJECTIVES

- To recognize the breadth, depth, and scope of chemistry
- To understand what is meant by the scientific method
- To be able to classify matter
- To understand the development of the atomic model
- To know the meaning of isotopes and atomic masses
- To become familiar with the periodic table and to be able to use it as a predictive tool

CHAPTER OUTLINE

1.1 Chemistry in the Modern World

1.2 The Scientific Method

1.3 A Description of Matter

1.4 A Brief History of Chemistry

1.5 The Atom

1.6 Isotopes and Atomic Masses

1.7 Introduction to the Periodic Table

1.8 Essential Elements

ESSENTIAL SKILLS 1

Units of Measurement

Scientific Notation

Significant Figures

Accuracy and Precision

As you begin your study of college chemistry, those of you who do not intend to become professional chemists may well wonder why you need to study chemistry. In fact, as you will soon discover, a basic understanding of chemistry is useful in a wide range of disciplines and career paths. You will also discover that an understanding of chemistry helps you make informed decisions about many issues that affect you, your community, and your world. A major goal of this text is to demonstrate the importance of chemistry in your daily life and in our collective understanding of both the physical world we occupy and the biological realm of which we are a part. The objective of this chapter is to introduce the breadth, the importance, and some of the challenges of modern chemistry, and to present some of the fundamental concepts and definitions you will need to understand how chemists think and work.

An atomic corral for electrons. A "corral" of 48 Fe atoms (yellow-orange) on a smooth Cu surface (cyan-purple) confine the electrons on the surface of the Cu, producing a pattern of "ripples" in the distribution of the electrons. Scientists assembled the 713-pm-diameter corral by individually positioning Fe atoms with the tip of a scanning tunneling microscope.

1.1 Chemistry in the Modern World

Chemistry is the study of matter and the changes that material substances undergo. Of all the scientific disciplines, it is perhaps the most extensively connected to other fields of study. Geologists who want to locate new mineral or oil deposits use chemical techniques to analyze and identify rock samples. Oceanographers use chemistry to track ocean currents, determine the flux of nutrients into the sea, and measure the rate of exchange of nutrients between ocean layers. Engineers consider the relationships between the structures and properties of substances when they specify materials for various uses. Physicists take advantage of the properties of substances to detect new subatomic particles. Astronomers use chemical signatures to determine the age and distance of stars and thus answer questions about how stars form and how old the universe is. The entire subject of environmental science depends on chemistry to explain the origin and impacts of phenomena such as air pollution, ozone layer depletion, and global warming.

The disciplines that focus on living organisms and their interactions with the physical world also rely heavily on chemistry. A living cell contains a large collection of complex molecules that carry out thousands of chemical reactions, including those that are necessary for the cell to reproduce. Biological phenomena such as vision, taste, smell, and movement are all the result of numerous series of chemical reactions. Thus, fields such as medicine, pharmacology, nutrition, and toxicology focus specifically on how the chemical substances that enter our body interact with the chemical components of the body to maintain our health and well-being. For example, in the specialized area of sports medicine, a knowledge of chemistry is needed to understand why muscles get sore after exercise as well as how prolonged exercise produces the euphoric feeling known as "runners' high."

Examples of the practical applications of chemistry are everywhere (Figure 1.1). Engineers need to understand the chemical properties of the substances they work with in order to design biologically compatible implants for joint replacements, or to design

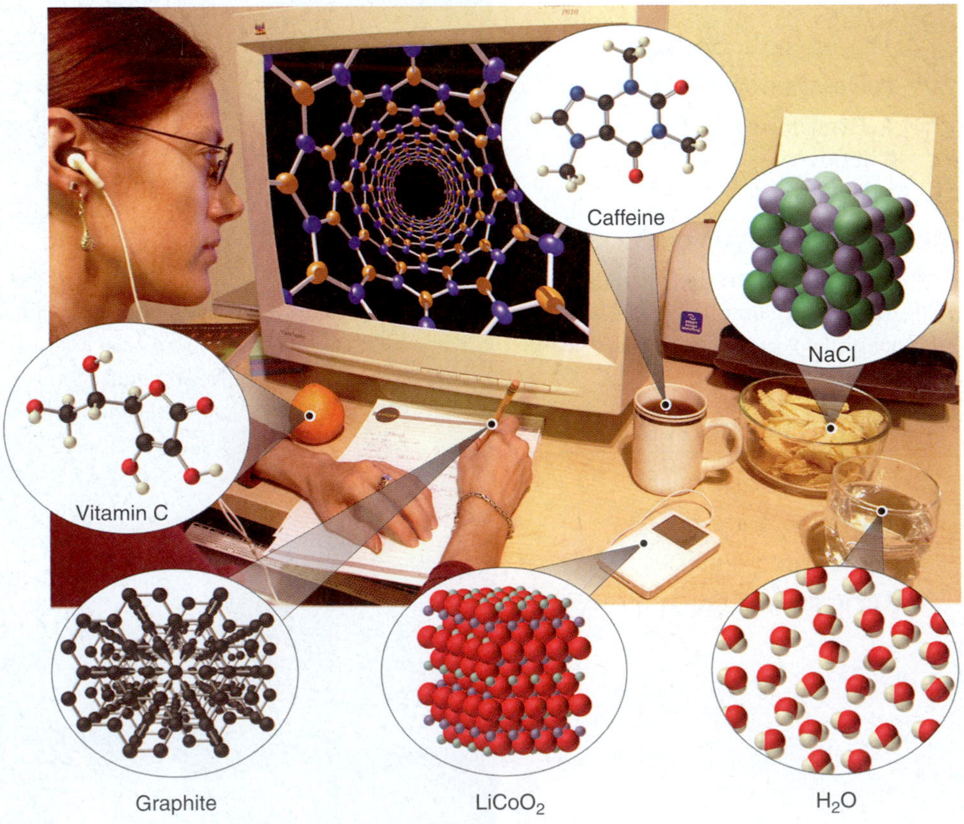

Figure 1.1 Chemistry in everyday life. Although most people do not recognize it, chemistry and chemical compounds are crucial ingredients in almost everything we eat, wear, and use.

roads, bridges, buildings, and nuclear reactors that do not collapse because of weakened structural materials such as steel and cement. Archaeology and paleontology rely on chemical techniques to date bones and artifacts and to identify their origins. Although law is not normally considered a field that is related to chemistry, forensic scientists use chemical methods to analyze blood, fibers, and other evidence as they investigate crimes. In particular, DNA matching—comparing of biological samples of genetic material to see whether they could have come from the same person—has been used to solve many high-profile criminal cases as well as to clear those individuals who have been erroneously convicted. Forensics is one of the most rapidly growing areas of applied chemistry. In addition, the proliferation of chemical and biochemical innovations in industry is producing rapid growth in the area of patent law. Ultimately, the dispersal of information in all the fields in which chemistry plays a part requires experts who are able to explain complex chemical issues to the public through television, print journalism, the Internet, and popular books.

By this point, it shouldn't surprise you to learn that chemistry was essential in explaining a pivotal event in the history of our planet: the disappearance of the dinosaurs. Although dinosaurs ruled Earth for more than 150 million years, fossil evidence suggests that they became extinct rather abruptly approximately 66 million years ago. Proposed explanations for their extinction have ranged from an epidemic caused by some deadly microbe or virus to more gradual phenomena such as massive climate changes. In 1978 Luis Alvarez (a Nobel-Prize-winning physicist), his son, the geologist Walter Alvarez, and their co-workers discovered a thin layer of sedimentary rock formed 66 million years ago that contained unusually high concentrations of iridium, a rather rare metal (Figure 1.2a). This layer was deposited at about the time dinosaurs disappeared from the fossil record. Although iridium is very rare in most rocks, accounting for only 0.0000001% of Earth's crust, it is much more abundant in comets and asteroids. Because corresponding samples of rocks at sites in Italy and Denmark contained high iridium concentrations, the Alvarezes suggested that the impact of a large asteroid with Earth brought about the extinction of the dinosaurs. When additional samples of 66-million-year-old sediments from sites around the world were analyzed by chemists, all were found to contain high levels of iridium. In addition, small grains of quartz in most of the iridium-containing layers exhibit microscopic cracks characteristic of high-intensity shock waves (Figure 1.2b). These grains apparently originated from terrestrial rocks at the impact site, which were pulverized upon impact and blasted into the upper atmosphere before they settled out all over the world.

Scientists calculate that a collision of Earth with a stony asteroid about 10 kilometers (6 miles) in diameter, traveling at 25 kilometers per second (~56,000 miles per hour), would almost instantaneously release energy equivalent to the explosion of about 100 million megatons of TNT. This is more energy than is stored in the entire nuclear arsenal of the world! The energy released by such an impact would set fire to vast areas of forest, and the smoke from the fires and the dust created by the impact would block the sun's light for months or years, eventually killing virtually all green plants and most organisms that depend on them. This could explain why about 70% of *all* species—not just dinosaurs—disappeared at the same time. Scientists also calculate that this impact would form a crater at least 125 kilometers (78 miles) in diameter. Recently, a partially submerged crater 180 kilometers (112 miles) in diameter was detected near the tip of Mexico's Yucatan Peninsula in the Gulf of Mexico and identified as the probable impact site (Figure 1.3). Thus, simple chemical measurements of the abundance of one element in rocks led to a new and dramatic explanation for the extinction of the dinosaurs. Though still controversial, this explanation is supported by additional evidence, much of it chemical.

This is only one example of how chemistry has been applied to an important scientific problem. Other chemical applications and explanations that we will discuss in this text include how astronomers determine the distance of galaxies and how fish can survive in subfreezing water under polar ice sheets. We will also

(a)

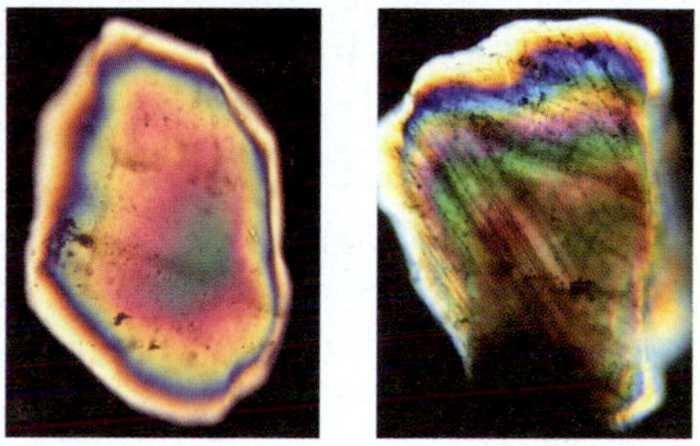

(b)

Figure 1.2 Evidence for the asteroid impact that may have caused the extinction of the dinosaurs. (a) Luis Alvarez and his son Walter in front of a rock formation in Italy that shows the thin white layer of iridium-rich clay deposited at the time the dinosaurs became extinct. The concentration of iridium is 30 times higher in this layer than in the rocks immediately above and below it. There are no significant differences between the clay layer and the surrounding rocks in the concentrations of any of 28 other elements examined. (b) Microphotographs of an unshocked quartz grain (left) and a quartz grain from the iridium-rich layer exhibiting microscopic cracks resulting from shock (right).

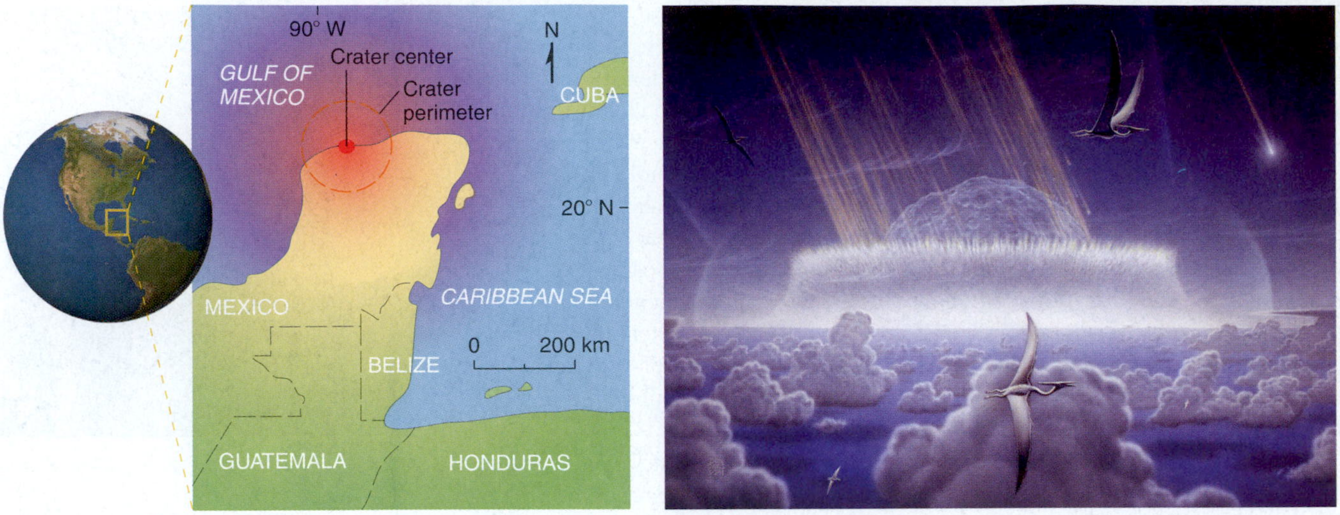

(a) (b)

Figure 1.3 Asteroid impact. (a) The location of the asteroid impact crater near what is now the tip of the Yucatan Peninsula in Mexico. (b) An artist's depiction of the asteroid impact that is believed by many scientists to have caused the extinction of the dinosaurs 66 million years ago.

consider ways in which chemistry affects our daily lives: the addition of iodine to table salt; the development of more effective drugs to treat diseases such as cancer, AIDS, and arthritis; the retooling of industry to use non-chlorine-containing refrigerants, propellants, and other chemicals in order to preserve Earth's ozone layer; the use of modern materials in engineering; current efforts to control the problems of acid rain and global warming; and the awareness that in order to function properly our bodies require small amounts of some chemical substances that are toxic when ingested in larger doses. By the time you finish this text, you will be able to discuss these kinds of topics knowledgeably, either as a beginning scientist who intends to spend your career studying such problems, or as an informed observer who is able to participate in the public debates that will certainly arise as society grapples with scientific issues.

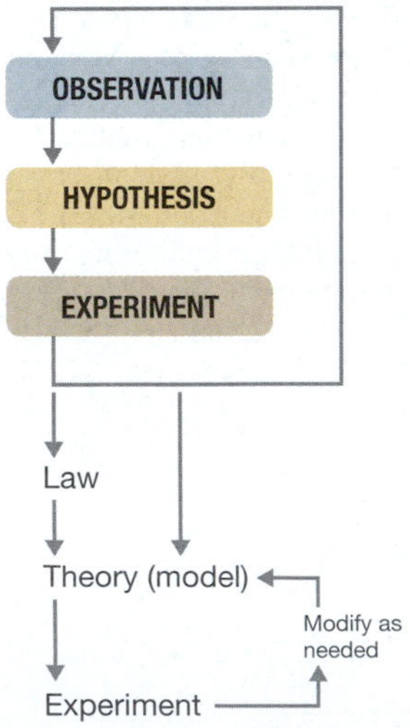

Figure 1.4 The scientific method. As depicted in this flowchart, the scientific method consists of making observations, formulating hypotheses, and designing experiments. A scientist may enter the cycle at any point.

1.2 The Scientific Method

Scientists search for answers to questions and solutions to problems by using a procedure called the **scientific method**. This procedure consists of making *observations*, formulating *hypotheses*, and designing *experiments*, which lead in turn to additional observations, hypotheses, and experiments in repeated cycles (Figure 1.4).

Observations can be qualitative or quantitative. *Qualitative observations* describe properties or occurrences in ways that do not rely on numbers. Examples are the observations that crystalline sulfur is yellow and that a penny dissolves in dilute nitric acid to form a blue solution and a brown gas. *Quantitative observations* are measurements, which by definition consist of both a *number* and a *unit*. Examples are the observations that the melting point (mp) of crystalline sulfur is 115.21°C and that 35.9 grams (g) of table salt (sodium chloride) dissolves in 100 g of water at 20°C. For the question of the dinosaurs' extinction, the initial observation was quantitative: Iridium concentrations in sediments dating to 66 million years ago were 20–160 times higher than normal.

After deciding to learn more about an observation or a set of observations, scientists generally begin an investigation by forming a **hypothesis**, a tentative explanation for the observation(s). The hypothesis may not be correct, but it puts the

MGC The Scientific Method

scientist's understanding of the system being studied into a form that can be tested. For example, the observation that we experience alternating periods of light and darkness corresponding to observed movements of the sun, moon, clouds, and shadows is consistent with either of two hypotheses: (1) Earth rotates on its axis every 24 hours, alternately exposing one side to the sun, or (2) the sun revolves around Earth every 24 hours. Suitable experiments can be designed to choose between these two alternatives. For the disappearance of the dinosaurs, the hypothesis was that the impact of a large extraterrestrial object caused their extinction. Unfortunately (or perhaps fortunately), this hypothesis is not amenable to direct testing by any obvious experiment, but scientists can collect additional data that either support or refute it.

After a hypothesis has been formed, scientists conduct experiments to test its validity. **Experiments** are systematic observations or measurements, preferably made under controlled conditions—that is, conditions in which the variable of interest is clearly distinguished from any others. A properly designed and executed experiment enables the scientist to decide whether the original hypothesis is valid. Often experiments demonstrate that the hypothesis is incorrect or that it must be modified. More experimental data are then collected and analyzed, at which point the scientist may begin to think that the results are sufficiently reproducible (that is, dependable) to merit being summarized in a **law**, a verbal or mathematical description of a phenomenon that allows for general predictions. A law simply says *what* happens; it does not address the question of *why*. For example, the **law of definite proportions**, which was discovered by the French scientist Joseph Proust (1754–1826), states that a chemical substance always contains the same proportions of elements by mass. Thus, sodium chloride (table salt) always contains 39.34% sodium and 60.66% chlorine by mass, and sucrose (table sugar) is always 42.11% carbon, 6.48% hydrogen, and 51.41% oxygen by mass.* (For a review of common units of measurement, see Essential Skills 1 at the end of this chapter.) The law of definite proportions may now seem obvious, but the head of the U.S. Patent Office did not accept it as a fact until the early 20th century!

Whereas a law states only *what* happens, a **theory** attempts to explain *why* nature behaves as it does. Laws are unlikely to change greatly over time unless a major experimental error is discovered. In contrast, a theory, by definition, is incomplete and imperfect, evolving with time to explain new facts as they are discovered. The theory developed to explain the extinction of the dinosaurs, for example, is that Earth occasionally encounters small to medium-sized asteroids and that these encounters have unfortunate implications for the continued existence of most species. This theory is by no means proven, but it is consistent with the bulk of the evidence amassed to date. Figure 1.5 summarizes the application of the scientific method in this case.

EXAMPLE 1.1

Classify each statement as a law, theory, experiment, hypothesis, qualitative observation, or quantitative observation. **(a)** Ice always floats on liquid water; **(b)** Birds evolved from dinosaurs; **(c)** Hot air is less dense than cold air, probably because the components of hot air are moving more rapidly; **(d)** When 10 grams (g) of ice was added to 100 milliliters (mL) of water at 25°C, the temperature of the water decreased to 15.5°C after the ice melted; **(e)** The ingredients of Ivory soap were analyzed to see whether it really is 99.44% pure, as advertised.

Strategy

Refer to the definitions in this section to decide which category best describes each statement.

---

* You will learn in Chapter 12 that some solid compounds do not strictly obey the law of definite proportions.

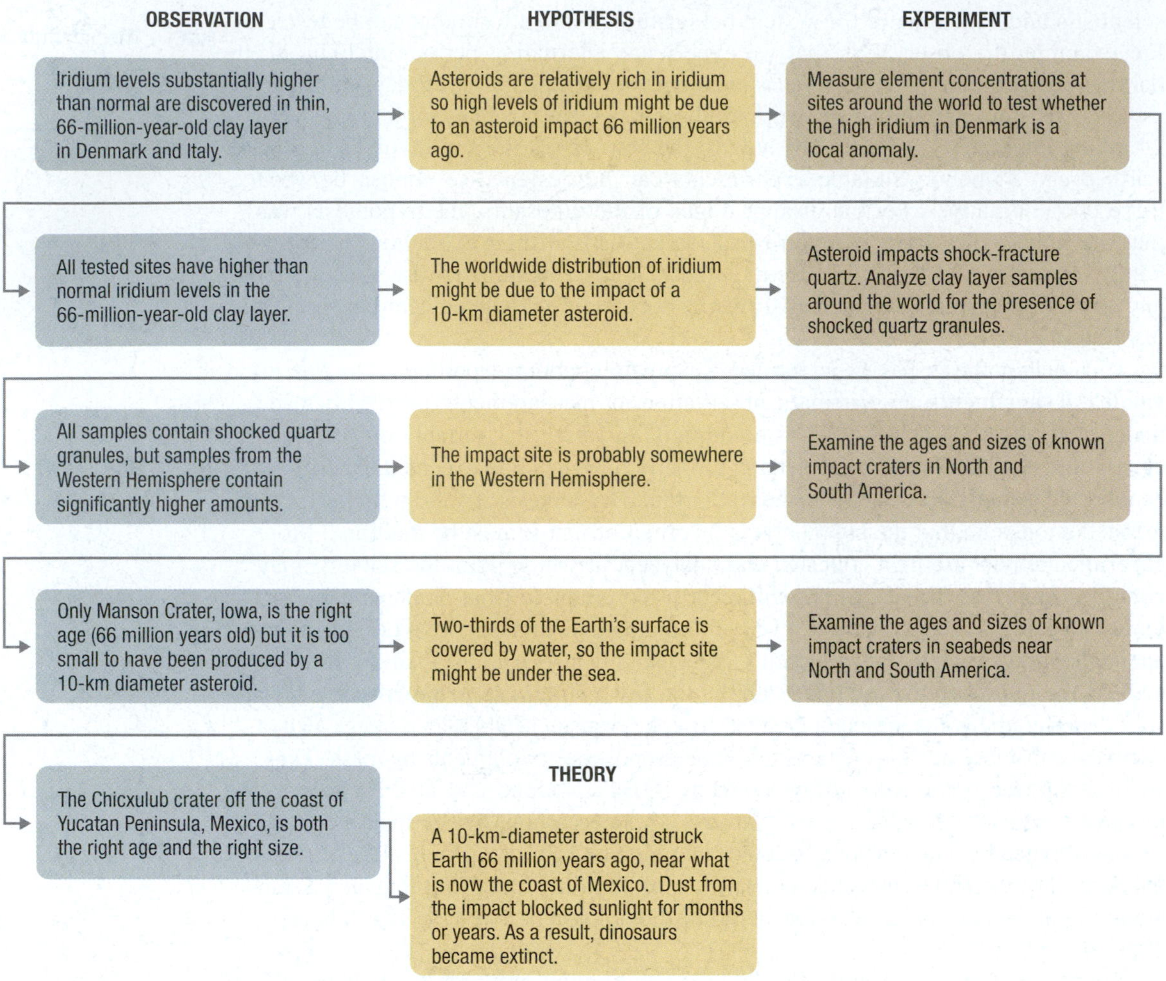

Figure 1.5 A summary of how the scientific method was used in developing the asteroid impact theory to explain the disappearance of the dinosaurs from Earth.

Solution

(a) This is a law, a general statement of a relationship between the properties of liquid and solid water.
(b) This is a hypothesis, a possible explanation for the origin of birds.
(c) This is a theory, a statement that tries to explain the relationship between the temperature and the density of air based on fundamental principles.
(d) This is a quantitative observation, the temperatures measured before and after a change is made in a system.
(e) This is an experiment, an analysis designed to test a hypothesis (in this case, the manufacturer's claim of purity).

EXERCISE 1.1

Classify each statement as a law, theory, experiment, hypothesis, qualitative observation, or quantitative observation. **(a)** Measured amounts of acid were added to a Rolaids tablet to see whether it really "consumes 47 times its weight in excess stomach acid"; **(b)** Heat always flows from hot objects to cooler ones, never in the other direction; **(c)** The universe was formed by a massive explosion that propelled matter into a vacuum; **(d)** Michael Jordan was the greatest pure shooter ever to play professional basketball; **(e)** Limestone is relatively insoluble in water but dissolves readily in dilute acid with the evolution of a gas; **(f)** Gas mixtures that contain more than 4% hydrogen in air are potentially explosive.

Answer **(a)** experiment; **(b)** law; **(c)** theory; **(d)** hypothesis; **(e)** qualitative observation; **(f)** quantitative observation

Because scientists can enter the cycle shown in Figure 1.4 at any point, the actual application of the scientific method to different topics can take many different forms. For example, a scientist may start with a hypothesis formed by reading about work done by others in the field, rather than by making direct observations.

It is important to remember that scientists are human. They have a tendency to formulate hypotheses in familiar terms, simply because it is difficult to propose something that has never been encountered or imagined before. Scientists may try to explain the unexpected by extending familiar concepts and models, rather than by looking objectively at the accumulated data. As a result, scientists sometimes discount or overlook unexpected findings that disagree with the basic assumptions behind the hypothesis or theory being tested. A mature theory can therefore be a two-edged sword: it can help us refine our understanding of a given system by focusing our thinking and experiments, but it can also blind us to other possible explanations that are equally likely. Fortunately, truly important findings are immediately subject to independent verification by scientists in other laboratories, and so science is a discipline that is largely self-correcting. When the Alvarezes originally suggested that an extraterrestrial impact caused the extinction of the dinosaurs, the response was almost universal skepticism and scorn. In only 20 years, however, the persuasive nature of the evidence has overcome the skepticism of many scientists, and their initial hypothesis has now evolved into a theory that has revolutionized paleontology and geology.

1.3 • A Description of Matter

MGC Mixtures, Solutions, Pure Substances, etc.

Chemists study the structures, physical properties, and chemical properties of material substances. *Matter* is anything that occupies space and possesses mass. Gold and iridium are matter, as are peanuts, people, and postage stamps. Smoke, smog, and laughing gas are matter. Energy, light, and sound, however, are not matter, nor are ideas and emotions.

The **mass** of an object is the quantity of matter it contains. Do not confuse an object's mass with its *weight*, which is a force caused by the gravitational attraction that operates on the object. Mass is a fundamental property of an object that does not depend on its location.* Weight, on the other hand, does depend on the location of an object. Thus, an astronaut whose mass is 95 kilograms (kg) weighs about 210 pounds on Earth but only about 35 pounds on the moon because the gravitational force he or she experiences on the moon is approximately one-sixth the force experienced on Earth. For practical purposes, weight and mass are often used interchangeably in laboratories. Because the force of gravity can be considered to be the same everywhere on Earth's surface, 2.2 pounds (a weight) equals 1.0 kg (a mass), regardless of the location of the laboratory on Earth.†

* In physical terms, the mass of an object is directly proportional to the force required to change its speed or direction.

† A more detailed discussion of the differences between weight and mass and the units used to measure them is included in Essential Skills 1 at the end of this chapter.

Figure 1.6 The three states of matter. Solids have a defined shape and volume. Liquids have a fixed volume but flow to assume the shape of their containers. Gases fill their container completely, regardless of volume.

Under normal conditions, there are three distinct *states of matter*: solids, liquids, and gases (Figure 1.6). **Solids** are relatively rigid and have fixed shapes and volumes. **Liquids** have fixed volumes but flow to assume the shape of their containers. **Gases** have neither fixed shapes nor fixed volumes and expand to fill their containers completely. The volume of gases depends strongly on their temperature and *pressure*, the amount of force exerted on a given area, whereas the volumes of liquids and solids are virtually independent of temperature and pressure.

Pure Substances and Mixtures

A *pure chemical substance* is any matter that has a fixed chemical composition and characteristic properties. Very few samples of matter consist of pure substances; instead, most are **mixtures**, combinations of two or more pure substances in variable proportions in which the individual substances retain their identity. Air, tap water, milk, blue cheese, bread, and dirt are all mixtures. If all portions of a material are in the same state, have no visible boundaries, and are uniform throughout, then the material is **homogeneous**. Examples of homogeneous mixtures are the air we breathe and the tap water we drink. Homogeneous mixtures are also called *solutions*. Thus, air is a solution of nitrogen, oxygen, water vapor, carbon dioxide, and several other gases; tap water is a solution of small amounts of several substances in water. The specific compositions of both of these solutions are not fixed, however, but depend on the source and location; for example, the compositions of tap water in Boise, Idaho, and Buffalo, New York, are *not* the same. Although most solutions we encounter are liquid, solutions can also be solid. The gray substance still used by some dentists to fill tooth cavities is a complex solid solution that contains 70% silver, 25% lead, 3% copper, and 2% mercury. Solid solutions of two or more metals are commonly called *alloys*.

If the composition of a material is not completely uniform, then it is **heterogeneous** (for example, chocolate chip cookie dough, blue cheese, and dirt). Often mixtures that appear to be homogeneous are found to be heterogeneous upon microscopic examination. Milk, for example, appears to be homogeneous, but when examined under the microscope, it clearly consists of tiny globules of fat and protein dispersed in water (Figure 1.7). The components of heterogeneous mixtures can usually be separated by simple means. Solid–liquid mixtures such as sand in water or tea leaves in tea are readily separated by *filtration*, which consists of passing the mixture through a barrier with holes or pores that are smaller than the solid particles. In principle, mixtures of two or more solids, such as sugar and salt, can be separated by microscopic inspection and sorting. More complex operations are usually necessary, though, such as separating gold nuggets from river gravel by panning. First solid material is filtered from river water, and then the solids are separated by inspection. If the gold is embedded in the rock, it may have to be isolated using chemical methods.

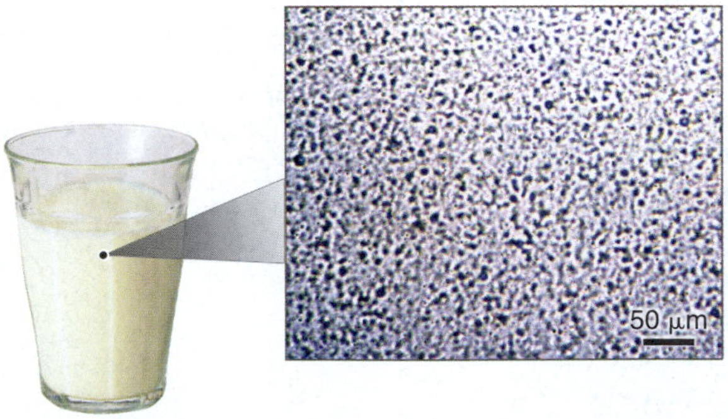

Figure 1.7 A heterogeneous mixture. Under the microscope, whole milk is seen to be a heterogeneous mixture composed of globules of fat and protein dispersed in water.

Homogeneous mixtures (solutions) can be separated into their component substances by physical processes that rely on differences in some physical property. Two of these processes are distillation and crystallization. *Distillation* makes use of differences in *volatility*, a measure of how easily a substance is converted to a gas at a given temperature. Figure 1.8 shows a simple distillation apparatus for separating a mixture of substances, at least one of which is a liquid. The most volatile component boils first and is condensed back to a liquid in the water-cooled

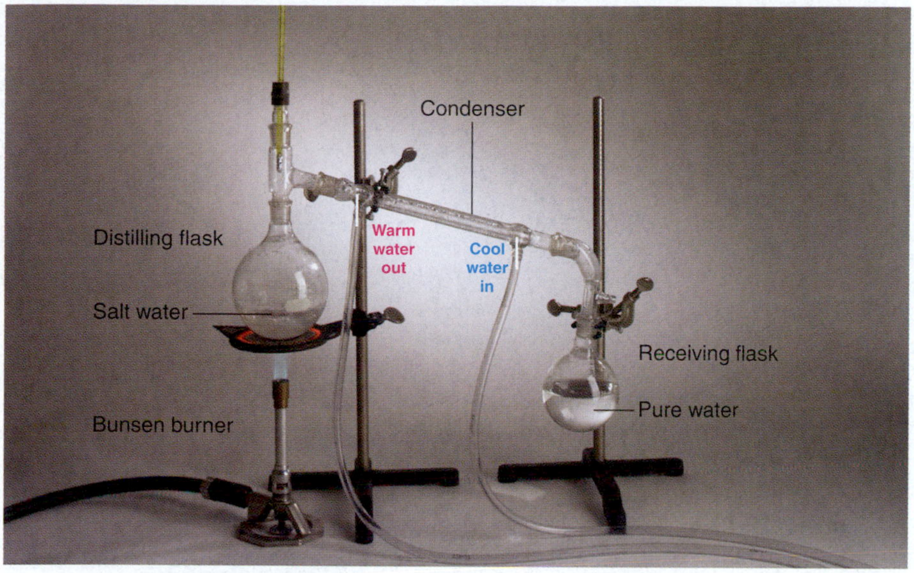

Figure 1.8 Distillation of a solution of table salt in water. The solution of salt in water is heated in the distilling flask until it boils. The resulting vapor is enriched in the more volatile component (water), which condenses to a liquid in the cold condenser and is then collected in the receiving flask.

condenser, from which it flows into the receiving flask. If a solution of salt and water is distilled, for example, the more volatile component, pure water, collects in the receiving flask, while the salt remains in the distillation flask.

Mixtures of two or more liquids with different boiling points (bp) can be separated with a more complex distillation apparatus. One example is the refining of crude petroleum into a range of useful products: aviation fuel, gasoline, kerosene, diesel fuel, and lubricating oil (in approximate order of decreasing volatility). Another example is the distillation of alcoholic spirits such as brandy or whiskey. This relatively simple procedure caused more than a few headaches for federal authorities in the 1920s during the era of Prohibition, when illegal stills proliferated in remote regions of the United States.

Figure 1.9 Crystallization of the compound sodium acetate from a concentrated solution of sodium acetate in water. Addition of a small "seed" crystal (left) causes the compound to form white crystals, which grow and eventually occupy most of the flask (right).

Crystallization separates mixtures based on differences in *solubility*, a measure of how much of a solid substance remains dissolved in a given amount of a specified liquid. Most substances are more soluble at higher temperatures, so a mixture of two or more substances can be dissolved at an elevated temperature and then allowed to cool slowly. Alternatively, the liquid, called the *solvent*, may be allowed to evaporate. In either case, the least soluble of the dissolved substances usually forms crystals first, and these crystals can be removed from the remaining solution by filtration. Figure 1.9 dramatically illustrates the process of crystallization.

Most mixtures can be separated into pure substances, which may be either elements or compounds. An **element** is a substance that cannot be broken down into simpler ones by chemical changes; a **compound** contains two or more elements and has chemical and physical properties that are usually different from those of the elements of which it is composed. With only a few exceptions, a particular compound has the same elemental composition (the same elements in the same proportions) regardless of its source or history. Currently, only about 115 elements are known, but millions of chemical compounds have been prepared. The known elements are listed alphabetically in the table inside the front cover of the text.

In general, compounds can be broken down into their elements by chemical processes. For example, water (a compound) can be decomposed into hydrogen and oxygen (both elements) by a process called *electrolysis*. In electrolysis, electricity provides the energy needed to separate the compound into its constituent elements

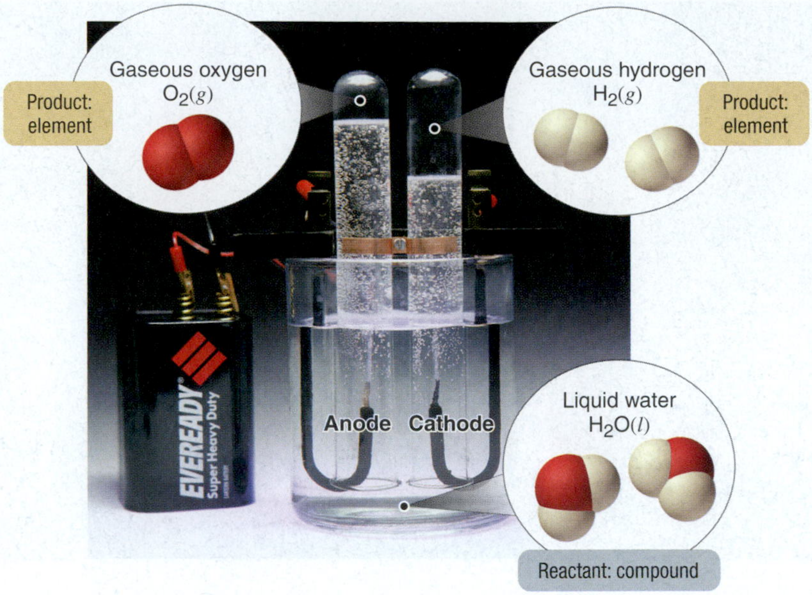

Figure 1.10 Decomposition of water to hydrogen and oxygen using electrical energy (electrolysis). Water is a chemical compound; hydrogen and oxygen are elements.

(Figure 1.10). A similar technique is used on a vast scale to obtain pure aluminum, an element, from its ores, which are mixtures of compounds. Because a great deal of energy is required for electrolysis, the cost of electricity is by far the greatest expense incurred in manufacturing pure aluminum. Thus, recycling aluminum is both cost-effective and ecologically sound.

The overall organization of matter and the methods used to separate mixtures are summarized in Figure 1.11.

EXAMPLE 1.2

Identify each substance as a compound, an element, a heterogeneous mixture, or a homogeneous mixture (solution). **(a)** Filtered tea; **(b)** Freshly squeezed orange juice; **(c)** A compact disk; **(d)** Aluminum oxide, a white powder that contains a 2:3 ratio of aluminum and oxygen atoms; **(e)** Selenium.

Given A chemical substance

Asked for Its classification

Strategy

A Decide whether the substance is chemically pure. If it is, the substance is either an element or a compound. If the substance can be separated into its elements, it is a compound.

B If the substance is not chemically pure, it is either a heterogeneous or homogeneous mixture. If its composition is uniform throughout, it is a homogeneous mixture.

Solution

(a) **A** Tea is a solution of carbon-containing compounds in water, so it is not chemically pure. It is usually separated from tea leaves by filtration. **B** Because the composition of the solution is uniform throughout, it is a homogeneous mixture.

(b) **A** Orange juice contains particles of solid (pulp) as well as liquid; it is not chemically pure. **B** Because its composition is not uniform throughout, orange juice is a heterogeneous mixture.

(c) **A** A compact disk is a solid material that contains more than one element, with regions of different compositions visible along its edge. Hence, a compact disk is not chemically pure. **B** The regions of different composition indicate that a compact disk is a heterogeneous mixture.

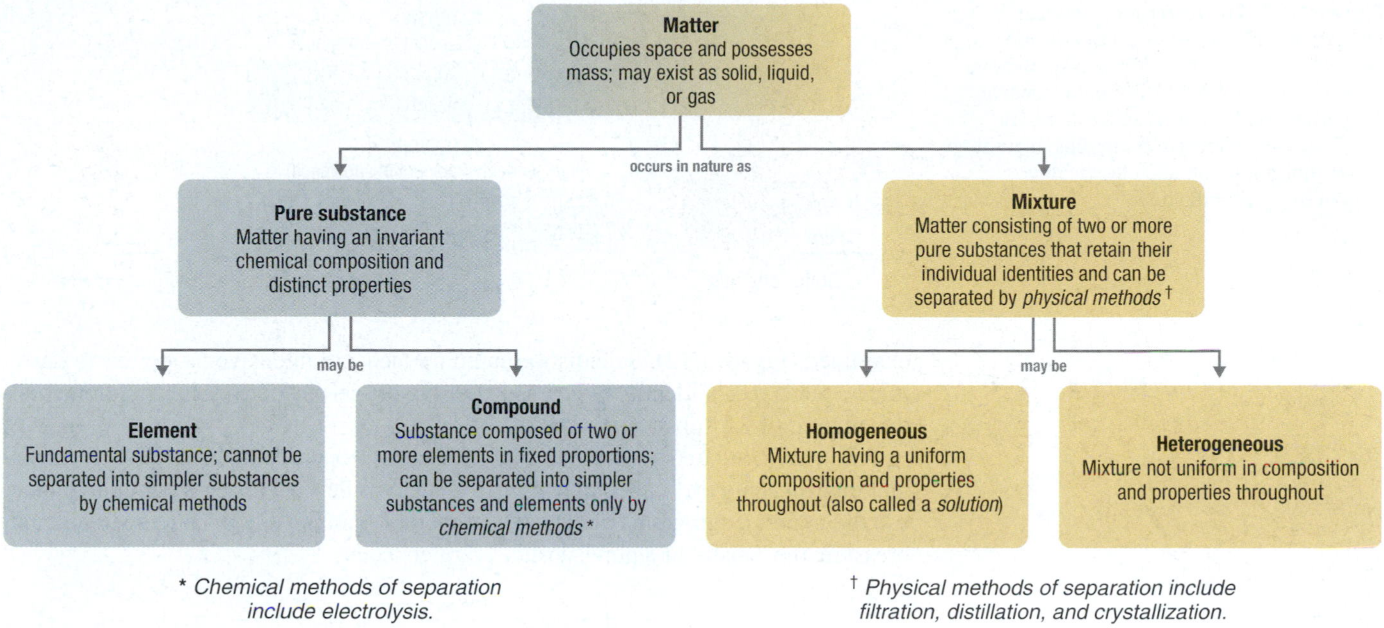

Figure 1.11 **Relationships between the types of matter and the methods used to separate mixtures.**

(d) Aluminum oxide is a single, chemically pure compound.
(e) Selenium is one of the known elements.

EXERCISE 1.2

Identify each substance as a compound, an element, a heterogeneous mixture, or a homogeneous mixture (solution): **(a)** White wine; **(b)** Mercury; **(c)** Ranch-style salad dressing; **(d)** Table sugar (sucrose).

Answer **(a)** solution, **(b)** element, **(c)** heterogeneous mixture, **(d)** compound

Properties of Matter

Physical properties are characteristics that scientists can measure without changing the composition of the sample under study, such as mass, color, and volume, the amount of space occupied by the sample. **Chemical properties** describe the characteristic ability of a substance to react to form new substances; they include its flammability and susceptibility to corrosion. Pure substances are characterized by the chemical and physical properties that are the same for all samples of the substance. For example, pure copper is always a reddish-brown solid (a physical property) and always dissolves in dilute nitric acid to produce a blue solution and a brown gas (a chemical property).

Physical properties can be extensive or intensive. **Extensive properties** vary with the amount of the substance and include mass, weight, and volume. **Intensive properties**, in contrast, do not depend on the amount of the substance; they include color, melting point, boiling point, electrical conductivity, and physical state at a given temperature. For example, elemental sulfur is a yellow crystalline solid that does not conduct electricity and has a melting point of 115.2°C, no matter what amount is

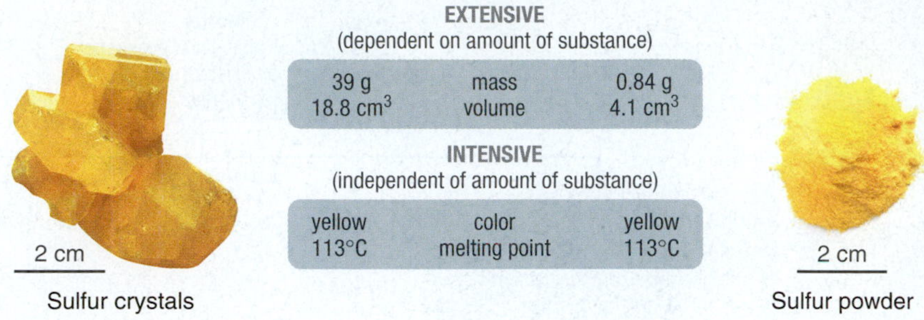

Figure 1.12 The difference between extensive and intensive properties of matter. Because they differ in size, the two samples of sulfur have different extensive properties, such as mass and volume. In contrast, their intensive properties, including color, melting point, and electrical conductivity, are identical.

examined (Figure 1.12). Scientists commonly measure intensive properties to determine a substance's identity, whereas extensive properties convey information about the amount of the substance in the sample.

Although mass and volume are both extensive properties, their ratio is an important intensive property called **density (*d*)**. Density is defined as mass per unit volume, usually expressed in grams per cubic centimeter (g/cm^3). At a given temperature and pressure, the density of a pure substance is a constant:

$$\text{density} = \frac{\text{mass}}{\text{volume}} \qquad d = \frac{m}{V} \tag{1.1}$$

Example 1.3 shows how density measurements can be used to identify pure substances.

EXAMPLE 1.3

The densities of some common liquids are listed in the table. Imagine that you have been given five bottles that contain colorless liquids (labeled *A*–*E*) and are asked to identify them by measuring their density. Using a pipet, a laboratory instrument for accurately measuring and transferring liquids, you carefully measure 25.00 mL of each liquid into five beakers of known mass (1 mL = 1 cm^3). You then weigh each sample on a laboratory balance. Use the tabulated data to calculate the density of each sample. Based solely on your results, can you unambiguously identify all five liquids?*
Masses of samples: *A*, 17.72 g; *B*, 19.75 g; *C*, 24.91 g; *D*, 19.65 g; *E*, 27.80 g

Substance	Density at 25°C, g/cm^3
Water	0.998
Ethanol ("alcohol" in beverages)	0.789
Methanol (wood alcohol)	0.792
Ethylene glycol (used in antifreeze)	1.113
Diethyl ether ("ether"; once widely used as an anesthetic)	0.708
Isopropanol (rubbing alcohol)	0.785

Given Volume and mass

Asked for Density

Strategy

A Calculate the density of each liquid from the volumes and masses given.
B Compare each calculated density with those given in the table. If the calculated density of a liquid is not significantly different from that of one of the liquids given in the table, then the unknown liquid is most likely the corresponding liquid.

* If necessary, review the use of significant figures in calculations in Essential Skills 1 at the end of this chapter prior to working this example.

C If none of the reported densities corresponds to the calculated density, then the liquid cannot be unambiguously identified.

Solution

A Density is mass per unit volume and is usually reported in g/cm^3 (or g/mL because 1 mL = 1 cm^3). The masses of the samples are given in grams, and the volume of all the samples is 25.00 mL (= 25.00 cm^3). The densities are calculated by dividing the mass of each sample by its volume (Equation 1.1). The density of sample *A* is

$$\frac{17.72 \text{ g}}{25.00 \text{ cm}^3} = 0.7088 \text{ g/cm}^3$$

Both the volume and the mass are given to four significant figures, so four significant figures are permitted in the result. The densities of the other samples (in g/cm^3) are *B*, 19.75 g/25.00 cm^3 = 0.7900; *C*, 0.9964; *D*, 0.7860; and *E*, 1.112.

B Comparing these results with the data given in the table shows that sample *A* is probably diethyl ether (0.708 g/cm^3 and 0.7088 g/cm^3 are not substantially different), sample *C* is probably water (0.998 g/cm^3 in the table versus 0.9964 g/cm^3 measured), and sample *E* is probably ethylene glycol (1.113 g/cm^3 in the table versus 1.112 g/cm^3 measured).

C Samples *B* and *D* are more difficult to identify for two reasons. First, both have similar densities (0.7900 and 0.7860 g/cm^3), so they may or may not be chemically identical. Second, within experimental error, the measured densities of *B* and *D* are indistinguishable from the densities of ethanol (0.789 g/cm^3), methanol (0.792 g/cm^3), and isopropanol (0.785 g/cm^3). Thus, some property other than density must be used to decide between them.

EXERCISE 1.3

Given the volumes and masses of five samples of compounds used in blending gasoline, together with the densities of several chemically pure liquids, identify as many of the samples as possible.

Sample	Volume, mL	Mass, g
A	337	250.0
B	972	678.1
C	243	190.9
D	119	103.2
E	499	438.7

Substance	Density, g/cm^3
Benzene	0.8787
Toluene	0.8669
m-Xylene	0.8684
Isooctane	0.6979
Methyl *t*-butyl ether	0.7405
t-Butyl alcohol	0.7856

Answer *A*, methyl *t*-butyl ether; *B*, isooctane; *C*, *t*-butyl alcohol; *D*, toluene or *m*-xylene; *E*, benzene

1.4 • A Brief History of Chemistry

MGC The Law of Definite Proportions

It was not until the era of the ancient Greeks that we have any record of how people tried to explain the chemical changes that they observed and utilized. At that time, natural objects were thought to consist of only four basic elements: earth, air, fire, and water. Then, in the fourth century B.C., two Greeks named Democritus and Leucippus suggested that matter was not infinitely divisible into smaller particles but instead consisted of fundamental, indivisible particles they called **atoms**. Unfortunately, these early philosophers did not have the technology to test their hypothesis. They

Figure 1.13 An alchemist at work. Alchemy was a form of chemistry that flourished during the Middle Ages and Renaissance. Although some alchemists were frauds, others made major contributions, including the discovery of several elements and the preparation of strong acids.

would have been unlikely to do so in any case, because the ancient Greeks did not conduct experiments or use the scientific method. They believed that the nature of the universe could be discovered by rational thought alone.

Over the next two millennia, the major advances in chemistry were achieved by *alchemists*, who engaged in a form of chemistry and speculative philosophy during the Middle Ages and Renaissance. Their major goal was to convert certain elements into others by a process they called *transmutation* (Figure 1.13). In particular, alchemists wanted to find a way to transform cheaper metals into gold. Although most alchemists did not approach chemistry systematically and many appear to have been outright frauds, alchemists in China, the Arab kingdoms, and medieval Europe made major contributions, including the discovery of elements such as quicksilver (mercury) and the preparation of several strong acids.

Modern Chemistry

The 16th and 17th centuries saw the beginnings of what we now recognize as modern chemistry. During this period, great advances were made in *metallurgy*, the extraction of metals from ores, and the first systematic quantitative experiments were carried out. In 1661, the Englishman Robert Boyle (1627–1691) published "The Skeptical Chymist," which described the relationship between the pressure and volume of air. More important, it was Boyle who defined an element as a substance that cannot be broken down into two or more simpler substances by chemical means. This led to the identification of a large number of elements, many of which were metals. Ironically, Boyle himself never thought that metals were elements.

During the 18th century, the English clergyman Joseph Priestley (1733–1804) discovered oxygen gas and found that many carbon-containing materials burn vigorously in an oxygen atmosphere, a process called *combustion*. Priestley also discovered that the gas produced by fermenting beer, which we now know to be carbon dioxide, is the same as one of the gaseous products of combustion. Priestley's studies of this gas did not continue as he would have liked, however. After he fell into a vat of fermenting beer, brewers prohibited him from working in their factories. Although Priestley did not understand its identity, he found that carbon dioxide dissolved in water to produce seltzer water. In essence, he may be considered the founder of the multi-billion-dollar carbonated soft drink industry.

Despite the pioneering studies of Priestley and others, a clear understanding of combustion remained elusive. In the late 18th century, however, the French scientist Antoine Lavoisier (1743–1794) showed that combustion is the reaction of a carbon-containing substance with oxygen to form carbon dioxide and water and that life depends on a similar reaction, which today we call *respiration*. Lavoisier also wrote the first modern chemistry text and is widely regarded as the father of modern chemistry. His most important contribution was the *law of conservation of mass*, which states that in any chemical reaction, the mass of the substances that react equals the mass of the products that are formed. Unfortunately, Lavoisier invested in a private corporation that collected taxes for the Crown, and royal tax collectors were not popular during the French Revolution. He was executed on the guillotine at age 51, prematurely terminating his contributions to chemistry.

The Atomic Theory of Matter

In 1803, the English schoolteacher John Dalton (1766–1844) expanded Proust's development of the law of definite proportions (Section 1.2) and Lavoisier's findings on the conservation of mass in chemical reactions to propose that elements consist of indivisible particles that he called *atoms* (taking the term from

Democritus and Leucippus). Dalton's *atomic theory of matter* contains four fundamental hypotheses:

1. All matter is composed of tiny indivisible particles called atoms.
2. All atoms of an element are identical in mass and chemical properties, whereas atoms of different elements differ in mass and fundamental chemical properties.
3. A chemical compound is a substance that always contains the same atoms in the same ratio.
4. In chemical reactions, atoms from one or more compounds or elements redistribute or rearrange in relation to other atoms to form one or more new compounds. *Atoms themselves do not undergo a change of identity in chemical reactions.*

This last hypothesis suggested that the alchemists' goal of transmuting other elements to gold was impossible, at least through chemical reactions. We now know that Dalton's atomic theory is essentially correct, with four minor modifications:

a. Not all atoms of an element must have precisely the same mass.
b. Atoms of one element can be transformed into another through nuclear reactions.
c. The compositions of many solid compounds are somewhat variable.
d. Under certain circumstances, some atoms can be divided (split into smaller particles).

These modifications illustrate the effectiveness of the scientific method: later experiments and observations were used to refine Dalton's original theory.

The Law of Multiple Proportions

Despite the clarity of his thinking, Dalton could not use his chemical theory to determine the elemental compositions of chemical compounds because he had no reliable scale of atomic masses; that is, he did not know the relative masses of elements such as carbon and oxygen. For example, he knew that the gas we now call carbon monoxide contained carbon and oxygen in the ratio 1:1.33 by mass and that a second compound, the gas we call carbon dioxide, contained carbon and oxygen in the ratio 1:2.66. Because 2.66/1.33 = 2.00, the second compound contained twice as many atoms of oxygen per atom of carbon as did the first. But what was the correct formula for each of these compounds? If the first compound consisted of particles that contain 1 carbon atom and 1 oxygen atom, the second must consist of particles that contain 1 carbon atom and 2 oxygen atoms. If the first compound had 2 carbon and 1 oxygen, the second must have 2 carbon and 2 oxygen. If the first had 1 carbon and 2 oxygen, the second would have 1 carbon and 4 oxygen, and so forth. Dalton had no way to distinguish among these or more complicated alternatives. However, these data led to a general statement that is now known as the *law of multiple proportions*: When two elements form a series of compounds, the ratios of the masses of the second element that are present per gram of the first element can almost always be expressed as the ratios of integers. (The same law holds for mass ratios of compounds forming a series that contains more than two elements.)

EXAMPLE 1.4

A chemist is studying a series of simple compounds of carbon and hydrogen. The table lists the masses of hydrogen that combine with 1 g of carbon to form the first three members of the series.

Compound	Mass of Hydrogen, g
A	0.0839
B	0.1678
C	0.2520
D	———

(a) Determine whether these data follow the law of multiple proportions; **(b)** Calculate the mass of hydrogen that would combine with 1 g of carbon to form the fourth compound in the series, *D*.

Given Mass of hydrogen per gram of carbon for three compounds

Asked for (a) Ratios of masses of hydrogen to carbon. (b) Mass of hydrogen per gram of carbon for fourth compound in series.

Strategy

A Select the lowest mass to use as the denominator, and then calculate the ratio of each of the other masses to that mass. Include other ratios if appropriate.
B If the ratios are small whole integers, the data follow the law of multiple proportions.
C Decide whether the ratios form a numerical series. If they do, then determine the next member of that series and predict the ratio corresponding to the next compound in the series.
D Use proportions to calculate the mass that corresponds to that compound.

Solution

A Compound *A* has the lowest mass of hydrogen, so we use it as the denominator. The ratios of the masses of hydrogen that combine with 1 g of carbon are therefore

$$\frac{C}{A} = \frac{0.2520\text{ g}}{0.0839\text{ g}} = 3.00 = \frac{3}{1} \qquad \frac{B}{A} = \frac{0.1678\text{ g}}{0.0839\text{ g}} = 2.00 = \frac{2}{1} \qquad \frac{C}{B} = \frac{0.2520\text{ g}}{0.1678\text{ g}} = 1.502 \approx \frac{3}{2}$$

(a) **B** The ratios of the masses of hydrogen that combine with 1 g of carbon are indeed small whole integers, as predicted by the law of multiple proportions.
(b) **C** The ratios *B/A* and *C/A* form the series 2/1, 3/1, so the next member of the series should be *D/A* = 4/1. **D** Thus, compound *D* should be formed by combining 4 × 0.0839 g = 0.336 g of hydrogen with 1 g of carbon. Such a compound does exist. It is *methane*, the major constituent of natural gas.

EXERCISE 1.4

Four compounds containing only sulfur and fluorine are known. The table lists the mass of fluorine that combines with 1 g of sulfur to form each compound.

Compound	Mass of Fluorine, g
A	3.54
B	2.96
C	2.36
D	0.59

(a) Determine the ratios of the masses of fluorine that combine with 1 g of sulfur in these compounds. Are these data consistent with the law of multiple proportions?; **(b)** Calculate the mass of fluorine that would combine with 1 g of sulfur to form the next compounds in the series, *E* and *F*.

Answer **(a)** *A/D* = 6.0 or 6/1; *B/D* ≈ 5.0, or 5/1; *C/D* = 4.0, or 4/1; yes; **(b)** Ratios of 3.0 and 2.0 give 1.8 g and 1.2 g of fluorine/g of sulfur, respectively. (Neither of these compounds is yet known.)

Avogadro's Hypothesis

The French chemist Joseph Gay-Lussac (1778–1850) carried out a series of experiments in an attempt to establish the formulas of chemical compounds.

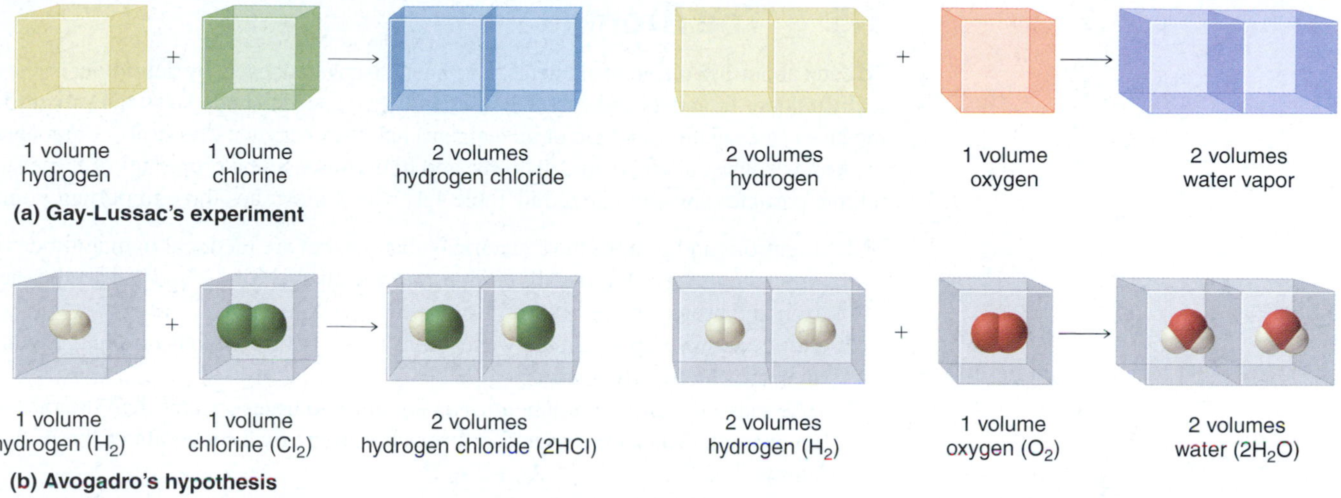

Figure 1.14 Gay-Lussac's experiments with chlorine gas and hydrogen gas. (a) One volume of chlorine gas reacted with 1 volume of hydrogen gas to produce 2 volumes of hydrogen chloride gas, and 1 volume of oxygen gas reacted with 2 volumes of hydrogen gas to produce 2 volumes of water vapor. (b) A summary of Avogadro's hypothesis, which interpreted Gay-Lussac's results in terms of atoms. Note that the simplest way for 2 molecules of hydrogen chloride to be produced is if hydrogen and chlorine each consist of molecules that contain 2 atoms of the element.

Under conditions of constant temperature and pressure, he carefully measured the *volumes* of gases that reacted to make a given chemical compound, together with the volumes of the products if they were gases. Gay-Lussac found, for example, that 1 volume of chlorine gas always reacted with 1 volume of hydrogen gas to produce 2 volumes of hydrogen chloride gas. Similarly, 1 volume of oxygen gas always reacted with 2 volumes of hydrogen gas to produce 2 volumes of water vapor (Figure 1.14a).

Gay-Lussac's results did not by themselves reveal the formulas for hydrogen chloride and water. The key insight that led to the exact formulas was provided by the Italian chemist Amadeo Avogadro (1776–1856), who proposed that *equal volumes of different gases contain equal numbers of gas particles* when measured at the same temperature and pressure. *Avogadro's hypothesis*, which explained Gay-Lussac's results, is summarized here and in Figure 1.14b:

1 (volume or particle of) hydrogen + 1 (volume or particle of) chlorine ⟶ 2 (volumes or particles of) hydrogen chloride

If Dalton's theory of atoms was correct, then each particle of hydrogen or chlorine had to contain *at least 2 atoms* of hydrogen or chlorine because 2 particles of hydrogen chloride were produced. The simplest, but not the only, explanation was that hydrogen and chlorine contained 2 atoms each (that is, they were *diatomic*) and that hydrogen chloride contained 1 atom each of hydrogen and chlorine. Applying this reasoning to Gay-Lussac's results with hydrogen and oxygen leads to the conclusion that water contains 2 atoms of hydrogen per atom of oxygen. Unfortunately, because no data supported Avogadro's hypothesis that equal volumes of gases contained equal numbers of particles, his explanations and formulas for simple compounds were not generally accepted for more than 50 years. Dalton and many others continued to believe that water particles contained 1 hydrogen atom and 1 oxygen atom, rather than 2 atoms of hydrogen and 1 of oxygen. The historical development of the concept of the atom is summarized in Figure 1.15.

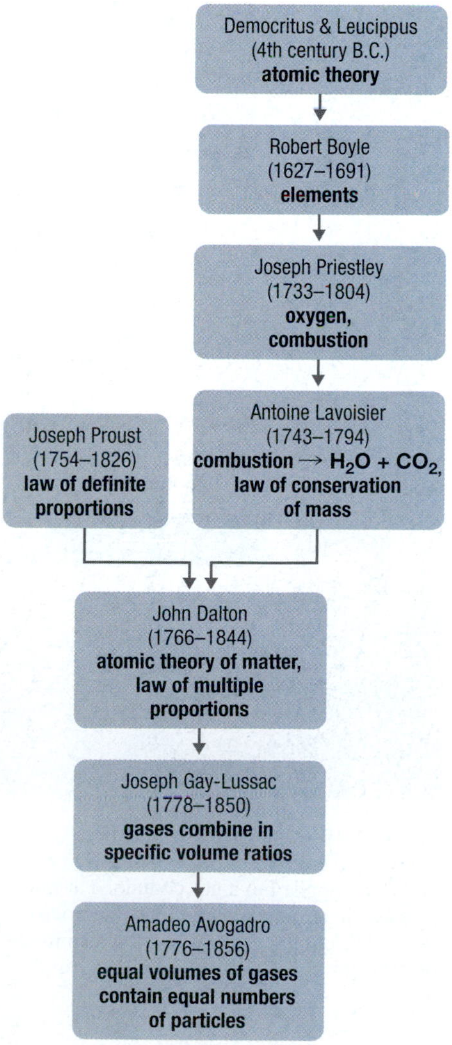

Figure 1.15 A summary of the historical development of the concept of the atom.

1.5 The Atom

MGC The Atom

To date, about 115 different elements have been discovered; each, by definition, is chemically unique. To understand why they are unique, you need to understand the structure of the atom (the smallest particle of an element) and the characteristics of its components.

Atoms consist of **electrons, protons**, and **neutrons**.* Some properties of these subatomic particles are summarized in Table 1.1, which illustrates three important points.

1. Electrons and protons have electrical charges that are identical in magnitude but opposite in sign. We usually assign *relative* charges of –1 and +1 to the electron and proton, respectively.
2. Neutrons have approximately the same mass as protons but no charge. Thus, they are electrically neutral.
3. The mass of a proton or a neutron is about 1836 times greater than the mass of an electron. Thus, protons and neutrons constitute by far the bulk of the mass of atoms.

The discovery of the electron and the proton was crucial to the development of the modern model of the atom and provides an excellent case study in the application of the scientific method. In fact, the elucidation of the atom's structure is one of the greatest detective stories in the history of science.

TABLE 1.1 Properties of subatomic particles [a]

Particle	Mass, g	Atomic Mass, amu	Electrical Charge, coulombs	Relative Charge
Electron	9.109×10^{-28}	0.0005486	-1.602×10^{-19}	−1
Proton	1.673×10^{-24}	1.007276	$+1.602 \times 10^{-19}$	+1
Neutron	1.675×10^{-24}	1.008665	0	0

[a] For a review of the use of scientific notation and units of measurement, see Essential Skills 1 at the end of this chapter.

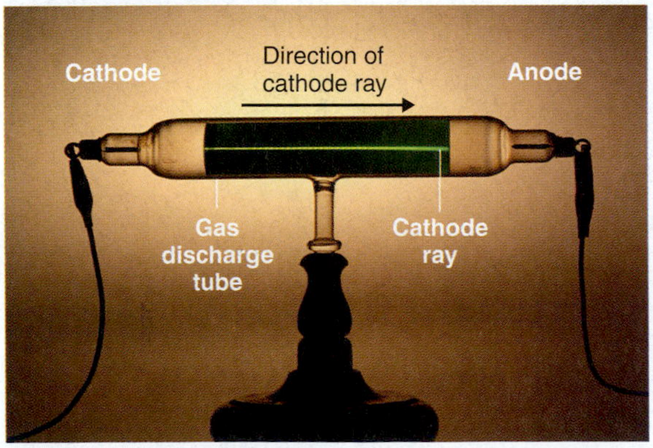

Figure 1.16 A gas discharge tube producing cathode rays. When a high voltage is applied to a gas contained at low pressure in a gas discharge tube, electricity flows through the gas and energy is emitted in the form of light.

The Electron

Long before the end of the 19th century, it was well known that applying a high voltage to a gas contained at low pressure in a sealed tube called a gas discharge tube caused electricity to flow through the gas, which then emitted light (Figure 1.16). Researchers trying to understand this phenomenon found that in addition an unusual form of energy was emitted from the *cathode*, or negatively charged electrode; hence, this form of energy was called *cathode rays*. In 1897, the British physicist J. J. Thomson (1856–1940) demonstrated that cathode rays could be deflected, or bent, by magnetic or electric fields, which indicated that the cathode rays consisted of charged particles (Figure 1.17). More important, by measuring the extent of the deflection of the cathode rays in magnetic or electric fields of various strengths, Thomson was able to calculate the *mass-to-charge ratio* of the particles. These particles were emitted by the negatively charged cathode and repelled by the negative terminal of an electric field. Because like charges repel each other and opposite charges attract, Thomson concluded that the particles had a net negative charge; we now call these particles *electrons*. Most important for chemistry, Thomson found that the mass-to-charge ratio of cathode rays was independent of the nature of the metal

* This is an oversimplification that ignores the other subatomic particles that have been discovered, but it is sufficient for our discussion of chemical principles.

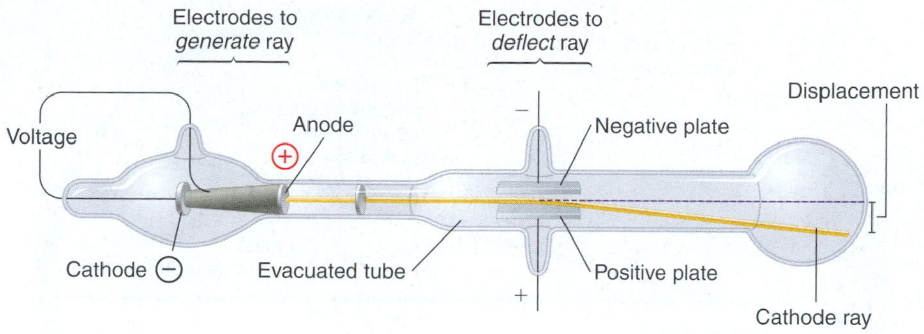

Figure 1.17 Deflection of cathode rays by an electric field. As the cathode rays travel toward the right, they are deflected toward the positive electrode (+), demonstrating that they are negatively charged.

electrodes or the gas, which suggested that electrons were fundamental components of all atoms.

Subsequently, the American scientist Robert Millikan (1868–1953) carried out a series of experiments using electrically charged oil droplets, which allowed him to calculate the charge on a single electron. With this information and Thomson's mass-to-charge ratio, Millikan determined the mass of an electron:

$$\frac{\text{mass}}{\cancel{\text{charge}}} \times \cancel{\text{charge}} = \text{mass}$$

It was at this point that two separate lines of investigation began to converge, both aimed at determining how and why energy is emitted by matter.

Radioactivity

The second line of investigation began in 1896, when the French physicist Henri Bequerel (1852–1908) discovered that certain minerals, such as uranium salts, emitted a new form of energy. Bequerel's work was greatly extended by Marie Curie (1867–1934) and her husband, Pierre (1854–1906); all three shared the Nobel Prize in physics in 1903. Marie Curie coined the term **radioactivity** (from the Latin *radius*, "ray") to describe the emission of energy rays by matter. She found that one particular uranium ore, pitchblende, was substantially more radioactive than most, which suggested that it contained one or more highly radioactive impurities. Starting with several tons of pitchblende, the Curies isolated two new radioactive elements after months of work: polonium, named for Marie's native Poland, and radium, named for its intense radioactivity. Pierre Curie carried a vial of radium in his coat pocket to demonstrate its greenish glow, a habit that caused him to become ill from radiation poisoning before he was run over by a horse-drawn wagon and killed instantly in 1906. Marie Curie, in turn, died of what was almost certainly radiation poisoning.

Radium bromide illuminated by its own radioactive glow. A 1922 photo, taken in the dark, in the Curie Laboratory.

Building on the Curies' work, the British physicist Ernest Rutherford (1871–1937) performed decisive experiments that led to the modern view of the structure of the atom. While working in J. J. Thomson's laboratory shortly after Thomson discovered the electron, Rutherford showed that compounds of uranium and other elements emitted at least two distinct types of radiation. One was readily absorbed by matter and seemed to consist of particles that had a positive charge and were massive compared to electrons. Because it was the first kind of radiation to be discovered, Rutherford called these substances α *particles*. Rutherford also showed that the particles in the second type of radiation, β *particles*, had the same charge and mass-to-charge ratio as Thomson's electrons; they are now known to be high-speed electrons. A third type of radiation, γ *rays*, was discovered somewhat later and found to be similar to a lower-energy form of radiation called X rays, now used to produce images of bones and teeth.

These three kinds of radiation—α particles, β particles, and γ rays—are readily distinguished by the way they are deflected by an electric field and by the degree to

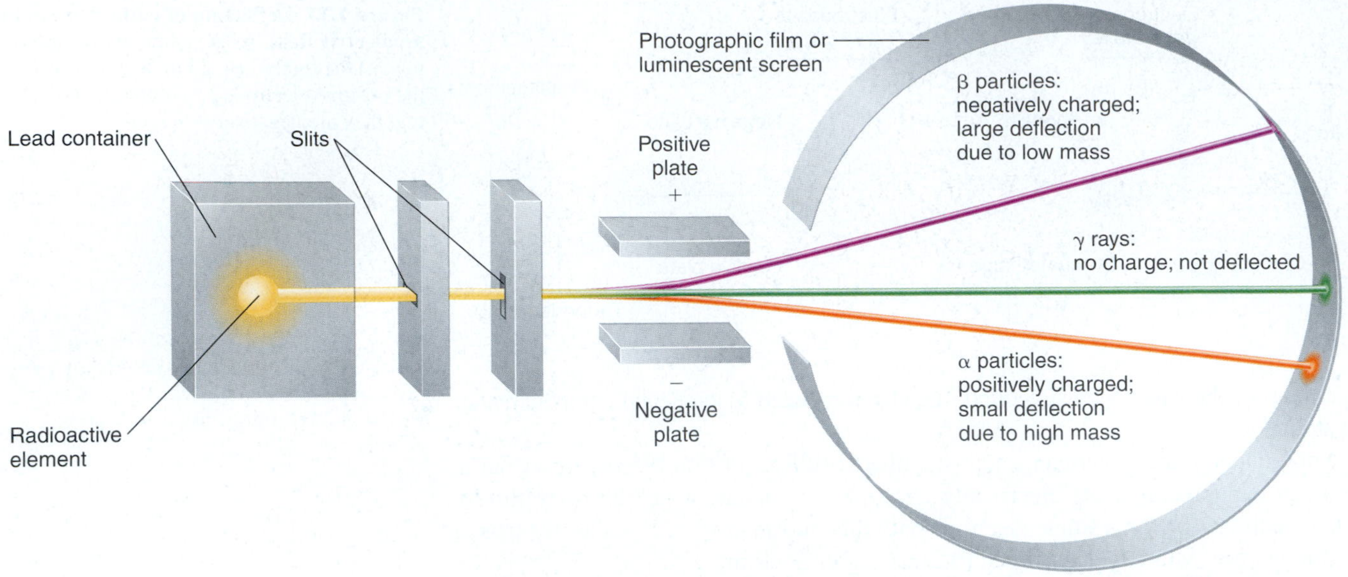

Figure 1.18 Effect of an electric field on α particles, β particles, and γ rays. Negatively charged β particles are deflected by a negative electrode, whereas positively charged α particles are deflected by a positive electrode. Uncharged γ rays are unaffected by an electric field. (Relative deflections are not shown to scale.)

which they penetrate matter. As Figure 1.18 illustrates, α particles and β particles are deflected in opposite directions; α particles are deflected to a much lesser extent because of their higher mass-to-charge ratio. In contrast, γ rays have no charge, so they are not deflected by electric or magnetic fields. Figure 1.19 shows that α particles have the least penetrating power and are stopped by a sheet of paper, whereas β particles can pass through thin sheets of metal but are absorbed by lead foil or even thick glass. In contrast, γ rays are able to penetrate matter readily; thick blocks of lead or concrete are needed to stop them.

The Atomic Model

Once scientists had concluded that all matter contains negatively charged electrons, it became clear that atoms, which are electrically neutral, must also contain positive charges to balance the negative ones. Thomson proposed that the electrons were embedded in a uniform sphere that contained both the positive charge and most of the mass of the atom, much like raisins in plum pudding or chocolate chips in a cookie (Figure 1.20).

In a single famous experiment, however, Rutherford showed unambiguously that Thomson's model of the atom could not be correct. Rutherford aimed a stream of α particles at a very thin gold foil target (Figure 1.21a) and examined how the α particles

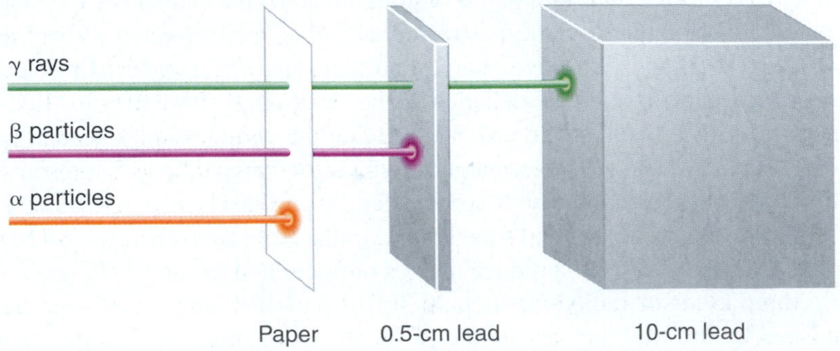

Figure 1.19 Relative penetrating power of the three types of radiation. Comparatively massive α particles are stopped by a sheet of paper, whereas β particles easily penetrate paper but are stopped by a thin piece of lead foil. Uncharged γ rays penetrate the paper and lead foil; a much thicker piece of lead or concrete is needed to absorb them.

were scattered by the foil. If Thomson's model of the atom were correct, the positively charged α particles should crash through the uniformly distributed mass of the gold target like cannonballs through the side of a wooden house. They might be moving a little slower when they emerged, but they should pass essentially straight through the target (Figure 1.21b). Instead, to Rutherford's amazement, a small fraction of the α particles were deflected at large angles, and a few were reflected directly back at the source (Figure 1.21c). According to Rutherford, "It was almost as incredible as if you fired a 15-inch shell at a piece of tissue paper and it came back and hit you."

Rutherford's results were *not* consistent with a model in which the mass and positive charge are distributed uniformly throughout the volume of the atom. Instead, they strongly suggested that both the mass and positive charge are concentrated in a tiny fraction of the volume of the atom, which Rutherford called the **nucleus**. It made sense that a small fraction of the α particles collided with the dense, positively charged nuclei in either a glancing fashion, resulting in large deflections, or almost head-on, causing them to be reflected straight back at the source.

Although Rutherford could not explain why repulsions between the positive charges in nuclei that contained more than one positive charge did not cause the nucleus to disintegrate, he reasoned that repulsions between negatively charged electrons would cause the electrons to be uniformly distributed throughout the atom's volume.* For this and other insights, Rutherford was awarded the Nobel Prize in

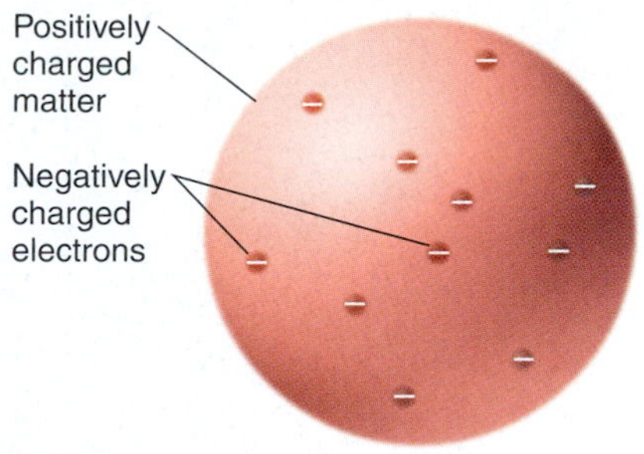

Figure 1.20 Thomson's "plum pudding" or "chocolate chip cookie" model of the atom. In this model, the electrons were embedded in a uniform sphere of positive charge.

Figure 1.21 A summary of Rutherford's experiments. (a) A representation of the apparatus he used to detect deflections in a stream of α particles aimed at a thin gold foil target. The particles were produced by a sample of radium. (b) If Thomson's model of the atom were correct, the α particles should have passed straight through the gold foil. (c) In fact, a small number of α particles were deflected in various directions, including right back at the source. This could be true only if the positive charge were much more massive than the α particle. It suggested that the mass of the gold atom is concentrated in a very small region of space, which he called the nucleus.

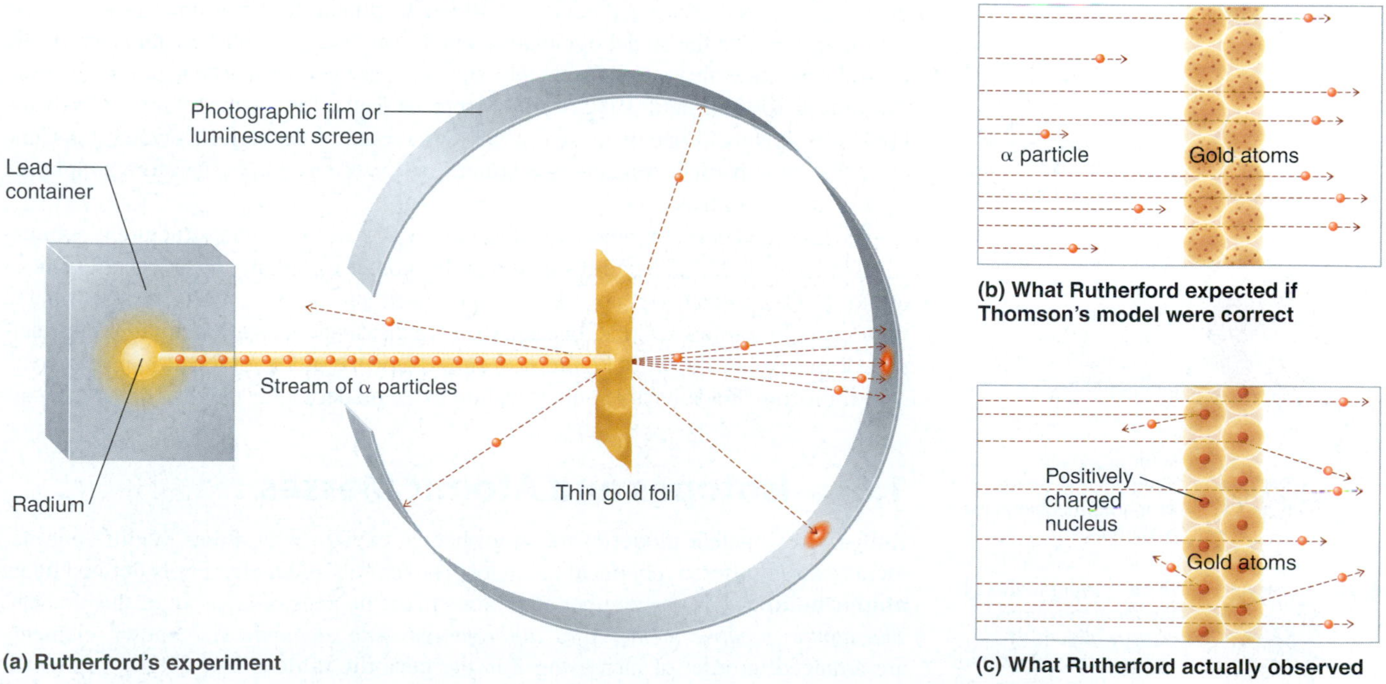

* Today we know that *strong nuclear forces*, which are much stronger than electrostatic interactions, hold the protons and neutrons together in the nucleus.

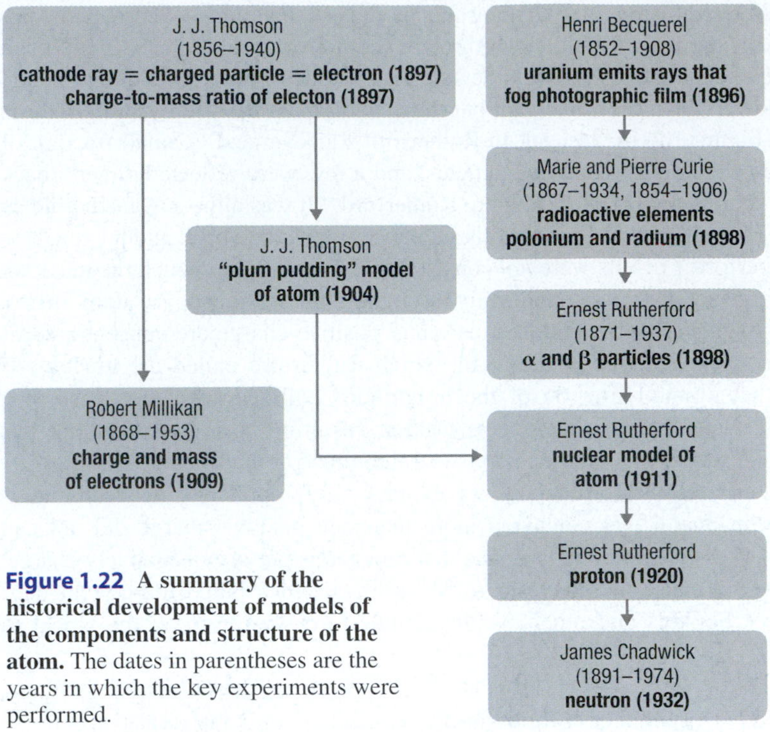

Figure 1.22 A summary of the historical development of models of the components and structure of the atom. The dates in parentheses are the years in which the key experiments were performed.

chemistry in 1908. Unfortunately, Rutherford would have preferred to receive the Nobel Prize in physics because he thought that physics was superior to chemistry. In his opinion, "All science is either physics or stamp collecting." (The authors of this text do *not* share Rutherford's view!)

Subsequently, Rutherford established that the nucleus of the hydrogen atom was indeed a positively charged particle, for which he coined the name *proton* in 1920. He also suggested that the nuclei of elements other than hydrogen must contain electrically neutral particles with approximately the same mass as the proton. The neutron, however, was not discovered until 1932, by a student of Rutherford named James Chadwick (1891–1974; Nobel Prize in physics, 1935). As a result of Rutherford's work, it became clear that an α particle contains two protons and neutrons and is therefore simply the nucleus of a helium atom.

The historical development of the different models of the atom's structure is summarized in Figure 1.22. Rutherford's model of the atom is essentially the same as the modern one, except that we now know that electrons are *not* uniformly distributed throughout the atom's volume. Instead, they are distributed according to a set of principles described in Chapter 6. Figure 1.23 shows how the model of the atom has evolved over time from the indivisible unit of Dalton to the modern view taught in classes today.

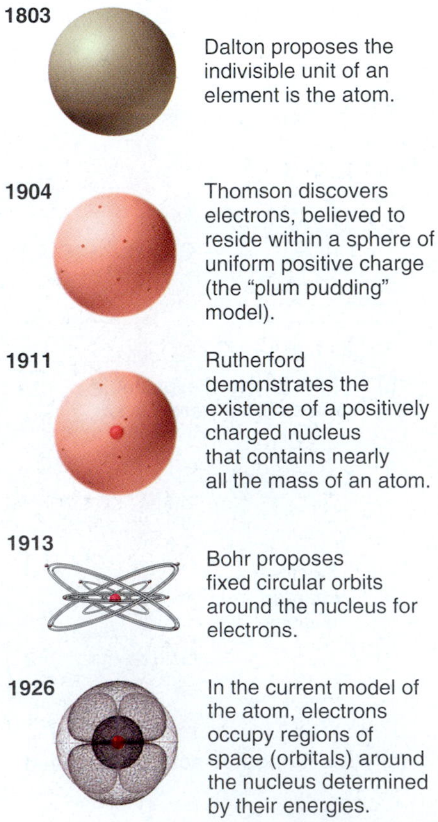

Figure 1.23 The evolution of atomic theory, as illustrated by models of the oxygen atom. Bohr's model and the current model are described in Chapter 6.

1.6 • Isotopes and Atomic Masses

Rutherford's nuclear model of the atom helped explain why atoms of different elements exhibit different chemical behavior. The identity of an element is defined by its **atomic number (*Z*)**, the number of protons in the nucleus of an atom of the element. *The atomic number is therefore different for each element.* The known elements are arranged in order of increasing *Z* in the **periodic table** (Figure 1.24),* in which each element is assigned a unique one-, two-, or three-letter symbol. The names of the elements are listed alphabetically in the table inside the front cover of the text, along

* We will explain the rationale for the peculiar format of the periodic table in Chapter 6.

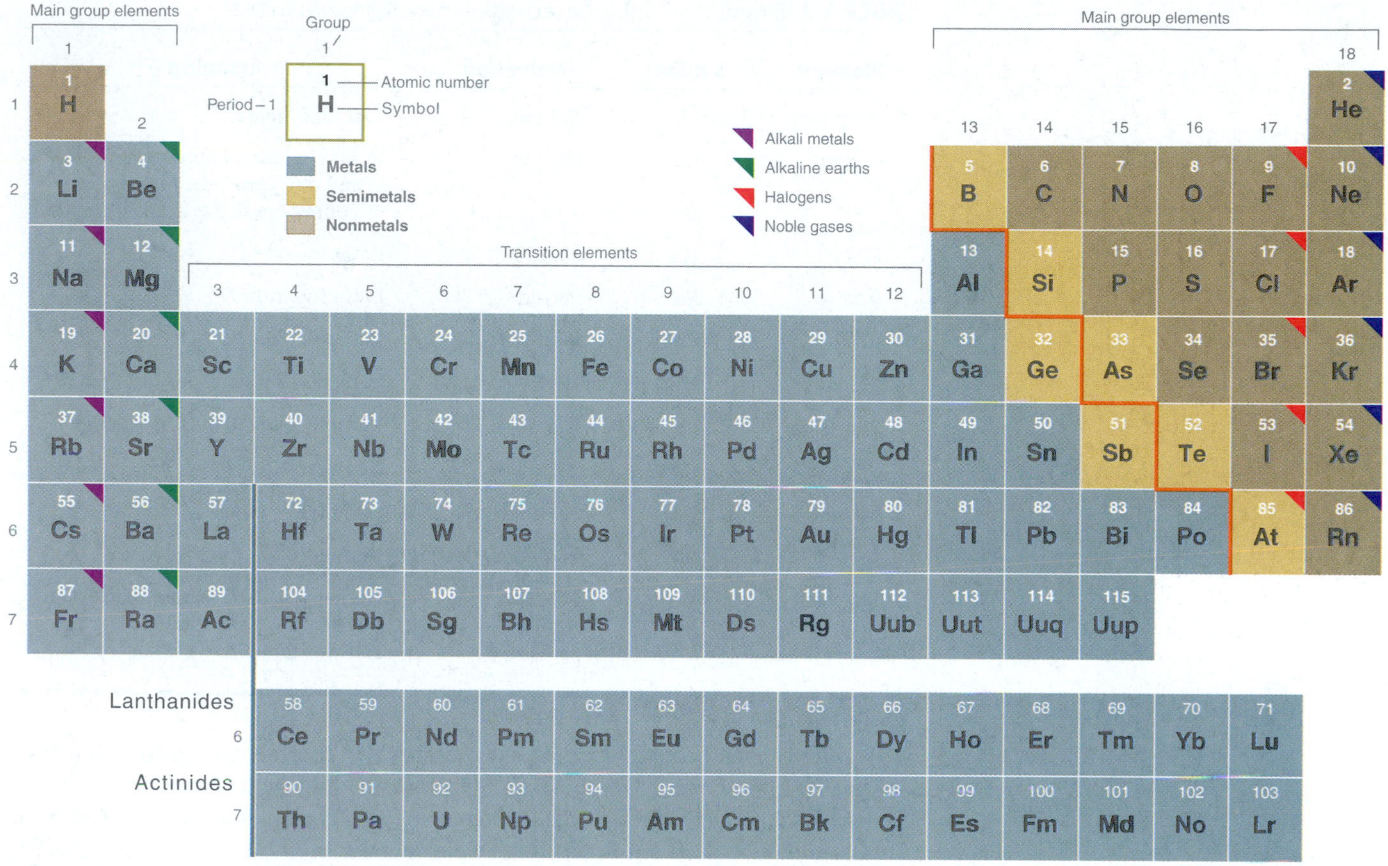

Figure 1.24 The periodic table, showing the elements in order of increasing *Z*. The chemistry of each element is determined by its number of protons and electrons. As described in Section 1.7, the metals are on the bottom left of the periodic table and the nonmetals are at the top right. The semimetals lie along a diagonal line separating the metals and nonmetals.

with their symbols, atomic numbers, and atomic masses. The chemistry of each element is determined by its number of protons and electrons; note that in a neutral atom the number of electrons equals the number of protons.

MGC Isotopes

In most cases, the symbols for the elements are derived directly from the element's name, such as C for carbon, U for uranium, Ca for calcium, and Po for polonium. Elements have also been named for their properties [such as radium (Ra) for its radioactivity], for the native country of the scientist(s) who discovered them [polonium (Po) for Poland], for eminent scientists [curium (Cm) for the Curies], for gods and goddesses [selenium (Se) for the Greek goddess of the moon, Selene], and for other poetic or historical reasons. Some of the symbols used for those elements that have been known since antiquity are derived from historical names that are no longer in use; only the symbols remain to remind us of their origin. Examples are Fe for iron, from the Latin *ferrum*; Na for sodium, from the Latin *natrium*; and W for tungsten, from the German *wolfram*. Examples are listed in Table 1.2. As you work through this text, you will encounter the names and symbols of the elements repeatedly, and much as you become familiar with characters in a play or a film, their names and symbols will become familiar.

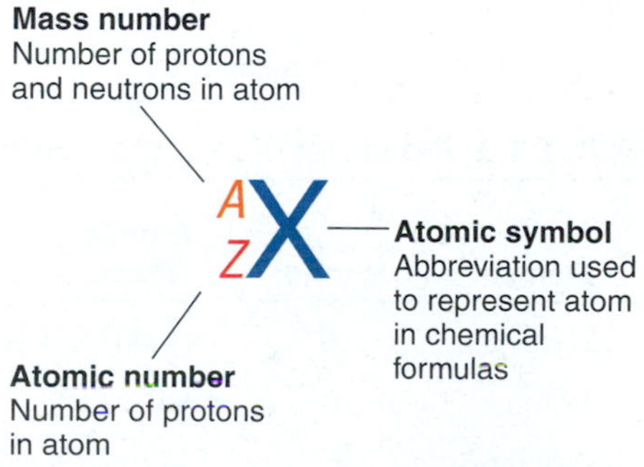

Recall from Section 1.5 that the nuclei of most atoms contain neutrons as well as protons. Unlike protons, the number of neutrons is not absolutely fixed for most elements. Atoms that have *the same number of protons*, and hence the same atomic number, but *different numbers of neutrons* are called **isotopes**. All isotopes of an element have the same number of protons and electrons, which means they exhibit the same chemistry. The isotopes of an element differ only in their atomic mass, which is given by the **mass number (*A*)**, the sum of the numbers of protons and neutrons.

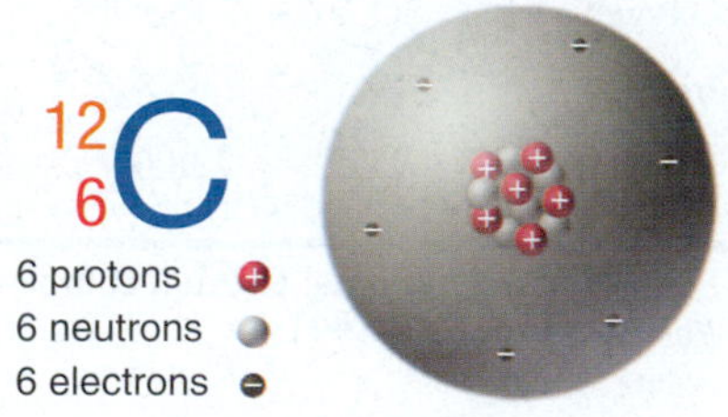

TABLE 1.2 Element symbols based on names no longer in use

Element	Symbol	Derivation	Meaning
Antimony	Sb	Stibium	Latin for "mark"
Copper	Cu	Cuprum	From *Cyprium*, Latin name for the island of Cyprus, the major source of copper ore in the Roman Empire
Gold	Au	Aurum	Latin for "gold"
Iron	Fe	Ferrum	Latin for "iron"
Lead	Pb	Plumbum	Latin for "heavy"
Mercury	Hg	Hydrargyrum	Latin for "liquid silver"
Potassium	K	Kalium	From the Arabic *al-qili*, "alkali"
Silver	Ag	Argentum	Latin for "silver"
Sodium	Na	Natrium	Latin for "sodium"
Tin	Sn	Stannum	Latin for "tin"
Tungsten	W	Wolfram	German for "wolf stone" because it interfered with the smelting of tin and was thought to devour the tin

The element carbon (C) has an atomic number Z of 6, which means that all neutral atoms of carbon contain 6 protons and 6 electrons. In a typical sample of carbon-containing material, 98.89% of the carbon atoms also contain 6 neutrons, so each has a mass number A of 12. An isotope of any element can be uniquely represented as ${}^{A}_{Z}X$, where X is the atomic symbol of the element. The isotope of carbon that has 6 neutrons is therefore ${}^{12}_{6}C$. The subscript indicating the atomic number Z is actually redundant because the atomic symbol already uniquely specifies Z. Consequently, ${}^{12}_{6}C$ is more often written as ${}^{12}C$, which is read as "carbon-12." Nevertheless, the value of Z is commonly included in the notation for *nuclear* reactions because these reactions involve changes in Z, as described in Chapter 20.

In addition to ${}^{12}C$, a typical sample of carbon contains 1.11% ${}^{13}_{6}C$ (${}^{13}C$), with 7 neutrons and 6 protons, and a trace of ${}^{14}_{6}C$ (${}^{14}C$), with 8 neutrons and 6 protons. The nucleus of ${}^{14}C$ is not stable, however, but undergoes a slow radioactive decay that is the basis of the carbon-14 dating technique used in archaeology (see Chapter 14). Many elements other than carbon have more than one stable isotope; tin, for example, has ten. The properties of some common isotopes are listed in Table 1.3.

TABLE 1.3 Properties of selected isotopes

Element	Symbol	Atomic Mass	Isotope Mass Number	Isotope Masses	Percent Abundances
Hydrogen	H	1.0079	1, 2	1.007825, 2.014102	99.9855, 0.0115
Boron	B	10.81	10, 11	10.012937, 11.009305	19.91, 80.09
Carbon	C	12.011	12, 13	12 (defined), 13.003355	99.89, 1.11
Oxygen	O	15.9994	16, 17, 18	15.994915, 16.999132, 17.999161	99.757, 0.0378, 0.205
Iron	Fe	55.845	54, 56, 57, 58	53.939611, 55.934938, 56.935394, 57.933276	5.82, 91.66, 2.19, 0.33
Uranium	U	238.03	234, 235, 238	234.040952, 235.043930, 238.050788	0.0054, 0.7204, 99.274

Sources of isotope data: G. Audi et al., *Nuclear Physics A* **729** (2003) 337–676; J. C. Kotz and K. F. Purcell, *Chemistry and Chemical Reactivity,* 2nd ed., 1991.

EXAMPLE 1.5

An element with three stable isotopes has 82 protons. The separate isotopes contain 124, 125, and 126 neutrons. Identify the element, and write symbols for the isotopes.

Given Number of protons and neutrons

Asked for Element and atomic symbol

Strategy

A Refer to the periodic table inside the front cover of this text, and use the number of protons to identify the element.
B Calculate the mass number A of each isotope by summing the numbers of protons and neutrons.
C Give the symbol of each isotope with the mass number A as the superscript and the number of protons Z as the subscript, both written to the left of the symbol of the element.

Solution

A The element with 82 protons (atomic number of 82) is lead, Pb. **B** For the first isotope, A = 82 protons + 124 neutrons = 206. Similarly, $A = 82 + 125 = 207$ and $A = 82 + 126 = 208$ for the second and third isotopes, respectively. **C** Their symbols are thus

$$^{206}_{82}\text{Pb} \qquad ^{207}_{82}\text{Pb} \qquad ^{208}_{82}\text{Pb}$$

which are usually abbreviated as

$$^{206}\text{Pb} \qquad ^{207}\text{Pb} \qquad ^{208}\text{Pb}$$

EXERCISE 1.5

Identify the element that contains 35 protons, and write the symbols for its isotopes with 44 and 46 neutrons.

Answer $^{79}_{35}$Br and $^{81}_{35}$Br or, more commonly, ^{79}Br and ^{81}Br

Although the masses of the electron, proton, and neutron are known to a high degree of precision (Table 1.1), the mass of any given atom is not simply the sum of the masses of its electrons, protons, and neutrons. For example, the ratio of the masses of ^{1}H (hydrogen) and ^{2}H (deuterium) is found to be 0.500384, rather than the value of 0.49979 that would be predicted from the numbers of neutrons and protons present. The difference is due to nuclear binding energy effects, which are discussed in Chapter 20. Although the difference in mass is small, it is extremely important because it is the source of the huge amounts of energy released in nuclear reactions.

Because atoms are much too small to measure individually and do not have a charge, there is no convenient way to accurately measure *absolute* atomic masses. Scientists can measure *relative* atomic masses very accurately, however, using an instrument called a *mass spectrometer*. The technique is conceptually similar to the one Thomson used to determine the mass-to-charge ratio of the electron. First, electrons are removed from or added to atoms or molecules, thus producing charged particles called **ions**. When an electric field is applied, the ions are accelerated into a separate chamber where they are deflected from their initial trajectory by a magnetic field, like the electrons in Thomson's experiment. The extent of the deflection depends on the mass-to-charge ratio of the ion. By measuring the relative deflection of ions that have the same charge, scientists can determine their relative masses

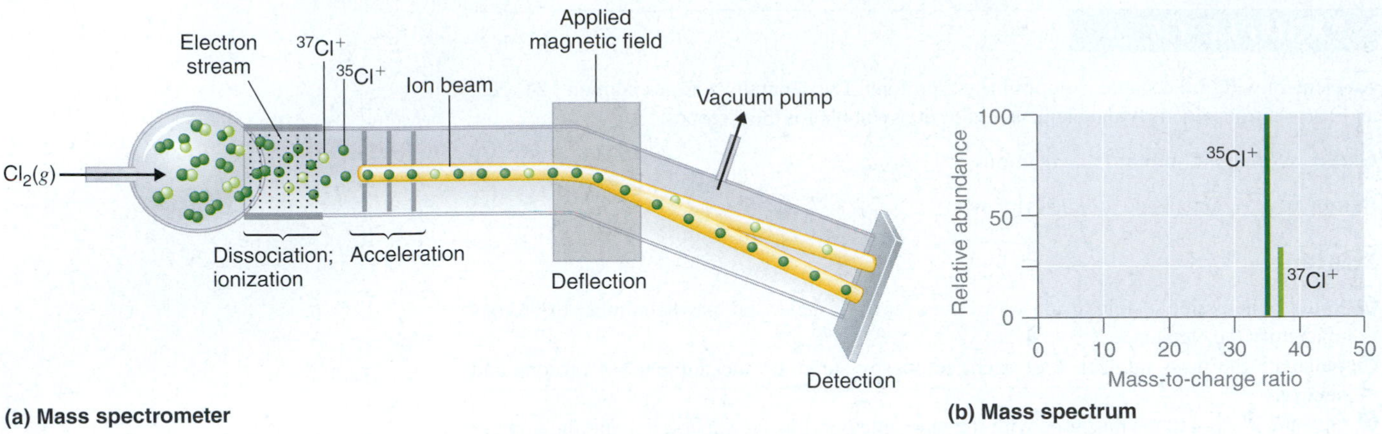

Figure 1.25 Determining relative atomic masses using a mass spectrometer. Chlorine consists of two isotopes, ^{35}Cl and ^{37}Cl, in approximately a 3:1 ratio. (a) When a sample of elemental chlorine is injected into the mass spectrometer, electrical energy is used to dissociate the Cl_2 molecules into Cl atoms and to convert the Cl atoms to Cl^+ ions. The ions are then accelerated into a magnetic field. The extent to which the ions are deflected by the magnetic field depends on their relative mass-to-charge ratios. Note that the lighter $^{35}Cl^+$ ions are deflected more than the heavier $^{37}Cl^+$ ions. By measuring the relative deflections of the ions, chemists can determine their mass-to-charge ratios and thus their masses. (b) Each peak in the mass spectrum corresponds to an ion with a particular mass-to-charge ratio (m/z). The abundance of the two isotopes can be determined from the heights of the peaks.

(Figure 1.25). Thus, it is not possible to calculate absolute atomic masses accurately by simply summing the masses of the electrons, protons, and neutrons, and *absolute* atomic masses cannot be measured, but *relative* masses can be measured very accurately. It is actually rather common in chemistry to encounter a quantity whose magnitude can be measured only relative to some other quantity, rather than absolutely. We will encounter a number of other examples later in this text. In such cases, chemists usually define a standard by arbitrarily assigning a numerical value to one of the quantities, which allows them to calculate numerical values for the rest.

The arbitrary standard that has been established for describing atomic mass is the **atomic mass unit (amu)**, defined as one-twelfth of the mass of one atom of ^{12}C. Because the masses of all other atoms are calculated relative to that of the ^{12}C standard, ^{12}C is the only atom listed in Table 1.3 for which the exact atomic mass is equal to the mass number A. Experiments have shown that 1 amu = 1.66×10^{-24} g.

Mass spectrometric experiments give a value of 0.167842 for the ratio of the mass of ^2H to the mass of ^{12}C, so the absolute mass of ^2H is

$$\frac{\text{mass of }^2\text{H}}{\cancel{\text{mass of }^{12}\text{C}}} \times \cancel{\text{mass of }^{12}\text{C}} = 0.167842 \times 12 \text{ amu} = 2.01410 \text{ amu}$$

The masses of the other elements are determined in a similar way.

The periodic table inside the front cover lists the atomic masses of all the common elements. If you compare these values with those given for some of the isotopes in Table 1.3, you can see that the atomic masses given in the periodic table never correspond exactly to those of any of the isotopes. Because most elements exist as mixtures of several stable isotopes, the **atomic mass** of an element is defined as the weighted average of the masses of the isotopes. For example, naturally occurring carbon is largely a mixture of two isotopes: 98.89% ^{12}C (mass = 12 amu by definition) and 1.11% ^{13}C (mass = 13.003355 amu). The percent abundance of ^{14}C is so low that it can be ignored in this calculation. The *average* atomic mass of carbon can be calculated as

$$(0.9889 \times 12 \text{ amu}) + (0.0111 \times 13.003355 \text{ amu}) = 12.01 \text{ amu}$$

Carbon is predominantly ^{12}C, so its average atomic mass should be close to 12 amu, in agreement with our calculation.

The value of 12.011 is shown under the symbol for C in the periodic table inside the front cover, although without the abbreviation "amu," which is customarily omitted. Thus, the tabulated *atomic mass* of carbon or any other element is the weighted average of the masses of the naturally occurring isotopes.

EXAMPLE 1.6

Naturally occurring bromine consists of the two isotopes listed in the table.

Isotope	Exact Mass, amu	Percent Abundance
^{79}Br	78.9183	50.69
^{81}Br	80.9163	49.31

Calculate the atomic mass of bromine.

Given Exact mass and percent abundance

Asked for Atomic mass

Strategy

A Convert the percent abundances to decimal form to obtain the mass fraction of each isotope.

B Multiply the exact mass of each isotope by its corresponding mass fraction (percent abundance ÷ 100%) to obtain its weighted mass.

C Sum the weighted masses to obtain the atomic mass of the element.

Solution

A The atomic mass is the weighted average of the masses of the isotopes. In general, we can write

atomic mass of element = (mass of isotope 1 in amu) (mass fraction of isotope 1) + (mass of isotope 2) (mass fraction of isotope 2) + . . .

Bromine has only two isotopes. Converting the percent abundances to mass fractions gives:

$$^{79}\text{Br}: \ \frac{50.69}{100} = 0.5069 \qquad ^{81}\text{Br}: \ \frac{49.31}{100} = 0.4931$$

B Multiplying the exact mass of each isotope by the corresponding mass fraction gives the isotope's weighted mass:

$$^{79}\text{Br}: \ (79.9183 \text{ amu})(0.5069) = 40.00 \text{ amu}$$
$$^{81}\text{Br}: \ (80.9163 \text{ amu})(0.4931) = 39.90 \text{ amu}$$

C The sum of the weighted masses is the atomic mass of bromine:

$$40.00 \text{ amu} + 39.90 \text{ amu} = 79.90 \text{ amu}$$

This value is about halfway between the masses of the two isotopes, as expected because the percent abundance of each is approximately 50%.

EXERCISE 1.6

Magnesium has the three isotopes listed in the table.

Isotope	Exact Mass, amu	Percent Abundance
^{24}Mg	23.98504	78.70
^{25}Mg	24.98584	10.13
^{26}Mg	25.98259	11.17

Use these data to calculate the atomic mass of magnesium.

Answer 24.31 amu

1.7 Introduction to the Periodic Table

MGC The Periodic Table

The periodic table (Figure 1.24) is probably the single most important learning aid in chemistry. It summarizes huge amounts of information about the elements in a way that permits you to predict many of their properties and chemical reactions.

The elements are arranged in seven horizontal rows, in order of increasing atomic number from left to right and from top to bottom. The rows are called **periods**, and they are numbered from 1 to 7. The elements are stacked in such a way that elements with similar chemical properties form vertical columns, called **groups**, numbered from 1 to 18 (older periodic tables use a system based on Roman numerals). Groups 1, 2, and 13–18 are the **main group elements**, listed as *A* in older tables. Groups 3–12 are in the middle of the periodic table and are the **transition elements**, listed as *B* in older tables. The two rows of 14 elements at the bottom of the periodic table are the *lanthanides* and *actinides*, whose positions in the periodic table are indicated in Group 3.

Metals, Nonmetals, and Semimetals

The heavy orange zigzag line running diagonally from the upper left to the lower right through Groups 13–16 in Figure 1.24 divides the elements into **metals** (in blue, below and to the left of the line) and **nonmetals** (in bronze, above and to the right). As you might expect, elements colored in gold that lie along the diagonal line exhibit properties intermediate between metals and nonmetals; they are called **semimetals**.

Note the pattern

Metals tend to lose electrons in chemical reactions.

The distinction between metals and nonmetals is one of the most fundamental in chemistry. Metals are good conductors of electricity and heat; they can be pulled into wires because they are *ductile*; they can be hammered or pressed into thin sheets or foils, being *malleable*; and most have a shiny appearance, so they are *lustrous*. In chemical reactions, as you will see, the overwhelming tendency of metals is to lose electrons to form positively charged ions. The vast majority of the known elements are metals. Of the metals, only mercury (Hg) is a liquid at room temperature and pressure; all the rest are solids.

Nonmetals, in contrast, are generally poor conductors of heat and electricity and are not lustrous. Nonmetals can be gases (like chlorine), liquids (like bromine), or solids (like iodine) at room temperature and pressure. Most solid nonmetals are *brittle*, so they break into small pieces when you hit them with a hammer or try to pull them into a wire. You will see that nonmetals tend to gain electrons in reactions with metals to form negatively charged ions or to share electrons in reactions with other nonmetals. As expected, semimetals exhibit properties intermediate between metals and nonmetals.

EXAMPLE 1.7

Based on its position in the periodic table, do you expect selenium (Se) to be a metal, nonmetal, or semimetal?

Given Element

Asked for Classification

Strategy

Find selenium in the periodic table shown in Figure 1.24 and then classify the element according to its location.

Solution

The atomic number of selenium is 34, which places it in Period 4 and Group 16. In Figure 1.24, selenium lies above and to the right of the diagonal line marking the boundary between metals and nonmetals, so it should be a nonmetal. Note, however, that because selenium is close to the metal/nonmetal dividing line, it would not be surprising if selenium were similar to a semimetal in some of its properties.

EXERCISE 1.7

Based on its location in the periodic table, do you expect indium (In) to be a nonmetal, metal, or semimetal?

Answer metal

Chemistry of the Groups

As we noted, the periodic table is arranged so that elements with similar chemical behavior are in the same group. Chemists often make general statements about the properties of the elements in a group using descriptive names with historical origins. For example, the elements of Group 1 are known as the **alkali metals**, those of Group 2 the **alkaline earths**, those of Group 17 the **halogens**, and those of Group 18 the **noble gases**.

The Alkali Metals

The alkali metals (Group 1) are lithium (Li), sodium (Na), potassium (K), rubidium (Rb), cesium (Cs), and francium (Fr). Hydrogen is unique in that it is generally placed in Group 1, but it is not a metal. All the Group 1 elements react readily with nonmetals to give ions with a +1 charge, such as Li^+ and Na^+. (The charge is indicated as a superscript.)

Note the pattern

The elements of Group 1 tend to form ions with a +1 charge.

Compounds of the alkali metals are common in nature and daily life. One example is table salt (sodium chloride), and lithium compounds are used in greases, in batteries, and as drugs to treat patients who exhibit manic-depressive, or bipolar, behavior. Although lithium, rubidium, and cesium are relatively rare in nature, and francium is so unstable and highly radioactive that it exists in only trace amounts, sodium and potassium are the seventh and eighth most abundant elements in Earth's crust, respectively.

The Alkaline Earths

Note the pattern

The elements of Group 2 tend to form ions with a +2 charge.

The alkaline earths (Group 2) are beryllium (Be), magnesium (Mg), calcium (Ca), strontium (Sr), barium (Ba), and radium (Ra). All are metals that react readily with nonmetals to give ions with a +2 charge, such as Mg^{2+} and Ca^{2+}. Beryllium, strontium, and barium are rather rare, and radium is unstable and highly radioactive. In contrast, calcium and magnesium are the fifth and sixth most abundant elements on Earth, respectively; they are found in huge deposits of limestone and other minerals.

> **Note the pattern**
> *The elements of Group 17 tend to form ions with a −1 charge.*

The Halogens

The halogens (Group 17) are fluorine (F), chlorine (Cl), bromine (Br), iodine (I), and astatine (At). The name *halogen* is derived from the Greek for "salt forming," which reflects that all the halogens react readily with metals to form compounds such as sodium chloride (also known as table salt or sea salt) and calcium chloride (used in some areas as road salt). In all of these reactions, the halogens form ions with a −1 charge (for example, Cl^- or I^-).

Compounds that contain F^- ions are added to toothpaste and the water supply to prevent dental cavities. Fluorine is also found in Teflon coatings on kitchen utensils. Although the "chlorofluorocarbon" propellants and refrigerants that are believed to lead to the depletion of Earth's ozone layer contain both fluorine and chlorine, the latter is responsible for the adverse effect on the ozone layer. Bromine and iodine are less abundant than chlorine, and astatine is so radioactive that it exists in only negligible amounts in nature.

The Noble Gases

The noble gases (Group 18) are helium (He), neon (Ne), argon (Ar), krypton (Kr), xenon (Xe), and radon (Rn). Because the noble gases are made up of single atoms, they are *monatomic*. At room temperature and pressure they are unreactive gases. Because of their lack of reactivity, for many years they were called "inert gases" or "rare gases." In 1962, however, the first chemical compounds containing the noble gases were prepared. Although the noble gases are relatively minor constituents of the atmosphere, natural gas contains substantial amounts of helium. Because of its low reactivity, argon is often used as an unreactive (*inert*) atmosphere for welding or in light bulbs. The red light emitted by neon in a gas discharge tube is used in neon lights.

1.8 ● Essential Elements

Of the approximately 115 elements known, only the 19 highlighted in purple in Figure 1.26 are absolutely required in the diets of humans. These elements—called **essential elements**—are restricted to the first four rows of the periodic table, with only two or three exceptions (Mo, I, and possibly Sn in the fifth row). Some other elements are essential for specific organisms. For example, boron is required for the growth of certain plants, bromine is widely distributed in marine organisms, and tungsten is necessary for some microorganisms.

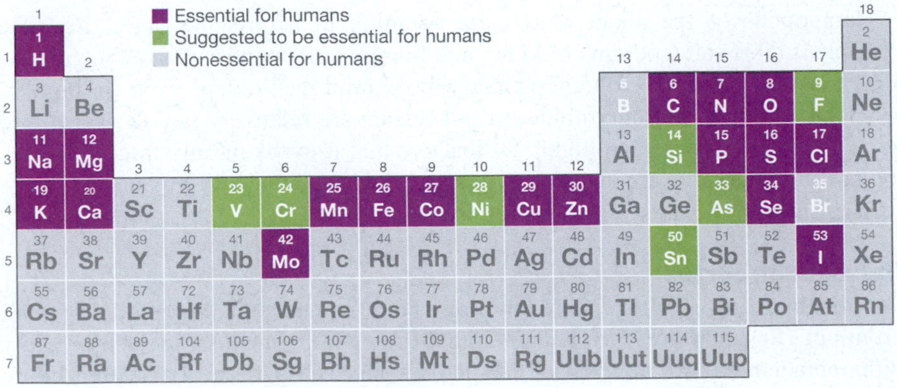

Figure 1.26 The periodic table, identifying the essential elements. Elements that are known to be essential for life in humans are shown in purple; elements that are suggested to be essential are shown in green. Elements not known to be essential are shown in gray.

What makes an element "essential"? By definition, an essential element is one that is required for life and whose absence results in death. Because of the experimental difficulties involved in producing deficiencies severe enough to cause death, especially for elements that are required in very low concentrations in the diet, a somewhat broader definition is generally used. An element is considered to be essential if a deficiency consistently causes abnormal development or functioning, and if dietary supplementation of that element, *and only that element*, prevents this adverse effect. Scientists determine whether an element is essential by raising rats, chicks, and other animals on a synthetic diet that has been carefully analyzed and supplemented with acceptable levels of all elements *except* the element of interest (E). Ultra-clean environments, in which plastic cages are used and dust from the air is carefully removed, minimize inadvertent contamination. If the animals grow normally on a diet that is as low as possible in E, then *either* E is not an essential element *or* the diet is not yet below the minimum required concentration. If the animals do not grow normally on a low-E diet, then their diet is supplemented with E until a level is reached at which the animals grow normally. This level is the *minimum required intake* of element E.

Classification of the Essential Elements

The approximate elemental composition of a healthy 70.0-kg (154-pound) adult human is listed in Table 1.4. Note that most living matter consists primarily of the so-called *bulk elements*—oxygen, carbon, hydrogen, nitrogen, and sulfur—which are the building blocks of the compounds that make up our organs and muscles. These five elements also constitute the bulk of our diet; tens of grams per day are required for humans. Six other elements—sodium, magnesium, potassium, calcium, chlorine, and phosphorus—are often referred to as *macrominerals* because they provide essential ions in body fluids and form the major structural components of the body. In addition, phosphorus is a key constituent of both DNA and RNA, the genetic building blocks of living organisms. The six macrominerals are present in the body in somewhat smaller amounts than the bulk elements, and thus correspondingly lower levels are required in the diet. The remaining essential elements—called *trace elements*—are present in very small amounts, ranging from a few grams to a few milligrams in an adult human. Finally, measurable levels of some elements are found in humans but are *not* required for growth or good health. Examples are rubidium (Rb) and strontium (Sr), whose chemistry is similar to that of the elements immediately above them in the periodic table (K and Ca, respectively, which are essential elements). Because the body's mechanisms for extracting potassium and calcium from foods are not 100% selective, small amounts of rubidium and strontium, which have no known biological function, are absorbed.

TABLE 1.4 Approximate elemental composition of a typical 70-kg human

Bulk Elements (kg)		**Macrominerals (g)**		**Trace Elements (mg)**					
Oxygen	44	Calcium	1700	Iron	5000	Manganese	70	Arsenic	~ 3
Carbon	12.6	Phosphorus	680	Silicon	3000	Iodine	70	Cobalt	~ 3
Hydrogen	6.6	Potassium	250	Zinc	1750	Aluminum	35	Chromium	~ 3
Nitrogen	1.8	Chlorine	115	Rubidium	360	Lead	35	Nickel	~ 3
Sulfur	0.1	Sodium	70	Copper	280	Barium	21	Selenium	~ 2
		Magnesium	42	Strontium	280	Molybdenum	14	Lithium	~ 2
				Bromine	140	Boron	14	Vanadium	~ 2
				Tin	140				

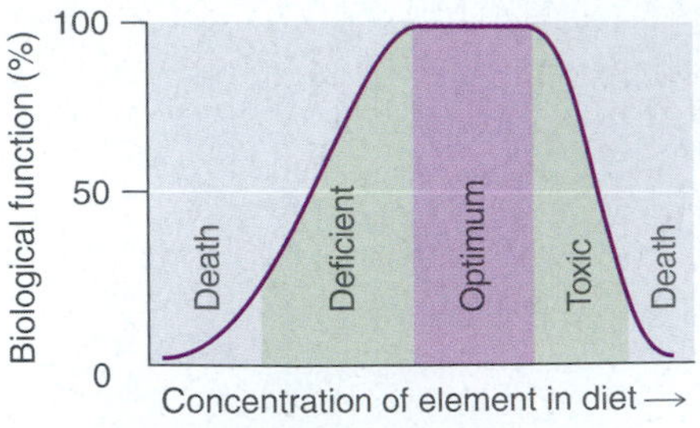

Figure 1.27 Possible concentrations of an essential element in the diet. The deficient, optimum, and toxic concentrations are different for different elements.

The Trace Elements

Because it is difficult to detect low levels of some of the essential elements, the trace elements were relatively slow to be recognized as essential. Iron was the first. In the 17th century, anemia was proved to be due to iron deficiency and often was cured by supplementing the diet with extracts of rusty nails. It was not until the 19th century, however, that trace amounts of iodine were found to eliminate goiter (an enlarged thyroid gland). This is why common table salt is "iodized"—a small amount of iodine is added. Copper was shown to be essential for humans in 1928, and manganese, zinc, and cobalt soon after that. Molybdenum was not known to be an essential element until 1953, and the need for chromium, selenium, vanadium, fluorine, and silicon was demonstrated only in the last 50 years. It seems likely that in the future other elements, possibly including tin, will be found to be essential at very low levels.

Many compounds of trace elements such as arsenic, selenium, and chromium are toxic and can even cause cancer, yet these elements are identified as essential elements in Figure 1.26. How can this be? First, the toxicity of an element often depends on its chemical form. For example, only certain compounds of chromium are toxic, whereas others are used in mineral supplements. Second, as shown in Figure 1.27, every element has three possible levels of dietary intake: *deficient, optimum*, and *toxic*, in order of increasing concentration in the diet. Very low intake levels lead to symptoms of deficiency. Over some range of higher intake levels, the organism is able to maintain its tissue concentrations of the element at a level that optimizes biological functions. Finally, at some higher intake level, the normal regulatory mechanisms are overloaded and toxic symptoms appear. Each element has its own characteristic curve. Both the width of the plateau and the specific concentration corresponding to the center of the plateau region differ by as much as several orders of magnitude for different elements. For example, the recommended daily dietary intake for an adult human is 10–18 mg of iron, 2–3 mg of copper, and less than 0.1 mg of chromium and selenium.

Amplification

How can elements that are present in such minuscule amounts have such large effects on the health of an organism? Our knowledge of the pathways by which each of the known trace elements affects health is far from complete, but certain general features are clear. The trace elements participate in an *amplification* mechanism; that is, they are essential components of larger biological molecules that are capable of interacting with or regulating the levels of relatively large amounts of other molecules. For example, vitamin B_{12} contains a single atom of cobalt, which is essential for its biological function. If the molecule whose level is controlled by the trace element can in turn regulate the level of another molecule, and so on, then the potential exists for extreme amplification of small variations in the level of the trace element. One goal of modern chemical research is to elucidate in detail the roles of the essential elements. In subsequent chapters, we will introduce some results of this research in order to demonstrate the biological importance of many of the elements and their compounds.

SUMMARY AND KEY TERMS

1.1 Chemistry in the Modern World (p. 2)

Chemistry is the study of matter and the changes material substances undergo. It is essential for understanding much of the natural world and central to many other scientific disciplines, including astronomy, geology, paleontology, biology, and medicine.

1.2 The Scientific Method (p. 4)

Chemists expand their knowledge through the process of making observations, carrying out **experiments**, and testing **hypotheses** in order to develop **laws** to summarize their results and **theories** to explain them. In doing so, they are using the **scientific method**.

1.3 A Description of Matter (p. 7)

Matter is anything that occupies space and has **mass**. The three states of matter are **gas, liquid**, and **solid**. Most matter consists of **mixtures** of pure substances, which can be **homogeneous** (uniform in composition) or **heterogeneous** (different regions possess different compositions and properties). Pure substances can be either chemical compounds or elements. **Compounds** can be broken down into elements by chemical reactions, but **elements** cannot be separated into simpler substances by chemical means. The properties used to describe material substances can be classified as either physical or chemical. Scientists can observe **physical properties** without changing the composition of the substance, whereas **chemical properties** describe the tendency of a substance to undergo chemical reactions that change its chemical composition. Physical properties can be intensive or extensive. **Intensive properties** are the same for all samples, do not depend on the sample size, and include color, physical state, and melting and boiling points. **Extensive properties** depend on the amount of material and include mass and volume. The ratio of two extensive properties, mass and volume, is an important intensive property called **density**.

1.4 A Brief History of Chemistry (p. 13)

The ancient Greeks first proposed that matter consisted of fundamental particles called **atoms**. Chemistry took its present scientific form in the 18th century, when careful quantitative experiments by Lavoisier, Proust, and Dalton resulted in the law of definite proportions, the law of conservation of mass, and the law of multiple proportions, which laid the groundwork for Dalton's atomic theory of matter. In particular, Avogadro's hypothesis provided the first link between the macroscopic properties of a substance (in this case, the volume of a gas) and the number of atoms or molecules present.

1.5 The Atom (p. 18)

Atoms, the smallest particles of an element that exhibit the properties of that element, consist of negatively charged **electrons** around a central **nucleus** composed of more massive positively charged **protons** and electrically neutral **neutrons**. **Radioactivity** is the emission of energetic particles and rays (radiation) by some substances. Three important kinds of radiation are α particles (helium nuclei), β particles (electrons traveling at high speed), and γ rays (similar to X rays but higher in energy).

1.6 Isotopes and Atomic Masses (p. 22)

Each atom of an element contains the same number of protons, which is the **atomic number** (Z). Neutral atoms have the same number of electrons and protons. Atoms of an element that contain different numbers of neutrons are called **isotopes**. Each isotope of a given element has the same atomic number but a different **mass number** (A), which is the sum of the numbers of protons and neutrons. The relative masses of atoms are reported using the **atomic mass unit (amu)**, which is defined as one-twelfth of the mass of one atom of carbon-12, with 6 protons, 6 neutrons, and 6 electrons. The **atomic mass** of an element is the weighted average of the masses of the naturally occurring isotopes. When one or more electrons are added to or removed from an atom or molecule, a charged particle called an **ion** is produced, whose charge is indicated by a superscript after the symbol.

1.7 Introduction to the Periodic Table (p. 28)

The **periodic table** is an arrangement of the elements in order of increasing atomic number. Elements that exhibit similar chemistry appear in vertical columns called **groups** (numbered 1–18 from left to right); the seven horizontal rows are called **periods**. Some of the groups have widely used common names, including the **alkali metals** (Group 1) and the **alkaline earths** (Group 2) on the far left, and the **halogens** (Group 17) and the **noble gases** (Group 18) on the far right. The elements can be broadly divided into **metals, nonmetals**, and **semimetals**. Semimetals exhibit properties intermediate between those of metals and nonmetals. Metals are located at the left of the periodic table, and nonmetals are located in the upper right. They are separated by a diagonal band of semimetals. Metals are lustrous, good conductors of electricity, and readily shaped (they are ductile and malleable), whereas solid nonmetals are generally brittle and poor electrical conductors. Metals and nonmetals exhibit distinctly different chemistry, with metals reacting to give up electrons and nonmetals reacting to gain or share electrons. Other important groupings of elements in the periodic table are the **main group elements, transition metals**, lanthanides, and actinides.

1.8 Essential Elements (p. 30)

About 19 of the approximately 115 known elements are essential for humans. An **essential element** is one whose absence results in abnormal biological function or development that is prevented by dietary supplementation with that element. Living organisms contain relatively large amounts of the bulk elements oxygen, carbon, hydrogen, nitrogen, and sulfur, and also contain the macrominerals sodium, magnesium, potassium, calcium, chlorine, and phosphorus. The other essential elements are the trace elements, which are present in very small quantities. Dietary intakes of elements range from deficient to optimum to toxic with increasing quantities; the optimum levels differ greatly for the essential elements.

QUESTIONS AND PROBLEMS

MGC For instructor-assigned homework, go to **www.masteringgenchem.com**

Please be sure you are familiar with the topics discussed in Essential Skills 1 at the end of this chapter before proceeding to the Questions and Problems.

Questions and Problems with colored numbers have answers in the Appendix and complete solutions in the Student Solutions Manual.

CONCEPTUAL

1.2 The Scientific Method

1. What are the three components of the scientific method? Is it necessary for an individual to conduct experiments in order to follow the scientific method?

2. Decide whether each statement is a theory or a law, and explain your reasoning.
   (a) The ratio of elements in a pure substance is constant.
   (b) An object appears black because it absorbs all the visible light that strikes it.
   (c) Energy cannot be created or destroyed.
   (d) Metals conduct electricity because their electrons are not tightly bound to a particular nucleus and are therefore free to migrate.
3. Which of the following statements are theories and which are laws? Explain your reasoning.
   (a) A pure chemical substance contains the same proportion of elements by mass.
   (b) The universe is expanding.
   (c) Oppositely charged particles attract each other.
   (d) Life exists on other planets.
4. Classify each statement as a qualitative or quantitative observation.
   (a) Mercury and bromine are the only elements that are liquids at room temperature.
   (b) An element is both malleable and ductile.
   (c) The density of iron is 7.87 g/cm^3.
   (d) Lead absorbs sound very effectively.
   (e) A meteorite contains 20% nickel by mass.
5. State whether each statement is a quantitative or qualitative observation.
   (a) Nickel deficiency in rats is associated with retarded growth.
   (b) Boron is a good conductor of electricity at high temperatures.
   (c) There are 1.4–2.3 g of zinc in an average 70-kg adult.
   (d) Certain osmium compounds found in air in concentrations as low as 10.7 $\mu g/m^3$ can cause lung cancer.

1.3 A Description of Matter

6. What is the difference between mass and weight? Is the mass of an object on Earth the same as the mass of the same object on Jupiter? Why or why not?
7. Is it accurate to say that a substance with a mass of 1 kg weighs 2.2 pounds? Why or why not?
8. What factor must be considered when reporting the weight of an object as opposed to its mass?
9. Construct a table with the headings "Solid," "Liquid," and "Gas." For any given substance, state what you expect for: (a) the relative densities of the three phases, (b) their physical shapes, (c) their volumes for the same mass of compound, (d) how sensitive the volume of each phase is to changes in temperature, and (e) how sensitive the volume is to changes in pressure.
10. Classify each substance as homogeneous or heterogeneous, and explain your reasoning: (a) platinum; (b) a carbonated beverage; (c) bronze; (d) wood; (e) natural gas; (f) styrofoam.
11. Classify each substance as homogeneous or heterogeneous: (a) snowflakes; (b) gasoline; (c) black tea; (d) plastic wrap; (e) blood; (f) containing ice cubes. Explain your answers.
12. Classify each of the following as a pure substance or a mixture, and explain your reasoning: (a) seawater; (b) coffee; (c) 14-karat gold; (d) diamond; (e) distilled water.
13. Classify each of the following as a pure substance or a mixture: (a) cardboard; (b) caffeine; (c) tin; (d) a vitamin tablet; (e) helium gas.
14. Classify each substance as an element or a compound: (a) sugar; (b) silver; (c) rust; (d) rubbing alcohol; (e) copper.
15. State whether each of the following is an element or a compound: (a) water; (b) palladium; (c) hydrogen gas; (d) glass; (e) nylon.
16. Describe the techniques you would use to separate (a) sugar and water from an aqueous solution of sugar; (b) a mixture of sugar and sand; and (c) a heterogeneous mixture of solids that have different solubilities.
17. Describe techniques that would effectively separate (a) solid calcium chloride from a solution of calcium chloride in water; (b) the components of a solution of vinegar in water; and (c) particulates from water in a fish tank.
18. Match the separation techniques in (a) with the physical/chemical property that each takes advantage of in (b).
    (a) Crystallization, distillation, filtration
    (b) Volatility, physical state, solubility
19. Classify each statement as describing an extensive or intensive property.
    (a) Carbon, in the form of diamond, is one of the hardest known materials.
    (b) A sample of crystalline silicon, a grayish solid, has a mass of 14.3 g.
    (c) Germanium has a density of 5.32 g/cm^3.
    (d) Gray tin converts to white tin at 13.2°C.
    (e) Lead is a bluish-white metal.
20. Classify each statement as describing a physical or a chemical property.
    (a) Fluorine etches glass.
    (b) Chlorine is a respiratory irritant.
    (c) Bromine is a reddish-brown liquid.
    (d) Iodine has a density of 11.27 g/L at 0°C.

1.4 A Brief History of Chemistry

21. Define *combustion* and discuss the contributions made by Priestley and Lavoisier toward understanding a combustion reaction.
22. Chemical engineers frequently use the concept of "mass balance" in their calculations, in which the mass of the reactants must equal the mass of the products. What law supports this practice?
23. Does the law of multiple proportions apply to both mass ratios and atomic ratios? Why or why not?
24. (a) What are the four hypotheses of the atomic theory of matter? (b) Much of the energy in France is provided by nuclear reactions. Are such reactions consistent with Dalton's hypotheses? Why or why not?
25. Does 1 L of air contain the same number of particles as 1 L of nitrogen gas? Explain your answer.

1.5 The Atom

26. Describe the experiment that provided evidence that the proton is positively charged.
27. What observation led Rutherford to propose the existence of the neutron?
28. What is the difference between Rutherford's model of the atom and the model chemists use today?
29. If cathode rays are not deflected when they pass through a region of space, what does this imply about the presence or absence of a magnetic field perpendicular to the path of the rays in that region?
30. Describe the outcome that would be expected from Rutherford's experiment if the charge on α particles had remained the same but the nucleus were negatively charged. If the nucleus were neutral, what would have been the outcome?

31. Describe the differences between an α particle, a β particle, and a γ ray. Which has the greatest ability to penetrate matter?

1.6 Isotopes and Atomic Masses

32. Complete the table for the missing elements, symbols, and numbers of electrons.

Element	Symbol	Number of Electrons
Molybdenum	___	___
________	___	19
Titanium	___	___
________	B	___
________	___	53
________	Sm	___
Helium	___	___
________	___	14

33. Complete the table.

Element	Symbol	Number of Electrons
Lanthanum	___	___
________	Ir	___
Aluminum	___	___
________	___	80
Sodium	___	___
________	Si	___
________	___	9
________	Be	___

34. Is the mass of an ion the same as the mass of its parent atom? Explain your answer.
35. What isotopic standard is used for determining the mass of an atom?
36. Give the symbol ${}^{A}_{Z}X$ for these elements, all of which exist as a single isotope: beryllium; Rh; phosphorus; Al; cesium; praseodymium; cobalt; Y; arsenic.
37. Give the symbol ${}^{A}_{Z}X$ for these elements, all of which exist as a single isotope: fluorine; He; Tb; I; aluminum; scandium; sodium; Nb; Mn.
38. Identify the elements, X, that have the given symbols:

$${}^{55}_{26}X;\ {}^{74}_{33}X;\ {}^{24}_{12}X;\ {}^{127}_{53}X;\ {}^{40}_{18}X;\ {}^{152}_{63}X$$

1.7 Introduction to the Periodic Table

39. Classify each of the elements in Problem 32 as a metal, nonmetal, or semimetal. If a metal, state whether it is an alkali metal, alkaline earth, or transition metal.
40. Classify each of the elements in Problem 33 as a metal, nonmetal, or semimetal. If a metal, state whether it is an alkali metal, alkaline earth, or transition metal.
41. Classify each element as a metal, semimetal, or nonmetal: Fe, Ta, S, Si, Cl, Ni, K, Rn, Zr. If the element is a metal, is it an alkali metal, alkaline earth, or transition metal?
42. Which of these sets of elements are all in the same period?
    (a) Potassium, vanadium, ruthenium
    (b) Lithium, carbon, chlorine
    (c) Sodium, magnesium, sulfur
    (d) Chromium, nickel, krypton
43. Which of these sets of elements are all in the same period?
    (a) Barium, tungsten, argon
    (b) Yttrium, zirconium, selenium
    (c) Potassium, calcium, zinc
    (d) Scandium, bromine, manganese
44. Which of these sets of elements are all in the same group?
    (a) Sodium, rubidium, barium
    (b) Nitrogen, phosphorus, bismuth
    (c) Copper, silver, gold
    (d) Magnesium, strontium, samarium
45. Which of these sets of elements are all in the same group?
    (a) Iron, ruthenium, osmium
    (b) Nickel, palladium, lead
    (c) Iodine, fluorine, oxygen
    (d) Boron, aluminum, gallium
46. Identify which of these elements are (a) transition metals, (b) halogens, or (c) noble gases: Mg; Ir; F; Xe; Li; C; Zn; Na; Tl; Zr; Sb; Cd.
47. Based on their locations in the periodic table, would you expect these elements to be malleable? (a) phosphorus; (b) chromium; (c) rubidium; (d) copper; (e) aluminum; (f) bismuth; (g) neodymium. Why or why not?
48. Based on their locations in the periodic table, would you expect these elements to be lustrous? (a) sulfur; (b) vanadium; (c) nickel; (d) arsenic; (e) strontium; (f) cerium; (g) sodium. Why or why not?

NUMERICAL

This section includes paired problems (marked by brackets) that require similar problem-solving skills.

1.3 A Description of Matter

49. If a person weighs 176 pounds on Earth, what is his or her mass on Mars, where the force of gravity is 37% that on Earth?
50. If a person weighs 135 pounds on Earth, what is his or her mass on Jupiter, where the force of gravity is 236% that of Earth?
51. Calculate the volume of 10.00 g of each element, and then arrange the elements in order of decreasing volume: (a) Cu (8.92); (b) Ca (1.54); (c) Ti (4.51); (d) Ir (22.85). The numbers in parentheses are densities (in g/cm^3).
52. Given 15.00 g of each of the following elements, calculate the volume of each, and then arrange the elements in order of increasing volume: (a) Au (19.32); (b) Pb (11.34); (c) Fe (7.87); (d) S (2.07). The numbers in parentheses are densities (in g/cm^3).
53. A silver bar has dimensions of 10.00 cm × 4.00 cm × 1.50 cm, and the density of silver is 10.49 g/cm^3. What is the mass of the bar?
54. Platinum has a density of 21.45 g/cm^3. What is the mass of a platinum bar measuring 3.00 cm × 1.50 cm × 0.500 cm?
55. Complete the table.

Density, g/cm^3	Mass, g	Volume, cm^3	Element
3.14	79.904	____	Br
3.51	______	3.42	C
____	39.1	45.5	K
11.34	207.2	____	Pb
____	107.868	10.28	Ag
6.51	______	14.0	Zr

56. Gold has a density of 19.30 g/cm^3. If a person who weighs 85.00 kg (1 kg = 1000 g) were given his or her weight in gold, what volume (in cm^3) would the gold occupy? Are we justified in using the SI unit of mass for the person's weight in this case?

57. An irregularly shaped piece of magnesium with a mass of 11.81 g was dropped into a graduated cylinder partially filled with water. The magnesium displaced 6.80 mL of water. What is the density of magnesium?

58. The density of copper is 8.92 g/cm^3. If a 10.00-g sample is placed into a graduated cylinder that contains 15.0 mL of water, what is the total volume that would be occupied?

59. At 20°C, the density of fresh water is 0.9982 kg/m^3 and the density of seawater is 1.025 kg/m^3. Will a ship float higher in fresh water or in seawater? Explain your reasoning.

1.4 A Brief History of Chemistry

60. Nitrogen and oxygen react to form three different compounds that contain 0.571, 1.143, and 2.285 g of oxygen/g of nitrogen, respectively. Is this consistent with the law of multiple proportions? Explain your answer.

61. One of the minerals found in soil has Al:Si:O atomic ratios of 0.2:0.2:0.5. Is this consistent with the law of multiple proportions? Why or why not? Is the ratio of elements consistent with Dalton's law of indivisible atoms?

62. Three binary compounds of vanadium and oxygen are known. The table gives the mass of oxygen that combines with 10.00 g of vanadium to form each compound.

Compound	Mass of Oxygen, g
A	4.71
B	6.27
C	7.84

(a) Determine the ratio of the masses of oxygen that combine with 10.00 g of vanadium in these compounds.
(b) Predict the mass of oxygen that would combine with 10.00 g of vanadium to form another compound in the series.

63. Three compounds containing titanium, magnesium, and oxygen are known. The table lists the masses of titanium and magnesium that react with 5.00 g of oxygen to form each compound.

Compound	Mass of Titanium, g	Mass of Magnesium, g
A	4.99	2.53
B	3.74	3.80
C	—	—

(a) Determine the ratios of the masses of titanium and magnesium that combine with 5.00 g of oxygen in these compounds.
(b) Predict the masses of titanium and magnesium that would combine with 5.00 g of oxygen to form another possible compound in the series, Compound C.

1.5 The Atom

64. Using the data in Table 1.1 and the periodic table inside the front cover, calculate the percentage of the mass of a helium atom that is due to (a) the electrons and (b) the protons.

65. Using the data in Table 1.1 and the periodic table, calculate the percentage of the mass of a silicon atom that is due to (a) the electrons and (b) the protons.

66. The radius of an atom is approximately 10^4 times larger than the radius of its nucleus. If the radius of the nucleus were 1.0 cm, what would be the radius of the atom in centimeters? In miles?

67. The total charge on an oil drop was found to be 3.84×10^{-18} coulombs. What is the total number of electrons contained in the drop?

1.6 Isotopes and Atomic Masses

68. Determine the number of protons, neutrons, and electrons in a neutral atom of each isotope: ^{97}Tc; ^{113}In; ^{63}Ni; ^{55}Fe.

69. The isotopes ^{131}I and ^{60}Co are commonly used in medicine. Determine the number of neutrons, protons, and electrons in a neutral atom of each.

70. The following isotopes are important in archaeological research. How many protons, neutrons, and electrons does each contain? (a) ^{207}Pb; (b) ^{16}O; (c) ^{40}K; (d) ^{137}Cs; (e) ^{40}Ar

71. Both technetium-97 and americium-240 are produced in nuclear reactors. Determine the number of protons, neutrons, and electrons in the neutral atoms of each.

72. Silicon consists of three isotopes with the following percent abundances:

Isotope	Percent Abundance	Atomic Mass, amu
^{28}Si	92.18	27.976926
^{29}Si	4.71	28.976495
^{30}Si	3.12	29.973770

Calculate the average atomic mass of silicon.

73. Copper, an excellent conductor of heat, has two isotopes, ^{63}Cu and ^{65}Cu. Use the following information to calculate the average atomic mass of copper:

Isotope	Percent Abundance	Atomic Mass, amu
^{63}Cu	69.09	62.9298
^{65}Cu	30.92	64.9278

74. Complete the table for neon; the average atomic mass of Ne is 20.1797 amu.

Isotope	Percent Abundance	Atomic Mass, amu
^{20}Ne	90.92	19.99244
^{21}Ne	0.257	20.99395
^{22}Ne		

75. Are $^{63}_{28}$X and $^{62}_{29}$X isotopes of the same element? Explain your answer.

76. Complete the table:

Isotope	^{57}Fe	^{40}X	^{36}S
Number of protons	___	20	___
Number of neutrons	___	___	___
Number of electrons	___	___	___

77. Complete the table:

Isotope	^{238}X	^{238}U	___
Number of protons	___	___	75
Number of neutrons	___	___	112
Number of electrons	95	___	___

78. The percent abundances of two of the three isotopes of oxygen are ^{16}O, 99.76%, and ^{18}O, 0.204%. Use the atomic mass of oxygen given in the periodic table and the following data to determine the mass of the third isotope, ^{17}O: ^{16}O, 15.994915 amu; ^{18}O, 17.999160 amu.

79. Using a mass spectrometer, a scientist determined the percent abundances of the isotopes of sulfur to be ^{32}S, 95.27%; ^{33}S, 0.51%; and ^{34}S, 4.22%. Use the atomic mass of sulfur from the periodic table and the following atomic masses to determine whether these data are accurate, assuming that these are the only isotopes of sulfur: ^{32}S, 31.972071 amu; ^{33}S, 32.971459 amu; ^{34}S, 33.967867 amu.

80. Which element has the higher proportion by mass in the compound KBr?

81. Which element has the higher proportion by mass in the compound NaI?

1.7 Introduction to the Periodic Table

82. Predict how many electrons are in each of these ions: (a) an oxygen ion that has a −2 charge; (b) a calcium ion that has a +2 charge; (c) a silver ion that has a +1 charge; (d) a selenium ion that has a +4 charge; (e) an iron ion that has a +2 charge; (f) a chlorine ion that has a –1 charge.

83. Predict how many protons and electrons are in each of these ions: (a) a copper ion that has a +2 charge, (b) a molybdenum ion that has a +4 charge, (c) an iodine ion that has a −1 charge, (d) a gallium ion that has a +3 charge, (e) an ytterbium ion that has a +3 charge, (f) a scandium ion that has a +3 charge.

APPLICATIONS

84. In 1953, James Watson and Francis Crick spent three days analyzing data to put together a model that was consistent with the known facts about the structure of DNA, the chemical substance that is the basis for life. For their work they were awarded the Nobel Prize. Based on this information, would you classify their proposed model for the structure of DNA as an experiment, a law, a hypothesis, or a theory? Explain your reasoning.

85. In each scenario, state the observation and the hypothesis.
(a) A recently discovered Neanderthal throat bone has been found to be similar in dimensions and appearance to that of modern humans; therefore, some scientists believe that Neanderthals could talk.
(b) Because DNA profiles from samples of human tissue are widely used in criminal trials, DNA sequences from plant residue on clothing can be used to place a person at the scene of a crime.

86. Small quantities of gold from far underground are carried to the surface by groundwater, where the gold can be taken up by certain plants and stored in their leaves. By identifying the kinds of plants that grow around existing gold deposits, one should be able to use this information to discover potential new gold deposits.
(a) State the observation.
(b) State the hypothesis.
(c) Devise an experiment to test the hypothesis.

87. Large amounts of nitrogen are used by the electronics industry to provide a gas blanket over a component during production. This ensures that undesired reactions with oxygen will not occur. Classify each statement as an extensive or intensive property of nitrogen.
(a) Nitrogen is a colorless gas.
(b) A volume of 22.4 L of nitrogen gas weighs 28 g at 0°C.
(c) Liquid nitrogen boils at 77.4 K.
(d) Nitrogen gas has a density of 1.25 g/L at 0°C.

88. Oxygen is the third most abundant element in the universe and makes up about two-thirds of the human body. Classify each statement as an extensive or intensive property of oxygen.
(a) Liquid oxygen boils at 90.2 K.
(b) Liquid oxygen is pale blue.
(c) A volume of 22.4 L of oxygen gas weighs 32 g at 0°C.
(d) Oxygen has a density of 1.43 g/L at 0°C.

89. One of the first high-temperature superconductors was found to contain elements in the ratio 1Y:2Ba:3Cu:6.8O. A material that contains elements in the ratio 1Y:2Ba:3Cu:6O, however, was not a high-temperature superconductor. Do these materials obey the law of multiple proportions? Is the ratio of elements in each compound consistent with Dalton's law of indivisible atoms?

90. There has been increased concern that human activities are causing changes in Earth's atmospheric chemistry. Recent research efforts have focused on atmospheric ozone (O_3) concentrations. The amount of ozone in the atmosphere is influenced by concentrations of gases that contain only nitrogen and oxygen, among others. The table gives the masses of nitrogen that combine with 1.00 g of oxygen to form three of these compounds.

Compound	Mass of Nitrogen, g
A	0.875
B	0.438
C	0.350

(a) Determine the ratios of the masses of nitrogen that combine with 1.00 g of oxygen in these compounds. Are these data consistent with the law of multiple proportions?
(b) Predict the mass of nitrogen that would combine with 1.00 g of oxygen to form another possible compound in the series.

91. Bromine has an average atomic mass of 79.904 amu. One of its two isotopes has an atomic mass of 80.9163 amu with a percent abundance of 49.46. What is the mass of the other isotope?

92. Earth's core is largely composed of iron, an element that is also a major component of black sands on beaches. Iron has four stable isotopes. Use the data to calculate the average atomic mass of iron.

Isotope	Percent Abundance	Atomic Mass, amu
^{54}Fe	5.82	53.9396
^{56}Fe	91.66	55.9349
^{57}Fe	2.19	56.9354
^{58}Fe	0.33	57.9333

93. Because ores are deposited during different geologic periods, lead ores from different mining regions of the world can contain different ratios of isotopes. Archaeologists use these differences to determine the origin of lead artifacts. For example, the table lists the percent abundances of three lead isotopes from one artifact recovered from Rio Tinto in Spain.

Isotope	Percent Abundance	Atomic Mass, amu
^{204}Pb	—	203.973028
^{206}Pb	24.41	205.974449
^{207}Pb	20.32	206.97580
^{208}Pb	50.28	207.976636

(a) If the only other lead isotope in the artifact is ^{204}Pb, what is its percent abundance?

(b) What is the average atomic mass of lead if the only other isotope in the artifact is ^{204}Pb?

(c) An artifact from Laurion, Greece, was found to have a $^{207}Pb : ^{206}Pb$ ratio of 0.8307. From the data given, can you decide whether the lead in the artifact from Rio Tinto came from the same source as the lead in the artifact from Laurion, Greece?

94. The macrominerals Na, Mg, K, Ca, Cl, and P are widely distributed in biological substances, although their distributions are far from uniform. Classify these elements by both their periods and their groups, and then state whether each is a metal, nonmetal, or semimetal. If a metal, is the element a transition metal?

95. The composition of fingernails is sensitive to exposure to certain elements, including Na, Mg, Al, Cl, K, Ca, Se, V, Cr, Mn, Fe, Co, Cu, Zn, Sc, As, and Sb. Classify these elements by both their periods and their groups, and then determine whether each is a metal, nonmetal, or semimetal. Of the metals, which are transition metals? Based on how you have classified these elements, predict other elements that could prove to be detectable in fingernails.

96. Mercury levels in hair have been used to identify individuals who have been exposed to toxic levels of mercury. Is mercury an essential element? A trace element?

97. Trace elements are usually present at levels of less than 50 mg/kg of body weight. Classify the essential trace elements by their groups and periods in the periodic table. Based on your classifications, would you predict that As, Cd, and Pb are potential essential trace elements?

ESSENTIAL SKILLS

1

TOPICS

- Units of Measurement
- Scientific Notation
- Significant Figures
- Accuracy and Precision

This section describes some of the fundamental mathematical skills you will need to complete the questions and problems in this text. For some of you, this discussion will serve as a review, whereas others may be encountering at least some of the ideas and techniques for the first time. We will introduce other mathematical skills in subsequent Essential Skills sections as the need arises. Be sure you are familiar with the topics discussed here before you start the Chapter 1 Questions and Problems.

Units of Measurement

A variety of instruments are available for making direct measurements of the macroscopic properties of a chemical substance. For example, we usually measure the *volume* of a liquid sample using pipets, burets, graduated cylinders, and volumetric flasks, whereas we usually measure the *mass* of a solid substance with a balance. Measurements on an atomic or molecular scale, in contrast, require specialized instrumentation, such as the mass spectrometer described in Chapter 1.

SI Units

All reported measurements must include an appropriate unit of measurement because to say that a substance has "a mass of 10," for example, does not tell whether the mass was measured in grams, pounds, tons, or some other unit. To establish worldwide standards for the consistent measurement of important physical and chemical properties, an international body called the General Conference on Weights and Measures devised the **Système Internationale d'Unités** (the International System of Units), abbreviated **SI**. This system is based on metric units and requires measurement to be expressed in decimal form. Table ES1.1 lists the seven base units of the SI system; all other SI units of measurement are derived from them.

The magnitudes of units are indicated by attaching prefixes to the base units; each prefix indicates that the base unit is multiplied by a specified power of 10. The prefixes, their symbols, and their numerical significance are listed in Table ES1.2. To study chemistry you need to know the information presented in Tables ES1.1 and ES1.2.

TABLE ES1.1 SI base units

Base Quantity	Unit Name	Abbreviation
Mass	kilogram	kg
Length	meter	m
Time	second	s
Temperature	kelvin	K
Electric current	ampere	A
Amount of substance	mole	mol
Luminous intensity	candela	cd

TABLE ES1.2 Prefixes used with SI units

Prefix	Symbol	Value	Power of 10	Word
tera	T	1,000,000,000,000	10^{12}	trillion
giga	G	1,000,000,000	10^{9}	billion
mega	M	1,000,000	10^{6}	million
kilo	k	1,000	10^{3}	thousand
hecto	h	100	10^{2}	hundred
deca	da	10	10^{1}	ten
—	—	1	10^{0}	one
deci	d	0.1	10^{-1}	tenth
centi	c	0.01	10^{-2}	hundredth
milli	m	0.001	10^{-3}	thousandth
micro	μ	0.000001	10^{-6}	millionth
nano	n	0.000000001	10^{-9}	billionth
pico	p	0.000000000001	10^{-12}	trillionth

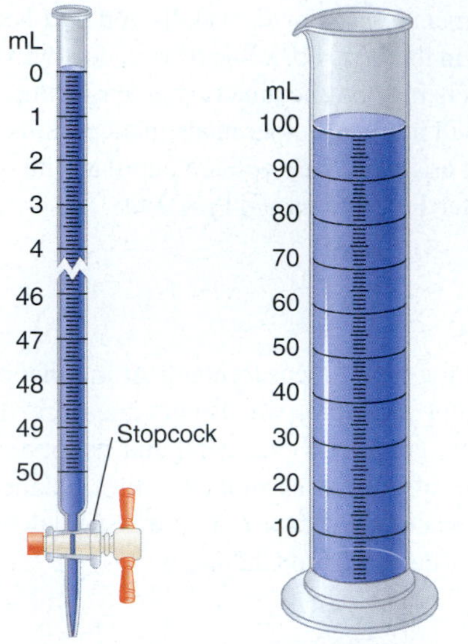

Buret Graduated cylinder

Graduated glassware: used to deliver variable volumes of liquid

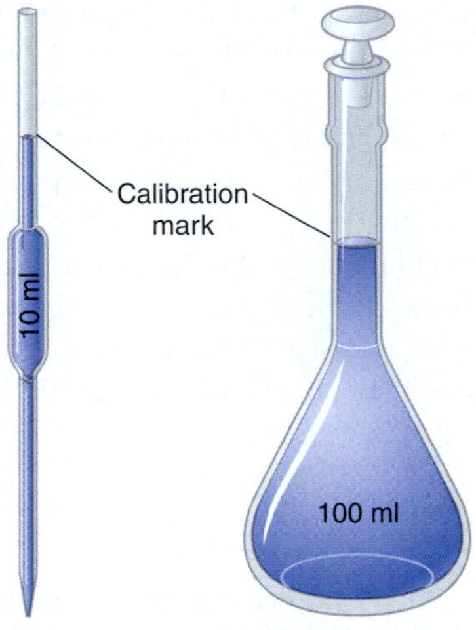

Pipet Volumetric flask

Volumetric glassware: used to deliver (pipet) or contain (volumetric flask) a single volume accurately when filled to the calibration mark

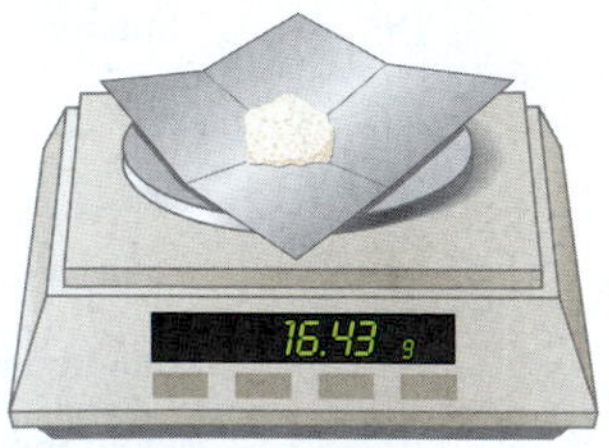

Balance: used to measure mass

Units of Mass, Volume, and Length

The units of measurement you will encounter most frequently in chemistry are those for mass, volume, and length. The basic SI unit for mass is the kilogram (kg), but in the laboratory mass is usually expressed in either grams or milligrams: 1000 grams = 1 kilogram (10^3 g, = 1 kg), 1000 milligrams = 1 gram (10^3 mg = 1 g), and 1,000,000 milligrams = 1 kilogram (10^6 mg = 1 kg). Units for volume are derived from the cube of the SI unit for length, which is the meter (m). Thus, the basic SI unit for volume is cubic meters (length × width × height = m^3). In chemistry, however, volumes are usually reported in cubic centimeters (cm^3) and cubic decimeters (dm^3) or in milliliters (mL) and liters (L), although the liter is not an SI unit of measurement. The relationships between these units are

$$1\ \text{L} = 1000\ \text{mL} = 1\ \text{dm}^3$$

$$1\ \text{mL} = 1\ \text{cm}^3$$

Notice that 1000 cm^3 = 1 L.

Scientific Notation

Chemists often work with numbers that are exceedingly large or small. For example, entering the mass in grams of a hydrogen atom into a calculator requires a display with at least 24 decimal places! A system called **scientific notation** avoids much of the tedium and awkwardness of manipulating numbers with large or small magnitudes. In scientific notation, these numbers are expressed in the form

$$N \times 10^n$$

where N is greater than or equal to 1 and less than 10 ($1 \leq N < 10$), and n is an integer that can be either positive or negative ($10^0 = 1$). The number 10 is called the *base* because it is this number that is being raised to the power n. Although a base number may have values other than 10, the base number in scientific notation is *always* 10.

A simple way to convert numbers to scientific notation is to move the decimal point as many places to the left or right as needed to give a number between 1 and 10 (N). The magnitude of n is then determined as follows: *If the decimal point had to be moved to the* left *n places, n is positive; if the decimal point had to be moved to the* right *n places, n is negative*. Another way to remember this is to recognize that as the number decreases in magnitude, the exponent increases, and vice versa. The application of this rule is illustrated in Skill Builder ES1.1.

SKILL BUILDER ES1.1

Convert each number to scientific notation: **(a)** 637.8; **(b)** 0.0479; **(c)** 7.86; **(d)** 12,378; **(e)** 0.00032.

Solution **(a)** To convert 637.8 to a number between 1 and 10, we move the decimal point two places to the *left*: **637.8**. Because the decimal point was moved two places to the left, $n = 2$. Thus, in scientific notation 637.8 = 6.378×10^2. **(b)** To convert 0.0479 to a number between 1 and 10, we move the decimal point two places to the *right*: **0.0479**. Because the decimal point was moved two places to the right, $n = -2$. Thus, in scientific notation 0.0479 = 4.79×10^{-2}. **(c)** 7.86×10^0, although this is usually expressed simply as 7.86. (Recall that 10^0 = 1.) **(d)** 1.2378×10^4. Because the decimal point was moved four places to the right, $n = 4$. **(e)** 3.2×10^{-4}. Because the decimal point was moved four places to the left, $n = -4$.

Addition and Subtraction

Before numbers expressed in scientific notation can be added or subtracted, they must be converted to a form in which all the exponents have the same value. The appropriate operation is then carried out on the values of N. Skill Builder ES1.2 illustrates how to do this.

SKILL BUILDER ES1.2

Carry out the appropriate operation on each number, and then express the answer in scientific notation: **(a)** $1.36 \times 10^2 + 4.73 \times 10^3$; **(b)** $6.923 \times 10^{-3} - 8.756 \times 10^{-4}$.

Solution **(a)** Both exponents must have the same value, so these numbers are converted to either $1.36 \times 10^2 + 47.3 \times 10^2$ or $0.136 \times 10^3 + 4.73 \times 10^3$. Choosing either alternative gives the same answer, reported to two decimal places:

$$1.36 \times 10^2 + 47.3 \times 10^2 = (1.36 + 47.3) \times 10^2 = 48.66 \times 10^2 = 4.87 \times 10^3$$
$$0.136 \times 10^3 + 4.73 \times 10^3 = (0.136 + 4.73) \times 10^3 = 4.87 \times 10^3$$

Notice that in converting 48.66×10^2 to scientific notation, n has become more positive by 1. **(b)** Converting the exponents to the same value gives either $6.923 \times 10^{-3} - 0.8756 \times 10^{-3}$ or $69.23 \times 10^{-4} - 8.756 \times 10^{-4}$. Completing the calculations gives the same answer, expressed to three decimal places:

$$6.923 \times 10^{-3} - 0.8756 \times 10^{-3} = (6.923 - 0.8756) \times 10^{-3} = 6.047 \times 10^{-3}$$
$$69.23 \times 10^{-4} - 8.756 \times 10^{-4} = (69.23 - 8.756) \times 10^{-4} = 60.474 \times 10^{-4} = 6.047 \times 10^{-3}$$

Multiplication and Division

When multiplying numbers expressed in scientific notation, we multiply the values of N and sum the values of n. Conversely, when dividing, we divide N in the dividend (the number being divided) by N in the divisor (the number by which we are dividing) and then subtract n in the divisor from n in the dividend. In contrast to addition and subtraction, the exponents do *not* have to be the same in multiplication and division. Examples of problems involving multiplication and division are shown in Skill Builder ES1.3.

SKILL BUILDER ES1.3

Perform the appropriate operation on each of the numbers, and express your answer in scientific notation: **(a)** $(6.022 \times 10^{23})(6.42 \times 10^{-2})$; **(b)** $(1.67 \times 10^{-24})/(9.12 \times 10^{-28})$; **(c)** $(6.63 \times 10^{-34})(6.0 \times 10)/(8.52 \times 10^{-2})$.

Solution **(a)** This is a multiplication problem, so we add the exponents:

$$(6.022 \times 10^{23})(6.42 \times 10^{-2}) = (6.022)(6.42) \times 10^{[23+(-2)]} = 38.7 \times 10^{21} = 3.87 \times 10^{22}$$

(b) In division, we subtract the exponents:

$$\frac{(1.67 \times 10^{-24})}{(9.12 \times 10^{-28})} = \frac{1.67}{9.12} \times 10^{[-24-(-28)]} = 0.183 \times 10^4 = 1.83 \times 10^3$$

(c) This problem has both multiplication *and* division:

$$\frac{(6.63 \times 10^{-34})(6.0 \times 10)}{(8.52 \times 10^{-2})} = \frac{(6.63)(6.0)}{8.52} \times 10^{[(-34)+1-(-2)]} = 4.7 \times 10^{-31}$$

Significant Figures

No measurement can be free from error. Error is introduced by the limitations of instruments and measuring devices (such as the size of the divisions on a graduated cylinder) and by the imperfection of human senses. Although errors in calculations can be enormous, they do *not* contribute to the uncertainty in measurements that we are discussing. Chemists describe the estimated degree of error in a measurement as the *uncertainty* of the measurement, and they are careful to report all measured values using only **significant figures**, numbers that describe the value without exaggerating the degree to which it is known to be accurate. Chemists report as significant all numbers known with absolute certainty, plus one more digit that is understood to contain some uncertainty. The uncertainty in the final digit is usually assumed to be ± 1, unless otherwise stated.

The following rules have been developed for counting the number of significant figures in a measurement or calculation:

1. A digit that is not zero is significant.
2. Any zero between nonzero digits is significant. The number 2005, for example, has four significant figures.
3. Any zero used as a placeholder preceding the first nonzero digit is not significant. Thus, 0.05 has one significant figure because the zeros are used to indicate the placement of the digit 5. In contrast, 0.050 has two significant figures because the last two digits correspond to the number 50; the last zero is not a placeholder. As a further example, the number 5.0 has two significant figures because the zero is used not to place the 5 but to indicate 5.0.
4. When a number does not contain a decimal point, zeros added after a nonzero number may or may not be significant. An example is the number 100, which may be interpreted as having one, two, or three significant figures. (Note: Treat all trailing zeros in exercises and problems in this textbook as significant unless you are specifically told otherwise.)
5. Integers obtained either by counting objects or from definitions are *exact numbers*, which are considered to have infinitely many significant figures. If we have counted four objects, for example, then the number 4 has an infinite number of significant figures (that is, it represents 4.000 . . .). Similarly, 1 foot is defined to contain 12 inches, so the number 12 in the equation below has infinitely many significant figures:

$$1 \text{ foot} = 12 \text{ inches}$$

An effective method for determining the number of significant figures is to convert the measured or calculated value to scientific notation because any zero used as a placeholder is eliminated in the conversion. An example is 0.0800, which is expressed in scientific notation as 8.00×10^{-2}. In this form, it is more readily apparent that the number has three significant figures rather than five; in scientific notation, the number preceding the exponential (N) determines the number of significant figures. Skill Builder ES1.4 provides practice with these rules.

SKILL BUILDER ES1.4

Give the number of significant figures in each: **(a)** 5.87; **(b)** 0.031; **(c)** 52.90; **(d)** 00.2001; **(e)** 500; **(f)** 6 atoms.

Solution **(a)** Three (rule 1); **(b)** Two (rule 3). In scientific notation, this number is represented as 3.1×10^{-2}, showing that it has two significant figures; **(c)** Four (rule 3); **(d)** Four (rule 2). This number is 2.001×10^{-1} in scientific notation, showing that it has four significant figures; **(e)** One, two, or three (rule 4); **(f)** Infinite (rule 5).

Mathematical operations are carried out using all the digits given and then rounding the final result to the correct number of significant figures to obtain a reasonable answer. This method avoids compounding inaccuracies by successively rounding intermediate calculations. After you complete a calculation, you may have to round the last significant figure up or down depending on the value of the digit that follows it. If the digit is 5 or greater, then the number is rounded up. For example, to three significant figures the number 5.215 is 5.22, whereas 5.213 is rounded to 5.21. Similarly, to three significant figures, 5.005 kg becomes 5.01 kg, whereas 5.004 kg becomes 5.00 kg. The procedures for dealing with significant figures are different for addition and subtraction versus multiplication and division, as described below.

When we are adding or subtracting measured values, the value with the fewest significant figures to the right of the decimal point determines the number of significant figures to the right of the decimal point in the answer. Drawing a vertical line to the right of the column corresponding to the smallest number of significant figures is a simple method of determining the proper length for the answer:

$$\begin{array}{r|l} 3240.7 & \\ +\quad 21.2 & 36 \\ \hline 3261.9 & 36 \end{array}$$

The line indicates that in the answer to this calculation, the digits 3 and 6 are not significant. Consequently, the answer is expressed as 3261.9, with five significant figures. The reason they are not is that values for the corresponding places in the other measurement are unknown (3240.7??). Again, numbers ≥ 5 are rounded up. If our second number in the calculation had been 21.256, then we would have needed to round 3261.956 to 3262.0 to complete our calculation.

When we are multiplying or dividing measured values, the answer is limited to the smallest number of significant figures in the calculation; thus, $42.9 \times 8.323 = 357.057 = 357$. Although the second number in the calculation has four significant figures, we are justified in reporting the answer to only three because the first number in the calculation has only three significant figures.

When you are solving a problem using a calculator, it is important to remember that the number shown in the calculator display often shows more digits than can be reported as significant in your answer. When a measurement reported as 5.0 kg is divided by 3.0 L, for example, the display may show 1.666666667 as the answer. We are justified in reporting an answer to only two significant figures, however, giving 1.7 kg/L as the answer, with the last digit understood to have some uncertainty.

In calculations involving several steps, slightly different answers can be obtained depending on how rounding is handled and if rounding is performed on intermediate results or if rounding is postponed until the last step, and rounding the final answer to the correct number of significant figures should be performed at the end. Rounding at intermediate steps is avoided, because rounding of intermediate results can sometimes cause the final answer to be significantly in error.

In practice, chemists generally work with a calculator, as described above. When working on paper, however, we often want to minimize the number of digits we have to write out. Because successive rounding can compound inaccuracies, intermediate roundings need to be handled correctly. When working on paper, always round an intermediate result so as to retain one more digit than can be justified, and carry this number into the next step in the calculation. The final answer is then rounded to the correct number of significant figures at the very end.

In the worked examples in this text, we will often show the results of intermediate steps in a calculation. In doing so, we will show the results to only the correct number of significant figures allowed *for that step*, in effect treating each step as a separate calculation. This is intended to make it easier for you to follow along, but in some cases this procedure may give a final answer that differs in the last digit from that obtained using a calculator and carrying all digits through to the last step as described above. Skill Builder ES1.5 provides practice with calculations using significant figures.

SKILL BUILDER ES1.5

Complete the calculations, and report your answers using the correct number of significant figures: **(a)** 87.25 mL + 3.0201 mL; **(b)** 26.843 g + 12.23 g; **(c)** 6×12.011; **(d)** 2(1.008) g + 15.99 g; **(e)** 137.3 + 2(35.45).

Solution **(a)** 90.27 mL; **(b)** 39.07 g; **(c)** 72.066 (See rule 5 under "Significant Figures."); **(d)** 2(1.008) g + 15.99 g = 2.016 g + 15.99 g = 18.01 g; **(e)** 137.3 + 2(35.45) = 137.3 + 70.90 = 208.2

Accuracy and Precision

Measurements may be **accurate**, meaning that the measured value is the same as the true value; they may be **precise**, meaning that multiple measurements give nearly identical values (*reproducible results*); or they may be both accurate and precise. The goal of scientists is to obtain measured values that are both accurate and precise.

Suppose, for example, that the mass of a sample of gold was measured on one balance and found to be 1.896 g. On a different balance, the same sample was found to have a mass of 1.125 g. Which was correct? Careful and repeated measurements, including measurements on a third balance, showed the sample to have a mass of 1.895 g. The masses obtained from the three balances are listed in the table.

Balance 1	Balance 2	Balance 3
1.896 g	1.125 g	1.893 g
1.895 g	1.158 g	1.895 g
1.894 g	1.067 g	1.895 g

Whereas the measurements obtained from balances 1 and 3 are reproducible (precise) and are close to the accepted value (accurate), those obtained from balance 2 are neither. Even if the measurements obtained from balance 2 had been precise (if, for example, they had been 1.125, 1.124, and 1.125), they still would not have been accurate. We can assess the precision of a set of measurements by calculating the *average deviation* of the measurements as follows:

1. Calculate the average value of all the measurements:

$$\text{average} = \frac{\text{sum of measurements}}{\text{number of measurements}}$$

2. Calculate the *deviation* of each measurement, which is the absolute value of the difference between each measurement and the average value:

$$\text{deviation} = |\text{measurement} - \text{average}|$$

where the symbol $|\ldots|$ means absolute value (in other words, convert any negative number to a positive number).

3. Add all of the deviations and divide by the number of measurements to obtain the average deviation:

$$\text{average deviation} = \frac{\text{sum of deviations}}{\text{number of measurements}}$$

Then we can express the precision as a percentage by dividing the average deviation by the average value of the measurements and multiplying the result by 100%:

$$\text{precision} = \frac{\text{average deviation}}{\text{average}} \times 100\%$$

In the case of balance 2, the average value is

$$\frac{1.125\text{ g} + 1.158 + 1.067\text{ g}}{3} = 1.117\text{ g}$$

The deviations are 1.125 g − 1.117 g = 0.008 g, 1.158 g − 1.117 g = 0.041 g, and |1.067 g − 1.117 g| = 0.050 g. Thus, the average deviation is

$$\frac{0.008\text{ g} + 0.041\text{ g} + 0.050\text{ g}}{3} = \frac{0.099\text{ g}}{3} = 0.033\text{ g}$$

The precision of this set of measurements is therefore

$$\frac{0.033\text{ g}}{1.117\text{ g}} \times 100\% = 3.0\%$$

When a series of measurements is precise but not accurate, the error is usually systematic. Systematic errors can be caused by faulty instrumentation or faulty technique. The difference between accuracy and precision is demonstrated in Skill Builder ES1.6.

SKILL BUILDER ES1.6

(a) A 1-carat diamond has a mass of 200.0 mg. When a jeweler repeatedly weighed a 2-carat diamond, he obtained measurements of 450.0 mg, 459.0 mg, and 463.0 mg. Were the jeweler's measurements accurate? Were they precise? **(b)** A single copper penny was tested three times to determine its composition. The first analysis gave a composition of 93.2% zinc and 2.8% copper, the second gave 92.9% zinc and 3.1% copper, and the third gave 93.5% zinc and 2.5% copper. The actual composition of the penny was 97.6% zinc and 2.4% copper. Were the results accurate? Were they precise?

Solution **(a)** The expected mass of a 2-carat diamond is 2 × 200.0 mg = 400.0 mg. The average of the three measurements is 457.3 mg, about 13% greater than the true mass. Hence, these measurements are not particularly accurate.

The deviations of the measurements are 0.0073 mg, 0.0017 mg, and 0.0057 mg, respectively, which give an average deviation of 0.0049 mg and a precision of

$$\frac{0.0049\text{ mg}}{457.3\text{ mg}} \times 100\% = 1.1\%$$

Thus, these measurements are rather precise. **(b)** The average values of the measurements are 93.2% zinc and 2.8% copper versus the true values of 97.6% zinc and 2.4% copper. Thus, these measurements are not very accurate, with errors of −4.5% and +20% for zinc and copper, respectively. (*Note*: The sum of the measured zinc and copper contents is only 96.0% rather than 100%, which tells us immediately that *either* there is a significant error in one or both measurements *or* some other element is present.)

The deviations of the measurements are 0.0%, 0.3%, and 0.3% for both zinc and copper, which give an average deviation of 0.2% for both metals. We might therefore conclude that the measurements are equally precise, but that is not the case. Recall that precision is the average deviation divided by the average value times 100%. Because the average value of the zinc measurements is much greater than the average value of the copper measurements (93.2% versus 2.8%), the copper measurements are much less precise.

$$\text{precision (Zn)} = \frac{0.2\%}{93.2\%} \times 100\% = 0.2\%$$

$$\text{precision (Cu)} = \frac{0.2\%}{2.8\%} \times 100\% = 7\%$$

2 Molecules, Ions, and Chemical Formulas

LEARNING OBJECTIVES

- To understand the differences between covalent and ionic bonding
- To be able to describe the composition of a chemical compound
- To be able to write an empirical formula
- To be able to name ionic compounds
- To be able to name covalent compounds that contain up to three elements
- To be able to identify and name common acids and bases

CHAPTER OUTLINE

Chapter 1 introduced some of the fundamental concepts of chemistry, with particular attention to the basic properties of atoms and elements. These entities are the building blocks of all the substances we encounter, yet most common substances do not consist of pure elements or of individual atoms. Instead, nearly all substances are chemical compounds or mixtures of chemical compounds. Although there are only about 115 elements (of which about 86 occur naturally), millions of chemical compounds are known, with a tremendous range of physical and chemical properties. Consequently, the emphasis of modern chemistry, and of this text, is on understanding the relationship between the structures and properties of chemical compounds.

In this chapter, you will learn how to describe the composition of chemical compounds. We introduce you to *chemical nomenclature*—the language of chemistry, which will enable you to recognize and name the most common

Using chemicals, catalysts, heat, and pressure, a petroleum refinery will separate, combine, and rearrange the structure and bonding patterns of the basic carbon–hydrogen molecules found in crude oil. The final products include petrol for car fuel, paraffin, diesel, lubricants, and bitumen.

kinds of compounds. Not only is an understanding of chemical nomenclature essential for your study of chemistry, but it also has other benefits. For example, you will be able understand the labels on products found in the supermarket and pharmacy. You will also be better equipped to understand many of the important environmental and medical issues that face society. By the end of this chapter, you will be able to describe what happens chemically when a doctor prepares a cast to stabilize a broken bone, and you will know the composition of common substances such as laundry bleach, the active ingredient in baking powder, and the foul-smelling compound responsible for the odor of spoiled fish. Finally, you will be able to explain the chemical differences between different grades of gasoline.

2.1 • Chemical Compounds

MGC Covalent Compounds vs. Ionic Compounds

The atoms in all substances that contain more than one atom are held together by *electrostatic interactions*, which are interactions between electrically charged particles such as protons and electrons. *Electrostatic attraction* between oppositely charged species (positive and negative) results in a force that causes them to move toward each other. In contrast, *electrostatic repulsion* between two species that have the same charge (either both positive or both negative) results in a force that causes them to repel each other. Atoms form chemical compounds when the attractive electrostatic interactions between them are stronger than the repulsive interactions. Collectively, we refer to the attractive interactions between atoms as **chemical bonds**.

Chemical bonds are generally divided into two fundamentally different kinds: ionic and covalent. In reality, however, the bonds in most substances are neither purely ionic nor purely covalent, but instead are closer to one extreme or the other. Although purely ionic and covalent bonds represent extreme cases that are seldom encountered in anything but very simple substances, a brief discussion of these two extremes helps us understand why substances that have different kinds of chemical bonds have very different properties. **Ionic compounds** consist of positively and negatively charged ions held together by strong electrostatic forces, whereas **covalent compounds** consist of **molecules**, which are groups of atoms in which one or more pairs of electrons are shared between bonded atoms. In a *covalent bond*, the atoms are held together by the electrostatic attraction between the positively charged nuclei of the bonded atoms and the negatively charged electrons they share. We begin our discussion of structures and formulas by describing covalent compounds.

Covalent Compounds

Just as an atom is the simplest unit that has the fundamental chemical properties of an element, a molecule is the simplest unit that has the fundamental chemical properties of a covalent compound. Each covalent compound is represented by a **molecular formula**, which gives the atomic symbol for each component element, in a prescribed order, accompanied by a subscript indicating the number of atoms of that element in the molecule. The subscript is written only if the number is greater than 1. For example, water, with 2 hydrogen atoms and 1 oxygen atom per molecule, is written as H_2O. Similarly, carbon dioxide, which contains 1 atom of carbon and 2 atoms of oxygen in each molecule, is written as CO_2.

Some pure elements also exist as covalent molecules. Hydrogen, nitrogen, oxygen, and the halogens occur naturally as the diatomic molecules H_2, N_2, O_2, F_2, Cl_2, Br_2, and I_2 (Figure 2.1a). Similarly, a few pure elements are *polyatomic* ("many atoms") *molecules*, such as elemental phosphorus and sulfur, which occur as P_4 and S_8 (Figure 2.1b).

Covalent compounds that contain predominantly carbon and hydrogen are called **organic compounds**. The convention for representing the formulas of organic compounds

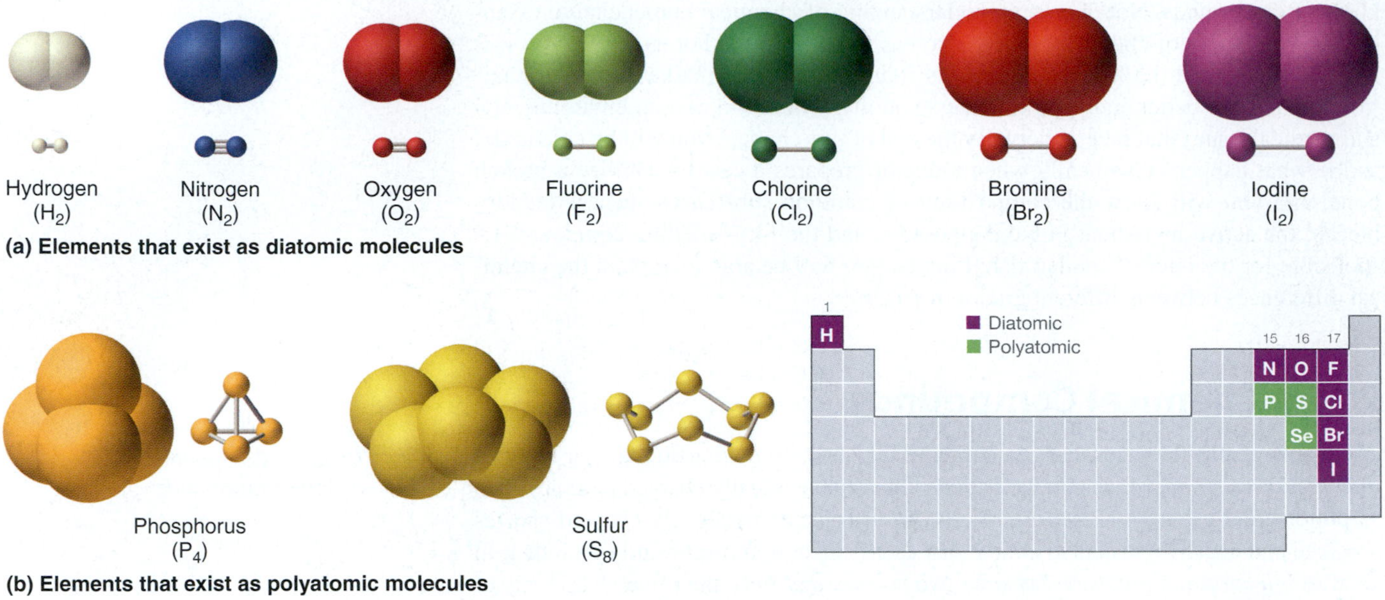

Figure 2.1 Elements that exist as covalent molecules. (a) Several elements naturally exist as diatomic molecules, in which two atoms (E) are joined by one or more covalent bonds to form a molecule with the general formula E_2. (b) A few elements naturally exist as polyatomic molecules, which contain more than two atoms. For example, phosphorus exists as P_4 tetrahedra, regular polyhedra with four triangular sides, with a phosphorus atom at each vertex. Elemental sulfur consists of a puckered ring of eight sulfur atoms connected by single bonds.

Note the pattern

For organic compounds: write C first, then H, then other elements in alphabetical order. For molecular inorganic compounds: start with element at far left; list elements in same group alphabetically.

is to write carbon first, followed by hydrogen and then by other elements in alphabetical order (CH_4O for methyl alcohol, for example, a fuel). Compounds that consist primarily of elements other than carbon and hydrogen are called **inorganic compounds**; they include both covalent and ionic compounds. In inorganic compounds, the component elements are listed beginning with the one farthest to the left in the periodic table; those in the same group are listed alphabetically. By convention, however, when an inorganic compound contains both hydrogen and an element from Groups 13–15, the latter is usually listed first in the formula. Examples are methane (CH_4) and ammonia (NH_3). Compounds such as water, whose compositions were established long before this convention was adopted, are always written with hydrogen first: Water, for example, is always written as H_2O, not OH_2. The conventions for inorganic acids, such as HCl and H_2SO_4, are described in Section 2.5.

EXAMPLE 2.1

Write the molecular formula of each compound: **(a)** The phosphorus–sulfur compound that is responsible for the ignition of so-called "strike anywhere" matches contains 4 phosphorus atoms and 3 sulfur atoms per molecule; **(b)** Ethyl alcohol, the "alcohol" of alcoholic beverages, contains 1 oxygen atom, 2 carbon atoms, and 6 hydrogen atoms per molecule; **(c)** Freon-11, once widely used in automobile air conditioners and implicated in damage to the ozone layer, contains 1 carbon atom, 3 chlorine atoms, and 1 fluorine atom per molecule.

Given Identity of elements present and number of atoms of each

Asked for Molecular formula

Strategy

A Identify the symbol for each element in the molecule. Then identify the substance as either an organic or inorganic compound.

B If the substance is an organic compound, arrange the elements in order beginning with carbon and hydrogen, and then list the other elements alphabetically. If it is an inorganic compound, list the elements beginning with the one farthest left in the periodic table. List elements in the same group alphabetically.

Ⓒ From the information given, add the subscript for each kind of atom to write the molecular formula

Solution

(a) Ⓐ✓ The molecule contains 4 P atoms and 3 S atoms. Because the compound does not contain mostly C and H, it is inorganic. Ⓑ✓ Phosphorus is in Group 15, and sulfur is in Group 16. Because phosphorus is to the left of sulfur, it is written first. Ⓒ✓ Writing the number of each kind of atom as a right-hand subscript gives P_4S_3 as the molecular formula.

(b) Ⓐ✓ Ethyl alcohol contains predominantly C and H, so it is an organic compound. Ⓑ✓ The formula for an organic compound is written with the number of C atoms first, the number of H atoms next, and the other atoms in alphabetical order: CHO. Ⓒ✓ Adding subscripts gives the molecular formula C_2H_6O.

(c) Ⓐ✓ Freon-11 contains C, Cl, and F. It can be viewed as either an inorganic or organic compound (in which fluorine has replaced hydrogen). The formula for Freon-11 can therefore be written using either of the two conventions. Ⓑ✓ According to the convention for inorganic compounds, carbon is written first because it is farther left in the periodic table. Fluorine and chlorine are in the same group, so they are listed alphabetically. We have CClF. Ⓒ✓ Adding subscripts gives the molecular formula CCl_3F. Ⓑ✓ Ⓒ✓ We obtain the same formula for Freon-11 using the convention for organic compounds. The number of carbon atoms is written first, followed by the number of H atoms (zero), and then the other elements in alphabetical order, also giving CCl_3F.

EXERCISE 2.1

Write the molecular formula for each compound: **(a)** Nitrous oxide, also called "laughing gas," contains 2 atoms of nitrogen and 1 of oxygen per molecule. Nitrous oxide is used as a mild anesthetic for minor surgery and as the propellant in cans of whipped cream; **(b)** Sucrose, also known as cane sugar, contains 12 carbon atoms, 11 oxygen atoms, and 22 hydrogen atoms; **(c)** The gas sulfur hexafluoride, used to pressurize "unpressurized" tennis balls and as a coolant in nuclear reactors, contains 6 fluorine atoms and 1 sulfur atom per molecule.

Answer **(a)** N_2O; **(b)** $C_{12}H_{22}O_{11}$; **(c)** SF_6

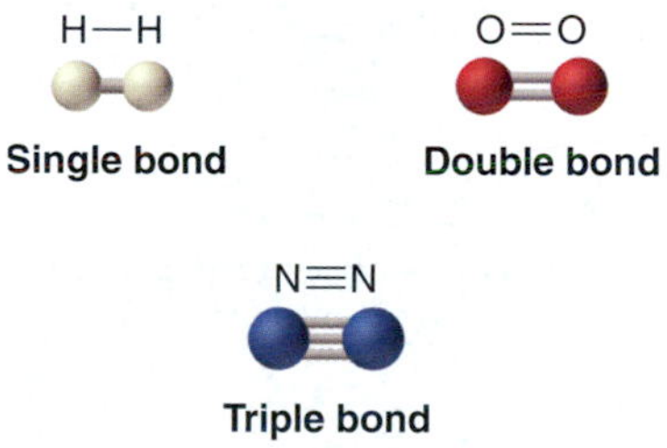

Figure 2.2 Molecules that contain single, double, and triple bonds between atoms. Hydrogen (H_2) has a single bond between atoms. Oxygen (O_2) has a double bond between atoms, indicated by two lines (═). Nitrogen (N_2) has a triple bond between atoms, indicated by three lines (≡).

Representations of Molecular Structures

Molecular formulas give only the elemental composition of molecules. In contrast, **structural formulas** show which atoms are bonded to one another and, in some cases, the approximate arrangement of the atoms in space. Knowing the structural formula of a compound enables chemists to create a three-dimensional model, which provides information about how that compound will behave physically and chemically.

The structural formula for the molecule H_2 can be drawn as H—H and that of I_2 as I—I, where the line indicates a single pair of shared electrons, a **single bond**. A **double bond**, in which two pairs of electrons are shared, is indicated by two lines (for example, O_2 is O═O). In a **triple bond**, shown by three lines, three pairs of electrons are shared (N_2 is N≡N) (Figure 2.2). Carbon is unique in the extent to which it forms single, double, and triple bonds to itself and to other elements. The number of bonds formed by an atom in its covalent compounds is *not* arbitrary. As you will learn in Chapter 8, hydrogen, oxygen, nitrogen, and carbon have a very strong tendency to form substances in which they have one, two, three, and four bonds to other atoms, respectively (Table 2.1).

TABLE 2.1 The number of bonds that selected atoms commonly form to other atoms

Atom	Number of Bonds
H (Group 1)	1
O (Group 16)	2
N (Group 15)	3
C (Group 14)	4

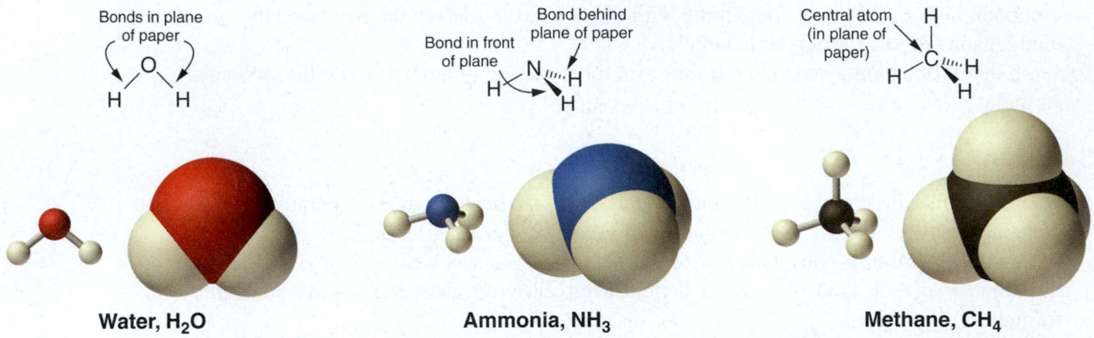

Figure 2.3 The three-dimensional structures of water (H_2O), ammonia (NH_3), and methane (CH_4). Water is a V-shaped molecule, in which all three atoms lie in a plane. In contrast, ammonia has a pyramidal structure, in which the three H atoms form the base of the pyramid and the N atom is at the vertex. The four H atoms of methane form a tetrahedron; the C atom lies in the center.

The structural formula for water can be drawn as H—O—H or H/O\H. Because the latter approximates the experimentally determined shape of the water molecule, it is more informative. Similarly, ammonia (NH_3) and methane (CH_4) are often written as

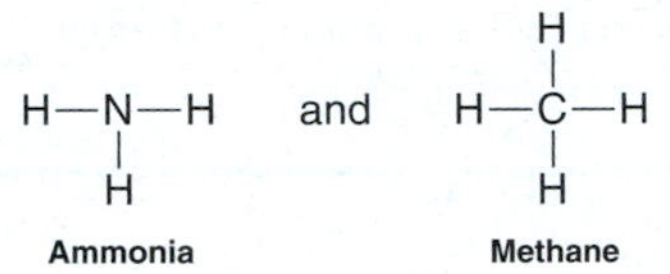

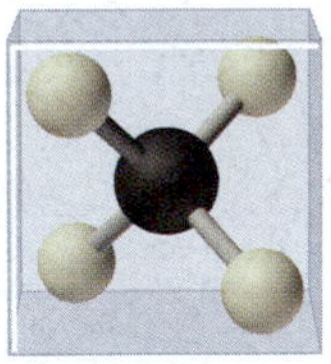

Tetrahedral structure of methane, CH_4

As shown in Figure 2.3, however, the actual three-dimensional structure of NH_3 looks like a pyramid with a triangular base of three H atoms. The structure of CH_4, with four H atoms arranged around a central C as shown in Figure 2.3, is *tetrahedral*. That is, the H atoms are positioned at every other vertex of a cube, as shown in the margin. Many compounds—carbon compounds, in particular—have four bonded atoms arranged around a central atom to form a tetrahedron.

Figures 2.1–2.3 illustrate different ways to represent the structures of molecules. It should be clear that there is no single "best" way to draw the structure of a molecule; the method you use depends on which aspect of the structure you want to emphasize, and how much time and effort you want to spend. Figure 2.4 shows some of the different ways to portray the structure of a slightly more complex molecule, methanol. These representations differ greatly in their information content. For example, the molecular

Figure 2.4 Different ways of representing the structure of a molecule. (a) The molecular formula for methanol gives only the number of each kind of atom present. (b) The structural formula shows which atoms are connected. (c) The ball-and-stick model shows the atoms as spheres and the bonds as sticks. (d) A perspective drawing (also called a *wedge-and-dash* representation) attempts to show the three-dimensional structure of the molecule. (e) The space-filling model shows the atoms in the molecule but not the bonds. (f) The condensed structural formula is by far the easiest and most common way to represent a molecule.

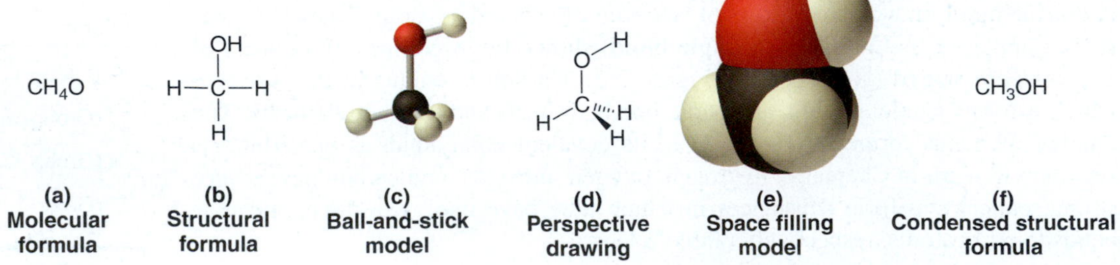

formula for methanol (Figure 2.4a) gives only the number of each kind of atom; writing methanol as CH_4O tells nothing about its structure. In contrast, the structural formula (Figure 2.4b) indicates how the atoms are connected, but it makes methanol look as if it is planar (which it is not). Both the ball-and-stick model (Figure 2.4c) and the perspective drawing (Figure 2.4d) show the three-dimensional structure of the molecule. The latter (also called a *wedge-and-dash* representation) is the easiest way to sketch the structure of a molecule in three dimensions. It shows which atoms are above and below the plane of the paper by using wedges and dashes, respectively; the central atom is always assumed to be in the plane of the paper. The space-filling model (Figure 2.4e) illustrates the approximate relative sizes of the atoms in the molecule, but it does not show the bonds between atoms. Also, in a space-filling model, atoms at the "back" of the molecule may be obscured by atoms in the front.

Although a structural formula, ball-and-stick model, perspective drawing, or space-filling model provides a significant amount of information about the structure of a molecule, each of these requires time and effort. Consequently, chemists often use a *condensed structural formula* (Figure 2.4f), which omits the lines representing bonds between atoms and simply lists the atoms bonded to a given atom next to it. Multiple groups attached to the same atom are shown in parentheses, followed by a subscript that indicates the number of such groups. For example, the condensed structural formula for methanol is CH_3OH, which tells that the molecule contains a CH_3 unit that looks like a fragment of methane (CH_4). Methanol can therefore be viewed either as a methane molecule in which one H atom has been replaced by an OH or as a water molecule in which one H atom has been replaced by a CH_3 fragment. Because of their ease of use and information content, we use condensed structural formulas for molecules throughout this text. Ball-and-stick models are used when needed to illustrate the three-dimensional structure of molecules, and space-filling models are used only when it is necessary to visualize the relative sizes of atoms or molecules in order to understand an important point.

EXAMPLE 2.2

Write the molecular formula for each compound. The structural formula is given: **(a)** Sulfur monochloride (also called disulfur dichloride) is a vile-smelling, corrosive yellow liquid used in the production of synthetic rubber; its structural formula is ClSSCl; **(b)** Ethylene glycol is the major ingredient in antifreeze; its structural formula is $HOCH_2CH_2OH$; **(c)** Trimethylamine is one of the substances responsible for the smell of spoiled fish; its structural formula is $(CH_3)_3N$.

Given Structural formula

Asked for Molecular formula

Strategy

A Identify every element in the structural formula, and then determine whether the compound is organic or inorganic.

B Use either the organic or inorganic convention to list the elements in order. Then add appropriate subscripts to indicate the number of atoms of each element present in the molecular formula.

Solution

The molecular formula lists the elements in the molecule and the number of atoms of each.

(a) **A** Sulfur monochloride contains 2 atoms each of S and Cl per molecule. It does not contain mostly carbon and hydrogen, so it is an inorganic compound. **B** Because sulfur lies to the left of chlorine in the periodic table, it is written first in the formula. Adding subscripts gives the molecular formula S_2Cl_2.

(b) **A** Counting the atoms in ethylene glycol, we get 6H, 2C, and 2O per molecule. The compound consists mostly of C and H atoms, so it is organic. **B** As with all organic compounds, C and H are written first in the molecular formula. Adding appropriate subscripts gives the molecular formula $C_2H_6O_2$.

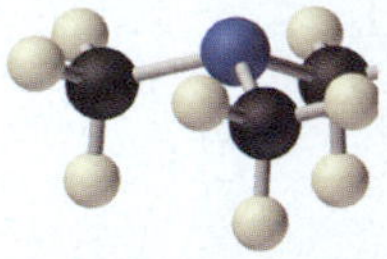

Trimethylamine

(c) The structural formula shows that trimethylamine contains $3CH_3$ units, so we have 1N, 3C, and 9H atoms per molecule. Because trimethylamine contains mostly carbon and hydrogen, it is an organic compound. According to the convention for organic compounds, C and H are written first, giving the molecular formula C_3H_9N.

EXERCISE 2.2

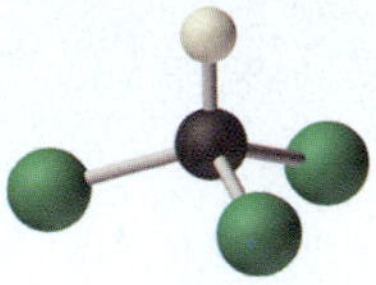

Chloroform

Write the molecular formula for each molecule: **(a)** Chloroform, which was one of the first anesthetics and was used in many cough syrups until recently, contains 1C atom, 1H atom, and 3Cl atoms; its structure is shown in the margin; **(b)** Hydrazine is used as a propellant in the attitude jets of the space shuttle; its structural formula is H_2NNH_2; **(c)** Putrescine is a pungent-smelling compound first isolated from extracts of rotting meat; its structural formula is $H_2NCH_2CH_2CH_2CH_2NH_2$.

Answer **(a)** $CHCl_3$; **(b)** N_2H_4; **(c)** $C_4H_{12}N_2$

Ionic Compounds

All the substances described in the preceding discussion are composed of molecules that are electrically neutral; that is, the number of positively charged protons in the nucleus is equal to the number of negatively charged electrons. In contrast, *ions* are atoms or assemblies of atoms that have a net electrical charge. Ions that contain fewer electrons than protons have a net positive charge and are called **cations**. Conversely, ions that contain more electrons than protons have a net negative charge and are called **anions**. *Ionic compounds* contain both cations and anions in a ratio that results in no net electrical charge.

In covalent compounds, electrons are shared between bonded atoms and are simultaneously attracted to more than one nucleus. In contrast, ionic compounds contain cations and anions rather than discrete neutral molecules. Ionic compounds are held together by the attractive electrostatic interactions between cations and anions. In an ionic compound, the cations and anions are arranged in space to form an extended three-dimensional array that maximizes the number of attractive electrostatic interactions and minimizes the number of repulsive electrostatic interactions (Figure 2.5). As shown in Equation 2.1, the electrostatic energy of the interaction

Figure 2.5 Covalent and ionic bonding. (a) In molecular hydrogen (H_2), two hydrogen atoms share two electrons to form a covalent bond. (b) The ionic compound NaCl forms when electrons from Na atoms are transferred to Cl atoms. The resulting Na^+ and Cl^- ions form a three-dimensional solid that is held together by attractive electrostatic interactions.

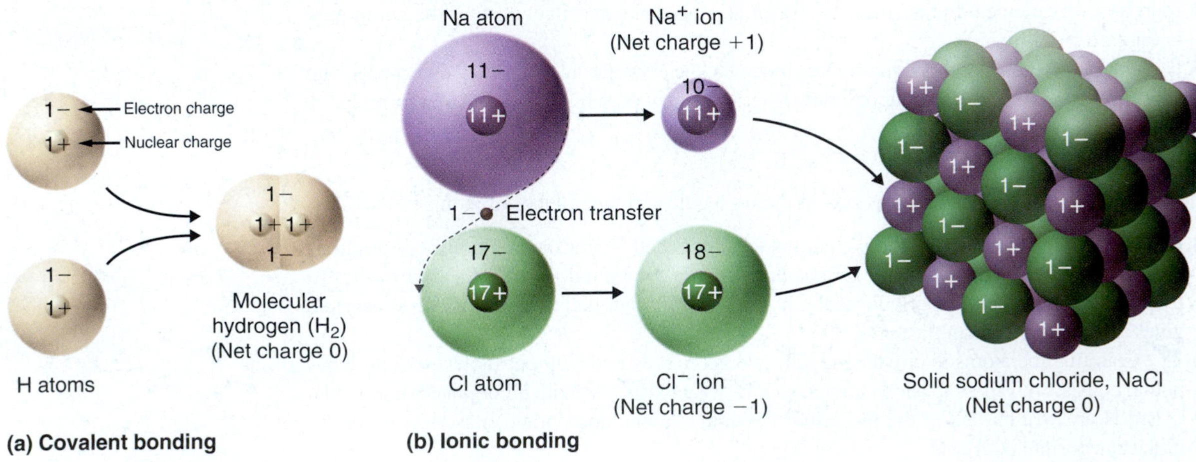

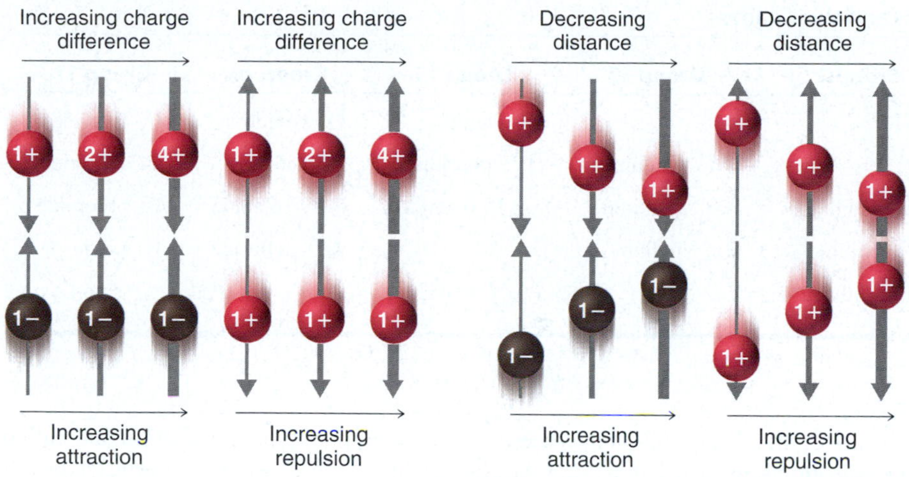

Figure 2.6 The effect of charge and distance on the strength of electrostatic interactions. As the charge on ions increases or the distance between ions decreases, so does the strength of the attractive (⊕···⊖) or repulsive (⊕···⊕ or ⊖···⊖) interactions.

between two charged particles is proportional to the product of the charges on the particles and inversely proportional to the distance between them:

$$\text{electrostatic energy} \propto \frac{(Q_1)(Q_2)}{r} \tag{2.1}$$

where Q_1 and Q_2 are the electrical charges on particles 1 and 2 and r is the distance between them. If the electrostatic energy is positive, the particles repel each other; if it is negative, the particles are attracted to each other. Thus, when Q_1 and Q_2 are both positive, corresponding to the charges on cations, the electrostatic energy is positive, and the cations repel each other. When Q_1 and Q_2 are both negative, corresponding to the charges on anions, the electrostatic energy is again positive and the anions repel each other. The electrostatic energy is negative only when the charges have opposite signs; thus, positively charged species are attracted to negatively charged species, and vice versa. As shown in Figure 2.6, the strength of the interaction is proportional to the *magnitude* of the charges and decreases as the *distance* between the particles increases.

Sodium chloride, NaCl, is an example of an ionic compound (Figure 2.7). Recall from Section 1.7 that the alkali metals tend to lose an electron to form ions with a +1 charge and the halogens tend to gain an electron to form ions with a −1 charge. Thus, when sodium and chlorine come into contact, each Na atom gives up an electron to become a Na^+ ion, with 11 protons in its nucleus but only 10 electrons, and each Cl atom gains an electron to become a Cl^- ion, with 17 protons in its nucleus and 18 electrons, as shown in Figure 2.5b. Solid sodium chloride contains equal numbers of cations (Na^+) and anions (Cl^-). Each Na^+ ion is surrounded by six Cl^- ions, and each Cl^- ion is surrounded by six Na^+ ions. Because of the large number of attractive $Na^+ \cdots Cl^-$ interactions, the total attractive electrostatic energy in NaCl is great.

Figure 2.7 NaCl, an ionic solid. The planes of an NaCl crystal reflect the regular three-dimensional arrangement of its Na^+ (purple) and Cl^- (green) ions.

Many elements have a tendency to gain or lose enough electrons to attain the same number of electrons as the noble gas closest to them in the periodic table. Thus, elements in Groups 1, 2, and 3 tend to *lose* one, two, and three electrons, respectively, to form cations such as Na^+ and Mg^{2+}, which like Ne have 10 electrons, or K^+, Ca^{2+}, and Sc^{3+}, which like Ar have 18 electrons. In addition, the elements in Groups 12 and 13 *lose* two or three electrons, resepectively, to form cations such as Zn^{2+}, Cd^{2+}, Al^{3+}, and Ga^{3+}. Because the lanthanides and actinides formally belong to Group 3, the most common ion formed by these elements is M^{3+}, where M represents the metal. Conversely, elements in Groups 17, 16, and 15 often react to *gain* one, two, and three electrons, respectively, to form ions such as

Note the pattern

Elements in Groups 1, 2, & 12, and 3 & 13 tend to form 1+, 2+, and 3+ ions, respectively; elements in Groups 15, 16, and 17 tend to form 3−, 2−, and 1− ions, respectively.

TABLE 2.2 Some common monatomic ions and their names

Group 1	Group 2	Group 3	Group 13	Group 15	Group 16	Group 17
Li^+ lithium	Be^{2+} beryllium			N^{3-} nitride	O^{2-} oxide	F^- fluoride
Na^+ sodium	Mg^{2+} magnesium		Al^{3+} aluminum	P^{3-} phosphide	S^{2-} sulfide	Cl^- chloride
K^+ potassium	Ca^{2+} calcium	Sc^{3+} scandium	Ga^{3+} gallium	As^{3-} arsenide	Se^{2-} selenide	Br^- bromide
Rb^+ rubidium	Sr^{2+} strontium	Y^{3+} yttrium	In^{3+} indium		Te^{2-} telluride	I^- iodide
Cs^+ cesium	Ba^{2+} barium	La^{3+} lanthanum				

Cl^-, S^{2-}, and P^{3-}. Ions such as these, which contain only a single atom, are called **monatomic ions**. You can predict the charges of most monatomic ions derived from the main group elements by simply looking at the periodic table and counting how many columns an element lies from the extreme left or right. Note that this method does not usually work for most of the transition metals, as you will learn in *Section 2.3*. Some common monatomic ions are listed in Table 2.2.

EXAMPLE 2.3

Predict the charges on the most common monatomic ion formed by each of these elements: **(a)** aluminum; **(b)** selenium; **(c)** yttrium.

Given Element

Asked for Ionic charge

Strategy

A Identify the group in the periodic table to which the element belongs. Based on its location in the periodic table, decide whether the element is a metal, which tends to lose electrons; a nonmetal, which tends to gain electrons; or a semimetal, which can do either.

B After locating the noble gas that is closest to the element, determine the number of electrons the element must gain or lose in order to have the same number of electrons as the nearest noble gas.

Solution

(a) **A** Aluminum, Al, is a metal in Group 13; consequently, it will tend to lose electrons. **B** The nearest noble gas to Al is Ne. Thus, aluminum will lose three electrons to form the Al^{3+} ion, which has the same number of electrons as Ne.

(b) **A** Selenium, Se, is a nonmetal in Group 16, so it will tend to gain electrons. **B** The nearest noble gas is Kr, so Se will gain two electrons to form the Se^{2-} ion, which has the same number of electrons as Kr.

(c) **A** Yttrium, Y, lies in Group 3 of the periodic table. Elements in this group are metals that tend to lose electrons. **B** The nearest noble gas to Y is Kr, so yttrium is predicted to lose three electrons to form Y^{3+}, which has the same number of electrons as Kr.

EXERCISE 2.3

Predict the charge on the most common monatomic ion formed by **(a)** calcium; **(b)** iodine; and **(c)** zirconium.

Answer **(a)** Ca^{2+}; **(b)** I^-; **(c)** Zr^{4+}

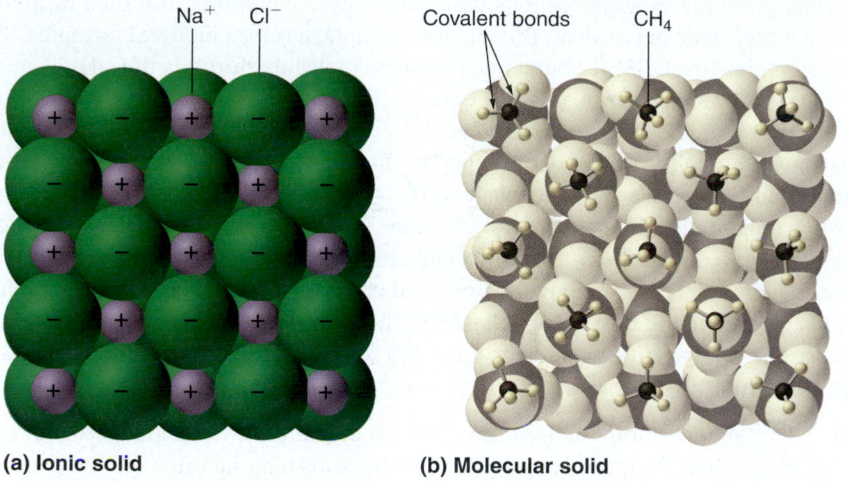

Figure 2.8 Interactions in ionic and covalent solids. (a) The positively and negatively charged ions in an ionic solid such as sodium chloride (NaCl) are held together by strong electrostatic interactions. (b) A representation of the packing of methane (CH_4) molecules in solid methane, a prototypical molecular solid. Note that the methane molecules are held together in the solid only by relatively weak intermolecular forces, even though the atoms within each methane molecule are held together by strong covalent bonds.

Physical Properties of Ionic and Covalent Compounds

In general, ionic and covalent compounds have different physical properties. Ionic compounds usually form hard crystalline solids that melt at rather high temperatures and are *very* resistant to evaporation. These properties stem from the characteristic internal structure of an ionic solid, illustrated schematically in Figure 2.8a, which shows the three-dimensional array of alternating positive and negative ions held together by strong electrostatic attractions. In contrast, as shown in Figure 2.8b, most covalent compounds consist of discrete molecules held together by comparatively weak *intermolecular* forces (the forces between molecules), even though the atoms within each molecule are held together by strong *intramolecular* covalent bonds (the forces within the molecule). Covalent substances can be gases, liquids, or solids at room temperature and pressure, depending on the strength of the intermolecular interactions. Covalent molecular solids tend to form soft crystals that melt at rather low temperatures and evaporate relatively easily.* The covalent bonds that hold the atoms together in the molecules are unaffected when covalent substances melt or evaporate, so a liquid or vapor of discrete, independent molecules is formed. For example, at room temperature, methane, the major constituent of natural gas, is a gas that is composed of discrete CH_4 molecules. The different physical properties of ionic and covalent molecular substances are summarized in Table 2.3.

TABLE 2.3 The physical properties of typical ionic and covalent compounds

Ionic Compounds	Covalent Molecular Substances
hard solids	gases, liquids, or soft solids
high melting points	low melting points
nonvolatile	volatile

2.2 • Chemical Formulas

MGC A Formula for Formulas

When chemists synthesize a new compound, they may not yet know its molecular or structural formula. In such cases, they usually begin by determining its **empirical formula**, the *relative* numbers of atoms of the elements in the compound, reduced to the smallest whole numbers. Because the empirical formula is based on experimental measurements of the numbers of atoms in a sample of the compound, it shows only the ratios of the numbers of the elements present. The difference between empirical and molecular formulas can be illustrated with butane, a covalent compound used as the fuel in disposable lighters. The molecular formula for butane is C_4H_{10}. Hence, the ratio of carbon atoms to hydrogen atoms in butane is 4:10, which can be reduced to 2:5. The empirical formula for butane is therefore C_2H_5. The **formula unit** is the

* Some covalent substances, however, are not molecular, but consist of infinite three-dimensional arrays of covalently bonded atoms and include some of the hardest materials known, such as diamond (Chapter 12).

absolute grouping of atoms or ions represented by the empirical formula of a compound, either ionic or covalent. Butane, for example, has the empirical formula C_2H_5, but it contains two C_2H_5 formula units, giving a molecular formula of C_4H_{10}.

Because ionic compounds do not contain discrete molecules, empirical formulas are used to indicate their compositions. All compounds, whether ionic or covalent, must be electrically neutral. Consequently, the positive and negative charges in a formula unit must exactly cancel each other. If the cation and the anion have charges of equal magnitude, such as Na^+ and Cl^-, then the compound must have a 1:1 ratio of cations to anions, and the empirical formula must be NaCl. If the charges are not the same magnitude, then a cation:anion ratio other than 1:1 is needed to produce a neutral compound. In the case of Mg^{2+} and Cl^-, for example, two Cl^- ions are needed to balance the two positive charges on each Mg^{2+} ion, giving an empirical formula of $MgCl_2$. Similarly, the formula for the ionic compound that contains Na^+ and O^{2-} ions is Na_2O.

Binary Ionic Compounds

An ionic compound that contains only two elements, one present as a cation and one as an anion, is called a *binary ionic compound*. The compound $MgCl_2$, for example, is a binary ionic compound. For such compounds, the subscripts in the empirical formula can also be obtained using the absolute value of the charge on one ion as the subscript for the other ion. This method is shown schematically in the drawing in the margin. When using this method, you will sometimes find it necessary to reduce the subscripts to their simplest ratio to write the empirical formula. Consider, for example, the compound formed by Mg^{2+} and O^{2-}. Using the absolute values of the charges on the ions as subscripts gives the formula Mg_2O_2:

Another method for obtaining subscripts in the empirical formula.

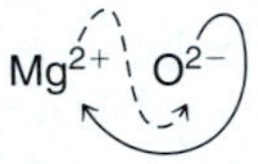

This simplifies to its correct empirical formula MgO. Hence, the empirical formula has one Mg^{2+} ion and one O^{2-} ion.

EXAMPLE 2.4

Write the empirical formulas for the simplest binary ionic compounds formed from these ions or elements: **(a)** Ga^{3+} and As^{3-}; **(b)** Eu^{3+} and O^{2-}; **(c)** calcium and chlorine.

Given Ions or elements

Asked for Empirical formula for binary ionic compound

Strategy

A If not given, determine the ionic charges based on the location of the elements in the periodic table.

B Use the absolute value of the charge on each ion as the subscript for the other ion. Reduce the subscripts to the lowest numbers to write the empirical formula. Check to make sure the empirical formula is electrically neutral.

Solution

(a) **B** ✓ Using the absolute values of the charges on the ions as the subscripts gives Ga_3As_3

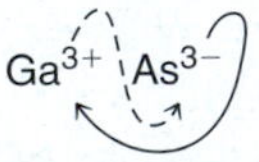

Reducing the subscripts to the smallest whole numbers gives the empirical formula GaAs, which is electrically neutral [$+3 + (-3) = 0$]. Alternatively, we could recognize that

Ga^{3+} and As^{3-} have charges of equal magnitude but opposite signs. Hence, one Ga^{3+} ion balances the charge on one As^{3-} ion, and a 1:1 compound will have no net charge. Since we write subscripts only if the number is greater than 1, the empirical formula is GaAs. GaAs is gallium arsenide, which is widely used in the electronics industry in transistors and other devices.

(b) Ⓑ Because Eu^{3+} has a charge of +3 and O^{2-} has a charge of −2, a 1:1 compound would have a net charge of +1. We must therefore find multiples of the charges that cancel. We use the absolute value of the charge on one ion as the subscript for the other ion:

Thus, the subscript for Eu^{3+} is 2 (from O^{2-}), and the subscript for O^{2-} is 3 (from Eu^{3+}), giving Eu_2O_3; the subscripts cannot be reduced further. The empirical formula contains a positive charge of 2(+3) = +6 and a negative charge of 3(−2) = −6, for a net charge of zero. Thus, the compound Eu_2O_3 is neutral, as it must be. Europium oxide is responsible for the red color in color television tubes.

(c) Ⓐ Because the charges on the ions are not given, we must first determine the charges expected for the most common ions derived from calcium and chlorine. Calcium (Ca) lies in Group 2, so it should lose two electrons to form Ca^{2+}. Chlorine (Cl) lies in Group 17, so it should gain one electron to form Cl^-. Ⓑ Two Cl^- ions are needed to balance the charge on one Ca^{2+} ion, which leads to the empirical formula $CaCl_2$. We could also use the absolute value of the charge on Ca^{2+} as the subscript for Cl and the absolute value of the charge on Cl^- as the subscript for Ca:

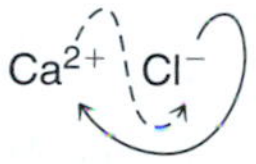

The subscripts in $CaCl_2$ cannot be reduced further. The empirical formula is electrically neutral [+2 + 2(−1) = 0]. This compound is calcium chloride, one of the substances used as "salt" to melt ice on roads and sidewalks in winter.

EXERCISE 2.4

Write the empirical formulas for the simplest binary ionic compounds formed from these ions or elements: **(a)** Li^+ and N^{3-}; **(b)** Al^{3+} and O^{2-}; **(c)** Li and O.

Answer **(a)** Li_3N; **(b)** Al_2O_3; **(c)** Li_2O

Polyatomic Ions

Polyatomic ions are groups of atoms that bear a net electrical charge, although the atoms that make up a polyatomic ion are held together by the same covalent bonds that hold atoms together in molecules. Just as there are many more kinds of molecules than simple elements, there are many more kinds of polyatomic ions than monatomic ions. Two examples of polyatomic cations are the ammonium (NH_4^+) and methylammonium ($CH_3NH_3^+$) ions. Polyatomic anions are much more numerous than polyatomic cations; some common examples are listed in Table 2.4.

The method we used to predict the empirical formulas for ionic compounds that contain monatomic ions can also be used for compounds that contain polyatomic ions. The overall charge on the cations must balance the overall charge on the anions in the formula unit. Thus, K^+ and NO_3^- ions combine in a 1:1 ratio to form KNO_3, known as potassium nitrate or saltpeter, a major ingredient in black gunpowder. Similarly, Ca^{2+} and SO_4^{2-} form $CaSO_4$, calcium sulfate, which combines with varying amounts of water to

TABLE 2.4 Common polyatomic ions and their names

Formula	Name of Ion
NH_4^+	ammonium
$CH_3NH_3^+$	methylammonium
OH^-	hydroxide
O_2^{2-}	peroxide
CN^-	cyanide
SCN^-	thiocyanate
NO_2^-	nitrite
NO_3^-	nitrate
CO_3^{2-}	carbonate
HCO_3^-	hydrogen carbonate, or bicarbonate
SO_3^{2-}	sulfite
SO_4^{2-}	sulfate
HSO_4^-	hydrogen sulfate, or bisulfate
PO_4^{3-}	phosphate
HPO_4^{2-}	hydrogen phosphate
$H_2PO_4^-$	dihydrogen phosphate
ClO^-	hypochlorite
ClO_2^-	chlorite
ClO_3^-	chlorate
ClO_4^-	perchlorate
MnO_4^-	permanganate
CrO_4^{2-}	chromate
$Cr_2O_7^{2-}$	dichromate
$C_2O_4^{2-}$	oxalate
HCO_2^-	formate
$CH_3CO_2^-$	acetate
$C_6H_5CO_2^-$	benzoate

form gypsum and plaster of Paris. The polyatomic ions NH_4^+ and NO_3^- form NH_4NO_3, (ammonium nitrate), which is a widely used fertilizer and, in the wrong hands, an explosive. One example of a compound in which the ions have charges of different magnitudes is calcium phosphate, which is composed of Ca^{2+} and PO_4^{3-} ions; it is a major component of bones. The compound is electrically neutral because the ions combine in a ratio of three Ca^{2+} ions $[3(+2) = +6]$ for every two PO_4^{3-} ions $[2(-3) = -6]$, giving an empirical formula of $Ca_3(PO_4)_2$; the parentheses around PO_4 in the empirical formula indicate that it is a polyatomic ion. Writing the formula for calcium phosphate as $Ca_3P_2O_8$ gives the correct number of each atom in the formula unit, but it obscures the fact that the compound contains readily identifiable PO_4^{3-} ions.

EXAMPLE 2.5

Write the empirical formulas for the compounds formed from these ions: **(a)** Na^+ and HPO_4^{2-}; **(b)** potassium cation and cyanide anion; **(c)** calcium cation and hypochlorite anion.

Given Ions

Asked for Empirical formula for ionic compound

Strategy

A If it is not given, determine the charge on a monatomic ion from its location in the periodic table. Use Table 2.4 to find the charge on a polyatomic ion.

B Use the absolute value of the charge on each ion as the subscript for the other ion. Reduce the subscripts to the smallest whole numbers when writing the empirical formula.

Solution

(a) **B** Because HPO_4^{2-} has a charge of -2 and Na^+ has a charge of $+1$, the empirical formula requires two Na^+ ions to balance the charge of the polyatomic ion, giving Na_2HPO_4. The subscripts are reduced to the lowest numbers, so the empirical formula is Na_2HPO_4. This compound is sodium hydrogen phosphate, which is used to provide texture in processed cheese, puddings, and "instant breakfasts."

(b) **A** The potassium cation is K^+, and the cyanide anion is CN^-. **B** Because the magnitude of the charge on each ion is the same, the empirical formula is KCN. Potassium cyanide is highly toxic, and at one time it was used as rat poison. This use has been discontinued, however, because too many people were being poisoned accidentally.

(c) **A** The calcium cation is Ca^{2+}, and the hypochlorite anion is ClO^-. **B** Two ClO^- ions are needed to balance the charge on one Ca^{2+} ion, giving $Ca(ClO)_2$. The subscripts cannot be reduced further, so the empirical formula is $Ca(ClO)_2$. This is calcium hypochlorite, the "chlorine" used to purify water in swimming pools.

EXERCISE 2.5

Write the empirical formulas for the compounds formed from these ions: **(a)** Ca^{2+} and $H_2PO_4^-$; **(b)** sodium cation and bicarbonate anion; **(c)** ammonium cation and sulfate anion.

Answer **(a)** $Ca(H_2PO_4)_2$ (calcium dihydrogen phosphate, one of the ingredients in baking powder); **(b)** $NaHCO_3$ (sodium bicarbonate, found in antacids and baking powder; in pure form, it is sold as baking soda); **(c)** $(NH_4)_2SO_4$ (ammonium sulfate, a common source of nitrogen in fertilizers)

Hydrates

Many ionic compounds occur as **hydrates**, compounds that contain specific ratios of loosely bound water molecules, called *waters of hydration*. Waters of hydration can often be removed simply by heating. For example, calcium dihydrogen phosphate can form a solid that contains one molecule of water per $Ca(H_2PO_4)_2$ unit. The empirical formula for the solid is $Ca(H_2PO_4)_2 \cdot H_2O$. In contrast, copper sulfate usually forms a blue solid that contains *five* waters of hydration per formula unit, with the empirical formula $CuSO_4 \cdot 5H_2O$. Upon heating, all five water molecules are lost, giving a white solid with the empirical formula $CuSO_4$ (Figure 2.9).

Compounds that differ only in the numbers of waters of hydration can have very different properties. For example, $CaSO_4 \cdot \frac{1}{2}H_2O$ is plaster of Paris, which is often used to make sturdy casts for broken arms or legs, whereas $CaSO_4 \cdot 2H_2O$ is the less dense, flakier gypsum, a mineral used in drywall panels for home construction. When a cast sets, a mixture of plaster of Paris (powdered $CaSO_4 \cdot \frac{1}{2}H_2O$) and water crystallizes to give solid $CaSO_4 \cdot 2H_2O$. Similar processes are used in the setting of cement and concrete.

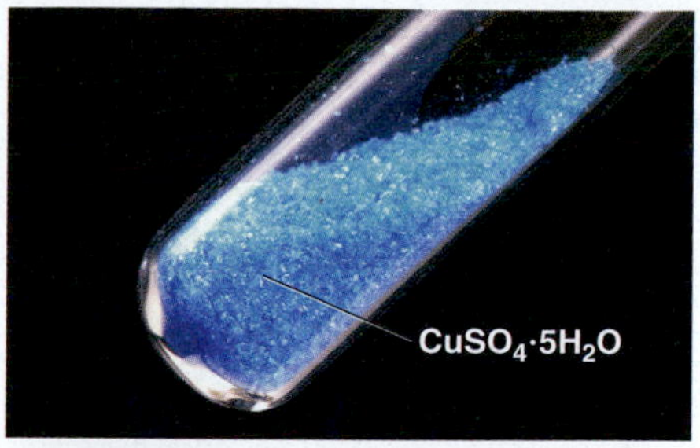

Figure 2.9 Loss of water from a hydrate upon heating. When blue $CuSO_4 \cdot 5H_2O$ is heated, two molecules of water are lost at 30°C, two more at 110°C, and the last at 250°C to give white $CuSO_4$.

2.3 • Naming Ionic Compounds

The empirical and molecular formulas discussed in the preceding section are precise and highly informative, but they have some disadvantages. First, they are inconvenient for routine verbal communication. For example, saying "C-A-three-P-O-four-two" for $Ca_3(PO_4)_2$ is much more difficult than saying "calcium phosphate." In addition, you will see in Section 2.4 that many compounds have the same empirical and molecular formulas but different arrangements of atoms, which result in very different chemical and physical properties. In such cases, it is necessary for the compounds to have different names that distinguish among the possible arrangements.

Many compounds, particularly those that have been known for a relatively long time, have more than one name: a *common* name (sometimes more than one) and a *systematic* name. Like the names of most of the elements, the common names of chemical compounds generally have historical origins, although they often appear to be unrelated to the compounds of interest. For example, a common name for KNO_3 (potassium nitrate) is "saltpeter."

In this text, we use a systematic nomenclature that has been developed in order to assign meaningful names to the millions of known substances. Unfortunately, some chemicals that are widely used in commerce and industry are still known almost exclusively by their common names; in such cases, you must be familiar with the common name as well as the systematic one. The objective of this and the next two sections is to teach you to write the formula for a simple inorganic compound from its name, and vice versa, and to introduce you to some of the more frequently encountered common names.

MGC Naming Ionic Compounds

We begin with *binary ionic compounds*, which contain only two elements. The procedure for naming such compounds is outlined in Figure 2.10 and uses the following steps:

1. ***Place the ions in their proper order: cation and then anion.*** The name of a binary ionic compound is the name of the cation followed by the name of the anion. For example, $CaCl_2$, which consists of Ca^+ and Cl^- ions, is calcium chloride.
2. ***Name the cation.***
   a. *Metals that form only one kind of positive ion.* As noted in Section 2.1, these metals are usually in Groups 1–3, 12, and 13. The name of the cation of a metal that forms only one kind of positive ion is the same as the name of the metal (with the word *ion* added if the cation is by itself). For example, Na^+ is the sodium ion, Ca^{2+} is the calcium ion, and Al^{3+} is the aluminum ion.
   b. *Metals that form more than one cation.* As shown in Figure 2.11, many metals can form more than one positively charged ion. This behavior is observed for most transition metals, many actinides, and the heaviest elements of Groups 13–15. In such cases, the positive charge on the metal is indicated by

Note the pattern

Cations are always named before anions.

Figure 2.10 Naming an ionic compound.

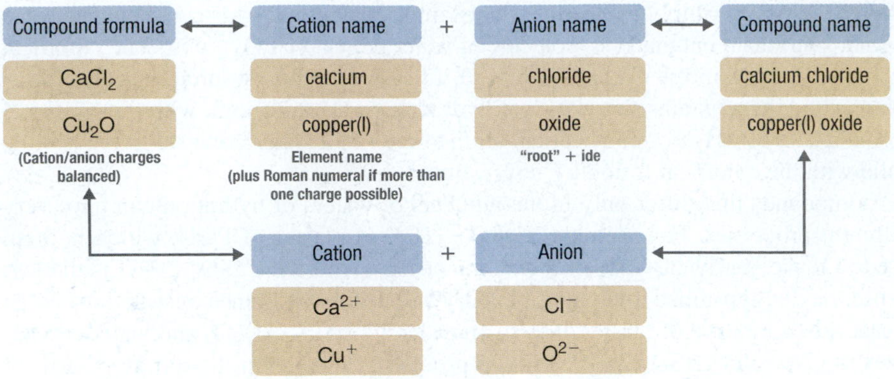

a Roman numeral in parentheses immediately following the name of the metal. Thus, Cu^+ is copper(I) (read as "copper one"), Fe^{2+} is iron(II), Fe^{3+} is iron(III), Sn^{2+} is tin(II), and Sn^{4+} is tin(IV).

An older system of nomenclature for such cations is still widely used, however. The name of the cation with the *higher* charge is formed from the root of the element's Latin name with the suffix *-ic* attached, and the name of the cation with the *lower* charge has the same root with the suffix *-ous*. The names of Fe^{3+}, Fe^{2+}, Sn^{4+}, and Sn^{2+} are therefore ferric, ferrous, stannic, and stannous, respectively. Even though this text uses the systematic names with Roman numerals, you should be able to recognize these common names because they are still used often. For example, on the label of your dentist's

Figure 2.11 Metals that form more than one positively charged ion and their locations in the periodic table. Note that, with only a few exceptions, these metals are usually transition metals or actinides.

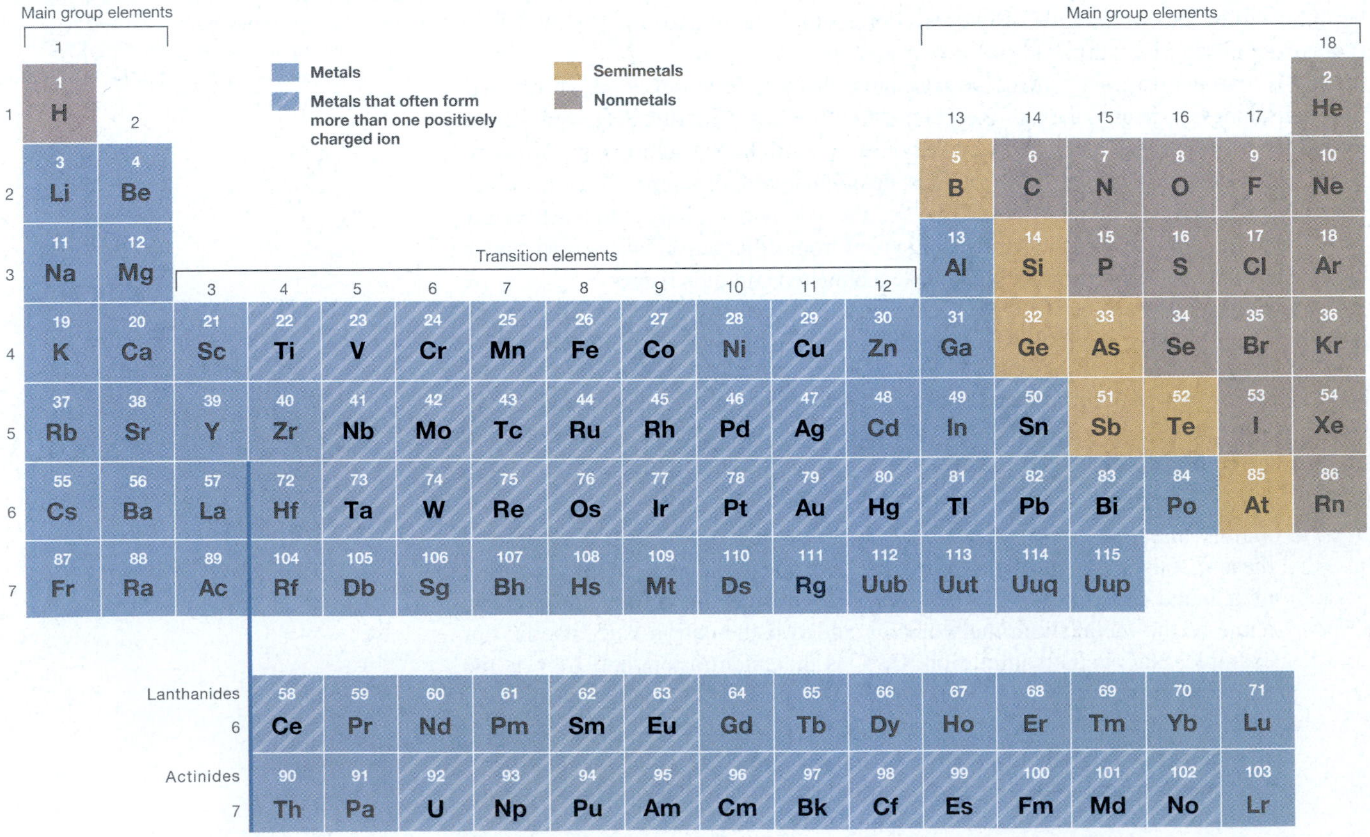

TABLE 2.5 Common cations of metals that form more than one ion

Cation	Systematic Name	Common Name	Cation	Systematic Name	Common Name
Cr^{2+}	chromium(II)	chromous	Cu^{2+}	copper(II)	cupric
Cr^{3+}	chromium(III)	chromic	Cu^{+}	copper(I)	cuprous
Mn^{2+}	manganese(II)	manganous[a]	Hg^{2+}	mercury(II)	mercuric
Mn^{3+}	manganese(III)	manganic[a]	$Hg_2{}^{2+}$	mercury(I)	mercurous[b]
Fe^{2+}	iron(II)	ferrous	Sn^{4+}	tin(IV)	stannic
Fe^{3+}	iron(III)	ferric	Sn^{2+}	tin(II)	stannous
Co^{2+}	cobalt(II)	cobaltous[a]	Pb^{4+}	lead(IV)	plumbic[a]
Co^{3+}	cobalt(III)	cobaltic[a]	Pb^{2+}	lead(II)	plumbous[a]

[a]Not widely used.
[b]The mercury(I) (mercurous) ion exists *only* as the $Hg_2{}^{2+}$ ion.

fluoride rinse, the compound chemists call tin(II) fluoride is usually listed as stannous fluoride.

Some examples of metals that form more than one cation are listed in Table 2.5 along with the names of the ions. Note that the simple Hg^{+} cation does not occur in chemical compounds. Instead, all compounds of mercury(I) contain a *dimeric* cation, $Hg_2{}^{2+}$, in which the two Hg atoms are bonded.

c. *Polyatomic cations.* The names of the common polyatomic cations that are relatively important in ionic compounds (such as $NH_4{}^{+}$, the ammonium ion) were listed in Table 2.4.

3. ***Name the anion.***

a. *Monatomic anions.* Monatomic anions are named by adding the suffix *-ide* to the root of the name of the parent element; thus, Cl^{-} is chloride, O^{2-} is oxide, P^{3-} is phosphide, and C^{4-} is carbide. Because the charges on these ions can be predicted from their position in the periodic table, it is *not* necessary to specify the charge in the name. Examples of monatomic anions were listed in Table 2.2.

b. *Polyatomic anions.* Polyatomic anions typically have common names that you must learn; some examples were listed in Table 2.4. Polyatomic anions that contain a single metal or nonmetal atom plus one or more oxygen atoms are called **oxoanions** (sometimes called *oxyanions*). In cases where only two oxoanions are known for an element, the name of the oxoanion with more oxygen atoms ends in *-ate*, and the name of the oxoanion with fewer oxygen atoms ends in *-ite*. For example, $NO_3{}^{-}$ is nitrate and $NO_2{}^{-}$ is nitrite.

The halogens and some of the transition metals form more extensive series of oxoanions with as many as four members. In the names of these oxoanions, the prefix *per-* is used to identify the oxoanion with the most oxygen atoms (so that $ClO_4{}^{-}$ is perchlorate and $ClO_3{}^{-}$ is chlorate), and the prefix *hypo-* is used to identify the anion with the fewest oxygen atoms ($ClO_2{}^{-}$ is chlorite and ClO^{-} is hypochlorite). The relationship between the names of oxoanions and the number of oxygen atoms present is diagrammed in Figure 2.12. Differentiating the oxoanions in such a series is no trivial matter. For example, the hypochlorite ion is the active ingredient in laundry bleach and swimming pool disinfectant, but compounds that contain the perchlorate ion can explode if they come into contact with organic substances.

4. ***Write the name of the compound as the name of the cation followed by the name of the anion.***

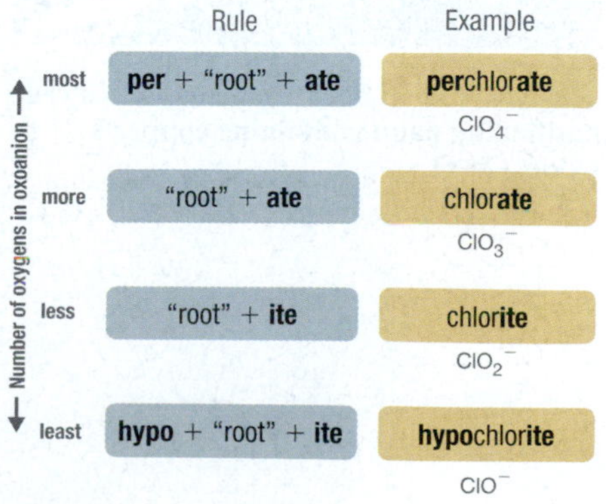

Figure 2.12 **The relationship between the names of oxoanions and the number of oxygen atoms present.**

It is *not* necessary to indicate the number of cations or anions present per formula unit in the name of an ionic compound because this information is implied by the charges on the ions. You must consider the charge of the ions when writing the formula for an ionic compound from its name, however. Because the charge on the chloride ion is −1 and the charge on the calcium ion is +2, for example, simple arithmetic tells you that calcium chloride must contain twice as many chloride ions as calcium ions, so the formula is $CaCl_2$. Similarly, calcium phosphate must be $Ca_3(PO_4)_2$ because the cation and the anion have charges of +2 and −3, respectively. The best way to learn how to name ionic compounds is to work through a few examples, referring to Figure 2.10, Tables 2.2, 2.4, and 2.5 as needed.

EXAMPLE 2.6

Write the systematic name of each ionic compound: **(a)** LiCl; **(b)** $MgSO_4$; **(c)** $(NH_4)_3PO_4$; **(d)** Cu_2O.

Given Empirical formula

Asked for Name

Strategy

Ⓐ If only one charge is possible for the cation, give its name, consulting Table 2.2 or 2.4 only if necessary. If the cation can have more than one charge (Table 2.5), specify the charge using Roman numerals.

Ⓑ If the anion does not contain oxygen, name it according to step 3a, using Tables 2.2 and 2.4 as references if necessary. For polyatomic anions that contain oxygen, use Table 2.4 and the appropriate prefix and suffix listed in step 3b.

Ⓒ Beginning with the cation, write the name of the compound.

The bottom of a boat protected with a red antifouling paint containing copper(I) oxide, Cu_2O.

Solution

(a) Ⓐ✓ Ⓑ✓ Lithium is in Group 1, so we know that it forms only the Li^+ cation, which is the lithium ion. Similarly, chlorine is in Group 7, so it forms the Cl^- anion, which is the chloride ion. Ⓒ✓ We begin with the cation. The name of this compound is lithium chloride, which is used medically as an antidepressant drug.

(b) Ⓐ✓ Ⓑ✓ The cation is the magnesium ion, and the anion, which contains oxygen, is sulfate. Ⓒ✓ We list the cation first, so the name of this compound is magnesium sulfate. A hydrated form of magnesium sulfate, $MgSO_4 \cdot 7H_2O$, is sold in drugstores as Epsom salts, a harsh but effective laxative.

(c) Ⓐ✓ Ⓑ✓ The cation is the ammonium ion (from Table 2.4), and the anion is phosphate. Ⓒ✓ The compound is therefore ammonium phosphate, which is widely used as a fertilizer. It is not necessary to specify that the formula unit contains three ammonium ions because three are required to balance the negative charge on phosphate.

(d) Ⓐ✓ Ⓑ✓ The cation is a transition metal that often forms more than one kind of ion (Table 2.5). We must therefore specify the positive charge on the cation in the name: copper(I) or, according to the older system, cuprous. The anion is oxide. Ⓒ✓ The name of this compound is copper(I) oxide or, in the older system, cuprous oxide. Copper(I) oxide is used as a red glaze on ceramics and in antifouling paints to prevent organisms from growing on the bottoms of boats.

EXERCISE 2.6

Name each compound: **(a)** $CuCl_2$; **(b)** $MgCO_3$; **(c)** $FePO_4$.

Answer **(a)** copper(II) chloride (or cupric chloride); **(b)** magnesium carbonate; **(c)** iron(III) phosphate (or ferric phosphate)

EXAMPLE 2.7

Write the formula for each compound from its name: **(a)** calcium dihydrogen phosphate; **(b)** aluminum sulfate; **(c)** chromium(III) oxide.

Given Systematic name

Asked for Formula

Strategy

Ⓐ Identify the cation and its charge using the location of the element in the periodic table and Tables 2.2, 2.4, and 2.5. If the cation is derived from a metal that can form cations with different charges, use the appropriate Roman numeral or suffix to indicate its charge.

Ⓑ Identify the anion using Tables 2.2 and 2.4. Beginning with the cation, write the compound's formula, and then determine the number of cations and anions needed to achieve electrical neutrality.

Solution

(a) Ⓐ✓ Calcium is in Group 2, so it forms only the Ca^{2+} ion. Ⓑ✓ Dihydrogen phosphate is the $H_2PO_4^-$ ion (Table 2.4). Two $H_2PO_4^-$ ions are needed to balance the positive charge on Ca^{2+}, to give $Ca(H_2PO_4)_2$. A hydrate of calcium dihydrogen phosphate, $Ca(H_2PO_4)_2{\cdot}H_2O$, is the active ingredient in baking powder.

(b) Ⓐ✓ Aluminum, toward the top of Group 13 in the periodic table, forms only one cation, Al^{3+} (Figure 2.11). Ⓑ✓ Sulfate is SO_4^{2-} (Table 2.4). To balance the electrical charges, we need two Al^{3+} cations and three SO_4^{2-} anions, giving $Al_2(SO_4)_3$. Aluminum sulfate is a compound used to tan leather and to purify drinking water.

(c) Ⓐ✓ Because chromium is a transition metal, it can form cations with different charges. The Roman numeral tells us that the positive charge in this case is +3, so the cation is Cr^{3+}. Ⓑ✓ Oxide is O^{2-}. Thus, two cations (Cr^{3+}) and three anions (O^{2-}) are required to give an electrically neutral compound, Cr_2O_3. This compound, chromium(III) oxide, is a common green pigment that has many uses, including camouflage coatings.

Chromium(III) oxide, Cr_2O_3, is a common pigment in dark green paints, such as camouflage paint.

EXERCISE 2.7

Write the formula for each compound: **(a)** barium chloride; **(b)** sodium carbonate; **(c)** iron(III) hydroxide.

Answer **(a)** $BaCl_2$; **(b)** Na_2CO_3; **(c)** $Fe(OH)_3$

2.4 ● Naming Covalent Compounds

MGC Naming Covalent Compounds

As with ionic compounds, the system that chemists have devised for naming covalent compounds enables us to write the molecular formula from the name, and vice versa. In this and the following section, we describe the rules for naming simple covalent compounds. We begin with inorganic compounds and then turn to simple organic compounds that contain only carbon and hydrogen.

Binary Inorganic Compounds

Binary covalent compounds—that is, covalent compounds that contain only two elements—are named using a procedure similar to that used to name simple ionic

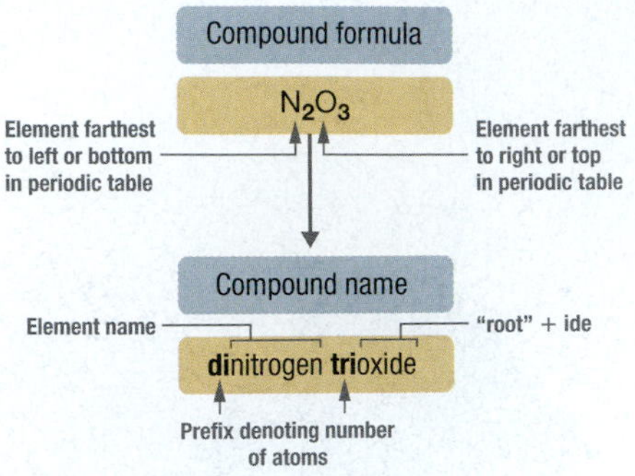

Figure 2.13 Naming a covalent inorganic compound.

compounds, but with prefixes added as needed to indicate the number of atoms of each kind. The procedure, diagrammed in Figure 2.13, uses the following steps:

1. ***Place the elements in their proper order.***
   a. The element farthest to the left in the periodic table is usually named first. If both elements are in the same group, the element closer to the bottom of the column is named first.
   b. The second element is named as if it were a monatomic anion in an ionic compound (even though it is not!), with the suffix *-ide* attached to the root of the element name.
2. ***Identify the number of each type of atom present.***
   a. Prefixes derived from Greek stems are used to indicate the number of each type of atom in the formula unit (Table 2.6). The prefix *mono-* ("one") is used only when absolutely necessary to avoid confusion, just as we omit the subscript 1 when writing molecular formulas.

      To demonstrate steps 1 and 2a, we name HCl as hydrogen chloride (because hydrogen is to the left of chlorine in the periodic table), and PCl_5 as phosphorus pentachloride. The order of the elements in the name of BrF_3, bromine trifluoride, is determined by the fact that bromine lies below fluorine in Group 17.
   b. If the molecule contains more than one atom of both elements, then prefixes are used for both. Thus, N_2O_3 is *di*nitrogen *tri*oxide, as shown in Figure 2.13.
   c. In some names, the final *a* or *o* of the prefix is dropped to avoid awkward pronunciation. Thus, OsO_4 is osmium tetroxide rather than osmium tetraoxide.
3. ***Write the name of the compound.***
   a. Binary compounds of the elements with oxygen are generally named as "element oxide," with prefixes that indicate the number of atoms of each element per formula unit. For example, CO is carbon monoxide. The only exception is binary compounds of oxygen with fluorine, which are named as oxygen fluorides. (The reasons for this convention will become clear in Chapters 7 and 8.)
   b. Certain compounds are *always* called by the common names that were assigned long ago when names rather than formulas were used. For example, H_2O is water (not dihydrogen oxide), NH_3 is ammonia, PH_3 is phosphine, SiH_4 is silane, and B_2H_6, a *dimer* of BH_3, is diborane. For many compounds, the systematic name and the common name are both used frequently, so you must be familiar with them. For example, the systematic name for NO is nitrogen monoxide, but it is much more commonly called nitric oxide. Similarly, N_2O is usually called nitrous oxide rather than dinitrogen monoxide. Notice that the suffixes *-ic* and *-ous* are the same ones used for ionic compounds.

Note the pattern

Start with the element at the far left and work to the right. If two or more elements are in the same column, start with the bottom element and work up.

TABLE 2.6 Prefixes for indicating numbers of atoms in chemical names

Prefix	Number
mono-	1
di-	2
tri-	3
tetra-	4
penta-	5
hexa-	6
hepta-	7
octa-	8
nona-	9
deca-	10
undeca-	11
dodeca-	12

EXAMPLE 2.8

Write the name of each binary covalent compound: **(a)** SF_6; **(b)** N_2O_4; **(c)** ClO_2.

Given Molecular formula

Asked for Name of compound

Strategy

A List the elements in order according to their positions in the periodic table. Identify the number of each type of atom in the chemical formula, and then use Table 2.6 to determine the prefixes needed.

B If the compound contains oxygen, follow step 3a. If not, decide whether to use the common or the systematic name.

Solution

(a) **A** Because S is to the left of F in the periodic table, sulfur is named first. Because there is only one S in the formula, no prefix is needed. **B** There are, however, six fluorines, so we use the prefix for "six," *hexa-* (Table 2.6). The compound is sulfur hexafluoride.

(b) **A** Because N is to the left of O in the periodic table, nitrogen is named first. Because more than one atom of each element is present, prefixes are needed to indicate the number of atoms of each. According to Table 2.6, the prefix for "two" is *di-* and the prefix for "four" is *tetra-*. **B** The compound is dinitrogen tetroxide (omitting the *a* in *tetra-* according to step 2c), used as a component of some rocket fuels.

(c) **A** Although oxygen lies to the left of chlorine in the periodic table, it is not named first because ClO_2 is an oxide of an element other than fluorine (see step 3a). Consequently, chlorine is named first, but a prefix is not necessary because each molecule has only one atom of chlorine. **B** Because there are two O atoms, the compound is a dioxide. Thus, the compound is chlorine dioxide. It is widely used as a substitute for chlorine in municipal water treatment plants because, unlike chlorine, it does not react with organic compounds in water to produce potentially toxic chlorinated compounds.

EXERCISE 2.8

Name each compound: **(a)** IF_7; **(b)** N_2O_5; **(c)** OF_2.

Answer **(a)** iodine heptafluoride; **(b)** dinitrogen pentoxide; **(c)** oxygen difluoride

EXAMPLE 2.9

Write the formula for each compound: **(a)** sulfur trioxide; **(b)** diiodine pentoxide.

Given Name of compound

Asked for Formula

Strategy

List the elements in the same order as in the formula, use Table 2.6 to identify the number of each type of atom present, and then indicate this quantity as a subscript to the right of that element when writing the formula.

Solution

(a) Sulfur has no prefix, which means that each molecule has only one S atom. The prefix *tri-* indicates that there are three O atoms. The formula is therefore SO_3. Sulfur trioxide is produced industrially in huge amounts as an intermediate in the synthesis of sulfuric acid.

(b) The prefix *di-* tells you that each molecule has two I atoms, and the prefix *penta-* indicates that there are five O atoms. The formula is thus I_2O_5, a compound used to remove carbon monoxide from air in respirators.

EXERCISE 2.9

Write the formula for each compound: **(a)** silicon tetrachloride; **(b)** disulfur decafluoride.

Answer **(a)** $SiCl_4$; **(b)** S_2F_{10}

The structures of some of the compounds mentioned in Examples and Exercises 2.8 and 2.9 are shown in Figure 2.14, along with the location of the "central atom" of each compound in the periodic table. It may seem that the compositions

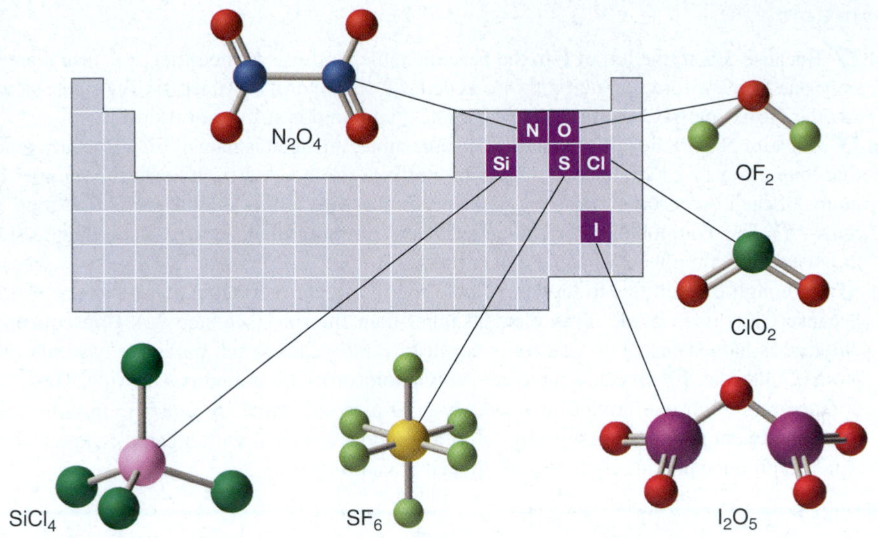

Figure 2.14 The structures of some covalent inorganic compounds and the locations of the "central atoms" in the periodic table. The compositions and structures of covalent inorganic compounds are not random. As you will learn in Chapters 7 and 8, they can be predicted from the locations of the component atoms in the periodic table.

and structures of such compounds are entirely random, but this is not true. By the time you have mastered the material in Chapters 7 and 8, you will be able to predict the compositions and structures of compounds of this type with a high degree of accuracy.

Hydrocarbons

Note the pattern

The number of bonds between carbon atoms in a hydrocarbon is indicated in the suffix: alkane: only C—C single bonds; alkene: at least one C=C double bond; alkyne: at least one C≡C triple bond

Approximately one-third of the compounds produced industrially are organic compounds. All living organisms are made of organic compounds, as is most of the food you consume and the medicines you take, the fibers in the clothes you wear, and the plastics in the materials you use. Section 2.1 introduced two organic compounds: methane (CH_4) and methanol (CH_3OH). These and other organic compounds appear frequently in discussions and examples throughout this text.

The simplest class of organic compounds is the **hydrocarbons**, which consist entirely of carbon and hydrogen. Petroleum and natural gas are complex, naturally occurring mixtures of many different hydrocarbons that furnish raw materials for the chemical industry. The four major classes of hydrocarbons are the **alkanes**, which contain only carbon–hydrogen and carbon–carbon single bonds; the **alkenes**, which contain at least one carbon–carbon double bond; the **alkynes**, which contain at least one carbon–carbon triple bond; and the *aromatics*, which usually contain rings of six carbon atoms that can be drawn with alternating single and double bonds. Alkanes are also called *saturated* hydrocarbons, whereas hydrocarbons that contain multiple bonds (alkenes, alkynes, and aromatics) are *unsaturated*.

Alkanes

The simplest alkane is CH_4 or *methane*, a colorless, odorless gas that is the major component of natural gas. In larger alkanes whose carbon atoms are joined in an unbranched chain (*straight-chain alkanes*), each carbon atom is bonded to at most two other carbon atoms. The structures of two simple alkanes are shown in Figure 2.15, and the names and condensed structural formulas for the first 10 straight-chain alkanes are listed in Table 2.7. Note that the names of all alkanes end in *-ane* and that their boiling points increase with more carbon atoms present.

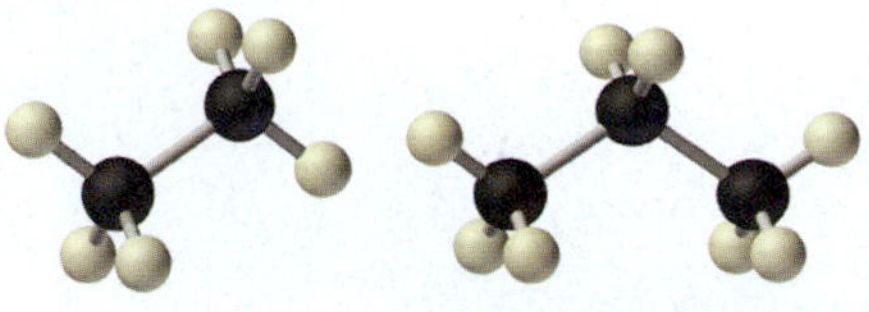

Ethane, C_2H_6 Propane, C_3H_8

Figure 2.15 Straight-chain alkanes that contain 2 and 3 carbon atoms.

Alkanes that contain four or more carbon atoms can have more than one arrangement of atoms. The carbon atoms can form a single unbranched chain, or the primary chain of carbon atoms can have one or more shorter chains that form branches. For example, *butane* (C_4H_{10}) has two possible structures. *Normal* butane (usually called *n*-butane) is $CH_3CH_2CH_2CH_3$, in which the carbon atoms form a single unbranched

TABLE 2.7 The first 10 straight-chain alkanes

Name	Number of Carbon Atoms	Molecular Formula	Condensed Structural Formula	Boiling Point (°C)	Uses
methane	1	CH_4	CH_4	−162	natural gas constituent
ethane	2	C_2H_6	CH_3CH_3	−89	natural gas constituent
propane	3	C_3H_8	$CH_3CH_2CH_3$	−42	bottled gas
butane	4	C_4H_{10}	$CH_3CH_2CH_2CH_3$ or $CH_3(CH_2)_2CH_3$	0	lighters, bottled gas
pentane	5	C_5H_{12}	$CH_3(CH_2)_3CH_3$	36	solvent, gasoline
hexane	6	C_6H_{14}	$CH_3(CH_2)_4CH_3$	69	solvent, gasoline
heptane	7	C_7H_{16}	$CH_3(CH_2)_5CH_3$	98	solvent, gasoline
octane	8	C_8H_{18}	$CH_3(CH_2)_6CH_3$	126	gasoline
nonane	9	C_9H_{20}	$CH_3(CH_2)_7CH_3$	151	gasoline
decane	10	$C_{10}H_{22}$	$CH_3(CH_2)_8CH_3$	174	kerosene

chain. In contrast, the condensed structural formula for *isobutane* is $(CH_3)_2CHCH_3$, in which the primary chain of three carbon atoms has a one-carbon chain branching at the central carbon. Three-dimensional representations of both structures are shown in the margin.

The systematic names for branched hydrocarbons use the lowest possible number to indicate the position of the branch along the longest straight carbon chain in the structure. Thus, the systematic name for isobutane is 2-methylpropane, which indicates that a methyl group (a branch consisting of $—CH_3$) is attached to the second carbon of a propane molecule. Similarly, you will learn in Section 2.6 that one of the major components of gasoline is commonly called isooctane; its structure is shown in the margin. As you can see, the compound has a chain of five carbon atoms, so it is a derivative of pentane. There are two methyl group branches at one carbon atom and one methyl group at another. Using the lowest possible numbers for the branches gives 2,2,4-trimethylpentane for the systematic name of this compound.

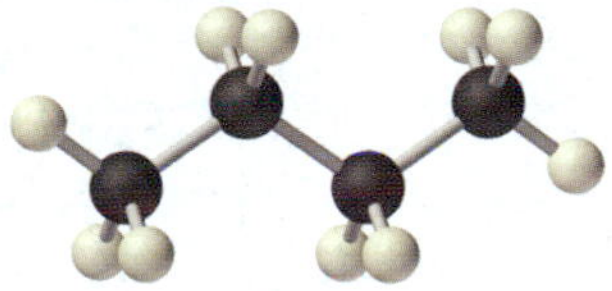

n-Butane, C_4H_{10}

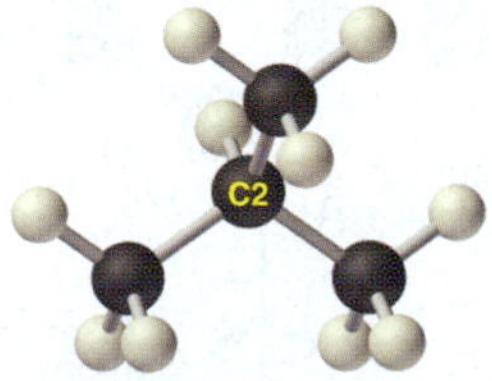

Isobutane (2-methylpropane), C_4H_{10}

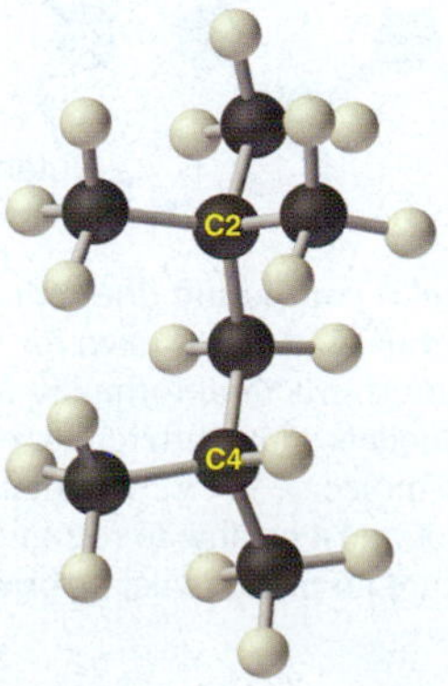

Isooctane (2,2,4-trimethylpentane)

Alkenes

The simplest alkenes are *ethylene*, C_2H_4 or $CH_2{=}CH_2$, and *propylene*, C_3H_6 or $CH_3CH{=}CH_2$ (Figure 2.16a). The names of alkenes that have more than three carbon atoms use the same stems as the names of the alkanes (Table 2.7) but end in *-ene* instead of *-ane*.

Once again, more than one structure is possible for alkenes that contain four or more carbons. For example, a four-carbon alkene has three possible structures. One is $CH_2{=}CHCH_2CH_3$ (1-butene), which has the double bond between the first and second carbons in the chain. The other two structures have the double bond between the second and third carbon atoms, and are forms of $CH_3CH{=}CHCH_3$ (2-butene). All four carbon atoms in 2-butane lie in the same plane, so there are two possible structures (Figure 2.16a). If the two methyl groups are on the same side of the double bond, the compound is *cis*-2-butene (from the Latin *cis*, "on the same side"). If the two methyl groups are on opposite sides of the double bond, the compound is *trans*-2-butene (from the Latin *trans*, "across").

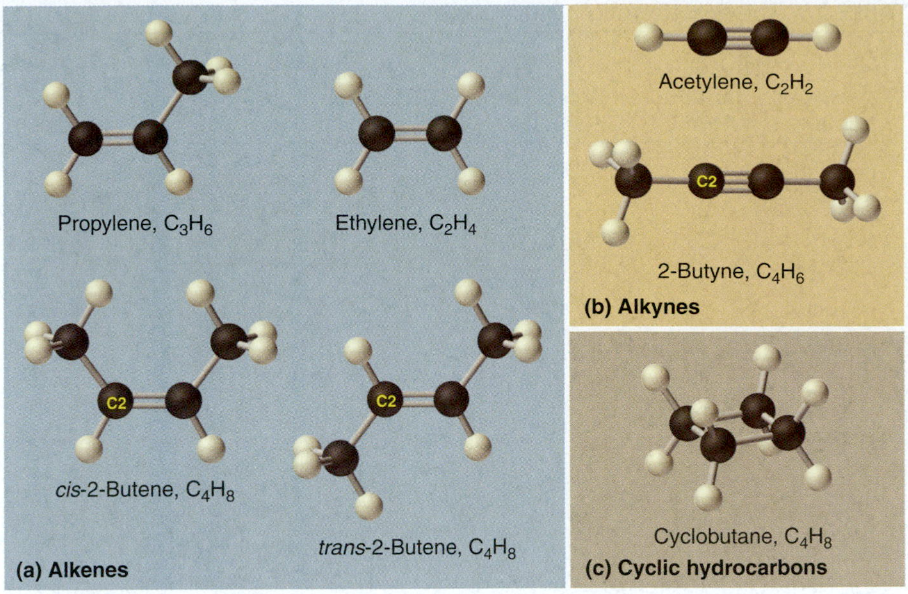

Figure 2.16 Some simple (a) alkenes, (b) alkynes, and (c) cyclic hydrocarbons. Positions of the carbon atoms (C) in the chain are indicated by C1 or C2.

Just as the positions of branches in an alkane are indicated by a number, the number in the name of an alkene specifies the position of the *first* carbon atom of the double bond. Again, the name is based on the lowest possible number starting from *either end* of the carbon chain, so $CH_3CH_2CH{=}CH_2$ is called 1-butene, *not* 3-butene. Note that $CH_2{=}CHCH_2CH_3$ and $CH_3CH_2CH{=}CH_2$ are just two different ways of writing the *same molecule* (1-butene) but in two different orientations as shown in the margin.

Note the pattern

The positions of groups or multiple bonds are always indicated by the lowest numbers possible.

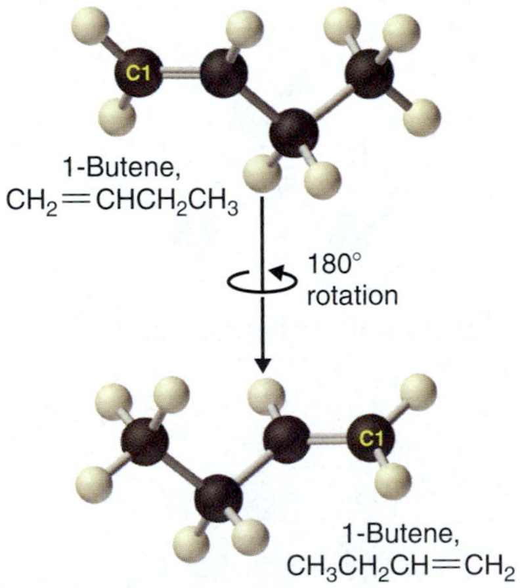

The name of a compound does *not* depend on its orientation. As illustrated for 1-butene, both condensed structural formulas and molecular models show different orientations of the same molecule. Don't let orientation fool you; you must be able to recognize the same structure no matter what its orientation.

Alkynes

The simplest alkyne is *acetylene*, C_2H_2 or $HC{\equiv}CH$ (Figure 2.16b). Because a mixture of acetylene and oxygen burns with a flame that is hot enough (>3000°C) to cut metals such as hardened steel, acetylene is widely used in cutting and welding torches. The names of other alkynes are similar to those of the corresponding alkanes but end in *-yne*. For example, $HC{\equiv}CCH_3$ is *propyne*, and $CH_3C{\equiv}CCH_3$ is *2-butyne* because the multiple bond begins on the second carbon.

Cyclic Hydrocarbons

In a **cyclic hydrocarbon**, the ends of a hydrocarbon chain are connected to form a ring of covalently bonded carbon atoms. Cyclic hydrocarbons are named by attaching the prefix *cyclo-* to the name of the alkane, alkene, or alkyne. The simplest cyclic alkanes are *cyclopropane*, C_3H_6, a flammable gas that is also a powerful anesthetic, and *cyclobutane*, C_4H_8 (Figure 2.16c). The most common way to draw the structures of cyclic alkanes is to sketch a polygon with the same number of vertices as there are carbon atoms in the ring; each vertex represents a CH_2 unit. The structures of the cycloalkanes that contain three to six carbon atoms are shown schematically in Table 2.8.

Aromatic Hydrocarbons

Alkanes, alkenes, alkynes, and cyclic hydrocarbons are generally called *aliphatic hydrocarbons*. The name comes from the Greek *aleiphar*, "oil," because the first examples were extracted from animal fats. In contrast, the first examples of **aromatic hydrocarbons**, also called **arenes**, were obtained by the distillation and degradation of highly scented (thus *aromatic*) resins from tropical trees.

The simplest aromatic hydrocarbon is *benzene*, C_6H_6, which was first obtained from a coal distillate. The word *aromatic* now refers to benzene and structurally

TABLE 2.8 The simple cyclic alkanes

Name	Molecular Formula	Structural Formula	Name	Molecular Formula	Structural Formula
cyclopropane	C_3H_6	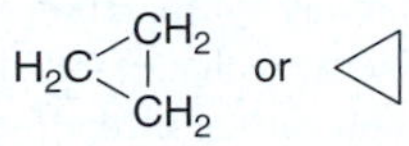	cyclopentane	C_5H_{10}	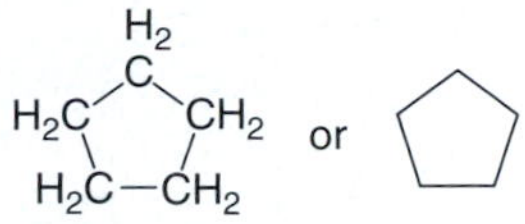
cyclobutane	C_4H_8	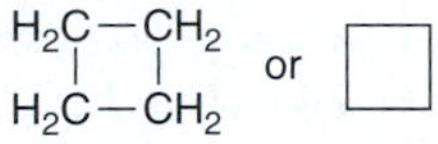	cyclohexane	C_6H_{12}	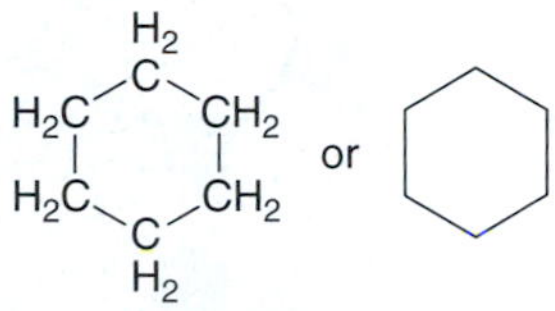

similar compounds. As shown in Figure 2.17 (left), it is possible to draw the structure of benzene in two different but equivalent ways, depending on which carbon atoms are connected by double bonds or single bonds. *Toluene* is similar to benzene, except that one H atom is replaced by a —CH_3 group; it has the formula C_7H_8 (Figure 2.17, right). As you will soon learn, the chemical behavior of aromatic compounds differs from the behavior of aliphatic compounds. Benzene and toluene are found in gasoline, and benzene is the starting material for preparing substances as diverse as aspirin and nylon.

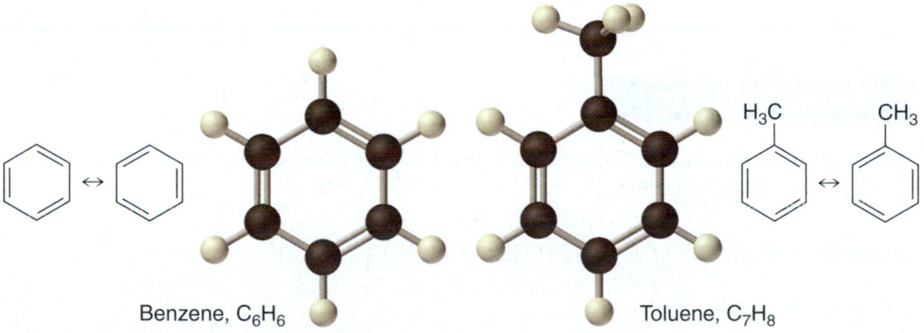

Figure 2.17 **Two aromatic hydrocarbons: benzene and toluene.**

Figure 2.18 illustrates two of the molecular structures possible for hydrocarbons that contain six carbon atoms. As you can see, compounds with the same molecular formula can have very different structures.

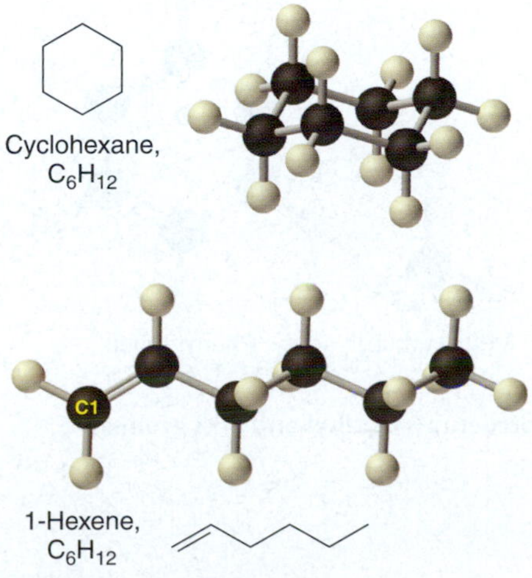

Figure 2.18 **Two hydrocarbons with the molecular formula C_6H_{12}.**

EXAMPLE 2.10

Write the condensed structural formula for each hydrocarbon: **(a)** *n*-heptane; **(b)** 2-pentene; **(c)** 2-butyne; **(d)** cyclooctene.

Given Name of hydrocarbon

Asked for Condensed structural formula

Strategy

A Use the prefix to determine the number of carbon atoms in the molecule and whether it is cyclic. From the suffix, determine whether multiple bonds are present.

B Identify the position of any multiple bonds from the number(s) in the name, and then write the condensed structural formula.

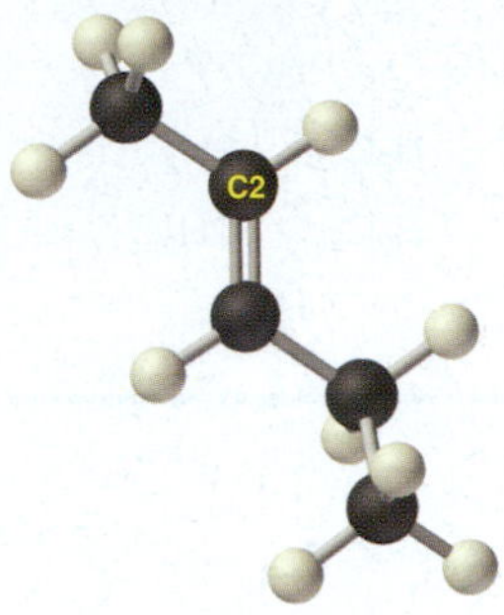

2-Pentene

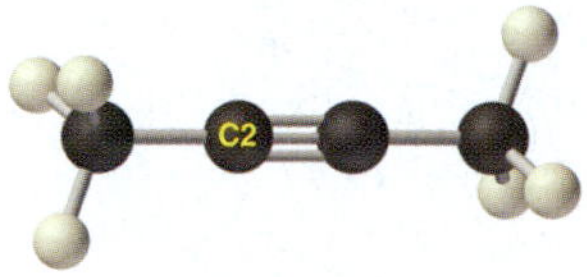

2-Butyne

Solution

(a) The prefix *hept-* tells us that this hydrocarbon contains seven C atoms, and *n-* indicates that the carbon atoms form a straight chain. The suffix *-ane* tells that it is an alkane, with no carbon–carbon double or triple bonds. The condensed structural formula is $CH_3CH_2CH_2CH_2CH_2CH_2CH_3$, which can also be written as $CH_3(CH_2)_5CH_3$.

(b) The prefix *pent-* tells us that this hydrocarbon contains five carbon atoms, and the suffix *-ene* indicates that it is an alkene, with a C=C double bond. The 2- tells us that the double bond begins on the second carbon of the 5-carbon chain. The condensed structural formula of the compound is therefore $CH_3CH{=}CHCH_2CH_3$.

(c) The prefix *but-* tells us that the compound contains a chain of four C atoms, and the suffix *-yne* indicates that it contains a C≡C triple bond. The 2- tells us that the triple bond begins on the second carbon of the 4-carbon chain. Hence, the condensed structural formula for the compound is $CH_3C{\equiv}CCH_3$.

(d) The prefix *cyclo-* tells us that this hydrocarbon has a ring structure, and *oct-* indicates that it contains eight carbon atoms, which we can draw as

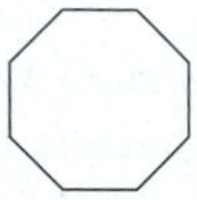

The suffix *-ene* tells that the compound contains a C=C double bond, but where in the ring do we place the double bond? Because all eight carbon atoms are identical, it doesn't matter. We can draw the structure of cyclooctene as

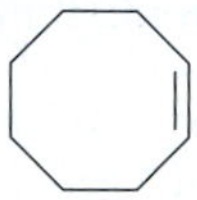

EXERCISE 2.10

Write condensed structural formulas for the hydrocarbons: **(a)** *n*-octane; **(b)** 2-hexene; **(c)** 1-heptyne; **(d)** cyclopentane.

Answer **(a)** $CH_3(CH_2)_6CH_3$; **(b)** $CH_3CH{=}CHCH_2CH_2CH_3$;

(c) $HC{\equiv}C(CH_2)_4CH_3$; **(d)** 

The general name for a group of atoms derived from an alkane is an *alkyl group*. The name of an alkyl group is derived from the name of the alkane by adding the suffix *-yl*. Thus, the CH_3— fragment is a *methyl* group, the CH_3CH_2— fragment is an *ethyl* group, and so on, where the dash represents a single bond to some other atom or group. Similarly, groups of atoms derived from aromatic hydrocarbons are *aryl groups*, which sometimes have unexpected names. For example, the C_6H_5— fragment is derived from benzene, but it is called a *phenyl* group. In general formulas and structures, alkyl and aryl groups are often abbreviated as **R**.

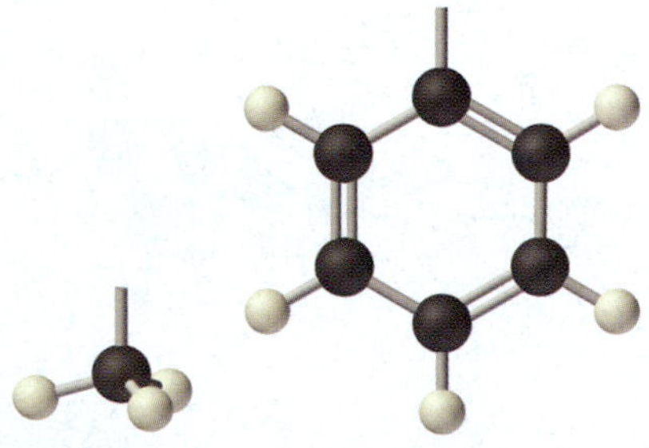

Methyl group, CH_3— Phenyl group, C_6H_5—

Structures of alkyl and aryl groups.

Alcohols

Replacing one or more hydrogen atoms of a hydrocarbon with an —OH group gives an **alcohol**, represented as R—OH. The simplest alcohol, CH_3OH, is called either *methanol* (its systematic name) or *methyl alcohol* (its common name) (Figure 2.4). Methanol is the antifreeze in automobile windshield washer fluids, and it is also used as an efficient fuel for racing cars, most notably in the Indianapolis 500. Ethanol, or ethyl alcohol (CH_3CH_2OH) is familiar as the alcohol in fermented or distilled beverages, such as beer, wine, and whiskey; it is also used as a gasoline additive (Section 2.6).

The simplest alcohol derived from an aromatic hydrocarbon is C_6H_5OH, *phenol* (shortened from *phenyl* alcohol), a potent disinfectant used in some sore throat medications and mouthwashes.

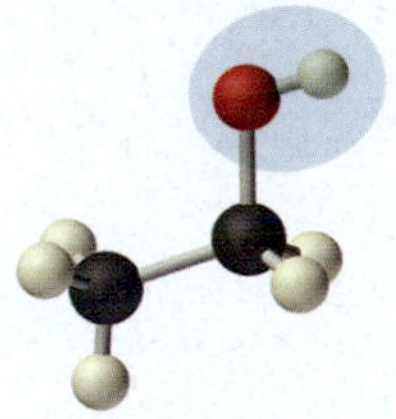

Ethanol, CH_3CH_2**OH**

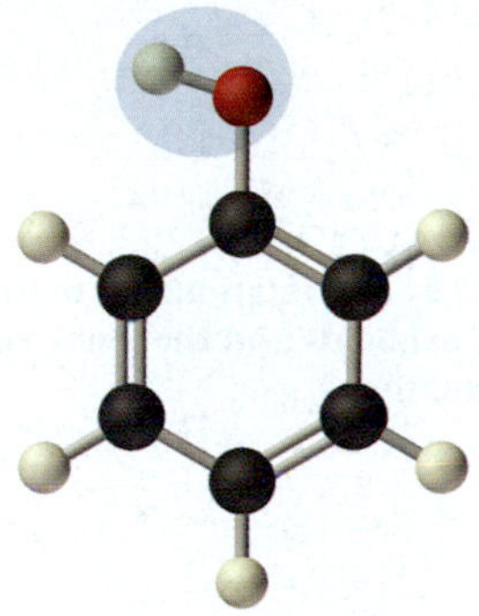

Phenol, C_6H_5**OH**

2.5 • Acids and Bases

For our purposes at this point in the text, we can define an **acid** as a substance with at least one hydrogen atom that can dissociate to form an anion and an H^+ ion (a proton) in aqueous solution, thereby forming an *acidic solution*. **Bases** are compounds that produce hydroxide ions (OH^-) and a cation when dissolved in water, thus forming a *basic solution*. Solutions that are neither basic nor acidic are *neutral*. We will discuss the chemistry of acids and bases in more detail in Chapters 4, 8, and 16, but in this chapter we describe the nomenclature of common acids and identify some important bases so that you can recognize them in future discussions. Pure acids and bases and their concentrated aqueous solutions are commonly encountered in the laboratory. They are usually highly corrosive, so they must be handled with care.

Acids

MGC Acid Names

The names of acids differentiate between (1) acids in which the H^+ ion is attached to an oxygen atom of a polyatomic anion (these are called **oxoacids**, or occasionally *oxyacids*), and (2) acids in which the H^+ ion is attached to some other element. In the latter case, the name of the acid begins with *hydro-* and ends in *-ic*, with the root of the name of the other element or ion in between. Recall that the name of the anion derived from this kind of acid always ends in *-ide*. Thus, hydrogen chloride (HCl) gas dissolves in water to form hydrochloric acid (which contains H^+ and Cl^- ions), hydrogen cyanide (HCN) gas forms hydrocyanic acid (which contains H^+ and CN^- ions), and so on (Table 2.9). Examples of this kind of acid are commonly encountered and very important. For instance, your stomach contains a dilute solution of hydrochloric acid to help digest food. When the mechanisms that prevent the stomach from digesting itself malfunction, the acid destroys the lining of the stomach and an ulcer forms.

If an acid contains one or more H^+ ions attached to oxygen, it is a derivative of one of the common oxoanions, such as sulfate (SO_4^{2-}) or nitrate (NO_3^-). These acids contain as many H^+ ions as are necessary to balance the negative charge on the anion, resulting in a neutral species such as H_2SO_4 and HNO_3.

The names of acids are derived from the names of anions according to the following rules:

1. ***If the name of the anion ends in -ate, then the name of the acid ends in -ic.*** For example, because NO_3^- is the nitrate ion, HNO_3 is nitric acid. Similarly,

TABLE 2.9 Some common acids that do not contain oxygen

Formula	Name in Aqueous Solution	Name of Gaseous Species
HF	hydrofluoric acid	hydrogen fluoride
HCl	hydrochloric acid	hydrogen chloride
HBr	hydrobromic acid	hydrogen bromide
HI	hydroiodic acid	hydrogen iodide
HCN	hydrocyanic acid	hydrogen cyanide
H_2S	hydrosulfuric acid	hydrogen sulfide

ClO_4^- is the perchlorate ion, so $HClO_4$ is perchloric acid. Two important acids are sulfuric acid, H_2SO_4, from the sulfate ion, SO_4^{2-}, and phosphoric acid, H_3PO_4, from the phosphate ion, PO_4^{3-}. Note that these two names use a slight variant of the root of the anion name: *sulf*ate becomes *sulfur*ic and *phosph*ate becomes *phosphor*ic.

2. ***If the name of the anion ends in -ite, then the name of the acid ends in -ous.*** For example, OCl^- is the hypochlorite ion and HOCl is hypochlorous acid, NO_2^- is the nitrite ion and HNO_2 is nitrous acid, and SO_3^{2-} is the sulfite ion and H_2SO_3 is sulfurous acid. The same roots are used whether the acid name ends in *-ic* or *-ous*; thus, *sulfite* becomes *sulfurous*.

The relationship between the names of oxoacids and the parent oxoanions is illustrated in Figure 2.19, and some common oxoacids are listed in Table 2.10.

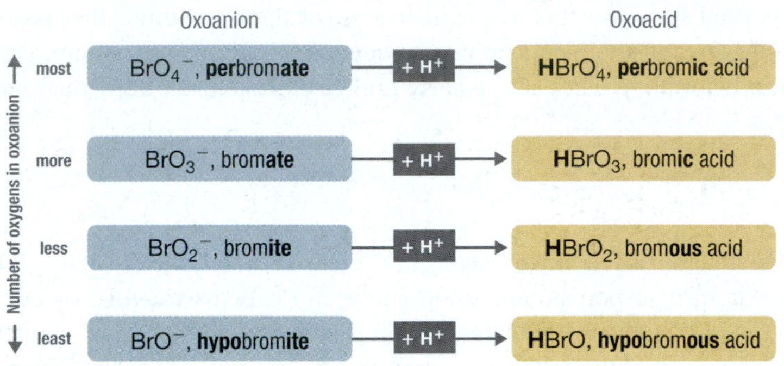

Figure 2.19 **The relationship between the names of oxoacids and the names of the parent oxoanions.**

EXAMPLE 2.11

Give the names of **(a)** the acid formed by adding a proton to the hypobromite ion, OBr^-, and **(b)** the acid formed by adding two protons to the selenate ion, SeO_4^{2-}.

Given Anion

Asked for Parent acid

Strategy

Refer to Tables 2.9 and 2.10 to find the name of the acid. If the acid is not listed, use the guidelines given above.

Solution

Neither of these species is listed in Table 2.9 or 2.10, so we must use the information given above to derive the name of the acid from the name of the polyatomic anion.

(a) The anion name, *hypobromite*, ends in *-ite*, so the name of the parent acid ends in *-ous*. The acid is therefore hypobromous acid, HOBr.

(b) *Selenate* ends in *-ate*, so the name of the parent acid ends in *-ic*. The acid is therefore selenic acid, H_2SeO_4.

EXERCISE 2.11

Give the names and formulas for **(a)** the acid formed by adding a proton to the perbromate ion, BrO_4^-, and **(b)** the acid formed by adding three protons to the arsenite ion, AsO_3^{3-}.

Answer **(a)** perbromic acid, $HBrO_4$; **(b)** arsenous acid, H_3AsO_3

TABLE 2.10 Some common oxoacids

Formula	Name
HNO_2	nitrous acid
HNO_3	nitric acid
H_2SO_3	sulfurous acid
H_2SO_4	sulfuric acid
H_3PO_4	phosphoric acid
H_2CO_3	carbonic acid
HClO	hypochlorous acid
$HClO_2$	chlorous acid
$HClO_3$	chloric acid
$HClO_4$	perchloric acid

TABLE 2.11 Some common carboxylic acids

Formula (Structure)	Name	Uses
HCO_2H	formic acid	tanning, dyeing
CH_3CO_2H	acetic acid	vinegar, food preservative
$CH_3CH_2CO_2H$	propionic acid	food preservative
$CH_3CH_2CH_2CO_2H$	butyric acid	varnishes
$C_6H_5CO_2H$	benzoic acid	food preservative, dyeing
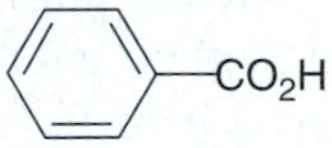		

Many organic compounds contain the *carbonyl group*, a carbon double-bonded to an oxygen. In **carboxylic acids**, an —OH is covalently bonded to the carbon atom of the carbonyl group. Their general formula is RCO_2H:

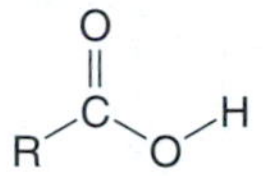

Carboxylic acid

where R can be an alkyl group, an aryl group, or a hydrogen atom. The simplest example, HCO_2H, is *formic acid*, so called because it is found in the secretions of stinging ants (from the Latin *formica*, "ant"). Another example is *acetic acid*, CH_3CO_2H, found in vinegar. Like many acids, carboxylic acids tend to have sharp odors. For example, butyric acid, $CH_3CH_2CH_2CO_2H$, is responsible for the smell of rancid butter, and the characteristic odor of sour milk and vomit is due to lactic acid, $CH_3CH(OH)CO_2H$. Some common carboxylic acids are listed in Table 2.11.

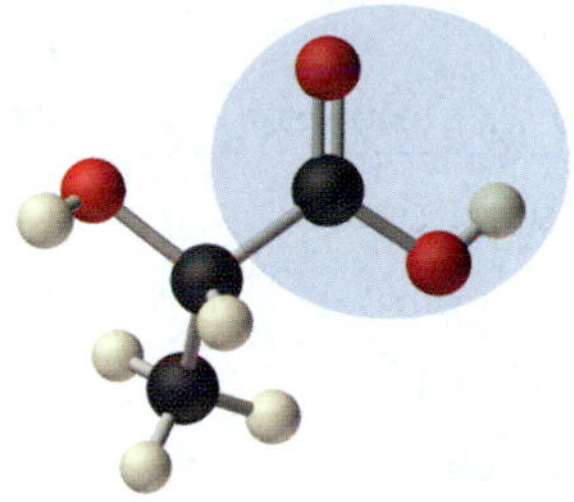
Lactic acid, $CH_3CH(OH)\mathbf{CO_2H}$

Although carboxylic acids are covalent compounds, when they dissolve in water they dissociate to produce H^+ ions (just like any other acid) and RCO_2^- ions. Note that *only the hydrogen attached to the oxygen atom of the carboxylic acid dissociates to form an H^+ ion*. In contrast, the hydrogen atom attached to the oxygen atom of an alcohol does *not* dissociate to form an H^+ ion when an alcohol is dissolved in water. The reasons for the difference in behavior between carboxylic acids and alcohols will be discussed in Chapter 8.

Bases

We will present more comprehensive definitions of bases in later chapters, but virtually all of the bases you encounter in the meantime will be ionic compounds such as sodium hydroxide, NaOH, and barium hydroxide, $Ba(OH)_2$, that contain the hydroxide ion and a metal cation. These have the general formula $M(OH)_n$. Because alcohols, ROH, are *covalent* compounds, they do *not* dissociate in water to form a basic solution (containing OH^- ions). When a base reacts with any of the acids we have discussed, it accepts a proton (H^+). For example, the hydroxide ion (OH^-) accepts a proton to form H_2O. Hence, bases are also referred to as *proton acceptors*.

Concentrated aqueous solutions of ammonia, NH_3, contain significant amounts of the hydroxide ion (even though the dissolved substance is *not* primarily "ammonium hydroxide," $NH_4^+OH^-$, as is often stated on the label). Hence, aqueous ammonia solution is also a common base. Replacing a hydrogen atom of NH_3 with an alkyl group results in an **amine**, RNH_2, which is also a base. Amines have pungent odors; for example, methylamine (CH_3NH_2) is one of the compounds responsible for the foul odor

Note the pattern

Metal hydroxides (MOH) yield OH^- ions and are bases; alcohols (ROH) do not yield OH^- or H^+ ions and are neutral; and carboxylic acids (RCO_2H) yield H^+ ions and are acids.

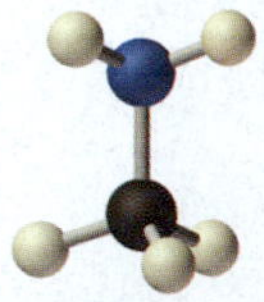

Methylamine

associated with spoiled fish. The physiological importance of amines is suggested in the word *vitamin*, which is derived from the phrase "vital amines." The word was coined to describe dietary substances that were effective at preventing scurvy, rickets, and other diseases, because these substances were assumed to be amines. Subsequently, some "vitamins" have indeed been confirmed to be amines.

2.6 • Industrially Important Chemicals

MGC Octane Rating

It isn't easy to comprehend the scale on which the chemical industry must operate in order to supply the huge amounts of chemicals required in modern industrial societies. Table 2.12 lists the names and formulas of the chemical industry's "Top 25" for 2002—the 25 chemicals produced in the largest quantity in the United States that year—along with the amounts produced, in *billions* of pounds. To put these numbers in perspective, consider that the 88.80 billion pounds of sulfuric acid produced in the United States in 2002 has a volume of 21.90 million cubic meters (2.19×10^7 m^3), enough to fill the Pentagon, probably the largest office building in the world, about 22 times!

TABLE 2.12 Top 25 chemicals produced in the United States in 2002*

Rank	Name	Formula (Structure)	Billions of Pounds
1	sulfuric acid	H_2SO_4	88.80
2	nitrogen	N_2	58.70
3	oxygen	O_2	42.38
4	ethylene	$H_2C{=}CH_2$	40.41
5	ammonia	NH_3	35.95
6	calcium oxide (lime)	CaO	34.72
7	phosphoric acid	H_3PO_4	25.36
8	sodium hydroxide	$NaOH$	24.02
9	propylene (propene)	$CH_3CH{=}CH_2$	22.60
10	chlorine	Cl_2	22.28
11	sodium carbonate	Na_2CO_3	20.89
12	urea	$(NH_2)_2C{=}O$ 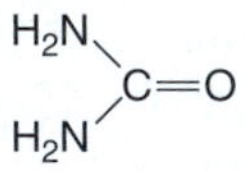	16.84
13	nitric acid	HNO_3	16.08
14	1,2-dichloroethane (ethylene dichloride)	$ClCH_2CH_2Cl$	15.94
15	ammonium nitrate	NH_4NO_3	15.33
16	vinyl chloride	$CH_2{=}CHCl$	13.23
17	benzene	C_6H_6	12.01
18	ethylbenzene	$C_6H_5CH_2CH_3$ 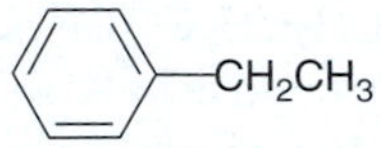	10.99
19	carbon dioxide	CO_2	10.91
20	methyl *t*-butyl ether	$CH_3OC(CH_3)_3$ 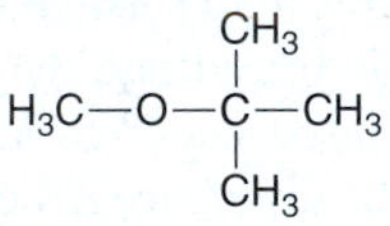	10.86
21	styrene	$C_6H_5CH{=}CH_2$ 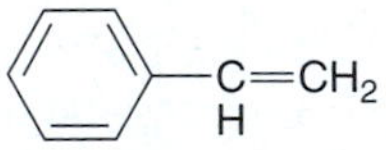	8.94
22	methanol	CH_3OH	8.73
23	formaldehyde	$H_2C{=}O$ 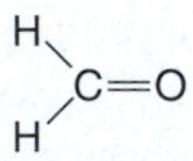	6.98
24	xylenes (mixture)[a]	$(CH_3)_2C_6H_4$ 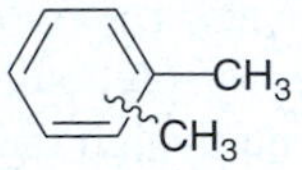	6.38
25	toluene	$C_6H_5CH_3$	6.03

* Nonlinear structures not previously introduced are shown with their respective formulas.
[a] The squiggly line indicates that the second $—CH_3$ group can be attached to any of the other positions of the ring.

According to Table 2.12, 4 of the top 5, 8 of the top 10, and 11 of the top 15 compounds produced in the United States are inorganic, and the total mass of inorganic chemicals produced is almost twice the mass of organic chemicals. Yet the diversity of organic compounds used in industry is such that over half of the top 25 compounds (13 out of 25) are organic.

Why are such huge quantities of chemical compounds produced annually? The reason is that they are used both directly as components of compounds and materials that we encounter on an almost daily basis and indirectly in the production of those compounds and materials. The single largest use of industrial chemicals is in the production of foods: 7 of the top 15 chemicals are either fertilizers (ammonia, urea, and ammonium nitrate) or used primarily in the production of fertilizers (sulfuric acid, nitric acid, nitrogen, and phosphoric acid). Many of the organic chemicals on the list are used primarily as ingredients in the plastics and related materials that are so prevalent in contemporary society. Ethylene (#4) and propylene (#9), for example, are used to produce polyethylene and polypropylene, which are made into plastic milk bottles, sandwich bags, indoor–outdoor carpets, and other common items. Vinyl chloride (#16), in the form of polyvinylchloride (PVC), is used in everything from pipes to floor tiles to trash bags. Though not listed in Table 2.12, butadiene (#36) and carbon black (#37) are used in the manufacture of synthetic rubber for tires, and phenol (#34) and formaldehyde (#23) are ingredients in plywood, fiberglass, and many hard plastic items.

We do not have the space in this text to consider the applications of all these compounds in any detail, but we will return to many of them after we have developed the concepts necessary to understand their underlying chemistry. Instead, we conclude this chapter with a brief discussion of petroleum refining as it relates to gasoline and octane ratings, and with a look at the production and utilization of the #1 industrial chemical, sulfuric acid.

Petroleum

The petroleum that is pumped out of the ground at locations around the world is a complex mixture of several thousand organic compounds, including straight-chain alkanes, cycloalkanes, alkenes, and aromatic hydrocarbons that contain from four to several hundred carbon atoms. The identities and relative abundances of the components vary depending on the source. Thus, Texas crude oil is somewhat different from Saudi Arabian crude. In fact, the analysis of petroleum from different deposits can produce a "fingerprint" of each, which is useful in tracking down the sources of spilled crude oil. For example, Texas crude oil is "sweet," meaning that it contains a small amount of sulfur-containing molecules, whereas Saudi Arabian crude oil is "sour," meaning that it contains a relatively large amount of sulfur-containing molecules.

Gasoline

Petroleum is converted to useful products such as gasoline in three steps: distillation, cracking, and reforming. Recall from Chapter 1 that distillation separates compounds on the basis of their relative volatility, which is usually inversely proportional to their boiling points. Figure 2.20a shows a cutaway drawing of a column used in the petroleum industry for separating the components of crude oil. The petroleum is heated to approximately 400°C (750°F), at which temperature it has become a mixture of liquid and vapor. This mixture, called the *feedstock*, is introduced into the refining tower. The most volatile components (those with the lowest boiling points) condense at the top of the column where it is cooler, while the less volatile components condense nearer the bottom. Some materials are so nonvolatile that they collect at the bottom without evaporating at all. Thus, the composition of the liquid condensing at each level is different. These different fractions, each of which usually consists of a mixture of compounds with similar numbers of carbon atoms, are drawn off separately.

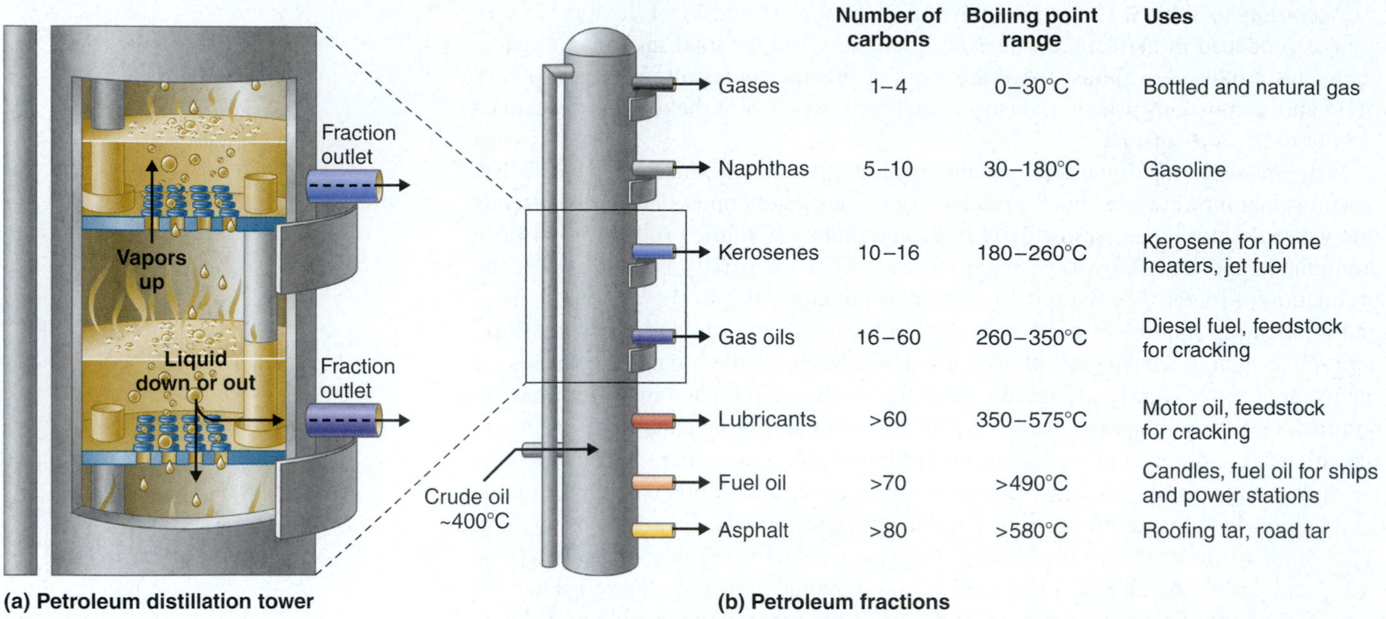

Figure 2.20 Distillation of petroleum. (a) Diagram of a distillation column used for separating petroleum fractions. (b) Petroleum fractions condense at different temperatures, depending on the number of carbon atoms in the molecules, and are drawn off from the column. The most volatile components (those with the lowest boiling points) condense at the top of the column, and the least volatile (those with the highest boiling points) condense at the bottom.

Figure 2.20b shows the typical fractions collected at refineries, the numbers of carbon atoms they contain, their boiling points, and their ultimate uses. These products range from gases used in natural and bottled gas to liquids used in fuels and lubricants to gummy solids used as tar on roads and roofs.

The economics of petroleum refining are complex. For example, the market demand for kerosene and lubricants is much lower than the demand for gasoline, yet all three fractions are obtained from the distillation column in comparable amounts. Furthermore, most gasolines and jet fuels are blends with very carefully controlled compositions that cannot vary as their original feedstocks did. To make petroleum refining more profitable, the less volatile, lower-value fractions must be converted to more volatile, higher-value mixtures that have carefully controlled formulas. The first process used to accomplish this transformation is **cracking**, in which the larger and heavier hydrocarbons in the kerosene and higher-boiling-point fractions are heated to temperatures as high as 900°C. High-temperature reactions cause the C—C bonds to break, which converts the compounds to lighter molecules similar to those in the gasoline fraction. Thus, in cracking, a straight-chain alkane with a number of carbon atoms corresponding to the kerosene fraction is converted to a mixture of hydrocarbons with a number of carbon atoms corresponding to the lighter gasoline fraction. The second process used to increase the amounts of valuable products is called **reforming**; it is the chemical conversion of straight-chain alkanes to either branched-chain alkanes or mixtures of aromatic hydrocarbons. The necessary chemical reactions are brought about by the use of metals such as platinum. The mixtures of products obtained from cracking and reforming are separated by fractional distillation.

Octane Ratings

The quality of a fuel is indicated by its **octane rating**, which is a measure of its ability to burn in a combustion engine without knocking or pinging. Knocking and pinging signal premature combustion (Figure 2.21), which can be caused either by an engine malfunction or by a fuel that burns too fast. In either case, the gasoline–air mixture detonates at the wrong point in the engine cycle, which reduces the power output and can damage valves,

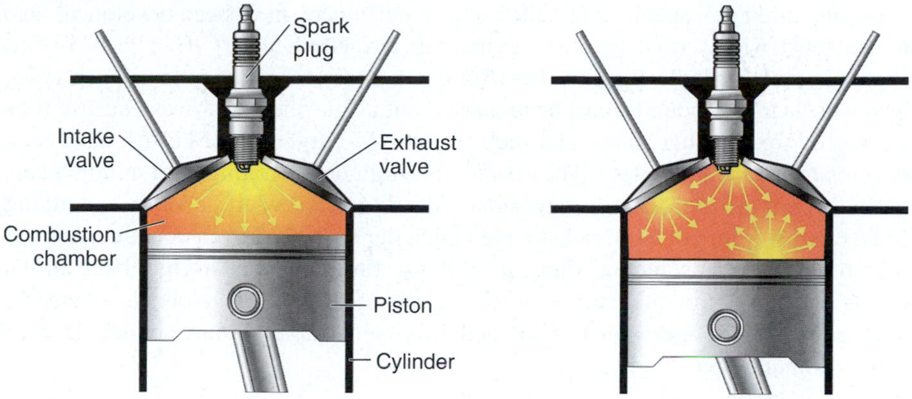

(a) Normal combustion **(b) Premature combustion**

Figure 2.21 Burning of gasoline in an internal combustion engine. (a) Normally the fuel is ignited by the spark plug and combustion spreads uniformly outward. (b) Gasoline with an octane rating that is too low for the engine can ignite prematurely, resulting in uneven burning that causes knocking and pinging.

pistons, bearings, and other engine components. The various gasoline formulations are designed to provide the mix of hydrocarbons least likely to cause knocking or pinging in a given type of engine performing at a particular level.

The octane scale was established in 1927 using a standard test engine and two pure compounds: *n*-heptane and isooctane (2,2,4-trimethylpentane). *n*-Heptane, which causes a great deal of knocking upon combustion, was assigned an octane rating of 0, whereas isooctane, a very smooth-burning fuel, was assigned an octane rating of 100. Chemists assign octane ratings to different blends of gasoline by burning a sample of each in a test engine and comparing the observed knocking with the amount of knocking caused by specific mixtures of *n*-heptane and isooctane. For example, the octane rating of a blend of 89% isooctane and 11% *n*-heptane is simply the average of the octane ratings of the components weighted by the relative amounts of each in the blend. Converting percentages to decimals, we obtain the octane rating of the mixture:

$$0.89(100) + 0.11(0) = 89$$

A gasoline that performs at the same level as a blend of 89% isooctane and 11% *n*-heptane is assigned an octane rating of 89; this represents an intermediate grade of gasoline. Regular gasoline typically has an octane rating of 87; premium has a rating of 93 or higher.

As shown in Table 2.13, many compounds that are now available have octane ratings higher than 100, which means they are better fuels than pure isooctane. In

The octane rating of gasoline is calibrated using the compounds *n*-heptane and isooctane as standards.

TABLE 2.13 Octane ratings of some hydrocarbons and common additives

Name	Condensed Structural Formula	Octane Rating	Name	Condensed Structural Formula	Octane Rating
n-heptane	$CH_3CH_2CH_2CH_2CH_2CH_2CH_3$	0	ethanol	CH_3CH_2OH	108
n-hexane	$CH_3CH_2CH_2CH_2CH_2CH_3$	25	*t*-butyl alcohol	$(CH_3)_3COH$	113
n-pentane	$CH_3CH_2CH_2CH_2CH_3$	62			
isooctane	$(CH_3)_3CCH_2CH(CH_3)_2$	100	*p*-xylene		116
benzene		106			
methanol	CH_3OH	107	methyl *t*-butyl ether	$H_3COC(CH_3)_3$	116
o-xylene	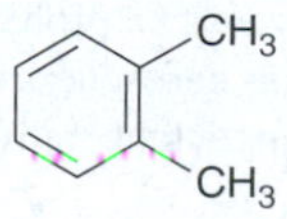	107	toluene	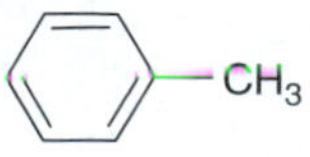	118

addition, antiknock agents, also called octane enhancers, have been developed. One of the most widely used for many years was tetraethyllead, $(C_2H_5)_4Pb$, which at approximately 3 g/gallon gives a 10–15-point increase in octane rating. Since 1975, however, lead compounds have been phased out as gasoline additives because they are highly toxic. Other enhancers, such as methyl *t*-butyl ether (MTBE), have been developed to take their place. They combine a high octane rating with minimal corrosion to engine and fuel system parts. Unfortunately, when gasoline containing MTBE leaks from underground storage tanks, the result has been contamination of the groundwater in some locations, resulting in limitations or outright bans on the use of MTBE in certain areas. As a result, the use of alternative octane enhancers such as ethanol, which can be obtained from renewable resources such as corn stalks, is increasing.

EXAMPLE 2.12

You have a crude (that is, unprocessed or *straight-run*) petroleum distillate consisting of 10% *n*-heptane, 10% *n*-hexane, and 80% *n*-pentane by mass, with an octane rating of 52. What percentage of methyl *t*-butyl ether (MTBE) by mass would you need to increase the octane rating of the distillate to that of regular-grade gasoline (a rating of 87), assuming that the octane rating is directly proportional to the amounts of the compounds present? Use the information presented in Table 2.13.

Given Composition of petroleum distillate, initial octane rating, and final octane rating

Asked for Percentage MTBE by mass in final mixture

Strategy

Ⓐ Define the unknown as the percentage of MTBE in the final mixture. Then subtract this unknown from 100% to obtain the percentage of petroleum distillate.

Ⓑ Multiply the percentage of MTBE and the percentage of petroleum distillate by their respective octane ratings, and then sum these values to obtain the overall octane rating of the new mixture.

Ⓒ Solve for the unknown to obtain the percentage of MTBE needed.

Solution

Ⓐ✓ The question asks what percentage of MTBE will give an overall octane rating of 87 when mixed with the straight-run fraction. From Table 2.13, we see that the octane rating of MTBE is 116. Let x be the percentage of MTBE, and let $100 - x$ be the percentage of petroleum distillate. Ⓑ✓ Multiplying the percentage of each component by its respective octane rating and setting the sum equal to the desired octane rating of the mixture (87) times 100% give

$$\begin{aligned}\text{final octane rating of mixture} = 87(100) &= 52(100 - x) + 116x \\ &= 5200 - 52x + 116x \\ &= 5200 + 64x\end{aligned}$$

Ⓒ✓ Solving the equation gives $x = 55\%$. Thus, the final mixture must contain 55% MTBE by mass.

To obtain a composition of 55% MTBE by mass, you would have to add more than an equal mass of MTBE (actually 0.55/0.45, or 1.2 times) to the straight-run fraction. This is 1.2 tons of MTBE per ton of straight-run gasoline, which would be prohibitively expensive. Hence, there are sound economic reasons for reforming the kerosene fractions to produce toluene and other aromatic compounds, which have high octane ratings and are much cheaper than MTBE.

EXERCISE 2.12

As shown in Table 2.13, toluene is one of the fuels suitable for use in automobile engines. How much toluene would have to be added to a blend of the petroleum fraction from Example 2.12 containing 15% MTBE by mass to increase the octane rating to that of premium gasoline (93)?

Answer The final blend is 56% toluene by mass, which requires a ratio of 56/44, or 1.3 tons of toluene per ton of blend.

Sulfuric Acid

Sulfuric acid is one of the oldest chemical compounds known. It was probably first prepared by alchemists who burned sulfate salts such as $FeSO_4 \cdot 7H_2O$, called green vitriol from its color and glassy appearance (the Latin word for "glass" is *vitrum*). Because pure sulfuric acid was found to be useful for dyeing textiles, enterprising individuals looked for ways to improve its production. Thus, by the mid-18th century, sulfuric acid was being produced in multiton quantities by the *lead-chamber process*, invented by John Roebuck in 1746. In this process, sulfur was burned in a large room lined with lead, and the resulting fumes were absorbed in water.

Production

Today the production of sulfuric acid is likely to start with elemental sulfur obtained through an ingenious technique called the *Frasch process*, which takes advantage of the low melting point of elemental sulfur (115.21°C). Large deposits of elemental sulfur are found in porous limestone rocks in the same geological formations that often contain petroleum. In the Frasch process, water at high temperature (160°C) and high pressure is pumped underground to melt the sulfur, and compressed air is used to force the liquid sulfur–water mixture to the surface (Figure 2.22). The material that emerges from the ground is more than 99% pure sulfur. After it solidifies, it is pulverized and shipped in railroad cars to the plants that produce sulfuric acid, as shown in the photograph in the margin.

A train carrying elemental sulfur through the White Canyon of the Thompson River in British Columbia, Canada.

An increasing number of sulfuric acid manufacturers have begun to use sulfur dioxide (SO_2) as a starting material instead of elemental sulfur. Sulfur dioxide is recovered from the burning of oil and gas, which contain small amounts of sulfur compounds. When not recovered, SO_2 is released into the atmosphere, where it is converted to an environmentally hazardous form that leads to *acid rain* (Chapter 4).

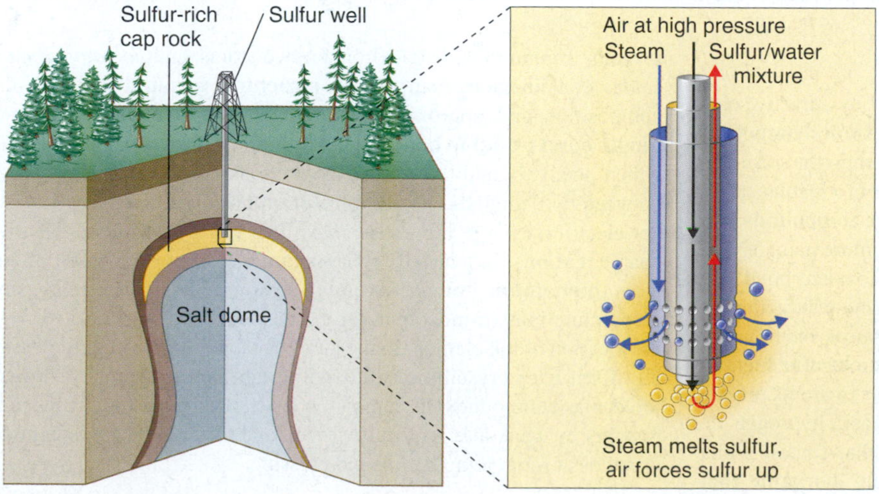

Figure 2.22 Extraction of elemental sulfur from underground deposits. In the Frasch process for extracting sulfur, very hot water at high pressure is injected into the sulfur-containing rock layer to melt the sulfur. The resulting mixture of liquid sulfur and hot water is forced up to the surface by compressed air.

If sulfur is the starting material, the first step in the production of sulfuric acid is the combustion of sulfur with oxygen to produce SO_2. Next, SO_2 is converted to SO_3 by the *contact process*, in which SO_2 and O_2 react in the presence of V_2O_5 to achieve about 97% conversion to SO_3. The SO_3 can then be treated with a small amount of water to produce sulfuric acid. Usually, however, the SO_3 is absorbed in concentrated sulfuric acid to produce *oleum*, a more potent form called fuming sulfuric acid. Because of its high SO_3 content (approximately 99% by mass), oleum is cheaper to ship than concentrated sulfuric acid. At the point of use, the oleum is diluted with water to give concentrated sulfuric acid (*very carefully* because dilution generates enormous amounts of heat). Because SO_2 is a pollutant, the small amounts of unconverted SO_2 are recovered and recycled to minimize the amount released into the air.

Uses

Two-thirds of the sulfuric acid produced in the United States is used to make fertilizers, most of which contain nitrogen, phosphorus, and potassium (in a form called potash). In earlier days, phosphate-containing rocks were simply ground up and spread on fields as fertilizer, but the extreme insolubility of many salts that contain the phosphate ion (PO_4^{3-}) limits the availability of phosphorus from these sources. Sulfuric acid serves as a source of protons (H^+ ions) that react with phosphate minerals to produce more soluble salts containing HPO_4^{2-} or $H_2PO_4^-$ as the anion, which are much more readily taken up by plants. In this context, sulfuric acid is used in two principal ways: (1) the phosphate rocks are treated with concentrated sulfuric acid to produce "superphosphate," a mixture of 32% $CaHPO_4$ and $Ca(H_2PO_4)_2{\cdot}H_2O$, 50% $CaSO_4{\cdot}2H_2O$, approximately 3% absorbed phosphoric acid, and other nutrients; and (2) sulfuric acid is used to produce phosphoric acid (H_3PO_4), which can then be used to convert phosphate rocks to "triple superphosphate," which is largely $Ca(H_2PO_4)_2{\cdot}H_2O$.

Sulfuric acid is also used to produce *potash*, one of the other major ingredients in fertilizers. The name *potash* originally referred to potassium carbonate (obtained by boiling wood *ash*es with water in iron *pot*s), but today it also refers to compounds such as potassium hydroxide (KOH) and potassium oxide (K_2O). The usual source of potassium in fertilizers is actually potassium sulfate (K_2SO_4), which is produced by several routes, including the reaction of concentrated sulfuric acid with solid KCl (obtained as the pure salt from mineral deposits).

SUMMARY AND KEY TERMS

2.1 Chemical Compounds (p. 47)

The atoms in chemical compounds are held together by attractive electrostatic interactions known as **chemical bonds**. **Ionic compounds** contain positively and negatively charged ions in a ratio that results in an overall charge of zero. The ions are held together in a regular spatial arrangement by electrostatic forces. Most **covalent compounds** consist of **molecules**, groups of atoms in which one or more pairs of electrons are shared by at least two atoms to form a covalent bond. The atoms in molecules are held together by the electrostatic attraction between the positively charged nuclei of the bonded atoms and the negatively charged electrons shared by the nuclei. The **molecular formula** of a covalent compound gives the types and numbers of atoms present. Compounds that contain predominantly carbon and hydrogen are called **organic compounds**, whereas compounds that consist primarily of elements other than carbon and hydrogen are **inorganic compounds**. *Diatomic molecules* contain two atoms, and *polyatomic molecules* contain more than two. A **structural formula** indicates the composition and approximate structure and shape of a molecule. **Single bonds, double bonds**, and **triple bonds** are covalent bonds in which one, two, and three pairs of electrons, respectively, are shared between two bonded atoms. Atoms or groups of atoms that possess a net electrical charge are called *ions*; they can have either a positive charge (**cations**) or a negative charge (**anions**). Ions can consist of one atom (**monatomic ions**) or several (**polyatomic ions**). The charges on monatomic ions of most main group elements can be predicted from the location of the element in the periodic table. Ionic compounds usually form hard crystalline solids with high melting points. Covalent molecular compounds, in contrast, consist of discrete molecules held together by weak intermolecular forces, and they can be gases, liquids, or solids at room temperature and pressure.

2.2 Chemical Formulas (p. 55)

An **empirical formula** gives the *relative* numbers of atoms of the elements in a compound, reduced to the lowest whole numbers. The **formula unit** is the *absolute* grouping represented by the empirical formula of a compound, either ionic or covalent. Empirical formulas are particularly useful for describing the composition of ionic compounds, which do not contain readily identifiable molecules. Some ionic compounds occur as **hydrates**, which contain specific ratios of loosely bound water molecules called *waters of hydration*.

2.3 Naming Ionic Compounds (p. 59)

Ionic compounds are named according to systematic procedures, although common names are widely used. Systematic nomenclature enables us to write the structure of any compound from its name, and vice versa. Ionic compounds are named by writing the cation first, followed by the anion. If a metal can form cations with more than one charge, the charge is indicated by Roman numerals in parentheses following the name of the metal.

2.4 Naming Covalent Compounds (p. 63)

Covalent inorganic compounds are named by a procedure similar to that used for ionic compounds, using prefixes to indicate the numbers of atoms in the molecular formula. The simplest organic compounds are the **hydrocarbons**, which contain *only* carbon and hydrogen. **Alkanes** contain only C—H and C—C single bonds, **alkenes** contain at least one carbon–carbon double bond, and **alkynes** contain one or more carbon–carbon triple bonds. Hydrocarbons can also be **cyclic**, with the ends of the chain connected to form a ring. Collectively, alkanes, alkenes, and alkynes are called *aliphatic hydrocarbons*. **Aromatic hydrocarbons**, or **arenes**, are another important class of hydrocarbons that contain rings of carbon atoms related to the structure of benzene, C_6H_6. A derivative of an alkane or an arene from which one hydrogen atom has been removed is called an *alkyl group* or an *aryl group*, respectively. **Alcohols** are another common class of organic compound, which contain an —OH group covalently bonded to either an alkyl or an aryl group.

2.5 Acids and Bases (p. 71)

Common **acids** and the polyatomic anions derived from them have their own names and rules for nomenclature. The nomenclature of acids differentiates between **oxoacids**, in which the H^+ ion is attached to an oxygen atom of a polyatomic ion, and acids in which the H^+ ion is attached to another element. **Carboxylic acids** are an important class of organic acids. Ammonia is an important **base**, as are its organic derivatives, the **amines**.

2.6 Industrially Important Chemicals (p. 74)

Many chemical compounds are prepared industrially in huge quantities and used to produce foods, fuels, or plastics and other such materials. Petroleum refining takes a complex mixture of naturally occurring hydrocarbons as a feedstock and, through a series of steps involving distillation, **cracking**, and **reforming**, converts them to mixtures of simpler organic compounds with desirable properties. A major use of petroleum is in the production of motor fuels such as gasoline. The performance of such fuels in engines is described by their **octane rating**, which depends on the identity of the compounds present and their relative abundance in the blend.

Sulfuric acid is the compound produced in the largest quantity in the industrial world. Much of the sulfur used in the production of sulfuric acid is obtained via the Frasch process, in which very hot water forces liquid sulfur out of the ground in nearly pure form. Sulfuric acid is produced by the reaction of sulfur dioxide with oxygen in the presence of vanadium(V) oxide (the contact process), followed by absorption of the sulfur trioxide in concentrated sulfuric acid to produce oleum. Most sulfuric acid is used to prepare fertilizers.

QUESTIONS AND PROBLEMS

MGC For instructor-assigned homework, go to **www.masteringgenchem.com**

Questions and Problems with colored numbers have answers in the Appendix and complete solutions in the Student Solutions Manual.

CONCEPTUAL

2.1 Chemical Compounds

1. Ionic and covalent compounds are both held together by electrostatic attractions between oppositely charged particles. Describe the differences in the nature of the attractions in ionic and covalent compounds. Which class of compounds contains pairs of electrons shared between bonded atoms?
2. Which contains fewer electrons than the neutral atom: the corresponding cation or anion?
3. What is the difference between an organic and an inorganic compound?
4. Give two pieces of information that we obtain from a structural formula that we cannot obtain from an empirical formula.
5. What is the advantage of writing a structural formula as a condensed formula?
6. The formulas of alcohols are often written as ROH rather than as empirical formulas. For example, methanol is generally written as CH_3OH rather than CH_4O. Explain why the ROH notation is preferred.
7. The compound dimethyl sulfide has the empirical formula C_2H_6S and the structural formula CH_3SCH_3. What information do we obtain from the structural formula that we do not get from the empirical formula? Write the condensed structural formula for the compound.
8. The majority of elements that exist as diatomic molecules are found in one group of the periodic table. Identify the group.
9. Discuss the differences between covalent and ionic compounds with regard to (a) the forces that hold the atoms together; (b) melting points: and (c) physical states at room temperature and pressure.
10. Benzene (C_6H_6) is an organic compounds and KCl is an ionic compound. The sum of the masses of the atoms in each empirical formula is approximately the same. How would you expect the two to compare with regard to (a) melting point; (b) type of

bonding; (c) rate of evaporation; and (d) structure? What species are present in benzene vapor?

11. Why do covalent compounds generally tend to have lower melting points than ionic compounds?

2.2 Chemical Formulas

12. What are the differences and similarities between a polyatomic ion and a molecule?
13. Classify each compound as ionic or covalent: (a) $Zn_3(PO_4)_2$; (b) $C_6H_5CO_2H$; (c) potassium dichromate; (d) CH_3CH_2SH; (e) ammonium bromide; (f) CCl_2F_2.
14. Classify each compound as ionic or covalent: (a) $CH_3CH_2CO_2H$; (b) $CaCl_2$; (c) $Y(NO_3)_3$; (d) H_2S; (e) sodium acetate. Which are organic and which are inorganic?
15. Generally, one cannot determine the molecular formula directly from an empirical formula. What other information is needed?
16. What is the correct formula for magnesium hydroxide: $MgOH_2$ or $Mg(OH)_2$? Why?
17. Magnesium cyanide is written as $Mg(CN)_2$, not $MgCN_2$. Why?
18. Name these cations: (a) K^+; (b) Al^{3+}; (c) $NH_4{}^+$; (d) Mg^{2+}; (e) Li^+.
19. Name these anions: (a) Br^-; (b) $CO_3{}^{2-}$; (c) S^{2-}; (d) $NO_3{}^-$; (e) $HCO_2{}^-$; (f) F^-; (g) ClO^-; (h) $C_2O_4{}^{2-}$.
20. Name these anions: (a) $PO_4{}^{3-}$; (b) Cl^-; (c) $SO_3{}^{2-}$; (d) $CH_3CO_2{}^-$; (e) $HSO_4{}^-$; (f) $ClO_4{}^-$; (g) $NO_2{}^-$; (h) O^{2-}.
21. Name these anions: (a) $SO_4{}^{2-}$; (b) CN^-; (c) $Cr_2O_7{}^{2-}$; (d) N^{3-}; (e) OH^-; (f) I^-; (g) $O_2{}^{2-}$.
22. Does a given always contain the same number of waters of hydration?

2.3 Naming Ionic Compounds

23. Name each compound: (a) $MgBr_2$; (b) NH_4CN; (c) CaO; (d) $KClO_3$; (e) K_3PO_4; (f) NH_4NO_2; (g) NaN_3.
24. Name each compound: (a) $NaNO_3$; (b) $Cu_3(PO_4)_2$; (c) NaOH; (d) Li_4C; (e) CaF_2; (f) NH_4Br; (g) $MgCO_3$.
25. Name each compound: (a) RbBr; (b) $Mn_2(SO_4)_3$; (c) NaClO; (d) $(NH_4)_2SO_4$; (e) NaBr; (f) KIO_3; (g) Na_2CrO_4.
26. Name each compound: (a) NH_4ClO_4; (b) $SnCl_4$; (c) $Fe(OH)_2$; (d) Na_2O; (e) $MgCl_2$; (f) K_2SO_4; (g) $RaCl_2$.
27. Give the name of each compound: (a) KCN; (b) LiOH; (c) $CaCl_2$; (d) $NiSO_4$; (e) NH_4ClO_2; (f) $LiClO_4$; (g) $La(CN)_3$.

2.4 Naming Covalent Compounds

28. Can an inorganic compound be classified as a hydrocarbon? Why or why not?
29. Is the compound $NaHCO_3$ a hydrocarbon? Why or why not?
30. Name each compound: (a) NiO; (b) TiO_2; (c) N_2O; (d) CS_2; (e) SO_3; (f) NF_3; (g) SF_6.
31. Name each compound: (a) $HgCl_2$; (b) IF_5; (c) N_2O_5; (d) Cl_2O; (e) HgS; (f) PCl_5.
32. Would you expect PCl_3 to be an ionic or a covalent compound? Explain your reasoning.
33. What distinguishes an aromatic hydrocarbon from an aliphatic hydrocarbon?
34. The following general formulas represent specific classes of hydrocarbons. Refer to Tables 2.7 and 2.8 and Figure 2.16 and identify the classes. (a) C_nH_{2n+2}; (b) C_nH_{2n}; (c) C_nH_{2n-2}
35. Using R to represent an alkyl group, show the general structure of (a) an alcohol and (b) an amine.

2.5 Acids and Bases

36. Name each acid: (a) HCl; (b) $HBrO_3$; (c) HNO_3; (d) H_2SO_4; (e) HIO_3.
37. Name each acid: (a) HBr; (b) H_2SO_3; (c) $HClO_3$; (d) HCN; (e) H_3PO_4.
38. Name the aqueous acid that corresponds to each gaseous species: (a) hydrogen bromide; (b) hydrogen cyanide; (c) hydrogen iodide.
39. When each of the following compounds is added to water, is the resulting solution acidic, neutral, or basic? (a) CH_3CH_2OH; (b) $Mg(OH)_2$; (c) $C_6H_5CO_2H$; (d) LiOH; (e) $C_3H_7CO_2H$; (f) H_2SO_4
40. Draw the structure of the simplest example of each: (a) alkane; (b) alkene; (c) alkyne; (d) aromatic compound; (e) alcohol; (f) carboxylic acid; (g) amine; (h) cycloalkane.
41. Identify the class of organic compound represented by each compound.

(a) 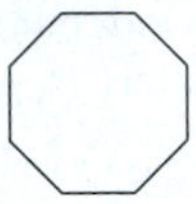 (b) CH_3CH_2OH (c) $HC{\equiv}CH$

(d)  (e) $C_3H_7NH_2$ (f) $CH_3CH{=}CHCH_2CH_3$

(g) 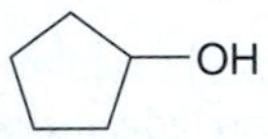

(h) 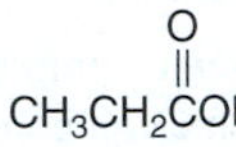

42. Which class of organic compound does each substance represent?

(a) 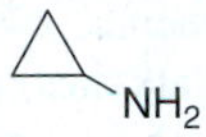

(b) 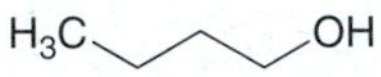

(c)  (d) 

(e) 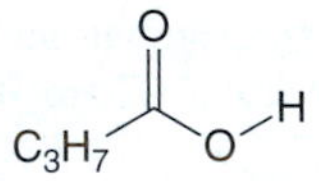

(f) $CH_3C{\equiv}CH$

(g) 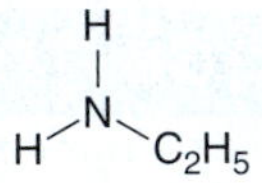

(h) 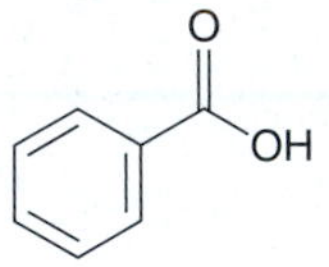

2.6 Industrially Important Chemicals

43. Describe the processes used for converting crude oil to transportation fuels.
44. If your automobile engine is knocking, is the octane rating of your gasoline too low or too high? Explain your answer.
45. Tetraethyllead is no longer used as a fuel additive to prevent knocking. Instead, fuel is now marketed as "unleaded." Why is tetraethyllead no longer used?
46. If you were to try to extract sulfur from an underground source, what process would you use? Describe briefly the essential features of this process.
47. Why are the phosphate-containing minerals used in fertilizers treated with sulfuric acid?

NUMERICAL

This section includes paired problems (marked by brackets) that require similar problem-solving skills.

2.1 Chemical Compounds

48. The structural formula for chloroform ($CHCl_3$) was shown in Exercise 2.2. Based on this information, draw the structural formula of dichloromethane (CH_2Cl_2).

49. How many electrons are present in each of these ions? (a) Ca^{2+}; (b) Se^{2-}; (c) In^{3+}; (d) Sr^{2+}; (e) As^{3+}; (f) N^{3-}; (g) Tl^{+}

50. How many electrons are present in each of these ions? (a) F^{-}; (b) Rb^{+}; (c) Ce^{3+}; (d) Zr^{4+}; (e) Zn^{2+}; (f) Kr^{2+}; (g) B^{3+}

51. Predict the charge on the most common monatomic ion formed by each element: (a) Na; (b) Se; (c) Ba; (d) Rb; (e) N; (f) Al.

52. Predict the charge on the most common monatomic ion formed by each element: (a) Cl; (b) P; (c) Sc; (d) Mg; (e) As; (f) O.

2.2 Chemical Formulas

53. Write the formula for each of the following compounds.
(a) Magnesium sulfate, which contains 1 magnesium atom, 4 oxygen atoms, and 1 sulfur atom
(b) Ethylene glycol (antifreeze), which contains 6 hydrogen atoms, 2 carbon atoms, and 2 oxygen atoms
(c) Acetic acid, which contains 2 oxygen atoms, 2 carbon atoms, and 4 hydrogen atoms
(d) Potassium chlorate, which contains 1 chlorine atom, 1 potassium atom, and 3 oxygen atoms
(e) Sodium hypochlorite pentahydrate, which contains 1 chlorine atom, 1 sodium atom, 6 oxygen atoms, and 10 hydrogen atoms

54. Write the formula for each of the following compounds.
(a) Cadmium acetate, containing 1 cadmium atom, 4 oxygen atoms, 4 carbon atoms, and 6 hydrogen atoms
(b) Barium cyanide, containing 1 barium atom, 2 carbon atoms, and 2 nitrogen atoms
(c) Iron(III) phosphate dihydrate, containing 1 iron atom, 1 phosphorus atom, 6 oxygen atoms, and 4 hydrogen atoms
(d) Manganese(II) nitrate hexahydrate, containing 1 manganese atom, 12 hydrogen atoms, 12 oxygen atoms, and 2 nitrogen atoms
(e) Silver phosphate, containing 1 phosphorus atom, 3 silver atoms, and 4 oxygen atoms

55. Complete the table by filling in the formula for the ionic compound formed by each cation–anion pair.

Ion	K^+	Fe^{3+}	NH_4^+	Ba^{2+}
Cl^-	KCl	___	___	___
SO_4^{2-}	___	___	___	___
PO_4^{3-}	___	___	___	___
NO_3^-	___	___	___	___
OH^-	___	___	___	___

56. Give the empirical formula for the binary compound formed by the most common monatomic ions formed by each pair of elements: (a) zinc and sulfur; (b) barium and iodine; (c) magnesium and chlorine; (d) silicon and oxygen; (e) sodium and sulfur.

57. Write the empirical formula for the binary compound formed by the most common monatomic ions formed by each pair of elements: (a) lithium and nitrogen; (b) cesium and chlorine; (c) germanium and oxygen; (d) rubidium and sulfur; (e) arsenic and sodium.

58. Give the empirical formula for each compound: (a) $Na_2S_2O_4$; (b) B_2H_6; (c) $C_6H_{12}O_6$; (d) P_4O_{10}; (e) $KMnO_4$.

59. Write the empirical formula for each compound: (a) Al_2Cl_6; (b) $K_2Cr_2O_7$; (c) C_2H_4; (d) $(NH_2)_2CNH$; (e) CH_3COOH.

2.3 Naming Ionic Compounds

60. Name the cation and anion, and give the charge on each ion in the following ionic compounds: (a) $Zn(NO_3)_2$; (b) CoS; (c) $BeCO_3$; (d) Na_2SO_4; (e) $K_2C_2O_4$; (f) NaCN.

61. Give the names of the cation and anion, and give the charge on each ion in the following ionic compounds: (a) BeO; (b) $Pb(OH)_2$; (c) BaS; (d) $Na_2Cr_2O_7$; (e) $ZnSO_4$; (f) KClO; (g) NaH_2PO_4.

62. Write the formula for (a) lead(II) nitrate; (b) ammonium phosphate; (c) silver sulfide; (d) barium sulfate; (e) cesium iodide; (f) sodium bicarbonate; and (g) potassium dichromate.

63. Write the formula for each compound: (a) magnesium carbonate; (b) aluminum sulfate; (c) potassium phosphate; (d) lead(IV) oxide; (e) silicon nitride; (f) sodium hypochlorite; (g) titanium (IV)chloride; (h) disodium ammonium phosphate.

64. Write the chemical formula for each of the following: (a) calcium fluoride; (b) sodium nitrate; (c) iron(III) oxide; (d) copper(II) acetate; (e) sodium nitrite.

65. Write the chemical formula for each compound: (a) zinc cyanide; (b) silver chromate; (c) lead(II) iodide; (d) benzene; (e) copper(II) perchlorate.

66. Write the chemical formula for (a) sodium chlorite; (b) potassium nitrite; (c) sodium azide; (d) calcium phosphide; (e) tin(II) chloride; (f) calcium hydrogen phosphate; and (g) iron(II) chloride dihydrate.

67. Write the formula for each compound: (a) sodium hydroxide; (b) calcium cyanide; (c) magnesium phosphate; (d) sodium sulfate; (e) nickel(II) bromide; (f) calcium chlorite; (g) titanium(IV) bromide.

68. Write the formula for each compound: (a) potassium carbonate; (b) chromium(III) sulfite; (c) cobalt(II) phosphate; (d) magnesium hypochlorite; (e) nickel(II) nitrate hexahydrate.

2.4 Naming Covalent Compounds

69. Write the chemical formula for (a) dinitrogen oxide; (b) silicon tetrafluoride; (c) boron trichloride; (d) nitrogen trifluoride; and (e) phosphorus tribromide.

70. Write the chemical formula for (a) dinitrogen trioxide; (b) iodine pentafluoride; (c) boron tribromide; (d) oxygen difluoride; and (e) arsenic trichloride.

71. Write the chemical formula for (a) thallium(I) selenide; (b) neptunium(IV) oxide; (c) iron(II) sulfide; (d) copper(I) cyanide; and (e) nitrogen trichloride.

72. Name each compound: (a) RuO_4; (b) PbO_2; (c) MoF_6; (d) $Hg_2(NO_3)_2{\cdot}2H_2O$; (e) WCl_4.

73. Name each compound: (a) NbO_2; (b) MoS_2; (c) P_4S_{10}; (d) Cu_2O; (e) ReF_5.

74. Draw the structure of each of the following: (a) the 3-carbon carboxylic acid, propionic acid; (b) the 2-carbon alcohol, ethanol; (c) the 6-carbon alkane, *n*-hexane; (d) the 3-carbon cycloalkane, cyclopropane; (e) the 6-carbon arene, benzene.

75. Draw the structure of each of the following compounds: (a) the 4-carbon alkene, 1-butene; (b) the 5-carbon alkyne, 2-pentyne; (c) the 7-carbon cycloalkane, cycloheptane; (d) the 4-carbon amine, diethylamine; (e) the 2-carbon carboxylic acid, acetic acid; (f) the 6-carbon aromatic alcohol, phenol.

2.5 Acids and Bases

76. Write the formula for each compound: (a) hydroiodic acid; (b) hydrogen sulfide; (c) phosphorous acid; (d) perchloric acid; (e) calcium azide.

77. Write the formula for each compound: (a) hypochlorous acid; (b) perbromic acid; (c) hydrobromic acid; (d) sulfurous acid; (e) sodium azide.

78. Name each compound: (a) H_2SO_4; (b) HNO_2; (c) K_2HPO_4; (d) H_3PO_3; (e) $Ca(H_2PO_4)_2{\cdot}H_2O$.

79. Name each compound: (a) HBr; (b) H_2SO_3; (c) HCN; (d) $HClO_4$; (e) $NaHSO_4$.

2.6 Industrially Important Chemicals

80. In Example 2.12, the crude petroleum had an overall octane rating of 52. What is the composition of a solution of MTBE and *n*-heptane that has this octane rating?

APPLICATIONS

81. Carbon tetrachloride (CCl_4) was used as a dry cleaning solvent until it was found to cause liver cancer. Based on the structure of chloroform given in Section 2.1, draw the structure of "carbon tet."

82. Ammonium nitrate and ammonium sulfate are used in fertilizers as a source of nitrogen. The ammonium cation is tetrahedral. Refer to Section 2.1 to draw the structure of the ammonium ion.

83. The white light in fireworks displays is produced by burning magnesium in air, which contains oxygen. What compound is formed?

84. Sodium hydrogen sulfite is made from sulfur dioxide, and it is used for bleaching and swelling leather and to preserve flavor in almost all commercial wines. What are the formulas for these two sulfur-containing compounds?

85. Carbonic acid is used in carbonated drinks. When combined with lithium hydroxide, it produces lithium carbonate, a compound used to increase the brightness of pottery glazes and as a primary treatment for depression and bipolar disorder. Write the formula for both of these carbon-containing compounds.

86. Vinegar is a dilute solution of acetic acid, an organic acid, in water. What grouping of atoms would you expect to find in the structural formula for acetic acid?

87. Sodamide, or sodium amide, is prepared from sodium metal and gaseous ammonia. Sodamide contains the amide ion, NH_2^-, which reacts with water to form the hydroxide anion by removing a proton from water. Sodium amide is also used in the preparation of sodium cyanide. (a) Write the formula for each of these sodium-containing compounds. (b) What are the products of the reaction of sodamide with water?

88. A mixture of isooctane, *n*-pentane, and *n*-heptane is known to have an octane rating of 87. Use the data in Table 2.13 to calculate how much isooctane and *n*-heptane are present if the mixture is known to contain 30% *n*-pentane?

89. A crude petroleum distillate consists of 60% *n*-pentane, 25% methanol, and the remainder *n*-hexane by mass (Table 2.13).
(a) What is the octane rating?
(b) How much MTBE would have to be added to increase the octane rating to 93?

90. Premium gasoline sold in much of the midwestern United States has an octane rating of 93 and contains 10% ethanol. What is the octane rating of the gasoline fraction before ethanol is added? (See Table 2.13.)

3 Chemical Reactions

LEARNING OBJECTIVES

- To know how to calculate the molecular mass of a covalent compound and the formula mass of an ionic compound
- To be able to use the concepts of the mole and molar mass to calculate the number of atoms, molecules, or formula units in a sample of a substance
- To be able to determine the empirical formula of a compound from its composition by mass
- To know how to derive the molecular formula of a compound from its empirical formula
- To be able to describe a chemical reaction
- To know how to calculate the quantities of compounds produced or consumed in a chemical reaction
- To identify fundamental types of chemical reactions
- To begin to predict the types of chemical reactions that substances undergo

CHAPTER OUTLINE

ESSENTIAL SKILLS 2

Chapter 2 introduced you to a wide variety of chemical compounds, many of which have interesting applications. For example, nitrous oxide, a mild anesthetic, is also used as the propellant in cans of whipped cream, while copper(I) oxide is used as both a red glaze for ceramics and in anti-fouling bottom paints for boats. In addition to the physical properties of substances, chemists are also interested in their **chemical reactions**, in which a substance is converted to one or more other substances with different compositions and properties. Our very existence depends on chemical reactions, such as those between the oxygen in the air we breathe and the nutrient molecules in the foods we eat. Other reactions cook those foods, heat our homes, and provide the energy to run our cars. Many of the materials and pharmaceuticals that we take for granted today, such as silicon nitride for the sharp edge of cutting tools and antibiotics like amoxicillin, were unknown only a few years ago. Their development required that chemists understand how substances combine in certain ratios and under specific conditions to produce a new substance with particular properties.

Sodium metal, the fourth most abundant alkali metal on earth, is a highly reactive element and is never found free in nature. When heated to 250°C, it bursts into flames if exposed to air.

We begin this chapter by describing the relationship between the mass of a sample of a substance and it composition. We then develop methods for determining the quantities of compounds produced or consumed in chemical reactions, and we describe some fundamental types of chemical reactions. By applying the concepts and skills introduced in this chapter, you will be able to explain what happens to the sugar in a candy bar you eat, what reaction occurs in the battery when you start your car, what may be causing the "ozone hole" over Antarctica, and how we might prevent its growth.

3.1 • The Mole and Molar Masses

MGC Counting Atoms

As you learned in Chapter 1, the atomic mass of an isotope is the sum of the numbers of protons and neutrons present in the nucleus, while the average atomic mass of an element is calculated from the relative masses and abundances of its naturally occurring isotopes. Because a molecule or polyatomic ion is an assembly of atoms whose identities are given in its molecular or ionic formula, we can calculate the average atomic mass of any molecule or polyatomic ion from its composition. The average mass of a monatomic ion is taken to be the same as the average mass of the element because the mass of electrons is negligible.

Molecular and Formula Masses

The **molecular mass** of a substance is the sum of the average masses of the atoms in one molecule of the substance. It is calculated by summing the atomic masses of the elements in the substance, each multiplied by its subscript (written or implied) in the molecular formula. Because the units of atomic mass are atomic mass units (amu), the units of molecular mass are also amu. The procedure for calculating molecular masses is illustrated in Example 3.1.

EXAMPLE 3.1

Calculate the molecular mass of ethanol, whose molecular formula is CH_3CH_2OH.

Given Molecule

Asked for Molecular mass

Strategy

Ⓐ Determine the number of atoms of each element in the molecule.
Ⓑ Obtain atomic masses from the table inside the front cover, and multiply the atomic mass of each element by the number of atoms of that element.
Ⓒ Sum the masses to give the molecular mass.

Solution

Ⓐ The molecular formula of ethanol may be written in three different ways: CH_3CH_2OH (which is given, and illustrates the presence of an ethyl group, CH_3CH_2- and an OH group), C_2H_5OH, and C_2H_6O; all show that ethanol contains two carbon atoms, six hydrogen atoms, and one oxygen atom. Ⓑ Taking atomic masses from the table inside the front cover, we obtain

$$2 \times \text{atomic mass of carbon} = 2\ \cancel{\text{atoms}} \left(12.011\ \frac{\text{amu}}{\cancel{\text{atom}}}\right) = 24.022\ \text{amu}$$

$$6 \times \text{atomic mass of hydrogen} = 6\ \cancel{\text{atoms}} \left(1.0079\ \frac{\text{amu}}{\cancel{\text{atom}}}\right) = 6.0474\ \text{amu}$$

$$1 \times \text{atomic mass of oxygen} = 1\ \cancel{\text{atom}} \left(15.9994\ \frac{\text{amu}}{\cancel{\text{atom}}}\right) = 15.9994\ \text{amu}$$

C Summing the masses gives the molecular mass:

$$24.022 \text{ amu} + 6.0474 \text{ amu} + 15.9994 \text{ amu} = 46.069 \text{ amu}$$

Alternatively, we could have used unit conversions to reach the result in one step, as described in Essential Skills 2 at the end of this chapter:

$$\left[2 \cancel{\text{atoms C}}\left(\frac{12.011 \text{ amu}}{1 \cancel{\text{atom C}}}\right)\right] + \left[6 \cancel{\text{atoms H}}\left(\frac{1.0079 \text{ amu}}{1 \cancel{\text{atom H}}}\right)\right] + \left[1 \cancel{\text{atom O}}\left(\frac{15.9994 \text{ amu}}{1 \cancel{\text{atom O}}}\right)\right] = 46.069 \text{ amu}$$

The same calculation can also be done in a tabular format, which is especially helpful for more complex molecules:

2 C	(2 atoms)(12.011 amu/atom)	= 24.022 amu
6 H	(6 atoms)(1.0079 amu/atom)	= 6.0474 amu
+ 1 O	(1 atom)(15.9994 amu/atom)	= 15.9994 amu
C_2H_6O	molecular mass of ethanol	= 46.069 amu

EXERCISE 3.1

Calculate the molecular mass of CCl_3F (trichlorofluoromethane, or Freon-11), which until recently was used as a refrigerant. The structure of a molecule of Freon-11 is shown in the margin.

Answer 137.368 amu

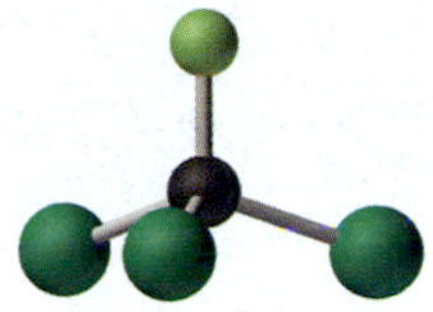

Freon-11, CCl_3F

Note the pattern

Atomic mass, molecular mass, and formula mass all have the same units: amu.

Unlike molecules, which are covalent compounds, ionic compounds do not have a readily identifiable molecular unit. Hence, for ionic compounds we use the *formula mass* (also called the *empirical formula mass*) of the compound rather than the molecular mass. The **formula mass** is the sum of the atomic masses of all the elements in the empirical formula, each multiplied by its subscript (written or implied). It is directly analogous to the molecular mass of a covalent compound. Once again, the units are amu.

EXAMPLE 3.2

Calculate the formula mass of calcium phosphate, $Ca_3(PO_4)_2$.

Given Ionic compound

Asked for Formula mass

Strategy

A Determine the number of atoms of each element in the empirical formula.
B Obtain atomic masses from the periodic table inside the front cover, and multiply the atomic mass of each element by the number of atoms of that element.
C Sum the masses to give the formula mass.

Solution

A The empirical formula, $Ca_3(PO_4)_2$, indicates that the simplest electrically neutral unit of calcium phosphate contains three Ca^{2+} ions and two PO_4^{3-} ions. The mass of this formula unit is calculated by summing the atomic masses of three calcium atoms, two phosphorus atoms, and eight oxygen atoms. B Taking atomic masses from the table inside the front cover, we obtain

$$3 \times \text{atomic mass of calcium} = 3 \cancel{\text{atoms}}\left(40.078 \frac{\text{amu}}{\cancel{\text{atom}}}\right) = 120.234 \text{ amu}$$

$$2 \times \text{atomic mass of phosphorus} = 2 \cancel{\text{atoms}}\left(30.973761 \frac{\text{amu}}{\cancel{\text{atom}}}\right) = 61.947522 \text{ amu}$$

$$8 \times \text{atomic mass of oxygen} = 8 \cancel{\text{atoms}}\left(15.9994 \frac{\text{amu}}{\cancel{\text{atom}}}\right) = 127.9952 \text{ amu}$$

Summing the masses gives the formula mass of $Ca_3(PO_4)_2$:

$$120.234 \text{ amu} + 61.947522 \text{ amu} + 127.9952 \text{ amu} = 310.177 \text{ amu}$$

We could also find the formula mass of $Ca_3(PO_4)_2$ in one step by using unit conversions or a tabular format:

$$\left[3 \cancel{\text{atoms Ca}}\left(\frac{40.078 \text{ amu}}{1 \cancel{\text{atom Ca}}}\right)\right] + \left[2 \cancel{\text{atoms P}}\left(\frac{30.9738 \text{ amu}}{1 \cancel{\text{atom P}}}\right)\right] + \left[8 \cancel{\text{atoms O}}\left(\frac{15.9994 \text{ amu}}{1 \cancel{\text{atom O}}}\right)\right] = 310.177 \text{ amu}$$

3 Ca	(3 atoms)(40.078 amu/atom)	= 120.234 amu
2 P	(2 atoms)(30.973761 amu/atom)	= 61.947522 amu
+8 O	(8 atoms)(15.9994 amu/atom)	= 127.9952 amu
$Ca_3P_2O_8$	formula mass of $Ca_3(PO_4)_2$	= 310.177 amu

EXERCISE 3.2

Calculate the formula mass of Si_3N_4, commonly called silicon nitride. It is an extremely hard and inert material that is used to make cutting tools for machining hard metal alloys.

Answer 140.29 amu

The Mole

In Chapter 1 we described Dalton's theory that each chemical compound contains a particular combination of atoms, and that the ratios of the *numbers* of atoms of the elements present are usually small whole numbers. We also described the law of multiple proportions, which states that the ratios of the *masses* of elements that form a series of compounds are small whole numbers. The problem for Dalton and other early chemists was to discover the quantitative relationship between the number of atoms in a chemical substance and its mass. Because the masses of individual atoms are so minuscule (on the order of 10^{-23} g/atom), chemists do not measure the mass of individual atoms or molecules. In the laboratory, for example, the masses of compounds and elements used by chemists typically range from milligrams to grams, while in industry, chemicals are bought and sold in kilograms and tons. In order to analyze the transformations that occur between atoms or molecules in a chemical reaction, it is therefore absolutely essential for chemists to know how many atoms or molecules are contained in a given mass of sample. The unit that provides this information is the *mole*, from the Latin *moles*, "pile" or "heap" (*not* from the small subterranean animal!).

Many familiar items are sold in numerical quantities that have unusual names. For example, cans of soda come in a six-pack, eggs are sold by the dozen (12), and pencils often come in a gross (12 dozen, or 144). Sheets of printer paper are packaged in reams of 500, a seemingly large number. Atoms are so small, however, that even 500 atoms are too small to see or measure by most common techniques. Any readily measurable mass of an element or compound contains an extraordinarily large number of atoms, molecules, or ions, so an extraordinarily large numerical unit is needed to count them. The mole is used for this purpose.

A **mole** (abbreviated mol) is defined as the amount of substance that contains the number of carbon atoms in exactly 12 g of isotopically pure carbon-12. According to the most recent experimental measurements, this mass of carbon-12 contains 6.022142×10^{23} atoms, but for most purposes 6.022×10^{23} provides an adequate number of significant figures. Just as a mole of atoms contains 6.022×10^{23} atoms, a mole of eggs contains 6.022×10^{23} eggs. The number in a mole is called **Avogadro's number**, after the 19th-century Italian scientist who first proposed a relationship between the volumes of gases and the numbers of particles they contain.

It is not obvious why eggs come in dozens rather than 10's or 14's, or why a ream of paper contains 500 sheets rather than 400 or 600. The definition of the mole—that is, the decision to base it on 12 g of carbon-12—is also arbitrary. The important point is that *one mole of carbon—or of anything else, whether atoms, compact disks, or houses—always contains the same number of objects*.

To appreciate the magnitude of Avogadro's number, consider a mole of pennies. Stacked vertically, a mole of pennies would be 4.5×10^{17} miles high, or almost exactly six times the diameter of the Milky Way galaxy. If a mole of pennies were distributed equally among the entire population of the earth, each person would get more than one trillion dollars. Clearly, the mole is so large that it is useful only for measuring very small objects, such as atoms.

The concept of the mole allows us to count out a specific number of individual atoms and molecules by weighing measurable quantities of elements and compounds. To obtain a mole of carbon-12 atoms, we would weigh out 12 g of isotopically pure carbon-12. Because each element has a different atomic mass, however, a mole of each element has a different mass, even though it contains the same number of atoms (6.022×10^{23}). This is analogous to the fact that a dozen extra large eggs weighs more than a dozen small eggs, or that the total weight of 50 adult humans is greater than the total weight of 50 children. Because of the way in which the mole is defined, for every element the number of grams of an element in a mole is the same as the number of amu in the atomic mass of the element. For example, the mass of 1 mole (mol) of magnesium (atomic mass = 24.305 amu) is 24.305 g. Note that because the atomic mass of magnesium (24.305 amu) is slightly more than twice that of a carbon-12 atom (12 amu), the mass of a mole of magnesium atoms (24.305 g) is a little more than twice that of a mole of carbon-12 (12 g). Similarly, the mass of 1 mol of helium (atomic mass = 4.002602 amu) is 4.002602 g, about one-third that of a mole of carbon-12.

Using the concept of the mole, we can now restate Dalton's theory: *One mole of a compound is formed by combining elements in amounts whose mole ratios are small whole numbers.* For example, 1 mol of water molecules (H_2O) contains 2 mol of H atoms and 1 mol of O atoms.

Molar Mass

The **molar mass** of a substance is defined as the mass in grams of 1 mol of that substance. One mole of isotopically pure carbon-12 has a mass of 12 g. For an element, the molar mass is the mass of a mole of atoms of that element; for a covalent molecular compound, it is the mass of a mole of molecules of that compound; for an ionic compound, it is the mass of a mole of formula units. That is, the molar mass of a substance is the mass (in grams per mole) of 6.022×10^{23} atoms, molecules, or formula units of that substance. In each case, the number of grams in a mole is the same as the number of atomic mass units that describe the atomic mass, molecular mass, or formula mass, respectively.

The table of atomic masses inside the front cover lists the atomic mass of carbon as 12.011 amu; the average molar mass of carbon—the mass of 6.022×10^{23} carbon atoms—is therefore 12.011 g/mol, as shown in the table in the margin. Note that the molar mass of naturally occurring carbon is different from that of carbon-12 because carbon occurs as a mixture of carbon-12, carbon-13, and carbon-14. One mole of carbon still contains 6.022×10^{23} atoms of carbon, but 98.89% of those atoms are carbon-12, 1.11% are carbon-13, and a trace (about 1 atom in 10^{12}) are carbon-14 (see Section 1.6). Similarly, the molar mass of uranium is 238.03 g/mol, and the molar mass of iodine is 126.90 g/mol. When we deal with elements such as iodine and sulfur, which occur as the diatomic molecule, I_2, and the polyatomic molecule, S_8, respectively, molar mass usually refers to the mass of 1 mol of *atoms* of the element, in this case I and S, *not* to the mass of 1 mol of *molecules* of the element (I_2 and S_8).

Note the pattern

The molar mass of any substance is its atomic mass, molecular mass, or formula mass in grams per mole.

Substance (formula)	Atomic, molecular, or formula mass, amu	Molar mass, g/mol
Carbon (C)	12.011 (atomic mass)	12.011
Ethanol (C_2H_5OH)	46.069 (molecular mass)	46.069
$Ca_3(PO_4)_2$	310.177 (formula mass)	310.177

Figure 3.1 Samples of 1 mol of some common substances.

The molar mass of ethanol is the mass of ethanol that contains 6.022×10^{23} ethanol molecules. As you calculated in Example 3.1, the molecular mass of ethanol is 46.069 amu. Because 1 mol of ethanol contains 2 mol of carbon atoms (2×12.011 g), 6 mol of hydrogen atoms (6×1.008 g), and 1 mol of oxygen atoms (1×15.999 g), its molar mass is 46.069 g/mol. Similarly, the formula mass of calcium phosphate, $Ca_3(PO_4)_2$, is 310.177 amu, so its molar mass is 310.177 g/mol. This is the mass of calcium phosphate that contains 6.022×10^{23} $Ca_3(PO_4)_2$ formula units. Figure 3.1 shows samples that contain precisely one molar mass of several common substances.

The mole is the basis of quantitative chemistry. It provides chemists with a way to convert easily between the mass of a substance and the number of atoms, molecules, or formula units of that substance. Conversely, it enables chemists to calculate the mass of a substance needed in order to obtain a desired number of atoms, molecules, or formula units. For example, to convert moles of a substance to mass, we use the relationship

$$(\text{moles})(\text{molar mass}) \longrightarrow \text{mass} \tag{3.1}$$

or, more specifically,

$$\cancel{\text{moles}}\left(\frac{\text{grams}}{\cancel{\text{mole}}}\right) = \text{grams}$$

Conversely, to convert the mass of a substance to moles, we use

$$\frac{\text{mass}}{\text{molar mass}} \longrightarrow \text{moles}$$
$$\frac{\text{grams}}{\text{grams/mole}} = \cancel{\text{grams}}\left(\frac{\text{mole}}{\cancel{\text{grams}}}\right) = \text{moles} \tag{3.2}$$

Be sure to pay attention to the units when converting between mass and moles!

Figure 3.2 shows a flowchart for converting between mass, number of moles, and number of atoms, molecules, or formula units. The use of these conversions is illustrated in Examples 3.3 and 3.4.

Figure 3.2 A flowchart for converting between mass, number of moles, and number of atoms, molecules, or formula units.

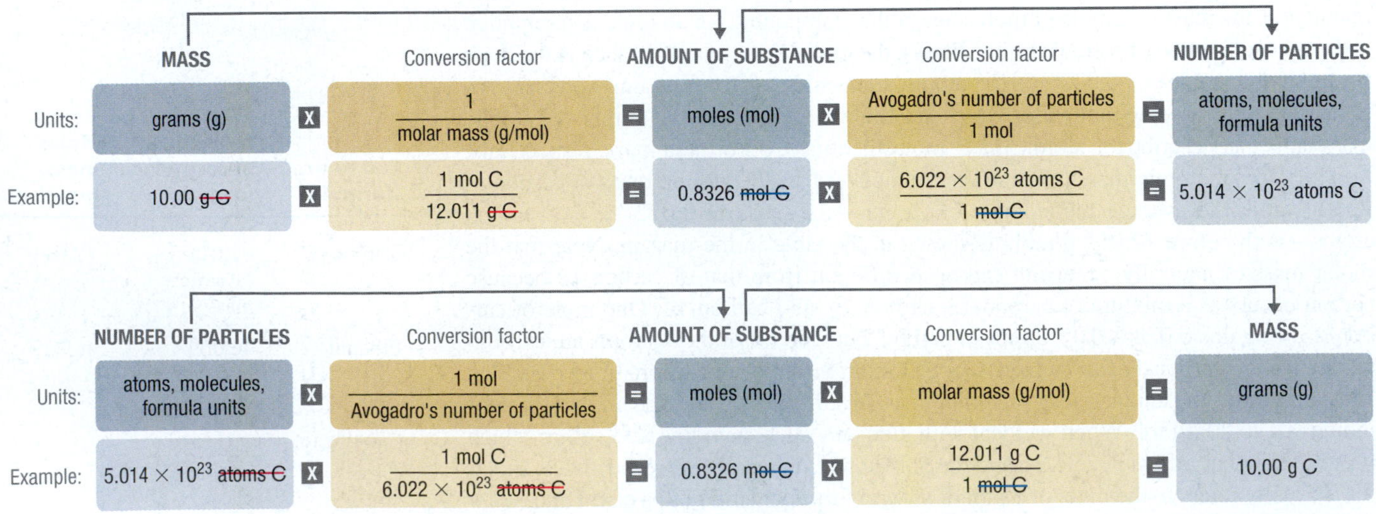

EXAMPLE 3.3

Calculate **(a)** the number of moles and **(b)** the number of molecules in 35.00 g of ethylene glycol, $HOCH_2CH_2OH$.

Given Mass and molecular formula

Asked for Number of moles and number of molecules

Strategy

A Use the molecular formula of the compound to calculate its molecular mass in grams per mole.
B Convert from mass to moles by dividing the mass given by the compound's molar mass.
C Convert from moles to molecules by multiplying the number of moles by Avogadro's number.

Solution

(a) **A** The molecular mass of ethylene glycol can be calculated from its molecular formula using the method illustrated in Example 3.1:

2 C	(2 atoms)(12.011 amu/atom)	= 24.022 amu
6 H	(6 atoms)(1.0079 amu/atom)	= 6.0474 amu
+2 O	(2 atoms)(15.9994 amu/atom)	= 31.9988 amu
$C_2H_6O_2$	molecular mass of ethylene glycol	= 62.068 amu

The molar mass of ethylene glycol is therefore 62.068 g/mol. **B** The number of moles of ethylene glycol present in 35.00 g can be calculated by dividing the mass (in g) by the molar mass (in g/mol):

$$\frac{\text{mass of ethylene glycol (g)}}{\text{molar mass (g/mol)}} = \text{moles ethylene glycol (mol)}$$

We obtain

$$35.00\ \cancel{\text{g ethylene glycol}}\left(\frac{1\ \text{mol ethylene glycol}}{62.068\ \cancel{\text{g ethylene glycol}}}\right) = 0.5639\ \text{mol ethylene glycol}$$

Note: It is always a good idea to estimate the answer ***before*** *you do the actual calculation.* In this case, the mass given (35.00 g) is *less* than the molar mass, so the answer should be less than 1 mol. The calculated answer (0.5639 mol) is indeed less than 1 mol, so we have probably not made a major error in the calculations.

(b) **C** To calculate the number of molecules in the sample, we multiply the number of moles by Avogadro's number:

$$\text{molecules of ethylene glycol} = 0.5639\ \cancel{\text{mol}}\left(\frac{6.022 \times 10^{23}\ \text{molecules}}{1\ \cancel{\text{mol}}}\right)$$
$$= 3.396 \times 10^{23}\ \text{molecules}$$

Again, because we are dealing with slightly more than $\frac{1}{2}$ mol of ethylene glycol, we expect the number of molecules present to be slightly more than $\frac{1}{2}$ of Avogadro's number, or a little more than 3×10^{23} molecules, which is indeed the case.

EXERCISE 3.3

Calculate **(a)** the number of moles and **(b)** the number of molecules in 75.0 g of CCl_3F (Freon-11).

Answer **(a)** 0.546 mol; **(b)** 3.29×10^{23} molecules

EXAMPLE 3.4

Calculate the mass of 1.75 mol of **(a)** S_2Cl_2 (common name: sulfur monochloride; systematic name: disulfur dichloride) and **(b)** $Ca(ClO)_2$ (calcium hypochlorite).

Given Number of moles and molecular or empirical formula

Asked for Mass

Strategy

A Calculate the molecular mass of the compound in grams from its molecular formula (if covalent) or empirical formula (if ionic).

B Convert from moles to mass by multiplying the moles of the compound given by its molar mass.

Solution

We begin by calculating the molecular mass of S_2Cl_2 and the formula mass of $Ca(ClO)_2$.

(a) **A** The molar mass of S_2Cl_2 is obtained from its molecular mass as follows:

2 S	(2 atoms)(32.065 amu/atom)	=	64.130 amu
+2 Cl	(2 atoms)(35.453 amu/atom)	=	70.906 amu
S_2Cl_2	molecular mass of S_2Cl_2	=	135.036 amu

The molar mass of S_2Cl_2 is thus 135.036 g/mol. **B** The mass of 1.75 mol of S_2Cl_2 is calculated as follows:

$$\text{moles } S_2Cl_2 \left[\text{molar mass}\left(\frac{\text{g}}{\text{mol}}\right)\right] \longrightarrow \text{mass of } S_2Cl_2 \text{ (g)}$$

$$1.75\ \cancel{\text{mol } S_2Cl_2}\left(\frac{135.036\text{ g } S_2Cl_2}{1\ \cancel{\text{mol } S_2Cl_2}}\right) = 236\text{ g } S_2Cl_2$$

(b) **A** The formula mass of $Ca(ClO)_2$ is obtained as follows:

1 Ca	(1 atom)(40.078 amu/atom)	=	40.078 amu
2 Cl	(2 atoms)(35.453 amu/atom)	=	70.906 amu
+2 O	(2 atoms)(15.9994 amu/atom)	=	31.9988 amu
$CaCl_2O_2$	formula mass of $Ca(ClO)_2$	=	142.983 amu

The molar mass of $Ca(ClO)_2$ is thus 142.983 g/mol. **B** The mass of 1.75 mol of $Ca(ClO)_2$ is calculated as

$$\text{moles } Ca(ClO)_2\left(\frac{\text{molar mass } Ca(ClO)_2}{1\text{ mol } Ca(ClO)_2}\right) = \text{mass } Ca(ClO)_2$$

$$1.75\ \cancel{\text{mol } Ca(ClO)_2}\left(\frac{142.983\text{ g } Ca(ClO)_2}{1\ \cancel{\text{mol } Ca(ClO)_2}}\right) = 250\text{ g } Ca(ClO)_2$$

Because 1.75 mol is a little less than 2 mol, the final quantity in grams in both cases should be slightly less than twice the molar mass, as it is.

EXERCISE 3.4

Calculate the mass of 0.0122 mol of **(a)** Si_3N_4 and **(b)** trimethylamine, $(CH_3)_3N$.

Answer **(a)** 1.71 g; **(b)** 0.721 g

3.2 • Determining Empirical and Molecular Formulas

When a new chemical compound such as a potential new pharmaceutical is synthesized in the laboratory or isolated from a natural source, chemists determine its elemental composition, its empirical formula, and its structure in order to understand its properties. In this section, we focus on how to determine the empirical formula of a compound and how to use it to determine the molecular formula if the molar mass of the compound is known.

MGC Empirical Formula

Calculating Mass Percentages

The law of definite proportions states that a chemical compound always contains the same proportion of elements by mass; that is, the *percent composition*—the percentage of each element present in a pure substance—is constant (although we now know there are exceptions to this law). For example, sucrose (cane sugar) is 42.11% carbon, 6.48% hydrogen, and 51.41% oxygen by mass. This means that 100.00 g of sucrose always contains 42.11 g of carbon, 6.48 g of hydrogen, and 51.41 g of oxygen. First we will use the molecular formula of sucrose, $C_{12}H_{22}O_{11}$, to calculate the mass percentage of each of the component elements, and then we will show how mass percentages can be used to determine an empirical formula.

According to its molecular formula, each molecule of sucrose contains 12 carbon atoms, 22 hydrogen atoms, and 11 oxygen atoms. A mole of sucrose molecules therefore contains 12 mol of carbon atoms, 22 mol of hydrogen atoms, and 11 mol of oxygen atoms. We can use this information to calculate the mass of each element in a mole of sucrose, which will give us the molar mass of sucrose. We can then use these masses to calculate the percent composition of sucrose.

$$\begin{aligned}
\text{mass of C/mol sucrose} &= 12\ \cancel{\text{mol C}} \times \frac{12.011\ \text{g C}}{1\ \cancel{\text{mol C}}} = 144.132\ \text{g C} \\
\text{mass of H/mol sucrose} &= 22\ \cancel{\text{mol H}} \times \frac{1.008\ \text{g H}}{1\ \cancel{\text{mol H}}} = 22.176\ \text{g H} \\
\text{mass of O/mol sucrose} &= 11\ \cancel{\text{mol O}} \times \frac{15.999\ \text{g O}}{1\ \cancel{\text{mol O}}} = 175.989\ \text{g O}
\end{aligned} \tag{3.3}$$

Thus, 1 mol of sucrose has a mass of 342.297 g; note that more than half of the mass (175.989 g) is oxygen, and almost half of the mass (144.132 g) is carbon.

The mass percentage of each element in sucrose is the mass of the element present in 1 mol of sucrose divided by the molar mass of sucrose, multiplied by 100% to give a percentage, shown to two decimal places:

$$\begin{aligned}
\text{mass \% C in sucrose} &= \frac{\text{mass of C/mol sucrose}}{\text{molar mass of sucrose}} \times 100\% \\
&= \frac{144.132\ \text{g C}}{342.297\ \text{g/mol sucrose}} \times 100\% = 42.12\% \\
\text{mass \% H in sucrose} &= \frac{\text{mass of H/mol sucrose}}{\text{molar mass of sucrose}} \times 100\% \\
&= \frac{22.176\ \text{g H}}{342.297\ \text{g/mol sucrose}} \times 100\% = 6.48\% \\
\text{mass \% O in sucrose} &= \frac{\text{mass of O/mol sucrose}}{\text{molar mass of sucrose}} \times 100\% \\
&= \frac{175.989\ \text{g O}}{342.297\ \text{g/mol sucrose}} \times 100\% = 51.41\%
\end{aligned}$$

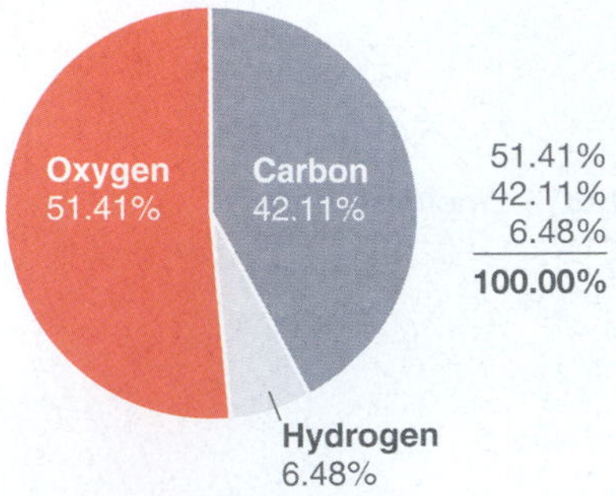

Percent composition (any amount sucrose)

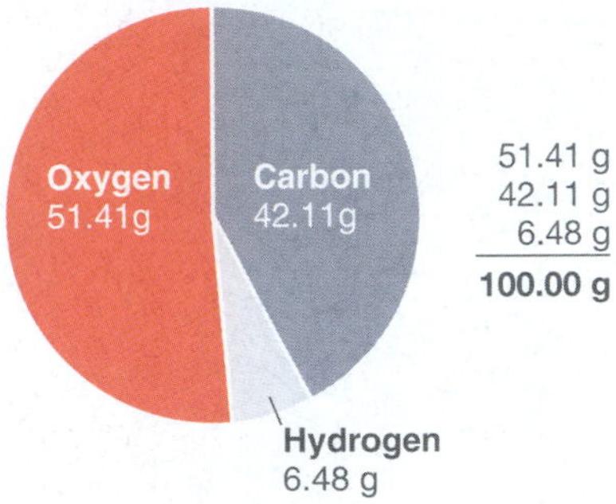

Composition (100 g sample sucrose)

You can check your work by verifying that the sum of the percentages of all the elements in the compound is 100%, as it must be: 42.11% + 6.48% + 51.41% = 100.00%. If the sum is not 100%, you have made an error in your calculations. (Rounding to the correct number of decimal places can, however, cause the total to be slightly different from 100%.) Thus, 100.00 g of sucrose contains 42.11 g of carbon, 6.48 g of hydrogen, and 51.41 g of oxygen, and, to two decimal places, the percent composition of sucrose is indeed 42.11% carbon, 6.48% hydrogen, and 51.41% oxygen, as stated previously.

We could also calculate the mass percentages using atomic masses and molecular masses, with units of amu. Because the answer we are seeking is a ratio, expressed as a percentage, the units of mass cancel whether they are grams (using molar masses) or amu's (using atomic and molecular masses).

EXAMPLE 3.5

Aspartame is the artificial sweetener sold as NutraSweet and Equal. Its molecular formula is $C_{14}H_{18}N_2O_5$. **(a)** Calculate the mass percentage of each element in aspartame. **(b)** Calculate the mass of carbon in a 1.00-g packet of Equal, assuming it is pure aspartame.

Given Molecular formula and mass of sample

Asked for Mass percentage of all elements, and mass of one element in sample

Strategy

A Use atomic masses from inside the front cover to calculate the molar mass of aspartame.
B Divide the mass of each element by the molar mass of aspartame; then multiply by 100% to obtain percentages.
C To find the mass of an element contained in a given mass of aspartame, multiply the mass of aspartame by the mass percentage of that element, expressed as a decimal.

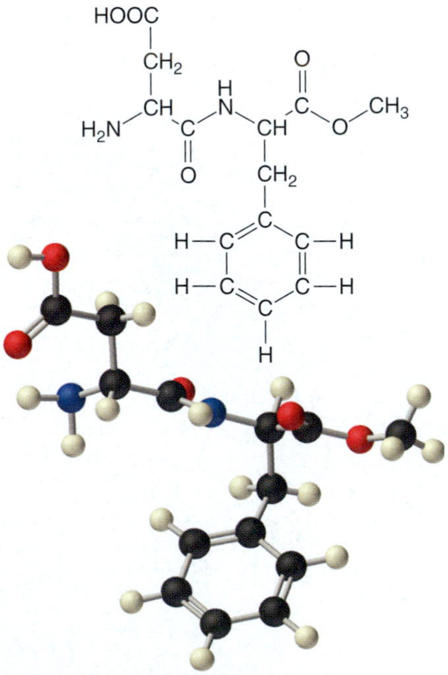

Aspartame

Solution

(a) **A** We calculate the mass of each element in 1 mol of aspartame and the molar mass of aspartame:

14 C	(14 mol C)(12.011g/mol C)	= 168.154 g
18 H	(18 mol H)(1.008 g/mol H)	= 18.114 g
2 N	(2 mol N)(14.007 g/mol N)	= 28.014 g
+ 5 O	(5 mol O)(15.999 g/mol O)	= 79.995 g
$C_{14}H_{18}N_2O_5$	molar mass of aspartame	= 294.277 g/mol

Thus, more than half the mass of 1 mol of aspartame (294.31 g) is carbon (168.154 g). **B** To calculate the mass percentage of each element, we divide the mass of each element in the compound by the molar mass of aspartame and then multiply by 100% to obtain percentages, here reported to two decimal places:

$$\text{mass \% C} = \frac{168.154 \text{ g C}}{294.277 \text{ g aspartame}} \times 100\% = 57.14\% \text{ C}$$

$$\text{mass \% H} = \frac{18.114 \text{ g H}}{294.277 \text{ g aspartame}} \times 100\% = 6.16\% \text{ H}$$

$$\text{mass \% N} = \frac{28.014 \text{ g N}}{294.277 \text{ g aspartame}} \times 100\% = 9.52\% \text{ N}$$

$$\text{mass \% O} = \frac{79.995 \text{ g O}}{294.277 \text{ g aspartame}} \times 100\% = 27.18\% \text{ O}$$

As a check, we can sum the percentages:

$$57.14\% + 6.16\% + 9.52\% + 27.18\% = 100.00\%$$

If you obtain a total that differs from 100% by more than about ±1%, there must be an error somewhere in the calculation.

(b) The mass of carbon in 1.00 g of aspartame is calculated as follows:

$$\text{mass of C} = 1.00 \text{ g aspartame} \times \frac{57.14 \text{ g C}}{100 \text{ g aspartame}} = 0.571 \text{ g C}$$

EXERCISE 3.5

Calculate the mass percentage of each element in aluminum oxide, Al_2O_3.

Answer 52.93% Al, 47.08% O

Determining the Empirical Formula of Penicillin

Just as we can use the empirical formula of a substance to determine its percent composition, we can use the percent composition of a sample to determine its empirical formula, which can then be used to determine its molecular formula. Such a procedure was actually used to determine the empirical and molecular formulas of the first antibiotic to be discovered, penicillin.

Antibiotics are chemical compounds that selectively kill microorganisms, many of which cause diseases. Although we may take antibiotics for granted today, penicillin was discovered barely 75 years ago. The subsequent development of a wide array of other antibiotics for treating many common diseases has contributed greatly to the substantial increase in life expectancy during the last 50 years. The discovery of penicillin is a historical detective story in which the use of mass percentages to determine empirical formulas played a key role.

In 1928, Alexander Fleming, a young microbiologist at the University of London, was working with a common bacterium that causes boils and other infections such as blood poisoning. For laboratory study, bacteria are commonly grown on the surface of a nutrient-containing gel in small, flat culture dishes. One day Fleming noticed that one of his cultures was contaminated by a bluish-green mold similar to the mold found on spoiled bread or fruit. Such accidents are rather common, and most laboratory workers would have simply thrown the cultures away. Fleming noticed, however, that the bacteria were growing everywhere on the gel *except* near the contaminating mold (Figure 3.3a), and he hypothesized that the mold must be producing a substance that either killed the bacteria or prevented their growth. To test this hypothesis, he grew the mold in a liquid and then filtered the liquid and added it to various bacteria cultures. The liquid killed not only the bacteria Fleming had originally been studying but a wide range of other disease-causing bacteria as well. Because the mold was a member of the

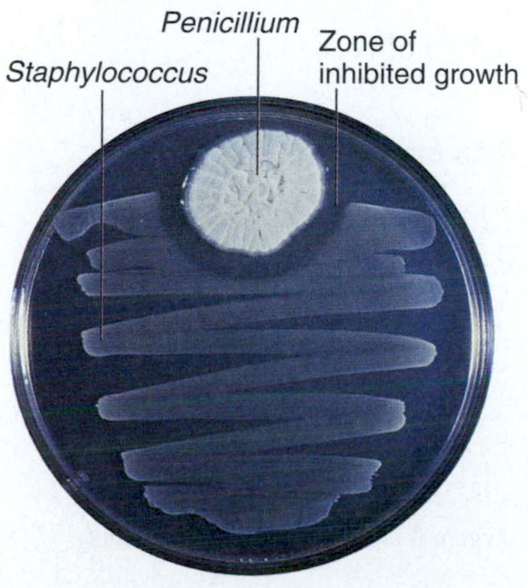

(a)

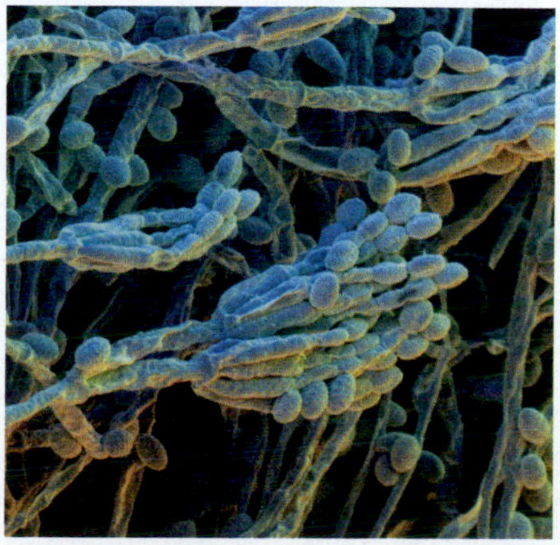

(b)

Figure 3.3 Penicillium. (a) Photograph of the *Penicillium* mold growing in a culture dish, showing its effect on bacterial growth. (b) Photomicrograph of *Penicillium*, showing its rod- and pencil-shaped branches. The name comes from the Latin *penicillus*, "paintbrush."

Penicillium family (named for their pencil-shaped branches under the microscope—Figure 3.3b), Fleming called the active ingredient in the broth *penicillin*.

Although Fleming was unable to isolate penicillin in pure form, the medical importance of his discovery stimulated researchers in other laboratories. Finally, in 1940, two chemists at Oxford University, Howard Florey (1898–1968) and Ernst Chain (1906–1979), were able to isolate an active product, which they called penicillin G. Within three years, penicillin G was in widespread use for treating pneumonia, gangrene, gonorrhea, and other diseases, and its use greatly increased the survival rate of wounded soldiers in World War II. As a result of their work, Fleming, Florey, and Chain shared the Nobel Prize in medicine in 1945.

As soon as they had succeeded in isolating pure penicillin G, Florey and Chain subjected the compound to a procedure called combustion analysis (described in the next section) to determine what elements were present and in what quantities. The results of such analyses are usually reported as mass percentages. They discovered that a typical sample of penicillin G contains 53.9% carbon, 4.8% hydrogen, 7.9% nitrogen, 9.0% sulfur, and 6.5% sodium by mass. Notice that these numbers add up to only 82.1% rather than 100.0%, which implies that there must be one or more additional elements. A reasonable candidate is oxygen, which is a common component of compounds that contain carbon and hydrogen;* for technical reasons, however, it is difficult to analyze for oxygen directly. If we assume that all of the missing mass is due to oxygen, then penicillin G contains (100.0% − 82.1%) = 17.9% oxygen. From these mass percentages, the empirical formula and eventually the molecular formula of the compound can be determined.

To determine the empirical formula from the mass percentages of the elements in a compound such as penicillin G, we need to convert the mass percentages to relative numbers of atoms. For convenience, we assume that we are dealing with a 100.0-g sample of the compound, even though the sizes of samples used for analyses are generally much smaller, on the order of milligrams. This assumption simplifies the arithmetic because a 53.9% mass percentage of carbon corresponds to 53.9 g of carbon in a 100.0-g sample of penicillin G; likewise, 4.8% hydrogen corresponds to 4.8 g of hydrogen in 100.0 g of penicillin G; and so on for the other elements. We can then divide each of these masses by the molar mass of the element to determine how many moles of each element are present in the 100.0-g sample:

$$\frac{\text{mass (g)}}{\text{molar mass (g/mol)}} = (\cancel{\text{g}})\left(\frac{\text{mol}}{\cancel{\text{g}}}\right) = \text{mol}$$

$$\begin{aligned}
53.9\ \cancel{\text{g C}}\left(\frac{1\ \text{mol C}}{12.011\ \cancel{\text{g C}}}\right) &= 4.49\ \text{mol C} \\
4.8\ \cancel{\text{g H}}\left(\frac{1\ \text{mol H}}{1.008\ \cancel{\text{g H}}}\right) &= 4.8\ \text{mol H} \\
7.9\ \cancel{\text{g N}}\left(\frac{1\ \text{mol N}}{14.007\ \cancel{\text{g N}}}\right) &= 0.56\ \text{mol N} \\
9.0\ \cancel{\text{g S}}\left(\frac{1\ \text{mol S}}{32.065\ \cancel{\text{g S}}}\right) &= 0.28\ \text{mol S} \\
6.5\ \cancel{\text{g Na}}\left(\frac{1\ \text{mol Na}}{22.990\ \cancel{\text{g Na}}}\right) &= 0.28\ \text{mol Na} \\
17.9\ \cancel{\text{g O}}\left(\frac{1\ \text{mol O}}{15.999\ \cancel{\text{g O}}}\right) &= 1.12\ \text{mol O}
\end{aligned} \tag{3.4}$$

---

* Do not assume that the "missing" mass is always due to oxygen. It could be any other element.

Thus, 100.0 g of penicillin G contains 4.49 mol of carbon, 4.8 mol of hydrogen, 0.56 mol of nitrogen, 0.28 mol of sulfur, 0.28 mol of sodium, and 1.12 mol of oxygen (assuming all the missing mass was oxygen). Note that the number of significant figures in the numbers of moles of elements varies between two and three because some of the analytical data were reported to only two significant figures.

These results tell us the ratios of the moles of the various elements in the sample (4.49 mol of carbon to 4.8 mol of hydrogen to 0.56 mol of nitrogen, and so on), but they are not the whole-number ratios we need for the empirical formula—the empirical formula expresses the *relative* numbers of atoms in the *smallest whole numbers possible*. To obtain whole numbers, we divide the numbers of moles of all the elements in the sample by the number of moles of the element present in the lowest relative amount, which in this example is sulfur or sodium. The results will be the subscripts of the elements in the empirical formula. To two significant figures, the results are

$$\text{C: } \frac{4.49}{0.28} = 16 \qquad \text{H: } \frac{4.8}{0.28} = 17 \qquad \text{N: } \frac{0.56}{0.28} = 2.0$$
$$\text{S: } \frac{0.28}{0.28} = 1.0 \qquad \text{Na: } \frac{0.28}{0.28} = 1.0 \qquad \text{O: } \frac{1.12}{0.28} = 4.0 \qquad (3.5)$$

The empirical formula of penicillin G is therefore $C_{16}H_{17}N_2NaO_4S$. Other experiments have shown that penicillin G is actually an ionic compound that contains Na^+ cations and $[C_{16}H_{17}N_2O_4S]^-$ anions in a 1:1 ratio. The complex structure of penicillin G (Figure 3.4) was not determined until 1948.

In some cases, one or more of the subscripts in a formula calculated using this procedure may not be integers. Does this mean that the compound of interest contains a nonintegral number of atoms? Not at all; rounding errors in the calculations as well as experimental errors in the data can result in nonintegral ratios. When this happens, you must exercise some judgment in interpreting the results, as illustrated in Example 3.6. In particular, ratios of 1.50, 1.33, or 1.25 suggest that you should multiply *all* subscripts in the formula by 2, 3, or 4, respectively. Only if the ratio is within 5% of an integral value should you consider rounding to the nearest integer.

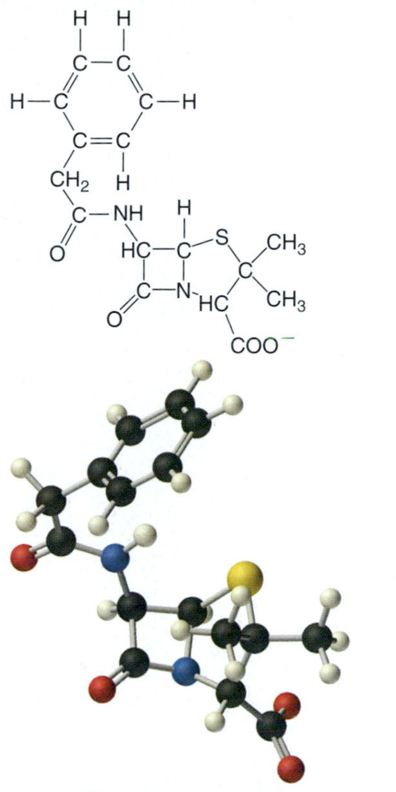

Figure 3.4 Structural formula and ball-and-stick model of the anion of penicillin G.

EXAMPLE 3.6

Calculate the empirical formula of the ionic compound calcium phosphate, a major component of fertilizer and a polishing agent in toothpastes. Elemental analysis indicates that it contains 38.77% Ca, 19.97% P, and 41.27% O.

Given Percent composition

Asked for Empirical formula

Strategy

Ⓐ Assume a 100-g sample and calculate the number of moles of each element in that sample.
Ⓑ Obtain the relative numbers of atoms of each element in the compound by dividing the number of moles of each element in the 100-g sample by the number of moles of the element present in the smallest amount.
Ⓒ If the atomic ratios are not integers, multiply all subscripts by the same number to give integral values.
Ⓓ Because this is an ionic compound, identify the anion and cation, and write the formula so that their charges balance.

Solution

A A 100-g sample of calcium phosphate contains 38.77 g of Ca, 19.97 g of P, and 41.27 g of O. Dividing the mass of each element in the 100-g sample by its molar mass gives the number of moles of each element in the sample:

$$\text{moles Ca} = 38.77\ \text{g Ca} \times \frac{1\ \text{mol Ca}}{40.078\ \text{g Ca}} = 0.9674\ \text{mol Ca}$$

$$\text{moles P} = 19.97\ \text{g P} \times \frac{1\ \text{mol P}}{30.9738\ \text{g P}} = 0.6447\ \text{mol P}$$

$$\text{moles O} = 41.27\ \text{g O} \times \frac{1\ \text{mol O}}{15.9994\ \text{g O}} = 2.5800\ \text{mol O}$$

B To obtain the *relative* numbers of atoms of each element in the compound, we need to divide the number of moles of each element in the 100-g sample by the number of moles of the element in the smallest amount, in this case P:

$$\text{P:}\ \frac{0.6447\ \text{mol P}}{0.6447\ \text{mol P}} = 1.000 \qquad \text{Ca:}\ \frac{0.9674}{0.6447} = 1.501 \qquad \text{O:}\ \frac{2.5800}{0.6447} = 4.002$$

C We could write the empirical formula of calcium phosphate as $Ca_{1.501}P_{1.000}O_{4.002}$, but the empirical formula should show the ratios of the elements as small whole numbers. To convert the result to integral form, we multiply all the subscripts by 2 to get $Ca_{3.002}P_{2.000}O_{8.004}$. The deviation from integral atomic ratios is small and can be attributed to minor experimental errors; the empirical formula is therefore $Ca_3P_2O_8$.

D The calcium ion (Ca^{2+}) is a cation, so in order to maintain electrical neutrality, P and O must form a polyatomic anion. We know from Chapter 2 that phosphorus and oxygen form PO_4^{3-}, the phosphate ion (see Table 2.4). Because there are two phosphorus atoms in the empirical formula, two phosphate ions must be present. Thus, we write the formula of calcium phosphate as $Ca_3(PO_4)_2$. When we encounter nonintegral atomic ratios, as in this case, it is because no atom has a subscript of 1 in the empirical formula.

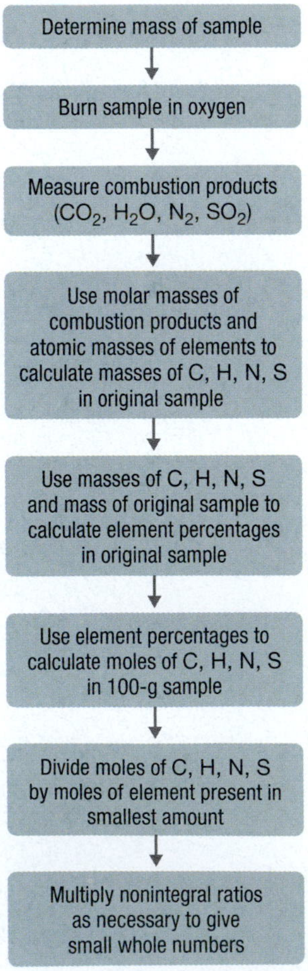

Figure 3.5 Steps for obtaining an empirical formula from combustion analysis.

EXERCISE 3.6

Calculate the empirical formula of ammonium nitrate, an ionic compound that contains 35.00% N, 5.04% H, and 59.96% O by mass; refer to Table 2.4 if necessary. Although ammonium nitrate is widely used as a fertilizer, it can be dangerously explosive. For example, it was a major component of the explosive used in the 1995 Oklahoma City disaster.

Answer $N_2H_4O_3 = NH_4^+NO_3^-$, written as NH_4NO_3

Combustion Analysis

One of the most common ways to determine the elemental composition of an unknown hydrocarbon is an analytical procedure called *combustion analysis*. A small, carefully weighed sample of an unknown compound that may contain carbon, hydrogen, nitrogen, and/or sulfur is burned in an oxygen atmosphere,* and the quantities of the resulting gaseous products (CO_2, H_2O, N_2, and SO_2, respectively) are determined by one of several possible methods. During the combustion reaction, 1 mol of C or S in the sample produces 1 mol of CO_2 or SO_2, respectively, while 2 mol of H or N in the sample produce 1 mol of H_2O or N_2, respectively. Because the percentages of C, H, N, and S in CO_2, H_2O, N_2, and SO_2 are known, these results can be used to determine the amounts of C, H, N, and S in the original sample. The procedure used in combustion analysis is outlined schematically in Figure 3.5, and a typical combustion analysis is illustrated in Example 3.7.

* Other elements, such as metals, can be determined by other methods.

EXAMPLE 3.7

Naphthalene, the active ingredient in one variety of mothballs, is an organic compound that contains only carbon and hydrogen. Complete combustion of a 20.10-mg sample of naphthalene in oxygen gave 69.00 mg of CO_2 and 11.30 mg of H_2O. Determine the empirical formula of naphthalene.

Given Mass of sample and mass of combustion products

Asked for Empirical formula

Strategy

A Use the masses and molar masses of the combustion products, CO_2 and H_2O, to calculate the masses of C and H present in the original sample of naphthalene.

B Use those masses and the molar masses of the elements to calculate the empirical formula of naphthalene.

Solution

A Upon combustion, 1 mol of CO_2 is produced for each mol of carbon in the original sample. Similarly, 1 mol of H_2O is produced for every 2 mol of hydrogen atoms present in the sample. The masses of carbon and hydrogen in the original sample can be calculated from these ratios, the masses of CO_2 and H_2O, and their molar masses. Because the units of molar mass are g/mol, we must first convert the masses from mg to g:

$$\text{mass of C} = 69.00\ \cancel{\text{mg CO}_2} \times \frac{1\ \text{g}}{1000\ \cancel{\text{mg}}} \times \frac{1\ \cancel{\text{mol CO}_2}}{44.010\ \cancel{\text{g CO}_2}} \times \frac{1\ \cancel{\text{mol C}}}{1\ \cancel{\text{mol CO}_2}} \times \frac{12.011\ \text{g}}{1\ \cancel{\text{mol C}}}$$
$$= 1.883 \times 10^{-2}\ \text{g C} = 18.83\ \text{mg C}$$

$$\text{mass of H} = 11.30\ \cancel{\text{mg H}_2\text{O}} \times \frac{1\ \text{g}}{1000\ \cancel{\text{mg}}} \times \frac{1\ \cancel{\text{mol H}_2\text{O}}}{18.015\ \cancel{\text{g H}_2\text{O}}} \times \frac{2\ \cancel{\text{mol H}}}{1\ \cancel{\text{mol H}_2\text{O}}} \times \frac{1.0079\ \text{g}}{1\ \cancel{\text{mol H}}}$$
$$= 1.264 \times 10^{-3}\ \text{g H} = 1.264\ \text{mg H}$$

B To obtain the *relative* numbers of atoms of both elements present, we need to calculate the number of moles of each and divide by the number of moles of the element present in the smallest amount:

$$\text{moles C} = 1.883 \times 10^{-2}\ \cancel{\text{g C}} \times \frac{1\ \text{mol C}}{12.011\ \cancel{\text{g C}}} = 1.568 \times 10^{-3}\ \text{mol C}$$

$$\text{moles H} = 1.264 \times 10^{-3}\ \cancel{\text{g H}} \times \frac{1\ \text{mol H}}{1.0079\ \cancel{\text{g H}}} = 1.254 \times 10^{-3}\ \text{mol H}$$

Dividing each number by the number of moles of the element present in the smaller amount (H) gives

$$\text{H:}\ \frac{1.254 \times 10^{-3}}{1.254 \times 10^{-3}} = 1.000 \qquad \text{C:}\ \frac{1.568 \times 10^{-3}}{1.254 \times 10^{-3}} = 1.250$$

Thus, naphthalene contains a 1.25:1 ratio of moles of carbon to moles of hydrogen: $C_{1.25}H_{1.0}$. Because the ratios of the elements in the empirical formula must be expressed as small whole numbers, we multiply both subscripts by 4, which gives C_5H_4 as the empirical formula of naphthalene. In fact, the molecular formula of naphthalene is $C_{10}H_8$, which is consistent with our results.

EXERCISE 3.7

(a) Xylene, an organic compound that is a major component of many gasoline blends, contains only carbon and hydrogen. Complete combustion of a 17.12-mg sample of xylene in oxygen gave 56.77 mg of CO_2 and 14.53 mg of H_2O. Determine the empirical formula of xylene. **(b)** The empirical formula of benzene is CH (its molecular formula is C_6H_6). If

10.00 mg of benzene is subjected to combustion analysis, what mass of CO_2 and H_2O will be produced?

Answer **(a)** The empirical formula is C_4H_5. (The molecular formula of xylene is actually C_8H_{10}.) **(b)** 33.81 mg of CO_2; 6.92 mg of H_2O

From Empirical Formula to Molecular Formula

The empirical formula gives only the *relative* numbers of atoms in a substance in the smallest possible ratio. For a covalent substance, we are usually more interested in the molecular formula, which gives the *actual* number of atoms of each kind present per molecule. Without additional information, however, it is impossible to know whether the formula of penicillin G, for example, is $C_{16}H_{17}N_2NaO_4S$ or an integral multiple, such as $C_{32}H_{34}N_4Na_2O_8S_2$, $C_{48}H_{51}N_6Na_3O_{12}S_3$, or in general $(C_{16}H_{17}N_2NaO_4S)_n$, where n is an integer. (The actual structure of penicillin G was shown in Figure 3.4.)

Consider glucose, the sugar that circulates in our blood to provide fuel for our bodies and especially for our brains. Results from combustion analysis of glucose report that glucose contains 39.68% C and 6.58% H. Because combustion occurs in the presence of O_2, it is impossible to determine the percent oxygen in a compound directly using combustion analysis; other more complex methods are necessary. If we assume that the remaining percentage is due to oxygen, then glucose would contain 53.79% O. A 100.0-g sample of glucose would therefore contain 39.68 g of C, 6.58 g of H, and 53.79 g of O. To calculate the number of moles of each element in the 100.0-g sample, we divide the mass of each element by its molar mass:

$$
\begin{aligned}
\text{moles C} &= 39.68\ \cancel{\text{g C}} \times \frac{1\ \text{mol C}}{12.011\ \cancel{\text{g C}}} = 3.304\ \text{mol C} \\
\text{moles H} &= 6.58\ \cancel{\text{g H}} \times \frac{1\ \text{mol H}}{1.0079\ \cancel{\text{g H}}} = 6.53\ \text{mol H} \\
\text{moles O} &= 53.79\ \cancel{\text{g O}} \times \frac{1\ \text{mol O}}{15.9994\ \cancel{\text{g O}}} = 3.362\ \text{mol O}
\end{aligned} \tag{3.6}
$$

Once again, we find the subscripts of the elements in the empirical formula by dividing the number of moles of each element by the number of moles of the element present in the smallest amount (3.304 mol of C):

$$
\text{C:}\ \frac{3.304}{3.304} = 1.000 \qquad \text{H:}\ \frac{6.53}{3.304} = 1.98 \qquad \text{O:}\ \frac{3.362}{3.304} = 1.018
$$

The O:C ratio is 1.018, or approximately 1, and the H:C ratio is approximately 2. The empirical formula of glucose is therefore CH_2O, but what is its molecular formula?

Many compounds are known that have the empirical formula CH_2O, including formaldehyde, which is used to preserve biological specimens and has properties that are very different from the sugar circulating in our blood. At this point, we cannot know whether glucose is CH_2O, $C_2H_4O_2$, or any other $(CH_2O)_n$. We can, however, use the experimentally determined molar mass of glucose (180 g/mol) to resolve this dilemma.

First, we calculate the *formula mass*, the molar mass of the formula unit, which is the sum of the atomic masses of the elements in the empirical formula multiplied by their respective subscripts. For glucose,

$$
\begin{aligned}
\text{formula mass of CH}_2\text{O} &= \left[\cancel{1\ \text{mol C}}\left(\frac{12.011\ \text{g}}{\cancel{1\ \text{mol C}}}\right)\right] + \left[2\ \cancel{\text{mol H}}\left(\frac{1.0079\ \text{g}}{1\ \cancel{\text{mol H}}}\right)\right] \\
&\quad + \left[\cancel{1\ \text{mol O}}\left(\frac{15.9994\ \text{g}}{\cancel{1\ \text{mol O}}}\right)\right] = 30.026\ \text{g}
\end{aligned} \tag{3.7}
$$

This is much smaller than the observed molar mass of 180 g/mol.

Second, we determine the number of formula units per mol. For glucose, we can calculate the number of (CH_2O) units—that is, the n in $(CH_2O)_n$—by dividing the molar mass of glucose by the formula mass of CH_2O:

$$n = \frac{180\text{ g}}{30.026\text{ g}/CH_2O} = 5.99 \approx 6\ CH_2O\text{ formula units} \tag{3.8}$$

Thus, each glucose contains six CH_2O formula units, which gives a molecular formula for glucose of $(CH_2O)_6$, more commonly written as $C_6H_{12}O_6$. The molecular structures of formaldehyde and glucose, both of which have the empirical formula CH_2O, are shown in Figure 3.6.

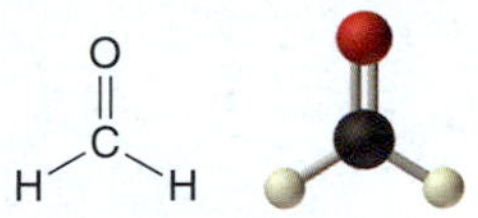

(a) Formaldehyde, CH_2O

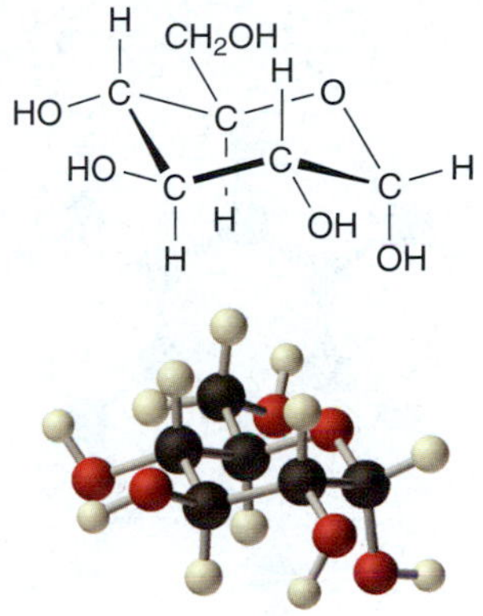

(b) Glucose, $(CH_2O)_6$

Figure 3.6 **Structural formulas and ball-and-stick models of (a) formaldehyde and (b) glucose, both of which have the empirical formula CH_2O.**

EXAMPLE 3.8

Calculate the molecular formula of caffeine, a compound found in coffee, tea, and cola drinks that has a marked stimulatory effect on mammals. Chemical analysis of caffeine shows that it contains 49.18% C, 5.39% H, 28.65% N, and 16.68% O by mass, and its experimentally determined molar mass is 196 g/mol.

Given Percent composition and molar mass

Asked for Molecular formula

Strategy

A Assume 100 g of compound. From the percentages given, use the procedure given in Example 3.6 to calculate the empirical formula of caffeine.

B Calculate the formula mass, and then divide the experimentally determined molar mass by the formula mass. This gives the number of formula units present.

C Multiply each subscript in the empirical formula by the number of formula units to give the molecular formula.

Solution

A We begin by dividing the mass of each element in 100.0 g of caffeine (49.18 g of C, 5.39 g of H, 28.65 g of N, 16.68 g of O) by its molar mass. This gives the number of moles of each element in 100 g of caffeine.

$$\text{moles C} = 49.18\ \cancel{\text{g C}} \times \frac{1\text{ mol C}}{12.011\ \cancel{\text{g C}}} = 4.095\text{ mol C}$$

$$\text{moles H} = 5.39\ \cancel{\text{g H}} \times \frac{1\text{ mol H}}{1.0079\ \cancel{\text{g H}}} = 5.35\text{ mol H}$$

$$\text{moles N} = 28.65\ \cancel{\text{g N}} \times \frac{1\text{ mol N}}{14.0067\ \cancel{\text{g N}}} = 2.045\text{ mol N}$$

$$\text{moles O} = 16.68\ \cancel{\text{g O}} \times \frac{1\text{ mol O}}{15.9994\ \cancel{\text{g O}}} = 1.043\text{ mol O}$$

To obtain the relative numbers of atoms of each element present, we divide the number of moles of each element by the number of moles of the element present in the least amount:

$$\text{O: } \frac{1.043}{1.043} = 1.000 \quad \text{C: } \frac{4.095}{1.043} = 3.926 \quad \text{H: } \frac{5.35}{1.043} = 5.13 \quad \text{N: } \frac{2.045}{1.043} = 1.960$$

These results are fairly typical of actual experimental data. Note that none of the atomic ratios is exactly integral but all are within 5% of integral values. Just as in Example 3.6, it is reasonable to assume that such small deviations from integral values are due to minor experimental errors and to round off to the nearest integer. The empirical formula of caffeine is thus $C_4H_5N_2O$.

B The *molecular* formula of caffeine could be $C_4H_5N_2O$, but it could also be any integral multiple of this. To determine the actual molecular formula, we must divide the

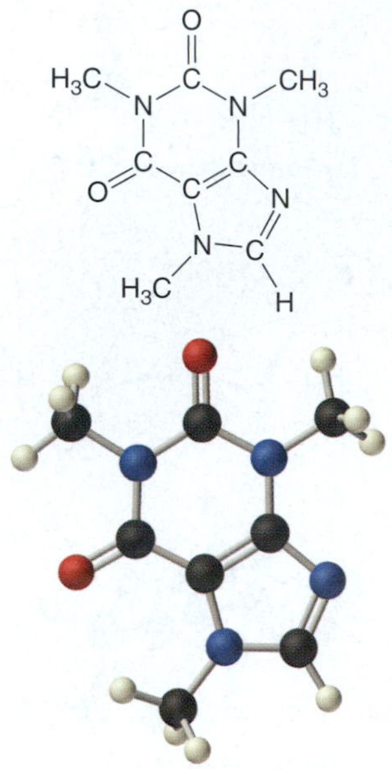

Caffeine

experimentally determined molar mass by the formula mass. The formula mass is calculated as follows:

4 C	(4 atoms C)(12.011 g/atom C)	= 48.044 g
5 H	(5 atoms H) (1.0079 g/atom H)	= 5.0395 g
2 N	(2 atoms N)(14.0067 g/atom N)	= 28.0134 g
+1 O	(1 atom O)(15.9994 g/atom O)	= 15.9994 g
$C_4H_5N_2O$	formula mass of caffeine	97.096 g

Dividing the measured molar mass of caffeine (196 g/mol) by the calculated formula mass gives

$$\frac{196 \text{ g/mol}}{97.096 \text{ g}/C_4H_5N_2O} = 2.02 \cong 2\ C_4H_5N_2O\text{/empirical formula units}$$

Thus, there are two $C_4H_5N_2O$ formula units in caffeine, and the molecular formula must be $(C_4H_5N_2O)_2 = C_8H_{10}N_4O_2$. The structure of caffeine is shown in the margin.

EXERCISE 3.8

Calculate the molecular formula of Freon-114, which contains 13.85% C, 41.89% Cl, and 44.06% F. The experimentally measured molar mass of this compound is 171 g/mol. Like Freon-11, Freon-114 is a commonly used refrigerant that has been implicated in the destruction of the ozone layer.

Answer $C_2Cl_2F_4$

3.3 Chemical Equations

MGC Balancing Chemical Equations

As shown in Figure 3.7, applying a small amount of heat to a pile of orange ammonium dichromate powder results in a vigorous reaction known as the "ammonium dichromate volcano." Heat, light, and gas are produced as a large pile of fluffy green chromium(III) oxide forms. We can describe this reaction with a **chemical equation**, an expression that gives the identities and quantities of the substances in a chemical reaction. Chemical formulas and other symbols are used to indicate the starting material(s), or

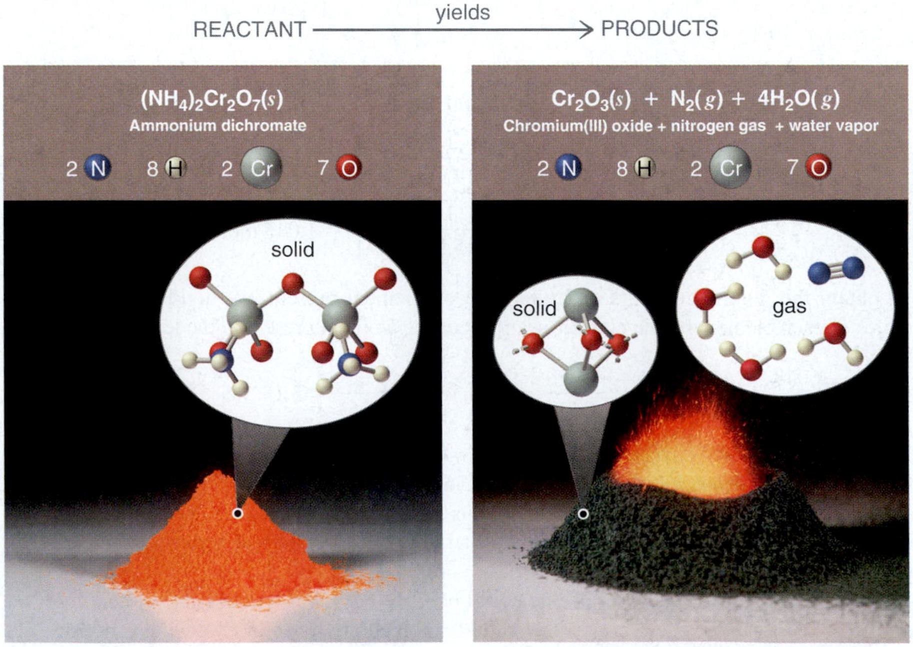

Figure 3.7 An ammonium dichromate volcano, illustrating the change in the distribution of atoms that occurs during a chemical reaction. The starting material *(left)* is solid ammonium dichromate. A chemical reaction *(right)* transforms it to solid chromium(III) oxide, nitrogen gas, and water vapor. (In addition, energy in the form of heat and light is released.) During the reaction, the distribution of atoms changes, but the *number* of atoms of each element does not change. Because the numbers of each type of atom are the same in the reactants and the products, the chemical equation is balanced.

reactant(s), which by convention are written on the left side of the equation, and the final compound(s), or **product(s)**, which are written on the right. An arrow points from the reactant to the products:

$$\underbrace{(NH_4)_2Cr_2O_7}_{\text{Reactant}} \longrightarrow \underbrace{Cr_2O_3 + N_2 + 4H_2O}_{\text{Products}} \quad (3.9)$$

The arrow is read as "yields" or "reacts to form." Thus, Equation 3.9 tells us that ammonium dichromate (the reactant) yields chromium(III) oxide, nitrogen, and water (the products).

The equation for this reaction is even more informative when written as

$$(NH_4)_2Cr_2O_7(s) \longrightarrow Cr_2O_3(s) + N_2(g) + 4H_2O(g) \quad (3.10)$$

Equation 3.10 is identical to Equation 3.9 except for the addition of abbreviations in parentheses to indicate the physical state of each species. The abbreviations are (s) for solid, (l) for liquid, (g) for gas, and (aq) for an *aqueous solution*, a solution of the substance in water.

Notice that the numbers of each type of atom are the same on both sides of Equations 3.9 and 3.10. As illustrated in Figure 3.7, each side has two Cr atoms, seven O atoms, two N atoms, and eight H atoms. In a *balanced chemical equation*, both the numbers of each type of atom and the total charge are the same on both sides. Thus, Equations 3.9 and 3.10 are both balanced equations. What is different on each side of the equation is how the atoms are arranged to make molecules or ions. Thus, *a chemical reaction represents a change in the distribution of atoms but not in the number of atoms*. In this reaction, and in most chemical reactions, bonds are broken in the reactants (in this case, Cr—O and N—H bonds) and new bonds are formed to create the products (O—H and N≡N bonds). If the numbers of each type of atom are different on the two sides of a chemical equation, then the equation is unbalanced, and it cannot correctly describe what happens during the reaction. In order to proceed, the equation must first be balanced.

Note the pattern

*A chemical reaction changes only the distribution of atoms, **not** the number of atoms.*

Interpreting Chemical Equations

In addition to providing qualitative information about the identities and physical states of the reactants and products, a balanced chemical equation provides *quantitative* information. Specifically, it tells the relative amounts of reactants and products consumed or produced in the reaction. The number of atoms, molecules, or formula units of a reactant or product in a balanced chemical equation is the **coefficient** of that species (for example, the "4" preceding "H_2O" in Equation 3.9). When no coefficient is written in front of a species, the coefficient is assumed to be 1. As illustrated in Figure 3.8, the coefficients allow us to interpret Equation 3.9 in any of the following ways:

1. Two NH_4^+ ions and one $Cr_2O_7^{2-}$ ion yield one formula unit of Cr_2O_3, one N_2 molecule, and four H_2O molecules.
2. One mole of $(NH_4)_2Cr_2O_7$ yields 1 mol of Cr_2O_3, 1 mol of N_2, and 4 mol of H_2O.

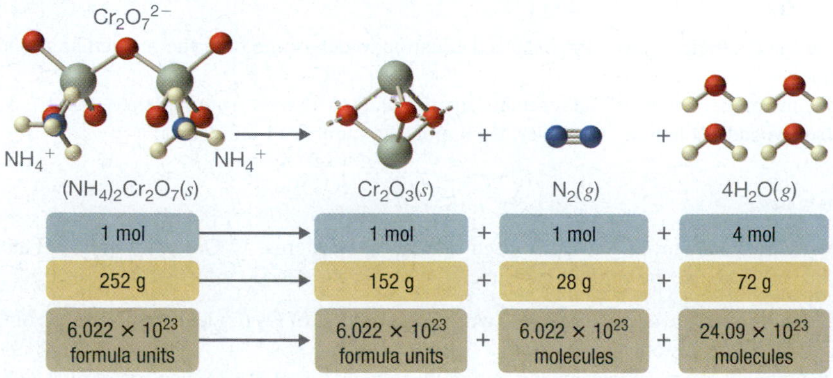

Figure 3.8 The relationships among moles, masses, and formula units of compounds in the balanced chemical reaction for the ammonium dichromate volcano.

3. 252 g of $(NH_4)_2Cr_2O_7$ yields 152 g of Cr_2O_3, 28 g of N_2, and 72 g of H_2O.
4. 6.022×10^{23} formula units of $(NH_4)_2Cr_2O_7$ yields 6.022×10^{23} formula units of Cr_2O_3, 6.022×10^{23} molecules of N_2, and 24.09×10^{23} molecules of H_2O.

These are all *chemically equivalent* ways of stating the information given in the balanced chemical equation, using the concepts of the mole, molar or formula mass, and Avogadro's number. The ratio of the number of moles of one substance to the number of moles of another is called the **mole ratio**. For example, the mole ratio of H_2O to N_2 in Equation 3.9 is 4:1. Notice that the total mass of reactants equals the total mass of products, as predicted by Dalton's law of conservation of mass: 252 g of $(NH_4)_2Cr_2O_7$ yields $152 + 28 + 72 = 252$ g of products. The chemical equation does *not*, however, show whether the reaction occurs rapidly, slowly, or not at all, or whether energy in the form of heat or light is given off. We will consider these issues in more detail in later chapters.

Antoine Lavoisier, the 18th-century French chemist, was interested in the chemistry of living organisms as well as simple chemical systems. In a classic series of experiments, he measured the CO_2 and heat produced by a guinea pig during respiration, in which organic compounds are used as fuel to produce energy, CO_2, and H_2O. Lavoisier found that the ratio of heat produced to CO_2 exhaled was similar to the ratio observed for the reaction of charcoal with oxygen in the air to produce carbon dioxide—a process chemists call *combustion*. Based on these experiments, he proposed that "Respiration is a combustion, slow it is true, but otherwise perfectly similar to that of charcoal." Lavoisier was correct, although the organic compounds consumed in respiration are substantially different from those found in charcoal. One of the most important fuels in the human body is glucose, $C_6H_{12}O_6$, which is virtually the only fuel used in the brain.

EXAMPLE 3.9

The balanced chemical equation for the combustion of glucose in the laboratory (or in the brain) is

$$C_6H_{12}O_6(s) + 6O_2(g) \longrightarrow 6CO_2(g) + 6H_2O(l)$$

Construct a table showing how to interpret the information in this equation in terms of: **(a)** A single molecule of glucose; **(b)** Moles of reactants and products; **(c)** Grams of reactants and products represented by 1 mol of glucose; **(d)** Numbers of molecules of reactants and products represented by 1 mol of glucose.

Given Balanced chemical equation

Asked for Molecule, mole, and mass relationships

Strategy

A Use the coefficients from the balanced equation to determine both the molecular and mole ratios.
B Use the molar masses of the reactants and products to convert from moles to grams.
C Use Avogadro's number to convert from moles to number of molecules.

Solution

This equation is balanced as written: each side has six C atoms, 18 O atoms, and 12 H atoms. We can therefore use the coefficients directly to obtain the desired information.

(a) **A** One molecule of glucose reacts with six molecules of O_2 to yield six molecules of CO_2 and six molecules of H_2O.
(b) One mole of glucose reacts with 6 mol of O_2 to yield 6 mol of CO_2 and 6 mol of H_2O.

(c) To interpret the equation in terms of masses of reactants and products, we need their molar masses and the mole ratios from part b. The molar masses in g/mol are: glucose, 180.16; O_2, 31.9988; CO_2, 44.010; and H_2O, 18.015. Thus,

$$\text{mass of reactants} = \text{mass of products}$$

$$\text{g glucose} + \text{g } O_2 = \text{g } CO_2 + \text{g } H_2O$$

$$\left[1\ \cancel{\text{mol glucose}}\left(\frac{180.16\text{ g}}{1\ \cancel{\text{mol glucose}}}\right)\right] + \left[6\ \cancel{\text{mol } O_2}\left(\frac{31.9988\text{ g}}{1\ \cancel{\text{mol } O_2}}\right)\right]$$

$$= \left[6\ \cancel{\text{mol } CO_2}\left(\frac{44.010\text{ g}}{1\ \cancel{\text{mol } CO_2}}\right)\right] + \left[6\ \cancel{\text{mol } H_2O}\left(\frac{18.015\text{ g}}{1\ \cancel{\text{mol } H_2O}}\right)\right]$$

$$372.15\text{ g} = 372.15\text{ g}$$

(d) One mole of glucose contains Avogadro's number (6.022×10^{23}) of glucose molecules. Thus, 6.022×10^{23} glucose molecules react with ($6 \times 6.022 \times 10^{23}$) = 3.613×10^{24} oxygen molecules to yield ($6 \times 6.022 \times 10^{23}$) = 3.613×10^{24} molecules each of CO_2 and H_2O.

In tabular form:

$C_6H_{12}O_6(s)$	+ $6O_2(g)$	$\longrightarrow$ $6CO_2(g)$	+ $6H_2O(l)$
Glucose	**O_2**	**CO_2**	**H_2O**
(a) 1 molecule	6 molecules	6 molecules	6 molecules
(b) 1 mol	6 mol	6 mol	6 mol
(c) 180.16 g	191.9928 g	264.06 g	108.09 g
(d) 6.022×10^{23} molecules	3.613×10^{24} molecules	3.613×10^{24} molecules	3.613×10^{24} molecules

EXERCISE 3.9

Ammonium nitrate is a common fertilizer, but under the wrong conditions it can be hazardous. In 1947, a ship loaded with ammonium nitrate caught fire while unloading and exploded, destroying the town of Texas City, Texas. The reaction that produced the explosion is

$$2NH_4NO_3(s) \longrightarrow 2N_2(g) + 4H_2O(g) + O_2(g)$$

Construct a table showing how to interpret the information in the equation in terms of: **(a)** Individual molecules and ions; **(b)** Moles of reactants and products; **(c)** Grams of reactants and products given 2 mol of ammonium nitrate; **(d)** Numbers of molecules or formula units of reactants and products given 2 mol of ammonium nitrate.

Answer

$2NH_4NO_3(s)$	$\longrightarrow$ $2N_2(g)$	+ $4H_2O(g)$	+ $O_2(g)$
NH_4NO_3	**N_2**	**H_2O**	**O_2**
(a) $2NH_4^+$ ions and $2NO_3^-$ ions	2 molecules	4 molecules	1 molecule
(b) 2 mol	2 mol	4 mol	1 mol
(c) 160.0864 g	56.0268 g	72.0608 g	31.9988 g
(d) 1.204×10^{24} formula units	1.204×10^{24} molecules	2.409×10^{24} molecules	6.022×10^{23} molecules

Ammonium nitrate can be hazardous. This aerial photograph of Texas City, Texas, shows the devastation caused by the explosion of a shipload of ammonium nitrate on April 16, 1947.

Balancing Simple Chemical Equations

When a chemist encounters a new reaction, it does not usually come with a label that shows the balanced chemical equation. Instead, the chemist must identify the reactants and products and then write them down in the form of a chemical equation that

Figure 3.9 An example of a combustion reaction. The wax in a candle is a high-molecular-mass hydrocarbon, which produces gaseous carbon dioxide and water vapor upon complete combustion. When the candle is allowed to burn inside a flask, drops of water, one of the products of combustion, condense at the top of the inner surface of the flask.

may or may not be balanced as first written. Consider, for example, the combustion of *n*-heptane, C_7H_{16}, an important component of gasoline:

$$C_7H_{16}(l) + O_2(g) \longrightarrow CO_2(g) + H_2O(g) \tag{3.11}$$

The complete combustion of any hydrocarbon with sufficient oxygen *always* yields carbon dioxide and water (Figure 3.9).

Equation 3.11 is not balanced: the number of each type of atom on the reactants side of the equation (7 C, 16 H, 2 O) is not the same as the number of each type of atom on the products side (1 C, 2 H, 3 O). Consequently, we must adjust the coefficients of the reactants and products to give the same number of atoms of each type on both sides of the equation. Because the identities of the reactants and products are fixed, we *cannot* balance the equation by changing the subscripts of any of the reactants or products. To do so would change the chemical identity of the species being described, as illustrated in Figure 3.10.

The simplest and most generally useful method for balancing chemical equations is "inspection," better known as trial and error. We present an efficient approach to balancing a chemical equation using this method.

Steps in Balancing a Chemical Equation

Step 1 Identify the most complex substance.

Step 2 Beginning with that substance, choose an element that appears in only one reactant and one product, if possible. Adjust the coefficients to obtain the same number of atoms of this element on both sides.

Step 3 Balance polyatomic ions (if present) as a unit.

Step 4 Balance the remaining atoms, usually ending with the least complex substance and using fractional coefficients if necessary. If a fractional coefficient has been used, multiply both sides of the equation by the denominator to obtain whole numbers for the coefficients.

Step 5 Count the numbers of atoms of each kind on both sides of the equation to be sure that the chemical equation is balanced.

To demonstrate this approach, let's use the combustion of *n*-heptane (Equation 3.11) as an example.

Step 1 **Identify the most complex substance.** The most complex substance is the one with the largest number of different atoms, which in Equation 3.11 is C_7H_{16}. We will assume initially that the final balanced equation contains one molecule or formula unit of this substance.

Step 2 **Adjust the coefficients.** We try to adjust the coefficients of the molecules on the other side of the equation to obtain the same numbers of atoms on both sides. Because one molecule of *n*-heptane contains seven C atoms, we need seven CO_2 molecules, each of which contains one C atom, on the right side:

$$C_7H_{16} + O_2 \longrightarrow 7CO_2 + H_2O \tag{3.12}$$

Step 3 **Balance polyatomic ions as a unit.** There are no polyatomic ions to be considered in this reaction.

Step 4 **Balance the remaining atoms.** Because one molecule of *n*-heptane contains 16 H atoms, we need eight H_2O molecules, each of which contains two H atoms, on the right side:

$$C_7H_{16} + O_2 \longrightarrow 7CO_2 + 8H_2O \tag{3.13}$$

The C and H atoms are now balanced, but we have 22 O atoms on the right side and only two O atoms on the left. We can balance the O atoms by adjusting the coefficient in front of the least complex substance, O_2, on the reactants side:

$$C_7H_{16}(l) + 11O_2(g) \longrightarrow 7CO_2(g) + 8H_2O(g) \tag{3.14}$$

Step 5 **Check your work.** The equation is now balanced and there are no fractional coefficients: there are seven C atoms, 16 H atoms, and 22 O atoms on each side, as you should verify. *Always check to be sure that a chemical equation is balanced!*

The assumption that the final balanced equation contains only one molecule or formula unit of the most complex substance is not always valid, but it is a good place to start. Consider, for example, a similar reaction, the combustion of isooctane, C_8H_{18}. Because the combustion of any hydrocarbon with oxygen produces carbon dioxide and water, the unbalanced chemical equation is

$$C_8H_{18}(l) + O_2(g) \longrightarrow CO_2(g) + H_2O(g) \qquad (3.15)$$

Step 1 **Identify the most complex substance.** Begin the balancing process by assuming that the final balanced equation contains a single molecule of isooctane.

Step 2 **Adjust the coefficients.** The first element that appears only once in the reactants is carbon: eight carbon atoms in isooctane means that there must be eight CO_2 molecules in the products:

$$C_8H_{18} + O_2 \longrightarrow 8CO_2 + H_2O \qquad (3.16)$$

Step 3 **Balance polyatomic ions as a unit.** Again, this step does not apply to this problem.

Step 4 **Balance the remaining atoms.** Eighteen hydrogen atoms in isooctane means that there must be nine H_2O molecules in the products:

$$C_8H_{18} + O_2 \longrightarrow 8CO_2 + 9H_2O \qquad (3.17)$$

The C and H atoms are now balanced, but we have 25 oxygen atoms on the right side and only two oxygen atoms on the left. We can balance the least complex substance, O_2, but because there are two oxygen atoms per O_2 molecule, we must use a fractional coefficient $\left(\frac{25}{2}\right)$ to balance the oxygen atoms:

$$C_8H_{18} + \frac{25}{2}O_2 \longrightarrow 8CO_2 + 9H_2O \qquad (3.18)$$

Equation 3.18 is now balanced, but we usually write equations with whole-number coefficients. We can eliminate the fractional coefficient by multiplying all coefficients on both sides of the chemical equation by 2:

$$2C_8H_{18}(l) + 25O_2(g) \longrightarrow 16CO_2(g) + 18H_2O(g) \qquad (3.19)$$

Step 5 **Check your work.** The balanced equation contains 16 carbon atoms, 36 hydrogen atoms, and 50 oxygen atoms on each side.

Balancing equations requires some practice on your part as well as some common sense. If you find yourself using very large coefficients or if you have spent several minutes without success, go back and make sure that you have written the formulas of the reactants and products correctly.

Figure 3.10 You cannot change subscripts in a chemical formula to balance a chemical equation; you can change only the coefficients. Changing subscripts changes the ratios of atoms in the molecule and the resulting chemical properties. For example, water (H_2O) and hydrogen peroxide (H_2O_2) are chemically distinct substances. H_2O_2 decomposes to liquid water and O_2 gas when it comes in contact with the metal platinum. (No such reaction occurs between water and platinum.)

EXAMPLE 3.10

Reaction of the mineral hydroxyapatite, $Ca_5(PO_4)_3(OH)$, with phosphoric acid and water gives $Ca(H_2PO_4)_2 \cdot H_2O$ (calcium dihydrogen phosphate monohydrate). Write and balance the equation for this reaction.

Given Reactants and product

Asked for Balanced chemical equation

Strategy

Ⓐ Identify the product and reactants, and then write the unbalanced chemical equation.
Ⓑ Follow the steps for balancing a chemical equation.

Solution

Ⓐ We must first identify the product and reactants and write an equation for the reaction. The formulas for hydroxyapatite and calcium dihydrogen phosphate monohydrate are given in the problem. Recall from Chapter 2 that phosphoric acid is H_3PO_4. The initial (unbalanced) equation is thus

$$Ca_5(PO_4)_3(OH)(s) + H_3PO_4(aq) + H_2O(l) \longrightarrow Ca(H_2PO_4)_2{\cdot}H_2O(s)$$

Step 1 Ⓑ **Identify the most complex substance.** We start by assuming that only one molecule or formula unit of the most complex substance, $Ca_5(PO_4)_3(OH)$, appears in the balanced equation.

Step 2 **Adjust the coefficients.** Because calcium is present in only one reactant and one product, we begin with it. One formula unit of $Ca_5(PO_4)_3(OH)$ contains five Ca atoms, so we need five $Ca(H_2PO_4)_2{\cdot}H_2O$ on the right side:

$$Ca_5(PO_4)_3(OH) + H_3PO_4 + H_2O \longrightarrow 5Ca(H_2PO_4)_2{\cdot}H_2O$$

Step 3 **Balance polyatomic ions as a unit.** It is usually easier to balance an equation if we recognize that certain combinations of atoms occur on both sides. In this equation, the polyatomic phosphate ion, PO_4^{3-}, shows up in three different places.* Thus, it is easier to balance PO_4 units rather than counting individual P and O atoms. There are 10 PO_4 units on the right side but only four on the left. The simplest way to balance the PO_4 units is to place a coefficient of 7 in front of H_3PO_4:

$$Ca_5(PO_4)_3(OH) + 7H_3PO_4 + H_2O \longrightarrow 5Ca(H_2PO_4)_2{\cdot}H_2O$$

Although OH^- is also a polyatomic ion, it does not appear on both sides of the equation. Hence, O and H must be balanced separately.

Step 4 **Balance the remaining atoms.** We now have 30 hydrogen atoms on the right side but only 24 on the left. We can balance the hydrogen atoms using the least complex substance, H_2O, by placing a coefficient of 4 in front of H_2O on the left side, giving a total of $4H_2O$:

$$Ca_5(PO_4)_3(OH)(s) + 7H_3PO_4(aq) + 4H_2O(l) \longrightarrow 5Ca(H_2PO_4)_2{\cdot}H_2O(s)$$

The equation is now balanced, as you should verify. Even though we have not explicitly balanced oxygen atoms, there are 45 oxygen atoms on each side.

Step 5 **Check your work.** Both sides of the equation contain five Ca atoms, seven P atoms, 30 H atoms, and 45 O atoms.

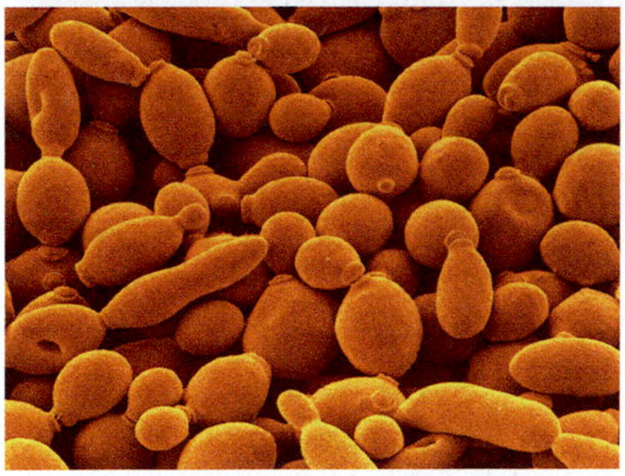

Commercial use of fermentation. Photograph of microbrewery vats used to prepare beer (*top*). Fermentation of glucose by yeast cells (*bottom*) is the reaction that makes beer production possible.

EXERCISE 3.10

Fermentation is the biochemical process that enables yeast cells to live in the absence of oxygen. Humans have exploited it for centuries to produce wine and beer and to cause bread to rise. In fermentation, sugars such as glucose are converted to ethanol and carbon dioxide. Write a balanced chemical reaction for the fermentation of glucose.

Answer $C_6H_{12}O_6(s) \longrightarrow 2C_2H_5OH(l) + 2CO_2(g)$

* In H_3PO_4 the phosphate ion is combined with three H^+ ions to make phosphoric acid (H_3PO_4), whereas in $Ca(H_2PO_4)_2{\cdot}H_2O$ it is combined with two H^+ ions to give the dihydrogen phosphate ion ($H_2PO_4^-$).

3.4 • Mass Relationships in Chemical Equations

MGC Stoichiometry

A balanced chemical equation gives the identity of the reactants and products as well as the accurate number of molecules or moles of each that are consumed or produced. **Stoichiometry** is a collective term for the quantitative relationships between the masses, numbers of moles, and numbers of particles (atoms, molecules, and ions) of the reactants and products in a balanced reaction. A **stoichiometric quantity** is the amount of product or reactant specified by the coefficients in a balanced chemical equation. In Section 3.3, for example, you learned how to express the stoichiometry of the reaction that is responsible for the ammonium dichromate volcano in terms of the atoms, ions, or molecules involved and the number of moles, grams, and formula units of each (recognizing, for instance, that 1 mol of ammonium dichromate produces 4 mol of water). This section describes how to use the stoichiometry of a reaction to answer questions like the following: How much O_2 is needed to ensure complete combustion of a given amount of isooctane? (This information is crucial to the design of nonpolluting and efficient automobile engines.) How many grams of pure gold can be obtained from a ton of low-grade gold ore? (The answer determines whether the ore deposit is worth mining.) If an industrial plant must produce a certain number of tons of sulfuric acid per week, how many tank cars of elemental sulfur must arrive by rail each week?

All these questions can be answered using the concepts of the mole and molar and formula masses, along with the coefficients in the appropriate balanced chemical equation.

Stoichiometry Problems

When we carry out a reaction in either an industrial setting or the laboratory, it is easier to work with *masses* of substances than with the numbers of molecules or moles. The general method for converting from the mass of any reactant or product to the mass of any other reactant or product using a balanced chemical equation is outlined in Figure 3.11 and described below.

Steps in Converting Between Masses of Reactant and Product

Step 1 Convert the mass of one substance (substance A) to the corresponding number of moles using its molar mass.

Step 2 From the balanced chemical equation, obtain the number of moles of another substance (B) from the number of moles of substance A using the appropriate mole ratio (the ratio of their coefficients).

Step 3 Convert the number of moles of substance B to mass using its molar mass.

Converting amounts of substances to moles, and vice versa, is the key to all stoichiometry problems, whether the amounts are given in units of mass (grams or kilograms), weight (pounds or tons), or volume (liters or gallons).

To illustrate this procedure, let's return to the combustion of glucose. We saw earlier that glucose reacts with oxygen to produce carbon dioxide and water according to the equation

$$C_6H_{12}O_6(s) + 6O_2(g) \longrightarrow 6CO_2(g) + 6H_2O(l) \qquad (3.20)$$

Figure 3.11 A flowchart for stoichiometric calculations involving pure substances. The molar masses of the reactants and products are used as conversion factors so that you can calculate the mass of product from the mass of reactant, and vice versa.

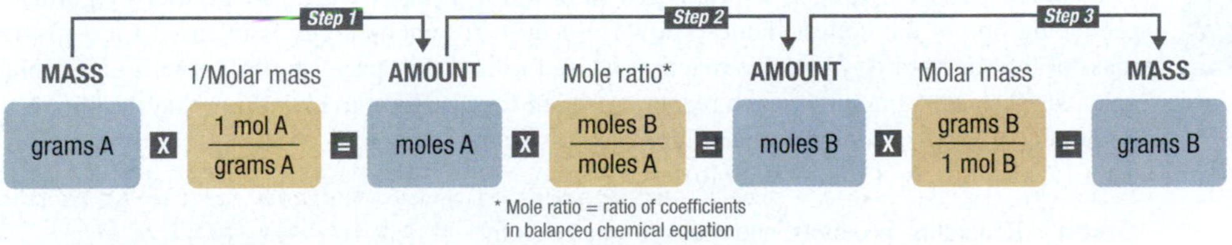

Suppose that, just before a chemistry exam, your friend reminds you that glucose is the major fuel used by the human brain. You therefore decide to eat a candy bar to make sure that your brain doesn't run out of energy during the exam (even though there is no direct evidence that consumption of candy bars improves performance on chemistry exams). If a typical $2\frac{1}{8}$-ounce candy bar contains the equivalent of 45.3 g of glucose and the glucose is completely converted to CO_2 during the exam, how many grams of CO_2 will you produce and exhale into the exam room?

The initial step in solving a problem of this type must be to write the balanced chemical equation for the reaction. Inspection of Equation 3.20 shows that it is balanced as written, so we can proceed to the strategy outlined in Figure 3.11, adapting it as follows:

Step 1 We use the molar mass of glucose (180.2 g/mol) to determine the number of moles of glucose in the candy bar:

$$\text{moles glucose} = 45.3\ \cancel{\text{g glucose}} \times \frac{1\ \text{mol glucose}}{180.2\ \cancel{\text{g glucose}}} = 0.251\ \text{mol glucose}$$

Step 2 According to the balanced equation, 6 mol of CO_2 is produced per mole of glucose; the mole ratio of CO_2 to glucose is therefore 6:1. The number of moles of CO_2 produced is thus

$$\begin{aligned}\text{moles } CO_2 &= \text{mol glucose} \times \frac{6\ \text{mol } CO_2}{1\ \text{mol glucose}} \\ &= 0.251\ \cancel{\text{mol glucose}} \times \frac{6\ \text{mol } CO_2}{1\ \cancel{\text{mol glucose}}} \\ &= 1.51\ \text{mol } CO_2\end{aligned}$$

Step 3 We use the molar mass of CO_2 (44.010 g/mol) to calculate the mass of CO_2 corresponding to 1.51 mol of CO_2:

$$\text{mass of } CO_2 = 1.51\ \cancel{\text{mol } CO_2} \times \frac{44.010\ \text{g } CO_2}{1\ \cancel{\text{mol } CO_2}} = 66.5\ \text{g } CO_2$$

We can summarize these operations as follows:

$$45.3\ \cancel{\text{g glucose}} \times \underbrace{\frac{1\ \cancel{\text{mol glucose}}}{180.16\ \cancel{\text{g glucose}}}}_{\text{Step 1}} \times \underbrace{\frac{6\ \cancel{\text{mol } CO_2}}{1\ \cancel{\text{mol glucose}}}}_{\text{Step 2}} \times \underbrace{\frac{44.010\ \text{g } CO_2}{1\ \cancel{\text{mol } CO_2}}}_{\text{Step 3}} = 66.4\ \text{g } CO_2$$

Discrepencies between the two values are attributed to rounding errors resulting from the use of stepwise calculations in Steps 1-3 (see Essential Skills 1). In Chapter 10, you will discover that this amount of gaseous CO_2 occupies an enormous volume—more than 33 liters! We could use similar methods to calculate the amount of O_2 consumed or the amount of H_2O produced.

In the preceding problem, we used the balanced equation to calculate the mass of product that is formed from a certain amount of reactant. We can also use the balanced chemical equation to determine the masses of reactants that are necessary to form a certain amount of product or, as shown in Example 3.11, the mass of one reactant that is required to consume a given mass of another reactant.

The U.S. space shuttle *Discovery* during liftoff. The large cylinder in the middle contains the oxygen and hydrogen that fuel the shuttle's main engine.

EXAMPLE 3.11

The combustion of hydrogen with oxygen to produce gaseous water is extremely vigorous, producing one of the hottest flames known. Because so much energy is released for a given mass of hydrogen or oxygen, this reaction is used to fuel the space shuttles. NASA engineers calculate the exact amount of each reactant needed for the flight to make sure that the shuttles do not carry excess fuel into orbit. Calculate how many tons of H_2 the shuttle must carry for each 1.00 ton of O_2 (1 ton = 2000 lb).

Given Reactants, products, and mass of one reactant

Asked for Mass of other reactant

Strategy

A Write the balanced chemical equation for the reaction.
B Convert mass of O_2 to moles. From the mole ratio in the balanced chemical equation, determine the number of moles of H_2 required. Then convert moles of H_2 to the equivalent mass in tons.

Solution

We use the same general strategy for solving stoichiometric calculations as in the preceding example. Because the amount of O_2 is given in tons rather than grams, however, we also need to convert tons to units of mass in grams. Another conversion is needed at the end to report the final answer in tons.

A We first use the information given to write a balanced chemical equation. Because we know the identity of both the reactants and the product, we can write the reaction as

$$H_2(g) + O_2(g) \longrightarrow H_2O(g)$$

This equation is not balanced as written because there are two O atoms on the left side and only one on the right. Assigning a coefficient of 2 to both H_2O and H_2 gives the balanced equation

$$2H_2(g) + O_2(g) \longrightarrow 2H_2O(g)$$

Thus, 2 mol of H_2 react with 1 mol of O_2 to produce 2 mol of H_2O.

Step 1 **B** To convert tons of oxygen to units of mass in grams, we multiply by the appropriate conversion factors:

$$\text{mass of } O_2 = 1.00 \cancel{\text{ton}} \times \frac{2000 \cancel{\text{lb}}}{\cancel{\text{ton}}} \times \frac{453.6 \text{ g}}{\cancel{\text{lb}}}$$
$$= 9.07 \times 10^5 \text{ g } O_2$$

Using the molar mass of O_2 (32.00 g/mol, to three significant figures), we can calculate the number of moles of O_2 contained in this mass of O_2:

$$\text{moles } O_2 = 9.07 \times 10^5 \cancel{\text{g } O_2} \times \frac{1 \text{ mol } O_2}{32.00 \cancel{\text{g } O_2}}$$
$$= 2.83 \times 10^4 \text{ mol } O_2$$

Step 2 We now use the coefficients in the balanced chemical equation to obtain the number of moles of H_2 needed to react with this number of moles of O_2:

$$\text{moles } H_2 = \text{mol } O_2 \times \frac{2 \text{ mol } H_2}{1 \text{ mol } O_2}$$
$$= 2.83 \times 10^4 \cancel{\text{mol } O_2} \times \frac{2 \text{ mol } H_2}{1 \cancel{\text{mol } O_2}}$$
$$= 5.66 \times 10^4 \text{ mol } H_2$$

Step 3 The molar mass of H_2 (2.016 g/mol) allows us to calculate the corresponding mass of H_2:

$$\text{mass of } H_2 = 5.66 \times 10^4 \cancel{\text{mol } H_2} \times \frac{2.016 \text{ g } H_2}{\cancel{\text{mol } H_2}}$$
$$= 1.14 \times 10^5 \text{ g } H_2$$

Finally, we need to convert the mass of H_2 to the desired units (tons) by using the appropriate conversion factors:

$$\text{tons } H_2 = 1.14 \times 10^5 \cancel{\text{g}} \text{ } H_2 \times \frac{1 \cancel{\text{lb}}}{453.6 \cancel{\text{g}}} \times \frac{1 \text{ ton}}{2000 \cancel{\text{lb}}} = 0.126 \text{ ton } H_2$$

Thus, the space shuttle must be designed to carry 0.126 ton of H_2 for each 1.00 ton of O_2. Note that even though 2 mol of H_2 is needed to react with each mole of O_2, the molar mass of H_2 is so much smaller than that of O_2 that only a relatively small mass of H_2 is needed compared to the mass of O_2.

EXERCISE 3.11

Alchemists produced elemental mercury by roasting the mercury-containing ore cinnabar (HgS) in air:

$$HgS(s) + O_2(g) \longrightarrow Hg(l) + SO_2(g)$$

The volatility and toxicity of mercury make this a hazardous procedure, which likely shortened the life span of many alchemists. Given 100 g of cinnabar, how much elemental mercury can be produced through this reaction?

Answer 86.2 g

Limiting Reactants

In all the examples discussed thus far, the reactants were assumed to be present in stoichiometric quantities. Consequently, none of the reactants was left over at the end of the reaction. This is often desirable, as in the case of the space shuttle, where excess oxygen or hydrogen is simply extra freight to be hauled into orbit as well as an explosion hazard. More often, however, reactants are present in mole ratios that are not the same as the ratio of the coefficients in the balanced chemical equation. As a result, one or more of them will not be used up completely but will be left over when the reaction is completed. In this situation, the amount of product that can be obtained is limited by the amount of only one of the reactants. The reactant that restricts the amount of product obtained is called the **limiting reactant**. The reactant that remains after a reaction has gone to completion was present *in excess*.

To be certain you understand these concepts, let's first consider a nonchemical example. Assume you are having some friends over for dinner and you want to bake brownies for dessert. You find two boxes of brownie mix in your pantry and see that each package requires two eggs. The balanced equation for brownie preparation is thus

$$\text{1 box mix} + \text{2 eggs} \longrightarrow \text{1 batch brownies} \tag{3.21}$$

If you have a dozen eggs, which ingredient will determine the number of batches of brownies that you can prepare? Because each box of brownie mix requires two eggs and you have two boxes, you need four eggs. Twelve eggs is eight more eggs than you need. Although the ratio of eggs to boxes in Equation 3.21 is 2:1, the ratio in your possession is 6:1. Hence, the eggs are the ingredient (reactant) that is present in excess, and the brownie mix is the limiting reactant (Figure 3.12). Even if you had a refrigerator full of eggs, you could make only two batches of brownies.

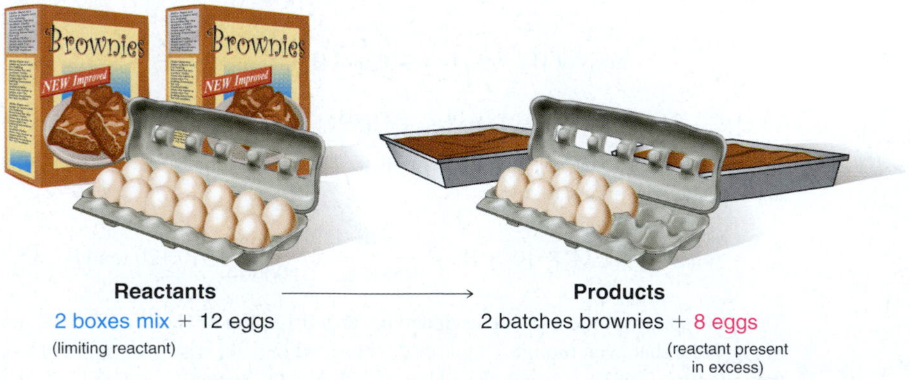

Figure 3.12 Illustration of the concept of a limiting reactant in the preparation of brownies.

Let's now turn to a chemical example of a limiting reactant: the production of pure titanium. This metal is fairly light (45% lighter than steel and only 60% heavier than aluminum) and has great mechanical strength (as strong as steel and twice as strong as aluminum). Because it is also highly resistant to corrosion and can withstand extreme temperatures, titanium has many applications in the aerospace industry. Titanium is also used in medical implants and portable computer housings because it is light and resistant to corrosion. Although titanium is the ninth most common element in the earth's crust, it is relatively difficult to extract from its ores. In the first step of the extraction process, titanium-containing oxide minerals react with solid carbon and chlorine gas to form titanium tetrachloride, $TiCl_4$, and CO_2. The titanium tetrachloride is then converted to metallic titanium by reaction with magnesium metal at high temperature:

$$TiCl_4(g) + 2Mg(l) \longrightarrow Ti(s) + 2MgCl_2(l) \qquad (3.22)$$

Because titanium ores, carbon, and chlorine are all rather inexpensive, the high price of titanium (about $1100 per kilogram) is largely due to the high cost of magnesium metal. Under these circumstances, magnesium metal is chosen as the limiting reactant in the production of metallic titanium.

Suppose you are given 1.00 kg of titanium tetrachloride and 200 g of magnesium metal. How much titanium metal can you produce according to Equation 3.22? Solving this type of problem requires that you carry out the following steps:

Step 1 Determine the number of moles of each reactant.
Step 2 Compare the mole ratio of the reactants with the ratio in the balanced chemical equation to determine which reactant is limiting.
Step 3 Calculate the number of moles of product that can be obtained from the limiting reactant.
Step 4 Convert the number of moles of product to mass of product.

Step 1: To determine the number of moles of reactants present, you must calculate or look up their molar masses: 189.679 g/mol for $TiCl_4$ and 24.305 g/mol for Mg. The number of moles of each is calculated as follows:

$$\text{moles TiCl}_4 = \frac{\text{mass TiCl}_4}{\text{molar mass TiCl}_4}$$
$$= 1000\ \cancel{\text{g TiCl}_4} \times \frac{1\ \text{mol TiCl}_4}{189.679\ \cancel{\text{g TiCl}_4}} = 5.272\ \text{mol TiCl}_4$$

$$\text{moles Mg} = \frac{\text{mass Mg}}{\text{molar mass Mg}}$$
$$= 200\ \cancel{\text{g Mg}} \times \frac{1\ \text{mol Mg}}{24.305\ \cancel{\text{g Mg}}} = 8.23\ \text{mol Mg}$$

Step 2: You have more moles of Mg than of $TiCl_4$, but the ratio is only

$$\frac{\text{mol Mg}}{\text{mol TiCl}_4} = \frac{8.23\ \text{mol}}{5.272\ \text{mol}} = 1.56$$

Because the ratio of the coefficients in the balanced chemical equation is

$$\frac{2\ \text{mol Mg}}{1\ \text{mol TiCl}_4} = 2$$

you do not have enough Mg to react with all of the $TiCl_4$. If this point is not clear from the mole ratio, you should calculate the number of moles of one reactant that is required for complete reaction of the other reactant. For example, you have 8.23 mol of Mg, and you need $(8.23 \div 2) = 4.12$ mol of $TiCl_4$ for complete reaction. Because you have 5.272 mol

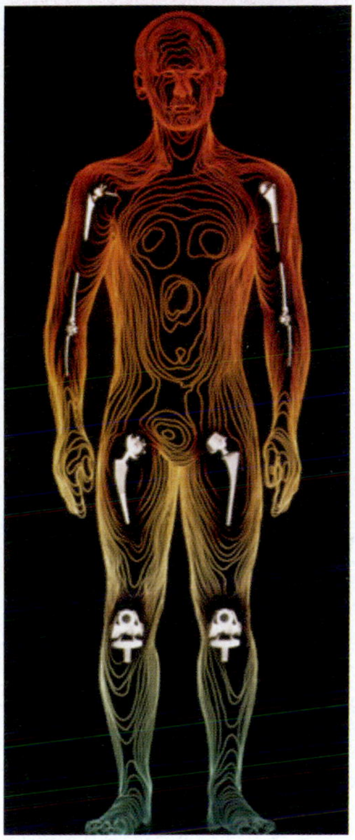

Medical use of titanium. Some places in the human body where titanium has been successfully used in joint replacement implants.

of $TiCl_4$, *$TiCl_4$ is present in excess.* Conversely, 5.272 mol of $TiCl_4$ requires $2 \times 5.272 =$ 10.54 mol of Mg, but you have only 8.23 mol. Thus, *Mg is the limiting reactant.*

Step 3: Because Mg is the limiting reactant, the number of moles of Mg determines the number of moles of Ti that can be formed:

$$\text{moles Ti} = 8.23\ \cancel{\text{mol Mg}} = \frac{1\ \text{mol Ti}}{2\ \cancel{\text{mol Mg}}} = 4.12\ \text{mol Ti}$$

Thus, only 4.12 mol of Ti can be formed.

Step 4: To calculate the mass of titanium metal that you can obtain, multiply the number of moles of Ti by the molar mass of Ti (47.867 g/mol):

$$\text{moles Ti} = \text{mass Ti} \times \text{molar mass Ti} = 4.12\ \cancel{\text{mol Ti}} \times \frac{47.867\ \text{g Ti}}{1\ \cancel{\text{mol Ti}}} = 197\ \text{g Ti}$$

Here is a simple and reliable way to identify the limiting reactant in any problem of this sort:

Step 1 Calculate the number of moles of each reactant present: 5.272 mol of $TiCl_4$ and 8.23 mol of Mg.

Step 2 Divide the actual number of moles of each reactant by its stoichiometric coefficient in the balanced chemical equation:

$$TiCl_4\text{:}\ \frac{5.272\ \text{mol (actual)}}{1\ \text{mol (stoich)}} = 5.272 \qquad \text{Mg:}\ \frac{8.23\ \text{mol (actual)}}{2\ \text{mol (stoich)}} = 4.12$$

Step 3 The reactant with the smallest mole:coefficient ratio is limiting. Thus, Mg, with a calculated stoichiometric mole ratio of 4.12, is the limiting reactant.

As you learned in Chapter 1, density is the mass per unit volume of a substance. If we are given the density of a substance, we can use it in stoichiometric calculations involving liquid reactants and/or products, as Example 3.12 demonstrates.

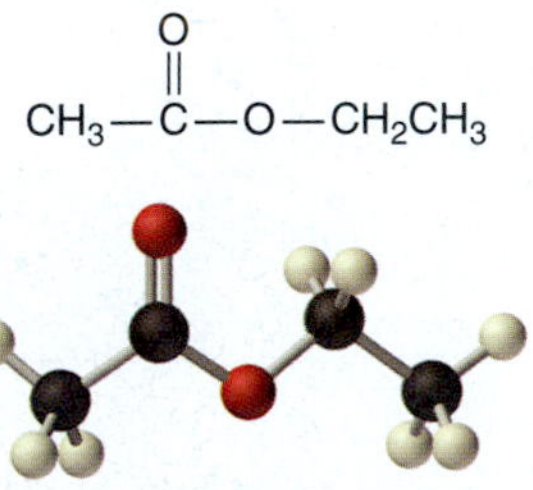

Ethyl acetate

EXAMPLE 3.12

Ethyl acetate, $CH_3CO_2C_2H_5$, is the solvent in many fingernail polish removers. It is prepared by the reaction of ethanol, C_2H_5OH, with acetic acid, CH_3CO_2H; the other product is water. A small amount of sulfuric acid is used to accelerate the reaction, but the sulfuric acid is not consumed and does not appear in the balanced equation. Given 10.0 mL each of acetic acid and ethanol, how many grams of ethyl acetate can be prepared using this reaction? The densities of acetic acid and ethanol are 1.0492 g/mL and 0.7893 g/mL, respectively.

Given Reactants, products, and volumes and densities of reactants

Asked for Mass of product

Strategy

A Balance the chemical equation for the reaction.

B Use the given densities to convert from volume to mass. Then use each molar mass to convert from mass to moles.

C Using mole ratios, determine which substance is the limiting reactant. Once the limiting reactant is identified, use mole ratios based on the number of moles of limiting reactant to determine the number of moles of product.

D Convert from moles of product to mass of product.

Solution

A We begin by writing the balanced chemical equation for the reaction:

$$C_2H_5OH(l) + CH_3CO_2H(aq) \longrightarrow CH_3CO_2C_2H_5(aq) + H_2O(l)$$

B We need to calculate the number of moles of ethanol and acetic acid that are present in 10.0 mL of each. Recall from Chapter 1 that the density of a substance is the mass divided by the volume:

$$\text{density} = \frac{\text{mass}}{\text{volume}}$$

Rearranging this expression gives mass = (density)(volume). We can replace mass by the product of the density and the volume to calculate the number of moles of each substance in 10.0 mL (1 mL = 1 cm^3):

$$\begin{aligned}
\text{moles } C_2H_5OH &= \frac{\text{mass } C_2H_5OH}{\text{molar mass } C_2H_5OH} \\
&= \frac{\text{volume } C_2H_5OH \times \text{density } C_2H_5OH}{\text{molar mass } C_2H_5OH} \\
&= 10.0 \cancel{\text{mL } C_2H_5OH} \times \frac{0.7893 \cancel{\text{g } C_2H_5OH}}{1 \cancel{\text{mL } C_2H_5OH}} \times \frac{1 \text{ mol } C_2H_5OH}{46.07 \cancel{\text{g } C_2H_5OH}} \\
&= 0.171 \text{ mol } C_2H_5OH
\end{aligned}$$

$$\begin{aligned}
\text{moles } CH_3CO_2H &= \frac{\text{mass } CH_3CO_2H}{\text{molar mass } CH_3CO_2H} \\
&= \frac{\text{volume } CH_3CO_2H \times \text{density } CH_3CO_2H}{\text{molar mass } CH_3CO_2H} \\
&= 10.0 \cancel{\text{mL } CH_3CO_2H} \times \frac{1.0492 \cancel{\text{g } CH_3CO_2H}}{1 \cancel{\text{mL } CH_3CO_2H}} \times \frac{1 \text{ mol } CH_3CO_2H}{60.05 \cancel{\text{g } CH_3CO_2H}} \\
&= 0.175 \text{ mol } CH_3CO_2H
\end{aligned}$$

C The number of moles of acetic acid exceeds the number of moles of ethanol. Because the reactants both have coefficients of 1 in the balanced chemical equation, the mole ratio is 1:1. We have 0.171 mol of ethanol and 0.175 mol of acetic acid, so ethanol is the limiting reactant, and acetic acid is present in excess. The coefficient in the balanced chemical equation for the product (ethyl acetate) is also 1, so the mole ratio of ethanol and ethyl acetate is also 1:1. This means that given 0.171 mol of ethanol, the amount of ethyl acetate produced must also be 0.171 mol:

$$\begin{aligned}
\text{moles ethyl acetate} &= \text{mol ethanol} \times \frac{1 \text{ mol ethyl acetate}}{1 \text{ mol ethanol}} \\
&= 0.171 \cancel{\text{mol } C_2H_5OH} \times \frac{1 \text{ mol } CH_3CO_2C_2H_5}{1 \cancel{\text{mol } C_2H_5OH}} \\
&= 0.171 \text{ mol } CH_3CO_2C_2H_5
\end{aligned}$$

D The final step in our strategy is to determine the mass of ethyl acetate that can be formed, which we do by multiplying the number of moles by the molar mass:

$$\begin{aligned}
\text{mass of ethyl acetate} &= \text{mol ethyl acetate} \times \text{molar mass ethyl acetate} \\
&= 0.171 \cancel{\text{mol } CH_3CO_2C_2H_5} \times \frac{88.11 \text{ g } CH_3CO_2C_2H_5}{1 \cancel{\text{mol } CH_3CO_2C_2H_5}} \\
&= 15.1 \text{ g } CH_3CO_2C_2H_5
\end{aligned}$$

Thus, 15.1 g of ethyl acetate can be prepared in this reaction. If necessary, you could use the density of ethyl acetate (0.9003 g/cm^3) to determine the volume of ethyl acetate that could be produced:

$$\begin{aligned}
\text{volume of ethyl acetate} &= 15.1 \cancel{\text{g } CH_3CO_2C_2H_5} \times \frac{1 \text{ mL } CH_3CO_2C_2H_5}{0.9003 \cancel{\text{g } CH_3CO_2C_2H_5}} \\
&= 16.8 \text{ mL } CH_3CO_2C_2H_5
\end{aligned}$$

EXERCISE 3.12

Under appropriate conditions, the reaction of elemental phosphorus and elemental sulfur produces the compound P_4S_{10}. How much P_4S_{10} can be prepared starting with 10.0 g of P_4 and 30.0 g of S_8?

Answer 35.9 g of P_4S_{10}

Percent Yields

You have learned that when reactants are not present in stoichiometric quantities, the limiting reactant determines the maximum amount of product that can be formed from the reactants. The amount of product calculated in this way is the **theoretical yield**, the amount you would obtain if the reaction occurred perfectly and your method of purifying the product were 100% efficient.

In reality, you almost always obtain less product than is theoretically possible because of mechanical losses (such as spilling), separation procedures that are not 100% efficient, competing reactions that form undesired products, and reactions that simply do not go all the way to completion, resulting in a mixture of products and reactants. This last possibility is a common occurrence and is the subject of Chapter 15. Thus, the **actual yield**, the measured mass of products obtained from a reaction, is almost always less than the theoretical yield (often much less). The **percent yield** of a reaction is the ratio of the actual yield to the theoretical yield, multiplied by 100% to give a percentage:

$$\text{percent yield} = \frac{\text{actual yield (g)}}{\text{theoretical yield (g)}} \times 100\% \tag{3.23}$$

The method used to calculate the percent yield of a reaction is illustrated in Example 3.13.

EXAMPLE 3.13

Procaine is a key component of Novocain, an injectable local anesthetic used in dental work and minor surgery. Procaine can be prepared in the presence of H_2SO_4 (indicated above the arrow) by the reaction

$$\underset{p\text{-Aminobenzoic acid}}{C_7H_7NO_2} + \underset{\text{2-Diethylaminoethanol}}{C_6H_{15}NO} \xrightarrow{H_2SO_4} \underset{\text{Procaine}}{C_{13}H_{20}N_2O_2} + H_2O$$

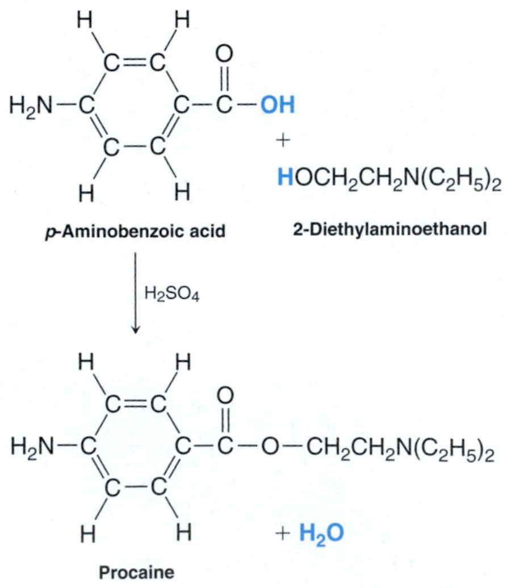

The reaction of *p*-aminobenzoic acid with 2-diethylaminoethanol to give procaine and water.

If we carried out this reaction using 10.0 g of *p*-aminobenzoic acid and 10.0 g of 2-diethylaminoethanol, and we isolated 15.7 g of procaine, what was the percent yield?

Given Masses of reactants and product

Asked for Percent yield

Strategy

A Write the balanced chemical equation.

B Convert from mass of reactants and product to moles using molar masses, and then use mole ratios to determine which is the limiting reactant. Based on the number of moles of the limiting reactant, use mole ratios to determine the theoretical yield.

C Calculate the percent yield by dividing the actual yield by the theoretical yield and multiplying by 100%.

Solution

A From the formulas given for the reactants and products, we see that the chemical equation is balanced as written. According to the equation, 1 mol of each reactant combines to give 1 mol of product plus 1 mol of water.

B To determine which reactant is limiting, we need to know their molar masses, which are calculated from their structural formulas: *p*-aminobenzoic acid, $C_7H_7NO_2$, 137.14 g/mol;

2-diethylaminoethanol, $C_6H_{15}NO$, 117.19 g/mol. Thus, the reaction used the following numbers of moles of reactants:

$$\text{moles } p\text{-aminobenzoic acid} = 10.0\ \cancel{\text{g}} \times \frac{1\ \text{mol}}{137.14\ \cancel{\text{g}}} = 0.0729\ \text{mol } p\text{-aminobenzoic acid}$$

$$\text{moles 2-diethylaminoethanol} = 10.0\ \cancel{\text{g}} \times \frac{1\ \text{mol}}{117.19\ \cancel{\text{g}}} = 0.0853\ \text{mol 2-diethylaminoethanol}$$

The reaction requires a 1:1 mole ratio of the two reactants, so *p*-aminobenzoic acid is the limiting reactant. Based on the coefficients in the balanced chemical equation, 1 mol of *p*-aminobenzoic acid yields 1 mol of procaine. We can therefore only obtain a maximum of 0.0729 mol of procaine. To calculate the corresponding mass of procaine, we use its structural formula, $C_{13}H_{20}N_2O_2$, to calculate its molar mass, 236.31 g/mol. Thus,

$$\text{theoretical yield of procaine} = 0.0729\ \cancel{\text{mol}} \times \frac{236.31\ \text{g}}{1\ \cancel{\text{mol}}} = 17.2\ \text{g}$$

The actual yield was only 15.7 g of procaine, so the percent yield was

$$\text{percent yield} = \frac{15.7\ \text{g}}{17.2\ \text{g}} \times 100\% = 91.3\%$$

(If the product were pure and dry, this yield would indicate that we have very good lab technique!)

EXERCISE 3.13

Lead was one of the earliest metals to be isolated in pure form. It occurs as concentrated deposits of a distinctive ore called galena (PbS), which is easily converted to lead oxide in 100% yield by roasting in air via the reaction $2PbS(s) + 3O_2(g) \longrightarrow 2PbO(s) + 2SO_2(g)$. The resulting PbO is then converted to the pure metal by reaction with charcoal. Because lead has such a low melting point (327°C), it runs out of the ore–charcoal mixture as a liquid that is easily collected. The reaction for the conversion of lead oxide to pure lead is

$$PbO(s) + C(s) \longrightarrow Pb(l) + CO(g)$$

If 93.3 kg of PbO is heated with excess charcoal and 77.3 kg of pure lead is obtained, what is the percent yield?

Answer 89.2%

Percent yields can range from 0% to 100%.* A 100% yield means that everything worked perfectly and you obtained all of the product that could have been produced. Anyone who has tried to do something as simple as fill a salt shaker or add oil to a car's engine without spilling knows how unlikely a 100% yield is. At the other extreme, a yield of 0% means that *no* product was obtained. A percent yield of 80–90% is usually considered good to excellent; a yield of 50% is only fair. In part because of the problems and costs of waste disposal, industrial production facilities face considerable pressures to optimize the yields of products and to make them as close to 100% as possible.

Crystalline galena (*top*) and a sample of lead (*bottom*). Pure lead is soft enough to be shaped easily with a hammer, unlike the brittle mineral galena (PbS), the main ore of lead.

3.5 • Classifying Chemical Reactions

The chemical reactions we have described are only a tiny sampling of the infinite number of chemical reactions possible. How do chemists cope with this overwhelming diversity? How do they predict which compounds will react with one another and what products will be formed? The key to success is to find useful ways to categorize

* In the laboratory, a student will occasionally obtain a yield that appears to be greater than 100%. This usually happens when the product is impure or is wet with a solvent such as water. If this is not the case, then the student *must* have made an error in weighing either the reactants or the products. The law of conservation of mass applies even to undergraduate chemistry laboratory experiments!

MGC Classifying Reactions

reactions. Familiarity with a few basic types of reactions will help you to predict the products that form when certain kinds of compounds or elements come in contact.

Most chemical reactions can be classified into one or more of only four basic types: *acid–base reactions, exchange reactions, condensation reactions* (and the reverse, *cleavage reactions*), and *oxidation–reduction reactions*. The general forms of these four kinds of reactions are summarized in Table 3.1, along with examples of each. It is important to note, however, that many reactions can be assigned to more than one classification, as you will see in our discussion. Remember that the classification is only for our convenience; the same reaction can be classified in different ways, depending on which of its characteristics is most important to us. Oxidation–reduction reactions, in which there is a net transfer of electrons from one atom to another, and condensation reactions are discussed below. Acid–base reactions and one kind of exchange reaction—the formation of an insoluble salt such as $BaSO_4$ when solutions of two soluble salts are mixed together—will be discussed in Chapter 4.

Oxidation–Reduction Reactions

The term **oxidation** was first used to describe reactions in which metals react with oxygen in air to produce metal oxides. When iron is exposed to air in the presence of water, for example, the iron turns to rust, an iron oxide. When exposed to air, aluminum metal develops a white powder on its surface, an aluminum oxide. In both cases, the metal acquires a positive charge by transferring electrons to the neutral oxygen atoms of an oxygen molecule. As a result, the oxygen atoms acquire a negative charge and form oxide ions (O^{2-}). Because the metals have lost electrons to oxygen, they have been *oxidized*; oxidation is therefore the loss of electrons. Conversely, because the oxygen atoms have gained electrons, they have been *reduced*; reduction is therefore the gain of electrons. The reaction of aluminum with oxygen to produce aluminum oxide is

$$4Al(s) + 3O_2(g) \longrightarrow 2Al_2O_3(s) \qquad (3.24)$$

For every oxidation there must be an associated reduction. Originally the term **reduction** referred to the decrease in mass observed when a metal oxide was heated with hydrogen, a reaction that was widely used to extract metals from their ores. When solid Cu_2O is heated with hydrogen, for example, its mass decreases because the formation of pure copper is accompanied by the loss of oxygen atoms as a volatile product (water). The reaction is

$$Cu_2O(s) + H_2(g) \longrightarrow 2Cu(s) + H_2O(g) \qquad (3.25)$$

Note the pattern

Any oxidation must be accompanied by a reduction, and vice versa.

Oxidation and reduction reactions are now characterized by a change in the oxidation states of one or more elements in the reactants. The **oxidation state** of each atom in a compound is the charge that atom would have if all of its bonding electrons

TABLE 3.1 Basic types of chemical reactions

Name of Reaction	General Form	Example
Oxidation–reduction (redox)	oxidant + reductant $\longrightarrow$ reduced oxidant + oxidized reductant	$C_7H_{16}(l) + 11O_2(g) \longrightarrow 7CO_2(g) + 8H_2O(g)$
Acid–base	acid + base $\longrightarrow$ salt	$NH_3(aq) + HNO_3(aq) \longrightarrow NH_4^+(aq) + NO_3^-(aq)$
Exchange	$AB + C \longrightarrow AC + B$ $AB + CD \longrightarrow AD + CB$	$CH_3Cl + OH^- \longrightarrow CH_3OH + Cl^-$ $BaCl_2(aq) + Na_2SO_4(aq) \longrightarrow BaSO_4(s) + 2NaCl(aq)$
Condensation	$A + B \longrightarrow AB$	$HBr + H_2C{=}CH_2 \longrightarrow CH_3CHBr$ $CO_2(g) + H_2O(l) \longrightarrow H_2CO_3(aq)$
Cleavage	$AB \longrightarrow A + B$	$CaCO_3(s) \longrightarrow CaO(s) + CO_2(g)$ $CH_3CH_2Cl \longrightarrow H_2C{=}CH_2 + HCl$

were transferred to the atom with the greater attraction for electrons. Atoms in their elemental form, such as O_2 or H_2, are assigned an oxidation state of zero. In Equation 3.24, each neutral oxygen atom gains two electrons and becomes negatively charged, forming an oxide ion; thus, oxygen has an oxidation state of −2 in the product so it has been reduced. Each neutral aluminum atom loses three electrons to produce an aluminum ion with an oxidation state of +3 in the product so aluminum has been oxidized. Thus, in the formation of Al_2O_3, electrons are transferred as follows (a superscript 0 emphasizes the oxidation state of the elements):

$$4Al^0 + 3O_2{}^0 \longrightarrow 4Al^{3+} + 6O^{2-} \quad (3.26)$$

Equation 3.24 is an example of an **oxidation–reduction reaction**, often called a **redox reaction**. In redox reactions there is a net transfer of electrons from one reactant to another. In any redox reaction, *the total number of electrons lost must equal the total number of electrons gained*. In Equation 3.26, for example, the total number of electrons lost by aluminum is equal to the total number gained by oxygen:

$$\text{electrons lost} = 4\ \cancel{\text{Al atoms}} \times \frac{3\ e^-\ \text{lost}}{\cancel{\text{Al atom}}} = 12\ e^-\ \text{lost}$$
$$\text{electrons gained} = 6\ \cancel{\text{O atoms}} \times \frac{2\ e^-\ \text{gained}}{\cancel{\text{O atom}}} = 12\ e^-\ \text{gained} \quad (3.27)$$

The same pattern is seen in all oxidation–reduction reactions: the number of electrons lost must equal the number of electrons gained.

Note the pattern

In all oxidation–reduction (redox) reactions, the number of electrons lost equals the number of electrons gained.

Assigning Oxidation States

Assigning oxidation states to the elements in binary ionic compounds is straightforward: the oxidation states of the elements are identical to the charges on the monatomic ions. In Chapter 2, you learned how to predict the formulas of simple ionic compounds based on the sign and magnitude of the charge on monatomic ions formed by the neutral elements. Examples of such compounds are NaCl (Figure 3.13), MgO, and $CaCl_2$. In covalent compounds, in contrast, electrons are shared by atoms. Oxidation states in covalent compounds are somewhat arbitrary, but they are useful bookkeeping devices to help you understand and predict many reactions.

A set of rules for assigning oxidation states to atoms in chemical compounds is shown in the margin. As we discuss atomic and molecular structure in Chapters 6–9, the principles underlying these rules will be described more fully.

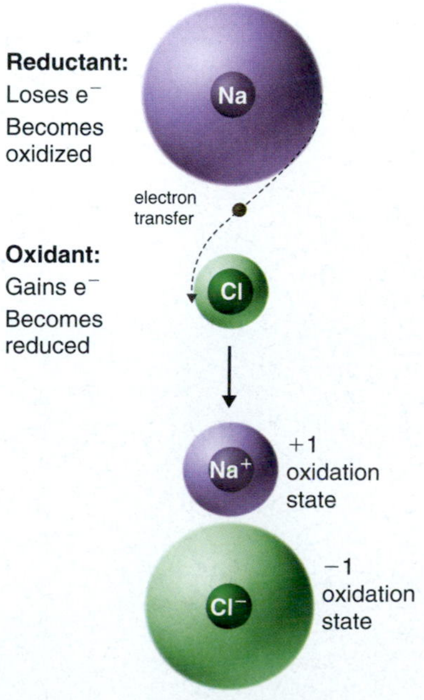

Figure 3.13 The reaction of a neutral sodium atom with a neutral chlorine atom. The result is the transfer of one electron from Na to Cl, forming the ionic compound NaCl.

Rules for assigning oxidation states[a]

1. The sum of the oxidation states of all the atoms in a neutral molecule or ion must equal the charge on the molecule or ion.
2. The oxidation state of an atom in any pure element, whether monatomic, diatomic, or polyatomic, is zero.
3. The oxidation state of a monatomic ion is the same as its charge; for example: $Na^+ = +1$, $Cl^- = -1$.
4. The oxidation state of fluorine in chemical compounds is always −1. Other halogens usually have oxidation states of −1 as well, except when combined with oxygen or other halogens.
5. Hydrogen is assigned an oxidation state of +1 in its compounds with nonmetals and −1 in its compounds with metals.
6. Oxygen is normally assigned an oxidation state of −2 in compounds, with two exceptions: in compounds that contain O—F or O—O bonds, the oxidation state of oxygen is determined by the oxidation states of the other elements present.

[a]Nonintegral oxidation states are encountered occasionally, they are usually due to the presence of two or more atoms of the same element with different oxidation states.

In any chemical reaction, the net charge must be conserved; that is, in a chemical reaction, the total number of electrons is constant, just like the total number of atoms. Consistent with this, rule 1 states that the sum of the individual oxidation states of the atoms in a molecule or ion must equal the net charge on that molecule or ion. In NaCl, for example, Na has an oxidation state of +1 and Cl is −1. The net charge is zero, as it must be for any compound.

Rule 4 is required because fluorine attracts electrons more strongly than any other element, for reasons you will discover in Chapter 6. Hence, fluorine provides a reference for calculating the oxidation states of other atoms in chemical compounds. Rule 5 reflects the difference in chemistry observed for compounds of hydrogen with nonmetals (such as chlorine) as opposed to compounds of hydrogen with metals (such as sodium). For example, NaH contains the H^- ion, whereas HCl forms H^+ and Cl^- ions when dissolved in water. Rule 6 is necessary because fluorine has a greater attraction for electrons than oxygen does; this rule also prevents violations of rule 1. Thus, the oxidation state of oxygen is +2 in OF_2 but $-\frac{1}{2}$ in KO_2. Note that an oxidation state of $-\frac{1}{2}$ for O in KO_2 is perfectly acceptable.

The reduction of copper(I) oxide shown in Equation 3.25 demonstrates how to apply these rules. Rule 2 states that atoms in their elemental form have an oxidation state of zero, which applies to H_2 and Cu. From rule 5, hydrogen in H_2O has an oxidation state of +1, and from rule 6, oxygen in both Cu_2O and H_2O has an oxidation state of −2. Rule 1 states that the sum of the oxidation states in a compound must equal the net charge on that compound. This means that each Cu atom in Cu_2O must have a charge of +1: 2(+1) + (−2) = 0. Thus, the oxidation states are

$$\underset{+1\ -2}{Cu_2O(s)} + \underset{0}{H_2(g)} \longrightarrow \underset{0}{2Cu(s)} + \underset{+1\ -2}{H_2O(g)} \tag{3.28}$$

Assigning oxidation states allows us to see that there has been a net transfer of electrons from hydrogen (0 ⟶ +1) to copper (+1 ⟶ 0). Thus, this is a redox reaction. Once again, the number of electrons lost equals the number of electrons gained, and there is a net conservation of charge:

$$\begin{aligned} \text{electrons lost} &= 2\ \cancel{\text{H atoms}} \times \frac{1\ e^-\ \text{lost}}{\cancel{\text{H atom}}} = 2\ e^-\ \text{lost} \\ \text{electrons gained} &= 2\ \cancel{\text{Cu atoms}} \times \frac{1\ e^-\ \text{gained}}{\cancel{\text{Cu atom}}} = 2\ e^-\ \text{gained} \end{aligned} \tag{3.29}$$

Remember that oxidation states are useful for visualizing the transfer of electrons in oxidation–reduction reactions, but the oxidation state of an atom and its actual charge are the same for only simple ionic compounds. Oxidation states are only a convenient way of assigning electrons to atoms, but they are useful for predicting the types of reactions that substances undergo.

EXAMPLE 3.14

Assign oxidation states to all the atoms in each compound: **(a)** SF_6; **(b)** CH_3OH; **(c)** $(NH_4)_2SO_4$; **(d)** Fe_3O_4; **(e)** CH_3CO_2H.

Given Molecular or empirical formula

Asked for Oxidation states

Strategy

Begin with atoms whose oxidation states can be determined unambiguously from the rules presented (such as F, other halogens, O, and monatomic ions). Determine the oxidation states of other atoms present using rule 1.

Solution

(a) We know from rule 4 that fluorine always has an oxidation state of -1 in its compounds. The six F atoms in sulfur hexafluoride give a total negative charge of -6. Because rule 1 requires that the sum of the oxidation states of all the atoms be zero in a neutral SF_6 molecule, the oxidation state of sulfur must be $+6$: $[(6 \text{ F atoms})(-1)] + [(1 \text{ S atom})(+6)] = 0$.

(b) According to rules 5 and 6, hydrogen and oxygen have oxidation states of $+1$ and -2, respectively. Because methanol has no net charge, carbon must have an oxidation state of -2 as well: $[(4 \text{ H atoms})(+1)] + [(1 \text{ O atom})(-2)] + [(1 \text{ C atom})(-2)] = 0$.

(c) Note that $(NH_4)_2SO_4$ is an ionic compound that consists of both a polyatomic cation (NH_4^+) and a polyatomic anion (SO_4^{2-}) (see Table 2.4). We assign oxidation states to the atoms in each polyatomic ion separately. For NH_4^+, hydrogen has an oxidation state of $+1$ (rule 5), so nitrogen must have an oxidation state of -3: $[(4 \text{ H atoms})(+1)] + [(1 \text{ N atom})(-3)] = +1$, the charge on the NH_4^+ ion. For SO_4^{2-}, oxygen has an oxidation state of -2 (rule 6), so sulfur must have an oxidation state of $+6$: $[(4 \text{ O atoms})(-2)] + [(1 \text{ S atom})(+6)] = -2$, the charge on the sulfate ion.

(d) Oxygen has an oxidation state of -2 (rule 6), giving an overall charge of -8 per formula unit. This must be balanced by the positive charge on three iron atoms, giving an oxidation state of $+\frac{8}{3}$ for iron: $\left[(4 \text{ O atoms})(-2)\right] + \left[\left(3 \text{ Fe atoms}\right)\left(+\frac{8}{3}\right)\right] = 0$. Fractional oxidation states are allowed because oxidation states are a somewhat arbitrary way of keeping track of electrons. In fact, Fe_3O_4 can be viewed as having two Fe^{3+} ions and one Fe^{2+} ion per formula unit, giving a net positive charge of $+8$ per formula unit. Fe_3O_4 is a magnetic iron ore commonly called magnetite. In ancient times magnetite was known as lodestone because it could be used to make primitive compasses that pointed toward Polaris (the North Star), which was called the "lodestar."

(e) The compound CH_3CO_2H is acetic acid. Initially, we assign oxidation states to its components in the same way as any other compound. Hydrogen and oxygen have oxidation states of $+1$ and -2 (rules 5 and 6, respectively), resulting in a total charge of $[(4 \text{ H atoms})(+1)] + [(2 \text{ O atoms})(-2)] = 0$ for H and O together. Hence, the oxidation state of carbon must also be zero (rule 1). This is, however, an *average* oxidation state for the two carbon atoms present. Because each carbon atom has a different set of atoms bonded to it, they are likely to have different oxidation states. To determine the oxidation states of the individual carbon atoms, we use the same rules as before but with the additional assumption that bonds between atoms of the same element do not affect the oxidation states of those atoms. The carbon atom of the methyl group (—CH_3) is bonded to three H atoms and one C atom. We know from rule 5 that H has an oxidation state of $+1$, and we have just said that the C—C bond can be ignored in calculating the oxidation state of the carbon atom. For the methyl group to be electrically neutral, its carbon atom must have an oxidation state of -3. Similarly, the carbon atom of the carboxylic acid group, —CO_2H, is bonded to one carbon atom and two oxygen atoms. Again ignoring the bonded carbon atom, we assign oxidation states of -2 and $+1$ to the O and H atoms, respectively, leading to a net charge of $[(2 \text{ O atoms})(-2)] + [(1 \text{ H atom})(+1)] = -3$. To obtain an electrically neutral carboxylic acid group, the charge on this carbon must be $+3$. The oxidation states of the individual atoms in acetic acid are thus

$$\overset{+1}{\underset{-3}{C}}\,\overset{}{H_3}\,\overset{+3}{C}\,\underset{-2}{O_2}\,\overset{+1}{H}$$

Note that the sum of the oxidation states of the two carbon atoms is indeed zero.

EXERCISE 3.14

Assign oxidation states to the atoms in each compound: **(a)** BaF_2; **(b)** CH_2O; **(c)** $K_2Cr_2O_7$; **(d)** CsO_2; **(e)** CH_3CH_2OH.

Answer **(a)** Ba, $+2$; F, -1; **(b)** C, 0; H, $+1$; O, -2; **(c)** K, $+1$; Cr, $+6$; O, -2; **(d)** Cs, $+1$; O, $-\frac{1}{2}$; **(e)** C, -3; H, $+1$; C, -1; H, $+1$; O, -2; H, $+1$

Oxidants and Reductants

Compounds that are capable of accepting electrons, such as O_2 or F_2, are called **oxidants**, or *oxidizing agents*, because they can oxidize other compounds. *In the process of accepting electrons, an oxidant is reduced.* Compounds that are capable of donating electrons, such as sodium metal or cyclohexane, C_6H_{12}, are called **reductants**, or *reducing agents*, because they can cause the reduction of another compound. *In the process of donating electrons, a reductant is oxidized.* These relationships are summarized in Equation 3.30:

$$\text{oxidant} + \text{reductant} \longrightarrow \text{oxidation} - \text{reduction} \qquad (3.30)$$

$$O_2(g) + 4Na(s) \longrightarrow 2Na_2O(s)$$

$$\underset{\substack{\text{Gains e}^-\\\text{(is reduced)}}}{9O_2(g)} + \underset{\substack{\text{Loses e}^-\\\text{(is oxidized)}}}{C_6H_{12}(l)} \longrightarrow \underset{\text{redox reaction}}{6CO_2(g) + 6H_2O(l)}$$

Some oxidants have a greater ability than others to remove electrons from other compounds. Thus, oxidants can range from very powerful, capable of oxidizing most compounds with which they come in contact, to rather weak. Both F_2 and Cl_2 are powerful oxidants: for example, F_2 will oxidize H_2O in a vigorous, potentially explosive reaction. In contrast, S_8 is a rather weak oxidant, and O_2 falls somewhere in between. Conversely, reductants vary in their tendency to donate electrons to other compounds. Reductants can also range from very powerful, capable of giving up electrons to almost anything, to weak. The alkali metals are powerful reductants, so they must be kept away from atmospheric oxygen to avoid a potentially hazardous redox reaction.

Combustion analysis was introduced in Section 3.3. A **combustion reaction** is an oxidation–reduction reaction in which the oxidant is O_2. Consider, for example, the combustion of cyclohexane, a typical hydrocarbon, in excess oxygen. The balanced chemical equation for the reaction, with the oxidation state shown for each atom, is

$$\overset{+1}{\underset{-2}{C_6H_{12}}} + \overset{0}{9O_2} \longrightarrow \overset{+4}{\underset{-2}{6CO_2}} + \overset{+1}{\underset{-2}{6H_2O}} \qquad (3.31)$$

If we compare the oxidation state of each element in the products and reactants, we see that hydrogen is the only element whose oxidation state does not change; it remains $+1$. Carbon, however, has an oxidation state of -2 in cyclohexane and $+4$ in CO_2; that is, each carbon atom loses six electrons during the reaction. Oxygen has an oxidation state of 0 in the reactants, but it gains electrons to have an oxidation state of -2 in CO_2 and H_2O. Because carbon has been oxidized, cyclohexane is the reductant; because oxygen has been reduced, it is the oxidant. All combustion reactions are therefore oxidation–reduction reactions.

Condensation Reactions

The reaction of bromine with ethylene to give 1,2-dibromoethane, used in agriculture to kill nematodes in the soil, is

$$C_2H_4(g) + Br_2(g) \longrightarrow BrCH_2CH_2Br(g) \qquad (3.32)$$

According to Table 3.1, this is a condensation reaction because it has the general form $A + B \longrightarrow AB$. This reaction, however, can also be viewed as an oxidation–reduction reaction, in which electrons are transferred from carbon ($-2 \longrightarrow -1$) to bromine ($0 \longrightarrow -1$). Another example of such a reaction is the one used for the industrial synthesis of ammonia:

$$3H_2(g) + N_2(g) \longrightarrow 2NH_3(g) \qquad (3.33)$$

This reaction also has the general form of a condensation reaction. Notice, however, that hydrogen has been oxidized ($0 \longrightarrow +1$) and nitrogen has been reduced ($0 \longrightarrow -3$), so it can also be classified as an oxidation–reduction reaction.

Not all condensation reactions are redox reactions. As shown in the margin, for example, the reaction of an amine with a carboxylic acid is a variant of a condensation reaction in which —OH from the carboxylic acid group and —H from the amine group are eliminated as H_2O. The reaction forms an *amide bond* (also called a *peptide bond*), which links the two fragments. Amide bonds are the essential structural unit linking the building blocks of proteins and many polymers together, as described in Chapter 12.

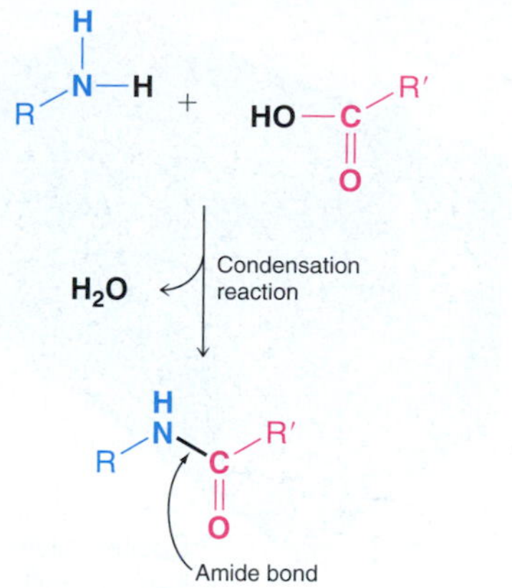

The reaction of an amine with a carboxylic acid proceeds by eliminating water and forming a new C–N (amide) bond.

EXAMPLE 3.15

The following reactions all have important industrial applications. Refer to Table 3.1 and classify each as an oxidation–reduction, acid–base, exchange, or condensation/cleavage reaction. For each redox reaction, identify the oxidant and reductant and specify which atoms are oxidized or reduced. (*Note*: Don't forget that some reactions can be placed into more than one category!): **(a)** $C_2H_4(g) + Cl_2(g) \longrightarrow ClCH_2CH_2Cl(g)$; **(b)** $AgNO_3(aq) + NaCl(aq) \longrightarrow AgCl(s) + NaNO_3(aq)$; **(c)** $CaCO_3(s) \longrightarrow CaO(s) + CO_2(g)$; **(d)** $Ca_5(PO_4)_3(OH)(s) + 7H_3PO_4(aq) + 4H_2O(l) \longrightarrow 5Ca(H_2PO_4)_2{\cdot}H_2O(s)$; **(e)** $Pb(s) + PbO_2(s) + 2H_2SO_4(aq) \longrightarrow 2PbSO_4(s) + 2H_2O(l)$.

Given Balanced chemical equation

Asked for Classification of chemical reaction

Strategy

A Determine the general form of the equation by referring to Table 3.1, and then classify the reaction.

B For redox reactions, assign oxidation states to all of the atoms present in the reactants and products. If the oxidation state of one or more atoms changes, then the reaction is a redox reaction. If not, the reaction must be one of the other types of reaction listed in Table 3.1.

Solution

(a) **A** This reaction is used to prepare 1,2-dichloroethane, number 14 of the top 25 industrial chemicals listed in Table 2.12. It has the general form A + B ⟶ AB, which is typical of a condensation reaction. **B** Because reactions may fit into more than one category, we need to look at the oxidation states of the atoms:

+1 0 −1 +1

$$C_2H_4 + Cl_2 \longrightarrow ClCH_2CH_2Cl$$

−2 −1 +1 −1 −1

The oxidation states show that chlorine is reduced from 0 to −1 and carbon is oxidized from −2 to −1, so this is a redox reaction as well as a condensation reaction. Ethylene is the reductant, and chlorine is the oxidant.

(b) **A** This reaction is used to prepare silver chloride for making photographic film. The equation has the general form AB + CD ⟶ AD + CB, so it is classified as an exchange reaction. **B** The oxidation states of the atoms are

+1 +5 +1 +1 +1 +5

$$AgNO_3 + NaCl \longrightarrow AgCl + NaNO_3$$

−2 −1 −1 −2

There is no change in the oxidation states, so this is not a redox reaction.

(c) **A** This reaction is used to prepare lime (CaO) from limestone ($CaCO_3$). The reaction has the general form AB ⟶ A + B. The equation's general form indicates that it can be classified as a cleavage reaction. **B** The oxidation states of the atoms are

+2 +4 +2 +4

$$CaCO_3 \longrightarrow CaO + CO_2$$

−2 −2 −2

Because the oxidation states of all the atoms are the same in the products and reactant, this is *not* a redox reaction.

(d) **A** This reaction is used to prepare "super triple phosphate" in fertilizer. One of the reactants is phosphoric acid, which transfers a proton (H^+) to the phosphate and hydroxide ions of hydroxyapatite, $Ca_5(PO_4)_3(OH)$, to form $H_2PO_4^-$ and H_2O, respectively. Thus, this is an acid–base reaction, in which H_3PO_4 is the acid (H^+ donor) and $Ca_5(PO_4)_3(OH)$ is the base (H^+ acceptor).

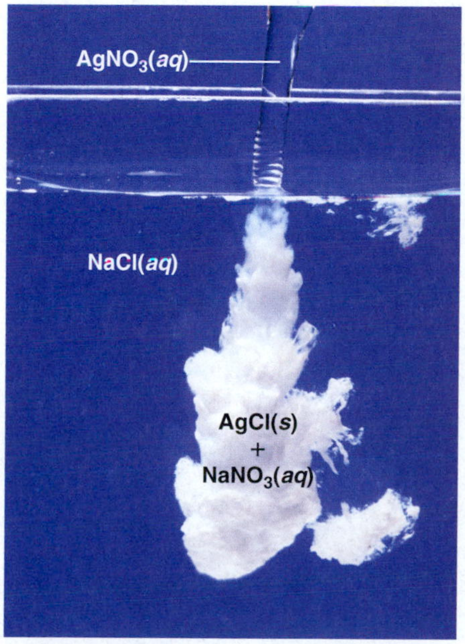

$AgCl(s)$ precipitates when solutions of $AgNO_3(aq)$ and $NaCl(aq)$ are mixed. $NaNO_3(aq)$ is in solution as Na^+ and NO_3^- ions.

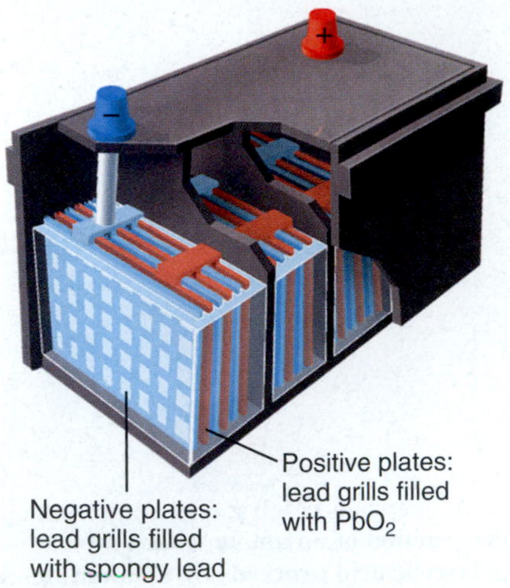

Schematic drawing of a 12-volt car battery. The locations of the reactants (lead metal in a spongy form with large surface area) and PbO_2 are shown. The product, $PbSO_4$, forms as a white solid between the plates.

To determine whether it is a redox reaction, we assign oxidation states to the atoms:

$$\overset{+2}{Ca}_5(\overset{+5}{P}\underset{-2}{O}_4)_3(\underset{-2}{O}\overset{+1}{H}) + 7\overset{+1}{H}_3\overset{+5}{P}\underset{-2}{O}_4 + 4\overset{+1}{H}_2\underset{-2}{O} \longrightarrow 5\overset{+2}{Ca}(\overset{+1}{H}_2\overset{+5}{P}\underset{-2}{O}_4)_2\cdot\overset{+1}{H}_2\underset{-2}{O}$$

Because there is no change in oxidation state, this is *not* a redox reaction.

(e) This reaction occurs in a conventional car battery every time the engine is started. An acid (H_2SO_4) is present and transfers protons to oxygen atoms of PbO_2 to form water during the reaction. The reaction can therefore be described as an acid–base reaction. The oxidation states are

$$\underset{0}{Pb} + \overset{+4}{Pb}\underset{-2}{O}_2 + 2\overset{+1}{H}_2\overset{+6}{S}\underset{-2}{O}_4 \longrightarrow 2\overset{+2}{Pb}\overset{+6}{S}\underset{-2}{O}_4 + 2\overset{+1}{H}_2\underset{-2}{O}$$

The oxidation state of lead changes from 0 in Pb and +4 in PbO_2 (both reactants) to +2 in $PbSO_4$. Thus, this is also a redox reaction, in which elemental lead is the reductant and PbO_2 is the oxidant. Which description is correct? Both.

EXERCISE 3.15

Using Table 3.1, classify each reaction as an oxidation–reduction, acid–base, exchange, or condensation/cleavage reaction. For each oxidation–reduction reaction, identify the oxidant and reductant and the atoms that are oxidized or reduced. **(a)** $Al(s) + OH^-(aq) + 3H_2O(l) \longrightarrow \frac{3}{2}H_2(g) + [Al(OH)_4]^-(aq)$; **(b)** $TiCl_4(l) + 2Mg(l) \longrightarrow Ti(s) + 2MgCl_2(l)$; **(c)** $MgCl_2(aq) + Na_2CO_3(aq) \longrightarrow MgCO_3(s) + 2NaCl(aq)$; **(d)** $CO(g) + Cl_2(g) \longrightarrow Cl_2CO(l)$; and **(e)** $H_2SO_4(l) + 2NH_3(g) \longrightarrow (NH_4)_2SO_4(s)$.

Answer **(a)** Redox reaction; reductant is Al, oxidant is H_2O; Al is oxidized, H is reduced. This is the reaction that occurs when Drano is used to clear a clogged drain. **(b)** Redox reaction; reductant is Mg, oxidant is $TiCl_4$; Mg is oxidized, Ti is reduced. **(c)** Exchange reaction. This reaction is responsible for the "scale" that develops in coffee makers in areas that have "hard water." **(d)** Both a condensation reaction and a redox reaction; reductant is CO, oxidant is Cl_2; C is oxidized, Cl is reduced. The product of this reaction is phosgene, a highly toxic gas used as a chemical weapon in World War I. Phosgene is now used to prepare polyurethanes, which are used in foams for bedding and furniture and in a variety of coatings. **(e)** Acid–base reaction.

Catalysts

Many chemical reactions, including some of those discussed above, occur more rapidly in the presence of a **catalyst**, a substance that participates in a reaction and cause it to occur more rapidly, but that can be recovered unchanged at the end of the reaction and reused. Because catalysts are not involved in the stoichiometry of the reaction, they are usually shown above the arrow in a net chemical equation. Chemical processes in industry rely heavily on the use of catalysts, which are usually added to a reaction mixture in trace amounts, and most biological reactions do not take place without a biological catalyst or **enzyme**. Examples of catalyzed reactions in industry are the use of platinum in petroleum cracking and reforming, the reaction of SO_2 and O_2 in the presence of V_2O_5 to produce SO_3 in the industrial synthesis of sulfuric acid, and the use of sulfuric acid in the synthesis of compounds such as ethyl acetate and procaine. Not only do catalysts greatly increase the rates of reactions but in some cases such as in petroleum refining, they also control which products are formed. The acceleration of a reaction by a catalyst is called **catalysis**.

Catalysts may be classified as homogeneous or heterogeneous. A **homogeneous catalyst** is uniformly dispersed throughout the reactant mixture to form a solution. Sulfuric acid, for example, is a homogeneous catalyst used in the synthesis of esters such as procaine (Example 3.13). An ester, whose general structure is shown in the margin, has a structure similar to that of a carboxylic acid, in which the H atom attached to oxygen has

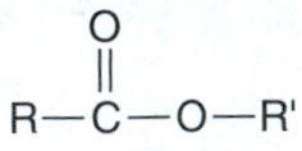

An ester

been replaced by an R group. They are responsible for the fragrances of many fruits, flowers, and perfumes. Other examples of homogeneous catalysts are the enzymes that allow our bodies to function. In contrast, a **heterogeneous catalyst** is in a different physical state than the reactants. For economic reasons, most industrial processes use heterogeneous catalysts in the form of solids that are added to solutions of the reactants. Because such catalysts often contain expensive precious metals such as platinum or palladium, it makes sense to formulate them as solids that can be easily separated from the liquid or gaseous reactant–product mixture and recovered. Examples of heterogeneous catalysts are the iron oxides used in the industrial synthesis of ammonia and the catalytic converters found on virtually all modern automobiles, which contain precious metals like palladium and rhodium. Catalysis will be discussed in more detail in Chapter 14 when we discuss reaction rates, but you will encounter the term frequently throughout the text.

A heterogeneous catalyst. This large circular gauze, woven from rhodium–platinum wire, is used as a heterogeneous catalyst in the commercial production of nitric acid by the oxidation of ammonia.

3.6 ∘ Chemical Reactions in the Atmosphere

Section 3.5 described different classes of chemical reactions. Of the many different chemical reactions that occur in the earth's atmosphere, some are important and controversial because they affect our quality of life and our health. The atmospheric reactions presented in this section provide examples of the various classes of reaction introduced in this chapter that are implicated in the destruction of earth's protective ozone layer.

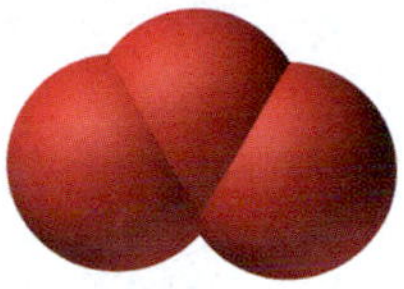

Ozone, O_3

Each year since the mid-1970s, scientists have noted a disappearance of approximately 70% of the ozone (O_3) layer above Antarctica during the Antarctic spring, creating what is commonly known as the "ozone hole." **Ozone** is an unstable form of oxygen that consists of three oxygen atoms bonded together. In September 2004, the Antarctic ozone hole reached 24.2 million square kilometers (9.3 million square miles), about the size of North America. The largest area ever recorded was in the year 2000, when the hole measured 29.2 million square kilometers and for the first time extended over a populated area—the city of Punta Arenas, Chile (population 120,000) (Figure 3.14). A less extensive zone of depletion has been detected over the Arctic as well. Years of study from the ground, from the air, and from satellites in space have shown that chlorine from industrial chemicals used in spray cans, foam packaging, and refrigeration materials is largely responsible for the catalytic depletion of ozone through a series of condensation, cleavage, and oxidation–reduction reactions.

MGC Catalysts

The Earth's Atmosphere and the Ozone Layer

The earth's atmosphere at sea level is an approximately 80:20 solution of nitrogen and oxygen gases, with small amounts of carbon dioxide, water vapor, and the noble

Figure 3.14 Satellite photos of earth reveal the sizes of the Antarctic ozone hole over time. Dark blue colors correspond to the thinnest ozone; light blue, green, yellow, orange, and red indicate progressively thicker ozone. In September 2000, the Antarctic ozone hole briefly approached a record 30 million square kilometers.

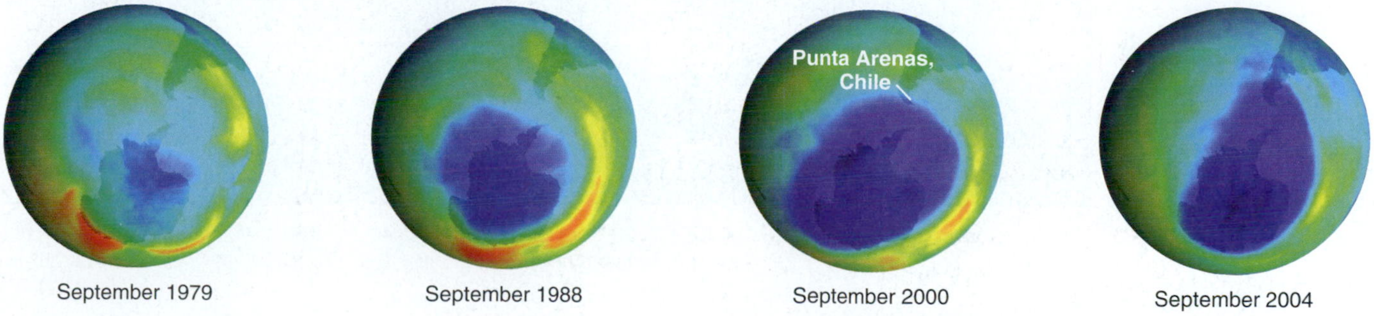

TABLE 3.2 The composition of the earth's atmosphere at sea level[a]

Gas	Formula	Percent by Volume
Nitrogen	N_2	78.084
Oxygen	O_2	20.948
Argon	Ar	0.934
Carbon dioxide[b]	CO_2	0.0314
Neon	Ne	0.00182
Helium	He	0.000524
Krypton	Kr	0.000114
Methane	CH_4	0.0002
Hydrogen	H_2	0.00005
Nitrous oxide	N_2O	0.00005
Xenon	Xe	0.0000087

[a]In addition, air contains as much as 7% water vapor, H_2O; 0.0001% sulfur dioxide, SO_2; 0.00007% ozone, O_3; 0.000002% carbon monoxide, CO; and 0.000002% nitrogen dioxide, NO_2.
[b]Carbon dioxide levels are highly variable; the typical range is 0.01–0.1%.

gases, and trace amounts of a variety of other compounds (Table 3.2). A key feature of the atmosphere is that its composition, temperature, and pressure vary dramatically with altitude. Consequently, scientists have divided the atmosphere into distinct layers, which interact differently with the continuous flux of solar radiation from the top and the land and ocean masses at the bottom. Some of the characteristic features of the layers of the atmosphere are illustrated in Figure 3.15.

The *troposphere* is the lowest layer of the atmosphere, extending from earth's surface to an altitude of about 11–13 km (7–8 miles). Above the troposphere lies the *stratosphere*, which extends from an altitude of 13 km (8 miles) to about 44 km (27 miles). As shown in Figure 3.15, the temperature of the troposphere decreases steadily with increasing altitude. Because "hot air rises," this temperature gradient leads to continuous mixing of the upper and lower regions *within* the layer. The thermally induced turbulence in the troposphere produces fluctuations in temperature and precipitation that we collectively refer to as "weather." In contrast, mixing *between* the layers of the atmosphere occurs relatively slowly, and as a result each layer has distinctive chemistry. We focus our attention on the stratosphere, which contains the highest concentration of ozone.

The sun's radiation is the major source of energy that initiates chemical reactions in the atmosphere. The sun emits many kinds of radiation, including **visible light**, which is radiation that the human eye can detect, and **ultraviolet light**, which is higher energy radiation that cannot be detected by the human eye. Ultraviolet light has a higher energy than visible light and can cause a wide variety of chemical reactions that are harmful to organisms. For example, ultraviolet light is used to sterilize items, and, as anyone who has ever suffered a severe sunburn knows, it can produce extensive tissue damage.

Light in the higher energy ultraviolet range is almost totally absorbed by oxygen molecules in the upper layers of the atmosphere, causing the O_2 molecules to dissociate into two oxygen atoms in a cleavage reaction:

$$O_2(g) \xrightarrow{\text{light}} 2O(g) \qquad (3.34)$$

Figure 3.15 Variation of temperature with altitude in the earth's atmosphere. Note the important chemical species present in each layer.

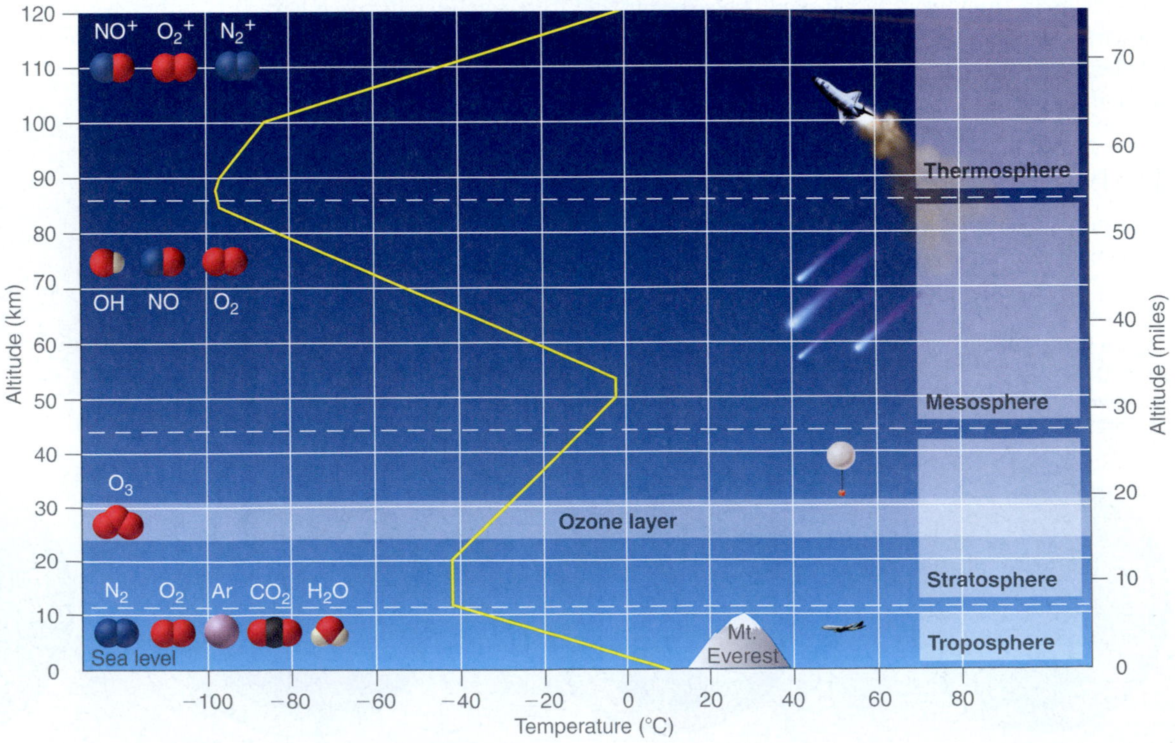

In Equation 3.34, "light" is written above the arrow to indicate that light is required for the reaction to occur. The oxygen atoms produced in Equation 3.34 can undergo a condensation reaction with O_2 molecules to form ozone:

$$O(g) + O_2(g) \longrightarrow O_3(g) \qquad (3.35)$$

Ozone is responsible for the pungent smell we associate with lightning discharges and electric motors. It is also toxic.

In the stratosphere, the ozone produced via Equation 3.35 has a major beneficial effect. Ozone absorbs the less-energetic range of ultraviolet light, undergoing a cleavage reaction in the process to give O_2 and O:

$$O_3(g) \xrightarrow{\text{light}} O_2(g) + O(g) \qquad (3.36)$$

The formation of ozone (Equation 3.35) and its decomposition (Equation 3.36) are normally in balance, resulting in essentially constant levels of about 10^{15} ozone molecules per liter in the stratosphere. This so-called **ozone layer** acts as a protective screen that absorbs ultraviolet light that would otherwise reach the surface of the earth.

In 1974, F. Sherwood Rowland and Michael Molina published a paper claiming that commonly used *chlorofluorocarbon compounds (CFCs)* were causing major damage to the ozone layer (Table 3.3). Chlorofluorocarbons had been used as refrigerants and propellants in aerosol cans for many years, releasing millions of tons of CFC molecules into the atmosphere. Because CFCs are volatile compounds that do not readily undergo chemical reactions, they persist in the atmosphere long enough to be carried to the top of the troposphere, where they eventually enter the stratosphere. There they are exposed to intense ultraviolet light and undergo a cleavage reaction to produce a chlorine atom, as shown below for Freon-11:

$$CCl_3F(g) \xrightarrow{\text{light}} CCl_2F(g) + Cl(g) \qquad (3.37)$$

The resulting chlorine atoms act as a homogeneous catalyst in two redox reactions (Equations 3.38 and 3.39):

$$Cl(g) + O_3(g) \longrightarrow ClO(g) + O_2(g) \qquad (3.38)$$
$$ClO(g) + O(g) \longrightarrow Cl(g) + O_2(g) \qquad (3.39)$$

Adding the two reactions in Equations 3.38 and 3.39 gives

$$Cl(g) + ClO(g) + O_3(g) + O(g) \longrightarrow 2O_2(g) + Cl(g) + ClO(g) \qquad (3.40)$$

Because Cl and ClO appear on both sides of the equation, they can be canceled to give the following net reaction:

$$O_3(g) + O(g) \longrightarrow 2O_2(g) \qquad (3.41)$$

TABLE 3.3 Common chlorofluorocarbons (CFCs) and related compounds

Name	Molecular Formula	Industrial Name
Trichlorofluoromethane	CCl_3F	CFC-11 (Freon-11)
Dichlorodifluoromethane	CCl_2F_2	CFC-12 (Freon-12)
Chlorotrifluoromethane	$CClF_3$	CFC-13 (Freon-13)
Bromotrifluoromethane	$CBrF_3$	Halon-1301[a]
Bromochlorodifluoromethane	$CBrClF_2$	Halon-1211

[a]Halons, compounds similar to CFCs that contain at least one bromine atom, are used as fire extinguishers in specific applications such as in the engine rooms of ships.

Thus, in the presence of Cl atoms, one O_3 molecule and one O atom react to give two O_2 molecules. Although Cl is necessary for the overall reaction to occur, it does not appear in the net equation. Thus, the Cl atoms are a catalyst that increases the rate at which ozone is converted to oxygen.

Because the stratosphere is relatively isolated from the layers of the atmosphere above and below it, once they enter the stratosphere chlorine-containing species remain there for long periods of time. Each Cl atom produced from a CFC molecule can lead to the destruction of large numbers of ozone molecules, thereby decreasing the concentration of ozone in the stratosphere. Eventually, however, the Cl atom reacts with a water molecule to form HCl, which is carried back into the troposphere and then washed out of the atmosphere in rainfall.

The Ozone Hole

Massive ozone depletions were first observed in 1975 over the Antarctic and more recently over the Arctic. Although the reactions in Equations 3.38 and 3.39 appear to account for most of the ozone destruction observed at low to middle latitudes, Equation 3.37 requires intense sunlight to generate Cl atoms, and sunlight is in very short supply during the polar winters. Thus, at high latitudes (near the poles), a different set of reactions must be responsible for the depletion.

Recent research has shown that, in the absence of O atoms, ClO can react with stratospheric NO_2 in a redox reaction to form $ClONO_2$ (chlorine nitrate). When chlorine nitrate is in the presence of trace amounts of HCl or adsorbed on ice particles in stratospheric clouds, additional redox reactions can occur in which chlorine nitrate produces Cl_2 or HOCl:

$$HCl(g) + ClONO_2(g) \longrightarrow Cl_2(g) + HNO_3(g) \quad (3.42)$$

$$H_2O(g) + ClONO_2(g) \longrightarrow HOCl(g) + HNO_3(g) \quad (3.43)$$

Both Cl_2 and HOCl undergo cleavage reactions by even weak sunlight to give reactive Cl atoms. Thus, when the sun finally rises after the long polar night, relatively large amounts of Cl_2 and HOCl are present and rapidly generate high levels of Cl atoms. The reactions shown in Equations 3.38 and 3.39 then cause ozone levels to fall dramatically.

Stratospheric ozone levels decreased about 2.5% from 1978 to 1988, which coincided with a fivefold increase in the widespread use of CFCs since the 1950s. If the trend were allowed to continue, the results could be catastrophic. Fortunately, many countries have banned the use of CFCs in aerosols. In 1987, representatives from 43 nations signed the Montreal Protocol, committing themselves to reducing CFC emissions by 50% by the year 2000. Later, representatives from a large number of countries, alarmed by data showing rapid depletion of stratospheric chlorine, agreed to phase out CFCs completely by the early 21st century; the United States banned their use in 1995. The projected effects of these agreements on atmospheric chlorine levels are shown in Figure 3.16. Because of the very slow rate at which CFCs are removed from the stratosphere, however, stratospheric chlorine levels will not fall to the level at which the Antarctic ozone hole was first observed until about 2050. Molina and Rowland's work was recognized by the scientific community in 1995, when they shared the Nobel Prize in chemistry.

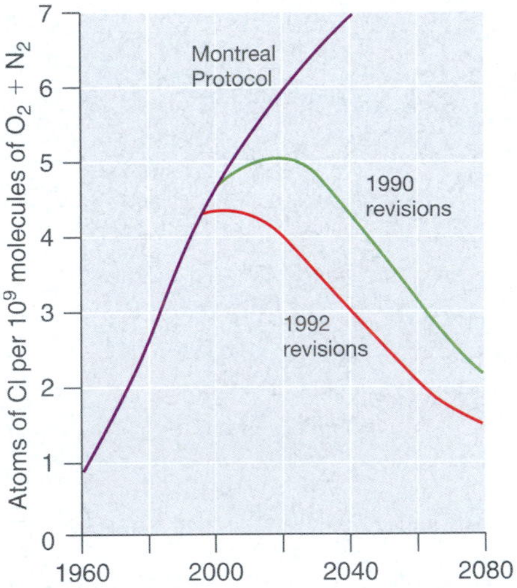

Figure 3.16 Projected effects of international agreements on atmospheric chlorine levels. The graph plots atmospheric chlorine content in atoms of Cl per 10^9 molecules of O_2 plus N_2 from 1960 to 1990 (actual data) and 1990 to 2080 (estimates for various schemes for regulating CFC emissions).

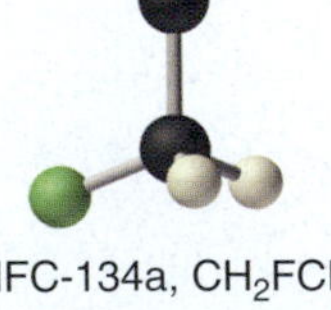

A hydrofluorocarbon (HFC) used as a replacement for chlorofluorocarbons (CFCs).

Manufacturing companies are now under great political and economic pressure to find alternatives to the CFCs used in the air-conditioning units of cars, houses, and commercial buildings. One approach is to use *hydrochlorofluorocarbons (HCFCs)*, hydrocarbons in which only some of the hydrogen atoms are replaced by chlorine or fluorine, and *hydrofluorocarbons (HFCs)*, which do not contain Cl (Table 3.4). The C–H bonds in HCFCs and HFCs act as "handles" that permit additional chemical reactions to occur. Consequently, these substances are degraded more rapidly, and most are washed out of the atmosphere before they can reach the stratosphere.

Nonetheless, the small fraction of HCFCs that reaches the stratosphere will deplete ozone levels just as CFCs do, so they are not the final answer. Indeed, the 1990 London amendment to the Montreal Protocol specifies that HCFCs must be phased

TABLE 3.4 Selected hydrochlorofluorocarbons (HCFCs) and hydrofluorocarbons (HFCs)

Name	Molecular Formula	Industrial Name
Chlorodifluoromethane	$CHClF_2$	HCFC-22 (Freon-22)
1-Chloro-1,1-difluoroethane	CH_3CClF_2	HCFC-141b
2,2-Dichloro-1,1,1-trifluoroethane	$CHCl_2CF_3$	HCFC-123
1,1,1,2-Tetrafluoroethane	CH_2FCF_3	HFC-134a

out by 2040. Finding a suitable replacement for use as a refrigerant is just one of the challenges facing chemists in the 21st century.

EXAMPLE 3.16

Nitric oxide, NO, may also be an important factor in the destruction of the ozone layer. One source of this compound is the combustion of hydrocarbons in jet engines. The fact that high-flying supersonic aircraft inject NO directly into the stratosphere was a major argument against the development of commercial supersonic transports. Do you agree with this decision? Why or why not?

Given Identity of compound

Asked for Assessment of likely role in ozone depletion

Strategy

Predict what reactions are likely to occur between NO and ozone, and then determine whether the reactions are likely to deplete ozone from the atmosphere.

Solution

Both NO and NO_2 are known oxides of nitrogen. Thus, NO is likely to react with ozone according to the chemical equation $NO(g) + O_3(g) \longrightarrow NO_2(g) + O_2(g)$, resulting in ozone depletion. If $NO_2(g)$ also reacts with atomic oxygen according to the equation $NO_2(g) + O(g) \longrightarrow NO(g) + O_2(g)$, then we would have a potential catalytic cycle for ozone destruction similar to that caused by Cl atoms. Based on these reactions, development of commercial supersonic transports is not recommended until the environmental impact has been further tested. (Although these reactions have been observed, they do not appear to be a major factor in ozone destruction.)

EXERCISE 3.16

An industrial manufacturer proposed that halons such as CF_3Br could be used as replacements for CFC propellants. Do you think that this is a reasonable suggestion, or is there a potential problem with such a use?

Answer Because the compound CF_3Br contains carbon, fluorine, and a bromine atom that is chemically similar to chlorine, it is likely that it too would be a catalyst for ozone destruction. There is therefore a potential problem with its use.

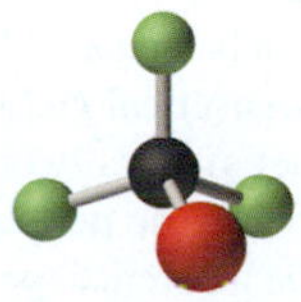

Halon–1301, $CBrF_3$

SUMMARY AND KEY TERMS

3.1 The Mole and Molar Masses (p. 86)

The **molecular mass** and the **formula mass** of a compound are obtained by summing the atomic masses of the atoms present in the molecular or empirical formula, respectively; the units of both are atomic mass units (amu). The **mole** is a unit used to measure the number of atoms, molecules, or (in the case of ionic compounds) formula units in a given mass of a substance. The mole is defined as the amount of substance that contains the number of carbon atoms in exactly 12 g of carbon-12 and consists of **Avogradro's number** (6.022×10^{23}) of atoms of carbon-12. The **molar mass** of a substance is defined as the mass of a mole of that substance, expressed in grams per mole, and is equal to the mass of 6.022×10^{23} atoms, molecules, or formula units of that substance.

3.2 Determining Empirical and Molecular Formulas (p. 92)

The empirical formula of a substance can be calculated from the experimentally determined *percent composition*, the percentage of each element present in a pure substance by mass. In many cases, these percentages can be determined by *combustion analysis*. If the molar mass of the compound is known, the molecular formula can be determined from the empirical formula.

3.3 Chemical Equations (p. 102)

In a **chemical reaction**, one or more substances are transformed to new substances. A chemical reaction is described by a **chemical equation**, an expression that gives the identities and quantities of the substances involved in a reaction. A chemical equation shows the starting compound(s)—the **reactants**—on the left and the final compound(s)—the **products**—on the right, separated by an arrow. In a *balanced chemical equation*, the numbers of atoms of each element and the total charge are the same on both sides of the equation. The number of atoms, molecules, or formula units of a reactant or product in a balanced chemical equation is the **coefficient** of that species. The **mole ratio** of two substances in a chemical reaction is the ratio of their coefficients in the balanced chemical equation.

3.4 Mass Relationships in Chemical Equations (p. 109)

The **stoichiometry** of a reaction describes the relative amounts of reactants and products in the balanced chemical equation. A **stoichiometric quantity** of a reactant is that amount necessary to react completely with the other reactant(s). If a quantity of a reactant remains unconsumed after complete reaction has occurred, it was originally present *in excess*. The reactant that is consumed first and limits the amount of product(s) that can be obtained is the **limiting reactant**. To identify the limiting reactant, we calculate the number of moles of each reactant present and compare their ratio to the mole ratio of the reactants in the balanced chemical equation. The maximum amount of product(s) that can be obtained in a reaction from a given amount of reactant(s) is the **theoretical yield** of the reaction. The **actual yield** is the amount of product(s) actually obtained in the reaction; it cannot exceed the theoretical yield. The **percent yield** of a reaction is the ratio of the actual yield to the theoretical yield, expressed as a percentage.

3.5 Classifying Chemical Reactions (p. 118)

Chemical reactions may be classified as acid–base, exchange, condensation or cleavage, and oxidation–reduction (or redox) reactions. To keep track of electrons in chemical reactions, oxidation states are assigned to atoms in compounds. The **oxidation state** is the charge an atom would have if all its bonding electrons were transferred completely to the atom that has the greater attraction for electrons. In an **oxidation–reduction reaction**, one atom must lose electrons and another must gain electrons. **Oxidation** is the loss of electrons, and an element whose oxidation state increases is said to be *oxidized*. **Reduction** is the gain of electrons, and an element whose oxidation state decreases is said to be *reduced*. **Oxidants** are compounds that are capable of accepting electrons from other compounds, so they are reduced during an oxidation–reduction reaction. In contrast, **reductants** are compounds that are capable of donating electrons to other compounds, so they are oxidized during an oxidation–reduction reaction. A **combustion reaction** is a redox reaction in which the oxidant is $O_2(g)$. An *amide bond* is formed from the condensation reaction between a carboxylic acid and an amine; it is the essential structural unit of proteins and many polymers. A **catalyst** is a substance that increases the rate of a chemical reaction without undergoing a net chemical change itself. A biological catalyst is called an **enzyme**. **Catalysis** is an acceleration in the rate of a reaction caused by the presence of a substance that does not appear in the chemical equation. A **homogeneous catalyst** is uniformly dispersed in a solution of the reactants, whereas a **heterogeneous catalyst** is present as a different phase, usually a solid.

3.6 Chemical Reactions in the Atmosphere (p. 125)

The earth's atmosphere consists of discrete layers that do not mix readily with one another. The sun emits radiation with a wide range of energies, including **visible light**, which can be detected by the human eye, and **ultraviolet light**, which is more energetic than visible light and cannot be detected by the human eye. In the stratosphere, ultraviolet light reacts with O_2 molecules to form atomic oxygen. Atomic oxygen then reacts with an O_2 molecule to produce **ozone**, O_3. As a result of this reaction, the stratosphere contains an appreciable concentration of ozone molecules that constitutes the **ozone layer**. The absorption of ultraviolet light in the stratosphere protects the earth's surface from the sun's harmful effects. Volatile organic compounds that contain chlorine and fluorine (*chlorofluorocarbons, CFCs*) are capable of reaching the stratosphere, where they can react with ultraviolet light to generate chlorine atoms and other chlorine-containing species that catalyze the conversion of ozone to O_2, thereby decreasing the amount of O_3 in the stratosphere. Replacing chlorofluorocarbons with *hydrochlorofluorocarbons (HCFCs)* or *hydrofluorocarbons (HFC's)* is one strategy that has been developed to minimize further damage to the earth's ozone layer.

QUESTIONS AND PROBLEMS

MGC For instructor-assigned homework, go to **www.masteringgenchem.com**

Please be sure you are familiar with the topics discussed in Essential Skills 2 at the end of this chapter before proceeding to the Questions and Problems. Questions and Problems with colored numbers have answers in the appendix and complete solutions in the Student Solutions Manual.

CONCEPTUAL

3.1 The Mole and Molar Masses

1. Describe the relationship between an atomic mass unit and a gram.

2. Is it correct to say that ethanol has a formula mass of 46? Why or why not?
3. If 2 mol of sodium react completely with 1 mol of chlorine to produce sodium chloride, does this mean that 2 g of sodium reacts completely with 1 g of chlorine to give the same product? Explain your answer.
4. Construct a flowchart to show how you would calculate the number of moles of silicon in a 37.0-g sample of orthoclase, $KAlSi_3O_8$, a mineral used in the manufacture of porcelain.
5. Construct a flowchart to show how you would calculate the number of moles of nitrogen in a 22.4-g sample of nitroglycerin that contains 18.5% nitrogen by mass.

3.2 Determining Empirical and Molecular Formulas

6. What is the relationship between an empirical formula and a molecular formula?
7. Construct an outline showing how you would determine the empirical formula of a compound from its percent composition.

3.3 Chemical Equations

8. How does a balanced equation agree with the law of definite proportions?
9. What is the difference between S_8 and 8S? Use this example to explain why subscripts in a formula must not be changed.
10. What factors determine whether a chemical equation is balanced?
11. What information can be obtained from a balanced chemical equation? Does a balanced chemical equation give information about the rate of the reaction?

3.4 Mass Relationships in Chemical Equations

12. Engineers use conservation of mass, called a "mass balance," to determine the amount of product that can be obtained from a chemical reaction. Mass balance assumes that the total mass of reactants is equal to the total mass of products. Is this a chemically valid practice? Explain your answer.
13. Given the equation $2H_2(g) + O_2(g) \longrightarrow 2H_2O(g)$, is it correct to say that 10 g of hydrogen will react with 10 g of oxygen to produce 20 g of water vapor?
14. What does it mean to say that a reaction is *stoichiometric*?
15. When sulfur is burned in air to produce sulfur dioxide, what is the limiting reactant? Explain your answer.
16. Is it possible for the percent yield to be greater than the theoretical yield? Justify your answer.

3.5 Classifying Chemical Reactions

17. What is a combustion reaction? How can it be distinguished from an exchange reaction?
18. What two products are formed in the combustion of an organic compound containing only C, H, and O? Is it possible to form only these two products from a reaction that is not a combustion reaction? Explain your answer.
19. What factors determine whether a reaction can be classified as a redox reaction?
20. Name three characteristics of a balanced redox reaction.
21. Does an oxidant accept electrons or donate them?
22. Does the oxidation state of a reductant become more positive or more negative during a redox reaction?
23. Nitrogen, hydrogen, and ammonia are known to have existed on primordial earth, yet mixtures of nitrogen and hydrogen do not usually react to give ammonia. What natural phenomenon would have enough energy to initiate a reaction between these two primordial gases?
24. Catalysts are not added to reactions in stoichiometric quantities. Why?
25. Decide whether each of the following uses a homogeneous or heterogeneous catalyst:
    (a) Pt metal is used in the catalytic converter of an automobile
    (b) Biological conversion of nitrogen to ammonia by an enzyme
    (c) Conversion of carbon monoxide and hydrogen to methane and water using a Ni catalyst
    (d) Using a dissolved Rh compound as a catalyst for the conversion of an alkene to an alkane
26. State whether each of the following uses a homogeneous or heterogeneous catalyst:
    (a) Pellets of ZSM-5, an aluminum- and silicon-containing mineral, are used to catalyze the conversion of methanol to gasoline
    (b) The conversion of glucose to a carboxylic acid by the enzyme glucose oxidase
    (c) Using metallic Rh for the conversion of CO and H_2O to CO_2 and H_2
27. Complete the table to describe some key differences between homogeneous and heterogeneous catalysis.

	Homogeneous	Heterogeneous
Number of phases	________	________
Ease of separation from product	________	________
Ease of recovery of catalyst	________	________

28. To increase the rate of a reaction, a scientist decided to use a catalyst. Unexpectedly, the scientist discovered that the catalyst actually decreased the yield of the desired product, rather than increasing it. What might have happened?

3.6 Chemical Reactions in the Atmosphere

29. Carbon monoxide is a toxic gas that can be produced from the combustion of wood in wood-burning stoves when excess oxygen is not present. Write a balanced equation showing how carbon monoxide is produced from carbon and suggest what might be done to prevent it from being a reaction product.
30. Explain why stratospheric ozone depletion has developed over the coldest part of the earth (the poles) and reaches a maximum at the beginning of the polar spring.
31. What type of reactions produce species that are believed to be responsible for catalytic depletion of ozone in the atmosphere?

NUMERICAL

3.1 The Mole and Molar Masses

This section includes "paired problems" (marked by brackets) that require similar problem-solving skills.

32. Derive an expression that relates the number of molecules in a sample of a substance to its mass and molecular mass.
33. Calculate the molecular mass or formula mass of each compound: (a) V_2O_4; (b) $CaSiO_3$; (c) BiOCl; (d) CH_3COOH; (e) Ag_2SO_4; (f) Na_2CO_3; (g) $(CH_3)_2CHOH$.

34. Calculate the molecular mass or formula mass of each compound: (a) KCl; (b) NaCN; (c) H_2S; (d) NaN_3; (e) H_2CO_3; (f) K_2O; (g) $Al(NO_3)_3$; (h) $Cu(ClO_4)_2$.

35. Calculate the number of moles in 5.00×10^2 g of each of the following substances: (a) lime, CaO; (b) chalk, $CaCO_3$; (c) sucrose (cane sugar), $C_{12}H_{22}O_{11}$; (d) bleach, NaOCl; (e) dry ice, CO_2. How many molecules or formula units are present in each sample?

36. Calculate the mass in grams of each sample: (a) 0.520 mol of N_2O_4; (b) 1.63 mol of $C_6H_4Br_2$; (c) 4.62 mol of $(NH_4)_2SO_3$.

37. Give the number of molecules or formula units in each sample: (a) 1.30×10^{-2} mol of SCl_2; (b) 1.03 mol of N_2O_5; (c) 0.265 mol of $Ag_2Cr_2O_7$.

38. Give the number of moles of compound in each sample: (a) 9.58×10^{26} molecules of Cl_2; (b) 3.62×10^{27} formula units of KCl; (c) 6.94×10^{28} formula units of $Fe(OH)_2$.

39. Solutions of iodine are used as antiseptics and disinfectants. How many iodine atoms correspond to 11.0 g of molecular iodine (I_2)?

40. What is the total number of atoms in each sample? (a) 0.980 mol of Na; (b) 2.35 mol of O_2; (c) 1.83 mol of Ag_2S; (d) 1.23 mol of propane (C_3H_8)

41. What is the total number of atoms in each sample? (a) 0.431 mol of Li; (b) 2.783 mol of methanol (CH_3OH); (c) 0.0361 mol of $CoCO_3$; (d) 1.002 mol of $SeBr_2O$.

42. What is the total number of atoms in each sample? (a) 2.48 g of HBr; (b) 4.77 g of CS_2; (c) 1.89 g of NaOH; (d) 1.46 g of SrC_2O_4

43. Decide whether each statement is true or false and explain your reasoning.
   (a) 0.5 mol of Cl_2 contains more molecules than does 0.5 mol of H_2.
   (b) One mole of H_2 contains 6.022×10^{23} hydrogen atoms.
   (c) The molecular mass of H_2O is 18.0 amu.
   (d) The formula mass of benzene is 78 amu.

44. Complete the table.

Substance	Mass, g	Number of Moles	Number of Molecules or Formula Units	Number of Atoms or Ions
$MgCl_2$	37.62	___	___	___
$AgNO_3$	___	2.84	___	___
BH_4Cl	___	___	8.93×10^{25}	___
K_2S	___	___	___	7.69×10^{26}
H_2SO_4	___	1.29	___	___
C_6H_{14}	11.84	___	___	___
$HClO_3$	___	___	2.45×10^{26}	___

45. Give the formula mass or molecular mass of each: (a) $MoCl_5$; (b) B_2O_3; (c) UO_2CO_3; (d) $NH_4UO_2AsO_4$.

46. Give the formula mass or molecular mass of each: (a) PbClF; (b) $Cu_2P_2O_7$; (c) $BiONO_3$; (d) Tl_2SeO_4.

3.2 Determining Empirical and Molecular Formulas

47. Determine the mass percentage of water in each hydrate: (a) $H_3AsO_4 \cdot 0.5H_2O$; (b) $NH_4NiCl_3 \cdot 6H_2O$; (c) $Al(NO_3)_3 \cdot 9H_2O$.

48. What is the mass percentage of water in each hydrate? (a) $CaSO_4 \cdot 2H_2O$; (b) $Fe(NO_3)_3 \cdot 9H_2O$; (c) $(NH_4)_3ZrOH(CO_3)_3 \cdot 2H_2O$

49. Which of the following contains the greatest mass percentage of oxygen: $KMnO_4$, $K_2Cr_2O_7$, or Fe_2O_3?

50. Which of the following contains the greatest mass percentage of oxygen: $ThOCl_2$, $MgCO_3$, or NO_2Cl?

51. Calculate the percent composition of the element shown in bold in each of the following compounds: (a) Sb**Br**$_3$; (b) **As**$_2$I$_4$; (c) Al**P**O$_4$; (d) **C**$_6$H$_{10}$O.

52. Give the percent composition of the element shown in bold in each of the following compounds: (a) **H**BrO$_3$; (b) **Cs**ReO$_4$; (c) **C**$_3$H$_8$O; (d) Fe**S**O$_4$.

53. A sample of a chromium compound has a molar mass of 151.99 g/mol. Elemental analysis of the compound shows that it contains 68.43% Cr and 31.57% O. What is the identity of the compound?

54. The percentages of Fe and O in the three most common binary compounds of iron and oxygen are given in the table below: Write the empirical formulas of these three compounds.

Compound	% Fe	% O	Empirical Formula
1	69.9	30.1	___
2	77.7	22.3	___
3	72.4	27.6	___

55. Calculate the mass percentage of water in each hydrate: (a) $LiCl \cdot H_2O$; (b) $MgSO_4 \cdot 7H_2O$; (c) $Sr(NO_3)_2 \cdot 4H_2O$.

56. Determine the percentage of H_2O in each hydrate: (a) $CaHPO_4 \cdot 2H_2O$; (b) $FeCl_2 \cdot 4H_2O$; (c) $Mg(NO_3)_2 \cdot 4H_2O$.

57. Two hydrates were weighed, heated to drive off waters of hydration, and then cooled. The residues were then reweighed. Based on the following results, what are the formulas of the hydrates?

Compound	Initial Mass, g	Mass after Cooling, g
$NiSO_4 \cdot xH_2O$	2.08	1.22
$CoCl_2 \cdot xH_2O$	1.62	0.88

58. Which contains the greatest mass percentage of sulfur: FeS_2, $Na_2S_2O_4$, or Na_2S?

59. Given equal masses of each, which contains the larger amount of sulfur by mass: $NaHSO_4$ or K_2SO_4?

60. Calculate the mass percentage of oxygen in each polyatomic ion: (a) bicarbonate; (b) chromate; (c) acetate; (d) sulfite.

61. Calculate the mass percentage of oxygen in each polyatomic ion: (a) oxalate; (b) nitrite; (c) dihydrogen phosphate; (d) thiocyanate.

62. The empirical formula of garnet, a gemstone, is $Fe_3Al_2Si_3O_{12}$. Analysis of a sample of garnet gave a value of 13.8% for the mass percentage of silicon. Is this consistent with the empirical formula?

63. A compound has the empirical formula C_2H_4O, and its formula mass is 88 amu. What is its molecular formula?

64. Mirex is an insecticide that contains 22.01% C and 77.99% Cl, and has a molecular mass of 545.59 amu. What is its empirical formula? What is its molecular formula?

65. Combustion of a 34.8-mg sample of benzaldehyde, which contains only carbon, hydrogen, and oxygen, produced 101 mg of CO_2 and 17.7 mg of H_2O.

(a) What was the mass of carbon and hydrogen in the sample?
(b) Assuming that the original sample contained only carbon, hydrogen, and oxygen, what was the mass of oxygen in the sample?
(c) What was the mass percentage of oxygen in the sample?
(d) What is the empirical formula of benzaldehyde?
(e) Benzaldehyde has a molar mass of 106.12 g/mol. What is its molecular formula?

66. Salicylic acid is used to make aspirin. It contains only carbon, oxygen, and hydrogen. Combustion of a 43.5-mg sample of this compound produced 97.1 mg of CO_2 and 17.0 mg of H_2O.
(a) What is the mass of oxygen in the sample?
(b) What is the mass percentage of oxygen in the sample?
(c) What is the empirical formula of salicylic acid?
(d) Salicylic acid has a molar mass of 138.12 g/mol. What is its molecular formula?

67. Given equal masses of the following acids, which contains the greatest amount of hydrogen that can dissociate to form H^+: nitric acid, hydroiodic acid, hydrocyanic acid, or chloric acid?

68. Calculate the formula or molecular mass of each of the following: (a) the 7-carbon carboxylic acid, heptanoic acid, (b) the 3-carbon alcohol, 2-propanol; (c) $KMnO_4$; (d) tetraethyllead; (e) sulfurous acid; (f) the 8-carbon arene, ethylbenzene.

69. Determine the formula or molecular mass of each of the following: (a) $MoCl_5$; (b) B_2O_3; (c) bromobenzene; (d) propene; (e) cyclohexene; (f) phosphoric acid; (g) ethylamine.

70. Given equal masses of butane, cyclobutane, and propene, which contains the greatest mass of carbon?

71. Given equal masses of urea $[(NH_2)_2CO]$ and ammonium sulfate, which contains the most nitrogen for use as a fertilizer?

3.3 Chemical Equations

72. Balance each chemical equation:
(a) $Be(s) + O_2(g) \longrightarrow BeO(s)$
(b) $N_2O_3(g) + H_2O(l) \longrightarrow HNO_2(aq)$
(c) $Na(s) + H_2O(l) \longrightarrow NaOH(aq) + H_2(g)$
(d) $CaO(s) + HCl(aq) \longrightarrow CaCl_2(aq) + H_2O(l)$
(e) $CH_3NH_2(g) + O_2(g) \longrightarrow H_2O(g) + CO_2(g) + N_2(g)$
(f) $Fe(s) + H_2SO_4(aq) \longrightarrow FeSO_4(aq) + H_2(g)$

73. Balance each chemical equation:
(a) $KI(aq) + Br_2(l) \longrightarrow KBr(aq) + I_2(s)$
(b) $MnO_2(s) + HCl(aq) \longrightarrow MnCl_2(aq) + Cl_2(g) + H_2O(l)$
(c) $Na_2O(s) + H_2O(l) \longrightarrow NaOH(aq)$
(d) $Cu(s) + AgNO_3(aq) \longrightarrow Cu(NO_3)_2(aq) + Ag(s)$
(e) $SO_2(g) + H_2O(l) \longrightarrow H_2SO_3(aq)$
(f) $S_2Cl_2(l) + NH_3(l) \longrightarrow S_4N_4(s) + S_8(s) + NH_4Cl(s)$

74. Balance each chemical equation:
(a) $H_2S(g) + O_2(g) \longrightarrow H_2O(l) + S_8(s)$
(b) $KCl(aq) + HNO_3(aq) + O_2(g) \longrightarrow KNO_3(aq) + Cl_2(g) + H_2O(l)$
(c) $NH_3(g) + O_2(g) \longrightarrow NO(g) + H_2O(g)$
(d) $CH_4(g) + O_2(g) \longrightarrow CO(g) + H_2(g)$
(e) $NaF(aq) + Th(NO_3)_4(aq) \longrightarrow NaNO_3(aq) + ThF_4(s)$
(f) $Ca_5(PO_4)_3F(s) + H_2SO_4(aq) + H_2O(l) \longrightarrow H_3PO_4(aq) + CaSO_4{\cdot}2H_2O(s) + HF(aq)$

75. Balance each chemical equation:
(a) $N_2O_5(g) \longrightarrow NO_2(g) + O_2(g)$
(b) $NaNO_3(s) \longrightarrow NaNO_2(s) + O_2(g)$
(c) $Al(s) + NH_4NO_3(s) \longrightarrow N_2(g) + H_2O(l) + Al_2O_3(s)$
(d) $C_3H_5N_3O_9(l) \longrightarrow CO_2(g) + N_2(g) + H_2O(g) + O_2(g)$
(e) Reaction of butane with excess oxygen
(f) $IO_2F(s) + BrF_3(l) \longrightarrow IF_5(l) + Br_2(l) + O_2(g)$

76. Balance each chemical equation:
(a) $NaCl(aq) + H_2SO_4(aq) \longrightarrow Na_2SO_4(aq) + HCl(g)$
(b) $K(s) + H_2O(l) \longrightarrow KOH(aq) + H_2(g)$
(c) Reaction of octane with excess oxygen
(d) $S_8(s) + Cl_2(g) \longrightarrow S_2Cl_2(l)$
(e) $CH_3OH(l) + I_2(s) + P_4(s) \longrightarrow CH_3I(l) + H_3PO_4(aq) + H_2O(l)$
(f) $(CH_3)_3Al(s) + H_2O(l) \longrightarrow CH_4(g) + Al(OH)_3(s)$

77. Write a balanced chemical equation for each reaction:
(a) Magnesium burns in oxygen.
(b) Carbon dioxide and sodium oxide react to produce sodium carbonate.
(c) Aluminum reacts with hydrochloric acid.
(d) An aqueous solution of silver nitrate reacts with a solution of potassium chloride.
(e) Methane burns in oxygen.
(f) Sodium nitrate and sulfuric acid react to produce sodium sulfate and nitric acid.

78. Write a balanced equation for each reaction:
(a) Aluminum reacts with bromine.
(b) Sodium reacts with chlorine.
(c) Aluminum hydroxide and acetic acid react to produce aluminum acetate and water.
(d) Ammonia and oxygen react to produce nitrogen monoxide and water.
(e) Nitrogen and hydrogen react at elevated temperature and pressure to produce ammonia.
(f) An aqueous solution of barium chloride reacts with a solution of sodium sulfate.

3.4 Mass Relationships in Chemical Equations

79. What is the formula mass of each compound? (a) ammonium chloride; (b) sodium cyanide; (c) magnesium hydroxide; (d) calcium phosphate; (e) lithium carbonate

80. Give the formula mass of each species: (a) potassium permanganate; (b) sodium sulfate; (c) hydrogen cyanide; (d) potassium thiocyanate; (e) ammonium oxalate; (f) hydrogen sulfite ion; (g) lithium acetate; (h) hydrogen peroxide.

81. Convert each mass to moles: (a) 10.76 g of Si; (b) 8.6 g of Pb; (c) 2.49 g of Mg; (d) 0.94 g of La; (e) 2.68 g of chlorine gas; (f) 0.089 g of As.

82. How many moles are contained in each of the following? (a) 8.6 g of CO_2; (b) 2.7 g of CaO; (c) 0.89 g of KCl; (d) 4.3 g of $SrBr_2$; (e) 2.5 g of NaOH; (f) 1.87 g of $Ca(OH)_2$

83. Convert the following to moles and millimoles: (a) 1.68 g of $Ba(OH)_2$; (b) 0.792 g of H_3PO_4; (c) 3.21 g of K_2S; (d) 0.8692 g of $Cu(NO_3)_2$; (e) 10.648 g of $Ba_3(PO_4)_2$; (f) 5.79 g of $(NH_4)_2SO_4$; (g) 1.32 g of $Pb(C_2H_3O_2)_2$; (h) 4.29 g of $CaCl_2{\cdot}6H_2O$.

84. How many moles and millimoles are contained in each of the following? (a) 0.089 g of silver nitrate; (b) 1.62 g of aluminum chloride; (c) 2.37 g of calcium carbonate; (d) 1.004 g of iron(II) sulfide; (e) 2.12 g of dinitrogen pentoxide; (f) 2.68 g of lead(II) nitrate; (g) 3.02 g of ammonium phosphate; (h) 5.852 g of sulfuric acid; (i) 4.735 g of potassium dichromate

85. Convert to grams and milligrams: (a) 5.68 mol of Ag; (b) 2.49 mol of Sn; (c) 0.0873 mol of Os; (d) 1.74 mol of Si; (e) 0.379 mol of H_2; (f) 1.009 mol of Zr.

86. What is the mass of each of the following substances (in grams and milligrams)? (a) 2.080 mol of CH_3OH; (b) 0.288 mol of P_4; (c) 3.89 mol of $ZnCl_2$; (d) 1.800 mol of $Fe(CO)_5$; (e) 0.798 mol of S_8; (f) 4.01 mol of NaOH

87. What is the mass of each of the following (in kilograms)? (a) 6.38 mol of P_4O_{10}; (b) 2.26 mol of $Ba(OH)_2$; (c) 4.35 mol of K_3PO_4; (d) 2.03 mol of $Ni(ClO_3)_2$; (e) 1.47 mol of NH_4NO_3; (f) 0.445 mol of $Co(NO_3)_3$

88. How many atoms are contained in each? (a) 2.32 mol of Bi; (b) 0.066 mol of V; (c) 0.267 mol of Ru; (d) 4.87 mol of C; (e) 2.74 g of I_2; (f) 1.96 g of Cs; (g) 7.78 g of O_2

89. Convert each to milligrams: (a) 5.89×10^{22} atoms of Pt; (b) 2.899×10^{21} atoms of Hg; (c) 4.826×10^{22} atoms of chlorine.

90. Write a balanced chemical equation for each reaction, and then determine which reactant is in excess.
(a) 2.46 g barium(*s*) + 3.89 g bromine(*l*) in water to give barium bromide
(b) 1.44 g bromine(*l*) + 2.42 g potassium iodide(*s*) in water to give potassium bromide and iodine
(c) 1.852 g of Zn metal is added to 3.62 g of sulfuric acid in water to give zinc sulfate and hydrogen gas
(d) 0.147 g of iron metal reacts with 0.924 g of silver acetate in water to give iron(II) acetate and silver metal.
(e) 3.142 g of ammonium phosphate reacts with 1.648 g of barium hydroxide in water to give ammonium hydroxide and barium phosphate

91. Under the proper conditions, ammonia and oxygen will react to form dinitrogen monoxide (nitrous oxide, also called laughing gas) and water. Write a balanced chemical equation for this reaction. Decide which reactant is in excess for each of the following combinations of reactants:
(a) 24.6 g of ammonia and 21.4 g of oxygen
(b) 3.8 mol of ammonia and 84.2 g of oxygen
(c) 3.6×10^{24} molecules of ammonia and 318 g of oxygen
(d) 2.1 mol of ammonia and 36.4 g of oxygen

92. When a piece of zinc metal is placed in aqueous hydrochloric acid, zinc chloride is produced and hydrogen gas is evolved. Write a balanced chemical equation for this reaction. Decide which reactant is in excess for each of the following combinations of reactants:
(a) 12.5 g of HCl and 7.3 g of Zn
(b) 6.2 mol of HCl and 100 g of Zn
(c) 2.1×10^{23} molecules of Zn and 26.0 g of HCl
(d) 3.1 mol of Zn and 97.4 g of HCl

93. Determine the mass of each reactant needed to give the indicated amount of product. Be sure that the equations are balanced.
(a) $NaI(aq) + Cl_2(g) \longrightarrow NaCl(aq) + I_2(s)$; 1.0 mol of NaCl
(b) $NaCl(aq) + H_2SO_4(aq) \longrightarrow HCl(g) + Na_2SO_4(aq)$; 0.50 mol of HCl
(c) $NO_2(g) + H_2O(l) \longrightarrow HNO_2(aq) + HNO_3(aq)$; 1.5 mol of HNO_3

94. Determine the mass of each reactant needed to give the indicated amount of product. Be sure that the equations are balanced.
(a) $AgNO_3(aq) + CaCl_2(s) \longrightarrow AgCl(s) + Ca(NO_3)_2(aq)$; 1.25 mol of AgCl
(b) $Pb(s) + PbO_2(s) + H_2SO_4(aq) \longrightarrow PbSO_4(s) + H_2O(l)$; 3.8 g of $PbSO_4$
(c) $H_3PO_4(aq) + MgCO_3(s) \longrightarrow Mg_3(PO_4)_2(s) + CO_2(g) + H_2O(l)$; 6.41 g of $Mg_3(PO_4)_2$

95. Determine the percent yield of each reaction. Be sure that the equations are balanced. Assume that any reactants for which amounts are not given are present in excess. (The symbol Δ indicates that the reactants are heated.)
(a) $KClO_3(s) \xrightarrow{\Delta} KCl(s) + O_2(g)$; 2.14 g of $KClO_3$ produces 0.67 g of O_2
(b) $Cu(s) + H_2SO_4(aq) \longrightarrow CuSO_4(aq) + SO_2(g) + H_2O(l)$; 4.00 g of copper gives 1.2 g of sulfur dioxide
(c) $AgC_2H_3O_2(aq) + Na_3PO_4(aq) \longrightarrow Ag_3PO_4(s) + NaC_2H_3O_2(aq)$; 5.298 g of silver acetate produces 1.583 g of silver phosphate

96. If each step of a four-step reaction has a yield of 95%, what is the percent yield for the overall reaction?

97. A three-step reaction had actual yields of 87% for the first step, 94% for the second, and 55% for the third. What was the percent yield of the overall reaction?

98. Give a general expression relating the theoretical yield (in grams) of product that can be obtained from *x* grams of B, assuming neither A nor B is limiting:

$$A + 3B \longrightarrow 2C$$

99. Under certain conditions, the reaction of hydrogen with carbon monoxide can produce methanol.
(a) Write a balanced chemical equation for this reaction.
(b) Calculate the percent yield if exactly 200 g of methanol is produced from exactly 300 g of carbon monoxide.

100. Chlorine dioxide is a bleaching agent used in the paper industry. It can be prepared by the reaction

$$NaClO_2(s) + Cl_2(g) \longrightarrow ClO_2(aq) + NaCl(aq)$$

(a) What mass of chlorine is needed for the complete reaction of 30.5 g of $NaClO_2$?
(b) Give a general equation for the conversion of *x* grams of sodium chlorite to chlorine dioxide.

101. The reaction of propane with chlorine gas produces two monochloride products, $CH_3CH_2CH_2Cl$ and $CH_3CHClCH_3$. The first is obtained in a 43% yield and the second in a 57% yield.
(a) If you began your reaction with 2.78 g of propane gas, how much chlorine gas would you need for the reaction to go to completion?
(b) How many grams of each product could theoretically be obtained from the reaction starting with 2.78 g of propane?
(c) Use the actual percent yields to calculate how many grams of each product would actually be obtained.

102. Protactinium (Pa), a highly toxic metal, is one of the rarest and most expensive elements. The following reaction is one method for preparing protactinium metal under relatively extreme conditions:

$$2PaI_5(s) \xrightarrow{\Delta} 2Pa(s) + 5I_2(s)$$

(a) Given 15.8 mg of reactant, how many milligrams of protactinium could be synthesized?
(b) If 3.4 mg of Pa was obtained, what was the percent yield of this reaction?
(c) If you obtained 3.4 mg of Pa and the percent yield was 78.6%, how many grams of PaI_5 were used in the preparation?

103. Aniline can be produced from chlorobenzene via the following reaction

$$\underset{\text{Chlorobenzene}}{C_6H_5Cl(l)} + 2NH_3(g) \longrightarrow \underset{\text{Aniline}}{C_6H_5NH_2(l)} + NH_4Cl(s)$$

Assume that 20.0 g of chlorobenzene at 92% purity is mixed with 8.30 g of ammonia.
(a) Which is the limiting reactant?
(b) Which reactant is present in excess?
(c) What is the theoretical yield of ammonium chloride in grams?
(d) If 4.78 g of NH_4Cl was recovered, what was the percent yield?
(e) Derive a general expression for the theoretical yield of ammonium chloride in terms of grams of chlorobenzene reactant, if ammonia is present in excess.

104. A stoichiometric amount of chlorine gas is added to an aqueous solution of NaBr to produce an aqueous solution of sodium chloride and liquid bromine. Write the chemical equation for this reaction. Then assume an 89% yield and calculate the mass of chlorine in the reactants given the following:
(a) 9.36×10^{24} formula units of NaCl
(b) 8.5×10^4 mol of Br_2
(c) 3.7×10^8 g of NaCl

3.5 Classifying Chemical Reactions

105. Classify each chemical reaction according to the types listed in Table 3.1:
(a) $12FeCl_2(s) + 3O_2(g) \longrightarrow 8FeCl_3(s) + 2Fe_2O_3(s)$
(b) $CaCl_2(aq) + K_2SO_4(aq) \longrightarrow CaSO_4(s) + 2KCl(aq)$
(c) $HCl(aq) + NaOH(aq) \longrightarrow NaCl(aq) + H_2O(l)$
(d) $Br_2(l) + C_2H_4(g) \longrightarrow BrCH_2CH_2Br(l)$

106. Classify each reaction according to the types listed in Table 3.1:
(a) $4FeO(s) + O_2(g) \longrightarrow 2Fe_2O_3(s)$
(b) $Ca_3(PO_4)_2(s) + 3H_2SO_4(aq) \longrightarrow 3CaSO_4(s) + 2H_3PO_4(aq)$
(c) $HNO_3(aq) + KOH(aq) \longrightarrow KNO_3(aq) + H_2O(l)$
(d) ethane(g) + oxygen(g) $\longrightarrow$ carbon dioxide(g) + water(g)

107. Assign oxidation states to all atoms in each of the following: (a) $(NH_4)_2S$; (b) the phosphate ion; (c) $[AlF_6]^{3-}$; (d) CuS; (e) HCO_3^-; (f) NH_4^+; (g) H_2SO_4; (h) formic acid; (i) *n*-butanol.

108. Give the oxidation states of all atoms in each of the following: (a) ClO_2; (b) HO_2^-; (c) sodium bicarbonate; (d) MnO_2; (e) PCl_5; (f) $[Mg(H_2O)_6]^{2+}$; (g) N_2O_4; (h) butanoic acid; (i) methanol.

109. Balance this chemical equation:

$$NaHCO_3(aq) + H_2SO_4(aq) \longrightarrow Na_2SO_4(aq) + CO_2(g) + H_2O(l)$$

What type of reaction is this? Justify your answer.

110. Give the oxidation state of each atom in the following: (a) Iron(III) nitrate; (b) Al_2O_3; (c) potassium sulfate; (d) Cr_2O_3; (e) sodium perchlorate; (f) Cu_2S; (g) hydrazine (N_2H_4); (h) NO_2; (i) *n*-pentanol.

111. Assign oxidation states to all atoms in the following: (a) calcium carbonate; (b) NaCl; (c) CO_2; (d) potassium dichromate; (e) $KMnO_4$; (f) ferric oxide; (g) $Cu(OH)_2$; (h) SO_4^{2-}; (i) *n*-hexanol.

112. For each redox reaction, determine the identities of the oxidant, the reductant, the species oxidized, and the species reduced:
(a) $H_2(g) + I_2(s) \longrightarrow 2HI(g)$
(b) $2Na(s) + 2H_2O(l) \longrightarrow 2NaOH(aq) + H_2(g)$
(c) $2F_2(g) + 2NaOH(aq) \longrightarrow OF_2(g) + 2NaF(aq) + H_2O(l)$

113. Identify the oxidant, the reductant, the species oxidized, and the species reduced in each chemical reaction:
(a) $2Na(s) + Cl_2(g) \longrightarrow 2NaCl(s)$
(b) $SiCl_4(l) + 2Mg(s) \longrightarrow 2MgCl_2(s) + Si(s)$
(c) $2H_2O_2(aq) \longrightarrow 2H_2O(l) + O_2(g)$

114. Balance each chemical equation. Then identify the oxidizing agent, the reducing agent, the species oxidized, and the species reduced (Δ indicates that the reaction requires heating).
(a) $H_2O(g) + CO(g) \longrightarrow CO_2(g) + H_2(g)$
(b) The reaction of aluminum oxide, carbon, and chlorine gas at 900°C to produce aluminum chloride and carbon monoxide
(c) $HgO(s) \xrightarrow{\Delta} Hg(l) + O_2(g)$

115. Balance each chemical equation. Then identify the oxidant, the reductant, the species oxidized, and the species reduced.
(a) The reaction of water and carbon at 800°C to produce hydrogen and carbon monoxide
(b) $Mn(s) + S_8(s) + CaO(s) \longrightarrow CaS(s) + MnO(s)$
(c) The reaction of ethylene and oxygen at elevated temperature in the presence of a silver catalyst to produce ethylene oxide

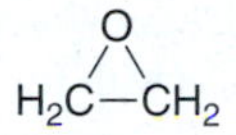

Ethylene oxide

(d) $ZnS(s) + H_2SO_4(aq) + O_2(g) \longrightarrow ZnSO_4(aq) + S_8(s) + H_2O(l)$

116. Silver is tarnished by hydrogen sulfide, an atmospheric contaminant, to form a thin layer of dark silver sulfide (Ag_2S) along with hydrogen gas.
(a) Write a balanced equation for this reaction.
(b) Which species has been oxidized and which has been reduced?
(c) Assuming 2.2 g of Ag has been converted to silver sulfide, construct a table showing the reaction in terms of the number of atoms in the reactants and products, moles of reactants and products, grams of reactants and products, and molecules of reactants and products.

117. The following reaction is used in the paper and pulp industry:

$$Na_2SO_4(aq) + C(s) + NaOH(aq) \longrightarrow Na_2CO_3(aq) + Na_2S(aq) + H_2O(l)$$

(a) Balance the chemical equation.
(b) Identify the oxidant and the reductant.
(c) How much carbon is needed to convert 2.8 kg of sodium sulfate to sodium sulfide?
(d) If the yield of the reaction were only 78%, how many kilograms of sodium carbonate would be produced from 2.80 kg of sodium sulfate?
(e) If 240 g of C and 2.80 kg of sodium sulfate were used in the reaction, what would be the limiting reactant (assuming an excess of sodium hydroxide)?

118. Methyl butyrate, an artificial apple flavor used in the food industry, is produced by the reaction of butanoic acid with methanol in the presence of an acid catalyst (H^+):

$$CH_3CH_2CH_2CO_2H(l) + CH_3OH(l) \xrightarrow{H^+} CH_3CH_2CH_2CO_2CH_3(l) + H_2O(l)$$

(a) Given 7.8 g of butanoic acid, how many grams of methyl butyrate would be synthesized, assuming a 100% yield?
(b) The reaction produced 5.5 g of methyl butyrate. What was the percent yield?
(c) Is the catalyst used in this reaction heterogeneous or homogeneous?

119. In the presence of a Pt catalyst, hydrogen and bromine react at elevated temperatures (300°C) to form hydrogen bromide (heat is indicated by Δ):

$$H_2(g) + Br_2(l) \xrightarrow[\Delta]{Pt} 2HBr(g)$$

Given the following, calculate the mass of hydrogen bromide produced:
(a) 8.23×10^{22} molecules of H_2
(b) 6.1×10^3 mol of H_2
(c) 1.3×10^5 g of H_2
(d) Is the catalyst used in this reaction heterogeneous or homogeneous?

3.6 Chemical Reactions in the Atmosphere

120. Sulfur dioxide and hydrogen sulfide are important atmospheric contaminants that have resulted in the deterioration of ancient objects. Sulfur dioxide combines with water to produce sulfurous acid, which then reacts with atmospheric oxygen to produce sulfuric acid. Sulfuric acid is known to attack many metals that were used by ancient cultures. Give the formulas for these four sulfur-containing species. What is the percentage of sulfur in each of these compounds? What is the percentage of oxygen in each?

APPLICATIONS

121. Hydrogen sulfide is a noxious and toxic gas produced from decaying organic matter that contains sulfur. A lethal concentration in rats corresponds to an inhaled dose of 715 molecules per million molecules of air. How many molecules does this correspond to per mole of air? How many moles of hydrogen sulfide does this correspond to per mole of air?

122. Bromine, sometimes produced from brines (salt lakes) and ocean water, can be used for bleaching fibers and silks. How many moles of bromine atoms are found in 8.0 g of molecular bromine (Br_2)?

123. Paris yellow is a lead compound that is used as a pigment; it contains 16.09% chromium, 19.80% oxygen, and 64.11% lead. What is the empirical formula of Paris yellow?

124. A particular chromium compound used for dyeing and waterproofing fabrics has the elemental composition 18.36% Cr, 13.81% K, 45.19% O, and 22.64% S. What is the empirical formula of this compound?

125. Compounds that contain aluminum and silicon are commonly found in the clay fractions of soils derived from volcanic ash. One of these compounds is *vermiculite*, which is formed in reactions caused by exposure to weather. Vermiculite has the following formula: $Ca_{0.7}[Si_{6.6}Al_{1.4}]Al_4O_{20}(OH)_4$. (Note that the content of Ca, Si, and Al are not shown as integers because the relative amounts of these elements vary from sample to sample.) What is the mass percent of each element in this sample of vermiculite?

126. Pheromones are chemical signals secreted by a member of one species to evoke a response in another member of the same species. One honeybee pheromone is an organic compound known as an alarm pheromone, which smells like bananas. It induces an aggressive attack by other honeybees, causing swarms of angry bees to attack the same aggressor. The composition of this alarm pheromone is 64.58% C, 10.84% H, and 24.58% O by mass, and its molecular mass is 130.2 amu.
(a) Calculate the empirical formula of this pheromone.
(b) Determine its molecular formula.
(c) Assuming a honeybee secretes 1.00×10^{-11} g of pure pheromone, how many molecules of pheromone are secreted?

127. Amoxicillin is a prescription drug used to treat a wide variety of bacterial infections, including infections of the middle ear and the upper and lower respiratory tracts. It destroys the cell walls of bacteria, which causes them to die. The elemental composition of amoxicillin is 52.59% C, 5.24% H, 11.50% N, 21.89% O, and 8.77% S by mass. What is its empirical formula?

128. Monosodium glutamate, or MSG (molar mass = 169 g/mol), is used as a flavor enhancer in food preparation. It is known to cause headaches and chest pains in some individuals, the so-called "Chinese food syndrome." Its composition was found to be 35.51% C, 4.77% H, 8.28% N, and 13.59% Na by mass. If the "missing" mass is oxygen, what is the empirical formula of MSG?

129. Ritalin is a mild central nervous system stimulant that is prescribed to treat attention deficit disorders and narcolepsy (an uncontrollable desire to sleep). Its chemical name is methylphenidate hydrochloride, and its empirical formula is $C_{14}H_{20}ClNO_2$. If you sent a sample of this compound to a commercial laboratory for elemental analysis, what results would you expect for the mass percentages of C, H, and N?

130. Fructose, a sugar found in fruit, contains only carbon, oxygen, and hydrogen. It is used in ice cream to prevent a sandy texture. Complete combustion of 32.4 mg of fructose in oxygen produced 47.6 mg of CO_2 and 19.4 mg of H_2O. What is the empirical formula of fructose?

131. Coniine, the primary toxin in hemlock, contains only carbon, nitrogen, and hydrogen. When ingested, it causes paralysis and eventual death. Complete combustion of 28.7 mg of coniine produced 79.4 mg of CO_2 and 34.4 mg of H_2O. What is the empirical formula of the coniine?

132. Copper–tin alloys (bronzes) with a high arsenic content were presumably used by Bronze Age metallurgists because bronze produced from arsenic-rich ores had superior casting and working properties. The compositions of some representative bronzes of this type are given below:

Origin	% Composition	
	Cu	As
Dead Sea	87.0	12.0
Central America	90.7	3.8

If ancient metallurgists had used the mineral As_2S_3 as their source of arsenic, how much As_2S_3 would have had to be processed with 100 g of cuprite (Cu_2O) bronzes with these compositions?

133. The phrase "mad as a hatter" refers to mental disorders caused by exposure to mercury(II) nitrate in the felt hat manufacturing trade during the 18th and 19th centuries. An even greater danger to humans, however, arises from alkyl derivatives of mercury.
(a) Give the empirical formula of mercury(II) nitrate.
(b) One alkyl derivative, dimethylmercury, is a highly toxic compound that can cause mercury poisoning in humans. How many molecules are contained in a 5.0-g sample of dimethylmercury?
(c) What is the percentage of mercury in the sample?

134. Magnesium carbonate, aluminum hydroxide, and sodium bicarbonate are commonly used as antacids. Give the empirical formulas and determine the molar masses of these compounds. Based on their formulas, suggest another compound that might be an effective antacid.

135. Nickel(II) acetate, lead(II) phosphate, zinc nitrate, and beryllium oxide have all been reported to induce cancers in experimental animals.
(a) Give the empirical formulas for these compounds.
(b) Calculate their formula masses.
(c) Based on the location of cadmium in the periodic table, would you predict that cadmium chloride might also induce cancer?

136. Methane, the major component of natural gas, is found in the atmospheres of Jupiter, Saturn, Uranus, and Neptune.
(a) Draw the structure of methane.
(b) Calculate the molecular mass of methane.
(c) Calculate the mass percentage of both elements present in methane.

137. Sodium saccharin, which is approximately 500 times sweeter than sucrose, is frequently used as a sugar substitute. What are the percentages of carbon, oxygen, and sulfur in this artificial sweetener?

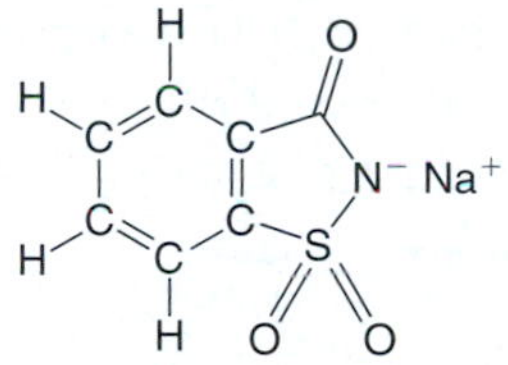

Sodium saccharin

138. Lactic acid, found in sour milk, dill pickles, and sauerkraut, has the functional groups of both an alcohol and a carboxylic acid. The empirical formula for this compound is CH_2O, and its molar mass is 90 g/mol. If this compound were sent to a laboratory for elemental analysis, what results would you expect for C, H, and O content?

139. The compound 2-nonenal is a cockroach repellant that is found in cucumbers, watermelon, and carrots. Determine its molecular mass.

$$CH_3CH_2CH_2CH_2CH_2CH_2CH{=}CH{-}\overset{\overset{\displaystyle O}{\|}}{C}{-}H$$

2-Nonenal

140. You have obtained a 720-mg sample of what you believe to be pure fructose, although it is possible that the sample has been contaminated with formaldehyde. Fructose and formaldehyde both have the empirical formula CH_2O. Could you use the results from combustion analysis to determine whether your sample is pure?

141. The booster rockets in the space shuttles use a mixture of aluminum metal and ammonium perchlorate for fuel. Upon ignition, this mixture can react according to the chemical equation

$$Al(s) + NH_4ClO_4(s) \longrightarrow Al_2O_3(s) + AlCl_3(g) + NO(g) + H_2O(g)$$

Balance the equation and construct a table showing how to interpret this information in terms of the following:
(a) Numbers of individual atoms, molecules, and ions
(b) Moles of reactants and products
(c) Grams of reactants and products
(d) Numbers of molecules of reactants and products given 1 mol of aluminum metal

142. One of the byproducts of the manufacturing of soap is glycerol. In 1847, it was discovered that the reaction of glycerol with nitric acid produced nitroglycerin according to the following unbalanced chemical equation:

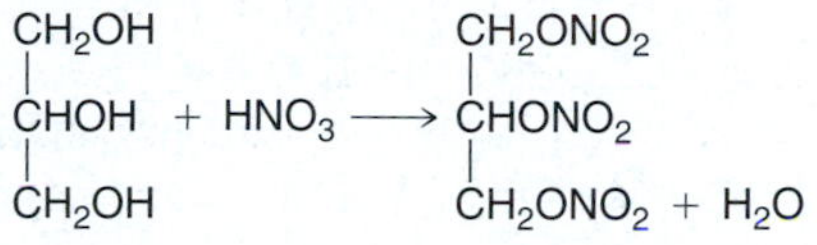

Glycerol **Nitroglycerin**

Nitroglycerine is both an explosive liquid and a blood vessel dilator that is used to treat a heart condition known as angina.
(a) Balance the chemical equation, and determine how many grams of nitroglycerine would be produced from 15.00 g of glycerol.
(b) If 9.00 g of nitric acid had been used in the reaction, which would be the limiting reactant?
(c) What is the theoretical yield in grams of nitroglycerin?
(d) If 9.3 g of nitroglycerin was produced from 9.0 g of nitric acid, what would be the percent yield?
(e) Given the data in part d, how would you rate the success of this reaction according to the criteria mentioned in the text?
(f) Derive a general expression for the theoretical yield of nitroglycerin in terms of x grams of glycerol.

143. A significant weathering reaction in geochemistry is hydration–dehydration. An example is the transformation of hematite to ferrihydrite as the relative humidity of the soil approaches 100%:

$$\underset{\textbf{Hematite}}{Fe_2O_3(s)} + H_2O(l) \longrightarrow \underset{\textbf{Ferrihydrite}}{Fe_{10}O_{15}{\cdot}9H_2O(s)}$$

This reaction occurs during advanced stages of the weathering process.
(a) Balance the chemical equation.
(b) Is this a redox reaction? Explain your answer.
(c) If 1 ton of hematite rock weathered in this manner, how many kilograms of ferrihydrite would be formed?

144. Hydrazine (N_2H_4) is used not only as a rocket fuel but also in industry to remove toxic chromates from waste water according to the chemical equation

$$4CrO_4^{2-}(aq) + 3N_2H_4(l) + 4H_2O(l) \longrightarrow 4Cr(OH)_3(s) + 3N_2(g) + 8OH^-(aq)$$

Identify the species that is oxidized and the species that is reduced. What mass of water is needed for the complete reaction of 15.0 kg of hydrazine? Write a general equation for the mass of $Cr(OH)_3$ produced from x grams of hydrazine.

145. *Corrosion* is a term for the deterioration of metals through chemical reaction with their environment. A particularly difficult problem for the archaeological chemist is the formation of CuCl, an unstable compound that is formed by corrosion of copper and its alloys. Although copper and bronze objects can survive burial for centuries without significant deterioration, exposure to air can cause cuprous chloride to react with atmospheric oxygen to form Cu_2O and cupric chloride. The cupric chloride then reacts with the free metal to produce cuprous chloride. Continued reaction of oxygen and water with cuprous chloride causes "bronze disease," which consists of spots of a pale green, powdery deposit of [$CuCl_2 \cdot 3Cu(OH)_2 \cdot H_2O$] on the surface of the object that continues to grow. Using the series of reactions described above, complete and balance the following equations, which together result in bronze disease:

Equation 1: _____ + $O_2 \longrightarrow$ _____ + _____

Equation 2: _____ + Cu $\longrightarrow$ _____

Equation 3: _____ + $O_2 + H_2O \longrightarrow$

$$CuCl_2 \cdot 3Cu(OH)_2 \cdot H_2O + CuCl_2$$

Bronze disease

(a) Which species are the oxidants and which are the reductants in each equation?
(b) If 8.0% by mass of a 350.0-kg copper statue consisted of CuCl, and the statue succumbed to bronze disease, how many pounds of the powdery green hydrate would be formed?
(c) What factors could affect the rate of deterioration of a recently excavated bronze artifact?

146. Iron submerged in seawater will react with dissolved oxygen, but when an iron object, such as a ship, sinks into the seabed where there is little or no free oxygen, the iron remains fresh until it is brought to the surface. Even in the seabed, however, iron can react with salt water according to the unbalanced equation

$$Fe(s) + NaCl(aq) + H_2O(l) \longrightarrow FeCl_2(s) + NaOH(aq) + H_2(g)$$

The ferrous chloride and water then form hydrated ferrous chloride according to the equation

$$FeCl_2(s) + 2H_2O(l) \longrightarrow FeCl_2 \cdot 2H_2O(s)$$

When the submerged iron object is removed from the seabed, the ferrous chloride dihydrate reacts with atmospheric moisture to form a solution that seeps outward, producing a characteristic "sweat" that may continue to emerge for many years. Oxygen from the air oxidizes the solution to ferric resulting in the formation of what is commonly referred to as rust (ferric oxide):

$$FeCl_2(aq) + O_2(g) \longrightarrow FeCl_3(aq) + Fe_2O_3(s)$$

The rust layer will continue to grow until arrested.
(a) Balance the chemical equations.
(b) Given a 10.0-ton ship of which 2.60% is now rust, how many kilograms of iron were converted to $FeCl_2$, assuming that the ship was pure iron?
(c) What mass of rust in grams would result?
(d) What is the overall change in the oxidation state of iron for this process?
(e) In the first equation given, what species has been reduced? What has been oxidized?

147. The glass industry uses lead oxide in the production of fine crystal glass such as crystal goblets. Lead oxide can be formed by the reaction:

$$PbS(s) + O_2(g) \longrightarrow PbO(s) + SO_2(g)$$

Balance the equation, and determine what has been oxidized and what has been reduced. How many grams of sulfur dioxide would be produced from 4.0×10^3 g of lead sulfide? Discuss some potential environmental hazards that stem from this reaction.

148. The Deacon process is one way to recover Cl_2 on-site in industrial plants where chlorination of hydrocarbons produces HCl. The reaction utilizes oxygen to oxidize HCl to chlorine, as shown.

$$HCl(g) + O_2(g) \longrightarrow Cl_2(g) + H_2O(g)$$

The reaction is frequently carried out in the presence of NO as a catalyst.
(a) Balance the chemical equation.
(b) Which compound is the oxidant, and which is the reductant?
(c) If 26 kg of HCl was produced during a chlorination reaction, how many kilograms of water would result from the Deacon process?

149. In 1834, Eilhardt Mitscherlich of the University of Berlin synthesized benzene by heating benzoic acid with calcium oxide according to this balanced equation:

$$C_6H_5CO_2H(s) + CaO(s) \xrightarrow{\Delta} C_6H_6(l) + CaCO_3(s)$$

Benzoic acid

(Heating is indicated by the symbol Δ.) How much benzene would you expect from the reaction of 16.9 g of benzoic acid and 18.4 g of calcium oxide? Which is the limiting reactant? How many grams of benzene would you expect to obtain from this reaction, assuming a 73% yield?

150. Aspirin is synthesized by the reaction of salicylic acid ($C_7H_6O_3$) with acetic anhydride ($C_4H_6O_3$) according to the equation

$$C_7H_6O_3(s) + C_4H_6O_3(l) \longrightarrow C_9H_8O_4(s) + H_2O(l)$$

Salicylic acid — Acetic anhydride — Aspirin

Balance the equation and find the limiting reactant given 10.0 g of acetic anhydride and 8.0 g of salicylic acid. How many grams of aspirin would you expect from this reaction, assuming an 83% yield?

151. Hydrofluoric acid etches glass because it dissolves silicon dioxide, as represented in the following chemical equation

$$SiO_2(s) + HF(aq) \longrightarrow SiF_6^{2-}(aq) + H^+(aq) + H_2O(l)$$

(a) Balance the equation.
(b) How many grams of silicon dioxide will react with 5.6 g of HF?
(c) How many grams of HF are needed to remove 80% of the silicon dioxide from a 4.0-kg piece of glass? (Assume that the glass is pure silicon dioxide.)

152. Lead sulfide and hydrogen peroxide react to form lead sulfate and water. This reaction is used to clean oil paintings that have blackened due to the reaction of the lead-based paints with atmospheric hydrogen sulfide.
(a) Write the balanced equation for the oxidation of lead sulfide by hydrogen peroxide.
(b) What mass of hydrogen peroxide would be needed to remove 3.4 g of lead sulfide?
(c) If the painting had originally been covered with 5.4 g of lead sulfide and you had 3.0 g of hydrogen peroxide, what percent of the lead sulfide could be removed?

153. It has been suggested that diacetylene (C_4H_2, HC≡C—C≡CH) may be the "ozone" of the outer planets. As the largest hydrocarbon yet identified in planetary atmospheres, diacetylene shields planetary surfaces from ultraviolet radiation and is itself reactive when exposed to light. One reaction of diacetylene is an important route for the formation of higher hydrocarbons, as shown in these chemical equations:

$$C_4H_2(g) + C_4H_2(g) \longrightarrow C_8H_3(g) + H(g)$$
$$C_8H_3(g) + C_4H_2(g) \longrightarrow C_{10}H_3(g) + C_2H_2(g)$$

Consider the second reaction shown.

(a) Given 18.4 mol of C_8H_3 and 1000 g of C_4H_2, which is the limiting reactant?

(b) Given 2.8×10^{24} molecules of C_8H_3 and 250 g of C_4H_2, which is the limiting reactant?

(c) Given 385 g of C_8H_3 and 200 g of C_4H_2, which is in excess? How many grams of excess reactant would remain?

(d) Suggest why this reaction might be of interest to scientists.

154. Glucose ($C_6H_{12}O_6$) can be converted to ethanol and carbon dioxide using certain enzymes. As alcohol concentrations are increased, however, catalytic activity is inhibited and alcohol production ceases.

(a) Write a balanced chemical equation for the conversion of glucose to ethanol and carbon dioxide.

(b) Given 12.6 g of glucose, how many grams of ethanol would be produced, assuming complete conversion?

(c) If 4.3 g of ethanol had been produced, what would be the percent yield for this reaction?

(d) Is a heterogeneous or homogeneous catalyst used in this reaction?

(e) You have been asked to find a way to increase the rate of this reaction given stoichiometric quantities of each reactant. How would you do this?

155. Early spacecraft developed by NASA for its manned missions used capsules that had a pure oxygen atmosphere. This practice was stopped when a spark from an electrical short in the wiring inside the capsule of the Apollo III spacecraft ignited its contents. The resulting explosion and fire killed the three astronauts on board within minutes. What chemical steps could have been taken to prevent this disaster?

ESSENTIAL SKILLS 2

TOPICS

- Proportions
- Percentages
- Unit Conversions

In Essential Skills 1, we introduced you to some of the fundamental mathematical operations you need to successfully manipulate mathematical equations in chemistry. Before proceeding to the Questions and Problems in Chapter 3, you should become familiar with the additional skills described in this section on proportions, percentages, and unit conversions.

Proportions

We can solve many problems in general chemistry by using ratios, or proportions. For example, if the ratio of some quantity A to some quantity B is known, and the relationship between these quantities is known to be constant, then any change in A (from A_1 to A_2) produces a proportional change in B (from B_1 to B_2), and vice versa. The relationship between A_1, B_1, A_2, and B_2 can be written as

$$\frac{A_1}{B_1} = \frac{A_2}{B_2} = \text{constant}$$

To solve this equation for A_2, we multiply both sides of the equality by B_2, thus canceling B_2 from the denominator:

$$B_2 \frac{A_1}{B_1} = \cancel{B_2} \frac{A_2}{\cancel{B_2}}$$

$$\frac{B_2 A_1}{B_1} = A_2$$

Similarly, we can solve for B_2 by multiplying both sides of the equality by $1/A_2$, thus canceling A_2 from the numerator:

$$\frac{1}{A_2}\frac{A_1}{B_1} = \frac{1}{\cancel{A_2}}\frac{\cancel{A_2}}{B_2}$$

$$\frac{A_1}{A_2 B_1} = \frac{1}{B_2}$$

If the values of A_1, A_2, and B_1 are known, then we can solve the left side of the equation and invert the answer to obtain B_2:

$$\frac{1}{B_2} = \text{numerical value}$$

$$B_2 = \frac{1}{\text{numerical value}}$$

If the value of A_1, A_2, or B_1 is unknown, however, we can solve for B_2 by inverting both sides of the equality:

$$B_2 = \frac{A_2 B_1}{A_1}$$

When you manipulate equations, remember that *any operation carried out on one side of the equality must be carried out on the other.*

Skill Builder E2.1 illustrates how to find the value of an unknown by using proportions.

SKILL BUILDER ES2.1

If 38.4 g of element A is needed to combine with 17.8 g of element B, then how many grams of A are needed to combine with 52.3 g of B?

Solution

We set up the proportions as follows:

$$A_1 = 38.4 \text{ g}$$
$$B_1 = 17.8 \text{ g}$$
$$A_2 = ?$$
$$B_2 = 52.3 \text{ g}$$
$$\frac{A_1}{B_1} = \frac{A_2}{B_2}$$
$$\frac{38.4 \text{ g}}{17.8 \text{ g}} = \frac{A_2}{52.3 \text{ g}}$$

Multiplying both sides of the equation by 52.3 g (B_2) gives

$$\frac{(38.4 \text{ g})(52.3 \not{\text{g}})}{17.8 \not{\text{g}}} = \frac{A_2(\cancel{52.3 \text{ g}})}{\cancel{52.3 \text{ g}}}$$
$$A_2 = 113 \text{ g}$$

Notice that grams cancel to leave us with an answer that is in the correct units. *Always check to make sure that your answer has the correct units!*

SKILL BUILDER ES2.2

Solve to find the indicated variables: **(a)** $\frac{16.4 \text{ g}}{41.2 \text{ g}} = \frac{x}{18.3 \text{ g}}$; **(b)** $\frac{2.65 \text{ m}}{4.02 \text{ m}} = \frac{3.28 \text{ m}}{y}$;
(c) $\frac{3.27 \times 10^{-3} \text{ g}}{x} = \frac{5.0 \times 10^{-1} \text{ g}}{3.2 \text{ g}}$; **(d)** $\frac{P_1}{P_2} = \frac{V_2}{V_1}$ Solve for V_1;
(e) $\frac{P_1V_1}{T_1} = \frac{P_2V_2}{T_2}$ Solve for T_1.

Solution

(a) Multiply both sides of the equality by 18.3 g to remove this measurement from the denominator:

$$\frac{16.4 \text{ g} \times 18.3 \not{\text{g}}}{41.2 \not{\text{g}}} = \frac{x(\cancel{18.3 \text{ g}})}{\cancel{18.3 \text{ g}}}$$
$$7.28 \text{ g} = x$$

(b) Multiply both sides of the equality by 1/3.28 m, solve the left side of the equation, and then invert to solve for y:

$$\frac{2.65 \not{\text{m}}}{4.02 \text{ m}(3.28 \not{\text{m}})} = \frac{\cancel{3.28 \text{ m}}}{y(\cancel{3.28 \text{ m}})} = \frac{1}{y}$$
$$y = \frac{(4.02)(3.29)}{2.65} = 4.99 \text{ m}$$

(c) Multiply both sides of the equality by $1/3.27 \times 10^{-3}$ g, solve the right side of the equation, and then invert to find x:

$$\frac{\cancel{3.27 \times 10^{-3} \text{ g}}}{(\cancel{3.27 \times 10^{-3} \text{ g}})x} = \frac{5.0 \times 10^{-1} \not{\text{g}}}{(3.27 \times 10^{-3} \not{\text{g}})(3.2 \text{ g})} = \frac{1}{x}$$
$$x = \frac{(3.2 \text{ g})(3.27 \times 10^{-3})}{5.0 \times 10^{-1}} = 2.1 \times 10^{-2} \text{ g}$$

(d) Multiply both sides of the equality by $1/V_2$, and then invert both sides to obtain V_1:

$$\frac{P_1}{P_2V_2} = \frac{\cancel{V_2}}{V_1\cancel{V_2}}$$

$$\frac{P_2V_2}{P_1} = V_1$$

(e) Multiply both sides of the equality by $1/P_1V_1$, and then invert both sides to obtain T_1:

$$\frac{\cancel{P_1V_1}}{T_1\cancel{P_1V_1}} = \frac{P_2V_2}{T_2P_1V_1}$$

$$T_1 = \frac{T_2P_1V_1}{P_2V_2}$$

Percentages

Because many measurements are reported as percentages, many chemical calculations require an understanding of how to manipulate such values. You may, for example, need to calculate the mass percentage of a substance, as described in Chapter 3, or determine the percentage of product obtained from a particular reaction mixture.

You can convert a percentage to decimal form by dividing the percentage by 100:

$$52.8\% = \frac{52.8}{100} = 0.528$$

Conversely, you can convert a decimal to a percentage by multiplying the decimal by 100%:

$$0.356 \times 100\% = 35.6\%$$

Suppose, for example, you want to determine the mass of substance A, one component of a sample with a mass of 27 mg, and you are told that the sample consists of 82% A. You begin the calculation by converting the percentage to decimal form:

$$82\% = \frac{82}{100} = 0.82$$

The mass of A can then be calculated from the mass of the sample:

$$0.82 \times 27 \text{ mg} = 22 \text{ mg}$$

The following exercises provide practice converting and using percentages.

SKILL BUILDER ES2.3

Convert each number to a percentage or to decimal form: **(a)** 29.4%; **(b)** 0.390; **(c)** 101%; **(d)** 1.023.

Solution

(a) $\frac{29.4}{100} = 0.294$

(b) $0.390 \times 100\% = 39.0\%$

(c) $\frac{101}{100} = 1.01$

(d) $1.023 \times 100\% = 102.3\%$

SKILL BUILDER ES2.4

Use percentages to answer the following questions, being sure to use the correct number of significant figures (see Essential Skills 1). Express your answer in scientific notation where appropriate. **(a)** What is the mass of hydrogen in 52.83 g of a compound that is 11.2% hydrogen? **(b)** What is the percentage of carbon in 28.4 g of a compound that contains 13.79 g of that element? **(c)** A compound that is 4.08% oxygen contains 194 mg of that element. What is the mass of the compound?

Solution

(a) $52.83 \text{ g} \times \frac{11.2}{100} = 52.83 \text{ g} \times 0.112 = 5.92 \text{ g}$

(b) $\left(\frac{13.79 \text{ g carbon}}{28.4 \text{ g}}\right) \times 100\% = 48.6\% \text{ carbon}$

(c) This problem can be solved by using a proportion:

$$\frac{4.08\% \text{ oxygen}}{100\% \text{ compound}} = \frac{194 \text{ mg}}{x \text{ mg}}$$

$$x = 4.75 \times 10^3 \text{ mg (or 4.75 g)}$$

Unit Conversions

As you learned in Essential Skills 1, all measurements must be expressed in the correct units to have any meaning. This sometimes requires converting between different units (see Table ES1.1). Conversions are carried out using conversion factors such as the ones listed on the inside of the back cover of this text. Conversion factors are ratios constructed from the relationships between different units or measurements. The relationship between milligrams and grams, for example, can be expressed as either 1 g/1000 mg or 1000 mg/1 g. When making unit conversions, use arithmetic steps accompanied by cancellation of units.

Suppose you have measured a mass in milligrams but you need to report the measurement in kilograms. In problems that involve SI units, you can use the definitions of the prefixes given in Table ES1.2 to get the necessary conversion factors. For example, you can convert milligrams to grams and then convert grams to kilograms:

$$\text{milligrams} \longrightarrow \text{grams} \longrightarrow \text{kilograms}$$
$$1000 \text{ mg} \longrightarrow 1 \text{ g}$$
$$1000 \text{ g} \longrightarrow 1 \text{ kilogram}$$

If you have measured 928 mg of a substance, you can convert that value to kilograms as follows:

$$928 \cancel{\text{mg}} \times \frac{1 \text{ g}}{1000 \cancel{\text{mg}}} = 0.928 \text{ g}$$

$$0.928 \cancel{\text{g}} \times \frac{1 \text{ kg}}{1000 \cancel{\text{g}}} = 0.000928 \text{ kg} = 9.28 \times 10^{-4} \text{ kg}$$

Notice that in each arithmetic step, units cancel as if they were algebraic variables, leaving us with an answer in kilograms. In the conversion to grams, we begin with milligrams in the numerator. Milligrams must therefore appear in the denominator of the conversion factor in order to produce an answer in grams. The individual steps may be connected as follows:

$$928 \cancel{\text{mg}} \times \frac{1 \cancel{\text{g}}}{1000 \cancel{\text{mg}}} \times \frac{1 \text{ kg}}{1000 \cancel{\text{g}}} = \frac{928 \text{ kg}}{10^6} = 928 \times 10^{-6} \text{ kg} = 9.28 \times 10^{-4} \text{ kg}$$

The next exercise provides practice converting between units.

SKILL BUILDER ES2.5

Use the information in Table ES1.2 and inside the back cover of this text to convert each measurement: **(a)** 59.2 cm to dm; **(b)** 3.7×10^5 mg to kg; **(c)** 270 mL to dm^3; **(d)** 2.04×10^3 g to tons; **(e)** 9.024×10^{10} seconds to years. Be sure that your answers contain the correct number of significant figures and that they are expressed in scientific notation where appropriate.

Solution

(a) $59.2\ \cancel{cm} \times \dfrac{1\ \cancel{m}}{100\ \cancel{cm}} \times \dfrac{10\ dm}{1\ \cancel{m}} = 5.92\ dm$

(b) $3.7 \times 10^5\ \cancel{mg} \times \dfrac{1\ \cancel{g}}{1000\ \cancel{mg}} \times \dfrac{1\ kg}{1000\ \cancel{g}} = 3.7 \times 10^{-1}\ kg$

(c) $270\ \cancel{mL} \times \dfrac{1\ \cancel{L}}{1000\ \cancel{mL}} \times \dfrac{1\ dm^3}{1\ \cancel{L}} = 270 \times 10^{-3}\ dm^3 = 2.70 \times 10^{-1}\ dm^3$

(d) $2.04 \times 10^3\ \cancel{g} \times \dfrac{1\ \cancel{lb}}{453.6\ \cancel{g}} \times \dfrac{1\ ton}{2{,}000\ \cancel{lb}} = 0.00225\ ton = 2.25 \times 10^{-3}\ tons$

(e) $9.024 \times 10^{10}\ \cancel{s} \times \dfrac{1\ \cancel{min}}{60\ \cancel{s}} \times \dfrac{1\ \cancel{h}}{60\ \cancel{min}} \times \dfrac{1\ \cancel{day}}{24\ \cancel{h}} \times \dfrac{1\ yr}{365\ \cancel{days}} = 2.86 \times 10^3\ yr$

4 Reactions in Aqueous Solution

LEARNING OBJECTIVES

- To understand how and why solutions form
- To describe the concentrations of solutions quantitatively
- To be able to balance equations that describe reactions in solution
- To solve quantitative problems involving the stoichiometry of reactions in solution
- To know the characteristic properties of acids and bases
- To know three important kinds of reactions that occur in aqueous solutions: precipitation, acid–base, and oxidation–reduction reactions
- To know how to use titration methods to analyze solutions quantitatively

CHAPTER OUTLINE

ESSENTIAL SKILLS 3

In Chapter 3, we described chemical reactions in general and introduced some techniques that are used to characterize them quantitatively. For the sake of simplicity, we only discussed situations in which the reactants and products of a given reaction were the only chemical species present. In reality, however, virtually all of the chemical reactions that take place within and around us, such as the oxidation of foods to generate energy or the treatment of an upset stomach with an antacid tablet, occur *in solution*. In fact, many reactions must be carried out in solution, and do not take place at all if the solid reactants are simply mixed.

As you learned in Chapter 1, a **solution** is a homogeneous mixture in which substances present in lesser amounts, called **solutes**, are dispersed uniformly throughout the substance in the greater amount, the **solvent**. An **aqueous**

The reaction of mercury(II) acetate with sodium iodide. When colorless aqueous solutions of each reactant are mixed, they produce a red precipitate, mercury(II) iodide, the result of a double displacement reaction.

solution is a solution in which the solvent is water, whereas in a *nonaqueous solution*, any substance other than water is the solvent. Familiar examples of nonaqueous solvents are ethyl acetate, used in nail polish removers, and turpentine, used to clean paint brushes. In this chapter, we focus on reactions that occur in aqueous solution.

There are many reasons for carrying out reactions in solution. For a chemical reaction to occur, individual atoms, molecules, or ions must collide, and collisions between two solids, which are not dispersed at the atomic, molecular, or ionic level, do not occur at a significant rate. In addition, when the amount of a substance required for a reaction is so small that it cannot be weighed accurately, the use of a solution of that substance, in which the solute is dispersed in a much larger mass of solvent, enables chemists to measure its quantity with great precision. Chemists can also control the amount of heat consumed or produced in a reaction more effectively when the reaction occurs in solution, and sometimes the nature of the reaction itself can be controlled by the choice of solvent.

This chapter introduces techniques for preparing and analyzing aqueous solutions, for balancing equations that describe reactions in solution, and for solving problems using solution stoichiometry. By the time you complete this chapter, you will know enough about aqueous solutions to explain what causes acid rain, why acid rain is harmful, and how a "breathalyzer" measures alcohol levels. You will also understand the chemistry of photographic development, be able to explain why rhubarb leaves are toxic, and learn about a possible chemical reason for the decline and fall of the Roman Empire.

4.1 Aqueous Solutions

MGC Solutions

Water, which makes up about 70% of the mass of the human body, is essential for life; many of the chemical reactions that keep us alive depend on the interaction of water molecules with dissolved compounds. Moreover, as we will discuss in Chapter 5, the presence of large amounts of water on the surface of our planet helps to maintain its surface temperature in a range suitable for life. In this section, we describe some of the interactions of water with various substances and introduce you to the characteristics of aqueous solutions.

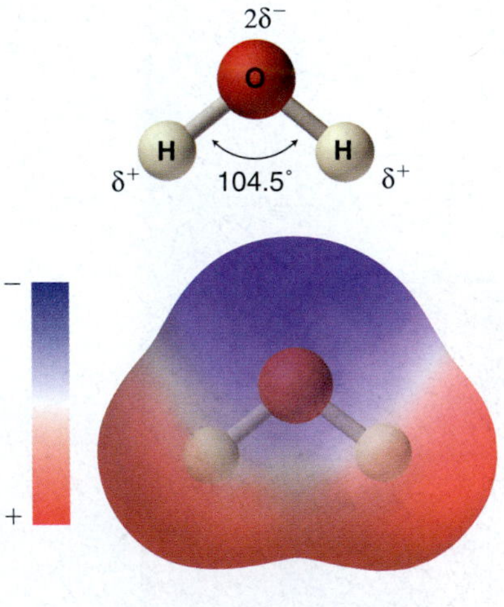

Figure 4.1 The polar nature of water. Each water molecule consists of two hydrogen atoms bonded to an oxygen atom in a bent (V-shaped) structure. Because the oxygen atom attracts electrons more strongly than hydrogen atoms do, the oxygen atom is partially negatively charged ($2\delta^-$; blue) and the hydrogen atoms are partially positively charged (δ^+; red). In order for the molecule to have a net charge of zero, the partial negative charge on oxygen must be twice as large as the partial positive charge on each hydrogen.

Polar Substances

As shown in Figure 4.1, the individual water molecule consists of two hydrogen atoms bonded to an oxygen atom in a bent (V-shaped) structure. As you will learn in Chapter 7, the oxygen atom in each O—H covalent bond attracts the electrons more strongly than the hydrogen atom does. Consequently, the O and H nuclei do not share the electrons equally. Instead, the hydrogen atoms are electron-poor compared with a neutral hydrogen atom and have a partial positive charge, indicated by the symbol δ^+. The oxygen atom, in contrast, is more electron-rich than a neutral oxygen atom, so the oxygen atom has a partial negative charge that is indicated by the symbol $2\delta^-$; its partial negative charge must be twice as large as the partial positive charge on each hydrogen in order for the molecule to have a net charge of zero. This unequal distribution of charge creates a **polar bond**. Because of the arrangement of polar bonds in a water molecule, water is described as a *polar* substance. As shown in Figure 4.1, one portion of the molecule carries a partial negative charge, while the other portion carries partial positive charges.

Because of the asymmetric charge distribution in the water molecule, adjacent water molecules are held together by attractive electrostatic ($\delta^+ \ldots \delta^-$) interactions between the partially negatively charged oxygen atom of one molecule and the partially positively charged hydrogen atoms of adjacent molecules (Figure 4.2). Energy is needed to overcome these electrostatic attractions; without them water would evaporate at a much lower temperature and neither we nor earth's oceans would exist! Ionic compounds such as NaCl are also held together by electrostatic

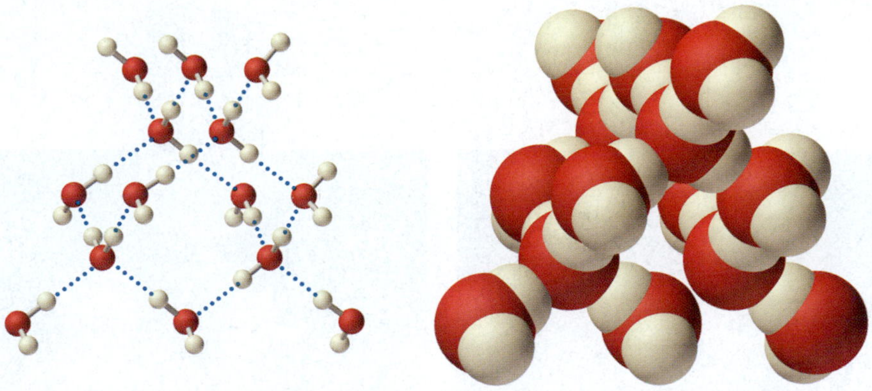

Figure 4.2 The structure of liquid water. A ball-and-stick structure is shown on the left, and the view on the right is a space-filling model. Water molecules are held together by electrostatic attractions (dotted lines) between the partially negatively charged oxygen atom of one molecule and the partially positively charged hydrogen atoms on adjacent molecules. As a result, the water molecules in liquid water form transient networks with structures similar to that shown. Because the interactions between water molecules are continually breaking and reforming, liquid water does not have a single fixed structure.

interactions—in this case, between oppositely charged ions in the highly ordered solid, where each ion is surrounded by ions of the opposite charge in a fixed arrangement. In contrast, the structure of liquid water is not completely ordered because the interactions between molecules in a liquid are constantly breaking and reforming.

Note the pattern

Polar liquids are good solvents for ionic compounds.

The unequal charge distribution in polar liquids such as water makes them good solvents for ionic compounds. When an ionic solid dissolves in water, the partially negatively charged oxygen atoms of the H_2O molecules surround the cations (Na^+ in the case of NaCl) (Figure 4.3). Furthermore, the partially positively charged hydrogen atoms in H_2O surround the anions (Cl^-) (Figure 4.3). The individual cations and anions, each surrounded by its own shell of water molecules, are called **hydrated ions**. We can describe the dissolution of NaCl in water as

$$NaCl(s) \xrightarrow{H_2O(l)} Na^+(aq) + Cl^-(aq) \tag{4.1}$$

where (aq) indicates that Na^+ and Cl^- are hydrated ions.

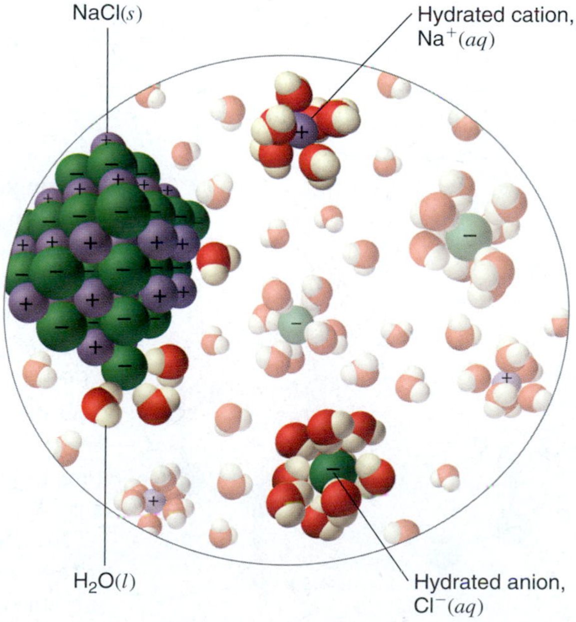

Figure 4.3 The dissolution of sodium chloride in water. An ionic solid such as NaCl dissolves in water because of the electrostatic attraction between the cations (Na^+) and the partially negatively charged oxygen atoms of water molecules, and between the anions (Cl^-) and the partially positively charged hydrogen atoms of water.

Electrolytes

When electricity, in the form of an *electrical potential*, is applied to a solution, ions in solution migrate toward the oppositely charged rod or plate to complete an electrical circuit, whereas neutral molecules in solution do not (Figure 4.4). Thus, solutions that contain ions conduct electricity, while solutions that contain only neutral molecules do not. Electrical current will flow through the circuit shown in Figure 4.4 and the bulb will glow *only* if ions are present. The lower the concentration of ions in solution, the weaker the current and the dimmer the glow. Pure water, for example, contains only very low concentrations of ions, so it is a poor electrical conductor.

An *electrolyte* is any compound that can form ions when it dissolves in water. When **strong electrolytes** dissolve, the constituent ions dissociate completely, producing aqueous solutions that conduct electricity very well (Figure 4.4). Examples

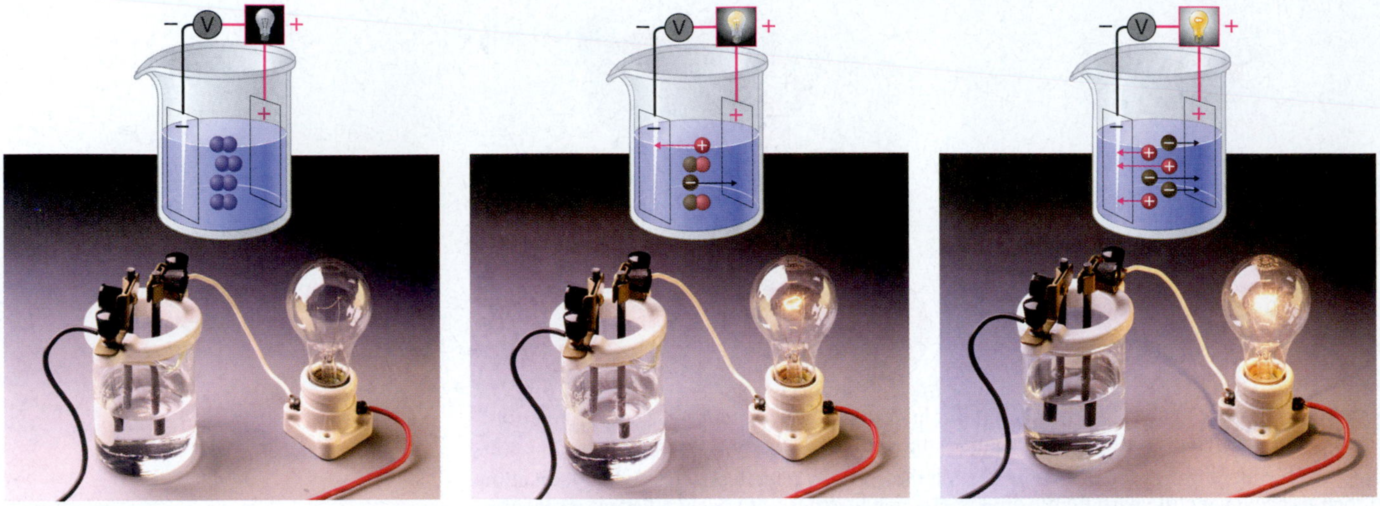

(a) Nonelectrolyte (b) Weak electrolyte (c) Strong electrolyte

Figure 4.4 The effect of ions on the electrical conductivity of water. An electrical current will flow and light the bulb only if the solution contains ions. (a) Pure water or an aqueous solution of a nonelectrolyte allows no current to flow, and the bulb does not light. (b) A weak electrolyte produces a few ions, allowing some current to flow and the bulb to glow dimly. (c) A strong electrolyte produces many ions, allowing more current to flow and the bulb to shine brightly.

include ionic compounds such as $BaCl_2$ and NaOH, which are both strong electrolytes and dissociate as follows:

$$BaCl_2(s) \xrightarrow{H_2O(l)} Ba^{2+}(aq) + 2Cl^-(aq) \quad (4.2)$$

$$NaOH(s) \xrightarrow{H_2O(l)} Na^+(aq) + OH^-(aq) \quad (4.3)$$

The single arrows from reactant to products in Equations 4.2 and 4.3 indicate that dissociation is complete.

When **weak electrolytes** dissolve, they produce relatively few ions in solution. This does *not* mean that the compounds do not dissolve readily in water; many weak electrolytes contain polar bonds and are therefore very soluble in a polar solvent such as water. They do not completely dissociate to form ions, however. Because very few of the dissolved particles are ions, aqueous solutions of weak electrolytes do not conduct electricity as well as solutions of strong electrolytes. One such compound is acetic acid, CH_3CO_2H, which contains the $—CO_2H$ unit. It is a weak acid and therefore also a weak electrolyte, whose behavior is described in more detail when we discuss acid–base reactions in Section 4.6.

Nonelectrolytes dissolve in water as neutral molecules, and thus have essentially no effect on conductivity. Examples of nonelectrolytes that are very soluble in water are ethanol, ethylene glycol, glucose, and sucrose, all of which contain the —OH group that is characteristic of alcohols. In Chapter 8, we will discuss why alcohols and carboxylic acids behave differently in aqueous solution, but for now, you can simply look for the presence of these groups when attempting to predict whether a substance is a strong electrolyte, a weak electrolyte, or a nonelectrolyte. In addition to alcohols, two other classes of organic compounds that are nonelectrolytes are *aldehydes* and *ketones*, whose general structures are shown in the margin. The distinctions between soluble and insoluble substances and between strong, weak, and nonelectrolytes are illustrated in Figure 4.5.

Note the pattern

Ionic substances and carboxylic acids are electrolytes; alcohols, aldehydes, and ketones are nonelectrolytes.

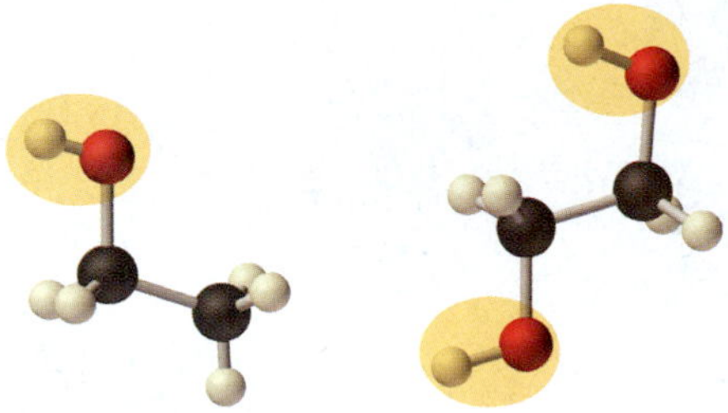

Ethanol CH_3CH_2OH Ethylene glycol CH_2OHCH_2OH

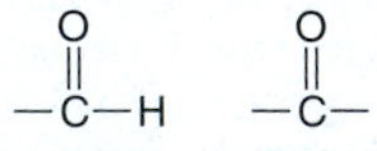

Aldehyde Ketone

Ball-and-stick model of the general structures of an aldehyde and ketone.

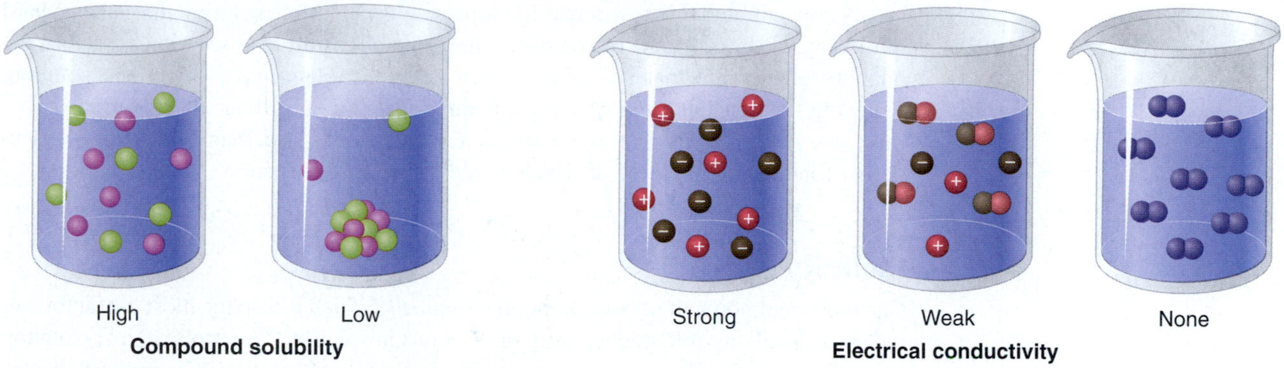

Figure 4.5 The difference between soluble and insoluble compounds (left) and strong, weak, and nonelectrolytes (right). When a soluble compound dissolves, its constituent atoms, molecules, or ions disperse throughout the solvent. In contrast, the constituents of an insoluble compound remain associated with one another in the solid. A soluble compound is a strong electrolyte if it dissociates completely into ions, a weak electrolyte if it dissociates only slightly into ions, and a nonelectrolyte if it dissolves to produce only neutral molecules.

EXAMPLE 4.1

Predict whether each compound is a strong electrolyte, a weak electrolyte, or a nonelectrolyte in water: (a) formaldehyde; (b) cesium chloride.

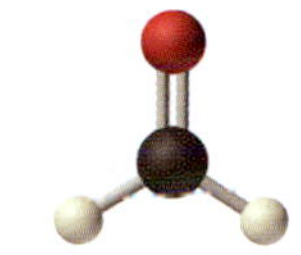

Formaldehyde
CH_2O

Given Compound

Asked for Relative ability to form ions in water

Strategy

Ⓐ Classify the compound as ionic or covalent.
Ⓑ If ionic, the compound is a strong electrolyte that, if it dissolves, will dissociate in water completely to produce a solution that conducts electricity well. If the compound is covalent and organic, determine whether it contains the carboxylic acid group. If the compound contains this group, it is a weak electrolyte. If not, it is a nonelectrolyte.

Solution

(a) Ⓐ✓ Formaldehyde is an organic compound, so it is covalent. Ⓑ✓ It contains only an aldehyde group, not a carboxylic acid group, so it should be a nonelectrolyte.
(b) Ⓐ✓ Cesium chloride, CsCl, is an ionic compound that consists of Cs^+ and Cl^- ions. Ⓑ✓ Like virtually all other ionic compounds that are soluble in water, cesium chloride will dissociate completely into $Cs^+(aq)$ and $Cl^-(aq)$ ions. Hence, CsCl should be a strong electrolyte.

EXERCISE 4.1

Predict whether each compound will act as a strong electrolyte or a nonelectrolyte in aqueous solution: (a) $(CH_3)_2CHOH$ (2-propanol); (b) ammonium sulfate.

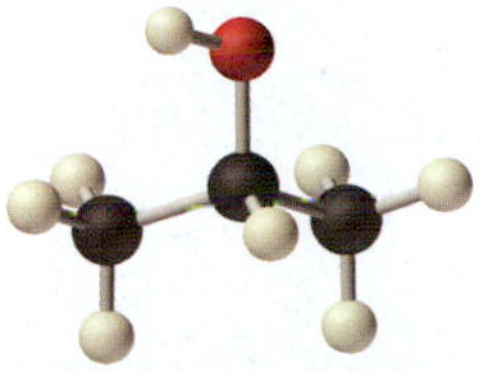

2-Propanol (isopropyl alcohol)
$(CH_3)_2CHOH$

Answer (a) nonelectrolyte; (b) strong electrolyte

4.2 • Solution Concentrations

All of us have a qualitative idea of what is meant by *concentration.* Anyone who has made instant coffee or lemonade knows that too much powder gives a strongly flavored,

MGC Molarity

highly concentrated drink, whereas too little results in a dilute solution that may be hard to distinguish from water. In chemistry, the **concentration** of a solution describes the quantity of a solute that is contained in a particular quantity of solvent or solution. Knowing the concentration of solutes is important in controlling the stoichiometry of reactants for reactions that occur in solution. Chemists use many different ways to define concentrations, some of which are described in this section.

Molarity

The most common unit of concentration is *molarity*, which is also the most useful for calculations involving the stoichiometry of reactions in solution. The **molarity** of a solution is the number of moles of solute present in exactly 1 L of solution. The molarity is also the number of millimoles of solute present in exactly 1 mL of solution:

$$\text{molarity} = \frac{\text{moles of solute}}{\text{liters of solution}} = \frac{\text{mmoles of solute}}{\text{mL of solution}} \tag{4.4}$$

The units of molarity are therefore moles per liter of solution (mol/L), abbreviated as M. Thus, an aqueous solution that contains 1 mol (342 g) of sucrose in enough water to give a final volume of 1.00 L has a sucrose concentration of 1.00 mol/L or 1.00 M. In chemical notation, the molar concentration of a solute is indicated by placing square brackets around the name or formula of the solute. Thus,

$$[\text{sucrose}] = 1.00\ M$$

is read as "the concentration of sucrose is 1.00 molar" or "the solution is 1.00 molar in sucrose." The relationships among volume, molarity, and moles may be expressed as either

$$V_{\text{L}}\, M_{\text{mol/L}} = \cancel{\text{L}}\left(\frac{\text{mol}}{\cancel{\text{L}}}\right) = \text{moles} \tag{4.5}$$

or

$$V_{\text{mL}}\, M_{\text{mmol/mL}} = \cancel{\text{mL}}\left(\frac{\text{mmol}}{\cancel{\text{mL}}}\right) = \text{mmoles} \tag{4.6}$$

The use of Equations 4.5 and 4.6 is illustrated in the following example.

EXAMPLE 4.2

Calculate the number of moles of NaOH contained in 2.50 L of 0.100 M NaOH.

Given Identity of solute and volume and molarity of solution

Asked for Amount of solute in moles

Strategy

Use either Equation 4.5. or 4.6, depending on the units given in the problem.

Solution

Because we are given the volume of the solution in liters and we are asked for the number of moles of substance, Equation 4.5 is more useful:

$$\begin{aligned}\text{moles NaOH} &= V_{\text{L}}\, M_{\text{mol/L}} \\ &= (2.50\ \cancel{\text{L}})\left(\frac{0.100\ \text{mol}}{\cancel{\text{L}}}\right) = 0.250\ \text{mol NaOH}\end{aligned}$$

EXERCISE 4.2

Calculate the number of millimoles of the compound alanine in 27.2 mL of a 1.53 *M* alanine solution.

Answer 41.6 mmol

Concentrations are often reported on a mass-to-mass basis (m/m) or on a mass-to-volume basis (m/v), particularly in clinical laboratories and engineering applications. A concentration expressed on a mass-to-mass basis is equal to the number of grams of solute per g of solution; a concentration on a mass-to-volume basis is the number of grams of solute per mL of solution. Each of these measurements can be expressed as a percentage by multiplying the ratio by 100% where the result is reported as percent m/m or percent m/v. The concentrations of very dilute solutions are often expressed in *parts per million* (*ppm*), grams of solute per 10^6 g of solution, or in *parts per billion* (*ppb*), grams of solute per 10^9 g of solution. For aqueous solutions at 20°C, one ppm corresponds to one microgram per mL, and one ppb corresponds to one nanogram per mL.

Preparation of Solutions

To prepare a solution that contains a specified concentration of a substance, it is necessary to dissolve the desired number of moles of solute in enough solvent to give the desired final volume of solution. Figure 4.6 illustrates this procedure for a solution of cobalt(II) chloride dihydrate in ethanol. Note that the volume of the *solvent* is not specified. Because the solute occupies space in the solution, the volume of the solvent that is needed is almost always *less* than the desired volume of solution. For example, if the desired volume were 1.00 L, it would be incorrect to add 1.00 L of water to 342 g of sucrose because that would produce more than 1.00 L of solution. As shown in Figure 4.7, for some substances this effect can be significant, especially in the preparation of concentrated solutions.

EXAMPLE 4.3

The solution shown in Figure 4.6 contains 10.0 g of cobalt(II) chloride dihydrate, $CoCl_2 \cdot 2H_2O$, in enough ethanol to make exactly 500 mL of solution. What is the molar concentration of $CoCl_2 \cdot 2H_2O$?

Figure 4.6 Preparation of a solution of known concentration using a solid solute.

(a) An amount of solute is weighed out on an analytical balance and then transferred to a volumetric flask.

(b) A portion of the solvent is added to the volumetric flask.

(c) The mixture is swirled until *all* of the solute is dissolved.

(d) Additional solvent is added up to the mark on the volumetric flask.

Given Mass of solute and volume of solution

Asked for Concentration (M)

Strategy

To find the number of moles of $CoCl_2{\cdot}2H_2O$, divide the mass of the compound by its molar mass. Calculate the molarity of the solution by dividing the number of moles of solute by the volume of the solution in liters.

Solution

The molar mass of $CoCl_2{\cdot}2H_2O$ is 165.87 g/mol. Therefore,

$$\text{moles } CoCl_2{\cdot}2H_2O = \frac{10.0 \cancel{\text{g}}}{165.87 \cancel{\text{g}}/\text{mol}} = 0.0603 \text{ mol}$$

The volume of the solution in liters is

$$\text{volume} = 500 \cancel{\text{mL}} \left(\frac{1 \text{ L}}{1000 \cancel{\text{mL}}}\right) = 0.500 \text{ L}$$

Molarity is the number of moles of solute per liter of solution, so the molarity of the solution is

$$\text{molarity} = \frac{0.0603 \text{ mol}}{0.500 \text{ L}} = 0.121\ M = \text{Conc. of } CoCl_2{\cdot}2H_2O$$

EXERCISE 4.3

The solution shown in Figure 4.7 contains 90.0 g of $(NH_4)_2Cr_2O_7$ in enough water to give a final volume of exactly 250 mL. What is the molar concentration of ammonium dichromate?

Answer Conc. of $(NH_4)_2Cr_2O_7$ = 1.43 M

To prepare a particular volume of a solution that contains a specified concentration of a solute, we first need to calculate the number of moles of solute in the desired volume of solution using the relationship shown in Equation 4.5. We then convert the number of moles of solute to the corresponding mass of solute needed. This procedure is illustrated in the next example.

EXAMPLE 4.4

The so-called D5W solution used for the intravenous replacement of body fluids contains 0.310 M glucose. [D5W is an approximately 5% solution of dextrose (the medical name for glucose) in

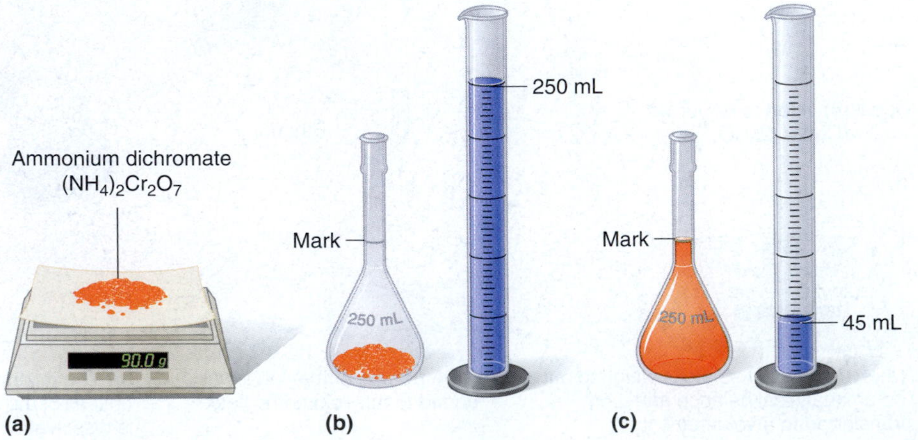

Figure 4.7 Preparation of 250 mL of a solution of $(NH_4)_2Cr_2O_7$ in water. The solute occupies space in the solution, so less than 250 mL of water is needed to make 250 mL of solution.

water.] Calculate the mass of glucose necessary to prepare a 500-mL pouch of D5W. Glucose has a molar mass of 180.16 g/mol.

Given Molarity, volume, and molar mass of solute

Asked for Mass of solute

Strategy

A Calculate the number of moles of glucose contained in the specified volume of solution by multiplying the volume of the solution by its molarity.
B Obtain the mass of glucose needed by multiplying the number of moles of the compound by its molar mass.

Solution

A We must first calculate the number of moles of glucose contained in 500 mL of a 0.310 *M* solution:

$$VM = \text{moles}$$

$$500\ \cancel{\text{mL}}\left(\frac{1\ \cancel{\text{L}}}{1000\ \cancel{\text{mL}}}\right)\left(\frac{0.310\ \text{mol glucose}}{1\ \cancel{\text{L}}}\right) = 0.155\ \text{mol glucose}$$

B We then convert the number of moles of glucose to the required mass of glucose:

$$\text{mass of glucose} = 0.155\ \cancel{\text{mol glucose}}\left(\frac{180.16\ \text{g glucose}}{1\ \cancel{\text{mol glucose}}}\right) = 27.9\ \text{g glucose}$$

EXERCISE 4.4

Another solution commonly used for intravenous injections is "normal saline," a 0.16 *M* solution of sodium chloride in water. Calculate the mass of sodium chloride needed to prepare 250 mL of normal saline solution.

Answer 2.3 g NaCl

A solution of a desired concentration can also be prepared by diluting a small volume of a more concentrated solution with additional solvent. A *stock solution*, which is a commercially prepared solution of known concentration, is often used for this purpose. Diluting a stock solution is preferred because the alternative method, weighing out tiny amounts of solute, is difficult to carry out with a high degree of accuracy. Dilution is also used to prepare solutions from substances that are sold as concentrated aqueous solutions, such as strong acids.

The procedure for preparing a solution of known concentration from a stock solution is shown in Figure 4.8. It requires calculating the number of moles of solute desired in the final volume of the more dilute solution, and then calculating the volume of the stock solution that contains this amount of solute. Remember that diluting a given quantity of stock solution with solvent does *not* change the number of moles of solute present. The relationship between the volume and concentration of the stock solution and the volume and concentration of the desired diluted solution is therefore

$$(V_s)(M_s) = \text{moles of solute} = (V_d)(M_d) \qquad (4.7)$$

where the subscripts s and d indicate the stock and dilute solutions, respectively. Example 4.5 demonstrates the calculations involved in diluting a concentrated stock solution.

EXAMPLE 4.5

What volume of a 3.00 *M* glucose stock solution is necessary to prepare 2500 mL of the D5W solution in Example 4.4?

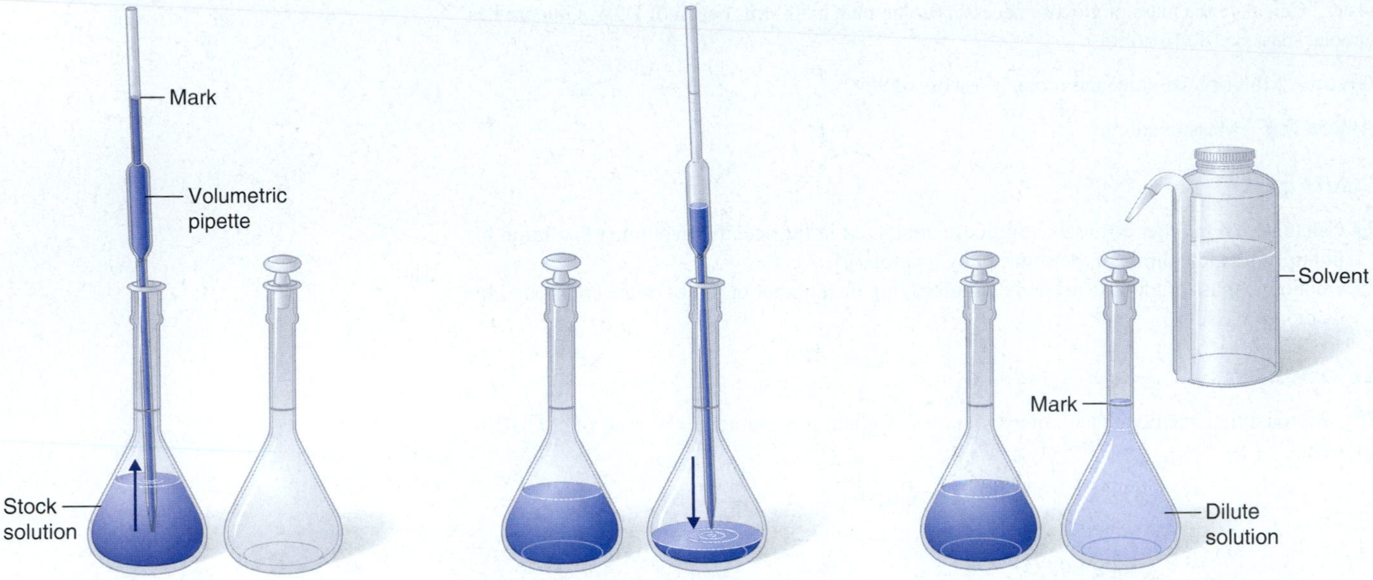

(a) A volume (V_s) containing the desired moles of solute (M_s) is measured from a stock solution of known concentration.

(b) The measured volume of stock solution is transferred to a second volumetric flask.

(c) The measured volume in the second flask is then diluted with solvent up to the volumetric mark [$(V_s)(M_s) = (V_d)(M_d)$].

Figure 4.8 Preparation of a solution of known concentration by dilution of a stock solution.

Given Volume and molarity of dilute solution

Asked for Volume of stock solution

Strategy

A Calculate the number of moles of glucose contained in the indicated volume of dilute solution by multiplying the volume of the solution by its molarity.

B To determine the volume of stock solution needed, divide the number of moles of glucose by the molarity of the stock solution.

Solution

A The D5W solution in Example 4.4 was 0.310 *M* glucose. Following our strategy, we begin by using Equation 4.7 to calculate the number of moles of glucose contained in 2500 mL of the solution:

$$\text{moles glucose} = 2500\ \cancel{\text{mL}}\left(\frac{1\ \cancel{\text{L}}}{1000\ \cancel{\text{mL}}}\right)\left(\frac{0.310\ \text{mol glucose}}{1\ \cancel{\text{L}}}\right)$$
$$= 0.775\ \text{mol glucose}$$

B We must now determine the volume of the 3.00 *M* stock solution that contains this amount of glucose:

$$\text{volume of stock soln} = 0.775\ \cancel{\text{mol glucose}}\left(\frac{1\ \text{L}}{3.00\ \cancel{\text{mol glucose}}}\right)$$
$$= 0.258\ \text{L or } 258\ \text{mL}$$

Notice that in determining the volume of stock solution that was needed, we had to divide the desired number of moles of glucose by the concentration of the stock solution to obtain the appropriate units. Also notice that the number of moles of solute in 258 mL of the stock solution is the same as the number of moles in 2500 mL of the more dilute solution; *only the amount of solvent has changed*. The answer we obtained makes sense: diluting the stock solution about 10-fold increases its volume by about a factor of 10 (258 mL $\longrightarrow$ 2500 mL). Consequently, the concentration of the solute must decrease by about a factor of 10, as it does (3.00 $M \longrightarrow$ 0.310 M).

We could also have solved this problem in a single step by solving Equation 4.7 for V_s and substituting the appropriate values:

$$V_s = \frac{(V_d)(M_d)}{M_s} = \frac{(2.500 \text{ L})(0.310 \, M)}{3.00 \, M} = 0.258 \text{ L}$$

As we have noted, there is often more than one correct way to solve a problem.

EXERCISE 4.5

Calculate the volume of a 5.0 *M* stock solution of NaCl needed to prepare 500 mL of normal saline solution (0.16 *M* NaCl).

Answer 16 mL

Ion Concentrations in Solution

In Exercise 4.3, you calculated that the concentration of a solution containing 90.00 g of ammonium dichromate in a final volume of 250 mL is 1.43 *M*. Let's consider in more detail exactly what that means. Ammonium dichromate is an ionic compound that contains two NH_4^+ ions and one $Cr_2O_7^{2-}$ ion per formula unit. Like other ionic compounds, it is a strong electrolyte that dissociates in aqueous solution to give hydrated NH_4^+ and $Cr_2O_7^{2-}$ ions:

$$(NH_4)_2Cr_2O_7(s) \xrightarrow{H_2O(l)} 2NH_4^+(aq) + Cr_2O_7^{2-}(aq) \tag{4.8}$$

Thus, 1 mol of ammonium dichromate formula units dissolves in water to produce 1 mol of $Cr_2O_7^{2-}$ anions and 2 mol of NH_4^+ cations (Figure 4.9).

When we carry out a chemical reaction using a solution of a salt such as ammonium dichromate, we need to know the concentration of each ion present in the solution. If a solution contains 1.43 *M* $(NH_4)_2Cr_2O_7$, then the concentration of $Cr_2O_7^{2-}$ must also be 1.43 *M* because there is one $Cr_2O_7^{2-}$ ion per formula unit. However, there are two NH_4^+ ions per formula unit, so the concentration of NH_4^+ ions is $2 \times 1.43 \, M = 2.86 \, M = [NH_4^+]$. Because each formula unit of $(NH_4)_2Cr_2O_7$ produces *three* ions when dissolved in water ($2 \, NH_4^+ + 1 \, Cr_2O_7^{2-}$), the *total* concentration of ions in the solution is $3 \times 1.43 \, M = 4.29 \, M$.

Figure 4.9 The dissolution of 1 mol of an ionic compound in water produces *more* than 1 mol of ions. In this case, dissolving 1 mol of $(NH_4)_2Cr_2O_7$ produces a solution that contains 1 mol of $Cr_2O_7^{2-}$ ions and 2 mol of NH_4^+ ions. (Water molecules omitted from molecular view of solution for clarity.)

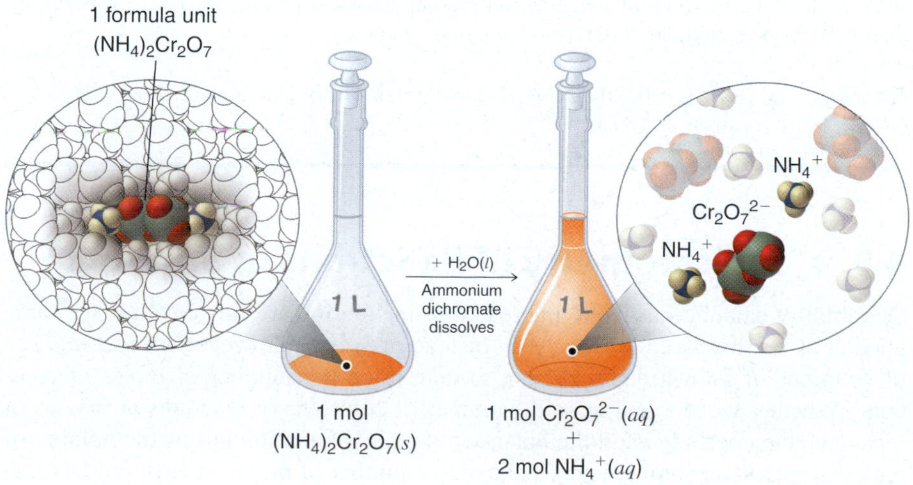

EXAMPLE 4.6

What are the concentrations of all species derived from the solutes in these aqueous solutions? **(a)** 0.21 M NaOH; **(b)** 3.7 M $(CH_3)CHOH$; **(c)** 0.032 M $In(NO_3)_3$

Given Molarity

Asked for Concentrations

Strategy

Ⓐ Classify each compound as either a strong electrolyte or a nonelectrolyte.
Ⓑ If the compound is a nonelectrolyte, its concentration is the same as the molarity of the solution. If the compound is a strong electrolyte, determine the number of each ion contained in one formula unit. Find the concentration of each species by multiplying the number of each ion by the molarity of the solution.

Solution

(a) Ⓐ✓ Sodium hydroxide is an ionic compound that is a strong electrolyte (and a strong base) in aqueous solution:

$$NaOH(s) \xrightarrow{H_2O(l)} Na^+(aq) + OH^-(aq)$$

Ⓑ✓ Because each formula unit of NaOH produces one Na^+ ion and one OH^- ion, the concentration of each ion is the same as the concentration of NaOH: $[Na^+] = 0.21\ M$ and $[OH^-] = 0.21\ M$.

(b) Ⓐ✓ The formula $(CH_3)_2CHOH$ represents 2-propanol (isopropyl alcohol) which contains the —OH group, so it is an alcohol. Recall from Section 4.1 that alcohols are covalent compounds that dissolve in water to give solutions of neutral molecules. Thus, alcohols are nonelectrolytes. Ⓑ✓ The only solute species in solution is therefore $(CH_3)_2CHOH$ molecules, so $[(CH_3)_2CHOH] = 3.7\ M$.

(c) Ⓐ✓ Indium nitrate is an ionic compound that contains In^{3+} ions and NO_3^- ions, so we expect it to behave like a strong electrolyte in aqueous solution:

$$In(NO_3)_3(s) \xrightarrow{H_2O(l)} In^{3+}(aq) + 3NO_3^-(aq)$$

Ⓑ✓ One formula unit of $In(NO_3)_3$ produces one In^{3+} ion and three NO_3^- ions, so a 0.032 M $In(NO_3)_3$ solution contains 0.032 M In^{3+} and $3 \times 0.032\ M = 0.096\ M$ NO_3^-. That is, $[In^{3+}] = 0.032\ M$ and $[NO_3^-] = 0.096\ M$.

EXERCISE 4.6

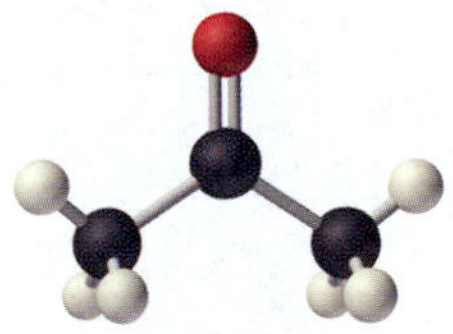

Acetone
$(CH_3)_2CO$

What are the concentrations of all the species present in these solutions? **(a)** 0.0012 M $Ba(OH)_2$; **(b)** 0.17 M Na_2SO_4; **(c)** 0.50 M $(CH_3)_2CO$, commonly known as acetone.

Answer **(a)** $[Ba^{2+}] = 0.0012\ M$, $[OH^-] = 0.0024\ M$; **(b)** $[Na^+] = 0.34\ M$, $[SO_4^{2-}] = 0.17\ M$; **(c)** $[(CH_3)_2CO] = 0.50\ M$

4.3 • Stoichiometry of Reactions in Solution

MGC Solution Stoichiometry

Quantitative calculations involving reactions in solution are carried out in the same manner as we discussed in Chapter 3. Instead of *masses*, however, we use *volumes* of solutions of known concentration to determine the number of moles of reactants. Whether we are dealing with volumes of solutions of reactants or masses of reactants, the coefficients in the balanced chemical equation tell us the number of moles of each reactant that is needed and the number of moles of each product that can be produced.

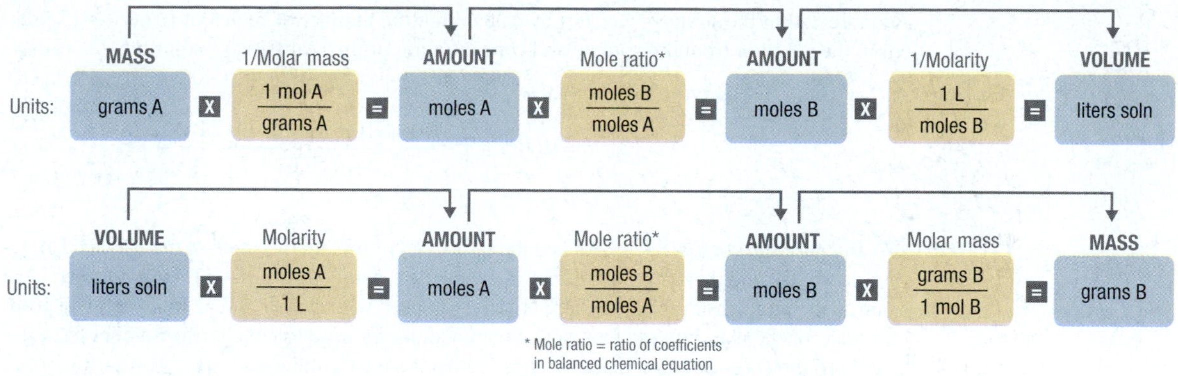

Figure 4.10 An expanded flowchart for stoichiometric calculations. Either the masses or the volumes of solutions of reactants and products can be used to determine the amounts of other species in the balanced chemical equation.

Calculating Moles from Volume

An expanded version of the flowchart for stoichiometric calculations illustrated in Figure 3.5 is shown in Figure 4.10. Notice that we can use the balanced equation for the reaction and *either* the masses of solid reactants and products *or* the volumes of solutions of reactants and products to determine the amounts of other species, as illustrated in Examples 4.7–4.9.

Note the pattern

The balanced equation for a reaction and either the masses of solid reactants and products or the volumes of solutions of reactants and products can be used in stoichiometric calculations.

EXAMPLE 4.7

Gold is extracted from its ores by treatment with aqueous cyanide solution, which causes a reaction that forms the soluble $[Au(CN)_2]^-$ ion. The gold is then recovered by reduction with metallic zinc according to the equation

$$Zn(s) + 2[Au(CN)_2]^-(aq) \longrightarrow [Zn(CN)_4]^{2-}(aq) + 2Au(s)$$

What mass of gold would you expect to recover from 400.0 L of a 3.30×10^{-4} *M* solution of $[Au(CN)_2]^-$?

Given Chemical equation and molarity and volume of reactant

Asked for Mass of product

Strategy

Ⓐ Check the chemical equation to make sure it is balanced as written. If it is not, balance the equation. Then calculate the number of moles of $[Au(CN)_2]^-$ present by multiplying the volume of the solution by its concentration.

Ⓑ From the balanced equation, use a mole ratio to calculate the number of moles of Au that can be obtained from the reaction. To calculate the mass of gold recovered, multiply the number of moles of gold by its molar mass.

Solution

Ⓐ The equation is balanced as written, so we can proceed to the stoichiometric calculation. We can adapt Figure 4.10 for this particular problem as follows:

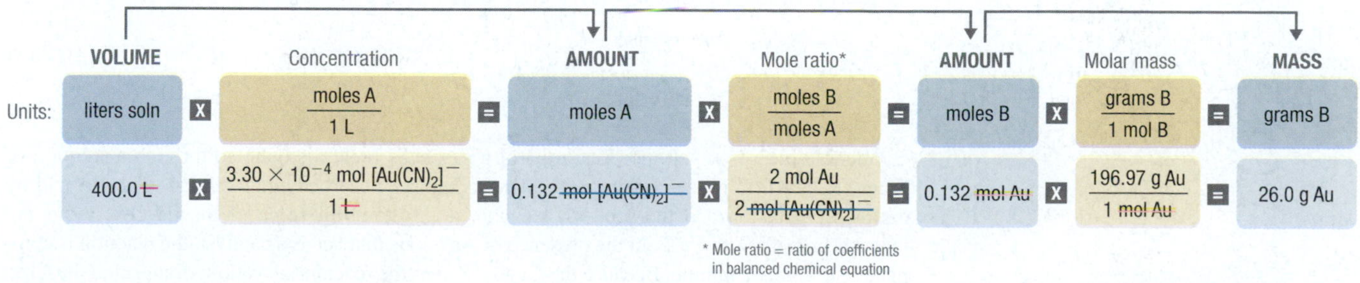

As indicated in the strategy, we start by calculating the number of moles of $[Au(CN)_2]^-$ present in the solution from the volume and concentration of the $[Au(CN)_2]^-$ solution:

$$\text{moles } [Au(CN)_2]^- = VM$$
$$= 400.0\ \cancel{L}\left(\frac{3.30 \times 10^{-4}\ \text{mol } [Au(CN)_2]^-}{1\ \cancel{L}}\right)$$
$$= 0.132\ \text{mol } [Au(CN)_2]^-$$

B✓ Because the coefficients of Au and the $[Au(CN)_2]^-$ ion are the same in the balanced equation, if we assume that Zn(s) is present in excess, the number of moles of Au produced is the same as the number of moles of $[Au(CN)_2]^-$ we started with (that is, 0.132 mol Au). The problem asks for the mass of gold that can be obtained, so we need to convert the number of moles of gold to the corresponding mass using the molar mass of gold:

$$\text{mass of Au} = (\text{moles Au})(\text{molar mass Au})$$
$$= 0.132\ \cancel{\text{mol Au}}\left(\frac{196.97\ \text{g Au}}{1\ \cancel{\text{mol Au}}}\right) = 26.0\ \text{g Au}$$

At a market price of \$300 per troy ounce (31.10 g), this amount of gold is worth

$$26.0\ \cancel{\text{g Au}} \times \frac{1\ \cancel{\text{troy oz}}}{31.10\ \cancel{\text{g}}} \times \frac{\$300}{1\ \cancel{\text{troy oz Au}}} = \$251$$

EXERCISE 4.7

What mass of solid lanthanum(III) oxalate nonahydrate, $La_2(C_2O_4)_3{\cdot}9H_2O$, can be obtained from 650 mL of a 0.0170 M aqueous solution of $LaCl_3$ by adding a stoichiometric amount of sodium oxalate?

Answer 3.89 g

Limiting Reactants in Solutions

The concept of limiting reactants applies to reactions that are carried out in solution as well as to reactions that involve pure substances. If all the reactants but one are present in excess, then the amount of the limiting reactant may be calculated as illustrated in the next example.

EXAMPLE 4.8

Because the consumption of alcoholic beverages adversely affects the performance of tasks that require skill and judgment, in most countries it is illegal to drive while under the influence of alcohol. In almost all U.S. states, a blood alcohol level of 0.08% by volume is considered legally drunk. Higher levels cause acute intoxication (0.20%), unconsciousness (~0.30%), and even death (~0.50%). The *breathalyzer* is a portable device that measures the ethanol concentration in a person's breath, which is directly proportional to the blood alcohol level. The reaction used in the breathalyzer is the oxidation of ethanol by the dichromate ion:

$$3CH_3CH_2OH(aq) + \underset{\text{Yellow-orange}}{2Cr_2O_7^{2-}(aq)} + 16H^+(aq) \xrightarrow[H_2SO_4(aq)]{Ag^+} 3CH_3CO_2H(aq) + \underset{\text{Green}}{4Cr^{3+}(aq)} + 11H_2O(l)$$

When a measured volume (52.5 mL) of a suspect's breath is bubbled through a solution of excess $K_2Cr_2O_7$ in dilute H_2SO_4, the ethanol is rapidly absorbed and oxidized to acetic acid by the dichromate ions. In the process, the chromium atoms in some of the $Cr_2O_7^{2-}$ ions are reduced from Cr^{6+} to Cr^{3+}. In the presence of Ag^+ ions that act as a catalyst, the reaction is complete in less than a minute. Because the $Cr_2O_7^{2-}$ ion (the reactant) is yellow-orange and the Cr^{3+}

ion (the product) forms a green solution, the amount of ethanol in the person's breath (the limiting reactant) can be determined quite accurately by comparing the color of the final solution with the colors of standard solutions prepared with known amounts of ethanol.

A typical breathalyzer ampule contains 3.0 mL of a 0.25 mg/mL solution of $K_2Cr_2O_7$ in 50% H_2SO_4 as well as a fixed concentration of $AgNO_3$ (typically 0.25 mg/mL is used for this purpose). How many grams of ethanol must be present in 52.5 mL of a person's breath to convert all of the Cr^{6+} to Cr^{3+}?

A breathalyzer ampule before (left) and after (right) ethanol is added. When a measured volume of a suspect's breath is bubbled through the solution, the ethanol is oxidized to acetic acid and the solution changes color from yellow-orange to green. The intensity of the green color indicates the amount of ethanol in the sample.

Given Volume and concentration of one reactant

Asked for Mass of other reactant needed for complete reaction

Strategy

A Calculate the number of moles of dichromate ion in 1 mL of the breathalyzer solution by dividing the mass of $K_2Cr_2O_7$ by its molar mass.

B Find the total number of moles of dichromate ion in the breathalyzer ampule by multiplying the number of moles contained in 1 mL by the total volume of the breathalyzer solution (3.0 mL).

C Use the mole ratios from the balanced equation to calculate the number of moles of ethanol needed to react completely with the number of moles of dichromate ions present. Then find the mass of ethanol needed by multiplying the number of moles of ethanol by its molar mass.

Solution

A In any stoichiometry problem, the first step is always to calculate the number of moles of reactants present. In this case, we are given the mass of $K_2Cr_2O_7$ in 1 mL of solution, which we can use to calculate the number of moles of $K_2Cr_2O_7$ contained in 1 mL:

$$\text{moles } K_2Cr_2O_7 / 1\text{ mL} = \frac{(0.25\ \cancel{\text{mg}}\ K_2Cr_2O_7)}{\text{mL}}\left(\frac{1\ \cancel{\text{g}}}{1000\ \cancel{\text{mg}}}\right)\left(\frac{1\text{ mol}}{294.18\ \cancel{\text{g}}\ K_2Cr_2O_7}\right)$$
$$= 8.5 \times 10^{-7}\text{ mol } K_2Cr_2O_7/\text{mL}$$

B Because 1 mol of $K_2Cr_2O_7$ produces 1 mol of $Cr_2O_7{}^{2-}$ when it dissolves, each milliliter of solution contains 8.5×10^{-7} mol of $Cr_2O_7{}^{2-}$. The total number of moles of $Cr_2O_7{}^{2-}$ in a 3.0-mL breathalyzer ampule is thus

$$\text{moles } Cr_2O_7{}^{2-} = \left(\frac{8.5 \times 10^{-7}\text{ mol}}{1\ \cancel{\text{mL}}}\right)(3.0\ \cancel{\text{mL}}) = 2.6 \times 10^{-6}\text{ mol } Cr_2O_7{}^{2-}$$

C The balanced equation tells us that 3 mol of ethanol is needed to consume 2 mol of dichromate ion, so the total number of moles of ethanol required for complete reaction is

$$\text{moles } C_2H_5OH = (2.6 \times 10^{-6}\ \cancel{\text{mol } Cr_2O_7{}^{2-}})\left(\frac{3\text{ mol } C_2H_5OH}{2\ \cancel{\text{mol } Cr_2O_7{}^{2-}}}\right) = 3.9 \times 10^{-6}\text{ mol } C_2H_5OH$$

As indicated in the strategy, this number can be converted to the mass of ethanol using its molar mass:

$$\text{mass of } C_2H_5OH = (3.9 \times 10^{-6}\ \cancel{\text{mol } C_2H_5OH})\left(\frac{46.07\text{ g}}{\cancel{\text{mol } C_2H_5OH}}\right) = 1.8 \times 10^{-4}\text{ g } C_2H_5OH$$

Thus, 1.8×10^{-4} g or 0.18 mg of ethanol must be present. Experimentally, it is found that this value corresponds to a blood alcohol level of 0.7%, which can be fatal.

EXERCISE 4.8

The compound *para*-nitrophenol (molar mass = 139 g/mol) reacts with sodium hydroxide in aqueous solution to generate a yellow anion via the reaction

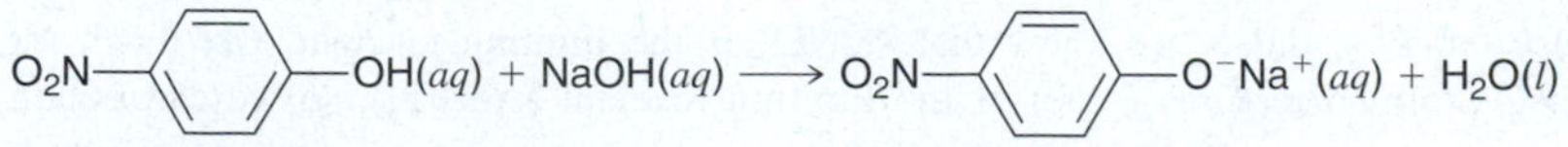

para-Nitrophenol Yellow

Because the amount of *para*-nitrophenol is easily estimated from the intensity of the yellow color that results when excess NaOH is added, reactions that produce *para*-nitrophenol are commonly used to measure the activity of enzymes, the catalysts in biological systems. What volume of 0.105 *M* NaOH must be added to 50.0 mL of a solution containing 7.20×10^{-4} g of *para*-nitrophenol to ensure that formation of the yellow anion is complete?

Answer 4.93×10^{-5} L or (49.3 μL)

In our examples, the identity of the limiting reactant has been apparent: $[Au(CN)_2]^-$, $LaCl_3$, ethanol, and *para*-nitrophenol. When the limiting reactant is not apparent, we can determine which reactant is limiting by comparing the molar amounts of the reactants with their coefficients in the balanced chemical equation, just as we did in Section 3.2. The only difference is that now we use the volumes and concentrations of solutions of reactants rather than the masses of reactants to calculate the number of moles of reactants, as illustrated in the next example.

EXAMPLE 4.9

When aqueous solutions of silver nitrate and potassium dichromate are mixed, $Ag_2Cr_2O_7$ is obtained as a red solid. The overall equation for the reaction is

$$2AgNO_3(aq) + K_2Cr_2O_7(aq) \longrightarrow Ag_2Cr_2O_7(s) + 2KNO_3(aq)$$

What mass of $Ag_2Cr_2O_7$ is formed when 500 mL of 0.17 *M* $K_2Cr_2O_7$ is mixed with 250 mL of 0.57 *M* $AgNO_3$?

Given Balanced chemical equation and volume and concentration of each reactant

Asked for Mass of product

Strategy

Ⓐ Calculate the number of moles of each reactant by multiplying the volume of each solution by its molarity.
Ⓑ Determine which of the reactants is limiting by dividing the number of moles of each reactant by its stoichiometric coefficient in the balanced chemical equation.
Ⓒ Use mole ratios to calculate the number of moles of product that can be formed from the limiting reactant. Multiply the number of moles of the product by its molar mass to obtain the corresponding mass of product.

Solution

Ⓐ The balanced chemical equation tells us that 2 mol of $AgNO_3(aq)$ reacts with 1 mol of $K_2Cr_2O_7(aq)$ to form 1 mol of $Ag_2Cr_2O_7$ (Figure 4.11). The first step is to calculate the number of moles of each reactant in the specified volumes:

$$\text{moles } K_2Cr_2O_7 = 500 \cancel{\text{mL}} \left(\frac{1 \cancel{\text{L}}}{1000 \cancel{\text{mL}}}\right)\left(\frac{0.17 \text{ mol } K_2Cr_2O_7}{1 \cancel{\text{L}}}\right) = 0.085 \text{ mol } K_2Cr_2O_7$$

$$\text{moles } AgNO_3 = 250 \cancel{\text{mL}} \left(\frac{1 \cancel{\text{L}}}{1000 \cancel{\text{mL}}}\right)\left(\frac{0.57 \text{ mol } AgNO_3}{1 \cancel{\text{L}}}\right) = 0.14 \text{ mol } AgNO_3$$

Ⓑ Now we can determine which of the reactants is limiting by dividing the number of moles of each reactant by its stoichiometric coefficient:

$$K_2Cr_2O_7\text{: } \frac{0.085 \text{ mol}}{1 \text{ mol}} = 0.085 \qquad AgNO_3\text{: } \frac{0.14 \text{ mol}}{2 \text{ mol}} = 0.070$$

Because $0.070 < 0.085$, we know that $AgNO_3$ is the limiting reactant. Ⓒ Each mole of $Ag_2Cr_2O_7$ formed requires 2 mol of the limiting reactant ($AgNO_3$), so we can obtain only

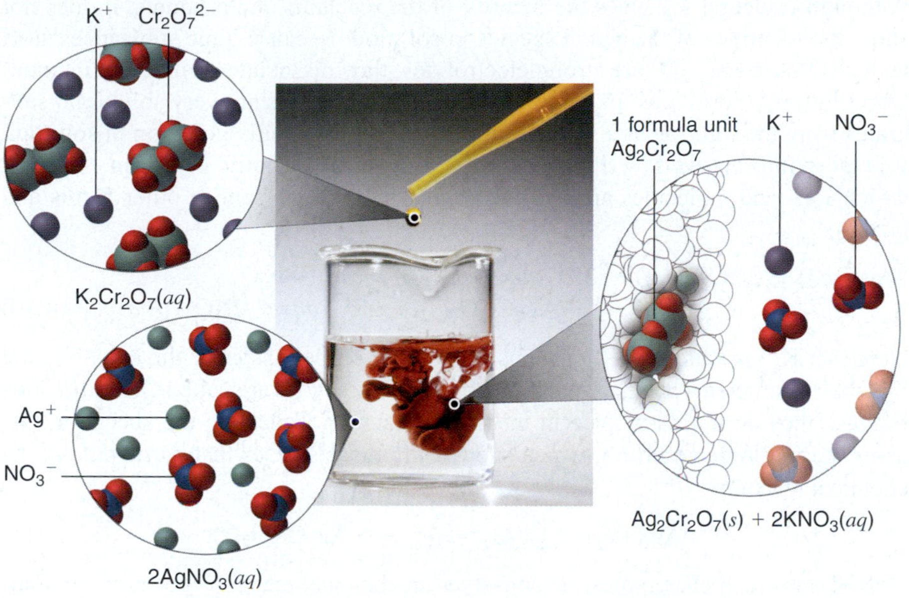

Figure 4.11 An illustration of what is happening at the molecular level when solutions of $AgNO_3$ and $K_2Cr_2O_7$ are mixed. The Ag^+ and $Cr_2O_7^{2-}$ ions form a red precipitate of solid $Ag_2Cr_2O_7$, while the K^+ and NO_3^- ions remain in solution. (Water molecules omitted from molecular views of solutions for clarity.)

0.14/2 = 0.070 mol of $Ag_2Cr_2O_7$. Finally, we convert the number of moles of $Ag_2Cr_2O_7$ to the corresponding mass:

$$\text{mass of } Ag_2Cr_2O_7 = 0.070 \text{ mol} \left(\frac{431.72 \text{ g}}{1 \text{ mol}}\right) = 30 \text{ g } Ag_2Cr_2O_7$$

EXERCISE 4.9

Aqueous solutions of sodium bicarbonate and sulfuric acid react to produce carbon dioxide according to this equation:

$$2NaHCO_3(aq) + H_2SO_4(aq) \longrightarrow 2CO_2(g) + Na_2SO_4(aq) + 2H_2O(l)$$

If 13.0 mL of 3.0 M H_2SO_4 is added to 732 mL of 0.112 M $NaHCO_3$, what mass of CO_2 is produced?

Answer 3.4 g

4.4 • Ionic Equations

MGC Net Ionic Equations

The chemical equations discussed in earlier chapters showed the identities of the reactants and products and gave the stoichiometries of the reactions, but they told us very little about what was occurring in solution. In contrast, equations that show only the hydrated species focus our attention on the chemistry that is taking place and allow us to see similarities between reactions that might not otherwise be apparent.

Let's consider the reaction of silver nitrate with potassium dichromate. As you learned in Example 4.9, when aqueous solutions of silver nitrate and potassium dichromate are mixed, $Ag_2Cr_2O_7$ forms as a red solid. The **overall equation** for the reaction shows all of the reactants and products as undissociated, electrically neutral compounds:

$$2AgNO_3(aq) + K_2Cr_2O_7(aq) \longrightarrow Ag_2Cr_2O_7(s) + 2KNO_3(aq) \quad (4.9)$$

Although Equation 4.9 gives the identity of the reactants and products, it does not show the identities of the actual species in solution. Because ionic substances such as $AgNO_3$ and $K_2Cr_2O_7$ are strong electrolytes, they dissociate completely in aqueous solution to form ions. In contrast, because $Ag_2Cr_2O_7$ is not very soluble, it separates from the solution as a solid. To find out what is actually going on in solution, it is more informative to write the reaction as a **complete ionic equation**, showing which ions and molecules are hydrated and which are present in other forms and phases:

$$2Ag^+(aq) + 2NO_3^-(aq) + 2K^+(aq) + Cr_2O_7^{2-}(aq) \longrightarrow Ag_2Cr_2O_7(s) + 2K^+(aq) + 2NO_3^-(aq) \quad (4.10)$$

Note that $K^+(aq)$ and NO_3^- (aq) ions are present on both sides of the equation and that their coefficients are the same on both sides. These ions are called **spectator ions** because they do not participate in the actual reaction. Canceling the spectator ions gives the **net ionic equation**, which shows only those species that participate in the chemical reaction:

$$2Ag^+(aq) + Cr_2O_7^{2-}(aq) \longrightarrow Ag_2Cr_2O_7(s) \quad (4.11)$$

Both mass and charge must be conserved in chemical reactions because the numbers of electrons and protons do not change. For charge to be conserved, the sum of the charges of the ions multiplied by their coefficients must be the same on both sides of the equation. In Equation 4.11, the charge on the left side is $2(+1) + 1(-2) = 0$, which is the same as the charge of a neutral $Ag_2Cr_2O_7$ formula unit.

By eliminating the spectator ions, we can focus on the chemistry that takes place in a solution. For example, the overall equation for the reaction between silver fluoride and ammonium dichromate is

$$2AgF(aq) + (NH_4)_2Cr_2O_7(aq) \longrightarrow Ag_2Cr_2O_7(s) + 2NH_4F(aq) \quad (4.12)$$

The complete ionic equation for the reaction is

$$2Ag^+(aq) + 2F^-(aq) + 2NH_4^+(aq) + Cr_2O_7^{2-}(aq) \longrightarrow Ag_2Cr_2O_7(s) + 2NH_4^+(aq) + 2F^-(aq) \quad (4.13)$$

Because $2NH_4^+(aq)$ and $2F^-(aq)$ ions appear on both sides of Equation 4.13, they are spectator ions. They can therefore be canceled to give the net ionic equation (Equation 4.14), which is identical to Equation 4.11:

$$2Ag^+(aq) + Cr_2O_7^{2-}(aq) \longrightarrow Ag_2Cr_2O_7(s) \quad (4.14)$$

If we look at net ionic equations, it becomes apparent that many different combinations of reactants can result in the same net chemical reaction. For example, we can predict that silver fluoride could be replaced by silver nitrate in the preceding reaction without affecting the outcome of the reaction.

EXAMPLE 4.10

Write the overall equation, the complete ionic equation, and the net ionic equation for the reaction of aqueous barium nitrate with aqueous sodium phosphate to give solid barium phosphate and a solution of sodium nitrate.

Given Reactants and products

Asked for Overall, complete ionic, and net ionic equations

Strategy

Balance the overall chemical equation. Write all soluble reactants and products in their dissociated form to give the complete ionic equation, and then cancel species that appear on both sides of the complete ionic equation to give the net ionic equation.

Solution

From the information given, we can write the unbalanced equation for the reaction:

$$Ba(NO_3)_2(aq) + Na_3PO_4(aq) \longrightarrow Ba_3(PO_4)_2(s) + NaNO_3(aq)$$

Because the product is $Ba_3(PO_4)_2$, which contains three Ba^{2+} ions and two PO_4^{3-} ions per formula unit, we can balance the equation by inspection:

$$3Ba(NO_3)_2(aq) + 2Na_3PO_4(aq) \longrightarrow Ba_3(PO_4)_2(s) + 6NaNO_3(aq)$$

This is the overall balanced equation for the reaction, showing the reactants and products in their undissociated form. To obtain the complete ionic equation, we write all the soluble reactants and products in their dissociated forms:

$$3Ba^{2+}(aq) + 6NO_3^{-}(aq) + 6Na^{+}(aq) + 2PO_4^{3-}(aq) \longrightarrow Ba_3(PO_4)_2(s) + 6Na^{+}(aq) + 6NO_3^{-}(aq)$$

The six $NO_3^{-}(aq)$ ions and the six $Na^{+}(aq)$ ions that appear on both sides of the equation are spectator ions that can be canceled to give the net ionic equation:

$$3Ba^{2+}(aq) + 2PO_4^{3-}(aq) \longrightarrow Ba_3(PO_4)_2(s)$$

EXERCISE 4.10

Write **(a)** the overall equation, **(b)** the complete ionic equation, and **(c)** the net ionic equation for the reaction of aqueous silver fluoride with aqueous sodium phosphate to give solid silver phosphate and a solution of sodium fluoride.

Answer

(a) $3AgF(aq) + Na_3PO_4(aq) \longrightarrow Ag_3PO_4(s) + 3NaF(aq)$

(b) $3Ag^{+}(aq) + 3F^{-}(aq) + 3Na^{+}(aq) + PO_4^{3-}(aq) \longrightarrow Ag_3PO_4(s) + 3Na^{+}(aq) + 3F^{-}(aq)$

(c) $3Ag^{+}(aq) + PO_4^{3-}(aq) \longrightarrow Ag_3PO_4(s)$

So far, we have always told you whether a reaction will occur when solutions are mixed and, if so, what products will form. As you advance in chemistry, however, it will be up to you to predict the results of mixing solutions of compounds, to anticipate what kind of reaction, if any, will occur, and to predict the identities of the products. Students tend to think that this means they are supposed to "just know" what will happen when two substances are mixed. Nothing could be further from the truth: an infinite number of chemical reactions are possible, and neither you nor anyone else could possibly memorize them all. Instead, you must begin by identifying the various reactions that *could* occur and then assessing which is the most probable (or least improbable) outcome.

The most important step in analyzing an unknown reaction is to *write down all the species—whether molecules or dissociated ions—that are actually present in the solution* (not forgetting the solvent itself), so that you can assess which of them are most likely to react with one another. The easiest way to make that kind of prediction is to attempt to place the reaction into one of several familiar classifications, refinements of the four general kinds of reactions introduced in Chapter 3 (acid–base, exchange, condensation or cleavage, and oxidation–reduction reactions). In the sections that follow, we discuss three of the most important kinds of reactions that occur in aqueous solutions: precipitation reactions, acid–base reactions, and oxidation–reduction reactions.

4.5 • Precipitation Reactions

A **precipitation reaction** is a reaction that yields an insoluble product—a **precipitate**—when two solutions are mixed. In Section 4.4, we described a precipitation reaction in which a colorless solution of silver nitrate was mixed with a yellow-orange

MGC Precipitation Reactions

solution of potassium dichromate to give a reddish precipitate of silver dichromate, $Ag_2Cr_2O_7$:

$$AgNO_3(aq) + K_2Cr_2O_7(aq) \longrightarrow Ag_2Cr_2O_7(s) + KNO_3(aq) \tag{4.15}$$

This equation has the general form of an exchange reaction:

$$AC + BD \longrightarrow \underset{\text{Insoluble}}{AD} + BC \tag{4.16}$$

Note the pattern

Precipitation reactions are a subclass of exchange reactions.

Thus, precipitation reactions are a subclass of exchange reactions that occur between ionic compounds when one of the products is insoluble. Because both components of each compound change partners, such reactions are sometimes called "double-displacement reactions." Important uses of precipitation reactions are to isolate metals that have been extracted from their ores and to recover precious metals for recycling.

Predicting Solubilities

Table 4.1 gives guidelines for predicting the solubility of a wide variety of ionic compounds. To decide whether a precipitation reaction will occur, we identify all of the species in the solution and then refer to Table 4.1 to see which, if any, combination(s) of cation and anion are likely to produce an insoluble salt. In doing so, it is important to recognize that *soluble* and *insoluble* are relative terms that span a wide range of actual solubilities. We will discuss solubilities in more detail in Chapter 17, where you will learn that very small amounts of the constituent ions remain in solution even after precipitation of an "insoluble" salt. For our purposes, however, we will assume that precipitation of an insoluble salt is complete.

Just as important as predicting the product of a reaction is knowing when a chemical reaction will *not* occur. Simply mixing solutions of two different chemical substances

TABLE 4.1 Guidelines for predicting the solubility of ionic compounds in water

Rule	Soluble		Insoluble
1	Most salts that contain an alkali metal cation (Li^+, Na^+, K^+, Rb^+, Cs^+) or ammonium (NH_4^+)		
2	Most salts that contain the nitrate (NO_3^-) anion		
3	Most salts of anions derived from monocarboxylic acids (such as $CH_3CO_2^-$)	except	silver acetate and salts of long-chain carboxylates
4	Most chloride, bromide, and iodide salts	except	salts of metal ions located in the right side of the metals, such as Cu^+, Ag^+, Pb^{2+}, and Hg_2^{2+}
5	Sulfate (SO_4^{2-}) salts that contain +1 cations, Mg^{2+}, and dipositive transition metal cations (such as Ni^{2+})	except	cations derived from main group elements with ≥ +2 charge, such as the metals of Groups 2 (except Mg^{2+}), 3 and 13, and Pb^{2+}
6	Most ionic compounds that contain the hydroxide (OH^-) and sulfide (S^{2-}) anions and alkali metal (Group 1), alkaline earth (Group 2), and ammonium cations	except	$Be(OH)_2$, $Mg(OH)_2$, and $Ca(OH)_2$
7	Carbonate (CO_3^{2-}) and phosphate (PO_4^{3-}) salts that contain alkali metal or ammonium cations	but not	most other metal cations

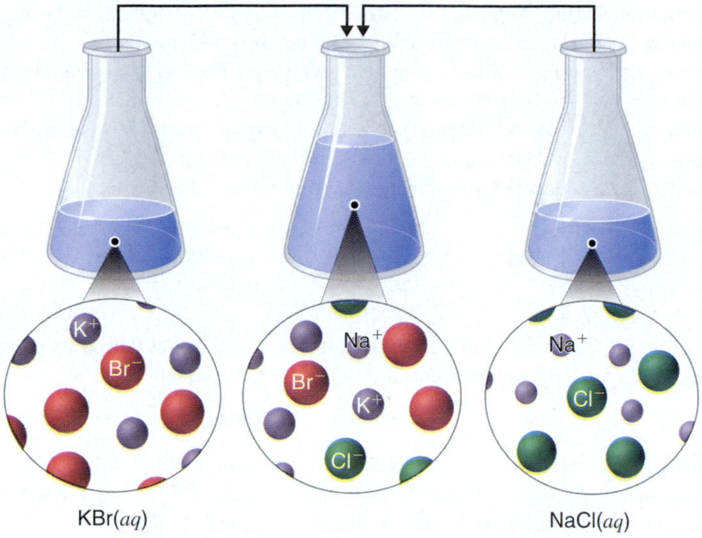

Figure 4.12 The effect of mixing aqueous KBr and NaCl solutions. Because no net reaction occurs, the only effect is to dilute each solution with the other. (Water molecules omitted from molecular views of solutions for clarity.)

does *not* guarantee that a reaction will take place. For example, if 500 mL of a 1.0 *M* aqueous NaCl solution is mixed with 500 mL of a 1.0 *M* aqueous KBr solution, the final solution has a volume of 1.00 L and contains 0.50 *M* $Na^+(aq)$, 0.50 *M* $Cl^-(aq)$, 0.50 *M* $K^+(aq)$, and 0.50 *M* $Br^-(aq)$. As you will see in the following sections, none of these species reacts with any of the others. Hence, when the solutions are mixed, the only effect is to dilute each solution with the other (Figure 4.12).

EXAMPLE 4.11

Using the information in Table 4.1, predict what will happen in each case involving strong electrolytes. Write the net ionic equation for any reaction that occurs. (a) Aqueous solutions of barium chloride and lithium sulfate are mixed. (b) Aqueous solutions of rubidium hydroxide and cobalt(II) chloride are mixed. (c) Aqueous solutions of strontium bromide and aluminum nitrate are mixed. (d) Solid lead(II) acetate is added to an aqueous solution of ammonium iodide.

Given Reactants

Asked for Reaction and net ionic equation

Strategy

A Identify the ions present in solution and write the products of all possible exchange reactions.
B Refer to Table 4.1 to determine which, if any, of the products is insoluble and will therefore form a precipitate. If a precipitate forms, write the net ionic equation for the reaction.

Solution

(a) A Because the reactants, barium chloride and lithium sulfate, are strong electrolytes, each dissociates completely in water to give a solution that contains the constituent anions and cations. Mixing the two solutions *initially* gives an aqueous solution that contains Ba^{2+}, Cl^-, Li^+, and SO_4^{2-} ions. The only possible exchange reaction is to form LiCl and $BaSO_4$:

$$Ba^{2+}(aq) + 2Cl^-(aq) + 2Li^+(aq) + SO_4^{2-}(aq)$$

B We now need to decide whether either of these products is insoluble. Table 4.1 shows that LiCl is soluble in water (rules 1 and 4) but $BaSO_4$ is not (rule 5). Hence, $BaSO_4$ will precipitate according to the net ionic equation

$$Ba^{2+}(aq) + SO_4^{2-}(aq) \longrightarrow BaSO_4(s)$$

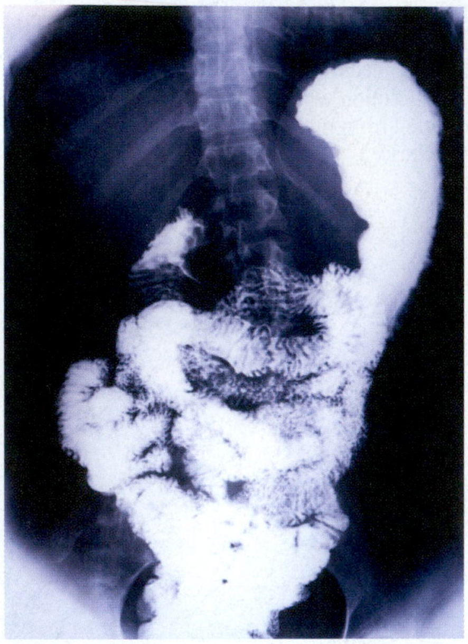

An X ray of the digestive organs of a patient who has swallowed a "barium milkshake." A barium milkshake is a suspension of very fine $BaSO_4$ particles in water; the high atomic mass of barium makes it opaque to X rays.

Although soluble barium salts are toxic, $BaSO_4$ is so insoluble that it can be used to diagnose stomach and intestinal problems without being absorbed into tissues. An outline of the digestive organs appears on X rays of patients who have been given a "barium milkshake" or a "barium enema"—a suspension of very fine $BaSO_4$ particles in water.

(b) A✓ Rubidium hydroxide and cobalt(II) chloride are strong electrolytes, so when their aqueous solutions are mixed, the resulting solution initially contains Rb^+, OH^-, Co^{2+}, and Cl^- ions. The possible products of an exchange reaction are RbCl and $Co(OH)_2$:

$$Rb^+(aq) + OH^-(aq) + Co^{2+}(aq) + 2Cl^-(aq)$$

B✓ According to Table 4.1, RbCl is soluble (rules 1 and 4) but $Co(OH)_2$ is not (rule 6). Hence, $Co(OH)_2$ will precipitate according to the net ionic equation

$$Co^{2+}(aq) + 2OH^-(aq) \longrightarrow Co(OH)_2(s)$$

(c) A✓ When aqueous solutions of strontium bromide and aluminum nitrate are mixed, we initially obtain a solution that contains Sr^{2+}, Br^-, Al^{3+}, and NO_3^- ions. The two possible products from an exchange reaction are $AlBr_3$ and $Sr(NO_3)_2$:

$$Sr^{2+}(aq) + 2Br^-(aq) + Al^{3+}(aq) + 3NO_3^-(aq)$$

B✓ According to Table 4.1, both $AlBr_3$ (rule 4) and $Sr(NO_3)_2$ (rule 2) are soluble. Thus, no net reaction will occur.

(d) A✓ According to Table 4.1, lead acetate is soluble (rule 3). Thus, solid lead acetate dissolves in water to give Pb^{2+} and $CH_3CO_2^-$ ions. Because the solution also contains NH_4^+ and I^- ions, the possible products of an exchange reaction are $NH_4CH_3CO_2$ and PbI_2:

$$NH_4^+(aq) + I^-(aq) + Pb^{2+}(aq) + 2CH_3CO_2^-(aq)$$

B✓ According to Table 4.1, ammonium acetate is soluble (rules 1 and 3) but PbI_2 is insoluble (rule 4). Thus, $Pb(CH_3CO_2)_2$ will dissolve and PbI_2 will precipitate. The net ionic equation is

$$Pb^{2+}(aq) + 2I^-(aq) \longrightarrow PbI_2(s)$$

EXERCISE 4.11

Using the information given in Table 4.1, write the net ionic equation for any reaction that occurs in each case: **(a)** An aqueous solution of strontium hydroxide is added to an aqueous solution of iron(II) chloride. **(b)** Solid potassium phosphate is added to an aqueous solution of mercury(II) perchlorate. **(c)** Solid sodium fluoride is added to an aqueous solution of ammonium formate. **(d)** Aqueous solutions of calcium bromide and cesium carbonate are mixed.

Answer

(a) $Fe^{2+}(aq) + 2OH^-(aq) \longrightarrow Fe(OH)_2(s)$
(b) $2PO_4^{3-}(aq) + 3Hg^{2+}(aq) \longrightarrow Hg_3(PO_4)_2(s)$
(c) NaF(s) dissolves; no net reaction
(d) $Ca^{2+}(aq) + CO_3^{2-}(aq) \longrightarrow CaCO_3(s)$

AgBr before exposure to light

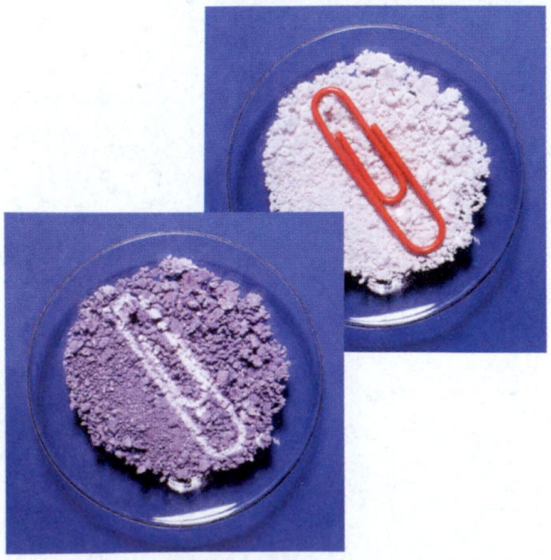

AgBr after exposure to light

Darkening of silver bromide crystals by exposure to light.

Precipitation Reactions in Photography

Precipitation reactions can be used to recover silver from solutions used to develop conventional photographic film. As shown in the margin, silver bromide is an off-white solid that turns black when it is exposed to light, due to the formation of small particles of silver metal. Black-and-white photography uses this reaction to capture images in

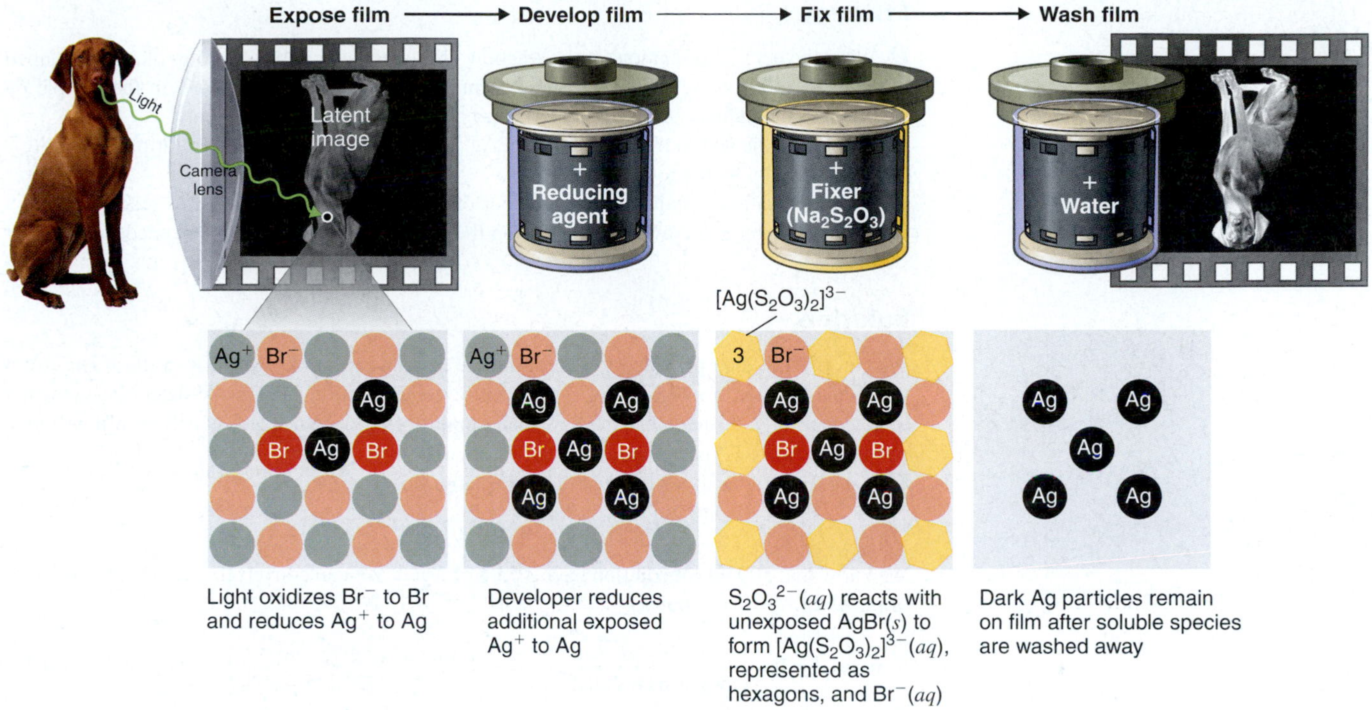

Figure 4.13 Outline of the steps involved in producing a black-and-white photograph.

shades of gray, with the darkest areas of the film corresponding to the areas that received the most light. The first step in film processing is to enhance the black/white contrast by using a developer to increase the amount of black. The developer is a reductant: because silver atoms catalyze the reduction reaction, grains of AgBr that have already been partially reduced by exposure to light react with the reductant much more rapidly than unexposed grains.

After the film is developed, any unexposed AgBr must be removed by a process called "fixing"; otherwise, the entire film would turn black upon further exposure to light. Although AgBr is insoluble in water, it is soluble in a dilute solution of sodium thiosulfate, $Na_2S_2O_3$ (called *photographer's hypo*), because of the formation of $[Ag(S_2O_3)_2]^{3-}$ ions. Thus, washing the film with thiosulfate solution dissolves unexposed AgBr and leaves a pattern of metallic silver granules that constitutes the negative. This procedure is summarized in Figure 4.13. The negative image is then projected onto paper coated with silver halides, and the developing and fixing processes are repeated to give a positive image. (Color photography works in much the same way, with a combination of silver halides and organic dyes superimposed in layers.) "Instant photo" operations can generate more than a hundred gallons of dilute silver waste solution per day. Recovery of silver from thiosulfate fixing solutions involves first removing the thiosulfate by oxidation and then precipitating Ag^+ ions with excess chloride ions.

EXAMPLE 4.12

A silver recovery unit can process 1500 L of photographic silver waste solution per day. Adding excess solid NaCl to a 500-mL sample of the waste (after removal of the thiosulfate, as described above) gives a white precipitate that, after filtration and drying, consists of 3.73 g of AgCl. What mass of NaCl must be added to the 1500 L of silver waste to ensure that all of the Ag^+ ions precipitate?

Given Volume of solution of one reactant, and mass of product from a sample of reactant solution

Asked for Mass of second reactant needed for complete reaction

Strategy

A Write the net ionic equation for the reaction. Calculate the number of moles of AgCl obtained from the 500-mL sample, and then determine the concentration of Ag^+ in the sample by dividing the number of moles of AgCl formed by the volume of solution.

B Determine the total number of moles of Ag^+ in the 1500-L solution by multiplying the Ag^+ concentration by the total volume.

C Use mole ratios to calculate the number of moles of chloride needed to react with Ag^+. Obtain the mass of NaCl by multiplying the number of moles of NaCl needed by its molar mass.

Solution

We can use the data provided to determine the concentration of Ag^+ ions in the waste, from which the number of moles of Ag^+ in the entire waste solution can be calculated. From the net ionic equation, we can determine how many moles of Cl^- are needed, which in turn will give us the mass of NaCl necessary.

A The first step is to write the net ionic equation for the reaction:

$$Cl^-(aq) + Ag^+(aq) \longrightarrow AgCl(s)$$

We know that 500 mL of solution gave 3.73 g of AgCl. We can convert this value to the number of moles of AgCl as follows:

$$\text{moles AgCl} = \frac{\text{grams AgCl}}{\text{molar mass AgCl}} = 3.73\ \cancel{\text{g AgCl}}\left(\frac{1\ \text{mol AgCl}}{143.32\ \cancel{\text{g AgCl}}}\right) = 0.0260\ \text{mol AgCl}$$

Therefore, the 500-mL sample of the solution contained 0.0260 mol of Ag^+. The Ag^+ concentration is

$$[Ag^+] = \frac{\text{moles } Ag^+}{\text{liters soln}} = \frac{0.0260\ \text{mol AgCl}}{0.500\ \text{L}} = 0.0520\ M$$

B The total number of moles of Ag^+ present in 1500 L of solution is

$$\text{moles } Ag^+ = 1500\ \cancel{\text{L}}\left(\frac{0.520\ \text{mol}}{1\ \cancel{\text{L}}}\right) = 78.1\ \text{mol } Ag^+$$

C According to the net ionic equation, one Cl^- ion is required for each Ag^+ ion. Thus, 78.1 mol of NaCl is needed to precipitate the silver. The corresponding mass of NaCl is

$$\text{mass of NaCl} = 78.1\ \cancel{\text{mol NaCl}}\left(\frac{58.44\ \text{g NaCl}}{1\ \cancel{\text{mol NaCl}}}\right) = 4560\ \text{g NaCl} = 4.56\ \text{kg NaCl}$$

Note that 78.1 mol of AgCl corresponds to 8.43 kg of metallic silver, which is worth about $1160 at current prices ($4.27 per troy ounce). Thus, silver recovery may be economically attractive as well as ecologically sound, although the procedure outlined is becoming increasingly obsolete with the growth of digital photography.

EXERCISE 4.12

Because of its toxicity, arsenic is the active ingredient in many pesticides. The arsenic content of a pesticide can be measured by oxidizing the arsenic compounds to the arsenate ion, AsO_4^{3-}, which forms an insoluble silver salt, Ag_3AsO_4. Suppose you are asked to assess the purity of technical grade sodium arsenite, $NaAsO_2$, the active ingredient in a pesticide used against termites. You dissolve a 10.00-g sample in water, oxidize it to arsenate, and dilute it with water to a final volume of 500 mL. You then add excess $AgNO_3$ solution to a 50.0-mL sample of the arsenate solution. The resulting precipitate of Ag_3AsO_4 has a mass of 3.24 g after drying. What is the percentage by mass of $NaAsO_2$ in the original sample?

Answer 91.0%

4.6 ◦ Acid–Base Reactions

Acid–base reactions are essential in both biochemistry and industrial chemistry. Moreover, many of the substances that we encounter in the household, the supermarket, and the pharmacy are acids or bases. For example, aspirin is an acid (acetylsalicylic acid) and antacids are bases. Every amateur chef who has prepared mayonnaise or squeezed a wedge of lemon to marinate a piece of fish has carried out an acid–base reaction. Before we discuss the characteristics of such reactions, let's first describe some of the properties of acids and bases.

MGC Acids, Bases, and Salts

Definitions of Acids and Bases

In Chapter 2, we defined *acids* as substances that dissolve in water to produce H^+ ions, whereas *bases* were defined as substances that dissolve in water to produce OH^- ions. In fact, this is only one possible set of definitions. Although the general properties of acids and bases have been known for more than a thousand years, the definitions of *acid* and *base* have changed dramatically as scientists have learned more about them. In ancient times, an acid was any substance that had a sour taste (such as vinegar or lemon juice), caused consistent color changes in dyes derived from plants (for example, turning blue litmus paper red), reacted with certain metals to produce hydrogen gas and a solution of a salt containing a metal cation, and dissolved carbonate salts such as limestone ($CaCO_3$) with the evolution of carbon dioxide. In contrast, a base was any substance that had a bitter taste, felt slippery to the touch, and caused color changes in plant dyes that differed diametrically from the changes caused by acids (for example, turning red litmus paper blue). Although these definitions were useful, they were entirely descriptive.

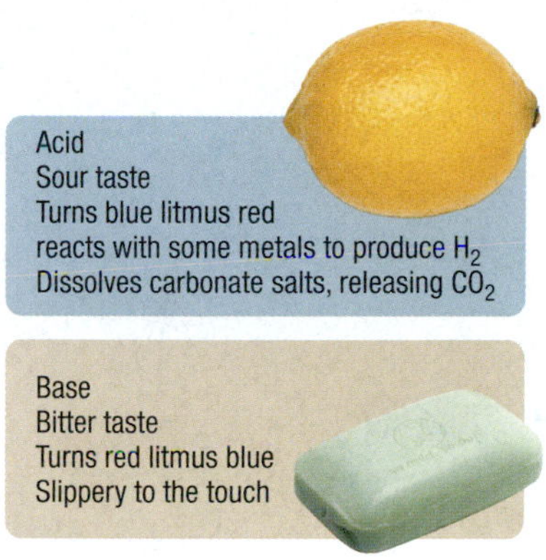

The Arrhenius Definition of Acids and Bases

The first person to define acids and bases in detail was the Swedish chemist Svante Arrhenius (1859–1927; Nobel Prize in chemistry, 1903). According to the *Arrhenius definition*, an acid is a substance like HCl that dissolves in water to produce H^+ ions (protons) (Equation 4.17), and a base is a substance like NaOH that dissolves in water to produce hydroxide (OH^-) ions (Equation 4.18):

$$\underset{\text{An Arrhenius acid}}{HCl(g)} \xrightarrow{H_2O(l)} H^+(aq) + Cl^-(aq) \tag{4.17}$$

$$\underset{\text{An Arrhenius base}}{NaOH(s)} \xrightarrow{H_2O(l)} Na^+(aq) + OH^-(aq) \tag{4.18}$$

According to Arrhenius, the characteristic properties of acids and bases are due exclusively to the presence of H^+ and OH^- ions, respectively, in solution.

Although Arrhenius's ideas were widely accepted, his definition of acids and bases had two major limitations. First, because acids and bases were defined in terms of ions obtained from water, the Arrhenius concept applied only to substances in aqueous solution. Second, and more important, the Arrhenius definition predicted that *only* substances that dissolve in water to produce H^+ and OH^- ions should exhibit the properties of acids and bases, respectively. For example, according to the Arrhenius definition, the reaction of ammonia, a base, with gaseous HCl, an acid, to give ammonium chloride (Equation 4.19) is not an acid–base reaction because it does not involve H^+ and OH^-:

$$NH_3(g) + HCl(g) \longrightarrow NH_4Cl(s) \tag{4.19}$$

The Brønsted–Lowry Definition of Acids and Bases

Because of the limitations of the Arrhenius definition, a more general definition of acids and bases was needed. One was proposed independently in 1923 by the Danish chemist

J. N. Brønsted (1879–1947) and the British chemist T. M. Lowry (1874–1936), who defined acid–base reactions in terms of the transfer of a proton (H^+ ion) from one substance to another.

According to Brønsted and Lowry, an **acid** is any substance that can donate a proton, and a **base** is any substance that can accept a proton. Thus, the Brønsted–Lowry definition of an acid is essentially the same as Arrhenius's definition, except that it is not restricted to aqueous solutions. The Brønsted–Lowry definition of a base, however, is far more general because the hydroxide ion is just one of many substances that can accept a proton. Ammonia, for example, reacts with a proton to form NH_4^+, so in Equation 4.19, NH_3 is a Brønsted–Lowry base and HCl is a Brønsted–Lowry acid. Because of its more general nature, the Brønsted–Lowry definition is used throughout this text unless otherwise specified. We will present a third definition—Lewis acids and bases—in Chapter 8 when we discuss molecular structure.

Polyprotic Acids

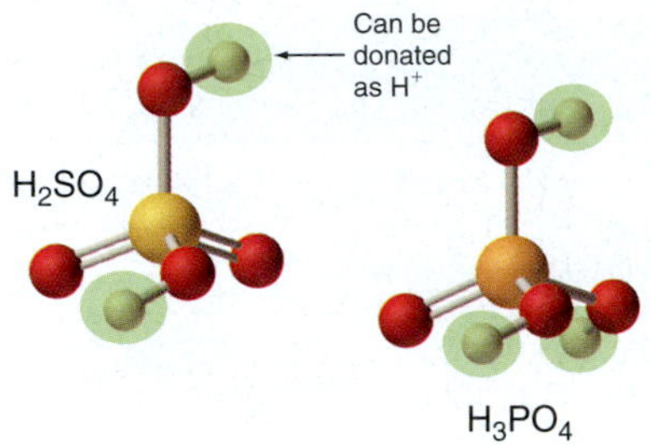

Acids differ in the number of hydrogen ions they can donate. For example, **monoprotic acids** are compounds that are capable of donating a single proton per molecule. Monoprotic acids include HF, HCl, HBr, HI, HNO_3, and HNO_2. All carboxylic acids that contain a single $-CO_2H$ group, such as acetic acid (CH_3CO_2H), are monoprotic acids, dissociating to form RCO_2^- and H^+ (Section 4.1). **Polyprotic acids** can donate more than one hydrogen ion per molecule. For example, H_2SO_4 can donate two hydrogen ions in separate steps, so it is a *diprotic acid*, and H_3PO_4, which is capable of donating three protons in successive steps, is a *triprotic acid* (Equations 4.20 a–c):

$$H_3PO_4(l) \xrightleftharpoons{H_2O(l)} H^+(aq) + H_2PO_4^-(aq) \qquad (4.20a)$$

$$H_2PO_4^-(aq) \rightleftharpoons H^+(aq) + HPO_4^{2-}(aq) \qquad (4.20b)$$

$$HPO_4^{2-}(aq) \rightleftharpoons H^+(aq) + PO_4^{3-}(aq) \qquad (4.20c)$$

In chemical equations such as these, a double arrow is used to indicate that both the forward and reverse reactions occur simultaneously, so the forward reaction does not go to completion. Instead, the solution contains significant amounts of both reactants and products. Over time the reaction reaches a state in which the concentration of each species in solution remains constant. The reaction is then said to be in *equilibrium*.

Strengths of Acids and Bases

We will not discuss the strengths of acids and bases quantitatively until Chapter 16. Qualitatively, however, we can state that **strong acids** react essentially completely with water to give H^+ and the corresponding anion. Similarly, **strong bases** dissociate essentially completely in water to give OH^- and the corresponding cation. Thus, strong acids and strong bases are both strong electrolytes. In contrast, only a fraction of the molecules of **weak acids** and **weak bases** react with water to produce ions, so weak acids and weak bases are also weak electrolytes. Typically less than 5% of a weak electrolyte dissociates into ions in solution, whereas more than 95% is present in undissociated form.

In practice, only a few strong acids are commonly encountered: HCl, HBr, HI, HNO_3, $HClO_4$, and H_2SO_4. The most common strong bases are ionic compounds that contain the hydroxide ion as the anion; three examples are NaOH, KOH, and $Ca(OH)_2$. Common weak acids include HCN, H_2S, HF, oxoacids such as HNO_2 and HClO, and carboxylic acids such as acetic acid. The ionization reaction of acetic acid is

$$CH_3CO_2H(l) \xrightleftharpoons{H_2O(l)} H^+(aq) + CH_3CO_2^-(aq) \qquad (4.21)$$

Thus, although acetic acid is *very* soluble in water, almost all of the acetic acid in solution exists in the form of neutral molecules (less than 1% dissociates), as we stated in Section 4.1. Sulfuric acid is unusual in that it is a strong acid when it donates its first proton (Equation 4.22a) but a weak acid when it donates its second proton (Equation 4.22b) as indicated by the single and double arrows, respectively:

$$\underset{\text{Strong acid}}{H_2SO_4(l)} \xrightarrow{H_2O(l)} H^+(aq) + HSO_4^-(aq) \quad (4.22a)$$

$$\underset{\text{Weak acid}}{HSO_4^-(aq)} \rightleftharpoons H^+(aq) + SO_4^{2-}(aq) \quad (4.22b)$$

Consequently, an aqueous solution of sulfuric acid contains $H^+(aq)$ ions and a mixture of $HSO_4^-(aq)$ and $SO_4^{2-}(aq)$ ions but no H_2SO_4 molecules.

The most common weak base is ammonia, which reacts with water to form small amounts of hydroxide ion:

$$NH_3(g) + H_2O(l) \rightleftharpoons NH_4^+(aq) + OH^-(aq) \quad (4.23)$$

Most of the ammonia (>99%) is present in the form of $NH_3(aq)$. Amines, which are organic analogues of ammonia, are also weak bases, as are ionic compounds that contain anions derived from weak acids (such as S^{2-}).

Some common strong acids and bases are listed in Table 4.2. Acids other than the six common strong acids are almost invariably weak acids. The only common strong bases are the hydroxides of the alkali metals and the heavier alkaline earths (Ca, Sr, and Ba); any other bases you encounter are most likely weak. Remember that *there is no correlation between solubility and whether a substance is a strong or a weak electrolyte!* Many weak acids and bases are extremely soluble in water.

EXAMPLE 4.13

Classify each of the following as a strong acid, weak acid, strong base, weak base, or none of these: **(a)** $CH_3CH_2CO_2H$; **(b)** CH_3OH; **(c)** $Sr(OH)_2$; **(d)** $CH_3CH_2NH_2$; **(e)** $HBrO_4$.

Given Compound

Asked for Acid or base strength

TABLE 4.2 Common strong acids and bases

Strong Acids		Strong Bases	
Hydrogen Halides	**Oxoacids**	**Group 1 Hydroxides**	**Hydroxides of the Heavier Group 2 Elements**
Hydrochloric acid, HCl	Nitric acid, HNO_3	LiOH	$Ca(OH)_2$
Hydrobromic acid, HBr	Sulfuric acid, H_2SO_4	NaOH	$Sr(OH)_2$
Hydroiodic acid, HI	Perchloric acid, $HClO_4$	KOH	$Ba(OH)_2$
		RbOH	
		CsOH	

Strategy

A Determine whether the compound is organic or inorganic.

B If inorganic, determine whether the compound is acidic or basic by the presence of dissociable H^+ or OH^- ions, respectively. If organic, identify the compound as a weak base or a weak acid by the presence of an amine or a carboxylic acid group, respectively. Recall that all polyprotic acids except H_2SO_4 are weak acids.

Solution

(a) A✓ This compound is propionic acid, which is organic. B✓ It contains a carboxylic acid group analogous to that in acetic acid, so it must be a weak acid.

(b) A✓ CH_3OH is methanol, an organic compound that contains the $-OH$ group. B✓ As a covalent compound, it does not dissociate to form the OH^- ion. Because it does not contain a carboxylic acid ($-CO_2H$) group, methanol cannot dissociate to form $H^+(aq)$ ions either. Thus, we predict that in aqueous solution methanol is neither an acid nor a base.

(c) A✓ $Sr(OH)_2$ is an inorganic compound that contains one Sr^{2+} and two OH^- ions per formula unit. B✓ We therefore expect it to be a strong base, similar to $Ca(OH)_2$.

(d) A✓ $CH_3CH_2NH_2$ is an amine (ethylamine), an organic compound. B✓ Consequently, we expect it to behave similarly to ammonia (Equation 4.23), reacting with water to produce small amounts of the OH^- ion. Ethylamine is therefore a weak base.

(e) A✓ $HBrO_4$ is perbromic acid, an inorganic compound. B✓ It is not listed in Table 4.2 as one of the common strong acids, but that does not necessarily mean that it is a weak acid. If you examine the periodic table, you can see that Br lies directly below Cl in Group 17. We might therefore expect that $HBrO_4$ is chemically similar to $HClO_4$, a strong acid—and, in fact, it is.

EXERCISE 4.13

Classify each of the following as a strong acid, weak acid, strong base, weak base, or none of these: (a) $Ba(OH)_2$; (b) $HBrO_4$; (c) $CH_3CH_2CH_2CO_2H$; (d) $(CH_3)_2NH$; (e) CH_2O.

Answer (a) strong base; (b) strong acid; (c) weak acid; (d) weak base; (e) none of these—formaldehyde is a neutral molecule.

The Hydronium Ion

Because isolated protons are *very* unstable and hence very reactive, an acid never simply "loses" an H^+ ion. Instead, the proton is always transferred to another substance, which acts as a base in the Brønsted–Lowry definition. Thus, in every acid–base reaction, one species acts as an acid and one species acts as a base. Occasionally the same substance performs both roles, as you will see below. When a strong acid dissolves in water, the proton that is released is transferred to a H_2O molecule that acts as a proton acceptor or base, as shown for the dissociation of sulfuric acid:

$$\underset{\substack{\text{Acid} \\ \text{(proton donor)}}}{H_2SO_4(l)} + \underset{\substack{\text{Base} \\ \text{(proton acceptor)}}}{H_2O(l)} \longrightarrow \underset{\text{Acid}}{H_3O^+(aq)} + \underset{\text{Base}}{HSO_4^-(aq)} \tag{4.24}$$

Hydronium ion
H_3O^+

Technically, therefore, it is imprecise to describe the dissociation of a strong acid as producing $H^+(aq)$ ions, as we have been doing. The resulting H_3O^+ ion, called the **hydronium ion**, is a more accurate representation of $H^+(aq)$. For the sake of brevity, however, in discussing acid dissociation reactions, we will often show the product as $H^+(aq)$ (as in Equation 4.21) with the understanding that the product is actually the $H_3O^+(aq)$ ion.

Conversely, bases that do not contain the hydroxide ion accept a proton from water, so small amounts of OH^- are produced, as in

$$\underset{\text{Base}}{NH_3(g)} + \underset{\text{Acid}}{H_2O(l)} \rightleftharpoons \underset{\text{Acid}}{NH_4^+(aq)} + \underset{\text{Base}}{OH^-(aq)} \tag{4.25}$$

Again, the double arrow indicates that the reaction does not go to completion but rather reaches a state of equilibrium. In this reaction, water acts as an acid by donating a proton to ammonia, and ammonia acts as a base by accepting a proton from water. Thus, water can act as either an acid or a base by donating a proton to a base or by accepting a proton from an acid. Substances that can behave as both an acid and a base are said to be *amphoteric*.

Notice that the products of an acid–base reaction are also an acid and a base. In Equation 4.24, for example, the products of the reaction are the hydronium ion, here an acid, and the HSO_4^- ion, here a weak base. In Equation 4.25, the products are NH_4^+, an acid, and OH^-, a base. The product NH_4^+ is called the *conjugate acid* of the base NH_3, and the product OH^- is called the *conjugate base* of the acid H_2O. Thus, all acid–base reactions actually involve two *conjugate acid–base pairs*; in Equation 4.25, they are NH_4^+/NH_3 and H_2O/OH^-. We will describe the relationship between conjugate acid–base pairs in more detail in Chapter 16.

Neutralization Reactions

A **neutralization reaction** is one in which an acid and a base react in stoichiometric amounts to produce water and a **salt**, the general term for any ionic substance that does not have OH^- as the anion or H^+ as the cation. If the base is a metal hydroxide, then the general formula for the reaction of an acid with a base is described as: *Acid plus base yields water plus salt*. For example, the reaction of equimolar amounts of HBr and NaOH to give water and a salt (NaBr) is a neutralization reaction:

Note the pattern

Acid plus base yields water plus salt.

$$\underset{\text{Acid}}{HBr(aq)} + \underset{\text{Base}}{NaOH(aq)} \longrightarrow \underset{\text{Water}}{H_2O(l)} + \underset{\text{Salt}}{NaBr(aq)} \tag{4.26}$$

If we write out the complete ionic equation for the reaction in Equation 4.26, we see that $Na^+(aq)$ and $Br^-(aq)$ are spectator ions and hence are not involved in the reaction:

$$H^+(aq) + \cancel{Br^-(aq)} + \cancel{Na^+(aq)} + OH^-(aq) \longrightarrow H_2O(l) + \cancel{Na^+(aq)} + \cancel{Br^-(aq)} \tag{4.27}$$

The overall reaction is therefore simply the combination of $H^+(aq)$ and $OH^-(aq)$ to produce H_2O, as shown in the net ionic equation:

$$H^+(aq) + OH^-(aq) \longrightarrow H_2O(l) \tag{4.28}$$

The net ionic equation for the reaction of any strong acid with any strong base is identical to Equation 4.28.

The strengths of the acid and the base generally determine whether the reaction goes to completion. The reaction of *any* strong acid with *any* strong base goes essentially to completion, as does the reaction of a strong acid with a weak base, and a weak acid with a strong base. Examples of the last two are

Note the pattern

The reaction of a strong acid with a strong base, a weak acid with a strong base, and a strong acid with a weak base go to completion.

$$\underset{\substack{\text{Strong}\\\text{acid}}}{HCl(aq)} + \underset{\substack{\text{Weak}\\\text{base}}}{NH_3(aq)} \longrightarrow \underset{\text{Salt}}{NH_4Cl(aq)} \tag{4.29}$$

$$\underset{\substack{\text{Weak}\\\text{acid}}}{CH_3CO_2H(aq)} + \underset{\substack{\text{Strong}\\\text{base}}}{NaOH(aq)} \longrightarrow \underset{\text{Salt}}{CH_3CO_2Na(aq)} + H_2O(l) \tag{4.30}$$

Notice that sodium acetate is written with the organic component first followed by the cation, as is usual for organic salts. Most reactions of a weak acid with a weak base also go essentially to completion. One example is the reaction of acetic acid with ammonia:

$$\underset{\substack{\text{Weak}\\\text{acid}}}{CH_3CO_2H(aq)} + \underset{\substack{\text{Weak}\\\text{base}}}{NH_3(aq)} \longrightarrow \underset{\text{Salt}}{CH_3CO_2NH_4(aq)} \tag{4.31}$$

An example of an acid–base reaction that does not go to completion is the reaction of a weak acid or a weak base with water, which is both an extremely weak acid

Figure 4.14 The reaction of dilute aqueous HNO_3 with a solution of Na_2CO_3. Note the vigorous formation of gaseous CO_2.

and an extremely weak base. We will discuss these reactions in more detail in Chapter 16.

In some cases, the reaction of an acid with an anion derived from a weak acid (such as HS^-) produces a gas (in this case, H_2S). Because the gaseous product escapes from solution in the form of bubbles, these reactions tend to be forced, or driven, to completion. Examples include reactions in which an acid is added to ionic compounds that contain the HCO_3^-, CN^-, or S^{2-} anions, all of which are driven to completion (Figure 4.14):

$$\begin{aligned} HCO_3^-(aq) + H^+(aq) &\longrightarrow H_2CO_3(aq) \\ H_2CO_3(aq) &\longrightarrow CO_2(g) + H_2O(l) \end{aligned} \tag{4.32}$$

$$CN^-(aq) + H^+(aq) \longrightarrow HCN(g) \tag{4.33}$$

$$\begin{aligned} S^{2-}(aq) + H^+(aq) &\longrightarrow HS^-(aq) \\ HS^-(aq) + H^+(aq) &\longrightarrow H_2S(g) \end{aligned} \tag{4.34}$$

The reactions in Equation 4.34 are responsible for the rotten egg smell that is produced when metal sulfides come in contact with acids.

EXAMPLE 4.14

Calcium propionate is used to inhibit the growth of molds in foods, tobacco, and some medicines. Write a balanced equation for the reaction of aqueous propionic acid, $CH_3CH_2CO_2H$, with aqueous calcium hydroxide to give calcium propionate. Do you expect this reaction to go to completion, making it a feasible method for the preparation of calcium propionate?

Given Reactants and product

Asked for Balanced chemical equation and whether the reaction will go to completion

Strategy

Write the balanced equation for the reaction of propionic acid with calcium hydroxide. Based on their acid and base strengths, predict whether the reaction will go to completion.

Solution

Propionic acid is $CH_3CH_2CO_2H$, an organic compound that is a weak acid, and calcium hydroxide is $Ca(OH)_2$, an inorganic compound that is a strong base. The balanced chemical equation is

$$2CH_3CH_2CO_2H(aq) + Ca(OH)_2(aq) \longrightarrow (CH_3CH_2CO_2)_2Ca(aq) + 2H_2O(l)$$

The reaction of a weak acid and a strong base will go to completion, so it is reasonable to prepare calcium propionate by mixing solutions of propionic acid and calcium hydroxide in a 2:1 mole ratio.

EXERCISE 4.14

Write a balanced equation for the reaction of solid sodium acetate with dilute sulfuric acid to give sodium sulfate.

Answer $2CH_3CO_2Na(s) + H_2SO_4(aq) \longrightarrow Na_2SO_4(aq) + 2CH_3CO_2H(aq)$

One of the most familiar and most heavily advertised applications of acid–base chemistry is *antacids*, which are bases that neutralize stomach acid. The human stomach contains an approximately 0.1 *M* solution of hydrochloric acid that helps digest foods. If the protective lining of the stomach breaks down, this acid can attack the stomach tissue, resulting in the formation of an *ulcer*. Because one factor that is believed to contribute to the formation of stomach ulcers is the production of excess acid in the stomach, many individuals routinely consume large quantities of antacids. The active ingredients in antacids include $NaHCO_3$ and $KHCO_3$ (Alka-Seltzer); a mixture of $Mg(OH)_2$ and $Al(OH)_3$ (Maalox, Mylanta); $CaCO_3$ (Tums); and a complex salt, $Na^+[Al(OH)_2CO_3]^-$ (original Rolaids). Each has certain advantages and disadvantages. For example, $Mg(OH)_2$ is a powerful laxative (it is the active ingredient in milk of magnesia), whereas $Al(OH)_3$ causes constipation. When mixed, each tends to counteract the unwanted effects of the other. Although all antacids contain both an anionic base (OH^-, CO_3^{2-}, or HCO_3^-) and an appropriate cation, they differ substantially in the amount of active ingredient in a given mass of product.

The reaction of an antacid tablet with 0.1 *M* HCl (approximately the concentration found in the human stomach).

EXAMPLE 4.15

Assume that the stomach of someone suffering from acid indigestion contains 75 mL of 0.20 *M* HCl. How many Tums tablets are required to neutralize 90% of the stomach acid, if each tablet contains 500 mg of $CaCO_3$? (Neutralizing all of the stomach acid is not desirable because that would shut down digestion completely.)

Given Volume and molarity of acid, and mass of base in an antacid tablet

Asked for Number of tablets required for 90% neutralization

Strategy

A Write the balanced equation for the reaction, and then decide whether the reaction will go to completion.

B Calculate the number of moles of acid present. Multiply the number of moles by the percentage to obtain the quantity of acid that must be neutralized. Using mole ratios, calculate the number of moles of base required to neutralize the acid.

C Calculate the number of moles of base contained in one tablet by dividing the mass of base by the corresponding molar mass. Calculate the number of tablets required by dividing the moles of base by the moles contained in one tablet.

Solution

A We first write the balanced equation for the reaction:

$$2HCl(aq) + CaCO_3(s) \longrightarrow CaCl_2(aq) + H_2CO_3(aq)$$

Each carbonate ion can react with 2 mol of H^+ to produce H_2CO_3, which rapidly decomposes to H_2O and CO_2. Because HCl is a strong acid and CO_3^{2-} is a weak base, the reaction will go to completion. **B** Next we need to determine the number of moles of HCl present:

$$75\ \cancel{mL}\left(\frac{1\ \cancel{L}}{1000\ \cancel{mL}}\right)\left(\frac{0.20\text{ mol HCl}}{\cancel{L}}\right) = 0.015\text{ mol HCl}$$

Because we want to neutralize only 90% of the acid present, we multiply the number of moles of HCl by 0.90:

$$(0.015\text{ mol HCl})(0.90) = 0.014\text{ mol HCl}$$

We know from the stoichiometry of the reaction that each mole of $CaCO_3$ reacts with 2 mol of HCl, so we need

$$\text{moles } CaCO_3 = 0.014\ \cancel{\text{mol HCl}}\left(\frac{1\text{ mol }CaCO_3}{2\ \cancel{\text{mol HCl}}}\right) = 0.0070\text{ mol }CaCO_3$$

C Each Tums tablet contains

$$\left(\frac{500\ \cancel{\text{mg }CaCO_3}}{1\text{ Tums tablet}}\right)\left(\frac{1\ \cancel{g}}{1000\ \cancel{\text{mg }CaCO_3}}\right)\left(\frac{1\text{ mol }CaCO_3}{100.1\ \cancel{g}}\right) = 0.00500\text{ mol }CaCO_3$$

Thus, we need $\dfrac{0.0070\ \cancel{\text{mol }CaCO_3}}{0.00500\ \cancel{\text{mol }CaCO_3}/\text{Tums}} = 1.4$ Tums tablets.

EXERCISE 4.15

Assume that as a result of overeating a person's stomach contains 300 mL of 0.25 *M* HCl. How many Rolaids tablets must be consumed to neutralize 95% of the acid, if each tablet contains 400 mg of $NaAl(OH)_2CO_3$? The neutralization reaction can be written as

$$NaAl(OH)_2CO_3(s) + 4HCl(aq) \longrightarrow AlCl_3(aq) + NaCl(aq) + CO_2(g) + 3H_2O(l)$$

Answer 6.4 tablets

The pH Scale

One of the key factors affecting reactions that occur in dilute solutions of acids and bases is the concentration of H^+ and OH^- ions. The **pH scale** provides a convenient way of expressing the hydrogen ion (H^+) concentration of a solution and enables us to describe acidity or basicity in quantitative terms.

Pure liquid water contains extremely low but measurable concentrations of $H_3O^+(aq)$ and $OH^-(aq)$ ions produced via an *autoionization reaction*, in which water acts simultaneously as an acid and as a base:

$$H_2O(l) + H_2O(l) \rightleftharpoons H_3O^+(aq) + OH^-(aq) \tag{4.35}$$

The concentration of hydrogen ions in pure water is only 1.0×10^{-7} *M* at 25°C. Because the autoionization reaction produces both a proton and a hydroxide ion, the OH^- concentration in pure water is also 1.0×10^{-7} *M*. Thus, pure water is a **neutral solution**, in which $[H^+] = [OH^-] = 1.0 \times 10^{-7}$ *M*.

The pH scale describes the hydrogen ion concentration of a solution in a way that avoids the use of exponential notation: **pH** is defined as the negative base-10 logarithm of the hydrogen ion concentration:*

$$pH = -\log[H^+] \tag{4.36}$$

Conversely,

$$[H^+] = 10^{-pH} \tag{4.37}$$

(If you are not familiar with logarithms or with using a calculator to obtain logarithms and antilogarithms, consult Essential Skills 3, which follows this chapter.)

Because the hydrogen ion concentration is 1.0×10^{-7} *M* in pure water at 25°C, the pH of pure liquid water (and, by extension, of any neutral solution) is

$$pH = -\log[1.0 \times 10^{-7}] = 7.00 \tag{4.38}$$

Note the pattern

pH decreases with increasing $[H^+]$.

Adding an acid to pure water increases the hydrogen ion concentration and decreases the hydroxide ion concentration because a neutralization reaction occurs, such as that shown in Equation 4.28. Because the negative exponent of $[H^+]$ becomes smaller as $[H^+]$ increases, the pH *decreases* with increasing $[H^+]$. For example, a 1.0 *M* solution of a strong monoprotic acid such as HCl or HNO_3 has a pH of 0.00:

$$pH = -\log[1.0] = 0.00 \tag{4.39}$$

Conversely, adding a base to pure water increases the hydroxide ion concentration and decreases the hydrogen ion concentration. Because the autoionization reaction of water does not go to completion, neither does the neutralization reaction. Thus, even a strongly

* The pH is actually defined as the negative base-10 logarithm of the hydrogen ion *activity*. As you will learn in a more advanced course, the activity of a substance in solution is related to its concentration. For dilute solutions such as those we are discussing, the activity and the concentration are approximately the same.

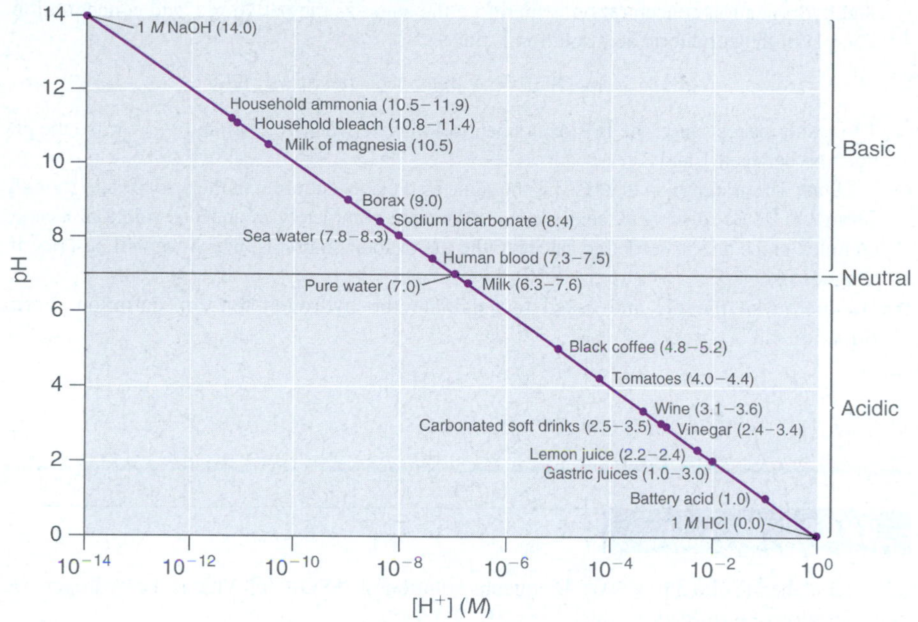

Figure 4.15 A plot of pH versus $[H^+]$ for some common aqueous solutions. While many substances exist in a range of pH values (indicated in parentheses) they are plotted using typical values. Interactive Graph

basic solution contains a detectable amount of H^+ ions. For example, a 1.0 *M* OH^- solution has $[H^+] = 1.0 \times 10^{-14}$ *M*. The pH of a 1.0 *M* NaOH solution is therefore

$$pH = -\log[1.0 \times 10^{-14}] = 14.00 \tag{4.40}$$

For practical purposes, the pH scale runs from pH = 0 (corresponding to 1 *M* H^+) to pH 14 (corresponding to 1 *M* OH^-) although pH values less than 0 or greater than 14 are possible.

We can summarize the relationships between acidity, basicity, and pH as follows:

If pH = 7.0, the solution is neutral.

If pH $<$ 7.0, the solution is acidic.

If pH $>$ 7.0, the solution is basic.

Keep in mind that the pH scale is logarithmic, so a change of 1.0 in the pH of a solution corresponds to a *10-fold* change in the hydrogen ion concentration. The foods and consumer products we encounter daily represent a wide range of pH values, as shown in Figure 4.15.

EXAMPLE 4.16

(a) What is the pH of a 2.1×10^{-2} *M* aqueous solution of $HClO_4$? **(b)** The pH of a vinegar sample is 3.80. What is its hydrogen ion concentration?

Given Molarity of acid or pH

Asked for pH or $[H^+]$

Strategy

Using the balanced equation for the acid dissociation reaction and Equation 4.38 or 4.37, determine $[H^+]$ and convert it to pH, or vice versa.

Solution

(a) $HClO_4$, perchloric acid, is one of the common strong acids, so it dissociates completely into hydrogen ions and perchlorate ions:

$$HClO_4(l) \longrightarrow H^+(aq) + ClO_4^-(aq)$$

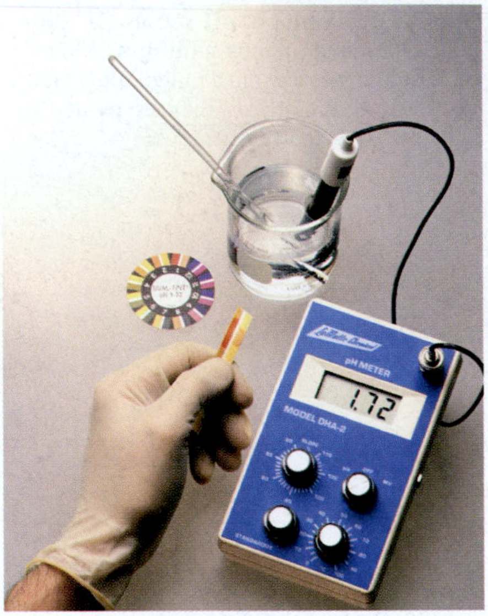

Figure 4.16 Two ways of measuring the pH of a solution: pH paper and a pH meter. Note that both show that the pH is 1.7, but the pH meter gives a more precise value.

The hydrogen ion concentration is therefore the same as the perchloric acid concentration. The pH of the perchloric acid solution is thus

$$pH = -\log[H^+] = -\log(2.1 \times 10^{-2}) = 1.68$$

The result makes sense: the H^+ ion concentration is between 10^{-1} *M* and 10^{-2} *M*, so the pH must be between 1 and 2.

Note: The assumption that $[H^+]$ is the same as the concentration of the acid is valid for *only* strong acids! Because weak acids do *not* dissociate completely in aqueous solution, a more complex procedure is needed to calculate the pH of their solutions, which we will describe in Chapter 16.

(b) We are given the pH and asked to calculate the hydrogen ion concentration. From Equation 4.37

$$10^{-pH} = [H^+]$$

Thus, $[H^+] = 10^{-3.80} = 1.6 \times 10^{-4}$ *M*.

EXERCISE 4.16

(a) What is the pH of a 3.0×10^{-5} *M* aqueous solution of HNO_3? **(b)** What is the hydrogen ion concentration of turnip juice, which has a pH of 5.41?

Answer **(a)** pH = 4.52; **(b)** $[H^+] = 3.9 \times 10^{-6}$ *M*

Tools have been developed that make the measurement of pH simple and convenient (Figure 4.16). For example, pH paper consists of strips of paper impregnated with one or more acid–base **indicators**, which are intensely colored organic molecules whose colors change dramatically depending on the pH of the solution. Placing a drop of a solution on a strip of pH paper and comparing its color with standards give the solution's approximate pH. A more accurate tool, the pH meter, makes use of a glass electrode, a device whose voltage depends on the hydrogen ion concentration.

4.7 • The Chemistry of Acid Rain

MGC pH Calculation

Acid–base reactions can have a strong environmental impact. For example, a dramatic increase in the acidity of rain and snow over the last 150 years is dissolving marble and limestone surfaces, accelerating the corrosion of metal objects, and decreasing the pH of natural waters. This environmental problem is called **acid rain** and has significant consequences for all of us. To understand acid rain requires an understanding of acid–base reactions in aqueous solution.

The term *acid rain* is actually somewhat misleading because even pure rainwater collected in areas remote from civilization is slightly acidic (pH ≈ 5.6) due to dissolved carbon dioxide, which reacts with water to give carbonic acid, a weak acid:

$$CO_2(g) + H_2O(l) \rightleftharpoons H_2CO_3(aq) \rightleftharpoons H^+(aq) + HCO_3^-(aq) \quad (4.41)$$

The English chemist Robert Angus Smith is generally credited with coining the phrase *acid rain* in 1872 to describe the increased acidity of the rain in British industrial centers (such as Manchester), which was apparently caused by the unbridled excesses of the early Industrial Revolution, although the connection was not yet understood. At that time, there was no good way to measure hydrogen ion concentrations, so it is difficult to know the actual pH of the rain observed by Smith. Typical pH values for rain in the continental United States now range from 4 to 4.5, with values as low as 2.0 reported for areas such as Los Angeles. Note from Figure 4.15 that rain with a pH of 2 is comparable in acidity to lemon juice, and even "normal" rain is now as acidic as tomato juice or black coffee.

What is the source of the increased acidity in rain and snow? Chemical analysis shows the presence of large quantities of sulfate (SO_4^{2-}) and nitrate (NO_3^-) ions, and a wide variety of evidence indicates that a significant fraction of these species come from nitrogen and sulfur oxides produced during the combustion of fossil fuels. At the high temperatures found in internal combustion engines and in lightning discharges, molecular nitrogen and molecular oxygen react to give nitric oxide:

$$N_2(g) + O_2(g) \longrightarrow 2NO(g) \tag{4.42}$$

Nitric oxide then reacts rapidly with excess oxygen to give nitrogen dioxide, the compound responsible for the brown color of smog:

$$2NO(g) + O_2(g) \longrightarrow 2NO_2(g) \tag{4.43}$$

When nitrogen dioxide dissolves in water, it forms a 1:1 mixture of nitrous acid and nitric acid:

$$2NO_2(g) + H_2O(l) \longrightarrow HNO_2(aq) + HNO_3(aq) \tag{4.44}$$

Because molecular oxygen eventually oxidizes nitrous acid to nitric acid, the overall reaction is

$$2N_2(g) + 5O_2(g) + 2H_2O(l) \longrightarrow 4HNO_3(aq) \tag{4.45}$$

Large amounts of sulfur dioxide have always been released into the atmosphere by natural sources such as volcanoes, forest fires, and the microbial decay of organic materials, but for most of our planet's recorded history the natural cycling of sulfur from the atmosphere into oceans and rocks kept the acidity of rain and snow in check. Unfortunately, the burning of fossil fuels seems to have tipped the balance. Many coals contain as much as 5–6% pyrite (FeS_2) by mass, and fuel oils typically contain at least 0.5% sulfur by mass. Since the mid-19th century, these fuels have been burned on a huge scale to supply the energy needs of our modern industrial society, releasing tens of millions of tons of additional SO_2 into the atmosphere annually. In addition, roasting sulfide ores to obtain metals such as zinc and copper produces large amounts of SO_2 via reactions such as

$$2ZnS(s) + 3O_2(g) \longrightarrow 2ZnO(s) + 2SO_2(g) \tag{4.46}$$

Regardless of the source, the SO_2 dissolves in rainwater to give sulfurous acid (Equation 4.47a), which is eventually oxidized by oxygen to sulfuric acid (Equation 4.47b):

$$SO_2(g) + H_2O(l) \longrightarrow H_2SO_3(aq) \tag{4.47a}$$

$$2H_2SO_3(aq) + O_2(g) \longrightarrow 2H_2SO_4(aq) \tag{4.47b}$$

Concerns about the harmful effects of acid rain have led to strong pressure on industry to minimize the release of SO_2 and NO. For example, coal-burning power plants now use SO_2 "scrubbers," which trap SO_2 by its reaction with lime (CaO) to produce calcium sulfite dihydrate ($CaSO_3 \cdot 2H_2O$) (Figure 4.17).

The damage that acid rain does to limestone and marble buildings and sculptures is due to a classic acid–base reaction. Marble and limestone both consist of calcium carbonate, $CaCO_3$, which is a salt derived from the weak acid H_2CO_3. As we saw in Section 4.6, the reaction of a strong acid with a salt of a weak acid goes to completion. Thus, we can write the reaction of limestone or marble with dilute sulfuric acid as

$$CaCO_3(s) + H_2SO_4(aq) \longrightarrow CaSO_4(s) + H_2O(l) + CO_2(g) \tag{4.48}$$

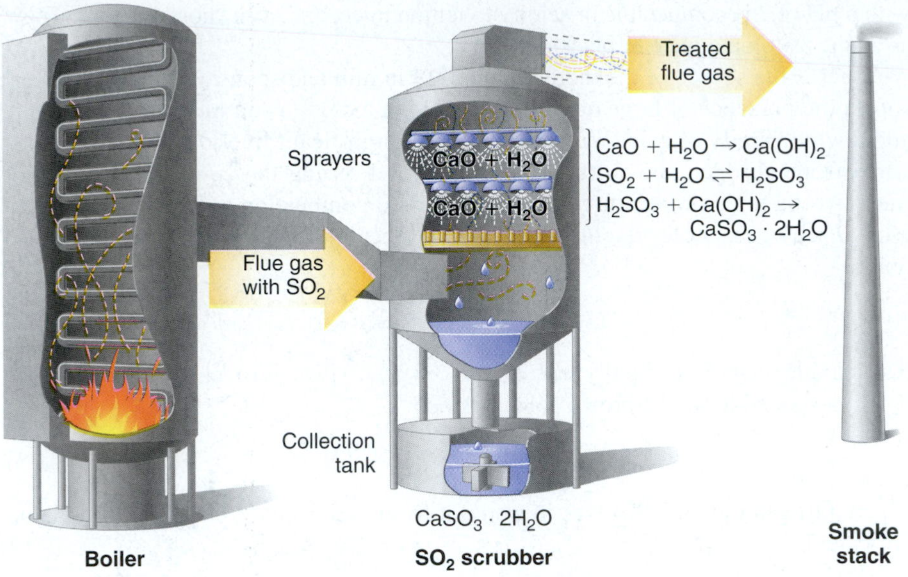

Figure 4.17 Schematic diagram of a wet scrubber system. In coal-burning power plants, SO_2 can be removed ("scrubbed") from exhaust gases by its reaction with a lime (CaO) and water spray to produce calcium sulfite dihydrate ($CaSO_3 \cdot 2H_2O$). Removing SO_2 from the gases prevents its conversion to SO_3 and subsequent reaction with rainwater ("acid rain"). Scrubbing systems are now commonly used to minimize the environmental effects of large-scale fossil fuel combustion.

Figure 4.18 Acid rain damage to a statue of George Washington. Both marble and limestone consist of $CaCO_3$, which reacts with acid rain in an acid–base reaction to produce $CaSO_4$. Because $CaSO_4$ is somewhat soluble in water, significant damage to the structure can result.

Because $CaSO_4$ is sparingly soluble in water, the net result of this reaction is to dissolve the marble or limestone. The Lincoln Memorial in Washington, DC, which was built in 1922, already shows significant damage from acid rain, and many older objects are exhibiting even greater damage (Figure 4.18). Metal objects can also suffer damage from acid rain through reactions of the kind discussed in Section 4.8.

The biological effects of acid rain are more complex. As indicated in Figure 4.15, biological fluids such as blood have a pH of 7–8. Organisms such as fish can maintain their internal pH in water that has a pH in the range of 6.5–8.5. If the external pH is too low, however, many aquatic organisms can no longer maintain their internal pH and they die. Thus, a pH of 4 or lower is fatal for virtually all fish, most invertebrate animals, and many microorganisms. As a result of acid rain, the pH of some lakes in Europe and the United States has dropped below 4. Recent surveys suggest that up to 6% of the lakes in the Adirondack Mountains of upstate New York and 4% of the lakes in Sweden and Norway are essentially dead and contain no fish. Neither location contains large concentrations of industry, but New York lies downwind of the industrial Midwest and Scandinavia is downwind of the most industrialized regions of western Europe. Hence, both regions appear to have borne the brunt of the pollution produced by their upwind neighbors. One possible way to counter the effects of acid rain in isolated lakes is by adding large quantities of finely ground limestone, which neutralizes the acid via the reaction shown in Equation 4.48 (see Problem 123).

A second major way in which acid rain can cause biological damage is less direct. Trees and many other plants are sensitive to the presence of aluminum and other metals in groundwater. Under normal circumstances, aluminum hydroxide, $Al(OH)_3$, which is present in some soils, is insoluble. At lower pH values, however, $Al(OH)_3$ dissolves via the reaction

$$Al(OH)_3(s) + 3H^+(aq) \longrightarrow Al^{3+}(aq) + 3H_2O(l) \quad (4.49)$$

The result is increased levels of Al^{3+} ions in groundwater. Because the Al^{3+} ion is toxic to plants, high concentrations can affect plant growth. Acid rain can also weaken the leaves and roots of plants so much that the plants are unable to withstand other stresses. The combination of the two effects can cause significant damage to established forests, such as the Black Forest in Germany and the forests of the northeastern United States and Canada (Figure 4.19).

Figure 4.19 Acid rain damage to a forest in the Czech Republic. Trees and many other plants are sensitive to aluminum and other metals in groundwater. Acid rain increases the concentration of Al^{3+} in groundwater, thereby adversely affecting plant growth. Large sections of established forests have been severely damaged.

4.8 ○ Oxidation–Reduction Reactions in Solution

MGC Balancing Redox Equations

We described the defining characteristics of oxidation–reduction, or redox, reactions in Chapter 3. Most of the reactions we considered there were relatively simple, and balancing them was straightforward. When oxidation–reduction reactions occur in aqueous solution, however, the equations are more complex and can be more difficult to balance by inspection. Because a balanced chemical equation is the most important prerequisite for solving any stoichiometry problem, we need a method for balancing oxidation–reduction reactions in aqueous solution that is generally applicable. One such method uses *oxidation states,* and a second is referred to as the *half-reaction* method. We show you how to balance redox equations using oxidation states in this section; the half-reaction method will be described in Chapter 19. (Instructors who prefer the half-reaction method may assign Section 19.2 at this point without disrupting the flow of the presentation.)

Balancing Redox Equations Using Oxidation States

To balance a redox equation using the **oxidation state method**, we conceptually separate the overall reaction into two parts: an oxidation—in which the atoms of one element lose electrons—and a reduction—in which the atoms of one element gain electrons. Consider, for example, the reaction of $Cr^{2+}(aq)$ with MnO_2 in the presence of dilute acid. Equation 4.50 is the net ionic equation for this reaction before balancing; the oxidation state of each element in each species has been assigned using the procedure described in Section 3.5 and is shown under the equation:

$$\underset{+2}{Cr^{2+}}(aq) + \underset{+4\ -2}{MnO_2}(s) + \underset{+1}{H^+}(aq) \longrightarrow \underset{+3}{Cr^{3+}}(aq) + \underset{+2}{Mn^{2+}}(aq) + \underset{+1\ -2}{H_2O}(l) \quad (4.50)$$

Notice that chromium is oxidized from the +2 to the +3 oxidation state, while manganese is reduced from the +4 to the +2 oxidation state. We can write an equation for this reaction that shows only the species that are oxidized and reduced:

$$Cr^{2+}(aq) + MnO_2(s) \longrightarrow Cr^{3+}(aq) + Mn^{2+}(aq) \quad (4.51)$$

The oxidation can be written as

$$Cr^{2+}(aq) \longrightarrow Cr^{3+}(aq) + e^- \quad (4.52)$$

and the reduction as

$$MnO_2(s) + 2e^- \longrightarrow Mn^{2+}(aq) \quad (4.53)$$

For the overall equation to be balanced, the number of electrons lost by the reductant must equal the number gained by the oxidant. We must therefore multiply the oxidation and the reduction equations by appropriate coefficients to give us the same number of electrons in both. Thus, in this example we must multiply the oxidation equation by 2 to give

$$2Cr^{2+}(aq) \longrightarrow 2Cr^{3+}(aq) + 2e^- \quad (4.54)$$

We then add the equations for the oxidation and the reduction and cancel the electrons on both sides of the equation:

$$\begin{array}{r} 2Cr^{2+}(aq) \longrightarrow 2Cr^{3+}(aq) + \cancel{2e^-} \\ MnO_2(s) + \cancel{2e^-} \longrightarrow Mn^{2+}(aq) \\ \hline 2Cr^{2+}(aq) + MnO_2(s) \longrightarrow 2Cr^{3+}(aq) + Mn^{2+}(aq) \end{array} \quad (4.55)$$

Notice that although the electrons cancel and the metal atoms are balanced, the total charge on the left side of the equation (+4) does not equal the charge on the right side (+8). Because the reaction is carried out in the presence of aqueous acid, however, we can add H^+ as necessary to either side of the equation to balance the charge. By the same token, if the reaction were carried out in the presence of aqueous base, we could balance the charge by adding OH^- as necessary to either side of the equation. In this case, adding four H^+ ions to the left side of the equation gives

$$2Cr^{2+}(aq) + MnO_2(s) + 4H^+(aq) \longrightarrow 2Cr^{3+}(aq) + Mn^{2+}(aq) \quad (4.56)$$

Although the charges are now balanced, we have two oxygen atoms on the left side of the equation and none on the right. Because the reaction takes place in aqueous solution, we can balance the oxygen atoms without affecting the overall charge balance by adding H_2O as necessary to either side of the equation. In this case, we add two H_2O molecules to the right side:

$$2Cr^{2+}(aq) + MnO_2(s) + 4H^+(aq) \longrightarrow 2Cr^{3+}(aq) + Mn^{2+}(aq) + 2H_2O(l) \quad (4.57)$$

Note that although we did not explicitly balance the hydrogen atoms, we can see by inspection that the overall equation is now balanced. All that remains is to check to make sure that we have not made a mistake. This procedure for balancing reactions is summarized in Table 4.3 and illustrated again in the following example.

EXAMPLE 4.17

Recall from Section 4.3 that the reaction of ethanol with dichromate ion in acidic solution yields acetic acid and Cr^{3+}. Balance the following equation for that reaction using oxidation states:

$$Cr_2O_7^{2-}(aq) + CH_3CH_2OH(aq) \xrightarrow{H^+} Cr^{3+}(aq) + CH_3CO_2H$$

Given Reactants and products in acidic solution

Asked for Balanced chemical equation using oxidation states

Strategy

Follow the procedure given in Table 4.3 for balancing a redox equation using oxidation states. When you are done, be certain to check that the equation is balanced.

TABLE 4.3 Procedure for balancing oxidation–reduction reactions by the oxidation state method

1. Write the unbalanced equation for the reaction, showing the reactants and products.
2. Assign oxidation states to all atoms in the reactants and products (see Section 3.5), and determine which atoms change oxidation state.
3. Write separate equations for oxidation and reduction, showing the atom(s) that is (are) oxidized and reduced plus the number of electrons accepted or donated.
4. Multiply the oxidation and reduction equations by appropriate coefficients so that both contain the same number of electrons.
5. Write the equations showing the actual chemical forms of the reactants and products, adjusting the coefficients as necessary.
6. Add the two equations and cancel the electrons.
7. Balance the charge by adding H^+ or OH^- ions as necessary for reactions in acidic or basic solution, respectively.
8. Balance the oxygen atoms by adding H_2O molecules to one side of the equation.
9. Check to make sure that the equation is balanced in both atoms and total charges.

Solution

1. *Write an equation showing the reactants and products.* Because we are given this information, we can skip this step.
2. *Assign oxidation states using the procedure described in Section 3.5, and identify those atoms that undergo a change in oxidation state.* The oxidation state of chromium in the dichromate ion is +6, and the oxidation state of the chromium ion is +3. Conversely, the carbon atom bonded to oxygen in ethanol is oxidized from the −1 oxidation state to +3 in acetic acid (the oxidation state of carbon in the CH_3 group remains unchanged, as do the oxidation states of hydrogen and oxygen):

$$\underset{+6}{Cr_2O_7^{2-}}(aq) + \underset{-1}{CH_3CH_2OH}(aq) \xrightarrow{H^+} \underset{+3}{Cr^{3+}}(aq) + \underset{+3}{CH_3CO_2H}(aq)$$

3. *Write separate equations for oxidation and reduction.* Each chromium atom (in $Cr_2O_7^{2-}$) is reduced from the +6 to the +3 oxidation state, which requires the addition of three electrons:

$$\text{Reduction:} \quad \underset{+6}{Cr^{6+}} + 3e^- \longrightarrow \underset{+3}{Cr^{3+}}$$

One carbon atom of ethanol is oxidized from −1 (in RCH_2OH) to +3 (in RCO_2H), which requires the loss of four electrons:

$$\text{Oxidation:} \quad \underset{-1}{RCH_2OH} \longrightarrow \underset{+3}{RCO_2H} + 4e^-$$

4. *Multiply the oxidation and reduction equations by appropriate coefficients to obtain equations with the same number of electrons.* The smallest integer that is divisible by both 3 and 4 is 12. Consequently, we multiply the reduction equation by 4 and the oxidation equation by 3 to obtain

$$\text{Reduction } (\times 4): \quad 4Cr^{6+} + 12e^- \longrightarrow 4Cr^{3+}$$
$$\text{Oxidation } (\times 3): \quad 3RCH_2OH \longrightarrow 3RCO_2H + 12e^-$$

5. *Write the oxidation and reduction equations showing the actual reactants and products, adjusting coefficients as necessary to give the numbers of atoms shown in step 4.* Because each $Cr_2O_7^{2-}$ ion contains two chromiums, we only need two dichromate ions to give us four Cr^{3+}. Inserting the actual chemical forms of chromium and adjusting the coefficients give

$$\text{Reduction:} \quad 2Cr_2O_7^{2-} + 12e^- \longrightarrow 4Cr^{3+}$$
$$\text{Oxidation:} \quad 3CH_3CH_2OH \longrightarrow 3CH_3CO_2H + 12e^-$$

6. *Add the two equations and cancel the electrons.* The sum of the two equations in step 5 is

$$2Cr_2O_7^{2-} + 3CH_3CH_2OH + 12e^- \longrightarrow 4Cr^{3+} + 3CH_3CO_2H + 12e^-$$

Canceling electrons gives

$$2Cr_2O_7^{2-} + 3CH_3CH_2OH \longrightarrow 4Cr^{3+} + 3CH_3CO_2H$$

7. *Add H^+ or OH^- ions as necessary to balance the charge.* Because the reaction is carried out in acidic solution, we can add H^+ ions to whichever side of the equation requires them in order to balance the charge. The overall charge on the left side is $2 \times (-2) = -4$, and the charge on the right side is $4 \times (+3) = +12$. Adding 16 H^+ ions to the left side gives a charge of +12 on both sides of the equation:

$$2Cr_2O_7^{2-} + 3CH_3CH_2OH + 16H^+ \longrightarrow 4Cr^{3+} + 3CH_3CO_2H$$

8. *Add H_2O molecules to one side of the equation as necessary to balance the O atoms without affecting the overall charge.* There are 17 O atoms on the left and only 6 on the right side of the equation above. Adding 11 H_2O molecules to the right side balances the O atoms:

$$2Cr_2O_7^{2-} + 3CH_3CH_2OH + 16H^+ \longrightarrow 4Cr^{3+} + 3CH_3CO_2H + 11H_2O$$

Although we have not explicitly balanced H atoms, each side of the equation has 34 H atoms.

9. *Check to make sure that the equation is balanced.* To guard against careless errors, it is important to check that both the total number of atoms of each element and the total charges are the same on both sides of the equation:

$$\text{Atoms:} \quad 4Cr + 6C + 17O + 34H = 4Cr + 6C + 17O + 34H$$

$$\text{Total charge:} \quad 2(-2) + 16(+1) = 4(+3)$$

$$+12 = +12$$

EXERCISE 4.17

Copper commonly occurs as the sulfide mineral, CuS. The first step in extracting copper from CuS is to dissolve the mineral in nitric acid, which oxidizes the sulfide to sulfate and reduces nitric acid to NO. Balance the equation for this reaction using oxidation states:

$$CuS(s) + H^+(aq) + NO_3^-(aq) \longrightarrow Cu^{2+}(aq) + NO(g) + SO_4^{2-}(aq)$$

Answer $3CuS(s) + 8H^+(aq) + 8NO_3^-(aq) \longrightarrow 3Cu^{2+}(aq) + 8NO(g) + 3SO_4^{2-}(aq) + 4H_2O(l)$

Reactions in basic solutions are balanced in exactly the same manner. To make sure you understand the procedure, consider the next example.

EXAMPLE 4.18

The commercial solid drain cleaner, Drano, contains a mixture of sodium hydroxide and powdered aluminum. The sodium hydroxide dissolves in standing water to form a strongly basic solution, capable of slowly dissolving organic substances, such as hair, that may be clogging the drain. The aluminum dissolves in the strongly basic solution to produce bubbles of hydrogen gas that agitate the solution to help break up the clogs. The reaction is

$$Al(s) + H_2O(aq) \longrightarrow [Al(OH)_4]^-(aq) + H_2(g)$$

Balance this equation using oxidation states.

Given Reactants and products in a basic solution

Asked for Balanced chemical equation

Strategy

Follow the procedure given in Table 4.3 for balancing a redox reaction using oxidation states. When you are done, check to be sure the equation is balanced.

Solution

We will apply the same procedure used in Example 4.17, but in a more abbreviated form.

1. The equation for the reaction is given, so we can skip this step.
2. The oxidation state of aluminum changes from 0 in metallic Al to +3 in $[Al(OH)_4]^-$. The oxidation state of hydrogen changes from +1 in H_2O to 0 in H_2. Thus, aluminum is oxidized, while hydrogen is reduced:

$$\underset{0}{Al}(s) + \underset{+1}{H_2}O(aq) \longrightarrow [\underset{+3}{Al}(OH)_4]^-(aq) + \underset{0}{H_2}(g)$$

3. Oxidation: $\underset{0}{Al^0} \longrightarrow \underset{+3}{Al^{3+}} + 3e^-$

   Reduction: $\underset{+1}{H^+} + e^- \longrightarrow \underset{0}{H^0}$ (in H_2)

4. Multiply the reduction equation by 3 to obtain an equation with the same number of electrons as the oxidation equation:

   Oxidation: $Al^0 \longrightarrow Al^{3+} + 3e^-$

   Reduction: $3H^+ + 3e^- \longrightarrow 3H^0$ (in H_2)

5. Insert the actual chemical forms of the reactants and products, adjusting the coefficients as necessary to obtain the correct numbers of atoms as in step 4. Because a molecule of H_2O contains two protons, $3H^+$ corresponds to $\frac{3}{2}H_2O$. Similarly, each molecule of hydrogen gas contains two atoms of hydrogen, so 3H corresponds to $\frac{3}{2}H_2$.

   Oxidation: $Al \longrightarrow [Al(OH)_4]^- + 3e^-$

   Reduction: $\frac{3}{2}H_2O + 3e^- \longrightarrow \frac{3}{2}H_2$

6. Add the equations and cancel electrons:

$$Al + \tfrac{3}{2}H_2O + \cancel{3e^-} \longrightarrow [Al(OH)_4]^- + \tfrac{3}{2}H_2 + \cancel{3e^-}$$
$$Al + \tfrac{3}{2}H_2O \longrightarrow [Al(OH)_4]^- + \tfrac{3}{2}H_2$$

   To remove fractional coefficients, multiply both sides of the equation by 2:

$$2Al + 3H_2O \longrightarrow 2[Al(OH)_4]^- + 3H_2$$

7. The right side of the equation has a total charge of −2, whereas the left side has a total charge of 0. Because the reaction is carried out in basic solution, we can balance the charge by adding two OH^- ions to the left side:

$$2Al + 2OH^- + 3H_2O \longrightarrow 2[Al(OH)_4]^- + 3H_2$$

8. The left side of the equation contains five O atoms, and the right side contains eight O atoms. We can balance the O atoms by adding three H_2O molecules to the left side:

$$2Al + 2OH^- + 6H_2O \longrightarrow 2[Al(OH)_4]^- + 3H_2$$

9. Be sure the equation is balanced:

   Atoms: $2Al + 8O + 14H = 2Al + 8O + 14H$

   Total charge: $(2 \times 0) + [2 \times (-1)] + (6 \times 0) = [2 \times (-1)] + (3 \times 0)$

$$-2 = -2$$

   The balanced equation is therefore

$$2Al(s) + 2OH^-(aq) + 6H_2O(l) \longrightarrow 2[Al(OH)_4]^-(aq) + 3H_2(g)$$

   Thus, 3 mol of H_2 gas is produced for every 2 mol of Al.

EXERCISE 4.18

The permanganate ion reacts with nitrite ion in basic solution to produce manganese(IV) oxide and nitrate ion. Write a balanced equation for the reaction.

Answer $2MnO_4^-(aq) + 3NO_2^-(aq) + H_2O(l) \longrightarrow 2MnO_2(s) + 3NO_3^-(aq) + 2OH^-(aq)$

Note the pattern

Species in high oxidation states act as oxidants, whereas species in low oxidation states act as reductants.

As suggested in Examples 4.17 and 4.18, a wide variety of redox reactions are possible in aqueous solution. Often the identity of the products obtained from a given set of reactants depends on both the ratio of oxidant to reductant, and whether the reaction is carried out in acidic or basic solution, which is one reason it can be difficult to predict the outcome of a reaction. Because oxidation–reduction reactions in solution are so common and so important, however, chemists have developed two general guidelines for predicting whether a redox reaction will occur and for predicting the identity of the products:

1. Compounds of elements in high oxidation states (such as ClO_4^-, NO_3^-, MnO_4^-, $Cr_2O_7^{2-}$, and UF_6) tend to act as *oxidants* and *become reduced* in chemical reactions.
2. Compounds of elements in low oxidation states (such as CH_4, NH_3, H_2S, and HI) tend to act as *reductants* and *become oxidized* in chemical reactions.

When an aqueous solution of a compound that contains an element in a high oxidation state is mixed with an aqueous solution of a compound that contains an element in a low oxidation state, an oxidation–reduction reaction is likely to occur.

Redox Reactions of Solid Metals in Aqueous Solution

A widely encountered class of oxidation–reduction reactions is the reaction of aqueous solutions of acids or metal salts with solid metals. An example is the corrosion of metal objects, such as the rusting of an automobile (Figure 4.20). Rust is formed from a complex oxidation–reduction reaction involving dilute acid solutions that contain Cl^- ions (effectively, dilute HCl), iron metal, and oxygen. When an object rusts, iron metal reacts with HCl(*aq*) to produce iron(II) chloride and hydrogen gas:

$$Fe(s) + 2HCl(aq) \longrightarrow FeCl_2(aq) + H_2(g) \qquad (4.58)$$

In subsequent steps, $FeCl_2$ oxidizes further to form a reddish-brown precipitate of $Fe(OH)_3$.

Many metals dissolve through reactions of this type, which have the general form

$$\text{metal} + \text{acid} \longrightarrow \text{salt} + \text{hydrogen} \qquad (4.59)$$

Some of these reactions have important consequences. For example, it has been proposed that one factor that contributed to the fall of the Roman Empire was the widespread use of lead in cooking utensils and in pipes that carried water. Rainwater, as we have seen, is slightly acidic, and foods such as fruits, wine, and vinegar contain organic acids. In the presence of these acids, lead dissolves:

$$Pb(s) + 2H^+(aq) \longrightarrow Pb^{2+}(aq) + H_2(g) \qquad (4.60)$$

Consequently, it has been speculated that both the water and the food consumed by Romans contained toxic levels of lead, which resulted in widespread lead poisoning and eventual madness. Perhaps this explains why the Roman Emperor Caligula appointed his favorite horse as consul!

Figure 4.20 Rust formation. The corrosion process involves an oxidation–reduction reaction in which metallic iron is converted to $Fe(OH)_3$, a reddish-brown solid.

Single-Displacement Reactions

Certain metals are oxidized by aqueous acid, whereas others are oxidized by aqueous solutions of various metal salts. Both types of reactions are called **single-displacement reactions**, in which the ion in solution is displaced through oxidation of the metal. Two examples of single-displacement reactions are the reduction of iron salts by zinc (Equation 4.61) and the reduction of silver salts by copper (Equation 4.62 and Figure 4.21):

$$Zn(s) + Fe^{2+}(aq) \longrightarrow Zn^{2+}(aq) + Fe(s) \quad (4.61)$$

$$Cu(s) + 2Ag^{+}(aq) \longrightarrow Cu^{2+}(aq) + 2Ag(s) \quad (4.62)$$

The reaction in Equation 4.61 is widely used to prevent (or at least postpone) the corrosion of iron or steel objects such as nails and sheet metal. The process of "galvanizing" consists of applying a thin coating of zinc to the iron or steel, thus protecting it from oxidation as long as any of the zinc remains on the object.

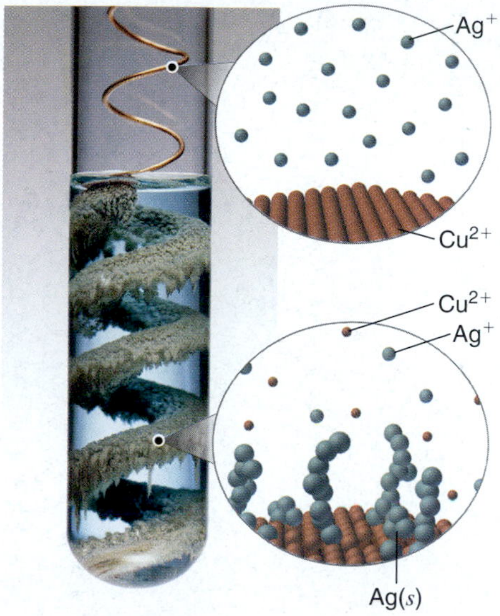

Figure 4.21 The single-displacement reaction of metallic copper with a solution of silver nitrate. When a copper coil is placed in a solution of silver nitrate, silver ions are reduced to metallic silver on the copper surface and some of the copper metal dissolves. Note the formation of a metallic silver precipitate on the copper coil and a blue color in the surrounding solution due to the presence of aqueous Cu^{2+} ions. (Water molecules omitted from molecular views of solution for clarity.)

The Activity Series

By observing what happens when samples of various metals are placed in contact with solutions of other metals, chemists have been able to arrange the metals according to the relative ease or difficulty with which they can be oxidized in a single-displacement reaction. For example, we saw in Equations 4.61 and 4.62 that metallic zinc reacts with iron salts and metallic copper reacts with silver salts. Experimentally, it is found that zinc reacts with both copper salts and silver salts, producing Zn^{2+}. Zinc therefore has a greater tendency to be oxidized than does iron, copper, or silver. Although zinc will not react with magnesium salts to give magnesium metal, magnesium metal will react with zinc salts to give zinc metal:

$$Zn(s) + Mg^{2+}(aq) \not\longrightarrow Zn^{2+}(aq) + Mg(s) \quad (4.63)$$

$$Mg(s) + Zn^{2+}(aq) \longrightarrow Mg^{2+}(aq) + Zn(s) \quad (4.64)$$

Thus, magnesium has a greater tendency to be oxidized than zinc does.

Pairwise reactions of this sort are the basis of the **activity series** (Table 4.4), which lists metals and hydrogen in order of their relative tendency to be oxidized. The metals

TABLE 4.4 The activity series

	Element	Oxidation Reaction
React with simple acids to form H_2 (Lithium–Lead); React with steam to form H_2 (Lithium–Iron); React vigorously with cold H_2O to form H_2 (Lithium–Sodium)	Lithium	$Li \longrightarrow Li^{+} + e^{-}$
	Potassium	$K \longrightarrow K^{+} + e^{-}$
	Barium	$Ba \longrightarrow Ba^{2+} + 2e^{-}$
	Calcium	$Ca \longrightarrow Ca^{2+} + 2e^{-}$
	Sodium	$Na \longrightarrow Na^{+} + e^{-}$
	Magnesium	$Mg \longrightarrow Mg^{2+} + 2e^{-}$
	Aluminum	$Al \longrightarrow Al^{3+} + 3e^{-}$
	Manganese	$Mn \longrightarrow Mn^{2+} + 2e^{-}$
	Zinc	$Zn \longrightarrow Zn^{2+} + 2e^{-}$
	Chromium	$Cr \longrightarrow Cr^{3+} + 3e^{-}$
	Iron	$Fe \longrightarrow Fe^{2+} + 2e^{-}$
	Cadmium	$Cd \longrightarrow Cd^{2+} + 2e^{-}$
	Cobalt	$Co \longrightarrow Co^{2+} + 2e^{-}$
	Nickel	$Ni \longrightarrow Ni^{2+} + 2e^{-}$
	Tin	$Sn \longrightarrow Sn^{2+} + 2e^{-}$
	Lead	$Pb \longrightarrow Pb^{2+} + 2e^{-}$
	Hydrogen	$H_2 \longrightarrow 2H^{+} + 2e^{-}$
Will not dissolve in simple acids (Copper–Gold)	Copper	$Cu \longrightarrow Cu^{2+} + 2e^{-}$
	Silver	$Ag \longrightarrow Ag^{+} + e^{-}$
	Mercury	$Hg \longrightarrow Hg^{2+} + 2e^{-}$
	Platinum	$Pt \longrightarrow Pt^{2+} + 2e^{-}$
	Gold	$Au \longrightarrow Au^{+} + e^{-}$

Increasing ease of oxidation (upward arrow, Gold to Lithium)

at the top of the series, which have the greatest tendency to lose electrons, are the alkali metals (Group 1), alkaline earths (Group 2), and Al (Group 13). In contrast, the metals at the bottom of the series, which have the lowest tendency to be oxidized, are the precious metals or coinage metals Pt, Au, Ag, and Cu as well as Hg, which are located in the lower right portion of the metals in the periodic table. You should be generally familiar with which kinds of metals are **active metals** (located at the top of the series) and which are **inert metals** (at the bottom of the series).

When using the activity series to predict the outcome of a reaction, keep in mind that *any element will reduce compounds of the elements below it in the series.* Because magnesium is above zinc in Table 4.4, magnesium metal will reduce zinc salts, but not vice versa. Similarly, the precious metals are at the bottom of the activity series, so virtually any other metal will reduce precious metal salts to the pure precious metals. Hydrogen is included in the series, and the tendency of a metal to react with an acid is indicated by its position relative to hydrogen in the activity series. *Only those metals that lie above hydrogen in the activity series dissolve in acids to produce H_2.* Because the precious metals lie below hydrogen, they do not dissolve in dilute acid and therefore do not corrode readily. The next example demonstrates how a familiarity with the activity series allows you to predict the products of many single-displacement reactions. We will return to the activity series when we discuss oxidation–reduction reactions in more detail in Chapter 19.

EXAMPLE 4.19

Using the activity series, predict what happens in each situation. If a reaction occurs, write the net ionic equation. (a) A strip of aluminum foil is placed in an aqueous solution of silver nitrate. (b) A few drops of liquid mercury are added to an aqueous solution of lead(II) acetate. (c) Some sulfuric acid from a car battery is accidentally spilled on the lead cable terminals.

Given Reactants

Asked for Overall reaction and net ionic equation

Strategy

Ⓐ Locate the reactants in the activity series in Table 4.4, and from their relative positions, predict whether a reaction will occur. If a reaction does occur, identify which metal is oxidized and which is reduced.

Ⓑ Write the net ionic equation for the redox reaction.

Solution

(a) Ⓐ Aluminum is an active metal that lies above silver in the activity series, so we expect a reaction to occur. According to their relative positions, aluminum will be oxidized and dissolve, and silver ions will be reduced to silver metal. Ⓑ The net ionic equation is

$$Al(s) + 3Ag^+(aq) \longrightarrow Al^{3+}(aq) + 3Ag(s)$$

Recall from our discussion of solubilities that most nitrate salts are soluble. In this case, therefore, the nitrate ions are spectator ions and are not involved in the reaction.

(b) Ⓐ Mercury lies below lead in the activity series, so no reaction will occur.

(c) Ⓐ Lead is above hydrogen in the activity series, so the lead terminals will be oxidized and the acid will be reduced to form H_2. Ⓑ From our discussion of solubilities, recall that Pb^{2+} and SO_4^{2-} form insoluble lead(II) sulfate. In this case, therefore, the sulfate ions are *not* spectator ions and the reaction is

$$Pb(s) + 2H^+(aq) + SO_4^{2-}(aq) \longrightarrow PbSO_4(s) + H_2(g)$$

Lead(II) sulfate is the white solid that forms on corroded battery terminals.

Corroded battery terminals. The white solid is lead(II) sulfate, formed from the reaction of solid lead with a solution of sulfuric acid.

EXERCISE 4.19

Using the activity series, write net ionic equations to describe what happens in each situation. If no reaction occurs, say so. **(a)** A strip of chromium metal is placed in an aqueous solution of aluminum chloride. **(b)** A strip of zinc is placed in an aqueous solution of chromium(III) nitrate. **(c)** A piece of aluminum foil is dropped into a glass that contains vinegar (active ingredient: acetic acid).

Answer **(a)** No reaction; **(b)** $3Zn(s) + 2Cr^{3+}(aq) \longrightarrow 3Zn^{2+}(aq) + 2Cr(s)$; **(c)** $2Al(s) + 6CH_3CO_2H(aq) \longrightarrow 2Al^{3+}(aq) + 6CH_3CO_2^-(aq) + 3H_2(g)$

4.9 • Quantitative Analysis Using Titrations

MGC Acid–Base Titration

To determine the amounts or concentrations of substances present in a sample, chemists use a combination of chemical reactions and stoichiometric calculations in a methodology called **quantitative analysis**. Suppose, for example, we know the identity of a certain compound in a solution but not its concentration. If the compound reacts rapidly and completely with another reactant, we may be able to use the reaction to determine the concentration of the compound of interest. In a **titration**, a carefully measured volume of a solution of known concentration, called the **titrant**, is added to a measured volume of a solution containing a compound whose concentration is to be determined (the *unknown*). The reaction used in a titration can be an acid–base, precipitation, or oxidation–reduction reaction. In all cases, the reaction chosen for the analysis must be fast, complete, and *specific*; that is, *only* the compound of interest should react with the titrant. The **equivalence point** is reached when the stoichiometric amount of the titrant has been added, that is the amount required to react completely with the unknown, the substance whose concentration is to be determined.

Determining the Concentration of an Unknown Solution Using a Titration

The chemical nature of the species present in the unknown dictates which type of reaction is most appropriate and also how to determine the equivalence point. The volume of titrant added, its concentration, and the coefficients from the balanced chemical equation for the reaction allow us to calculate the total number of moles of the unknown in the original solution. Because we have measured the volume of the solution that contains the unknown, we can calculate the molarity of the unknown substance. This procedure is summarized graphically below:

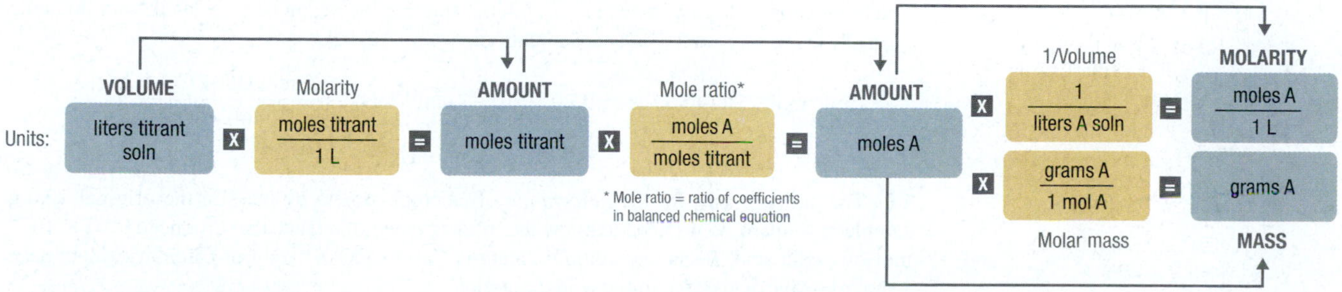

EXAMPLE 4.20

The calcium salt of oxalic acid, $Ca(O_2CCO_2)$, is found in the sap and leaves of some vegetables, including spinach and rhubarb, and in many ornamental plants. Because oxalic acid and its salts are toxic, when a food such as rhubarb is processed commercially, the leaves must be removed and the oxalate content carefully monitored.

Figure 4.22 The titration of oxalic acid with permanganate. As small volumes of permanganate solution are added to the oxalate solution, a transient purple color appears and then disappears as the permanganate is consumed (left). As more permanganate is added, eventually all of the oxalate has been oxidized, and a faint purple color from the presence of excess permanganate appears, marking the endpoint (right).

The reaction of MnO_4^- with oxalic acid, HO_2CCO_2H, in acidic aqueous solution produces Mn^{2+} and CO_2:

$$\underset{\text{Purple}}{MnO_4^-(aq)} + HO_2CCO_2H(aq) + H^+(aq) \longrightarrow \underset{\text{Colorless}}{Mn^{2+}(aq)} + CO_2(g) + H_2O(l)$$

Because this reaction is rapid and goes to completion, potassium permanganate, $KMnO_4$, is widely used as a reactant for determining the concentration of oxalic acid.

Suppose that you stirred a 10.0-g sample of canned rhubarb with enough dilute $H_2SO_4(aq)$ to obtain 127 mL of colorless solution. Because the added permanganate is rapidly consumed, adding small volumes of a 0.0247 *M* $KMnO_4$ solution, which has a deep purple color, to the rhubarb extract does not initially change the color of the extract. When a total of 15.4 mL of the permanganate solution has been added, however, the solution becomes faint purple due to the presence of a slight excess of permanganate (Figure 4.22). If we assume that oxalic acid is the only species in solution that reacts with permanganate, what percentage of the mass of the original sample was calcium oxalate?

Given Equation, mass of sample, volume of solution, and molarity and volume of titrant

Asked for Mass percentage of unknown in sample

Strategy

A Balance the chemical equation for the reaction using oxidation states.

B Calculate the number of moles of permanganate consumed by multiplying the volume of the titrant by its molarity. Then calculate the number of moles of oxalate in the solution by multiplying by the ratio of the coefficients in the balanced equation. Because calcium oxalate contains a 1:1 ratio of Ca^{2+}:$^-O_2CCO_2^-$, the number of moles of oxalate in the solution is the same as the number of moles of calcium oxalate in the original sample.

C Find the mass of calcium oxalate by multiplying the number of moles of calcium oxalate in the sample by its molar mass. Divide the mass of calcium oxalate by the mass of the sample and convert to a percentage to calculate the percentage by mass of calcium oxalate in the original sample.

Solution

A As in all other problems of this type, the first requirement is a balanced equation for the reaction. Using oxidation states gives

$$2MnO_4^-(aq) + 5HO_2CCO_2H(aq) + 6H^+(aq) \longrightarrow 2Mn^{2+}(aq) + 10CO_2(g) + 8H_2O(l)$$

Thus, each mole of MnO_4^- added consumes 2.5 mol of oxalic acid.

B Because we know the concentration of permanganate (0.0247 *M*) and the volume of permanganate solution that was needed to consume all of the oxalic acid (15.4 mL), we can calculate the number of moles of MnO_4^- consumed:

$$15.4\ \cancel{\text{mL}}\left(\frac{1\ \cancel{\text{L}}}{1000\ \cancel{\text{mL}}}\right)\left(\frac{0.0247\ \text{mol } MnO_4^-}{1\ \cancel{\text{L}}}\right) = 3.80 \times 10^{-4}\ \text{mol } MnO_4^-$$

The number of moles of oxalic acid, and thus oxalate, present can be calculated from the mole ratio of the reactants in the balanced equation:

$$\begin{aligned}\text{moles } HO_2CCO_2H &= 3.80 \times 10^{-4}\ \cancel{\text{mol } MnO_4^-}\left(\frac{5\ \text{mol } HO_2CCO_2H}{2\ \cancel{\text{mol } MnO_4^-}}\right)\\ &= 9.50 \times 10^{-4}\ \text{mol } HO_2CCO_2H\end{aligned}$$

C The problem asks for the *percentage* of calcium oxalate by mass in the original 10.0-g sample of rhubarb, so we need to know the mass of calcium oxalate that produced 9.50×10^{-4} mol of oxalic acid. Because calcium oxalate is $Ca(O_2CCO_2)$, 1 mol of calcium oxalate gave 1 mol of oxalic acid in the initial acid extraction:

$$Ca(O_2CCO_2)(s) + 2H^+(aq) \longrightarrow Ca^{2+}(aq) + HO_2CCO_2H(aq)$$

The mass of calcium oxalate originally present was thus

$$\begin{aligned}\text{mass of } CaC_2O_4 &= 9.50 \times 10^{-4}\ \cancel{\text{mol } HO_2CCO_2H}\left(\frac{1\ \cancel{\text{mol } CaC_2O_4}}{1\ \cancel{\text{mol } HO_2CCO_2H}}\right)\left(\frac{128.10\ \text{g } CaC_2O_4}{1\ \cancel{\text{mol } CaC_2O_4}}\right)\\ &= 0.122\ \text{g } CaC_2O_4\end{aligned}$$

The original sample contained 0.122 g of calcium oxalate per 10.0 g of rhubarb. The percentage of calcium oxalate by mass was thus

$$\% \ CaC_2O_4 = \frac{0.122 \text{ g}}{10.0 \text{ g}} \times 100\% = 1.22\%$$

Because the problem asked for the percentage by mass of calcium oxalate in the original sample rather than for the *concentration* of oxalic acid in the extract, we do not need to know the volume of the oxalic acid solution for this calculation.

EXERCISE 4.20

Glutathione is a low-molecular-weight compound found in living cells; its structure is shown in the margin. It is found in two forms: one abbreviated as GSH (indicating the presence of an —SH group) and the other as GSSG (the *disulfide* form, in which an S—S bond links two glutathione units). The GSH form is easily oxidized to GSSG by elemental iodine:

$$2\text{GSH}(aq) + I_2(aq) \longrightarrow \text{GSSG}(aq) + 2\text{HI}(aq)$$

A small amount of soluble starch is added as an indicator. Because starch reacts with excess I_2 to give an intense blue color, the appearance of a blue color indicates that the equivalence point of the reaction has been reached.

Adding small volumes of a 0.0031 *M* aqueous solution of I_2 to 194 mL of a solution that contains glutathione and a trace of soluble starch initially causes no change. After 16.3 mL of iodine solution has been added, however, a permanent pale blue color appears because of the formation of the starch-iodine complex. What is the concentration of glutathione in the original solution?

Answer 5.2×10^{-4} *M* glutathione

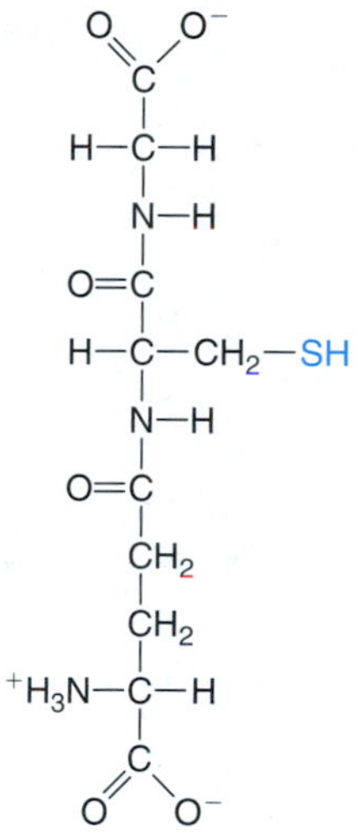

Glutathione (GSH)

Standard Solutions

In Exercise 4.20, the concentration of the titrant, I_2, was accurately known. The accuracy of any titration analysis depends in large part on an accurate knowledge of the concentration of the titrant. Thus, most titrants are first *standardized*; that is, their concentration is measured by titration with a **standard solution**, a solution whose concentration is known precisely. Only pure crystalline compounds that do not react with water or CO_2 are suitable for use in preparing a standard solution. One such compound is potassium hydrogen phthalate (KHP), a weak monoprotic acid suitable for standardizing solutions of bases such as NaOH. As shown in the margin, the reaction of KHP with NaOH is a simple acid–base reaction. If the concentration of the KHP solution is known accurately and the titration of a NaOH solution with the KHP solution is carried out carefully, then the concentration of the NaOH solution can be calculated precisely. The standardized NaOH solution can then be used to titrate a solution of an acid whose concentration is unknown.

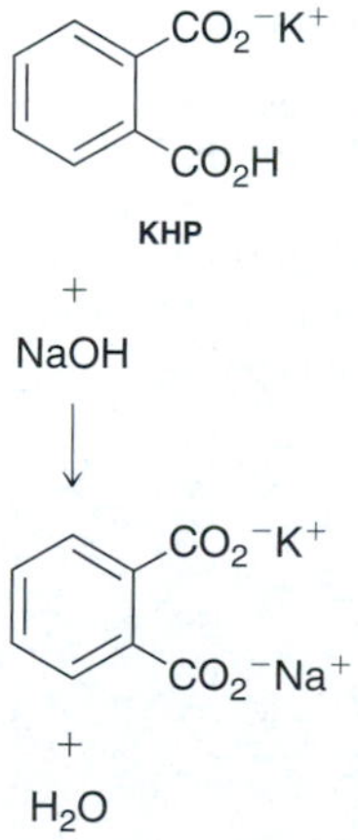

The reaction of KHP with NaOH.

Acid–Base Titrations

Because most common acids and bases are not intensely colored, a small amount of an *acid–base indicator* is usually added to detect the equivalence point in an acid–base titration. The point in the titration at which an indicator changes color is called the **endpoint**. The procedure is illustrated in the next example.

EXAMPLE 4.21

The structure of vitamin C (ascorbic acid, abbreviated as HAsc to indicate that it acts as a monoprotic acid) is shown in the margin on page 192. An absence of vitamin C in the diet leads to the disease known as scurvy, a breakdown of connective tissue throughout the body

Figure 4.23 The titration of ascorbic acid with a solution of NaOH. The solution, containing bromothymol blue as an indicator, is initially yellow (*left*). The addition of a trace excess of NaOH causes the solution to turn green at the endpoint (*right*) and then blue.

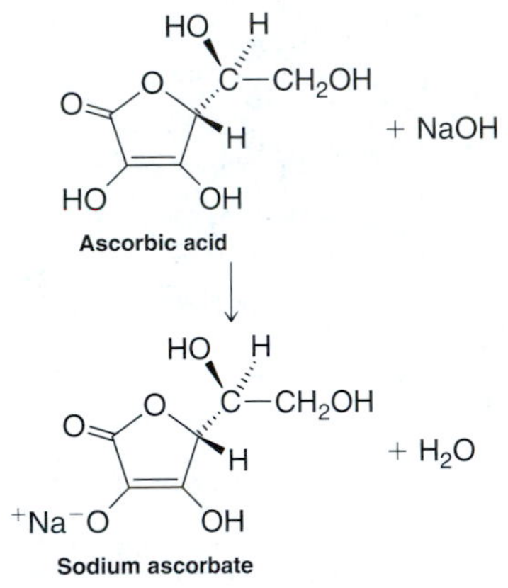

The reaction of ascorbic acid (vitamin C) with NaOH. In this three-dimensional representation, hatched lines indicate bonds that are behind the plane of the paper, and wedged lines indicate bonds that are out of the plane of the paper.

and of dentin in the teeth. Because fresh fruits and vegetables rich in vitamin C are readily available in developed countries today, scurvy is not a major problem. In the days of slow voyages in wooden ships, however, scurvy was common. Ferdinand Magellan, the first person to sail around the world, lost more than 90% of his crew, many to scurvy. Although a diet rich in fruits and vegetables contains more than enough vitamin C to prevent scurvy, many people take supplemental doses of vitamin C, hoping that the extra amounts will help prevent colds and other illness.

Suppose a tablet advertised as containing 500 mg of vitamin C is dissolved in 100.0 mL of distilled water that contains a small amount of the acid–base indicator bromothymol blue, an indicator that is yellow in acid solution and blue in basic solution, to give a yellow solution. The addition of 53.5 mL of a 0.0520 *M* solution of NaOH results in a change to green at the endpoint, due to a mixture of the blue and yellow forms of the indicator (Figure 4.23). What is the actual mass of vitamin C in the tablet? (The molar mass of ascorbic acid is 176.13 g/mol.)

Given Reactant, volume of sample solution, and volume and molarity of titrant

Asked for Mass of unknown

Strategy

Ⓐ Write the balanced chemical equation for the reaction, and calculate the number of moles of base needed to neutralize the ascorbic acid.

Ⓑ Using mole ratios, determine the amount of ascorbic acid consumed. Calculate the mass of vitamin C by multiplying the number of moles of ascorbic acid by its molar mass.

Solution

Ⓐ Because ascorbic acid acts as a monoprotic acid, we can write the balanced equation for the reaction as

$$\mathrm{HAsc}(aq) + \mathrm{OH^-}(aq) \longrightarrow \mathrm{Asc^-}(aq) + \mathrm{H_2O}(l)$$

where HAsc is ascorbic acid and $\mathrm{Asc^-}$ is ascorbate. The number of moles of $\mathrm{OH^-}$ ions needed to neutralize the ascorbic acid is

$$\text{moles OH}^- = 53.5\ \cancel{\text{mL}}\left(\frac{1\ \cancel{\text{L}}}{1000\ \cancel{\text{mL}}}\right)\left(\frac{0.0520\ \text{mol OH}^-}{1\ \cancel{\text{L}}}\right) = 2.78 \times 10^{-3}\ \text{mol OH}^-$$

Ⓑ The mole ratio of the base added to the acid consumed is 1:1, so the number of moles of $\mathrm{OH^-}$ added equals the number of moles of ascorbic acid present in the tablet. Thus,

$$\text{mass of ascorbic acid} = 2.78 \times 10^{-3} \cancel{\text{mol HAsc}} \left(\frac{176.13 \text{ g HAsc}}{1 \cancel{\text{mol HAsc}}} \right) = 0.490 \text{ g HAsc}$$

The tablet therefore contains about 2% less vitamin C than advertised.

EXERCISE 4.21

Vinegar is essentially a dilute solution of acetic acid in water. Vinegar is usually produced in a concentrated form and then diluted with water to give a final concentration of 4–7% acetic acid; that is, a 4% (m/v) solution contains 4.00 g of acetic acid per 100 mL of solution. If a drop of bromothymol blue indicator is added to 50.0 mL of concentrated vinegar stock and 31.0 mL of 2.51 *M* NaOH is needed to turn the solution from yellow to green, what is the percentage of acetic acid in the vinegar stock? (Assume that the density of the vinegar solution is 1.00 g/mL.)

Answer 9.35%

KEY EQUATIONS

Definition of molarity	$\text{molarity} = \dfrac{\text{moles of solute}}{\text{liters of solution}} = \dfrac{\text{mmoles of solute}}{\text{mL of solution}}$	(4.4)
Relationships among volume, molarity, and moles	$V_L\, M_{\text{mol/L}} = \text{L}\left(\dfrac{\text{mol}}{\text{L}}\right) = \text{moles}$	(4.5)
Relationship between volume and concentration of stock and dilute solutions	$(V_s)(M_s) = \text{moles of solute} = (V_d)(M_d)$	(4.7)
Definition of pH	$\text{pH} = -\log[\text{H}^+]$	(4.36)
	$[\text{H}^+] = 10^{-\text{pH}}$	(4.37)

SUMMARY AND KEY TERMS

4.1 Aqueous Solutions (p. 146)

Most chemical reactions are carried out in **solutions**, homogeneous mixtures of two or more substances. In a solution, a **solute**, the substance present in the lesser amount, is dispersed in a **solvent**, the substance present in the greater amount. **Aqueous solutions** contain water as the solvent, whereas *nonaqueous solutions* have solvents other than water.

Polar substances, such as water, contain asymmetric arrangements of **polar bonds**, in which electrons are shared unequally between bonded atoms. Polar substances and ionic compounds tend to be most soluble in water because they interact favorably with its structure. In aqueous solution, dissolved ions become **hydrated**; that is, they are surrounded by a shell of water molecules.

Substances that dissolve in water may be categorized according to whether the resulting aqueous solutions conduct electricity. **Strong electrolytes** dissociate completely into ions to produce solutions that conduct electricity well. **Weak electrolytes** produce a relatively small number of ions, resulting in solutions that conduct electricity only poorly. **Nonelectrolytes** dissolve as uncharged molecules and have no effect on the electrical conductivity of water.

4.2 Solution Concentrations (p. 149)

The **concentration** of a substance is the quantity of solute present in a given quantity of solution. Concentrations are usually expressed as **molarity**, the number of moles of solute in 1 L of solution. Solutions of known concentration can be prepared either by dissolving a known mass of solute in a solvent and diluting to a desired final volume or by diluting the appropriate volume of a more concentrated solution (a **stock solution**) to the desired final volume.

4.3 Stoichiometry of Reactions in Solution (p. 156)

Quantitative calculations that involve the stoichiometry of reactions in solution use volumes of solutions of known concentration instead of masses of reactants or products. The coefficients in the balanced chemical equation tell how many moles of reactants are needed and how many moles of product can be produced.

4.4 Ionic Equations (p. 161)

The chemical equation for a reaction in solution can be written in three ways. The **overall equation** shows all of the substances present in their undissociated form; the **complete ionic equation** shows all of the substances present in the form in which they actually exist in solution; and the **net ionic equation** is derived from the complete ionic equation by omitting all **spectator ions**, ions that occur on both sides of the equation with the same coefficients. Net ionic equations demonstrate that many different combinations of reactants can give the same net chemical reaction. The three common kinds of reactions that occur in aqueous solution are acid–base, precipitation, and oxidation–reduction reactions.

4.5 Precipitation Reactions (p. 163)

In a **precipitation reaction**, a subclass of exchange reactions, an insoluble material (a **precipitate**) forms when solutions of two substances are mixed. To predict the product of a precipitation reaction, all species initially present in the solutions are identified, as are any combinations likely to produce an insoluble salt.

4.6 Acid–Base Reactions (p. 169)

Acid–base reactions require both an acid and a base. In Brønsted–Lowry terms, an **acid** is a substance that can donate a hydrogen ion (a proton), and a **base** is one that can accept a hydrogen ion (a proton). All acid–base reactions contain two acid–base pairs, the reactants and the products. Acids can donate one hydrogen ion (**monoprotic acids**), two hydrogen ions (**diprotic acids**), or three hydrogen ions (*triprotic acids*). Compounds that are capable of donating more than one proton are generally called *polyprotic acids*. Acids also differ in their tendency to donate a hydrogen ion, a measure of their acid strength. **Strong acids** react completely with water to produce $H_3O^+(aq)$ (the **hydronium ion**), whereas **weak acids** dissociate only partially in water. Conversely, **strong bases** react completely with water to produce the hydroxide ion, whereas **weak bases** react only partially with water to form hydroxide ions. The reaction of a strong acid with a strong base is a **neutralization reaction**, which produces water plus a salt.

The acidity or basicity of an aqueous solution is described quantitatively using the **pH scale**. The **pH** of a solution is the negative logarithm of the hydrogen ion concentration and typically ranges from 0 for strongly acidic solutions to 14 for strongly basic ones. Because of the *autoionization reaction* of water, which produces small amounts of hydronium ions and hydroxide ions, a **neutral solution** of water contains $1 \times 10^{-7}\ M\ H^+$ ions and has a pH of 7.0. An **indicator** is an intensely colored organic substance whose color is pH dependent; it is used to determine the pH of a solution.

4.7 The Chemistry of Acid Rain (p. 178)

Acid rain is rainfall whose pH is less than 5.6, the value typically observed due to the presence of dissolved carbon dioxide. Acid rain is caused by nitrogen oxides and sulfur dioxide produced both by natural processes and by the combustion of fossil fuels. Eventually, these oxides react with oxygen and water to give nitric acid and sulfuric acid.

4.8 Oxidation–Reduction Reactions in Solution (p. 181)

In oxidation–reduction reactions, electrons are transferred from one substance or atom to another. We can balance oxidation–reduction reactions in solution using the **oxidation state method** (Table 4.3), in which the overall reaction is separated into an oxidation equation and a reduction equation. **Single-displacement reactions** are reactions of metals with either acids or another metal salt that result in dissolution of the first metal and precipitation of a second (or evolution of hydrogen gas). The outcome of these reactions can be predicted using the **activity series** (Table 4.4), which arranges metals and H_2 in decreasing order of their tendency to be oxidized. Any metal will reduce metal ions below it in the activity series. **Active metals** lie at the top of the activity series, whereas **inert metals** such as gold, silver, and platinum are at the bottom of the activity series.

4.9 Quantitative Analysis Using Titrations (p. 189)

The concentration of a species in solution can be determined by **quantitative analysis**. One such method is a **titration**, in which a measured volume of a solution of one substance, the **titrant**, is added to a solution of another substance to determine its concentration. The **equivalence point** in a titration is the point at which exactly enough reactant has been added for the reaction to go to completion. A **standard solution**, a solution whose concentration is known precisely, is used to determine the concentration of the titrant. Many titrations, especially those that involve acid–base reactions, rely on an indicator. The point at which a color change is observed is the **endpoint**, which is close to the equivalence point if the indicator is chosen properly.

QUESTIONS AND PROBLEMS

Please be sure you are familiar with the topics discussed in Essential Skills 3 at the end of this chapter before proceeding to the Questions and Problems. Questions and Problems with colored numbers have answers in the appendix and complete solutions in the Student Solutions Manual.

CONCEPTUAL

4.1 Aqueous Solutions

1. What are the advantages to carrying out a reaction in solution rather than simply mixing the pure reactants?

2. What types of compounds dissolve in polar solvents?
3. Describe the charge distribution in liquid water. How does this distribution affect its physical properties?
4. Must a molecule have an asymmetric charge distribution in order to be polar? Explain your answer.
5. Why are many ionic substances soluble in water?
6. Explain the phrase "Like dissolves like."
7. What kinds of covalent compounds are soluble in water?
8. Why do most aromatic hydrocarbons have only limited solubility in water? Would you expect their solubility to be higher, lower, or the same in ethanol compared with water? Why?
9. Predict whether each compound will form a homogeneous mixture in water, and explain why: (a) toluene; (b) acetic acid; (c) sodium acetate; (d) butanol; (e) pentanoic acid.
10. Decide whether each compound will dissolve in water, and explain why: (a) ammonium chloride; (b) 2-propanol; (c) heptane; (d) potassium dichromate; (e) 2-octanol.
11. Given water and toluene, choose the better solvent for each compound, and explain your reasoning: (a) sodium cyanide; (b) benzene; (c) acetic acid; (d) sodium ethoxide (CH_3CH_2ONa).
12. Of water and toluene, predict which is the better solvent for each compound, and explain why: (a) *t*-butanol; (b) calcium chloride; (c) sucrose; (d) cyclohexene.
13. Compound A is divided into three equal samples. The first sample does not dissolve in water, the second sample dissolves only slightly in ethanol, and the third sample dissolves completely in toluene. What does this suggest about the polarity of A?
14. You are given a mixture of three solid compounds, A, B, and C, and are told that A is a polar compound, B is slightly polar, and C is nonpolar. Suggest a method for separating these three compounds.
15. A laboratory technician is given a sample that contains only sodium chloride, sucrose, and cyclodecanone, a ketone. You must tell the technician how to separate these three compounds from the mixture. What would you suggest?
16. Many over-the-counter drugs are sold as ethanol/water solutions rather than as purely aqueous solutions. Give a plausible reason for this practice.
17. What distinguishes a weak electrolyte from a strong electrolyte?
18. Which organic groups result in aqueous solutions that conduct electricity?
19. It is considered highly dangerous to splash barefoot in puddles during a lightning storm. Why?
20. Which solution would you expect to conduct electricity well? (a) an aqueous solution of sodium chloride; (b) a solution of ethanol in water; (c) a solution of calcium chloride in water; (d) a solution of sucrose in water. Explain your reasoning.
21. Determine whether each solution will conduct electricity well: (a) an aqueous solution of acetic acid; (b) an aqueous solution of potassium hydroxide; (c) a solution of ethylene glycol in water; (d) a solution of ammonium chloride in water. Explain your reasoning.
22. Which of the following is a strong electrolyte, weak electrolyte, or nonelectrolyte in an aqueous solution? (a) potassium hydroxide; (b) ammonia; (c) calcium chloride; (d) butanoic acid
23. Decide whether each compound is a strong electrolyte, weak electrolyte, or nonelectrolyte in an aqueous solution: (a) magnesium hydroxide; (b) butanol; (c) ammonium bromide; (d) pentanoic acid.
24. Predict whether each compound will act as a strong electrolyte, weak electrolyte, or nonelectrolyte in aqueous solution, and explain why: (a) H_2SO_4; (b) diethylamine; (c) 2-propanol; (d) ammonium chloride; (e) propanoic acid.
25. Would you expect a 1.0 *M* solution of $CaCl_2$ to be a better conductor of electricity than a 1.0 *M* solution of NaCl? Why or why not?

4.2 Solution Concentrations

26. An alternative way to define the concentration of a solution is *molality*, abbreviated as *m*. Molality is defined as the number of moles of solute in 1 kg of *solvent*. How is this different from molarity? Would you expect a 1 *M* solution of sucrose to be more or less concentrated than a 1 *m* solution of sucrose? Explain your answer.
27. What are the advantages of using solutions for quantitative calculations?

4.3 Stoichiometry of Reactions in Solution

28. What information is required to determine the mass of solute in a solution if you know the molar concentration of the solution?
29. Is it possible for one reactant to be limiting in a reaction that does not go to completion?

4.4 Ionic Equations

30. What information can be obtained from a complete ionic equation that cannot be obtained from the overall equation?

4.5 Precipitation Reactions

31. Predict whether mixing each of the following pairs of solutions will result in the formation of a precipitate. If so, identify the precipitate.
(a) $FeCl_3(aq) + Na_2S(aq)$
(b) $NaOH(aq) + H_3PO_4(aq)$
(c) $ZnCl_2(aq) + (NH_4)_2S(aq)$
32. Predict which pairs of solutions will form a precipitate when mixed, and predict the identity of the precipitate.
(a) $KOH(aq) + H_3PO_4(aq)$
(b) $K_2CO_3(aq) + BaCl_2(aq)$
(c) $Ba(NO_3)_2(aq) + Na_2SO_4(aq)$

4.6 Acid–Base Reactions

33. Why was it necessary to expand on the Arrhenius definition of an acid and a base? What specific point does the Brønsted–Lowry definition address?
34. State whether each of the following compounds is an acid, a base, or a salt: (a) $CaCO_3$; (b) $NaHCO_3$; (c) H_2SO_4; (d) $CaCl_2$; (e) $Ba(OH)_2$.
35. Decide whether each of the following compounds is an acid, a base, or a salt: (a) NH_3; (b) NH_4Cl; (c) H_2CO_3; (d) CH_3COOH; (e) NaOH.
36. Classify each compound as a strong acid, weak acid, strong base, or weak base in aqueous solution: (a) sodium hydroxide; (b) acetic acid; (c) magnesium hydroxide; (d) tartaric acid; (e) sulfuric acid; (f) ammonia; (g) hydroxylamine (NH_2OH); (h) hydrocyanic acid.
37. Decide whether each compound forms an aqueous solution that is strongly acidic, weakly acidic, strongly basic, or

weakly basic: (a) propanoic acid; (b) hydrobromic acid; (c) methylamine; (d) lithium hydroxide; (e) citric acid; (f) sodium acetate; (g) ammonium chloride; (h) barium hydroxide.

38. What is the relationship between the strength of an acid and the strength of the conjugate base derived from that acid? Would you expect the $CH_3CO_2^-$ ion to be a strong or weak base? Why? Is the hydronium ion a strong or weak acid? Explain your answer.

39. What are the products of an acid–base reaction? Under what circumstances is one of the products a gas?

40. Explain how an aqueous solution that is strongly basic can have a pH, which is a measure of the *acidity* of a solution.

4.7 The Chemistry of Acid Rain

41. Why is it recommended that marble countertops not be used in kitchens? Marble is composed mostly of $CaCO_3$.

42. Explain why desulfurization of fossil fuels is an area of intense research.

43. What is the role of NO_x in the formation of acid rain?

4.8 Oxidation–Reduction Reactions in Solution

44. Which elements in the periodic table tend to be good oxidants? Which tend to be good reductants?

45. If two compounds are mixed, one containing an element that is a poor oxidant and one with an element that is a poor reductant, do you expect a redox reaction to occur? Explain your answer. What do you predict if one is a strong oxidant and the other is a weak reductant? Why?

46. In each redox reaction, determine which species is oxidized and which is reduced:

(a) $Zn(s) + H_2SO_4(aq) \longrightarrow ZnSO_4(aq) + H_2(g)$

(b) $Cu(s) + 4HNO_3(aq) \longrightarrow Cu(NO_3)_2(aq) + 2NO_2(g) + 2H_2O(l)$

(c) $BrO_3^-(aq) + 2MnO_2(s) + H_2O(l) \longrightarrow Br^-(aq) + 2MnO_4^-(aq) + 2H^+(aq)$

47. Single-displacement reactions are a subset of redox reactions. In this subset, what is oxidized and what is reduced? Give an example of a redox reaction that is *not* a single-displacement reaction.

48. Of the following elements, which would you expect to have the greatest tendency to be oxidized: Zn, Li, or S? Explain your reasoning.

49. Of these elements, which would you expect to be easiest to reduce: Se, Sr, or Ni? Explain your reasoning.

50. Which of these metals produce H_2 in acidic solution? (a) Ag; (b) Cd; (c) Ca; (d) Cu

51. Using the activity series, predict what happens in each situation. If a reaction occurs, write the net ionic equation.

(a) $Mg(s) + Cu^{2+}(aq) \longrightarrow$

(b) $Au(s) + Ag^+(aq) \longrightarrow$

(c) $Cr(s) + Pb^{2+}(aq) \longrightarrow$

(d) $K(s) + H_2O(l) \longrightarrow$

(e) $Hg(l) + Pb^{2+}(aq) \longrightarrow$

4.9 Quantitative Analysis Using Titrations

52. Explain how to determine the concentration of a substance using a titration.

53. The titration procedure is an application of the use of limiting reactants. Explain why this is so.

NUMERICAL

This section includes "paired problems" (marked by brackets) that require similar problem-solving skills.

4.2 Solution Concentrations

54. Calculate the number of grams of solute in 1.000 L of each solution: (a) 0.1065 *M* BaI_2; (b) 1.135 *M* Na_2SO_4; (c) 1.428 *M* NH_4Br; (d) 0.889 *M* sodium acetate.

55. Give the number of grams of solute in 1.000 L of each solution: (a) 0.2593 *M* $NaBrO_3$; (b) 1.592 *M* KNO_3; (c) 1.559 *M* acetic acid; (d) 0.943 *M* potassium iodate.

56. If all solutions are of the same solute, which solution contains the greater mass of solute?

(a) 1.40 L of a 0.334 *M* solution or 1.10 L of a 0.420 *M* solution

(b) 25.0 mL of a 0.134 *M* solution or 10.0 mL of a 0.295 *M* solution

(c) 250 mL of a 0.489 *M* solution or 150 mL of a 0.769 *M* solution

57. Complete the table for 500 mL of solution.

Compound	Mass (g)	Moles	Concentration (*M*)
Calcium sulfate	4.86	____	____
Acetic acid	____	3.62	____
Hydrogen iodide dihydrate	____	____	1.273
Barium bromide	3.92	____	____
Glucose	____	____	0.983
Sodium acetate	____	2.42	____

58. What are the concentrations of each of the species present in the following aqueous solutions?

(a) 0.324 mol of K_2MoO_4 in 250 mL of solution

(b) 0.528 mol of potassium formate in 300 mL of solution

(c) 0.477 mol of $KClO_3$ in 900 mL of solution

(d) 0.378 mol of potassium iodide in 750 mL of solution

59. Determine the concentration of each species in the following aqueous solutions:

(a) 0.489 mol of $NiSO_4$ in 600 mL of solution

(b) 1.045 mol of magnesium bromide in 500 mL of solution

(c) 0.146 mol of glucose in 800 mL of solution

(d) 0.479 mol of $CeCl_3$ in 700 mL of solution

60. What is the molar concentration of each solution?

(a) 12.8 g of sodium hydrogen sulfate in 400 mL of solution

(b) 7.5 g of potassium hydrogen phosphate in 250 mL of solution

(c) 11.4 g of barium chloride in 350 mL of solution

(d) 4.3 g of tartaric acid ($C_4H_6O_6$) in 250 mL of solution

61. Determine the molar concentration of each solution:

(a) 8.7 g of calcium bromide in 250 mL of solution

(b) 9.8 g of lithium sulfate in 300 mL of solution

(c) 12.4 g of sucrose ($C_{12}H_{22}O_{11}$) in 750 mL of solution

(d) 14.2 g of iron(III) nitrate hexahydrate in 300 mL of solution

62. Give the concentration of each reactant in the following equations, assuming 20.0 g of each and a solution volume of 250 mL for each reactant:

(a) $BaCl_2(aq) + Na_2SO_4(aq) \longrightarrow$

(b) $Ca(OH)_2(aq) + H_3PO_4(aq) \longrightarrow$

(c) $Al(NO_3)_3(aq) + H_2SO_4(aq) \longrightarrow$
(d) $Pb(NO_3)_2(aq) + CuSO_4(aq) \longrightarrow$
(e) $Al(CH_3CO_2)_3(aq) + NaOH(aq) \longrightarrow$

63. An experiment required 200.0 mL of a 0.330 *M* solution of Na_2CrO_4. A stock solution of Na_2CrO_4 containing 20.0% solute by mass with a density of 1.19 g/cm^3 was used to prepare this solution. Describe how to prepare 200.0 mL of a 0.330 *M* solution of Na_2CrO_4 using the stock solution.

64. Refer to the breathalyzer test described in Example 4.8. How much ethanol must be present in 89.5 mL of a person's breath to consume all of the potassium dichromate in a breathalyzer ampule containing 3.0 mL of a 0.40 mg/mL solution of potassium dichromate?

65. Calcium hypochlorite, $Ca(OCl)_2$, is an effective disinfectant for clothing and bedding. If a solution has a $Ca(OCl)_2$ concentration of 3.4 g per 100 mL of solution, what is the molarity of hypochlorite?

66. Phenol, C_6H_5OH, is often used as an antiseptic in mouthwashes and throat lozenges. If a mouthwash has a phenol concentration of 1.5 g per 100 mL of solution, what is the molarity of phenol?

67. If a tablet containing 100 mg of caffeine ($C_8H_{10}N_4O_2$) is dissolved in water to give 10.0 fluid ounces of solution, what is the molar concentration of caffeine in the solution?

68. A certain drug label carries instructions to add 10.0 mL of sterile water, stating that each milliliter of the resulting solution will contain 0.500 g of medication. If a patient has a prescribed dose of 900.0 mg, how many milliliters of the solution should be administered?

4.3 Stoichiometry of Reactions in Solution

69. Barium chloride and sodium sulfate react to produce sodium chloride and barium sulfate. If 50.00 mL of a 2.55 *M* solution of barium chloride is used in the reaction, how many grams of sodium sulfate are needed for the reaction to go to completion?

70. Phosphoric acid and magnesium hydroxide react to produce magnesium phosphate and water. If 45.00 mL of a 1.50 *M* solution of phosphoric acid is used in the reaction, how many grams of magnesium hydroxide are needed for the reaction to go to completion?

71. How many grams of ammonium bromide are produced from the reaction of 50.00 mL of 2.08 *M* iron(II) bromide with a stoichiometric amount of ammonium sulfide? What is the second product? How many grams of the second product are produced?

72. How many grams of sodium phosphate are obtained from the reaction of 75.00 mL of 2.80 *M* sodium carbonate with a stoichiometric amount of phosphoric acid? A second product is water; what is the third product? How many grams of the third product are obtained?

73. Silver nitrate and sodium chloride react to produce sodium nitrate and silver chloride. If 2.60 g of AgCl was obtained by adding excess NaCl to 100 mL of $AgNO_3$, what was the molarity of the silver nitrate solution?

74. Lead(II) nitrate and hydroiodic acid react to produce lead(II) iodide and nitric acid. If 3.25 g of lead(II) iodide was obtained by adding excess HI to 150.0 mL of lead(II) nitrate, what was the molarity of the lead(II) nitrate solution?

4.5 Precipitation Reactions

75. What mass of precipitate would you expect to obtain by mixing 250 mL of a solution containing 4.88 g of Na_2CrO_4 with 200 mL of a solution containing 3.84 g of $AgNO_3$? What is the final nitrate ion concentration?

76. Adding 10.0 mL of a dilute solution of zinc(II) nitrate to 246 mL of a 2.00 *M* solution of sodium sulfide produced 0.279 g of a precipitate. How many grams of zinc(II) nitrate and sodium sulfide were consumed to produce this quantity of product? What was the concentration of each ion in the original solutions? What is the concentration of the sulfide ion in solution after the precipitation reaction, assuming no further reaction?

4.6 Acid–Base Reactions

77. Derive an equation to relate the hydrogen ion concentration to the molarity of a solution of a strong monoprotic acid.

78. Referring to the periodic table, derive an equation to relate the hydroxide ion concentration to the molarity of a solution of (a) a Group I hydroxide and (b) a Group II hydroxide.

79. Given the following salts, identify the acid and the base in the neutralization reactions and then write the complete ionic equation: (a) barium sulfate; (b) lithium nitrate; (c) sodium bromide; (d) calcium perchlorate.

80. What is the pH of each solution? (a) 5.8×10^{-3} mol of HNO_3 in 257 mL of water; (b) 0.0079 mol of HI in 750 mL of water; (c) 0.011 mol of $HClO_4$ in 500 mL of water; (d) 0.257 mol of HBr in 5.00 L of water

81. What is the hydrogen ion concentration of each of the following in the indicated pH range?
(a) black coffee (pH 5.10)
(b) milk (pH 6.30–7.60)
(c) tomatoes (pH 4.00–4.40)

82. Give the range of H^+ ion concentrations for each:
(a) orange juice (pH 3–4)
(b) fresh egg white (pH 7.60–7.80)
(c) lemon juice (pH 2.20–2.40)

83. What is the pH of a solution prepared by diluting 25.00 mL of 0.879 *M* HCl to a volume of 555 mL?

84. Vinegar is primarily an aqueous solution of acetic acid. Commercial vinegar typically contains 5.0 g of acetic acid in 95.0 g of water. What is the concentration of commercial vinegar? If only 3.1% of the acetic acid dissociates to $CH_3CO_2^-$ and H^+, what is the pH of the solution? (Assume the density of the solution is 1.00 g/mL.)

85. If a typical household cleanser is 0.50 *M* in strong base, what volume of 0.998 *M* strong monoprotic acid is needed to neutralize 50.0 mL of the cleanser?

86. A 25.00-mL sample of a 0.9005 *M* solution of HCl is diluted to 500.0 mL. What is the molarity of the final solution? How many milliliters of 0.223 *M* NaOH are needed to neutralize 25.00 mL of this final solution?

87. If 20.0 mL of 0.10 *M* NaOH is needed to neutralize 15.0 mL of gastric fluid, what is the molarity of HCl in the fluid? (Assume all of the acidity is due to the presence of HCl.) What other base might be used in this titration?

88. Malonic acid, $C_3H_4O_4$, is a diprotic acid used in the manufacture of barbiturates. How many grams of malonic acid are in a 25.00-mL sample that requires 32.68 mL of 1.124 *M* KOH for complete neutralization to occur? Malonic acid is a dicarboxylic acid; propose a structure for malonic acid.

89. Describe how you would prepare 500 mL of a 1.00 *M* stock solution of HCl from an HCl solution that is 12.11 *M*. Using your stock solution, how would you prepare 500 mL of a solution that is 0.012 *M* in HCl?

90. Given a stock solution that is 8.52 *M* in HBr, describe how you would prepare 500 mL solutions with these concentrations: (a) 2.50 *M*; (b) 4.00×10^{-3} *M*; (c) 0.989 *M*.

91. How many moles of solute are contained in each? (a) 25.00 mL of 1.86 *M* NaOH; (b) 50.00 mL of 0.0898 *M* HCl; (c) 13.89 mL of 0.102 *M* HBr

92. A chemist needed a solution that was approximately 0.5 *M* in HCl but could measure only 10.00-mL samples into a 50.00-mL volumetric flask. Propose a method for making up the solution. (Assume that concentrated HCl is 12.0 *M*.)

93. Give the balanced equation for the reaction of (a) perchloric acid with potassium hydroxide and (b) nitric acid with calcium hydroxide.

94. Write a balanced equation for the reaction of (a) solid strontium hydroxide with hydrobromic acid and (b) aqueous sulfuric acid with solid sodium hydroxide.

95. A neutralization reaction gives calcium nitrate as one of the two products. Identify the acid and the base in this reaction. What is the second product? If the product had been cesium iodide, what would have been the acid and the base? What is the complete ionic equation for each of these reactions?

4.8 Oxidation–Reduction Reactions in Solution

96. Balance each redox reaction under the conditions indicated:
(a) $MnO_4^-(aq) + S_2O_3^{2-}(aq) \longrightarrow Mn^{2+}(aq) + SO_4^{2-}(aq)$; acidic solution
(b) $Fe^{2+}(aq) + Cr_2O_7^{2-}(aq) \longrightarrow Fe^{3+}(aq) + Cr^{3+}(aq)$; acidic solution
(c) $Fe(s) + CrO_4^{2-}(aq) \longrightarrow Fe_2O_3(s) + Cr_2O_3(s)$; basic solution
(d) $Cl_2(aq) \longrightarrow ClO_3^-(aq) + Cl^-(aq)$; acidic solution
(e) $CO_3^{2-}(aq) + N_2H_4(aq) \longrightarrow CO(g) + N_2(g)$; basic solution

97. Balance each of the following redox reactions:
(a) $CuS(s) + NO_3^-(aq) \longrightarrow Cu^{2+}(aq) + SO_4^{2-}(aq) + NO(g)$; acidic solution
(b) $Ag(s) + HS^-(aq) + CrO_4^{2-}(aq) \longrightarrow Ag_2S(s) + Cr(OH)_3(s)$; basic solution
(c) $Zn(s) + H_2O(l) \longrightarrow Zn^{2+}(aq) + H_2(g)$; acidic solution
(d) $O_2(g) + Sb(s) \longrightarrow H_2O_2(aq) + SbO_2^-(aq)$; basic solution
(e) $UO_2^{2+}(aq) + Te(s) \longrightarrow U^{4+}(aq) + TeO_4^{2-}(aq)$; acidic solution

98. Using the activity series, predict what happens when each experiment is performed. If a reaction occurs, write the net ionic equation for the reaction; then give the complete ionic equation.
(a) A few drops of $NiBr_2$ are dropped onto a piece of iron.
(b) A strip of zinc is placed into a solution of HCl.
(c) Copper is dipped into a solution of $ZnCl_2$.
(d) A solution of silver nitrate is dropped onto an aluminum plate.

99. Predict what happens when each experiment is carried out. If a reaction occurs, write the net ionic equation and the complete ionic equation for the reaction.
(a) Platinum wire is dipped into hydrochloric acid.
(b) Manganese metal is added to a solution of iron(II) chloride.
(c) Tin is heated with steam.
(d) Hydrogen gas is bubbled through a solution of lead(II) nitrate.

100. Dentists occasionally use metallic mixtures called *amalgams* for fillings. If an amalgam contains zinc, however, water can contaminate the amalgam as it is being manipulated, producing hydrogen gas under basic conditions. As the filling hardens, the gas can be released, causing pain and cracking the tooth. Write a balanced equation for this reaction.

101. Copper metal readily dissolves in dilute aqueous nitric acid to form blue $Cu^{2+}(aq)$ and nitric oxide gas.
(a) What has been oxidized and what has been reduced?
(b) Balance the equation.

102. Classify each reaction as an acid–base, precipitation, or redox reaction, or state if there is no reaction, and then complete and balance the equation:
(a) $Zn(s) + HCl(aq) \longrightarrow$
(b) $HNO_3(aq) + AlCl_3(aq) \longrightarrow$
(c) $K_2CrO_4(aq) + Ba(NO_3)_2(aq) \longrightarrow$
(d) $Zn(s) + Ni^{2+}(aq) \longrightarrow Zn^{2+}(aq) + Ni(s)$

103. Determine whether each reaction is an acid–base, precipitation, or redox reaction, and then complete and balance the equation:
(a) $Pt^{2+}(aq) + Ag(s) \longrightarrow$
(b) $HCN(aq) + NaOH(aq) \longrightarrow$
(c) $Fe(NO_3)_3(aq) + NaOH(aq) \longrightarrow$
(d) $CH_4(g) + O_2(g) \longrightarrow$

4.9 Quantitative Analysis Using Titrations

104. A 10.00-mL sample of a 1.07 *M* solution of potassium hydrogen phthalate (KHP, MM = 204.22 g/mol) is diluted to 250.0 mL. What is the molarity of the final solution? How many grams of KHP are in the 10.00-mL sample?

105. What volume of a 0.978 *M* solution of NaOH must be added to 25.0 mL of 0.583 *M* HCl to completely neutralize the acid? How many moles of NaOH are needed for the neutralization?

106. A student was titrating 25.00 mL of a basic solution with an HCl solution that was 0.281 *M*. The student ran out of the HCl solution after having added 32.46 mL, so she borrowed an HCl solution that was labeled as 0.317 *M*. An additional 11.5 mL of the second solution was needed to complete the titration. What was the concentration of the basic solution?

APPLICATIONS

107. Acetaminophen (molar mass = 151 g/mol) is an analgesic used as a substitute for aspirin. If a child's dose contains 80.0 mg of acetaminophen/5.00 mL of ethanol/water, what is the molar concentration? Acetaminophen is frequently packaged as an ethanol/water solution rather than as an aqueous one. Why?

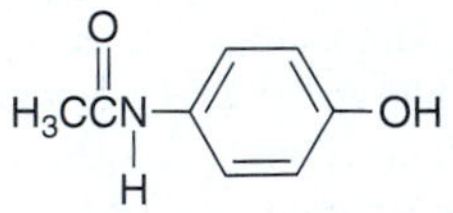

Acetaminophen

108. Lead may have been the first metal ever recovered from its ore by humans. Its cation, Pb^{2+}, forms a precipitate with Cl^- according to the equation

$$Pb^{2+}(aq) + 2Cl^-(aq) \longrightarrow PbCl_2(s)$$

When $PbCl_2$ is dissolved in hot water, its presence can be confirmed by its reaction with CrO_4^{2-}, with which it forms a yellow precipitate:

$$Pb^{2+}(aq) + CrO_4^{2-}(aq) \longrightarrow PbCrO_4(s)$$

The precipitate is used as a rust inhibitor and in pigments.
(a) What type of reaction does each of the above equations represent?
(b) If 100 mL of a Pb^{2+} solution produces 1.65 g of lead chromate, what was the concentration of the lead solution?
(c) What volume of a potassium chromate solution containing 0.503 g of solute per 250.0 mL is needed for this reaction?
(d) If all the $PbCrO_4$ originated from $PbCl_2$, what volume of a 1.463 *M* NaCl solution was needed for the initial reaction?
(e) Why is there environmental concern over the use of $PbCrO_4$?

109. Reactions that affect buried marble artifacts present a problem for archaeological chemists. Groundwater dissolves atmospheric carbon dioxide to produce an aqueous solution of carbonic acid:

$$CO_2(g) + H_2O(l) \longrightarrow H_2CO_3(aq)$$

This weakly acidic carbonic acid solution dissolves marble, converting it to soluble calcium bicarbonate:

$$CaCO_3(s) + H_2CO_3(aq) \longrightarrow Ca(HCO_3)_2(aq)$$

Evaporation of water causes carbon dioxide to be driven off. The result is the precipitation of calcium carbonate:

$$Ca(HCO_3)_2(aq) \longrightarrow CaCO_3(s) + H_2O(l) + CO_2(g)$$

The reprecipitated calcium carbonate forms a hard scale, or incrustation, on the surface of the object.
(a) If 8.5 g of calcium carbonate was obtained by evaporating 250 mL of a solution of calcium bicarbonate followed by drying, what was the molarity of the initial calcium bicarbonate solution, assuming complete reaction?
(b) If the overall reaction sequence was 75% efficient, how many grams of carbonic acid were initially dissolved in the 250 mL to produce the calcium bicarbonate?

110. How many Maalox tablets are needed to neutralize 5.00 mL of 0.100 *M* HCl stomach acid if each tablet contains 200 mg $Mg(OH)_2$ + 200 mg $Al(OH)_3$? Each Rolaids tablet contains 412 mg $CaCO_3$ + 80.0 mg $Mg(OH)_2$. How many Rolaids tablets are needed? Suggest another formula (and approximate composition) for an effective antacid tablet.

111. Citric acid ($C_6H_8O_7$, molar mass = 192.12 g/mol) is a triprotic acid extracted from citrus fruits and pineapple waste that is used in beverages and jellies to provide tartness. How many grams of citric acid are contained in a 25.00-mL sample that requires 38.43 mL of 1.075 *M* NaOH for neutralization to occur? Give the formula of the calcium salt of this compound.

112. A method for determining the molarity of a strongly acidic solution has been developed based on the fact that a standard solution of potassium iodide and potassium iodate yields iodine when treated with acid:

$$IO_3^- + 5I^- + 6H^+ \longrightarrow 3I_2 + 3H_2O$$

Starch is used as the indicator in this titration because starch reacts with iodine in the presence of iodide to form an intense blue complex. The amount of iodine produced from this reaction can be determined by subsequent titration with thiosulfate:

$$2S_2O_3^{2-} + I_2 \longrightarrow S_4O_6^{2-} + 2I^-$$

The endpoint is reached when the solution becomes colorless.
(a) The thiosulfate solution was determined to be 1.023 *M*. If 37.63 mL of thiosulfate solution was needed to titrate a 25.00-mL sample of an acid, what was the H^+ ion concentration of the acid?
(b) If the 25.00-mL sample that was titrated had been produced by dilution of a 10.00-mL sample of acid, what was the molarity of the acid in the original solution?
(c) Why might this be an effective method for determining the molarity of a strong acid, such as H_2SO_4?

113. Sewage processing occurs in three stages. Primary treatment removes suspended solids, secondary treatment involves biological processes that decompose organic matter, and tertiary treatment removes specific pollutants that arise from secondary treatment (generally phosphates). Phosphate can be removed by treating the HPO_4^{2-} solution produced in the second stage with lime (CaO) to precipitate hydroxyapatite, $Ca_5(PO_4)_3OH$.
(a) Write a balanced equation for the reaction that occurs in the tertiary treatment process.
(b) What has been neutralized in this process?
(c) Four pounds of hydroxyapatite precipitated from the water. What mass of lime was used in the reaction?
(d) Assuming a volume of water of 30 m^3, what was the hydrogen phosphate anion concentration in the water?

114. $Ca(OH)_2$ and $CaCO_3$ are effective in neutralizing the effects of acid rain on lakes. Suggest other compounds that might be effective in treating lakes. Give a plausible reason to explain why $Ca(OH)_2$ and $CaCO_3$ are used.

115. Approximately 95% of the chlorine produced industrially comes from the electrolysis of sodium chloride solutions (brine). This reaction is

$$NaCl(aq) + H_2O(l) \longrightarrow Cl_2(g) + H_2(g) + NaOH(aq)$$

Chlorine is a respiratory irritant whose presence is easily detected by its penetrating odor and greenish-yellow color. One use for the chlorine produced is in the manufacture of hydrochloric acid.
(a) In the chemical equation shown, what has been oxidized and what has been reduced?
(b) Write the oxidation and reduction equations for this reaction.
(c) Balance the net ionic equation.
(d) Name another salt that might produce chlorine by electrolysis, and give the expected products. Identify those products as gases, liquids, or solids.

116. The lead/acid battery used in automobiles consists of six cells that produce a 12-volt electrical system. During discharge, lead(IV) oxide, lead, and aqueous sulfuric acid react to form lead(II) sulfate and water.
(a) What has been oxidized and what has been reduced?
(b) Write and balance the equation for the reaction.
(c) What is the net ionic equation?
(d) What is the complete ionic equation?
(e) What hazard is associated with handling automobile batteries?

117. The use of iron, an element abundant in the earth's crust, goes back to prehistoric times. In fact, it is believed that the Egyptians used iron implements approximately 5000 years ago. One method for quantifying the iron concentration in a sample involves three steps. The first step is to dissolve a portion of the sample in concentrated hydrochloric acid to produce ferric chloride; the second is to reduce Fe^{3+} to Fe^{2+} using zinc metal; and the third is to titrate Fe^{2+} with permanganate, producing $Mn^{2+}(aq)$ and ferric iron in the form of Fe_2O_3.
(a) Write equations for all three steps.
(b) Write the net ionic equations for these three reactions.
(c) If 27.64 mL of a 1.032 *M* solution of permanganate is required to titrate 25.00 mL of Fe^{2+} in the third step, how many grams of Fe were in the original sample?
(d) Based on your answer to part c, if the original sample weighed 50.32 g, what was the percentage of iron?

118. Baking powder, which is a mixture of tartaric acid and sodium bicarbonate, is used in baking cakes and bread. Why does bread rise when you use baking powder? What type of reaction is involved?

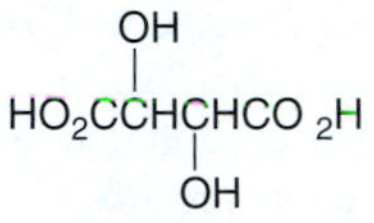

Tartaric acid

119. An activity series exists for the halogens, which is based on the ease of *reduction* of the diatomic halogen molecule, X_2, to X^-. Experimentally, it is found that fluorine is the easiest halogen to reduce (that is, F_2 is the best oxidant), and iodine is the hardest halogen to reduce (I_2 is the worst oxidant). Consequently, the addition of any diatomic halogen, Y_2, to solutions containing a halide ion, X^-, that lies below Y in the periodic table will result in the reduction of Y_2 to Y^- and the oxidation of X^-, Describe what you would expect to occur when (a) chlorine is added to an aqueous solution of bromide and (b) iodine crystals are added to a solution of potassium bromide. Bromide is present in naturally occurring salt solutions called *brines*. Based on your answers, propose an effective method to remove bromide from brine.

120. Marble is composed of mostly calcium carbonate. Assuming that acid rain contains 4.0×10^{-5} *M* H_2SO_4, approximately what volume of rain is necessary to dissolve a 250-pound marble statue?

121. One of the "first-aid" measures used to neutralize lakes whose pH has dropped to critical levels is to spray them with slaked lime ($Ca(OH)_2$) or limestone ($CaCO_3$). (A slower but effective alternative is to add limestone boulders.) How much slaked lime would be needed to neutralize the acid in a lake that contains 4.0×10^{-5} *M* H_2SO_4 and has a volume of 1.2 cubic miles (5.0×10^{12} L)?

ESSENTIAL SKILLS 3

TOPICS

- Base-10 Logarithms
- Calculations Using Common Logarithms

Essential Skills 1 and 2 described some fundamental mathematical operations used for solving problems in chemistry. This section introduces you to base-10 logarithms, a topic with which you must be familiar to do the Questions and Problems at the end of Chapter 4. We will return to the subject of logarithms in Essential Skills 6 at the end of Chapter 11.

Base-10 (Common) Logarithms

Essential Skills 1 introduced exponential notation, in which a base number is multiplied by itself the number of times indicated in the exponent. The number 10^3, for example, is the base 10 multiplied by itself 3 times ($10 \times 10 \times 10 = 1000$). Now suppose that we do not know what the exponent is—that we are given only a base of 10 and the final number. If our answer is 1000, the problem can be expressed as

$$10^a = 1000$$

We can determine the value of a by using an operation called the *base-10 logarithm*, or *common logarithm*, abbreviated as *log*, which represents the power to which 10 is raised to give the number to the right of the equals sign. This relationship is stated as $\log 10^a = a$. In this case, the logarithm is 3 because $10^3 = 1000$:

$$\log 10^3 = 3$$
$$\log 1000 = 3$$

Now suppose you are asked to find a when the final number is 659. The problem can be solved as follows (remember that any operation applied to one side of an equality must also be applied to the other side):

$$10^a = 659$$
$$\log 10^a = \log 659$$
$$a = \log 659$$

If you enter the number 659 into your calculator and press "log," you get an answer of 2.819, which means that $a = 2.819$ and $10^{2.819} = 659$. Conversely, if you enter the value 2.819 into your calculator and press "10^x," you get an answer of 659.

You can decide whether your answer is reasonable by comparing it with the results you get when $a = 2$ and $a = 3$:

$$a = 2: \quad 10^2 = 100$$
$$a = 2.819: \quad 10^{2.819} = 659$$
$$a = 3: \quad 10^3 = 1000$$

Because the number 659 is between 100 and 1000, a must be between 2 and 3, which is indeed the case.

Table ES3.1 lists some base-10 logarithms, their numerical values, and their exponential forms.

TABLE ES3.1 Relationships in base-10 logarithms

Numerical Value	Exponential Form	Logarithm (a)
1000	10^3	3
100	10^2	2
10	10^1	1
1	10^0	0
0.1	10^{-1}	−1
0.01	10^{-2}	−2
0.001	10^{-3}	−3

Base-10 logarithms may also be expressed as $\log_{10}$, in which the base is indicated as a subscript. Thus, we can write $\log 10^a = a$ in either of two ways:

$$\log 10^a = a$$
$$\log_{10} 10^a = a$$

The second equation explicitly indicates that we are solving for the base-10 logarithm of 10^a.

The number of significant figures in a logarithmic value is the same as the number of digits *after* the decimal point in its logarithm, so log 62.2, a value with three significant figures, is 1.794, with three significant figures after the decimal point; that is, $10^{1.794} = 62.2$, *not* 62.23. The following exercises provide practice converting a value to its exponential form and then calculating its logarithm.

SKILL BUILDER ES3.1

Express each number as a power of 10, and then find the common logarithm: **(a)** 10,000; **(b)** 0.00001; **(c)** 10.01; **(d)** 2.87; **(e)** 0.134.

Solution

(a) $10{,}000 = 1 \times 10^4$
$\log 1 \times 10^4 = 4.0$

(b) $0.00001 = 1 \times 10^{-5}$
$\log 1 \times 10^{-5} = -5.0$

(c) $10.01 = 1.001 \times 10$
$\log 10.01 = 1.0004$ (Enter 10.01 into your calculator and press "log.")
Thus, $10^{1.0004} = 10.01$.

(d) $2.87 = 2.87 \times 10^0$
$\log 2.87 = 0.458$ (Enter 2.87 into your calculator and press "log.")
Thus, $10^{0.458} = 2.87$.

(e) $0.134 = 1.34 \times 10^{-1}$
$\log 0.134 = -0.873$ (Enter 0.134 into your calculator and press "log.")
Thus, $10^{-0.873} = 0.134$.

SKILL BUILDER ES3.2

Convert each base-10 logarithm to its numerical value: **(a)** 3; **(b)** −2.0; **(c)** 1.62; **(d)** −0.23; **(e)** −4.872.

Solution **(a)** 10^3 **(b)** 10^{-2} **(c)** $10^{1.62} = 42$; **(d)** $10^{-0.23} = 0.59$; **(e)** $10^{-4.872} = 1.34 \times 10^{-5}$

Calculations Using Common Logarithms

Because logarithms are exponents, all of the properties of exponents you learned in Essential Skills 1 apply to logarithms as well, as summarized in Table ES3.2. The logarithm of 4.08×20.67, for example, can be computed as

$$\begin{aligned}\log(4.08 \times 20.67) &= \log 4.08 + \log 20.67\\ &= 0.611 + 1.3153\\ &= 1.926\end{aligned}$$

We can be sure that this answer is correct by checking that $10^{1.926}$ is equal to 4.08×20.67, and it is ($84.3 = 84.3$).

In an alternative approach, we multiply the two values before computing the logarithm:

$$\begin{aligned}4.08 \times 20.67 &= 84.3\\ \log 84.3 &= 1.926\end{aligned}$$

We could also have expressed 84.3 as a power of 10 and then calculated the logarithm:

$$\log 84.3 = \log(8.43 \times 10) = \log 8.43 + \log 10 = 0.926 + 1 = 1.926$$

TABLE ES3.2 Properties of logarithms

Operation	Exponential Form	Logarithm
Multiplication	$(10^a)(10^b) = 10^{a+b}$	$\log(ab) = \log a + \log b$
Division	$\dfrac{10^a}{10^b} = 10^{a-b}$	$\log\left(\dfrac{a}{b}\right) = \log a - \log b$

As you can see, *there may be more than one way to solve a problem correctly.*

We can use the properties of exponentials and logarithms to show that the logarithm of the inverse of a number ($1/B$) is the negative logarithm of that number ($-\log B$):

$$\log\left(\frac{1}{B}\right) = -\log B$$

If we use the formula for division given Table ES3.2 and recognize that $\log 1 = 0$, then the logarithm of $1/B$ is

$$\log\left(\frac{1}{B}\right) = \log 1 - \log B = 0 - \log B = -\log B$$

SKILL BUILDER ES3.3

Convert each number to exponential form, and then calculate the logarithm (assume all trailing zeros on whole numbers are not significant): **(a)** 100×1000; **(b)** $0.100 \div 100$; **(c)** 1000×0.010; **(d)** 200×3000; **(e)** $20.5 \div 0.026$.

Solution

(a) $100 \times 1000 = (1 \times 10^2)(1 \times 10^3)$
$\log[(1 \times 10^2)(1 \times 10^3)] = 2.0 + 3.0 = 5.0$
Alternatively, $(1 \times 10^2)(1 \times 10^3) = 1 \times 10^{2+3} = 1 \times 10^5$
$\log(1 \times 10^5) = 5.0$

(b) $0.100 \div 100 = (1.00 \times 10^{-1}) \div (1 \times 10^2)$
$\log[(1.00 \times 10^{-1}) \div (1 \times 10^2)] = -1.000 - 2.0 = -3.0$
Alternatively, $(1.00 \times 10^{-1}) \div (1 \times 10^2) = 1 \times 10^{[(-1)-2]} = 1 \times 10^{-3}$
$\log(1 \times 10^{-3}) = -3.0$

(c) $1000 \times 0.010 = (1 \times 10^3)(1.0 \times 10^{-2})$
$\log[(1 \times 10^3)(1 \times 10^{-2})] = 3.0 + (-2.0) = 1.0$
Alternatively, $(1 \times 10^3)(1.0 \times 10^{-2}) = 1 \times 10^{[3+(-2)]} = 1 \times 10^1$
$\log(1 \times 10^1) = 1.0$

(d) $200 \times 3000 = (2 \times 10^2)(3 \times 10^3)$
$$\begin{aligned}\log[(2 \times 10^2)(3 \times 10^3)] &= \log(2 \times 10^2) + \log(3 \times 10^3)\\ &= (\log 2 + \log 10^2) + (\log 3 + \log 10^3)\\ &= 0.30 + 2 + 0.48 + 3 = 5.8\end{aligned}$$
Alternatively, $(2 \times 10^2)(3 \times 10^3) = 6 \times 10^{2+3} = 6 \times 10^5$
$\log(6 \times 10^5) = \log 6 + \log 10^5 = 0.78 + 5 = 5.8$

(e) $20.5 \div 0.026 = (2.05 \times 10) \div (2.6 \times 10^{-2})$
$$\begin{aligned}\log[(2.05 \times 10) \div (2.6 \times 10^{-2})] &= (\log 2.05 + \log 10) - (\log 2.6 + \log 10^{-2})\\ &= (0.3118 + 1) - [0.415 + (-2)]\\ &= 1.3118 + 1.585 = 2.90\end{aligned}$$
Alternatively, $(2.05 \times 10) \div (2.6 \times 10^{-2}) = 0.788 \times 10^{[1-(-2)]} = 0.788 \times 10^3$
$\log(0.79 \times 10^3) = \log 0.79 + \log 10^3 = -0.102 + 3 = 2.90$

SKILL BUILDER ES3.4

Convert each number to exponential form, and then calculate its logarithm (assume all trailing zeros on whole numbers are not significant): **(a)** $10 \times 100{,}000$; **(b)** $1000 \div 0.10$; **(c)** $25{,}000 \times 150$; **(d)** $658 \div 17$.

Solution

(a) $(1 \times 10)(1 \times 10^5)$; logarithm $= 6.0$
(b) $(1 \times 10^3) \div (1.0 \times 10^{-1})$; logarithm $= 4.00$
(c) $(2.5 \times 10^4)(1.50 \times 10^2)$; logarithm $= 6.57$
(d) $(6.58 \times 10^2) \div (1.7 \times 10)$; logarithm $= 1.59$

5 Energy Changes in Chemical Reactions

LEARNING OBJECTIVES

- To understand the concept of energy and its various forms
- To know the relationship between energy, work, and heat
- To know the key features of a state function
- To be able to use Hess's law and thermochemical cycles to calculate enthalpy changes of chemical reactions
- To know how to use calorimetric data to calculate enthalpy changes
- To know how to use thermochemical concepts to solve nutritional and environmental problems

CHAPTER OUTLINE

ESSENTIAL SKILLS 4

Temperature

Unit Conversions: Dimensional Analysis

In Chapter 3 you learned that applying a small amount of heat to solid ammonium dichromate initiates a vigorous reaction that produces chromium(III) oxide, nitrogen gas, and water vapor. Those are not the only products of this reaction that interest chemists, however; the reaction also releases energy in the form of heat and light. Hence, our description of this reaction was incomplete. A complete description of a chemical reaction includes not only the identity, amount, and chemical form of the reactants and products, but also the quantity of energy produced or consumed. In combustion reactions, heat is always a product; in other reactions, heat may be produced or consumed.

This chapter introduces you to **thermochemistry**, the branch of chemistry that describes the energy changes that occur during chemical reactions. In some situations, the energy produced by chemical reactions is actually of greater interest to chemists than the material products of the reaction. For

The highly exothermic and dramatic thermite reaction is thermodynamically spontaneous. Reactants of aluminum and a metal oxide, usually iron, which are stable at room temperature, are ignited either in the presence of heat or by the reaction of potassium permanganate and glycerin. The resulting products are aluminum oxide, free and molten elemental metal, and a great deal of heat, which makes this an excellent method for on-site welding. Because this reaction has its own oxygen supply, it can be used for underwater welding as well.

example, the controlled combustion of organic molecules, primarily sugars and fats, within our cells provides the energy for physical activity, thought, and all of the other complex chemical transformations that occur in our bodies. Similarly, our energy-intensive society extracts energy from the combustion of fossil fuels, such as coal, petroleum, and natural gas, to manufacture clothing and furniture, to heat your room in winter and cool it in summer, and to power the car or bus that gets you to class and to the movies. By the end of this chapter, you will know enough about thermochemistry to explain why ice cubes cool a glass of soda, how instant cold packs and hot packs work, and why swimming pools and waterbeds are heated. You will also understand what factors determine the caloric content of your diet and why even "nonpolluting" uses of fossil fuels may be affecting the environment.

5.1 Energy and Work

Because energy takes many forms, only some of which can be seen or felt, it is defined by its effect on matter. For example, microwave ovens produce the energy to cook our food, but we cannot see that energy. In contrast, we can see the energy produced by a light bulb when we switch on a lamp. In this section we describe forms of energy and discuss the relationship between energy, heat, and work.

MGC Kinetic Energy

Forms of Energy

The forms of energy include radiant, thermal, chemical, nuclear, and electrical energy (Figure 5.1). **Thermal energy** results from atomic and molecular motion; the faster the motion, the greater the thermal energy. The **temperature** of an object is a measure of its thermal energy content. *Radiant energy* is the energy carried by light, microwaves, and radio waves. Objects left in bright sunshine or exposed to microwaves become warm because much of the radiant energy they absorb is converted to thermal energy. *Electrical energy* results from the flow of electrically charged particles. When the

Figure 5.1 Forms of energy. (a) *Radiant energy* (from the sun, for example) is the energy in light, microwaves, and radio waves. (b) *Thermal energy* results from atomic and molecular motion; molten steel at 2000°C has a very high thermal energy content. (c) *Chemical energy* results from the particular arrangement of atoms in a chemical compound; the heat and light produced in this reaction are due to energy released during the breaking and reforming of chemical bonds. (d) *Nuclear energy* is released when particles in the nucleus of the atom are rearranged. (e) Lightning is an example of *electrical energy*, which is due to the flow of electrically charged particles.

Figure 5.2 Interconversion of forms of energy. When a swimmer steps off the platform to dive into the water, potential energy is converted to kinetic energy. As the swimmer climbs back up to the top of the diving platform, chemical energy is converted to mechanical work.

ground and a cloud develop a separation of charge, for example, the resulting flow of electrons from one to the other produces lightning, a natural form of electrical energy. *Nuclear energy* is stored in the nucleus of an atom, and *chemical energy* is stored within a chemical compound because of a particular arrangement of atoms.

Electrical, nuclear, and chemical energy are different forms of **potential energy**, which is energy stored in an object because of the relative positions or orientations of its components. A brick lying on the window sill of a tenth-floor office has a great deal of potential energy, but until its position changes by falling, the energy is contained. In contrast, **kinetic energy** is energy due to the *motion* of an object. When the brick falls, its potential energy is transformed to kinetic energy, which is then transferred to the object on the ground that it strikes. The electrostatic attraction between oppositely charged particles is a form of potential energy, which is converted to kinetic energy when the charged particles move toward each other.

Energy can be converted from one form to another (Figure 5.2) or, as we saw with the brick, transferred from one object to another. For example, when you climb a ladder to a high diving board, your body uses chemical energy produced by the combustion of organic molecules. As you climb, the chemical energy is converted to *mechanical work* to overcome the force of gravity. When you stand on the end of the diving board, your potential energy is greater than it was before you climbed the ladder: the greater the distance from the water, the greater the potential energy. When you then dive into the swimming pool, your potential energy is converted to kinetic energy as you fall, and when you hit the surface, some of that energy is transferred to the water, causing it to splash into the air. Chemical energy can also be converted to radiant energy; one common example is the light emitted by fireflies, which is produced from a chemical reaction.

Although energy can be converted from one form to another, *the total amount of energy in the universe remains constant*. This is known as the **law of conservation of energy**.* Thus, *energy cannot be created or destroyed.*

Energy, Heat, and Work

Figure 5.3 An example of mechanical work. One form of energy is mechanical work, the energy required to move an object of mass *m* a distance *d* when opposed by a force *F*, such as gravity.

One definition of **energy** is the capacity to do work. The easiest form of work to visualize is **mechanical work** (Figure 5.3), which is the energy required to move an object a distance *d* when opposed by a force *F*, such as gravity:

$$\text{work} = \text{force} \times \text{distance} \qquad (5.1)$$
$$w = Fd$$

Because *F*, the force that opposes the action, is equal to the mass of the object times its acceleration, we can also write Equation 5.1 as†

$$\text{work} = \text{mass} \times \text{acceleration} \times \text{distance} \qquad (5.2)$$
$$w = mad$$

Consider the mechanical work required for you to travel from the first floor of a building to the second. Whether you take an elevator or escalator, trudge upstairs, or leap up the stairs two at a time, energy is expended to overcome the force of gravity. The amount of work done (w) and thus the energy required depends on the height of the second floor (the distance d); on your mass, which must be raised that distance against the downward acceleration due to gravity; and on the path, as you will learn in Section 5.2.

---

* As you will learn in Chapter 18, the law of conservation of energy is also known as the first law of thermodynamics.

† Recall from Chapter 1 that weight is a force caused by the gravitational attraction between two masses, such as you and the earth.

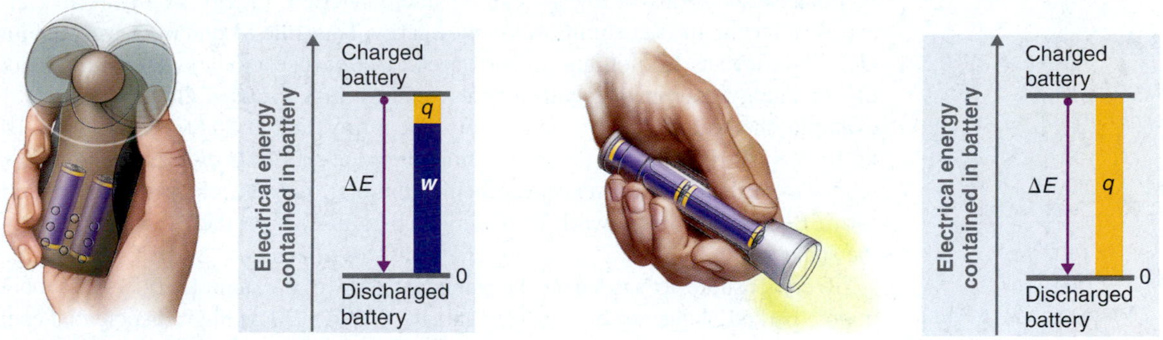

(a) Energy (*E*) released as work (*w*) and heat (*q*) **(b) Energy (*E*) released as heat (*q*) and light**

Figure 5.4 Energy can be transferred as heat, work, or any combination of the two. Discharging a fully charged battery releases the same amount of energy whether the battery is used to run a fan (a) or to illuminate a light bulb (b). In (a), most of the energy is used to perform work, which turns the blades of the fan and thus moves the air; only a small portion of the energy is released as heat by the electric motor. In (b), all of the energy is released as heat and light; no work is done.

In contrast, **heat (*q*)** is thermal energy that can be transferred from an object at one temperature to an object at another temperature. The net transfer of thermal energy stops when the two objects reach the same temperature.

The energy of an object can be changed by only the transfer of energy to or from another object in the form of heat,* work performed on or by the object, or some combination of heat and work. Consider, for example, the energy stored in a fully charged battery. As shown in Figure 5.4, this energy can be used primarily to perform work, such as running an electric fan, or to generate light and heat, such as illuminating a light bulb. When the battery is fully discharged in either case, the total change in energy is the same, even though the fraction released as work or heat varies greatly. Thus, the sum of the heat produced and the work performed equals the change in energy, ΔE:

$$\begin{aligned} \text{energy change} &= \text{heat} + \text{work} \\ \Delta E &= q + w \end{aligned} \tag{5.3}$$

Energy is an extensive property of matter (Chapter 1); for example, the amount of thermal energy in an object is proportional to both its mass and its temperature. Thus, a water heater that holds 150 L of water at 50°C contains much more thermal energy than does a 1-L pan of water at 50°C. Similarly, a bomb contains much more chemical energy than does a firecracker.

Kinetic and Potential Energy

The kinetic energy, *KE*, of an object is related to its mass *m* and velocity *v*:

$$KE = \frac{1}{2}mv^2 \tag{5.4}$$

For example, the kinetic energy of a 1360-kg (~3000-lb) automobile traveling at a velocity of 26.8 m/s (~60 mi/h) is

$$KE = \frac{1}{2}(1360\ \text{kg})\left(\frac{26.8\ \text{m}}{\text{s}}\right)^2 = 4.88 \times 10^5\ \frac{\text{kg}\cdot\text{m}^2}{\text{s}^2} \tag{5.5}$$

* As you will learn in the next chapter, hot objects can also lose energy as radiant energy, such as heat or light. This energy is converted to heat when it is absorbed by another object. Hence, radiant energy is equivalent to heat.

Note the pattern

The units of energy are the same for all forms of energy.

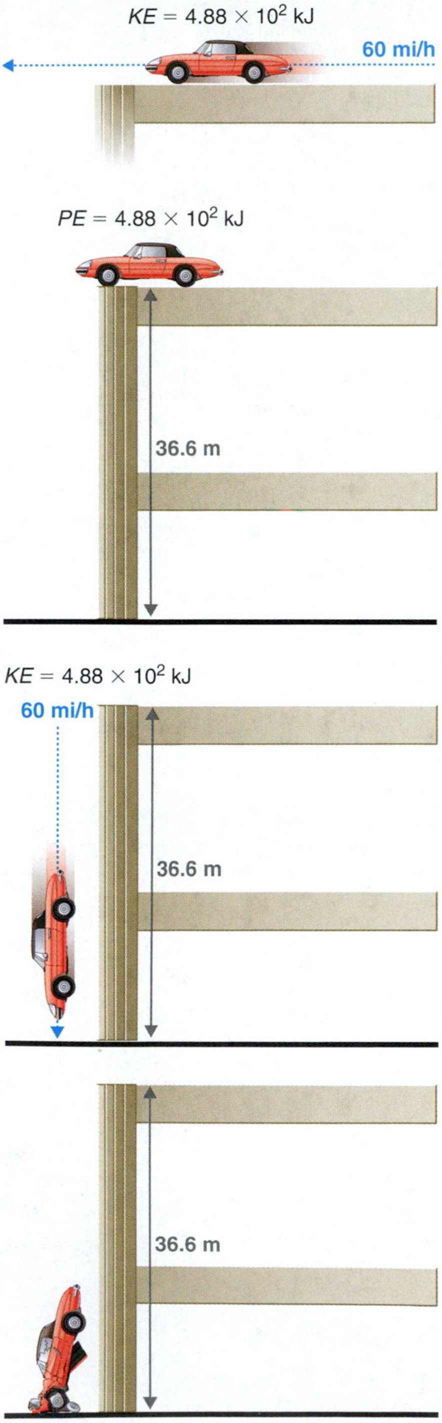

$PE = 0$ kJ, $KE = 0$
(see Eq. 5.5, and 5.6)

Because all forms of energy can be interconverted, energy in any form can be expressed using the same units as kinetic energy. Thus, the SI unit of energy, the **joule (J)**,* is defined as 1 (kilogram · meter^2)/second^2. Because a joule is such a small quantity of energy, chemists usually express energy in kilojoules (1 kJ = 10^3 J). For example, the kinetic energy of the 1360-kg car traveling at 26.8 m/s is 4.88×10^5 J or 4.88×10^2 kJ. It is important to remember that *the units of energy are the same regardless of the form of energy*, whether thermal, radiant, chemical, or any other form. Because heat and work result in changes in energy, their units must also be the same.

To demonstrate, let's calculate the potential energy of the same 1360-kg automobile if it were parked on the top of a parking garage 36.6 m (120 ft) high. Its potential energy is equivalent to the amount of work required to raise the vehicle from street level to the top of the parking garage, which is given by Equation 5.1 ($w = Fd$). According to Equation 5.2, the force F exerted by gravity on any object is equal to its mass m (in this case, 1360 kg) times the acceleration a due to gravity, or g (9.81 m/s^2 at the earth's surface). The distance d is the height above street level: $h = 36.6$ m. Thus, the potential energy, PE, of the car is

$$PE = Fd = mad = mgh$$
$$PE = (1360\ \text{kg})\left(\frac{9.81\ \text{m}}{\text{s}^2}\right)(36.6\ \text{m}) = 4.88 \times 10^5\ \frac{\text{kg} \cdot \text{m}^2}{\text{s}^2} \tag{5.6}$$
$$= 4.88 \times 10^5\ \text{J} = 4.88 \times 10^2\ \text{kJ}$$

Notice that the units of potential energy are the same as the units of kinetic energy. Notice also that in this case the potential energy of the stationary automobile at the top of a 36.6-m-high parking garage is the same as its kinetic energy at 60 mi/h. If the vehicle fell from the roof of the parking garage, its potential energy would be converted to kinetic energy, and it is reasonable to infer that the vehicle would be traveling at 60 mi/h just before it hit the ground, neglecting air resistance. After the car hit the ground, its potential and kinetic energy would both be zero. Last, potential energy is usually defined relative to an arbitrary standard position (in this case, the street was assigned an elevation of zero). As a result, we usually calculate only differences in potential energy: in this case, the difference between the potential energy of the car on the roof of the parking garage and the potential energy of the same car in the street at the base of the garage.

Energy can also be expressed in the non-SI units of **calories**, where 1 calorie (cal) was originally defined as the amount of energy needed to raise the temperature of exactly 1 g of water from 14.5°C to 15.5°C.† The name is derived from the Latin *calor*, "heat." Although energy may be expressed as either calories or joules, calories were defined in terms of heat whereas joules were defined in terms of motion. Because calories and joules are both units of energy, however, the calorie is now defined in terms of the joule:

$$\begin{aligned} 1\ \text{cal} &= 4.184\ \text{J exactly} \\ 1\ \text{J} &= 0.2390\ \text{cal} \end{aligned} \tag{5.7}$$

In this text, we will use the SI units—joules (J) and kilojoules (kJ)—exclusively, except when we deal with nutritional information, as in Section 5.4.

* After British physicist James Joule (1818–1889), an early worker in the field of energy.

† We specify the exact temperatures because the amount of energy needed to raise the temperature of 1 g of water 1°C varies slightly with elevation. To three significant figures, however, this amount is 1.00 cal over the temperature range 0–100°C.

EXAMPLE 5.1

(a) If the mass of a baseball is 149 g, what is the kinetic energy of a fastball clocked at 100 mi/h? **(b)** A batter hits a pop fly, and the baseball (with a mass of 149 g) reaches an altitude of 250 ft. If we assume that the ball was 3 ft above home plate when hit by the batter, what is the increase in its potential energy?

Given Mass and velocity or height

Asked for Kinetic and potential energy

Strategy

Use Equation 5.4 to calculate the kinetic energy and Equation 5.6 to calculate the potential energy, as appropriate.

Solution

(a) The kinetic energy of an object is given by $\frac{1}{2}mv^2$. In this case, we know both the mass and the velocity, but we must convert the velocity to SI units:

$$v = \left(\frac{100\ \cancel{\text{mi}}}{1\ \cancel{\text{h}}}\right)\left(\frac{1\ \cancel{\text{h}}}{60\ \cancel{\text{min}}}\right)\left(\frac{1\ \cancel{\text{min}}}{60\ \text{s}}\right)\left(\frac{1.61\ \cancel{\text{km}}}{1\ \cancel{\text{mi}}}\right)\left(\frac{1000\ \text{m}}{1\ \cancel{\text{km}}}\right) = 44.7\ \text{m/s}$$

The kinetic energy of the baseball is therefore

$$KE = \frac{1}{2}mv^2 = \frac{1}{2}(149\ \cancel{\text{g}})\left(\frac{1\ \text{kg}}{1000\ \cancel{\text{g}}}\right)\left(\frac{44.7\ \text{m}}{\text{s}}\right)^2 = 1.49 \times 10^2\ \frac{\text{kg}\cdot\text{m}^2}{\text{s}^2} = 1.49 \times 10^2\ \text{J}$$

(b) The increase in potential energy is the same as the amount of work required to raise the ball to its new altitude, which is $(250 - 3) = 247$ feet above its initial position. Thus,

$$PE = mgh = (149\ \cancel{\text{g}})\left(\frac{1\ \text{kg}}{1000\ \cancel{\text{g}}}\right)\left(\frac{9.81\ \text{m}}{\text{s}^2}\right)(247\ \cancel{\text{ft}})\left(\frac{0.3048\ \text{m}}{1\ \cancel{\text{ft}}}\right)$$
$$= 1.10 \times 10^2\ \frac{\text{kg}\cdot\text{m}^2}{\text{s}^2} = 1.10 \times 10^2\ \text{J}$$

EXERCISE 5.1

(a) In a bowling alley, the distance from the foul line to the head pin is 59 ft, $10\frac{13}{16}$ in. (18.26 m). If a 16-lb (7.3-kg) bowling ball takes 2.0 s to reach the head pin, what is its kinetic energy at impact? (Assume its speed is constant.) **(b)** What is the potential energy of a 16-lb bowling ball held 3.0 ft above your foot?

Answer **(a)** 3.0×10^2 J; **(b)** 65 J

5.2 ○ Enthalpy

MGC Enthalpy

To study the flow of energy during a chemical reaction, we need to distinguish between the **system,** the small, well-defined part of the universe in which we are interested (such as a chemical reaction), and the **surroundings,** the rest of the universe, including the container in which the reaction is carried out (Figure 5.5). In the discussion that follows, the mixture of chemical substances that undergoes a reaction is always the system, and the flow of heat can be from the system to the surroundings, or vice versa.

Three kinds of systems are important in chemistry. An **open system** can exchange both matter and energy with its surroundings. A pot of boiling water is an open system, because energy in the form of heat is being supplied by the burner

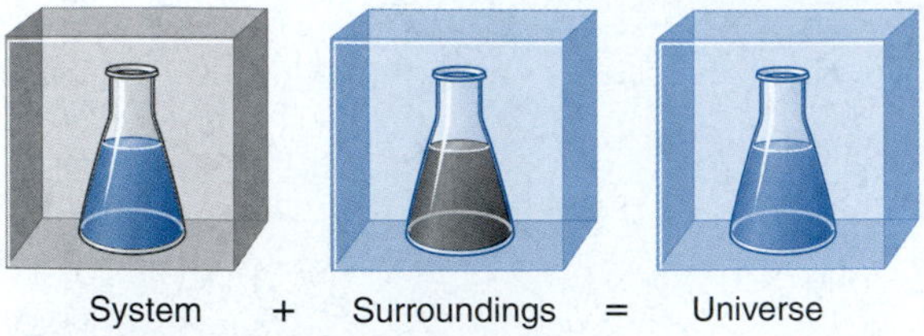

Figure 5.5 A system and its surroundings. The system is that part of the universe we are interested in studying, such as a chemical reaction inside a flask. The surroundings are the rest of the universe, including the container in which the reaction is carried out.

and matter in the form of water vapor is being lost as the water boils. A **closed system** can exchange energy but not matter with its surroundings. The sealed pouch of a ready-made dinner that is dropped into a pot of boiling water is a closed system, because thermal energy is transferred to the system from the boiling water but no matter is exchanged (unless the pouch leaks, in which case it is no longer a closed system). An **isolated system** exchanges neither energy nor matter with the surroundings. Energy is always exchanged between a system and its surroundings, although this process may take place very slowly. Thus, a truly isolated system does not actually exist. An insulated thermos containing hot coffee approximates an isolated system, but eventually the coffee cools as heat is transferred to the surroundings. In all cases, the amount of heat lost by the system is equal to the amount of heat gained by the surroundings, and vice versa. That is, *the total energy of the system plus the surroundings is constant*, which must be true if *energy is conserved*.

The *state* of a system is a complete description of the system at a given time, including its temperature and pressure, the amount of matter it contains, its chemical composition, and the physical state of the matter. A **state function** is a property of a system whose magnitude depends only on the present state of the system and not on its previous history. Temperature, pressure, volume, and potential energy are all state functions. Heat and work, on the other hand, are not state functions because they are *path dependent*. For example, a car sitting on top of a parking garage has the same potential energy whether it was lifted by a crane, set there by a helicopter, driven up, or pushed up by a group of students (Figure 5.6). The amount of work expended to get it there, however, can differ greatly depending on the path chosen. If the students decided to carry the car to the top of the ramp, they would perform a great deal more work than if they simply pushed the car up the ramp (unless, of course, they neglected to release the parking brake, in which case the work expended would increase substantially!). The potential energy of the car is the same, however, no matter which path they choose.

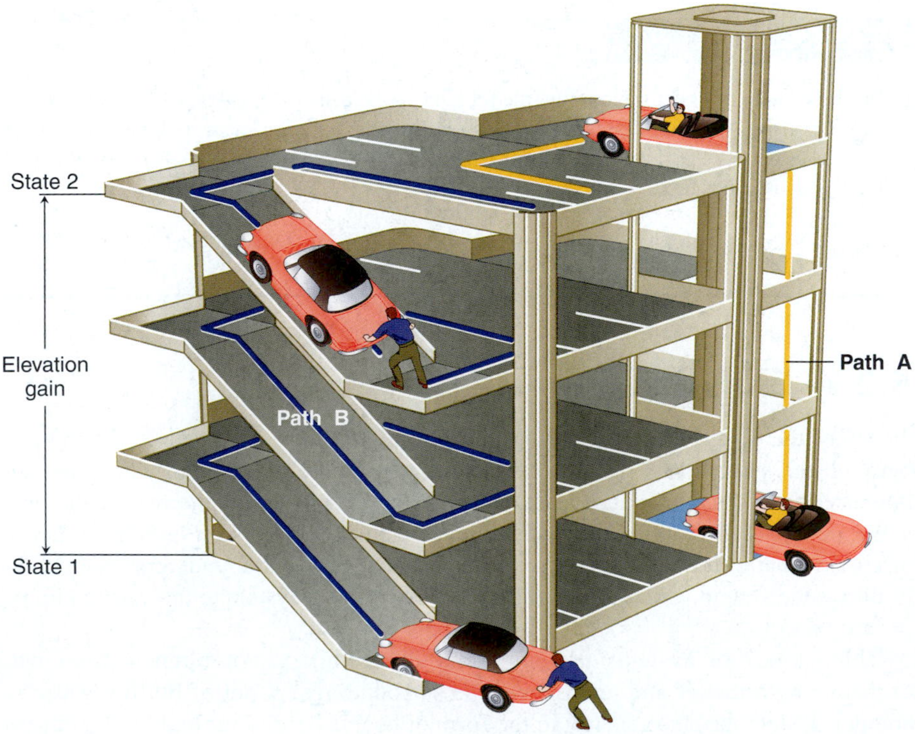

Figure 5.6 Elevation as an example of a state function. The change in elevation between state 1 (at the bottom of the parking garage) and state 2 (at the top of the parking garage) is the same for both paths A and B; it does not depend on which path is taken from the bottom to the top. In contrast, the distance traveled and the work needed to reach the top do depend on which path is taken. Thus, elevation is a state function, but distance and work are *not*.

Direction of Heat Flow

The reaction of powdered aluminum with iron(III) oxide, known as the thermite reaction, generates an enormous amount of heat—enough, in fact, to melt steel (see chapter opening image). The balanced chemical equation for the reaction is

$$2Al(s) + Fe_2O_3(s) \longrightarrow 2Fe(s) + Al_2O_3(s) \quad (5.8)$$

We can also write this equation as

$$2Al(s) + Fe_2O_3(s) \longrightarrow 2Fe(s) + Al_2O_3(s) + \text{heat} \quad (5.9)$$

to indicate that heat is one of the products. Chemical equations in which heat is shown as either a reactant or a product are called *thermochemical equations*. In this reaction, the system consists of Al, Fe, and O atoms; everything else, including the container, makes up the surroundings. During the reaction, so much heat is produced that the iron liquefies. Eventually, the system cools and the iron solidifies as heat is transferred to the surroundings. A process in which heat (q) is transferred *from* the system *to* the surroundings is described as **exothermic**. By convention, $q < 0$ for an exothermic reaction.

When you hold an ice cube in your hand, heat from the surroundings (including your hand) is transferred to the system (the ice), causing the ice to melt and your hand to become cold. We can describe this process by the equation

$$\text{heat} + H_2O(s) \longrightarrow H_2O(l) \quad (5.10)$$

When heat is transferred *to* the system *from* the surroundings, the process is **endothermic.** By convention, $q > 0$ for an endothermic reaction.

Enthalpy of Reaction

We have stated that the change in energy, ΔE, is equal to the sum of the heat produced and the work performed. More precisely, the symbol E represents the *internal energy* of a system, which is the sum of the kinetic and potential energy of all of its components. To measure the energy changes that occur in chemical reactions, however, chemists usually use a related quantity called *enthalpy, H* (from the Greek *enthalpein*, "to warm"). The **enthalpy** of a system is defined as the sum of its internal energy E and the product of its pressure P and volume V:

$$H = E + PV \quad (5.11)$$

Because internal energy, pressure, and volume are all state functions, enthalpy is also a state function.

Most chemical processes, including combustion reactions, such as those that generate energy in our bodies, and all reactions carried out in an open beaker or flask occur under conditions of constant pressure (approximately atmospheric pressure). Measuring changes in enthalpy for such processes is straightforward because at constant pressure, the change in enthalpy is equal to the heat flow. Consider, for example, a reaction that produces a gas, such as dissolving a piece of copper in concentrated nitric acid. The chemical equation for this reaction is

$$Cu(s) + 4HNO_3(aq) \longrightarrow Cu(NO_3)_2(aq) + 2H_2O(l) + 2NO_2(g) \quad (5.12)$$

If the reaction is carried out in a closed system that is maintained at constant pressure by a movable piston, the piston will rise as nitrogen dioxide gas is formed (Figure 5.7). Thus, the system is performing work by lifting the piston against the downward force exerted by the atmosphere (atmospheric pressure). This kind of work is called *pressure-volume work*, and it will be discussed in more detail in Chapter 18. For now, we can state that the amount of work performed by the system on its surroundings is

$$w = -P\Delta V \quad (5.13)$$

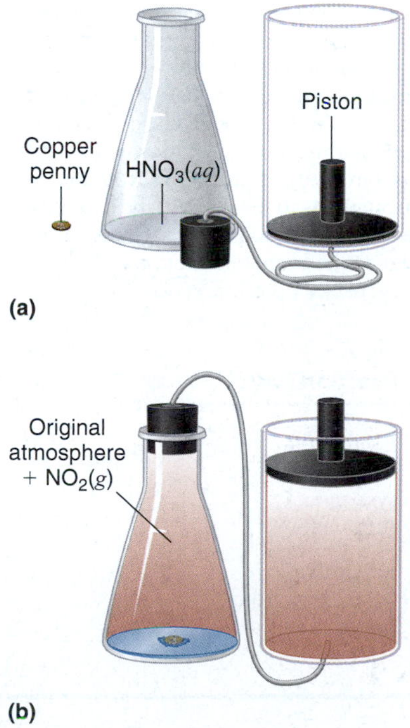

Figure 5.7 An example of work performed by a reaction carried out at constant pressure. (a) Initially the system (a copper penny and concentrated nitric acid) is at atmospheric pressure. (b) When the penny is added to the nitric acid, the volume of NO_2 gas that is formed causes the piston to move upward to maintain the system at atmospheric pressure. In doing so, the system is performing work on the surroundings.

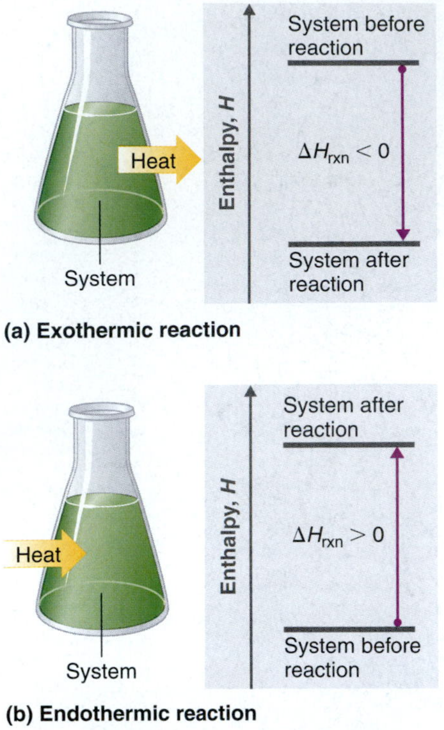

Figure 5.8 The enthalpy of reaction. Energy changes in chemical reactions are usually measured as changes in enthalpy. (a) If heat flows from the system to the surroundings, the enthalpy of the system decreases, ΔH_{rxn} is negative, and the reaction is exothermic; it is energetically downhill. (b) Conversely, if heat flows from the surroundings to the system, the enthalpy of the system increases, ΔH_{rxn} is positive, and the reaction is endothermic; it is energetically uphill. Interactive Graph

Reaction Type	q	ΔH_{rxn}
*Exo*thermic (heat flows from system to surroundings)	<0	<0
*Endo*thermic (heat flows from surroundings to system)	>0	>0

According to Equation 5.13, if the volume increases at constant pressure ($\Delta V > 0$), the work done by the system is negative, indicating that the system has lost energy by performing work on the surroundings. Conversely, if the volume decreases ($\Delta V < 0$), the work done by the system is positive, which means that the surroundings has performed work on the system, thereby increasing its energy.

From Equation 5.14, if a chemical change occurs at constant pressure ($\Delta P = 0$), the change in enthalpy, ΔH, defined as $H_{\text{final}} - H_{\text{initial}}$, is

$$H_{\text{final}} - H_{\text{initial}} = \Delta H = \Delta(E + PV) = \Delta E + \Delta(PV) = \Delta E + P\Delta V \quad (5.14)$$

Substituting $q + w$ for ΔE (Equation 5.3) and $-w$ for $P\Delta V$ (Equation 5.13) we obtain:

$$\Delta H = \Delta E + P\Delta V = q_p + w - w = q_p \quad (5.15)$$

Where the subscript p is used to emphasize that this equation is true only for a process that occurs at constant pressure.

Thus, at constant pressure the heat transferred from the surroundings to the system (or vice versa) is identical to the **change in enthalpy, ΔH** of the system.

$$q_p = H_{\text{final}} - H_{\text{initial}} = \Delta H \quad \text{(at constant pressure)} \quad (5.16)$$

Because enthalpy is a state function, the magnitude of ΔH depends on only the initial and final states of the system, not on the path taken. Most important, the enthalpy change is the same even if the process does *not* occur at constant pressure.

When we study energy changes in chemical reactions, the most important quantity is usually the **enthalpy of reaction (ΔH_{rxn})**, the change in enthalpy that occurs during the reaction (such as the reaction describing the dissolution of the piece of copper in nitric acid). If heat flows from the system to the surroundings, the enthalpy of the system decreases, and by convention ΔH_{rxn} is negative. Conversely, if heat flows from the surroundings to the system, the enthalpy of the system increases, and ΔH_{rxn} is positive. Thus, $\Delta H_{rxn} < 0$ *for an exothermic reaction*, and $\Delta H_{rxn} > 0$ *for an endothermic reaction*. The sign conventions for heat flow and enthalpy changes are summarized in the table in the margin. Note that if ΔH_{rxn} is negative, then the enthalpy of the products is less than the enthalpy of the reactants; that is, *an exothermic reaction is enthalpically downhill* (Figure 5.8a). Conversely, if ΔH_{rxn} is positive, then the enthalpy of the products is greater than the enthalpy of the reactants; thus, *an endothermic reaction is enthalpically uphill* (Figure 5.8b). Two important characteristics of enthalpy and changes in enthalpy are summarized in the following discussion.

Reversing a reaction or a process changes the sign of ΔH. Ice *absorbs* heat when it melts, so liquid water must *release* heat when it freezes:

$$\text{heat} + H_2O(s) \longrightarrow H_2O(l) \qquad \Delta H > 0 \quad (5.17a)$$

$$H_2O(l) \longrightarrow H_2O(s) + \text{heat} \qquad \Delta H < 0 \quad (5.17b)$$

Note that the *magnitude* of the enthalpy change is the same in both cases; only the *sign* is different.

Enthalpy is an extensive property (like mass). The magnitude of ΔH for a reaction is proportional to the amounts of the substances that react. For example, a large fire produces more heat than a single match, even though the chemical reaction—the combustion of wood—is the same in both cases. For this reason, the enthalpy change for a reaction is usually given in units of kilojoules per mole of a particular reactant or product. Consider Equation 5.18, which describes the reaction of Al with Fe_2O_3 at constant pressure. According to the reaction stoichiometry, 2 mol of Fe, 1 mol of

Al_2O_3, and 851.5 kJ of heat are produced for every 2 mol of Al and 1 mol of Fe_2O_3 consumed:

$$2Al(s) + Fe_2O_3(s) \longrightarrow 2Fe(s) + Al_2O_3(s) + 851.5\text{ kJ} \quad (5.18)$$

Thus, $\Delta H = -851.5$ kJ/mol of Fe_2O_3. We can also describe ΔH for the reaction as -425.8 kJ/mol of Al: because 2 mol of Al is consumed in the balanced reaction: we divide 851.5 kJ by 2. When a value for ΔH, in units of kJ rather than kJ/mol, is written after the reaction, as in Equation 5.19, it is the value of ΔH corresponding to the reaction of the molar quantities of reactants as given in the balanced equation:

$$2Al(s) + Fe_2O_3(s) \longrightarrow 2Fe(s) + Al_2O_3(s) \qquad \Delta H_{rxn} = -851.5\text{ kJ} \quad (5.19)$$

If 4 mol of Al and 2 mol of Fe_2O_3 react, the change in enthalpy is $2 \times (-851.5\text{ kJ}) = -1703$ kJ. We can summarize the relationship between the amount of each substance and the enthalpy change for this reaction as follows:

$$-\frac{851.5\text{ kJ}}{2\text{ mol Al}} = -\frac{425.8\text{ kJ}}{1\text{ mol Al}} = -\frac{1703\text{ kJ}}{4\text{ mol Al}} \quad (5.20)$$

The relationship between the magnitude of the enthalpy change and the mass of reactants is illustrated in Example 5.2.

EXAMPLE 5.2

Certain parts of the world, such as southern California and Saudi Arabia, are short of freshwater for drinking. One possible solution to the problem is to tow icebergs from Antarctica and then melt them as needed. If ΔH is 6.01 kJ/mol for the reaction $H_2O(s) \longrightarrow H_2O(l)$ at 0°C and constant pressure, how much energy would be required to melt a moderately large iceberg with a mass of 1.00 million metric tons (1.00×10^6 metric tons)? (A metric ton is 1000 kg.)

Given Energy per mole of ice and mass of iceberg

Asked for Energy required to melt iceberg

Strategy

A Calculate the number of moles of ice contained in 1 million metric tons (1.00×10^6 tons) of ice.
B Calculate the energy needed to melt the ice by multiplying the number of moles of ice in the iceberg by the amount of energy required to melt 1 mol of ice.

Solution

A Because enthalpy is an extensive property, the amount of energy required to melt ice depends on the amount of ice present. We are given ΔH for the process—that is, the amount of energy needed to melt 1 mol (or 18.015 g) of ice—so we need to calculate the number of moles of ice in the iceberg and multiply that number by ΔH (+6.01 kJ/mol):

$$\text{moles } H_2O = 1.00 \times 10^6 \cancel{\text{tons } H_2O}\left(\frac{1000 \cancel{\text{kg}}}{1 \cancel{\text{ton}}}\right)\left(\frac{1000 \cancel{\text{g}}}{1 \cancel{\text{kg}}}\right)\left(\frac{1 \text{ mol } H_2O}{18.015 \cancel{\text{g } H_2O}}\right)$$
$$= 5.55 \times 10^{10} \text{ mol } H_2O$$

B The energy needed to melt the iceberg is thus

$$\left(\frac{6.01\text{ kJ}}{\cancel{\text{mol } H_2O}}\right)(5.55 \times 10^{10} \cancel{\text{mol } H_2O}) = 3.34 \times 10^{11}\text{ kJ}$$

Possible sources of the approximately 3.34×10^{11} kJ needed to melt a 1.00×10^{6}-metric-ton iceberg
Combustion of 3.8×10^{3} ft^3 of natural gas
Combustion of 68,000 barrels of oil
Combustion of 15,000 tons of coal
1.1×10^{8} kilowatt-hours of electricity

Because so much energy is needed to melt the iceberg, this plan would require a relatively inexpensive source of energy to be practical. To give you some idea of the scale of such an operation, the amounts of different energy sources equivalent to the amount of energy needed to melt the iceberg are shown in the table in the margin.

EXERCISE 5.2

If 17.3 g of powdered aluminum is allowed to react with excess Fe_2O_3, how much heat is produced?

Answer 273 kJ

Hess's Law

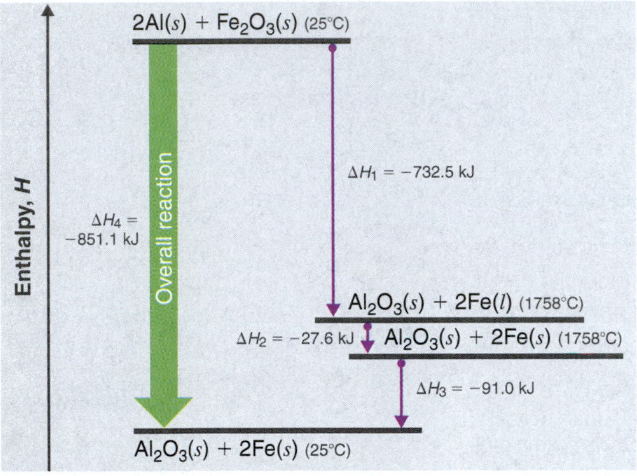

Figure 5.9 Energy changes accompanying the thermite reaction. Because enthalpy is a state function, the overall enthalpy change for the reaction of 2 mol of Al(*s*) with 1 mol of $Fe_2O_3(s)$ is −851.1 kJ, whether the reaction occurs in a single step (ΔH_4) or in three hypothetical steps that involve the successive formation of solid Al_2O_3 and liquid iron (ΔH_1), solid iron at 1758°C (ΔH_2), and solid iron at 25°C (ΔH_3). Thus, $\Delta H_4 = \Delta H_1 + \Delta H_2 + \Delta H_3$, as stated by Hess's law.

Because enthalpy is a state function, the enthalpy change for a reaction depends *only* on the masses of the reacting substances and on the physical states of the reactants and products, and *not* on the path by which reactants are converted to products. If you climbed a mountain, for example, the altitude change would not depend on whether you climbed the entire way without stopping or you stopped many times to take a break. If you stopped often, the overall change in altitude would be the sum of the changes in altitude for each of the short stretches climbed. Similarly, when we add two or more balanced chemical equations to obtain a net chemical equation, *ΔH for the net reaction is the sum of the ΔH values for the individual reactions.* This principle is called **Hess's law**, after the Swiss-born Russian chemist Germain Hess (1802–1850), a pioneer in the study of thermochemistry. Hess's law allows us to calculate ΔH values for reactions that are difficult to carry out directly by summing known ΔH values for individual steps that give the overall reaction when added, even though the overall reaction may not actually occur via those steps.

We can illustrate Hess's law using the thermite reaction. The overall reaction shown in Equation 5.19 can be viewed as occurring in three distinct steps with known ΔH values. As shown in Figure 5.9, the first reaction produces 1 mol of solid aluminum(III) oxide and 2 mol of liquid iron at its melting point of 1758°C (Equation 5.21a); the enthalpy change for this reaction is −732.5 kJ/mol of Fe_2O_3. The second reaction is the conversion of 2 mol of liquid iron at 1758°C to 2 mol of solid iron at 1758°C (Equation 5.21b); the enthalpy change for this reaction is 13.8 kJ/mol of Fe (−27.6 kJ per 2 mol Fe). In the third reaction, 2 mol of solid iron at 1758°C is converted to 2 mol of solid iron at 25°C (Equation 5.21c); the enthalpy change for this reaction is −45.5 kJ/mol of Fe (−91.0 kJ per 2 mol Fe). As you can see in Figure 5.9, the overall reaction is given by the longest arrow, which is the sum of the three shorter arrows. Thus, adding Equations 5.21a, 5.21b, and 5.21c gives the overall reaction, Equation 5.21d:

$$
\begin{array}{llll}
2Al(s) + Fe_2O_3(s) \longrightarrow \cancel{2Fe(l, 1758°C)} + Al_2O_3(s) & \Delta H = & -732.5\ \text{kJ} & (5.21a)\\
\cancel{2Fe(l, 1758°C)} \longrightarrow \cancel{2Fe(s, 1758°C)} & \Delta H = & -27.6\ \text{kJ} & (5.21b)\\
\cancel{2Fe(s, 1758°C)} \longrightarrow 2Fe(s, 25°C) & \Delta H = & -91.0\ \text{kJ} & (5.21c)\\
\hline
2Al(s) + Fe_2O_3(s) \longrightarrow Al_2O_3(s) + 2Fe(s)\ \text{(all at 25°C)} & \Delta H_{rxn} = & -851.1\ \text{kJ} & (5.21d)
\end{array}
$$

The net reaction in Equation 5.21d is identical to Equation 5.19. By Hess's law, the enthalpy change for Equation 5.21d is the sum of the enthalpy changes for Equations 5.21a, 5.21b, and 5.21c. In essence, Hess's law enables us to calculate the enthalpy change for the sum of a series of reactions without having to draw a diagram like that in Figure 5.9.

Comparing Equations 5.21a and 5.21d also illustrates an important point: *The magnitude of ΔH for a reaction depends on the physical states of the reactants and products (gas, liquid, solid, or solution).* Note that when the product is liquid iron at its melting point (Equation 5.21a), only 732.5 kJ of heat is released to the surroundings compared with 852 kJ when the product is solid iron at 25°C (Equation 5.21d). The difference, 120 kJ, is the amount of energy that is released when 2 mol of liquid iron solidifies and cools to 25°C. Thus, it is important to specify the physical state of all reactants and products when writing a thermochemical equation.

When using Hess's law to calculate the value of ΔH for a reaction, follow this procedure:

1. Identify the equation whose ΔH value is unknown, and write individual reactions with known ΔH values that, when summed, will give the desired equation.
2. Arrange the chemical equations so that the reaction of interest is the sum of the individual reactions.
3. If a reaction must be reversed, change the sign of ΔH for that reaction. Additionally, if a reaction must be multiplied by a factor to obtain the correct number of moles of a substance, multiply its ΔH value by that same factor.
4. Sum the individual reactions and their corresponding ΔH values to obtain the reaction of interest and the unknown ΔH.

In the next example we illustrate how to use this procedure.

EXAMPLE 5.3

When carbon is burned with limited amounts of oxygen gas, carbon monoxide is the main product:

$$2C(s) + O_2(g) \longrightarrow 2CO(g) \qquad \Delta H_1 = -221.0 \text{ kJ} \tag{1}$$

When carbon is burned in excess oxygen, carbon dioxide is produced:

$$C(s) + O_2(g) \longrightarrow CO_2(g) \qquad \Delta H_2 = -393.5 \text{ kJ} \tag{2}$$

Use this information to calculate the enthalpy change per mole of CO for the reaction of carbon monoxide with oxygen to give carbon dioxide.

Given Two balanced chemical equations and their ΔH values

Asked for Enthalpy change for a third reaction

Strategy

A After balancing the chemical equation for the overall reaction, write two equations whose ΔH values are known and that, when added together, give the equation for the overall reaction. (Reverse the direction of one or more of the equations as necessary, making sure to also reverse the sign of ΔH.)

B Multiply the equations by appropriate factors to ensure that they give the desired overall equation when added together. To obtain the enthalpy change per mole of CO, write the resulting equations as a sum, along with the enthalpy change for each.

Solution

A We begin by writing the balanced chemical equation for the reaction of interest:

$$CO(g) + \frac{1}{2}O_2(g) \longrightarrow CO_2(g) \qquad \Delta H_{rxn} = ? \tag{3}$$

There are at least two ways to solve this problem using Hess's law and the data provided. The simplest is to write two equations that can be added together to give the desired equation and for which the enthalpy changes are known. Observing that CO, a reactant in Equation 3, is a product in Equation 1, we can reverse Equation 1 to give

$$2CO(g) \longrightarrow 2C(s) + O_2(g) \qquad \Delta H = +221.0 \text{ kJ}$$

Because we have reversed the direction of the reaction, the sign of ΔH is changed. We can use Equation 2 as written because its product, CO_2, is the product we want in Equation 3:

$$C(s) + O_2(g) \longrightarrow CO_2(g) \qquad \Delta H_2 = -393.5 \text{ kJ}$$

B Adding these two equations together does not give the desired reaction, however, because the numbers of C(*s*) on the left and right sides do not cancel. According to our strategy, we can multiply the second equation by 2 to obtain 2 mol of C(*s*) as the reactant:

$$2C(s) + 2O_2(g) \longrightarrow 2CO_2(g) \qquad \Delta H = -787.0 \text{ kJ}$$

Writing the resulting equations as a sum, along with the enthalpy change for each, gives

$$\begin{array}{ll} 2CO(g) \longrightarrow \cancel{2C(s)} + \cancel{O_2(g)} & \Delta H = -\Delta H_1 = +221.0 \text{ kJ} \\ \cancel{2C(s)} + \cancel{2}O_2(g) \longrightarrow 2CO_2(g) & \Delta H = 2\Delta H_2 = -787.0 \text{ kJ} \\ \hline 2CO(g) + O_2(g) \longrightarrow 2CO_2(g) & \Delta H = -566.0 \text{ kJ} \end{array}$$

Note that the overall equation and the enthalpy change for the reaction are both for the reaction of 2 mol of CO with oxygen, and the problem asks for the amount *per mole of CO*. Consequently, we must divide both sides of the final equation *and the magnitude of* ΔH by 2:

$$CO(g) + \frac{1}{2}O_2(g) \longrightarrow CO_2(g) \qquad \Delta H_{rxn} = -283.0 \text{ kJ/mol CO}$$

An alternative and equally valid way to solve this problem is to write the two given equations as occurring in steps. Note that we have multiplied the equations by the appropriate factors to allow us to cancel terms:

$$\begin{array}{ll} \text{(A)} \quad 2C(s) + O_2(g) \longrightarrow \cancel{2CO(g)} & \Delta H_A = \Delta H_1 = -221.0 \text{ kJ} \\ \text{(B)} \quad \cancel{2CO(g)} + O_2(g) \longrightarrow 2CO_2(g) & \Delta H_B = ? \\ \hline \text{(C)} \quad 2C(s) + 2O_2(g) \longrightarrow 2CO_2(g) & \Delta H_C = 2 \times \Delta H_2 = 2 \times (-393.5 \text{ kJ}) = -787.0 \text{ kJ} \end{array}$$

The sum of reactions A and B is reaction C, which corresponds to the combustion of 2 mol of carbon to give CO_2. From Hess's law, $\Delta H_A + \Delta H_B = \Delta H_C$, and we are given ΔH for reactions A and C. Substituting the appropriate values gives

$$\begin{aligned} -221.0 \text{ kJ} + \Delta H_B &= -787.0 \text{ kJ} \\ \Delta H_B &= -566.0 \text{ kJ} \end{aligned}$$

This is again the enthalpy change for the conversion of 2 mol of CO to CO_2. The enthalpy change for the conversion of 1 mol of CO to CO_2 is therefore $-566.0 \div 2 = -283.0$ kJ/mol of CO, which is the same result we obtained earlier. As you can see, *there may be more than one correct way to solve a problem.*

EXERCISE 5.3

The reaction of acetylene (C_2H_2) with hydrogen can produce either ethylene (C_2H_4) or ethane (C_2H_6):

$$\begin{array}{ll} C_2H_2(g) + H_2(g) \longrightarrow C_2H_4(g) & \Delta H = -175.7 \text{ kJ/mol } C_2H_2 \\ C_2H_2(g) + 2H_2(g) \longrightarrow C_2H_6(g) & \Delta H = -312.0 \text{ kJ/mol } C_2H_2 \end{array}$$

What is ΔH for the reaction of ethylene with hydrogen to form ethane?

Answer -136.3 kJ/mol of C_2H_4

Enthalpies of Formation and Reaction

Chapters 2–4 presented a wide variety of chemical reactions, and you learned how to write balanced chemical equations that include all the reactants and products except heat. One way to report the heat absorbed or released would be to compile a massive set of reference tables that list the enthalpy changes for all possible chemical reactions, which would require an incredible amount of effort. Fortunately, Hess's law allows us to calculate the enthalpy change for virtually any conceivable chemical reaction using a relatively small set of tabulated data, such as the following:

Enthalpy of combustion, ΔH_{comb}: Enthalpy changes have been measured for the combustion of virtually any substance that will burn in O_2; these values are usually reported as the enthalpy of combustion per mole of substance.

Enthalpy of fusion, ΔH_{fus}: The enthalpy change that accompanies the melting, or *fusion*, of 1 mol of a substance; these values have been measured for almost all of the elements and for most simple compounds.

Enthalpy of vaporization, ΔH_{vap}: The enthalpy change that accompanies the vaporization of 1 mol of a substance; these values have also been measured for nearly all the elements and for most volatile compounds.

Enthalpy of solution, ΔH_{soln}: The enthalpy change when a specified amount of solute dissolves in a given quantity of solvent.

Enthalpy of formation, ΔH_f: The enthalpy change for the formation of 1 mol of a compound from its component elements, such as the formation of CO_2 from C and O_2. The corresponding relationship is

$$\text{elements} \longrightarrow \text{compound} \qquad \Delta H_{rxn} = \Delta H_f \tag{5.22}$$

For example,

$$C(s) + O_2(g) \longrightarrow CO_2(g) \qquad \Delta H_{rxn} = \Delta H_f\,[CO_2(g)]$$

The sign convention for ΔH_f is the same as for any enthalpy change: $\Delta H_f < 0$ if heat is released when elements combine to form a compound and $\Delta H_f > 0$ if heat is absorbed. Values of ΔH_{vap} and ΔH_{fus} for some common substances are listed in Table 5.1. These values are used in enthalpy calculations when any of the substances undergoes a change of physical state during a reaction.

TABLE 5.1 Enthalpies of vaporization and fusion for selected substances at their boiling points and melting points

Substance	ΔH_{vap}, kJ/mol	ΔH_{fus}, kJ/mol
Argon (Ar)	6.3	1.3
Methane (CH_4)	9.2	0.84
Ethanol (CH_3CH_2OH)	39.3	7.6
Benzene (C_6H_6)	31.0	10.9
Water (H_2O)	40.7	6.0
Mercury (Hg)	59.0	2.29
Iron (Fe)	340	14

Note the pattern

The sign convention is the same for all enthalpy changes: negative if heat is released by the system, positive if heat is absorbed by the system.

Standard Enthalpies of Formation

The magnitude of ΔH for a reaction depends on the physical states of the reactants and products (gas, liquid, solid, or solution), the pressure of any gases present, and the temperature at which the reaction is carried out. To avoid confusion caused by differences in reaction conditions and to ensure uniformity of data, the scientific community has selected a specific set of conditions under which enthalpy changes are measured. These standard conditions serve as a reference point for measuring differences in enthalpy, much as sea level is the reference point for measuring the height of a mountain or for reporting the altitude of an airplane.

The **standard conditions** for which most thermochemical data are tabulated are a *pressure* of 1 atmosphere (atm) for all gases and a *concentration* of 1 *M* for all species in solution (1 mol/L). In addition, each pure substance must be in its **standard state.** This is usually its most stable form at a pressure of 1 atm at a specified temperature. We assume a temperature of 25°C (298 K) for all enthalpy changes given in this text, unless otherwise indicated. Enthalpies of formation measured under these conditions are called **standard enthalpies of formation,** ΔH°_f (pronounced "delta H eff naught"). *The standard enthalpy of formation of any element in its standard state is zero by definition.* For example, although oxygen can exist as ozone (O_3), atomic oxygen (O), and molecular oxygen (O_2), O_2 is the most stable form of oxygen at 1 atm pressure and 25°C. Similarly, hydrogen is $H_2(g)$, not atomic hydrogen (H). Graphite and diamond are both forms of elemental carbon, but graphite is more stable at 1 atm pressure and 25°C. So, the standard state of carbon is graphite (Figure 5.10). Therefore, $O_2(g)$, $H_2(g)$, and graphite have ΔH°_f values of zero.

Figure 5.10 Elemental carbon. Although graphite and diamond are both forms of elemental carbon, graphite is more stable at 1 atm pressure and 25°C. Given enough time, diamond will revert to graphite under these conditions. Hence, graphite is the standard state of carbon.

The standard enthalpy of formation of glucose from the elements at 25°C is the enthalpy change for the reaction

$$6C(\text{graphite}) + 6H_2(g) + 3O_2(g) \longrightarrow C_6H_{12}O_6(s) \qquad \Delta H^\circ_f = -1273.3 \text{ kJ} \qquad (5.23)$$

It is not possible to measure the value of ΔH°_f for glucose, −1273.3 kJ/mol, by simply mixing appropriate amounts of graphite, O_2, and H_2 and measuring the heat evolved as glucose is formed; the reaction shown in Equation 5.23 does not occur at a measurable rate under any known conditions. Glucose is not unique; most compounds cannot be prepared by the chemical equations that define their standard enthalpies of formation. Instead, values of ΔH°_f are obtained using Hess's law and standard enthalpy changes that have been measured for other reactions, such as combustion reactions. Values of ΔH°_f for an extensive list of compounds are given in Appendix A. Note that ΔH°_f values are always reported in units of kilojoules per mole of the substance of interest. Also notice in Appendix A that the standard enthalpy of formation of $O_2(g)$ is zero because it is the most stable form of oxygen in its standard state.

EXAMPLE 5.4

Write balanced chemical equations for the reactions corresponding to the standard enthalpy of formation of these compounds: **(a)** $HCl(g)$; **(b)** $MgCO_3(s)$; **(c)** $CH_3(CH_2)_{14}CO_2H(s)$ (palmitic acid).

Given Compound

Asked for Balanced chemical equation for its formation from elements in standard states

Strategy

Identify the standard state for each element by using Appendix A. Write a chemical equation that describes the formation of the compound from the elements in their standard states, and then balance it so that one mole of product is made.

Solution

To calculate the standard enthalpy of formation of a compound, we must start with the elements in their standard states. The standard state of an element can be identified in Appendix A by a ΔH°_f value of 0 kJ/mol.

(a) Hydrogen chloride contains one atom of hydrogen and one of chlorine. Because the standard states of elemental hydrogen and elemental chlorine are $H_2(g)$ and $Cl_2(g)$, respectively, the unbalanced equation is

$$H_2(g) + Cl_2(g) \longrightarrow HCl(g)$$

Note that fractional coefficients are required in this case because ΔH°_f values are reported for *1 mol* of the product, HCl. Thus, multiplying both $H_2(g)$ and $Cl_2(g)$ by a coefficient of $\frac{1}{2}$ balances the equation:

$$\frac{1}{2}H_2(g) + \frac{1}{2}Cl_2(g) \longrightarrow HCl(g)$$

(b) The standard states of the elements in this compound are $Mg(s)$, $C(s, \text{graphite})$, and $O_2(g)$. The unbalanced equation is thus

$$Mg(s) + C(s, \text{graphite}) + O_2(g) \longrightarrow MgCO_3(s)$$

This equation can be balanced by inspection to give

$$Mg(s) + C(s, \text{graphite}) + \frac{3}{2}O_2(g) \longrightarrow MgCO_3(s)$$

(c) Palmitic acid contains H, C, and O, so the unbalanced equation for its formation from the elements in their standard states is

$$C(s, \text{graphite}) + H_2(g) + O_2(g) \longrightarrow CH_3(CH_2)_{14}CO_2H(s)$$

There are 16 C atoms and 32 H atoms in 1 mol of palmitic acid, so the balanced equation is

$$16C(s, \text{graphite}) + 16H_2(g) + O_2(g) \longrightarrow CH_3(CH_2)_{14}CO_2H(s)$$

EXERCISE 5.4

Write balanced chemical equations for the reactions corresponding to the standard enthalpy of formation of these compounds: **(a)** $NaCl(s)$; **(b)** $H_2SO_4(l)$; **(c)** $CH_3CO_2H(l)$ (acetic acid).

Answer

(a) $Na(s) + \frac{1}{2}Cl_2(g) \longrightarrow NaCl(s)$
(b) $H_2(g) + \frac{1}{8}S_8(s) + 2O_2(g) \longrightarrow H_2SO_4(l)$
(c) $2C(s) + O_2(g) + 2H_2(g) \longrightarrow CH_3CO_2H(l)$

Standard Enthalpies of Reaction

Tabulated values of standard enthalpies of formation can be used to calculate enthalpy changes for *any* reaction involving substances whose ΔH_f° values are known. The **standard enthalpy of reaction, ΔH_{rxn}°**, is the enthalpy change that occurs when a reaction is carried out with all reactants and products in their standard states. Consider the general reaction

$$aA + bB \longrightarrow cC + dD \tag{5.24}$$

where A, B, C, and D are chemical substances and a, b, c, and d are their stoichiometric coefficients. The magnitude of ΔH_{rxn}° is the sum of the standard enthalpies of formation of the products, each multiplied by its appropriate coefficient, minus the sum of the standard enthalpies of formation of the reactants, also multiplied by their coefficients:

$$\Delta H_{rxn}^\circ = \underbrace{[c\Delta H_f^\circ(C) + d\Delta H_f^\circ(D)]}_{\text{Products}} - \underbrace{[a\Delta H_f^\circ(A) + b\Delta H_f^\circ(B)]}_{\text{Reactants}} \tag{5.25}$$

More generally, we can write

$$\Delta H_{rxn}^\circ = \Sigma m\Delta H_f^\circ(\text{product}) - \Sigma n\Delta H_f^\circ(\text{reactants}) \tag{5.26}$$

where the symbol Σ means "sum of" and m and n are the stoichiometric coefficients of each of the products and reactants, respectively. "Products minus reactants" summations such as Equation 5.26 arise from the fact that enthalpy is a state function. Because many other thermochemical quantities are also state functions, "products minus reactants" summations are very common in chemistry; we will encounter many others in subsequent chapters.

Note the pattern

"Products minus reactants" summations are typical of state functions.

To demonstrate the use of tabulated ΔH_f° values, we will use them to calculate ΔH_{rxn}° for the combustion of glucose, the reaction that provides energy for your brain:

$$C_6H_{12}O_6(s) + 6O_2(g) \longrightarrow 6CO_2(g) + 6H_2O(l) \tag{5.27}$$

Using Equation 5.26, we write

$$\begin{aligned}\Delta H_{rxn}^\circ = {} & \{6\Delta H_f^\circ[CO_2(g)] + 6\Delta H_f^\circ[H_2O(l)]\} \\ & - \{\Delta H_f^\circ[C_6H_{12}O_6(s)] + 6\Delta H_f^\circ[O_2(g)]\}\end{aligned} \tag{5.28}$$

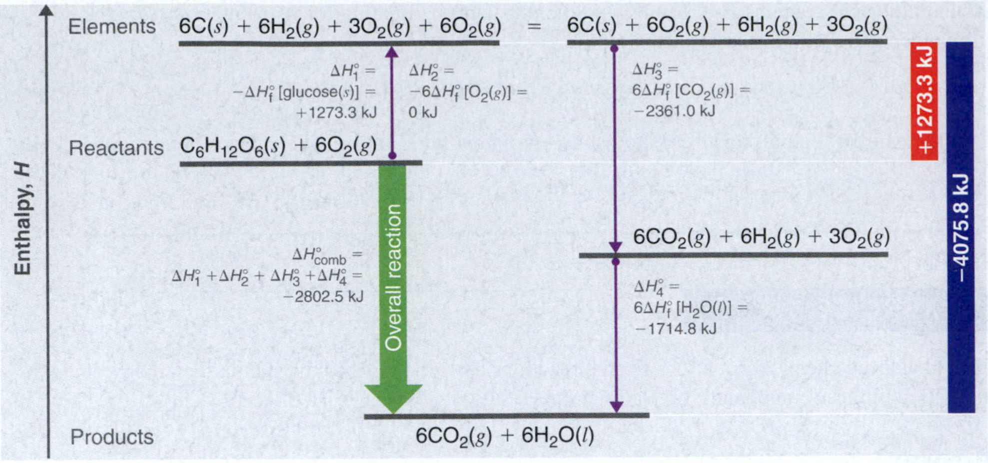

Figure 5.11 A thermochemical cycle for the combustion of glucose. Two hypothetical pathways are shown from the reactants to the products. The combustion reaction is indicated by the green arrow labeled $\Delta H^\circ_{\text{comb}}$. Alternatively, we could first convert the reactants to the elements via the reverse of the equations that define their standard enthalpies of formation, labeled ΔH°_1 and ΔH°_2. Then we could convert the elements to the products via the equations used to define their standard enthalpies of formation, labeled ΔH°_3 and ΔH°_4. Because enthalpy is a state function, $\Delta H^\circ_{\text{comb}}$ is equal to the sum of the enthalpy changes $\Delta H^\circ_1 + \Delta H^\circ_2 + \Delta H^\circ_3 + \Delta H^\circ_4$. Interactive Graph

From Appendix A, the relevant ΔH°_f values are $\Delta H^\circ_f[CO_2(g)] = -393.5$ kJ/mol, $\Delta H^\circ_f[H_2O(l)] = -285.8$ kJ/mol, and $\Delta H^\circ_f[C_6H_{12}O_6(s)] = -1273.3$ kJ/mol. Because $O_2(g)$ is a pure element in its standard state, $\Delta H^\circ_f[O_2(g)] = 0$ kJ/mol. Inserting these values into Equation 5.28 and changing the subscript to indicate that this is a combustion reaction, we obtain.

$$\begin{aligned}\Delta H^\circ_{\text{comb}} &= [6(-393.5\text{ kJ/mol}) + 6(-285.8\text{ kJ/mol})] \\ &\quad - [-1273.3\text{ kJ/mol} + 6(0\text{ kJ/mol})] = -2802.5\text{ kJ/mol}\end{aligned} \tag{5.29}$$

As illustrated in Figure 5.11, we can use Equation 5.28 to calculate $\Delta H^\circ_{\text{comb}}$ for glucose because enthalpy is a state function. The figure shows two pathways from reactants (middle left) to products (bottom). The more direct pathway is the downward green arrow labeled $\Delta H^\circ_{\text{comb}}$. The alternative hypothetical pathway consists of four separate reactions that convert the reactants *to the elements* in their standard states (upward purple arrow at left), and then convert the elements into the desired products (downward purple arrows at right). The reactions that convert the reactants to the elements are the reverse of the equations that define the ΔH°_f values of the reactants. Consequently, the enthalpy changes are

$$\Delta H^\circ_1 = \Delta H^\circ_f[\text{glucose}(s)] = -1\ \cancel{\text{mol glucose}}\left(\frac{-1273.3\text{ kJ}}{1\ \cancel{\text{mol glucose}}}\right) = +1273.3\text{ kJ}$$

$$\Delta H^\circ_2 = 6\Delta H^\circ_f[O_2(g)] = -6\ \cancel{\text{mol } O_2}\left(\frac{0\text{ kJ}}{1\ \cancel{\text{mol } O_2}}\right) = 0\text{ kJ}$$

(Recall that when we reverse a reaction, we must also reverse the sign of the accompanying enthalpy change.) The overall enthalpy change for conversion of the reactants (1 mol of glucose and 6 mol of O_2) to the elements is therefore +1273.3 kJ.

The reactions that convert the elements to final products (downward purple arrows in Figure 5.11) are identical to those used to define the ΔH°_f values of the products. Consequently, the enthalpy changes (from Appendix A) are

$$\Delta H^\circ_3 = 6\Delta H^\circ_f[CO_2(g)] = 6\ \cancel{\text{mol } CO_2}\left(\frac{-393.5\ \text{kJ}}{1\ \cancel{\text{mol } CO_2}}\right) = -2361.0\ \text{kJ}$$

$$\Delta H^\circ_4 = 6\Delta H^\circ_f[H_2O(l)] = -6\ \cancel{\text{mol } H_2O}\left(\frac{-285.8\ \text{kJ}}{1\ \cancel{\text{mol } H_2O}}\right) = -1714.8\ \text{kJ}$$

The overall enthalpy change for conversion of the elements to products (6 mol of carbon dioxide and 6 mol of liquid water) is therefore −4075.8 kJ. Because enthalpy is a state function, the difference in enthalpy between an initial state and a final state can be computed using *any* pathway that connects the two. Thus, the enthalpy change for the combustion of glucose to carbon dioxide and water is the sum of the enthalpy changes for the conversion of glucose and oxygen to the elements (+1273.3 kJ) and for the conversion of the elements to carbon dioxide and water (−4075.8 kJ):

$$\Delta H^\circ_{comb} = +1273.3\ \text{kJ} + (-4075.8\ \text{kJ}) = -2802.5\ \text{kJ} \qquad (5.30)$$

This is the same result we obtained using the "products minus reactants" rule (Equation 5.26) and ΔH°_f values. The two results must be the same because Equation 5.28 is just a more compact way of describing the thermochemical cycle shown in Figure 5.11.

EXAMPLE 5.5

Long-chain fatty acids such as palmitic acid, $CH_3(CH_2)_{14}CO_2H$, are one of the two major sources of energy in our diet ($\Delta H^\circ_f = -891.5$ kJ/mol). Use the data in Appendix A to calculate ΔH°_{comb} for the combustion of palmitic acid. Based on the energy released upon combustion *per gram*, which is the better fuel: glucose or palmitic acid?

Given Compound and ΔH°_f values

Asked for ΔH°_{comb} per mole and per gram

Strategy

A After writing the balanced chemical equation for the reaction, use Equation 5.26 and ΔH°_f values from Appendix A to calculate ΔH°_{comb}, the energy released by the combustion of 1 mol of palmitic acid.

B Divide this value by the molar mass of palmitic acid to find the energy released from the combustion of 1 g of palmitic acid. Compare this value with the value calculated in Equation 5.29 for the combustion of glucose to determine which is the better fuel.

Solution

A To determine the energy released by the combustion of palmitic acid, we need to calculate its ΔH°_{comb}. As always, the first requirement is a balanced chemical equation:

$$C_{16}H_{32}O_2(s) + 23O_2(g) \longrightarrow 16CO_2(g) + 16H_2O(l)$$

Using Equation 5.26 ("products minus reactants") with ΔH°_f values from Appendix A (and omitting the physical states of the reactants and products to save space) gives

$$\begin{aligned}\Delta H^\circ_{comb} &= \Sigma m\Delta H^\circ_f\ \text{(products)} - \Sigma n\Delta H^\circ_f\ \text{(reactants)}\\ &= [16(-393.5\ \text{kJ/mol } CO_2) + 16(-285.8\ \text{kJ/mol } H_2O]\\ &\quad - [(-891.5\ \text{kJ/mol } C_{16}H_{32}O_2) + 23(0\ \text{kJ/mol } O_2)]\\ &= -9977.3\ \text{kJ/mol}\end{aligned}$$

This is the energy released by the combustion of 1 mol of palmitic acid.

B ✓ The energy released by the combustion of 1 g of palmitic acid is

$$\Delta H^\circ_{comb} \text{ per gram} = \left(\frac{-9977.3\ \text{kJ}}{\cancel{1\ \text{mol}}}\right)\left(\frac{\cancel{1\ \text{mol}}}{256.42\ \text{g}}\right) = -38.910\ \text{kJ/g}$$

As calculated in Equation 5.29, ΔH°_{comb} of glucose is -2802.5 kJ/mol. The energy released by the combustion of 1 g of glucose is therefore

$$\Delta H^\circ_{comb} \text{ per gram} = \left(\frac{-2802.5\ \text{kJ}}{\cancel{1\ \text{mol}}}\right)\left(\frac{\cancel{1\ \text{mol}}}{180.16\ \text{g}}\right) = -15.556\ \text{kJ/g}$$

Thus, the combustion of fats such as palmitic acid releases more than twice as much energy per gram as the combustion of sugars such as glucose. This is one reason many people try to minimize the fat content of their diet in order to lose weight.

EXERCISE 5.5

Use the data in Appendix A to calculate ΔH°_{rxn} for the *water–gas shift reaction*, which is used industrially on an enormous scale to obtain $H_2(g)$:

$$CO(g) + H_2O(g) \longrightarrow CO_2(g) + H_2(g)$$

Water–gas shift reaction

Answer -41.2 kJ/mol

We can also measure the enthalpy change for another reaction, such as a combustion reaction, and then use it to calculate a compound's ΔH°_f, which we could not obtain otherwise. This procedure is illustrated in the next example.

EXAMPLE 5.6

Beginning in 1923, tetraethyllead, $(C_2H_5)_4Pb$, was used as an antiknock additive in gasoline in the United States. Its use was completely phased out in 1986 because of the health risks associated with chronic lead exposure. Tetraethyllead is a highly poisonous, colorless liquid that burns in air to give an orange flame with a green halo. The combustion products are $CO_2(g)$, $H_2O(l)$, and red $PbO(s)$. What is the standard enthalpy of formation of tetraethyllead, given that ΔH°_{comb} is -19.29 kJ/g for the combustion of tetraethyllead and ΔH°_f of red $PbO(s)$ is -219.0 kJ/mol?

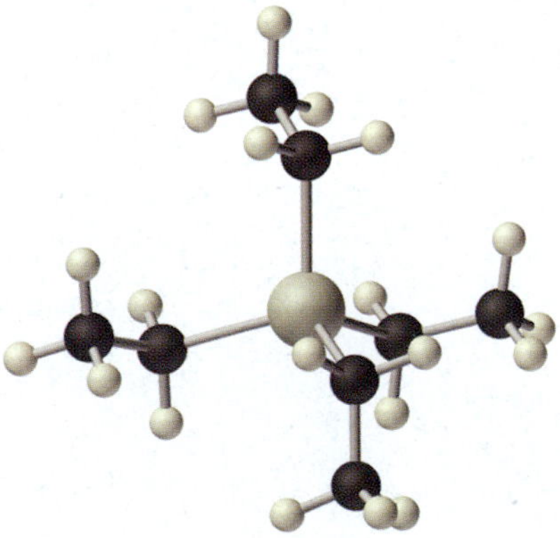

Tetraethyllead

Given Reactant, products, and ΔH°_{comb} and ΔH°_f values

Asked for ΔH°_f of reactant

Strategy

A Write the balanced equation for the combustion of tetraethyllead. Then insert the appropriate quantities into Equation 5.26 to get the equation for ΔH°_f of tetraethyllead.

B Convert ΔH°_{comb} per gram given in the problem to ΔH°_{comb} per mole by multiplying ΔH°_{comb} per gram by the molar mass of tetraethyllead.

C Use Appendix A to obtain values of ΔH°_f for the other reactants and products. Insert these values into the equation for ΔH°_f of tetraethyllead and solve the equation.

Solution

A ✓ The balanced equation for the combustion reaction is

$$2(C_2H_5)_4Pb(l) + 27O_2(g) \longrightarrow 2PbO(s) + 16CO_2(g) + 20H_2O(l)$$

Using Equation 5.26 gives

$$\Delta H^\circ_{comb} = [2\Delta H^\circ_f(PbO) + 16\Delta H^\circ_f(CO_2) + 20\Delta H^\circ_f(H_2O)] - \{2\Delta H^\circ_f[(C_2H_5)_4Pb] + 27\Delta H^\circ_f(O_2)\}$$

Solving for $\Delta H^\circ_f[(C_2H_5)_4Pb]$ gives

$$\Delta H^\circ_f[(C_2H_5)_4Pb] = \Delta H^\circ_f(PbO) + 8\Delta H^\circ_f(CO_2) + 10\Delta H^\circ_f(H_2O) - \frac{27}{2}\Delta H^\circ_f(O_2) - \frac{\Delta H^\circ_{comb}}{2}$$

The values of all terms other than $\Delta H^\circ_f[(C_2H_5)_4Pb]$ and ΔH°_{comb} are given in Appendix A.

B✓ The magnitude of ΔH°_{comb} is given in the problem in units of kilojoules per *gram* of tetraethyllead. We must therefore multiply this value by the molar mass of tetraethyllead (323.44 g/mol) to get ΔH°_{comb} for 1 mol of tetraethyllead:

$$\Delta H^\circ_{comb} = \left(\frac{-19.29\text{ kJ}}{\cancel{\text{g}}}\right)\left(\frac{323.44\ \cancel{\text{g}}}{\text{mol}}\right) = -6239\text{ kJ/mol}$$

Note that because the balanced equation contains 2 mol of tetraethyllead, ΔH°_{rxn} is

$$\Delta H^\circ_{rxn} = 2\ \cancel{\text{mol }(C_2H_5)_4Pb}\left(\frac{-6239\text{ kJ}}{1\ \cancel{\text{mol }(C_2H_5)_4Pb}}\right) = -12{,}480\text{ kJ}$$

C✓ Inserting the appropriate values into the equation for $\Delta H^\circ_f[(C_2H_5)_4Pb]$ gives

$$\begin{aligned}\Delta H^\circ_f[(C_2H_5)_4Pb] &= (1\ \cancel{\text{mol PbO}})(-219.0\text{ kJ/}\cancel{\text{mol PbO}})\\ &\quad + (8\ \cancel{\text{mol CO}_2})(-393.5\text{ kJ/}\cancel{\text{mol CO}_2})\\ &\quad + (10\ \cancel{\text{mol H}_2\text{O}})(-285.8\text{ kJ/}\cancel{\text{mol H}_2\text{O}})\\ &\quad - \left(\tfrac{27}{2}\ \cancel{\text{mol O}_2}\right)(0\text{ kJ/}\cancel{\text{mol O}_2})\\ &\quad - \left[\tfrac{-12{,}480}{2}\text{ kJ/mol }(C_2H_5)_4Pb\right]\\ &= -219.0\text{ kJ} - 3148\text{ kJ} - 2858\text{ kJ} - 0\text{ kJ} + 6240\text{ kJ/mol}\\ &= 15\text{ kJ/mol}\end{aligned}$$

EXERCISE 5.6

Ammonium sulfate, $(NH_4)_2SO_4$, is used as a fire retardant and wood preservative; it is prepared industrially by the highly exothermic reaction of gaseous ammonia with sulfuric acid:

$$2NH_3(g) + H_2SO_4(aq) \longrightarrow (NH_4)_2SO_4(s)$$

The value of ΔH°_{rxn} is -2805 kJ/g H_2SO_4. Use the data in Appendix A to calculate the standard enthalpy of formation of ammonium sulfate (in kJ/mol).

Answer -1181 kJ/mol

Enthalpies of Solution and Dilution

Physical changes, such as melting or vaporization, and chemical reactions, in which one substance is converted to another, are accompanied by changes in enthalpy. Two other kinds of changes that are accompanied by changes in enthalpy are the dissolution of solids and the dilution of concentrated solutions.

The dissolution of a solid can be described by the equation

$$\text{solute}(s) + \text{solvent}(l) \longrightarrow \text{solution}(l) \qquad (5.31)$$

TABLE 5.2 Enthalpies of solution at 25°C of selected ionic compounds in water (in kJ/mol)

	Anion								
Cation	**Fluoride**	**Chloride**	**Bromide**	**Iodide**	**Hydroxide**	**Nitrate**	**Acetate**	**Carbonate**	**Sulfate**
Lithium	4.7	−37.0	−48.8	−63.3	−23.6	−2.5	—	−18.2	−29.8
Sodium	0.9	3.9	−0.6	−7.5	−44.5	20.5	−17.3	−26.7	2.4
Potassium	−17.7	17.2	19.9	20.3	−57.6	34.9	−15.3	−30.9	23.8
Ammonium	−1.2	14.8	16.8	13.7	—	25.7	−2.4	—	6.6
Silver	−22.5	65.5	84.4	112.2	—	22.6	—	22.6	17.8
Magnesium	−17.7	−160.0	−185.6	−213.2	2.3	−90.9	—	−25.3	−91.2
Calcium	11.5	−81.3	−103.1	−119.7	−16.7	−19.2	—	−13.1	−18.0

The values of ΔH_{soln} for some common substances are given in Table 5.2. The sign and the magnitude of ΔH_{soln} depend on specific attractive and repulsive interactions between the solute and the solvent, factors that will be discussed in Chapter 13. When substances dissolve, the process can be either exothermic ($\Delta H_{soln} < 0$) or endothermic ($\Delta H_{soln} > 0$), as you can see from the data in Table 5.2.

Substances with large positive or negative enthalpies of solution have commercial applications as instant cold or hot packs. Single-use versions of these products are based on the dissolution of either calcium chloride ($CaCl_2$, $\Delta H_{soln} = -81.3$ kJ/mol) or ammonium nitrate (NH_4NO_3, $\Delta H_{soln} = +25.7$ kJ/mol). Both types consist of a plastic bag that contains about 100 mL of water plus a dry chemical (40 g of $CaCl_2$ or 30 g of NH_4NO_3) in a separate plastic pouch. When the pack is twisted or struck sharply, the inner plastic bag of water ruptures and the salt dissolves in the water. If the salt is $CaCl_2$, heat is released to produce a solution with a temperature of about 90°C; hence, the product is an "instant hot compress." If the salt is NH_4NO_3, heat is absorbed when it dissolves, and the temperature drops to about 0° for an "instant cold pack."

A similar product based on the *precipitation* of sodium acetate, not its dissolution, is marketed as a reusable hand warmer (Figure 5.12). At high temperatures, sodium acetate forms a highly concentrated aqueous solution. Upon cooling, an unstable *supersaturated* solution is formed containing excess solute. When the pack is agitated, sodium acetate trihydrate, $CH_3CO_2Na \cdot 3H_2O$, crystallizes and heat is evolved:

$$Na^+(aq) + CH_3CO_2^-(aq) + 3H_2O(l) \longrightarrow CH_3CO_2Na \cdot 3H_2O(s)$$
$$\Delta H = -\Delta H_{soln} = -19.7 \text{ kJ/mol} \quad (5.32)$$

Thus, a bag of concentrated sodium acetate solution can be carried until heat is needed, at which time vigorous agitation induces crystallization and heat is released. The pack can be reused after it is immersed in hot water until the sodium acetate redissolves.

The amount of heat released or absorbed when a substance is dissolved is not a constant, but depends on the final concentration of the solute. The values of ΔH_{soln} given above and in Table 5.2, for example, were obtained by measuring the enthalpy changes at various concentrations and extrapolating the data to infinite dilution.

Because ΔH_{soln} depends on the concentration of the solute, diluting a solution can produce a change in enthalpy. If the initial dissolution process is exothermic ($\Delta H < 0$), then the dilution process is exothermic as well. This phenomenon is particularly relevant for strong acids and bases, which are often sold or stored as concentrated aqueous solutions. If water is added to a concentrated solution of sulfuric acid (which is 98% H_2SO_4

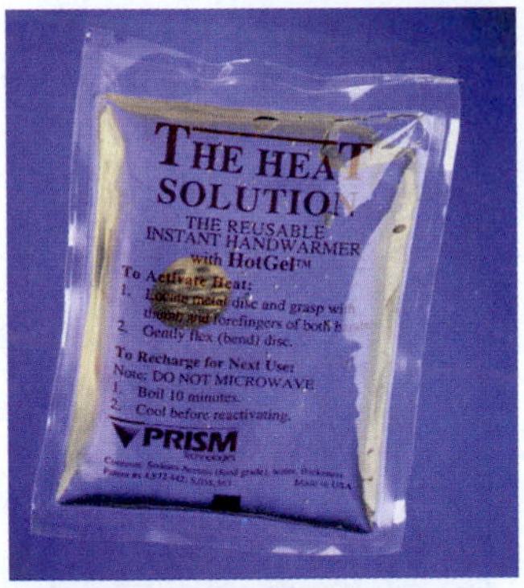

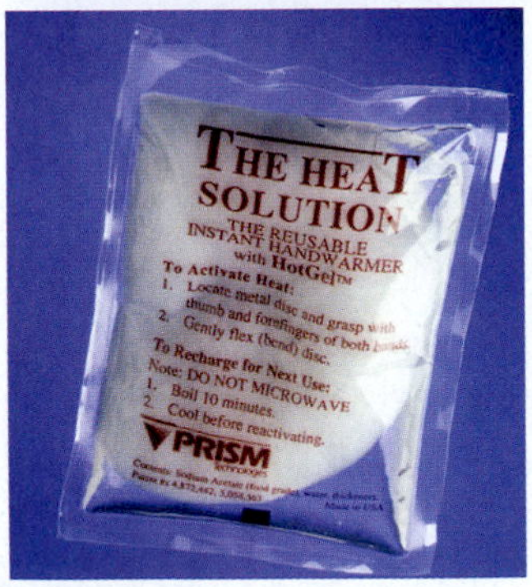

Figure 5.12 An instant hot pack based on the crystallization of sodium acetate. The hot pack is at room temperature prior to agitation (top). Because the sodium acetate is in solution, you can see the metal disc inside the pack. After the hot pack has been agitated, the sodium acetate crystallizes (bottom) to release heat. Because of the mass of white sodium acetate that has crystallized, the metal disc is no longer visible.

and only 2% H_2O) or sodium hydroxide, the heat released by the large negative ΔH can cause the solution to boil. Dangerous spattering of strong acid or base can be avoided if the concentrated acid or base is slowly added to water, so that the heat liberated is largely dissipated by the water. Thus, you should **never add water to a strong acid or base**; a useful way to avoid the danger is to remember: *Add water to acid and get blasted!*

5.3 ● Calorimetry

MGC Calorimetry

Thermal energy itself cannot be measured easily, but the temperature change caused by the flow of thermal energy between objects or substances can be measured. The term **calorimetry** describes the set of techniques employed to measure enthalpy changes in chemical processes using devices called *calorimeters*.

To have any meaning, the quantity that is actually measured in a calorimetric experiment, the change in the temperature of the device, must be related to the heat evolved or consumed in a chemical reaction. We begin this section by explaining how the flow of thermal energy affects the temperature of an object.

Heat Capacity

We have seen that the temperature of an object changes when it absorbs or loses thermal energy. The *magnitude* of the temperature change depends on both the *amount* of thermal energy transferred (q) and the *heat capacity* of the object. Its **heat capacity (*C*)** is the amount of energy needed to raise the temperature of the object exactly 1°C; the units of C are joules per degree Celsius (J/°C). The change in temperature, ΔT is

$$\Delta T = \frac{q}{C} \qquad (5.33)$$

where q is the amount of heat (in units of J), C is the heat capacity (in units of J/°C), and ΔT is $T_{\text{final}} - T_{\text{initial}}$ (in units of °C). Note that ΔT is *always* written as the final temperature minus the initial temperature. The value of C is intrinsically a positive number, but ΔT and q can be either positive or negative; they must have the *same* sign, however. If ΔT and q are positive, then *heat flows from the surroundings into the object*. If ΔT and q are negative, then *heat flows from the object into the surroundings*.

The heat capacity of an object depends on both its *mass* and its *composition*. For example, doubling the mass of an object doubles its heat capacity. Consequently, the amount of substance must be indicated when the heat capacity of the substance is reported. The **molar heat capacity (C_p)** is the amount of energy needed to increase the temperature of 1 mol of a substance by 1°C; the units of C_p are thus J/(mol · °C).* The **specific heat (C_s)** is the amount of energy needed to increase the temperature of 1 g of a substance by 1°C; its units are thus J/(g · °C). We can relate the quantity of a substance, the amount of heat transferred, its heat capacity, and the temperature change in two ways:

$$q = nC_p\Delta T \quad \text{where } n = \text{number of moles of substance} \qquad (5.34)$$

$$q = mC_s\Delta T \quad \text{where } m = \text{mass of substance in grams} \qquad (5.35)$$

The specific heats of some common substances are given in Table 5.3. Note that the specific heat values of most solids are less than 1 J/(g · °C), whereas those of most liquids are about 2 J/(g · °C). Water in its solid and liquid states is an exception. The heat capacity of ice is twice as high as that of most solids; the heat capacity of liquid water, 4.184 J/(g · °C), is one of the highest known.

The high specific heat of liquid water has important implications for life on earth. A given mass of water releases more than five times as much heat for a 1°C temperature

TABLE 5.3 Specific heats (C_s) of selected substances at 25°C

Compound	Specific Heat, J/(g · °C)
$H_2O(l)$	4.184
$H_2O(g)$	2.062
CH_3OH (methanol)	2.531
CH_3CH_2OH (ethanol)	2.438
n-C_6H_{14} (n-hexane)	2.270
C_6H_6 (benzene)	1.745
C(s) (graphite)	0.709
C(s) (diamond)	0.509
Al(s)	0.897
Fe(s)	0.449
Cu(s)	0.385
Au(s)	0.129
Hg(l)	0.140
NaCl(s)	0.864
MgO(s)	0.921
$SiO_2(s)$ (quartz)	0.742
$CaCO_3(s)$ (calcite)	0.915

* The subscript p indicates that the value was measured at constant pressure.

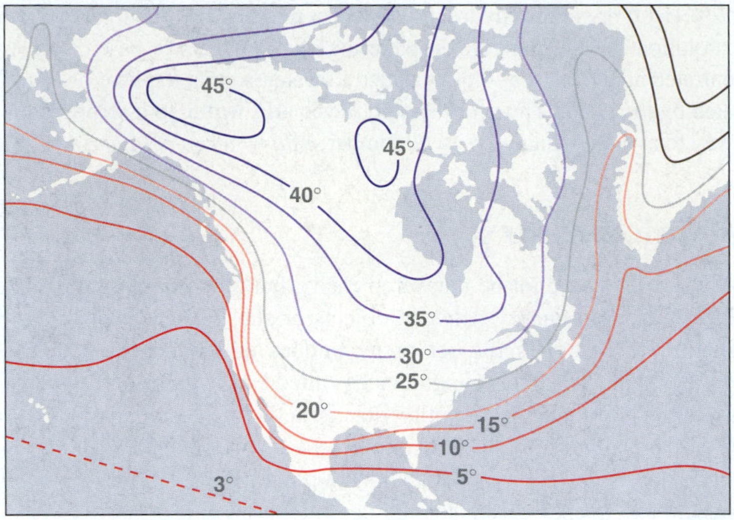

Figure 5.13 **The high specific heat of liquid water has major effects on climate.** Regions that are near very large bodies of water, such as oceans or lakes, tend to have smaller temperature differences between summer and winter months than regions in the center of a continent. The contours on this map show the difference between January and July monthly mean surface temperatures (in °C).

change as does the same mass of limestone or granite. Consequently, coastal regions of our planet tend to have less variable climates than the regions in the center of a continent. After absorbing large amounts of thermal energy from the sun in summer, the water slowly releases the energy during the winter, thus keeping the coastal areas warmer than otherwise would be expected (Figure 5.13). Water's capacity to absorb large amounts of energy without undergoing a large increase in temperature also explains why swimming pools and waterbeds are usually heated. Heat must be applied to raise the temperature of the water to a comfortable level for swimming or sleeping and to maintain that level as heat is exchanged with the surroundings. Moreover, because the human body is about 70% water by mass, a great deal of energy is required to change its temperature by even 1°C. Consequently, the mechanism for maintaining our body temperature at about 37°C does not have to be as finely tuned as would be necessary if our bodies were primarily composed of a substance with a lower specific heat.

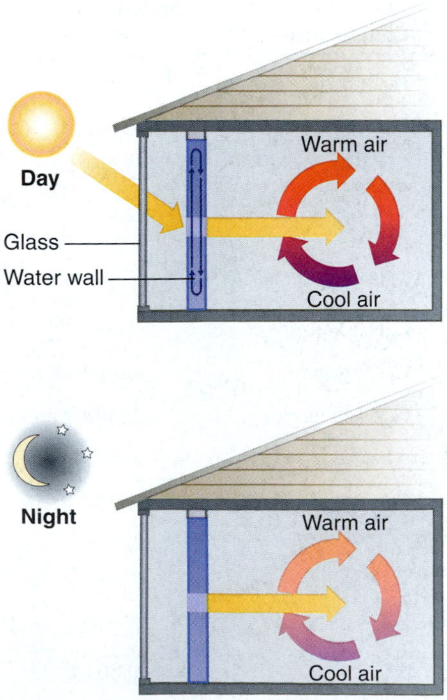

Passive solar system. (top) During the day, sunlight is absorbed by the water circulating in the water wall. (bottom) At night, heat stored in the water wall continues to warm the air inside the house.

EXAMPLE 5.7

A home solar energy storage unit uses 400 L of water for storing thermal energy. On a sunny day, the initial temperature of the water is 22.0°C. During the course of the day, the temperature of the water rises to 38.0°C as it circulates through the water wall. How much energy has been stored in the water? (Note that the density of water at 22.0°C is 0.998 g/mL.)

Given Volume and density of water and initial and final temperatures

Asked for Amount of energy stored

Strategy

A Use the density of water at 22.0°C to obtain the mass of water, m, that corresponds to 400 L of water. Then compute ΔT for the water.

B Determine the amount of heat absorbed by substituting values for m, C_s, and ΔT into Equation 5.35.

Solution

A ✓ The mass of water is

$$\text{mass of } H_2O = 400\ \cancel{\text{L}}\left(\frac{1000\ \cancel{\text{mL}}}{1\ \cancel{\text{L}}}\right)\left(\frac{0.998\ \text{g}}{1\ \cancel{\text{mL}}}\right) = 3.99 \times 10^5\ \text{g}\ H_2O$$

The temperature change, ΔT, is (38.0°C − 22.0°C) = +16.0°C.

B From Table 5.3, the specific heat of water is 4.184 J/(g · °C). From Equation 5.35, the heat absorbed by the water is thus

$$q = mC_s\Delta T = (3.99 \times 10^5\ \cancel{g})\left(\frac{4.184\ J}{\cancel{g} \cdot \cancel{°C}}\right)(16.0\ \cancel{°C}) = 2.67 \times 10^7\ J = 2.67 \times 10^4\ kJ$$

Note that both q and ΔT are positive, consistent with the fact that energy has been absorbed by the water.

EXERCISE 5.7

Some solar energy devices used in homes circulate air over a bed of rocks that absorb thermal energy from the sun. If a house uses a solar heating system that contains 2500 kg of sandstone rocks, what amount of energy is stored if the temperature of the rocks increases from 20.0°C to 34.5°C during the day? Assume that the specific heat of sandstone is the same as that of quartz (SiO_2) in Table 5.3.

Answer 2.7×10^4 kJ (Note that even though the mass of sandstone is more than six times the mass of the water in Example 5.7, the amount of thermal energy stored is the same to two significant figures.)

When two objects at different temperatures are placed in contact, heat flows from the warmer object to the cooler one until the temperature of both objects is the same. The law of conservation of energy says that the total energy cannot change during this process:

$$q_{cold} + q_{hot} = 0 \qquad (5.36)$$

The equation implies that the amount of heat that flows *from* the warmer object is the same as the amount of heat that flows *into* the cooler object. Because the direction of heat flow is opposite for the two objects, the sign of the heat flow values must be opposite:

$$q_{cold} = -q_{hot} \qquad (5.37)$$

Thus, heat is conserved in any such process, consistent with the law of conservation of energy.

Substituting for q from Equation 5.35 gives

$$[mC_s\Delta T]_{hot} + [mC_s\Delta T]_{cold} = 0 \qquad (5.38)$$

which can be rearranged to give

$$[mC_s\Delta T]_{cold} = -[mC_s\Delta T]_{hot} \qquad (5.39)$$

When two objects initially at different temperatures are placed in contact, we can use Equation 5.39 to calculate the final temperature if we know the chemical composition and mass of the objects.

Note the pattern

The amount of heat lost by the warmer object equals the amount of heat gained by the cooler object.

EXAMPLE 5.8

If a 30.0-g piece of copper pipe at 80.0°C is placed in 100.0 g of water at 27.0°C, what is the final temperature? Assume that no heat is transferred to the surroundings.

Given Mass and initial temperature of two objects

Asked for Final temperature

Strategy

Using Equation 5.39 and writing ΔT as $T_{final} - T_{initial}$ for both the copper and the water, substitute the appropriate values of m, C_s, and $T_{initial}$ into the equation and solve for T_{final}.

Solution

We can adapt Equation 5.39 to solve this problem, remembering that ΔT is defined as $T_{\text{final}} - T_{\text{initial}}$:

$$[mC_s(T_{\text{final}} - T_{\text{initial}})]_{\text{Cu}} + [mC_s(T_{\text{final}} - T_{\text{initial}})]_{\text{H}_2\text{O}} = 0$$

Substituting the data provided in the problem and Table 5.3 gives

$$\{[30.0\text{ g}][0.385\text{ J/(g}\cdot{}^\circ\text{C)}][T_{\text{final}} - 80.0^\circ\text{C}]\} + \{[100.0\text{ g}][4.184\text{ J/(g}\cdot{}^\circ\text{C)}][T_{\text{final}} - 27.0^\circ\text{C}]\} = 0$$

$$T_{\text{final}}(11.6\text{ J/}^\circ\text{C}) - 924\text{ J} + T_{\text{final}}(418.4\text{ J/}^\circ\text{C}) - 11{,}300\text{ J} = 0$$

$$T_{\text{final}}(430\text{ J/}^\circ\text{C}) = 12{,}224\text{ J}$$

$$T_{\text{final}} = 28.4^\circ\text{C}$$

EXERCISE 5.8a

If a 14.0-g chunk of gold at 20.0°C is dropped into 25.0 g of water at 80.0°C, what is the final temperature if no heat is transferred to the surroundings?

Answer 80.0°C

EXERCISE 5.8b

A 28.0-g chunk of aluminum is dropped into 100.0 g of water with an initial temperature of 20.0°C. If the final temperature of the water is 24.0°C, what was the initial temperature of the aluminum? (Assume that no heat is transferred to the surroundings.)

Answer 90.6°C

Measuring Heat Flow

In Example 5.7, radiant energy from the sun was used to raise the temperature of water. A calorimetric experiment uses essentially the same procedure, except that the thermal energy change accompanying a chemical reaction is responsible for the change in temperature that takes place in a calorimeter. If the reaction releases heat ($q_{\text{rxn}} < 0$), then heat is absorbed by the calorimeter ($q_{\text{calorimeter}} > 0$) and its temperature increases. Conversely, if the reaction absorbs heat ($q_{\text{rxn}} > 0$), then heat is transferred from the calorimeter to the system ($q_{\text{calorimeter}} < 0$) and the temperature of the calorimeter decreases. In both cases, *the amount of heat absorbed or released by the calorimeter is equal in magnitude and opposite in sign to the amount of heat produced or consumed by the reaction.* The heat capacity of the calorimeter or of the reaction mixture may be used to calculate the amount of heat released or absorbed by the chemical reaction. The amount of heat released or absorbed per gram or mole of reactant can then be calculated from the mass of the reactants.

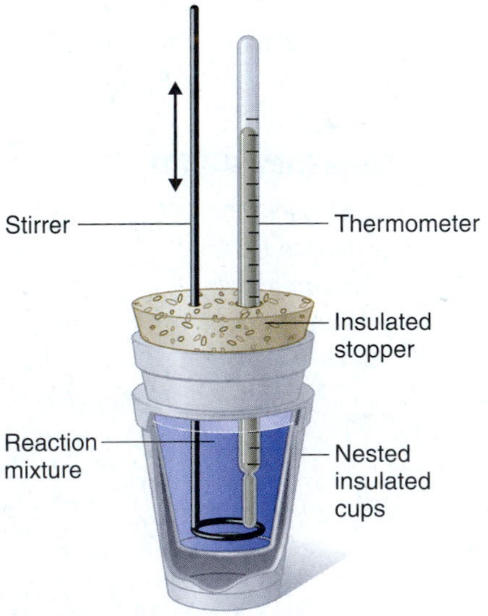

Figure 5.14 A coffee-cup calorimeter. This simplified version of a constant-pressure calorimeter consists of two Styrofoam cups nested and sealed with an insulated stopper to thermally isolate the system (the solution being studied) from the surroundings (the air and the laboratory bench). Two holes in the stopper allow the use of a thermometer to measure the temperature and a stirrer to mix the reactants.

Constant-Pressure Calorimetry

Because ΔH is defined as the heat flow at constant pressure, measurements made using a **constant-pressure calorimeter** give ΔH values directly. This device is particularly well suited to studying reactions carried out in solution at a constant atmospheric pressure. A "student" version, called a "coffee-cup calorimeter" (Figure 5.14), is often encountered in general chemistry laboratories. Commercial calorimeters operate on the same principle, but they can be used with smaller volumes of solution, have better thermal insulation, and can detect a change in temperature as small as several millionths of a degree (10^{-6}°C). Because the heat released or absorbed at constant pressure is equal to the change in enthalpy, the relationship between heat and ΔH_{rxn} is

$$\Delta H_{rxn} = q_{rxn} = -q_{calorimeter} = -mC_s\Delta T \tag{5.40}$$

The use of a constant-pressure calorimeter is illustrated in the next example.

EXAMPLE 5.9

When 5.03 g of solid potassium hydroxide is dissolved in 100.0 mL of distilled water in a coffee-cup calorimeter, the temperature of the liquid increases from 23.0°C to 34.7°C. The density of water in this temperature range averages 0.9969 g/cm^3. What is ΔH_{soln} (in kJ/mol)? Assume that the calorimeter absorbs a negligible amount of heat and, because of the large volume of water, the specific heat of the solution is the same as the specific heat of pure water.

Given Mass of substance, volume of solvent, and initial and final temperatures

Asked for ΔH_{soln}

Strategy

A Calculate the mass of the solution from its volume and density, and calculate the temperature change of the solution.
B Find the heat flow that accompanies the dissolution reaction by substituting the appropriate values into Equation 5.40.
C Use the molar mass of KOH to calculate the molar enthalpy of solution, ΔH_{soln}.

Solution

A To calculate ΔH_{soln}, we must first determine the amount of heat released in the calorimetry experiment. The mass of the solution is

$$(100.0 \cancel{\text{mL}}\ H_2O)(0.9969\ \text{g}/\cancel{\text{mL}}) + 5.03\ \text{g KOH} = 104.72\ \text{g}$$

The temperature change is (34.7°C − 23.0°C) = +11.7°C.
B Because the solution is not very concentrated (~0.9 *M*), we assume that the specific heat of the solution is the same as that of water. The heat flow that accompanies dissolution is thus

$$q_{calorimeter} = mC_s\Delta T = (104.72\ \cancel{\text{g}})\left(\frac{4.184\ \text{J}}{\cancel{\text{g}}\cdot\cancel{°\text{C}}}\right)(11.7\cancel{°\text{C}}) = 5130\ \text{J} = 5.13\ \text{kJ}$$

The temperature of the solution increased because heat was absorbed by the solution ($q > 0$). Where did this heat come from? It was released by KOH dissolving in water. From Equation 5.40, we see that

$$\Delta H_{rxn} = -q_{calorimeter} = -5.13\ \text{kJ}$$

This experiment tells us that dissolving 5.03 g of KOH in water is accompanied by the *release* of 5.13 kJ of energy. Because the temperature of the solution increased, the dissolution of KOH in water must be exothermic.
C The last step is to use the molar mass of KOH to calculate the molar enthalpy of solution—that is, the heat released upon dissolving 1 mol of KOH:

$$\Delta H_{soln} = \left(\frac{-5.13\ \text{kJ}}{5.03\ \cancel{\text{g}}}\right)\left(\frac{56.11\ \cancel{\text{g}}}{1\ \text{mol}}\right) = -57.2\ \text{kJ/mol}$$

EXERCISE 5.9

A coffee-cup calorimeter contains 50.0 mL of distilled water at 22.7°C. Solid ammonium bromide (3.14 g) is added and the solution is stirred, giving a final temperature of 20.3°C. Using the same assumptions as in Example 5.9, find ΔH_{soln} for NH_4Br (in kJ/mol).

Answer 16.6 kJ/mol

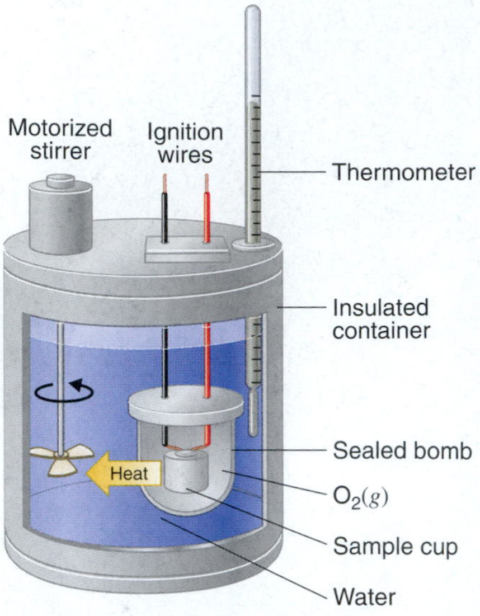

Figure 5.15 A bomb calorimeter. After the temperature of the water in the insulated container has reached a constant value, the combustion reaction is initiated by passing an electric current through a wire embedded in the sample. Because this calorimeter operates at constant volume, the heat released is not precisely the same as the enthalpy change for the reaction.

Constant-Volume Calorimetry

Constant-pressure calorimeters are not very well suited for studying reactions in which one or more of the reactants is a gas, such as a combustion reaction. The enthalpy changes that accompany combustion reactions are therefore measured using a constant-volume calorimeter, such as the **bomb calorimeter** shown schematically in Figure 5.15. The reactant is placed in a steel cup inside a steel vessel with a fixed volume (the "bomb"). The bomb is then sealed, filled with excess oxygen gas, and placed inside an insulated container that holds a known amount of water. Because combustion reactions are exothermic, the temperature of the bath and the calorimeter increases upon combustion. If the heat capacity of the bomb and the mass of water are known, the heat released can be calculated.

Because the volume of the system (the inside of the bomb) is fixed, the combustion reaction occurs under conditions in which the volume is constant, but not the pressure. As you will learn in Chapter 18, the heat released by a reaction carried out at constant volume is identical to the change in *internal energy*, ΔE, rather than the enthalpy change, ΔH; ΔE is related to ΔH by an expression that depends on the change in the number of moles of gas during the reaction. The difference between the heat flow measured at constant volume and the enthalpy change is usually quite small, however (on the order of a few percent). Assuming that $\Delta E \approx \Delta H$, the relationship between the measured temperature change and ΔH_{comb} is given in Equation 5.41, where C_{bomb} is the total heat capacity of the steel bomb and the water surrounding it:

$$\Delta H_{comb} \approx q_{comb} = -q_{calorimeter} = -C_{bomb}\Delta T \tag{5.41}$$

To measure the heat capacity of the calorimeter, we first burn a carefully weighed mass of a standard compound whose enthalpy of combustion is accurately known. Benzoic acid, $C_6H_5CO_2H$, is often used for this purpose because it is a crystalline solid that can be obtained in high purity. The combustion of benzoic acid in a bomb calorimeter releases 26.38 kJ of heat per gram (that is, its $\Delta H_{comb} = -26.38$ kJ/g). This value and the measured increase in temperature of the calorimeter can be used in Equation 5.41 to determine C_{bomb}. The use of a bomb calorimeter to measure the enthalpy of combustion of a substance is illustrated in the next example.

EXAMPLE 5.10

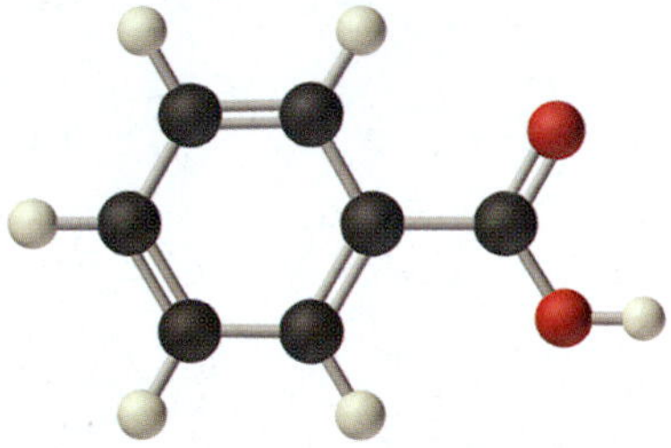

Benzoic acid

The combustion of 0.579 g of benzoic acid in a bomb calorimeter caused a 2.08°C increase in the temperature of the calorimeter. The chamber was then emptied and recharged with 1.732 g of glucose and excess oxygen. Ignition of the glucose resulted in a temperature increase of 3.64°C. What is the molar enthalpy of combustion of glucose?

Given Mass and ΔT for combustion of standard and sample

Asked for ΔH_{comb} of glucose

Strategy

A Calculate the value of q_{rxn} for benzoic acid by multiplying the mass of benzoic acid by its ΔH_{comb}. Then use Equation 5.41 to determine the heat capacity of the calorimeter, C_{bomb}, from q_{comb} and ΔT.

B Calculate the amount of heat released during the combustion of glucose by multiplying the heat capacity of the bomb by the temperature change. Determine the molar enthalpy of combustion of glucose, ΔH_{comb}, by multiplying the amount of heat released per gram by the molar mass of glucose.

Solution

A The first step is to use Equation 5.41 and the information obtained from the combustion of benzoic acid to calculate the heat capacity of the calorimeter, C_{bomb}. We are given ΔT, and we can calculate q_{comb} from the mass of benzoic acid:

$$q_{comb} = (0.579\ \cancel{g})(-26.38\ \text{kJ}/\cancel{g}) = -15.3\ \text{kJ}$$

Thus, from Equation 5.41,

$$-C_{\text{bomb}} = \frac{q_{\text{comb}}}{\Delta T} = \frac{-15.3\ \text{kJ}}{2.08°\text{C}} = -7.34\ \text{kJ/°C}$$

B According to the strategy, we can now use the heat capacity of the bomb to calculate the amount of heat released during the combustion of glucose:

$$q_{\text{comb}} = -C_{\text{bomb}}\,\Delta T = (-7.34\ \text{kJ/}\cancel{°\text{C}})(3.64\cancel{°\text{C}}) = -26.7\ \text{kJ}$$

Because the combustion of 1.732 g of glucose released 26.7 kJ of energy, the molar enthalpy of combustion of glucose is

$$\Delta H_{\text{comb}} = \left(\frac{-26.7\ \text{kJ}}{1.732\ \cancel{\text{g}}}\right)\left(\frac{180.16\ \cancel{\text{g}}}{\text{mol}}\right) = -2780\ \text{kJ/mol} = -2.78 \times 10^3\ \text{kJ/mol}$$

This result is in good agreement ($<1\%$ error) with the value of $\Delta H_{\text{comb}} = -2803$ kJ/mol that we calculated in Section 5.2 using enthalpies of formation.

EXERCISE 5.10

When 2.123 g of benzoic acid is ignited in a bomb calorimeter, a temperature increase of 4.75°C is observed. When 1.932 g of methylhydrazine, CH_3NHNH_2, is ignited in the same calorimeter, the temperature increase is 4.64°C. Calculate the molar enthalpy of combustion of methylhydrazine, the fuel used in the maneuvering jets of the U.S. space shuttle.

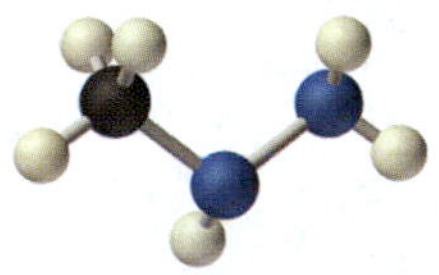
Methylhydrazine

Answer -1.30×10^3 kJ/mol

5.4 Thermochemistry and Nutrition

MGC Nutritional Energy

The thermochemical quantities that you probably encounter most often are the caloric values of food. Food supplies the raw materials that your body needs to replace cells and the energy that keeps those cells functioning. About 80% of this energy is released as heat to maintain your body temperature at a sustainable level to keep you alive.

The nutritional **Calorie** (with a capital C) that you see on food labels is equal to 1 kilocalorie (kcal). The caloric content of a food is determined from its enthalpy of combustion per gram, as measured in a bomb calorimeter, using the general reaction

$$\text{food} + \text{excess } O_2(g) \longrightarrow CO_2(g) + H_2O(l) + N_2(g) \tag{5.42}$$

There are two important differences, however, between the caloric values reported for foods and the enthalpies of combustion of the same foods burned in a calorimeter. First, enthalpies of combustion described in joules (or kilojoules) are negative for all substances that can be burned. In contrast, the caloric content of a food is always expressed as a *positive* number because it is *stored energy*. Therefore,

$$\text{caloric content} = -\Delta H_{\text{comb}} \tag{5.43}$$

Second, when foods are burned in a calorimeter, any nitrogen they contain (largely from proteins, which are rich in nitrogen) is transformed to N_2. In the body, however, the nitrogen from foods is converted to urea, $(H_2N)_2C{=}O$, rather than N_2 before it is excreted. The enthalpy of combustion of urea measured by bomb calorimetry is $\Delta H_{\text{comb}} = -632.0$ kJ/mol. Consequently, the enthalpy change measured by calorimetry for any nitrogen-containing food is greater than the amount of energy the body would obtain from it. The difference in the values is equal to the enthalpy of combustion of urea multiplied by the number of moles of

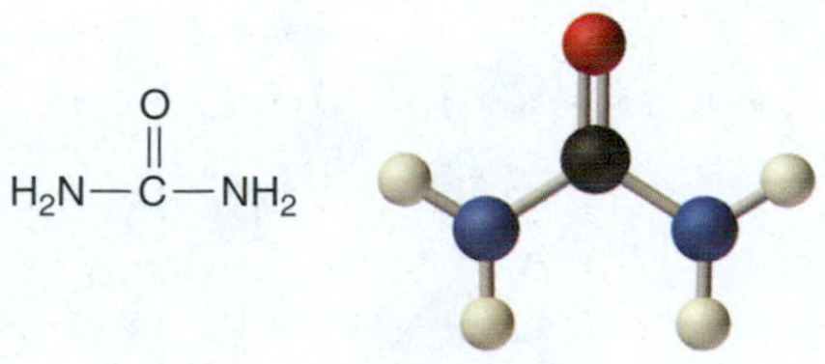

Urea

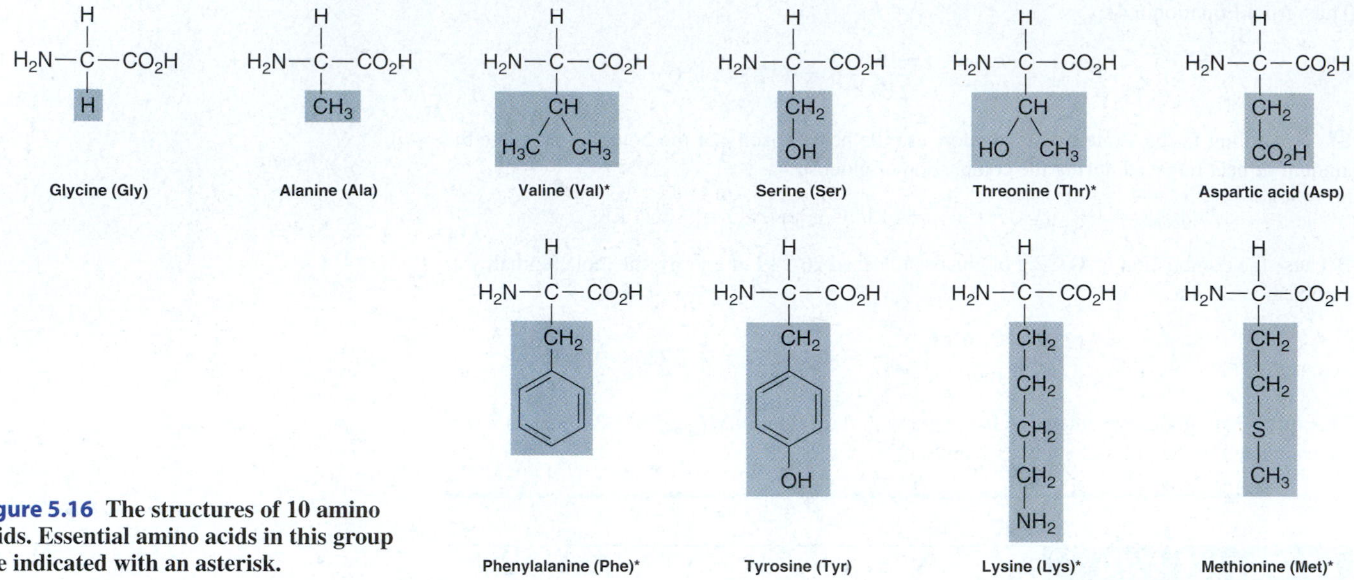

Figure 5.16 The structures of 10 amino acids. Essential amino acids in this group are indicated with an asterisk.

urea formed when the food is broken down. This point is illustrated schematically in the following equations:

$$\begin{aligned}
&\text{food} + \text{excess } O_2(g) \longrightarrow CO_2(g) + H_2O(l) + \cancel{(H_2N)_2C{=}O(s)} && \Delta H_1 < 0 \\
&\cancel{(H_2N)_2C{=}O(s)} + \text{excess } \tfrac{3}{2}O_2(g) \longrightarrow CO_2(g) + 2H_2O(l) + N_2(g) && \Delta H_2 = -632.0 \text{ kJ/mol} \quad (5.44) \\
\hline
&\text{food} + \text{excess } O_2(g) \longrightarrow 2CO_2(g) + 3H_2O(l) + N_2(g) && \Delta H_3 = \Delta H_1 + \Delta H_2 < 0
\end{aligned}$$

All three ΔH values are negative and, by Hess's law, $\Delta H_3 = \Delta H_1 + \Delta H_2$. Thus, the magnitude of ΔH_1 must be less than ΔH_3, the calorimetrically measured ΔH_{comb} for a food. By producing urea rather than N_2, therefore, humans are excreting some of the energy that was stored in their food.

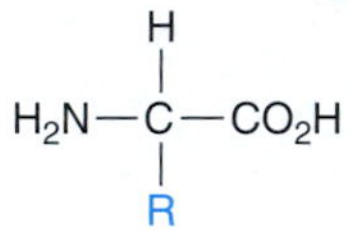

General structure of an amino acid, which contains an amine group ($-NH_2$) and a carboxylic acid group ($-CO_2H$).

Because of their different chemical compositions, foods vary widely in caloric content. As we saw in Example 5.5, for instance, a fatty acid such as palmitic acid produces about 39 kJ/g upon combustion, while a sugar such as glucose produces 15.6 kJ/g. Fatty acids and sugars are the building blocks of fats and carbohydrates, respectively, two of the major sources of energy in the diet. Nutritionists typically assign average values of 38 kJ/g (about 9 Cal/g) and 17 kJ/g (about 4 Cal/g) for fats and carbohydrates, respectively, although the actual values for specific foods vary because of differences in composition. Proteins, the third major source of calories in the diet, vary as well. Proteins are composed of amino acids, which have the general structure shown in the margin.

In addition to their amine and carboxylic acid components, amino acids may contain a wide range of other functional groups: R can be hydrogen ($-H$), an alkyl group (such as $-CH_3$), an aryl group (for example, $-CH_2C_6H_5$), or a substituted alkyl group that contains an amine, an alcohol, or a carboxylic acid (Figure 5.16). Of the 20 naturally occurring amino acids, 10 are required in the human diet; these 10 are called *essential amino acids* because our bodies are unable to synthesize them from other compounds. Because R can be any of several different groups, each amino acid has a different value of ΔH_{comb}. Proteins are usually estimated to have an average ΔH_{comb} of 17 kJ/g (about 4 Cal/g).

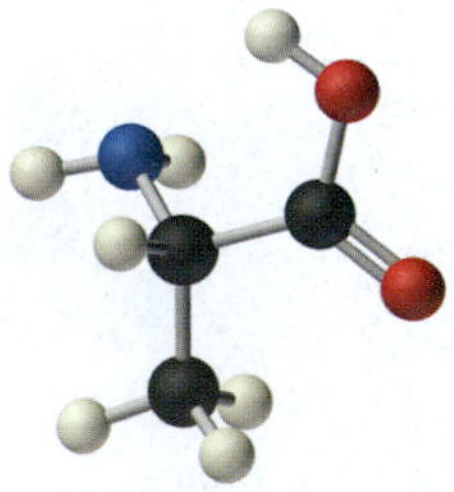

Alanine

EXAMPLE 5.11

Calculate the amount of available energy obtained from the biological oxidation of 1.000 g of the amino acid alanine. Remember that the nitrogen-containing product is urea, not N_2, so biological oxidation of alanine will yield *less* energy than will combustion. The value of ΔH_{comb} for alanine is -1577 kJ/mol.

Given Amino acid and ΔH_{comb} per mole

Asked for Caloric content per gram

Strategy

A Write balanced equations for the oxidation of alanine to CO_2, H_2O, and urea; the combustion of urea; and the combustion of alanine. Multiply both sides of the equations by appropriate factors, and then rearrange them to cancel urea from both sides when the equations are added.

B Use Hess's law to obtain an expression for ΔH for the oxidation of alanine to urea in terms of ΔH_{comb} of alanine and urea. Substitute the appropriate values of ΔH_{comb} into the equation, and solve for ΔH for the oxidation of alanine to CO_2, H_2O, and urea.

C Calculate the amount of energy released per gram by dividing the value of ΔH by the molar mass of alanine.

Solution

The actual energy available biologically from alanine is less than its enthalpy of combustion because of the production of urea rather than N_2. We know ΔH_{comb} values for alanine and for urea, so we can use Hess's law to calculate ΔH for the oxidation of alanine to CO_2, H_2O, and urea.

A We begin by writing balanced equations for (1) the oxidation of alanine to carbon dioxide, water, and urea; (2) the combustion of urea; and (3) the combustion of alanine. Because alanine contains only a single nitrogen atom whereas urea and N_2 each contain two nitrogen atoms, it is easier to balance Equations 1 and 3 if we write them for the oxidation of 2 mol of alanine:

$$2C_3H_7NO_2(s) + 6O_2(g) \longrightarrow 5CO_2(g) + 5H_2O(l) + (H_2N)_2C{=}O(s) \qquad (1)$$

$$(H_2N)_2C{=}O(s) + \frac{3}{2}O_2(g) \longrightarrow CO_2(g) + 2H_2O(l) + N_2(g) \qquad (2)$$

$$2C_3H_7NO_2(s) + \frac{15}{2}O_2(g) \longrightarrow 6CO_2(g) + 7H_2O(l) + N_2(g) \qquad (3)$$

Adding Equations 1 and 2 and canceling urea from both sides give the overall equation directly:

$$\begin{array}{lll} (1) & 2C_3H_7NO_2(s) + 6O_2(g) \longrightarrow 5CO_2(g) + 5H_2O(l) + \cancel{(H_2N)_2C{=}O(s)} & \Delta H_1 \\ (2) & \cancel{(H_2N)_2C{=}O(s)} + \frac{3}{2}O_2(g) \longrightarrow CO_2(g) + 2H_2O(l) + N_2(g) & \Delta H_2 \\ \hline (3) & 2C_3H_7NO_2(s) + \frac{15}{2}O_2(g) \longrightarrow 6CO_2(g) + 7H_2O(l) + N_2(g) & \Delta H_3 \end{array}$$

B By Hess's law, $\Delta H_3 = \Delta H_1 + \Delta H_2$. We know that $\Delta H_3 = 2\Delta H_{comb}$ (alanine), $\Delta H_2 = \Delta H_{comb}$ (urea), and $\Delta H_1 = 2\Delta H$ (alanine $\longrightarrow$ urea). Rearranging and substituting the appropriate values gives

$$\begin{aligned} \Delta H_1 = \Delta H_3 - \Delta H_2 &= 2\Delta H_{comb}(\text{alanine}) - \Delta H_{comb}\,(\text{urea}) \\ &= 2(-1577 \text{ kJ/mol}) - (-632.0 \text{ kJ/mol}) = -2522 \text{ kJ/(2 mol alanine)} \end{aligned}$$

Thus, ΔH (alanine $\longrightarrow$ urea) $= -2522$ kJ/(2 mol of alanine) $= -1261$ kJ/mol of alanine. Oxidation of alanine to urea rather than to nitrogen therefore results in about a 20% decrease in the amount of energy released (-1261 kJ/mol versus -1577 kJ/mol).

C The energy released per gram by the biological oxidation of alanine is

$$\left(\frac{-1261 \text{ kJ}}{1 \cancel{\text{mol}}}\right)\left(\frac{1 \cancel{\text{mol}}}{89.094 \text{ g}}\right) = -14.15 \text{ kJ/g}$$

This is equal to -3.382 Cal/g.

EXERCISE 5.11

Calculate the energy released per gram from the oxidation of the amino acid valine to carbon dioxide, water, and urea. Report your answer to three significant figures. The value of ΔH_{comb} for valine is -2922 kJ/mol.

Answer -22.2 kJ/g (-5.31 Cal/g)

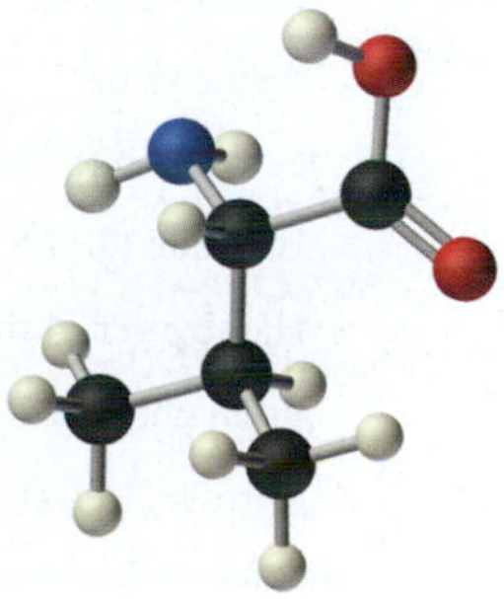

Valine

TABLE 5.4 Approximate composition and fuel value of an 8-ounce slice of roast beef

Composition		Calories
97.5 g of water	×	0 Cal/g = 0
58.7 g of protein	×	4 Cal/g = 235
69.3 g of fat	×	9 Cal/g = 624
0 g of carbohydrates	×	4 Cal/g = 0
1.5 g of minerals	×	0 Cal/g = 0
Total mass: 227.0 g		**Total calories: ~900 Cal**

The reported caloric content of foods does not include enthalpies of combustion for those components that are not digested, such as fiber. For example, meats and fruits are 50−70% water, but water cannot be oxidized by O_2 to obtain energy. Thus, water contains no calories. Some foods contain large amounts of fiber, which is primarily composed of sugars. Although fiber can be burned in a calorimeter just like glucose to give carbon dioxide, water, and heat, humans lack the enzymes needed to break fiber down into smaller molecules that can be oxidized. Hence, fiber does not contribute to the caloric content of food either.

We can determine the caloric content of foods in two ways. The most precise method is to dry a carefully weighed sample and carry out a combustion reaction in a bomb calorimeter. The more typical approach, however, is to analyze the food for protein, carbohydrate, fat, water, and "minerals" (everything that doesn't burn), and then calculate the caloric content using the average values for each component that produces energy (9 Cal/g for fats and 4 Cal/g for carbohydrates and proteins). An example of this approach is shown in Table 5.4 for a slice of roast beef. The compositions and caloric contents of some common foods are given in Table 5.5.

Because the Calorie represents such a large amount of energy, a few of them go a long way. An average 73-kg (160-lb) person needs about 67 Cal/h (1600 Cal/day) just to fuel the basic biochemical processes that keep that person alive. This energy is required to maintain body temperature, to keep the heart beating, to power the muscles used for breathing, to carry out chemical reactions in cells, and to send the nerve impulses that control those automatic functions. Physical activity increases the amount of energy required but not by as much as many of us hope (Table 5.6). A moderately active individual requires about 2500−3000 Cal/day, and athletes or others engaged in strenuous activity can burn 4000 Cal/day. Any excess caloric intake is stored by the body for future use, usually in the form of fat, which is the most compact way to store energy. When more energy is needed than the diet supplies, stored fuels are mobilized and oxidized. We usually exhaust the supply of stored carbohydrates before turning to fats, which accounts in part for the popularity of low-carbohydrate diets.

TABLE 5.5 Approximate compositions and fuel values of some common foods

Food (quantity)	Approximate Composition, %				Food Value, Cal/g	Calories
	Water	Carbohydrate	Protein	Fat		
Beer (12 oz)	92	3.6	0.3	0	0.4	150
Coffee (6 oz)	100	~0	~0	~0	~0	~0
Milk (1 cup)	88	4.5	3.3	3.3	0.6	150
Egg (1 large)	75	2	12	12	1.6	80
Butter (1 tbsp)	16	~0	~0	79	7.1	100
Apple (8 oz)	84	15	~0	0.5	0.6	125
Bread, white (2 slices)	37	48	8	4	2.6	130
Brownie (40 g)	10	55	5	30	4.8	190
Hamburger (4 oz)	54	0	24	21	2.9	326
Fried chicken (1 drumstick)	53	8.3	22	15	2.7	195
Carrots (1 cup)	87	10	1.3	~0	0.4	70

EXAMPLE 5.12

What is the minimum number of Calories expended by a 160-lb person who climbs a 30-story building? (Assume each flight of stairs is 14 ft high.) How many grams of glucose are required to supply this amount of energy? (The energy released upon combustion of glucose was calculated in Example 5.5.)

Given Mass, height, and energy released by combustion of glucose

Asked for Calories expended and mass of glucose needed

Strategy

A Convert the mass and height to SI units, and then substitute these values into Equation 5.6 to calculate the change in potential energy (in kJ). Divide the calculated energy by 4.184 Cal/kJ to convert the potential energy change to Calories.

B Use the value obtained in Example 5.5 for the combustion of glucose to calculate the mass of glucose needed to supply this amount of energy.

Solution

The energy needed to climb the stairs equals the difference between the person's potential energy (PE) at the top of the building and at ground level.

A Recall from Section 5.1 that $PE = mgh$. Because m and h are given in non-SI units, we must convert them to kilograms and meters, respectively:

$$m = (160\ \cancel{\text{lb}})\left(\frac{454\ \cancel{\text{g}}}{1\ \cancel{\text{lb}}}\right)\left(\frac{1\ \text{kg}}{1000\ \cancel{\text{g}}}\right) = 72.6\ \text{kg}$$

$$h = (30\ \cancel{\text{stories}})\left(\frac{14\ \cancel{\text{ft}}}{1\ \cancel{\text{story}}}\right)\left(\frac{12\ \cancel{\text{in.}}}{1\ \cancel{\text{ft}}}\right)\left(\frac{2.54\ \cancel{\text{cm}}}{1\ \cancel{\text{in.}}}\right)\left(\frac{1\ \text{m}}{100\ \cancel{\text{cm}}}\right) = 128\ \text{m}$$

Thus,

$$PE = (72.6\ \text{kg})(9.81\ \text{m/s}^2)(128\ \text{m}) = 8.55 \times 10^4\ (\text{kg} \cdot \text{m}^2)/\text{s}^2 = 91.2\ \text{kJ}$$

To convert to Calories, we divide by 4.184 kJ/kcal:

$$PE = (91.2\ \cancel{\text{kJ}})\left(\frac{1\ \text{kcal}}{4.184\ \cancel{\text{kJ}}}\right) = 21.8\ \text{kcal} = 21.8\ \text{Cal}$$

B Because the combustion of glucose produces 15.6 kJ/g (Example 5.5), the mass of glucose needed to supply 85.5 kJ of energy is

$$(91.2\ \cancel{\text{kJ}})\left(\frac{1\ \text{g glucose}}{15.6\ \cancel{\text{kJ}}}\right) = 5.85\ \text{g glucose}$$

This mass corresponds to only about a teaspoonful of sugar! Because the body is only about 30% efficient in using the energy in glucose, the actual amount of glucose required would be higher: (100%/30%) × 5.85 g = 19.5 g. Nonetheless, this calculation illustrates the difficulty many people have in trying to lose weight by exercise alone.

EXERCISE 5.12

Calculate how many times a 160-lb person would have to climb the tallest building in the United States, the 110-story Sears Tower in Chicago, to burn off 1.0 pound of stored fat. Assume that each story of the building is 14 feet high, and use a calorie content of 9.0 kcal/g of fat.

Answer About 55 times

TABLE 5.6 Approximate energy expenditure by a 160-lb person engaged in various activities

Activity	Cal/h
Sleeping	80
Driving a car	120
Standing	140
Eating	150
Walking 2.5 mi/h	210
Mowing lawn	250
Swimming 0.25 mi/h	300
Roller skating	350
Tennis	420
Bicycling 13 mi/h	660
Running 10 mi/h	900

The calculations in Example 5.12 and Exercise 5.12 ignore factors such as how fast the person is climbing. Although the rate is irrelevant in calculating the change in potential energy, it is very relevant to the amount of energy actually required to ascend the stairs. The calculations also ignore the fact that the body's conversion of chemical energy to mechanical work is significantly less than 100% efficient. According to the average energy expended for various activities listed in Table 5.6, a person must run more than $4\frac{1}{2}$ hours at 10 miles per hour or bicycle for 6 hours at 13 miles per hour to burn off 1 pound of fat (1.0 lb × 454 g/lb × 9.0 Cal/g = 4100 Cal). Note, however, that riding a bicycle at 13 miles per hour for only 1 hour per day six days a week will burn off 50 pounds of fat in the course of a year (assuming, of course, the cyclist doesn't increase his or her intake of calories to compensate for the exercise).

5.5 • Energy Sources and the Environment

MGC Law of Conservation of Energy

Our contemporary society requires the constant expenditure of huge amounts of energy to heat our homes, to provide telephone and cable service, to transport us from one location to another, to provide light when it is dark outside, and to run the machinery that manufactures material goods. The United States alone consumes almost 10^6 kJ per person per day, which is about a hundred times the normal required energy content of the human diet. This figure is about 30% of the world's total energy usage, although only about 5% of the total population of the world lives in the United States. In contrast, the average energy consumption elsewhere in the world is about 10^5 kJ per person per day, although actual values vary widely depending on a country's level of industrialization. In this section we describe various sources of energy and their impact on the environment.

Fuels

According to the law of conservation of energy, energy can never actually be "consumed"; it can only be changed from one form to another. What *is* consumed on a huge scale, however, are resources that can be readily converted to a form of energy that is useful for doing work. As you will see in Chapter 18, energy that is not used to perform work is either stored as potential energy for future use or transferred to the surroundings as heat.

A major reason for the huge consumption of energy by our society is the low *efficiency* of most machines in transforming stored energy into work. Efficiency can be defined as the ratio of useful work accomplished to energy expended. Automobiles, for example, are only about 20% efficient in converting the energy stored in gasoline to mechanical work; the rest of the energy is released as heat, either emitted in the exhaust or produced by friction in bearings and tires. The production of electricity by coal- or oil-powered steam turbines is significantly more efficient (Figure 5.17): about 38% of the energy released from combustion is converted to electricity. In comparison, modern nuclear power plants can be more than 50% efficient.

In general, it is more efficient to use primary sources of energy directly (such as natural gas or oil) than to transform them to a secondary source such as electricity prior to their use. For example, if a furnace is well maintained, heating a house with natural gas is about 70% efficient. In contrast, burning the natural gas in a remote power plant, converting it to electricity, transmitting it long distances through wires, and heating the house by electric baseboard heaters have an overall efficiency of less than 35%.

The total expenditure of energy in the world each year is about 3×10^{17} kJ. More than 80% of this energy is provided by the combustion of fossil fuels: oil,

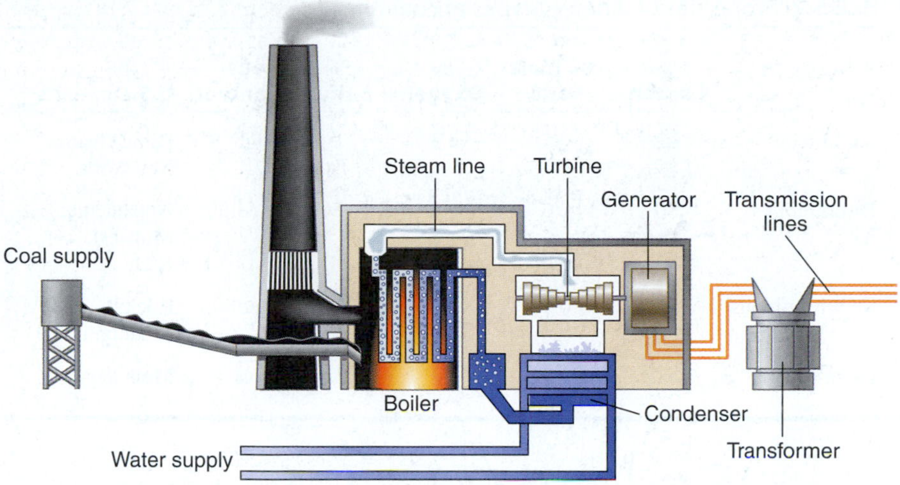

Figure 5.17 Electricity from coal. A coal-powered electric power plant uses the combustion of coal to produce steam, which drives a turbine to produce electricity.

coal, and natural gas. (The sources of the energy consumed in the United States in 2003 are shown in Figure 5.18.) Natural gas and petroleum, whose compositions were described in Chapter 2, are the preferred fuels because they or products derived from them are gases or liquids that are readily transported, stored, and burned. Natural gas and petroleum are derived from the remains of marine creatures that died hundreds of millions of years ago and were buried beneath layers of sediment. As the sediment turned to rock, the tremendous heat and pressure inside the earth transformed the organic components of the buried sea creatures to petroleum and natural gas.

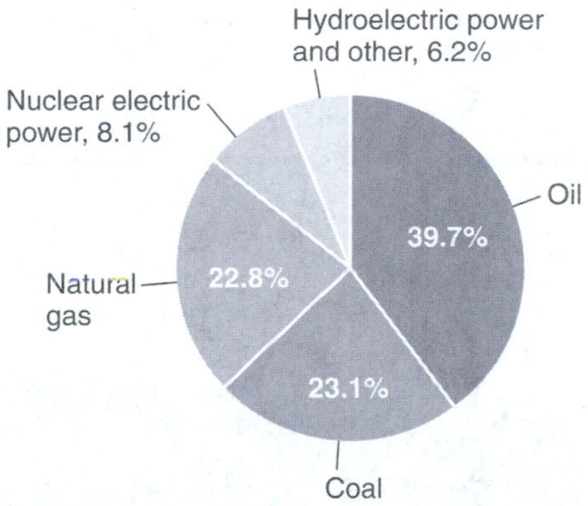

Figure 5.18 Energy consumption in the United States by source, 2003. More than 80% of the total energy expended is provided by the combustion of fossil fuels, such as oil, coal, and natural gas.

Coal

Coal is a complex solid material derived primarily from plants that died and were buried hundreds of millions of years ago and were subsequently subjected to high temperatures and pressures. Because plants contain large amounts of *cellulose*, derived from linked glucose units, the structure of coal is more complex than that of petroleum (Figure 5.19). In particular, coal contains a large number of oxygen atoms that link parts of the structure together, in addition to the basic framework of carbon–carbon bonds. It is impossible to draw a single structure for coal, but because of the prevalence of rings of carbon atoms (due to the original high cellulose content), coal is more similar to an aromatic hydrocarbon than to an aliphatic one.

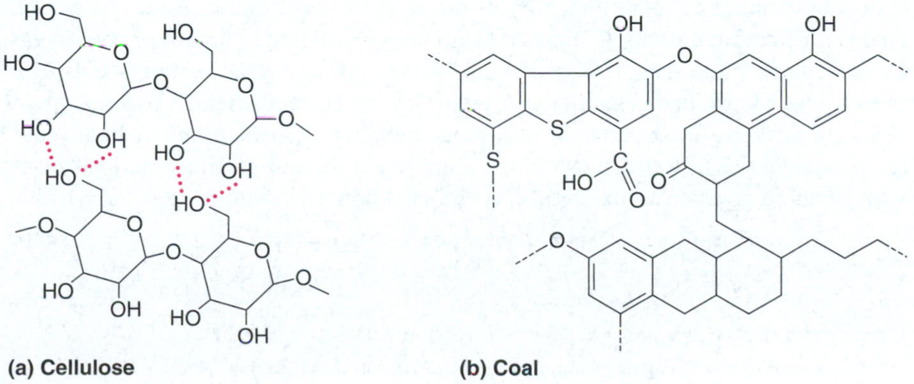

Figure 5.19 The structures of cellulose and coal. (a) Cellulose consists of long chains of cyclic glucose molecules linked by hydrogen bonds. (b) When cellulose is subjected to high pressures and temperatures for long periods of time, water is eliminated and bonds are formed between the rings, eventually producing coal. This drawing shows some of the common structural features of coal; note the presence of many different kinds of ring structures.

TABLE 5.8 Enthalpies of combustion of common fuels and selected organic compounds

Fuel	ΔH_{comb}, kJ/g
Dry wood	−15
Peat	−20.8
Bituminous coal	**−28.3**
Charcoal	−35
Kerosene	−37
C_6H_6 (benzene)	−41.8
Crude oil	−43
Natural gas	**−50**
C_2H_2 (acetylene)	−50.0
CH_4 (methane)	−55.5
Gasoline	**−84**
Hydrogen	−143

TABLE 5.7 Properties of different types of coal

Type	% Carbon	H/C Mole Ratio	% Oxygen	% Sulfur	Heat Content	U.S. Deposits
Anthracite	92	0.5	3	1	High	Pennsylvania, New York
Bituminous	80	0.6	8	5	Medium	Appalachia, Midwest, Utah
Subbituminous	77	0.9	16	1	Medium	Rocky Mountains
Lignite	71	1.0	23	1	Low	Montana

There are four distinct classes of coal (Table 5.7); their hydrogen and oxygen contents depend on the length of time the coal has been buried and the pressures and temperatures to which it has been subjected. Lignite, with an H/C ratio of about 1.0 and a high oxygen content, has the lowest enthalpy of combustion. Anthracite, in contrast, with an H/C ratio of about 0.5 and the lowest oxygen content, has the highest enthalpy of combustion and is the highest grade of coal. The most abundant form in the United States is bituminous coal, which has a high sulfur content because of the presence of small particles of FeS_2 (pyrite). As discussed in Chapter 4, the combustion of coal releases the sulfur in FeS_2 as SO_2, which is a major contributor to acid rain. Table 5.8 compares the enthalpies of combustion per gram of oil, natural gas, and coal with those of selected organic compounds.

Figure 5.20 **A peat bog.** Peat is a smoky fuel that burns poorly and produces little heat, but it has been used as a fuel since ancient times.

Peat, a precursor to coal, is the partially decayed remains of plants that grow in swampy areas. It is removed from the ground in the form of soggy bricks of mud that will not burn until they have been dried. Even though peat is a smoky, poor-burning fuel that gives off relatively little heat, humans have burned it since ancient times (Figure 5.20). If a peat bog were buried under many layers of sediment for a few million years, the peat could eventually be compressed and heated enough to become lignite, the lowest grade of coal; given enough time and heat, lignite would eventually become anthracite, a much better fuel.

Converting Coal to Gaseous and Liquid Fuels

Oil and natural gas resources are limited. Current estimates suggest that the known reserves of petroleum will be exhausted in about 60 years, and supplies of natural gas are estimated to run out in about 120 years. Coal, on the other hand, is relatively abundant, making up more than 90% of the world's fossil fuel reserves. As a solid, coal is much more difficult to mine and ship than petroleum (a liquid) or natural gas. Consequently, more than 75% of the coal produced each year is simply burned in power plants to produce electricity. A great deal of current research focuses on developing methods to convert coal to gaseous fuels (*coal gasification*) or liquid fuels (*coal liquefaction*). In the most common approach to coal gasification, coal reacts with steam to produce a mixture of CO and H_2 known as *synthesis gas*, or *syngas*:*

$$C(s) + H_2O(g) \longrightarrow CO(g) + H_2(g) \qquad \Delta H = 131 \text{ kJ} \qquad (5.45)$$

* Because coal is 70–90% carbon by mass, it is approximated as C in Equation 5.45.

Converting coal to syngas removes any sulfur present and produces a clean-burning mixture of gases.

Syngas is also used as a reactant to produce methane and methanol. A promising approach is to convert coal directly to methane through a series of reactions:

$$\begin{array}{lll}
 & 2C(s) + 2H_2O(g) \longrightarrow \cancel{2CO(g)} + \cancel{2H_2(g)} & \Delta H_1 = 262\text{ kJ} \\
 & \cancel{CO(g)} + \cancel{H_2O(g)} \longrightarrow CO_2(g) + \cancel{H_2(g)} & \Delta H_2 = -41\text{ kJ} \\
 & \cancel{CO(g)} + \cancel{3H_2(g)} \longrightarrow CH_4(g) + \cancel{H_2O(g)} & \Delta H_3 = -206\text{ kJ} \\
\hline
\text{Overall:} & 2C(s) + 2H_2O(g) \longrightarrow CH_4(g) + CO_2(g) & \Delta H_T = 15\text{ kJ}
\end{array} \quad (5.46)$$

The energy consumed by these reactions is provided by burning a small amount of coal or methane. Unfortunately, methane produced by this process is currently significantly more expensive than natural gas. As supplies of natural gas become depleted, however, the coal-based process may well become competitive in cost.

The standard industrial unit of measure for crude oil is a 42-gallon barrel.

Similarly, the techniques available for converting coal to liquid fuels are not yet economically competitive with the production of liquid fuels from petroleum. Current approaches to coal liquefaction use a catalyst to break the complex network structure of coal into more manageable fragments. The products are then treated with hydrogen (from syngas or other sources) under high pressure to produce a liquid more like petroleum. Subsequent distillation, cracking, and reforming can be used to create products similar to those obtained from petroleum. The total yield of liquid fuels is about 5.5 barrels of crude liquid per ton of coal (one barrel is 42 gallons or 160 liters). Although the economics of coal liquefaction are currently even less attractive than for coal gasification, liquid fuels based on coal are likely to become economically competitive as supplies of petroleum are consumed.

EXAMPLE 5.13

If bituminous coal is converted to methane by the process in Equations 5.46, what is the ratio of the enthalpy of combustion of the methane produced to the enthalpy of the coal consumed to produce the methane? (Note that 1 mol of CH_4 is produced for every 2 mol of carbon in coal.)

Given Chemical reaction and enthalpies of combustion (Table 5.8)

Asked for Ratio of ΔH_{comb} of methane produced to coal consumed

Strategy

A Write a balanced equation for the conversion of coal to methane. Referring to Table 5.8, calculate the enthalpy of combustion of methane and carbon.

B Calculate the ratio of the energy released by combustion of the methane to the energy released by combustion of the carbon.

Solution

A The balanced equation for the conversion of coal to methane is

$$2C(s) + 2H_2O(g) \longrightarrow CH_4(g) + CO_2(g)$$

Thus, 1 mol of methane is produced for every 2 mol of carbon consumed. The enthalpy of combustion of 1 mol of methane is

$$1\ \cancel{\text{mol CH}_4}\left(\frac{16.043\ \cancel{\text{g}}}{1\ \cancel{\text{mol CH}_4}}\right)\left(\frac{-55.5\ \text{kJ}}{\cancel{\text{g}}}\right) = -890\ \text{kJ}$$

The enthalpy of combustion of 2 mol of carbon (as coal) is

$$2\ \cancel{\text{mol C}}\left(\frac{12.011\ \cancel{\text{g}}}{1\ \cancel{\text{mol C}}}\right)\left(\frac{-28.3\ \text{kJ}}{\cancel{\text{g}}}\right) = -680\ \text{kJ}$$

B✓ The ratio of the energy released from the combustion of methane to the energy released from the combustion of carbon is

$$\frac{-890\ \cancel{\text{kJ}}}{-680\ \cancel{\text{kJ}}} = 1.31$$

Thus, the energy released from the combustion of the product (methane) is 131% of that of the reactant (coal). The fuel value of coal is actually *increased* by the process!

How is this possible, since the law of conservation of energy states that energy cannot be created? Note that the reaction consumes 2 mol of water ($\Delta H_f^\circ = -285.8$ kJ/mol) but produces only 1 mol of CO_2 ($\Delta H_f^\circ = -393.5$ kJ/mol). Part of the difference in potential energy between the two (approximately 180 kJ/mol) is stored in CH_4 and can be released upon combustion.

EXERCISE 5.13

Using the data in Table 5.8, calculate the mass of hydrogen necessary to provide as much energy upon combustion as one barrel of crude oil (density approximately 0.75 g/mL).

Answer 36 kg

The Carbon Cycle and the Greenhouse Effect

Even if carbon-based fuels could be burned with 100% efficiency, producing only $CO_2(g)$ and $H_2O(g)$, doing so could still potentially damage the environment when carried out on the vast scale required by an industrial society. The amount of CO_2 released is so large and is increasing so rapidly that it is apparently overwhelming the natural ability of the planet to remove CO_2 from the atmosphere. In turn, the elevated levels of CO_2 are thought to be affecting the temperature of the planet through a mechanism known as the "greenhouse effect." As you will see, there is little doubt that atmospheric CO_2 levels are increasing and that the major reason for the increase is the combustion of fossil fuels. There is substantially less agreement, however, on whether the increased CO_2 levels are responsible for a significant increase in temperature.

The Global Carbon Cycle

Figure 5.21 illustrates the global **carbon cycle**, the distribution and flow of carbon throughout the planet. Normally, the fate of atmospheric CO_2 is either to dissolve in the oceans and eventually precipitate as carbonate rocks or to be taken up by plants. The rate of uptake of CO_2 by the ocean is limited by its surface area and the rate at which gases dissolve, which are approximately constant. The rate of uptake of CO_2 by plants, representing about 60 billion metric tons of carbon per year, depends in part on how much of the earth's surface is covered by vegetation. Unfortunately, rapid deforestation for agriculture is reducing the overall amount of vegetation, and about 60 billion metric tons of carbon are released annually as CO_2 from animal respiration and plant decay.

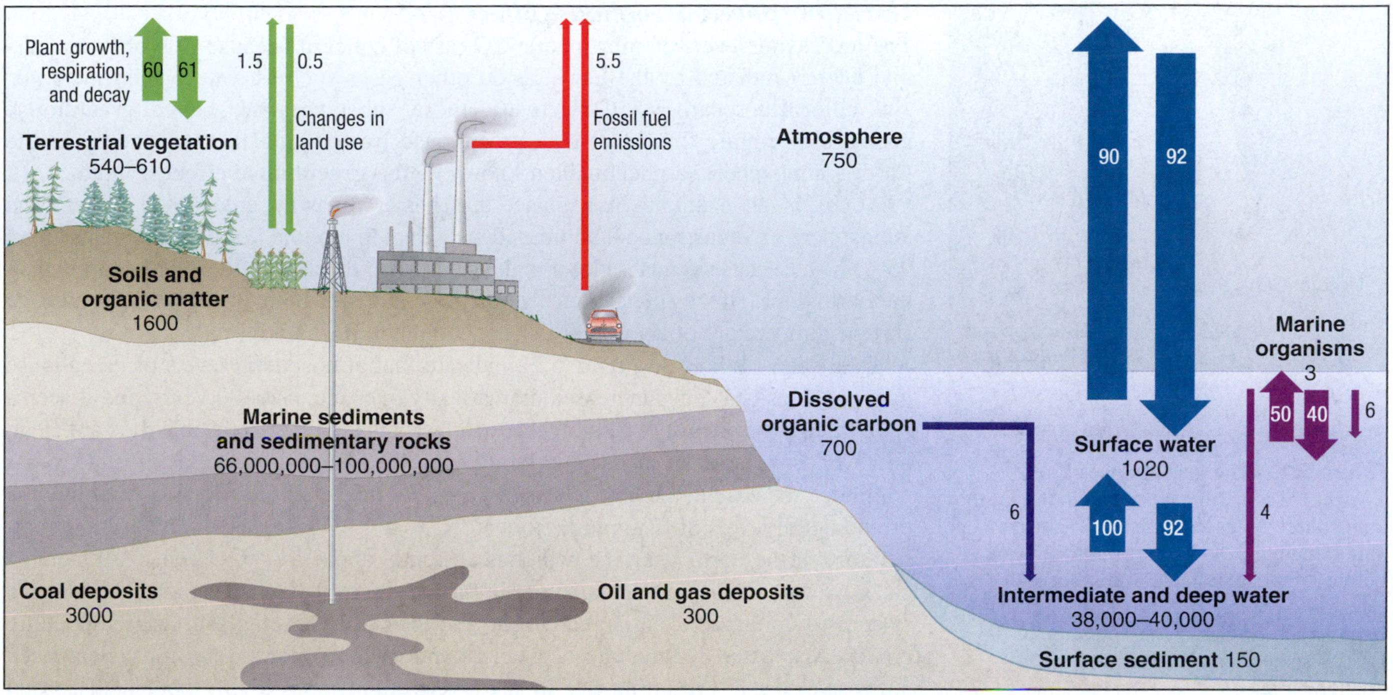

Figure 5.21 The global carbon cycle. Most of the earth's carbon is found in the crust, where it is stored as calcium and magnesium carbonate in sedimentary rocks. The oceans also contain a large reservoir of carbon, primarily as the bicarbonate ion, HCO_3^-. Green plants consume about 60 billion metric tons of carbon per year as CO_2 during photosynthesis, and about the same amount of carbon is released as CO_2 annually from animal and plant respiration and decay. The combustion of fossil fuels releases about 5.5 billion metric tons of carbon per year as CO_2.

The amount of carbon released as CO_2 every year by fossil fuel combustion is estimated to be about 5.5 billion metric tons. The net result is a system that is slightly out of balance, experiencing a slow but steady increase in atmospheric CO_2 levels (Figure 5.22). As a result, average CO_2 levels have increased by about 30% since 1850.

Figure 5.22 Changes in atmospheric CO_2 levels. (a) Average worldwide CO_2 levels have increased by about 30% since 1850. (b) Atmospheric CO_2 concentrations measured at Mauna Loa in Hawaii show seasonal variations caused by the removal of CO_2 from the atmosphere by green plants during the growing season along with a general increase in CO_2 levels.

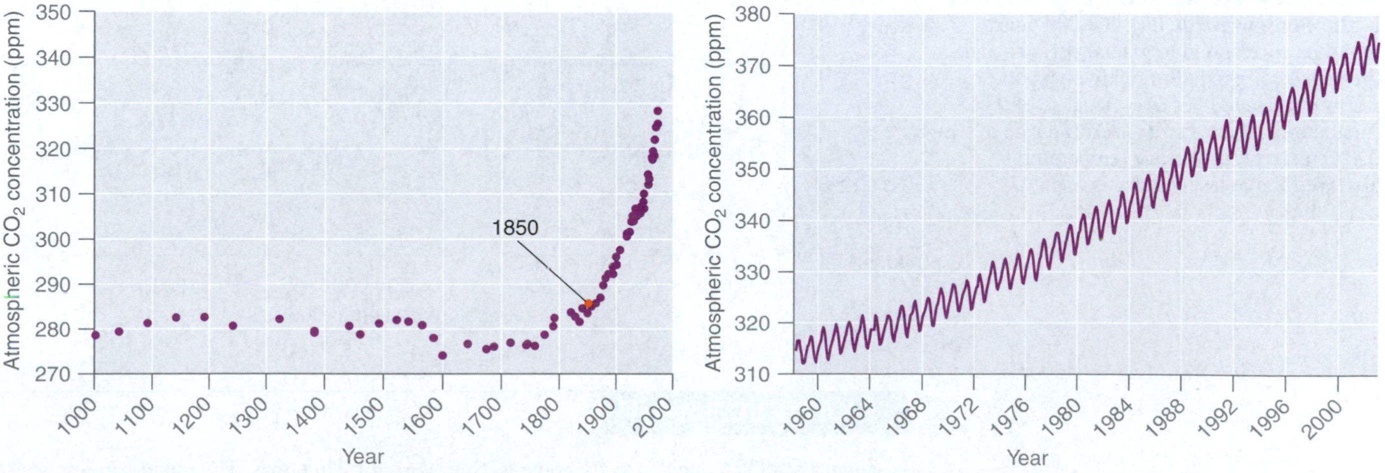

(a) Records from Antarctic ice cores (1006-1969 A.D.)

(b) Records from monthly air samples, Mauna Loa Observatory, Hawaii (1958-2002)

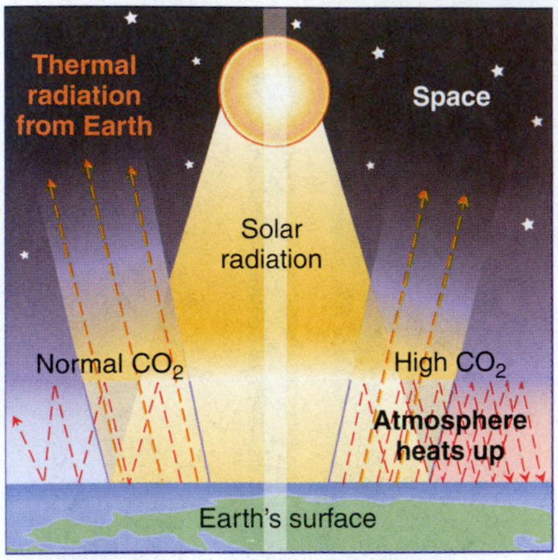

Figure 5.23 The greenhouse effect. Thermal energy can be trapped in the Earth's atmosphere by gases such as CO_2, water vapor, methane, and chlorofluorocarbons before it can be radiated into space—like the effect of a greenhouse. It is not clear how large an increase in the temperature of the Earth's surface can be attributed to this phenomenon.

The Atmospheric Greenhouse Effect

The increasing levels of atmospheric CO_2 are of concern because CO_2 absorbs thermal energy radiated by the Earth, as do other gases such as water vapor, methane, and chlorofluorocarbons. Collectively, these substances are called **greenhouse gases**; they mimic the effect of a greenhouse by trapping thermal energy in the Earth's atmosphere, a phenomenon known as the **greenhouse effect** (Figure 5.23).

Venus is an example of a planet that has a runaway greenhouse effect. The atmosphere of Venus is about 95 times denser than that of the Earth and contains about 95% CO_2. Because Venus is closer to the sun, it also receives more solar radiation than the Earth does. The result of increased solar radiation and high CO_2 levels is an average surface temperature of about 450°C—hot enough to melt lead!

Data such as those in Figure 5.22 indicate that atmospheric levels of greenhouse gases such as CO_2 have increased dramatically over the past 100 years, and it seems clear that the heavy use of fossil fuels by industry is largely responsible. It is not clear, however, how large an increase in temperature (*global warming*) may result from a continued increase in the levels of these gases. Estimates of the effects of doubling the preindustrial levels of CO_2 range from a 0°C to a 4.5°C increase in the average temperature of the Earth's surface, which is currently about 14.4°C. Even small increases, however, could cause major perturbations in our planet's delicately balanced systems. For example, an increase of 5°C in the Earth's average surface temperature could cause extensive melting of glaciers and of the Antarctic ice cap. It has been suggested that the resulting rise in sea levels could flood highly populated coastal areas such as New York City, Calcutta, Tokyo, Rio de Janeiro, and Sydney.

The increase in CO_2 levels is only one of many trends that can affect the Earth's temperature, however. In fact, geologic evidence shows that the average temperature of the Earth has fluctuated significantly over the past 400,000 years, with a series of glacial periods (during which the temperature was 10–15°C *lower* than it is now and large glaciers covered much of the globe) interspersed with relatively short, warm interglacial periods (Figure 5.24). Thus, although average temperatures appear to have increased by 0.5°C in the last century, the statistical significance of this increase is open to question, as is the existence of a cause-and-effect relationship between the temperature change and CO_2 levels. Despite the lack of solid scientific evidence, however, many people believe that we should take steps now to limit CO_2 emissions and explore alternative sources of energy, such as solar, geothermal energy from volcanic steam, and nuclear energy, to avoid even the possibility of creating major perturbations in the Earth's environment.

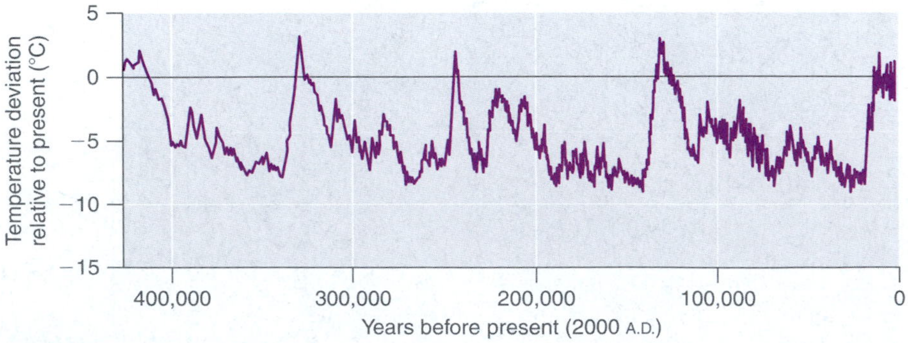

Figure 5.24 Average surface temperature of the earth over the last 400,000 years. The dips correspond to glacial periods, and the peaks correspond to relatively short, warm interglacial periods. Because of these fluctuations, the statistical significance of the 0.5°C increase in average temperatures observed in the last century is open to question.

EXAMPLE 5.14

A student at UCLA decided to fly home to New York for Christmas. The round-trip was 4500 air miles, and part of the cost of her ticket went to buy the 100 gallons of jet fuel necessary to transport her and her baggage. Assuming that jet fuel is primarily *n*-dodecane, $C_{12}H_{26}$, with a density of 0.75 g/mL, how much energy was expended and how many tons of CO_2 were emitted into the upper atmosphere to get her home and back?

Given Volume and density of reactant in combustion reaction

Asked for Energy expended and mass of CO_2 emitted

Strategy

Ⓐ After writing a balanced chemical equation for the reaction, calculate ΔH°_{comb} using Equation 5.26 and ΔH°_f values given in Appendix A.
Ⓑ Determine the number of moles of dodecane in 100 gallons by using the density and molar mass of dodecane and the appropriate conversion factors.
Ⓒ Obtain the amount of energy expended by multiplying ΔH°_{comb} by the number of moles of dodecane. Calculate the amount of CO_2 emitted in tons by using mole ratios from the balanced chemical equation and the appropriate conversion factors.

Solution

Ⓐ✓ We first need to write a balanced chemical equation for the reaction:

$$2C_{12}H_{26}(l) + 37O_2(g) \longrightarrow 24CO_2(g) + 26H_2O(l)$$

We can calculate ΔH°_{comb} for this reaction from Equation 5.26 and the ΔH°_f values in Appendix A corresponding to each substance in the specified phase (phases are not shown for simplicity):

$$\begin{aligned}\Delta H^\circ_{comb} &= \Sigma m\Delta H^\circ_f\,(\text{products}) - \Sigma n\Delta H^\circ_f\,(\text{reactants})\\ &= [24\Delta H^\circ_f(CO_2) + 26\Delta H^\circ_f(H_2O)] - [37\Delta H^\circ_f(O_2) + 2\Delta H^\circ_f(C_{12}H_{26})]\\ &= [24(-393.5\text{ kJ/mol }CO_2) + 26(-285.8\text{ kJ/mol }H_2O)] - [37(0\text{ kJ/mol }O_2)\\ &\quad + 2(-350.9\text{ kJ/mol }C_{12}H_{26})]\\ &= -16{,}173.0\text{ kJ}\end{aligned}$$

According to the balanced equation for the reaction, this value is ΔH° for the combustion of 2 mol of *n*-dodecane. Hence, we must divide by 2 to obtain ΔH°_{comb} per mole of *n*-dodecane:

$$\Delta H^\circ_{comb} = -8086.5\text{ kJ/mol }C_{12}H_{26}$$

Ⓑ✓ The number of moles of dodecane in 100 gallons can be calculated as follows, using the density, molar mass, and appropriate conversion factors:

$$100\,\cancel{\text{gal}}\left(\frac{3.785\,\cancel{\text{L}}}{1\,\cancel{\text{gal}}}\right)\left(\frac{1000\,\cancel{\text{mL}}}{\cancel{\text{L}}}\right)\left(\frac{0.75\,\cancel{\text{g}}}{\cancel{\text{mL}}}\right)\left(\frac{1\text{ mol}}{170.34\,\cancel{\text{g}}}\right) = 1.7\times10^3\text{ mol }C_{12}H_{26}$$

Ⓒ✓ The total energy released is

$$\Delta H^\circ_{comb} = (-8086.5\text{ kJ/}\cancel{\text{mol}})(1.7\times10^3\,\cancel{\text{mol}}) = -1.4\times10^7\text{ kJ}$$

From the balanced equation for the reaction, we see that each mole of dodecane forms 12 mol of CO_2 upon combustion. Hence, the amount of CO_2 emitted is

$$1.7\times10^3\,\cancel{\text{mol }C_{12}H_{26}}\left(\frac{\frac{24}{2}\,\cancel{\text{mol }CO_2}}{1\,\cancel{\text{mol }C_{12}H_{26}}}\right)\left(\frac{44.0\,\cancel{\text{g}}}{1\,\cancel{\text{mol }CO_2}}\right)\left(\frac{1\,\cancel{\text{lb}}}{454\,\cancel{\text{g}}}\right)\left(\frac{1\text{ ton}}{2000\,\cancel{\text{lb}}}\right) = 0.99\text{ ton}$$

EXERCISE 5.14

Suppose the student in Example 5.14 couldn't afford the plane fare, so she decided to drive home instead. Assume that the round-trip distance by road was 5572 miles, her fuel consumption averaged 31 miles per gallon, and her fuel was pure isooctane, C_8H_{18} (density 0.69 g/mL). How much energy was expended and how many tons of CO_2 were produced during her trip?

Answer 2.2×10^7 kJ; 1.6 tons of CO_2 (about twice as much as is released by flying)

KEY EQUATIONS

General definition of work	$w = Fd$ $w = mad$	(5.1) (5.2)
Relationship between energy, heat, and work	$\Delta E = q + w$	(5.3)
Kinetic energy	$KE = \frac{1}{2}mv^2$	(5.4)
Potential energy in a gravitational field	$PE = mgh$	(5.6)
Definition of enthalpy	$H = E + PV$	(5.11)
Pressure-volume work	$w = -P\Delta V$	(5.13)
Enthalpy change at constant pressure	$\Delta H = \Delta E + P\Delta V$ $\Delta H = q_p$	(5.14) (5.15)
Relationship between ΔH°_{rxn} and ΔH°_f	$\Delta H^\circ_{rxn} = \Sigma m\Delta H^\circ_f(\text{products}) - \Sigma n\Delta H^\circ_f(\text{reactants})$	(5.26)
Relationship of quantity of a substance, heat capacity, heat flow, and temperature change	$q = nC_p\Delta T$ $q = mC_s\Delta T$	(5.34) (5.35)
Constant-pressure calorimetry	$\Delta H_{rxn} = q_{rxn} = -q_{calorimeter} = -mC_s\Delta T$	(5.40)
Constant-volume calorimetry	$\Delta H_{comb} \approx q_{comb} = -q_{calorimeter} = -C_{bomb}\Delta T$	(5.41)

SUMMARY AND KEY TERMS

5.1 Energy and Work (p. 205)

Thermochemistry is the branch of chemistry that qualitatively and quantitatively describes the energy changes that occur during chemical reactions. **Energy** is the capacity to do work. **Mechanical work** is the amount of energy required to move an object a given distance when opposed by a force. **Thermal energy** is due to the random motions of atoms, molecules, or ions in a substance. The **temperature** of an object is a measure of the amount of thermal energy it contains. **Heat (q)** is the transfer of thermal energy from a hotter object to a cooler one. Energy can take many forms; most are different varieties of **potential energy**, energy caused by the relative position or orientation of an object. **Kinetic energy** is the energy an object possesses due to its motion. Energy can be converted from one form to another, but the **law of conservation of energy** states that energy can be neither created nor destroyed. The most common units of energy are the **joule (J)**, defined as 1 (kg · m^2)/s^2, and the **calorie**, defined as the amount of energy needed to raise the temperature of 1 g of water by 1°C (1 cal = 4.184 J).

5.2 Enthalpy (p. 209)

In chemistry, the small part of the universe that we are studying is the **system**, and the rest of the universe is the **surroundings**. **Open systems** can exchange both matter and energy with their surroundings, **closed systems** can exchange energy but not matter with their surroundings, and **isolated systems** can exchange neither matter nor energy with their surroundings. A **state function** is a property of a system that depends only on its present state, not on its history. A reaction or process in which heat is transferred from the system to the surroundings is **exothermic**. A reaction or process in which heat is transferred to the system from the surroundings is **endothermic**. **Enthalpy** is a state function used to measure the heat transferred from a system to its surroundings, or vice versa, at constant pressure. Only the **change in enthalpy, ΔH**, can be measured. A negative ΔH means that heat flows from the system to the surroundings; a positive ΔH means that heat flows into the system from the surroundings. For a chemical reaction, the **enthalpy of reaction, ΔH_{rxn}**, is the difference in enthalpy between products and reactants; the units of ΔH_{rxn} are kilojoules per mole. Reversing a chemical reaction reverses the sign of ΔH_{rxn}. The magnitude of ΔH_{rxn} also depends on the physical state of the reactants and products because processes such as melting solids or vaporizing liquids are accompanied by enthalpy changes as well: the **enthalpy of fusion, ΔH_{fus}**, and the **enthalpy of vaporization, ΔH_{vap}**, respectively. The overall enthalpy change for a series of reactions is the sum of the enthalpy changes for the individual reactions, which is **Hess's law**. The **enthalpy of combustion, ΔH_{comb}**, is the enthalpy change that occurs when a substance is burned in excess oxygen. The **enthalpy of formation, ΔH_f**, is the enthalpy change that accompanies the formation of a compound from its elements. **Standard enthalpies of formation, ΔH°_f**, are determined under **standard conditions**: a pressure of 1 atm for gases and a concentration of 1 M for species in solution, with all pure substances present in their **standard states** (their most stable forms at 1 atm pressure and the temperature of the measurement). The standard heat of formation of any element in its most stable form is defined to be zero. The **standard enthalpy of reaction, ΔH°_{rxn},** can be calculated from the sum of the standard enthalpies of formation of the products (each multiplied by its stoichiometric coefficient) minus the sum of the standard enthalpies of formation of the reactants (each multiplied by its stoichiometric coefficient)—the "products minus reactants" rule. The **enthalpy of solution, ΔH_{soln},** is the heat released or absorbed when a specified amount of a solute dissolves in a certain quantity of solvent at constant pressure.

5.3 Calorimetry (p. 225)

Calorimetry is the set of techniques used to measure enthalpy changes during chemical processes. It uses devices called *calorimeters*, which measure the change in temperature when a chemical reaction is carried out. The magnitude of the temperature change depends on the amount of heat released or absorbed and on the heat capacity of the system. The **heat capacity, (C)** of an object is the amount of energy needed to raise its temperature by 1°C; its units are J/°C. The **specific heat (C_s)** of a substance is the amount of energy needed to raise the temperature of 1 g of the substance by 1°C, and the **molar heat capacity (C_p)** is the amount of energy needed to raise the temperature of 1 mol of a substance by 1°C. Liquid water has one of the highest specific heats known. Heat flow measurements can be made using either a **constant-pressure calorimeter**, which gives ΔH values directly, or a **bomb calorimeter**, which operates at constant volume and is particularly useful for measuring enthalpies of combustion.

5.4 Thermochemistry and Nutrition (p. 231)

The nutritional **Calorie** is equivalent to 1 kcal (4.184 kJ). The caloric content of a food is its enthalpy of combustion per gram. Combustion of nitrogen-containing substances produces $N_2(g)$, but the biological oxidation of such substances produces urea. Hence, the actual energy available from nitrogen-containing substances such as proteins is less than the enthalpy of combustion by the enthalpy of combustion of urea multiplied by the number of moles of urea produced. Typical caloric contents for food are 9 Cal/g for fats and 4 Cal/g for carbohydrates and proteins.

5.5 Energy Sources and the Environment (p. 236)

More than 80% of the energy used by modern society (about 3×10^{17} kJ/yr) is provided by combustion of fossil fuels. Because of their availability, ease of transport, and facile conversion to convenient fuels, natural gas and petroleum are currently the preferred fuels. Supplies of **coal,** a complex solid material derived from plants that lived long ago, are much greater, but the difficulty in transporting and burning a solid makes it less attractive as a fuel. Coal releases the smallest amount of energy per gram upon combustion of any fossil fuel, and natural gas the greatest amount. Combustion of fossil fuels releases large amounts of CO_2 that upset the balance of the **carbon cycle** and result in a steady increase in atmospheric CO_2 levels. Because CO_2 is a **greenhouse gas,** which absorbs heat before it can be radiated from the Earth into space, CO_2 in the atmosphere can result in increased surface temperatures (the **greenhouse effect**). The temperature increases caused by increased CO_2 levels because of human activities are, however, superimposed upon much larger variations in the Earth's temperature that have produced phenomena such as the ice ages and are still poorly understood.

QUESTIONS AND PROBLEMS

MGC For instructor-assigned homework, go to **www.masteringgenchem.com**

Please be sure you are familiar with the topics discussed in Essential Skills 4 at the end of this chapter before proceeding to the Questions and Problems. Questions and Problems with colored numbers have answers in the Appendix and complete solutions in the Student Solutions Manual.

CONCEPTUAL

5.1 Energy and Work

1. What is the relationship between mechanical work and energy?
2. Does a person with a mass of 50 kg climbing a height of 15 m do work? Explain your answer. Does that same person do work while descending a mountain?
3. If a person exerts a force on an immovable object, does that person do work? Explain your answer.
4. Explain the differences between electrical energy, nuclear energy, and chemical energy.
5. The chapter describes thermal, radiant, electrical, nuclear, and chemical energy. Which form(s) are represented by each of the following: (a) sunlight; (b) the energy produced by a cathode ray tube, such as that found in a television; (c) the energy emitted from radioactivity; (d) the energy emitted from a burning candle; (e) the energy associated with a steam engine; (f) the energy emitted by a cellular phone; (g) the energy associated with a stick of dynamite
6. Describe the various forms of energy that are interconverted when a flashlight is switched on.
7. Describe the forms of energy that are interconverted when the space shuttle lifts off.
8. Categorize each as representing kinetic energy or potential energy:
   (a) The energy associated with a laptop computer sitting on a desk
   (b) Shoveling snow
   (c) Water pouring out of a fire hydrant
   (d) The energy released by an earthquake
   (e) The energy in a volcano about to erupt
   (f) The energy associated with a coiled spring
9. Are the units for potential energy the same as the units for kinetic energy? Can an absolute value for potential energy be obtained? Explain your answer.

10. Classify each as representing either potential energy or kinetic energy:
(a) Water cascading over Niagara Falls
(b) A beaker balanced on the edge of a sink
(c) The energy released during a mudslide
(d) Rollerblading
(e) The energy in a block of ice on a rooftop before a thaw

11. Why does hammering a piece of sheet metal cause the metal to heat up?

5.2 Enthalpy

12. Heat implies the flow of energy from one object to another. Describe the energy flow in (a) an exothermic reaction and (b) an endothermic reaction.

13. When a thermometer is suspended in an insulated thermos that contains a block of ice, the temperature recorded on the thermometer drops. Describe the direction of heat flow.

14. In each of these processes, the system is defined as the mixture of chemical substances that undergoes a reaction. State whether each process is endothermic or exothermic.
(a) Water is added to NaOH pellets, and the flask becomes hot.
(b) Glucose is metabolized by the body to CO_2 and H_2O.
(c) Ammonium nitrate crystals are dissolved in water, causing the solution to become cool.

15. In each scenario, the system is defined as the mixture of chemical substances that undergoes a reaction or change of state. Determine whether each process is endothermic or exothermic.
(a) Concentrated acid is added to water in a flask, and the flask becomes warm.
(b) Water evaporates from your skin, causing you to shiver.
(c) A container of ammonium nitrate detonates.

16. Is the earth's environment an isolated system, an open system, or a closed system? Explain your answer.

17. Why is it impossible to measure the absolute magnitude of the enthalpy of an object or a compound?

18. Determine whether energy is consumed or released in each scenario: (a) a leaf falls from a tree; (b) a motorboat maneuvers against a current; (c) a child jumps rope; (d) dynamite detonates; (e) a jogger sprints down a hill. Explain your reasoning.

19. The chapter states that enthalpy is an extensive property. Why? Describe a situation that illustrates this fact.

20. The enthalpy of a system is affected by the physical states of the reactants and products. Explain why.

21. Is the distance a person travels on a trip a state function? Why or why not?

22. Describe how Hess's law can be used to calculate the enthalpy change of a reaction that cannot be observed directly.

23. When you apply Hess's law, what enthalpy values do you need to account for each change in physical state?
(a) The melting of a solid
(b) The conversion of a gas to a liquid
(c) Solidification of a liquid
(d) Dissolution of a solid into water

24. The structure of coal is quite different from the structure of gasoline. How do their structural differences affect their enthalpies of combustion? Explain your answer.

25. What is the difference between ΔH_f° and ΔH_f?

26. In their elemental form, A_2 and B_2 exist as diatomic molecules. Given the following reactions, each with an associated ΔH°, describe how you would calculate ΔH_f° for the compound AB_2.

$$2AB \longrightarrow A_2 + B_2 \qquad \Delta H_1^\circ$$
$$3AB \longrightarrow AB_2 + A_2B \qquad \Delta H_2^\circ$$
$$2A_2B \longrightarrow 2A_2 + B_2 \qquad \Delta H_3^\circ$$

27. How can ΔH_f° of a compound be determined if the compound cannot be prepared by the reactions used to define its standard enthalpy of formation?

28. Write a balanced chemical equation for the reaction associated with the standard enthalpy of formation of each compound: (a) HBr; (b) CH_3OH; (c) $NaHCO_3$.

29. Describe the distinction between ΔH_{soln} and ΔH_f.

30. Does adding water to concentrated acid result in an endothermic or an exothermic process?

31. The table lists ΔH_{soln}° values for some ionic compounds. If 1 mol of each solute is dissolved in 100 mL of water, rank the resulting solutions from warmest to coldest.

Compound	ΔH_{soln}°, kJ/mol
KOH	−57.61
$LiNO_3$	−2.51
$KMnO_4$	43.56
$NaC_2H_3O_2$	−17.32

5.3 Calorimetry

32. Can an object have a negative heat capacity? Why or why not?

33. What two factors determine the heat capacity of an object? Does the specific heat also depend on these two factors? Explain your answer.

34. Explain why regions along seacoasts have a more moderate climate than inland regions do.

35. Although soapstone is more expensive than brick, soapstone is frequently the building material of choice for fireplaces, particularly in northern climates with harsh winters. Propose an explanation for this.

5.4 Thermochemistry and Nutrition

36. Can water be considered a food? Explain your answer.

37. Describe how you would determine the caloric content of a bag of popcorn using a calorimeter.

38. Why do some people initially feel cold after eating a meal and then begin to feel warm?

39. In humans, one of the biochemical products of the combustion/digestion of amino acids is urea. What effect does this have on the energy available from these reactions? Speculate why conversion to urea is preferable to generation of N_2.

5.5 Energy Sources and the Environment

40. Why is it preferable to convert coal to syngas before use rather than burning the coal as a solid fuel?

41. What is meant by the term *greenhouse gases*? List three greenhouse gases that have been implicated in global warming.

42. Name three factors that determine the rate of planetary CO_2 uptake.

NUMERICAL

This section includes "paired problems" (marked by brackets) that require similar problem-solving skills.

5.1 Energy and Work

43. Describe the mathematical relationship between (a) the thermal energy stored in an object and that object's mass, and (b) the thermal energy stored in an object and that object's temperature.

44. How much energy (in kJ) is released or stored when each of the following occurs?
(a) A 230-lb football player is lifted to a height of 4.00 ft.
(b) An 11.8-lb cat jumps from a height of 6.50 ft.
(c) A 3.75-lb book falls off of a shelf that is 5.50 ft high.

45. Calculate how much energy (in kJ) is released or stored when each of the following occurs:
(a) A 130-lb ice skater is lifted 7.50 ft off the ice.
(b) A 48-lb child jumps from a height of 4.0 ft.
(c) An 18.5-lb light fixture falls from a 10.0-ft ceiling.

46. A car weighing 1438 kg falls off a bridge that is 211 ft high. Ignoring air resistance, how much energy is released when the car hits the water?

47. A 1-ton roller coaster filled with passengers reaches a height of 28 m before accelerating downhill. How much energy is released when the roller coaster reaches the bottom of the hill? Assume no energy is lost due to friction.

48. Approximately 810 kJ of energy is needed to evaporate water from the leaves of a 9.2-m tree in one day. What mass of water is evaporated from the tree?

5.2 Enthalpy

49. Using Appendix A, calculate ΔH°_{rxn} for each chemical reaction:
(a) $2Mg(s) + O_2(g) \longrightarrow 2MgO(s)$
(b) $CaCO_3(s, \text{calcite}) \longrightarrow CaO(s) + CO_2(g)$
(c) $AgNO_3(s) + NaCl(s) \longrightarrow AgCl(s) + NaNO_3(s)$

50. Using Appendix A, determine ΔH°_{rxn} for each chemical reaction:
(a) $2Na(s) + Pb(NO_3)_2(s) \longrightarrow 2NaNO_3(s) + Pb(s)$
(b) $Na_2CO_3(s) + H_2SO_4(l) \longrightarrow Na_2SO_4(s) + CO_2(g) + H_2O(l)$
(c) $2KClO_3(s) \longrightarrow 2KCl(s) + 3O_2(g)$

51. Find ΔH°_{rxn} for each chemical equation. Notice that not all equations are balanced.
(a) $Fe(s) + CuCl_2(s) \longrightarrow FeCl_2(s) + Cu(s)$
(b) $(NH_4)_2SO_4(s) + Ca(OH)_2(s) \longrightarrow CaSO_4(s) + NH_3(g) + H_2O(l)$
(c) $Pb(s) + PbO_2(s) + H_2SO_4(l) \longrightarrow PbSO_4(s) + H_2O(l)$

52. Calculate ΔH°_{rxn} for each reaction:
(a) $4HBr(g) + O_2(g) \longrightarrow 2H_2O(l) + 2Br_2(l)$
(b) $2KBr(s) + H_2SO_4(l) \longrightarrow K_2SO_4(s) + 2HBr(g)$
(c) $4Zn(s) + 9HNO_3(l) \longrightarrow 4Zn(NO_3)_2(s) + NH_3(g) + 3H_2O(l)$

53. Use the data in Appendix A to calculate ΔH°_f of $HNO_3(l)$ if $\Delta H^\circ_{rxn} = -320.0$ kJ for the reaction $Sn(s, \text{white}) + 4HNO_3(l) \longrightarrow SnO_2(s) + 4NO_2(g) + 2H_2O(l)$.

54. Use the data in Appendix A to calculate ΔH°_f of P_4O_{10} if $\Delta H^\circ_{rxn} = -362.1$ kJ for the reaction $P_4O_{10}(s) + 6H_2O(l) \longrightarrow 4H_3PO_4(l)$.

55. How much heat is released or required in the reaction of 0.50 mol of HBr(*g*) with 1.0 mol of chlorine gas to produce bromine gas?

56. Is the decomposition of gaseous dinitrogen pentoxide to gaseous nitrogen dioxide and oxygen exothermic or endothermic? How much energy is released or consumed if 10.0 g of N_2O_5 is completely decomposed as described?

57. In the mid-1700s, a method was devised for preparing chlorine gas using the reaction

$$NaCl(s) + H_2SO_4(l) + MnO_2(s) \longrightarrow Na_2SO_4(s) + MnCl_2(s) + H_2O(l) + Cl_2(g)$$

Calculate ΔH°_{rxn} for this reaction. Is the reaction exothermic or endothermic?

58. Would you expect heat to be evolved during the reaction of (a) solid sodium oxide with gaseous sulfur dioxide to give solid sodium sulfite; (b) solid aluminum chloride reacting with water to give solid aluminum oxide and hydrogen chloride gas?

59. How much heat is released in the preparation of an aqueous solution containing 6.3 g of calcium chloride and one containing 2.9 g of potassium carbonate, and then when the two solutions are mixed together to produce potassium chloride and calcium carbonate?

60. Methanol is used as a fuel in Indianapolis 500 race cars. Use the table to determine whether methanol releases more energy per liter upon combustion than 2,2,4-trimethylpentane (isooctane).

Fuel	ΔH°_{comb}, kJ/mol	Density, g/mL
Methanol	−726.1	0.791
2,2,4-Trimethylpentane	−5461.4	0.692

61. (a) Use the enthalpies of combustion given in the table to determine which organic compound releases the greatest amount of energy per gram upon combustion.

Fuel	ΔH°_{comb}, kJ/mol
Methanol	−726.1
1-Ethyl-2-methylbenzene	−5210.2
n-Octane	−5470.5

(b) Calculate the standard enthalpy of formation of 1-ethyl-2-methylbenzene.

62. Given the enthalpies of combustion which organic compound is the best fuel per gram?

Fuel	ΔH_f°, kJ/mol
Ethanol	−1366.8
Benzene	−3267.6
Cyclooctane	−5434.7

63. A 60-watt lightbulb is burned for 6 hours. If we assume an efficiency of 38% in the conversion of energy from oil to electricity, how much oil must be consumed to supply the electrical energy needed to light the bulb? (1 watt = 1 J/s)

64. How many liters of cyclohexane need to be burned to release as much energy as burning 10.0 lb of pine logs? The density of cyclohexane is 0.7785 g/mL, and its $\Delta H_{comb} = -46.6$ kJ/g.

5.3 Calorimetry

65. Using Equations 5.34 and 5.35, derive a mathematical relationship betwen C_s and C_p.

66. Complete the table for 28.0 g of each element at an initial temperature of 22.0°C.

Element	*q*, J	C_p, J/(mol·K)	Final *T*, °C
Nickel	137	26.07	____
Silicon	____	19.789	3.0
Zinc	603	____	77.5
Mercury	137	____	57

67. Using Table 5.3, how much heat is needed to raise the temperature of a 2.5-g piece of copper wire from 20°C to 80°C? How much heat is needed to increase the temperature of an equivalent mass of aluminum by the same amount? If you were using one of these metals to channel heat away from electrical components, which metal would you use? Once heated, which metal will cool faster? Give the specific heat for each metal.

68. Gold has a molar heat capacity of 25.418 J/(mol · K), and silver has a molar heat capacity of 23.350 J/(mol · K).
(a) If you put silver and gold spoons of equal mass into a cup of hot liquid and wait until the temperature of the liquid is constant, which spoon will take longer to cool down when removed from the hot liquid?
(b) If 8.00-g spoons of each metal at 20.0°C are placed in an insulated mug with 50.0 g of water at 97.0°C, what will be the final temperature of the water after the system has equilibrated? (Assume that no heat is transferred to the surroundings.)

69. In an exothermic reaction, how much heat would need to be evolved in order to raise the temperature of 150 mL of water 7.5°C? Explain how this process illustrates the law of conservation of energy.

70. How much heat must be evolved by a reaction to raise the temperature of 8.0 fluid oz of water 5.0°C? What mass of lithium iodide would be needed to be dissolved in this volume of water to produce this temperature change?

71. A solution is made by dissolving 3.35 g of an unknown salt in 150 mL of water, and the temperature of the water rises 3.0°C. Addition of a silver nitrate solution results in a precipitate. Assuming that the heat capacity of the solution is the same as that of pure water, use the information in Table 5.2 and solubility rules to identify the salt.

72. Using the data in Table 5.8, calculate the change in temperature of a calorimeter with a heat capacity of 1.78 kJ/°C when 3.0 g of charcoal is burned in the calorimeter. If the calorimeter is in a 2-L bath of water at an initial temperature of 21.5°C, what will be the final temperature of the water after the combustion reaction (assuming no heat is lost to the surroundings)?

73. A 3.00-g sample of TNT ($C_7H_5N_3O_6$) is placed in a bomb calorimeter with a heat capacity of 1.93 kJ/°C; the enthalpy of combustion of TNT is -3403.5 kJ/mol. If the initial temperature of the calorimeter is 19.8°C, what will be the final temperature of the calorimeter after the combustion reaction (assuming no heat is lost to the surroundings)? What is the enthalpy of formation of TNT?

5.4 Thermochemistry and Nutrition

74. Determine the amount of energy available from the biological oxidation of 1.50 g of the amino acid leucine ($\Delta H_{comb} = -3581.7$ kJ/mol).

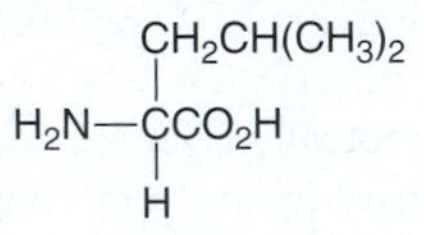

Leucine

75. Calculate the energy released (in kJ) from the metabolism of 1.5 fluid oz of vodka that is 62% water and 38% ethanol by volume, assuming that the total volume is equal to the sum of the volume of the two components. The density of ethanol is 0.824 g/mL. What is this enthalpy change in nutritional Calories?

76. While exercising, a person lifts an 80-lb barbell 7 ft off the ground. Assuming that the transformation of chemical energy to mechanical energy is only 35% efficient, how many Calories would the person utilize to accomplish this task? From Figure 5.11, how many grams of glucose would be needed to provide the energy to accomplish this task?

77. A 30-g sample of potato chips is placed in a bomb calorimeter with a heat capacity of 1.80 kJ/°C, and the bomb calorimeter is immersed in 1.5 L of water. Calculate the energy contained in the food per gram if, upon combustion of the chips, the temperature of the calorimeter increases to 58.6°C from an initial temperature of 22.1°C.

5.5 Energy Sources and the Environment

78. One of the side reactions that occurs during the burning of fossil fuels is

$$4FeS_2(s) + 11O_2(g) \longrightarrow 2Fe_2O_3(s) + 8SO_2(g)$$

(a) How many kilojoules of energy are released during the combustion of 10 lb of FeS_2?
(b) How many pounds of SO_2 are released into the atmosphere?
(c) Discuss the potential environmental impacts of this combustion reaction.

79. How many kilograms of CO_2 are released during the combustion of 16 gallons of gasoline? Assume gasoline is pure isooctane with a density of 0.6919 g/mL. If this combustion was used to heat 4.5×10^3 L of water from an initial temperature of 11.0°C, what would be the final temperature of the water assuming 42% efficiency in the energy transfer?

APPLICATIONS

80. Palm trees grow on the coast of southern England even though the latitude is the same as that of Winnipeg, Canada. What is a plausible explanation for this phenomenon? (Hint: the Gulf Stream current is a factor.)

81. During intense exercise, your body cannot provide enough oxygen to allow the complete combustion of glucose to CO_2. Under these conditions, an alternative means of obtaining energy from glucose is used in which glucose is converted to lactic acid. The equation for this reaction is

$$\underset{\text{Glucose}}{C_6H_{12}O_6} \longrightarrow \underset{\text{Lactic acid}}{2C_3H_5O_3H}$$

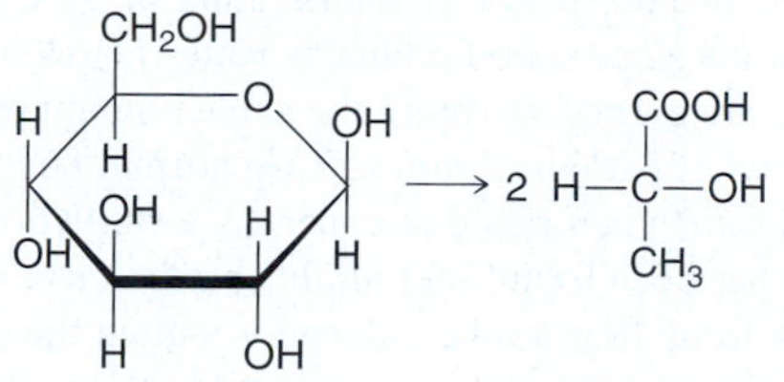

Glucose Lactic acid

(a) Calculate the energy yield for this reaction per mole of glucose.
(b) How does this energy yield compare with that obtained per mole of glucose for the combustion reaction?
(c) Muscles become sore after intense exercise. Propose a chemical explanation for this.

82. During the late spring, icebergs in the North Atlantic pose a hazard to shipping. To avoid them, ships travel routes that are about 30% longer. Many attempts have been made to destroy icebergs, including using explosives, torpedoes, and bombs. How much heat must be generated to melt 15% of a 1.9×10^8-kg iceberg? How many kilograms of TNT ($C_7H_5N_3O_6$) would be needed to provide enough energy to melt the ice? (The enthalpy change for explosive decomposition of TNT is -1035.8 kJ/mol.)

83. Many biochemical processes occur through sequences of reactions called *pathways*. The total energy released by many of these pathways is much more than the energy a cell could handle if it were all released in a single step. For example, the combustion of glucose in a single step would release enough energy to kill a cell. By using a series of smaller steps that release less energy per reaction, however, the cell can extract the maximum energy from glucose without being destroyed. Referring to Equation 5.29, calculate how many grams of glucose would need to be metabolized to raise the temperature of a liver cell from an average body temperature of 37°C to 100°C, if the cell has a volume of 5000 cubic micrometers. Although the cell is only 69% water, assume that the density of the cell is 1.00 and that its C_s is the same as that of water.

84. During smelting, naturally occurring metal oxides are reduced by carbon at high temperature. For copper(II) oxide, this process includes a series of equations:

$$CuO(s) + C(s) \longrightarrow Cu(l) + CO(g)$$
$$2C(s) + O_2(g) \longrightarrow 2CO(g)$$
$$CuO(l) + CO(g) \longrightarrow Cu(s) + CO_2(g)$$

The final products are CO_2 and Cu. The discovery of this process led to the increasing use of ores as sources of metals in ancient cultures. In fact, between 3000 and 2000 B.C., the smelting of Cu was well established, and beads made from copper are some of the earliest known metal artifacts.
(a) Write a balanced equation for the overall reaction of CuO, C, and O_2 to give *only* CO_2 and Cu.
(b) Using Hess's law and the data in Appendix A, calculate ΔH_{rxn} for the smelting of CuO ore to give Cu and CO_2.
(c) Assuming complete reaction, how much heat was released if 23 g of Cu metal was produced from its ore?

85. The earliest known Egyptian artifacts made from tin metal date back to approximately 1400 B.C. If the smelting process for SnO_2 occurs via the reaction sequence

$$SnO_2(s) + 2C(s) \longrightarrow Sn(s) + 2CO(g)$$
$$SnO_2(s) + 2CO(g) \longrightarrow Sn(s) + 2CO_2(g)$$

what is ΔH_{rxn} for the conversion of SnO_2 to Sn(white) by this smelting process? How much heat was released or required if 28 g of Sn metal was produced from its ore, assuming complete reaction?

86. An average American consumes approximately 10^6 kJ of energy per day. The average life expectancy of an American is 71 years.
(a) How much coal would need to be burned to provide enough energy to meet a person's energy demands if the efficiency of energy production from coal is 38%?
(b) If the coal contains 0.6% by mass FeS_2, how many kilograms of sulfuric acid are produced during the time in part (a)?

87. Several theories propose that life on earth evolved in the absence of oxygen. One theory is that primitive organisms used fermentation processes, in which sugars are decomposed in an oxygen-free environment, to obtain energy. Many kinds of fermentation processes are possible, including the conversion of glucose (a) to lactic acid, (b) to CO_2 and ethanol, and (c) to ethanol and acetic acid:
(a) $C_6H_{12}O_6(s) \longrightarrow 2C_3H_6O_3(s)$
(b) $C_6H_{12}O_6(s) \longrightarrow 2C_2H_5OH(l) + 2CO_2(g)$
(c) $C_6H_{12}O_6(s) + H_2O(l) \longrightarrow$
$C_2H_5OH(l) + CH_3CO_2H(l) + 2H_2(g) + 2CO_2(g)$
The first reaction occurs in rapidly exercising muscle cells, the second in yeast, and the third in intestinal bacteria. Using Appendix A, calculate which reaction gives the greatest energy yield (most negative ΔH°_{rxn}) per mole of glucose.

88. A 70-kg person expends 85 Cal/h watching television. If the person eats 8 cups of popcorn that contains 55 Cal per cup, how many kilojoules of energy from the popcorn will have been burned during a 2-h movie? After the movie, the person goes outside to play tennis and burns approximately 500 Cal/h. How long will that person have to play tennis to work off all of the residual energy from the popcorn?

89. Photosynthesis in higher plants is a complex process in which glucose is synthesized from atmospheric CO_2 and water in a sequence of reactions that uses light as an energy source. The overall reaction is

$$6CO_2(g) + 6H_2O(l) \xrightarrow{\text{light}} C_6H_{12}O_6(s) + 6O_2(g)$$

The glucose may then be used to produce the complex carbohydrates, such as cellulose, that make up the plant tissues.
(a) Is the reaction shown endothermic or exothermic?
(b) How many grams of glucose are produced per kilogram of CO_2?
(c) A 2.5-lb sweet potato is approximately 73% water by mass. If the remaining mass is made up of carbohydrates derived from glucose, how much CO_2 was needed to grow this sweet potato?
(d) How many kilojoules of energy are stored in the potato?
(e) Which releases more energy: digestion of the potato or combustion of the potato?

90. Adipose (fat) tissue consists of cellular protoplasm, which is mostly water, and fat globules. Nearly all of the energy stored in adipose tissue comes from the chemical energy of its fat globules, totaling approximately 3500 Calories per pound of tissue. How many kilojoules of energy are stored in 10 g of adipose tissue? How many 50 g brownies would you need to consume to generate 10 lb of fat? (Refer to Table 5.5 for the necessary caloric data.)

91. If a moderate running pace of 5 min/km expends energy at a rate of about 400 kJ/km, how many 8-oz apples would a person have to eat in order to have enough energy to run 5 mi? How many 8-oz hamburgers? (Refer to Table 5.5 for the necessary caloric data.)

92. Proteins contain approximately 4 Cal/g, carbohydrates approximately 4 Cal/g, and fat approximately 9 Cal/g. How many kilojoules of energy are available from the consumption of one serving (8 oz) of each food in the table? (Data are shown per serving.)

Food	Protein, g	Fat, g	Carbohydrates, g
Sour cream	7	48	10
Banana	2	1	35
Cheeseburger	60	31	40
Green peas	8	1	21

93. When you eat a bowl of cereal with 500 g of milk, how many Calories must your body burn to warm the milk from 4°C to a normal body temperature of 37°C? (Assume milk has the same specific heat as water.) How many Calories are burned warming the same amount of milk in a 32°C bowl of oatmeal from 32°C to normal body temperature? In some countries that experience starvation conditions, it has been found that infants don't starve even though the milk from their mothers doesn't contain the number of Calories thought necessary to sustain them. Propose an explanation for this.

94. If a person's fever is caused by an increase in the temperature of water inside the body, how much additional energy is needed if a 70-kg person with a normal body temperature of 37°C runs a temperature of 39.5°C? (A person's body is approximately 79% water.) The old adage "feed a fever" may contain some truth in this case. How many hamburgers would the person need to consume to cause this change? (Refer to Table 5.5 for the necessary caloric data.)

ESSENTIAL SKILLS

4

TOPICS

- Temperature
- Unit Conversions: Dimensional Analysis

The previous Essential Skills sections introduced some fundamental operations that you need to successfully manipulate mathematical equations in chemistry. This section describes how to convert between temperature scales and further develops the topic of unit conversions begun in Essential Skills 2.

Temperature

The concept of temperature may seem familiar to you, but many people confuse temperature with heat. Temperature is a measure of how hot or cold an object is relative to another object (its thermal energy content), whereas heat is the flow of thermal energy between objects with different temperatures.

Three different scales are commonly used to measure temperature: Fahrenheit (expressed as °F), Celsius (°C), and Kelvin (K). Thermometers measure temperature by using materials that expand or contract when heated or cooled. Mercury or alcohol thermometers, for example, have a reservoir of liquid that expands when heated and contracts when cooled, so the liquid column lengthens or shortens as the temperature of the liquid changes.

The Fahrenheit Scale

The Fahrenheit temperature scale was developed in 1717 by the German physicist Gabriel Fahrenheit, who designated the temperature of a bath of ice melting in a solution of salt as the zero point on his scale. Such a solution was commonly used in the 18th century to carry out low-temperature reactions in the laboratory. The scale was measured in increments of 12; its upper end, designated as 96 degrees, was based on the armpit temperature of a healthy person—in this case, Fahrenheit's wife. Later, the number of increments shown on a thermometer increased as measurements became more precise. The upper point is based on the boiling point of water, designated as 212 degrees in order to maintain the original magnitude of a Fahrenheit degree, whereas the melting point of ice is designated as 32 degrees.

The Celsius (Centigrade) Scale

The Celsius scale was developed in 1742 by the Swedish astronomer Anders Celsius. It is based on the melting and boiling points of water under normal atmospheric conditions. The current scale is an inverted form of the original scale, which was divided into 100 increments. Because of these 100 divisions, the Celsius scale is also called the centigrade scale.

The Kelvin Scale

Lord Kelvin, working in Scotland, developed the Kelvin scale in 1848. His scale uses molecular energy to define the extremes of hot and cold. Absolute zero, or 0 K, corresponds to the point at which molecular energy is at a minimum. The Kelvin scale is preferred in scientific work, although the Celsius scale is also commonly used. Temperatures measured on the Kelvin scale are reported simply as K, not °K. Figure ES4.1 compares the three scales.

Converting Between Scales

The kelvin is the same size as the Celsius degree, so measurements are easily converted from one to the other. The freezing point of water is 0°C = 273.15 K; the boiling point of water is 100°C = 373.15 K. Thus, the Kelvin and Celsius scales are related as follows:

$$T\text{ (in °C)} + 273.15 = T\text{ (in K)}$$
$$T\text{ (in K)} - 273.15 = T\text{ (in °C)}$$

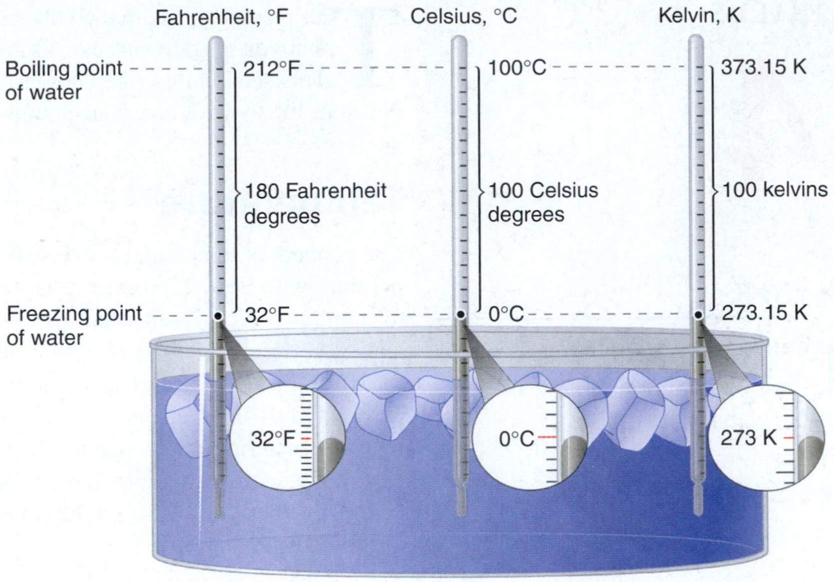

Figure ES4.1 A comparison of the Fahrenheit, Celsius, and Kelvin temperature scales. Because the difference between the freezing point and the boiling point of water is 100 degrees on both the Celsius and Kelvin scales, the size of a degree Celsius (°C) and a kelvin (K) are precisely the same. In contrast, both a degree Celsius and a kelvin are 9/5 the size of a degree Fahrenheit (°F).

The degrees on the Fahrenheit scale, however, are based on an English tradition of using 12 divisions, just as 1 foot = 12 inches. The relationship between degrees Fahrenheit and degrees Celsius is

$$T\,(\text{in °F}) = \frac{9\text{°F}}{5\text{°}\cancel{\text{C}}}\,[T\,(\text{°}\cancel{\text{C}})] + 32\text{°F}$$

$$T\,(\text{in °C}) = \frac{5\text{°C}}{9\text{°}\cancel{\text{F}}}\,[T\,(\text{°}\cancel{\text{F}}) - 32\text{°F}]$$

where the coefficient for degrees Fahrenheit is exact. (Most calculators have a function that allows you to convert directly between °F and °C.) There is only one temperature for which the numerical value is the same on both the Fahrenheit and Celsius scales: −40°C = −40°F.

SKILL BUILDER ES4.1

(a) Convert the temperature of the surface of the sun (5800 K) and the boiling points of gold (3080 K) and liquid nitrogen (77.36 K) to °C and °F. **(b)** A student is ill with a temperature of 103.5°F. What is her temperature in °C and K?

Solution

(a) Sun: $5800\text{ K} = (5800 - 273.15)\text{°C} = 5527\text{°C}$

$$\text{°F} = \frac{9\text{°F}}{5\text{°}\cancel{\text{C}}}(5527\text{°}\cancel{\text{C}}) + 32\text{°F} = 9981\text{°F}$$

Gold: $3080\text{ K} = (3080 - 273.15)\text{°C} = 2807\text{°C}$

$$\text{°F} = \frac{9\text{°F}}{5\text{°}\cancel{\text{C}}}(2807\text{°}\cancel{\text{C}}) + 32\text{°F} = 5085\text{°F}$$

N_2: $77.36\text{ K} = (77.36 - 273.15)\text{°C} = -195.79\text{°C}$

$$\text{°F} = \frac{9\text{°F}}{5\text{°}\cancel{\text{C}}}(-195.79\text{°}\cancel{\text{C}}) + 32\text{°F} = -320\text{°F}$$

(b) $$\text{°C} = \frac{5\text{°C}}{9\text{°}\cancel{\text{F}}}(103.5\text{°}\cancel{\text{F}} - 32\text{°}\cancel{\text{F}}) = 40\text{°C}$$

$K = 40\text{°C} + 273.15 = 313\text{ K}$

Unit Conversions: Dimensional Analysis

In Essential Skills 2 you learned a convenient way of converting between units of measure, such as from grams to kilograms or seconds to hours. The use of units in a calculation to ensure that we obtain the final proper units is called dimensional analysis. For example, if we observe experimentally that an object's potential energy is related to its mass, its height from the ground, and to a gravitational force, then when multiplied, the units of mass, height, and the force of gravity must give us units corresponding to those of energy.

Energy is typically measured in joules, calories, or electron volts (abbreviated eV), defined by the following expressions:

$$1\text{ J} = 1(\text{kg}\cdot\text{m}^2)/\text{s}^2 = 1\text{ coulomb}\cdot\text{volt}$$
$$1\text{ cal} = 4.184\text{ J}$$
$$1\text{ eV} = 1.602\times 10^{-19}\text{ J}$$

To illustrate the use of dimensional analysis to solve energy problems, let us calculate the kinetic energy in joules of a 320-g object traveling at 123 cm/s. To obtain an answer in joules, which have dimensions of $(\text{kg}\cdot\text{m}^2)/\text{s}^2$, we must convert grams to kilograms and centimeters to meters. Using Equation 5.4, the calculation may be set up as follows:

$$\text{KE} = \frac{1}{2}mv^2 = \frac{1}{2}(\text{g})\left(\frac{\text{kg}}{\text{g}}\right)\left[\left(\frac{\text{cm}}{\text{s}}\right)\left(\frac{\text{m}}{\text{cm}}\right)\right]^2 = (\cancel{\text{g}})\left(\frac{\text{kg}}{\cancel{\text{g}}}\right)\left(\frac{\cancel{\text{cm}^2}}{\text{s}^2}\right)\left(\frac{\text{m}^2}{\cancel{\text{cm}^2}}\right) = \frac{\text{kg}\cdot\text{m}^2}{\text{s}^2}$$
$$= \frac{1}{2}320\ \cancel{\text{g}}\left(\frac{1\text{ kg}}{1000\ \cancel{\text{g}}}\right)\left[\left(\frac{123\ \cancel{\text{cm}}}{1\text{ s}}\right)\left(\frac{1\text{ m}}{100\ \cancel{\text{cm}}}\right)\right]^2 = \frac{0.320\text{ kg}}{2}\left[\frac{123\text{ m}}{\text{s}(100)}\right]^2$$
$$= \frac{1}{2}0.320\text{ kg}\left[\frac{(123)^2\text{m}^2}{\text{s}^2(100)^2}\right] = 0.242(\text{kg}\cdot\text{m}^2)/\text{s}^2 = 0.242\text{ J}$$

Alternatively, the conversions may be carried out in a stepwise manner:

$$320\ \cancel{\text{g}}\left(\frac{1\text{ kg}}{1000\ \cancel{\text{g}}}\right) = 0.320\text{ kg}$$
$$123\ \cancel{\text{cm}}\left(\frac{1\text{ m}}{100\ \cancel{\text{cm}}}\right) = 1.23\text{ m}$$
$$= \frac{1}{2}0.320\text{ kg}\left(\frac{1.23\text{ m}}{\text{s}}\right)^2 = \frac{1}{2}0.320\text{ kg}\left(\frac{1.513\text{ m}^2}{\text{s}^2}\right) = 0.242(\text{kg}\cdot\text{m}^2)/\text{s}^2 = 0.242\text{ J}$$

Notice, however, that this second method involves an additional step.

Now suppose you wish to report the number of kilocalories of energy contained in a 7.00-ounce piece of chocolate in units of kilojoules per gram. To obtain an answer in kilojoules, we must convert 7.00 ounces to grams, and kilocalories to kilojoules. Food reported to contain a value in Calories actually contains that same value in kilocalories. Thus, if the chocolate wrapper lists the caloric content as 120 Calories, the chocolate contains 120 kcal of energy. If we choose to use multiple steps to obtain our answer, we can begin with the conversion of kilocalories to kilojoules:

$$120\ \cancel{\text{kcal}}\left(\frac{1000\ \cancel{\text{cal}}}{\cancel{\text{kcal}}}\right)\left(\frac{4.184\ \cancel{\text{J}}}{1\ \cancel{\text{cal}}}\right)\left(\frac{1\text{ kJ}}{1000\ \cancel{\text{J}}}\right) = 502\text{ kJ}$$

We next convert the 7.00 ounces of chocolate to grams:

$$7.00\ \cancel{\text{oz}}\left(\frac{28.35\text{ g}}{1\ \cancel{\text{oz}}}\right) = 199\text{ g}$$

The number of kilojoules per gram is therefore

$$\frac{502\text{ kJ}}{199\text{ g}} = 2.52\text{ kJ/g}$$

Alternatively, we could solve the problem in one step:

$$\left(\frac{120\ \cancel{\text{kcal}}}{7.00\ \cancel{\text{oz}}}\right)\left(\frac{1000\ \cancel{\text{cal}}}{\cancel{\text{kcal}}}\right)\left(\frac{4.184\ \cancel{\text{J}}}{1\ \cancel{\text{cal}}}\right)\left(\frac{1\ \text{kJ}}{1000\ \cancel{\text{J}}}\right)\left(\frac{1\ \cancel{\text{oz}}}{28.35\ \text{g}}\right) = 2.53\ \text{kJ/g}$$

The discrepancy between the two answers is attributable to rounding to the correct number of significant figures for each step when carrying out the calculation in a stepwise manner. Recall that all digits in the calculator should be carried forward when carrying out a calculation using multiple steps. In this problem, notice that we first converted kcalories to kilojoules and then converted ounces to grams. The following exercises allow you to practice making multiple conversions between units in a single step.

SKILL BUILDER ES4.2

(a) Write a single equation to show how to convert cm/min to km/h; cal/oz to J/g; lb/in.^{2} to kg/m^2; and °C/s to K/h. **(b)** How many Calories are contained in an 8.0-oz serving of green beans if their fuel value is 1.5 kJ/g? **(c)** Gasoline has a fuel value of 48 kJ/g. How much energy in joules can be obtained by filling an automobile's 16.3-gallon tank with gasoline, assuming gasoline has a density of 0.70 g/mL?

Solution

(a)

$$\left(\frac{\cancel{\text{cm}}}{\cancel{\text{min}}}\right)\left(\frac{1\ \cancel{\text{m}}}{100\ \cancel{\text{cm}}}\right)\left(\frac{1\ \text{km}}{1000\ \cancel{\text{m}}}\right)\left(\frac{60\ \cancel{\text{min}}}{\text{h}}\right) = \text{km/h}$$

$$\left(\frac{\cancel{\text{cal}}}{\cancel{\text{oz}}}\right)\left(\frac{4.184\ \text{J}}{\cancel{\text{cal}}}\right)\left(\frac{16\ \cancel{\text{oz}}}{\cancel{\text{lb}}}\right)\left(\frac{\cancel{\text{lb}}}{453.59\ \text{g}}\right) = \text{J/g}$$

$$\left(\frac{\cancel{\text{lb}}}{\cancel{\text{in.}^2}}\right)\left(\frac{16\ \cancel{\text{oz}}}{\cancel{\text{lb}}}\right)\left(\frac{28.35\ \cancel{\text{g}}}{\cancel{\text{oz}}}\right)\left(\frac{\text{kg}}{1000\ \cancel{\text{g}}}\right)\left[\frac{(36\ \cancel{\text{in.}})^2}{(1\ \cancel{\text{yd}})^2}\right]\left[\frac{(1.09\ \cancel{\text{yd}})^2}{(1\ \text{m})^2}\right] = \text{kg/m}^2$$

$$\left(\frac{°\text{C}}{\cancel{\text{s}}}\right)\left(\frac{60\ \cancel{\text{s}}}{1\ \cancel{\text{min}}}\right)\left(\frac{60\ \cancel{\text{min}}}{\text{h}}\right) + 273.15 = \text{K/h}$$

(b) Our goal is to convert 1.5 kJ/g to Cal in 8 oz:

$$\left(\frac{1.5\ \cancel{\text{kJ}}}{\cancel{\text{g}}}\right)\left(\frac{1000\ \cancel{\text{J}}}{\cancel{\text{kJ}}}\right)\left(\frac{1\ \cancel{\text{cal}}}{4.184\ \cancel{\text{J}}}\right)\left(\frac{\text{Cal}}{1000\ \cancel{\text{cal}}}\right)\left(\frac{28.35\ \cancel{\text{g}}}{1\ \cancel{\text{oz}}}\right)(8.0\ \cancel{\text{oz}}) = 81\ \text{Cal}$$

(c) Our goal is to use the energy content, 48 kJ/g, and the density, 0.70 g/mL, to obtain the number of joules in 16.3 gal of gasoline:

$$\left(\frac{48\ \cancel{\text{kJ}}}{\cancel{\text{g}}}\right)\left(\frac{1000\ \text{J}}{\cancel{\text{kJ}}}\right)\left(\frac{0.70\ \cancel{\text{g}}}{\cancel{\text{mL}}}\right)\left(\frac{1000\ \cancel{\text{mL}}}{\cancel{\text{L}}}\right)\left(\frac{3.79\ \cancel{\text{L}}}{\cancel{\text{gal}}}\right)(16.3\ \cancel{\text{gal}}) = 2.1 \times 10^9\ \text{J}$$

6 The Structure of Atoms

LEARNING OBJECTIVES

- To know the characteristics of electromagnetic energy
- To understand how energy is quantized
- To know the relationship between atomic spectra and the electronic structure of atoms
- To understand the wave–particle duality of matter
- To be able to apply the results of quantum mechanics to chemistry
- To be able to write the electron configuration of any element and to relate its electron configuration to its position in the periodic table

CHAPTER OUTLINE

Part I introduced a wide variety of chemical substances and some of the most fundamental concepts in chemistry, providing you with general descriptions of chemical bonding, mass relationships in chemical equations, and the energy changes associated with chemical reactions. You learned that the atoms of each element contain a unique number of positively charged protons in the nucleus, that a neutral atom has the same number of electrons as protons, that the protons and neutrons constitute most of the mass of the atom, and that the electrons occupy most of the volume of an atom. These facts do not, however, explain the stoichiometries and the structures of chemical compounds; a deeper understanding of the electronic structure of atoms is required.

Helium, neon, argon, krypton, and xenon, all monatomic elements, are five of the six known noble gases. When atoms of these gases are excited by the flow of electrons in the discharge tubes, they emit visible light photons of their characteristic spectral color. All noble gases have the maximum number of electrons possible in their outer shell (2 for helium, 8 for all others), so they do not form chemical compounds easily.

In this chapter, we describe how electrons are arranged in atoms and how the spatial arrangements of electrons are related to their energies. We also explain how knowing the arrangement of electrons in an atom enables chemists to predict and explain the chemistry of an element and how the shape of the periodic table reflects the electronic arrangements of elements. In this and subsequent chapters, we build on this information to explain why certain chemical changes occur and others do not.

After reading this chapter, you will know enough about the theory of the electronic structure of atoms to explain what causes the characteristic colors of neon signs, how laser beams are created, and why gemstones and fireworks have such brilliant colors. In later chapters, we will develop the concepts introduced here to explain why the only compound formed by sodium and chlorine is NaCl, an ionic compound, whereas neon and argon form no stable compounds at all, and why carbon and hydrogen combine to form an almost endless array of covalent compounds, including CH_4, C_2H_2, C_2H_4, and C_2H_6. You will discover that knowing how to use the periodic table is the single most important skill you can acquire to understand the incredible chemical diversity of the elements.

(a) Circular waves

(b) Linear waves

Figure 6.1 Waves in water. (a) When a drop of water falls onto a smooth water surface, it generates a set of waves that travel outward in a circular direction. (b) Wind at sea generates linear waves that travel long distances and can reach heights of more than 10 m.

6.1 • Waves and Electromagnetic Radiation

Scientists discovered much of what we know about the structure of the atom by observing the interaction of atoms with various forms of radiant, or transmitted, energy, such as the energy associated with the visible light we detect with our eyes, the infrared radiation we feel as heat, the ultraviolet light that causes sunburn, and the X rays that produce images of our teeth or bones. All these forms of radiant energy are probably familiar to you. We begin our discussion of the development of our current atomic model with a description of the properties of waves and the various forms of electromagnetic radiation.

Properties of Waves

A **wave** is a periodic oscillation that transmits energy through space. Anyone who has visited a beach or dropped a stone into a puddle has observed waves traveling through water (Figure 6.1). These waves are produced when wind or a stone or some other disturbance, such as a passing boat, transfers energy to the water, causing the surface to oscillate up and down as the energy travels outward from its point of origin. As a wave passes a particular point on the surface of the water, anything floating there moves up and down.

Waves have characteristic properties (Figure 6.2). As you may have noticed in Figure 6.1, waves are **periodic**; that is, they repeat regularly in both space and time.

Figure 6.2 Important properties of waves. (a) Wavelength (λ), frequency (v, labeled in Hz), and amplitude are indicated on this drawing of a wave. (b) The wave with the shortest wavelength (smallest λ) has the greatest number of wavelengths per unit time (highest frequency, v). If two waves have the same frequency and speed, the one with the greater amplitude has the higher energy.

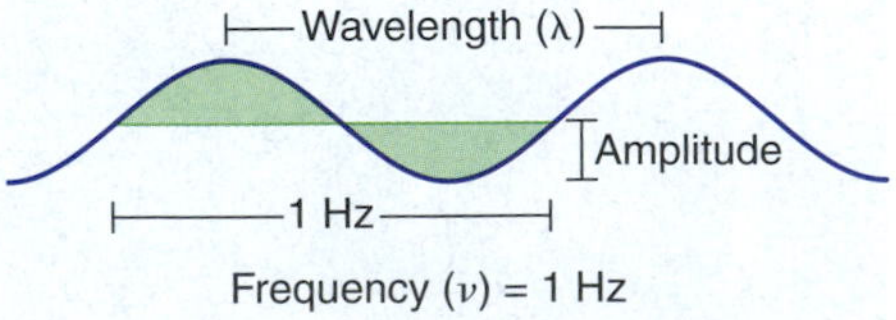

(a)

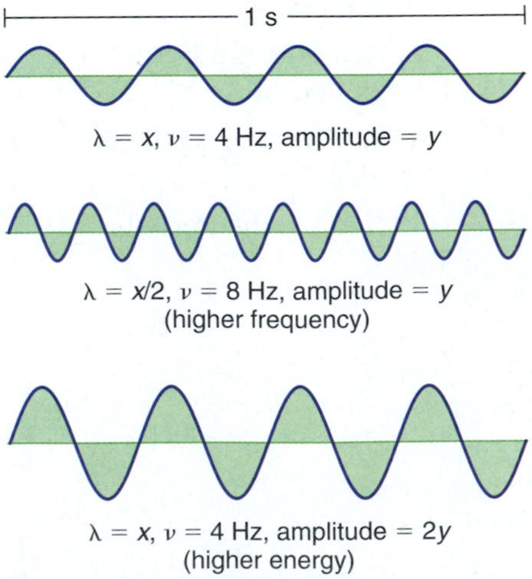

(b)

The distance between two corresponding points in a wave—between the midpoints of two peaks, for example, or two troughs—is the **wavelength**, symbolized by λ (lowercase Greek lambda). Wavelengths are described by any appropriate unit of distance, such as meters. The **frequency** of a wave, which is the number of oscillations that pass a particular point in a given period of time, is represented by ν (lowercase Greek nu). The usual units are oscillations per second or $1/s = s^{-1}$, which in the SI system is called the hertz (Hz).* The **amplitude**, or vertical height, of a wave is defined as half the peak-to-trough height; as the amplitude of a wave with a given frequency increases, so does its energy. As you can see in Figure 6.2, two waves can have the same amplitude but different wavelengths, and vice versa. The distance traveled by a wave per unit time is its **speed** (v), which is typically measured in meters per second (m/s). The speed of a wave is equal to the product of its wavelength and frequency:

MGC Properties of Waves

$$\text{(wavelength)(frequency)} = \text{speed}$$
$$\lambda\nu = v \tag{6.1}$$
$$\left(\frac{\text{meters}}{\cancel{\text{wave}}}\right)\left(\frac{\cancel{\text{waves}}}{\text{second}}\right) = \frac{\text{meters}}{\text{second}}$$

Water waves are slow compared to sound waves, which can travel through solids, liquids, and gases. Whereas water waves may travel a few meters per second, the speed of sound in dry air at 20°C is 343.5 m/s. Ultrasonic waves, which travel at a higher speed yet (>1500 m/s) and have a greater frequency, are used in such diverse applications as locating underwater objects and the medical imaging of internal organs.

Electromagnetic Radiation

Water waves transmit energy through space by the periodic oscillation of matter (the water). In contrast, energy that is transmitted, or radiated, through space in the form of periodic oscillations of an electric and a magnetic field is known as **electromagnetic radiation** (Figure 6.3). Some forms of electromagnetic radiation are shown in Figure 6.4. In a vacuum, all forms of electromagnetic radiation—whether microwaves, visible light, or gamma rays—travel at the **speed of light**, a fundamental physical constant with a value of 2.99792458×10^8 m/s ($\sim 3.00 \times 10^8$ m/s or 1.86×10^5 mi/s). This is about a million times faster than the speed of sound! The speed of light in a vacuum is represented by the symbol c.

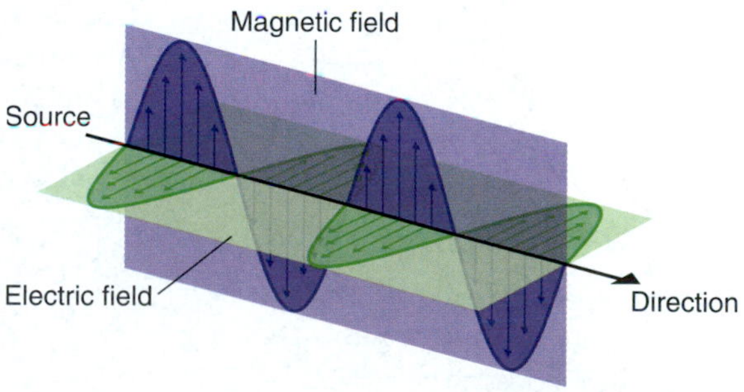

Figure 6.3 The nature of electromagnetic radiation. All forms of electromagnetic radiation consist of perpendicular oscillating electric and magnetic fields.

Because the various kinds of electromagnetic radiation all have the same speed (c), they differ only in wavelength and frequency. As shown in Figure 6.4 and Table 6.1, the wavelengths of familiar electromagnetic radiation range from 10^1 m for radio waves to 10^{-12} m for gamma rays, which are emitted by nuclear reactions. By inserting c in place of v in Equation 6.1, we can show that the frequency of electromagnetic radiation is inversely proportional to the wavelength:

$$c = \lambda\nu$$
$$\nu = \frac{c}{\lambda} \tag{6.2}$$

For example, the frequency of radio waves is about 10^8 Hz, whereas the frequency of gamma rays is about 10^{20} Hz. Visible light, which is electromagnetic radiation that can

* Named after German physicist Heinrich Hertz (1857–1894), a pioneer in the field of electromagnetic radiation.

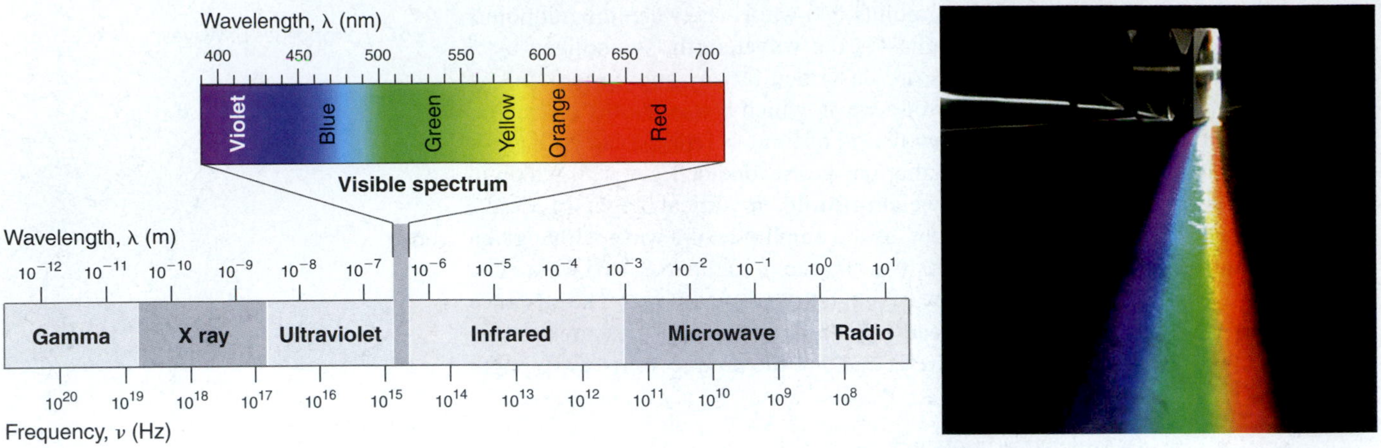

(a) (b)

Figure 6.4 The electromagnetic spectrum. (a) The wavelength and frequency ranges of electromagnetic radiation. The visible portion of the electromagnetic spectrum is the narrow region with wavelengths between about 400 and 700 nm. (b) When white light is passed through a prism, it is split into light of different wavelengths, whose colors correspond to the visible spectrum.

be detected by the human eye, has wavelengths between about 7×10^{-7} m (700 nm, corresponding to a frequency of 4.3×10^{14} Hz) and 4×10^{-7} m (400 nm, or 7.5×10^{14} Hz). Within this range, the eye perceives radiation of different wavelengths (or frequencies) as light of different colors, ranging from red to violet in order of decreasing wavelength. The components of white light, which is a mixture of all the frequencies of visible light, can be separated by a prism, as shown in Figure 6.4b. A similar phenomenon creates a rainbow, where water droplets suspended in the air act as tiny prisms.

As you will soon see, the energy of electromagnetic radiation is directly proportional to its frequency and inversely proportional to its wavelength:

$$E \propto \nu \tag{6.3}$$

$$E \propto \frac{1}{\lambda} \tag{6.4}$$

Note the pattern

The energy of electromagnetic radiation increases *with* increasing frequency *and* decreasing wavelength.

Thus, whereas visible light is essentially harmless to our skin, ultraviolet light, with wavelengths of ≤400 nm, has enough energy to cause severe damage to our skin in the form of sunburn. Because the ozone layer described in Chapter 3 absorbs sunlight

TABLE 6.1 Common wavelength units for electromagnetic radiation

Unit	Symbol	Wavelength, m	Type of Radiation
Picometer	pm	10^{-12}	Gamma ray
Ångstrom	Å	10^{-10}	X-ray
Nanometer	nm	10^{-9}	X-ray
Micrometer	μm	10^{-6}	Infrared
Millimeter	mm	10^{-3}	Infrared
Centimeter	cm	10^{-2}	Microwave
Meter	m	10^{0}	Radio

with wavelengths less than 350 nm, it protects us from the damaging effects of highly energetic ultraviolet radiation.

EXAMPLE 6.1

Your favorite FM station for jazz, W-XYZ, broadcasts at a frequency of 101.1 MHz. What is the wavelength of this radiation?

Given Frequency

Asked for Wavelength

Strategy

Substitute the value for the speed of light in meters per second into Equation 6.2 to calculate the wavelength in meters.

Solution

From Equation 6.2, we know that the product of the wavelength and the frequency is the speed of the wave, which for electromagnetic radiation is 2.998×10^8 m/s:

$$\lambda \nu = c = 2.998 \times 10^8 \text{ m/s}$$

Thus, the wavelength λ is given by

$$\lambda = \frac{c}{\nu} = \left(\frac{2.998 \times 10^8 \text{ m/}\cancel{\text{s}}}{101.1 \cancel{\text{MHz}}}\right)\left(\frac{1 \cancel{\text{MHz}}}{10^6 \cancel{\text{s}}^{-1}}\right) = 2.965 \text{ m}$$

EXERCISE 6.1

While the officer was writing up your speeding ticket, she mentioned that she was using a state-of-the-art radar gun operating at 35.5 GHz. What is the wavelength of the radiation emitted by the radar gun?

Answer 8.45 mm

In the next two sections, we describe how scientists developed our current understanding of the structure of atoms using the scientific method described in Chapter 1. You will discover why scientists had to rethink their classical understanding of the nature of electromagnetic energy, which clearly distinguished between the particulate behavior of matter and the wavelike nature of energy.

6.2 • The Quantization of Energy

By the late 19th century, many physicists thought their discipline was well on the way to explaining most natural phenomena. They could calculate the motions of material objects using Newton's laws of classical mechanics, and they could describe the properties of radiant energy using mathematical relationships known as Maxwell's equations, developed in 1873 by James Clerk Maxwell, a Scottish physicist. The universe appeared to be a simple and orderly place: it contained matter, which consisted of particles that had mass and whose location and motion could be accurately described, and it contained electromagnetic radiation, which was viewed as having no mass and whose exact position in space could not be fixed. Thus, matter and energy were considered distinct and unrelated phenomena. Soon, however, scientists began to look more closely at a few inconvenient phenomena that could *not* be explained by the theories available at the time.

MGC The Photoelectric Effect

Blackbody Radiation

One phenomenon that seemed to contradict the theories of classical physics was **blackbody radiation**, the energy emitted by an object when it is heated. The wavelength of energy emitted depends only on the temperature of the object and not on its surface or composition. Hence, an electric stove burner or the filament of a space heater glows dull red or orange when heated, whereas the much hotter tungsten wire in an incandescent light bulb gives off a yellowish light (Figure 6.5).

Figure 6.5 Blackbody radiation. When heated, all objects emit electromagnetic radiation whose wavelength (and color) depends on the temperature of the object. A relatively low-temperature object such as an electric stove element on a low setting appears red, whereas a higher-temperature object such as the filament of an incandescent light bulb appears yellow or white.

The *intensity* of radiation is a measure of the energy emitted per unit area. A plot of the intensity of blackbody radiation as a function of wavelength for an object at various temperatures is shown in Figure 6.6. One of the major assumptions of classical physics was that energy increased or decreased in a smooth, continuous manner. For example, classical physics predicted that as wavelength decreased, the intensity of the radiation an object emits should increase without limit at *all* temperatures, as shown by the broken line for 6000 K in Figure 6.6. Thus, classical physics could not explain the sharp *decrease* in the intensity of radiation emitted at shorter wavelengths (primarily in the ultraviolet region of the spectrum), which we now refer to as the "ultraviolet catastrophe." In 1900, however, the German physicist Max Planck (1858–1947) explained the "ultraviolet catastrophe" by assuming that the energy of electromagnetic waves is *quantized* rather than continuous. This meant that energy could be gained or lost only in integral multiples of some smallest unit of energy, a **quantum**.

Although quantization may seem to be an unfamiliar concept, we all encounter it frequently. For example, U.S. money is integral multiples of pennies. Similarly, musical instruments like a piano or a trumpet can produce only certain musical notes, such as C or F sharp. Because these instruments cannot produce a continuous range of frequencies,

Figure 6.6 Relationship between the temperature of an object and the spectrum of blackbody radiation it emits. At relatively low temperatures, most of the radiation is emitted at wavelengths longer than 700 nm, in the infrared portion of the spectrum. The dull red glow of the electric stove element in Figure 6.5 is due to the small amount of radiation emitted at wavelengths less than 700 nm, which the eye can detect. As the temperature of the object increases, the maximum intensity shifts to shorter wavelengths, successively resulting in orange, yellow, and finally white light. At high temperatures, all wavelengths of visible light are emitted with approximately equal intensities. The white light spectrum shown for the object at 6000 K closely approximates the spectrum of light emitted by the sun (Figure 6.14). Note the sharp decrease in the intensity of radiation emitted at wavelengths below 400 nm, which constituted the "ultraviolet catastrophe." The classical prediction fails to fit the experimental curves entirely and does not have a maximum intensity.

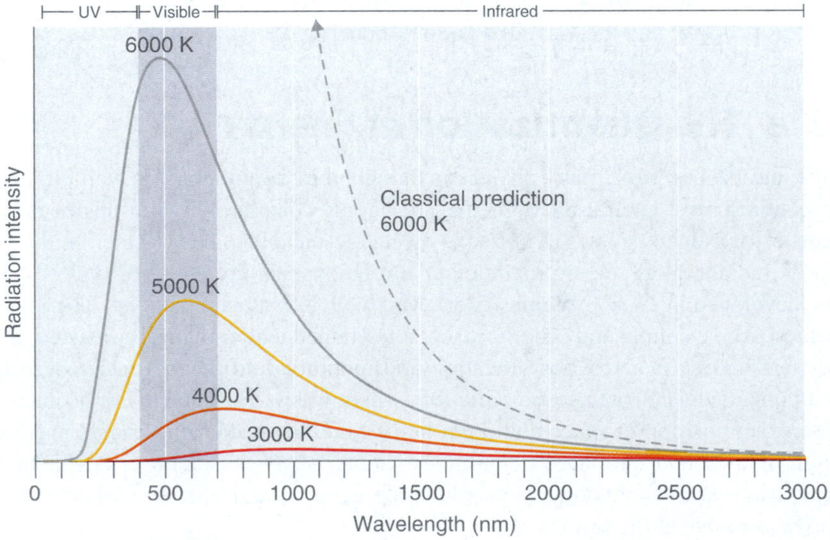

their frequencies are quantized. Even electrical charge is quantized: an ion may have a charge of −1 or −2, but *not* −1.33.

Planck postulated that the energy of a particular quantum of radiant energy could be described explicitly by the equation

$$E = h\nu \qquad (6.5)$$

where the proportionality constant h is today called Planck's constant, one of the most accurately known fundamental constants in science. For our purposes, its value to four significant figures is generally sufficient:

$$h = 6.626 \times 10^{-34} \text{ joule} \cdot \text{second (J} \cdot \text{s)}$$

Thus, as the frequency of electromagnetic radiation increases, the magnitude of the associated quantum of radiant energy increases. By assuming that energy can be emitted by an object only in integral multiples of $h\nu$, Planck devised an equation that fit the experimental data shown in Figure 6.6. We can understand Planck's explanation of the "ultraviolet catastrophe" qualitatively as follows: At low temperatures, only radiation with relatively low frequencies is emitted, corresponding to low-energy quanta. As the temperature of an object increases, there is an increased probability of emitting radiation with higher frequencies, corresponding to higher-energy quanta. At any temperature, however, it is simply more probable for an object to lose energy by emitting n lower-energy quanta than a single very high-energy quantum that corresponds to ultraviolet radiation. The result is a maximum in the plot of intensity of emitted radiation versus wavelength, as shown in Figure 6.6, and a shift in the position of the maximum to lower wavelength (higher frequency) with increasing temperature.

At the time he proposed his radical hypothesis, Planck could not explain *why* energies should be quantized. Initially, his hypothesis explained only one set of experimental data—blackbody radiation. If quantization were observed for a large number of different phenomena, then quantization would become a law (as defined in Chapter 1). In time, a theory might be developed to explain that law. As things turned out, Planck's hypothesis was the seed from which modern physics grew.

Max Planck (1858–1947). In addition to being a physicist, Planck was a gifted pianist, who at one time considered music as a career. During the 1930s, Planck felt it was his duty to remain in Germany, despite his open opposition to the Nazi government's policies. One of his sons was executed in 1944 for his part in an unsuccessful attempt to assassinate Hitler, and Planck's home was destroyed by bombing during the last weeks of the war.

The Photoelectric Effect

Only five years after he proposed it, Planck's quantization hypothesis was used to explain a second phenomenon that conflicted with the accepted laws of classical physics. When certain metals are exposed to light, electrons are ejected from their surface (Figure 6.7). Classical physics predicted that the number of electrons emitted and their kinetic energy should depend only on the intensity of the light, *not* on its frequency. In fact, however, each metal was found to have a characteristic threshold frequency of light; below that frequency, *no* electrons are emitted regardless of the light's intensity. Above the threshold frequency, the number of electrons emitted was found to be proportional to the intensity of the light, and their kinetic energy proportional to its frequency. This phenomenon was called the **photoelectric effect**.

Albert Einstein (1879–1955, Nobel Prize in physics, 1921) quickly realized that Planck's hypothesis about the quantization of radiant energy could also explain the photoelectric effect. The key feature of Einstein's hypothesis was the assumption that radiant energy arrives at the metal surface in "particles" that we now call **photons**, each possessing a particular energy E given by Equation 6.5. Einstein postulated that each metal has a particular electrostatic attraction for its electrons that must be overcome before an electron can be emitted from its surface ($E_o = h\nu_o$). If photons of light with energy less than E_o strike a metal surface, no single photon has enough energy to eject an electron, and consequently no electrons are emitted regardless of the intensity

Albert Einstein (1879–1955). In 1900, Einstein was working in the Swiss patent office in Bern. He had been born in Germany, and throughout his childhood his parents and teachers had worried that he might be retarded. The patent office job was a low-level civil service position that was not very demanding, but it did allow Einstein to spend a great deal of time reading and thinking about physics. In 1905, he published his paper on the photoelectric effect, for which he received the Nobel Prize in 1921.

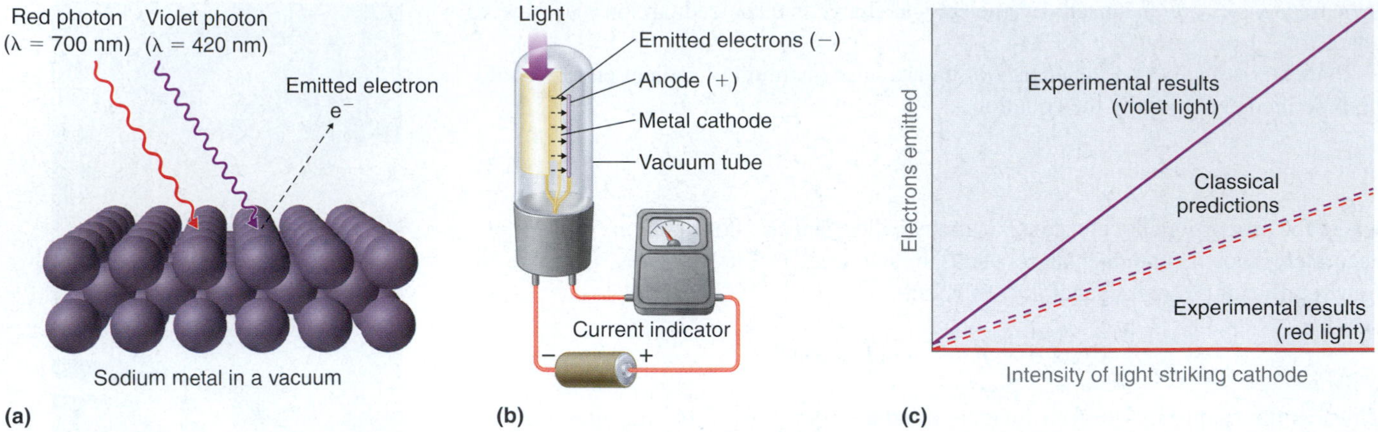

Figure 6.7 The photoelectric effect. (a) Irradiating a metal surface with photons of sufficiently high energy causes electrons to be ejected from the metal. (b) A photocell that uses the photoelectric effect, similar to those found in automatic door openers. When light strikes the metal cathode, electrons are emitted and attracted to the anode, resulting in a flow of electrical current. If the incoming light is interrupted by, for example, a passing person, the current drops to zero. (c) In contrast to predictions using classical physics, no electrons are emitted when photons of light with energy less than E_o, such as red light, strike the cathode. The energy of violet light is above the threshold frequency, so the number of emitted photons is proportional to the light's intensity.

of the light. If a photon with energy greater than E_o strikes the metal, then part of its energy is used to overcome the forces that hold the electron to the metal surface, and the excess energy appears as the kinetic energy of the ejected electron:

$$\text{kinetic energy of ejected electron} = E - E_o = h\nu - h\nu_o = h(\nu - \nu_o) \quad (6.6)$$

When a metal is struck by light with energy above the threshold energy E_o, the *number* of emitted electrons is proportional to the *intensity* of the light beam, which corresponds to the number of photons per square centimeter, but the *kinetic energy* of the emitted electrons is proportional to the *frequency* of the light. Thus, Einstein showed that the energy of the emitted electrons depended on the frequency of the light, contrary to the prediction of classical physics.

Planck's and Einstein's postulate that energy is quantized is in many ways similar to Dalton's description of atoms. Both theories are based on the existence of simple building blocks, atoms in one case and quanta of energy in the other. The work of Planck and Einstein thus suggested a connection between the quantized nature of energy and the properties of individual atoms. In fact, Einstein's Nobel Prize was awarded for his work on the photoelectric effect (*not* for his more famous equation $E = mc^2$), demonstrating its fundamental importance.

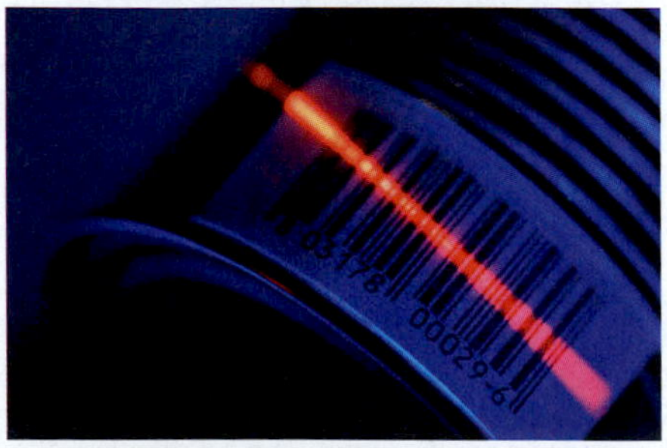

Figure 6.8 A beam of red light emitted by a ruby laser. Ruby lasers, which emit red light at a wavelength of 694.3 nm, are used to read bar codes. When used for commercial applications, such lasers are generally designed to emit radiation over a narrow range of wavelengths to reduce their cost.

EXAMPLE 6.2

A ruby laser, a device that produces light in a narrow range of wavelengths (Section 6.3), emits red light at a wavelength of 694.3 nm (Figure 6.8). What is the energy in joules of **(a)** a single photon and **(b)** a mole of photons?

Given Wavelength

Asked for Energy of single photon and mole of photons

Strategy

A Use Equations 6.2 and 6.5 to calculate the energy in joules.

B Multiply the energy of a single photon by Avogadro's number to obtain the energy in a mole of photons.

Solution

(a) A✓ The energy of a single photon is given by $E = h\nu = hc/\lambda$. Thus,

$$E = \frac{(6.626 \times 10^{-34}\ \text{J}\cdot\cancel{\text{s}})(2.998 \times 10^{8}\ \cancel{\text{m}}\cdot\cancel{\text{s}}^{-1})}{694.3 \times 10^{-9}\ \cancel{\text{m}}}$$
$$= 2.861 \times 10^{-19}\ \text{J}$$

(b) B✓ To calculate the energy in a mole of photons, we multiply the energy of a single photon by the number of photons in a mole (Avogadro's number). If we write the energy of a photon as 2.86×10^{-19} J/photon, we obtain the energy of a mole of photons with wavelength 694.3 nm:

$$E = \left(\frac{2.861 \times 10^{-19}\ \text{J}}{\cancel{\text{photon}}}\right)\left(\frac{6.022 \times 10^{23}\ \cancel{\text{photon}}}{\text{mol}}\right)$$
$$= 1.723 \times 10^{5}\ \text{J/mol}$$
$$= 172.3\ \text{kJ/mol}$$

This energy is of the same magnitude as some of the enthalpies of reaction that we considered in Chapter 5, and, as you will see in Chapter 8, it is comparable to the strength of many chemical bonds. As a result, light can be used to initiate chemical reactions. In fact, an entire area of chemistry called *photochemistry* is devoted to studying such processes. In the phenomenon of *photosynthesis*, green plants use the energy of visible light to convert carbon dioxide and water into sugars such as glucose.

EXERCISE 6.2

An X-ray generator, such as those used in hospitals, emits radiation with a wavelength of 1.544 Å. What is the energy in joules of **(a)** a single photon and **(b)** a mole of photons? **(c)** How many times more energetic is a single X-ray photon of this wavelength than a photon emitted by a ruby laser?

Answer **(a)** 1.287×10^{-15} J/photon; **(b)** 7.748×10^{8} J/mol $= 7.748 \times 10^{5}$ kJ/mol; **(c)** 4497 times

6.3 • Atomic Spectra and Models of the Atom

MGC The Bohr Equation

The photoelectric effect provided indisputable evidence for the existence of the photon and thus the particle-like behavior of electromagnetic radiation. The concept of the photon, however, emerged from experimentation using thermal radiation, which produces a continuous spectrum of energies. More direct evidence was needed to verify the quantized nature of electromagnetic radiation. In this section, we describe how experimentation using visible light provided this evidence.

Line Spectra

Although objects at high temperature emit a continuous spectrum of electromagnetic radiation (Figure 6.6), a different kind of spectrum is observed when pure samples of individual elements are heated. For example, when a high-voltage electrical discharge is passed through a sample of H_2 gas at low pressure, the hydrogen atoms emit a red light. Unlike blackbody radiation, the color of the light emitted by hydrogen atoms does not depend greatly on the temperature of the gas in the tube. When the emitted light is passed through a prism, only a few narrow lines, called a **line spectrum**, are seen (Figure 6.9), rather than a continuous range of colors. The light emitted by hydrogen is red because the most intense lines in its spectrum are in the red portion of the visible spectrum, at 656 nm. With sodium, however, we observe a yellow color because the most intense lines in its spectrum are in the yellow portion of the spectrum, at about 589 nm.

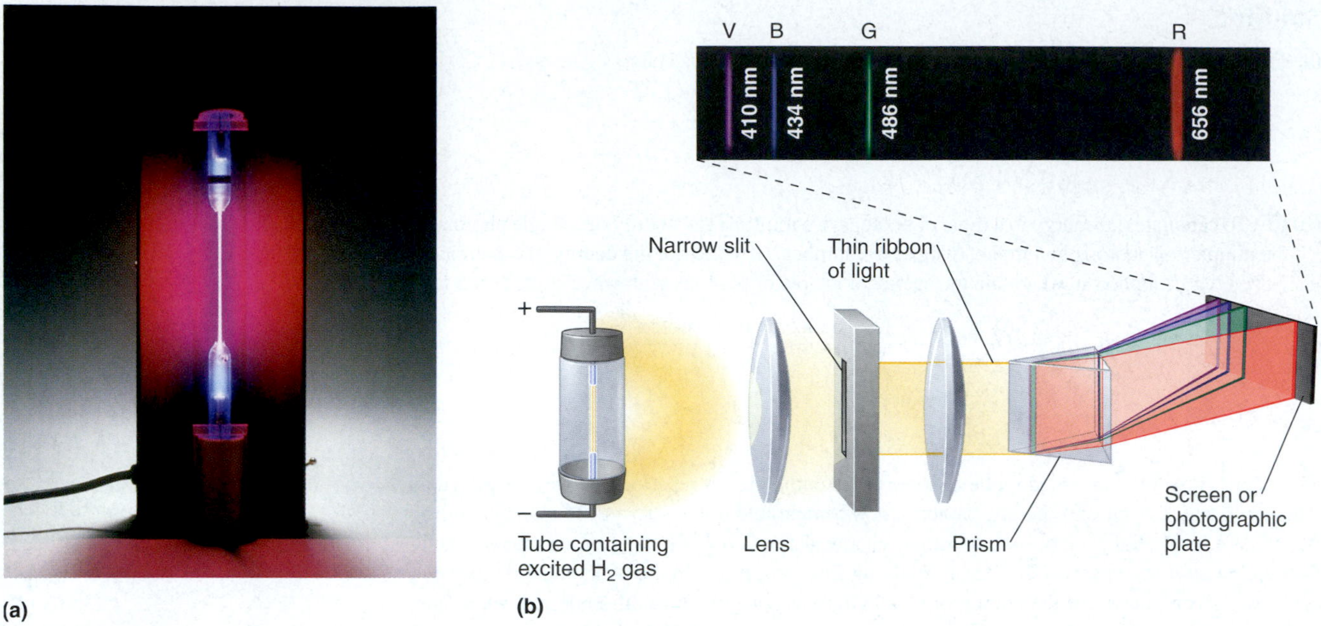

Figure 6.9 Emission of light by hydrogen atoms. (a) A sample of excited hydrogen gas emits a characteristic red light. (b) When the light emitted by a sample of excited hydrogen gas is split into its component wavelengths by a prism, four characteristic violet, blue, green, and red emission lines can be observed, the most intense of which is at 656 nm.

Such *emission spectra* were observed for many other elements in the late 19th century, which presented a major challenge because classical physics was unable to explain them. Part of the explanation is provided by Planck's equation (Equation 6.5): the observation of only a few values of λ (or ν) in the line spectrum meant that only a few values of E were possible. Thus, *the energy levels of a hydrogen atom had to be quantized*; in other words, only states that had certain values of energy were possible, or *allowed*. If a hydrogen atom could have *any* value of energy, then a continuous spectrum would have been observed, similar to blackbody radiation.

In 1885, a Swiss mathematics teacher, Johann Balmer (1825–1898), showed that the frequencies of the lines observed in the visible region of the spectrum of hydrogen fit a simple equation that can be expressed as:

$$\nu = \text{constant}\left(\frac{1}{2^2} - \frac{1}{n^2}\right) \tag{6.7}$$

where $n = 3, 4, 5, 6$. As a result, these lines are known as the *Balmer series*. The Swedish physicist Johannes Rydberg (1854–1919) subsequently restated and expanded Balmer's result in the *Rydberg equation*:

$$\frac{1}{\lambda} = \mathcal{R}\left(\frac{1}{n_1^2} - \frac{1}{n_2^2}\right) \tag{6.8}$$

where n_1 and n_2 are positive integers, $n_2 > n_1$, and $\mathcal{R}$, the *Rydberg constant*, has a value of $1.09737 \times 10^7\ \text{m}^{-1}$.

Like Balmer's, Rydberg's simple equation described the wavelengths of the visible lines in the emission spectrum of hydrogen (with $n_1 = 2$, $n_2 = 3, 4, 5, \ldots$). More important, Rydberg's equation also described the wavelengths of other series of lines that were to be observed in the emission spectrum of hydrogen: one in the ultraviolet ($n_1 = 1$, $n_2 = 2, 3, 4, \ldots$) and one in the infrared ($n_1 = 3$, $n_2 = 4, 5, 6$). Unfortunately, scientists had not yet developed any theoretical justification for an equation of this form.

Johann Balmer (1825–1898). A mathematics teacher at a secondary school for girls in Switzerland, Balmer was 60 years old when he wrote the paper on the spectral lines of hydrogen that made him famous. He published only one other paper on the topic, which appeared when he was 72.

The Bohr Model

In 1913, a Danish physicist, Niels Bohr (1885–1962, Nobel Prize in physics, 1922), proposed a theoretical model for the hydrogen atom that explained its emission spectrum. Bohr's model required only one assumption: *The electron moves around the nucleus in circular orbits that can have only certain allowed radii.* Rutherford's earlier model of the atom had also assumed that the electrons moved in circular orbits around the nucleus, and that the atom was held together by the electrostatic attraction between the positively charged nucleus and the negatively charged electron (Chapter 1). Although we now know that the assumption of circular orbits was incorrect, Bohr's insight was to propose that *the electron could occupy only certain regions of space.*

Using classical physics, Bohr showed that the energy of an electron in a particular orbit is given by

$$E_n = -\frac{\mathfrak{R}hc}{n^2} \tag{6.9}$$

where $\mathfrak{R}$ is the Rydberg constant, h is Planck's constant, c is the speed of light, and n is a positive integer corresponding to the number assigned to the orbit, with $n = 1$ corresponding to the orbit closest to the nucleus.* Thus, the orbit with $n = 1$ is the lowest in energy. Because a hydrogen atom with its one electron in this orbit has the lowest possible energy, this is the **ground state**, the most stable arrangement for a hydrogen atom. As n increases, the radius of the orbit increases and the energy of that orbit becomes less negative (Figure 6.10). A hydrogen atom with an electron in an orbit with $n > 1$ is therefore in an **excited state**: its energy is higher than the energy of the ground state. When an atom in an excited state undergoes a transition to the ground state, called *decay*, it loses energy by emitting a photon whose energy corresponds to the difference in energy between the two states (Figure 6.11).

In this description, the difference in energy, ΔE, between any two orbits or energy levels is given by $\Delta E = E_{n_1} - E_{n_2}$, where n_1 is the final orbit and n_2 the initial orbit. Substituting from Bohr's equation (Equation 6.9) for each energy value gives

$$\Delta E = E_{\text{final}} - E_{\text{initial}} \quad \Delta E = -\frac{\mathfrak{R}hc}{n_1^2} - \left(-\frac{\mathfrak{R}hc}{n_2^2}\right) = -\mathfrak{R}hc\left(\frac{1}{n_1^2} - \frac{1}{n_2^2}\right) \tag{6.10}$$

If $n_2 > n_1$, the transition is from a higher energy state (larger-radius orbit) to a lower energy state (smaller-radius orbit), as shown by the dashed arrow in Figure 6.11a. Substituting hc/λ for ΔE gives

$$\Delta E = \frac{hc}{\lambda} = -\mathfrak{R}hc\left(\frac{1}{n_1^2} - \frac{1}{n_2^2}\right) \tag{6.11}$$

Canceling hc on both sides gives

$$\frac{1}{\lambda} = -\mathfrak{R}\left(\frac{1}{n_1^2} - \frac{1}{n_2^2}\right) \tag{6.12}$$

Except for the negative sign, this is the same equation that Rydberg obtained experimentally. The negative sign in Equations 6.11 and 6.12 indicates that energy is released as the electron moves from orbit n_2 to orbit n_1 because orbit n_2 is at a higher energy than orbit n_1. Bohr calculated the value of $\mathfrak{R}$ independently and obtained a value of 1.0974×10^7 m^{-1}, the same number Rydberg had obtained by analyzing the emission spectra.

Niels Bohr (1885–1962). During the Nazi occupation of Denmark in World War II, Bohr escaped to the United States, where he became associated with the Atomic Energy Project. In his final years, he devoted himself to the peaceful application of atomic physics and to resolving political problems arising from the development of atomic weapons.

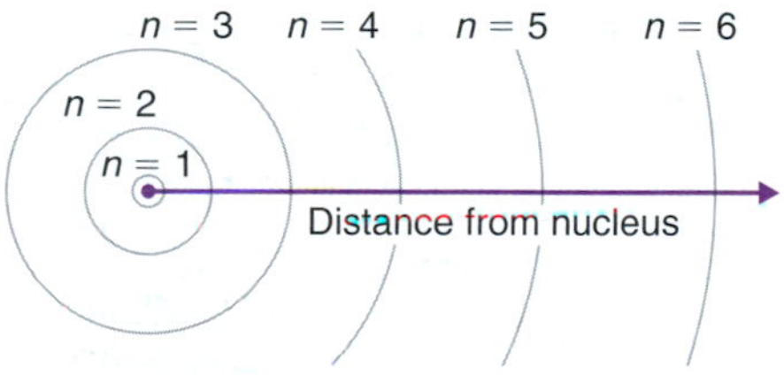

(a) Hydrogen orbits (Bohr model). Orbits are not drawn to scale.

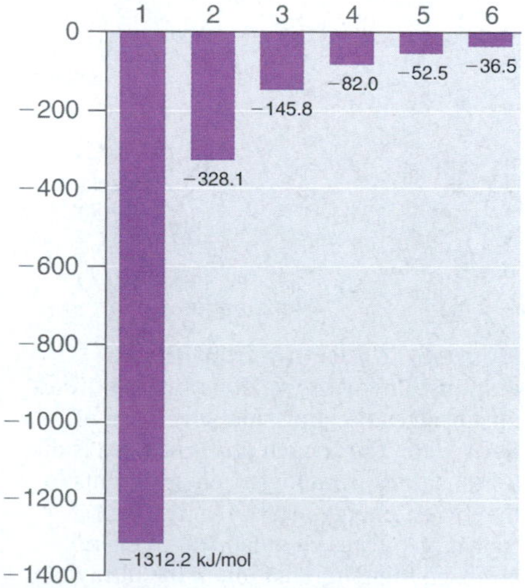

(b) Hydrogen orbit energy (kJ/mol) vs. *n*

Figure 6.10 Bohr model of the hydrogen atom. (a) The distance of the orbit from the nucleus increases with increasing n. (b) The energy of the orbit becomes increasingly less negative with increasing n.

* The negative sign in Equation 6.9 is a convention indicating that the electron–nucleus pair is more stable (has a lower energy) when they are near each other than when they are infinitely far apart, corresponding to $n = \infty$. The latter condition is arbitrarily assigned an energy of zero.

Figure 6.11 Emission of light by a hydrogen atom in an excited state. (a) Light is emitted when the electron undergoes a transition from an orbit with a higher value of n (at a higher energy) to an orbit with a lower value of n (at lower energy). (b) The Balmer series of emission lines is due to transitions from orbits with $n \geq 3$ to the orbit with $n = 2$. The differences in energy between these levels corresponds to light in the visible portion of the electromagnetic spectrum.

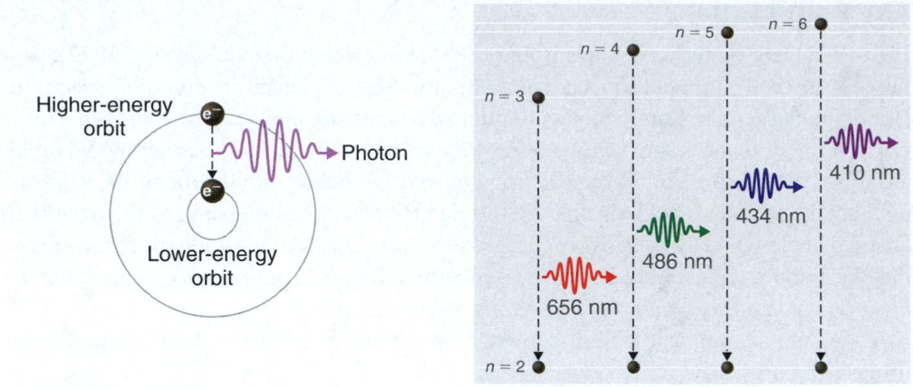

(a) Electronic emission transition **(b) Balmer series transitions**

We can now understand the physical basis for the Balmer series of lines in the emission spectrum of hydrogen (Figure 6.9b). As shown in Figure 6.11b, the lines in this series correspond to transitions from higher-energy orbits ($n > 2$) to the *second* orbit ($n = 2$). Thus, the hydrogen atoms in the sample have absorbed energy from the electrical discharge and decayed from a higher-energy excited state ($n > 2$) to a lower-energy state ($n = 2$) by emitting a photon of electromagnetic radiation whose energy corresponds exactly to the *difference* in energy between the two states (Figure 6.11a). Thus, the $n = 3$ to $n = 2$ transition gives rise to the line at 656 nm (red), the $n = 4$ to $n = 2$ transition to the line at 486 nm (green), the $n = 5$ to $n = 2$ transition to the line at 434 nm (blue), and the $n = 6$ to $n = 2$ transition to the line at 410 nm (violet). Because a sample of hydrogen contains a large number of atoms, the intensity of the various lines in a line spectrum depends on the number of atoms in each excited state. At the temperature in the gas discharge tube, more atoms are in the $n = 3$ than the $n \geq 4$ levels. Consequently, the $n = 3$ to $n = 2$ transition is the most intense line, producing the characteristic red color of a hydrogen discharge (Figure 6.9a). Other families of lines are produced by transitions from excited states with $n > 1$ to the orbit with $n = 1$ or to orbits with $n \geq 3$. These transitions are shown schematically in Figure 6.12.

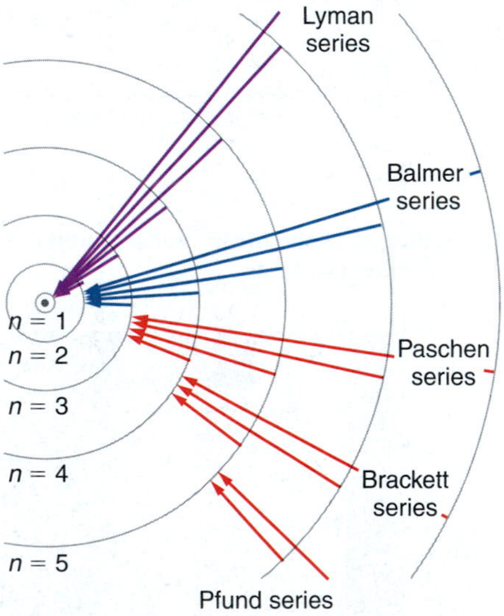

Figure 6.12 Electron transitions responsible for the various series of lines observed in the emission spectrum of hydrogen. The Lyman series of lines is due to transitions from higher-energy orbits to the lowest-energy orbit ($n = 1$); these transitions release a great deal of energy, corresponding to radiation in the ultraviolet portion of the electromagnetic spectrum. The Paschen, Brackett, and Pfund series of lines are due to transitions from higher-energy orbits to orbits with $n = 3$, 4, and 5, respectively; these transitions release substantially less energy, corresponding to infrared radiation. (Not to scale.)

EXAMPLE 6.3

The so-called Lyman series of lines in the emission spectrum of hydrogen corresponds to transitions from various excited states to the $n = 1$ orbit. Calculate the wavelength of the lowest-energy line in the Lyman series to three significant figures. In what region of the electromagnetic spectrum does it occur?

Given Lowest-energy orbit in the Lyman series

Asked for Wavelength of the lowest-energy Lyman line and corresponding region of the spectrum

Strategy

A Substitute the appropriate values into Equation 6.8 (the Rydberg equation) and solve for λ.
B Use Figure 6.4 to locate the region of the electromagnetic spectrum corresponding to the calculated wavelength.

Solution

We can use the Rydberg equation to calculate the wavelength:

$$\frac{1}{\lambda} = \mathcal{R}\left(\frac{1}{n_1^2} - \frac{1}{n_2^2}\right)$$

A For the Lyman series, $n_1 = 1$. The lowest-energy line is due to a transition from the $n = 2$ to $n = 1$ orbit because they are the closest in energy. Thus,

$$\frac{1}{\lambda} = \mathcal{R}\left(\frac{1}{n_1^2} - \frac{1}{n_2^2}\right) = 1.097 \times 10^7 \text{ m}^{-1}\left(\frac{1}{1} - \frac{1}{4}\right) = 8.228 \times 10^6 \text{ m}^{-1}$$

and

$$\lambda = 1.215 \times 10^{-7} \text{ m} = 122 \text{ nm}$$

This wavelength is in the ultraviolet region of the spectrum.

EXERCISE 6.3

The Pfund series of lines in the emission spectrum of hydrogen corresponds to transitions from higher excited states to the $n = 5$ orbit. Calculate the wavelength of the *second* line in the Pfund series to three significant figures. In which region of the spectrum does it lie?

Answer 4.65×10^3 nm; infrared

Bohr's model of the hydrogen atom gave an exact explanation for its observed emission spectrum. Here are his key contributions to our understanding of atomic structure:

1. Electrons can occupy only certain regions of space, called *orbits*.
2. Orbits closer to the nucleus are more stable (that is, they are at lower energy levels).
3. Electrons can move from one orbit to another by absorbing or emitting energy, giving rise to characteristic spectra.

Unfortunately, Bohr could not explain *why* the electron should be restricted to particular orbits. Also, despite a great deal of tinkering, such as assuming that orbits could be ellipses rather than circles, his model could not quantitatively explain the emission spectra of any element other than hydrogen (Figure 6.13). In fact, Bohr's model worked only for species that contained just one electron: H, He^+, Li^{2+}, and so on. Scientists needed a fundamental change in their way of thinking about the electronic structure of atoms to advance beyond the Bohr model.

Thus far we have explicitly considered only the emission of light by atoms in excited states, which produces an **emission spectrum**. The converse, absorption of light by ground-state atoms to produce an excited state, can also occur, producing an **absorption spectrum**. Because each element has characteristic emission and absorption spectra, scientists can use such spectra to analyze the composition of matter, as we describe in the next section.

Note the pattern

When an atom emits light, it decays to a lower energy state; when an atom absorbs light, it is excited to a higher energy state.

Figure 6.13 Emission spectra of elements compared with hydrogen. (a) Hydrogen gas, which is atomized to H in the discharge tube, (b) neon, and (c) mercury.

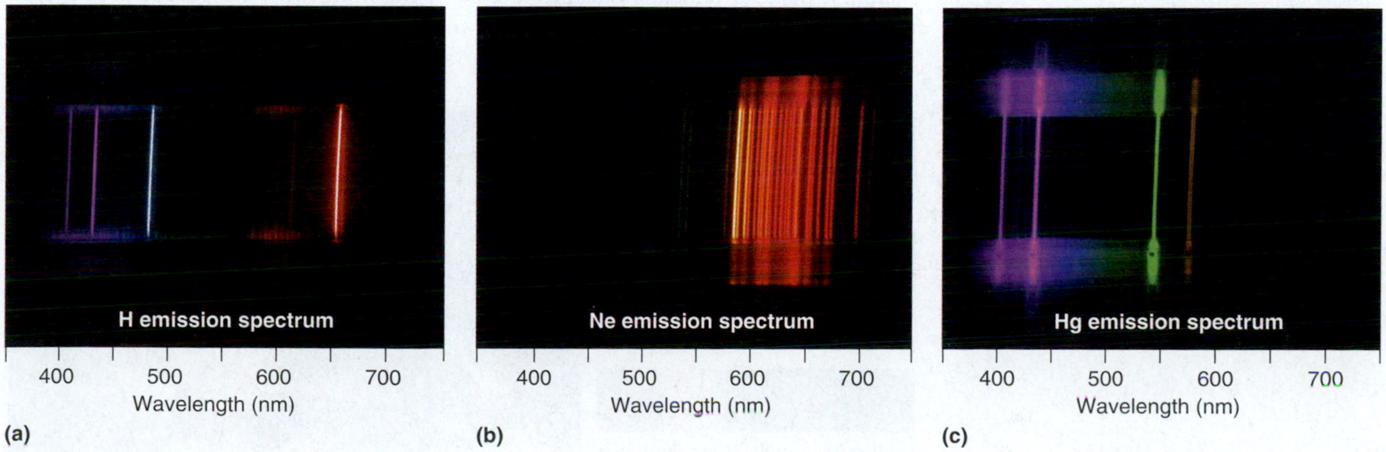

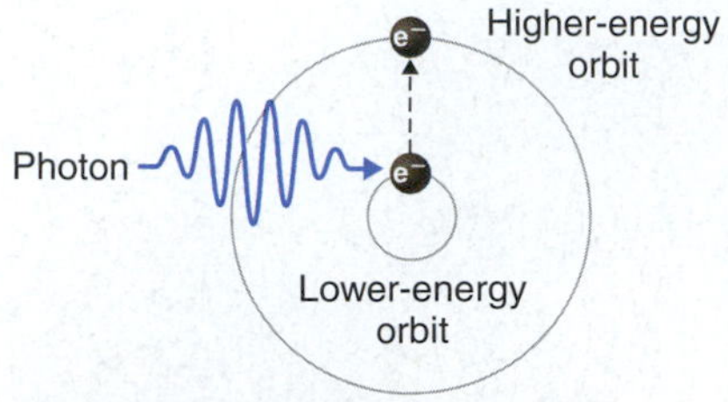

(a) Electronic absorption transition

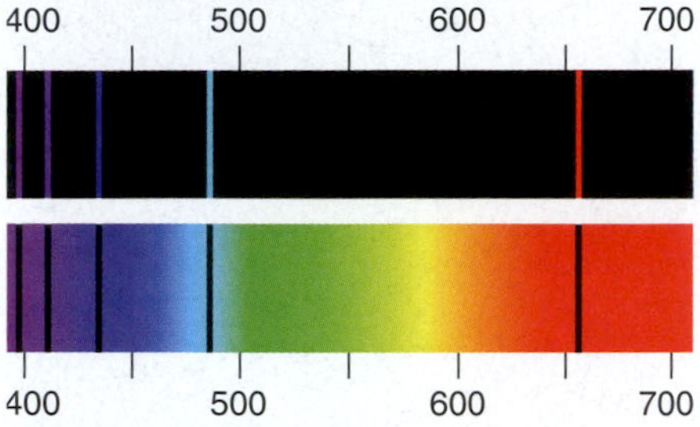

(b) Emission spectrum (top), absorption spectrum (bottom).

Absorption of light by a hydrogen atom. (a) When a hydrogen atom absorbs a photon of light, an electron is excited to an orbit that has a higher energy and larger value of *n*. (b) Emission vs. absorption spectra of hydrogen.

Uses of Emission and Absorption Spectra

If white light is passed through a sample of hydrogen, hydrogen atoms absorb energy as an electron is excited to higher energy levels (orbits with $n \geq 2$). If the light that emerges is passed through a prism, it forms a continuous spectrum with *black* lines (corresponding to no light passing through the sample) at 656, 468, 434, and 410 nm. These wavelengths correspond to the $n = 2$ to $n = 3$, $n = 2$ to $n = 4$, $n = 2$ to $n = 5$, and $n = 2$ to $n = 6$ transitions. Any given element therefore has both a characteristic emission spectrum and a characteristic absorption spectrum, which are essentially complementary images.

Emission and absorption spectra form the basis of *spectroscopy*, which uses spectra to provide information about the structure and composition of a substance or an object. In particular, astronomers use emission and absorption spectra to determine the composition of stars and interstellar matter. As an example, consider the spectrum of sunlight shown in Figure 6.14. Because the sun is very hot, the light it emits is in the form of a continuous emission spectrum. Superimposed upon it, however, is a series of dark lines due primarily to the absorption of specific frequencies of light by cooler atoms in the outer atmosphere of the sun. By comparing these lines with the spectra of elements measured on earth, we now know that the sun contains large amounts of hydrogen, iron, and carbon, along with smaller amounts of other elements. During the solar eclipse of 1868, the French astronomer Pierre Janssen (1824–1907) observed a set of lines that did not match those of any known element. He suggested that they were due to the presence of a new element, which he named *helium*, from the Greek *helios*, "sun." Helium was finally discovered in uranium ores on earth in 1895.

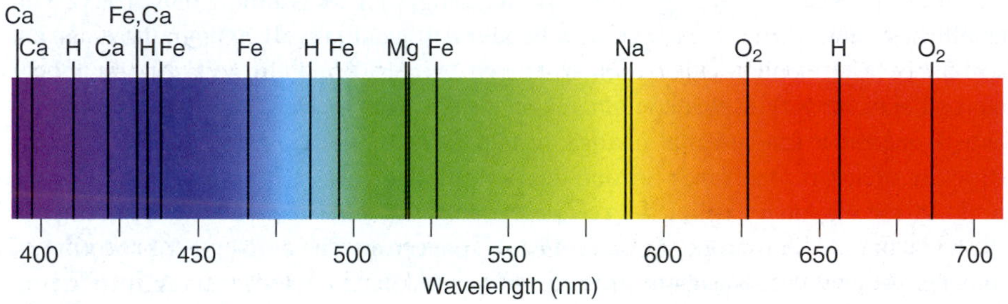

Figure 6.14 The visible spectrum of sunlight. The characteristic dark lines are mostly due to the absorption of light by elements that are present in the cooler outer part of the sun's atmosphere; specific elements are indicated by the labels. The lines at 628 and 687 nm, however, are due to the absorption of light by O_2 molecules in the earth's atmosphere.

The familiar red color of the "neon signs" used in advertising is due to the emission spectrum of neon shown in Figure 6.13b. Similarly, the blue and yellow colors of certain street lights are caused, respectively, by mercury and sodium discharges. In all these cases, an electrical discharge is used to excite neutral atoms to a higher energy state, and light is emitted when the atoms decay to the ground state. In the case of mercury, most of the emission lines are below 450 nm, which produces a blue light (Figure 6.13c). In the case of sodium, the most intense emission lines are at 589 nm, which produces an intense yellow light.

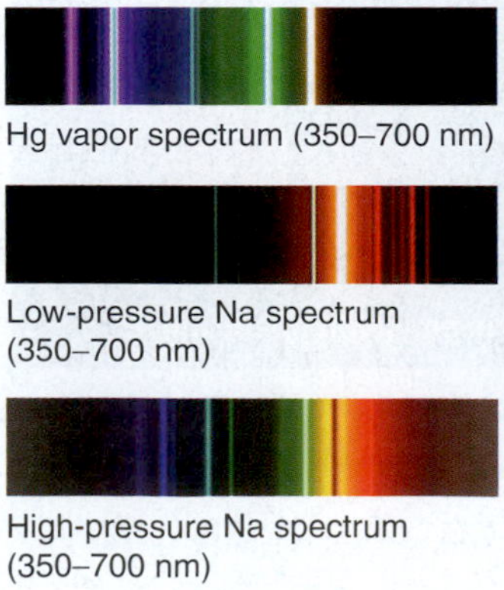

Sodium and mercury street lights. Many street lights use bulbs that contain sodium or mercury vapor. Due to the very different emission spectra of these elements, they emit light of different colors.

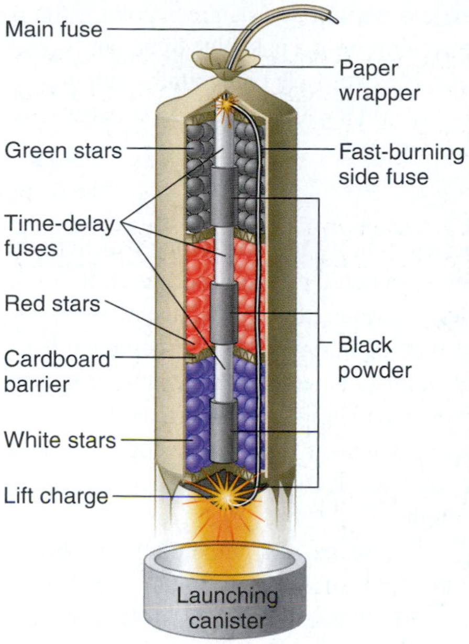

Figure 6.15 Chemistry of fireworks. (a) Diagram of a "multibreak" shell used for fireworks. The chambers contain mixtures of fuels and oxidizers plus compounds for special effects ("stars") connected by time-delay fuses so that the chambers explode in stages. (b) The finale of a fireworks display usually consists of many shells fired simultaneously to give a dazzling multicolor display. The labels indicate the substances that are responsible for the colors of some of the fireworks shown.

The Chemistry of Fireworks

The colors of fireworks are also due to atomic emission spectra. As diagrammed in Figure 6.15a, a typical shell used in a fireworks display contains gunpowder to propel the shell into the air and a fuse to initiate a variety of redox reactions that produce heat and small explosions. The thermal energy excites the atoms to higher energy states, and as they decay to lower energy states, the atoms emit light that gives the familiar colors. Typical oxidant/reductant mixtures include potassium perchlorate ($KClO_4$)/sulfur and aluminum, ammonium perchlorate/charcoal, and potassium chlorate/magnesium. When these mixtures are ignited, a flash of white or yellow light is produced along with a loud bang. Achieving the colors shown in Figure 6.15b requires adding a small amount of a substance that has an emission spectrum in the desired portion of the visible spectrum. For example, sodium is used for yellow because of its 589-nm emission lines. The intense yellow color of sodium would mask most other colors, so potassium and ammonium salts, rather than sodium salts, are usually used as oxidants to produce other colors, which explains the preponderance of such salts in Table 6.2. Strontium salts, which are

TABLE 6.2 Common chemicals used in the manufacture of fireworks[a]

Oxidizers	Fuels (Reductants)	Special Effects
Ammonium perchlorate	Aluminum	*Blue flame*: copper carbonate, copper sulfate, or copper oxide
Barium chlorate	Antimony sulfide	*Red flame*: strontium nitrate or strontium carbonate
Barium nitrate	Charcoal	*White flame*: magnesium or aluminum
Potassium chlorate	Magnesium	*Yellow flame*: sodium oxalate or cryolite (Na_3AlF_6)
Potassium nitrate	Sulfur	*Green flame*: barium nitrate or barium chlorate
Potassium perchlorate	Titanium	*White smoke*: potassium nitrate plus sulfur
Strontium nitrate		*Colored smoke*: potassium chlorate and sulfur, plus organic dye
		Whistling noise: potassium benzoate or sodium salicylate
		White sparks: aluminum, magnesium, or titanium
		Gold sparks: iron fillings or charcoal

[a] Almost any combination of an oxidizer and a fuel may be used along with the compounds needed to produce a desired special effect.

also used in highway flares, emit red light, whereas barium gives a green color. Blue is one of the most difficult colors to achieve. Copper salts emit a pale blue light, but copper is dangerous to use because it forms highly unstable explosive compounds with anions such as chlorate. As you might guess, preparing fireworks with the desired properties is a complex, challenging, and potentially hazardous process.

Lasers

Most light emitted by atoms is *polychromatic*, containing more than one wavelength. In contrast, *lasers* (from *l*ight *a*mplification by *s*timulated *e*mission of *r*adiation) emit *monochromatic* light, consisting of only a single wavelength. Lasers have many applications in fiber-optic telecommunications, the reading and recording of compact disks and DVDs, steel cutting, and supermarket checkout scanners. Laser beams are generated by the same general phenomenon that gives rise to emission spectra, with one difference: only a single excited state is produced, which in principle results in only a single frequency of emitted light. In practice, inexpensive commercial lasers actually emit light with a very narrow range of wavelengths.

The operation of a ruby laser, the first type of laser used commercially, is shown schematically in Figure 6.16. Ruby is an impure form of Al_2O_3, in which some Al^{3+} ions are replaced by Cr^{3+}. The red color of the gem is caused by the absorption of light in the blue region of the visible spectrum by Cr^{3+} ions, which leaves only the longer wavelengths to be reflected back to the eye. One end of a ruby bar is coated with a fully reflecting mirror, and the mirror on the other end is only partially reflecting. When the Cr^{3+} ions are excited by flashes of white light from a flash lamp, they initially decay to a relatively long-lived excited state and can subsequently decay to the ground state by emitting a photon of red light. Some of these photons are reflected back and forth by the mirrored surfaces. As shown in Figure 6.16b, each time a photon interacts with an excited Cr^{3+} ion, it can stimulate that ion to emit another photon that has the same wavelength and is synchronized (*in phase*) with the first wave. This process produces a *cascade* of photons traveling back and forth, until the intense beam emerges through the partially reflecting mirror. Ruby is only one substance that is used to produce a laser; the choice of material determines the wavelength of light emitted, from infrared to ultraviolet, and the light output can be either continuous or pulsed.

When used in a DVD or CD player, light emitted by a laser passes through a transparent layer of plastic on the compact disk and is reflected by an underlying aluminum layer, which contains pits or flat regions that were created when the compact disk was recorded.

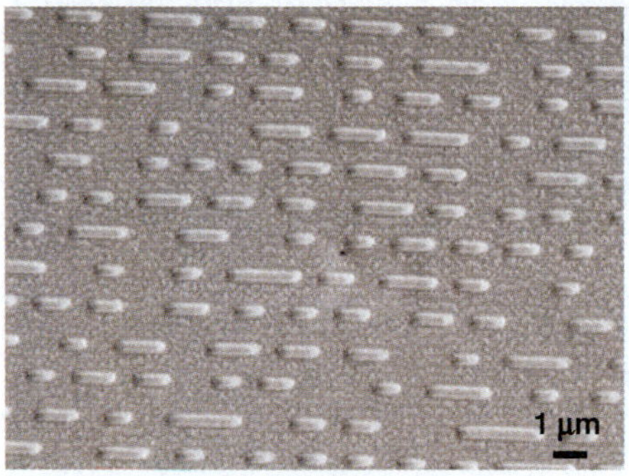

How a CD player uses a laser to read a CD. Inside a CD is a flat, light-reflecting layer called "land." On the land are many "pits" recorded in a spiral-shaped track. (From the read laser's point of view, pits are actually the "bumps" shown here because the master disc with pits is duplicated negatively, turning the pits into bumps.) Pits have the same light-reflecting surface as land, but there are differences in the frequencies of the reflected light in pit and land, making light reflected by pits relatively dark compared with light reflected by land.

Figure 6.16 A ruby laser. (a) Cutaway view of a ruby laser, showing the ruby rod, the flash lamp used to excite the Cr^{3+} ions in the ruby, and the totally and partially reflective mirrors. (b) A schematic drawing illustrating how light from the flash lamp is used to excite the Cr^{3+} ions to a short-lived excited state, which is followed by decay to a longer-lived excited state that is responsible for the stimulated in-phase emission of light by the laser.

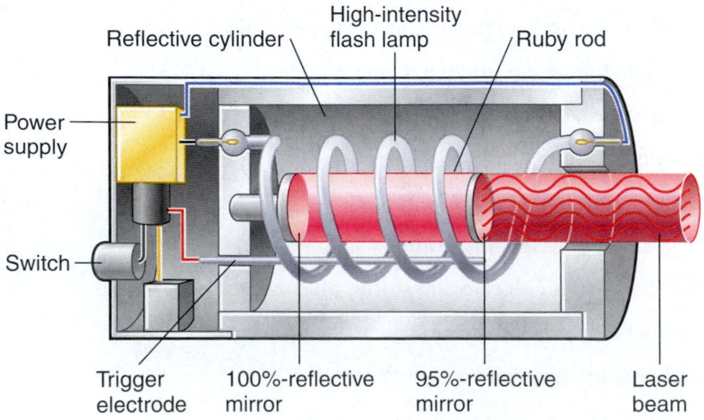

(a)

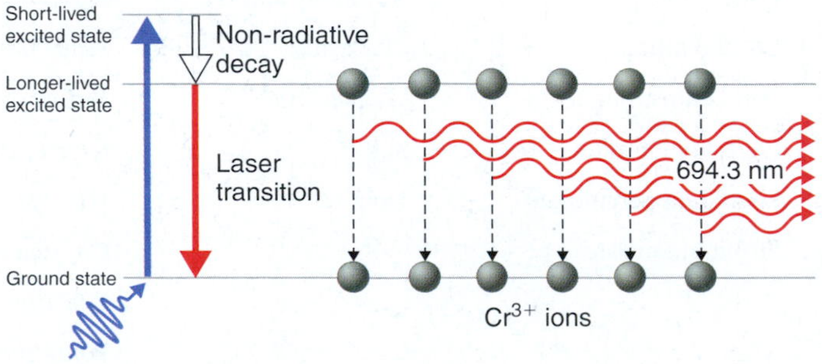

(b)

Differences in the frequencies of the transmitted and reflected light are detected by light-sensitive equipment that converts these differences into binary code, a series of 1's and 0's, which is translated electronically into recognizable sounds and images.

6.4 The Relationship Between Energy and Mass

MGC Relationship Between Energy and Mass

Einstein's photons of light were individual packets of energy that had many of the characteristics of particles. Recall that the collision of an electron (a particle) with a sufficiently energetic photon can eject a *photoelectron* from the surface of a metal. Any excess energy is transferred to the electron and is converted to the kinetic energy of the ejected electron. Einstein's hypothesis that energy is concentrated in localized bundles, however, was in sharp contrast to the classical notion that energy is spread out uniformly in a wave. We now describe Einstein's theory of the relationship between energy and mass, a theory that others built upon to develop our current model of the atom.

The Wave Character of Matter

Einstein initially assumed that photons had zero mass, which made them a peculiar sort of particle indeed. In 1905, however, he published his special theory of relativity, which related energy and mass by the equation

$$E = h\nu = \frac{hc}{\lambda} = mc^2 \tag{6.13}$$

According to this theory, a photon of wavelength λ and frequency ν has a nonzero mass, which is given as

$$m = \frac{E}{c^2} = \frac{h\nu}{c^2} = \frac{h}{\lambda c} \tag{6.14}$$

That is, light, which had always been regarded as a wave, also has properties typical of particles, a condition known as ***wave-particle duality***. Depending on conditions, light could be viewed as being either a wave or a particle.

In 1922, the American physicist Arthur Compton (1892–1962) reported the results of experiments involving the collision of X rays and electrons that supported the particle nature of light. At about the same time, a young French physics student named Louis de Broglie (1892–1972) began to wonder whether the converse was true: Could particles exhibit the properties of waves? In his Ph.D. dissertation submitted to the Sorbonne in 1924, de Broglie proposed that a particle such as an electron could be described by a wave whose wavelength is given by

$$\lambda = \frac{h}{mv} \tag{6.15}$$

where h is Planck's constant, m is the mass of the particle, and v is the velocity of the particle. This revolutionary idea was quickly confirmed by American physicists Clinton Davisson and Lester Germer, who showed that beams of electrons, regarded as particles, were diffracted by a sodium chloride crystal in the same manner as X rays, which were regarded as waves. Thus, it was proven experimentally that electrons do indeed exhibit the properties of waves! For his work, de Broglie received the Nobel Prize in Physics in 1929.

If particles exhibit the properties of waves, why had no one observed them before? The answer lies in the numerator of de Broglie's equation, which is an extremely small number. As you will calculate in the next example, Planck's constant (6.63×10^{-34} J·s) is so small that the wavelength of a particle with a large mass is too short (less than the diameter of an atomic nucleus) to be noticeable.

EXAMPLE 6.4

Calculate the wavelength of a baseball, which has a mass of 149 g and a speed of 100 mi/h.

Given Mass and speed of object

Asked for Wavelength

Strategy

A Convert the speed of the baseball to the appropriate SI units, meters per second.
B Substitute values into Equation 6.15 and solve for the wavelength.

Solution

A The wavelength of a particle is given by $\lambda = h/mv$. We know that $m = 0.149$ kg, so all we need to find is the speed of the baseball:

$$v = \left(\frac{100\ \cancel{\text{mi}}}{\cancel{\text{h}}}\right)\left(\frac{1\ \cancel{\text{h}}}{60\ \cancel{\text{min}}}\right)\left(\frac{1\ \cancel{\text{min}}}{60\ \text{s}}\right)\left(\frac{1.609\ \cancel{\text{km}}}{\cancel{\text{mi}}}\right)\left(\frac{1000\ \text{m}}{\cancel{\text{km}}}\right) = 44.69\ \text{m/s}$$

B Recall that the joule is a derived unit, whose units are $(\text{kg}\cdot\text{m}^2)/\text{s}^2$. Thus, the wavelength of the baseball is

$$\lambda = \frac{6.626 \times 10^{-34}\ \text{J}\cdot\text{s}}{(0.149\ \text{kg})\ (44.69\ \text{m}\cdot\text{s}^{-1})} = \frac{6.626 \times 10^{-34}\ \cancel{\text{kg}}\cdot\text{m}^{\cancel{2}}\cdot\cancel{\text{s}^{-2}}\cdot\cancel{\text{s}}}{(0.149\ \cancel{\text{kg}})(44.69\ \cancel{\text{m}}\cdot\cancel{\text{s}^{-1}})} = 9.95 \times 10^{-35}\ \text{m}$$

(You should verify that units cancel to give the wavelength in meters.) Given that the diameter of the nucleus of an atom is approximately 10^{-14} m, the wavelength of the baseball is almost unimaginably small.

EXERCISE 6.4

Calculate the wavelength of a neutron that is moving at 3.00×10^3 m/s.

Answer 1.32 Å, or 132 pm

As you calculated in Example 6.4, objects such as a baseball or a neutron have such short wavelengths that they are best regarded primarily as particles. In contrast, objects with very small masses (such as photons) have large wavelengths and can be viewed primarily as waves. Objects with intermediate masses, such as electrons, exhibit the properties of both particles and waves. Although we still usually think of electrons as particles, the wave nature of electrons is utilized in an instrument called the *electron microscope*, which has revealed most of what we know about the microscopic structure of living organisms and materials. Because the wavelength of an electron beam is much shorter than the wavelength of a beam of visible light, this instrument can resolve smaller details than a light microscope can (Figure 6.17).

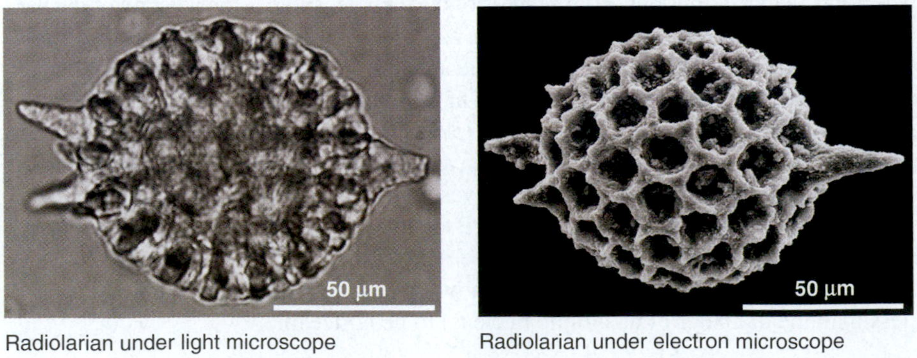

Figure 6.17 Comparison of images obtained using a light microscope and an electron microscope. Because of their shorter wavelength, high-energy electrons have a higher resolving power than visible light. Consequently, the electron microscope is able to resolve finer details than the light microscope. (Radiolaria are unicellular planktonic organisms.)

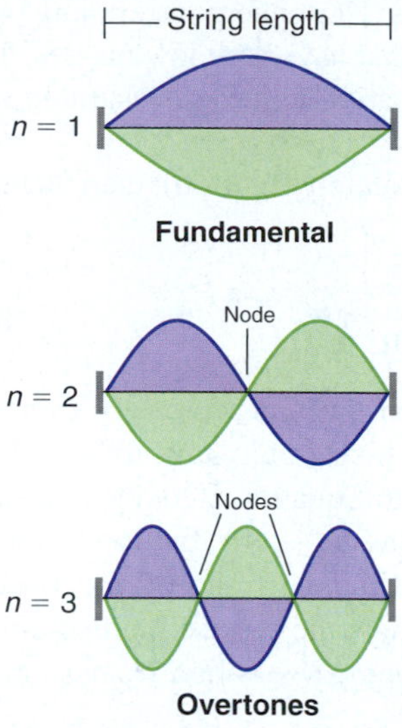

Figure 6.18 Standing waves in a vibrating string. The vibration with $n = 1$ is the fundamental and contains no nodes. Vibrations with higher values of n are called overtones; they contain $n - 1$ nodes.

Standing Waves

De Broglie also investigated why only certain orbits were allowed in Bohr's model of the hydrogen atom. He hypothesized that the electron behaves like a **standing wave**, a wave that does not travel in space. An example of a standing wave is the motion of a string of a violin or guitar. When the string is plucked, it vibrates at certain fixed frequencies because it is fastened at both ends (Figure 6.18). If the length of the string is L, then the lowest-energy vibration (the *fundamental*) has wavelength

$$\frac{\lambda}{2} = L \quad \lambda = 2L \tag{6.16}$$

Higher-energy vibrations (*overtones*) are produced when the string is plucked more strongly; they have wavelengths given by

$$\lambda = \frac{2L}{n} \tag{6.17}$$

where n has any integral value. Thus, the vibrational energy of the string is quantized, and only certain wavelengths and frequencies are possible. Notice in Figure 6.18 that all overtones have one or more **nodes**, points where the string does not move. The amplitude of the wave at a node is zero.

Quantized vibrations and overtones containing nodes are not restricted to one-dimensional systems such as strings. A two-dimensional surface such as a drumhead also has quantized vibrations. Similarly, when the ends of a string are joined to form a circle, the only allowed vibrations are those with wavelength

$$2\pi r = n\lambda \tag{6.18}$$

where r is the radius of the circle. De Broglie argued that Bohr's allowed orbits could be understood if the electron behaved like a *standing circular wave* (Figure 6.19). The standing wave could exist only if the circumference of the circle was an integral multiple of the wavelength causing *constructive interference*. Otherwise, the wave would be out of phase with itself on successive orbits and would cancel out, causing *destructive interference*. De Broglie's idea explained Bohr's allowed orbits and energy levels nicely: in the lowest energy level, corresponding to $n = 1$ in Equation 6.18, one complete wavelength would close the circle. Higher energy levels would have successively higher values of n with a corresponding number of nodes.

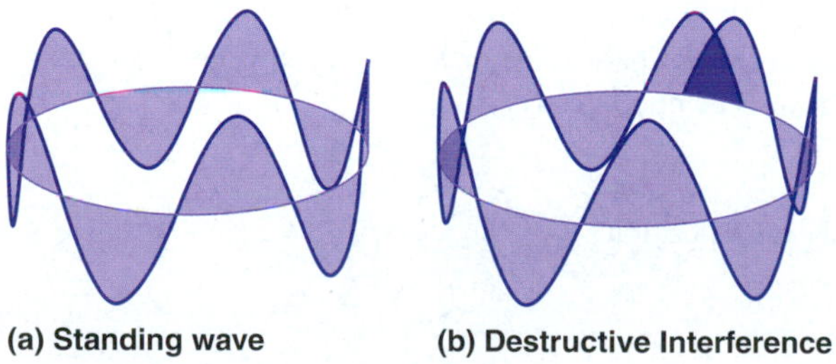

Figure 6.19 Standing circular wave and destructive interference. (a) A standing circular wave with $n = 5$; note that the circumference of the circle corresponds to exactly five wavelengths, which results in constructive interference of the wave with itself when overlapping occurs. (b) If the circumference of the circle is not equal to an integral multiple of wavelengths, then the wave does not overlap exactly with itself and the resulting destructive interference will result in cancellation of the wave. Consequently, a standing wave cannot exist under these conditions.

As you will see, several of de Broglie's ideas are retained in the modern theory of the electronic structure of the atom: the wave behavior of the electron, the concept of standing waves, and the presence of nodes that increase in number as the energy level increases. Unfortunately, his explanation also contains one major feature that we know to be incorrect: in the currently accepted model, the electron in a given orbit is *not* always at the same distance from the nucleus.

The Heisenberg Uncertainty Principle

Because a wave is a disturbance that travels in space, it has no fixed position. One might therefore expect that it would also be hard to specify the exact position of a *particle* that exhibits wavelike behavior. This situation was described mathematically by the German physicist Werner Heisenberg (1901–1976, Nobel Prize 1932). The **Heisenberg uncertainty principle** states that the uncertainty in the position of a particle (Δx) multiplied by the uncertainty in its momentum [$\Delta(mv)$] is greater than or equal to Planck's constant divided by 4π:

$$(\Delta x)[\Delta(mv)] \geq \frac{h}{4\pi} \tag{6.19}$$

Because Planck's constant is a very small number, the Heisenberg uncertainty principle is important only for particles such as electrons that have very low masses. These are the same particles predicted by de Broglie's equation to have measurable wavelengths.

If the precise position x of a particle is known absolutely ($\Delta x = 0$), then the uncertainty in its momentum, $\Delta(mv)$, must be infinite:

$$\Delta(mv) = \frac{h}{4\pi(\Delta x)} = \frac{h}{4\pi(0)} = \infty \qquad (6.20)$$

Because the mass of the electron at rest, m, is both constant and accurately known, the uncertainty in $\Delta(mv)$ must be due to the Δv term, which would have to be infinitely large for $\Delta(mv)$ to equal infinity. That is, according to Equation 6.20, the more accurately we know the exact position of the electron (as $\Delta x \rightarrow 0$), the less accurately we know the speed and hence the kinetic energy of the electron ($1/2\ mv^2$) because $\Delta(mv) \rightarrow \infty$. Conversely, the more accurately we know the precise momentum (and hence the energy) of the electron [as $\Delta(mv) \rightarrow 0$], then $\Delta x \rightarrow \infty$ and we have no idea where the electron is.

Bohr's model of the hydrogen atom violated the Heisenberg uncertainty principle by trying to specify simultaneously both the position (an orbit of a particular radius) and the energy (a quantity related to the momentum) of the electron. Moreover, given its mass and wavelike nature, the electron in the hydrogen atom could not possibly orbit the nucleus in a well-defined circular path as predicted by Bohr. You will see, however, that the *most probable radius* of the electron in the hydrogen atom is exactly the one predicted by Bohr's model.

EXAMPLE 6.5

Calculate the minimum uncertainty in the position of the pitched baseball from Example 6.4 that has a mass of exactly 149 g and a speed of 100 ± 1 mi/h.

Given Mass and speed of object

Asked for Minimum uncertainty in its position

Strategy

A Rearrange the inequality that describes the Heisenberg uncertainty principle (Equation 6.19) to solve for the minimum uncertainty in the position of an object, Δx.

B Find Δv by converting the velocity of the baseball to the appropriate SI units, meters per second.

C Substitute the appropriate values into the expression for the inequality and solve for Δx.

Solution

A The Heisenberg uncertainty principle tells us that $(\Delta x)[\Delta(mv)] \geq h/4\pi$. Rearranging the inequality gives

$$\Delta x \geq \frac{h}{4\pi} \cdot \frac{1}{\Delta(mv)}$$

B We know that $h = 6.626 \times 10^{-34}$ J·s and $m = 0.149$ kg. Because there is no uncertainty in the mass of the baseball, $\Delta(mv) = m\Delta v$ and $\Delta v = \pm 1$ mi/h. We have

$$\Delta v = \left(\frac{1\ \cancel{\text{mi}}}{\cancel{\text{h}}}\right)\left(\frac{1\ \cancel{\text{h}}}{60\ \cancel{\text{min}}}\right)\left(\frac{1\ \cancel{\text{min}}}{60\ \text{s}}\right)\left(\frac{1.609\ \cancel{\text{km}}}{\cancel{\text{mi}}}\right)\left(\frac{1000\ \text{m}}{\cancel{\text{km}}}\right) = 0.4469\ \text{m/s}$$

C Therefore,

$$\Delta x \geq \left(\frac{6.626 \times 10^{-34}\ \text{J}\cdot\text{s}}{4\,(3.1416)}\right)\left(\frac{1}{(0.149\ \text{kg})\,(0.4469\ \text{m/s})}\right)$$

Inserting the definition of a joule ($1\ \text{J} = 1\ \text{kg}\cdot\text{m}^2\cdot\text{s}^{-2}$) gives

$$\Delta x \geq \left(\frac{6.626 \times 10^{-34}\ \cancel{\text{kg}}\cdot\text{m}^{\cancel{2}}\cdot\cancel{\text{s}}}{4\,(3.1416)(\text{s}^{\cancel{2}})}\right)\left(\frac{\cancel{1\ \text{s}}}{(0.149\ \cancel{\text{kg}})\ (0.4469\ \cancel{\text{m}})}\right)$$
$$\geq 7.92 \times 10^{-34}\ \text{m}$$

This is equal to 3.12×10^{-32} inches! We can safely say that if a batter misjudges the speed of a fastball by 1 mi/h (about 1%), he will not be able to blame Heisenberg's uncertainty principle for striking out.

EXERCISE 6.5

Calculate the minimum uncertainty in the position of an electron traveling at one-third the speed of light, if the uncertainty in its speed is ±0.1%. Assume its mass to be equal to its rest mass.

Answer 6×10^{-10} m, or 0.6 nm (about the diameter of a benzene molecule)

Erwin Schrödinger (1887–1961). Schrödinger's unconventional approach to atomic theory was typical of his unconventional approach to life. He was notorious for his intense dislike of memorizing data and learning from books. When Hitler came to power in Germany, Schrödinger escaped to Italy. He then worked at Princeton University in the United States, but eventually moved to the Institute for Advanced Studies in Dublin, Ireland, where he remained until his retirement in 1955.

6.5 • Atomic Orbitals and Their Energies

The paradox described by Heisenberg's uncertainty principle and the wavelike nature of subatomic particles such as the electron made it impossible to use the equations of classical physics to describe the motion of electrons in atoms. Scientists needed a new approach that took the wave behavior of the electron into account. In 1926, an Austrian physicist, Erwin Schrödinger (1887–1961, Nobel Prize 1931), developed *wave mechanics*, a mathematical technique to describe the relationship between the motion of a particle that exhibits wavelike properties (such as an electron) and its allowed energies. In doing so, Schrödinger developed the theory of **quantum mechanics**, which is used today to describe the energies and spatial distributions of electrons in atoms and molecules.

Although quantum mechanics makes use of sophisticated mathematics, you do not need to understand the mathematical details in order to follow our discussion of its general conclusions. We focus on the properties of the *wave functions* that are the solutions of Schrödinger's equations.

MGC Quantum Numbers

Wave Functions

A **wave function** is a mathematical function that relates the location of an electron at a given point in space (identified by x, y, z coordinates) to the amplitude of its wave, which corresponds to its energy. Thus, each wave function Ψ (capital Greek letter psi) is associated with a particular energy E. The properties of wave functions derived from quantum mechanics are summarized here:

1. ***A wave function uses three variables to describe the position of an electron.*** A fourth variable is usually required to fully describe the location of objects in motion. Three specify the position in space (as with the Cartesian coordinates x, y, z), and one specifies the time at which the object is at the specified location. For example, if you wanted to intercept an enemy submarine, you would need to know its latitude, longitude, and depth, as well as the time at which it was going to be at this position (Figure 6.20). For electrons, we can ignore the time dependence because we will be using standing waves, which by definition do not change with time, to describe the position of the electron.
2. ***The magnitude of the wave function at a particular point in space is proportional to the amplitude of the wave at that point.*** Many wave functions, Ψ, are complex functions, which is a mathematical term indicating that they

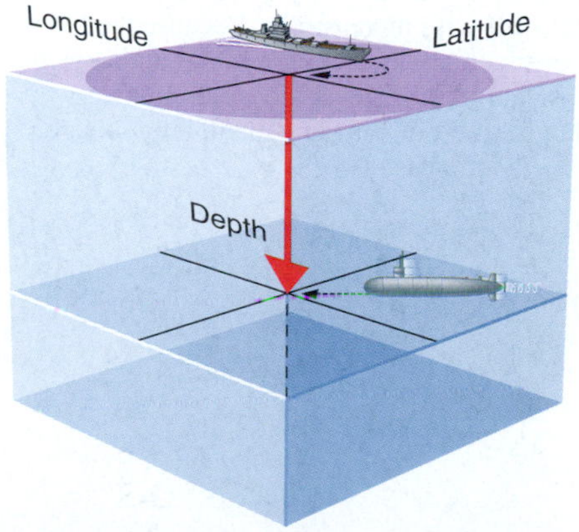

Figure 6.20 The four variables (latitude, longitude, depth, and time) required to precisely locate an object such as a submarine. If you are the captain of a ship trying to intercept an enemy submarine, you need to deliver your depth charge to the right location at the right time.

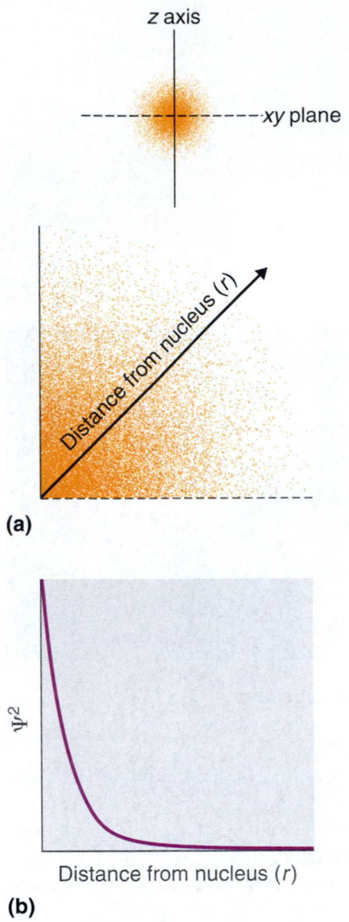

Figure 6.21 Probability of finding the electron in the ground state of the hydrogen atom at different points in space. (a) The electron probability is shown by the density of the dots. (b) A plot of Ψ^2 versus r for the ground state of the hydrogen atom. Note that the electron probability density is greatest at $r = 0$ (the nucleus) and falls off with increasing r. Because the line never actually reaches the horizontal axis, the probability of finding the electron at very large values of r is very small, but *not* zero.

contain $\sqrt{-1}$, represented as i. Hence, the amplitude of the the wave has no real physical significance. The sign of the wave function (either positive or negative) corresponds to the phase of the wave, which will be important in our discussion of chemical bonding in Chapter 9. The sign of the wave function should *not* be confused with a positive or negative electrical charge.

3. ***The square of the wave function at a given point is proportional to the probability of finding an electron at that point, which leads to a distribution of probabilities in space.*** The square of the wave function, Ψ^2, is always a real quantity [recall that $(\sqrt{-1})^2 = -1$] that is proportional to the probability of finding an electron at a given point.* We use probabilities because, according to Heisenberg's uncertainty principle, we cannot specify the position of the electron precisely. The probability of finding the electron at any point in space depends on several factors, including the distance from the nucleus and, in many cases, the atomic equivalent of latitude and longitude. In one way of representing the probability distribution graphically, the probability of finding the electron is indicated by the density of colored dots, as shown for the ground state of the hydrogen atom in Figure 6.21.
4. ***Describing the electron distribution as a standing wave leads to sets of "quantum numbers" characteristic of each wave function.*** From the patterns of one- and two-dimensional standing waves shown in Figures 6.18 and 6.19, you might expect (correctly) that the patterns of three-dimensional standing waves would be complex. Fortunately, however, in the 18th century, a French mathematician, Adrien Legendre (1752–1783), developed a set of equations to describe the motion of tidal waves on the surface of a flooded planet. Schrödinger incorporated Legendre's equations into his wave functions. The requirement that the waves must be in phase with one another to avoid cancellation and produce a standing wave results in a limited number of solutions (wave functions), each of which is specified by a set of numbers called **quantum numbers**.
5. ***Each wave function Ψ is associated with a particular energy.*** Thus, as in Bohr's model, the energy of an electron in an atom is quantized; it can have only certain allowed values. The major difference between Bohr's model and Schrödinger's is that Bohr had to impose the idea of quantization arbitrarily, whereas in Schrödinger's approach, quantization is a natural consequence of describing the electron as a standing wave.

Quantum Numbers

Schrödinger's approach uses three quantum numbers (n, l, and m_l) to specify any wave function. The quantum numbers provide information about the spatial distribution of the electron. Although n can be any positive integer, only certain values of l and m_l are allowed for a given value of n.

The Principal Quantum Number

The quantum number n is the **principal quantum number**, which tells the average relative distance of the electron from the nucleus:

$$n = 1, 2, 3, 4, \ldots \quad (6.21A)$$

As n increases for a given atom, so does the average distance of the electron from the nucleus. A negatively charged electron that is, on average, closer to the positively

* More accurately, the probability is given by the product of the wave function Ψ and its complex conjugate Ψ^*, in which all terms that contain i are replaced by $-i$.

charged nucleus is attracted to the nucleus more strongly than an electron that is farther out in space. This means that electrons with higher values of n are easier to remove from an atom. All wave functions that have the same value of n are said to constitute a **principal shell** because those electrons have similar average distances from the nucleus. As you will see, the principal quantum number n corresponds to the n used by Bohr to describe electron orbits and by Rydberg to describe atomic energy levels.

The Azimuthal Quantum Number

The second quantum number is l, often called the **azimuthal quantum number**. The value of l describes the *shape* of the region of space occupied by the electron. The allowed values of l depend on the value of n and can range from 0 to $n - 1$:

$$l = 0, 1, 2, \ldots, n - 1 \qquad (6.21B)$$

For example, if $n = 1$, l can be only 0; if $n = 2$, l can be 0 or 1; and so on. For a given atom, all wave functions that have the same values of both n and l form a **subshell**. The regions of space occupied by electrons in the same subshell usually have the same shape, but they are oriented differently in space.

The Magnetic Quantum Number

The third quantum number is m_l, the **magnetic quantum number**. The value of m_l describes the *orientation* of the region in space occupied by the electron with respect to an applied magnetic field. The allowed values of m_l depend on the value of l: m_l can range from $-l$ to l in integral steps:

$$m_l = -l, -l + 1, \ldots, 0, \ldots, l - 1, l \qquad (6.21C)$$

For example, if $l = 0$, m_l can be only 0; if $l = 1$, m_l can be -1, 0, or $+1$; and if $l = 2$, m_l can be -2, -1, 0, $+1$, or $+2$.

Each wave function with an allowed combination of n, l, and m_l values describes an **atomic orbital**, a particular spatial distribution for an electron. Thus, for a given set of quantum numbers, each principal shell contains a fixed number of subshells, and each subshell contains a fixed number of orbitals.

EXAMPLE 6.6

How many subshells and orbitals are contained within the principal shell with $n = 4$?

Given Value of n

Asked for Number of subshells and orbitals in principal shell

Strategy

A Given $n = 4$, calculate the allowed values of l. From these allowed values, count the number of subshells.

B For each allowed value of l, calculate the allowed values of m_l. Sum the number of orbitals in each subshell to obtain the number of orbitals contained within the principal shell.

Solution

A We know that l can have all integral values from 0 to $n - 1$. If $n = 4$, then l can equal 0, 1, 2, or 3. Because the shell has four values of l, it has four subshells, each of which will contain a different number of orbitals, depending on the allowed values of m_l. **B** For $l = 0$, m_l can be only 0, and thus the $l = 0$ subshell contains only one orbital. For $l = 1$, m_l can be 0 or ±1; thus, the $l = 1$ subshell contains three orbitals. For $l = 2$, m_l can be 0, ±1, or ±2, and thus there are five orbitals in the $l = 2$ subshell. The last allowed value of l is $l = 3$, for which m_l can be 0, ±1, ±2, or ±3, resulting in seven orbitals in the $l = 3$ subshell. The total number of orbitals in the $n = 4$ principal shell is the sum of the number of orbitals in each of the subshells and is equal to n^2:

$$\underset{(l=0)}{1} + \underset{(l=1)}{3} + \underset{(l=2)}{5} + \underset{(l=3)}{7} = 16 \text{ orbitals} = (4 \text{ principal shells})^2$$

EXERCISE 6.6

How many subshells and orbitals are contained within the principal shell with $n = 3$?

Answer three subshells; nine orbitals

Rather than specifying all the values of n and l every time we refer to a subshell or an orbital, chemists use an abbreviated system with lowercase letters to denote the value of l for a particular subshell or orbital:

	$l =$ 0	1	2	3
Designation	*s*	*p*	*d*	*f*

The principal quantum number is named first, followed by the letter *s*, *p*, *d*, or *f* as appropriate. Thus, a 1*s* orbital has $n = 1$ and $l = 0$; a 2*p* subshell has $n = 2$ and $l = 1$ (and contains three 2*p* orbitals, corresponding to $m_l = -1, 0$, and $+1$); a 3*d* subshell has $n = 3$ and $l = 2$ (and contains five 3*d* orbitals, corresponding to $m_l = -2, -1, 0, +1$, and $+2$); and so on.

We can summarize the relationships between the quantum numbers and the number of subshells and orbitals as follows (Table 6.3):

Note the pattern

Each principal shell contains n subshells, and each subshell contains 2l + 1 orbitals.

1. Each principal shell contains n subshells. For $n = 1$, only a single subshell is possible (1*s*); for $n = 2$, there are two subshells (2*s* and 2*p*); for $n = 3$, there are three subshells (3*s*, 3*p*, and 3*d*); and so on. Notice that every shell contains an *ns* subshell, any shell with $n \geq 2$ also contains an *np* subshell, and any shell with $n \geq 3$ also contains an *nd* subshell. Because a 2*d* subshell would require both $n = 2$ and $l = 2$, which is not an allowed value of l for $n = 2$, a 2*d* subshell does not exist.
2. Each subshell contains $2l + 1$ orbitals. This means that all *ns* subshells contain a single *s* orbital, all *np* subshells contain three *p* orbitals, all *nd* subshells contain five *d* orbitals, and all *nf* subshells contain seven *f* orbitals.

TABLE 6.3 Values of n, l, and m_l through $n = 4$

n	l	Subshell Designation	m_l	Number of Orbitals in Subshell	Number of Orbitals in Shell
1	0	1*s*	0	1	1
2	0	2*s*	0	1	4
	1	2*p*	−1, 0, 1	3	
3	0	3*s*	0	1	9
	1	3*p*	−1, 0, 1	3	
	2	3*d*	−2, −1, 0, 1, 2	5	
4	0	4*s*	0	1	16
	1	4*p*	−1, 0, 1	3	
	2	4*d*	−2, −1, 0, 1, 2	5	
	3	4*f*	−3, −2, −1, 0, 1, 2, 3	7	

Orbital Shapes

An orbital is the quantum mechanical refinement of Bohr's orbit. In contrast to his concept of a simple circular orbit with a fixed radius, orbitals are mathematically derived regions of space with different *probabilities* of containing the electron.

One way of representing electron probability distributions was illustrated in Figure 6.21 for the 1*s* orbital of hydrogen. Because Ψ^2 gives the probability of finding the electron in a given volume of space (such as a cubic picometer), a plot of Ψ^2 versus distance from the nucleus, *r*, is a plot of the *probability density*. The 1*s* orbital is spherically symmetrical, so the probability of finding a 1*s* electron at any given point depends *only* on its distance from the nucleus. The probability density is greatest at $r = 0$ (at the nucleus) and decreases steadily with increasing distance. At very large values of *r*, the electron probability density is very small, but *not* zero.

In contrast, we can calculate the *radial probability* (the probability of finding a 1*s* electron at a distance *r* from the nucleus) by summing the probabilities of the electron being at all points on a series of *x* spherical shells of radius $r_1, r_2, r_3, \ldots, r_{x-1}, r_x$. In effect, we are dividing the atom into very thin concentric shells, much like the layers of an onion (Figure 6.22a), and calculating the probability of finding the electron on each of the spherical shells. Recall that the electron probability density is greatest at $r = 0$ (Figure 6.22b), and so the density of dots is greatest for the smallest spherical shells in Figure 6.22a. In contrast, the surface area of each spherical shell is equal to $4\pi r^2$, which increases very rapidly with increasing *r* (Figure 6.22c). Because the surface area of the spherical shells increases more rapidly with increasing *r* than the electron probability density decreases, the plot of radial probability has a maximum at a particular distance, as shown in Figure 6.22d. Most important, when *r* is very small, the surface area of a

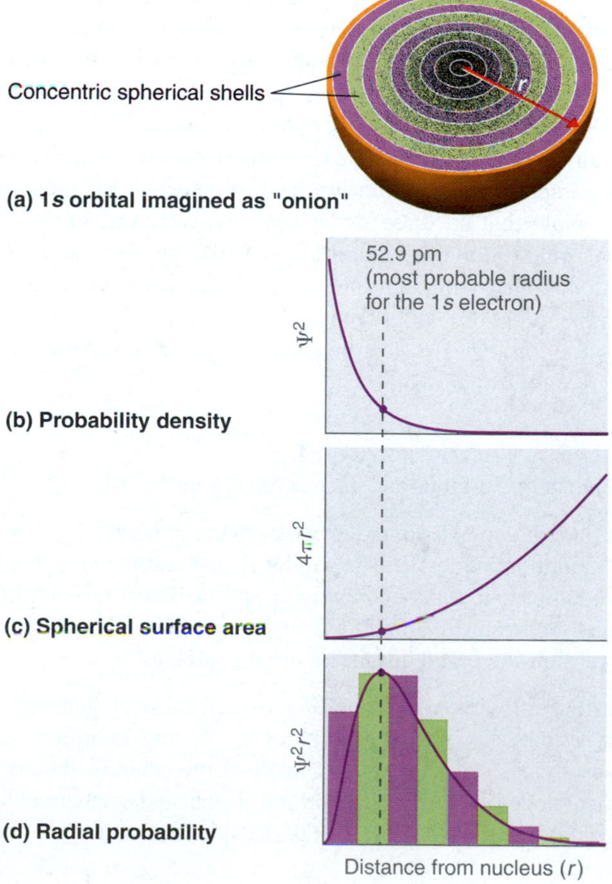

Figure 6.22 Most probable radius for the electron in the ground state of the hydrogen atom. (a) Imagine dividing the atom's total volume into very thin concentric shells as shown in the "onion" drawing. (b) A plot of electron probability density Ψ^2 versus *r* shows that the electron probability density is greatest at $r = 0$ and falls off smoothly with increasing *r*. The density of the dots is therefore greatest in the innermost shells of the "onion." (c) The surface area of each shell, given by $4\pi r^2$, increases rapidly with increasing *r*. (d) If we count the number of dots in each spherical shell, we obtain the total probability of finding the electron at a given value of *r*. Because the surface area of each shell increases more rapidly with increasing *r* than the electron probability density decreases, a plot of electron probability versus *r* (the *radial probability*) shows a peak. This peak corresponds to the most probable radius for the electron, 52.9 picometers (pm), exactly the radius predicted by Bohr's model of the hydrogen atom.

Interactive Graph

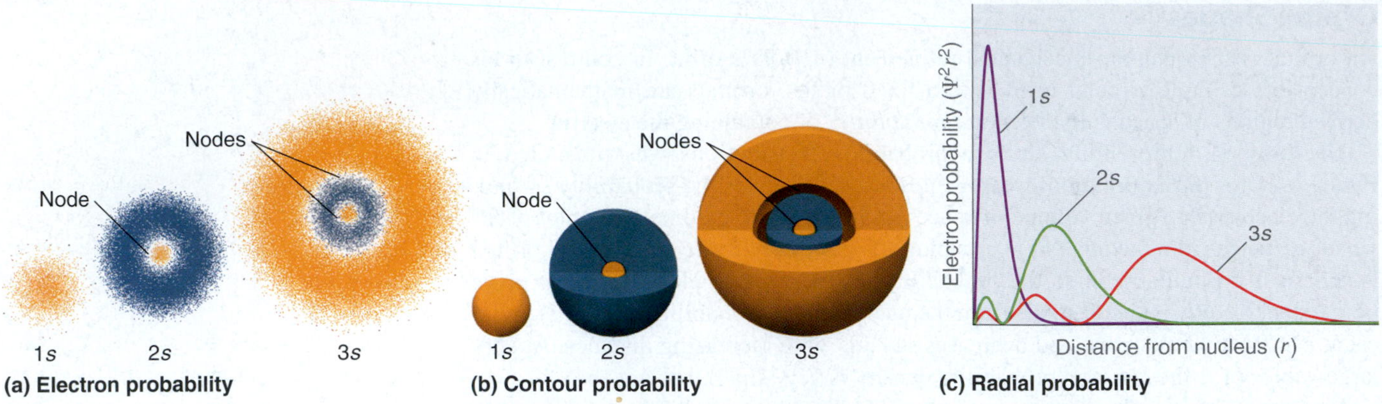

Figure 6.23 Probability densities for the 1*s*, 2*s*, and 3*s* orbitals of the hydrogen atom. (a) The electron probability density in any plane that contains the nucleus. Note the presence of circular regions where the probability density is zero (nodes). (b) Contour surfaces enclosing 90% of the electron probability, illustrating the different sizes of the 1*s*, 2*s*, and 3*s* orbitals. The cutaway drawings give partial views of the internal spherical nodes. The orange color corresponds to regions of space where the phase of the wave function Ψ is positive, and the blue color corresponds to regions of space where the phase of the wave function is negative. (c) Plots of electron probability as a function of distance from the nucleus, r, in all directions (radial probability). Note that the most probable radius increases as n increases but the 2*s* and 3*s* orbitals have regions of significant electron probability at small values of r.

spherical shell is so small that the *total* probability of finding an electron close to the nucleus is very low, and at the nucleus the electron probability vanishes (Figure 6.22d).

For the hydrogen atom, the peak in the radial probability plot occurs at $r = 0.529$ Å (52.9 pm), which is exactly the radius calculated by Bohr for the $n = 1$ orbit. Thus, the *most probable radius* obtained from quantum mechanics is identical to the radius calculated by classical mechanics. In Bohr's model, however, the electron was assumed to be at this distance 100% of the time, whereas in the Schrödinger model, it is at this distance only some of the time. The difference between the two models is attributable to the wavelike behavior of the electron and the Heisenberg uncertainty principle.

Figure 6.23 compares the electron probability densities for the hydrogen 1*s*, 2*s*, and 3*s* orbitals. Note that all three are spherically symmetrical. For the 2*s* and 3*s* orbitals, however (and for all other *s* orbitals as well), the electron probability density does not fall off smoothly with increasing r. Instead, a series of minima and maxima are observed in the radial probability plots (Figure 6.23c). The minima correspond to spherical nodes (regions of zero electron probability), which alternate with spherical regions of nonzero electron probability.

s Orbitals

Three things happen to *s* orbitals as n increases (Figure 6.23):

1. They become larger, extending farther from the nucleus.
2. They contain more nodes. This is similar to a standing wave that has regions of significant amplitude separated by nodes, points with zero amplitude.
3. For a given atom, the *s* orbitals also become higher in energy as n increases because of their increased distance from the nucleus.

Orbitals are generally drawn as three-dimensional surfaces that enclose 90% of the electron density, as was shown for the hydrogen 1*s*, 2*s*, and 3*s* orbitals in Figure 6.23b. Although such drawings show the relative sizes of the orbitals, they do not normally show the spherical nodes in the 2*s* and 3*s* orbitals because the spherical nodes lie inside the 90% surface. Fortunately, the positions of the spherical nodes are not important for chemical bonding.

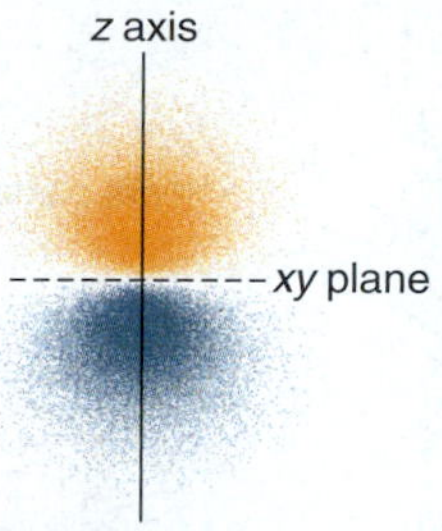

Figure 6.24 Electron probability distribution for a hydrogen 2*p* orbital. Note the nodal plane of zero electron density separating the two lobes of the 2*p* orbital. As in Figure 6.23, the colors correspond to regions of space where the phase of the wave function is positive (orange) and negative (blue).

p Orbitals

Only *s* orbitals are spherically symmetrical. As the value of l increases, the number of orbitals in a given subshell increases and the shapes of the orbitals become more complex. Because the 2*p* subshell has $l = 1$, with three values of m_l ($-1, 0, +1$), there are three 2*p* orbitals.

The electron probability distribution for one of the hydrogen 2*p* orbitals is shown in Figure 6.24. Because this orbital has two lobes of electron density arranged along the *z* axis, with an electron density of zero in the *xy* plane (that is, the *xy* plane is a nodal plane), it is a $2p_z$ orbital. As shown in Figure 6.25, the other two 2*p* orbitals have identical shapes, but they lie along the *x* axis ($2p_x$) and *y* axis ($2p_y$), respectively. Note that each *p* orbital has just one nodal plane. In each case, the phase of the wave function for each of the 2*p* orbitals is positive for the lobe that points along the positive axis, and negative for the lobe that points along the negative axis. It is important to emphasize that these signs correspond to the *phase* of the wave that describes the electron motion, *not* to positive or negative charges!

Just as with the *s* orbitals, the size and complexity of the *p* orbitals for any atom increase as the principal quantum number *n* increases. The shapes of the 90% probability surfaces of the 3*p*, 4*p*, and higher-energy *p* orbitals are, however, essentially the same as those shown in Figure 6.25.

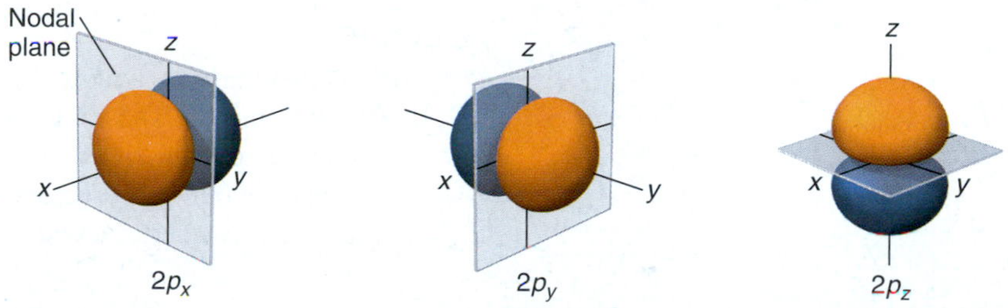

Figure 6.25 The three equivalent 2*p* orbitals of the hydrogen atom. The surfaces shown enclose 90% of the total electron probability for the $2p_x$, $2p_y$, and $2p_z$ orbitals. Note that each of the orbitals is oriented along the axis indicated by the subscript and that each 2*p* orbital is bisected by a nodal plane that is perpendicular to that axis. The phase of the wave function is positive (orange) in the region of space where *x*, *y*, or *z* is positive, and negative (blue) where *x*, *y*, or *z* is negative.

d Orbitals

Subshells with $l = 2$ contain five *d* orbitals; the first principal shell to have a *d* subshell corresponds to $n = 3$. The five *d* orbitals have m_l values of $-2, -1, 0, +1$, and $+2$.

The hydrogen 3*d* orbitals, shown in Figure 6.26, have more complex shapes than the 2*p* orbitals. All five 3*d* orbitals contain two nodal surfaces, as compared to one for each *p* orbital and zero for each *s* orbital. In three of the *d* orbitals, the lobes of electron density are oriented between the *x* and *y*, *x* and *z*, and *y* and *z* planes; these orbitals are referred to as the $3d_{xy}$, $3d_{xz}$, and $3d_{yz}$ orbitals, respectively. A fourth *d* orbital has lobes lying along the *x* and *y* axes; this is the $3d_{x^2-y^2}$ orbital. The fifth 3*d* orbital, called the $3d_{z^2}$ orbital, has a unique shape: it looks like a $2p_z$ orbital combined with an additional doughnut of electron probability lying in the *xy* plane. Despite its peculiar shape, the $3d_{z^2}$ orbital is mathematically equivalent to the other four and has the same energy. In contrast to *p* orbitals, for *d* orbitals the phase of the wave function is the same for opposite pairs of lobes. As shown in Figure 6.26, the phase of the wave function is positive for the two lobes of the d_{z^2} orbital that lie along the *z* axis, whereas the phase of the wave function is negative for the "doughnut" of electron density in the *xy* plane. Like the *s* and *p* orbitals, as *n* increases, the size of the *d* orbitals increases but the overall shapes remain similar to those depicted in Figure 6.26.

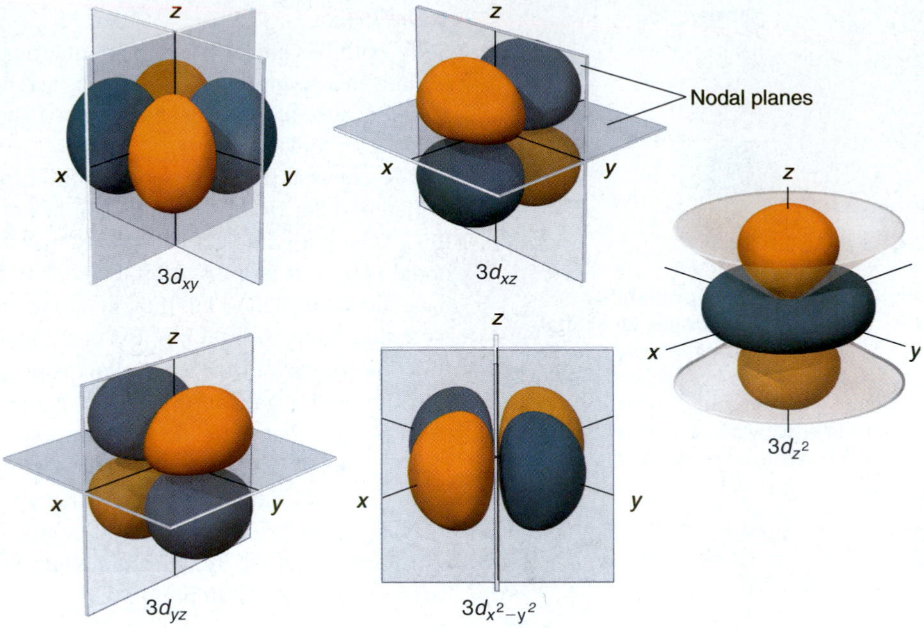

Figure 6.26 The five equivalent 3*d* orbitals of the hydrogen atom. The surfaces shown enclose 90% of the total electron probability for the five hydrogen 3*d* orbitals. Note that four of the five 3*d* orbitals consist of four lobes arranged in a plane that is intersected by two perpendicular nodal planes. These four orbitals have the same shape but different orientations. The fifth 3*d* orbital, $3d_{z^2}$, has a distinct shape even though it is mathematically equivalent to the others. The phase of the wave function for the different lobes is indicated by color: orange for positive and blue for negative.

f Orbitals

Principal shells with $n \geq 4$ can have subshells with $l = 3$ and m_l values of -3, -2, -1, 0, $+1$, $+2$, and $+3$. These subshells consist of seven *f* orbitals. Each *f* orbital has three nodal surfaces, so their shapes are complex. Because *f* orbitals are not particularly important for our purposes, we do not discuss them further, and orbitals with higher values of *l* are not discussed at all.

Orbital Energies

Although we have discussed the shapes of orbitals, we have said little about their comparative energies. We begin our discussion of orbital energies by considering atoms or ions that contain only a single electron (such as H or He^+).

The relative energies of the atomic orbitals with $n \leq 4$ for a hydrogen atom are plotted in Figure 6.27; note that the orbital energies depend *only* on the principal quantum number *n*. Consequently, the energies of the 2*s* and 2*p* orbitals of hydrogen are the same; the energies of the 3*s*, 3*p*, and 3*d* orbitals are the same; and so forth. The orbital energies obtained for hydrogen using quantum mechanics are exactly the same as the allowed energies calculated by Bohr. In contrast to Bohr's model, however, which allowed only one orbit for each energy level, quantum mechanics predicts that there are *4* orbitals with different electron density distributions in the $n = 2$ principal shell (one 2*s* and three 2*p* orbitals), *9* in the $n = 3$ principal shell, and *16* in the $n = 4$ principal shell.* As we have just seen, however, quantum mechanics also predicts that in the hydrogen atom, all orbitals with the same value of *n* (for example, the three 2*p* orbitals) are **degenerate**, meaning that they have the same energy. Figure 6.27 shows that the energy levels become closer and closer together as the value of *n* increases, as expected because of the $1/n^2$ dependence of orbital energies.

The energies of the orbitals in any species that contain only one electron can be calculated by a minor variation of Bohr's equation (Equation 6.9), which can be extended

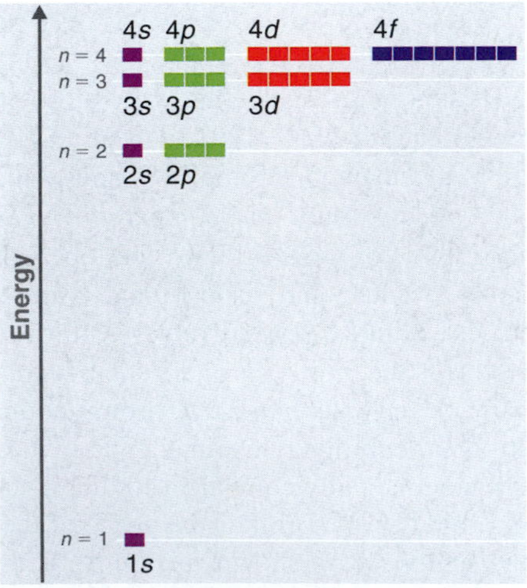

Figure 6.27 Orbital energy level diagram for the hydrogen atom. Each box corresponds to one orbital. Note that the difference in energy between orbitals decreases rapidly with increasing values of *n*.

* The different values of *l* and m_l for the individual orbitals within a given principal shell are not important for understanding the emission or absorption spectrum of the hydrogen atom under most conditions, but they do explain the splittings of the main lines that are observed when hydrogen atoms are placed in a magnetic field.

to other single-electron species by incorporating the nuclear charge Z (the number of protons in the nucleus):

$$E = -\frac{Z^2}{n^2}\mathfrak{R}hc \tag{6.22}$$

In general, both energy and radius decrease as the nuclear charge increases. Thus, the most stable orbitals (those with the lowest energy) are those that are closest to the nucleus. For example, in the ground state of the hydrogen atom, the single electron is in the $1s$ orbital, whereas in the first excited state, the atom has absorbed energy and the electron has been promoted to one of the $n = 2$ orbitals. In ions that contain only one electron, the energy of a given orbital depends only on n, and all subshells within a principal shell such as the p_x, p_y, and p_z orbitals, are degenerate.

Effective Nuclear Charges

For an atom or an ion that has only a single electron, we can calculate the potential energy by considering only the electrostatic attraction between the positively charged nucleus and the negatively charged electron. When more than one electron is present, however, the total energy of the atom or ion depends not only on attractive electron–nucleus interactions but also on repulsive electron–electron interactions. When there are two electrons, the repulsive interactions depend on the positions of *both* electrons at a given instant and as we cannot specify the exact positions of the electrons, it is impossible to calculate the repulsive interactions exactly. Consequently, we must use approximate methods to deal with the effect of electron–electron repulsions on orbital energies.

On one hand, if an electron is far from the nucleus (that is, if the distance r between the nucleus and the electron is large), then at any given moment most of the other electrons will be *between* that electron and the nucleus (Figure 6.28a). Hence, the electrons will cancel a portion of the positive charge of the nucleus and thereby decrease the attractive interaction between it and the electron farther away. As a result, the electron farther away experiences an **effective nuclear charge**, Z_{eff}, that is *less* than the actual nuclear charge Z. This effect is called **electron shielding**. As the distance between an electron and the nucleus approaches infinity, Z_{eff} approaches a value of 1 because all of the other $(Z - 1)$ electrons in the neutral atom are, on the average, between it and the nucleus. If, on the other hand, an electron is very close to the nucleus, then at any given moment most of the other electrons are farther from the nucleus and do not shield the nuclear charge (Figure 6.28b). Thus, at $r \approx 0$, the positive charge experienced by an electron is approximately the full nuclear charge, or $Z_{eff} \approx Z$. At intermediate values of

Note the pattern

Due to electron shielding, Z_{eff} increases more rapidly going across a row of the periodic table than going down a column.

Figure 6.28 Electron shielding in multielectron atoms. (a) If an electron is very far from the nucleus, at any given time most of the other electrons lie between it and the nucleus, leading to a small effective nuclear charge ($Z_{eff} \approx 1$). (b) If an electron is very close to the nucleus, at any given time it is between the nucleus and most of the other electrons. Consequently, it experiences a large effective nuclear charge ($Z_{eff} \approx Z$).

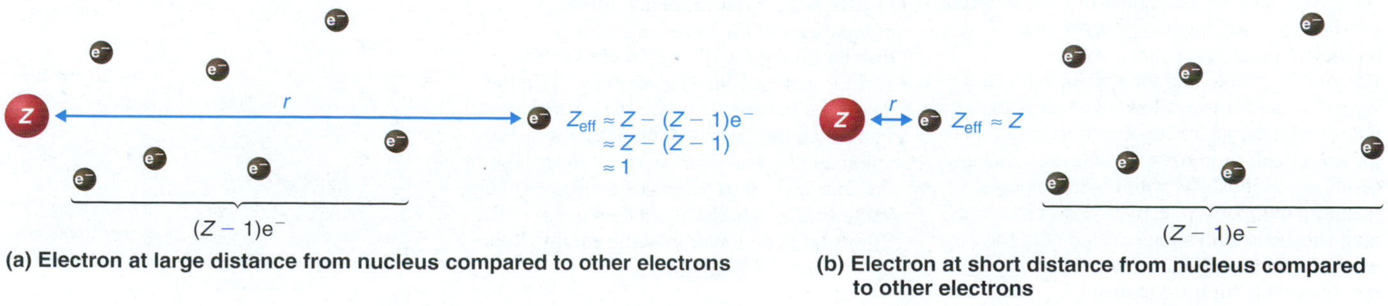

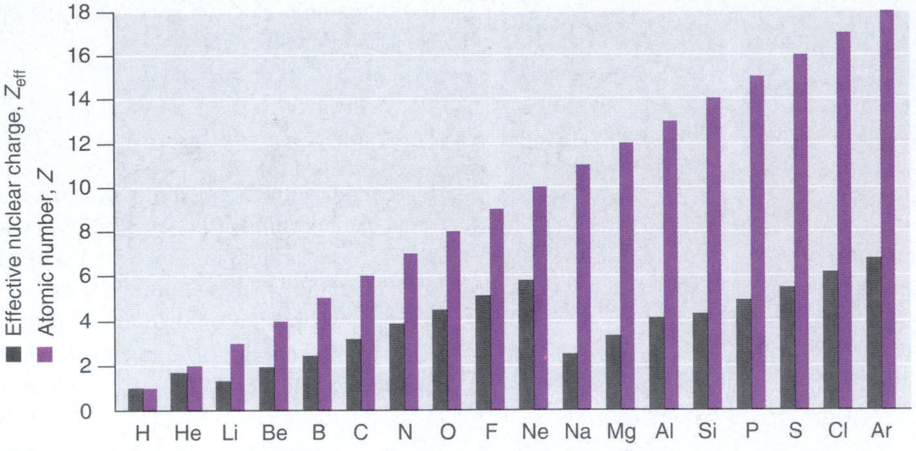

Figure 6.29 Relationship between effective nuclear charge Z_{eff} and atomic number Z for the outer electrons of the elements of the first three rows of the periodic table. Notice that, except for H, Z_{eff} is *always* less than Z and that Z_{eff} increases from left to right as you go across a row.

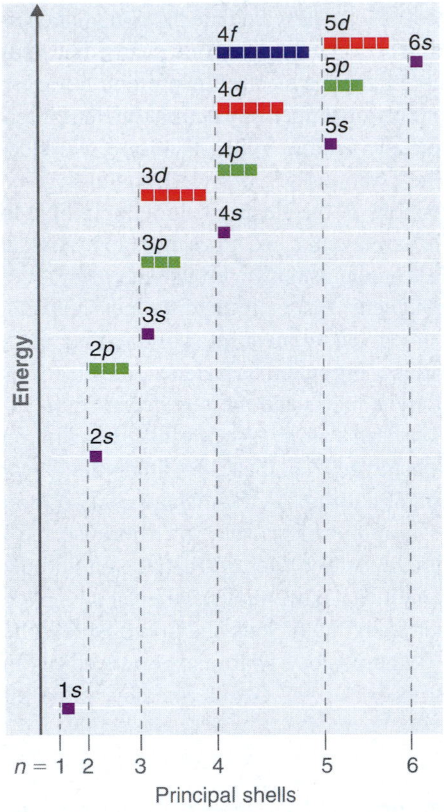

Figure 6.30 Orbital energy level diagram for a typical multielectron atom. Because of the effects of shielding and the different radial distributions of orbitals that have the same value of n but different values of l, the different subshells are no longer degenerate in a multielectron atom. (Compare with Figure 6.27.) Note that for a given value of n, the ns orbital is always lower in energy than the np orbitals, which are lower in energy than the nd orbitals, and so forth. Note also that as a result, some subshells with higher principal quantum numbers are actually lower in energy than subshells with a lower value of n; for example, the 4s orbital is lower in energy than the 3d orbitals for most atoms.

r, the effective nuclear charge is somewhere between 1 and Z: $1 \leq Z_{eff} \leq Z$. Thus, the actual Z_{eff} experienced by an electron in a given orbital depends not only on the spatial distribution of the electron in that orbital but also on the distribution of all of the other electrons present. This leads to large differences in Z_{eff} for different elements, as shown in Figure 6.29 for the elements of the first three rows of the periodic table. Notice that only for hydrogen does $Z_{eff} = Z$, and only for helium are Z_{eff} and Z comparable in magnitude.

The energies of the different orbitals for a typical multielectron atom are shown in Figure 6.30. Within a given principal shell of a multielectron atom, the orbital energies increase with increasing l. Thus, an ns orbital always lies below the corresponding np orbital, which in turn lies below the nd orbital. These energy differences are caused by the effects of shielding and *penetration*, the extent to which a given orbital lies inside other filled orbitals. As shown in Figure 6.31, for example, an electron in the 2s orbital penetrates inside a filled 1s orbital more than an electron in a 2p orbital does. Hence, in an atom with a filled 1s orbital, the effective nuclear charge experienced by a 2s electron is greater than the charge of a 2p electron. Consequently, the 2s electron is more tightly bound to the nucleus and has a lower energy, consistent with the order of energies shown in Figure 6.30.

Notice in Figure 6.30 that the difference in energies between subshells can be so large that the energies of orbitals from different principal shells can become approximately equal. For example, the energy of the 3d orbitals in most atoms is actually *between* the energies of the 4s and the 4p orbitals.

Figure 6.31 Orbital penetration. A comparison of the radial probability distribution of the 2s and 2p orbitals for various states of the H atom shows that the 2s orbital penetrates inside the 1s orbital more than the 2p orbital does. Consequently, when an electron is in the small inner lobe of the 2s orbital, it experiences a relatively large value of Z_{eff}, which causes the energy of the 2s orbital to be lower than the energy of the 2p orbital. Interactive Graph

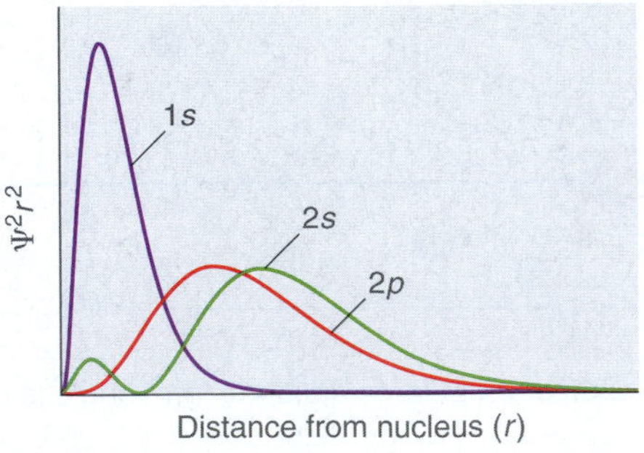

6.6 • Building Up the Periodic Table

MGC Orbital Diagrams

Now you can use the information you learned in Section 6.5 to determine the electronic structure of every element in the periodic table. The process of describing each atom's electronic structure consists, essentially, of beginning with hydrogen and adding one proton and one electron at a time to create the next heavier element in the table.* Before demonstrating how to do this, however, we must introduce the concept of electron spin and the Pauli principle.

Electron Spin: The Fourth Quantum Number

When scientists analyzed the emission and absorption spectra of the elements more closely, they saw that for elements having more than one electron, nearly all the lines in the spectra were actually *pairs* of very closely spaced lines. Because each line represents an energy level available to electrons in the atom, there are twice as many energy levels available as would be predicted solely based on the quantum numbers n, l, and m_l. Scientists also discovered that applying a magnetic field causes the lines in the pairs to split farther apart. In 1925, two graduate students in physics in the Netherlands, George Uhlenbeck and Samuel Goudsmit, proposed that the splittings were caused by an electron spinning about its axis, much as the earth spins about its axis. When an electrically charged object spins, it produces a magnetic moment parallel to the axis of rotation, making it behave like a magnet. Although the electron cannot be viewed solely as a particle, spinning or otherwise, it is indisputable that it does have a magnetic moment. This magnetic moment is called **electron spin**.

In an external magnetic field, the electron has two possible orientations (Figure 6.32). These are described by a fourth quantum number, m_s, which for any electron can have only two possible values, designated $+\frac{1}{2}$ (up) and $-\frac{1}{2}$ (down) to indicate that the two orientations are opposites; the subscript s is for spin. Thus, an electron behaves like a magnet that has one of two possible orientations, aligned either with the magnetic field or against it.

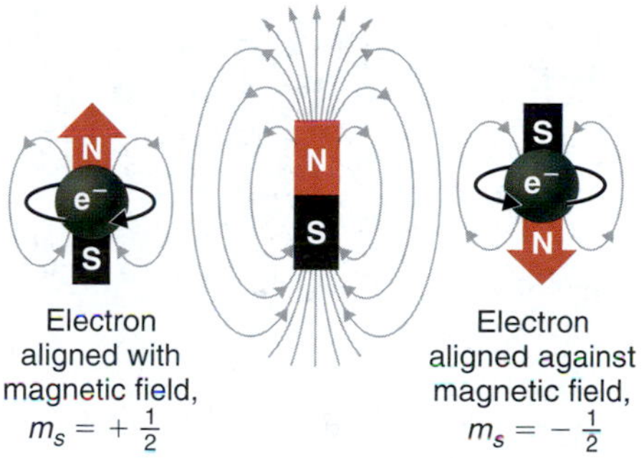

Figure 6.32 Electron spin. In a magnetic field, an electron has two possible orientations with different energies, one with spin up, aligned with the magnetic field, and one with spin down, aligned against it. All other orientations are forbidden.

The Pauli Principle

The implications of electron spin for chemistry were recognized almost immediately by an Austrian physicist, Wolfgang Pauli (1900–1958, Nobel Prize 1945), who determined that each orbital can contain no more than two electrons. He developed the **Pauli exclusion principle**: *No two electrons in an atom can have the same values of all four quantum numbers* (n, l, m_l, m_s).

By giving the values of n, l, and m_l, we also specify a particular orbital (for example, 1s with $n = 1$, $l = 0$, $m_l = 0$). Because m_s has only two possible values ($+\frac{1}{2}$ or $-\frac{1}{2}$), two electrons, *and only two electrons*, can occupy any given orbital, one with spin up and one with spin down. With this information, we can proceed to construct the entire periodic table, which, as you learned in Chapter 1, was originally based on the physical and chemical properties of the known elements.

EXAMPLE 6.7

List all allowed combinations of the four quantum numbers (n, l, m_l, m_s) for electrons in a 2p orbital, and predict the maximum number of electrons the 2p subshell can accommodate.

Given Orbital

Asked for Allowed quantum numbers and maximum number of electrons in orbital

* All stable nuclei other than hydrogen also contain one or more neutrons. Because neutrons have no electrical charge, however, they can be ignored in the following discussion.

Strategy

A List the quantum numbers (n, l, m_l) that correspond to an n = 2p orbital. List all allowed combinations of (n, l, m_l).
B Build upon these combinations to list all allowed combinations of (n, l, m_l, m_s).
C Sum the number of combinations to predict the maximum number of electrons the 2p subshell can accommodate.

Solution

A For a 2p orbital, we know that $n = 2$, $l = n - 1 = 1$, and $m_l = -l, (-l+1), \ldots, (l-1), l$. Thus, there are only three possible combinations of (n, l, m_l): (2, 1, 1), (2, 1, 0), and (2, 1, −1). **B** Because m_s is independent of the other quantum numbers and can have values of only $+\frac{1}{2}$ and $-\frac{1}{2}$, there are six possible combinations of (n, l, m_l, m_s): $(2, 1, 1, +\frac{1}{2})$, $(2, 1, 1, -\frac{1}{2})$, $(2, 1, 0, +\frac{1}{2})$, $(2, 1, 0, -\frac{1}{2})$, $(2, 1, -1, +\frac{1}{2})$, and $(2, 1, -1, -\frac{1}{2})$. **C** Hence, the 2p subshell, which consists of three 2p orbitals (2p_x, 2p_y, and 2p_z), can contain a total of six electrons, two in each orbital.

EXERCISE 6.7

List all allowed combinations of the four quantum numbers (n, l, m_l, m_s) for a 6s orbital, and predict the total number of electrons it can contain.

Answer $(6, 0, 0, +\frac{1}{2})$, $(6, 0, 0, -\frac{1}{2})$; two electrons

Electron Configurations of the Elements

The **electron configuration** of an element is the arrangement of its electrons in its atomic orbitals. By knowing the electron configuration of an element, we can predict and explain a great deal of its chemistry.

The Aufbau Principle

We construct the periodic table by following the **aufbau principle** (German for "building up"). First we determine the number of electrons in the atom, and then we add electrons one at a time to the lowest-energy orbital available *without violating the Pauli principle*. We use the orbital energy diagram of Figure 6.30, recognizing that each of the orbitals can hold two electrons, one with spin up ↑, corresponding to $m_s = +\frac{1}{2}$, which is arbitrarily written first, and one with spin down ↓, corresponding to $m_s = -\frac{1}{2}$. A filled orbital is indicated by ↑↓, in which the electron spins are said to be *paired*. Here is a schematic orbital diagram for a hydrogen atom in its ground state:

H: 2p __ __ __
2s __
1s ↑

From the orbital diagram, we can write the electron configuration in an abbreviated form in which the occupied orbitals are identified by their principal quantum number n and their value of l (s, p, d, or f), with the number of electrons in the subshell indicated by a superscript. For hydrogen, therefore, the single electron is placed in the 1s orbital, which is the orbital lowest in energy (Figure 6.30), and the electron configuration is written as $1s^1$, read as "one-s-one."

A neutral helium atom, with an atomic number of 2 ($Z = 2$), contains two electrons. We place one electron in the orbital that is lowest in energy, the 1s orbital. From the Pauli principle, we know that an orbital can contain two electrons with opposite spin, so we place the second electron in the same orbital as the first but pointing down, so that the electrons are paired. The orbital diagram for the helium atom is therefore

He: 2p __ __ __
2s __
1s ↑↓

written as $1s^2$, where the superscript 2 implies the pairing of spins. Otherwise, our configuration would violate the Pauli principle.

The next element is lithium, with $Z = 3$ and three electrons in the neutral atom. We know that the 1s orbital can hold two of the electrons with their spins paired. Figure 6.30 tells us that the next lowest energy orbital is 2s, so the orbital diagram for lithium is

Li: 2p __ __ __
2s ↑
1s ↑↓

This electron configuration is written as $1s^2 2s^1$.

The next element is beryllium, with $Z = 4$ and four electrons. We fill both the 1s and 2s orbitals to achieve a $1s^2 2s^2$ electron configuration:

Be: 2p __ __ __
2s ↑↓
1s ↑↓

When we reach boron, with $Z = 5$ and five electrons, we must place the fifth electron in one of the 2p orbitals. Because all three 2p orbitals are degenerate, it doesn't matter which one we select. The electron configuration of boron is $1s^2 2s^2 2p^1$:

B: 2p ↑ __ __
2s ↑↓
1s ↑↓

At carbon, with $Z = 6$ and six electrons, we are faced with a choice. Should the sixth electron be placed in the same 2p orbital that already contains an electron, or should it go in one of the empty 2p orbitals? If it goes in an empty 2p orbital, will the sixth electron have its spin aligned with or be opposite to the spin of the fifth? In short, which of the following three orbital diagrams is correct for carbon, remembering that the 2p orbitals are degenerate?

	(a)	(b)	(c)
2p	↑↓ __ __	↑ ↓ __	↑ ↑ __
2s	↑↓	↑↓	↑↓
1s	↑↓	↑↓	↑↓

Because of electron–electron repulsions, it is more favorable energetically for an electron to be in an unoccupied orbital than in one that is already occupied; hence, we can eliminate choice (a). Similarly, experiments have shown that choice (b) is slightly higher in energy (less stable) than choice (c) because electrons in degenerate orbitals prefer to line up with their spins parallel; thus, we can eliminate choice (b). Choice (c) illustrates **Hund's rule** (named after the German physicist F. H. Hund), which today says that the lowest-energy electron configuration for an atom is the one that has the maximum number of electrons with parallel spins in degenerate orbitals. By Hund's rule, the electron configuration of carbon, which is $1s^2 2s^2 2p^2$, is understood to correspond to the orbital diagram shown in (c). Experimentally, it is found that the ground state of a neutral carbon atom does indeed contain two unpaired electrons.

When we get to nitrogen ($Z = 7$, with seven electrons), Hund's rule tells us that the lowest-energy arrangement is

N: 2*p* ↑ ↑ ↑
2*s* ↑↓
1*s* ↑↓

with three unpaired electrons. The electron configuration of nitrogen is thus $1s^2 2s^2 2p^3$.

At oxygen, with $Z = 8$ and eight electrons, we have no choice. One electron must be paired with another in one of the 2*p* orbitals, which gives us two unpaired electrons and a $1s^2 2s^2 2p^4$ electron configuration. Because all the 2*p* orbitals are degenerate, it doesn't matter which one contains the pair of electrons.

O: 2*p* ↑↓ ↑ ↑
2*s* ↑↓
1*s* ↑↓

Similarly, fluorine has the electron configuration $1s^2 2s^2 2p^5$:

F: 2*p* ↑↓ ↑↓ ↑
2*s* ↑↓
1*s* ↑↓

When we reach neon, with $Z = 10$, we have filled the 2*p* subshell, giving a $1s^2 2s^2 2p^6$ electron configuration:

Ne: 2*p* ↑↓ ↑↓ ↑↓
2*s* ↑↓
1*s* ↑↓

Notice that for neon, as for helium, all of the orbitals through the 2*p* level are completely filled. This fact is very important in dictating both the chemical reactivity and the bonding of helium and neon, as you will see.

Valence Electrons

As we continue through the periodic table in this way, writing the electron configurations of larger and larger atoms, it becomes tedious to keep copying the configurations of the filled inner subshells. In practice, chemists simplify the notation by using a bracketed noble gas symbol to represent the configuration of the noble gas from the preceding row because all the orbitals in a noble gas are filled. For example, [Ne] represents the $1s^2 2s^2 2p^6$ electron configuration of neon ($Z = 10$), so the electron configuration of sodium, with $Z = 11$, which is $1s^2 2s^2 2p^6 3s^1$, is written as [Ne] $3s^1$:

Neon	$Z = 10$	$1s^2 2s^2 2p^6$
Sodium	$Z = 11$	$1s^2 2s^2 2p^6 3s^1$ = [Ne] $3s^1$

Because electrons in filled inner orbitals are closer to the nucleus and more tightly bound to it, they are rarely involved in chemical reactions. This means that the chemistry of an atom depends mostly on the electrons in its outermost shell, which are called the **valence electrons**. The simplified notation allows us to see the valence-electron configuration more easily. Using this notation to compare the electron configurations of sodium and lithium, we have:

Sodium	$1s^2 2s^2 2p^6 3s^1$ = [Ne] $3s^1$
Lithium	$1s^2 2s^1$ = [He] $2s^1$

It is readily apparent that both sodium and lithium have one *s* electron in their valence shell. We would therefore predict that sodium and lithium have very similar chemistry, which is indeed the case.

As we continue to build the eight elements of Period 3, the 3*s* and 3*p* orbitals are filled, one electron at a time. This row concludes with the noble gas argon, which has the electron configuration [Ne] $3s^2 3p^6$, corresponding to a filled valence shell.

EXAMPLE 6.8

Draw an orbital diagram and use it to derive the electron configuration of phosphorus, $Z = 15$. What is its valence-electron configuration?

Given Atomic number

Asked for Orbital diagram and valence-electron configuration for phosphorus

Strategy

A Locate the nearest noble gas preceding phosphorus in the periodic table. Then subtract its number of electrons from those in phosphorus to obtain the number of valence electrons in phosphorus.

B Referring to Figure 6.30, draw an orbital diagram to represent those valence orbitals. Following Hund's rule, place the valence electrons in the available orbitals, beginning with the orbital that is lowest in energy. Write the electron configuration from your orbital diagram.

C Ignore the inner orbitals (those that correspond to the electron configuration of the nearest noble gas), and write the valence-electron configuration for phosphorus.

Solution

A Because phosphorus is in the third row of the periodic table, we know that it contains a [Ne] closed shell with 10 electrons. We begin by subtracting 10 electrons from the 15 in phosphorus. **B** The additional five electrons are placed in the next available orbitals, which Figure 6.30 tells us are the 3*s* and 3*p* orbitals:

P: [Ne] 3*p* __ __ __
3*s* __

Because the 3*s* orbital is lower in energy than the 3*p* orbitals, we fill it first:

P: [Ne] 3*p* __ __ __
3*s* ↑↓

Hund's rule tells us that the remaining three electrons will occupy the degenerate 3*p* orbitals separately but with their spins aligned:

P: [Ne] 3*p* ↑ ↑ ↑
3*s* ↑↓

The electron configuration is [Ne] $3s^2 3p^3$. **C** We obtain the valence-electron configuration by ignoring the inner orbitals, which for phosphorus means that we ignore the [Ne] closed shell. This gives a valence-electron configuration of $3s^2 3p^3$.

EXERCISE 6.8

Derive the electron configuration of chlorine, $Z = 17$. What is its valence-electron configuration?

Answer [Ne] $3s^2 3p^5$; $3s^2 3p^5$

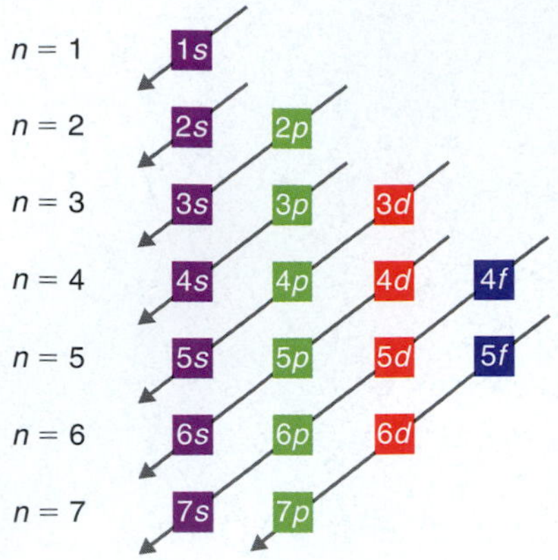

Figure 6.33 Diagram for predicting the order in which orbitals are filled in multielectron atoms. If you write the subshells for each value of the principal quantum number on successive lines, the observed order in which they are filled is indicated by a series of diagonal lines running from upper right to lower left, as shown.

The general order in which orbitals are filled is depicted in Figure 6.33. Subshells corresponding to each value of *n* are written from left to right on successive horizontal lines, where each row represents a row in the periodic table. The order in which these orbitals are filled is indicated by the diagonal lines running from upper right to lower left. Accordingly, the 4*s* orbital is filled prior to the 3*d* orbital because of shielding and penetration effects. Consequently, the electron configuration of potassium, which begins the fourth period, is $[Ar]4s^1$, and the configuration of calcium is $[Ar]4s^2$. Five 3*d* orbitals are filled by the next 10 elements, the transition metals, followed by three 4*p* orbitals. Notice that the last member of this row is the noble gas krypton ($Z = 36$, $[Ar]\,4s^2 3d^{10} 4p^6 = [Kr]$), which has filled 4*s*, 3*d*, and 4*p* orbitals. The fifth row of the periodic table is essentially the same as the fourth, except that the 5*s*, 4*d*, and 5*p* orbitals are filled sequentially.

Notice that the sixth row of the periodic table will be different from the preceding two because the 4*f* orbitals, which can hold 14 electrons, are filled between the 6*s* and the 5*d* orbitals. The elements that contain 4*f* orbitals in their valence shell are the lanthanides. When the 6*p* orbitals are finally filled, we have reached the next (and last known) noble gas, radon ($Z = 86$, $[Xe]\,6s^2 4f^{14} 5d^{10} 6p^6 = [Rn]$). In the last row, the 5*f* orbitals are filled between the 7*s* and the 6*d* orbitals, which gives the 14 actinide elements. Because the large number of protons makes their nuclei unstable, all of the actinides are radioactive.

EXAMPLE 6.9

Write the electron configuration of mercury ($Z = 80$), showing all the inner orbitals.

Given Atomic number

Asked for Complete electron configuration

Strategy

Using the orbital diagram in Figure 6.33 and the periodic table as a guide, fill the orbitals until all 80 electrons have been placed.

Solution

By placing the electrons in orbitals following the order shown in Figure 6.33 and using the periodic table as a guide, we obtain

$1s^2$	row 1	2 electrons
$2s^2 2p^6$	row 2	8 electrons
$3s^2 3p^6$	row 3	8 electrons
$4s^2 3d^{10} 4p^6$	row 4	18 electrons
$5s^2 4d^{10} 5p^6$	row 5	18 electrons
	rows 1–5	54 electrons

After filling the first five rows, we still have $80 - 54 = 26$ more electrons to accommodate. According to Figure 6.33, we need to fill the 6*s* (2 electrons), 4*f* (14 electrons), and 5*d* (10 electrons) orbitals. The result is Hg:

$$1s^2 2s^2 2p^6 3s^2 3p^6 4s^2 3d^{10} 4p^6 5s^2 4d^{10} 5p^6 6s^2 4f^{14} 5d^{10} = \text{Hg} = [\text{Xe}]\,6s^2 4f^{14} 5d^{10}$$

with a filled 5*d* subshell, a $6s^2 4f^{14} 5d^{10}$ valence-shell configuration, and a total of 80 electrons. (You should always check to be sure that the total number of electrons equals the atomic number.)

EXERCISE 6.9

Although element 114 is not stable enough to occur in nature, two isotopes of element 114 were created for the first time in a nuclear reactor in 1999 by a team of Russian and American scientists. Write the complete electron configuration for element 114.

Answer $1s^2\,2s^2\,2p^6\,3s^2\,3p^6\,4s^2\,3d^{10}\,4p^6\,5s^2\,4d^{10}\,5p^6\,6s^2\,4f^{14}\,5d^{10}\,6p^6\,7s^2\,5f^{14}\,6d^{10}\,7p^2$

The electron configurations of the elements are presented in Table 6.4, which lists the orbitals in the order in which they are filled. Notice that in several cases, the ground-state electron configurations are different from those predicted by Figure 6.33. Some of these anomalies occur as the 3*d* orbitals are filled. For example, the observed ground-state electron configuration of chromium is [Ar] $4s^1\,3d^5$ rather than the predicted [Ar] $4s^2\,3d^4$. Similarly, the observed electron configuration of copper is [Ar] $4s^1\,3d^{10}$ instead of [Ar] $4s^2\,3d^9$. The actual electron configuration may be rationalized in terms of an added stability associated with a half-filled (ns^1, np^3, nd^5, nf^7) or filled (ns^2, np^6, nd^{10}, nf^{14}) subshell. Given the small differences between higher energy levels, this added stability is enough to shift an electron from one orbital to another. In heavier elements, other more complex effects can also be important, leading to some of the additional anomalies indicated in Table 6.4. For example, cerium has an electron configuration of [Xe] $6s^2\,4f^1\,5d^1$, which is impossible to rationalize in simple terms. In most cases, however, these apparent anomalies do not have important chemical consequences.

Blocks in the Periodic Table

As you have learned, the electron configurations of the elements explain the otherwise peculiar shape of the periodic table. Although the table was originally organized on the basis of physical and chemical similarities between the elements within groups, these similarities are ultimately attributable to orbital energy levels and the Pauli principle, which cause the individual subshells to be filled in a particular order. As a result, the periodic table can be divided into "blocks" corresponding to the type of subshell that is being filled, as illustrated in Figure 6.34. For example, the two columns on the left, known as the ***s*-block elements**, consist of elements in which the *ns* orbitals are being filled. The six columns on the right, elements in which the *np* orbitals are being filled, constitute the ***p* block**. In between are the 10 columns of the ***d* block**, elements in which the $(n - 1)d$ orbitals are filled. At the bottom lie the 14 columns of the ***f* block**, elements in which the $(n - 2)f$ orbitals are filled. Because two electrons can be accommodated per orbital, the number of columns in each block is the same as the maximum electron capacity of the subshell: 2 for *ns*, 6 for *np*, 10 for $(n - 1)d$, and 14 for $(n - 2)f$. Within each column, all of the elements have the same valence-electron configuration—for example, ns^1 (Group 1) or $ns^2\,np^1$ (Group 13). As you will see, this is reflected in important similarities in chemical reactivity and bonding for the elements in each column.

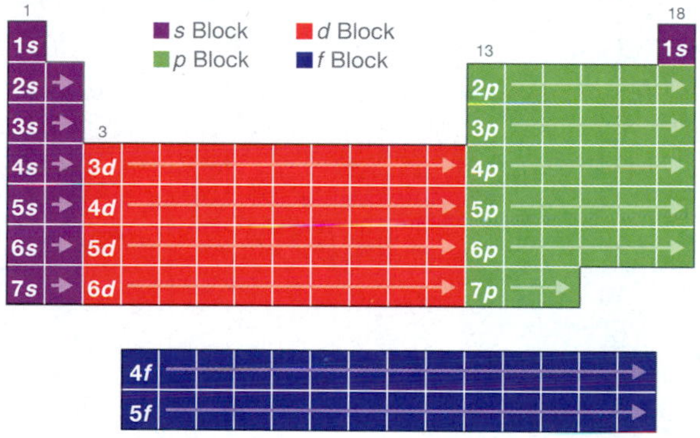

Figure 6.34 The periodic table, showing how the elements are grouped according to the kind of subshell (*s*, *p*, *d*, *f*) that is being filled with electrons in the valence shell of each element. The electron configurations of the elements are listed in Table 6.4.

Hydrogen and helium are placed somewhat arbitrarily. Although hydrogen is not an alkali metal, its $1s^1$ electron configuration suggests a similarity to lithium ([He] $2s^1$) and the other elements in the first column. Although helium, with a filled *ns* subshell, should be similar chemically to other elements with an ns^2 electron configuration, the closed principal shell dominates its chemistry, justifying its placement above neon on the right. In Chapter 7, we will examine how electron configurations affect the properties and reactivity of the elements.

Note the pattern

Because each orbital can contain a maximum of 2 electrons, there are 2 columns in the s block, 6 columns in the p block, 10 columns in the d block, and 14 columns in the f block.

EXAMPLE 6.10

Use the periodic table to predict the valence-electron configuration of all the elements of Group 2 (Be, Mg, Ca, Sr, Ba, Ra).

Given Series of elements

Asked for Valence-electron configurations

TABLE 6.4 Electron configurations of the elements*

Atomic number	Symbol	Electron configuration	Atomic number	Symbol	Electron configuration	Atomic number	Symbol	Electron configuration
1	H	$1s^1$	37	Rb	[Kr]$5s^1$	73	Ta	[Xe]$6s^2 4f^{14} 5d^3$
2	He	$1s^2$	38	Sr	[Kr]$5s^2$	74	W	[Xe]$6s^2 4f^{14} 5d^4$
3	Li	[He]$2s^1$	39	Y	[Kr]$5s^2 4d^1$	75	Re	[Xe]$6s^2 4f^{14} 5d^5$
4	Be	[He]$2s^2$	40	Zr	[Kr] $5s^2 4d^2$	76	Os	[Xe]$6s^2 4f^{14} 5d^6$
5	B	[He]$2s^2 2p^1$	41	Nb	[Kr]$5s^1 4d^4$	77	Ir	[Xe]$6s^2 4f^{14} 5d^7$
6	C	[He]$2s^2 2p^2$	42	Mo	[Kr]$5s^1 4d^5$	78	Pt	[Xe]$6s^1 4f^{14} 5d^9$
7	N	[He]$2s^2 2p^3$	43	Tc	[Kr]$5s^2 4d^5$	79	Au	[Xe]$6s^1 4f^{14} 5d^{10}$
8	O	[He]$2s^2 2p^4$	44	Ru	[Kr]$5s^1 4d^7$	80	Hg	[Xe]$6s^2 4f^{14} 5d^{10}$
9	F	[He]$2s^2 2p^5$	45	Rh	[Kr]$5s^1 4d^8$	81	Tl	[Xe]$6s^2 4f^{14} 5d^{10} 6p^1$
10	Ne	[He]$2s^2 2p^6$	46	Pd	[Kr]$4d^{10}$	82	Pb	[Xe]$6s^2 4f^{14} 5d^{10} 6p^2$
11	Na	[Ne]$3s^1$	47	Ag	[Kr]$5s^1 4d^{10}$	83	Bi	[Xe]$6s^2 4f^{14} 5d^{10} 6p^3$
12	Mg	[Ne]$3s^2$	48	Cd	[Kr]$5s^2 4d^{10}$	84	Po	[Xe]$6s^2 4f^{14} 5d^{10} 6p^4$
13	Al	[Ne]$3s^2 3p^1$	49	In	[Kr]$5s^2 4d^{10} 5p^1$	85	At	[Xe]$6s^2 4f^{14} 5d^{10} 6p^5$
14	Si	[Ne]$3s^2 3p^2$	50	Sn	[Kr]$5s^2 4d^{10} 5p^2$	86	Rn	[Xe]$6s^2 4f^{14} 5d^{10} 6p^6$
15	P	[Ne]$3s^2 3p^3$	51	Sb	[Kr]$5s^2 4d^{10} 5p^3$	87	Fr	[Rn]$7s^1$
16	S	[Ne]$3s^2 3p^4$	52	Te	[Kr]$5s^2 4d^{10} 5p^4$	88	Ra	[Rn]$7s^2$
17	Cl	[Ne]$3s^2 3p^5$	53	I	[Kr]$5s^2 4d^{10} 5p^5$	89	Ac	[Rn]$7s^2 6d^1$
18	Ar	[Ne]$3s^2 3p^6$	54	Xe	[Kr]$5s^2 4d^{10} 5p^6$	90	Th	[Rn]$7s^2 6d^2$
19	K	[Ar]$4s^1$	55	Cs	[Xe]$6s^1$	91	Pa	[Rn]$7s^2 5f^2 6d^1$
20	Ca	[Ar]$4s^2$	56	Ba	[Xe]$6s^2$	92	U	[Rn]$7s^2 5f^3 6d^1$
21	Sc	[Ar]$4s^2 3d^1$	57	La	[Xe]$6s^2 5d^1$	93	Np	[Rn]$7s^2 5f^4 6d^1$
22	Ti	[Ar]$4s^2 3d^2$	58	Ce	[Xe]$6s^2 4f^1 5d^1$	94	Pu	[Rn]$7s^2 5f^6$
23	V	[Ar]$4s^2 3d^3$	59	Pr	[Xe]$6s^2 4f^3$	95	Am	[Rn]$7s^2 5f^7$
24	Cr	[Ar]$4s^1 3d^5$	60	Nd	[Xe]$6s^2 4f^4$	96	Cm	[Rn]$7s^2 5f^7 6d^1$
25	Mn	[Ar]$4s^2 3d^5$	61	Pm	[Xe]$6s^2 4f^5$	97	Bk	[Rn]$7s^2 5f^9$
26	Fe	[Ar]$4s^2 3d^6$	62	Sm	[Xe]$6s^2 4f^6$	98	Cf	[Rn]$7s^2 5f^{10}$
27	Co	[Ar]$4s^2 3d^7$	63	Eu	[Xe]$6s^2 4f^7$	99	Es	[Rn]$7s^2 5f^{11}$
28	Ni	[Ar]$4s^2 3d^8$	64	Gd	[Xe]$6s^2 4f^7 5d^1$	100	Fm	[Rn]$7s^2 5f^{12}$
29	Cu	[Ar]$4s^1 3d^{10}$	65	Tb	[Xe]$6s^2 4f^9$	101	Md	[Rn]$7s^2 5f^{13}$
30	Zn	[Ar]$4s^2 3d^{10}$	66	Dy	[Xe]$6s^2 4f^{10}$	102	No	[Rn]$7s^2 5f^{14}$
31	Ga	[Ar]$4s^2 3d^{10} 4p^1$	67	Ho	[Xe]$6s^2 4f^{11}$	103	Lr	[Rn]$7s^2 5f^{14} 6d^1$
32	Ge	[Ar]$4s^2 3d^{10} 4p^2$	68	Er	[Xe]$6s^2 4f^{12}$	104	Rf	[Rn]$7s^2 5f^{14} 6d^2$
33	As	[Ar]$4s^2 3d^{10} 4p^3$	69	Tm	[Xe]$6s^2 4f^{13}$	105	Db	[Rn]$7s^2 5f^{14} 6d^3$
34	Se	[Ar]$4s^2 3d^{10} 4p^4$	70	Yb	[Xe]$6s^2 4f^{14}$	106	Sg	[Rn]$7s^2 5f^{14} 6d^4$
35	Br	[Ar]$4s^2 3d^{10} 4p^5$	71	Lu	[Xe]$6s^2 4f^{14} 5d^1$	107	Bh	[Rn]$7s^2 5f^{14} 6d^5$
36	Kr	[Ar]$4s^2 3d^{10} 4p^6$	72	Hf	[Xe]$6s^2 4f^{14} 5d^2$	108	Hs	[Rn]$7s^2 5f^{14} 6d^6$
						109	Mt	[Rn]$7s^2 5f^{14} 6d^7$
						110	Ds	[Rn]$7s^1 5f^{14} 6d^9$
						111	Rg	[Rn]$7s^1 5f^{14} 6d^{10}$

*The electron configurations of elements indicated in red are exceptions due to the added stability associated with half-filled and filled subshells. The electron configurations of the elements indicated in blue are also anomalous, but the reasons for the observed configurations are more complex. For elements after No, electron configurations are tentative.

Strategy

Ⓐ Identify the block in the periodic table to which the Group 2 elements belong. Locate the nearest noble gas preceding each element, and identify the principal quantum number of the valence shell of each element.

Ⓑ Write the valence-electron configuration of each element by first indicating the filled inner shells using the symbol for the nearest preceding noble gas and then listing the principal quantum number of its valence shell, its valence orbitals, and the number of valence electrons in each orbital as superscripts.

Solution

Ⓐ✓ The Group 2 elements are in the *s* block of the periodic table, and as Group 2 elements, they all have two valence electrons. Beginning with Be, we see that its nearest preceding noble gas is He and that the principal quantum number of its valence shell is $n = 2$.

Ⓑ✓ Thus, Be has a [He]$2s^2$ electron configuration. The next element down, magnesium, is expected to have exactly the same arrangement of electrons in the $n = 3$ principal shell: [Ne]$3s^2$. By extrapolation, we expect all of the Group 2 elements to have an ns^2 electron configuration.

EXERCISE 6.10

Use the periodic table to predict the characteristic valence-electron configuration of the halogens, Group 17.

Answer All have an ns^2np^5 electron configuration, one electron short of a noble-gas electron configuration. (Note that the heavier halogens also have filled $(n-1)d^{10}$ subshells as well as an $(n-2)f^{14}$ subshell for Rn; these do not, however, affect their chemistry in any significant way.)

SUMMARY AND KEY TERMS

6.1 Waves and Electromagnetic Radiation (p. 256)

A basic knowledge of the electronic structure of atoms requires an understanding of the properties of waves and electromagnetic radiation. A **wave** is a periodic oscillation by which energy is transmitted through space. All waves are **periodic**, repeating regularly in both space and time. Waves are characterized by several interrelated properties: **wavelength**, the distance between successive waves; **frequency**, the number of waves that pass a fixed point per unit time; **speed**, the rate at which the wave propagates through space; and **amplitude**, the magnitude of the oscillation about the mean position. The speed of a wave is equal to the product of its wavelength and frequency. **Electromagnetic radiation** consists of two perpendicular waves, one electric and one magnetic, propagating at the **speed of light**, abbreviated *c*. Electromagnetic radiation is radiant energy that includes radio waves, microwaves, visible light, X rays, and gamma rays, which differ only in their frequencies and wavelengths.

6.2 The Quantization of Energy (p. 259)

The properties of **blackbody radiation**, the radiation emitted by hot objects, could not be explained using classical physics. Max Planck postulated that energy was quantized and could be emitted or absorbed only in integral multiples of a small unit of energy, a **quantum**. The energy of a quantum is proportional to the frequency of the radiation; the proportionality constant *h* is a fundamental constant (Planck's constant). Albert Einstein used Planck's concept of the quantization of energy to explain the **photoelectric effect**, the ejection of electrons from certain metals upon exposure to light. Einstein postulated the existence of what today we call **photons**, particles of light with a particular energy, $E = h\eta$. Thus, both energy and matter have fundamental building blocks: quanta and atoms, respectively.

6.3 Atomic Spectra and Models of the Atom (p. 263)

Atoms of individual elements emit light at only specific wavelengths, producing a **line spectrum** rather than the continuous spectrum of all wavelengths produced by a hot object. Niels Bohr explained the line spectrum of the hydrogen atom by assuming that the electron moved in circular orbits and that only orbits with certain radii were allowed. Lines in the spectrum were due to transitions in which an electron moved from a higher-energy orbit with larger radius to a lower-energy orbit with smaller radius. The orbit closest to the nucleus represented the **ground state** of the atom and was most stable; orbits farther away were higher-energy **excited states**. Transitions from an excited state to a lower-energy state resulted in the emission of light with only a limited number of wavelengths. Bohr's model could not, however, explain the spectra of atoms heavier than hydrogen.

Most light is polychromatic, containing light of many wavelengths. Light that contains only a single wavelength is monochromatic and is produced by devices called lasers, which utilize transitions between two atomic energy levels to produce light in a very narrow range of wavelengths. Atoms can also absorb light of certain energies, resulting in a transition from the ground state or a lower-energy excited state to a higher-energy excited state. This produces an **absorption spectrum**, which has dark lines in the same position as the bright lines in the **emission spectrum** of an element.

6.4 The Relationship Between Energy and Mass (p. 271)

The modern model for the electronic structure of the atom is based on the recognition that an electron possesses properties of both a particle and a wave, the so-called **wave–particle duality**. Louis de Broglie showed that the wavelength of a particle is equal to Planck's constant divided by the mass times the velocity of the particle. The electron in Bohr's circular orbits could thus be described as a **standing wave**, one that does not move through space. Standing waves are familiar from music: the lowest-energy standing wave is the **fundamental** vibration, and higher-energy vibrations are **overtones** and have successively more **nodes**, points where the amplitude of the wave is always zero. Werner Heisenberg's **uncertainty principle** states that it is impossible to describe precisely both the location and the speed of particles that exhibit wavelike behavior.

6.5 Atomic Orbitals and Their Energies (p. 275)

Because of the wave–particle duality, scientists must deal with the probability of an electron being at a particular point in space. To do so required the development of **quantum mechanics**, which uses **wave functions** to describe the mathematical relationship between the motion of electrons in atoms and molecules and their energies. Wave functions have five important properties: (1) the wave function Ψ uses three variables (Cartesian axes x, y, z) to describe the position of an electron; (2) the magnitude of the wave function Ψ is proportional to the intensity of the wave; (3) the probability of finding an electron at a given point is proportional to the square of the wave function at that point, leading to a distribution of probabilities in space that is often portrayed as an **electron density** plot; (4) describing electron distributions as standing waves leads naturally to the existence of sets of **quantum numbers** characteristic of each wave function; and (5) each spatial distribution of the electron described by a wave function with a given set of quantum numbers has a particular energy.

Quantum numbers provide important information about the energy and spatial distribution of an electron. The **principal quantum number** n can be any positive integer; as n increases for an atom, so does the average distance of the electron from the nucleus. All wave functions with the same value of n constitute a **principal shell** in which the electrons have similar average distances from the nucleus. The **azimuthal quantum number** l can have integral values between 0 and $n - 1$; it describes the shape of the electron distribution. Wave functions that have the same values of both n and l constitute a **subshell**, corresponding to electron distributions that usually differ in orientation rather than in shape or in average distance from the nucleus. The **magnetic quantum number** m_l can have $2l + 1$ integral values, ranging from $-l$ to $+l$, and describes the orientation of the electron distribution. Each wave function with a given set of values of n, l, and m_l describes a particular spatial distribution of an electron in an atom, an **atomic orbital**.

The four chemically important types of atomic orbital correspond to values of $l = 0$, 1, 2, and 3. Orbitals with $l = 0$ are s orbitals and are spherically symmetrical, with the greatest probability of finding the electron occurring at the nucleus. All orbitals with values of $n > 1$ and $l \geq 0$ contain one or more nodes. Orbitals with $l = 1$ are p orbitals and contain a nodal plane that includes the nucleus, giving rise to a "dumbbell" shape. Orbitals with $l = 2$ are d orbitals and have more complex shapes with at least two nodal surfaces. Orbitals with $l = 3$ are f orbitals, which are still more complex.

Because the energy of an electron is determined by its average distance from the nucleus, each atomic orbital with a given set of quantum numbers has a particular energy associated with it, the **orbital energy**. In atoms or ions that contain only a single electron, all orbitals with the same value of n have the same energy (they are **degenerate**), and the energies of the principal shells increase smoothly as n increases. An atom or ion with the electron(s) in the lowest-energy orbital(s) is said to be in its ground state, whereas an atom or ion in which one or more electrons occupy higher-energy orbitals is said to be in an excited state. Calculation of orbital energies in atoms or ions that contain more than one electron (multielectron atoms or ions) is complicated by repulsive interactions between the electrons. The concept of **electron shielding**, in which intervening electrons act to reduce the positive nuclear charge experienced by an electron, allows the use of hydrogen-like orbitals and an **effective nuclear charge** to describe electron distributions in more complex atoms or ions. The degree to which orbitals with different values of l and the same value of n overlap or penetrate filled inner shells results in slightly different energies for different subshells in the same principal shell in most atoms.

6.6 Building Up the Periodic Table (p. 285)

In addition to the three quantum numbers (n, l, m_l) dictated by quantum mechanics, a fourth quantum number is required to explain certain properties of atoms. This is the **electron spin** quantum number m_s, which can have values of $+\frac{1}{2}$ or $-\frac{1}{2}$ for any electron, corresponding to the two possible orientations of an electron in a magnetic field. The concept of electron spin has important consequences for chemistry because the **Pauli exclusion principle** implies that no orbital can contain more than two electrons (with opposite spin). Based on the Pauli principle and a knowledge of orbital energies obtained using hydrogen-like orbitals, it is possible to construct the periodic table by filling up the available orbitals beginning with the lowest-energy orbitals (the **aufbau principle**), which gives rise to a particular arrangement of electrons for each element (its **electron configuration**). **Hund's rule** says that the lowest-energy arrangement of electrons is the one that places them in degenerate orbitals with their spins parallel. For chemical purposes, the most important electrons are those in the outermost principal shell, the **valence electrons**. The arrangement of atoms in the periodic table results in blocks corresponding to filling of the ns, np, nd, and nf orbitals to produce the distinctive chemical properties of the ***s*-block elements**, ***p*-block elements**, ***d*-block elements**, and ***f* block elements**, respectively.

KEY EQUATIONS

Relationship between wavelength, frequency, and speed of a wave	$\lambda\nu = v$	(6.1)
Relationship between wavelength, frequency, and speed of electromagnetic radiation	$c = \lambda\nu$	(6.2)
Quantization of energy	$E = h\nu$	(6.5)
Rydberg equation	$\frac{1}{\lambda} = \mathcal{R}\left(\frac{1}{n_1^2} - \frac{1}{n_2^2}\right)$	(6.8)
Einstein's relationship between mass and energy	$E = h\nu = \frac{hc}{\lambda} = mc^2$	(6.13)
de Broglie's relationship between mass, speed, and wavelength	$\lambda = \frac{h}{mv}$	(6.15)
Heisenberg's uncertainty principle	$(\Delta x)[\Delta(mv)] \geq \frac{h}{4\pi}$	(6.19)
Energy of hydrogen-like orbitals	$E = -\frac{Z^2}{n^2}\mathcal{R}hc$	(6.22)

QUESTIONS AND PROBLEMS

MGC For instructor-assigned homework, go to **www.masteringgenchem.com**

Questions and Problems with colored numbers have answers in the Appendix and complete solutions in the Student Solutions Manual.

CONCEPTUAL

6.1 Radiant Energy and Electromagnetic Radiation

1. What are the characteristics of a wave? What is the relationship between electromagnetic radiation and wave energy?
2. At constant wavelength, what effect does increasing the frequency of a wave have on its speed? On its amplitude?
3. List the following forms of electromagnetic radiation in order of increasing wavelength: X rays, radio waves, infrared waves, microwaves, ultraviolet waves, visible waves, gamma rays. List them in order of increasing frequency. Which has the highest energy?
4. A large industry is centered around developing skin-care products, such as suntan lotions and cosmetics, that cannot be penetrated by ultraviolet radiation. How does the wavelength of visible light compare with the wavelength of ultraviolet light? How does the energy of visible light compare with the energy of ultraviolet light? Why is this industry focused on blocking ultraviolet light rather than visible light?

6.2 The Quantization of Energy

5. Describe the relationship between the energy of a photon and its frequency.
6. How was the "ultraviolet catastrophe" explained?
7. If electromagnetic radiation with a continuous range of frequencies above the threshold frequency of a metal is allowed to strike a metal surface, is the kinetic energy of the ejected electrons continuous or quantized? Explain your answer.
8. The vibrational energy of a plucked guitar string is said to be *quantized*. What do we mean by this? Are the sounds emitted from the 88 keys on a piano also quantized?
9. Which of the following exhibit quantized behavior: a human voice; the speed of a car; a harp; colors of light; automobile tire sizes; waves from a speedboat; Bohr's model of the atom?

6.3 Atomic Spectra and Models of the Atom

10. Is the spectrum of the light emitted by isolated atoms of an element discrete or continuous? How do these spectra differ from those obtained by heating a bulk sample of the solid element? Explain your answers.
11. Explain why each element has a characteristic emission and absorption spectrum. If spectral emissions had been found to be continuous rather than discrete, what would have been the implications for Bohr's model of the atom?
12. Explain the differences between a ground state and an excited state. Describe what happens in the spectrum of a species when an electron moves from a ground state to an excited state. What happens in the spectrum when the electron falls from an excited state to a ground state?
13. What phenomenon causes a neon sign to have a characteristic color? If the emission spectrum of an element is constant, why do some neon signs have more than one color?
14. How is light from a laser different from the light emitted by a light source such as a light bulb? Describe how a laser produces light.
15. Explain what is meant by each term: (a) standing wave; (b) fundamental; (c) overtone; (d) node. Illustrate each with a sketch.

6.4 The Relationship Between Energy and Mass

16. How does Einstein's theory of relativity illustrate the wave–particle duality of light? What properties of light can be explained by a wave model? By a particle model?
17. In the modern theory of the electronic structure of the atom, which of de Broglie's ideas have been retained? Which proved to be incorrect?

18. According to Bohr, what is the relationship between an atomic orbit and the energy of an electron in that orbit? Is Bohr's model of the atom consistent with Heisenberg's uncertainty principle? Explain your answer.

19. The development of ideas frequently builds on the work of predecessors. Complete the chart by filling in the names of those responsible for each theory shown:

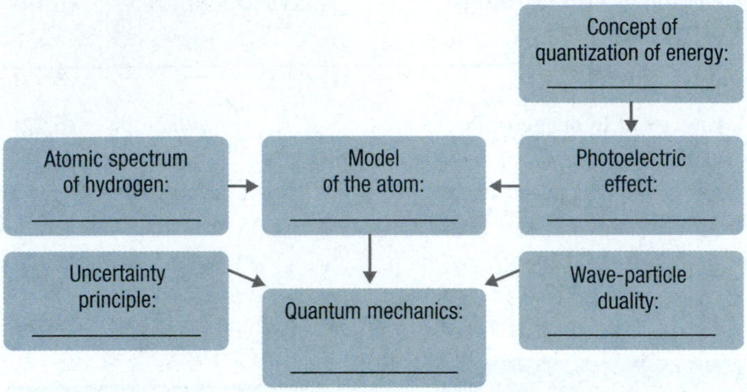

6.5 Atomic Orbitals and Their Energies

20. Why is an electron in an orbital with $n = 1$ in a hydrogen atom more stable than a free electron ($n = \infty$)?

21. What four variables are required to fully describe the position of any object in space? In quantum mechanics, one of these variables is not explicitly considered. Which one, and why?

22. Chemists generally refer to the square of the wave function rather than to the wave function itself. Why?

23. Orbital energies of species that contain only one electron are defined by only one of the quantum numbers. Which one? In such a species, is the energy of an orbital with $n = 2$ greater than, less than, or equal to the energy of an orbital with $n = 4$? Justify your answer.

24. In each pair of subshells for a hydrogen atom, which has the higher energy? Give the principle and the azimuthal quantum number for each pair. (a) 1*s*, 2*p*; (b) 2*p*, 2*s*; (c) 2*s*, 3*s*; (d) 3*d*, 4*s*

25. What is the relationship between the energy of an orbital and its average radius? If an electron made a transition from an orbital with an average radius of 846.4 pm to an orbital with an average radius of 476.1 pm, would an emission spectrum or an absorption spectrum be produced? Why?

26. In making a transition from an orbital with a principle quantum number of 4 to an orbital with a principle quantum number of 7, does the electron of a hydrogen atom emit or absorb a photon of energy? What would be the energy of the photon? To what region of the electromagnetic spectrum does this energy correspond?

27. What quantum number defines each of the following? (a) the overall shape of an orbital; (b) the orientation of an electron with respect to a magnetic field; (c) the orientation of an orbital in space; (d) the average energy and distance of an electron from the nucleus

28. In an attempt to explain the properties of the elements, Niels Bohr initially proposed electronic structures for several elements that utilized orbits that held a certain number of electrons, some of which are listed in the table:

Element	Number of Electrons	Electrons in Orbits with $n = 4\ 3\ 2\ 1$
H	1	1
He	2	2
Ne	10	8 2
Ar	18	8 8 2
Li	3	1 2
Na	11	1 8 2
K	19	1 8 8 2
Be	4	2 2

(a) Draw the electron configuration of each atom based only on the information given in the table. What are the differences between Bohr's initially proposed structures and those accepted today?
(b) Using Bohr's model, what are the implications for the reactivity of each of these elements?
(c) Give the actual electron configuration of each element listed.

29. What happens to the energy of a given orbital as the nuclear charge *Z* of a species increases? In a multielectron atom and for a given nuclear charge, the effective nuclear charge experienced by an electron depends on its value of *l*. Why?

30. The electron density of a particular atom is divided into two regions. Name these two regions, and describe what each of them defines.

31. As the principal quantum number increases, the energy difference between successive energy levels decreases. Why? What would happen to the electron configurations of the transition metals if this decrease did not occur?

32. Describe the relationship between electron shielding and the effective nuclear charge on the outermost electrons of an atom. Predict how chemical reactivity is affected by a decreased effective nuclear charge.

33. If a given atom or ion contains a single electron in each of the following subshells, which electron is easier to remove? (a) 2*s*, 3*s*; (b) 3*p*, 4*d*; (c) 2*p*, 1*s*; (d) 3*d*, 4*s*

6.6 Building Up the Periodic Table

34. A set of four quantum numbers specifies each wave function. What information is given by each of the four quantum numbers? What does the specified wave function describe?

35. List two pieces of evidence to support the statement that electrons have a spin.

36. The periodic table is divided into blocks. Identify each block, and explain the principle behind the divisions. Which quantum number distinguishes the horizontal rows?

37. Identify the element with each ground-state electron configuration: [He] $2s^2 2p^3$; [Ar] $4s^2 3d^1$; [Kr] $5s^2 4d^{10} 5p^3$; [Xe] $6s^2 4f^6$

38. Which element has each ground-state electron configuration? [He] $2s^2 2p^1$; [Ar] $4s^2 3d^8$; [Kr] $5s^2 4d^{10} 5p^4$; [Xe] $6s^2$

39. Propose an explanation as to why the noble gases are inert.

NUMERICAL

This section includes paired problems (marked by brackets) that require similar problem-solving skills.

6.1 Waves and Electromagnetic Energy

40. The human eye is sensitive to what fraction of the electromagnetic spectrum, assuming a spectral range of 10^4 to 10^{20} Hz? If we came from the planet Krypton and had X-ray vision (that is, if our eyes were sensitive to X rays in addition to visible light), how would this fraction be changed?

41. What is the frequency of light with these wavelengths? (a) 5.8×10^{-7} m; (b) 3.7×10^{-10} m; (c) 8.6×10^7 m; (d) 6.2 mm; (e) 3.7 nm; (f) 2.3 Å

42. What is the frequency in megahertz corresponding to each wavelength? (a) 755 m; (b) 6.73 nm; (c) 1.77×10^3 km; (d) 9.88 Å

43. Given the following characteristic wavelengths for each species, identify the spectral region (UV, visible, etc.) in which the following line spectra will occur. Given 1.00 mol of each compound and the wavelength of absorbed or emitted light, how much energy does this correspond to? (a) NH_3, 1.0×10^{-2} m; (b) CH_3CH_2OH, 9.0 μm; (c) Mo atom, 7.1 Å

44. What is the speed of a wave in m/s that has a wavelength of 1250 m and a frequency of 2.36×10^5 s^{-1}?

45. A wave travels at 3.70 m/s with a frequency of 4.599×10^7 Hz and an amplitude of 1.0 m. What is the wavelength in nm?

46. An FM station broadcasts with a wavelength corresponding to 3.21 m. What is the broadcast frequency of the station in megahertz? An FM radio typically has a broadcast range of 82–112 MHz. What is the corresponding wavelength range in meters for this reception?

47. An AM station broadcasts with a wavelength of 248.0 m. What is the broadcast frequency of the station in kilohertz?

48. A microwave oven operates at a frequency of approximately 2450 MHz. What is the corresponding wavelength? Water, with its polar molecules, absorbs electromagnetic radiation primarily in the infrared portion of the spectrum. Given this fact, why are microwave ovens used for cooking food?

6.2 The Quantization of Energy

49. What is the energy of a photon of light with each of the following wavelengths: (a) 4.33×10^5 m; (b) 0.065 nm; (c) 786 pm? To which region of the electromagnetic spectrum does each wavelength belong?

50. How much energy is contained in each of the following: (a) 250 photons with a wavelength of 3.0 m; (b) 4.2×10^6 photons with a wavelength of 92 μm; (c) 1.78×10^{22} photons with a wavelength of 2.1 Å? To which region of the electromagnetic spectrum does each wavelength belong?

51. A mole of photons is found to have an energy of 225 kJ. What is the wavelength of the radiation?

52. Use the data in Table 6.1 to calculate how much more energetic a single gamma-ray photon is than a radio-wave photon. How many photons from a radio source operating at a frequency of 8×10^5 Hz would be required to provide the same amount of energy as a single gamma-ray photon with a frequency of 3×10^{19} Hz?

53. Using the data in Table 6.1, calculate the range in energy differences between a single X-ray photon and a photon of ultraviolet light.

54. A radio station has a transmitter that broadcasts at a frequency of 100.7 MHz with a power output of 50 kilowatts (kW). If 1 W = 1 J/s, how many photons are emitted by the transmitter each second?

6.3 Atomic Spectra and Models of the Atom

55. Using a Bohr model and the transition from $n = 2$ to $n = 3$ in an atom with a single electron, describe the mathematical relationship between an emission spectrum and an absorption spectrum. What is the energy of this transition? What does the sign of the energy value represent in this case? What range of light is associated with this transition?

56. If a hydrogen atom is excited from an $n = 1$ state to an $n = 3$ state, how much energy does this correspond to? Is this an absorption or an emission? What is the wavelength of the photon involved in this process? To what region of the electromagnetic spectrum does this correspond?

57. The hydrogen atom emits a photon with a 486-nm wavelength, corresponding to an electron dropping from the $n = 4$ level to which other level? What is the color of the emission?

58. An electron in a hydrogen atom can drop from the $n = 3$ level to $n = 2$. What is the color of the emitted light? What is the energy of this transition?

59. Calculate the wavelength and energy of the photon that gives rise to the third line in the Lyman series in the emission spectrum of hydrogen. In what region of the spectrum does this wavelength occur? Describe qualitatively what the absorption spectrum looks like.

60. The wavelength of one of the lines in the Lyman series of hydrogen is 121 nm. In what region of the spectrum does this occur? To which electronic transition does this correspond?

61. The emission spectrum of helium is shown here. What difference in energy in kJ/mol, ΔE, gives rise to each line?

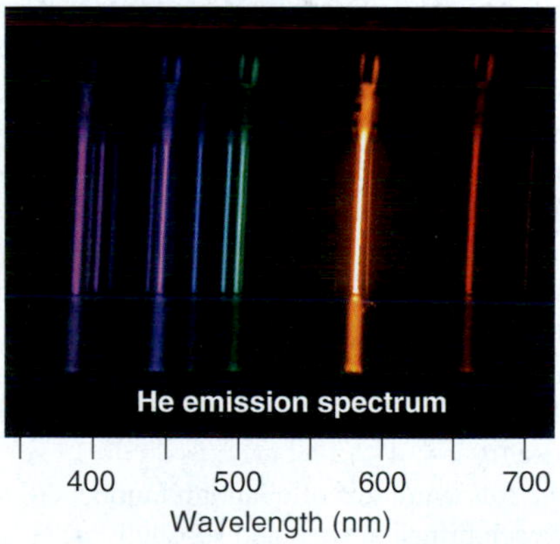

62. Removing an electron from solid potassium requires 222 kJ/mol. Would you expect to observe a photoelectric effect for potassium using a photon of blue light ($\lambda = 485$ nm)? What is the maximum wavelength of energy needed to observe a photoelectric effect for potassium? What is the corresponding color of light of this wavelength?

63. The binding energy of an electron is the energy needed to remove an electron from its lowest energy state. According to Bohr's postulates, calculate the binding energy of an electron in a hydrogen atom in kJ/mol. What wavelength in nm is required to remove such an electron?

64. As a radioastronomer, you have observed spectral lines for hydrogen corresponding to a state with $n = 320$, and you would like to produce these lines in the laboratory. Is this feasible? Why or why not?

6.4 The Relationship Between Energy and Mass

65. How much heat is generated by shining a CO_2 laser with a wavelength of 1.065 microns on a 68.95-kg sample of water if 1.000 mol of photons is absorbed and converted to heat? Is this enough heat to raise the temperature of the water 4°C?

66. Show the mathematical relationship between energy and mass. Between wavelength and mass. What is the effect of (a) doubling the mass of an object on its energy; (b) doubling the mass of an object on its wavelength; (c) doubling the frequency on its mass?

67. What is the de Broglie wavelength of a bullet that has a mass of 39 g and is traveling at a speed of 1020 m/s ± 10 m/s? What is the minimum uncertainty in the bullet's position?

68. What is the de Broglie wavelength of an aircraft carrier that has a mass of 6800 tons and is traveling at a speed of 18 ± 0.1 knots (1 knot = 1.15 mi/h)? What is the minimum uncertainty in its position?

69. Calculate the mass of a particle if it is traveling at 2.2×10^6 m/s and has a frequency of 6.67×10^7 Hz. If the uncertainty in the velocity is known to be 0.1%, what is the minimum uncertainty in the position of the particle?

70. Determine the wavelength of an automobile that has a mass of 2800 lb and is traveling at a speed of 80 mi/h ± 3%. How does this compare with the diameter of the nucleus of an atom? You are standing 3 in. from the edge of the highway. What is the minimum uncertainty in the position of the automobile in inches?

6.5 Atomic Orbitals and Their Energies

71. How many subshells are possible for $n = 3$? What are they?

72. For $n = 5$, how many subshells are possible? What are they?

73. What value of l corresponds to a d subshell? How many orbitals are contained in this subshell?

74. Give the value of l that corresponds to an f subshell. How many orbitals are contained in this subshell?

75. How many electrons can occupy each subshell? (a) $2s$; (b) $3p$; (c) $4d$; (d) $6f$

76. State the number of orbitals and electrons that can occupy each subshell: (a) $1s$; (b) $4p$; (c) $5d$; (d) $4f$.

77. How many orbitals and subshells are found within the principal shell with $n = 6$? How do these orbital energies compare with those for $n = 4$?

78. Given the following sets of quantum numbers (n, l, m_l, m_s), identify each principal shell and subshell: (a) $[1, 0, 0, \frac{1}{2}]$; (b) $[2, 1, 0, \frac{1}{2}]$; (c) $[3, 2, 0, \frac{1}{2}]$; (d) $[4, 3, 3, \frac{1}{2}]$.

79. How many nodes would you expect a $4p$ orbital to have? A $5s$ orbital?

80. A p orbital is found to have one node in addition to the nodal plane that bisects the lobes. What would you predict to be the value of n? If an s orbital has two nodes, what is the value of n?

6.6 Building Up the Periodic Table

81. How many magnetic quantum numbers are possible for a $4p$ subshell? For a $3d$ subshell? How many orbitals are in these subshells?

82. How many magnetic quantum numbers are possible for a $6s$ subshell? For a $4f$ subshell? How many orbitals does each subshell contain?

83. If $l = 2$ and $m_l = 2$, give all allowed combinations for the four quantum numbers for electrons in the corresponding $3d$ subshell.

84. Give all allowed combinations of the four quantum numbers (n, l, m_l, m_s) for electrons in a $4d$ subshell. How many electrons can the $4d$ orbital accommodate? How would this differ from a situation in which there were only three quantum numbers (n, l, m)?

85. Is each set of quantum numbers allowed? Explain your answers.
    (a) $n = 2; l = 1; m_l = 2; m_s = +\frac{1}{2}$
    (b) $n = 3, l = 0; m_l = -1; m_s = -\frac{1}{2}$
    (c) $n = 2; l = 2; m_l = 1; m_s = +\frac{1}{2}$
    (d) $n = 3; l = 2; m_l = 2; m_s = +\frac{1}{2}$

86. List the set of quantum numbers for each of the electrons in the valence shell of these elements: (a) C; (b) Mg; (c) Br.

87. List the set of quantum numbers for the electrons in the valence shell of each of these elements: (a) Be; (b) Xe; (c) Li; (d) F; (e) S.

88. Sketch the shape of the periodic table if there were three possible values of m_s for each electron ($+\frac{1}{2}$, $-\frac{1}{2}$, and 0); assume that the Pauli principle is still valid.

89. Predict the shape of the periodic table if eight electrons could occupy the p subshell.

90. If the electron could have only spin $+\frac{1}{2}$ and the Pauli principle were still valid, what would the periodic table look like?

91. If three electrons could occupy each s orbital, what would be the electron configuration of each species? (a) Na; (b) Ti; (c) F; (d) Ca

92. If Hund's rule was not followed and maximum pairing occurred, how many unpaired electrons would each species have: (a) P; (b) I; (c) Mn? How do these numbers compare with the number found using Hund's rule?

93. Write electron configurations for these elements in their ground states: (a) B; (b) Rb; (c) Br; (d) Ge; (e) V; (f) Pd; (g) Bi; (h) Eu.

94. What is the ground-state configuration for each element? (a) Al; (b) Ca; (c) S; (d) Sn; (e) Ni; (f) W; (g) Nd; (h) Am

95. Give the complete electron configuration for each element: Sn; Cu; F; Hg; Th; Y.

96. Give the complete electron configuration for each element: Mg; K; Ti; Se; I; U; Ge.

97. Write the valence-electron configuration for (a) Sm; (b) Pr; (c) B; (d) Co.

98. Using the Pauli exclusion principle and Hund's rule, draw valence orbital diagrams for Cl, Si, and Sc.

99. Following the Pauli exclusion principle and Hund's rule, draw valence orbital diagrams for Ba, Nd, and I.

100. How many unpaired electrons does each species contain? (a) He; (b) O; (c) Bi; (d) Ag; (e) B

101. How many unpaired electrons does each species contain? (a) Pb; (b) Cs; (c) Cu; (d) Si; (e) Se

102. For each element, give the complete electron configuration, draw the valence electron configuration, and give the number of unpaired electrons present: (a) Li; (b) Mg; (c) Si; (d) Cs; (e) Pb.

103. Use an orbital diagram to illustrate the aufbau principle, the Pauli exclusion principle, and Hund's rule using (a) carbon and (b) sulfur.

APPLICATIONS

104. The lamps in street lights use emission of light from excited states of atoms to produce a characteristic glow. Light is generated by electron bombardment of a metal vapor. Of calcium and strontium, which metal vapor would you use to produce yellow light? Which metal would you use to

produce red light? Calculate the energy associated with each transition and propose an explanation for the colors of the emitted light.

105. Lasers have useful medical applications because their light is directional (permitting tight focus of the laser beam for precise cutting), monochromatic, and intense. Typically CO_2 lasers, emitting at a wavelength of 1.06×10^4 nm, are used in surgery.
(a) What are the frequency and energy (in kJ/mol) of a photon from a CO_2 laser?
(b) Why is monochromatic light desirable in a surgical procedure?
(c) Biological tissue consists primarily of water, which absorbs electromagnetic radiation in the infrared region of the spectrum. Suggest a plausible reason for using CO_2 lasers in surgery.

106. An excimer laser (*excimer* stands for "excited dimer") emits light in the ultraviolet region of the spectrum. An example of such a laser is KrF, which emits light at a wavelength of 248 nm. What is the energy in joules of a mole of photons emitted from the KrF laser? How much more energetic is a single photon of this wavelength than a photon from a CO_2 laser used in surgery (10,600 nm)?

107. Wavelengths less than 10 nm are needed to "see" objects on an atomic or molecular scale. Such imaging can be accomplished by an electron microscope, which uses an electric or magnetic field to focus and accelerate a beam of electrons to a high velocity. Electron microscopy is now a powerful tool in chemical research. What electron velocity is needed to produce electrons with a wavelength of 4×10^{-3} nm, which is sufficient to produce an image of an atom? If electromagnetic radiation were used, what region of the electromagnetic spectrum would this correspond to?

108. Microwave ovens operate by emitting microwave radiation, which is primarily absorbed by water molecules in food. The absorbed radiation is converted to heat through rapid oscillations of polar water molecules, which cooks the food and warms beverages. If 7.2×10^{28} photons are needed to heat 150.0 g of water from 20.0°C to 100.0°C in a microwave oven, what is the frequency of the microwaves? Metal objects should not be placed in a microwave oven because they cause sparks. Why does this cause sparks?

109. The magnitude of the energy gap between an excited state and a ground state determines the color of visible light that is absorbed. The observed color of an object is not the color of the light it absorbs but rather the complement of that color. The accompanying rosette, first developed by Isaac Newton, shows the colors increasing in energy from red to violet. Any two colors that are opposite each other are said to be *complementary* (for example, red and green are complementary).

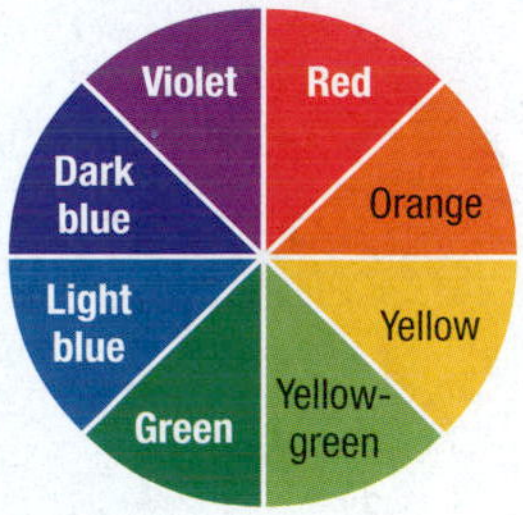

(a) Given the following absorption spectra and table, what are the colors of the objects that produce spectra A, B, and C?

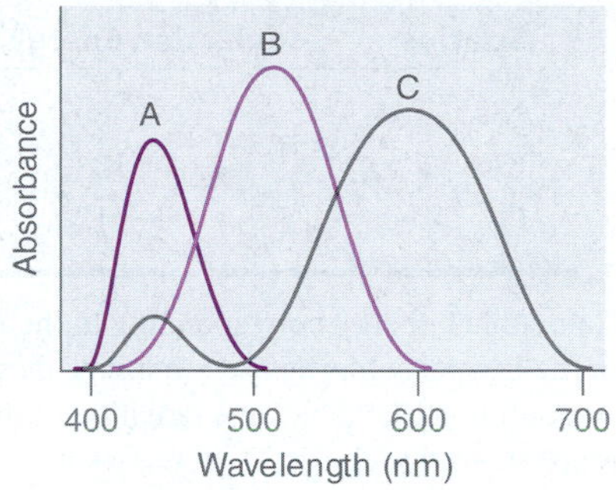

Wavelength, nm	Color of Light
390–453	Violet
453–492	Blue
492–577	Green
577–597	Yellow
597–622	Orange
622–780	Red

(b) The decomposition of a pigment depends on many factors, but the rate of decomposition often depends on the energy of the radiation absorbed by the pigment. Which of the above compounds, A, B, or C, will likely fade fastest?
(c) A painter has a sample of yellow paint. What would you expect the absorption spectrum of the paint to look like? What would the absorption spectrum look like if the paint had been contaminated with traces of a violet pigment?

110. Photodegradation of atmospheric ozone occurs via the reaction $O_3 + h\nu \longrightarrow O_2 + O$; the maximum absorption occurs at approximately 255 nm. In what region of the electromagnetic spectrum does this occur? Based on this information, what would be the effect of depleting the ozone layer of earth's atmosphere?

111. A microscope's resolution (its ability to distinguish between two points separated by a given distance) depends on the wavelength of light used to illuminate an object. The resolution R is given by the equation $R = \lambda/2N$, where N is a constant related to the aperture. If a microscope has an aperture constant of 0.25, what is the smallest distance between two objects that can be resolved using the following light sources? (a) red light ($E = 171$ kJ/mol); (b) UV light ($E = 1.20 \times 10^3$ kJ/mol); (c) X rays ($E = 1.20 \times 10^5$ kJ/mol)

112. Silver bromide is the photosensitive material in 35-mm photographic film. When monochromatic light falls on film, the photons are recorded if they contain sufficient energy to react with AgBr in the film. Given that the minimum energy needed to do this is approximately 57.9 kJ/mol, explain why red light is used to light a darkroom. What happens when the door to the darkroom is opened, allowing yellow light to enter?

113. A lighting system has recently been developed that uses a quartz bulb the size of a golf ball filled with an inert gas and a small amount of sulfur. When irradiated by microwaves, the bulb puts out as much light as hundreds of high-intensity mercury vapor lamps. Because 1000 kJ/mol is needed to ionize sulfur, can this process occur simply by irradiating sulfur atoms with microwaves? Explain your answer.

114. The table lists the ionization energies of some common atmospheric species:

Species	Ionization Energy, kJ/mol
NO	897
CO_2	1330
O_2	1170

(a) Can radiant energy corresponding to the lowest energy line in the Lyman series be used to ionize these molecules?
(b) According to the table, can radiation of this energy be transmitted through an O_2 atmosphere?

115. An artist used a pigment that has a significant absorption peak at 450 nm, with a trace absorption at 530 nm. Based on the color chart and table in Problem 109, what was the color of the paint? Draw the absorption pattern. What would the absorption spectrum have looked like if the artist had wanted green? Using absorption spectra, explain why an equal combination of red and yellow paints produces orange.

116. You live in a universe where the electron has 4 different spins $\left(m_s = +\frac{1}{2}, +\frac{1}{4}, -\frac{1}{2}, -\frac{1}{4}\right)$ and the periodic table goes up to only $Z = 36$. Which elements would be noble gases? What would the periodic table look like? (Assume that the Pauli principle is still valid.)

117. If you were living on a planet where there were three quantum numbers (n, l, m) instead of four, what would be the allowed combinations for an electron in a $3p$ orbital? How many electrons would this orbital contain assuming the Pauli principle were still in effect? How does this compare with the actual number of allowed combinations found on earth?

118. X rays are frequently used to project images of the human body. Recently, however, a superior technique called magnetic resonance imaging (MRI) has been developed that uses *proton* spin to image tissues in spectacular detail. In MRI, spinning hydrogen nuclei in an organic material are irradiated with photons that contain enough energy to flip the protons to the opposite orientation. If 33.121 kJ/mol of energy is needed to flip a proton, what is the resonance frequency required to produce an MRI spectrum? Suggest why this frequency of electromagnetic radiation would be preferred over X rays.

119. Vanadium has been found to be a key component in a biological catalyst that reduces nitrogen to ammonia. What is the valence-electron configuration of vanadium? What are the quantum numbers for each of the valence electrons? How many unpaired electrons does vanadium have?

120. Tellurium, a metal used in semiconductor devices, is also used as a coloring agent in porcelains and enamels. Illustrate the aufbau principle, the Pauli exclusion principle, and Hund's rule using tellurium metal.

7 The Periodic Table and Periodic Trends

LEARNING OBJECTIVES

- To become familiar with the history of the periodic table
- To understand periodic trends in atomic radii
- To be able to predict relative ionic sizes within an isoelectronic series
- To correlate ionization energies and electron affinities with the chemistry of the elements
- To become familiar with the relationship between the electronegativity of an element and its chemistry
- To understand the correlation between the chemical properties and reactivity of the elements and their positions in the periodic table
- To be able to describe some of the roles of trace elements in biological systems

CHAPTER OUTLINE

7.1 The History of the Periodic Table
7.2 Sizes of Atoms and Ions
7.3 Energetics of Ion Formation
7.4 The Chemical Families
7.5 Trace Elements in Biological Systems

In Chapter 6, we presented the contemporary quantum mechanical model of the atom. In using this model to describe the electronic structures of the elements in order of increasing atomic number, we saw that periodic similarities in electron configuration correlate with periodic similarities in properties, which is the basis for the structure of the periodic table. For example, all of the noble gases have what is often called filled or closed-shell valence electron configurations. These closed shells are actually filled *s* and *p* subshells containing a total of eight electrons, called *octets*; helium is an exception, with a closed 1*s* shell containing two electrons. Because of their filled valence shells, the noble gases are generally unreactive. In contrast, all of the alkali metals have a single valence electron outside a closed shell and readily lose this electron to elements that require electrons to achieve an

Crookes's spiral periodic table, 1888. Created by Sir William Crookes (1832–1919), the spiral represents the relationships between the elements and the order of evolution of the elements from what he believed to be primal matter.

octet, such as the halogens. Thus, because of their periodic similarities in electron configuration, atoms in the same column of the periodic table tend to form compounds with the same oxidation states and stoichiometries. Chapter 6 ended with the observation that, because all the elements in a column have the same valence electron configuration, the periodic table can be used to find the electron configuration of most of the elements at a glance.

In this chapter, we explore the relationship between the electron configurations of the elements, as reflected in their arrangement in the periodic table, and their physical and chemical properties. In particular, we focus on the similarities between elements in the same column and on the trends in properties that are observed across horizontal rows or down vertical columns. By the end of this chapter, your understanding of these trends and relationships will provide you with clues to why argon is used in incandescent light bulbs, why coal and wood burst into flames when they come in contact with pure F_2, why aluminum was discovered so late despite being the third most abundant element in the earth's crust, and why lithium is commonly used in batteries. We begin the chapter by expanding on the brief discussion of the history of the periodic table presented in Chapter 1, and describing how it came to be created many years before electrons had even been discovered, much less discussed in terms of shells, subshells, orbitals, and electron spin.

John Newlands (1838–1898). Newlands noticed that elemental properties repeated every seventh (or multiple of seven) element, as musical notes repeat every eighth note.

7.1 The History of the Periodic Table

The modern periodic table has evolved through a long history of attempts by chemists to arrange the elements according to their properties as an aid in predicting chemical behavior. One of the first to suggest such an arrangement was the German chemist Johannes Dobereiner (1780–1849), who had noticed that many of the known elements could be grouped in **triads**, sets of three elements that have similar properties: chlorine, bromine, and iodine, for example, or copper, silver, and gold. Dobereiner proposed that all elements could be grouped in such triads, but subsequent attempts to expand his concept were unsuccessful. We now know that portions of the periodic table—the *d* block in particular—do contain triads of elements with substantial similarities. The middle three members of most of the other columns, such as sulfur, selenium, and tellurium in Group 16, or aluminum, gallium, and indium in Group 13, also have remarkably similar chemistry.

MGC Periodic Trends of the Elements

Dimitri Mendeleev (1834–1907). When his family's glass factory was destroyed by fire, Mendeleev moved to St. Petersburg, Russia, to study science. He became ill and was not expected to recover, but finished his Ph.D. with the help of his professors and fellow students. In addition to the periodic table, another contribution to science by Mendeleev was an outstanding textbook, which was used for many years.

By the mid-19th century, the atomic masses of many of the elements had been determined. The English chemist John Newlands (1838–1898), hypothesizing that the chemistry of the elements might be related to their masses, arranged the known elements in order of increasing atomic mass and discovered that every seventh element had similar properties (Figure 7.1). (The noble gases were still unknown.) Newlands therefore suggested that the elements could be classified into **octaves**, corresponding to the *horizontal* rows in the main group elements. Unfortunately, Newlands's "law of octaves" did not seem to work for elements heavier than calcium, and his idea was publicly ridiculed. At one scientific meeting, Newlands was asked why he didn't arrange the elements in alphabetical order instead of by atomic mass, since that would make just as much sense! Actually, Newlands was on the right track—with only a few exceptions, atomic mass does increase with atomic number, and similar properties occur every time a set of $ns^2\,np^6$ subshells is filled. Despite the fact that Newlands's table had no logical place for the *d*-block elements, he was honored for his idea by the Royal Society of London in 1887.

The periodic table achieved its modern form through the work of German chemist Julius Lothar Meyer (1830–1895) and Russian chemist Dimitri Mendeleev (1834–1907), both of whom focused on the relationships between atomic mass and various physical and chemical properties. In 1869, they independently proposed essentially identical arrangements of the elements. Meyer aligned the elements in his table according to periodic variations in simple atomic properties such as "atomic volume" (Figure 7.2), which

No.	No.	No.	No.	No.	No.	No.	No.
H 1	F 8	Cl 15	Co & Ni 22	Br 29	Pd 36	I 42	Pt & Ir 50
Li 2	Na 9	K 16	Cu 23	Rb 30	Ag 37	Cs 44	Tl 51
G 3	Mg 10	Ca 17	Zn 25	Sr 31	Bd [sic-Cd] 38	Ba & V 45	Pb 54
Bo 4	Al 11	Cr 19	Y 24	Ce & La 33	U 40	Ta 46	Th 56
C 5	Si 12	Ti 18	In 26	Zr 32	Sn 39	W 47	Hg 52
N 6	**P 13**	Mn 20	**As 27**	Di & Mo 34	**Sb 41**	Nb 48	**Bi 55**
O 7	S 14	Fe 21	Se 28	Ro & Ru 35	Te 43	Au 49	Os 51

NOTE. -- Where two elements happen to have the same equivalent, both are designated by the same number.

Figure 7.1 The arrangement of the elements into "octaves" as proposed by Newlands. The table above accompanied a letter from a 27-year-old Newlands to the editor of the journal *Chemical News* in which he wrote: "If the elements are arranged in the order of their equivalents, with a few slight transpositions, as in the accompanying table, it will be observed that elements belonging to the same group usually appear on the same horizontal line. It will also be seen that the numbers of analogous elements generally differ either by 7 or by some multiple of seven; in other words, members of the same group stand to each other in the same relation as the extremities of one or more octaves in music. Thus, in the nitrogen group, between nitrogen and phosphorus there are 7 elements; between phosphorus and arsenic, 14; between arsenic and antimony, 14; and lastly, between antimony and bismuth, 14 also. This peculiar relationship I propose to provisionally term the *Law of Octaves*. I am, &c. John A. R. Newlands, F.C.S. Laboratory, 19, Great St. Helen's, E.C., August 8, 1865."

he obtained by dividing the atomic mass (molar mass) in grams per mole by the density of the element in grams per cubic centimeter. This property is equivalent to what is today defined as **molar volume** (measured in cm^3/mol):

$$\frac{\text{molar mass (g/mol)}}{\text{density (g/cm}^3)} = \text{molar volume (cm}^3\text{/mol)} \tag{7.1}$$

As shown in Figure 7.2, the alkali metals have the highest molar volumes of the solid elements. In Meyer's plot of atomic volume versus atomic mass, the nonmetals occur on the rising portion of the graph, and metals occur at the peaks, in the valleys, and on the downslopes.

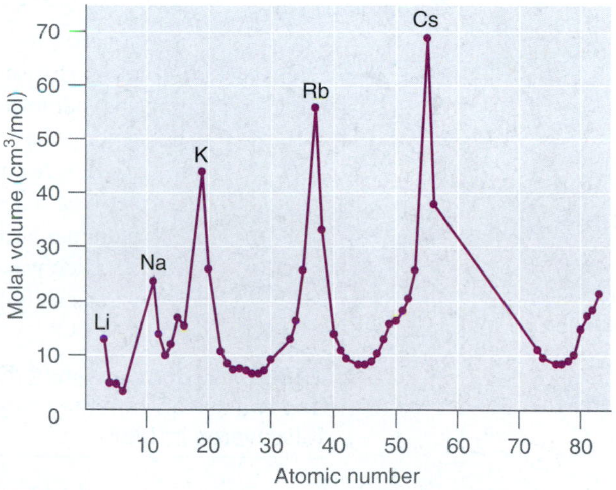

Figure 7.2 Variation of atomic volume with atomic number, adapted from Meyer's plot of 1870. Note the periodic increase and decrease in the atomic volume. Because the noble gases had not yet been discovered at the time this graph was formulated, the peaks correspond to the alkali metals (Group 1).

Mendeleev's Periodic Table

Mendeleev, who first published his periodic table in 1869 (Figure 7.3), is usually credited with the origin of the modern periodic table. The key difference between his arrangement of the elements and that of Meyer and others is that Mendeleev did not assume that all of the elements had been discovered (actually, only about two-thirds of the naturally occurring elements were known at the time). Instead, he deliberately left blanks in his table at atomic masses 44, 68, 72, and 100, in the expectation that elements with those atomic masses would be discovered. Those blanks correspond to the elements we now know as scandium, gallium, germanium, and technetium.

The most convincing evidence in support of Mendeleev's arrangement of the elements was the discovery of two previously unknown elements whose properties closely corresponded with his predictions (Table 7.1). Two of the blanks Mendeleev had left in his original table were situated below aluminum and silicon, awaiting the discovery of two as-yet-unknown elements, *eka*-aluminum and *eka*-silicon (from the Sanskrit *eka*, "one," as in "one beyond aluminum"). The observed properties of gallium and germanium matched those of *eka*-aluminum and *eka*-silicon so well that once they were discovered, Mendeleev's periodic table rapidly gained acceptance.

Reihen	Gruppe I. — R^2O	Gruppe II. — RO	Gruppe III. — R^2O^3	Gruppe IV. RH^4 RO^2	Gruppe V. RH^3 R^2O^5	Gruppe VI. RH^2 RO^3	Gruppe VII. RH R^2O^7	Gruppe VIII. — RO^4
1	H=1							
2	Li=7	Be=9,4	B=11	C=12	N=14	O=16	F=19	
3	Na=23	Mg=24	Al=27,3	Si=28	P=31	S=32	Cl=35,5	
4	K=39	Ca=40	— =44	Ti=48	V=51	Cr=52	Mn=55	Fe=56, Co=59, Ni=59, Cu=63.
5	(Cu=63)	Zn=65	— =68	— =72	As=75	Se=78	Br=80	
6	Rb=85	Sr=87	?Yt=88	Zr=90	Nb=94	Mo=96	— =100	Ru=104, Rh=104, Pd=106, Ag=108.
7	(Ag=108)	Cd=112	In=113	Sn=118	Sb=122	Te=125	J=127	
8	Cs=133	Ba=137	?Di= 138	?Ce= 40	—	—	—	— — — —
9	(—)	—	—	—	—	—	—	
10	—	—	?Er=178	?La=180	Ta=182	W=184	—	Os=195, Ir=197, Pt=198, Au=199.
11	(Au=199)	Hg=200	Tl=204	Pb=207	Bi=208	—	—	
12	—	—	—	Th=231	—	U=240	—	— — — —

Figure 7.3 Mendeleev's periodic table, as published in the German journal *Annalen der Chemie und Pharmacie*, 1872. Column headings "Reihen" and "Gruppe" are German for "row" and "group." Formulas indicate the type of compounds formed by each group, with "R" standing for "any element" and superscripts used where we now use subscripts. Atomic masses are shown after equal signs and increase across each row from left to right.

When the chemical properties of an element suggested that it might have been assigned the wrong place in earlier tables, Mendeleev carefully reexamined its atomic mass. He discovered, for example, that the atomic masses previously reported for beryllium, indium, and uranium were incorrect. The atomic mass of indium had originally been reported as 75.6, based on an assumed stoichiometry of InO for its oxide. If this atomic mass were correct, then indium would have to be placed in the middle of the nonmetals, between arsenic (atomic mass 75) and selenium (atomic mass 78). Because elemental indium is a silvery-white *metal*, however, Mendeleev postulated that the stoichiometry of its oxide was really In_2O_3 rather than InO. This would mean that indium's atomic mass was actually 113, placing the element between two other metals, cadmium and tin.

TABLE 7.1 Comparison of the properties predicted by Mendeleev in 1869 for *eka*-aluminum and *eka*-silicon with the properties of gallium (discovered in 1875) and germanium (discovered in 1886)

Property	*eka*-Aluminum (predicted)	Gallium (observed)	*eka*-Silicon (predicted)	Germanium (observed)
Atomic mass	68	69.723	72	72.64
Element	Metal	Metal	Dirty-gray metal	Gray-white metal
	low melting point	mp 29.8°C	high melting point	mp 938°C
	d = 5.9 g/cm^3	d = 5.91 g/cm^3	d = 5.5 g/cm^3	d = 5.323 g/cm^3
Oxide	E_2O_3	Ga_2O_3	EO_2	GeO_2
	d = 5.5 g/cm^3	d = 6.0 g/cm^3	d = 4.7 g/cm^3	d = 4.25 g/cm^3
Chloride	ECl_3	$GaCl_3$	ECl_4	$GeCl_4$
	volatile	mp 78°C	bp <100°C	bp 87°C
		bp 201°C		

One group of elements that is absent from Mendeleev's table is the noble gases, all of which were discovered more than 20 years later, between 1894 and 1898, by Sir William Ramsay (1852–1916, Nobel Prize 1904). Initially, Ramsay did not know where to place these elements in the periodic table. Argon, the first to be discovered, had an atomic mass of 40. This was higher than chlorine's and comparable to that of potassium, so Ramsay, using the same kind of reasoning as Mendeleev, decided to place the noble gases between the halogens and the alkali metals.

H. G. J. Moseley (1887–1915). This young scientist served in the British army as an engineer and was killed in World War I.

The Role of Atomic Number in the Periodic Table

Despite its usefulness, Mendeleev's periodic table was based entirely on empirical observation supported by very little understanding. It was not until 1913, when a young British physicist, H. G. J. Moseley (1887–1915), analyzed the frequencies of X rays emitted by the elements, that the underlying foundation of the order of the elements was discovered to be *atomic number*, not atomic mass. Moseley hypothesized that the placement of each element in his series corresponded to its atomic number Z, which is the number of positive charges (protons) in its nucleus. Argon, for example, with an atomic mass higher than that of potassium (39.9 amu versus 39.1 amu, respectively), was placed *before* potassium in the periodic table. While analyzing the frequencies of the emitted X rays, Moseley noticed that the atomic number of argon is 18, whereas that of potassium is 19, which indicated that they were placed correctly. Moseley also noticed three gaps in his table of X-ray frequencies, so he predicted the existence of three unknown elements: technetium ($Z = 43$), discovered in 1937; promethium ($Z = 61$), discovered in 1945; and rhenium ($Z = 75$), discovered in 1925.

EXAMPLE 7.1

Before its discovery in 1999, some theoreticians believed that an element with a Z of 114 existed in nature. Use Mendeleev's reasoning to name element 114 as *eka*-______, thus identifying the known element whose chemistry you predict to be most similar to that of element 114.

Given Atomic number

Asked for Name using prefix *eka*-

Strategy

A Using the periodic table on the inside cover of this text, locate the $n = 7$ row. Identify the location of the unknown element with $Z = 114$; then identify the known element that is directly above this location.

B Name the unknown element by using the prefix *eka*- before the name of the known element.

Solution

A The $n = 7$ row can be filled in by assuming the existence of elements with atomic numbers higher than 112, which is underneath mercury (Hg). Counting three boxes to the right gives element 114, which lies directly below lead (Pb). **B** Thus, if Mendeleev were alive today, he would call element 114 *eka*-lead.

EXERCISE 7.1

Use the *eka*-______ nomenclature to name element 112, and thus identify the element whose chemistry you predict to be most similar to that of element 112.

Answer *eka*-mercury

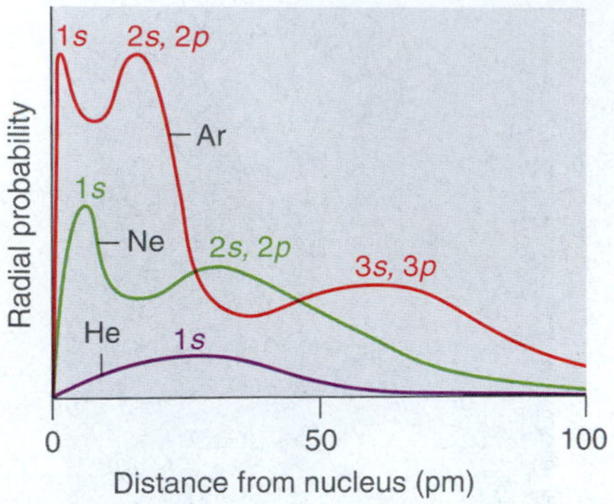

Figure 7.4 Plots of radial probability as a function of distance from the nucleus for He, Ne, and Ar. In He, the 1*s* electrons have a maximum radial probability at around 30 pm from the nucleus. In Ne, the 1*s* electrons have a maximum at around 8 pm, and the 2*s* and 2*p* electrons combine to form another maximum at around 35 pm (the $n = 2$ shell). In Ar, the 1*s* electrons have a maximum at around 2 pm, the 2*s* and 2*p* electrons combine to form a maximum at around 18 pm, and the 3*s* and 3*p* electrons combine to form a maximum at around 70 pm. Interactive Graph

MGC Atomic Radii

7.2 Sizes of Atoms and Ions

Although some people fall into the trap of visualizing atoms and ions as small, hard spheres similar to miniature Ping-Pong balls or marbles, the quantum mechanical model tells us that their shapes and boundaries are much less definite than those images suggest. As a result, atoms and ions cannot be said to have exact sizes. In this section, we discuss how atomic and ion "sizes" are defined and obtained.

Atomic Radii

Recall from Chapter 6 that the probability of finding an electron in the various available orbitals falls off slowly as the distance from the nucleus increases. This point is illustrated in Figure 7.4, which shows a plot of total electron density for all occupied orbitals for three noble gases as a function of their distance from the nucleus. Electron density diminishes gradually with increasing distance, which makes it impossible to draw a sharp line marking the boundary of the atom.

Figure 7.4 also shows that there are distinct peaks in the total electron density at particular distances, and that these peaks occur at different distances from the nucleus for each element. Each peak in a given plot corresponds to electron density in a given principal shell. Because helium has only a single filled shell ($n = 1$), it shows only a single peak. In contrast, neon, with filled $n = 1$ and 2 principal shells, has two peaks. Argon, with filled $n = 1$ principal shells and 2 and 3*s* and 3*p* subshells, has three peaks. The peak for the filled $n = 1$ shell occurs at successively shorter distances for neon ($Z = 10$) and argon ($Z = 18$) because, with a greater number of protons, their nuclei are more positively charged than that of helium. Since the $1s^2$ shell is closest to the nucleus, its electrons are very poorly shielded by electrons in filled shells with larger values of n. Consequently, the two electrons in the $n = 1$ shell experience nearly the full nuclear charge, resulting in a strong electrostatic interaction between the electrons and the nucleus. The energy of the $n = 1$ shell also decreases tremendously (the filled 1*s* orbital becomes more stable) as the nuclear charge increases. For similar reasons, the filled $n = 2$ shell in argon is located closer to the nucleus and has a lower energy than the $n = 2$ shell in neon.

Figure 7.4 illustrates the difficulty of measuring the dimensions of an individual atom. Because distances between the nuclei in pairs of covalently bonded atoms can be measured quite precisely, however, chemists use these distances as a basis for describing the approximate sizes of atoms. For example, the internuclear distance in the diatomic Cl_2 molecule is known to be 198 pm. We assign half of this distance to each chlorine atom, giving chlorine a **covalent atomic radius** of 99 pm or 0.99 Å (Figure 7.5a).*

* Atomic radii are often measured in angstroms (Å), a non-SI unit: $1\ \text{Å} = 1 \times 10^{-10}\ \text{m} = 100\ \text{pm}$.

Figure 7.5 Definitions of atomic radius. (a) The covalent atomic radius, r_{cov}, is half the distance between the nuclei of two like atoms joined by a covalent bond in the same molecule, such as Cl_2. (b) The metallic atomic radius, r_{met}, is half the distance between the nuclei of two adjacent atoms in a pure solid metal, such as aluminum. (c) The van der Waals atomic radius, r_{vdW}, is half the distance between the nuclei of two like atoms, such as argon, that are closely packed but not bonded. (d) Covalent versus van der Waals radii of chlorine.

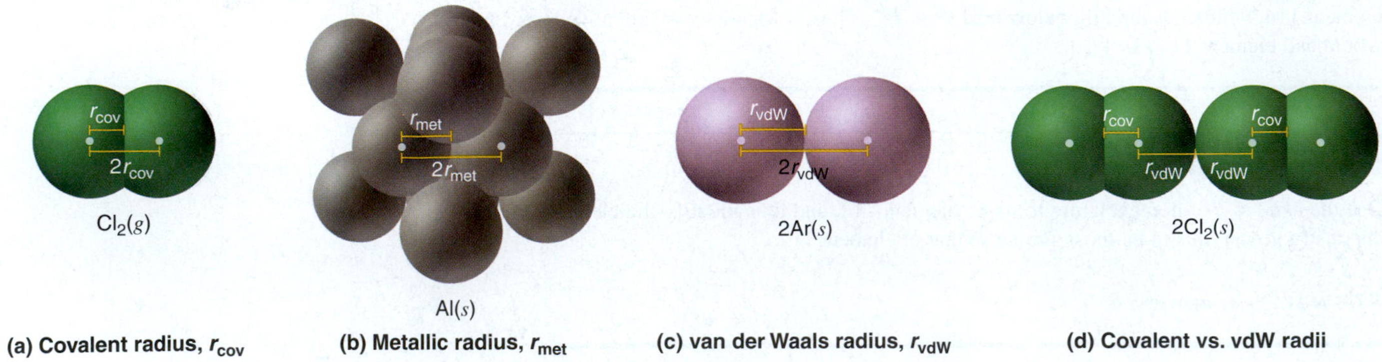

In a similar approach, we can use the lengths of carbon–carbon single bonds in organic compounds, which are remarkably uniform at 154 pm, to assign a value of 77 pm as the covalent atomic radius of carbon. If these values do indeed reflect the actual sizes of the atoms, then we should be able to predict the lengths of covalent bonds formed between different elements by adding them. For example, we would predict a carbon–chlorine distance of 77 pm + 99 pm = 176 pm for a C—Cl bond, which is very close to the average value observed in many organochlorine compounds.*

Covalent atomic radii can be determined for most of the nonmetals, but how do chemists obtain atomic radii for elements that do not form covalent bonds? For these elements, a variety of other methods have been developed. With a metal, for example, the *metallic atomic radius* is defined as half the distance between the nuclei of two adjacent metal atoms (Figure 7.5b). For elements such as the noble gases, most of which form no stable compounds, we can use what is called the *van der Waals atomic radius*, half the internuclear distance between two non-bonded atoms in the solid (Figure 7.5c). Note that an atom such as chlorine has both a covalent radius (the distance between the two atoms in a Cl_2 molecule) and a van der Waals radius (the distance between two Cl atoms in different molecules in, for example, $Cl_2(s)$ at low temperatures), and that these radii are generally not the same (Figure 7.5d).

Periodic Trends in Atomic Radii

Because it is impossible to measure the sizes of both metallic and nonmetallic elements using any one method, chemists have developed a self-consistent way of calculating atomic radii using the quantum mechanical functions described in Chapter 6. Although the radii values obtained by such calculations are not identical to any of the experimentally measured sets of values, they do provide a way to compare the intrinsic sizes of all the elements and clearly show that atomic size varies in a periodic fashion (Figure 7.6). In the periodic table, atomic radii decrease from left to right across a row, and increase from top to bottom down a column. Because of these two trends, the largest atoms are found in the lower left corner of the periodic table, and the smallest in the upper right corner (Figure 7.7).

Note the pattern

Atomic radii decrease from left to right across a row, and increase from top to bottom down a column.

Trends in atomic size result from differences in the effective nuclear charges experienced by electrons in the outermost orbitals of the elements. As we described in Chapter 6, for all elements except H, the effective nuclear charge is always *less* than the actual nuclear charge because of shielding effects. The greater the effective nuclear charge, the more strongly the outermost electrons are attracted to the nucleus, and the smaller the atomic radius.

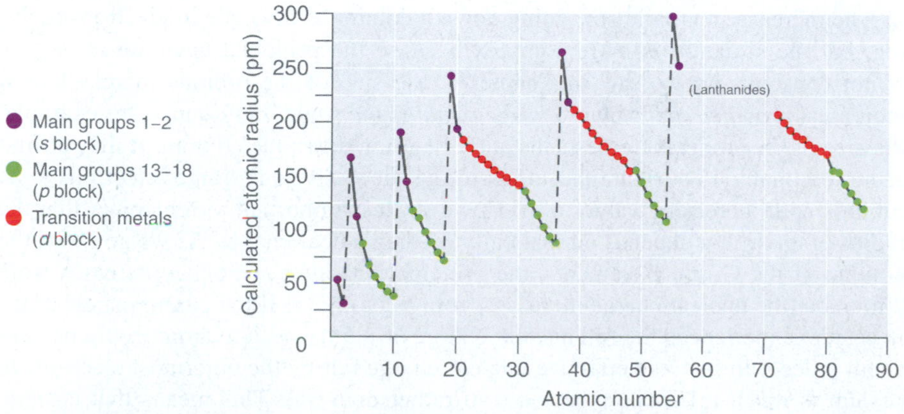

Figure 7.6 Plot of periodic variation of atomic radius with atomic number for the first six rows of the periodic table. Note the similarity to the plot of atomic volume versus atomic number (Figure 7.2)—a variation of Meyer's early plot. Interactive Graph

* A similar approach for measuring the size of ions is discussed below.

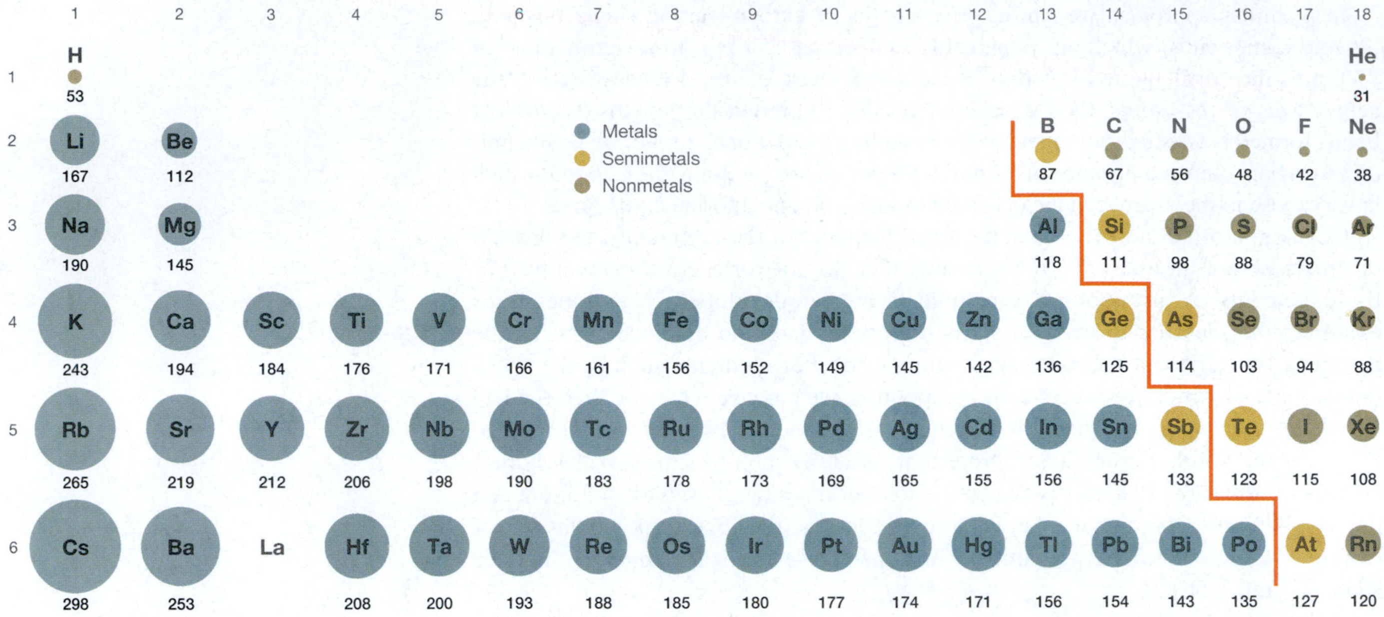

Figure 7.7 Calculated atomic radii (in pm) of the *s*-, *p*-, and *d*-block elements. The sizes of the circles illustrate the relative sizes of the atoms. Calculated values based on quantum mechanical wave functions. (Source of data: www.webelements.com.) Interactive Graph

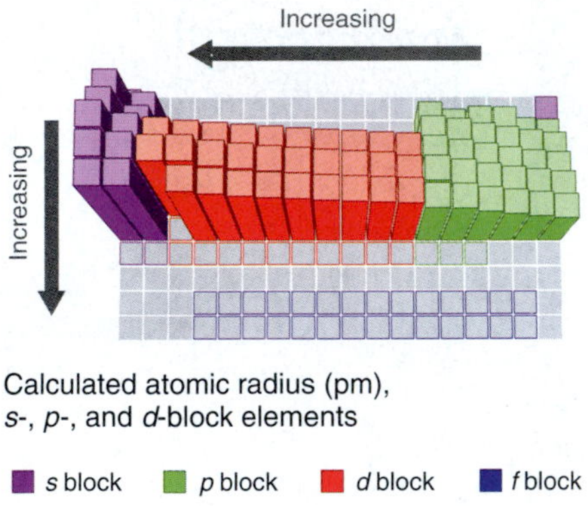

The atoms in the second row of the periodic table (Li through Ne) illustrate the effect of electron shielding (see Figure 6.29). All have a filled $1s^2$ inner shell, but as we go from left to right across the row, the nuclear charge increases from +3 to +10. Although electrons are being added to the 2*s* and 2*p* orbitals, *electrons in the same principal shell are not very effective at shielding one another from the nuclear charge*. Thus, the single 2*s* electron in Li experiences an effective nuclear charge of approximately +1 because the electrons in the filled $1s^2$ shell effectively neutralize two of the three positive charges in the nucleus. (More detailed calculations give a value of $Z_{eff} = +1.26$ for Li.) In contrast, the two 2*s* electrons in beryllium do not shield each other very well, although the filled $1s^2$ shell effectively neutralizes two of the four positive charges in the nucleus. This means that the effective nuclear charge experienced by the 2*s* electrons in beryllium is between +1 and +2 (the calculated value is +1.66). Consequently, beryllium is significantly smaller than lithium. Similarly, as we proceed across the row, the increasing nuclear charge is not effectively neutralized by the electrons being added to the 2*s* and 2*p* orbitals. The result is a steady increase in effective nuclear charge and a steady decrease in atomic size.

The increase in atomic size going down a column is also due to electron shielding, but the situation is more complex because the principal quantum number *n* is not constant. As we saw in Chapter 6, the size of the orbitals increases as *n* increases, *provided the nuclear charge remains the same*. In Group 1, for example, the size of the atoms increases substantially going down the column. It may at first seem reasonable to attribute this effect to the successive addition of electrons to *ns* orbitals with increasing values of *n*. However, it is important to remember that the radius of an orbital depends dramatically on the nuclear charge. As we go down the column of the Group 1 elements, the principal quantum number *n* increases from 2 to 6, but the nuclear charge increases from +3 to +55! If the outermost electrons in cesium experienced the full nuclear charge of +55, a cesium atom would be very small indeed. In fact, the effective nuclear charge felt by the outermost electrons in cesium is much less than expected (≈6 rather than 55). This means that cesium, with a $6s^1$ valence-electron configuration, is much larger than lithium, with a $2s^1$ valence-electron configuration. The effective nuclear charge changes relatively little from lithium to cesium because *electrons in filled inner shells are highly effective at shielding electrons in outer shells from the nuclear charge*. Even though cesium has a nuclear charge of +55, it has 54 electrons in its filled $1s^2 2s^2 2p^6 3s^2 3p^6 4s^2 3d^{10} 4p^6 5s^2 4d^{10} 5p^6$ shells, which effectively neutralize most of the 55 positive charges in the nucleus. The same dynamic is responsible for the steady increase in size observed as we go down the other columns of the periodic table. Irregularities can usually be explained by variations in effective nuclear charge.

EXAMPLE 7.2

On the basis of their positions in the periodic table, arrange these elements in order of increasing atomic radius: aluminum, carbon, silicon.

Given Three elements

Asked for Arrange in order of increasing atomic radius

Strategy

A Identify the location of the elements in the periodic table. Determine the relative sizes of elements located in the same column from their principal quantum number *n*. Then determine the order of elements in the same row from their effective nuclear charges. If the elements are not in the same column or row, use pairwise comparisons.

B List the elements in order of increasing atomic radius.

Solution

A These elements are not all in the same column or row, so we must use pairwise comparisons. Carbon and silicon are both in Group 14 with carbon lying above, so carbon is smaller than silicon (C < Si). Aluminum and silicon are both in the third row with aluminum lying to the left, so silicon is smaller than aluminum (Si < Al) because its effective nuclear charge is greater. **B** Combining the two inequalities gives the overall order: C < Si < Al.

EXERCISE 7.2

Based on their positions in the periodic table, arrange these elements in order of increasing size: oxygen, phosphorus, potassium, sulfur.

Answer O < S < P < K

Ionic Radii and Isoelectronic Series

As you learned in Chapter 2, ionic compounds consist of regular repeating arrays of alternating cations and anions. Although it is not possible to measure an ionic radius directly for the same reason it is not possible to measure an atom's radius directly, it *is* possible to measure the distance between the nuclei of a cation and an adjacent anion in an ionic compound in order to determine the **ionic radius** of one or both. As illustrated in Figure 7.8, the internuclear distance corresponds to the *sum* of the radii of the cation and anion. A variety of methods have been developed to divide the experimentally measured distance proportionally between the smaller cation and larger anion. These methods produce sets of ionic radii that are internally consistent from one ionic compound to another, although each method gives slightly different values. For example, the radius of the Na^+ ion is essentially the same in NaCl and Na_2S, as long as the same method is used to measure it. Thus, despite minor differences due to methodology, certain trends can be observed.

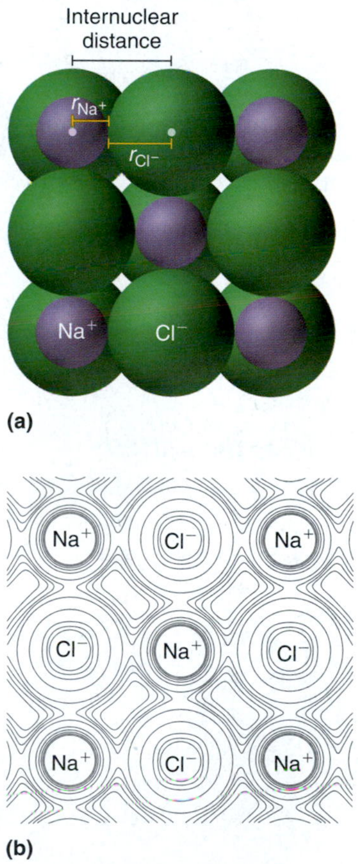

Figure 7.8 Definition of ionic radius. (a) Internuclear distance is apportioned between adjacent cations and anions in the ionic structure, as shown here for Na^+ and Cl^- in sodium chloride. (b) Electron density contours for a single plane of atoms in the NaCl structure. The lines connect points of equal electron density. Note the relative sizes of the electron density contour lines around Cl^- and Na^+.

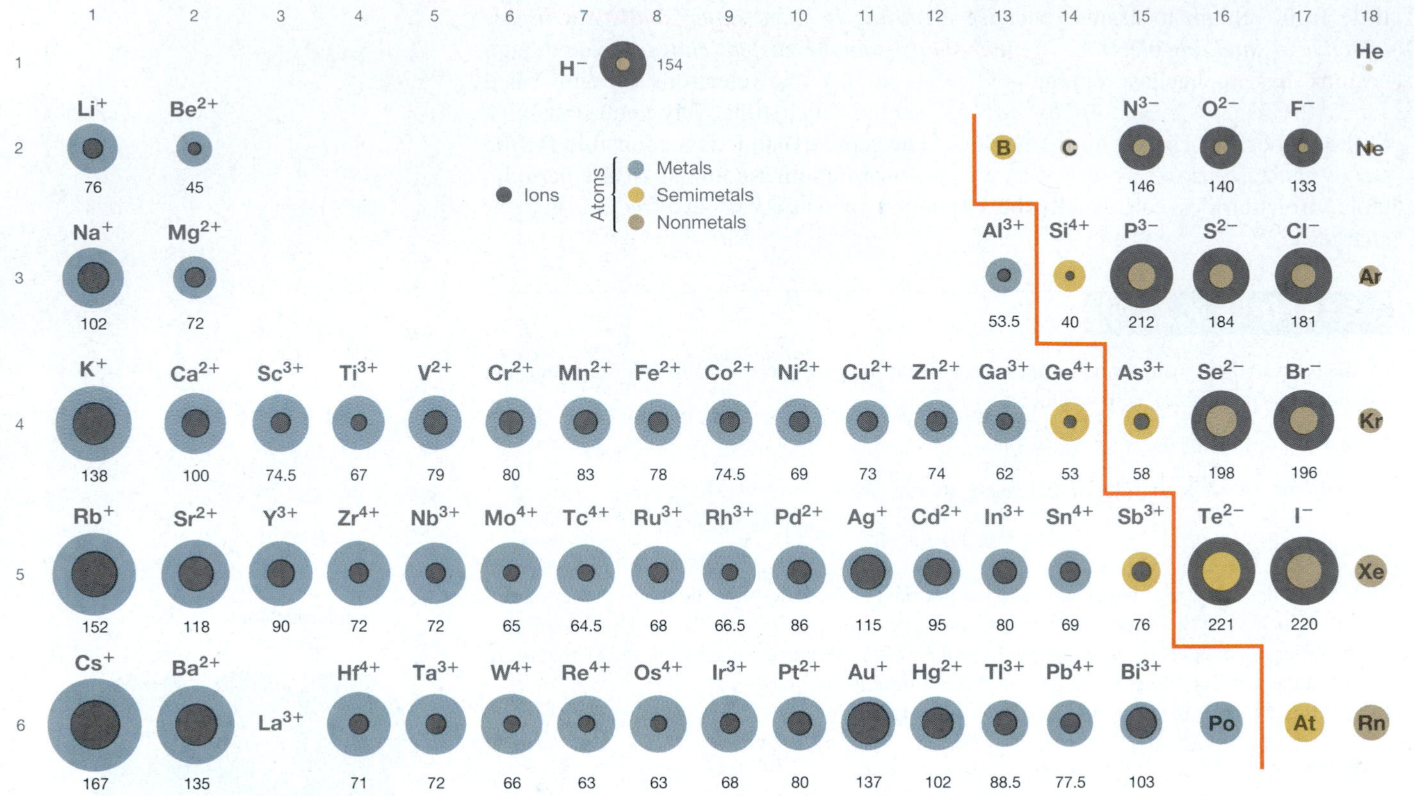

Figure 7.9 Ionic radii (in pm) of the most common oxidation states of the *s*-, *p*-, and *d*-block elements. Gray circles indicate the sizes of the ions shown; colored circles indicate the sizes of the neutral atoms, previously shown in Figure 7.7. [Source of ionic radius data: R. D. Shannon, *Acta Crystallographica* (1976).]

> **Note the pattern**
>
> *Cations are always smaller than the neutral atom, and anions are always larger.*

Comparison of ionic radii with atomic radii (Figure 7.9) shows that a *cation is always smaller than its parent neutral atom, and an anion is always larger than the parent neutral atom.* When one or more electrons is removed from a neutral atom, two things happen: (1) repulsions between electrons in the same principal shell decrease, because fewer electrons are present, and (2) the effective nuclear charge felt by the remaining electrons increases, because there are fewer electrons to shield one another from the nucleus. Consequently, the size of the region of space occupied by electrons decreases (compare Li at 167 pm with Li^+ at 76 pm). If different numbers of electrons can be removed to produce ions with different charges, the ion with the greatest positive charge is the smallest (compare Fe^{2+} at 78 pm with Fe^{3+} at 64.5 pm). Conversely, adding one or more electrons to a neutral atom causes electron–electron repulsions to increase and the effective nuclear charge to decrease, so that the size of the probability region increases (compare F at 42 pm with F^- at 133 pm).

Because most elements form either a cation or an anion but not both, there are few opportunities to compare the sizes of a cation and an anion derived from the same neutral atom. A few compounds of sodium, however, contain the Na^- ion, allowing comparison of its size with that of the far more familiar Na^+ ion, which is found in *many* compounds. The radius of sodium in each of its three known oxidation states is given in Table 7.2. All three species have a nuclear charge of +11, but they contain 10 (Na^+), 11 (Na^0), and 12 (Na^-) electrons. The Na^+ ion is significantly smaller than the neutral Na atom because the $3s^1$ electron has been removed to give a closed shell with principal quantum number 2. The Na^- ion is larger than the parent Na atom because the additional electron produces a $3s^2$ valence-electron configuration, while the nuclear charge remains the same.

TABLE 7.2 Experimentally measured values for the radius of sodium in its three known oxidation states

	Na^+	Na^0	Na^-
Electron Configuration	$1s^2 2s^2 2p^6$	$1s^2 2s^2 2p^6 3s^1$	$1s^2 2s^2 2p^6 3s^2$
Radius (pm)	102	154[a]	202[b]

[a]The metallic radius measured for Na(*s*).
[b]Source: M. J. Wagner and J. L. Dye, *Annu. Rev. Mater. Sci.*, **23**, 225–253 (1993).

Ionic radii follow the same vertical trend as atomic radii; that is, for ions with the same charge, the ionic radius increases going down a column. The reason is the same as for atomic radii: shielding by filled inner shells produces little change in the effective nuclear charge felt by the outermost electrons. Again, principal shells with larger values of n lie at successively greater distances from the nucleus.

Because elements in different columns tend to form ions with different charges, it is not possible to compare ions of the same charge across a row of the periodic table. Instead, elements that are next to each other tend to form ions with the same number of electrons but with different overall charges because of their different atomic numbers. Such a set of species is known as an **isoelectronic series**. For example, the isoelectronic series of species with the neon closed-shell configuration ($1s^2 2s^2 2p^6$) is shown in Table 7.3. Notice that the sizes of the ions in this series decrease smoothly from N^{3-} to Al^{3+}. All six of the ions contain 10 electrons in the $1s$, $2s$, and $2p$ orbitals, but the nuclear charge varies from +7 (N) to +13 (Al). As the positive charge of the nucleus increases while the number of electrons remains the same, there is a greater electrostatic attraction between the electrons and the nucleus, which causes a decrease in radius. Consequently, the ion with the greatest nuclear charge (Al^{3+}) is the smallest, and the ion with the smallest nuclear charge (N^{3-}) is the largest. Notice that one member of this isoelectronic series is not listed in Table 7.3: the neon atom. Because neon forms no covalent or ionic compounds, its radius is difficult to measure.

TABLE 7.3 Radius of ions with the neon closed-shell electron configuration[a]

Ion	Radius, pm	Atomic Number
N^{3-}	146	7
O^{2-}	140	8
F^-	133	9
Na^+	102	11
Mg^{2+}	72	12
Al^{3+}	53.5	13

[a]Source of data: R.D. Shannon, *Acta Crystallogaphica* (1976).

EXAMPLE 7.3

Based on their positions in the periodic table, arrange these ions in order of increasing radius: Cl^-, K^+, S^{2-}, Se^{2-}.

Given Four ions

Asked for Order by increasing radius

Strategy

Ⓐ Determine which ions form an isoelectronic series. Of those ions, predict their relative sizes based on their nuclear charges. For ions that do not form an isoelectronic series, locate their positions in the periodic table.

Ⓑ Determine the relative sizes of the ions based on their principal quantum numbers n and their locations within a row.

Solution

Ⓐ We see that S and Cl are at the right of the third row, while K and Se are at the far left and right ends of the fourth row, respectively. Notice that K^+, Cl^-, and S^{2-} form an isoelectronic series with the [Ar] closed-shell electron configuration; that is, all three ions contain 18 electrons but have different nuclear charges. Because K^+ has the greatest nuclear charge ($Z = 19$), its radius is smallest, and S^{2-} with $Z = 16$ has the largest radius. Because selenium is directly below sulfur, we expect the Se^{2-} ion to be even larger than S^{2-}. Ⓑ The order must therefore be $K^+ < Cl^- < S^{2-} < Se^{2-}$.

EXERCISE 7.3

Arrange these ions in order of increasing size: Br^-, Ca^{2+}, Rb^+, Sr^{2+}.

Answer $Ca^{2+} < Sr^{2+} < Rb^+ < Br^-$

7.3 • Energetics of Ion Formation

MGC Ionization Energy

We have seen that when elements react, they often gain or lose enough electrons to achieve the valence electron configuration of the nearest noble gas. In this section, we develop a more quantitative approach to predicting such reactions by examining periodic trends in the energy changes that accompany ion formation.

Ionization Energies

Because atoms do not spontaneously lose electrons, energy is required to remove an electron from an atom to form a cation. Chemists define the **ionization energy** (I) of an element as the amount of energy needed to remove an electron from the gaseous atom E in its ground state. I is therefore the energy required for the reaction

$$E(g) \rightarrow E^+(g) + e^- \qquad \text{energy required} = I \tag{7.2}$$

Because an input of energy is required, the ionization energy is always positive ($I > 0$) for the reaction as written in Equation 7.2. Larger values of I mean that the electron is more tightly bound to the atom and harder to remove. Typical units for ionization energies are kilojoules/mole (kJ/mol) or electron volts (eV):

$$1 \text{ eV/atom} = 96.49 \text{ kJ/mol}$$

Note the pattern

Successive ionization energies for an element increase steadily.

If an atom possesses more than one electron, the amount of energy needed to remove successive electrons increases steadily. We can define a first ionization energy (I_1), a second ionization energy (I_2), and in general an nth ionization energy (I_n) according to the following reactions:

$$E(g) \rightarrow E^+(g) + e^- \qquad I_1 = \text{1st ionization energy} \tag{7.3}$$

$$E^+(g) \rightarrow E^{2+}(g) + e^- \qquad I_2 = \text{2nd ionization energy} \tag{7.4}$$

$$E^{(n-1)+}(g) \rightarrow E^{n+}(g) + e^- \qquad I_n = n\text{th ionization energy} \tag{7.5}$$

Values for the ionization energies of Li and Be listed in Table 7.4 show that successive ionization energies for an element increase steadily; that is, it takes more energy to

TABLE 7.4 Ionization energies, I (in kJ/mol), for removing successive electrons from lithium and beryllium[a]

Reaction	I	Reaction	I
$\underset{1s^2 2s^1}{Li(g)} \rightarrow \underset{1s^2}{Li^+(g)} + e^-$	$I_1 = 520.2$	$\underset{1s^2 2s^2}{Be(g)} \rightarrow \underset{1s^2 2s^1}{Be^+(g)} + e^-$	$I_1 = 899.5$
$\underset{1s^2}{Li^+(g)} \rightarrow \underset{1s^1}{Li^{2+}(g)} + e^-$	$I_2 = 7298.2$	$\underset{1s^2 2s^1}{Be^+(g)} \rightarrow \underset{1s^2}{Be^{2+}(g)} + e^-$	$I_2 = 1757.1$
$\underset{1s^1}{Li^{2+}(g)} \rightarrow \underset{1s^0}{Li^{3+}(g)} + e^-$	$I_3 = 11,815.0$	$\underset{1s^2}{Be^{2+}(g)} \rightarrow \underset{1s^1}{Be^{3+}(g)} + e^-$	$I_3 = 14,848.8$
		$\underset{1s^1}{Be^{3+}(g)} \rightarrow \underset{1s^0}{Be^{4+}(g)} + e^-$	$I_4 = 21,006.6$

[a]Source of data: *CRC Handbook of Chemistry and Physics* (2004).

remove the second electron from an atom than the first, and so on. There are two reasons for this trend. First, the second electron is being removed from a positively charged species rather than a neutral one, so in accordance with Coulomb's law, more energy is required. Second, removing the first electron reduces the repulsive forces among the remaining electrons, so the attraction of the remaining electrons to the nucleus is stronger.

The most important consequence of the values listed in Table 7.4 is that the chemistry of Li is dominated by the Li^+ ion, while the chemistry of Be is dominated by the +2 oxidation state. Note that the energy required to remove the *second* electron from Li

$$Li^+(g) \rightarrow Li^{2+}(g) + e^-$$

is more than 10 times greater than the energy needed to remove the first electron. Similarly, the energy required to remove the *third* electron from Be

$$Be^{2+}(g) \rightarrow Be^{3+}(g) + e^-$$

is about 15 times greater than the energy needed to remove the first electron and around 8 times greater than the energy required to remove the second electron. Both Li^+ and Be^{2+} have $1s^2$ closed-shell configurations, and much more energy is required to remove an electron from the $1s^2$ core than from the $2s$ valence orbital of the same element. The chemical consequences are enormous: lithium (and all the alkali metals) forms compounds that contain the 1+ ions but not the 2+ or 3+ ions. Similarly, beryllium (and all the alkaline earths) forms compounds that contain the 2+ ion but not the 3+ or 4+ ions. *The energy required to remove electrons from a filled core is prohibitively large and simply cannot be achieved in normal chemical reactions.*

Ionization Energies of s- and p-Block Elements

Ionization energies of the elements in the third row of the periodic table exhibit the same pattern as those of Li and Be (Table 7.5): successive ionization energies increase steadily as electrons are removed from the valence orbitals ($3s$ or $3p$, in this case), followed by an especially large increase in ionization energy when electrons are removed from filled core levels as indicated by the bold diagonal line in Table 7.5. Thus, in the third row of the periodic table, the largest increase in ionization energy corresponds to removing the fourth electron from Al, the fifth electron from Si, and so on—that is, to removing an electron from an ion that has the valence-electron configuration of the preceding noble gas. This pattern explains why the chemistry of the elements normally involves valence electrons only. Too much energy is required to either remove or share the inner electrons.

TABLE 7.5 Successive ionization energies, *I* (in kJ/mol), for the elements in the third row of the periodic table[a]

Element	I_1	I_2	I_3	I_4	I_5	I_6	I_7
Na	495.8	**4562.4**[b]	—	—	—	—	—
Mg	737.7	1450.7	**7732.7**	—	—	—	—
Al	577.5	1816.7	2744.8	**11,577.5**	—	—	—
Si	786.5	1577.1	3231.6	4355.5	**16,090.6**	—	—
P	1011.8	1907.5	2914.1	4963.6	6274.0	**21,267.4**	—
S	999.6	2251.8	3357	4556.2	7004.3	8495.8	**27,107.4**
Cl	1251.2	2297.7	3822	5158.6	6540	9362	11,018.2
Ar	1520.6	2665.9	3931	5771	7238	8781.0	11,995.3

[a]Source of data: *CRC Handbook of Chemistry and Physics* (2004).
[b]Inner-shell electron

EXAMPLE 7.4

From their locations in the periodic table, predict which of these elements has the highest fourth ionization energy: B, C, or N.

Given Three elements

Asked for Element with highest fourth ionization energy

Strategy

Ⓐ List the electron configuration of each element.

Ⓑ Determine whether electrons are being removed from a filled or partially filled valence shell. Predict which element has the highest fourth ionization energy, recognizing that the highest energy corresponds to the removal of electrons from a filled electron core.

Solution

Ⓐ✓ These elements all lie in the second row of the periodic table and have the following electron configurations:

$$\text{B:}\quad [\text{He}]\,2s^2\,2p^1$$
$$\text{C:}\quad [\text{He}]\,2s^2\,2p^2$$
$$\text{N:}\quad [\text{He}]\,2s^2\,2p^3$$

Ⓑ✓ The fourth ionization energy of an element (I_4) is defined as the energy required to remove the fourth electron:

$$\text{E}^{3+}(g) \rightarrow \text{E}^{4+}(g) + \text{e}^-$$

Because carbon and nitrogen have four and five valence electrons, respectively, their fourth ionization energies correspond to removing an electron from a partially filled valence shell. The fourth ionization energy for boron, however, corresponds to removing an electron from the filled $1s^2$ subshell. This should require much more energy. The actual values are: B, 25,026 kJ/mol; C, 6223 kJ/mol; N, 7475 kJ/mol.

EXERCISE 7.4

On the basis of their locations in the periodic table, which of these elements would you expect to have the lowest second ionization energy: Sr, Rb, or Ar?

Answer Sr

The first column of data in Table 7.5 shows that first ionization energies tend to increase across the third row of the periodic table. This is because the valence electrons do not screen each other very well, allowing the effective nuclear charge to increase steadily across the row. The valence electrons are therefore attracted more strongly to the nucleus, so atomic sizes decrease and ionization energies increase. These effects represent two sides of the same coin: stronger electrostatic interactions between the electrons and the nucleus further increase the energy required to remove the electrons.

Notice, however, that the first ionization energy decreases at Al ([Ne] $3s^2\,3p^1$) and at S ([Ne] $3s^2\,3p^4$). The electrons in aluminum's filled $3s^2$ subshell are better at screening the $3p^1$ electron than they are at screening each other from the nuclear charge, so the *s* electrons penetrate closer to the nucleus than the *p* electron does. The decrease at S occurs because the two electrons in the same *p* orbital repel each other. This makes the S atom slightly less stable than would otherwise be expected, as is true of all of the Group 16 elements.

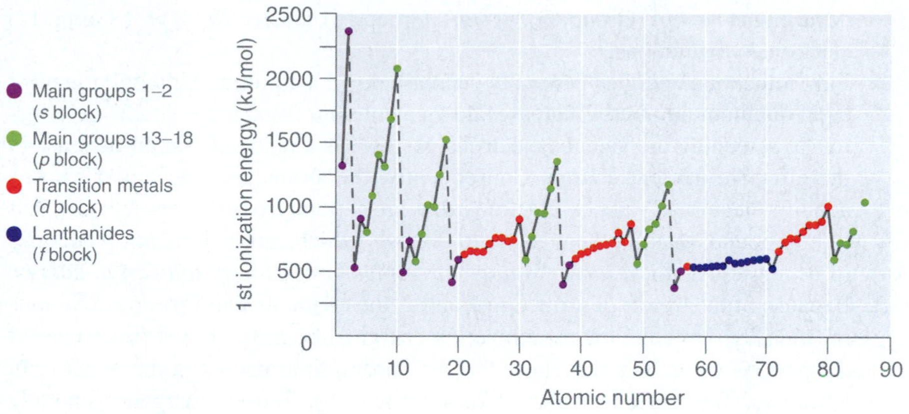

Figure 7.10 Plot of periodic variation of first ionization energy with atomic number for the first six rows of the periodic table. Note the decrease in ionization energy within a group (most easily seen here for Groups 1 and 18).

The first ionization energies of the elements in the first six rows of the periodic table are plotted in Figure 7.10 and presented numerically and graphically in Figure 7.11. These figures illustrate three important trends:

1. The changes seen in the second (Li to Ne), fourth (K to Kr), fifth (Rb to Xe), and sixth (Cs to Rn) rows of the *s* and *p* blocks follow a pattern similar to the pattern described above for the third row of the periodic table. The transition metals are included in the fourth, fifth, and sixth rows, however, and the lanthanides are included in the sixth row. Notice that the first ionization energies of the transition metals are somewhat similar to one another, as are those of the lanthanides. Ionization energies increase from left to right across each row, with discrepancies

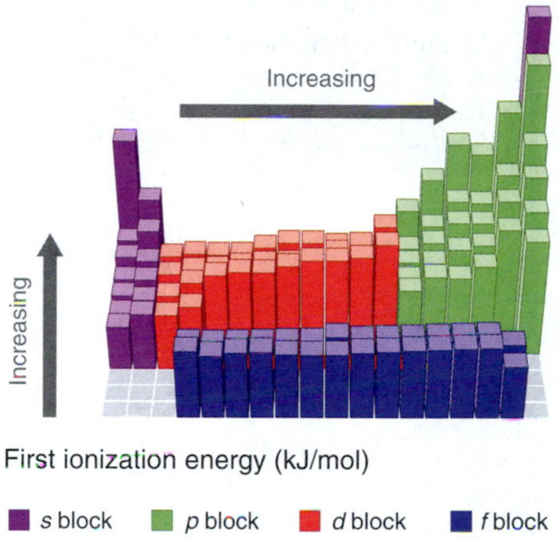

Figure 7.11 First ionization energies of the *s*-, *p*-, *d*-, and *f*-block elements. The darkness of the shading inside the cells of the table indicates the relative magnitudes of the ionization energies. Elements in gray have undetermined first ionization energies. [Source of data: *CRC Handbook of Chemistry and Physics* (2004).]

375.7 kJ/mol — 2372.3 kJ/mol

	1	2	3	4	5	6	7	8	9	10	11	12	13	14	15	16	17	18
1	H 1312.0																	He 2372.3
2	Li 520.2	Be 899.5											B 800.6	C 1086.5	N 1402.3	O 1313.9	F 1681.0	Ne 2080.7
3	Na 495.8	Mg 737.7											Al 577.5	Si 786.5	P 1011.8	S 999.6	Cl 1251.2	Ar 1520.6
4	K 418.8	Ca 589.8	Sc 633.1	Ti 658.8	V 650.9	Cr 652.9	Mn 717.3	Fe 762.5	Co 760.4	Ni 737.1	Cu 745.5	Zn 906.4	Ga 578.8	Ge 762.2	As 944.5	Se 941.0	Br 1139.9	Kr 1350.8
5	Rb 403.0	Sr 549.5	Y 599.9	Zr 640.1	Nb 652.1	Mo 684.3	Tc 702	Ru 710.2	Rh 719.7	Pd 804.4	Ag 731.0	Cd 867.8	In 558.3	Sn 708.6	Sb 830.6	Te 869.3	I 1008.4	Xe 1170.3
6	Cs 375.7	Ba 502.9	La 538.1	Hf 658.5	Ta 728.4	W 758.8	Re 755.8	Os 814.2	Ir 865.2	Pt 864.4	Au 890.1	Hg 1007.1	Tl 589.4	Pb 715.6	Bi 703.0	Po 812.1	At	Rn 1037.1
7	Fr 393.0	Ra 509.3	Ac 498.8	Rf 580	Db	Sg	Bh	Hs	Mt	Ds	Rg	Uub	Uut	Uuq	Uup			

Lanthanides 6	Ce 534.4	Pr 528.1	Nd 533.1	Pm 538.6	Sm 544.5	Eu 547.1	Gd 593.4	Tb 565.8	Dy 573.0	Ho 581.0	Er 589.3	Tm 596.7	Yb 603.4	Lu 523.5
Actinides 7	Th 608.5	Pa 568	U 597.6	Np 604.5	Pu 581.4	Am 576.4	Cm 578.1	Bk 598.0	Cf 606.1	Es 619	Fm 627	Md 635	No 642	Lr 472.8

occurring at $ns^2\,np^1$ (Group 13), $ns^2\,np^4$ (Group 16), and $ns^2\,(n-1)d^{10}$ (Group 12) electron configurations.

2. First ionization energies generally decrease down a column. Although the principal quantum number n increases down a column, filled inner shells are effective at screening the valence electrons, so there is a relatively small increase in the effective nuclear charge. Consequently, the atoms become larger as they acquire electrons. Valence electrons that are farther from the nucleus are less tightly bound, making them easier to remove, which causes ionization energies to decrease. Again, *a larger radius corresponds to a lower ionization energy.*
3. Because of the trends described in 1 and 2, the elements that form positive ions most easily (have the lowest ionization energies) lie in the lower left corner of the periodic table, whereas those that are hardest to ionize lie in the upper right corner of the periodic table. Consequently, ionization energies generally increase diagonally from lower left (Cs) to upper right (He).

Note the pattern

Generally, I_1 increases diagonally from the lower left of the periodic table to the upper right.

Gallium (Ga), which is the first element following the first row of transition metals, has an [Ar] $4s^2\,3d^{10}\,4p^1$ electron configuration. Its first ionization energy is significantly lower than that of the immediately preceding element, Zn, because the filled $3d^{10}$ subshell lies inside the $4p$ subshell, screening the single $4p$ electron from the nucleus. Experiments have revealed something of even greater interest: the second and third electrons that are removed when gallium is ionized come from the $4s^2$ orbital, *not* the $3d^{10}$ subshell. The chemistry of gallium is dominated by the resulting Ga^{3+} ion, with its [Ar]$3d^{10}$ electron configuration. This and similar electron configurations are particularly stable and are often encountered in the heavier p-block elements. They are sometimes referred to as **pseudo noble gas configurations**. In fact, for elements that exhibit these configurations, *no chemical compounds are known in which electrons are removed from the $(n-1)d^{10}$ filled subshell.*

Ionization Energies of Transition Metals and Lanthanides

As we noted, the first ionization energies of the transition metals and lanthanides change very little across each row. Differences in their second and third ionization energies are also rather small, in sharp contrast to the pattern seen with the s- and p-block elements. The reason for these similarities is that the transition metals and lanthanides form cations by losing the ns electrons before the $(n-1)d$ or $(n-2)f$ electrons, respectively. This means that transition metal cations have $(n-1)d^n$ valence electron configurations, and lanthanide cations have $(n-2)f^n$ valence electron configurations. Because the $(n-1)d$ and $(n-2)f$ shells are closer to the nucleus than the ns shell, the $(n-1)d$ and $(n-2)f$ electrons screen the ns electrons quite effectively, reducing the effective nuclear charge felt by the ns electrons. As Z increases, the increasing positive charge is largely canceled by the electrons added to the $(n-1)d$ or $(n-2)f$ orbitals.

That the ns electrons are removed before the $(n-1)d$ or $(n-2)f$ electrons may surprise you because the orbitals were filled in the reverse order (Chapter 6). In fact, the ns, $(n-1)d$, and $(n-2)f$ orbitals are so close to one another in energy, and interpenetrate one another so extensively, that very small changes in the effective nuclear charge can change the order of their energy levels. As the d orbitals are filled, the effective nuclear charge causes the $3d$ orbitals to be slightly lower in energy than the $4s$ orbitals. The [Ar] $3d^2$ electron configuration of Ti^{2+} tells us that the $4s$ electrons of titanium are lost before the $3d$ electrons; this is confirmed by experiment. A similar pattern is seen with the lanthanides, producing cations with an $(n-2)f^n$ valence electron configuration.

Because their first, second, and third ionization energies change so little across a row, these elements have important *horizontal* similarities in chemical properties in addition to the expected vertical similarities. For example, all the first-row transition metals except Sc form stable compounds as M^{2+} ions, whereas the lanthanides primarily form compounds in which they exist as M^{3+} ions.

EXAMPLE 7.5

Use their locations in the periodic table to predict which of the following elements has the lowest first ionization energy: Ca, K, Mg, Na, Rb, or Sr.

Given Six elements

Asked for Element with lowest first ionization energy

Strategy

Locate the elements in the periodic table. Based on trends in ionization energies across a row and down a column, identify the element with the lowest first ionization energy.

Solution

These six elements form a rectangle in the two far-left columns of the periodic table. Because we know that ionization energies increase from left to right in a row and from bottom to top of a column, we can predict that the element at the bottom left of the rectangle will have the lowest first ionization energy: Rb.

EXERCISE 7.5

Which of these elements would you predict to have the highest first ionization energy: As, Bi, Ge, Pb, Sb, Sn?

Answer As

Electron Affinities

The **electron affinity (*EA*)** of an element E is defined as the energy change that occurs when an electron is added to a gaseous atom:

$$E(g) + e^- \rightarrow E^-(g) \qquad \text{energy change} = EA \tag{7.6}$$

Unlike ionization energies, which are always positive for a neutral atom because energy is required to remove an electron, electron affinities can be negative (in which case energy is released when an electron is added), or they can be positive (in which case energy must be added to the system to produce an anion), or zero (the process is energetically neutral). Notice that this sign convention is consistent with our discussion of energy changes in Chapter 5, where a negative value corresponded to the energy change for an exothermic process, which is one in which heat is released.

Chlorine has the most negative electron affinity of any element, which means that more energy is released when an electron is added to a gaseous chlorine atom than to an atom of any other element:

$$Cl(g) + e^- \rightarrow Cl^-(g) \qquad EA = -348.6 \text{ kJ/mol} \tag{7.7}$$

In contrast, beryllium does not form a stable anion, so it has an effective electron affinity of ≥ 0:

$$Be(g) + e^- \rightarrow Be^-(g) \qquad EA \geq 0 \tag{7.8}$$

Nitrogen is unique in that it has an electron affinity of approximately zero. Adding an electron neither releases nor requires a significant amount of energy:

$$N(g) + e^- \rightarrow N^-(g) \qquad EA \geq 0 \tag{7.9}$$

Note the pattern

Generally, EAs become more negative across a row of the periodic table.

Figure 7.12 Plot of periodic variation of electron affinity with atomic number for the first six rows of the periodic table.

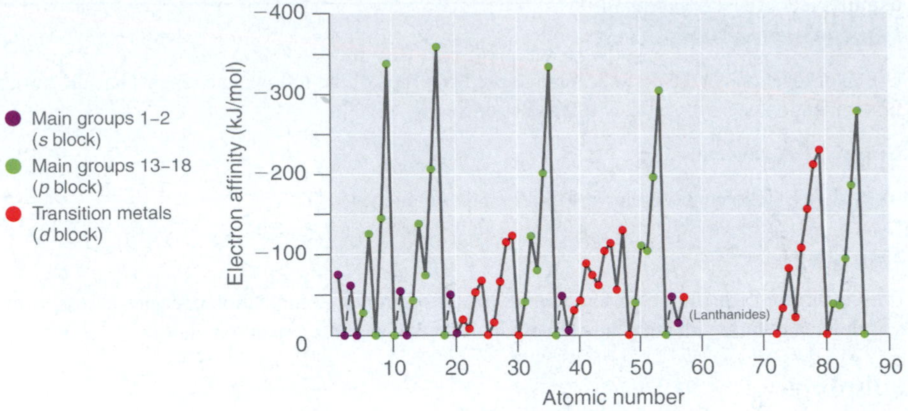

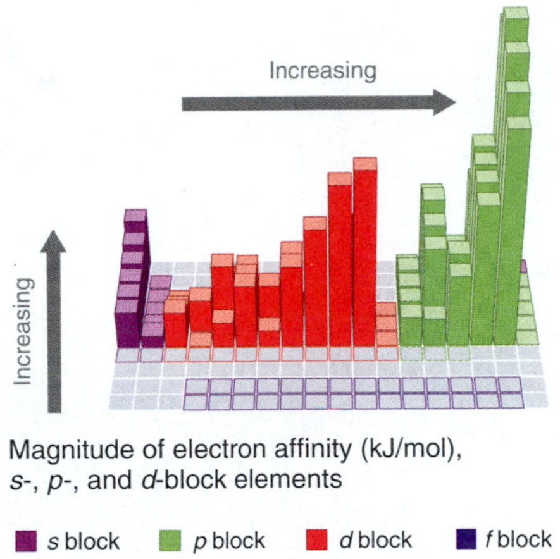

Electron affinities for the first six rows of the periodic table are plotted in Figure 7.12 and presented numerically and graphically in Figure 7.13. Both figures show that the halogens, with their $ns^2 np^5$ valence-electron configuration, have the most negative electron affinities. In general, electron affinities become more negative as we go across a row of the periodic table. This pattern corresponds to the increased effective nuclear charge felt by the valence electrons across a row, which leads to increased electrostatic attraction between the added electron and the nucleus (a more negative *EA*). The trend, however, is not as uniform as the one observed for ionization energies. Some of the alkaline earths (Group 2), the elements of Group 12, and all of the noble gases (Group 18) have effective electron affinities of ≥ 0, while the electron affinities for the elements of Group 15 are

Figure 7.13 Electron affinities (in kJ/mol) of the *s*-, *p*-, and *d*-block elements. Notice that there are many more exceptions to the trends across rows and down columns than with first ionization energies. Elements that do not form stable ions, such as the noble gases, are assigned an effective electron affinity of ≥0. Elements for which no data are available are shown in gray. [Source of data: *J. Phys. Chem. Ref. Data,* Vol 28, No. 6 (1999).] Interactive Graph

≥0 kJ/mol — −348.6 kJ/mol

	1	2	3	4	5	6	7	8	9	10	11	12	13	14	15	16	17	18
1	H -72.8																	He ≥0
2	Li -59.6	Be ≥0											B -27.0	C -121.8	N ≥0	O -141.0	F -328.2	Ne ≥0
3	Na -52.9	Mg ≥0											Al -41.8	Si -134.1	P -72.0	S -200.4	Cl -348.6	Ar ≥0
4	K -48.4	Ca -2.4	Sc -18	Ti -8	V -51	Cr -65.2	Mn ≥0	Fe -15	Co -64.0	Ni -111.7	Cu -119.2	Zn ≥0	Ga -40	Ge -118.9	As -78	Se -195.0	Br -324.5	Kr ≥0
5	Rb -46.9	Sr -5.0	Y -30	Zr -41	Nb -86	Mo -72.1	Tc -60	Ru -101.0	Rh -110.3	Pd -54.2	Ag -125.9	Cd ≥0	In -39	Sn -107.3	Sb -101.1	Te -190.2	I -295.2	Xe ≥0
6	Cs -45.5	Ba -14.0	La -45	Hf ≥0	Ta -31	W -79	Re -20	Os -104.0	Ir -150.9	Pt -205.0	Au -222.7	Hg ≥0	Tl -37	Pb -35	Bi -90.9	Po -180	At -270	Rn ≥0
7	Fr	Ra	Ac	Rf	Db	Sg	Bh	Hs	Mt	Ds	Rg	Uub	Uut	Uuq	Uup			

Lanthanides 6	Ce	Pr	Nd	Pm	Sm	Eu	Gd	Tb	Dy	Ho	Er	Tm	Yb	Lu
Actinides 7	Th	Pa	U	Np	Pu	Am	Cm	Bk	Cf	Es	Fm	Md	No	Lr

usually less negative than those for the Group 14 elements. These exceptions can be explained by the groups' electron configurations. Both the alkaline earths and the noble gases have valence electron shells with filled subshells (ns^2 and $ns^2\,np^6$, respectively). In each case, the added electron must enter a higher-energy orbital, requiring an input of energy. All of the Group 15 elements have an $ns^2\,np^3$ valence-electron configuration, in which each of the three *p* orbitals contains a single electron, according to Hund's rule; hence, the added electron must enter an already occupied *p* orbital. The resulting electron–electron repulsions destabilize the anion, causing the electron affinities of these elements to be less negative than we would otherwise expect. In the case of nitrogen, the 2*p* orbital is quite small, and the electron–electron repulsions are so strong that nitrogen has approximately zero affinity for an extra electron. In the heavier elements, however, the effect is relatively small because they have larger valence *p* orbitals.

In general, electron affinities of the main-group elements become less negative as we proceed down a column. This is because as *n* increases, the extra electrons enter orbitals that are increasingly far from the nucleus. Atoms with the largest radii, which have the lowest ionization energies (affinity for their own valence electrons), also have the lowest affinity for an added electron. There are, however, two major exceptions to this trend:

1. The electron affinities of elements B through F in the second row of the periodic table are *less* negative than those of the elements immediately below them in the third row. Apparently, the increased electron–electron repulsions experienced by electrons confined to the relatively small 2*p* orbitals overcome the increased electron–nucleus attraction at short nuclear distances. Fluorine, therefore, has a lower affinity for an added electron than does chlorine. Consequently, the elements of the *third* row ($n = 3$) have the most negative electron affinities. Farther down a column, the attraction for an added electron decreases because the electron is entering an orbital more distant from the nucleus. Electron–electron repulsions also decrease because the valence electrons occupy a greater volume of space. These effects tend to cancel one another, so the changes in electron affinity within a family are much smaller than the changes in ionization energy.
2. The electron affinities of the alkaline earths become more negative from Be to Ba. As you learned in Chapter 6, the energy separation between the filled ns^2 and the empty *np* subshells decreases with increasing *n*, so that formation of an anion from the heavier elements becomes energetically more favorable.

The equations for second and higher electron affinities are analogous to those for second and higher ionization energies:

$$E(g) + e^- \rightarrow E^-(g) \qquad \text{energy change} = EA_1 \tag{7.10}$$

$$E^-(g) + e^- \rightarrow E^{2-}(g) \qquad \text{energy change} = EA_2 \tag{7.11}$$

As we have seen, the first electron affinity can be ≥0 or negative, depending on the electron configuration of the atom. In contrast, the second electron affinity is *always* positive because the increased electron–electron repulsions in a dianion are far greater than the attraction of the nucleus for the extra electrons. For example, the first electron affinity of oxygen is −141 kJ/mol, but the second electron affinity is +744 kJ/mol:

$$O(g) + e^- \rightarrow O^-(g) \qquad EA_1 = -141\ \text{kJ/mol} \tag{7.12}$$

$$O^-(g) + e^- \rightarrow O^{2-}(g) \qquad EA_2 = +744\ \text{kJ/mol} \tag{7.13}$$

Thus, the formation of a gaseous oxide (O^{2-}) ion is energetically quite unfavorable:

$$O(g) + 2e^- \rightarrow O^{2-}(g) \qquad EA_1 + EA_2 = +603\ \text{kJ/mol} \tag{7.14}$$

Similarly, the formation of all common dianions (such as S^{2-}) or trianions (such as P^{3-}) is energetically unfavorable in the gas phase.

If energy is required to form both positively charged ions and monatomic polyanions, why do ionic compounds such as MgO, Na_2S, and Na_3P form at all? The key factor in the formation of stable ionic compounds is the favorable electrostatic interactions between the cations and the anions *in the crystalline salt*. We will describe the energetics of ionic compounds in more detail in Chapter 8.

EXAMPLE 7.6

Based on their positions in the periodic table, which of these elements would you predict to have the most negative electron affinity: Sb, Se, Te?

Given Three elements

Asked for Element with most negative electron affinity

Strategy

Ⓐ Locate the elements in the periodic table. Use the trends in electron affinities going down a column for elements in the same group. Similarly, use the trends in electron affinities from left to right for elements in the same row.

Ⓑ Place the elements in order, listing the element with the most negative electron affinity first.

Solution

Ⓐ✓ We know that electron affinities become less negative going down a column (except for the anomalously low electron affinities of the elements of the second row), so we can predict that the electron affinity of Se is more negative than that of Te. We also know that electron affinities become more negative from left to right across a row, and that the Group 15 elements tend to have values that are less negative than expected. Because Sb is located to the left of Te and belongs to Group 15, we predict that the electron affinity of Te is more negative than that of Sb. Ⓑ✓ The overall order is Se < Te < Sb, so Se has the most negative electron affinity among the three elements.

EXERCISE 7.6

Based on their positions in the periodic table, predict which of these elements is most likely to form a gaseous anion: Rb, Sr, or Xe?

Answer Rb

Electronegativity

The elements with the highest ionization energies are generally those with the most negative electron affinities, which are located toward the upper right corner of the periodic table (compare Figures 7.11 and 7.13). Conversely, the elements with the lowest ionization energies are generally those with the least negative electron affinities and are located in the lower left corner of the periodic table.

Because the tendency of an element to gain or lose electrons is so important in determining its chemistry, various methods have been developed to describe this tendency quantitatively. The most important method uses a measurement called **electronegativity** (represented by the Greek letter *chi*, χ, pronounced "ky" as in "sky"), defined as the *relative* ability of an atom to attract electrons to itself *in a chemical compound.* Elements with high electronegativities tend to acquire electrons in chemical reactions, and are found in the upper right corner of the periodic table. Elements with low electronegativities tend to lose electrons in chemical reactions, and are found in the lower left corner of the periodic table.

Note the pattern

Electronegativity increases diagonally from the lower left of the periodic table to the upper right.

Unlike ionization energy or electron affinity, the electronegativity of an atom is not a simple, fixed property that can be directly measured in a single experiment. In fact, an atom's electronegativity should depend to some extent on its chemical environment because the properties of an atom are influenced by its neighbors in a chemical compound. Nevertheless, when different methods for measuring the electronegativity of an

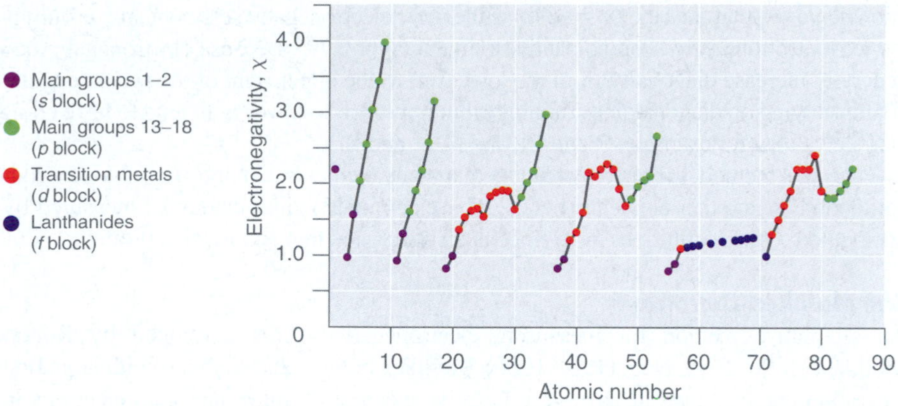

Figure 7.14 **Plot of periodic variation of electronegativity with atomic number for the first six rows of the periodic table.**

atom are compared, they all tend to assign similar relative values to a given element. For example, all scales predict that fluorine has the highest electronegativity and cesium the lowest of the stable elements, which suggests that all the methods are measuring the same fundamental property.

The Pauling Electronegativity Scale

The original electronegativity scale, developed in the 1930s by Linus Pauling (1901–1994) was based on measurements of the strengths of covalent bonds between different elements. Pauling arbitrarily set the electronegativity of fluorine at 4.0 (although today it has been refined to 3.98), thereby creating a scale in which all elements have values between 0 and 4.0.

Periodic variations in Pauling's electronegativity values are illustrated in Figures 7.14 and 7.15. If we ignore the inert gases and elements for which no stable isotopes are

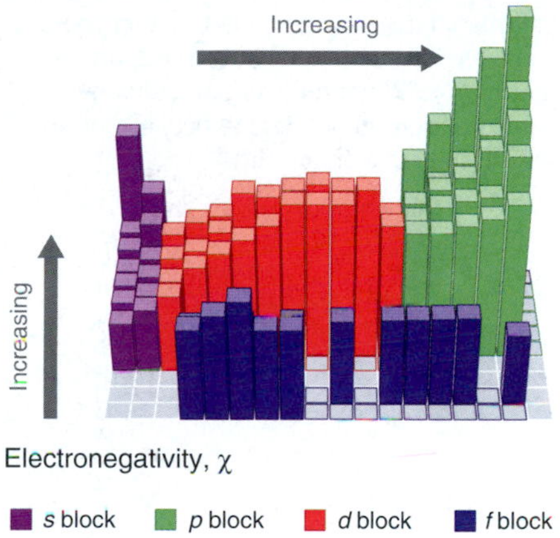

Figure 7.15 **Pauling electronegativity values of the *s*-, *p*-, *d*-, and *f*-block elements.** Values for most actinides are approximate. Elements for which no data are available are shown in gray. [Source of data: L. Pauling, *The Nature of the Chemical Bond,* 3d Edition (1960).] Interactive Graph

Scale: 0.7 – 3.98

Period	1	2	3	4	5	6	7	8	9	10	11	12	13	14	15	16	17	18
1	H 2.20																	He
2	Li 0.98	Be 1.57											B 2.04	C 2.55	N 3.04	O 3.44	F 3.98	Ne
3	Na 0.93	Mg 1.31											Al 1.61	Si 1.90	P 2.19	S 2.58	Cl 3.16	Ar
4	K 0.82	Ca 1.00	Sc 1.36	Ti 1.54	V 1.63	Cr 1.66	Mn 1.55	Fe 1.83	Co 1.88	Ni 1.91	Cu 1.90	Zn 1.65	Ga 1.81	Ge 2.01	As 2.18	Se 2.55	Br 2.96	Kr
5	Rb 0.82	Sr 0.95	Y 1.22	Zr 1.33	Nb 1.6	Mo 2.16	Tc 2.10	Ru 2.2	Rh 2.28	Pd 2.20	Ag 1.93	Cd 1.69	In 1.78	Sn 1.96	Sb 2.05	Te 2.1	I 2.66	Xe
6	Cs 0.79	Ba 0.89	La 1.10	Hf 1.3	Ta 1.5	W 1.7	Re 1.9	Os 2.2	Ir 2.2	Pt 2.2	Au 2.4	Hg 1.9	Tl 1.8	Pb 1.8	Bi 1.9	Po 2.0	At 2.2	Rn
7	Fr 0.7	Ra 0.9	Ac 1.1	Rf	Db	Sg	Bh	Hs	Mt	Ds	Rg	Uub	Uut	Uuq	Uup			

Lanthanides 6	Ce 1.12	Pr 1.13	Nd 1.14	Pm	Sm 1.17	Eu	Gd 1.20	Tb	Dy 1.22	Ho 1.23	Er 1.24	Tm 1.25	Yb	Lu 1.0
Actinides 7	Th 1.3	Pa 1.5	U 1.7	Np 1.3	Pu 1.3	Am	Cm	Bk	Cf	Es	Fm	Md	No	Lr

Linus Pauling (1901–1994). Pauling won two Nobel Prizes, one for chemistry in 1954 and one for peace in 1962. When he was nine Pauling's father died, and his mother tried to convince him to quit school to support the family. He did not quit school, but was denied a high school degree because of his refusal to take a civics class.

known, we see that fluorine ($\chi = 3.98$) is the most electronegative element and cesium is the least electronegative nonradioactive element ($\chi = 0.79$). Because electronegativities generally increase diagonally from the lower left to the upper right of the periodic table, elements lying on diagonal lines running from upper left to lower right tend to have comparable values (for example, O and Cl, and N, S, and Br).

Pauling's method is limited by the fact that many elements do not form stable covalent compounds with other elements; hence, their electronegativities cannot be measured by his method. Other definitions have since been developed that address this problem.

The Mulliken Definition

An alternative method for measuring electronegativity was developed by Robert Mulliken (1896–1986; Nobel Prize 1966). Mulliken noticed that elements with large first ionization energies tend to have very negative electron affinities and gain electrons in chemical reactions. Conversely, elements with small first ionization energies tend to have slightly negative (or even positive) electron affinities and lose electrons in chemical reactions. Mulliken recognized that an atom's tendency to gain or lose electrons could therefore be described quantitatively by the average of the values of its first ionization energy and the absolute value of its electron affinity. Using our definition of electron affinity, we can write Mulliken's original expression for electronegativity in the following way:*

$$\chi = \frac{I + |EA|}{2} \tag{7.15}$$

Elements with a large first ionization energy and a very negative electron affinity have a large positive value in the numerator of Equation 7.15, so their electronegativity is high. Elements with a small first ionization energy and a small electron affinity have a small positive value for the numerator in Equation 7.15, so they have a low electronegativity.†

As noted previously, all electronegativity scales give essentially the same results for one element relative to another. Thus, even though the Mulliken scale is based on the properties of individual *atoms* and the Pauling scale is based on the properties of atoms in *molecules*, they both apparently measure the same basic property of an element. In the following discussion, we will focus on the relationship between electronegativity and the tendency of atoms to form positive or negative ions. We will therefore be implicitly using the Mulliken definition of electronegativity. Because of the parallels between the Mulliken and Pauling definitions, however, the conclusions are likely to apply to atoms in molecules as well.

Electronegativity Differences Between Metals and Nonmetals

An element's electronegativity provides us with a single value that we can use to characterize the chemistry of an element. Elements with a high electronegativity ($\chi \geq 2.2$) have very negative affinities and large ionization potentials, hence they are generally nonmetals and electrical insulators that tend to gain electrons in chemical reactions (they are *oxidants*). In contrast, elements with a low electronegativity ($\chi \leq 1.8$) have electron affinities that have either positive or small negative values and small ionization potentials; hence, they are generally metals and good electrical conductors that tend to lose their valence electrons in chemical reactions (they are *reductants*). In between the metals

* Mulliken's definition used the *magnitude* of the ionization energy and the electron affinity. By definition, the magnitude of a quantity is a positive number. Our definition of electron affinity produces negative values for the electron affinity for most elements, so vertical lines indicating absolute value are needed in Equation 7.15 to make sure that we are adding two positive numbers in the numerator.

† Inserting the appropriate data from Figure 7.11 and Figure 7.13 into Equation 7.15 gives a Mulliken electronegativity value for fluorine of 1004.6 (in units of kJ/mol). In order to compare Mulliken's electronegativity values with those obtained by Pauling, Mulliken's values are divided by 252.4 kJ/mol, which gives Pauling's value (3.98).

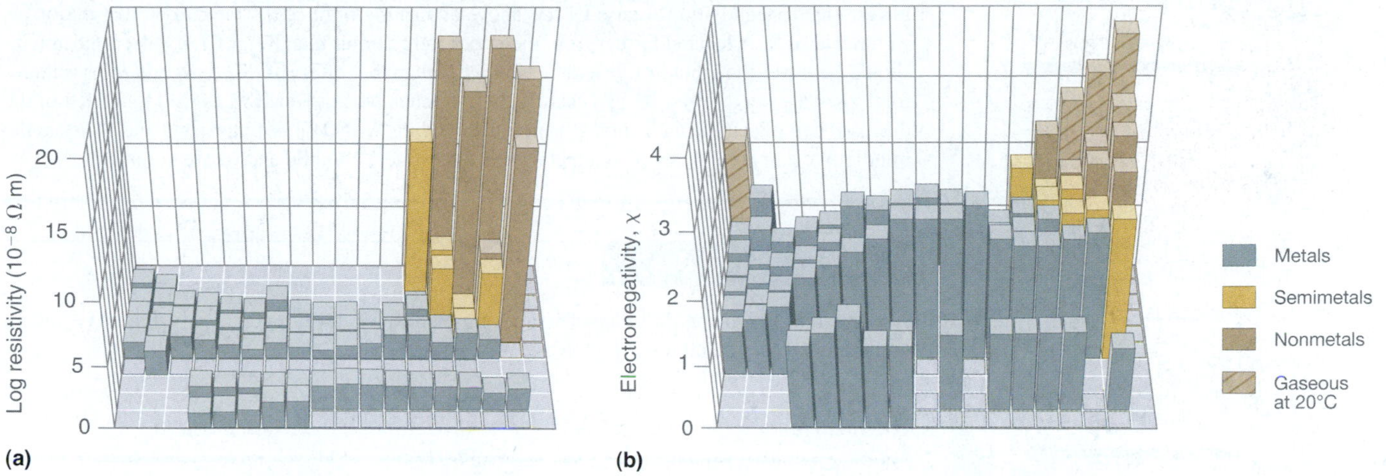

Figure 7.16 Three-dimensional plots demonstrating the relationship between electronegativity and the metallic/nonmetallic character of the elements. (a) Plot of electrical resistivity (measured resistivity to electron flow) at or near room temperature. Substances with high resistivity (little to no measured electron flow) are electrical insulators, whereas substances with low resistivity (high measured electron flow) are metals. (b) Plot of Pauling electronegativities for a like set of elements. High electronegativity values (≥ about 2.2) correlate with high electrical resistivities (insulators). Low electronegativity values (≤ about 2.2) correlate with low resistivities (metals). Note: Because electrical resistivity is typically measured only for solids and liquids, the gaseous elements do not appear in part (a). Interactive Graph

and nonmetals, along the heavy diagonal line running from B to At in Figure 1.24, is a group of elements with intermediate electronegativities ($\chi \sim 2.0$). These are the semimetals, elements that have some of the chemical properties of both nonmetals and metals. The distinction between metals and nonmetals is one of the most fundamental we can make in categorizing the elements and predicting their chemical behavior. Figure 7.16 shows the strong correlation between electronegativity values, metallic versus nonmetallic character, and location in the periodic table.

The rules for assigning oxidation states that were introduced in Chapter 3 are based on the relative electronegativities of the elements; the more electronegative element in a binary compound is assigned a negative oxidation state. As we shall see, electronegativity values are also used to predict bond energies, bond polarities, and the kinds of reactions that compounds undergo.

Note the pattern

Electronegativity values increase from lower left to upper right in the periodic table.

EXAMPLE 7.7

On the basis of their positions in the periodic table, arrange the following elements in order of increasing electronegativity and identify each as a metal, nonmetal, or semimetal: Cl, Se, Si, Sr.

Given Four elements

Asked for Order by increasing electronegativity, and classification

Strategy

Ⓐ Locate the elements in the periodic table. From their diagonal positions from lower left to upper right, predict their relative electronegativities.

Ⓑ Arrange the elements in order of increasing electronegativity.

Ⓒ Classify each element as a metal, nonmetal, or semimetal according to its location about the diagonal belt of semimetals running from B to At.

Solution

Ⓐ Electronegativity increases from lower left to upper right in the periodic table (Figure 7.15). Because Sr lies far to the left of the other elements given, we can predict that it will have the

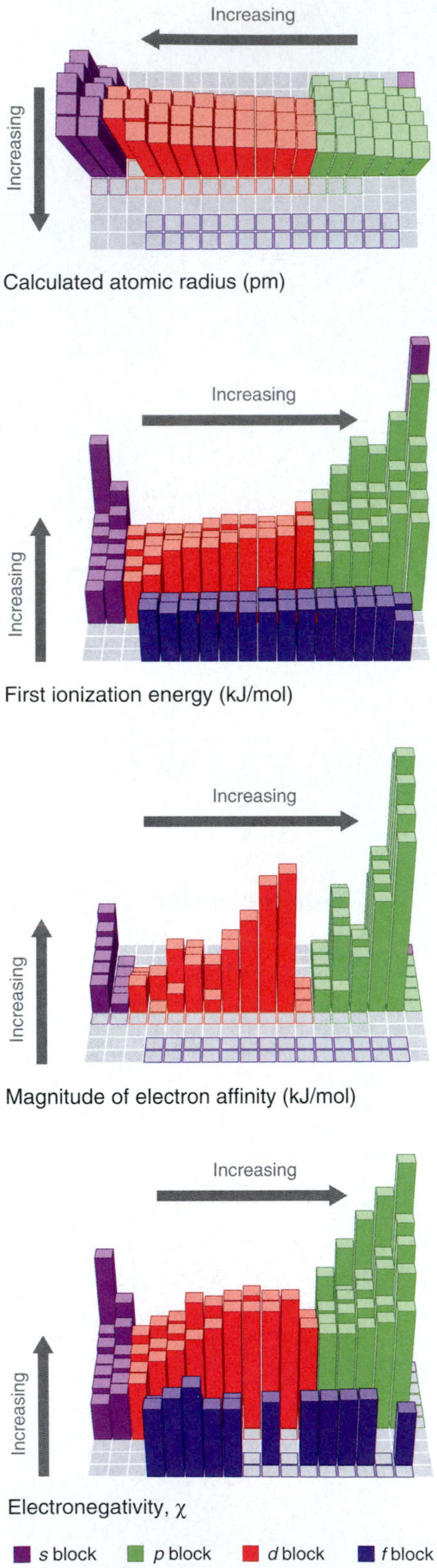

Figure 7.17 Summary of major periodic trends. Notice that the general trends for first ionization energy, electron affinity, and electronegativity are opposite to the general trend for covalent atomic radius. Interactive Graph

lowest electronegativity. Because Cl lies above and to the right of Se, we can predict that $\chi_{Cl} > \chi_{Se}$. Because Si is located farther from the upper right corner than Se or Cl, its electronegativity should be lower than those of Se and Cl but higher than that of Sr. B The overall order is therefore: $\chi_{Sr} < \chi_{Si} < \chi_{Se} < \chi_{Cl}$. C To classify the elements, we note that Sr lies well to the left of the diagonal belt of semimetals running from B to Te; while Se and Cl lie to the right and Si lies in the middle. We can predict that Sr is a metal, Si is a semimetal, and Se and Cl are nonmetals.

EXERCISE 7.7

Arrange these elements in order of increasing electronegativity and classify each as a metal, nonmetal, or semimetal: Ge, N, O, Rb, Zr.

Answer Rb < Zr < Ge < N < O. Metals (Rb, Zr); semimetal (Ge); nonmetal (N, O).

The trends in periodic properties are summarized in Figure 7.17. As discussed, atomic radii decrease from lower left to upper right in the periodic table; while ionization energies become more positive, electron affinities become more negative, and electronegativities increase from the lower left to the upper right.

7.4 The Chemical Families

Periodic trends in properties such as atomic and ionic size, ionization energy, electron affinity, and electronegativity illustrate the strong connection between the chemical properties and reactivity of the elements and their positions in the periodic table. In this section, we explore that connection by focusing on two periodic properties that correlate strongly with the chemical behavior of the elements: valence electron configurations and Mulliken electronegativities.

The Main Group Elements

We have said that elements with the same valence electron configuration (that is, elements located in the same column of the periodic table) often have similar chemistry. This correlation is particularly evident for the elements of Groups 1, 2, 3, 13, 16, 17, and 18. The intervening families in the *p* block (Groups 14 and 15) straddle the diagonal line separating metals from nonmetals. The lightest members of these two families are nonmetals, so they react differently compared to the heaviest members, which are metals. We begin our survey with the noble gases (Group 18), which have full valence electron shells, and end with the alkali metals (Group 1), which contain only a single electron outside a noble gas electron configuration.

Group 18, the Noble Gases

The noble gases are helium, neon, argon, krypton, xenon, and radon. All have filled valence-electron configurations and therefore are unreactive elements found in nature as monatomic gases. The noble gases were long referred to as either "rare gases" or "inert gases," but they are neither rare nor inert. Argon constitutes about 1% of the atmosphere, which also contains small amounts of the lighter Group 18 elements, and helium is found in large amounts in many natural gas deposits. The group's perceived "rarity" stems in part from the fact that the noble gases were the last major family of elements to be discovered.

The noble gases have $EA \geq 0$, so they do not form compounds in which they have negative oxidation states. Because ionization energies decrease down the column, the only noble gases that form compounds in which they have positive oxidation states are Kr, Xe, and Rn. Of these three elements, only xenon forms an extensive series of compounds. The chemistry of radon is severely limited by its extreme radioactivity, and the chemistry of krypton is limited by its high ionization energy (1350.8 kJ/mol versus 1170.4 kJ/mol for xenon). In essentially all its compounds, Xe is bonded to highly electronegative atoms such as fluorine or oxygen. In fact, the only significant reaction of xenon is with elemental fluorine, which can give XeF_2, XeF_4, or XeF_6 (shown in the margin). Oxides such as XeO_3 are produced when xenon fluorides react with water, and oxidation with ozone produces the perxenate ion, XeO_6^{4-}, in which Xe acquires a +8 oxidation state by formally donating all eight of its valence electrons to the more electronegative oxygen atoms. In all of its stable compounds, xenon has a positive, even-numbered oxidation state: +2, +4, +6, or +8. The actual stability of these compounds varies greatly. For example, XeO_3 is a shock-sensitive, white crystalline solid with an explosive power comparable to that of TNT, whereas another compound, Na_2XeF_8, is stable up to 300°C.

MGC Chemical Families

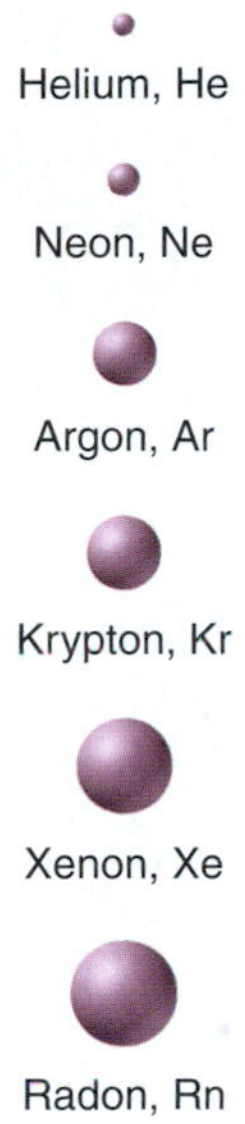

Group 18, noble gases

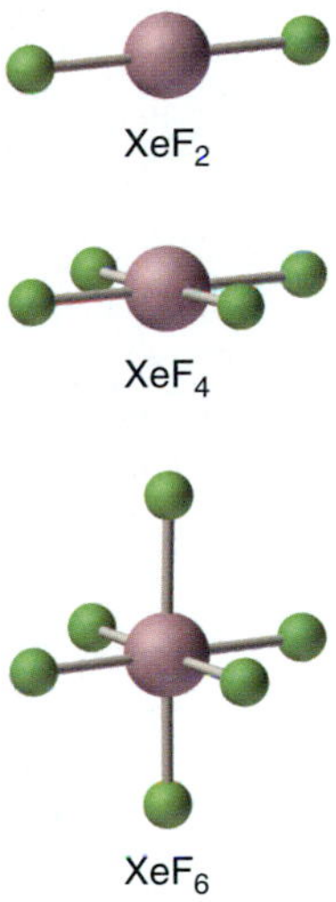

Although none of the noble gas compounds is commercially significant, the elements themselves have important applications. For example, argon is used in incandescent light bulbs, where it provides an inert atmosphere that protects the tungsten filament from oxidation. It is also used in arc welding and in the manufacture of reactive elements, such as titanium, or of ultrapure products, such as the silicon used by the electronics industry. Helium, with a boiling point of only 4.2 K, is used as a liquid for studying the properties of substances at very low temperatures. It is also combined in an 80:20 mixture with oxygen that is used by scuba divers instead of compressed air when they descend to great depths. Because He is less soluble in water than N_2—a component of compressed air—replacing N_2 with He prevents the formation of bubbles in blood vessels, a condition called "the bends" that can occur during rapid ascents. Neon is familiar to all of us as the gas responsible for the red glow in neon lights.

Group 17, the Halogens

The term *halogen*, derived from the Greek *háls*, "salt," and *genes*, "producing," was first applied to chlorine because of its tendency to react with metals to form salts. All of the halogens have an ns^2np^5 valence-electron configuration, and all but astatine are diatomic molecules in which the two halogen atoms share a pair of electrons. Diatomic F_2 and Cl_2 are pale yellow-green and pale green gases, respectively, while Br_2 is a red liquid and I_2 is a purple solid. The halogens were not isolated until the 18th and 19th centuries.

Because of their relatively high electronegativities, the halogens are nonmetallic and generally react by gaining one electron per atom to attain a noble gas electron configuration and an oxidation state of −1. *Halides* are produced according to the following equation, in which X denotes a halogen:

$$E + \frac{n}{2}X_2 \rightarrow EX_n \tag{7.16}$$

If the element E has a low electronegativity (as does Na), the product is typically an ionic halide (NaCl). If the element E is highly electronegative (as P is), the product is typically a covalent halide (PCl_5). Ionic halides tend to be nonvolatile substances with high melting points, whereas covalent halides tend to be volatile substances with low melting points. Elemental fluorine (F_2) is the most reactive of the halogens and iodine (I_2) the least, which is consistent with their relative electronegativities.* In fact, F_2 reacts with nearly all of the elements at room temperature. Under more extreme conditions, it combines with all elements except He, Ne, and Ar.

Group 17, halogens

The halogens react with hydrogen to form the hydrogen halides (HX):

$$H_2(g) + X_2(g, l, s) \rightarrow 2HX(g) \tag{7.17}$$

* As we shall see in subsequent chapters, however, factors such as bond strengths are also important in dictating the reactivities of these elements.

Fluorine is so reactive that any substance containing hydrogen, including coal, wood, and even water, will burst into flames if it comes into contact with pure F_2.

Because it is the most electronegative element known, fluorine never has a positive oxidation state in any compound. In contrast, the other halogens (Cl, Br, I) form compounds in which their oxidation states are +1, +3, +5, and +7, as in the *oxoanions*, XO_n^-, where n = 1–4. Because oxygen has the second highest electronegativity of any element, it stabilizes the positive oxidation states of the halogens in these ions.

All of the halogens except astatine (which is radioactive) are commercially important. The NaCl in salt water is purified for use as table salt. Chlorine and hypochlorite (OCl^-) salts are used to sanitize public water supplies, swimming pools, and waste water, and hypochlorite salts are also used as bleaches because they oxidize colored organic molecules. Organochlorine compounds are used as drugs and pesticides. Fluoride (usually in the form of NaF) is added to many municipal water supplies to help prevent tooth decay, and bromine (in AgBr) is a component of the light-sensitive coating on photographic film. Because iodine is essential to life—it is a key component of the hormone produced by the thyroid gland—small amounts of KI are added to table salt to produce "iodized salt," which prevents thyroid hormone deficiencies.

Group 16, the Chalcogens

Group 16, chalcogens

The Group 16 elements are often referred to as the **chalcogens**—from the Greek *chalkós*, "copper," and *genes*, "producing"—because the most ancient copper ore, copper sulfide, is also rich in two other Group 16 elements, selenium and tellurium. Recall that the trend in most groups is for the lightest member to have properties that are quite different from those of the heavier members. Consistent with this trend, oxygen has unique properties. For example, in its most common pure form, it is a diatomic gas, O_2, whereas sulfur is a volatile solid that contains S_8 rings, selenium and tellurium are gray or silver solids that have chains of atoms, and polonium is a silvery metal with a regular array of atoms. Like astatine and radon, the metal polonium is a highly radioactive element.

All of the chalcogens have ns^2np^4 valence-electron configurations. Hence, their chemistry is dominated by three oxidation states: −2, in which two electrons are added to achieve the closed-shell electron configuration of the *next* noble gas; +6, in which all six valence electrons are lost to give the closed-shell electron configuration of the *preceding* noble gas; and +4, in which only the four np electrons are lost to give a filled ns^2 subshell. Oxygen has the second highest electronegativity of any element, and so its chemistry is dominated by the −2 oxidation state (as in MgO and H_2O). No compounds of oxygen in the +4 or +6 oxidation state are known. In contrast, sulfur can form compounds in all three oxidation states. Sulfur accepts electrons from less electronegative elements to give H_2S and Na_2S, for example, and it donates electrons to more electronegative elements to give compounds such as SO_2, SO_3, and SF_6. Selenium and tellurium, near the diagonal line between nonmetals and semimetals, behave similarly to sulfur but are somewhat more likely to be found in positive oxidation states.

Oxygen, the second most electronegative element in the periodic table, was not discovered until the late 18th century, even though it constitutes 20% of the atmosphere and is the most abundant element in the earth's crust. Oxygen is essential for life because our metabolism is based on the oxidation of organic compounds by O_2 to produce CO_2 and H_2O. Commercially, oxygen is used in the conversion of pig iron to steel, as the oxidant in oxyacetylene torches for cutting steel, as a fuel for the U.S. space shuttle, and in hospital respirators.

Sulfur is the "brimstone" in "fire and brimstone" from ancient times. Partly as a result of its long history, it is employed in a wide variety of commercial products and processes. In fact, as you learned in Chapter 2, more sulfuric acid is produced worldwide than any other compound. Sulfur is used to cross-link the polymers in rubber in a process called *vulcanization*, which was discovered by Charles Goodyear in the 1830s and commercialized by Benjamin Goodrich in the 1870s. Vulcanization gives rubber its unique combination of strength, elasticity, and stability.

Selenium, the only other commercially important chalcogen, was discovered in 1817, and today it is widely used in light-sensitive applications. For example, photocopying, or xerography, from the Greek *xèrós*, "dry," and *graphia*, "writing," uses selenium films to transfer an image from one piece of paper to another, while compounds such as cadmium selenide are used to measure light in photographic light meters and automatic street lights.

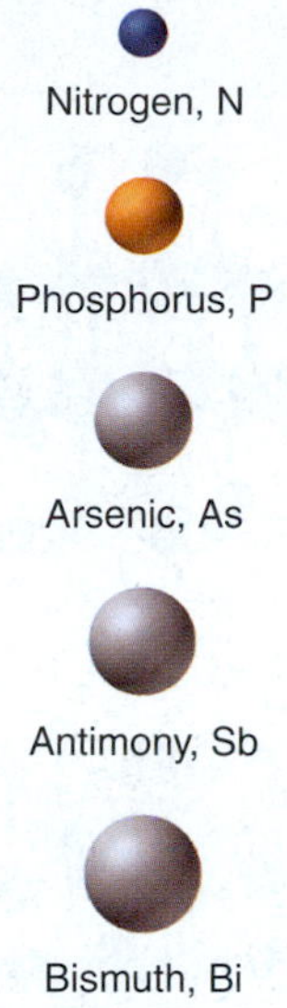

Group 15, pnicogens

Group 15, the Pnicogens

This is the first time we encounter a set of *five* stable elements; one isotope of bismuth (^{209}Bi) is nonradioactive and is the heaviest nonradioactive isotope of any element. The Group 15 elements are called the **pnicogens**—from the Greek *pnigein*, "to choke," and *genes*, "producing"—ostensibly because of the noxious fumes that many nitrogen and phosphorus compounds produce. Once again, the lightest member of the family has unique properties. Thus, although both nitrogen and phosphorus are nonmetals, nitrogen under standard conditions is a diatomic gas, N_2, whereas phosphorus consists of three allotropes: white, a volatile, low-melting solid consisting of P_4 tetrahedra; red, a solid comprised of P_8, P_9, and P_{10} cages linked by P_2 units; and black, which consists of layers of corrugated phosphorus sheets. The next two elements, arsenic and antimony, are semimetals with extended three-dimensional network structures, and bismuth is a silvery metal with a pink tint.

All of the pnicogens have ns^2np^3 valence-electron configurations, leading to three common oxidation states: −3, in which three electrons are added to give the closed-shell electron configuration of the next noble gas; +5, in which all five valence electrons are lost to give the closed-shell electron configuration of the preceding noble gas; and +3, in which only the three *np* electrons are lost to give a filled ns^2 subshell. Because the electronegativity of nitrogen is similar to that of chlorine, nitrogen accepts electrons from most elements to form compounds in the −3 oxidation state (such as in NH_3). Nitrogen has only positive oxidation states when combined with highly electronegative elements, such as oxygen and the halogens (for example, HNO_3, NF_3). Although phosphorus and arsenic can combine with active metals and hydrogen to produce compounds in which they have a −3 oxidation state (PH_3, for example), they typically attain oxidation states of +3 and +5 when combined with more electronegative elements, such as PCl_3 and H_3PO_4. Antimony and bismuth are relatively unreactive metals, but form compounds with oxygen and the halogens in which their oxidation states are +3 and +5 (as in Bi_2O_3 and SbF_5).

Although it is present in most biological molecules, nitrogen was the last pnicogen to be discovered. Nitrogen compounds such as ammonia, nitric acid, and their salts are used agriculturally in huge quantities; nitrates and nitrites are used as preservatives in meat products such as ham and bacon, and nitrogen is a component of nearly all explosives.

Phosphorus, too, is essential for life, and phosphate salts are used in fertilizers, toothpaste, and baking powder. A phosphorus sulfide, P_4S_3, is used to ignite modern safety matches. Arsenic, in contrast, is toxic; its compounds are used as pesticides and poisons. Antimony and bismuth are primarily used in metal alloys, but a bismuth compound is the active ingredient in the popular antacid medication Pepto-Bismol.

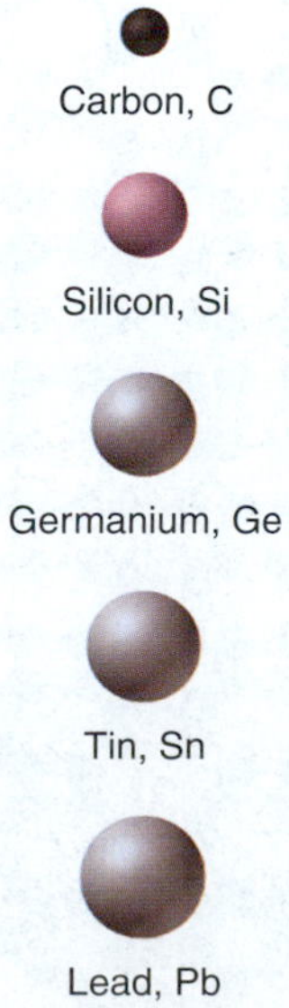

Group 14

Group 14

The Group 14 elements straddle the diagonal line that divides nonmetals from metals. Thus, carbon is a nonmetal, silicon and germanium are semimetals, and tin and lead are metals. As with Groups 15 and 16, the structures of the pure elements vary greatly.

The $ns^2\,np^2$ valence-electron configurations of Group 14 gives rise to three oxidation states: −4, in which four electrons are added to achieve the closed-shell electron configuration of the next noble gas; +4, in which all four valence electrons are lost to give the closed-shell electron configuration of the preceding noble gas; and +2, in which the loss of two np^2 electrons gives a filled ns^2 subshell.

The electronegativities of the Group 14 elements are lower than those of Groups 15–17. That of carbon is only 2.5, which is in the middle of the electronegativity range, so carbon forms covalent compounds with a wide variety of elements, and is the basis of all organic compounds. All of the Group 14 elements form compounds in the +4 oxidation state, so that all of them are able to form dioxides (from CO_2 to PbO_2) and tetrachlorides (CCl_4 and $PbCl_4$). Only the two metallic elements, Sn and Pb, form an extensive series of compounds in the +2 oxidation state. Stannous fluoride, SnF_2, was once used as an additive in toothpaste, and $PbSO_4$ is formed when your car battery discharges.

Carbon has at least four allotropes that are stable at room temperature: graphite; diamond; a group of related cage structures called **fullerenes** (such as C_{60}); and **nanotubes**, which are cylinders of carbon atoms (Figure 7.18). Graphite consists of extended planes of covalently bonded hexagonal rings; because the planes are not linked by covalent bonds, they can slide across one another easily. This makes graphite ideally suited as a lubricant and as the "lead" in lead pencils. Graphite also provides the black color in inks and tires, and graphite fibers are used in high-tech items such as golf clubs, tennis rackets, airplanes, and sailboats because of their light weight, strength, and stiffness.

In contrast to the layered structure of graphite, each carbon atom in diamond is bonded to four others to form a rigid three-dimensional array, making diamond the hardest substance known; consequently, it is used in industry as a cutting tool. Fullerenes, on the other hand, are spherical or ellipsoidal molecules that contain six- and five-membered rings of carbon atoms; they are volatile substances that dissolve in organic solvents. Fullerenes of extraterrestrial origin have been found in meteorites. Carbon nanotubes, intermediate in structure between graphite and the fullerenes, can be described as sheets of graphite that have been rolled up into a cylinder or, alternatively, fullerene cages that have been stretched in one direction. Currently, carbon nanotubes are being studied for use in the construction of molecular electronic devices and computers.

Silicon and germanium have strong, three-dimensional network structures similar to that of diamond. Silicon is the second most abundant element in the earth's crust. For example, sand is primarily SiO_2, which is used commercially to make glass and to prevent caking in food products. Complex compounds of silicon and oxygen with elements such as aluminum are used in detergents and talcum powder and as industrial catalysts. Because silicon-chip technology laid the foundation for the modern electronics industry, the San Jose region of California, where many of the most important advances in electronics and computers were developed, has been nicknamed "Silicon Valley."

Elemental tin and lead are metallic solids. Tin is primarily used to make alloys such as bronze, which consists of tin and copper; solder, which is tin and lead; and pewter, which is tin, antimony, and copper. In ancient times, lead was used for everything from pipes to cooking pots because it is easily hammered into different shapes. In fact, the term *plumbing* is derived from the Latin name for lead, *plumbum*. Until recently, lead compounds were used as pigments in paints and tetraethyllead was an important antiknock

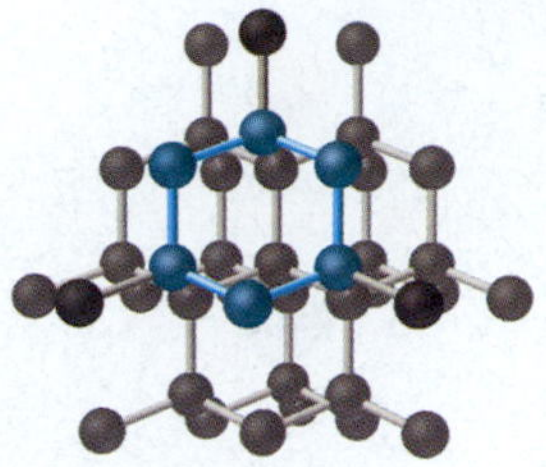

Diamond

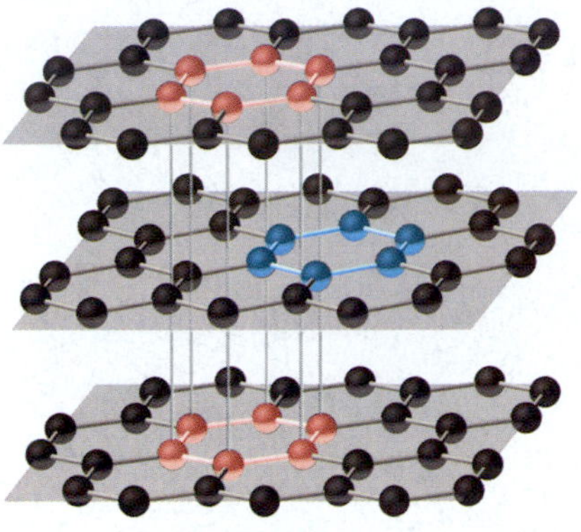

Graphite

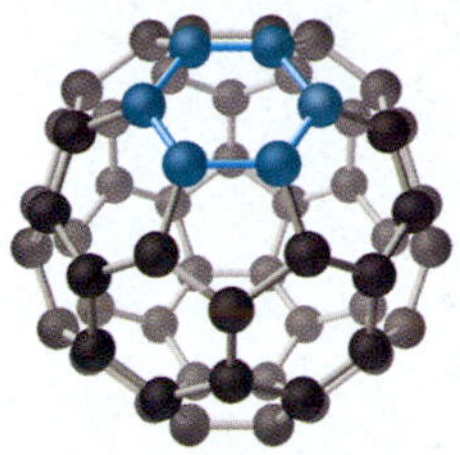

Buckminsterfullerene, C_{60}

Carbon nanotube

Figure 7.18 Four allotropes of carbon. Diamond consists of a rigid three-dimensional array of carbon atoms, making it the hardest substance known. In contrast, graphite forms from extended planes of covalently bonded hexagonal rings of carbon atoms that can slide across one another easily. Fullerenes are spherical or ellipsoidal molecules that contain 6- and 5-membered rings of carbon atoms, and nanotubes are sheets of graphite rolled up into a cylinder.

agent in gasoline. Now, however, lead has been banned from many uses because of its toxicity, although it is still widely used in lead storage batteries for automobiles.

Group 13

Of the Group 13 elements, only the lightest, boron, lies on the diagonal line that separates nonmetals and metals. Thus, boron is a semimetal, whereas the rest of the Group 13 elements are metals (as are the rest of the elements in the periodic table, except for hydrogen). Elemental boron has an unusual structure consisting of B_{12} icosahedra covalently bonded to one another (shown in the margin); the other elements are typical metallic solids.

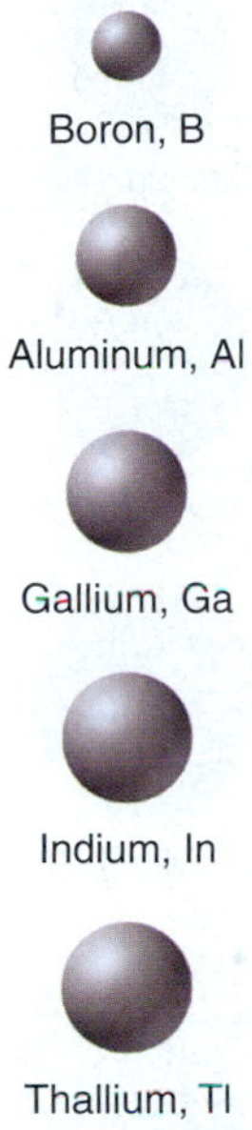

Group 13

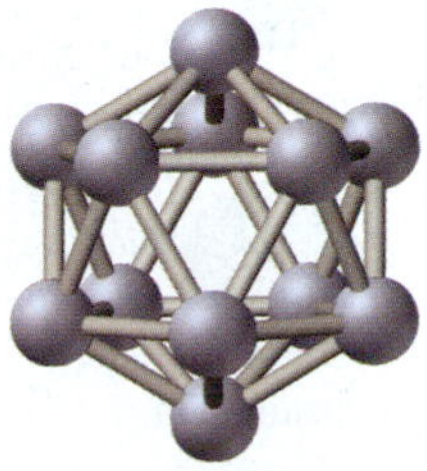

B_{12} icosahedron in elemental boron

No Group 13 elements were known in ancient times, not because they are scarce—Al is the third most abundant element in the earth's crust—but because they are highly reactive and form extremely stable compounds with oxygen. Potent reducing agents and careful handling were needed to isolate the pure elements.

The elements of Group 13 have ns^2np^1 valence-electron configurations. Consequently, two oxidation states are important: +3, from losing three valence electrons to give the closed-shell electron configuration of the preceding noble gas; and +1, from losing the single electron in the *np* subshell. Because these elements have small, negative electron affinities (boron's is only −27.0 kJ/mol), they are unlikely to acquire five electrons to reach the next noble-gas configuration. In fact, the chemistry of these elements is almost exclusively characterized by +3. Only the heaviest element, Tl, has extensive chemistry in the +1 oxidation state. It loses the single 6*p* electron to produce the TlX monohalides and the oxide, Tl_2O.

In the 19th century, aluminum was considered a precious metal. Thus, aluminum knives and forks were reserved for the French Emperor Louis Napoleon III, while his less important guests had to be content with gold or silver cutlery. Dedication of the Washington Monument in 1885 was celebrated by placing a 100-ounce chunk of pure aluminum at the top. In contrast, today aluminum is used on an enormous scale, with copper, manganese, magnesium, and silicon in aircraft, automobile engines, armor, cookware, and beverage containers. It is valued for its combination of low density, high strength, and corrosion resistance. In addition, aluminum compounds are the active ingredients in most antiperspirant deodorants.

Compounds of boron, such as one form of BN, are hard, have a high melting point, and are resistant to corrosion. They are particularly useful in materials that are exposed to extreme conditions, such as aircraft turbines, brake linings, and polishing compounds. Boron is also a major component of many kinds of glasses, and sodium perborate, $Na_2B_2O_4(OH)_4$, is the active ingredient in many so-called "color-safe" laundry bleaches.

Gallium, indium, and thallium are less widely used, but gallium arsenide is the red light-emitting diode (LED) in digital readouts in calculators, and $MgGa_2O_4$ produces the green light emitted in many xerographic machines. Compounds of thallium(I) are extremely toxic. Thus, although Tl_2SO_4 is an excellent rat or ant poison, it is so toxic to humans that it is no longer used for this purpose.

A solid pyramid of aluminum caps the Washington Monument. Installed in 1884, at the time of the monument's completion, the pyramid was part of the original lightning protection system. In the 1880s, aluminum was a rare metal, and used primarily for jewelry. Measuring 9-inches high and weighing 100 ounces, this was the largest single piece of aluminum of its day.

Group 2, the Alkaline Earth Metals

The elements of Group 2 are collectively referred to as the *alkaline earths*, a name that originated in the Middle Ages, when an "earth" was defined as a substance that did not melt and was not transformed by fire. *Alkali* (from the Arabic *al-qili*, meaning ashes of the saltwort plant from salt marshes) was a general term for substances derived from wood ashes, all of which possessed a bitter taste and were able to neutralize acids. Alkalis that did not melt easily were called "alkaline earths."

As in the other families, the properties of the lightest element—in this case, beryllium—tend to be different from those of its heavier *congeners,* the other members of the group. Beryllium is relatively unreactive but forms many covalent compounds, whereas the other group members are much more reactive metals and form ionic compounds. As is the case in Groups 16–18, the heaviest element,

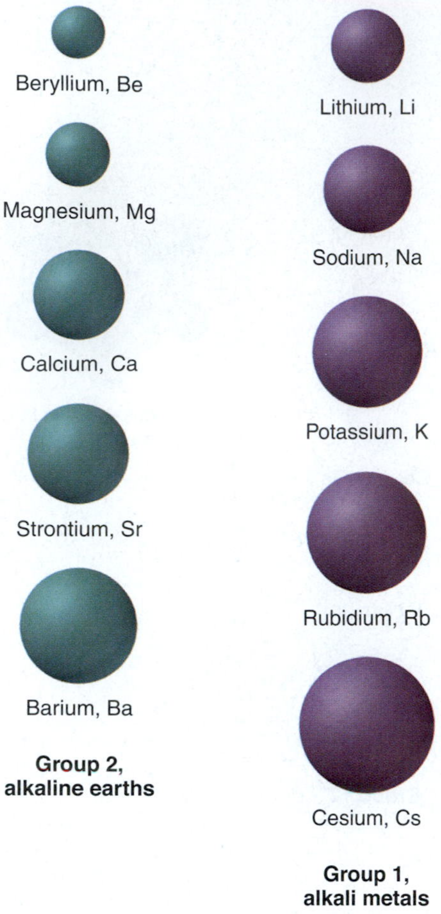

radium, is highly radioactive, making its size difficult to measure. Radium was discovered in 1902 by Marie Curie [1867–1934; Nobel Prize 1903 (physics); Nobel Prize 1911 (chemistry)], who, with her husband, Pierre, isolated 120 mg of radium chloride from tons of residues from uranium mining.

All the alkaline earth elements have ns^2 valence-electron configurations, and all have electronegativities less than 1.6. This means that they behave chemically as metals (although beryllium compounds are covalent) and lose the two valence electrons to form compounds in the +2 oxidation state. Examples include the dihalides, EX_2, and the oxides, EO.

Compounds of the Group 2 elements have been commercially important since Egyptian and Roman times, when blocks of limestone or marble, which are both $CaCO_3$, were used as building materials, and gypsum ($CaSO_4 \cdot 2H_2O$) or lime (CaO) was used as mortar. Calcium sulfate is still used in Portland cement and plaster of Paris. Magnesium and beryllium form lightweight, high-strength alloys that are used in the aerospace, automotive, and other high-tech industries. As you learned in Chapter 6, one of the most impressive uses of these elements is in fireworks; strontium and barium salts, for example, give red or green colors, respectively. Except for beryllium, which is highly toxic, the Group 2 elements are also important biologically. Bone is largely hydroxyapatite, $Ca_5(PO_4)_3OH$, mollusc shells are calcium carbonate, magnesium is part of the chlorophyll molecule in green plants, and calcium is important in hormonal and nerve signal transmission. Because $BaSO_4$ is so insoluble, it is used in "barium milk shakes" to obtain X rays of the gastrointestinal tract.

Group 1, the Alkali Metals

The elements of Group 1 are called the *alkali metals*. Their oxides, too, were obtained from wood ashes, paralleling the history of the alkaline earths, but the alkali metals had lower melting points. Potassium and sodium were first isolated in 1807 by the British chemist Sir Humphry Davy (1778–1829) by passing an electrical current through molten samples of potash (K_2CO_3) and soda ash (Na_2CO_3). The potassium burst into flames as soon as it was produced because it reacts readily with oxygen at the higher temperature. Also like the Group 2 elements, the Group 1 elements become less reactive with air or water as their atomic number decreases. Once again, the heaviest element (francium), is so radioactive that studying its chemistry is very difficult; it was not discovered until 1939.

The alkali metals have ns^1 valence-electron configurations and the lowest electronegativity of any group; hence they are often referred to as being *electropositive* elements. As a result, they have a strong tendency to lose their single valence electron to form compounds in the +1 oxidation state, producing the EX monohalides and the E_2O oxides.

Because they are so reactive, pure Group 1 elements are powerful reducing agents that are used in lithium batteries and cardiac pacemakers. Sodium salts such as common table salt (NaCl), baking soda ($NaHCO_3$), soda ash (Na_2CO_3), and caustic soda (NaOH) are important industrial chemicals. Other compounds of the alkali metals are important in biology. Because potassium is required for plant growth, its compounds are used in fertilizers, and lithium salts are used to treat manic-depressive, or bipolar, disorders.

Potassium burning. A piece of potassium dropped in a beaker of water will burn as it skips across the top of the water.

The Transition Metals, Lanthanides, and Actinides

As expected for elements with the same valence-electron configuration, the elements in each column of the *d* block have vertical similarities in chemical behavior. In contrast to the *s*- and *p*-block elements, however, elements in the *d* block also display strong *horizontal* similarities. The horizontal trends compete with the vertical trends. In further contrast to the *p*-block elements, which tend to have stable oxidation states that are separated by *two* electrons, the transition metals have multiple oxidation states that are separated by only *one* electron.

The Group 6 elements, chromium, molybdenum, and tungsten, illustrate the competition that occurs between these horizontal and vertical trends. For example, the maximum oxidation state for all elements in Group 6 is +6, achieved by losing all six valence electrons (recall that Cr has a $4s^1\,3d^5$ valence-electron configuration), yet nearly all the elements in the first row of the transition metals, including Cr, form compounds that contain the dication, M^{2+}, and many also form a trication, M^{3+}. As a result, the transition metals in Group 6 have very different tendencies to achieve their maximum oxidation state. The most common oxidation state for chromium is +3, whereas the most common oxidation state for molybdenum and tungsten is +6.

Note the pattern

The d-block elements display both strong vertical and horizontal similarities in chemistry.

Groups 3 (scandium, lanthanum, actinium), 11 (copper, silver, gold), and 12 (zinc, cadmium, mercury) are the only transition metal groups in which the oxidation state predicted by the valence-electron configuration dominates the chemistry of the group. The elements of Group 3 have three valence electrons outside an inner closed shell, so their chemistry is almost exclusively that of the M^{3+} ions produced by losing all three valence electrons. The elements of Group 11 have 11 valence electrons, and so all three lose a single electron to form a monocation, M^+, with a closed $(n-1)d^{10}$ electron configuration. Consequently, compounds of Cu^+, Ag^+, and Au^+ are very common, although there is also a great deal of chemistry involving Cu^{2+}. Similarly, the elements of Group 12 all have an $ns^2(n-1)d^{10}$ electron configuration, so they lose two electrons to form M^{2+} ions with an $(n-1)d^{10}$ electron configuration; indeed, the most important ions for these elements are Zn^{2+}, Cd^{2+}, and Hg^{2+}. Mercury, however, also forms the dimeric mercurous ion, Hg_2^{2+}, because of a subtle balance between the energies needed to remove additional electrons and the energy released when bonds are formed. The +3 oxidation state is the most important for the lanthanides and for most of the actinides.

EXAMPLE 7.8

Based on the information given below, determine the most likely identities for elements D and E.

(a) Element D is a shiny gray solid that conducts electricity only moderately; it forms two oxides, DO_2 and DO_3.

(b) Element E is a reddish metallic substance that is an excellent conductor of electricity; it forms two oxides, EO and E_2O, and two chlorides, ECl and ECl_2.

Given Physical and chemical properties of two elements

Asked for Identities

Strategy

A Based on the conductivity of the elements, determine whether each is a metal, nonmetal, or semimetal. Confirm your prediction from its physical appearance.

B From the compounds each element forms, determine its common oxidation states.

C If the element is a nonmetal, it must be located in the *p* block of the periodic table. If a semimetal, it must lie along the diagonal line of semimetals from B to At. Transition metals can have two oxidation states separated by one electron.

D From your classification, the oxidation states of the element, and its physical appearance, deduce its identity.

Solution

(a) **A** The moderate electrical conductivity of element D tells us that it is a semimetal. It must lie in the *p* block of the periodic table because all of the semimetals are located there. **B** The stoichiometry of the oxides tells us that two common oxidation states for D are +4 and +6. **C** The element D must be located in Group 16 because the common oxidation states for the chalcogens (Group 16) include +6 (by losing all six valence elections) and +4 (by losing the four electrons from the *p* subshell). Thus, D is likely to be Se or Te. **D** Additional information is needed to distinguish between the two.

(b) A✓ Element E is an excellent electrical conductor, so it is a metal. B✓ The stoichiometry of the oxides and chlorides, however, tells us that common oxidation states for E are +2 and +1. C✓ Metals that can have two oxidation states separated by one electron are usually transition metals. The +1 oxidation state is characteristic of only one group: Group 11. Within Group 11, copper is the only element with common oxidation states of +1 and +2. D✓ Copper also has a reddish hue. Thus, E is probably copper.

EXERCISE 7.8

Based on the following information, identify elements G and J.

(a) Element G is a red liquid that does not conduct electricity. It forms three compounds with fluorine, GF, GF_3, and GF_5, and one with sodium, NaG.

(b) Element J is a soft, dull gray solid that conducts electricity well and forms two oxides, JO and JO_2.

Answer **(a)** Br; **(b)** Sn or Pb

TABLE 7.6 Relative abundance of some essential elements in the earth's crust and oceans[a]

Element	Crust, ppm (average)	Seawater, mg/L = ppm
O[b]	461,000	857,000
Si	282,000	2.2
Al	82,300	0.002
Fe	56,300	0.002
Ca	41,500	412
Na	23,600	10,800
Mg	23,300	1290
K	20,900	399
H	1400	108,000
P	1050	0.06
Mn	950	0.0002
F	585	1.3
S	350	905
C	200	28
Cl	145	19,400
V	120	0.0025
Cr	102	0.0003
Ni	84	0.00056
Zn	70	0.0049
Cu	60	0.00025
Co	25	0.00002
Li	20	0.18
N	19	0.5
Br	2.4	67.3
Mo	1.2	0.01
I	0.45	0.06
Se	0.05	0.0002

[a]Source of data: *CRC Handbook of Chemistry and Physics* (2004).
[b]Elements in boldface are known to be essential to humans.

7.5 • Trace Elements in Biological Systems

Of the more than 100 known elements, approximately 28 are known to be **essential** for the growth of at least one biological species, and only 19 are essential to humans (see Figure 1.26). What makes some elements essential to an organism and the rest nonessential? There are at least two reasons:

1. The element must have some unique chemical property that an organism can use to its advantage and without which it cannot survive.
2. Adequate amounts of the element must be available in the environment in an easily accessible form.

As you can see in Table 7.6, many of the elements that are abundant in the earth's crust are nevertheless not found in an easily accessible form (for example, as ions dissolved in seawater). Instead, they tend to form insoluble oxides, hydroxides, or carbonate salts. For example, although silicon is the second most abundant element in the earth's crust, SiO_2 and many silicate minerals are insoluble, so they are not easily absorbed by living tissues. This is also the case for iron and aluminum, which form insoluble hydroxides. Many organisms have therefore developed elaborate strategies to obtain iron from their environment. In contrast, molybdenum and iodine, though not particularly abundant, are highly soluble—molybdenum as molybdate, MoO_4^{2-}, and iodine as iodide, I^-, and iodate, IO_3^-—and thus are more abundant in seawater than iron. Not surprisingly, both molybdenum and iodine are used by many organisms.

Fortunately, many of the elements essential to life are necessary in only small amounts (Table 1.4 lists trace elements in humans). Even so, elements that are present in trace amounts can exert large effects on the health of an organism. Such elements function as part of an **amplification mechanism**, in which a molecule containing a trace element is an essential part of a larger molecule that acts in turn to regulate the concentrations of other molecules, and so forth. The amplification mechanism enables small variations in the concentration of the trace element to have large biochemical effects.

Essential trace elements in mammals can have four general roles: they can behave as macrominerals, participate in the catalysis of group-transfer reactions or oxidation–reduction reactions, or serve as structural components.

Macrominerals

MGC Biological Trace Elements

The **macrominerals**—Na, Mg, K, Ca, Cl, and P—are found in large amounts in biological tissues and are present as inorganic compounds, either dissolved or precipitated. All form monatomic ions (Na^+, Mg^{2+}, K^+, Ca^{2+}, Cl^-) except for phosphorus, which is found as the phosphate ion, PO_4^{3-}. Recall that calcium salts are used by many organisms as structural materials, such as in bone [hydroxyapatite—$Ca_5(PO_4)_3OH$], sea shells and egg shells ($CaCO_3$), and in plants calcium oxalate is used as a repository for Ca^{2+}.

The body fluids of all multicellular organisms contain relatively high concentrations of these ions. Some ions (Na^+, Ca^{2+}, and Cl^-) are located primarily in *extra*cellular fluids such as blood plasma, whereas K^+, Mg^{2+}, and phosphate are located primarily in *intra*cellular fluids. Substantial amounts of energy are required to selectively transport these ions across cell membranes. The selectivity of these **ion pumps** is based on differences in ionic radius (Section 7.2) and ionic charge.

Maintaining optimum levels of macrominerals is important because temporary changes in their concentration within a cell affect biological functions. For example, nerve impulse transmission requires a sudden, reversible increase in the amount of Na^+ that flows into the nerve cell. Similarly, when hormones bind to a cell, they can cause Ca^{2+} ions to enter that cell. In a complex series of reactions, the Ca^{2+} ions trigger events such as muscle contraction, release of neurotransmitters, or secretion of hormones. When people who exercise vigorously for long periods of time overhydrate with water, low blood salt levels can result in a condition known as *hyponatremia*, which causes nausea, fatigue, weakness, seizures, and even death. For this reason, athletes should hydrate with a sports drink that contains salts, not just water.

Group-Transfer Reactions

Trace metal ions also play crucial roles in many biological **group-transfer reactions**. In these reactions, a recognizable functional group, such as a phosphoryl unit ($-PO_3^-$), is transferred from one molecule to another. In this example,

$$ROPO_3^{2-} + H_2O \rightarrow ROH + HOPO_3^{2-} \qquad (7.18)$$

a PO_3^- unit is transferred from an alkoxide, RO^-, to hydroxide, OH^-. In order to neutralize the negative charge on the molecule that is undergoing the reaction, many biological reactions of this type require the presence of metal ions, such as Zn^{2+}, Mn^{2+}, Ca^{2+}, or Mg^{2+}, and occasionally Ni^{2+} or Fe^{3+}. The effectiveness of the metal ion depends largely on its charge and radius.

Zinc is an important component of enzymes that catalyze the hydrolysis of proteins, the addition of water to CO_2 to produce HCO_3^- and H^+, and most of the reactions involved in DNA and RNA synthesis, repair, and replication. Consequently, zinc deficiency has serious adverse effects, including abnormal growth and sexual development and loss of the sense of taste.

Biological Oxidation–Reduction Reactions

A third important role of trace elements is to transfer electrons in biological oxidation–reduction reactions. Iron and copper, for example, are found in proteins and enzymes that participate in O_2 transport, reduction of O_2, oxidation of organic molecules, and the conversion of atmospheric N_2 to NH_3. These metals usually transfer one electron per metal ion by alternating between oxidation states 3+/2+ (Fe) or 2+/1+ (Cu).

Because most transition metals have multiple oxidation states separated by only one electron, they are uniquely suited to transfer multiple electrons one at a time. Examples include molybdenum (+6/+5/+4), which is widely used for two-electron oxidation–reduction reactions, and cobalt (+3/+2/+1), which is found in vitamin B_{12}. In contrast, many of the *p*-block elements are well suited for transferring *two* electrons at once. Selenium (+4/+2), for example, is found in the enzyme that

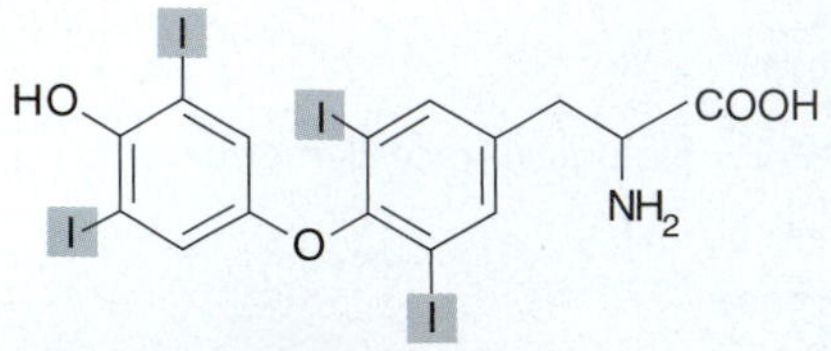

Thyroid hormone, thyroxine

An individual with goiter. In the United States, "iodized salt" prevents the occurrence of goiter.

catalyzes the oxidation of glutathione (whose function is described in Chapter 4): $2GSH + H_2O_2 \rightarrow 2H_2O + GSSG$ (GSH = glutathione).

Structural Components

Trace elements also act as essential structural components of biological tissues or molecules. In many systems where trace elements do not change oxidation states or otherwise participate directly in biochemical reactions, it is often assumed, though frequently with no direct evidence, that the element stabilizes a particular three-dimensional structure of the biomolecule in which it is found. One example is a sugar-binding protein containing Mn^{2+} and Ca^{2+} that is a part of the biological defense system of certain plants. Other examples include enzymes that require a Zn^{2+} at one site in order for activity to occur at a different site on the molecule. Some nonmetallic elements, such as F^-, also appear to have structural roles. Fluoride, for example, displaces hydroxide ion from hydroxyapatite in bone and teeth to form fluoroapatite, $Ca_5(PO_4)_3F$. Fluoroapatite is less soluble in acid and provides increased resistance to tooth decay. Another example of a nonmetal that plays a structural role is iodine, which in humans is found in only one molecule, the thyroid hormone *thyroxine*, whose structure is shown in the margin. When a person's diet does not contain sufficient iodine, the thyroid glands in their neck become greatly enlarged, leading to a condition called *goiter*. Because iodine is found primarily in ocean fish and seaweed, many of the original settlers of the American Midwest developed goiter due to the lack of seafood in their diet. Today most table salt contains small amounts of "iodine" (actually potassium iodide, KI) to prevent this problem.

EXAMPLE 7.9

There is some evidence that tin is an essential element in mammals. Based solely on what you know about the chemistry of tin and its position in the periodic table, predict a likely biological function for this element.

Given Element and data in Table 1.4

Asked for Likely biological function

Strategy

From the position of tin in the periodic table, its common oxidation states, and the data in Table 1.4, predict a likely biological function for the element.

Solution

From its position in the lower part of Group 14, we know that tin is a metallic element whose most common oxidation states are +4 and +2. Given the low levels of tin in mammals (140 mg/70-kg human), tin is unlikely to function as a macromineral. Although a role in catalyzing group-transfer reactions or as an essential structural component cannot be ruled out, the most likely role for tin would be in catalyzing oxidation–reduction reactions that involve two-electron transfers. This would take advantage of the ability of tin to have two oxidation states separated by two electrons.

EXERCISE 7.9

Based on what you know of its chemistry so far, predict a plausible role for vanadium in biological systems. Be sure to consider its position in the periodic table.

Answer Catalysis of oxidation–reduction reactions. Because vanadium is a first-row transition metal, it is likely to have multiple oxidation states.

SUMMARY AND KEY TERMS

7.1 The History of the Periodic Table (p. 302)

The periodic table arranges the elements according to their electron configurations, such that elements in the same column have the same valence electron configurations. Periodic variations in size and chemical properties are important factors in dictating the types of chemical reaction the elements undergo and the kinds of chemical compounds they form. The modern periodic table was based on empirical correlations of properties such as atomic mass; early models using limited data noted the existence of **triads** and **octaves** of elements with similar properties. The periodic table achieved its current form through the work of Dimitri Mendeleev and Julius Lothar Meyer, who both focused on the relationship between atomic mass and chemical properties. Meyer arranged the elements by their atomic volume, which today is equivalent to the **molar volume**, defined as the molar mass divided by molar density. The correlation with the electronic structure of atoms was made when H. G. J. Moseley showed that the periodic arrangement of the elements was determined by atomic number, not atomic mass.

7.2 Sizes of Atoms and Ions (p. 306)

A variety of methods have been established to measure the size of a single atom or ion. The **covalent atomic radius** is half the internuclear distance in a molecule that contains two identical atoms bonded to each other, whereas the **metallic atomic radius** is defined as half the distance between the nuclei of two adjacent atoms in a metallic element. The **van der Waals radius** of an element is half the internuclear distance between two non-bonded atoms in a solid. Atomic radii decrease from left to right across a row because of the increase in effective nuclear charge due to poor electron screening by other electrons in the same principal shell. Moreover, atomic radii increase from top to bottom down a column because the effective nuclear charge remains relatively constant as the principal quantum number increases. The **ionic radii** of cations and anions are always smaller or larger, respectively, than the parent atom due to changes in electron-electron repulsions, and the trends in ionic radius parallel those in atomic size. Comparison of the dimensions of atoms or ions that have the same number of electrons but different nuclear charges, called an **isoelectronic series**, shows a clear correlation between increasing nuclear charge and decreasing size.

7.3 Energetics of Ion Formation (p. 312)

The tendency of an element to lose or gain electrons is one of the most important factors in determining the kind of compounds it forms. Periodic behavior is most evident for **ionization energy** (***I***), the energy required to remove an electron from a gaseous atom. The energy required to remove successive electrons from an atom increases steadily, with a substantial increase occurring with the removal of an electron from a filled inner shell. Consequently, only valence electrons can be removed in chemical reactions, leaving the filled inner shell intact. Ionization energies explain the common oxidation states observed for the elements. Ionization energies increase diagonally from the lower left of the periodic table to the upper right. Minor deviations from this trend can be explained in terms of particularly stable electronic configurations, called **pseudo-noble gas configurations,** in either the parent atom or the resulting ion. The **electron affinity** (***EA***) of an element is the energy change that occurs when an electron is added to a gaseous atom to give an anion. In general, elements with the most negative electron affinities (the highest affinity for an added electron) are those with the smallest size and highest ionization energies, and are located in the upper right corner of the periodic table. The **electronegativity** (χ) of an element is the relative ability of an atom to attract electrons to itself in a chemical compound, and increases diagonally from the lower left of the periodic table to the upper right. The Pauling electronegativity scale is based on measurements of the strengths of covalent bonds between different atoms, whereas the **Mulliken electronegativity of an element** is the average of its first ionization energy and the absolute value of its electron affinity. Elements with a high electronegativity are generally non-metals and electrical insulators, and tend to behave as oxidants in chemical reactions. Conversely, elements with a low electronegativity are generally metals and good electrical conductors, and tend to behave as reductants in chemical reactions.

7.4 The Chemical Families (p. 324)

The **chemical families** consist of elements that have the same valence electron configuration and tend to have similar chemistry. The noble gases (Group 18) are monatomic gases that are chemically quite unreactive due to the presence of a filled shell of electrons. The halogens (Group 17) all have ns^2np^5 valence electron configurations, and are diatomic molecules that tend to react chemically by accepting a single electron. The **chalcogens** (Group 16) have ns^2np^4 valence electron configurations and react chemically by either gaining two electrons or by formally losing four or six electrons. The **pnicogens** (Group 15) all have ns^2np^3 valence electron configurations; hence, they form compounds in oxidation states ranging from -3 to $+5$. Elements in Group 14 have ns^2np^2 valence electron configurations, but exhibit a variety of chemical behaviors because they range from a nonmetal (carbon) to metals (tin/lead). Carbon, the basis of organic compounds, has at least four allotropes with distinct structures: diamond, graphite, **fullerenes**, and carbon **nanotubes**. Group 13 elements have ns^2np^1 valence electron configurations, and have an overwhelming tendency to form compounds in the $+3$ oxidation state. The alkaline earths (Group 2) have ns^2 valence electron configurations and form M^{2+} ions, while the alkali metals (Group 1) have ns^1 valence electron configurations and form M^+ ions. The **transition metals** (Groups 3–10) contain partially filled sets of *d* orbitals, and the **lanthanides** and **actinides** are those groups in which *f* orbitals are being filled. These groups exhibit strong horizontal similarities in behavior. Many of the transition metals form M^{2+} ions, whereas the chemistry of the lanthanides and actinides is dominated by M^{3+} ions.

7.5 Trace Elements in Biological Systems (p. 332)

Many of the elements in the periodic table are **essential trace elements** that are required for the growth of most organisms. Although they are present in only small quantities they have important biological effects because of their participation in an **amplification mechanism**. **Macrominerals** are present in larger amounts and play structural roles or act as electrolytes whose distribution in cells is tightly controlled. These ions are selectively transported across cell membranes by **ion pumps**. Other trace elements catalyze **group-transfer reactions** or biological oxidation-reduction reactions, while others yet are essential structural components of biological molecules.

QUESTIONS AND PROBLEMS

For instructor-assigned homework, go to www.masteringgenchem.com

Questions and Problems with colored numbers have answers in the Appendix and complete solutions in the Student Solutions Manual.

CONCEPTUAL

7.1 The History of the Periodic Table

1. Johannes Dobereiner is credited with developing the concept of chemical triads. Which of the Group 15 elements would you expect to make up a triad? Would you expect B, Al, and Ga to act as a triad? Justify your answers.
2. Despite the fact that Dobereiner, Newlands, Meyer, and Mendeleev all contributed to the development of the modern periodic table, Mendeleev is credited with its origin. Why was Mendeleev's periodic table accepted so rapidly?
3. How did Moseley's contribution to the development of the periodic table explain the location of the noble gases?
4. The *eka-* naming scheme devised by Mendeleev was used to describe undiscovered elements. (a) Use this naming method to predict the atomic number of each: *eka*-mercury, *eka*-astatine, *eka*-thallium, *eka*-hafnium. (b) Using the *eka*-prefix, identify the elements with these atomic numbers: 79, 40, 51, 117, 121.

7.2 Sizes of Atoms and Ions

5. The electrons of the 1*s* shell have a stronger electrostatic attraction to the nucleus than electrons in the 2*s* shell. Give two reasons for this.
6. Predict whether Na or Cl has the more stable $1s^2$ shell, and explain your rationale.
7. Arrange these elements in order of decreasing atomic radius: K, F, Ba, Pb, B, I.
8. Arrange these elements in order of increasing atomic radius: Ag, Pt, Mg, C, Cu, Si.
9. Using the periodic table, arrange the following elements in order of increasing atomic radius: Li, Ga, Ba, Cl, Ni.
10. Element M is a metal that forms compounds of the type MX_2, MX_3, and MX_4, where X is a halogen. What is the expected trend in the ionic radius of M in these compounds? Arrange these compounds in order of decreasing ionic radius of M.
11. The atomic radii of Na and Cl are 190 and 79 pm, respectively. However, the distance between sodium and chlorine in NaCl is 282 pm. Explain this discrepancy.
12. Are shielding effects on the atomic radius more pronounced across a row or down a group? Why?
13. What two factors influence the size of an ion relative to the size of its parent atom? Would you expect the ionic radius of S^{2-} to be the same in both MgS and Na_2S? Why or why not?
14. Arrange these ions in order of increasing radius: Br^-, Al^{3+}, Sr^{2+}, F^-, O^{2-}, I^-.
15. Arrange these ions in order of decreasing ionic radius: P^{3-}, N^{3-}, Cl^-, In^{3+}, S^{2-}.
16. How is an isoelectronic series different from a series of ions with the same charge? Do the cations in magnesium-, strontium-, and potassium sulfate form an isoelectronic series? Why or why not?
17. What isoelectronic series arises from the elements fluorine, nitrogen, magnesium, and carbon? Arrange the ions in this series by (a) increasing nuclear charge and (b) increasing size.
18. What would be the charge and electron configuration of an ion formed from calcium that is isoelectronic with (a) a chloride ion, and (b) Ar^+ ?

7.3 Energetics of Ion Formation

19. Answer each of the following true or false questions with regard to ionization energies. Support each answer with a short explanation.
    (a) Ionization energies increase with atomic radius.
    (b) Ionization energies decrease down a group.
    (c) Ionization energies increase with an increase in the magnitude of the electron affinity.
    (d) Ionization energies decrease diagonally across the periodic table from He to Cs.
    (e) Ionization energies depend upon electron configuration.
    (f) Ionization energies decrease across a row.
20. Based on electronic configurations, explain why the first ionization energies of the Group 16 elements are lower than those of the Group 15 elements, which is contrary to the general trend.
21. The first through third ionization energies do not vary greatly across the lanthanides. Why? How does the effective nuclear charge experienced by the n*s* electron change upon going from left to right (with increasing atomic number) in this series?
22. Most of the first row transition metals can form at least two stable cations, for example iron(II) and iron(III). In contrast, scandium and zinc each form only a single cation, the Sc^{3+} and Zn^{2+} ions, respectively. Use the electron configuration of these elements to provide an explanation.
23. Which of the following readily form(s) +3 ions: Nd, Al, or Ar? Why?
24. Orbital energies can reverse when an element is ionized. For which of these ions would you expect this reversal to occur? B^{3+}, Ga^{3+}, Pr^{3+}, Cr^{3+}, As^{3+}. Explain your reasoning.
25. The periodic trends in electron affinities are not as regular as periodic trends in ionization energies, even though the processes are essentially the converse of one another. Why are there so many more exceptions to the trends in electron affinities compared to ionization energies?
26. Elements lying on a lower right to upper left diagonal line cannot be arranged in order of increasing electronegativity according to where they occur in the periodic table. Why?
27. Why do ionic compounds form, if energy is required to form gaseous cations?
28. Why is Pauling's definition of electronegativity considered to be somewhat limited?
29. Based on their positions in the periodic table, arrange the following elements in order of increasing electronegativity: Sb, O, P, Mo, K, H.

30. Based on their positions in the periodic table, arrange these elements in order of decreasing electronegativity: V, F, B, In, Na, S.

7.4 The Chemical Families

31. Of the Group 1 elements, which would you expect to be the best reductant? Why? Would you expect boron to be a good reductant? Why or why not?
32. Classify each element as a metal, nonmetal, or semimetal: Hf, I, Tl, S, Si, He, Ti, Li, Sb. Which would you expect to be good electrical conductors, and why?
33. Identify each element as a metal, nonmetal, or semimetal: Au, Bi, P, Kr, V, Na, Po. Which would you expect to be good electrical insulators, and why?
34. Of the elements Kr, Xe, and Ar, why does only xenon form an extensive series of compounds? Would you expect Xe^{2+} to be a good oxidant? Why or why not?
35. Answer each of the following true or false questions about the halogens. Support each answer with a short explanation.
    (a) Halogens have filled valence-electron configurations.
    (b) Halogens tend to form salts with metals.
    (c) As the free elements, halogens are monatomic.
    (d) Halogens have appreciable nonmetallic character.
    (e) Halogens tend to have an oxidation state of −1.
    (f) Halogens are good reductants.
36. Nitrogen forms compounds in the +5, +4, +3, +2, and −3 oxidation states, whereas Bi forms ions only in the +5 and +3 oxidation states. Propose an explanation for the differences in behavior.
37. Which of these elements would you expect to form covalent halides: Mg, Al, O, P, Ne? Why? How do the melting points of covalent halides compare with those of ionic halides?
38. Which of these elements would you expect to form ionic halides upon reaction with Cl_2: Li, Ga, As, Xe? Explain your reasoning. Which are usually more volatile, ionic or covalent halides? Why?
39. Predict the relationship between the oxidative strength of the oxoanions of bromine, BrO_n^- (n = 1–4), and the number of oxygen atoms present (n). Explain your reasoning.
40. The stability of the binary hydrides of the chalcogens decreases in the order $H_2O > H_2S > H_2Se > H_2Te$. Why?
41. With which of the following elements will nitrogen form a compound in a positive oxidation state? O, Al, H, Cl. Write a reasonable chemical formula for an example of a binary compound with each element.
42. How do you explain the differences in chemistry observed for the Group 14 elements as you go down the column? Classify each of the Group 14 elements as a metal, nonmetal, or semimetal. Do you expect the Group 14 elements to form covalent or ionic compounds? Explain your reasoning.
43. Why is the chemistry of the Group 13 elements less varied than the chemistry of the Group 15 elements? Would you expect the chemistry of the Group 13 elements to be more or less varied than that of the Group 17 elements? Explain your reasoning.
44. If you needed to design a substitute for $BaSO_4$, the barium milkshake used to examine the large and small intestine by X-rays, would $BeSO_4$ be an inappropriate substitute? Explain your reasoning.
45. The alkali metals have an ns^1 valence electron configuration, and consequently they tend to lose an electron to form ions with +1 charge. Based on their valence electron configuration, what other kind of ion can the alkali metals form? Explain your answer.
46. Would Mo or W be the more appropriate biological substitute for Cr? Explain your reasoning.

7.5 Trace Elements in Biological Systems

47. Give at least one criterion for essential elements involved in biological oxidation–reduction reactions. Which region of the periodic table contains elements that are very well suited for this role? Explain your reasoning.
48. What are the general biological roles of trace elements that do not have two or more accessible oxidation states?

NUMERICAL

This section includes paired problems (marked by backets) that require similar problem-solving skills.

7.1 The History of the Periodic Table

49. Based on the data given, complete the table.

Species	Molar Mass, g/mol	Density, g/cm³	Molar Volume, cm³/mol
A	40.078	____	25.85
B	39.09	0.856	____
C	32.065	____	16.35
D	____	1.823	16.98
E	26.98	____	9.992
F	22.98	0.968	____

Plot molar volume versus molar mass for these substances. According to Meyer, which would be considered metals and which would be considered nonmetals?

7.2 Sizes of Atoms and Ions

50. Plot the ionic charge versus ionic radius using the following data for Mo: Mo^{3+}, 69 pm; Mo^{4+}, 65 pm; and Mo^{5+}, 61 pm. Then use this plot to predict the ionic radius of Mo^{6+}. Is the observed trend consistent with the general trends discussed in the chapter? Why or why not?
51. Internuclear distances for selected ionic compounds are given in the table. (a) If the ionic radius of Li^+ is 76 pm, what is the ionic radius of each of the anions? (b) What is the ionic radius of Na^+?

	LiF	LiCl	LiBr	LiI
Distance, pm	209	257	272	296

	NaF	NaCl	NaBr	NaI
Distance, pm	235	282	298	322

52. Arrange the following gaseous species in order of increasing radius, and justify your decisions: Mg^{2+}, P^{3-}, Br^-, S^{2-}, F^-, N^{3-}.

7.3 Energetics of Ion Formation

53. The table gives values of the first and third ionization energies (I, kJ/mol) for selected elements:

Number of Electrons	Element	I_1 ($E \rightarrow E^+ + e^-$)	Element	I_3 ($E^{2+} \rightarrow E^{3+} + e^-$)
11	Na	495.9	Al	2744.8
12	Mg	737.8	Si	3231.6
13	Al	577.6	P	2914.1
14	Si	786.6	S	3357
15	P	1011.9	Cl	3822
16	S	999.6	Ar	3931
17	Cl	1251.2	K	4419.6
18	Ar	1520.6	Ca	4912.4

Plot the ionization energies versus number of electrons. Explain why the slopes of the I_1 and I_3 plots are different, even though the species in each row of the table have the same electron configurations.

54. Would you expect the third ionization energy of iron, corresponding to the removal of an electron from a gaseous Fe^{2+} ion, to be larger or smaller than the fourth ionization energy, corresponding to removal of an electron from a gaseous Fe^{3+} ion? Why? How would these ionization energies compare to the first ionization energy of Ca?
55. Which would you expect to have the highest first ionization energy: Mg, Al, or Si? Which would you expect to have the highest third ionization energy. Why?
56. Use the values of the first ionization energies given in Figure 7.11 to construct plots of first ionization energy versus atomic number for (a) boron through oxygen in the second period; and (b) oxygen through tellerium in Group 16. Which plot shows more variation? Explain the reason for the variation in first ionization energies for this group of elements.
57. Arrange these elements in order of increasing first ionization energies: Ga, In, Zn. Would the order be the same for second and third ionization energies? Explain your reasoning.
58. Arrange each set of elements in order of increasing magnitude of electron affinity: (a) Pb, Bi, Te; (b) Na, K, Rb; (c) P, C, Ge.
59. Arrange each set of elements in order of decreasing magnitude of electron affinity: (a) As, Bi, N; (b) O, F, Ar; (c) Cs, Ba, Rb.
60. Which of these species has the highest electron affinity? F, O^-, Al^{3+}, Li^+. Explain your reasoning.
61. Which of the following has the highest electron affinity? O^-, N^{2-}, Hg^{2+}, H^+. Which has the lowest electron affinity? Justify your answers.
62. The Mulliken electronegativity of element A is 542 kJ/mol. If the electron affinity of A is –72 kJ/mol, what is the first ionization energy of element A? Use the data in the table below as a guideline to decide if A is a metal, nonmetal, or semimetal. If one gram of A contains 4.85×10^{21} molecules, what is element A?

	Na	Al	Si	S	Cl
EA (kJ/mol)	−59.6	−41.8	−134.1	−200.4	−348.6
Ionization Energy (kJ/mol)	495.8	577.5	786.5	999.6	1251.2

63. Based on their valence electron configurations, classify these elements as either electrical insulators, electrical conductors, or substances with intermediate conductivity: S, Ba, Fe, Al, Te, Be, O, C, P, Sc, W, Na, B, Rb.
64. Using the data in Problem 62, what conclusions can you draw with regard to the relationship between electronegativity and electrical properties? Estimate the approximate electronegativity of a pure element that is very dense, lustrous, and malleable.
65. Which, if any, of these elements is a good reductant? Al, Mg, O_2, Ti, I_2, H_2. Explain your reasoning.
66. Which of the following, if any, would you expect to be a good oxidant: Zn, B, Li, Se, Co, Br_2? Explain your reasoning.
67. Determine whether each species is a good oxidant, a good reductant, or neither: Ba, Mo, Al, Ni, O_2, Xe.
68. Determine whether each species is a good oxidant, a good reductant, or neither: Ir, Cs, Be, B, N, Po, Ne.
69. Choose which of the following you would expect to be a good oxidant. Then justify your answer. I_2, O^-, Zn, Sn^{2+}, K^+.
70. Based on the valence electron configuration of the noble gases, would you expect them to have positive or negative electron affinities? What does this imply about their most likely oxidation states? About their reactivity?

7.4 The Chemical Families

71. Write a balanced equation for formation of XeO_3 from elemental Xe and O_2. What is the oxidation state of Xe in XeO_3? Would you expect Ar to undergo an analogous reaction? Why or why not?
72. Which of the *p*-block elements exhibit the greatest variation in oxidation states? Why? Based on their valence electron configurations, identify these oxidation states.
73. Based on its valence electron configuration, what are the three common oxidation states of selenium? In a binary compound, what atoms bonded to Se will stabilize the highest oxidation state? The lowest oxidation state?
74. Would you expect sulfur to be readily oxidized by HCl? Why or why not? Would you expect phosphorus to be readily oxidized by sulfur? Why or why not?
75. What are the most common oxidation states for the pnicogens? What factors determine the relative stabilities of these oxidation states for the lighter and the heavier pnicogens? What is likely to be the most common oxidation state for phosphorus and arsenic? Why?
76. Which would be the least stable: NF_3, NCl_3, or NI_3? Explain your answer. Which of the following would be the least stable: BrO^-, ClO^-, FO^-? Explain your answer.
77. In an attempt to explore the chemistry of the superheavy element ununquadium, Z = 114, you isolated two distinct salts by exhaustively oxidizing metal samples with chlorine gas. These salts are found to have the formulas MCl_2 and MCl_4. What would be the name of ununquadium using Mendeleev's *eka*-notation?
78. Would you expect the compound CCl_2 to be stable? $SnCl_2$? Why or why not?
79. A newly discovered element, Z, is a good conductor of electricity and reacts only slowly with oxygen. Reaction of one gram of Z with oxygen under three different sets of conditions gives products with masses of 1.333 g, 1.668 g, and 1.501 g, respectively. To what family of elements does Z belong? What is the atomic mass of the element?

80. An unknown element, Z, is a dull, brittle powder that reacts with oxygen at high temperatures. Reaction of 0.665 gram of Z with oxygen under two different sets of conditions forms gaseous products with masses of 1.328 g and 1.660 g. To which family of elements does Z belong? What is the atomic mass of the element?

81. Why are the alkali metals such powerful reductants? Would you expect Li to be able to reduce H_2? Would Li reduce V? Why or why not?

82. What do you predict to be the most common oxidation state for (a) Au; (b) Sc; (c) Ag; (d) Zn? Give the valence electron configuration for each of these elements in its most stable oxidation state.

83. Complete the table.

	Mg	C	Ne	Fe	Br
Valence Electron Configuration	—	—	—	—	—
Common Oxidation States	—	—	—	—	—
Oxidizing Strength	—	—	—	—	—

84. Use the following information to identify elements T, X, Y, and Z. Element T reacts with oxygen to form at least three compounds: TO, T_2O_3, and TO_2. Element X reacts with oxygen to form XO_2, but X is also known to form compounds in the +2 oxidation state. Element Y forms Y_2O_3, and element Z reacts vigorously and forms Z_2O. Electrical conductivity measurements showed that element X exhibited electrical conductivity intermediate between metals and insulators, while T, Y, and Z were good conductors of electricity. Element T is a hard, lustrous, silvery metal, element X is a blue-gray metal, element Y is a light, silvery metal, and element Z is a soft, low-melting metal.

85. Predict whether each of these elements will react with oxygen: Cs, F_2, Al, He. If a reaction will occur, identify the products.

86. Predict whether each of these elements will react with Cl_2: K, Ar, O, Al. If a reaction will occur, identify the products.

87. Use the following information to identify elements X, Y, and Z.

(a) Element X is a soft, white metal that is flammable in air and reacts vigorously with water. Its first ionization energy is less than 500 kJ/mol, but the second ionization energy is greater than 3,000 kJ/mol.

(b) Element Y is a gas that reacts with F_2 to form a series of fluorides ranging from YF_2 to YF_6. It is inert to most other chemicals.

(c) Element Z is a deep red liquid that reacts with fluorine to form ZF_3 and with chlorine to form ZCl and ZCl_3, and with iodine to form ZI. Element Z also reacts with the alkali metals and alkaline earths.

88. Adding a reactive metal to water in the presence of oxygen results in a fire. In the absence of oxygen, addition of 551 mg of the metal to water produces 6.4 mg of hydrogen gas. Treatment of 2.00 gram of this metal with 6.3 grams of Br_2 results in the formation of 3.86 grams of an ionic solid. To which chemical family does this element belong? What is the identity of the element? Write and balance the chemical equation for the reaction of water with the metal to form hydrogen gas.

APPLICATIONS

89. Most plants, animals and bacteria use oxygen as the terminal oxidant in their respiration process. In a few locations, however, whole ecosystems have developed in the absence of oxygen, containing creatures that can utilize sulfur compounds instead of oxygen. List the common oxidation states of sulfur, provide an example of any oxoanions containing sulfur in these oxidation states, and name the ions. Which ion would be the best oxidant?

90. Titanium is currently used in the aircraft industry and is being studied for its use in ships, which operate in a highly corrosive environment. This interest is due to the fact that titanium is strong, "light", and corrosion-resistant. The densities of selected elements are given below. Why can't an element with an even lower density such as calcium be used to produce an even lighter structural material?

Element	Density (g/cm^3)	Element	Density (g/cm^3)
K	0.865	Cr	7.140
Ca	1.550	Mn	7.470
Sc	2.985	Fe	7.874
Ti	4.507	Co	8.900
V	6.110	Ni	8.908

91. The compound Fe_3O_4 was called lodestone in ancient times, because it responds to Earth's magnetic field and can be used to construct a primitive compass. Today Fe_3O_4 is commonly called magnetite, because it contains both Fe^{2+} and Fe^{3+}, and the unpaired electrons on these ions align to form tiny magnets. How many unpaired electrons does each ion have? Would you expect to observe magnetic behavior in compounds containing Zn^{2+}? Why or why not? Would you expect Fe or Zn to have the lower third ionization energy? Why?

92. Understanding trends in periodic properties allows us to predict the properties of individual elements. For example, if we need to know whether Fr is a liquid at room temperature (approximately 20°C), we could obtain this information by plotting the melting points of the other alkali metals versus atomic number. Based on the data in the table below, would you predict Fr to be a solid, a liquid, or a gas at 20°C?

	Li	Na	K	Rb	Cs
Melting point, °C	180	97.8	63.7	39.0	28.5

Francium is found in minute traces in uranium ores. Is this consistent with your conclusion? Why or why not? Why would Fr be found in these ores, but only in small quantities?

8 Structure and Bonding I:

Ionic Versus Covalent Bonding

LEARNING OBJECTIVES

- To be able to quantitatively describe the energetic factors involved in the formation of an ionic bond
- To understand the relationship between the lattice energy and physical properties of an ionic compound
- To be able to calculate lattice energy using a Born–Haber cycle
- To be able to predict the number of bonds an element will form using Lewis dot symbols
- To be able to use Lewis dot symbols to explain the stoichiometry of a compound
- To understand the concept of resonance structures
- To be able to identify Lewis acids and bases
- To understand the relationships among bond order, bond length, and bond energy
- To be able to calculate the percent ionic character of a covalent polar bond

In Chapter 7, we described the relationship between the chemical properties and reactivity of an element and its position in the periodic table. In this chapter and the next, we describe the interactions that hold atoms together in chemical substances, and we examine the factors that determine how the atoms of a substance are arranged in space. Our goal is to understand how the properties of the component atoms in a chemical compound determine the structure and reactivity of the compound.

The properties described in Chapters 6 and 7 were properties of isolated atoms, yet most of the substances in our world consist of atoms held together in molecules, ionic compounds, or metallic solids. The properties of these substances depend not only on the characteristics of the component atoms but also on how those atoms are bonded to one another.

Although both Group-14 elements, carbon and silicon, form bonds with oxygen, how they form those bonds results in a vast difference in physical properties. Because of its simple molecular bond, carbon dioxide is a gas that exists only as a volatile molecular solid, "dry ice," at temperatures of –78ºC and below. Silicon dioxide is a giant covalent structure, whose strong bonds in three dimensions make it a hard, high-melting-point solid, such as quartz.

What you learn in this chapter about chemical bonding and molecular structure will help you understand how different substances that contain the same atoms can have vastly different physical and chemical properties. For example, oxygen gas (O_2) is essential for life, yet ozone (O_3) is toxic to cells, although as you learned in Chapter 3, ozone in the upper atmosphere shields us from harmful ultraviolet light. Moreover, you saw in Chapter 7 that diamond is a hard, transparent solid that is a gemstone, graphite is a soft, black solid that is a lubricant, and fullerenes are molecular species with carbon cage structures, yet all are made of carbon. As you learn about bonding, you will also discover why, although carbon and silicon both have ns^2np^2 valence-electron configurations and form dioxides, CO_2 is normally a gas that condenses into the volatile molecular solid known as dry ice, whereas SiO_2 is a nonvolatile solid with a network structure that can take several forms, including beach sand and quartz crystals.

CHAPTER OUTLINE

MGC Chemical Bonding

8.1 • An Overview of Chemical Bonding

In Chapter 2, we defined a *chemical bond* as the force that holds atoms together in a chemical compound. We also introduced two idealized types of bonding: *covalent bonding*, in which electrons are shared between atoms in a molecule or polyatomic ion, and *ionic bonding*, in which positively and negatively charged ions are held together by electrostatic forces. The concepts of covalent and ionic bonding were developed to explain the properties of different kinds of chemical substances. Ionic compounds, for example, typically dissolve in water to form aqueous solutions that conduct electricity (see Chapter 4). In contrast, most covalent compounds that dissolve in water form solutions that do not conduct electricity. Furthermore, many covalent compounds are volatile, whereas ionic compounds are not.

Despite the differences in the distribution of electrons between these two idealized types of bonding, all models of chemical bonding have three features in common:

1. Atoms interact with one another to form aggregates such as molecules, compounds, and crystals because doing so lowers the total energy of the system; that is, the aggregates are more stable than the isolated atoms.
2. Energy is required to dissociate bonded atoms or ions into isolated atoms or ions. For ionic solids, in which the ions form a three-dimensional array called a *lattice*, this energy is called the **lattice energy**, the enthalpy change that occurs when a solid ionic compound is transformed into gaseous ions. For covalent compounds, this energy is called the **bond energy**, the enthalpy change that occurs when a given bond in a gaseous molecule is broken.
3. Each chemical bond is characterized by a particular optimal internuclear distance r_0, called the **bond distance**.

We explore these characteristics further, after a brief description of the energetic factors involved in the formation of an ionic bond.

8.2 • Ionic Bonding

Chapter 2 explained that ionic bonds are formed when positively and negatively charged ions are held together by electrostatic forces. You learned that the energy of the electrostatic attraction (E), a measure of the force's strength, is inversely proportional to the internuclear distance between the charged particles (r):

$$E \propto \frac{Q_1Q_2}{r} \qquad E = k\frac{Q_1Q_2}{r} \tag{8.1}$$

where each ion's charge is represented by the symbol Q. The proportionality constant k is equal to 2.31×10^{-28} J·m.* If Q_1 and Q_2 have opposite signs (as in NaCl, for example, where Q_1 is +1 for Na^+ and Q_2 is −1 for Cl^-), then E is negative, which means that energy is released when oppositely charged ions are brought together from an infinite distance to form an isolated ion pair. As shown by the green curve in the lower half of Figure 8.1, Equation 8.1 predicts that the maximum energy is released when the ions are infinitely close to each other, at $r = 0$. Because ions occupy space, however, they cannot be infinitely close together. At very short distances, repulsive electron–electron interactions between electrons on adjacent ions become stronger than the attractive interactions between ions with opposite charges, as shown by the red curve in the upper half of Figure 8.1. The total energy of the system is a balance between the attractive and repulsive interactions. The purple curve in Figure 8.1 shows that the total energy of the system reaches a minimum at r_0, the point where the electrostatic repulsions and attractions are exactly balanced. This distance is the same as the experimentally measured bond distance.

Let's consider the energy released when a gaseous Na^+ ion and a gaseous Cl^- ion are brought together from $r = \infty$ to $r = r_0$. Given that the observed gas-phase internuclear distance is 236 pm, the energy change associated with the formation of an ion pair from an $Na^+(g)$ ion and a $Cl^-(g)$ ion is

$$E = k\frac{Q_1Q_2}{r_0} = (2.31 \times 10^{-28}\ \text{J}\cdot\cancel{\text{m}})\frac{(+1)(-1)}{236\ \cancel{\text{pm}} \times 10^{-12}\ \cancel{\text{m}}/\cancel{\text{pm}}} \tag{8.2}$$
$$= -9.79 \times 10^{-19}\ \text{J/ion pair}$$

The negative value indicates that energy is released. To calculate the energy change in the formation of a mole of NaCl pairs, we need to multiply the energy per ion pair by Avogadro's number:

$$E = (-9.79 \times 10^{-19}\ \text{J}/\cancel{\text{ion pair}})(6.022 \times 10^{23}\ \cancel{\text{ion pair}}/\text{mol}) \tag{8.3}$$
$$= -589\ \text{kJ/mol}$$

This is the energy released when a mole of gaseous ion pairs is formed, *not* when a mole of positive and negative ions condense to form a crystalline lattice. Although

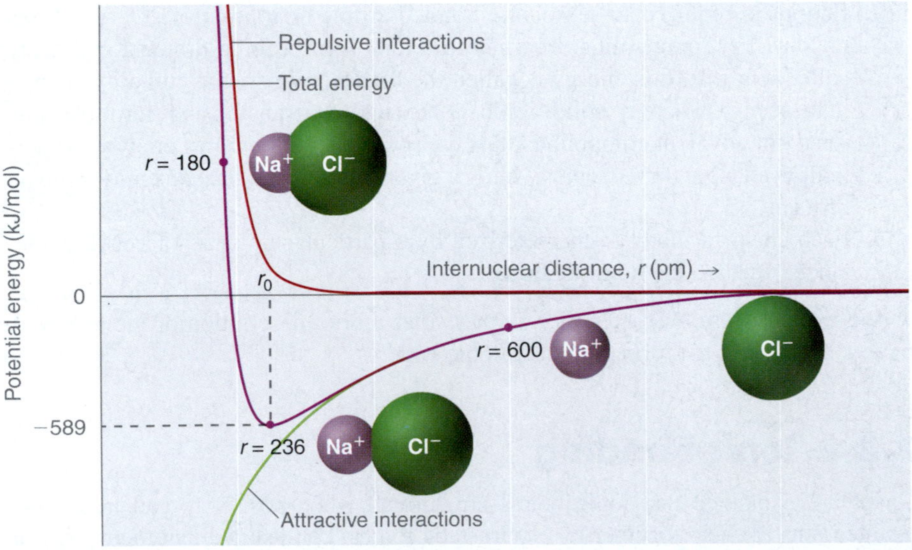

Figure 8.1 A plot of potential energy vs. internuclear distance for the interaction between a gaseous Na^+ ion and a gaseous Cl^- ion. Note that the energy of the system reaches a minimum at a particular distance, r_0, where the attractive and repulsive interactions are balanced. Interactive Graph

* This value of k includes the charge of a single electron (1.6022×10^{-19} C) for each ion. The equation can also be written using the charge of each ion, expressed in coulombs (C) incorporated into the constant. In this case, the proportionality constant, k, equals 8.999×10^9 J · m/C^2. Thus, in the example given, $Q_1 = +1$ (1.6022×10^{-19} C) and $Q_2 = -1$ (1.6022×10^{-19} C).

this energy does not correspond directly to the lattice energy of the crystalline solid because of long-range interactions in the lattice structure, the large negative value indicates that bringing positive and negative ions together is energetically very favorable, whether an ion pair or a crystalline lattice is formed.

We summarize the important points about ionic bonding:

At r_0, the ions are more stable (have a lower potential energy) than they are at an infinite internuclear distance. When oppositely charged ions are brought together from $r = \infty$ to $r = r_0$, the energy of the system is lowered (energy is released).

Because of the low potential energy at r_0, energy must be added to the system to separate the ions. The amount of energy needed is the bond energy.

The energy of the system reaches a minimum at a particular internuclear distance (the bond distance).

EXAMPLE 8.1

Calculate the energy released when a mole of gaseous Li^+F^- ion pairs is formed from the separated ions. The observed internuclear distance in the gas phase is 156 pm.

Given Cation and anion, amount, internuclear distance

Asked for Energy released from formation of gaseous ion pairs

Strategy

Substitute the appropriate values into Equation 8.1 to obtain the energy released in the formation of a single ion pair, and then multiply this value by Avogadro's number to obtain the energy released per mole.

Solution

Inserting the values for Li^+F^- into Equation 8.1 (where $Q_1 = +1$, $Q_2 = -1$, and $r = 156$ pm), we find that the energy associated with the formation of a single pair of Li^+F^- ions is

$$E = k\frac{Q_1Q_2}{r_0} = (2.31 \times 10^{-28}\ \text{J}\cdot\cancel{\text{m}})\frac{(+1)(-1)}{156\ \cancel{\text{pm}} \times 10^{-12}\ \cancel{\text{m}}/\cancel{\text{pm}}} = -1.48 \times 10^{-18}\ \text{J/ion pair}$$

Then the energy released per mole of Li^+F^- ion pairs is

$$E = (-1.48 \times 10^{-18}\ \text{J}/\cancel{\text{ion pair}})(6.022 \times 10^{23}\ \cancel{\text{ion pair}}/\text{mol}) = -891\ \text{kJ/mol}$$

Because Li^+ and F^- are smaller than Na^+ and Cl^- (see Figure 7.9), the internuclear distance in LiF is shorter than in NaCl. Consequently, in accordance with Equation 8.1, much more energy is released when a mole of gaseous Li^+F^- ion pairs is formed (−891 kJ/mol) than when a mole of gaseous Na^+Cl^- ion pairs is formed (−589 kJ/mol).

EXERCISE 8.1

Calculate the amount of energy released when a mole of gaseous MgO ion pairs is formed. The internuclear distance in the gas phase is 175 pm.

Answer −3180 kJ/mol = -3.18×10^3 kJ/mol

8.3 • Lattice Energies in Ionic Solids

MGC Coulomb's Law and Lattice Energy

As you have learned, ionic compounds are usually rigid, brittle, crystalline substances with flat surfaces that intersect at characteristic angles. They are not easily deformed, and they melt at relatively high temperatures (NaCl, for example, melts at 801°C).

These properties result from the regular arrangement of the ions in the crystalline lattice and from the strong electrostatic attractive forces between ions with opposite charges. Recall from Chapter 2 that the reaction of a metal with a nonmetal produces an ionic compound; that is, electrons are transferred from the metal (the *reductant*) to the nonmetal (the *oxidant*).

Equation 8.1 has demonstrated that the formation of ion pairs from isolated ions releases large amounts of energy. Even more energy is released, however, when these ion pairs condense to form an ordered three-dimensional array (Figure 7.8). This is because each cation in the lattice is surrounded by more than one anion (typically four, six, or eight) and vice versa, an arrangement that is more stable than a system consisting of separate pairs of ions, in which there is only one cation–anion interaction in each pair. Note that r_0 may differ between the gas-phase dimer and the lattice.

TABLE 8.1 Representative calculated lattice energies

Substance	*U*, kJ/mol
NaI	682
CaI_2	1971
MgI_2	2293
NaOH	887
Na_2O	2481
$NaNO_3$	755
$Ca_3(PO_4)_2$	10,602
$CaCO_3$	2804

Source: *CRC Handbook of Chemistry and Physics* (2004).

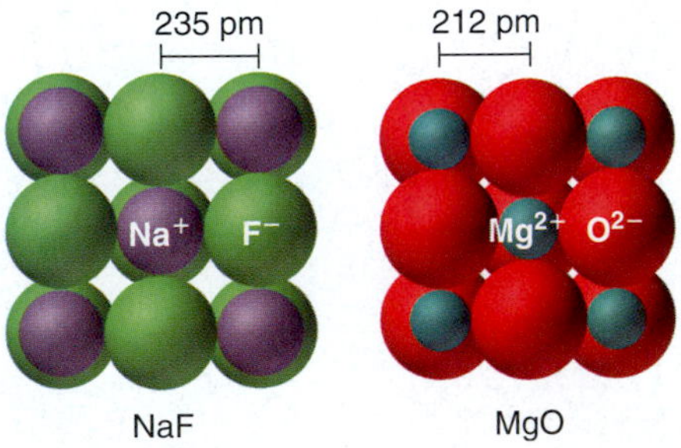

Calculating Lattice Energies

The lattice energy U of nearly any ionic solid can be calculated rather accurately using a modified form of Equation 8.1:

$$U = -k'\frac{Q_1Q_2}{r_0} \qquad U > 0 \tag{8.4}$$

U, which is always a positive number, represents the amount of energy required to dissociate a mole of an ionic solid into the gaseous ions:

$$\mathrm{MX}(s) \longrightarrow \mathrm{M}^{+n}(g) + \mathrm{X}^{-n}(g) \qquad \Delta H = U \tag{8.5}$$

As before, Q_1 and Q_2 are the charges on the ions and r_0 is the internuclear distance. Thus, the lattice energy is directly related to the product of the ion charges and inversely related to the internuclear distance. The value of the constant k' depends on the specific arrangement of ions in the solid lattice and on their valence-electron configurations, topics that will be discussed in more detail in Chapter 12. Representative values for calculated lattice energies, which range from about 600 to 10,000 kJ/mol, are listed in Table 8.1. Energies of this magnitude can be decisive in determining the chemistry of the elements.

Because the lattice energy depends on the *product* of the charges of the ions, a salt that contains a metal cation with a +2 charge, M^{2+}, and a nonmetal anion with a −2 charge, Y^{2-}, will have a lattice energy four times higher than one that contains M^+ and X^-, assuming the ions are of comparable size (and thus have similar internuclear distances). For example, the calculated value of U for NaF is 910 kJ/mol, whereas U for MgO (containing Mg^{2+} and O^{2-} ions) is 3795 kJ/mol.

Because the lattice energy is *inversely* related to the internuclear distance r_0, it is also inversely proportional to the size of the ions. This effect is illustrated in Figure 8.2, which shows that the lattice energy decreases for the series LiX, NaX, and KX as the radius of X^- increases. Because r_0 in Equation 8.4 is the sum of the ionic radii of the cation and the anion ($r_0 = r_+ + r_-$), r_0 increases as the cation becomes larger in the series, and therefore the magnitude of U decreases. A similar effect is seen when the anion becomes larger in a series of compounds that contain the same cation.

Note the pattern

Lattice energies are highest for substances that contain small, highly charged ions.

EXAMPLE 8.2

Arrange these compounds in order of increasing lattice energy: GaP, BaS, CaO, and RbCl.

Given Four compounds

Asked for Order of increasing lattice energy

Strategy

Using Equation 8.4, predict the order of the lattice energies based on the charges on the ions. For compounds that contain ions with the same charge, use the relative sizes of the ions to make this prediction.

Solution

The first compound, GaP, which is used in semiconductor electronics, contains Ga^{3+} and P^{3-} ions, the compound BaS contains Ba^{2+} and S^{2-} ions, CaO contains Ca^{2+} and O^{2-} ions, and RbCl has Rb^{+} and Cl^{-} ions. We know from Equation 8.4 that the lattice energy is directly proportional to the product of the ionic charges. Consequently, we expect RbCl, with a $(-1)(+1)$ term in the numerator, to have the lowest lattice energy, and GaP, with a $(+3)(-3)$ term, the highest. To decide whether BaS or CaO has the higher lattice energy, we need to consider the relative sizes of the ions because both compounds contain a $+2$ metal ion and a -2 chalcogenide ion. Because Ba^{2+} lies below Ca^{2+} in the periodic table, Ba^{2+} is larger than Ca^{2+}. Similarly, S^{2-} is larger than O^{2-}. Because the cation and the anion in BaS are both larger than the corresponding ions in CaO, the internuclear distance r_0 is greater in BaS and its lattice energy will be lower than that of CaO. The order of increasing lattice energy is RbCl < BaS < CaO < GaP.

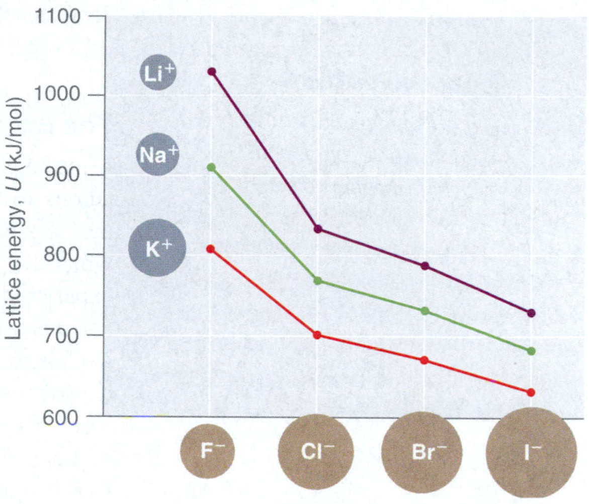

Figure 8.2 A plot of lattice energy vs. identity of the halide for the lithium, sodium, and potassium halides. Because the ionic radii of the cations decrease in the order $K^{+} > Na^{+} > Li^{+}$, for a given halide ion, the lattice energy decreases smoothly from Li^{+} to K^{+}. Conversely, for a given alkali metal ion, the fluoride salt always has the highest lattice energy and the iodide salt the lowest. Interactive Graph

EXERCISE 8.2

Arrange these compounds in order of decreasing lattice energy: InAs, KBr, LiCl, SrSe, ZnS.

Answer InAs > ZnS > SrSe > LiCl > KBr

The Relationship Between Lattice Energies and Physical Properties

The magnitude of the forces that hold an ionic substance together has a dramatic effect on many of its properties. The **melting point**, for example, is the temperature at which the individual ions have enough kinetic energy to overcome the attractive forces that hold them in place. At the melting point, the ions can move freely and the substance becomes a liquid. Thus, melting points vary with lattice energies for ionic substances that have similar structures. The melting points of the sodium halides (Figure 8.3), for example, decrease smoothly from NaF to NaI, following the same trend as seen for their lattice energies (Figure 8.2). Similarly, the melting point of MgO is 2825°C, compared with 996°C for NaF, reflecting the higher lattice energies associated with higher charges on the ions. In fact, because of its high melting point MgO is used as an electrical insulator in heating elements for electric stoves.

The **hardness** of ionic materials—that is, their resistance to scratching or abrasion—is also related to their lattice energies. Hardness is directly related to how tightly the ions are held together electrostatically, which, as we saw, is also reflected in the lattice energy. As an example, MgO is harder than NaF, which is consistent with its higher lattice energy.

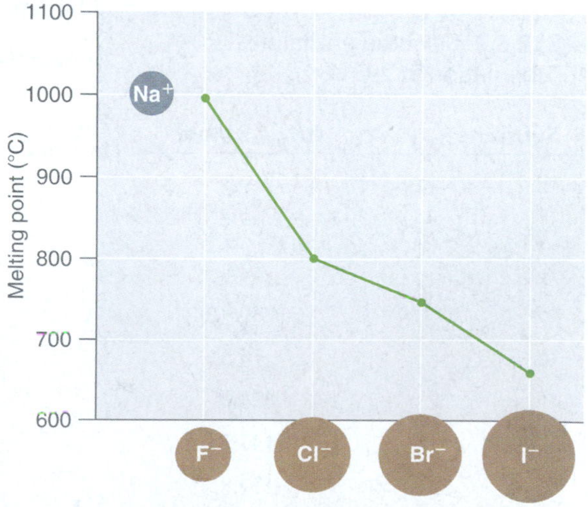

Figure 8.3 A plot of melting point vs. identity of the halide for the sodium halides. The melting points follow the same trend as the magnitude of the lattice energies in Figure 8.2.

In addition to determining melting point and hardness, lattice energies affect the solubilities of ionic substances in water. In general, *the higher the lattice energy, the less soluble the compound in water*. For example, the solubility of NaF in water at

Note the pattern

High lattice energies lead to hard, insoluble compounds with high melting points.

25°C is 4.13 g/100 mL, but under the same conditions, the solubility of MgO is only 0.65 mg/100 mL, meaning that it is essentially insoluble.

The Born–Haber Cycle

In principle, lattice energies could be measured by combining gaseous cations and anions to form an ionic solid and then measuring the heat evolved. Unfortunately, measurable quantities of gaseous ions have never been obtained under conditions where the heat flow can be measured. Instead, lattice energies are found using the experimentally determined enthalpy changes for other chemical processes, Hess's law, and a *thermochemical cycle*, called the **Born–Haber cycle**, similar to those introduced in Chapter 5. Developed by Max Born and Fritz Haber in 1919, the Born–Haber cycle describes a process in which an ionic solid is conceptually formed from its component elements in a stepwise manner.

Let's examine the Born–Haber cycle used to determine the lattice energy of CsF(*s*). CsF is a nearly ideal ionic compound because Cs is the least electronegative element that is not radioactive and F is the most electronegative element. To construct a thermochemical cycle for the formation of CsF, we need to know its enthalpy of formation ΔH_f, which is defined by the reaction

$$Cs(s) + \frac{1}{2}F_2(g) \longrightarrow CsF(s) \qquad (8.6)$$

Recall that enthalpy is a state function (see Chapter 5); thus, the overall ΔH for a series of reactions is the sum of the values of ΔH for the individual reactions (see Hess's law in Chapter 5). We can therefore use a thermochemical cycle to determine the enthalpy change that accompanies the formation of solid CsF from the parent *elements* (not ions).

A Born–Haber cycle for calculating the lattice energy of cesium fluoride is shown in Figure 8.4. This particular cycle consists of six reactions, Equation 8.6 plus the five reactions listed below:

Reaction 1: $Cs(s) \longrightarrow Cs(g) \qquad \Delta H_1 = \Delta H_{sub} = 76.5 \text{ kJ/mol}$

This equation describes the **sublimation** of elemental cesium, the conversion of the solid directly to a gas. The accompanying enthalpy change is called the **enthalpy of sublimation** (Table 8.2) and is *always* positive (energy is required to sublime a solid).

Reaction 2: $Cs(g) \longrightarrow Cs^+(g) + e^- \qquad \Delta H_2 = I_1 = 375.7 \text{ kJ/mol}$

TABLE 8.2 Selected enthalpies of sublimation at 298 K

Substance	ΔH_{sub}, kJ/mol
Li	159.3
Na	107.5
K	89.0
Rb	80.9
Cs	76.5
Be	324.0
Mg	147.1
Ca	177.8
Sr	164.4
Ba	180.0

Source: *CRC Handbook of Chemistry and Physics* (2004).

Figure 8.4 **A Born–Haber cycle illustrating the enthalpy changes involved in the formation of solid cesium fluoride from the elements.**

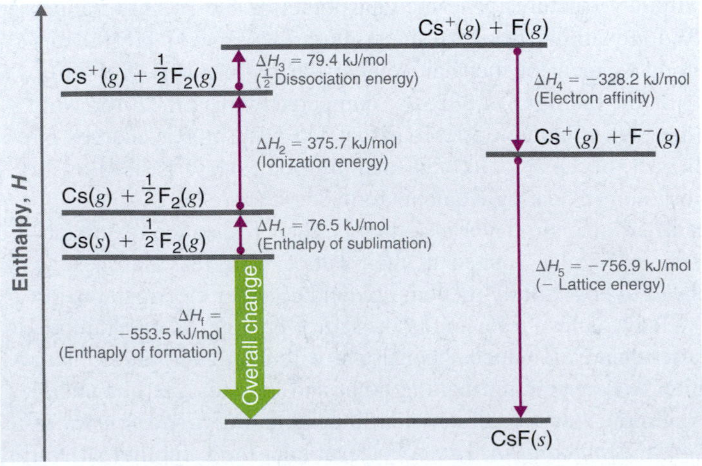

This equation describes the ionization of cesium, so the enthalpy change is the first ionization energy of cesium. Recall from Chapter 7 that energy is needed to ionize any neutral atom. Hence, regardless of the compound, the enthalpy change for this portion of the Born–Haber cycle is always positive.

Reaction 3: $\frac{1}{2}F_2(g) \longrightarrow F(g) \qquad \Delta H_3 = \frac{1}{2}D = 79.4 \text{ kJ/mol}$

This equation describes the dissociation of fluorine molecules into fluorine atoms, where D is the energy required for dissociation to occur (Table 8.3). Notice that we need to dissociate only $\frac{1}{2}$ mol of $F_2(g)$ molecules to obtain 1 mol of $F(g)$ atoms. The ΔH for this reaction, too, is always positive because energy is required to dissociate any stable diatomic molecule into the component atoms.

Reaction 4: $F(g) + e^- \longrightarrow F^-(g) \qquad \Delta H_4 = EA = -328.2 \text{ kJ/mol}$

This equation describes the formation of a gaseous fluoride ion from a fluorine atom; the enthalpy change is the electron affinity of fluorine. Recall from Chapter 7 that electron affinities can be positive, negative, or zero. In this case, ΔH is negative because of the highly negative electron affinity of fluorine.

Reaction 5: $Cs^+(g) + F^-(g) \longrightarrow CsF(s) \qquad \Delta H_5 = -U$

This equation describes the formation of the ionic solid from the gaseous ions. Because Reaction 5 is the reverse of the equation used to define lattice energy and U is defined to be a *positive* number, ΔH_5 is always *negative*.

If the enthalpy of formation of CsF from the elements is known ($\Delta H_f = -553.5$ kJ/mol at 298 K), then the thermochemical cycle shown in Figure 8.4 has only one unknown, the quantity $\Delta H_5 = -U$. From Hess's law, we can write

$$\Delta H_f = \Delta H_1 + \Delta H_2 + \Delta H_3 + \Delta H_4 + \Delta H_5 \qquad (8.7)$$

We can rearrange Equation 8.7 to give

$$-\Delta H_5 = \Delta H_1 + \Delta H_2 + \Delta H_3 + \Delta H_4 - \Delta H_f \qquad (8.8)$$

Substituting for the individual ΔH's, we obtain

$$U = \Delta H_{sub}(Cs) + I_1(Cs) + \frac{1}{2}D(F_2) + EA(F) - \Delta H_f(CsF) \qquad (8.9)$$

Substituting the appropriate values into Equation 8.9 gives

$$\begin{aligned} U &= 76.5 \text{ kJ/mol} + 375.7 \text{ kJ/mol} + 79.4 \text{ kJ/mol} \\ &\quad + (-328.2 \text{ kJ/mol}) - (-553.5 \text{ kJ/mol}) \\ &= 756.9 \text{ kJ/mol} \end{aligned} \qquad (8.10)$$

Notice that U is larger in magnitude than any of the other quantities in Equation 8.10. The process we have used to arrive at this value is summarized in Table 8.4.

TABLE 8.3 Selected bond dissociation enthalpies at 298 K

Substance	D, kJ/mol
$H_2(g)$	436.0
$N_2(g)$	945.3
$O_2(g)$	498.4
$F_2(g)$	158.8
$Cl_2(g)$	242.6
$Br_2(g)$	192.8
$I_2(g)$	151.1

Source: *CRC Handbook of Chemistry and Physics* (2004).

Predicting the Stability of Ionic Compounds

Equation 8.7 may be used as a tool for predicting which ionic compounds are likely to form from particular elements. As we have noted, ΔH_1 (ΔH_{sub}), ΔH_2 (I), and ΔH_3 (D) are always positive numbers, and ΔH_2 can be quite large. In contrast, ΔH_4 (EA) is comparatively small and can be positive, negative, or zero. Thus, the first three terms in Equation 8.7 make the formation of an ionic substance energetically unfavorable, and the fourth term contributes little either way. The formation of an ionic compound will be exothermic ($\Delta H_f < 0$) if and only if ΔH_5 ($-U$) is a large negative number. This means that *the lattice energy is the most important*

TABLE 8.4 Summary of reactions in the Born–Haber cycle for the formation of CsF(s)

Reaction	Enthalpy Change, kJ/mol
(1) $Cs(s) \longrightarrow Cs(g)$	$\Delta H_{sub} = 76.5$
(2) $Cs(g) \longrightarrow Cs^+(g) + e^-$	$I_1 = 375.7$
(3) $\frac{1}{2}F_2(g) \longrightarrow F(g)$	$\frac{1}{2}D = 79.4$
(4) $F(g) + e^- \longrightarrow F^-(g)$	$EA = -328.2$
(5) $Cs^+(g) + F^-(g) \longrightarrow CsF(s)$	$-U = -756.9$
$Cs(s) + \frac{1}{2}F_2(g) \longrightarrow CsF(s)$	$\Delta H_f = -553.5$

factor in determining the stability of an ionic compound. Another case in point is the formation of BaO:

$$Ba(s) + \frac{1}{2}O_2(g) \longrightarrow BaO(s) \qquad \Delta H = \Delta H_f \tag{8.11}$$

whose Born–Haber cycle is compared with that for the formation of CsF in Figure 8.5.

Reaction 1: $Ba(s) \longrightarrow Ba(g) \qquad \Delta H_1 = \Delta H_{sub} = 180.0$ kJ/mol

More than twice as much energy is required to sublime barium metal (180.0 kJ/mol) as is required to sublime cesium (76.5 kJ/mol).

Reaction 2: $Ba(g) \longrightarrow Ba^{2+}(g) + 2e^- \qquad \Delta H_2 = I_1 + I_2 = 1468.1$ kJ/mol

Nearly four times the energy is needed to form Ba^{2+} ions ($I_1 = 502.9$ kJ/mol, $I_2 = 965.2$ kJ/mol, $I_1 + I_2 = 1468.1$ kJ/mol) as Cs^+ ions ($I_1 = 375.7$ kJ/mol).

Reaction 3: $\frac{1}{2}O_2(g) \longrightarrow O(g) \qquad \Delta H_3 = \frac{1}{2}D = 249.2$ kJ/mol

Because the bond energy of $O_2(g)$ is 498.4 kJ/mol compared with 158.8 kJ/mol for $F_2(g)$, more than three times the energy is needed to form oxygen atoms from O_2 molecules as is required to form fluorine atoms from F_2.

Reaction 4: $O(g) + 2e^- \longrightarrow O^{2-}(g) \qquad \Delta H_4 = EA_1 + EA_2 = 603$ kJ/mol

Forming gaseous oxide (O^{2-}) ions is energetically unfavorable. Even though adding one electron to an oxygen atom is exothermic ($EA_1 = -141$ kJ/mol), adding a second electron to an $O^-(g)$ ion is energetically unfavorable ($EA_2 = +744$ kJ/mol)—so much so that the overall cost of forming $O^{2-}(g)$ from $O(g)$ is energetically prohibitive ($EA_1 + EA_2 = +603$ kJ/mol).

If the first four terms in the Born–Haber cycle are all substantially more positive for BaO than for CsF, why does BaO even form? The answer is the formation of the ionic solid from the gaseous ions (Reaction 5):

Reaction 5: $Ba^{2+}(g) + O^{2-}(g) \longrightarrow BaO(s) \qquad \Delta H_5 = -U$

Remember from Equation 8.4 that lattice energies are directly proportional to the product of the charges on the ions and inversely proportional to the internuclear distance. Although the internuclear distances are not significantly different for BaO and CsF (275 and 300 pm, respectively), the larger ionic charges in BaO produce a much

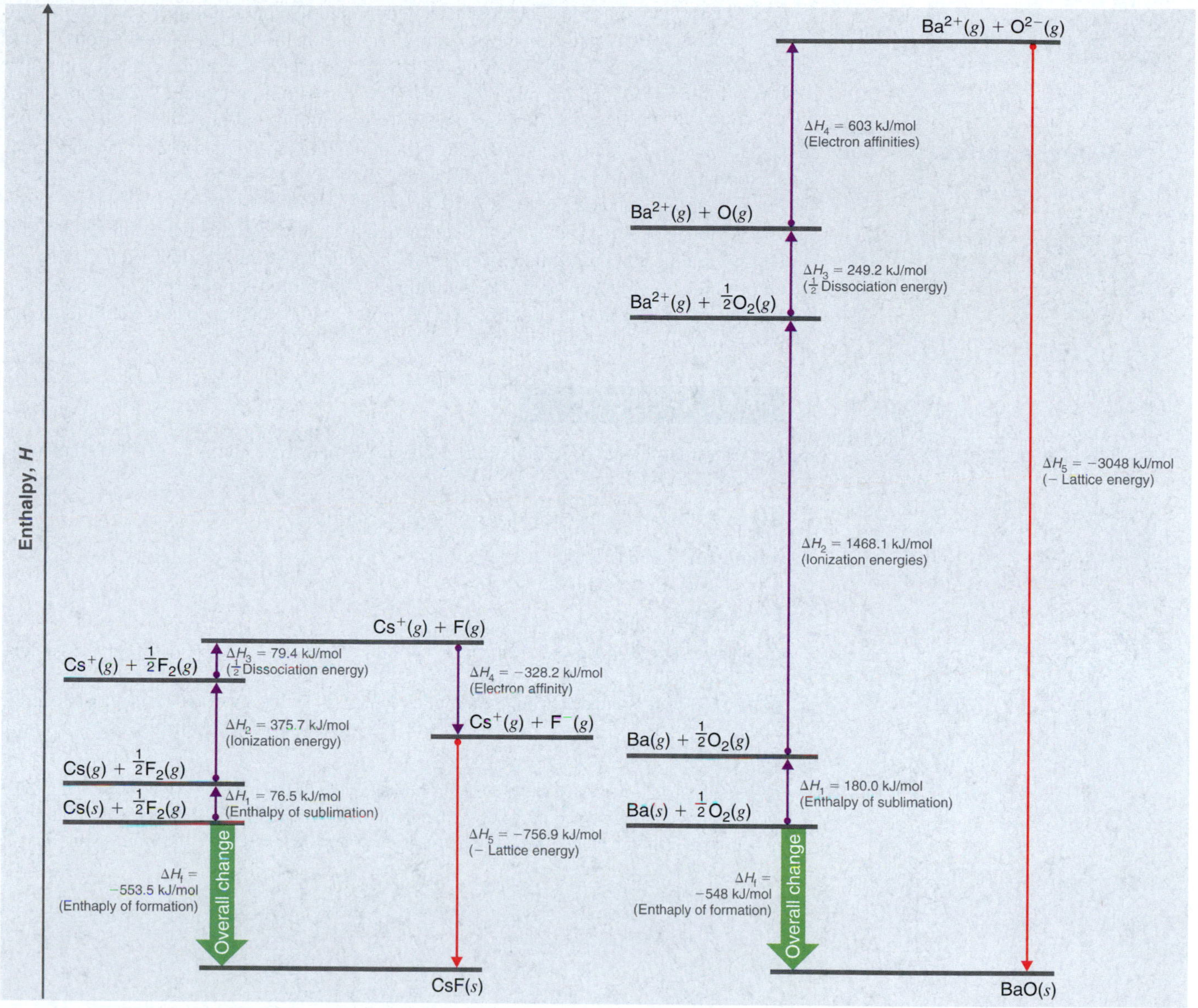

Figure 8.5 Comparison of the enthalpy changes involved in the formation of solid CsF and BaO from the elements. Note how the lattice energy of BaO, with a dipositive cation and a dinegative anion, dominates the Born–Haber cycle.

higher lattice energy. Substituting values for BaO ($\Delta H_f = -548.0$ kJ/mol) into Equation 8.9 gives

$$\begin{aligned} U &= \Delta H_{sub}(\text{Ba}) + [I_1(\text{Ba}) + I_2(\text{Ba})] + \frac{1}{2}D(\text{O}_2) \\ &\quad + [EA_1(\text{O}) + EA_2(\text{O})] - \Delta H_f(\text{BaO}) \\ &= 180.0 \text{ kJ/mol} + 1468.1 \text{ kJ/mol} + 249.2 \text{ kJ/mol} \\ &\quad + 603 \text{ kJ/mol} - (-548.0 \text{ kJ/mol}) \\ &= 3048 \text{ kJ/mol} \end{aligned}$$

Thus, U for BaO is slightly more than four times higher than U for CsF. The extra energy released when BaO forms from its ions more than compensates for the additional energy required to form $Ba^{2+}(g)$ and $O^{2-}(g)$ ions from Ba(s) and $\frac{1}{2}O_2(g)$.

If the formation of ionic lattices that contain multiply charged ions is so energetically favorable, why does CsF contain Cs^+ and F^- ions rather than Cs^{2+} and F^{2-} ions? If we assume that U for a $Cs^{2+}F^{2-}$ salt would be approximately the same as U for BaO, the formation of a lattice containing Cs^{2+} and F^{2-} ions would release 2291 kJ/mol (3048 kJ/mol − 756.9 kJ/mol) more energy than one containing Cs^+ and F^- ions. To form the Cs^{2+} ion from Cs^+, however, would require removing a $5p$ electron from a filled inner shell, which calls for a great deal of energy: I_2 = 2234.4 kJ/mol for Cs. Furthermore, forming an F^{2-} ion is expected to be even more energetically unfavorable than forming an O^{2-} ion. Not only is an electron being added to an already negatively charged ion, but because the F^- ion has a filled $2p$ subshell, the added electron would have to occupy an empty high-energy $3s$ orbital. Cesium fluoride, therefore, is not $Cs^{2+}F^{2-}$ because the energy cost of forming the doubly charged ions would be higher than the additional lattice energy that would be gained.

Note the pattern

The lattice energy is usually the most important energy factor in determining the stability of an ionic compound.

EXAMPLE 8.3

Use data from Figure 7.13, Tables 7.5, 8.2, and 8.3, and Appendix A to calculate the lattice energy of MgH_2.

Given Chemical compound, data from figures and tables

Asked for Lattice energy

Strategy

A Write a series of stepwise reactions for forming MgH_2 from the elements via the gaseous ions.

B Use Hess's law and data from the specified figures and tables to calculate the lattice energy.

Solution

A Hess's law allows us to use a thermochemical cycle (a Born–Haber cycle) to calculate the lattice energy for a given compound. We begin by writing reactions in which we form the component ions from the elements in a stepwise manner and then assemble the ionic solid:

(1)	$Mg(s) \longrightarrow Mg(g)$	$\Delta H_1 = \Delta H_{sub}(Mg)$
(2)	$Mg(g) \longrightarrow Mg^{2+}(g) + 2e^-$	$\Delta H_2 = I_1(Mg) + I_2(Mg)$
(3)	$H_2(g) \longrightarrow 2H(g)$	$\Delta H_3 = D(H_2)$
(4)	$2H(g) + 2e^- \longrightarrow 2H^-(g)$	$\Delta H_4 = 2EA(H)$
(5)	$Mg^{2+}(g) + 2H^-(g) \longrightarrow MgH_2(s)$	$\Delta H_5 = -U$
	$Mg(s) + H_2(g) \longrightarrow MgH_2(s)$	$\Delta H = \Delta H_f$

B Table 7.5 lists first and second ionization energies for the Period-3 elements [$I_1(Mg)$ = 737.7 kJ/mol, $I_2(Mg)$ = 1450.7 kJ/mol]. First electron affinities for all elements are given in Figure 7.13 [$EA(H)$ = −72.8 kJ/mol]. Table 8.2 lists selected enthalpies of sublimation [$\Delta H_{sub}(Mg)$ = 147.1 kJ/mol]. Table 8.3 lists selected bond dissociation energies [$D(H_2)$ = 436.0 kJ/mol]. Enthalpies of formation (ΔH_f = −75.3 kJ/mol for MgH_2) are listed in Appendix A. From Hess's law, ΔH_f is equal to the sum of the enthalpy changes for Reactions 1–5:

$$\begin{aligned}\Delta H_f &= \Delta H_1 + \Delta H_2 + \Delta H_3 + \Delta H_4 + \Delta H_5 \\ &= \Delta H_{sub}(Mg) + [I_1(Mg) + I_2(Mg)] + D(H_2) + 2EA(H) - U \\ -75.3\text{ kJ/mol} &= 147.1\text{ kJ/mol} + (737.7\text{ kJ/mol} + 1450.7\text{ kJ/mol}) + 436.0\text{ kJ/mol} \\ &\quad + 2(-72.8\text{ kJ/mol}) - U \\ U &= 2701.2\text{ kJ/mol}\end{aligned}$$

For MgH_2, U = 2701.2 kJ/mol. Once again, the lattice energy provides the driving force for forming this compound because $\Delta H_1, \Delta H_2, \Delta H_3 > 0$. When solving this type of problem, be sure to write the chemical equation for each step and double check that the enthalpy value used for each step has the correct sign *for the reaction in the direction it is written.*

EXERCISE 8.3

Use data from Figures 7.11 and 7.13, Tables 8.2 and 8.3, and Appendix A to calculate the lattice energy of Li_2O. Remember that the second electron affinity for oxygen $[O^-(g) + e^- \longrightarrow O^{2-}(g)]$ is *positive* (+744 kJ/mol; see Equation 7.13).

Answer 2809 kJ/mol

8.4 • Introduction to Lewis Electron Structures

At the beginning of the 20th century, the American chemist G. N. Lewis (1875–1946) devised a system of symbols—now called **Lewis dot symbols**—that can be used for predicting the number of bonds formed by most elements in their compounds (Figure 8.6). Each Lewis dot symbol consists of the chemical symbol for an element surrounded by dots that represent its valence electrons. Cesium, for example, has the electron configuration $[Xe]6s^1$, which indicates one valence electron outside a closed shell. In the Lewis dot symbol, this single electron is represented as a single dot: Cs·.

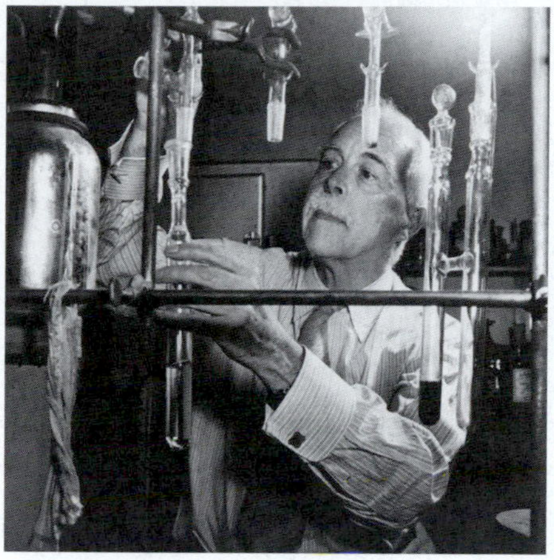

(a)

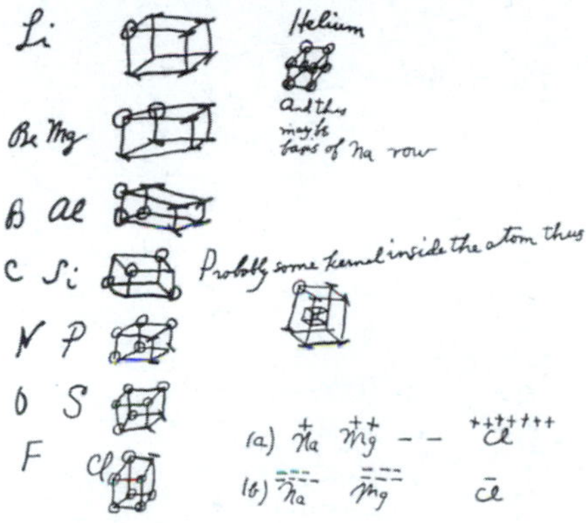

(b)

Figure 8.6 G. N. Lewis and the octet rule. (a) Lewis in the laboratory. (b) Lewis's original sketch for the octet rule. Note that Lewis initially placed the electrons at the corners of a cube rather than placing them as we do now.

MGC Introduction to Lewis Structures

Creating a Lewis Dot Symbol

To write an element's Lewis dot symbol, we place the dots representing its valence electrons, one at a time, around the element's chemical symbol. Up to four dots are placed above, below, to the left, and to the right of the symbol (in any order, as long as elements with four or fewer valence electrons have no more than one dot in each position). The next dots, for elements that have more than four valence electrons, are again distributed one at a time, each paired with one of the first four. Fluorine, for example, with the electron configuration $[He]2s^2 2p^5$, has seven valence electrons, so its Lewis dot symbol is constructed as follows:

Ḟ ·Ḟ ·Ḟ· ·F̈· ·F̈· :F̈· :F̤̈·

Notice that the number of dots in the Lewis dot symbol is the same as the number of valence electrons, which is the same as the last digit of the element's group number in the periodic table. Lewis dot symbols for the elements of Period 2 are given in Table 8.5.

Lewis used the unpaired dots to predict the number of bonds that an element will form in a compound. Consider the symbol for nitrogen shown in Table 8.5. The Lewis dot symbol explains why nitrogen, with three unpaired valence electrons, tends to form compounds in which it shares the unpaired electrons to form three bonds. Boron, which also has three unpaired valence electrons in its Lewis electron structure, also tends to form compounds with three bonds, whereas carbon, with four unpaired valence electrons in its Lewis electron structure, tends to share all of its unpaired valence electrons by forming compounds in which it has four bonds.

The Octet Rule

Lewis's major contribution to bonding theory was to recognize that atoms tend to lose, gain, or share electrons to reach a total of eight valence electrons, called an *octet*. This so-called **octet rule** explains the stoichiometry of most compounds in the

TABLE 8.5 Lewis dot symbols for the elements of Period 2

Group	1	2	13	14	15	16	17	18
Electron Configuration	$[He]2s^1$	$[He]2s^2$	$[He]2s^2 2p^1$	$[He]2s^2 2p^2$	$[He]2s^2 2p^3$	$[He]2s^2 2p^4$	$[He]2s^2 2p^5$	$[He]2s^2 2p^6$
Lewis Dot Symbol	Li·	·Be·	·Ḃ·	·Ċ· (dot below)	·N̈· (dot below)	:Ö· (dot below)	:F̈· (two dots below)	:N̈e: (two dots below)

s and *p* blocks of the periodic table. We now know from quantum mechanics that the number eight corresponds to one *ns* and three *np* valence orbitals, which together can accommodate a total of eight electrons. Remarkably, though, Lewis's insight was made nearly a decade before Rutherford proposed the nuclear model of the atom. An exception to the octet rule is helium, whose $1s^2$ electron configuration gives it a full $n = 1$ shell, and hydrogen, which tends to gain or share its one electron to achieve the electron configuration of helium.

Lewis dot symbols can also be used to represent the ions in ionic compounds. The reaction of cesium with a fluorine atom, for example, to produce the ionic compound CsF can be written as

$$\mathrm{Cs}\cdot + :\underset{\cdot\cdot}{\ddot{\mathrm{F}}}\cdot \longrightarrow \mathrm{Cs}^+[:\underset{\cdot\cdot}{\ddot{\mathrm{F}}}:]^- \qquad (8.12)$$

No dots are shown on Cs^+ in the product because cesium has lost its single valence electron to fluorine. The transfer of this electron produces the Cs^+ ion, which has the valence electron configuration of Xe, and the F^- ion, which has a total of eight valence electrons (an octet) and the Ne electron configuration. This description is consistent with the statement in Chapter 7 that among the main-group elements, ions in simple binary ionic compounds generally have the electron configurations of the nearest noble gas. The charge of each ion is written in the product, and the anion and its electrons are enclosed in brackets. This notation emphasizes that the ions are associated electrostatically; no electrons are shared between the two elements.

As you might expect for such a qualitative approach to bonding, there are exceptions to the octet rule, which we describe in Section 8.6. In the next section, however, we explain how to form molecular compounds by completing octets.

8.5 Lewis Structures and Covalent Bonding

MGC Lewis Structures

We begin our discussion of the relationship between structure and bonding in covalent compounds by describing the interaction between two identical neutral atoms—for example, the H_2 molecule, which contains a purely covalent bond. Each hydrogen atom in H_2 contains one electron and one proton, with the electron attracted to the proton by electrostatic forces. As the two hydrogen atoms are brought together, additional interactions must be considered (Figure 8.7):

The electrons in the two atoms repel each other because they have the same charge ($E > 0$).

Similarly, the protons in adjacent atoms repel each other ($E > 0$).

The electron in one atom is attracted to the oppositely charged proton in the other atom, and vice versa ($E < 0$).*

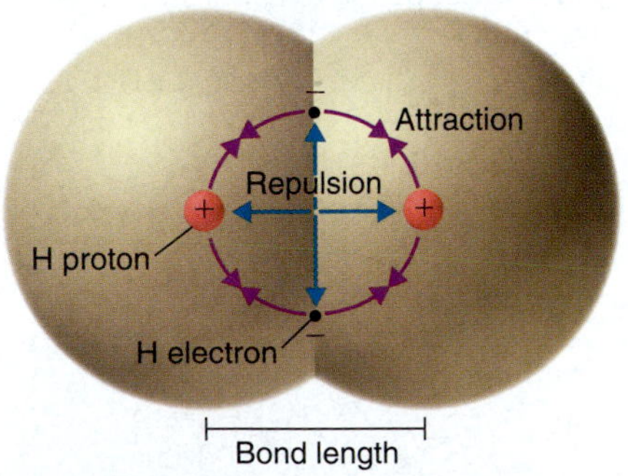

Figure 8.7 Attractive and repulsive interactions between electrons and nuclei in the hydrogen molecule. Electron–electron and proton–proton interactions are repulsive; electron–proton interactions are attractive. At the observed bond distance, the repulsive and attractive interactions are balanced.

A plot of the potential energy of the system as a function of the internuclear distance (Figure 8.8) shows that the system becomes more stable (the energy of the system decreases) as two hydrogen atoms move toward each other from $r = \infty$, until the energy reaches a minimum at $r = r_0$ (the observed internuclear distance in H_2, 74 pm). Thus, at intermediate distances, proton–electron attractive interactions dominate, but as the distance becomes very short, electron–electron and proton–proton repulsive interactions cause the energy of the system to increase rapidly. Notice the similarity between Figure 8.8 and Figure 8.1, which described a system containing two oppositely charged *ions*. The shapes of the energy versus distance curves in the two figures are similar because they both result from attractive and repulsive forces between charged entities.

---

* Recall from Chapter 6 that it is impossible to specify precisely the position of the electron in either hydrogen atom. Hence, the quantum mechanical probability distributions must be used.

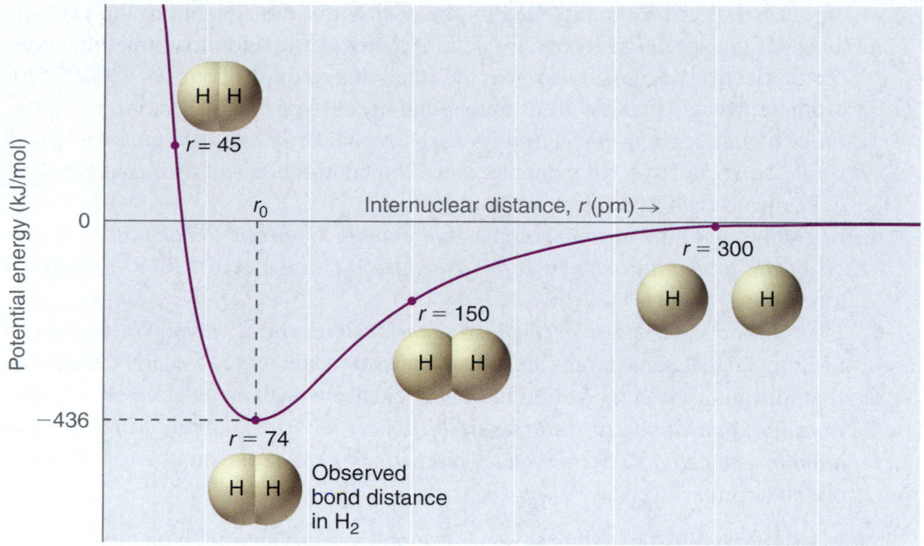

Figure 8.8 A plot of potential energy vs. internuclear distance for the interaction between two gaseous hydrogen atoms. At long distances, both attractive and repulsive interactions are small. As the distance between the atoms decreases, the attractive electron–proton interactions dominate and the energy of the system decreases. At the observed bond distance, the repulsive electron–electron and proton–proton interactions just balance the attractive interactions, preventing a further decrease in the internuclear distance. At very short internuclear distances, the repulsive interactions dominate, making the system less stable than the isolated atoms.

Interactive Graph

Using Lewis Electron Structures to Describe Covalent Bonding

The valence-electron configurations of the constituent atoms of a covalent compound are important factors in determining its structure, stoichiometry, and properties. For example, chlorine, with seven valence electrons, is one electron short of an octet. If two chlorine atoms share their unpaired electrons by making a covalent bond and forming Cl_2, they can each complete their valence shell:

$$:\ddot{\underset{..}{Cl}}\cdot + \cdot\ddot{\underset{..}{Cl}}: \longrightarrow :\ddot{\underset{..}{Cl}}:\ddot{\underset{..}{Cl}}: \tag{8.13}$$

Each chlorine atom now has an octet. The electron pair being shared by the atoms is called a **bonding pair**; the other three pairs of electrons on each chlorine atom are called **lone pairs**. Lone pairs are not involved in covalent bonding. If both electrons in a covalent bond come from the same atom, the bond is called a **coordinate covalent bond**. Examples of this type of bonding are presented in Section 8.6, when we discuss atoms that contain fewer than an octet of electrons.

We can illustrate the formation of a water molecule from two hydrogen atoms and an oxygen atom using Lewis dot symbols:

$$H\cdot + \cdot\ddot{\underset{..}{O}}\cdot + \cdot H \longrightarrow H:\ddot{\underset{..}{O}}:H \tag{8.14}$$

The structure on the right is the *Lewis electron structure*, or *Lewis structure*, for H_2O. With two bonding pairs and two lone pairs, the oxygen atom has now completed its octet. Moreover, by sharing a bonding pair with oxygen, each hydrogen atom now has a full valence shell of two electrons. Chemists usually indicate a bonding pair by a single line, as shown below for our two examples:

$$:\ddot{\underset{..}{Cl}}-\ddot{\underset{..}{Cl}}: \qquad H-\ddot{\underset{..}{O}}-H$$

The following procedure can be used to construct Lewis electron structures for more complex molecules and ions:

1. *Arrange the atoms to show specific connections*. When there is a central atom, it is usually the least electronegative element in the compound. Chemists usually list this central atom first in the chemical formula (as in CCl_4 and CO_3^{2-}, both with C as the central atom), which is another clue to the compound's structure. Hydrogen and the halogens are almost always connected to only one other atom, so they are usually *terminal* rather than central.
2. *Determine the total number of valence electrons in the molecule or ion*. Add up the valence electrons from each atom. (Recall from Chapter 6 that the number

of valence electrons is indicated by the position of the element in the periodic table.) If the species is a polyatomic ion, remember to add or subtract the number of electrons necessary to give the total charge on the ion. For CO_3^{2-}, for example, we add two electrons to the total because of the -2 charge.

3. *Place a bonding pair of electrons between each pair of adjacent atoms to give a single bond*. In H_2O, for example, there is a bonding pair of electrons between oxygen and each hydrogen.
4. Beginning with the terminal atoms, *add enough electrons to each atom to give all of the atoms an octet (two for hydrogen)*. These electrons will usually be lone pairs.
5. *If any electrons are left over, place them on the central atom*. We explain in Section 8.6 that some atoms are able to accommodate more than eight electrons.
6. If at this point the central atom has fewer electrons than an octet, *use lone pairs from terminal atoms to form multiple (double or triple) bonds to the central atom in order to achieve an octet*. This will not change the number of electrons on the terminal atoms.

Now let's apply this procedure to some examples, beginning with one we have already discussed.

H_2O

1. Because H atoms are almost always terminal, the arrangement within the molecule must be HOH.
2. Each H atom (Group 1) has one valence electron, and the O atom (Group 16) has six, for a total of eight valence electrons.
3. Placing one bonding pair of electrons between the O atom and each H atom gives H:O:H, with four electrons left over.
5. Adding the remaining four electrons to the oxygen (as two lone pairs) gives

$$\mathrm{H}:\ddot{\underset{\cdot\cdot}{\mathrm{O}}}:\mathrm{H}$$

This is the Lewis structure we drew earlier. Because it gives oxygen an octet and each hydrogen two electrons, we do not need to use step 6.

OCl^-

1. With only two atoms in the molecule, there is no central atom.
2. Oxygen (Group 16) has six valence electrons, and chlorine (Group 17) has seven; we must add one more for the negative charge on the ion, giving a total of 14 valence electrons.
3. Placing a bonding pair of electrons between O and Cl gives O:Cl, with 12 electrons left over.
4. If we place six electrons (as three lone pairs) on each atom, we obtain

$$\left[:\ddot{\underset{\cdot\cdot}{\mathrm{O}}}-\ddot{\underset{\cdot\cdot}{\mathrm{Cl}}}:\right]^-$$

Each atom now has an octet of electrons, so steps 5 and 6 are not needed. Notice that the Lewis electron structure is drawn within brackets as is customary for an ion, with the overall charge indicated outside of the brackets, and that the bonding pair of electrons is indicated by a solid line. This species, OCl^-, is the hypochlorite ion, the active ingredient in chlorine laundry bleach and swimming pool disinfectant.

CH_2O

1. Because carbon is less electronegative than oxygen and hydrogen is normally terminal, C must be the central atom. One possible arrangement is

$$\begin{array}{c}\mathrm{O}\\ \mathrm{H\;C\;H}\end{array}$$

2. Each hydrogen atom (Group 1) has one valence electron, carbon (Group 14) has four, and oxygen (Group 16) has six, for a total of [6 + 4 + (2)(1)] = 12 electrons.
3. Placing a bonding pair of electrons between each pair of bonded atoms gives

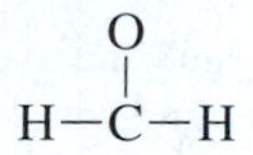

Six electrons are used and six are left over.

4. Adding all six remaining electrons to oxygen (as three lone pairs) gives

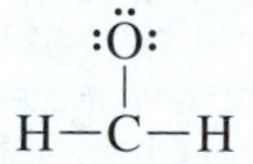

Although oxygen now has an octet and each hydrogen has two electrons, carbon has only six electrons.

5. There are no electrons left to place on the central atom.
6. To give carbon an octet of electrons, we use one of the lone pairs of electrons on oxygen to form a carbon–oxygen double bond:

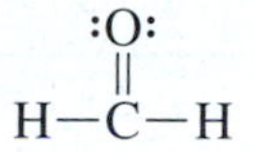

Both the oxygen and the carbon now have an octet of electrons, so this is an acceptable Lewis electron structure. Note that O has two bonding pairs and two lone pairs, and C has four bonding pairs. This is the structure of formaldehyde, which is used in embalming fluid.

An alternative structure can be drawn with one H bonded to O. *Formal charges*, discussed later in this section, suggest that such a structure is less stable than that shown above.

EXAMPLE 8.4

Write Lewis electron structures for these species: **(a)** NCl_3; **(b)** S_2^{2-}; **(c)** NOCl.

Given Chemical species

Asked for Lewis electron structures

Strategy

Use the procedure given above to write the Lewis electron structure for each compound.

Solution

(a) Nitrogen is less electronegative than chlorine, and halogen atoms are usually terminal, so nitrogen is the central atom. The nitrogen atom (Group 15) has five valence electrons and each chlorine atom (Group 17) has seven, for a total of 26. Using two electrons for each N—Cl bond and adding three lone pairs to each Cl account for $(3 \times 2) + (3 \times 2 \times 3) = 24$ electrons. Rule 5 leads us to place the remaining two electrons on the central N:

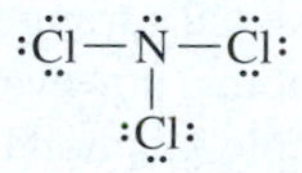

Nitrogen trichloride

Nitrogen trichloride is an unstable oily liquid once used to bleach flour; this use is now prohibited in the United States.

(b) In a diatomic molecule or ion, we do not need to worry about a central atom. Each sulfur atom (Group 16) contains six valence electrons, and we need to add two electrons for the −2 charge, giving a total of 14. Using two electrons for the S—S bond, we arrange the remaining 12 electrons as three lone pairs on each sulfur, giving each S atom an octet of electrons:

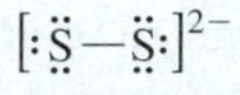

(c) Because nitrogen is less electronegative than oxygen or chlorine, it is the central atom. The N atom (Group 15) has five valence electrons, the O atom (Group 16) has six, and the Cl

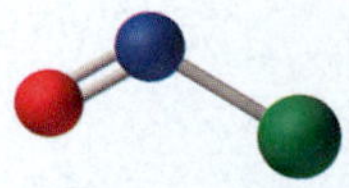

Nitrosyl chloride

atom (Group 17) has seven, giving a total of 18 valence electrons. Placing one bonding pair of electrons between each pair of bonded atoms uses four electrons and gives

O—N—Cl

Adding three lone pairs each to oxygen and to chlorine uses 12 more electrons, leaving two electrons to place as a lone pair on nitrogen:

:Ö—N—C̈l:

Because this Lewis structure has only six electrons around the central nitrogen, a lone pair of electrons on a terminal atom must be used to form a bonding pair. We could use a lone pair on either O or Cl. Because we have seen many structures in which O forms a double bond but none with a double bond to Cl, we select a lone pair from O to give

Ö=N̈—C̈l:

All atoms now have octet configurations. This is the Lewis electron structure of nitrosyl chloride, a highly corrosive reddish-orange gas.

EXERCISE 8.4

Write Lewis electron structures for CO_2 and SCl_2, a vile-smelling, unstable red liquid that is used in the manufacture of rubber.

Answer **(a)** Ö=C=Ö **(b)** :C̈l—S̈—C̈l:

Carbon dioxide

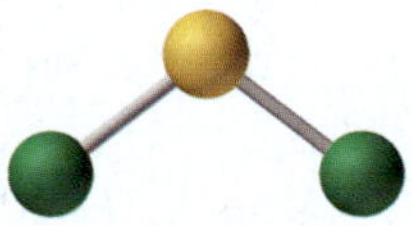

Sulfur dichloride

Using Lewis Electron Structures to Explain Stoichiometry

Lewis dot symbols provide a simple rationalization of why elements form compounds with the observed stoichiometries. In the Lewis model, the number of bonds formed by an element in a neutral compound is the same as the number of unpaired electrons it must share with other atoms to complete its octet of electrons. For the elements of Group 17 (the halogens), this number is one; for the elements of Group 16 (the chalcogens), it is two; for Group 15, three; and for Group 14, four. These requirements are illustrated by the Lewis structures shown in the margin for the hydrides of the lightest members of each group.

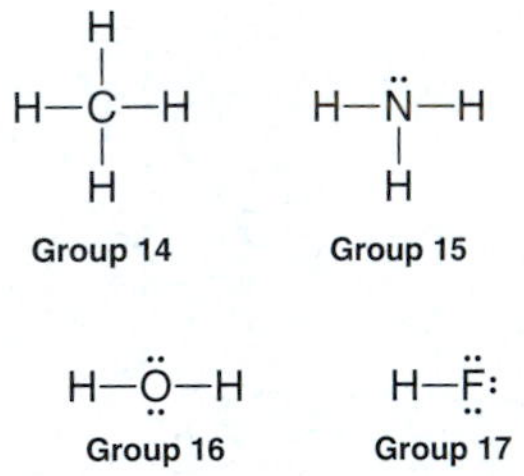

Number of bonds formed

Elements may form multiple bonds to complete an octet. In ethylene, for example, whose Lewis electron structure is shown in the margin, each carbon contributes two electrons to the double bond, giving each carbon an octet (2 electrons/bond × 4 bonds = 8 electrons). Neutral structures with fewer or more bonds than those indicated above exist, but they are unusual and violate the octet rule.

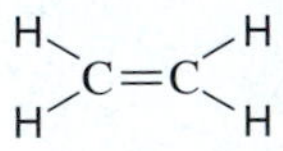

Ethylene

Allotropes of an element can have very different physical and chemical properties because of different three-dimensional arrangements of the atoms; the number of bonds formed by the component atoms, however, is always the same. As noted at the beginning of the chapter, for example, diamond is a hard, transparent solid; graphite is a soft, black solid, and fullerenes have open cage structures. Despite these differences, the carbon atoms in all three allotropes form four bonds, in accordance with the octet rule. Elemental phosphorus also exists in three forms: white phosphorus, a toxic, waxy substance that initially glows and then spontaneously ignites on contact with air; red phosphorus, an amorphous substance that is used commercially in safety matches, fireworks, and smoke bombs; and black phosphorus, an unreactive crystalline solid with a texture similar to graphite (Figure 8.9). Nonetheless, the phosphorus atoms in all three forms obey the octet rule and form three bonds per phosphorus atom.

Note the pattern

Lewis structures explain why the elements of Groups 14–17 form neutral compounds with four, three, two, and one bonded atom(s), respectively.

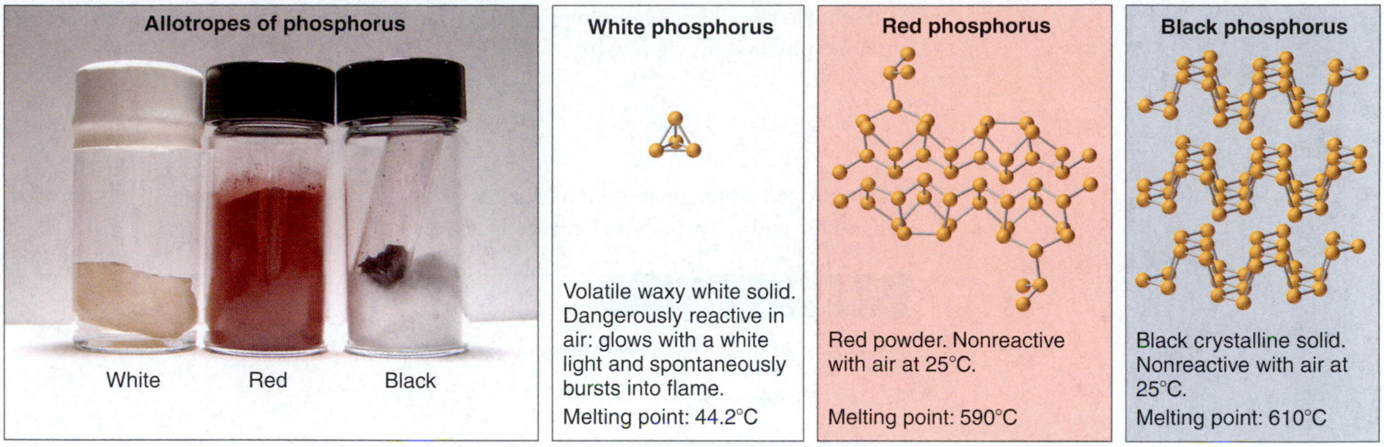

Figure 8.9 The three allotropes of phosphorus: white, red, and black. All three forms contain only phosphorus atoms, but they differ in the arrangement and connectivity of their atoms. White phosphorus contains P_4 tetrahedra, red phosphorus is a network of linked P_8 and P_9 units, and black phosphorus forms sheets of six-membered rings. As a result, their physical and chemical properties differ dramatically.

Formal Charges

It is sometimes possible to write more than one Lewis structure for a substance that does not violate the octet rule, although not all of the Lewis structures may be equally reasonable. In these situations, we can choose the most stable Lewis structure by considering the **formal charge** on the atoms, which is the difference between the number of valence electrons in the free atom and the number assigned to it in the Lewis electron structure. The formal charge is a way of computing the charge distribution within a Lewis structure; the sum of the formal charges on the atoms within a molecule or an ion must equal the overall charge on the molecule or ion. A formal charge does *not* represent a true charge on an atom in a covalent bond, but is simply used to predict the most likely structure when a compound has more than one valid Lewis structure.

To calculate formal charges, we assign electrons in the molecule to individual atoms according to these rules:

Nonbonding electrons are assigned to the atom on which they are located.

Bonding electrons are divided equally between the bonded atoms.

For each atom, we then compute a formal charge:

$$\text{formal charge} = \underbrace{\text{valence e}^-}_{\text{Free atom}} - \underbrace{\left(\text{nonbonding e}^- + \frac{\text{bonding e}^-}{2}\right)}_{\text{Atom in Lewis structure}} \qquad (8.15)$$

To illustrate this method, let's calculate the formal charge on the atoms in ammonia, NH_3, whose Lewis electron structure is shown in the margin.

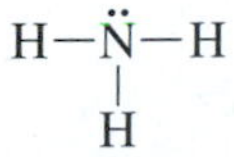

Ammonia

A neutral nitrogen atom contains five valence electrons because it is in Group 15. From its Lewis electron structure, the nitrogen atom in ammonia has one lone pair and shares three bonding pairs with hydrogen atoms, so nitrogen is assigned a total of five electrons [2 nonbonding e^- + (6 bonding e^-/2)]. Substituting into Equation 8.15, we obtain

$$\text{formal charge (N)} = 5 \text{ valence e}^- - \left(2 \text{ nonbonding e}^- + \frac{6 \text{ bonding e}^-}{2}\right) = 0 \quad (8.16)$$

A neutral hydrogen atom has one valence electron. Each hydrogen atom in the molecule shares one pair of bonding electrons and is therefore assigned one electron

[0 nonbonding e$^-$ + (2 bonding e$^-$/2)]. Using Equation 8.15 to calculate the formal charge on hydrogen, we obtain

$$\text{formal charge (H)} = 1 \text{ valence e}^- - \left(0 \text{ nonbonding e}^- + \frac{2 \text{ bonding e}^-}{2}\right) = 0 \quad (8.17)$$

The hydrogen atoms in ammonia have the same number of electrons as neutral hydrogen atoms, and so their formal charge is also zero.

EXAMPLE 8.5

Calculate the formal charges on all atoms in the NH_4^+ ion.

Given Chemical species

Asked for Formal charges

Strategy

Identify the number of valence electrons in each atom in the NH_4^+ ion. Use the Lewis electron structure of NH_4^+ to identify the number of bonding and nonbonding electrons associated with each atom, and then use Equation 8.15 to calculate the formal charge on each atom.

Solution

The Lewis electron structure for the NH_4^+ ion is

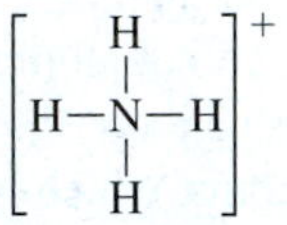

The nitrogen atom shares four bonding pairs of electrons, and a neutral nitrogen atom has five valence electrons. Using Equation 8.15, the formal charge on the nitrogen atom is therefore

$$\text{formal charge (N)} = 5 - \left(0 + \frac{8}{2}\right) = +1$$

Each hydrogen atom in NH_4^+ has one bonding pair. The formal charge on each hydrogen atom is therefore

$$\text{formal charge (H)} = 1 - \left(0 + \frac{2}{2}\right) = 0$$

The formal charges on the atoms in the NH_4^+ ion are thus

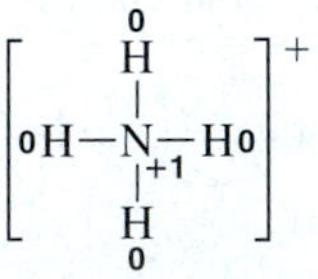

Summing the formal charges on the atoms should give us the total charge on the molecule or ion. In this case, the sum of the formal charges is 0 + 1 + 0 + 0 + 0 = +1.

EXERCISE 8.5

Write the formal charges on all atoms in BH_4^-.

Answer

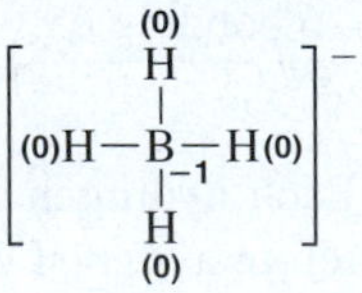

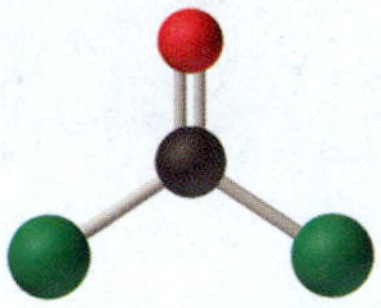

Carbonyl chloride

Notice that if an atom in a molecule or ion has the number of bonds that is typical for that atom (four bonds for carbon, for example), its formal charge is zero.

Using Formal Charges to Distinguish Between Lewis Structures

As an example of how formal charges can be used to determine the most stable Lewis structure for a substance, let's compare two possible structures for carbon dioxide. Both structures conform to the rules for Lewis electron structures.

CO_2

1. C is less electronegative than O, so it is the central atom.
2. C has four valence electrons and each O has six, for a total of 16 valence electrons.
3. Placing one electron pair between the C and each O gives O—C—O, with 12 electrons left over.
4. Dividing the remaining electrons between the O atoms gives three lone pairs on each atom:

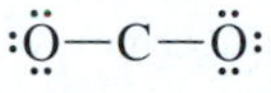

This structure has an octet of electrons around each O atom but only four electrons around the C atom.

5. No electrons are left for the central atom.
6. To give the carbon atom an octet of electrons, we can convert two of the lone pairs on the oxygens to bonding electron pairs. There are, however, two ways to do this. We can either take one electron pair from each oxygen to form a symmetrical structure or take both electron pairs from a single oxygen atom to give an asymmetrical structure:

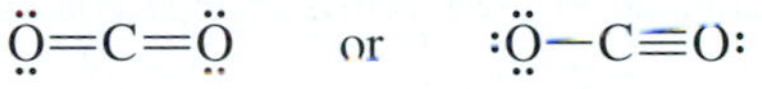

Both of these Lewis electron structures give all three atoms an octet. How do we decide between these two possibilities? The structure with the lower formal charges on the atoms is likely to be the more stable. Moreover, in cases where there are positive or negative formal charges on various atoms, stable structures generally have negative formal charges on the more electronegative atoms, and positive formal charges on the less electronegative atoms. The formal charges for the two Lewis electron structures of CO_2 are

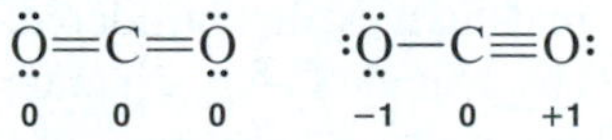

Both Lewis structures have a net formal charge of zero, but the structure on the right has a +1 charge on the more electronegative atom (O). Thus, the symmetrical Lewis structure on the left is predicted to be more stable, and it is, in fact, the structure observed experimentally. Remember, though, that formal charges do *not* represent the actual charges on atoms in a molecule or ion. They are used simply as a bookkeeping method for predicting the most stable Lewis structure for a compound.

Note the pattern

The Lewis structure that gives the lowest formal charges is usually the most stable.

EXAMPLE 8.6

The thiocyanate ion, SCN^-, used in printing and as a corrosion inhibitor against acidic gases, has at least two possible Lewis electron structures. Draw two possible structures, assign formal charges on all atoms in both, and decide which is the preferred arrangement of electrons.

Given Chemical species

Asked for Lewis electron structures, formal charges, preferred arrangement

Strategy

A Use the procedure given above to write two plausible Lewis electron structures for the thiocyanate ion.

B Calculate the formal charge on each atom using Equation 8.15.
C Predict which structure is preferred based on the formal charge on each atom and its electronegativity relative to the other atoms present.

Solution

A Possible Lewis structures for the SCN^- ion are

$[:\ddot{\underset{..}{S}}-C\equiv N:]^-$ (a) $[\ddot{\underset{..}{S}}=C=\ddot{\underset{..}{N}}]^-$ (b) $[:S\equiv C-\ddot{\underset{..}{N}}:]^-$ (c)

B We must calculate the formal charges on each atom in order to identify the more stable structure. If we begin with carbon, we notice that the carbon atom in each of these structures shares four bonding pairs, the number of bonds typical for carbon, so it has a formal charge of zero. Continuing with sulfur, we observe that in (a) the sulfur atom shares one bonding pair and has three lone pairs, and has a total of six valence electrons. The formal charge on the sulfur atom is therefore $6 - \left(6 + \frac{2}{2}\right) = -1$. In (b), the sulfur atom has two bonding pairs and two lone pairs, giving it a formal charge of zero. In (c), sulfur has a formal charge of +1. Completing our calculations with nitrogen, in (a) the nitrogen atom has three bonding pairs, giving it a formal charge of zero. In (b), the nitrogen atom has two lone pairs and shares two bonding pairs, giving it a formal charge of $5 - \left(4 + \frac{4}{2}\right) = -1$. In (c), nitrogen has a formal charge of −2. **C** Which structure is preferred? Structure (b) because the negative charge is on the more electronegative atom, N, and it has lower formal charges on each atom as compared to structure (c) (0, −1 versus +1, −2).

EXERCISE 8.6

Salts containing the fulminate ion, CNO^-, are used in explosive detonators. Draw two Lewis electron structures for fulminate, and use formal charges to predict which is more stable. (Note: N is the central atom.)

Answer $[:\underset{..}{C}=N=\underset{..}{O}:]^-$ (−2 +1 0) or $[:C\equiv N-\ddot{\underset{..}{O}}:]^-$ (−1 +1 −1) or $[:\ddot{\underset{..}{C}}-N\equiv O:]^-$ (−3 +1 +1); the second structure

Resonance Structures

Sometimes, even when formal charges are considered, the bonding in some molecules or ions cannot be described by a single Lewis structure. Such is the case for ozone, O_3, an allotrope of oxygen with a V-shaped structure and an O—O—O angle of 117.5°.

O_3

1. We know that ozone is V-shaped, so one O atom is "central":

O
O O

2. Each O atom has six valence electrons, for a total of 18.
3. Assigning one bonding pair of electrons to each oxygen–oxygen bond gives

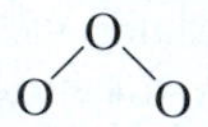

with 14 electrons left over.
4. If we place three lone pairs of electrons on each terminal oxygen, we obtain

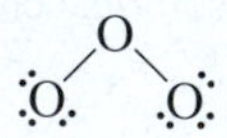

and we have two electrons left over.

5. At this point, both terminal oxygen atoms have octets of electrons. We therefore place the last two electrons on the central atom:

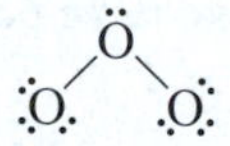

6. Note that the central oxygen has only six electrons. We must convert one lone pair on a terminal oxygen atom to a bonding pair of electrons, but which one? Depending on which one we choose, we obtain either

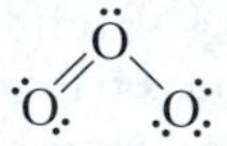

or

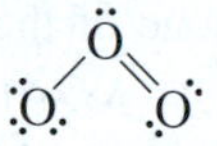

Which is correct? In fact, neither. Both predict one O—O single bond and one O═O double bond. As you will learn in Section 8.8, if the bonds were of different types (one single and one double, for example), they would have different lengths. It turns out, however, that both O—O bond distances are identical, 127.2 pm, which is shorter than a typical O—O single bond (148 pm) and longer than the O═O double bond in O_2 (120.8 pm).

Equivalent Lewis dot structures, such as those of ozone, are called **resonance structures**. The position of the *atoms* is the same in the various resonance structures of a compound, but the position of the *electrons* is different. The different resonance structures of a compound are linked by double-headed arrows:

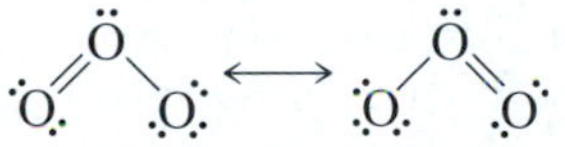

The double-headed arrow indicates that the actual electronic structure is an average of those shown, not that the molecule oscillates between the two structures.

Note the pattern

When it is possible to write more than one equivalent resonance structure for a molecule or ion, the actual structure is the average of the resonance structures.

CO_3^{2-}

Like ozone, the electronic structure of the carbonate ion cannot be described by a single Lewis electron structure. Unlike O_3, though, the actual structure of CO_3^{2-} is an average of *three* resonance structures.

1. Because carbon is the least electronegative element, we place it in the central position:

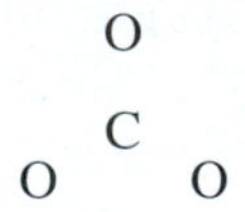

2. Carbon has four valence electrons, each oxygen has six, and there are two more for the −2 charge. This gives $4 + (3 \times 6) + 2 = 24$ valence electrons.
3. Six electrons are used to form 3 bonding pairs between the oxygens and the carbon:

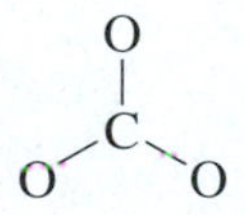

4. We divide the remaining 18 electrons equally among the three oxygen atoms by placing three lone pairs on each:

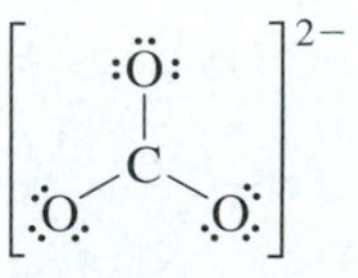

5. No electrons are left for the central atom.

6. At this point, the carbon atom has only six valence electrons, so we must take one lone pair from an oxygen and use it to form a carbon–oxygen double bond. In this case, however, there are *three* possible choices:

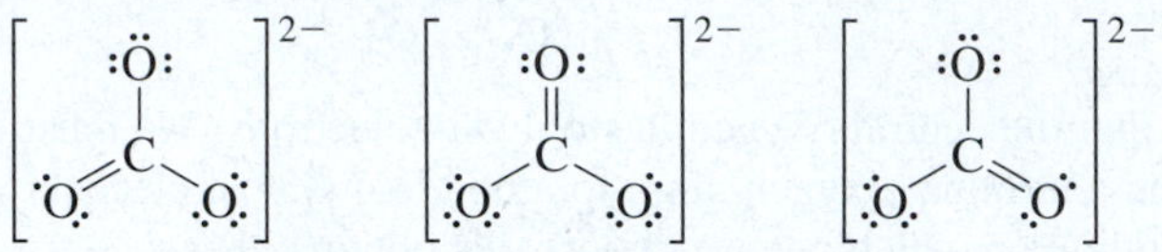

As with ozone, none of these structures describes the bonding exactly. Each predicts one carbon–oxygen double bond and two carbon–oxygen single bonds, but experimentally all C—O bond lengths are identical. We can write resonance structures (in this case, three of them) for the carbonate ion:

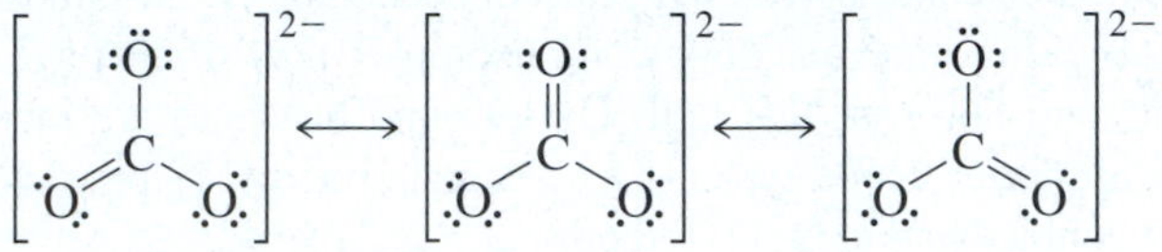

The actual structure is an average of these three resonance structures.

EXAMPLE 8.7

Benzene is a common organic solvent that was previously used in gasoline; it is no longer used for this purpose, however, because it is now known to be a carcinogen. The benzene molecule, C_6H_6, consists of a regular hexagon of carbon atoms, each of which is also bonded to a hydrogen atom. Use resonance structures to describe the bonding in benzene.

Given Molecular formula, molecular geometry

Asked for Resonance structures

Strategy

(A) Draw a structure for benzene illustrating the bonded atoms. Then calculate the number of valence electrons used in this drawing.

(B) Subtract this number from the total number of valence electrons in benzene, and then locate the remaining electrons such that each atom in the structure reaches an octet.

(C) Draw the resonance structures for benzene.

Solution

(A) Each hydrogen atom contributes one valence electron and each carbon atom contributes four, for a total of $(6 \times 1) + (6 \times 4) = 30$. If we place a single bonding electron pair between each pair of carbon atoms and between each carbon and a hydrogen atom, we obtain

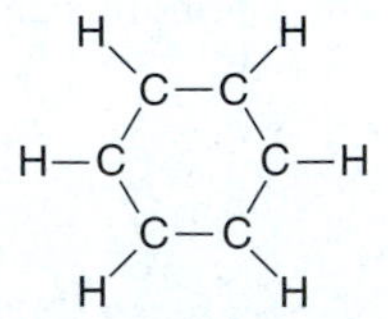

Each carbon atom in this structure has only six electrons and has a formal charge of +1, but we have used only 24 of the 30 valence electrons. (B) If the difference, six electrons, are uniformly distributed pairwise on alternate carbon atoms, we obtain

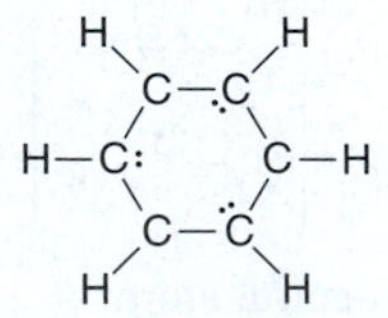

Three carbon atoms now have an octet configuration and a formal charge of −1, while three carbon atoms have only six electrons and a formal charge of +1. We can convert each lone pair to a bonding electron pair, which gives each atom an octet of electrons and a formal charge of 0, by making three C═C double bonds. There are, however, two ways to do this:

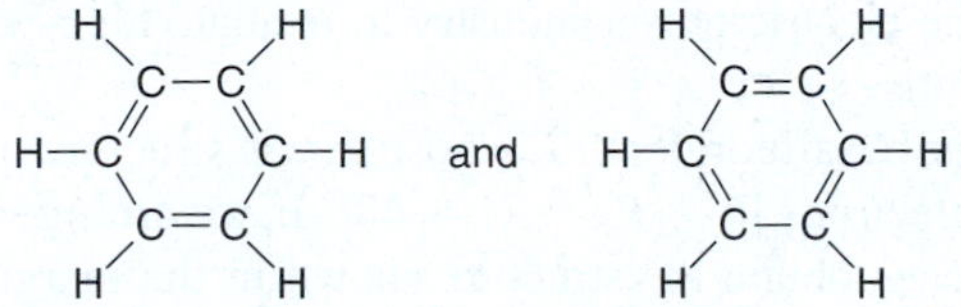

Each of these structures has alternating double and single bonds, but experiment shows that all of the carbon–carbon bonds in benzene are identical, with bond lengths (139.9 pm) intermediate between those typically found for a C—C single bond (154 pm) and a C═C double bond (134 pm). We can describe the bonding in benzene using the two resonance structures, but the actual electronic structure is an average of the two. The existence of multiple resonance structures for aromatic hydrocarbons like benzene is often indicated by drawing either a circle or dashed lines inside the hexagon as shown in the margin.

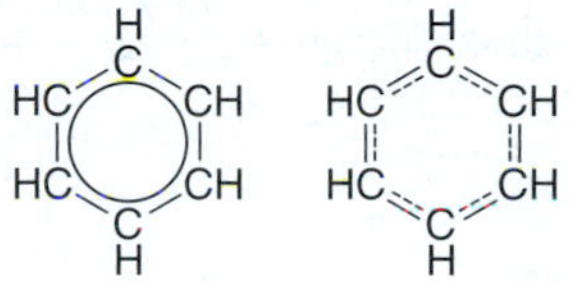

Benzene

EXERCISE 8.7

The sodium salt of nitrite is used to relieve muscle spasms. Draw two resonance structures for the nitrite ion, NO_2^-.

Answer $[:\ddot{O}-\ddot{N}=\ddot{O}]^- \longleftrightarrow [\ddot{O}=\ddot{N}-\ddot{O}:]^-$

Resonance structures are particularly common in oxoanions of the *p*-block elements, such as sulfate and phosphate, and in aromatic hydrocarbons, such as benzene and naphthalene.

8.6 ● Exceptions to the Octet Rule

MGC Exceptions to the Octet Rule

Lewis dot structures provide a simple model for rationalizing the bonding in most known compounds. However, there are three general exceptions to the octet rule: (1) molecules, such as NO, that have an *odd* number of electrons; (2) molecules in which one or more atoms possess *more* than eight electrons, such as SF_6; and (3) molecules such as BCl_3, in which one or more atoms possess *fewer* than eight electrons.

Odd Number of Electrons

Because most molecules or ions that consist of *s*- and *p*-block elements contain even numbers of electrons, their bonding can be described using a model that assigns every electron to either a bonding pair or a lone pair.* There are, however, a few molecules containing only *p*-block elements that have an odd number of electrons. Some important examples are nitric oxide, NO, whose biochemical importance was described in earlier chapters; nitrogen dioxide, NO_2, an oxidizing agent in rocket propulsion; and chlorine dioxide, ClO_2, used in water purification plants. Consider NO, for example: with 5 + 6 = 11 valence electrons, there is no way to draw a Lewis structure that gives each atom an octet of electrons. Molecules such as NO, NO_2, and ClO_2 require a more sophisticated treatment of bonding, which will be developed in Chapter 9.

* Molecules or ions that contain *d*-block elements frequently contain an odd number of electrons, and their bonding cannot adequately be described using the simple approach we have developed so far. Bonding in these compounds will be discussed in Chapter 23.

More Than an Octet of Electrons

The most common exception to the octet rule is a molecule or ion with at least one atom that possesses more than an octet of electrons. Such compounds are found for elements of Period 3 and beyond. Examples from the *p*-block elements include SF_6, a substance used by the electric power industry to insulate high-voltage lines, and the SO_4^{2-} and PO_4^{3-} ions.

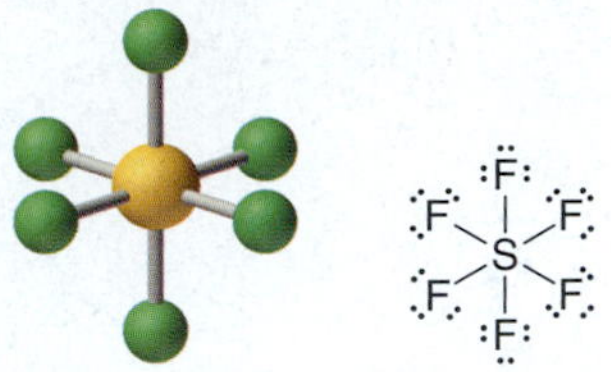

Sulfur hexafluoride

Let's look at sulfur hexafluoride, SF_6, whose Lewis structure must accommodate a total of 48 valence electrons $[6 + (6 \times 7) = 48]$. If we arrange the atoms and electrons symmetrically, we obtain the structure shown in the margin with six bonds to sulfur; that is, it is *six-coordinate*. Each fluorine atom has an octet, but the sulfur atom has 12 electrons surrounding it rather than eight.*

The octet rule is based on the fact that each valence orbital (typically, one *ns* and three *np* orbitals) can accommodate only two electrons. To accommodate more than eight electrons, sulfur must be using not only the *ns* and *np* valence orbitals but additional orbitals as well. Sulfur has an $[Ne]3s^2 3p^4 3d^0$ electron configuration, so in principle it could accommodate more than eight valence electrons by using one or more *d* orbitals. Thus, species such as SF_6 are often called **expanded-valence molecules**. Whether or not such compounds really do use *d* orbitals in bonding is controversial, but this model explains why compounds exist with more than an octet of electrons around an atom.

There is no correlation between the stability of a molecule or ion and whether or not it has an expanded valence shell. Some species with expanded valences, such as PF_5, are highly reactive, whereas others, such as SF_6, are very unreactive. In fact, SF_6 is so inert that it has many commercial applications. In addition to its use as an electrical insulator, it is used as the coolant in some nuclear power plants, and it is the pressurizing gas in "unpressurized" tennis balls.

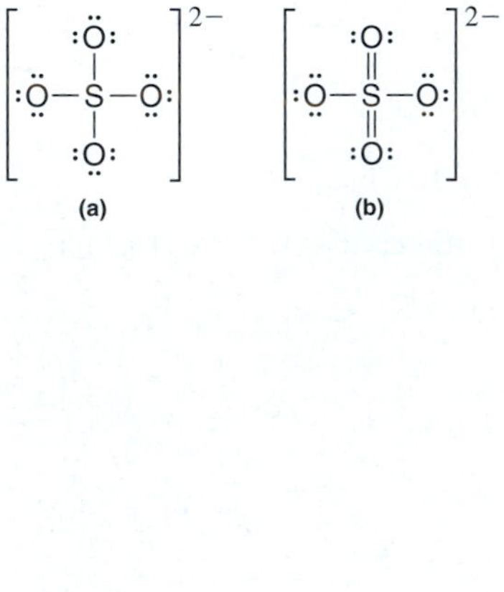

(a) (b)

An expanded valence shell is often written for oxoanions of the heavier *p*-block elements, such as sulfate (SO_4^{2-}) and phosphate (PO_4^{3-}). Sulfate, for example, has a total of 32 valence electrons $[6 + (4 \times 6) + 2]$. If we use a single pair of electrons to connect the sulfur and each oxygen, we obtain the four-coordinate Lewis structure (a) shown in the margin. We know that sulfur can accommodate more than eight electrons by using its empty valence *d* orbitals, just as in SF_6. An alternative structure (b) can be written that contains S═O double bonds, making the sulfur again six-coordinate. We can draw five other resonance structures equivalent to (b) that vary only in the arrangement of the single and double bonds. In fact, experimental data show that the S-to-O bonds in the SO_4^{2-} ion are intermediate in length between single and double bonds, as expected for a system whose resonance structures all contain two S—O single bonds and two S═O double bonds. When calculating the formal charges on structures (a) and (b), we see that the S atom in (a) has a formal charge of +2, whereas the S atom in (b) has a formal charge of 0. Thus, a +2 formal charge on S can be eliminated through the use of an expanded octet.

Fewer Than an Octet of Electrons

Molecules with atoms that possess fewer than an octet of electrons generally contain the lighter *s*- and *p*-block elements, especially beryllium and boron. These two elements tend to form compounds with four and six electrons, respectively, around the central atom. One example, boron trichloride (BCl_3), is used to produce fibers for reinforcing high-tech tennis rackets and golf clubs. The compound has 24 valence electrons and the Lewis structure

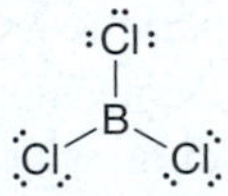

* The third step in our procedure for writing Lewis electron structures, in which we place an electron pair between each pair of bonded atoms, requires that an atom have more than eight electrons whenever it is bonded to more than four other atoms.

Boron trichloride

The boron atom has only six valence electrons, while each chlorine atom has eight. A reasonable solution might be to use a lone pair from one of the chlorine atoms to form a B-to-Cl double bond:

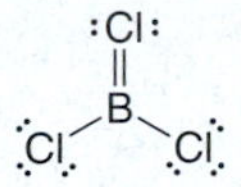

This resonance structure, however, results in a formal charge of +1 on the doubly bonded Cl atom and −1 on the B atom. The high electronegativity of Cl makes this separation of charge unlikely and suggests that this is not the most important resonance structure for BCl_3, which is shown to be true based on B—Cl bond lengths. Electron-deficient compounds have a strong tendency to gain an additional pair of electrons by reacting with species containing a lone pair of electrons.

EXAMPLE 8.8

Draw Lewis dot structures for **(a)** $BeCl_2$ gas, a compound used to produce beryllium, which in turn is used to produce structural materials for missiles and communication satellites, and **(b)** SF_4, a compound that reacts violently with water. Include resonance structures where appropriate.

Given Two compounds

Asked for Lewis electron structures

Strategy

A Use the procedure given earlier to write a Lewis electron structure for each compound. If necessary, place any remaining valence electrons on the element most likely to be able to accommodate more than an octet.

B After all valence electrons have been placed, decide whether you have drawn an acceptable Lewis structure.

Solution

(a) **A** Because it is the least electronegative element, Be is the central atom. The molecule has 16 valence electrons (two from Be and seven from each Cl). Drawing two Be–Cl bonds and placing three lone pairs on each Cl give

:Cl—Be—Cl:

B Although this arrangement gives beryllium only four electrons, it is an acceptable Lewis structure for $BeCl_2$. Beryllium is known to form compounds in which it is surrounded by fewer than an octet of electrons.

(b) **A** Sulfur is the central atom because it is less electronegative than fluorine. The molecule has 34 valence electrons (six from S and seven from each F). The S—F bonds use eight electrons, and another 24 are placed around the F atoms:

The only place to put the remaining two electrons is on the sulfur, giving sulfur 10 valence electrons:

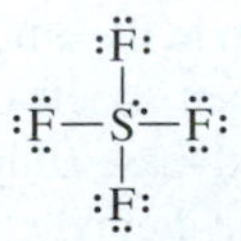

B Sulfur can accommodate more than an octet, so this is an acceptable Lewis structure.

EXERCISE 8.8

Draw Lewis dot structures for XeF_4.

Answer 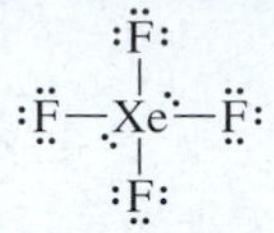

8.7 Lewis Acids and Bases

As you learned in Chapter 4, the Brønsted–Lowry concept of acids and bases defines a base as any species that can accept a proton, and an acid as any substance that can donate a proton. Lewis proposed an alternative definition that focuses on *pairs of electrons* instead.

A **Lewis base** is defined as any species that can *donate* a pair of electrons, and a **Lewis acid** is any species that can *accept* a pair of electrons. All Brønsted–Lowry bases (proton acceptors), such as OH^-, H_2O, and NH_3, are also electron-pair donors. Consequently, the Lewis definition of acids and bases does not contradict the Brønsted–Lowry definition. Rather, it expands the definition of acids to include substances other than the H^+ ion.

Note the pattern

Electron-deficient molecules (those with fewer than an octet of electrons) are Lewis acids.

As described in the preceding section, **electron-deficient molecules**, such as BCl_3, contain fewer than an octet of electrons around one atom and have a strong tendency to gain an additional pair of electrons by reacting with substances that possess a lone pair of electrons. Lewis's definition, which is less restrictive than either the Brønsted–Lowry or Arrhenius definition, grew out of his observation of this tendency.

A general Brønsted–Lowry acid–base reaction can be depicted in Lewis electron symbols as

$$\underset{\substack{\text{Lewis acid}\\ \text{Electron-pair}\\ \text{acceptor}}}{H^+} + \underset{\substack{\text{Lewis base}\\ \text{Electron-pair}\\ \text{donor}}}{:B} \longrightarrow H:B^+ \qquad (8.18)$$

The proton (H^+), which has no valence electrons, is a Lewis acid because it accepts a lone pair of electrons on the base to form a bond. The proton, however, is just one of many electron-deficient species that are known to react with bases. For example, neutral compounds of boron, aluminum, and the other Group-13 elements, which possess only six valence electrons, have a very strong tendency to gain an additional electron pair. Such compounds are therefore potent Lewis acids that react with an electron-pair donor such as ammonia to form an acid–base **adduct**, a new covalent bond, as shown here for boron trifluoride:

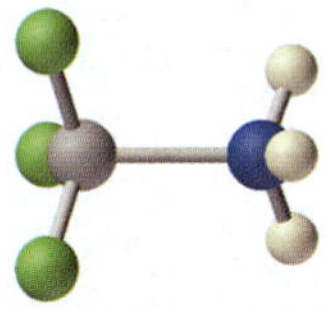

Boron trifluoride-ammonia adduct

$$\underset{\substack{\text{Lewis}\\ \text{acid}}}{BF_3} + \underset{\substack{\text{Lewis}\\ \text{base}}}{:NH_3} \longrightarrow \underset{\substack{\text{Acid-base}\\ \text{adduct}}}{F_3B:NH_3} \qquad (8.19)$$

The bond formed between the Lewis acid and the Lewis base is called a *coordinate covalent bond* because both electrons are provided by only one of the atoms (N, in the case of $F_3B:NH_3$). After it is formed, however, a coordinate covalent bond behaves like any other covalent single bond.

Species that are very weak Brønsted–Lowry bases can be relatively strong Lewis bases. For example, many of the Group-13 trihalides are highly soluble in ethers (R—O—R′) because the oxygen atom in the ether contains two lone pairs of electrons, just as in H_2O. Hence, the predominant species in solutions of electron-deficient trihalides in ether solvents is a Lewis acid–base adduct. A reaction of this type is shown in Equation 8.20 for boron trichloride and diethyl ether:

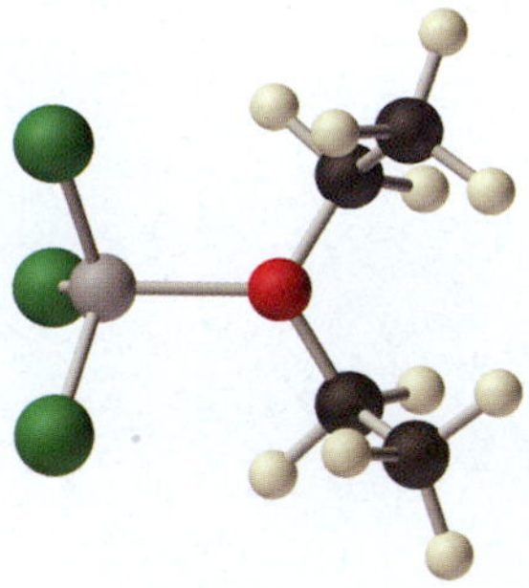

Boron trichloride-diethyl ether adduct

$$\underset{\substack{\text{Lewis}\\ \text{acid}}}{BCl_3} + \underset{\substack{\text{Lewis}\\ \text{base}}}{:\ddot{O}(CH_2CH_3)_2} \longrightarrow \underset{\text{Acid-base adduct}}{Cl_3B:\ddot{O}(CH_2CH_3)_2} \qquad (8.20)$$

Many molecules that contain multiple bonds can act as Lewis acids. In these cases, the Lewis base typically donates a pair of electrons to form a bond to the central atom of the molecule, while a pair of electrons displaced from the multiple bond becomes a lone pair on a terminal atom. A typical example is the reaction of hydroxide ion with carbon dioxide to give the bicarbonate ion, as shown in Equation 8.21. The highly electronegative oxygen atoms pull electron density away from carbon, so the carbon atom acts as a Lewis acid. Arrows indicate the direction of electron flow.

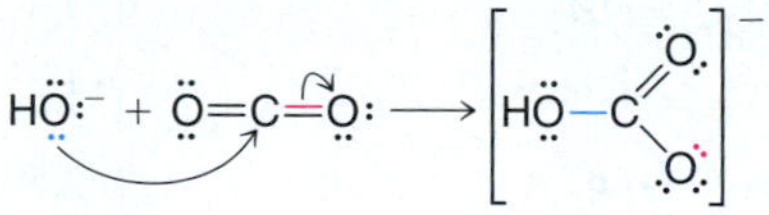

(8.21)

EXAMPLE 8.9

Identify the acid and the base in each Lewis acid–base reaction:

(a) $BH_3 + (CH_3)_2S \longrightarrow H_3B{:}S(CH_3)_2$
(b) $CaO + CO_2 \longrightarrow CaCO_3$
(c) $BeCl_2 + 2Cl^- \longrightarrow BeCl_4^{2-}$

Given Reactants and products

Asked for Identity of Lewis acid and Lewis base

Strategy

In each equation, identify the reactant that is electron deficient and the reactant that is an electron-pair donor. The electron-deficient compound is the Lewis acid, whereas the other is the Lewis base.

Solution

(a) In BH_3, boron has only six valence electrons. It is therefore electron deficient and can accept a lone pair. Like oxygen, the sulfur atom in $(CH_3)_2S$ has two lone pairs. Thus, $(CH_3)_2S$ donates an electron pair on sulfur to the boron atom of BH_3. The Lewis base is $(CH_3)_2S$, and the Lewis acid is BH_3.
(b) As in the reaction shown in Equation 8.21, CO_2 accepts a pair of electrons from the O^{2-} ion in CaO to form the carbonate ion. The oxygen in CaO is an electron-pair donor, so CaO is the Lewis base. Carbon accepts a pair of electrons, so CO_2 is the Lewis acid.
(c) The chloride ion contains four lone pairs. In this reaction, each chloride ion donates one lone pair to $BeCl_2$, which has only four electrons around Be. Thus, the chloride ions are Lewis bases, and $BeCl_2$ is the Lewis acid.

EXERCISE 8.9

Identify the acid and the base in each Lewis acid–base reaction:

(a) $(CH_3)_2O + BF_3 \longrightarrow (CH_3)_2O{:}BF_3$
(b) $H_2O + SO_3 \longrightarrow H_2SO_4$

Answer **(a)** Lewis base: $(CH_3)_2O$, Lewis acid: BF_3; **(b)** Lewis base: H_2O, Lewis acid: SO_3

8.8 • Properties of Covalent Bonds

In proposing his theory that octets can be completed by two atoms sharing electron pairs, Lewis provided scientists with the first description of covalent bonding. In this section, we expand on this and describe some of the properties of covalent bonds.

TABLE 8.6 Bond lengths and bond dissociation energies for bonds with different bond orders in selected gas-phase molecules at 298 K

Compound	Bond Order	Bond Length, pm	Bond Dissociation Energy, kJ/mol	Compound	Bond Order	Bond Length, pm	Bond Dissociation Energy, kJ/mol
$H_3C—CH_3$	1	153.5	376	$H_3C—NH_2$	1	147.1	331
$H_2C=CH_2$	2	133.9	728	$H_2C=NH$	2	127.3	644
$HC\equiv CH$	3	120.3	965	$HC\equiv N$	3	115.3	937
$H_2N—NH_2$	1	144.9	275.3	$H_3C—OH$	1	142.5	377
$HN=NH$	2	125.2	456	$H_2C=O$	2	120.8	732
$N\equiv N$	3	109.8	945.3	$O=C=O$	2	116.0	799
$HO—OH$	1	147.5	213	$C\equiv O$	3	112.8	1076.5
$O=O$	2	120.7	498.4				

Sources: *CRC Handbook of Chemistry and Physics* (2004); *Lange's Handbook of Chemistry* (2005); srdata.nist.gov/cccbdb.

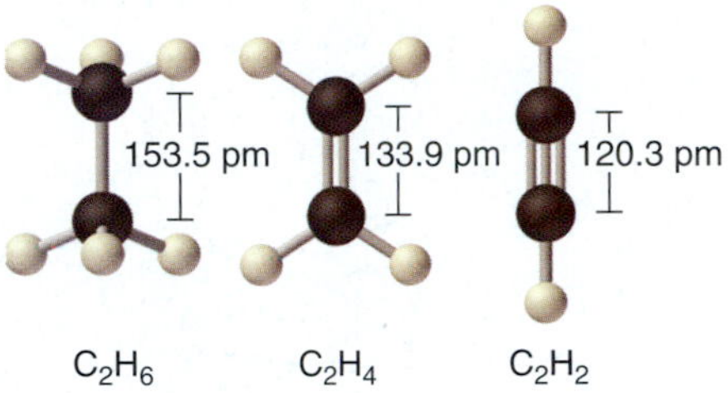

Bond Order

When we draw Lewis structures, we place one, two, or three pairs of electrons between adjacent atoms. In the Lewis bonding model, the number of electron pairs that hold two atoms together is called the **bond order**. For a single bond, such as the C—C bond in $H_3C—CH_3$, the bond order is one. For a double bond (such as $H_2C=CH_2$), the bond order is two. For a triple bond, such as $HC\equiv CH$, the bond order is three.

When analogous bonds in similar compounds are compared, bond length decreases as bond order increases. The bond length data in Table 8.6, for example, show that the C—C distance in $H_3C—CH_3$ (153.5 pm) is longer than the distance in $H_2C=CH_2$ (133.9 pm), which in turn is longer than that in $HC\equiv CH$ (120.3 pm). Additionally, as noted in Section 8.5, molecules or ions whose bonding must be described using resonance structures usually have bond distances that are intermediate between those of single and double bonds, as we demonstrated with the C—C distances in benzene. The relationship between bond length and bond order is not linear, however. A double bond is not half as long as a single bond, nor is the length of a C=C bond the average of the lengths of C≡C and C—C bonds. Nevertheless, as bond orders increase, bond lengths generally decrease.

The Relationship Between Bond Order and Bond Energy

As shown in Table 8.6, not only are triple bonds between like atoms shorter than double bonds, but because more energy is required to completely break all three bonds than to completely break two, a triple bond is also stronger than a double bond. Similarly, double bonds between like atoms are stronger and shorter than single bonds. Bonds of the same order between *different* atoms show a wide range of bond energies, however. Table 8.7 lists the average values for some commonly encountered bonds. Although the values shown vary widely, we can observe four trends:

1. Bonds between hydrogen and atoms in the same column decrease in strength as we go down the column. Thus, an H—F bond is stronger than an H—I bond, C—H is stronger than Si—H, N—H is stronger than P—H, O—H is stronger than S—H, and so forth. The reason for this is that the region of space in which electrons are shared between two atoms becomes proportionally smaller as one of the atoms becomes larger (Figure 8.10a).
2. Bonds between like atoms usually become *weaker* as we go down a column (important exceptions are noted below). For example, the C—C single bond is stronger than the Si—Si single bond, which is stronger than the Ge—Ge bond,

TABLE 8.7 Average bond energies (kJ/mol) for commonly encountered bonds at 273 K

Single Bonds										Multiple Bonds	
H—H	432	C—C	346	N—N	~167	O—O	~142	F—F	155	C=C	602
H—C	411	C—Si	318	N—O	201	O—F	190	F—Cl	249	C≡C	835
H—Si	318	C—N	305	N—F	283	O—Cl	218	F—Br	249	C=N	615
H—N	386	C—O	358	N—Cl	313	O—Br	201	F—I	278	C≡N	887
H—P	~322	C—S	272	N—Br	243	O—I	201	Cl—Cl	240	C=O	749
H—O	459	C—F	485	P—P	201	S—S	226	Cl—Br	216	C≡O	1072
H—S	363	C—Cl	327			S—F	284	Cl—I	208	N=N	418
H—F	565	C—Br	285			S—Cl	255	Br—Br	190	N≡N	942
H—Cl	428	C—I	213			S—Br	218	Br—I	175	N=O	607
H—Br	362	Si—Si	222					I—I	149	O=O	494
H—I	295	Si—O	452							S=O	532

Source: J. E. Huheey, E. A. Keiter, and R. L. Keiter, *Inorganic Chemistry,* 4th edition (1993).

and so forth. As two bonded atoms become larger, the region between them occupied by bonding electrons becomes *proportionally* smaller, as illustrated in Figure 8.10b. Noteworthy exceptions are single bonds between the Period-2 atoms of Groups 15, 16, and 17 (that is, N, O, F), which are unusually weak compared with single bonds between their larger congeners. It is likely that the N—N, O—O, and F—F single bonds are weaker than might be expected due to strong repulsive interactions between lone pairs of electrons on *adjacent* atoms. The trend in bond energies for the halogens is therefore

$$\text{Cl—Cl} > \text{Br—Br} > \text{F—F} > \text{I—I}$$

Similar effects are also seen for the O—O versus S—S and for N—N versus P—P single bonds.

3. Because elements in Periods 3 and 4 rarely form multiple bonds with themselves, their multiple bond energies are not accurately known. Nonetheless, they are presumed to be significantly weaker than multiple bonds between lighter atoms of the same families. Compounds containing an Si=Si double bond, for example, have only recently been prepared, whereas compounds containing C=C double bonds are one of the best-studied and most important classes of organic compounds.

Figure 8.10 The strength of covalent bonds depends on the overlap between the valence orbitals of the bonded atoms. The relative sizes of the region of space in which electrons are shared between (a) a hydrogen atom and lighter (smaller) vs. heavier (larger) atoms in the same periodic group; and (b) two lighter vs. two heavier atoms in the same group. Note that, although the absolute amount of shared space increases in both cases on going from a light to a heavy atom, the amount of space *relative to the size of the bonded atom* decreases. That is, the *percentage* of total orbital volume decreases with increasing size. Hence, the strength of the bond decreases.

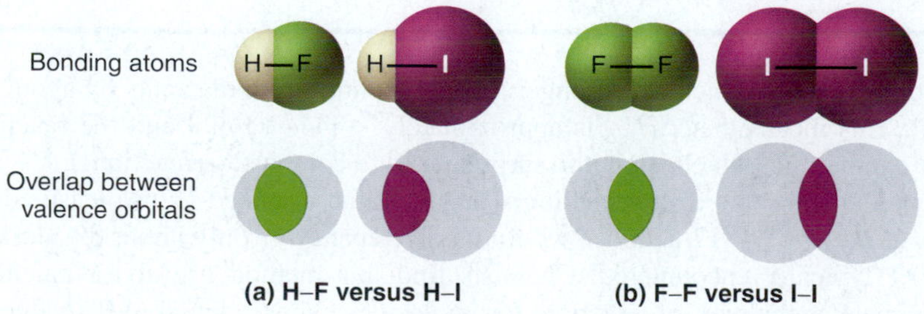

Note the pattern

Bond strengths increase as bond order increases, while bond distances decrease.

4. Multiple bonds between carbon, oxygen, or nitrogen and a Period-3 element such as phosphorus or sulfur tend to be unusually strong. In fact, multiple bonds of this type dominate the chemistry of the Period-3 elements of Groups 15 and 16. Multiple bonds to phosphorus or sulfur occur as a result of *d*-orbital interactions, as we discussed for the SO_4^{2-} ion in Section 8.6. In contrast, silicon in Group 14 has little tendency to form discrete silicon–oxygen double bonds. Consequently, SiO_2 has a three-dimensional network structure in which each silicon atom forms four Si—O single bonds, which makes the physical and chemical properties of SiO_2 very different from those of CO_2.

The Relationship Between Molecular Structure and Bond Energy

Bond energy is defined as the energy required to break a particular bond in a molecule in the gas phase. Its value depends not only on the identity of the bonded atoms but also on their environment. Thus, the bond energy of a C—H single bond is *not* the same in all organic compounds. For example, the energy required to break a C—H bond in methane varies by as much as 25% depending on how many other bonds in the molecule have already been broken (Table 8.8). That is, the C—H bond energy depends on its molecular environment. Except for diatomic molecules, the bond energies listed in Table 8.7 are *average* values for all bonds of a given type in a range of molecules. Even so, they are not likely to differ from the actual value of a given bond by more than about 10%.

TABLE 8.8 Energies for the dissociation of successive C—H bonds in methane

Reaction	*D*, kJ/mol
$CH_4(g) \longrightarrow CH_3(g) + H(g)$	439
$CH_3(g) \longrightarrow CH_2(g) + H(g)$	462
$CH_2(g) \longrightarrow CH(g) + H(g)$	424
$CH(g) \longrightarrow C(g) + H(g)$	338

Source: *CRC Handbook of Chemistry and Physics* (2004).

We can estimate the enthalpy change for a chemical reaction by summing the average energies of the bonds broken in the reactants and the average energies of the bonds formed in the products, and then calculating the difference between the two. If the bonds formed in the products are stronger than those broken in the reactants, then energy will be released in the reaction ($\Delta H_{rxn} < 0$):

$$\Delta H_{rxn} \approx \Sigma(\text{bond energies of bonds broken}) - \Sigma(\text{bond energies of bonds formed}) \quad (8.22)$$

The ≈ sign is used because we are summing *average* bond energies; hence, this approach does not give exact values for ΔH_{rxn}.

Let's consider the reaction of 1 mol of *n*-heptane, C_7H_{16}, with oxygen gas to give carbon dioxide and water. This is one of the reactions that occurs during the combustion of gasoline:

$$CH_3(CH_2)_5CH_3(l) + 11O_2(g) \rightarrow 7CO_2(g) + 8H_2O(g) \quad (8.23)$$

In this reaction, six C—C bonds, 16 C—H bonds, and 11 O=O bonds are broken per mole of *n*-heptane, while 14 C=O bonds (two for each CO_2) and 16 O—H bonds (two for each H_2O) are formed. The energy changes can be tabulated as follows:

Bonds Broken, kJ/mol				Bonds Formed, kJ/mol			
6 C—C	346	=	2,076	14 C=O	799	=	11,186
16 C—H	411	=	6,576	16 O—H	459	=	7,344
11 O=O	494	=	5,434		**Total**	=	18,530
	Total	=	14,086				

The bonds in the products are stronger than the bonds in the reactants by about 4444 kJ/mol. This means that ΔH_{rxn} is approximately −4444 kJ/mol and the reaction is highly exothermic (which is not too surprising for a combustion reaction!).

If we compare this approximation with the value obtained from measured ΔH_f° values ($\Delta H_{rxn} = -4817$ kJ/mol), we find a discrepancy of only about 8%, less than the 10% typically encountered. Chemists find this method useful for calculating approximate enthalpies of reaction for molecules whose actual ΔH_f° values are unknown. These approximations can be important for predicting whether a reaction is exothermic or endothermic, and to what degree.

EXAMPLE 8.10

When RDX, a more powerful explosive than dynamite, is detonated, it produces gaseous products and heat according to the reaction shown below. Use the approximate bond energies in Table 8.7 to estimate the ΔH_{rxn} per mole of RDX.

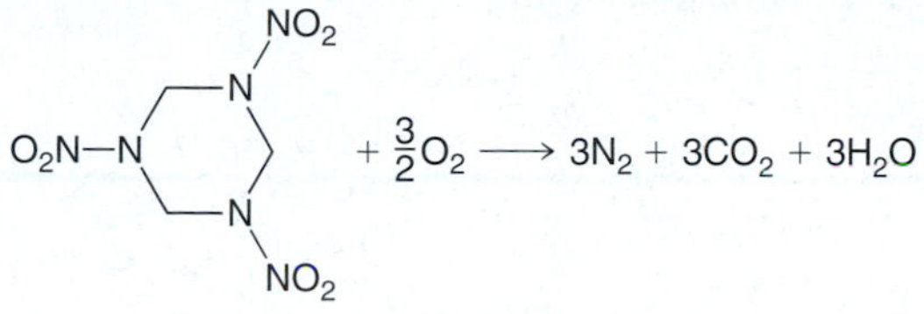

RDX

Given Chemical reaction, structure of reactant, Table 8.7

Asked for ΔH_{rxn} per mole

Strategy

A List the types of bonds broken in RDX, along with the bond energy required to break each type. Multiply the number of each type by the energy required to break one bond of that type and then sum the energies. Repeat this procedure for the bonds formed in the reaction.

B Use Equation 8.22 to calculate the amount of energy consumed or released upon reaction, ΔH_{rxn}.

Solution

We must sum the energies of the bonds in the reactants and compare that quantity with the sum of the energies of the bonds in the products. A nitro group (—NO_2) can be viewed as having one N—O single bond and one N=O double bond, as shown:

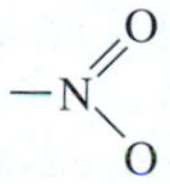

In fact, however, both N—O distances are usually the same because of the presence of two equivalent resonance structures.

A We can organize our data by constructing a table:

Bonds Broken, kJ/mol			
6 C—N	305	=	1830
6 C—H	411	=	2466
3 N—N	167	=	501
3 N—O	201	=	603
3 N=O	607	=	1821
1.5 O=O	494	=	741
	Total	=	7962

Bonds Formed, kJ/mol			
3 N≡N	942	=	2,826
6 C=O	799	=	4,794
6 O—H	459	=	2,754
	Total	=	10,374

B From Equation 8.22, we have

$$\Delta H_{rxn} \approx \Sigma(\text{bond energies of bonds broken}) - \Sigma(\text{bond energies of bonds formed})$$
$$= 7962 \text{ kJ/mol} - 10{,}374 \text{ kJ/mol}$$
$$= -2412 \text{ kJ/mol}$$

Thus, this reaction is also highly exothermic.

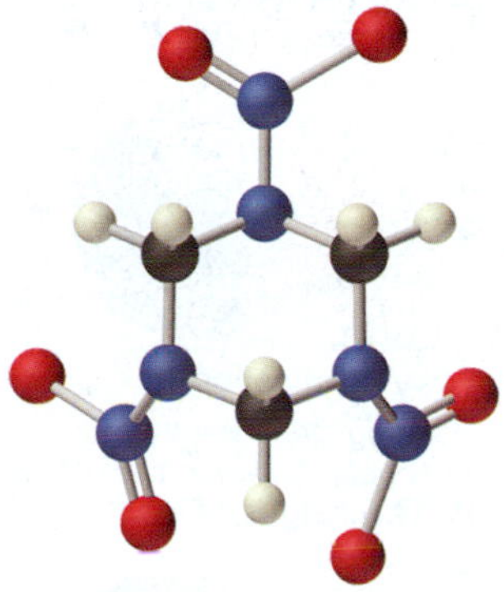

RDX

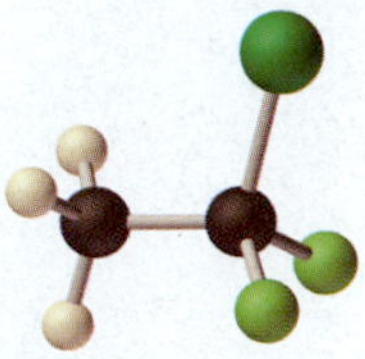

HCFC-142b

EXERCISE 8.10

The molecule HCFC-142b, a hydrochlorofluorocarbon used in place of CFCs such as Freons, can be prepared by adding HCl to 1,1-difluoroethylene:

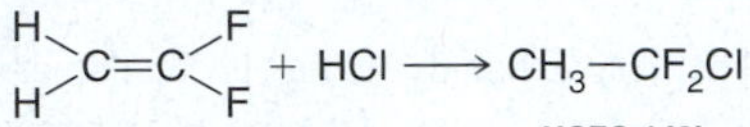

HCFC-142b

Use tabulated bond energies to calculate ΔH_{rxn}.

Answer −54 kJ/mol

8.9 Polar Covalent Bonds

In Chapter 2 and Section 8.1, we described the two idealized extremes of chemical bonding: (1) ionic bonding, in which one or more electrons are transferred completely from one atom to another, and the resulting ions are held together by purely electrostatic forces, and (2) covalent bonding, in which electrons are shared equally between two atoms. Most compounds, however, have **polar covalent bonds**, which means that electrons are shared *unequally* between the bonded atoms. Figure 8.11 compares the electron distribution in a polar covalent bond with those in an ideally covalent and an ideally ionic bond. Recall from Section 4.1 that a lowercase Greek delta (δ) is used to indicate that a bonded atom possesses a partial positive charge, indicated by δ^+, or a partial negative charge, indicated by δ^-, and that a bond between two atoms that possess partial charges is a *polar bond*.

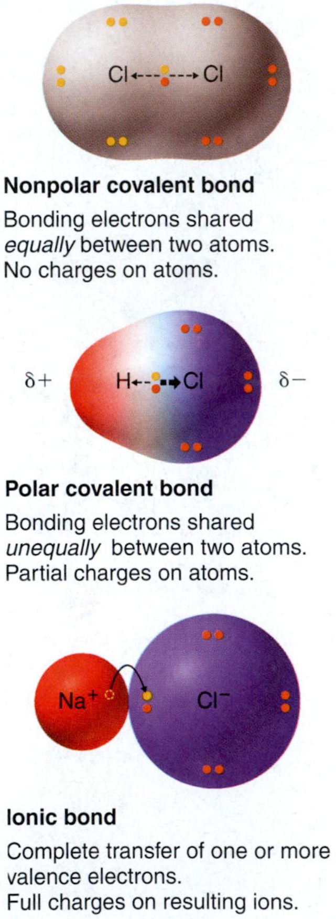

Figure 8.11 The electron distribution in a nonpolar covalent bond, a polar covalent bond, and an ionic bond using Lewis electron structures. In a purely covalent bond (top), the bonding electrons are shared equally between the atoms. In a purely ionic bond (bottom), an electron has been transferred completely from one atom to the other. A polar covalent bond (center) is intermediate between the two extremes: the bonding electrons are shared unequally between the two atoms, and the electron distribution is asymmetrical with the electron density being greater around the more electronegative atom. Electron-rich (negatively charged) regions are shown in blue; electron-poor (positively charged) regions are shown in red.

Bond Polarity

The polarity of a bond—the extent to which it is polar—is determined largely by the relative electronegativities of the bonded atoms. In Chapter 7, electronegativity (χ) was defined as the ability of an atom in a molecule or ion to attract electrons to itself. Thus, there is a direct correlation between electronegativity and bond polarity. A bond is *nonpolar* if the bonded atoms have equal electronegativities. If the electronegativities of the bonded atoms are not equal, however, the bond is *polarized* toward the more electronegative atom. A bond in which the electronegativity of B (χ_B) is greater than the electronegativity of A (χ_A), for example, is indicated with the partial negative charge on the more electronegative atom:

$$\overset{\delta^+}{\text{A}}\text{—}\overset{\delta^-}{\text{B}}$$

Less electronegative (A) — More electronegative (B)

One way of estimating the ionic character of a bond—that is, the magnitude of the charge separation in a polar covalent bond—is to calculate the difference in electronegativity between the two atoms: $\Delta\chi = \chi_B - \chi_A$.

To predict the polarity of the bonds in Cl_2, HCl, and NaCl, for example, we look at the electronegativities of the relevant atoms: $\chi_{Cl} = 3.16$, $\chi_H = 2.20$, and $\chi_{Na} = 0.93$ (see Figure 7.14). Cl_2 must be nonpolar because the electronegativity difference, $\Delta\chi$, is zero; hence, the bonding electrons are shared equally by the two chlorine atoms. In sodium chloride, $\Delta\chi$ is 2.23. This high value is typical of an ionic compound ($\Delta\chi \geq \approx 1.5$) and means that the valence electron of sodium has been completely transferred to chlorine to form Na^+ and Cl^- ions. In hydrogen chloride, however, $\Delta\chi$ is only 0.96. The bonding electrons are more strongly attracted to the more electronegative chlorine atom, and so the charge distribution is

$$\underset{\delta^+}{\text{H}}\text{—}\underset{\delta^-}{\text{Cl}}$$

Remember that electronegativities are difficult to measure precisely and different definitions produce slightly different numbers. In practice, the polarity of a bond is usually estimated rather than calculated.

As with bond energies, the electronegativity of an atom depends to some extent on its chemical environment. It is therefore unlikely that the reported electronegativities of a chlorine atom in NaCl, Cl_2, ClF_5, and $HClO_4$ would be exactly the same.

Note the pattern

Bond polarity and ionic character increase with increasing difference in electronegativity.

Dipole Moments

The asymmetrical charge distribution in a polar substance such as HCl produces a **dipole moment**, abbreviated by the Greek letter mu (μ). The dipole moment is defined as the product of the partial charge Q on the bonded atoms and the distance r between the partial charges:

MGC Bond Polarity

$$\mu = Qr \tag{8.24}$$

where Q is measured in coulombs (C) and r in meters. The unit for dipole moments is the debye (D):

$$1\ \text{D} = 3.3356 \times 10^{-30}\ \text{C}\cdot\text{m} \tag{8.25}$$

When a molecule with a dipole moment is placed in an electric field, it tends to orient itself with the electric field because of its asymmetrical charge distribution (Figure 8.12).

We can measure the partial charges on the atoms in a molecule such as HCl using Equation 8.24. If the bonding in HCl were purely ionic, an electron would be transferred from H to Cl, so there would be a full +1 charge on the H atom and a full −1 charge on the Cl atom. The dipole moment of HCl is 1.109 D, as determined by measuring the extent of its alignment in an electric field, and the reported gas-phase H—Cl distance is 127.5 pm. Hence, the charge on each atom is

$$\begin{aligned} Q = \frac{\mu}{r} &= 1.109\ \cancel{\text{D}}\left(\frac{3.3356 \times 10^{-30}\ \text{C}\cdot\cancel{\text{m}}}{1\ \cancel{\text{D}}}\right)\left(\frac{1}{127.5\ \cancel{\text{pm}}}\right)\left(\frac{1\ \cancel{\text{pm}}}{10^{-12}\ \cancel{\text{m}}}\right) \\ &= 2.901 \times 10^{-20}\ \text{C} \end{aligned} \tag{8.26}$$

By dividing this calculated value by the charge on a single electron (1.6022×10^{-19} C), we find that the charge on the Cl atom of an HCl molecule is about −0.18, corresponding to about 0.18 e^-:

$$\frac{2.901 \times 10^{-20}\ \cancel{\text{C}}}{1.6022 \times 10^{-19}\ \cancel{\text{C}}/e^-} = 0.1811\ e^- \tag{8.27}$$

To form a neutral compound, the charge on the H atom must be equal but opposite. Thus, the measured dipole moment of HCl indicates that the H—Cl bond has approximately 18% ionic character (0.1811 × 100%), or 82% covalent character. Instead of writing HCl as $\overset{\delta+}{\text{H}}—\overset{\delta-}{\text{Cl}}$, we can therefore indicate the charge separation quantitatively as

$$\overset{0.18\delta^+}{\text{H}}—\overset{0.18\delta^-}{\text{Cl}}$$

Our calculated results are in agreement with the electronegativity difference between hydrogen and chlorine ($\chi_H = 2.20$; $\chi_{Cl} = 3.16$, $\chi_{Cl} - \chi_H = 0.96$), a value well within the range for polar covalent bonds. We indicate the dipole moment by writing an arrow above the molecule.* In HCl, for example, the dipole moment is indicated as

$$\overset{+\!\!\longrightarrow}{\text{H—Cl}}$$

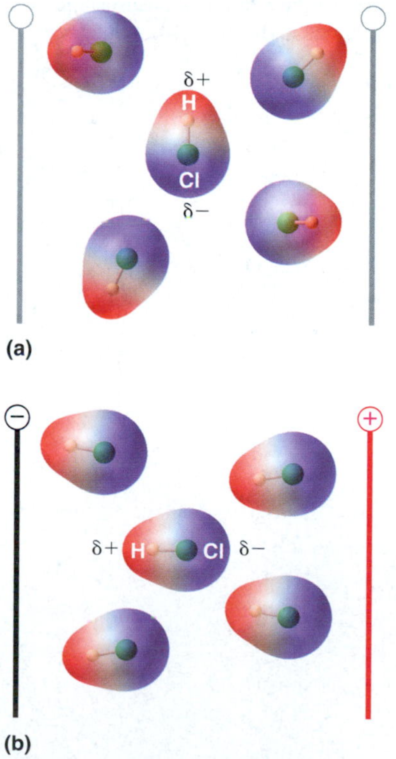

Figure 8.12 Molecules that possess a dipole moment partially align themselves with an applied electric field. In the absence of a field (a), the HCl molecules are randomly oriented. When an electric field is applied (b), the molecules tend to align themselves with the field, such that the positive end of the molecular dipole points toward the negative terminal, and vice versa.

* Mathematically, dipole moments are vectors, and they possess both a magnitude and a direction. The dipole moment of a molecule is the vector sum of the dipoles of the individual bonds.

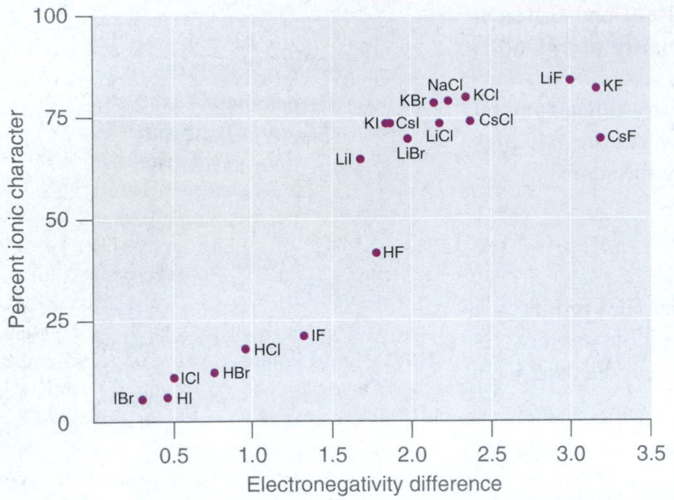

Figure 8.13 A plot of the percent ionic character of a bond as determined from measured dipole moments vs. the difference in electronegativity of the bonded atoms. In the gas phase, even CsF, which has the largest possible difference in electronegativity between atoms, is not 100% ionic. Solid CsF, however, is best viewed as 100% ionic because of the additional electrostatic interactions in the lattice. Interactive Graph

The arrow shows the direction of electron flow by pointing toward the more electronegative atom.

The charge on the atoms of many substances in the gas phase can be calculated using measured dipole moments and bond distances. Figure 8.13 shows a plot of the percent ionic character versus the difference in electronegativity of the bonded atoms for several substances. According to the graph, the bonding in species such as NaCl(g) and CsF(g) is substantially less than 100% ionic in character. As the gas condenses into a solid, however, dipole–dipole interactions between polarized species increase the charge separations. Hence, in the crystal, an electron is transferred from the metal to the nonmetal, and these substances behave like classic ionic compounds. The data in Figure 8.13 show that diatomic species with an electronegativity difference of <1.5 are less than 50% ionic in character, which is consistent with our earlier description of these species as containing polar covalent bonds. The use of dipole moments to determine the ionic character of a polar bond is illustrated in the next example.

EXAMPLE 8.11

In the gas phase, NaCl has a dipole moment of 9.001 D and an Na—Cl distance of 236.1 pm. Calculate the percent ionic character in NaCl.

Given Chemical species, dipole moment, internuclear distance

Asked for Percent ionic character

Strategy

A Compute the charge on each atom using the information given and Equation 8.24.

B Find the percent ionic character from the ratio of the actual charge to the charge of a single electron.

Solution

A The charge on each atom is given by

$$Q = \frac{\mu}{r} = 9.001\ \cancel{\text{D}}\left(\frac{3.3356 \times 10^{-30}\ \text{C} \cdot \cancel{\text{m}}}{1\ \cancel{\text{D}}}\right)\left(\frac{1}{236.1\ \cancel{\text{pm}}}\right)\left(\frac{1\ \cancel{\text{pm}}}{10^{-12}\ \cancel{\text{m}}}\right)$$
$$= 1.272 \times 10^{-19}\ \text{C}$$

Thus, NaCl behaves as if it had charges of 1.272×10^{-19} C on each atom separated by 236.1 pm.

B The percent ionic character is given by the ratio of the actual charge to the charge of a single electron (the charge expected for the complete transfer of one electron):

$$\%\ \text{ionic character} = \left(\frac{1.272 \times 10^{-19}\ \cancel{\text{C}}}{1.6022 \times 10^{-19}\ \cancel{\text{C}}}\right)(100) = 79.39\% \simeq 79\%$$

EXERCISE 8.11

In the gas phase, silver chloride, AgCl, has a dipole moment of 6.08 D and an Ag—Cl distance of 228.1 pm. What is the percent ionic character in silver chloride?

Answer 55.5%

SUMMARY AND KEY TERMS

8.1 An Overview of Chemical Bonding (p. 341)

Chemical bonding is the general term used to describe the forces that hold atoms together in molecules and ions. Two idealized types of bonding are ionic bonding, in which positively and negatively charged ions are held together by electrostatic forces, and covalent bonding, in which electron pairs are shared between atoms. All models of chemical bonding have three common features: atoms form bonds because the products are more stable than the isolated atoms; bonding interactions are characterized by a particular energy (the **bond energy** or **lattice energy**), which is the amount of energy required to dissociate the substance into its components; and bonding interactions have an optimal internuclear distance, the **bond distance**.

8.2 Ionic Bonding (p. 341)

The strength of the electrostatic attraction between ions with opposite charges is directly proportional to the magnitude of the charges on the ions and inversely proportional to the internuclear distance. The total energy of the system is a balance between the repulsive interactions between electrons on adjacent ions and the attractive interactions between ions with opposite charges.

8.3 Lattice Energies in Ionic Solids (p. 343)

Ionic compounds have strong electrostatic attractions between oppositely charged ions in a regular array. The lattice energy U of an ionic substance is defined as the energy required to dissociate the solid into gaseous ions; U can be calculated from the charges on the ions, the arrangement of the ions in the solid, and the internuclear distance. Because U depends on the product of the ionic charges, substances that contain di- or tripositive cations and/or di- or trinegative anions tend to have higher lattice energies than their singly charged counterparts. Higher lattice energies typically result in higher **melting points** and increased **hardness** because more thermal energy is needed to overcome the forces that hold the ions together. Lattice energies cannot be measured directly but are obtained from a thermochemical cycle called a **Born–Haber cycle**, in which Hess's law is used to calculate the lattice energy from the measured enthalpy of formation of the ionic compound, along with other thermochemical data. The Born–Haber cycle can be used to predict which ionic compounds are likely to form. **Sublimation**, the conversion of a solid directly to a gas, has an accompanying enthalpy change called the **enthalpy of sublimation**.

8.4 Introduction to Lewis Electron Structures (p. 351)

One convenient way to predict the number and basic arrangement of bonds in compounds is by using **Lewis dot symbols**, which consist of the chemical symbol for an element surrounded by dots that represent its valence electrons, grouped into pairs often placed above, below, and to the left and right of the symbol. The structures reflect the fact that the elements in Period 2 and beyond tend to gain, lose, or share electrons to reach a total of eight valence electrons in their compounds, the so-called **octet rule**. Hydrogen, with only two valence electrons, does not obey the octet rule.

8.5 Lewis Structures and Covalent Bonding (p. 352)

A plot of the overall energy of a covalent bond as a function of internuclear distance is identical to a plot of an ionic pair because both result from attractive and repulsive forces between charged entities. In Lewis electron structures we encounter **bonding pairs**, which are shared by two atoms, and **lone pairs**, which are not shared between atoms. If both electrons in a covalent bond come from the same atom, the bond is called a **coordinate covalent bond**. Lewis structures are an attempt to rationalize why certain stoichiometries are commonly observed for the elements of particular families. Neutral compounds of Group-14 elements typically contain four bonds around each atom (a double bond counts as two, a triple bond as three), whereas neutral compounds of Group-15 elements typically contain three bonds. In cases where it is possible to write more than one Lewis electron structure with octets around all the nonhydrogen atoms of a compound, the **formal charge** on each atom in alternative structures must be considered to decide which of the valid structures can be excluded and which is the most reasonable. The formal charge is the difference between the number of valence electrons of the free atom and the number of electrons assigned to it in the compound, where bonding electrons are divided equally between the bonded atoms. The Lewis structure with the lowest formal charges on atoms is almost always the most stable one. Some molecules have two or more chemically equivalent Lewis electron structures, called **resonance structures**. These structures are written with a double-headed arrow between them, indicating that *none* of the Lewis structures accurately describes the bonding but that the actual structure is an *average* of the individual resonance structures.

8.6 Exceptions to the Octet Rule (p. 363)

Molecules with an odd number of electrons are relatively rare in the *s* and *p* blocks but rather common among the *d*- and *f*-block elements. Compounds that contain more than an octet of electrons around an atom are called **expanded-valence molecules**. One model to explain their existence uses one or more *d* orbitals in bonding in addition to the valence *ns* and *np* orbitals. Such species are known for only atoms in Period 3 or below, which contain *nd* subshells in their valence shell. **Electron-deficient molecules**, which contain fewer than an octet of electrons around one atom, are relatively common. They tend to acquire an octet electron configuration by reacting with an atom that contains a lone pair of electrons.

8.7 Lewis Acids and Bases (p. 366)

A **Lewis acid** is a compound with a strong tendency to *accept* an additional pair of electrons from a **Lewis base**, which can *donate* a pair of electrons. Such an acid–base reaction forms an **adduct**, which is a compound that contains a *coordinate covalent bond* in which both electrons are provided by only one of the atoms.

8.8 Properties of Covalent Bonds (p. 367)

Bond order is the number of electron pairs that hold two atoms together. Single bonds have a bond order of one, and multiple bonds with bond orders of two (a double bond) and three (a triple bond) are quite common. In closely related compounds that contain bonds between the same kinds of atoms, the bond with the highest bond order is both the shortest and the strongest. In bonds with the same bond order between different atoms, trends are observed that, with few exceptions, result in the strongest single bonds being formed between the smallest atoms. Tabulated values of average bond energies can be used to calculate the enthalpy change of many chemical reactions. If the bonds in the products are stronger than those in the reactants, the reaction is exothermic, and vice versa.

8.9 Polar Covalent Bonds (p. 372)

Compounds with **polar covalent bonds** have electrons that are shared unequally between the bonded atoms. The polarity of such a bond is determined largely by the relative electronegativites of the bonded atoms. The asymmetrical charge distribution in a polar substance produces a **dipole moment**, which is the product of the partial charges on the bonded atoms and the distance between them.

KEY EQUATIONS

Lattice energy	$U = -k'\frac{Q_1Q_2}{r_0}$	(8.4)
Formal charge on an atom	$\text{formal charge} = \text{valence e}^- - \left(\text{nonbonding e}^- + \frac{\text{bonding e}^-}{2}\right)$	(8.15)
Enthalpy of reaction	$\Delta H_{rxn} = \Sigma(\text{bond energies of bonds broken}) - \Sigma(\text{bond energies of bonds formed})$	(8.22)
Dipole moment	$\mu = Qr$	(8.24)

QUESTIONS AND PROBLEMS

Questions and Problems with colored numbers have answers in the Appendix and complete solutions in the Student Solutions Manual.

CONCEPTUAL

8.1 An Overview of Chemical Bonding

1. Describe the differences between covalent bonding and ionic bonding. Which best describes the bonding in $MgCl_2$ and PF_5?
2. What three features do all chemical bonds have in common?

8.2 Ionic Bonding

3. Describe the differences in behavior between NaOH and CH_3OH in aqueous solution. Which solution would be a better conductor of electricity? Explain your reasoning.
4. What is the relationship between the strength of the electrostatic attraction between oppositely charged ions and the distance between the ions? How does the strength of the electrostatic interactions change as the size of the ions increases?
5. Which will result in the release of more energy: the interaction of a gaseous sodium ion with a gaseous oxide ion or the interaction of a gaseous sodium ion with a gaseous bromide ion? Why?
6. Which will result in the release of more energy: the interaction of a gaseous chloride ion with a gaseous sodium ion or a gaseous potassium ion? Explain your answer.
7. What are the predominant interactions when oppositely-charged ions are: (a) far apart; (b) at internuclear distances close to r_0; (c) when they are very close together (at a distance that is *less* than the sum of the ionic radii)?
8. A number of factors contribute to the stability of ionic compounds. Describe one type of interaction that *destabilizes* ionic compounds. Describe the interactions that *stabilize* ionic compounds.
9. What is the relationship between the electrostatic attractive energy between charged particles and the distance between the particles?

8.3 Lattice Energies in Ionic Solids

10. If a great deal of energy is required to form gaseous ions, why do ionic compounds form at all?
11. What are the general physical characteristics of ionic compounds?
12. Ionic compounds consist of crystalline lattices rather than discrete ion pairs. Why?
13. What factors affect the magnitude of the lattice energy of an ionic compound? What is the relationship between ionic size and lattice energy?
14. Which would have the larger lattice energy: an ionic compound consisting of a large cation and a large anion or one consisting of a large anion and a small cation? Explain your answer and the assumptions you made in responding.
15. How would the lattice energy of an ionic compound consisting of a monovalent cation and a divalent anion compare with the lattice energy of an ionic compound containing a monovalent cation and a monovalent anion, if the internuclear distance was the same in both compounds? Explain your answer.
16. Which would have the larger lattice energy: $CrCl_2$ or $CrCl_3$, assuming similar arrangements of ions in the lattice? Explain your answer.
17. Which cation in each of the following pairs would be expected to form a chloride salt with the larger lattice energy, assuming similar arrangements of ions in the lattice: (a) Na^+, Mg^{2+}; (b) Li^+, Cs^+; (c) Cu^+, Cu^{2+}? Explain your reasoning.
18. Which cation in each of the following pairs would be expected to form an oxide with the higher melting point, assuming

similar arrangements of ions in the lattice: (a) Mg^{2+}, Sr^{2+}; (b) Cs^{+}, Ba^{2+}; (c) Fe^{2+}, Fe^{3+}? Explain your reasoning.

19. How can a thermochemical cycle be used to determine lattice energies? Which steps in such a cycle require an input of energy?
20. Although NaOH and CH_3OH have similar formulas and molecular masses, the materials have radically different properties. One has a high melting point, and the other is a liquid at room temperature. Which compound is which, and why?

8.4 Introduction to Lewis Electron Structures

21. The Lewis electron system is a simplified approach for understanding bonding in covalent and ionic compounds. Why do chemists still find it useful?
22. Is a Lewis dot symbol an exact representation of the valence electrons in an atom or ion? Explain your answer.
23. How can the Lewis electron system help to predict the stoichiometry of a compound and its chemical and physical properties?
24. How is a Lewis dot representation consistent with the quantum mechanical model of the atom described in Chapter 6? How is it different?

8.5 Lewis Structures and Covalent Bonding

25. Compare and contrast covalent and ionic compounds with regard to: (a) volatility, (b) melting point, (c) electrical conductivity, and (d) physical appearance.
26. What are the similarities between plots of the overall energy versus internuclear distance for an ionic compound and a covalent compound? Why are the plots so similar?
27. Which atom do you expect to be the central atom in each of the following species: SO_4^{2-}, NH_4^+, BCl_3, SO_2Cl_2?
28. Which atom is the central atom in each of the following species: PCl_3, $CHCl_3$, SO_2, IF_3?
29. What is the relationship between the number of bonds typically formed by the second-period elements in Groups 14, 15, and 16 and their Lewis electron structures?
30. Although formal charges do not represent actual charges on atoms in molecules or ions, they are still useful. Why?
31. What evidence is there that resonance structures are important?
32. In what types of compounds are resonance structures particularly common?

8.6 Exceptions to the Octet Rule

33. What regions of the periodic table contain elements that frequently form molecules with an odd number of electrons? Explain your answer.
34. How can atoms "expand" their valence shell? What is the relationship between an expanded valence shell and the stability of an ion or molecule?
35. What elements are known to form compounds with fewer than an octet of electrons? Why do "electron-deficient" compounds form?
36. List three elements that form compounds that do not obey the octet rule. Describe the factors that are responsible for the stability of these compounds.

8.7 Lewis Acids and Bases

37. Construct a table comparing how OH^-, NH_3, H_2O, and BCl_3 are classified according to the Arrhenius, Brønsted–Lowry, and Lewis definitions of acids and bases.
38. Describe how the proton, H^+, can simultaneously behave as an Arrhenius acid, a Brønsted–Lowry acid, and a Lewis acid.
39. Would you expect aluminum to form compounds with covalent bonds or coordinate covalent bonds? Explain your answer.
40. Classify each of the following compounds as a Lewis acid or a Lewis base, and justify your choice: (a) $AlCl_3$; (b) $(CH_3)_3N$; (c) IO_3^-.
41. Explain how a carboxylate ion (RCO_2^-) can act as both a Brønsted–Lowry base and a Lewis base.

8.8 Properties of Covalent Bonds

42. Which would you expect to be stronger: an S—S bond or an Se—Se bond? Why?
43. Which element forms the strongest multiple bond with oxygen: nitrogen, phosphorus, or arsenic? Why?
44. Why do multiple bonds between oxygen and third period elements tend to be unusually strong?
45. What can bond energies tell you about reactivity?
46. Bond energies are typically reported as average values for a range of bonds in a molecule rather than as specific values for a single bond? Why?
47. If the bonds in the products are weaker than those in the reactants, is a reaction exothermic or endothermic? Explain your answer.
48. A student assumed that because heat was required to initiate a particular reaction, the reaction product would be stable. Instead, the product exploded. What information might have allowed the student to predict this outcome?

8.9 Polar Covalent Bonds

49. Why do ionic compounds such as KI exhibit substantially less than 100% ionic character in the gas phase?
50. Which would you expect to behave more like a classical ionic compound: LiI or LiF? Explain your answer. Which would have the greater dipole moment in the gas phase, LiI or LiF?

NUMERICAL

This section includes paired problems (marked by brackets) that require similar problem-solving skills.

8.2 Ionic Bonding

51. How does the energy of the electrostatic interaction between ions with charges +1 and −1 compare to the interaction between ions with charges +3 and −1, if the distance between the ions is the same in both cases? How does this compare with the magnitude of the interaction between ions with +3 and −3 charges?
52. How many grams of gaseous $MgCl_2$ are needed to give the same electrostatic attractive energy as 0.5 mol of gaseous LiCl? The ionic radii are $Li^+ = 76$ pm, $Mg^{+2} = 72$ pm and $Cl^- = 181$ pm.
53. Sketch a diagram showing the relationship between potential energy and internuclear distance (from $r = \infty$ to $r = 0$) for the interaction of a bromide ion and a potassium ion to form gaseous KBr. Explain why the energy of the system increases as the distance between the ions decreases from $r = r_0$ to $r = 0$.
54. Calculate the magnitude of the electrostatic attractive energy, E (in kJ), for 85.0 g of gaseous SrS ion pairs. The observed internuclear distance in the gas phase is 244.05 pm.
55. What is the electrostatic attractive energy, E(in kJ), for 130 g of gaseous HgI_2? The internucular distance is 255.3 pm.

8.3 Lattice Energies in Ionic Solids

56. Compare BaO and MgO with respect to each of the following properties: (a) enthalpy of sublimation, (b) ionization energy of the metal, (c) lattice energy, (d) enthalpy of formation.
57. Arrange the following compounds in order of decreasing lattice energy: SrO, PbS, PrI_3.

58. Would you expect the formation of SrO from its component elements to be exothermic or endothermic? Why or why not? How does the valence-electron configuration of the component elements help you determine this?

59. Use a thermochemical cycle and data from Figure 7.13, Tables 7.5, 8.2, and 8.3, and Appendix A to calculate the lattice energy, U, of magnesium chloride ($MgCl_2$).

60. Use a thermochemical cycle and data from Tables 8.2, 8.3, and Appendix A to calculate the lattice energy of calcium oxide. The first and second ionization energies of calcium are 589.8 kJ/mol and 1145.4 kJ/mol.

61. Using the information in questions 59 and 60, predict which is likely to have the higher melting point, CaO or $MgCl_2$.

8.5 Lewis Structures and Covalent Bonding

62. Give the electron configuration and the Lewis dot symbol for each of the following atoms: Na, Br, Ne, C, Ga, How many more electrons can each of these atoms accommodate?

63. Give the electron configuration and the Lewis dot symbol for each of the following atoms: Se, Kr, Li, Sr, H? How many more electrons can each of these atoms accommodate?

64. Based on Lewis dot symbols, predict the preferred oxidation state of each of the following elements: Br, Rb, O, Si, Sr.

65. Based on their Lewis dot symbols, predict the preferred oxidation state for each of the following elements: Be, F, B, C, Cs.

66. Based on their Lewis dot symbols, predict how many bonds gallium, silicon, and selenium will form in their neutral compounds.

67. Determine the total number of valence electrons in each of the following species: Ag, Pt^{2+}, H_2S, OH^-, I_2, CH_4, SO_4^{2-}, NH_4^+.

68. Give the total number of valence electrons in each of the following species: Cr, Cu^+, NO^+, XeF_2, Br_2, CH_2Cl_2, NO_3^-, H_3O^+.

69. Draw Lewis electron structures for Br_2, CH_3Br, SO_4^{2-}, O_2, S_2^{2-}, and BF_3.

70. Draw Lewis electron structures for: F_2, SO_2, $AlCl_4^-$, SO_3^{2-}, BrCl, XeF_4, NO^+, PCl_3.

71. Draw Lewis electron structures for CO_2, NO_2^-, SO_2, NO_2^+. From your diagram, predict which pair(s) of compounds have similar electronic structures.

72. Write Lewis dot symbols for the following pairs of elements: (a) Li, F; (b) Cs, Br; (c) Ca, Cl; (d) B, F. For a reaction between each pair of elements, predict which element is the oxidant, and which element is the reductant, as well as the final stoichiometry of the compound formed.

73. Write Lewis dot symbols for the following pairs of elements: (a) K, S; (b) Sr, Br; (c) Al, O; (d) Mg, Cl. For a reaction between each pair of elements, predict which element is the oxidant, and which element is the reductant, as well as the final stoichiometry of the compound formed.

74. Use Lewis dot symbols to predict whether ICl and NO_4^{1-} are chemically reasonable formulas.

75. Draw a reasonable Lewis electron structure for a compound with the molecular formula CH_4O.

76. Draw a plausible Lewis electron structure for a compound with the molecular formula Cl_3PO.

77. A student proposed the Lewis structure shown below for acetaldehyde. Why is this not a feasible structure? Draw an acceptable Lewis structure for acetaldehyde. Show the formal charges of all nonhydrogen atoms in both the correct and incorrect structures.

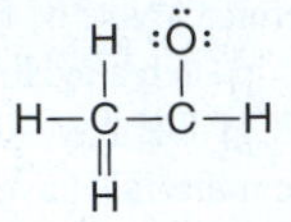

78. While reviewing her notes, a student noticed that she had drawn the structure shown below in her notebook for acetic acid:

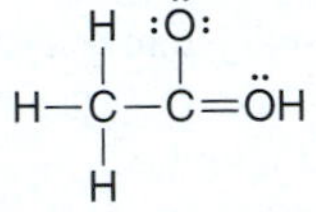

Why is this structure incorrect? Draw an acceptable Lewis structure for this compound. Give the formal charges of all nonhydrogen atoms in both the correct and incorrect structures.

79. Draw the most plausible Lewis structure for NO_3^-. Does this ion have any other resonance structures? Draw at least one other Lewis structure for nitrate that is not plausible based on formal charges.

80. Draw the most likely structure for HCN based on formal charges, showing the formal charge on each atom in your structure. Does this compound have any plausible resonance structures? If so, draw one.

81. Using arguments based on formal charges, explain why the most feasible Lewis structure for SO_4^{2-} has two sulfur–oxygen double bonds.

82. At least two Lewis structures can be drawn for BCl_3. Using arguments based on formal charges, explain why the most likely structure is the one with three B—Cl single bonds.

83. At least two distinct Lewis structures can be drawn for N_3^-. Use arguments based on formal charges to explain why the most likely structure contains a nitrogen–nitrogen double bond.

84. Is H—O=C—H a reasonable structure for a compound with the formula CH_2O? Use Lewis electron dot structures to justify your answer.

85. Is H—O—N=O a reasonable structure for the compound HNO_2? Justify your answer using Lewis electron dot structures.

86. Explain why the following Lewis structure for SO_3^{2-} is or is not reasonable.

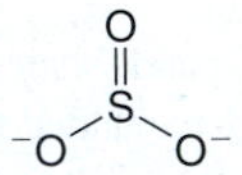

87. Draw all of the resonance structures for: (a) the HSO_4^- ion; and (b) the HSO_3^- ion.

8.6 Exceptions to the Octet Rule

88. What is the major weakness of the Lewis system in predicting the electron structures of PCl_6^- and other species containg atoms from the third period and beyond?

89. The compound aluminum trichloride consists of Al_2Cl_6 molecules with the following structure (lone pairs of electrons removed for clarity):

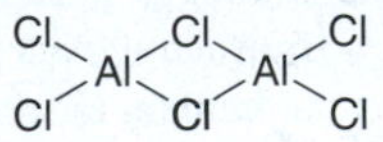

Does this structure satisfy the octet rule? What is the formal charge on each atom? Given the chemical similarity between aluminum and boron, what is a plausible explanation for the fact that aluminum trichloride forms a dimeric structure rather than the monomeric trigonal planar structure of BCl_3?

90. Draw Lewis electron structures for ICl_3, Cl_3PO, Cl_2SO, and AsF_6^-.

91. Draw Lewis electron structures for ClO_4^-, IF_5, $SeCl_4$, and SbF_5.

92. Draw an acceptable Lewis structure for PCl_5, a compound used in manufacturing a form of cellulose. What is the formal charge of the central atom? What is the oxidation number of the central atom?

93. Draw plausible Lewis structures for the phosphate ion, including resonance structures. What is the formal charge on each atom in your structures?

94. Draw an acceptable Lewis structure for xenon trioxide, XeO_3, including all resonance structures.

95. Using Lewis structures, draw all of the resonance structures for the BrO_3^- ion.

8.7 Lewis Acids and Bases

96. Use Lewis dot symbols to depict the reaction of BCl_3 with dimethyl ether, $(CH_3)_2O$. How is this reaction similar to the protonation of ammonia?

97. In each of the following reactions, identify the Lewis acid and the Lewis base, and complete the reaction: (a) $(CH_3)_2O$ + $AlCl_3$; (b) $SnCl_4 + 2\ Cl^-$.

8.8 Properties of Covalent Bonds

98. What is the carbon–carbon bond order in the following compounds: ethylene (C_2H_4), BrH_2CCH_2Br, and FCCH? Arrange the compounds in order of increasing C—C bond distance. Which would you expect to have the largest C—C bond energy? Why?

99. What is the bond order about the central atom(s) of hydrazine (N_2H_4), nitrogen, and diimide (N_2H_2)? Draw Lewis electron structures for each compound, and then arrange these compounds in order of increasing N—N bond distance. Which of these compounds would you expect to have the largest N—N bond energy? Explain your answer.

100. Which of the following pairs of elements has the greater bond strength: (a) P—P, Sb—Sb; (b) Cl—Cl, I—I; (c) O—O, Se—Se; (d) S—S, Cl—Cl; (e) Al—Cl, B—Cl. Explain your choice in each case.

101. Which of the following pairs of elements has the greater bond strength: (a) Te—Te, S—S; (b) C—H, Ge—H; (c) Si—Si, P—P; (d) Cl—Cl, F—F; (e) Ga—H, Al—H? Explain your choice in each case.

102. Approximately how much energy/mol is required to completely dissociate acetone [$(CH_3)_2CO$] and urea [$(NH_2)_2CO$] into their constituent atoms?

103. Approximately how much energy/mol is required to completely dissociate each of the following compounds into its constituent atoms: ethanol, formaldehyde, and hydrazine?

104. Is the reaction of diimine (N_2H_2) with oxygen to produce nitrogen and water exothermic or endothermic? Quantify your answer.

8.9 Polar Covalent Bonds

105. Predict whether each of the following compounds are purely covalent, purely ionic, or polar covalent: RbCl, S_8, $TiCl_2$, $SbCl_3$, LiI, and Br_2.

106. Based on relative electronegativities, classify the bonding in each of the following as ionic, covalent, or polar covalent: NO, HF, MgO, $AlCl_3$, SiO_2, and the C═O bond in acetone. Indicate the direction of the bond dipole for each polar covalent bond.

107. Based on relative electronegativities, classify the bonding in each of the following as ionic, covalent, or polar covalent: NaBr, OF_2, BCl_3, the S—S bond in $CH_3CH_2SSCH_2CH_3$, the C—Cl bond in CH_2Cl_2, the O—H bond in CH_3OH, and $PtCl_4^{2-}$. Indicate the direction of the bond dipole for each polar covalent bond.

108. Classify each of the following compounds as having 0–40% ionic character, 40–60% ionic character, or 60–100% ionic character based on the type of bonding you would expect: CaO, S_8, $AlBr_3$, ICl, Na_2S, SiO_2, and LiBr. Justify your reasoning.

109. If the bond distance in HCl (dipole moment = 1.109 D) were double the actual value of 127.46 pm, what would be the effect on the charge localized on each atom? What would be the percent negative charge on Cl? At the actual bond distance, how would doubling the charge on each atom affect the dipole moment? Would this represent more ionic or covalent character?

110. Calculate the percent ionic character of HF (dipole moment = 1.826 D) if the H—F bond distance is 92 pm.

111. Calculate the percent ionic character of CO (dipole moment = 0.110 D) if the C—O distance is 113 pm.

112. Calculate the percent ionic character of PbS and PbO in the gas phase, given the following information: for PbS, r = 228.69 pm and μ = 3.59 D; for PbO, r = 192.18 pm and μ = 4.64 D. Would you classify these compounds as having covalent or polar covalent bonds in the solid state?

APPLICATIONS

113. Until recently, benzidine (structure shown below) was used in forensic medicine to detect the presence of human blood; when mixed with human blood, benzidine turns a characteristic blue color. Because benzidine has recently been identified as a carcinogen, it has been replaced by other indicators. Draw the complete Lewis dot structure for benzidine. Would you expect this compound to behave as a Lewis acid or Lewis base?

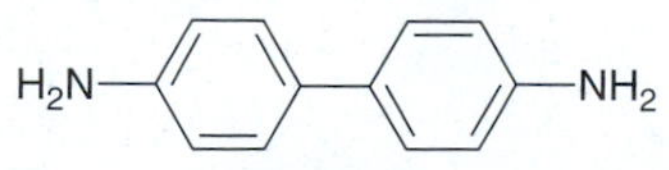

Benzidine

114. There are three possible ways to connect carbon, nitrogen, and oxygen to form a monoanion: CNO^-, CON^-, and OCN^-. One is the cyanate ion, a common and stable species; one is the fulminate ion, salts of which are used as explosive detonators; and one is so unstable that it has never been isolated. Use Lewis electron structures and the concept of formal charge to determine which isomer is cyanate, which is the fulminate, and which is the least stable.

115. The colorless gas N_2O_4 is a deadly poison that has been used as an oxidizing agent in rocket fuel. The compound has a single N—N bond, with a formal charge of +1 on each nitrogen atom. Draw resonance structures for this molecule.

116. Naphthalene is an organic compound that is commonly used in veterinary medicine as the active ingredient in dusting powders; it is also used internally as an intestinal antiseptic. From its chemical structure and the ΔH_f of CO_2 and H_2O, estimate the molar enthalpy of combustion and the enthalpy of formation of naphthalene.

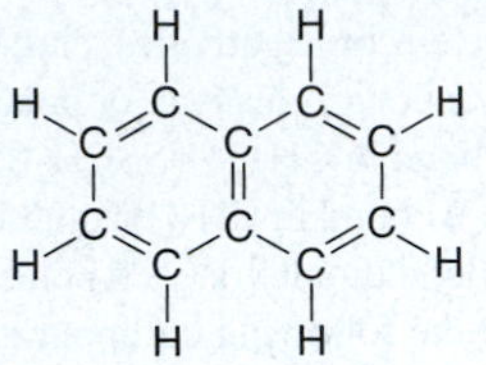

Naphthalene

117. Compare the combustion of hydrazine, N_2H_4, which produces nitrogen and water, with the combustion of methanol. Use the chemical structures to estimate which has a higher heat of combustion. Given equal volumes of hydrazine (d = 1.004 g/ml) and methanol (d = 0.791 g/ml), which is the better fuel (in other words, which provides more energy per unit volume upon combustion)? Can you think of a reason why hydrazine is not used in internal combustion engines?

118. Race car drivers frequently prefer methanol to isooctane as a fuel. Is this justified, based on enthalpies of combustion? If you had a choice between 10 gallons of methanol (d = 0.791) and the same volume of isooctane (d = 0.688), whch fuel would you prefer?

119. An atmospheric reservoir species is a molecule that is rather unreactive, but which contains elements that can be converted to reactive forms. For example, chlorine nitrate, $ClONO_2$, is a reservoir species for both chlorine and nitrogen dioxide. In fact, most of the chlorine in the atmosphere is usually bound up in chlorine nitrate as a result of the reaction of ClO with NO_2.

(a) Write a balanced equation for this reaction.

(b) Draw Lewis electron structures for each species in this reaction. What difficulty is associated with the structure of the reactants? How does this affect the reactivity of the compounds?

Chlorine nitrate can react in a surface reaction with water to form HClO and nitric acid.

(c) Draw Lewis electron structures to describe this reaction.

(d) Identify the Lewis and Brønsted–Lowry acids.

120. Aniline is an oily liquid used in the preparation of organic dyes, in varnishes, and in black shoe polishes.

(a) Draw a complete Lewis structure for the molecule (including the nitrogen atom).

(b) The —NH_2 bound to the ring contains a lone pair of electrons that can participate in resonance in the following way:

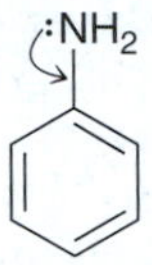

Draw a second Lewis structure for aniline that takes this interaction into account.

(c) Calculate the formal charge on each non-hydrogen atom in both Lewis structures.

(d) What other resonance structures can be drawn for aniline that satisfy the octet rule?

9 Structure and Bonding II:
Molecular Geometry and Models of Covalent Bonding

LEARNING OBJECTIVES

- To be able to use the VSEPR model to predict molecular geometries
- To predict whether a molecule has a dipole moment
- To be able to describe the bonding in simple compounds using valence bond theory
- To be able to use molecular orbital theory to predict bond orders
- To be able to explain resonance structures using molecular orbitals
- To understand the strengths and limitations of the various approaches to bonding

CHAPTER OUTLINE

In Chapter 8, we described the interactions that hold atoms together in chemical substances, focusing on the lattice energy of ionic compounds and the bond energy of covalent compounds. In the process, we introduced Lewis electron structures, which provide a simple method for predicting the number of bonds in common substances. As you learned in Chapter 8, the dots in Lewis structures represent the valence electrons of the constituent atoms and are paired according to the octet rule. As you will soon discover, however, the bonding in more complex molecules, such as those that contain multiple bonds or odd numbers of electrons, cannot be explained using this simple approach.

An experimental image of a covalent bond. This image shows that the bonding electrons on the copper atom in Cu_2O occupy d_{z^2} orbitals that point toward the oxygen atoms located at the center and corners of a cube.

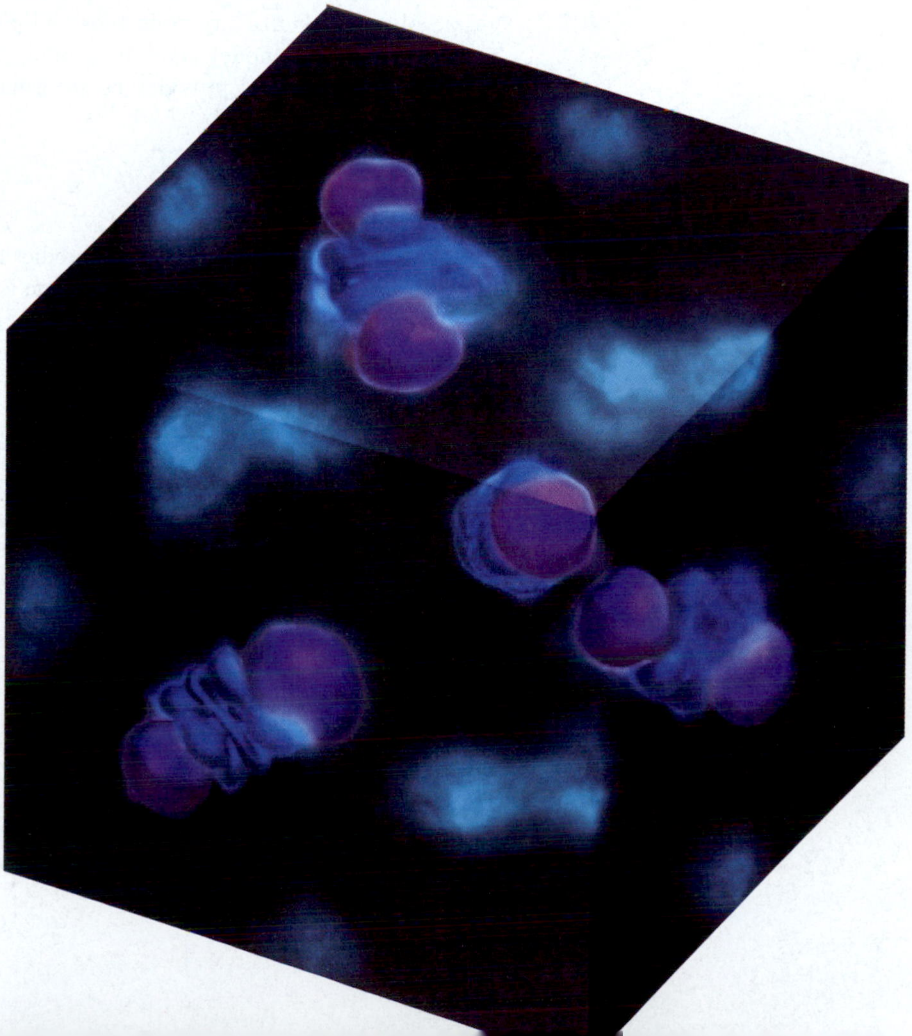

The purpose of this chapter is to introduce you to conceptual models used by chemists to describe the bonding in more complex compounds.

We begin with a general method for predicting the structures of simple covalent molecules and polyatomic ions, and then we discuss the actual distribution of electrons in covalent bonds. We make use of two distinct approaches for describing covalent bonds: a localized model to describe bonding in molecules that contain two or more atoms attached to a central atom, and a delocalized model to explain and predict which diatomic species are able to exist and which are not. We conclude by combining the two approaches to describe more complex molecules and ions that contain multiple bonds. The tools you acquire in this chapter will enable you to explain why Ca_2 is too unstable to exist in nature, and why the unpaired electrons on O_2 are crucial to the existence of life as we know it. You will also discover why carbon, the basic component of all organic compounds, forms four bonds despite having only two unpaired electrons in its valence-electron configuration, and how the structure of retinal, the key light-sensing component in our eyes, allows us to detect visible light.

9.1 Predicting the Geometry of Molecules and Polyatomic Ions

MGC Molecular Geometry

The Lewis electron-pair approach described in Chapter 8 can be used to predict the number and types of bonds between the atoms in a substance, and it indicates which atoms have lone pairs of electrons. This approach gives no information about the actual arrangement of atoms in space, however. We continue our discussion of structure and bonding by introducing the **valence-shell electron-pair repulsion (VSEPR) model** (pronounced "vesper"), which can be used to predict the shapes of many molecules and polyatomic ions. Keep in mind, however, that the VSEPR model, like any model, is a limited representation of reality; in this case, the model provides no information about bond lengths or the presence of multiple bonds.

The VSEPR Model

The VSEPR model is able to predict the structure of nearly any molecule or polyatomic ion in which the central atom is a nonmetal, as well as the structures of many compounds that contain a central metal atom. The VSEPR model is *not* a theory; it does not attempt to explain observations. Instead, it is a counting procedure that accurately predicts the structures of a large number of compounds whose structures cannot be predicted using the Lewis electron-pair approach.

In discussions of the structures of molecules or polyatomic ions, it is convenient to classify species according to the number of atoms (n) of one type (B) attached to the central atom (A) using the notation AB_n. Not all AB_n species with the same value of n have the same structure, however. For example, carbon dioxide (CO_2) and sulfur dioxide (SO_2) are both AB_2 species, but CO_2 is linear and SO_2 is bent. Similarly, ammonia (NH_3), boron trichloride (BCl_3), and bromine trifluoride (BrF_3) are all AB_3 species, yet each has a different molecular geometry (Figure 9.1). Although we cannot account for these differences using the Lewis electron-pair approach, we can explain them using the VSEPR model.

The VSEPR model assumes that the electron pairs around the central atom of a Lewis structure occupy space, whether they are bonding pairs or lone pairs. Because electrons repel each other electrostatically, the most stable arrangement of electron pairs (that is, the one with the lowest energy) is the one that minimizes repulsions between the electrons. The VSEPR model distinguishes between the *electron-pair*

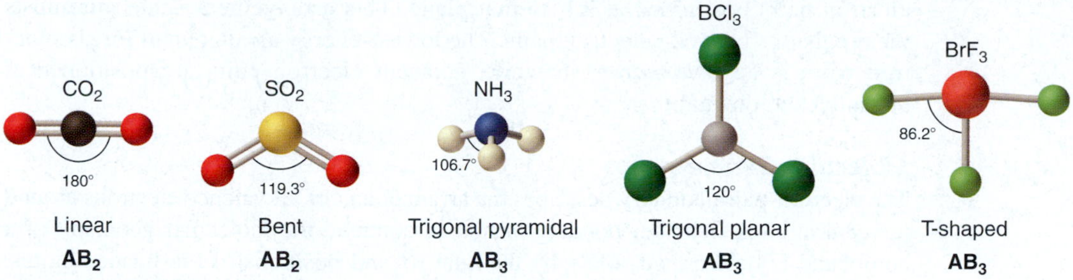

Figure 9.1 Common structures for molecules and polyatomic ions that consist of a central A atom bonded to two or three B atoms.

geometry, which is the three-dimensional arrangement of *electron pairs* around the central atom, and the **molecular geometry**, the arrangement of the *bonded atoms* in a molecule or polyatomic ion.

Electron-Pair Geometry

Because the strength of the electrostatic repulsion between two particles that have the same charge depends on the distance between them, the repulsions between electron pairs depend strongly on the angle between them: the smaller the angle, the closer they are on average and the greater the electrostatic repulsion. For compounds with two electron pairs around the central atom, the lowest-energy arrangement has the electron pairs on opposite sides of the central atom with a 180° angle between them, as shown in Table 9.1. This *linear* arrangement minimizes the electrostatic repulsions. Any angle smaller than 180° places the electron pairs closer together, which increases the electrostatic repulsions.

For compounds with three electron pairs around the central atom, the lowest-energy arrangement is *trigonal planar*, with the electron pairs at 120° angles from each other. This arrangement again minimizes the electrostatic repulsions between electrons. In compounds with four electron pairs around the central atom, the electron pairs point toward the vertices of a *tetrahedron*, a structure already familiar to you, with 109.5° angles between adjacent electron pairs. In compounds with five electron pairs, the electron pairs have a *trigonal bipyramidal* arrangement, which has two sets of angles between adjacent electron pairs: one set contains three electron pairs at 120° angles to one another in a plane, and a second set contains two

TABLE 9.1 Electron-pair geometries for species with two to six electron pairs

Electron Pairs	2	3	4	5	6
Geometry	Linear	Trigonal planar	Tetrahedral	Trigonal bipyramidal	Octahedral
		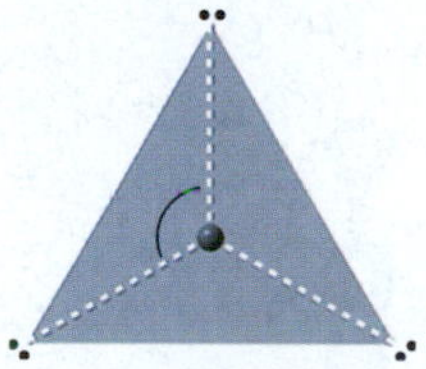	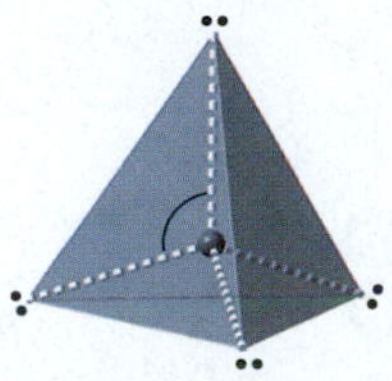	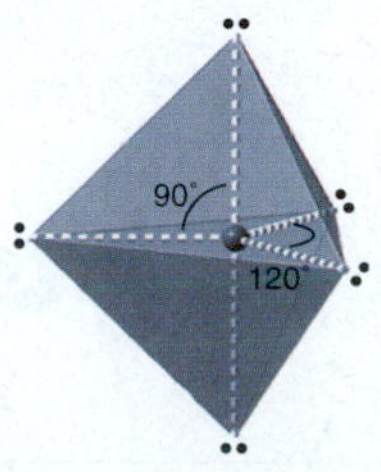	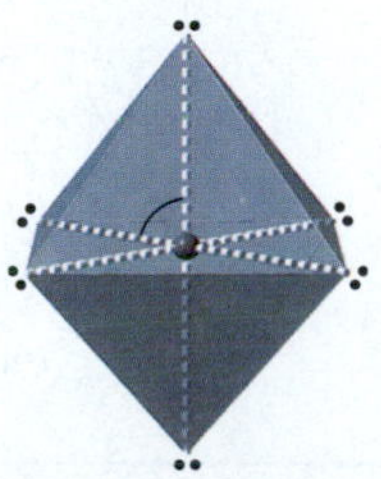
Predicted Bond Angles	180°	120°	109.5°	90°, 120°	90°

electron pairs positioned at 90° to that plane. This arrangement again minimizes the repulsions between electron pairs. The lowest-energy arrangement for six electron pairs is an *octahedron*, in which adjacent electron pairs are positioned at 90° angles to one another.

Molecular Geometry

The electron-pair geometry describes the arrangement of all valence electrons around the central atom, whether bonding or not. In contrast, the molecular geometry of a compound is determined solely by the number and positions of the bonded atoms, which share one or more pairs of electrons with the central atom. Thus, the molecular geometry is related to the electron-pair geometry but not necessarily identical to it, as we shall see. The relative positions of the atoms are given by the bond lengths and the angles between the bonds, the **bond angles**. Common three-dimensional molecular geometries of various AB_n molecules are illustrated in Table 9.2. The idealized bond angles characteristic of each structure are also listed.

From Table 9.2, we discover that the geometry of an AB_2 molecular species can be either *linear* (BAB bond angle = 180°) or *bent* (BAB bond angle ≠ 180°). Two common geometries for AB_3 species are *trigonal planar* and *trigonal pyramidal*. Two common geometries for AB_4 compounds are *square planar* and *tetrahedral*. In contrast, only one structure is commonly found for AB_5 (*trigonal bipyramidal*) and

TABLE 9.2 Common molecular geometries for species with two to six bonded atoms[a]

AB_n Notation	AB_2	AB_2	AB_3	AB_3
Geometry	Linear	Bent (V-shaped)	Trigonal planar	Trigonal pyramidal
	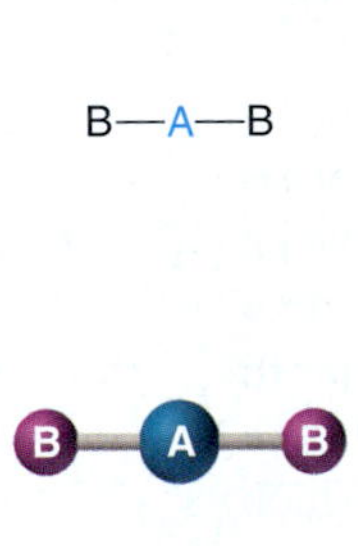	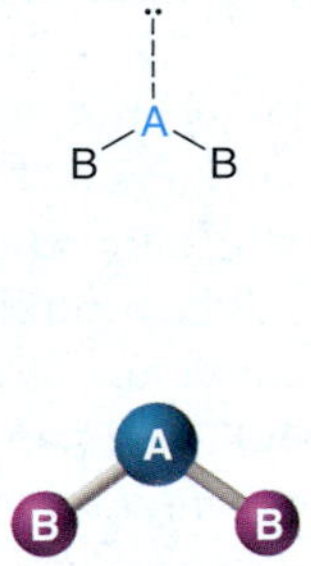	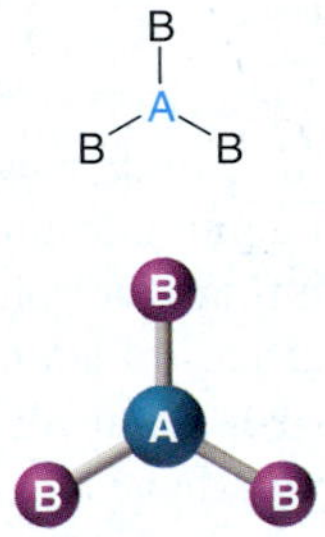	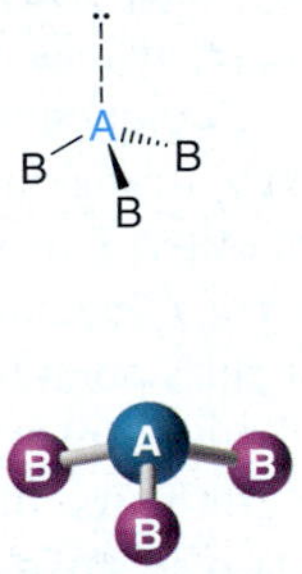
Idealized Bond Angles	180°	<180°	120°	<120°
AB_n Notation	AB_4	AB_4	AB_5	AB_6
Geometry	Square planar	Tetrahedral	Trigonal bipyramidal	Octahedral
	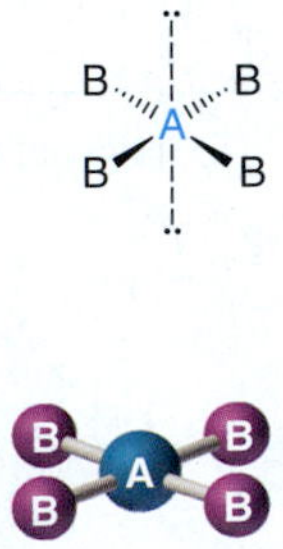	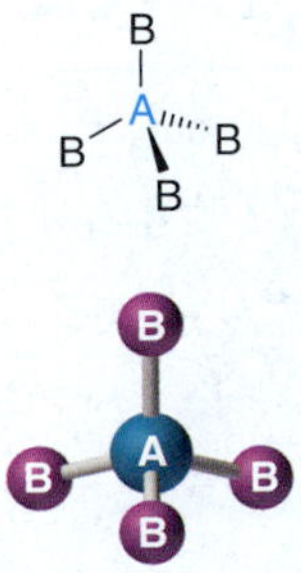	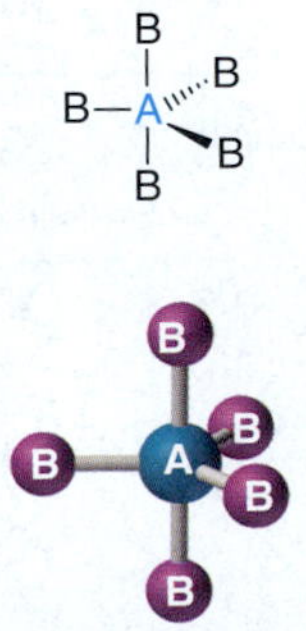	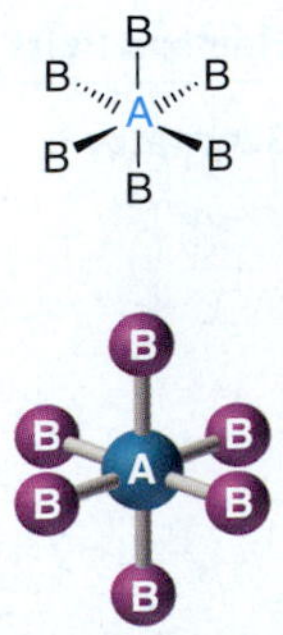
Idealized Bond Angles	90°	109.5°	90°, 120°	90°

[a] Lone pairs are shown using a dashed line.

AB_6 (*octahedral*) species. The AB_n molecules shown in Table 9.2 are the simplest to consider. As we shall see, many other species can be viewed as derived from these AB_n structures as one or more of the B atoms are replaced by other atoms or groups. In general, this has only modest effects on the structure predicted by the VSEPR model.

Using the VSEPR Model

We can use the VSEPR model to predict the geometry of most polyatomic molecules and ions by focusing on only the number of electron pairs around the *central atom*, and ignoring all other valence electrons present. We begin by drawing the Lewis electron structure of the molecule or ion. Because experiments have shown that the presence of multiple bonds has only a minor effect on the arrangement of atoms in space, *we treat all multiple bonds in a Lewis structure as single electron pairs*. After the electron-pair geometry is determined, we can predict the molecular geometry of a molecule or polyatomic ion by using the electron pairs around the A atom to form bonds between the central A atom and the B atoms. We will use the following procedure:

Step 1 Draw the Lewis electron structure of the molecule or polyatomic ion.

Step 2 Count the number of valence-electron pairs around the atom of interest, treating any multiple bonds or single unpaired electrons as single electron pairs. This number determines the electron-pair geometry around the central atom according to Table 9.1.

Step 3 Identify each electron pair as a bonding pair (BP) or lone (nonbonding) pair (LP). Remember that the VSEPR model counts a single unpaired electron as a lone pair.

Step 4 To determine the molecular geometry, arrange the bonded atoms around the central atom to minimize repulsions between electron pairs. Identify the molecular geometry from Table 9.2.

Molecules with No Lone Pairs of Electrons

We will illustrate the use of this procedure with several examples that have the general formula AB_n (n = 2–6), beginning with molecules that have no lone pairs of electrons around the central atom.

AB_2: BeH_2

Step 1 The central atom, beryllium, contributes two valence electrons, and each hydrogen atom contributes one. The Lewis electron structure is

H:Be:H or H—Be—H

Lewis structure

Step 2 There are two electron pairs around the central atom. The electron-pair geometry that minimizes the repulsions between the electron pairs is therefore *linear*:

Electron-pair geometry (linear)

Step 3 Both electron pairs around the central atom are bonding pairs (BP).

Step 4 With two bonding pairs and no lone pairs, the molecular geometry of BeH_2 is the same as the electron-pair geometry—that is, linear:

H—Be—H

Molecular geometry (linear)

AB_3: BCl_3

Step 1 The central atom, boron, contributes three valence electrons, and each chlorine atom contributes seven. The Lewis electron structure is

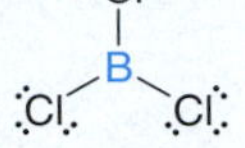

Lewis structure

As you learned in Chapter 8, this compound is electron deficient because boron has less than an octet of electrons.

Step 2 There are three electron pairs around the central atom. According to Table 9.1, the electron-pair geometry is therefore *trigonal planar*:

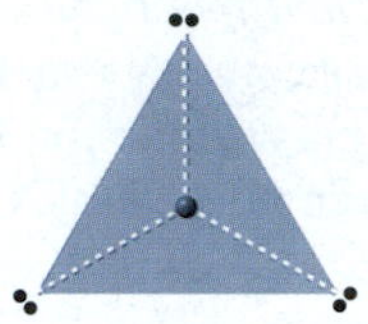

Electron-pair geometry (trigonal planar)

Step 3 With three electron pairs and three bonded Cl atoms, all electron pairs around the central atom must be bonding pairs (BP).

Step 4 Because each electron pair is a bonding pair, the molecular geometry is trigonal planar. The electron-pair geometry and the molecular geometry are the same:

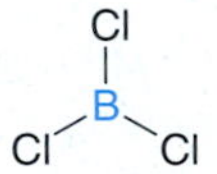

Molecular geometry (trigonal planar)

We now examine another AB_3 species, the carbonate ion.

AB_3: CO_3^{2-}

Step 1 The central atom, carbon, has four valence electrons, and each oxygen atom has six. As you discovered in Chapter 8, the Lewis electron structure can be represented as

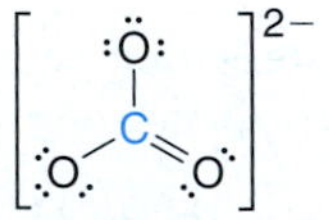

Lewis structure

Step 2 Because the VSEPR model treats multiple bonds as single electron pairs, according to the VSEPR model there are three electron pairs around carbon in the carbonate ion. The electron-pair geometry that minimizes electron pair–electron pair repulsions in CO_3^{2-} is again *trigonal planar*, as in BCl_3:

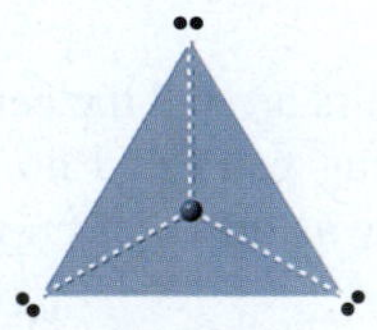

Electron-pair geometry (trigonal planar)

Step 3 Because there are three electron pairs and three bonded O atoms around the central atom, all electron pairs must be bonding pairs (BP).

Step 4 Each O atom shares one electron pair to give a trigonal planar molecular geometry:

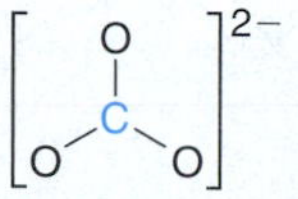

Molecular geometry (trigonal planar)

If we compare the representation shown in step 4 with the Lewis electron structure shown in step 1, we see that the VSEPR model specifies *only* the molecular geometry. It indicates nothing about multiple bonds, nor does it focus on octets of electrons. The Lewis electron structure gives a better picture of the bonding in carbonate, with one C═O double bond and two C—O single bonds, but it does not show that the carbonate ion is trigonal planar. Consequently, the two approaches are complementary: one predicts the structure of the species, while the other describes its bonding. Linear and planar systems such as BeH_2, BCl_3, and CO_3^{2-} are relatively straight-forward to visualize. We now examine methane (CH_4), for which a two-dimensional representation is not adequate.

AB_4: CH_4

Step 1 The central carbon atom contributes four valence electrons, and each hydrogren atom has one electron, so the full Lewis electron structure is

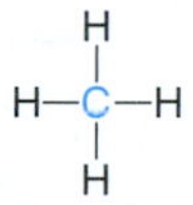

Lewis structure

Step 2 There are four electron pairs around the central atom. From Table 9.1 we see that the electron-pair geometry that minimizes electron–electron repulsions is *tetrahedral*:

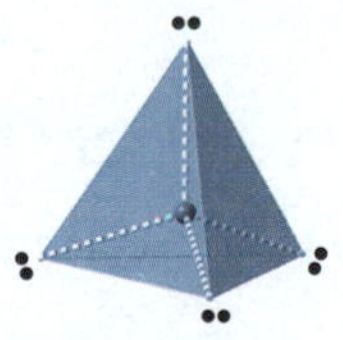

Electron-pair geometry (tetrahedral)

Step 3 Because there are four electron pairs and four bonded H atoms, all electron pairs must be bonding pairs (BP).

Step 4 Connecting each hydrogen atom to carbon using one of the bonding pairs gives a tetrahedral arrangement of hydrogen atoms, thereby minimizing BP–BP repulsions:

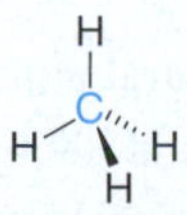

Molecular geometry (tetrahedral)

The electron-pair geometry and the molecular geometry are the same.

We now describe an example in which the central atom has an expanded valence, which, as you learned in Section 8.6, is not easily understood in terms of an octet configuration.

AB_6: SF_6

Step 1 The central atom, S, contributes six valence electrons, and each F atom has seven, so the Lewis electron structure is

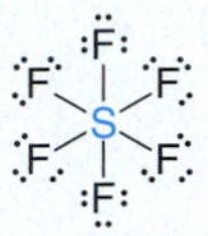

Lewis structure

With an expanded valence, we know from Section 8.6 that this species is an exception to the octet rule.

Step 2 There are six electron pairs around the central atom. Table 9.1 shows that the electron-pair geometry that minimizes electron–electron repulsions is *octahedral*:

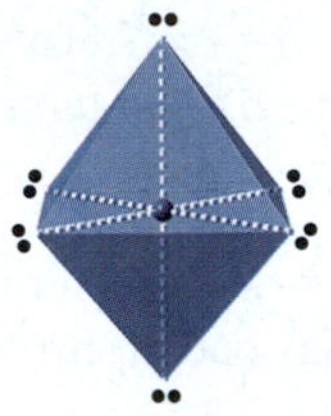

Electron-pair geometry (octahedral)

Step 3 Because there are six electron pairs around S and six bonded F atoms, all electron pairs must be bonding pairs (BP).

Step 4 Connecting a fluorine atom to sulfur using each of the bonding pairs gives an octahedral arrangement of atoms:

Molecular geometry (octahedral)

Hence, the electron-pair geometry and the molecular geometry are again the same.

Molecules with Lone Pairs of Electrons

In each of the preceding examples, the number of bonded atoms is the same as the number of electron pairs. As a result, the electron-pair geometry and the molecular geometry are the same. This is not always the case, however. In compounds that contain one or more lone pairs of electrons, the electron-pair geometry and the molecular geometry are *not* identical, as illustrated in the next examples.

Note the pattern

The molecular geometry and the electron-pair geometry are the same only if the central atom has no lone pairs of electrons.

AB_3: NH_3

Step 1 In ammonia, the central atom, N, has five valence electrons and each H donates one, producing the Lewis structure

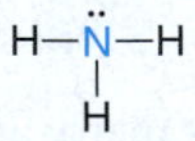

Lewis structure

Step 2 There are four electron pairs around N, so the electron-pair geometry is *tetrahedral*:

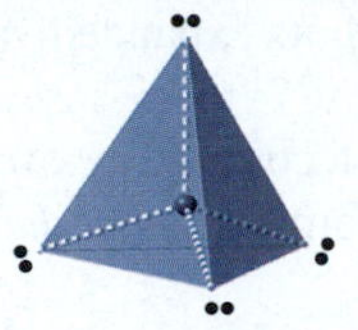

Electron-pair geometry (tetrahedral)

Step 3 With four electron pairs and only three bonded H atoms, one electron pair must be a lone pair.

Step 4 Because molecular geometries are determined by the arrangement of the *bonded* atoms, the molecular geometry of NH_3 is described as *trigonal pyramidal*:

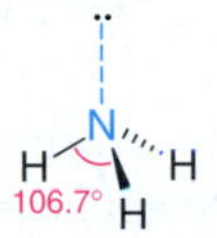

Molecular geometry (trigonal pyramidal)

The actual H—N—H bond angle in ammonia is 106.7°, which is less than the ideal value of 109.5° for a tetrahedron. The smaller bond angle suggests that the lone pair actually occupies a larger region of space around the central atom than the bonding pairs. The H—N—H bond angles are therefore slightly smaller than the angles in an idealized tetrahedral structure.

In CH_4, all four electron pairs are bonding pairs. In NH_3, three are bonding pairs and one is a lone pair. As you discovered in Chapter 8, in H_2O, two electron pairs are bonding pairs and two are lone pairs. All of these molecules have the same electron-pair geometry (tetrahedral) because they all contain four electron pairs. In the case of methane, however, the electron-pair geometry and the molecular geometry are the same, but in ammonia, the molecular geometry is trigonal pyramidal. Similarly, in water, the electron-pair geometry is tetrahedral but the molecular geometry is bent.

The angles in NH_3 and H_2O deviate significantly from the angles of a perfect tetrahedron, decreasing steadily from the tetrahedral value of 109.5° in CH_4 to 106.7° in NH_3 and to 104.5° in H_2O:

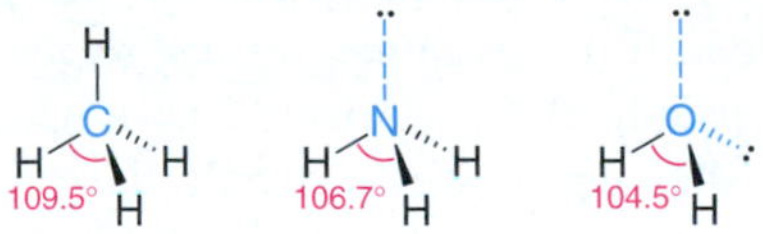

Notice that as the number of lone pairs increases, the angle between the bonding electron pairs decreases. This is because lone pairs occupy more space around the central atom than bonding electron pairs (Figure 9.2). A bonding pair is attracted simultaneously to *two* atomic nuclei, so the electron density is concentrated along the internuclear axis. In contrast, a lone pair of electrons is attracted to only a single nucleus, and as a result it tends to occupy a larger region of space. Thus, the angle between bonded atoms decreases as the space occupied by lone pairs of electrons increases. The following examples further illustrate the structural effect of lone pairs of electrons.

AB_5: BrF_5

Step 1 The central atom, Br, has seven valence electrons. Each F also has seven, so the Lewis electron structure is

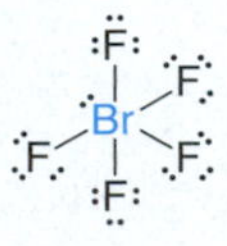

Lewis structure

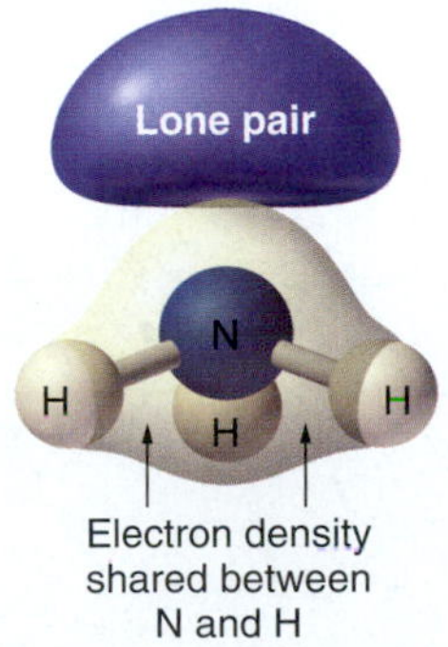

Figure 9.2 The difference in the space occupied by a lone pair of electrons and by a bonding pair. This composite model of electron distribution and negative electrostatic potential in ammonia shows that a lone pair of electrons occupies a larger region of space around the nitrogen atom than does a bonding pair of electrons that is shared with a hydrogen atom.

Once again, with its expanded valence, this species is an exception to the octet rule.

Step 2 There are six electron pairs around Br, so according to Table 9.1, the electron-pair geometry is *octahedral*:

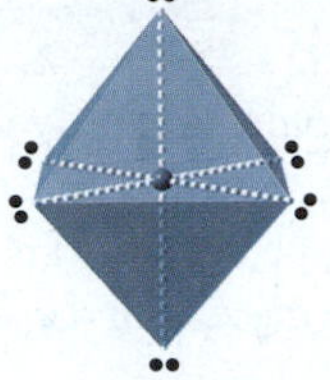

Electron-pair geometry (octahedral)

Step 3 Because there are six electron pairs but only five bonded F atoms, one electron pair must be a lone pair.

Step 4 Placing five F atoms around Br while minimizing BP–BP and LP–BP repulsions gives the following structure:

Molecular geometry (square pyramidal)

The BrF_5 structure has four F atoms in a plane in an *equatorial* position and one F atom located below the center of the plane in an *axial* position, so its molecular structure is *square pyramidal*. Note that the F—Br—F angle between the unique axial F atom and any of the other four equatorial F atoms is only 85.1°, smaller than the 90° angle expected for an octahedral arrangement of electron pairs. This effect is attributable to the lone pair, which occupies more space than the bonding electron pairs.

The molecules in the preceding two examples both contained a single lone pair on the central atom. It did not matter where we placed the lone pair of electrons because all positions were equivalent. If, for example, we had selected one of the other electron pairs to be the lone pair in BrF_5, we would have generated exactly the same molecular geometry. Our drawing, however, would have been rotated in space by either 180° or 90°:

180° rotation 90° rotation

In some cases, however, the location of the lone pairs is important. The following examples illustrate this point.

AB_4: SF_4

Step 1 The S atom has six valence electrons and each F has seven, so the Lewis electron structure is

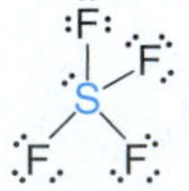

Lewis structure

With an expanded valence, this species is an exception to the octet rule.

Step 2 There are five electron pairs around S, so the electron-pair geometry is *trigonal bipyramidal* (Table 9.1):

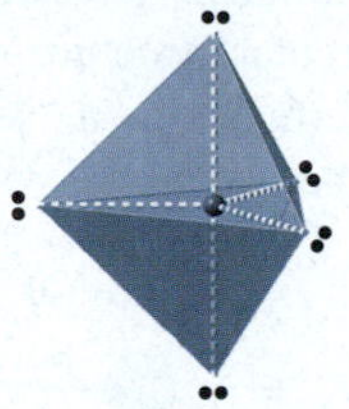

Electron-pair geometry (trigonal bipyramidal)

Step 3 Because there are five electron pairs but only four bonded F atoms, there must be one lone pair.

Step 4 In a trigonal bipyramid, the five positions are *not* chemically equivalent. There are three equatorial positions separated by 120° from one another, and two axial positions at 90° to the equatorial plane:

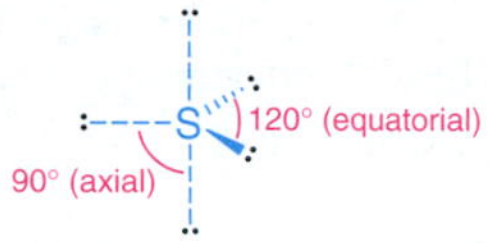

We can place the lone electron pair in either the axial or the equatorial position, as shown below. How do we decide which of these structures is more stable? According to the VSEPR model, we must choose the structure that minimizes electrostatic repulsions. Lone pair–lone pair repulsions increase rapidly as the angle between the lone pairs decreases from 180° to 90° (Figure 9.3). Because lone pairs occupy more space than bonding pairs, 90° lone pair–lone pair (LP–LP) repulsions are greater than 90° lone pair–bonding pair (LP–BP) repulsions, which in turn are greater than 90° bonding pair–bonding pair (BP–BP) repulsions. The LP–LP and LP–BP interactions found in SF_4 can be tabulated as follows:

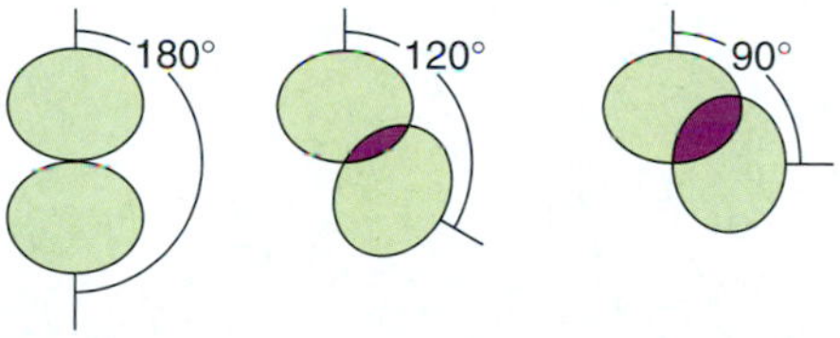

Figure 9.3 Illustration of the area shared by two electron pairs versus the angle between them. At 90°, the two electron pairs share a relatively large region of space, which leads to strong repulsive electron–electron interactions.

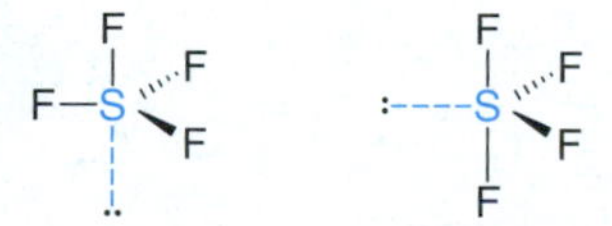

	Lone pair axial	Lone pair equatorial
Interactions		
90° LP–LP	0	0
90° LP–BP	3	2

With only one lone pair on the central atom, neither structure has any 90° LP–LP interactions. The structure with an axial lone pair has three 90° LP–BP interactions, however, whereas the structure with an equatorial lone pair has only two. Consequently, the most stable structure for SF_4 is the one with the lone pair in an equatorial position.* The actual structure of SF_4 is

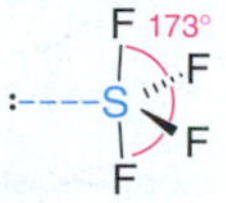

* Because 90° BP–BP interactions are less important than 90° LP–LP and LP–BP interactions, it is necessary to consider them *only* if the numbers of 90° LP–LP and LP–BP interactions are the same for two or more possible structures.

The axial F atoms are bent slightly toward the equatorial F atoms (the F_{axial}—S—F_{axial} angle is 173° rather than 180°), which indicates that the lone pair of electrons, occupying more space than a bonding pair, is located in the equatorial plane. Although this bent structure is derived from a trigonal bipyramid, the molecular structure of SF_4 is *not* trigonal pyramidal. Instead, it is usually referred to as a "seesaw" structure, for reasons that become apparent when the structure is rotated 90°:

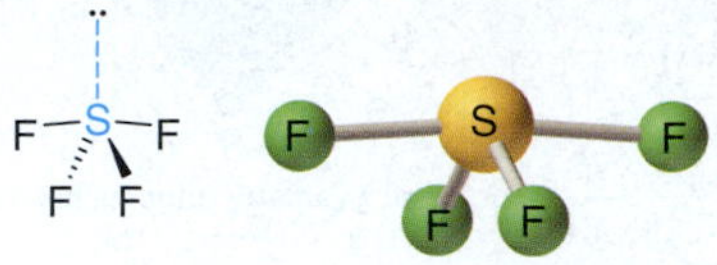

Molecular geometry (seesaw)

In the next example, we examine a compound that contains more than one lone pair on the central atom.

AB_3: BrF_3

Step 1 The Br atom has seven valence electrons, and each F has seven, so the Lewis electron structure is

Lewis structure

This compound has an expanded valence, so it is an exception to the octet rule.

Step 2 There are five electron pairs around the central atom, so the electron-pair geometry is again *trigonal bipyramidal*:

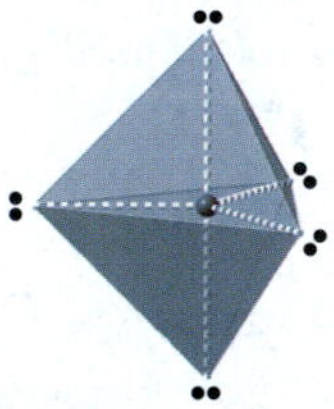

Electron-pair geometry (trigonal bipyramidal)

Step 3 Because there are five electron pairs and only three bonded F atoms, the structure must have two lone pairs.

Step 4 Because the axial and equatorial positions in a trigonal bipyramid are not equivalent, we must select the structure with the fewest electron-pair repulsions. Is it the one with both lone pairs axial, both equatorial, or one axial and one equatorial?

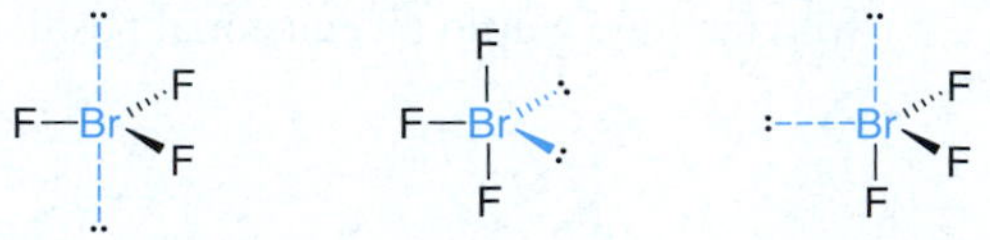

	Lone pairs axial (a)	Lone pairs equatorial (b)	Lone pairs axial and equatorial (c)
Interactions			
90° LP–LP	0	0	1
90° LP–BP	6	4	3

Structure c has one 90° LP–LP interaction, and the other two structures have none. Consequently, structure c can be eliminated from consideration. Structure a, with both lone pairs in axial positions, has six 90° LP–BP interactions, and structure b, with both lone pairs in equatorial positions, has only four 90° LP–BP interactions. We therefore predict that BrF_3 has structure b, although it is somewhat distorted because of the additional space occupied by the lone pairs of electrons. In fact, the axial F atoms are bent back slightly toward the single equatorial F atom (F_{axial}—Br—F_{axial} angle = 172°). The BrF_3 structure is usually referred to as "T-shaped", again for reasons that become apparent when the structure is rotated 90°:

Note the pattern

Because lone pairs occupy more space around the central atom than bonding pairs, lone pairs experience greater electrostatic repulsion.

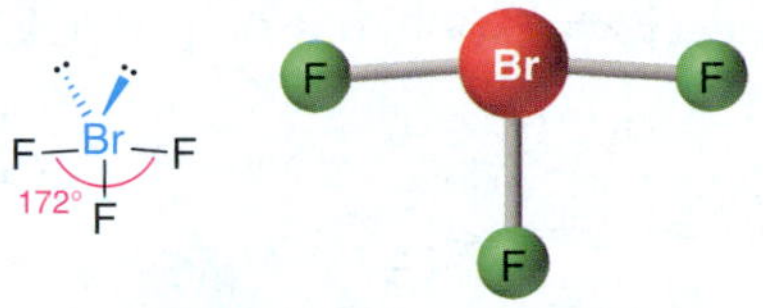

Molecular geometry (T-shaped)

By combining Lewis electron structures with the VSEPR model, we can describe the bonding in a molecule or polyatomic ion and predict its three-dimensional geometry. As you discovered for the carbonate ion, for example, the VSEPR model correctly predicts a trigonal planar structure. In contrast, any of the three resonance structures from the Lewis electron-pair approach predicts a C═O double bond and two C—O single bonds. Because the VSEPR model ignores multiple bonds, it does not address the concept of resonance.

Like lone pairs of electrons, multiple bonds cause other bond angles to be somewhat smaller than expected. Two examples are formaldehyde, CH_2O, and ethylene, C_2H_4, whose structures are shown here:

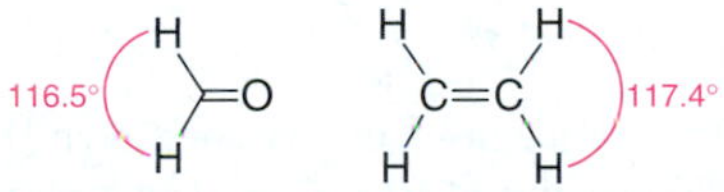

According to the VSEPR model, each carbon is trigonal planar. The H—C—H angle in both molecules is smaller than 120°, however, because the presence of multiple bonds increases the electron density between the bonded C and O atoms in formaldehyde and the two C atoms in ethylene. Thus, multiple bonds occupy more space than single bonds.

EXAMPLE 9.1

Using the VSEPR model, predict the electron-pair geometry and the molecular structure of each molecule or ion: **(a)** PF_5; **(b)** H_3O^+.

Given Two chemical species

Asked for Electron-pair geometry, molecular structure

Strategy

A Draw the Lewis electron structure of the molecule. From Table 9.1, select the structure that corresponds to the number of valence-electron pairs. This is the electron-pair geometry.

B Determine the number of bonding pairs of electrons. Then subtract the number of bonding pairs from the total number of electron pairs to obtain the number of lone pairs.

Ⓒ Use the number of bonding pairs and Table 9.2 to determine the molecular geometry. If the compound has lone pairs of electrons, minimize the number of repulsive interactions to obtain the correct molecular structure.

Solution

(a) Ⓐ✓ The central P atom of PF_5 has five valence electrons, and each F atom has seven, so the Lewis electron structure is

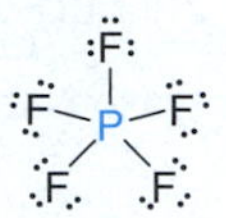

showing five electron pairs about P. This means that the electron-pair geometry of PF_5 is trigonal bipyramidal:

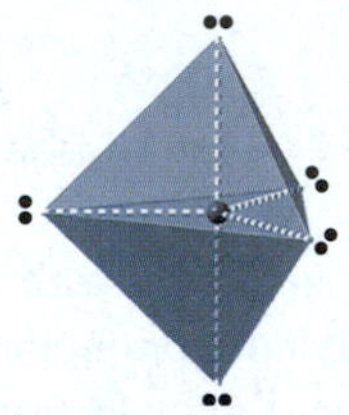

Ⓑ✓ The compound has five bonding pairs and no lone pairs. Ⓒ✓ With no lone pairs, the molecular geometry is the same as the electron-pair geometry. Phosphorus pentafluoride is used as a catalyst in certain organic reactions.

(b) Ⓐ✓ The central O atom has six valence electrons, and each H atom contributes one. Subtracting one electron for the positive charge gives a total of eight electrons, so the Lewis electron structure is:

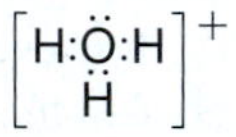

With four electron pairs, the electron-pair geometry is tetrahedral. Ⓑ✓ There are three bonding pairs, leaving one lone pair. Ⓒ✓ With three bonded H atoms and one lone pair, the molecular geometry of H_3O^+ is trigonal pyramidal (analogous to that of ammonia).

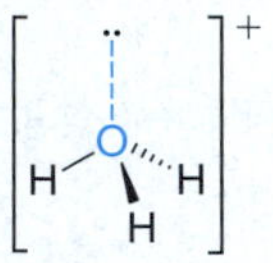

EXERCISE 9.1

Predict the molecular structure of each substance using the VSEPR model: **(a)** XeO_3; **(b)** PF_6^-; **(c)** NO_2^+.

Answer **(a)** trigonal pyramidal; **(b)** octahedral; **(c)** linear

EXAMPLE 9.2

Predict the electron-pair geometry and the molecular structure of (a) XeF_2 and (b) SO_2.

Given Two chemical compounds

Asked for Electron-pair geometry, molecular structure

Strategy

Use the strategy given in Example 9.1.

Solution

(a) A Xenon contributes eight electrons and each fluorine seven, so the Lewis electron structure is

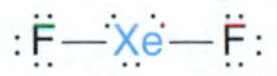

with five electron pairs around the central atom. This gives an electron-pair geometry that is trigonal bipyramidal. B There are five electron pairs of which two are bonding pairs, giving three lone pairs. C With only two terminal F atoms, there are three possible structures to choose from: both F atoms can be axial, one can be axial and one equatorial, or both can be equatorial. We begin by tabulating the LP–LP interactions:

Interactions			
90° LP–LP	0	2	2

With both fluorine atoms in axial positions, structure a minimizes lone pair–lone pair repulsions. Thus, the molecular structure of XeF_2 is linear.

(b) A Sulfur donates six valence electrons, and the oxygen atoms contribute six each. With 18 valence electrons, one possible Lewis electron structure is shown below (several other resonance structures can be drawn).

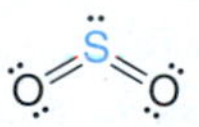

Because we count double bonds as single bonding pairs in the VSEPR model, there are three electron pairs around the central atom. Hence, the electron-pair geometry is trigonal planar. B C With three electron pairs and two bonded atoms, there are two bonding pairs and one lone pair. The molecular geometry must be bent, with an O—S—O angle of about 120°.

EXERCISE 9.2

Predict the molecular structure of (a) SO_3 and (b) XeF_4.

Answer (a) trigonal planar; (b) square planar

The relationship between the number of electron pairs around a central atom, the number of lone pairs, and the molecular geometry is summarized in Table 9.3. For molecules that have no lone pairs on the central atom, the table shows that the molecular geometry is always the same as the electron-pair geometry. For molecules and

TABLE 9.3 Overview of molecular geometries

Electron Pairs	2	3	4	5	6
Electron Pair Geometry 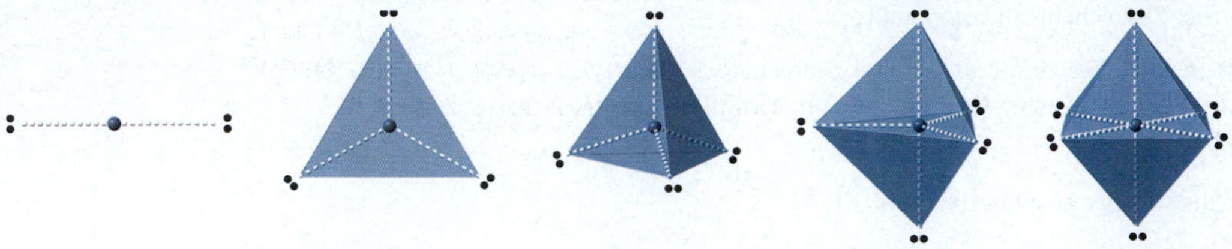	Linear	Trigonal planar	Tetrahedral	Trigonal bipyramidal	Octahedral
Molecular Geometry: Zero Lone Pairs 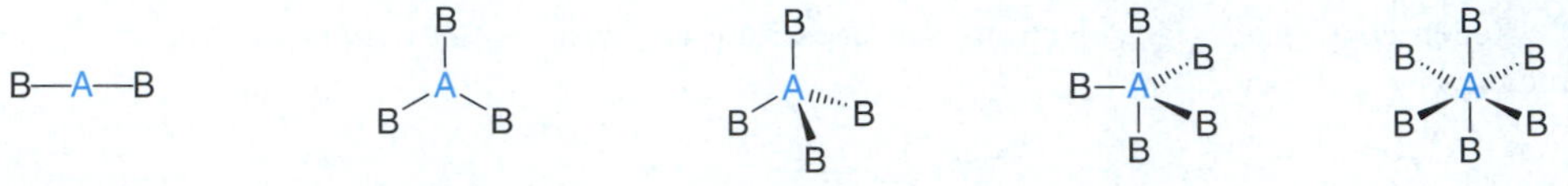	Linear AB_2	Trigonal planar AB_3	Tetrahedral AB_4	Trigonal bipyramidal AB_5	Octahedral AB_6
Molecular Geometry: One Lone Pair 		Bent (V-shaped) AB_2	Trigonal pyramidal AB_3	Seesaw AB_4	Square pyramidal AB_5
Molecular Geometry: Two Lone Pairs 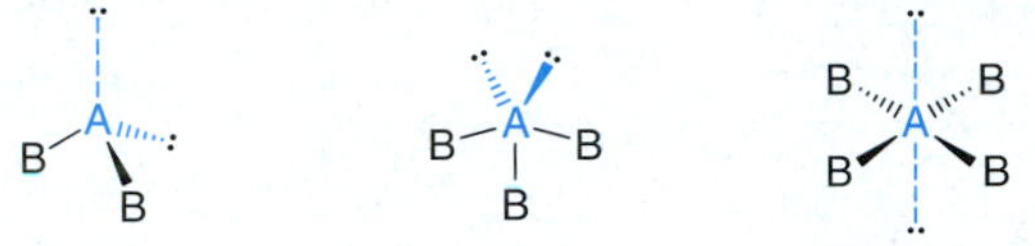			Bent (V-shaped) AB_2	T-shaped AB_3	Square planar AB_4
Molecular Geometry: Three Lone Pairs 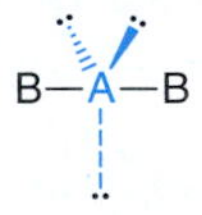				Linear AB_2	

polyatomic ions that have one or more lone pairs on the central atom, however, the molecular geometry is *not* the same as the electron-pair geometry but derived from it. For example, a bent (V-shaped) molecule can result from a trigonal planar arrangement of three electron pairs (with one lone pair) or from a tetrahedral arrangement of four electron pairs (with two lone pairs). The tetrahedral electron-pair geometry can also produce a pyramidal AB_3 molecular structure with one lone pair on the central atom. The trigonal bipyramidal electron-pair geometry can produce seesaw (AB_4), T-shaped (AB_3), and linear (AB_2) molecular geometries with one, two, and three lone pairs on the central atom, respectively. The octahedral electron-pair geometry can produce square pyramidal AB_5 or square planar AB_4 molecular geometries, with one or two lone pairs on the central atom, respectively.

Molecules with No Well-Defined Central Atom

The VSEPR model can be used to predict the structure of somewhat more complex molecules by treating them as linked AB_n fragments. We will demonstrate with methyl isocyanate (CH_3—N=C=O), a volatile and highly toxic molecule used to produce the pesticide Sevin. In 1984, large quantities of Sevin were accidentally released in Bhopal, India, when water leaked into storage tanks. The resulting highly exothermic reaction caused a rapid increase in pressure that ruptured the tanks, releasing large amounts of methyl isocyanate that killed approximately 3800 people and wholly or partially disabled about 50,000 others. Moreover, significant damage was done to livestock and crops.

We can treat methyl isocyanate as linked AB_n fragments beginning with the carbon atom at the left, which is connected to three H atoms and one N atom by single bonds. The four bonds around carbon mean that it must be surrounded by four bonding electron pairs in a configuration similar to AB_4. We can therefore predict the CH_3—N portion of the molecule to be roughly tetrahedral, similar to methane:

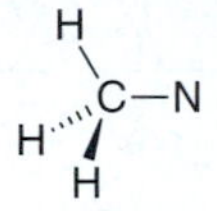

The nitrogen atom is connected to one carbon by a single bond and to the other carbon by a double bond, producing a total of three bonds, C—N=C. For nitrogen to have an octet of electrons, it must also have a lone pair:

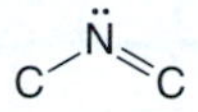

Because multiple bonds are not shown in the VSEPR model, the nitrogen is effectively surrounded by three electron pairs. Thus, according to the VSEPR model, the C—N=C fragment should be bent with an angle less than 120°.

The carbon in the —N=C=O fragment is doubly bonded to both nitrogen and oxygen, which in the VSEPR model gives carbon a total of two electron pairs. The N=C=O angle should therefore be 180°, or linear. The three fragments combine to give

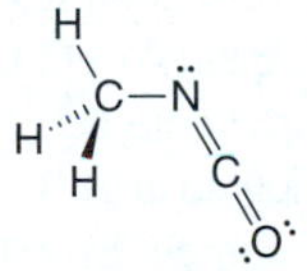

We predict that all four nonhydrogen atoms lie in a single plane, with a C—N—C angle of approximately 120°. The experimentally determined structure of methyl isocyanate confirms our prediction (Figure 9.4).

Certain patterns are seen in the structures of moderately complex molecules. For example, carbon atoms that have four bonds (such as the carbon on the left in our drawing of methyl isocyanate) are generally tetrahedral. Similarly, the carbon atom on the right has two double bonds that are similar to those in CO_2 (AB_2), so its geometry, like that of CO_2, is linear. Recognizing similarities to simpler molecules will help you predict molecular structures of more complex molecules.

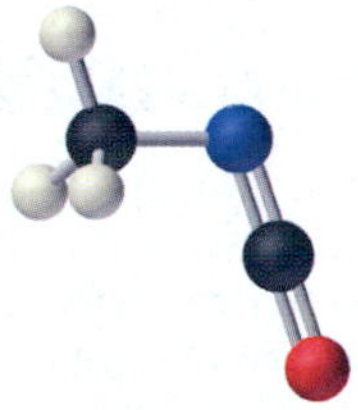

Figure 9.4 The experimentally determined structure of methyl isocyanate.

EXAMPLE 9.3

Use the VSEPR model to predict the molecular structure of propyne, $H_3C—C≡CH$, a gas with some anesthetic properties.

Given Chemical compound

Asked for Molecular structure

Strategy

Count the number of electron pairs around each carbon, recognizing that in the VSEPR model, a multiple bond counts as a single bond. Use Table 9.2 to determine the molecular geometry around each carbon atom, and then deduce the structure of the molecule as a whole.

Solution

Because the carbon atom on the left is bonded to four other atoms, we know that it is approximately tetrahedral. The next two carbon atoms share a triple bond, and each has an additional single bond. Because a multiple bond is counted as a single bond in the VSEPR model, each of the carbon atoms behaves as if it had two electron pairs. This means that both of these carbons are linear, with C—C≡C and C≡C—H angles of 180°.

EXERCISE 9.3

Predict the geometry of allene, $H_2C{=}C{=}CH_2$, a compound with narcotic properties that is used in making more complex organic molecules.

Answer The terminal carbon atoms are trigonal planar, the central carbon is linear, and the C—C—C angle is 180°.

Molecular Dipole Moments

In Chapter 8, you learned how to calculate the *dipole moments* of simple diatomic molecules. In more complex molecules that contain polar covalent bonds, the three-dimensional geometry and the compound's symmetry determine whether there is a net dipole moment. Mathematically, dipole moments are *vectors*; they possess both a *magnitude* and a *direction*. The dipole moment of a molecule is therefore the *vector sum* of the dipole moments of the individual bonds in the molecule. If the individual bond dipole moments cancel one another, there is no net dipole moment. Such is the case for CO_2, a linear molecule (Figure 9.5a). Each C—O bond in CO_2 is polar, yet experiments show that the CO_2 molecule has no dipole moment. Because the two C—O bond dipoles in CO_2 are equal in magnitude and oriented at 180° to each other, they cancel. As a result, the CO_2 molecule has no *net* dipole moment even though it has a substantial separation of charge. In contrast, the H_2O molecule is not linear (Figure 9.5b); it is bent in three-dimensional space, so the dipole moments do not cancel each other. Thus, a molecule such as H_2O has a net dipole moment. This concentration of negative charge on the oxygen and positive charge on the two hydrogens allows H_2O to hydrogen-bond to other polarized or charged species, including other water molecules (Section 4.1).

Other examples of molecules that contain polar bonds are shown in Figure 9.6. Notice that in molecular structures that are highly symmetrical (most notably tetrahedral and square planar AB_4, trigonal bipyramidal AB_5, and octahedral AB_6), individual bond dipole moments completely cancel, and there is no net dipole moment. Although a molecule like $CHCl_3$ is best described as tetrahedral, the atoms bonded to carbon are not identical. Consequently, the bond dipole moments cannot cancel one another, and the molecule has a dipole moment. Due to the arrangement of the bonds in molecules

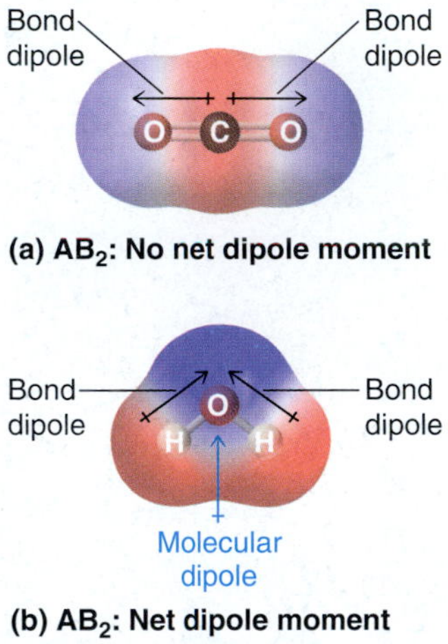

Figure 9.5 How individual bond dipole moments add to give an overall molecular dipole moment for two AB_2 molecules with different structures, CO_2 and H_2O. (a) In CO_2, the C—O bond dipoles are equal in magnitude but oriented in opposite directions (at 180°). Their vector sum is zero, and CO_2 therefore has no net dipole. (b) In H_2O, the O—H bond dipoles are also equal in magnitude, but they are oriented at 104.5° to each other. Hence, the vector sum is not zero, and H_2O has a net dipole moment.

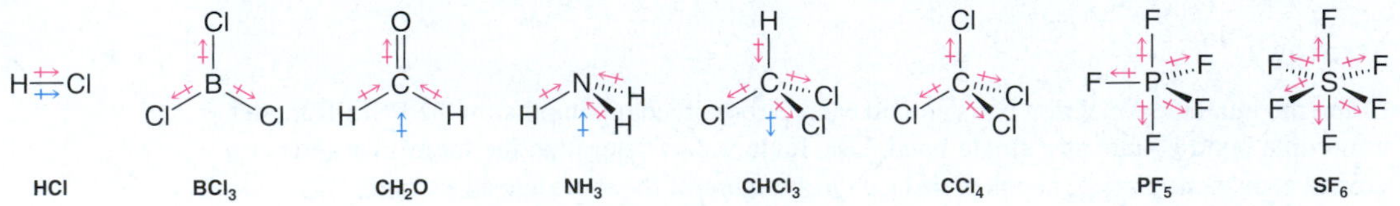

Figure 9.6 Examples of molecules that contain polar bonds. Individual bond dipole moments are indicated in red. Due to their different three-dimensional structures, some molecules that contain polar bonds have a net dipole moment (HCl, CH_2O, NH_3, $CHCl_3$), indicated in blue, whereas others do not because the bond dipole moments cancel (BCl_3, CCl_4, PF_5, SF_6).

that have V-shaped, trigonal pyramidal, seesaw, T-shaped, and square pyramidal geometries, the bond dipole moments cannot cancel one another. Consequently, molecules with these geometries always have a nonzero dipole moment.

EXAMPLE 9.4

Which of these molecules has a net dipole moment? (a) H_2S; (b) NH_3; (c) BF_3

Given Three chemical compounds

Asked for Net dipole moment

Strategy

For each three-dimensional molecular structure, predict whether the bond dipoles cancel. If they do not, then the molecule has a net dipole moment.

Solution

(a) The total number of electrons around the central atom, S, is eight, which gives four electron pairs. Two of these electron pairs are bonding pairs and two are lone pairs, so the molecular structure of H_2S is bent. The bond dipoles cannot cancel one another, so the molecule has a net dipole moment.

(b) Ammonia has a tetrahedral electron-pair geometry and a trigonal pyramidal molecular geometry. Because of the lone pair of electrons on nitrogen, the molecule is not symmetrical and the bond dipoles of NH_3 cannot cancel one another. This means that NH_3 has a net dipole moment.

(c) The electron-pair geometry and the molecular geometry of BF_3 are trigonal planar. Because all the B—F bonds are equal and the molecule is highly symmetrical, the dipoles cancel one another in three-dimensional space. Thus, BF_3 has a net dipole moment of zero:

EXERCISE 9.4

Which of these species has a net dipole moment? (a) CH_3Cl; (b) SO_3; (c) XeO_3

Answer CH_3Cl and XeO_3

9.2 ● Localized Bonding and Hybrid Atomic Orbitals

MGC Hybrid Orbitals

Although the VSEPR model is a simple and useful method for qualitatively predicting the structures of a wide range of compounds, it is *not* infallible. It predicts, for example, that H_2S and PH_3 should have structures similar to those of H_2O and NH_3, respectively. In fact, structural studies have shown that the H—S—H and H—P—H angles are more than 12° smaller than the corresponding bond angles in H_2O and NH_3. More disturbing, the VSEPR model predicts that the simple Group-2 halides (MX_2), which have four valence electrons, should all be two-electron-pair AB_2 molecules with linear X—M—X geometries. Instead, many of these species, including SrF_2 and BaF_2, are significantly bent. A more sophisticated treatment of bonding is needed for systems such as these. In this section, we present a quantum mechanical description of bonding, in which bonding electrons are viewed as being localized

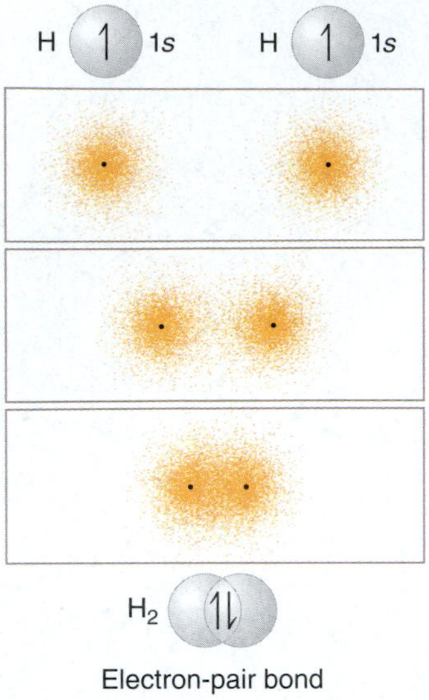

Figure 9.7 Overlap of two singly occupied hydrogen 1*s* atomic orbitals produces an H—H bond in H_2. The formation of H_2 from two hydrogen atoms, each containing a single electron in a 1*s* orbital, occurs as the electrons are shared to form an electron-pair bond, as indicated schematically by the gray spheres and black arrows. The orange electron density distributions show that the formation of an H_2 molecule increases the electron density in the region between the two positively charged nuclei.

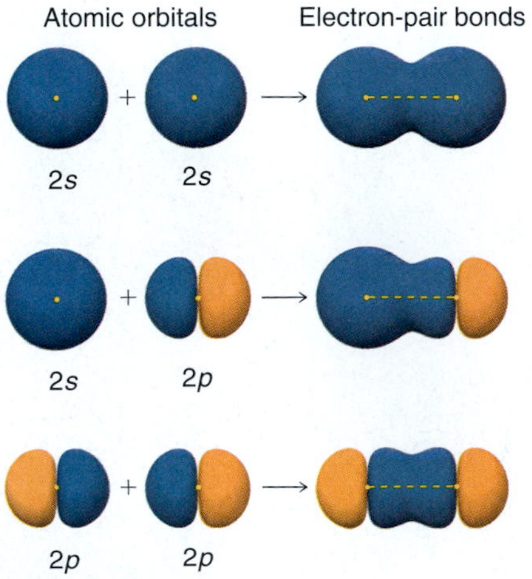

Figure 9.8 Three different ways to form an electron-pair bond. An electron-pair bond can be formed by the overlap of any of the following combinations of two singly occupied atomic orbitals: two *ns* atomic orbitals (top), an *ns* and an *np* atomic orbital (middle), and two *np* atomic orbitals (bottom) where $n = 2$. The positive lobe is indicated in yellow and the negative lobe in blue.

between the nuclei of the bonded atoms. The overlap of bonding orbitals is substantially increased through a process called *hybridization*, which results in the formation of stronger bonds.

Valence Bond Theory: A Localized Bonding Approach

In Chapter 8, you learned that as two hydrogen atoms approach each other from an infinite distance, the energy of the system reaches a minimum. This region of minimum energy in the energy diagram corresponds to the formation of a covalent bond between the two atoms at an H—H distance of 74 pm (Figure 8.8). According to quantum mechanics, bonds form between atoms because their atomic orbitals overlap, with each region of overlap accommodating a maximum of two electrons with opposite spin, in accordance with the Pauli principle. In this case, a bond forms between the two hydrogen atoms when the singly-occupied 1*s* atomic orbital of one hydrogen atom overlaps with the singly-occupied 1*s* atomic orbital of a second hydrogen atom. Electron density between the nuclei is increased because of this orbital overlap and results in a *localized electron-pair bond* (Figure 9.7).

Although Lewis and VSEPR structures also contain localized electron-pair bonds, neither description uses an atomic orbital approach to predict the stability of the bond. Doing so forms the basis for a description of chemical bonding known as **valence bond theory**, which is built on two assumptions:

1. The strength of a covalent bond is proportional to the amount of overlap between atomic orbitals. That is, the greater the overlap, the more stable the bond.
2. An atom can use different combinations of atomic orbitals to maximize the overlap of orbitals used by bonded atoms.

Figure 9.8 shows an electron-pair bond formed by the overlap of two *ns* atomic orbitals, two *np* atomic orbitals, and an *ns* and an *np* orbital where $n = 2$. Maximum overlap occurs between orbitals that have the same spatial orientation and similar energies.

Let's examine the bonds in BeH_2, for example. According to the VSEPR model, BeH_2 is a linear compound that contains four valence electrons and two Be—H bonds. Its bonding can also be described using an atomic orbital approach. Beryllium has a $1s^2\,2s^2$ electron configuration, and each H atom has a $1s^1$ electron configuration. Because the Be atom has a filled 2*s* subshell, however, it has no singly occupied orbitals available to overlap with the singly occupied 1*s* orbitals on the H atoms. If a singly occupied 1*s* orbital on hydrogen were to overlap with a filled 2*s* orbital on beryllium, the resulting bonding orbital would contain *three* electrons, but the maximum allowed by quantum mechanics is *two*. How then is beryllium able to bond to two hydrogen atoms? One way would be to add enough energy to excite one of its 2*s* electrons into an empty 2*p* orbital and reverse its spin, in a process called **promotion**:

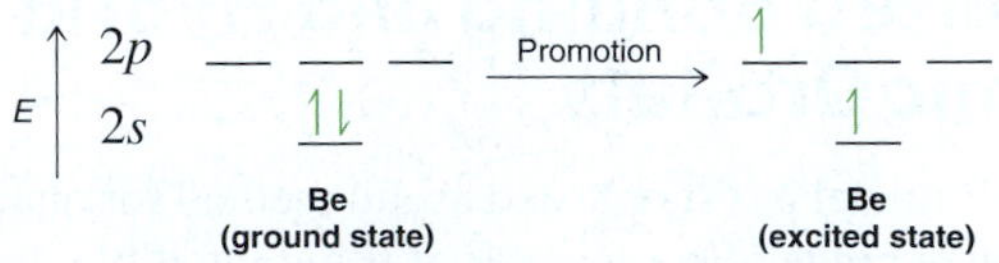

In this excited state, the Be atom would contain two singly occupied atomic orbitals (the 2*s* and one of the 2*p* orbitals), each of which could overlap with a singly occupied 1*s* orbital of an H atom to form an electron-pair bond. Although this would produce BeH_2, the two Be—H bonds would not be equivalent: the 1*s* orbital of one hydrogen atom would overlap with a Be 2*s* orbital, and the 1*s* orbital of the other hydrogen atom would overlap with an orbital of a different energy, a Be 2*p* orbital. Experimental evidence indicates, however, that the two Be—H bonds have identical energies. To resolve this discrepancy and explain how molecules such as BeH_2 form, scientists developed the concept of hybridization.

Hybridization of *s* and *p* Orbitals

The localized bonding approach uses a process called **hybridization**, in which atomic orbitals that are similar in energy but not equivalent are combined mathematically to produce sets of equivalent orbitals that are properly oriented to form bonds. These new combinations are called **hybrid atomic orbitals** because they are produced by combining (*hybridizing*) two or more atomic orbitals from the same atom.

In BeH_2, we can generate two equivalent orbitals by combining the 2*s* orbital of beryllium and any one of the three degenerate 2*p* orbitals. By taking the sum and the difference of Be 2*s* and $2p_z$ atomic orbitals, for example,

$$sp = \frac{1}{\sqrt{2}}(2s + 2p_z) \quad \text{and} \quad sp = \frac{1}{\sqrt{2}}(2s - 2p_z) \qquad (9.1)$$

we produce two new orbitals with major and minor lobes oriented along the *z* axes, as shown in Figure 9.9.* The value $1/\sqrt{2}$ is needed to indicate that the 2*s* and 2*p* orbitals contribute equally to each hybrid orbital.

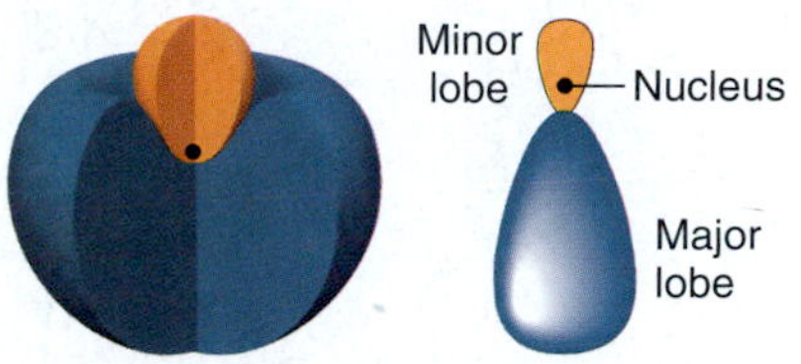

The position of the atomic nucleus with respect to an *sp* hybrid orbital. Note that the nucleus is actually located slightly inside the minor lobe, not at the node separating the major and minor lobes.

Note that the nucleus resides just inside the minor lobe of each orbital. In this case, the new orbitals are called *sp hybrids* because they are formed from one *s* and one *p* orbital. The two new orbitals are equivalent in energy, and their energy is between the energy values associated with pure *s* and *p* orbitals, as illustrated in this diagram:

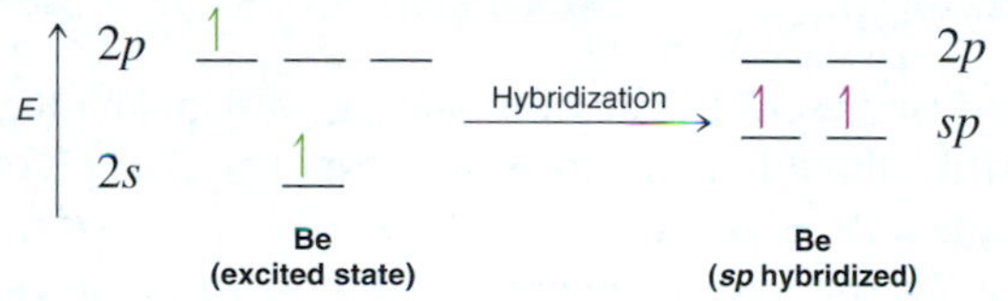

Each singly occupied *sp* hybrid orbital can now form an electron-pair bond with the singly occupied 1*s* atomic orbital of one of the H atoms. As shown in Figure 9.10, each *sp* orbital on Be has the correct orientation for the major lobes to overlap with the 1*s* atomic orbital of an H atom. The formation of two energetically equivalent Be—H bonds produces a linear BeH_2 molecule. Thus, valence bond theory does what neither the Lewis electron structure nor the VSEPR model is able to do: explain why the bonds in BeH_2 are equivalent in energy, and why BeH_2 has a linear geometry.

Because both promotion and hybridization require an input of energy, the formation of a set of singly occupied hybrid atomic orbitals is energetically uphill. The overall process of forming a compound with hybrid orbitals will be energetically favorable *only* if the amount of energy released by the formation of covalent bonds is

Figure 9.9 Formation of *sp* hybrid orbitals. Taking the mathematical sum and difference of an *ns* and an *np* atomic orbital where $n = 2$ gives two equivalent *sp* hybrid orbitals oriented at 180° to each other.

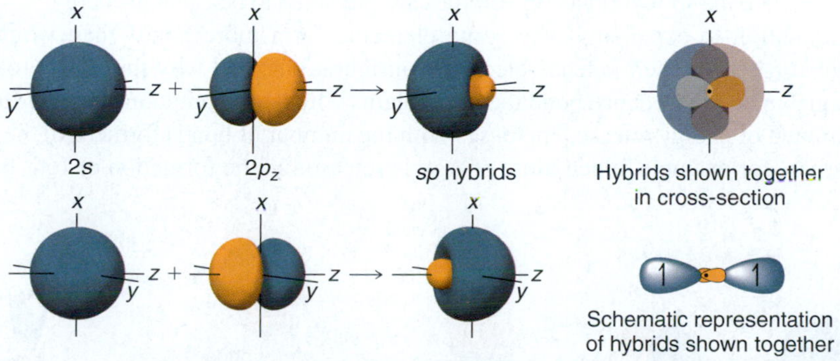

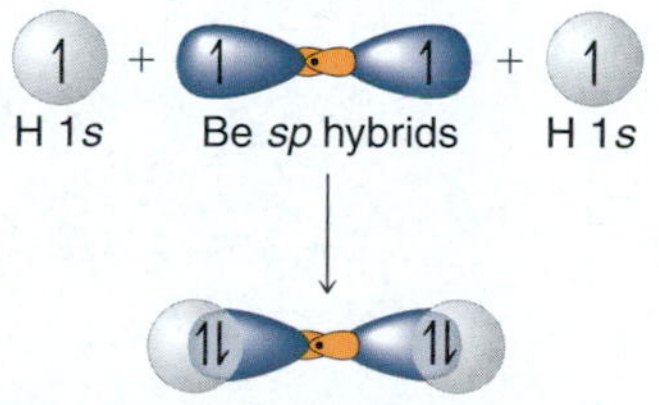

Figure 9.10 Explanation of the bonding in BeH_2 using *sp* hybrid orbitals. Each singly occupied *sp* hybrid orbital on beryllium can form an electron-pair bond with the singly occupied 1*s* orbital of a hydrogen atom. Because the two *sp* hybrid orbitals are oriented at a 180° angle, the BeH_2 molecule is linear.

* Because the difference A − B can also be written as A + (−B), in Figure 9.9 and subsequent figures we have reversed the phase(s) of the orbital being subtracted, which is the same as multiplying it by −1 and adding.

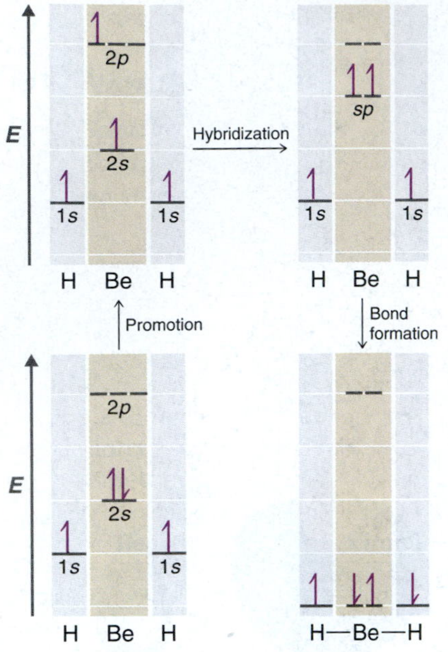

Figure 9.11 A hypothetical stepwise process for the formation of BeH_2 from a gaseous Be atom and two gaseous H atoms. Note that promotion of an electron from the 2*s* orbital of beryllium to one of the 2*p* orbitals is energetically uphill. The overall process of forming a BeH_2 molecule from a Be atom and two H atoms will therefore be energetically favorable *only* if the amount of energy released by the formation of the two Be—H bonds is greater than the amount of energy required for promotion and hybridization.

greater than the amount of energy used to form the hybrid orbitals (Figure 9.11). As we will see, some compounds are highly unstable or do not exist because the amount of energy required to form hybrid orbitals is greater than the amount of energy that would be released by the formation of additional bonds.

The concept of hybridization also explains why boron, with a $2s^2 2p^1$ valence-electron configuration, forms three bonds with fluorine to produce BF_3, as predicted by the Lewis and VSEPR approaches. With only a single unpaired electron in its ground state, boron should form only a single covalent bond. By the promotion of one of its 2*s* electrons to an unoccupied 2*p* orbital, however, followed by the hybridization of the three singly occupied orbitals (the 2*s* and two 2*p* orbitals), boron acquires a set of three equivalent hybrid orbitals containing one electron each, as shown in the diagram below. The hybrid orbitals are degenerate and are oriented at 120° angles to each other (Figure 9.12). Because the hybrid atomic orbitals are formed from one *s* and two *p* orbitals, boron is said to be sp^2 *hybridized* (pronounced "s-p-two" or "s-p-squared").

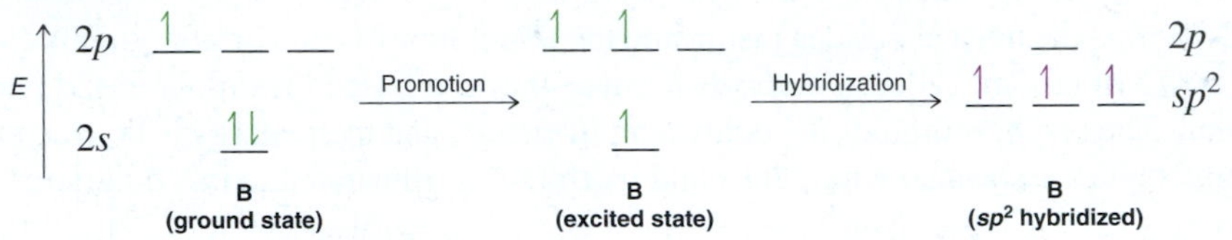

The singly occupied sp^2 hybrid atomic orbitals can overlap with the singly occupied orbitals on each of the three F atoms to form a trigonal planar structure with three energetically equivalent B—F bonds.

Looking at the $2s^2 2p^2$ valence-electron configuration of carbon, we might expect carbon to use its two unpaired 2*p* electrons to form compounds with only two covalent bonds. We know, however, that carbon typically forms compounds that contain four covalent bonds. We can explain this apparent discrepancy by the hybridization of the 2*s* orbital and the three 2*p* orbitals on carbon to give a set of four degenerate sp^3 ("s-p-three" or "s-p-cubed") hybrid orbitals, each of which contains a single electron:

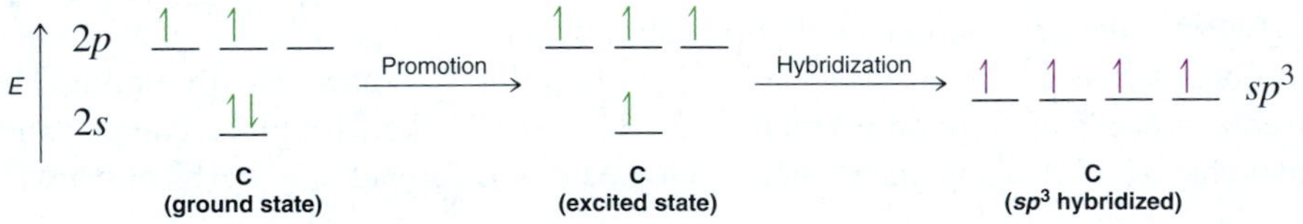

The large lobes of the hybridized orbitals are oriented toward the vertices of a tetrahedron, with 109.5° angles between them (Figure 9.13). Like all of the hybridized orbitals discussed earlier, the sp^3 hybrid atomic orbitals are predicted to be equal in energy.

In addition to explaining why some elements form more bonds than would be expected based on their valence-electron configurations, and why the bonds formed are equal in energy, valence bond theory explains why these compounds are so stable: the amount of energy released increases with the number of bonds formed. In the case of carbon, for example, much more energy is released in the formation of four bonds

Figure 9.12 Formation of sp^2 hybrid orbitals. Combining one *ns* and two *np* atomic orbitals gives three equivalent sp^2 hybrid orbitals in a trigonal planar arrangement; that is, oriented at 120° to one another.

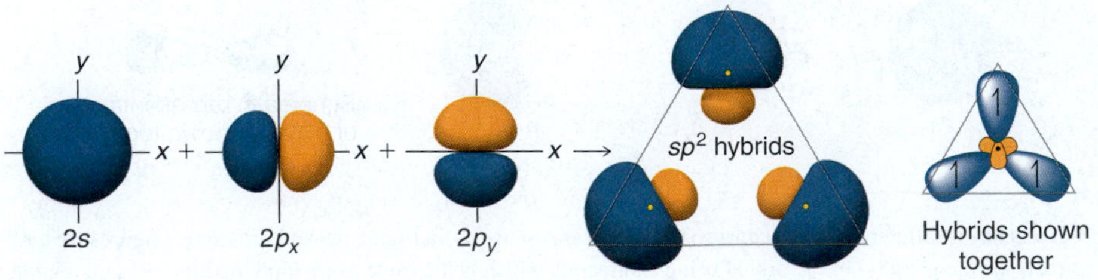

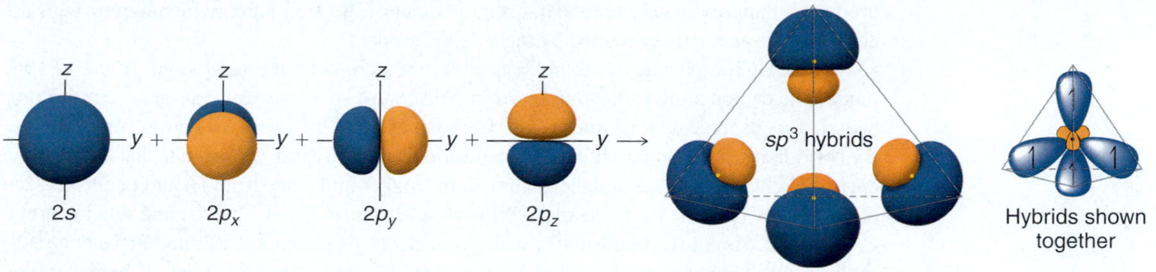

Figure 9.13 Formation of sp^3 hybrid orbitals. Combining one ns and three np atomic orbitals results in four sp^3 hybrid orbitals oriented at 109.5° to one another in a tetrahedral arrangement.

than two, so compounds of carbon with four bonds tend to be more stable than those with only two. Carbon does form compounds with only two covalent bonds (such as CH_2 or CF_2), but these species are highly reactive, unstable intermediates that form in only certain chemical reactions.

The bonding in molecules such as NH_3 or H_2O, which have lone pairs on the central atom, can also be described in terms of hybrid atomic orbitals. In NH_3, for example, N, with a $2s^22p^3$ valence-electron configuration, can hybridize its $2s$ and $2p$ orbitals to produce four sp^3 hybrid orbitals. Placing five valence electrons in the four hybrid orbitals, we obtain three that are singly occupied and one that contains a pair of electrons:

$$\underline{\uparrow\downarrow}\ \ \underline{\uparrow}\ \ \underline{\uparrow}\ \ \underline{\uparrow}\ \ sp^3$$

The three singly occupied sp^3 lobes can form bonds with three H atoms, while the fourth orbital accommodates the lone pair of electrons. Similarly, water has an sp^3 hybridized oxygen atom that uses two singly occupied sp^3 lobes to bond to two H atoms, and two to accommodate the two lone pairs predicted by the VSEPR model. Such descriptions explain the approximately tetrahedral distribution of electron pairs on the central atom in NH_3 and H_2O. Unfortunately, however, recent experimental evidence indicates that in CH_4 and NH_3, the hybridized orbitals are *not* entirely equivalent in energy.

EXAMPLE 9.5

Use the VSEPR model to predict the electron-pair and molecular geometry, and then describe the hybridization and bonding of all atoms except hydrogen in **(a)** H_2S and **(b)** $CHCl_3$.

Given Two chemical compounds

Asked for Electron-pair and molecular geometry, hybridization, bonding

Strategy

Ⓐ Using the approach from Example 9.1, determine the electron-pair and molecular geometry of the molecule.

Ⓑ From the valence-electron configuration of the central atom, predict the number and type of hybrid orbitals that can be produced. Fill these hybrid orbitals with the total number of valence electrons around the central atom, and describe the hybridization.

Solution

(a) Ⓐ H_2S has four electron pairs around the sulfur atom with two bonded atoms, so the VSEPR model predicts an electron-pair geometry that is tetrahedral and a molecular geometry that is bent. Ⓑ Sulfur has a $3s^23p^4$ valence-electron configuration containing six electrons, but by hybridizing its $3s$ and $3p$ orbitals, it can produce four sp^3 hybrids. If the six valence electrons are placed in these orbitals, two contain electron pairs and two are singly occupied. The two sp^3 hybrid orbitals that are singly occupied are used to form S—H bonds, whereas the other two contain lone pairs of electrons. Together, the four sp^3 hybrid orbitals produce an approximately tetrahedral arrangement of electron pairs, which agrees with the electron-pair geometry predicted by the VSEPR model.

(b) Ⓐ The $CHCl_3$ molecule has four valence electrons around the central atom. In the VSEPR model, the carbon atom forms four electron-pair bonds, so its electron-pair geometry is tetrahedral, as is its molecular geometry. Ⓑ Carbon has a $2s^2 2p^2$ valence-electron configuration. By hybridizing its 2*s* and 2*p* orbitals, it can form four sp^3 hybridized orbitals that are equal in energy. Eight electrons around the central atom (four from C, one from H, and one from each of the three Cl atoms) fill three sp^3 hybrid orbitals to form C—Cl bonds, and one to form a C—H bond. Similarly, the Cl atoms, with seven electrons each in their 3*s* and 3*p* valence subshells, can be viewed as sp^3 hybridized. Each Cl atom uses a singly occupied sp^3 hybrid orbital to form a C—Cl bond and three hybrid orbitals to accommodate lone pairs.

EXERCISE 9.5

What are the hybridization and bonding of all atoms in **(a)** the BF_4^- ion, and **(b)** hydrazine, $H_2N—NH_2$?

Answer **(a)** B is sp^3 hybridized; F is also sp^3 hybridized so it can accommodate one B—F electron-pair bond and three lone pairs. **(b)** Each N atom is sp^3 hybridized and uses one sp^3 hybrid orbital to form the N—N bond, two to form N—H bonds, and one to accommodate a lone pair.

Note the pattern

The number of hybrid orbitals used by the central atom is the same as the number of electron pairs around the central atom.

Hybridization Using *d* Orbitals

Hybridization is not restricted to *ns* and *np* atomic orbitals. The bonding in compounds that have central atoms in the third period and below can also be described using hybrid atomic orbitals. In these cases, the central atom can use its valence *nd* orbitals as well as its *ns* and *np* orbitals to form hybrid atomic orbitals, which allows it to accommodate five or more bonded atoms (as in PF_5 and SF_6). Using the *ns* orbital, all three *np* orbitals, and one *nd* orbital gives a set of five sp^3d hybrid orbitals that point toward the vertices of a trigonal bipyramid (Figure 9.14a). In this case, the five hybrid orbitals are *not* all equivalent: three form a triangular array oriented at 120° angles, and the other two are oriented at 90° to the first three and at 180° to each other.

Similarly, the combination of the *ns* orbital, all three *np* atomic orbitals, and *two nd* orbitals gives a set of six equivalent sp^3d^2 hybrid orbitals oriented toward the vertices of an octahedron (Figure 9.14b). In the VSEPR model, PF_5 and SF_6 are predicted to be trigonal bipyramidal and octahedral, respectively, which agrees with a valence bond description in which sp^3d or sp^3d^2 hybrid orbitals are used for bonding.

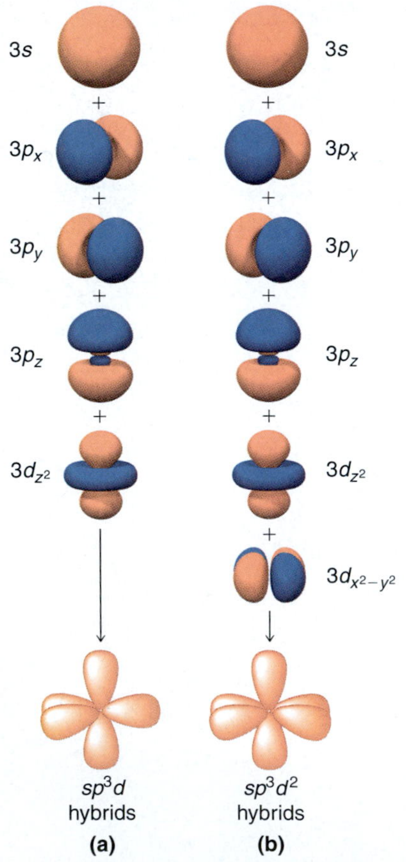

Figure 9.14 Hybrid orbitals involving *d* orbitals. The formation of a set of (a) five sp^3d hybrid orbitals and (b) six sp^3d^2 hybrid orbitals from *ns*, *np*, and *nd* atomic orbitals where $n = 4$.

EXAMPLE 9.6

What is the hybridization of the central atom in **(a)** XeF_4, **(b)** SO_4^{2-}, and **(c)** SF_4? Describe the bonding in each species.

Given Three chemical species

Asked for Hybridization of central atom

Strategy

Ⓐ Determine the electron-pair geometry of the molecule using the strategy in Example 9.1. From the valence-electron configuration of the central atom and the electron-pair geometry, determine the hybridization.

Ⓑ Place the total number of electrons around the central atom in the hybrid orbitals and describe the bonding.

Solution

(a) Ⓐ Using the VSEPR model, we find that Xe in XeF_4 has six electron pairs. With four bonded atoms, the structure is square planar and has two lone pairs. Six electron pairs form an octahedral arrangement, so the Xe must be sp^3d^2 hybridized. Ⓑ With 12 electrons around Xe, four of the six sp^3d^2 hybrid orbitals form Xe—F bonds, and two are occupied by lone pairs of electrons.

(b) Ⓐ The S in the SO_4^{2-} ion contains four electron pairs and four bonded atoms, so the structure is tetrahedral. The sulfur must be sp^3 hybridized in order to generate four S—O bonds. Ⓑ Filling the sp^3 hybrid orbitals with eight electrons from four bonds produces four filled sp^3 hybrid orbitals.

(c) Ⓐ The S atom in SF_4 contains five electron pairs and four bonded atoms. The molecule has a seesaw structure with one lone pair:

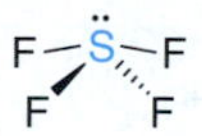

To accommodate five electron pairs, the sulfur atom must be sp^3d hybridized. Ⓑ Filling these orbitals with 10 electrons gives four sp^3d hybrid orbitals forming S—F bonds and one containing a lone pair of electrons.

EXERCISE 9.6

What is the hybridization of the central atom in **(a)** PCl_4^+, **(b)** BrF_3, and **(c)** SO_3?

Answer **(a)** sp^3; **(b)** sp^3d; **(c)** sp^2

Hybridization using *d* orbitals is a model that allows chemists to explain the structures and properties of many molecules and ions. Like most such models, however, it is not universally accepted. Nonetheless, it does explain a fundamental difference between the chemistry of the elements in the second period (C, N, and O) and those in the third period and below (such as Si, P, and S).

Second-period elements do not form compounds in which the central atom is covalently bonded to five or more atoms, although such compounds are common for the heavier elements. Thus, whereas carbon and silicon both form tetrafluorides (CF_4 and SiF_4), only SiF_4 reacts with F^- to give a stable hexafluoro dianion, SiF_6^{2-}. Because there are no 2*d* atomic orbitals, the formation of octahedral CF_6^{2-} would require the use of hybrid orbitals created from 2*s*, 2*p*, and 3*d* atomic orbitals. The 3*d* orbitals of carbon are so high in energy that the amount of energy needed to form a set of sp^3d^2 hybrid orbitals cannot be equaled by the energy released in the formation of two additional C—F bonds. These additional bonds are expected to be weak because the carbon atom (and other atoms in the second period) is so small that it cannot accommodate five or six F atoms at normal C—F bond lengths due to repulsions between electrons on adjacent fluorine atoms. Consequently, species such as CF_6^{2-} have never been prepared.

EXAMPLE 9.7

What is the hybridization of the oxygen atom in OF_4? Is OF_4 likely to exist?

Given Chemical compound

Asked for Hybridization, stability

Strategy

Ⓐ Predict the electron-pair geometry of OF_4 using the VSEPR model.

Ⓑ From the number of electron pairs around O in OF_4, predict the hybridization of O. Compare the number of hybrid orbitals with the number of electron pairs to decide whether the molecule is likely to exist.

Solution

A✓ The VSEPR model predicts that OF_4 will have five electron pairs, resulting in a trigonal bipyramidal electron-pair geometry with four bonding pairs and one lone pair. B✓ To accommodate five electron pairs, the O atom would have to be sp^3d hybridized. The only d orbital available for forming a set of sp^3d hybrid orbitals is a $3d$ orbital, which is *much* higher in energy than the $2s$ and $2p$ valence orbitals of oxygen. As a result, the OF_4 molecule is unlikely to exist. In fact, it has not been detected.

EXERCISE 9.7

What is the hybridization of the boron atom in BF_6^{3-}? Is this ion likely to exist?

Answer sp^3d^2 hybridization; no

9.3 Delocalized Bonding and Molecular Orbitals

None of the approaches we have described so far can adequately explain why some compounds are colored and others are not, why some substances with unpaired electrons are stable, and why others are effective semiconductors (Chapter 12). These approaches also cannot describe the nature of resonance. Such limitations led to the development of a new approach to bonding in which electrons are *not* viewed as being localized between the nuclei of bonded atoms, but are instead delocalized throughout the entire molecule. Just as with the valence bond theory, the approach we are about to discuss is based on a quantum mechanical model.

In Chapter 6, we described the electrons in isolated atoms as having certain spatial distributions, called *orbitals*, each with a particular *orbital energy*. Just as the positions and energies of electrons in *atoms* can be described in terms of *atomic orbitals* (AOs), the positions and energies of electrons in *molecules* can be described in terms of *molecular orbitals*. A **molecular orbital** (MO) is a spatial distribution of electrons *in a molecule* that is associated with a particular orbital energy. As the name suggests, molecular orbitals are not localized on a single atom but extend over the entire molecule. Consequently, the molecular orbital approach, called **molecular orbital theory**, is a *delocalized* approach to bonding.

Molecular Orbital Theory: A Delocalized Bonding Approach

Although the molecular orbital theory is computationally demanding, the principles on which it is based are similar to those we used to determine electron configurations for atoms. The key difference is that in molecular orbitals, the electrons are allowed to interact with more than one atomic nucleus at a time. Just as with atomic orbitals, we create an energy-level diagram by listing the molecular orbitals in order of increasing energy. We then fill the orbitals with the required number of valence electrons according to the Pauli principle. This means that each molecular orbital can accommodate a maximum of two electrons with opposite spins.

Molecular Orbitals Involving Only ns *Atomic Orbitals*

We begin our discussion of molecular orbitals with the simplest molecule, H_2, formed from two isolated hydrogen atoms, each with a $1s^1$ electron configuration. As we explained in Chapter 6, electrons can behave like waves. In the molecular orbital approach, the overlapping atomic orbitals are described by mathematical equations called *wave functions*. The $1s$ atomic orbitals on the two hydrogen atoms interact to

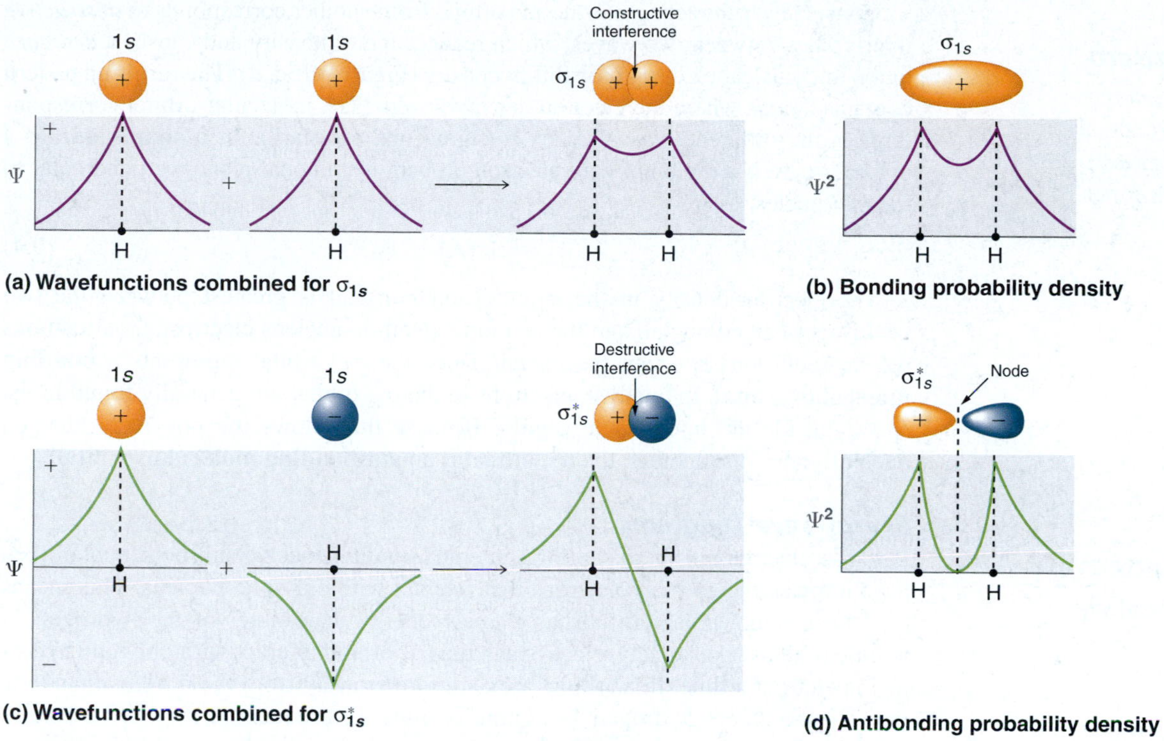

Figure 9.15 Molecular orbitals for the H_2 molecule. (a) Formation of a bonding σ_{1s} molecular orbital for H_2 as the sum of the wave functions (Ψ) of two H 1*s* atomic orbitals. (b) A plot of the square of the wave function, Ψ^2, for the bonding σ_{1s} molecular orbital, illustrating the increased electron probability density between the two hydrogen nuclei. (Recall from Chapter 6 that the probability density is proportional to the *square* of the wave function.) (c) Formation of an antibonding σ^*_{1s} molecular orbital for H_2 as the difference of the wave functions (Ψ) of two H 1*s* atomic orbitals. (d) A plot of the square of the wave function, Ψ^2, for the antibonding σ^*_{1s} molecular orbital, illustrating the node corresponding to zero electron probability density between the two hydrogen nuclei.

form two new molecular orbitals, one produced by taking the *sum* of the two H 1*s* wave functions, and the other produced by taking their *difference*:

$$\begin{aligned} \text{MO}(1) &= \text{AO}(\text{atom A}) + \text{AO}(\text{atom B}) \\ \text{MO}(2) &= \text{AO}(\text{atom A}) - \text{AO}(\text{atom B}) \end{aligned} \tag{9.2}$$

The molecular orbitals created from Equation 9.2 are called **linear combinations of atomic orbitals** (LCAOs). A molecule must have as many molecular orbitals as there are atomic orbitals.

Note the pattern

A molecule must have as many molecular orbitals as there are atomic orbitals.

Adding two atomic orbitals corresponds to *constructive* interference between two waves, thus reinforcing their intensity; the internuclear electron probability density is *increased*. The molecular orbital corresponding to the sum of the two H 1*s* orbitals is called a σ_{1s} combination (pronounced "sigma one ess") (see Figures 9.15a, b). In a **sigma (σ) orbital**, the electron density along the internuclear axis and between the nuclei has cylindrical symmetry. That is, all cross-sections perpendicular to the internuclear axis are circles. The subscript 1*s* denotes the atomic orbitals from which the molecular orbital was derived:*

$$\sigma_{1s} \approx 1s(\text{A}) + 1s(\text{B}) \tag{9.3}$$

* The ≈ sign is used rather than an = sign because we are ignoring certain constants that are not important to our argument.

Note the pattern

Antibonding orbitals contain a node perpendicular to the internuclear axis; bonding orbitals do not.

Conversely, subtracting one atomic orbital from another corresponds to *destructive* interference between two waves, which reduces their intensity and causes a *decrease* in the internuclear electron probability density (Figure 9.15c, d). The resulting pattern contains a *node* where the electron density is zero. The molecular orbital corresponding to the difference is called σ^*_{1s} ("sigma one ess star"). In a **sigma star (σ^*) orbital**, there is a region of zero electron probability, a nodal plane, perpendicular to the internuclear axis:

$$\sigma^*_{1s} \approx 1s(\text{A}) - 1s(\text{B}) \tag{9.4}$$

The electron density in the σ_{1s} molecular orbital is greatest between the two positively charged nuclei, and the resulting electron–nucleus electrostatic attractions reduce repulsions between the nuclei. Thus, the σ_{1s} orbital represents a **bonding molecular orbital**. In contrast, electrons in the σ^*_{1s} orbital are generally found in the space outside the internuclear region. Because this allows the positively charged nuclei to repel one another, the σ^*_{1s} orbital is an **antibonding molecular orbital**.

Energy-Level Diagrams

Note the pattern

A bonding MO is always lower in energy (more stable) than the component atomic orbitals, whereas an antibonding MO is always higher in energy (less stable).

Because electrons in the σ_{1s} orbital interact simultaneously with both nuclei, they have a lower energy than electrons that interact with only one nucleus. This means that the σ_{1s} molecular orbital has a *lower* energy than either of the hydrogen 1*s* atomic orbitals. Conversely, electrons in the σ^*_{1s} orbital interact with only one hydrogen nucleus at a time. In addition, they are farther away from the nucleus than they were in the parent hydrogen 1*s* atomic orbitals. Consequently, the σ^*_{1s} molecular orbital has a *higher* energy than either of the hydrogen 1*s* atomic orbitals. The σ_{1s} (bonding) molecular orbital is *stabilized* relative to the 1*s* atomic orbitals, and the σ^*_{1s} (antibonding) molecular orbital is *destabilized*. The relative energy levels of these orbitals are shown in the *energy-level diagram* in Figure 9.16.

To describe the bonding in a **homonuclear diatomic molecule** such as H_2 using molecular orbitals; that is, a molecule in which two identical atoms interact, we insert the total number of valence electrons into the energy-level diagram (Figure 9.16). We fill the orbitals according to the Pauli principle and Hund's rule: each orbital can accommodate a maximum of two electrons with opposite spins, and the orbitals are filled in order of increasing energy. Because each H atom contributes one valence electron, the resulting two electrons are exactly enough to fill the σ_{1s} bonding molecular orbital. The two electrons enter an orbital whose energy is lower than that of the parent atomic orbitals, so the H_2 molecule is more stable than the two isolated hydrogen atoms. Thus, molecular orbital theory correctly predicts that H_2 is a stable molecule. Because bonds form when electrons are concentrated in the space between the nuclei, this approach is also consistent with our earlier discussion of electron-pair bonds.

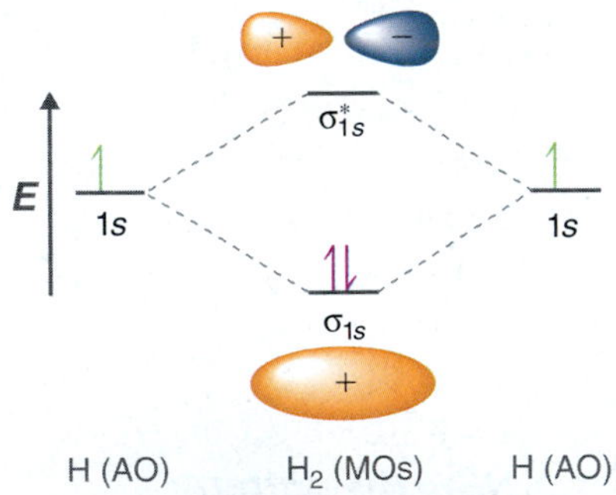

Figure 9.16 Molecular orbital energy-level diagram for H_2. Note that the two available electrons (one from each H atom) in this diagram fill the bonding σ_{1s} molecular orbital. Because the energy of the σ_{1s} molecular orbital is lower than that of the two H 1*s* atomic orbitals, the H_2 molecule is more stable (at a lower energy) than the two isolated H atoms.

Bond Order in Molecular Orbital Theory

In the Lewis electron structures described in Chapter 8, the number of electron pairs holding two atoms together was called the *bond order*. In the molecular orbital approach, **bond order** is defined as one-half the *net* number of bonding electrons:

$$\text{bond order} = \frac{\text{number of bonding electrons} - \text{number of antibonding electrons}}{2} \tag{9.5}$$

To calculate the bond order of H_2, we see from Figure 9.16 that the σ_{1s} (bonding) molecular orbital contains two electrons, while the σ^*_{1s} (antibonding) molecular orbital is empty. The bond order of H_2 is therefore

$$\text{bond order} = \frac{2 - 0}{2} = 1 \tag{9.6}$$

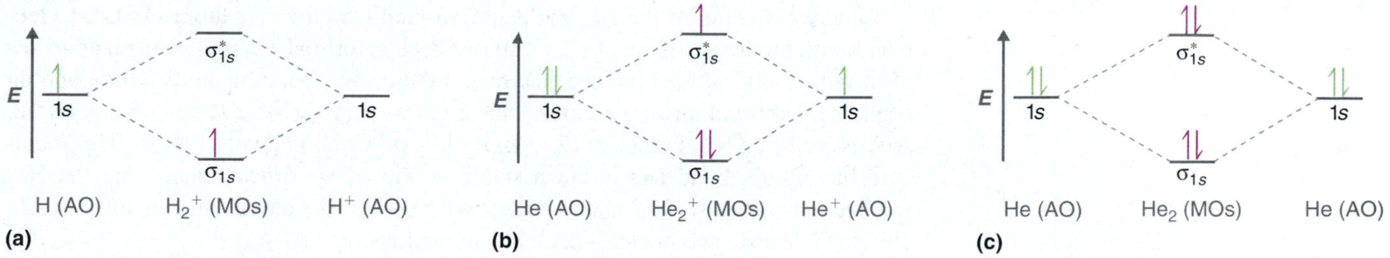

Figure 9.17 Molecular orbital energy-level diagrams for diatomic molecules that contain only 1*s* atomic orbitals. (a) The H_2^+ ion; (b) the He_2^+ ion; and (c) the He_2 molecule.

This result corresponds to the single covalent bond predicted by Lewis electron symbols. Thus, molecular orbital theory and the Lewis electron-pair approach agree that a single bond containing two electrons has a bond order of 1. Double and triple bonds contain four or six electrons, respectively, and correspond to bond orders of 2 and 3.

We can use energy-level diagrams such as the one in Figure 9.16 to describe the bonding in other pairs of atoms and ions where $n = 1$, such as the H_2^+ ion, the He_2^+ ion, and the He_2 molecule. Again, we fill the lowest-energy molecular orbitals first while being sure not to violate the Pauli principle or Hund's rule.

Figure 9.17a shows the energy-level diagram for the H_2^+ ion, which contains two protons and only one electron. The single electron occupies the σ_{1s} bonding molecular orbital, giving a $(\sigma_{1s})^1$ electron configuration. The number of electrons in an orbital is indicated by a superscript. In this case, the bond order is $(1 - 0) \div 2 = \frac{1}{2}$. Because the bond order is greater than zero, the H_2^+ ion should be more stable than an isolated H atom and a proton. We can therefore use a molecular orbital energy-level diagram and the calculated bond order to predict the relative stability of species such as H_2^+. With a bond order of only $\frac{1}{2}$, the bond in H_2^+ should be weaker and the H—H bond should be longer than in the H_2 molecule. As shown in Table 9.4, these predictions agree with the experimental data.

Figure 9.17b is the molecular orbital energy-level diagram for He_2^+. This ion has a total of three valence electrons. Because the first two electrons completely fill the σ_{1s} molecular orbital, the Pauli principle states that the third electron must be in the σ^*_{1s} antibonding orbital, giving a $(\sigma_{1s})^2(\sigma^*_{1s})^1$ electron configuration. This electron configuration gives a bond order of $(2 - 1) \div 2 = \frac{1}{2}$. As with H_2^+, the He_2^+ ion should be stable, but the He—He bond should be weaker and longer than in H_2. In fact, the He_2^+ ion can be prepared, and its properties are consistent with our predictions (Table 9.4).

TABLE 9.4 Molecular orbital electron configurations, bond orders, bond lengths, and bond energies for some simple homonuclear diatomic molecules and ions

Molecule or Ion	Electron Configuration	Bond Order	Bond Length, pm	Bond Energy, kJ/mol
H_2^+	$(\sigma_{1s})^1$	$\frac{1}{2}$	106	269
H_2	$(\sigma_{1s})^2$	1	74	436
He_2^+	$(\sigma_{1s})^2(\sigma^*_{1s})^1$	$\frac{1}{2}$	108	251
He_2	$(\sigma_{1s})^2(\sigma^*_{1s})^2$	0	Not observed	Not observed

Finally, we examine the He_2 molecule, formed from two He atoms with $1s^2$ electron configurations. Figure 9.17c is the molecular orbital energy-level diagram for He_2. With a total of four valence electrons, both the σ_{1s} bonding and σ^*_{1s} antibonding orbitals must contain two electrons. This gives a $(\sigma_{1s})^2\,(\sigma^*_{1s})^2$ electron configuration, with a predicted bond order of $(2 - 2) \div 2 = 0$, which indicates that the He_2 molecule has no net bond and is not a stable species. Experiments show that the He_2 molecule is actually *less* stable than two isolated He atoms due to unfavorable electron–electron and nucleus–nucleus interactions.

In molecular orbital theory, *electrons in antibonding orbitals effectively cancel the stabilization resulting from electrons in bonding orbitals.* Consequently, any system that has equal numbers of bonding and antibonding electrons will have a bond order of 0, and it is predicted to be unstable and therefore not to exist in nature. Notice that in contrast to Lewis electron structures and the valence bond approach, molecular orbital theory is able to accommodate systems that contain an odd number of electrons, such as the H_2^+ ion.

EXAMPLE 9.8

Use a molecular orbital energy-level diagram, such as those in Figure 9.17, to predict the bond order in the He_2^{2+} ion. Is this a stable species?

Given Chemical species

Asked for Molecular orbital energy-level diagram, bond order, stability

Strategy

A Combine the two He valence atomic orbitals to produce bonding and antibonding molecular orbitals. Draw the molecular orbital energy-level diagram for the system.

B Determine the total number of valence electrons in the He_2^{2+} ion. Fill the molecular orbitals in the energy-level diagram beginning with the orbital with the lowest energy. Be sure to obey the Pauli principle and Hund's rule while doing so.

C Calculate the bond order, and predict whether the species is stable.

Solution

A Two He $1s$ atomic orbitals combine to give two molecular orbitals: a σ_{1s} bonding orbital at lower energy than the atomic orbitals and a σ^*_{1s} antibonding orbital at higher energy. The bonding in any diatomic molecule that contains two He atoms can be described using the following molecular orbital diagram:

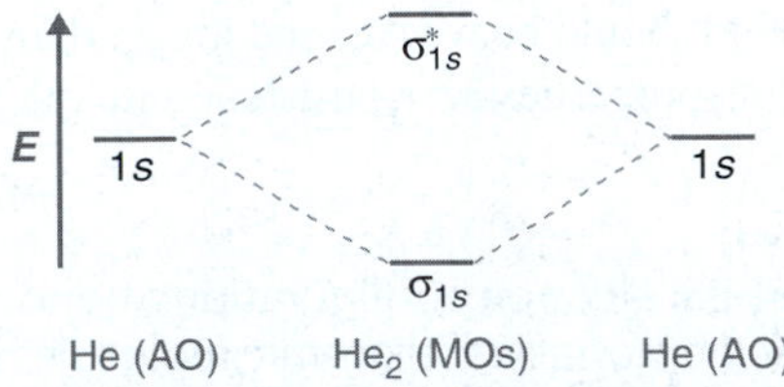

B The He_2^{2+} ion contains only two valence electrons (two from each He atom minus two for the +2 charge). We can also view He_2^{2+} as being formed from two He^+ ions, each of which has a single valence electron in the $1s$ atomic orbital. We can now fill the molecular orbital diagram:

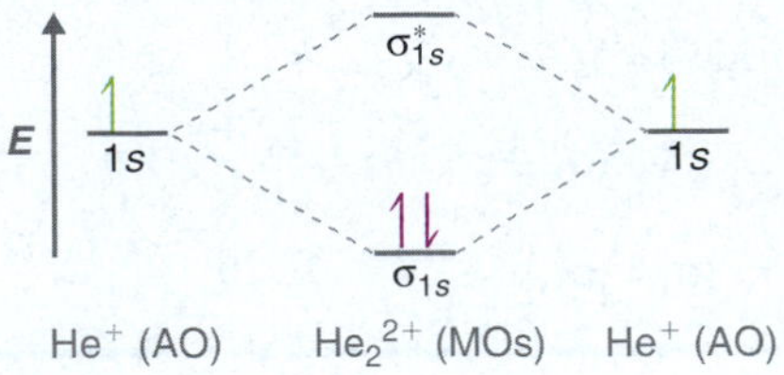

The two electrons occupy the lowest-energy molecular orbital, which is the bonding (σ_{1s}) orbital, giving a $(\sigma_{1s})^2$ electron configuration. To avoid violating the Pauli principle, the electron spins must be paired. The bond order is

$$\text{bond order} = \frac{2-0}{2} = 1$$

He_2^{2+} is therefore predicted to contain a single He—He bond. Thus, it should be a stable species.

EXERCISE 9.8

Use a molecular orbital energy-level diagram to predict the valence-electron configuration of the H_2^{2-} ion, its bond order, and its stability.

Answer H_2^{2-} has a valence-electron configuration of $(\sigma_{1s})^2(\sigma_{1s}^*)^2$ with a bond order of 0. It is therefore predicted to be unstable.

So far, our discussion of molecular orbitals has been confined to the interaction of valence orbitals, which tend to lie farthest from the nucleus. When two atoms are close enough for their valence orbitals to overlap significantly, the filled inner electron shells are largely unperturbed; hence, they do not need to be considered in a molecular orbital scheme. Also, when the inner orbitals are completely filled, they contain exactly enough electrons to completely fill both the bonding and antibonding molecular orbitals that arise from their interaction. Thus, the interaction of filled shells always gives a bond order of 0, and filled shells are therefore not a factor when predicting the stability of a species. This means that we can focus our attention on the molecular orbitals derived from valence atomic orbitals.

A molecular orbital diagram that can be applied to any homonuclear diatomic molecule that contains two identical alkali metal atoms (Li_2 and Cs_2, for example) is shown in Figure 9.18a, where M represents the metal atom. Only two energy levels are important for describing the valence-electron molecular orbitals of these species: a σ_{ns} bonding molecular orbital and a σ_{ns}^* antibonding molecular orbital. Because each alkali metal, M, has an ns^1 valence-electron configuration, the M_2 molecule contains two valence electrons, which fill the σ_{ns} bonding orbital. As a result, a bond order of 1 is predicted for all homonuclear diatomic species formed from the alkali metals (Li_2, Na_2, K_2, Rb_2, and Cs_2). The general features of these M_2 diagrams are identical to the diagram for the H_2 molecule in Figure 9.16. Experimentally, all are found to be stable in the gas phase, and some are even stable in solution.

Similarly, the molecular orbital diagrams for homonuclear diatomic compounds of the alkaline earths (such as Be_2), in which each metal atom has an ns^2 valence-electron configuration, resemble the diagram for the He_2 molecule in Figure 9.17c. As shown in

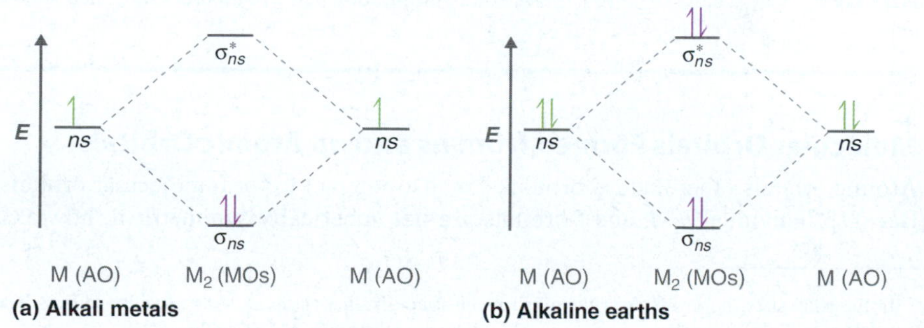

Figure 9.18 Molecular orbital energy-level diagrams for alkali metal and alkaline earth diatomic (M_2) molecules. (a) For alkali metal diatomic molecules, the two valence electrons are enough to fill the σ_{ns} (bonding) level, giving a bond order of 1. (b) For alkaline earth diatomic molecules, the four valence electrons fill both the σ_{ns} (bonding) and the σ_{ns}^* (nonbonding) levels, leading to a predicted bond order of 0.

Figure 9.18b, this is indeed the case. All the homonuclear alkaline earth diatomic molecules contain four valence electrons, which fill both the σ_{ns} bonding orbital and the σ^*_{ns} antibonding orbital and give a bond order of 0. Thus, Be_2, Mg_2, Ca_2, Sr_2, and Ba_2 are all expected to be unstable, in agreement with experiment.*

EXAMPLE 9.9

Use a qualitative molecular orbital energy-level diagram to predict the valence-electron configuration, bond order, and likely existence of the Na_2^- ion.

Given Chemical species

Asked for Molecular orbital energy-level diagram, valence-electron configuration, bond order, stability

Strategy

Ⓐ Combine the two sodium valence atomic orbitals to produce bonding and antibonding molecular orbitals. Draw the molecular orbital energy-level diagram for this system.
Ⓑ Determine the total number of valence electrons in the Na_2^- ion. Fill the molecular orbitals in the energy-level diagram beginning with the orbital with the lowest energy. Be sure to obey the Pauli principle and Hund's rule while doing so.
Ⓒ Calculate the bond order, and predict whether the species is stable.

Solution

Ⓐ Because sodium has a [Ne]$3s^1$ electron configuration, the molecular orbital energy-level diagram is qualitatively identical to the diagram for the interaction of two $1s$ atomic orbitals, as shown below. Ⓑ The Na_2^- ion has a total of three valence electrons (one from each Na atom and one for the negative charge), resulting in a filled σ_{3s} molecular orbital, a half-filled σ^*_{3s} molecular orbital, and a $(\sigma_{3s})^2(\sigma^*_{3s})^1$ electron configuration.

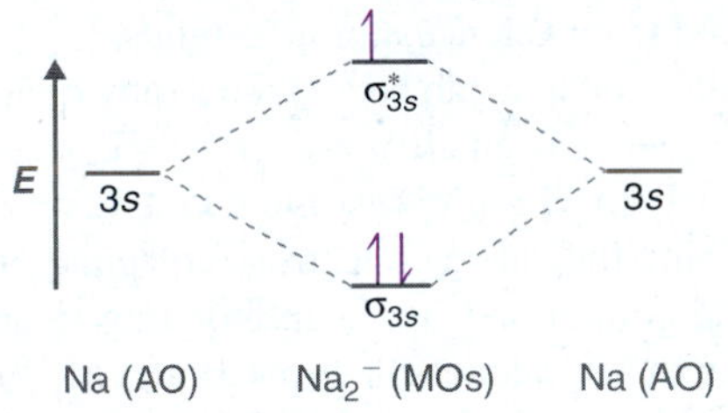

Ⓒ The bond order is $(2 - 1) \div 2 = \frac{1}{2}$. With a fractional bond order, we predict that the Na_2^- ion exists but that it is highly reactive.

EXERCISE 9.9

Use a qualitative molecular orbital energy-level diagram to predict the valence-electron configuration, bond order, and likely existence of the Ca_2^+ ion.

Answer Ca_2^+ has a $(\sigma_{4s})^2(\sigma^*_{4s})^1$ electron configuration and a bond order of $\frac{1}{2}$, and should exist.

Molecular Orbitals Formed from *ns* and *np* Atomic Orbitals

Atomic orbitals other than *ns* orbitals can also interact to form molecular orbitals. Because individual *p*, *d*, and *f* orbitals are not spherically symmetrical, however,

* In the solid state, however, all of the alkali metals and alkaline earths exist as extended lattices held together by metallic bonding (see Chapter 12). At low temperatures, Be_2 is stable.

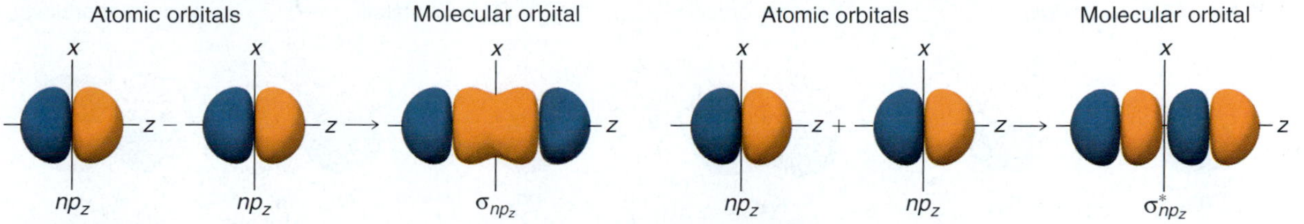

Figure 9.19 Formation of σ molecular orbitals from np_z atomic orbitals on adjacent atoms. (a) By convention, in a linear molecule or ion, the z axis always corresponds to the internuclear axis, with $+z$ to the right. As a result, the signs of the lobes of the np_z atomic orbitals on the two atoms alternate $- + - +$, from left to right. In this case, the σ (bonding) molecular orbital corresponds to the mathematical *difference*, in which the overlap of lobes with the same sign results in increased probability density between the nuclei. (b) In contrast, the σ^* (antibonding) molecular orbital corresponds to the mathematical *sum*, in which the overlap of lobes with opposite signs results in a nodal plane of zero probability density perpendicular to the internuclear axis.

we need to define a coordinate system so we know which lobes are interacting in three-dimensional space. Recall that for each *np* subshell, for example, there are np_x, np_y, and np_z orbitals (Figure 6.25). All have the same energy and are therefore degenerate, but they have different spatial orientations.

Just as with *ns* orbitals, we can form molecular orbitals from *np* orbitals by taking their mathematical sum and difference. When two positive lobes with the appropriate spatial orientation overlap, as illustrated for two np_z atomic orbitals in Figure 9.19a, it is the mathematical *difference* of their wave functions that results in *constructive* interference, which in turn increases the electron probability density between the two atoms. The difference therefore corresponds to a molecular orbital called a σ_{np_z} *bonding molecular orbital* because, just as with the σ orbitals discussed above, it is symmetrical about the internuclear axis (in this case, the z axis):

$$\sigma_{np_z} = np_z\,(A) - np_z\,(B) \tag{9.7}$$

The other possible combination of the two np_z orbitals is the mathematical sum:

$$\sigma^*_{2np_z} = np_z\,(A) + np_z\,(B) \tag{9.8}$$

In this combination, shown in Figure 9.19b, the positive lobe of one np_z atomic orbital overlaps the negative lobe of the other, leading to *destructive* interference of the two waves and creating a node between the two atoms. Hence, this is an antibonding molecular orbital. Because it, too, is symmetrical about the internuclear axis, this molecular orbital is called a $\sigma^*_{np_z}$ *antibonding molecular orbital*. Whenever orbitals combine, *the bonding combination is always lower in energy* (more stable) than the atomic orbitals from which it was derived, and *the antibonding combination is higher in energy* (less stable).

Note the pattern

Overlap of atomic orbital lobes with the same sign produces a bonding molecular orbital, regardless of whether it corresponds to the sum or the difference of the atomic orbitals.

The remaining *p* orbitals on each of the two atoms, np_x and np_y, do not point directly toward each other. Instead, they are perpendicular to the internuclear axis. If we arbitrarily label the axes as shown in Figure 9.20, we see that we have two pairs of *np* orbitals: the two np_x orbitals lying in the plane of the page, and two np_y orbitals perpendicular to the plane. Although these two pairs are equivalent in energy, the np_x orbital on one atom can interact with only the np_x orbital on the other, and the np_y orbital on one atom can interact with only the np_y on the other. Notice that these interactions are side-to-side rather than the head-to-head interactions characteristic of σ orbitals. Each pair of overlapping atomic orbitals again forms two molecular orbitals: one corresponds to the arithmetic sum of the two atomic orbitals, and one to the difference. The sum of these side-to-side interactions increases the electron probability in the region above and below a line connecting the nuclei, so it is a bonding molecular orbital that is called a **pi (π) orbital**. The difference results in the overlap of orbital lobes with opposite signs, which

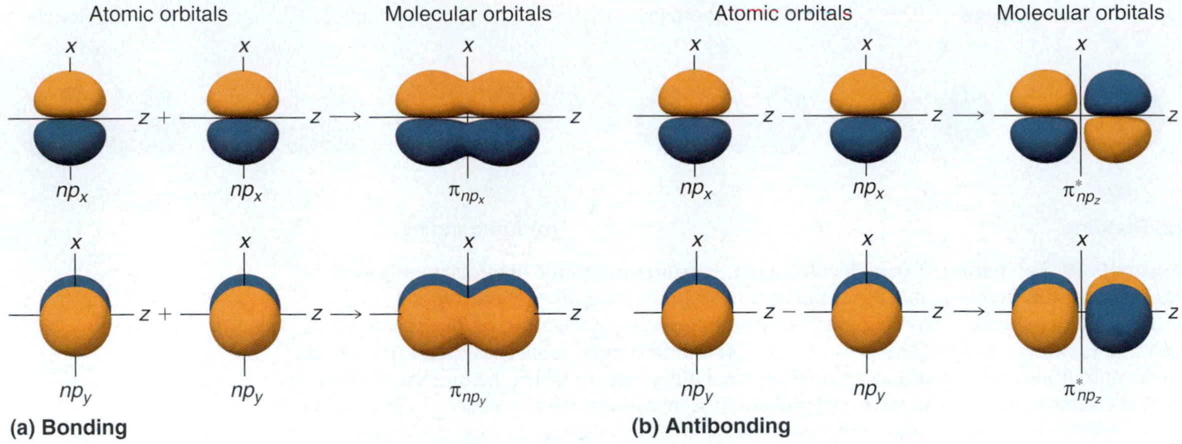

Figure 9.20 Formation of π molecular orbitals from $np_{x,y}$ atomic orbitals on adjacent atoms. (a) Because the signs of the lobes of both the np_x and the np_y atomic orbitals on adjacent atoms are the same, in both cases the mathematical sum corresponds to a π (bonding) molecular orbital. (b) In contrast, in both cases, the mathematical difference corresponds to a π* (antibonding) molecular orbital, with a nodal plane of zero probability density perpendicular to the internuclear axis.

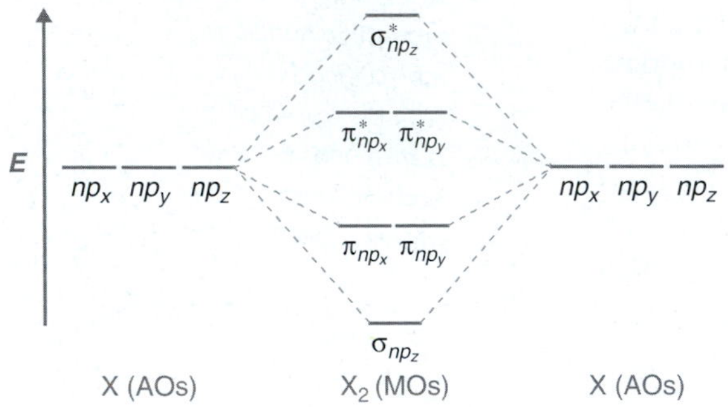

Figure 9.21 The relative energies of the σ and π molecular orbitals derived from $np_{x,y,z}$ orbitals on identical adjacent atoms. Because the two np_z orbitals point directly at each other, their orbital overlap is greater, so the difference in energy between the σ and σ^* molecular orbitals is greater than the energy difference between the π and π^* orbitals.

produces a nodal plane perpendicular to the internuclear axis; hence, it is an antibonding molecular orbital, called a **pi star (π^*) orbital**.

$$\pi_{np_x} = np_x\,(A) + np_x\,(B) \qquad (9.9)$$

$$\pi^*_{np_x} = np_x\,(A) - np_x\,(B) \qquad (9.10)$$

The two np_y orbitals can also combine using side-to-side interactions to produce a bonding π_{np_y} molecular orbital and an antibonding $\pi^*_{np_y}$ molecular orbital. Because the np_x and np_y atomic orbitals interact in the same way (side-to-side) and have the same energy, the π_{np_x} and π_{np_y} molecular orbitals are a degenerate pair, as are the $\pi^*_{np_x}$ and $\pi^*_{np_y}$ molecular orbitals.

Figure 9.21 is an energy-level diagram that can be applied to two identical interacting atoms that have three *np* atomic orbitals each. There are six degenerate *p* atomic orbitals (three from each atom) that combine to form six molecular orbitals, three bonding and three antibonding. The bonding molecular orbitals are lower in energy than the atomic orbitals because of the increased stability associated with the formation of a bond. Conversely, the antibonding molecular orbitals are higher in energy, as shown. Notice that the energy difference between the σ and σ^* molecular orbitals is significantly greater than the difference between the two π and π^* sets. The reason for this is that the atomic orbital overlap and thus the strength of the interaction are greater for a σ bond than a π bond, which means that the σ molecular orbital is more stable (lower in energy) than the π molecular orbitals.

Although many combinations of atomic orbitals form molecular orbitals, we will discuss only one other interaction: an *ns* atomic orbital on one atom with an np_z atomic orbital on another. As shown in Figure 9.22, the sum of the two atomic wave functions ($ns + np_z$) produces a σ bonding molecular orbital. Their difference ($ns - np_z$) produces a σ^* antibonding molecular orbital, which has a nodal plane of zero probability density perpendicular to the internuclear axis.

Molecular Orbital Diagrams for Second-Period Homonuclear Diatomic Molecules

We now describe examples of systems involving second-period homonuclear diatomic molecules, such as N_2, O_2, and F_2. When we draw a molecular orbital diagram for a molecule, there are four key points to remember:

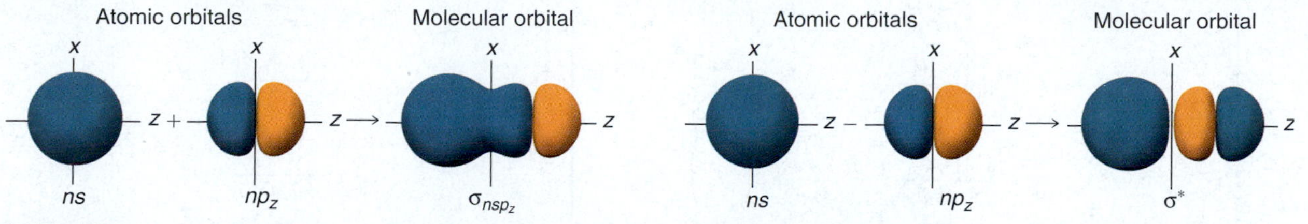

Figure 9.22 Formation of molecular orbitals from an *ns* atomic orbital on one atom and an np_z atomic orbital on an adjacent atom. (a) The mathematical sum results in a σ (bonding) molecular orbital, with increased probability density between the nuclei. (b) The mathematical difference results in a σ^* (antibonding) molecular orbital, with a nodal plane of zero probability density perpendicular to the internuclear axis.

1. The number of molecular orbitals produced is the same as the number of atomic orbitals used to create them ("the law of conservation of orbitals").
2. As the overlap between two atomic orbitals increases, the difference in energy between the resulting bonding and antibonding molecular orbitals increases.
3. When two atomic orbitals combine to form a pair of molecular orbitals, the bonding molecular orbital is stabilized about as much as the antibonding molecular orbital is destabilized.
4. The interaction between atomic orbitals is greatest when they have the same energy.

Note the pattern

The number of molecular orbitals is always equal to the total number of atomic orbitals we started with.

We illustrate how to use these points by constructing a molecular orbital energy-level diagram for F_2. We use the diagram in Figure 9.23a; the $n = 1$ orbitals (σ_{1s} and σ^*_{1s}) are located well below those of the $n = 2$ level and are not shown. As illustrated in the diagram, the σ_{2s} and σ^*_{2s} molecular orbitals are much lower in energy than the molecular orbitals derived from the 2*p* atomic orbitals because of the large difference in energy between the 2*s* and 2*p* atomic orbitals of fluorine. The lowest-energy molecular orbital derived from the three 2*p* orbitals on each F is σ_{2p_z}, and the next most stable are the two degenerate orbitals, π_{2p_x} and π_{2p_y}. Notice that for each bonding orbital in the diagram there is an antibonding orbital, and that the antibonding orbital is destabilized by about as much as the corresponding bonding orbital is stabilized. As a result, the $\sigma^*_{2p_z}$ orbital is higher in energy than either of the degenerate $\pi^*_{2p_x}$ and $\pi^*_{2p_y}$ orbitals. We can now fill the orbitals, beginning with the one that is lowest in energy.

Each F has seven valence electrons, so there are a total of 14 valence electrons in the F_2 molecule. Starting at the lowest energy level, the electrons are placed in the orbitals according to the Pauli principle and Hund's rule. Two electrons each fill the σ_{2s} and σ^*_{2s} orbitals, two fill the σ_{2p_z} orbital, four fill the two degenerate π orbitals, and four fill the two degenerate π^* orbitals, for a total of 14 electrons. To determine what type of bonding the molecular orbital approach predicts F_2 to have, we must calculate the bond order. According to our diagram, there are eight bonding electrons and six antibonding electrons, giving a bond order of $(8 - 6) \div 2 = 1$. Thus, F_2 is predicted to have a stable F—F single bond, in agreement with experiment.

We now turn to a molecular orbital description of the bonding in O_2. It so happens that the molecular orbital description of this molecule provided an explanation for a long-standing puzzle that could not be explained using other bonding models. To obtain the molecular orbital energy-level diagram for O_2, we need to place 12 valence electrons (six from each O atom) in the energy-level diagram shown in

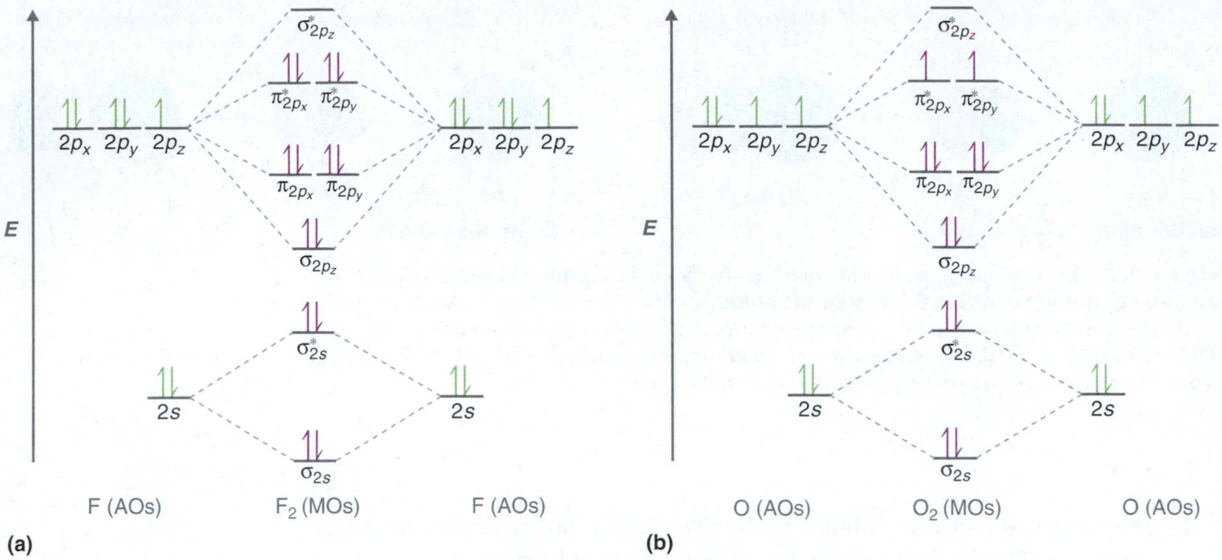

Figure 9.23 Molecular orbital energy-level diagrams for homonuclear diatomic molecules. (a) For F_2, with 14 valence electrons (seven from each F atom), all of the energy levels except the highest, $\sigma^*_{2p_z}$, are filled. This diagram shows eight electrons in bonding orbitals and six in antibonding orbitals, resulting in a bond order of 1. (b) For O_2, with 12 valence electrons (six from each O atom), there are only two electrons to place in the ($\pi^*_{2p_x}$, $\pi^*_{2p_y}$) pair of orbitals. Hund's rule dictates that one electron occupies each orbital and that their spins are parallel, giving the O_2 molecule two unpaired electrons. This diagram shows eight electrons in bonding orbitals and four in antibonding orbitals, resulting in a predicted bond order of 2.

Figure 9.23b. We again fill the orbitals according to Hund's rule and the Pauli principle, beginning with the orbital that is lowest in energy. Two electrons each are needed to fill the σ_{2s} and σ^*_{2s} orbitals, two more to fill the σ_{2p_z} orbital, and four to fill the degenerate π_{2p_x} and π_{2p_y} orbitals. According to Hund's rule, the last two electrons must be placed in separate π^* orbitals with their spins parallel, giving two unpaired electrons. This leads to a predicted bond order of $(8 - 4) \div 2 = 2$, which corresponds to a double bond, in agreement with experimental data: the O—O bond length is 120.7 pm, and the bond energy is 498.4 kJ/mol at 298 K.

None of the other bonding models can predict the presence of two unpaired electrons in O_2. Chemists had long wondered why, unlike most other substances, liquid O_2 is attracted into a magnetic field. As shown in Figure 9.24, it actually remains suspended between the poles of a magnet until the liquid boils away. The only way to explain this behavior was for O_2 to contain unpaired electrons, making it paramagnetic, exactly as predicted by molecular orbital theory. This result was one of the earliest triumphs of molecular orbital theory over the other approaches we have discussed.

Figure 9.24 Liquid O_2 suspended between the poles of a magnet. Because the O_2 molecule contains two unpaired electrons, it is paramagnetic. Consequently, it is attracted into a magnetic field, which allows it to remain suspended between the poles of a powerful magnet until it evaporates.

The magnetic properties of O_2 are not just a laboratory curiosity; they are absolutely crucial to the existence of life. Because Earth's atmosphere contains 20% oxygen, all organic compounds, including those that make up our body tissues, should react rapidly with air to form H_2O, CO_2, and N_2 in an exothermic reaction. Fortunately for us, however, this reaction is very, very slow. The reason for the unexpected stability of organic compounds in an oxygen atmosphere is that virtually all organic compounds, as well as H_2O, CO_2, and N_2, contain only paired electrons, whereas oxygen has two unpaired electrons. Thus, the reaction of O_2 with organic compounds to give H_2O, CO_2, and N_2 would require that at least one of the electrons on O_2 change its spin during the reaction. This would require a large input of energy, an obstacle that chemists call a *spin barrier*. Consequently, reactions of this type are usually exceedingly slow. If they were not so

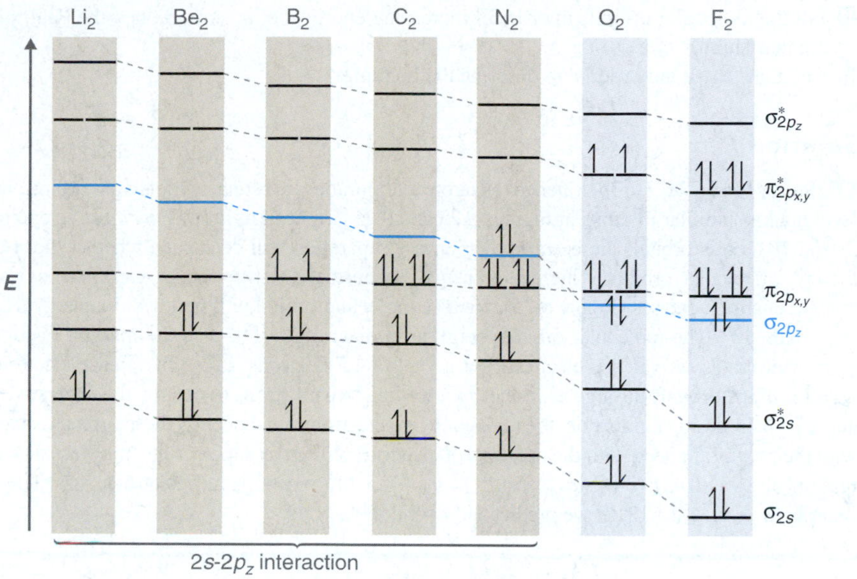

Figure 9.25 Molecular orbital energy-level diagrams for the diatomic molecules of the second-period elements. Unlike earlier diagrams, only the molecular orbital energy levels for the molecules are shown here. For simplicity, the atomic orbital energy levels for the component *atoms* have been omitted. Note that for Li_2 through N_2, the σ_{2p_z} orbital is higher in energy than the $\pi_{2p_{x,y}}$ orbitals. In contrast, the σ_{2p_z} orbital is *lower* in energy than the $\pi_{2p_{x,y}}$ orbitals for O_2 and F_2 due to the increase in the energy difference between the $2s$ and $2p$ atomic orbitals as the nuclear charge increases across the row.

slow, all organic substances, including this book and you, the reader, would disappear in a puff of smoke!

For second-period diatomic molecules to the left of N_2 in the periodic table, a slightly different molecular orbital energy-level diagram is needed because the σ_{2p_z} molecular orbital is slightly *higher* in energy than the degenerate π_{2p_x} and π_{2p_y} orbitals. The difference in energy between the $2s$ and $2p$ atomic orbitals increases from Li_2 to F_2 due to increasing nuclear charge and poor screening of the $2s$ electrons by electrons in the $2p$ subshell. The bonding interaction between the $2s$ orbital on one atom and the $2p_z$ orbital on the other is most important when the two orbitals have similar energies. This interaction decreases the energy of the σ_{2s} orbital and increases the energy of the σ_{2p_z} orbital. Thus, for Li_2, Be_2, B_2, C_2, and N_2, the σ_{2p_z} orbital is higher in energy than the $\pi_{2p_{x,y}}$ orbitals, as shown in Figure 9.25. Experimentally, it is found that the energy gap between the *ns* and *np* atomic orbitals *increases* as the nuclear charge increases (Figure 9.25). Thus, for example, the σ_{2p_z} molecular orbital is at a lower energy than the $\pi_{2p_{x,y}}$ pair.

Completing the diagram for N_2 in the same manner as demonstrated above, we find that the 10 valence electrons result in eight bonding electrons and two antibonding electrons, for a predicted bond order of 3, a triple bond. Experiments show that the N—N bond is significantly shorter than the F—F bond (109.8 pm in N_2 versus 141.2 pm in F_2), and the bond energy is much greater for N_2 than for F_2 (945.3 kJ/mol versus 158.8 kJ/mol, respectively). Thus, the N_2 bond is much shorter and stronger than the F_2 bond, consistent with what we would expect when comparing a triple bond with a single bond.

EXAMPLE 9.10

Use a qualitative molecular orbital energy-level diagram to describe the bonding in S_2, a bright blue gas at high temperatures. Predict the bond order and the number of unpaired electrons.

Given Chemical species

Asked for Molecular orbital energy-level diagram, bond order, number of unpaired electrons

Strategy

A Write the valence-electron configuration of sulfur, and determine the type of molecular orbitals formed in S_2. Predict the relative energies of the molecular orbitals based on how close in energy the valence atomic orbitals are to one another.

B Draw the molecular orbital energy-level diagram for this system, and determine the total number of valence electrons in S_2.

(C) Fill the molecular orbitals in order of increasing energy, being sure to obey the Pauli principle and Hund's rule.
(D) Calculate the bond order, and describe the bonding.

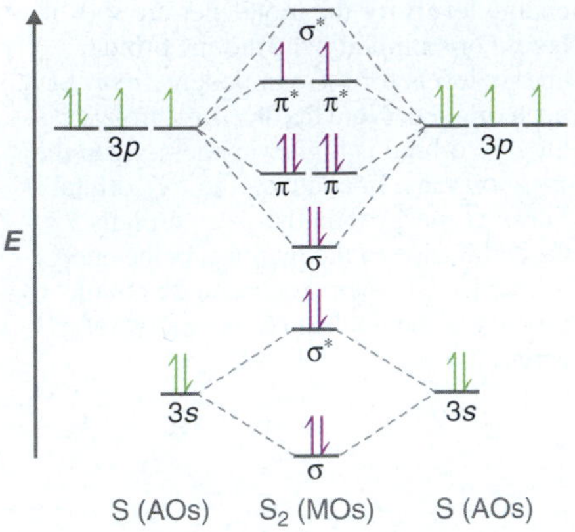

Solution

(A) Sulfur has a [Ne] $3s^2\,3p^4$ valence- electron configuration. To create a molecular orbital energy-level diagram similar to those in Figures 9.23 and 9.25, we need to know how close in energy the 3*s* and 3*p* atomic orbitals are because their energy separation will determine whether the $\pi_{3p_{x,y}}$ or the σ_{3p_z} molecular orbital is higher in energy. Because the *ns*–*np* energy gap *increases* as the nuclear charge increases (Figure 9.25), the σ_{3p_z} molecular orbital will be lower in energy than the $\pi_{3p_{x,y}}$ pair. (B) The molecular orbital energy-level diagram is shown in the margin. Each sulfur atom contributes six valence electrons, for a total of 12 electrons. (C) Ten valence electrons are used to fill the orbitals through π_{3p_x} and π_{3p_y}, leaving two electrons to occupy the degenerate $\pi^*_{3p_x}$ and $\pi^*_{3p_y}$ pair. From Hund's rule, the remaining two electrons must occupy these orbitals separately with their spins aligned. With the numbers of electrons written as superscripts, the electron configuration of S_2 is $(\sigma_{3s})^2(\sigma^*_{3s})^2(\sigma_{3p_z})^2(\pi_{3p_{x,y}})^4(\pi^*_{3p_{x,y}})^2$ with two unpaired electrons. (D) The bond order is $(8 - 4) \div 2 = 2$, so we predict an S=S double bond.

EXERCISE 9.10

Use a qualitative molecular orbital energy-level diagram to predict the electron configuration, bond order, and number of unpaired electrons in the peroxide ion, O_2^{2-}.

Answer O_2^{2-}: $(\sigma_{2s})^2(\sigma^*_{2s})^2(\sigma_{2p_z})^2(\pi_{2p_{x,y}})^4(\pi^*_{2p_{x,y}})^4$; bond order 1; no unpaired electrons

Note the pattern

An molecular orbital energy-level diagram is always skewed toward the more electronegative atom.

Molecular Orbitals for Heteronuclear Diatomic Molecules

Diatomic molecules that contain two different atoms are called **heteronuclear diatomic molecules**. When two nonidentical atoms interact to form a chemical bond, the interacting atomic orbitals do not have the same energy. If, for example, element B is more electronegative than element A ($\chi_B > \chi_A$), the net result is a "skewed" molecular orbital energy-level diagram, such as the one shown for a hypothetical A—B molecule in Figure 9.26. The atomic orbitals of element B are uniformly lower in energy than the corresponding atomic orbitals of element A because of the enhanced stability of the electrons in element B. The molecular orbitals are no longer symmetrical, and the energies of the bonding molecular orbitals are more similar to those of the atomic orbitals of B. Hence, the electron density of bonding electrons is likely to be closer to the more electronegative atom. In this way, molecular orbital theory can describe a polar covalent bond.

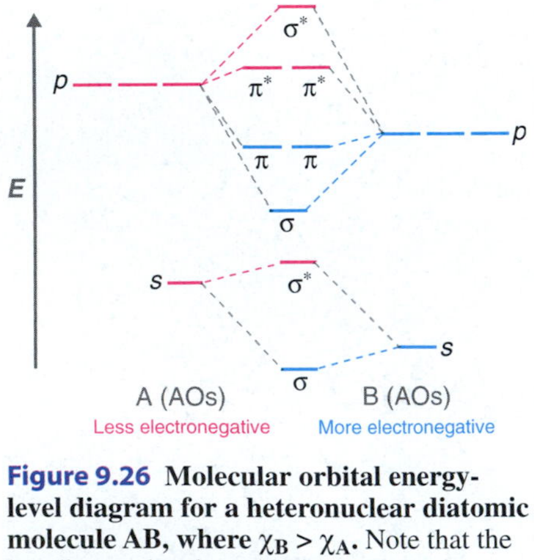

Figure 9.26 Molecular orbital energy-level diagram for a heteronuclear diatomic molecule AB, where $\chi_B > \chi_A$. Note that the bonding molecular orbitals are closer in energy to the atomic orbitals of the more electronegative B atom. Consequently, the electrons in the bonding orbitals are not shared equally between the two atoms. On average, they are closer to the B atom, resulting in a polar covalent bond.

An Odd Number of Valence Electrons: NO

Nitric oxide (NO) is an example of a heteronuclear diatomic molecule. The reaction of O_2 with N_2 at high temperatures in internal combustion engines forms nitric oxide, which undergoes a complex reaction with O_2 to produce the NO_2 that is responsible for the brown color we associate with air pollution. Recently, however, nitric oxide has also been recognized to be a vital biological messenger involved in regulating blood pressure and long-term memory in mammals.

Because NO has an odd number of valence electrons (five from N and six from O, for a total of 11), its bonding and properties cannot be successfully explained by either the Lewis electron-pair approach or valence bond theory. The molecular orbital energy-level diagram for NO (Figure 9.27) shows that the general pattern is similar to that for the O_2 molecule. Because 10 electrons are sufficient to fill all the bonding molecular orbitals derived from 2*p* atomic orbitals, the 11th electron must occupy one of the degenerate π^* orbitals. The predicted bond order for NO is therefore

$(8 - 3) \div 2 = 2\frac{1}{2}$. Experimental data, showing an N—O bond length of 115 pm and N—O bond energy of 631 kJ/mol, are consistent with this description. These values lie between those of the N_2 and O_2 molecules, which have triple and double bonds, respectively. As we stated earlier, molecular orbital theory can therefore explain the bonding in molecules that contain an odd number of electrons, such as NO, whereas Lewis electron structures cannot.

Molecular orbital theory can also tell us something about the *chemistry* of NO. As indicated in the energy-level diagram in Figure 9.27, NO contains a single electron in a relatively high-energy molecular orbital. We might therefore expect it to have similar reactivity as alkali metals such as Li and Na with their single valence electrons. In fact, NO is easily oxidized to the NO^+ cation, which is isoelectronic with N_2 and has a bond order of 3, corresponding to an N≡O triple bond.

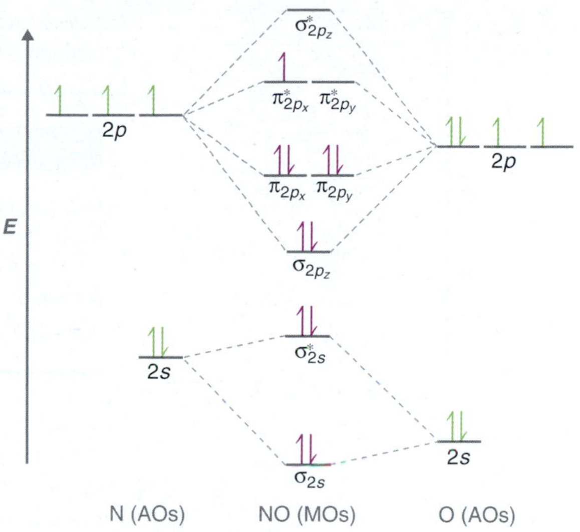

Figure 9.27 Molecular orbital energy-level diagram for NO. Because NO has 11 valence electrons, it is paramagnetic, with a single electron occupying the $(\pi^*_{2p_x}, \pi^*_{2p_y})$ pair of orbitals.

Nonbonding Molecular Orbitals

Molecular orbital theory is also able to explain the presence of lone pairs of electrons. Consider, for example, the HCl molecule, whose Lewis electron structure has three lone pairs of electrons on the Cl atom. Using the molecular orbital approach to describe the bonding in HCl, we can see from Figure 9.28 that the 1*s* orbital of H is closest in energy to the 3*p* orbitals of Cl. Consequently, the filled Cl 3*s* atomic orbital is not involved in bonding to any appreciable extent, and the only important interactions are those between the H 1*s* and Cl 3*p* orbitals. Of the three *p* orbitals, only one, designated as $3p_z$, can interact with the H 1*s* orbital. The $3p_x$ and $3p_y$ atomic orbitals have no net overlap with the 1*s* orbital on H, so they are not involved in bonding. Because the energies of the Cl 3*s*, $3p_x$, and $3p_y$ orbitals do not change when HCl forms, they are called **nonbonding molecular orbitals** (*nb*). A nonbonding molecular orbital occupied by a pair of electrons is the molecular orbital equivalent of a lone pair of electrons. By definition, electrons in nonbonding orbitals have no effect on bond order, so they are not counted in the calculation of bond order. Thus, the predicted bond order of HCl is $(2 - 0) \div 2 = 1$. Because the σ bonding molecular orbital is closer in energy to the Cl $3p_z$ than to the H 1*s* atomic orbital, the electrons in the σ orbital are concentrated closer to the Cl atom than to H. A molecular orbital approach to bonding can therefore be used to describe the polarization of the H—Cl bond to give $H^{\delta+}$—$Cl^{\delta-}$ as described in Chapter 8.

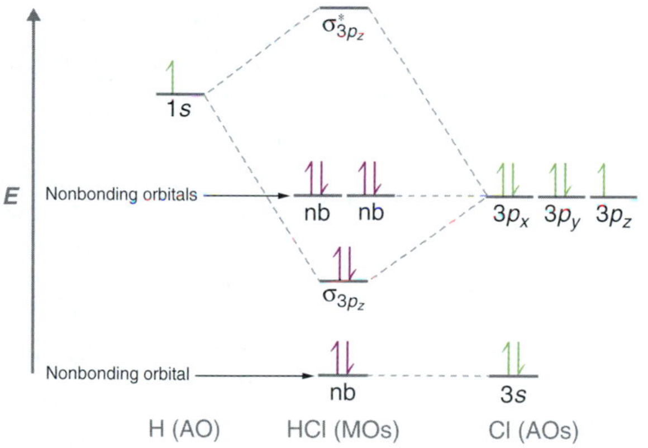

Figure 9.28 Molecular orbital energy-level diagram for HCl. The hydrogen 1*s* atomic orbital interacts most strongly with the $3p_z$ orbital on chlorine, producing a bonding/antibonding pair of molecular orbitals. The other electrons on Cl are best viewed as nonbonding. As a result, only the bonding σ orbital is occupied by electrons, giving a bond order of 1.

Note the pattern

Electrons in nonbonding molecular orbitals have no effect on bond order.

EXAMPLE 9.11

Use a "skewed" molecular orbital energy-level diagram like the one in Figure 9.26 to describe the bonding in the cyanide ion, CN^-. What is the bond order?

Given Chemical species

Asked for "Skewed" molecular orbital energy-level diagram, bonding description, bond order

Strategy

Ⓐ Calculate the total number of valence electrons in CN^-. Then place these electrons in a molecular orbital energy-level diagram like Figure 9.26 in order of increasing energy. Be sure to obey the Pauli principle and Hund's rule while doing so.

Ⓑ Calculate the bond order, and describe the bonding in CN^-.

Solution

Ⓐ The CN^- ion has a total of 10 valence electrons: four from C, five from N, and one for the −1 charge. Placing these electrons in an energy-level diagram like Figure 9.26 fills the five lowest-energy orbitals, as shown in the margin. Because $\chi_N > \chi_C$, the atomic orbitals of N (on the right) are lower in energy than those of C. Ⓑ The resulting valence-electron configuration

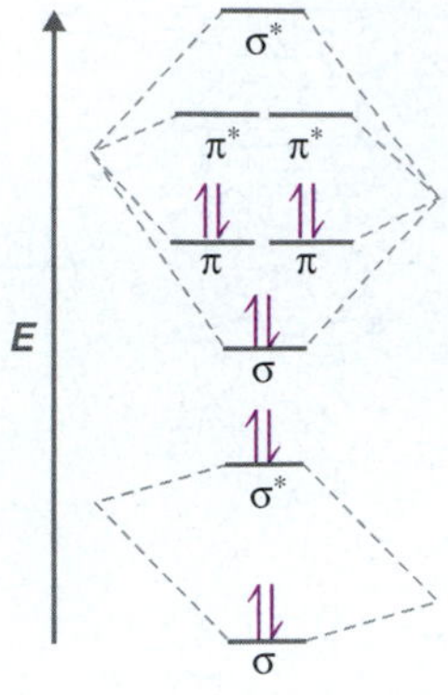

CN⁻ (MOs)

gives a predicted bond order of (8 − 2) ÷ 2 = 3, indicating that the CN^- ion has a triple bond, analogous to that in N_2.

EXERCISE 9.11

Use a qualitative molecular orbital energy-level diagram to describe the bonding in the hypochlorite ion, OCl^-. What is the bond order?

Answer All molecular orbitals except the highest-energy σ^* are filled, giving a bond order of 1.

Although the molecular orbital approach reveals a great deal about the bonding in a given molecule, the procedure quickly becomes computationally intensive for molecules of even moderate complexity. Furthermore, because the computed molecular orbitals extend over the entire molecule, they are often difficult to represent in a way that is easy to visualize. Therefore we do not use a pure molecular orbital approach to describe the bonding in molecules or ions that contain more than two atoms. Instead, we combine a valence bond approach with a molecular orbital approach, which allows us to explain, among other things, the concept of resonance, which cannot adequately be explained using other approaches.

9.4 Combining the Valence Bond and Molecular Orbital Approaches

So far in our molecular orbital descriptions we have not dealt with polyatomic systems that contain multiple bonds. To do so, we can use a combined approach in which we describe σ bonding using localized electron-pair bonds formed by hybrid atomic orbitals, and π bonding using molecular orbitals formed by unhybridized *np* atomic orbitals.

Multiple Bonds

We begin our discussion by considering the bonding in ethylene, C_2H_4. Experimentally, we know that the H—C—H and H—C—C angles in ethylene are approximately 120°. This angle suggests that the carbon atoms are sp^2 hybridized, which means that a singly occupied sp^2 orbital on one carbon overlaps with a singly occupied *s* orbital on each H and a singly occupied sp^2 lobe on the other C. Thus, each carbon forms a set of three σ bonds: two C—H ($sp^2 + s$) and one C—C ($sp^2 + sp^2$) (Figure 9.29a). The sp^2 hybridization can be represented as follows:

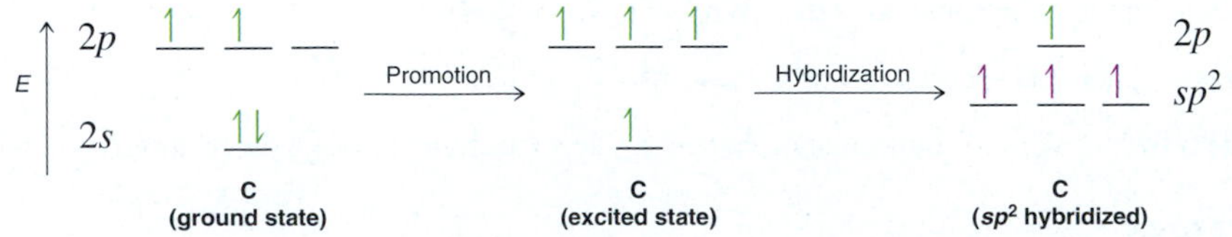

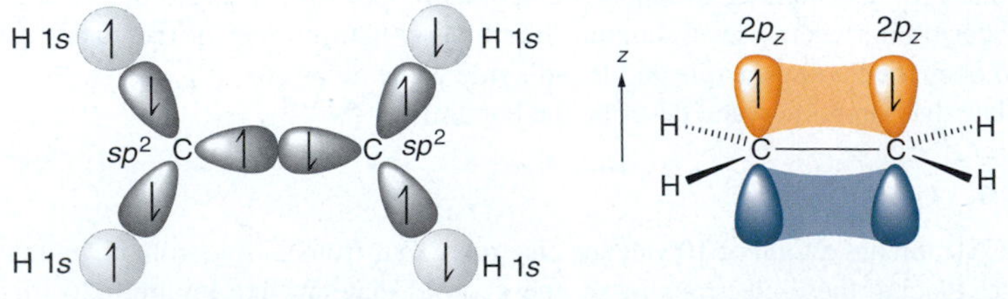

(a) C_2H_4 sigma-bonded framework (b) C_2H_4 pi bonding

Figure 9.29 Bonding in ethylene. (a) The σ-bonded framework is formed by the overlap of two sets of singly occupied carbon sp^2 hybrid orbitals and four singly occupied hydrogen 1*s* orbitals to form electron-pair bonds. This uses 10 of the 12 valence electrons to form a total of five σ bonds (four C—H bonds and one C—C bond). (b) One singly occupied unhybridized $2p_z$ orbital remains on each carbon atom to form a carbon–carbon π bond. (*Note:* By convention, in planar molecules the axis perpendicular to the molecular plane is always the *z* axis.)

After hybridization, each carbon still has one unhybridized $2p_z$ orbital that is perpendicular to the hybridized lobes and contains a single electron (Figure 9.29b). The two singly occupied $2p_z$ orbitals can overlap to form a π bonding orbital and a π^* antibonding orbital, which produces the energy-level diagram shown in Figure 9.30. With the formation of a π bonding orbital, the electron density increases in the plane between the carbon nuclei. The π^* orbital lies outside the internuclear region and has a nodal plane perpendicular to the internuclear axis. Because each $2p_z$ orbital contains a single electron, there are only two electrons, enough to fill only the bonding (π) level, leaving the π^* orbital empty. Consequently, the C—C bond in ethylene consists of a σ bond and a π bond, which together give a C═C double bond. Our model is supported by the facts that the measured carbon–carbon bond is shorter than that in ethane (133.9 pm versus 153.5 pm) and the bond is stronger (728 kJ/mol versus 376 kJ/mol in ethane). The two CH_2 fragments are coplanar, which maximizes the overlap of the two singly occupied $2p_z$ orbitals.

Figure 9.30 Molecular orbital energy-level diagram for π bonding in ethylene. As in the diatomic molecules discussed previously, the singly occupied $2p_z$ orbitals in ethylene can overlap to form a bonding/antibonding pair of π molecular orbitals. The two electrons remaining are enough to fill only the bonding π orbital. With one σ bond plus one π bond, the carbon–carbon bond order in ethylene is 2.

Triple bonds, as in acetylene (C_2H_2), can also be explained using a combination of hybrid atomic orbitals and molecular orbitals. The four atoms of acetylene are collinear, which suggests that each carbon is *sp* hybridized. If one *sp* lobe on each carbon atom is used to form a C—C σ bond and one is used to form the C—H σ bond, then each carbon will still have two unhybridized 2*p* orbitals (a $2p_{x,y}$ pair), each of which contains one electron (Figure 9.31a).

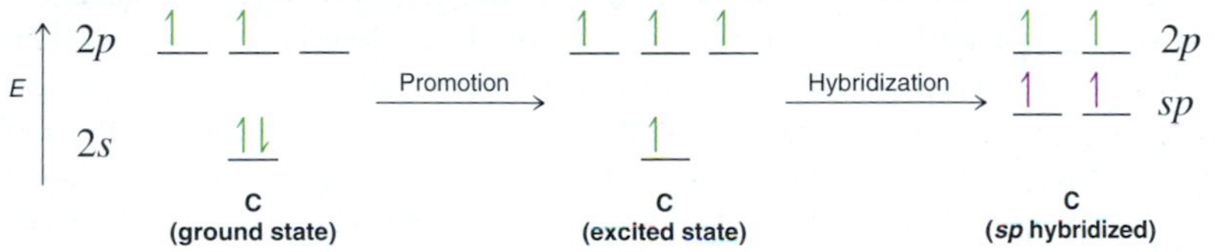

The two 2*p* orbitals on each carbon can align with the corresponding 2*p* orbitals on the adjacent carbon to simultaneously form a *pair* of π bonds (Figure 9.31b). Because each of the unhybridized 2*p* orbitals contains a single electron, four electrons are available for π bonding, which is enough to occupy only the bonding molecular orbitals. Acetylene must therefore contain a carbon–carbon triple bond, which consists of a C—C σ bond and two mutually perpendicular π bonds. Acetylene does in fact have a shorter carbon–carbon bond (120.3 pm) and a higher bond energy (965 kJ/mol) than ethane and ethylene, as we would expect for a triple bond.

Note the pattern

In complex molecules, use hybrid orbitals and valence bond theory to describe σ bonding, and use unhybridized p orbitals and molecular orbital theory to describe π bonding.

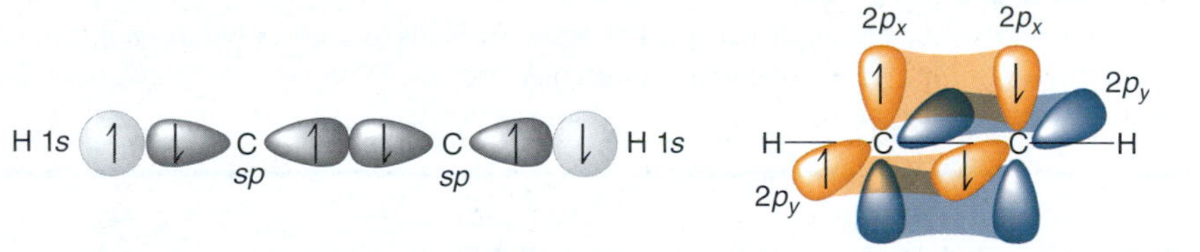

(a) C_2H_2 sigma-bonded framework **(b) C_2H_2 pi bonding**

Figure 9.31 Bonding in acetylene. (a) Formation of the σ-bonded framework by overlap of two sets of singly occupied carbon *sp* hybrid orbitals and two singly occupied hydrogen 1*s* orbitals. (b) Formation of two carbon–carbon π bonds in acetylene by overlap of two singly occupied unhybridized $2p_{x,y}$ orbitals on each carbon atom. With one σ bond plus two π bonds, the carbon–carbon bond order in acetylene is 3.

EXAMPLE 9.12

Describe the bonding in HCN using a combination of hybrid atomic orbitals and molecular orbitals. The HCN molecule is linear.

Given Chemical compound, molecular geometry

Asked for Bonding description using hybrid atomic orbitals and molecular orbitals

Strategy

A From the geometry given, predict the hybridization in HCN. Use the hybrid orbitals to form the σ-bonded framework of the molecule, and determine the number of valence electrons that are used for σ bonding.

B Determine the number of remaining valence electrons. Use any remaining unhybridized p orbitals to form π and π^* orbitals.

C Fill the orbitals with the remaining electrons in order of increasing energy. Describe the bonding in HCN.

Solution

A✓ Because HCN is a linear molecule, it is likely that the bonding can be described in terms of *sp* hybridization at carbon. Because the nitrogen atom can also be described as *sp* hybridized, we can use one *sp* hybrid on each atom to form a C—N σ bond. This leaves one *sp* hybrid on each atom to either bond to hydrogen (C) or hold a lone pair of electrons (N). Of ten valence electrons (five from N plus four from C plus one from H), four are used for σ bonding:

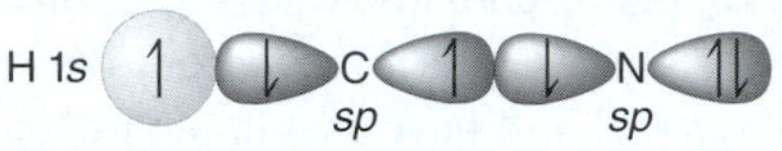

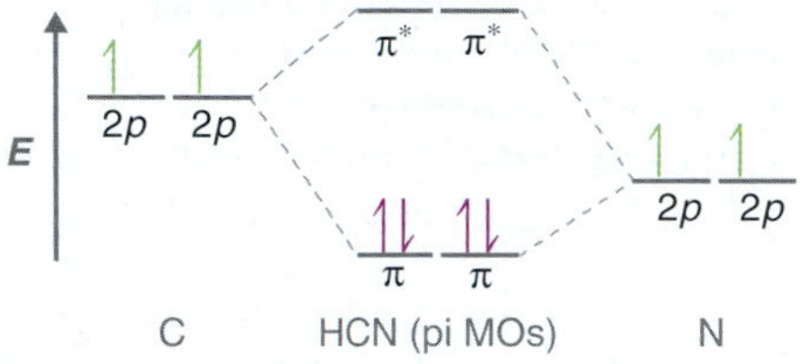

B✓ We are now left with two electrons on N (five valence electrons minus one bonding electron minus two electrons in the lone pair) and two on C (four valence electrons minus two bonding electrons). We have two unhybridized $2p$ atomic orbitals left on carbon and two on nitrogen, each occupied by a single electron. These four $2p$ atomic orbitals can be combined to give four molecular orbitals: two π (bonding) orbitals and two π^* (antibonding) orbitals. **C✓** With four electrons available, only the π orbitals are filled. The overall result is a triple bond (1 σ and 2 π) between C and N.

EXERCISE 9.12

Describe the bonding in formaldehyde ($H_2C{=}O$) in terms of a combination of hybrid atomic orbitals and molecular orbitals.

Answer

Sigma-bonding framework: Carbon and oxygen are sp^2 hybridized. Two sp^2 hybrid orbitals on oxygen contain lone pairs, two sp^2 hybrid orbitals on carbon form C—H bonds, and one sp^2 hybrid orbital on C and O forms a C—O σ bond.

Pi bonding: Unhybridized, singly occupied $2p$ atomic orbitals on carbon and oxygen interact to form π (bonding) and π^* (antibonding) molecular orbitals. With two electrons, only the π (bonding) orbital is occupied.

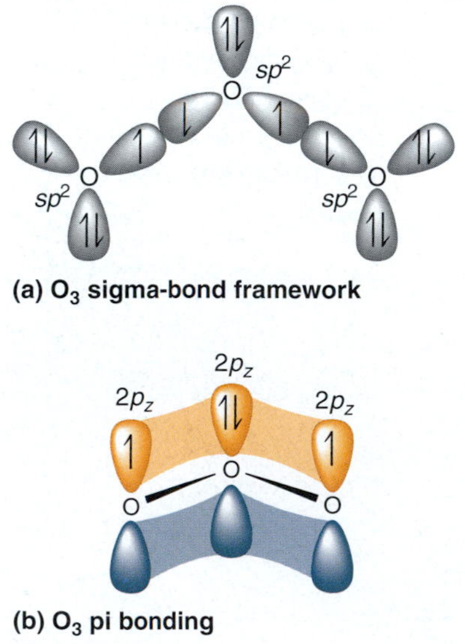

Figure 9.32 Bonding in ozone. (a) Formation of the σ-bonded framework by overlap of three sets of oxygen sp^2 hybrid orbitals to give two O—O σ bonds and five lone pairs, two on each terminal O and one on the central O. The σ bonds and lone pairs account for 14 of the 18 valence electrons of O_3. (b) One unhybridized $2p_z$ orbital remains on each oxygen atom that is available for π bonding. The unhybridized $2p_z$ orbital on each terminal O atom contains a single electron, whereas the unhybridized $2p_z$ orbital on the central O atom contains two electrons.

Molecular Orbitals and Resonance Structures

In Chapter 8, we used resonance structures to describe the bonding in molecules such as ozone, O_3, and the nitrite ion, NO_2^-. We showed that ozone can be represented by either of these Lewis dot structures:

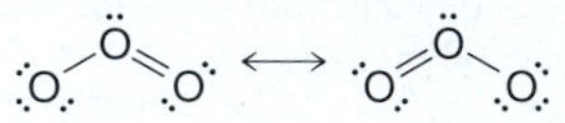

Although the VSEPR model correctly predicts that both species are bent, it gives no information about their bond orders.

Experimental evidence indicates that ozone has a bond angle of 117.5°. Because this angle is close to 120°, it is likely that the central oxygen atom in ozone is trigonal planar and sp^2 hybridized. If we assume that the terminal oxygen atoms are also sp^2 hybridized, then we obtain the σ-bonded framework shown in Figure 9.32. Two of the three sp^2 lobes on the central O are used to form O—O sigma bonds, and the third contains a lone pair of electrons. Each terminal oxygen atom has two lone pairs of electrons that are also in sp^2 lobes. In addition, each oxygen atom contains one unhybridized $2p$ orbital perpendicular to the molecular plane. The σ bonds and lone pairs account for a total of 14 electrons (five lone pairs and two σ bonds, each containing two electrons). Each oxygen atom in ozone has six valence electrons, so O_3 has a total of 18 valence electrons. Subtracting 14 electrons from the total gives us four electrons that must occupy the three unhybridized $2p$ orbitals.

With a molecular orbital approach to describe the π bonding, three $2p$ atomic orbitals give us three molecular orbitals, as shown in Figure 9.33. One of the molecular orbitals is a π bonding molecular orbital, which is shown as a banana-shaped region of electron density above and below the molecular plane. This region has *no* nodes perpendicular to the O_3 plane. The molecular orbital with the highest energy has two nodes that bisect the O—O σ bonds; it is a π^* antibonding orbital. The third molecular orbital contains a single node that is perpendicular to the O_3 plane and passes through the central O atom; it is a nonbonding molecular orbital. Because electrons in nonbonding orbitals are neither bonding nor antibonding, they are ignored in calculating bond orders.

We can now place the remaining four electrons in the three energy levels shown in Figure 9.33, thereby filling the π bonding and the nonbonding levels. Notice that the result is a single π bond holding *three* oxygen atoms together, or $\frac{1}{2}\pi$ bond per O—O. We therefore predict the overall O—O bond order to be $1\frac{1}{2}$ ($\frac{1}{2}\pi$ bond plus 1 σ bond), just as predicted using resonance structures. The molecular orbital approach, however, shows that the π nonbonding orbital is localized on the *terminal* O atoms, which suggests that they are more electron rich than the central O atom. The reactivity of ozone is consistent with the predicted charge localization.

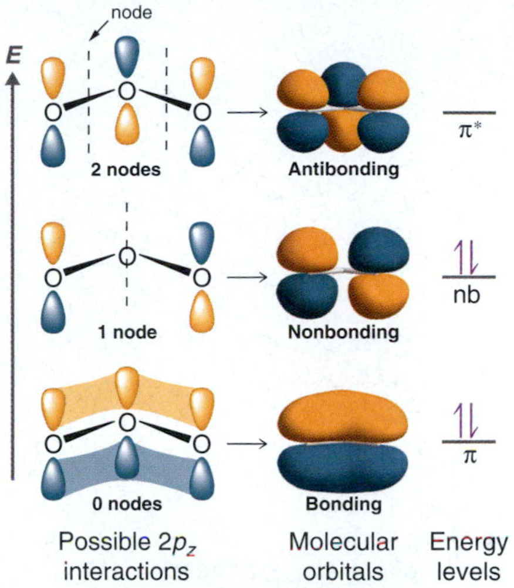

Figure 9.33 Pi bonding in ozone. The three unhybridized $2p_z$ atomic orbitals interact with one another to form three molecular orbitals: one π bonding orbital at lower energy, one π^* antibonding orbital at higher energy, and a nonbonding orbital in between. Placing four electrons in this diagram fills the bonding and nonbonding orbitals. With one filled π bonding orbital holding three atoms together, the net π bond order is $\frac{1}{2}$ per O—O bond. The combined σ/π bond order is thus $1\frac{1}{2}$ for each O—O bond.

EXAMPLE 9.13

Describe the bonding in the nitrite ion in terms of a combination of hybrid atomic orbitals and molecular orbitals. Lewis dot structures and the VSEPR model predict that the NO_2^- ion is bent.

Given Chemical species, molecular geometry

Asked for Bonding description using hybrid atomic orbitals and molecular orbitals

Strategy

Ⓐ Calculate the number of valence electrons in NO_2^-. From the structure, predict the type of atomic orbital hybridization in the ion.

Ⓑ Predict the number and type of molecular orbitals that form during bonding. Use valence electrons to fill these orbitals, and then calculate the number of electrons that remain.

Ⓒ If there are unhybridized orbitals, place the remaining electrons in these orbitals in order of increasing energy. Calculate the bond order, and describe the bonding.

Solution

Ⓐ The lone pair of electrons on nitrogen and a bent structure suggest that the bonding in NO_2^- is similar to the bonding in ozone. This conclusion is supported by the fact that nitrite also contains 18 valence electrons (five from N and six from each O, plus one for the −1 charge). The bent structure implies that the nitrogen is sp^2 hybridized. Ⓑ If we assume that the oxygen atoms are sp^2 hybridized as well, then we can use two sp^2 hybrid orbitals on each oxygen and one sp^2 hybrid orbital on nitrogen to accommodate the five lone pairs of electrons. Two sp^2 hybrid orbitals on nitrogen form σ bonds with the remaining sp^2 hybrid orbital on each oxygen. The σ bonds and lone pairs account for 14 electrons. We are left with three unhybridized $2p$ orbitals, one on each atom, perpendicular to the plane of the molecule, and four electrons. Just as with ozone, these three $2p$ orbitals interact to form bonding, nonbonding, and antibonding π molecular orbitals. The bonding molecular orbital is spread over the nitrogen and both oxygen atoms. Ⓒ Placing four electrons in the energy-level diagram shown in the margin fills both the bonding and nonbonding molecular orbitals, and gives a π bond order of $\frac{1}{2}$ per N—O bond. The overall N—O bond order is thus $1\frac{1}{2}$, consistent with a resonance structure.

Note the pattern

Resonance structures are a crude way of describing molecular orbitals that extend over more than two atoms.

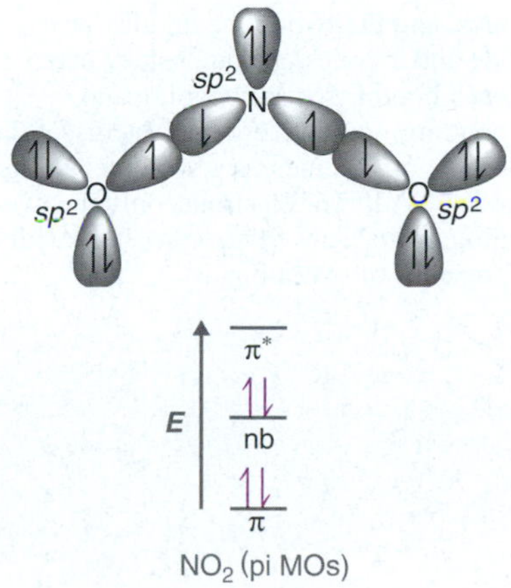

EXERCISE 9.13

Describe the bonding in the formate ion, HCO_2^-, in terms of a combination of hybrid atomic orbitals and molecular orbitals.

Answer Like nitrite, formate is a planar polyatomic ion that contains 18 valence electrons. The σ-bonding framework can be described in terms of sp^2 hybridized carbon and oxygen, which account for 14 electrons. The three unhybridized $2p$ orbitals (on C and both O atoms) form three π molecular orbitals, and the remaining four electrons occupy both the bonding and nonbonding π molecular orbitals. The overall C—O bond order is therefore $1\frac{1}{2}$.

The Chemistry of Vision

Note the pattern

As the number of interacting atomic orbitals increases, the energy separation between the resulting molecular orbitals steadily decreases.

Hydrocarbons in which two or more carbon–carbon double bonds are directly linked by carbon–carbon single bonds are generally more stable than expected because of resonance. Because the double bonds are close enough to interact electronically with one another, the π electrons are shared over all the carbon atoms, as illustrated for 1,3-butadiene in Figure 9.34. As the number of interacting atomic orbitals increases, the number of molecular orbitals increases, the energy spacing between molecular orbitals decreases, and the systems become more stable (Figure 9.35). Thus, as a chain of alternating double and single bonds becomes longer, the energy required to excite an electron from the highest-energy occupied (bonding) orbital to the lowest-energy unoccupied (antibonding) orbital decreases. If the chain is long enough, the amount of energy required to excite an electron corresponds to the energy of visible light. For example, vitamin A is yellow because its chain of five alternating double bonds is able to absorb violet light. Many of the colors we associate with dyes result from this same phenomenon; most dyes are organic compounds that contain alternating double bonds.

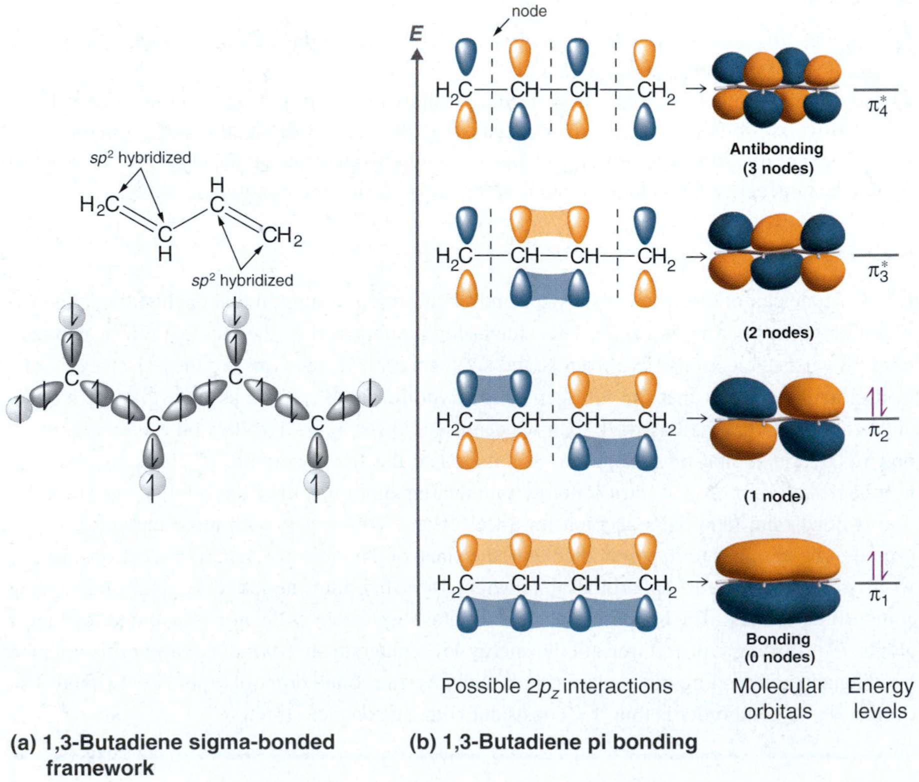

Figure 9.34 Pi bonding in 1,3-butadiene. (a) If each carbon atom is assumed to be sp^2 hybridized, we can construct a σ-bonded framework that accounts for the C-H and C-C single bonds, leaving *four* singly occupied $2p_z$ orbitals, one on each carbon atom. (b) As in ozone, these orbitals can interact, in this case to form *four* molecular orbitals. The molecular orbital at lowest energy is a bonding orbital with 0 nodes, the one at highest energy is antibonding with 3 nodes, and the two in the middle have 1 node and 2 nodes, and are somewhere between bonding or antibonding and nonbonding, respectively. The energy of the molecular orbital increases with the number of nodes. With four electrons, only the two bonding orbitals are filled, consistent with the presence of two π bonds.

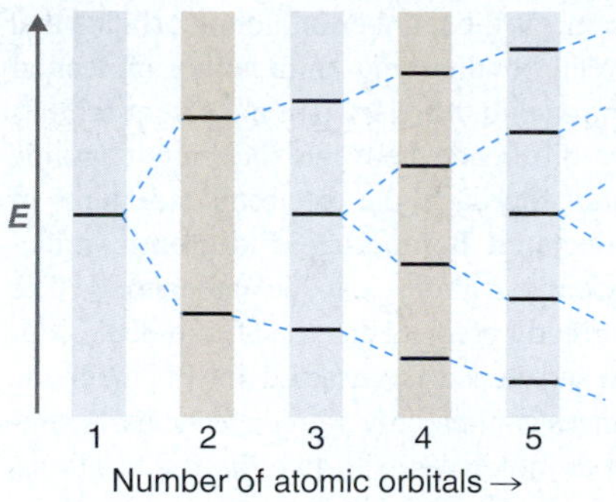

Figure 9.35 Molecular orbital energy-level diagrams for a chain of *n* interacting like orbitals ($n \leq 5$). Note that as the number of atomic orbitals increases, the difference in energy between the resulting molecular orbital energy levels decreases, which allows light of lower energy to be absorbed. As a result, organic compounds that contain long chains of carbon atoms with alternating single and double bonds tend to become more deeply colored as the number of double bonds increases.

A derivative of vitamin A called *retinal* is used by the human eye to detect light and has a structure with alternating C═C double bonds. When visible light strikes retinal, the energy separation between molecular orbitals is sufficiently close that the energy absorbed corresponds to the energy required to change one double bond in the molecule from *cis* to *trans,* initiating a process that causes a signal to be sent to the brain. If this mechanism is defective, we lose our vision in dim light. Once again, a molecular orbital approach to bonding explains a process that cannot be explained using any of the other approaches we have described.

SUMMARY AND KEY TERMS

9.1 Predicting the Geometry of Molecules and Polyatomic Ions (p. 384)

Lewis electron structures give no information about **molecular geometry**, the arrangement of bonded atoms in a molecule or polyatomic ion, which is crucial to understanding the chemistry of a molecule. The **valence-shell electron-pair repulsion (VSEPR) model** allows us to predict which of the possible structures is actually observed in most cases. It is based on the assumption that pairs of electrons occupy space, and hence the lowest-energy structure is the one that minimizes electron pair–electron pair repulsions. The VSEPR model distinguishes between the molecular geometry and the three-dimensional arrangement of electron pairs around the central atom, called the **electron-pair geometry**, but it tells nothing about the bonding. The molecular geometry of an AB_n molecule is determined by the A—B bond lengths and the **bond angles**, the angles formed by two A—B bonds. Only a small number of structures are known for AB_n molecules: AB_2, *linear* or *bent*; AB_3, *trigonal planar, trigonal pyramidal*, or *T-shaped*; AB_4, *tetrahedral, square planar* or *see-saw*; AB_5, *trigonal bipyramidal* or *square pyramidal*; and AB_6, *octahedral.* A combination of VSEPR and a bonding model such as Lewis electron structures, however, is necessary to understand the presence of multiple bonds.

Molecules that contain polar covalent bonds can have a *dipole moment*, an asymmetrical distribution of charge that results in a tendency for molecules to align themselves in an applied electric field. Any diatomic molecule that contains a polar covalent bond has a dipole moment, but in polyatomic molecules, the presence or absence of a net dipole moment depends on the structure. For some highly symmetrical AB_n structures, the individual bond dipole moments cancel one another, giving a dipole moment of zero.

9.2 Localized Bonding and Hybrid Atomic Orbitals (p. 401)

The *localized bonding* model (called **valence bond theory**) assumes that covalent bonds are formed when atomic orbitals overlap and that the strength of a covalent bond is proportional to the amount of overlap. It also assumes that atoms use combinations of atomic orbitals (*hybrids*) to maximize the overlap with adjacent atoms. The formation of **hybrid atomic orbitals** can be viewed as occurring via **promotion** of an electron from a filled ns^2 subshell to an empty np or $(n-1)d$ valence orbital, followed by **hybridization**, the combination of the orbitals to give a new set of (usually) equivalent orbitals that are oriented properly to form bonds. The combination of an ns and an np orbital gives rise to two equivalent *sp hybrids* oriented at 180°, whereas the combination of an ns and two or three np orbitals produces three equivalent sp^2 *hybrids* or four equivalent sp^3 *hybrids,* respectively. The bonding in molecules that contain more than an octet of electrons around a central atom can be explained by invoking the participation of one or two $(n-1)d$ orbitals to give sets of five sp^3d or six sp^3d^2 *hybrid* orbitals, capable of forming five or six bonds, respectively. The spatial orientation of the hybrid atomic orbitals is consistent with the geometries predicted using the VSEPR model.

9.3 Delocalized Bonding and Molecular Orbitals (p. 408)

A **molecular orbital** (MO) is an allowed spatial distribution of electrons in a molecule that is associated with a particular orbital energy. Unlike an atomic orbital (AO), which is centered on a single atom, a molecular orbital extends over all the atoms in a molecule or ion. Hence, the **molecular orbital theory** of bonding is a *delocalized* approach. Molecular orbitals are constructed using **linear combinations of atomic orbitals** (LCAOs), which are usually the mathematical sums and differences of wave functions that describe overlapping atomic orbitals. Atomic orbitals interact to form three types of molecular orbitals. (1) Orbitals or orbital lobes with the same sign interact to give increased electron probability along the plane of the internuclear axis because of *constructive* reinforcement of the wave functions. Consequently, electrons in such molecular orbitals help to hold the positively charged nuclei together. Such orbitals are **bonding molecular orbitals**, and they are always lower in energy than the parent atomic orbitals. (2) Orbitals or orbital lobes with opposite signs interact to give decreased electron probability density between the nuclei because of *destructive* interference of the wave functions. Consequently, electrons in such molecular orbitals are primarily located outside the internuclear region, leading to increased repulsions between the positively charged nuclei. These orbitals are called **antibonding molecular orbitals**, and they are always higher in energy than the parent atomic orbitals. (3) Some atomic orbitals interact only very weakly, and the resulting molecular orbitals give essentially no change in the electron probability density between the nuclei. Hence, electrons in such orbitals have no effect on the bonding in a molecule or ion. These orbitals are **nonbonding molecular orbitals**, and they have approximately the same energy as the parent atomic orbitals. A completely bonding molecular orbital contains no nodes (regions of zero electron probability) perpendicular to the internuclear axis, whereas a completely antibonding molecular orbital contains at least one node perpendicular to the internuclear axis. A **sigma (σ) orbital** (bonding) or a **sigma star (σ^*) orbital** (antibonding) is symmetrical about the internuclear axis. Hence, all cross-sections perpendicular to that axis are circular. Both a **pi (π) orbital** (bonding) and a **pi star (π^*) orbital** (antibonding) possess a nodal plane that contains the nuclei, with electron density localized on both sides of the plane.

The energies of the molecular orbitals versus those of the parent atomic orbitals can be shown schematically in an *energy-level diagram*. The electron configuration of a molecule is shown by placing the correct number of electrons in the appropriate energy-level diagram, starting with the lowest-energy orbital and obeying the Pauli principle; that is, placing only two electrons with opposite spin in each orbital. From the completed energy-level diagram, we can calculate a **bond order**, defined as one-half the net number of bonding electrons. In bond orders, electrons in antibonding molecular orbitals cancel electrons in bonding molecular orbitals, while electrons in nonbonding orbitals have no effect and are not counted. Bond orders of 1, 2, and 3 correspond to single, double, and triple bonds, respectively. Molecules with predicted bond orders of 0 are generally less stable than the isolated atoms and do not normally exist.

Molecular orbital energy-level diagrams for diatomic molecules can be created if the electron configuration of the parent atoms is known, following a few simple rules. Most important, the number of molecular orbitals in a molecule is the same as the number of atomic orbitals that interact. The difference between bonding and antibonding molecular orbital combinations is proportional to the overlap of the parent orbitals and decreases as the energy difference between the parent atomic orbitals increases. With such an approach, the electronic structures of virtually all commonly encountered **homonuclear diatomic molecules**, molecules with two identical atoms, can be understood. The molecular orbital approach correctly predicts that the O_2 molecule contains two unpaired electrons and hence is attracted into a magnetic field. In contrast, most substances contain only paired electrons. A similar procedure can be applied to molecules with two dissimilar atoms, called **heteronuclear diatomic molecules**. Using a molecular orbital energy-level diagram that is skewed or tilted toward the more electronegative element. Molecular orbital theory is able to describe the bonding in a molecule with an odd number of electrons such as NO and even to predict something about its chemistry.

9.4 Combining the Valence Bond and Molecular Orbital Approaches (p. 422)

To describe the bonding in more complex molecules that contain multiple bonds, we can use an approach that combines hybrid atomic orbitals to describe the σ bonding and molecular orbitals to describe the π bonding. In this approach, unhybridized *np* orbitals on atoms bonded to one another are allowed to interact to produce bonding, antibonding, or nonbonding combinations. For π bonds between two atoms (as in ethylene or acetylene), the resulting molecular orbitals are virtually identical to the π molecular orbitals in diatomic molecules such as O_2 and N_2. Application of the same approach to π bonding between three or four atoms requires combining three or four unhybridized *np* orbitals on adjacent atoms to generate π bonding, antibonding, and nonbonding molecular orbitals extending over all of the atoms. Filling the resulting energy-level diagram with the appropriate number of electrons explains the bonding in molecules or ions that previously required the use of resonance structures in the Lewis electron-pair approach.

KEY EQUATION

Bond order

$$\text{bond order} = \frac{\text{number of bonding electrons} - \text{number of antibonding electrons}}{2} \qquad (9.5)$$

QUESTIONS AND PROBLEMS

Questions and problems with colored numbers have answers in the Appendix and complete solutions in the Student Solutions Manual.

CONCEPTUAL

9.1 Predicting the Geometry of Molecules and Polyatomic Ions

1. What are the differences between molecular geometry and electron-pair geometry? Can two molecules with the same electron-pair geometry have different molecular geometries? Can two molecules with the same molecular geometry have different electron-pair geometries? In each case, support your answer with an example.
2. What is the main difference between the VSEPR model and Lewis electron structures?
3. How does the VSEPR model deal with the presence of multiple bonds?
4. Three molecules have the following generic formulas: AB_2 with no lone pairs, AB_2 with one lone pair, and AB_2 with two lone pairs. Predict the molecular geometry

of each, and arrange them in order of increasing B—A—B angle.

5. Which has the smaller B—A—B angle, H_2S or SiH_4? Why? Do the Lewis electron structures of these molecules predict which has the smaller angle?
6. Discuss in your own words why lone pairs of electrons occupy more space than bonding pairs. How does the presence of lone pairs affect the molecular geometry?
7. When using VSEPR to predict molecular geometry, the importance of repulsions between electron pairs decreases in the following order: lone pair–lone pair, lone pair–bonded pair, bonded pair–bonded pair. Explain this order. Draw structures of real molecules that separately show each of these interactions.
8. How do multiple bonds affect molecular geometry? Does a multiple bond take up more or less space around an atom than a single bond? A lone pair?
9. Straight-chain alkanes do not have linear structures, but are "kinked." Using *n*-hexane as an example, explain why this is so. Compare the geometry of 1-hexene to that of *n*-hexane.
10. How is molecular geometry related to the presence or absence of a molecular dipole moment?
11. How are molecular geometry and dipole moments related to physical properties such as melting point and boiling point?
12. What two features of a molecule's structure and bonding are required for a molecule to be considered polar? Is COF_2 likely to have a significant dipole moment? Explain your answer.
13. When a chemist says that a molecule is *polar,* what does this mean? What are the general physical properties of polar molecules?
14. Use the VSPER model and your knowledge of bonding and dipole moments to predict which of the following molecules will be liquids or solids at room temperature and which will be gases: CH_3Cl, PCl_3, CO, SF_6, IF_5, CH_3OCH_3, CCl_3H, and H_3COH. Explain your rationale for each choice. Justify your answers.
15. The idealized molecular geometry of BrF_5 is square pyramidal, with one lone pair. What effect does the lone pair have on the actual molecular geometry of BrF_5? If lone pair–bonding pair repulsions were *weaker* than bonding pair–bonding pair repulsions, what would be the effect on the molecular geometry of BrF_5?
16. Which has the smallest B—A—B bond angle: H_2S, H_2Se, or H_2Te? The largest? Justify your answers.
17. Which of these molecular geometries *always* results in a molecule with a net dipole moment: linear, bent, trigonal planar, tetrahedral, seesaw, trigonal pyramidal, square pyramidal, and octahedral? For the geometries that do not always produce a net dipole moment, what factor(s) will result in a net dipole moment?

9.2 Localized Bonding and Hybrid Atomic Orbitals

18. Arrange the following hybrid orbitals in order of increasing strength of the bond formed to a hydrogen atom: sp, sp^3, sp^2. Explain your reasoning.
19. What atomic orbitals are combined to form each of the following sets of hybrid orbitals: sp^3, sp, sp^3d^2, sp^3d? What is the maximum number of electron-pair bonds that can be formed using each of these sets of hybrid orbitals?
20. Why is it incorrect to say that an atom with sp^2 hybridization will form only three bonds? The carbon atom in the carbonate anion is sp^2 hybridized. How many bonds to carbon are present in the carbonate ion? Which orbitals on carbon are used to form each bond?
21. If hybridization did not occur, how many bonds would each of the following elements form in a neutral molecule, and what would be the approximate molecular geometry: N, O, C, B?
22. How are hybridization and molecular geometry related? Which has a stronger correlation: molecular geometry and hybridization, or electron-pair geometry and hybridization?
23. In the valence bond approach to bonding in BeF_2, which step(s) require(s) an energy input, and which release(s) energy?
24. The energies of hybrid orbitals are intermediate between the energies of the atomic orbitals from which they are formed. Why?
25. How are lone pairs on the central atom treated using hybrid orbitals?
26. Since nitrogen bonds to only three hydrogen atoms in ammonia, why doesn't the nitrogen atom use sp^2 hybrid orbitals instead of sp^3 hybrids?
27. Using arguments based on orbital hybridization, explain why the CCl_6^{2-} ion does not exist.
28. Species such as NF_5^{2-} and OF_4^{2-} are unknown. If $3d$ atomic orbitals were much lower energy, low enough to be involved in hybrid orbital formation, what effect would this have on the stability of such species? Why? What molecular geometry, electron-pair geometry, and hybridization would be expected for each molecule?

9.3 Delocalized Bonding and Molecular Orbitals

29. What is the distinction between an atomic orbital and a molecular orbital? How many electrons can a molecular orbital accommodate?
30. Why is the molecular orbital approach to bonding called a *delocalized* approach?
31. How is the energy of an electron affected by interacting with more than one positively charged atomic nucleus at a time? Does the energy of the system increase, decrease, or remain unchanged? Why?
32. Constructive and destructive interference of waves can be used to understand how bonding and antibonding molecular orbitals are formed from atomic orbitals. Does constructive interference of waves result in increased or decreased electron probability density between the nuclei? Is the result of constructive interference best described as a bonding molecular orbital or an antibonding molecular orbital?
33. What is a "node" in molecular orbital theory? How is it similar to the nodes found in atomic orbitals?
34. What is the difference between an s orbital and a σ orbital? How are the two similar?
35. Why is a σ_{1s} molecular orbital lower in energy than the two s atomic orbitals from which it is derived? Why is a σ^*_{1s} molecular orbital higher in energy than the two s atomic orbitals from which it is derived?
36. What is meant by the term *bond order* in molecular orbital theory? How is the bond order determined from molecular orbital theory different from the bond order obtained using Lewis electron structures? How is it similar?
37. What is the effect of placing an electron in an antibonding orbital on the bond order; the stability of the molecule; and the reactivity of a molecule?

38. How can the molecular orbital approach to bonding be used to predict a molecule's stability? What advantages does this method have over the Lewis electron-pair approach to bonding?

39. What is the relationship between bond length and bond order? What effect do antibonding electrons have on bond length? On bond strength?

40. Draw a diagram that illustrates how atomic *p* orbitals can form both σ and π molecular orbitals. Which type of molecular orbital typically results in a stronger bond?

41. What is the minimum number of nodes in each of these molecular orbitals: σ, π, σ^*, π^*? How are the nodes in bonding orbitals different from the nodes in antibonding orbitals?

42. It is possible to form both σ and π molecular orbitals with the overlap of a *d* orbital with a *p* orbital, yet it is only possible to form σ molecular orbitals between *s* and *d* orbitals. Illustrate why this is so with a figure showing the three types of overlap between this set of orbitals. Include a fourth image that shows why *s* and *d* orbitals cannot combine to form a π molecular orbital.

43. Is it possible for an np_x orbital on one atom to interact with an np_y orbital on another atom to produce molecular orbitals? Why or why not? Can the same be said of np_y and np_z orbitals on adjacent atoms?

44. What is meant by *degenerate orbitals* in molecular orbital theory? Is it possible for σ molecular orbitals to form a degenerate pair? Explain your answer.

45. Why are bonding molecular orbitals lower in energy than the parent atomic orbitals? Why are antibonding molecular orbitals higher in energy than the parent atomic orbitals?

46. What is meant by "the law of conservation of orbitals"?

47. Atomic orbitals on different atoms have different energies. When atomic orbitals from non-identical atoms are combined to form molecular orbitals, what is the effect of this difference in energy on the resulting molecular orbitals?

48. If two atomic orbitals have different energies, how does this affect the orbital overlap and the molecular orbitals formed by combining the atomic orbitals?

49. Are the Al—Cl bonds in $AlCl_3$ stronger, the same strength, or weaker than the Al—Br bonds in $AlBr_3$? Why?

50. Are the Ga—Cl bonds in $GaCl_3$ stronger, the same strength, or weaker than the Sb—Cl bonds in $SbCl_3$? Why?

51. What is meant by a *nonbonding* molecular orbital, and how is it formed? How does the energy of a nonbonding orbital compare with the energy of bonding or antibonding molecular orbitals derived from the same atomic orbitals?

52. Many features of molecular orbital theory have analogs in Lewis electron structures. How do Lewis electron structures represent
(a) nonbonding electrons, and
(b) electrons in bonding molecular orbitals?

53. How does electron screening affect the energy difference between the 2*s* and 2*p* atomic orbitals of the second-row elements? How does the energy difference between the 2*s* and 2*p* atomic orbitals depend upon the effective nuclear charge?

54. For each of the following combinations of atomic orbitals of similar energy, which of the resulting molecular orbitals is lower in energy: σ vs. π; π vs. σ^*; σ^* vs. π^*?

55. The energy of a σ molecular orbital is usually lower than the energy of a π molecular orbital derived from the same set of atomic orbitals. Under specific conditions, however, the order can be reversed. What causes this reversal? In which portion of the periodic table is this kind of orbital energy reversal most likely to be observed?

56. Is the σ_{2p_z} molecular orbital stabilized or destabilized by interaction with the σ_{2s} molecular orbital in N_2? In O_2? In which molecule is this interaction most important?

57. Explain how the Lewis electron-pair approach and molecular orbital theory differ in their treatment of bonding in O_2.

58. Why is it crucial to our existence that O_2 is paramagnetic?

59. Will NO or CO react more quickly with O_2? Explain your answer.

60. How is the energy-level diagram of a heteronuclear diatomic molecule, such as CO, different from that of a homonuclear diatomic molecule, such as N_2?

61. How does molecular orbital theory describe the existence of polar bonds? How is this apparent in the molecular orbital diagram of HCl?

9.4 Combining the Valence Bond and Molecular Orbital Approaches

62. What information is obtained by using the molecular orbital approach to bonding in O_3 that is not obtained using the VSEPR model? Can this information be obtained using a Lewis electron-pair approach?

63. How is resonance explained using the molecular orbital approach?

64. Indicate what information can be obtained by each method:

	Lewis Electron Structures	VSEPR Model	Valence Bond Theory	Molecular Orbital Theory
Geometry	______	______	______	______
Resonance	______	______	______	______
Orbital hybridization	______	______	______	______
Reactivity	______	______	______	______
Expanded valences	______	______	______	______
Bond order	______	______	______	______

NUMERICAL

This section includes paired problems (marked by brackets) that require similar problem-solving skills.

9.1 Predicting the Geometry of Molecules and Polyatomic Ions

65. Give the electron-pair geometry for each of the following molecules: BF_3, PCl_3, XeF_2, $AlCl_4^-$, CH_2Cl_2. Classify the electron pairs in each species as bonding pairs or lone pairs.

66. Determine the electron-pair geometry for each of the following species: ICl_3, CCl_3^+, H_2Te, XeF_4, NH_4^+. Identify the electron pairs in each species as bonding pairs or lone pairs.

67. Give the electron-pair geometry and the molecular geometry for: HCl, NF_3, ICl_2^+, N_3^-, and H_3O^+. For structures that are not linear, draw three-dimensional representations, clearly showing the positions of the lone pairs of electrons.

68. Give the electron-pair geometry and the molecular geometry of the following molecules: SO_3, NH_2^-, NO_3^-, I_3^-, and OF_2. Draw three-dimensional representations of all structures that are not linear, clearly showing the positions of the lone pairs of electrons.

69. What is the molecular geometry of ClF_3? Draw a three-dimensional representation of its structure, and explain the effect of any lone pairs on the idealized geometry.

70. Predict the molecular geometry of ICl_3, AsF_5, NO_2^-, and $TeCl_4$.

71. Predict whether each of the following molecules has a net dipole moment: NO, HF, PCl_3, CO_2, SO_2, SF_4, XeO_4. Justify your answers, and indicate the direction of any bond dipoles.

72. Predict whether each of the following molecules has a net dipole moment: OF_2, BCl_3, CH_2Cl_2, TeF_4, CH_3OH. Justify your answers, and indicate the direction of any bond dipoles.

73. Which of the following molecules has a net dipole moment: Cl_2CCCl_2, IF_3, and SF_6? Explain your reasoning.

74. Which of the following molecules is likely to have a net dipole moment: SO_3, XeF_4, and H_2CCCl_2? Explain your reasoning.

9.2 Localized Bonding and Hybrid Atomic Orbitals

75. Draw an energy-level diagram showing promotion and hybridization to describe the bonding in CH_3^-. How does your diagram compare with that for methane? What is the electron-pair geometry for this species? What is the molecular geometry?

76. Illustrate promotion and hybridization with an energy-level diagram for the bonding in CH_3^+. How does your diagram compare with that for methane? What is the electron-pair geometry for this species? What is the molecular geometry?

77. Draw the molecular structure, including any lone pairs on the central atom, state the hybridization of the central atom, and determine the molecular geometry for each molecule: BBr_3, PCl_3, NO_3^-.

78. Draw the molecular structure, show any lone pairs on the central atom, state the hybridization of the central atom, and determine the molecular geometry of each of the following species: $AsBr_3$, CF_3^+, and H_2O.

79. What is the hybridization of the central atom in CF_4, CCl_2^{2-}, IO_3^-, and SiH_4?

80. Give the hybridization of the central atom in CCl_3^+, CBr_2O, CO_3^{2-}, and IBr_2^-.

81. What is the hybridization of the central atom in PF_6^-. Is this ion likely to exist? Why or why not? What would be the shape of the molecule?

82. Give the hybridization of the central atom in SF_5^-. Is this ion likely to be a stable species? Why or why not? What would be the molecular geometry?

9.3 Delocalized Bonding and Molecular Orbitals

83. Use a qualitative molecular orbital energy-level diagram to describe the bonding in S_2^{2-}. What is the bond order? How many unpaired electrons does it have?

84. Use a qualitative molecular orbital energy-level diagram to describe the bonding in F_2^{2+}. What is the bond order? How many unpaired electrons does it have?

85. If three atomic orbitals combine to form molecular orbitals, how many molecular orbitals are generated? How many molecular orbitals result from the combination of four atomic orbitals? From five?

86. If two atoms interact to form a bond, and each atom has four atomic orbitals, how many molecular orbitals will form?

87. Sketch the possible ways of combining two 1*s* orbitals on adjacent atoms. How many molecular orbitals can be formed by this combination? Be sure to indicate any nodal planes.

88. Sketch the *four* possible ways of combining two 2*p* orbitals on adjacent atoms. How many molecular orbitals can be formed by this combination? Be sure to indicate any nodal planes.

89. If a diatomic molecule has a bond order of 2 and six bonding electrons, how many antibonding electrons must it have? What would be the corresponding Lewis electron structure (disregarding lone pairs)? What would be the effect of a one-electron reduction on the bond distance?

90. What is the bond order of a diatomic molecule that has six bonding electrons and no antibonding electrons? If an analogous diatomic molecule has six bonding electrons and four antibonding electrons, which has the stronger bond? The shorter bond distance? If the highest occupied molecular orbital in both molecules is bonding, how will a one-electron oxidation affect the bond length?

91. Qualitatively discuss how the bond distance in a diatomic molecule would be affected by:
(a) adding an electron to an antibonding orbital, and
(b) adding an electron to a bonding orbital.

92. Explain why oxidation of O_2 decreases the bond distance, whereas oxidation of N_2 *increases* the N—N distance. Could Lewis electron structures be utilized to answer this problem?

93. Draw a molecular orbital energy-level diagram for Na_2^+. What is the bond order in this ion? Is it likely to be a stable species? If not, would you recommend an oxidation or a reduction to improve stability? Explain your answer. Based on your answers, which is likely to be the most stable: Na_2^+, Na_2, or Na_2^-?

94. Draw a molecular orbital energy-level diagram for Xe_2^+, showing only the valence orbitals and electrons. What is the bond order? Is this ion likely to be a stable species? Explain your answer. Based on your answers, which is likely to be most stable: Xe_2^{2+}, Xe_2^+, or Xe_2?

95. Draw a molecular orbital energy-level diagram to predict the valence electron configuration, bond order, and stability of O_2^{2-}.

96. Construct a molecular orbital energy-level diagram for C_2^{2-}, and predict its valence electron configuration, bond order, and stability.

97. If all the *p* orbitals in the valence shells of two atoms interact, how many molecular orbitals are formed? Why is it not possible to form three π orbitals (and the corresponding antibonding orbitals) from the set of six π orbitals?

98. Draw a complete energy-level diagram for B_2. Determine the bond order and whether the molecule is paramagnetic or diamagnetic. Explain your rationale for the order of the molecular orbitals.

99. Sketch a molecular orbital energy-level diagram for
(a) NO^+ and
(b) NO^-.
Based on your diagram, what is the bond order of each species?

100. The diatomic molecule BN has never been detected. Assume that its MO diagram would be similar to that shown for CN^- in Section 9.3, but that the σ_{2p_z} molecular orbital is higher in energy than the $\pi_{2p_{x,y}}$ molecular orbitals.
(a) Sketch a MO diagram for BN.
(b) Based on your diagram, what would be the bond order of this molecule?
(c) Would you expect BN to be stable? Why or why not?

101. Which of the following species are isoelectronic: BN, CO, C_2, and N_2?

102. Which of the following species are isoelectronic: CN^-, NO^+, B_2^{2-}, and O_2^+?

9.4 Combining the Valence Bond and Molecular Orbital Approaches

103. Using a combined hybrid atomic orbital and molecular orbital approach, describe the bonding in BCl_3 and CS_3^{2-}.

104. Use a combined hybrid atomic orbital and molecular orbital approach to describe the bonding in CO_2 and N_3^-.

APPLICATIONS

105. Sulfur hexafluoride, SF_6, is a very stable gas that is used in a wide range of applications, because it is non-toxic, non-flammable, and non-corrosive. Unfortunately, it is also a very powerful "greenhouse gas" that is about 22,000 times more effective at causing global warming than the same mass of CO_2.
(a) Draw the Lewis electron structure of SF_6, and determine the electron-pair geometry, the molecular geometry, and the hybridization of the central atom.
(b) Suggest a reason for the extremely high stability of SF_6.
(c) Despite its rather high molecular mass (146.06 g/mol) and highly polar S—F bonds, SF_6 is a gas at room temperature (bp –63°C). Why?

106. The elevated concentrations of chlorine monoxide (ClO) that accompany ozone depletions in Earth's atmosphere can be explained by a sequence of reactions. In the first step, chlorine gas is split into chlorine atoms by sunlight. Each chlorine atom then catalyzes the decomposition of ozone through a chlorine monoxide intermediate.
(a) Write balanced equations showing this sequence of reactions.
(b) Sketch the molecular orbital energy-level diagram of ClO.
(c) Does ClO contain any unpaired electrons?
(d) Based on your molecular orbital diagram, is ClO likely to be a stable species? Explain your answer.

107. Saccharin is an artificial sweetener that was discovered in 1879. For several decades, it was used by people who had to limit their intake of sugar for medical reasons. Because it was implicated as a carcinogen in 1977, however, warning labels are now required on foods and beverages that contain saccharin. The structure of this sweetener is:

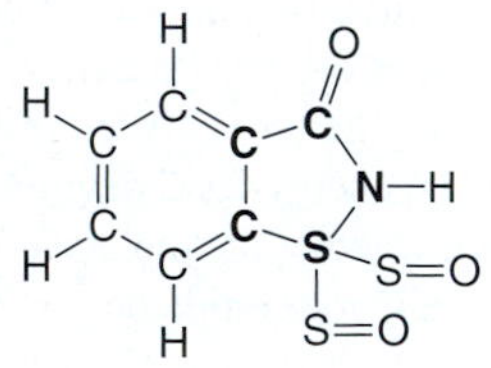

(a) Give the hybridization of all five atoms shown in bold in the structure. Note: all five atoms in the five-membered ring are coplanar.
(b) The carbon–oxygen bond is drawn as a double bond. If the nitrogen and carbon attached to the C═O group each contribute one electron to the bonding, use both a Lewis electron structure and a hybrid orbital approach to explain the presence of the double bond.
(c) If the sulfur and the carbon each contribute one electron to the nitrogen, how many lone pairs are present on the nitrogen?
(d) What is the geometry of the sulfur atom?
(e) The Lewis electron structure supports a single bond between the carbon and nitrogen and a double bond between the carbon and oxygen. In actuality, the C—O bond is longer than expected for a double bond, and the C—N bond is shorter. The nitrogen is also planar. Based on this information, what is the likely hybridization of the nitrogen? Using the concepts of molecular orbital theory, propose an explanation for this observation.

108. Pheromones are chemical signals used for communication between members of the same species. For example, the bark beetle uses an aggregation pheromone to signal other bark beetles to congregate at a particular site in a tree. Bark beetle infestations can cause severe damage, because the beetles carry a fungal infection that spreads rapidly and can kill the tree. One of the components of this aggregation pheromone has the following structure:

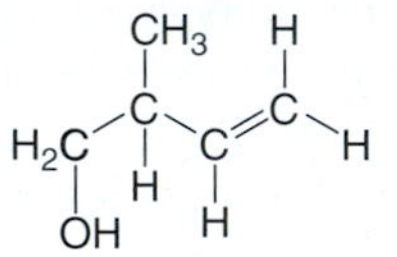

(a) Give the hybridization of all atoms except hydrogen in this pheromone.
(b) How many σ bonds are present in this molecule? How many π bonds are there?
(c) Describe the bonding in this molecule using a combination of the localized and delocalized approaches.

109. Carbon monoxide is highly poisonous, because it binds more strongly than O_2 to the iron in red blood cells, which transport oxygen in the blood. Consequently, a victim of CO poisoning suffocates from lack of oxygen. Draw a molecular orbital energy-level diagram for CO. What is the highest occupied molecular orbital? Are any of the molecular orbitals degenerate? If so, which ones?

10 Gases

LEARNING OBJECTIVES

- To be able to describe the characteristics of a gas
- To understand the relationships between the pressure, temperature, amount, and volume of a gas
- To be able to use the ideal gas law to describe the behavior of a gas
- To know how to determine the contribution of each component gas to the total pressure of a gaseous mixture
- To understand the significance of the kinetic molecular theory of gases
- To recognize the differences between the behaviors of an ideal gas and a real gas

CHAPTER OUTLINE

In Part II we focused on so-called *microscopic* properties of matter—the properties of individual atoms, ions, and molecules—and on how the electronic structures of atoms and ions determine the stoichiometry and three-dimensional geometry of the compounds they form. In Part III, we will focus on *macroscopic* properties—the behavior of aggregates that contain large numbers of atoms, ions, or molecules. An understanding of macroscopic properties is central to an understanding of chemistry. Why, for example, are many compounds gases under normal pressures and temperatures (1.0 atm, 25°C), whereas others are liquids or solids? We will examine each of the three forms of matter—gases, liquids, and solids—as well as the nature of the forces, such as hydrogen-bonding and electrostatic interactions, that hold molecular and ionic compounds together in these three states.

Hot-air balloons being prepared for flight. As the air inside each balloon is heated, the volume of the air increases, filling the balloon. The lower density of the air in the balloons allows the balloons to ascend through the substance with higher density, the cooler air.

In Chapter 10, we explore the relationships between the pressure, temperature, volume, and amount of a gas. You will learn how to use these relationships to describe the physical behavior of a sample of both a pure gaseous substance and mixtures of gases. By the end of this chapter, your understanding of the gas laws and the model used to explain the behavior of gases will allow you to explain how straws and hot-air balloons work, why hand pumps cannot be used in wells beyond a certain depth, why helium-filled balloons deflate so rapidly, and how a gas can be liquefied for use in preserving biological tissue.

10.1 Gaseous Elements and Compounds

The three common phases (or states) of matter are gas, liquid, and solid. Gases have the lowest density of the three, are highly compressible, and completely fill any container in which they are placed. Gases behave this way because their intermolecular forces are relatively weak, so the molecules are constantly moving independently of the other molecules present. Solids, in contrast, are relatively dense, rigid, and incompressible because their intermolecular forces are so strong that the molecules are essentially locked in place. Liquids are relatively dense and incompressible, like solids, but they flow readily to adapt to the shape of the container, like gases. We can therefore conclude that the sum of the intermolecular forces in liquids are between those of gases and solids. Figure 10.1 compares the three states of matter and illustrates the differences at the molecular level.

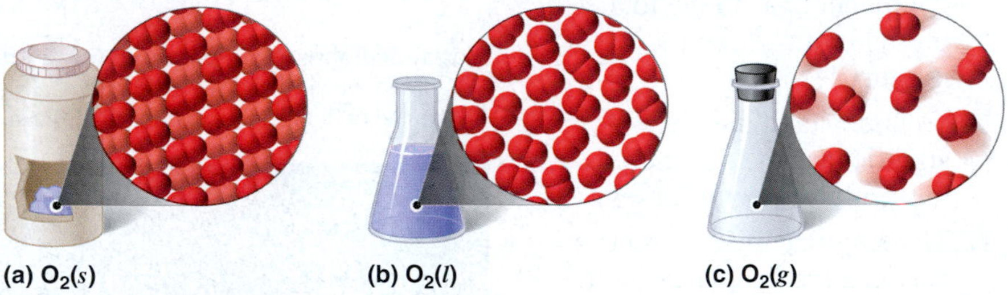

Figure 10.1 A diatomic substance (O_2) in the solid, liquid, and gaseous states. (a) Solid O_2 has a fixed volume and shape, and the molecules are packed tightly together. (b) Liquid O_2 conforms to the shape of its container but has a fixed volume; it contains relatively densely packed molecules. (c) Gaseous O_2 fills its container completely, regardless of the container's size or shape, and consists of widely separated molecules.

The state of a given substance depends strongly on conditions. For example, H_2O is commonly found in all three states: solid ice, liquid water, and water vapor, its gaseous form. Under most conditions, we encounter water as the liquid that is essential for life; we drink it, cook with it, and bathe in it. When the temperature is cold enough to transform the liquid to ice, we can ski or skate on it, pack it into a snowball or snowcone, and even build dwellings with it. Water vapor* is a component of the air we breathe, and it is produced whenever we heat water for cooking food or making coffee. Water vapor at temperatures higher than 100°C is called steam, and steam is used to drive large machinery, including turbines that generate electricity. Properties of the three states of water are summarized in Table 10.1.

TABLE 10.1 Properties of water at 1.0 atm

Temperature	State	Density, g/cm^3
≤0°C	Solid (ice)	0.9167 (0.0°C)
0–100°C	Liquid (water)	0.9997 (4.0°C)
≥100°C	Vapor (steam)	0.005476 (127°C)

* The distinction between a gas and a vapor is subtle: the term *vapor* is used to refer to the gaseous form of a substance that is a liquid or a solid under "normal" conditions (25°C, 1.0 atm). Nitrogen (N_2) and oxygen (O_2) are thus referred to as gases, but gaseous water in the atmosphere is called water vapor.

The geometric structure and the physical and chemical properties of atoms, ions, and molecules usually do *not* depend on the physical state; the individual water molecules in ice, liquid water, and steam, for example, are all identical. In contrast, the macroscopic properties of a substance depend strongly on its physical state, which is determined by intermolecular forces and conditions such as temperature and pressure.

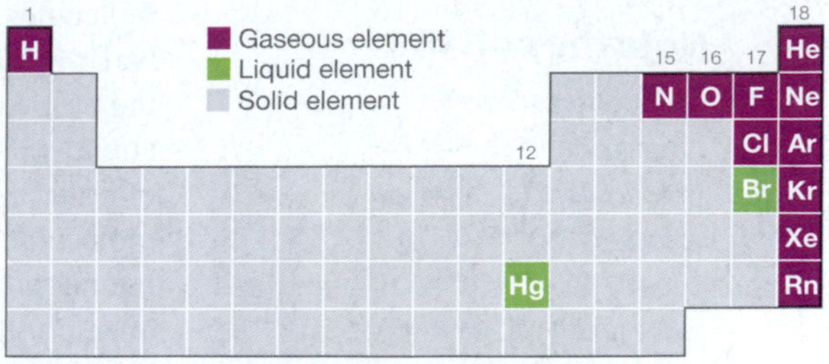

Figure 10.2 Elements that occur naturally as gases, liquids, and solids at 25°C and 1 atm. All the noble gases and mercury occur as monatomic species, whereas all the other gases and bromine are diatomic molecules.

Figure 10.2 shows the locations in the periodic table of those elements that are commonly found in the gaseous, liquid, and solid states. Except for hydrogen, the elements that occur naturally as gases are on the right side of the periodic table. Of these, all the noble gases (Group 18) are monatomic gases, whereas the other gaseous elements are diatomic molecules (H_2, N_2, O_2, F_2, and Cl_2). Oxygen can also form a second allotrope, the highly reactive triatomic molecule ozone (O_3), which is a gas. In contrast, bromine (as Br_2) and mercury (Hg) are liquids under normal conditions (25°C and 1.0 atm, commonly referred to as "room temperature and pressure"). Gallium (Ga), which melts at only 29.76°C, can be converted to a liquid simply by holding a container of it in your hand or keeping it in a non–air-conditioned room on a hot summer day. The rest of the elements are all solids under normal conditions.

Many of the elements and compounds we have encountered in this text are typically found as gases; some of the more common ones are listed in Table 10.2. Gaseous substances include many binary hydrides, such as the hydrogen halides (HX); hydrides of the chalcogens; hydrides of the Group-15 elements N, P, and As; hydrides of the Group-14 elements C, Si, and Ge; and diborane (B_2H_6). In addition, many of the simple covalent oxides of the nonmetals are gases, such as CO, CO_2, NO, NO_2, SO_2, SO_3, and ClO_2. Many low-molecular-mass organic compounds are gases as well, including all the hydrocarbons that have four or fewer carbon atoms and simple molecules such as dimethyl ether [$(CH_3)_2O$], methyl chloride (CH_3Cl), formaldehyde (CH_2O), and acetaldehyde (CH_3CHO). Finally, most of the commonly used refrigerants, such as the chlorofluorocarbons (CFCs) and the hydrochlorofluorocarbons (HCFCs) discussed in Chapter 3, are gases.

Dimethyl ether

Note that all of the gaseous substances mentioned above (other than the monatomic noble gases) contain covalent or polar covalent bonds and are nonpolar or polar

TABLE 10.2 Some common substances that are gases at 25°C and 1.0 atm

Elements	Compounds	
He (helium)	HF (hydrogen fluoride)	C_2H_4 (ethylene)
Ne (neon)	HCl (hydrogen chloride)	C_2H_2 (acetylene)
Ar (argon)	HBr (hydrogen bromide)	C_3H_8 (propane)
Kr (krypton)	HI (hydrogen iodide)	C_4H_{10} (butane)
Xe (xenon)	HCN (hydrogen cyanide)[a]	CO (carbon monoxide)
Rn (radon)	H_2S (hydrogen sulfide)	CO_2 (carbon dioxide)
H_2 (hydrogen)	NH_3 (ammonia)	NO (nitric oxide)
N_2 (nitrogen)	PH_3 (phosphine)	N_2O (nitrous oxide)
O_2 (oxygen)	CH_4 (methane)	NO_2 (nitrogen dioxide)
O_3 (ozone)	C_2H_6 (ethane)	SO_2 (sulfur dioxide)
F_2 (fluorine)		
Cl_2 (chlorine)		

[a]HCN boils at 26°C at 1 atm, so it is included in this table.

Note the pattern

Gaseous substances often: contain covalent or polar covalent bonds, exist as nonpolar or slightly polar molecules, have relatively low molecular masses, and contain five or fewer atoms from the second row or above.

molecules. In contrast, the strong electrostatic attractions in ionic compounds, such as NaBr (boiling point 1390°C) or LiF (boiling point 1673°C), prevent them from existing as gases at room temperature and pressure. Note too that the lightest members of any given family of compounds are most likely to be gases, and that the boiling points of polar compounds are generally higher than those of nonpolar compounds of similar molecular mass. Therefore, in a given series of compounds, the lightest and least polar members are the ones most likely to be gases. With relatively few exceptions, however, compounds that contain more than about five atoms from the second period or below are too heavy to exist as gases under normal conditions.

EXAMPLE 10.1

Which of the following compounds is most likely to be a gas at room temperature and pressure? **(a)** cyclohexene; **(b)** lithium carbonate; **(c)** cyclobutane; **(d)** vanadium(III) oxide; **(e)** benzoic acid, $C_6H_5CO_2H$

Given Compounds

Asked for Physical state

Strategy

A Decide whether each compound is ionic or covalent. An ionic compound is most likely to be a solid at room temperature and pressure, whereas a covalent compound may be a solid, liquid, or gas.

B Among the covalent compounds, those that are relatively nonpolar and have low molecular masses are most likely to be gases at room temperature and pressure.

Solution

A Lithium carbonate is Li_2CO_3, containing Li^+ and CO_3^{2-} ions, and vanadium(III) oxide is V_2O_3, containing V^{3+} and O^{2-} ions. Consequently, both are primarily ionic compounds that are expected to be solids. The remaining three compounds are all covalent. **B** Benzoic acid has more than four carbon atoms and is polar, so it is not likely to be a gas. Both cyclohexene and cyclobutane are essentially nonpolar molecules, but cyclobutane (C_4H_8) has a significantly lower molecular mass than cyclohexene (C_6H_{10}), which again has more than four carbon atoms. We therefore predict that cyclobutane is most likely to be a gas at room temperature and pressure, and cyclohexene a liquid. In fact, with a boiling point of only 12°C, compared to 83°C for cyclohexene, cyclobutane is indeed a gas at room temperature and pressure.

EXERCISE 10.1

Which of the following compounds would you predict to be gases under normal conditions? **(a)** *n*-butanol; **(b)** ammonium fluoride, NH_4F; **(c)** ClF; **(d)** ethylene oxide, $H_2C{-}CH_2$ (with O bridging the two carbons); **(e)** $HClO_4$

Answer ClF and ethylene oxide

10.2 Gas Pressure

At the macroscopic level, a complete physical description of a sample of a gas requires four quantities: *temperature* (expressed in K), *volume* (expressed in liters), *amount* (expressed in moles), and *pressure* (in atmospheres). As we explain in the next two sections, these variables are *not* independent. If we know the values of any *three* of these quantities, we can calculate the fourth and thereby obtain a full physical description of the gas. Temperature, volume, and amount have been discussed in previous chapters. We now discuss pressure and its units of measurement.

Units of Pressure

Any object, whether it is this text, a person, or a sample of gas, exerts a force on any surface with which it comes in contact. The air in a balloon, for example, exerts a force against the interior surface of the balloon, and a liquid injected into a mold exerts a force against the interior surface of the mold, just as a chair exerts a force against the floor because of its mass and the effects of gravity. If the air in a balloon is heated, the increased kinetic energy of the gas eventually causes the balloon to burst because of the increased **pressure** of the gas, the force *F* per unit area *A* of surface:

$$P = \frac{F}{A} \tag{10.1}$$

Pressure is dependent on *both* the force exerted *and* the size of the area to which the force is applied. We know from Equation 10.1 that applying the same force to a smaller area produces higher pressure. When we use a hose to wash a car, for example, we can increase the pressure of the water by reducing the size of the opening of the hose with a thumb.

The units of pressure are derived from the units used to measure force and area. In the English system, the units of force are pounds and the units of area are square inches, so we often see pressure expressed in pounds per square inch, $lb/in.^2$ (or psi). For scientific measurements, however, the SI units for force are preferred. The SI unit for pressure, derived from the SI units for force (newtons) and area (square meters), is the newton per square meter, N/m^2, which is called the **pascal (Pa)**, after the French mathematician Blaise Pascal (1623–1662):

$$1 \text{ pascal (Pa)} = 1 \text{ newton/meter}^2 \text{ (N/m}^2\text{)} \tag{10.2}$$

Blaise Pascal (1623–1662). In addition to his talents in mathematics (he invented modern probability theory), Pascal did research in physics and was an author and a religious philosopher as well. His accomplishments include invention of the first syringe and the first digital calculator, and development of the principle of hydraulic pressure transmission now used in brake systems and hydraulic lifts.

EXAMPLE 10.2

Assuming a paperback textbook has a mass of 2.00 kg and measures 27.0 cm × 21.0 cm × 4.5 cm, what pressure does it exert on a surface if it is (a) lying flat and (b) standing on edge in a bookcase?

Given Mass and dimensions of object

Asked for Pressure

Strategy

Ⓐ Calculate the force exerted by the book, and then compute the area that is in contact with a surface.

Ⓑ Substitute these two values into Equation 10.1 to find the pressure exerted on the surface in each orientation.

Solution

The force exerted by the book does *not* depend on its orientation. Recall from Chapter 5 that the force exerted by an object is $F = ma$, where *m* is its mass and *a* is its acceleration. In Earth's gravitational field, the acceleration is due to gravity (9.8067 m/s^2 at the Earth's surface). In SI units, the force exerted by the book is therefore

$$F = ma = (2.00 \text{ kg})(9.8067 \text{ m/s}^2) = 19.6 \text{ (kg}\cdot\text{m)/s}^2 = 19.6 \text{ N}$$

(a) Ⓐ✓ We calculated the force as 19.6 N. When the book is lying flat, the area is (0.270 m)(0.210 m) = 0.0567 m^2. Ⓑ✓ The pressure exerted by the text lying flat is thus

$$P = \frac{19.6 \text{ N}}{0.0567 \text{ m}^2} = 3.46 \times 10^2 \text{ Pa}$$

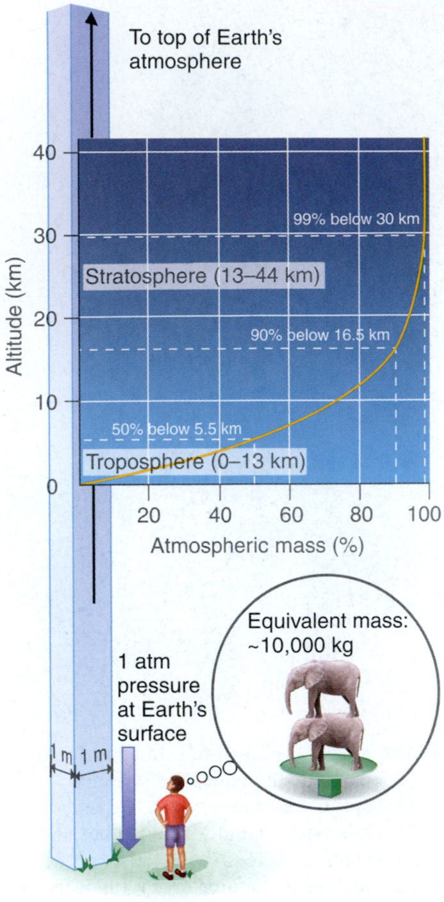

Figure 10.3 Atmospheric pressure. Each square meter of the Earth's surface supports a column of air that is more than 200 km high and weighs about 10,000 kg at the Earth's surface, resulting in a pressure at the surface of 1.01×10^5 N/m^2. This corresponds to a pressure of 101 kPa = 760 mmHg = 1 atm.

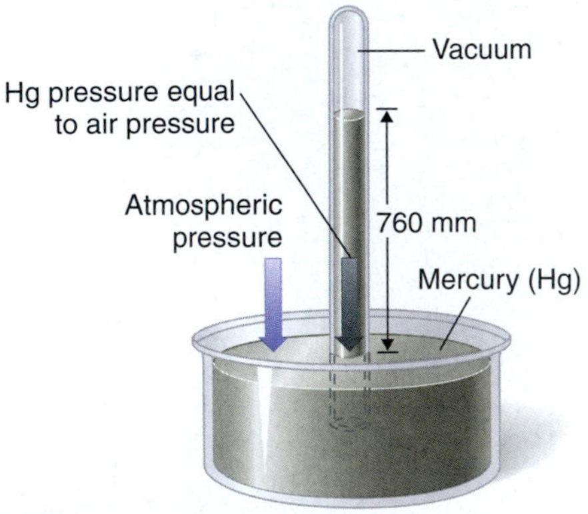

Figure 10.4 A mercury barometer. The pressure exerted by the atmosphere on the surface of the pool of mercury supports a column of mercury in the tube that is about 760 mm tall. Because the boiling point of mercury is quite high (356.73°C), there is very little mercury vapor in the space above the mercury column.

(b) Ⓐ✓ If the book is standing on end, the force remains the same but the area decreases:

$$(21.0\text{ cm})(4.5\text{ cm}) = (0.210\text{ m})(0.045\text{ m}) = 9.5 \times 10^{-3}\text{ m}^2$$

Ⓑ✓ The pressure exerted by the text in this position is thus

$$P = \frac{19.6\text{ N}}{9.5 \times 10^{-3}\text{ m}^2} = 2.1 \times 10^3\text{ Pa}$$

Thus, the *pressure* exerted by the text varies by a factor of about six depending on its orientation, although the *force* exerted by the text does not vary.

EXERCISE 10.2

Calculate the pressure exerted on the floor by a 60.0-kg student **(a)** standing flat-footed in the laboratory in a pair of tennis shoes (surface area of the soles is approximately 180 cm^2) and **(b)** as she steps heel-first onto a dance floor wearing high-heeled shoes (area of the heel = 1.0 cm^2).

Answer **(a)** 3.27×10^4 Pa (4.74 lb/in.^{2}); **(b)** 5.9×10^6 Pa (8.5×10^2 lb/in.^{2})

Atmospheric Pressure

Just as we exert pressure on a surface because of gravity, so does our atmosphere. We live at the bottom of an ocean of gases that becomes progressively less dense with increasing altitude. Approximately 99% of the mass of the atmosphere lies within 30 km of the surface of the Earth, and half of it is within the first 5.5 km (Figure 10.3). Every point on Earth's surface experiences a net pressure called *atmospheric pressure*. The pressure exerted by the atmosphere is considerable: a 1.0-m^2 column, measured from sea level to the top of the atmosphere, has a mass of about 10,000 kg, which gives a pressure of about 100 kPa:

$$\text{pressure} = \frac{(1.0 \times 10^4\text{ kg})(9.807\ \cancel{\text{m}}/\text{s}^2)}{1.0\ \text{m}^{\cancel{2}}} = 0.98 \times 10^5\text{ Pa} = 98\text{ kPa} \qquad (10.3)$$

In English units, this is about 14 lb/in.^{2}, but we are so accustomed to living under this pressure that we never notice it. Instead, what we notice are *changes* in the pressure, such as when our ears pop in fast elevators in skyscrapers or in airplanes during rapid changes in altitude. We make use of atmospheric pressure in many ways. We can use a drinking straw because sucking on it removes air and thereby reduces the pressure inside the straw. The atmospheric pressure pushing down on the liquid in the glass then forces the liquid up the straw.

Atmospheric pressure can be measured using a **barometer**, a device invented in 1643 by one of Galileo's students, Evangelista Torricelli (1608–1647). A barometer may be constructed from a long glass tube that is closed at one end. It is filled with mercury and placed upside down in a dish of mercury without allowing any air to enter the tube. Some of the mercury will run out of the tube, but a relatively tall column remains inside (Figure 10.4). Why doesn't all the mercury run out? Gravity is certainly exerting a downward force on the mercury in the tube, but it is opposed by the pressure of the atmosphere pushing down on the surface of the mercury in the dish, which has the net effect of pushing the mercury up into the tube. Because there is no air above the mercury inside the tube in a properly filled barometer (it contains a *vacuum*), there is no pressure pushing down on the column. Thus, the mercury runs out of the tube until the pressure exerted by the mercury column itself exactly balances the pressure of the atmosphere. Under normal weather conditions at sea level, the two forces are balanced when the top of the mercury column is approximately 760 mm above the level of the mercury in the dish, as shown in Figure 10.4. This value varies with meteorological conditions and altitude. In Denver,

Colorado, for example, at an elevation of about 1 mile, or 1609 m (5280 ft), the height of the mercury column is 630 mm rather than 760 mm.

Mercury barometers have been used to measure atmospheric pressure for so long that they have their own unit for pressure: the **millimeter of mercury (mmHg)**, often called the **torr**, after Torricelli. **Standard atmospheric pressure** is the atmospheric pressure required to support a column of mercury exactly 760 mm tall; this pressure is also referred to as 1 **atmosphere (atm)**. These units are also related to the pascal:

$$1 \text{ atm} = 760 \text{ mmHg} = 760 \text{ torr} = 1.01325 \times 10^5 \text{ Pa} = 101.325 \text{ kPa} \qquad (10.4)$$

Thus, a pressure of 1 atm equals 760 mmHg exactly and is approximately equal to 100 kPa.

EXAMPLE 10.3

One of the authors visited Rocky Mountain National Park several years ago. After departing from an airport at sea level in the eastern United States, he arrived in Denver (altitude 5280 ft), rented a car, and drove to the top of the highway outside Estes Park (elevation 14,000 ft). He noticed that even slight exertion was very difficult at this altitude, where the atmospheric pressure is only 454 mmHg. Convert this pressure to **(a)** atmospheres and **(b)** kilopascals.

Given Pressure in mmHg

Asked for Pressure in atm and kPa

Strategy

Use the conversion factors in Equation 10.4 to convert from mmHg to atm and kPa.

Solution

From Equation 10.4, we have 1 atm = 760 mmHg = 101.325 kPa. The pressure at 14,000 ft in atm is thus

$$P = (454 \cancel{\text{mmHg}})\left(\frac{1 \text{ atm}}{760 \cancel{\text{mmHg}}}\right) = 0.597 \text{ atm}$$

The pressure in kPa is given by

$$P = (0.597 \cancel{\text{atm}})\left(\frac{101.325 \text{ kPa}}{1 \cancel{\text{atm}}}\right) = 60.5 \text{ kPa}$$

EXERCISE 10.3

Mt. Everest, at 29,028 ft above sea level, is the world's tallest mountain. The normal atmospheric pressure at this altitude is about 0.308 atm. Convert this pressure to **(a)** mmHg and **(b)** kPa.

Answer **(a)** 234 mmHg; **(b)** 31.2 kPa

Manometers

Barometers measure atmospheric pressure, but **manometers** measure the pressures of samples of gases contained in an apparatus. The key feature of a manometer is a U-shaped tube containing mercury (or occasionally another nonvolatile liquid). A closed-end manometer is shown schematically in Figure 10.5a. When the bulb contains no gas (that is, when its interior is a near vacuum), the heights of the two columns of mercury are the same because the space above the mercury on the left is a near vacuum (it contains only a very small amount of mercury vapor). If a gas is

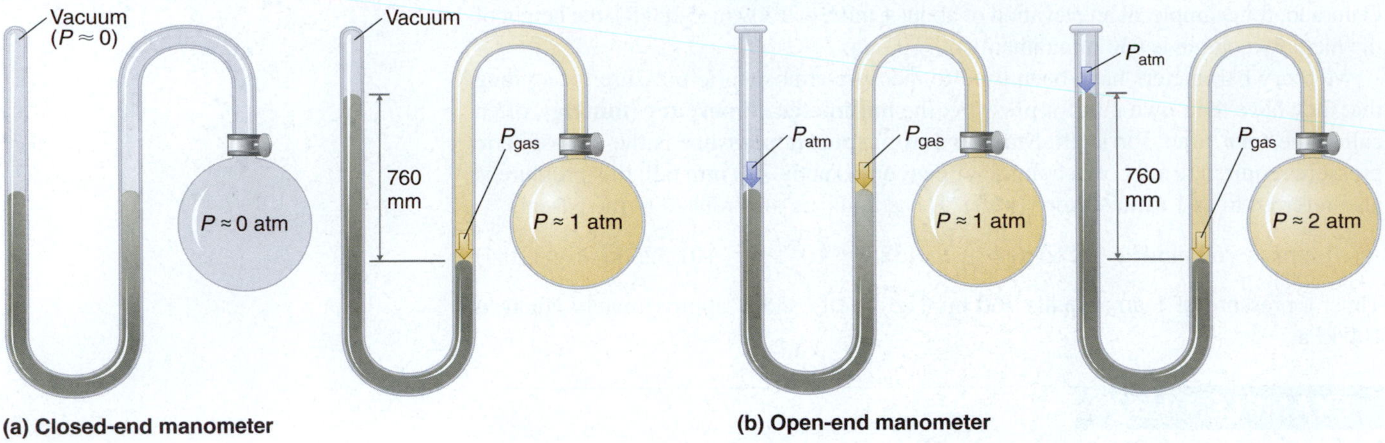

Figure 10.5 The two types of manometer. (a) In a closed-end manometer, the space above the mercury column on the left (the reference arm) is essentially a vacuum ($P \approx 0$), and the difference in the heights of the two columns gives the pressure of the gas contained in the bulb directly. (b) In an open-end manometer, the left (reference) arm is open to the atmosphere ($P \approx 1$ atm), and the difference in the heights of the two columns gives the *difference* between atmospheric pressure and the pressure of the gas in the bulb.

released into the bulb on the right, it will exert a pressure on the mercury in the right column, and the two columns of mercury will no longer be the same height. The *difference* between the heights of the two columns is equal to the pressure of the gas.

If the tube is open to the atmosphere instead of closed, in an open-end manometer (Figure 10.5b), then the two columns of mercury have the same height only if the gas in the bulb has a pressure equal to the atmospheric pressure. If the gas in the bulb has a *higher* pressure, the mercury in the open tube will be forced up by the gas pushing down on the mercury in the other arm of the U-tube. The pressure of the gas in the bulb is therefore the sum of the atmospheric pressure (measured with a barometer) and the difference in the heights of the two columns. If the gas in the bulb has a pressure *less* than that of the atmosphere, then the height of the mercury will be greater in the arm attached to the bulb. In this case, the pressure of the gas in the bulb is the atmospheric pressure minus the difference in the heights of the two columns.

EXAMPLE 10.4

Suppose you want to construct a closed-end manometer to measure gas pressures in the range 0.000–0.200 atm. Because of the toxicity of mercury, you decide to use water rather than mercury. How tall a column of water do you need? (The density of water is 0.9970 g/cm^3; the density of mercury is 13.53 g/cm^3.)

Given Pressure range, densities of water and mercury

Asked for Column height

Strategy

Ⓐ Calculate the height of a column of mercury corresponding to 0.200 atm in mmHg. This is the height needed for a mercury-filled column.

Ⓑ From the given densities, use a proportion to compute the height needed for a water-filled column.

Solution

Ⓐ A gas pressure of 0.200 atm equals

$$P = (0.200\ \cancel{\text{atm}})\left(\frac{760\ \text{mmHg}}{1\ \cancel{\text{atm}}}\right) = 152\ \text{mmHg}$$

Using a mercury manometer, you would need a mercury column at least 152 mm high. Because water is less dense than mercury, you need a *taller* column of water to achieve the same pressure as a given column of mercury. The height needed for a water-filled column corresponding to a pressure of 0.200 atm is proportional to the ratio of the density of mercury (d_{Hg}) to the density of water (d_{H_2O}):

$$(\text{height}_{H_2O})(d_{H_2O}) = (\text{height}_{Hg})(d_{Hg})$$

$$\text{height}_{H_2O} = (\text{height}_{Hg})\left(\frac{d_{Hg}}{d_{H_2O}}\right)$$

$$= (152 \text{ mm})\left(\frac{13.53 \cancel{\text{g/cm}^3}}{0.9970 \cancel{\text{g/cm}^3}}\right)$$

$$= 2.06 \times 10^3 \text{ mm } H_2O = 2.06 \text{ m } H_2O$$

This answer makes sense: it takes a taller column of a less dense liquid to achieve the same pressure.

EXERCISE 10.4

Suppose you want to design a barometer to measure atmospheric pressure in an environment that is always hotter than 30°C. To avoid using mercury, you decide to use gallium, which melts at 29.76°C; the density of liquid gallium is 6.114 g/cm^3. How tall a column of gallium do you need if $P = 1.00$ atm?

Answer 1.68 m

The answer to Example 10.4 also tells us the maximum depth of a farmer's well if a simple suction pump will be used to get the water out. If a column of water 2.06 m high corresponds to 0.200 atm, then 1.00 atm corresponds to a column height of

$$\frac{h}{2.06 \text{ m}} = \frac{1.00 \cancel{\text{atm}}}{0.200 \cancel{\text{atm}}}$$

$$h = 10.3 \text{ m}$$

A suction pump is just a more sophisticated version of a straw: it creates a vacuum above a liquid and relies on atmospheric pressure to force the liquid up a tube. If 1 atm pressure corresponds to a 10.3-m (33.8-ft) column of water, then it is physically impossible for atmospheric pressure to raise the water in a well higher than this. Until electric pumps were invented to push water mechanically from greater depths, this factor greatly limited where people could live because obtaining water from wells deeper than about 33 ft was difficult.

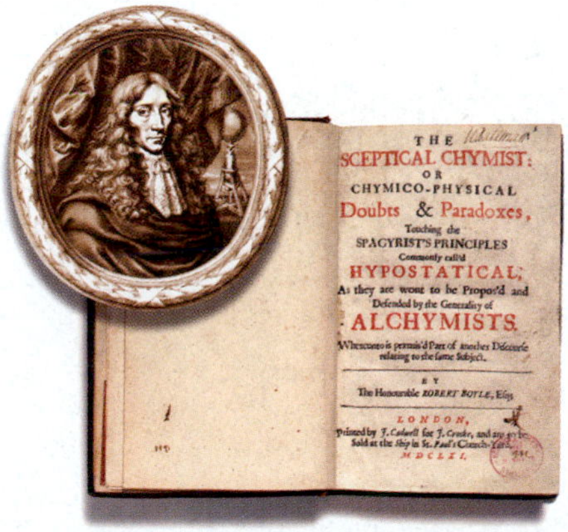

Title page of *The Sceptical Chymist* by Robert Boyle (1627–1691). Boyle, the youngest (and 14th!) child of the Earl of Cork, was an important early figure in chemistry whose views were often at odds with accepted wisdom. Boyle's studies of gases are reported to have utilized a very tall J-tube that he set up in the entryway of his house, which was several stories tall. He is known for the gas law that bears his name and for his book, *The Sceptical Chymist,* which was published in 1661 and influenced chemists for many years after his death. In addition, one of Boyle's early essays on morals is said to have inspired Jonathan Swift to write *Gulliver's Travels*.

10.3 Relationships Between Pressure, Temperature, Amount, and Volume

Early scientists explored the relationships between the pressure of a gas (P) and its temperature (T), volume (V), and amount (n) by holding two of the four variables constant (amount and temperature, for example), varying a third (such as pressure), and measuring the effect of the change on the fourth (in this case, volume). The history of their discoveries provides several excellent examples of the scientific method as presented in Chapter 1.

The Relationship Between Pressure and Volume

As the pressure on a gas increases, the volume of the gas decreases because the gas particles are forced closer together. Conversely, as the pressure on a gas decreases, the gas volume

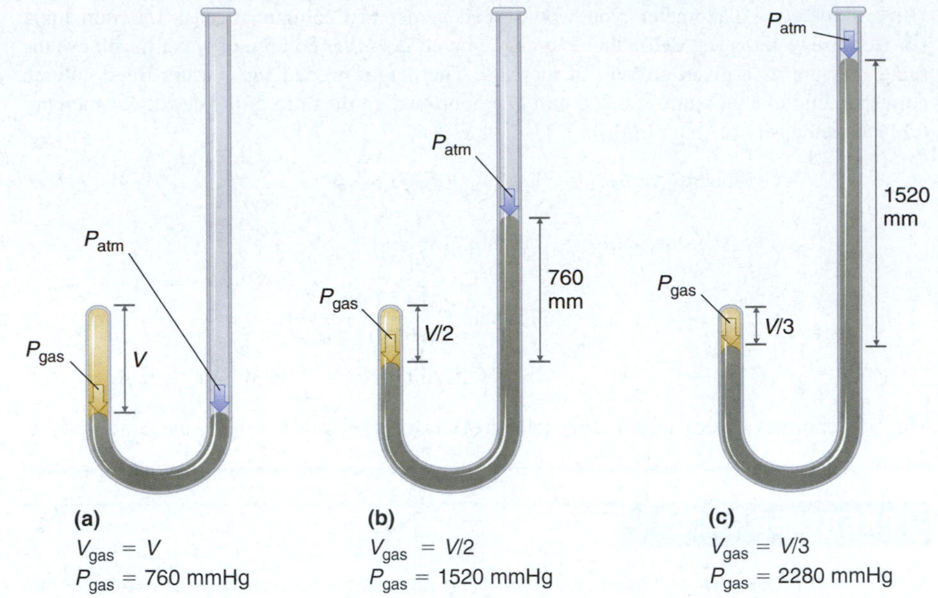

Figure 10.6 Boyle's experiment using a J-shaped tube to determine the relationship between gas pressure and volume. (a) Initially the gas is at a pressure of 1 atm = 760 mmHg (notice that the mercury is at the same height in both the arm containing the sample and the arm open to the atmosphere); its volume is V. (b) If enough mercury is added to the right side to give a difference in height of 760 mmHg between the two arms, the pressure of the gas is 760 mmHg (atmospheric pressure) plus 760 mmHg = 1520 mmHg and the volume is $V/2$. (c) If an additional 760 mmHg is added to the column on the right, the total pressure on the gas increases to 2280 mmHg and the volume of the gas decreases to $V/3$.

MGC Charles' Law

increases because the gas particles can now move farther apart. Weather balloons get larger as they rise through the atmosphere to regions of lower pressure because the volume of the gas has increased. That is, the atmospheric gas exerts less pressure on the surface of the balloon, so the interior gas expands until the internal and external pressures are equal.

The Irish chemist Robert Boyle (1627–1691) carried out some of the earliest experiments that determined the quantitative relationship between the pressure and volume of a gas. Boyle used a J-shaped tube partially filled with mercury, as shown in Figure 10.6. In these experiments, a small amount of a gas or air is trapped above the mercury column, and its volume is measured at atmospheric pressure and constant temperature. More mercury is then poured into the open arm to increase the pressure on the gas sample. The pressure on the gas is atmospheric pressure *plus* the difference in the heights of the mercury columns, and the resulting volume is measured. This process is repeated until either there is no more room in the open arm or the volume of the gas is too small to be measured accurately. Data such as those from one of Boyle's own experiments may be plotted in several ways (Figure 10.7). A simple plot of V

Figure 10.7 Plots of Boyle's data. (a) Actual data from a typical experiment conducted by Boyle. Note that Boyle used non-SI units to measure the volume (in^3 rather than cm^3) and the pressure (inHg rather than mmHg). (b) The plot of pressure vs. volume is a hyperbola. Because PV is a constant, decreasing the pressure by a factor of two results in a twofold increase in volume, and vice versa. (c) A plot of volume vs. 1/pressure for the same data shows the inverse linear relationship between the two quantities, as expressed by the equation V = constant/P. Interactive Graph

P (inHg)	V (in^3)	PV
12.0	117.5	1410
16.0	87.2	1400
20.0	70.7	1410
24.0	58.8	1410
32.0	44.2	1410
40.0	35.3	1410
48.0	29.1	1400

(a) Data from Boyle's experiment

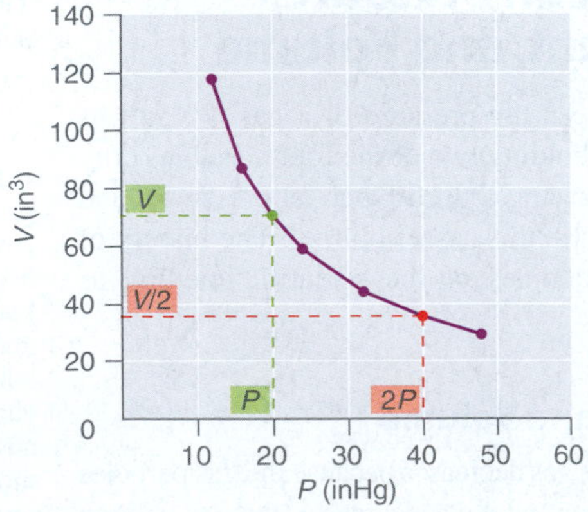

(b) Volume vs. pressure

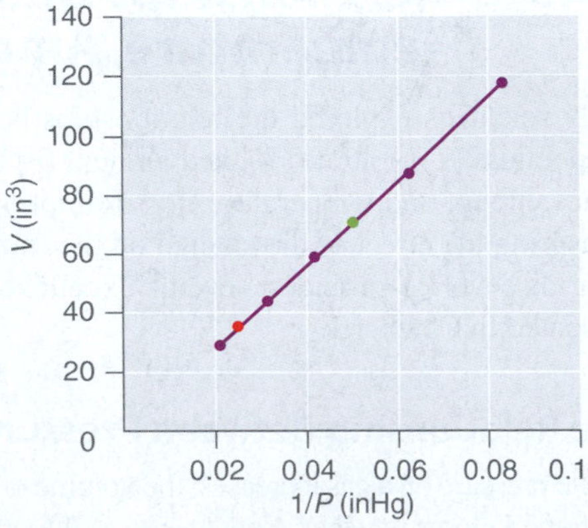

(c) Volume vs. 1/pressure

versus P gives a curve called a *hyperbola* and reveals an *inverse* relationship between pressure and volume: as the pressure is doubled, the volume decreases by a factor of two. This relationship between the two quantities is described by the equation

$$PV = \text{constant} \tag{10.5}$$

Dividing both sides by P gives an equation illustrating the inverse relationship between P and V:

$$V = \frac{\text{constant}}{P} = \text{constant}\left(\frac{1}{P}\right) \quad \text{or} \quad V \propto \frac{1}{P} \tag{10.6}$$

where the $\propto$ symbol is read "is proportional to." A plot of V versus $1/P$ is thus a straight line whose slope is equal to the constant in Equations 10.5 and 10.6. Dividing both sides of Equation 10.5 by V instead of P gives a similar relationship between P and $1/V$. The numerical value of the constant depends on the amount of gas used in the experiment and on the temperature at which the experiments are carried out. This relationship between pressure and volume is known as *Boyle's law*, after its discoverer, and can be stated as follows: *At constant temperature, the volume of a fixed amount of a gas is inversely proportional to its pressure.*

The Relationship Between Temperature and Volume

Hot air rises, which is why hot-air balloons ascend through the atmosphere and why warm air collects near the ceiling and cooler air collects at ground level. Because of this behavior, heating registers are placed on or near the floor, and vents for air-conditioning are placed on or near the ceiling. The fundamental reason for this behavior is that gases expand when they are heated. Because the same amount of substance now occupies a greater volume, hot air is less dense than cold air. The substance with the lower density—in this case hot air—rises through the substance with the higher density, the cooler air.

The first experiments to quantify the relationship between the temperature and volume of a gas were carried out in 1783 by an avid balloonist, French chemist Jacques Alexandre César Charles (1746–1823). Charles's initial experiments showed that a plot of the volume of a given sample of gas versus temperature (in °C) at constant pressure is a straight line. Similar but more precise studies were carried out by another balloon enthusiast, Frenchman Joseph-Louis Gay-Lussac (1778–1850), who showed that a plot of V versus T was a straight line that could be extrapolated to a point at zero volume, a theoretical condition, now known to correspond to −273.15°C (Figure 10.8).* Note from Figure 10.8a that the *slope* of the plot of V versus T varies for the same gas at different pressures, but that the *intercept* remains constant at −273.15°C. Similarly, as shown in Figure 10.8b, plots of V versus T for different amounts of varied gases are straight lines with different slopes but the *same* intercept on the T axis.

The significance of the invariant T intercept in plots of V versus T was recognized in 1848 by the British physicist William Thomson (1824–1907), later named Lord Kelvin. He postulated that −273.15°C was the *lowest possible temperature* that could theoretically be achieved, for which he coined the term **absolute zero (0 K)**.

We can state Charles's and Gay-Lussac's findings in simple terms: *At constant pressure, the volume of a fixed amount of gas is directly proportional to its absolute temperature (in K).* This relationship, illustrated in Figure 10.8b, is often referred to as *Charles's law* and is stated mathematically as

$$V = (\text{constant})[T \text{ (in K)}] \quad \text{or} \quad V \propto T \text{ (in K, at constant } P) \tag{10.7}$$

Jacques Alexandre César Charles (1746–1823) and Joseph-Louis Gay-Lussac (1778–1850). In 1783, Charles filled a balloon ("aerostatic globe") with hydrogen (generated by the reaction of iron with more than 200 kg of acid over several days) and flew successfully for almost an hour. When the balloon descended in a nearby village, however, the terrified townspeople destroyed it. In 1804, Gay-Lussac managed to ascend to 23,000 ft (more than 7000 m) to collect samples of the atmosphere to analyze its composition as a function of altitude. In the process, he had trouble breathing and nearly froze to death, but he set an altitude record that endured for decades. (To put Gay-Lussac's achievement in perspective, recall that modern jetliners cruise at only 35,000 ft!)

* A sample of gas cannot *really* have a volume of zero because any sample of matter must have *some* volume. Furthermore, at 1 atm pressure all gases liquefy at temperatures well above −273.15°C.

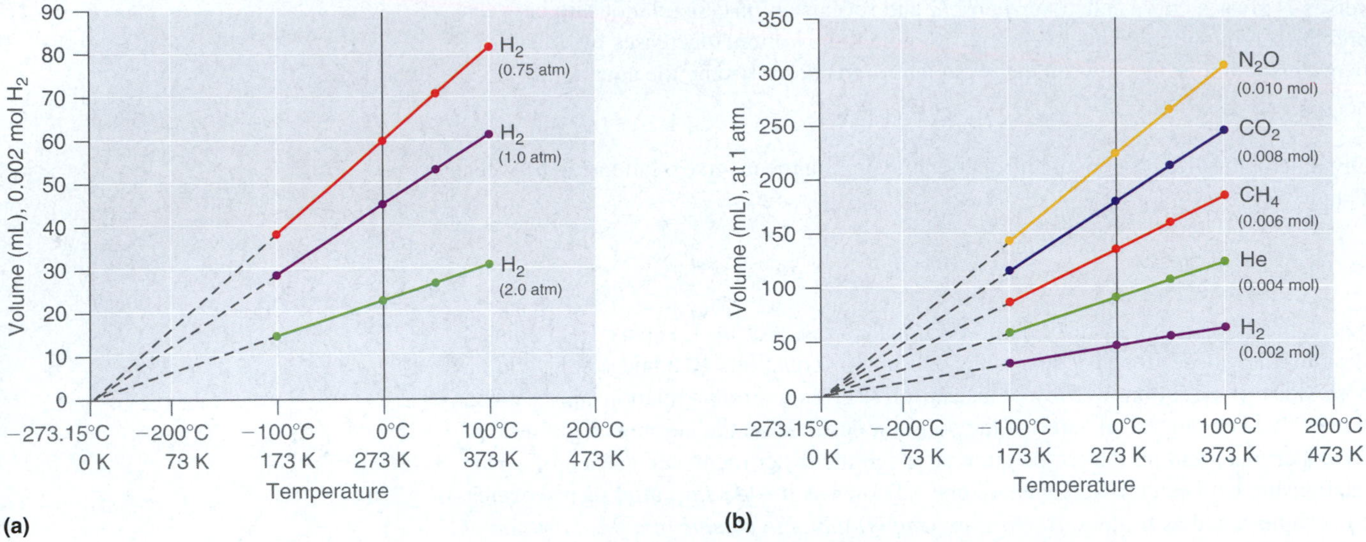

Figure 10.8 The relationship between volume and temperature. (a) Plots of volume vs. temperature for equal-sized samples of H_2 at three different pressures. The solid lines show the experimentally measured data down to −100°C, and the broken lines show the extrapolation of the data to $V = 0$. Note that the temperature scale is given in both °C and K. Although the slopes of the lines decrease with increasing pressure, all of the lines extrapolate to the same temperature at $V = 0$ (−273.15°C = 0 K). (b) Plots of volume vs. temperature for different amounts of selected gases at 1 atm pressure. Regardless of the identity or the amount of the gas, all the plots extrapolate to a value of $V = 0$ at −273.15°C. Interactive Graph

Charles's law is valid for virtually all gases at temperatures well above their boiling points. *Note that the temperature must be expressed in K,* not *in degrees Celsius!*

The Relationship Between Amount and Volume

We can demonstrate the relationship between the volume and amount of a gas by filling a balloon; as we add more gas, the balloon gets larger. The specific quantitative relationship was discovered by the Italian chemist Amedeo Avogadro, who recognized the importance of Gay-Lussac's work on combining volumes of gases. In 1811, Avogadro postulated that, at the same temperature and pressure, equal volumes of gases contain the same number of gaseous particles (Figure 10.9). (This is the historic "Avogadro's hypothesis" introduced in Chapter 1.) A logical corollary, sometimes called *Avogadro's law,* describes the relationship between volume and amount of gas: *At constant temperature and pressure, the volume of a sample of gas is directly proportional to the number of moles of gas in the sample*. Stated mathematically,

Note the pattern

For a sample of gas:

V↑ as P↓ (and vice versa)

V↑ as T↑ (and vice versa)

V↑ as n↑ (and vice versa)

$$V = (\text{constant})(n) \quad \text{or} \quad V \propto n \text{ (at constant } T \text{ and } P) \tag{10.8}$$

This relationship is valid for most gases at relatively low pressures, but deviations from strict linearity are observed at elevated pressures.

The relationships between the volume of a gas and its pressure, temperature, and amount are summarized in Figure 10.10. The volume *increases* with increasing temperature or amount, but *decreases* with increasing pressure.

10.4 • The Ideal Gas Law

MGC The Ideal Gas Law

In Section 10.3, you learned how the volume of a gas changes when its pressure, temperature, or amount is changed, as long as the other two variables are held constant. In this section, we describe how these relationships can be combined to give a general expression that describes the behavior of a gas.

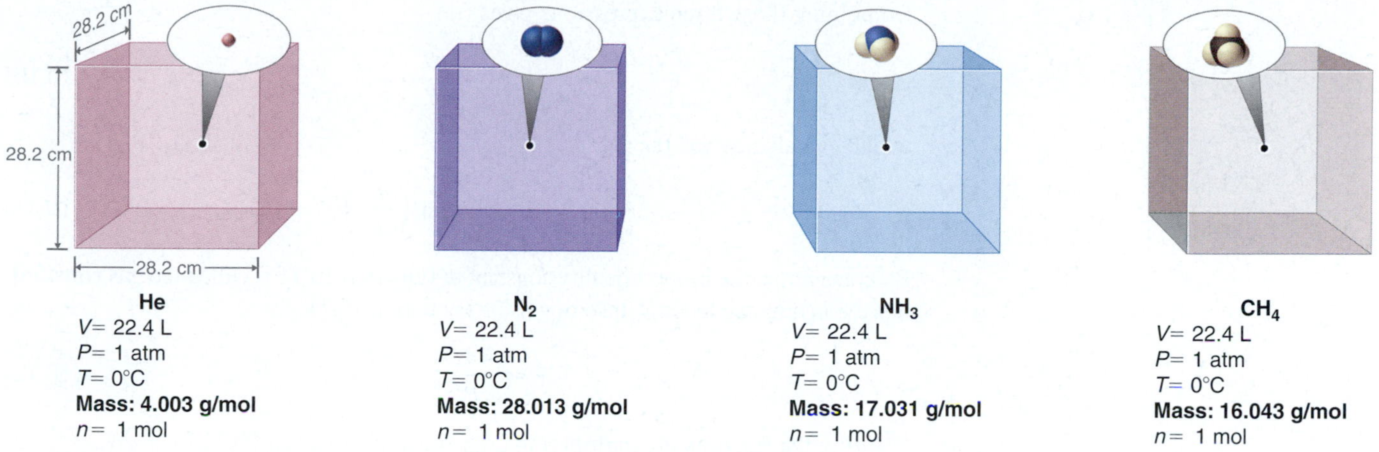

Figure 10.9 Avogadro's hypothesis. Equal volumes of four different gases at the same temperature and pressure contain the same number of gaseous particles. Because the molar mass of each of the gases is different, the *mass* of each gas sample is different even though all contain 1 mole of gas.

Deriving the Ideal Gas Law

Any set of relationships between a single quantity (such as V) and several other variables (P, T, n) can be combined into a single expression that describes all the relationships simultaneously. The three individual expressions derived in Section 10.3 are

$$V \propto \frac{1}{P} \quad \text{(at constant } n, T) \tag{10.9a}$$

$$V \propto T \quad \text{(at constant } n, P) \tag{10.9b}$$

$$V \propto n \quad \text{(at constant } T, P) \tag{10.9c}$$

Figure 10.10 The empirically determined relationships between gas pressure, volume, temperature, and amount. The thermometer and pressure gauge indicate the temperature and pressure qualitatively, the level in the flask indicates the volume, and relative amounts are indicated by the number of particles in each flask.

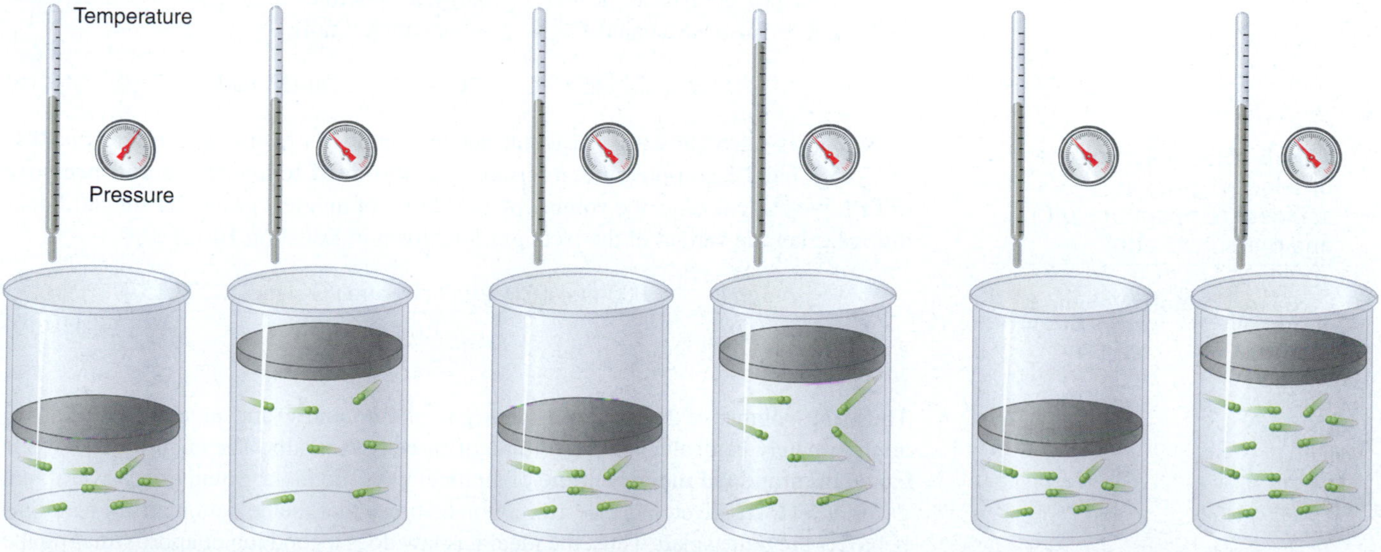

(a) *P* vs. *V* (Boyle's law): At constant temperature and amount of gas, pressure decreases as volume increases (and vice versa).

(b) *V* vs. *T* (Charles's law): At constant pressure and amount of gas, volume increases as temperature increases (and vice versa).

(c) *V* vs. *n* (Avogadro's hypothesis): At constant pressure and temperature, volume increases as amount of gas increases (and vice versa).

Combining these three expressions gives

$$V \propto \frac{nT}{P} \tag{10.10}$$

which can also be written as

$$V = (\text{constant})\left(\frac{nT}{P}\right) \tag{10.11}$$

By convention, the proportionality constant in Equation 10.11 is called the **gas constant**, represented by the letter R. Inserting R into Equation 10.11 gives

$$V = \frac{RnT}{P} = \frac{nRT}{P} \tag{10.12}$$

Clearing the fractions by multiplying both sides of Equation 10.12 by P gives

$$PV = nRT \tag{10.13}$$

This equation is known as the **ideal gas law**.

An *ideal gas* is defined as a hypothetical gaseous substance whose behavior is independent of attractive and repulsive forces and can be completely described by the ideal gas law. In reality, there is no such thing as an ideal gas, but an ideal gas is a useful conceptual model that allows us to understand how gases respond to changing conditions. As we shall see, under many conditions, most real gases exhibit behavior that closely approximates that of an ideal gas. The ideal gas law can therefore be used to predict the behavior of real gases under most conditions. As you will learn in Section 10.8, the ideal gas law does not work well at very low temperatures or very high pressures, where deviations from ideal behavior are most commonly observed.

Before we can use the ideal gas law, however, we need to know the value of the gas constant R. Its form depends on the units used for the other quantities in the expression. If V is expressed in liters (L), P in atmospheres (atm), T in kelvins (K), and n in moles (mol), then

$$R = 0.082057\ (\text{L}\cdot\text{atm})/(\text{K}\cdot\text{mol}) \tag{10.14}$$

Because the product PV has the units of energy, as described in Section 5.1 and Essential Skills 4, R can also have units of J/(K · mol) or cal/(K · mol):

$$R = 8.3145\ \text{J}/(\text{K}\cdot\text{mol}) = 1.9872\ \text{cal}/(\text{K}\cdot\text{mol}) \tag{10.15}$$

Scientists have chosen a particular set of conditions to use as a reference: 0°C (273.15 K) and 1 atm pressure, referred to as **standard temperature and pressure (STP)**. We can calculate the volume of 1.000 mol of an ideal gas under standard conditions using the variant of the ideal gas law given in Equation 10.12:

$$\begin{aligned} V = \frac{nRT}{P} &= \frac{(1.000\ \cancel{\text{mol}})[0.082057\ (\text{L}\cdot\cancel{\text{atm}})/(\cancel{\text{K}}\cdot\cancel{\text{mol}})](273.15\ \cancel{\text{K}})}{1.000\ \cancel{\text{atm}}} \\ &= 22.41\ \text{L} \end{aligned} \tag{10.16}$$

Thus, the volume of 1 mol of an ideal gas at 0°C and 1 atm pressure is 22.41 L, approximately equivalent to the volume of three basketballs. The quantity 22.41 L is called the **standard molar volume** of an ideal gas. The molar volumes of several real gases at STP are given in Table 10.3, which shows that the deviations from ideal gas behavior are quite small. Thus, the ideal gas law does a good job of approximating the behavior of real gases at STP. The relationships described in Section 10.3 as Boyle's, Charles's, and Avogadro's laws are simply special cases of the ideal gas law in which two of the four parameters (P, V, T, n) are held fixed.

TABLE 10.3 Molar volumes of selected gases at standard temperature (0°C) and pressure (1 atm)

Gas	Molar Volume, L
He	22.434
Ar	22.397
H_2	22.433
N_2	22.402
O_2	22.397
CO_2	22.260
NH_3	22.079

If n, R, and T are all constant in Equation 10.12, the equation reduces to

$$V = (\text{constant})\left(\frac{1}{P}\right) \quad \text{or} \quad V \propto \frac{1}{P} \tag{10.17}$$

which is exactly the same as Boyle's law in Equation 10.6.

Similarly, Charles's law states that the volume of a fixed quantity of gas is directly proportional to its temperature at constant pressure. If n and P in Equation 10.12 are fixed, then

$$V = \frac{nRT}{P} = (\text{constant})(T) \quad \text{or} \quad V \propto T \tag{10.18}$$

which is exactly the same as Equation 10.7.

Applying the Ideal Gas Law

The ideal gas law allows us to calculate the value of the fourth variable for a gaseous sample if we know the values of any three of the four variables (P, V, T, n). It also allows us to predict the *final state* of a sample of a gas (that is, its final temperature, pressure, volume, and amount) following any changes in conditions if the parameters (P, V, T, n) are specified for an *initial state.* Some applications are illustrated in the following examples. The approach used throughout is always to start with the same equation—the ideal gas law—and then decide which quantities are given and which need to be calculated. Let's begin with simple cases in which we are given three of the four parameters needed for a complete physical description of a gaseous sample.

EXAMPLE 10.5

The balloon that Charles used for his initial flight in 1783 was destroyed, but we can estimate that its volume was 31,150 L (1100 ft^3), given the dimensions recorded at the time. If the temperature at ground level was 86°F (30°C) and the atmospheric pressure was 745 mmHg, how many moles of hydrogen gas were needed to fill the balloon?

Given Volume, temperature, pressure

Asked for Amount of gas

Strategy

A Solve the ideal gas law for the unknown quantity, in this case n.

B Make sure that all quantities are given in units that are compatible with the units of the gas constant. If they are not, convert them to the appropriate units, insert them into the equation you have derived, and then calculate the number of moles of hydrogen gas needed.

Solution

A We are given values for P, T, and V and asked to calculate n. If we solve the ideal gas law (Equation 10.13) for n, we obtain

$$n = \frac{PV}{RT}$$

B P and T are given in units that are not compatible with the units of the gas constant [$R = 0.082057$ (L · atm)/(K · mol)]. We must therefore convert the temperature to degrees K and the pressure to atm:

$$P = (745\ \cancel{\text{mmHg}})\left(\frac{1\ \text{atm}}{760\ \cancel{\text{mmHg}}}\right) = 0.980\ \text{atm}$$

$$V = 31{,}150\ \text{L} \quad \text{(given)}$$

$$T = 30 + 273 = 303\ \text{K}$$

Substituting these values into the expression we derived for n, we obtain

$$n = \frac{PV}{RT} = \frac{(0.980 \text{ atm})(31{,}150 \text{ L})}{[0.082057 \text{ (L} \cdot \text{atm)/(K} \cdot \text{mol)}](303 \text{ K})} = 1.23 \times 10^3 \text{ mol } H_2$$

EXERCISE 10.5

Suppose that an "empty" aerosol spray-paint can has a volume of 0.406 L and contains 0.025 mol of a propellant gas such as CO_2. What is the pressure of the gas at 25°C?

Answer 1.5 atm

In Example 10.5 and Exercise 10.5, we were given three of the four parameters needed to describe a gas under a particular set of conditions, and we were asked to calculate the fourth. We can also use the ideal gas law to calculate the effect of *changes* in any of the specified conditions on any of the other parameters, as shown in Example 10.6.

EXAMPLE 10.6

Suppose that Charles had changed his plans and carried out his initial flight not in August but on a cold day in January, when the temperature at ground level was −10°C (14°F). How large a balloon would he have needed to contain the same amount of hydrogen gas at the same pressure as in Example 10.5?

Given Temperature, pressure, amount, and volume in August, temperature in January

Asked for Volume in January

Strategy

A Use the results from Example 10.5 for August as initial conditions and then calculate the *change in volume* due to the change in temperature from 86°F to 14°F. Begin by constructing a table showing the initial and final conditions.

B Rearrange the ideal gas law to isolate those quantities that differ between the initial and final states on one side of the equation, in this case V and T.

C Equate the ratios of those terms that change for the two sets of conditions. Making sure to use the appropriate units, insert the quantities and solve for the unknown parameter.

Solution

A To see exactly which parameters have changed and which are constant, make a table of the initial and final conditions:

	August (initial)	January (final)
T	30°C = 303 K	−10°C = 263 K
P	0.980 atm	0.980 atm
n	1.23×10^3 mol H_2	1.23×10^3 mol H_2
V	31,150 L	?

Thus, we are asked to calculate the effect of a change in temperature on the volume of a fixed amount of gas at constant pressure. **B** Recall that we can rearrange the ideal gas law to give

$$V = \left(\frac{nR}{P}\right)(T)$$

And because n and P are the same in both cases, dividing both sides by T gives

$$\frac{V}{T} = \frac{nR}{P} = \text{constant}$$

This is the relationship first noted by Charles. We see from this expression that under conditions where the amount (n) of gas and the pressure (P) do not change, the ratio V/T also does not change. If we have two sets of conditions for the same amount of gas at the same pressure, we can therefore write

$$\frac{V_1}{T_1} = \frac{V_2}{T_2}$$

where the subscripts 1 and 2 refer to the initial and final conditions, respectively. Solving for V_2 and inserting the given quantities in the appropriate units, we obtain

$$V_2 = \frac{V_1 T_2}{T_1} = \frac{(31{,}150\ \text{L})(263\ \cancel{\text{K}})}{303\ \cancel{\text{K}}} = 2.70 \times 10^4\ \text{L}$$

It is important to check your answer to be sure that it makes sense, just in case you have accidentally inverted a quantity or multiplied rather than divided. In this case, the temperature of the gas decreases. Because we know that gas volume decreases with decreasing temperature, the final volume *must* be less than the intial volume, so the answer makes sense. We could have calculated the new volume by plugging all the given numbers into the ideal gas law, but in general it is much easier and faster to focus on only the quantities that change.

EXERCISE 10.6

At a laboratory party, a helium-filled balloon with a volume of 2.00 L at 22°C is dropped into a large container of liquid nitrogen ($T = -196$°C). What is the final volume of the gas in the balloon?

Answer 0.52 L

Example 10.6 illustrates the relationship originally observed by Charles. We could work through similar examples illustrating the inverse relationship between pressure and volume noted by Boyle ($PV =$ constant) and the relationship between volume and amount observed by Avogadro ($V/n =$ constant). We will not do so, however, because it is more important to note that the historically important gas laws are only special cases of the ideal gas law in which two quantities are varied while the other two remain fixed. The method used in Example 10.6 can be applied in *any* such case, as we demonstrate in Example 10.7 (which also shows why heating a closed container of a gas, such as a butane lighter cartridge or an aerosol can, may cause an explosion).

EXAMPLE 10.7

Aerosol cans are prominently labeled with a warning such as "Do not incinerate this container when empty." Assume that you did not notice this warning and tossed the "empty" aerosol can in Exercise 10.5 (0.025 mol in 0.406 L, initially at 25°C and 1.5 atm internal pressure) into a fire at 750°C. What would be the pressure inside the can (if it did not explode)?

Given Initial volume, amount, temperature, and pressure, final temperature

Asked for Final pressure

Strategy

Follow the strategy outlined in Example 10.6.

Solution

Set up a table to determine which parameters change and which are held constant:

	Initial	Final
V	0.406 L	0.406 L
n	0.025 mol	0.025 mol
T	25°C = 298 K	750°C = 1023 K
P	1.5 atm	?

Once again, two parameters are constant while one is varied, and we are asked to calculate the fourth. As before, we begin with the ideal gas law and rearrange it as necessary to get all the constant quantities on one side. In this case, because V and n are constant, we rearrange to obtain

$$P = \left(\frac{nR}{V}\right)(T) = (\text{constant})(T)$$

Dividing both sides by T, we obtain an equation analogous to the one in Example 10.6, $P/T = nR/V =$ constant. Thus, the ratio of P to T does not change if the amount and volume of a gas are held constant. We can thus write the relationship between any two sets of values of P and T for the same sample of gas at the same volume as

$$\frac{P_1}{T_1} = \frac{P_2}{T_2}$$

In this example, $P_1 = 1.5$ atm, $T_1 = 298$ K, and $T_2 = 1023$ K, and we are asked to find P_2. Solving for P_2 and substituting the appropriate values, we obtain

$$P_2 = \frac{P_1 T_2}{T_1} = \frac{(1.5\text{ atm})(1023\text{ }\cancel{\text{K}})}{298\text{ }\cancel{\text{K}}} = 5.1\text{ atm}$$

This pressure is more than enough to rupture a thin sheet metal container and cause an explosion!

EXERCISE 10.7

Suppose that a fire extinguisher, filled with CO_2 to a pressure of 20.0 atm at 21°C at the factory, is accidentally left in the sun in a closed automobile in Tucson, Arizona in July. The interior temperature of the car rises to 160°F (71.1°C). What is the internal pressure in the fire extinguisher?

Answer 23.4 atm

In Examples 10.6 and 10.7, two of the four parameters (P, V, T, n) were fixed while one was allowed to vary, and we were interested in the effect on the value of the fourth. In fact, we often encounter cases where two of the variables P, V, and T are allowed to vary for a given sample of gas (hence, n is constant), and we are interested in the change in the value of the third under the new conditions. If we rearrange the ideal gas law so that P, V, and T, the quantities that change, are on one side and the constant terms (R and n for a given sample of gas) are on the other, we obtain

$$\frac{PV}{T} = nR = \text{constant} \qquad (10.19)$$

Thus, the quantity PV/T is constant if the total amount of gas is constant. We can therefore write the relationship between any two sets of parameters for a sample of gas as

$$\frac{P_1 V_1}{T_1} = \frac{P_2 V_2}{T_2} \qquad (10.20)$$

This equation can be solved for any of the quantities P_2, V_2, or T_2 if the initial conditions are known, as shown in Example 10.8.

EXAMPLE 10.8

We saw in Example 10.5 that Charles used a balloon with a volume of 31,150 L for his initial ascent, and that the balloon contained 1.23×10^3 mol of H_2 gas initially at 30°C and 745 mmHg. Suppose that Gay-Lussac had also used this balloon for his record-breaking ascent to 23,000 ft, and that the pressure and temperature at that altitude were 312 mmHg and −30°C, respectively. To what volume would the balloon have had to expand to hold the same amount of hydrogen gas at the higher altitude?

Given Initial pressure, temperature, amount, and volume, final pressure and temperature

Asked for Final volume

Strategy

Follow the strategy outlined in Example 10.6.

Solution

Begin by setting up a table of the two sets of conditions:

	Initial	Final
P	745 mmHg = 0.980 atm	312 mmHg = 0.411 atm
T	30°C = 303 K	−30°C = 243 K
n	1.23×10^3 mol H_2	1.23×10^3 mol H_2
V	31,150 L	?

Thus, all the quantities except V_2 are known. Solving Equation 10.20 for V_2 and substituting the appropriate values give

$$V_2 = V_1\left(\frac{P_1T_2}{P_2T_1}\right) = (31{,}150\text{ L})\left[\frac{(0.980\ \cancel{\text{atm}})(243\ \cancel{\text{K}})}{(0.411\ \cancel{\text{atm}})(303\ \cancel{\text{K}})}\right] = 5.96 \times 10^4\text{ L}$$

Does this answer make sense? Two opposing factors are at work in this problem: decreasing the pressure tends to *increase* the volume of the gas, while decreasing the temperature tends to *decrease* the volume of the gas. Which do we expect to predominate? The pressure drops by more than a factor of two, while the absolute temperature drops by only about 20%. Because the volume of a gas sample is directly proportional to both T and $1/P$, the variable that changes the most will have the greatest effect on V. In this case, the effect of decreasing pressure predominates, and we expect the volume of the gas to increase, as we found in our calculation.

We could also have solved this problem by solving the ideal gas law for V and then substituting the relevant parameters for an altitude of 23,000 ft:

$$V = \frac{nRT}{P} = \frac{(1.23 \times 10^3\ \cancel{\text{mol}})[0.082057\ (\text{L}\cdot\cancel{\text{atm}})/(\cancel{\text{K}}\cdot\cancel{\text{mol}})](243\ \cancel{\text{K}})}{0.411\ \cancel{\text{atm}}} = 5.97 \times 10^4\text{ L}$$

Except for a difference caused by rounding to the last significant figure, this is the same result we obtained previously. Thus, *there is often more than one "right" way to solve chemical problems!*

EXERCISE 10.8

A steel cylinder of compressed argon with a volume of 0.400 L was filled to a pressure of 145 atm at 10°C. At 1.00 atm pressure and 25°C, how many 15.0-mL incandescent light bulbs could be filled from this cylinder? (*Hint*: Find the number of moles of argon in each container.)

Answer 4.07×10^3

Using the Ideal Gas Law to Calculate Gas Densities and Molar Masses

The ideal gas law can also be used to calculate molar masses of gases from experimentally measured gas densities. To see how this is possible, we first rearrange the ideal gas law to obtain

$$\frac{n}{V} = \frac{P}{RT} \tag{10.21}$$

The left side has the units of moles per unit volume, mol/L. The number of moles of a substance equals its mass (in grams) divided by its molar mass (M, in grams per mole):

$$n \text{ (in moles)} = \frac{m \text{ (in grams)}}{M \text{ (in grams/mole)}} \tag{10.22}$$

Substituting this expression for n into Equation 10.21 gives

$$\frac{m}{MV} = \frac{P}{RT} \tag{10.23}$$

Because m/V is the density d of a substance, we can replace m/V by d and rearrange to give

$$d = \frac{PM}{RT} \tag{10.24}$$

The distance between molecules in gases is large compared to the size of the molecules, so their densities are much lower than the densities of liquids and solids. Consequently, gas density is usually measured in grams per liter (g/L) rather than grams per milliliter (g/mL).

EXAMPLE 10.9

Calculate the density of butane at 25°C and a pressure of 750 mmHg.

Given Compound, temperature, pressure

Asked for Density

Strategy

A Calculate the molar mass of butane, and convert all quantities to appropriate units for the value of the gas constant.

B Substitute these values into Equation 10.24 to obtain the density.

Solution

A The molar mass of butane (C_4H_{10}) is

$$(4)[12.011 \text{ (C)}] + (10)[1.0079 \text{ (H)}] = 58.123 \text{ g/mol}$$

Using 0.082057 (L · atm)/(K · mol) for R means that we need to convert the temperature from °C to K ($T = 25 + 273 = 298$ K) and the pressure from mmHg to atm:

$$(750 \cancel{\text{mmHg}})\left(\frac{1 \text{ atm}}{760 \cancel{\text{mmHg}}}\right) = 0.987 \text{ atm}$$

B Substituting these values into Equation 10.24 gives

$$d = \frac{PM}{RT} = \frac{(0.987 \cancel{\text{atm}})(58.123 \text{ g}/\cancel{\text{mol}})}{[0.082057 (\text{L} \cdot \cancel{\text{atm}})/(\cancel{\text{K}} \cdot \cancel{\text{mol}})](298 \cancel{\text{K}})} = 2.35 \text{ g/L}$$

EXERCISE 10.9

Radon (Rn) is a radioactive gas formed by the decay of naturally occurring uranium in rocks such as granite. It tends to collect in the basements of houses and poses a significant health risk if present in indoor air. Many states now require that houses be tested for radon before they are sold. Calculate the density of radon at 1.00 atm pressure and 20°C and compare it with the density of nitrogen gas, which constitutes 80% of the atmosphere, under the same conditions to see why radon is found in basements rather than in attics.

Answer radon, 9.23 g/L; N_2, 1.17 g/L

A common use of Equation 10.24 is to determine the molar mass of an unknown gas by measuring its density at a known temperature and pressure. This method is particularly useful in identifying a gas that has been produced in a reaction, and it is not difficult to carry out. A flask or glass bulb of known volume is carefully dried, evacuated, sealed, and weighed empty. It is then filled with a sample of a gas at a known temperature and pressure, and reweighed. The difference in mass between the two readings is the mass of the gas. The volume of the flask is usually determined by weighing the flask when empty and when filled with a liquid of known density such as water. The use of density measurements to calculate molar masses is illustrated in Example 10.10.

EXAMPLE 10.10

The reaction of a copper penny with nitric acid results in the formation of a red-brown gaseous compound containing nitrogen and oxygen. A sample of the gas at a pressure of 727 mmHg and a temperature of 18°C weighs 0.289 g in a flask with a volume of 157.0 mL. Calculate the molar mass of the gas, and suggest a reasonable chemical formula for the compound.

Given Pressure, temperature, mass, and volume

Asked for Molar mass and chemical formula

Strategy

A Solve Equation 10.24 for the molar mass of the gas, and then calculate the density of the gas from the information given.

B Convert all known quantities to the appropriate units for the gas constant being used. Substitute the known values into your equation and solve for the molar mass.

C Propose a reasonable empirical formula using the atomic masses of nitrogen and oxygen and the calculated molar mass of the gas.

Solution

A Solving Equation 10.24 for the molar mass gives

$$M = \frac{dRT}{P}$$

The density is the mass of the gas divided by its volume:

$$d = \frac{m}{V} = \frac{0.289 \text{ g}}{0.157 \text{ L}} = 1.84 \text{ g/L}$$

B We must convert the other quantities to the appropriate units before inserting them into the equation:

$$T = 18 + 273 = 291 \text{ K}$$

$$P = (727 \cancel{\text{mmHg}})\left(\frac{1 \text{ atm}}{760 \cancel{\text{mmHg}}}\right) = 0.957 \text{ atm}$$

The molar mass of the unknown gas is thus

$$M = \frac{dRT}{P} = \frac{(1.84\ \text{g/}\cancel{\text{L}})[0.082057\ (\cancel{\text{L}}\cdot\cancel{\text{atm}})/(\cancel{\text{K}}\cdot\text{mol})](291\ \cancel{\text{K}})}{0.957\ \cancel{\text{atm}}} = 45.9\ \text{g/mol}$$

The atomic masses of N and O are approximately 14 and 16, respectively, so we can construct a table showing the masses of possible combinations:

$$NO = 14 + 16 = 30\ \text{g/mol}$$
$$N_2O = (2)(14) + 16 = 44\ \text{g/mol}$$
$$NO_2 = 14 + (2)(16) = 46\ \text{g/mol}$$

The most likely choice is NO_2 which is in agreement with the data. The red-brown color of smog also results from the presence of NO_2 gas.

EXERCISE 10.10

You are in charge of interpreting the data from an unmanned space probe that has just landed on Venus and sent back a report on its atmosphere. The data are: pressure, 90 atm; temperature, 557°C; density, 58 g/L. The major constituent of the atmosphere (>95%) contains carbon. Calculate the molar mass of the major gas present and identify it.

Answer 44 g/mol; CO_2

10.5 • Mixtures of Gases

MGC Partial Pressures

In our use of the ideal gas law thus far, we have focused entirely on the properties of pure gases that contain only a single chemical species. But what happens when two or more gases are mixed? In this section, we show you how to determine the contribution of each gas present to the total pressure of the mixture.

Partial Pressures

The ideal gas law *assumes* that all gases behave identically and that their behavior is independent of attractive and repulsive forces. If the volume and temperature are held constant, the ideal gas equation can be rearranged to show that the pressure of a sample of gas is directly proportional to the number of moles of gas present:

$$P = n\left(\frac{RT}{V}\right) = n(\text{constant}) \quad (10.25)$$

Nothing in the equation depends on the *nature* of the gas, only on the amount.

With this assumption, let's suppose that we have a mixture of two ideal gases that are present in equal amounts. What is the total pressure of the mixture? Because the pressure depends on only the total number of particles of gas present, the total pressure of the mixture will simply be twice the pressure of either component. More generally, the total pressure exerted by a mixture of gases at a given temperature and volume is the sum of the pressures exerted by each of the gases alone. Furthermore, if we know the volume, temperature, and number of moles of each gas in a mixture, then we can calculate the pressure exerted by each gas individually, which is its **partial pressure**, the pressure the gas would exert if it were the only one present (at the same temperature and volume).

To summarize, *the total pressure exerted by a mixture of gases is the sum of the partial pressures of component gases*. This law was first discovered by John Dalton,

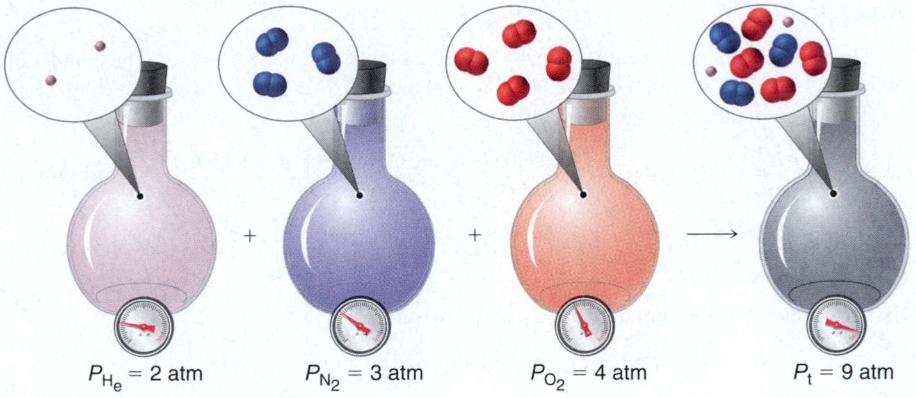

Figure 10.11 Dalton's law. The total pressure of a mixture of gases is the sum of the partial pressures of the individual gases.

the father of the atomic theory of matter. It is now known as *Dalton's law of partial pressures*. We can write it mathematically as

$$P_t = P_1 + P_2 + P_3 \cdots + P_i \tag{10.26}$$

where P_t is the total pressure and the other terms are the partial pressures of the individual gases (Figure 10.11).

For a mixture of two ideal gases, A and B, we can write an expression for the total pressure:

$$P_t = P_A + P_B = n_A\left(\frac{RT}{V}\right) + n_B\left(\frac{RT}{V}\right) = (n_A + n_B)\left(\frac{RT}{V}\right) \tag{10.27}$$

More generally, for a mixture of i components, the total pressure is given by

$$P_t = (n_1 + n_2 + n_3 + \cdots + n_i)\left(\frac{RT}{V}\right) \tag{10.28}$$

Equation 10.28 restates Equation 10.27 in a more general form and makes it explicitly clear that, at constant temperature and volume, the pressure exerted by a gas depends on *only* the total number of moles of gas present, whether the gas is a single chemical species or a mixture of dozens or even hundreds of gaseous species. For Equation 10.28 to be valid, the identity of the molecules present cannot have an effect. Thus, an ideal gas must be one whose properties are not affected by either the size of the molecules or the intermolecular interactions because both will vary from one gas to another. The calculation of total and partial pressures for mixtures of gases is illustrated in Example 10.11.

EXAMPLE 10.11

For reasons that we will examine in Chapter 13, deep-sea divers must use special gas mixtures in their tanks, rather than compressed air, to avoid serious problems, most notably a condition called "the bends." At depths of about 350 ft, divers are subject to a pressure of approximately 10 atm. A typical gas cylinder used for such depths contains 51.2 g of O_2 and 326.4 g of He and has a volume of 10.0 L. What is the partial pressure of each gas at 20.00°C, and what is the total pressure in the cylinder at this temperature?

Given Masses of components, total volume, and temperature

Asked for Partial pressures, total pressure

Strategy

Ⓐ Calculate the number of moles of He and O_2 present.

Ⓑ Use the ideal gas law to calculate the partial pressure of each gas. Then sum the partial pressures to obtain the total pressure of the gaseous mixture.

Solution

A The number of moles of He is (326.4 g)(1.000 mol/4.0026 g) = 81.55 mol. The number of moles of O_2 is (51.2 g)(1.000 mol/32.00 g) = 1.60 mol. B We can now use the ideal gas law to calculate the partial pressure of each:

$$P_{O_2} = n_{O_2}\left(\frac{RT}{V}\right) = (1.60\ \cancel{\text{mol}})\left\{\frac{[0.082057\ (\cancel{\text{L}}\cdot\text{atm})/(\cancel{\text{K}}\cdot\cancel{\text{mol}})](293.15\ \cancel{\text{K}})}{10.0\ \cancel{\text{L}}}\right\} = 3.85\ \text{atm}$$

$$P_{He} = n_{He}\left(\frac{RT}{V}\right) = (81.55\ \cancel{\text{mol}}])\left\{\frac{[0.082057\ (\cancel{\text{L}}\cdot\text{atm})/(\cancel{\text{K}}\cdot\cancel{\text{mol}})](293.15\ \cancel{\text{K}})}{10.0\ \cancel{\text{L}}}\right\} = 196\ \text{atm}$$

The total pressure is the sum of the two partial pressures:

$$P_t = P_{O_2} + P_{He} = 3.85\ \text{atm} + 196\ \text{atm} = 200\ \text{atm}$$

EXERCISE 10.11

A cylinder of compressed natural gas has a volume of 20.0 L and contains 1813 g of methane and 336 g of ethane. Calculate the partial pressure of each gas at 22.0°C and the total pressure in the cylinder.

Answer $P_{CH_4} = 137$ atm; $P_{C_2H_6} = 13.4$ atm; $P_t = 151$ atm

---

Mole Fractions of Gas Mixtures

The composition of a gas mixture can be described by the mole fractions of the gases present. The **mole fraction** (X) of any component of a mixture is the ratio of the number of moles of that component to the total number of moles of all the species present in the mixture (n_t):

$$\text{mole fraction of A} = X_A = \frac{\text{moles A}}{\text{total moles}} = \frac{n_A}{n_t} \tag{10.29}$$

The mole fraction is a dimensionless quantity between 0 and 1. If $X_A = 1.0$, then the sample is pure A, not a mixture. If $X_A = 0$, then no A is present in the mixture. The sum of the mole fractions of all the components present must equal 1.

To see how mole fractions can help us understand the properties of gas mixtures, let's evaluate the ratio of the pressure of a gas A to the total pressure of the gas mixture that contains A. We can use the ideal gas law to describe the pressures of both gas A and the mixture: $P_A = n_ART/V$ and $P_t = n_tRT/V$. The ratio of the two is thus

$$\frac{P_A}{P_t} = \frac{n_ART/V}{n_tRT/V} = \frac{n_A}{n_t} = X_A \tag{10.30}$$

Rearranging this equation gives

$$P_A = X_AP_t \tag{10.31}$$

That is, the partial pressure of any gas in a mixture is the total pressure multiplied by the mole fraction of that gas. This conclusion is a direct result of the ideal gas law, which assumes that all gas molecules behave identically. Consequently, the pressure of a gas in a mixture depends on only the percentage of molecules in the mixture that are of that type, and not on their specific physical or chemical properties. Recall from Chapter 3 that by volume, the Earth's atmosphere is about 78% N_2, 21% O_2, and 0.9% Ar, with traces of gases such as CO_2, H_2O, and others. This means that 78% of the molecules present in the atmosphere are N_2; hence, the mole fraction of N_2 is 78%/100% = 0.78. Similarly, the mole fractions of O_2 and Ar are 0.21 and 0.009, respectively. Using

Equation 10.31, we therefore know that the partial pressure of N_2 is 0.78 atm (assuming an atmospheric pressure of exactly 760 mmHg) and, similarly, the partial pressures of O_2 and Ar are 0.21 and 0.009 atm, respectively.

EXAMPLE 10.12

We have just calculated the partial pressures of the major gases in the air we inhale. Experiments that measure the composition of the air we *exhale* yield different results, however. The table gives the measured pressures of the major gases in both inhaled and exhaled air. Calculate the mole fractions of the gases in exhaled air.

	Inhaled Air, mmHg	Exhaled Air, mmHg
P_{N_2}	597	568
P_{O_2}	158	116
P_{CO_2}	0.3	28
P_{H_2O}	5	48
P_{Ar}	8	8
P_t	767	767

Given Pressures of gases in inhaled and exhaled air

Asked for Mole fractions of gases in exhaled air

Strategy

Calculate the mole fraction of each gas using Equation 10.31.

Solution

The mole fraction of any gas A is given by $X_A = P_A/P_t$, where P_A is the partial pressure of A and P_t is the total pressure. In this case, P_t = (767 mmHg)(1 atm/760 mmHg) = 1.01 atm. The table gives the values of P_A and X_A for exhaled air.

	P_A	X_A
P_{N_2}	$(568\ \cancel{\text{mmHg}})\left(\frac{1\ \text{atm}}{760\ \cancel{\text{mmHg}}}\right) = 0.747\ \text{atm}$	$\frac{0.747\ \cancel{\text{atm}}}{1.01\ \cancel{\text{atm}}} = 0.740$
P_{O_2}	$(116\ \cancel{\text{mmHg}})\left(\frac{1\ \text{atm}}{760\ \cancel{\text{mmHg}}}\right) = 0.153\ \text{atm}$	$\frac{0.153\ \cancel{\text{atm}}}{1.01\ \cancel{\text{atm}}} = 0.151$
P_{CO_2}	$(28\ \cancel{\text{mmHg}})\left(\frac{1\ \text{atm}}{760\ \cancel{\text{mmHg}}}\right) = 0.037\ \text{atm}$	$\frac{0.037\ \cancel{\text{atm}}}{1.01\ \cancel{\text{atm}}} = 0.037$
P_{H_2O}	$(48\ \cancel{\text{mmHg}})\left(\frac{1\ \text{atm}}{760\ \cancel{\text{mmHg}}}\right) = 0.063\ \text{atm}$	$\frac{0.063\ \cancel{\text{atm}}}{1.01\ \cancel{\text{atm}}} = 0.062$
P_{Ar}	$(8\ \cancel{\text{mmHg}})\left(\frac{1\ \text{atm}}{760\ \cancel{\text{mmHg}}}\right) = 0.011\ \text{atm}$	$\frac{0.011\ \cancel{\text{atm}}}{1.01\ \cancel{\text{atm}}} = 0.011$

EXERCISE 10.12

We saw in Exercise 10.10 that Venus is an inhospitable place, with a surface temperature of 560°C and a surface pressure of 90 atm. The atmosphere consists of about 96% CO_2 and 3% N_2, with traces of other gases, including water, sulfur dioxide, and sulfuric acid. Calculate the partial pressures of CO_2 and N_2.

Answer P_{CO_2} = 86 atm; P_{N_2} = 3 atm

10.6 • Gas Volumes and Stoichiometry

MGC Gas Law Stoichiometry

With the ideal gas law, we can use the relationship between the amounts of gases (in moles) and their volumes (in liters) to calculate the stoichiometry of reactions involving gases, if the pressure and temperature are known. This is important for several reasons. Many reactions that are carried out in the laboratory involve the formation or reaction of a gas, so chemists must be able to quantitatively treat gaseous products and reactants as readily as they quantitatively treat solids or solutions. Furthermore, many, if not most, industrially important reactions are carried out in the gas phase for practical reasons. Gases mix readily, are easily heated or cooled, and can be transferred from one place to another in a manufacturing facility via simple pumps and plumbing. As a chemical engineer said to one of the authors, "Gases always go where you want them to, liquids sometimes do, but solids almost never do."

EXAMPLE 10.13

Sulfuric acid, the industrial chemical produced in greatest quantity (almost 45 million tons per year in the United States alone), is prepared by the combustion of sulfur in air to give SO_2, followed by the reaction of SO_2 with O_2 in the presence of a catalyst to give SO_3, which reacts with water to give H_2SO_4. The overall equation is

$$\mathrm{S} + \frac{3}{2}\mathrm{O_2} + \mathrm{H_2O} \longrightarrow \mathrm{H_2SO_4}$$

What volume of O_2 (in liters) at 22°C and 745 mmHg pressure is required to produce 1.00 ton of H_2SO_4?

Given Reaction, temperature, pressure, mass of one product

Asked for Volume of gaseous reactant

Strategy

Ⓐ Calculate the number of moles of H_2SO_4 in 1.00 ton. From the stoichiometric coefficients in the balanced chemical equation, calculate the number of moles of O_2 required.

Ⓑ Use the ideal gas law to determine the volume of O_2 required under the given conditions. Be sure that all quantities are expressed in the appropriate units.

Solution

We can see from the stoichiometry of the reaction that $\frac{3}{2}$ mol of O_2 is required to produce 1 mol of H_2SO_4. Thus, this is a standard stoichiometry problem of the type presented in Chapter 3, except this problem asks for the volume of one of the reactants (O_2) rather than its mass. We proceed exactly as in Chapter 3, using the strategy

$$\text{mass of } \mathrm{H_2SO_4} \longrightarrow \text{moles } \mathrm{H_2SO_4} \longrightarrow \text{moles } \mathrm{O_2} \longrightarrow \text{liters } \mathrm{O_2}$$

Ⓐ We begin by calculating the number of moles of H_2SO_4 in 1.00 ton:

$$\text{moles } \mathrm{H_2SO_4} = (1.00\ \cancel{\text{ton } \mathrm{H_2SO_4}})\left(\frac{2000\ \cancel{\text{lb}}}{1\ \cancel{\text{ton}}}\right)\left(\frac{453.6\ \cancel{\text{g}}}{1\ \cancel{\text{lb}}}\right)\left(\frac{1\ \text{mol } \mathrm{H_2SO_4}}{98.08\ \cancel{\text{g}}}\right)$$

$$= 9250\ \text{mol } \mathrm{H_2SO_4}$$

We next calculate the number of moles of O_2 required:

$$\text{moles } \mathrm{O_2} = (9250\ \cancel{\text{mol } \mathrm{H_2SO_4}})\left(\frac{1.5\ \text{mol } \mathrm{O_2}}{1\ \cancel{\text{mol } \mathrm{H_2SO_4}}}\right) = 1.39 \times 10^4\ \text{mol } \mathrm{O_2}$$

B After converting all quantities to the appropriate units, we can use the ideal gas law to calculate the volume of O_2:

$$T = 273 + 22 = 295\ \text{K}$$

$$P = (745\ \text{mmHg})\left(\frac{1\ \text{atm}}{760\ \text{mmHg}}\right) = 0.980\ \text{atm}$$

$$\text{Volume of } O_2 = n_{O_2}\left(\frac{RT}{P}\right) = (1.39 \times 10^4\ \text{mol } O_2)\left\{\frac{[0.08206\ (\text{L}\cdot\text{atm})/(\text{K}\cdot\text{mol})](295\ \text{K})}{0.980\ \text{atm}}\right\}$$

$$= 3.43 \times 10^5\ \text{L } O_2$$

The answer means that more than 300,000 L of oxygen gas is needed to produce 1 ton of sulfuric acid. These numbers may give you some appreciation for the magnitude of the engineering and plumbing problems faced in industrial chemistry.

EXERCISE 10.13

In Example 10.5, we saw that Charles used a balloon containing approximately 31,150 L of H_2 for his initial flight in 1783. The hydrogen gas was produced by the reaction of metallic iron with dilute hydrochloric acid according to the equation

$$Fe(s) + 2HCl(aq) \longrightarrow H_2(g) + FeCl_2(aq)$$

How much iron (in kg) was needed to produce this volume of H_2 if the temperature was 30°C and the atmospheric presssure was 745 mmHg?

Answer 68.6 kg of Fe (~150 lb)

Many of the advances made in chemistry during the 18th and 19th centuries were the result of careful experiments done to determine the identity and quantity of gases produced in chemical reactions. For example, in 1774, Joseph Priestley was able to isolate oxygen gas by the thermal decomposition of mercuric oxide (HgO). In the 1780s, Antoine Lavoisier conducted experiments that showed that combustion reactions, which require oxygen, produce what we now know to be carbon dioxide. Both sets of experiments required the scientists to collect and manipulate gases produced in chemical reactions, and both used a simple technique that is still used in chemical laboratories today: collecting a gas by the displacement of water. As shown in Figure 10.12, the gas produced in a reaction can be channeled through a tube into inverted bottles filled with water. Because the gas is less dense than liquid water, it bubbles to the top of the bottle, displacing the water. Eventually, all the water is forced out and the bottle contains only gas. If a calibrated bottle is used (that is, one with markings to indicate the volume of the gas) and the bottle is raised or lowered until the level of the water is the same both inside and outside, then the pressure within the bottle will exactly equal the atmospheric pressure measured separately with a barometer.

The only gases that cannot be collected using this technique are those that dissolve readily in water (for example, NH_3, H_2S, and CO_2) and those that react rapidly with water (such as F_2 and NO_2). Remember, however, when calculating the amount of gas formed in the reaction, that the gas collected inside the bottle is *not* pure. Instead, it is a mixture of the product gas and water vapor. As we will discuss in Chapter 11, all liquids (including water) have a measurable amount of vapor in equilibrium with the liquid because molecules of the liquid are continuously escaping from the liquid's surface, while other molecules from the vapor phase collide with the surface and return to the liquid. The vapor thus exerts a pressure above the liquid, which is called the liquid's *vapor pressure*. In the case shown in Figure 10.12, the bottle is therefore actually filled with a mixture of O_2

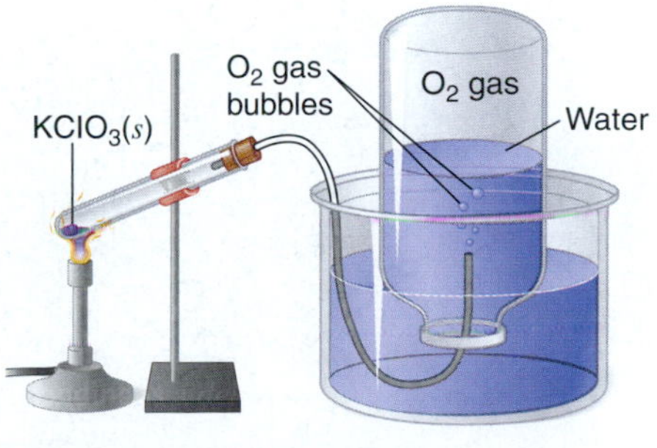

Figure 10.12 An apparatus for collecting gases by the displacement of water. When $KClO_3(s)$ is heated, O_2 is produced according to the equation $KClO_3(s) \longrightarrow KCl(s) + \frac{3}{2}O_2(g)$. The oxygen gas travels through the tube, bubbles up through the water, and is collected in a bottle as shown.

TABLE 10.4 Vapor pressure of water (in mmHg) at various temperatures (°C)

T	*P*	*T*	*P*	*T*	*P*	*T*	*P*
0	4.58	21	18.66	35	42.2	92	567.2
5	6.54	22	19.84	40	55.4	94	611.0
10	9.21	23	21.08	45	71.9	96	657.7
12	10.52	24	22.39	50	92.6	98	707.3
14	11.99	25	23.77	55	118.1	100	760.0
16	13.64	26	25.22	60	149.5	102	815.8
17	14.54	27	26.75	65	187.7	104	875.1
18	15.48	28	28.37	70	233.8	106	937.8
19	16.48	29	30.06	80	355.3	108	1004.2
20	17.54	30	31.84	90	525.9	110	1074.4

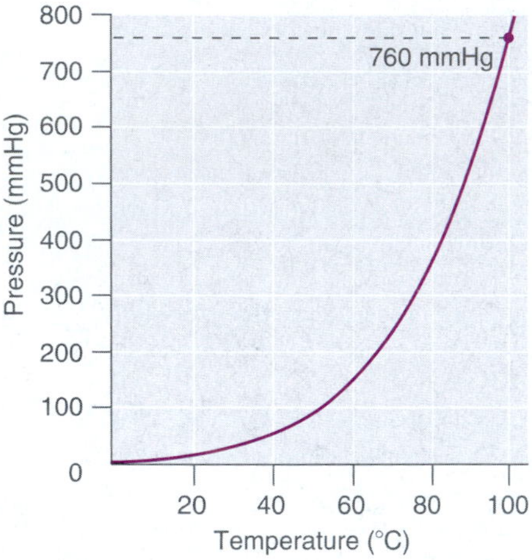

Figure 10.13 A plot of the vapor pressure of water vs. temperature. Note that the vapor pressure is very low (but not zero) at 0°C and reaches 1 atm = 760 mmHg at the normal boiling point, 100°C.

and water vapor, and the total pressure is, by Dalton's law of partial pressures, the sum of the pressures of the two components:

$$P_t = P_{O_2} + P_{H_2O}$$

If we want to know the pressure of the gas generated in the reaction in order to calculate the amount of gas formed, we must first subtract the pressure due to water vapor from the total pressure. This is done by referring to tabulated values of the vapor pressure of water as a function of temperature (Table 10.4). As shown in Figure 10.13, the vapor pressure of water increases rapidly with increasing temperature, and at the normal boiling point (100°C), the vapor pressure is exactly 1 atm. The methodology is illustrated in Example 10.14.

EXAMPLE 10.14

Sodium azide (NaN_3) decomposes to form sodium metal and nitrogen gas according to the equation

$$2NaN_3(s) \xrightarrow{\Delta} 2Na(s) + 3N_2(g)$$

This is the reaction used to inflate the air bags that cushion passengers during automobile collisions. The reaction is initiated in air bags by an electrical impulse and results in the rapid evolution of gas. If the N_2 gas that results from the decomposition of a 5.00-g sample of NaN_3 could be collected by displacing water from an inverted flask, as in Figure 10.12, what volume of gas would be produced at a temperature of 22°C and a pressure of 762 mmHg?

Given Reaction, mass of compound, temperature, and pressure

Asked for Volume of nitrogen gas produced

Strategy

A Calculate the number of moles of N_2 gas produced. From the data in Table 10.4, determine the partial pressure of N_2 gas in the flask.

B Use the ideal gas law to find the volume of N_2 gas produced.

Solution

A Because we know the mass of the reactant and the stoichiometry of the reaction, our first step is to calculate the number of moles of N_2 gas produced:

$$\text{moles } N_2 = (5.00 \cancel{\text{ g NaN}_3})\left(\frac{1 \cancel{\text{ mol NaN}_3}}{65.01 \cancel{\text{ g NaN}_3}}\right)\left(\frac{3 \text{ mol } N_2}{2 \cancel{\text{ mol NaN}_3}}\right) = 0.115 \text{ mol } N_2$$

The pressure given (762 mmHg) is the *total* pressure in the flask, which is the sum of the pressures due to the N_2 gas and to the water vapor present. Table 10.4 tells us that the vapor pressure of water is 19.84 mmHg at 22°C (295 K), so the partial pressure of the N_2 gas in the flask is only 762 − 19.84 = 742 mmHg = 0.976 atm. B Solving the ideal gas law for *V* and substituting the other quantities (in the appropriate units), we get

$$V = \frac{nRT}{P} = \frac{(0.115 \cancel{\text{mol N}_2})[0.082057\ (\text{L}\cdot\cancel{\text{atm}})/(\cancel{\text{K}}\cdot\cancel{\text{mol}})](295\ \cancel{\text{K}})}{0.976\ \cancel{\text{atm}}} = 2.85\ \text{L}$$

EXERCISE 10.14

A 1.00-g sample of zinc metal is added to a solution of dilute hydrochloric acid. It dissolves to produce H_2 gas according to the equation $Zn(s) + 2HCl(aq) \longrightarrow H_2(g) + ZnCl_2(aq)$. The resulting H_2 gas is collected in a water-filled bottle at 30°C and an atmospheric pressure of 760 mmHg. What volume does it occupy?

Answer 0.397 L

10.7 • The Kinetic Molecular Theory of Gases

The laws that describe the behavior of gases were well established long before anyone had developed a coherent model of the properties of gases. In this section, we introduce a theory that describes why gases behave the way they do. The theory we introduce can also be used to derive laws such as the ideal gas law from fundamental principles and the properties of individual molecules.

MGC The Kinetic Molecular Theory of Gases

A Molecular Description

The **kinetic molecular theory of gases** explains the laws that describe the behavior of gases. Developed during the mid-19th century by several physicists, including Austrian Ludwig Boltzmann (1844–1906), German Rudolf Clausius (1822–1888) and an Englishman, James Clerk Maxwell (1831–1879), who is also known for his contributions to electricity and magnetism, this theory is based on the properties of individual molecules as defined for an ideal gas and on fundamental concepts of physics. Thus, the kinetic molecular theory of gases provides a molecular explanation for the observations that led to the development of the ideal gas law. The kinetic molecular theory of gases is based on the following postulates:

1. A gas is composed of a large number of particles called molecules (whether monatomic or polyatomic) that are in constant random motion.
2. Because the distance between gas molecules is much greater than the size of the molecules, the volume of the molecules is negligible.
3. Intermolecular interactions, whether repulsive or attractive, are so weak that they are also negligible.
4. Gas molecules collide with one another and with the walls of the container, but these collisions are perfectly elastic; that is, they do not change the average kinetic energy of the molecules.
5. The average kinetic energy of the molecules of any gas depends on only the temperature, and at a given temperature, all gaseous molecules have exactly the same average kinetic energy.

Although the molecules of real gases have nonzero volumes and exert both attractive and repulsive forces on one another, for the moment we will focus on how the kinetic molecular theory of gases relates to the properties of gases that we have been discussing. In Section 10.8, we explain how this theory must be modified to account for the behavior of real gases.

Postulates 1 and 4 state that molecules are in constant motion and collide frequently with the walls of their container. The collision of molecules with the walls of the container results in a force exerted by the gas on the walls, just as a bowling ball exerts a force on the pins it strikes. Anything that increases the frequency with which the molecules strike the walls or increases the momentum of the gas molecules (in other words, how hard they hit the walls) increases the pressure; anything that decreases that frequency or the momentum of the molecules decreases the pressure.

Because volumes and intermolecular interactions are negligible, postulates 2 and 3 state that all gaseous particles behave identically, regardless of the chemical nature of their component molecules. This is the essence of the ideal gas law, which treats all gases as collections of particles that are identical in all respects except mass. Postulate 2 also explains why it is relatively easy to compress a gas; you simply decrease the distance between the gas molecules.

Postulate 5 provides a molecular explanation for the temperature of a gas. Recall from Chapter 5 that the kinetic energy of an object is given by $KE = \frac{1}{2}mv^2$, where m is the mass of the object and v is its velocity, or speed. Postulate 5 refers to the *average* kinetic energy of the molecules of a gas, which we will call $\overline{KE}$, and states that at a given temperature, all gases have the same value of $\overline{KE}$. The average kinetic energy of the molecules of a gas is

$$\overline{KE} = \frac{1}{2}m\overline{v^2} \tag{10.32}$$

where $\overline{v^2}$ is the average of the squares of the speeds of the particles. For n particles,

$$\overline{v^2} = \frac{v_1^2 + v_2^2 + v_3^2 + \cdots + v_n^2}{n} \tag{10.33}$$

The square root of $\overline{v^2}$ is the **root mean square (rms) speed (v_{rms})**:

$$v_{rms} = \sqrt{\frac{v_1^2 + v_2^2 + v_3^2 + \cdots + v_n^2}{n}} \tag{10.34}$$

Compare this with the formula used to calculate the average speed:

$$\overline{v} = \frac{v_1 + v_2 + v_3 + \cdots + v_n}{n} \tag{10.35}$$

Note the pattern

At a given temperature, all gaseous particles have the same v_{rms} but not the same v.

The root mean square speed and the average speed do not differ greatly (typically by less than 10%). The distinction is important, however, because the root mean square speed is the speed of a gas particle that has average kinetic energy. Particles of different gases at the same temperature have the same average kinetic energy, *not* the same average speed. In contrast, the most probable speed, v_p, is the speed at which the greatest number of particles are moving. If the average kinetic energy of the particles of a gas increases linearly with increasing temperature, then Equation 10.34 tells us that the root mean square speed must increase with temperature as well because the mass of the particles is constant. At higher temperatures, therefore, the molecules of a gas move more rapidly than at lower temperatures, and v_p increases.

EXAMPLE 10.15

The speeds of eight particles were found to be 1.0, 4.0, 4.0, 6.0, 6.0, 6.0, 8.0, and 10.0 m/s. Calculate their average speed ($\overline{v}$), root mean square speed (v_{rms}), and most probable speed (v_p).

Given Particle speeds

Asked for Average speed ($\bar{v}$), root mean square speed (v_{rms}), most probable speed (v_p)

Strategy

Use Equation 10.35 to calculate the average speed and Equation 10.34 to calculate the root mean square speed. Find the most probable speed by determining the speed at which the greatest number of particles are moving.

Solution

The average speed is the sum of the speeds divided by the number of particles:

$$\bar{v} = \frac{v_1 + v_2 + v_3 + \cdots + v_n}{n}$$

$$= \frac{1.0 + 4.0 + 4.0 + 6.0 + 6.0 + 6.0 + 8.0 + 10.0}{8} = 5.6 \text{ m/s}$$

The root mean square speed is the square root of the sum of the squared speeds divided by the number of particles:

$$v_{rms} = \sqrt{\frac{v_1^2 + v_2^2 + v_3^2 + \cdots + v_n^2}{n}}$$

$$= \sqrt{\frac{1.0^2 + 4.0^2 + 4.0^2 + 6.0^2 + 6.0^2 + 6.0^2 + 8.0^2 + 10.0^2}{8}} = 6.2 \text{ m/s}$$

The most probable speed is the speed at which the greatest number of particles are moving. Of the eight particles, three have speeds of 6.0 m/s, two have speeds of 4.0 m/s, and the other three particles have different speeds. Hence, $v_p = 6.0$ m/s. Note that the v_{rms} of the particles, which is related to the average kinetic energy, is greater than their average speed.

EXERCISE 10.15

Ten particles were found to have speeds of 0.1, 1.0, 2.0, 3.0, 3.0, 3.0, 4.0, 4.0, 5.0, and 6.0 m/s. Calculate the average speed, the root mean square speed, and the most probable speed of these particles.

Answer $\bar{v} = 3.1$ m/s; $v_{rms} = 3.5$ m/s; $v_p = 3.0$ m/s

Boltzmann Distributions

At any given time, what fraction of the molecules in a particular sample have a given speed? Some of the molecules will be moving more slowly than average, and some will be moving faster than average, but how many? Answers to questions such as these can have a substantial effect on the amount of product formed during a chemical reaction, as you will learn in Chapter 14. This problem was solved mathematically by Maxwell in 1866; he used statistical analysis to obtain an equation that describes the distribution of molecular speeds at a given temperature. Typical curves showing the distributions of speeds of molecules at several temperatures are displayed in Figure 10.14. Increasing the temperature has two effects. First, the peak of the curve moves to the right because the most probable speed increases. Second, the curve becomes broader because of the increased spread of the speeds. Thus, increased temperature increases the *value* of the most probable speed but decreases the relative number of molecules that have that speed. Although the mathematics behind curves such as those in Figure 10.14 was first worked out by Maxwell, the curves are almost universally referred to as **Boltzmann distributions**, after one of the other major figures responsible for the kinetic molecular theory of gases.

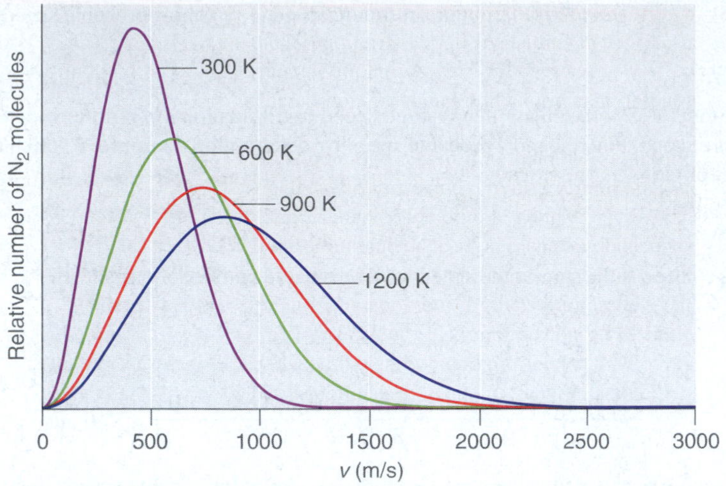

Figure 10.14 The distributions of molecular speeds for a sample of nitrogen gas at various temperatures. Increasing the temperature increases both the most probable speed (given at the peak of the curve) and the width of the curve. Interactive Graph

The Relationships Between Pressure, Volume, and Temperature

We now describe how the kinetic molecular theory of gases explains some of the important relationships we have discussed in previous sections.

Pressure Versus Volume

At constant temperature, the kinetic energy of the molecules of a gas and hence the root mean square speed remain unchanged. If a given gas sample is allowed to occupy a larger volume, then the speed of the molecules doesn't change, but the density of the gas (number of particles per unit volume) decreases and the average distance between the molecules increases. Hence, the molecules must, on average, travel farther between collisions. They therefore collide with one another and with the walls of the container less often, leading to a decrease in pressure. Conversely, increasing the pressure forces the molecules closer together and increases the density, until the collective impact of the collisions of the molecules with the walls of the container just balances the applied pressure.

Volume Versus Temperature

Raising the temperature of a gas increases the average kinetic energy and therefore the root mean square speed (and the average speed) of the gas molecules. Hence, as the temperature increases, the molecules collide with the walls of the container more frequently and with greater force. This increases the pressure, *unless* the volume increases to reduce the pressure, as we have just seen. Thus, an increase in temperature must be offset by an increase in volume for the net impact (pressure) of the gas molecules on the container walls to remain unchanged.

Pressure of Gas Mixtures

Postulate 3 of the kinetic molecular theory states that gas molecules exert no attractive or repulsive forces on one another. If the gaseous molecules do not interact, then the presence of one gas in a gas mixture will have no effect on the pressure exerted by another, and Dalton's law of partial pressures holds.

EXAMPLE 10.16

The temperature of a 4.75-L container of N_2 gas is increased from 0°C to 117°C. What is the qualitative effect of this change on the following: **(a)** the average kinetic energy of the N_2 molecules; **(b)** the rms speed of the N_2 molecules; **(c)** the average speed of the N_2 molecules;

(d) the impact of each N_2 molecule on the wall of the container during a collision with the wall; **(e)** the total number of collisions per second of N_2 molecules with the walls of the entire container; **(f)** the number of collisions per second of N_2 molecules with each square centimeter of the container wall; and **(g)** the pressure of the N_2 gas?

Given Temperatures, volume

Asked for Effect of increase in temperature

Strategy

Use the relationships between pressure, volume, and temperature to predict the qualitative effect of an increase in the temperature of the gas.

Solution

(a) Increasing the temperature increases the average kinetic energy of the N_2 molecules.
(b) An increase in average kinetic energy can only be due to an increase in the rms speed of the gas particles.
(c) If the rms speed of the N_2 molecules increases, the average speed increases, too.
(d) If, on average, the particles are moving faster, then they strike the container walls with more energy.
(e) Because the particles are moving faster, they collide with the walls of the container more often per unit time.
(f) The number of collisions per second of N_2 molecules with each square centimeter of container wall increases because the total number of collisions has increased, but the volume occupied by the gas and hence the total area of the walls are unchanged.
(g) As a result of (d) and (f), the pressure exerted by the N_2 gas increases when the temperature is increased at constant volume, as predicted by the ideal gas law.

EXERCISE 10.16

A sample of helium gas is confined in a cylinder with a gas-tight sliding piston. The initial volume is 1.34 L, and the temperature is 22°C. The piston is moved to allow the gas to expand to 2.12 L at constant temperature. What is the qualitative effect of this change on the following: **(a)** the average kinetic energy of the He atoms; **(b)** the rms speed of the He atoms; **(c)** the average speed of the He atoms; **(d)** the impact of each He atom on the wall of the container during a collision with the wall; **(e)** the total number of collisions per second of He atoms with the walls of the entire container; **(f)** the number of collisions per second of He atoms with each square centimeter of the container wall; and **(g)** the pressure of the helium gas?

Answer **(a)** no change; **(b)** no change; **(c)** no change; **(d)** no change; **(e)** decreases; **(f)** decreases; **(g)** decreases

Diffusion and Effusion

As you have learned, the molecules of a gas are *not* stationary but in constant motion. If someone opens a bottle of perfume in the next room, for example, you are likely to be aware of it soon. Your sense of smell relies on molecules of the aromatic substance coming into contact with specialized olfactory cells in your nasal passages, which contain specific receptors (protein molecules) that recognize the substance. How do the molecules responsible for the aroma get from the perfume bottle to your nose? You might think that they are blown by drafts, but, in fact, molecules can move from one place to another even in a draft-free environment. Figure 10.15 shows white fumes of solid ammonium chloride (NH_4Cl) forming when containers of aqueous ammonia and HCl are placed near each other, even with no draft to stir the air.

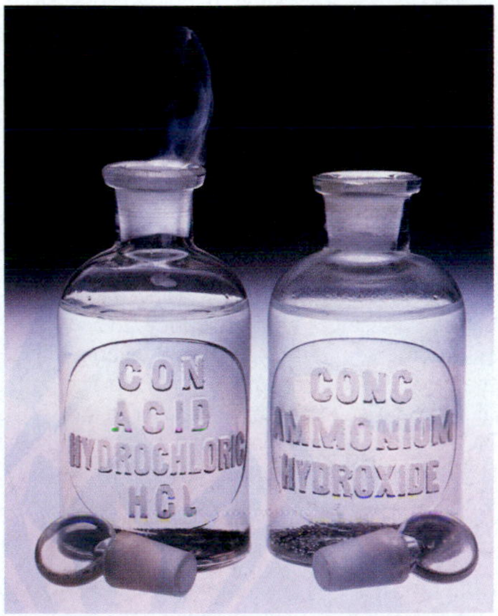

Figure 10.15 The diffusion of gaseous molecules. When open containers of aqueous NH_3 and HCl are placed near each other in a draft-free environment, molecules of the two substances diffuse, collide, and react to produce white fumes of solid ammonium chloride (NH_4Cl).

This phenomenon suggests that NH_3 and HCl molecules (as well as the more complex organic molecules responsible for the aromas of pizza and perfumes) move without assistance.

Diffusion is the gradual mixing of gases due to the motion of their component particles even in the absence of mechanical agitation such as stirring. The result is a gas mixture with uniform composition. As we shall see in Chapters 11–13, diffusion is also a property of the particles in liquids and liquid solutions and, to a lesser extent, of solids and solid solutions. We can describe the phenomenon shown in Figure 10.15 by saying that the molecules of HCl and NH_3 are able to diffuse away from their containers, and that NH_4Cl is formed where the two gases come into contact. Similarly, we say that a perfume or an aroma diffuses throughout a room or a house. The related process, **effusion**, is the escape of gaseous molecules through a small (usually microscopic) hole, such as a hole in a balloon, into an evacuated space.

The phenomenon of effusion had been known for thousands of years, but it was not until the early 19th century that quantitative experiments related the rate of effusion to molecular properties. The rate of effusion of a gaseous substance is inversely proportional to the square root of its molar mass. This relationship, rate $\propto 1/\sqrt{M}$, is referred to as *Graham's law*, after Scottish chemist Thomas Graham (1805–1869). The ratio of the effusion rates of two gases is the square root of the inverse ratio of their molar masses. If r is the effusion rate and M is the molar mass, then

$$\frac{r_1}{r_2} = \sqrt{\frac{M_2}{M_1}} \tag{10.36}$$

Note the pattern

At a given temperature, heavier molecules move more slowly than light molecules.

Although diffusion and effusion are different phenomena, the rate of diffusion is closely approximated using Equation 10.36. That is, if $M_1 < M_2$, then gas #1 will diffuse more rapidly than gas #2. This point is illustrated by the experiment shown in Figure 10.16, which is a more quantitative version of the case shown in Figure 10.15. The reaction is the same ($NH_3 + HCl \longrightarrow NH_4Cl$), but in this experiment, two cotton balls containing aqueous ammonia and HCl are placed along a meter stick in a draft-free environment, and the position at which the initial NH_4Cl fumes appear is noted. The white cloud forms much nearer the HCl-containing ball than the NH_3-containing ball. Because ammonia (M = 17.0 g/mol) diffuses much faster than HCl (M = 36.5 g/mol),

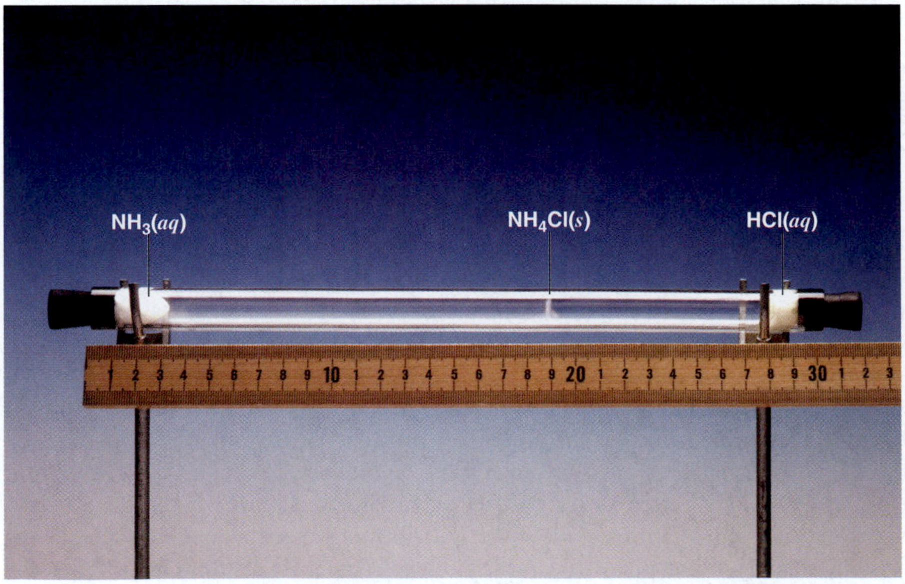

Figure 10.16 A simple experiment to measure the relative rates of diffusion of two gases. Cotton balls containing aqueous NH_3 (left) and HCl (right) are placed a measured distance apart in a draft-free environment, and the position at which white fumes of NH_4Cl first appear is noted. The puff of white NH_4Cl forms much closer to the HCl-containing ball than to the NH_3-containing ball. The left edge of the white puff marks where the reaction was first observed. The position of the white puff (18.8 − 3.3 = 15.5 cm from the NH_3, 28.0 − 18.8 = 9.2 cm from the HCl, giving a ratio of distances of 15.5/9.2 = 1.68) is approximately the location predicted by Graham's law based on the square root of the inverse ratio of the molar masses of the reactants (1.47).

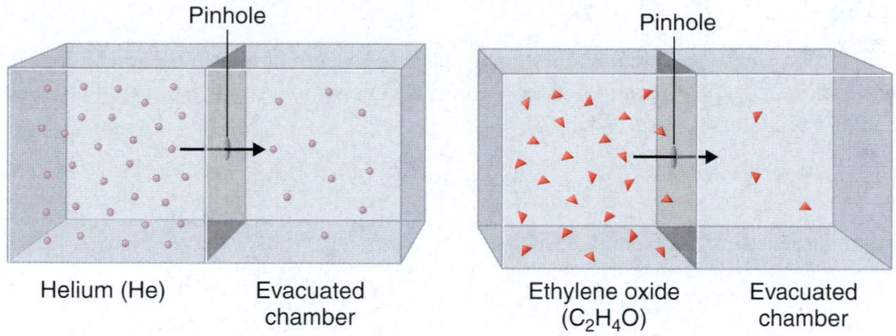

Figure 10.17 The relative rates of effusion of two gases with different masses. The lighter He atoms (M = 4.00 g/mol) effuse through the small hole more rapidly than the heavier ethylene oxide (C_2H_4O) molecules (M = 44.0 g/mol), as predicted by Graham's law.

the NH_4Cl fumes form closer to HCl because the HCl molecules travel a shorter distance. The ratio of the distances traveled by NH_3 and HCl in Figure 10.16 is about 1.7, in good agreement with the ratio of 1.47 predicted by their molar masses $[(36.5/17.0)^{1/2} = 1.47]$.

Heavy molecules effuse through a porous material more slowly than light molecules, as illustrated schematically in Figure 10.17 for ethylene oxide and helium. Helium (M = 4.00 g/mol) effuses much more rapidly than ethylene oxide (M = 44.0 g/mol). Thus, because helium is less dense than air, helium-filled balloons "float" at the end of a tethering string. Unfortunately, rubber balloons filled with helium soon lose their buoyancy along with much of their volume. In contrast, rubber balloons filled with air tend to retain their shape and volume for a much longer time. Because helium has a molar mass of 4.00 g/mol, whereas air has an average molar mass of about 29 g/mol, pure helium effuses through the microscopic pores in the rubber balloon $\sqrt{\frac{29}{4.00}} = 2.7$ times faster than air. For this reason, high-quality helium-filled balloons are usually made of Mylar, a dense, strong, opaque material with a high molecular mass that forms films that have many fewer pores than rubber. Mylar balloons can retain their helium for days.

EXAMPLE 10.17

During World War II, scientists working on the first atomic bomb were faced with the challenge of finding a way to obtain large amounts of ^{235}U. Naturally occurring uranium is only 0.720% ^{235}U, whereas most of the rest (99.275%) is ^{238}U, which is not only not fissionable (that is, it will not break apart to release nuclear energy) but actually poisons the fission process. Because both isotopes of uranium have the same reactivity, they cannot be separated chemically. Instead, a process of gaseous effusion was developed using the volatile compound UF_6 (boiling point 56°C). **(a)** Calculate the ratio of the rates of effusion of $^{235}UF_6$ and $^{238}UF_6$ for a single step in which UF_6 is allowed to pass through a porous barrier. (The atomic mass of ^{235}U is 235.04, and the atomic mass of ^{238}U is 238.05.) **(b)** If n identical successive separation steps are used, the overall separation is given by the separation in a single step (in this case, the ratio of effusion rates) raised to the nth power. How many effusion steps are needed to obtain 99.0% pure $^{235}UF_6$?

Given Isotopic content of naturally occurring uranium, atomic masses of ^{235}U and ^{238}U

Asked for Ratio of rates of effusion, number of effusion steps needed to obtain 99.0% pure $^{235}UF_6$

Strategy

A Calculate the molar masses of $^{235}UF_6$ and $^{238}UF_6$, and then use Graham's law to determine the ratio of the effusion rates. Use this value to determine the isotopic content of $^{235}UF_6$ after a single effusion step.

B Divide the final purity by the initial purity to obtain a value for the number of separation steps needed to achieve the desired purity. Use a logarithmic expression to compute the number of separation steps required.

Solution

(a) The first step is to calculate the molar mass of UF_6 containing ^{235}U and ^{238}U. Luckily for the success of the separation method, fluorine consists of a single isotope of atomic mass 18.998. The molar mass of $^{235}UF_6$ is

$$235.04 + (6)(18.998) = 349.03 \text{ g/mol}$$

The molar mass of $^{238}UF_6$ is

$$238.05 + (6)(18.998) = 352.04 \text{ g/mol}$$

The difference is only 3.01 g/mol (less than 1%). The ratio of the effusion rates can be calculated from Graham's law using Equation 10.36:

$$\frac{\text{rate } ^{235}UF_6}{\text{rate } ^{238}UF_6} = \sqrt{\frac{352.04}{349.03}} = 1.0043$$

Thus, passing UF_6 containing a mixture of the two isotopes through a single porous barrier gives an enrichment of 1.0043, so after one step the isotopic content is (0.720%)(1.0043) = 0.723% $^{235}UF_6$.

(b) To obtain 99.0% pure $^{235}UF_6$ requires many steps. We can set up an equation that relates the initial and final purity to the number of times the separation process is repeated:

$$\text{final purity} = (\text{initial purity})(\text{separation})^n$$

In this case, $0.990 = (0.00720)(1.0043)^n$, which can be rearranged to give

$$\frac{0.990}{0.00720} = (1.0043)^n = 138$$

Taking the logarithm of both sides gives

$$\log 138 = n \log 1.0043$$

$$n = \frac{\log 138}{\log 1.0043}$$

$$n = 1150 = 1.15 \times 10^3$$

Thus, at least a thousand effusion steps are necessary to obtain highly enriched ^{235}U. The photo in Figure 10.18 shows a small part of a system that is used to prepare enriched uranium on a large scale.

Figure 10.18 A portion of a plant for separating uranium isotopes by effusion of UF_6. The large cylindrical objects (note the human for scale) are so-called "diffuser" (actually effuser) units, in which gaseous UF_6 is pumped through a porous barrier to partially separate the isotopes. The UF_6 must be passed through multiple units to become substantially enriched in ^{235}U.

EXERCISE 10.17

Helium consists of two isotopes: ^3He (natural abundance 0.000134%) and ^4He (natural abundance 99.999866%). Their atomic masses are 3.01603 and 4.00260, respectively. Helium-3 has unique physical properties and is used in the study of ultralow temperatures. It is separated from the more abundant ^4He by a process of gaseous effusion. **(a)** Calculate the ratio of the effusion rates of ^3He and ^4He and thus the enrichment possible in a single effusion step. **(b)** How many effusion steps are necessary to yield 99.0% pure ^3He?

Answer **(a)** Ratio of effusion rates = 1.15200; one step gives 0.000154% ^3He; **(b)** 96 steps

Rates of Diffusion or Effusion

Graham's law is an empirical relationship that states that the ratio of the rates of diffusion or effusion of two gases is the square root of the inverse ratio of their molar masses. The

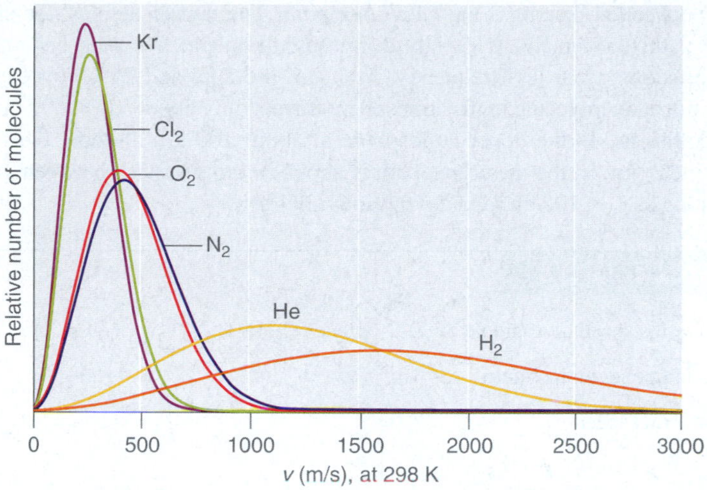

Figure 10.19 The wide variation in molecular speeds observed at 298 K for gases with different molar masses. The lightest gases have a wider distribution of speeds and the highest average speeds. Interactive Graph

relationship is based on the postulate that all gases at the same temperature have the same average kinetic energy. We can write the expression for the average kinetic energy of two gases with different molar masses:

$$\overline{KE} = \frac{1}{2}M_1v^2_{rms_1} = \frac{1}{2}M_2v^2_{rms_2} \quad (10.37)$$

Multiplying both sides by 2 and rearranging give

$$\frac{v^2_{rms_2}}{v^2_{rms_1}} = \frac{M_1}{M_2} \quad (10.38)$$

Taking the square root of both sides gives

$$\frac{v_{rms_2}}{v_{rms_1}} = \sqrt{\frac{M_1}{M_2}} \quad (10.39)$$

Thus, the rate at which a molecule, or a mole of molecules, diffuses or effuses is directly related to the speed at which it moves. Equation 10.39 shows that Graham's law is a direct consequence of the fact that gaseous molecules at the same temperature have the same average kinetic energy.

Note the pattern

Molecules with lower masses have a wider distribution of speeds.

Typically, gaseous molecules have a speed of hundreds of meters per second (hundreds of miles per hour). The effect of molar mass on these speeds is dramatic, as illustrated in Figure 10.19 for some common gases. Because all the gases have the same average kinetic energy, according to the Boltzman distribution, molecules with lower masses, such as hydrogen and helium, have a wider distribution of speeds. The postulates of the kinetic molecular theory lead to the following equation, which directly relates molar mass, temperature, and rms speed:

$$v_{rms} = \sqrt{\frac{3RT}{M}} \quad (10.40)$$

In this equation, v_{rms} has units of m/s; consequently, the units of molar mass M are kg/mol, temperature T is expressed in K, and the ideal gas constant R has the value 8.3145 J/(K·mol).

Gas molecules do not diffuse nearly as rapidly as their very high speeds might suggest. If molecules actually moved through a room at hundreds of miles per hour, we would detect odors faster than we hear sound. Instead, it can take several minutes for us to detect an aroma because molecules are traveling in a medium that contains other gas molecules. Because gas molecules collide as often as 10^{10} times per second, changing direction and speed with each collision, they do not diffuse across a room in a straight line, as illustrated schematically in Figure 10.20. The average distance traveled by a

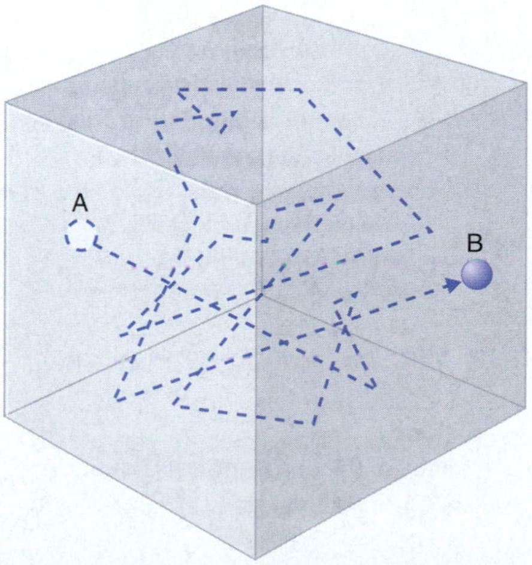

Figure 10.20 The path of a single particle in a gas sample. The frequent changes in direction are the result of collisions with other gas molecules and with the walls of the container.

Note the pattern

The denser the gas, the shorter the mean free path.

molecule between collisions is the *mean free path*. The denser the gas, the shorter the mean free path; conversely, as density decreases, the mean free path becomes longer because collisions occur less frequently. At 1 atm pressure and 25°C, for example, an oxygen or nitrogen molecule in the atmosphere travels only about 6.0×10^{-8} m (60 nm) between collisions. In the upper atmosphere at about 100 km altitude, where the gas density is much lower, the mean free path is about 10 cm; in space between galaxies, it can be as long as 1×10^{10} m (about 6 million miles!).

EXAMPLE 10.18

Calculate the rms speed of a sample of *cis*-2-butene (C_4H_8) at 20°C.

Given Compound, temperature

Asked for rms speed

Strategy

Calculate the molar mass of *cis*-2-butene. Be certain that all quantities are expressed in the appropriate units, and then use Equation 10.40 to calculate the rms speed of the gas.

Solution

To use Equation 10.40, we need to calculate the molar mass of *cis*-2-butene and make sure that each of the quantities is expressed in the appropriate units. Butene is C_4H_8, so its molar mass is 56.11 g/mol. Thus,

$$M = 56.11 \text{ g/mol} = 56.11 \times 10^{-3} \text{ kg/mol}$$

$$T = 20 + 273 = 293 \text{ K}$$

$$R = 8.3145 \text{ J/(K}\cdot\text{mol)} = 8.3144 \text{ (kg}\cdot\text{m}^2\text{)/(s}^2\cdot\text{K}\cdot\text{mol)}$$

$$v_{\text{rms}} = \sqrt{\frac{3RT}{M}} = \sqrt{\frac{(3)[8.3145 \text{ (}\cancel{\text{kg}}\cdot\text{m}^2\text{)/(s}^2\cdot\cancel{\text{K}}\cdot\cancel{\text{mol}}\text{)}](293 \cancel{\text{K}})}{56.11 \times 10^{-3} \cancel{\text{kg/mol}}}} = 3.61 \times 10^2 \text{ m/s}$$

or approximately 810 mi/h.

EXERCISE 10.18

Calculate the rms speed of a sample of radon gas at 23°C.

Answer 1.82×10^2 m/s (about 410 mi/h)

The kinetic molecular theory of gases demonstrates how a successful theory can explain previously observed empirical relationships (laws) in an intuitively satisfying way. Unfortunately, the actual gases that we encounter are not "ideal," although their behavior usually approximates that of an ideal gas. In Section 10.8, we explore how the behavior of real gases differs from that of ideal gases.

10.8 • The Behavior of Real Gases

MGC The van der Waals Equation

The postulates of the kinetic molecular theory of gases ignore both the volume occupied by the molecules of a gas and all interactions between molecules, whether attractive or repulsive. In reality, however, all gases have nonzero molecular volumes. Furthermore, the molecules of real gases interact with one another in ways that depend on the structure of the molecules and therefore differ for each gaseous substance. In this section, we consider the properties of real gases and how and why they differ from the predictions of the ideal gas law. We also examine liquefaction, a key property of real gases that is not predicted by the kinetic molecular theory.

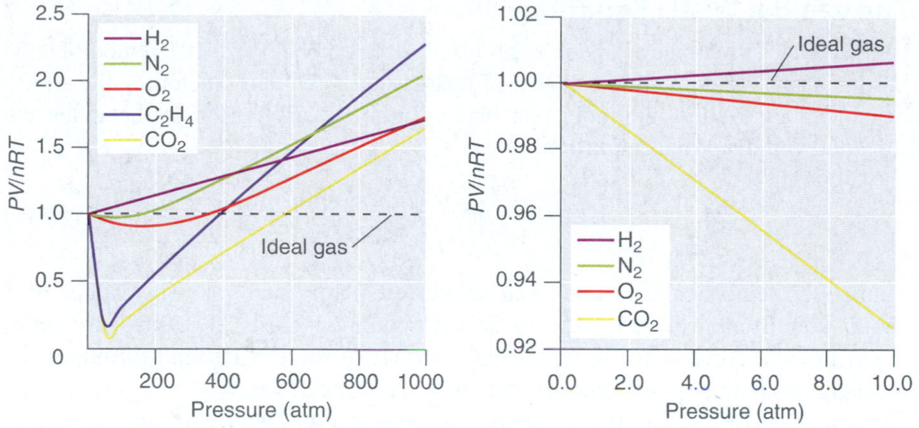

(a) ***PV/nRT*** **at high pressures** **(b)** ***PV/nRT*** **at low pressures**

Figure 10.21 Real gases do not obey the ideal gas law, especially at high pressures. (a) Plots of *PV/nRT* vs. *P* at 273 K for several common gases. The large negative deviations observed for C_2H_4 and CO_2 are because they liquefy at relatively low pressures. (b) These plots illustrate the relatively good agreement between experimental data for real gases and the ideal gas law at low pressures. Interactive Graph

Pressure, Volume, and Temperature Relationships in Real Gases

For an ideal gas, a plot of *PV/nRT* versus *P* gives a horizontal line with an intercept of 1 on the *PV/nRT* axis. Real gases, however, show significant deviations from the behavior expected for an ideal gas, particularly at high pressures (Figure 10.21a). Only at relatively low pressures (less than 1 atm) do real gases approximate ideal gas behavior (Figure 10.21b). Real gases also approach ideal gas behavior more closely at higher temperatures, as shown in Figure 10.22 for N_2. Why do real gases behave so differently from ideal gases at high pressures and low temperatures? Under these conditions, the two basic assumptions behind the ideal gas law—namely, that gas molecules have negligible volume and that intermolecular interactions are negligible—are no longer valid.

Because the molecules of an ideal gas are assumed to have zero volume, the volume available to them for motion is always the same as the volume of the container. In contrast, the molecules of a real gas have small but measurable volumes. At low pressures, the gaseous molecules are relatively far apart, but as the pressure of the gas increases, the intermolecular distances become smaller and smaller (Figure 10.23). As a result, the volume occupied by the molecules becomes significant compared with the volume of the container. Consequently, the total volume occupied by the gas is greater than the volume predicted by the ideal gas law. Thus, at very high pressures, the experimentally measured value of *PV/nRT* is greater than the value predicted by the ideal gas law.

Moreover, all molecules are attracted to one another by a combination of forces. These forces become particularly important for gases at low temperatures and high pressures, where intermolecular distances are shorter. Attractions between molecules reduce the number of collisions with the container wall, an effect that becomes more pronounced as the number of attractive interactions increases. Because the average distance between molecules decreases, the pressure exerted by the gas on the container wall decreases, and the observed pressure is *less* than expected (Figure 10.24). Thus, as shown in Figure 10.22, at low temperatures, the ratio of *PV/nRT* is lower than predicted for an ideal gas, an effect that becomes particularly evident for complex gases and for simple gases at low temperatures. At very high pressures, the effect of nonzero molecular volume predominates. The competition between these effects is responsible for the minimum observed in the *PV/nRT* versus *P* plot for many gases.

At high temperatures, the molecules have sufficient kinetic energy to overcome the intermolecular attractive forces, and the effects of nonzero molecular volume predominate. Conversely, as the temperature is lowered, the kinetic energy of the gas molecules decreases. Eventually, a point is reached where the molecules can no longer overcome the intermolecular attractive forces, and the gas liquefies (condenses to a liquid).

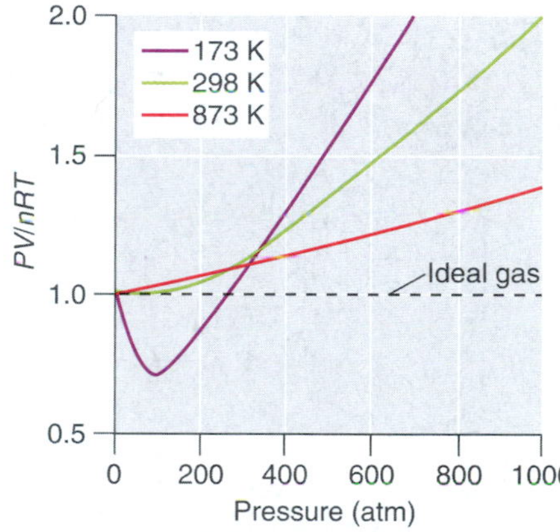

Figure 10.22 The effect of temperature on the behavior of real gases. A plot of *PV/nRT* vs. *P* for nitrogen gas at three temperatures shows that the approximation to ideal gas behavior becomes better as the temperature increases.

Note the pattern

Nonzero molecular volume makes the actual volume greater than predicted at high pressures; intermolecular attractions make the pressure less than predicted.

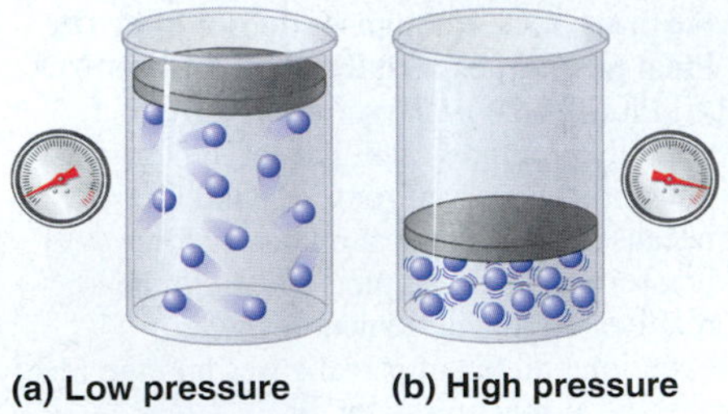

Figure 10.23 The effect of nonzero volume of gas particles on the behavior of gases at low and high pressures. (a) At low pressures, the volume occupied by the molecules themselves is small compared with the volume of the container. (b) At high pressures, the molecules occupy a large portion of the volume of the container, resulting in significantly decreased space in which the molecules can move.

The van der Waals Equation

Dutch physicist Johannes van der Waals (1837–1923, Nobel Prize 1910) modified the ideal gas law to describe the behavior of real gases by explicitly including the effects of molecular size and intermolecular forces. In his description of gas behavior, the so-called **van der Waals equation**,

$$\left(P + \frac{an^2}{V^2}\right)(V - nb) = nRT \tag{10.41}$$

a and b are empirical constants that are different for each gas. The values of a and b are listed in Table 10.5 for several common gases. The pressure term, $P + (an^2/V^2)$, corrects for intermolecular attractive forces that tend to reduce the pressure from that predicted by the ideal gas law. Here, n^2/V^2 represents the concentration of the gas (n/V) squared because it takes two particles to engage in the pairwise intermolecular interactions of the type shown in Figure 10.24. The volume term, $V - nb$, corrects for the volume occupied by the gaseous molecules.

The correction for volume is negative, but the correction for pressure is positive to reflect the effect of each of the factors on V and P, respectively. Because nonzero molecular volumes produce a measured volume that is *larger* than that predicted by the ideal gas law, we must subtract the molecular volumes to obtain the actual volume available. Conversely, attractive intermolecular forces produce a pressure that is *less* than that expected based on the ideal gas law, so the an^2/V^2 term must be added to the measured pressure to correct for these effects.

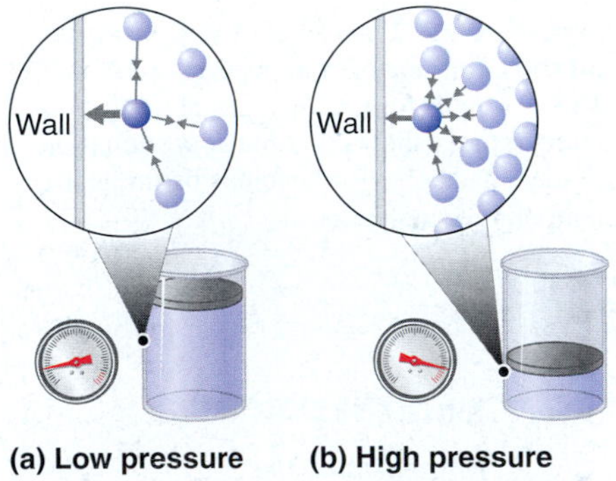

Figure 10.24 The effect of intermolecular attractive forces on the pressure a gas exerts on the container walls. (a) At low pressures, there are relatively few attractive intermolecular interactions to lessen the impact of the molecule striking the wall of the container, and the pressure is close to that predicted by the ideal gas law. (b) At high pressures, with the average intemolecular distance relatively small, the effect of intermolecular interactions is to lessen the impact of a given molecule striking the container wall, resulting in a lower pressure than predicted by the ideal gas law.

EXAMPLE 10.19

You are in charge of the manufacture of cylinders of compressed gas at a small company. Your company president would like to offer a 4.0-L cylinder containing 500 g of chlorine in the new catalog. The cylinders you have on hand have a rupture pressure of 40 atm. Use both the ideal gas law and the van der Waals equation to calculate the pressure in a cylinder at 25°C. Is this cylinder likely to be safe against sudden rupture (which would be disastrous because chlorine gas is highly toxic)?

Given Volume of cylinder, mass of compound, pressure, temperature

Asked for Safety

Strategy

A Use the molar mass of chlorine to calculate the amount of chlorine in the cylinder. Then calculate the pressure of the gas using the ideal gas law.

B Obtain a and b values for Cl_2 from Table 10.5. Use the van der Waals equation to solve for the pressure of the gas. Based on the value obtained, predict whether the cylinder is likely to be safe against sudden rupture.

Solution

A We begin by calculating the amount of chlorine in the cylinder using the molar mass of chlorine (70.906 g/mol):

$$(500\ \cancel{\text{g}})\left(\frac{1\ \text{mol}}{70.906\ \cancel{\text{g}}}\right) = 7.05\ \text{mol Cl}$$

Using the ideal gas law and the temperature in K (298 K), we calculate the pressure:

$$P = \frac{nRT}{V} = \frac{(7.05\ \cancel{\text{mol}})[0.08206\ (\cancel{\text{L}}\cdot\text{atm})/(\cancel{\text{K}}\cdot\cancel{\text{mol}})](298\ \cancel{\text{K}})}{4.0\ \cancel{\text{L}}} = 43\ \text{atm}$$

If chlorine behaves like an ideal gas, you have a real problem! Now let's use the van der Waals equation with the a and b values for Cl_2 from Table 10.5. Solving for P gives

$$P = \frac{nRT}{V - nb} - \frac{an^2}{V^2}$$

$$= \frac{(7.05\ \text{mol})(0.08206\ \text{L} \cdot \text{atm/K} \cdot \text{mol})(298\ \text{K})}{4.0\ \text{L} - (7.05\ \text{mol})(0.0542\ \text{L/mol})} - \frac{(6.260\ \text{L}^2 \cdot \text{atm/mol}^2)(7.05\ \text{mol})^2}{(4.0\ \text{L})^2}$$

$$= 47.7\ \text{atm} - 19.4\ \text{atm} = 28\ \text{atm (to two significant figures)}$$

This pressure is well within the safety limits of the cylinder. Note that the ideal gas law predicts a pressure 15 atm higher than the van der Waals equation!

TABLE 10.5 van der Waals constants for some common gases

Gas	a, $(L^2 \cdot atm)/mol^2$	b, L/mol
He	0.03410	0.0238
Ne	0.205	0.0167
Ar	1.337	0.032
H_2	0.2420	0.0265
N_2	1.352	0.0387
O_2	1.364	0.0319
Cl_2	6.260	0.0530
NH_3	4.170	0.0371
CH_4	2.273	0.0430
CO_2	3.610	0.0429

EXERCISE 10.19

A 10.0-L cylinder contains 500 g of methane. Calculate its pressure to two significant figures at 27°C using **(a)** the ideal gas law and **(b)** the van der Waals equation.

Answer **(a)** 77 atm; **(b)** 67 atm

Liquefaction of Gases

Liquefaction of gases is the condensation of gases into a liquid form, which is neither anticipated nor explained by the kinetic molecular theory of gases. Both the theory and the ideal gas law predict that gases compressed to very high pressures and cooled to very low temperatures should still behave like gases, albeit cold, dense ones. As gases are compressed and cooled, however, they invariably condense to form liquids, although very low temperatures are needed to liquefy light elements such as helium (for He, 4.2 K at 1 atm pressure).

Liquefaction can be viewed as an extreme deviation from ideal gas behavior. It occurs when the molecules of a gas are cooled to the point where they no longer possess sufficient kinetic energy to overcome the intermolecular attractive forces. The precise combination of temperature and pressure needed to liquefy a gas depends strongly on its molar mass and structure, with heavier and more complex molecules usually liquefying at higher temperatures. In general, substances with large van der Waals a coefficients are relatively easy to liquefy because large a coefficients indicate relatively strong intermolecular attractive interactions. Conversely, small molecules that contain only light elements have small a coefficients, indicating weak intermolecular interactions, and they are relatively difficult to liquefy. Gas liquefaction is used on a massive scale to separate O_2, N_2, Ar, Ne, Kr, and Xe. After a sample of air is liquefied, the mixture is warmed and the gases are separated according to their boiling points. In Chapter 11, we will consider in more detail the nature of the intermolecular forces that allow gases to liquefy.

The ultracold liquids formed from the liquefaction of gases are called **cryogenic liquids**, from the Greek *kryo* (cold) and *genes* (producing). They have applications as refrigerants in both industry and biology. For example, under carefully controlled conditions, the very cold temperatures afforded by liquefied gases such as nitrogen (boiling point 77 K at 1 atm) can preserve biological materials, such as semen for the artificial insemination of cows and other farm animals. These liquids can also be used in a specialized type of surgery called *cryosurgery,* which selectively destroys tissues with minimal loss of blood by the use of extreme cold.

Moreover, liquefaction of gases is tremendously important in the storage and shipment of fossil fuels (Figure 10.25). Liquefied natural gas (LNG) and liquefied petroleum gas (LPG) are liquefied forms of hydrocarbons produced from natural gas or

Figure 10.25 **Photo of liquid natural gas transport ship.**

petroleum reserves. LNG consists mostly of methane, with small amounts of heavier hydrocarbons; it is prepared by cooling natural gas to below about −162°C. It can be stored in double-walled, vacuum-insulated containers at or slightly above atmospheric pressure. Because LNG occupies only about 1/600 the volume of natural gas, it is easier and more economical to transport. LPG is typically a mixture of propane, propene, butane, and butenes and is primarily used as a fuel for home heating. It is also used as a feedstock for chemical plants and as an inexpensive and relatively nonpolluting fuel for automobiles.

SUMMARY AND KEY TERMS

10.1 Gaseous Elements and Compounds (p. 432)

Bulk matter can exist in three states: gas, liquid, and solid. Gases have the lowest density of the three, are highly compressible, and fill their container completely. Elements that exist as gases at room temperature and pressure are clustered on the right side of the periodic table; they occur as either monatomic gases (the noble gases) or diatomic molecules (some halogens, N_2, O_2). Many inorganic and organic compounds that contain four or fewer nonhydrogen atoms are also gases at room temperature and pressure. All gaseous substances are characterized by weak interactions between the constituent molecules or atoms.

10.2 Gas Pressure (p. 434)

Four quantities must be known for a complete physical description of a sample of a gas: *temperature, volume, amount*, and *pressure*. **Pressure** is force per unit area of surface; the SI unit for pressure is the **pascal (Pa)**, defined as 1 newton per square meter (N/m^2). The pressure exerted by an object is proportional to the force it exerts and inversely proportional to the area upon which the force is exerted. The pressure exerted by Earth's atmosphere, called *atmospheric pressure,* is about 101 kPa or 14.7 $lb/in.^2$ at sea level. Atmospheric pressure can be measured with a **barometer**, a closed, inverted tube filled with mercury. The height of the mercury column is proportional to the atmospheric pressure, which is often reported in units of **millimeters of mercury (mmHg)**, also called **torr. Standard atmospheric pressure**, the pressure required to support a column of mercury 760 mm tall, is yet another unit of pressure: 1 **atmosphere (atm)**. A **manometer** is an apparatus used to measure the pressure of a sample of a gas.

10.3 Relationships Between Pressure, Temperature, Amount, and Volume (p. 439)

Boyle showed that the volume of a sample of a gas is inversely proportional to its pressure (*Boyle's law*), Charles and Gay-Lussac demonstrated that the volume of a gas is directly proportional to its temperature (in K) at constant pressure (*Charles's law*), and Avogadro postulated that the volume of a gas is directly proportional to the number of moles of gas present (*Avogadro's law*). Plots of the volume of gases versus temperature extrapolate to zero volume at −273.15°C, which is **absolute zero,** the lowest temperature possible. Charles's law implies that the volume of a gas is directly proportional to its absolute temperature.

10.4 The Ideal Gas Law (p. 442)

The empirical relationships between volume, temperature, pressure, and amount of a gas can be combined into the **ideal gas law**, $PV = nRT$. The proportionality constant, R, is called the **gas constant** and has the value 0.08206 (L·atm)/(K·mol), 8.3145 J/(K·mol), or 1.9872 cal/(K·mol), depending on the units used. The ideal gas law describes the behavior of an *ideal gas,* a hypothetical substance whose behavior can be explained quantitatively by the ideal gas law and the kinetic molecular theory of gases (see below). **Standard temperature and pressure (STP)** is 0°C and 1 atm. The volume of 1 mol of an ideal gas at STP is 22.41 L, the **standard molar volume**. All of the empirical gas relationships are special cases of the ideal gas law in which two of the four parameters are held constant. The ideal gas law allows us to calculate the value of the fourth quantity (P, V, T, or n) needed to describe a gaseous sample when the others are known, and also to predict the value of all of these quantities following a change in conditions if the original conditions (values of P, V, T, and n) are known. The ideal gas law can also be used to calculate the density of a gas if its molar mass is known or, conversely, the molar mass of an unknown gas sample if its density is measured.

10.5 Mixtures of Gases (p. 452)

The pressure exerted by each gas in a gas mixture (its **partial pressure**) is independent of the pressure exerted by all other gases present. Consequently, the total pressure exerted by a mixture of gases is the sum of the partial pressures of the components (*Dalton's law of partial pressures*). The amount of gas present in a mixture may be described by its partial pressure or its mole fraction. The **mole fraction** of any component of a mixture is the ratio of the number of moles of that substance to the total number of moles of all substances present. In a mixture of gases, the partial pressure of each gas is the product of the total pressure and the mole fraction of that gas.

10.6 Gas Volumes and Stoichiometry (p. 456)

The relationship between the amounts of products and reactants in a chemical reaction can be expressed in units of moles or masses of pure substances, of volumes of solutions, or of volumes of gaseous substances. The ideal gas law can be used to calculate the volume of gaseous products or reactants as needed. In the laboratory, gases produced in a reaction are often collected by the displacement of water from filled vessels; the amount of gas can then be calculated from the volume of water displaced and the atmospheric pressure. A gas collected in such a way is not pure, however, but contains a significant amount of water vapor. The measured pressure must therefore be corrected for the vapor pressure of water, which depends strongly on the temperature.

10.7 The Kinetic Molecular Theory of Gases (p. 459)

The behavior of ideal gases is explained by the **kinetic molecular theory of gases**. Molecular motion, which leads to collisions between molecules and the container walls, explains pressure, and the large intermolecular distances in gases explain their high compressibility. Although all gases have the same average kinetic energy at a given temperature, they do not all possess the same **root mean square (rms) speed** (v_{rms}). The actual values of speed and kinetic energy are not the same for all particles of a gas but are given by a **Boltzmann distribution,** in which some molecules have higher or lower speeds (and kinetic energies) than average. **Diffusion** is the gradual mixing of gases to form a sample of uniform composition even in the absence of mechanical agitation. In contrast, **effusion** is the escape of a gas from a container through a tiny opening into an evacuated space. The rate of effusion of a gas is inversely proportional to the square root of its molar mass (*Graham's law*), a relationship that closely approximates the rate of diffusion. As a result, light gases tend to diffuse and effuse much more rapidly than heavier gases.

10.8 The Behavior of Real Gases (p. 468)

No real gas exhibits ideal gas behavior, although many real gases approximate it over a range of conditions. Deviations from ideal gas behavior can be seen in plots of PV/nRT versus P at a given temperature; for an ideal gas, PV/nRT vs. $P = 1$ under all conditions. At high pressures, most real gases exhibit larger PV/nRT values than predicted by the ideal gas law, whereas at low pressures, most real gases exhibit PV/nRT values close to those predicted by the ideal gas law. Gases most closely approximate ideal gas behavior at high temperatures and low pressures. Deviations from ideal gas law behavior can be described by the **van der Waals equation**, which includes empirical constants to correct for the actual volume of the gaseous molecules and to quantify the reduction in pressure due to intermolecular attractive forces. If the temperature of a gas is decreased sufficiently, **liquefaction** occurs, in which the gas condenses into a liquid form. Liquefied gases have many commercial applications, including the transport of large amounts of gases in small volumes and the uses of ultracold **cryogenic liquids**.

KEY EQUATIONS

Definition of pressure	$P = \frac{F}{A}$	(10.1)
Ideal gas law	$PV = nRT$	(10.13)
Relationship between initial and final conditions	$\frac{P_1V_1}{T_1} = \frac{P_2V_2}{T_2} \quad n = \text{constant}$	(10.20)
Density of a gas	$d = \frac{PM}{RT}$	(10.24)
Dalton's law of partial pressures	$P_t = (n_1 + n_2 + n_3 + \cdots + n_i)\left(\frac{RT}{V}\right)$	(10.28)
Mole fraction	$X_A = \frac{\text{moles A}}{\text{total moles}} = \frac{n_A}{n_t}$	(10.29)
Relationship between partial pressure and mole fraction	$P_A = X_A P_t$	(10.31)
Average kinetic energy	$\overline{KE} = \frac{1}{2}m\overline{v^2}$	(10.32)
Root mean square (rms) speed	$v_{rms} = \sqrt{\frac{v_1^2 + v_2^2 + v_3^2 + \cdots + v_n^2}{n}}$	(10.34)
Graham's law for diffusion and effusion	$\frac{r_1}{r_2} = \sqrt{\frac{M_2}{M_1}}$	(10.36)
Kinetic molecular theory	$v_{rms} = \sqrt{\frac{3RT}{M}}$	(10.40)
van der Waals equation	$\left(P + \frac{an^2}{V^2}\right)(V - nb) = nRT$	(10.41)

QUESTIONS AND PROBLEMS

Questions and problems with colored numbers have answers in the Appendix and complete solutions in the Student Solutions Manual.

CONCEPTUAL

10.1 Gaseous Elements and Compounds

1. Explain the differences between microscopic and macroscopic properties of matter. Is the boiling point of a compound a microscopic or macroscopic property? Molecular mass? Why?
2. Determine whether each of the following is a macroscopic or microscopic property of matter, and explain your reasoning: melting point, dipole moment, and electrical conductivity.
3. How do the microscopic properties of matter influence the macroscopic properties? Can you relate molecular mass to boiling point? Why or why not? Can polarity be related to boiling point?
4. For a substance that has gas, liquid, and solid phases, arrange these phases in order of increasing: (a) density; (b) strength of intermolecular interactions; (c) compressibility; (d) molecular motion; and (e) order in the arrangement of the molecules or atoms.
5. Explain what is wrong with this statement: *The state of matter largely determines the molecular properties of a substance.*
6. Describe the most important factors that determine the state of a given compound. What external conditions influence whether a substance exists in any one of the three states of matter?
7. Which elements of the periodic table exist as gases at room temperature and pressure? Of these, which are diatomic molecules and which are monatomic? Which elements are liquids at room temperature and pressure? Which portion of the periodic table contains elements whose binary hydrides are most likely to be gases at room temperature?
8. Is the following observation correct? "*Almost all nonmetal binary hydrides are gases at room temperature, but metal hydrides are all solids.*" Explain your reasoning.
9. Is the following observation correct? "*All the hydrides of the chalcogens are gases at room temperature and pressure except the binary hydride of oxygen, which is a liquid.*" Explain your reasoning. Would you expect 1-chloropropane to be a gas? Iodopropane? Why?
10. Explain why ionic compounds are not gases under normal conditions.

10.2 Gas Pressure

11. What four quantities must be known to completely describe a sample of a gas? What units are commonly used for each of these quantities?
12. If the applied force is constant, how does the pressure exerted by an object change as the area upon which the force is exerted decreases? In the real world, how does this relationship apply to the ease of driving a small nail versus a large nail?
13. As the force on a fixed area increases, does the pressure increase or decrease? With this in mind, would you expect a heavy person to need smaller or larger snowshoes than a lighter person? Explain.
14. What do we mean by *atmospheric pressure*? Is the atmospheric pressure at the summit of Mt. Rainier higher or lower than the pressure in Miami, Florida? Why?
15. Which has the highest atmospheric pressure: a cave in the Himalayas, a mine in South Africa, or a beach house in Florida? Which has the lowest?
16. Mars has an average atmospheric pressure of 0.007 atm. Would it be easier or harder to drink liquid from a straw on Mars than on Earth? Explain your answer.
17. Is the pressure exerted by a 1.0-kg mass on a 2.0-m^2 area greater or less than the pressure exerted by a 1.0-kg mass on a 1.0-m^2 area? What is the difference, if any, between the pressure of the atmosphere exerted on a 1.0-m^2 piston and a 2.0-m^2 piston?
18. If you used water in a barometer instead of mercury, what would be the major difference in the instrument?

10.3 Relationships Between Pressure, Temperature, Amount, and Volume

19. Sketch a graph of the volume of a gas versus the pressure on the gas. What would the graph of V versus P look like if volume was directly proportional to pressure?
20. What properties of a gas are described by Boyle's Law? By Charles's law? By Avogadro's law? In each of these cases, what quantities are held constant? Why does the constant in Boyle's law depend on the amount of gas used and on the temperature at which the experiments are carried out?
21. Using Charles's law, explain why cooler air sinks.
22. Use Boyle's law to explain why it is dangerous to heat even a small quantity of water in a sealed container.

10.4 The Ideal Gas Law

23. For an ideal gas, is volume directly proportional to temperature or inversely proportional? What is the volume of an ideal gas at a temperature of absolute zero?
24. What is meant by the acronym STP? If a gas is at STP, what further information is required to completely describe the state of the gas?
25. For a given amount of a gas, the volume, temperature, and pressure under any one set of conditions are related to the volume, temperature, and pressure under any other set of conditions by the equation $P_1V_1/T_1 = P_2V_2/T_2$. Derive this equation from the ideal gas law. At constant temperature, this equation reduces to one of the laws discussed in Section 10.3; which one? At constant pressure, this equation reduces to one of the laws discussed in Section 10.3; which one?
26. Predict the effect of each of the following changes on one variable if the other variables are held constant:
   (a) If the number of moles of gas increases, what is the effect on the temperature of the gas?
   (b) If the temperature of a gas decreases, what is the effect on the pressure?
   (c) If the volume of the gas increases, what is the effect on the temperature of the gas?
   (d) If the pressure of the gas increases, what is the effect on the number of moles of gas.

27. What would the ideal gas law be if the following were true? (a) volume were proportional to pressure; (b) temperature were proportional to amount; (c) pressure were inversely proportional to temperature; (d) volume were inversely proportional to temperature; (e) both pressure and volume were inversely proportional to temperature?
28. Given the following initial and final values, what additional information is needed to solve the problem using the ideal gas law?

Given	Solve for
V_1, T_1, T_2, n_1	n_2
P_1, P_2, T_2, n_2	n_1
T_1, T_2	V_2
P_1, n_1	P_2

29. Given the following information and using the ideal gas law, what equation would you use to solve the problem?

Given	Solve for
P_1, P_2, T_1	T_2
V_1, n_1, n_2	V_2
T_1, T_2, V_1, V_2, n_2	n_1
P_1, P_2, V_1, n_1, n_2	V_2

30. Using the ideal gas law as a starting point, derive the relationship between the density of a gas and its molecular mass. Which would you expect to be denser: nitrogen or oxygen? Why does radon gas accumulate in basements and mine shafts?
31. Use the ideal gas law to derive an equation that relates the remaining variables for a sample of an ideal gas if the following are held constant: (a) amount and volume; (b) pressure and amount; (c) temperature and volume; (d) temperature and amount; (e) pressure and temperature.
32. Tennis balls that are made for Denver, Colorado, feel soft and do not bounce well at lower altitudes. Use the ideal gas law to explain this observation. Will a tennis ball designed to be used at sea level be harder or softer and bounce better or worse at high altitude?

10.5 Mixtures of Gases

33. Dalton's law of partial pressures makes one key assumption about the nature of the intermolecular interactions in a mixture of gases; what is it?
34. What is the relationship between the partial pressure of a gas and its mole fraction in a mixture?

10.6 Gas Volumes and Stoichiometry

35. Why are so many industrially important reactions carried out in the gas phase?
36. The volume of gas produced during a chemical reaction can be measured by collecting the gas in an inverted container filled with water. The gas forces the water out of the container, and the volume of liquid displaced is a measure of the volume of gas. What additional information must be considered to determine the number of moles of gas produced? The volume of some gases cannot be measured using this method; what property of a gas precludes the use of this method?
37. Equal masses of two solid compounds (**A** and **B**) are placed in separate sealed flasks filled with air at one atmosphere pressure and heated to 50°C for ten hours. After cooling to room temperature, the pressure in the flask containing **A** was 1.5 atm. In contrast, the pressure in the flask containing **B** was 0.87 atm. Suggest an explanation for these observations. Would the masses of samples **A** and **B** still be equal after the experiment? Why or why not?

10.7 The Kinetic Molecular Theory of Gases

38. Which of the following processes represents effusion, and which represents diffusion: helium escaping from a hole in a balloon; vapor escaping from the surface of a liquid; and gas escaping through a membrane?
39. Which postulate of the kinetic molecular theory of gases most readily explains the observation that a helium-filled balloon is round?
40. Why is it relatively easy to compress a gas? How does the compressibility of a gas compare with that of a liquid? Of a solid? Why? Which of the postulates of the kinetic molecular theory of gases most readily explains these observations?
41. What happens to the average kinetic energy of a gas if the rms speed of its particles increases by a factor of 2? How is the rms speed different from the average speed?
42. Which gas has a higher average kinetic energy at 100°C: radon or helium? Which has a higher average speed? Why? Which postulate of the kinetic molecular theory of gases most readily supports your answer?
43. What is the relationship between the average speed of a gas particle and the temperature of the gas? What happens to the distribution of molecular speeds if the temperature of a gas is increased? If the temperature of the gas is decreased?
44. Qualitatively explain the relationship between the number of collisions of gas particles with the walls of a container and the pressure of a gas. How does increasing the temperature affect the number of collisions?
45. What happens to the average kinetic energy of a gas at constant temperature if (a) the volume of the gas is increased, and (b) the pressure of the gas is increased?
46. What happens to the density of a gas at constant temperature if (a) the volume of the gas is increased, and (b) the pressure of the gas is increased?
47. Use the kinetic molecular theory to describe how a decrease in volume produces an increase in pressure at constant temperature. Similarly, explain how a decrease in temperature leads to a decrease in volume at constant pressure.
48. Graham's law is valid only if the two gases are at the same temperature. Why?
49. If we lived in a helium atmosphere rather than in air, would we detect odors more or less rapidly than we do now? Explain your reasoning. Would we detect odors more or less rapidly at sea level or at high altitude? Why?

10.8 The Behavior of Real Gases

50. What factors cause deviations from ideal gas behavior? Use a sketch to explain your answer based on interactions at the molecular level.
51. Explain the effect of non-zero atomic volume on the ideal gas law at high pressure. Sketch a graph of volume versus $1/P$ for an ideal gas and for a real gas.

52. For an ideal gas, the product of pressure and volume should be constant, regardless of the pressure. Experimental data for methane, however, show that the value of *PV* decreases significantly over the pressure range 0 to 120 atm at 0°C. The decrease in PV over the same pressure range is much smaller at 100°C. Explain why PV decreases with increasing temperature, and why the decrease is less significant at higher temperatures.

53. What is the effect of intermolecular forces on liquefaction of a gas? At constant pressure and volume, does it become easier or harder to liquefy a gas as its temperature increases? Explain your reasoning. What is the effect of increasing the pressure on the liquefaction temperature?

54. Describe qualitatively what the two empirical constants, a and b, in the van der Waals equation represent.

55. In the van der Waals equation, why is the term that corrects for volume negative and the term that corrects for pressure positive? Why is n/V squared?

56. Liquefaction of a gas depends strongly on two factors—what are they? As the temperature is decreased, which gas will liquefy first: ammonia, methane, or carbon monoxide? Why?

57. What is a cryogenic liquid? Describe three uses of cryogenic liquids.

58. Air consists primarily of O_2, N_2, Ar, Ne, Kr, and Xe. Use the concepts discussed in this chapter to propose two methods by which air can be separated into its components. Which component of air will be isolated first?

59. How can gas liquefaction facilitate the storage and transport of fossil fuels? What are potential drawbacks to these methods?

NUMERICAL

This section includes paired problems (marked by brackets) that require similar problem-solving skills.

10.2 Gas Pressure

60. Calculate the pressure in pascals and in atmospheres exerted by a carton of milk that weighs 1.5 kg and has a base of 7.0 cm × 7.0 cm. If the carton were lying on its side (height = 25 cm), would it exert more or less pressure? Explain your reasoning.

61. Calculate the pressure in atmospheres and kilopascals exerted by a fish tank that is 2.0 ft long, 1.0 ft wide, and 2.5 ft high and contains 25.0 gal of water in a room that is at 20°C; the tank itself weighs 15 lb (d_{H_2O} = 1.00 g/cm^3 at 20°C). If the tank were 1 ft long, 1 ft wide, and 5 ft high, would it exert the same pressure? Explain your answer.

62. If atmospheric pressure at sea level is 1.0 × 10^5 Pa, what is the mass of air in kilograms above a 1.0-cm^2 area of your skin as you lie on the beach? If atmospheric pressure is 8.2 × 10^4 Pa on a mountaintop, what is the mass of air in kilograms above a 4.0-cm^2 patch of skin?

63. Complete the table:

atm	kPa	mmHg	torr
1.40	_____	_____	_____
_____	_____	723	_____
_____	43.2	_____	_____

64. The SI unit of pressure is the pascal, which is equal to one N/m^2. Show how the product of the mass of an object and the acceleration due to gravity result in a force that, when exerted on a given area, leads to a pressure in the correct SI units. What mass in kilograms applied to a 1.0-cm^2 area is required to produce a pressure of: (a) 1.0 atm, (b) 1.0 torr, (c) 1 mmHg, and (d) 1 kPa?

65. If you constructed a manometer to measure gas pressures over the range 0.60–1.40 atm using the liquids given below, how tall a column would you need for each liquid? The density of mercury is 13.5 g/cm^3. Based on your results, explain why mercury is still used in barometers, despite its toxicity.

	Liquid Density (20°C)	Column Height, m
Isopropanol	0.785	_____
Coconut oil	0.924	_____
Glycerine	1.259	_____

10.3 Relationships Between Pressure, Temperature, Amount, and Volume

66. A 1.00-mol sample of gas is at a temperature of 300 K and a pressure of 4.11 atm. What is the volume of the gas under these conditions? The sample is compressed to 6.0 atm at constant temperature, giving a volume of 3.99 L. Is this result consistent with Boyle's law?

67. A 1.00-mol sample of gas at STP has an initial volume of 22.4 L. Calculate the results of the following changes, assuming all other conditions remain constant.
(a) The pressure is changed to 85.7 mmHg. How many milliliters does the gas occupy?
(b) The volume is reduced to 275 mL. What is the pressure in mmHg?
(c) The pressure is increased to 25.3 atm. What is the temperature in °C?
(d) The sample is heated to 25°C. What is the volume in liters?
(e) The sample is compressed to 1255 mL, and the pressure is increased to 2555 torr. What is the temperature of the gas in K?

10.4 The Ideal Gas Law

68. Determine the number of moles in each sample at STP: (a) 2200 cm^3 of CO_2; (b) 1200 cm^3 of N_2; (c) 3800 mL of SO_2; (d) 13.75 L of NH_3.

69. Calculate the number of moles in each sample at STP: (a) 1580 mL of NO_2; (b) 847 cm^3 of HCl; (c) 4.792 L of H_2; (d) a 15.0 cm × 6.7 cm × 7.5 cm container of ethane.

70. Calculate the mass of each of the following at STP: (a) 3.2 L of N_2O; (b) 65 cm^3 of Cl_2; (c) 3600 mL of HBr.

71. Determine the mass of each of the following at STP: (a) 36 mL of HI; (b) 550 L of H_2S; (c) 1380 cm^3 of CH_4.

72. Calculate the volume in liters of each of the following at STP: (a) 3.2 g of Xe; (b) 465 mg of CS_2; (c) 5.34 kg of acetylene, C_2H_2.

73. Determine the volume in liters of each of the following at STP: (a) 1.68 g of Kr; (b) 2.97 kg of propane, C_3H_8; (c) 0.643 mg of $(CH_3)_2O$.

74. Determine the volume of each of the following gas samples at STP: (a) 2.30 L at 23°C and 740 mmHg; (b) 320 mL at 13°C and 97.2 kPa; (c) 100.5 mL at 35°C and 1.4 atm.

75. Determine the volume of each of the following gas samples at STP: (a) 1.7 L at 28°C and 96.4 kPa; (b) 38.0 mL at 17°C and 103.4 torr; (c) 650 mL at −15°C and 723 mmHg.

76. At a constant temperature, what pressure in atmospheres is needed to compress 14.2 L of gas initially at 25.2 atm to a volume of 12.4 L? What pressure is needed to compress 27.8 L of gas to 20.6 L under similar conditions?

77. A 8.60-L tank of nitrogen gas at a pressure of 455 mmHg is connected to an empty tank with a volume of 5.35 L. What is the final pressure in the system after the valve connecting the two tanks is opened? Assume that the temperature is constant.

78. An 3.50-g sample of acetylene is burned in excess oxygen according to the reaction

$$2C_2H_2(g) + 5O_2(g) \longrightarrow 4CO_2(g) + 2H_2O(l)$$

At STP, what volume of $CO_2(g)$ is produced?

79. One method for preparing hydrogen gas is to pass HCl gas over hot aluminum; the other product of the reaction is $AlCl_3$. If you wanted to use this reaction to fill a balloon with a volume of 28,500 L at sea level and a temperature of 78°F, what mass of aluminum would you need? What volume of HCl at STP would you need?

80. Determine the density of O_2 under each of the following sets of conditions: (a) 42 g at 1.1 atm and 25°C; (b) 0.87 mol at 820 mmHg and 45°C; and (c) 16.7 g at 2.4 atm and 67°C.

81. Calculate the density of ethylene, C_2H_4, under each of the following sets of conditions: (a) 7.8 g at 0.89 atm and 26°C; (b) 6.3 mol at 102.6 kPa and 38°C; and (c) 9.8 g at 3.1 atm and −45°C.

82. At 140°C, the pressure of a diatomic gas in a 3.0-L flask is 635 kPa. The mass of the gas is 88.7 g. What is the most likely identity of the gas?

83. What must be the volume of a balloon that can hold 313.0 g of helium gas and ascend from sea level to an elevation of 1.5 km, where the temperature is 10.0°C and the pressure is 635.4 mmHg?

84. What volume must a balloon have to be able to hold 6.20 kg of H_2 for an ascent from sea level to an elevation of 20,320 ft, where the temperature is −37°C and the pressure is 369 mmHg?

85. A typical automobile tire is inflated to a pressure of 28.0 $lb/in.^2$ Assume that the tire is inflated when the air temperature is 20°C, and the car is then driven at high speeds, which increases the temperature of the tire to 43°C. What is the pressure in the tire? If the volume of the tire had increased by 8% at the higher temperature, what would the pressure be?

86. The average respiratory rate for adult humans is 20 breaths per minute. If each breath has a volume of 310 mL of air at 20°C and 0.997 atm, how many moles of air does a person inhale each day? If the density of air is 1.19 kg/m^3, what is the average molecular mass of air?

87. Kerosene has a self-ignition temperature of 255°C. It is a common accelerant used by arsonists, but its presence is easily detected in fire debris by a variety of methods. If a 1.0-L glass bottle containing a mixture of air and kerosene vapor at an initial pressure of 1 atm and an initial temperature of 23°C is pressurized, at what pressure would the kerosene vapor ignite?

10.5 Mixtures of Gases

88. Determine the partial pressure of each gas in the following 3.0-L mixtures at 37°C, as well as the total pressure: (a) 0.128 mol of SO_2 and 0.098 mol of CH_4; (b) 3.40 g of acetylene, C_2H_2, and 1.54 g of He; and (c) 0.267 g of NO, 4.3 g of Ar, and 0.872 g of SO_2.

89. What is the partial pressure of each gas if the following amounts of substances are placed in a 25.0 L container at 25°C: (a) 1.570 mol of methane, CH_4, and 0.870 mol of carbon dioxide; (b) 2.63 g of CO and 1.24 g of NO_2; and (c) 1.78 kg of CH_3Cl and 0.92 kg of SO_2? What is the total pressure of each mixture?

90. A 2.00-L flask originally contains 1.00 g of ethane, C_2H_6, and 32.0 g of oxygen at 21°C. Upon ignition, the ethane reacts completely with oxygen to produce CO_2 and water vapor, and the temperature of the flask increases to 200°C. Determine the total pressure and the partial pressure of each gas before and after the reaction.

91. In a mixture of helium, oxygen, and methane in a 2.00-L container, the partial pressures of He and O_2 are 13.6 kPa and 29.2 kPa, respectively, and the total pressure inside the container is 95.4 kPa. What is the partial pressure of methane? If the methane is ignited to initiate its combustion with oxygen and the system is then cooled to the original temperature of 30°C, what is the final pressure inside the container (in kPa)?

92. If a 20.0-L cylinder at 19°C is charged with 5.0 g each of sulfur dioxide and oxygen, what is the partial pressure of each gas? The sulfur dioxide is ignited in the oxygen to produce sulfur trioxide gas, and the mixture is allowed to cool to 19°C at constant pressure. What is the final volume of the cylinder? What is the partial pressure of each gas in the piston?

93. The highest point on the continent of Europe is Mt. Elbrus in Russia, with an elevation of 18,476 ft. The highest point on the continent of South America is Mt. Aconcagua in Argentina, with an elevation of 22,841 ft.

(a) The table shows the variation of atmospheric pressure with elevation. Use the data in the table to construct a plot of pressure versus elevation.

Elevation, km	Pressure in Summer, mmHg	Pressure in Winter, mmHg
0.0	760.0	760.0
1.0	674.8	670.6
1.5	635.4	629.6
2.0	598.0	590.8
3.0	528.9	519.7
5.0	410.6	398.7
7.0	314.9	301.6
9.0	237.8	224.1

(b) Use your graph to estimate the pressures in mmHg during the summer and the winter at the top of both mountains in both atmospheres and kilopascals.

(c) Given that air is 20.95% O_2 by volume, what is the partial pressure of oxygen in atmospheres during the summer at each location?

10.6 Gas Volumes and Stoichiometry

94. Balance each of the following chemical equations, and then determine the volume of the indicated reactant at STP required for complete reaction. Assuming complete reaction, what is the volume of the products?

(a) $SO_2(g) + O_2(g) \longrightarrow SO_3(g)$ given 2.4 mol of O_2
(b) $H_2(g) + Cl_2(g) \longrightarrow HCl(g)$ given 0.78 g of H_2
(c) $C_2H_6(g) + O_2(g) \longrightarrow CO_2(g) + H_2O(g)$ given 1.91 mol of O_2

95. Complete decomposition of a sample of potassium chlorate produced 1.34 g of potassium chloride and oxygen gas. Determine:
(a) the mass of $KClO_3$ in the original sample;
(b) the mass of oxygen produced;
(c) the volume of oxygen produced at STP.

96. During the smelting of iron, carbon reacts with oxygen to produce carbon monoxide, which then reacts with iron(III) oxide to produce iron metal and carbon dioxide. If 1.82 L of CO_2 at STP is produced, calculate: (a) the mass of CO consumed; (b) the volume of CO at STP consumed; (c) the volume of O_2 at STP utilized; (d) the mass of carbon consumed; (e) the mass of iron metal produced.

97. Combustion of a 300.0-mg sample of an antidepressant in excess oxygen produced 326 ml of CO_2 and 164 ml of H_2O vapor at STP. Separate analysis shows that the sample contained 23.28% oxygen. If the sample is known to contain only C, H, O, and N, determine the percent composition and the empirical formula of the antidepressant.

98. Combustion of a 100.0-mg sample of an herbicide in excess oxygen produced 83.16 ml of CO_2 and 72.9 ml of H_2O vapor at STP. Separate analysis shows that the sample contained 16.44 mg of chlorine. If the sample is known to contain only C, H, Cl, and N, determine the percent composition and empirical formula of the herbicide.

10.7 The Kinetic Molecular Theory of Gases

99. At a given temperature, what is the ratio of the rms speed of the atoms of Ar gas to the rms speed of H_2 gas?

100. At a given temperature, what is the ratio of the rms speed of molecules of CO to the rms speed of molecules of H_2S gas?

101. What is the ratio of the rms speeds of argon and oxygen at any temperature? Which diffuses more rapidly?

102. What is the ratio of the rms speeds of Kr and NO at any temperature? Which diffuses more rapidly?

103. Tritium, T, is a heavy isotope of hydrogen with an atomic mass of 3.016 amu. The natural abundance of tritium is 0.000138%. Effusion of hydrogen gas (containing a mixture of H_2, HD, and HT molecules) through a porous membrane can be used to obtain samples of hydrogen that are enriched in tritium. How many membrane passes are necessary to give a sample of hydrogen gas in which 1% of the hydrogen molecules are HT?

104. Samples of HBr gas and NH_3 gas are placed at opposite ends of a 1-m tube. If the two gases are allowed to diffuse through the tube toward one another, at what distance from each end of the tube will the gases meet and form solid NH_4Br?

10.8 The Behavior of Real Gases

105. The van der Waals constants for xenon are $a = 4.19$ $(L^2 \cdot atm)/mol^2$ and $b = 0.0510$ L/mol. If a 0.250-mol sample of xenon with a volume of 3.65 L is cooled to $-90°C$, what is the pressure of the sample assuming ideal gas behavior? What would be the *actual* pressure under these conditions?

106. The van der Waals constants for water vapor are $a = 5.46$ $(L^2{\cdot}atm)/mol^2$ and $b = 0.0305$ L/mol. If a 20.0-g sample of water is heated to 120°C in a container with a volume of 5.0 L, what is the pressure of the sample, assuming ideal gas behavior? What would be the *actual* pressure under these conditions?

APPLICATIONS

107. Oxalic acid, $C_2H_2O_4$, is a metabolic product of many molds. Although oxalic acid is toxic to humans if ingested, many plants and vegetables contain significant amounts of oxalic acid or oxalate salts. In solution, oxalic acid can be oxidized by air via the equation

$$H_2C_2O_4(aq) + O_2(g) \longrightarrow H_2O_2(l) + 2CO_2(g)$$

If a plant metabolized enough oxalic acid to produce 3.2 L of CO_2 on a day when the temperature was 29°C and the pressure was 752 mmHg, how many grams of oxalic acid were converted to CO_2? Given that air is 21% oxygen, what volume of air was needed for the oxidation?

108. The decomposition of iron oxide is used to produce gas during the manufacture of porous, expanded materials. These materials have very low densities due to the swelling that occurs during the initial rapid heating. Consequently, they are used as additives to provide insulation in concrete, road building, and other construction materials. Iron oxide decomposes at 1150°C according to the equation

$$6Fe_2O_3(s) \longrightarrow 4Fe_3O_4(s) + O_2(g)$$

(a) Explain how this reaction could cause materials containing Fe_2O_3 to swell upon heating.
(b) If you begin with 15.4 g of Fe_2O_3, what volume of O_2 gas at STP is produced?
(c) What is the volume of the same amount of O_2 gas at 1200°C, assuming a constant pressure of 1.0 atm?
(d) If 6.3 L of gas was produced at 1200°C and 1 atm, how many kilograms of Fe_2O_3 were initially used in the reaction? How many kilograms of Fe_3O_4 are produced?

109. A 70-kg man expends 480 kcal of energy per hour shoveling snow. The oxidation of organic nutrients such as glucose during metabolism liberates approximately 3.36 kcal of energy per gram of oxygen consumed. If air is 21% oxygen, what volume of air at STP is needed to produce enough energy for the man to clear snow from a walkway that requires 35 minutes of shoveling?

110. Calcium carbonate is an important filler in the processing industry. Its many uses include a reinforcing agent for rubber and improving the whiteness and hiding power of paints. When calcium nitrate is used as a starting material in the manufacture of fertilizers, calcium carbonate is produced according to the following reaction:

$$Ca(NO_3)_2 + 2NH_3 + CO_2 + H_2O \longrightarrow CaCO_3 + 2NH_4NO_3$$

(a) What volume of CO_2 at STP is needed to react completely with 28.0 g of calcium nitrate?
(b) What volume of ammonia at STP is needed to react with the amount of calcium nitrate?
(c) If this reaction were carried out in Denver, Colorado (pressure = 630 mmHg), what volumes of CO_2 and NH_3 at room temperature (20°C) would be needed?

111. Calcium nitrate used in the process described in Problem 110 is produced by the reaction of fluoroapatite, $Ca_5[(PO_4)_3(F)]$, with nitric acid:

$$Ca_5[(PO_4)_3(F)] + 10HNO_3 \longrightarrow 5Ca(NO_3)_2 + HF + 3H_3PO_4$$
$$Ca(NO_3)_2 + 2NH_3 + CO_2 + H_2O \longrightarrow CaCO_3 + 2NH_4NO_3$$

(a) Your lab is in Denver, Colorado, and you have a cylinder that contains 8.40 L of CO_2 at room temperature (20°C) and 4.75 atm pressure. What would be the volume of the CO_2 gas at a pressure of 630 mmHg? How many grams of fluoroapatite could be converted to $CaCO_3$ using the two reactions and this amount of CO_2?

(b) If 3.50-L of a 0.753 *M* HF solution was produced during the conversion of fluoroapatite to calcium nitrate, how many grams of calcium carbonate could you produce, assuming 100% efficiency?

(c) How many liters of ammonia gas at this higher altitude and 20°C are required to convert the calcium nitrate to calcium carbonate *and* to neutralize all of the HF solution?

(d) If the reaction of fluoroapatite with nitric acid takes place at 330 atm and 80°C, what volumes of NH_3 and CO_2 are needed to convert 20.5 kg of fluoroapatite to calcium carbonate?

112. Mars has an average temperature of −47°C, a surface pressure of 500 Pa, and an atmosphere that is 95% carbon dioxide, 3% nitrogen, and 2% argon by mass, with traces of other gases. What is the partial pressure (in atm) of each gas in this atmosphere? A 5.0-L sample is returned to Earth and stored in a laboratory at 19°C and one atmosphere. What is the volume of this sample?

113. Chlorofluorocarbons (CFCs) are inert substances that were long used as refrigerants. Because CFCs are inert, when they are released into the atmosphere they are not rapidly destroyed in the lower atmosphere. Instead, they are carried into the stratosphere, where they cause ozone depletion. A method for destroying CFC stockpiles passes the CFC through packed sodium oxalate, $Na_2C_2O_4$, powder at 270°C. The reaction for Freon-12, CF_2Cl_2, is

$$CF_2Cl_2(g) + 2Na_2C_2O_4(s) \longrightarrow 2NaF(s) + 2NaCl + C(s) + 4CO_2(g)$$

(a) If this reaction produced 11.4 L of CO_2 gas at 21°C and 752 mmHg, what mass of sodium oxalate was consumed in the reaction?

(b) If you had to design a reactor to carry out the reaction at a maximum safe pressure of 10.0 atm while destroying 1.0 kg of Freon-12, what volume reactor would you need?

(c) A 2.50-L reaction vessel is charged with 20.0 atm of Freon-12 and excess sodium oxalate at 20°C. The temperature is increased to 270°C, and the pressure is monitored as the reaction progresses. What is the initial pressure at 270°C, and what is the final pressure when the reaction has gone to completion?

114. The exhaust from a typical six-cylinder car contains the following average compositions of carbon monoxide and carbon dioxide under different conditions (data reported as percent by volume; rpm = rotations per minute):

Species	Idling, 1000 rpm	Accelerating, 4000 rpm	Decelerating, 800 rpm
CO	1.0	1.2	0.60
CO_2	0.80	0.40	0.40

(a) What are the mole fractions of CO and CO_2 under each set of conditions?

(b) If the engine takes in 4.70 L of air at STP mixed with fine droplets of gasoline, C_8H_{18}, with each rotation, how many grams of gasoline are burned per minute?

115. Automobile airbags inflate by the decomposition of sodium azide, NaN_3, which produces sodium metal and nitrogen gas according to the equation

$$2NaN_3 \longrightarrow 2Na(s) + 3N_2(g)$$

How many grams of sodium azide are needed to inflate a 15.0-L airbag at 20°C and 760 mmHg? The density of NaN_3 is 1.847 g/cm^3. What is the volume of the gas produced compared to the solid reactant?

116. Under basic conditions, the reaction of hydrogen peroxide, H_2O_2, and potassium permanganate, $KMnO_4$, produces oxygen and MnO_2. During a laboratory exercise, you carefully weighed out your sample of $KMnO_4$. Unfortunately, however, you lost your data just before mixing the $KMnO_4$ with an H_2O_2 solution of unknown concentration. Devise a method to determine the mass of your sample of $KMnO_4$ using excess H_2O_2.

117. Carbonated beverages are pressurized with CO_2. In an attempt to produce another bubbly soda beverage, an intrepid chemist attempted to use three other gases: He, N_2, and Xe. Rank the four beverages in order of how fast the drink would go "flat," and explain your reasoning. Which beverage would have the shortest shelf life (that is, how long will an *unopened* bottle still be good)? Explain your answer.

118. Urea is synthesized industrially by the reaction of ammonia and carbon dioxide to produce ammonium carbamate, followed by dehydration of ammonium carbamate to give urea and water. This process is shown in the following set of equations:

$$2NH_3(g) + CO_2(g) \longrightarrow NH_2CO_2NH_4(s)$$
$$NH_2CO_2NH_4(s) \longrightarrow NH_2CONH_2(s) + H_2O(g)$$

A 50.0-L reaction vessel is charged with 2.5 atm of ammonia and 2.5 atm of CO_2 at 50°C, and the vessel is then heated to 150°C. What is the pressure in the vessel when the reaction has gone to completion? If the vessel is then cooled to 20°C, what is the pressure? An aqueous solution of urea and water is drained from the reaction vessel. What is the molarity of the urea solution? Industrially, this reaction is actually carried out at pressures ranging from 130 to 260 atm and temperatures of approximately 180°C. Give a plausible reason for using these extreme conditions.

119. Explain what happens to the temperature, volume, or pressure of a gas during each of the following operations, and give the direction of heat flow, if any.

(a) The gas is allowed to expand from V_1 to V_2; a heat transfer occurs to maintain a constant gas temperature.

(b) The gas is allowed to expand from V_2 to V_3 with no concomitant heat transfer.

(c) The gas is compressed from V_3 to V_4; a heat transfer occurs to maintain a constant gas temperature.

(d) The gas is compressed from V_4 to V_1; no heat transfer occurs.

These four processes constitute the cycle used in refrigeration, in which a gas such as Freon is alternately compressed and allowed to expand in the piston of a compressor. Which step eventually causes the food in a refrigerator to cool? Where does the thermal energy go that was removed in the cooling process?

ESSENTIAL SKILLS

5

TOPICS

- Preparing a Graph
- Interpreting a Graph

Previous Essential Skills sections presented the fundamental mathematical operations you need to know to solve problems by manipulating chemical equations. This section describes how to prepare and interpret graphs, two other skills that chemistry students must have to understand concepts and solve problems.

Preparing a Graph

A *graph* is a pictorial representation of a mathematical relationship. It is an extremely effective tool for understanding and communicating the relationship between two or more variables. Each axis is labeled with the name of the variable to which it corresponds, along with the unit in which the variable is measured, and each axis is divided by tic marks or grid lines into segments that represent those units (or multiples). The scale of the divisions should be chosen so that the plotted points are distributed across the entire graph. Whenever possible, data points should be combined with a bar that intersects the data point and indicates the range of error of the measurement, although for simplicity the bars are frequently omitted in undergraduate textbooks. Lines or curves that represent theoretical or computational results are drawn using a "best-fit" approach. That is, data points are not connected as a series of straight-line segments; rather, a smooth line or curve is drawn that provides the best fit to the plotted data.

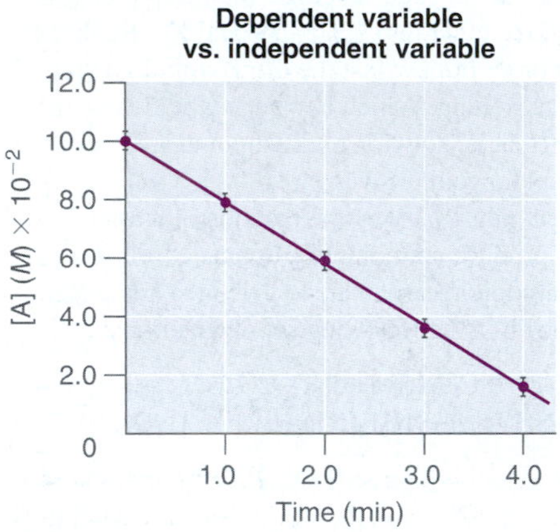

Interactive Graph

The independent variable is usually assigned to the horizontal, or x, axis (also called the *abscissa*), and the dependent variable to the vertical, or y, axis (called the *ordinate*). Let's examine, for example, an experiment in which we are interested in plotting the change in the concentration of compound A with time. Because time does not depend on the concentration of A but the concentration of A does depend on the amount of time that has passed during the reaction, time is the independent variable and concentration is the dependent variable. In this case, the time is assigned to the horizontal axis and the concentration of A to the vertical axis.

We may plot more than one dependent variable on a graph, but the lines or curves corresponding to each set of data must be clearly identified with labels, different types of lines (a dashed line, for example), or different symbols for the respective data points (a triangle versus a circle). When words are used to label a line or curve, either a key identifying the different sets of data or a label placed next to each line or curve is used.

Interpreting a Graph

Two types of graphs are frequently encountered in beginning chemistry courses: linear and log-linear. We describe each of these types.

Linear graphs

In a *linear graph*, the plot of the relationship between the variables is a straight line and thus can be expressed by the equation for a straight line:

$$y = mx + b$$

where m is the slope of the line and b is the point where the line intersects the vertical axis (where $x = 0$), called the *y intercept*. The slope is calculated using the formula

$$m = \frac{y_2 - y_1}{x_2 - x_1} = \frac{\Delta y}{\Delta x} = \frac{\text{rise}}{\text{run}}$$

For accuracy, two widely separated points should be selected for use in the formula to minimize the effects of any reporting or measurement errors that may have occurred in any given region of the graph. For example, when concentrations are measured, limitations in the sensitivity of an instrument as well as human error may result in measurements being less accurate for samples with low concentrations than for those that are more concentrated. The graph of the change in the concentration of A with time is an example of a linear graph. The key features of a linear plot are shown on the generic example in the margin.

It is important to remember that when a graphical procedure is used to calculate a slope, the scale on each axis must be of the same order (they must have the same exponent). For example, although acceleration is a change in velocity over time ($\Delta v/\Delta t$), the slope of a linear plot of velocity versus time only gives the correct value for acceleration (m/s^2) if the average acceleration over the interval and the instantaneous acceleration are identical. That is, the acceleration must be constant over the same interval.

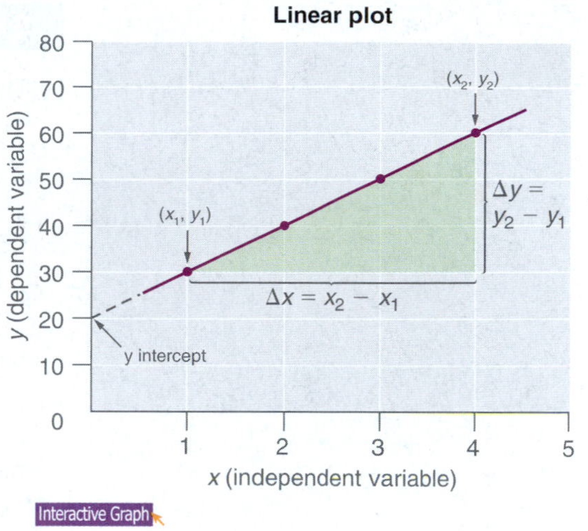

Log-Linear graphs

A log-linear plot is a representation of the following general mathematical relationship:

$$y = Ac^{mx}$$

Here, y is equal to some value Ac when $x = 0$. As described in Essential Skills 3, taking the logarithm of both sides produces

$$\log y = \log A + mx \log c = (m \log c)\, x + \log A$$

When expressed in this form, the equation is that of a straight line ($y = mx + b$), where the plot of y is on a logarithmic axis and $(m \log c)x$ is on a linear axis. This type of graph is known as a *log-linear plot*. Log-linear plots are particularly useful for graphing changes in pH versus changes in the concentration of another substance. One example of a log-linear plot, where $y = [HA]$ and $x = pH$, is shown in the margin.

From the linear equation, notice that a log-linear plot has a y intercept of $\log A$, and so the value of A may be obtained directly from the plot *if the x axis begins at 0* (note that in this case it does not). Using our example, however, we can calculate [HA] at the y intercept first by calculating the slope of the line using any two points and the equation $(\log[HA]_2 - \log[HA]_1)/(pH_2 - pH_1)$:

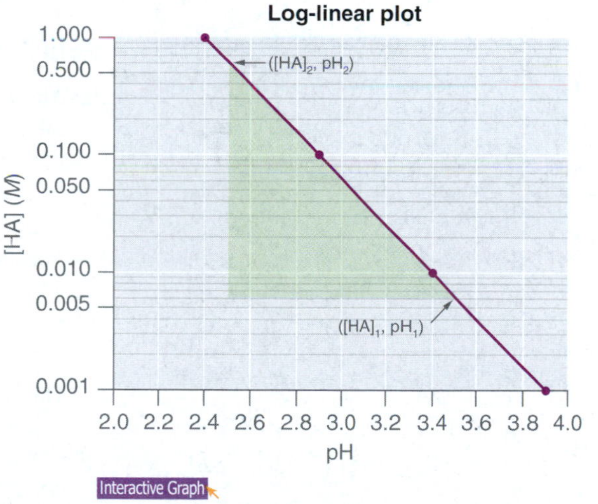

$$[HA_2] = 0.600 \qquad [HA_1] = 0.006$$
$$\log[HA_2] = -0.222 \qquad \log[HA_1] = -2.222$$
$$pH_2 = 2.5 \qquad pH_1 = 3.5$$
$$m \log c = \frac{-0.222 - (-2.222)}{2.5 - 3.5} = \frac{2.000}{-1.0} = -2.0$$

Using any point along the line (for example, $[HA] = 0.100$, $pH = 2.9$), we can then calculate the y intercept ($pH = 0$):

$$\log[HA] = -2.0\text{pH} + b$$
$$-1.00 = [(-2.0)(2.9)] + b$$
$$b = 4.8$$

Thus, at a pH of 0.0, $\log[HA] = 4.8$ and $[HA] = 6.31 \times 10^4\ M$. The exercises provide practice in drawing and interpreting graphs.

SKILL BUILDER ES5.1

(a) The absorbance of light by various aqueous solutions of phosphate was measured and tabulated as follows:

Absorbance (400 nm)	$[PO_4^{3-}]$, mol/L
0.16	3.2×10^{-5}
0.38	8.4×10^{-5}
0.62	13.8×10^{-5}
0.88	19.4×10^{-5}

Graph the data with the dependent variable on the *y* axis and the independent variable on the *x* axis, and then calculate the slope. If a sample has an absorbance of 0.45, what is the phosphate concentration in the sample?

(b) The table lists the conductivity of three aqueous solutions with varying concentrations. Create a plot from these data, and then predict the conductivity of each sample at a concentration of 15.0×10^3 ppm.

	0 ppm	5.00×10^3 ppm	10.00×10^3 ppm
K_2CO_3	0.0	7.0	14.0
Seawater	0.0	8.0	15.5
Na_2SO_4	0.0	6.0	11.8

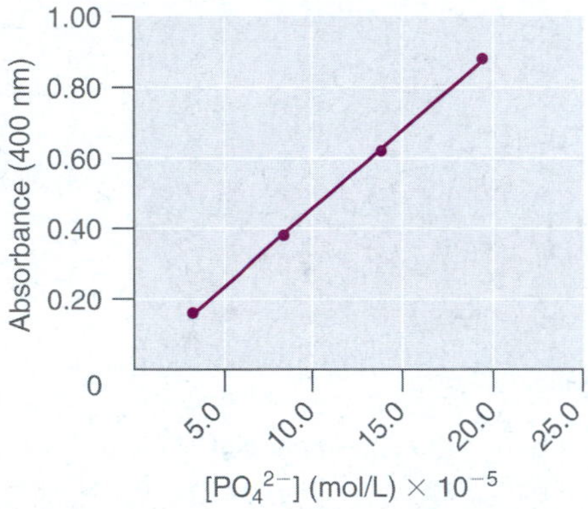

(a)

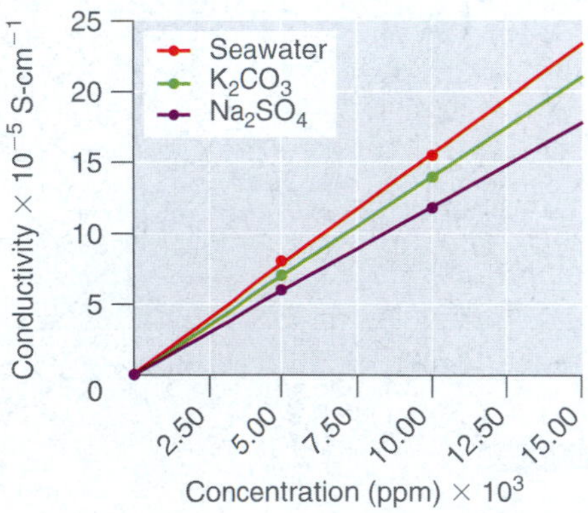

(b)

Solution

(a) Absorbance is the dependent variable, and concentration is the independent variable. We calculate the slope using two widely separated data points:

$$m = \frac{\Delta y}{\Delta x} = \frac{0.62 - 0.16}{13.8 \times 10^{-5} - 3.2 \times 10^{-5}} = 4.3 \times 10^3$$

According to our graph, the *y*-intercept, *b*, is 0.00. Thus, when $y = 0.45$,

$$0.45 = (4.3 \times 10^3)(x) + 0.00$$
$$x = 10 \times 10^{-5}$$

This is in good agreement with a graphical determination of the phosphate concentration at an absorbance of 0.45, which gives a value of 10.2×10^{-5} mol/L.

(b) Conductivity is the dependent variable, and concentration is the independent variable. From our graph, at 15.0×10^3 ppm, the conductivity of K_2CO_3 is predicted to be 21; that of seawater, 24; and that of Na_2SO_4, 18.

11 Liquids

LEARNING OBJECTIVES

- To be familiar with the kinetic molecular description of liquids
- To be able to describe the intermolecular forces in liquids
- To be able to describe the unique properties of liquids
- To know how and why the vapor pressure of a liquid varies with temperature
- To be able to calculate the energy changes that accompany phase changes
- To know what is meant by the critical temperature and pressure of a liquid
- To understand the general features of a phase diagram

CHAPTER OUTLINE

In Chapter 10, you learned that attractive intermolecular forces cause most gases to condense to a liquid at high pressure, low temperature, or both. Many substances occur normally as liquids, and are held together by exactly the same forces that are responsible for the liquefaction of gases. One such substance is water, the solvent in which all biochemical reactions take place. Among its many important roles, water, because of its thermal properties, modulates the temperature of Earth, maintaining a temperature range suitable for life. Other liquids are used to manufacture objects that we use every day; a solid material is converted to a liquid, the liquid is injected into a mold, and it is then solidified into complex shapes under conditions that are carefully controlled. In order to understand such processes, our study of the macroscopic properties of matter must include an understanding of the properties of liquids and of the interconversion of the three states of matter: gases, liquids, and solids.

Water beading up on the surface of a freshly waxed car. The waxed, nonpolar surface does not interact strongly with the polar water molecules. The absence of attractive interactions causes the water to form round beads.

In this chapter, we look more closely at the intermolecular forces that are responsible for the properties of liquids, describe some of the unique properties of liquids compared with the other states of matter, and then consider changes in state between liquids and gases or solids. By the end of the chapter, you will understand what is happening at the molecular level when you dry yourself with a towel, why you feel cold when you come out of the water, why ice is slippery, and how it is possible to decaffeinate coffee without removing important flavor components. You will also learn how LCD devices in calculators and digital watches function, and how adhesive strips used to measure body temperature change color to indicate a fever.

11.1 The Kinetic Molecular Description of Liquids

MGC Phases of Matter

The kinetic molecular theory described in Chapter 10 gives a reasonably accurate description of the behavior of gases. A similar model can be applied to liquids, but it must take into account the nonzero volumes of particles and the presence of strong intermolecular attractive forces.

In a gas, the distance between molecules is very large compared with the size of the molecules; thus, gases have a low density and are highly compressible. In contrast, the molecules in liquids are very close together, with essentially no empty space between them. As in gases, however, the molecules in liquids are in constant motion, and their kinetic energy (and hence their speed) depends on their temperature. We begin our discussion by examining some of the characteristic properties of liquids to see how each is consistent with a modified kinetic molecular description.

Density

The molecules of a liquid are packed relatively close together. Consequently, liquids are much denser than gases. The density of a liquid is typically about the same as the density of the solid state of the substance. Densities of liquids are therefore more commonly measured in units of grams per cubic centimeter (g/cm^3) or grams per milliliter (g/mL) than in grams per liter (g/L), the units commonly used for gases.

Molecular Order

Liquids exhibit short-range order because strong intermolecular attractive forces cause the molecules to pack together rather tightly. Because of their higher kinetic energy compared to the molecules in a solid, however, the molecules in a liquid move rapidly with respect to one another. Thus, unlike the ions in the ionic solids discussed in Section 8.2, the molecules in liquids are not arranged in a repeating three-dimensional array. Unlike the molecules in gases, however, the arrangement of the molecules in a liquid is not completely random.

Compressibility

Liquids have so little empty space between the component molecules that they cannot be readily compressed. Compression would force the atoms on adjacent molecules to occupy the same region of space.

Thermal Expansion

The intermolecular forces in liquids are strong enough to keep them from expanding significantly when heated (typically only a few percent over a 100°C temperature range). Thus, the volumes of liquids are somewhat fixed.

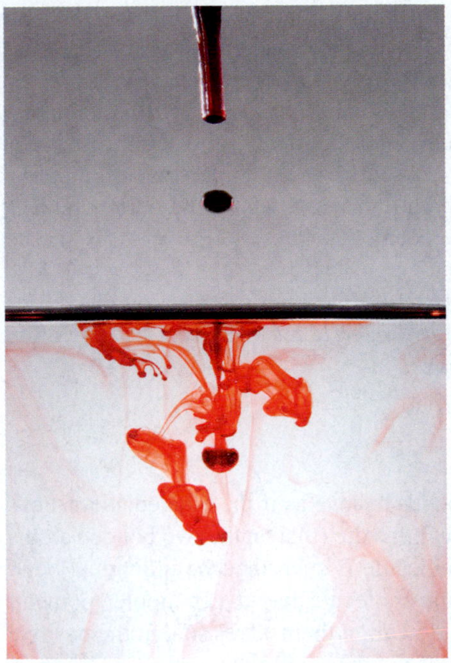

Figure 11.1 Molecular diffusion in a liquid. A small amount of an aqueous solution containing a marker dye has been poured into a larger volume of water. Note how the color of the dye is fainter at the edges as it diffuses.

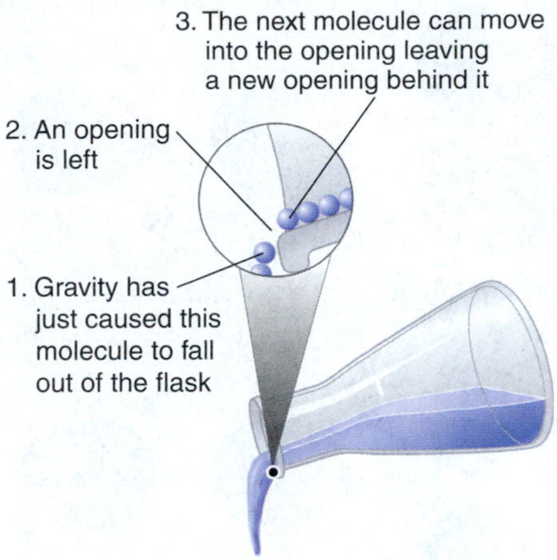

Figure 11.2 Why liquids flow. Molecules in a liquid are in constant motion. Consequently, when the flask is tilted, molecules move to the left and down due to the force of gravity, and the openings are occupied by other molecules. The result is a net flow of liquid out of the container.

Diffusion

Molecules in liquids diffuse because they are in constant motion (Figure 11.1). A molecule in a liquid cannot move far before colliding with another molecule, however, so the mean free path in liquids is very short and the rate of diffusion much slower than in gases.

Fluidity

Liquids can flow, adjusting to the shape of the container, because their molecules are free to move. This freedom of motion and their close spacing allow the molecules in a liquid to move rapidly into the openings left by other molecules, in turn generating more openings, and so on (Figure 11.2).

11.2 ○ Intermolecular Forces

MGC Intermolecular Forces in Liquids

The properties of liquids are intermediate between those of gases and solids, but more similar to solids. In contrast to *intra*molecular forces, such as the covalent bonds that hold atoms together in molecules and polyatomic ions, *inter*molecular forces hold molecules together in a liquid or solid. Intermolecular forces are generally much weaker than covalent bonds. For example, it requires 927 kJ to overcome the intramolecular forces and break both O—H bonds in 1 mol of water, but it takes only about 41 kJ to overcome the intermolecular attractions and convert 1 mol of liquid water to water vapor at 100°C. (Despite this seemingly low value, the intermolecular forces in liquid water are among the strongest such forces known!) Due to the large difference in the strengths of intra- and intermolecular forces, changes between the solid, liquid, and gaseous states almost invariably occur for molecular substances *without breaking covalent bonds.*

Note the pattern

The properties of liquids are intermediate between those of gases and solids, but more similar to solids.

Intermolecular forces determine bulk properties such as the melting points of solids and the boiling points of liquids. Liquids boil when the molecules have enough thermal energy to overcome the intermolecular attractive forces that hold them together, thereby forming bubbles of vapor within the liquid. Similarly, solids melt when the molecules acquire enough thermal energy to overcome the intermolecular forces that lock them into place in the solid.

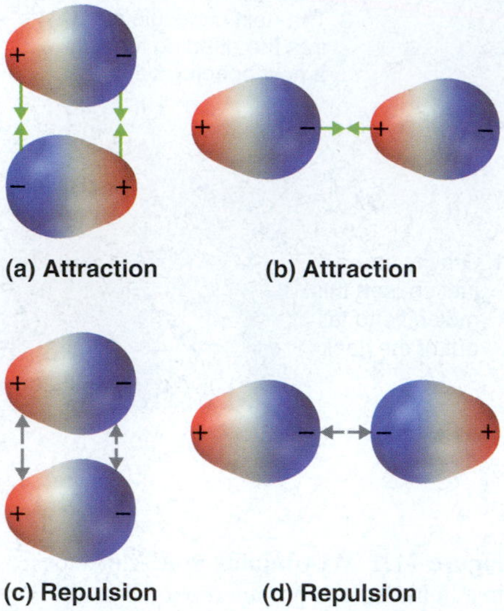

Figure 11.3 Attractive and repulsive dipole–dipole interactions. (a, b) Molecular orientations in which the positive end of one dipole is near the negative end of another (and vice versa) produce attractive interactions. (c, d) Molecular orientations that juxtapose the positive or negative ends of the dipoles on adjacent molecules produce repulsive interactions.

Intermolecular forces are electrostatic in nature; that is, they arise from the interaction between positively and negatively charged species. Like covalent and ionic bonds, intermolecular interactions are the sum of both attractive and repulsive components. Because electrostatic interactions fall off rapidly with increasing distance between molecules, intermolecular interactions are most important for solids and liquids, where the molecules are close together. These interactions become important for gases only at very high pressures, where they are responsible for the observed deviations from the ideal gas law at high pressures (see Section 10.8).

We explicitly consider three kinds of intermolecular interactions:* *dipole–dipole interactions*, *London dispersion forces*, and *hydrogen bonds*. The first two are often described collectively as **van der Waals forces**.

Dipole–Dipole Interactions

Recall from Chapter 9 that polar covalent bonds behave as if the bonded atoms have localized fractional charges that are equal but opposite (that is, the two bonded atoms generate a *dipole*). If the structure of the molecule is such that the individual bond dipoles do not cancel one another, then the molecule has a net dipole moment. Molecules that have net dipole moments tend to align themselves so that the positive end of one dipole is near the negative end of another, and vice versa, as shown in Figure 11.3a, b. These arrangements are more stable than arrangements in which two positive or two negative ends are adjacent (Figure 11.3c, d). Hence, **dipole–dipole interactions** such as those in Figure 11.3a, b are *attractive intermolecular interactions*, whereas those in Figure 11.3c, d are *repulsive intermolecular interactions*. Because molecules in a liquid move freely and continuously, molecules always experience both attractive and repulsive dipole–dipole interactions simultaneously, as shown in Figure 11.4. On average, however, the attractive interactions dominate.

Because each end of a dipole possesses only a fraction of the charge of an electron, dipole–dipole interactions are substantially weaker than the interactions between two ions, each of which has a charge of at least ± 1, or between a dipole and an ion, in which one of the species has at least a full positive or negative charge. In addition, the attractive interaction between dipoles falls off much more rapidly with increasing distance than do the ion–ion interactions we considered in Chapter 8. Recall that the attractive energy between two ions is proportional to $1/r$, where r is the distance between the ions. Doubling the distance ($r \longrightarrow 2r$) decreases the attractive energy by one-half. In contrast, the energy of the interaction of two dipoles is proportional to $1/r^6$, so doubling the distance between the dipoles decreases the strength of the interaction by 2^6, or 64-fold. Thus, a substance such as HCl, which is held together partially by dipole–dipole interactions, is a gas at room temperature and 1 atm pressure, whereas NaCl, which is held together by interionic interactions, is a high-melting-point solid. Within a series of compounds of similar molar mass, the strength of the intermolecular interactions increases as the dipole moment of the molecules increases, as shown in Table 11.1. Using what we learned in Chapter 9 about predicting relative bond polarities from the electronegativities of the bonded atoms, we can make educated guesses about the relative boiling points of similar molecules.

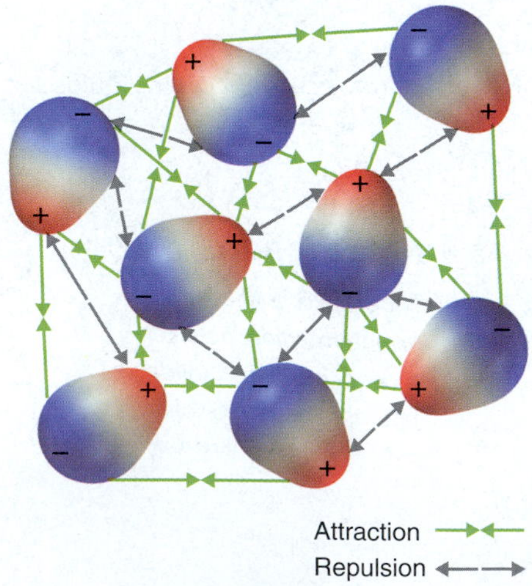

Figure 11.4 Both attractive and repulsive dipole–dipole interactions occur in a liquid sample that contains many molecules.

* There are two additional types of electrostatic interaction that you are already familiar with: the ion–ion interactions that are responsible for ionic bonding (see Chapter 8) and the ion–dipole interactions that occur when ionic substances dissolve in a polar substance such as water (see Chapters 4 and 5).

TABLE 11.1 Relationship between dipole moment and boiling point for organic compounds of similar molar mass

Compound	Molar Mass, g/mol	Dipole Moment (μ), D	Boiling Point, K
C_3H_6 (cyclopropane)	42	0	240
CH_3OCH_3 (dimethyl ether)	46	1.30	248
CH_3CN (acetonitrile)	41	3.9	355

EXAMPLE 11.1

Arrange the following compounds in order of increasing boiling points: ethyl methyl ether, $CH_3OCH_2CH_3$; 2-methylpropane [isobutane, $(CH_3)_2CHCH_3$]; acetone, CH_3COCH_3. Their structures are shown in the margin.

Given Compounds

Asked for Order of increasing boiling points

Strategy

Compare the molar masses and the polarities of the compounds. Compounds that have higher molar masses and are polar will have the highest boiling points.

Solution

All of the compounds have essentially the same molar mass (58–60 g/mol), so we must look at differences in polarity to predict the strength of the intermolecular dipole–dipole interactions and thus the boiling points of the compounds. Isobutane contains only C—H bonds, which are not very polar because C and H have similar electronegativities. Isobutane, therefore, should have a very small (but nonzero) dipole moment and quite a low boiling point. Ethyl methyl ether has a structure similar to H_2O, and contains two polar C—O single bonds oriented at about a 109° angle to each other, in addition to relatively nonpolar C—H bonds. As a result, the C—O bond dipoles partially reinforce one another and generate a significant dipole moment that should give a moderately high boiling point. Acetone contains a polar C=O double bond oriented at about 120° to two methyl groups that have nonpolar C—H bonds. The C—O bond dipole therefore corresponds to the molecular dipole, which should result in both a rather large dipole moment and a high boiling point. Thus, we predict the following order of boiling points: isobutane < ethyl methyl ether < acetone. This result is in good agreement with the data: isobutane, bp −11.7°C (the dipole moment, μ, is 0.13 D); methyl ethyl ether, bp 7.4°C, μ = 1.17 D; acetone, bp 56.1°C, μ = 2.88 D.

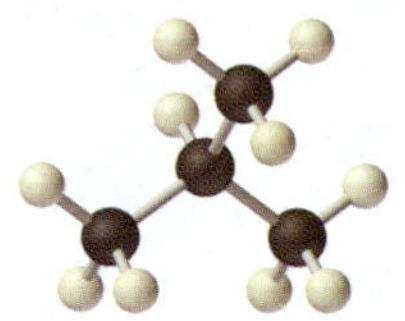

Isobutane

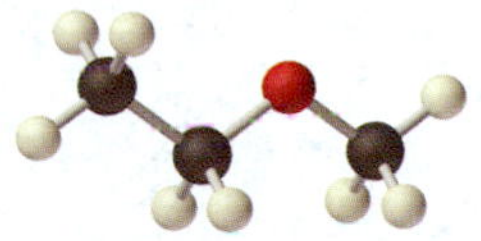

Ethyl methyl ether

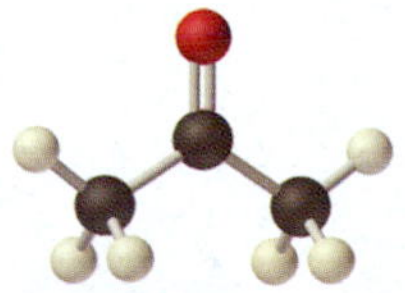

Acetone

EXERCISE 11.1

Arrange these compounds in order of decreasing boiling points: carbon tetrafluoride, CF_4; ethyl methyl sulfide, $CH_3SC_2H_5$; dimethyl sulfoxide, $(CH_3)_2S{=}O$; 2-methylbutane (isopentane), $(CH_3)_2CHCH_2CH_3$.

Answer dimethyl sulfoxide (bp 189.9°C) > ethyl methyl sulfide (bp 67°C) > 2-methylbutane (bp 27.8°C) > carbon tetrafluoride (bp −128°C)

London Dispersion Forces

Thus far we have considered only interactions between polar molecules, but other factors must be considered to explain why many nonpolar molecules, such as bromine,

TABLE 11.2 Normal melting and boiling points of some elements and nonpolar compounds

Substance	Molar Mass, g/mol	Melting Point, °C	Boiling Point, °C
Ar	40	−189.4	−185.9
Xe	131	−111.8	−108.1
N_2	28	−210	−195.8
O_2	32	−218.8	−183.0
F_2	38	−219.7	−188.1
I_2	254	113.7	184.4
CH_4	16	−182.5	−161.5

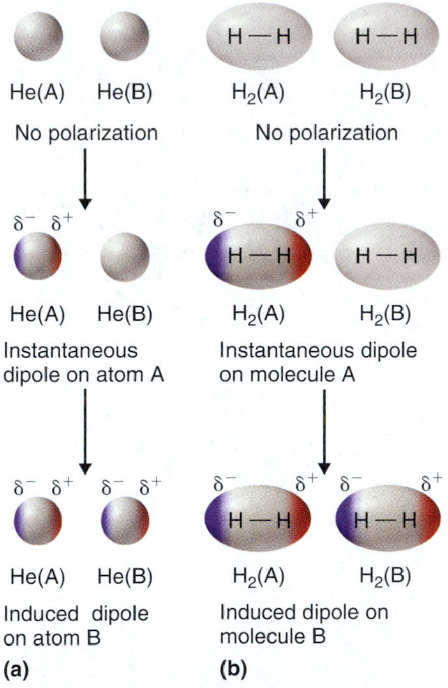

Figure 11.5 Instantaneous dipole moments. The formation of an instantaneous dipole moment on one He atom (a) or H_2 molecule (b) results in the formation of an induced dipole on an adjacent atom or molecule.

Note the pattern

For similar substances, London dispersion forces get stronger with increasing molecular size.

benzene, and hexane, are liquids at room temperature, and others, such as iodine and naphthalene, are solids. Even the noble gases can be liquefied or solidified at low temperatures, high pressures, or both (Table 11.2). What kind of attractive forces can exist between nonpolar molecules or atoms? This question was answered by Fritz London (1900–1954), a German physicist who later worked in the United States. In 1930, London proposed that temporary fluctuations in the electron distributions within atoms and nonpolar molecules could result in the formation of short-lived **instantaneous dipole moments**, which produce attractive forces called **London dispersion forces** between otherwise nonpolar substances.

Consider a pair of adjacent He atoms, for example. On average, the two electrons in each He atom are uniformly distributed around the nucleus. Because the electrons are in constant motion, however, their distribution in one atom is likely to be asymmetrical at any given instant, resulting in an instantaneous dipole moment. As shown in Figure 11.5a, the instantaneous dipole moment on one atom can interact with the electrons in an adjacent atom, pulling them toward the positive end of the instantaneous dipole or repelling them from the negative end. The net effect is that the first atom causes the temporary formation of a dipole, called an **induced dipole**, in the second. Interactions between these temporary dipoles cause atoms to be attracted to one another. These attractive interactions are weak and fall off rapidly with increasing distance. London was able to show quantum mechanically that the attractive energy between molecules due to temporary dipole–induced dipole interactions falls off as $1/r^6$. Doubling the distance therefore decreases the attractive energy by 2^6, or 64-fold.

Instantaneous dipole–induced dipole interactions between nonpolar molecules can produce intermolecular attractions just as they produce interatomic attractions in monatomic substances like xenon. This effect, illustrated for two H_2 molecules in Figure 11.5b, tends to become more pronounced as atomic and molecular masses increase (Table 11.2). For example, Xe boils at −108.1°C, whereas He boils at −269°C. The reason for this trend is that the strength of London dispersion forces is related to the ease with which the electron distribution in a given atom can be perturbed. In small atoms such as He, the two 1*s* electrons are held close to the nucleus in a very small volume, and electron–electron repulsions are strong enough to prevent significant asymmetry in their distribution. In larger atoms such as Xe, however, the outer electrons are much less strongly attracted to the nucleus because of filled intervening shells (see Chapter 7 for a discussion of shielding). As a result, it is relatively easy to temporarily deform the electron distribution to generate an instantaneous or induced dipole. The ease of deformation of the electron distribution in an atom or molecule is called its *polarizability*. Because the electron distribution is more easily perturbed in large, heavy species than in small, light ones, we say that heavier substances tend to be much more *polarizable* than lighter ones.

The polarizability of a substance also determines how it interacts with ions and with species that possess permanent dipoles, as we shall see when we discuss solutions in Chapter 13. Thus, London dispersion forces are responsible for the general trend toward higher boiling points with increased molecular mass and greater surface area in a homologous series of compounds such as the alkanes (Figure 11.6a). The strengths of London dispersion forces also depend significantly on molecular shape because shape determines how much of one molecule can interact with its neighboring molecules at any given time. For example, Figure 11.6b shows 2,2-dimethylpropane (neopentane) and *n*-pentane, both of which have the empirical formula C_5H_{12}. Neopentane is almost spherical, with a small surface area for intermolecular interactions, whereas *n*-pentane has an extended conformation that enables it to come into close contact with other *n*-pentane molecules. As a result, the boiling point of neopentane (9.5°C) is more than 25°C lower than the boiling point of *n*-pentane (36.1°C).

All molecules, whether polar or nonpolar, are attracted to one another by London dispersion forces in addition to any other attractive forces that may be present. In general, however, dipole–dipole interactions in small polar molecules are so much stronger than London dispersion forces that the former predominate.

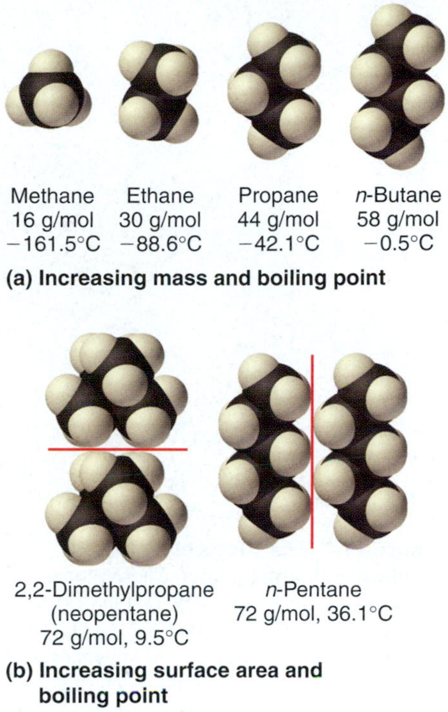

Figure 11.6 Mass and surface area affect the strength of London dispersion forces. (a) In this series of four simple alkanes, larger molecules have stronger London forces between them than smaller molecules and consequently higher boiling points. (b) Linear *n*-pentane molecules have a larger surface area and stronger intermolecular forces than spherical neopentane molecules. As a result, neopentane is a gas at room temperature, whereas *n*-pentane is a volatile liquid.

EXAMPLE 11.2

Arrange the following nonpolar compounds in order of increasing boiling points: *n*-butane, propane, 2-methylpropane [isobutane, $(CH_3)_2CHCH_3$], *n*-pentane.

Given Compounds

Asked for Order of increasing boiling points

Strategy

Determine the intermolecular forces in the compounds, and then arrange the compounds according to the strength of those forces. The substance with the weakest forces will have the lowest boiling point.

Solution

All of these compounds are alkanes and are nonpolar, so London dispersion forces are the only important intermolecular forces. These forces are generally stronger with increasing molecular mass, so propane should have the lowest boiling point and *n*-pentane should have the highest, with the two butane isomers falling in between. Of the two butane isomers, 2-methylpropane is more compact and *n*-butane has the more extended shape. Consequently, we expect intermolecular interactions for *n*-butane to be stronger due to its larger surface area, resulting in a higher boiling point. The overall order is thus propane (−42.1°C) < 2-methylpropane (−11.7°C) < *n*-butane (−0.5°C) < *n*-pentane (36.1°C).

EXERCISE 11.2

Arrange these compounds in order of decreasing boiling points: GeH_4, $SiCl_4$, SiH_4, CH_4, $GeCl_4$.

Answer $GeCl_4$ (87°C) > $SiCl_4$ (57.6°C) > GeH_4 (−88.5°C) > SiH_4 (−111.8°C) > CH_4(−161°C)

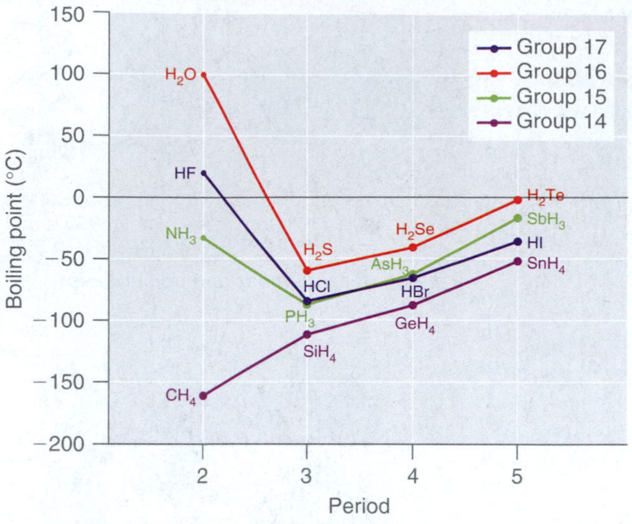

Figure 11.7 The effects of hydrogen bonding on boiling points. These plots of the boiling points of the covalent hydrides of the elements of Groups 14–17 show that the boiling points of the lightest members of each series for which hydrogen bonding is possible (HF, NH_3, H_2O) are anomalously high for compounds with such low molecular masses. Interactive Graph

Hydrogen Bonds

Molecules with hydrogen atoms bonded to electronegative atoms such as O, N, F, and to a much lesser extent Cl and S, tend to exhibit unusually strong intermolecular interactions. These result in much higher boiling points than are observed for substances in which London dispersion forces dominate, as illustrated for the covalent hydrides of elements of Groups 14–17 in Figure 11.7. Methane and its heavier congeners in Group 14 form a series whose boiling points increase smoothly with increasing molar mass. This is the expected trend in nonpolar molecules, for which London dispersion forces are the exclusive forces. In contrast, the hydrides of the lightest members of Groups 15–17 have boiling points that are more than 100°C higher than predicted based on their molar masses. The effect is most dramatic for water: if we extend the straight line connecting the points for H_2Te and H_2Se to the line for Period 2, we obtain an estimated boiling point of −130°C for water! Imagine the implications for life on Earth if water boiled at −130°C rather than 100°C.

What is the reason for the strong intermolecular forces that produce such anomalously high boiling points and other unusual properties, such as high enthalpies of vaporization and high melting points? The answer lies in the highly polar nature of the bonds between hydrogen and very electronegative elements such as O, N, and F. The large difference in electronegativity results in a large partial positive charge on hydrogen and a correspondingly large partial negative charge on the O, N, or F atom. Consequently, H—O, H—N, and H—F bonds have very large bond dipoles that can interact strongly with one another. Because a hydrogen atom is so small, these dipoles can also approach one another more closely than most other dipoles. The combination of large bond dipoles and short dipole–dipole distances results in very strong dipole–dipole interactions called **hydrogen bonds**, as shown for ice in Figure 11.8. A hydrogen bond is usually indicated by a dotted line between the hydrogen atom attached to O, N, or F (the *hydrogen bond donor*) and the atom that has the lone pair of electrons (the *hydrogen bond acceptor*). Because each water molecule contains two hydrogen atoms and two lone pairs, a tetrahedral arrangement maximizes the number of hydrogen bonds that can be formed. In the structure of ice, each oxygen atom is surrounded by a distorted tetrahedron of hydrogen atoms that form bridges to the oxygen atoms of adjacent water molecules. The bridging hydrogen atoms are *not* equidistant from the two oxygen atoms they connect, however. Instead, each hydrogen atom is 101 pm from one oxygen and 174 pm from the other. In contrast, each oxygen atom is bonded to two H atoms at the shorter distance and two at the longer distance, corresponding to two O—H covalent bonds and two O···H hydrogen bonds from adjacent water molecules, respectively. The resulting open, cagelike structure of ice means that the solid is actually slightly less dense than the liquid, which is the reason ice floats on water rather than sinks.

Note the pattern

Hydrogen bond formation requires both a hydrogen bond donor and a hydrogen bond acceptor.

Because ice is less dense than liquid water, rivers, lakes, and oceans freeze from the top down. In fact, the ice forms a protective surface layer that insulates the rest of the water, allowing organisms such as fish to survive in the lower levels of a frozen lake or sea. If ice were denser than the liquid, the ice formed at the surface in cold weather would sink as fast as it formed. Bodies of water would freeze from the bottom up, which would be lethal for most aquatic creatures. The expansion of water when freezing also explains why automobile or boat engines must be protected by "antifreeze" (we will discuss how antifreeze works in Chapter 13), and why unprotected pipes in houses break if they are allowed to freeze.

Although hydrogen bonds are significantly weaker than covalent bonds, with typical dissociation energies of only 15–25 kJ/mol, they have a significant influence on the physical properties of a compound. Compounds such as HF can form only two hydrogen bonds at a time as can, on average, pure liquid NH_3. Consequently, even though their molecular masses are similar to that of water, their boiling points are significantly lower than the boiling point of water, which forms *four* hydrogen bonds at a time.

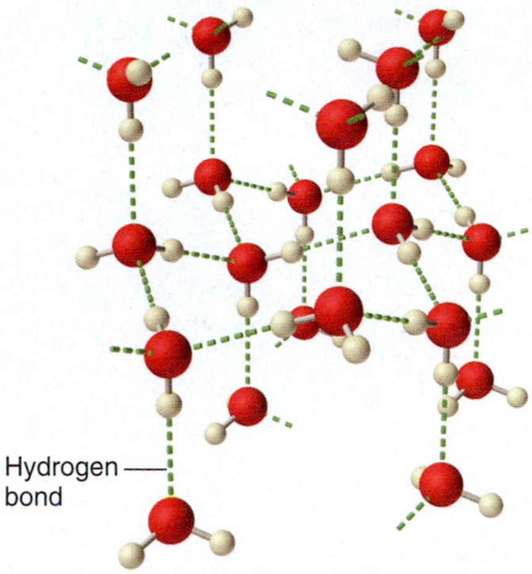

Figure 11.8 The hydrogen-bonded structure of ice. Each water molecule accepts two hydrogen bonds from two other water molecules and donates two hydrogen atoms to form hydrogen bonds with two more water molecules, producing an open, cagelike structure. The structure of liquid water is very similar, but in the liquid, the hydrogen bonds are continually broken and formed because of rapid molecular motion.

EXAMPLE 11.3

Determine which of these substances can form hydrogen bonds with themselves, and draw the hydrogen-bonded structures: CH_3OH, C_2H_6, Xe, $(CH_3)_3N$.

Given Compounds

Asked for Formation of hydrogen bonds, structure

Strategy

A Identify the compounds that contain a hydrogen atom attached to O, N, or F. These are likely to be able to act as hydrogen bond donors.

B Of the compounds that can act as hydrogen bond donors, identify those that also contain lone pairs of electrons, which allow them to be hydrogen bond acceptors. If the substance is both a hydrogen donor and an acceptor, draw a structure showing the hydrogen bonding.

Solution

A Of the species listed, xenon, Xe, ethane, C_2H_6, and trimethylamine, $(CH_3)_3N$, do not contain a hydrogen atom attached to O, N, or F; hence, they cannot act as hydrogen bond donors.
B The one compound that can act as a hydrogen bond donor, methanol, CH_3OH, contains both a hydrogen atom attached to O (making it a hydrogen bond donor) and two lone pairs of electrons on O (making it a hydrogen bond acceptor); methanol can thus form hydrogen bonds by acting as either a hydrogen bond donor or an acceptor. The hydrogen-bonded structure of methanol is

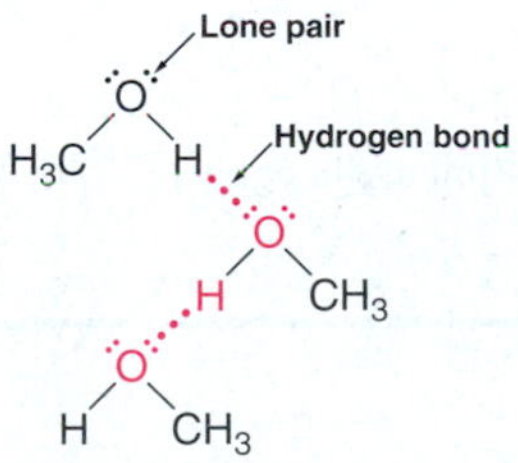

Hydrogen bonding in methanol

EXERCISE 11.3

Predict which of these substances can form hydrogen bonds, and draw the hydrogen-bonded structures: CH_3CO_2H, $(CH_3)_3N$, NH_3, CH_3F.

Answer CH_3CO_2H and NH_3

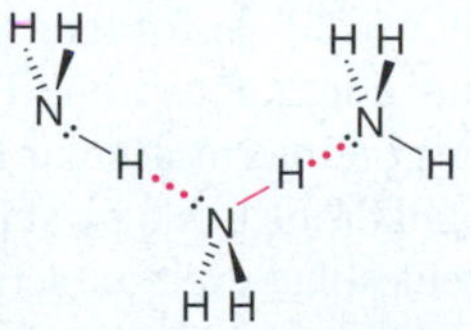

Hydrogen bonding in ammonia

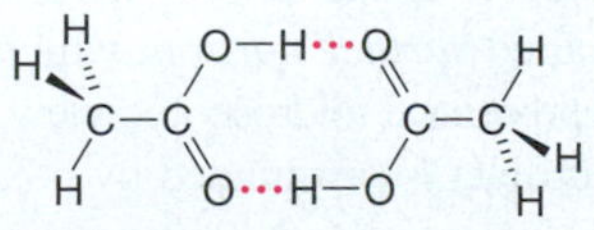

Hydrogen bonding in acetic acid

EXAMPLE 11.4

Arrange the following compounds in order of increasing boiling points: C_{60} (buckminsterfullerene, which has a cage structure), NaCl, He, Ar, N_2O.

Given Compounds

Asked for Order of increasing boiling points

Strategy

Identify the intermolecular forces in each compound, and then arrange the compounds according to the strength of those forces. The substance with the weakest forces will have the lowest boiling point.

Solution

Electrostatic interactions are strongest for an ionic compound, so we expect NaCl to have the highest boiling point. To predict the relative boiling points of the other compounds, we have to consider their polarity (for dipole–dipole interactions), their ability to form hydrogen bonds, and their molar mass (for London dispersion forces). Helium is nonpolar and by far the lightest, so it should have the lowest boiling point. Argon and N_2O have very similar molar masses (40 and 44 g/mol, respectively), but N_2O is polar while Ar is not. Consequently, N_2O should have a higher boiling point. A C_{60} molecule is nonpolar, but its molar mass is 720 g/mol, much greater than that of Ar or N_2O. Because the boiling points of nonpolar substances increase rapidly with molecular mass, C_{60} should boil at a higher temperature than the other nonionic substances. The predicted order is thus as follows, with actual boiling points in parentheses: He (−269°C) < Ar (−185.7°C) < N_2O (−88.5°C) < C_{60} (>280°C) < NaCl (1465°C).

EXERCISE 11.4

Arrange these compounds in order of decreasing boiling points: 2,4-dimethylheptane, Ne, CS_2, Cl_2, KBr.

Answer KBr (1435°C) > 2,4-dimethylheptane (132.9°C) > CS_2 (46.6°C) > Cl_2 (−34.6°C) > Ne (−246°C)

MGC Properties of Liquids

11.3 • Unique Properties of Liquids

Although you have been introduced to some of the interactions that hold molecules together in a liquid, we have not yet discussed the consequences of those interactions for the bulk properties of liquids. We now turn our attention to three unique properties of liquids that depend intimately on the nature of intermolecular interactions: surface tension, capillary action, and viscosity.

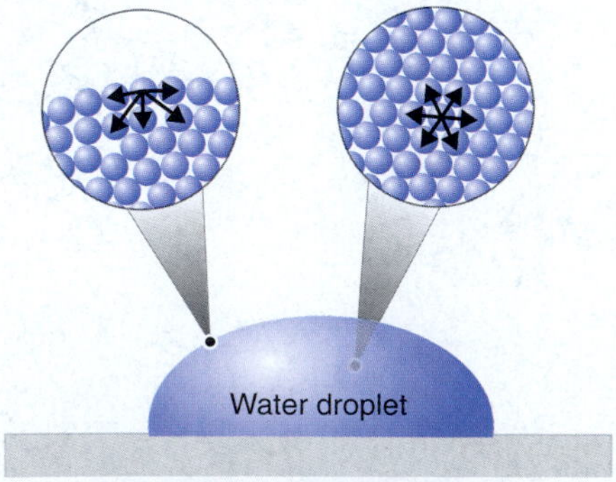

Figure 11.9 A representation of surface tension in a liquid. Molecules at the surface of the liquid experience a net attraction into the liquid, whereas those in the interior do not.

Surface Tension

We stated in Section 11.1 that liquids tend to adopt the shapes of their containers. Why, then, do small amounts of water on a freshly waxed car form raised droplets, as shown in the chapter opener photo, instead of a thin, continuous film? The answer lies in a property called *surface tension*, which depends on intermolecular forces.

Figure 11.9 presents a microscopic view of a liquid droplet. A typical molecule in the *interior* of the droplet is surrounded by other molecules that exert attractive forces from all directions. Consequently, there is no *net* force on the molecule that would cause it to move in a particular direction. In contrast, a molecule on the *surface* experiences a net attraction toward the drop because there are no molecules on the outside to balance the forces exerted by adjacent molecules in the interior. Because a sphere has the smallest possible surface area for a given volume, the absence of intermolecular attractive interactions on the surface away from the drop causes the droplet to adopt a spherical shape. Hence, raindrops are almost spherical, and drops of water on a waxed (nonpolar) surface, which does not interact strongly with water, form round beads (see chapter opener photo). A dirty car is covered with a mixture of substances, some of which are polar and able to interact with the water. These polar interactions cause the water to spread and prevent it from forming beads.

The same phenomenon holds molecules together at the surface of a bulk sample of water, almost as if they formed a skin. When filling a glass with water, the glass can be overfilled so that the level of the liquid actually extends *above* the rim. Similarly, a sewing needle or a paper clip can be placed on the surface of a glass of water where it "floats," even though steel is much denser than water (Figure 11.10a). Many insects take advantage of this property to walk on the surface of puddles or ponds without sinking (Figure 11.10b).

All of these phenomena are manifestations of **surface tension**, which is defined as the energy required to increase the surface area of a liquid by a specific amount. Surface tension is therefore measured as energy per unit area, such as joules per square meter (J/m^2). Values of the surface tension of some representative liquids are listed in Table 11.3. Note the correlation between the surface tension of a liquid and the strength of the intermolecular forces: the stronger the intermolecular forces, the higher the surface tension. For example, water, with its strong intermolecular hydrogen bonding, has one of the highest surface tension values of any liquid, whereas low-boiling-point organic molecules, which have relatively weak intermolecular forces, have much lower surface tensions. Mercury is an apparent anomaly, but its very high surface tension is due to the presence of strong metallic bonding, which we will discuss in more detail in Chapter 12.

The surface tension of water can be reduced by the addition of soaps and detergents that disrupt the intermolecular attractions between adjacent water molecules.

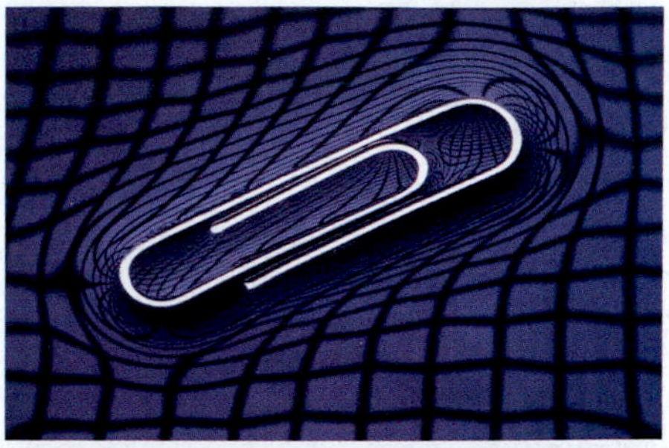

(a)

(b)

Figure 11.10 Effects of the high surface tension of liquid water. (a) A paper clip can "float" on water because of surface tension. (b) Surface tension also allows insects such as this water strider to "walk on water."

TABLE 11.3 Surface tension, viscosity, vapor pressure (at 25°C unless otherwise indicated), and normal boiling points of common liquids

Substance	Surface Tension, $\times 10^{-3}$ J/m^2	Viscosity, mPa · s	Vapor Pressure, mmHg	Normal Boiling Point, °C
		Organic compounds		
Diethyl ether	17	0.22	531	34.6
n-Hexane	18	0.30	149	68.7
Acetone	23	0.31	227	56.5
Ethanol	22	1.07	59	78.3
Ethylene glycol	48	16.1	~0.08	198.9
		Liquid elements		
Bromine	41	0.94	218	58.8
Mercury	486	1.53	0.0020	357
		Water		
(0°C)	75.6	1.79	4.6	
(20°C)	72.8	1.00	17.5	
(60°C)	66.2	0.47	149	
(100°C)	58.9	0.28	760	

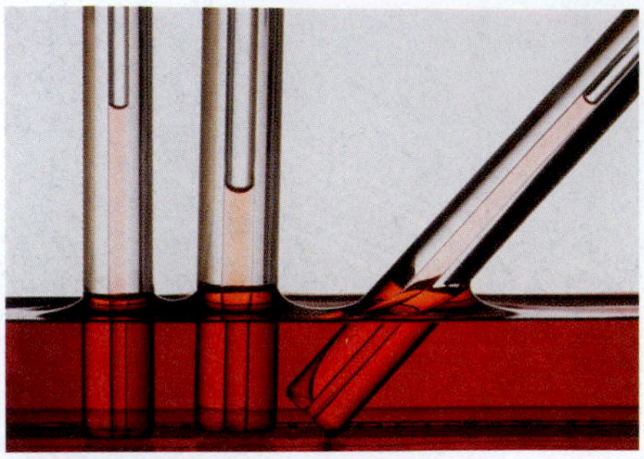

Figure 11.11 The phenomenon of capillary action. When a glass capillary is placed in liquid water, the water rises up into the capillary. The smaller the diameter of the capillary, the higher the water rises. Note that the height of the water does *not* depend on the angle at which the capillary is tilted.

Because they affect the surface properties of a liquid, such substances are called surface-active agents, or **surfactants**. In the 1960s, U.S. Navy researchers developed a method of fighting fires aboard aircraft carriers using "foams," which are aqueous solutions of fluorinated surfactants. The surfactants reduce the surface tension of water below that of fuel, so the fluorinated solution is able to spread across the burning surface and extinguish the fire. Such foams are now used universally to fight large-scale fires of organic liquids.

Capillary Action

Intermolecular forces also cause a phenomenon called **capillary action**, which is the tendency of a polar liquid to rise against gravity into a small-diameter tube (a *capillary*), as shown in Figure 11.11. When a glass capillary is put into a dish of water, the water is drawn up into the tube. The height to which the water rises depends on the diameter of the tube (the smaller the diameter, the higher the liquid rises) and the temperature of the water, but *not* on the angle at which the tube enters the water.

Capillary action is the net result of two opposing sets of forces: **cohesive forces**, which are the intermolecular forces that hold the liquid together, and **adhesive forces**, which are the attractive forces between the liquid and the substance that makes up the capillary. Water has both strong adhesion to glass, which contains polar SiOH groups, and strong intermolecular cohesion. When a glass capillary is put into water, the surface tension due to cohesive forces constricts the surface area of the water within the tube, while adhesion between the water and the glass creates an upward force that maximizes the amount of glass surface in contact with the water. If the adhesive forces are stronger than the cohesive forces, as is the case for water, then the liquid in the capillary rises to the level where the downward force of gravity exactly balances this upward force. If, however, the cohesive forces are stronger than the adhesive forces, as is the case for mercury and glass, the liquid pulls itself down into the capillary below the surface of the bulk liquid to minimize contact with the glass (Figure 11.12a). The upper surface of a liquid in a tube is called the **meniscus**, and the shape of the meniscus depends on the relative strengths of the cohesive and adhesive forces. In liquids such as water, the meniscus is concave; in liquids such as mercury, however, which have very strong cohesive forces and weak adhesion to glass, the meniscus is convex (Figure 11.12b).

Note the pattern

Polar substances are drawn up a glass capillary and have a concave meniscus.

Fluids and nutrients are transported up the stems of plants or the trunks of trees by capillary action. Plants contain tiny rigid tubes composed of cellulose, to which water has strong adhesion. Because of the strong adhesive forces, nutrients can be transported from the roots to the tops of trees that are more than 50 m tall. Cotton towels are also made of cellulose; they absorb water because the tiny tubes act like capillaries and "wick" the water away from your skin. The moisture is absorbed by the entire fabric, not just the layer in contact with your body.

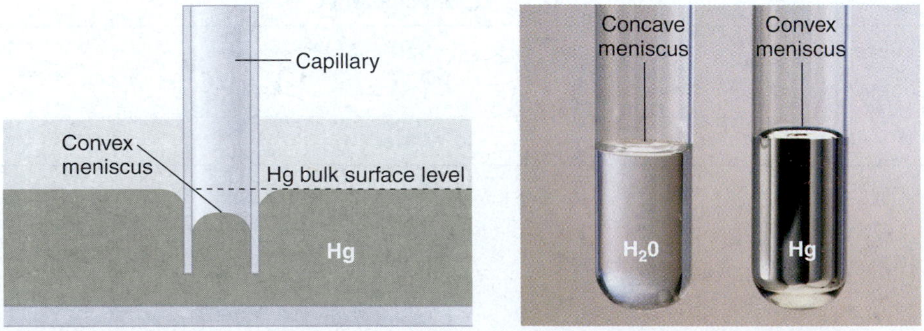

(a) (b)

Figure 11.12 The effects of capillary action. (a) Drawing illustrating the shape of the meniscus and the relative height of the mercury column when a glass capillary is put into liquid mercury. Note that the meniscus is convex and that the surface of the liquid inside the tube is *lower* than the level of the liquid outside the tube. (b) Because water adheres strongly to the polar surface of glass, it has a concave meniscus, whereas mercury, which does not adhere to the glass, has a convex meniscus.

Viscosity

Viscosity (η) is the resistance of a liquid to flow. Some liquids, such as gasoline, ethanol, and water, flow very readily and hence have a *low viscosity*. Others, such as motor oil, molasses, and maple syrup, flow very slowly and so have a *high viscosity*. The two most common methods for evaluating the viscosity of a liquid are to measure the time it takes a quantity of liquid to flow through a narrow vertical tube, and to measure the time it takes steel balls to fall through a given volume of the liquid. The higher the viscosity, the slower the liquid flows through the tube and the steel balls fall. Viscosity is expressed in units of the poise (mPa · s); the higher the number, the higher the viscosity. The viscosities of some representative liquids are listed in Table 11.3 and show a correlation between viscosity and intermolecular forces. Because a liquid can flow only if the molecules can move past one another with minimal resistance, strong intermolecular attractive forces make it more difficult for molecules to move with respect to one another. The addition of a second hydroxyl group to ethanol, for example, which produces ethylene glycol ($HOCH_2CH_2OH$), increases the viscosity 15-fold. This effect is due to the increased number of hydrogen bonds that can form between hydroxyl groups in adjacent molecules, resulting in dramatically stronger intermolecular attractive forces.

Ethanol

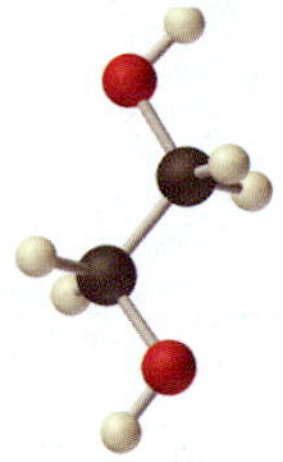
Ethylene glycol

There is also a correlation between viscosity and molecular shape. Liquids consisting of long, flexible molecules tend to have higher viscosities than those composed of more spherical or shorter-chain molecules. The longer the molecules, the easier it is for them to become "tangled" with one another, making it more difficult for them to move past one another. London dispersion forces also increase with chain length. Due to a combination of these two effects, long-chain hydrocarbons (such as motor oils) are highly viscous.

Motor oils and other lubricants demonstrate the practical importance of controlling viscosity. The oil in an automobile engine must effectively lubricate under a wide range of conditions, from subzero starting temperatures to the 200°C that oil can reach in an engine in the heat of the Mohave Desert in August. Viscosity decreases rapidly with increasing temperatures because the kinetic energy of the molecules increases, and higher kinetic energy enables the molecules to overcome the attractive forces that prevent the liquid from flowing. As a result, an oil that is thin enough to be a good lubricant in a cold engine will become too "thin" (have too low a viscosity) to be effective at high temperatures. The viscosity of motor oils is described by an SAE (Society of Automotive Engineers) rating ranging from SAE 5 to SAE 50 for engine oils: the lower the number, the lower the viscosity. So-called *single-grade oils* can cause major problems. If they are viscous enough to work at high operating temperatures (SAE 50, for example), then at low temperatures, they can be so viscous that the car is difficult to start or the engine is not properly lubricated. Consequently, most modern oils are *multigrade*, with designations such as SAE 20W/50 (a grade used in high-performance sports cars), in which case the oil has the viscosity of an SAE 20 oil at subzero temperatures (hence the W for winter) and the viscosity of an SAE 50 oil at high temperatures. These properties are achieved by a careful blend of additives that modulate the intermolecular interactions in the oil, thereby controlling the temperature dependence of the viscosity. Many of the commercially available oil additives "for improved engine performance" are highly viscous materials that increase the viscosity and effective SAE rating of the oil, but overusing these additives can cause the same problems experienced with highly viscous single-grade oils.

> ***Note the pattern***
> *Viscosity increases as intermolecular interactions or molecular size increases.*

EXAMPLE 11.5

Based on the nature and strength of the intermolecular cohesive forces and the probable nature of the liquid–glass adhesive forces, predict what will happen when a glass capillary is put into a beaker of SAE 20 motor oil. Will the oil be pulled up into the tube by capillary action or pushed down below the surface of the liquid in the beaker? What will be the shape of the meniscus (convex or concave)? *Hint*: The surface of glass is lined with Si—OH groups.

Given Substance, composition of glass surface

Asked for Behavior of oil, shape of meniscus

Strategy

Ⓐ Identify the cohesive forces in the motor oil.
Ⓑ Determine whether the forces interact with the surface of glass. From the strength of this interaction, predict the behavior of the oil and the shape of the meniscus.

Solution

Ⓐ Motor oil is a nonpolar liquid consisting largely of hydrocarbon chains. The cohesive forces responsible for its high boiling point are almost solely London dispersion forces between the hydrocarbon chains. Ⓑ Such a liquid cannot form strong interactions with the polar Si—OH groups of glass, so the surface of the oil inside the capillary will be lower than the level of the liquid in the beaker. The oil will have a convex meniscus similar to that of mercury.

EXERCISE 11.5

Predict what will happen when a glass capillary is put into a beaker of ethylene glycol.

Answer Capillary action will pull the ethylene glycol up into the capillary. The meniscus will be concave.

11.4 • Vapor Pressure

MGC Vapor Pressure

Nearly all of us have heated a pan of water with the lid in place and shortly thereafter heard the sounds of the lid rattling and hot water spilling onto the stovetop. When the liquid is heated, its molecules obtain sufficient kinetic energy to overcome the forces holding them in the liquid and they escape into the gaseous phase. By doing so, they generate a population of molecules in the vapor phase above the liquid that produces a pressure, the **vapor pressure** of the liquid. In the situation we described, enough pressure was generated to move the lid, which allowed the vapor to escape. If the vapor is contained in a sealed vessel, however, such as an unvented flask, and the vapor pressure becomes too high, the flask will explode (as many students have discovered!). In this section, we describe vapor pressure in more detail and explain how to quantitatively determine the vapor pressure of a liquid.

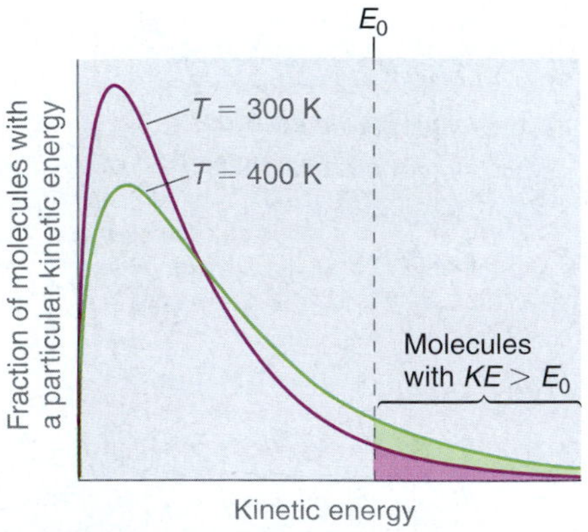

Figure 11.13 The distribution of the kinetic energies of the molecules of a liquid at two temperatures. Just as with gases, increasing the temperature shifts the peak to a higher energy and broadens the curve. Only molecules with kinetic energy greater than E_0 can escape from the liquid to enter the vapor phase, and the proportion of molecules with $KE > E_0$ is greater at the higher temperature. Interactive Graph

Evaporation and Condensation

Because the molecules of a liquid are in constant motion, we can plot the fraction of molecules that have a given kinetic energy against their kinetic energy to obtain the kinetic energy distribution of the molecules in the liquid (Figure 11.13), just as we did for a gas (Figure 10.19). Note that, as for gases, increasing the temperature increases both the average kinetic energy of the particles in a liquid and the range of kinetic energy of the individual molecules. If we assume that a minimum amount of energy, E_0, is needed to overcome the intermolecular attractive forces that hold the liquid together, then some fraction of molecules in the liquid always has a kinetic energy greater than E_0. The fraction of molecules with kinetic energy greater than this minimum value increases with increasing temperature. Any molecule with kinetic energy greater than E_0 has enough energy to overcome the forces holding it in the liquid and escape into the vapor phase. Before it can do so, however, the molecule must also be at the surface of the liquid, where it is physically possible for the molecule to leave the liquid surface. That is, only molecules at the surface can undergo **evaporation**, or **vaporization**, where atoms or molecules gain sufficient energy to enter a gaseous state above the liquid's surface, thereby creating a vapor pressure.

To understand the causes of vapor pressure, consider the apparatus shown in Figure 11.14. When a liquid is introduced into an evacuated chamber (Figure 11.14a), the initial pressure above the liquid is approximately zero because there are as yet no molecules in the vapor phase. Some molecules at the surface, however, will have sufficient kinetic energy to escape from the liquid and form a vapor, thus increasing the pressure inside the container. As long as the temperature of the liquid is held constant, the fraction of molecules with $KE > E_0$ will not change, and the rate at which molecules escape from the liquid into the vapor phase will depend only on the surface area of the liquid phase.

As soon as some vapor has formed, a fraction of the molecules in the vapor phase will collide with the surface of the liquid and reenter the liquid phase in a process known as **condensation** (Figure 11.14b). As the number of molecules in the vapor phase increases, the number of collisions between vapor-phase molecules and the surface will also increase. Eventually, a *steady state* will be reached in which exactly as many molecules per unit time leave the surface of the liquid (vaporize) as collide with it (condense). At this point, the pressure over the liquid stops increasing and remains constant at a particular value that is characteristic of the liquid at a given temperature. The rates of evaporation and condensation over time for a system such as this are shown graphically in Figure 11.15.

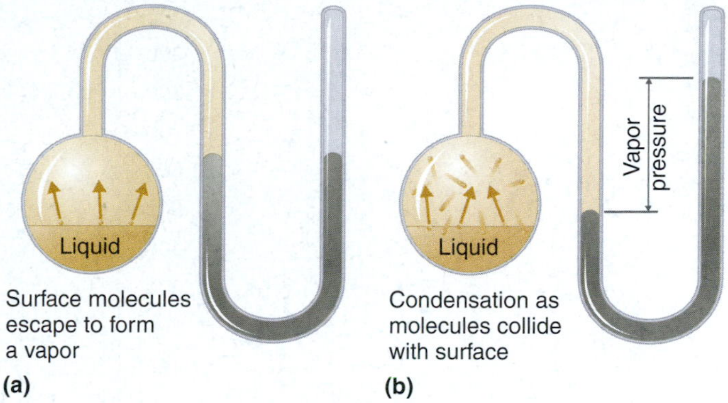

Figure 11.14 Vapor pressure. (a) When a liquid is introduced into an evacuated chamber, molecules with sufficient kinetic energy escape from the surface and enter the vapor phase, causing the pressure in the chamber to increase. (b) When sufficient molecules are in the vapor phase, the rate of condensation equals the rate of evaporation (a steady state is reached), and the pressure in the container becomes constant.

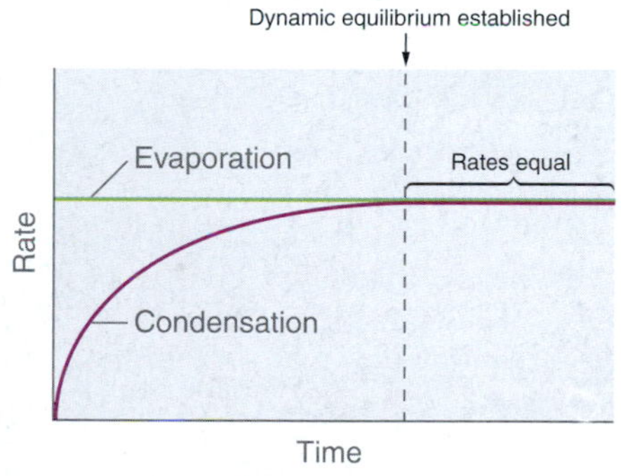

Figure 11.15 The relative rates of evaporation and condensation as a function of time after a liquid is introduced into a sealed chamber. The rate of evaporation depends only on the surface area of the liquid and is essentially constant. The rate of condensation depends on the number of molecules in the vapor phase and increases steadily until it equals the rate of evaporation.

Equilibrium Vapor Pressure

Two opposing processes (such as evaporation and condensation) that occur at the same rate, thus producing no *net* change in the system, constitute a **dynamic equilibrium**. In the case of a liquid enclosed in a chamber, the molecules continuously evaporate and condense, but the amounts of liquid and vapor do not change with time. The pressure exerted by a vapor in dynamic equilibrium with a liquid is the **equilibrium vapor pressure** of the liquid.

If a liquid is placed in an *open* container, however, most of the molecules that escape into the vapor phase will *not* collide with the surface of the liquid and return to the liquid phase. Instead, they will diffuse through the gas phase away from the container, and an equilibrium will never be established. Under these conditions, the liquid will continue to evaporate until it has "disappeared." The speed with which this occurs depends on the vapor pressure of the liquid and the temperature. **Volatile liquids** have relatively high vapor pressures and tend to evaporate readily; **nonvolatile liquids** have low vapor pressures and evaporate more slowly. Although the dividing line between volatile and nonvolatile liquids is not clear-cut, as a general guideline we can say that substances with vapor pressures higher than that of water (Table 11.3) are relatively volatile, whereas those with vapor pressures lower than water are relatively nonvolatile. Thus, diethyl ether (ethyl ether), acetone, and gasoline are volatile, but mercury, ethylene glycol, and motor oil are nonvolatile.

The equilibrium vapor pressure of a substance at a particular temperature is a characteristic of the material, like its molecular mass, melting point, and boiling point (Table 11.3). It does *not* depend on the amount of liquid as long as at least a tiny amount of liquid is present in equilibrium with the vapor. The equilibrium vapor pressure does, however, depend very strongly on the temperature and on the intermolecular forces present, as shown for several substances in Figure 11.16. Molecules that can hydrogen bond, such as ethylene glycol, for example, have a much lower equilibrium vapor pressure than those that cannot, such as octane. The nonlinear increase in vapor

Note the pattern

Volatile substances have low boiling points and relatively weak intermolecular interactions; nonvolatile substances have high boiling points and relatively strong intermolecular interactions.

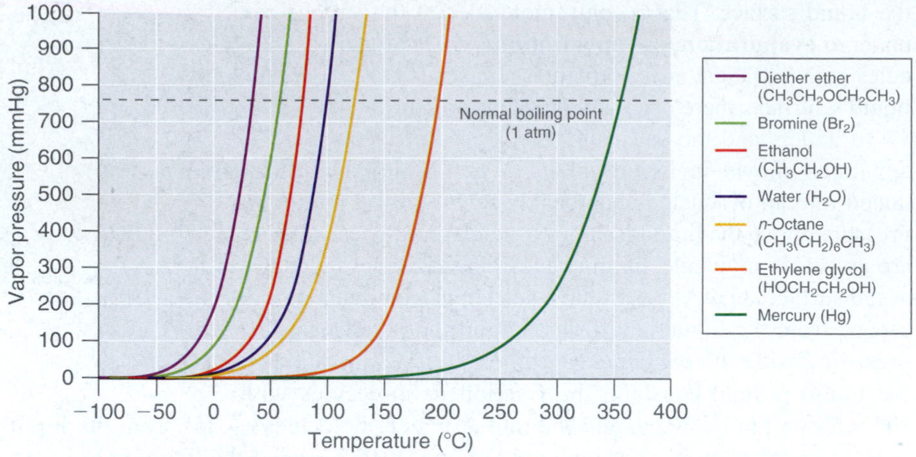

Figure 11.16 Vapor pressures of several liquids as a function of temperature. The point at which the vapor pressure curve crosses the $P = 1$ atm line (dashed) is the normal boiling point of the liquid. Interactive Graph

pressure with increasing temperature is *much* steeper than the increase in pressure expected for an ideal gas over the corresponding temperature range. The temperature dependence is so strong because the vapor pressure depends on the fraction of molecules that have a kinetic energy higher than that needed to escape from the liquid, and this fraction increases exponentially with temperature. As a result, sealed containers of volatile liquids are potential bombs if subjected to large increases in temperature. The gas tanks on automobiles are vented, for example, so that the car won't explode when parked in the sun. Similarly, the small cans (1–5 gallons) used to transport gasoline are required by law to have a pop-off pressure release.

The exponential rise in vapor pressure with increasing temperature in Figure 11.16 allows us to use natural logarithms to express the nonlinear relationship as a linear one.*

$$\ln P = \frac{-\Delta H_{vap}}{R}\left(\frac{1}{T}\right) + C \qquad (11.1)$$

Equation for a straight line: *y* = *m* *x* + *b*

where $\ln P$ is the natural logarithm of the vapor pressure, ΔH_{vap} is the enthalpy of vaporization, R is the universal gas constant [8.314 J/(mol · K)], T is the temperature in kelvins, and C is the y intercept, which is a constant for any given line. A plot of $\ln P$ versus the inverse of the absolute temperature ($1/T$) is a straight line with a slope of $-\Delta H_{vap}/R$. Equation 11.1, called the *Clausius–Clapeyron equation*, can be used to calculate the enthalpy of vaporization of a liquid from its measured vapor pressure at two or more temperatures. The simplest way to determine ΔH_{vap} is to measure the vapor pressure of a liquid at two temperatures and insert the values of P and T for these points into Equation 11.2, which is derived from the Clausius–Clapeyron equation:

$$\ln\left(\frac{P_2}{P_1}\right) = \frac{-\Delta H_{vap}}{R}\left(\frac{1}{T_2} - \frac{1}{T_1}\right) \qquad (11.2)$$

Conversely, if we know ΔH_{vap} and the vapor pressure P_1 at any temperature T_1, we can use Equation 11.2 to calculate the vapor pressure P_2 at any other temperature T_2, as shown in the next example.

* For a review of natural logarithms, refer to Essential Skills 6 at the end of this chapter.

EXAMPLE 11.6

The experimentally measured vapor pressures of liquid Hg at four temperatures are listed in the table:

***T*, °C**	80.0	100	120	140
***P*, torr**	0.0888	0.2729	0.7457	1.845

From these data, calculate the enthalpy of vaporization, ΔH_{vap}, of mercury, and predict the vapor pressure of the liquid at 160°C. (*Safety note:* Mercury is highly toxic, and when it is spilled, its vapor pressure generates hazardous levels of mercury vapor.)

Given Vapor pressures at four temperatures

Asked for Enthalpy of vaporization, ΔH_{vap}, of mercury; vapor pressure at 160°C

Strategy

A Use Equation 11.2 to obtain ΔH_{vap} directly from two of the pairs of values in the table, making sure to convert all values to the appropriate units.

B Substitute the calculated value of ΔH_{vap} into Equation 11.2 to obtain the unknown pressure, P_2.

Solution

A The table gives the measured vapor pressures of liquid Hg for four temperatures. Although one way to proceed would be to plot the data using Equation 11.1 and find the value of ΔH_{vap} from the slope of the line, a second, alternative approach is to use Equation 11.2 to obtain ΔH_{vap} directly from two of the pairs of values listed in the table, assuming no errors in our measurement. We therefore select two sets of values from the table and convert the temperatures from °C to K because the equation requires absolute temperatures. Substituting the values measured at 80.0°C (T_1) and 120.0°C (T_2) into Equation 11.2 gives

$$\ln\left(\frac{0.7457\ \cancel{\text{torr}}}{0.0888\ \cancel{\text{torr}}}\right) = \frac{-\Delta H_{vap}}{8.314\ \text{J/(mol}\cdot\text{K)}}\left[\frac{1}{(120+273)\ \text{K}} - \frac{1}{(80.0+273)\ \text{K}}\right]$$

$$\ln(8.398) = \frac{-\Delta H_{vap}}{8.314\ \text{J}\cdot\text{mol}^{-1}\cdot\cancel{\text{K}^{-1}}}(-2.88\times10^{-4}\ \cancel{\text{K}^{-1}})$$

$$2.13 = -\Delta H_{vap}(-0.346\times10^{-4})\ \text{J}^{-1}\cdot\text{mol}$$

$$\Delta H_{vap} = 61{,}400\ \text{J/mol} = 61.4\ \text{kJ/mol}$$

B We can now use this value of ΔH_{vap} to calculate the vapor pressure of the liquid (P_2) at 160.0°C (T_2):

$$\ln\left(\frac{P_2}{0.0888\ \text{torr}}\right) = \frac{-61{,}400\ \cancel{\text{J}}\cdot\cancel{\text{mol}^{-1}}}{8.314\ \cancel{\text{J}}\cdot\cancel{\text{mol}^{-1}}\cdot\text{K}^{-1}}\left[\frac{1}{(160+273)\ \text{K}} - \frac{1}{(80.0+273)\ \text{K}}\right] = 3.86$$

Using the relationship $e^{\ln x} = x$, we have

$$\ln\left(\frac{P_2}{0.0888\ \text{torr}}\right) = 3.86$$

$$\frac{P_2}{0.0888\ \text{torr}} = e^{3.86} = 47.5$$

$$P_2 = 4.21\ \text{torr}$$

Thus, at 160°C, liquid Hg has a vapor pressure of 4.21 torr, substantially greater than the pressure at 80.0°C, as we would expect.

EXERCISE 11.6

The vapor pressure of liquid Ni at 1606°C is 0.100 torr, whereas at 1805°C, its vapor pressure is 1.000 torr. At what temperature does the liquid have a vapor pressure of 2.500 torr?

Answer 1896°C

Boiling Points

As the temperature of a liquid increases, the vapor pressure of the liquid increases until it equals the external pressure, or the atmospheric pressure in the case of an open container. Bubbles of vapor begin to form throughout the liquid, and the liquid begins to boil. The temperature at which a liquid boils at exactly 1 atm pressure is the **normal boiling point** of the liquid. For water, the normal boiling point is exactly 100°C. The normal boiling points of the other liquids in Figure 11.16 are represented by the points at which the vapor pressure curves cross the line corresponding to a pressure of 1 atm. Although we usually cite the "normal" boiling point of a liquid, the *actual* boiling point depends on the pressure. At a pressure higher than 1 atm, water boils at a temperature higher than 100°C as the increased pressure forces vapor molecules to condense. Hence, the molecules must have greater kinetic energy to escape from the surface. Conversely, at pressures lower than 1 atm, water boils below 100°C.

Typical variations in atmospheric pressure at sea level are relatively small, causing only minor changes in the boiling point of water. For example, the highest recorded atmospheric pressure at sea level is 813 mmHg, recorded during a Siberian winter; the lowest sea-level pressure ever measured was 658 mmHg in a Pacific typhoon. At these pressures, the boiling point of water changes minimally, to 102°C and 96°C, respectively. At high altitudes, on the other hand, the dependence of the boiling point of water on pressure becomes significant. Table 11.4 lists the boiling points of water at several locations with different altitudes. At an elevation of only 5000 ft, for example, the boiling point of water is already lower than the lowest ever recorded at sea level. The lower boiling point of water has major consequences for cooking everything from soft-boiled eggs (a "three-minute egg" may well take four or more minutes in the Rockies and even longer in the Himalayas) to cakes (cake mixes are often sold with separate "high-altitude" instructions). Conversely, pressure cookers, which have a seal that allows the pressure inside them to exceed 1 atm, are used to cook food more rapidly by raising the boiling point of water and thus the temperature at which the food is being cooked.

Note the pattern

As pressure increases, the boiling point of a liquid increases, and vice versa.

TABLE 11.4 The boiling points of water at various locations on earth

Place	Altitude Above Sea Level, ft	Atmospheric Pressure, mmHg	Boiling Point of Water, °C
Mt. Everest, Nepal/Tibet	29,028	240	70
Bogota, Colombia	11,490	495	88
Denver, Colorado	5,280	633	95
Washington, DC	25	759	100
Dead Sea, Israel/Jordan	−1,312	799	101.4

EXAMPLE 11.7

Use Figure 11.16 to estimate **(a)** the boiling point of water in a pressure cooker operating at 1000 mmHg, and **(b)** the pressure required for mercury to boil at 250°C.

Given Data in Figure 11.16, pressure, boiling point

Asked for Corresponding boiling point and pressure

Strategy

A To estimate the boiling point of water at 1000 mmHg, refer to Figure 11.16 and find the point where the vapor pressure curve of water intersects the line corresponding to a pressure of 1000 mmHg.

B To estimate the pressure required for mercury to boil at 250°C, find the point where the vapor pressure curve of mercury intersects the line corresponding to a temperature of 250°C.

Solution

(a) **A** The vapor pressure curve of water intersects the P = 1000 mmHg line at about 110°C; this is therefore the boiling point of water at 1000 mmHg.

(b) **B** The vertical line corresponding to a temperature of 250°C intersects the vapor pressure curve of mercury at $P \approx 75$ mmHg. Hence, this is the pressure required for mercury to boil at 250°C.

EXERCISE 11.7

Use the data in Figure 11.16 to estimate **(a)** the normal boiling point of ethylene glycol, and **(b)** the pressure required for diethyl ether to boil at 20°C.

Answer **(a)** 200°C; **(b)** 450 mmHg

11.5 • Changes of State

MGC Heating and Cooling Curves

We take advantage of changes between the gas, liquid, and solid states to cool a drink with ice cubes (solid to liquid), cool our bodies by perspiration (liquid to gas), and cool food inside a refrigerator (gas to liquid and vice versa). We use dry ice, which is solid CO_2, as a refrigerant (solid to gas), and we make artificial snow for skiing and snowboarding by transforming a liquid to a solid. In this section, we examine what happens when any of the three forms of matter is converted to either of the other two. These changes of state are often called **phase changes**. The six most common phase changes are indicated in Figure 11.17.

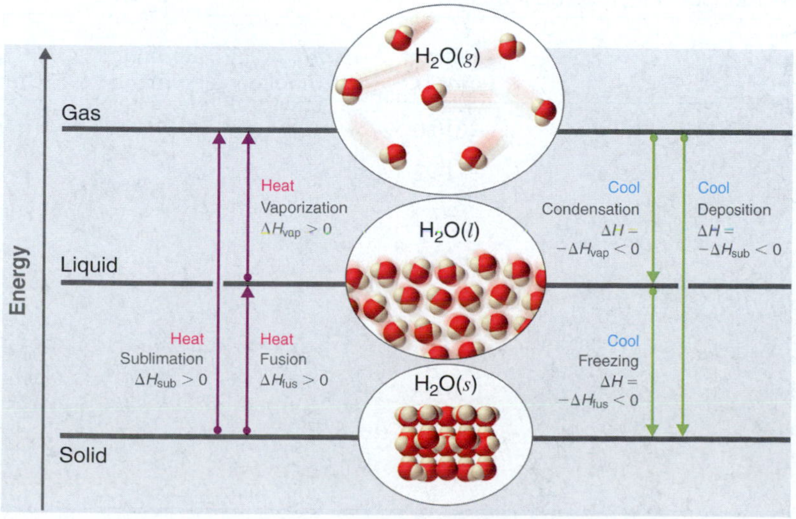

Figure 11.17 The three phases of matter and the processes that interconvert them when the temperature is changed. Enthalpy changes that accompany phase transitions are indicated by purple and green arrows.

Energy Changes That Accompany Phase Changes

Note the pattern

ΔH is positive for any transition from a more ordered to a less ordered state, and negative for a transition from a less ordered to a more ordered state.

Phase changes are *always* accompanied by a change in the energy of the system. For example, converting a liquid, in which the molecules are close together, to a gas, in which the molecules are, on the average, far apart, requires an input of energy (heat), in order to give the molecules enough kinetic energy to allow them to overcome the intermolecular attractive forces. The stronger the attractive forces, the more energy is needed to overcome them. Solids, which are highly ordered, have the strongest intermolecular interactions, whereas gases, which are very disordered, have the weakest. Thus, any transition from a more ordered to a less ordered state (solid to liquid, liquid to gas, or solid to gas) requires an input of energy—it is *endothermic*. Conversely, any transition from a less ordered to a more ordered state (liquid to solid, gas to liquid, or gas to solid) releases energy—it is *exothermic*. The energy change associated with each of the common phase changes is indicated in Figure 11.17.

In Chapter 5, we defined the enthalpy changes associated with various chemical and physical processes. The melting points and molar **enthalpies of fusion, ΔH_{fus},** the energy required to convert from a solid to a liquid, as well as the normal boiling points and enthalpies of vaporization, ΔH_{vap}, of selected compounds are listed in Table 11.5. Notice that the substances with the highest melting points usually have the highest enthalpies of fusion; they tend to be ionic compounds that are held together by very strong electrostatic interactions. Note too that substances with high boiling points are those with strong intermolecular interactions that must be overcome to convert a liquid to a gas, resulting in high enthalpies of vaporization. The enthalpy of vaporization of a given substance is much higher than its enthalpy of fusion because it takes more energy to completely separate molecules (conversion from liquid to gas) than to enable them only to move past one another freely (conversion from solid to liquid).

Figure 11.18 Sublimation of solid iodine. When solid iodine is heated at ordinary atmospheric pressure, it sublimes. When the I_2 vapor comes in contact with a cold surface, it deposits I_2 crystals.

The direct conversion of a solid to a gas, without an intervening liquid phase, is called **sublimation**. The amount of energy required to sublime 1 mol of a pure solid is the **enthalpy of sublimation, ΔH_{sub}**. Common substances that sublime at STP include CO_2 (dry ice), iodine (Figure 11.18), naphthalene, a substance used to protect woolen clothing against moths, and 1,4-dichlorobenzene. As shown in Figure 11.17, the enthalpy of sublimation of a substance is the sum of its enthalpies of fusion and vaporization provided all values are at the same T; this is an application of Hess's law (see Chapter 5):

$$\Delta H_{sub} = \Delta H_{fus} + \Delta H_{vap} \tag{11.3}$$

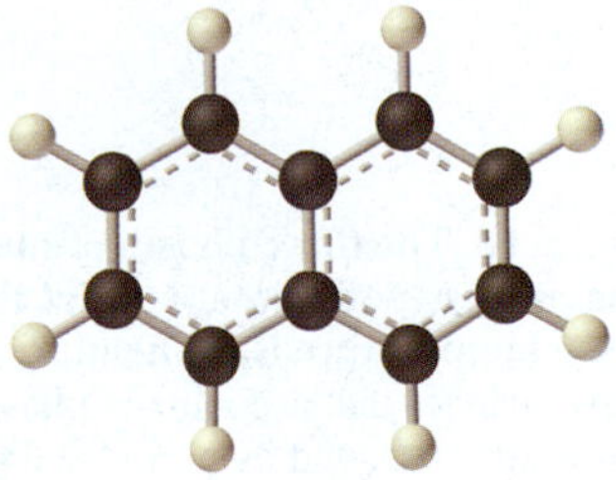

Naphthalene

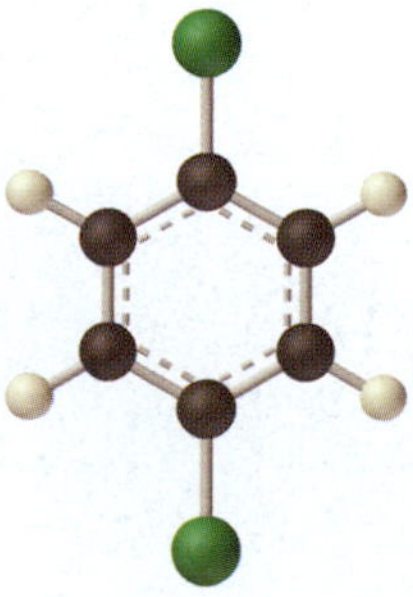

1,4-Dichlorobenzene

TABLE 11.5 Melting and boiling points and enthalpies of fusion and vaporization for selected substances

Substance	Melting Point, °C	ΔH_{fus}, kJ/mol	Boiling Point, °C	ΔH_{vap}, kJ/mol
N_2	−210.0	0.71	−195.8	5.6
HCl	−114.2	2.00	−85.1	16.2
Br_2	−7.2	10.6	58.8	30.0
CCl_4	−22.6	2.56	76.8	29.8
CH_3CH_2OH (ethanol)	−114.1	4.93	78.3	38.6
$CH_3(CH_2)_4CH_3$ (*n*-hexane)	−95.4	13.1	68.7	28.9
H_2O	0	6.01	100	40.7
Na	97.8	2.6	883	97.4
NaF	996	33.4	1704	176.1

Fusion, vaporization, and sublimation are endothermic processes; they occur only with the absorption of heat. Anyone who has ever stepped out of a swimming pool on a cool, breezy day has felt the heat loss that accompanies evaporation of water from the skin. Our bodies use this same phenomenon to maintain a constant temperature: we perspire continuously, even when at rest, losing about 600 mL of water daily by evaporation from the skin. We also lose about 400 mL of water as water vapor in the air we exhale, which also contributes to cooling. Refrigerators and air conditioners operate on a similar principle: heat is absorbed from the object or area to be cooled and used to vaporize a low-boiling-point liquid, such as ammonia or the CFCs and HCFCs discussed in Chapter 3 in connection with the ozone layer. The vapor is then transported to a different location and compressed, thus releasing and dissipating the heat. Likewise, ice cubes efficiently cool a drink not because of their low temperature, but because heat is required to convert ice at 0°C to liquid water at 0°C, as demonstrated later in Example 11.8.

Temperature Curves

The processes on the right side of Figure 11.17—freezing, condensation, and deposition, which are the reverse of fusion, sublimation, and vaporization—are exothermic. Thus, heat pumps that use refrigerants are essentially air conditioners running in reverse. Heat from the environment is used to vaporize the refrigerant, which is then condensed to a liquid in coils within a house to provide heat. The energy changes that occur during phase changes can be quantified by using a heating or cooling curve, a plot of the temperature of a sample versus time or heat added or removed.

Heating Curves

Figure 11.19 shows a **heating curve**, a plot of temperature versus heating time, for a 75-g sample of water. The sample is initially ice at −23°C, but, as heat is added, the temperature of the ice increases linearly with time. The slope of the line depends on both the mass of the ice and the specific heat of ice, which is the number of joules required to raise the temperature of 1 g of ice by 1°C. As the temperature of the ice increases, the water molecules in the ice crystal absorb more and more energy and vibrate more vigorously. At the melting point, they have enough kinetic energy to overcome attractive forces and to move with respect to one another. As more heat is added, the temperature of the system does *not* increase further, but remains constant at 0°C until all the ice has melted. Once all the ice has been converted to liquid water, the temperature of the water again begins to increase. Now, however, the temperature increases more slowly than before

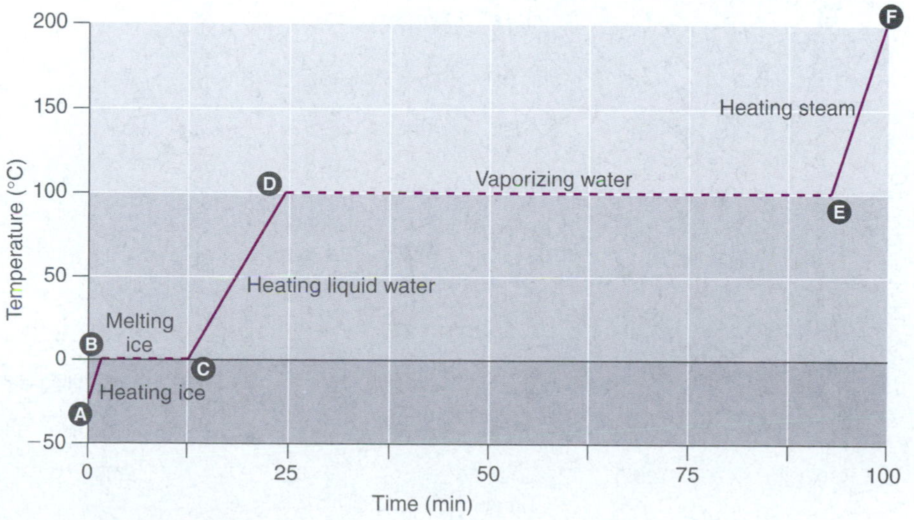

Figure 11.19 A heating curve for water. A plot of temperature as heat is added at a constant rate to a 75-g sample of ice initially at −23°C. A–B: heating solid ice; B–C: melting ice; C–D: heating liquid water; D–E: vaporizing water; E–F: heating steam.

Interactive Graph

because the specific heat capacity, C_s, of water is *greater* than that of ice. When the temperature of the water reaches 100°C, the water begins to boil. Here, too, the temperature remains constant at 100°C until all the water has been converted to steam. At this point, the temperature again begins to rise, but at a *faster* rate than seen in the other phases because the heat capacity of steam is *less* than that of ice or water.

Thus, *the temperature of the system does not change during a phase change*. In this example, as long as even a tiny amount of ice is present, the temperature of the system remains at 0°C during the melting process, and as long as even a small amount of liquid water is present, the temperature of the system remains at 100°C during the boiling process. The rate at which heat is added does *not* affect the temperature of the ice/water or water/steam mixture because the added heat is being used exclusively to overcome the attractive forces that hold the more condensed phase together. Many cooks think that food will cook faster if the heat is turned up higher so that the water boils more rapidly. Instead, the pot of water will boil to dryness sooner, but the temperature of the water does not depend on how vigorously it boils.

Note the pattern

The temperature of a sample does not change during a phase change.

If heat is added at a constant rate, as in Figure 11.19, then the length of the horizontal lines, which represents the time during which the temperature does not change, is directly proportional to the magnitude of the enthalpies associated with the phase changes. In Figure 11.19, the horizontal line at 100°C is much longer than the line at 0°C because the enthalpy of vaporization of water is several times higher than the enthalpy of fusion.

A **superheated liquid** is a sample of a liquid at the temperature and pressure at which it should be a gas. Superheated liquids are not stable; the liquid will eventually boil, sometimes violently. The phenomenon of superheating causes "bumping" when a liquid is heated in the laboratory. When a test tube containing water is heated over a Bunsen burner, for example, one portion of the liquid can easily become too hot. When the superheated liquid converts to a gas, it can push or "bump" the rest of the liquid out of the test tube. Placing a stirring rod or a small piece of ceramic (a "boiling chip") in the test tube allows bubbles of vapor to form and allows the liquid to boil instead of becoming superheated. Superheating is the reason a liquid heated in a smooth cup in a microwave oven may not boil until the cup is moved, when the motion of the cup allows bubbles to form.

Cooling Curves

The **cooling curve** in Figure 11.20 plots temperature versus time as a 75-g sample of steam, initially at 200°C, is cooled. Although we might expect the cooling curve to be the mirror image of the heating curve in Figure 11.19, the cooling curve is *not* the identical

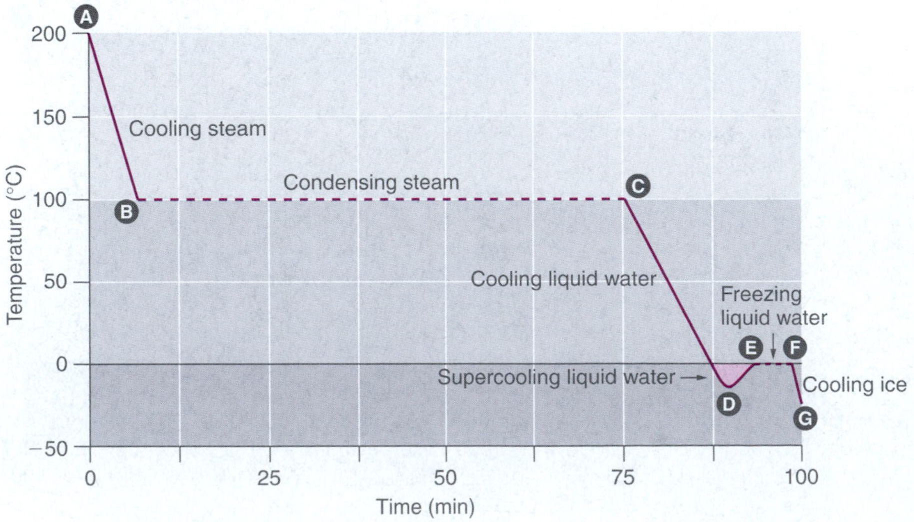

Figure 11.20 A cooling curve for water. A plot of temperature as heat is removed at a constant rate from a 75-g sample of steam initially at 200°C. A–B: cooling steam; B–C: condensing steam; C–D: cooling liquid water to give a supercooled liquid; D–E: warming the liquid as it begins to freeze; E–F: freezing liquid water; F–G: cooling ice.

Interactive Graph

mirror image. As heat is removed from the steam, the temperature falls until it reaches 100°C. At this temperature, the steam begins to condense to liquid water. No further temperature change occurs until all the steam is converted to the liquid, but then the temperature again decreases as the water is cooled. We might expect to reach another plateau at 0°C, where the water is converted to ice, but in reality this does not always occur. Instead, the temperature often drops below the freezing point for some time, as shown by the little dip in the cooling curve below 0°C. This region corresponds to an unstable form of the liquid, a **supercooled liquid**. If the liquid is allowed to stand, if cooling is continued, or if a small crystal of the solid phase is added (a *seed crystal*), the supercooled liquid will convert to a solid, sometimes quite suddenly. As the water freezes, the temperature increases slightly due to the heat evolved during the freezing process, and then holds constant at the melting point as the rest of the water freezes. Subsequently, the temperature of the ice decreases again as more heat is removed from the system.

Supercooling effects have a huge impact on Earth's climate. For example, supercooling of water droplets in clouds can prevent the clouds from releasing precipitation over regions that are persistently arid as a result. Clouds consist of tiny droplets of water, which in principle should be dense enough to fall as rain. In fact, however, the droplets must aggregate to reach a certain size before they can fall to the ground. Usually a small particle (a *nucleus*) is required for the droplets to aggregate; the nucleus can be a dust particle, an ice crystal, or a particle of silver iodide dispersed in a cloud during *seeding* (a method of inducing rain). Unfortunately, the small droplets of water generally remain as a supercooled liquid down to about −10°C, rather than freezing into ice crystals that are more suitable nuclei for raindrop formation. One approach to producing rainfall from an existing cloud is to cool the water droplets so that they crystallize to provide nuclei around which raindrops can grow. This is best done by dispersing small granules of solid CO_2 (dry ice) into the cloud from an airplane. Solid CO_2 sublimes directly to the gas at pressures of 1 atm or lower, and the enthalpy of sublimation is substantial (25.3 kJ/mol). Thus, as the CO_2 sublimes, it absorbs heat from the cloud, often with the desired results.

EXAMPLE 11.8

If a 50.0-g ice cube at 0.0°C is added to 500 mL of tea at 20.0°C, what is the temperature of the tea when the ice cube has just melted? Assume that no heat is transferred to or from the surroundings. The density of water (and iced tea) is 1.00 g/mL over the range 0–20°C, the specific heats of liquid water and ice are 4.184 J/(g·°C) and 2.062 J/(g·°C), respectively, and the enthalpy of fusion of ice is 6.01 kJ/mol.

Given Mass, volume, initial temperature, density, specific heats, ΔH_{fus}

Asked for Final temperature

Strategy

Substitute the values given into the general equation relating heat gained to heat lost (Equation 5.39) to obtain the final temperature of the mixture.

Solution

Recall from Chapter 5 that when two substances or objects at different temperatures are brought into contact, heat will flow from the warmer one to the cooler. The amount of heat that flows is given by

$$q = mC_s\Delta T$$

where q is heat, m is mass, C_s is the specific heat, and ΔT is the temperature change. Eventually, the temperatures of the two substances will become equal at a value somewhere between their initial temperatures. Calculating the temperature of iced tea after adding an ice cube is slightly more complicated. The general equation relating heat gained and heat lost is still valid, but in this case we also have to take into account the amount of heat required to melt the ice cube from ice at 0.0°C to liquid water at 0.0°C:

$$q_{\text{lost}} = -q_{\text{gained}}$$

$$(m_{\text{iced tea}})[C_s(H_2O)](\Delta T_{\text{iced tea}}) = -\{(m_{\text{ice}})[C_s(H_2O)](\Delta T_{\text{ice}}) + (\text{mol}_{\text{ice}})\Delta H_{\text{fus}}(\text{ice})\}$$

$$(500\ \text{g})[4.184\ \text{J/(g}\cdot{}^\circ\text{C)}](T_f - 20.0^\circ\text{C}) = -\ [(50.0\ \text{g})[4.184\ \text{J/(g}\cdot{}^\circ\text{C)}](T_f - 0.0^\circ\text{C}) + \left(\frac{50.0\ \text{g}}{18.0\ \text{g/mol}}\right)(6.01 \times 10^3\ \text{J/mol})$$

$$(2090\ \text{J/}^\circ\text{C})(T_f) - 4.18 \times 10^4\ \text{J} = -[(209\ \text{J/}^\circ\text{C})(T_f) + 1.67 \times 10^4\ \text{J}]$$

$$2.53 \times 10^4\ \text{J} = (2310\ \text{J/}^\circ\text{C})\ T_f \quad \text{or} \quad T_f = 11.0^\circ\text{C}$$

EXERCISE 11.8

Suppose you are overtaken by a blizzard while ski touring and you take refuge in a tent. You are thirsty, but you forgot to bring liquid water. You have a choice of eating a few handfuls of snow (say 400 g) at −5.0°C immediately to quench your thirst, or setting up your propane stove, melting the snow, and heating the water to body temperature before drinking it. You recall that the survival guide you leafed through at the hotel said something about not eating snow, but you can't remember why—after all, it's just frozen water. To understand the guide's recommendation, calculate the amount of heat that your body will have to supply to bring 400 g of snow at −5.0°C to 37°C. Use the data in Example 11.7.

Answer 200 kJ (4.1 kJ to bring the ice from −5.0°C to 0.0°C, 133.6 kJ to melt the ice at 0.0°C, and 61.9 kJ to bring the water from 0.0°C to 37°C), energy that would not have been expended had you first melted the snow.

11.6 • Critical Temperature and Pressure

Note the pattern

High-boiling-point, nonvolatile liquids have high critical temperatures, and vice versa.

In Section 10.8, we saw that a combination of high pressure and low temperature allows gases to be liquefied. As we increase the temperature of a gas, liquefaction becomes more and more difficult because higher and higher pressures are required to overcome the increased kinetic energy of the molecules. In fact, for every substance, there is some temperature above which the gas can no longer be liquefied, regardless of pressure. This temperature is the **critical temperature (T_c)**, the highest temperature at which a substance can exist as a liquid. Above the critical temperature, the molecules have too much kinetic energy for the intermolecular attractive forces to hold them together in a separate liquid phase. Instead, the substance forms a single phase that completely occupies the volume of the container. Substances with strong intermolecular forces tend to form a liquid phase over a very large temperature range and therefore have high critical temperatures. Conversely, substances with weak intermolecular interactions have relatively low critical temperatures. Each substance also has a **critical pressure (P_c)**, the minimum pressure needed to liquefy it at the critical temperature. The combination of critical temperature and critical pressure is called the **critical point** of a substance. The critical temperatures and pressures of several common substances are listed in Table 11.6.

TABLE 11.6 Critical temperatures and pressures of some simple substances

Substance	T_c, °C	P_c, atm
NH_3	132.4	113.5
CO_2	31.0	73.8
CH_3CH_2OH (ethanol)	240.9	61.4
He	−267.96	2.27
Hg	1477	1587
CH_4	−82.6	46.0
N_2	−146.9	33.9
H_2O	374.0	217.7

Supercritical Fluids

To understand what happens at the critical point consider the effects of temperature and pressure on the densities of liquids and gases, respectively. As the temperature of a liquid increases, its density decreases. As the pressure of a gas increases, its density increases. At the critical point, the liquid and gas phases have exactly the same

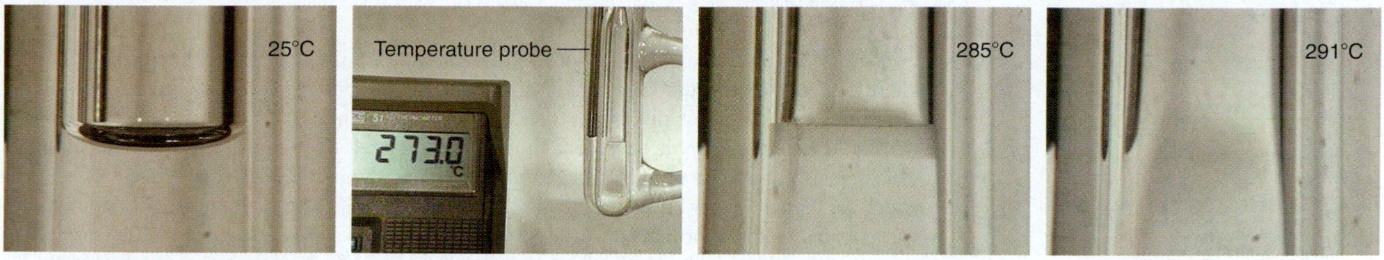

Figure 11.21 Supercritical benzene. Below the critical temperature of benzene (T_c = 289°C), the meniscus between the liquid and gas phases is apparent. At the critical temperature, the meniscus disappears because the density of the vapor is equal to the density of the liquid. Above T_c, a dense homogeneous fluid fills the tube.

density, and only a single phase exists. This single phase is called a **supercritical fluid**, which exhibits many of the properties of a gas but has a density more typical of a liquid. For example, the density of water at its critical point (T = 374°C, P = 217.7 atm) is 0.32 g/mL, about one-third that of liquid water at room temperature but much higher than that of water vapor under most conditions. The transition between a liquid/gas mixture and a supercritical phase is demonstrated for a sample of benzene in Figure 11.21. At the critical temperature, the meniscus separating the liquid and gas phases disappears.

In the last few years, supercritical fluids (SCFs) have evolved from laboratory curiosities to substances with important commercial applications. For example, carbon dioxide has a low critical temperature (31°C), a comparatively low critical pressure (73 atm), and low toxicity, making it easy to contain and relatively safe to manipulate. Because many substances are quite soluble in supercritical CO_2, commercial processes that use it as a solvent are now well established in the oil industry, the food industry, and others. Supercritical CO_2 is pumped into oil wells that are no longer producing much oil in order to dissolve the residual oil in the underground reservoirs. The less-viscous solution is then pumped to the surface, where the oil can be recovered by evaporation (and recycling) of the CO_2. In the food, flavor, and fragrance industry, supercritical CO_2 is used to extract components from natural substances for use in perfumes, to remove objectionable organic acids from hops prior to making beer, and to selectively extract caffeine from whole coffee beans without removing important flavor components. The latter process was patented in 1974, and now virtually all decaffeinated coffee is produced this way. The earlier method used volatile organic solvents such as methylene chloride (dichloromethane), CH_2Cl_2 (bp 40°C), which is difficult to remove completely from the beans and is known to cause cancer in laboratory animals at high doses.

EXAMPLE 11.9

Arrange these substances in order of increasing critical temperatures: methanol, *n*-butane, *n*-pentane, N_2O.

Given Compounds

Asked for Order of increasing critical temperatures

Strategy

A Identify the intermolecular forces in each molecule, and then assess the strengths of those forces.

B Arrange the compounds in order of increasing critical temperatures.

Solution

A The critical temperature depends on the strength of the intermolecular interactions that hold a substance together as a liquid. In N_2O, a slightly polar substance, weak dipole–dipole interactions and London dispersion forces are important. Butane and pentane are larger, nonpolar molecules that exhibit only London dispersion forces. Methanol, in contrast, should have substantial intermolecular hydrogen bonding interactions. Because hydrogen bonds are stronger than the other intermolecular forces, methanol will have the highest T_c. London forces are more important for pentane (C_5H_{12}) than for butane (C_4H_{10}) because of its larger size, so *n*-pentane will have a higher T_c than *n*-butane. The only remaining question is whether N_2O is polar enough to have stronger intermolecular interactions than pentane or butane. Because the electronegativities of O and N are quite similar, the answer is probably no, so N_2O should have the lowest T_c. **B** We predict the order of increasing critical temperatures as N_2O < *n*-butane < *n*-pentane < methanol. The actual values are N_2O (36.9°C) < *n*-butane (152.0°C) < *n*-pentane (196.9°C) < methanol (239.9°C). This is the same order as their normal boiling points—N_2O (−88.7°C) < *n*-butane (−0.2°C) < *n*-pentane (36.0°C) < methanol (65°C)—because both critical temperature and boiling point depend on the relative strengths of the intermolecular interactions.

EXERCISE 11.9

Arrange these substances in order of increasing critical temperatures: ethanol, methanethiol (CH_3SH), ethane, *n*-hexanol.

Answer ethane (32.3°C) < methanethiol (196.9°C) < ethanol (240.9°C) < *n*-hexanol (336.9°C)

Molten Salts and Ionic Liquids

Heating a salt to its melting point produces a **molten salt**. If we heated a sample of solid NaCl to its melting point of 801°C, for example, it would melt to give a stable liquid that conducts electricity. Other characteristics of molten salts are high heat capacity, ability to attain very high temperatures (over 700°C) as a liquid, and utility as solvents because of their relatively low toxicity.

Molten salts have many uses in industry and in the laboratory. For example, in solar power towers in the desert of California, mirrors collect and focus sunlight to melt a mixture of sodium nitrite and sodium nitrate. The heat stored in the molten salt is used to produce steam that drives a steam turbine and generator, thereby producing electricity from the sun for southern California.

Due to their low toxicity and high thermal efficiency, molten salts have also been used in nuclear reactors to operate at temperatures higher than 750°C. One prototype reactor tested in the 1950s used a fuel and coolant consisting of molten fluoride salts, including NaF, ZrF_4, and UF_4. Molten salts are also useful in catalytic processes such as coal gasification, in which carbon and water react at high temperatures to form CO and H_2.

More recently, chemists have been studying the characteristics of **ionic liquids**, ionic substances that are liquid at room temperature. These substances consist of small, symmetrical anions, such as $[PF_6]^-$ and $[BF_4]^-$, combined with larger, asymmetrical organic cations that prevent the formation of a highly organized structure, resulting in a low melting point. By varying the cation and anion, chemists can tailor the liquid to specific needs, such as for use as a solvent in a given reaction or for extracting specific molecules from a solution. For example, an ionic liquid consisting of a bulky cation and anions that bind metal contaminants such as mercury and cadmium ions can remove those toxic metals from the environment. A similar approach has been applied to the removal of uranium and americium from water contaminated by nuclear waste.

Initial interest in ionic liquids centered on their use as a low-temperature alternative to molten salts in batteries for missiles, nuclear warheads, and space probes. Further research revealed that ionic liquids had other useful properties: for example, some could dissolve the black rubber of discarded tires, allowing it to be recovered for recycling. Others could be used to produce commercially important organic compounds with high molecular mass, such as Styrofoam and Plexiglas, at rates 10 times faster than traditional methods.

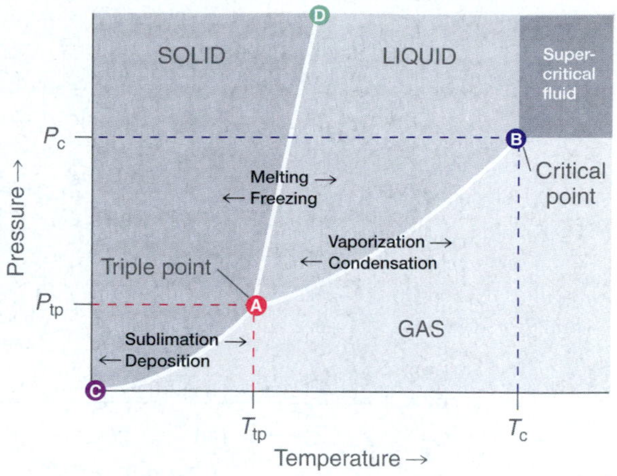

Figure 11.22 A typical phase diagram for a substance that exhibits three phases: solid, liquid, and gas, and a supercritical region. Interactive Graph

MGC Phase Diagrams

11.7 • Phase Diagrams

The state exhibited by a given sample of matter depends on the identity, temperature and pressure of the sample. A **phase diagram** is a graphic summary of the physical state of the substance as a function of temperature and pressure in a closed system.

A typical phase diagram, shown in Figure 11.22, consists of discrete regions that represent the different phases exhibited by the substance. Each region corresponds to the range of combinations of temperature and pressure over which that phase is stable. Note that the combination of high pressure and low temperature (upper left of Figure 11.22) corresponds to the solid phase, whereas the gas phase is favored at high temperature and low pressure (lower right). The combination of high temperature and high pressure (upper right) corresponds to a supercritical fluid.

Note the pattern

The solid phase is favored at low temperature and high pressure; the gas phase is favored at high temperature and low pressure.

General Features of a Phase Diagram

The lines in a phase diagram correspond to the combinations of temperature and pressure at which two phases can coexist in equilibrium. In Figure 11.22, the line that connects points A and D separates the solid and liquid phases and shows how the melting point of the solid varies with pressure. The solid and liquid phases are in equilibrium all along this line; crossing the line horizontally corresponds to melting or freezing. The line that connects points A and B is the vapor pressure curve of the liquid, which we discussed in Section 11.4. It ends at the critical point beyond which the substance exists as a supercritical fluid. The line that connects points A and C is the vapor pressure curve of the *solid* phase. Along this line, the solid is in equilibrium with the vapor phase through sublimation and deposition. Finally, point A, where the solid/liquid, liquid/gas, and solid/gas lines intersect, is the **triple point**; it is the *only* combination of temperature and pressure at which all three phases (solid, liquid, and gas) are in equilibrium and can therefore exist simultaneously. Because no more than three phases can ever coexist, a phase diagram can never have more than three lines intersecting at a single point.

Remember that a phase diagram, such as the one in Figure 11.22, is for a single pure substance in a closed system, not for a liquid in an open beaker in contact with air at 1 atm pressure. In practice, however, the conclusions reached about the behavior of a substance in a closed system can usually be extrapolated to an open system without a great deal of error.

The Phase Diagram of Water

The phase diagram of water, shown in Figure 11.23, illustrates that the triple point of water occurs at 0.01°C and 0.00604 atm (4.59 mmHg). Far more reproducible than the melting point of ice, which depends on the amount of dissolved air and the atmospheric pressure, the triple point (273.16 K) is used to define the absolute (Kelvin) temperature scale. The triple point also represents the lowest pressure at which a

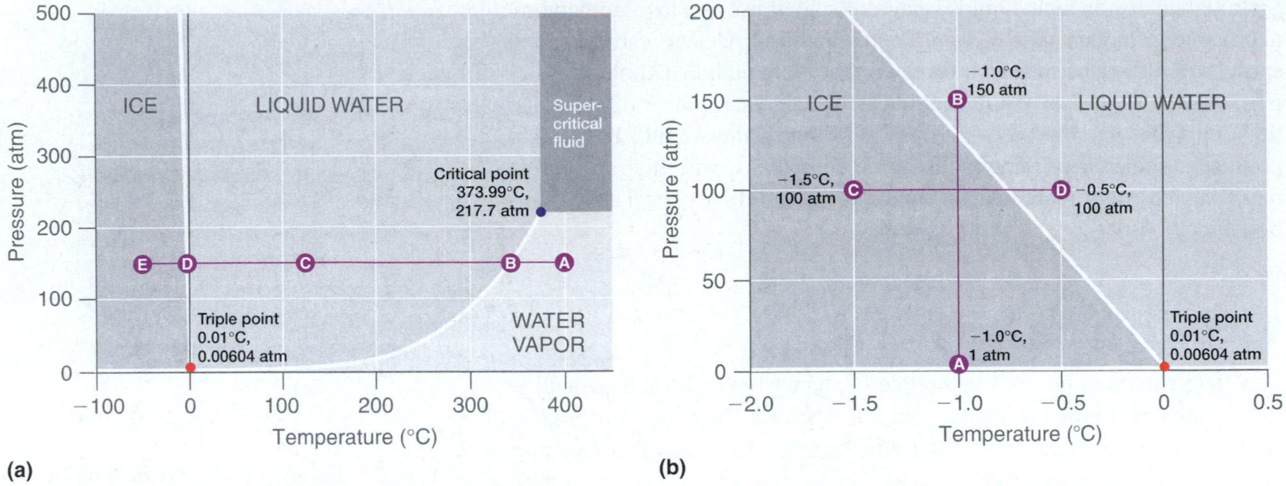

Figure 11.23 Two versions of the phase diagram of water. (a) A graph with linear temperature and pressure axes, in which the boundary between ice and liquid water is almost vertical. (b) A graph with an expanded scale, illustrating the decrease in melting point with increasing pressure. (The letters refer to points discussed in Example 11.10 and Exercise 11.10.) Interactive Graph

liquid phase can exist in equilibrium with the solid or vapor. At pressures below 0.00604 atm, therefore, ice does not melt to a liquid as the temperature increases, but instead sublimes directly to water vapor. Sublimation of water at low temperature and pressure can be used to "freeze-dry" foods and beverages. The food or beverage is first cooled to subzero temperatures and placed in a container in which the pressure is maintained below 0.00604 atm. Then, as the temperature is increased, the water sublimes, leaving the dehydrated food (such as that used by backpackers or astronauts) or the powdered beverage (as with freeze-dried coffee).

The phase diagram for water illustrated in Figure 11.23b shows the boundary between ice and water on an expanded scale. Notice that the melting curve of ice slopes up and slightly to the left rather than up and to the right as in Figure 11.22. That is, the melting point of ice *decreases* with increasing pressure; at 100 MPa (987 atm), ice melts at −9°C. Water behaves this way because it is one of the few known substances for which the crystalline solid is *less dense* than the liquid (others include antimony and bismuth). Increasing the pressure of ice that is in equilibrium with water at 0°C and 1 atm tends to push some of the molecules closer together, thus decreasing the volume of the sample. The decrease in volume (and corresponding increase in density) is smaller for a solid or liquid than for a gas, but it is sufficient to melt some of the ice.

In Figure 11.23b, point A is located at $P = 1$ atm and $T = -1.0$°C, within the solid (ice) region of the phase diagram. As the pressure increases to 150 atm while the temperature remains the same, the line from point A crosses the ice/water boundary to point B, which lies in the liquid water region. Consequently, applying a pressure of 150 atm will melt ice at −1.0°C. We have already indicated that the pressure dependence of the melting point of water is of vital importance. If the solid/liquid boundary in the phase diagram of water were to slant up and to the right rather than to the left, ice would be denser than water, ice cubes would sink, water pipes would not burst when they freeze, and antifreeze would be unnecessary in automobile engines.

Until recently, many textbooks described ice skating as being possible because the pressure generated by the skater's blade is high enough to melt the ice under the blade, thereby creating a lubricating layer of liquid water that enables the blade to slide across the ice. Although this explanation is intuitively satisfying, it is incorrect, as we can show by a simple calculation. Recall from Chapter 10 that pressure P is the force F applied per unit area A:

$$P = \frac{F}{A} \tag{11.4}$$

To calculate the pressure an ice skater exerts on the ice, we need to calculate only the force exerted and the area of the skate blade. If we assume a 75.0-kg (165-lb) skater, then the force F exerted by the skater on the ice due to gravity is

$$F = mg \tag{11.5}$$

where m is the mass and g is the acceleration due to Earth's gravity (9.81 m/s^2). Thus, the force is

$$F = (75.0 \text{ kg})(9.81 \text{ m/s}^2) = 736 \text{ (kg·m)/s}^2 = 736 \text{ N} \tag{11.6}$$

If we assume that the skate blades are 2.0 mm wide and 25 cm long, then the area A of the bottom of each blade is

$$A = (2.0 \times 10^{-3} \text{ m})(25 \times 10^{-2} \text{ m}) = 5.0 \times 10^{-4} \text{ m}^2 \tag{11.7}$$

If the skater is gliding on one foot, the pressure P exerted on the ice is

$$P = \frac{736 \text{ N}}{5.0 \times 10^{-4} \text{ m}^2} = 1.5 \times 10^6 \text{ N/m}^2 = 1.5 \times 10^6 \text{ Pa} = 15 \text{ atm} \tag{11.8}$$

The pressure is much lower than the pressure needed to decrease the melting point of ice by even 1°C, and experience indicates that it is possible to skate even when the temperature is well below freezing. Hence, pressure-induced melting of the ice cannot explain the low friction that enables skaters (and hockey pucks!) to glide. Recent research indicates that the surface of ice, where the ordered array of water molecules meets the air, consists of one or more layers of almost liquid water. These layers, together with melting induced by friction as the skater pushes forward, appear to account for both the ease with which a skater glides and the fact that skating becomes more difficult below about −7°C, when the number of lubricating surface water layers decreases.

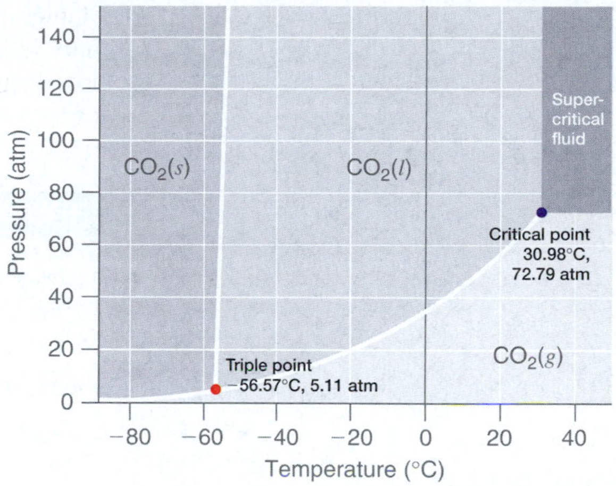

Figure 11.24 The phase diagram of carbon dioxide. Note the critical point, the triple point, and the normal sublimation temperature. Interactive Graph

The Phase Diagram of Carbon Dioxide

In contrast to the phase diagram of water, the phase diagram of CO_2 (Figure 11.24) has a more typical melting curve, sloping up and to the right. The triple point is −56.6°C and 5.11 atm, which means that liquid CO_2 cannot exist at pressures lower than 5.11 atm. At 1 atm, therefore, solid CO_2 sublimes directly to the vapor while maintaining a temperature of −78.5°C, the normal sublimation temperature. Solid CO_2 is generally known as "dry ice" because it is a cold solid with no liquid phase observed when it is warmed.

The phase diagrams for water and carbon dioxide shown here are relatively simple because each contains only one solid, liquid, and gas phase. As we shall see in Chapter 12, however, many substances exhibit more than one stable structure in the solid phase, depending on temperature and pressure, and some substances exhibit multiple liquid or liquid-like phases as well. For example, carbon has three solid phases, as shown in Figure 11.25. Graphite is the stable form of carbon at low temperatures and pressures, but at higher than about 10,000 atm pressure, the more compact diamond structure is favored. (In addition, a third phase, carbon-III, is stable only at extremely high pressures.) Consequently,

Figure 11.25 The phase diagram of carbon. Note the three different solid phases, each with a different structure. [After F.P. Bundy et al., The pressure-temperature phase and transformation diagram for carbon, updated through 1994, *carbon* 1996, 34(2): 141–153; and J.M. Zazula, LHC Project Note 78, 1997, Geneva: CERN.]

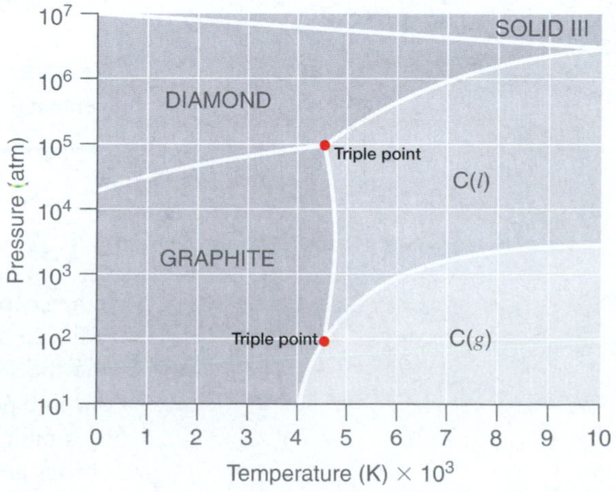

samples of graphite that are maintained at pressures around 100,000 atm and temperatures higher than 1000°C gradually transform into diamond. When the pressure is removed and the sample is cooled, the diamond phase does not convert back to graphite at a measurable rate.

EXAMPLE 11.10

Referring to the phase diagram of water in Figure 11.23, **(a)** predict the physical form of a sample of water at 400°C and 150 atm, and **(b)** describe the changes that occur as the sample in part a is slowly allowed to cool to −50°C at a constant pressure of 150 atm.

Given Phase diagram, temperature, pressure

Asked for Physical form, physical changes

Strategy

Ⓐ Identify the region of the phase diagram corresponding to the initial conditions, and identify the phase that exists in this region.

Ⓑ Draw a line corresponding to the given pressure. Move along that line in the appropriate direction (in this case cooling), and describe the phase changes.

Solution

(a) Ⓐ Locate the starting point on the phase diagram in Figure 11.23a. The initial conditions correspond to point A, which lies in the region of the phase diagram representing water vapor. Thus, water at $T = 400°C$ and $P = 150$ atm is a gas.

(b) Ⓑ Cooling the sample at constant pressure corresponds to moving left along the horizontal line in Figure 11.23a. At about 340°C (point B), we cross the vapor pressure curve, at which point water vapor will begin to condense and the sample will consist of a mixture of vapor and liquid. When all of the vapor has condensed, the temperature drops further, and we enter the region corresponding to liquid water (indicated by point C). Further cooling brings us to the melting curve, the line that separates the liquid and solid phases at a little below 0°C (point D), at which point the sample will consist of a mixture of liquid and solid water (ice). When all of the water has frozen, cooling the sample to −50°C takes us along the horizontal line to point E, which lies within the region corresponding to solid water. At $P = 150$ atm and $T = -50°C$, therefore, the sample is solid ice.

EXERCISE 11.10

Referring again to the phase diagram of water, predict what happens as the pressure on a sample of water at −0.0050°C is gradually increased from 1.0 mmHg to 218 atm.

Answer The sample is initially a gas, condenses to a solid as the pressure increases, and then melts when the pressure is increased further to give a liquid.

11.8 • Liquid Crystals

MGC Liquid Crystals

When cooled, most liquids undergo a simple phase transition to an ordered crystalline solid. In the phase diagrams for these liquids, there are no regions between the liquid and the solid phases. Thousands of substances are known, however, that exhibit one or more phases intermediate between the liquid state, in which the molecules are free to tumble and move past one another, and the solid state, in which the molecules or ions are rigidly locked into place. In these intermediate phases, the molecules have an ordered arrangement and yet can still flow like a liquid. Hence, they are called **liquid crystals**, and their unusual properties have found a wide range of commercial applications. They are used, for example, in the LCDs (liquid crystal displays) in digital watches, calculators, and computer and video displays.

The first documented example of a liquid crystal was reported by the Austrian Frederick Reinitzer in 1888. Reinitzer was studying the properties of a cholesterol derivative, cholesteryl benzoate, and noticed that it behaved strangely as it melted. The white solid first formed a cloudy white liquid phase at 145°C, which reproducibly transformed into a clear liquid at 179°C (Figure 11.26). The transitions were completely reversible: cooling molten cholesteryl benzoate below 179°C caused the clear liquid to revert to a milky one, which then crystallized at the melting point of 145°C.

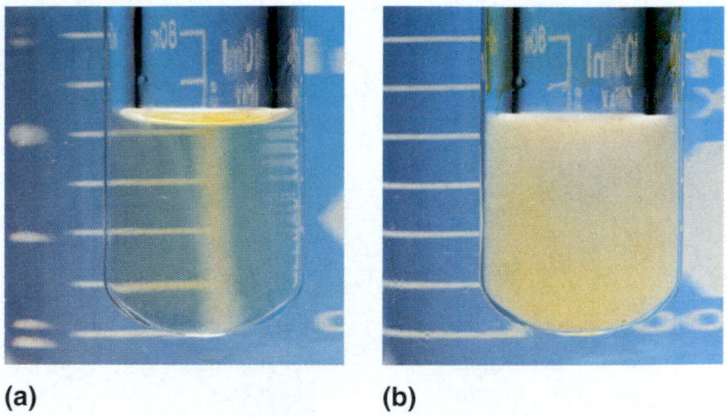

Figure 11.26 Cholesteryl benzoate (a) above 179°C and (b) between 145°C and 179°C. In (a), the substance is an isotropic liquid through which letters can be read. In (b), the substance is in the cholesteric liquid crystalline phase and is an opaque, milky liquid.

In a normal liquid, the molecules possess enough thermal energy to overcome the intermolecular attractive forces and tumble freely. This arrangement of the molecules is described as **isotropic**, which means that it is equally disordered in all directions. Liquid crystals, in contrast, are **anisotropic**: their properties depend on the direction in which they are viewed. Hence, liquid crystals are not as disordered as a liquid because the molecules have some degree of alignment.

Most substances that exhibit the properties of liquid crystals consist of long, rigid rod- or disc-shaped molecules that are easily polarizable and can orient themselves in one of three different ways, as shown in Figure 11.27. In the *nematic* phase, the molecules are not layered but they are pointed in the same direction. As a result, the molecules are free to rotate or to slide past one another. In the *smectic* phase, the molecules maintain the general order of the nematic phase, but in addition they are aligned in layers. Several variants of the smectic phase are known, depending on the angle formed between the molecular axes and the planes of molecules. The simplest such structure is the so-called smectic A phase, in which the molecules can rotate about their long axes within a given plane, but they cannot readily slide past one another. In the *cholesteric* phase, the molecules are directionally oriented and stacked in a helical pattern, with each layer rotated at a slight angle to the ones above and below it. As the degree of molecular ordering increases from the nematic to the cholesteric phase, the liquid becomes more opaque, although direct comparisons are somewhat difficult because most compounds form only one of these liquid crystal phases when the solid is melted or the liquid is cooled.

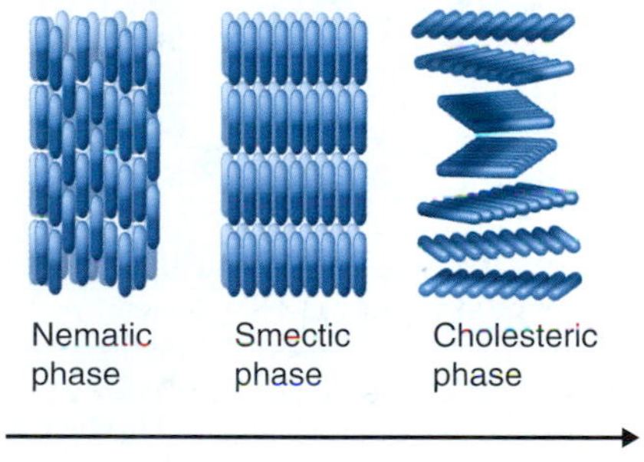

Figure 11.27 The arrangement of molecules in nematic, smectic, and cholesteric liquid crystal phases. In the *nematic* phase, only the long axes of the molecules are parallel and the ends are staggered at random intervals. In the *smectic* phase, the long axes of the molecules are parallel and the molecules are also arranged in planes. Finally, in the *cholesteric* phase, the molecules are arranged in layers; each layer is rotated with respect to the ones above and below it to give a spiral structure. Note the increase in molecular order on going from the nematic to the smectic to the cholesteric phase, which become increasingly opaque.

Molecules that form liquid crystals tend to be rigid molecules with polar groups that exhibit relatively strong dipole–dipole or dipole–induced dipole interactions, hydrogen bonds, or some combination of both. Some examples of substances that form liquid crystals are listed in Table 11.7. In most cases, the intermolecular interactions are due to the presence of polar or polarizable groups. Notice that aromatic rings and multiple bonds between carbon and nitrogen or oxygen are especially common. Moreover, many liquid crystals are made up of molecules that contain two similar halves connected by a unit containing a multiple bond.

Because of their anisotropic structures, liquid crystals exhibit unusual optical and electrical properties. The intermolecular forces are rather weak and can be perturbed by an applied electric field. Because the molecules are polar, they interact with the electric field, which causes them to change their orientation slightly. Nematic liquid crystals, for example, tend to be relatively translucent, but many of them become opaque when an electric field is applied and the molecular orientation changes. This behavior is ideal for producing dark images on a light or opalescent background, and it is used in the LCDs in digital watches, handheld calculators, flat-screen monitors, and car, ship, and aircraft instrumentation. Although each of these applications differs in the details of its construction and operation, the basic principles are similar, as illustrated in Figure 11.28.

Note the pattern

Long, rigid molecules that contain polar groups tend to form liquid crystals.

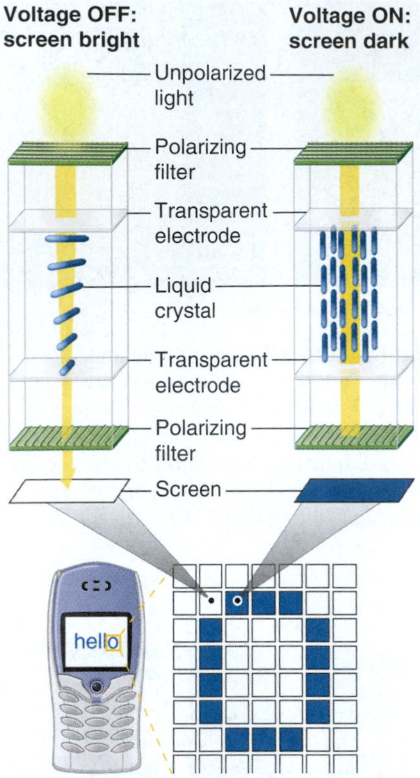

Figure 11.28 Schematic drawing of a liquid crystal display (LCD) device, showing the various layers. Applying a voltage to selected segments of the device will produce any of the numbers. The device is a sandwich that contains a number of very thin layers, consisting of (from top to bottom) a sheet of polarizer to produce polarized light, a transparent electrode, a thin layer of a liquid crystalline substance, a second transparent electrode, a second polarizer, and a screen. Applying an electrical voltage to the liquid crystal changes its orientation slightly, which rotates the plane of the polarized light and makes the area appear dark.

TABLE 11.7 Structures of typical molecules that form liquid crystals[a]

Structure	Liquid Crystal Phase	Liquid Crystalline Temperature Range, °C
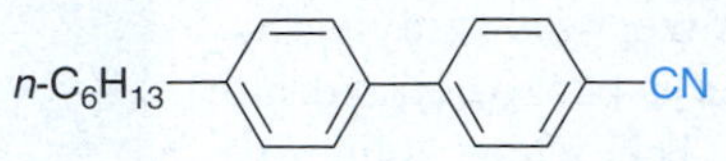	Nematic	14–28
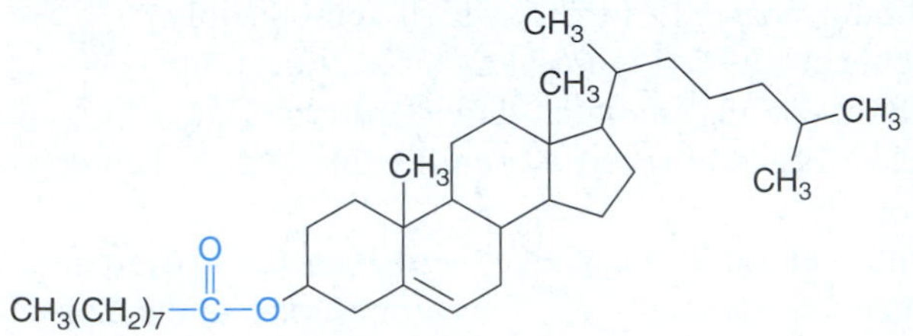	Cholesteric	78–90
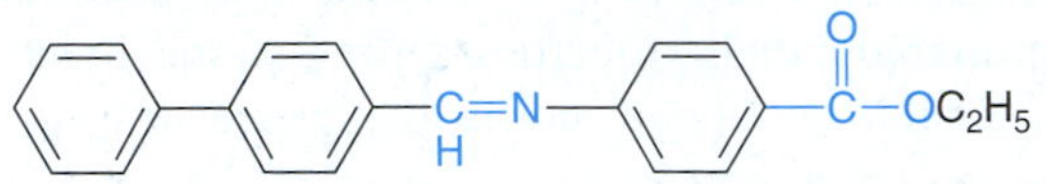	Smectic	121–131

[a]Polar or polarizable groups are indicated in blue.

Changes in molecular orientation that are dependent on temperature result in an alteration of the wavelength of reflected light. Changes in reflected light produce a change in color, which can be customized by using either a single type of liquid crystalline material or mixtures. It is therefore possible to build a liquid crystal thermometer that indicates the temperature by color (Figure 11.29), and to use liquid crystals in heat-sensitive films to detect flaws in electronic board connections where overheating can occur.

With only molecular structure as a guide, one cannot predict precisely which of the various liquid crystalline phases a given compound will actually form. One can, however, identify molecules containing the kinds of structural features that tend to result in liquid crystalline behavior, as demonstrated in the next example.

Figure 11.29 An inexpensive fever thermometer that uses liquid crystals. Each section contains a liquid crystal sample with a different liquid crystalline range. The section whose liquid crystalline range corresponds to the temperature of the body becomes translucent, indicating the temperature.

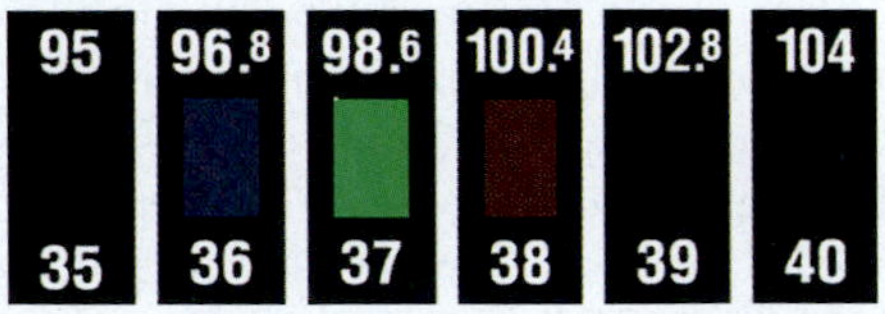

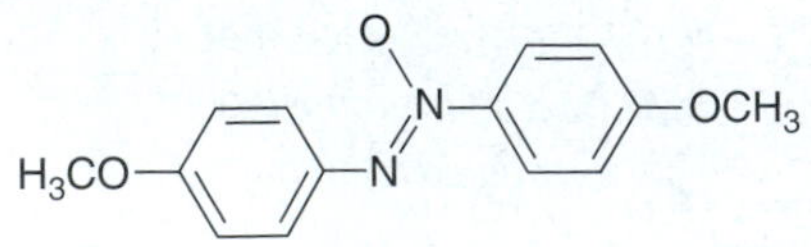

p-Azoxyanisole

EXAMPLE 11.11

Which of the following molecules is most likely to form a liquid crystalline phase as the isotropic liquid is cooled? **(a)** isooctane (2,2,4-trimethylpentane) **(b)** ammonium thiocyanate [$NH_4(SCN)$] **(c)** *p*-azoxyanisole (shown in margin) **(d)** sodium decanoate {$Na[CH_3(CH_2)_8CO_2]$}

Given Compounds

Asked for Liquid crystalline behavior

Strategy

Determine which compounds have a rigid structure and contain polar groups. Those that do are likely to exhibit liquid crystal behavior.

Solution

(a) Isooctane is not long and rigid and contains no polar groups, so it is unlikely to exhibit liquid crystal behavior.

(b) Ammonium thiocyanate is ionic, and ionic compounds tend to have high melting points, so it should not form a liquid crystalline phase. In fact, ionic compounds that form liquid crystals are very rare indeed.

(c) *p*-Azoxyanisole combines two planar phenyl rings linked through a multiply bonded unit, and it contains polar groups. The combination of a long, rigid shape and polar groups makes it a reasonable candidate for a liquid crystal.

(d) Sodium decanoate is the sodium salt of a straight-chain carboxylic acid. The *n*-alkyl chain is long, but it is flexible rather than rigid, so the compound is probably not a liquid crystal.

EXERCISE 11.11

Which one of the following compounds is least likely to form a liquid crystal phase?

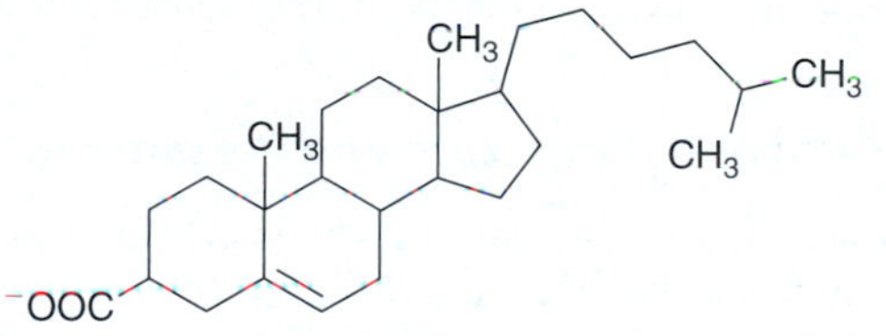

(a) Cholesterol acetate

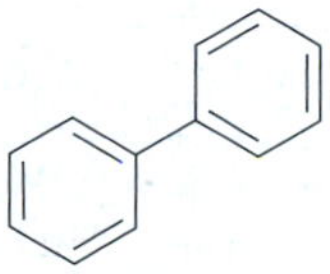

(b) Biphenyl

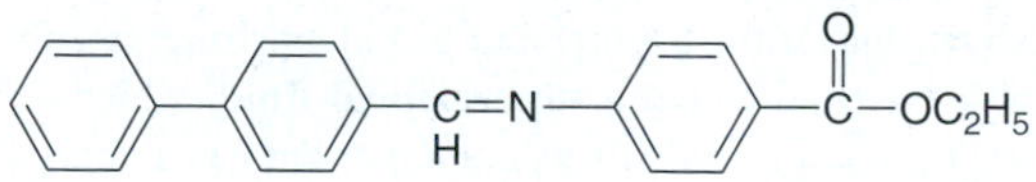

(c)

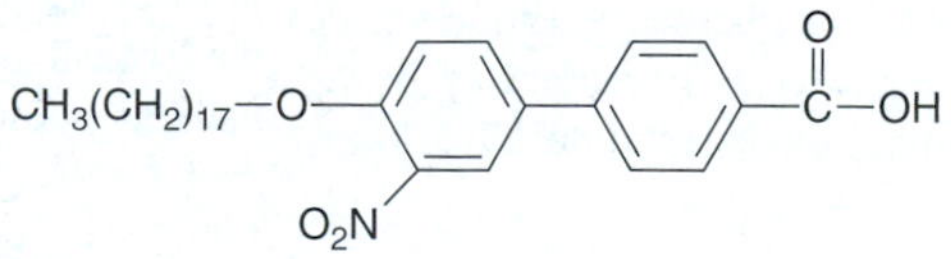

(d)

Answer Biphenyl; although it is rather long and rigid, it lacks any polar substituents.

SUMMARY AND KEY TERMS

11.1 The Kinetic Molecular Description of Liquids (p. 484)

The properties of liquids can be explained using a modified version of the kinetic molecular theory described in Chapter 10. This model explains the higher density, greater order, and lower compressibility of liquids versus gases, the thermal expansion of liquids, why they diffuse, and why they adopt the shape (but not the volume) of their containers.

11.2 Intermolecular Forces (p. 485)

Molecules in liquids are held to other molecules by intermolecular interactions, which are weaker than the intramolecular interactions that hold the atoms together within molecules and polyatomic ions. Transitions between the solid and liquid or the liquid and gas phases are due to changes in intermolecular interactions but do not affect intramolecular interactions. The three major types of intermolecular interactions are dipole–dipole interactions, London dispersion forces (these two are often referred to collectively as **van der Waals forces**), and hydrogen bonds. **Dipole–dipole interactions** arise from the electrostatic interactions of the positive and negative ends of molecules that contain permanent dipole moments; their strength is proportional to the magnitude of the dipole moment and to $1/r^6$, where r is the distance between dipoles. **London dispersion forces** are due to the formation of **instantaneous dipole moments** in polar or nonpolar molecules as a result of short-lived fluctuations of electron charge distribution, which in turn cause the temporary formation of an **induced dipole** in adjacent molecules. Like dipole–dipole interactions, their energy falls off as $1/r^6$. Larger atoms tend to be more *polarizable* than smaller ones because their outer electrons are less tightly bound and are therefore more easily perturbed. **Hydrogen bonds** are especially strong dipole–dipole interactions between molecules that contain hydrogen bonded to a highly electronegative atom such as O, N, or F. The resulting partially positively charged H atom on one molecule (the *hydrogen bond donor*) can interact strongly with a lone pair of electrons of a partially negatively charged O, N, or F atom on adjacent molecules (the *hydrogen bond acceptor*). Because of strong $O \cdots H$ hydrogen bonding between water molecules, water has an unusually high boiling point, and ice has an open, cagelike structure that is less dense than liquid water.

11.3 Unique Properties of Liquids (p. 492)

Surface tension is the energy required to increase the surface area of a liquid by a given amount. The stronger the intermolecular interactions, the higher the surface tension. **Surfactants** are molecules such as soaps and detergents that reduce the surface tension of polar liquids like water. **Capillary action** is the phenomenon in which liquids rise up into a narrow tube called a capillary. It results when **cohesive forces**, the intermolecular forces in the liquid, are weaker than **adhesive forces**, the attraction between the liquid and the surface of the capillary. The shape of the **meniscus**, the upper surface of a liquid in a tube, also reflects the balance between adhesive and cohesive forces. The **viscosity** of a liquid is its resistance to flow. Liquids that have strong intermolecular forces tend to have high viscosities.

11.4 Vapor Pressure (p. 496)

Because the molecules of a liquid are in constant motion and possess a wide range of kinetic energies, at any moment some fraction of them has enough energy to escape from the surface of the liquid to enter the gas or vapor phase. This process, called **vaporization** or **evaporation**, generates a **vapor pressure** above the liquid. Molecules in the gas phase can collide with the liquid surface and reenter the liquid via **condensation**. Eventually, a *steady state* is reached in which the number of molecules evaporating and condensing per unit time is the same, and the system is in a state of **dynamic equilibrium**. Under these conditions, a liquid exhibits a characteristic **equilibrium vapor pressure** that depends only on the temperature. We can express the nonlinear relationship between vapor pressure and temperature as a linear relationship using the Clausius–Clapeyron equation. This equation can be used to calculate the enthalpy of vaporization of a liquid from its measured vapour pressure at two or more temperatures. **Volatile liquids** are liquids with high vapor pressures, which tend to evaporate readily from an open container; **nonvolatile liquids** have low vapor pressures. When the vapor pressure equals the external pressure, bubbles of vapor form within the liquid, and it boils. The temperature at which a substance boils at a pressure of 1 atm is its **normal boiling point**.

11.5 Changes of State (p. 501)

Changes of state are examples of **phase changes**, or **phase transitions**. All phase changes are accompanied by changes in the energy of the system. Changes from a more-ordered to a less-ordered state (such as liquid to gas) are *endothermic*. Changes from a less-ordered to a more-ordered state (such as liquid to solid) are always *exothermic*. The conversion of a solid to a liquid is called melting or **fusion**. The energy required to melt 1 mol of the substance is its enthalpy of fusion, ΔH_{fus}. The energy change required to vaporize 1 mol of a substance is the enthalpy of vaporization, ΔH_{vap}. The direct conversion of a solid to a gas is **sublimation**. The amount of energy needed to sublime 1 mol of a substance is its **enthalpy of sublimation, ΔH_{sub}**, and is the sum of the enthalpies of fusion and vaporization. Plots of the temperature of a substance versus heat added or versus heating time at a constant rate of heating are called **heating curves**. Heating curves relate temperature changes to phase transitions. A *superheated liquid*, a liquid at a temperature and pressure at which it should be a gas, is not stable. A **cooling curve** is not exactly the reverse of the heating curve because many liquids do not freeze at the expected temperature. Instead, they form a **supercooled liquid**, a metastable liquid phase that exists below the normal melting point. Supercooled liquids usually crystallize upon standing, or crystallization can be induced by the addition of a *seed crystal* of the same or another substance.

11.6 Critical Temperature and Pressure (p. 506)

A substance cannot form a liquid above its **critical temperature**, regardless of the applied pressure. Above the critical temperature, the molecules have enough kinetic energy to overcome the intermolecular attractive forces. The minimum pressure needed to liquefy a substance at its critical temperature is its **critical pressure**. The combination of the critical temperature and critical pressure of a substance is its **critical point**. Above the critical temperature and pressure, a substance exists as a dense fluid called a **supercritical fluid**, which resembles a gas in that it completely fills its container but has a density comparable to that of a liquid. A **molten salt** is a salt heated to its melting point, giving a stable liquid that conducts electricity. **Ionic liquids** are ionic substances that are liquid at room temperature. Their disorganized structure results in a low melting point.

11.7 Phase Diagrams (p. 509)

The states of matter exhibited by a substance under different temperatures and pressures can be summarized graphically in a **phase diagram**, a plot of pressure versus temperature. Phase diagrams contain discrete regions corresponding to the solid, liquid, and gaseous phases. The solid and liquid regions are separated by the melting curve of the substance, and the liquid and gas regions are separated by its vapor pressure curve, which ends at the critical point. Within a given region, only a single phase is stable, but along the lines that separate the regions, two phases are in equilibrium at a given temperature and pressure. The lines separating the three phases intersect at a single point, the **triple point**, which is the only combination of temperature and pressure at which all three phases can coexist in equilibrium. Water has an unusual phase diagram: its melting point decreases with increasing pressure because ice is less dense than liquid water. The phase diagram of carbon dioxide shows that liquid carbon dioxide cannot exist at atmospheric pressure. Consequently, solid carbon dioxide sublimes directly to the gas.

11.8 Liquid Crystals (p. 512)

Many substances exhibit phases that have properties intermediate between those of a crystalline solid and a normal liquid. These substances, which possess long-range molecular order but still flow like liquids, are called **liquid crystals**. Liquid crystals are typically long, rigid molecules that can interact strongly with one another; they do not have **isotropic** structures, which are completely disordered, but rather have **anisotropic** structures, which exhibit different properties when viewed from different directions. In the *nematic* phase, only the long axes of the molecules are aligned, whereas in the *smectic* phase, the long axes of the molecules are parallel and the molecules are arranged in planes. In the *cholesteric* phase, the molecules are arranged in planes, but each layer is rotated by a certain amount with respect to those above and below it, giving a helical structure.

KEY EQUATIONS

Clausius–Clapeyron equation	$\ln P = \dfrac{-\Delta H_{vap}}{R}\left(\dfrac{1}{T}\right) + C$	(11.1)
Using vapor pressure at two temperatures to calculate ΔH_{vap}	$\ln\left(\dfrac{P_2}{P_1}\right) = \dfrac{-\Delta H_{vap}}{R}\left(\dfrac{1}{T_2} - \dfrac{1}{T_1}\right)$	(11.2)

QUESTIONS AND PROBLEMS

MGC For instructor-assigned homework, go to **www.masteringgenchem.com**

Please be sure you are familiar with the topics discussed in Essential Skills 6 at the end of this chapter before proceeding to the Questions and Problems. Questions and Problems with colored numbers have answers in the Appendix and complete solutions in the Student Solutions Manual.

CONCEPTUAL

11.1 The Kinetic Molecular Description of Liquids

1. A liquid, unlike a gas, is virtually *incompressible*. Explain what this means using macroscopic and microscopic descriptions. What general physical properties do liquids share with solids? What properties do liquids share with gases?
2. Discuss the differences and similarities between liquids and gases with regard to thermal expansion using a kinetic molecular approach.
3. Discuss the differences and similarities between liquids and gases with regard to fluidity using a kinetic molecular approach.
4. Discuss the differences and similarities between liquids and gases with regard to diffusion using a kinetic molecular approach.
5. How must the ideal gas law be altered to apply the kinetic molecular theory of gases to liquids? Explain.
6. Why are the rms speeds of molecules in liquids less than the rms speeds of molecules in gases?

11.2 Intermolecular Forces

7. What is the main difference between intramolecular interactions and intermolecular interactions? Which is typically stronger? How are changes of state affected by these different kinds of interactions?
8. Describe the three major kinds of intermolecular interactions discussed in this chapter and their major features. The hydrogen bond is actually an example of one of the other two types of interaction. Identify the kind of interaction that includes hydrogen bonds, and explain why hydrogen bonds fall into this category.
9. Which are stronger, dipole–dipole interactions or London dispersion forces? Which are likely to be more important in a molecule that contains heavy atoms? Explain your answers.
10. Explain why hydrogen bonds are unusually strong compared to other dipole–dipole interactions. How does the strength of hydrogen bonds compare with the strength of covalent bonds?
11. Liquid water is essential for life as we know it, but based on its molecular mass, water should be a gas under standard conditions. Why is water a liquid rather than a gas under standard conditions?
12. Describe the effect of each of the following on the melting point and boiling point of a substance: polarity, molecular mass, and hydrogen bonding.
13. Why are intermolecular interactions more important for liquids and solids than for gases? Under what conditions must these interactions be considered for gases?
14. Using acetic acid as an example, illustrate both attractive and repulsive intermolecular interactions. How does the boiling point of a substance depend on the magnitude of the repulsive intermolecular interactions?
15. In Group 17, elemental fluorine and chlorine are gases, whereas bromine is a liquid and iodine is a solid. Why?
16. The boiling points of the anhydrous hydrogen halides are: HF, 19°C; HCl, −85°C; HBr, −67°C; HI, −34°C. Explain any trends in the data, as well as any deviations from that trend.
17. Identify the most important intermolecular interaction in each of the following: SO_2, HF, CO_2, CCl_4, and CH_2Cl_2.

18. Identify the most important intermolecular interaction in each of the following compounds: LiF, I_2, ICl, NH_3, NH_2Cl?

19. Would you expect London dispersion forces to be more important for Xe or Ne? Why? The atomic radius of Ne is 38 pm, whereas that of Xe is 108 pm.

20. Arrange the following species in order of increasing polarizability: Kr, Cl_2, H_2, N_2, Ne, and O_2. Explain your reasoning.

21. Both water and methanol have anomalously high boiling points due to hydrogen bonding, but the boiling point of water is higher than that of methanol despite its lower molecular mass. Why? Draw the structures of these two compounds, including any lone pairs, and indicate potential hydrogen bonds.

22. The structures of ethanol, ethylene glycol, and glycerin are shown below. Arrange these compounds in order of increasing boiling point. Explain your rationale.

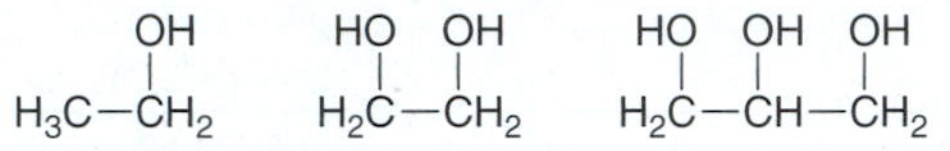

Ethanol Ethylene glycol Glycerin

23. Do you expect the boiling point of H_2S to be higher or lower than that of H_2O? Justify your answer.

24. Ammonia (NH_3), methyl amine (CH_3NH_2), and ethyl amine ($CH_3CH_2NH_2$) are gases at room temperature, while propyl amine ($CH_3CH_2CH_2NH_2$) is a liquid at room temperature. Explain these observations.

25. Why is it not advisable to freeze a sealed glass bottle that is completely filled with water? Use both macroscopic and microscopic models to explain your answer. Is a similar consideration required for a bottle that contains pure ethanol? Why or why not?

26. Which compound in the following pairs will have the higher boiling point: (a) NH_3 or PH_3; (b) ethylene glycol ($HOCH_2CH_2OH$) or ethanol; (c) 2,2-dimethylpropanol ($CH_3C(CH_3)_2CH_2OH$) or *n*-butanol ($CH_3CH_2CH_2OH$)? Explain your reasoning.

27. Some recipes call for vigorous boiling, while others call for gentle simmering. What is the difference in the temperature of the cooking liquid between boiling and simmering? What is the difference in energy input?

28. Use the melting of a metal such as lead to explain the process of melting in terms of what is happening at the molecular level. As a piece of lead melts, the temperature of the metal remains constant, even though energy is being added continuously. Why?

29. How does the O—H distance in a hydrogen bond in liquid water compare with the O—H distance in the covalent O—H bond in the H_2O molecule? What effect does this have on the structure and density of ice?

30. (a) Explain why the hydrogen bonds in liquid HF are stronger than the corresponding intermolecular H ··· I interactions in liquid HI.
(b) In which substance are the individual hydrogen bonds stronger: HF or H_2O? Explain your reasoning.
(c) For which substance will hydrogen bonding have the greater effect on the boiling point: HF or H_2O? Explain your reasoning.

11.3 Unique Properties of Liquids

31. Why is a water droplet round?

32. How is the environment of molecules on the surface of a liquid droplet different from that of molecules in the interior of the droplet? How is this difference related to the concept of surface tension?

33. Explain the role of intermolecular and intramolecular forces in surface tension.

34. A mosquito is able to walk across water without sinking, but if a few drops of detergent are added to the water, the insect will sink. Why?

35. Explain how soaps or surfactants decrease the surface tension of a liquid. How does the meniscus of an aqueous solution in a capillary change if a surfactant is added? Illustrate your answer with a sketch.

36. Which of the following liquids has the lowest viscosity? Which has the highest surface tension? Explain your reasoning in each case: CH_2Cl_2, hexane, ethanol.

37. At 25°C, cyclohexanol has a surface tension of 32.92 mN/m^2, whereas the surface tension of cyclohexanone, which is very similar chemically, is only 25.45 mN/m^2. Why is the surface tension of cyclohexanone so much less than that of cyclohexanol?

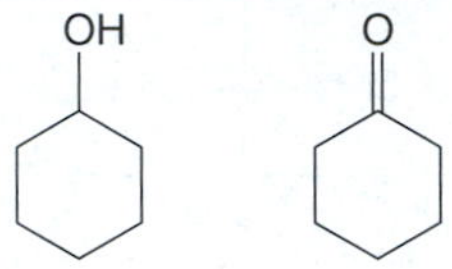

cyclohexanol cyclohexanone

38. What is the relationship between: (a) surface tension and temperature; (b) viscosity and temperature? Explain your answers in terms of a microscopic picture.

39. What two opposing forces are responsible for capillary action? How do these forces determine the shape of the meniscus?

40. Which of the following liquids will have a concave meniscus in a glass capillary: (a) pentane, (b) diethyleneglycol ($HOCH_2CH_2OCH_2CH_2OH$), or (c) carbon tetrachloride? Explain your reasoning.

41. How does viscosity depend upon molecular shape? What molecular features make liquids highly viscous?

11.4 Vapor Pressure

42. What is the relationship between the boiling point of a substance, its vapor pressure, temperature, and atmospheric pressure?

43. What is the difference between a volatile and a nonvolatile liquid? Suppose that two liquid substances have the same molecular mass, but one is volatile and the other is nonvolatile. What differences in the molecular structures of the two substances could account for the differences in volatility?

44. An "old wives' tale" states that applying ethanol to the wrists of a child with a very high fever will help to reduce the fever, because blood vessels in the wrists are close to the skin. Is there a scientific basis for this recommendation? Would water be as effective as ethanol?

45. Why is the air over a strip of grass significantly cooler than the air over a sandy beach only a few feet away?

46. If gasoline is allowed to sit in an open container, it often feels much colder than the surrounding air. Explain this observation. Describe the flow of heat into or out of the system, as well as any transfer of mass that occurs. Would the temperature of a sealed can of gasoline be higher, lower, or the same as that of the open can? Explain your answer.

47. What is the relationship between the vapor pressure of a liquid and: (a) its temperature; (b) the surface area of the liquid; (c) the pressure of other gases on the liquid; and (d) its viscosity.

48. At 25°C, benzene has a vapor pressure of 12.5 kPa, whereas the vapor pressure of acetic acid is 2.1 kPa. Which is more volatile? Based on the intermolecular interactions in the two liquids, explain why acetic acid has the lower vapor pressure.

11.5 Changes of State

49. In extremely cold climates, snow can disappear with no evidence of its melting. How can this happen? What change(s) in state are taking place? Would you expect this phenomenon to be more common at high or low altitudes? Explain your answer.
50. Why do car manufacturers recommend that an automobile should not be left standing in subzero temperatures if its radiator contains only water? Car manufacturers also warn car owners that they should only check the fluid level in a radiator when the engine is cool. What is the basis for this warning? What is likely to happen if it is ignored?
51. Use Hess's law and a thermochemical cycle to show that, for any solid, the enthalpy of sublimation is equal to the sum of the enthalpy of fusion of the solid and the enthalpy of vaporization of the resulting liquid.
52. Three distinct processes occur when an ice cube at −10°C is used to cool a glass of water at 20°C. What are they? Which causes the greatest temperature change in the water?
53. When frost forms on a piece of glass, crystals of ice are deposited from water vapor in the air. How is this process related to sublimation? Describe the energy changes that take place as the water vapor is converted to frost.
54. What phase changes are involved in each of the following processes: (a) ice melting; (b) distillation; (c) condensation forming on a window; (d) the use of dry ice to create a cloud for a theatrical production? Which of these processes are exothermic, and which are endothermic?
55. Describe the phase change involved in each of the following processes, and state whether each process is endothermic or exothermic: (a) evaporation of methanol; (b) crystallization; (c) liquefaction of natural gas; (d) the use of naphthalene crystals to repel moths.
56. Why do substances that have high enthalpies of fusion tend to have high melting points?
57. Why is the enthalpy of vaporization of a compound invariably much larger than its enthalpy of fusion?
58. What is the opposite of each of the following processes: fusion, sublimation, and condensation? Describe the phase change in each pair of opposing processes, and state whether each phase change is exothermic or endothermic.
59. Sketch a typical heating curve (temperature versus amount of heat added at a constant rate) for conversion of a solid to a liquid and then to a gas. What causes some regions of the plot to have a positive slope? What is happening in the regions of the plot where the curve is horizontal, meaning that the temperature does not change even though heat is being added?
60. If you know the mass of a sample of a substance, how could you use a heating curve to calculate the specific heat of the substance, as well as the change in enthalpy associated with a phase change?
61. Sketch the heating curve for a liquid that has been superheated; how does this differ from a normal heating curve for a liquid? Sketch the cooling curve for a liquid that has been supercooled; how does this differ from a normal cooling curve for a liquid?

11.6 Critical Temperature and Pressure

62. Describe the changes that take place when a liquid is heated above its critical temperature. How does this affect the physical properties?
63. What is meant by the term *critical pressure*? What is the effect of increasing the pressure on a gas to above its critical pressure? Would it make any difference if the temperature of the gas was higher than its critical temperature?
64. Do you expect the physical properties of a supercritical fluid to be more like those of the gas or the liquid phase? Explain. Can an ideal gas form a supercritical fluid? Why or why not?
65. What are the limitations in using supercritical fluids to extract organic materials? What are the advantages?
66. Describe the differences between a molten salt and an ionic liquid. Under what circumstances would an ionic liquid be preferred over a molten salt?

11.7 Phase Diagrams

67. A phase diagram is a graphic representation of the stable phase of a substance at any combination of temperature and pressure. What do the lines separating different regions in a phase diagram indicate? What information does the slope of a line in a phase diagram convey about the physical properties of the phases it separates? Can a phase diagram have more than one point where three lines intersect?
68. If the slope of the line corresponding to the solid/liquid boundary in the phase diagram of water were positive rather than negative, what would be the effect on aquatic life during periods of subzero temperatures? Explain your answer.

11.8 Liquid Crystals

69. Describe the common structural features of molecules that form liquid crystals. What kind of intermolecular interactions are most likely to result in a long-chain molecule that exhibits liquid crystalline behavior? Does an electrical field affect these interactions?
70. What is the difference between an *isotropic* liquid and an *anisotropic* liquid? Which is more anisotropic, a cholesteric liquid crystal or a nematic liquid crystal?

NUMERICAL

This section includes paired problems (marked by brackets) that require similar problem-solving skills.

11.3 Unique Properties of Liquids

71. The viscosities of five liquids at 25°C are given in the table. Explain the observed trends in viscosity.

Compound	Molecular Formula	Viscosity, mPa · s
Benzene	C_6H_6	0.604
Aniline	$C_6H_5NH_2$	3.847
1,2-Dichloroethane	$C_2H_4Cl_2$	0.779
Heptane	C_7H_{16}	0.357
1-Heptanol	$C_7H_{15}OH$	5.810

72. The following table gives values for the viscosity, boiling point, and surface tension of four substances. Examine these data carefully to see if the data for each compound are internally

consistent, and point out any obvious errors or inconsistencies. Explain your reasoning.

Compound	Viscosity, mPa · s at 20°C	Boiling Point, °C	Surface Tension, dyn/cm at 25°C
A	0.41	61	27.16
B	0.55	65	22.55
C	0.92	105	36.76
D	0.59	110	28.53

73. Surface tension data (in dyn/cm) for propionic acid, $C_3H_6O_2$, and isopropanol, C_3H_8O, as a function of temperature are given in the following table. Plot the data for each compound, and explain the differences between the two graphs. Based on these data, which molecule is more polar?

Compound	25°C	50°C	75°C
Propanoic acid	26.20	23.72	21.23
2-Propanol	20.93	18.96	16.98

11.4 Vapor Pressure

74. The table gives the vapor pressure of water at various temperatures. Plot the data, and use your graph to estimate the vapor pressure of water at 25°C and at 75°C. What is the vapor pressure of water at 110°C? Use these data to determine the value of ΔH_{vap} for water.

T, °C	0	10	30	50	60	80	100
P, mmHg	4.6	9.2	31.8	92.6	150	355	760

75. Acetylene, C_2H_2, which is used for industrial welding, is transported in pressurized cylinders. Its vapor pressure at various temperatures is given in the following table. Plot the data, and use your plot to predict the vapor pressure of acetylene at 293 K. Then use your plot to determine the value of ΔH_{vap} for acetylene. How much energy is required to vaporize 2.00 g of acetylene at 250 K?

T, K	145	155	175	200	225	250	300
P, mmHg	1.3	7.8	32.2	190	579	1370	5093

76. The normal boiling point of sodium is 883°C. If ΔH_{vap} is 97.4 kJ/mol, what is the vapor pressure (in mmHg) of liquid sodium at 300°C?

77. The enthalpy of vaporization of carbon tetrachloride is 29.8 kJ/mol, and its normal boiling point is 76.8°C. What would the boiling point be at a pressure of 0.100 atm?

78. An unknown liquid has a boiling point of 75.8°C at 0.910 atm and a boiling point of 57.2°C at 0.430 atm. Use the data in Table 11.5 to identify the liquid.

79. An unknown liquid has a vapor pressure of 0.860 atm at 63.7°C and a vapor pressure of 0.330 atm at 35.1°C. Use Table 11.5 to identify the liquid.

80. If the vapor pressure of a liquid is 0.799 atm at 99.0°C and 0.842 atm at 111°C, what is the normal boiling point of the liquid?

81. If the vapor pressure of a liquid is 0.850 atm at 20°C and 0.897 atm at 25°C, what is the normal boiling point of the liquid?

82. The vapor pressure of liquid SO_2 is 33.4 torr at −63.4°C and 100.0 torr at −47.7 K.
(a) Determine the enthalpy of vaporization of SO_2.
(b) What is its vapor pressure at a temperature of −24.5 K?
(c) At what temperature is the vapor pressure equal to 220 torr?

83. The vapor pressure of CO_2 at various temperatures is given in the following table:

T, °C	−120	−110	−100	−90
P, torr	9.81	34.63	104.81	279.5

(a) What is ΔH_{vap} over this temperature range?
(b) What is the vapor pressure of CO_2 at −70°C?
(c) At what temperature does CO_2 have a vapor pressure of 310 torr?

11.5 Changes of State

84. The density of propane at 1 atm and various temperatures is given in the following table. Plot the data, and use your graph to predict the normal boiling point of propane.

T, K	100	125	150	175	200	225	250	275
d, mol/L	16.3	15.7	15.0	14.4	13.8	13.2	0.049	0.044

85. The table lists the density of oxygen at various temperatures and at 1 atm. Graph the data, and then predict the boiling point of oxygen.

T, K	60	70	80	90	100	120	140
d, mol/L	40.1	38.6	37.2	35.6	0.123	0.102	0.087

86. Propionic acid has a melting point of −20.8°C and a boiling point of 141°C. Sketch a heating curve showing the temperature versus time as heat is added at a constant rate to show the behavior of a sample of propionic acid as it is heated from −50°C to its boiling point. What happens above 141°C?

87. Sketch the cooling curve for a sample of the vapor of a compound that has a melting point of 34°C and a boiling point of 77°C as it is cooled from 100°C to 0°C.

88. A 2.0-L sample of gas at 210°C and 0.762 atm condenses to give 1.20 mL of liquid; 476 J of heat is released during the process. What is the enthalpy of vaporization of the compound?

89. A 0.542-g sample of I_2 requires 96.1 J of energy to be converted to vapor. What is the enthalpy of sublimation of I_2?

90. How much energy is released upon freezing 100.0 g of dimethyl disulfide, $C_2H_6S_2$, initially at a temperature of 20°C? Use the following information: mp = −84.7°C, ΔH_{fus} = 9.19 kJ/mol, C_p = 118.1 J/(mol · K).

91. One fuel used for jet engines and rockets is aluminum borohydride, $Al(BH_4)_3$, a liquid that readily reacts with water to produce hydrogen. The liquid has a boiling point of 44.5°C. How much energy is needed to vaporize 1.0 kg of aluminum borohydride at 20°C, given a ΔH_{vap} of 30 kJ/mol and a molar heat capacity (C_p) of 194.6 J/(mol · K)?

Problems 92–95 use the following information: ΔH_{fus} *(H_2O) = 6.01 kJ/mol;* ΔH_{vap} *(H_2O) = 40.66 kJ/mol;* $C_{p(s)}$ *(crystalline H_2O) = 38.02 J/(K · mol);* $C_{p(l)}$ *(liquid H_2O) = 75.35 J/(K · mol);* $C_{p(g)}$ *(H_2O gas) = 33.60 J/(K · mol).*

92. How much heat must be applied to convert a 1.00-g piece of ice at −10°C to steam at 120°C?

93. How much heat is released in the conversion of 1.00 L of steam at 21.9 atm and 200°C to ice at −6.0°C and 1 atm?

94. How many grams of ice at −5.0°C must be added to 150.0 g of water at 22°C to give a final temperature of 15°C?

95. How many grams of boiling water must be added to a glass that contains 25.0 g of ice at −3°C to obtain a liquid with a temperature of 45°C?

11.7 Phase Diagrams

96. Naphthalene, $C_{10}H_8$, is the key ingredient in mothballs. It has a normal melting point of 81°C and a normal boiling point of 218°C. The triple point of naphthalene is 80°C at 1000 Pa. Use these data to construct a phase diagram for naphthalene; label all the regions of your diagram.

97. Argon, an inert gas used in welding, has normal boiling and freezing points of 87.3 K and 83.8 K, respectively. The triple point of argon is 83.8 K at 0.68 atm. Use these data to construct a phase diagram for argon, and label all regions of your diagram.

APPLICATIONS

98. During cold periods, workers in the citrus industry often spray water on orange trees to prevent them from being damaged, even though ice forms on the fruit. (a) Explain the scientific basis for this practice. (b) To illustrate why the production of ice prevents damage to the fruit during cold weather, calculate the heat released by formation of ice from 1000 L of water at 10°C.

99. Relative humidity is the ratio of the actual partial pressure of water in the air to the vapor pressure of water at that temperature (that is, if the air was completely saturated with water vapor), multiplied by 100% to give a percentage. On a summer day in the Chesapeake, when the temperature was recorded as 35°C, the partial pressure of water was reported to be 33.9 mmHg. (a) Refer to the data in Problem 74, and calculate the relative humidity. (b) Why does it seem "drier" in the winter, even though the relative humidity may be the same as in the summer?

100. Liquids are frequently classified according to their physical properties, such as surface tension, vapor pressure, and boiling point. Such classifications are useful when substitutes are needed for a liquid that might not be available. (a) Draw the structure of each of the following compounds: methanol, benzene, pentane, toluene, cyclohexane, 1-butanol, trichloroethylene, acetic acid, acetone, chloroform. (b) Identify the most important kind of intermolecular interaction in each. (c) Sort the compounds into three groups with similar characteristics. (d) If you needed a substitute for trimethylpentane, from which group would you make your selection?

101. In the process of freeze drying, which is used as a preservation method and to aid in the shipping or storage of fruit and biological samples, a sample is cooled and then placed in a compartment in which a very low pressure is maintained, ~0.01 atm. (a) Explain how this process removes water and "dries" the sample. (b) Identify the phase change that occurs during this process. (c) Using the Clausius–Clapeyron equation, show why it is possible to remove water and still maintain a low temperature at this pressure.

102. Many industrial processes for preparing compounds use "continuous-flow reactors," which are chemical reaction vessels in which the reactants are mixed and allowed to react as they flow along a tube. The products are removed at a certain distance from the starting point, when the reaction is nearly complete. The key operating parameters in a continuous-flow reactor are temperature, reactor volume, and reactant flow rate. As an industrial chemist, you think you have successfully modified a particular process to produce a higher product yield by substituting one reactant for another. The viscosity of the new reactant is, however, higher than that of the initial reactant.
(a) Which of the operating parameters will be most greatly affected by this change?
(b) What other parameter could be changed to compensate for the substitution?
(c) Predict the possible effects on your reactor and your process if you do not compensate for the substitution.

ESSENTIAL SKILLS 6

TOPICS

- Natural Logarithms
- Calculations Using Natural Logarithms

Essential Skills 3 introduced the common, or base-10, logarithms and showed how to use the properties of exponents to perform logarithmic calculations. In this section, we describe natural logarithms, their relationship to common logarithms, and how to do calculations with them using the same properties of exponents.

Natural Logarithms

Many natural phenomena exhibit an exponential rate of increase or decrease. Population growth is an example of an exponential rate of increase, whereas a runner's performance may show an exponential decline if initial improvements are substantially greater than those that occur at later stages of training. Exponential changes are represented logarithmically by e^x, where e is an irrational number whose value is approximately 2.7183. The *natural logarithm*, abbreviated as *ln*, is the power x to which e must be raised to obtain a particular number. The natural logarithm of e is 1 ($\ln e = 1$).

Some important relationships between base-10 logarithms and natural logarithms are

$$10^1 = 10 = e^{2.303}$$
$$\ln e^x = x$$
$$\ln 10 = \ln(e^{2.303}) = 2.303$$
$$\log 10 = \ln e = 1$$

According to these relationships, $\ln 10 = 2.303$ and $\log 10 = 1$. Because multiplying by 1 does not change an equality,

$$\ln 10 = 2.303 \log 10$$

Substituting any value y for 10 gives

$$\ln y = 2.303 \log y$$

Other important relationships are

$$\log A^x = x \log A$$
$$\ln(e^x) = x \ln e = x = e^{\ln x}$$

Entering a value x, such as 3.86, into your calculator and pressing the "ln" key gives the value of $\ln(x)$, which is 1.35 for $x = 3.86$. Conversely, entering the value 1.35 and pressing "e^x" gives an answer of 3.86.* Hence,

$$e^{\ln 3.86} = e^{1.35} = 3.86$$
$$\ln(e^{3.86}) = 3.86$$

SKILL BUILDER ES6.1

Calculate the natural logarithm of each number, and express each as a power of the base e: **(a)** 0.523; **(b)** 1.63.

Solution

(a) $\ln(0.523) = -0.648$; $e^{-0.648} = 0.523$
(b) $\ln(1.63) = 0.489$; $e^{0.489} = 1.63$

* On some calculators, pressing [INV] and then [ln x] is equivalent to pressing [e^x].

SKILL BUILDER ES6.2

What number is each of the following the natural logarithm of? **(a)** 2.87; **(b)** 0.030; **(c)** -1.39

Solution

(a) $\ln x = 2.87$; $x = e^{2.87} = 17.6 = 18$ to two significant figures
(b) $\ln x = 0.030$; $x = e^{0.030} = 1.03 = 1.0$ to two significant figures
(c) $\ln x = -1.39$; $x = e^{-1.39} = 0.249 = 0.25$

Calculations Using Natural Logarithms

Like common logarithms, natural logarithms make use of the properties of exponents. We can expand the table in Essential Skills 3 to include natural logarithms:

Operation	Exponential Form	Logarithm
Multiplication	$(10^a)(10^b) = 10^{a+b}$	$\log(AB) = \log A + \log B$
	$(e^x)(e^y) = e^{x+y}$	$\ln(e^x e^y) = \ln(e^x) + \ln(e^y) = x + y$
Division	$\dfrac{10^a}{10^b} = 10^{a-b}$	$\log\left(\dfrac{A}{B}\right) = \log A - \log B$
	$\dfrac{e^x}{e^y} = e^{x-y}$	$\ln\left(\dfrac{X}{Y}\right) = \ln X - \ln Y$
		$\ln\left(\dfrac{e^x}{e^y}\right) = \ln(e^x) - \ln(e^y) = x - y$
Inverse		$\log\left(\dfrac{1}{A}\right) = -\log A$
		$\ln\left(\dfrac{1}{x}\right) = -\ln x$

The number of significant figures in a number is the same as the number of digits after the decimal point in its logarithm. For example, the natural logarithm of 18.45 is 2.9151, which means that $e^{2.9151}$ is equal to 18.45.

SKILL BUILDER ES6.3

Calculate the natural logarithm of each number: **(a)** 22×18.6; **(b)** $\dfrac{0.51}{2.67}$; **(c)** 0.079×1.485; **(d)** $\dfrac{20.5}{0.026}$.

Solution

(a) $\ln(22 \times 18.6) = \ln(22) + \ln(18.6) = 3.09 + 2.923 = 6.01$. Alternatively, $22 \times 18.6 = 410$; $\ln(410) = 6.02$.

(b) $\ln\left(\dfrac{0.51}{2.67}\right) = \ln(0.51) - \ln(2.67) = -0.67 - 0.982 = -1.65$. Alternatively, $\dfrac{0.51}{2.67} = 0.19$; $\ln(0.19) = -1.66$.

(c) $\ln(0.079 \times 1.485) = \ln(0.079) + \ln(1.485) = -2.54 + 0.395 = -2.15$. Alternatively, $0.079 \times 1.485 = 0.12$; $\ln(0.12) = -2.12$.

(d) $\ln\left(\frac{20.5}{0.026}\right) = \ln(20.5) - \ln(0.026) = 3.0204 - (-3.65) = 6.67$. Alternatively,

$\frac{20.5}{0.026} = 790$; $\ln(790) = 6.67$.

Note that the answers obtained using the two methods may differ slightly due to rounding errors.

SKILL BUILDER ES6.4

Calculate the natural logarithm of each number: **(a)** 34×16.5; **(b)** $\frac{2.10}{0.052}$; **(c)** 0.402×3.930; **(d)** $\frac{0.164}{10.7}$.

Solution

(a) 6.33
(b) 3.70
(c) 0.457
(d) −4.178

12 Solids

LEARNING OBJECTIVES

- To be able to recognize the unit cell of a crystalline solid
- To be able to calculate the density of a solid given its unit cell
- To understand the origin and nature of defects in crystals
- To understand how X rays are diffracted by crystalline solids
- To understand the correlation between bonding and the properties of solids
- To be able to describe the electrical properties of a solid using band theory
- To be familiar with the properties of superconductors
- To understand the differences between synthetic and biological polymers
- To be familiar with the properties of some contemporary materials

CHAPTER OUTLINE

In this chapter, we turn our attention to the structures and properties of solids. The solid state is distinguished from the gas and liquid states by a rigid structure in which the component atoms, ions, or molecules are usually locked into place. In many solids, the components are arranged in extended three-dimensional patterns, producing a wide range of properties that can often be tailored to specific functions. Thus, diamond, one allotrope of elemental carbon, is the hardest material known, yet graphite, another allotrope of carbon, is a soft, slippery material used in pencil lead and as a lubricant. Metallic sodium is soft enough to be cut with a dull knife, but crystalline sodium chloride turns into a fine powder when struck with a hammer.

Solids, also called *materials*, are so important in today's technology that the subdisciplines of solid-state chemistry and materials science are among the

A drawing by M. C. Escher showing two possible choices for the repeating unit. M. C. Escher's "Symmetry Drawing E128"

most active and exciting areas of modern chemical research. After presenting a basic survey of the structures of solids, we will examine how the properties of solids are determined by their composition and structure. We will also explore the principles underlying the electrical properties of metals, insulators, semiconductors (which are at the heart of the modern electronics industry), and superconductors. By the end of the chapter, you will know why some metals "remember" their shape after being bent and why ceramics are used in jet engines. You will also understand why carbon- or boron-fiber materials are used in high-performance golf clubs and tennis rackets, why nylon is used to make parachutes, and how solid electrolytes improve the performance of high-capacity batteries.

12.1 ○ Crystalline and Amorphous Solids

Crystalline faces. The faces of crystals can intersect at right angles, as in galena (PbS) and pyrite (FeS_2), or at other angles, as in quartz.

With few exceptions, the particles that make up a solid material, whether ionic, molecular, covalent, or metallic, are held in place by strong attractive forces between them. When we discuss solids, therefore, we consider the *positions* of the atoms, molecules, or ions, which are essentially fixed in space, rather than their motions (which are more important in liquids and gases). The constituents of a solid can be arranged in two general ways: they can form a regular repeating three-dimensional structure called a **crystal lattice**, thus producing a **crystalline solid**, or they can aggregate with no particular order, in which case they form an **amorphous solid** (from the Greek *ámorphos*, "shapeless").

Crystalline solids, or *crystals*, have distinctive internal structures that in turn lead to distinctive flat surfaces, or *faces*. The faces intersect at angles that are characteristic of the substance. Each structure also produces a distinctive pattern when exposed to X-rays that can be used to identify the material (see Section 12.3). The characteristic angles do not depend on the size of the crystal, but instead reflect the regular repeating arrangement of the component atoms, molecules, or ions in space. When an ionic crystal is cleaved (Figure 12.1), for example, repulsive interactions cause it to break along fixed planes to produce new faces that intersect at the same angles as those in the original crystal. In a covalent solid such as a cut diamond, the angles at which the faces meet are also not arbitrary but are determined by the arrangement of the carbon atoms in the crystal.

Crystals tend to have relatively sharp, well-defined melting points because all of the component atoms, molecules, or ions are the same distance from the same number and type of neighbors; that is, the regularity of the crystalline lattice creates local environments that are the same. Thus, the intermolecular forces holding the solid together are uniform, and the same amount of thermal energy is needed to break all of the interactions simultaneously.

Cleavage surfaces of an amorphous solid. Obsidian, a volcanic glass with the same chemical composition as granite (typically $KAlSi_3O_8$), tends to have curved, irregular surfaces when cleaved.

Amorphous solids have two characteristic properties. When cleaved or broken, they produce fragments with irregular, often curved surfaces; and they have poorly defined patterns when exposed to X-rays because their components are not arranged in a regular array. An amorphous, translucent solid is called a *glass*. Almost any substance can solidify in amorphous form if the liquid phase is cooled rapidly enough. Some solids, however, are intrinsically amorphous, either because their components

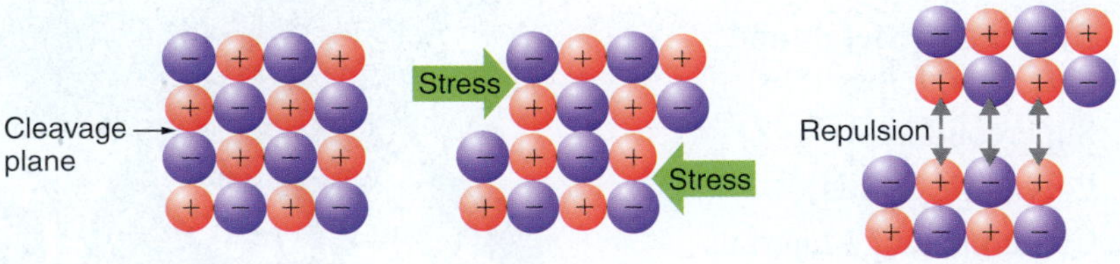

Figure 12.1 Cleaving a crystal of an ionic compound along a plane of ions. Deformation of the ionic crystal causes one plane of atoms to slide along another. The resulting repulsive interactions between ions with like charges cause the layers to separate.

cannot fit together well enough to form a stable crystalline lattice or because they contain impurities that disrupt the lattice. For example, although the chemical composition and the basic structural units of a quartz crystal and quartz glass are the same—both are SiO_2 and both consist of linked SiO_4 tetrahedra—the arrangements of the atoms in space are not. As shown in the margin, crystalline quartz contains a highly ordered arrangement of silicon and oxygen atoms, but in quartz glass the atoms are arranged almost randomly. When molten SiO_2 is cooled rapidly (4 K/min), it forms quartz glass, whereas the large, perfect quartz crystals sold in mineral shops have had cooling times of thousands of years. In contrast, aluminum crystallizes much more rapidly. Amorphous aluminum forms only when the liquid is cooled at the extraordinary rate of 4×10^{13} K/s, which prevents the atoms from arranging themselves into a regular array.

In an amorphous solid, the local environment, including both the distances to the neighboring units and the numbers of neighbors, varies throughout the material. Different amounts of thermal energy are needed to overcome these different interactions. Consequently, amorphous solids tend to soften slowly over a wide temperature range rather than having a well-defined melting point like a crystalline solid. If an amorphous solid is maintained at a temperature just below its melting point for long periods of time, the component molecules, atoms, or ions can gradually rearrange into a more highly ordered crystalline form.

MGC Unit Cells and Crystal Packing

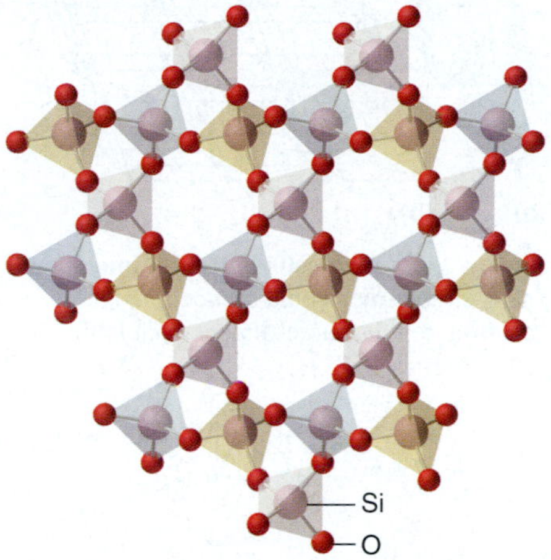

The regular arrangement of atoms in the lattice of crystalline quartz (SiO_2).

Note the pattern

Crystals have sharp, well-defined melting points; amorphous solids do not.

12.2 • The Arrangement of Atoms in Crystalline Solids

Because a crystalline solid consists of repeating patterns of its components in three dimensions (a crystal lattice), we can represent the entire crystal by drawing the structure of the smallest identical units that, when stacked together, form the crystal. This basic repeating unit is called a **unit cell**. For example, the "unit cell" of a sheet of identical postage stamps is a single stamp, and the "unit cell" of a stack of bricks is a single brick. In this section, we describe the arrangements of atoms in various unit cells.

Unit cells are easiest to visualize in two dimensions. In many cases, more than one unit cell can be used to represent a given structure, as shown for the Escher drawing on the opening page of this chapter and for a two-dimensional crystal lattice in Figure 12.2a–c. Usually the smallest unit cell that completely describes the order is chosen. The only requirement for a valid unit cell is that repeating it in space must produce the regular lattice. Thus, the unit cell shown in Figure 12.2d is not a valid choice because repeating it in space does not produce the desired lattice (there are triangular holes). The concept of unit cells is extended to a three-dimensional lattice in the schematic drawing in Figure 12.3.

MGC Cubic Structures

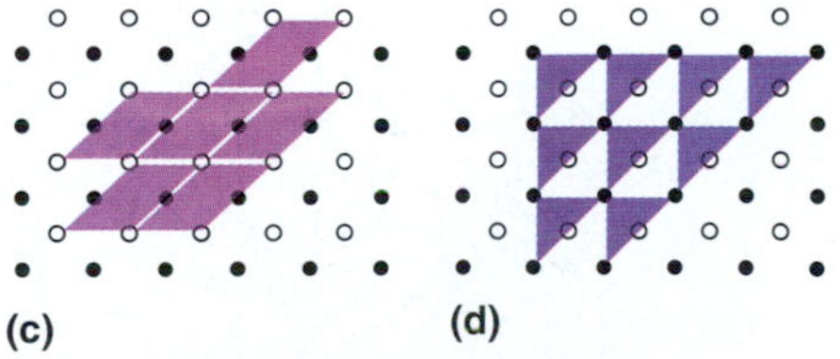

Figure 12.2 Unit cells in two dimensions. Two-dimensional lattices illustrating the possible choices of the unit cell. (a–c) The unit cells differ in their relative locations or orientations within the lattice, but are all valid choices because repeating them in any direction fills the overall pattern of dots. (d) The triangle is not a valid unit cell because repeating it in space fills only half of the space in the pattern.

The Unit Cell

There are only seven fundamentally different kinds of unit cell, which differ in the relative lengths of the edges and the angles between them (Figure 12.4). Each unit cell has six sides, and each side is a parallelogram. We focus primarily on the cubic unit cells, in which all sides have the same length and all angles are 90°, but the concepts that we introduce also apply to substances whose unit cells are not cubic.

If the cubic unit cell consists of eight component atoms, molecules, or ions located at the corners of the cube, then it is called **simple cubic** (Figure 12.5a). If the unit cell also contains an identical component in the center of the cube, then it is **body-centered cubic (bcc)** (Figure 12.5b). If there are components in the center of each face in addition to those at the corners of the cube, then the unit cell is **face-centered cubic (fcc)** (Figure 12.5c).

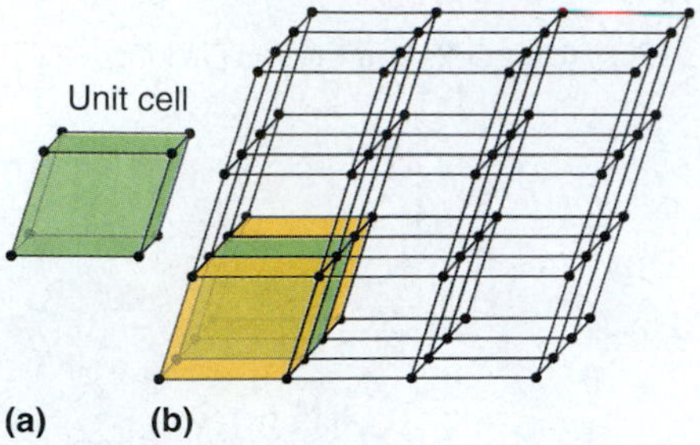

Figure 12.3 Unit cells in three dimensions. (a) A three-dimensional unit cell; (b) the resulting regular three-dimensional lattice.

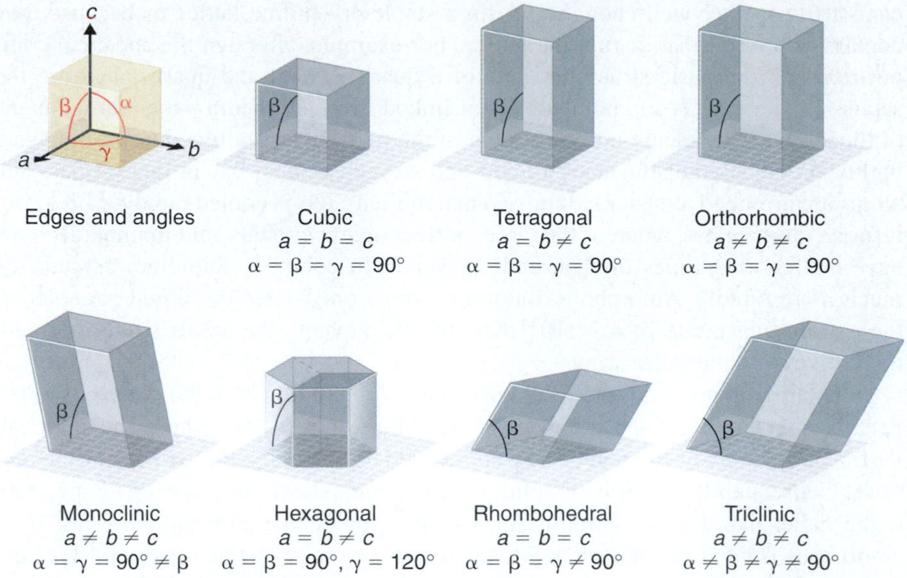

Figure 12.4 The general features of the seven basic unit cells. The lengths of the edges of the unit cells are indicated by a, b, and c, and the angles are defined as follows: α, the angle between b and c; β, the angle between a and c; and γ, the angle between a and b.

As indicated in Figure 12.5, a solid consists of a large number of unit cells arrayed in three dimensions. Any intensive property of the bulk material, such as its density, must therefore also be related to its unit cell. Because density is the mass of substance per unit volume, we can calculate the density of the bulk material from the density of a single unit cell. To do this, we need to know the size of the unit cell (to obtain its volume), the molar mass of its components, and the number of components per unit cell. When we count atoms or ions in a unit cell, however, those lying on a face, an edge, or a corner contribute to more than one unit cell, as shown in Figure 12.5. For

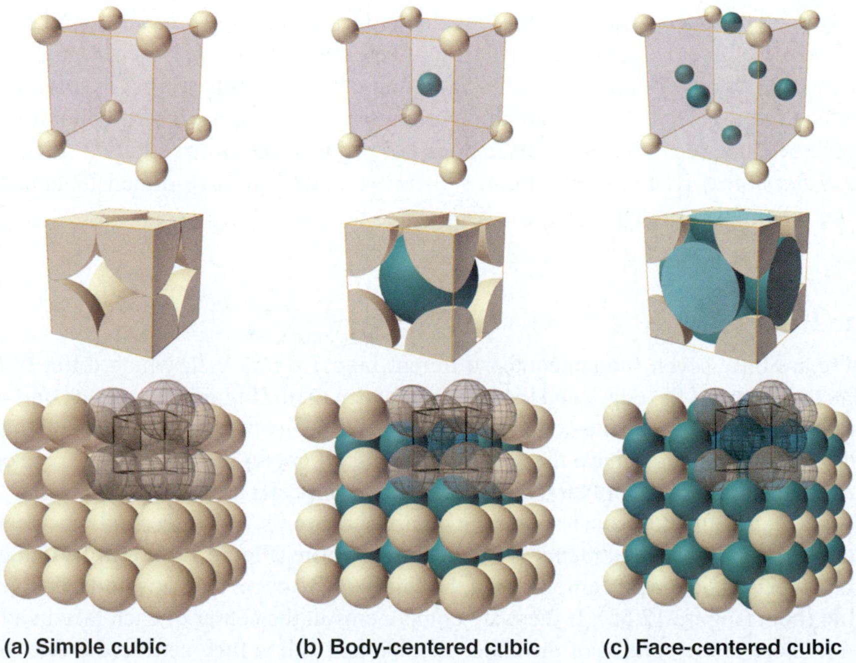

Figure 12.5 The three kinds of cubic unit cell. (a) Simple cubic, (b) body-centered cubic, and (c) face-centered cubic. Three representations are shown for each: a ball-and-stick model, a space-filling cutaway model that shows the portion of each atom that lies within the unit cell, and an aggregate of several unit cells.

example, an atom that lies on a face of a unit cell is shared by two adjacent unit cells and is therefore counted as $\frac{1}{2}$ atom per unit cell. Similarly, an atom that lies on the edge of a unit cell is shared by four adjacent unit cells, so it contributes $\frac{1}{4}$ atom to each. An atom at a corner of a unit cell is shared by all eight adjacent unit cells and therefore contributes $\frac{1}{8}$ atom to each.* In contrast, atoms that lie entirely within a unit cell, such as the atom in the center of a body-centered cubic unit cell, belong to only that one unit cell.

Note the pattern

For all unit cells except hexagonal: atoms on faces contribute $\frac{1}{2}$ atom to each unit cell; atoms on edges contribute $\frac{1}{4}$ atom to each unit cell; and atoms on corners contribute $\frac{1}{8}$ atom to each unit cell.

EXAMPLE 12.1

Metallic gold has a face-centered cubic unit cell. How many Au atoms are in each unit cell?

Given Unit cell

Asked for Number of atoms per unit cell

Strategy

Using Figure 12.5, identify the positions of the Au atoms in a face-centered cubic unit cell, and then determine how much each Au atom contributes to the unit cell. Add the contributions of all the Au atoms to obtain the total number of Au atoms in a unit cell.

Solution

As shown in Figure 12.5, a face-centered cubic unit cell has eight atoms at the corners of the cube and six atoms on the faces. Because atoms on a face are shared by two unit cells, each counts as $\frac{1}{2}$ atom per unit cell, giving $6 \times \frac{1}{2} = 3$ Au atoms per unit cell. Atoms on a corner are shared by eight unit cells and hence contribute only $\frac{1}{8}$ atom per unit cell, giving $8 \times \frac{1}{8} = 1$ Au atom per unit cell. The total number of Au atoms in each unit cell is thus $3 + 1 = 4$.

EXERCISE 12.1

Metallic iron has a body-centered cubic unit cell. How many atoms of iron are in each unit cell?

Answer two

Now that we know how to count atoms in unit cells, we can use unit cells to calculate the densities of simple compounds. Note, however, that we are assuming a solid consists of a perfect regular array of unit cells, whereas real substances contain impurities and defects that affect many of their bulk properties, including the density. Consequently, the results of our calculations will be close but not identical to the experimentally obtained values.

EXAMPLE 12.2

Calculate the density of metallic iron, which has a body-centered cubic unit cell with an edge length of 286.6 pm.

Given Unit cell, edge length

* The statement that atoms lying on an edge or a corner of a unit cell count as $\frac{1}{4}$ or $\frac{1}{8}$ atom per unit cell, respectively, is true for all unit cells *except* the hexagonal one, in which three unit cells share each vertical edge and six share each corner (Figure 12.4), leading to values of $\frac{1}{3}$ and $\frac{1}{6}$ atom per unit cell, respectively, for atoms in these positions.

Asked for Density

Strategy

Ⓐ Determine the number of iron atoms per unit cell.
Ⓑ Calculate the mass of iron atoms in the unit cell from the molar mass and Avogadro's number. Then divide the mass by the volume of the cell.

Solution

Ⓐ✓ We know from Exercise 12.1 that each unit cell of metallic iron contains two Fe atoms.
Ⓑ✓ The molar mass of iron is 55.85 g/mol. Because density is mass per unit volume, we need to calculate the mass of the iron atoms in the unit cell from the molar mass and Avogadro's number and then divide the mass by the volume of the cell (making sure to use suitable units to get density in g/cm^3):

$$\text{mass of Fe} = (2\ \cancel{\text{atoms}}\ \text{Fe})\left(\frac{1\ \cancel{\text{mol}}}{6.022 \times 10^{23}\ \cancel{\text{atoms}}}\right)\left(\frac{55.85\ \text{g}}{\cancel{\text{mol}}}\right) = 1.855 \times 10^{-22}\ \text{g}$$

$$\text{volume} = \left[(286.6\ \cancel{\text{pm}})\left(\frac{10^{-12}\ \cancel{\text{m}}}{\cancel{\text{pm}}}\right)\left(\frac{10^{2}\ \text{cm}}{\cancel{\text{m}}}\right)\right]^3 = 2.354 \times 10^{-23}\ \text{cm}^3$$

$$\text{density} = \frac{1.855 \times 10^{-22}\ \text{g}}{2.354 \times 10^{-23}\ \text{cm}^3} = 7.880\ \text{g/cm}^3$$

This result compares well with the tabulated experimental value of 7.874 g/cm^3.

EXERCISE 12.2

Calculate the density of gold, which has a face-centered cubic unit cell with an edge length of 407.8 pm.

Answer 19.29 g/cm^3

Packing of Spheres

Our discussion of the three-dimensional structures of solids has considered only substances in which all the components are identical. As we shall see, such substances can be viewed as consisting of identical spheres packed together in space, and the way the components are packed together produces the different unit cells. Most of the substances with structures of this type are metals.

Simple Cubic Structure

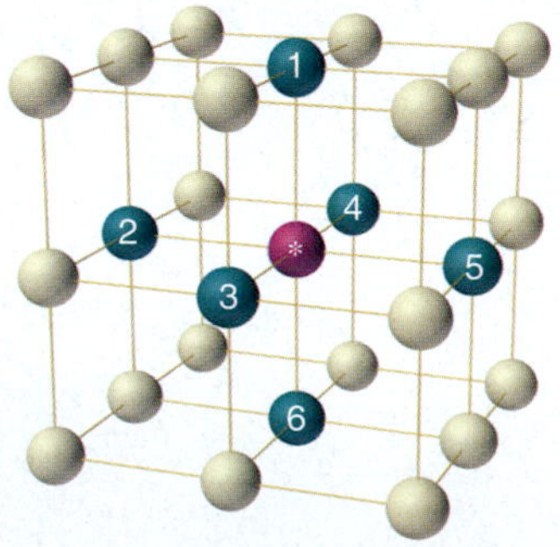

Octahedral arrangement of six nearest neighbors in a simple cubic lattice.

The arrangement of the atoms in a solid that has a simple cubic unit cell was shown in Figure 12.5a. Note that each atom in the lattice has only six nearest neighbors in an octahedral arrangement. Consequently, the simple cubic lattice is an inefficient way to pack atoms together in space: only 52% of the total space is filled by the atoms. The only element that crystallizes in a simple cubic unit cell is polonium (Po). Simple cubic unit cells are, however, common among binary ionic compounds, where each cation is surrounded by six anions, and vice versa, as shown in the margin.

Body-Centered Cubic (bcc) Structure

The body-centered cubic unit cell is a more efficient way to pack spheres together and it is much more common among pure elements. Each atom has eight nearest neighbors in the unit cell, and 68% of the volume is occupied by the atoms. As shown in Figure 12.5b, the body-centered cubic structure consists of a single layer of spheres in contact with each other and aligned so that their centers are at the corners of a square; a second layer of spheres occupies the square-shaped "holes" above the spheres in the first layer. The third layer of spheres occupies the square holes formed by the second layer, so that each lies directly above a sphere in the first layer, and so on. All the alkali metals, barium, radium, and several of the transition metals have body-centered cubic structures.

Hexagonal Close-Packed (hcp) and Cubic Close-Packed (ccp) Structures

The most efficient way to pack spheres is the close-packed arrangement, which has two variants. A single layer of close-packed spheres is shown in Figure 12.6a. Note that each sphere is surrounded by six others in the same plane to produce a hexagonal arrangement. Above any set of seven spheres are six depressions arranged in a hexagon. In principle, all six of the sites are the same, and any one of them could be occupied by an atom in the next layer. Actually, however, these six sites can be divided into two sets, labeled B and C in Figure 12.6a. Sites B and C differ because as soon as we place a sphere at a B position, we can no longer place a sphere in any of the three C positions adjacent to A, and vice versa.

If we place the second layer of spheres at the B positions in Figure 12.6a, we obtain the two-layered structure shown in Figure 12.6b. There are now two alternatives for placing the first atom of the third layer: we can place it directly over one of the atoms in the first layer (an A position), or we can place it at one of the C positions, corresponding to the positions that we did *not* use for the atoms in the first or second layers (Figure 12.6c). If we choose the first arrangement and repeat the pattern in succeeding layers, the positions of the atoms alternate from layer to layer in the pattern ABABAB . . . , resulting in a **hexagonal close-packed (hcp) structure** (Figure 12.7a). If we choose the second arrangement and repeat the pattern indefinitely, the positions of the atoms alternate as ABCABC . . . , giving a **cubic close-packed (ccp) structure** (Figure 12.7b). Because the ccp structure contains hexagonally packed layers, it does not look particularly cubic. As shown in Figure 12.7b, however, simply rotating the structure reveals its cubic nature, which is identical to a face-centered cubic (fcc) structure. The hcp and ccp structures differ only in the way their layers are stacked. Both structures have an overall packing efficiency of 74%, and in both each atom has 12 nearest neighbors (6 in the same plane plus 3 in each of the planes immediately above and below).

Table 12.1 compares the packing efficiency and the number of nearest neighbors for the different cubic and close-packed structures; the number of nearest neighbors is

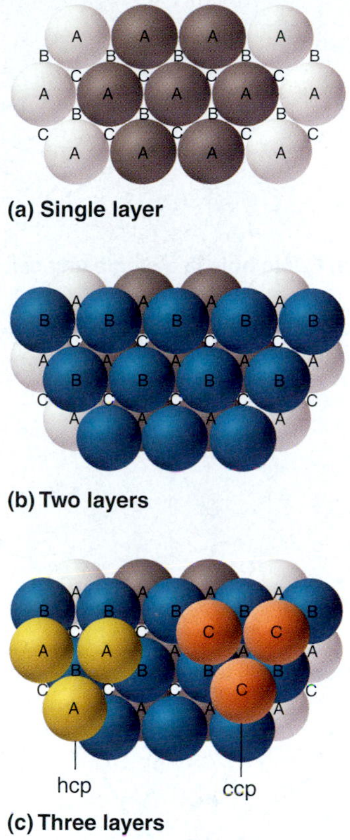

Figure 12.6 Close-packed layers of spheres. (a) A single layer of close-packed spheres. Note that each sphere is surrounded by six others in a hexagonal arrangement. (b) Placing an atom at a B position prohibits placing an atom at *any* of the adjacent C positions and results in all the atoms in the second layer occupying the B positions. (c) Placing the atoms in the third layer over the atoms at A positions in the first layer gives the hexagonal close-packed structure. Placing the third-layer atoms over the C positions gives the cubic close-packed structure.

Figure 12.7 Close-packed structures: (a) hexagonal (hcp) and (b) cubic (ccp). (a) An exploded view, a side view, and a top view of the structure. The simple hexagonal unit cell is outlined in the side and top views; note the similarity to the hexagonal unit cell shown in Figure 12.4. (b) An exploded view, a side view, and a rotated view of the structure. The latter emphasizes the face-centered cubic nature of the unit cell (outlined). The line that connects the atoms in the first and fourth layers of the ccp structure is the body diagonal of the cube.

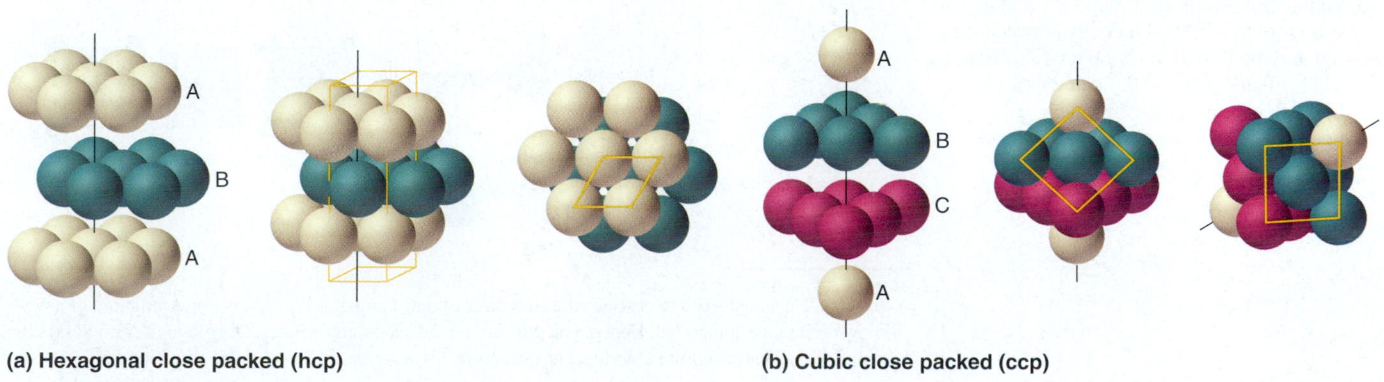

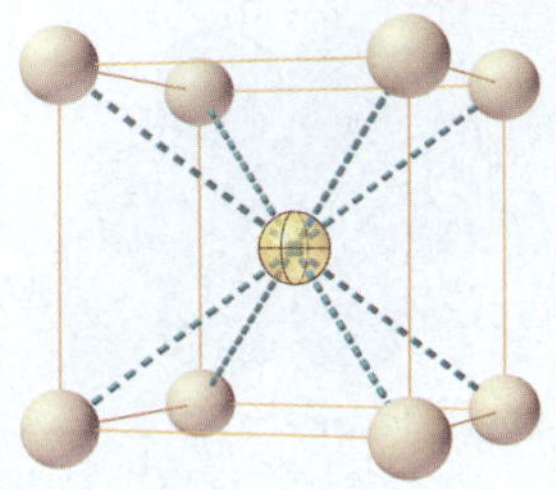

(a) Cubic hole in a single unit cell

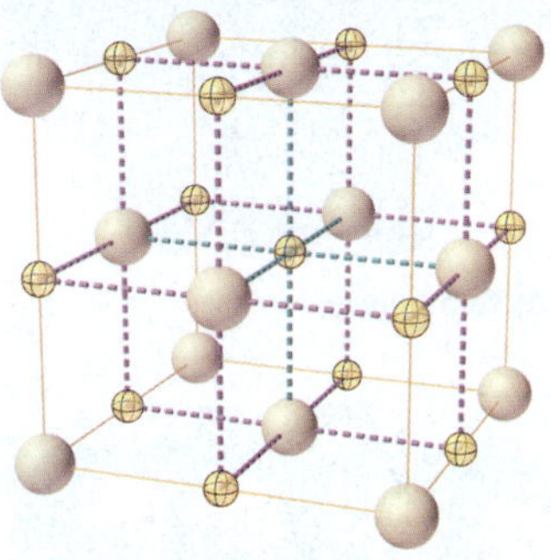

(b) Octahedral holes in a single unit cell

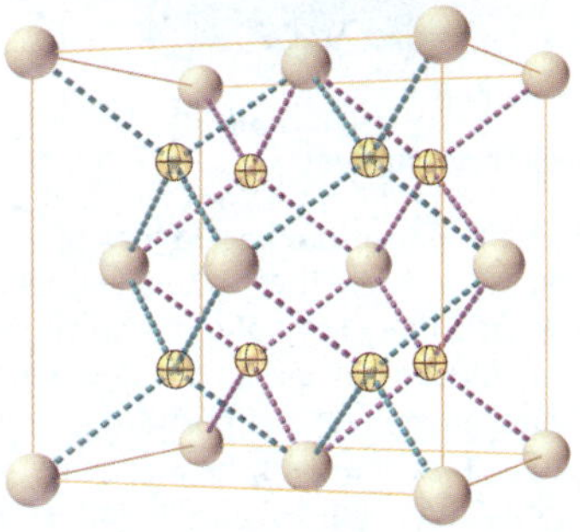

(c) Tetrahedral holes in a single unit cell

Figure 12.8 Holes in cubic lattices. (a) The cubic hole in the center of a simple cubic lattice of anions. (b) The locations of the octahedral holes in a face-centered cubic lattice of anions. (c) The locations of the tetrahedral holes in a face-centered cubic lattice of anions.

Figure 12.9 The CsCl structure. The Cs^+ ion occupies the cubic hole in the center of a cube of Cl^- ions. The drawings at the right are horizontal cross-sections through the unit cell at the bottom ($z = 0$) and halfway between the bottom and top ($z = 0.5$). A top cross-section ($z = 1$) is identical to $z = 0$. Such cross-sections often help us visualize the arrangement of atoms or ions in the unit cell more easily.

TABLE 12.1 Properties of the common structures of metals

Structure	Percentage of Space Occupied by Atoms	Coordination Number
Simple cubic	52	6
Body-centered cubic	68	8
Hexagonal close packed (hcp)	74	12
Cubic close packed (ccp) (identical to face-centered cubic)	74	12

called the **coordination number**. Most metals have hcp, ccp, or bcc structures, although a number of metals exhibit both hcp and ccp structures, depending on the temperature and pressure.

12.3 Structures of Simple Binary Compounds

The structures of most binary compounds can be described using the packing schemes we have just discussed for metals. To do so, we generally focus on the arrangement in space of the largest species present. In ionic solids, this generally means the anions (see Chapter 7), which are usually arranged in a simple cubic, bcc, fcc, or hcp lattice. Often, however, the anion lattices are not truly "close packed"; because the cations are large enough to prop them apart somewhat, the anions are not actually in contact with one another. In ionic compounds, the cations usually occupy the "holes" between the anions, thus balancing the negative charge. The ratio of cations to anions within a unit cell is required to achieve electrical neutrality and corresponds to the bulk stoichiometry of the compound.

Common Structures of Binary Compounds

As shown in Figure 12.8a, a simple cubic lattice of anions contains only one kind of hole, located in the center of the unit cell. Because this hole is equidistant from all eight atoms at the corners of the unit cell, it is called a *cubic hole*. An atom or ion in a cubic hole therefore has a coordination number of 8. Many ionic compounds that contain relatively large cations and a 1:1 cation:anion ratio have this structure, which is called the **cesium chloride structure** (Figure 12.9) because CsCl is a common example.* The unit cell in CsCl contains a single Cs^+ ion as well as $8 \times \frac{1}{8} Cl^- = 1\ Cl^-$ ion, for an overall stoichiometry of CsCl. The CsCl structure is most common for

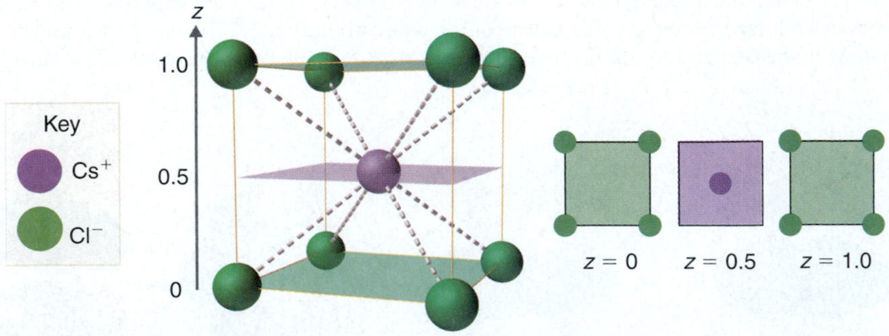

* Solid-state chemists tend to describe the structures of new compounds in terms of the structure of a well-known reference compound. Hence, you will often read statements such as "Compound X possesses the cesium chloride (or sodium chloride, etc.) structure" to describe the structure of compound X.

TABLE 12.2 Relationship between cation : anion radius ratio and the site occupied by the cations

Approximate Range of Cation : Anion Radius Ratio	Hole Occupied by Cation	Cation Coordination Number
0.225–0.414	Tetrahedral	4
0.414–0.732	Octahedral	6
0.732–1.000	Cubic	8

ionic substances that contain relatively large cations, in which the ratio of the radius of the cation to the radius of the anion is in the range shown in Table 12.2.

In contrast, a face-centered cubic array of atoms or anions contains two types of holes: *octahedral holes*, one in the center of the unit cell plus a shared one in the middle of each edge (Figure 12.8b), and *tetrahedral holes*, located between an atom at a corner and the three atoms at the centers of the adjacent faces (Figure 12.8c). As shown in Table 12.2, the ratio of the radius of the cation to the radius of the anion is the most important determinant of whether cations occupy the cubic holes in a cubic anion lattice or the octahedral or tetrahedral holes in an fcc lattice of anions. Very large cations occupy cubic holes in a cubic anion lattice, cations of intermediate size tend to occupy the octahedral holes in an fcc anion lattice, and relatively small cations tend to occupy the tetrahedral holes in an fcc anion lattice. In general, larger cations have higher coordination numbers than small cations.

Note the pattern

Very large cations occupy cubic holes, cations of intermediate size occupy octahedral holes, and small cations occupy tetrahedral holes in the anion lattice.

The most common structure based on a face-centered cubic lattice is the **sodium chloride structure** (Figure 12.10), which contains an fcc array of Cl^- ions with Na^+ ions in all the octahedral holes. We can understand the sodium chloride structure by recognizing that filling all the octahedral holes in an fcc lattice of Cl^- ions with Na^+ ions gives a total of 4 Cl^- ions (one on each face gives $6 \times \frac{1}{2} = 3$ plus one on each corner gives $8 \times \frac{1}{8} = 1$, for a total of 4) and 4 Na^+ ions (one on each edge gives $12 \times \frac{1}{4} = 3$ plus one in the middle, for a total of 4). The result is an electrically neutral unit cell and a stoichiometry of NaCl. As shown in Figure 12.10, the Na^+ ions in the NaCl structure also form an fcc lattice. The sodium chloride structure is favored for substances that contain two atoms or ions in a 1:1 ratio and in which the ratio of the radius of the cation to the radius of the anion is between 0.414 and 0.732. It is observed in many compounds, including MgO and TiC.

MGC Structures of Binary Ionic Crystals

The structure shown in Figure 12.11 is called the **zinc blende structure**, from the common name of the mineral ZnS. It results when the cation in a substance that has a 1:1 cation:anion ratio is much smaller than the anion (if the cation : anion radius ratio is less than about 0.414). For example, ZnS contains an fcc lattice of S^{2-} ions, and the cation:anion radius ratio is only about 0.40, so we predict that the cation would occupy either a tetrahedral or octahedral hole. In fact, the relatively small Zn^{2+} cations occupy the tetrahedral holes in the lattice. If all 8 tetrahedral holes in the unit

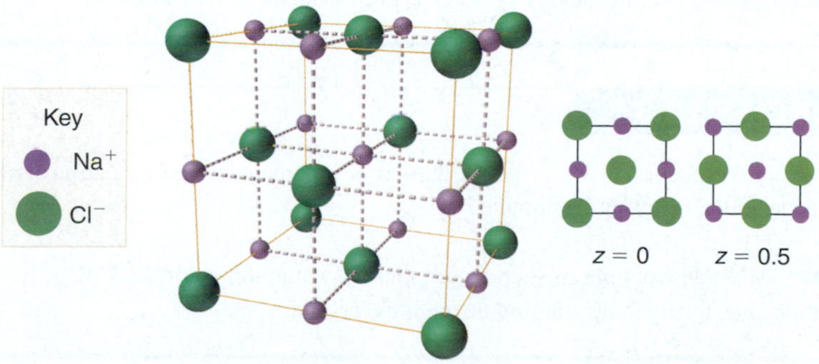

Figure 12.10 The NaCl structure. In NaCl, the Na^+ ions occupy the octahedral holes in a face-centered cubic lattice of Cl^- ions, resulting in a face-centered cubic array of Na^+ ions as well.

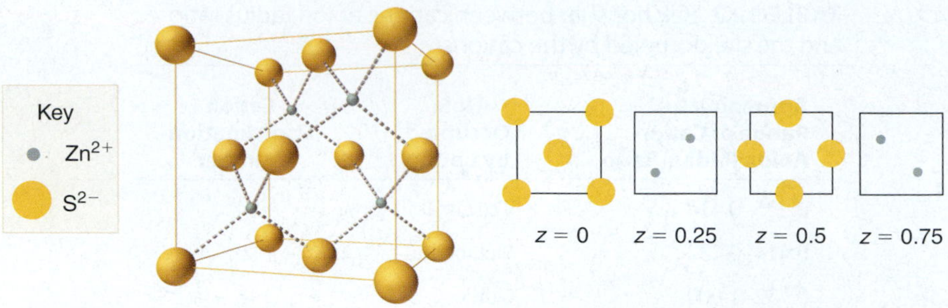

Figure 12.11 The zinc blende structure. Zn^{2+} ions occupy every other tetrahedral hole in the face-centered cubic array of S^{2-} ions. Each Zn^{2+} ion is surrounded by four S^{2-} ions in a tetrahedral arrangement.

cell were occupied by Zn^{2+} ions, however, the unit cell would contain 4 S^{2-} and 8 Zn^{2+} ions, giving a formula of Zn_2S and a net charge of +4 per unit cell. Consequently, the Zn^{2+} ions occupy *every other* tetrahedral hole, as shown in Figure 12.11, giving a total of 4 Zn^{2+} and 4 S^{2-} ions per unit cell and a formula of ZnS. Note that the zinc blende structure results in a coordination number of 4 for each Zn^{2+} ion and a tetrahedral arrangement of the four S^{2-} ions around each Zn^{2+} ion.

EXAMPLE 12.3

(a) If all the tetrahedral holes in an fcc lattice of anions are occupied by cations, what is the stoichiometry of the resulting compound? **(b)** Use the ionic radii given in Figure 7.9 to identify a plausible oxygen-containing compound with this stoichiometry and structure.

Given Lattice, occupancy of tetrahedral holes, ionic radii

Asked for Stoichiometry and identity

Strategy

Ⓐ Use Figure 12.8 to determine the number and location of the tetrahedral holes in an fcc unit cell of anions, and place a cation in each.

Ⓑ Determine the total number of cations and anions in the unit cell; their ratio is the stoichiometry of the compound.

Ⓒ From the stoichiometry, suggest reasonable charges for the cation and anion. Use the data in Figure 7.9 to identify a cation–anion combination that has a cation:anion radius ratio within a reasonable range.

Solution

(a) Ⓐ Figure 12.8 shows that the tetrahedral holes in an fcc unit cell of anions are located entirely within the unit cell, for a total of eight (one near each corner). Ⓑ Because the tetrahedral holes are located entirely within the unit cell, there are eight cations per unit cell. We calculated above that an fcc unit cell of anions contains a total of four anions per unit cell. The stoichiometry of the compound is therefore M_8Y_4 or, reduced to the smallest whole numbers, M_2Y. **(b)** Ⓒ The M_2Y stoichiometry is consistent with a lattice composed of M^+ ions and Y^{2-} ions. If the anion is O^{2-} (ionic radius 140 pm), we need a monocation with a radius no larger than about $140 \times 0.414 = 58$ pm to fit into the tetrahedral holes. According to Figure 7.9, none of the monocations has such a small radius; therefore, the most likely possibility is Li^+ at 76 pm. Thus, we expect Li_2O to have a structure that is an fcc array of O^{2-} anions with Li^+ cations in all the tetrahedral holes.

EXERCISE 12.3

If only half the octahedral holes in an fcc lattice of anions are filled by cations, what is the stoichiometry of the resulting compound?

Answer MX_2; an example of such a compound is cadmium chloride, $CdCl_2$, in which the empty cation sites form planes running through the crystal.

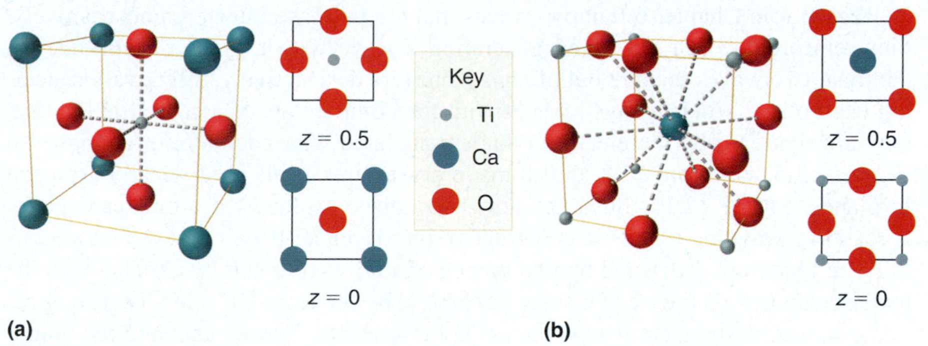

Figure 12.12 The perovskite structure of $CaTiO_3$. Two equivalent views are shown: (a) a view with the Ti atom at the center and (b) an alternative view with the Ca atom at the center.

We examine only one other structure of the many that are known, the **perovskite structure.** *Perovskite* is the generic name for oxides that contain two different kinds of metal and have the general formula $MM'O_3$, such as $CaTiO_3$. The structure is a body-centered array of two metal ions, with one M (Ca in this case) located at the corners of the cube, and the other M′ (in this case Ti) in the centers of the cube. The oxides are in the centers of the square faces (Figure 12.12a). The stoichiometry predicted from the unit cell shown in Figure 12.12a agrees with the general formula; each unit cell contains $8 \times \frac{1}{8} = 1$ Ca, 1 Ti, and $6 \times \frac{1}{2} = 3$ O atoms. The Ti and Ca atoms have coordination numbers of 6 and 12, respectively. We will return to the perovskite structure when we discuss high-temperature superconductors in Section 12.7.

X-Ray Diffraction

As you learned in Chapter 6, the wavelengths of X rays are approximately the same magnitude as the distances between atoms in molecules or ions. Consequently, X rays are a useful tool for obtaining information about the structures of crystalline substances. In a technique called **X-ray diffraction**, a beam of X rays is aimed at a sample of a crystalline material and the X rays are diffracted by layers of atoms in the crystalline lattice (Figure 12.13a). When the beam strikes photographic film, it produces an X-ray diffraction pattern, which consists of dark spots on a light background (Figure 12.13b). In 1912, the German physicist Max von Laue (1879–1960; Nobel Prize, 1914) predicted that X rays should be diffracted by crystals, and his prediction was rapidly confirmed. Within a year, two British physicists, William Henry Bragg (1862–1942) and his son, William Lawrence Bragg (1890–1972), had worked out the mathematics that allows X-ray diffraction to be used to measure interatomic distances in crystals. The Braggs shared the Nobel Prize in 1915, when the son was only 25 years old. Virtually everything we know today about the detailed structures of solids and molecules in solids is due to the X-ray diffraction technique.

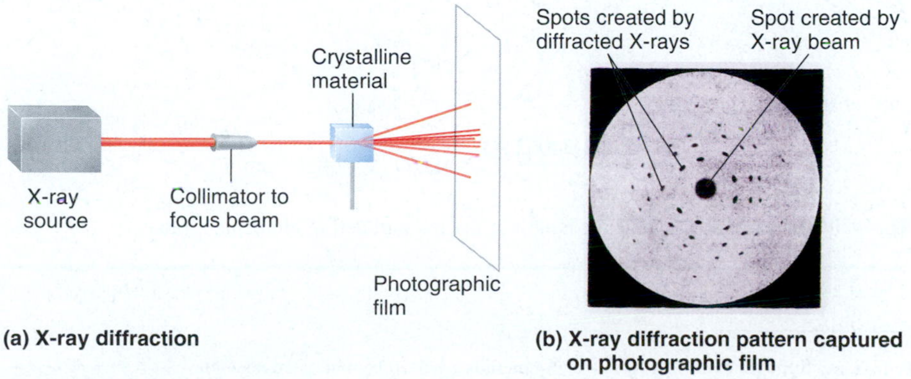

Figure 12.13 X-ray diffraction. (a) Schematic drawing of X-ray diffraction. (b) The X-ray diffraction pattern of a zinc blende crystalline solid captured on photographic film.

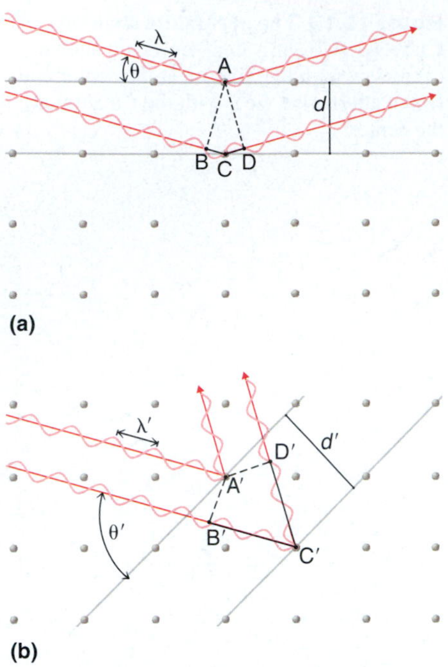

Figure 12.14 **Reflection of X rays from two adjacent planes of atoms can result in constructive interference of the X rays.** (a) The X ray diffracted by the lower layer of atoms must travel a distance that is longer by $2d \sin \theta$ than the distance traveled by the X ray diffracted by the upper layer of atoms. Only if this distance (BC plus CD) equals an integral number of wavelengths of the X rays (that is, only if $n\lambda = 2d \sin \theta$) will the X rays arrive at the detector in phase. (b) In a solid, many different sets of planes of atoms can diffract X rays. Each has a different interplanar distance and therefore diffracts the X rays at a different angle θ, which produces a characteristic pattern of spots.

Recall from Chapter 6 that two waves that are in phase interfere constructively, thus reinforcing each other and generating a wave with a greater amplitude. In contrast, two waves that are out of phase interfere destructively, effectively canceling each other. When X rays interact with the components of a crystalline lattice, they are scattered by the electron clouds associated with each atom. As shown in Figures 12.5, 12.7, and 12.8, the atoms in crystalline solids are typically arranged in planes. Figure 12.14 illustrates how two adjacent planes of atoms can scatter X rays in a way that results in constructive interference. If two X rays that are initially in phase are diffracted by two planes of atoms separated by a distance d, the lower beam travels the extra distance indicated by the lines BC and CD. The *angle of incidence*, designated as θ, is the angle between the X-ray beam and the planes in the crystal. Because BC = CD = $d \sin \theta$, the extra distance that the lower beam in Figure 12.14 must travel compared with the upper beam is $2d \sin \theta$. In order for these two X rays to arrive at a detector in phase, the extra distance traveled must be an integral multiple n of the wavelength λ:

$$2d \sin \theta = n\lambda \tag{12.1}$$

Equation 12.1 is the **Bragg equation**. The structures of crystalline substances that contain both small molecules and ions or very large biological molecules, with molecular masses in excess of 100,000 amu, can now be determined accurately and routinely using X-ray diffraction and the Bragg equation. The next example illustrates how to use the Bragg equation to calculate the distance between planes of atoms in crystals.

EXAMPLE 12.4

X rays from a copper X-ray tube (λ = 1.54062 Å or 154.062 pm)* are diffracted at an angle of 10.89° from a sample of crystalline gold. Assuming that $n = 1$, what is the distance between the planes that gives rise to this reflection? Give your answer in angstroms and picometers to four significant figures.

Given Wavelength, diffraction angle, number of wavelengths

Asked for Distance between planes

Strategy

Substitute the given values into the Bragg equation and solve to obtain the distance between planes.

Solution

We are given n, θ, and λ and asked to solve for d, so this is a straightforward application of the Bragg equation. For an answer in angstroms, we do not even have to convert units. Solving the Bragg equation for d gives

$$d = \frac{n\lambda}{2 \sin \theta}$$

and substituting values gives

$$d = \frac{(1)(1.54062\ \text{Å})}{2 \sin 10.89°} = 4.077\ \text{Å} = 407.7\ \text{pm}$$

This value corresponds to the edge length of the fcc unit cell of elemental gold.

* In X-ray diffraction, the angstrom (Å) is generally used as the unit of wavelength.

EXERCISE 12.4

X rays from a molybdenum X-ray tube ($\lambda = 0.709300$ Å) are diffracted from a sample of metallic iron at an angle of 7.11°. Assuming that $n = 1$, what is the distance between the planes responsible for the reflection? Give your answer in angstroms and picometers. Report your answer to four significant figures.

Answer 2.87 Å or 287 pm (corresponding to the edge length of the bcc unit cell of elemental iron)

12.4 • Defects in Crystals

MGC Defects in Metals

The crystal lattices we have described represent an idealized, simplified system that can be used to understand many of the important principles governing the behavior of solids. In contrast, real crystals contain large numbers of *defects* (typically more than 10^4 per milligram) ranging from variable amounts of impurities to missing or misplaced atoms or ions. These defects occur for three main reasons:

1. It is impossible to obtain any substance in 100% pure form. Some impurities are always present.
2. Even if a substance were 100% pure, forming a perfect crystal would require cooling the liquid phase infinitely slowly to allow all atoms, ions, or molecules to find their proper positions. Cooling at more realistic rates usually results in one or more components being trapped in the "wrong" place in a lattice or in areas where two lattices that grew separately intersect.
3. Applying an external stress to a crystal, such as a hammer blow, can cause microscopic regions of the lattice to move with respect to the rest, thus resulting in imperfect alignment.

In this section, we discuss how defects determine some of the properties of solids. We begin with solids that consist of neutral atoms, specifically metals, and then turn to ionic compounds.

Defects in Metals

A **point defect** is any defect that involves only a single particle (a lattice point) or sometimes a very small set of points. A **line defect** is restricted to a row of lattice points, and a **plane defect** involves an entire plane of lattice points in a crystal. A **vacancy** occurs where an atom is missing from the normal crystalline array; it constitutes a tiny vacuum in the middle of a solid (Figure 12.15). We focus primarily on point and plane defects in our discussion because they are encountered most frequently.

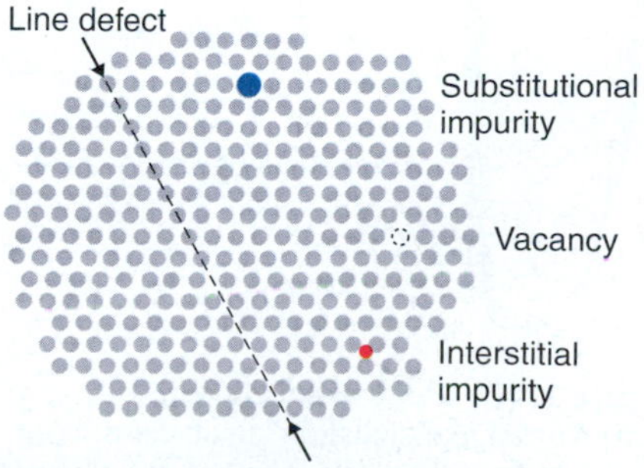

Figure 12.15 Common defects in crystals. A two-dimensional representation of a crystal lattice containing substitutional and interstitial impurities, vacancies, and line defects. The position of the line defect is indicated by a dashed line and arrows.

Impurities

An **interstitial impurity** is usually a smaller atom (typically about 45% smaller than the host) that can fit into the octahedral or tetrahedral holes in the metal lattice (Figure 12.15). Steels consist of iron with a specific amount of carbon atoms present as interstitial impurities (Table 12.3). The inclusion of one or more transition metals or semimetals often improves the corrosion resistance of steel.

In contrast, a **substitutional impurity** is a different atom of about the same size that simply replaces one of the atoms that make up the host lattice (Figure 12.15). Substitutional impurities are usually chemically similar to the substance that constitutes the bulk of the sample, and they generally have atomic radii that are with-

TABLE 12.3 Compositions, properties, and uses of some types of steel

Name of Steel	Typical Composition[a]	Properties	Applications
Low-carbon	<0.15% C	Soft, ductile	Wire
Mild carbon	0.15–0.25% C	Malleable, ductile	Cables, chains, nails
High-carbon	0.60–1.5% C	Hard, brittle	Knives, cutting tools, drill bits, springs
Stainless	15–20% Cr, 1–5% Mn, 5–10% Ni, 1–3% Si, 1% C, 0.05% P	Corrosion resistant	Cutlery, instruments, marine fittings
Invar	36% Ni	Low coefficient of thermal expansion	Measuring tapes, meter sticks
Manganese	10–20% Mn	Hard, wear resistant	Armor plate, safes, rails
High-speed	14–20% W	Retains hardness at high temperatures	High-speed cutting tools
Silicon	1–5% Si	Hard, strong, highly magnetic	Magnets in electric motors and transformers

[a] In addition to enough iron to bring the total percentage up to 100%, most steels contain small amounts of carbon (0.5–1.5%) and manganese (<2%).

in about 15% of the radius of the host. For example, strontium and calcium are chemically similar and have similar radii, and as a result, strontium is a common impurity in crystalline calcium, with the Sr atoms randomly occupying sites normally occupied by Ca.

Dislocations, Deformations, and Work Hardening

Inserting an extra plane of atoms into a crystal lattice produces an **edge dislocation**. A familiar example of an edge dislocation occurs when an ear of corn contains an extra row of kernels between the other rows (Figure 12.16). An edge dislocation in a crystal causes the planes of atoms in the lattice to become kinked where the extra plane of atoms begins (Figure 12.16). This is the core of the dislocation, which is the most distorted region of the solid and frequently determines whether the solid will deform and fail under stress.

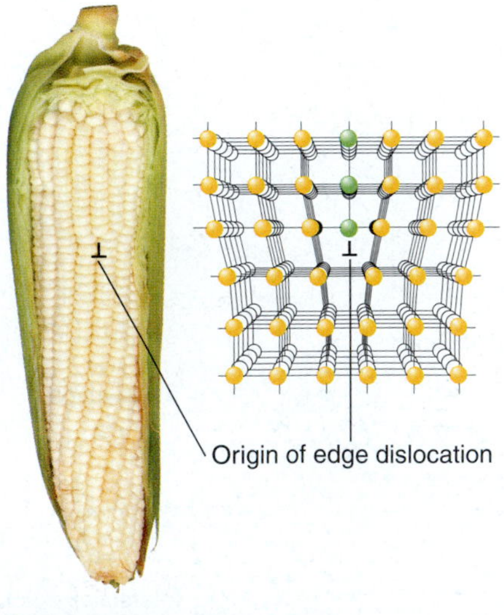

Figure 12.16 Edge dislocations. (left) An "edge dislocation" in an ear of corn. (right) A three-dimensional representation of an edge dislocation in a solid, illustrating how an edge dislocation can be viewed as a simple line defect arising from the insertion of an extra set of atoms into the lattice. In both cases, the origin of the edge dislocation is indicated by the symbol ⊥.

To shape a solid without shattering it, adjacent planes of atoms must move past one another to a new position that is energetically equivalent to the old one. The plane along which motion occurs is called a *slip plane*. The deformation process begins when an edge dislocation, perhaps resulting from a vacancy formed during crystallization or by an impact, moves through the crystal. As the dislocation moves, only one set of contacts is broken at a time, but the net result is that the atoms in one half of the lattice move with respect to the other half.

To illustrate the process of deformation, suppose you have a heavy rug that is lying a few inches off-center on a nonskid pad. To move the rug to its proper place, you could pick up one end and pull it. Because of the large area of contact between the rug and the pad, however, they will probably move as a unit. Alternatively, you could pick up the rug and try to set it back down exactly where you want it, but that requires a great deal of effort (and probably at least one extra person). An easier solution is to create a small wrinkle at one end of the rug and gradually push the wrinkle across, resulting in a net movement of the rug as a whole (Figure 12.17a). Moving the wrinkle requires only a small amount of energy because only a small part of the rug is actually moving at any one time. The wrinkle is analogous to an edge dislocation in a solid, where the contacts between layers are broken in only one place at a time yet facilitate the net deformation of the solid.

If the rug we have just described has a *second* wrinkle at a different angle, however, it is very difficult to move the first one where the two wrinkles intersect (Figure 12.17b); this process is called **pinning**. Similarly, intersecting dislocations in a solid prevent them from moving, thereby increasing the mechanical strength of the material. In fact, one of the major goals of materials science is to find ways to pin dislocations in order to strengthen or harden a material.

Pinning can also be achieved by introducing selected impurities in appropriate amounts. Substitutional impurities that are a mismatch in size to the host prevent dislocations from migrating smoothly along a plane, with higher concentrations of impurities generally producing stronger materials. For example, bronze, which contains about 20% tin and 80% copper by mass, produces a much harder and sharper weapon than does either pure tin or pure copper. Similarly, pure gold is too soft to make durable jewelry, so most gold jewelry contains 75% (18 carat) or 58% (14 carat) gold by mass, with the remainder consisting of copper, silver, or both.

If an interstitial impurity forms polar covalent bonds to the host atoms, these bonds can prevent layers from sliding past one another, even when only a small amount of the impurity is present. For example, because iron forms polar covalent bonds to carbon, the strongest steels contain about 1% carbon by mass (Table 12.3).

Most materials are polycrystalline, which means they consist of many microscopic individual crystals called *grains* that are randomly oriented with respect to one another. The place where two grains intersect is called a **grain boundary**, and each boundary can be viewed as a two-dimensional dislocation. Deformations are usually limited to the movement of dislocations within a single grain, and their movement through a solid tends to stop at a grain boundary. Consequently, controlling the grain size in solids is critical for obtaining desirable mechanical properties; fine-grained materials are usually much stronger than coarse-grained ones.

Work hardening is the introduction of a dense network of dislocations throughout a solid, which makes it very tough and hard. If all the linear defects in a single 1-cm^3 sample of a work-hardened material were laid end to end, their total length could be 10^6 km! The legendary blades of the Japanese and Moorish swordsmiths owed much of their strength to repeated work hardening of the steel. As the density of defects increases, however, the metal becomes more brittle (less malleable). For example, bending a paper clip back and forth several times causes the wire to break because of the increased brittleness resulting from work hardening.

(a) Dislocation

(b) Pinning

Figure 12.17 The role of dislocation in the motion of one planar object across another. (a) Pushing the wrinkle across the rug results in a net movement of the rug with relatively little expenditure of energy because at any given time only a very small amount of the rug is not in contact with the floor. (b) A second intersecting wrinkle prevents movement of the first by "pinning" it.

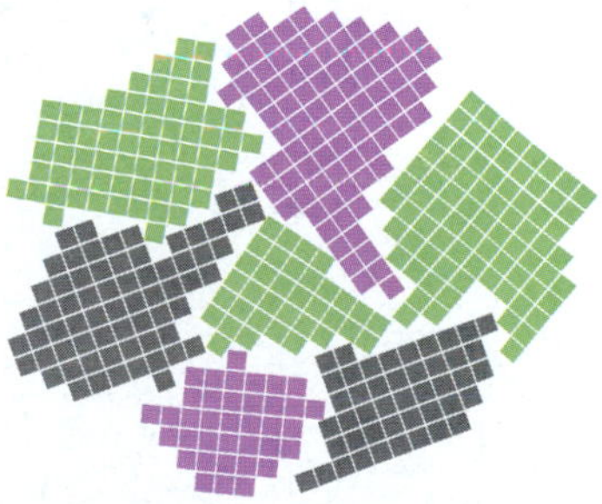

Grain boundaries. As a polycrystalline material solidifies, grains form that have irregular shapes. The interfaces between grains constitute grain boundaries. (Squares represent unit cells within grains.)

Memory Metal

The compound NiTi, popularly known as "memory metal" or nitinol (nickel–titanium Naval Ordinance Laboratory, after the site where it was first prepared), illustrates the importance of deformations. If a sample of this metal is warmed from room temperature to higher than about 50°C, it will revert to a shape in which it has been previously set. For example, if a straight piece of NiTi wire is wound into a spiral, it will remain in the spiral shape indefinitely, *unless* it is warmed to 50–60°C, at which point it will spontaneously straighten out again. The chemistry behind the temperature-induced change in shape is moderately complex, but for our purposes it is sufficient to know that NiTi can exist in two different solid phases.

The high-temperature phase has the cubic CsCl structure, in which a Ti atom is embedded in the center of a cube of Ni atoms (or vice versa). The low-temperature phase has a related but kinked structure, in which one of the angles of the unit cell is no longer 90°. Bending an object made of the low-temperature phase creates defects that change the pattern of kinks within the structure. If the object is heated to a temperature higher than about 50°C, the material undergoes a transition to the cubic high-temperature phase, causing the object to return to its original shape. The shape of the object above 50°C is controlled by a complex set of defects and dislocations that can be relaxed or changed only by the thermal motion of the atoms within individual NiTi grains and at grain boundaries. These changes require much higher temperatures than the phase transition itself. As a result, the shape of a NiTi object can be changed by holding it in the desired position with a mold or other device and maintaining it at about 500°C for 15 minutes. The rigid mold is needed to prevent the object from reverting to its original shape during the initial heating. In addition to the application shown in the margin, memory metal has many practical applications, such as temperature-sensitive springs that open and close valves in the automatic transmissions of cars. Because NiTi can also undergo pressure- or tension-induced phase transitions, it is used to make wires for straightening teeth in orthodontic braces and for surgical staples that change shape at body temperature to hold broken bones together.

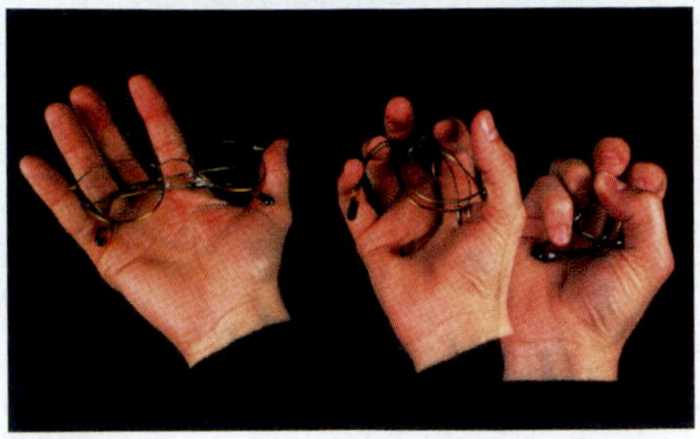

Memory metal. Flexon is a fatigue-resistant alloy of Ti and Ni that is used as a frame for glasses because of its durability and corrosion resistance.

Another flexible, fatigue-resistant alloy composed of titanium and nickel is Flexon. Originally discovered by metallurgists who were creating titanium-based alloys for use in missile heat shields, Flexon is now used as a durable, corrosion-resistant frame for glasses, among other uses.

EXAMPLE 12.5

Because steels that contain at least 4% chromium are much more corrosion resistant than iron, they are collectively sold as "stainless steel." Referring to the composition of stainless steel in Table 12.3 and, if needed, the atomic radii in Figure 7.7, predict which type of impurity is represented by each of the elements, excluding iron, that are present in at least 0.05% by mass.

Given Composition of stainless steel, atomic radii

Asked for Type of impurity

Strategy

Using the data listed in Table 12.3 and the atomic radii given in Chapter 7, determine whether the impurities listed are similar in size to an iron atom. Then determine whether each impurity is chemically similar to Fe. If similar in both size and chemistry, the impurity is likely to be a substitutional impurity. If not, it is likely to be an interstitial impurity.

Solution

According to Table 12.3, a typical stainless steel contains about 1% carbon, 1–5% manganese, 0.05% phosphorus, 1–3% silicon, and up to 18–20% nickel and chromium. The three transition elements (Mn, Ni, and Cr) lie near Fe in the periodic table, so they should be similar to Fe in chemical properties and atomic size (atomic radius 125 pm). Hence, they almost certainly will substitute for iron in the Fe lattice. Carbon is a second-period element that is nonmetallic and much smaller (atomic radius 77 pm) than iron. Carbon will therefore tend to occupy interstitial sites in the iron lattice. Phosphorus and silicon are chemically quite different from iron (phosphorus is a nonmetal and silicon is a semimetal), even though they are similar in size (atomic radii of 106 and 111 pm, respectively). Thus, they are unlikely to be substitutional impurities in the iron lattice or to fit into interstitial sites, but they could aggregate into layers that would constitute plane defects.

EXERCISE 12.5

Consider nitrogen, vanadium, zirconium, and uranium impurities in a sample of titanium metal. Which is most likely to form an interstitial impurity? A substitutional impurity?

Answer N; V

Defects in Ionic and Molecular Crystals

All the defects and impurities described for metals are seen in ionic and molecular compounds as well. Because ionic compounds contain both cations and anions rather than only neutral atoms, however, they exhibit additional types of defects that are not possible in metals.

The most straightforward variant is a substitutional impurity in which a cation or anion is replaced by another of similar charge and size. For example, Br^- can substitute for Cl^-, so tiny amounts of Br^- are usually present in a chloride salt such as $CaCl_2$ or $BaCl_2$. If the substitutional impurity and the host have *different* charges, however, the situation becomes more complicated. Suppose, for example, that Sr^{2+} (ionic radius 118 pm) substitutes for K^+ (ionic radius 138 pm) in KCl. Because the ions are approximately the same size, Sr^{2+} should fit nicely into the face-centered cubic lattice of KCl. The difference in charge, however, must somehow be compensated for in order to preserve electrical neutrality. The simplest way is for a second K^+ ion to be lost elsewhere in the crystal, producing a vacancy. Thus, substitution of K^+ by Sr^{2+} in KCl results in the introduction of *two* defects: a site in which an Sr^{2+} ion occupies a K^+ position and a vacant cation site. Substitutional impurities whose charges do not match the host's are often introduced intentionally to produce compounds with specific properties (Section 12.7).

Virtually all of the colored gems used in jewelry are due to substitutional impurities in simple oxide structures. For example, α-Al_2O_3, a hard white solid called *corundum* that is used as an abrasive in fine sandpaper, is the primary component, or *matrix*, of a wide variety of gems. Because many trivalent transition metal ions have ionic radii only a little larger than the radius of Al^{3+} (ionic radius 53.5 pm), they can replace Al^{3+} in the octahedral holes of the oxide lattice. Substituting small amounts of Cr^{3+} ions (ionic radius 75 pm) for Al^{3+} gives the deep red color of ruby, and a mixture of impurities (Fe^{2+}, Fe^{3+}, and Ti^{4+}) gives the deep blue of sapphire. True amethyst contains small amounts of Fe^{3+} in an SiO_2 (quartz) matrix. The same metal ion substituted into different mineral lattices can produce very different colors. For example, Fe^{3+} ions are responsible for the yellow color of topaz and the violet color of amethyst. The distinct environments cause differences in *d* orbital energies, enabling the Fe^{3+} ions to absorb light of different frequencies, a topic we describe in more detail in Chapter 23.

The same cation in different environments. An Fe^{3+} substitutional impurity produces substances with strikingly different colors.

Substitutional impurities are also observed in molecular crystals if the structure of the impurity molecule is similar to the host, and they can have major effects on the properties of the crystal. In pure anthracene for example, an electrical conductor, the transfer of electrons through the molecule is much slower if the anthracene crystal contains even very small amounts of tetracene despite the strong structural similarities.

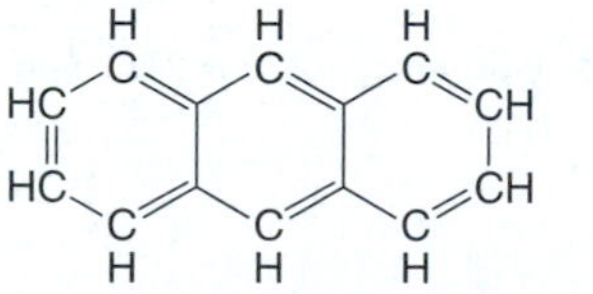

Anthracene

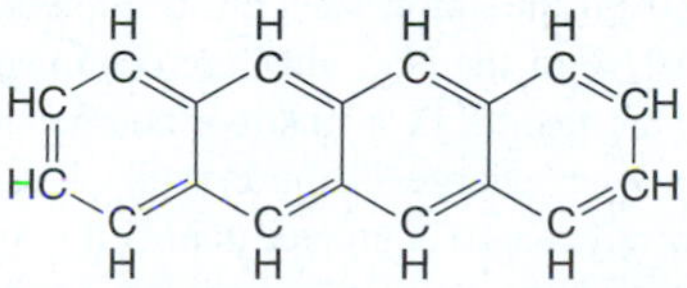

Tetracene

If a cation or anion is simply missing, leaving a vacant site in an ionic crystal, then the requirement that a crystal be electrically neutral implies that there must also be a corresponding vacancy of the ion with the opposite charge somewhere in the crystal. In compounds such as KCl, the charges are equal but opposite, so one anion vacancy

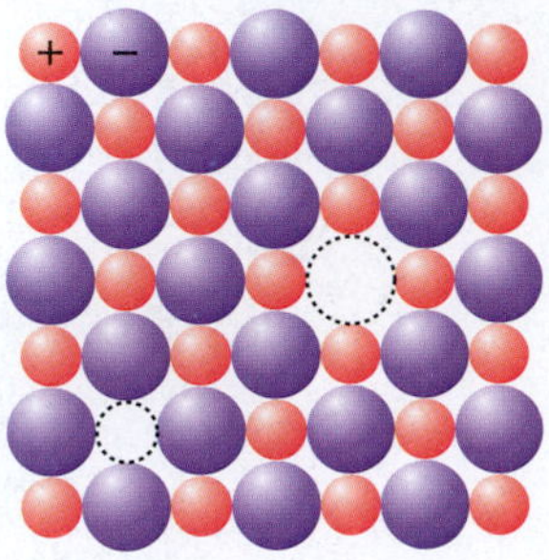

(a) Schottky defect

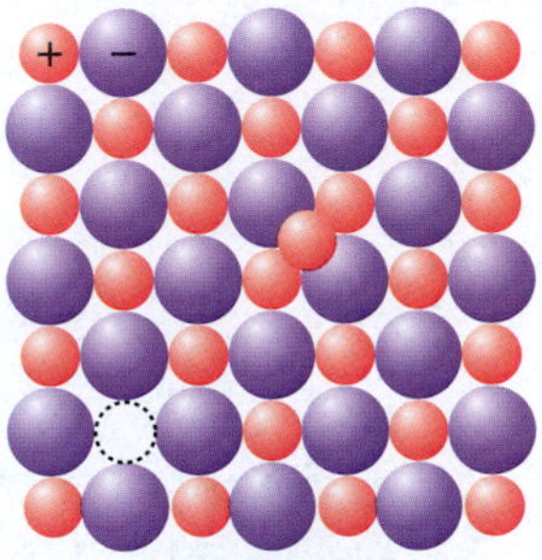

(b) Frenkel defect

Figure 12.18 The two most common defects in ionic solids. (a) A Schottky defect in KCl, showing the missing cation/anion pair. (b) A Frenkel defect in AgI, showing a misplaced Ag^+ cation.

is sufficient to compensate for each cation vacancy. In compounds such as $CaCl_2$, however, *two* Cl^- anion sites must be vacant to compensate for each missing Ca^{2+} cation. These pairs (or sets) of vacancies are called **Schottky defects** and are particularly common in simple alkali metal halides such as KCl (Figure 12.18a). Many microwave diodes, which are devices that allow a current to flow in a single direction, are composed of materials with Schottky defects.

Occasionally one of the ions in an ionic lattice is simply in the wrong position. An example of this phenomenon, called a **Frenkel defect**, is a cation that occupies a tetrahedral hole rather than an octahedral hole in the anion lattice (Figure 12.18b). Frenkel defects are most common in salts that have a large anion and a relatively small cation. To preserve electrical neutrality, one of the normal cation sites, usually octahedral, must be vacant.

Frenkel defects are particularly common in the silver halides AgCl, AgBr, and AgI, which combine a rather small cation (Ag^+, ionic radius 115 pm) with large, polarizable anions. Certain more complex salts that contain a second cation in addition to Ag^+ and Br^- or I^- have so many Ag^+ ions in tetrahedral holes that they are good electrical conductors in the solid state; hence, they are called **solid electrolytes**. (As you learned in Chapter 4, most ionic compounds do not conduct electricity in the solid state, although they do conduct electricity when molten or dissolved in a solvent that separates the ions, allowing them to migrate in response to an applied electric field.) In response to an applied voltage, the cations in solid electrolytes can diffuse rapidly through the lattice via octahedral holes, creating Frenkel defects as the cations migrate. Sodium–sulfur batteries use a solid Al_2O_3 electrolyte that contains small amounts of solid Na_2O. Because the electrolyte cannot leak, it cannot cause corrosion, which gives a battery that uses a solid electrolyte a significant advantage over one with a liquid electrolyte.

EXAMPLE 12.6

In a sample of NaCl, one of every 10,000 sites normally occupied by Na^+ is occupied instead by Ca^{2+}. Assuming that all of the Cl^- sites are fully occupied, what is the stoichiometry of the sample?

Given Ionic solid, number and type of defect

Asked for Stoichiometry

Strategy

A Identify the unit cell of the host compound. Compute the stoichiometry if 0.01% of the Na^+ sites are occupied by Ca^{2+}. If the overall charge is greater than 0, then the stoichiometry must be incorrect.

B If incorrect, adjust the stoichiometry of the Na^+ ion to compensate for the additional charge.

Solution

A Pure sodium chloride has a 1:1 ratio of Na^+ and Cl^- ions arranged in a face-centered cubic lattice (the NaCl structure). If all the anion sites are occupied by Cl^-, the negative charge is -1.00 per formula unit. If 0.01% of the Na^+ sites are occupied by Ca^{2+} ions, the cation stoichiometry is $Na_{0.99}Ca_{0.01}$. This results in a positive charge of $(0.99)(+1) + (0.01)(+2) = +1.01$ per formula unit, for a net charge in the crystal of $+1.01 + (-1.00) = +0.01$ per formula unit. Because the overall charge is greater than 0, this stoichiometry must be incorrect. **B** The most plausible way for the solid to adjust its composition to become electrically neutral is for some of the Na^+ sites to be vacant. If one Na^+ site is vacant for each site that contains a Ca^{2+} cation, then the cation stoichiometry is $Na_{0.98}Ca_{0.01}$. This results in a positive charge of $(0.98)(+1) + (0.01)(+2) = +1.00$ per formula unit, which exactly neutralizes the negative charge. The stoichiometry of the solid is thus $Na_{0.98}Ca_{0.01}Cl_{1.00}$.

EXERCISE 12.6

In a sample of MgO that has the NaCl structure, 0.02% of the Mg^{2+} ions are replaced by Na^{+} ions. If the cation sites are completely occupied, what is the stoichiometry of the sample?

Answer If the formula of the compound is $Mg_{0.98}Na_{0.02}O_{1-x}$, then x must equal 0.01 to preserve electrical neutrality. The formula is thus $Mg_{0.98}Na_{0.02}O_{0.99}$.

Nonstoichiometric Compounds

Nonstoichiometric compounds are solids that have intrinsically variable stoichiometries. Such compounds appear to violate the law of multiple proportions (see Chapter 1), which states that chemical compounds contain fixed integral ratios of atoms. In fact, nonstoichiometric compounds contain large numbers of defects, usually vacancies, which give rise to stoichiometries that can depart significantly from simple integral ratios *without affecting the fundamental structure of the crystal.* Nonstoichiometric compounds frequently consist of transition metals, lanthanides, and actinides, with polarizable anions such as oxide (O^{2-}) and sulfide (S^{2-}). Some common examples are listed in Table 12.4, along with their basic structure type. These compounds are nonstoichiometric because their constituent metals can exist in various oxidation states in the solid, which preserves electrical neutrality.

One example is iron(II) oxide (ferrous oxide), which produces the black color in clays and is used as an abrasive. Its stoichiometry is *not* FeO because it always contains fewer than 1.00 Fe per O^{2-} (typically 0.90–0.95). This is possible because Fe can exist in both the +2 and +3 oxidation states. Thus, the difference in charge created by a vacant Fe^{2+} site can be balanced by two Fe^{2+} sites that contain Fe^{3+} ions instead [+2 vacancy = (3 − 2) + (3 − 2)]. The crystal lattice is able to accommodate this relatively high fraction of substitutions and vacancies with no significant change in structure.

TABLE 12.4 Some nonstoichiometric compounds

Compound	Observed Range of x
Oxides[a]	
Fe_xO	0.85–0.95
Ni_xO	0.97–1.00
TiO_x	0.75–1.45
VO_x	0.9–1.20
NbO_x	0.9–1.04
Sulfides	
Cu_xS	1.77–2.0
Fe_xS	0.80–0.99
ZrS_x	0.9–1.0

[a] All the oxides listed have the NaCl structure.

Note the pattern

Because a crystal must be electrically neutral, any defect that affects the number or charge of the cations must be compensated by a corresponding defect in the number or charge of the anions.

12.5 Correlation Between Bonding and the Properties of Solids

Based on the nature of the forces that hold the component atoms, molecules, or ions together, solids may be classified as *ionic, molecular, covalent (network)*, or *metallic*. The variation in the relative strengths of these four types of interactions correlates nicely with their wide variation in properties.

MGC Attractive Forces Within Solids

Ionic Solids

You learned in Chapter 8 that **ionic solids** consist of positively and negatively charged ions held together by electrostatic forces. The strength of the attractive forces depends on the charge and size of the ions that make up the lattice and determines many of the physical properties of the crystal.

The *lattice energy*, the energy required to separate 1 mol of the crystalline ionic solid into its component ions in the gas phase, is directly proportional to the product of the ionic charges and inversely proportional to the sum of the radii of the ions. For example, NaF and CaO both crystallize in the fcc NaCl structure, and the sizes of their component ions are about the same: Na^{+} (102 pm) versus Ca^{2+} (100 pm), and F^{-} (133 pm) versus O^{2-} (140 pm). Because of the higher charge on the ions in CaO, however, the lattice energy of CaO is almost four times higher than that of NaF (3401 versus 923 kJ/mol). Because the forces that hold Ca and O together in CaO are much stronger than those that hold Na and F together in NaF, the heat of fusion of CaO is almost twice that of NaF (59 versus 33.4 kJ/mol), and the melting point of CaO is

2927°C versus 996°C for NaF. In both cases, however, the values are large; that is, simple ionic compounds have high melting points and are relatively hard (and brittle) solids.

Molecular Solids

The structure of solid benzene. In solid benzene, the molecules are not arranged with their planes parallel to one another but at 90° angles.

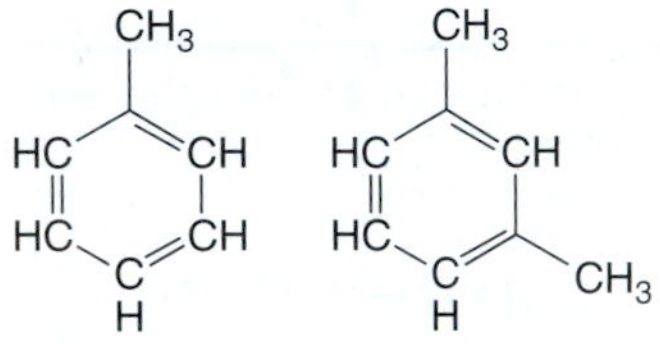

The methyl groups attached to the phenyl ring in toluene and *m*-xylene prevent the rings from packing together as in solid benzene.

Molecular solids consist of atoms or molecules held to each other by dipole–dipole interactions, London dispersion forces, or hydrogen bonds, or any combination of these, all discussed in Chapter 11. The arrangement of the molecules in solid benzene is shown in the margin. Because the intermolecular interactions in a molecular solid are relatively weak compared with ionic and covalent bonds, molecular solids tend to be soft, low melting, and easily vaporized (ΔH_{fus} and ΔH_{vap} are low). For similar substances, the strength of the London dispersion forces increases smoothly with increasing molecular mass. For example, the melting points of benzene (C_6H_6), naphthalene ($C_{10}H_8$), and anthracene ($C_{14}H_{10}$), with one, two, and three fused aromatic rings, are 5.5°C, 80.2°C, and 215°C, respectively. The enthalpies of fusion also increase smoothly within the series: benzene (9.95 kJ/mol) < naphthalene (19.1 kJ/mol) < anthracene (28.8 kJ/mol). If the molecules have shapes that cannot pack together efficiently in the crystal, however, then the melting points and the enthalpies of fusion tend to be unexpectedly low because the molecules are unable to arrange themselves to optimize intermolecular interactions. Thus, toluene ($C_6H_5CH_3$) and *m*-xylene [m-$C_6H_4(CH_3)_2$] have melting points of −95°C and −48°C, respectively, which are significantly lower than the melting point of the lighter but more symmetrical analog, benzene.

Covalent Solids

Covalent solids are formed by networks or chains of atoms or molecules held together by covalent bonds. A perfect single crystal of a covalent solid is therefore a single giant molecule. For example, the structure of diamond, shown in Figure 12.19a, consists of sp^3 hybridized carbon atoms, each bonded to four other carbon atoms in a tetrahedral array to create a giant network. Note that the carbon atoms form six-membered rings.

The unit cell of diamond can be described as an fcc array of carbon atoms with four additional carbon atoms inserted into four of the tetrahedral holes. It thus has the zinc

Figure 12.19 The structures of diamond and graphite. (a) Diamond consists of sp^3 hybridized carbon atoms, each bonded to four other carbon atoms. The tetrahedral array forms a giant network in which carbon atoms form six-membered rings. (b) Side (left) and top (right) views of the graphite structure, showing the layers of fused six-membered rings and the arrangement of atoms in alternate layers of graphite. Note that the rings in alternate layers are staggered, such that every other carbon atom in one layer lies directly under (and above) the center of a six-membered ring in an adjacent layer.

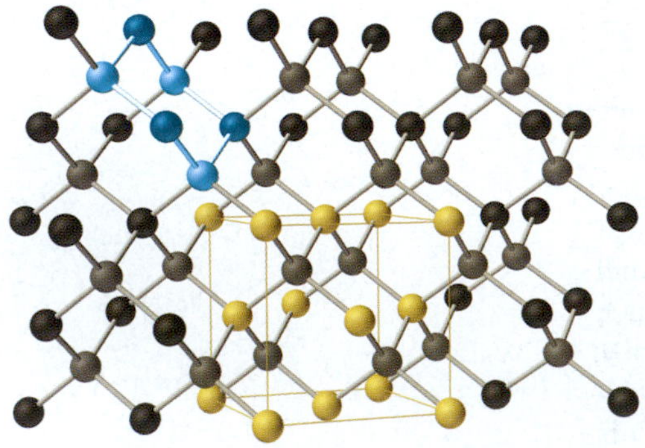

(a) Diamond

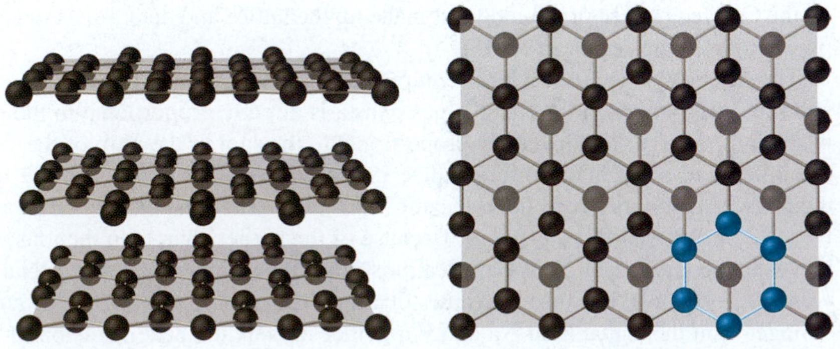

(b) Graphite

blende structure described in Section 12.3, except that in zinc blende the atoms that make up the fcc array are sulfur and the atoms in the tetrahedral holes are zinc. Elemental silicon has the same structure, as does silicon carbide (SiC), which has alternating C and Si atoms. The structure of crystalline quartz, SiO_2, shown in Section 12.1, can be viewed as being derived from the structure of silicon by inserting an oxygen atom between each pair of silicon atoms.

All compounds with the diamond and related structures are hard, high-melting-point solids that are not easily deformed. Instead, they tend to shatter when subjected to large stresses, and they usually do not conduct electricity very well. In fact, diamond (mp 3500°C at 63.5 atm) is the hardest substance known, and silicon carbide (mp 2986°C) is used commercially as an abrasive in sandpaper and grinding wheels. It is difficult to deform or melt these and related compounds because strong covalent (C—C or Si—Si) or polar covalent (Si—C or Si—O) bonds must be broken, which requires a large input of energy.

Other covalent solids have very different structures. For example, graphite, the other common allotrope of carbon, has the structure shown in Figure 12.19b. It contains planar networks of six-membered rings of sp^2 hybridized carbon atoms in which each carbon is bonded to three others. This leaves a single electron in an unhybridized $2p_z$ orbital that can be used to form C═C double bonds, resulting in a ring with alternating double and single bonds. Because of its resonance structures, the bonding in graphite is best viewed as consisting of a network of C—C single bonds with one-third of a π bond holding the carbons together, similar to the bonding in benzene.

To completely describe the bonding in graphite, we need a molecular orbital approach similar to the one used for benzene in Chapter 9. In fact, the C—C distance in graphite (141.5 pm) is slightly longer than the distance in benzene (139.5 pm), consistent with a net carbon–carbon bond order of 1.33. In graphite, the two-dimensional planes of carbon atoms are stacked to form a three-dimensional solid; only London dispersion forces hold the layers together. As a result, graphite exhibits properties typical of both covalent and molecular solids. Due to the strong covalent bonding within the layers, graphite has a very high melting point, as expected for a covalent solid (it actually sublimes at about 3915°C). It is also very soft because the layers can slide past one another easily because of the weak interlayer interactions. Consequently, graphite is used as a lubricant and as the "lead" in pencils; the friction between graphite and a piece of paper is sufficient to leave a thin layer of carbon on the paper. Graphite is unusual among covalent solids in that its electrical conductivity is very high parallel to the planes of carbon atoms because of delocalized C—C π bonding. Last, graphite is black because it contains an immense number of alternating double bonds, which results in a very small energy difference between the individual molecular orbitals. Consequently, light of virtually all wavelengths is absorbed. Diamond, on the other hand, is colorless when pure because it has no delocalized electrons.

Metallic Solids

Metals are characterized by their ability to reflect light, called *luster*, their high electrical and thermal conductivity, their high heat capacity, and their malleability and ductility. Every lattice point in a pure metallic element is occupied by an atom of the same metal. The packing efficiency in metallic crystals tends to be high, so the resulting **metallic solids** are dense, with each atom having as many as 12 nearest neighbors.

Bonding in metallic solids is quite different from the bonding in the other kinds of solids we have discussed. Because all the atoms are the same, there can be no ionic bonding, yet metals always contain too few electrons or valence orbitals to form covalent bonds with each of their neighbors. Instead, the valence electrons are delocalized throughout the crystal, providing a strong cohesive force that holds the metal atoms together.

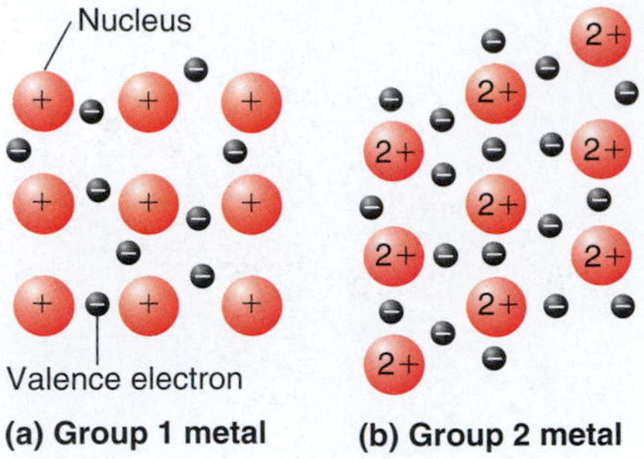

Figure 12.20 The electron-sea model of bonding in metals. Fixed, positively charged metal nuclei from Group 1 (a) or Group 2 (b) are surrounded by a "sea" of mobile valence electrons. Because a Group-2 metal has twice the number of valence electrons as a Group-1 metal, it should have a higher melting point.

The strength of metallic bonds varies dramatically. For example, cesium melts at 28.4°C and mercury is a liquid at room temperature, whereas tungsten melts at 3680°C. Metallic bonds tend to be weakest for elements that have nearly empty (as in Cs) or nearly full (Hg) valence subshells, and strongest for elements with approximately half-filled valence shells (as in W). As a result, the melting points of the metals increase to a maximum around Group 6 and then decrease again from left to right across the *d* block. Other properties related to the strength of metallic bonds, such as enthalpies of fusion, boiling points, and hardness, have similar periodic trends.

A somewhat oversimplified way to describe the bonding in a metallic crystal is to depict the crystal as consisting of positively charged nuclei in an *electron sea* (Figure 12.20). In this model, the valence electrons are not tightly bound to any one atom but are distributed uniformly throughout the structure. Very little energy is needed to remove electrons from a solid metal because they are not bound to a single nucleus. When an electrical potential is applied, the electrons can migrate through the solid toward the positive electrode, thus producing high electrical conductivity. The ease with which metals can be deformed under pressure is attributed to the ability of the metal ions to change positions within the electron sea without breaking any specific bonds. The transfer of energy through the solid by successive collisions between the metal ions also explains the high thermal conductivity of metals. This model does not, however, explain many of the other properties of metals, such as their metallic luster and the observed trends in bond strength as reflected in melting points or enthalpies of fusion. A more complete description of metallic bonding is presented in Section 12.6.

Substitutional Alloys

An **alloy** is a mixture of metals with metallic properties that differ from those of its constituent elements. Brass (Cu and Zn in a 2:1 ratio) and bronze (Cu and Sn in a 4:1 ratio) are examples of **substitutional alloys**, which are metallic solids that contain large numbers of substitutional impurities. In contrast, small numbers of interstitial impurities, such as carbon in the iron lattice of steel, give an **interstitial alloy**. Because scientists can combine two or more metals in varying proportions to tailor the properties of the material for particular applications, most of the metallic substances we encounter are actually alloys. Examples include the low-melting-point alloys used in solder (Pb and Sn in a 2:1 ratio) and in fuses and fire sprinklers (Bi, Pb, Sn, and Cd in a 4:2:1:1 ratio).

The compositions of most alloys can vary over wide ranges. In contrast, **intermetallic compounds** consist of certain metals that combine in only specific proportions. Their compositions are largely determined by the relative sizes of their component atoms and the ratio of the total number of valence electrons to the number of atoms present. The structures and physical properties of intermetallic compounds are frequently quite different from those of their constituent elements. For example, Cr_3Pt is an intermetallic compound used to coat razor blades advertised as "platinum coated"; it is very hard and dramatically lengthens the useful life of the razor blade.

Some general properties of the four major classes of solids are summarized in Table 12.5.

EXAMPLE 12.7

Classify each of these solids as ionic, molecular, covalent, or metallic, and arrange them in order of increasing melting points: Ge, RbI, $C_6(CH_3)_6$, Zn.

Given Compounds

Asked for Classification and order of melting points

TABLE 12.5 Properties of the major classes of solids

Ionic Solids	Molecular Solids	Covalent Solids	Metallic Solids
Poor conductors of heat and electricity	Poor conductors of heat and electricity	Poor conductors of heat and electricity[a]	Good conductors of heat and electricity
Relatively high melting point	Low melting point	High melting point	Melting points depend strongly on electron configuration
Hard but brittle; shatter under stress	Soft	Very hard and brittle	Easily deformed under stress; ductile and malleable
Relatively dense	Low density	Low density	Usually high density
Dull surface	Dull surface	Dull surface	Lustrous

[a] Many exceptions exist. For example, graphite has a relatively high electrical conductivity within the carbon planes, and diamond has the highest thermal conductivity of any known substance.

Strategy

A Locate the component element(s) in the periodic table. Based on their positions, predict whether each solid is ionic, molecular, covalent, or metallic.
B Arrange the solids in order of increasing melting points based on your classification, beginning with molecular solids.

Solution

A Germanium lies in the *p* block just under Si, along the diagonal line of semimetallic elements, which suggests that elemental Ge is likely to have the same structure as Si (the diamond structure). Thus, Ge is probably a covalent solid. RbI contains a metal from Group 1 and a nonmetal from Group 17, so it is an ionic solid containing Rb^+ and I^- ions. The compound $C_6(CH_3)_6$ is a hydrocarbon (hexamethylbenzene), which consists of isolated molecules that stack to form a molecular solid with no covalent bonds between them. Zn is a *d*-block element, so it is a metallic solid. **B** Arranging these substances in order of increasing melting points is straightforward, with one exception. We expect the molecular solid $[C_6(CH_3)_6]$ to have the lowest melting point and the covalent solid (Ge) to have the highest melting point, with the ionic solid (RbI) somewhere between. The melting points of metals, however, are difficult to predict based on the models presented thus far. Because Zn has a filled valence shell, it should not have a particularly high melting point, so a reasonable guess is $C_6(CH_3)_6$ < Zn ~ RbI < Ge. The actual melting points are $C_6(CH_3)_6$ (166°C); Zn (419°C); RbI (642°C); and Ge (938°C), in agreement with our prediction.

Note the pattern

The general order of increasing strength of interactions in a solid is: molecular solids < ionic solids ≈ metallic solids < covalent solids.

EXERCISE 12.7

Classify each substance as an ionic, covalent, molecular, or metallic solid, and then arrange them in approximate order of increasing melting points: C_{60}, $BaBr_2$, GaAs, AgZn.

Answer C_{60} (molecular) < AgZn (metallic) ~ $BaBr_2$ (ionic) < GaAs (covalent). The actual melting points are C_{60} (about 300°C), AgZn (about 700°C), $BaBr_2$ (856°C), and GaAs (1238°C).

12.6 • Bonding in Metals and Semiconductors

MGC Introduction to Band Theory

To explain the observed properties of metals, a more sophisticated approach is needed than the electron-sea model described in Section 12.5. The molecular orbital theory we used in Chapter 9 to explain the delocalized π bonding in polyatomic ions and molecules such as NO_2^-, ozone, and 1,3-butadiene can be adapted to accommodate the much higher number of atomic orbitals that interact with one another simultaneously in metals.

Band Theory

In a 1-mol sample of a metal, there can be more than 10^{24} orbital interactions to consider. In our molecular orbital description of metals, however, we begin by considering a simple one-dimensional example: a linear arrangement of n metal atoms, each of which contains a single electron in an s orbital. We use this example to describe an approach to metallic bonding called **band theory**, which assumes that the valence orbitals of the atoms in a solid interact, generating a set of molecular orbitals that extend throughout the solid.

One-Dimensional Systems

If the distance between the metal atoms is short enough for the orbitals to interact, they produce bonding, antibonding, and nonbonding molecular orbitals. The left portion of Figure 12.21, which is the same as the molecular orbital diagram in Figure 9.35, shows the pattern of molecular orbitals that results from the interaction of n s orbitals as n increases from 2 to 5.

As we saw in Chapter 9, the lowest-energy orbital is the completely bonding molecular orbital, whereas the highest-energy orbital is the completely antibonding molecular orbital. Molecular orbitals of intermediate energy have fewer nodes than the totally antibonding molecular orbital. The energy separation between adjacent orbitals decreases as the number of interacting orbitals increases. For $n = 30$, there are still discrete, well-resolved energy levels, but as n increases from 30 to a number close to Avogadro's number, the spacing between adjacent energy levels becomes almost infinitely small. The result is essentially a continuum of energy levels, as shown on the right in Figure 12.21, each of which corresponds to a particular molecular orbital extending throughout the linear array of metal atoms. The levels that are lowest in energy correspond to mostly bonding combinations of atomic orbitals, those highest in energy correspond to mostly antibonding combinations, and those in the middle correspond to essentially nonbonding combinations.

The continuous set of allowed energy levels shown on the right in Figure 12.21 is called an **energy band**. The difference in energy between the highest and lowest

Figure 12.21 The molecular orbital energy-level diagram for a linear arrangement of *n* atoms, each of which contains a singly occupied *s* orbital. This is the same diagram as Figure 9.35, with the addition of the far right-hand portion, corresponding to $n = 30$ and $n = \infty$. As n becomes very large, the energy separation between adjacent levels becomes so small that a single continuous band of allowed energy levels results. The lowest-energy molecular orbital corresponds to positive overlap between all the atomic orbitals to give a totally bonding combination, whereas the highest-energy molecular orbital contains a node between each pair of atoms and is thus totally antibonding.

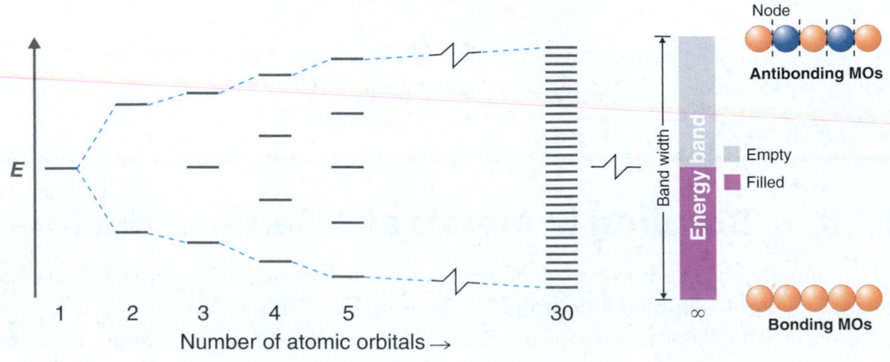

energy levels is the **bandwidth**, and it is proportional to the strength of the interaction between orbitals on adjacent atoms: the stronger the interaction, the larger the bandwidth. Because the band contains as many energy levels as molecular orbitals, and the number of molecular orbitals is the same as the number of interacting atomic orbitals, the band in Figure 12.21 contains n energy levels corresponding to the combining of s orbitals from n metal atoms. Each of the original s orbitals could contain a maximum of two electrons, so the band can accommodate a total of $2n$ electrons. Recall, however, that each of the metal atoms we started with contained only a single electron in each s orbital, so there are only n electrons to place in the band. Just as with atomic orbitals or molecular orbitals, the electrons occupy the lowest energy levels available. Consequently, only the lower half of the band is filled. This corresponds to filling all of the *bonding* molecular orbitals in the linear array of metal atoms and results in the strongest possible bonding.

Multidimensional Systems

The example discussed above was a one-dimensional array of atoms that contained only s orbitals. To extrapolate to two- or three-dimensional systems and to atoms that contain electrons in p and d orbitals is straightforward in principle, even though in practice the mathematics becomes more complex and the resulting molecular orbitals are more difficult to visualize. The resulting energy-level diagrams are essentially the same as the diagram of the one-dimensional example in Figure 12.21, with the following exception: they contain as many bands as there are different types of interacting orbitals. Because different atomic orbitals interact differently, each band will have a different bandwidth and will be centered at a different energy, corresponding to the energy of the parent atomic orbital of an isolated atom.

Band Gap

Because the $1s$, $2s$, and $2p$ orbitals of a third-period atom are filled core levels, they do not interact strongly with the corresponding orbitals on adjacent atoms. Hence, they form rather narrow bands that are well separated in energy (Figure 12.22). These bands are completely filled (both the bonding and antibonding levels are completely populated), so they do not make a net contribution to bonding in the solid. The energy difference between the highest level of one band and the lowest level of the next is the **band gap**. It represents a set of forbidden energies that do not correspond to any allowed combinations of atomic orbitals.

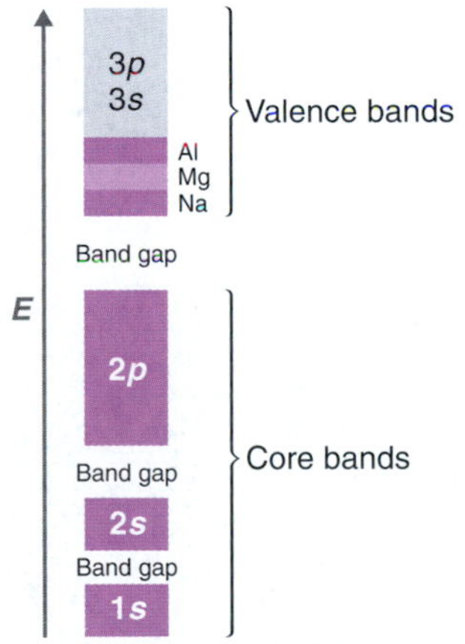

Figure 12.22 The band structures of the third-period metals Na, Mg, and Al. The $3s$ and $3p$ valence bands overlap in energy to form a continuous set of energy levels that can hold a maximum of eight electrons per atom.

Because they extend farther from the nucleus, the valence orbitals of adjacent atoms ($3s$ and $3p$ in Figure 12.22) interact much more strongly with one another than do the filled core levels, and as a result the valence bands have a larger bandwidth. In fact, the bands derived from the $3s$ and $3p$ atomic orbitals are wider than the energy gap between them, so the result is **overlapping bands**. These have molecular orbitals derived from two or more valence orbitals with similar energies. As the valence band is filled with one, two, or three electrons per atom for Na, Mg, and Al, respectively, the combined band that arises from the overlap of the $3s$ and $3p$ bands is also filling up; it has a total capacity of eight electrons per atom (two electrons for each $3s$ orbital and six electrons for each set of $3p$ orbitals). With Na, therefore, which has one valence electron, the combined valence band is one-eighth filled; with Mg (two valence electrons), it is one-fourth filled; and with Al, it is three-eighths filled, as indicated in Figure 12.22. The partially filled valence band is absolutely crucial for explaining metallic behavior because it guarantees that there are unoccupied energy levels at an infinitesimally small energy above the highest occupied level.

Band theory can explain virtually all the properties of metals. Metals conduct electricity, for example, because only a very small amount of energy is required to excite an electron from a filled level to an empty one, where it is free to migrate rapidly throughout the crystal in response to an applied electric field. Similarly, metals have high heat capacities (as you no doubt remember from the last time a doctor or a nurse placed a stethoscope on your skin!) because the electrons in the valence band can absorb thermal energy by being excited to the low-lying empty energy levels. Finally, metals are lustrous because light of various wavelengths can be absorbed, causing the valence electrons to be excited into any of the empty energy levels above the highest occupied level. When the electrons decay back to low-lying empty levels, they emit light of different wavelengths. Because electrons can be excited from many different filled levels in a metallic solid and can then decay back to any of many empty levels, light of varying wavelengths is absorbed and reemitted, which results in the characteristic shiny appearance that we associate with metals.

Requirements for Metallic Behavior

Note the pattern
Metallic behavior requires a partially occupied band.

For a solid to exhibit metallic behavior, it must have a set of delocalized orbitals forming a band of allowed energy levels, and the resulting band must be only partially filled (10–90%) with electrons. Without a set of delocalized orbitals there is no pathway by which electrons can move through the solid.

Band theory explains the correlation between the valence-electron configuration of a metal and the strength of metallic bonding. The valence electrons of transition metals occupy either their valence *ns*, $(n - 1)d$, and *np* orbitals (with a total capacity of 18 electrons per metal atom) or their *ns* and $(n - 1)d$ orbitals (a total capacity of 12 electrons per metal atom). These atomic orbitals are close enough in energy that the derived bands overlap, so the valence electrons are not confined to a specific orbital. Metals with six to nine valence electrons (which correspond to Groups 6 to 9) are those most likely to fill the valence bands approximately halfway. Those electrons therefore occupy the highest possible number of bonding levels, while the number of antibonding levels occupied is minimal. Not coincidentally, the elements of these groups exhibit physical properties consistent with the presence of the strongest metallic bonding, such as very high melting points.

Insulators

In contrast to metals, **electrical insulators** are materials that conduct electricity poorly because their valence bands are full. The energy gap between the highest filled levels and the lowest empty levels is so large that the empty levels are inaccessible: thermal energy cannot excite an electron from a filled level to an empty one. The valence-band structure of diamond, for example, is shown in Figure 12.23a. Because diamond has only 4 bonded neighbors rather than the 6 to 12 typical of metals, the carbon 2*s* and 2*p* orbitals combine to form two bands in the solid, with the one at lower energy representing bonding molecular orbitals and the one at higher energy representing antibonding molecular orbitals. Each band can contain four electrons per atom, so only the lower band is occupied. Because the energy gap between the filled band and the empty band is very large (530 kJ/mol), at normal temperatures thermal energy cannot excite electrons from the filled level into the empty band. Thus, there is no pathway by which

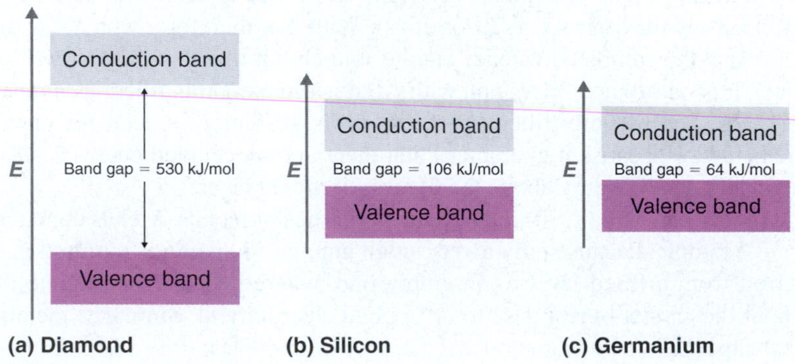

Figure 12.23 Energy-band diagrams for diamond, silicon, and germanium. Note that the band gap gets smaller from C ⟶ Ge.
Interactive Graph

electrons can move through the solid, and as a result diamond has one of the lowest electrical conductivities known.

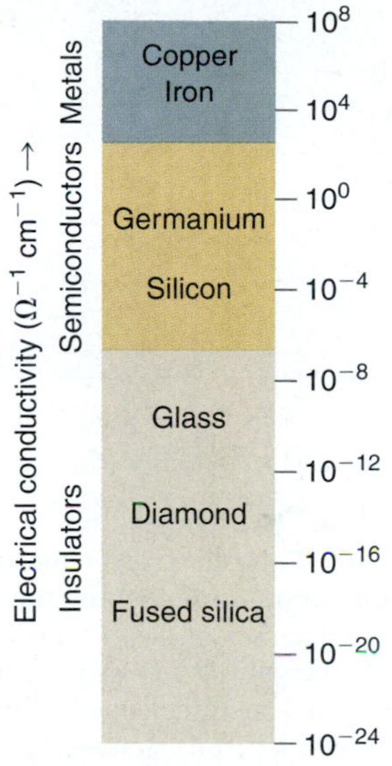

Figure 12.24 A logarithmic scale illustrating the enormous range of electrical conductivities of solids.

Semiconductors

What if the difference in energy between the highest occupied level and the lowest empty level is intermediate between those of electrical conductors and insulators? Such is the case for silicon and germanium, which have the same structure as diamond. Because Si—Si and Ge—Ge bonds are substantially weaker than C—C bonds (see Chapter 7), the energy gap between the filled and empty bands becomes much smaller as we go down Group 14 (Figure 12.23b and c). Consequently, thermal energy is able to excite a small number of electrons from the filled valence band of Si and Ge into the empty band above it, which is called the **conduction band**.

Exciting electrons from the filled valence band to the empty conduction band causes an increase in electrical conductivity for two reasons. First, the electrons in the previously vacant conduction band are free to migrate through the crystal in response to an applied electric field. Second, excitation of an electron from the valence band produces a "hole" in the valence band that is equivalent to a *positive* charge. The hole in the valence band can migrate through the crystal in the direction opposite that of the electron in the conduction band by means of a "bucket brigade" mechanism in which an adjacent electron fills the hole, thus generating a hole where the second electron had been, and so forth. Consequently, Si is a much better electrical conductor than diamond, and Ge is even better, although both are still much poorer conductors than a typical metal (Figure 12.24). Substances such as Si and Ge that have conductivities between those of metals and insulators are called **semiconductors**. Many binary compounds of the main group elements exhibit semiconducting behavior similar to that of Si and Ge. For example, gallium arsenide (GaAs) is isoelectronic with Ge and has the same crystalline structure, with alternating Ga and As atoms; not surprisingly, it is also a semiconductor. The electronic structure of semiconductors is compared with the structures of metals and insulators in Figure 12.25.

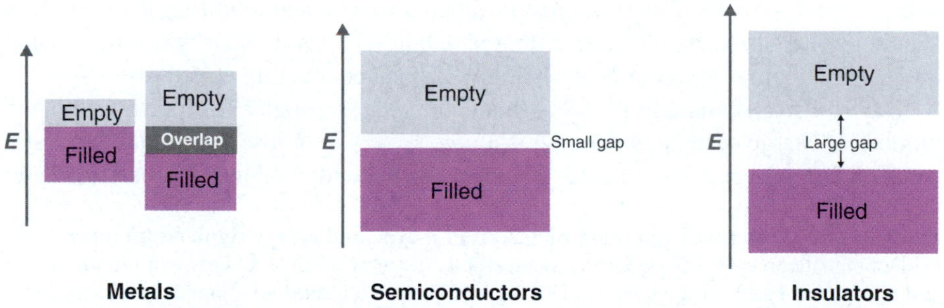

Figure 12.25 Comparison of the key features of the band structures of metals, semiconductors, and insulators. Note that metallic behavior can arise either from the presence of a single partially filled band or from two overlapping bands (one full and one empty). Interactive Graph

Temperature and Conductivity

Because thermal energy can excite electrons across the band gap in a semiconductor, increasing the temperature increases the number of electrons that have sufficient kinetic energy to be promoted into the conduction band. The electrical conductivity of a semiconductor therefore increases rapidly with increasing temperature, in contrast to the behavior of a purely metallic crystal. In a metal, as an electron travels through the crystal in response to an applied electrical potential, it cannot travel very far before it encounters and collides with a metal nucleus. The more often such encounters occur, the slower the *net* motion of the electron through the crystal, and the *lower* the conductivity. As the temperature of the solid increases, the metal atoms in the lattice acquire more and more kinetic energy. Because their positions are fixed in the lattice, however, the increased kinetic energy increases only the extent to which they vibrate about their fixed positions. At higher temperatures, therefore, the metal nuclei collide with the mobile

Note the pattern

The electrical conductivity of a semiconductor increases with increasing temperature, whereas the electrical conductivity of a metal decreases with increasing temperature.

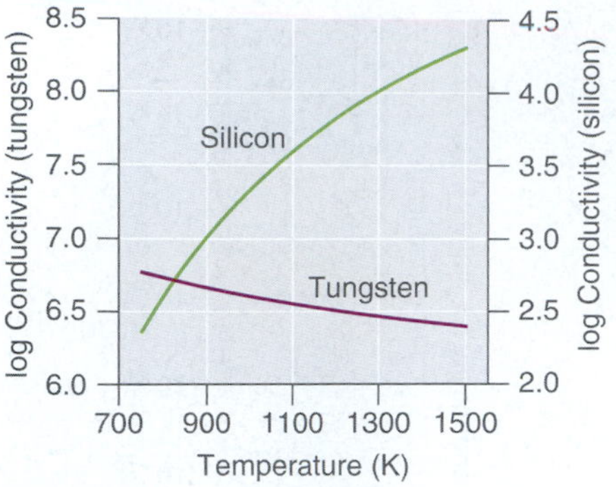

Figure 12.26 The temperature dependence of the electrical conductivity of a metal versus a semiconductor. Note that the conductivity of the metal (tungsten) decreases relatively slowly with increasing temperature, whereas the conductivity of the semiconductor (silicon) increases much more rapidly.

electrons more frequently and with greater energy, thus *decreasing* the conductivity. This effect is, however, substantially smaller than the increase in conductivity with temperature exhibited by semiconductors. For example, the conductivity of a tungsten wire decreases by a factor of only about two over the temperature range 750–1500 K, whereas the conductivity of silicon increases approximately 100-fold over the same temperature range. These trends are illustrated in Figure 12.26.

n- and *p*-Type Semiconductors

Doping is a process that is used to tune the electrical properties of commercial semiconductors by deliberately introducing small amounts of impurities. If an impurity contains *more* valence electrons than the atoms of the host lattice (for example, when small amounts of a Group-15 atom are introduced into a crystal of a Group-14 element), then the doped solid has more electrons available to conduct current than the pure host has. As shown in Figure 12.27a, adding an impurity such as phosphorus to a silicon crystal creates occasional electron-rich sites in the lattice. The electronic energy of these sites lies between those of the filled valence band and the empty conduction band, but closer to the conduction band. Because the atoms that were introduced are surrounded by host atoms, and the electrons associated with the impurity are close in energy to the conduction band, those extra electrons are relatively easily excited into the empty conduction band of the host. Such a substance is called an ***n*-type semiconductor**, with the *n* indicating that the added charge carriers are negative (they are electrons).

Note the pattern

n-Type semiconductor: negative charge carriers; impurity has more valence electrons than host.

p-Type semiconductor: positive charge carriers; impurity has fewer valence electrons than host.

If the impurity atoms contain *fewer* valence electrons than the atoms of the host (for example, when small amounts of a Group-13 atom are introduced into a crystal of a Group-14 element), then the doped solid has fewer electrons than the pure host. Perhaps unexpectedly, this also results in *increased* conductivity because the impurity atoms generate holes in the valence band. As shown in Figure 12.27b, adding an impurity such as gallium to a silicon crystal creates isolated electron-deficient sites in the host lattice. The electronic energy of these empty sites also lies between those of the filled valence band and the empty conduction band of the host, but much closer to the filled valence band. It is therefore relatively easy to excite electrons from the valence band of the host to the isolated impurity atoms, thus forming holes in the valence band. This kind of substance is called a ***p*-type semiconductor**, with the *p* standing for positive charge carrier (that is, a hole). Holes in what was a filled band are just as effective as electrons in an empty band at conducting electricity.

The electrical conductivity of a semiconductor is roughly proportional to the number of charge carriers, so doping is a precise way to adjust the conductivity of a semiconductor over a wide range. The entire semiconductor industry is built upon

Figure 12.27 Structures and band diagrams of *n*-type and *p*-type semiconductors. (a) Doping silicon with a Group-15 element results in a new *filled* level between the valence and conduction bands of the host. (b) Doping silicon with a Group-13 element results in a new *empty* level between the valence and conduction bands of the host. In both cases, the effective band gap is substantially decreased, and the electrical conductivity at a given temperature increases dramatically.

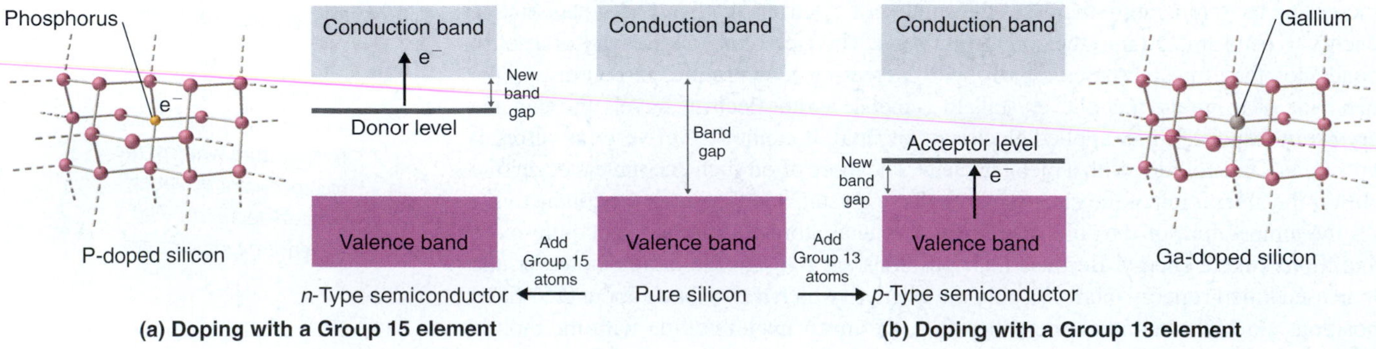

methods for preparing samples of Si, Ge, or GaAs doped with precise amounts of desired impurities, and for assembling silicon chips and other complex devices that contain junctions between *n*- and *p*-type semiconductors in varying numbers and arrangements.

EXAMPLE 12.8

A crystalline solid has the band structure shown in the margin, with the purple areas representing regions occupied by electrons. Thus, the lower band is completely occupied by electrons and the upper level is about one-third filled with electrons. **(a)** Predict the electrical properties of this solid. **(b)** What would happen to the electrical properties if all of the electrons were removed from the upper band? Would you use a chemical oxidant or reductant agent to effect this change? **(c)** What would happen to the electrical properties if enough electrons were added to completely fill the upper band? Would you use a chemical oxidizing or reducing agent to effect this change?

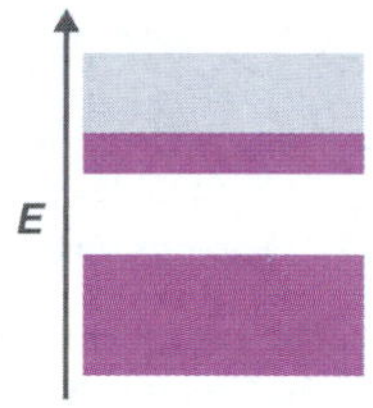

Given Band structure

Asked for Variations in electrical properties with conditions

Strategy

A Based on the occupancy of the lower and upper bands, predict whether the substance will be an electrical conductor. Then predict how its conductivity will change with temperature.

B After all the electrons are removed from the upper band, predict how the band gap would affect the electrical properties of the material. Determine whether you would use a chemical oxidant or reductant to remove electrons from the upper band.

C Predict the effect of a filled upper band on the electrical properties of the solid. Then decide whether you would use an oxidant or a reductant to fill the upper band.

Solution

(a) **A** The material has a partially filled band, which is critical for metallic behavior. The solid will therefore behave like a metal, with high electrical conductivity that decreases slightly with increasing temperature. **(b)** **B** Removing all of the electrons from the partially filled upper band would create a solid with a filled lower band and an empty upper band, separated by an energy gap. If the band gap is large, the material will be an electrical insulator. If the gap is relatively small, the substance will be a semiconductor whose electrical conductivity increases rapidly with increasing temperature. Removing the electrons would require an oxidizing agent because oxidizing agents accept electrons. **(c)** **C** Adding enough electrons to completely fill the upper band would produce an electrical insulator. Without another empty band relatively close in energy above the filled band, semiconductor behavior would be impossible. Adding electrons to the solid would require a reductant because reductants are electron donors.

EXERCISE 12.8

A substance has the band structure shown in the margin, in which the lower band is half-filled with electrons (purple area) and the upper band is empty. **(a)** Predict the electrical properties of the solid. **(b)** What would happen to the electrical properties if all of the electrons were removed from the lower band? Would you use a chemical oxidizing agent or reducing agent to effect this change? **(c)** What would happen to the electrical properties if enough electrons were added to completely fill the lower band? Would you use a chemical oxidizing or reducing agent to effect this change?

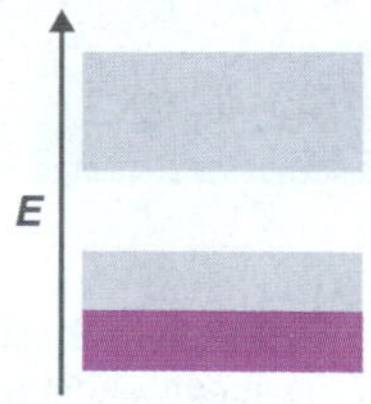

Answer **(a)** The solid has a partially filled band, so it has the electrical properties of a conductor. **(b)** Removing all electrons from the lower band would produce an electrical insulator with two empty bands. An oxidant is required. **(c)** Adding enough electrons to completely fill the lower level would result in an electrical insulator if the energy gap between the upper and lower bands is relatively large, or a semiconductor if the band gap is relatively small. A reductant is required.

12.7 • Superconductors

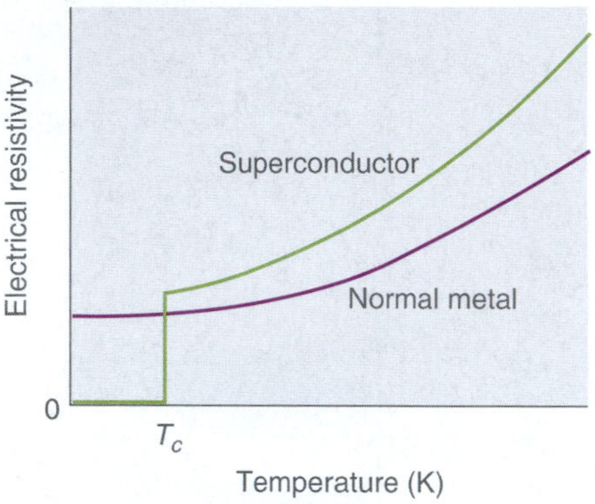

Figure 12.28 The temperature dependence of the electrical resistivity of a normal metal and a superconductor. The superconducting transition temperature, T_c, is the temperature at which the resistivity of a superconductor drops to zero.

The phenomenon of superconductivity was discovered by the Danish physicist H. Kamerlingh Onnes (1853–1926, Nobel Prize 1913), who had found a way to liquefy helium, which boils at 4.2 K and 1 atm pressure. To exploit the very low temperatures made possible by this new cryogenic fluid, he began a systematic study of the properties of metals, especially their electrical properties. Because the electrical resistance of a sample is technically easier to measure than its conductivity, Onnes measured the *resistivity* of his samples. The resistivity and conductivity of a material are inversely proportional:

$$\text{conductivity} = \frac{1}{\text{resistivity}} \tag{12.2}$$

In 1911, Onnes discovered that at about 4 K, the resistivity of metallic mercury (mp 234 K) decreased suddenly to essentially zero, rather than continuing to decrease only slowly with decreasing temperature as expected (Figure 12.28). He called this phenomenon **superconductivity** because a resistivity of zero means that an electrical current can flow forever. Onnes soon discovered that many other metallic elements exhibit superconductivity at very low temperatures. Each of these **superconductors** has a characteristic **superconducting transition temperature, T_c**, at which its resistivity drops to zero. At temperatures lower than their T_c, superconductors also completely expel a magnetic field from their interior (Figure 12.29a). This phenomenon is called the **Meissner effect** after one of its discoverers, the German physicist W. Meissner, who described the phenomenon in 1933. Due to the Meissner effect, a superconductor will actually "float" over a magnet, as shown in Figure 12.29b.

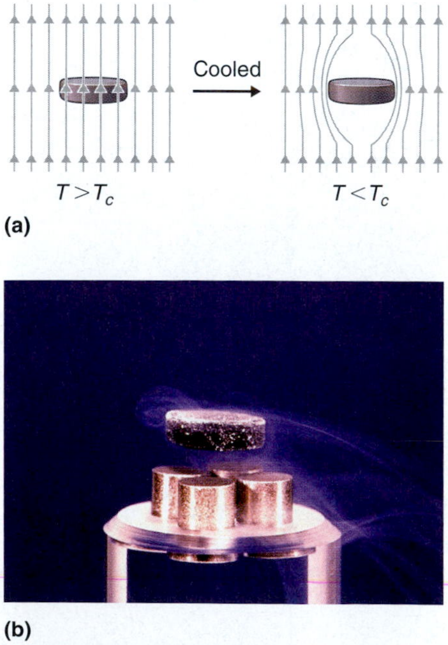

Figure 12.29 The Meissner effect. (a) Below its T_c, a superconductor completely expels magnetic lines of force from its interior. (b) Magnetic levitation, in which a small magnet "floats" over a disk of a high-temperature superconducting material ($YBa_2Cu_3O_{7-x}$) cooled in liquid nitrogen.

BCS Theory

For many years, the phenomenon of superconductivity could not be satisfactorily explained by the laws of conventional physics. In the early 1950s, however, American physicists J. Bardeen, L. Cooper, and J. R. Schrieffer formulated a theory for superconductivity (called the *BCS theory* from their initials) that earned them the Nobel Prize in physics in 1972. According to **BCS theory**, electrons are able to travel through a solid with zero resistance because of an attractive interaction between two electrons that are at some distance from each other. As one electron moves through the lattice, the surrounding nuclei are attracted to it. The motion of the nuclei can create a transient (short-lived) hole that pulls the second electron in the same direction as the first. The nuclei then return to their original positions to avoid colliding with the second electron as it approaches. The pairs of electrons, called *Cooper pairs*, migrate through the crystal as a unit. The electrons in Cooper pairs change partners frequently, like dancers in a ballet.

According to BCS theory, as the temperature of the solid increases, the vibrations of the atoms in the lattice increase continuously, until eventually the electrons cannot avoid colliding with them. The collisions result in the loss of superconductivity at higher temperatures.

The phenomenon of superconductivity suggested many exciting technological applications. For example, using superconducting wires in power cables would result in zero power losses, even over distances of hundreds of miles. Additionally, because superconductors expel magnetic fields, a combination of magnetic rails and superconducting wheels (or vice versa) could be used to produce *magnetic levitation* of, for example, a train over the track, resulting in friction-free transportation.

Unfortunately, for many years the only superconductors known had serious limitations, especially the need for very low temperatures, which required the use of expensive cryogenic fluids such as liquid He. In addition, the superconducting properties of many substances are destroyed by large electrical currents or even moderately large magnetic fields, making them useless for applications in power cables or high-field magnets. The ability of materials such as NbTi, NbSn, Nb_3Si, and Nb_3Ge to tolerate rather high magnetic fields, however, has led to a number of commercial applications of superconductors, including high-field magnets for nuclear magnetic resonance (NMR) spectrometers and magnetic resonance imaging (MRI) instruments in medicine, which, unlike X rays, can detect small changes in soft tissues in the body.

High-Temperature Superconductors

Because of these limitations, scientists continued to search for materials that exhibited superconductivity at temperatures higher than 77 K (the temperature of liquid nitrogen, the least expensive cryogenic fluid). In 1986, J. G. Bednorz and K. A. Müller, working for IBM in Zurich, showed that certain mixed-metal oxides containing La, Ba, and Cu exhibited superconductivity above 30 K. These compounds had been prepared by French workers as potential solid catalysts some years earlier, but their electrical properties had never been examined at low temperatures. Although initially the scientific community was extremely skeptical, the compounds were so easy to prepare that the results were confirmed within a few weeks. These **high-temperature superconductors** earned Bednorz and Müller the Nobel Prize in physics in 1987. Subsequent research has produced new compounds with related structures that are superconducting at temperatures as high as 95 K. The best known of these was discovered by P. Chu and M. K. Wu, Jr., and is called the "Chu–Wu phase" or the 1-2-3 superconductor.

The formula for the 1-2-3 superconductor is $YBa_2Cu_3O_{7-x}$, where x is about 0.1 for superconducting samples. If $x \approx 1.0$, giving a formula of $YBa_2Cu_3O_6$, the material is an electrical insulator. The superconducting phase is thus a nonstoichiometric compound, with a fixed ratio of metal atoms but a variable oxygen content. The reactants are heated to about 950°C for 12–24 hours, cooled, and ground into a fine powder that is then pressed into pellets and heated in oxygen at 800–900°C for several hours. The resulting shiny black pellets become superconducting at about 95 K. The overall equation for the synthesis of this material is

$$Y_2O_3(s) + 4BaCO_3(s) + 6CuO(s) + \tfrac{1}{2}O_2(g) \xrightarrow{\Delta} 2YBa_2Cu_3O_7(s) + 4CO_2(g) \tag{12.3}$$

If we assume that the superconducting phase is really stoichiometric $YBa_2Cu_3O_7$, then the average oxidation states of O, Y, Ba, and Cu are -2, $+3$, $+2$, and $+\frac{7}{3}$, respectively. The simplest way to view the average oxidation state of Cu is to assume that two Cu atoms per formula unit are present as Cu^{2+} and one is present as the rather unusual Cu^{3+}. In $YBa_2Cu_3O_6$, the insulating form, the oxidation state of Cu is $+\frac{5}{3}$, so there are two Cu^{2+} and one Cu^{+} per formula unit.

As shown in Figure 12.30, the unit cell of the 1-2-3 superconductor is related to the unit cell of the simple perovskite structure (Figure 12.12b). The only difference between the superconducting and insulating forms of the compound is that an O atom has been removed from between the Cu^{3+} ions, which destroys the chains of Cu atoms and leaves the Cu in the center of the unit cell as Cu^{+}. The chains of Cu atoms are crucial to the formation of the superconducting state.

Table 12.6 lists the ideal compositions of some of the known high-temperature superconductors that have been discovered in recent years. Engineers have learned how to process the brittle polycrystalline 1-2-3 and related compounds into wires, tapes, and films that can carry enormous electrical currents. Commercial applications include their use in infrared sensors and in analog signal processing and microwave devices.

TABLE 12.6 The composition of various superconductors

Compound	T_c, K
$Ba(Pb_{1-x}Bi_x)O_3$	13.5
$(La_{2-x}Sr_x)CuO_4$	35
$YBa_2Cu_3O_{7-x}$	95
$Bi_2(Sr_{2-x}Ca_x)CuO_6$[a]	80
$Bi_2Ca_2Sr_2Cu_3O_{10}$[a]	110
$Tl_2Ba_2Ca_2Cu_3O_{10}$[a]	125
$HgBa_2Ca_2Cu_3O_8$[a]	133
K_3C_{60}	18
Rb_3C_{60}	30

[a] Nominal compositions only. Oxygen deficiencies or excesses are common in these compounds.

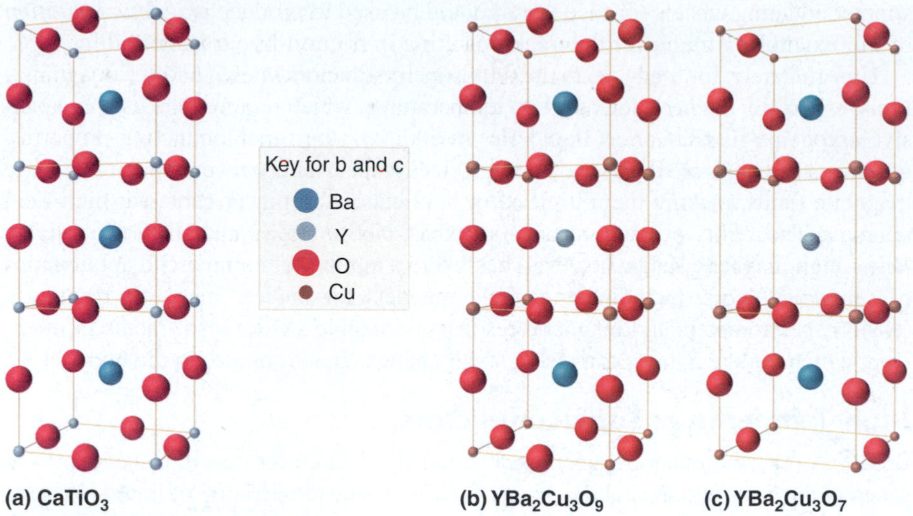

Figure 12.30 Relationship of the structure of a superconductor consisting of Y-Ba-Cu-O to a simple perovskite structure. (a) Stacking three unit cells of the Ca-centered $CaTiO_3$ perovskite structure (Figure 12.12b) together with (b) replacement of all Ti atoms by Cu, replacement of Ca in the top and bottom cubes by Ba, and replacement of Ca in the central cube by Y gives a $YBa_2Cu_3O_9$ stoichiometry. (c) Removal of two oxygen atoms per unit cell gives the nominal $YBa_2Cu_3O_7$ stoichiometry of the superconducting material.

EXAMPLE 12.9

Calculate the average oxidation state of Cu in a sample of $YBa_2Cu_3O_{7-x}$ with $x = 0.5$. How do you expect its structure to differ from those shown in Figure 12.30 for $YBa_2Cu_3O_9$ and $YBa_2Cu_3O_7$?

Given Stoichiometry

Asked for Average oxidation state, structure

Strategy

A Based on the oxidation states of the other component atoms, calculate the average oxidation state of Cu that would make an electrically neutral compound.

B Compare the stoichiometry of the structures shown in Figure 12.30 with the stoichiometry of the given compound to predict how the structures differ.

Solution

A The net negative charge is $(7.0 - 0.5)\ \text{O} \times (-2) = -13$, and the sum of the charges on the Y and Ba atoms is $[1 \times (+3)] + [2 \times (+2)] = +7$. This leaves a net charge of -6 per unit cell, which must be compensated for by the three Cu atoms, for a net charge of $+\frac{6}{3} = +2$ per Cu. **B** The most likely structure would be one in which every other O atom between the Cu atoms in the Cu chains of $YBa_2Cu_3O_7$ has been removed.

EXERCISE 12.9

Calculate the average oxidation state of Cu in a sample of $HgBa_2Ca_2Cu_3O_8$. Assume that Hg is present as Hg^{2+}.

Answer +2

12.8 • Polymeric Solids

Most of the reactions discussed in this text have involved molecules or ions with low molecular masses, ranging from tens to hundreds of amu. Many of the molecular materials in consumer goods today, however, have very high molecular masses, ranging from thousands to millions of amu, and are formed from a carefully controlled series of condensation or addition reactions that produce giant molecules called

polymers (from the Greek *poly* and *meros*, "many parts"). Polymers are used in corrective eye lenses, plastic containers, clothing and textiles, and medical implant devices. They consist of basic structural units called **monomers**, which are repeated many times in each molecule. As shown schematically in Figure 12.31, *polymerization* is the process by which monomers are connected into chains or networks by covalent bonds. Polymers can form via a condensation reaction, in which two monomer molecules are joined by a new covalent bond and a small molecule such as water is eliminated, or by an addition reaction, where the components of a species A-B are added to adjacent atoms of a multiple bond. Many people confuse the terms *plastics* and *polymers*. **Plastic** is the property of a material that allows it to be molded into almost any shape. Although many plastics are polymers, many polymers are not plastics. In this section, we introduce the reactions that produce naturally occurring and synthetic polymers.

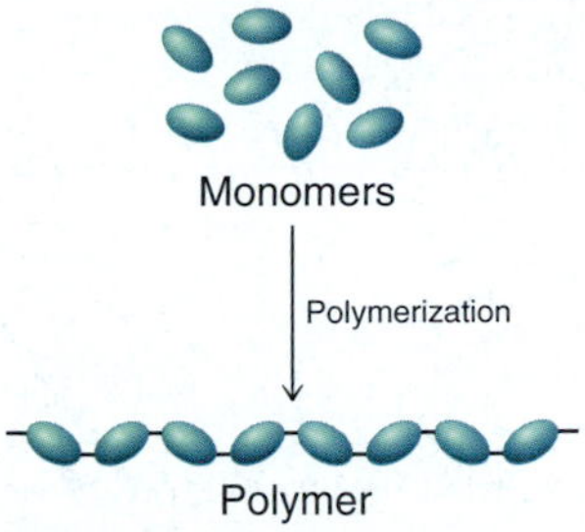

Figure 12.31 Polymer formation. During a polymerization reaction, a large number of monomers become connected by covalent bonds to form a single long molecule, a polymer.

Biological Polymers: Peptides and Proteins

Naturally occurring, or biological, polymers are crucial components of all organisms and form the fabric of our lives. Hair, silk, skin, feathers, muscle, and connective tissue are all primarily composed of proteins, the most familiar kind of biological polymer. The monomers of many biological polymers are the amino acids introduced in Section 5.4, each called an *amino acid residue*. The residues are linked together by amide bonds, also called **peptide bonds**, via a condensation reaction where H_2O is eliminated:

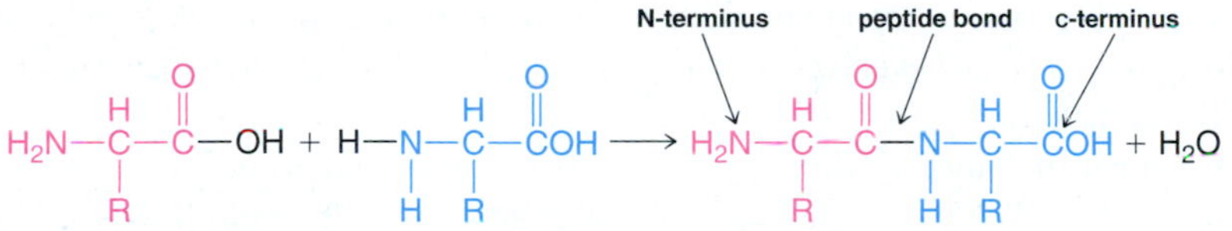

(12.4)

In Equation 12.4 R represents an alkyl or aryl group, or hydrogen, depending on the amino acid. We write the structural formula of the product with the free amino group on the left (the *N-terminus*) and the free carboxylate group on the right (the *C-terminus*). For example, the structural formula for the product formed from glycine and valine (glycyl-valine) is:

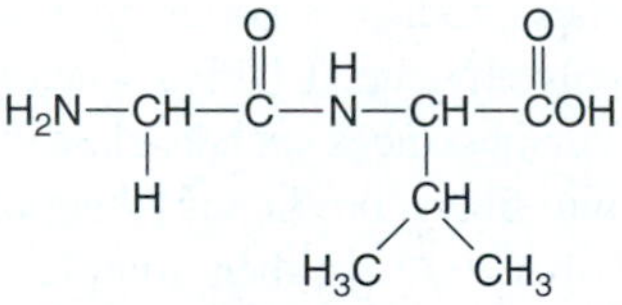

Glycyl-valine

The most important difference between synthetic and biological polymers is that the former usually contain very few different monomers, whereas proteins can have as many as 20 different kinds of amino acid residues arranged in many different orders.

Chains that contain fewer than about 50 amino acid residues are called **peptides**, whereas those that contain more than about 50 amino acid residues are called **proteins**. Many proteins are **enzymes**, catalysts that increase the rate of a biological reaction.

Many small peptides have potent physiological activities. The *endorphins*, for example, are powerful, naturally occurring painkillers found in the brain. Other important peptides are the hormones vasopressin and oxytocin. Although their structures and amino acid sequences are similar, vasopressin is a blood pressure regulator, whereas oxytocin induces labor in pregnant women and milk production in nursing mothers. Oxytocin was the first biologically active peptide to be prepared in the laboratory by Vincent du Vigneaud (1901–1978), who was awarded the Nobel Prize for his accomplishment in 1955.

Synthetic Polymers

Many of the synthetic polymers we use, such as plastics and rubbers, have commercial advantages over naturally occurring polymers because they can be produced inexpensively. Moreover, many synthetic polymers are actually more desirable than their natural counterparts because scientists can select monomer units to tailor the physical properties of the resulting polymer for particular purposes. For example, in many applications, wood has been replaced by plastics that are more durable, lighter, and easier to shape and to maintain. Polymers are also increasingly used in engineering applications where weight reduction and corrosion resistance are required. Steel rods used to support concrete structures, for example, are often coated with a polymeric material when the structures are near ocean environments where steel is vunerable to corrosion (see Chapter 19). The use of polymers in engineering applications is an active area of research.

Probably the best-known example of a synthetic polymer is *nylon* (Figure 12.32). Its monomers are linked by amide bonds (called peptide bonds in biological polymers), so its physical properties are similar to those of some proteins because of their common structural unit, the amide group. Nylon is easily drawn into silky fibers, particles that are more than a hundred times longer than they are wide, which can be woven into fabrics. Nylon fibers are so light and strong that during World War II, all available nylon was commandeered for use in parachutes, ropes, and other military items. With polymer chains that are fully extended and run parallel to the fiber axis, nylon fibers resist stretching, just like naturally occurring silk fibers, although the structures of nylon and silk are otherwise different. Replacing the flexible —CH_2— units in nylon by aromatic rings produces a stiffer and stronger polymer, such as the very strong polymer known as Kevlar. Kevlar fibers are so strong and rigid that they are used in lightweight army helmets, bulletproof vests, and even sailboat and canoe hulls, all of which contain multiple layers of Kevlar fabric.

Not all synthetic polymers are linked by amide bonds; for example, *polyesters* contain monomers that are linked by ester bonds. Polyesters are sold under trade names such as Dacron, Kodel, and Fortrel, which are used in clothing, and Mylar, which is used in magnetic tape, helium-filled balloons, and high-tech sails for sailboats. Although the fibers are flexible, properly prepared Mylar films are almost as strong as steel. Polymers based on skeletons that contain only carbon are all synthetic. Most of these are formed from ethylene ($CH_2{=}CH_2$), a two-carbon building block, and its derivatives. The relative lengths of the chains and any branches control the properties of polyethylene. For example, higher numbers of branches produce a softer, more flexible, lower-melting-point polymer called low-density polyethylene (LDPE), whereas high-density polyethylene (HDPE) contains few branches. Substances such as glass that melt at relatively low temperatures can also be formed into fibers, producing *fiberglass*.

Because most synthetic fibers are neither soluble nor low melting, multistep processes are required to manufacture them and to form them into objects. Graphite

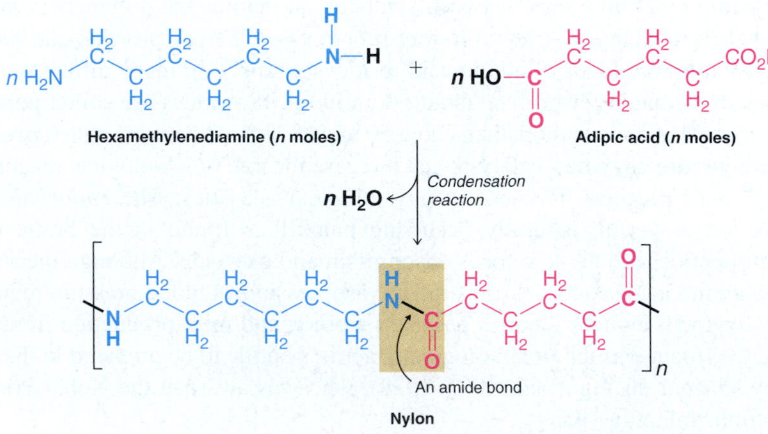

Figure 12.32 The synthesis of nylon. Nylon is a synthetic condensation polymer created by the reaction of a dicarboxylic acid and a diamine to form amide bonds and water.

fibers are formed by heating a precursor polymer at high temperatures to decompose it, a process called **pyrolysis**. The usual precursor for graphite is polyacrylonitrile, better known by its trade name, Orlon. A similar approach is used to prepare fibers of silicon carbide using a organosilicon precursor such as polydimethylsilane, $[—(CH_3)_2Si—]_n$.

Because there are no good polymer precursors for elemental boron or boron nitride, these fibers have to be prepared by indirect methods that are time consuming and costly. Thus, even though boron fibers are about eight times stronger than metallic aluminum and 10% lighter, they are significantly more expensive. Consequently unless an application requires boron's greater resistance to oxidation, these fibers cannot compete with less costly graphite fibers.

EXAMPLE 12.10

Polyethylene is used in a wide variety of products, including beach balls and the hard plastic bottles used to store solutions in a chemistry laboratory. Which of these products is formed from the more highly branched polyethylene?

Given Type of polymer

Asked for Application

Strategy

Determine whether the polymer is low-density polyethylene (LDPE), used in applications that require flexibility, or high-density polyethylene (HDPE), used for its strength and rigidity.

Solution

A highly branched polymer is less dense and less rigid than a relatively unbranched polymer. Hence, hard, strong polyethylene objects such as bottles are made of HDPE with relatively few branches. In contrast, a beach ball has to be flexible so it can be inflated. It is therefore made of highly branched LDPE.

EXERCISE 12.10

Which products are manufactured from low-density polyethylene and which from high-density polyethylene? **(a)** lawn chair frames; **(b)** rope; **(c)** disposable syringes; **(d)** automobile protective covers

Answer **(a)**, **(c)** HDPE; **(b)**, **(d)** LDPE

12.9 • Contemporary Materials

In addition to polymers, other materials, such as ceramics, high-strength alloys, and composites, play a major role in almost every aspect of our lives. Until relatively recently, steel was used for any application that required an especially strong and durable material, such as bridges, automobiles, airplanes, golf clubs, and tennis rackets. In the last 15 to 20 years, however, graphite or boron fiber golf clubs and tennis rackets have made wood and steel obsolete for these items. Likewise, a modern jet engine now contains virtually 0% Fe by weight (Table 12.7). The percentage of iron in wings and fuselages is similarly low, which indicates the extent to which other materials have supplanted steel. The Chevrolet Corvette introduced in 1953 was unusual because its body was fiberglass rather than steel, and by 1992, Jaguar was producing an all-aluminum limited-edition vehicle. Current models of many automobiles have engines that are made mostly of aluminum rather than steel. In this section, we describe some of the chemistry behind these new materials.

TABLE 12.7 Approximate elemental composition of a modern jet engine

Element	Percentage by Mass
Titanium	38
Nickel	37
Chromium	12
Cobalt	6
Aluminum	3
Niobium	1
Tantalum	0.025

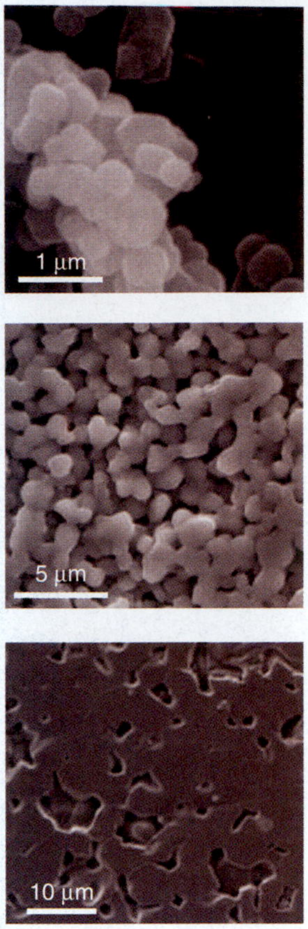

Figure 12.33 Sintering. Electron micrographs of magnesium oxide grains (top); the microstructure of the ceramic after sintering for two hours at 1250°C (center); and the microstructure after sintering for two hours at 1450°C (bottom). During the sintering process, the grains fuse, forming a dense and strong material.

Ceramics

A **ceramic** is any nonmetallic, inorganic solid that is strong enough for use in structural applications. Traditional ceramics, based on metal silicates or aluminosilicates, are the materials used to make pottery, china, bricks, and concrete. Modern ceramics contain a much wider range of components and can be classified as either *ceramic oxides*, which are based on metal oxides such as alumina (Al_2O_3), zirconia (ZrO_2), and beryllia (BeO), or *nonoxide ceramics*, which are based on metal carbides such as silicon carbide (carborundum, SiC) and tungsten carbide (WC), or nitrides like silicon nitride (Si_3N_4) and boron nitride (BN).

All modern ceramics are hard, lightweight, and stable at very high temperatures. Unfortunately, however, they are also rather brittle, tending to crack or break under stresses that would cause metals to bend or dent. Thus, a major challenge for materials scientists is to take advantage of the desirable properties of ceramics, such as their thermal and oxidative stability, chemical inertness, and toughness, while finding ways to decrease their brittleness in order to use them in new applications. Few metals can be used in jet engines for example, because most lose mechanical strength and react with oxygen at the very high operating temperatures inside the engines (~2000°C). In contrast, ceramic oxides such as Al_2O_3 cannot react with oxygen regardless of the temperature because aluminum is already in its highest possible oxidation state (Al^{3+}). Even nonoxide ceramics such as silicon and boron nitrides and silicon carbide are essentially unreactive in air up to about 1500°C. Producing a high-strength ceramic for use in such applications involves a process called *sintering*, which fuses the grains into a dense and strong material (Figure 12.33).

One of the most widely used raw materials for making ceramics is clay. Clay minerals consist of hydrated alumina (Al_2O_3) and silica (SiO_2) that contain a broad range of impurities, including barium, calcium, sodium, potassium, and iron. Although the structures of clay minerals are complicated, they all contain layers of metal atoms linked by oxygen atoms. Water molecules fit between the layers to form a thin film of water. When hydrated, clays can be easily molded, but during high-temperature heat treatment, called *firing*, a dense and strong ceramic is produced.

Because ceramics are so hard, they are easily contaminated by the material used to grind them. In fact, the ceramic often grinds the metal surface of the mill almost as fast as the mill grinds the ceramic! The **sol-gel process** for manufacturing ceramics, developed to address this problem, is also used in nature to create opal gemstones. A water-soluble precursor species, usually a metal or semimetal alkoxide, $M(OR)_n$, undergoes a hydrolysis reaction to form a cloudy aqueous dispersion called a *sol*. The sol contains particles of the metal or semimetal hydroxide, $M(OH)_n$, which are typically 1–100 nm in diameter. As the reaction proceeds, molecules of water are eliminated from between the $M(OH)_n$ units in a condensation reaction and the particles fuse together, producing oxide bridges, M-O-M. Eventually, the particles become linked in a three-dimensional network that causes the solution to form a *gel*, similar to a gelatin dessert. Heating the gel to 200–500°C causes more water to be eliminated, thus forming small particles of metal oxide that can be amazingly uniform in size. This chemistry starts with highly pure $SiCl_4$ and proceeds via the following reactions:

$$SiCl_4(s) + 4CH_3CH_2OH(l) + 4NH_3(g) \longrightarrow Si(OCH_2CH_3)_4(s) + 4NH_4Cl(s) \quad (12.5)$$

Alkoxide formation

$$Si(OCH_2CH_3)_4(s) + 4H_2O(l) \longrightarrow (HO)_3Si{-}OH(s) + 4CH_3CH_2OH(aq) \quad (12.6)$$

Hydrolysis of the alkoxide (sol)

$$(HO)_3Si{-}OH(s) + n\,HO{-}Si(OH)_3(s) \longrightarrow (HO)_3Si({-}O{-}Si(OH)_3)n(s) + nH_2O(l) \quad (12.7)$$

Condensation

Superalloys

Superalloys are high-strength alloys, often with a complex composition, that are used in systems requiring mechanical strength, high surface stability (minimal flaking or pitting), and resistance to high temperatures. The aerospace industry, for example, requires materials that have high strength-to-weight ratios to improve the fuel efficiency of advanced propulsion systems, and these systems must operate safely at temperatures higher than 1000°C.

Although most superalloys are based on nickel, cobalt, or iron, other metals are used as well. Pure nickel or cobalt is relatively easily oxidized, but adding small amounts of other metals (Al, Co, Cr, Mo, Nb, Ti, and W) results in an alloy that has superior properties. Consequently, most of the internal parts of modern gas turbine jet engines are now made of superalloys based on either nickel (used in blades and disks) or cobalt (used in vanes, combustion chamber liners, and afterburners). The cobalt-based superalloys are not as strong as the nickel-based ones, but they have excellent corrosion resistance at high temperatures.

Other alloys, such as aluminum–lithium and alloys based on titanium, also have applications in the aerospace industry. Because aluminum–lithium alloys are lighter, stiffer, and more resistant to fatigue at high temperatures than aluminum itself, they are used in engine parts and in the metal "skins" that cover wings and bodies. The high strength, corrosion resistance, and light weight of titanium are equally desirable for applications where minimizing weight is important (as in airplanes). Unfortunately, however, metallic titanium reacts rapidly with air at high temperatures to form TiN and TiO_2. Welding of titanium or any similar processes must therefore be carried out in an argon or inert gas atmosphere, which adds significantly to the cost. Initially, titanium and its alloys were primarily used in military applications, but more recently, they have been used as components of the airframes of commercial planes, in ship structures, and in biological implants.

Composite Materials

Composite materials have at least two distinct components: the matrix (which constitutes the bulk of the material) and fibers or granules that are embedded within the matrix and limit the growth of cracks by pinning defects in the bulk material (Figure 12.34). The resulting material is stronger, tougher, stiffer, and more resistant to corrosion than either component alone. Composites are thus the nanometer-scale equivalent of reinforced concrete, in which steel rods greatly increase the mechanical strength of the cement matrix. Three distinct types of composite material are generally recognized, distinguished by the nature of the matrix: polymer-matrix composites, metal-matrix composites, and ceramic-matrix composites.

Fiberglass is a **polymer-matrix composite** that consists of glass fibers embedded in a polymer, forming tapes that are then arranged in layers impregnated with epoxy. The result is a strong, stiff, lightweight material that is resistant to chemical degradation. It is not strong enough, however, to resist cracking or puncturing on impact. Stronger, stiffer polymer-matrix composites contain fibers of carbon (graphite), boron, or polyamides such as Kevlar. High-tech tennis rackets and golf clubs as well as the skins of modern military aircraft such as the "stealth" F-117A fighters and B-2 bombers are made from both carbon fiber–epoxy and boron fiber–epoxy composites. Compared with metals, these materials are 25–50% lighter. Similarly, the space shuttle payload bay doors and panels are made of a carbon fiber–epoxy composite.

Metal-matrix composites consist of metals or metal alloys reinforced with fibers. They offer significant advantages for high-temperature applications but pose major manufacturing challenges. For example, obtaining a uniform distribution and alignment of the reinforcing fibers can be difficult, and, because organic polymers cannot survive the high temperatures of molten metals, only fibers

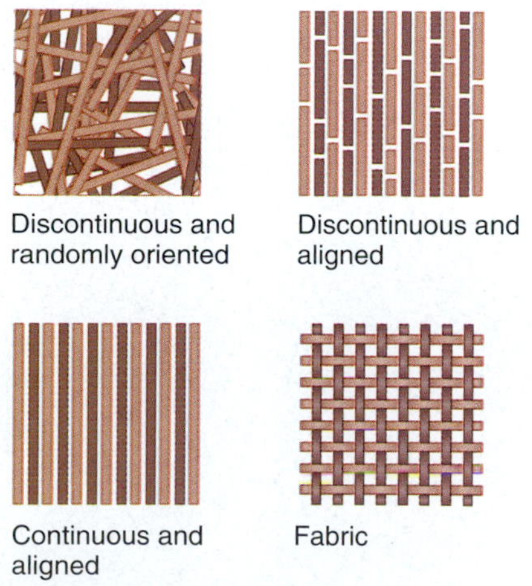

Figure 12.34 Some possible arrangements of fibers in fiber-reinforced composite materials. The arrangements shown range from discontinuous and randomly oriented to continuous and aligned. The fibers limit the growth of cracks by pinning defects within the matrix.

composed of boron, carbon, or ceramic (such as silicon carbide) can be used. Aluminum alloys reinforced with boron fibers are used in the aerospace industry, where their strength and light weight make up for their relatively high cost. The skins of hypersonic aircraft and structural units in the space shuttle are made of metal-matrix composites.

Ceramic-matrix composites contain ceramic fibers in a ceramic matrix material. A typical example is alumina reinforced with silicon carbide fibers. Combining the two very high-melting-point materials results in a composite that has excellent thermal stability, great strength, and corrosion resistance, while the SiC fibers reduce brittleness and cracking. Consequently, these materials are used in very high-temperature applications, such as the leading edge of wings of hypersonic airplanes and jet engine parts. They are also used in the protective ceramic tiles on the space shuttle, which contain short fibers of pure SiO_2 mixed with fibers of an aluminum–boron–silicate ceramic. These tiles are excellent thermal insulators and extremely light (their density is only ~0.2 g/cm^3). Although their surface reaches a temperature of about 1250°C during reentry, the temperature of the underlying aluminum alloy skin stays below 200°C.

EXAMPLE 12.11

An engineer is assigned to design a jet ski hull. What material is most suited to this application, and why?

Given Design objective

Asked for Most suitable material

Strategy

Determine under what conditions the design will be used. Then decide what type of material is most appropriate.

Solution

A jet ski hull must be lightweight to maximize speed and fuel efficiency. Because of its use in a marine environment, it must also be resistant to impact and corrosion. A ceramic material provides rigidity but is brittle and therefore tends to break or crack under stress, such as when it impacts waves at high speeds. Superalloys provide strength and stability, but a superalloy is probably too heavy for this application. Depending on the selection of metals, it might not be resistant to corrosion in a marine environment either. Composite materials, however, provide strength, stiffness, and corrosion resistance, and in addition they are lightweight. This is not a high-temperature application, so we do not need a metal-matrix or ceramic-matrix composite. The best choice of material is a polymer-matrix composite that contains Kevlar fibers to increase the strength of the composite on impact.

EXERCISE 12.11

In designing a new generation of space shuttle, NASA engineers are considering thermal-protection devices to protect the skin of the craft. Among the materials being considered are titanium- or nickel-based alloys and silicon carbide ceramic reinforced with carbon fibers. Why are these materials suitable for this application?

Answer Ti- or Ni-based alloys have a high strength-to-weight ratio, resist corrosion, and are safe at high temperatures. Reinforced ceramic is lightweight, has high thermal and oxidative stability, and is chemically inert, tough, and impact resistant.

SUMMARY AND KEY TERMS

12.1 Crystalline and Amorphous Solids (p. 526)

Solids are characterized by an extended three-dimensional arrangement of atoms, ions, or molecules in which the components are generally locked into their positions. The components can be arranged in a regular repeating three-dimensional array (a **crystal lattice**), which results in a **crystalline solid**, or more or less randomly to produce an **amorphous solid**. Crystalline solids have well-defined edges and faces, diffract X rays, and tend to have sharp melting points. In contrast, amorphous solids have irregular or curved surfaces, do not give well-resolved X-ray diffraction patterns, and melt over a wide range of temperatures.

12.2 The Arrangement of Atoms in Crystalline Solids (p. 527)

The smallest repeating unit of a crystal lattice is the **unit cell**. The **simple cubic** unit cell contains only eight atoms, molecules, or ions at the corners of a cube. A **body-centered cubic (bcc)** unit cell contains one additional component in the center of the cube. A **face-centered cubic (fcc)** unit cell contains a component in the center of each face in addition to those at the corners of the cube. Simple cubic and body-centered cubic arrangements fill only 52% and 68% of the available space with atoms, respectively. The **hexagonal close-packed (hcp) structure** has an ABABAB . . . repeat arrangement, and the **cubic close-packed (ccp) structure** has an ABCABC . . . repeat pattern; the latter is identical to a face-centered cubic lattice. The hcp and ccp arrangements fill 74% of the available space and have a **coordination number** of 12 for each atom in the lattice, the number of nearest neighbors. The simple cubic and body-centered cubic lattices have coordination numbers of 6 and 8, respectively.

12.3 Structures of Simple Binary Compounds (p. 532)

The structures of most binary compounds are dictated by the packing arrangement of the largest species present (the anions), with the smaller species (the cations) occupying appropriately sized holes in the anion lattice. A simple cubic lattice of anions contains a single cubic hole in the center of the unit cell. Placing a cation in the cubic hole results in the **cesium chloride structure**, with a 1:1 cation:anion ratio and a coordination number of 8 for both the cation and the anion. A face-centered cubic array of atoms or ions contains both octahedral holes and tetrahedral holes. If the octahedral holes in an fcc lattice of anions are filled with cations, the result is a **sodium chloride structure**. It also has a 1:1 cation:anion ratio, and each ion has a coordination number of 6. Occupation of half the tetrahedral holes by cations results in the **zinc blende structure**, with a 1:1 cation:anion ratio and a coordination number of 4 for the cations. More complex structures are possible if there are more than two kinds of atoms in a solid. One example is the **perovskite structure**, in which the two metal ions form an alternating body-centered array with the anions in the centers of the square faces. Because the wavelength of X-ray radiation is comparable to the interatomic distances in most solids, **X-ray diffraction** can be used to provide information about the structures of crystalline solids. X-rays diffracted from different planes of atoms in a solid reinforce one another if they are in phase, which occurs only if the extra distance they travel corresponds to an integral number of wavelengths. This relationship is described by the **Bragg equation**: $2d \sin \theta = n\lambda$.

12.4 Defects in Crystals (p. 537)

Real crystals contain large numbers of defects. Defects may affect only a single point in the lattice (a **point defect**), a row of lattice points (a **line defect**), or a plane of atoms (a **plane defect**). A point defect can be an atom missing from a site in the crystal (a **vacancy**) or an impurity atom that occupies either a normal lattice site (a **substitutional impurity**) or a hole in the lattice between atoms (an **interstitial impurity**). In an **edge dislocation**, an extra plane of atoms is inserted into part of the crystal lattice. Multiple defects can be introduced into materials so that the presence of one defect prevents the motion of another, in a process called **pinning**. Because defect motion tends to stop at **grain boundaries**, controlling the size of the grains in a material controls its mechanical properties. In addition, a process called **work hardening** introduces defects to toughen metals. **Schottky defects** are a coupled pair of vacancies, one cation and one anion, that maintains electrical neutrality. A **Frenkel defect** is an ion that occupies an incorrect site in the lattice. Cations in such compounds are often able to move rapidly from one site in the crystal to another, resulting in high electrical conductivity in the solid material. Such compounds are called **solid electrolytes**. **Nonstoichiometric compounds** have variable stoichiometries over a given range with no dramatic change in crystal structure. This behavior is due to a large number of vacancies or substitutions of one ion by another ion with a different charge.

12.5 Correlation Between Bonding and the Properties of Solids (p. 543)

The major types of solids are ionic, molecular, covalent, and metallic. **Ionic solids** consist of positively and negatively charged ions held together by electrostatic forces; the strength of the bonding is reflected in the lattice energy. Ionic solids tend to have high melting points and to be rather hard. **Molecular solids** are held together by relatively weak forces, such as dipole–dipole interactions, hydrogen bonds, and London dispersion forces. As a result, they tend to be rather soft and have low melting points, which depend on their molecular structure. **Covalent solids** consist of two- or three-dimensional networks of atoms held together by covalent bonds; they tend to be very hard and have high melting points. **Metallic solids** have unusual properties: in addition to having high thermal and electrical conductivity and being malleable and ductile, they exhibit luster, a shiny surface that reflects light. An **alloy** is a mixture of metals that has bulk metallic properties different from those of its constituent elements. Alloys can be formed by substituting one metal atom for another of similar size in the lattice (**substitutional alloys**), by inserting smaller atoms into holes in the metal lattice (**interstitial alloys**), or by a combination. Although the elemental composition of most alloys can vary over wide ranges, certain metals combine in only fixed proportions to form **intermetallic compounds** with unique properties.

12.6 Bonding in Metals and Semiconductors (p. 547)

Band theory assumes that the valence orbitals of the atoms in a solid interact to generate a set of molecular orbitals that extend throughout the solid; the continuous set of allowed energy levels is an **energy band**. The difference in energy between the highest and lowest allowed levels within a given band is the **bandwidth**, and the difference in energy between the highest level of one band and the lowest level of the band above it is the **band gap**. If the width of adjacent bands is larger than the energy gap between them, **overlapping bands** result, in which molecular orbitals derived from two or more kinds of valence orbitals have similar energies. Metallic properties depend on a partially occupied band corresponding to a set of molecular orbitals that extend throughout the solid to form a band of energy levels. If a solid has a filled valence band with a relatively low-lying empty band above it (a **conduction band**), then electrons can be excited by thermal energy from the filled band into the vacant band where they can then migrate through the crystal, resulting in electrical conductivity. **Electrical insulators** are poor conductors because their valence bands are full. **Semiconductors** have electrical conductivities intermediate between those of insulators and metals. The electrical conductivity of semiconductors increases rapidly with increasing temperature, whereas the electrical conductivity of metals decreases slowly with increasing temperature. The properties of semiconductors can be modified by **doping**, or introducing impurities. Adding an element with more valence electrons than the atoms of the host populates the conduction band, resulting in an ***n*-type semiconductor** with increased electrical conductivity. Adding an element with fewer valence electrons than the atoms of the host generates holes in the valence band, resulting in a ***p*-type semiconductor** that also exhibits increased electrical conductivity.

12.7 Superconductors (p. 554)

Superconductors are solids that at low temperatures exhibit zero resistance to the flow of electrical current, a phenomenon known as **superconductivity**. The temperature at which the electrical resistance of a substance drops to zero is its **superconducting transition temperature, T_c**. Superconductors also expel a magnetic field from their interior, a phenomenon known as the **Meissner effect**. Superconductivity can be explained by the **BCS theory**, which says that electrons are able to travel through a solid with no resistance because they couple to form pairs of electrons (Cooper pairs). **High-temperature superconductors** have T_c values higher than 30 K.

12.8 Polymeric Solids (p. 556)

Polymers are giant molecules that consist of long chains of units called **monomers** connected by covalent bonds. *Polymerization* is the process of linking monomers together to form a polymer. **Plastic** is the property of a material that allows it to be molded. Biological polymers formed from amino acid residues are called **peptides** or **proteins**, depending on their size. **Enzymes** are proteins that catalyze a biological reaction. A particle that is more than a hundred times longer than it is wide is a **fiber**, which can be formed by a high-temperature decomposition reaction called **pyrolysis**.

12.9 Contemporary Materials (p. 559)

Ceramics are nonmetallic, inorganic solids that are typically strong with high melting points but brittle. The two major classes of modern ceramics are ceramic oxides, and nonoxide ceramics, composed of nonmetal carbides or nitrides. The production of ceramics generally involves pressing a powder of the material into the desired shape and sintering at a temperature just below its melting point. The necessary fine powders of ceramic oxides with uniformly sized particles can be produced by the **sol-gel process**. **Superalloys** are new metal phases based on cobalt, nickel, or iron that exhibit unusually high temperature stability and resistance to oxidation. **Composite materials** consist of at least two phases: a matrix that constitutes the bulk of the material, and fibers or granules that act as a reinforcement. **Polymer-matrix composites** have reinforcing fibers embedded in a polymer matrix. **Metal-matrix composites** have a metal matrix and fibers of boron, graphite, or ceramic. **Ceramic-matrix composites** use reinforcing fibers, usually also ceramic, to make the matrix phase less brittle.

KEY EQUATION

Bragg equation $$2d \sin \theta = n\lambda \quad (12.1)$$

QUESTIONS AND PROBLEMS

MGC For instructor-assigned homework, go to **www.masteringgenchem.com**

Questions and Problems with colored numbers have answers in the Appendix and complete solutions in the Student Solutions Manual.

CONCEPTUAL

12.1 Crystalline and Amorphous Solids

1. Compare the solid and liquid states in terms of (a) rigidity of structure; (b) long-range order; and (c) short-range order.
2. How do amorphous solids differ from crystalline solids in: (a) rigidity of structure; (b) long-range order; and (c) short-range order? Which of the two types of solid is most similar to a liquid?
3. Why is the arrangement of the constituent atoms or molecules more important in determining the properties of a solid than a liquid or gas?
4. Why are the structures of solids usually described in terms of the positions of the constituent atoms rather than their motion?
5. What physical characteristics distinguish a crystalline solid from an amorphous solid? Describe at least two ways to determine experimentally whether a material is crystalline or amorphous.
6. Explain why each of the following would or would not favor formation of an amorphous solid: (a) slow cooling of pure

molten material; (b) impurities in the liquid from which the solid is formed; and (c) weak intermolecular attractive forces.

7. A student obtained a solid product in a laboratory synthesis. To verify the identity of the solid, she measured its melting point, and found that the material melted over a 12°C range. After it had cooled, she measured the melting point of the same sample again, and found that this time the solid had a sharp melting point at the temperature that is characteristic of the desired product. Why were the two melting points different? What was responsible for the change in the melting point?

12.2 The Arrangement of Atoms in Crystalline Solids

8. Why is it valid to represent the structure of a crystalline solid by the structure of its unit cell? What are the most important constraints in selecting a unit cell?
9. All unit cell structures have six sides. Can crystals of a solid have more than six sides? Explain your answer.
10. Explain how the intensive properties of a material are reflected in the unit cell. Are all the properties of a bulk material the same as those of its unit cell? Explain your answer.
11. The experimentally measured density of a bulk material is slightly *higher* than expected based on the structure of the pure material. Propose two explanations for this observation.
12. The experimentally determined density of a material is *lower* than expected based on the arrangement of the atoms in the unit cell, the formula mass, and the size of the atoms. What conclusion(s) can you draw about the material?
13. Only one element (polonium) crystallizes with a simple cubic unit cell. Why is polonium the only example of an element with this structure?
14. What is meant by the term "coordination number" in the structure of a solid? How does the coordination number depend upon the structure of the metal?
15. Arrange the three types of cubic unit cell in order of increasing packing efficiency. What is the difference in packing efficiency between the hexagonal close-packed structure and the cubic close-packed structure?
16. The structures of many metals depends upon the pressure and temperature. Which structure would be more likely in a given metal at very high pressures: bcc or hcp? Explain your reasoning.
17. A metal has two crystalline phases. The "transition temperature," the temperature at which one phase is converted to the other, is 95°C at 1 atm and 135°C at 1000 atm. Sketch a phase diagram for this substance. The metal is known to have either a cubic close-packed structure or a simple cubic structure; label the regions in your diagram appropriately, and justify your selection for the structure of each phase.

12.3 Structures of Simple Binary Compounds

18. Using circles or spheres, sketch a unit cell containing an octahedral hole. Which of the basic structural types possess octahedral holes? If an ion were placed in an octahedral hole, what would its coordination number be?
19. Using circles or spheres, sketch a unit cell containing a tetrahedral hole. Which of the basic structural types possess tetrahedral holes? If an ion were placed in a tetrahedral hole, what would its coordination number be?
20. How many octahedral holes are there in each unit cell of the NaCl structure? Potassium fluoride contains an fcc lattice of F^- ions that is identical to the arrangement of Cl^- ions in the NaCl structure. Do you expect K^+ ions to occupy the tetrahedral or octahedral holes in the fcc lattice of F^- ions?
21. The unit cell of cesium chloride consists of a cubic array of chloride ions with a cesium ion in the center. Why then is cesium chloride described as having a simple cubic structure rather than a body-centered cubic structure? The unit cell of iron also consists of a cubic array of iron atoms with an iron atom in the center of the cube. Is this a body-centered cubic or a simple cubic unit cell? Explain your answer.
22. Why are X rays used to determine the structure of crystalline materials? Could gamma rays also be used to determine crystalline structures? Why or why not?
23. X rays are higher in energy than most other forms of electromagnetic radiation, including visible light. Why can't you use visible light to determine the structure of a crystalline material?
24. When X rays interact with the atoms in a crystal lattice, what relatonship between the distances between planes of atoms in the crystal structure and the wavelength of the X rays results in the scattered X rays being exactly in phase with one another? What difference in structure between amorphous materials and crystalline materials makes it difficult to determine the structures of amorphous materials by X-ray diffraction?
25. It is possible to use different X-ray sources to generate X rays with different wavelengths. Use the Bragg equation to predict how the diffraction angle would change if a molybdenum X-ray source (X-ray wavelength = 70.93 pm) were used instead of a copper source (X-ray wavelength = 154.1 pm).
26. Based on the Bragg equation, if crystal A has larger spacing in its diffraction pattern than crystal B, what conclusion can you draw about the spacing between layers of atoms in A compared with B?

12.4 Defects in Crystals

27. How are defects and impurities in a solid related? Can a pure, crystalline compound be free of defects? How can a substitutional impurity produce a vacancy?
28. Why does applying a mechanical stress to a covalent solid cause it to fracture? Use an atomic level description to explain why a metal is ductile under conditions that cause a covalent solid to fracture.
29. How does work hardening increase the strength of a metal? How does work hardening affect the physical properties of a metal?
30. Work-hardened metals and covalent solids such as diamonds are both susceptible to cracking when stressed. Explain how such different materials can both exhibit this property.
31. Suppose you want to produce a ductile material with improved properties. Would impurity atoms of similar or dissimilar atomic size be better at maintaining the ductility of a metal? Why? How would introducing an impurity that forms polar covalent bonds with the metal atoms affect the ductility of the metal? Explain your reasoning.
32. Substitutional impurities are often used to tune the properties of material. Why are substitutional impurities generally more effective at high concentrations, whereas interstitial impurities are usually effective at low concentrations?
33. If an O^{2-} ion (radius = 132 pm) is substituted for an F^- ion (radius = 133 pm) in an ionic crystal, what structural changes in the ionic lattice will result?

34. How will the introduction of a metal ion with a different charge as an impurity induce formation of oxygen vacancies in an ionic metal-oxide crystal?

35. Many nonstoichiometric compounds are transition metal compounds. How can such compounds exist, given that their nonintegral cation/anion ratios apparently contradict one of the basic tenets of Dalton's atomic theory of matter?

36. If you wanted to induce formation of oxygen vacancies in an ionic crystal, which would you introduce as substitutional impurities: cations with a higher or lower positive charge than the cations in the parent structure? Explain your reasoning.

12.5 Correlation Between Bonding and the Properties of Solids

37. Four vials labeled A–D contain sucrose, zinc, quartz, and sodium chloride, although not necessarily in that order. The table summarizes the results of the series of analyses you have performed on the contents:

	A	B	C	D
Melting Point	high	high	high	low
Thermal Conductivity	poor	poor	good	poor
Electrical Conductivity in Solid State	moderate	poor	high	poor
Electrical Conductivity in Liquid State	high	poor	high	poor
Hardness	hard	hard	soft	soft
Luster	none	none	high	none

Match each vial with its contents.

38. Do ionic solids generally have higher or lower melting points than molecular solids? Why? Do ionic solids generally have higher or lower melting points than covalent solids? Explain your reasoning.

39. The strength of London dispersion forces in molecular solids tends to increase with molecular mass, causing a smooth increase in melting points. Some molecular solids, however, have significantly lower melting points than predicted by their molecular masses. Why?

40. Suppose you want to synthesize a solid that is both heat-resistant and a good electrical conductor. What specific types of bonding and molecular interactions would you want in your starting materials?

41. Explain the differences between an interstitial alloy and a substitutional alloy. Given an alloy in which the identity of one metallic element is known, how could you determine whether it is a substitutional alloy or an interstitial alloy?

42. How are intermetallic compounds different from interstitial or substitutional alloys?

12.6 Bonding in Metals and Semiconductors

43. Can band theory be applied to metals that have two electrons in their valence *s* orbitals? To metals that have no electrons in their valence *s* orbitals? Why or why not?

44. Given a sample of a metal that contains 10^{20} atoms, how does the width of the band arising from *p* orbital interactions compare with the width of the band arising from *s* orbital interactions? With the width of the band arising from *d* orbital interactions?

45. Diamond has one of the lowest electrical conductivities known. Based on this fact, do you expect diamond to be colored? Why? How do you account for the fact that some diamonds are colored (such as "pink" diamond or "green" diamond)?

46. Why do silver halides, used in the photographic industry, have band gaps typical of semiconducting materials, whereas alkali metal halides have very large band gaps?

47. As the ionic character of a compound increases, does its band gap increase or decrease? Why?

48. Why is silicon used in the semiconductor industry, rather than carbon or germanium?

49. Carbon is an insulator, and silicon and germanium are semiconductors. Explain the relationship between the valence-electron configuration of each of these elements and their band structures. Which will have the higher electrical conductivity at room temperature, silicon or germanium?

50. How does doping affect the electrical conductivity of a semiconductor? Sketch the effect of doping on the energy levels of the valence band and the conduction band for both an *n*-type and a *p*-type semiconductor.

12.7 Superconductors

51. Why does BCS theory predict that superconductivity is not possible at temperatures above approximately 30 K?

52. How does the formation of Cooper pairs lead to superconductivity?

12.8 Polymers

53. How are amino acids and proteins related to monomers and polymers? Draw the general structure of an amide bond linking two amino acid residues.

54. Although proteins and synthetic polymers (such as nylon) both contain amide bonds, different terms are used to describe the two types of polymer. Compare and contrast the terminology used for (a) the smallest repeating unit, and (b) the covalent bond connecting the units.

12.9 Contemporary Materials

55. Can a titanium oxide–based compound qualify as a ceramic material? Explain your answer.

56. What features make ceramic materials attractive for use under extreme conditions? What are some potential drawbacks of ceramics?

57. How do composite materials differ from the other classes of materials discussed in this chapter? What advantages do composites have versus other materials?

58. How does the matrix control the properties of a composite material? What is the role of an additive in determining the properties of a composite material?

NUMERICAL

This section includes paired problems (marked by brackets) that require similar problem-solving skills.

12.2 The Arrangement of Atoms in Crystalline Solids

59. Metallic nickel has an fcc unit cell. How many atoms of nickel does each unit cell contain?

60. Chromium has a structure that contains two atoms per unit cell. Is the structure of this metal simple cubic, bcc, fcc, or hcp?

61. The density of nickel is 8.908 g/cm^3. If the metallic radius of nickel is 125 pm, what is the structure of metallic nickel?

62. The density of tungsten is 19.3g/cm^3. If the metallic radius of tungsten is 139 pm, what is the structure of metallic tungsten?

63. An element has a density of 10.25 g/cm^3 and a metallic radius of 136.3 pm. The metal crystallizes in a bcc lattice. Identify the element.

64. A 21.64-g sample of a nonreactive metal is placed in a flask containing 12.00-mL of water; the final volume is 13.81 mL. If the length of the edge of the unit cell is 387 pm and the metallic radius is 137 pm, determine the packing arrangement, and identify the element.

65. A sample of an alkali metal that has a bcc unit cell is found to have a mass of 1.000 g and a volume of 1.0298 cm^3. The metal is allowed to react with excess water, producing 539.29 mL of hydrogen gas at 0.980 atm and 23°C. Identify the metal, determine the unit cell dimensions, and give the approximate size of the atom in pm.

66. A sample of an alkaline earth metal that has a bcc unit cell is found to have a mass 5.000 g and a volume of 1.392 cm^3. Complete reaction with chlorine gas requires 848.3 mL of chlorine gas at 1.050 atm and 25°C. Identify the metal, determine the unit cell dimensions, and give the approximate size of the atom in pm.

67. Lithium crystallizes in a bcc structure with an edge length of 3.509 Å. Calculate its density. What is the approximate metallic radius of lithium in pm?

68. Vanadium is used in the manufacture of rust-resistant vanadium steel. It forms bcc crystals with a density of 6.11 g/cm^3 at 18.7°C. What is the length of the edge of the unit cell? What is the approximate metallic radius of the vanadium in pm?

69. A simple cubic cell contains one metal atom with a metallic radius of 100 pm.
(a) Determine the volume of the atom(s) contained in one unit cell [the volume of a sphere $= \left(\frac{4}{3}\right)\pi r^3$].
(b) What is the length of one edge of the unit cell? (*Hint:* There is no empty space between atoms.)
(c) Calculate the volume of the unit cell.
(d) Determine the packing efficiency for this structure.

70. Use the steps in Problem 69 to calculate the packing efficiency for a bcc unit cell with a metallic radius of 1.00Å.

12.3 Structures of Simple Binary Compounds

71. Thallium bromide crystallizes in the CsCl structure. This bcc structure contains a Tl^+ ion in the center of the cube with Br^- ions at the corners. Sketch an alternative unit cell for this compound.

72. Potassium fluoride has a lattice identical to that of sodium chloride. The potassium ions occupy octahedral holes in an fcc lattice of fluoride ions. Propose an alternative unit cell that can also represent the structure of KF.

73. Calcium fluoride is used to fluoridate drinking water to promote dental health. Crystalline CaF_2 (d = 3.1805 g/cm^3) has a structure in which calcium ions are located at each of the corners and the middle of each edge of the unit cell, which contains eight fluoride ions per unit cell. The length of the edge of this unit cell is 5.463 Å. Use this information to determine Avogadro's number.

74. Zinc and oxygen form a compound that is used both as a semiconductor and as a paint pigment. This compound has the following structure:

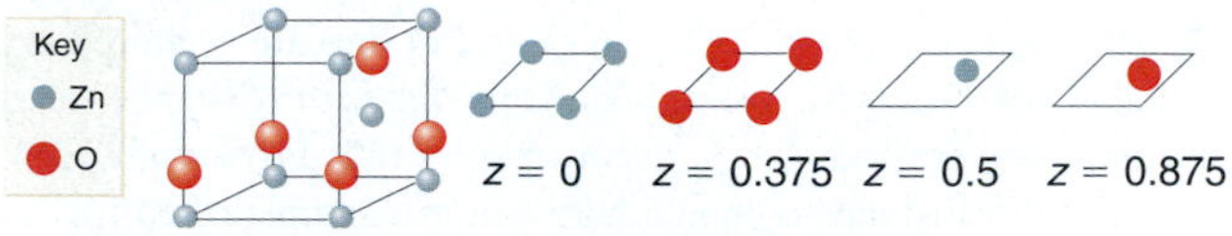

What is the empirical formula of this compound?

75. Shown below are two representations of the perovskite structure. Are they identical? What is the empirical formula corresponding to each representation?

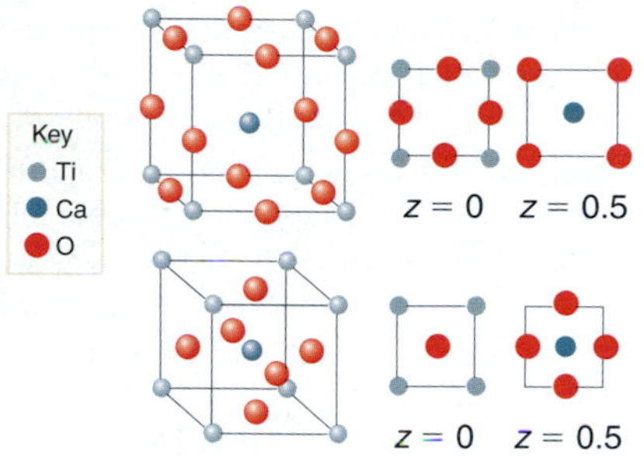

76. The salt MX_2 has a ccp structure in which all the tetrahedral holes are filled by cations. What is the coordination number of M? Of X?

77. A compound has a structure based on simple cubic packing of the anions, and the cations occupy half of the cubic holes. What is the empirical formula of this compound? What is the coordination number of the cation?

78. Barium and fluoride form a compound that crystallizes in the fluorite structure, in which the fluoride ions occupy all of the tetrahedral holes in a ccp array of barium ions. This particular compound is used in embalming fluid. What is its empirical formula?

79. Cadmium chloride is used in paints as a yellow pigment. Is the following structure consistent with an empirical formula of $CdCl_2$? If not, what is the empirical formula of the structure shown?

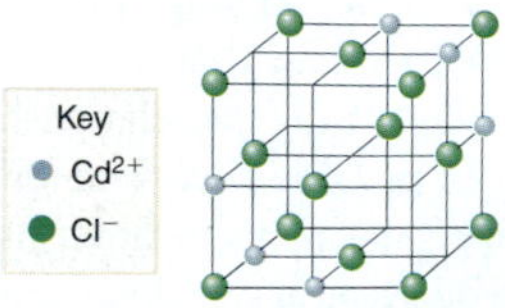

80. Use the information in the table to decide whether the cation will occupy a tetrahedral, octahedral, or cubic hole in each case:

Cation Radius, pm	Anion Radius, pm
78.0	132
165	133
81	174

81. Calculate the angle of diffraction when X rays from a copper tube (wavelength = 154 pm) are diffracted by planes of atoms parallel to the faces of the cubic unit cell for the following metals: Mg (260 pm), Zn (247 pm), Ni (216 pm). The length on one edge of the unit cell is given in parentheses: assume first-order diffraction ($n = 1$).

82. If X rays from a copper target ($\lambda = 154$ pm) are scattered at an angle of 17.23° by a sample of Mg, what is the distance (in pm) between the planes responsible for this diffraction? How does this distance compare with that in a sample of Ni for which $\theta = 20.88°$?

12.4 Defects in Crystals

83. The ionic radius of K^+ is 133 pm, whereas that of Na^+ is 98 pm. Do you expect K^+ to be a common substitutional impurity in compounds that contain Na^+? Why or why not?

84. Given the following atomic radii, which atom is most likely to act as an interstitial impurity in a Sn lattice ($\text{radius}_{Sn} = 141$ pm): Cs (262 pm), Tl (171 pm), or B (88 pm)? Why?

85. Iron is the second most abundant metal in the Earth's crust after aluminum. The silvery-white, ductile metal has a bcc unit cell with an edge length of 286.65 pm.
(a) Use this information to calculate the density of iron.
(b) What would the density of iron be if 0.15% of the iron sites were vacant?
(c) How does the mass of 1.00 cm^3 of iron without defects compare with the mass of 1.00 cm^3 of iron with 0.15% vacancies?

86. Certain ceramic materials are good electrical conductors due to high mobility of oxide ions resulting from the presence of oxygen vacancies. Zirconia, ZrO_2, can be doped with yttrium by the adding of Y_2O_3. If 0.35 g of Y_2O_3 can be incorporated into 25.0 g of ZrO_2 while maintaining the zirconia structure, what is the percentage of oxygen vacancies in the structure?

87. Which of the following ions is most effective at inducing an O^{2-} vacancy in crystal of CaO? Ionic radii are: O^{2-} = 132 pm; Ca^{2+}, 100 pm; Sr^{2+}, 127 pm; F^-, 133 pm; La^{3+}, 104 pm; K^+, 133 pm. Explain your reasoning.

12.5 Correlation Between Bonding and the Properties of Solids

88. Will the melting point of lanthanum(III) oxide be higher or lower than that of ferrous bromide? Relevant ionic radii are: La^{3+}, 104 pm; O^{2-}, 132 pm; Fe^{2+}, 83 pm; and Br^-, 196 pm. Explain your reasoning.

89. Which has the higher melting point: Os or Hf? SnO_2 or ZrO_2? Al_2O_3 or SiO_2? Explain your reasoning in each case.

90. Sketch a graph showing the relationship between the electrical conductivity of metallic silver and temperature.

91. Sketch a graph showing the relationship between the electrical conductivity of a typical semiconductor and temperature.

12.6 Bonding in Metals and Semiconductors

92. Which of these elements would convert pure silicon into a *p*-type semiconductor upon doping: Ca, N, B, or Ge? Explain your reasoning.

93. Which of these elements would convert pure germanium into an *n*-type semiconductor upon doping: Ga, Si, Br, or P? Explain your reasoning.

APPLICATIONS

94. Cadmium selenide, CdSe, is a semiconductor used in photoconductors and photoelectric cells that conduct electricity when illuminated. In a related process, a CdSe crystal can absorb enough energy to excite electrons from the valence band to the conduction band, and the excited electrons can return to the valence band by emitting light. The relative intensity and peak wavelength of the emitted light in one experiment are shown in the following table:

Relative Intensity, %	Wavelength, nm	Temperature, °C
100	720	23
50	725	45
10	730	75

(a) Explain why the emitted light shifts to longer wavelength at higher temperatures. (*Hint:* Consider the expansion of the crystal and resulting changes in orbital interactions upon heating.)
(b) Why does the relative intensity of the emitted light decrease as the temperature increases?

95. A large fraction of electrical energy is currently lost as heat during transmission due to the electrical resistance of transmission wires. How could superconducting technology improve the transmission of electrical power? What are some potential drawbacks of this technology?

96. Light-emitting diodes (LEDs) are semiconductor-based devices that are used in consumer electronics products ranging from digital clocks to fiber-optic telephone transmission lines. The color of the emitted light is determined in part by the band gap of the semiconductor. Electrons can be promoted to the conduction band and return to the valence band by emitting light, or by increasing the magnitude of atomic vibrations in the crystal, which increases its temperature. If you wanted to increase the efficiency of an LED display, and thereby the intensity of the emitted light, would you increase or decrease the operating temperature of the LED? Explain your answer.

97. Strips of pure Au and Al are often used in close proximity to each other on circuit boards. As the boards became warm during use, however, the metals can diffuse, forming a purple alloy known as "the purple plague" between the strips. Because the alloy is electrically conductive, the board short-circuits. Structural analysis of the purple alloy showed that its structure contained an fcc lattice of atoms of one element, with atoms of the other element occupying tetrahedral holes. What type of alloy is this? Which element is most likely to form the fcc lattice? Which element is most likely to occupy the tetrahedral holes? Explain your answers. What is the empirical formula of the "purple plague"?

98. Glasses are mixtures of oxides, the main component of which is silica, SiO_2. Silica is called the glass *former*, while additives are referred to as glass *modifiers*. The crystalline lattice of the glass former breaks down upon heating, producing the random atomic arrangements typical of a liquid. Adding a modifier and cooling the melt rapidly produces a glass. How does the three-dimensional structure of the glass differ from that of the crystalline glass former? Would you expect the melting point of a glass to be higher or lower than that of pure SiO_2? Lead glass, a particular favorite of the Romans, was formed by adding lead oxide as the modifier. Would you expect lead glass to be more or less dense than soda-lime glass formed by adding sodium and potassium salts as modifiers?

99. Many glasses eventually crystallize, rendering them brittle and opaque. Modifying agents such as TiO_2 are frequently added to molten glass to reduce their tendency to crystallize. Why does the addition of small amounts of TiO_2 stabilize the amorphous structure of the glass?

100. The carbon–carbon bond distances in polyacetylene (—CH=CH—)$_n$ alternate between short and long, resulting in the following band structure:

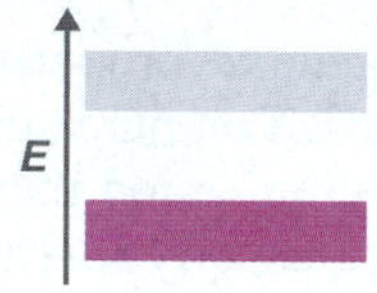

(a) Is polyacetylene a metal, a semiconductor, or an insulator?
(b) Based on its band structure, how would treating polyacetylene with potent oxidant affect its electrical conductivity? What would be the effect of treating polyacetylene with small amounts of a powerful reductant? Explain your answer.

101. Enkephalins are pentapeptides, short biopolymers that are synthesized by humans to control pain. Enkephalins bind to certain receptors in brain cells, which are also known to bind morphine and heroin. One enkephalin has the structure tyrosine–glycine–glycine–phenylalanine–methionine. Draw its structure.

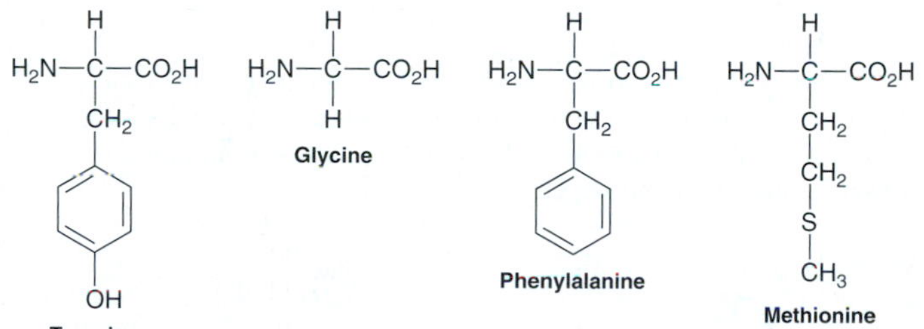

102. A polymerization reaction is used to synthesize Saran, a flexible material used in packaging film and seat covers. The monomeric unit for Saran is 1,1-dichloroethylene (CH_2=CCl_2), also known as vinylidene chloride. Draw a reasonable structure for the polymer. Why do pieces of Saran "cling" to one another when they are brought in contact?

103. Polymers are often amorphous solids. Like other materials, polymers can also undergo phase changes. For example, many polymers are flexible above a certain temperature, called the glass-transition temperature (T_g). Below the glass transition temperature, the polymer becomes hard and brittle. Biomedical devices that replace or augment parts of the human body often contain a wide variety of materials whose properties must be carefully controlled.
(a) Polydimethylsiloxane has a T_g of −123°C, whereas poly(methylmethacrylate) has a T_g of 105°C. Which of these polymers is likely to be used in dentures, and which is likely to be used for soft-tissue replacement?
(b) If you were designing biomedical devices, which class of biomaterials (alloys, ceramics, polymers) would you consider for each of the following applications: finger joint replaccments, eyeball replacements, windpipe replacements, shoulder joint replacements, and bridging bone fractures?

13 Solutions

LEARNING OBJECTIVES

- To understand how enthalpy and entropy changes affect solution formation
- To understand the relationship between solubility and molecular structure
- To be able to describe the concentration of a solution in the way that is most appropriate for a particular problem or application
- To understand the relationships between temperature, pressure, and solubility
- To describe the relationship between solute concentration and the colligative properties of a solution
- To distinguish between true solutions and solutions that contain aggregate particles

CHAPTER OUTLINE

In Chapters 10–12, we explored the general properties of gases, liquids, and solids. Most of the discussion focused on pure substances that contain a single kind of atom, molecule, or cation–anion pair. The substances we encounter in our daily lives, however, are usually mixtures rather than pure substances. Some are *heterogeneous mixtures*, which consist of at least two phases that are not uniformly dispersed on a microscopic scale; others are *homogeneous mixtures*, consisting of a single phase in which the components are uniformly distributed (see Section 1.3). Homogeneous mixtures are also called **solutions**; they include the air we breathe, the gas we use to cook and heat our homes, the water we drink, the gasoline or diesel fuel that powers our cars, and the gold and silver jewelry we wear.

Many of the concepts that we will use in our discussion of solutions were introduced in earlier chapters. In Chapter 4, for example, we described

Beads of oil in water. When a nonpolar liquid (oil) is dispersed in a polar solvent (water), it does not dissolve, but forms spherical beads. Oil is insoluble in water because the intermolecular interactions *within* the solute (oil) and the solvent (water) are stronger than the intermolecular interactions *between* oil and water.

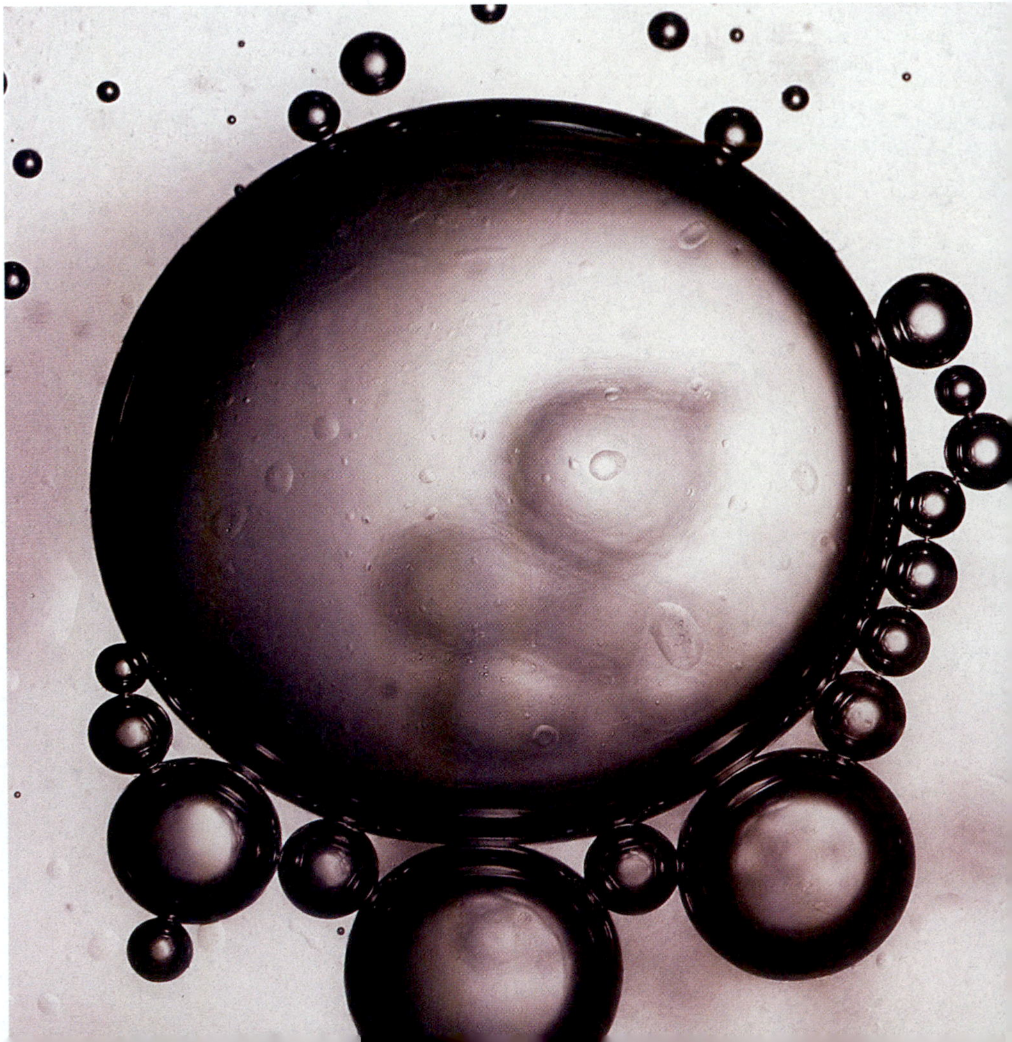

reactions that occur in aqueous solution and how to use *molarity* to describe concentrations. In Chapters 4, 7, and 11, we introduced the principles that govern ion–ion and molecule–molecule interactions in pure substances; similar interactions also occur in solutions. Now we use the principles developed in those chapters to understand the factors that determine how much of one substance can dissolve in another, and how the properties of a solution differ from those of its components.

The properties of mixtures of gases were described in Chapter 10, and the properties of certain types of solid solutions, such as alloys and doped semiconductors, were discussed in Chapter 12. This chapter focuses on liquid solutions, aqueous or otherwise. By the end of this chapter, your understanding of solutions will enable you to explain why the radiator in your car must contain ethylene glycol to avoid damage to the engine on cold winter nights, why salt is spread on icy roads in the winter (and why it isn't effective when the temperature is too low), why certain vitamins accumulate in your body at toxic levels while others are rapidly excreted, and how salt can be removed from seawater to provide drinking water.

13.1 Factors Affecting Solution Formation

In all solutions, whether gaseous, liquid, or solid, the substance present in the greatest amount is the *solvent*, and the substance or substances present in lesser amounts are the *solute(s)*. The solute does not have to be in the same physical state as the solvent, but the physical state of the solvent usually determines the state of the solution. As long as the solute and solvent combine to give a homogeneous solution, the solute is said to be *soluble* in the solvent. Table 13.1 lists some common examples of gaseous, liquid, and solid solutions and identifies the physical states of the solute and solvent in each.

Forming a Solution

The formation of a solution from a solute and a solvent is a physical process, not a chemical one. That is, both solute and solvent can be recovered in chemically unchanged form using appropriate separation methods. For example, zinc nitrate dissolves in water to form an aqueous solution of zinc nitrate:

$$\mathrm{Zn(NO_3)_2}(s) \xrightarrow{\mathrm{H_2O}(l)} \mathrm{Zn^{2+}}(aq) + 2\mathrm{NO_3^-}(aq) \qquad (13.1)$$

Because $Zn(NO_3)_2$ can be recovered easily by evaporating the water, this is a physical process. In contrast, metallic zinc *appears* to dissolve in aqueous hydrochloric acid.

TABLE 13.1 Types of solutions

Solution	Solute	Solvent	Examples
Gas	Gas	Gas	Air, natural gas
Liquid	Gas	Liquid	Seltzer water (CO_2 gas in water)
Liquid	Liquid	Liquid	Alcoholic beverage (ethanol in water), gasoline
Liquid	Solid	Liquid	Tea, salt water
Solid	Gas	Solid	H_2 in Pd (used for H_2 storage)
Solid	Solid	Liquid	Mercury in silver or gold (amalgam often used in dentistry)
Solid	Solid	Solid	14-carat gold (copper in gold), brass (zinc in copper), solder (lead in tin)

Note the pattern

Dissolution of a solute in a solvent to form a solution does not involve a chemical transformation.

In fact, however, the two substances undergo a chemical reaction to form an aqueous solution of zinc chloride with evolution of hydrogen gas:

$$Zn(s) + 2H^+(aq) + 2Cl^-(aq) \longrightarrow Zn^{2+}(aq) + 2Cl^-(aq) + H_2(g) \quad (13.2)$$

When the solution evaporates, we do not recover metallic zinc, so we cannot say that metallic zinc is *soluble* in aqueous hydrochloric acid because it is chemically transformed when it dissolves. *The dissolution of a solute in a solvent to form a solution does not involve a chemical transformation.*

Substances that form a single homogeneous phase in all proportions are said to be completely **miscible** in one another. Ethanol and water are miscible, just as mixtures of gases are miscible. If two substances are essentially insoluble in each other, they are *immiscible*, such as oil and water. Examples of gaseous solutions that we have already discussed include Earth's atmosphere (see Chapter 3) and natural gas (see Chapter 10).

The Role of Enthalpy in Solution Formation

As we saw in Chapters 10–12, energy is required to overcome the intermolecular interactions in a solute. This energy can be supplied only by the new interactions that occur in the solution, when each solute particle is surrounded by particles of the solvent in a process called **solvation**, or **hydration** when the solvent is water. Thus, all of the solute–solute interactions and many of the solvent–solvent interactions must be disrupted for a solution to form. In this section we describe the role of enthalpy in this process.

Because enthalpy is a state function (see Section 5.2), we can use the same type of thermochemical cycle described in Chapter 5 to analyze the energetics of solution formation. The process occurs in three discrete steps, indicated by ΔH_1, ΔH_2, and ΔH_3 in Figure 13.1. The overall enthalpy change in the formation of the solution, ΔH_{soln}, is the sum of the enthalpy changes in the three steps:

$$\Delta H_{soln} = \Delta H_1 + \Delta H_2 + \Delta H_3 \quad (13.3)$$

Figure 13.1 Enthalpy changes that accompany the formation of a solution. Solvation can be an exothermic or endothermic process depending on the nature of the solute and solvent. In both cases, step 1, separation of the solvent particles, is energetically uphill ($\Delta H_1 > 0$), as is step 2, separation of the solute particles ($\Delta H_2 > 0$). In contrast, energy is released in step 3 ($\Delta H_3 < 0$) because of interactions between the solute and solvent. (**a**) When ΔH_3 is larger in magnitude than the sum of ΔH_1 and ΔH_2, the overall process is exothermic ($\Delta H_{soln} < 0$), as shown in the thermochemical cycle. (**b**) When ΔH_3 is smaller in magnitude than the sum of ΔH_1 and ΔH_2, the overall process is endothermic ($\Delta H_{soln} > 0$).

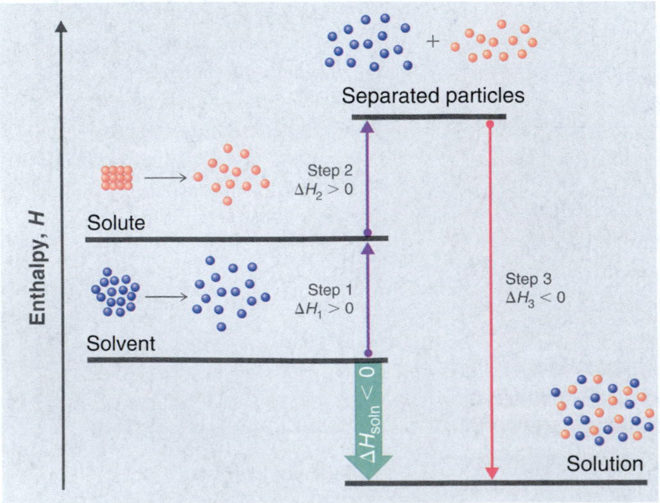

(a) Exothermic solution formation

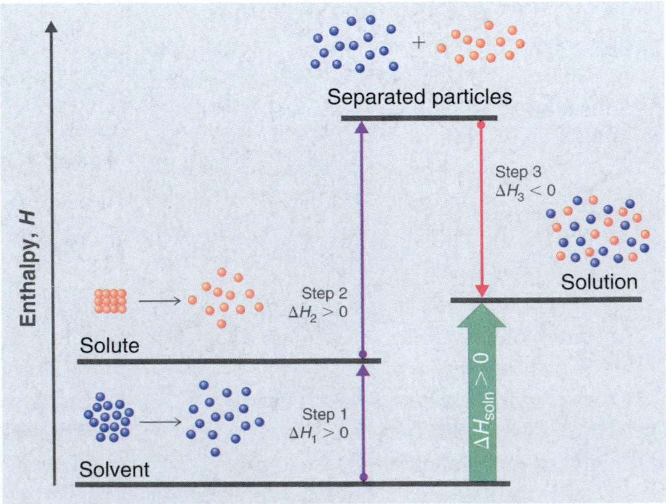

(b) Endothermic solution formation

Steps 1 and 2 are both endothermic because energy is required to overcome the intermolecular interactions in the solvent (ΔH_1) and the solute (ΔH_2). Because ΔH is positive for both steps 1 and 2, in order for the dissolution process to be exothermic ($\Delta H_{soln} < 0$), the solute–solvent interactions (ΔH_3) must be stronger than the solute–solute and solvent–solvent interactions they replace. When the solute is an ionic solid, ΔH_2 corresponds to the lattice energy that must be overcome to form a solution and, as you learned in Chapter 12, the higher the charge of the ions in an ionic solid, the higher the lattice energy. Consequently, solids that have very high lattice energies, such as MgO (−3791 kJ/mol), are generally insoluble in all solvents.

As you will see in Chapter 18, a positive value for ΔH_{soln} does not mean that a solution will not form. Whether a given process, including formation of a solution, occurs spontaneously depends on whether the *total* energy of the system is lowered as a result, and enthalpy is only one of the contributing factors. If a substance's ΔH_{soln} is high, however, then the substance is usually not very soluble. Instant cold packs used commercially to treat athletic injuries, for example, take advantage of the large positive ΔH_{soln} of ammonium nitrate during dissolution (+25.7 kJ/mol), which produces temperatures lower than 0°C (Figure 13.2).

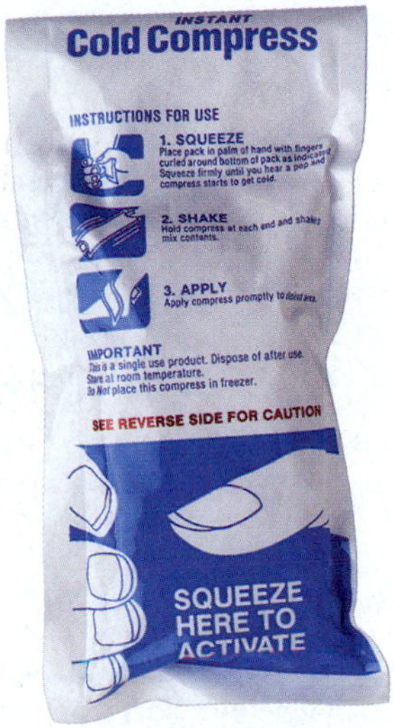

Figure 13.2 Commercial cold packs for treating injuries. The packs contain solid NH_4NO_3 and water in separate compartments. When the seal between the compartments is broken, the NH_4NO_3 dissolves in the water. Because ΔH_{soln} for NH_4NO_3 is much greater than zero, heat is absorbed by the cold pack during the dissolution process, producing local temperatures lower than 0°C.

Entropy and Solution Formation

The enthalpy change that accompanies a process is important because processes that release substantial amounts of energy tend to occur spontaneously, but a second property of any system, its *entropy*, is also important in helping us determine whether a given process occurs spontaneously. We will discuss entropy in more detail in Chapter 18, but now we can state that **entropy (*S*)** is a thermodynamic property of all substances that is proportional to their degree of disorder. A perfect crystal at 0 K, whose atoms are regularly arranged in a perfect lattice and are motionless, is arbitrarily assigned an entropy of zero. In contrast, gases have large positive entropies because their molecules are highly disordered and in constant motion at high speeds.

The formation of a solution disperses molecules, atoms, or ions of one kind throughout a second substance, which generally increases the disorder and results in an increase in the entropy of the system. Thus, entropic factors almost always favor formation of a solution. In contrast, a change in enthalpy may or may not favor solution formation. The London dispersion forces that hold cyclohexane and *n*-hexane together in pure liquids, for example, are similar in nature and strength. Consequently, ΔH_{soln} should be approximately zero, as is observed experimentally. Mixing the two liquids, however, produces a solution in which the *n*-hexane and cyclohexane molecules are uniformly distributed over approximately twice the initial volume. In this case, the driving force for solution formation is not a negative ΔH_{soln} but rather the increase in entropy due to the increased disorder in the mixture. *All spontaneous processes with $\Delta H \geq 0$ are characterized by an increase in entropy.* In other cases, such as mixing oil with water, salt with gasoline, or sugar with hexane, the enthalpy of solution is large and positive, and the increase in entropy resulting from solution formation is not enough to overcome it. Thus, in these cases a solution does not form.

Note the pattern

All spontaneous processes with $\Delta H \geq 0$ are characterized by an increase in entropy.

Table 13.2 summarizes how enthalpic factors affect solution formation for four general cases. The column on the far right uses the relative magnitudes of the enthalpic contributions to predict whether a solution will form from each of the four. Keep in mind that in each case, entropy favors solution formation. Note that in two of them, the enthalpy of solution is expected to be relatively small and can be either positive or negative. Thus, the entropic contribution dominates, and we expect a solution to form readily. In the other two cases, however, the enthalpy of solution is expected to be large and positive; the entropic contribution, though favorable, is

TABLE 13.2 Relative changes in enthalpies for different solute–solvent combinations[a]

ΔH_1 (separation of solvent molecules)	ΔH_2 (separation of solute particles)	ΔH_3 (solute–solvent interactions)	ΔH_{soln} ($\Delta H_1 + \Delta H_2 + \Delta H_3$)	Result of Mixing Solute and Solvent[b]
Large; positive	Large; positive	Large; negative	Small; positive or negative	Solution will usually form
Small; positive	Large; positive	Small; negative	Large; positive	Solution will not form
Large; positive	Small; positive	Small; negative	Large; positive	Solution will not form
Small; positive	Small; positive	Small; negative	Small; positive or negative	Solution will usually form

[a] ΔH_1, ΔH_2, and ΔH_3 refer to the processes indicated in the thermochemical cycle shown in Figure 13.1.
[b] In all four cases, entropy increases.

usually too small to overcome the unfavorable enthalpy term. Hence, we expect that solutions will not form readily.

In contrast to liquid solutions, the intermolecular interactions in gases are weak (they are considered to be nonexistent in ideal gases). Hence, mixing gases is usually a thermally neutral process ($\Delta H_{soln} \approx 0$), and the entropic factor due to the increase in disorder is dominant (Figure 13.3). Consequently, all gases dissolve readily in one another in all proportions to form solutions. We will return to a discussion of enthalpy and entropy in Chapter 18, where we treat their relationship quantitatively.

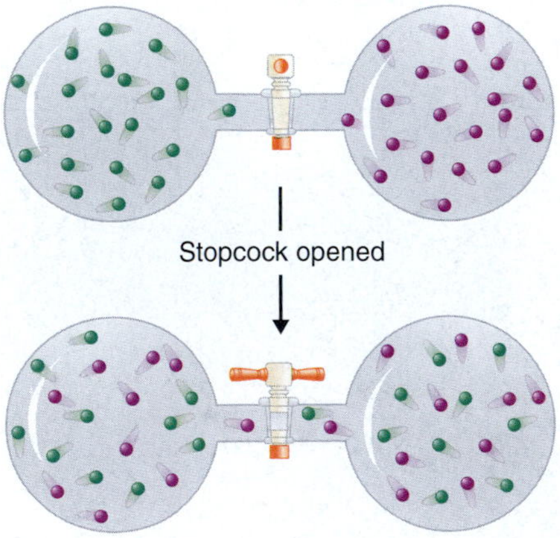

Figure 13.3 Formation of a solution of two gases. (*Top*) Pure samples of two different gases are in separate bulbs. (*Bottom*) When the connecting stopcock is opened, diffusion causes the two gases to mix together and form a solution. Even though ΔH_{soln} is zero for the process, the increased entropy of the solution (the increased disorder) versus that of the separate gases favors solution formation.

EXAMPLE 13.1

Predict which of these substances will be most soluble and which will be least soluble in water: LiCl, benzoic acid ($C_6H_5CO_2H$), naphthalene.

Given Three compounds

Asked for Relative solubilities in water

Strategy

Assess the relative magnitude of the enthalpy change for each step in the process shown in Figure 13.1. Then use Table 13.2 to predict the solubility of each compound in water, and arrange them in order of decreasing solubility.

Solution

The first substance, LiCl, is an ionic compound, so a great deal of energy is required to separate its anions and cations and overcome the lattice energy ($\Delta H_2 >> 0$ in Equation 13.3). Because water is a polar substance, however, the interactions between both Li^+ and Cl^- ions and water should be favorable and strong. Thus, we expect $\Delta H_3 << 0$, making LiCl soluble in water. In contrast, naphthalene is a nonpolar compound, with only London dispersion forces holding the molecules together in the solid state. We therefore expect ΔH_2 to be small and positive. We also expect the interaction between polar water molecules and nonpolar naphthalene molecules to be weak ($\Delta H_3 \approx 0$). Hence, we do not expect naphthalene to be very soluble in water, if at all. Benzoic acid has a polar carboxylic acid group and a nonpolar aromatic ring. We therefore expect that the energy required to separate solute molecules (ΔH_2) will be higher than for naphthalene and lower than for LiCl. The strength of the interaction of benzoic acid with water should also be intermediate between those of LiCl and naphthalene. Hence, benzoic acid is expected to be more soluble in water than naphthalene but less soluble than LiCl. We thus predict LiCl to be the most soluble in water and naphthalene to be the least soluble.

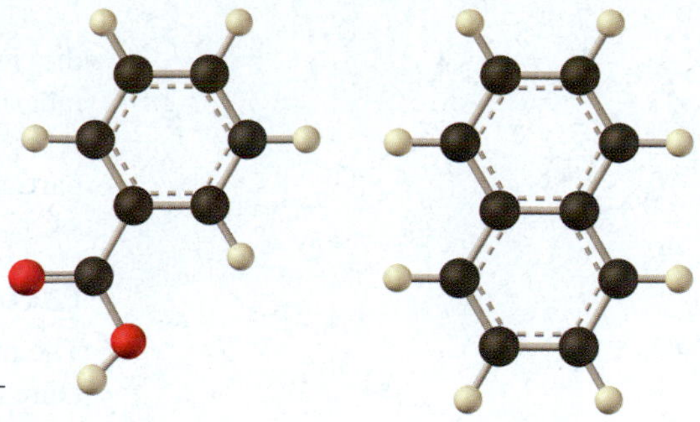

Benzoic acid Naphthalene

EXERCISE 13.1

Predict which of these solutes will be most soluble and which will be least soluble in benzene: ammonium chloride, cyclohexane, ethylene glycol ($HOCH_2CH_2OH$).

Answer The most soluble is cyclohexane; the least soluble is ammonium chloride.

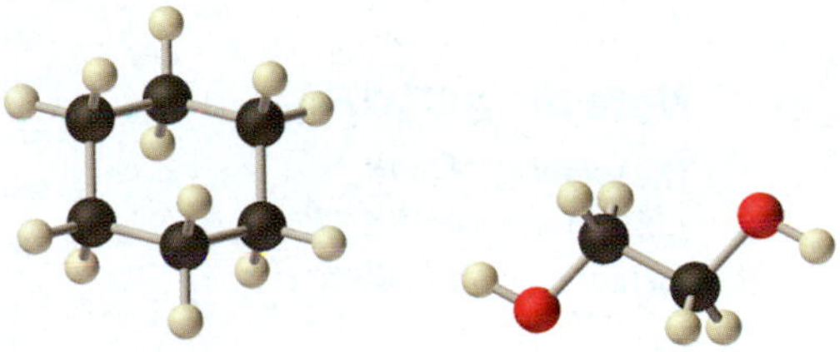

Cyclohexane Ethylene glycol

13.2 ○ Solubility and Molecular Structure

When a solute dissolves, its individual atoms, molecules or ions interact with the solvent, become solvated, and are able to diffuse independently throughout the solution (Figure 13.4a). This is not, however, a unidirectional process. If the molecule or ion happens to collide with the surface of a particle of the undissolved solute, it may adhere to the particle in a process called *crystallization*. Dissolution and crystallization continue as long as excess solid is present, resulting in a *dynamic equilibrium* analogous to the equilibrium that maintains the vapor pressure of a liquid (see Chapter 11). We can represent these opposing processes as

$$\text{solute} + \text{solvent} \underset{\text{crystallization}}{\overset{\text{dissolution}}{\rightleftharpoons}} \text{solution} \tag{13.4}$$

Figure 13.4 Dissolution and precipitation. (a) When a solid is added to a solvent in which it is soluble, solute particles leave the surface of the solid and become solvated by the solvent, initially forming an unsaturated solution. **(b)** When the maximum possible amount of solute has dissolved, the solution becomes saturated. If excess solute is present, the rate at which solute particles leave the surface of the solid equals the rate at which they return to the surface of the solid. **(c)** An unstable supersaturated solution can usually be formed from a saturated solution by filtering off the excess solute and lowering the temperature. **(d)** When a seed crystal of the solute is added to a supersaturated solution, solute particles leave the solution and form a crystalline precipitate.

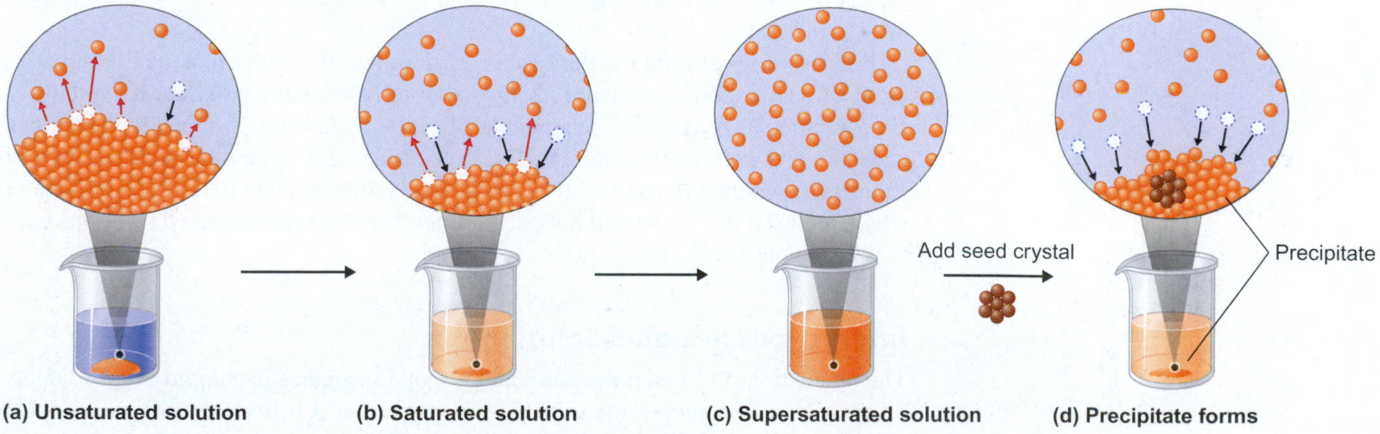

Although the terms *precipitation* and *crystallization* are both used to describe the separation of solid solute from a solution, *crystallization* refers to the formation of a solid with a well-defined crystalline structure, whereas *precipitation* refers to the formation of any solid phase, often one that contains very small particles.

Factors Affecting Solubility

The maximum amount of a solute that can dissolve in a solvent at a specified temperature and pressure is its **solubility**. Solubility is often expressed as the mass of solute per volume (g/L) or mass of solute per mass of solvent (g/g), or as the moles of solute per volume (mol/L). Even for very soluble substances, however, there is usually a limit to how much solute can dissolve in a given quantity of solvent. In general, the solubility of a substance depends not only on the energetic factors we have discussed but also on the temperature and, for gases, the pressure. At 20°C, for example, 177 g of NaI, 91.2 g of NaBr, 35.9 g of NaCl, and only 4.1 g of NaF dissolve in 100 g of water. At 70°C, however, the solubilities increase to 295 g of NaI, 119 g of NaBr, 37.5 g of NaCl, and 4.8 g of NaF. As you learned in Chapter 12, the lattice energies of the sodium halides increase from NaI to NaF. The fact that the solubilities decrease as the lattice energy increases suggests that the ΔH_2 term in Figure 13.1 dominates for this series of compounds.

Note the pattern

The solubility of most substances increases with increasing temperature.

A solution that contains the maximum possible amount of solute is **saturated**. If a solution contains less than the maximum amount of solute, it is *unsaturated*. When a solution is saturated and excess solute is present, the rate of dissolution is exactly equal to the rate of crystallization (Figure 13.4b). Using the value just stated, a saturated aqueous solution of NaCl, for example, contains 35.9 g of NaCl per 100 mL of water at 20°C. We can prepare a homogeneous saturated solution by adding excess solute (in this case, more than 35.9 g of NaCl) to the solvent (water), stirring until the maximum possible amount of solute has dissolved, and then removing undissolved solute by filtration.

Because the solubility of many substances generally increases with increasing temperature, a saturated solution that was prepared at a higher temperature usually contains more dissolved solute than it would contain at a lower temperature. When the solution is cooled, it can therefore become **supersaturated** (Figure 13.4c). Like a supercooled or superheated liquid (see Chapter 11), a supersaturated solution is unstable. Consequently, adding a small particle of the solute, a *seed crystal*, will usually cause the excess solute to rapidly precipitate or crystallize, sometimes with spectacular results, as was shown in Figure 1.9. The rate of crystallization in Equation 13.4 is greater than the rate of dissolution, so crystals or a precipitate form (Figure 13.4d). In contrast, adding a seed crystal to a saturated solution reestablishes the dynamic equilibrium, and the *net* quantity of dissolved solute no longer changes.

Because crystallization is the reverse of dissolution, a substance that requires an input of heat to form a solution ($\Delta H_{soln} > 0$) releases that heat when it crystallizes from solution ($\Delta H_{crys} < 0$). The amount of heat released is proportional to the amount of solute that exceeds its solubility. Two substances that have a positive enthalpy of solution are sodium thiosulfate ($Na_2S_2O_3$) and sodium acetate (CH_3CO_2Na), both of which are used in commercial hot packs, small bags of supersaturated solutions used to warm hands (see Figure 5.13).

Interactions in Liquid Solutions

The interactions that determine the solubility of a substance in a liquid depend largely on the chemical nature of the solute (such as whether it is ionic or molecular) rather than on its physical state (solid, liquid, or gas). We will first describe the general case of forming a solution of a molecular species in a liquid solvent, and then describe the formation of a solution of an ionic compound.

Solutions of Molecular Substances in Liquids

The London dispersion forces, dipole–dipole interactions, and hydrogen bonds that hold molecules to other molecules are generally weak. Even so, energy is required to disrupt these interactions. As we described in Section 13.1, unless some of that energy is recovered in the formation of new, favorable solute–solvent interactions, the increase in entropy on solution formation is not enough for a solution to form.

For solutions of gases in liquids, we can safely ignore the energy required to separate the solute molecules ($\Delta H_2 = 0$) because the molecules are already separated. Thus, we need to consider only the energy required to separate the solvent molecules (ΔH_1) and the energy released by new solute–solvent interactions (ΔH_3). Nonpolar gases such as N_2, O_2, and Ar have no dipole moment and cannot engage in dipole–dipole interactions or hydrogen bonding. Consequently, the only way they can interact with a solvent is by means of London dispersion forces, which may be weaker than the solvent–solvent interactions in a polar solvent. It is not surprising, then, that nonpolar gases are most soluble in nonpolar solvents. In this case, ΔH_1 and ΔH_3 are both small and of similar magnitude. In contrast, for a solution of a nonpolar gas in a polar solvent, $\Delta H_1 >> \Delta H_3$. As a result, nonpolar gases are less soluble in polar solvents than in nonpolar solvents. For example, the concentration of N_2 in a saturated solution of N_2 in water, a polar solvent, is only 7.07×10^{-4} *M* compared with 4.5×10^{-3} *M* for a saturated solution of N_2 in benzene, a nonpolar solvent.

The solubilities of nonpolar gases in water generally increase as the molecular mass of the gas increases, as shown in Table 13.3. This is precisely the trend expected: as the gas molecules become larger, the strength of the solvent–solute interactions due to London dispersion forces increases, approaching the strength of the solvent–solvent interactions.

Virtually all common organic liquids, whether polar or not, are miscible: the strengths of the intermolecular attractions are comparable, thus the enthalpy of solution is expected to be small ($\Delta H_{soln} \approx 0$), and the increase in entropy drives the formation of a solution. If the predominant intermolecular interactions in two liquids are very different from one another, however, they may be immiscible. For example, organic liquids such as benzene, hexane, CCl_4, and CS_2 (S=C=S) are nonpolar and have no ability to act as hydrogen bond donors or acceptors with hydrogen bonding solvents such as H_2O, HF, and NH_3; hence, they are immiscible in these solvents. When shaken with water, they form separate phases or layers separated by an *interface* (Figure 13.5), the region between the two layers. Just because two liquids are immiscible, however, does *not* mean that they are completely insoluble in each other. For example, 188 mg of benzene dissolves in 100 mL of water at 23.5°C. Adding more benzene results in the separation of an upper layer consisting of benzene that contains a small amount of dissolved water (the solubility of water in benzene is only 178 mg/100 mL of benzene).

The solubilities of simple alcohols in water are given in Table 13.4. Notice that only the three lightest alcohols (methanol, ethanol, and *n*-propanol) are completely miscible with water. As the molecular mass of the alcohol increases, so does the proportion of hydrocarbon in the molecule. Correspondingly, the importance of hydrogen bonding and dipole–dipole interactions in the pure alcohol decreases, while the importance of London dispersion forces increases, which leads to progressively fewer favorable electrostatic interactions with water. Organic liquids such as acetone, ethanol, and tetrahydrofuran (see the structure in the margin) are sufficiently polar to be completely miscible with water yet sufficiently nonpolar to be completely miscible with all organic solvents.

The same principles govern the solubilities of molecular solids in liquids. For example, elemental sulfur is a solid consisting of cyclic S_8 molecules that have no

TABLE 13.3 Solubilities of selected gases in water at 20°C and 1 atm pressure

Gas	Solubility (*M*)
He	3.90×10^{-4}
Ne	4.65×10^{-4}
Ar	15.2×10^{-4}
Kr	27.9×10^{-4}
Xe	50.2×10^{-4}
H_2	8.06×10^{-4}
N_2	7.07×10^{-4}
CO	10.6×10^{-4}
O_2	13.9×10^{-4}
N_2O	281×10^{-4}
CH_4	15.5×10^{-4}

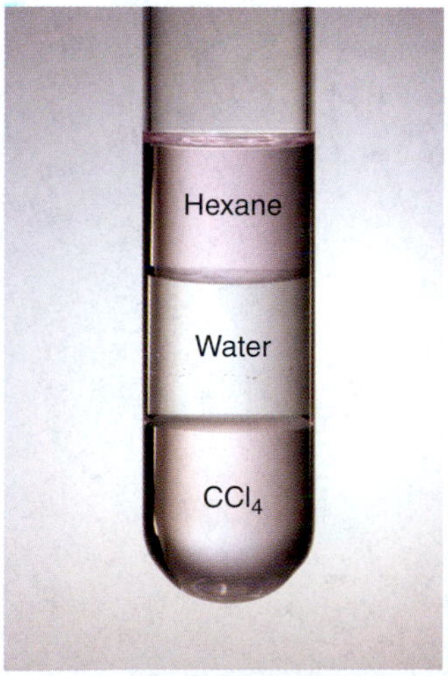

Figure 13.5 Immiscible liquids. Water is immiscible with both CCl_4 and hexane. When all three liquids are mixed, they separate into three distinct layers. Because CCl_4 is denser than water, the water layer floats on the CCl_4. In contrast, hexane is less dense than water, so the hexane floats on the water layer. Because I_2 is intensely purple and quite soluble in both CCl_4 and hexane, but insoluble in water, a small amount of I_2 has been added to help identify the hexane and CCl_4 layers.

TABLE 13.4 Solubilities of straight-chain organic alcohols in water at 20°C

Alcohol	Solubility, mol/100 g of H_2O
Methanol	Completely miscible
Ethanol	Completely miscible
n-Propanol	Completely miscible
n-Butanol	0.11
n-Pentanol	0.030
n-Hexanol	0.0058
n-Heptanol	0.0008

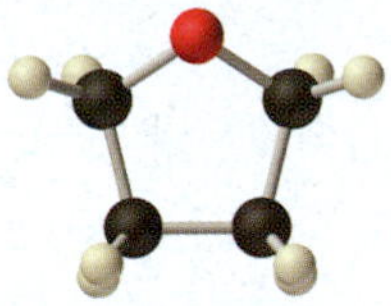

Tetrahydrofuran (THF)

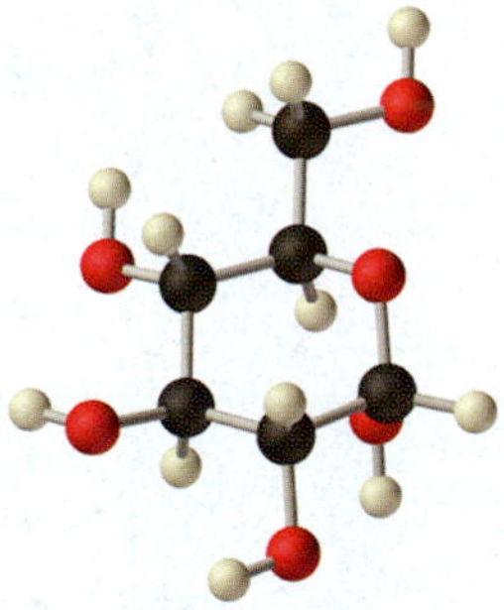

D-Glucose

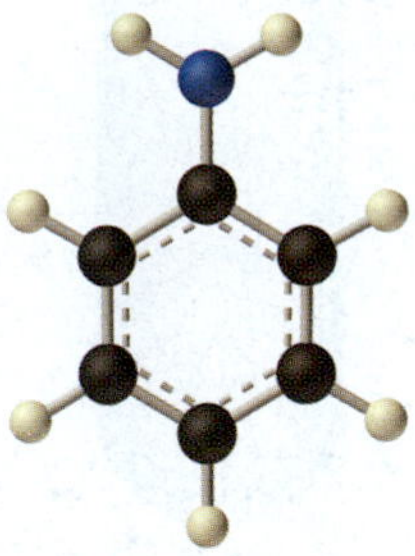

Aniline

dipole moment. Because the S_8 rings in solid sulfur are held to other rings by London dispersion forces, elemental sulfur is insoluble in water. It is, however, soluble in nonpolar solvents that have comparable London dispersion forces, such as CS_2 (23 g/100 mL). In contrast, glucose, whose structure is shown in the margin, contains five —OH groups that can form hydrogen bonds. Consequently, glucose is very soluble in water (91 g/120 mL of water) but essentially insoluble in nonpolar solvents such as benzene.

Low molecular mass hydrocarbons with highly electronegative and polarizable halogen atoms, such as chloroform ($CHCl_3$) and methylene chloride (CH_2Cl_2), have both significant dipole moments and relatively strong London dispersion forces. They are therefore powerful solvents for a wide range of polar and nonpolar compounds. Thus, both naphthalene, which is nonpolar, and phenol (C_6H_5OH), which is polar, are very soluble in chloroform. In contrast, the solubility of ionic compounds is largely determined not by the polarity of the solvent, but rather by its *dielectric constant*, a measure of its ability to separate ions in solution, as you will soon see.

EXAMPLE 13.2

Identify the most important solute–solvent interactions in these solutions: **(a)** iodine in benzene; **(b)** aniline ($C_6H_5NH_2$) in dichloromethane (CH_2Cl_2); **(c)** iodine in water.

Given Components of solutions

Asked for Predominant solute–solvent interactions

Strategy

Identify all possible intermolecular interactions for both the solute and the solvent: London dispersion forces, dipole–dipole interactions, or hydrogen bonding. Determine which is likely to be the most important factor in solution formation.

Solution

(a) Benzene and I_2 are both nonpolar molecules. The only possible attractive forces are London dispersion forces. **(b)** Aniline is a polar molecule that contains an —NH_2 group, which can act as a hydrogen bond donor. Dichloromethane is also polar, but it has no obvious hydrogen bond acceptor. Therefore, the most important interactions between aniline and CH_2Cl_2 are likely to be London interactions. **(c)** Water is a highly polar molecule that engages in extensive hydrogen bonding, whereas I_2 is a nonpolar molecule that cannot act as a hydrogen bond donor or acceptor. Thus, the slight solubility of I_2 in water (1.3×10^{-3} mol/L at 25°C) is due to London dispersion forces.

EXERCISE 13.2

Identify the most important interactions in each solution: **(a)** ethylene glycol ($HOCH_2CH_2OH$) in acetone; **(b)** acetonitrile ($CH_3C{\equiv}N$) in acetone; **(c)** *n*-hexane in benzene.

Answer **(a)** hydrogen bonding; **(b)** London interactions; **(c)** London dispersion forces

Hydrophilic and Hydrophobic Solutes

MGC Hydrophilic vs. Hydrophobic

A solute can be classified as **hydrophilic** (literally, "water loving"), meaning that there is an electrostatic attraction to water, or **hydrophobic** ("water fearing"), meaning that it repels water. A hydrophilic substance is polar and often contains

O—H or N—H groups that can form hydrogen bonds to water. For example, glucose with its five O—H groups is hydrophilic. In contrast, a hydrophobic substance may be polar but usually contains C—H bonds that do not interact favorably with water, as is the case with naphthalene and *n*-octane. Hydrophilic substances tend to be very soluble in water and other strongly polar solvents, whereas hydrophobic substances are essentially insoluble in water and soluble in nonpolar solvents such as benzene and cyclohexane.

The difference between hydrophilic and hydrophobic substances has substantial consequences in biological systems. For example, vitamins can be classified as either *fat soluble* or *water soluble*. Fat-soluble vitamins, such as vitamin A, are mostly nonpolar, hydrophobic molecules. As a result, they tend to be absorbed into fatty tissues and stored there. In contrast, water-soluble vitamins, such as vitamin C, are polar, hydrophilic molecules that circulate in the blood and intracellular fluids, which are primarily aqueous. Water-soluble vitamins are therefore excreted much more rapidly from the body and must be replenished in our daily diet. Comparison of the chemical structures of vitamin A and vitamin C, shown in the margin, quickly reveals why one is hydrophobic and the other hydrophilic.

Because water-soluble vitamins are rapidly excreted, the risk of consuming them in excess is relatively small. Eating a dozen oranges a day is likely to make you tired of oranges long before you suffer any ill effects due to their high vitamin C content. In contrast, fat-soluble vitamins constitute a significant health hazard if consumed in large amounts. For example, the livers of polar bears and other large animals that live in cold climates contain large amounts of vitamin A, which have occasionally proven fatal to unwary people who have eaten them.

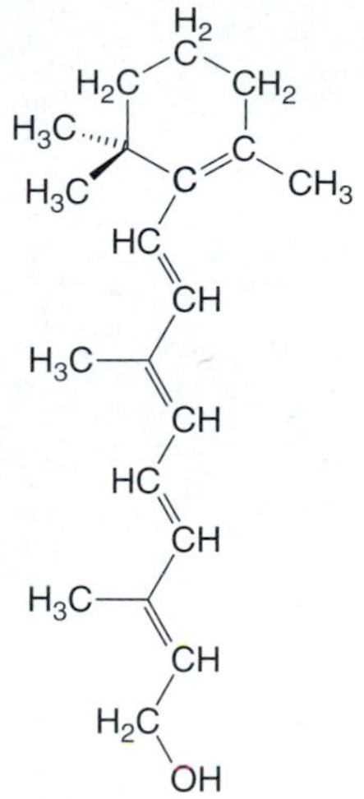

Vitamin A
(*trans*-Retinol)

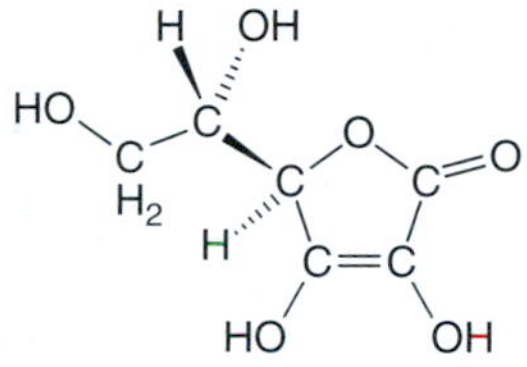

Vitamin C
(L-Ascorbic acid)

EXAMPLE 13.3

The substances whose structures are shown below are essential components of the human diet. Using what you know of hydrophilic and hydrophobic solutes, classify each as water soluble or fat soluble, and predict which are likely to be required in the diet on a daily basis: **(a)** arginine; **(b)** pantothenic acid; **(c)** oleic acid ($C_{17}H_{33}CO_2H$).

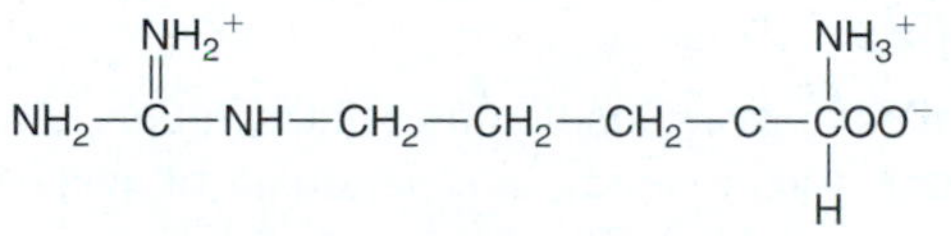

Arginine

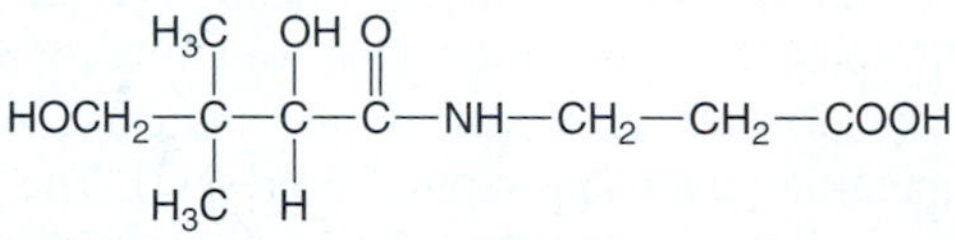

Pantothenic acid

$CH_3(CH_2)_7CH{=}CH(CH_2)_7COOH$

Oleic acid

Given Chemical structures

Asked for Classification as water soluble or fat soluble; dietary requirement

Strategy

Based on the structure of each compound, decide whether it is hydrophilic or hydrophobic. If it is hydrophilic, it is likely to be required on a daily basis.

Solution

(a) Arginine is a highly polar molecule that contains two positively charged groups and one negatively charged group, all of which can form hydrogen bonds with water. As a result, it is hydrophilic and required in our daily diet. **(b)** Although pantothenic acid contains a hydrophobic hydrocarbon portion, it also contains several polar functional groups (—OH and —CO_2H) that should interact strongly with water. Because of the latter, it is likely to be water soluble and therefore required in the diet. (In fact, pantothenic acid is almost always a component of multiple-vitamin tablets.) **(c)** Oleic acid is a hydrophobic molecule with a single polar group at one end. It should be fat soluble and not required each day.

EXERCISE 13.3

These compounds are consumed by humans: caffeine, acetaminophen, and vitamin D. Identify each as primarily hydrophilic (water soluble) or hydrophobic (fat soluble), and predict whether each is likely to be excreted from the body rapidly or slowly.

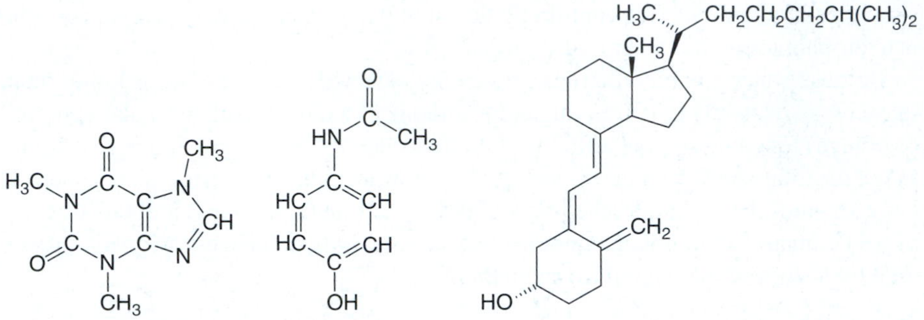

Caffeine Acetaminophen Vitamin D (cholecalciferol)

Answer Caffeine and acetaminophen are water soluble and rapidly excreted; whereas vitamin D is fat soluble and slowly excreted.

Solutions of Solids

Solutions are not limited to gases and liquids; solid solutions also exist. For example, **amalgams**, which are usually solids, are solutions of metals in liquid mercury. Because most metals are soluble in mercury, amalgams are used in gold mining, dentistry, and many other applications. In mining gold, the major difficulty is separating very small particles of pure gold from tons of crushed rock. One way to accomplish this is to agitate a suspension of the crushed rock with liquid mercury, which dissolves the gold (as well as any metallic silver that might be present). The very dense liquid gold–mercury amalgam is then isolated, and the mercury is distilled away to leave the gold.

Network solids such as diamond, graphite, and SiO_2 are insoluble in all solvents with which they do not react chemically. The covalent bonds that hold the network or lattice together are simply too strong to be broken under normal conditions, and they are certainly much stronger than any conceivable combination of intermolecular interactions that might occur in solution. Most metals are insoluble in virtually all solvents for the same reason. That is, the delocalized metallic bonding is much stronger than any favorable metal atom–solvent interactions. Many metals do react with solutions such as aqueous acid or base to produce a solution. However, as we saw in Section 13.1, in these instances the metal undergoes a chemical transformation that cannot be reversed by simply removing the solvent.

Note the pattern

Solids that have very strong intermolecular bonding interactions tend to be insoluble.

Solubilities of Ionic Substances in Liquids

Ionic substances are generally most soluble in polar solvents, and the higher the lattice energy, the more polar the solvent must be to overcome the lattice energy and dissolve the substance. Because of its high polarity, water is the most common solvent for ionic compounds. Many ionic compounds are soluble in other polar solvents, however, such as liquid ammonia, liquid hydrogen fluoride, and methanol. Because all these solvents consist of molecules that have relatively large dipole moments, they can interact favorably with the dissolved ions.

The interaction of water with Na^+ and Cl^- ions in an aqueous solution of NaCl was illustrated in Figure 4.3. The ion–dipole interactions between Li^+ ions and acetone molecules in a solution of LiCl in acetone are shown in Figure 13.6. The energetically favorable Li^+–acetone interactions make ΔH_3 in Figure 13.1 sufficiently negative to overcome the positive ΔH_1 and ΔH_2. Because the dipole moment of acetone (2.88 D), and thus its polarity, is actually larger than that of water (1.85 D), one might even expect that LiCl would be more soluble in acetone than in water. In fact, however, the opposite is true: 83 g of LiCl dissolves in 100 mL of water at 20°C, but only about 4.1 g of LiCl dissolves in 100 mL of acetone. This apparent contradiction arises from the fact that the dipole moment is a property of a single molecule in the gas phase. A more useful measure of the ability of a solvent to dissolve ionic compounds is its **dielectric constant (ε)**, which is the ability of a bulk substance to decrease the electrostatic forces between two charged particles. By definition, the dielectric constant of a vacuum is 1. In essence, a solvent with a high dielectric constant causes the charged particles to behave as if they have been moved farther apart. At 25°C, the dielectric constant of water is 80.1, one of the highest known, and that of acetone is only 21.0. Hence, water is better able to decrease the electrostatic attraction between Li^+ and Cl^- ions, so LiCl is more soluble in water than in acetone. This behavior is in contrast to that of molecular substances, for which polarity is the dominant factor governing solubility.

It is also possible to dissolve ionic compounds in organic solvents using **crown ethers**, cyclic compounds with the general formula $(OCH_2CH_2)_n$. Crown ethers are named using both the total number of atoms in the ring and the number of oxygen atoms. Thus, 18-crown-6 is an 18-membered ring with six oxygen atoms (Figure 13.7a). The cavity in the center of the crown ether molecule is lined with oxygen atoms and is large enough to be occupied by a cation, such as K^+. The cation is stabilized by interacting with lone pairs of electrons on the surrounding oxygen atoms. Thus, crown ethers solvate cations inside a hydrophilic cavity, whereas the outer shell, consisting of C—H bonds, is hydrophobic. Crown ethers are therefore useful for dissolving ionic substances such as $KMnO_4$ in organic

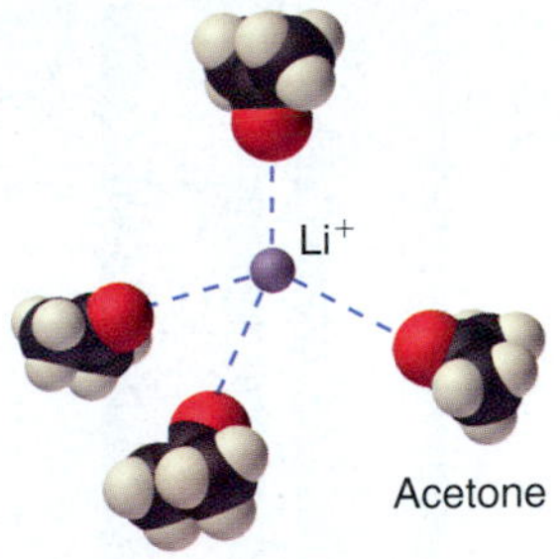

Figure 13.6 The ion–dipole interactions in the solvation of Li^+ ions by acetone, a polar solvent.

Note the pattern

Dielectric constants are the most useful measure of the ability of a solvent to dissolve ionic compounds.

Figure 13.7 Crown ethers and cryptands. (a) The potassium complex of the crown ether 18-crown-6. Note how the cation is nestled within the central cavity of the molecule and interacts with lone pairs of electrons on the oxygen atoms. **(b)** The potassium complex of 2,2,2-cryptand, showing how the cation is almost hidden by the cryptand. Cryptands solvate cations via lone pairs of electrons on both oxygen and nitrogen atoms.

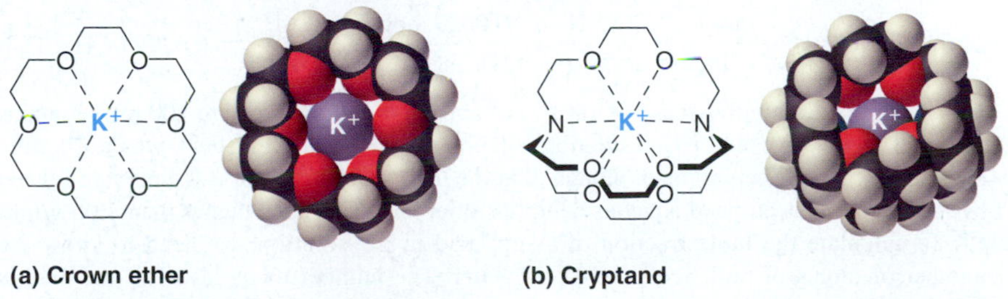

solvents such as isopropanol [$(CH_3)_2CHOH$] (Figure 13.8). The availability of crown ethers with cavities of different sizes allows specific cations to be solvated with a high degree of selectivity.

Cryptands (from the Greek *kryptós*, "hidden") are compounds that can completely surround a cation with lone pairs of electrons on oxygen and nitrogen atoms (Figure 13.7b). The number in the name of the cryptand is the number of oxygen atoms in each strand of the molecule. Like crown ethers, cryptands can be used to prepare solutions of ionic compounds in solvents that are otherwise too nonpolar to dissolve them.

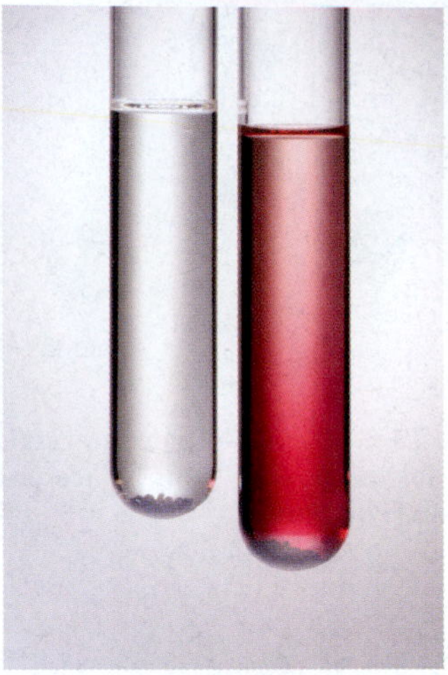

Figure 13.8 The effect of a crown ether on the solubility of $KMnO_4$ in isopropanol (2-propanol). (*Left*) Normally $KMnO_4$, which is intensely purple, is completely insoluble in isopropanol, which has a relatively low dielectric constant. (*Right*) In the presence of a small amount of 18-crown-6, $KMnO_4$ dissolves in isopropanol, as shown by the reddish-purple color due to the presence of permanganate ions in solution.

MGC Units of Concentration

13.3 • Units of Concentration

There are several different ways to quantitatively describe the **concentration** of a solution, which is the amount of solute in a given quantity of solution. For example, molarity was introduced in Chapter 4 as a useful way to describe solution concentrations for reactions that are carried out in solution. Mole fractions, introduced in Chapter 10, are used not only to describe gas concentrations but also to determine the vapor pressures of mixtures of similar liquids. The next example reviews the methods for calculating the molarity and mole fraction of a solution when the masses of its components are known.

EXAMPLE 13.4

Commercial vinegar is essentially a solution of acetic acid in water. Calculate (a) the molarity and (b) the mole fraction of acetic acid in a bottle of vinegar that contains 3.78 g of acetic acid per 100.0 g of solution. Assume that the density of the solution is 1.00 g/mL.

Given Mass of substance and mass and density of solution

Asked for Molarity and mole fraction

Strategy

Ⓐ Calculate the number of moles of acetic acid in the sample. Then calculate the number of liters of solution from its mass and density. Use these results to determine the molarity of the solution.

Ⓑ Determine the mass of the water in the sample, and calculate the number of moles of water. Then determine the mole fraction of acetic acid by dividing the number of moles of acetic acid by the total number of moles of substances in the sample.

Solution

(a) Ⓐ The molarity is the number of moles of acetic acid per liter of solution. We can calculate the number of moles of acetic acid as its mass divided by its molar mass. The volume of the solution equals its mass divided by its density. The calculations follow:

$$\text{moles } CH_3CO_2H = \frac{3.78\ \cancel{g}\ CH_3CO_2H}{60.05\ \cancel{g}/\text{mol}} = 0.0629 \text{ mol}$$

$$\text{volume} = \frac{\text{mass}}{\text{density}} = \frac{100.0\ \cancel{g}\text{ solution}}{1.00\ \cancel{g}/\text{mL}} = 100 \text{ mL}$$

$$\text{molarity of } CH_3CO_2H = \frac{\text{moles } CH_3CO_2H}{\text{liter solution}} = \frac{0.0629 \text{ mol } CH_3CO_2H}{(100\ \cancel{\text{mL}})(1 \text{ L}/1000\ \cancel{\text{mL}})} = 0.629\ M\ CH_3CO_2H$$

This result makes intuitive sense. If 100.0 g of aqueous solution (equal to 100 mL) contains 3.78 g of acetic acid, then 1 L of solution will contain 37.8 g of acetic acid, which is a little more than $\frac{1}{2}$ mole. Keep in mind, though, that the mass and volume of a solution are related by its density; concentrated aqueous solutions often have densities higher than 1.00 g/mL.

(b) Ⓑ To calculate the mole fraction of acetic acid in the solution, we need to know the number of moles of both acetic acid and water. The number of moles of acetic acid is 0.0629 mol, as calculated in part (a). We know that 100.0 g of vinegar contains 3.78 g of acetic acid; hence, the solution also contains (100.0 g − 3.78 g) = 96.2 g of water. We have

$$\text{moles } H_2O = \frac{96.2 \text{ g } H_2O}{18.02 \text{ g/mol}} = 5.34 \text{ mol } H_2O$$

The mole fraction X of acetic acid is the ratio of the number of moles of acetic acid to the total number of moles of substances present:

$$X_{CH_3CO_2H} = \frac{\text{moles } CH_3CO_2H}{\text{moles } CH_3CO_2H + \text{moles } H_2O} = \frac{0.0629 \text{ mol}}{0.0629 \text{ mol} + 5.34 \text{ mol}} = 0.0116$$
$$= 1.16 \times 10^{-2}$$

This answer makes sense, too. There are approximately 100 times as many moles of water as moles of acetic acid, so the ratio should be approximately 0.01.

EXERCISE 13.4

Calculate **(a)** the molarity and **(b)** the mole fraction of a solution of HCl in water (sold commercially as "muriatic acid") that contains 20.22 g of HCl per 100.0 g of solution. The density of the solution is 1.10 g/mL.

Answer **(a)** 6.10 M HCl; **(b)** $X_{HCl} = 0.111$

The concentration of a solution can also be described by its **molality (*m*)**, the number of moles of solute per *kilogram* of solvent:

$$\text{molality } (m) = \frac{\text{moles solute}}{\text{kilogram solvent}} \tag{13.5}$$

Molality, therefore, has the same numerator as molarity (the number of moles of solute) but a different denominator (kilogram of solvent rather than liter of solution). For dilute aqueous solutions, the molality and molarity are nearly the same because dilute solutions are mostly solvent. Thus, because the density of water under standard conditions is very close to 1.0 g/mL, the volume of 1.0 kg of H_2O under these conditions is very close to 1.0 L, and a 0.50 M solution of KBr in water, for example, has approximately the same concentration as a 0.50 m solution.

Another common way of describing concentration is as the ratio of the mass of the solute to the total mass of the solution. The result can be expressed as **mass percentage**, **parts per million (ppm)**, or **parts per billion (ppb)**:

$$\text{mass percentage} = \left(\frac{\text{mass of solute}}{\text{mass of solution}}\right)(100\%) \tag{13.6}$$

$$\text{parts per million (ppm)} = \left(\frac{\text{mass of solute}}{\text{mass of solution}}\right)(10^6) \tag{13.7}$$

$$\text{parts per billion (ppb)} = \left(\frac{\text{mass of solute}}{\text{mass of solution}}\right)(10^9) \tag{13.8}$$

The labels on bottles of commercial reagents often describe the contents in terms of mass percentage. Sulfuric acid, for example, is sold as a 95% aqueous solution, or 95 g of H_2SO_4 per 100 g of solution. Parts per million (ppm) and parts per billion (ppb) are used to describe concentrations of highly dilute solutions. These measurements correspond to milligrams (mg) and micrograms (μg) of solute per kilogram of solution, respectively. For dilute aqueous solutions, this is equal to mg and μg of solute per liter of solution (assuming a density of 1.0 g/mL).

EXAMPLE 13.5

Several years ago, millions of bottles of mineral water were contaminated with benzene at ppm levels. This incident received a great deal of attention because the lethal concentration of benzene in rats is 3.8 ppm. Calculate **(a)** the molarity and **(b)** the mass of benzene in a 250-mL sample of mineral water that contains 12.7 ppm of benzene. Because the contaminated mineral water is a very dilute aqueous solution, we can assume that its density is approximately 1.00 g/mL.

Given Volume of sample, solute concentration, and density of solution

Asked for Molarity of solute and mass of solute in 250 mL

Strategy

Ⓐ Use the concentration of the solute in ppm to calculate the molarity.
Ⓑ Use the concentration of the solute in ppm to calculate the mass of the solute in the specified volume of solution.

Solution

(a) Ⓐ To calculate the molarity of benzene, we need to determine the number of moles of benzene in 1 L of solution. We know that the solution contains 12.7 ppm of benzene. Because 12.7 ppm is equivalent to 12.7 mg/1000 g of solution and the density of the solution is 1.00 g/mL, the solution contains 12.7 mg of benzene per liter (1000 mL). The molarity is therefore

$$\text{molarity} = \frac{\text{moles}}{\text{liter solution}} = \frac{(12.7\ \cancel{\text{mg}})\left(\dfrac{1\ \cancel{\text{g}}}{1000\ \cancel{\text{mg}}}\right)\left(\dfrac{1\ \text{mol}}{78.114\ \cancel{\text{g}}}\right)}{1.00\ \text{L}} = 1.63 \times 10^{-4}\ M$$

(b) Ⓑ We are given that there are 12.7 mg of benzene per 1000 g of solution, which is equal to 12.7 mg per liter of solution. Hence, the mass of benzene in 250 mL (250 g) of solution is

$$\text{mass of benzene} = \frac{(12.7\ \text{mg benzene})(250\ \cancel{\text{mL}})}{1000\ \cancel{\text{mL}}} = 3.18\ \text{mg} = 3.18 \times 10^{-3}\ \text{g benzene}$$

EXERCISE 13.5

The maximum allowable concentration of lead in drinking water is 9.0 ppb. What is the molarity of lead in a 9.0-ppb aqueous solution?

Answer $4.3 \times 10^{-8}\ M$

How do chemists decide which units of concentration to use for a particular application? Although molarity is commonly used to express concentrations for reactions in solution or for titrations, it does have one drawback—molarity is the number of moles of solute divided by the volume of the solution, and the volume of a solution depends on its density, which is a function of temperature. Because

TABLE 13.5 Different units for expressing the concentrations of solutions[a]

Unit	Definition	Application
Molarity (*M*)	Moles of solute/liter of solution (mol/L)	Used for quantitative reactions in solution and titrations; mass and molecular mass of solute and volume of solution are known
Mole fraction (*X*)	Moles of solute/total moles present (mol/mol)	Used for partial pressures of gases and vapor pressures of some solutions; mass and molecular mass of each component are known
Molality (*m*)	Moles of solute/kg of solvent (mol/kg)	Used in determining how colligative properties vary with solute concentration; masses and molecular mass of solute are known
Mass percentage (%)	[Mass of solute (g)/mass of solvent (g)] × 100%	Useful when masses are known but molecular masses are unknown
Parts per million (ppm)	[Mass of solute/mass of solution] × 10^6 (mg solute/kg solution)	Used for trace quantities; masses are known but molecular masses may be unknown
Parts per billion (ppb)	[Mass of solute/mass of solution] × 10^9 (μg solute/kg solution)	Used for trace quantities; masses are known but molecular masses may be unknown

[a] The molarity of a solution is temperature dependent, but the other units shown in this table are independent of temperature.

volumetric glassware is calibrated at a particular temperature, typically 20°C, the molarity may differ from the original value by several percent if a solution is prepared or used at a significantly different temperature, such as 40°C or 0°C. For many applications this may not be a problem, but for precise work these errors can become important. In contrast, mole fraction, molality, and mass percentage depend on only the masses of the solute and solvent, which are independent of temperature.

Mole fraction is not very useful for experiments that involve quantitative reactions, but it is convenient for calculating the partial pressure of gases in mixtures, as we saw in Chapter 10. As you will learn in Section 13.5, mole fractions are also useful for calculating the vapor pressures of certain types of solutions. Molality is particularly useful for determining how properties such as the freezing or boiling point of a solution vary with solute concentration. Mass percentage and parts per million or billion enable us to express the concentrations of substances even if their molecular mass is unknown because these are simply different ways of expressing the ratios of the mass of a solute to the mass of the solution. Units of ppb or ppm are also used to express very low concentrations, such as those of residual impurities in foods or of pollutants in environmental studies.

Table 13.5 summarizes the different units of concentration and typical applications for each. If the molar mass of the solute and the density of the solution are known, then with a little practice it becomes relatively easy to interconvert between any of the units of concentration we have discussed, as illustrated in the next example.

EXAMPLE 13.6

Vodka is essentially a solution of pure ethanol in water. A typical vodka is sold as "80 proof," which means that it contains 40.0% ethanol by volume. The density of pure ethanol is 0.789 g/mL at 20°C. If we assume that the volume of the solution is the sum of the volumes of the components (which is not strictly correct), what are **(a)** the mass percentage; **(b)** the mole fraction; **(c)** the molarity; and **(d)** the molality of ethanol in "80 proof" vodka?

Given Volume percent and density

Asked for Mass percentage, mole fraction, molarity, and molality

Strategy

A Use the density of the solute to calculate the mass of the solute in 100.0 mL of solution. Calculate the mass of water in 100.0 mL of solution.

B Determine the mass percentage of solute by dividing the mass of ethanol by the mass of the solution and multiplying by 100%.

C Convert grams of solute and solvent to moles of solute and solvent. Calculate the mole fraction of solute by dividing the moles of solute by the total number of moles of substances present in solution.

D Calculate the molarity of the solution: moles of solute per liter of solution. Determine the molality of the solution by dividing the number of moles of solute by the kilograms of solvent.

Solution

The key to this problem is to use the density of pure ethanol to determine the mass of ethanol (CH_3CH_2OH), abbreviated as EtOH, that is present in a given volume of solution. From this, we can calculate the number of moles of ethanol and the concentration of ethanol in any of the required units. **A** Because we are given a percentage by volume, we assume that we have 100.0 mL of solution. The volume of ethanol will thus be 40.0% of 100.0 mL, or 40.0 mL of ethanol, and the volume of water will be 60.0% of 100.0 mL, or 60.0 mL of water. The mass of ethanol is obtained from its density:

$$\text{mass of EtOH} = (40.0\ \cancel{\text{mL}})\left(\frac{0.789\ \text{g}}{\cancel{\text{mL}}}\right) = 31.6\ \text{g EtOH}$$

If we assume the density of water is 1.00 g/mL, the mass of water is 60.0 g. We now have all the information we need to calculate the concentration of ethanol in the solution.

(a) **B** The mass percentage of ethanol is the ratio of the mass of ethanol to the total mass of the solution, expressed as a percentage:

$$\%\ \text{EtOH} = \left(\frac{\text{mass of EtOH}}{\text{mass of solution}}\right)(100\%) = \left(\frac{31.6\ \cancel{\text{g}}\ \text{EtOH}}{31.6\ \cancel{\text{g}}\ \text{EtOH} + 60.0\ \cancel{\text{g}}\ \text{H}_2\text{O}}\right)(100\%) = 34.5\%$$

(b) **C** The mole fraction of ethanol is the ratio of the number of moles of ethanol to the total number of moles of substances in the solution. Because 40.0 mL of ethanol has a mass of 31.6 g, we can use the molar mass of ethanol (46.07 g/mol) to determine the number of moles of ethanol in 40.0 mL:

$$\text{moles EtOH} = (31.6\ \cancel{\text{g EtOH}})\left(\frac{1\ \text{mol}}{46.07\ \cancel{\text{g EtOH}}}\right) = 0.686\ \text{mol CH}_3\text{CH}_2\text{OH}$$

Similarly, the number of moles of water is

$$\text{moles H}_2\text{O} = (60.0\ \cancel{\text{g H}_2\text{O}})\left(\frac{1\ \text{mol H}_2\text{O}}{18.02\ \cancel{\text{g H}_2\text{O}}}\right) = 3.33\ \text{mol H}_2\text{O}$$

The mole fraction of ethanol is thus

$$X_{\text{EtOH}} = \frac{0.686\ \cancel{\text{mol}}}{0.686\ \cancel{\text{mol}} + 3.33\ \cancel{\text{mol}}} = 0.171$$

(c) **D** The molarity of the solution is the number of moles of ethanol per liter of solution. We already know the number of moles of ethanol per 100.0 mL of solution, so the molarity is

$$M_{\text{EtOH}} = \left(\frac{0.686\ \text{mol}}{100\ \cancel{\text{mL}}}\right)\left(\frac{1000\ \cancel{\text{mL}}}{\text{L}}\right) = 6.86\ M$$

(d) The molality of the solution is the number of moles of ethanol per kilogram of solvent. Because we know the number of moles of ethanol in 60.0 g of water, the calculation is again straightforward:

$$m_{\text{EtOH}} = \left(\frac{0.686 \text{ mol EtOH}}{60.0 \text{ g } H_2O}\right)\left(\frac{1000 \text{ g}}{\text{kg}}\right) = \frac{11.4 \text{ mol EtOH}}{\text{kg } H_2O} = 11.4\ m$$

EXERCISE 13.6

Calculate **(a)** mass percentage, **(b)** mole fraction, **(c)** molarity, and **(d)** molality of toluene in a solution prepared by mixing 100.0 mL of toluene with 300.0 mL of benzene. The densities of toluene and benzene are 0.867 g/mL and 0.874 g/mL, respectively. Assume that the volume of the solution is the sum of the volumes of the components.

Answer **(a)** mass percentage toluene = 24.8%; **(b)** $X_{\text{toluene}} = 0.219$; **(c)** 2.35 M toluene; **(d)** 3.59 m toluene

13.4 • Effects of Temperature and Pressure on Solubility

Experimentally, it is found that the solubility of most compounds depends strongly on temperature and, if a gas, on pressure as well. As we shall see, the ability to manipulate the solubility by changing the temperature and pressure has several important consequences.

MGC Henry's Law

Effect of Temperature on the Solubility of Solids

Figure 13.9 shows plots of the solubilities of several organic and inorganic compounds in water as a function of temperature. Although the solubility of a substance generally increases with increasing temperature, there is no simple relationship between the structure of a substance and the temperature dependence of its solubility. Many compounds (such as glucose and CH_3CO_2Na) exhibit a dramatic increase in

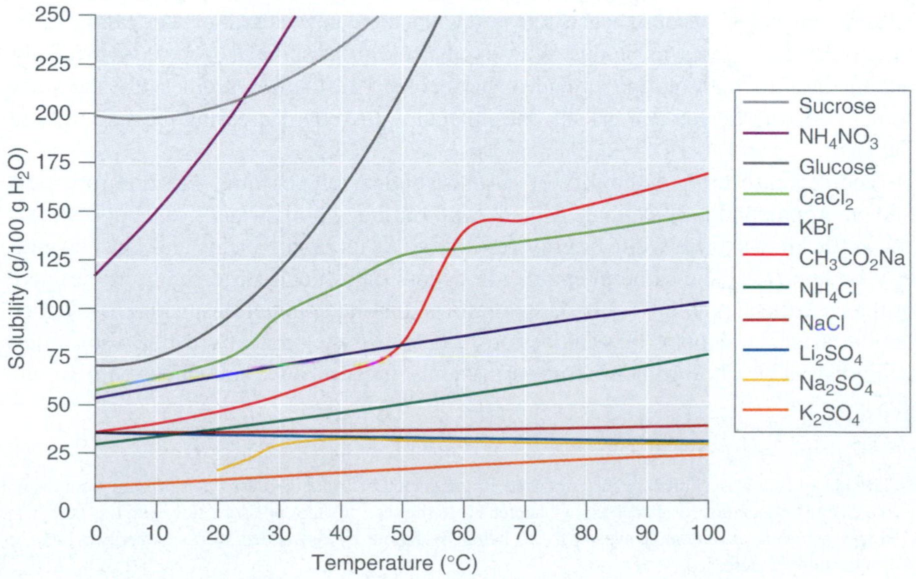

Figure 13.9 The solubilities of several inorganic and organic solids in water as a function of temperature. Solubility may increase or decrease with temperature; the magnitude of this temperature dependence varies widely among compounds.

solubility with increasing temperature. Others (such as NaCl and K_2SO_4) exhibit little variation, and still others (such as $CaSO_4$ and Li_2SO_4) become less soluble with increasing temperature.

Notice in particular the curves for NH_4NO_3 and $CaCl_2$. The dissolution of ammonium nitrate in water is endothermic ($\Delta H_{soln} = +25.7$ kJ/mol), whereas the dissolution of calcium chloride is exothermic ($\Delta H_{soln} = -68.2$ kJ/mol), yet Figure 13.9 shows that the solubility of both compounds increases sharply with increasing temperature. In fact, the magnitudes of the changes in both enthalpy and entropy for dissolution are temperature dependent. Because the solubility of a compound is ultimately determined by relatively small differences between large numbers, there is generally no good way to predict how it will vary with temperature.

The variation of solubility with temperature has been measured for a wide range of compounds, and the results are published in many standard reference books. Chemists are often able to use this information to separate the components of a mixture by **fractional crystallization**, the separation of compounds based on their solubilities in a given solvent. For example, if we have a mixture of 150 g of sodium acetate (CH_3CO_2Na) and 50 g of KBr, we can separate the two compounds by dissolving the mixture in 100 g of water at 80°C and then cooling the solution slowly to 0°C. According to the temperature curves in Figure 13.9, both compounds dissolve in water at 80°C, and all 50 g of KBr remains in solution at 0°C. Only about 36 g of CH_3CO_2Na is soluble in 100 g of water at 0°C, however, so approximately (150 g − 36 g) = 114 g of CH_3CO_2Na crystallizes out upon cooling. The crystals can then be separated by filtration. Thus, fractional crystallization allows us to recover about 75% of the original CH_3CO_2Na in essentially pure form in only one step.

Fractional crystallization is a common technique for purifying compounds as diverse as those shown in Figure 13.9 to antibiotics and enzymes. For the technique to work properly, the compound of interest must be more soluble at high temperature than at low temperature, so that lowering the temperature causes it to crystallize out of solution. In addition, the impurities must be *more* soluble than the compound of interest (as was KBr in this example) and preferably present in relatively small amounts.

Effect of Temperature on the Solubility of Gases

The solubility of gases in liquids decreases with increasing temperature, as shown in Figure 13.10. Attractive intermolecular interactions in the gas phase are essentially zero for most substances. When a gas dissolves, it does so because its molecules interact with solvent molecules. Because heat is released when these new attractive interactions form, dissolving most gases in liquids is an exothermic process ($\Delta H_{soln} < 0$). Conversely, adding heat to the solution provides thermal energy that overcomes the attractive forces between the gas and the solvent molecules, thereby decreasing the solubility of the gas.*

The decrease in the solubility of gases at higher temperatures has both practical and environmental implications. Anyone who routinely boils water in a teapot or electric kettle knows that a white or gray deposit builds up on the inside and must eventually be removed. The same phenomenon occurs on a much larger scale in the giant boilers used to supply hot water or steam for industrial applications, where it is called "boiler scale," a deposit that can seriously decrease the capacity of hot water pipes (Figure 13.11). The problem is not a uniquely modern one; aqueducts built by the

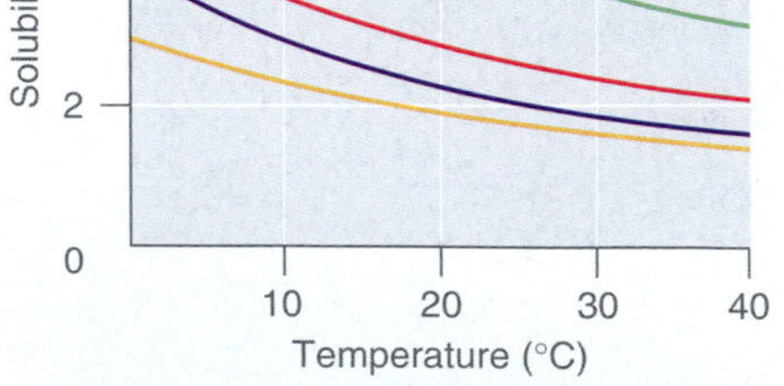

Figure 13.10 The solubilities of several common gases in water as a function of temperature at a partial pressure of 1 atm. The solubility of all gases decreases with increasing temperature. Interactive Graph

* The phenomenon is similar to that involved in the increase in vapor pressure of a pure liquid with increasing temperature, as discussed in Chapter 11. In the case of vapor pressure, however, it is attractive forces between *solvent* molecules that are being overcome by the added thermal energy when the temperature is increased.

Romans 2000 years ago to carry cold water from alpine regions to warmer, drier regions in southern France were also clogged by similar deposits. The chemistry behind the formation of these deposits is moderately complex and will be described in more detail in Chapter 17, but the driving force is the loss of dissolved CO_2 from solution. Hard water contains dissolved Ca^{2+} and HCO_3^- (bicarbonate) ions. Calcium bicarbonate [$Ca(HCO_3)_2$] is rather soluble in water, but calcium carbonate ($CaCO_3$) is quite insoluble. A solution of bicarbonate ions can react to form carbon dioxide, carbonate ion, and water:

$$2HCO_3^-(aq) \longrightarrow CO_3^{2-}(aq) + H_2O(l) + CO_2(aq) \qquad (13.9)$$

Heating the solution decreases the solubility of CO_2, which escapes into the gas phase above the solution. In the presence of calcium ions, the carbonate ions precipitate as insoluble calcium carbonate, the major component of boiler scale.

In *thermal pollution*, lake or river water that is used to cool an industrial reactor or a power plant is returned to the environment at a higher temperature than normal. Because of the reduced solubility of O_2 at higher temperatures (Figure 13.10), the warmer water contains less dissolved oxygen than the water did when it entered the plant. Fish and other aquatic organisms that need dissolved oxygen to live can literally suffocate if the oxygen concentration of their habitat is too low. Because the warm, oxygen-depleted water is less dense, it tends to float on top of the cooler, denser, more oxygen-rich water in the lake or river, forming a barrier that prevents atmospheric oxygen from dissolving. Eventually, even deep lakes can be suffocated if the problem is not corrected. Additionally, most fish and other nonmammalian aquatic organisms are cold-blooded, which means that their body temperature is the same as the temperature of their environment. Temperatures substantially higher than the normal range can lead to severe stress or even death. Given these considerations, cooling systems for power plants and other such facilities have to be designed to minimize the adverse effect on the temperature of surrounding bodies of water.

Figure 13.11 Boiler scale in a water pipe. Calcium carbonate ($CaCO_3$) deposits in hot water pipes can significantly reduce pipe capacity. These deposits, called boiler scale, form when dissolved CO_2 is driven into the gas phase at high temperatures.

Effect of Pressure on the Solubility of Gases: Henry's Law

External pressure has very little effect on the solubility of liquids and solids. In contrast, the solubility of gases increases as the partial pressure of the gas above a solution increases. This point is illustrated in Figure 13.12, which shows the effect of increased

Figure 13.12 A model that explains why the solubility of a gas increases as the partial pressure increases at constant temperature. (**a**) When a gas comes in contact with a pure liquid, some of the gas molecules (purple spheres) collide with the surface of the liquid and dissolve. When the concentration of dissolved gas molecules has increased so that the rate at which gas molecules escape into the gas phase is the same as the rate at which they dissolve, a dynamic equilibrium has been established, as depicted here. This equilibrium is entirely analogous to the one that maintains the vapor pressure of a liquid (see Chapter 11). (**b**) Increasing the pressure of the gas increases the number of molecules of gas per unit volume, which increases the rate at which gas molecules collide with the surface of the liquid and dissolve. (**c**) As additional gas molecules dissolve at the higher pressure, the concentration of dissolved gas increases until a new dynamic equilibrium is established.

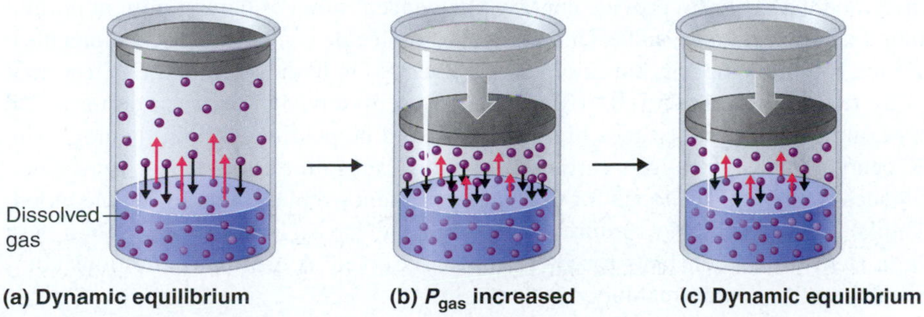

TABLE 13.6 Henry's law constants for selected gases in water at 20°C

Gas	Henry's Law Constant (*k*), mol/(L · atm)
He	3.9×10^{-4}
Ne	4.7×10^{-4}
Ar	15×10^{-4}
H_2	8.1×10^{-4}
N_2	7.1×10^{-4}
O_2	14×10^{-4}
CO_2	392×10^{-4}

pressure on the dynamic equilibrium that is established between the dissolved gas molecules in solution and the molecules in the gas phase above the solution. Because the concentration of molecules in the gas phase increases with increasing pressure, the concentration of dissolved gas molecules in the solution at equilibrium is also higher at higher pressures.

The relationship between pressure and the solubility of a gas is described quantitatively by **Henry's law**, named for its discoverer, the English physician and chemist, William Henry (1775–1836):

$$C = kP \tag{13.10}$$

where C is the concentration of dissolved gas at equilibrium, P is the partial pressure of the gas, and k is the *Henry's law constant*, which must be determined experimentally for each combination of gas, solvent, and temperature. Although the gas concentration may be expressed in any convenient units, we will use molarity exclusively. The units of the Henry's law constant are therefore mol/(L · atm) = M/atm. Values of the Henry's law constants for solutions of several gases in water at 20°C are listed in Table 13.6.

As the data in Table 13.6 demonstrate, the concentration of a dissolved gas in water at a given pressure depends strongly on its physical properties. For a series of related substances, London dispersion forces increase as molecular mass increases. Thus, among the elements of Group 18, the Henry's law constants increase smoothly from He to Ne to Ar. The table also shows that O_2 is almost twice as soluble as N_2. Although London dispersion forces are too weak to explain such a large difference, O_2 is paramagnetic and hence more polarizable than N_2, which explains its high solubility.

Gases that react chemically with water, such as HCl and the other hydrogen halides, H_2S, and NH_3, do not obey Henry's law; all of these gases are much more soluble than predicted by Henry's law. For example, HCl reacts with water to give $H^+(aq)$ and $Cl^-(aq)$, *not* dissolved HCl molecules, and its dissociation into ions results in a much higher solubility than expected for a neutral molecule.

Henry's law has important applications. For example, bubbles of CO_2 form as soon as a carbonated beverage is opened because the drink was bottled under CO_2 at a pressure higher than 1 atm. When the bottle is opened, the pressure of CO_2 above the solution drops rapidly, and some of the dissolved gas escapes from the solution as bubbles. Henry's law also explains why scuba divers have to be careful to ascend to the surface slowly after a dive if they are breathing compressed air. At the higher pressures under water, more N_2 from the air dissolves in the diver's internal fluids. If the diver ascends too quickly, the rapid pressure change causes small bubbles of N_2 to form throughout the body, a condition known as "the bends." These bubbles can block the flow of blood through the small blood vessels, causing great pain and even proving fatal in some cases.

Due to the low Henry's law constant for O_2 in water, the levels of dissolved oxygen in water are too low to support the energy needs of multicellular organisms, including humans. To increase the O_2 concentration in internal fluids, organisms synthesize highly soluble carrier molecules that bind O_2 reversibly. For example, human red blood cells contain a protein called hemoglobin that specifically binds O_2 and facilitates its transport from the lungs to the tissues, where it is used to oxidize food molecules to provide energy. The concentration of hemoglobin in normal blood is about 2.2 mM, and each hemoglobin molecule can bind four O_2 molecules. Hence, although the concentration of dissolved O_2 in blood serum at 37°C (normal body temperature) is only 0.010 mM, the total dissolved O_2 concentration is 8.8 mM, almost a thousand times higher than would be possible without hemoglobin. Recently, synthetic oxygen carriers, or "blood substitutes," based on fluorinated alkanes have been developed for use as an emergency replacement for whole blood. Unlike blood, they do not require refrigeration and have a long shelf life. Their very high Henry's law constants for O_2 result in dissolved oxygen concentrations comparable to those in normal blood.

EXAMPLE 13.7

The Henry's law constant for O_2 in water at 25°C is 1.27×10^{-3} *M*/atm, and the mole fraction of O_2 in the atmosphere is 0.21. Calculate the solubility of O_2 in water at 25°C at an atmospheric pressure of 1.00 atm.

Given Henry's law constant, mole fraction of O_2, and pressure

Asked for Solubility

Strategy

A Use Dalton's law of partial pressures (see Section 10.5) to calculate the partial pressure of oxygen.
B Use Henry's law to calculate the solubility, expressed as the concentration of dissolved gas.

Solution

A According to Dalton's law, the partial pressure of O_2 is proportional to the mole fraction of O_2:

$$P_A = X_A P_t = (0.21)(1.00 \text{ atm}) = 0.21 \text{ atm}$$

B From Henry's law, the concentration of dissolved oxygen under these conditions is

$$C_{O_2} = kP_{O_2} = (1.27 \times 10^{-3}\ M/\cancel{\text{atm}})(0.21\ \cancel{\text{atm}}) = 2.7 \times 10^{-4}\ M$$

EXERCISE 13.7

To understand why soft drinks "fizz" and then go "flat" after being opened, calculate the concentration of dissolved CO_2 in a soft drink **(a)** bottled under a pressure of 5.0 atm of CO_2, and **(b)** in equilibrium with the normal partial pressure of CO_2 in the atmosphere (approximately 3×10^{-4} atm). The Henry's law constant for CO_2 in water at 25°C is 3.4×10^{-2} *M*/atm.

Answer: **(a)** 0.17 *M*; **(b)** 1×10^{-5} *M*

13.5 • Colligative Properties of Solutions

MGC Boiling Point Elevation and Freezing Point Depression

Many of the physical properties of solutions differ significantly from those of the pure substances discussed in earlier chapters, and these differences have important consequences. For example, the limited temperature range of liquid water (0–100°C) severely limits its use. Aqueous solutions have both a lower freezing point and a higher boiling point than pure water. Probably one of the most familiar applications of this phenomenon is the addition of ethylene glycol ("antifreeze") to the water in an automobile radiator to lower the freezing point of the water and prevent the engine from cracking in very cold weather from the expansion of water on freezing. Antifreeze also enables the cooling system to operate at temperatures higher than 100°C without generating enough pressure to explode.

Changes in the freezing point and boiling point of a solution depend primarily on the *number* of solute particles present rather than the *kind* of particles. Such properties of solutions are called **colligative properties** (from the Latin *colligatus*, "bound together" as in a quantity). As we will see, the vapor pressure and osmotic pressure of solutions are also colligative properties.

When we determine the number of particles in a solution, it is important to remember that not all solutions with the same molarity contain the same concentration of solute particles. Consider, for example, 0.01 *M* aqueous solutions of sucrose, NaCl, and $CaCl_2$. Because sucrose dissolves to give a solution of neutral molecules, the concentration of solute particles in a 0.01 *M* sucrose solution is 0.01 *M*. In contrast, both NaCl and $CaCl_2$ are ionic

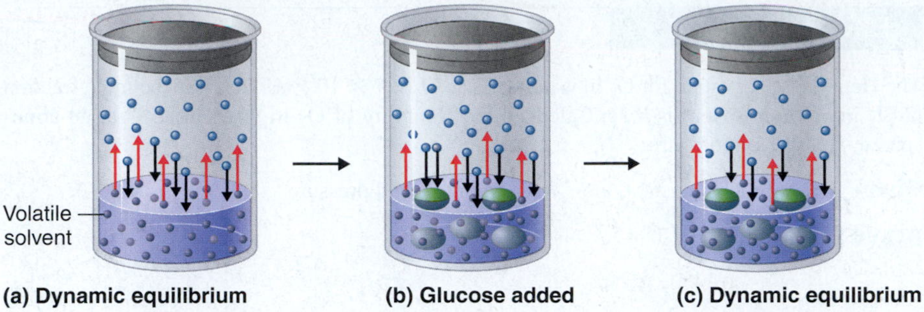

Figure 13.13 A model that explains why the vapor pressure of a solution of glucose is lower than the vapor pressure of pure water. (**a**) When water or any volatile solvent is in a closed container, water molecules move into and out of the liquid phase at the same rate in a dynamic equilibrium. (**b**) If a nonvolatile solute such as glucose is added, some fraction of the surface area is occupied by solvated solute molecules. As a result, the rate at which water molecules evaporate is decreased, although initially their rate of condensation is unchanged. (**c**) When the glucose solution reaches equilibrium, the concentration of water molecules in the vapor phase, and hence the vapor pressure, is lower than that of pure water.

compounds that dissociate in water to yield ions. As a result, a 0.01 *M* aqueous solution of NaCl contains 0.01 *M* Na^+ ions *and* 0.01 *M* Cl^- ions, for a total particle concentration of 0.02 *M*. Similarly, the $CaCl_2$ solution contains 0.01 *M* Ca^{2+} ions and 0.02 *M* Cl^- ions, for a total particle concentration of 0.03 *M*.* The sum of the concentrations of the various dissolved solute particles dictates the physical properties of a solution. In the following discussion, we must therefore keep the chemical nature of the solute firmly in mind.

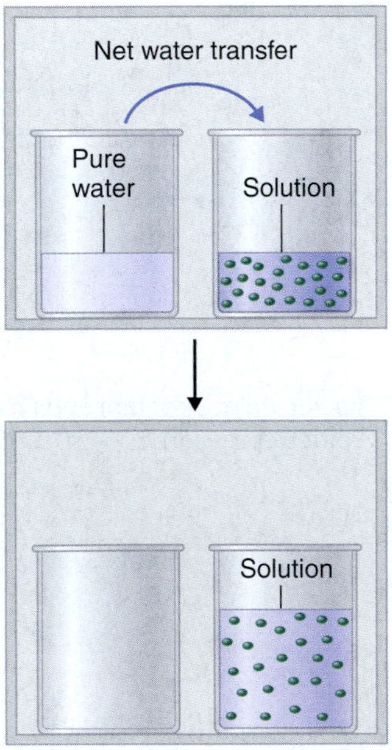

Figure 13.14 Transfer of water to a beaker containing a glucose solution. (*Top*) One beaker contains an aqueous solution of glucose, and the other contains pure water. If they are placed in a sealed chamber, the lower vapor pressure of water in the glucose solution results in a net transfer of water from the beaker containing pure water to the beaker containing the glucose solution. (*Bottom*) Eventually, all of the water is transferred to the beaker that contains the glucose solution.

Vapor Pressure of Solutions and Raoult's Law

Adding a nonvolatile solute, one whose vapor pressure is too low to measure readily, to a volatile solvent decreases the vapor pressure of the solvent. We can understand this phenomenon qualitatively by examining Figure 13.13, which is a schematic diagram of the surface of a solution of glucose in water. In an aqueous solution of glucose, a portion of the surface area is occupied by nonvolatile glucose molecules rather than by volatile water molecules. As a result, fewer water molecules can enter the vapor phase per unit time, even though the surface water molecules have the same kinetic energy distribution as they would in pure water. At the same time, the rate at which water molecules in the vapor phase collide with the surface and reenter the solution is unaffected. The net effect is to shift the dynamic equilibrium between water in the vapor and the liquid phases, decreasing the vapor pressure of the solution compared with the vapor pressure of the pure solvent.

Figure 13.14 shows two beakers, one containing pure water and one containing an aqueous glucose solution, in a sealed chamber. We can view the system as having two competing equilibria: water vapor will condense in both beakers at the same rate, but water molecules will evaporate more slowly from the glucose solution because fewer water molecules are at the surface. Eventually, all of the water will evaporate from the beaker containing the liquid with the higher vapor pressure (pure water) and condense in the beaker containing the liquid with the lower vapor pressure (the glucose solution). If the system consisted of only a beaker of water inside a sealed container, equilibrium between the liquid and vapor would be achieved rather rapidly, and the amount of liquid water in the beaker would remain constant.

---

* These values are correct for dilute solutions, where the dissociation of the compounds to form separately solvated ions is complete. At higher concentrations (typically >1 *M*), especially with salts of small, highly charged ions (such as Mg^{2+} or Al^{3+}), or in solutions in less polar solvents, dissociation to give separate ions is often incomplete (see Figure 13.21).

If the particles of a solute are essentially the same size as those of the solvent and both solute and solvent have roughly equal probabilities of being at the surface of the solution, then the effect of a solute on the vapor pressure of the solvent is proportional to the number of sites occupied by solute particles at the surface of the solution. Doubling the concentration of a given solute causes twice as many surface sites to be occupied by solute molecules, resulting in twice the decrease in vapor pressure. The relationship between solution composition and vapor pressure is therefore

$$P_A = X_A P_A^0 \tag{13.11}$$

where P_A is the vapor pressure of component A of the solution (in this case the solvent), X_A is the mole fraction of A in solution, and P_A^0 is the vapor pressure of pure A. Equation 13.11 is known as **Raoult's law**, after the French chemist who developed it. If the solution contains only a single nonvolatile solute (B), then $X_A + X_B = 1$, and we can substitute $X_A = 1 - X_B$ to obtain

$$P_A = (1 - X_B)P_A^0 \tag{13.12}$$

Rearranging and defining $\Delta P = P_A^0 - P_A$, we obtain a relationship between the decrease in vapor pressure and the mole fraction of nonvolatile solute:

$$P_A^0 - P_A = \Delta P = X_B P_A^0 \tag{13.13}$$

We can solve vapor pressure problems in either of two ways: by using Equation 13.11 to calculate the actual vapor pressure above a solution of a nonvolatile solute, or by using Equation 13.13 to calculate the decrease in vapor pressure caused by a specified amount of a nonvolatile solute.

EXAMPLE 13.8

Ethylene glycol ($HOCH_2CH_2OH$), the major ingredient in commercial automotive antifreeze, increases the boiling point of radiator fluid by lowering its vapor pressure. At 100°C, the vapor pressure of pure water is 760 mmHg. Calculate the vapor pressure of an aqueous solution that contains 30.2% ethylene glycol by mass, a concentration commonly used in climates that do not get extremely cold in winter.

Given Identity of solute, percentage by mass, and vapor pressure of pure solvent

Asked for Vapor pressure of solution

Strategy

A Calculate the number of moles of ethylene glycol in an arbitrary quantity of water, and then calculate the mole fraction of water.

B Use Raoult's law to calculate the vapor pressure of the solution.

Solution

A A 30.2% solution of ethylene glycol contains 302 g of ethylene glycol per kilogram of solution; the remainder (698 g) is water. To use Raoult's law to calculate the vapor pressure of the solution, we must know the mole fraction of water. Thus, we must first calculate the number of moles of both ethylene glycol (EG) and water present:

$$\text{moles EG} = (302\ \cancel{g})\left(\frac{1\ \text{mol}}{62.07\ \cancel{g}}\right) = 4.87\ \text{mol EG}$$

$$\text{moles water} = (698\ \cancel{g})\left(\frac{1\ \text{mol}}{18.02\ \cancel{g}}\right) = 38.7\ \text{mol water}$$

The mole fraction of water is thus

$$X_{H_2O} = \frac{38.7\ \cancel{\text{mol}}\ \text{water}}{38.7\ \cancel{\text{mol}}\ \text{water} + 4.87\ \cancel{\text{mol}}\ \text{EG}} = 0.888$$

From Raoult's law (Equation 13.11), the vapor pressure of the solution is

$$P_{H_2O} = (X_{H_2O})\,(P^0_{H_2O}) = (0.888)(760 \text{ mmHg}) = 675 \text{ mmHg}$$

Alternatively, we could solve this problem by calculating the mole fraction of ethylene glycol and then using Equation 13.13 to calculate the resulting decrease in vapor pressure:

$$X_{EG} = \frac{4.87 \text{ \sout{mol} EG}}{4.87 \text{ \sout{mol} EG} + 38.7 \text{ \sout{mol} water}} = 0.112$$

$$\Delta P = (X_{EG})(P^0_{H_2O}) = (0.112)(760 \text{ mmHg}) = 85.1 \text{ mmHg}$$

$$P_{H_2O} = P^0_{H_2O} - \Delta P = 760 \text{ mmHg} - 85.1 \text{ mmHg} = 675 \text{ mmHg}$$

The same result is obtained using either method.

EXERCISE 13.8

Seawater is an approximately 3.0% aqueous solution of NaCl by mass that also contains about 0.5% of other salts by mass. Calculate the decrease in the vapor pressure of water at 25°C caused by this concentration of NaCl, remembering that 1 mol of NaCl produces 2 mol of solute particles. The vapor pressure of pure water at 25°C is 23.8 mmHg.

Answer 0.45 mmHg. (This may seem like a small amount, but it constitutes about a 2% decrease in the vapor pressure of water and accounts in part for the higher humidity in the north-central United States near the Great Lakes, which are freshwater lakes. The decrease therefore has important implications for climate modeling.)

Even if the solute is volatile, meaning that it has a measurable vapor pressure, we can still use Raoult's law. In this case, we calculate the vapor pressure of each component separately. The total vapor pressure of the solution, P_T, is the sum of the vapor pressures of the components:

$$P_T = P_A + P_B = X_A P^0_A + X_B P^0_B \tag{13.14}$$

Because $X_B = 1 - X_A$ for a two-component system,

$$P_T = X_A P^0_A + (1 - X_A) P^0_B \tag{13.15}$$

Thus, we need to specify the mole fraction of only one of the components in a two-component system. Consider, for example, the vapor pressure of solutions of benzene and toluene of various compositions. At 20°C, the vapor pressures of pure benzene and toluene are 74.7 and 22.3 mmHg, respectively. The vapor pressure of benzene in a benzene–toluene solution is

$$P_{C_6H_6} = X_{C_6H_6} P^0_{C_6H_6} \tag{13.16}$$

and the vapor pressure of toluene in the solution is

$$P_{C_6H_5CH_3} = X_{C_6H_5CH_3} P^0_{C_6H_5CH_3} \tag{13.17}$$

Notice that Equations 13.16 and 13.17 are both in the form of the equation for a straight line: $y = mx + b$, where $b = 0$. Plots of the vapor pressures of both components versus the mole fractions are therefore straight lines that pass through the origin, as shown in Figure 13.15. Furthermore, a plot of the total vapor pressure of the solution versus the mole fraction is a straight line that represents the sum of the vapor pressures of the pure components. Thus, the vapor pressure of the solution is always greater than the vapor pressure of either component.

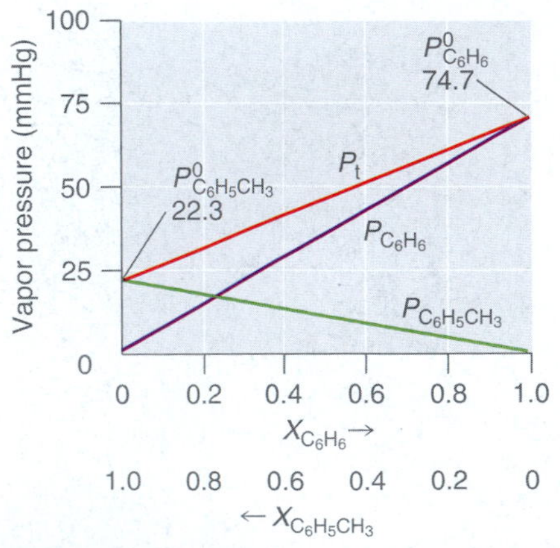

Figure 13.15 Vapor pressures of benzene–toluene solutions. Plots of the vapor pressures of benzene (C_6H_6) and toluene ($C_6H_5CH_3$) versus the mole fractions at 20°C are straight lines. For a solution like this, which approximates an ideal solution, the total vapor pressure of the solution (P_t) is the sum of the vapor pressures of the components. Interactive Graph

A solution of two volatile components that behaves like the solution in Figure 13.15 is an **ideal solution**, which is defined as a solution that obeys Raoult's law. Like an ideal gas, an ideal solution is a hypothetical system whose properties can be described in terms of a simple model. Mixtures of benzene and toluene approximate an ideal solution because the intermolecular forces in the two pure liquids are almost identical in both kind and magnitude. Consequently, the change in enthalpy on solution formation is essentially zero ($\Delta H_{soln} \approx 0$), which is one of the defining characteristics of an ideal solution.

Note the pattern

Ideal solutions and ideal gases are both simple models that ignore intermolecular interactions.

Most real solutions, however, do not obey Raoult's law precisely, just as most real gases do not obey the ideal gas law exactly. Real solutions generally deviate from Raoult's law because the intermolecular interactions between the two components A and B differ. We can distinguish between two general kinds of behavior, depending on whether the intermolecular interactions between molecules A and B are stronger or weaker than the A–A and B–B interactions in the pure components. If the A–B interactions are *stronger* than the A–A and B–B interactions, each component of the solution exhibits a *lower* vapor pressure than expected for an ideal solution, as does the solution as a whole. The favorable A–B interactions effectively stabilize the solution compared with the vapor. This kind of behavior is called a *negative deviation from Raoult's law*. Systems stabilized by hydrogen bonding interactions between the two molecules, such as acetone and ethanol, exhibit negative deviations from Raoult's law. Conversely, if the A–B interactions are *weaker* than the A–A and B–B interactions yet the entropy increase is enough to allow the solution to form, both A and B have an increased tendency to escape from the solution into the vapor phase. The result is a *higher* vapor pressure than expected for an ideal solution, producing a *positive deviation from Raoult's law*. In a solution of CCl_4 and methanol, for example, the nonpolar CCl_4 molecules interrupt the extensive hydrogen bonding network in methanol, and the lighter methanol molecules have weaker London dispersion forces than the heavier CCl_4 molecules. Consequently, solutions of CCl_4 and methanol exhibit positive deviations from Raoult's law.

EXAMPLE 13.9

For each of the following systems, compare the intermolecular interactions in the pure liquids and in the solution to decide whether the vapor pressure will be higher than that predicted by Raoult's law (positive deviation), approximately equal to that predicted by Raoult's law (an ideal solution), or lower than the pressure predicted by Raoult's law (negative deviation): **(a)** cyclohexane and ethanol; **(b)** methanol and acetone; **(c)** *n*-hexane and isooctane (2,2,4-trimethylpentane).

Given Identity of pure liquids

Asked for Predicted deviation from Raoult's law

Strategy

Identify whether each liquid is polar or nonpolar, and then predict the type of intermolecular interactions that occur in solution.

Solution

(a) Liquid ethanol contains an extensive hydrogen bonding network, and cyclohexane is nonpolar. Because the cyclohexane molecules cannot interact favorably with the polar ethanol molecules, they will disrupt the hydrogen bonding. Hence, the A–B interactions will be weaker than the A–A and B–B interactions, leading to a higher vapor pressure than predicted by Raoult's law (a positive deviation). **(b)** Methanol contains an extensive hydrogen bonding network, but in this case the polar acetone molecules create A–B interactions that are stronger than the A–A or B–B interactions, leading to a negative enthalpy of solution and a lower vapor pressure than predicted by Raoult's law (a negative deviation).

(c) Hexane and isooctane are both nonpolar molecules (isooctane actually has a very small dipole moment, but it is so small that it can be ignored). Hence, the predominant intermolecular forces in both liquids are London dispersion forces. We expect the A–B interactions to be comparable in strength to the A–A and B–B interactions, leading to a vapor pressure in good agreement with that predicted by Raoult's law (an ideal solution).

EXERCISE 13.9

For each of the following systems, compare the intermolecular interactions in the pure liquids with those in the solution to decide whether the vapor pressure will be higher than that predicted by Raoult's law (positive deviation), approximately equal to that predicted by Raoult's law (an ideal solution), or lower than the pressure predicted by Raoult's law (negative deviation): **(a)** benzene and *n*-hexane; **(b)** ethylene glycol and CCl_4.

Answer **(a)** approximately equal; **(b)** positive deviation (vapor pressure higher than predicted)

Boiling-Point Elevation

Note the pattern

The boiling point of a solution that contains a nonvolatile solute is always higher than the boiling point of the pure solvent.

Recall from Chapter 11 that the normal boiling point of a substance is the temperature at which the vapor pressure equals 1 atm. If a nonvolatile solute lowers the vapor pressure of the solvent, it must also affect the boiling point. Because the vapor pressure of the solution at a given temperature is *lower* than the vapor pressure of the pure solvent, achieving a vapor pressure of 1 atm for the solution requires a *higher* temperature than the normal boiling point of the solvent. Thus, *the boiling point of a solution is always higher than that of the pure solvent.* We can see why this must be true by comparing the phase diagram for an aqueous solution with the phase diagram for pure water (Figure 13.16). Note that the vapor pressure of the solution is lower than that of pure water *at all temperatures.* Consequently, the liquid-vapor curve for the solution crosses the horizontal line corresponding to $P = 1$ atm at a higher temperature than does the curve for pure water.

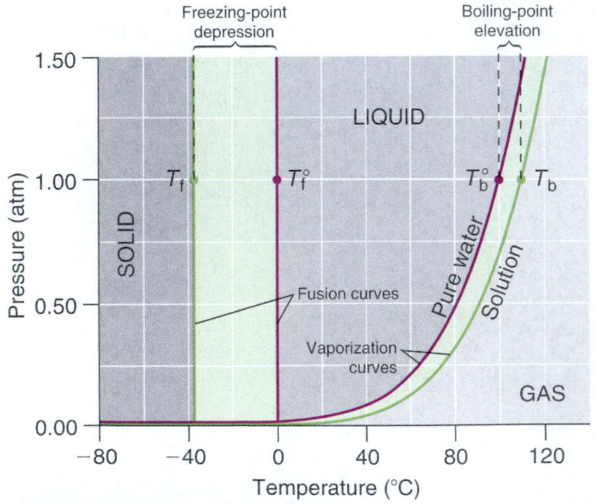

Figure 13.16 Phase diagrams of pure water and an aqueous solution of a nonvolatile solute. Note that the vaporization curve for the solution lies below the curve for pure water at all temperatures, which results in an increase in the boiling point and a decrease in the freezing point of the solution. Interactive Graph

The magnitude of the increase in the boiling point is related to the magnitude of the decrease in the vapor pressure. As we have just discussed, the decrease in the vapor pressure is proportional to the concentration of the solute in the solution. Hence, the magnitude of the increase in the boiling point must also be proportional to the concentration of the solute (Figure 13.17). We can define the **boiling point elevation (ΔT_b)** as the difference between the boiling points of the solution and the pure solvent:

$$\Delta T_b = T_b - T_b^0 \tag{13.18}$$

where T_b is the boiling point of the solution and T_b^0 is the boiling point of the pure solvent. We can express the relationship between ΔT_b and concentration as follows:

$$\Delta T_b = mK_b \tag{13.19}$$

where m is the concentration of the solute expressed in molality, and K_b is the *molal boiling-point elevation constant* of the solvent, which has units of °C/*m*. Table 13.7 lists characteristic K_b values for several commonly used solvents.

The concentration of the solute is typically expressed as molality rather than mole fraction or molarity for two reasons. First, because the density of a solution changes

TABLE 13.7 Boiling-point elevation constants (K_b) and freezing-point depression constants (K_f) for some solvents

Solvent	Boiling Point, °C	K_b, (°C/*m*)	Freezing Point, °C	K_f, (°C/*m*)
Acetic acid	117.90	3.22	16.64	3.63
Benzene	80.09	2.64	5.49	5.07
d-(+)-Camphor	207.4	4.91	178.8	37.8
Carbon disulfide	46.2	2.42	−112.1	3.74
Carbon tetrachloride	76.8	5.26	−22.62	31.4
Chloroform	61.17	3.80	−63.41	4.60
Nitrobenzene	210.8	5.24	5.70	6.87
Water	100.00	0.51	0.00	1.86

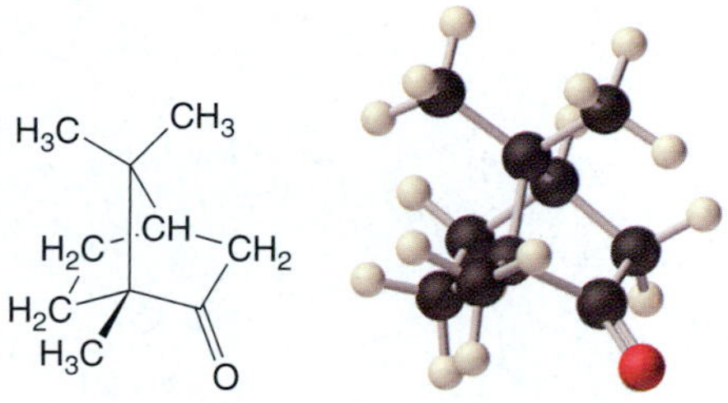

d-(+)-Camphor

with temperature, the value of molarity also varies with temperature. If the boiling point depends on the solute concentration, then by definition the system is not maintained at a constant temperature. Second, molality and mole fraction are proportional for relatively dilute solutions, but molality has a larger numerical value (a mole fraction can be only between zero and one). Using molality allows us to eliminate nonsignificant zeros.

According to Table 13.7, the molal boiling-point elevation constant for water is 0.51°C/*m*. Thus, a 1.00 *m* aqueous solution of a nonvolatile molecular solute such as glucose or sucrose will have an increase in boiling point of 0.51°C, to give a boiling point of 100.51°C at 1.00 atm. The increase in the boiling point of a 1.00 *m* aqueous NaCl solution will be approximately twice as large as that of the glucose or sucrose solution because 1 mol of NaCl produces 2 mol of dissolved ions. Hence, a 1.00 *m* NaCl solution will have a boiling point of about 101.02°C.

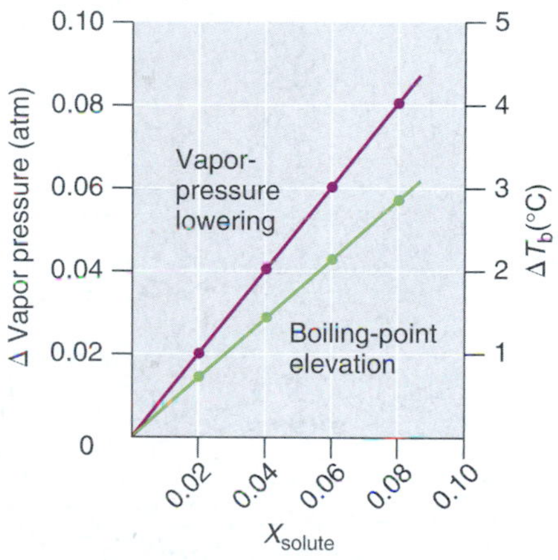

Figure 13.17 Vapor-pressure decrease and boiling-point increase as a function of the mole fraction of a nonvolatile solute. For relatively dilute solutions, the magnitude of both properties is proportional to the solute concentration. Interactive Graph

EXAMPLE 13.10

In Example 13.8, we calculated that the vapor pressure of a 30.2% aqueous solution of ethylene glycol at 100°C is 85.1 mmHg less than the vapor pressure of pure water. We stated (without offering proof) that this should result in a higher boiling point for the solution compared with pure water. Now that we have seen why this assertion is correct, calculate the boiling point of the aqueous ethylene glycol solution.

Given Composition of solution

Asked for Boiling point

Strategy

Calculate the molality of ethylene glycol in the 30.2% solution. Then use Equation 13.19 to calculate the increase in boiling point.

Solution

From Example 13.8, we know that a 30.2% solution of ethylene glycol in water contains 302 g of ethylene glycol (4.87 mol) per 698 g of water. The molality of the solution is thus

$$\text{molality of ethylene glycol} = \left(\frac{4.87\ \text{mol}}{698\ \cancel{\text{g}}\ \text{H}_2\text{O}}\right)\left(\frac{1000\ \cancel{\text{g}}}{1\ \text{kg}}\right) = 6.98\ m$$

From Equation 13.19, the increase in boiling point is therefore

$$\Delta T_b = mK_b = (6.98\ \cancel{m})(0.51°\text{C}/\cancel{m}) = 3.6°\text{C}$$

The boiling point of the solution is thus predicted to be 104°C. With a solute concentration of almost 7 m, however, the assumption of a dilute solution used to obtain Equation 13.19 may not be valid.

EXERCISE 13.10

Assume that a tablespoon (5.00 g) of NaCl is added to 2.00 L of water at 20.0°C, which is then brought to a boil to cook spaghetti. At what temperature will the water boil?

Answer 100.04°C, or 100°C to three significant figures. (Recall that 1 mol of NaCl produces 2 mol of dissolved particles. The small increase in temperature means that adding salt to the water used to cook pasta has essentially no effect on the cooking time.)

Freezing-Point Depression

The phase diagram in Figure 13.16 shows that dissolving a nonvolatile solute in water not only raises the boiling point of the water but also lowers its freezing point. The solid–liquid curve for the solution crosses the line corresponding to $P = 1$ atm at a *lower* temperature than the curve for pure water.

We can understand this result by imagining that we have a sample of water at the normal freezing-point temperature, where there is a dynamic equilibrium between solid and liquid. That is, water molecules are continuously colliding with the ice surface and entering the solid phase at the same rate that water molecules are leaving the surface of the ice and entering the liquid phase. If we dissolve a nonvolatile solute such as glucose in the liquid, the dissolved glucose molecules will reduce the number of collisions per unit time between water molecules and the ice surface because some of the molecules colliding with the ice will be glucose. Glucose, though, has a very different structure than water, and it cannot fit into the ice lattice. Consequently, the presence of glucose molecules in the solution can only decrease the rate at which water molecules in the liquid collide with the ice surface and solidify. Meanwhile, the rate at which the water molecules leave the surface of the ice and enter the liquid phase is unchanged. The net effect is to cause the ice to melt. The only way to reestablish a dynamic equilibrium between solid and liquid water is to lower the temperature of the system, which decreases the rate at which water molecules leave the surface of the ice crystals until it equals the rate at which water molecules in the solution collide with the ice.

By analogy to our treatment of boiling-point elevation, the **freezing-point depression (ΔT_f)** is defined as the difference between the freezing point of the pure solvent and the freezing point of the solution:

$$\Delta T_f = T_f^0 - T_f \qquad (13.20)$$

where T_f^0 is the freezing point of the pure solvent and T_f is the freezing point of the solution. Note that the order of the terms is reversed compared with Equation 13.18 in order to express the freezing-point depression as a positive number. The relationship between ΔT_f and the solute concentration is given by an equation analogous to Equation 13.19:

$$\Delta T_f = mK_f \qquad (13.21)$$

where m is the molality of the solution and K_f is the *molal freezing-point depression constant* for the solvent (in units of °C/m). Like K_b, each solvent has a characteristic value of K_f (see Table 13.7). As with boiling-point elevation, freezing-point depression depends on the total number of dissolved non-volatile solute particles. Thus, an aqueous NaCl solution has twice as large a freezing-point depression as a glucose solution of the same molality.

People who live in cold climates use freezing-point depression to their advantage in many ways. For example, salt is used to melt ice and snow on roads and sidewalks, ethylene glycol is added to automotive antifreeze to prevent the car motor from being destroyed, and methanol is added to windshield washer fluid to prevent the fluid from freezing.

Freezing plant and animal tissues produces ice crystals that rip cells apart, causing severe frostbite and degrading the quality of fish or meat. Consequently, we might expect that exposing any living organism to freezing temperatures would be fatal. In fact, however, many organisms that live in cold climates are able to survive at temperatures well below freezing by synthesizing their own chemical antifreeze in concentrations that prevent freezing. Such substances are typically small organic molecules with multiple —OH groups analogous to ethylene glycol.

Note the pattern

The decrease in vapor pressure, increase in boiling point, and decrease in freezing point of a solution versus a pure liquid all depend on the total number of dissolved non-volatile solute particles.

EXAMPLE 13.11

In colder regions of the United States, NaCl or $CaCl_2$ is often sprinkled on icy roads in winter to melt the ice and make driving safer. Use the data in Figure 13.9 to estimate the concentrations of two saturated solutions at 0°C, one of NaCl and one of $CaCl_2$, and calculate the freezing points of both solutions to see which salt is likely to be more effective at melting ice.

Given Solubilities of two compounds

Asked for Concentrations and freezing points

Strategy

A Estimate the solubility of each salt in 100 g of water from Figure 13.9. Determine the number of moles of each in 100 g, and calculate the molalities.

B Determine the concentrations of the dissolved species in the solutions. Substitute these values into Equation 13.21 to calculate the freezing-point depressions of the solutions.

Solution

A From Figure 13.9, we can estimate the solubility of NaCl and $CaCl_2$ to be about 36 g and 60 g per 100 g of water at 0°C, respectively. The corresponding concentrations in molality are

$$m_{\text{NaCl}} = \left(\frac{36\ \cancel{\text{g NaCl}}}{100\ \cancel{\text{g}}\ \text{H}_2\text{O}}\right)\left(\frac{1\ \text{mol NaCl}}{58.44\ \cancel{\text{g NaCl}}}\right)\left(\frac{1000\ \cancel{\text{g}}}{1\ \text{kg}}\right) = 6.2\ m$$

$$m_{\text{CaCl}_2} = \left(\frac{60\ \cancel{\text{g CaCl}_2}}{100\ \cancel{\text{g}}\ \text{H}_2\text{O}}\right)\left(\frac{1\ \text{mol CaCl}_2}{110.98\ \cancel{\text{g CaCl}_2}}\right)\left(\frac{1000\ \cancel{\text{g}}}{1\ \text{kg}}\right) = 5.4\ m$$

The lower formula mass of NaCl more than compensates for its lower solubility, resulting in a saturated solution that has a slightly higher concentration than $CaCl_2$. **B** Because these salts are ionic compounds that dissociate in water to yield two and three ions per formula unit of NaCl and $CaCl_2$, respectively, the actual concentrations of the dissolved species in the two saturated solutions are $2 \times 6.2\ m = 12\ m$ for NaCl and $3 \times 5.4\ m = 16\ m$ for $CaCl_2$. The resulting freezing-point depressions can be calculated using Equation 13.21:

$$\text{NaCl:}\quad \Delta T_f = mK_f = (12\ \cancel{m})(1.86°\text{C}/\cancel{m}) = 22°\text{C}$$

$$\text{CaCl}_2\text{:}\quad \Delta T_f = mK_f = (16\ \cancel{m})(1.86°\text{C}/\cancel{m}) = 30°\text{C}$$

Because the freezing point of pure water is 0°C, the actual freezing points of the solutions are −22°C and −30°C, respectively. Note that $CaCl_2$ is substantially more effective at lowering the freezing point of water because its solutions contain three ions per formula unit. In fact, $CaCl_2$ is the salt usually sold for home use, and it is also often used on highways.

Because the solubility of both salts decreases with decreasing temperature, the freezing point can be depressed by only a certain amount, regardless of how much salt is spread on an icy road. If the temperature is significantly below the minimum temperature at which one of these salts will cause ice to melt (say −35°C), there is no point in using salt until it gets warmer.

EXERCISE 13.11

Calculate the freezing point of the 30.2% solution of ethylene glycol in water whose vapor pressure and boiling point we calculated in Examples 13.8 and 13.10.

Answer −13.0°C

EXAMPLE 13.12

Arrange these aqueous solutions in order of decreasing freezing points: 0.1 *m* KCl, 0.1 *m* glucose, 0.1 *m* $SrCl_2$, 0.1 *m* ethylene glycol, 0.1 *m* benzoic acid, 0.1 *m* HCl.

Given Molalities of six solutions

Asked for Relative freezing points

Strategy

Ⓐ Identify each solute as a strong, weak, or nonelectrolyte, and use this information to determine the actual number of solute particles produced.

Ⓑ Multiply this number by the concentration of the solution to obtain the effective concentration of solute particles. The solution with the highest effective concentration of solute particles has the largest freezing-point depression.

Solution

Ⓐ✓ Because the molal concentrations of all six solutions are the same, we must focus on which of the substances are strong electrolytes, which are weak electrolytes, and which are nonelectrolytes in order to determine the actual numbers of particles in solution. KCl, $SrCl_2$, and HCl are strong electrolytes, producing two, three, and two ions per formula unit, respectively. Benzoic acid is a weak electrolyte (approximately one particle per molecule), and glucose and ethylene glycol are both nonelectrolytes (one particle per molecule). Ⓑ✓ The actual molalities of the solutions in terms of the total particles of solute are therefore: KCl and HCl, 0.2 *m*; $SrCl_2$, 0.3 *m*; glucose and ethylene glycol, 0.1 *m*; and benzoic acid, between 0.1 and 0.2 *m*. Because the magnitude of the decrease in freezing point is proportional to the concentration of dissolved particles, the order of freezing points of the solutions is: glucose and ethylene glycol (highest freezing point, smallest freezing-point depression) > benzoic acid > HCl = KCl > $SrCl_2$.

EXERCISE 13.12

Arrange these aqueous solutions in order of increasing freezing points: 0.2 *m* NaCl, 0.3 *m* acetic acid, 0.1 *m* $CaCl_2$, 0.2 *m* sucrose.

Answer 0.2 *m* NaCl (lowest freezing point) < 0.3 *m* acetic acid ≈ 0.1 *m* $CaCl_2$ < 0.2 *m* sucrose (highest freezing point)

Colligative properties can also be used to determine the molar mass of an unknown compound. One method that can be carried out in the laboratory with minimal equipment is to measure the freezing point of a solution that contains a known mass of solute, which is accurate for dilute solutions (≤1% by mass). Changes in the freezing point are usually large enough to measure accurately and precisely, even for dilute solutions of small molecules. By comparing K_b and K_f values in Table 13.7, we see that changes in the boiling point are smaller than changes in the freezing point for a given solvent. Boiling point elevations are thus more difficult to measure precisely. For this reason, freezing-point depression is more commonly used to determine molar mass than is boiling-point elevation. Because of its very large value of K_f (37.8°C/*m*), *d*-(+)-camphor (Table 13.7) is often used to determine the molar mass of organic compounds by this method.

EXAMPLE 13.13

A 7.08 g sample of elemental sulfur is dissolved in 75.0 g of CS_2 to give a solution whose freezing point is −112.90°C. Use these data to calculate the molar mass of elemental sulfur and thus the formula of the dissolved S_n molecules (that is, what is the value of *n*?).

Given Masses of solute and solvent, freezing point

Asked for Molar mass and number of S atoms per molecule

Strategy

A Use Equation 13.20, the measured freezing point of the solution, and the freezing point of CS_2 from Table 13.7 to calculate the freezing-point depression. Then use Equation 13.21 and the value of K_f from Table 13.7 to calculate the molality of the solution.
B From the calculated molality, determine the number of moles of solute present.
C Use the mass and number of moles of the solute to calculate the molar mass of sulfur in solution. Divide the result by the molar mass of atomic sulfur to obtain *n*, the number of sulfur atoms per mole of dissolved sulfur.

Solution

A The first step is to calculate the freezing-point depression using Equation 13.20:

$$\Delta T_f = T_f^0 - T_f = -111.5°\text{C} - (-112.90°\text{C}) = 1.4°\text{C}$$

Then Equation 13.21 gives

$$m = \frac{\Delta T_f}{K_f} = \frac{1.4°\cancel{\text{C}}}{3.74°\cancel{\text{C}}/m} = 0.37\ m$$

B The total number of moles of solute present in the solution is thus

$$\text{moles solute} = \left(\frac{0.37\ \text{mol}}{\cancel{\text{kg}}}\right)(75.0\ \text{g})\left(\frac{1\ \cancel{\text{kg}}}{1000\ \text{g}}\right) = 0.028\ \text{mol}$$

C We now know that 0.708 g of elemental sulfur corresponds to 0.028 mol of solute. The molar mass of dissolved sulfur is thus

$$\text{molar mass} = \frac{7.08\ \text{g}}{0.028\ \text{mol}} = 260\ \text{g/mol}$$

The molar mass of atomic sulfur is 32 g/mol, so there must be 260/32 = 8.1 sulfur atoms per mole, corresponding to a formula of S_8.

EXERCISE 13.13

One of the byproducts formed during the synthesis of C_{60} is a deep red solid that contains only carbon. A solution of 205 mg of this compound in 10.0 g of CCl_4 has a freezing point of −23.38°C. What are the molar mass and most probable formula of the substance?

Answer 847 g/mol; C_{70}

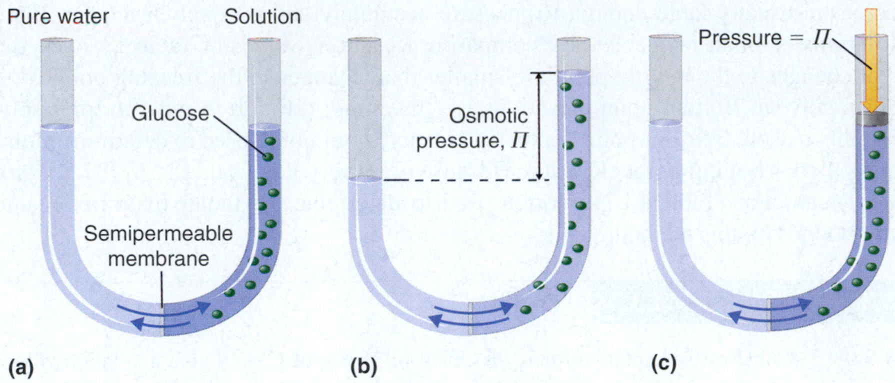

Figure 13.18 Osmotic pressure. (**a**) A dilute solution of glucose in water is placed in the right arm of a U-tube, and the left arm is filled to the same height with pure water; the two arms are separated by a semipermeable membrane. Because the flow of pure solvent through the membrane from left to right (from pure water to the solution) is greater than the flow of solvent in the reverse direction, the level of liquid in the right tube rises. (**b**) At equilibrium, the pressure differential, equal to the osmotic pressure of the solution (π_{soln}), equalizes the flow rate of solvent in both directions. (**c**) Applying an external pressure equal to the osmotic pressure of the original glucose solution to the liquid in the right arm reverses the flow of solvent and restores the original situation.

Osmotic Pressure

Osmotic pressure is a colligative property of solutions that is observed using a **semipermeable membrane**, a barrier with pores small enough to allow solvent molecules to pass through, but not solute molecules or ions. The net flow of solvent through a semipermeable membrane is called **osmosis** (from the Greek *osmós*, "push"). The direction of net solvent flow is *always* from the side with the lower concentration of solute to the side with the higher concentration.

Osmosis can be demonstrated using a U-tube like the one shown in Figure 13.18, which contains pure water in the left arm and a dilute aqueous solution of glucose in the right arm. A net flow of water through the membrane occurs until the levels in the arms eventually stop changing, which indicates that equilibrium has been reached. The **osmotic pressure** of the glucose solution is the difference in the pressure between the two sides, in this case the heights of the two columns. Although the semipermeable membrane allows water molecules to flow through in either direction, the *rate* of flow is not the same in both directions because the concentration of water is not the same in the two arms. The net flow of water through the membrane can be prevented by applying a pressure to the right arm that is equal to the osmotic pressure of the glucose solution.

Just as with any other colligative property, the osmotic pressure of a solution depends on the concentration of dissolved solute particles. Abbreviated as Π, osmotic pressure obeys a law that resembles the ideal gas equation:

$$\Pi = \frac{nRT}{V} = MRT \tag{13.22}$$

where M is the number of moles of solute per unit volume of solution (that is, the *molarity* of the solution), R is the ideal gas constant, and T is the absolute temperature. As shown in the next example, osmotic pressures tend to be quite high, even for rather dilute solutions.

EXAMPLE 13.14

When placed in a concentrated salt solution, certain yeasts are able to produce high internal concentrations of glycerol to counteract the osmotic pressure of the surrounding medium. Suppose that the yeast cells are placed in an aqueous solution that contains 4.0% NaCl by mass; the solution density is 1.02 g/mL at 25°C. (a) Calculate the osmotic pressure of a 4.0% aqueous NaCl solution at 25°C. (b) If the normal osmotic pressure inside a yeast cell is 7.3 atm, corresponding to a total concentration of dissolved particles of 0.30 *M*, what concentration of glycerol must the cells synthesize in order to exactly balance the external osmotic pressure at 25°C?

Given Concentration, density, and temperature of NaCl solution; internal osmotic pressure of cell

Asked for Osmotic pressure of NaCl solution; concentration of glycerol needed

Strategy

A Calculate the molarity of the NaCl solution using the formula mass of the solute and the density of the solution. Then calculate the total concentration of dissolved particles.
B Use Equation 13.22 to calculate the osmotic pressure of the solution.
C Subtract the normal osmotic pressure of the cells from the osmotic pressure of the salt solution to obtain the additional pressure needed to balance the two. Use Equation 13.22 to calculate the molarity of glycerol needed to create this osmotic pressure.

Solution

(a) A The solution contains 4.0 g of NaCl per 100 g of solution. Using the formula mass of NaCl (58.44 g/mol) and the density of the solution (1.02 g/mL), we can calculate the molarity:

$$M_{\text{NaCl}} = \frac{\text{moles NaCl}}{\text{liter solution}} = \left(\frac{4.0\ \cancel{\text{g}}\ \text{NaCl}}{58.44\ \cancel{\text{g}}/\text{mol NaCl}}\right)\left(\frac{1}{100\ \cancel{\text{g solution}}}\right)\left(\frac{1.02\ \cancel{\text{g solution}}}{1.00\ \cancel{\text{mL}}\ \text{solution}}\right)\left(\frac{1000\ \cancel{\text{mL}}}{\text{L}}\right) = 0.70\ M\ \text{NaCl}$$

Because 1 mol of NaCl produces 2 mol of particles in solution, the total concentration of dissolved particles in the solution is (2)(0.70 *M*) = 1.4 *M*.

B Now we can use Equation 13.22 to calculate the osmotic pressure of the solution:

$$\Pi = MRT = (1.4\ \cancel{\text{mol}/\text{L}})[0.0821\ (\cancel{\text{L}} \cdot \text{atm})/(\cancel{\text{K}} \cdot \cancel{\text{mol}})](298\ \cancel{\text{K}}) = 34\ \text{atm}$$

(b) C If the yeast cells are to exactly balance the external osmotic pressure, they must produce enough glycerol to give an additional internal pressure of (34 atm − 7.3 atm) = 27 atm. Glycerol is a nonelectrolyte, so we can solve Equation 13.22 for the molarity corresponding to this osmotic pressure:

$$M = \frac{\Pi}{RT} = \frac{27\ \cancel{\text{atm}}}{[0.0821\ (\text{L} \cdot \cancel{\text{atm}})/(\cancel{\text{K}} \cdot \text{mol})](298\ \cancel{\text{K}})} = 1.1\ M\ \text{glycerol}$$

In solving this problem, we could also have recognized that the only way the osmotic pressures can be the same inside the cells and in the solution is if the concentrations of dissolved particles are the same. We are given that the normal concentration of dissolved particles in the cells is 0.3 *M*, and we calculated that the NaCl solution is effectively 1.4 *M* in dissolved particles. The yeast cells must therefore synthesize enough glycerol to increase the internal concentration of dissolved particles from 0.3 *M* to 1.4 *M*—that is, an additional 1.1 *M* concentration of glycerol.

EXERCISE 13.14

Assume that the fluids inside a sausage are approximately 0.80 *M* in dissolved particles due to the salt and sodium nitrite used to prepare them. Calculate the osmotic pressure inside the sausage at 100°C to learn why experienced cooks pierce the semipermeable skin of sausages *before* boiling them.

Answer 24 atm

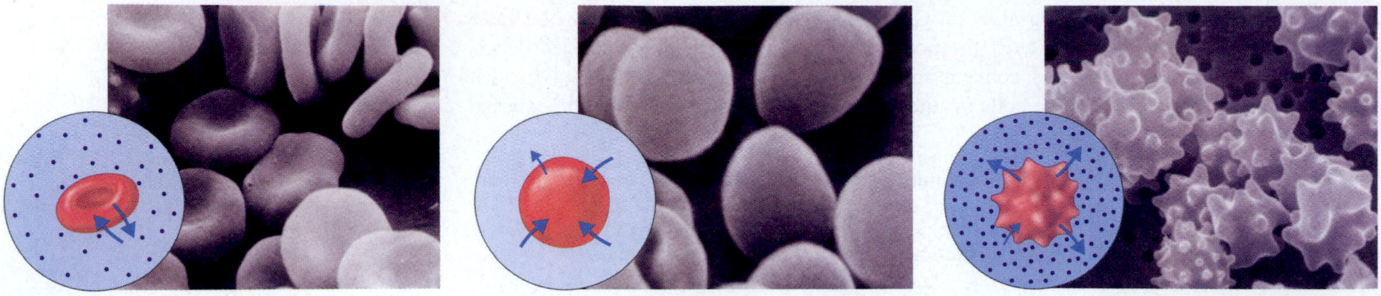

Figure 13.19 Effect of the osmotic pressure of the surrounding solution on red blood cells. (**a**) When red blood cells are placed in dilute salt solution, whose osmotic pressure is the same as that of the intracellular fluid, the rate of flow of water into and out of the cells is the same, and their shape does not change. (**b**) If cells are placed in distilled water, whose osmotic pressure is lower than that of the intracellular fluid, the rate of flow of water into the cells is greater than the rate of flow out of the cells. The cells swell and eventually burst. (**c**) If cells are placed in a concentrated salt solution, whose osmotic pressure is higher than that of the intracellular fluid, the rate of flow of water out of the cells is greater than the rate of flow into the cells. The cells shrivel and become so deformed that they cannot function.

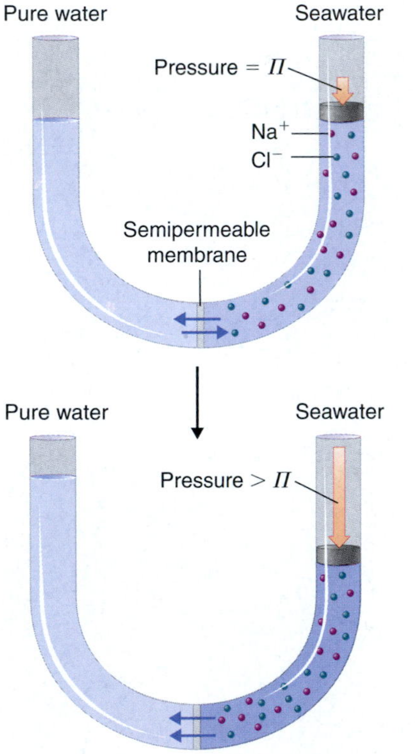

Figure 13.20 Desalinization of seawater by reverse osmosis. (*Top*) When the pressure applied to seawater equals its osmotic pressure (Π_{soln}), there is no net flow of water across the semipermeable membrane. (*Bottom*) The application of pressure higher than the osmotic pressure of seawater forces water molecules to flow through the membrane, leaving behind a concentrated salt solution. In desalinization plants, seawater is continuously introduced under pressure and pure water is collected, so the process continues indefinitely.

Because of the large magnitude of osmotic pressures, osmosis is extraordinarily important in biochemistry, biology, and medicine. Virtually every barrier that separates an organism or cell from its environment acts like a semipermeable membrane, permitting the flow of water but not solutes. The same is true of the compartments inside an organism or cell. Some specialized barriers, such as those in your kidneys, are slightly more permeable and use a related process called **dialysis**, which permits both water and small molecules to pass through, but not large molecules such as proteins.

The same principle has long been used to preserve fruits and their essential vitamins over the long winter. High concentrations of sugar are used in jams and jellies not for sweetness alone but because they greatly increase the osmotic pressure. Thus, any bacteria not killed in the cooking process are dehydrated, which keeps them from multiplying in an otherwise rich medium for bacterial growth. A similar process using salt prevents bacteria from growing in ham, bacon, salt pork, salt cod, and other preserved meats. The effect of osmotic pressure is dramatically illustrated in Figure 13.19, which shows what happens when red blood cells are placed in a solution whose osmotic pressure is much lower or much higher than the internal pressure of the cells.

Trees use osmotic pressure to transport water and other nutrients from the roots to the upper branches. Evaporation of water from the leaves results in a local increase in the salt concentration, which generates an osmotic pressure that pulls water up the trunk of the tree to the leaves.

Finally, a process called *reverse osmosis* can be used to produce pure water from seawater. As shown schematically in Figure 13.20, applying high pressure to seawater forces water molecules to flow through a semipermeable membrane that separates pure water from the solution, leaving the dissolved salt behind. Large-scale desalinization plants that can produce hundreds of thousands of gallons of freshwater per day are common in the desert lands of the Middle East, where they supply a large proportion of the freshwater needed by the population. Similar facilities are now being used to supply freshwater in southern California. Small, hand-operated reverse osmosis units can produce approximately 5 L of freshwater per hour, enough to keep 25 people alive, and are now standard equipment on U.S. Navy lifeboats.

Colligative Properties of Electrolyte Solutions

Thus far we have assumed that we could simply multiply the molar concentration of a solute by the number of ions per formula unit to obtain the actual concentration of dissolved particles in an electrolyte solution. We have used this simple model to predict such properties as freezing points, melting points, vapor pressure, and osmotic pressure. If this model were perfectly correct, we would expect the freezing-point depression of a 0.10 *m* solution of sodium chloride, with 2 mol of ions per mole of NaCl in solution, to be exactly twice that of a 0.10 *m* solution of glucose, with only 1 mol of molecules per mole of glucose in solution. In reality, this is not always the case. The observed change in freezing points for 0.10 *m* aqueous solutions of NaCl and KCl are significantly less than expected (−0.348°C and −0.344°C, respectively, rather than −0.372°C), which suggests that fewer particles than expected are present in solution.

The relationship between the actual number of moles of solute added to form a solution and the apparent number as determined by colligative properties is called the **van't Hoff factor (*i*)** and is defined as*

$$i = \frac{\text{apparent number of particles in solution}}{\text{number of moles of solute dissolved}} \tag{13.23}$$

As the data in Table 13.8 show, the van't Hoff factors for ionic compounds are somewhat lower than expected; that is, their solutions apparently contain fewer particles than predicted by the number of ions per formula unit. The deviation from the expected value increases as the concentration of the solute increases because ionic compounds generally do *not* totally dissociate in aqueous solution. Instead, some of the ions exist as **ion pairs**, a cation and an anion that for a brief time are associated with each other without an intervening shell of water molecules (Figure 13.21). Each of these temporary units behaves like a single dissolved particle until it dissociates. Highly charged ions such as Mg^{2+}, Al^{3+}, SO_4^{2-}, and PO_4^{3-} have a greater tendency to form ion pairs because of their strong electrostatic interactions. The actual number of solvated ions present in a solution can be determined by measuring a colligative property at several solute concentrations.

TABLE 13.8 van't Hoff factors for 0.0500 *M* aqueous solutions of selected compounds at 25°C

Compound	*i* (measured)	*i* (ideal)
Glucose	1.0	1.0
Sucrose	1.0	1.0
NaCl	1.9	2.0
HCl	1.9	2.0
$MgCl_2$	2.7	3.0
$AlCl_3$	3.2	4.0
$FeCl_3$	3.4	4.0
$MgSO_4$	1.4	2.0
$Ca(NO_3)_2$	2.5	3.0

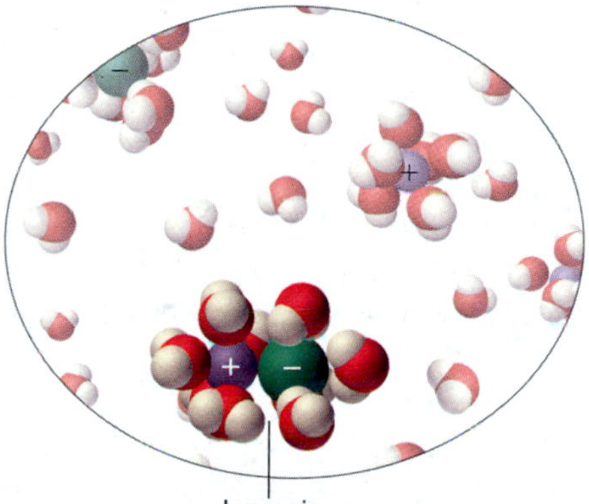

Figure 13.21 Ion pairs. In concentrated solutions of electrolytes like NaCl, some of the ions form neutral ion pairs that are not separated by solvent and diffuse as single particles.

EXAMPLE 13.15

A 0.0500 *M* aqueous solution of $FeCl_3$ has an osmotic pressure of 4.15 atm at 25°C. Calculate the van't Hoff factor *i* for the solution.

Given Solute concentration, osmotic pressure, and temperature

Asked for van't Hoff factor

Strategy

Ⓐ Use Equation 13.22 to calculate the expected osmotic pressure of the solution based on the effective concentration of dissolved particles in the solvent.

Ⓑ Calculate the ratio of the observed osmotic pressure to the expected value. Multiply this number by the number of ions of solute per formula unit, and then use Equation 13.23 to calculate the van't Hoff factor.

Solution

Ⓐ If $FeCl_3$ dissociated completely in aqueous solution, it would produce four ions per formula unit [$Fe^{3+}(aq)$ plus 3 $Cl^-(aq)$] for an effective concentration of dissolved particles of 4 × 0.0500 *M* = 0.200 *M*. The osmotic pressure would be

$$\Pi = MRT = (0.200\ \cancel{\text{mol}}/\cancel{\text{L}})[0.0821\ (\cancel{\text{L}} \cdot \text{atm})/(\cancel{\text{K}} \cdot \cancel{\text{mol}})](298\ \cancel{\text{K}}) = 4.89\ \text{atm}$$

* Named for Jacobus Hendricus van't Hoff (1852–1911), a Dutch chemistry professor at the University of Amsterdam who won the first Nobel Prize in chemistry (1901) for his work on thermodynamics and solutions.

B The observed osmotic pressure is only 4.15 atm, presumably due to ion pair formation. The ratio of the observed osmotic pressure to the calculated value is 4.15 atm/4.89 atm = 0.849, which indicates that the solution contains (0.849)(4) = 3.40 particles per mole of $FeCl_3$ dissolved. Alternatively, we can calculate the observed particle concentration from the osmotic pressure of 4.15 atm:

$$4.15\ \cancel{\text{atm}} = M\,[0.0821\ (\text{L}\cdot\cancel{\text{atm}})/(\cancel{\text{K}}\cdot\text{mol})](298\ \cancel{\text{K}})$$
$$M = 0.170\ \text{mol/L}$$

The ratio of this value to the expected value of 0.200 M is 0.170 M/0.200 M = 0.850, which again gives us (0.850)(4) = 3.40 particles per mole of $FeCl_3$ dissolved. From Equation 13.23, the van't Hoff factor for the solution is

$$i = \frac{3.39\ \text{particles observed}}{1\ \text{formula unit FeCl}_3} = 3.39$$

EXERCISE 13.15

Calculate the van't Hoff factor for a 0.050 m aqueous solution of $MgCl_2$ that has a measured freezing point of −0.25°C.

Answer 2.7

13.6 • Aggregate Particles in Aqueous Solution

MGC Solutions, Suspensions, and Colloids

Suspensions and colloids are two common types of mixtures whose properties are in many ways intermediate between those of true solutions and heterogeneous mixtures. A **suspension** is a heterogeneous mixture of particles with diameters of about 1 μm (1000 nm) that are distributed throughout a second phase. Common suspensions include paint, blood, and hot chocolate, which are solid particles in a liquid, and aerosol sprays, which are liquid particles in a gas. If the suspension is allowed to stand, the two phases will separate, which is why paints must be thoroughly stirred or shaken before use. A **colloid** is also a heterogeneous mixture, but the particles of a colloid are typically smaller than those of a suspension, generally in the range of 2 to about 500 nm in diameter. Colloids include fog and clouds (liquid particles in a gas), milk (solid particles in a liquid), and butter (solid particles in a solid). Unlike a suspension, the particles in a colloid do *not* separate into two phases on standing. The only combination of substances that cannot produce a suspension or a colloid is a mixture of two gases because their particles are so small that they always form true solutions. The properties of suspensions, colloids, and solutions are summarized in Table 13.9.

TABLE 13.9 Properties of liquid solutions, colloids, and suspensions

Type of Mixture	Approximate Size of Particles, nm	Characteristic Properties	Examples
Solution	< 2	Not filterable; does not separate on standing; does not scatter visible light	Air, wine, gasoline, salt water
Colloid	2–500	Scatters visible light; translucent or opaque; not filterable; does not separate on standing	Smoke, fog, ink, milk, butter, cheese
Suspension	500–1000	Cloudy or opaque; filterable; separates on standing	Muddy water, hot cocoa, blood, paint

Colloids and Suspensions

Figure 13.22 Tyndall effect, the scattering of light by colloids. Both cylinders contain a solution of red food coloring in water, but a small amount of gelatin has been added to the cylinder on the right to form a colloidal suspension of gelatin particles. Note that the beam of light goes straight through the true solution on the left, but the light beam is scattered by the colloid on the right.

Colloids were first characterized in about 1860 by Thomas Graham, who also gave us Graham's law of diffusion and effusion. Although some substances, such as starch, gelatin, and glue, appear to dissolve in water to produce solutions, Graham found that they diffuse very slowly or not at all compared with solutions of substances such as salt and sugar. Graham coined the word *colloid* (from the Greek *kólla*, "glue") to describe these substances, as well as the words *sol* and *gel* to describe certain types of colloids in which all of the solvent has been absorbed by the solid particles, thus preventing the mixture from flowing readily, as we see in Jell-O®. Two other important types of colloids are **aerosols**, which are dispersions of solid or liquid particles in a gas, and **emulsions**, which are dispersions of one liquid in another liquid with which it is immiscible.

Colloids share many properties with solutions. For example, the particles in both are invisible without a powerful microscope, do not settle on standing, and pass through most filters. However, the particles in a colloid scatter a beam of visible light, a phenomenon known as the **Tyndall effect**,* whereas the particles of a solution do not. The Tyndall effect is responsible for the way the beams from automobile headlights are clearly visible from the side on a foggy night but cannot be seen from the side on a clear night. It is also responsible for the colored rays of light seen in many sunsets, where the sun's light is scattered by water droplets and dust particles high in the atmosphere. An example of the Tyndall effect is shown in Figure 13.22.

Although colloids and suspensions can have particles similar in size, the two differ in stability: the particles of a colloid remain dispersed indefinitely unless the temperature or chemical composition of the dispersing medium is changed. The chemical explanation for the stability of colloids depends on whether the colloidal particles are hydrophilic or hydrophobic.

Most proteins, including those responsible for the properties of gelatin and glue, are hydrophilic because their exterior surface is largely covered with polar or charged groups. Starch, a long-branched polymer of glucose molecules, is also hydrophilic. A hydrophilic colloid particle interacts strongly with water, resulting in a shell of tightly bound water molecules that prevents the particles from aggregating when they collide. Heating such a colloid can cause aggregation because the particles collide with greater energy and disrupt the protective shell of solvent. Moreover, heat causes protein structures to unfold, exposing previously buried hydrophobic groups that can now interact with other hydrophobic groups and cause the particles to aggregate and precipitate from solution. When an egg is boiled, for example, the egg white, which is primarily a colloidal suspension of a protein called *albumin*, unfolds and exposes its hydrophobic groups, which aggregate and cause the albumin to precipitate as a white solid.

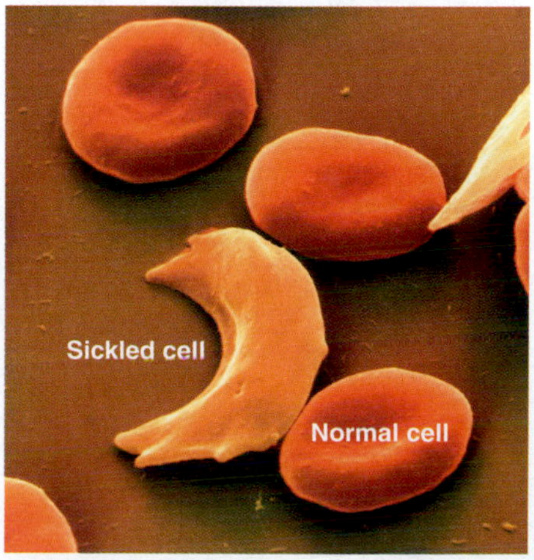

Figure 13.23 Sickle-cell anemia. The characteristic shape of sickled red blood cells is the result of fibrous aggregation of hemoglobin molecules inside the cell.

In some cases, a stable colloid can be transformed to an aggregated suspension by a minor chemical modification. Consider, for example, the behavior of hemoglobin, a major component of red blood cells. Hemoglobin molecules normally form a colloidal suspension inside red blood cells, which typically have a "donut" shape and are easily deformed, allowing them to squeeze through the capillaries to deliver oxygen to tissues. In a common inherited disease called sickle-cell anemia, one of the amino acids in hemoglobin that has a hydrophilic carboxylic acid side chain (glutamate) is replaced by another amino acid that has a hydrophobic side chain (valine). Under some conditions, the abnormal hemoglobin molecules can aggregate to form long, rigid fibers that cause the red blood cells to deform, adopting a characteristic sickle shape that prevents them from passing through the capillaries (Figure 13.23). The reduction in blood flow

* The effect is named after its discoverer, John Tyndall, an English physicist (1820–1893).

Figure 13.24 Formation of new land by the destabilization of a colloid suspension. This satellite photograph shows the Mississippi River delta from New Orleans (*top*) to the Gulf of Mexico (*bottom*). Where seawater mixes with freshwater from the Mississippi River, colloidal clay particles in the river water precipitate (*tan area*).

results in severe cramps, swollen joints, and liver damage. Until recently, many patients with sickle-cell anemia died before the age of 30 from infection, blood clots, or heart or kidney failure, although individuals with the sickle-cell genetic trait are more resistant to malaria than are those with "normal" hemoglobin.

Aggregation and precipitation can also result when the outer, charged layer of a particle is neutralized by ions with the opposite charge. In inland waterways, clay particles, which have a charged surface, form a colloidal suspension. High salt concentrations in seawater neutralize the charge on the particles, causing them to precipitate and form land at the mouths of large rivers, as seen in the satellite view in Figure 13.24. Charge neutralization is also an important strategy for precipitating solid particles from gaseous colloids such as smoke, and it is widely used to reduce particulate emissions from power plants that burn fossil fuels.

Emulsions

Emulsions are colloids formed by the dispersion of a hydrophobic liquid in water, thereby bringing two mutually insoluble liquids, such as oil and water, in close contact. Various agents have been developed to stabilize emulsions, the most successful being molecules that combine a relatively long hydrophobic "tail" with a hydrophilic "head":

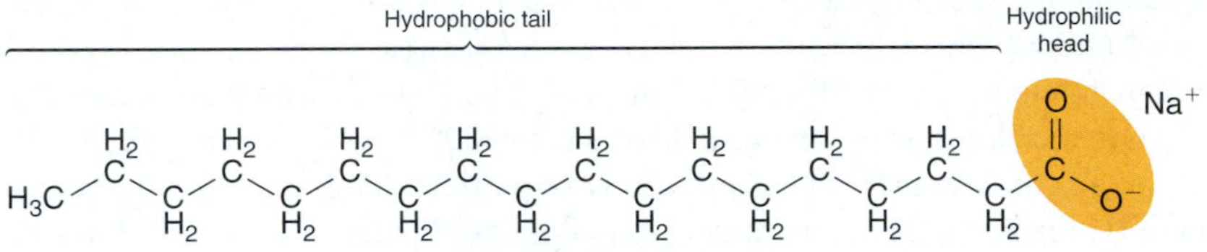

Sodium stearate: a soap

Examples of such emulsifying agents include soaps, which are salts of long-chain carboxylic acids, such as sodium stearate [$CH_3(CH_2)_{16}CO_2^-Na^+$], and detergents, such as sodium dodecyl sulfate [$CH_3(CH_2)_{11}OSO_3^-Na^+$], whose structures are shown in the margin.

Sodium stearate: a soap

Sodium dodecyl sulfate: a detergent

When you wash your laundry, the hydrophobic tails of soaps and detergents interact with hydrophobic particles of dirt or grease through dispersion forces, dissolving in the interior of the hydrophobic particle. The hydrophilic group is then exposed at the surface of the particle, which enables it to interact with water through ion–dipole forces and hydrogen bonding. This causes the particles of dirt or grease to disperse in the wash water and allows them to be removed by rinsing. Similar agents are used in the food industry to stabilize emulsions such as mayonnaise.

A related mechanism allows us to absorb and digest the fats in buttered popcorn and French fries. To solubilize the fats so that they can be absorbed, the gall bladder secretes a fluid called *bile* into the small intestine. Bile contains a variety of *bile salts,* detergent-like molecules that emulsify the fats.

Micelles

Detergents and soaps are surprisingly soluble in water in spite of their hydrophobic tails. The reason for their solubility is that they do not, in fact, form simple solutions. Instead, above a certain concentration, they spontaneously form **micelles**, which are spherical or cylindrical aggregates that minimize contact between the hydrophobic tails and water. In a micelle, only the hydrophilic heads are in direct contact with water, and the hydrophobic tails are in the interior of the aggregate (Figure 13.25a).

A large class of biological molecules called **phospholipids** consist of detergent-like molecules that contain a hydrophilic head and *two* hydrophobic tails, as can be seen in the molecule of phosphatidylcholine shown in the margin. The additional tail

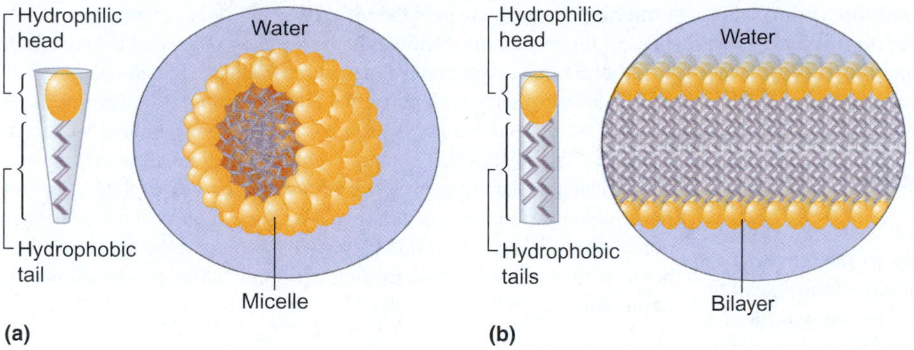

Figure 13.25 Micelles and a phospholipid bilayer. (**a**) Soaps and detergents, which contain a single hydrophobic tail on each molecule, form spherical micelles with the intertwined tails in the interior and the hydrophilic head groups on the exterior. (**b**) Phospholipids, which have two hydrophobic tails, tend to form extended double layers in which the hydrophobic tails are sandwiched between the hydrophilic head groups.

results in a cylindrical shape that prevents phospholipids from forming a spherical micelle. Consequently, phospholipids form *bilayers*, extended sheets consisting of a double layer of molecules. As shown in Figure 13.25b, the hydrophobic tails are in the center of the bilayer, where they are not in contact with water, and the hydrophilic heads are on the two surfaces, in contact with the surrounding aqueous solution.

A **cell membrane** is essentially a mixture of phospholipids that form a phospholipid bilayer. One definition of a **cell** is a collection of molecules surrounded by a phospholipid bilayer that is capable of reproducing itself. The simplest cells are bacteria, which consist of only a single compartment surrounded by a single membrane. Animal and plant cells are much more complex, however, and contain many different kinds of compartments, each surrounded by a membrane and able to carry out specialized tasks.

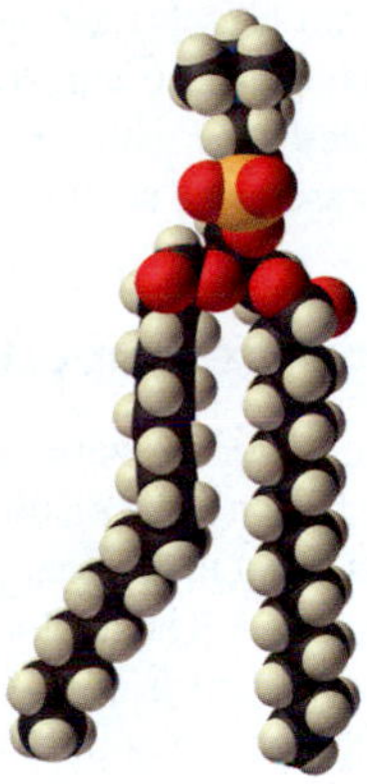

Phosphatidylcholine: a phospholipid

SUMMARY AND KEY TERMS

13.1 Factors Affecting Solution Formation (p. 571)

Solutions are homogeneous mixtures of two or more substances whose components are uniformly distributed on a microscopic scale. The component present in the greatest amount is the *solvent*, and the components present in lesser amounts are the *solute(s)*. The formation of a solution from a solute and a solvent is a physical process, not a chemical one. Substances that are **miscible**, such as gases, form a single phase in all proportions when mixed. Substances that form separate phases are *immiscible*. **Solvation** is the process in which solute particles are surrounded by solvent molecules. When the solvent is water, the process is called **hydration**. The overall enthalpy change that accompanies the formation of a solution, ΔH_{soln}, is the sum of the enthalpy change for breaking the intermolecular interactions in both the solvent and the solute and the enthalpy change for the formation of new solute–solvent interactions. Exothermic ($\Delta H_{\text{soln}} < 0$) processes favor solution formation. In addition, the change in **entropy**, the degree of disorder of the system, must be considered when predicting whether a solution will form. An increase in entropy (a decrease in order) favors dissolution.

13.2 Solubility and Molecular Structure (p. 575)

The **solubility** of a substance is the maximum amount of a solute that can dissolve in a given quantity of solvent; it depends on the chemical nature of both the solute and the solvent and on the temperature and pressure. If a solution contains the maximum amount of solute that can dissolve under a given set of conditions, it is **saturated**. Otherwise, it is *unsaturated*. **Supersaturated** solutions, which contain more dissolved solute than allowed under particular conditions, are not stable; the addition of a seed crystal, a small particle of solute, will usually cause the excess solute to crystallize. A system in which crystallization and dissolution occur at the same rate is in *dynamic equilibrium*. The solubility of a substance in a liquid is determined by intermolecular interactions, which also determine whether two liquids are miscible. Solutes can be classified as **hydrophilic** (water loving) or **hydrophobic** (water fearing). Vitamins with hydrophilic structures are water soluble, whereas those with hydrophobic structures are fat soluble. Many metals dissolve in liquid mercury to form **amalgams**. Covalent network solids and most metals are insoluble in nearly all solvents. The solubility of ionic compounds is largely determined by the **dielectric constant (ε)** of the solvent, a measure of its ability to decrease the electrostatic forces between charged particles. Solutions of many ionic compounds in organic solvents can be dissolved using **crown ethers**, cyclic polyethers large enough to accommodate a metal ion in the center, or **cryptands**, compounds that completely surround a cation.

13.3 Units of Concentration (p. 582)

The **concentration** of a solution is the quantity of solute in a given quantity of solution. It can be expressed in several ways: molarity (moles of solute per liter of solution); mole fraction, the ratio of the number of moles of solute to the total number of moles of substances present; **mass percentage**, the ratio of the mass of the solute to the mass of the solution times 100%; **parts per million (ppm)**, milligrams of solute per kilogram of solution; **parts per billion (ppb)**, micrograms of solute per kilogram of solution; and **molality (*m*)**, the number of moles of solute per kilogram of solvent.

13.4 Effects of Temperature and Pressure on Solubility (p. 587)

The solubility of most substances depends strongly on the temperature and, in the case of gases, on the pressure. The solubility of most solid or liquid solutes increases with increasing temperature. The components of a mixture can often be separated using **fractional crystallization**, which separates compounds according to their relative solubilities. The solubility of gases decreases with increasing temperature. The relationship between the pressure and solubility of a gas is described by **Henry's law**.

13.5 Colligative Properties of Solutions (p. 591)

The **colligative properties** of a solution depend on only the total number of dissolved particles in solution, not on their chemical identity. Colligative properties include vapor pressure, boiling point, freezing point, and osmotic pressure. The addition of a *nonvolatile solute* (one without a measurable vapor pressure) decreases the vapor pressure of the solvent. The vapor pressure of the solution is proportional to the mole fraction of solvent in the solution, a relationship known as **Raoult's law**. Solutions that obey Raoult's law are called **ideal solutions**. Most real solutions exhibit positive or negative deviations from Raoult's law. The **boiling-point elevation (ΔT_b)** and **freezing-point depression (ΔT_f)** of a solution are defined as the differences between the boiling and freezing points, respectively, of the solution and the pure solvent. Both are proportional to the molality of the solute. When a solution and a pure solvent are separated by a **semipermeable membrane**, a barrier that allows solvent molecules but not solute molecules to pass through, the flow of solvent in opposing directions is unequal and produces an **osmotic pressure**, which is the difference in pressure between the two sides of the membrane. **Osmosis** is the net flow of solvent through such a membrane due to different solute concentrations. **Dialysis** uses a semipermeable membrane with pores that allow only small solute molecules and solvent molecules to pass through. In more concentrated solutions, or in solutions of salts that contain highly charged ions, the cations and anions can associate to form **ion pairs**, which decreases their effect on the colligative properties of the solution. The extent of ion pair formation is given by the **van't Hoff factor (*i*)**, the ratio of the apparent number of particles in solution to the number predicted by the stoichiometry of the salt.

13.6 Aggregate Particles in Aqueous Solution (p. 606)

A **suspension** is a heterogeneous mixture of particles of one substance distributed throughout a second phase; the dispersed particles separate from the dispersing phase on standing. In contrast, the particles in a **colloid** are smaller and do not separate on standing. A colloid can be classified as a **sol**, a dispersion of solid particles in a liquid or solid; an **aerosol**, a dispersion of solid or liquid particles in a gas; or an **emulsion**, a dispersion of one liquid phase in another. A colloid can be distinguished from a true solution by its ability to scatter a beam of light, known as the **Tyndall effect**. Hydrophilic colloids contain an outer shell of groups that interact favorably with water, whereas hydrophobic colloids have an outer surface with little affinity for water. **Emulsions** are prepared by dispersing a hydrophobic liquid in water. In the absence of a dispersed hydrophobic liquid phase, solutions of detergents in water form organized spherical aggregates called **micelles**. **Phospholipids** are a class of detergent-like molecules that have two hydrophobic tails attached to a hydrophilic head. **Cells** are collections of molecules that are surrounded by a phospholipid bilayer called a **cell membrane** and are able to reproduce themselves.

KEY EQUATIONS

Molality	$\text{molality } (m) = \dfrac{\text{moles solute}}{\text{kilogram solvent}}$	(13.5)
Mass percentage	$\text{mass percentage} = \left(\dfrac{\text{mass of solute}}{\text{mass of solution}}\right)(100\%)$	(13.6)
Parts per million	$\text{ppm} = \left(\dfrac{\text{mass of solute}}{\text{mass of solution}}\right)(10^6)$	(13.7)
Parts per billion	$\text{ppb} = \left(\dfrac{\text{mass of solute}}{\text{mass of solution}}\right)(10^9)$	(13.8)
Henry's law	$C = kP$	(13.10)
Raoult's law	$P_A = X_A P_A^0$	(13.11)
Vapor-pressure lowering	$P_A^0 - P_A = \Delta P = X_B P_A^0$	(13.13)
Vapor pressure of a system containing two volatile components	$P_T = X_A P_A^0 + (1 - X_A) P_B^0$	(13.15)
Boiling-point elevation	$\Delta T_b = mK_b$	(13.19)
Freezing-point depression	$\Delta T_f = mK_f$	(13.21)
Osmotic pressure	$\Pi = \dfrac{nRT}{V} = MRT$	(13.22)
van't Hoff factor	$i = \dfrac{\text{apparent number of particles in solution}}{\text{number of moles of solute dissolved}}$	(13.23)

QUESTIONS AND PROBLEMS

For instructor-assigned homework, go to www.masteringgenchem.com

Questions and Problems with colored numbers have answers in the Appendix and complete solutions in the Student Solutions Manual.

CONCEPTUAL

13.1 Factors Affecting Solution Formation

1. Classify each of the following as a heterogeneous or homogeneous mixture: aqueous ammonia, liquid decongestant, vinegar, seawater, gasoline, fog. Explain your rationale in each case.
2. Solutions and heterogeneous mixtures are at the extreme ends of the solubility scale. Name one type of mixture that is intermediate on this scale. How are the properties of the mixture you have chosen different from those of a solution or a heterogeneous mixture?
3. Classify each process as simple dissolution or a chemical reaction: (a) a naphthalene moth ball dissolving in benzene; (b) a sample of Drano® (a mixture of NaOH crystals and Al chunks) dissolving in water to give H_2 gas and an aqueous solution of Na^+, OH^-, and Al^{3+} ions; (c) an iron ship anchor slowly dissolving in seawater; (d) sodium metal dissolving in liquid ammonia.
4. Classify each process as simple dissolution or a chemical reaction: (a) a sugar cube dissolving in a cup of hot tea; (b) SO_3 gas dissolving in water to produce sulfuric acid; (c) calcium oxide dissolving in water to produce a basic solution; (d) metallic gold dissolving in a small quantity of liquid mercury.
5. You notice that a gas is evolved as you are dissolving a solid in a liquid. Will you be able to recover your original solid by evaporation? Why or why not?
6. Why is heat evolved when sodium hydroxide pellets are dissolved in water? Does this process correspond to simple dissolution or a chemical reaction? Justify your answer.
7. Which of the following processes is simple formation of a solution, and which involves a chemical reaction: mixing an aqueous solution of NaOH with an aqueous solution of HCl; bubbling HCl gas through water; adding iodine crystals to CCl_4; adding sodium metal to ethanol to produce sodium ethoxide ($Na^+C_2H_5O^-$) and hydrogen gas?
8. Using thermochemical arguments, explain why some substances that do not form a solution at room temperature will form a solution when heated. Explain why a solution can form even if ΔH_{soln} is positive.
9. If you wanted to formulate a new compound that could be used in an instant cold pack, would you select a compound with a positive or negative value of ΔH_{soln} in water? Justify your answer.
10. Why is entropy the dominant factor in the formation of solutions of two or more gases? Is it possible for two gases to be immiscible? Why or why not?

13.2 Solubility and Molecular Structure

11. If a compound is only slightly soluble in a particular solvent, what are the relative strengths of the solvent–solvent and solute–solute interactions versus the solute–solvent interactions?
12. Predict whether each of the following sets of conditions favors formation of a solution:

Intermolecular Attractive Forces, Solute	Intermolecular Attractive Forces, Solvent	ΔH_{soln}
London dispersion	Hydrogen bonding	Slightly positive
Dipole–dipole	Hydrogen bonding	Very negative
Ionic	Dipole–dipole	Slightly positive
Ionic	London dispersion	Positive

13. Arrange the following liquids in order of increasing solubility in water: *t*-butanol [$(CH_3)_3COH$], benzene, ammonia, and heptane. Justify your answer.
14. Which compound in each of the following pairs will be more soluble in water: toluene (C_7H_8) or ethyl ether ($C_2H_5OC_2H_5$); chloroform ($CHCl_3$) or acetone (CH_3COCH_3); carbon tetrachloride (CCl_4) or tetrahydrofuran (C_4H_8O); $CaCl_2$ or CH_2Cl_2? Explain your reasoning in each case.
15. Which compound in each of the following pairs will be more soluble in benzene: cyclohexane or methanol; I_2 or $MgCl_2$; methylene chloride (CH_2Cl_2) or acetic acid? Explain your reasoning in each case.
16. Two water-insoluble compounds, *n*-decylamine [$CH_3(CH_2)_9NH_2$] and *n*-decane, can be separated by the following procedure: The compounds are dissolved in a solvent such as toluene that is immiscible with water. Upon addition of an aqueous HCl solution to the mixture and stirring vigorously, the HCl reacts with one of the compounds to produce a salt. When the stirring is stopped and the mixture is allowed to stand, two layers are formed. At this point, each layer contains only one of the two original compounds. After the layers are separated, adding aqueous NaOH to the aqueous layer liberates one of the original compounds, which can then be removed by stirring with a second portion of toluene to extract it from the water.
    (a) Identify the compound that is present in each layer following the addition of HCl. Explain your reasoning.
    (b) How can the original compounds be recovered from the toluene solution?
17. Bromine and iodine are both soluble in CCl_4, but bromine is much more soluble. Why?
18. A solution is made by mixing 50.0 mL of liquid A with 75.0 mL of liquid B. Which is the solute, and which is the solvent? Is it valid to assume that the volume of the resulting solution will be 125 mL? Explain your answer.
19. The compounds NaI, NaBr, and NaCl are far more soluble in water than NaF, a substance used to fluoridate drinking water. In fact, at 25°C the solubility of NaI is 184 g/100 mL of water, versus only 4.2 g/100 mL of water for NaF. Why is

sodium iodide so much more soluble in water? Do you expect KCl to be more soluble or less soluble in water than NaCl?

20. When water is mixed with a solvent with which it is immiscible, the two liquids usually form two separate layers. If the density of the nonaqueous solvent is 1.75 g/mL at room temperature, sketch the appearance of the heterogeneous mixture in a beaker and label which layer is which. If you were not sure of the density and the identity of the other liquid, how might you be able to identify which is the aqueous layer?

21. When two liquids are immiscible, the addition of a third liquid can occasionally be used to induce the formation of a homogeneous solution containing all three. (a) For example, ethylene glycol ($HOCH_2CH_2OH$) and hexane are immiscible, but adding acetone [$(CH_3)_2CO$] produces a homogeneous solution. Why does adding a third solvent produce a homogeneous solution? (b) Methanol and *n*-hexane are immiscible. Which of the following solvents would you add to create a homogeneous solution: water, *n*-butanol, or cyclohexane? Justify your choice.

22. Some proponents of vitamin therapy for combating illness encourage the consumption of large amounts of fat-soluble vitamins. Why can this be dangerous? Would it be as dangerous to consume large amounts of water-soluble vitamins? Why or why not?

23. Why are most metals insoluble in virtually all solvents?

24. Because sodium reacts violently with water, it is difficult to weigh out small quantities of sodium metal for a reaction due to its rapid reaction with small amounts of moisture in the air. Would a Na/Hg amalgam be as sensitive to moisture as metallic sodium? Why or why not? A Na/K alloy is a liquid at room temperature. Will it be more or less sensitive to moisture than solid Na or K?

25. Dental amalgams often contain high concentrations of Hg, which is highly toxic. Why isn't dental amalgam toxic?

26. Arrange the following compounds in order of increasing dipole moment: 2,2,3-trimethylpentane, 1-propanol, toluene (C_7H_8), and dimethyl sulfoxide [$(CH_3)_2S{=}O$]. Explain your reasoning.

27. Arrange the following compounds in order of increasing dielectric constant: acetone, chloroform, cyclohexane, and 2-butanol. Explain your reasoning.

28. Dissolving a white crystalline compound in ethanol gave a blue solution. Evaporating the ethanol from the solution gave a bluish-crystalline product, which slowly transformed into the original white solid upon standing in the air for several days. Explain what happened. How does the mass of the initial bluish solid compare to the mass of the white solid finally recovered?

29. You have been asked to develop a new drug that could be used to bind Fe^{3+} ions in patients who suffer from iron toxicity, allowing the bound iron to be excreted in the urine. Would you consider a crown ether or a cryptand to be a reasonable candidate for such a drug? Explain your answer.

30. Describe two different situations in which fractional crystallization will not work as a separation technique when attempting to isolate a single compound from a mixture.

31. You have been given a mixture of two compounds, A and B, and have been told to isolate pure A. You know that pure A has a lower solubility than pure B and that the solubilities of both A and B increase with temperature. Outline a procedure to isolate pure A. If B had the lower solubility, could you use the same procedure to isolate pure A? Why or why not?

13.3 Units of Concentration

32. Does the molality have the same numerical value as the molarity for a highly concentrated aqueous solution of fructose ($C_6H_{12}O_6$) (approximately 3.2 *M*)? Why or why not?

33. Explain why the molality and molarity of an aqueous solution are not always numerically identical. Will the difference between the two be greater for a dilute or a concentrated solution? Explain your answer.

34. Under what conditions are molality and molarity likely to be equal? Is the difference between the two greater for water as the solvent or for solvents other than water? Why?

35. What is the key difference between using mole fraction or molality versus molarity to describe the concentration of a solution? Which unit(s) of concentration is most appropriate for experiments that must be carried out at several different temperatures?

36. An experiment that relies upon very strict control of the reaction stoichiometry calls for adding 50.0 mL of a 0.95 *M* solution of A to 225 mL of a 1.01 *M* solution of B, followed by heating for 1 h at 60°C. To save time, a student decided to heat solution B to 60°C *before* measuring out 225 mL of solution B, transferring it to the flask containing solution A, and proceeding normally. This change in procedure caused the yield of product to be lower than usual. How could such an apparently minor change in procedure have resulted in a decrease in the yield?

13.4 Effects of Temperature and Pressure on Solubility

37. Use the kinetic molecular theory of gases discussed in Chapter 10 to explain why the solubility of virtually all gases in liquids decreases with increasing temperature.

38. An industrial plant uses water from a nearby stream to cool its reactor, and returns the water to the stream after use. Over a period of time, dead fish start to appear downstream from the plant, but there is no evidence for any leaks of potentially toxic chemicals into the stream. What other factor might be causing the fish to die?

39. One manufacturer's instructions for setting up an aquarium specify that if boiled water is used, the water must be cooled to room temperature and allowed to stand overnight before fish are added. Why is it necessary for the water to stand?

40. Using a carbonated beverage as an example, discuss the effect of temperature on the "fizz." How does the "foaminess" of a carbonated beverage differ between Los Angeles, California, and Denver, Colorado?

41. A common laboratory technique for degassing a solvent is to place it in a flask that is sealed to the atmosphere, and then evacuate the flask to remove any gases above the liquid. Why is this procedure effective? Why does the temperature of the solvent usually decrease substantially during this process?

13.5 Colligative Properties of Solutions

42. Why does the vapor pressure of a solvent decrease upon addition of a nonvolatile solute?

43. Does seawater boil at the same temperature as distilled water? If not, which has the higher boiling point? Explain your answer.

44. Which will be more soluble in benzene, O_2 or HCl? Will H_2S or HCl be more soluble in water? Explain your reasoning in each case.

45. Will the vapor pressure of a solution of hexane and heptane have an ideal vapor pressure curve (in other words, obey Raoult's law)? Explain your answer. What properties of two liquids determine whether a solution of the two will exhibit an ideal behavior?

46. Predict whether the following mixtures will exhibit negative, zero, or positive deviations from Raoult's law: carbon tetrachloride and heptane; methanol and tetrahydrofuran (C_4H_8O); acetone [$(CH_3)_2C{=}CO$] and dichloromethane; hexane and methanol. Explain your reasoning in each case.

47. Why are deviations from the ideal behavior predicted by Raoult's law more common for solutions of liquids than are deviations from the ideal behavior predicted by the ideal gas law for solutions of gases?

48. Boiling-point elevation is proportional to the molal concentration of the solute. Is it also proportional to the molar concentration of the solution? Why or why not?

49. Many packaged foods in sealed bags are cooked by placing the bag in boiling water. How could you reduce the time required to cook the contents of the bag using this cooking method?

50. If the costs per kilogram of ethylene glycol and of ethanol were the same, which would be the more cost-effective antifreeze?

51. Many people get thirsty after eating foods such as ice cream or potato chips, which have a high sugar or salt content, respectively. Suggest an explanation for this phenomenon.

52. When two aqueous solutions with identical concentrations are separated by a semipermeable membrane, no net movement of water occurs. What happens when a solute that cannot penetrate the membrane is added to one of the solutions? Why?

53. Solutions injected into blood vessels must have electrolyte concentrations that are nearly identical to that found in blood plasma. Why? What would happen if red blood cells were placed in distilled water? What would happen to red blood cells if they were placed in a solution that had twice the electrolyte concentration of blood plasma?

54. If you were stranded on a desert island, why would drinking seawater lead to an increased rate of dehydration, eventually causing you to die of thirst?

55. What is the relationship between the van't Hoff factor for a compound and its lattice energy?

13.6 Aggregate Particles in Aqueous Solution

56. How does a colloid differ from a suspension? Which has a greater effect on solvent properties, such as vapor pressure?

57. Is homogenized milk a colloid or a suspension? Is human plasma a colloid or a suspension? Justify your answers.

58. How would you separate the components of an emulsion of fat dispersed in an aqueous solution of sodium chloride?

NUMERICAL

This section includes paired problems (marked by brackets) that require similar problem-solving skills.

13.3 Units of Concentration

59. Complete the following table for aqueous solutions of the compounds given:

Compound	Molarity, *M*	Solution, Density, g/mL	Mole Fraction, *X*
H_2SO_4	18.0	1.84	
CH_3COOH		1.00	7.21×10^{-3}
KOH	3.60	1.16	

60. Complete the following table for each compound given:

Compound	Mass, g	Volume of Solution, mL	Molarity, *M*
Na_2SO_4	7.80	225	
KNO_3		125	1.27
NaO_2CCH_3	18.64		0.95

61. How would you prepare 100.0 mL of an aqueous solution that contains 0.40 *M* KI? A solution that contains 0.65 *M* NaCN?

62. Calculate the molality of a solution that contains 775 mg of NaCl in 500.0 g of water. Do you expect the molarity to be the same as the molality? Why or why not?

63. What is the molarity of each of the following solutions: 12.8 g of glucose ($C_6H_{12}O_6$) in water, total volume 150.0 mL; 9.2 g of Na_3PO_4 in water, total volume 200.0 mL; 843 mg of I_2 in EtOH, total volume 150.0 mL?

64. An aqueous solution that is 4.61% NaOH by mass has a density of 1.06 g/mL. Determine (a) the molarity, (b) the molality, and (c) the mole fraction of NaOH.

65. A solution of concentrated phosphoric acid contains 85.0% H_3PO_4 by mass and has a density of 1.684 g/mL. Calculate (a) the molarity, (b) the mole fraction of H_3PO_4, and (c) the molality of the solution.

66. A solution of commercial concentrated nitric acid is 16 *M* HNO_3 and has a density of 1.42 g/mL. What is the percentage of HNO_3 in the solution by mass? What is the molality?

67. Calculate (a) the molarity and (b) the mole fraction of ammonia in a commercial aqueous ammonia solution that is 28.0% NH_3 by mass. The density of the solution is 0.899 g/mL.

68. Concentrated, or glacial, acetic acid is pure acetic acid and has a density of 1.053 g/mL. It is widely used in organic syntheses, in the manufacture of rayon and plastics, as a preservative in foods, and occasionally to treat warts. What volume of glacial acetic acid is required to prepare 5.00 L of a 1.75 *M* solution of acetic acid in ethanol?

69. Solutions of sodium carbonate decahydrate, washing soda, are used as skin cleansers. The solubility of this compound in cold water is 21.52 g/100 mL. If a saturated solution has a density of 1.20 g/mL, what is its molarity? What is the mole fraction of sodium carbonate decahydrate in this solution?

70. Hydrogen peroxide, H_2O_2, is usually sold over the counter as an aqueous solution that is 3% by mass. Assuming a solution density of 1.01 g/mL, what is the molarity of hydrogen peroxide? What is the molar concentration of a solution that is 30% hydrogen peroxide by mass (density = 1.112 g/mL)? How would you prepare 100.0 mL of a 3% solution from the 30% solution?

71. Determine the concentration of a solution that contains 825 mg of Na_2HPO_4 dissolved in 450.0 mL of H_2O at 20°C in molarity, molality, mole fraction, and parts per million. Assume that the density of the solution is the same as that of water. Which unit of concentration is most convenient for calculating vapor pressure changes? Why?

72. How many moles of Cl^- are there in 25.0 mL of a 0.15 *M* $CaCl_2$ solution?

73. How many moles of Na^+ are there in 25.0-g of a 1.33×10^{-3} *m* Na_2HPO_4 solution?

74. How many grams of copper are there in 30.0-mL of a 0.100 *M* $CuSO_4$ solution?

75. How many grams of nitrate ion are there in 75.0-g of a 1.75×10^{-4} *m* $Pb(NO_3)_2$ solution?

76. How many milliliters of a 0.750 *M* solution of K_2CrO_4 are required to deliver 250 mg of chromate ion?

77. How many milliliters of a 1.95×10^{-6} *M* solution of Ag_3PO_4 are required to deliver 14.0 mg of Ag^+?

78. Iron reacts with bromine according to the following equation:

$$2Fe(s) + 3Br_2(aq) \longrightarrow 2FeBr_3(aq)$$

How many milliliters of a 5.0×10^{-2} *M* solution of bromine in water are required to react completely with 750.0 mg of iron metal?

79. Aluminum reacts with HCl according to the following equation:

$$2Al(s) + 6HCl(aq) \longrightarrow 2AlCl_3(aq) + 3H_2(g)$$

If 25.0 mL of a solution of HCl in water is required to react completely with 1.05 g of aluminum metal, what is the molarity of the HCl solution?

80. The precipitation of silver chloride is a diagnostic test for the presence of chloride ion. If 25.0 mL of 0.175 *M* $AgNO_3$ was required to completely precipitate the chloride ions from 10.0 mL of an NaCl solution, what was the original concentration of NaCl?

81. Barium sulfate is virtually insoluble. If a 10.0-mL solution of 0.333 *M* $Ba(NO_3)_2$ is stirred with 40.0 mL of a 0.100 *M* Na_2SO_4, how many grams of barium sulfate will precipitate? Which reactant is present in excess? What is its final concentration?

13.4 Effects of Temperature and Pressure on Solubility

82. The solubility of O_2 in 100 g of H_2O at varying temperatures and a pressure of 1 atm is given in the following table:

Solubility, g	Temperature, °C
0.0069	0
0.0054	10
0.0043	20

(a) What is the value of the Henry's law constant at each temperature?

(b) Does Henry's law constant increase or decrease with increasing temperature?

(c) At what partial pressure of O_2 would the concentration of O_2 in water at 0°C be the same as the concentration in water at 20°C at a partial pressure of 1 atm?

(d) Assuming that air is 20% O_2 by volume, at what atmospheric pressure would the O_2 concentration be the same at 20°C as it is at atmospheric pressure and 0°C?

83. The solubility of CO_2 in water at 0°C and 1 atm is 0.335 g/100 g of H_2O. At 20°C and 1 atm, the solubility of CO_2 in water is 0.169 g/100 g of H_2O. What volume of CO_2 would be released by warming 750 g of water saturated with CO_2 from 0°C to 20°C? What is the value of the Henry's law constant for CO_2 under each set of conditions?

13.5 Colligative Properties of Solutions

84. To determine the molar mass of the antifreeze protein from the Arctic right-eye flounder, the osmotic pressure of a solution containing 13.2 mg of protein per mL was measured and found to be 21.2 mmHg at 10°C. What is the molar mass of the protein?

85. Hemoglobin is the protein that is responsible for the red color of blood and for transporting oxygen from the lungs to the tissues. A solution that contains 11.2 mg of hemoglobin per mL has an osmotic pressure of 2.9 mmHg at 5°C. What is the molecular mass of hemoglobin?

86. A solution of $NaNO_3$ is generated by dissolving 1.25 g of $NaNO_3$ in enough water to give a final volume of 25.0 mL. What would be the osmotic pressure of this sample at 25.0°C?

87. What is the osmotic pressure at 21.0°C of 13.5 mL of a solution that contains 1.77 g of sucrose ($C_{12}H_{22}O_{11}$)?

88. What is the total particle concentration expected for each of the following aqueous solutions: 0.35 *M* KBr, 0.11 *M* $MgSO_4$, 0.26 *M* $MgCl_2$, 0.24 *M* glucose ($C_6H_{12}O_6$)? Which would produce the highest osmotic pressure?

89. Which would have the lower vapor pressure: an aqueous solution that is 0.12 *M* in glucose or one that is 0.12 *M* in $CaCl_2$? Why?

90. Calculate the boiling point of a solution of sugar prepared by dissolving 8.4 g of glucose ($C_6H_{12}O_6$) in 250 g of water.

91. The boiling point of an aqueous solution of sodium chloride is 100.37°C. What is the molality of the solution? How many grams of NaCl are present in 125 g of the solution?

92. How many grams of KCl must be added to reduce the vapor pressure of 500.0 g of H_2O from 17.5 mmHg to 16.0 mmHg at 20.0°C?

93. At 37°C, the vapor pressure of 300.0 g of water was reduced from 0.062 atm to 0.058 atm by the addition of NaBr. How many grams of NaBr were added?

94. You have three solutions with the following compositions: 12.5 g of KCl in 250 mL of water; 12.5 g of glucose in 400 mL of water; and 12.5 g of $MgCl_2$ in 350 mL of water. Which will have the highest boiling point?

95. How much NaCl would you have to add to 2.0 L of water at a mountain lodge at an elevation of 7350 ft, where the pressure is 0.78 atm and the boiling point of water is 94°C, to get the water to boil at the same temperature as in New Orleans, Louisiana, where the pressure is 1.00 atm?

96. How many grams of KNO_3 must be added to water to produce the same boiling-point elevation as a solution of 2.03 g of $MgCl_2$ in a total volume of 120.0 mL of solution, assuming complete dissociation? If the van't Hoff factor for $MgCl_2$ at this concentration is 2.73, how much KNO_3 would be needed?

97. Assuming the price per kilogram is the same, which is a better salt to use for deicing wintry roads: NaCl or $MgCl_2$? Why? Would magnesium chloride be an effective deicer at a temperature of −8°C?

98. Calculate the quantity of each compound that would need to be added to lower the freezing point of 500.0 mL of water by

1.0°C: KBr, ethylene glycol, $MgBr_2$, ethanol. Assume that the density of water is 1.00 g/cm^3.

99. Four solutions of urea in water were prepared, with concentrations of 0.32 *m*, 0.55 *m*, 1.52 *m*, and 3.16 *m*. The freezing points of these solutions were found to be −0.595°C, −1.02°C, −2.72°C, and −5.71°C, respectively. Graphically determine the freezing-point depression constant for water. A fifth solution made by dissolving 6.22 g of urea in 250.0 g of water has a freezing point of −0.75°C. Use these data to determine the molar mass of urea.

100. The melting point depression of biphenyl (mp 69.0°C) can be used to determine the molecular mass of organic compounds. A mixture of 100.0 g of biphenyl and 2.67 g of naphthalene ($C_{10}H_8$) has a melting point of 68.50°C. If a mixture of 1.00 g of an unknown compound with 100.0 g of biphenyl has a melting point of 68.86°C, what is the molar mass of the unknown compound?

101. The term *osmolarity* has been used to describe the total solute concentration of a solution (generally water), where 1 osmole is equal to 1 mol of an ideal, non-ionizing molecule.

(a) What is the osmolarity of a 1.5 *M* solution of glucose? Of a 1.5 *M* solution of NaCl? Of a 1.5 *M* solution of $CaCl_2$?

(b) What is the relationship between osmolarity and the concentration of water?

(c) What would be the direction of flow of water through a semi-permeable membrane separating a 0.1 *M* solution of NaCl and a 0.1 *M* solution of $CaCl_2$?

102. At 40°C, the vapor pressures of pure CCl_4 and cyclohexane are 0.2807 atm and 0.2429 atm, respectively. Assuming ideal behavior, what is the vapor pressure of a solution with a CCl_4 mole fraction of 0.475? What is the mole fraction of cyclohexane in the vapor phase? The boiling points of CCl_4 and cyclohexane are 76.8°C and 80.7°C, respectively.

103. A benzene/toluene solution with a mole fraction of benzene of 0.6589 boils at 88°C at 1 atm. The vapor pressures of pure benzene and toluene at this temperature are 1.259 atm and 0.4993 atm, respectively. What is the composition of the vapor above the boiling solution at this temperature?

104. Plot the vapor pressure of the solution versus composition for the system CCl_4–CH_3CN at 45°C, given the following experimental data [adapted from L. Brown and W. Foch, *Aust. J. Chem.*, 9:180 (1956)]:

X_{CCl_4} liquid	0.035	0.375	0.605	0.961
X_{CCl_4} vapor	0.180	0.543	0.594	0.800
Total *P*, atm	0.326	0.480	0.488	0.414

Does your diagram show behavior characteristic of an ideal solution? Explain your answer.

APPLICATIONS

105. Scuba divers utilize high-pressure gas in their tanks to allow them to breathe under water. At depths as shallow as 100 ft (30 m), the pressure exerted by water is 4.0 atm. At 25°C the values of Henry's law constants for N_2, O_2, and He in blood are: $N_2 = 6.5 \times 10^{-4}$ mol/(L · atm), $O_2 = 1.28 \times 10^{-3}$ mol/(L · atm), and He = 3.7×10^{-4} mol/(L · atm). What would be the concentration of nitrogen and oxygen in blood at sea level where the air is 21% oxygen and 79% nitrogen? At a depth of 30 m, assuming that the diver is breathing compressed air?

106. Many modern batteries take advantage of lithium ions dissolved in suitable electrolytes. Typical batteries have lithium concentrations of 0.10 *M*. Which aqueous solution has the higher concentration of ion pairs: 0.08 *M* LiCl or 1.4 *M* LiCl? Why? Does an increase in the number of ion pairs correspond to a higher or lower van't Hoff factor? Batteries rely on a high concentration of unpaired Li^+ ions. Why is using a more concentrated solution not an ideal strategy in this case?

107. Hydrogen sulfide, which is extremely toxic to humans, can be detected at a concentration of 2.0 ppb. At this level, headaches, dizziness, and nausea occur. At higher concentrations, however, the sense of smell is lost, and the lack of warning can result in coma and death can result. What is the concentration of H_2S in mg/L at the detection level? The lethal dose of hydrogen sulfide by inhalation for rats is 7.13×10^{-4} g/L. What is this lethal dose in ppm? The density of air is 1.2929 g/L.

108. One class of antibiotics consists of cyclic polyethers that can bind alkali metal cations in aqueous solution. Given the following antibiotics and cation selectivities, what conclusion can you draw regarding the relative sizes of the cavities?

Antibiotic	Cation Selectivity
Nigericin	$K^+ > Rb^+ > Na^+ > Cs^+ > Li^+$
Lasalocid	$Ba^{2+} >> Cs^+ > Rb^+, K^+ > Na^+, Ca^{2+}, Mg^{2+}$

109. Phenylpropanolamine hydrochloride is a common nasal decongestant. An aqueous solution of phenylpropanolamine hydrochloride that is sold commercially as a children's decongestant has a concentration of 6.67×10^{-3} *M*. If a common dose is 1.0 mL/12 lb of body weight, how many moles of the decongestant should be given to a 26-lb child?

110. The "freeze-thaw" method is often used to remove dissolved oxygen from solvents in the laboratory. In this technique, a liquid is placed in a flask that is then sealed to the atmosphere, the liquid is frozen, and the flask is evacuated to remove any gas and solvent vapor in the flask. The connection to the vacuum pump is closed, the liquid is warmed to room temperature and then refrozen, and the process is repeated. Why is this technique effective for degassing a solvent?

111. Suppose that, on a planet in a galaxy far, far away, a species has evolved whose biological processes require even more oxygen than we do. The partial pressure of oxygen on this planet, however, is much lower than on Earth. The chemical composition of the "blood" of this species is also different. Do you expect their "blood" to have a higher or lower value of the Henry's law constant for oxygen at standard temperature and pressure? Justify your answer.

112. A car owner who had never taken general chemistry decided that he needed to put some ethylene glycol antifreeze in his car's radiator. After reading the directions on the container, however, he decided that "more must be better." Instead of using the recommended mixture (30% ethylene glycol/70% water), he decided to reverse the amounts and used a 70% ethylene glycol/30% water mixture instead. Serious engine problems developed. Why?

113. The ancient Greeks produced "Attic" ware, pottery with a characteristic black and red glaze. In order to separate smaller clay particles from larger ones, the powdered clay was suspended in water and allowed to settle. This process yielded clay fractions with coarse, medium, and fine particles, and one of these fractions was used for painting. Which size of clay particles form a suspension, which form a precipitate, and which form a colloidal dispersion? Would the colloidal dispersion be better characterized as an emulsion? Why or why not? Which fraction of clay particles was used for painting?

114. The Tyndall effect is often observed in movie theaters, where it makes the beam of light from the projector clearly visible. What conclusions can you draw about the quality of the air in a movie theater where you observe a large Tyndall effect?

115. Aluminum sulfate is the active ingredient in styptic pencils, which are used to stop bleeding from small cuts. The Al^{3+} ions induce aggregation of colloids in the blood, which facilitates formation of a blood clot. How can Al^{3+} ions induce aggregation of a colloid? What is the probable charge on the colloidal particles in blood?

116. The liver secretes bile, which is essential for the digestion of fats. As discussed in Chapter 5, fats are biomolecules that contain long hydrocarbon chains. The globules of fat released by partial digestion of food particles in the stomach and lower intestine are too large to be absorbed by the intestine unless they are emulsified by bile salts, such as glycocholate (structure shown below). Explain why a molecule like glycocholate is effective at creating an aqueous dispersion of fats in the digestive tract.

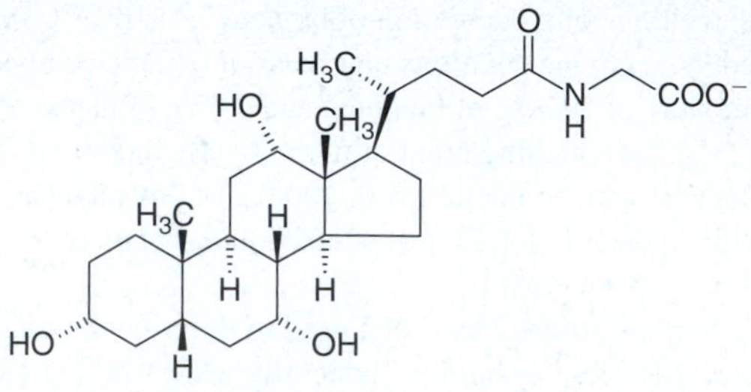

Glycocholate (a bile salt)

14 Chemical Kinetics

LEARNING OBJECTIVES

- To understand the factors that affect the rate of chemical reactions
- To be able to determine a reaction rate
- To understand the meaning of a rate law
- To be able to determine the concentration dependence of the rate of a chemical reaction
- To be able to predict the individual steps of a simple reaction
- To begin to understand why chemical reactions occur

CHAPTER OUTLINE

The gases, liquids, and solids you learned to describe quantitatively in Part III were systems whose chemical composition did not change with time. Now we will present a quantitative description of a far more common situation in which the chemical composition of a system is not constant with time. An example of such a system is the stratosphere, where chemicals rising from the ground level initiate reactions that lead to a decrease in the concentration of stratospheric ozone—the so-called ozone hole (see Chapter 3). Another example involves the production of polyethylene, in which the properties of the plastic are determined by the relative speeds of events that occur during the polymerization reaction (see Chapter 12). The techniques you are about to learn will enable you to describe the speed of many such changes and to predict how the composition of each system will change in response to changing conditions.

The Belousov-Zhabotinsky reaction, a chemical reaction that oscillates in time and space. When a very thin layer of an acidic solution containing $KBrO_3$, $(NH_4)_2Ce(NO_3)_6$, malonic acid ($HO_2CCH_2CO_2H$), and an indicator is poured into a shallow dish, local fluctuations in the concentration of the reactants and a complex series of reactions cause striking geometric patterns of concentric circles and spirals to propagate across the dish.

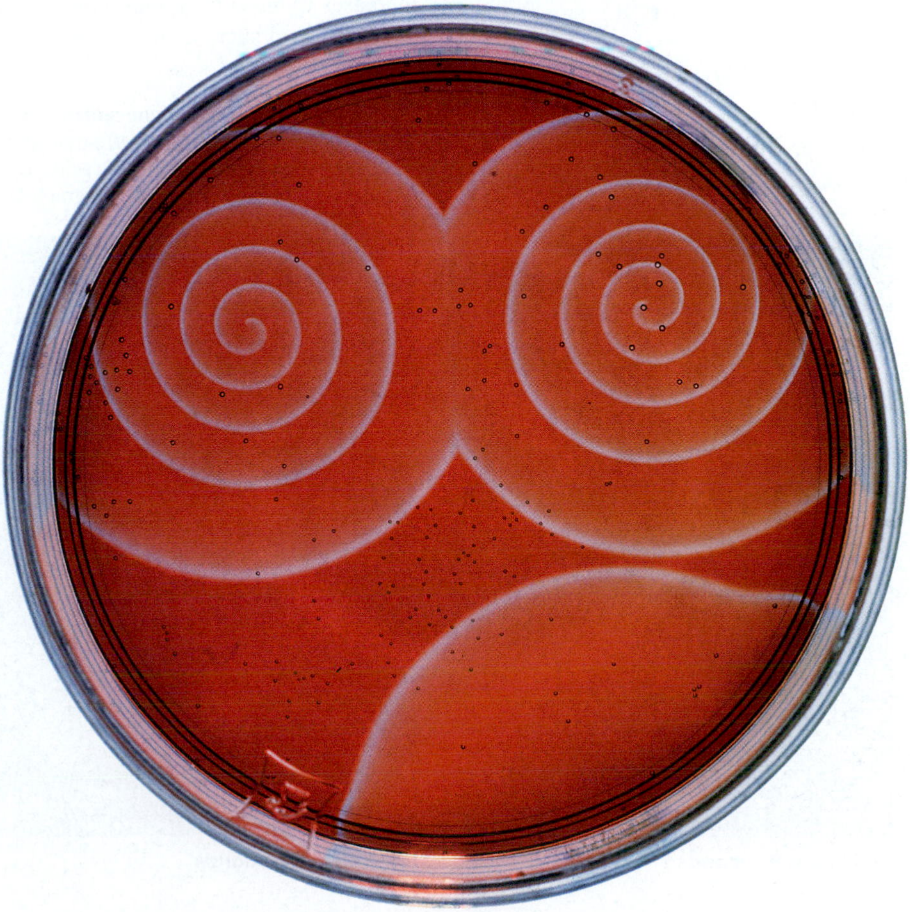

We begin Part IV with a discussion of **chemical kinetics**, the study of **reaction rates**, or the changes in the concentrations of reactants and products with time. As you learn about the factors that affect reaction rates, the methods chemists use for reporting and calculating those rates, and the clues that reaction rates provide about events at the molecular level, you will also discover the answers to questions such as these: How can normally stable substances such as flour and coal cause devastating explosions? How do archaeologists use isotopic composition to estimate the ages of ancient artifacts? How do the catalysts used in catalytic converters, some laundry detergents, and meat tenderizers work?

14.1 • Factors That Affect Reaction Rates

Although a balanced chemical equation for a reaction describes the quantitative relationships between the amounts of reactants present and the amounts of products that can be formed, it gives us no information about whether or how fast a given reaction will occur. This information is obtained by studying the chemical kinetics of a reaction, which depend on various factors: reactant concentrations, temperature, physical states and surface areas of reactants, as well as solvent and catalyst properties if either are present. By studying the kinetics of a reaction, chemists gain insights into how to control reaction conditions to achieve a desired outcome.

Concentration Effects

Two substances cannot possibly react with each other unless their constituent particles (molecules, atoms, or ions) come into contact. If there is no contact, the rate of reaction will be zero. Conversely, the more reactant particles that collide per unit time, the more often a reaction between them can occur. Consequently, the rate of a reaction usually increases as the concentration of the reactants increases. One example of this effect is the reaction of sucrose (table sugar) with sulfuric acid, shown in Figure 14.1.

Figure 14.1 The effect of concentration on reaction rates. Mixing sucrose with *dilute* sulfuric acid in a beaker (a, *right*) produces a simple solution. Mixing the same amount of sucrose with *concentrated* sulfuric acid (a, *left*) results in a dramatic reaction (b) that eventually produces a column of black porous graphite (c) and an intense smell of burning sugar.

Temperature Effects

You learned in Chapter 10 that increasing the temperature of a system increases the average kinetic energy of its constituent particles. As the average kinetic energy increases, the particles move faster, so they collide more frequently per unit time and possess greater energy when they collide. Both of these factors increase the rate of the reaction. Hence, the rate of virtually all reactions increases with increasing temperature. Conversely, the rate of virtually all reactions decreases with decreasing temperature. For example, refrigeration retards the rate of growth of bacteria in foods by decreasing the rates of the biochemical reactions that enable bacteria to reproduce. Figure 14.2 shows how temperature affects the light emitted by two chemiluminescent light sticks.

Figure 14.2 The effect of temperature on reaction rates. At high temperature, the reaction that produces light in a chemiluminescent light stick occurs more rapidly, producing more photons of light per unit time. Consequently, the light glows brighter in hot water *(left)* than in ice water *(right)*.

In systems where more than one reaction is possible, the same reactants can produce different products under different reaction conditions. For example, in the presence of dilute sulfuric acid and at temperatures around 100°C, ethanol is converted to diethyl ether:

$$2CH_3CH_2OH \xrightarrow{H_2SO_4} CH_3CH_2OCH_2CH_3 + H_2O \qquad (14.1)$$

At 180°C, however, a completely different reaction occurs, producing ethylene rather than diethyl ether as the major product:

$$CH_3CH_2OH \xrightarrow{H_2SO_4} C_2H_4 + H_2O \qquad (14.2)$$

Phase and Surface Area Effects

When two reactants are in the same fluid phase, their particles collide more frequently than when one or both of the reactants is solid (or when they are in different fluids that do not mix). If the reactants are uniformly dispersed in a single homogeneous solution, then the number of collisions per unit time depends on concentration and temperature, as we have just seen. If the reaction is heterogeneous, however, the reactants are in two different phases, and collisions between the reactants can occur only at interfaces between phases. The number of collisions between reactants per unit time is substantially reduced relative to the homogeneous case, and, hence, so is the reaction rate. The rate of a heterogeneous reaction depends on the surface area of the more condensed phase.

Automobile engines use surface area effects to increase reaction rates. Gasoline is injected into each cylinder, where it combusts upon ignition by a spark from the spark plug. The gasoline is injected in the form of microscopic droplets because in that form it has a much larger surface area and can burn much more rapidly than if it were fed into the cylinder as a stream. Similarly, a pile of finely divided flour burns slowly (or not at all), but spraying finely divided flour into a flame produces a vigorous reaction (Figure 14.3). Similar phenomena are partially responsible for the dust explosions that occasionally destroy grain elevators or coal mines.

Figure 14.3 The effect of surface area on reaction rates. A pile of flour is only scorched by a flame *(right)*, but when the same flour is sprayed into the flame, it burns rapidly *(left)*.

Solvent Effects

The nature of the solvent can also affect the reaction rates of solute particles. For example, a sodium acetate solution reacts with methyl iodide in an exchange reaction to give methyl acetate and sodium iodide.

$$CH_3CO_2Na(soln) + CH_3I(l) \longrightarrow CH_3CO_2CH_3(soln) + NaI(soln) \qquad (14.3)$$

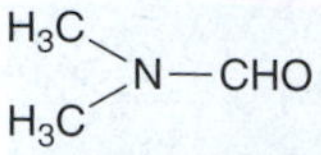

DMF
(dimethylformamide)

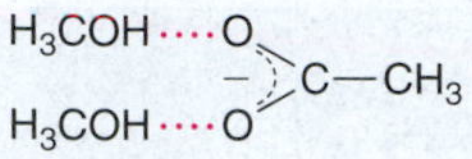

Methanol Acetate ion

This reaction occurs 10 million times more rapidly in the organic solvent DMF [$(CH_3)_2NCHO$] than it does in methanol (CH_3OH). Although both are organic solvents with similar dielectric constants (36.7 for DMF versus 32.6 for methanol), methanol is able to hydrogen bond with acetate ions, whereas DMF cannot. Hydrogen bonding reduces the reactivity of the oxygen atoms of the acetate ion.

Solvent viscosity is also important in determining reaction rates. In highly viscous solvents, dissolved particles diffuse much more slowly than in less viscous solvents and can collide less frequently per unit time. Thus, the rates of most reactions decrease rapidly with increasing solvent viscosity.

Figure 14.4 The effect of catalysts on reaction rates. A solution of hydrogen peroxide (H_2O_2) decomposes in water so slowly that the change is not noticeable *(left).* Iodide ion acts as a catalyst for the decomposition of H_2O_2, producing oxygen gas. The solution turns brown because of the reaction of H_2O_2 with I^-, which generates small amounts of I_3^- *(center).* The enzyme catalase is about 3 *billion* times more effective than iodide as a catalyst. Even in the presence of very small amounts of enzyme, the decomposition is vigorous *(right).*

Catalyst Effects

You learned in Chapter 3 that a catalyst is a substance that participates in a chemical reaction and increases the rate of the reaction without undergoing a net chemical change itself. Consider, for example, the decomposition of hydrogen peroxide in the presence and absence of different catalysts (Figure 14.4). Because most catalysts are highly selective, they often determine the product of a reaction by accelerating only one of several possible reactions that could occur.

Most of the bulk chemicals produced in industry are formed through catalyzed reactions. Recent estimates indicate that about 30% of the gross national product of the United States and other industrialized nations relies either directly or indirectly on the use of catalysts.

14.2 • Reaction Rates and Rate Laws

The factors discussed in Section 14.1 affect the rate of a chemical reaction, which in turn may determine whether a desired product is formed. In this section we will show you how to quantitatively determine the rate of a reaction.

Reaction Rates

MGC Reaction Rates

Reaction rates are usually expressed as the concentration of reactant consumed or the concentration of product formed per unit time. The units are thus moles per liter per unit time, written as *M*/s, *M*/min, or *M*/h. To measure reaction rates, chemists initiate the reaction, measure the concentration of the reactant or product at different times as the reaction progresses, perhaps plot the concentration as a function of time on a graph, and then calculate the *change* in the concentration per unit time.

The progress of a simple reaction ($A \longrightarrow B$) is shown in Figure 14.5, where the beakers are snapshots of the composition of the solution at 10-s intervals. The numbers of molecules of reactant (A) and product (B) are plotted as a function of time in the graph. Each point in the graph corresponds to one beaker in Figure 14.5. The reaction rate is the change in the concentration of either the reactant or the product over a period of time.

$$\text{rate} = \frac{\Delta[\text{B}]}{\Delta t} = -\frac{\Delta[\text{A}]}{\Delta t} \tag{14.4}$$

Just as in Chapters 4 and 5, square brackets indicate molar concentrations and the capital Greek delta (Δ) means "change in." The concentration of A decreases with time, while the concentration of B increases with time. Because chemists follow the

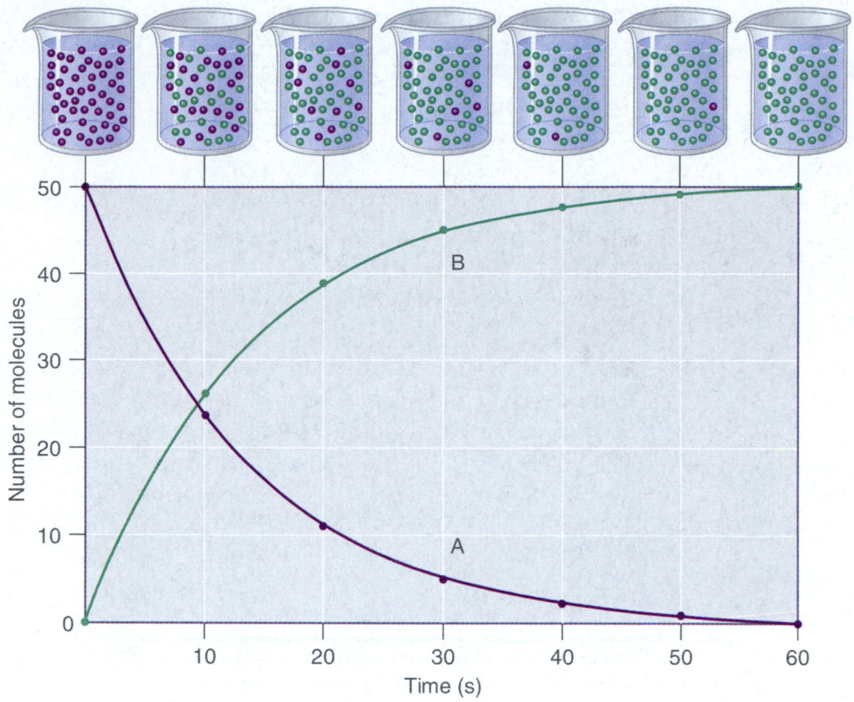

Figure 14.5 The progress of reaction A ⟶ B over a 1-min period *(top)*. Note that the mixture initially contains only A molecules (purple). With increasing time the number of A molecules decreases and more B molecules (green) are formed.

A plot of the progress of the reaction *(bottom)*. This graph shows the change in the numbers of A and B molecules in the reaction depicted above as a function of time. Interactive Graph

convention of expressing all reaction rates as positive numbers, however, a negative sign is inserted in front of $\Delta[A]/\Delta t$ to convert that expression to a positive number. Notice that the rate we would calculate for reaction A ⟶ B using Equation 14.4 would be different for each interval (this is not true for every reaction, as you will see). A much greater change occurs in [A] and [B] during the first 10-s interval, for example, than during the last, which means that the rate of the reaction is fastest at first. This is consistent with the concentration effects described in the preceding section because the concentration of A is greatest at the beginning of the reaction.

Determining the Rate of Hydrolysis of Aspirin

We can use Equation 14.4 to determine the rate of hydrolysis of aspirin, probably the most commonly used drug in the world. (More than 25 million kilograms are produced annually worldwide.) Aspirin (acetylsalicylic acid) reacts with water (such as water in body fluids) to give salicylic acid and acetic acid.

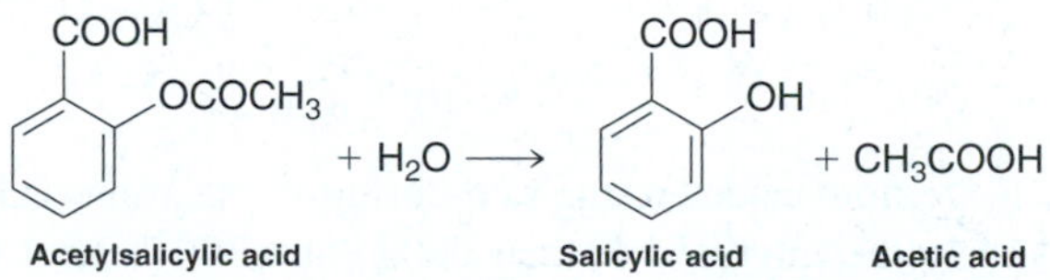

(14.5)

Because salicylic acid is the actual substance that relieves pain and reduces fever and inflammation, a great deal of research has focused on understanding this reaction and the factors that affect its rate. Data for the hydrolysis of a sample of aspirin are listed in Table 14.1 and are shown in the graph in Figure 14.6. These data were obtained by removing samples of the reaction mixture at the indicated times and analyzing them for the concentrations of the reactant (aspirin) and one of the products (salicylic acid).

We can calculate the **average reaction rate** for a given time interval from the concentrations of either the reactant or one of the products at the beginning of the interval (time $= t_0$) and at the end of the interval (t_1). Using salicylic acid, for example, we

TABLE 14.1 Data for the hydrolysis of aspirin in aqueous solution at pH 7.0 and 37°C[a]

Time (h)	[Aspirin], M	[Salicylic Acid], M
0	5.55×10^{-3}	0×10^{-3}
2.0	5.51×10^{-3}	0.040×10^{-3}
5.0	5.45×10^{-3}	0.10×10^{-3}
10	5.35×10^{-3}	0.20×10^{-3}
20	5.15×10^{-3}	0.40×10^{-3}
30	4.96×10^{-3}	0.59×10^{-3}
40	4.78×10^{-3}	0.77×10^{-3}
50	4.61×10^{-3}	0.94×10^{-3}
100	3.83×10^{-3}	1.72×10^{-3}
200	2.64×10^{-3}	2.91×10^{-3}
300	1.82×10^{-3}	3.73×10^{-3}

[a] Note that the reaction at pH 7.0 is very slow. It is *much* faster under acidic conditions, such as those found in the stomach.

find the rate of the reaction for the interval between $t = 0$ h and $t = 2.0$ h (recall that Δ is always calculated as "final minus initial"):

$$\text{rate}_{(t=0-2.0\text{ h})} = \frac{[\text{salicylic acid}]_2 - [\text{salicylic acid}]_0}{2.0\text{ h} - 0\text{ h}}$$

$$= \frac{0.040 \times 10^{-3}\,M - 0\,M}{2.0\text{ h}} = 2.0 \times 10^{-5}\,M/\text{h}$$

We can also calculate the rate of the reaction from the concentrations of aspirin at the beginning and the end of the same interval, remembering to insert a negative sign, because its concentration decreases:

$$\text{rate}_{(t=0-2.0\text{ h})} = -\frac{[\text{aspirin}]_2 - [\text{aspirin}]_0}{2.0\text{ h} - 0\text{ h}}$$

$$= -\frac{5.51 \times 10^{-3}\,M - 5.55 \times 10^{-3}\,M}{2.0\text{ h}}$$

$$= 2 \times 10^{-5}\,M/\text{h}$$

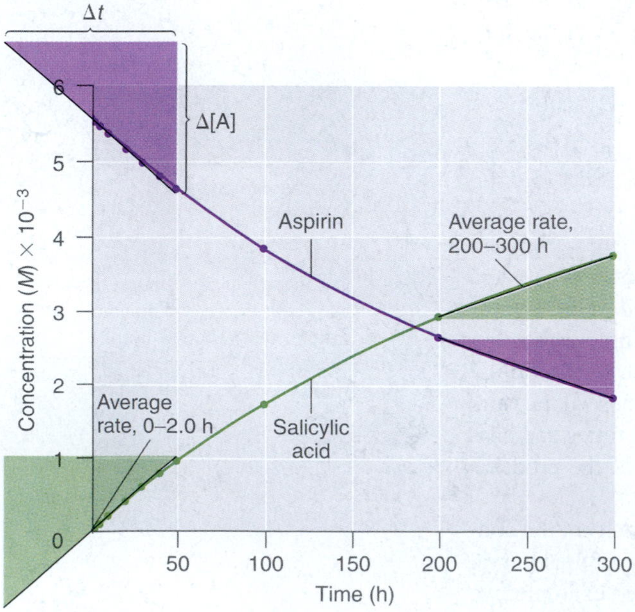

Figure 14.6 The hydrolysis of aspirin. This graph shows the concentrations of aspirin and salicylic acid as a function of time, based on the hydrolysis data in Table 14.1. The time dependence of the concentration of the other product, acetate, is not shown, but based on the stoichiometry of the reaction it is identical to the data for salicylic acid. Interactive Graph

If we now calculate the rate during the last interval given in Table 14.1 (the interval between 200 h and 300 h after the start of the reaction), we find that the rate of the reaction is significantly slower than it was during the first interval ($t = 0–2.0$ h):

$$\text{rate}_{(t=200-300\text{ h})} = \frac{[\text{salicylic acid}]_{300} - [\text{salicylic acid}]_{200}}{300\text{ h} - 200\text{ h}}$$

$$= \frac{3.73 \times 10^{-3}\,M - 2.91 \times 10^{-3}\,M}{100\text{ h}}$$

$$= 8.2 \times 10^{-6}\,M/\text{h}$$

(You should verify from the data in Table 14.1 that you get the same rate using the concentrations of aspirin measured at 200 h and 300 h.)

Calculating the Rate of Fermentation of Sucrose

Notice in the preceding example that the stoichiometric coefficients in the balanced chemical equation are the same for all reactants and products; that is, the reactants and products all have the coefficient 1. Let us look at a reaction in which the coefficients are *not* all the same: the fermentation of sucrose to ethanol and carbon dioxide, which we encountered in Chapter 3.

$$\underset{\text{Sucrose}}{C_{12}H_{22}O_{11}(aq)} + H_2O(l) \longrightarrow 4C_2H_5OH(aq) + 4CO_2(g) \qquad (14.6)$$

The coefficients show us that the reaction produces four molecules of ethanol and four molecules of carbon dioxide for every one molecule of sucrose that is consumed. As before, we can find the rate of the reaction by looking at the change in the concentration of any reactant or product. In this particular case, however, a chemist would probably use the concentration of either sucrose or ethanol because gases are usually measured as volumes and, as you learned in Chapter 10, the volume of CO_2 gas formed will depend on the total volume of the solution being studied and on the solubility of the gas in the solution, not just on the concentration of sucrose. The coefficients in the balanced equation tell us that the rate at which ethanol is formed is always four times faster than the rate at which sucrose is consumed:

$$\frac{\Delta[C_2H_5OH]}{\Delta t} = -\frac{4\Delta[\text{sucrose}]}{\Delta t} \qquad (14.7)$$

Note again that the concentration of the reactant—in this case sucrose—*decreases* with increasing time, so the value of Δ[sucrose] is negative. Consequently, a minus sign is inserted in front of Δ[sucrose] in Equation 14.7 so that the rate of change of the sucrose concentration is expressed as a positive value. Conversely, the ethanol concentration *increases* with increasing time, so its rate of change is automatically expressed as a positive value.

Often the rate of the reaction is expressed in terms of the reactant or product that has the smallest coefficient in the balanced equation. The smallest coefficient in the sucrose fermentation reaction (Equation 14.6) corresponds to sucrose, so the rate is generally defined as

$$\text{rate} = -\frac{\Delta[\text{sucrose}]}{\Delta t} = \frac{1}{4}\left(\frac{\Delta[C_2H_5OH]}{\Delta t}\right) \qquad (14.8)$$

EXAMPLE 14.1

Consider the thermal decomposition of gaseous N_2O_5 to NO_2 and O_2 via the equation

$$2N_2O_5(g) \xrightarrow{\Delta} 4NO_2(g) + O_2(g)$$

Write expressions for the reaction rate in terms of the rates of change in the concentrations of the reactant and each product with time.

Given Balanced chemical equation

Asked for Reaction rate expressions

Strategy

Ⓐ Choose the species in the equation that has the smallest coefficient. Then write an expression for the rate of change of that species with time.

Ⓑ For the remaining species in the equation, use molar ratios to obtain equivalent expressions for the rate of the reaction.

Solution

A Because O_2 has the smallest coefficient in the balanced equation for the reaction, we define the rate of the reaction as the rate of change in the concentration of O_2 and write that expression. **B** We know from the balanced equation that 2 mol of N_2O_5 must decompose for each 1 mol of O_2 produced and that 4 mol of NO_2 are produced for every 1 mol of O_2 produced. The molar ratios of O_2 to N_2O_5 and to NO_2 are thus 1:2 and 1:4, respectively. This means that we divide the rate of change of $[N_2O_5]$ and of $[NO_2]$ by its stoichiometric coefficient to obtain equivalent expressions for the rate of the reaction. For example, because NO_2 is produced at four times the rate of O_2, we must divide the rate of production of NO_2 by 4. The reaction rate expressions are thus

$$\text{rate} = \frac{\Delta[O_2]}{\Delta t} = \frac{\Delta[NO_2]}{4\Delta t} = -\frac{\Delta[N_2O_5]}{2\Delta t}$$

EXERCISE 14.1

The key step in the industrial production of sulfuric acid is the reaction of SO_2 with O_2 to produce SO_3.

$$2SO_2(g) + O_2(g) \longrightarrow 2SO_3(g)$$

Write expressions for the reaction rate in terms of the rate of change of the concentration of each species.

Answer $\text{rate} = -\dfrac{\Delta[O_2]}{\Delta t} = -\dfrac{\Delta[SO_2]}{2\Delta t} = \dfrac{\Delta[SO_3]}{2\Delta t}$

EXAMPLE 14.2

Using the reaction shown in Example 14.1, calculate the rate of the reaction from the following data taken at 56°C:

$$2N_2O_5\,(g) \longrightarrow 4NO_2\,(g) + O_2(g)$$

Time, s	$[N_2O_5]$, M	$[NO_2]$, M	$[O_2]$, M
240	0.0388	0.0314	0.00792
600	0.0197	0.0699	0.0175

Given Balanced chemical equation and concentrations at specific times

Asked for Rate of the reaction

Strategy

A Using the equations in Example 14.1, subtract the initial concentration of a species from its final concentration, and substitute that value into the equation for that species.

B Substitute the value for the time interval into the equation. Check to make sure your units are consistent.

Solution

A We are asked to calculate the rate of the reaction in the interval between $t_1 = 240\ s$ and $t_2 = 600\ s$. From Example 14.1, we see that we can evaluate the rate of the reaction using any of three expressions:

$$\text{rate} = \frac{\Delta[O_2]}{\Delta t} = \frac{\Delta[NO_2]}{4\Delta t} = -\frac{\Delta[N_2O_5]}{2\Delta t}$$

Subtracting the initial concentration from the final concentration of N_2O_5 and inserting the corresponding time interval into the rate expression for N_2O_5 gives

$$\text{rate} = -\frac{\Delta[N_2O_5]}{2\Delta t} = -\frac{[N_2O_5]_{600} - [N_2O_5]_{240}}{2(600\text{ s} - 240\text{ s})}$$

Substituting actual values into the expression gives

$$\text{rate} = -\frac{0.0197\,M - 0.0388\,M}{2(360\text{ s})} = 2.65 \times 10^{-5}\,M/\text{s}$$

Similarly, we can use NO_2 to calculate the rate:

$$\text{rate} = \frac{\Delta[NO_2]}{4\Delta t} = \frac{[NO_2]_{600} - [NO_2]_{240}}{4(600\text{ s} - 240\text{ s})} = \frac{0.0699\,M - 0.0314\,M}{4(360\text{ s})} = 2.67 \times 10^{-5}\,M/\text{s}$$

If we allow for experimental error, this is the same rate we obtained using the data for N_2O_5, as it should be because the rate of the reaction should be the same no matter which concentration is used. We can also use the data for O_2:

$$\text{rate} = \frac{\Delta[O_2]}{\Delta t} = \frac{[O_2]_{600} - [O_2]_{240}}{600\text{ s} - 240\text{ s}} = \frac{0.0175\,M - 0.00792\,M}{360\text{ s}} = 2.66 \times 10^{-5}\,M/\text{s}$$

Again, this is the same value we obtained from the N_2O_5 and NO_2 data. Thus, the rate of the reaction does not depend on which reactant or product is used to measure it.

EXERCISE 14.2

Using the data in the table, calculate the rate of the reaction of $SO_2(g)$ with $O_2(g)$ to give $SO_3(g)$.

$$2SO_2(g) + O_2(g) \longrightarrow 2SO_3(g)$$

Time, s	$[SO_2]$, M	$[O_2]$, M	$[SO_3]$, M
300	0.0270	0.0500	0.0072
720	0.0194	0.0462	0.0148

Answer $9.0 \times 10^{-6}\,M/\text{s}$

Instantaneous Rates of Reaction

So far we have determined average reaction rates over particular intervals of time. We can also determine the **instantaneous rate** of a reaction, which is the rate at any given point in time. As the period of time used to calculate an average rate of a reaction becomes shorter and shorter, the average rate approaches the instantaneous rate.*

Think of the distinction between the instantaneous and average rates of a reaction as being similar to the distinction between the actual speed of a car at any given time on a trip and the average speed of the car for the entire trip. Although you may travel for a long time at 65 mi/h on an interstate during a long trip, there may be times when you travel only 25 mi/h in construction zones, or 0 mi/h if you stop for meals or gas. Thus, your average speed on the trip may be only 50 mi/h, whereas your instantaneous speed on the interstate at a given moment may be 65 mi/h. Whether you are able to stop the car in time to avoid an accident depends on your instantaneous speed, not on your average speed. There are important differences between the speed of a car during a trip

* If you have studied calculus, you may recognize that the instantaneous rate of a reaction at a given time corresponds to the slope of a line tangent to the concentration-versus-time curve at that point—that is, the derivative of concentration with respect to time.

and the speed of a chemical reaction, however. The speed of a car may vary unpredictably over the length of the trip, and the initial part of the trip is often one of the slowest. In a chemical reaction, the initial interval normally has the fastest rate (though this is not always the case), and the rate generally changes smoothly over time.

In chemical kinetics, we generally focus on one particular instantaneous rate, $t = 0$, which is the initial rate of the reaction. Initial rates are determined by measuring the rate of the reaction at various times and then extrapolating a plot of rate versus time to $t = 0$.

Rate Laws

In the preceding section you learned that reaction rates generally decrease with time because reactant concentrations decrease as reactants are converted to products. You also learned that reaction rates generally increase when reactant concentrations are increased. We now examine the mathematical expressions called **rate laws**, which describe the relationships between reactant rates and reactant concentrations. Rate laws are laws as defined in Chapter 1; that is, they are mathematical descriptions of experimentally verifiable data.

Rate laws may be written from either of two different, but related, perspectives. A **differential rate law** expresses the rate of a reaction in terms of *changes* in the concentration of one or more reactants, $\Delta[R]$, over a specific time interval, Δt. In contrast, an **integrated rate law** describes the rate of a reaction in terms of the *initial* concentration, $[R]_0$, and the *measured* concentration of one or more reactants, [R], after a given amount of time, t; we will discuss integrated rate laws in the next section. The integrated rate law can be found by using calculus to integrate the differential rate law; the method of doing so is beyond the scope of this text. *Whether you use a differential or integrated rate law, always check to make sure that the rate law gives the proper units for the rate, usually* M/s.

Reaction Orders

For a reaction with the general equation

$$aA + bB \longrightarrow cC + dD \tag{14.9}$$

the experimentally determined rate law usually has the form

$$\text{rate} = k[A]^m[B]^n \tag{14.10}$$

The *proportionality constant, k*, is called the **rate constant**, and its value is characteristic of the reaction and reaction conditions. A given reaction has a particular value of the rate constant under a given set of conditions, such as temperature, pressure, and solvent; varying the temperature or the solvent usually changes the value of the rate constant. The numerical value of k, however, does *not* change as the reaction progresses under a given set of conditions.

Thus, the rate of a reaction depends on the rate constant for the given set of reaction conditions and on the concentration of A and B, raised to the powers m and n, respectively. The values of m and n are derived from experimental measurements of the changes in reactant concentrations over time and indicate the **reaction order**, the degree to which the rate of the reaction depends on the concentration of each reactant; m and n need not be integers. For example, Equation 14.10 tells us that Reaction 14.9 is mth order in reactant A and nth order in reactant B. It is important to remember that n and m are *not* related to the stoichiometric coefficients a and b in the balanced chemical equation but must be determined experimentally. The *overall reaction order* is the sum of all the exponents in the rate law, or $m + n$.

Although differential rate laws are generally used to describe what is occurring on a molecular level during a reaction, integrated rate laws are used for determining the reaction order and the value of the rate constant from experimental measurements.

(We present general forms for integrated rate laws in Section 14.3.) To illustrate how chemists interpret a differential rate law, we turn to the experimentally derived rate law for the hydrolysis of *t*-butyl bromide in 70% aqueous acetone. This reaction produces *t*-butanol according to the equation

$$(CH_3)_3CBr(soln) + H_2O(soln) \longrightarrow (CH_3)_3COH(soln) + HBr(soln) \qquad (14.11)$$

Combining the rate expression in Equations 14.4 and 14.10 gives us a general expression for the differential rate law:

$$rate = -\frac{\Delta[A]}{\Delta t} = k[A]^m[B]^n \qquad (14.12)$$

Inserting the identities of the reactants into Equation 14.12 gives the following expression for the differential rate law for the reaction:

$$\text{rate} = -\frac{\Delta[(CH_3)_3CBr]}{\Delta t} = k[(CH_3)_3CBr]^m[H_2O]^n \qquad (14.13)$$

Experiments done to determine the rate law for the hydrolysis of *t*-butyl bromide show that the rate of the reaction is directly proportional to the concentration of $(CH_3)_3CBr$ but is independent of the concentration of water. Thus, *m* and *n* in Equation 14.13 are 1 and 0, respectively, and

$$\text{rate} = k[(CH_3)_3CBr]^1[H_2O]^0 = k[(CH_3)_3CBr] \qquad (14.14)$$

Because the exponent for the reactant is 1, the reaction is *first order* in $(CH_3)_3CBr$. It is *zeroth order* in water because the exponent for $[H_2O]$ is 0 (recall that anything raised to the zeroth power equals 1). Thus, the overall reaction order is $1 + 0 = 1$. What the reaction orders tell us in practical terms is that doubling the concentration of $(CH_3)_3CBr$ doubles the rate of the hydrolysis reaction, halving the concentration of $(CH_3)_3CBr$ halves the rate, and so on. Conversely, increasing or decreasing the concentration of water has *no effect* on the rate of the reaction. (Again, remember when you work with rate laws that there is no simple correlation between the stoichiometry of the reaction and the rate law. The values of *k*, *m*, and *n* in the rate law *must* be determined experimentally.) Experimental data show that *k* has the value $5.15 \times 10^{-4}\ s^{-1}$ at 25°C. Note that the rate constant has units of reciprocal seconds (s^{-1}) because the rate of the reaction is defined in units of concentration per unit time (*M*/s). The units of a rate constant depend on the rate law for a particular reaction.

Under conditions identical to those for the *t*-butyl bromide reaction, the experimentally derived differential rate law for the hydrolysis of methyl bromide (CH_3Br) is

$$\text{rate} = -\frac{\Delta[CH_3Br]}{\Delta t} = k'[CH_3Br] \qquad (14.15)$$

This reaction also has an overall reaction order of 1, but the rate constant in Equation 14.15 is approximately 10^6 times smaller than that for *t*-butyl bromide. Thus, methyl bromide hydrolyzes about 1 million times more slowly than *t*-butyl bromide, and this information tells chemists how the reactions differ on a molecular level.

Frequently, changes in reaction conditions also produce changes in a rate law. In fact, chemists often change reaction conditions to obtain clues about what is occurring during a reaction. For example, when *t*-butyl bromide is hydrolyzed in aqueous acetone solution containing OH^- ions rather than in aqueous acetone alone, the differential rate law for the hydrolysis reaction does not change. For methyl bromide, in contrast, the differential rate law becomes rate $= k''[CH_3Br][OH^-]$, with an overall reaction order of 2. Thus, although the two reactions proceed similarly in neutral solution, they proceed very differently in the presence of a base, which again provides clues as to how the reactions differ on a molecular level.

EXAMPLE 14.3

We present three reactions and their experimentally determined differential rate laws. For each reaction: (a) give the *units* of the rate constant; (b) give the *order* of the reaction with respect to each reactant; (c) give the *overall order* of the reaction; and (d) predict what happens to the rate when the concentration of the first species in each chemical equation is doubled.

1. $2HI(g) \xrightarrow{Pt} H_2(g) + I_2(g)$ $\quad$ $\text{Rate} = -\frac{1}{2}\left(\frac{\Delta[HI]}{\Delta t}\right) = k[HI]^2$
2. $2N_2O(g) \xrightarrow{\Delta} 2N_2(g) + O_2(g)$ $\quad$ $\text{Rate} = -\frac{1}{2}\frac{\Delta[N_2O]}{\Delta t} = k$
3. Cyclopropane$(g) \longrightarrow$ propane(g) $\quad$ $\text{Rate} = -\frac{\Delta[\text{cyclopropane}]}{\Delta t} = k[\text{cyclopropane}]$

Given Balanced chemical equations and differential rate laws

Asked for Units of rate constant, reaction orders, and effect of doubling reactant concentration

Strategy

A Express the rate as mol/(L·s), or *M*/s. Then determine the units of each chemical species in the rate law. Divide the units for the rate of the reaction by the units for all species in the rate law to obtain the units for the rate constant.

B Identify the exponent of each species in the rate law to determine the order of the reaction with respect to that species. Sum all exponents to obtain the overall order of the reaction.

C Use the mathematical relationships as expressed in the rate law to determine the effect of doubling the concentration of a single species on the rate of the reaction.

Solution

1. **A** **(a)** Note that $[HI]^2$ will give units of M^2. For the reaction rate to have units of *M*/s, the rate constant must have units of $1/(M \cdot s)$:

$$kM^2 = \frac{M}{s} \qquad k = \frac{M/s}{M^2} = \frac{1}{M \cdot s} = M^{-1} \cdot s^{-1}$$

B **(b)** The exponent in the rate law is 2, so the reaction is second order in HI. **(c)** Because HI is the only reactant and the only species that appears in the rate law, the reaction is also second order overall. **C** **(d)** If the concentration of HI is doubled, the rate will increase from $k[HI]_0^2$ to $k(2[HI])_0^2 = 4k[HI]_0^2$. The rate will therefore quadruple.

2. **A** **(a)** Because no concentration term appears in the rate law, the rate constant must have units of *M*/s in order for the rate to have units of *M*/s. **B** **(b, c)** The rate law tells us that the rate is constant and is independent of the N_2O concentration. That is, the reaction is zeroth order in N_2O and zeroth order overall. **C** **(d)** Because the rate is independent of the N_2O concentration, doubling the concentration will have no effect on the rate.

3. **A** **(a)** The rate law contains only one concentration term raised to the first power. Hence, the rate constant must have units of reciprocal seconds (s^{-1}) to have units of *M*/s for the rate: $M \cdot s^{-1} = M/s$. **B** **(b)** The only concentration in the rate law is that of cyclopropane, and its exponent is 1. This means that the reaction is first order in cyclopropane. **(c)** Cyclopropane is the only species that appears in the rate law, so the reaction is also first order overall. **C** **(d)** Doubling the initial cyclopropane concentration will increase the rate from $k[\text{cyclopropane}]_0$ to $2k[\text{cyclopropane}]_0$. This doubles the rate.

EXERCISE 14.3

Given the following two reactions and their experimentally determined differential rate laws: (a) determine the *units* of the rate constant if time is in seconds; (b) determine the *order* of the reaction with respect to each reactant; (c) give the *overall order* of the reaction; and (d) predict what will happen to the rate when the concentration of the first species in each equation is doubled.

1. $CH_3N{=}NCH_3(g) \longrightarrow C_2H_6(g) + N_2(g)$

$$\text{Rate} = -\frac{\Delta[CH_3N{=}NCH_3]}{\Delta t} = k[CH_3N{=}NCH_3]$$

2. $2NO_2(g) + F_2(g) \longrightarrow 2NO_2F(g)$

$$\text{Rate} = -\frac{\Delta[F_2]}{\Delta t} = -\frac{1}{2}\left(\frac{\Delta[NO_2]}{\Delta t}\right) = k[NO_2][F_2]$$

Answer

1. **(a)** s^{-1}; **(b)** first order in $CH_3N{=}NCH_3$; **(c)** first order overall; **(d)** doubling $[CH_3N{=}NCH_3]$ will double the rate
2. **(a)** $M^{-1} \cdot s^{-1}$; **(b)** first order in NO_2, first order in F_2; **(c)** second order overall; **(d)** doubling $[NO_2]$ will double the rate

14.3 • Methods of Determining Reaction Orders

MGC Reaction Order

In the examples in this text, the exponents in the rate law are almost always the positive integers 1 and 2 or even 0. Thus, the reactions are zeroth, first, or second order in each of the reactants. The common patterns used to identify the order of a reaction are described in this section, where we focus on characteristic types of differential and integrated rate laws and on how to determine reaction orders from experimental data.

Zeroth-Order Reactions

A **zeroth-order reaction** is one whose rate is independent of concentration; its differential rate law is rate = k. We refer to these reactions as *zeroth order* because we could also write their rate in a form such that the exponent of the reactant in the rate law is 0:

$$\text{rate} = -\frac{\Delta[A]}{\Delta t} = k[\text{reactant}]^0 = k(1) = k \tag{14.16}$$

Because the rate is independent of reactant concentration, a graph of the concentration of any reactant as a function of time is a straight line with a slope of $-k$. The value of k is negative because the concentration of the reactant decreases with time. Conversely, a graph of the concentration of any product as a function of time is a straight line with a slope of k, a positive value.

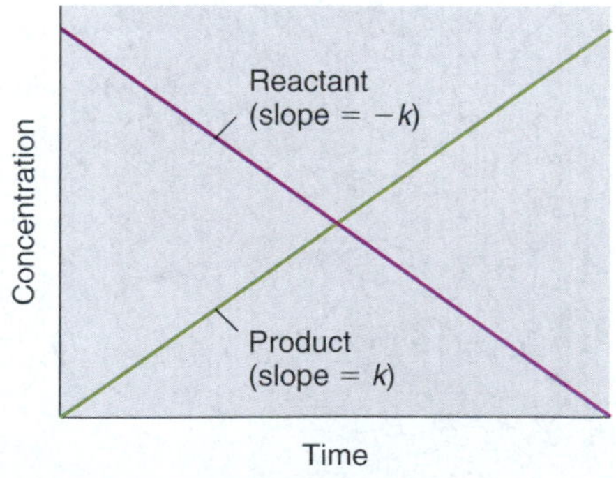

Shapes of zeroth-order curves. The change in concentration of reactant and product with time in a zeroth-order reaction.

The integrated rate law for a zeroth-order reaction also produces a straight line and has the general form

$$[A] = [A]_0 - kt \tag{14.17}$$

where $[A]_0$ is the initial concentration of reactant A. (Notice that Equation 14.17 has the form of the algebraic equation for a straight line, $y = mx + b$, with $y = [A]$, $mx = -kt$, and $b = [A]_0$.) In a zeroth-order reaction, the rate constant must have the same units as the rate of the reaction, typically M/s.

Although it may seem counterintuitive for the rate of a reaction to be independent of the reactant concentration(s), such reactions are rather common. They occur most often when the reaction rate is determined by available surface area. An example is the decomposition of N_2O on a platinum surface to produce N_2 and O_2, which occurs at temperatures ranging from 200°C to 400°C:

$$2N_2O(g) \xrightarrow{\text{Pt}} 2N_2(g) + O_2(g) \tag{14.18}$$

Without a platinum surface, the reaction requires temperatures higher than 700°C, but between 200°C and 400°C the only factor that determines how rapidly N_2O decomposes is the amount of Pt *surface* available (not the *amount* of Pt). As long as there is

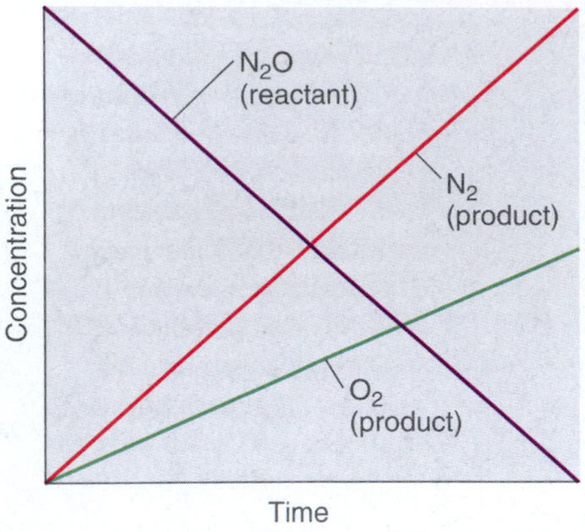

Figure 14.7 A zeroth-order reaction. This graph shows the concentrations of reactants and products vs. time for the zeroth-order catalyzed decomposition of N_2O to N_2 and O_2 on a platinum surface. Note that the change in the concentrations of all species with time is linear. Interactive Graph

Note the pattern

If a plot of reactant concentration versus time is linear, then the reaction is zeroth order *in that reactant.*

enough N_2O to react with the entire Pt surface, doubling or quadrupling the N_2O concentration will have no effect on the rate.* The rate of the reaction is

$$\text{rate} = -\frac{1}{2}\left(\frac{\Delta[N_2O]}{\Delta t}\right) = \frac{1}{2}\left(\frac{\Delta[N_2]}{\Delta t}\right) = \frac{\Delta[O_2]}{\Delta t} = k[N_2O]^0 = k \qquad (14.19)$$

Thus, the rate at which N_2O is consumed and the rates at which N_2 and O_2 are produced are independent of concentration. As shown in Figure 14.7, the change in the concentrations of all species with time is linear. Most important, the exponent (0) corresponding to the N_2O concentration in the experimentally derived rate law is *not* the same as the reactant's stoichiometric coefficient in the balanced equation (2). For this reaction, as for all others, *the rate law must be determined experimentally.*

A zeroth-order reaction that takes place in the human liver is the oxidation of ethanol (from alcoholic beverages) to acetaldehyde, catalyzed by the enzyme alcohol dehydrogenase. At high ethanol concentrations, this reaction is also a zeroth-order reaction. The overall reaction equation is

$$CH_3CH_2OH + NAD^+ \xrightarrow[\text{dehydrogenase}]{\text{Alcohol}} CH_3\overset{\overset{\displaystyle O}{\|}}{C}H + NADH + H^+ \qquad (14.20)$$

where NAD^+ and NADH are the oxidized and reduced forms of a species used by all organisms to transport electrons. When an alcoholic beverage is consumed, the ethanol is rapidly absorbed into the blood. Its concentration then decreases at a constant rate until it reaches zero (Figure 14.8a). An average 70-kg person typically takes about 2.5 h to oxidize the 15 mL of ethanol contained in a single 12-oz can of beer, a 5-oz glass of wine, or a shot of distilled spirits, such as whiskey or brandy. The actual rate, however, varies a great deal from person to person, depending on body size and the amount of alcohol dehydrogenase in the liver. The rate of the reaction does not increase if a greater quantity of alcohol is consumed over the same period of time because the rate is determined only by the amount of enzyme present in the liver.† When the ethanol has been completely oxidized and its concentration drops to essentially zero, the rate of oxidation also drops rapidly (Figure 14.8b).

These examples illustrate two important points: (1) *In a zeroth-order reaction, the reaction rate does not depend on the reactant concentration;* and (2) *a linear change in concentration with time is a clear indication of a zeroth-order reaction.*

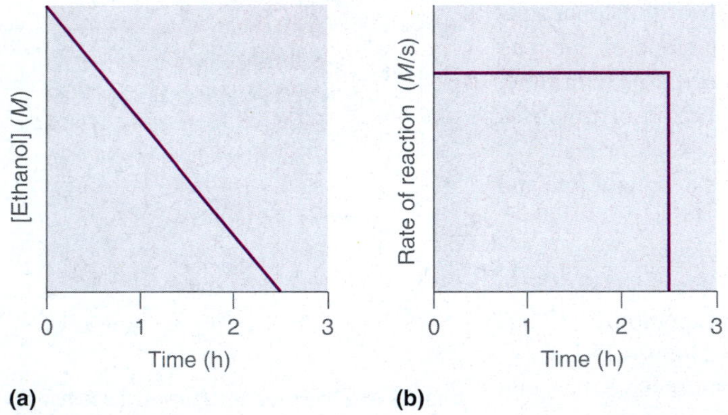

Figure 14.8 The catalyzed oxidation of ethanol. **(a)** The concentration of ethanol in human blood decreases linearly with time, which is typical of a zeroth-order reaction. **(b)** The rate at which ethanol is oxidized is constant until the ethanol concentration reaches essentially zero, at which point the rate of the reaction drops to zero. Interactive Graph

First-Order Reactions

In a **first-order reaction**, the reaction rate is directly proportional to the concentration of one of the reactants. First-order reactions often have the general form A $\longrightarrow$ products. The differential rate for a first-order reaction is

$$\text{rate} = -\frac{\Delta[A]}{\Delta t} = k[A] \qquad (14.21)$$

Thus, if the concentration of A is doubled, the rate of the reaction doubles; if the concentration of A is increased by a factor of 10, the rate increases by a factor of 10; and so forth. Because the units of the rate of the reaction are *M*/s, as always, the units of a first-order rate constant are inverse seconds, s^{-1}.

* At very low concentrations of N_2O, where there are not enough molecules present to occupy the entire available Pt surface, the rate of the reaction is dependent on the N_2O concentration.

† Contrary to popular belief, the caffeine in coffee is ineffective at catalyzing the oxidation of ethanol.

The integrated rate law for a first-order reaction can be written in two different ways: one using exponentials and one using logarithms. The exponential form is

$$[A] = [A]_0 e^{-kt} \quad (14.22)$$

where $[A]_0$ is the initial concentration of reactant A at $t = 0$, k is the rate constant, and e is the base of the natural logarithms which has the value 2.718 (see Essential Skills 6), to three decimal places. Recall that an integrated rate law gives the relationship between reactant concentration and time. Equation 14.22 predicts that the concentration of A will decrease in a smooth exponential curve over time. By taking the natural logarithm of each side of Equation 14.22 and rearranging, we obtain an alternative logarithmic expression of the relationship between the concentration of A and t:

$$\ln[A] = \ln[A]_0 - kt \quad (14.23)$$

Because Equation 14.23 has the form of the algebraic equation for a straight line, $y = mx + b$, with $y = \ln[A]$ and $b = \ln[A]_0$, a plot of $\ln[A]$ versus t for a first-order reaction should give a straight line with a slope of $-k$ and an intercept of $\ln[A]_0$. Either the differential rate law (Equation 14.21) or the integrated rate law (Equation 14.23) can be used to determine whether a particular reaction is first order.

First-order reactions are very common. In this chapter we have already encountered two examples of first-order reactions: the hydrolysis of aspirin (Equation 14.5) and the reaction of *t*-butyl bromide with water to give *t*-butanol (Equation 14.11). Another reaction that exhibits apparent first-order kinetics is the hydrolysis of the anticancer drug cisplatin.

Cisplatin, the first "inorganic" anticancer drug to be discovered, is unique in its ability to cause complete remission of the relatively rare but deadly cancers of the reproductive organs in young adults. The structures of cisplatin and its hydrolysis product are

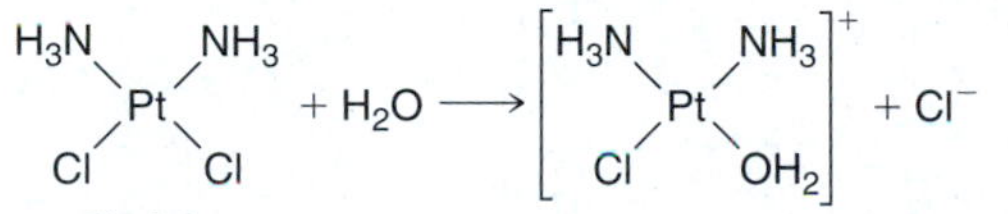

(14.24)

Cisplatin

Both platinum compounds have four groups arranged in a square plane around a Pt(II) ion. The reaction shown in Equation 14.24 is important because cisplatin, the form in which the drug is administered, is *not* the form in which the drug is active. Instead, at least one chloride ion must be replaced by water to produce a species that reacts with DNA to prevent cell division and tumor growth. Consequently, the kinetics of the reaction in Equation 14.24 have been studied extensively to find ways of maximizing the concentration of the active species.

The rate law and reaction order of the hydrolysis of cisplatin are determined from experimental data such as those displayed in Table 14.2. The table lists initial rate data for four experiments in which the reaction was run at pH 7.0 and 25°C but with different initial concentrations of cisplatin. Because the rate increases with increasing cisplatin concentration, we know this cannot be a zeroth-order reaction. Comparing Experiments 1 and 2 in Table 14.2 shows that the rate of the reaction doubles

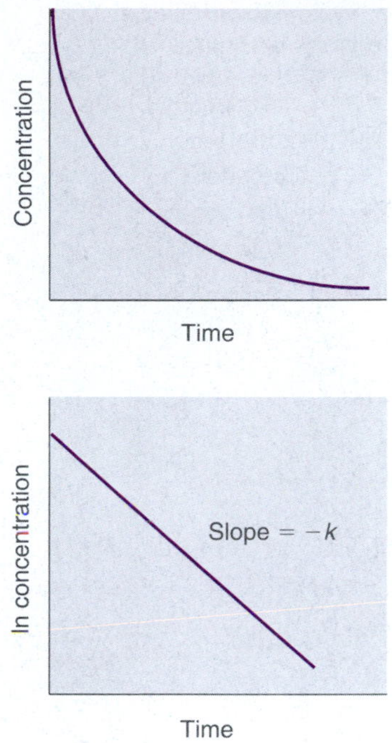

Shapes of first-order curves. The expected shapes of the curves for plots of reactant concentration vs. time *(top)* and the natural logarithm of reactant concentration vs. time *(bottom)* for a first-order reaction.

Note the pattern

If a plot of reactant concentration versus time is not *linear but a plot of the natural logarithm of reactant concentration versus time is* linear, *then the reaction is* first order.

TABLE 14.2 Rates of hydrolysis of cisplatin as a function of concentration at pH 7.0 and 25°C

Experiment	$[\text{Cisplatin}]_0$, *M*	Initial Rate, *M*/min
1	0.0060	9.0×10^{-6}
2	0.012	1.8×10^{-5}
3	0.024	3.6×10^{-5}
4	0.030	4.5×10^{-5}

[(1.8×10^{-5} M/min) ÷ (9.0×10^{-6} M/min) = 2.0] when the concentration of cisplatin is doubled (from 0.0060 M to 0.012 M). Similarly, comparing Experiments 1 and 4 shows that the rate of the reaction increases by a factor of 5 [(4.5×10^{-5} M/min) ÷ (9.0×10^{-6} M/min) = 5.0] when the concentration of cisplatin is increased by a factor of 5 (from 0.0060 M to 0.030 M). Because the rate of the reaction is directly proportional to the concentration of the reactant, the exponent of the cisplatin concentration in the rate law must be 1, and so the rate law is rate = $k[\text{cisplatin}]^1$. Thus, the reaction is first order. Knowing this, we can calculate the rate constant using the differential rate law for a first-order reaction and the data in any line of Table 14.2. For example, substituting the values for Experiment 3 into Equation 14.21 gives

$$3.6 \times 10^{-5}\ M/\text{min} = k(0.024\ M)$$
$$1.5 \times 10^{-3}\ \text{min}^{-1} = k$$

Knowing the rate constant for the hydrolysis of cisplatin and the rate constants for subsequent reactions that produce species that are highly toxic enables hospital pharmacists to provide patients with solutions that contain only the desired form of the drug.

EXAMPLE 14.4

At high temperatures, ethyl chloride produces HCl and ethylene by the reaction

$$CH_3CH_2Cl(g) \xrightarrow{\Delta} HCl(g) + C_2H_4(g)$$

Using the rate data for the reaction at 650°C presented in the table, calculate the reaction order with respect to the concentration of ethyl chloride, and determine the rate constant for the reaction.

Experiment	$[CH_3CH_2Cl]_0$, M	Initial Rate, M/s
1	0.010	1.6×10^{-8}
2	0.015	2.4×10^{-8}
3	0.030	4.8×10^{-8}
4	0.040	6.4×10^{-8}

Given Balanced chemical equation, initial concentrations of reactant, and initial rates of reaction

Asked for Reaction order and rate constant

Strategy

A Compare the data from two experiments to determine the effect of changing the concentration of a species on the rate of the reaction.
B Compare the observed effect with behavior characteristic of zeroth- and first-order reactions to determine the order of the reaction. Write the rate law for the reaction.
C Use measured concentrations and rate data from any of the experiments to find the rate constant.

Solution

The order of the reaction with respect to ethyl chloride is determined by examining the effect of changes in the ethyl chloride concentration on the rate of the reaction. **A** Comparing Experiments 2 and 3 shows that doubling the concentration doubles the rate, so the rate is proportional to $[CH_3CH_2Cl]$. Similarly, comparing Experiments 1 and 4 shows that quadrupling the concentration quadruples the rate, again indicating that the rate is directly proportional to $[CH_3CH_2Cl]$. **B** This behavior is characteristic of a first-order reaction, for which the rate law is rate = $k[CH_3CH_2Cl]$. **C** We can calculate the rate constant, k, using any line in the table. Selecting Experiment 1 gives

$$1.60 \times 10^{-8}\ M/\text{s} = k(0.010\ M)$$
$$1.6 \times 10^{-6}\ \text{s}^{-1} = k$$

EXERCISE 14.4

Sulfuryl chloride, SO_2Cl_2, decomposes to SO_2 and Cl_2 by the reaction

$$SO_2Cl_2(g) \longrightarrow SO_2(g) + Cl_2(g)$$

Data for the reaction at 320°C are listed in the table. Calculate the order of the reaction with regard to sulfuryl chloride, and determine the rate constant for the reaction.

Experiment	$[SO_2Cl_2]_0$, M	Initial Rate, M/s
1	0.0050	1.10×10^{-7}
2	0.0075	1.65×10^{-7}
3	0.0100	2.20×10^{-7}
4	0.0125	2.75×10^{-7}

Answer First order; $k = 2.2 \times 10^{-5}\ s^{-1}$

We can also use the integrated rate law to determine the reaction rate for the hydrolysis of cisplatin. To do this, we examine the change in the concentration of the reactant or the product as a function of time at a single initial cisplatin concentration. Figure 14.9a shows plots for a solution that originally contained 0.0100 M cisplatin and was maintained at pH 7 and 25°C. The concentration of cisplatin decreases smoothly with time, and the concentration of chloride ion increases in a similar way. When we plot the natural logarithm of the concentration of cisplatin versus time, we obtain the plot shown in Figure 14.9b. The straight line is consistent with the behavior of a system that obeys a first-order rate law. We can use any two points on the line to calculate the slope of the line, which gives us the rate constant for the reaction. Thus, taking the points from Figure 14.9a for t = 100 min ([cisplatin] = 0.0086 M) and t = 1000 min ([cisplatin] = 0.0022 M) gives

$$\text{slope} = \frac{\ln[\text{cisplatin}]_{1000} - \ln[\text{cisplatin}]_{100}}{1000\text{ min} - 100\text{ min}}$$

$$-k = \frac{\ln 0.0022 - \ln 0.0086}{1000\text{ min} - 100\text{ min}} = \frac{-6.12 - (-4.76)}{900\text{ min}} = -1.51 \times 10^{-3}\text{ min}^{-1}$$

$$k = 1.5 \times 10^{-3}\text{ min}^{-1}$$

Notice that the slope is negative because we are calculating the rate of disappearance of cisplatin. Also notice that the rate constant has units of min^{-1} because the times plotted on the horizontal axes in Figures 14.9a and 14.9b are in minutes rather than seconds.

The order of the reaction and the magnitude of the rate constant we obtain using the integrated rate law are exactly the same as those we calculated earlier using the differential rate law. This must be true if the experiments were carried out under the same conditions.

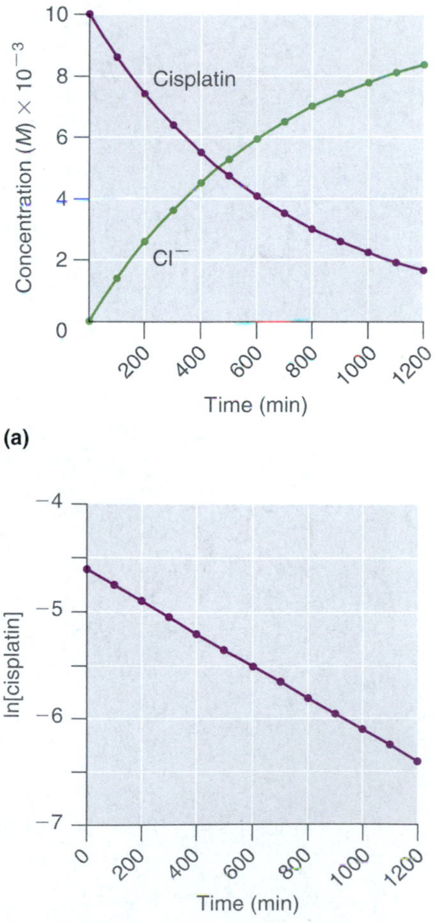

Figure 14.9 The hydrolysis of cisplatin, a first-order reaction. These plots show hydrolysis of cisplatin at pH 7.0 and 25°C as **(a)** the experimentally determined concentrations of cisplatin and chloride ions vs. time and **(b)** the natural logarithm of the cisplatin concentration vs. time. The straight line in (b) is expected for a first-order reaction. Interactive Graph

EXAMPLE 14.5

Refer back to Example 14.4. If a sample of ethyl chloride with an initial concentration of 0.0200 M is heated at 650°C, what is the concentration of ethyl chloride after 10 h? How many hours at 650°C must elapse in total for the concentration to decrease to 0.0050 M? (Recall that we calculated the rate constant for this reaction in Example 14.4.)

Given Initial concentration, rate constant, and time interval

Asked for Concentration at specified time and time required to obtain particular concentration

Strategy

A Substitute values for the initial concentration, $[A]_0$, and the calculated rate constant for the reaction, k, into the integrated rate law for a first-order reaction. Calculate the concentration, [A], at the given time, t.

B Given a concentration [A], solve the integrated rate law for time t.

Solution

The exponential form of the integrated rate law for a first-order reaction (Equation 14.22) is $[A] = [A]_0 e^{-kt}$. **A** Having been given the initial concentration of ethyl chloride, $[A]_0$, and having calculated the rate constant, k, in Example 14.4 ($k = 1.6 \times 10^{-6}\ s^{-1}$), we can use the rate law to calculate the concentration of the reactant at a given time t. Substituting the known values into the integrated rate law gives

$$\begin{aligned}[CH_3CH_2Cl]_{10\,h} &= [CH_3CH_2Cl]_0\, e^{-kt} \\ &= 0.0200\, M\{e^{-(1.6\times10^{-6}\,s^{-1})[(10\,h)(60\,min/h)(60\,s/min)]}\} \\ &= 0.0189\, M\end{aligned}$$

We could also have used the logarithmic form of the integrated rate law (Equation 14.23), which gives

$$\begin{aligned}\ln[CH_3CH_2Cl]_{10\,h} &= \ln[CH_3CH_2Cl]_0 - kt \\ &= \ln 0.0200 - (1.6 \times 10^{-6}\, s^{-1})[(10\,h)(60\,min/h)(60\,s/min)] \\ &= -3.912 - 0.0576 = -3.970 \\ [CH_3CH_2Cl]_{10\,h} &= e^{-3.970}\, M \\ &= 0.0189\, M\end{aligned}$$

B To calculate the amount of time required to reach a given concentration, we must solve the integrated rate law for t. Equation 14.23 gives

$$\ln[CH_3CH_2Cl]_t = \ln[CH_3CH_2Cl]_0 - kt$$

$$kt = \ln[CH_3CH_2Cl]_0 - \ln[CH_3CH_2Cl]_t = \ln\frac{[CH_3CH_2Cl]_0}{[CH_3CH_2Cl]_t}$$

$$t = \frac{1}{k}\left(\ln\frac{[CH_3CH_2Cl]_0}{[CH_3CH_2Cl]_t}\right) = \frac{1}{1.6\times10^{-6}\,s^{-1}}\left(\ln\frac{0.0200\,M}{0.0050\,M}\right)$$

$$= \frac{\ln 4.0}{1.6\times10^{-6}\,s^{-1}} = 8.7\times10^5\,s = 240\,h = 2.4\times10^2\,h$$

EXERCISE 14.5

In Exercise 14.4 you found that the decomposition of sulfuryl chloride (SO_2Cl_2) is first order, and you calculated the rate constant at 320°C. Use the form(s) of the integrated rate law to find the amount of SO_2Cl_2 that remains after 20 h if a sample with an original concentration of 0.123 M is heated at 320°C. How long would it take for 90% of the SO_2Cl_2 to decompose?

Answer 0.0252 M; 29 h

Second-Order Reactions

The simplest kind of **second-order reaction** is one whose rate is proportional to the square of the concentration of one reactant. These generally have the form 2A ⟶ products. A second kind of second-order reaction has a rate that is proportional to the *product* of the concentrations of two reactants. Such reactions generally have the form A + B ⟶ products. An example of the former is a *dimerization reaction,* in which two smaller molecules, each called a *monomer*, combine to form a larger molecule (a *dimer*).

The differential rate law for the simplest second-order reaction in which 2A ⟶ products is

$$\text{rate} = -\frac{\Delta[\text{A}]}{2\Delta t} = k[\text{A}]^2 \qquad (14.25)$$

Consequently, doubling the concentration of A quadruples the rate of the reaction. In order for the units of the rate to be M/s, the units of a second-order rate constant must be $M^{-1} \cdot s^{-1}$. Because molarity is expressed as mol/L, the unit of the rate constant can also be written as L(mol · s).

For the reaction 2A ⟶ products, the concentration of the reactant at a given time can be described by the following integrated rate law:

$$\frac{1}{[\text{A}]} = \frac{1}{[\text{A}]_0} + kt \qquad (14.26)$$

Because Equation 14.26 has the form of an algebraic equation for a straight line, $y = mx + b$, with $y = 1/[\text{A}]$ and $b = 1/[\text{A}]_0$, a plot of 1/[A] versus t for a simple second-order reaction is a straight line with a slope of k and an intercept of $1/[\text{A}]_0$.

Simple second-order reactions are common. In addition to dimerization reactions, two other examples are the decomposition of NO_2 to NO and O_2 and the decomposition of HI to I_2 and H_2. Most examples involve simple inorganic molecules, but there are organic examples, too. We can follow the progress of the reaction described in the next paragraph by monitoring the decrease in the intensity of the red color of the reaction mixture.

Many cyclic organic compounds that contain two carbon–carbon double bonds undergo a dimerization reaction to give complex structures. One example is

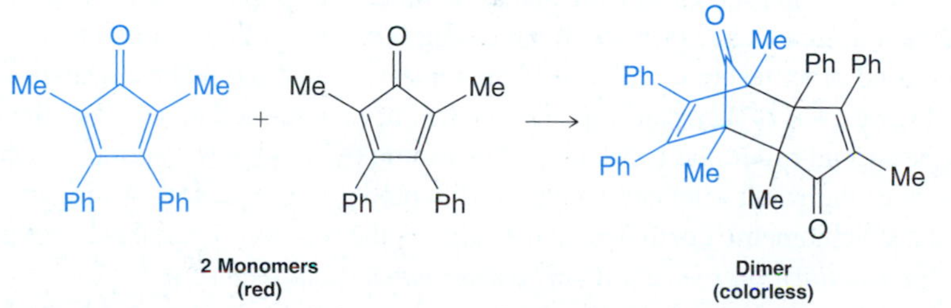

(14.27)

For simplicity, we will refer to this reactant and product as "monomer" and "dimer," respectively.* Because the monomers are the same, the general equation for this reaction is 2A ⟶ product. This reaction represents an important class of organic reactions used in the pharmaceutical industry to prepare complex carbon skeletons for the synthesis of drugs. Like the first-order reactions studied in the preceding section, it can be analyzed using either the differential rate law (Equation 14.25) or the integrated rate law (Equation 14.26).

To determine the differential rate law for the reaction, we need data on how the rate of the reaction varies as a function of monomer concentrations, which are provided in Table 14.3. From the data, we see that the reaction rate is not independent of the monomer concentration, so this is not a zeroth-order reaction. We also see that the reaction rate is not proportional to the monomer concentration, so the reaction is not first order. Comparing the data in the second and fourth lines shows that the rate decreases by a factor of 2.8 when the monomer concentration decreases by a factor of 1.7:

$$\frac{5.0 \times 10^{-5}\ M/\text{min}}{1.8 \times 10^{-5}\ M/\text{min}} = 2.8 \quad \text{and} \quad \frac{3.4 \times 10^{-3}\ M}{2.0 \times 10^{-3}\ M} = 1.7$$

Because $(1.7)^2 = 2.9 \approx 2.8$, the rate of the reaction is approximately proportional to the *square* of the monomer concentration.

$$\text{rate} \propto [\text{monomer}]^2$$

---

* The systematic name of the monomer is 2,5-dimethyl-3,4-diphenylcyclopentadienone.

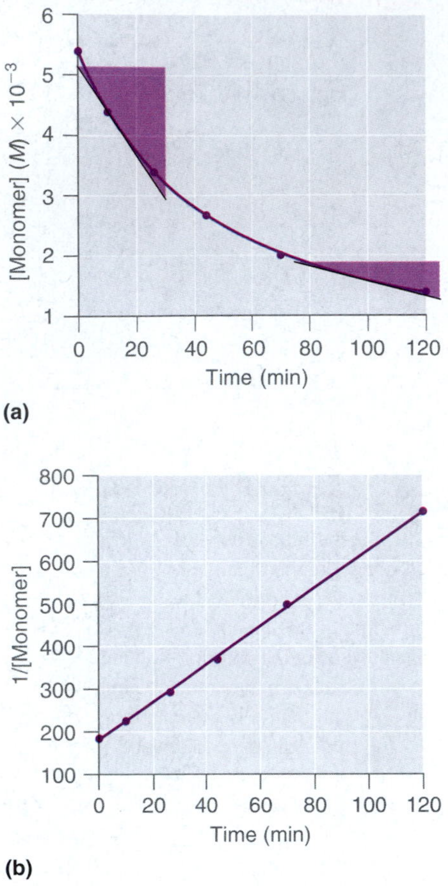

Figure 14.10 Dimerization of a monomeric compound, a second-order reaction. These plots correspond to dimerization of the monomer in Equation 14.27 as **(a)** the experimentally determined concentration of monomer vs. time and **(b)** 1/[monomer] vs. time. The straight line in (b) is expected for a simple second-order reaction. Interactive Graph

TABLE 14.3 Rates of reaction as a function of monomer concentration for an initial monomer concentration of 0.0054 *M*

Time, min	[Monomer], *M*	Instantaneous Rate, *M*/min
10	0.0044	8.0×10^{-5}
26	0.0034	5.0×10^{-5}
44	0.0027	3.1×10^{-5}
70	0.0020	1.8×10^{-5}
120	0.0014	8.0×10^{-6}

This means that the reaction is second order in the monomer. Using Equation 14.25 and the data from any line in Table 14.3, we can calculate the rate constant. Substituting values at time = 10 min, for example, gives

$$\text{rate} = k[\text{A}]^2$$
$$8.0 \times 10^{-5}\ M/\text{min} = k(4.4 \times 10^{-3}\ M)^2$$
$$4.1\ M^{-1} \cdot \text{min}^{-1} = k$$

We can also determine the reaction order using the integrated rate law. To do so, we use the decrease in the concentration of the monomer as a function of time for a single reaction, plotted in Figure 14.10a. The measurements show that the concentration of the monomer (initially $5.4 \times 10^{-3}\ M$) decreases with increasing time. This graph also shows that the *rate* of the reaction decreases smoothly with increasing time. According to the integrated rate law for a second-order reaction, a plot of 1/[monomer] versus *t* should be a straight line, as shown in Figure 14.10b. Any pair of points on the line can be used to calculate the slope, which is the second-order rate constant. In this example, $k = 4.1\ M^{-1} \cdot \text{min}^{-1}$, which is consistent with the result obtained using the differential rate equation. Although in this example the stoichiometric coefficient is the same as the reaction order, this is not always the case. *The reaction order must always be determined experimentally.*

Note that for two or more reactions of the *same order,* the reaction with the largest rate constant is the fastest. Because the units of the rate constants for zeroth-, first-, and second-order reactions are different, however, we cannot compare the magnitudes of rate constants for reactions that have different orders. The differential and integrated rate laws for zeroth-, first-, and second-order reactions and their corresponding graphs are shown in Table 14.6 in Section 14.4.

Note the pattern

If a plot of reactant concentration versus time is not linear but a plot of 1 over reaction concentration versus time is linear, then the reaction is second order.

EXAMPLE 14.6

At high temperatures, nitrogen dioxide decomposes to nitric oxide and oxygen.

$$2NO_2(g) \xrightarrow{\Delta} 2NO(g) + O_2(g)$$

Experimental data for the reaction at 300°C and four initial concentrations of NO_2 are listed in the table.

Experiment	$[NO_2]_0$, *M*	Initial Rate, *M*/s
1	0.015	1.22×10^{-4}
2	0.010	5.40×10^{-5}
3	0.0080	3.46×10^{-5}
4	0.0050	1.35×10^{-5}

Determine the order of the reaction and the rate constant.

Given Balanced chemical equation, initial concentrations, and initial rates

Asked for Reaction order and rate constant

Strategy

Ⓐ From the experiments, compare the changes in the initial rates of reaction with the corresponding changes in the initial concentrations. Determine whether the changes are characteristic of zeroth-, first-, or second-order reactions.
Ⓑ Determine the appropriate rate law. Using this rate law and data from any experiment, solve for the rate constant, *k*.

Solution

Ⓐ We can determine the order of the reaction with respect to nitrogen dioxide by comparing the changes in NO_2 concentrations with the corresponding reaction rates. Comparing Experiments 2 and 4 shows that doubling the concentration quadruples the rate $[(5.40 \times 10^{-5}) \div (1.35 \times 10^{-5}) = 4.0]$, which means that the rate is proportional to $[NO_2]^2$. Similarly, comparing Experiments 1 and 4 shows that tripling the concentration increases the rate by a factor of 9, again indicating that the rate is proportional to $[NO_2]^2$. This behavior is characteristic of a second-order reaction. Ⓑ We have rate $= k\,[NO_2]^2$. We can calculate the rate constant, *k*, using data from any experiment in the table. Selecting Experiment 2, for example, gives

$$\text{rate} = k[NO_2]^2$$
$$5.40 \times 10^{-5}\ M/\text{s} = k(0.010\ M)^2$$
$$0.54\ M^{-1} \cdot s^{-1} = k$$

EXERCISE 14.6

When the highly reactive species HO_2 forms in the atmosphere, one important reaction that then removes it from the atmosphere is

$$2HO_2(g) \longrightarrow H_2O_2(g) + O_2(g)$$

The kinetics of this reaction have been studied in the laboratory, and some initial rate data at 25°C are listed in the table.

Experiment	$[HO_2]_0$, *M*	Initial Rate, *M*/s
1	1.1×10^{-8}	1.7×10^{-7}
2	2.5×10^{-8}	8.8×10^{-7}
3	3.4×10^{-8}	1.6×10^{-6}
4	5.0×10^{-8}	3.5×10^{-6}

Determine the order of the reaction and the rate constant.

Answer Second order in HO_2; $k = 1.4 \times 10^9\ M^{-1} \cdot s^{-1}$

EXAMPLE 14.7

If a flask that initially contains 0.056 *M* NO_2 is heated at 300°C, what will be the concentration of NO_2 after 1.0 h? How long will it take for the concentration of NO_2 to decrease to 10% of the initial concentration? Use the integrated rate law for a second-order reaction (Equation 14.26) and the rate constant calculated in Example 14.6.

Given Balanced chemical equation, rate constant, time interval, and initial concentration

Asked for Final concentration and time required to reach specified concentration

Strategy

A Given k, t, and $[A]_0$, use the integrated rate law for a second-order reaction to calculate [A].
B Setting [A] equal to one-tenth $[A]_0$, use the same equation to solve for time t.

Solution

A We know k and $[NO_2]_0$, and we are asked to determine $[NO_2]$ at t = 1 h (3600 s). Substituting the appropriate values into Equation 14.26 gives

$$\frac{1}{[NO_2]_{3600}} = \frac{1}{[NO_2]_0} + kt = \frac{1}{0.056\ M} + [(0.54\ M^{-1} \cdot s^{-1})(3600\ s)]$$
$$= 2.0 \times 10^3\ M^{-1}$$

Thus, $[NO_2]_{3600} = 5.1 \times 10^{-4}\ M$. **B** In this case, we know k and $[NO_2]_0$, and we are asked to calculate at what time $[NO_2] = 0.1[NO_2]_0 = 0.1(0.056\ M) = 0.0056\ M$. To do this, we solve Equation 14.26 for t, using the concentrations given.

$$t = \frac{(1/[NO_2]) - (1/[NO_2]_0)}{k} = \frac{(1/0.0056\ M) - (1/0.056\ M)}{0.54\ M^{-1} \cdot s^{-1}} = 3 \times 10^2\ s = 5.0\ \text{min}$$

Notice that NO_2 decomposes very rapidly; under these conditions, the reaction is 90% complete in only 5 min.

EXERCISE 14.7

In Exercise 14.6 you calculated the rate constant for the decomposition of HO_2 as $k = 1.4 \times 10^9\ M^{-1} \cdot s^{-1}$. This high rate constant means that HO_2 decomposes rapidly under the reaction conditions given in the problem. In fact, the HO_2 molecule is so reactive that it is virtually impossible to obtain in high concentrations. Given a 0.0010 M sample of HO_2, calculate the concentration of HO_2 that remains after 1.0 h at 25°C. How long will it take for 90% of the HO_2 to decompose? Use the integrated rate law for a second-order reaction (Equation 14.26) and the rate constant calculated in Exercise 14.6.

Answer $2.0 \times 10^{-13}\ M$; 6.4×10^{-6} s

In addition to the simple second-order reaction and rate law we have just described, another very common second-order reaction has the general form A + B ⟶ products, in which the reaction is first order in A and first order in B. The differential rate law for this reaction is

$$\text{rate} = -\frac{\Delta[A]}{\Delta t} = -\frac{\Delta[B]}{\Delta t} = k[A][B] \qquad (14.28)$$

Because the reaction is first order both in A and in B, it has an overall reaction order of 2. (The integrated rate law for this reaction is rather complex, so we will not describe it.) We can recognize second-order reactions of this sort because the rate is proportional to the concentrations of each of two reactants. We presented one example at the end of Section 14.2, the reaction of CH_3Br with OH^- to produce CH_3OH.

Determining the Rate Law of a Reaction

The number of fundamentally different mechanisms (sets of steps in a reaction) is actually rather small compared to the large number of chemical reactions that can occur. Thus, understanding reaction mechanisms can simplify what might seem to be a confusing variety of chemical reactions. The first step in discovering the mechanism of a reaction is to determine the reaction's rate law. This can be done by designing experiments that measure the concentration(s) of one or more of the reactants or

TABLE 14.4 Rate data for a hypothetical reaction of the form $A + B \longrightarrow$ products

Experiment	[A], *M*	[B], *M*	Initial Rate, *M*/min
1	0.50	0.50	8.5×10^{-3}
2	0.75	0.50	19×10^{-3}
3	1.00	0.50	34×10^{-3}
4	0.50	0.75	8.5×10^{-3}
5	0.50	1.00	8.5×10^{-3}

products as a function of time. For the reaction $A + B \longrightarrow$ products, for example, we need to determine the value of k and the exponents m and n in the equation

$$\text{rate} = k[A]^m[B]^n \tag{14.29}$$

To do this, we might keep the initial concentration of B constant while varying the initial concentration of A and calculating the initial rate of the reaction. This information would permit us to deduce the order of the reaction with respect to A. Similarly, we could determine the order of the reaction with respect to B by studying the initial reaction rate when the initial concentration of A is kept constant while the initial concentration of B is varied. In earlier examples we determined the order of a reaction with respect to a given reactant by comparing the different rates obtained when only the concentration of the reactant in question was changed. An alternative way of determining reaction orders is to set up a proportion using the rate laws for two different experiments.

Rate data for a hypothetical reaction of the type $A + B \longrightarrow$ products are given in Table 14.4. The general rate law for the reaction is given in Equation 14.29. We can obtain the value of m or n directly by using a proportion of the rate laws for two experiments in which the concentration of one of the reactants is the same, such as Experiments 1 and 3 in Table 14.4.

$$\frac{\text{rate}_1}{\text{rate}_3} = \frac{k[A_1]^m[B_1]^n}{k[A_3]^m[B_3]^n}$$

Inserting the appropriate values from Table 14.4 gives

$$\frac{8.5 \times 10^{-3}\ \cancel{M/\text{min}}}{34 \times 10^{-3}\ \cancel{M/\text{min}}} = \frac{\cancel{k}[0.50\ \cancel{M}]^m\cancel{[0.50\ M]^n}}{\cancel{k}[1.00\ \cancel{M}]^m\cancel{[0.50\ M]^n}}$$

Noting that 1.00 to any power is 1, we can cancel like terms to give $0.25 = [0.50]^m$, which can also be written as $1/4 = [1/2]^m$. Thus, we can conclude that $m = 2$ and that the reaction is second order in A. Note that by selecting two experiments in which the concentration of B is the same, we were able to solve for the value of m.

Conversely, by selecting two experiments in which the concentration of A is the same (for example, Experiments 5 and 1), we can solve for the value of n.

$$\frac{\text{rate}_1}{\text{rate}_5} = \frac{k[A_1]^m[B_1]^n}{k[A_5]^m[B_5]^n}$$

Substituting the appropriate values from Table 14.4 gives

$$\frac{8.5 \times 10^{-3}\ \cancel{M/\text{min}}}{8.5 \times 10^{-3}\ \cancel{M/\text{min}}} = \frac{\cancel{k[0.50\ M]^m}[0.50\ \cancel{M}]^n}{\cancel{k[0.50\ M]^m}[1.00\ \cancel{M}]^n}$$

Canceling leaves $1.0 = [0.50]^n$, which gives $n = 0$; that is, the reaction is zeroth order in B. The experimentally determined rate law is therefore

$$\text{rate} = k[A]^2[B]^0 = k[A]^2$$

We can now calculate the rate constant by inserting the data from any line of Table 14.4 into the experimentally determined rate law and solving for k. Using Experiment 2, we obtain

$$19 \times 10^{-3}\ M/\text{min} = k(0.75\ M)^2$$
$$3.4 \times 10^{-2}\ M^{-1} \cdot \text{min}^{-1} = k$$

You should verify that using data from any other line of Table 14.4 gives the same rate constant. This must be true as long as the experimental conditions, such as temperature and solvent, are the same.

EXAMPLE 14.8

Nitric oxide is produced in the body by several different enzymes and acts as a signal that controls blood pressure, long-term memory, and other critical functions. The major route for the removal of NO from biological fluids is via reaction with O_2 to give NO_2, which then reacts rapidly with water to give nitrous acid and nitric acid:

$$2NO + O_2 \longrightarrow 2NO_2 \xrightarrow{H_2O} HNO_2 + HNO_3$$

These reactions are important in maintaining steady levels of NO. The table lists kinetics data for the reaction of NO with O_2 at 25°C:

$$2NO(g) + O_2(g) \longrightarrow 2NO_2(g)$$

Determine the rate law for the reaction, and calculate the rate constant.

Experiment	$[NO]_0$, M	$[O_2]_0$, M	Initial Rate, M/s
1	0.0235	0.0125	7.98×10^{-3}
2	0.0235	0.0250	15.9×10^{-3}
3	0.0470	0.0125	32.0×10^{-3}
4	0.0470	0.0250	63.5×10^{-3}

Given Balanced chemical equation, initial concentrations, and initial rates

Asked for Rate law and rate constant

Strategy

A Compare the changes in initial concentrations with the corresponding changes in rates of reaction to determine the order of the reaction for each species. Write the rate law for the reaction.

B Using data from any experiment, substitute appropriate values into the rate law. Solve the rate equation for k, the rate constant.

Solution

A Comparing Experiments 1 and 2 shows that as $[O_2]$ is doubled at a constant value of $[NO_2]$, the rate approximately doubles. Thus, the rate of the reaction is proportional to $[O_2]^1$, so the reaction is first order in O_2. Comparing Experiments 1 and 3 shows that the rate of the reaction essentially quadruples when [NO] is doubled and $[O_2]$ is held constant. That is, the reaction rate is proportional to $[NO]^2$, which indicates that the reaction is second order in NO. Using these relationships, we can write the rate law for the reaction:

$$\text{rate} = k[NO]^2[O_2]$$

B The data in any line can be used to calculate the rate constant. Using Experiment 1, for example, gives

$$k = \frac{\text{rate}}{[NO]^2[O_2]} = \frac{7.98 \times 10^{-3}\ M/s}{(0.0235\ M)^2\,(0.0125\ M)} = 1.16 \times 10^3\ M^{-2} \cdot s^{-1}$$

The overall reaction order $(m + n)$ is 3, so this is a third-order reaction. Notice that the units of the rate constant become more complex as the overall reaction order increases.

EXERCISE 14.8

The peroxydisulfate ion, $S_2O_8^{2-}$, is a potent oxidizing agent that reacts rapidly with iodide ion in water:

$$S_2O_8^{2-}(aq) + 3I^-(aq) \longrightarrow 2SO_4^{2-}(aq) + I_3^-(aq)$$

The table lists kinetics data for this reaction at 25°C. Determine the rate law, and calculate the rate constant.

Experiment	$[S_2O_8^{2-}]_0$, M	$[I^-]_0$, M	Initial Rate, M/s
1	0.27	0.38	2.05
2	0.40	0.38	3.06
3	0.40	0.22	1.76

Answer Rate $= k[S_2O_8^{2-}][I^-]$; $k = 20\ M^{-1} \cdot s^{-1}$

14.4 Using Graphs to Determine Rate Laws, Rate Constants, and Reaction Orders

In Section 14.3 you learned that the integrated rate law for each of the common types of reaction (zeroth, first, or second order in a single reactant) can be plotted as a straight line. The use of these plots offers an alternative to the methods described in that section for showing how reactant concentration changes with time and for determining the order of the reaction.

We will illustrate the use of these graphs by considering the thermal decomposition of NO_2 gas at elevated temperatures, which occurs according to the equation

$$2NO_2(g) \xrightarrow{\Delta} 2NO(g) + O_2(g) \qquad (14.30)$$

Experimental data for this reaction at 330°C are listed in Table 14.5; they are provided as $[NO_2]$, $\ln[NO_2]$, and $1/[NO_2]$ versus time to correspond to the integrated rate laws for zeroth-, first-, and second-order reactions, respectively. The actual concentrations of NO_2 are plotted versus time in Figure 14.11a. Because the plot of $[NO_2]$

TABLE 14.5 Concentration of NO_2 as a function of time at 330°C

Time, s	$[NO_2]$, M	$\ln[NO_2]$	$1/[NO_2]$, M^{-1}
0	1.00×10^{-2}	−4.605	100
60	6.83×10^{-3}	−4.986	146
120	5.18×10^{-3}	−5.263	193
180	4.18×10^{-3}	−5.477	239
240	3.50×10^{-3}	−5.655	286
300	3.01×10^{-3}	−5.806	332
360	2.64×10^{-3}	−5.937	379

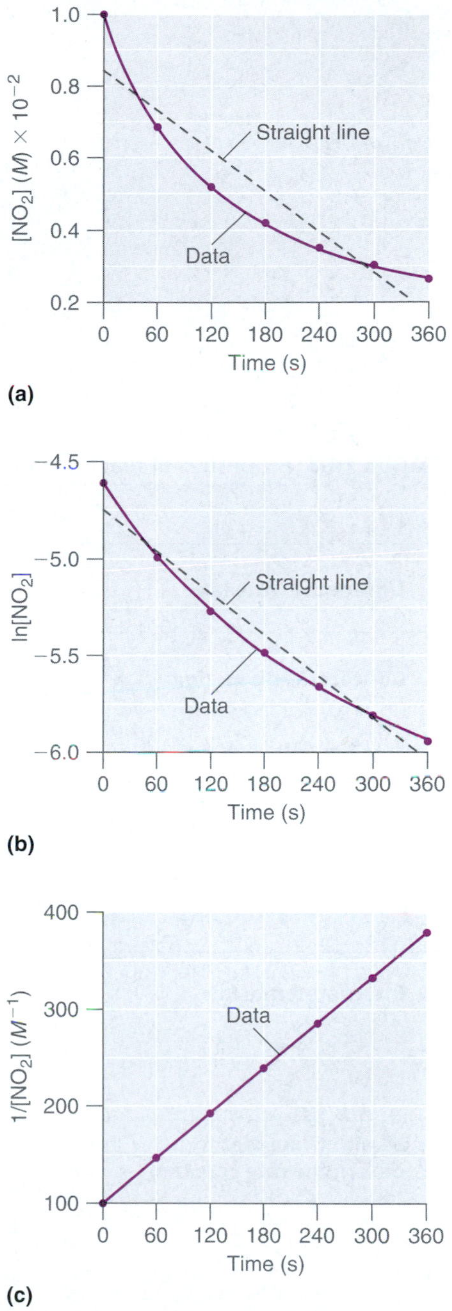

Figure 14.11 The decomposition of NO_2. These plots show the decomposition of a sample of NO_2 at 330°C as **(a)** the concentration of NO_2 vs. time, **(b)** the natural logarithm of $[NO_2]$ vs. time, and **(c)** $1/[NO_2]$ vs. time.

MGC Using Graphs to Determine Rate

versus t is not a straight line, we know the reaction is not zeroth order in NO_2. A plot of $\ln[NO_2]$ versus t (Figure 14.11b) shows us that the reaction is not first order in NO_2 either because a first-order reaction would give a straight line. Having eliminated zeroth-order and first-order behavior, we construct a plot of $1/[NO_2]$ versus t (Figure 14.11c). This plot is a straight line, indicating that the reaction is second order in NO_2.

Notice that we have just determined the order of a reaction using data from a single experiment by plotting the concentration of the reactant as a function of time. Because of the characteristic shapes of the lines shown in Table 14.6, the graphs can be used to determine the reaction order of an unknown reaction. In contrast, the method described in Section 14.3 required multiple experiments at different NO_2 concentrations as well as accurate initial rates of reaction, which can be difficult to obtain for rapid reactions.

TABLE 14.6 Properties of reactions that obey zeroth-, first-, and second-order rate laws

	Zeroth Order	**First Order**	**Second Order**
Differential rate law	$\text{Rate} = -\dfrac{\Delta[A]}{\Delta t} = k$	$\text{Rate} = -\dfrac{\Delta[A]}{\Delta t} = k[A]$	$\text{Rate} = -\dfrac{\Delta[A]}{\Delta t} = k[A]^2$
Concentration vs. time	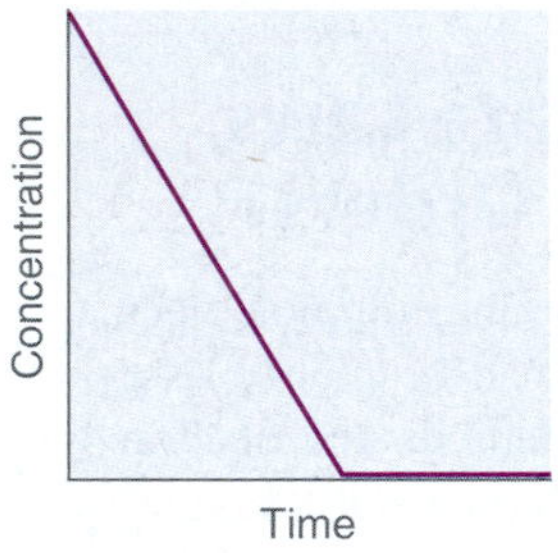	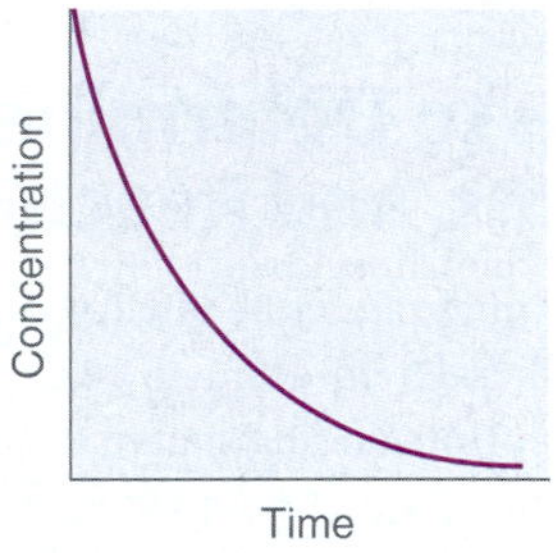	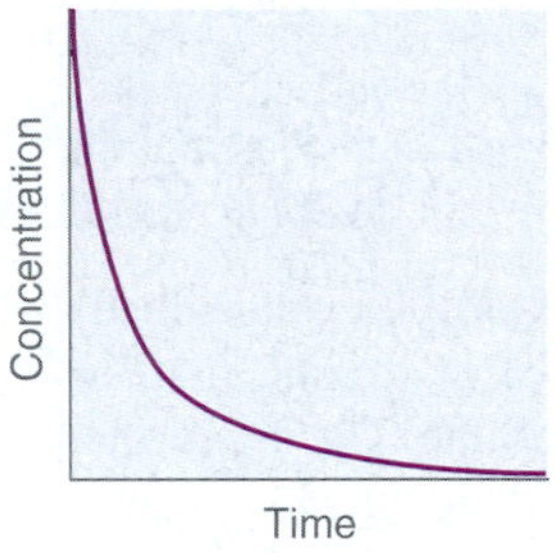
Integrated rate law	$[A] = [A]_0 - kt$	$[A] = [A]_0 e^{-kt}$ or $\ln[A] = \ln[A]_0 - kt$	$\dfrac{1}{[A]} = \dfrac{1}{[A]_0} + kt$
Straight-line plot to determine rate constant	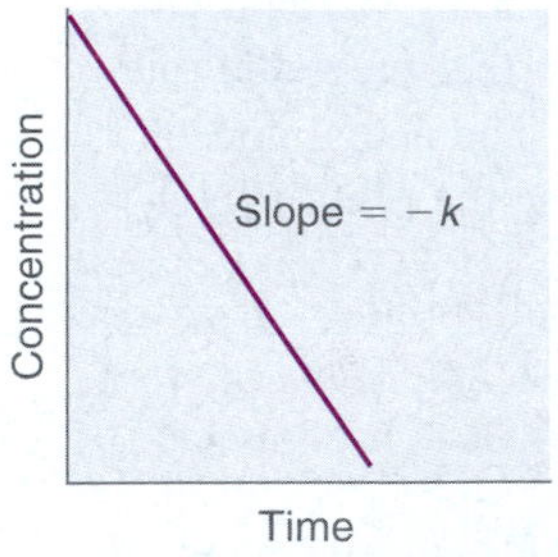	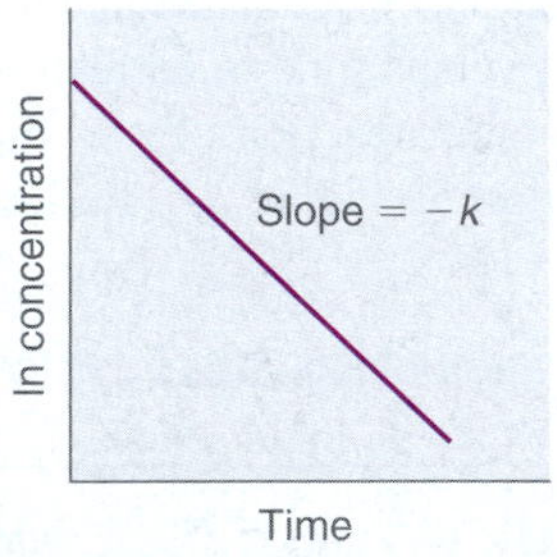	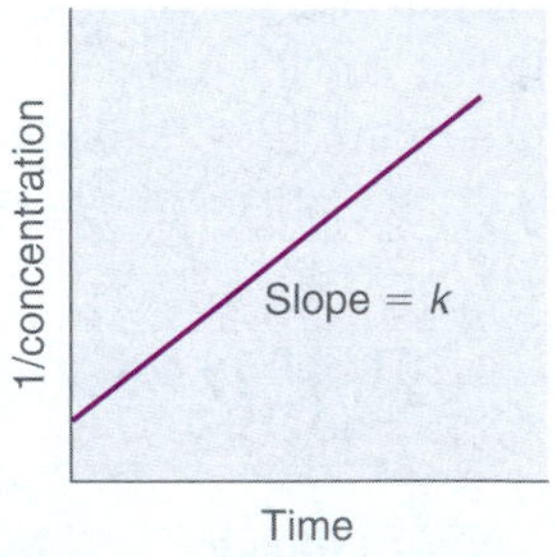
Relative rate vs. concentration	[A] (M) — Rate (M/s) 1 — 1 2 — 1 3 — 1	[A] (M) — Rate (M/s) 1 — 1 2 — 2 3 — 3	[A] (M) — Rate (M/s) 1 — 1 2 — 4 3 — 9
Half-life	$t_{1/2} = \dfrac{[A]_0}{2k}$	$t_{1/2} = \dfrac{0.693}{k}$	$t_{1/2} = \dfrac{1}{k[A]_0}$
Units of *k*, rate constant	M/s	1/s	$M^{-1} \cdot s^{-1}$

EXAMPLE 14.9

Dinitrogen pentoxide (N_2O_5) decomposes to NO_2 and O_2 at relatively low temperatures in the reaction

$$2N_2O_5(soln) \longrightarrow 4NO_2(soln) + O_2(g)$$

This reaction is carried out in a CCl_4 solution at 45°C. The concentrations of N_2O_5 as a function of time are listed in the table, together with the natural logarithms and reciprocal N_2O_5 concentrations. Plot a graph of the concentration versus *t*, ln concentration versus *t*, and 1/concentration versus *t*, and then determine the rate law and calculate the value of the rate constant.

Time, s	$[N_2O_5]$, *M*	$\ln[N_2O_5]$	$1/[N_2O_5]$, M^{-1}
0	0.0365	−3.310	27.4
600	0.0274	−3.597	36.5
1200	0.0206	−3.882	48.5
1800	0.0157	−4.154	63.7
2400	0.0117	−4.448	85.5
3000	0.00860	−4.756	116
3600	0.00640	−5.051	156

Note the pattern

For a zeroth-order reaction, a plot of reactant concentration vs time is a straight line. For a first-order reaction, a plot of the natural logarithm of reactant concentration is a straight line. For a second-order reaction, a plot of 1/reactant concentration vs time is a straight line.

Given Balanced chemical equation, reaction times, and concentrations

Asked for Graph of data, rate law, and rate constant

Strategy

A Use the data in the table to separately plot concentration, the natural logarithm of the concentration, and the reciprocal concentration (the vertical axis) versus time (the horiztonal axis). Compare the graphs with those shown in Table 14.6 to determine the order of the reaction.

B Write the rate law for the reaction. Using the appropriate data from the table and the linear graph corresponding to the rate law for the reaction, calculate the slope of the plotted line to obtain the rate constant for the reaction.

Solution

A Here are plots of $[N_2O_5]$ versus *t*, $\ln[N_2O_5]$ versus *t*, and $1/[N_2O_5]$ versus *t*:

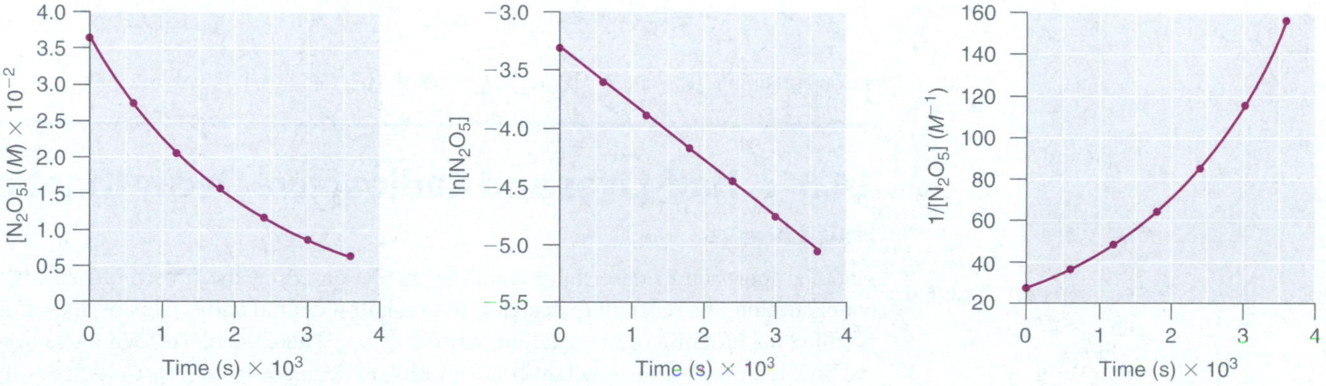

The plot of $\ln[N_2O_5]$ versus *t* gives a straight line, whereas the plots of $[N_2O_5]$ versus *t* and $1/[N_2O_5]$ versus *t* do not. This means that the decomposition of N_2O_5 is first order in $[N_2O_5]$.

B The rate law for the reaction is therefore

$$\text{rate} = k[N_2O_5]$$

Calculating the rate constant is straightforward because we know that the slope of the plot of ln[A] versus *t* for a first-order reaction is $-k$. We can calculate the slope using any two points that lie on the line in the plot of $\ln[N_2O_5]$ versus t. Using the points for $t = 0$ and 3000 s gives

$$\text{slope} = \frac{\ln[N_2O_5]_{3000} - \ln[N_2O_5]_0}{3000\ \text{s} - 0\ \text{s}} = \frac{(-4.756) - (-3.310)}{3000\ \text{s}} = -4.820 \times 10^{-4}\ \text{s}^{-1}$$

Thus, $k = 4.820 \times 10^{-4}\ \text{s}^{-1}$.

EXERCISE 14.9

1,3-Butadiene ($CH_2{=}CH{-}CH{=}CH_2$) is a volatile and reactive organic molecule used in the production of rubber. Above room temperature it reacts slowly to form products. Concentrations of 1,3-butadiene (C_4H_6) as a function of time at 326°C are listed in the table along with $\ln[C_4H_6]$ and the reciprocal concentrations. Graph the data as concentration versus t, In concentration versus t, and 1/concentration versus t. Then determine the order of the reaction in C_4H_6, the rate law, and the rate constant for the reaction.

Time, s	$[C_4H_6]$, M	$\ln[C_4H_6]$	$1/[C_4H_6]$, M^{-1}
0	1.72×10^{-2}	−4.063	58.1
900	1.43×10^{-2}	−4.247	69.9
1800	1.23×10^{-2}	−4.398	81.3
3600	9.52×10^{-3}	−4.654	105
6000	7.30×10^{-3}	−4.920	137

Answer

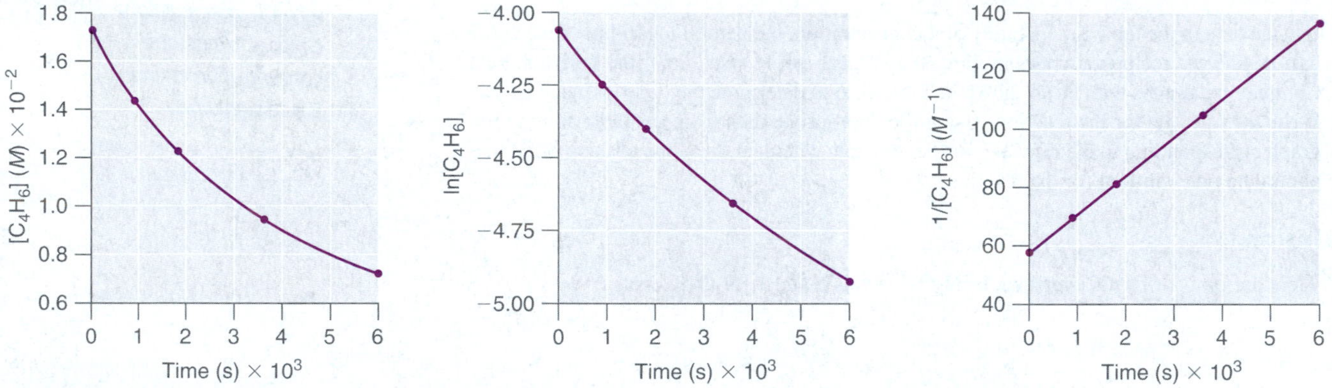

Second order in C_4H_6; rate $= k[C_4H_6]^2$; $k = 1.3 \times 10^{-2}\ M^{-1} \cdot \text{s}^{-1}$

14.5 Half-Lives and Radioactive Decay Kinetics

Half-Lives

MGC Half-life

Another approach to describing reaction rates is based on the time required for the concentration of a reactant to decrease to one-half its initial value. This period of time is called the **half-life** of the reaction, written as $t_{1/2}$. Thus, the half-life of a reaction is the time required for the reactant concentration to decrease from $[A]_0$ to $[A]_0/2$. If two reactions have the same order, the faster reaction will have a shorter half-life and the slower reaction will have a longer half-life.

The half-life of a first-order reaction under a given set of reaction conditions is a constant. This is not true for zeroth- or second-order reactions. Most important, the half-life of a first-order reaction is *independent of the concentration of the reactants.* This becomes evident when we rearrange the integrated rate law for a first-order reaction (Equation 14.23) to produce the following equation:

$$\ln\frac{[A]_0}{[A]} = kt \qquad (14.31)$$

Substituting $[A]_0/2$ for [A] and $t_{1/2}$ for t (to indicate a half-life) into Equation 14.31 gives

$$\ln\frac{[A]_0}{[A]_0/2} = \ln 2 = kt_{1/2}$$

The natural logarithm of 2 (to three decimal places) is 0.693. Substituting this value into the equation, we obtain the expression for the half-life of a first-order reaction:

$$t_{1/2} = \frac{0.693}{k} \qquad (14.32)$$

Thus, for a first-order reaction, each successive half-life is the same length of time, as shown in Figure 14.12, and is independent of [A].

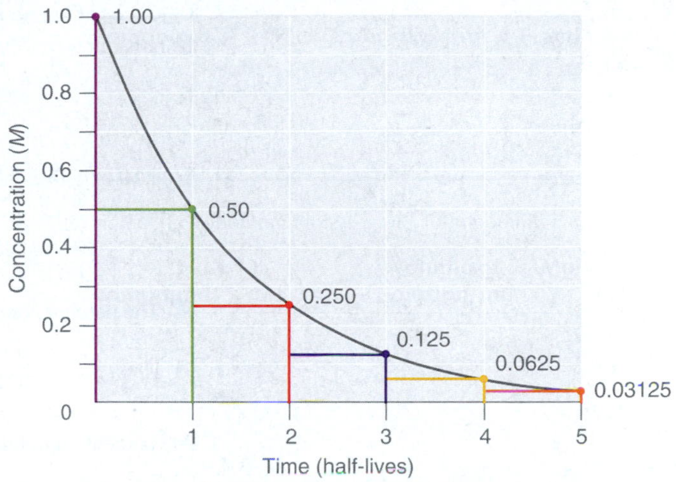

Figure 14.12 The half-life of a first-order reaction. This plot shows the concentration of the reactant in a first-order reaction as a function of time and identifies a series of half-lives, intervals in which the reactant concentration decreases by a factor of 2. In a first-order reaction, every half-life is the same length. Interactive Graph

If we know the rate constant for a first-order reaction, then we can use half-lives to predict how much time is needed for the reaction to reach a certain percent completion.

Number of Half-Lives	% Reactant Remaining	
1	$\frac{100\%}{2} = 50\%$	$\frac{1}{2}(100\%) = 50\%$
2	$\frac{50\%}{2} = 25\%$	$\frac{1}{2}\left(\frac{1}{2}\right)(100\%) = 25\%$
3	$\frac{25\%}{2} = 12.5\%$	$\frac{1}{2}\left(\frac{1}{2}\right)\left(\frac{1}{2}\right)(100\%) = 12.5\%$
n	$\frac{100\%}{2^n}$	$\left(\frac{1}{2}\right)^n(100\%) = \left(\frac{1}{2}\right)^n\%$

As you can see from the table, the amount of reactant left after n half-lives of a first-order reaction is $(1/2)^n$ times the initial concentration.

Note the pattern

For a first-order reaction, the concentration of the reactant decreases by a constant with each half-life and is independent of [A].

EXAMPLE 14.10

The anticancer drug cisplatin hydrolyzes in water with a rate constant of 1.5×10^{-3} min^{-1} at pH 7.0 and 25°C. Calculate the half-life for the hydrolysis reaction under these conditions. If a freshly prepared solution of cisplatin has a concentration of 0.053 *M*, what will be the concentration of cisplatin after five half-lives? After 10 half-lives? What is the percent completion of the reaction after five half-lives? After 10 half-lives?

Given Rate constant, initial concentration, and number of half-lives

Asked for Half life, final concentrations, and percent completion

Strategy

A Use Equation 14.32 to calculate the half-life of the reaction.
B Multiply the initial concentration by (0.50) to the power corresponding to the number of half-lives to obtain the remaining concentrations after those half-lives.
C Subtract the remaining concentration from the initial concentration. Then divide by the initial concentration, multiplying the fraction by 100 to obtain the percent completion.

Solution

A We can calculate the half-life of the reaction using Equation 14.32:

$$t_{1/2} = \frac{0.693}{k} = \frac{0.693}{1.5 \times 10^{-3}\ \text{min}^{-1}} = 4.6 \times 10^2\ \text{min}$$

Thus, it takes almost 8 h for half of the cisplatin to hydrolyze. B✓ After five half-lives (about 38 h), the remaining concentration of cisplatin will be

$$\frac{0.053\ M}{2^5} = \frac{0.053\ M}{32} = 0.0017\ M$$

After 10 half-lives (77 h), the remaining concentration of cisplatin will be

$$\frac{0.053\ M}{2^{10}} = \frac{0.053\ M}{1024} = 5.2 \times 10^{-5}\ M$$

C✓ The percent completion after five half-lives will be

$$\text{percent completion} = \frac{(0.053\ M - 0.0017\ M)\,(100)}{0.053\ M} = 97\%$$

The percent completion after 10 half-lives will be

$$\text{percent completion} = \frac{(0.053\ M - 5.2 \times 10^{-5}\ M)\,(100)}{0.053\ M} = 100\%$$

Thus, a first-order chemical reaction is 97% complete after five half-lives and to two significant figures is 100% complete after 10 half-lives.

EXERCISE 14.10

In Example 14.4 you found that ethyl chloride decomposes to ethylene and HCl in a first-order reaction that has a rate constant of $1.6 \times 10^{-6}\ s^{-1}$ at 650°C. What is the half-life for the reaction under these conditions? If a flask that originally contains 0.077 *M* ethyl chloride is heated at 650°C, what is the concentration of ethyl chloride after four half-lives?

Answer 4.3×10^5 s = 120 h = 5.0 days; $4.8 \times 10^{-3}\ M$

Radioactive Decay Rates

As you learned in Chapter 1, radioactivity, or radioactive decay, is the emission of a particle or a photon that results from the spontaneous decomposition of the unstable nucleus of an atom. The rate of radioactive decay is an intrinsic property of each radioactive isotope, independent of the chemical and physical form of the radioactive isotope. The rate is also independent of temperature. In this section we will describe radioactive decay rates and how half-lives can be used to monitor radioactive decay processes.

In any sample of a given radioactive substance, the number of atoms of the radioactive isotope must decrease with time as their nuclei decay to nuclei of a more stable isotope. Using N to represent the number of atoms of the radioactive isotope, we can define the **rate of decay** of the sample, which is also called its **activity**, A, as the decrease in the number of the radioisotope's nuclei per unit time:

$$A = -\frac{\Delta N}{\Delta t} \tag{14.33}$$

Activity is usually measured in *disintegrations per second (dps) or disintegrations per minute (dpm).*

The activity of a sample is directly proportional to the number of atoms of the radioactive isotope in the sample:

$$A = kN \tag{14.34}$$

Here, the symbol k is the *radioactive decay constant*, which has units of inverse time (for example, s^{-1}, yr^{-1}) and has a characteristic value for each radioactive isotope. If

TABLE 14.7 Half-lives and applications of some radioactive isotopes

Radioactive Isotope	Half-Life	Typical Uses
Hydrogen-3 (tritium)	12.32 yr	Biochemical tracer
Carbon-11	20.33 min	PET scans (biomedical imaging)
Carbon-14	5.70×10^3 yr	Dating of artifacts
Sodium-24	14.951 hr	Cardiovascular system tracer
Phosphorus-32	14.26 days	Biochemical tracer
Potassium-40	1.248×10^9 yr	Dating of rocks
Iron-59	44.495 days	Red blood cell lifetime tracer
Cobalt-60	5.2712 yr	Radiation therapy for cancer
Technetium-99*m*[a]	6.006 h	Biomedical imaging
Iodine-131	8.0207 days	Thyroid studies tracer
Radium-226	1.600×10^3 yr	Radiation therapy for cancer
Uranium-238	4.468×10^9 yr	Dating of rocks and Earth's crust
Americium-241	432.2 yr	Smoke detectors

[a]The *m* denotes metastable, where an excited state nucleus decays to the ground state of the same isotope.

we combine Equations 14.33 and 14.34, we obtain the relationship between the number of decays per unit time and the number of atoms of the isotope in a sample:

$$-\frac{\Delta N}{\Delta t} = kN \qquad (14.35)$$

Equation 14.35 is the same as the equation for the rate of a first-order reaction (Equation 14.21), except that it uses numbers of atoms instead of concentrations. In fact, radioactive decay *is* a first-order process and can be described in terms of either the differential rate law (Equation 14.35) or the integrated rate law:

$$N = N_0\, e^{-kt}$$

$$\ln \frac{N}{N_0} = -kt \qquad (14.36)$$

Because radioactive decay is a first-order process, the time required for half of the nuclei in any sample of a radioactive isotope to decay is a constant, called the *half-life of the isotope.* The half-life tells us how radioactive an isotope is (the number of decays per unit time); thus, it is the most commonly cited property of any radioisotope. For a given number of atoms, isotopes with shorter half-lives decay more rapidly, undergoing a greater number of radioactive decays per unit time than do isotopes with longer half-lives. The half-lives of several isotopes are listed in Table 14.7, along with some of their applications.

Radioisotope Dating Techniques

In our earlier discussion, we used the half-life of a first-order reaction to calculate how long the reaction had been occurring. Because nuclear decay reactions follow first-order kinetics and have a rate constant that is independent of temperature and chemical or physical environment, we can perform similar calculations using the half-lives of isotopes to estimate the ages of geological and archaeological artifacts. The techniques that have been developed for this application are known as *radioisotope dating* techniques.

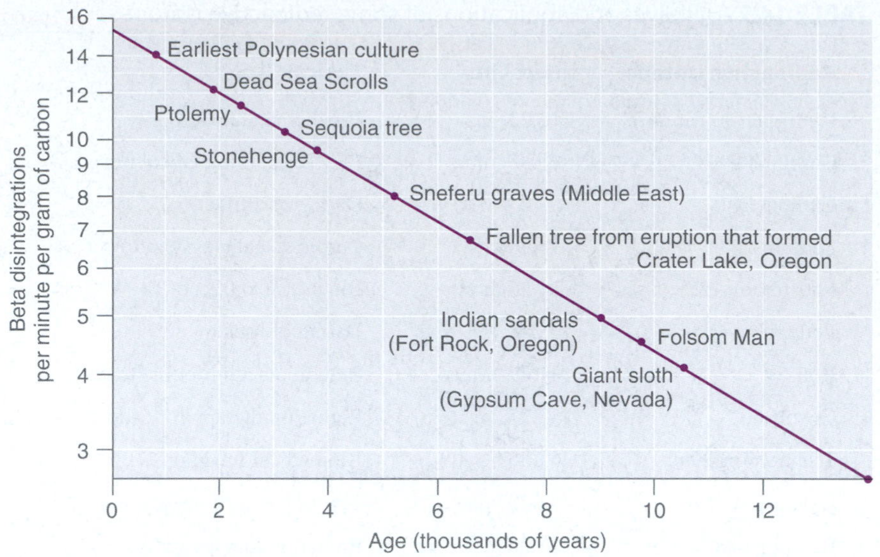

Figure 14.13 Radiocarbon dating. A plot of the specific activity of ^{14}C vs. age for a number of archaeological samples shows an inverse linear relationship between ^{14}C content (a log scale) and age (a linear scale).

The most common method for measuring the age of ancient objects is *carbon-14 dating*. The carbon-14 isotope, created continuously in the upper regions of Earth's atmosphere, reacts with atmospheric oxygen or ozone to form $^{14}CO_2$. As a result, the CO_2 that plants use as a carbon source for synthesizing organic compounds always includes a certain proportion of $^{14}CO_2$ molecules as well as nonradioactive $^{12}CO_2$ and $^{13}CO_2$. Any animal that eats a plant ingests a mixture of organic compounds that contains approximately the same proportions of carbon isotopes as the atmosphere. When the animal or plant dies, the carbon-14 nuclei in its tissues decay to nitrogen-14 nuclei by a radioactive process known as *beta decay*, which releases low-energy electrons (β particles) that can be detected and measured:

$$^{14}C \longrightarrow {}^{14}N + \beta^- \tag{14.37}$$

The half-life for this reaction is 5700 $\pm$ 30 yr.

The $^{14}C/^{12}C$ ratio in living organisms is 1.3×10^{-12}, with a decay rate of 15 dpm per gram of carbon (dpm/g carbon) (Figure 14.13). Comparing the disintegrations per minute per gram of carbon from an archaeological sample with those from a recently living sample enables scientists to estimate the age of the artifact, as illustrated in Example 14.11.*

EXAMPLE 14.11

In 1990 the remains of an apparently prehistoric man were found in a melting glacier in the Italian Alps. Analysis of the ^{14}C content of samples of wood from his tools gave a decay rate of 8.0 dpm/g carbon. How long ago did the man die?

Given Isotope and final activity

Asked for Elapsed time

Strategy

Ⓐ Use Equation 14.34 to calculate N_0/N. Then substitute the value for the half-life of ^{14}C into Equation 14.32 to find the rate constant for the reaction.

* Using this method implicitly assumes that the $^{14}CO_2/^{12}CO_2$ ratio in the atmosphere is constant, which is not strictly correct. Other methods, such as tree-ring dating, have been used to calibrate the dates obtained by radiocarbon dating, and all radiocarbon dates reported are now corrected for minor changes in the $^{14}CO_2/^{12}CO_2$ ratio over time.

Ⓑ Using the values obtained for N_0/N and the rate constant, solve Equation 14.36 to obtain the elapsed time, *t*.

Solution

We know the initial activity from the isotope's identity (15 dpm/g), the final activity (8.0 dpm/g), and the half-life, so we can use the integrated rate law for a first-order nuclear reaction (Equation 14.36) to calculate the elapsed time, *t* (the amount of time elapsed since the wood for the tools was cut and began to decay).

$$\ln\frac{N}{N_0} = -kt$$

$$t = \frac{\ln(N/N_0)}{k}$$

Ⓐ✓ From Equation 14.34, we know that $A = kN$. We can therefore use the initial and final activities ($A_0 = 15$ dpm and $A = 8.0$ dpm) to calculate N_0/N:

$$\frac{A_0}{A} = \frac{\cancel{k}N_0}{\cancel{k}N} = \frac{N_0}{N} = \frac{15}{8.0}$$

Now we need only calculate the rate constant for the reaction from its half-life (5730 yr) using Equation 14.32:

$$t_{1/2} = \frac{0.693}{k}$$

This equation can be rearranged to give

$$k = \frac{0.693}{t_{1/2}} = \frac{0.693}{5730\ \text{yr}} = 1.22 \times 10^{-4}\ \text{yr}^{-1}$$

Ⓑ✓ Substituting into the equation for *t* gives

$$t = \frac{\ln(N_0/N)}{k} = \frac{\ln(15/8.0)}{1.22 \times 10^{-4}\ \text{yr}^{-1}} = 5.2 \times 10^3\ \text{yr}$$

From our calculations, the man died 5200 years ago.

EXERCISE 14.11

It is believed that humans first arrived in the Western Hemisphere during the last Ice Age, presumably by traveling over an exposed land bridge between Siberia and Alaska. Archaeologists have estimated that this occurred about 11,000 yr ago, but some argue that recent discoveries in several sites in North and South America suggest a much earlier arrival. Analysis of a sample of charcoal from a fire in one such site gave a ^{14}C decay rate of 0.4 dpm/g of carbon. What is the approximate age of the sample?

Answer 30,000 yr

14.6 • Reaction Rates—A Microscopic View

MGC Mechanisms and Molecularity

One of the major reasons for studying chemical kinetics is to use measurements of the macroscopic properties of a system, such as the rate of change in the concentration of reactants or products with time, to discover the sequence of events that occur at the molecular level during a reaction. This molecular description is the *mechanism* of the reaction; it describes how individual atoms, ions, or molecules interact to form particular products. The stepwise changes are called the **reaction mechanism**.

In an internal combustion engine, for example, isooctane reacts with oxygen to give carbon dioxide and water.

$$2C_8H_{18}(l) + 25O_2(g) \longrightarrow 16CO_2(g) + 18H_2O(g) \quad (14.38)$$

For this reaction to occur in a single step, 25 dioxygen molecules and 2 isooctane molecules would have to collide simultaneously and be converted to 34 molecules of product, which is very unlikely. It is more likely that a complex series of reactions takes place in a stepwise fashion. Each of the individual reactions, called **elementary reactions,** involves one, two, or (rarely) three atoms, molecules, or ions. The overall sequence of elementary reactions is the mechanism of the reaction. *The sum of the individual steps, or elementary reactions, in the mechanism must give the balanced chemical equation for the overall reaction.*

Molecularity and the Rate-Determining Step

To demonstrate how the analysis of elementary steps helps us determine the overall mechanism of a reaction, we will examine the much simpler reaction of carbon monoxide with nitrogen dioxide.

$$NO_2(g) + CO(g) \longrightarrow NO(g) + CO_2(g) \quad (14.39)$$

From the balanced chemical equation, one might expect the reaction to occur via a collision of one molecule of NO_2 with a molecule of CO that results in the transfer of an oxygen atom from nitrogen to carbon. The experimentally determined rate law for the reaction, however, is

$$\text{rate} = k[NO_2]^2 \quad (14.40)$$

The fact that the reaction is second order in $[NO_2]$ and independent of [CO] tells us that it does not occur by the simple collision of an NO_2 molecule with a CO molecule. If it did, its predicted rate law would be rate $= k[NO_2][CO]$.

The following two-step mechanism is consistent with the rate law if Step 1 is much slower than Step 2:

Step 1	$NO_2 + NO_2 \xrightarrow{\text{slow}} NO_3 + NO$	Elementary reaction
Step 2	$NO_3 + CO \longrightarrow NO_2 + CO_2$	Elementary reaction
Sum	$NO_2 + CO \longrightarrow NO + CO_2$	Overall reaction

According to this mechanism, the overall reaction occurs in two steps, or elementary reactions. Notice that summing Steps 1 and 2 and canceling on both sides of the equation gives the overall balanced chemical equation for the reaction. The NO_3 molecule is an **intermediate** in the reaction, a species that does not appear in the balanced equation for the overall reaction. It is formed as a product of the first step but then is consumed in the second step.

Using Molecularity to Describe a Rate Law

The **molecularity** of an elementary reaction is the number of molecules that collide during that step in the mechanism. If there is only a single reactant molecule in an elementary reaction, that step is designated as *unimolecular*; if there are two reactant molecules, it is *bimolecular*; and if there are three reactant molecules (a relatively rare situation), it is *termolecular*. Elementary reactions that involve the simultaneous collision of more than three molecules are highly improbable and have never been observed experimentally. (To understand why, try to make three or more marbles or pool balls collide with one another simultaneously!)

Writing the rate law for an elementary reaction is straightforward because we know how many molecules must collide simultaneously for the elementary reaction to occur; hence, the order of the *elementary reaction* is the same as its molecularity

TABLE 14.8 Common types of elementary reactions and their rate laws

Elementary Reaction	Molecularity	Rate Law	Reaction Order
$A \longrightarrow$ products	Unimolecular	Rate = $k[A]$	First
$2A \longrightarrow$ products	Bimolecular	Rate = $k[A]^2$	Second
$A + B \longrightarrow$ products	Bimolecular	Rate = $k[A][B]$	Second
$2A + B \longrightarrow$ products	Termolecular	Rate = $k[A]^2[B]$	Third
$A + B + C \longrightarrow$ products	Termolecular	Rate = $k[A][B][C]$	Third

(Table 14.8). In contrast, the rate law for the reaction cannot be determined from the balanced equation for the overall reaction. The general rate law for a unimolecular elementary reaction ($A \longrightarrow$ products) is rate = $k[A]$. For bimolecular reactions, the reaction rate depends on the number of collisions per unit time, which is proportional to the product of the concentrations of the reactants, as shown in Figure 14.14. Hence, for a bimolecular elementary reaction of the form $A + B \longrightarrow$ products, the general rate law is rate = $k[A][B]$.

Identifying the Rate-Determining Step

Note the important difference between writing rate laws for the elementary steps of a reaction and for the balanced chemical equation of the overall reaction. Because the balanced chemical equation does not necessarily reveal the individual elementary reactions by which the reaction occurs, we *cannot* obtain the rate law for a reaction from the overall balanced equation alone. In fact, the rate law for the overall reaction is the same as the rate law for the slowest step in the reaction mechanism, the **rate-determining step**.* The reason for this is that any process that occurs through a sequence of steps can take place no faster than the slowest step in the sequence. In an automotive assembly line, for example, a component cannot be used faster than it is produced. Similarly, blood pressure is regulated by the flow of blood through the smallest passages, the capillaries. Because movement through capillaries constitutes

Figure 14.14 **The basis for writing rate laws of elementary reactions.** This diagram illustrates how the number of possible collisions per unit time between two reactant species, A and B, depends on the number of A and B particles present. Note that the number of collisions between A and B particles increases as the *product* of the number of particles, not as the sum. This is why the rate law for an elementary reaction depends on the product of the concentrations of the species that collide in that step.

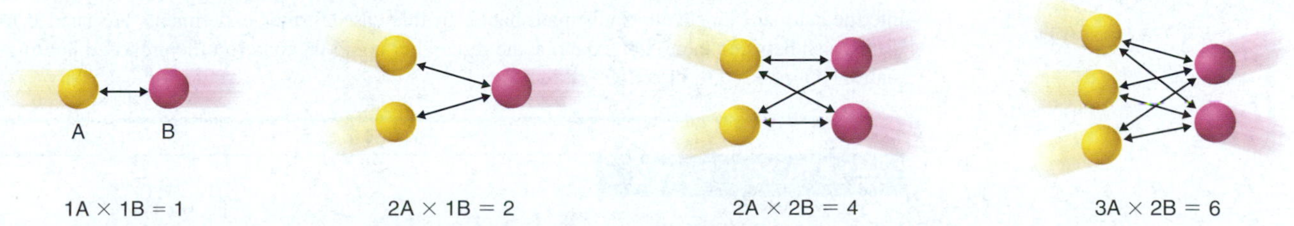

* This statement is true if one step is substantially slower than all the others, typically by a factor of 10 or more. If two or more slow steps have comparable rates, the experimentally determined rate laws can become complex. Our discussion is limited to reactions in which one step can be identified as being substantially slower than any other.

Rate-determining step. The phenomenon of a rate-determining step can be compared to a succession of funnels. The smallest-diameter funnel controls the rate at which the bottle is filled, whether it is the first or the last in the series. Pouring liquid into the first funnel faster than it can drain through the smallest results in an overflow.

the rate-determining step in blood flow, blood pressure can be regulated by medications that cause the capillaries to contract or dilate. A chemical reaction that occurs via a series of elementary reactions can take place no faster than the slowest step in the series of reactions.

Take a look at the rate laws for each elementary reaction in our example as well as for the overall reaction.

Step 1	$NO_2 + NO_2 \xrightarrow{k_1} NO_3 + NO$	Rate = $k_1[NO_2]^2$ (predicted)
Step 2	$NO_3 + CO \xrightarrow{k_2} NO_2 + CO_2$	Rate = $k_2[NO_3][CO]$ (predicted)
Sum	$NO_2 + CO \xrightarrow{k} NO + CO_2$	Rate = $k[NO_2]^2$ (observed)

The experimentally determined rate law for the reaction of NO_2 with CO is the same as the predicted rate law for Step 1. This tells us that the first elementary reaction is the rate-determining step, so k for the overall reaction must equal k_1. That is, NO_3 is formed slowly in Step 1, but once it is formed it reacts very rapidly with CO in Step 2.

Sometimes chemists are able to propose two or more mechanisms that are consistent with the available data. If a proposed mechanism predicts the wrong experimental rate law, however, the mechanism must be incorrect.

EXAMPLE 14.12

In an alternative mechanism for the reaction of NO_2 with CO, N_2O_4 appears as an intermediate.

Step 1	$NO_2 + NO_2 \xrightarrow{k_1} N_2O_4$
Step 2	$N_2O_4 + CO \xrightarrow{k_2} NO + NO_2 + CO_2$
Sum	$NO_2 + CO \longrightarrow NO + CO_2$

Write the rate law for each of the elementary steps. Is this mechanism consistent with the experimentally determined rate law (rate = $k[NO_2]^2$)?

Given Elementary steps of reaction

Asked for Rate law for each elementary step and overall rate law

Strategy

A Determine the rate law for each elementary step in the reaction.

B Determine which of these rate laws corresponds to the experimentally determined rate law for the reaction. This rate law is the one for the rate-determining step.

Solution

A The rate law for Step 1 is rate = $k_1[NO_2]^2$; for Step 2 it is rate = $k_2[N_2O_4][CO]$. **B** If Step 1 is slow (and therefore the rate-determining step), then the overall rate law for the reaction will be the same: rate = $k_1[NO_2]^2$. This is the same as the experimentally determined rate law. Hence, this mechanism, with N_2O_4 as an intermediate, and the one described previously, with NO_3 as an intermediate, are kinetically indistinguishable. In this case, further experiments are needed to distinguish between them. For example, the researcher could try to detect the proposed intermediates, NO_3 and N_2O_4, directly.

EXERCISE 14.12

Iodine monochloride, ICl, reacts with H_2 as follows:

$$2ICl(l) + H_2(g) \longrightarrow 2HCl(g) + I_2(s)$$

The experimentally determined rate law is rate = $k[ICl][H_2]$. Write a two-step mechanism for this reaction using only bimolecular elementary steps, and show that it is consistent with the experimental rate law. (*Hint:* HI is an intermediate.)

Answer

Step 1	$ICl + H_2 \longrightarrow HCl + HI$	Rate $= k_1[ICl][H_2]$ (slow)
Step 2	$HI + ICl \longrightarrow HCl + I_2$	Rate $= k_2[HI][ICl]$ (fast)
Sum	$2ICl + H_2 \longrightarrow 2HCl + I_2$	

This mechanism is consistent with the experimental rate law *if* the first step is the rate-determining step.

Chain Reactions

Many reaction mechanisms, like those discussed so far, consist of only two or three elementary reactions. Many others consist of long series of elementary reactions. The most common mechanisms are **chain reactions**, in which one or more elementary reactions that contain a highly reactive species repeat again and again during the reaction process. Chain reactions occur in fuel combustion, in explosions, in the formation of many polymers, and in the tissue changes associated with aging. They are also important in the chemistry of the atmosphere.

Chain reactions are described as having three stages. The first is *initiation*, a step that produces one or more reactive intermediates. Often these intermediates are **radicals**, species that have an unpaired valence electron. In the second stage, *propagation*, reactive intermediates are continuously consumed and regenerated while products are formed. Intermediates are also consumed but not regenerated in the final stage of a chain reaction, *termination*, usually by forming stable products.

Let us look at the reaction of methane with chlorine at elevated temperatures (400–450°C), a chain reaction used in industry to manufacture methyl chloride (CH_3Cl), dichloromethane (CH_2Cl_2), chloroform ($CHCl_3$), and carbon tetrachloride (CCl_4):

$$CH_4 + Cl_2 \longrightarrow CH_3Cl + HCl$$
$$CH_3Cl + Cl_2 \longrightarrow CH_2Cl_2 + HCl$$
$$CH_2Cl_2 + Cl_2 \longrightarrow CHCl_3 + HCl$$
$$CHCl_3 + Cl_2 \longrightarrow CCl_4 + HCl$$

Direct chlorination generally produces a mixture of all four carbon-containing products, which must then be separated by distillation. In our discussion we will examine only the chain reactions that lead to the preparation of CH_3Cl.

In the initiation stage of this reaction, the relatively weak Cl—Cl bond cleaves at temperatures of about 400°C to produce chlorine atoms ($Cl\cdot$):

$$Cl_2 \longrightarrow 2Cl\cdot$$

During propagation, a chlorine atom removes a hydrogen atom from a methane molecule to give HCl and $CH_3\cdot$, the methyl radical:

$$Cl\cdot + CH_4 \longrightarrow CH_3\cdot + HCl$$

The methyl radical then reacts with a chlorine molecule to form methyl chloride and another chlorine atom, $Cl\cdot$:

$$CH_3\cdot + Cl_2 \longrightarrow CH_3Cl + Cl\cdot$$

Note that the sum of the propagation reactions is the same as the overall balanced equation for the reaction:

$$\cancel{Cl} + CH_4 \longrightarrow \cancel{CH_3}\cdot + HCl$$
$$\cancel{CH_3} + Cl_2 \longrightarrow CH_3Cl + \cancel{Cl}$$
$$\overline{Cl_2 + CH_4 \longrightarrow CH_3Cl + HCl}$$

Without a chain-terminating reaction, the propagation reactions would continue until either the methane or the chlorine was consumed. Because radical species react rapidly with almost anything, however, including each other, they eventually form neutral compounds, thus terminating the chain reaction in any of three ways:

$$CH_3\cdot + Cl\cdot \longrightarrow CH_3Cl$$
$$CH_3\cdot + CH_3\cdot \longrightarrow H_3CCH_3$$
$$Cl\cdot + Cl\cdot \longrightarrow Cl_2$$

Here is the overall chain reaction, with the desired product (CH_3Cl) highlighted:

Initiation
$$Cl_2 \longrightarrow 2Cl\cdot$$

Propagation
$$Cl\cdot + CH_4 \longrightarrow CH_3\cdot + HCl$$
$$CH_3\cdot + Cl_2 \longrightarrow CH_3Cl + Cl\cdot$$

Termination
$$CH_3\cdot + Cl\cdot \longrightarrow CH_3Cl$$
$$CH_3\cdot + CH_3\cdot \longrightarrow H_3CCH_3$$
$$Cl\cdot + Cl\cdot \longrightarrow Cl_2$$

The chain reactions responsible for explosions generally have an additional feature: one or more *chain branching steps*, in which one radical reacts to produce two or more radicals, each of which can then go on to start a new chain reaction. Repetition of the branching step has a cascade effect such that a single initiation step generates large numbers of chain reactions. The result is a very rapid reaction or an explosion. The reaction of H_2 and O_2, used to propel rockets, is an example of a chain branching reaction:

Initiation
$$H_2 + O_2 \longrightarrow HO_2\cdot + H\cdot$$

Propagation
$$HO_2\cdot + H_2 \longrightarrow H_2O + OH\cdot$$
$$OH\cdot + H_2 \longrightarrow H_2O + H\cdot$$

Branching
$$H\cdot + O_2 \longrightarrow OH\cdot + \cdot O\cdot$$
$$\cdot O\cdot + H_2 \longrightarrow OH\cdot + H\cdot$$

Termination reactions occur when the extraordinarily reactive $H\cdot$ or $OH\cdot$ radicals react with a third species. The complexity of a chain reaction makes it unfeasible to write a rate law for the overall reaction.

14.7 • The Collision Model of Chemical Kinetics

MGC The Arrhenius Equation

In Section 14.6 you saw that it is possible to use kinetics studies of a chemical system, such as the effect of changes in reactant concentrations, to deduce events that occur on a microscopic scale, such as collisions between individual particles. Such studies have led to the *collision model of chemical kinetics*, which is a useful tool for understanding the behavior of reacting chemical species. According to the collision model, a chemical reaction can occur only when the reactant molecules, atoms, or ions collide with more than a certain amount of kinetic energy and in the proper orientation. The collision model explains why, for example, most collisions between molecules do *not* result in a chemical reaction. Nitrogen and oxygen molecules in a single liter of air at room temperature and 1 atm of pressure collide about 10^{30} times per second. If every collision produced two molecules of NO, the atmosphere would have been converted to NO and then NO_2 a long time ago. Instead, in most collisions the molecules simply bounce off one another without reacting, much as marbles bounce off each other when they collide. The collision model also explains why such chemical reactions occur more rapidly at higher temperatures. For example, the rates of many reactions that occur at room temperature approximately double with a temperature increase of only 10°C. In this section we will use the collision model to analyze this relationship between temperature and reaction rates.

Activation Energy

In Chapter 10 we discussed the kinetic molecular theory of gases, which showed that the average kinetic energy of the particles of a gas increases with increasing temperature. Because the speed of a particle is proportional to the square root of its kinetic energy, increasing the temperature will also increase the number of collisions between molecules per unit time. What the kinetic molecular theory of gases does not explain is why the rate of most reactions approximately doubles with a 10°C temperature increase. This result is surprisingly large considering that a 10°C increase in the temperature of a gas from 300 K to 310 K increases the kinetic energy of the particles by only about 4%, leading to an increase in molecular speed of only about 2% and a correspondingly small increase in the number of bimolecular collisions per unit time.

The collision model of chemical kinetics explains this behavior by introducing the concept of activation energy. We will define this concept using the reaction of NO with ozone, which plays an important role in the depletion of ozone in the ozone layer:

$$NO(g) + O_3(g) \longrightarrow NO_2(g) + O_2(g) \tag{14.41}$$

Increasing the temperature from 200 K to 350 K causes the rate constant for this particular reaction to increase by a factor of more than 10, whereas the increase in the frequency of bimolecular collisions over this temperature range is only 30%. Thus, something other than an increase in the collision rate must be affecting the reaction rate.

The rate of a reaction will vary with concentrations but not the rate constant. However, the rate constant does vary with temperature. Figure 14.15 shows a plot of the rate constant of the reaction of NO with O_3 at various temperatures. Notice that the relationship is not linear, but instead resembles the relationships seen in graphs of vapor pressure versus temperature (Chapter 11) and of conductivity versus temperature (Chapter 12). In all three cases, the shape of the plots results from a distribution of kinetic energy over a population of particles (electrons in the case of conductivity; molecules in the case of vapor pressure; and molecules, atoms, or ions in the case of reaction rates). Only a fraction of the particles have sufficient energy to overcome an energy barrier.

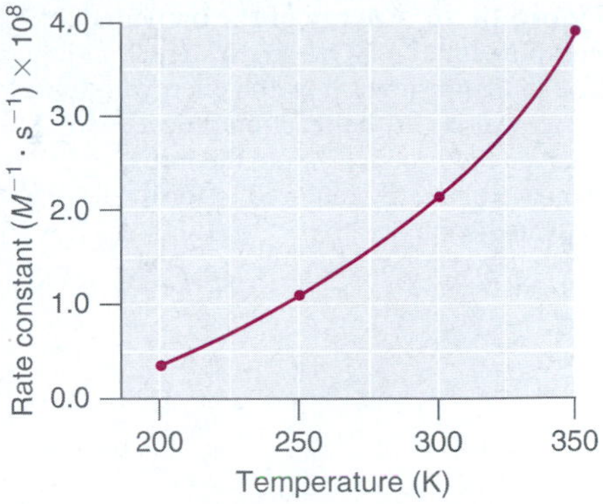

Figure 14.15 Rate constant vs. temperature for the reaction of NO with O_3. The nonlinear shape of the curve is caused by a distribution of kinetic energy over a population of molecules. Only a fraction of the particles have enough energy to overcome an energy barrier, but as the temperature is increased, the size of that fraction increases. Interactive Graph

In the case of vapor pressure, particles must overcome an energy barrier to escape from the liquid to the gas phase. This barrier corresponds to the energy of the intermolecular forces that hold the molecules together in the liquid. In conductivity, the barrier is the energy gap between the filled and empty bands. In chemical reactions, the energy barrier corresponds to the amount of energy the particles must have in order to react when they collide. This energy threshold, called the **activation energy (E_a)**, was first postulated in 1888 by the Swedish chemist Svante Arrhenius (1859–1927, Nobel Prize 1903). It is the minimum amount of energy needed for a reaction to occur. Reacting molecules must have enough energy to overcome electrostatic repulsion, and a minimum amount of energy is required to break chemical bonds so that new ones may be formed. Molecules that collide with less than the threshold energy bounce off one another chemically unchanged, with only their direction of travel and their speed altered by the collision. Molecules that are able to overcome the energy barrier, and so are able to react, form an arrangement of atoms called the **activated complex** or the **transition state** of the reaction. The activated complex is *not* a reaction intermediate; it does not last long enough to be detected readily.

Note the pattern

Any phenomenon that depends on the distribution of thermal energy in a population of particles has a nonlinear temperature dependence.

Graphing the Energy Changes During a Reaction

We can graph the energy of a reaction by plotting the potential energy of the system as the reaction progresses. Figure 14.16 shows a plot for the NO—O_3 system in which the vertical axis is potential energy and the horizontal axis is the *reaction coordinate*, indicating the progress of the reaction with time. The activated complex is shown in

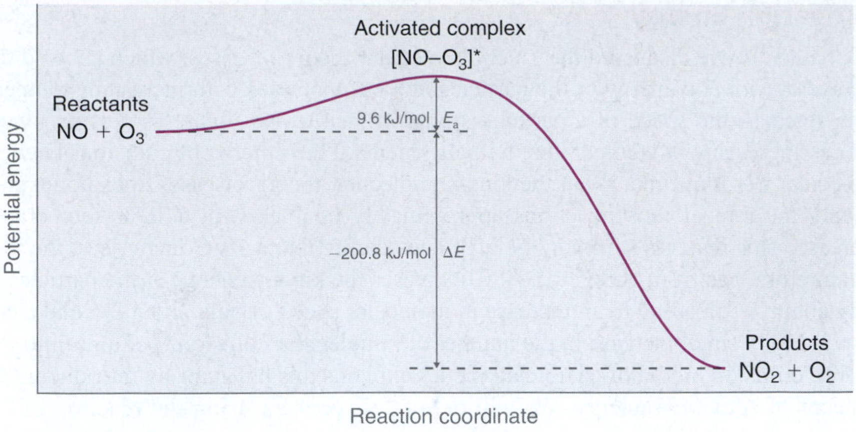

Figure 14.16 Energy of the activated complex for the $NO—O_3$ system. The diagram shows how the energy of this system varies as the reaction proceeds from reactants to products. Note the initial increase in energy required to form the activated complex.

brackets with an asterisk. The overall change in potential energy for the reaction (ΔE) is negative, which means that the reaction releases energy. (In this case ΔE is -200.8 kJ/mol.) In order to react, however, the molecules must overcome the energy barrier to reaction (E_a is 9.6 kJ/mol). That is, 9.6 kJ/mol must be put into the system as the activation energy. Below this threshold the particles do not have enough energy for the reaction to occur.

Figure 14.17a illustrates the general situation in which the products have a lower potential energy than the reactants. In contrast, Figure 14.17b illustrates the case in which the products have a higher potential energy than the reactants, so the overall reaction requires an input of energy; that is, it is energetically uphill and $\Delta E > 0$. Although the energy changes that result from a reaction can be positive, negative, or even zero, in all cases an energy barrier must be overcome before the reaction can occur. This means that the activation energy, E_a, is always positive.

Whereas ΔE is related to the tendency of a reaction to occur spontaneously (one of the central topics of Chapter 18), E_a gives us information about the rate of a reaction and how rapidly the rate changes with temperature. *For two similar reactions under comparable conditions, the reaction with the smallest E_a will occur more rapidly.*

Even when the energy of collisions between two reactant species is greater than E_a, however, most collisions do not produce a reaction. The probability of a reaction occurring depends not only on the collision energy but also on the spatial orientation of the molecules when they collide. For NO and O_3 to produce NO_2 and O_2, a terminal oxygen atom of O_3 must collide with the nitrogen atom of NO at an angle that allows O_3 to transfer an oxygen atom to NO to produce NO_2 (Figure 14.18). All other collisions produce no reaction. Because fewer than 1% of all possible orientations of NO and O_2 result in a reaction at kinetic energies greater than E_a, most collisions of

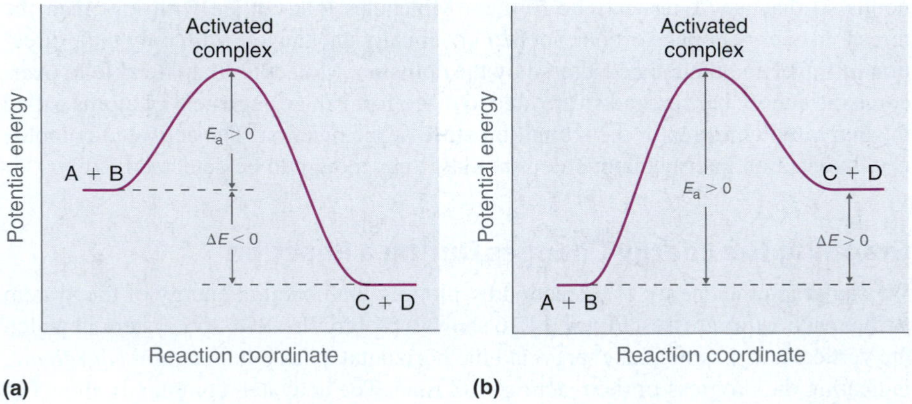

Figure 14.17 Differentiating between E_a and ΔE. The potential energy diagrams for a reaction with **(a)** $\Delta E < 0$ and **(b)** $\Delta E > 0$ illustrate the change in the potential energy of the system as reactants are converted to products. Note that E_a is always positive. For a reaction such as the one shown in **(b)**, E_a must be *greater* than ΔE. Interactive Graph

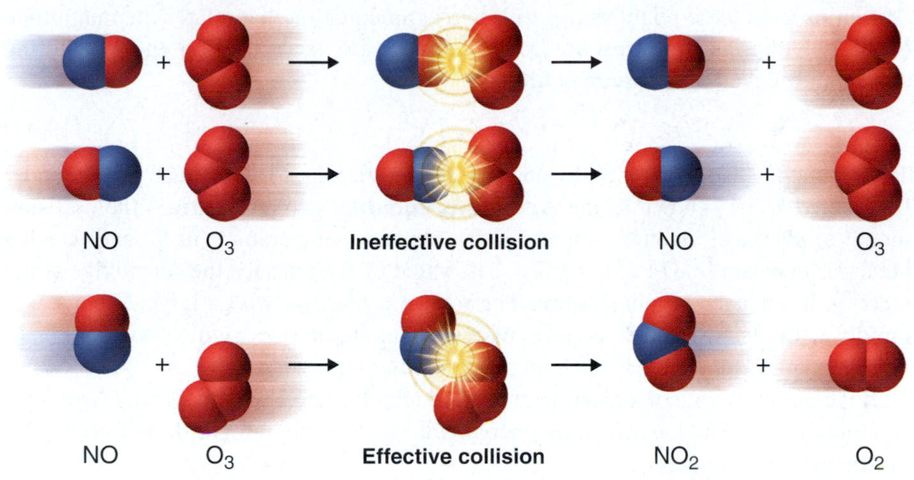

Figure 14.18 The effect of molecular orientation on reaction of NO and O_3. Most collisions of NO and O_3 molecules occur with an incorrect orientation for a reaction to occur. Only those collisions in which the N atom of NO collides with one of the terminal O atoms of O_3 are likely to produce NO_2 and O_2, even if the molecules collide with $E > E_a$.

NO and O_3 are unproductive. The fraction of orientations that result in a reaction is called the **steric factor**, ***p***, and, in general, its value can range from 0 (no orientations of molecules result in reaction) to 1 (all orientations result in reaction).

The Arrhenius Equation

Figure 14.19 shows both the kinetic energy distributions and a potential energy diagram for a reaction. The shaded areas show that at the lower temperature (300 K) only a small fraction of molecules collide with kinetic energy greater than E_a; however, at the higher temperature (500 K) a much larger fraction of molecules collide with kinetic energy greater than E_a. Consequently, the rate of the reaction is much slower at the lower temperature because only a relatively few molecules collide with enough energy to overcome the potential energy barrier.

For an A + B elementary reaction, all the factors that affect the reaction rate can be summarized in a single series of relationships:

$$\text{rate} = (\text{collision frequency})(\text{steric factor})(\text{fraction of collisions with } E > E_a)$$

where

$$\text{rate} = k[\text{A}][\text{B}] \tag{14.42}$$

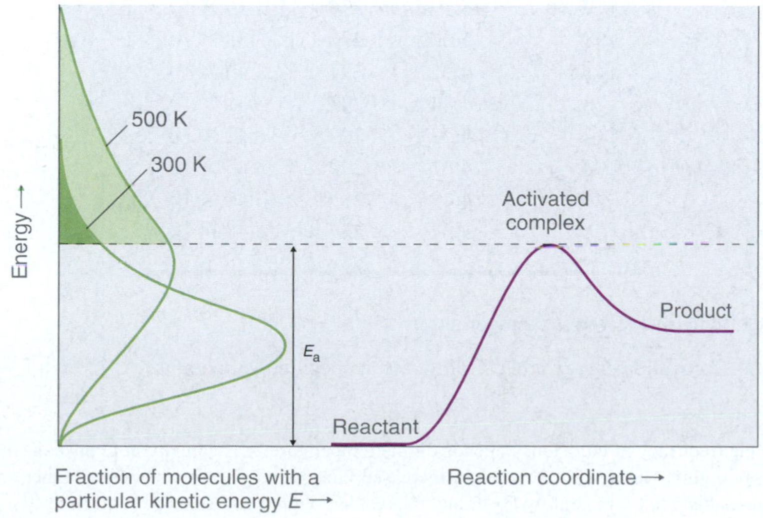

Figure 14.19 Surmounting the energy barrier to a reaction. This chart juxtaposes the energy distributions of lower-temperature (300 K) and higher-temperature (500 K) samples of a gas against the potential energy diagram for a reaction. Only those molecules in the shaded region of the energy distribution curve have $E > E_a$ and are therefore able to cross the energy barrier separating reactants and products. The fraction of molecules with $E > E_a$ is much greater at 500 K than at 300 K, and so the reaction will occur much more rapidly at 500 K. Interactive Graph

Arrhenius used these relationships to arrive at an equation that relates the magnitude of the rate constant for a reaction to the temperature, the activation energy, and the constant, A, called the **frequency factor**:

$$k = Ae^{-E_a/RT} \tag{14.43}$$

The frequency factor is used to convert concentrations to collisions per second.* Equation 14.43 is known as the **Arrhenius equation** and summarizes the collision model of chemical kinetics, where T is the absolute temperature (in K) and R is the ideal gas constant [8.314 J/(K · mol)]. The value of E_a indicates the sensitivity of the reaction to changes in temperature. The rate of a reaction with a large E_a increases rapidly with increasing temperature, whereas the rate of a reaction with a smaller E_a increases much more slowly with increasing temperature.

If we know the rate of a reaction at various temperatures, we can use the Arrhenius equation to calculate the activation energy. Taking the natural logarithm of both sides of Equation 14.43 gives

$$\ln k = \ln A + \left(-\frac{E_a}{RT}\right) = \ln A + \left[\left(-\frac{E_a}{R}\right)\left(\frac{1}{T}\right)\right] \tag{14.44}$$

Equation 14.44 is the equation of a straight line, $y = mx + b$, where $y = \ln k$ and $x = 1/T$. This means that a plot of ln k versus $1/T$ is a straight line with a slope of $-E_a/R$ and an intercept of ln A. In fact, we need to measure the rate of a reaction at only two temperatures to estimate E_a.

Knowing the value of E_a at one temperature allows us to predict the rate of a reaction at other temperatures. This is important in cooking and food preservation, for example, as well as in controlling industrial reactions to prevent potential disasters. The procedure for determining E_a from reaction rates measured at several temperatures is illustrated in the next example.

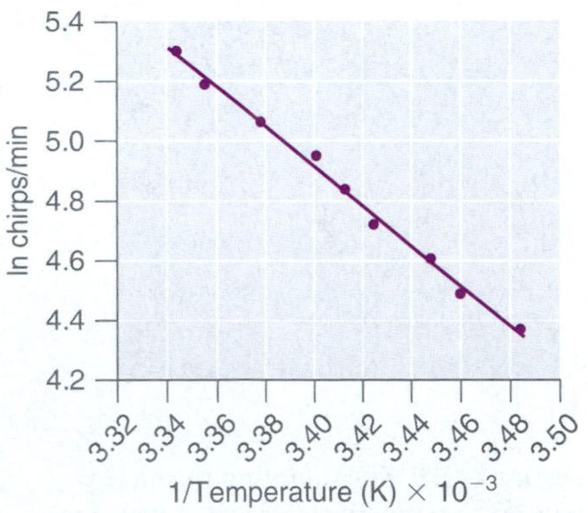

Figure 14.20 Graphical determination of E_a for tree cricket chirping. When the natural logarithm of the rate of tree cricket chirping is plotted vs. $1/T$, a straight line results. The slope of the line suggests that the chirping rate is controlled by a single reaction with an E_a of 55 kJ/mol.

EXAMPLE 14.13

Many people believe that the rate of a tree cricket's chirping is related to the temperature. To see whether this is true, biologists have carried out accurate measurements of the rate of tree cricket chirping (f) as a function of temperature. Use the data in the table, along with the graph of ln[chirping rate] versus $1/T$ in Figure 14.20, to calculate E_a for the biochemical reaction that controls cricket chirping. Then predict the chirping rate on a very hot night, when the temperature is 308 K.

Frequency, f, chirps/min	ln f	T, K	$1/T$, K
200	5.30	299	3.34×10^{-3}
179	5.19	298	3.36×10^{-3}
158	5.06	296	3.38×10^{-3}
141	4.95	294	3.40×10^{-3}
126	4.84	293	3.41×10^{-3}
112	4.72	292	3.42×10^{-3}
100	4.61	290	3.45×10^{-3}
89	4.49	289	3.46×10^{-3}
79	4.37	287	3.48×10^{-3}

Given Chirping rate at various temperatures

Asked for Activation energy and chirping rate at specified temperature

* Because the frequency of collisions depends on the temperature, A is actually *not* constant. Instead, A increases slightly with temperature as the increased kinetic energy of molecules at higher temperatures causes them to move slightly faster and thus undergo more collisions per unit time.

Strategy

A From the plot of $\ln f$ versus $1/T$ in Figure 14.20, calculate the slope of the line ($-E_a/R$); then solve for the activation energy.
B Express Equation 14.44 in terms of k_1 and T_1, and then in terms of k_2 and T_2.
C Subtract the two equations, then rearrange the result to describe k_2/k_1 in terms of T_2 and T_1.
D Using measured data from the table, solve the equation to obtain the ratio k_2/k_1. Using the value listed in the table for k_1, solve for k_2.

Solution

A If cricket chirping is controlled by a reaction that obeys the Arrhenius equation, then a plot of $\ln f$ versus $1/T$ should give a straight line (Figure 14.20). Also, the slope of the plot of $\ln f$ versus $1/T$ should be equal to $-E_a/R$. We can use the two endpoints in Figure 14.20 to estimate the slope:

$$\text{slope} = \frac{\Delta \ln f}{\Delta(1/T)} = \frac{5.30 - 4.37}{3.34 \times 10^{-3}\text{ K}^{-1} - 3.48 \times 10^{-3}\text{ K}^{-1}}$$

$$= \frac{0.93}{-0.14 \times 10^{-3}\text{ K}^{-1}} = -6.6 \times 10^3\text{ K}$$

A computer best-fit line through all the points has a slope of -6.67×10^3, so our estimate is very close. We now use it to solve for the activation energy:

$$E_a = -(\text{slope})(R) = -(-6.6 \times 10^3\text{ K})\left(\frac{8.314\text{ J}}{\text{K} \cdot \text{mol}}\right)\left(\frac{1\text{ KJ}}{1000\text{ J}}\right) = \frac{55\text{ kJ}}{\text{mol}}$$

B If the activation energy of a reaction and the rate constant at one temperature are known, then we can calculate the rate at any other temperature. We can use Equation 14.44 to express the known rate constant, k_1, at the first temperature, T_1, as

$$\ln k_1 = \ln A - \frac{E_a}{RT_1}$$

Similarly, we can express the unknown rate constant, k_2, at the second temperature, T_2, as

$$\ln k_2 = \ln A - \frac{E_a}{RT_2}$$

C These two equations contain four known quantities (E_a, T_1, T_2, and k_1) and two unknowns (A and k_2). We can eliminate the term A by subtracting the first equation from the second:

$$\ln k_2 - \ln k_1 = \left(\ln A - \frac{E_a}{RT_2}\right) - \left(\ln A - \frac{E_a}{RT_1}\right) = -\frac{E_a}{RT_2} + \frac{E_a}{RT_1}$$

Then

$$\ln \frac{k_2}{k_1} = \frac{E_a}{R}\left(\frac{1}{T_1} - \frac{1}{T_2}\right)$$

D To obtain the best prediction of chirping rate at 308 K (T_2), we try to choose for T_1 and k_1 the measured rate constant and corresponding temperature in the data table that is closest to the best-fit line in the graph. Choosing data for $T_1 = 296$ K, where $f = 158$, and using the value of E_a calculated above gives

$$\ln \frac{k_{T_2}}{k_{T_1}} = \frac{E_a}{R}\left(\frac{1}{T_1} - \frac{1}{T_2}\right) = \frac{55\text{ kJ/mol}}{8.314\text{ J/(K} \cdot \text{mol)}}\left(\frac{1000\text{ J}}{1\text{ kJ}}\right)\left(\frac{1}{296\text{ K}} - \frac{1}{308\text{ K}}\right) = 0.87$$

Thus, $k_{308}/k_{296} = 2.4$ and $k_{308} = (2.4)(158) = 380$, and the chirping rate on a night when the temperature is 308 K is predicted to be 380 chirps per minute.

EXERCISE 14.13

The equation for the decomposition of NO_2 to NO and O_2 is second order in NO_2:

$$2NO_2(g) \longrightarrow 2NO(g) + O_2(g)$$

Data for the rate of the reaction as a function of temperature are listed in the table. Calculate E_a for the reaction and the rate constant at 700 K.

T, K	k, $M^{-1} \cdot s^{-1}$
592	522
603	755
627	1700
652	4020
656	5030

Answer $E_a = 114$ kJ/mol; $k_{700} = 13{,}600\ M^{-1} \cdot s^{-1} = 1.9 \times 10^4\ M^{-1} \cdot s^{-1}$.

What value of E_a results in a doubling of the reaction rate with a 10°C increase in temperature from 20° to 30°C?

Answer About 51 kJ/mol.

14.8 Catalysis

MGC Activation Energy and Catalysts

Chapter 3 described **catalysts** as substances that increase the rate of a chemical reaction without being consumed in the process. A catalyst, therefore, does not appear in the overall stoichiometry of the reaction it catalyzes, but it *must* appear in at least one of the elementary steps in the mechanism for the catalyzed reaction. The catalyzed pathway has a lower E_a, but the *net* change in energy that results from the reaction (the difference between the energy of the reactants and the energy of the products) is *not* affected by the presence of a catalyst (Figure 14.21). Nevertheless, because of its lower E_a, the rate of a catalyzed reaction is faster than the rate of the uncatalyzed reaction at the same temperature. Note that because a catalyst decreases the height of the energy barrier, its presence increases the rates of *both* the forward *and* the reverse reactions by the same amount. In this section we will examine the three major classes of catalysts: heterogeneous catalysts, homogeneous catalysts, and enzymes.

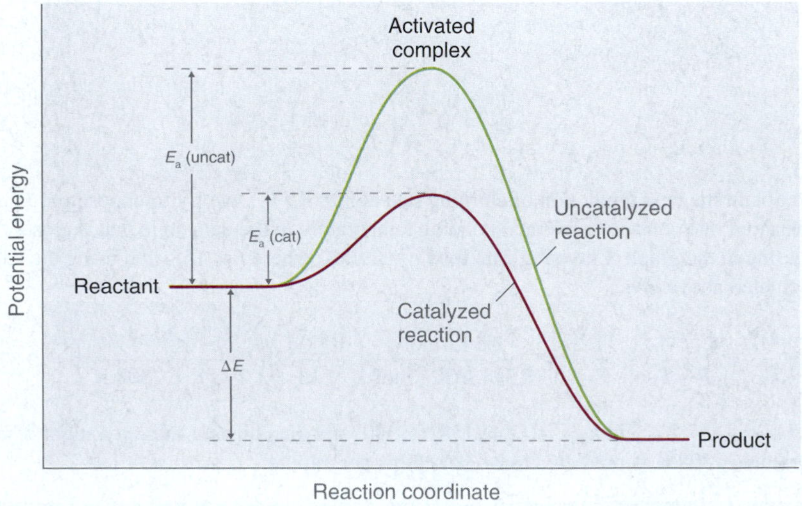

Figure 14.21 Lowering the activation energy of a reaction by a catalyst. This graph compares potential energy diagrams for a single step reaction in the presence and absence of a catalyst. Note that the only effect of the catalyst is to lower the activation energy of the reaction. The catalyst does not affect the energy of the reactants or products (and thus does not affect ΔE). Interactive Graph

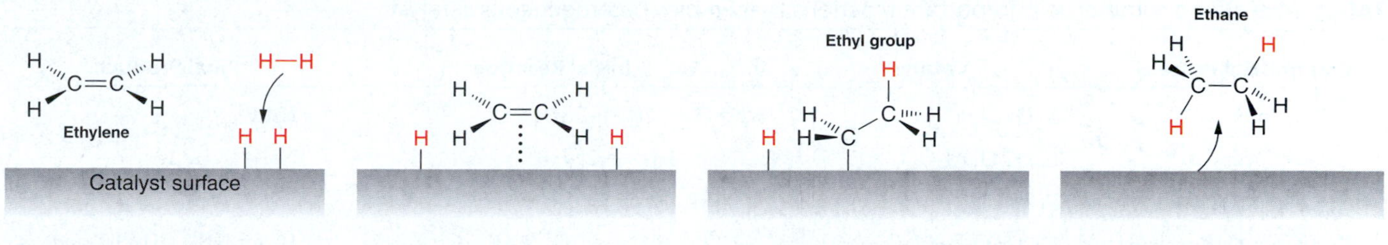

(a) Hydrogen (H_2) adsorbs to the catalyst surface (M) to form adsorbed H atoms.

(b) Ethylene (C_2H_4) adsorbs to the catalyst surface.

(c) Adsorbed ethylene reacts with an adsorbed H atom to give a bound ethyl group (C_2H_5).

(d) The ethyl group reacts with a second adsorbed H atom to give the product ethane (C_2H_6).

Figure 14.22 Hydrogenation of ethylene on a heterogeneous catalyst. When a molecule of hydrogen absorbs to the catalyst surface, the H—H bond breaks and new M—H bonds are formed. The individual H atoms are more reactive than gaseous H_2. When a molecule of ethylene interacts with the catalyst surface, it reacts with the H atoms in a stepwise process to eventually produce ethane, which is released.

Heterogeneous Catalysis

In **heterogeneous catalysis** the catalyst is in a different phase from the reactants. At least one of the reactants interacts with the solid surface in a physical process called **adsorption** in such a way that a chemical bond in the reactant becomes weak and then breaks. *Poisons* are substances that bind irreversibly to catalysts, preventing reactants from adsorbing and thus reducing or destroying the catalyst's efficiency.

An example of heterogeneous catalysis is the interaction of hydrogen gas with the surface of a metal such as Ni, Pd, or Pt. As shown in Figure 14.22a, the hydrogen–hydrogen bonds break and produce individual adsorbed hydrogen atoms on the surface of the metal. Because the adsorbed atoms can move around on the surface, two hydrogen atoms can collide and form a molecule of hydrogen gas that can then leave the surface in the reverse process, called *desorption.* Adsorbed H atoms on a metal surface are substantially more reactive than a hydrogen molecule. Because the relatively strong H—H bond (dissociation energy = 432 kJ/mol) has already been broken, the energy barrier for most reactions of H_2 is substantially lower on the catalyst surface.

Figure 14.22 shows a process called *hydrogenation*, in which hydrogen atoms are added to the double bond of an alkene, such as ethylene, to give a product that contains C—C single bonds, in this case ethane. Hydrogenation is used in the food industry to convert vegetable oils, which consist of long chains of alkenes, to more commercially valuable solid derivatives that contain alkyl chains. Hydrogenation of some of the double bonds in polyunsaturated vegetable oils, for example, produces margarine, a product with a melting point, texture, and other physical properties similar to those of butter.

Several important examples of industrial heterogeneous catalytic reactions are listed in Table 14.9. Although the mechanisms of these reactions are considerably more complex than the simple hydrogenation reaction described above, they all involve adsorption of the reactants onto a solid catalytic surface, chemical reaction of the adsorbed species (sometimes via a number of intermediate species), and finally desorption of the products from the surface.

Homogeneous Catalysis

In **homogeneous catalysis**, the catalyst is in the same phase as the reactant(s). The number of collisions between reactants and catalyst is at a maximum because the catalyst is uniformly dispersed throughout the reaction mixture. Many homogeneous catalysts in industry are transition metal compounds (Table 14.10), but recovering these expensive catalysts from solution has been a major challenge. As an added barrier to their widespread commercial use, many homogeneous catalysts can be used only at

TABLE 14.9 Some commercially important reactions that employ heterogeneous catalysts

Commercial Process	Catalyst	Initial Reaction	Final Product
Contact process	V_2O_5 or Pt	$2SO_2 + O_2 \longrightarrow 2SO_3$	H_2SO_4
Haber process	Fe, K_2O, Al_2O_3	$N_2 + 3H_2 \longrightarrow 2NH_3$	NH_3
Ostwald process	Pt and Rh	$4NH_3 + 5O_2 \longrightarrow 4NO + 6H_2O$	HNO_3
Water–gas shift reaction	Fe, Cr_2O_3, or Cu	$CO + H_2O \longrightarrow CO_2 + H_2$	H_2 for NH_3, CH_3OH, and other fuels
Steam reforming	Ni	$CH_4 + H_2O \longrightarrow CO + 3H_2$	H_2
Methanol synthesis	ZnO and Cr_2O_3	$CO + 2H_2 \longrightarrow CH_3OH$	CH_3OH
SOHIO process	Bismuth phosphomolybdate	$CH_2{=}CHCH_3 + NH_3 + \frac{3}{2}O_2 \longrightarrow CH_2{=}CHCN + 3H_2O$	$CH_2{=}CHCN$ **Acrylonitrile**
Catalytic hydrogenation	Ni, Pd, or Pt	$RCH{=}CHR' + H_2 \longrightarrow RCH_2{-}CH_2R'$	Partially hydrogenated oils for margarine, etc.

relatively low temperatures, and even then they tend to decompose slowly in solution. Despite these problems, a number of commercially viable processes have been developed in recent years. High-density polyethylene and polypropylene are produced by homogeneous catalysis.

Enzymes

Enzymes, catalysts that occur naturally in living organisms, are almost all protein molecules with typical molecular masses of 20,000–100,000 amu. Some are homogeneous catalysts that react in aqueous solution within a cellular compartment of an organism. Others are heterogeneous catalysts embedded within the membranes that separate cells and cellular compartments from their surroundings. The reactant in an enzyme-catalyzed reaction is called a **substrate**.

Because enzymes can increase reaction rates by enormous factors (up to 10^{17} times the uncatalyzed rate!) and tend to be very specific, typically producing only a single product in quantitative yield, they are the focus of active research. At the same time, enzymes are usually expensive to obtain, they often cease functioning at temperatures higher than 37°C, they have limited stability in solution, and they have such high specificity that they are confined to turning one particular set of reactants into one particular

TABLE 14.10 Some commercially important reactions that employ homogeneous catalysts

Commercial Process	Catalyst	Reactants	Final Product
Union Carbide	$[Rh(CO)_2I_2]^-$	$CO + CH_3OH$	CH_3CO_2H
Hydroperoxide process	Mo(VI) complexes	$CH_3CH{=}CH_2 + R{-}C{-}O{-}O{-}H$	$CH_3\underset{\diagdown O \diagup}{CH{-}CH_2} + ROH$ **Propylene oxide**
Hydroformylation	Rh/PR_3 complexes	$RCH{=}CH_2 + CO + H_2$	RCH_2CH_2CHO
Adiponitrile process	Ni/PR_3 complexes	$2HCN + CH_2{=}CHCH{=}CH_2$	$NCCH_2CH_2CH_2CH_2CN$ used to synthesize nylon
Olefin polymerization	$(RC_5H_5)_2ZrCl_2$	$CH_2{=}CH_2$	$\text{-(}CH_2CH_2\text{)-}_n$ High-density polyethylene

product. This means that separate processes using different enzymes must be developed for chemically similar reactions, which is time-consuming and expensive. Thus far, enzymes have found only limited industrial applications, although they are used as ingredients in laundry detergents, contact lens cleaners, and meat tenderizers. The enzymes in these applications tend to be *proteases*, which are able to cleave the amide bonds that hold amino acids together in proteins. Meat tenderizers, for example, contain a protease called papain, which is isolated from papaya juice. It cleaves some of the long, fibrous protein molecules that make inexpensive cuts of beef tough, producing a tenderer piece of meat. Some insects, like the bombadier beetle, carry an enzyme capable of catalyzing the decomposition of hydrogen peroxide to water (Figure 14.23).

Figure 14.23 A catalytic defense mechanism. The scalding, foul-smelling spray emitted by this bombardier beetle is produced by the catalytic decomposition of H_2O_2.

Enzyme inhibitors cause a decrease in the rate of an enzyme-catalyzed reaction by binding to a specific portion of an enzyme and thus slowing or preventing a reaction from occurring. Irreversible inhibitors are therefore the equivalent of poisons in heterogeneous catalysis. One of the oldest and most widely used commercial enzyme inhibitors is aspirin, which selectively inhibits one of the enzymes involved in the synthesis of molecules that trigger inflammation. The design and synthesis of related molecules that are more effective, more selective, and less toxic than aspirin are important objectives of biomedical research.

SUMMARY AND KEY TERMS

14.0 Introduction (p. 617)

Chemical kinetics is the study of **reaction rates**, changes in the concentrations of reactants and products with time.

14.1 Factors That Affect Reaction Rates (p. 618)

Factors that influence the rates of chemical reactions include concentration of reactants, temperature, physical state of reactants and their dispersion, solvent, and presence of a catalyst.

14.2 Reaction Rates and Rate Laws (p. 620)

Reaction rates are reported either as the **average rate** over a period of time or as the **instantaneous rate** at a single time.

The **rate law** for a reaction is a mathematical relationship between the reaction rate and the concentrations of species in solution. Rate laws can be expressed either as a **differential rate law**, describing the change in reactant or product concentrations as a function of time, or as an **integrated rate law**, describing the actual concentrations of reactants or products as a function of time.

The **rate constant**, k, of a rate law is a constant of proportionality between reaction rate and reactant concentration. The power to which a concentration is raised in a rate law indicates the **reaction order**, the degree to which the rate of the reaction depends on the concentration of a particular reactant.

14.3 Methods of Determining Reaction Orders (p. 629)

The rate of a **zeroth-order reaction** is independent of the concentration of the reactants. The rate of a **first-order reaction** is directly proportional to the concentration of one of the reactants. The rate of a simple **second-order reaction** is proportional to the square of the concentration of one of the reactants.

14.4 Using Graphs to Determine Rate Laws, Rate Constants, and Reaction Orders (p. 641)

For a zeroth-order reaction, a plot of the concentration of any reactant versus time is a straight line with a slope of $-k$. For a first-order reaction, a plot of the natural logarithm of the concentration of a reactant versus time is a straight line with a slope of $-k$. For a second-order reaction, a plot of the inverse of the concentration of a reactant versus time is a straight line with a slope of k.

14.5 Half-Lives and Radioactive Decay Kinetics (p. 644)

The **half-life** of a reaction is the time required for the reactant concentration to decrease to one-half its initial value. The half-life of a first-order reaction is a constant that is related to the rate constant for the reaction: $t_{1/2} = 0.693/k$.

Radioactive decay reactions are first-order reactions. The **rate of decay**, or **activity**, of a sample of a radioactive substance is the decrease in the number of radioactive nuclei per unit time.

14.6 Reaction Rates—A Microscopic View (p. 649)

A **reaction mechanism** is the microscopic path by which reactants are transformed into products. Each step is an **elementary reaction**. Species that are formed in one step and consumed in another are **intermediates**. Each elementary step can be described in terms of its **molecularity**, the number of molecules that collide in that step.

The slowest step in a reaction mechanism is the **rate-determining step**. **Chain reactions** consist of three kinds of reactions: initiation, propagation, and termination. Intermediates in chain reactions are often **radicals**, species that have an unpaired valence electron.

14.7 The Collision Model of Chemical Kinetics (p. 654)

A minimum energy (**activation energy, E_a**) is required for a collision between molecules to result in a chemical reaction. Plots of potential energy for a system versus the reaction coordinate show an energy barrier that must be overcome for the reaction to occur. The arrangement of atoms at the highest point of this barrier is the **activated complex**, or **transition state**, of the reaction. At a given temperature, the higher the E_a, the slower the reaction. The fraction of orientations that result in a reaction is the **steric factor**. The **frequency factor**, steric factor, and activation energy are related to the rate constant in the **Arrhenius equation**: $k = Ae^{-E_a/RT}$. A plot of the natural logarithm of k versus $1/T$ is a straight line with a slope of $-E_a/R$.

14.8 Catalysis (p. 660)

Catalysts participate in a chemical reaction and increase its rate. They do not appear in the reaction's net equation, and they are not consumed during the reaction. Catalysts allow a reaction to proceed via a pathway that has a lower activation energy than the uncatalyzed reaction. **Heterogeneous catalysts** provide a surface to which reactants bind in a process of **adsorption**. **Homogeneous catalysts** are in the same phase as the reactants. **Enzymes** are biological catalysts that produce large increases in reaction rates and tend to be specific for certain reactants and products. The reactant in an enzyme-catalyzed reaction is called a **substrate**. **Enzyme inhibitors** cause a decrease in the rate of an enzyme-catalyzed reaction.

KEY EQUATIONS

General definition of rate for A ⟶ B	$\text{rate} = \frac{\Delta[B]}{\Delta t} = -\frac{\Delta[A]}{\Delta t}$	(14.4)
General form of rate law when A and B are reactants	$\text{rate} = k[A]^m[B]^n$	(14.10)
Zeroth-order reaction	$\text{rate} = -\frac{\Delta[A]}{\Delta t} = k$	(14.16)
	$[A] = [A]_0 - kt$	(14.17)
First-order reaction	$\text{rate} = -\frac{\Delta[A]}{\Delta t} = k[A]$	(14.21)
	$[A] = [A]_0e^{-kt}$	(14.22)
	$\ln[A] = \ln[A]_0 - kt$	(14.23)
Second-order reaction	$\text{rate} = -\frac{\Delta[A]}{\Delta t} = k[A]^2$	(14.25)
	$\frac{1}{[A]} = \frac{1}{[A]_0} + kt$	(14.26)
Half-life of first-order reaction	$t_{1/2} = \frac{0.693}{k}$	(14.32)
Radioactive decay	$A = kN$	(14.34)
Arrhenius equation	$k = Ae^{-E_a/RT}$	(14.43)

QUESTIONS AND PROBLEMS

MGC For instructor-assigned homework, go to **www.masteringgenchem.com**

Questions and Problems with colored numbers have answers in the Appendix and complete solutions in the Student Solutions Manual.

CONCEPTUAL

14.1 Factors That Affect Reaction Rates

1. What information can you obtain by studying the chemical kinetics of a reaction? Does a balanced chemical equation provide the same information? Why or why not?
2. If you were given the task of determining whether to proceed with a particular reaction in an industrial facility, why would studying the chemical kinetics of the reaction be important to you?
3. What is the relationship between each of these factors and the rate of a reaction: reactant concentration, temperature of the reaction, physical properties of the reactants, physical and chemical properties of the solvent, and the presence of a catalyst?
4. A slurry is a mixture of a substance that is only sparingly soluble in a liquid. As you prepare a reaction, you notice that one of your reactants forms a slurry with the solvent. What effect will this have on the reaction rate? What steps can you take to try to solve the problem?
5. Why does the rate of virtually all reactions increase with an increase in temperature? If you were to make a glass of sweetened iced tea the old-fashioned way, by adding sugar and ice cubes to a glass of hot tea, which would you add first?
6. In a typical laboratory setting, a reaction is carried out in a ventilated hood with air circulation provided by outside air. A student noticed that a reaction that gave a high yield of a

product during the winter months gave a low yield of that same product in the summer, even though his technique did not change and the reagents and concentrations used were identical. What is a plausible explanation for the different yields?

7. A very active area of chemical research involves the development of solubilized catalysts that are not made inactive during the reaction process. Such catalysts are expected to increase reaction rates significantly relative to the same reaction run in the presence of a heterogeneous catalyst. What is the reason for anticipating that the relative rate will increase?
8. Water has a dielectric constant more than two times greater than that of methanol (dielectric constant = 80.1 for H_2O and 33.0 for CH_3OH). Which would be your solvent of choice for a substitution reaction between an ionic compound and a polar reagent, both of which are soluble in either methanol or water? Why?

14.2 Reaction Rates and Rate Laws

9. Explain why the rate of a reaction is generally fastest at early time intervals. For the second-order $A + B \rightarrow C$, what would the plot of the concentration of C versus time look like during the course of the reaction?
10. Explain the differences between a differential rate law and an integrated rate law. What two components do they have in common? Which form is preferred for obtaining a reaction order and a rate constant? Why?
11. Diffusion-controlled reactions have rates that are determined only by the rate at which two reactant molecules can diffuse together. These reactions are rapid, with second order rate constants typically on the order of 10^{10} L/mol · s. Would you expect the reactions to be faster or slower in solvents that have a low viscosity? Why? Consider the reactions $H_3O^+ + OH^- \rightarrow 2H_2O$ and $H_3O^+ + N(CH_3)_3 \rightarrow H_2O + HN(CH_3)_3{}^+$ in aqueous solution. Which would have the higher rate constant? Why?
12. What information can you get from the reaction order? What correlation does the reaction order have with the stoichiometry of the overall equation?
13. During the hydrolysis $A + H_2O \rightarrow B + C$, the concentration of A decreases much more rapidly in a polar solvent than in one that is not polar. How do you expect this effect to be reflected in the overall reaction order?

14.3 Methods of Determining Reaction Orders

14. What are the characteristics of a zeroth-order reaction? Experimentally, how would you determine whether a reaction is zeroth order?
15. Predict whether the following reactions are zeroth order: (a) a substitution reaction of an alcohol with HCl to form an alkyl halide and water; (b) catalytic hydrogenation of an alkene; (c) hydrolysis of an alkyl halide to an alcohol; and (d) enzymatic conversion of nitrate to nitrite in a soil bacterium. Explain your reasoning.
16. In a first-order reaction, what is the advantage of using the integrated rate law expressed in natural logarithms over the rate law expressed in exponential form?
17. If the rate of a reaction is directly proportional to the concentration of a reactant, what does this tell you about (a) the order of the reaction with respect to the reactant, and (b) the overall order of the reaction?
18. The reaction of NO with O_2 is found to be second order with respect to NO and first order with respect to O_2. What is the overall reaction order? What is the effect of doubling the concentration of each reagent on the rate?

14.4 Using Graphs to Determine Rate Laws, Rate Constants, and Reaction Orders

19. Compare first-order differential and integrated rate laws with respect to (a) the magnitude of the rate constant, (b) the information needed to determine the order, and (c) the shape of the graphs. Is there any information that can be obtained from the integrated rate law that cannot be obtained from the differential rate law?
20. In the single step second order reaction $2A \rightarrow$ products, how would a graph of concentration of A vs. time compare to a plot of 1/A vs. time? Which of these would be the most similar to the same set of graphs for A during the single step second order reaction $A + B \rightarrow$ products? Explain.
21. For reactions of the same order, what is the relationship between the magnitude of the rate constant and reaction rate? If you were comparing reactions with different orders, could the same arguments be made? Why?

14.5 Half-Lives and Radioactive Decay Kinetics

22. What do chemists mean when they refer to the *half-life* of a reaction?
23. If a sample of one isotope undergoes more disintegrations per second than the same number of atoms of another isotope, how do their half-lives compare?

14.6 Reaction Rates—A Microscopic View

24. How does the term *molecularity* relate to the elementary reactions? How does it relate to the overall balanced chemical equation?
25. What is the relationship between the order of a rate law and the molecularity of a reaction? What is the relationship between the rate law and the balanced chemical equation?
26. When you determine the rate law for a given reaction, why is it valid to assume that the concentration of an intermediate does not change with time during the course of the reaction?
27. If you know the rate law for an overall reaction, how would you determine which elementary step is rate determining? If an intermediate is contained in the rate-determining step, how can the experimentally determined rate law for the reaction be derived from this step?
28. Give the rate-determining step for each case: (a) traffic is backed up on a highway because two lanes merge into one; (b) gas flows from a pressurized cylinder fitted with a gas regulator and then is bubbled through a solution; and (c) a document containing text and graphics is downloaded from the Internet.
29. Before being sent on an assignment, an aging James Bond was sent off to a health farm where part of the program's focus was to purge his body of radicals. Why was this goal considered important to his health?

14.7 The Collision Model of Chemical Kinetics

30. Although an increase in temperature results in an increase in kinetic energy, this increase in kinetic energy is not sufficient to explain the relationship between temperature and reaction rates. How does the activation energy relate to the chemical kinetics of a reaction? Why does an increase in temperature increase the reaction rate despite the fact that the average kinetic energy is still lower than the activation energy?
31. For any given reaction, what is the relationship between the energy of activation and (a) electrostatic repulsions, (b) bond formation, and (c) the nature of the activated complex?

32. If you are concerned with whether a reaction will occur rapidly, why would you be more interested in knowing the magnitude of the activation energy than the change in potential energy for the reaction?

33. The product C in the reaction $A + B \longrightarrow C + D$ can be separated easily from the reaction mixture. You have been given pure A and pure B and are told to determine the activation energy for this reaction to determine whether the reaction is suitable for the industrial synthesis of C. How would you do this? Why do you need to know the magnitude of the activation energy to make a decision about feasibility?

34. Above E_a, molecules collide with enough energy to overcome the energy barrier for a reaction. Is it possible for a reaction to occur at a temperature lower than that needed to reach E_a? Explain your answer.

35. What is the relationship between *A*, activation energy, and temperature? How does an increase in the frequency factor affect the rate of reaction?

36. Of two highly exothermic reactions with different values of E_a, which would need to be monitored more carefully: the one with the smaller value or the one with the higher value? Why?

14.8 Catalysis

37. What effect does a catalyst have on the energy of activation of a reaction? What effect does it have on the frequency factor, *A*? What effect does it have on the change in potential energy for the reaction?

38. How is it possible to affect the product distribution of a reaction by using a catalyst?

39. A heterogeneous catalyst works by interacting with a reactant in a process called *adsorption*. What occurs during the process of adsorption? Explain how this can lower the activation energy.

40. What effect does increasing the surface area of a heterogeneous catalyst have on a reaction? Does increasing the surface area affect the energy of activation? Explain your answer.

41. What are the differences between a heterogeneous catalyst and a homogeneous catalyst in terms of (a) ease of recovery, (b) collision frequency, (c) temperature sensitivity, and (d) cost?

42. An area of intensive chemical research involves the development of homogeneous catalysts, even though homogeneous catalysts generally have a number of operational difficulties. Propose one or two reasons why a homogenous catalyst may be preferred.

43. Consider the reaction between cerium(IV) and thallium(I) ions

$$2Ce^{4+} + Tl^{+} \longrightarrow 2Ce^{3+} + Tl^{3+}$$

This reaction is slow, but it is catalyzed by Mn^{2+}, as shown in the following mechanism:

$$Ce^{4+} + Mn^{2+} \longrightarrow Ce^{3+} + Mn^{3+}$$
$$Ce^{4+} + Mn^{3+} \longrightarrow Ce^{3+} + Mn^{4+}$$
$$Mn^{4+} + Tl^{+} \longrightarrow Tl^{3+} + Mn^{2+}$$

In what way does Mn^{2+} increase the reaction rate?

44. The text lists several factors that limit the industrial applications of enzymes. Still, there is keen interest in understanding how enzymes work in order to design catalysts for industrial applications. Why?

45. Most enzymes have an optimal pH range; however, care must be taken when determining pH effects on enzyme activity. A decrease in activity could be due to the effects of changes in pH on groups at the catalytic center or to the effects on groups located elsewhere in the enzyme. Both examples are observed in chymotrypsin, a digestive enzyme that is a protease which hydrolyzes polypeptide chains. Explain how a change in pH could affect the catalytic activity due to (a) effects at the catalytic center and (b) effects elsewhere in the enzyme. (*Hint:* Remember that enzymes are composed of functional amino acids.)

NUMERICAL

This section includes "paired problems" (marked by brackets) that require similar problem-solving skills.

14.2 Reaction Rates and Rate Laws

46. The rate of a particular reaction in which A and B react to make C is rate $= -\Delta[A]/\Delta t = \frac{1}{2}\,\Delta[C]/\Delta t$. Write a reaction equation that is consistent with this rate law. What is the rate expression with respect to time if 2A is converted to 3C?

47. While commuting to work, a person drove for 12 min at 35 mi/h, then stopped at an intersection for 2 min, continued the commute at 50 mi/h for 28 min, drove slowly through traffic at 38 mi/h for 18 min, and then spent 1 min pulling into a parking space at 3 mi/h. What was the average rate of the commute? What was the instantaneous rate at 13 min? At 28 min?

48. Why do most studies of chemical reactions use the initial rates of reaction to generate a rate law? How is this initial rate determined? Given the following instantaneous rate data, what is the reaction order? Estimate.

Time, s	[A], (*M*)
120	0.158
240	0.089
360	0.062

49. Predict how the rate of each reaction will be affected by doubling the concentration of the first species in each equation.

(a) $C_2H_5I \longrightarrow C_2H_4 + HI$	Rate $= k[C_2H_5I]$
(b) $SO + O_2 \longrightarrow SO_2 + O$	Rate $= k[SO][O_2]$
(c) $2CH_3 \longrightarrow C_2H_6$	Rate $= k[CH_3]^2$
(d) $ClOO \longrightarrow Cl + O_2$	Rate $= k$

50. Three chemical processes occur at an altitude of approximately 100 km in Earth's atmosphere.

$$N_2^{+} + O_2 \xrightarrow{k_1} N_2 + O_2^{+}$$
$$O_2^{+} + O \xrightarrow{k_2} O_2 + O^{+}$$
$$O^{+} + N_2 \xrightarrow{k_3} NO^{+} + N$$

Write a rate law for each of the elementary steps. If the rate law for the overall reaction were found to be rate $= k[N_2^{+}][O_2]$, which one of the steps is rate limiting?

51. Cleavage of C_2H_6 to produce two $CH_3\cdot$ radicals is a gas-phase reaction that occurs at 700°C. This reaction is first order, with $k = 5.46 \times 10^{-4}\ s^{-1}$. How long will it take for the reaction to go to 15% completion? To 50% completion?

52. The oxidation of aqueous iodide by arsenic acid to give I_3^- and arsenious acid proceeds via the reaction

$$H_3AsO_4(aq) + 3I^-(aq) + 2H^+(aq) \underset{k_r}{\overset{k_f}{\rightleftharpoons}} H_3AsO_3(aq) + I_3^-(aq) + H_2O(l)$$

When the rate of the forward reaction is equal to the rate of the reverse reaction, what is $k_f/k_r = [H_3AsO_3][I_3^-]/[H_3AsO_4][I^-]^3[H^+]^2$? Then write an expression for $\Delta[I_3^-]/\Delta t$.

14.3 Methods of Determining Reaction Orders

53. Iodide reduces Fe(III) according to the reaction

$$2Fe^{3+}(soln) + 2I^-(soln) \longrightarrow 2Fe^{2+}(soln) + I_2(soln)$$

Experimentally, it was found that doubling the concentration of Fe(III) doubled the rate, and doubling the iodide concentration increased the reaction rate by a factor of 4. What is the order of the reaction with respect to each species? What is the overall rate law? What is the overall reaction order?

54. Benzoyl peroxide is a medication used to treat acne. Its rate of thermal decomposition at several concentrations was determined experimentally, and the data were tabulated as follows:

Experiment	[Benzoyl Peroxide]$_0$, *M*	Initial Rate, *M*/s
1	1.00	2.22×10^{-4}
2	0.70	1.64×10^{-4}
3	0.50	1.12×10^{-4}
4	0.25	0.59×10^{-4}

What is the order of this reaction with respect to benzoyl peroxide? What is the rate law for this reaction?

55. 1-Bromopropane is a colorless liquid that reacts with $S_2O_3^{2-}$ according to the equation

$$C_3H_7Br + S_2O_3^{2-} \longrightarrow C_3H_7S_2O_3^- + Br^-$$

The reaction is first order in 1-bromopropane and first order in $S_2O_3^{2-}$, with a rate constant of $8.05 \times 10^{-4}\ M^{-1} \cdot s^{-1}$. If you began a reaction with 40 mmol/100 mL of C_3H_7Br and an equivalent concentration of $S_2O_3^{2-}$, what would the initial rate of reaction be? If you were to decrease the concentration of each reactant to 20 mmol/100 mL, what would the initial rate of reaction be?

56. The experimental rate law for the reaction $3A + 2B \rightarrow C + D$ was found to be $\Delta[C]/\Delta t = k[A]^2[B]$ for an overall reaction that is third order. Because graphical analysis is difficult beyond second-order reactions, explain the procedure for determining the rate law experimentally.

14.4 Using Graphs to Determine Rate Laws, Rate Constants, and Reaction Orders

57. One method of using graphs to determine reaction order is to use relative rate information. Plotting the log of the relative rate vs. log of relative concentration provides information about the reaction. Here is an example of data from a zeroth-order reaction:

Relative [A], *M*	Relative Rate, *M*/s
1	1
2	1
3	1

Varying [A] does not alter the rate of the reaction. Using relative rates as shown in the table, generate plots of log (rate) vs. log (concentration) for zeroth-, first- and second order reactions. What does the slope of each line represent?

58. The table below follows the decomposition of N_2O_5 gas by examining the partial pressure of the gas as a function of time at 45°C. What is the order of the reaction? What is the rate constant? How long would it take for the pressure to reach 105 mm Hg at 45°C?

Time, s	Pressure, mm Hg
0	348
400	276
1600	156
3200	69
4800	33

14.5 Half-Lives and Radioactive Decay Kinetics

59. Half-lives for the reaction $A + B \rightarrow C$ were calculated at three values of $[A]_0$, and the concentration of B was the same in all cases. The data are listed in the table.

$[A]_0$, *M*	$t_{1/2}$, s
0.50	420
0.75	280
1.0	210

Does this reaction follow first-order kinetics? On what do you base your answer?

60. Ethyl-2-nitrobenzoate ($NO_2C_6H_4CO_2C_2H_5$) hydrolyzes under basic conditions. A plot of [ethyl-2-nitrobenzoate] versus *t* was used to calculate $t_{1/2}$, with the following results:

[Ethyl-2-nitrobenzoate], mol/cm^3	$t_{1/2}$, s
0.050	240
0.040	300
0.030	400

Is this a first-order reaction? Explain your reasoning.

61. Azomethane($CH_3N_2CH_3$) decomposes at 600 K to C_2H_6 and N_2. The decomposition is first order in azomethane. Calculate $t_{1/2}$ from the data in the table.

Time, s	$P_{CH_3N_2CH_3}$, atm
0	8.2×10^{-2}
2000	3.99×10^{-2}
4000	1.94×10^{-2}

How long will it take for the decomposition to be 99.9% complete?

62. The first-order decomposition of hydrogen peroxide has a half-life of 10.7 hr at 20°C. What is the rate constant (expressed in s^{-1}) for this reaction? If you started with a solution that was $7.5 \times 10^{-3}\ M$ H_2O_2, what would be the initial rate of decomposition (*M*/s)? What would be the concentration of H_2O_2 after 3.3 hr?

14.6 Reaction Rates—A Microscopic View

63. Cyclopropane, a mild anesthetic, rearranges to propylene via a collision that produces and destroys an energized species. Here are the important steps in this rearrangement.

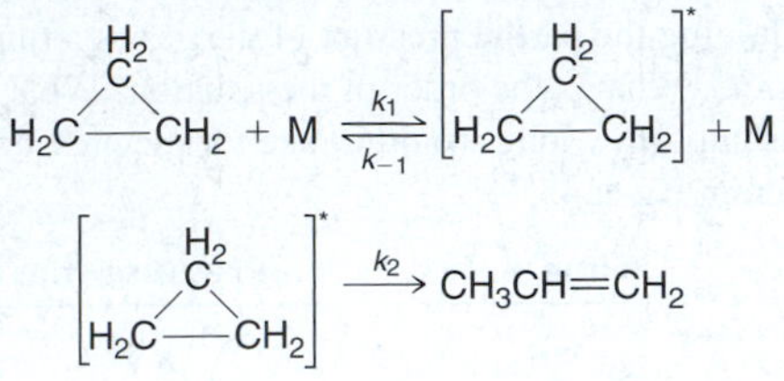

where M is any molecule, including cyclopropane. Only those cyclopropane molecules with sufficient energy (denoted with an asterisk) can rearrange to propylene. Which step determines the rate constant of the overall reaction?

64. Above approximately 500 K, the reaction between NO_2 and CO to produce CO_2 and NO follows the second-order rate law $\Delta[CO_2]/\Delta t = k[NO_2][CO]$. At lower temperatures, however, the rate law is $\Delta[CO_2]/\Delta t = k'[NO_2]^2$, for which it is known that NO_3 is an intermediate in the mechanism. Propose a complete low-temperature mechanism for the reaction based on this rate law. Which step is the slowest?

65. Nitramide (O_2NNH_2) decomposes in aqueous solution to N_2O and H_2O. What is the experimental rate law, $\Delta[N_2O]/\Delta t$, for the decomposition of nitramide if the mechanism for the decomposition is

$$O_2NNH_2 \underset{k_{-1}}{\overset{k_1}{\rightleftharpoons}} O_2NNH^- + H^+ \quad \text{(fast)}$$

$$O_2NNH^- \xrightarrow{k_2} N_2O + OH^- \quad \text{(slow)}$$

$$H^+ + OH^- \xrightarrow{k_3} H_2O \quad \text{(fast)}$$

66. Given the reactions

$$A + B \underset{k_{-1}}{\overset{k_1}{\rightleftharpoons}} C + D$$

$$D + E \xrightarrow{k_2} F$$

what is the relationship between the relative magnitudes of k_{-1} and k_2 if these reactions have the rate law $\Delta[F]/\Delta t = k[A][B][E]/[C]$? How does the magnitude of k_1 compare to that of k_2? Under what conditions would you expect the rate law to be $\Delta[F]/\Delta t = k'[A][B]$?

14.7 The Collision Model of Chemical Kinetics

67. What happens to the approximate rate of a reaction when the temperature of the reaction is increased from 20°C to 30°C? What happens to the reaction rate when the temperature is raised to 70°C? For a given reaction at room temperature (20°C), what is the shape of a plot of reaction rate versus temperature as the temperature is increased to 70°C?

68. Acetaldehyde, used in silvering mirrors and in some perfumes, undergoes a second-order decomposition between 700 and 840 K. From the data in the table, would you say that acetaldehyde follows the general rule that each 10 K increase in temperature doubles the reaction rate?

T, K	k_2, *M*/s
720	0.024
740	0.051
760	0.105
800	0.519

69. Bromoethane reacts with hydroxide ion in water to produce ethanol. The energy of activation for this reaction is 90 kJ/mol. If the rate of reaction is 3.6×10^{-5} *M*/s at 25°C, what would the rate of reaction be at (a) 15°C, (b) 30°C, (c) 45°C.

70. An enzyme-catalyzed reaction has an activation energy of 15 kcal/mol. How would the value of the rate constant differ between 20°C and 30°C? If the enzyme reduced the value of E_a from 25 kcal/mol to 15 kcal/mol, by what factor has the enzyme increased the reaction rate at each temperature?

71. The data in the table are the rate constants as a function of temperature for the dimerization of 1,3-butadiene. What is the energy of activation for this reaction?

T, K	*k*, $M^{-1} \cdot min^{-1}$
529	1.4
560	3.7
600	25
645	82

72. The rate of a reaction at 25°C is 1.0×10^{-4} *M*/s. Increasing the temperature to 75°C causes the rate to increase to 7.0×10^{-2} *M*/s. Estimate the activation energy for this process. If E_a were 25 kJ/mol and the rate at 25°C is 1.0×10^{-4} *M*/s, what would be the rate at 75°C?

14.8 Catalysis

73. At some point during an enzymatic reaction, the concentration of the activated complex, called an enzyme–substrate complex (ES), and other intermediates involved in the reaction is nearly constant. When a single substrate is involved, the reaction can be represented by the following sequence of equations:

$$\text{enzyme (E)} + \text{substrate (S)} \rightleftharpoons$$
$$\text{enzyme–substrate complex (ES)} \rightleftharpoons$$
$$\text{enzyme} + \text{product (P)}$$

This can also be shown as

$$E + S \underset{k_{-1}}{\overset{k_1}{\rightleftharpoons}} ES \underset{k_{-2}}{\overset{k_2}{\rightleftharpoons}} E + P$$

Using molar concentrations and rate constants, write an expression for the rate of disappearance of the enzyme–substrate complex. Typically, enzyme concentrations are small and substrate concentrations are high. If you were determining the rate law by varying the substrate concentrations under these conditions, what would be your apparent reaction order?

74. A particular reaction was found to proceed via the mechanism

$$A + B \longrightarrow C + D$$
$$2C \longrightarrow E$$
$$E + A \longrightarrow 3B + F$$

What is the overall reaction? Is this reaction catalytic, and if so, what species is the catalyst? Identify the intermediates.

75. A particular reaction has two accessible pathways (A and B), each of which favors conversion of *X* to a different product (*Y* and *Z*, respectively). Under uncatalyzed conditions pathway A is favored, but in the presence of a catalyst pathway B is favored. Pathway B is reversible, whereas pathway A is not. Which product is favored in the presence of a catalyst? Without a catalyst? Draw a diagram illustrating what is occurring with and without the catalyst.

76. The kinetics of an enzyme-catalyzed reaction can be analyzed by plotting the rate of reaction versus the substrate concentration. This type of analysis is referred to as a Michaelis–Menten treatment. At low substrate concentrations the plot shows behavior characteristic of first-order kinetics, but at very high substrate concentrations the behavior shows zeroth-order kinetics. Explain this phenomenon.

APPLICATIONS

77. Atmospheric chemistry in the region below the clouds of Venus appears to be dominated by reactions of sulfur and carbon-containing compounds. Included in representative elementary reaction steps are the following:

$$SO_2 + CO \longrightarrow SO + CO_2$$
$$SO + CO \longrightarrow S + CO_2$$
$$SO + SO_2 \longrightarrow S + SO_3$$

For each elementary reaction, write an expression for the reaction rate in terms of changes in the concentrations of each reactant and product with time.

78. In acid, nitriles hydrolyze to produce a carboxylic acid and ammonium ion. For example, acetonitrile, a substance used to extract fatty acids from fish liver oils, is hydrolyzed to acetic acid via the reaction

$$CH_3C{\equiv}N(l) + 2H_2O(l) + H^+(aq) \longrightarrow CH_3\overset{\overset{\displaystyle O}{\|}}{C}OH(aq) + NH_4{}^+(aq)$$

Express the rate of the reaction in terms of changes in the concentrations of each reactant and each product with time.

79. Ozone production occurs at lower altitudes according to the elementary reaction $O + O_2 \longrightarrow O_3$, with an estimated rate of ozone production of 4.86×10^{31} molecules $\cdot$ s^{-1} worldwide. What is the overall order of this reaction? If the rate of loss of O is 0.89×10^{31} molecules $\cdot$ s^{-1}, and 0.06×10^{31} molecules $\cdot$ s^{-1} of ozone is transported to other atmospheric regions, is ozone being produced faster than it is being destroyed? Measurements show that ozone concentrations are not increasing rapidly. What conclusion can you draw from these data?

80. The water in a fishery became polluted when toxic waste was dumped into its pond, causing the fish population to substantially decline. The percents of fish that survived are recorded in the table.

Day	1	2	3	4	5
% Survival	79	55	38	31	19

What is the order of the reaction live fish $\longrightarrow$ dead fish? What is the rate constant? If the fish continue to die at this rate, how many fish will be alive after 10 days?

81. Until 200 years ago, manufactured iron contained charcoal produced from freshly cut wood that was added during the smelting process. As a result of this practice, older samples of iron can be dated accurately using the carbon-14 method. An archaeologist found a cast iron specimen that she believed dated to the period between 480 and 221 B.C. in Hunan, China. Radiocarbon dating of the sample indicated a 24% reduction in carbon-14 content. Was the archaeologist correct?

82. Because of its short half-life, ^{32}P-labeled compounds must be shipped as quickly as possible so that they can be used as radioactive tags in biological studies. A 50-g sample that contained 0.60 % ^{32}P by mass was shipped at 11 A.M. on Monday morning. The package was delivered to a chemist via an overnight delivery service such that it arrived the next day.
(a) What would be the mass of ^{32}P remaining in the sample if he received the package on Tuesday afternoon but was unable to use it until 9 A.M. on Wednesday?
(b) What would be the mass of ^{32}P present in the sample if the shipper had not delivered the sample until Friday afternoon and then it sat on a loading dock until 9 A.M. Monday morning?
(c) The late shipment was used immediately on Monday morning, but the biological samples were not analyzed until Thursday at 5 P.M. What percentage of the sample still consists of ^{32}P?

83. Tritium (^{3}H) is a radioactive isotope that is commonly used to follow biochemical reactions. (a) Using the data in Table 14.7, calculate the radioactive decay constant, *k*, for tritium. (b) Use the value of *k* to determine mass of tritium that is still present in a 5.00 g sample of NaB^3H_4 that is 17.57 years old.

84. L-Aspartic acid is an amino acid found in fossil bone. It can convert to a geometrically different form (D-aspartic acid) at 20°C, with a half-life corresponding to the conversion of L $\rightarrow$ D of 14,000–20,000 years. If the temperature of an archaeological site is constant, then the extent of the conversion can be used to date fossils. In one such case, archaeologists dated the arrival of humans on the North American continent to be 20,000 years ago, but the conversion of L-aspartic acid to D-aspartic acid in human fossils indicated that Paleo-Indians were living in California at least 48,000 years ago. What would be the relative concentrations of the L- and D-forms that produced this result? Carbon-14 has a half-life of approximately 5700 years. What percent of the carbon-14 originally present would have been found in the bones?

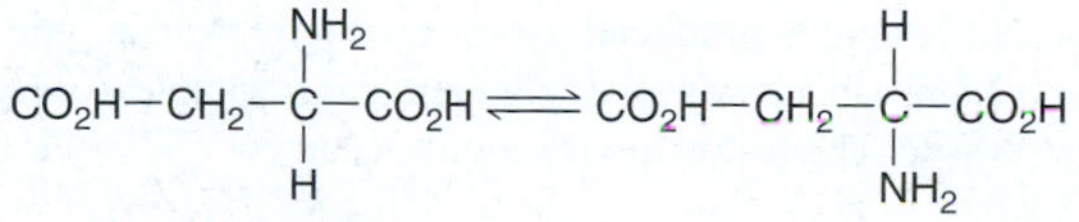

L-Aspartic acid   D-Aspartic acid

The technique described is frequently used in conjunction with radiocarbon dating. In cases where the results from the two techniques are in gross disagreement, what information can you get by comparing the two results?

85. Peroxides are able to initiate the radical polymerization of alkenes. Polyethylene, for example, is a high-molecular-weight polymer used as a film in packaging, as kitchenware, and as tubing. It is produced by heating ethylene at high pressure in the presence of oxygen or peroxide. It is formed by this radical process:

$$RO{:}OR \xrightarrow{\Delta} 2RO\cdot$$

$$RO\cdot + CH_2{=}CH_2 \longrightarrow RO{-}CH_2{-}CH_2\cdot$$

$$RO{-}CH_2{-}CH_2\cdot + CH_2{=}CH_2 \longrightarrow RO{-}CH_2{-}CH_2{-}CH_2{-}CH_2\cdot$$

(a) Label the steps that correspond to initiation and propagation.
(b) Show all available chain-terminating steps.
(c) The polymerization of styrene ($C_6H_5CH{=}CH_2$) occurs by a similar process to produce polystyrene, also used as a packaging material. Draw the structure of the polymer that results from five propagation cycles.

86. Lucite and Plexiglas are transparent polymers used as a glass substitute when a plastic material is preferred for safety. The compound used to synthesize Lucite and Plexiglas is methyl methacrylate, shown here. During the polymerization reaction, light produces a radical initiator from hydrogen peroxide ($H_2O_2 \rightarrow HO\cdot$). Show the mechanism for the polymerization, being sure to include the initiation and propagation steps.

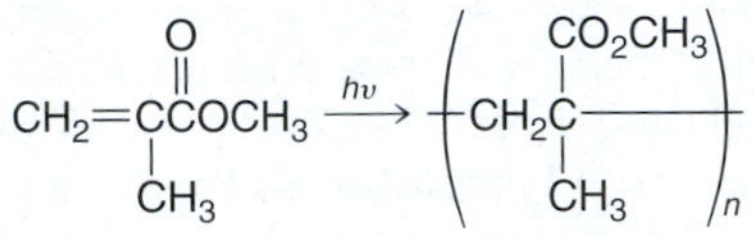

Methyl methacrylate Plexiglas

87. At higher altitudes ozone is converted to O_2 by the reaction $O + O_3 \rightarrow 2O_2$, with a rate constant at 220 K of 6.8×10^{-16} $cm^3 \cdot molecule^{-1} \cdot s^{-1}$.
(a) What is the overall reaction order?
(b) What is E_a for this reaction if $A = 8 \times 10^{-12}$ $cm^3 \cdot molecule^{-1} \cdot s^{-1}$?
If Cl is present, the rate constant at 220 K becomes 3.7×10^{-11} $cm^3 \cdot molecule^{-1} \cdot s^{-1}$, with $A = 4.7 \times 10^{-11}$ $cm^3 \cdot molecule^{-1} \cdot s^{-1}$.
(c) Calculate E_a for the depletion of ozone in the presence of Cl.
(d) Show an energy-level diagram for these two processes, clearly labeling reactants, products, and energies of activation.
(e) If you were an environmental scientist using these data to explain the effects of Cl on ozone concentration, what would be your conclusions?

88. Nitric acid is produced commercially by the catalytic oxidation of ammonia by air over platinum gauze at approximately 900°C. The following reactions occur:

$$NH_3(g) + \tfrac{5}{4}O_2(g) \longrightarrow NO(g) + \tfrac{3}{2}H_2O(g) \qquad \Delta H° = -226.3 \text{ kJ/mol}$$

$$NO(g) + \tfrac{1}{2}O_2(g) \longrightarrow NO_2(g) \qquad \Delta H° = -57.1 \text{ kJ/mol}$$

$$3NO_2(g) + H_2O(l) \longrightarrow 2HNO_3(l) + NO(g) \qquad \Delta H° = -71.7 \text{ kJ/mol}$$

Why is platinum gauze rather than platinum wire used for the initial reaction? The reaction $4NH_3(g) + 3O_2(g) \longrightarrow 2N_2(g) + 6H_2O(g)$ has $\Delta H° = -316.6$ kJ/mol. What would occur if the catalyst were not present? If the gas leaving the catalyst is not free of NH_3, the following reaction takes place: $6NO(g) + 4NH_3(g) \longrightarrow 5N_2(g) + 6H_2O(g)$. If this occurs, what will be the overall reaction?

89. Figure 14.22 illustrates the mechanism for the reduction of ethylene on a platinum surface to produce ethane. Industrially important silanes are synthesized using a related mechanism and are used to increase adhesion between layers of glass fiber and between layers of silicone rubber. Predict the products of these reactions:

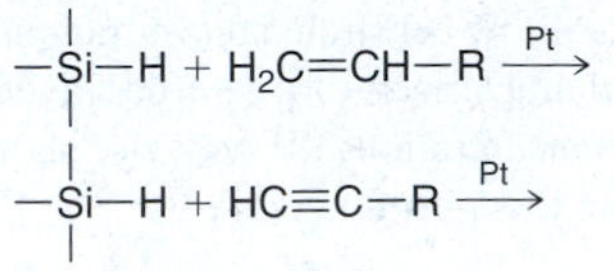

90. In catalysis, if a molecule forms strong bonds to the catalyst, then the catalyst may become poisoned. Experiments on various catalysts showed these results.
i. Fe, Ru, and Os form weak bonds with N_2; however, O_2, alkynes, alkenes, CO, H_2, and CO_2 interact more strongly.
ii. CO_2 and H_2 form weak bonds with a Co or Ni surface.
iii. Rh, Pd, Ir, and Pt form weak bonds with H_2 but do not bond at all with CO_2.
iv. Cu, Ag, and Au form weak bonds with CO and ethylene.
(a) Explain why Fe was chosen as a catalyst to convert nitrogen and hydrogen to ammonia. Why is Fe more suitable than Ru or Os?
(b) Because alkenes generally interact more strongly with metal surfaces than does H_2, what catalyst would you choose for hydrogenation of an alkene such as ethylene?
(c) Although platinum is used in catalytic converters for automobile exhaust, it was not found to be a particularly effective catalyst for the reaction of H_2 with a mixture of carbon monoxide and carbon dioxide to produce methane. Why?
(d) If you were interested in developing a catalyst to reversibly bind ethylene, which of the catalysts listed here would you choose?

91. Nonstoichiometric metal oxides can be effective catalysts for oxidation–reduction reactions. One such catalyst is $Ni_{1-x}O$, found to be effective for converting CO to CO_2 when oxygen is present. Why is it so effective?

92. The chemical reactions in an organism can be controlled by regulating the activity of certain enzymes. Efficient regulation results in an enzyme being active only when it is needed. For example, if a cell needed histidine, the nine enzymes needed to synthesize histidine would all be active. If the cell had adequate histidine, however, those enzymes would be inactive. The following diagram illustrates a situation in which three amino acids (D, F, H) are all synthesized from a common species, A. The numbers above the arrows refer to the enzymes that catalyze each step. Which enzymes would need to be regulated to produce (a) D, (b) F, (c) H?

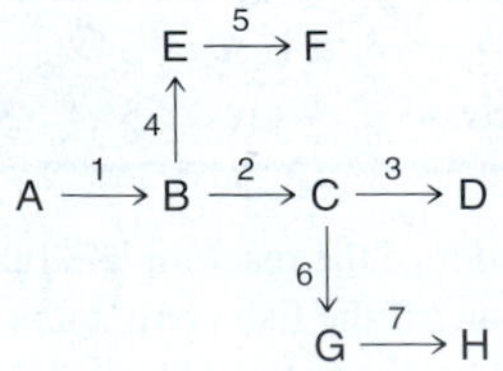

93. Because phosphorus-32 is incorporated into DNA, it can be used to detect DNA fragments. Consequently, it is used extensively in biological research, including the Human Genome Project, whose goal is to determine the complete sequence of human DNA. If you were to start with a 20 g sample of phosphorus that contained 10% ^{32}P by mass, converted it into DNA via several chemical steps that had an overall yield of 75% and took 25 days, and then incorporated it into bacteria and allowed them to grow for 5 more days, what mass of ^{32}P would be available for analysis at the end of this time?

94. The enzyme urease contains two atoms of nickel and catalyzes the hydrolysis of urea by the reaction

$$H_2NC(O)NH_2 + H_2O \longrightarrow 2NH_3 + CO_2$$

Urease is one of the most powerful catalysts known. It lowers the activation energy for the hydrolysis of urea from 137 kJ/mol to only 37 kJ/mol. Calculate the ratio of the rate of the catalyzed reaction to the rate of the uncatalyzed reaction at 37°C. Assume that the frequency factor is the same for both reactions.

95. As noted in Section 14.8, the rate for the hydrogenation of ethylene to give ethane can be increased by heterogeneous catalysts such as Pt or Ni:

$$H_2(g) + H_2C{=}CH_2(g) \xrightarrow{\text{Pt,Ni}} H_3C{-}CH_3(g)$$

The activation energy for the uncatalyzed reaction is large (188 kJ/mol), so the reaction is very slow at room temperature. In the presence of finely divided metallic Ni, the activation energy is only 84 kJ/mol. Calculate the ratio of the rate of the catalyzed reaction to the rate of the uncatalyzed reaction at 75°C.

15 Chemical Equilibrium

LEARNING OBJECTIVES

- To understand what is meant by chemical equilibrium
- To know the relationship between the equilibrium constant and the rate constants for the forward and reverse reactions
- To be able to write an equilibrium constant expression for any reaction
- To be able to solve quantitative problems involving chemical equilibrium
- To predict the direction of reaction
- To predict the effects of stresses on a system at equilibrium
- To understand different ways to control the products of a reaction

CHAPTER OUTLINE

ESSENTIAL SKILLS 7

In Chapter 14, we discussed the principles of chemical kinetics, which deal with the *rate of change*, or how quickly a given chemical reaction occurs. We now turn our attention to the *extent* to which a reaction occurs and how reaction conditions affect the final concentrations of reactants and products. For most of the reactions that we have discussed so far, you may have assumed that once reactants are converted to products, they are likely to remain that way. In fact, however, virtually all chemical reactions are *reversible* to some extent. That is, an opposing reaction occurs in which the products react, to a greater or lesser degree, to re-form the reactants. Eventually, the rates of the forward and reverse reactions become the same, and the system reaches **chemical equilibrium**, the point at which the composition of the system no longer changes with time.

A smoggy sunset in Shenzhen, China. The reaction of O_2 with N_2 at high temperature in an internal combustion engine produces small amounts of NO, which in turn reacts with atmospheric O_2 to form NO_2, an important component of smog. The reddish-brown color of NO_2 is responsible for the characteristic color of smog, as shown in this true-color photo.

We introduced the concept of equilibrium in Chapter 11, where you learned that a liquid and a vapor are in equilibrium when the number of molecules evaporating from the surface of the liquid per unit time is the same as the number of molecules condensing out of the vapor phase. Vapor pressure is an example of a *physical equilibrium* because only the physical form of the substance changes. Similarly, in Chapter 13, we discussed saturated solutions, another example of a physical equilibrium, in which the rate of dissolution of a solute is the same as the rate at which it crystallizes from solution.

In this chapter, we describe the methods chemists use to quantitatively describe the composition of chemical systems at equilibrium, and we discuss how factors such as temperature and pressure influence the equilibrium composition. As you study these concepts, you will also learn how urban smog forms, and how reaction conditions can be altered to produce H_2 rather than the combustion products CO_2 and H_2O from the methane in natural gas. You will discover how to control the composition of the gases emitted in automobile exhaust, and how synthetic polymers such as the polyacrylonitrile used in sweaters and carpets are produced on an industrial scale.

15.1 The Concept of Chemical Equilibrium

Chemical equilibrium is a dynamic process that consists of a forward reaction, in which reactants are converted to products, and a reverse reaction, in which products are converted to reactants. At equilibrium, the forward and reverse reactions proceed at equal rates. Consider, for example, a simple system that contains only one reactant and one product, the reversible dissociation of dinitrogen tetroxide, N_2O_4, to nitrogen dioxide, NO_2. You may recall from Chapter 14 that NO_2 is responsible for the brown color we associate with smog. When a sealed tube containing solid N_2O_4 (mp = −9.3°C, bp = 21.2°C) is heated from −78.4°C to 25°C, the red-brown color of NO_2 appears (Figure 15.1). The reaction can be followed visually because the product, NO_2, is colored, whereas the reactant, N_2O_4, is colorless:

$$\underset{\text{Colorless}}{N_2O_4(g)} \rightleftharpoons \underset{\text{Red-brown}}{2NO_2(g)} \qquad (15.1)$$

The double arrow indicates that both the forward and reverse reactions are occurring simultaneously; it is read "is in equilibrium with."

Figure 15.1 The $N_2O_4(g) \rightleftharpoons 2NO_2$ system at different temperatures. (*left*) At dry ice temperature (−78.4°C), the system contains essentially pure solid N_2O_4, which is colorless. (*center*) As the system is warmed above the melting point of N_2O_4 (−9.3°C), the N_2O_4 melts and then evaporates, and some of the vapor dissociates to red-brown NO_2. (*right*) Eventually the sample reaches room temperature, and a mixture of gaseous N_2O_4 and NO_2 is present. The composition of the mixture and hence the color do not change further with time: the system has reached equilibrium at the new temperature.

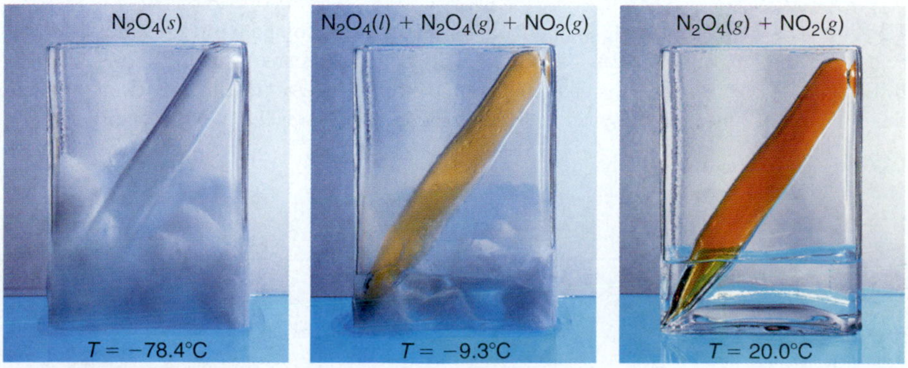

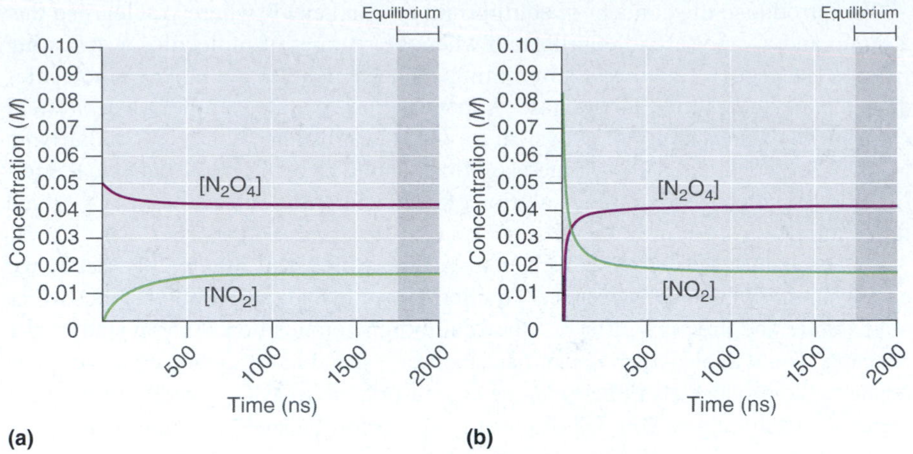

Figure 15.2 The composition of N_2O_4/NO_2 mixtures as a function of time at room temperature. (a) Initially, this idealized system contains 0.0500 *M* gaseous N_2O_4 and no gaseous NO_2. The concentration of N_2O_4 decreases with time as the concentration of NO_2 increases. **(b)** Initially, this system contains 0.1000 *M* NO_2 and no N_2O_4. The concentration of NO_2 decreases with time as the concentration of N_2O_4 increases. In both cases, the final concentrations of the substances are the same: $[N_2O_4] = 0.0422$ *M* and $[NO_2] = 0.0156$ *M* at equilibrium. Interactive Graph

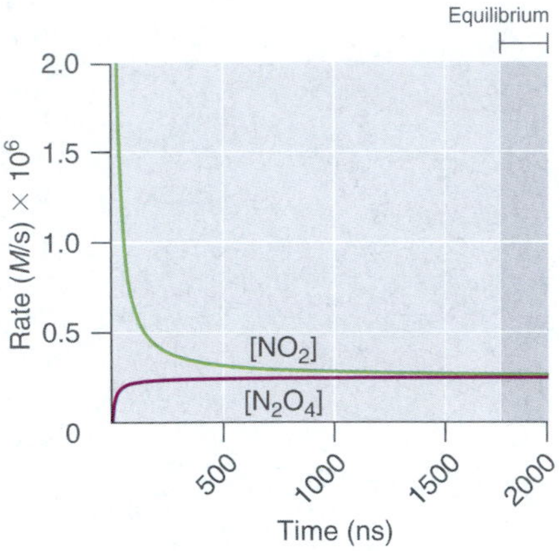

Figure 15.3 The rates of the forward and reverse reactions as a function of time for the $N_2O_4(g) \rightleftharpoons 2NO_2(g)$ system shown in Figure 15.2b. The rate of dimerization of NO_2 (reverse reaction) decreases rapidly with time, as expected for a second-order reaction. Because the initial concentration of N_2O_4 is zero, the rate of the dissociation reaction (forward reaction) at $t = 0$ is also zero. As the dimerization reaction proceeds, the N_2O_4 concentration increases, and its rate of dissociation also increases. Eventually the rates of the two reactions are equal: chemical equilibrium has been reached, and the concentrations of N_2O_4 and NO_2 no longer change. Interactive Graph

Figure 15.2 shows how the composition of this system would vary as a function of time at a constant temperature. If the initial concentration of NO_2 were zero, then it increases as the concentration of N_2O_4 decreases. Eventually the composition of the system stops changing with time; thus, chemical equilibrium has been achieved. Conversely, if we start with a sample that contains no N_2O_4 but an initial NO_2 concentration twice the initial concentration of N_2O_4 in Figure 15.2a, in accordance with the stoichiometry of the reaction, we reach exactly the same equilibrium composition, as shown in Figure 15.2b. Thus, equilibrium can be approached from *either direction* in a chemical reaction.

Figure 15.3 shows the rates of the forward and reverse reactions for a sample that initially contains pure NO_2. Because the initial concentration of N_2O_4 is zero, the rate of the forward reaction (dissociation of N_2O_4) is initially zero as well. In contrast, the rate of the reverse reaction (dimerization of NO_2) is initially very high (2.0×10^6 *M*/s), but it decreases rapidly as the concentration of NO_2 decreases. (Recall from Chapter 14 that the rate of the dimerization reaction is expected to decrease rapidly because the reaction is second order in NO_2: rate $= k_r[NO_2]^2$, where k_r is the rate constant for the reverse reaction shown in Equation 15.1.) As the concentration of N_2O_4 increases, the rate of dissociation of N_2O_4 increases, but more slowly than the dimerization of NO_2 because the reaction is only first order in N_2O_4 (rate $= k_f[N_2O_4]$, where k_f is the rate constant for the forward reaction in Equation 15.1). Eventually, the rates of the forward and reverse reactions become identical, and the system has reached chemical equilibrium. If the forward and reverse reactions occur at *different* rates, then the system is *not* at equilibrium.

Note the pattern

At equilibrium, the rate of the forward reaction is equal to the rate of the reverse reaction.

EXAMPLE 15.1

The three reaction systems (1, 2, and 3) depicted in the accompanying illustration can all be described by the equation $2A \rightleftharpoons B$, where the blue circles are A and the purple ovals are B. Each set of panels shows the changing composition of one of the three reaction mixtures as a function of time. Which system has taken the longest to reach chemical equilbrium?

Given Three reaction systems

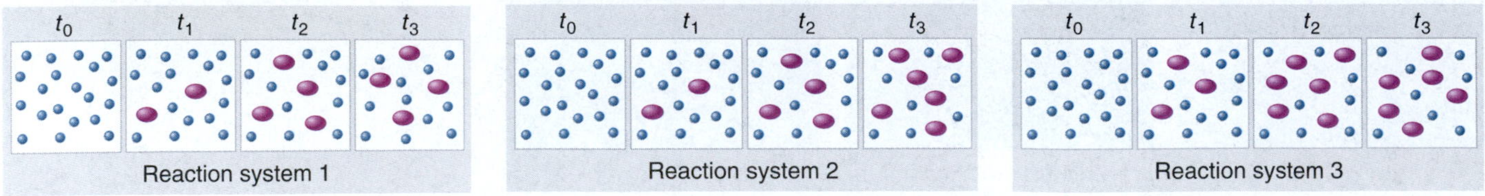

Asked for Relative time to reach chemical equilibrium

Strategy

Compare the concentrations of A and B at different times. The system whose composition takes the longest to stabilize has taken the longest to reach chemical equilibrium.

Solution

In both systems 1 and 3, the concentration of A decreases from t_0 through t_2 but is the same at both t_2 and t_3. Thus, systems 1 and 3 are at equilibrium by t_3. In system 2, the concentrations of A and B are still changing between t_2 and t_3, so system 2 may not yet have reached equilibrium by t_3. Thus, system 2 has taken the longest to reach chemical equilibrium.

EXERCISE 15.1

In the illustration, A is represented by blue circles, B by purple squares, and compound C by orange ovals; the equation for the reaction is A + B $\rightleftharpoons$ C. The sets of panels represent the compositions of three reaction mixtures as a function of time. Which, if any, of the systems shown has reached equilibrium?

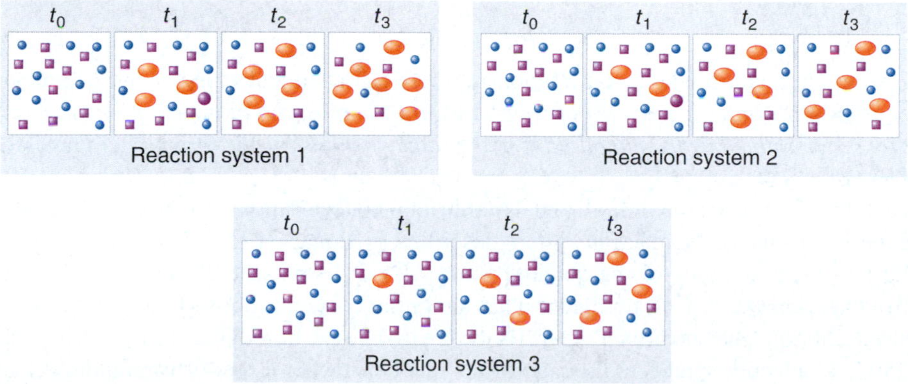

Answer System 2

15.2 ○ The Equilibrium Constant

MGC The Equilibrium Constant Expression

Because an equilibrium state is achieved when the rate of the forward reaction equals the rate of the reverse reaction, under a given set of conditions there must be a relationship between the composition of the system at equilibrium and the kinetics of a reaction (represented by rate constants). We can show this relationship using the system described in Equation 15.1, the decomposition of N_2O_4 to NO_2. Both the forward and reverse reactions for this system consist of a single elementary step, so the rates are

$$\text{forward rate} = k_f[N_2O_4] \tag{15.2a}$$

$$\text{reverse rate} = k_r[NO_2]^2 \tag{15.2b}$$

At equilibrium, the forward rate is equal to the reverse rate:

$$k_f[N_2O_4] = k_r[NO_2]^2 \tag{15.2c}$$

TABLE 15.1 Initial and equilibrium concentrations for NO_2/N_2O_4 mixtures at 25°C

	Initial Concentrations		Concentrations at Equilibrium		
Experiment	**$[N_2O_4]$, *M***	**$[NO_2]$, *M***	**$[N_2O_4]$, *M***	**$[NO_2]$, *M***	**$K = [NO_2]^2/[N_2O_4]$**
1	0.0500	0.0000	0.0417	0.0165	6.54×10^{-3}
2	0.0000	0.1000	0.0417	0.0165	6.54×10^{-3}
3	0.0750	0.0000	0.0647	0.0206	6.56×10^{-3}
4	0.0000	0.0750	0.0304	0.0141	6.54×10^{-3}
5	0.0250	0.0750	0.0532	0.0186	6.50×10^{-3}

so

$$\frac{k_f}{k_r} = \frac{[NO_2]^2}{[N_2O_4]} \tag{15.2d}$$

The ratio of the rate constants gives us a new constant, the **equilibrium constant (*K*)**, which is defined as

$$K = \frac{k_f}{k_r} \tag{15.3}$$

Note the pattern

The equilibrium constant K is equal to the rate constant for the forward reaction divided by the rate constant for the reverse reaction.

Hence, there is a fundamental relationship between chemical kinetics and chemical equilibrium: *Under a given set of conditions, the composition of the equilibrium mixture is determined by the magnitudes of the rate constants for the forward and reverse reactions.*

Table 15.1 lists the initial and equilibrium concentrations from five different experiments using the reaction system described by Equation 15.1. Note that at equilibrium the magnitude of the quantity $[NO_2]^2/[N_2O_4]$ is essentially the same for all five experiments. In fact, no matter what the initial concentrations of NO_2 and N_2O_4, at equilibrium the quantity $[NO_2]^2/[N_2O_4]$ will *always* be $6.53 \pm 0.03 \times 10^{-3}$ at 25°C, which corresponds to the ratio of the rate constants for the forward and reverse reactions. That is, at a given temperature, the equilibrium constant *K* for a reaction always has the same value, even though the specific concentrations of the reactants and products vary depending on their initial concentrations.

Developing an Equilibrium Constant Expression

In 1864, the Norwegian chemists Cato Guldberg (1836–1902) and Peter Waage (1833–1900) carefully measured the compositions of many reaction systems at equilibrium. They discovered that for *any* reversible reaction of the general form

$$aA + bB \rightleftharpoons cC + dD \tag{15.4}$$

where A and B are reactants, C and D are products, and *a*, *b*, *c*, and *d* are the stoichiometric coefficients in the balanced equation for the reaction, the ratio of the product of the equilibrium concentrations of the products (raised to their coefficients in the balanced equation) to the product of the equilibrium concentrations of the reactants (raised to their coefficients in the balanced equation) is always a constant under a given set of conditions. This relationship is known as the **law of mass action** and can be stated as

$$K = \frac{[C]^c[D]^d}{[A]^a[B]^b} \tag{15.5}$$

TABLE 15.2 Equilibrium constants for selected reactions[a]

Reaction	Temperature, K	Equilibrium Constant (*K*)
$S(s) + O_2(g) \rightleftharpoons SO_2(g)$	300	4.4×10^{53}
$2H_2(g) + O_2(g) \rightleftharpoons 2H_2O(g)$	500	2.4×10^{47}
$H_2(g) + Cl_2(g) \rightleftharpoons 2HCl(g)$	300	1.6×10^{33}
$H_2(g) + Br_2(g) \rightleftharpoons 2HBr(g)$	300	4.1×10^{18}
$2NO(g) + O_2(g) \rightleftharpoons 2NO_2(g)$	300	4.2×10^{13}
$3H_2(g) + N_2(g) \rightleftharpoons 2NH_3(g)$	300	2.7×10^{8}
$H_2(g) + D_2(g) \rightleftharpoons 2HD(g)$	100	1.92
$H_2(g) + I_2(g) \rightleftharpoons 2HI(g)$	300	2.9×10^{-1}
$Br_2(g) \rightleftharpoons 2Br(g)$	1000	4.0×10^{-7}
$I_2(g) \rightleftharpoons 2I(g)$	800	4.6×10^{-7}
$Cl_2(g) \rightleftharpoons 2Cl(g)$	1000	1.8×10^{-9}
$F_2(g) \rightleftharpoons 2F(g)$	500	7.4×10^{-13}

[a]Equilibrium constants vary with temperature. The *K* values shown are for systems at the indicated temperatures.

where *K* is the equilibrium constant for the reaction. Equation 15.4 is called the **equilibrium equation**, and the right side of Equation 15.5 is called the **equilibrium constant expression**. The relationship shown in Equation 15.5 is true for *any* pair of opposing reactions regardless of the mechanism of the reaction or of the number of steps in the mechanism.

The equilibrium constant can vary over a wide range of values. The values of *K* shown in Table 15.2, for example, vary by 60 orders of magnitude. Because products are in the numerator of the equilibrium constant expression and reactants are in the denominator, values of *K* greater than 10^3 indicate a strong tendency for reactants to form products. In this case, chemists say that equilibrium lies to the right as written, favoring the formation of products. An example is the reaction between H_2 and Cl_2 to produce HCl, which has an equilibrium constant of 1.6×10^{33} at 300 K. Because H_2 is a good reductant and Cl_2 is a good oxidant, the reaction proceeds essentially to completion. In contrast, values of *K* less than 10^{-3} indicate that the ratio of products to reactants at equilibrium is very small. That is, reactants do not tend to form products readily, and the equilibrium lies to the left as written, favoring the formation of reactants.

You will also notice in Table 15.2 that equilibrium constants have no units, even though Equation 15.5 suggests that the units of concentration might not always cancel because the exponents may vary. In fact, equilibrium constants are calculated using "effective concentrations", or *activities*, of reactants and products, which are the ratios of the measured concentrations to a standard state of 1 *M*.* As shown in Equation 15.6, the units of concentration cancel, which makes *K* unitless as well:

$$\frac{[A]_{\text{measured}}}{[A]_{\text{standard state}}} = \frac{\cancel{M}}{\cancel{M}} = \frac{\cancel{\text{mol/L}}}{\cancel{\text{mol/L}}} \tag{15.6}$$

Many reactions have equilibrium constants between 1000 and 0.001 ($10^3 \geq K \geq 10^{-3}$), neither very large nor very small. At equilibrium, these systems tend to

* As you will discover in more advanced chemistry courses, the effective concentration (the activity) is unitless. The standard state is assigned an activity of 1.

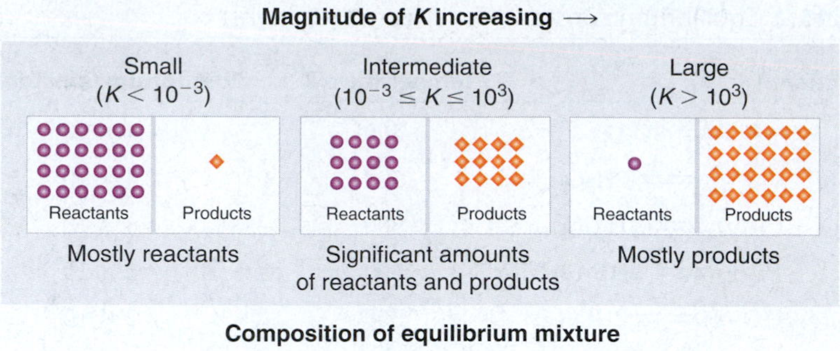

Figure 15.4 The relationship between the composition of the mixture at equilibrium and the magnitude of the equilibrium constant. The larger the value of *K*, the farther the reaction proceeds to the right before equilibrium is reached, and the higher the ratio of products to reactants at equilibrium.

contain significant amounts of both products and reactants, indicating that there is not a strong tendency to form either products from reactants or reactants from products. An example of this type of system is the reaction of gaseous hydrogen and deuterium, a component of high-stability fiber-optic light sources used in ocean studies, to form HD:

$$H_2(g) + D_2(g) \rightleftharpoons 2HD(g) \quad (15.7)$$

The equilibrium constant expression for this reaction is $[HD]^2/[H_2][D_2]$, and the value of K is between 1.9 and 4 over a wide temperature range (100–1000 K). Thus, an equilibrium mixture of H_2, D_2, and HD contains significant concentrations of both product and reactants.

Note the pattern

A large value of the equilibrium constant K means that products predominate at equilibrium; a small value means that reactants predominate at equilibrium.

Figure 15.4 summarizes the relationship between the magnitude of K and the relative concentrations of reactants and products at equilibrium for a general reaction, written as reactants $\rightleftharpoons$ products. Because there is a direct relationship between the kinetics of a reaction and the equilibrium concentrations of products and reactants (Equations 15.3 and 15.5), when $k_f \gg k_r$, K is a large number, and the concentration of products at equilibrium predominate. This corresponds to an essentially irreversible reaction. Conversely, when $k_f \ll k_r$, K is a very small number, and the reaction produces almost no products as written. Systems for which $k_f \approx k_r$ have significant concentrations of both reactants and products at equilibrium.

EXAMPLE 15.2

Write equilibrium constant expressions for these reactions:

(a) $N_2(g) + 3H_2(g) \rightleftharpoons 2NH_3(g)$
(b) $CO(g) + \frac{1}{2}O_2(g) \rightleftharpoons CO_2(g)$
(c) $2CO_2(g) \rightleftharpoons 2CO(g) + O_2(g)$

Given Balanced chemical equations

Asked for Equilibrium constant expressions

Strategy

Refer to Equation 15.5. Place the arithmetic product of the concentrations of the products (raised to their stoichiometric coefficients) in the numerator, and the product of the concentrations of the reactants (raised to their stoichiometric coefficients) in the denominator.

Solution

(a) The only product is ammonia, which has a coefficient of 2. For the reactants, N_2 has a coefficient of 1 and H_2 has a coefficient of 3. The equilibrium constant expression is therefore

$$\frac{[NH_3]^2}{[N_2][H_2]^3}$$

(b) The only product is carbon dioxide, which has a coefficient of 1. The reactants are CO, with a coefficient of 1, and O_2, with a coefficient of $\frac{1}{2}$. Thus, the equilibrium constant expression is

$$\frac{[CO_2]}{[CO][O_2]^{1/2}}$$

(c) This reaction is the reverse of the reaction in part b, with all coefficients multiplied by 2 to remove the fractional coefficient for O_2. The equilibrium constant expression is therefore the inverse of the expression in part b, with all exponents multiplied by 2:

$$\frac{[CO]^2[O_2]}{[CO_2]^2}$$

EXERCISE 15.2

Write equilibrium constant expressions for these reactions:

(a) $N_2O(g) \rightleftharpoons N_2(g) + \frac{1}{2}O_2(g)$
(b) $2C_8H_{18}(g) + 25O_2(g) \rightleftharpoons 16CO_2(g) + 18H_2O(g)$

Answer **(a)** $\frac{[N_2][O_2]^{1/2}}{[N_2O]}$; **(b)** $\frac{[CO_2]^{16}[H_2O]^{18}}{[C_8H_{18}]^2[O_2]^{25}}$

EXAMPLE 15.3

Predict which of the following systems at equilibrium will **(a)** contain essentially only products; **(b)** contain essentially only reactants; and **(c)** contain appreciable amounts of both products and reactants:

1. $H_2(g) + I_2(g) \rightleftharpoons 2HI(g)$, $K_{(700K)} = 54$
2. $2CO_2(g) \rightleftharpoons 2CO(g) + O_2(g)$, $K_{(1200K)} = 3.1 \times 10^{-18}$
3. $PCl_5(g) \rightleftharpoons PCl_3(g) + Cl_2(g)$, $K_{(613K)} = 97$
4. $2O_3(g) \rightleftharpoons 3O_2(g)$, $K_{(298K)} = 5.9 \times 10^{55}$

Given Systems and values of K

Asked for Composition of systems at equilibrium

Strategy

Use the value of the equilibrium constant to determine whether the equilibrium mixture will contain essentially only products, essentially only reactants, or significant amounts of both.

Solution

(a) Only system 4 has $K \gg 10^3$, so at equilibrium it will consist of essentially only products.
(b) System 2 has $K \ll 10^{-3}$, so the reactants have little tendency to form products under the conditions specified; thus, at equilibrium the system will contain essentially only reactants.
(c) Both systems 1 and 3 have equilibrium constants in the range $10^3 \geq K \geq 10^{-3}$, indicating that the equilibrium mixtures will contain appreciable amounts of both products and reactants.

EXERCISE 15.3

Hydrogen and nitrogen react to form ammonia according to this balanced chemical equation:

$$3H_2(g) + N_2(g) \rightleftharpoons 2NH_3(g)$$

Values of the equilibrium constant at various temperatures were reported as $K_{25°C} = 3.3 \times 10^8$, $K_{177°C} = 2.6 \times 10^3$, and $K_{327°C} = 4.1$. **(a)** At which temperature would you expect to find the highest proportion of H_2 and N_2 in the equilibrium mixture? **(b)** Assuming that the reaction rates are fast enough so that equilibrium is reached quickly, at what temperature would you design a commercial reactor to operate to maximize the yield of ammonia?

Answer **(a)** 327°C, where K is smallest; **(b)** 25°C

Variations in the Form of the Equilibrium Constant Expression

Because equilibrium can be approached from either direction in a chemical reaction, the equilibrium constant expression and thus the magnitude of the equilibrium constant depend on the form in which the chemical reaction is written. For example, if we write the reaction described in Equation 15.4 in reverse, we obtain

$$cC + dD \rightleftharpoons aA + bB \tag{15.8}$$

The corresponding equilibrium constant K' is

$$K' = \frac{[A]^a[B]^b}{[C]^c[D]^d} \tag{15.9}$$

This expression is the inverse of the expression for the original equilibrium constant K, so $K' = 1/K$. That is, when we write a reaction in the reverse direction, the equilibrium constant expression is inverted. For instance, the equilibrium constant K for the reaction $N_2O_4 \rightleftharpoons 2NO_2$ is

$$K = \frac{[NO_2]^2}{[N_2O_4]} \tag{15.10}$$

but for the opposite reaction, $2NO_2 \rightleftharpoons N_2O_4$, the equilibrium constant K' is given by the inverse expression:

$$K' = \frac{[N_2O_4]}{[NO_2]^2} \tag{15.11}$$

Consider another example, the formation of water: $2H_2(g) + O_2(g) \rightleftharpoons 2H_2O(g)$. Because H_2 is a good reductant and O_2 is a good oxidant, this reaction has a very large equilibrium constant ($K = 2.4 \times 10^{47}$at 500 K). Consequently, the equilibrium constant for the reverse reaction, the decomposition of water to form O_2 and H_2, is very small: $K' = 1/K = 1/(2.4 \times 10^{47} = 4.2 \times 10^{-48}$. As suggested by the very small equilibrium constant, and fortunately for life as we know it, a substantial amount of energy is indeed needed to dissociate water into H_2 and O_2.

Writing an equation in different but chemically equivalent forms also causes both the equilibrium constant expression and the magnitude of the equilibrium constant to be different. For example, we could write the equation for the reaction $2NO_2 \rightleftharpoons N_2O_4$ as $NO_2 \rightleftharpoons \frac{1}{2}N_2O_4$, for which the equilibrium constant K'' is

$$K'' = \frac{[N_2O_4]^{1/2}}{[NO_2]} \tag{15.12}$$

Note the pattern

The equilibrium constant for a reaction written in reverse is the inverse of the original equilibrium constant.

The values for K' (Equation 15.11) and K'' are related as follows:

$$K'' = (K')^{1/2} = \sqrt{K'} \tag{15.13}$$

In general, if all the coefficients in a balanced chemical equation are subsequently multiplied by n, then the new equilibrium constant is the original equilibrium constant raised to the nth power.

EXAMPLE 15.4

At 745 K, the value of K is 0.118 for the reaction

$$N_2(g) + 3H_2(g) \rightleftharpoons 2NH_3(g)$$

What are the equilibrium constants for these related reactions at 745 K?

(a) $2NH_3(g) \rightleftharpoons N_2(g) + 3H_2(g)$

(b) $\frac{1}{2}N_2(g) + \frac{3}{2}H_2(g) \rightleftharpoons NH_3(g)$

Given Balanced equilibrium equation, value of K at a given temperature, and equations of related reactions

Asked for Values of K for related reactions

Strategy

Write the equilibrium constant expression for the given reaction and for each related reaction. From these expressions, calculate K for each reaction.

Solution

The equilibrium constant expression for the given reaction of $N_2(g)$ with $H_2(g)$ to produce $NH_3(g)$ at 745 K is

$$K = \frac{[NH_3]^2}{[N_2][H_2]^3} = 0.118$$

(a) This reaction is the reverse of the one given, so its equilibrium constant expression is

$$K' = \frac{1}{K} = \frac{[N_2][H_2]^3}{[NH_3]^2} = \frac{1}{0.118} = 8.47$$

(b) In this reaction, all the stoichiometric coefficients of the given reaction are divided by 2, so the equilibrium constant is calculated as

$$K'' = \frac{[NH_3]}{[N_2]^{1/2}[H_2]^{3/2}} = K^{1/2} = \sqrt{K} = \sqrt{0.118} = 0.344$$

EXERCISE 15.4

At 527°C, the equilibrium constant for the reaction

$$2SO_2(g) + O_2(g) \rightleftharpoons 2SO_3(g)$$

is 7.9×10^4. Calculate the equilibrium constant for the following reaction at that same temperature:

$$SO_3(g) \rightleftharpoons SO_2(g) + \tfrac{1}{2}O_2(g)$$

Answer 3.6×10^{-3}

Equilibrium Constant Expressions for Systems That Contain Gases

For reactions that involve species in solution, the concentrations used in equilibrium calculations are usually expressed in moles/liter. For gases, however, the concentrations are usually expressed in terms of partial pressures rather than molarity, where the standard state is 1 atm of pressure. The symbol K_p is used to denote equilibrium constants calculated from partial pressures. For the general reaction

aA + bB $\rightleftharpoons$ cC + dD in which all the components are gases, we can write the equilibrium constant expression as the ratio of the partial pressures of the products and reactants (each raised to its coefficient in the chemical equation):

$$K_p = \frac{(P_A)^c\,(P_C)^d}{(P_B)^a\,(P_D)^b} \tag{15.14}$$

Thus, K_p for the decomposition of N_2O_4 (Equation 15.1) is

$$K_p = \frac{(P_{NO_2})^2}{P_{N_2O_4}} \tag{15.15}$$

Like K, K_p is a unitless quantity because the quantity that is actually used to calculate it is an "effective pressure," the ratio of the measured pressure to a standard state of 1 bar (approximately 1 atm) which produces a unitless quantity.*

Because partial pressures are usually expressed in atmospheres or mmHg, the molar concentration of a gas and its partial pressure do not have the same numerical value. Consequently, the numerical values of K and K_p are usually different. They are, however, related by the ideal gas constant R and the temperature T:

$$K_p = K(RT)^{\Delta n} \tag{15.16}$$

where K is the equilibrium constant expressed in units of concentration and Δn is the difference between the numbers of moles of gaseous products and gaseous reactants ($n_p - n_r$). The temperature is expressed as the absolute temperature in kelvins. According to Equation 15.16, $K_p = K$ only if the numbers of moles of gaseous products and gaseous reactants are the same (that is, $\Delta n = 0$). For the decomposition of N_2O_4, there are 2 mol of gaseous product and 1 mol of gaseous reactant, so $\Delta n = 1$. Thus, for this reaction, $K_p = K(RT)^1 = KRT$.

EXAMPLE 15.5

The equilibrium constant K for the reaction of nitrogen and hydrogen to give ammonia is 0.118 at 745 K:

$$N_2(g) + 3H_2(g) \rightleftharpoons 2NH_3(g)$$

What is K_p for this reaction at the same temperature?

Given Equilibrium equation, equilibrium constant, and temperature

Asked for Value of K_p

Strategy

Use the coefficients in the balanced equation to calculate Δn. Then use Equation 15.16 to calculate K from K_p.

Solution

This reaction has 2 mol of gaseous product and 4 mol of gaseous reactants, so $\Delta n = (2 - 4) = -2$. We know the value of K, and $T = 745$ K. Thus, from Equation 15.16, we have

$$K_p = K(RT)^{-2} = \frac{K}{(RT)^2} = \frac{0.118}{\{[0.08206\ (\text{L} \cdot \text{atm})/(\text{mol} \cdot \text{K})][745\ \text{K}]\}^2} = 3.16 \times 10^{-5}$$

Because K_p is a unitless quantity, the answer is $K_p = 3.16 \times 10^{-5}$.

---

* The "effective pressure" is called the *fugacity*, just as activity is the effective concentration.

EXERCISE 15.5

Calculate K_p for the reaction $2SO_2(g) + O_2(g) \rightleftharpoons 2SO_3(g)$ at 527°C, if $K = 7.9 \times 10^4$ at this temperature.

Answer $K_p = 1.2 \times 10^3$

Homogeneous and Heterogeneous Equilibria

When the products and reactants of an equilibrium reaction form a single phase, whether gas or liquid, the system is a **homogeneous equilibrium**. In such situations, the concentrations of the reactants and products can vary over a wide range. In contrast, a system whose reactants, products, or both are in more than one phase is a **heterogeneous equilibrium**, such as the reaction of a gas with a solid or liquid.

Because the molar concentrations of pure liquids and solids normally do not vary greatly with temperature, their concentrations are treated as constants, which allows us to simplify equilibrium constant expressions that involve pure solids or liquids.* (Recall from Chapter 11, for example, that the density of water, and thus its volume, changes by only a few percent between 0°C and 100°C.)

Consider the following reaction, which is used in the final firing of some types of pottery to produce brilliant metallic glazes:

$$CO_2(g) + C(s) \rightleftharpoons 2CO(g) \qquad (15.17)$$

The glaze is created when metal oxides are reduced to metals by the product, carbon monoxide. The equilibrium constant expression for this reaction is

$$K = \frac{[CO]^2}{[CO_2][C]} \qquad (15.18)$$

Because graphite is a solid, however, its molar concentration, determined from its density and molar mass, is essentially constant and has the following value:

$$[C] = \frac{2.26 \cancel{g}/\cancel{cm^3}}{12.01 \cancel{g}/mol} \times 1000 \cancel{cm^3}/L = 188 \text{ mol/L} = 188\ M \qquad (15.19)$$

We can rearrange Equation 15.18 so that the constant terms are on one side:

$$K[C] = K(188) = \frac{[CO]^2}{[CO_2]} \qquad (15.20)$$

Incorporating the constant value of [C] into the equilibrium equation for the reaction in Equation 15.17 gives

$$K' = \frac{[CO]^2}{[CO_2]} \qquad (15.21)$$

The equilibrium constant for this reaction can also be written in terms of the partial pressures of the gases:

$$K_p = \frac{(P_{CO})^2}{P_{CO_2}} \qquad (15.22)$$

Incorporating all the constant values into K' or K_p allows us to focus on the substances whose concentrations change during the reaction.

* The reference states for pure solids and liquids are those forms stable at 1 bar (approximately 1 atm), which are assigned an activity of 1.

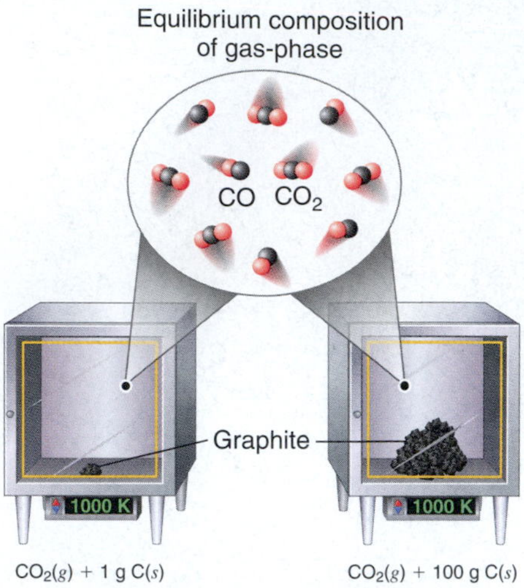

Figure 15.5 Effect of the amount of solid present on equilibrium in a heterogeneous solid–gas system. In the $CO_2(g) + C(s) \rightleftharpoons 2CO(g)$ system, the equilibrium composition of the gas phase at a given temperature, 1000 K in this case, is the same whether a small amount of solid carbon is present (*left*) or a large amount (*right*).

Although the concentrations of pure liquids or solids are not written explicitly in the equilibrium constant expression, these substances must be present in the reaction mixture for chemical equilibrium to occur. Whatever the concentrations of CO and CO_2, the system described in Equation 15.17 will reach chemical equilibrium only if a stoichiometric amount of solid carbon or excess solid carbon has been added so that some is still present once the system has reached equilibrium. As shown in Figure 15.5, it does not matter whether 1 g or 100 g of solid carbon is present; in either case, the composition of the gaseous components of the system will be the same at equilibrium.

EXAMPLE 15.6

Write expressions for K, incorporating all constants, and K_p for these equilibrium reactions:

(a) $PCl_3(l) + Cl_2(g) \rightleftharpoons PCl_5(s)$
(b) $Fe_3O_4(s) + 4H_2(g) \rightleftharpoons 3Fe(s) + 4H_2O(g)$

Given Balanced equilibrium equations

Asked for Expressions for K and K_p

Strategy

Find K by writing each equilibrium constant expression as the ratio of the concentrations of the products and reactants, each raised to its coefficient in the chemical equation. Then express K_p as the ratio of the partial pressures of the products and reactants, each also raised to its coefficient in the chemical equation.

Solution

(a) This reaction contains a pure solid (PCl_5) and a pure liquid (PCl_3). Their concentrations do not appear in the equilibrium constant expression because they do not change significantly. Thus,

$$K = \frac{1}{[Cl_2]} \quad \text{and} \quad K_p = \frac{1}{P_{Cl_2}}$$

(b) This reaction contains two pure solids (Fe_3O_4 and Fe), which do not appear in the equilibrium constant expressions. The two gases do, however, appear in the expressions:

$$K = \frac{[H_2O]^4}{[H_2]^4} \quad \text{and} \quad K_p = \frac{(P_{H_2O})^4}{(P_{H_2})^4}$$

EXERCISE 15.6

Write expressions for K and K_p for these reactions:

(a) $CaCO_3(s) \rightleftharpoons CaO(s) + CO_2(g)$
(b) $C_6H_{12}O_6(s) + 6O_2(g) \rightleftharpoons 6CO_2(g) + 6H_2O(g)$
Glucose

Answer **(a)** $K = [CO_2]$, $K_p = P_{CO_2}$; **(b)** $K = \dfrac{[CO_2]^6[H_2O]^6}{[O_2]^6}$, $K_p = \dfrac{(P_{CO_2})^6(P_{H_2O})^6}{(P_{O_2})^6}$

Note the pattern

The concentrations of pure solids, pure liquids, and solvents are omitted from equilibrium constant expressions because they don't change significantly during reactions when enough is present to reach equilibrium.

For reactions carried out in solution, the concentration of the solvent is omitted from the equilibrium constant expression even when the solvent appears in the balanced equation for the reaction. The concentration of the solvent is also typically much higher than the concentration of the reactants or products (recall that pure water is about 55.5 *M*, and pure ethanol is about 17 *M*). Consequently, the solvent concentration is essentially constant during chemical reactions, and the solvent is therefore treated as a pure liquid. The equilibrium constant expression for a reaction contains only those species whose concentrations could change significantly during the reaction.

Equilibrium Constant Expressions for the Sums of Reactions

Chemists frequently need to know the equilibrium constant for a reaction that has not been studied before. In such cases, the desired reaction can often be written as the sum of other reactions for which the equilibrium constants are known. The equilibrium constant for the unknown reaction can then be calculated from the tabulated values for the other reactions.

To illustrate this procedure, let's consider the reaction of N_2 with O_2 to give NO_2. As we stated in Section 15.1, this reaction is an important source of the NO_2 that gives urban smog its typical brown color. The reaction normally occurs in two distinct steps. In the first reaction [shown below as (1)], N_2 reacts with O_2 at the high temperatures inside an internal combustion engine to give NO. The released NO then reacts with additional O_2 to give NO_2 [shown below as (2)]. The equilibrium constant for each reaction at 100°C is also given.

$$(1)\ N_2(g) + O_2(g) \rightleftharpoons 2NO(g) \qquad K_1 = 2.0 \times 10^{-25}$$
$$(2)\ 2NO(g) + O_2(g) \rightleftharpoons 2NO_2(g) \qquad K_2 = 6.4 \times 10^{9}$$

Summing reactions (1) and (2) gives the overall reaction of N_2 with O_2:

$$(3)\ N_2(g) + 2O_2(g) \rightleftharpoons 2NO_2(g) \qquad K_3 = ?$$

The equilibrium constant expressions for the reactions are

$$K_1 = \frac{[NO]^2}{[N_2][O_2]} \qquad K_2 = \frac{[NO_2]^2}{[NO]^2[O_2]} \qquad K_3 = \frac{[NO_2]^2}{[N_2][O_2]^2}$$

What is the relationship between K_1, K_2, and K_3, all at 100°C? Note that the expression for K_1 has $[NO]^2$ in the numerator, the expression for K_2 has $[NO]^2$ in the denominator, and $[NO]^2$ does not appear at all in the expression for K_3. Multiplying K_1 by K_2 and canceling the $[NO]^2$ terms give

$$K_1K_2 = \frac{\cancel{[NO]^2}}{[N_2][O_2]} \times \frac{[NO_2]^2}{\cancel{[NO]^2}[O_2]} = \frac{[NO_2]^2}{[N_2][O_2]^2} = K_3$$

> **Note the pattern**
> To determine K for a reaction that is the sum of two or more reactions, add reactions, but multiply equilibrium constants.

Thus, the product of the equilibrium constant expressions for K_1 and K_2 is the same as the equilibrium constant expression for K_3:

$$K_3 = K_1K_2 = (2.0 \times 10^{-25})(6.4 \times 10^{9}) = 1.3 \times 10^{-15}$$

The equilibrium constant for a reaction that is the sum of two or more reactions is equal to the product *of the equilibrium constants for the individual reactions.* In contrast, recall that according to Hess's Law, ΔH for the sum of two or more reactions is the *sum* of the ΔH values for the individual reactions.

EXAMPLE 15.7

The following reactions occur at 1200°C:

$$(1)\ CO(g) + 3H_2(g) \rightleftharpoons CH_4(g) + H_2O(g) \qquad K_1 = 9.17 \times 10^{-2}$$
$$(2)\ CH_4(g) + 2H_2S(g) \rightleftharpoons CS_2(g) + 4H_2(g) \qquad K_2 = 3.3 \times 10^{4}$$

Calculate the equilibrium constant for this reaction at the same temperature:

$$(3)\ CO(g) + 2H_2S(g) \rightleftharpoons CS_2(g) + H_2O(g) + H_2(g) \qquad K_3 = ?$$

Given Two balanced equilibrium equations, values of *K*, and an equilibrium equation for the overall reaction

Asked for Equilibrium constant for the overall reaction

Strategy

Arrange the equations so that their sum produces the overall equation. If an equation had to be reversed, invert the value of K for that equation. Calculate K for the overall equation by multiplying the equilibrium constants for the individual equations.

Solution

The key to solving this problem is to recognize that reaction (3) is the sum of reactions (1) and (2):

$$CO(g) + \cancel{3H_2(g)} \rightleftharpoons \cancel{CH_4(g)} + H_2O(g)$$
$$\cancel{CH_4(g)} + 2H_2S(g) \rightleftharpoons CS_2(g) + \cancel{3H_2(g)} + H_2(g)$$
$$\overline{CO(g) + 2H_2S(g) \rightleftharpoons CS_2(g) + H_2O(g) + H_2(g)}$$

The values for K_1 and K_2 are given, so it is straightforward to calculate K_3:

$$K_3 = K_1K_2 = (9.17 \times 10^{-2})(3.3 \times 10^4) = 3.03 \times 10^3$$

EXERCISE 15.7

In the first of two steps in the industrial synthesis of sulfuric acid, elemental sulfur reacts with oxygen to produce sulfur dioxide. In the second step, sulfur dioxide reacts with additional oxygen to form sulfur trioxide. The reaction for each step is shown, as is the value of the corresponding equilibrium constant at 25°C. Calculate the equilibrium constant for the overall reaction at this same temperature.

$$(1)\ \frac{1}{8}S_8(s) + O_2(g) \rightleftharpoons SO_2(g) \qquad K_1 = 4.4 \times 10^{53}$$

$$(2)\ SO_2(g) + \frac{1}{2}O_2(g) \rightleftharpoons SO_3(g) \qquad K_2 = 2.6 \times 10^{12}$$

$$(3)\ \frac{1}{8}S_8(s) + \frac{3}{2}O_2(g) \rightleftharpoons SO_3(g) \qquad K_3 = ?$$

Answer $K_3 = 1.1 \times 10^{66}$

MGC Calculating Equilibrium Constants

15.3 • Solving Equilibrium Problems

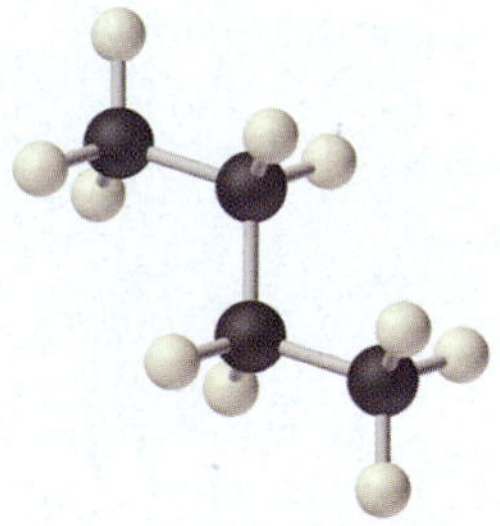

n-Butane

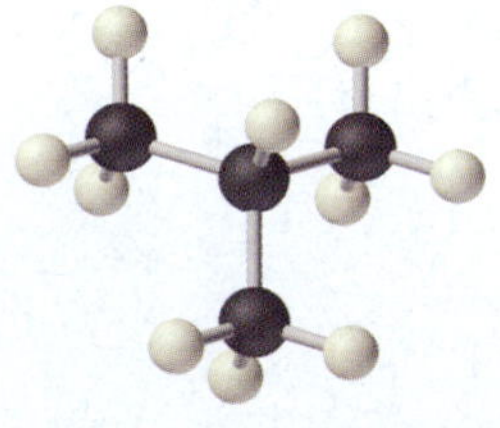

Isobutane (2-methylpropane)

There are two fundamental kinds of equilibrium problems: (1) those in which we are given the concentrations of the reactants and products at equilibrium (or, more often, information that allows us to calculate these concentrations) and we are asked to calculate the equilibrium constant for the reaction; and (2) those in which we are given the equilibrium constant and the initial concentrations of reactants and we are asked to calculate the concentration of one or more substances at equilibrium. In this section, we describe methods for solving both kinds of calculations.

Calculating an Equilibrium Constant from Equilibrium Concentrations

We saw in Exercise 15.6 that the equilibrium constant for the decomposition of $CaCO_3(s)$ to $CaO(s)$ and $CO_2(g)$ is $K = [CO_2]$. At 800°C, the concentration of CO_2 in equilibrium with solid $CaCO_3$ and CaO is 2.5×10^{-3} M. Thus, the value of K at 800°C is 2.5×10^{-3} (remember that equilibrium constants are unitless).

A more complex example of this type of problem is the conversion of *n*-butane, an additive used to increase the volatility of gasoline, to isobutane (2-methylpropane). This reaction can be written as

$$n\text{-butane}(g) \rightleftharpoons \text{isobutane}(g) \qquad (15.23)$$

for which the equilibrium constant $K = [i\text{-but}]/[n\text{-but}]$. At equilibrium, a mixture of n-butane and isobutane at room temperature was found to contain 0.041 M isobutane and 0.016 M n-butane. Substituting these concentrations into the equilibrium constant expression gives

$$K = \frac{[i\text{-but}]}{[n\text{-but}]} = \frac{0.041\ \cancel{M}}{0.016\ \cancel{M}} = 2.6 \tag{15.24}$$

Thus, the equilibrium constant K for the reaction as written is 2.6.

EXAMPLE 15.8

The reaction between gaseous sulfur dioxide and oxygen is a key step in the industrial synthesis of sulfuric acid:

$$2SO_2(g) + O_2(g) \rightleftharpoons 2SO_3(g)$$

A mixture of SO_2 and O_2 was maintained at a temperature of 800 K until the system reached equilibrium. The equilibrium mixture contained 5.0×10^{-2} M SO_3, 3.5×10^{-3} M O_2, and 3.0×10^{-3} M SO_2. Calculate the equilibrium constant, K and K_p, at this temperature.

Given Balanced equilibrium equation and composition of equilibrium mixture

Asked for Equilibrium constant

Strategy

Write the equilibrium constant expression for the reaction. Then substitute the appropriate equilibrium concentrations into this equation to obtain K.

Solution

Substituting the appropriate equilibrium concentrations into the equilibrium constant expression gives

$$K = \frac{[SO_3]^2}{[SO_2]^2[O_2]} = \frac{(5.0 \times 10^{-2})^2}{(3.0 \times 10^{-3})^2(3.5 \times 10^{-3})} = 7.9 \times 10^4$$

To solve for K_p we use Equation 16.16, where $\Delta n = 2 - 3 = -1$:

$$K_p = K(RT)^{\Delta n}$$
$$K_p = 7.9 \times 10^4\,[(0.08206\ \text{L} \cdot \text{atm/mol} \cdot \text{K})(800\ \cancel{K})]^{-1}$$
$$= 1.2 \times 10^3$$

EXERCISE 15.8

Hydrogen gas and iodine react to form hydrogen iodide via the reaction

$$H_2(g) + I_2(g) \rightleftharpoons 2HI(g)$$

A mixture of H_2 and I_2 was maintained at a temperature of 740 K until the system reached equilibrium. The equilibrium mixture contained 0.0137 M HI, 6.47×10^{-3} M H_2, and 5.94×10^{-4} M I_2. Calculate K and K_p for this reaction.

Answer $K = 48.8$, $K_p = 48.8$

Often chemists are not given the concentrations of all the substances or they are not likely to measure the equilibrium concentrations of all the relevant substances for a particular system. In such cases, we can obtain the equilibrium concentrations from the initial concentrations of the reactants and the balanced equation for the reaction, as long as the equilibrium concentration of one of the substances is known. The next example shows one way to do this.

EXAMPLE 15.9

A 1.00-mol sample of NOCl was placed in a 2.00-L reactor and heated to 227°C until the system reached equilibrium. The contents of the reactor were then analyzed and found to contain 0.056 mol of Cl_2. Calculate *K* at this temperature. The equation for the decomposition of NOCl to NO and Cl_2 is

$$2NOCl(g) \rightleftharpoons 2NO(g) + Cl_2(g)$$

Given Balanced equilibrium equation, amount of reactant, volume, and amount of one product at equilibrium

Asked for Value of *K*

Strategy

A Write the equilibrium constant expression for the reaction. Construct a table showing the initial concentrations, the changes in concentrations, and the final concentrations (as initial concentrations plus changes in concentrations).

B Calculate all possible initial concentrations from the data given and insert them in the table.

C Use the coefficients in the balanced equation to obtain the changes in concentration of all other substances in the reaction. Insert those concentration changes in the table.

D Obtain the final concentrations by summing the columns. Calculate the equilibrium constant for the reaction.

Solution

A The first step in any such problem is to balance the equation for the reaction (if not already balanced) and use it to derive the equilibrium constant expression. In this case, the equation is already balanced, and the equilibrium constant expression is thus

$$K = \frac{[NO]^2[Cl_2]}{[NOCl]^2}$$

To obtain the concentrations of NOCl, NO, and Cl_2 at equilibrium, we construct a table showing what is known and what needs to be calculated. We begin by writing the balanced equation at the top of the table, followed by three lines corresponding to the initial concentrations, the changes in concentrations required to get from the initial to the final state, and the final concentrations.

	$2NOCl(g) \rightleftharpoons$	$2NO(g)$ +	$Cl_2(g)$
	[NOCl]	**[NO]**	**[Cl_2]**
Initial	____	____	____
Change	____	____	____
Final	____	____	____

B Initially, the system contains 1.00 mol of NOCl in a 2.00-L container. Thus, $[NOCl]_i$ = 1.00 mol/2.00 L = 0.500 *M*. The initial concentrations of NO and Cl_2 are 0 M because initially no products are present. Moreover, we are told that at equilibrium the system contains 0.056 mol of Cl_2 in a 2.00-L container, so $[Cl_2]_f$ = 0.056 mol/2.00 L = 0.028 *M*. We insert these values into the table.

	$2NOCl(g) \rightleftharpoons$	$2NO(g)$ +	$Cl_2(g)$
	[NOCl]	**[NO]**	**[Cl_2]**
Initial	0.500	0	0
Change	____	____	____
Final	____	____	0.028

C We use the stoichiometric relationships given in the balanced equation to find the change in the concentration of Cl_2, the substance for which initial and final concentrations are known:

$$\Delta[Cl_2] = [0.028\ M\ (\text{final}) - 0.00\ M\ (\text{initial})] = +0.028\ M$$

According to the coefficients in the balanced equation, 2 mol of NO is produced for every 1 mol of Cl_2, so the change in the NO concentration is

$$\Delta[NO] = \left(\frac{0.028\ \cancel{\text{mol Cl}_2}}{\text{L}}\right)\left(\frac{2\ \text{mol NO}}{1\ \cancel{\text{mol Cl}_2}}\right) = 0.056\ M$$

Similarly, 2 mol of NOCl are consumed for every 1 mol of Cl_2 produced, so the change in the NOCl concentration is

$$\Delta[NOCl] = \left(\frac{0.028\ \cancel{\text{mol Cl}_2}}{\text{L}}\right)\left(\frac{-2\ \text{mol NOCl}}{1\ \cancel{\text{mol Cl}_2}}\right) = -0.056\ M$$

We insert these values into the table.

	$2NOCl(g)$ $\rightleftharpoons$	$2NO(g)$ +	$Cl_2(g)$
	[NOCl]	**[NO]**	**[Cl_2]**
Initial	0.500	0	0
Change	−0.056	+0.056	+0.028
Final	______	______	0.028

D We sum the numbers in the [NOCl] and [NO] columns to obtain the final concentrations of NO and NOCl:

$$[NO]_f = 0.000\ M + 0.056\ M = 0.056\ M$$
$$[NOCl]_f = 0.500\ M + (-0.056\ M) = 0.444\ M$$

We can now complete the table.

	$2NOCl(g)$ $\rightleftharpoons$	$2NO(g)$ +	$Cl_2(g)$
	[NOCl]	**[NO]**	**[Cl_2]**
Initial	0.500	0	0
Change	−0.056	+0.056	+0.028
Final	0.444	0.056	0.028

We can now calculate the equilibrium constant for the reaction:

$$K = \frac{[NO]^2[Cl_2]}{[NOCl]^2} = \frac{(0.056)^2(0.028)}{(0.444)^2} = 4.5 \times 10^{-4}$$

EXERCISE 15.9

The German chemist Fritz Haber (1868–1934, Nobel prize 1918) was able to synthesize ammonia, NH_3, by the reaction of 0.1248 $M\ H_2$ and 0.0416 $M\ N_2$ at about 500°C. At equilibrium, the mixture contained 0.00272 $M\ NH_3$. What is the equilibrium constant K for the reaction $N_2 + 3H_2 \rightleftharpoons 2NH_3$ at this temperature? What is K_p?

Answer $K = 0.105$, $K_p = 2.61 \times 10^{-5}$

The original laboratory apparatus designed by Fritz Haber and Robert Le Rossignol in 1908 for synthesizing ammonia from its elements. A metal catalyst bed, where ammonia was produced, is in the large cylinder at the left. The Haber–Bosch process used for the industrial production of ammonia uses essentially the same process and components, but on a much larger scale. Unfortunately, Haber's process enabled Germany to prolong World War I when German supplies of nitrogen compounds, which were used for explosives, had been exhausted in 1914.

Calculating Equilibrium Concentrations from the Equilibrium Constant

To describe how to calculate equilibrium concentrations from an equilibrium constant, we first consider a system that contains only a single product and a single reactant, the conversion of *n*-butane to isobutane (Equation 15.23), for which $K = 2.6$ at 25°C. If we begin with a 1.00 *M* sample of *n*-butane, we can determine the concentration of *n*-butane and isobutane at equilibrium by constructing a table showing what is known and what needs to be calculated, just as we did in Example 15.9.

$$n\text{-butane}(g) \rightleftharpoons \text{isobutane}(g)$$

	[*n*-Butane]	**[Isobutane]**
Initial	______	______
Change	______	______
Final	______	______

The initial concentrations of the reactant and product are both known: $[n\text{-but}]_i = 1.00\ M$ and $[i\text{-but}]_i = 0\ M$. We need to calculate the equilibrium concentrations of both *n*-butane and isobutane. Because it is generally difficult to calculate final concentrations directly, we focus on the *change* in the concentrations of the substances between the initial and final (equilibrium) conditions. If, for example, we define the change in the concentration of isobutane, $\Delta[i\text{-but}]$, as $+x$, then the change in the concentration of *n*-butane is $\Delta[n\text{-but}] = -x$. This is because the balanced equation for the reaction tells us that 1 mol of *n*-butane is consumed for every 1 mol of isobutane produced. We can then express the final concentrations in terms of the initial concentrations and the changes they have undergone.

$$n\text{-butane}(g) \rightleftharpoons \text{isobutane}(g)$$

	[*n*-Butane]	**[Isobutane]**
Initial	1.00	0
Change	$-x$	$+x$
Final	$(1.00 - x)$	$(0 + x) = x$

Substituting the expressions for the final concentrations of *n*-butane and isobutane from the table into the equilibrium equation gives

$$K = \frac{[i\text{-but}]}{[n\text{-but}]} = \frac{x}{1.00 - x} = 2.6$$

Rearranging and solving for x give

$$x = 2.6(1.00 - x) = 2.6 - 2.6x$$
$$x + 2.6x = 2.6$$
$$x = 0.72$$

We obtain the final concentrations by substituting this value of x into the expressions for the final concentrations of *n*-butane and isobutane listed in the table:

$$[n\text{-but}]_f = (1.00 - x)\ M = (1.00 - 0.72)\ M = 0.28\ M$$
$$[i\text{-but}]_f = (0.00 + x)\ M = (0.00 + 0.72)\ M = 0.72\ M$$

We can check the results by substituting them back into the equilibrium constant expression to see whether they give the same value of K that we used in the calculation:

$$K = \frac{[i\text{-but}]}{[n\text{-but}]} = \frac{0.72\ \cancel{M}}{0.28\ \cancel{M}} = 2.6 \qquad \text{(to two significant figures)}$$

This is the same value of K we were given, so we can be confident of our results.

The next examples illustrate some common types of equilibrium problems you are likely to encounter.

EXAMPLE 15.10

The *water–gas shift reaction* is important in a number of chemical processes, such as the production of H_2 for fuel cells. This reaction can be written as

$$H_2(g) + CO_2(g) \rightleftharpoons H_2O(g) + CO(g) \qquad K = 0.106 \text{ at } 700 \text{ K}$$

If a mixture of gases that initially contains 0.0150 M H_2 and 0.0150 M CO_2 is allowed to equilibrate at 700 K, what are the final concentrations of all substances present?

Given Balanced equilibrium equation, value of K, and initial concentrations

Asked for Final concentrations

Strategy

A Construct a table showing what is known and what needs to be calculated. Define x as the change in the concentration of one substance. Then use the reaction stoichiometry to express the changes in the concentrations of the other substances in terms of x. From the values in the table, calculate the final concentrations.

B Write the equilibrium equation for the reaction. Substitute appropriate values from the table to obtain the value of x.

C Calculate the final concentrations of all species present. Check your answers by substituting these values into the equilibrium constant expression to obtain K.

Solution

A The initial concentrations of the reactants are $[H_2]_i = [CO_2]_i = 0.0150\ M$. Just as before, we will focus on the *change* in the concentrations of the various substances between the initial and final states. If we define the change in the concentration of H_2O as x, then $\Delta[H_2O] = +x$. We can use the stoichiometry of the reaction to express the changes in the concentrations of the other substances in terms of x. For example, 1 mol of CO is produced for every 1 mol of H_2O, so the change in the CO concentration can be expressed as $\Delta[CO] = +x$. Similarly, for every 1 mol of H_2O produced, 1 mol each of H_2 and CO_2 is consumed, so the change in the concentration of the reactants is $\Delta[H_2] = \Delta[CO_2] = -x$. We enter the values in the table and calculate the final concentrations.

	$H_2(g)$ +	$CO_2(g)$ $\rightleftharpoons$	$H_2O(g)$ +	$CO(g)$
	$[H_2]$	$[CO_2]$	$[H_2O]$	$[CO]$
Initial	0.0150	0.0150	0	0
Change	$-x$	$-x$	$+x$	$+x$
Final	$(0.0150 - x)$	$(0.0150 - x)$	x	x

B We can now use the equilibrium equation and the value of K given to solve for x:

$$K = \frac{[H_2O][CO]}{[H_2][CO_2]} = \frac{(x)(x)}{(0.0150 - x)(0.0150 - x)} = \frac{x^2}{(0.0150 - x)^2} = 0.106$$

We could solve this equation using the quadratic formula (see Essential Skills 7 at the end of this chapter), but it is far easier to solve for x by recognizing that the left side of the equation is a perfect square; that is,

$$\frac{x^2}{(0.0150 - x)^2} = \left(\frac{x}{0.0150 - x}\right)^2 = 0.106$$

Taking the square root of the middle and right terms gives

$$\frac{x}{(0.0150 - x)} = (0.106)^{1/2} = 0.326$$

$$x = (0.326)(0.0150) - 0.326x$$

$$1.326x = 0.00489$$

$$x = 0.00369 = 3.69 \times 10^{-3} \quad \text{(to three significant figures)}$$

C The final concentrations of all species in the reaction mixture are

$$[H_2]_f = [H_2]_i + \Delta[H_2] = (0.0150 - 0.00369)\ M = 0.0113\ M$$

$$[CO_2]_f = [CO_2]_i + \Delta[CO_2] = (0.0150 - 0.00369)\ M = 0.0113\ M$$

$$[H_2O]_f = [H_2O]_i + \Delta[H_2O] = (0 + 0.00369)\ M = 0.00369\ M$$

$$[CO]_f = [CO]_i + \Delta[CO] = (0 + 0.00369)\ M = 0.00369\ M$$

We can check our work by inserting the calculated values back into the equilibrium constant expression:

$$K = \frac{[H_2O][CO]}{[H_2][CO_2]} = \frac{(0.00369)^2}{(0.0113)^2} = 0.107$$

To 2 significant figures, this value of K is the same as the value given in the problem, so our answer is confirmed.

EXERCISE 15.10

Hydrogen gas reacts with iodine vapor to give hydrogen iodide according to the chemical equation

$$H_2(g) + I_2(g) \rightleftharpoons 2HI(g) \qquad K = 54 \text{ at } 425°C$$

If 0.172 M H_2 and I_2 are injected into a reactor and maintained at 425°C until the system equilibrates, what is the final concentration of each substance in the reaction mixture?

Answer $[HI]_f = 0.270\ M$; $[H_2]_f = [I_2]_f = 0.037\ M$

In Example 15.10, the initial concentrations of the reactants were the same, which gave us an equation that was a perfect square and simplified our calculations. Often, however, the initial concentrations of the reactants are not the same, or one or more of the products may be present when the reaction starts. Under these conditions, there is usually no way to simplify the problem, and we must determine the equilibrium concentrations using other means. Such a case is described in the next example.

EXAMPLE 15.11

In the water–gas shift reaction shown in Example 15.10, a sample containing 0.632 M CO_2 and 0.570 M H_2 is allowed to equilibrate at 700 K. At this temperature $K = 0.106$. What is the composition of the reaction mixture at equilibrium?

Given Balanced equilibrium equation, concentrations of reactants, and value of K

Asked for Composition of reaction mixture at equilibrium

Strategy

A Write the equilibrium equation. Construct a table showing the initial concentrations of all substances in the mixture. Complete the table showing the changes in the concentrations (x) and the final concentrations.

B Write the equilibrium constant expression for the reaction. Substitute the known value of K and the final concentrations to solve for x.

C Calculate the final concentration of each substance in the reaction mixture. Check your answers by substituting these values into the equilibrium constant expression to obtain K.

Solution

A The initial concentrations of CO_2 and H_2 are given: $[CO_2]_i = 0.632\ M$ and $[H_2]_i = 0.570\ M$. Again, x is defined as the change in the concentration of H_2O: $\Delta[H_2O] = +x$. Because 1 mol of CO is produced for every 1 mol of H_2O, the change in the concentration of CO is the same as the change in the concentration of H_2O, so $\Delta[CO] = +x$. Similarly, because 1 mol of both H_2 and CO_2 is consumed for every 1 mol of H_2O produced, $\Delta[H_2] = \Delta[CO_2] = -x$. The final concentrations are the sums of the initial concentrations and the changes in concentrations at equilibrium.

	$H_2(g)$ +	$CO_2(g) \rightleftharpoons$	$H_2O(g)$ +	$CO(g)$
	$[H_2]$	$[CO_2]$	$[H_2O]$	$[CO]$
Initial	0.570	0.632	0	0
Change	$-x$	$-x$	$+x$	$+x$
Final	$(0.570 - x)$	$(0.632 - x)$	x	x

B We can now use the equilibrium equation and the known value of K to solve for x:

$$K = \frac{[H_2O][CO]}{[H_2][CO_2]} = \frac{x^2}{(0.570 - x)(0.632 - x)} = 0.106$$

In contrast to Example 15.10, however, there is no obvious way to simplify this expression. Thus, we must expand the expression and multiply both sides by the denominator:

$$x^2 = 0.106(0.360 - 1.202x + x^2)$$

Collecting terms on one side of the equation gives

$$0.894x^2 + 0.127x - 0.0382 = 0$$

This equation can be solved using the quadratic formula (see Essential Skills 7):

$$x = \frac{-b \pm \sqrt{b^2 - 4ac}}{2a} = \frac{-0.127 \pm \sqrt{(0.127)^2 - 4(0.894)(-0.0382)}}{2(0.894)} = 0.148, -0.290$$

Only the answer with the positive value has any physical significance, so $\Delta[H_2O] = \Delta[CO] = +0.148\ M$, and $\Delta[H_2] = \Delta[CO_2] = -0.148\ M$. C The final concentrations of all species in the reaction mixture are

$$[H_2]_f = [H_2]_i + \Delta[H_2] = 0.570\ M - 0.148\ M = 0.422\ M$$
$$[CO_2]_f = [CO_2]_i + \Delta[CO_2] = 0.632\ M - 0.148\ M = 0.484\ M$$
$$[H_2O]_f = [H_2O]_i + \Delta[H_2O] = 0\ M + 0.148\ M = 0.148\ M$$
$$[CO]_f = [CO]_i + \Delta[CO] = 0\ M + 0.148\ M = 0.148\ M$$

We can check our work by substituting these values into the equilibrium constant expression:

$$K = \frac{[H_2O][CO]}{[H_2][CO_2]} = \frac{(0.148)^2}{(0.422)(0.484)} = 0.107$$

Because the value of K is essentially the same as the value given in the problem, our calculations are confirmed.

EXERCISE 15.11

Exercise 15.8 showed the reaction of hydrogen and iodine vapor to form hydrogen iodide, for which $K = 54$ at 425°C. If a sample containing 0.200 M H_2 and 0.0450 M I_2 is allowed to equilibrate at 425°C, what is the final concentration of each substance in the reaction mixture?

Answer $[HI]_f = 0.0882\ M$; $[H_2]_f = 0.156\ M$; $[I_2]_f = 9.2 \times 10^{-4}\ M$

In many situations it is not necessary to solve a quadratic (or higher-order) equation. Most of these cases involve reactions for which the equilibrium constant is either very small ($K \leq 10^{-3}$) or very large ($K \geq 10^{3}$), which means that the change in the concentration (defined as x) is essentially negligible compared with the initial concentration of a substance. Knowing this simplifies the calculations dramatically, as illustrated in the next example.

EXAMPLE 15.12

Atmospheric nitrogen and oxygen react to form nitric oxide:

$$N_2(g) + O_2(g) \rightleftharpoons 2NO(g) \qquad K_p = 2.0 \times 10^{-31} \text{ at } 25^\circ\text{C}$$

What is the partial pressure of NO in equilibrium with N_2 and O_2 in the atmosphere (at 1 atm, $P_{N_2} = 0.78$ atm and $P_{O_2} = 0.21$ atm)?

Given Balanced equilibrium equation and values of K_p, P_{O_2}, and P_{N_2}

Asked for Partial pressure of NO

Strategy

Ⓐ Construct a table and enter the initial partial pressures, the changes in the partial pressures that occur during the course of the reaction, and the final partial pressures of all substances.
Ⓑ Write the equilibrium equation for the reaction. Then substitute values from the table to solve for the change in concentration, x.
Ⓒ Calculate the partial pressure of NO. Check your answer by substituting values into the equilibrium equation and solving for K.

Solution

Ⓐ Because we are given K_p and partial pressures are reported in atmospheres, we will use partial pressures. The initial partial pressure of O_2 is 0.21 atm, and that of N_2 is 0.78 atm. If we define the change in the partial pressure of NO as $2x$, then the change in the partial pressure of O_2 and of N_2 is $-x$ because 1 mol each of N_2 and of O_2 is consumed for every 2 mol of NO produced. Each substance has a final partial pressure equal to the sum of the initial pressure and the change in that pressure at equilibrium.

	$N_2(g)$	+	$O_2(g)$	$\rightleftharpoons$	$2NO(g)$
	P_{N_2}, atm		P_{O_2}, atm		P_{NO}, atm
Initial *P*	0.78		0.21		0
Change in *P*	$-x$		$-x$		$+2x$
Final *P*	$(0.78 - x)$		$(0.21 - x)$		$2x$

Ⓑ Substituting these values into the equation for the equilibrium constant gives

$$K_p = \frac{(P_{NO})^2}{(P_{N_2})(P_{O_2})} = \frac{(2x)^2}{(0.78 - x)(0.21 - x)} = 2.0 \times 10^{-31}$$

In principle, we could multiply out the terms in the denominator, rearrange, and solve the resulting quadratic equation. In practice, it is far easier to recognize that an equilibrium constant of this magnitude means that the extent of the reaction will be very small; therefore, the value of x will be negligible compared with the initial concentrations. If this assumption is correct, then to two significant figures, $(0.78 - x) = 0.78$ and $(0.21 - x) = 0.21$. Substituting these expressions into our original equation gives

$$\frac{(2x)^2}{(0.78)(0.21)} = 2.0 \times 10^{-31}$$

$$\frac{4x^2}{0.16} = 2.0 \times 10^{-31}$$

$$x^2 = \frac{0.33 \times 10^{-31}}{4}$$

$$x = 9.1 \times 10^{-17}$$

C Substituting this value of x into our expressions for the final partial pressures of the substances gives

$$P_{NO} = 2x \text{ atm} = 1.8 \times 10^{-16} \text{ atm}$$
$$P_{N_2} = (0.78 - x) \text{ atm} = 0.78 \text{ atm}$$
$$P_{O_2} = (0.21 - x) \text{ atm} = 0.21 \text{ atm}$$

From these calculations, we see that our initial assumption regarding x was correct: given two significant figures, 2.0×10^{-16} is certainly negligible compared with 0.78 and 0.21. When can we make such an assumption? As a general rule, if x is less than about 5% of the total, or $10^{-3} > K > 10^3$, then the assumption is justified. Otherwise, we must use the quadratic formula or some other approach. The results we have obtained agree with the general observation that toxic NO, an ingredient of smog, does not form from atmospheric concentrations of N_2 and O_2 to a substantial degree at 25°C. We can verify our results by substituting them into the original equilibrium equation:

$$K_p = \frac{(P_{NO})^2}{(P_{N_2})(P_{O_2})} = \frac{(1.8 \times 10^{-16})^2}{(0.78)(0.21)} = 2.0 \times 10^{-31}$$

The final value of K_p agrees with the value given at the beginning of this example.

EXERCISE 15.12

Under certain conditions, oxygen will react to form ozone, as shown in the equation

$$3O_2(g) \rightleftharpoons 2O_3(g) \qquad K_p = 2.5 \times 10^{-59} \text{at } 25^\circ\text{C}$$

What ozone partial pressure is in equilibrium with oxygen in the atmosphere ($P_{O_2} = 0.21$ atm)?

Answer 4.8×10^{-31} atm

Another type of problem that can be simplified by assuming that changes in concentration are negligible is one in which the equilibrium constant is very large ($K \geq 10^3$). A large equilibrium constant implies that reactants are converted almost entirely to products, so we can assume that the reaction proceeds 100% to completion. When we solve this type of problem, we view the system as equilibrating from the *products* side of the reaction rather than the reactants side. This approach is illustrated in the next example.

EXAMPLE 15.13

The chemical equation for the reaction of hydrogen with ethylene (C_2H_4) to give ethane (C_2H_6) is

$$H_2(g) + C_2H_4(g) \overset{\text{Ni}}{\rightleftharpoons} C_2H_6(g) \qquad K = 9.6 \times 10^{18} \text{ at } 25^\circ\text{C}$$

If a mixture of 0.200 M H_2 and 0.155 M C_2H_4 is maintained at 25°C in the presence of a powdered nickel catalyst, what is the equilibrium concentration of each substance in the mixture?

Given Balanced chemical equation, value of K, and initial concentrations of reactants

Asked for Equilibrium concentrations

Strategy

A Construct a table showing initial concentrations, concentrations that would be present if the reaction were to go to completion, changes in concentrations, and final concentrations.

B Write the equilibrium constant expression for the reaction. Then substitute values from the table into the expression to solve for x, the change in concentration.
C Calculate the equilibrium concentrations. Check your answers by substituting these values into the equilibrium equation.

Solution

A From the magnitude of the equilibrium constant, we see that the reaction goes essentially to completion. Because the initial concentration of ethylene (0.155 M) is less than the concentration of hydrogen (0.200 M), ethylene is the limiting reactant; that is, no more than 0.155 M ethane can be formed from 0.155 M ethylene. If the reaction were to go to completion, the concentration of ethane would be 0.155 M and the concentration of ethylene would be 0 M. Because the concentration of hydrogen is greater than is needed for complete reaction, the concentration of unreacted hydrogen in the reaction mixture would be 0.200 M − 0.155 M = 0.045 M. The equilibrium constant for the forward reaction is very large, so the equilibrium constant for the reverse reaction must be very small. The problem then is identical to the ones described in Example 15.12 and Exercise 15.12. If we define $-x$ as the change in the ethane concentration for the reverse reaction, then the change in the ethylene and hydrogen concentrations is $+x$. The final equilibrium concentrations are the sums of the concentrations for the forward and reverse reactions.

	$H_2(g)$ +	$C_2H_4(g)$ $\overset{Ni}{\rightleftharpoons}$	$C_2H_6(g)$
	$[H_2]$	**$[C_2H_4]$**	**$[C_2H_6]$**
Initial	0.200	0.155	0
Assuming 100% reaction	0.045	0	0.155
Change	$+x$	$+x$	$-x$
Final	$(0.045 + x)$	$(0 + x)$	$(0.155 - x)$

B Substituting values into the equilibrium constant expression gives:

$$K = \frac{[C_2H_6]}{[H_2][C_2H_4]} = \frac{0.155 - x}{(0.045 + x)x} = 9.6 \times 10^{18}$$

Once again, the magnitude of the equilibrium constant tells us that the equilibrium will lie far to the right as written, so the reverse reaction is negligible. Thus, x is likely to be very small compared with either 0.155 M or 0.045 M, and the equation can be simplified $[(0.045 + x) = 0.045$ and $(0.155 - x) = 0.155]$ to give

$$K = \frac{0.155}{0.045x} = 9.6 \times 10^{18}$$

$$x = 3.6 \times 10^{-19}$$

C The small value of x indicates that our assumption concerning the reverse reaction is correct, and we can therefore calculate the final concentrations by evaluating the expressions from the last line of the table:

$$[C_2H_6]_f = (0.155 - x)\,M = 0.155\,M$$
$$[C_2H_4]_f = x\,M = 3.6 \times 10^{-19}\,M$$
$$[H_2]_f = (0.045 + x)\,M = 0.045\,M$$

We can verify our calculations by substituting the final concentrations into the equilibrium constant expression:

$$K = \frac{[C_2H_6]}{[H_2][C_2H_4]} = \frac{0.155}{(0.045)(3.6 \times 10^{-19})} = 9.6 \times 10^{18}$$

This value of K agrees with our initial value at the beginning of the example.

EXERCISE 15.13

Hydrogen reacts with chlorine gas to form hydrogen chloride:

$$H_2(g) + Cl_2(g) \rightleftharpoons 2HCl(g) \qquad K_p = 4.0 \times 10^{31} \text{ at } 47°C$$

If a mixture of 0.257 M H_2 and 0.392 M Cl_2 is allowed to equilibrate at 47°C, what is the equilibrium composition of the mixture?

Answer $[H_2]_f = 4.8 \times 10^{-32}$ M; $[Cl_2]_f = 0.135$ M; $[HCl]_f = 0.514$ M

15.4 Nonequilibrium Conditions

In Section 15.3, we saw that knowing the magnitude of the equilibrium constant under a given set of conditions allows chemists to predict the extent of a reaction. Often, however, chemists must decide whether a system has reached equilibrium or the composition of the mixture will continue to change with time. In this section, we describe how to analyze the composition of a reaction mixture quantitatively in order to make this determination.

MGC The Reaction Quotient

The Reaction Quotient (Q)

To determine whether a system has reached equilibrium, chemists use a quantity called the **reaction quotient** (**Q**). The expression for the reaction quotient has precisely the same form as the equilibrium constant expression, except that Q may be derived from a set of values measured at *any* time during the reaction of *any* mixture of the reactants and products, regardless of whether the system is at equilibrium. Therefore, for the general reaction

$$aA + bB \rightleftharpoons cC + dD$$

the reaction quotient is defined as

$$Q = \frac{[C]^c[D]^d}{[A]^a[B]^b} \qquad (15.25)$$

A related quantity Q_p, analogous to K_p, can be written for any reaction that involves gases by using the partial pressures of the components.

To understand how information is obtained using a reaction quotient, consider the dissociation of dinitrogen tetroxide to nitrogen dioxide, $N_2O_4(g) \rightleftharpoons 2NO_2(g)$, for which $K = 4.65 \times 10^{-3}$ at 298 K. We can write Q for this reaction as

$$Q = \frac{[NO_2]^2}{[N_2O_4]} \qquad (15.26)$$

The table lists data from three experiments in which samples of the reaction mixture were obtained and analyzed at equivalent time intervals, and the corresponding values of Q calculated for each. Note that each experiment begins with different proportions of product and reactant:

Experiment	$[NO_2]$, M	$[N_2O_4]$, M	$Q = [NO_2]^2/[N_2O_4]$
1	0	0.0400	$\frac{0^2}{0.0400} = 0$
2	0.0600	0	$\frac{(0.0600)^2}{0} = \text{undefined}$
3	0.0200	0.0600	$\frac{(0.0200)^2}{0.0600} = 6.67 \times 10^{-3}$

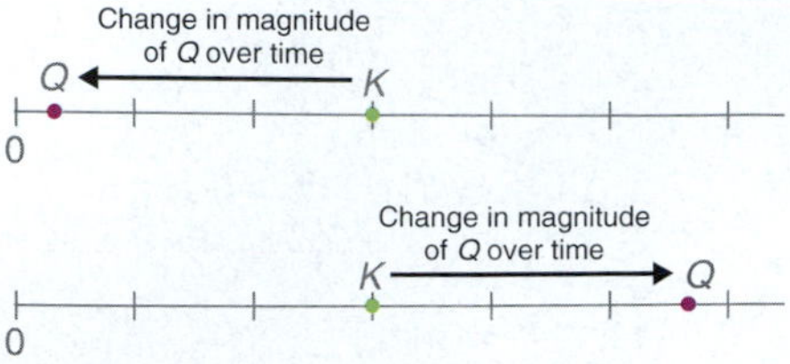

(a) *Q* versus *K*, t_n

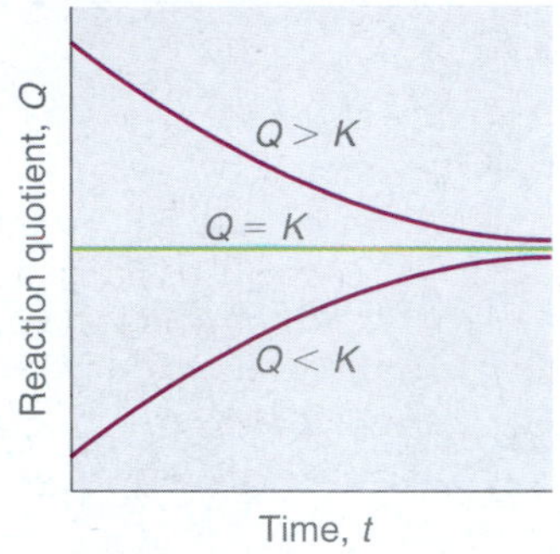

(b) *Q* versus *K*, t_0 to t_∞

Figure 15.6 Two different ways of illustrating how the composition of a system will change depending on the relative values of Q and K. **(a)** Both Q and K are plotted as points along a number line: the system will always react in the way that causes Q to approach K. **(b)** The change in the composition of a system with time is illustrated for systems with initial values of $Q > K$, $Q < K$, and $Q = K$. Interactive Graph

As these calculations demonstrate, Q can have any numerical value between 0 and infinity (undefined); that is, the value of Q can be greater than, less than, or equal to K.

Comparing the magnitudes of Q and K enables us to determine whether a reaction mixture is already at equilibrium and, if it is not, to predict how its composition will change with time in order to reach equilibrium (that is, whether the reaction will proceed to the right or to the left as written). All you need to remember is that the composition of a system that is not at equilibrium will change in a way that makes the value of Q approach the value of K. If $Q = K$, for example, then the system is already at equilibrium, and no further change in the composition of the system will occur unless the conditions are changed. If $Q < K$, then the ratio of the concentrations of products to the concentrations of reactants is *less* than the ratio at equilibrium. Therefore, the reaction will proceed to the right as written, forming products at the expense of reactants. Conversely, if $Q > K$, then the ratio of the concentrations of products to the concentrations of reactants is *greater* than at equilibrium, so the reaction will proceed to the left as written, forming reactants at the expense of products. These points are illustrated graphically in Figure 15.6.

Note the pattern

If $Q < K$, the reaction will proceed to the right as written. If $Q > K$, the reaction will proceed to the left as written. If $Q = K$, then the system is at equilibrium.

EXAMPLE 15.14

At elevated temperatures, methane (CH_4) reacts with water to produce hydrogen and carbon monoxide in what is known as a *steam-reforming* reaction:

$$CH_4(g) + H_2O(g) \rightleftharpoons CO(g) + 3H_2(g) \qquad K = 2.4 \times 10^{-4} \text{ at } 900 \text{ K}$$

Huge amounts of hydrogen are produced from natural gas in this way and are then used for the industrial synthesis of ammonia. If 1.2×10^{-2} mol of CH_4, 8.0×10^{-3} mol of H_2O, 1.6×10^{-2} mol of CO, and 6.0×10^{-3} mol of H_2 are placed in a 2.0-L steel reactor and heated to 900 K, will the reaction be at equilibrium or will it proceed to the right to produce CO and H_2 or to the left to form CH_4 and H_2O?

Given Balanced equation, value of K, amounts of reactants and products, and volume

Asked for Direction of reaction

Strategy

A Calculate the molar concentrations of the reactants and products.

B Use Equation 15.25 to determine Q. Compare the values of Q and K to decide in which direction the reaction will proceed.

Solution

A We must first determine the initial concentrations of the substances present. For example, we have 1.2×10^{-2} mol of CH_4 in a 2.0-L container, so

$$[CH_4] = \frac{1.2 \times 10^{-2} \text{ mol}}{2.0 \text{ L}} = 6.0 \times 10^{-3}\, M$$

We can calculate the other concentrations in a similar way: $[H_2O] = 4.0 \times 10^{-3}\, M$, $[CO] = 8.0 \times 10^{-3}\, M$, and $[H_2] = 3.0 \times 10^{-3}\, M$. **B** We now compute the value of Q and compare it with the value of K:

$$Q = \frac{[CO][H_2]^3}{[CH_4][H_2O]} = \frac{(8.0 \times 10^{-3})(3.0 \times 10^{-3})^3}{(6.0 \times 10^{-3})(4.0 \times 10^{-3})} = 9.0 \times 10^{-6}$$

Because $K = 2.4 \times 10^{-4}$, we see that $Q < K$. Thus, the ratio of the concentrations of products to the concentrations of reactants is less than the ratio for an equilibrium mixture. The reaction will therefore proceed to the right as written, forming H_2 and CO at the expense of H_2O and CH_4.

EXERCISE 15.14

In the water–gas shift reaction introduced in Example 15.10, CO produced by steam reforming of methane (Example 15.14) reacts with steam at elevated temperatures to produce more hydrogen:

$$CO(g) + H_2O(g) \rightleftharpoons CO_2(g) + H_2(g) \qquad K = 0.64 \text{ at } 900 \text{ K}$$

If 0.010 mol of both CO and H_2O, 0.0080 mol of CO_2, and 0.012 mol of H_2 are injected into a 4.0-L reactor and heated to 900 K, will the reaction proceed to the left or to the right as written?

Answer $Q = 0.96$ ($Q > K$), so the reaction will proceed to the left and CO and H_2O will form.

Predicting the Direction of Reaction Using a Graph

By graphing a few equilibrium concentrations for a system at a given temperature and pressure, we can readily see the range of reactant and product concentrations that correspond to equilibrium conditions, for which $Q = K$. Such a graph allows us to predict what will happen to a reaction when conditions change so that Q no longer equals K, such as when a reactant or product concentration is increased or decreased.

Lead carbonate decomposes to lead oxide and carbon dioxide according to the equation

$$PbCO_3(s) \rightleftharpoons PbO(s) + CO_2(g) \qquad (15.27)$$

Because $PbCO_3$ and PbO are solids, the equilibrium constant is simply $K = [CO_2]$. At a given temperature, therefore, any system that contains solid $PbCO_3$ and solid PbO will have exactly the same concentration of CO_2 at equilibrium, regardless of the ratio or the amounts of the solids present. This situation is represented in Figure 15.7, which shows a plot of $[CO_2]$ versus the amount of $PbCO_3$ added. Initially, the added $PbCO_3$ decomposes completely to CO_2 because the amount of $PbCO_3$ is not sufficient to give a CO_2 concentration equal to K. Thus, the left portion of the graph represents a system that is *not* at equilibrium because it contains only $CO_2(g)$ and $PbO(s)$. In contrast, when just enough $PbCO_3$ has been added to give $[CO_2] = K$, the system has reached equilibrium, and adding more $PbCO_3$ has no effect on the CO_2 concentration: the graph is a horizontal line. Thus, any CO_2 concentration that is not on the horizontal line represents a nonequilibrium state, and the system will adjust its composition to achieve equilibrium provided enough $PbCO_3$ and PbO are present. For example, the point labeled A in Figure 15.7 lies above the horizontal line, so it corresponds to a $[CO_2]$ that is greater than the equilibrium concentration of CO_2 ($Q > K$). To reach equilibrium, the system must decrease $[CO_2]$, which it can do only by the reaction of CO_2 with solid PbO to form solid $PbCO_3$. Thus, the reaction in Equation 15.27 will proceed to the left as written, until $[CO_2] = K$. Conversely, the point labeled B in Figure 15.7 lies below the horizontal line, so it corresponds to a $[CO_2]$ that is less than the equilibrium concentration of CO_2 ($Q < K$). To reach equilibrium, the system must increase $[CO_2]$, which it can do only by the decomposition of solid $PbCO_3$ to form CO_2 and solid PbO. The reaction in Equation 15.27 will therefore proceed to the right as written, until $[CO_2] = K$.

In contrast, for the reduction of cadmium oxide by hydrogen to give metallic cadmium and water vapor:

$$CdO(s) + H_2(g) \rightleftharpoons Cd(s) + H_2O(g) \qquad (15.28)$$

the equilibrium constant K is $[H_2O]/[H_2]$. Thus, if $[H_2O]$ is doubled at equilibrium, then $[H_2]$ must also be doubled for the system to remain at equilibrium. A plot of $[H_2O]$ versus $[H_2]$ at equilibrium is a straight line with a slope of K (Figure 15.8). Again, only those pairs of concentrations of H_2O and H_2 that lie on the line correspond to equilibrium states. Any point representing a pair of concentrations that does

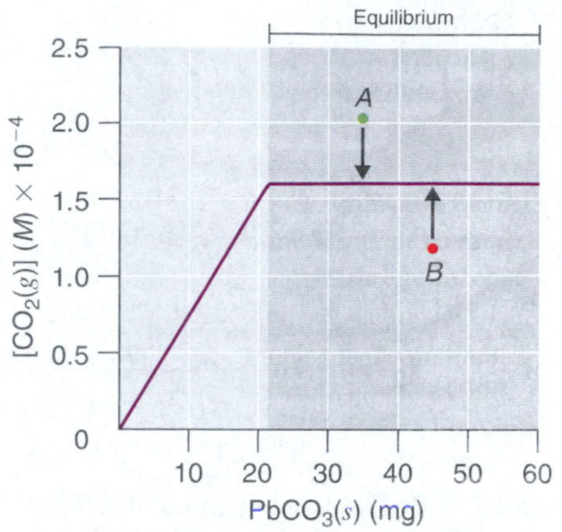

Figure 15.7 The concentration of gaseous CO_2 in a closed system at equilibrium as a function of the amount of solid $PbCO_3$ added. Initially the concentration of $CO_2(g)$ increases linearly with the amount of solid $PbCO_3$ added, as $PbCO_3$ decomposes to $CO_2(g)$ and solid PbO. Once the CO_2 concentration reaches the value that corresponds to the equilibrium concentration, however, adding more solid $PbCO_3$ has no effect on $[CO_2]$, as long as the temperature remains constant. Interactive Graph

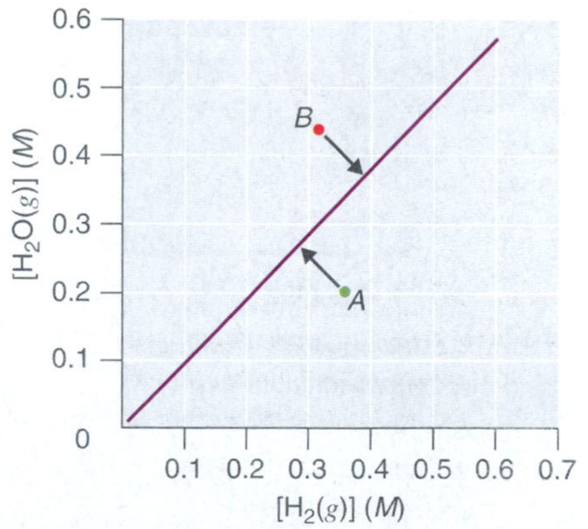

Figure 15.8 The concentration of water vapor versus the concentration of hydrogen for the $CdO(s) + H_2(g) \rightleftharpoons Cd(s) + H_2O(g)$ system at equilibrium. For any equilibrium concentration of $H_2O(g)$, there is only one equilibrium concentration of $H_2(g)$. Because the magnitudes of the two concentrations are directly proportional, a large $[H_2O]$ at equilibrium requires a large $[H_2]$, and vice versa. In this case, the slope of the line is equal to the value of the equilibrium constant K. Interactive Graph

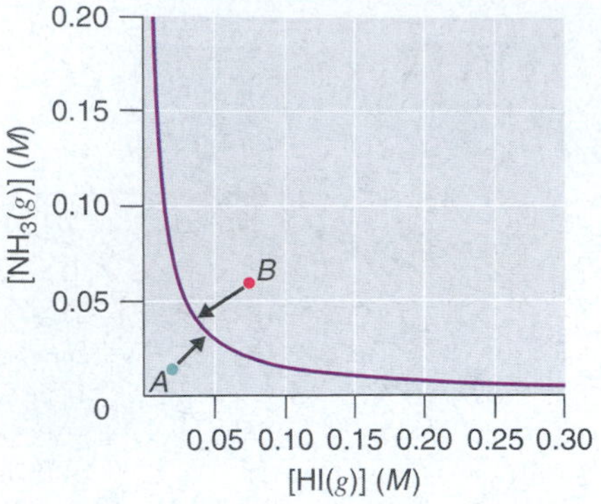

Figure 15.9 The concentration of $NH_3(g)$ versus the concentration of $HI(g)$ for the $NH_4I(s) \rightleftharpoons NH_3(g) + HI(g)$ system at equilibrium. Only one equilibrium concentration of $NH_3(g)$ is possible for any given equilibrium concentration of $HI(g)$. In this case, the two are inversely proportional. Thus, a large [HI] at equilibrium requires a small $[NH_3]$ at equilibrium, and vice versa.

Interactive Graph

not lie on the line corresponds to a nonequilibrium state. In such cases, the reaction in Equation 15.28 will proceed in whichever direction causes the composition of the system to move toward the equilibrium line. For example, point *A* in Figure 15.8 lies below the line, indicating that the $[H_2O]/[H_2]$ ratio is lower than the ratio of an equilibrium mixture ($Q < K$). Thus, the reaction in Equation 15.28 will proceed to the right as written, consuming H_2 and producing H_2O, which causes the concentration ratio to move up and to the left toward the equilibrium line. Conversely, point *B* in Figure 15.8 lies above the line, indicating that the $[H_2O]/[H_2]$ ratio is higher than the ratio of an equilibrium mixture ($Q > K$). Thus, the reaction in Equation 15.28 will proceed to the left as written, consuming H_2O and producing H_2, which causes the concentration ratio to move down and to the right toward the equilibrium line.

In another example, solid ammonium iodide dissociates to gaseous ammonia and hydrogen iodide at elevated temperatures:

$$NH_4I(s) \rightleftharpoons NH_3(g) + HI(g) \qquad (15.29)$$

For this system, the equilibrium constant *K* is equal to the product of the concentrations of the two products, $[NH_3] \times [HI]$. Thus, if we double the concentration of NH_3, the concentration of HI must decrease by approximately a factor of 2 to maintain equilibrium, as shown in Figure 15.9. As a result, for a given concentration of either HI or NH_3, only a *single* equilibrium composition that contains equal concentrations of both NH_3 and HI is possible, for which $[NH_3] = [HI] = K^{1/2}$. Any point that lies below and to the left of the equilibrium curve (such as point *A* in Figure 15.9) corresponds to $Q < K$, and the reaction in Equation 15.29 will therefore proceed to the right as written, causing the composition of the system to move toward the equilibrium line. Conversely, any point that lies above and to the right of the equilibrium curve (such as point *B* in Figure 15.9) corresponds to $Q > K$, and the reaction in Equation 15.29 will therefore proceed to the left as written, again causing the composition of the system to move toward the equilibrium line. Thus, by graphing equilibrium concentrations for a given system at a given temperature and pressure, we can predict the direction of reaction of that mixture when the system is not at equilibrium.

Le Châtelier's Principle

When a system at equilibrium is perturbed in some way, the effects of the perturbation can be predicted qualitatively using **Le Châtelier's principle** (named after the French chemist Henri Louis Le Châtelier, 1850–1936).* This principle can be stated as follows: *If a stress is applied to a system at equilibrium, the composition of the system will change to counteract the applied stress.* Stress occurs when any change in the system affects the magnitude of *Q* or *K*. In Equation 15.29, for example, increasing $[NH_3]$ produces a stress on the system that requires a decrease in [HI] for the system to return to equilibrium. Consider esters, which are one of the products of an equilibrium reaction between a carboxylic acid and an alcohol. Esters are responsible for the scents we associate with fruits (such as oranges and bananas), and they are also used as scents in perfumes. Applying a stress to the reaction of a carboxylic acid and an alcohol will change the composition of the system, leading to an increase or a decrease in the amount of ester produced. In the next section, we explore factors that affect equilibrium concentrations.

Note the pattern

If a stress is applied to a system at equilibrium, the composition of the system will change to counteract the applied stress (Le Châtelier's principle).

EXAMPLE 15.15

Write an equilibrium constant expression for each of the following reactions, and use this expression to predict what will happen to the concentration of the substance in red type when the indicated change is made if the system is to maintain equilibrium.

* The name is pronounced "Luh SHOT-lee-ay."

(a) $2HgO(s) \rightleftharpoons 2Hg(l) + O_2(g)$; the amount of HgO is doubled.
(b) $NH_4HS(s) \rightleftharpoons NH_3(g) + H_2S(g)$; the concentration of H_2S is tripled.
(c) n-butane(g) $\rightleftharpoons$ isobutane(g); the concentration of isobutane is halved.

Given Equilibrium systems and changes

Asked for Equilibrium constant expressions and effects of changes

Strategy

Write the equilibrium constant expression, remembering that pure liquids and solids do not appear in the expression. From this expression, predict the change that must occur to maintain equilibrium when the indicated changes is made.

Solution

(a) Because $HgO(s)$ and $Hg(l)$ are pure substances, they do not appear in the equilibrium constant expression. Thus, for this reaction, $K = [O_2]$. The equilibrium concentration of O_2 is a constant and does not depend on the amount of HgO present. Hence, adding more HgO will not affect the equilibrium concentration of O_2, so no compensatory change is necessary.
(b) The reactant, NH_4HS, does not appear in the equilibrium constant expression because it is a solid. Thus, $K = [NH_3][H_2S]$, which means that the concentrations of the products are inversely proportional. If the H_2S concentration is tripled by adding H_2S, for example, then the NH_3 concentration must decrease by about a factor of 3 for the system to remain at equilibrium so that the product of the concentrations is equal to K.
(c) For this reaction, $K = [i\text{-bu}]/[n\text{-bu}]$, so halving the concentration of isobutane means that the n-butane concentration must also decrease by about half if the system is to maintain equilibrium.

EXERCISE 15.15

Write an equilibrium constant expression for each reaction. What must happen to the concentration of the substance in red type when the indicated change occurs if the system is to maintain equilibrium?

(a) $HBr(g) + NaH(s) \rightleftharpoons NaBr(s) + H_2(g)$; the concentration of HBr is decreased by a factor of 3.
(b) $6Li(s) + N_2(g) \rightleftharpoons 2Li_3N(s)$; the amount of Li is tripled.
(c) $SO_2(g) + Cl_2(g) \rightleftharpoons SO_2Cl_2(l)$; the concentration of Cl_2 is doubled.

Answer **(a)** $[H_2]$ must decrease by about a factor of 3; **(b)** solid lithium does not appear in the equilibrium constant expression, so no compensatory change is necessary; **(c)** $[SO_2]$ must decrease by about half.

15.5 • Factors That Affect Equilibrium

MGC Le Châtelier's Principle

Chemists use various strategies to increase the yield of the desired products of reactions. When synthesizing an ester, for example, how can a chemist control the reaction conditions to obtain the maximum amount of the desired product? Only three types of stresses can change the composition of an equilibrium mixture: (1) a change in the concentrations (or partial pressures) of the components by the addition or removal of reactants or products, (2) a change in the total pressure or volume, and (3) a change in the temperature of the system. In this section, we explore how changes in reaction conditions can affect the equilibrium composition of a system and will explore each of these possibilities in turn.

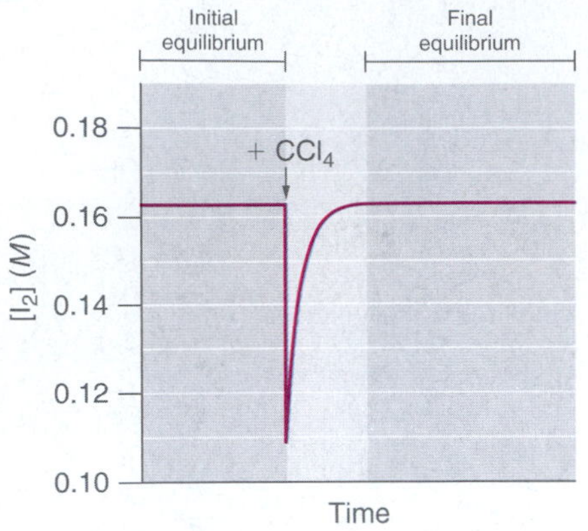

Figure 15.10 The concentration of dissolved I_2 as a function of time following the addition of more solvent to a saturated solution in contact with excess solid I_2. The concentration of I_2 decreases initially due to dilution but returns to its original value as long as solid I_2 is present. Interactive Graph

Changes in Concentration

If we add a small volume of carbon tetrachloride (CCl_4) solvent to a flask containing crystals of iodine, we obtain a saturated solution of I_2 in CCl_4, along with undissolved crystals:

$$I_2(s) \xrightleftharpoons{\text{solvent}} I_2\,(soln) \tag{15.30}$$

The system reaches equilibrium, with $K = [I_2]$. If we add more CCl_4, thereby diluting the solution, the equilibrium quotient Q is now less than K. Le Châtelier's principle tells us that the system will react to relieve the stress, but how? Adding solvent stressed the system by decreasing the concentration of dissolved I_2. Hence, more crystals will dissolve, thereby increasing the concentration of dissolved I_2 until the system again reaches equilibrium if enough solid I_2 is available (Figure 15.10). By adding solvent, we drove the reaction shown in Equation 15.30 to the right as written.

We encounter a more complex system in the reaction of hydrogen and nitrogen to form ammonia:

$$N_2(g) + 3H_2(g) \rightleftharpoons 2NH_3(g) \tag{15.31}$$

The K_p for this reaction is 2.14×10^{-2} at about 540 K. Under one set of equilibrium conditions, the partial pressure of ammonia is $P_{NH_3} = 0.454$ atm, that of hydrogen is $P_{H_2} = 2.319$ atm, and that of nitrogen is $P_{N_2} = 0.773$ atm. If an additional 1 atm of hydrogen is added to the reactor to give $P_{H_2} = 3.319$ atm, how will the system respond? Because the stress is an increase in P_{H_2}, the system must respond in some way that decreases the partial pressure of hydrogen to counteract the stress. The reaction will therefore proceed to the right as written, consuming H_2 and N_2 and forming additional NH_3. Initially, the partial pressures of H_2 and N_2 will decrease, and the partial pressure of NH_3 will increase until the system eventually reaches a new equilibrium composition, which will have a net increase in P_{H_2}.

We can confirm that this is indeed what will happen by evaluating the reaction quotient Q_p under the new conditions and comparing its value with the value of K_p. The equations used to evaluate K_p and Q_p have the same form: substituting the values after the addition of hydrogen into the expression for Q_p gives

$$Q_p = \frac{(P_{NH_3})^2}{(P_{N_2})(P_{H_2})^3} = \frac{(0.454)^2}{(0.773)(2.319 + 1.00)^3} = 7.29 \times 10^{-3}$$

Thus, $Q_p < K_p$, which tells us that the ratio of products to reactants is lower than at equilibrium. To reach equilibrium, the reaction must proceed to the right as written: the partial pressures of the products will increase, and the partial pressures of the reactants will decrease. Q_p will thereby increase until it equals K_p, and the system will once again be at equilibrium. Changes in the partial pressures of the various substances in the reaction mixture (Equation 15.31) as a function of time are shown in Figure 15.11.

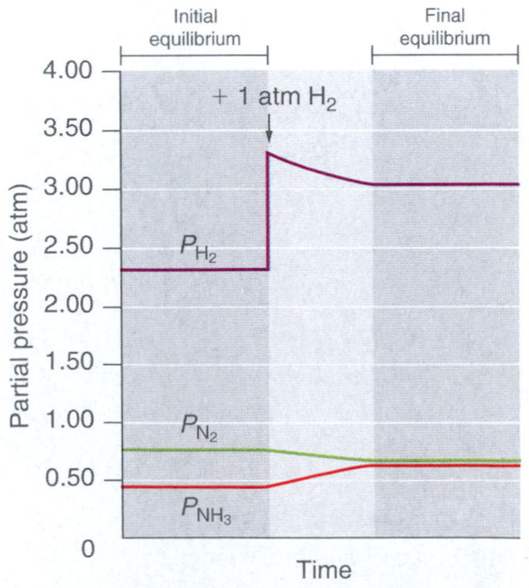

Figure 15.11 The partial pressures of H_2, N_2, and NH_3 as a function of time following the addition of more H_2 to an equilibrium mixture. Some of the added hydrogen is consumed by reaction with nitrogen to produce more ammonia, allowing the system to reach a new equilibrium composition. Interactive Graph

We can force a reaction to go essentially to completion, regardless of the magnitude of K, by continually removing one of the products from the reaction mixture. Consider, for example, the *methanation* reaction, in which hydrogen reacts with carbon monoxide to form methane and water:

$$CO(g) + 3H_2(g) \rightleftharpoons CH_4(g) + H_2O(g) \tag{15.32}$$

This reaction is used for the industrial production of methane, whereas the reverse reaction is used for the production of H_2 (Example 15.14). The reaction quotient Q has the form

$$Q = \frac{[CH_4][H_2O]}{[CO][H_2]^3} \tag{15.33}$$

Regardless of the magnitude of K, if either H_2O or CH_4 can be removed from the reaction mixture so that $[H_2O]$ or $[CH_4]$ is approximately zero, then $Q \approx 0$. In other words, when product is removed, the system is stressed ($Q << K$), and more product will form to counter the stress. Because water (bp = 100°C) is much less volatile than methane, hydrogen, or carbon monoxide (all of which have boiling points below −100°C), passing the gaseous reaction mixture through a cold coil will cause the water vapor to condense to a liquid that can be drawn off. Continuing to remove water from the system forces the reaction to the right as the system attempts to equilibrate, thus enriching the reaction mixture in methane. This technique, referred to as *driving a reaction to completion*, can be used to force a reaction to completion even if K is relatively small. For example, esters are usually synthesized by removing water, the other product of the condensation reaction. In Chapter 19, we will describe the thermodynamic basis for the change in the equilibrium position caused by changes in the concentrations of reaction components.

EXAMPLE 15.16

For each equilibrium system, predict the effect of the indicated stress on the specified quantity:

(a) $2SO_2(g) + O_2(g) \rightleftharpoons 2SO_3(g)$: (1) the effect of removing O_2 on P_{SO_2}, (2) the effect of removing O_2 on P_{SO_3}

(b) $CaCO_3(s) \rightleftharpoons CaO(s) + CO_2(g)$: (1) the effect of removing CO_2 on the amount of $CaCO_3$, (2) the effect of adding $CaCO_3$ on P_{CO_2}

Given Balanced chemical equations and changes

Asked for Effects of indicated stresses

Strategy

Use Q and K to predict the effect of the stress on each reaction.

Solution

(a) (1) Removing O_2 will decrease P_{O_2}, thereby decreasing the denominator in the reaction quotient and making $Q_p > K_p$. The reaction will proceed to the left as written, increasing the partial pressures of SO_2 and O_2 until Q_p once again equals K_p. (2) Removing O_2 will decrease P_{O_2} and thus increase Q_p, so the reaction will proceed to the left. The partial pressure of SO_3 will decrease.

(b) K_p and Q_p are both equal to P_{CO_2}. (1) Removing CO_2 from the system causes more $CaCO_3$ to react to produce CO_2, which increases P_{CO_2} to the partial pressure required by K_p. (2) Adding (or removing) solid $CaCO_3$ has no effect on P_{CO_2} because it does not appear in the expression for K_p (or Q_p).

EXERCISE 15.16

For each equilibrium system, predict the effect that the indicated stress will have upon the specified quantity:

(a) $H_2(g) + CO_2(g) \rightleftharpoons H_2O(g) + CO(g)$: (1) the effect of adding CO on $[H_2]$, (2) the effect of adding CO_2 on $[H_2]$

(b) $CuO(s) + CO(g) \rightleftharpoons Cu(s) + CO_2(g)$: (1) the effect of adding CO on the amount of Cu, (2) the effect of adding CO_2 on [CO]

Answer **(a)** (1) $[H_2]$ increases, (2) $[H_2]$ decreases; **(b)** (1) the amount of Cu increases, (2) [CO] increases

Changes in Total Pressure or Volume

Because liquids are relatively incompressible, changing the pressure above a liquid solution has little effect on the concentrations of dissolved substances. Consequently, changes in external pressure have very little effect on equilibrium systems

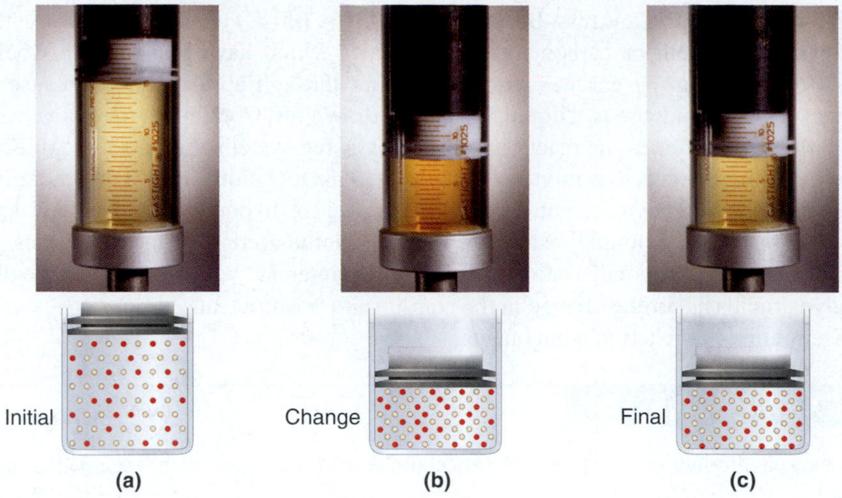

Figure 15.12 The effect of changing the volume (and thus the pressure) of an equilibrium mixture of N_2O_4 and NO_2 at constant temperature. **(a)** The syringe with a total volume of 15 mL contains an equilibrium mixture of N_2O_4 and NO_2; the red-brown color is proportional to the NO_2 concentration. **(b)** If the volume is rapidly decreased by a factor of 2 to 7.5 mL, the initial effect is to double the concentrations of all species present, including NO_2. Hence, the color becomes more intense. **(c)** With time, the system adjusts its composition in response to the stress as predicted by Le Châtelier's principle, forming colorless N_2O_4 at the expense of red-brown NO_2, which decreases the intensity of the color of the mixture.

that contain only solids or liquids. In contrast, because gases are highly compressible, their concentrations vary dramatically with pressure. From the ideal gas law, $PV = nRT$, described in Chapter 11, the concentration C of a gas is related to its pressure as follows:

$$C = \frac{n}{V} = \frac{P}{RT} \tag{15.34}$$

Hence, the concentration of any gaseous reactant or product is directly proportional to the applied pressure P and inversely proportional to the total volume V. Consequently, the equilibrium compositions of systems that contain gaseous substances are quite sensitive to changes in pressure, volume, and T.

These principles can be illustrated using the reversible dissociation of gaseous N_2O_4 to gaseous NO_2 (Equation 15.1). The syringe shown in Figure 15.12 initially contains an equilibrium mixture of colorless N_2O_4 and red-brown NO_2. Decreasing the volume by 50% causes the mixture to become darker because all concentrations have doubled. Decreasing the volume also constitutes a stress, however, as we can see by examining the effect of a change in volume on the reaction quotient Q. At equilibrium, $Q = K = [NO_2]^2/[N_2O_4]$ (Equation 15.10). If the volume is decreased by half, the concentrations of the substances in the mixture are doubled, so the new reaction quotient is

$$Q = \frac{[NO_2]_f^2}{[N_2O_4]_i} = \frac{2([NO_2]_i)^2}{2[N_2O_4]_i} = \frac{4([NO_2]_i)^2}{2[N_2O_4]_i} = 2K \tag{15.35}$$

Because Q is now greater than K, the system is no longer at equilibrium. The stress can be relieved if the reaction proceeds to the left, consuming 2 mol of NO_2 for every 1 mol of N_2O_4 produced. This will decrease the concentration of NO_2 and increase the concentration of N_2O_4, causing Q to decrease until it is once again equal to K. Thus, as shown in Figure 15.12c, the intensity of the brown color due to NO_2 decreases with time following the change in volume.

In general, *if a balanced reaction contains different numbers of gaseous reactant and product molecules, the equilibrium will be sensitive to changes in volume or pressure*. Increasing the pressure on a system (or decreasing the volume) will favor the side of the reaction that has fewer gaseous molecules, and vice versa.

Note the pattern

Increasing the pressure of a system (or decreasing the volume) favors the side of the reaction that has fewer gaseous molecules, and vice versa.

EXAMPLE 15.17

For each equilibrium system, predict the direction of the reaction if the pressure is decreased by a factor of 2 (for example, if the volume is doubled) at constant temperature:

(a) $N_2(g) + 3H_2(g) \rightleftharpoons 2NH_3(g)$
(b) $C_2H_2(g) + C_2H_6(g) \rightleftharpoons 2C_2H_4(g)$
(c) $2NO_2(g) \rightleftharpoons 2NO(g) + O_2(g)$

Given Balanced chemical equations

Asked for Direction of reaction if pressure is halved

Strategy

Use Le Châtelier's principle to predict the effect of the stress.

Solution

(a) Two moles of gaseous products are formed from 4 mol of gaseous reactants. Decreasing the pressure will cause the reaction to shift to the left because that side contains the larger number of moles of gas. Thus, the pressure increases, counteracting the stress.
(b) Two moles of gaseous products form from 2 mol of gaseous reactants. Decreasing the pressure will have no effect on the equilibrium composition because both sides of the balanced equation have the same number of moles of gas.
(c) Three moles of gaseous products are formed from 2 mol of gaseous reactants. Decreasing the pressure will favor the side that contains more moles of gas, so the reaction will shift toward the products to increase the pressure.

EXERCISE 15.17

For each equilibrium system, predict the direction in which the system will shift if the pressure is increased by a factor of 2 (for example, if the volume is halved) at constant temperature:

(a) $H_2O(g) + CO(g) \rightleftharpoons H_2(g) + CO_2(g)$
(b) $H_2(g) + C_2H_4(g) \rightleftharpoons C_2H_6(g)$
(c) $2SO_2(g) + O_2(g) \rightleftharpoons 2SO_3(g)$

Answer (a) no effect; (b) to the right; (c) to the right

Changes in Temperature

In all the cases we have considered so far, the magnitude of the equilibrium constant, K or K_p, was constant. Changes in temperature can, however, change the value of the equilibrium constant without immediately affecting the reaction quotient ($Q \neq K$). In this case, the system is no longer at equilibrium; hence, the composition of the system will change until Q equals K at the new temperature.

To predict how an equilibrium system will respond to a change in temperature, we must know something about the enthalpy change of the reaction, ΔH_{rxn}. As you learned in Chapter 5, heat is released to the surroundings in an exothermic reaction ($\Delta H_{rxn} < 0$), and heat is absorbed from the surroundings in an endothermic reaction ($\Delta H_{rxn} > 0$). We can express these changes in the following way:

$$\text{Exothermic:} \quad \text{reactants} \rightleftharpoons \text{products} + \text{heat} \quad (\Delta H < 0) \tag{15.36}$$

$$\text{Endothermic:} \quad \text{reactants} + \text{heat} \rightleftharpoons \text{products} \quad (\Delta H > 0) \tag{15.37}$$

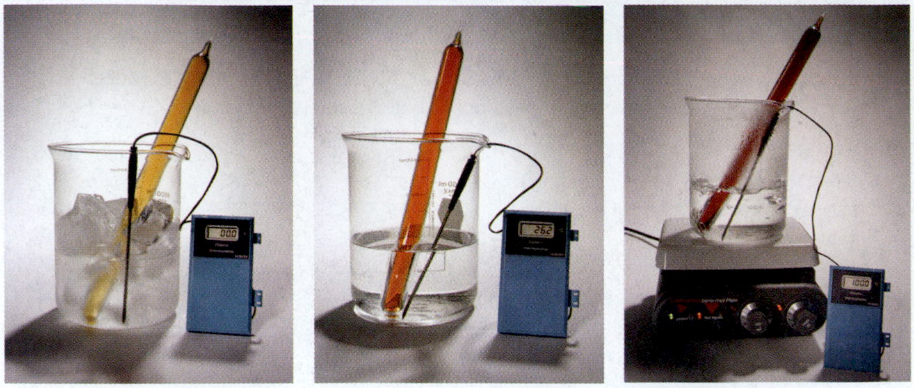

Figure 15.13 The effect of temperature on the equilibrium between gaseous N_2O_4 and NO_2. (*center*) A tube containing a mixture of N_2O_4 and NO_2 in the same proportion at room temperature is red-brown due to the NO_2 present. (*left*) Immersing the tube in ice water causes the mixture to become lighter in color due to a shift in the equilibrium composition toward colorless N_2O_4. (*right*) In contrast, immersing the same tube in boiling water causes the mixture to become darker due to a shift in the equilibrium composition toward the highly colored NO_2.

Thus, heat can be thought of as a product in an exothermic reaction and as a reactant in an endothermic reaction. Increasing the temperature of a system corresponds to adding heat. Le Châtelier's principle predicts that an exothermic reaction will shift to the left (toward the reactants) if the temperature of the system is increased (heat is added). Conversely, an endothermic reaction will shift to the right (toward the products) if the temperature of the system is increased. If a reaction is thermochemically neutral ($\Delta H_{rxn} = 0$), then a change in temperature will not affect the equilibrium composition.

We can examine the effects of temperature on the dissociation of N_2O_4 to NO_2, for which $\Delta H = +58$ kJ/mol. This reaction can be written as

$$58\text{ kJ} + N_2O_4(g) \rightleftharpoons 2NO_2(g) \tag{15.38}$$

Increasing the temperature (adding heat to the system) is a stress that will drive the reaction to the right, as illustrated in Figure 15.13. Thus, increasing the temperature increases the ratio of NO_2 to N_2O_4 at equilibrium, which increases the value of K.

The effect of increasing the temperature on a system at equilibrium can be summarized in the following way: *Increasing the temperature increases the magnitude of the equilibrium constant for an endothermic reaction, decreases the equilibrium constant for an exothermic reaction, and has no effect on the equilibrium constant for a thermally neutral reaction.* Table 15.3 shows the temperature dependence of the equilibrium constants for the synthesis of ammonia from hydrogen and nitrogen, which is an exothermic reaction with $\Delta H^0 = -91.8$ kJ/mol. Note that the values of both K and K_p decrease dramatically with increasing temperature, as predicted for an exothermic reaction.

Note the pattern

Increasing the temperature causes endothermic reactions to favor products and exothermic reactions to favor reactants.

TABLE 15.3 Temperature dependence of the equilibrium constants, K and K_p, for the reaction $N_2(g) + 3H_2(g) \rightleftharpoons 2NH_3(g)$

Temperature, K	K	K_p
298	3.3×10^8	5.6×10^5
400	3.9×10^4	3.6×10^1
450	2.6×10^3	1.9
500	1.7×10^2	1.0×10^{-1}
550	2.6×10^1	1.3×10^{-2}
600	4.1	1.7×10^{-3}

EXAMPLE 15.18

For each equilibrium reaction, predict the effect of decreasing the temperature:

(a) $N_2(g) + 3H_2(g) \rightleftharpoons 2NH_3(g)$, $\Delta H_{rxn} = -91.8$ kJ/mol
(b) $CaCO_3(s) \rightleftharpoons CaO(s) + CO_2(g)$, $\Delta H_{rxn} = 178$ kJ/mol

Given Balanced chemical equations and values of ΔH_{rxn}

Asked for Effects of decreasing temperature

Strategy

Use Le Châtelier's principle to predict the effect of decreasing the temperature on each reaction.

Solution

(a) The formation of NH_3 is exothermic, so we can view heat as one of the products:

$$N_2(g) + 3H_2(g) \rightleftharpoons 2NH_3(g) + 91.8 \text{ kJ}$$

If the temperature of the mixture is decreased, heat (one of the products) is being removed from the system, which causes the equilibrium to shift to the right. Hence, the formation of ammonia is favored at lower temperatures.

(b) The decomposition of calcium carbonate is endothermic, so heat can be viewed as one of the reactants:

$$CaCO_3(s) + 178 \text{ kJ} \rightleftharpoons CaO(s) + CO_2(g)$$

If the temperature of the mixture is decreased, heat (one of the reactants) is being removed from the system, which causes the equilibrium to shift to the left. Hence, the thermal decomposition of calcium carbonate is less favored at lower temperatures.

EXERCISE 15.18

For each equilibrium system, predict the effect of increasing the temperature on the reaction mixture:

(a) $2SO_2(g) + O_2(g) \rightleftharpoons 2SO_3(g)$, $\Delta H_{rxn} = -198$ kJ/mol
(b) $N_2(g) + O_2(g) \rightleftharpoons 2NO(g)$, $\Delta H_{rxn} = +181$ kJ/mol

Answer (a) reaction shifts to the *left*; (b) reaction shifts to the *right*

15.6 • Controlling the Products of Reactions

MGC Effects of Temperature and Pressure on Gaseous Reactions

Whether in the synthetic laboratory or in industrial settings, one of the primary goals of modern chemistry is to control the identity and quantity of the products of chemical reactions. For example, a process aimed at synthesizing ammonia is designed to maximize the amount of ammonia produced using a given amount of energy. Alternatively, other processes may be designed to minimize the creation of undesired products, such as pollutants emitted from an internal combustion engine. To achieve these goals, chemists must consider the competing effects of the reaction conditions that they can control.

One way to get a high yield of a desired compound is to make the rate of the desired reaction much faster than the rates of any other possible reactions that might occur in the system. Altering reaction conditions to control reaction rates, thereby obtaining a single product or set of products, is called **kinetic control**. A second approach, called **thermodynamic control**, consists of adjusting conditions so that at equilibrium only the desired products are present in significant quantities.

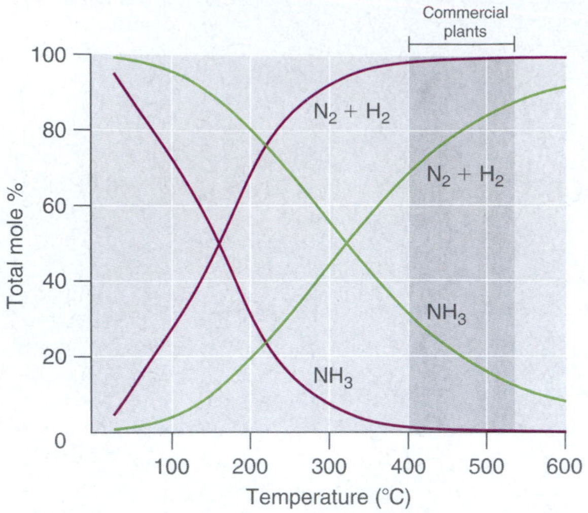

Figure 15.14 Effect of temperature and pressure on the equilibrium composition of two systems that originally contained a 3:1 mixture of hydrogen and nitrogen. At all temperatures, the total pressure in the systems was initially either 4 atm (purple curves) or 200 atm (green curves). Note the dramatic *decrease* in the proportion of NH_3 at equilibrium at higher temperatures in both cases, as well as the large *increase* in the proportion of NH_3 at equilibrium at any temperature for the system at higher pressure (green) versus lower pressure (purple). Commercial plants that use the Haber–Bosch process to synthesize ammonia on an industrial scale operate at temperatures of 400–530°C (indicated by the darker gray band) and total pressures of 130–330 atm.

Interactive Graph

An example of thermodynamic control is the Haber–Bosch process* used to synthesize ammonia via the reaction

$$N_2(g) + 3H_2(g) \rightleftharpoons 2NH_3(g) \qquad \Delta H_{rxn} = -91.8 \text{ kJ/mol} \qquad (15.39)$$

Because the reaction converts 4 mol of gaseous reactants to only 2 mol of gaseous product, Le Châtelier's principle predicts that the formation of NH_3 will be favored when the pressure is increased. The reaction is exothermic, however, ($\Delta H_{rxn} = -91.8$ kJ/mol), so the equilibrium constant decreases with increasing temperature, which causes an equilibrium mixture to contain only relatively small amounts of ammonia at high temperatures (Figure 15.14). Taken together, these considerations suggest that the maximum yield of NH_3 will be obtained if the reaction is carried out at as low a temperature and as high a pressure as possible. Unfortunately, at temperatures lower than approximately 300°C, where the equilibrium yield of ammonia would be relatively high, the reaction is too *slow* to be of any commercial use. The industrial process therefore uses a mixed oxide (Fe_2O_3/K_2O) catalyst that enables the reaction to proceed at a significant rate at temperatures of 400–530°C, where the formation of ammonia is less unfavorable than at higher temperatures.

Because of the low value of the equilibrium constant at high temperatures (for example, $K = 0.039$ at 800 K), there is no way to produce an equilibrium mixture that contains large proportions of ammonia at high temperature. We can, however, control the temperature and the pressure while using a catalyst to convert a fraction of the N_2 and H_2 in the reaction mixture to NH_3, as is done in the Haber–Bosch process. This process also makes use of the fact that the product, ammonia, is less volatile than the reactants. Because NH_3 is a liquid at room temperature at pressures higher than 10 atm, cooling the reaction mixture causes NH_3 to condense from the vapor as liquid ammonia, which is easily separated from unreacted N_2 and H_2. The unreacted gases are recycled until complete conversion of hydrogen and nitrogen to ammonia is eventually achieved. Figure 15.15 is a simplified layout of a Haber–Bosch process plant.

The Sohio acrylonitrile process, in which propene and ammonia react with oxygen to form acrylonitrile, is an example of a kinetically controlled reaction:

$$CH_2{=}CHCH_3(g) + NH_3(g) + \frac{3}{2}O_2(g) \rightleftharpoons CH_2{=}CHC{\equiv}N(g) + 3H_2O(g) \qquad (15.40)$$

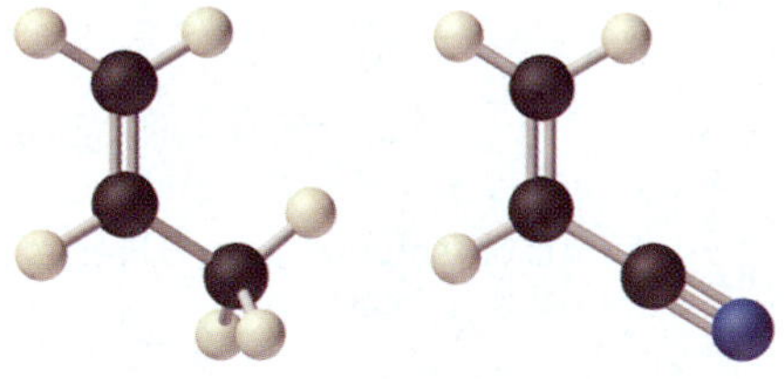

Like most oxidation reactions of organic compounds, this reaction is highly exothermic ($\Delta H^0 = -519$ kJ/mol) and has a very large equilibrium constant ($K = 1.2 \times 10^{94}$). Nonetheless, the reaction shown in Equation 15.40 is not the reaction a chemist would expect to occur when propene or ammonia is heated in the presence of oxygen. Competing combustion reactions that produce CO_2 and N_2 from the reactants, such as those shown below, are even more exothermic and have even larger equilibrium constants, thereby reducing the yield of the desired product, acrylonitrile:

$$CH_2{=}CHCH_3(g) + \frac{9}{2}O_2(g) \rightleftharpoons 3CO_2(g) + 3H_2O(g) \qquad \Delta H^0 = -1926.1 \text{ kJ/mol}, K = 4.5 \times 10^{338} \qquad (15.41)$$

$$2NH_3(g) + 3O_2(g) \rightleftharpoons N_2(g) + 6H_2O(g) \qquad \Delta H^0 = -1359.2 \text{ kJ/mol}, K = 4.4 \times 10^{234} \qquad (15.42)$$

* Karl Bosch (1874–1940) was a German chemical engineer who was responsible for designing the process that took advantage of Fritz Haber's discoveries regarding the $N_2 + H_2/NH_3$ equilibrium to make ammonia synthesis via this route cost effective. He received the Nobel Prize in chemistry in 1931 for his work. The industrial process is called either the Haber process or the Haber–Bosch process.

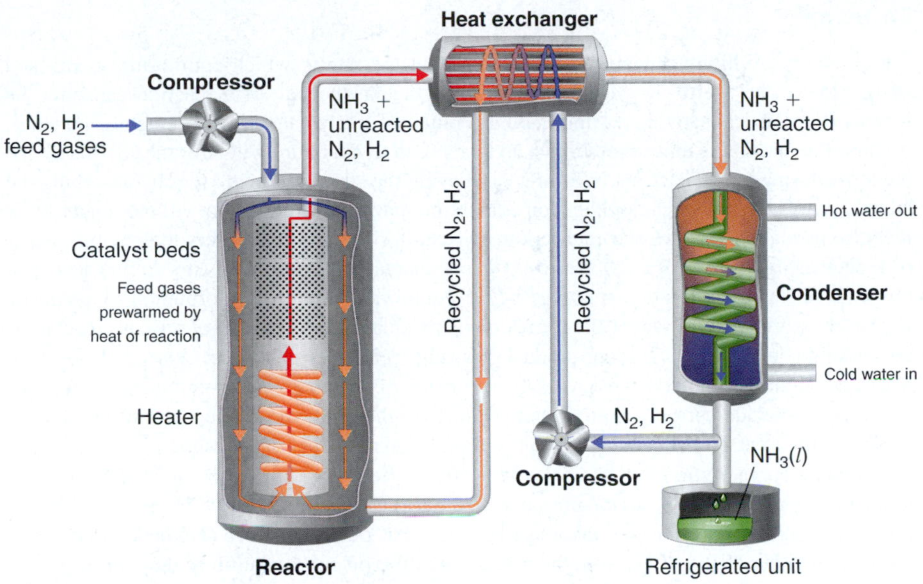

Figure 15.15 A schematic diagram of an industrial plant for the production of ammonia via the Haber–Bosch process. A 3:1 mixture of gaseous H_2 and N_2 is compressed to 130–330 atm, heated to 400–530°C, and passed over an Fe_2O_3/K_2O catalyst, which results in partial conversion to gaseous NH_3. The resulting mixture of gaseous NH_3, H_2, and N_2 is passed through a heat exchanger, which uses the hot gases to prewarm recycled N_2 and H_2, and a condensor to cool the NH_3, giving a liquid that is readily separated from unreacted N_2 and H_2. (Although the normal boiling point of NH_3 is −33°C, the boiling point increases rapidly with increasing pressure, to 20°C at 8.5 atm and 126°C at 100 atm.) The unreacted N_2 and H_2 are recycled to form more NH_3.

In fact, the formation of acrylonitrile in Equation 15.40 is accompanied by the release of approximately 760 kJ/mol of heat due to partial combustion of propene during the reaction.

The Sohio process uses a catalyst that selectively accelerates the rate of formation of acrylonitrile without significantly affecting the rates of the competing combustion reactions. Consequently, acrylonitrile is formed more rapidly than CO_2 and N_2 under the optimized reaction conditions (approximately 1.5 atm and 450°C). The reaction mixture is rapidly cooled to prevent further oxidation or combustion of acrylonitrile, which is then washed out of the vapor with a liquid water spray. Thus, controlling the kinetics of the reaction causes the desired product to be formed under conditions where equilibrium is not established. In industry, this reaction is carried out on an enormous scale. Acrylonitrile is the building block of the polymer called *polyacrylonitrile*, found in all the products referred to collectively as *acrylics*, whose wide range of uses includes the synthesis of fibers woven into clothing and carpets.

EXAMPLE 15.19

Recall that methanation is the reaction of hydrogen with carbon monoxide to form methane and water:

$$CO(g) + 3H_2(g) \rightleftharpoons CH_4(g) + H_2O(g)$$

Notice that this reaction is the reverse of the steam reforming of methane described in Example 15.14. The reaction is exothermic ($\Delta H^0 = -206$ kJ/mol), with an equilibrium constant at room temperature of $K_p = 7.6 \times 10^{24}$. Unfortunately, however, CO and H_2 do not react at an appreciable rate at room temperature. What conditions would you select to maximize the amount of methane formed per unit time by this reaction?

Given Balanced chemical equation and values of ΔH^0 and K

Asked for Conditions to maximize yield of product

Strategy

Consider the effect of changes in temperature and pressure and the addition of an effective catalyst on the rate and equilibrium of the reaction. Determine which combination of reaction conditions will result in the maximum production of methane.

Solution

The products are highly favored at equilibrium, but the rate at which equilibrium is reached is too slow to be useful. You learned in Chapter 14 that the rate of a reaction can often be increased dramatically by increasing the temperature of the reactants. Unfortunately, however, because the reaction is quite exothermic, an increase in temperature will shift the equilibrium to the left, causing more reactants to form and relieving the stress on the system by absorbing the added heat. Thus, if we increase the temperature too much, the equilibrium will no longer favor methane formation. (In fact, the equilibrium constant for this reaction is very temperature sensitive, decreasing to only 1.9×10^{-3} at 1000°C.) To increase the rate, we can try to find a catalyst that will operate at lower temperatures where equilibrium favors the formation of products. Higher pressures will also favor the formation of products because 4 mol of gaseous reactant is converted to only 2 mol of gaseous product. Very high pressures should not be needed, however, because the equilibrium constant favors the formation of products. Thus, optimal conditions for the reaction include carrying it out at temperatures higher than room temperature (but not too high), adding a catalyst, and using pressures higher than atmospheric pressure.

Industrially, catalytic methanation is typically carried out at pressures of 1–100 atm and temperatures of 250–450°C in the presence of a nickel catalyst. (At 425°C, K_p is 3.7×10^3, so the formation of products is still favored.) The synthesis of methane can also be favored by the removal of either H_2O or CH_4 from the reaction mixture by condensation as they form.

EXERCISE 15.19

As you learned in Example 15.10, the water–gas shift reaction is

$$H_2(g) + CO_2(g) \rightleftharpoons H_2O(g) + CO(g) \qquad K_p = 0.106 \text{ at } 700\text{ K}$$

At 700 K, $\Delta H = 41.2$ kJ/mol. What reaction conditions would you use to maximize the yield of carbon monoxide?

Answer high temperatures to increase the reaction rate and favor product formation, a catalyst to increase the reaction rate, and atmospheric pressure because the equilibrium will not be greatly affected by pressure

SUMMARY AND KEY TERMS

15.1 The Concept of Chemical Equilibrium (p. 673)

Chemical equilibrium is a dynamic process consisting of forward and reverse reactions that proceed at equal rates. At equilibrium, the composition of the system no longer changes with time. The composition of an equilibrium mixture is independent of the direction from which equilibrium is approached.

15.2 The Equilibrium Constant (p. 675)

The ratio of the rate constants for the forward and reverse reactions at equilibrium is the **equilibrium constant (K)**, a unitless quantity. The composition of the equilibrium mixture is therefore determined by the magnitudes of the forward and reverse rate constants at equilibrium. Under a given set of conditions, a reaction will always have the same value of K. For a system at equilibrium, the **law of mass action** relates K to the ratio of the equilibrium concentrations of the products to the concentrations of the reactants raised to their respective exponents to match the coefficients in the **equilibrium equation**. The ratio is called the **equilibrium constant expression**. When a reaction is written in the reverse direction, K and the equilibrium constant expression are inverted. For gases, the equilibrium constant expression can be written as the ratio of the partial pressures of the products to the partial pressures of the reactants, each raised to its coefficient in the chemical equation. An equilibrium constant calculated from partial pressures, K_p, is related to K by the ideal gas constant R, the temperature T, and the change in the number of moles of gas during the reaction. An equilibrium system that contains products and reactants in a single phase is a **homogeneous equilibrium**, whereas a system whose reactants, products, or both are in more than one phase is a **heterogeneous equilibrium**. When a reaction can be expressed as the sum of two or more reactions, its equilibrium constant is equal to the product of the equilibrium constants for the individual reactions.

15.3 Solving Equilibrium Problems (p. 686)

When an equilibrium constant is calculated from equilibrium concentrations, molar concentrations or partial pressures are substituted into the equilibrium constant expression for the reaction. Equilibrium constants can be used to calculate the equilibrium concentrations of reactants and products by using the quantities or concentrations of the reactants, the stoichiometry of the balanced equation for the reaction, and a tabular format to obtain the final concentrations of all species at equilibrium.

15.4 Nonequilibrium Conditions (p. 697)

The **reaction quotient** (**Q or Q_p**) has the same form as the equilibrium constant expression, but it is derived from concentrations obtained at any time. When a reaction system is at equilibrium, $Q = K$. Graphs derived by plotting a few equilibrium concentrations for a system at a given temperature and pressure can be used to predict the direction in which a reaction will proceed. Points that do not lie on the line or curve represent nonequilibrium states, and the system will adjust, if it can, to achieve equilibrium. **Le Châtelier's principle** states that if a stress is applied to a system at equilibrium, the composition of the system will adjust to counteract the stress.

15.5 Factors That Affect Equilibrium (p. 701)

Three types of stresses can alter the composition of an equilibrium system: adding or removing reactants or products, changing the total pressure or volume, and changing the temperature of the system. A reaction with an unfavorable equilibrium constant can be driven to completion by continually removing one of the products of the reaction. Equilibria that contain different numbers of gaseous reactant and product molecules are sensitive to changes in volume or pressure; higher pressures favor the side with fewer gaseous molecules. Removing heat from an exothermic reaction favors the formation of products, whereas removing heat from an endothermic reaction favors the formation of reactants.

15.6 Controlling the Products of Reactions (p. 707)

Changing conditions to affect the *rates* of reactions in order to obtain a single product is called **kinetic control** of the system. In contrast, **thermodynamic control** is adjusting conditions to ensure that at equilibrium only the desired product or products are present in significant concentrations.

KEY EQUATIONS

Definition of equilibrium constant in terms of forward and reverse rate constants	$K = \frac{k_f}{k_r}$	(15.3)
Equilibrium constant expression (law of mass action)	$K = \frac{[C]^c\,[D]^d}{[A]^a[B]^b}$	(15.5)
Equilibrium constant expression for reactions involving gases using partial pressures	$K_p = \frac{(P_C)^c\,(P_D)^d}{(P_A)^a\,(P_B)^b}$	(15.14)
Relationship between K_p and K	$K_p = K(RT)^{\Delta n}$	(15.16)
Reaction quotient	$Q = \frac{[C]^c[D]^d}{[A]^a[B]^b}$	(15.25)

QUESTIONS AND PROBLEMS

Please be sure you are familiar with the topics discussed in Essential Skills 7 at the end of this chapter before proceeding to the Questions and Problems.

Questions and Problems with colored numbers have answers in the Appendix and complete solutions in the Student Solutions Manual.

CONCEPTUAL

15.1 The Concept of Chemical Equilibrium

1. What is meant when a reaction is described as "having reached equilibrium"? What does this statement mean regarding the rates of the forward and reverse reactions? What does this statement mean regarding the concentrations or amounts of the reactants and products?
2. Is it correct to say that the reaction has "stopped" when it has reached equilibrium? Explain your answer and support it with a specific example.
3. Why is chemical equilibrium described as a dynamic process? Describe this process in the context of a saturated solution of NaCl in water. What is occurring on a microscopic level? What is happening on a macroscopic level?
4. Which of these systems exist(s) in a state of chemical equilibrium? (a) oxygen and hemoglobin in the human circulatory system; (b) iodine crystals in an open beaker; (c) the combustion of wood; (d) the amount of ^{14}C in a decomposing organism

15.2 The Equilibrium Constant

5. For an equilibrium reaction, what effect does reversing the reactants and products have on the value of the equilibrium constant?

6. Which of the following equilibria are homogeneous and which are heterogeneous?
(a) $2HF(g) \rightleftharpoons H_2(g) + F_2(g)$
(b) $C(s) + 2H_2(g) \rightleftharpoons CH_4(g)$
(c) $H_2C{=}CH_2(g) + H_2(g) \rightleftharpoons C_2H_6(g)$
(d) $2Hg(l) + O_2(g) \rightleftharpoons 2HgO(s)$

7. Classify each equilibrium system as being either homogeneous or heterogeneous:
(a) $NH_4NH_2CO_2(s) \rightleftharpoons 2NH_3(g) + CO_2(g)$
(b) $C(s) + O_2(g) \rightleftharpoons CO_2(g)$
(c) $2Mg(s) + O_2(g) \rightleftharpoons 2MgO(s)$
(d) $AgCl(s) \rightleftharpoons Ag^+(aq) + Cl^-(aq)$

8. If an equilibrium reaction is endothermic, what happens to the equilibrium constant if the temperature of the reaction is increased? If the temperature is decreased?

9. Industrial production of NO by the reaction $N_2(g) + O_2(g) \rightleftharpoons 2NO(g)$ is carried out at elevated temperatures in order to drive the reaction toward the formation of product. After sufficient product has formed, the reaction mixture is quickly cooled. Why?

10. How would you differentiate between a system that has reached chemical equilibrium and one that is reacting so slowly that changes in concentration are difficult to observe?

11. What is the relationship among the equilibrium constant K, the concentration of each component of the system, and the rate constants for the forward and reverse reactions?

12. Write expressions for K and K_p for each of these reactions:
(a) $CO(g) + H_2O(g) \rightleftharpoons CO_2(g) + H_2(g)$
(b) $PCl_3(g) + Cl_2(g) \rightleftharpoons PCl_5(g)$
(c) $2O_3(g) \rightleftharpoons 3O_2(g)$

13. Write expressions for K and K_p for each reaction:
(a) $2NO(g) + O_2(g) \rightleftharpoons 2NO_2(g)$
(b) $\frac{1}{2}H_2(g) + \frac{1}{2}I_2(g) \rightleftharpoons HI(g)$
(c) *cis*-stilbene $\rightleftharpoons$ *trans*-stilbene

14. Why is it incorrect to state that pure liquids, pure solids, and solvents are not part of an equilibrium constant expression?

15. Write expressions for K and K_p for each equilibrium reaction:
(a) $2S(s) + 3O_2(g) \rightleftharpoons 2SO_3(g)$
(b) $C(s) + CO_2(g) \rightleftharpoons 2CO(g)$
(c) $2ZnS(s) + 3O_2(g) \rightleftharpoons 2ZnO(s) + 2SO_2(g)$

16. Write expressions for K and K_p for each equilibrium reaction:
(a) $2HgO(s) \rightleftharpoons 2Hg(l) + O_2(g)$
(b) $H_2(g) + I_2(s) \rightleftharpoons 2HI(g)$
(c) $NH_4CO_2NH_2(s) \rightleftharpoons 2NH_3(g) + CO_2(g)$

17. At room temperature, the equilibrium constant of the reaction $2A(g) \rightleftharpoons B(g)$ is 1. What does this indicate about the concentrations of A and B at equilibrium? Would you expect K and K_p to vary significantly from each other? If so, how would their difference be affected by temperature?

18. For a certain series of reactions, if $[OH^-][HCO_3^-]/[CO_3^{2-}] = K_1$ and $[OH^-][H_2CO_3]/[HCO_3^-] = K_2$, what is the equilibrium expression for the overall reaction? Write the overall equilibrium equation.

19. In the equation for an enzymatic reaction, ES represents the complex formed between the substrate S and the enzyme protein E. In the final step of the oxidation reaction shown below, the product P dissociates from the ESO2 complex, which regenerates the active enzyme:

$$E + S \rightleftharpoons ES \qquad K_1$$
$$ES + O_2 \rightleftharpoons ESO_2 \qquad K_2$$
$$ESO_2 \rightleftharpoons E + P \qquad K_3$$

Give the overall reaction equation, and show that $K = K_1 \times K_2 \times K_3$.

15.3 Solving Equilibrium Problems

20. Calculations involving systems with very small or very large equilibrium constants can be dramatically simplified by making certain assumptions about the concentrations of products and reactants. What are these assumptions (a) when K is very large and (b) when K is very small? Illustrate this technique using the system $A + 2B \rightleftharpoons C$ for which you are to calculate the concentration of the product at equilibrium starting with only A and B. Under what circumstances should simplifying assumptions not be used?

15.4 Nonequilibrium Conditions

21. During a set of experiments, graphs were drawn of [reactants] versus [products] at equilibrium. Using Figures 15.8 and 15.9 as your guide, sketch the shape of each graph using appropriate labels.
(a) $H_2O(l) \rightleftharpoons H_2O(g)$
(b) $2Mg(s) + O_2(g) \rightleftharpoons 2MgO(s)$
(c) $2O_3(g) \rightleftharpoons 3O_2(g)$
(d) $2PbS(s) + 3O_2(g) \rightleftharpoons 2PbO(s) + 2SO_2(g)$

22. Write an equilibrium constant expression for each reaction system. Given the indicated changes, how must the concentration of the species in red type change if the system is to maintain equilibrium?
(a) $2NaHCO_3(s) \rightleftharpoons Na_2CO_3(s) + CO_2(g) + H_2O(g)$; $[CO_2]$ is doubled.
(b) $N_2F_4(g) \rightleftharpoons 2NF_2(g)$; $[NF_2]$ is decreased by a factor of 2.
(c) $H_2(g) + I_2(g) \rightleftharpoons 2HI(g)$; $[I_2]$ is doubled.

23. Write an equilibrium constant expression for each reaction system. Given the indicated changes, how must the concentration of the species in red type change if the system is to maintain equilibrium?
(a) $CS_2(g) + 4H_2(g) \rightleftharpoons CH_4(g) + 2H_2S(g)$; $[CS_2]$ is doubled.
(b) $PCl_5(g) \rightleftharpoons PCl_3(g) + Cl_2(g)$; $[Cl_2]$ is decreased by a factor of 2.
(c) $4NH_3(g) + 5O_2(g) \rightleftharpoons 4NO(g) + 6H_2O(g)$; [NO] is doubled.

15.5 Factors That Affect Equilibrium

24. If an equilibrium reaction is endothermic in the forward direction, what is the expected change in the concentration of each component of the system if the temperature of the reaction is increased? If the temperature is decreased?

25. Write the equilibrium equation for the system $4NH_3(g) + 5O_2(g) \rightleftharpoons 4NO(g) + 6H_2O(g)$. Would you expect equilibrium to shift toward the products or reactants with an increase in pressure? Why?

26. The rate of a reaction approximately doubles for every 10°C rise in temperature. What happens to K?

27. The formation of $A_2B_2(g)$ via the equilibrium reaction $2AB(g) \rightleftharpoons A_2B_2(g)$ is exothermic. What happens to the ratio k_f/k_r if the temperature is increased? If both temperature and pressure are increased?

28. In each system, predict the effect that the indicated change will have on the specified quantity at equilibrium:
(a) $H_2(g) + I_2(g) \rightleftharpoons 2HI(g)$
H_2 is removed; what is the effect on P_{I_2}?
(b) $2NOBr(g) \rightleftharpoons 2NO(g) + Br_2(g)$
Br_2 is removed; what is the effect on P_{NOBr}?
(c) $2NaHCO_3(s) \rightleftharpoons Na_2CO_3(s) + CO_2(g) + H_2O(g)$
CO_2 is removed; what is the effect on P_{NaHCO_3}?

29. What effect will the indicated change have on the specified quantity at equilibrium?
(a) $NH_4Cl(s) \rightleftharpoons NH_3(g) + HCl(g)$
NH_4Cl is increased; what is the effect on P_{HCl}?
(b) $H_2O(g) \rightleftharpoons 2H_2(g) + O_2(g)$
O_2 is added; what is the effect on P_{H_2}?
(c) $PCl_3(g) + Cl_2(g) \rightleftharpoons PCl_5(g)$
Cl_2 is removed; what is the effect on P_{PCl_5}?

15.6 Controlling the Products of Reactions

30. A reaction mixture will produce either product A or B depending on the reaction pathway. In the absence of a catalyst, product A is formed; in the presence of a catalyst, product B is formed. What conclusions can you draw about the forward and reverse rates of the reaction that produces A versus the reaction that produces B in (a) the absence of a catalyst and (b) the presence of a catalyst?

31. Describe how you would design an experiment to determine the equilibrium constant for the synthesis of ammonia: $N_2(g) + 3H_2(g) \rightleftharpoons 2NH_3(g)$. The forward reaction is exothermic ($\Delta H^\circ = -91.8$ kJ). What effect would an increase in temperature have on the equilibrium constant?

32. What effect does a catalyst have on the equilibrium position of a reaction? What effect does a catalyst have on the rate at which equilibrium is reached? What effect does a catalyst have on the equilibrium constant?

33. How can the ratio Q/K be used to determine in which direction a reaction will proceed to reach equilibrium?

34. Industrial reactions are frequently run under conditions in which competing reactions can occur. Explain how a catalyst can be used to achieve reaction selectivity. Does the ratio Q/K for the selected reaction change in the presence of a catalyst?

NUMERICAL

This section includes "paired problems" (marked by brackets) that require similar problem-solving skills.

15.2 The Equilibrium Constant

35. Explain what each of the following values for K tells you about the relative concentrations of reactants versus products in a given equilibrium reaction: $K = 0.892$; $K = 3.25 \times 10^8$; $K = 5.26 \times 10^{-11}$. Are products or reactants favored at equilibrium?

36. Write the equilibrium constant expression for each reaction:
(a) $N_2O_4(g) \rightleftharpoons 2NO_2(g)$
(b) $\frac{1}{2}N_2O_4(g) \rightleftharpoons NO_2(g)$
Are these equilibrium constant expressions equivalent? Explain.

37. Write the equilibrium constant expression for each reaction:
(a) $\frac{1}{2}N_2(g) + \frac{3}{2}H_2(g) \rightleftharpoons NH_3(g)$
(b) $\frac{1}{3}N_2(g) + H_2(g) \rightleftharpoons \frac{2}{3}NH_3(g)$
How are these two expressions mathematically related to the equilibrium constant expression for
$N_2(g) + 3H_2(g) \rightleftharpoons 2NH_3(g)$?

38. Write an equilibrium constant expression for each reaction:
(a) $C(s) + 2H_2O(g) \rightleftharpoons CO_2(g) + 2H_2(g)$
(b) $SbCl_3(g) + Cl_2(g) \rightleftharpoons SbCl_5(g)$
(c) $2O_3(g) \rightleftharpoons 3O_2(g)$

39. Give an equilibrium constant expression for each reaction:
(a) $2NO(g) + O_2(g) \rightleftharpoons 2NO_2(g)$
(b) $\frac{1}{2}H_2(g) + \frac{1}{2}I_2(g) \rightleftharpoons HI(g)$
(c) $CaCO_3(s) + 2HOCl(aq) \rightleftharpoons Ca^{2+}(aq) + 2OCl^-(aq) + 2H_2O(l) + CO_2(g)$

40. Calculate K and K_p for each reaction:
(a) $2NOBr(g) \rightleftharpoons 2NO(g) + Br_2(g)$; at 727°C, the equilibrium concentration of NO is 1.29 M, Br_2 is 10.52 M, and NOBr is 0.423 M.
(b) $C(s) + CO_2(g) \rightleftharpoons 2CO(g)$; at 1200 K, a 2.00-L vessel at equilibrium has partial pressures of 93.5 atm CO_2 and 76.8 atm CO, and the vessel contains 3.55 g of carbon.

41. Calculate K and K_p for each reaction:
(a) $N_2O_4(g) \rightleftharpoons 2NO_2(g)$; at the equilibrium temperature of −40°C, a 0.150 M sample of N_2O_4 undergoes a 0.456% decomposition.
(b) $CO(g) + 2H_2(g) \rightleftharpoons CH_3OH(g)$; an equilibrium is reached at 227°C in a 15.5-L reaction vessel with a total pressure of 6.71×10^2 atm and is found to contain 37.8 g of hydrogen gas, 457.7 g of carbon monoxide, and 7193 g of methanol.

42. Determine the equilibrium constant expression, K, and K_p (where applicable) for each reaction:
(a) $2H_2S(g) \rightleftharpoons 2H_2(g) + S_2(g)$; at 1065°C, an equilibrium mixture consists of 1.00×10^{-3} M H_2, 1.20×10^{-3} M S_2, and 3.32×10^{-3} M H_2S.
(b) $Ba(OH)_2(s) \rightleftharpoons 2OH^-(aq) + Ba^{2+}(aq)$; at 25°C, a 250-mL beaker contains 0.330 mol of barium hydroxide in equilibrium with 0.0267 mol of barium ions and 0.0537 mol of hydroxide ions.

43. Determine the equilibrium constant expression, K, and K_p for each reaction:
(a) $2NOCl(g) \rightleftharpoons 2NO(g) + Cl_2(g)$; at 500 K, a 24.3-m$M$ sample of NOCl has decomposed, leaving an equilibrium mixture that contains 72.7% of the original amount of NOCl.
(b) $Cl_2(g) + PCl_3(g) \rightleftharpoons PCl_5(g)$; at 250°C, a 500-mL reaction vessel contains 16.9 g of chlorine gas, 0.500 g of PCl_3, and 10.2 g of PCl_5 at equilibrium.

44. The equilibrium constant expression for a reaction is $[CO_2]^2/[SO_2]^2[O_2]$. What is the balanced chemical equation for the overall reaction if one of the reactants is $Na_2CO_3(s)$?

45. The equilibrium constant expression for a reaction is $[NO][H_2O]^{3/2}/[NH_3][O_2]^{5/4}$. What is the balanced chemical equation for the overall reaction?

46. Given $K = k_f/k_r$, what happens to the magnitude of the equilibrium constant if the rate of the forward reaction is doubled? If the rate of the reverse reaction for the overall reaction is decreased by a factor of 3?

47. The value of the equilibrium constant for the reaction $2H_2(g) + S_2(g) \rightleftharpoons 2H_2S(g)$ is 1.08×10^7 at 700°C.

What is the value of the equilibrium constant for the following related reactions?
(a) $H_2(g) + \frac{1}{2}S_2(g) \rightleftharpoons H_2S(g)$
(b) $4H_2(g) + 2S_2(g) \rightleftharpoons 4H_2S(g)$
(c) $H_2S(g) \rightleftharpoons H_2(g) + \frac{1}{2}S_2(g)$

15.3 Solving Equilibrium Problems

48. In the equilibrium reaction $A + B \rightleftharpoons C$, what happens to the value of the equilibrium constant K if the concentrations of the reactants are doubled? If they are tripled? Can the same be said about the equilibrium reaction $2A \rightleftharpoons B + C$?

49. The table shows reported values of the equilibrium P_{O_2} at three temperatures for the reaction $Ag_2O(s) \rightleftharpoons 2Ag(s) + \frac{1}{2}O_2(g)$, for which $\Delta H^0 = 31$ kJ/mol. Are these data consistent with what you would expect to occur? Why or why not?

T, °C	P_{O_2}, mmHg
150	182
184	143
191	126

50. Given the equilibrium system $N_2O_4(g) \rightleftharpoons 2NO_2(g)$, what happens to K_p if the initial concentration of N_2O_4 is doubled? If K_p is 1.7×10^{-1} at 2300°C, and the system initially contains 100% N_2O_4 at a pressure of 2.6×10^2 atm, what is the equilibrium pressure of each component?

51. At 430°C, 4.20 mol of HI in a 9.60-L reaction vessel reaches equilibrium according to the equation $H_2(g) + I_2(g) \rightleftharpoons 2HI(g)$. At equilibrium, $[H_2] = 0.047\ M$ and $[HI] = 0.345\ M$. What are the values of K and K_p for this reaction?

52. Methanol, a liquid used as an automobile fuel additive, is commercially produced from carbon monoxide and hydrogen at 300°C according to the reaction $CO(g) + 2H_2(g) \rightleftharpoons CH_3OH(g)$, for which $K_p = 1.3 \times 10^{-4}$. If 56.0 g of CO is mixed with excess hydrogen in a 250-mL flask at this temperature, and the hydrogen pressure is continuously maintained at 100 atm, what would be the maximum percent yield of methanol? What pressure of hydrogen would be required to obtain a minimum yield of methanol of 95% under these conditions?

53. If the total equilibrium pressure is 0.969 atm for the reaction $A(s) \rightleftharpoons 2B(g) + C(g)$, what is the value of K_p?

54. The decomposition of ammonium carbamate to ammonia and CO_2 at 40°C is written as $NH_4CO_2NH_2(s) \rightleftharpoons 2NH_3(g) + CO_2(g)$. If the partial pressure of NH_3 at equilibrium is 0.242 atm, what is the equilibrium partial pressure of CO_2? What is the total gas pressure of the system? What is K_p?

55. At 375 K, K_p for the reaction $SO_2Cl_2(g) \rightleftharpoons SO_2(g) + Cl_2(g)$ is 2.4, with pressures expressed in atmospheres. At 303 K, K_p is 2.9×10^{-2}.
(a) What is the equilibrium constant, K, for the reaction at each temperature?
(b) If a sample at 375 K has 0.100 M Cl_2 and 0.200 M SO_2 at equilibrium, what is the concentration of SO_2Cl_2?
(c) If the sample given in part b is cooled to 303 K, what is the pressure inside the bulb?

56. For the gas-phase reaction $aA \rightleftharpoons bB$, show that $K_p = K(RT)^{\Delta n}$ assuming ideal gas behavior.

57. For the gas-phase reaction $I_2 \rightleftharpoons 2I$, show that the total pressure is related to the equilibrium concentration by the equation $P_T = \sqrt{K_p P_{I_2}} + P_{I_2}$.

58. Experimental data on the system $Br_2(l) \rightleftharpoons Br_2(aq)$ are given in the table. Graph the moles of $Br_2(l)$ present versus $[Br_2]$, and then write the equilibrium equation and determine the value of K.

Grams Br_2 in 100 mL water	$[Br_2]$, M
1.0	0.0626
2.5	0.156
3.0	0.188
4.0	0.219
4.5	0.219

59. Data accumulated for the reaction n-butane $\rightleftharpoons$ isobutane at equilibrium are shown in the table. What is the equilibrium constant for this conversion? If 1 mol of n-butane is allowed to equilibrate under the same reaction conditions, what is the final number of moles each of n-butane and isobutane?

Moles n-butane	Moles isobutane
0.5	1.25
1.0	2.5
1.50	3.75

60. Solid ammonium carbamate, $NH_4CO_2NH_2$, dissociates completely to ammonia and carbon dioxide when it vaporizes:

$$NH_4CO_2NH_2(s) \rightleftharpoons 2NH_3(g) + CO_2(g)$$

At 25°C, the total pressure of the gases in equilibrium with the solid is 0.116 atm. What is the equilibrium partial pressure of each gas? What is K_p? If the concentration of CO_2 is doubled, and then equilibrates to its initial equilibrium partial pressure $+x$ atm, what change in the NH_3 concentration is necessary for the system to restore equilibrium?

61. The equilibrium constant for the reaction $COCl_2(g) \rightleftharpoons CO(g) + Cl_2(g)$ is $K_p = 2.2 \times 10^{-10}$ at 100°C. If the initial concentration of $COCl_2$ is 3.05×10^{-3} M, what is the partial pressure of each gas at equilibrium at 100°C? What assumption can be made to simplify your calculations?

62. Aqueous dilution of IO_4^- results in the reaction $IO_4^-(aq) + 2H_2O(l) \rightleftharpoons H_4IO_6^-(aq)$, for which $K = 3.5 \times 10^{-2}$. If you begin with 50 mL of a 0.896 M solution of IO_4^- that is diluted to 250 mL with water, how many moles of $H_4IO_6^-$ are formed at equilibrium?

63. Iodine and bromine react to form IBr, which then sublimes. At 184.4°C, the overall reaction proceeds according to the equation

$$I_2(g) + Br_2(g) \rightleftharpoons 2IBr(g) \qquad K_p = 1.2 \times 10^2$$

If you begin the reaction with 7.4 g of I_2 vapor and 6.3 g of Br_2 vapor in a 1.00 L container, what is the concentration of $IBr(g)$ at equilibrium? What is the partial pressure of each gas at equilibrium? What is the total pressure of the system?

64. For the reaction $C(s) + \frac{1}{2}N_2(g) + \frac{5}{2}H_2(g) \rightleftharpoons CH_3NH_2(g)$, $K = 1.8 \times 10^{-6}$. If you begin the reaction with 1.0 mol of N_2, 2.0 mol of H_2, and sufficient $C(s)$ in a 2.00-L container, what are the concentrations of N_2 and CH_3NH_2 at equilibrium? What happens to K if the concentration of H_2 is doubled?

15.4 Nonequilibrium Conditions

65. The data listed in the table were collected at 450°C for the reaction $N_2(g) + 3H_2(g) \rightleftharpoons 2NH_3(g)$:

	Equilibrium Partial Pressure, atm		
***P*, atm**	**NH_3**	**N_2**	**H_2**
30 (equilibrium)	1.740	6.588	21.58
100	15.20	19.17	65.13
600	321.6	56.74	220.8

The reaction equilibrates at 30 atm pressure. The pressure on the system is first increased to 100 atm and then to 600 atm. Is the system at equilibrium at each of these higher pressures? If not, in which direction will the reaction proceed in order to reach equilibrium?

66. For the reaction $2A \rightleftharpoons B + 3C$, K at 200°C is 2.0. A 6.00-L flask was used to carry out the reaction at this temperature. Given the experimental data in the table, all at 200°C, when the data for each experiment were collected, was the reaction at equilibrium? If it was not at equilibrium, in which direction will the reaction proceed?

Experiment	A	B	C
1	2.50 *M*	2.50 *M*	2.50 *M*
2	1.30 atm	1.75 atm	14.15 atm
3	12.61 mol	18.72 mol	6.51 mol

67. The following two reactions are carried out at 823 K:

$$CoO(s) + H_2(g) \rightleftharpoons Co(s) + H_2O(g) \qquad K = 67$$
$$CoO(s) + CO(g) \rightleftharpoons Co(s) + CO_2(g) \qquad K = 490$$

(a) Write the equilibrium expression for each reaction.
(b) Calculate the partial pressure of both gaseous components at equilibrium in each of the two reactions, if a 1.00 L reaction vessel initially contains 0.316 mol of H_2 or CO plus 0.500 mol CoO.
(c) Using the information provided, calculate K_p for the reaction $H_2(g) + CO_2(g) \rightleftharpoons CO(g) + H_2O(g)$
(d) Describe the shape of the graphs of [reactants] versus [products] as the amount of CoO changes.

68. Hydrogen iodide, HI, is synthesized via the reaction $H_2(g) + I_2(g) \rightleftharpoons 2HI(g)$, for which $K_p = 54.5$ at 425°C. Given a 2.0-L vessel containing 1.12×10^{-2} mol of H_2 and 1.8×10^{-3} mol of I_2 at equilibrium, what is the concentration of HI? Excess hydrogen is added to the vessel, so that the vessel now contains 3.64×10^{-1} mol of H_2. Calculate Q and then predict the direction in which the reaction will proceed. What are the new equilibrium concentrations?

15.5 Factors That Affect Equilibrium

69. For each equilibrium reaction, describe how Q and K change when (1) the pressure is increased; (2) the temperature is increased; (3) the volume of the system is increased; and (4) the concentration(s) of the reactant(s) is increased.
(a) $A(g) \rightleftharpoons B(g)$ $\qquad \Delta H = -20.6$ kJ/mol
(b) $2A(g) \rightleftharpoons B(g)$ $\qquad \Delta H = 0.3$ kJ/mol
(c) $A(g) + B(g) \rightleftharpoons 2C(g)$ $\qquad \Delta H = 46$ kJ/mol

70. For each reaction, describe how Q and K change when (1) the pressure in decreased; (2) the temperature is increased; (3) the volume of the system is decreased; (d) the concentration(s) of the reactant(s) is increased.
(a) $2A(g) \rightleftharpoons B(g)$ $\qquad \Delta H = -80$ kJ/mol
(b) $A(g) \rightleftharpoons 2B(g)$ $\qquad \Delta H = -18$ kJ/mol
(c) $2A(g) \rightleftharpoons 2B(g) + C(g)$ $\qquad \Delta H = 1.2$ kJ/mol

71. Le Châtelier's principle states that a system will change its composition to counteract stress. For the system $CO(g) + Cl_2(g) \rightleftharpoons COCl_2(g)$, write the equilibrium constant expression, K_p. What changes in the values of Q and K would you anticipate when (a) the volume is doubled; (b) the pressure is increased by a factor of 2; and (c) $COCl_2$ is removed from the system?

72. For the equilibrium system $3O_2(g) \rightleftharpoons 2O_3(g)$, $\Delta H^0 = 284$ kJ, write the equilibrium constant expression, K_p. What happens to the values of Q and K if the reaction temperature is increased? What happens to these values if both the temperature and pressure are increased?

73. Carbon and oxygen react to form CO_2 gas via the reaction $C(s) + O_2(g) \rightleftharpoons CO_2(g)$, for which $K = 1.2 \times 10^{69}$. Would you expect K to increase or decrease if the volume of the system were tripled? Why?

74. The reaction $COCl_2(g) \rightleftharpoons CO(g) + Cl_2(g)$ has $K = 2.2 \times 10^{-10}$ at 100°C. Starting with an initial P_{COCl_2} of 1.0 atm, you determine the following values of P_{CO} at three successive time intervals: 6.32×10^{-6} atm, 1.78×10^{-6} atm, and 1.02×10^{-5} atm. Based on these data, in which direction will the reaction proceed after each measurement? If chlorine gas is added to the system, what will be the effect on Q?

75. The table lists experimentally determined partial pressures at three temperatures for the reaction $Br_2(g) \rightleftharpoons 2Br(g)$.

T, K	1123	1173	1273
P_{Br_2} (atm)	3.000	0.3333	6.755×10^{-2}
P_{Br} (atm)	3.477×10^{-2}	2.159×10^{-2}	2.191×10^{-2}

Is this an endothermic or an exothermic reaction? Explain your reasoning.

76. The dissociation of water vapor proceeds according to the reaction $H_2O(g) \rightleftharpoons \frac{1}{2}O_2(g) + H_2(g)$. At 1300 K, there is 0.0027% dissociation, whereas at 2155 K, the dissociation is 1.18%. Calculate K and K_p. Is this an endothermic or an exothermic reaction? How do the magnitudes of the two equilibria compare? Would increasing the pressure improve the yield of H_2 gas at either temperature? (*Hint:* Assume that the system initially contains 1.00 mol of H_2O in a 1.00-L container.)

77. When 1.33 mol of CO_2 and 1.33 mol of H_2 are mixed in a 0.750-L container and are heated to 395°C, they react according to the equation $CO_2(g) + H_2(g) \rightleftharpoons CO(g) + H_2O(g)$. If $K = 0.802$, what are the equilibrium concentrations of each component of the equilibrium mixture? What happens to the value of K if H_2O is removed during the course of the reaction?

78. The equilibrium reaction $H_2(g) + Br_2(g) \rightleftharpoons 2HBr(g)$ has an equilibrium constant of 2.2×10^9. If you begin with 2.0 mol of

Br_2 and 2.0 mol of H_2 in a 5.0-L container, what is the partial pressure of HBr at equilibrium? What is the partial pressure of H_2 at equilibrium? If H_2 is removed from the system, what is the effect on the partial pressure of Br_2?

79. Iron(II) oxide reacts with carbon monoxide according to the equation $FeO(s) + CO(g) \rightleftharpoons Fe(s) + CO_2(g)$. At 800°C, $K = 0.34$; at 1000°C, $K = 0.40$.
(a) A 20.0-L container is charged with 800.0 g of carbon dioxide, 1436 g of FeO, and 1120 g of iron. What are the equilibrium concentrations of all components of the mixture at each temperature?
(b) What are the partial pressures of the gases at each temperature?
(c) If CO were removed, what would be the effect on P_{CO_2} at each temperature?

80. The equilibrium constant K for the reaction $C(s) + CO_2(g) \rightleftharpoons 2CO(g)$ is 1.9 at 1000 K and 0.133 at 298 K.
(a) If excess C is allowed to react with 25.0 g of CO_2 in a 3.00-L flask, how many grams of CO are produced at each temperature?
(b) What are the partial pressures of each gas at 298 K? At 1000 K?
(c) Would you expect K to increase or decrease if the pressure were increased at constant temperature and volume?

81. Data for the oxidation of methane, $CH_4(g) + 2O_2(g) \rightleftharpoons CO_2(g) + 2H_2O(g)$, in a closed 5.0 L vessel are listed in the table. Fill in the blanks in the following table, and determine the missing values of Q and K (indicated by ?) as the reaction is driven to completion.

	CH_4	O_2	CO_2	H_2O	Q	K
Initial (moles)	0.45	0.90	0	0	?	
At equilibrium	___	___	___	___		1.29
Add 0.50 mol of methane	0.95	___	___	___	?	
New equilibrium	___	___	___	___		?
Remove water	___	___	___	0	?	
New equilibrium	___	___	___	___		1.29

15.6 Controlling the Products of Reactions

82. The oxidation of carbon monoxide via the reaction $CO(g) + \frac{1}{2}O_2(g) \rightleftharpoons CO_2(g)$ has $\Delta H^0 = -283$ kJ. If you were interested in maximizing the yield of CO_2, what general conditions would you select with regard to temperature, pressure, and volume?

83. The oxidation of acetylene via the reaction $2C_2H_2(g) + 5O_2(g) \rightleftharpoons 4CO_2(g) + 2H_2O(l)$ has $\Delta H^0 = -2600$ kJ. What strategy would you use with regard to temperature, volume, and pressure to maximize the yield of product?

84. You are interested in maximizing the product yield of the system

$$2SO_2(g) + O_2(g) \rightleftharpoons 2SO_3(g) \qquad K = 280,\ \Delta H^0 = -158 \text{ kJ}$$

What general conditions would you select with regard to temperature, pressure, and volume? If SO_2 has an initial concentration of 0.200 M and the amount of O_2 is stoichiometric, what amount of SO_3 is produced at equilibrium?

APPLICATIONS

85. The total concentrations of dissolved Al in a soil sample represent the sum of "free" Al^{3+} and bound forms of Al that are stable enough to be considered definite chemical species. The distribution of aluminum among its possible chemical forms can be described using equilibrium constants such as these:

$$K_1 = [AlOH^{2+}]/[Al^{3+}][OH^-] = 1.0 \times 10^9$$
$$K_2 = [AlSO_4^{\ +}]/[Al^{3+}][SO_4^{\ 2-}] = 1.0 \times 10^3$$
$$K_3 = [AlF^{2+}]/[Al^{3+}][F^-] = 1.0 \times 10^7$$

Write an equilibrium equation for each expression. Which anion has the highest affinity for Al^{3+}: OH^-, $SO_4^{\ 2-}$, or F^-? Explain your reasoning. A 1.0 M solution of Al^{3+} is mixed with a 1.0 M solution of each of the anions. Which mixture has the lowest Al^{3+} concentration?

86. Many hydroxy acids undergo reactions forming lactones (cyclic esters) that contain a five- or six-member ring. Common hydroxy acids found in nature are glycolic acid, a constituent of cane-sugar juice; lactic acid, which has the characteristic odor and taste of sour milk; and citric acid, found in fruit juices. The general reaction is written as hydroxy acid $\rightleftharpoons$ lactone + H_2O:

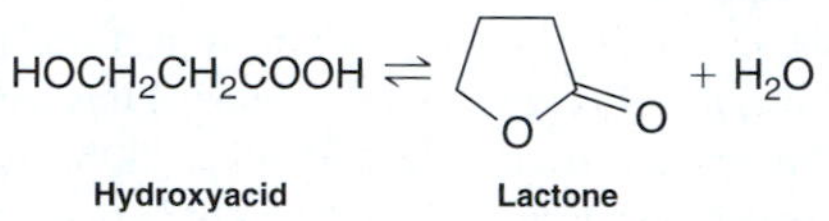

Given the information in the table, calculate the equilibrium constant for this reaction for each of the hydroxy acids indicated, and determine which ring size is most stable.

		At Equilibrium	
Hydroxy Acid Formula	**Size of Lactone Ring (atoms)**	**Hydroxy Acid, *M***	**Lactone, *M***
$HOCH_2CH_2COOH$	4	4.99×10^{-3}	5.00×10^{-5}
$HOCH_2CH_2CH_2COOH$	5	8.10×10^{-5}	2.19×10^{-4}
$HOCH_2CH_2CH_2CH_2COOH$	6	5.46×10^{-2}	5.40×10^{-9}
$HOCH_2CH_2CH_2CH_2CH_2COOH$	7	9.90×10^{-3}	1.00×10^{-4}

87. Phosphorus pentachloride, an important reagent in organic chemistry for converting alcohols to alkyl chlorides (ROH $\rightarrow$ RCl), is hydrolyzed in water to form phosphoric acid and hydrogen chloride. In the gaseous state, however, PCl_5 can decompose at 250°C according to the reaction $PCl_5(g) \rightleftharpoons PCl_3(g) + Cl_2(g)$, for which $K = 0.0420$.
(a) Are products or reactants favored in the decomposition of $PCl_5(g)$?
(b) If a 2.00-L flask containing 104.1 g PCl_5 is heated to 250°C, what is the equilibrium concentration of each species in this reaction?
(c) What effect would an increase in pressure have on the equilibrium position? Why?
(d) If a 1.00×10^3 L vessel containing 2.00×10^3 kg of PCl_3 with a constant chlorine pressure of 2.00 atm is allowed to reach equilibrium, how many kilograms of PCl_5 are produced? What is the percent yield of PCl_5?

88. Carbon disulfide is used in the manufacture of rayon and in electronic vacuum tubes. However, extreme caution must be used when handling CS_2 in its gaseous state because it is extremely toxic and can cause fatal convulsions. Chronic toxicity is marked by psychic disturbances and tremors. Carbon disulfide is used to synthesize H_2S at elevated T via the reaction $CS_2(g) + 4H_2(g) \rightleftharpoons CH_4(g) + 2H_2S(g)$; $K = 3.3 \times 10^4$.

(a) If the equilibrium concentration of methane in this reaction is $2.5 \times 10^{-2} M$ and the initial concentration of each reactant is 0.1635 M, what is the concentration of H_2S at equilibrium?

(b) Exposure to CS_2 concentrations greater than 300 ppm for several hours can start to produce adverse effects. After working for several hours in a laboratory that contains large quantities of carbon disulfide, you notice that the fume hoods were off and there was not enough ventilation to remove any carbon disulfide vapor. Given the equilibrium $CS_2(l) \rightleftharpoons CS_2(g)$, where $T = 20°C$ and $K_p = 0.391$, determine whether you are in any danger.

89. Chloral hydrate, a sedative commonly referred to as "knockout drops," is in equilibrium with trichloroacetaldehyde in aqueous solution:

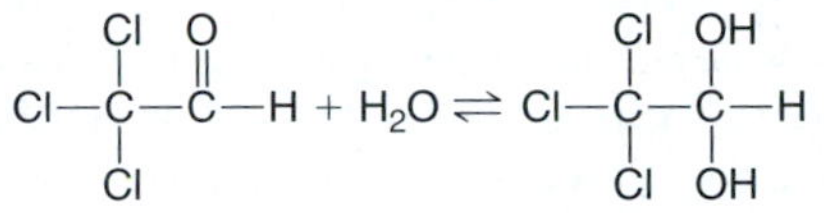

Trichloroacetaldehyde   Chloral hydrate

The equilibrium constant for this reaction as written is 3×10^4. Are the products or the reactants favored? Write an equilibrium expression for this reaction. How could you drive this reaction to completion?

90. Hydrogen cyanide is commercially produced in the United States by the reaction $CH_4(g) + NH_3(g) + \frac{3}{2}O_2(g) \rightleftharpoons HCN(g) + 3H_2O(g)$, where HCN is continuously removed from the system. This reaction is carried out at approximately 1100°C in the presence of a catalyst; however, the high temperature causes other reactions to occur. Why is it necessary to run this reaction at such an elevated temperature? Does the presence of the catalyst affect the equilibrium position?

91. Hemoglobin (Hb) transports oxygen from the lungs to the capillaries, and consists of four subunits, each capable of binding a single molecule of O_2. In the lungs, P_{O_2} is relatively high (100 mmHg), so Hb becomes nearly saturated with O_2. In the tissues, however, P_{O_2} is relatively low (40 mmHg), and Hb releases about half of its bound oxygen. Myoglobin, a protein in muscle, resembles a single subunit of hemoglobin. The plots show the percent O_2 saturation versus P_{O_2} for Hb and myoglobin. Based on these plots, which molecule has the higher affinity for oxygen? What advantage does Hb have over myoglobin as the oxygen transporter? Why is it advantageous to have myoglobin in muscle tissue? Use equilibrium to explain why it is more difficult to exercise at high altitudes where the partial pressure of oxygen is lower.

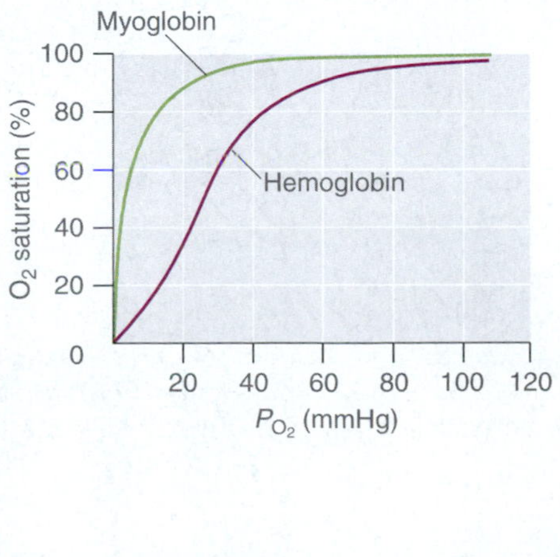

92. Sodium sulfate is widely used in the recycling industry as well as in the detergent and glass industries. This compound combines with H_2SO_4 via the reaction $Na_2SO_4(s) + H_2SO_4(g) \rightleftharpoons 2NaHSO_4(s)$. Sodium hydrogen sulfate is used as a cleaning agent because it is water soluble and acidic.

(a) Write an expression for the equilibrium constant K for this reaction.

(b) Relate this equilibrium constant to the equilibrium constant for the related reaction $2Na_2SO_4(s) + 2H_2SO_4(g) \rightleftharpoons 4NaHSO_4(s)$.

(c) The dissolution of Na_2SO_4 in water produces the equilibrium reaction $SO_4^{2-}(aq) + H_2O(l) \rightleftharpoons HSO_4^-(aq) + OH^-(aq)$ with $K = 8.33 \times 10^{-13}$. What is the concentration of OH^- in a solution formed from the dissolution of 1.00 g of sodium sulfate to make 150.0 mL of aqueous solution? Neglect the autoionization of water in your answer.

93. One of the Venera orbiter satellites measured S_2 concentrations at the surface of Venus. The resulting thermochemical data suggest that S_2 formation at the planet's surface occurs via the equilibrium reaction $4CO(g) + 2SO_2(g) \rightleftharpoons 4CO_2(g) + S_2(g)$. Write an expression for the equilibrium constant K for this reaction, and then relate this expression to those for the following reactions:

(a) $2CO(g) + SO_2(g) \rightleftharpoons 2CO_2(g) + \frac{1}{2}S_2(g)$

(b) $CO(g) + \frac{1}{2}SO_2(g) \rightleftharpoons CO_2(g) + \frac{1}{4}S_2(g)$

(c) At 450°C, the equilibrium pressure of CO_2 is 85.0 atm, SO_2 is 1.0 atm, CO is 1.0 atm, and S_2 is 3.0×10^{-8} atm. What are K and K_p at this temperature? What is the concentration of S_2?

94. Until the early part of the 20th century, commercial production of sulfuric acid was carried out by the "lead-chamber" process, in which SO_2 was oxidized to H_2SO_4 in a lead-lined room. This process may be summarized by the following sequence of reactions:

(1) $NO(g) + NO_2(g) + 2H_2SO_4(l) \rightleftharpoons 2NOHSO_4(s) + H_2O(l)$ K_1

(2) $2NOHSO_4(s) + SO_2(g) + 2H_2O(l) \rightleftharpoons 3H_2SO_4(l) + 2NO(g)$ K_2

(a) Write the equilibrium constant expressions for reactions 1 and 2 and the sum of the reactions (reaction 3).

(b) Show that $K_3 = K_1 \times K_2$.

(c) If insufficient water is added in reaction 2 such that the reaction becomes $NOHSO_4(s) + \frac{1}{2}SO_2(g) + H_2O(l) \rightleftharpoons \frac{3}{2}H_2SO_4(l) + NO(g)$, does K'_3 still equal $K_1 \times K_2$?

(d) Based on part c, write the equilibrium constant expression for K_2.

95. Phosgene (carbonic dichloride, $COCl_2$) is a colorless, highly toxic gas with an odor similar to that of moldy hay. Used as a lethal gas in war, phosgene can be immediately fatal; inhalation can cause either pneumonia or pulmonary edema. For the equilibrium reaction $COCl_2(g) \rightleftharpoons CO(g) + Cl_2(g)$, K_p is 0.680 at −10°C. If the initial pressure of $COCl_2$ is 0.681 atm what is the partial pressure of each component of this equilibrium system? Is the formation of products or reactant favored in this reaction?

96. British bituminous coal has a high sulfur content and produces much smoke when burned. In 1952, burning of this coal in London led to elevated levels of smog containing high concentrations of sulfur dioxide, a lung irritant, and more than 4000 people died. Sulfur dioxide emissions can be converted to SO_3 and ultimately to H_2SO_4, the cause of acid rain. The initial reaction is $2SO_2(g) + O_2(g) \rightleftharpoons 2SO_3(g)$, for which $K_p = 44$.
(a) Given this value of K_p, are products or reactant favored in this reaction?
(b) What is the partial pressure of each species under equilibrium conditions if the initial pressure of each reactant is 0.50 atm?
(c) Would an increase in pressure favor the formation of products or reactant? Why?

97. Oxyhemoglobin is the oxygenated form of hemoglobin, the oxygen-carrying pigment of red blood cells. Hemoglobin is built from α and β protein chains. Assembly of the oxygenated (oxy) and deoxygenated (deoxy) β-chains has been studied with the following results:

$$4\beta\,(\text{oxy}) \rightleftharpoons \beta_4\,(\text{oxy}) \qquad K = 9.07 \times 10^{15}$$
$$4\beta\,(\text{deoxy}) \rightleftharpoons \beta_4\,(\text{deoxy}) \qquad K = 9.20 \times 10^{13}$$

Is it more likely that hemoglobin β chains assemble in an oxygenated or deoxygenated state? Explain your answer.

98. Inorganic weathering reactions can turn silicate rocks, such as diopside ($CaMgSi_2O_6$), to carbonate via the reaction

$$CaMgSi_2O_6(s) + CO_2(g) \rightleftharpoons MgSiO_3(s) + CaCO_3(s) + SiO_2(s)$$

Write an expression for the equilibrium constant. Although this reaction occurs on both Earth and Venus, the high surface temperature of Venus causes the reaction to be driven in one direction on that planet. Predict whether the reaction will be driven to the right or the left by high temperatures, and then justify your answer. The estimated partial pressure of carbon dioxide on Venus is 85 atm due to the dense Venusian atmosphere. How does this pressure influence the reaction?

99. Silicon and its inorganic compounds are widely used to manufacture textile glass fibers, cement, ceramic products, and synthetic fillers. Two of the most important industrially utilized silicon halides are $SiCl_4$ and $SiHCl_3$, formed by reaction of elemental silicon with HCl at temperatures higher than 300°C:

$$Si(s) + 4HCl(g) \rightleftharpoons SiCl_4(g) + 2H_2(g)$$
$$Si(s) + 3HCl(g) \rightleftharpoons SiHCl_3(g) + H_2(g)$$

Which of these two reactions is favored by increasing the HCl concentration? By decreasing the volume of the system?

100. The first step in the utilization of glucose in humans is the conversion of glucose to glucose-6-phosphate via the transfer of a phosphate group from ATP (adenosine triphosphate), which produces glucose-6-phosphate and ADP (adenosine diphosphate):

$$\text{glucose} + \text{ATP} \rightleftharpoons \text{glucose-6-phosphate} + \text{ADP} \qquad K = 680 \text{ at } 25^\circ\text{C}$$

(a) Is the formation of products or reactant favored in this reaction?
(b) Would the value of K increase, decrease, or remain the same if the glucose concentration were doubled?
(c) If $-RT \ln K = -RT' \ln K$, what would the value of K be if the temperature were decreased to 0°C?
(d) Is the formation of products favored by an increase or a decrease in the temperature of the system?

101. In the presence of O_2, the carbon atoms of glucose can be fully oxidized to CO_2 with a release of free energy almost 20 times greater than that possible under conditions in which O_2 is not present. In many animal cells, the TCA cycle (*Tri*C*arboxylic* *A*cid cycle) is the second stage in the complete oxidation of glucose. One reaction in the TCA cycle is the conversion of citrate to isocitrate, for which $K = 0.08$ in the forward direction. Speculate as to why the cycle continues despite this unfavorable value of K. What happens if the citrate concentration increases?

102. Soil is an open system, subject to natural inputs and outputs that may change its chemical composition. Aqueous-phase, adsorbed, and solid-phase forms of Al(III) are of critical importance in controlling the acidity of soils, although industrial effluents, such as sulfur and nitrogen oxide gases, and fertilizers containing nitrogen can also have a large effect. Dissolution of the mineral gibbsite, which contains Al^{3+} in the form $Al(OH)_3(s)$, occurs in soil according to the reaction

$$Al(OH)_3(s) + 3H^+(aq) \rightleftharpoons Al^{3+}(aq) + 3H_2O(l)$$

When gibbsite is in a highly crystalline state, $K = 9.35$ for this reaction at 298 K. In the microcrystalline state, $K = 8.11$. Is this change consistent with the increased surface area of the microcrystalline state?

ESSENTIAL SKILLS 7

TOPICS

- The Quadratic Formula

Previous Essential Skills sections introduced many of the mathematical operations you need to solve chemical problems. We now introduce the quadratic formula, a mathematical relationship involving sums of powers in a single variable that you will need to apply to solve some of the problems in this chapter.

The Quadratic Formula

Mathematical expressions that involve a sum of powers in one or more variables (x, for example) multiplied by coefficients (such as a) are called *polynomials*. Polynomials of a single variable have the general form

$$a_n x^n + \cdots + a_2 x^2 + a_1 x + a_0$$

The highest power to which the variable in a polynomial is raised is called its *order*. Thus, the polynomial shown above is of the *nth* order. For example, if n were 3, the polynomial would be third order.

A *quadratic equation* is a second-order polynomial equation in a single variable x:

$$ax^2 + bx + c = 0$$

According to the fundamental theorem of algebra, a second-order polynomial equation has two solutions, called *roots*, which can be found using a method called *completing the square*. In this method, we first add $-c$ to both sides of the quadratic equation and then divide both sides by a:

$$x^2 + \frac{b}{a}x = -\frac{c}{a}$$

We can convert the left side of this equation to a perfect square by adding $b^2/4a^2$, which is equal to $(b/2a)^2$:

Left side: $$x^2 + \frac{b}{a}x + \frac{b^2}{4a^2} = \left(x + \frac{b}{2a}\right)^2$$

Having added a value to the left side, we must now add that same value to the right side:

$$\left(x + \frac{b}{2a}\right)^2 = -\frac{c}{a} + \frac{b^2}{4a^2}$$

The common denominator on the right side is $4a^2$. Rearranging the right side, we obtain

$$\left(x + \frac{b}{2a}\right)^2 = \frac{b^2 - 4ac}{4a^2}$$

Taking the square root of both sides and solving for x give us

$$x + \frac{b}{2a} = \frac{\pm\sqrt{b^2 - 4ac}}{2a}$$

$$x = \frac{-b \pm \sqrt{b^2 - 4ac}}{2a}$$

This equation, known as the *quadratic formula*, has two roots:

$$x = \frac{-b + \sqrt{b^2 - 4ac}}{2a} \quad \text{and} \quad x = \frac{-b - \sqrt{b^2 - 4ac}}{2a}$$

Thus, we can obtain the solutions to a quadratic equation by substituting the values of the coefficients (a, b, c) into the quadratic formula.

When you apply the quadratic formula to obtain solutions to a quadratic equation, it is important to remember that one of the two solutions may not make sense, or neither may make sense. There may be times, for example, when a negative solution is not reasonable, or when both solutions require that a square root be taken of a negative number. In such cases, we simply discard any solution that is unreasonable, and only report a solution that is reasonable. The next exercise gives you practice using the quadratic formula.

SKILL BUILDER ES15.1

Use the quadratic formula to solve for x in each equation. Report your answers to three significant figures.

(a) $x^2 + 8x - 5 = 0$
(b) $2x^2 - 6x + 3 = 0$
(c) $3x^2 - 5x - 4 = 6$
(d) $2x(-x + 2) + 1 = 0$
(e) $3x(2x + 1) - 4 = 5$

Solution

(a) $$x = \frac{-8 + \sqrt{8^2 - 4(1)(-5)}}{2(1)} = 0.583 \text{ and}$$

$$x = \frac{-8 - \sqrt{8^2 - 4(1)(-5)}}{2(1)} = -8.58$$

(b) $$x = \frac{-(-6) + \sqrt{(-6)^2 - 4(2)(3)}}{2(2)} = 2.37 \text{ and}$$

$$x = \frac{-(-6) - \sqrt{(-6)^2 - 4(2)(3)}}{2(2)} = 0.634$$

(c) $$x = \frac{-(-5) + \sqrt{(-5)^2 - 4(3)(-10)}}{2(3)} = 2.84 \text{ and}$$

$$x = \frac{-(-5) - \sqrt{(-5)^2 - 4(3)(-10)}}{2(3)} = -1.17$$

(d) $$x = \frac{-4 + \sqrt{4^2 - 4(-2)(1)}}{2(-2)} = -0.225 \text{ and}$$

$$x = \frac{-4 - \sqrt{4^2 - 4(-2)(1)}}{2(-2)} = 2.22$$

(e) $$x = \frac{-3 + \sqrt{3^2 - 4(6)(-9)}}{2(6)} = 1.00 \text{ and}$$

$$x = \frac{-3 - \sqrt{3^2 - 4(6)(-9)}}{2(6)} = -1.50$$

16 Aqueous Acid–Base Equilibria

LEARNING OBJECTIVES

- To understand the autoionization reaction of liquid water
- To know the relationship among pH, pOH, and pK_w
- To understand the concept of conjugate acid–base pairs
- To know the relationship between acid or base strength and the magnitude of K_a, K_b, pK_a, and pK_b
- To understand the leveling effect
- To be able to predict whether reactants or products are favored in an acid–base equilibrium
- To understand how molecular structure determines acid and base strengths
- To be able to use K_a and K_b values to calculate the percent ionization and pH of a solution of an acid or a base
- To be able to calculate the pH at any point in an acid–base titration
- To understand how the addition of a common ion affects the position of an acid–base equilibrium
- To understand how a buffer works and how to use the Henderson–Hasselbalch equation to calculate the pH of a buffer

Many vital chemical and physical processes take place exclusively in aqueous solution, including the complex biochemical reactions that occur in living organisms and the reactions that rust and corrode steel objects, such as bridges, ships, and automobiles. Among the most important reactions in aqueous solution are those that can be categorized as acid–base, precipitation, and complexation reactions. So far, our discussions of these reactions have been largely qualitative. In this chapter and the next, however, we take a more quantitative approach to understanding such reactions, using the concept of chemical equilibrium that we developed in Chapter 15 for simple gas-phase reactions. We will begin by revisiting acid–base reactions in a qualitative fashion and then develop quantitative methods to describe acid–base equilibria. In Chapter 17, we will use the same approach to describe the equilibria involved in the dissolution of sparingly soluble solids and the formation of metal complexes.

Indicaters are used to monitor changes in pH. The pH of a solution can be monitored using an acid–base indicator, a substance that undergoes a color change within a specific pH range that is characteristic of that indicator. The color changes for seven commonly used indicators over a pH range of 1–10 are shown here.

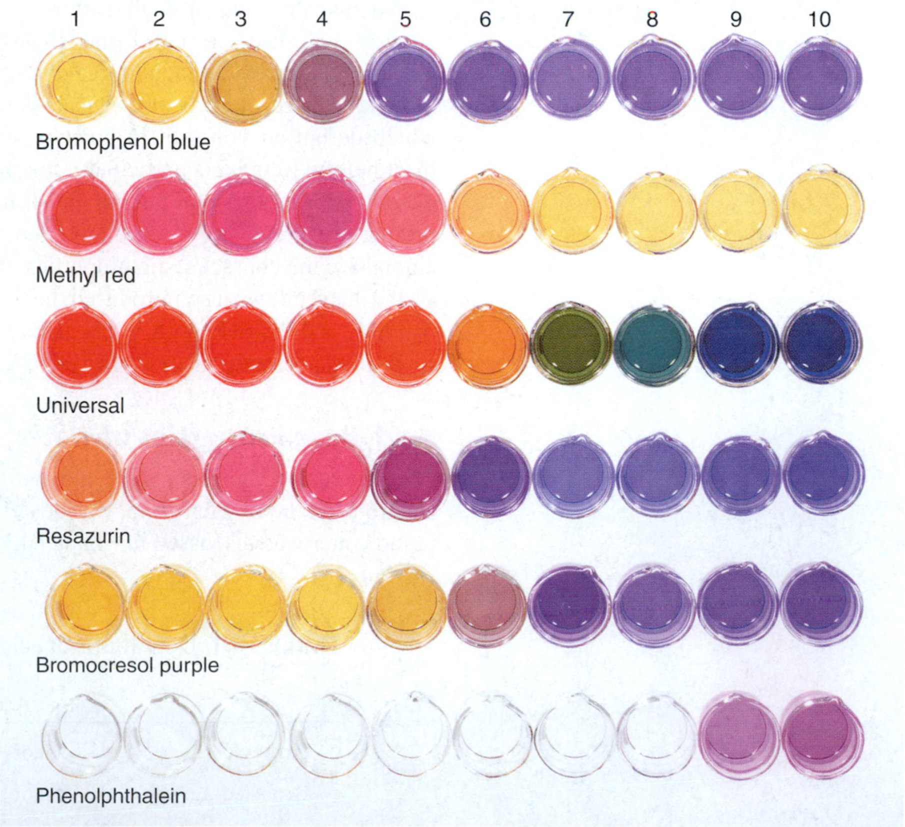

CHAPTER OUTLINE

In Chapter 4, we described how acid rain can adversely affect the survival of marine life and plant growth. Many significant phenomena such as acid rain can be understood only in terms of the acid–base behavior of chemical species. As we expand our discussion of acid–base behavior in this chapter, you will learn why lemon slices are served with fish, why the strengths of acids and bases can vary over many orders of magnitude, and why rhubarb leaves are toxic to humans. You will also understand how the pH of your blood is kept constant, even though you produce large amounts of acid when you exercise.

16.1 • The Autoionization of Water

As you learned in Chapters 4 and 8, acids and bases can be defined in several different ways. Recall that the Arrhenius definition of an acid is a substance that dissociates in water to produce H^+ ions (protons), and an Arrhenius base is a substance that dissociates in water to produce OH^- ions (hydroxide). According to this view, an acid–base reaction involves the reaction of a proton with a hydroxide ion to form water. Although Brønsted and Lowry defined an acid similarly to Arrhenius, as any substance that can *donate* a proton, the Brønsted–Lowry definition of a base is much more general than the Arrhenius definition. In Brønsted–Lowry terms, a base is any substance that can *accept* a proton and therefore is not limited to just a hydroxide ion. Recall from Chapter 4 that for every Brønsted–Lowry acid, there exists a corresponding conjugate base that contains one fewer proton. Consequently, all Brønsted–Lowry acid–base reactions actually involve *two* conjugate acid–base pairs and the transfer of a *proton* from one substance (the acid) to another (the base). In contrast, the Lewis definition of acids and bases, discussed in Chapter 8, focuses on accepting or donating *pairs of electrons* rather than protons. A Lewis base is an electron-pair donor, and a Lewis acid is an electron-pair acceptor. The definitions of acids and bases are summarized in Table 16.1.

Because this chapter deals with acid–base equilibria in *aqueous solution*, our discussion will use primarily the Brønsted–Lowry definitions and nomenclature. Remember, however, that all three definitions are just different ways of looking at the same kind of reaction: a proton is an acid and the hydroxide ion is a base, no matter which definition you use. In practice, chemists tend to use whichever definition is most helpful to make a particular point or to understand a given system. If, for example, we refer to a base as having one or more lone pairs of electrons that can accept a proton, we are simply combining the Lewis and Brønsted–Lowry definitions to emphasize the characteristic properties of a base.

In Chapter 4, we also introduced the acid–base properties of water, its *autoionization reaction,* and the definition of pH. The purpose of this section is to review those concepts and to describe them in terms of chemical equilibrium.

Acid–Base Properties of Water

The structure of the water molecule, with its polar O—H bonds and two lone pairs of electrons on the oxygen atom, was described in Chapters 4 and 8, and the structure of liquid water was discussed in Chapter 13. Recall that its highly polar structure allows

TABLE 16.1 Definitions of acids and bases

	Acids	Bases
Arrhenius	H^+ donor	OH^- donor
Brønsted–Lowry	H^+ donor	H^+ acceptor
Lewis	Electron-pair acceptor	Electron-pair donor

liquid water to act as either an acid (by donating a proton to a base) or a base (by using a lone pair of electrons to accept a proton). For example, when a strong acid such as HCl dissolves in water, it dissociates into chloride ions (Cl^-) and protons (H^+). As you learned in Chapter 4, the proton, in turn, reacts with a water molecule to form the **hydronium ion**, H_3O^+:

$$\underset{\text{Acid}}{HCl(aq)} + \underset{\text{Base}}{H_2O(l)} \longrightarrow \underset{\text{Acid}}{H_3O^+(aq)} + \underset{\text{Base}}{Cl^-(aq)} \qquad (16.1)$$

In this reaction, HCl is the acid and water is acting as a base by accepting an H^+ ion. The reaction in Equation 16.1 is often written in a simpler form, with the hydronium ion represented by H^+ (remember that free H^+ ions do not exist in liquid water):

$$HCl(aq) \longrightarrow H^+(aq) + Cl^-(aq) \qquad (16.2)$$

Conversely, in Equation 16.3, water acts as an acid by donating a proton to NH_3, which acts as a base:

$$\underset{\text{Acid}}{H_2O(l)} + \underset{\text{Base}}{NH_3(aq)} \rightleftharpoons \underset{\text{Acid}}{NH_4^+(aq)} + \underset{\text{Base}}{OH^-(aq)} \qquad (16.3)$$

Thus, water is **amphiprotic**, meaning that it can behave as either an acid or a base, depending on the nature of the other reactant. Notice in Equation 16.3 that the reaction is not complete, as indicated by the double arrow.

The Ion-Product Constant of Liquid Water

Because water is amphiprotic, one water molecule can react with another to form an OH^- ion and an H_3O^+ ion in an autoionization process:

$$2H_2O(l) \rightleftharpoons H_3O^+(aq) + OH^-(aq) \qquad (16.4)$$

The equilibrium constant K for this reaction can be written as

$$K = \frac{[H_3O^+][OH^-]}{[H_2O]^2} \qquad (16.5)$$

When pure liquid water is in equilibrium with hydronium and hydroxide ions at 25°C, the concentrations of hydronium ion and hydroxide ion are equal: $[H_3O^+] = [OH^-] = 1.003 \times 10^{-7}\ M$ (to three decimal places). At 25°C, the density of liquid water is 0.997 g/mL. Hence, the concentration of liquid water at 25°C is

$$[H_2O] = \frac{\text{mol}}{\text{L}} = \left(\frac{0.997\ \cancel{\text{g}}}{\cancel{\text{mL}}}\right)\left(\frac{1\ \text{mol}}{18.02\ \cancel{\text{g}}}\right)\left(\frac{1000\ \cancel{\text{mL}}}{\text{L}}\right) = 55.3\ M \qquad (16.6)$$

Because the number of dissociated water molecules is very small (approximately 2 ppb), the equilibrium of the autoionization reaction (Equation 16.4) lies far to the left. Consequently, the concentration of water is essentially unchanged by the autoionization reaction [$55.3\ M - (1.00 \times 10^{-7}\ M)$] and can be treated as a constant.

By treating $[H_2O]$ as a constant, we can rearrange Equation 16.5 to define a new equilibrium constant, the **ion-product constant of liquid water (K_w)**:

$$K[H_2O]^2 = [H_3O^+][OH^-]$$
$$K_w = [H_3O^+][OH^-] \qquad (16.7)$$

Substituting the values for $[H_3O^+]$ and $[OH^-]$ at 25°C into this expression gives

$$K_w = (1.00 \times 10^{-7})(1.00 \times 10^{-7}) = 1.01 \times 10^{-14} \qquad (16.8)$$

Like any other equilibrium constant, the value of K_w varies with temperature, ranging from 1.15×10^{-15} at 0°C to 4.99×10^{-13} at 100°C.

In pure water, the concentrations of the hydronium ion and the hydroxide ion are the same; hence, the solution is neutral. If $[H_3O^+] > [OH^-]$, however, the solution is acidic. Conversely, if $[H_3O^+] < [OH^-]$, the solution is basic. For an aqueous solution, the H_3O^+ concentration is a quantitative measure of acidity: the higher the H_3O^+ concentration, the more acidic the solution. Conversely, the higher the OH^- concentration, the more basic the solution. In most situations that you will encounter, the H_3O^+ and OH^- concentrations from the dissociation of water are so small ($\leq 10^{-7}$ *M*) that they can be ignored in calculating the H_3O^+ or OH^- concentrations of solutions of acids and bases, but this is not always the case.

The Relationship Among pH, pOH, and pK_w

The *pH scale* is a concise way of describing the H_3O^+ concentration, and hence the acidity or basicity of a solution. Recall from Chapter 4 that pH and the H^+ (actually H_3O^+) concentration are related as follows:

$$\mathrm{pH} = -\log_{10}[\mathrm{H^+}] \tag{16.9a}$$

$$[\mathrm{H^+}] = 10^{-\mathrm{pH}} \tag{16.9b}$$

Note the pattern

For any neutral solution, pH + pOH = 14.00 (at 25°C) and pH = $\frac{1}{2}$pK_w.

Recall also that the pH of a neutral solution ($[H_3O^+] = 1.00 \times 10^{-7}$ *M*)* is 7.00, whereas acidic solutions have pH < 7 (corresponding to $[H_3O^+] > 1.00 \times 10^{-7}$) and basic solutions have pH > 7 (corresponding to $[H_3O^+] < 1.00 \times 10^{-7}$). Because the scale is logarithmic, a pH difference of *one* between two solutions corresponds to a difference of a factor of *ten* in their hydronium ion concentrations. (Refer to Essential Skills 3 at the end of Chapter 4 if you need to refresh your memory about how to use logarithms.)

Similar notation systems are used to describe many other chemical quantities that contain a large negative exponent. For example, chemists use an analogous *pOH scale* to describe the hydroxide ion concentration of a solution. The pOH and $[OH^-]$ are related as follows:

$$\mathrm{pOH} = -\log_{10}[\mathrm{OH^-}] \tag{16.10a}$$

$$[\mathrm{OH^-}] = 10^{-\mathrm{pOH}} \tag{16.10b}$$

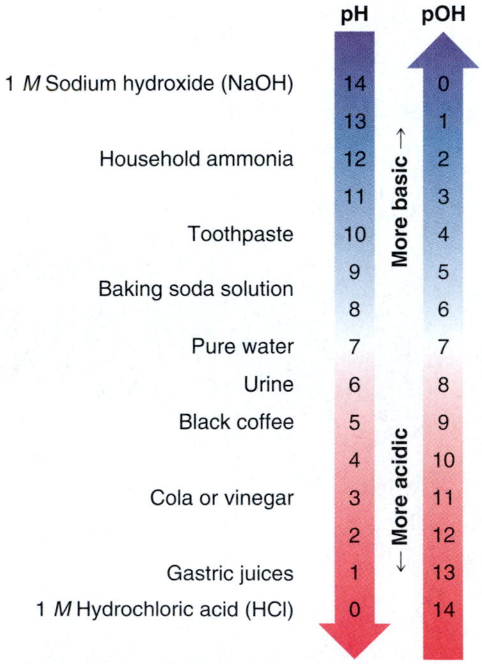

Figure 16.1 The inverse relationship between the pH and pOH scales. As pH decreases, $[H^+]$ and the acidity increase. As pOH increases, $[OH^-]$ and the basicity decrease. Common substances have pH values that range from extremely acidic to extremely basic.

The constant K_w can also be expressed using this notation, where p$K_w = -\log K_w = 14$.

Because a neutral solution has $[OH^-] = 1.00 \times 10^{-7}$, the pOH of a neutral solution is 7.00. Consequently, the sum of the pH and the pOH for a neutral solution at 25°C is 7.00 + 7.00 = 14.00.

We can show that the sum of pH and pOH is equal to 14.00 for *any* aqueous solution at 25°C by taking the negative logarithm of both sides of Equation 16.7:

$$\begin{aligned} -\log K_w = \mathrm{p}K_w &= -\log([\mathrm{H_3O^+}][\mathrm{OH^-}]) \\ &= (-\log[\mathrm{H_3O^+}]) + (-\log[\mathrm{OH^-}]) \\ &= \mathrm{pH} + \mathrm{pOH} \end{aligned} \tag{16.11}$$

At any temperature, pH + pOH = pK_w. Thus, at 25°C, where $K_w = 1.01 \times 10^{-14}$, pH + pOH = 14.00. More generally, the pH of any neutral solution is just half the value of pK_w at that temperature. The relationship among pH, pOH, and the acidity or basicity of a solution is summarized graphically in Figure 16.1 over the common pH range of 0 to 14. Notice the inverse relationship between the pH and pOH scales.

---

* Because most pH meters used in the laboratory read to only two decimal places, we will use only two decimal places for hydronium and hydroxide concentrations and K_w values in the rest of the chapter.

EXAMPLE 16.1

The value of K_w for water at 100°C is 4.99×10^{-13}. Calculate **(a)** the value of pK_w for water at this temperature and **(b)** the values of pH and pOH for a neutral aqueous solution at 100°C.

Given K_w

Asked for pK_w, pH, and pOH

Strategy

A Calculate the value of pK_w by taking the negative logarithm of K_w.
B For a neutral aqueous solution, $[H_3O^+] = [OH^-]$. Use this relationship and Equation 16.7 to calculate $[H_3O^+]$ and $[OH^-]$. Then determine the values of pH and pOH for the solution.

Solution

(a) **A** Because pK_w is the negative logarithm of K_w, we can write

$$pK_w = -\log K_w = -\log(4.99 \times 10^{-13}) = 12.302$$

The answer is reasonable: K_w is between 10^{-13} and 10^{-12}, so pK_w must be between 12 and 13.

(b) **B** Equation 16.7 shows that $K_w = [H_3O^+][OH^-]$. Because $[H_3O^+] = [OH^-]$ in a neutral solution, we can let $x = [H_3O^+] = [OH^-]$ to give

$$K_w = [H_3O^+][OH^-] = (x)(x) = x^2$$
$$x = \sqrt{K_w} = \sqrt{4.99 \times 10^{-13}} = 7.06 \times 10^{-7}\ M$$

Because x is equal to both $[H_3O^+]$ and $[OH^-]$,

$$\text{pH} = \text{pOH} = -\log(7.06 \times 10^{-7}) = 6.15 \text{ (to two decimal places)}$$

We could obtain the same answer more easily (without using logarithms) by using the value of pK_w. In this case, we know that $pK_w = 12.302$, and from Equation 16.11, we know that $pK_w = \text{pH} + \text{pOH}$. Because pH = pOH in a neutral solution, we can use Equation 16.11 directly, setting pH = pOH = y. Solving to two decimal places we obtain:

$$pK_w = \text{pH} + \text{pOH} = y + y = 2y$$
$$y = \frac{pK_w}{2} = \frac{12.302}{2} = 6.15 = \text{pH} = \text{pOH}$$

EXERCISE 16.1

Humans maintain an internal temperature of about 37°C. At this temperature, $K_w = 3.55 \times 10^{-14}$. Calculate pK_w and the pH and pOH of a neutral solution at 37°C.

Answer $pK_w = 13.49$; pH = pOH = 6.73

16.2 • A Qualitative Description of Acid–Base Equilibria

We now turn our attention to acid–base reactions to see how the concepts of chemical equilibrium and equilibrium constants can deepen our understanding of this kind of chemical behavior. We begin with a qualitative description of acid–base equilibria in terms of the Brønsted–Lowry model, and then we proceed to a quantitative description in Section 16.4.

Conjugate Acid–Base Pairs

We discussed the concept of conjugate acid–base pairs in Chapter 4, using the reaction of ammonia, the base, with water, the acid, as an example. In aqueous solutions, acids and bases can be defined in terms of the transfer of a proton from an acid to a base. Thus, for every acidic species in an aqueous solution, there exists a species derived from the acid by the loss of a proton. These two species that differ by only a proton constitute a **conjugate acid–base pair**. For example, in the reaction of HCl with water (Equation 16.1), HCl, the *parent acid*, donates a proton to a water molecule, the *parent base*, thereby forming Cl^-. Thus, HCl and Cl^- constitute a conjugate acid–base pair. (By convention, we always write a conjugate acid–base pair as the acid followed by its conjugate base.) In the reverse reaction, the Cl^- ion in solution acts as a base to accept a proton from H_3O^+, forming H_2O and HCl. Thus, H_3O^+ and H_2O constitute a second conjugate acid–base pair. In general, any acid–base reaction must contain *two* conjugate acid–base pairs, which in this case are HCl/Cl^- and H_3O^+/H_2O:

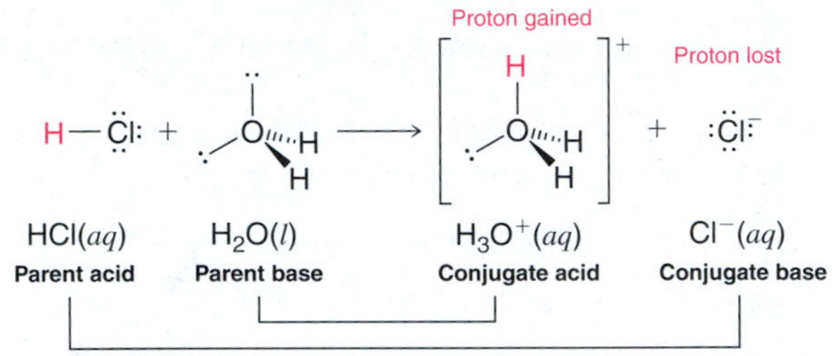

Similarly, in the reaction of acetic acid with water, acetic acid donates a proton to water, which acts as the base. In the reverse reaction, H_3O^+ is the acid that donates a proton to the acetate ion, which acts as the base. Once again, we have two conjugate acid–base pairs: the parent acid and its conjugate base, $CH_3CO_2H/CH_3CO_2^-$, and the parent base and its conjugate acid, written as H_3O^+/H_2O.

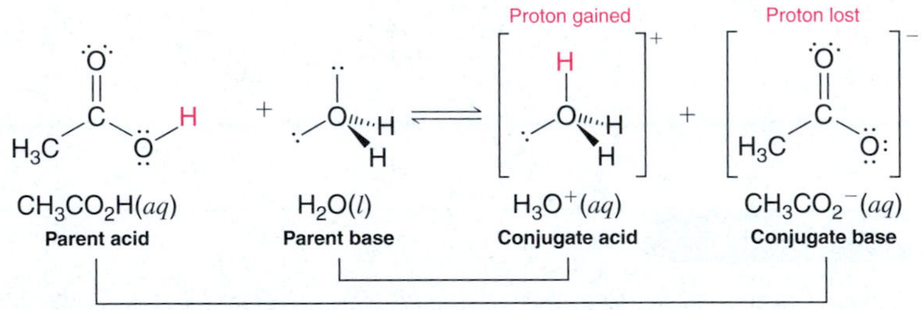

Note the pattern

All acid–base reactions contain two conjugate acid–base pairs.

In the reaction of ammonia with water to give ammonium ions and hydroxide ions (Equation 16.3), ammonia acts as a base by accepting a proton from a water molecule, which in this case is acting as an acid. In the reverse reaction, an ammonium ion acts as an acid by donating a proton to a hydroxide ion, which acts as a base. The conjugate acid–base pairs for this reaction are NH_4^+/NH_3 and H_2O/OH^-:

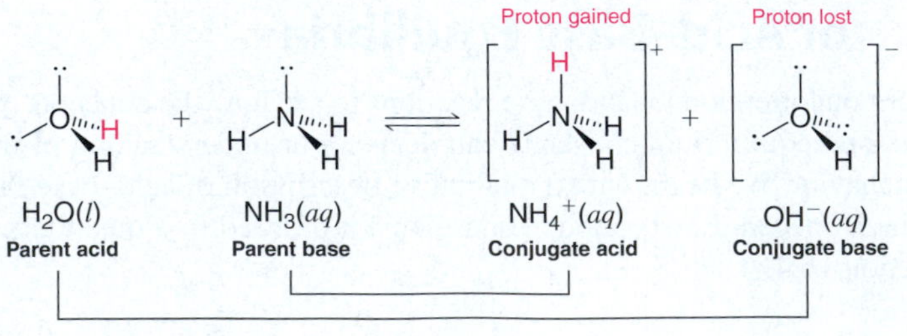

Acid–Base Equilibrium Constants: K_a, K_b, pK_a, and pK_b

MGC pK_a, pK_b, and Acid and Base Strength

The magnitude of the equilibrium constant for an ionization reaction can be used to determine the relative strengths of acids and bases. For example, the general equation for the ionization of a weak acid in water, where HA is the parent acid and A^- is its conjugate base, is

$$HA(aq) + H_2O(l) \rightleftharpoons H_3O^+(aq) + A^-(aq) \quad (16.12)$$

The equilibrium constant for this dissociation is

$$K = \frac{[H_3O^+][A^-]}{[H_2O][HA]} \quad (16.13)$$

As we noted earlier, the concentration of water is essentially constant for all reactions in aqueous solution, so $[H_2O]$ in Equation 16.13 can be incorporated into a new quantity, the **acid ionization constant (K_a)**, also called the *acid dissociation constant*:

$$K_a = K[H_2O] = \frac{[H_3O^+][A^-]}{[HA]} \quad (16.14)$$

Thus, the numerical values of K and K_a differ by the concentration of water (55.3 M). Again, for simplicity, H_3O^+ can be written as H^+ in Equation 16.14. Keep in mind, though, that free H^+ does not exist in aqueous solutions and that a proton is transferred to H_2O in all acid ionization reactions to form H_3O^+. Note that the larger the value of K_a, the stronger the acid and the higher the H^+ concentration at equilibrium.* Values of K_a for a number of common acids are given in Table 16.2.

TABLE 16.2 Values of K_a, pK_a, K_b, and pK_b for selected acids (HA) and their conjugate bases (A^-)

Acid	HA	K_a	pK_a	A^-	K_b	pK_b
Hydroiodic acid	HI	2×10^9	−9.3	I^-	5.0×10^{-24}	23.3
Sulfuric acid (1)[a]	H_2SO_4	1×10^2	−2	HSO_4^-	1.0^{-16}	16
Nitric acid	HNO_3	2.3×10^1	−1.37	NO_3^-	4.3×10^{-16}	15.37
Hydronium ion	H_3O^+	1.00	0.00	H_2O	1.0×10^{-14}	14.00
Sulfuric acid (2)[a]	HSO_4^-	1.0×10^{-2}	1.99	SO_4^{2-}	9.8×10^{-13}	12.01
Hydrofluoric acid	HF	6.3×10^{-4}	3.20	F^-	1.6×10^{-11}	10.80
Nitrous acid	HNO_2	5.6×10^{-4}	3.25	NO_2^-	1.8×10^{-11}	10.75
Formic acid	HCO_2H	1.78×10^{-4}	3.75	HCO_2^-	5.6×10^{-11}	10.25
Benzoic acid	$C_6H_5CO_2H$	6.3×10^{-5}	4.20	$C_6H_5CO_2^-$	1.6×10^{-10}	9.80
Acetic acid	CH_3CO_2H	1.7×10^{-5}	4.76	$CH_3CO_2^-$	5.8×10^{-10}	9.24
Pyridinium ion	$C_5H_5NH^+$	5.9×10^{-6}	5.23	C_5H_5N	1.7×10^{-9}	8.77
Hypochlorous acid	HOCl	4.0×10^{-8}	7.40	OCl^-	2.5×10^{-7}	6.60
Hydrocyanic acid	HCN	6.2×10^{-10}	9.21	CN^-	1.6×10^{-5}	4.79
Ammonium ion	NH_4^+	5.6×10^{-10}	9.25	NH_3	1.8×10^{-5}	4.75
Water	H_2O	1.0×10^{-14}	14.00	OH^-	1.00	0.00
Acetylene	C_2H_2	1×10^{-26}	26	HC_2^-	1×10^{12}	−12
Ammonia	NH_3	1×10^{-35}	35	NH_2^-	1×10^{21}	−21

[a] The number in parentheses indicates the ionization step referred to for a polyprotic acid.

* Like all equilibrium constants, acid–base ionization constants are actually measured in terms of the *activities* of H^+ or OH^-, making them unitless.

Weak bases react with water to produce the hydroxide ion, as shown in the following general equation, where B is the parent base and BH^+ is its conjugate acid:

$$B(aq) + H_2O(l) \rightleftharpoons BH^+(aq) + OH^-(aq) \tag{16.15}$$

The equilibrium constant for this reaction is the **base ionization constant (K_b)**, also called the *base dissociation constant*:

$$K_b = K[H_2O] = \frac{[BH^+][OH^-]}{[B]} \tag{16.16}$$

Once again, the concentration of water is constant, so it does not appear in the equilibrium constant expression; instead, it is included in the value of K_b. The larger the value of K_b, the stronger the base and the higher the OH^- concentration at equilibrium. Values of K_b for a number of common weak bases are given in Table 16.3.

There is a simple relationship between the magnitude of K_a for an acid and K_b for its conjugate base. Consider, for example, the ionization of hydrocyanic acid (HCN) in water to produce an acidic solution, and the reaction of CN^- with water to produce a basic solution:

$$HCN(aq) \rightleftharpoons H^+(aq) + CN^-(aq) \tag{16.17}$$

$$CN^-(aq) + H_2O(l) \rightleftharpoons OH^-(aq) + HCN(aq) \tag{16.18}$$

The equilibrium constant expression for the ionization of HCN is

$$K_a = \frac{[H^+][CN^-]}{[HCN]} \tag{16.19a}$$

and the corresponding expression for the reaction of cyanide with water is

$$K_b = \frac{[OH^-][HCN]}{[CN^-]} \tag{16.19b}$$

If we add Equations 16.17 and 16.18, we obtain the following (recall from Chapter 15 that the equilibrium constant for the *sum* of two reactions is the *product* of the equilibrium constants for the individual reactions):

$$\begin{array}{lll} \cancel{HCN(aq)} \rightleftharpoons H^+(aq) + \cancel{CN^-(aq)} & & K_a = [H^+]\cancel{[CN^-]}/\cancel{[HCN]} \\ \cancel{CN^-(aq)} + H_2O(l) \rightleftharpoons OH^-(aq) + \cancel{HCN(aq)} & & K_b = [OH^-]\cancel{[HCN]}/\cancel{[CN^-]} \\ \hline H_2O(l) \rightleftharpoons H^+(aq) + OH^-(aq) & & K = K_a \times K_b = [H^+][OH^-] \end{array}$$

TABLE 16.3 Values of K_b, pK_b, K_a, and pK_a for selected weak bases (B) and their conjugate acids (BH^+)

Base	**B**	K_b	pK_b	BH^+	K_a	pK_a
Hydroxide ion	OH^-	1.00	0.00	H_2O	1.0×10^{-14}	14.00
Phosphate ion	PO_4^{3-}	2.1×10^{-2}	1.68	HPO_4^{2-}	4.8×10^{-13}	12.32
Dimethylamine	$(CH_3)_2NH$	5.4×10^{-4}	3.27	$(CH_3)_2NH_2^+$	1.9×10^{-11}	10.73
Methylamine	CH_3NH_2	4.6×10^{-4}	3.34	$CH_3NH_3^+$	2.2×10^{-11}	10.66
Trimethylamine	$(CH_3)_3N$	6.3×10^{-5}	4.20	$(CH_3)_3NH^+$	1.6×10^{-10}	9.80
Ammonia	NH_3	1.8×10^{-5}	4.75	NH_4^+	5.6×10^{-10}	9.25
Pyridine	C_5H_5N	1.7×10^{-9}	8.77	$C_5H_5NH^+$	5.9×10^{-6}	5.23
Aniline	$C_6H_5NH_2$	7.4×10^{-10}	9.13	$C_6H_5NH_3^+$	1.3×10^{-5}	4.87
Water	H_2O	1.0×10^{-14}	14.00	H_3O^+	1.00	0.00

In this case, the sum of the reactions described by K_a and K_b is the equation for the autoionization of water, and the product of the two equilibrium constants is K_w:

$$K_aK_b = K_w \tag{16.20}$$

Thus, if we know the value of either K_a for an acid or K_b for its conjugate base, we can calculate the magnitude of the other equilibrium constant for any conjugate acid–base pair.

Just as with pH, pOH, and pK_w, we can use negative logarithms to avoid exponential notation in writing acid and base ionization constants by defining pK_a as

$$pK_a = -\log_{10}K_a \tag{16.21a}$$

$$K_a = 10^{-pK_a} \tag{16.21b}$$

and pK_b as

$$pK_b = -\log_{10}K_b \tag{16.22a}$$

$$K_b = 10^{-pK_b} \tag{16.22b}$$

Values of pK_a and pK_b are given for several common acids and bases in Tables 16.2 and 16.3, respectively, and a more extensive set of data is provided in Appendixes C and D. Because of the use of negative logarithms, *smaller values of* pK_a *correspond to larger acid ionization constants and hence stronger acids*. For example, nitrous acid (HNO_2), with a pK_a of 3.25, is about a thousand times stronger an acid than hydrocyanic acid (HCN), with a pK_a of 9.21. Conversely, *smaller values of* pK_b *correspond to larger base ionization constants and hence stronger bases*. Similarly, Equation 16.20, which expresses the relationship between K_a and K_b, can be written in logarithmic form as

$$pK_a + pK_b = pK_w \tag{16.23a}$$

At 25°C, this becomes

$$pK_a + pK_b = 14.00 \tag{16.23b}$$

The relative strengths of some common acids and their conjugate bases are shown graphically in Figure 16.2. The conjugate acid–base pairs are listed in order (from top to bottom) of increasing acid strength, which corresponds to decreasing values of pK_a. This order corresponds to decreasing strength of the conjugate base, or increasing values of pK_b. At the bottom left of Figure 16.2 are the common strong acids, and at the top right are the most common strong bases. Notice the inverse relationship between the strength of the parent acid and the strength of the conjugate base. Thus, the conjugate base of a strong acid is a very weak base, and the conjugate base of a very weak acid is a strong base.

We can use the relative strengths of acids and bases to predict the direction of an acid–base reaction by following a single rule: *An acid-base equilibrium always favors the side with the weaker acid and base*:

$$\text{stronger acid} + \text{stronger base} \rightleftharpoons \text{weaker acid} + \text{weaker base}$$

Thus, in an acid–base reaction, the proton always reacts with the stronger base.

For example, hydrochloric acid is a strong acid that ionizes essentially completely in dilute aqueous solution to produce H_3O^+ and Cl^-; only negligible amounts of HCl molecules remain undissociated. Hence, the ionization equilibrium lies virtually all the way to the right, as represented by a single arrow:

$$HCl(aq) + H_2O(l) \longrightarrow H_3O^+(aq) + Cl^-(aq) \tag{16.24}$$

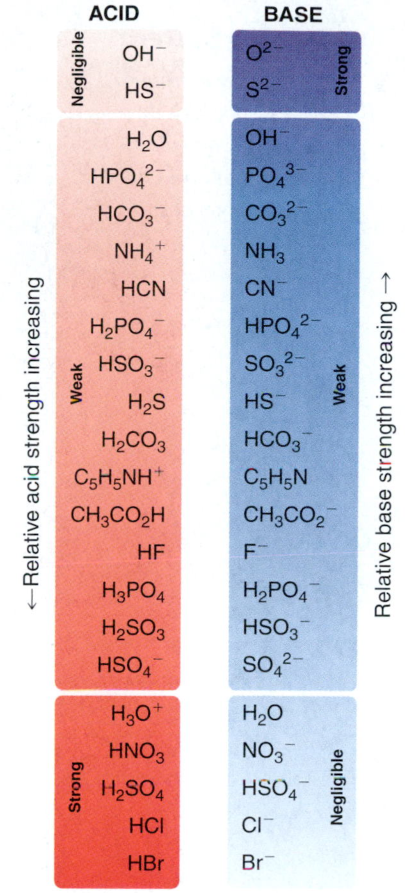

Figure 16.2 The relative strengths of some common conjugate acid–base pairs. The strongest acids are at the bottom left, and the strongest bases are at the top right. Note that the conjugate base of a strong acid is a very weak base and, conversely, the conjugate acid of a strong base is a very weak acid.

Note the pattern

The conjugate base of a strong acid is a weak base, and vice versa.

Note the pattern

All acid–base equilibria favor the side with the weaker acid and base. Thus, the proton is bound to the stronger base.

In contrast, acetic acid is a weak acid and water is a weak base. Consequently, aqueous solutions of acetic acid contain mostly acetic acid molecules in equilibrium with a small concentration of H_3O^+ and acetate ions, and the ionization equilibrium lies far to the left, as represented by these arrows:

$$CH_3CO_2H(aq) + H_2O(l) \rightleftharpoons H_3O^+(aq) + CH_3CO_2^-(aq) \qquad (16.25)$$

Similarly, in the reaction of ammonia with water, the hydroxide ion is a strong base and ammonia is a weak base, whereas the ammonium ion is a stronger acid than water. Hence, this equilibrium also lies to the left:

$$H_2O(l) + NH_3(aq) \rightleftharpoons NH_4^+(aq) + OH^-(aq) \qquad (16.26)$$

EXAMPLE 16.2

(a) Calculate the base ionization constant K_b and the pK_b of the butyrate ion, $CH_3CH_2CH_2CO_2^-$. The pK_a of butyric acid at 25°C is 4.83; butyric acid is responsible for the foul smell of rancid butter. **(b)** Calculate the acid ionization constant K_a and the pK_a of the dimethylammonium ion, $(CH_3)_2NH_2^+$. The base ionization constant K_b of dimethylamine, $(CH_3)_2NH$, is 5.4×10^{-4} at 25°C.

Given pK_a and K_b

Asked for Corresponding K_b and pK_b, K_a and pK_a

Strategy

The constants K_a and K_b are related as shown in Equation 16.20. The pK_a and pK_b for an acid and its conjugate base are related as shown in Equations 16.23a and b. Use the relationships $pK = -\log K$ and $K = 10^{-pK}$ (Equations 16.21a, b and 16.22a, b) to convert between K_a and pK_a or K_b and pK_b.

Solution

(a) We are given the value of pK_a for butyric acid and asked to calculate K_b and pK_b for its conjugate base, the butyrate ion. Because the pK_a value cited is for a temperature of 25°C, we can use Equation 16.23b: $pK_a + pK_b = pK_w = 14.00$. Substituting the value of pK_a and solving for pK_b give

$$4.83 + pK_b = 14.00$$
$$pK_b = 14.00 - 4.83 = 9.17$$

Because $pK_b = -\log K_b$, K_b is $10^{-9.17} = 6.8 \times 10^{-10}$.

(b) In this case, we are given K_b for a base (dimethylamine) and asked to calculate K_a and pK_a for its conjugate acid, the dimethylammonium ion. Because the initial quantity given is K_b rather than pK_b, we can use Equation 16.20: $K_aK_b = K_w$. Substituting the values of K_b and K_w at 25°C and solving for K_a give

$$K_a(5.4 \times 10^{-4}) = 1.01 \times 10^{-14}$$
$$K_a = 1.9 \times 10^{-11}$$

Because $pK_a = -\log K_a$, we have $pK_a = -\log(1.9 \times 10^{-11}) = 10.72$. We could also have converted K_b to pK_b to obtain the same answer:

$$pK_b = -\log(5.4 \times 10^{-4}) = 3.27$$
$$pK_a + pK_b = 14.00$$
$$pK_a = 10.73$$
$$K_a = 10^{-pKa} = 10^{-10.73} = 1.9 \times 10^{-11}$$

Notice that, if we are given any one of these four quantities for an acid or a base (K_a, pK_a, K_b, or pK_b), we can calculate the other three.

EXERCISE 16.2

Lactic acid, $CH_3CH(OH)CO_2H$, is responsible for the pungent taste and smell of sour milk; it is also thought to produce soreness in fatigued muscles. Its pK_a is 3.86 at 25°C. Calculate the value of K_a for lactic acid, and pK_b and K_b for the lactate ion.

Answer $K_a = 1.4 \times 10^{-4}$ for lactic acid; $pK_b = 10.14$ and $K_b = 7.2 \times 10^{-11}$ for the lactate ion.

Solutions of Strong Acids and Bases: The Leveling Effect

You will notice in Table 16.2 that acids like H_2SO_4 and HNO_3 lie *above* the hydronium ion, meaning that they have pK_a values less than zero and are stronger acids than the H_3O^+ ion.* In fact, all six of the common strong acids that we first encountered in Chapter 4 have pK_a values less than zero, which means that they have a greater tendency to lose a proton than does the H_3O^+ ion. Conversely, the conjugate bases of these strong acids are weaker bases than water. Consequently, the proton-transfer equilibria for these strong acids lie far to the right, and the addition of any of the common strong acids to water results in an essentially stoichiometric reaction of the acid with water to form a solution of the H_3O^+ ion and the conjugate base of the acid. Although, for example, K_a for HI is about 10^8 greater than K_a for HNO_3, the reaction of either HI or HNO_3 with water gives an essentially stoichiometric solution of H_3O^+ and I^- or NO_3^-. In fact, a 0.1 *M* aqueous solution of any strong acid actually contains 0.1 *M* H_3O^+, *regardless of the identity of the strong acid*. This phenomenon is called the **leveling effect**: Any species that is a stronger acid than the conjugate acid of water (H_3O^+) is *leveled* to the strength of H_3O^+ in aqueous solution because H_3O^+ is the strongest acid that can exist in equilibrium with water. Consequently, it is impossible to distinguish between the strengths of acids such as HI and HNO_3 in aqueous solution, and an alternative approach must be used to determine their relative acid strengths.

One method is to use a solvent such as anhydrous acetic acid. Since acetic acid is a stronger acid than water, it must also be a weaker base, with a lesser tendency to accept a proton than H_2O. Measurements of the conductivity of 0.1 *M* solutions of both HI and HNO_3 in acetic acid show that HI is completely dissociated but HNO_3 is only partially dissociated, behaving like a weak acid *in this solvent*. This result clearly tells us that HI is a stronger acid than HNO_3. The relative order of acid strengths and approximate K_a and pK_a values for the strong acids at the top of Table 16.2 were determined using measurements like this and different nonaqueous solvents.

The leveling effect applies to solutions of strong bases as well: In aqueous solution, any base stronger than OH^- is *leveled* to the strength of OH^- because OH^- is the strongest base that can exist in equilibrium with water. Salts such as K_2O, $NaOCH_3$ (sodium methoxide), and $NaNH_2$ (sodamide, or sodium amide), whose anions are the conjugate bases of species that would lie below water in Table 16.3, are all strong bases that react essentially completely (and often violently) with water, accepting a proton to give a solution of OH^- and the corresponding cation:

$$K_2O(s) + H_2O(l) \longrightarrow 2OH^-(aq) + 2K^+(aq) \tag{16.27a}$$

$$NaOCH_3(s) + H_2O(l) \longrightarrow OH^-(aq) + Na^+(aq) + CH_3OH(aq) \tag{16.27b}$$

$$NaNH_2(s) + H_2O(l) \longrightarrow OH^-(aq) + Na^+(aq) + NH_3(aq) \tag{16.27c}$$

* Recall from Chapter 4 that the acidic proton(s) in virtually all oxoacids is bonded to one of the oxygen atoms of the oxoanion. Thus, nitric acid should properly be written as $HONO_2$. Unfortunately, however, the formulas of oxoacids are almost always written with hydrogen on the left and oxygen on the right, giving HNO_3 instead.

In general, any substance whose anion is the conjugate base of a compound that is a weaker acid than water is a strong base that reacts quantitatively with water to form hydroxide ion. Other examples that you may encounter are potassium hydride (KH) and organometallic compounds such as methyl lithium (CH_3Li).

Polyprotic Acids and Bases

As you learned in Chapter 4, *polyprotic acids* such as H_2SO_4, H_3PO_4, and H_2CO_3 contain more than one ionizable proton, and the protons are lost in a stepwise manner. The fully protonated species is always the strongest acid because it is easier to remove a proton from a neutral molecule than from a negatively charged ion. Thus, acid strength decreases with the loss of subsequent protons, and, correspondingly, the pK_a increases. Consider H_2SO_4, for example:

$$H_2SO_4(aq) \rightleftharpoons HSO_4^-(aq) + H^+(aq) \qquad pK_a \approx -2 \qquad (16.28a)$$

$$HSO_4^-(aq) \rightleftharpoons SO_4^{2-}(aq) + H^+(aq) \qquad pK_a = 1.99 \qquad (16.28b)$$

The equilibrium in the first reaction lies far to the right, consistent with H_2SO_4 being a strong acid. In contrast, in the second reaction, appreciable quantities of both HSO_4^- and SO_4^{2-} are present at equilibrium.

Notice that HSO_4^- is both the conjugate base of H_2SO_4 and the conjugate acid of SO_4^{2-}. Just like water, HSO_4^- can therefore act as *either* an acid or a base, depending on whether the other reactant is a stronger acid or a stronger base. Conversely, the sulfate ion, SO_4^{2-}, is a *polyprotic base* that is capable of accepting two protons in a stepwise manner:

$$SO_4^{2-}(aq) + H_2O(aq) \rightleftharpoons HSO_4^-(aq) + OH^-(aq) \qquad (16.29a)$$

$$HSO_4^-(aq) + H_2O(aq) \rightleftharpoons H_2SO_4(aq) + OH^-(aq) \qquad (16.29b)$$

Like any other conjugate acid–base pair, the strengths of the conjugate acids and bases are related by pK_a + pK_b = pK_w. Consider, for example, the HSO_4^-/SO_4^{2-} conjugate acid–base pair. From Table 16.2, we see that the pK_a of HSO_4^- is 1.99. Hence, the pK_b of SO_4^{2-} is 14.00 − 1.99 = 12.01. Thus, sulfate is a rather weak base, whereas OH^- is a strong base, so the equilibrium shown in Equation 16.29a lies to the left. The HSO_4^- ion is also a very weak base [pK_a of H_2SO_4 = −2, pK_b of HSO_4^- = 14 − (−2) = 16], consistent with what we expect for the conjugate base of a strong acid. Thus, the equilibrium shown in Equation 16.29b also lies almost completely to the left. Once again, equilibrium favors formation of the weaker acid–base pair.

EXAMPLE 16.3

Predict whether the equilibrium for each reaction lies to the left or the right as written:

(a) $NH_4^+(aq) + PO_4^{3-}(aq) \rightleftharpoons NH_3(aq) + HPO_4^{2-}(aq)$
(b) $CH_3CH_2CO_2H(aq) + CN^-(aq) \rightleftharpoons CH_3CH_2CO_2^-(aq) + HCN(aq)$

Given Balanced chemical equation

Asked for Equilibrium position

Strategy

Identify the conjugate acid–base pairs in each reaction. Then refer to Tables 16.2 and 16.3 to determine which is the stronger acid and base. Equilibrium always favors the formation of the weaker acid–base pair.

Solution

(a) The conjugate acid–base pairs are NH_4^+/NH_3 and HPO_4^{2-}/PO_4^{3-}. According to Tables 16.2 and 16.3, NH_4^+ is a stronger acid ($pK_a = 9.25$) than HPO_4^{2-} ($pK_a = 12.32$), and PO_4^{3-} is a stronger base ($pK_b = 1.68$) than NH_3 ($pK_b = 4.75$). The equilibrium will therefore lie to the right, favoring the formation of the weaker acid–base pair:

$$\underset{\text{Stronger acid}}{NH_4^+(aq)} + \underset{\text{Stronger base}}{PO_4^{3-}(aq)} \rightleftharpoons \underset{\text{Weaker base}}{NH_3(aq)} + \underset{\text{Weaker acid}}{HPO_4^{2-}(aq)}$$

(b) The conjugate acid–base pairs are $CH_3CH_2CO_2H/CH_3CH_2CO_2^-$ and HCN/CN^-. According to Table 16.2, HCN is a weak acid ($pK_a = 9.21$) and CN^- is a moderately weak base ($pK_b = 4.79$). Propionic acid, $CH_3CH_2CO_2H$, is not listed in Table 16.2, however. In a situation like this, the best approach is to look for a similar compound whose acid–base properties *are* listed. For example, propionic acid and acetic acid are identical except for the groups attached to the carbon atom of the carboxylic acid ($-CH_2CH_3$ versus $-CH_3$), so we might expect the two compounds to have similar acid–base properties. In particular, we would expect the pK_a of propionic acid to be similar in magnitude to the pK_a of acetic acid. (In fact, the pK_a of propionic acid is 4.87, compared to 4.76 for acetic acid, which makes propionic acid a slightly weaker acid than acetic acid.) Thus, propionic acid should be a significantly stronger acid than HCN. Because the stronger acid forms the weaker conjugate base, we predict that cyanide will be a stronger base than propionate. The equilibrium will therefore lie to the right, favoring the formation of the weaker acid–base pair:

$$\underset{\text{Stronger acid}}{CH_3CH_2CO_2H(aq)} + \underset{\text{Stronger base}}{CN^-(aq)} \rightleftharpoons \underset{\text{Weaker base}}{CH_3CH_2CO_2^-(aq)} + \underset{\text{Weaker acid}}{HCN(aq)}$$

EXERCISE 16.3

Predict whether the equilibrium for each reaction lies to the left or the right as written:

(a) $H_2O(l) + HS^-(aq) \rightleftharpoons OH^-(aq) + H_2S(aq)$
(b) $HCO_2^-(aq) + HSO_4^-(aq) \rightleftharpoons HCO_2H(aq) + SO_4^{2-}(aq)$

Answer **(a)** left; **(b)** left, $K_{a(HSO_4^-)} > K_{a(HCO_2H)}$

Acid–Base Properties of Solutions of Salts

MGC Acid–Base Properties of Salt Solutions

We can also use the relative strengths of conjugate acid–base pairs to understand the acid–base properties of solutions of salts. In Chapter 4, you learned that a *neutralization reaction* can be defined as the reaction of an acid and a base to produce a salt and water. That is, another cation, such as Na^+, replaces the proton on the acid. An example is the reaction of CH_3CO_2H, a weak acid, with NaOH, a strong base:

$$\underset{\text{Acid}}{CH_3CO_2H(l)} + \underset{\text{Base}}{NaOH(s)} \xrightarrow{H_2O} \underset{\text{Salt}}{CH_3CO_2Na(aq)} + \underset{\text{Water}}{H_2O(l)} \quad (16.30)$$

Depending on the acid–base properties of its component ions, however, a salt can dissolve in water to produce a neutral solution, a basic solution, or an acidic solution.

When a salt such as NaCl dissolves in water, it produces $Na^+(aq)$ and $Cl^-(aq)$ ions. Using a Lewis approach, the Na^+ ion can be viewed as an acid because it is an electron pair acceptor, although its low charge and relatively large radius make it a very weak acid. The Cl^- ion is the conjugate base of the strong acid HCl, so it has essentially no basic character. Consequently, dissolving NaCl in water has no effect on the pH of a solution, and the solution remains neutral.

In contrast, aqueous solutions of potassium cyanide and sodium acetate are basic. Again, the cations (K^+ and Na^+) have essentially no acidic character, but the anions

(CN^- and $CH_3CO_2^-$) are the conjugate bases of the weak acids HCN and acetic acid, respectively. Hence, they are weak bases that can react with water:

$$CN^-(aq) + H_2O(l) \rightleftharpoons HCN(aq) + OH^-(aq) \quad (16.31)$$
$$CH_3CO_2^-(aq) + H_2O(l) \rightleftharpoons CH_3CO_2H(aq) + OH^-(aq) \quad (16.32)$$

Neither reaction proceeds very far to the right as written because HCN and acetic acid are stronger acids than water, and hydroxide is a stronger base than either acetate or cyanide. Thus, in both cases, the equilibrium lies to the left, favoring the weaker acid–base combination. Nonetheless, each of these reactions generates enough hydroxide ions to produce a basic solution. For example, the pH of a 0.1 *M* solution of sodium acetate or potassium cyanide at 25°C is 8.8 or 11.1, respectively. From Table 16.2 and Figure 16.2, we can see that CN^- is a stronger base ($pK_b = 4.79$) than acetate ($pK_b = 9.24$), which is consistent with KCN producing a more basic solution than sodium acetate at the same concentration.

In contrast, the conjugate acid of a weak base should be a weak acid. For example, ammonium chloride and pyridinium chloride are salts produced by the reaction of ammonia and pyridine, respectively, with HCl. As you already know, the chloride ion is such a weak base that it does not react with water. In contrast, the cations of the two salts are weak acids that react with water as follows:

$$NH_4^+(aq) + H_2O(l) \rightleftharpoons NH_3(aq) + H_3O^+(aq) \quad (16.33)$$
$$C_5H_5NH^+(aq) + H_2O(l) \rightleftharpoons C_5H_5N(aq) + H_3O^+(aq) \quad (16.34)$$

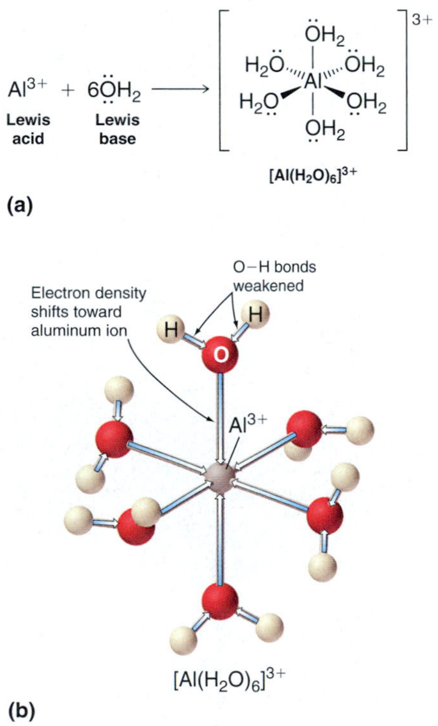

Figure 16.3 Effect of a metal ion on the acidity of water. **(a)** Reaction of the metal ion Al^{3+} with water to form the hydrated metal ion is an example of a Lewis acid–base reaction. **(b)** The positive charge on the aluminum ion attracts electron density from the oxygen atoms, which in turn shifts electron density away from the O—H bonds. The decrease in electron density weakens the O—H bonds in the water molecules and makes it easier for them to lose a proton.

Figure 16.2 shows that H_3O^+ is a stronger acid than either NH_4^+ or $C_5H_5NH^+$. Conversely, ammonia and pyridine are both stronger bases than water. The equilibrium will therefore lie far to the left in both cases, favoring the weaker acid–base pair. The H_3O^+ concentration produced by the reactions is great enough, however, to decrease the pH of the solution significantly: the pH of a 0.1 *M* solution of ammonium chloride or pyridinium chloride at 25°C is 5.13 or 3.12, respectively. This is consistent with the information shown in Figure 16.2, indicating that the pyridinium ion is more acidic than the ammonium ion.

What happens with aqueous solutions of a salt such as ammonium acetate, where both the cation and the anion can react separately with water to produce an acid and a base, respectively? According to Equation 16.33, the ammonium ion will lower the pH, while according to Equation 16.32, the acetate ion will raise the pH. This particular case is unusual, in that the cation is as strong an acid as the anion is a base ($pK_a \approx pK_b$). Consequently, the two effects cancel, and the solution remains neutral. With salts in which the cation is a stronger acid than the anion is a base, the final solution has pH $<$ 7. Conversely, if the cation is a weaker acid than the anion is a base, the final solution has pH $>$ 7.

Solutions of simple salts of metal ions can also be acidic, even though a metal ion cannot donate a proton directly to water to produce H_3O^+. Instead, a metal ion can act as a Lewis acid and interact with water, a Lewis base, by coordinating to a lone pair of electrons on the oxygen atom to form a hydrated metal ion (Figure 16.3a), as discussed in Chapter 4. A water molecule coordinated to a metal ion is more acidic than a free water molecule for two reasons. First, repulsive electrostatic interactions between the positively charged metal ion and the partially positively charged hydrogen atoms of the coordinated water molecule make it easier for the coordinated water to lose a proton:

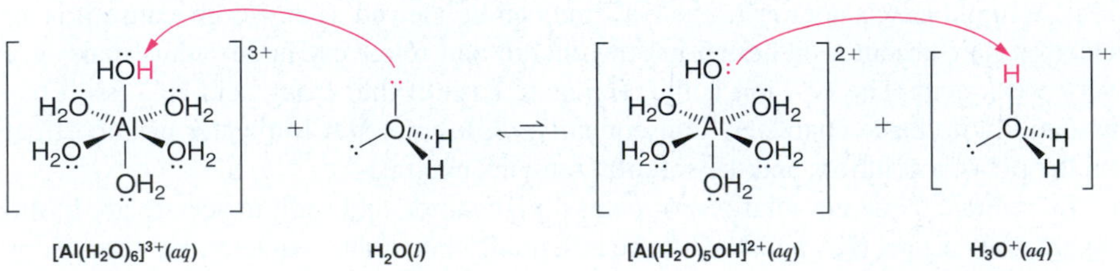

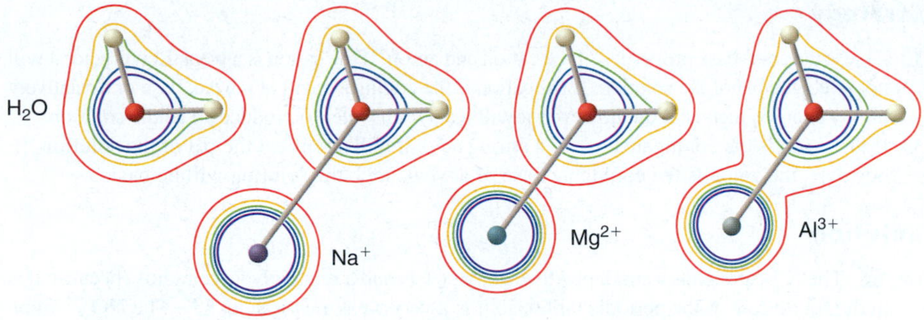

Figure 16.4 The effect of the charge and radius of a metal ion on the acidity of a coordinated water molecule. The contours show the electron density on the O and H atoms in both a free water molecule (*left*) and water molecules coordinated to Na^+, Mg^{2+}, and Al^{3+} ions. These contour maps demonstrate that the smallest, most highly charged metal ion (Al^{3+}) causes the greatest decrease in electron density of the O—H bonds of the water molecule. Due to this effect, the acidity of hydrated metal ions increases as the charge on the metal ion increases and its radius decreases.

Second, the positive charge on the Al^{3+} ion attracts electron density from the oxygen atoms of the water molecules, which in turn decreases the electron density in the O—H bonds, as shown in Figure 16.3b. With less electron density between the O and the H atoms, the O—H bonds are weaker than in a free H_2O molecule, making it easier to lose a H^+ ion.

The magnitude of this effect depends on these two factors (Figure 16.4):

- The *charge* on the metal ion: A divalent ion (M^{2+}) has approximately twice as strong an effect on the electron density in a coordinated water molecule as a monovalent ion (M^+) of the same radius.
- The *radius* of the metal ion: For metal ions with the same charge, the smaller the ion, the shorter the internuclear distance to the oxygen atom of the water molecule, and the greater the effect of the metal on the electron density distribution in the water molecule.

Thus, aqueous solutions of small, highly charged metal ions, such as Al^{3+} and Fe^{3+}, are acidic:

$$[Al(H_2O)_6]^{3+}(aq) \rightleftharpoons [Al(H_2O)_5(OH)]^{2+}(aq) + H^+(aq) \qquad (16.35)$$

The $[Al(H_2O)_6]^{3+}$ ion has a pK_a of 5.0, making it almost as strong an acid as acetic acid. The most important parameter for predicting the effect of a metal ion on the acidity of coordinated water molecules is actually the *charge-to-radius ratio* of the metal ion. A number of pairs of metal ions that lie on a diagonal line in the periodic table, such as Li^+ and Mg^{2+} or Ca^{2+} and Y^{3+}, have different sizes and charges but similar charge-to-radius ratios. As a result, these pairs of metal ions have similar effects on the acidity of coordinated water molecules, and they often exhibit other significant similarities in chemistry as well.

Reactions such as those discussed in this section, in which a salt reacts with water to give an acidic or basic solution, are often called **hydrolysis reactions**. The use of a separate name for this type of reaction is unfortunate because it suggests that they are somehow different. In fact, hydrolysis reactions are just acid–base reactions in which the acid is a cation or the base is an anion; they obey the same principles and rules as all other acid–base reactions.

EXAMPLE 16.4

Predict whether aqueous solutions of these compounds are acidic, basic, or neutral: **(a)** KNO_3; **(b)** $CrBr_3{\cdot}6H_2O$; **(c)** Na_2SO_4.

Given Compound

Asked for Acidity or basicity of aqueous solution

Strategy

A Assess the acid–base properties of the cation and anion. If the cation is a weak Lewis acid, it will not affect the pH of the solution. If the cation is the conjugate acid of a weak base or a relatively highly charged metal cation, however, it will react with water to produce an acidic solution.
B If the anion is the conjugate base of a strong acid, it will not affect the pH of the solution. If, however, the anion is the conjugate base of a weak acid, the solution will be basic.

Solution

(a) **A** The K^+ cation has a small positive charge (+1) and a relatively large radius (because it is in the fourth row of the periodic table), so it is a very weak Lewis acid. **B** The NO_3^- anion is the conjugate base of a strong acid, so it has essentially no basic character (Table 16.1). Hence, neither the cation nor the anion will react with water to produce H^+ or OH^-, and the solution will be neutral.

(b) **A** The Cr^{3+} ion is a relatively highly charged metal cation that should behave similarly to the Al^{3+} ion and form the $[Cr(H_2O)_6]^{3+}$ complex, which will behave as a weak acid:

$$[Cr(H_2O)_6]^{3+}(aq) \rightleftharpoons [Cr(H_2O)_5(OH)]^{2+}(aq) + H^+(aq)$$

B The Br^- anion is a very weak base (it is the conjugate base of the strong acid HBr), so it does not affect the pH of the solution. Hence, the solution will be acidic.

(c) **A** The Na^+ ion, like K^+, is a very weak acid, so it should not affect the acidity of the solution. **B** In contrast, SO_4^{2-} is the conjugate base of HSO_4^-, which is a weak acid. Hence, the SO_4^{2-} ion will react with water as shown in Equation 16.29a to give a slightly basic solution.

EXERCISE 16.4

Predict whether aqueous solutions of the following are acidic, basic, or neutral: **(a)** KI; **(b)** $Mg(ClO_4)_2$; **(c)** NaHS.

Answer **(a)** neutral; **(b)** acidic; **(c)** basic (due to the reaction of HS^- with water to form H_2S and OH^-)

16.3 • Molecular Structure and Acid–Base Strength

We have seen that the strengths of acids and bases vary over many orders of magnitude. In this section, we explore some of the structural and electronic factors that control the acidity or basicity of a molecule.

Bond Strengths

In general, the stronger the A—H or B—H^+ bond, the less likely the bond is to break to form H^+ ions, and thus the less acidic the substance. This effect can be illustrated using the hydrogen halides:

Relative Acid Strength	HF	<	HCl	<	HBr	<	HI
H—X Bond Energy, kJ/mol	570		432		366		298
pK_a	3.20		−6.1		−8		−9

The trend in bond energies is due to a steady decrease in overlap between the 1*s* orbital of hydrogen and the valence orbital of the halogen atom as the size of the halogen increases. The larger the atom to which H is bonded, the weaker the bond. Thus, the bond between

H and a large atom in a given family, such as I or Te, is weaker than the bond between H and a smaller atom in the same family, such as F or O. As a result, *acid strengths of binary hydrides increase as we go down a column of the periodic table*. For example, the order of acidity for the binary hydrides of Group 16 is the following, with pK_a values in parentheses: H_2O (14.00 = pK_w) < H_2S (7.05) < H_2Se (3.89) < H_2Te (2.6).

Note the pattern

Acid strengths of binary hydrides increase as we go down a column or from left to right across a row of the periodic table.

Stability of the Conjugate Base

Whether we write an acid–base reaction as $AH \rightleftharpoons A^- + H^+$ or as $BH^+ \rightleftharpoons B + H^+$, the conjugate base ($A^-$ or B) contains one more lone pair of electrons than the parent acid (AH or BH^+). *Any factor that stabilizes the lone pair on the conjugate base favors dissociation of H^+ and makes the parent acid a stronger acid.* Let's see how this explains the relative acidity of the binary hydrides of the elements in the second row of the periodic table. The observed order of increasing acidity is the following, with pK_a values in parentheses: CH_4 (~50) ≪ NH_3 (~36) < H_2O (14.00) < HF (3.20). Consider, for example, the compounds at both ends of this series: methane and hydrogen fluoride. The conjugate base of CH_4 is CH_3^-, and the conjugate base of HF is F^-. Because fluorine is much more electronegative than carbon, fluorine can better stabilize the negative charge in the F^- ion than carbon can stabilize the negative charge in the CH_3^- ion. Consequently, HF has a greater tendency to dissociate to form H^+ and F^- than does methane to form H^+ and CH_3^-, making HF a much stronger acid than CH_4.

The same trend is predicted by analysis of the properties of the conjugate acids. For a series of compounds of the general formula HE, as the electronegativity of E increases, the E—H bond becomes more polar, favoring dissociation to form E^- and H^+. Thus, due to both the increasing stability of the conjugate base and the increasing polarization of the E—H bond in the conjugate acid, *acid strengths of binary hydrides increase as we go from left to right across a row of the periodic table.*

Inductive Effects

Atoms or groups of atoms in a molecule other than those to which H is bonded can induce a change in the distribution of electrons within the molecule. This is called an *inductive effect*, and, much like the coordination of water to a metal ion, it can have a major effect on the acidity or basicity of the molecule. For example, the hypohalous acids (general formula HOX, with X representing a halogen) all have a hydrogen atom bonded to an oxygen atom. In aqueous solution, they all produce the following equilibrium:

$$HOX(aq) \rightleftharpoons H^+(aq) + OX^-(aq) \tag{16.36}$$

The acidities of these acids vary by about three orders of magnitude, however, due to the difference in electronegativity of the halogen atoms:

HOX	Electronegativity of X	pK_a
HOCl	3.0	7.40
HOBr	2.8	8.55
HOI	2.5	10.5

As the electronegativity of X increases, the distribution of electron density within the molecule changes: the electrons are drawn more strongly toward the halogen atom, and in turn away from the H in the O—H bond, thus weakening the O—H bond and allowing dissociation of hydrogen as H^+.

The acidity of oxoacids, with the general formula $HOXO_n$ (n = 0–3), depends strongly on the number of terminal oxygen atoms attached to the central atom X. As shown in Figure 16.5, the K_a values of the oxoacids of chlorine increase by a factor

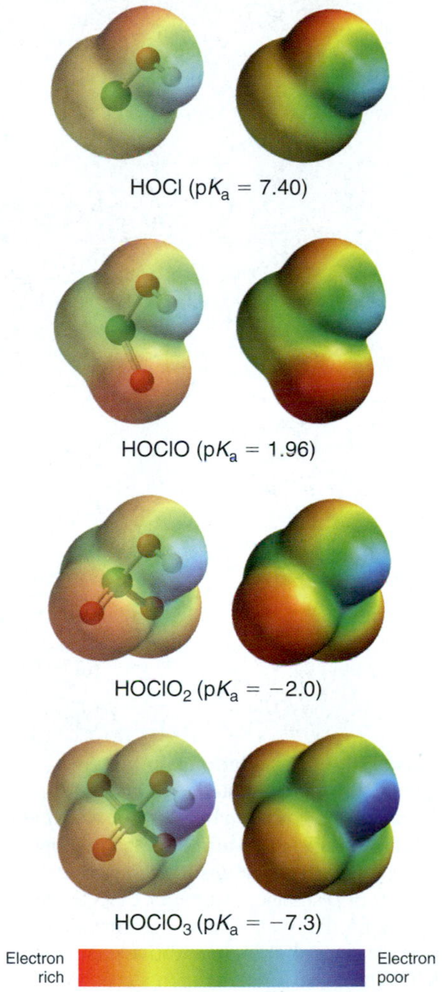

Figure 16.5 The relationship between the acid strengths of the oxoacids of chlorine and the electron density on the O—H unit. These electrostatic potential maps show how the electron density on the O—H unit decreases as the number of terminal oxygen atoms increases. Blue corresponds to low electron densities, whereas red corresponds to high electron densities. The oxygen atom in the O—H unit becomes steadily less red from HClO to $HClO_4$ (also written as $HOClO_3$), while the H atom becomes steadily bluer, indicating that the electron density on the O—H unit is decreasing. (Chlorine oxoacids pK_a values from J. R. Bowser, *Inorganic Chemistry*, 1993.)

of about 10^4 to 10^6 with each oxygen as successive oxygen atoms are added. The increase in acid strength with increasing number of terminal oxygen atoms is due to both an inductive effect and increased stabilization of the conjugate base.

Because oxygen is the second most electronegative element, adding terminal oxygen atoms causes electrons to be drawn away from the O—H bond, making it weaker and thereby increasing the strength of the acid. The colors in Figure 16.5 show how the electrostatic potential, a measure of the strength of the interaction of a point charge at any place on the surface of the molecule, changes as the number of terminal oxygen atoms increases. In Figures 16.5 and 16.6, blue corresponds to low electron densities, while red corresponds to high electron densities. Note that the oxygen atom in the O—H unit becomes steadily less red from HClO to $HClO_4$ (also written as $HOClO_3$), while the H atom becomes steadily bluer, indicating that the electron density on the O—H unit decreases as the number of terminal oxygen atoms increases. The decrease in electron density in the O—H bond weakens it, making it easier to lose hydrogen as H^+ ions and increasing the strength of the acid.

At least as important, however, is the effect of delocalization of the negative charge in the conjugate base. As shown in Figure 16.6, the number of resonance structures that can be written for the oxoanions of chlorine increases as the number of terminal oxygen atoms increases, allowing the single negative charge to be delocalized over successively more oxygen atoms. The electrostatic potential plots in Figure 16.6 demonstrate that the electron density on the terminal oxygen atoms decreases steadily as their number increases. Note that the oxygen atom in ClO^- is red, indicating that it is electron rich, and that the color of oxygen progressively changes to green in ClO_4^-, indicating that the oxygen atoms are becoming steadily less electron rich through the series. For example, in the perchlorate ion (ClO_4^-), the single negative charge is delocalized over all four oxygen atoms, whereas in the hypochlorite ion (OCl^-), the negative charge is largely localized on a single oxygen atom (Figure 16.6). As a result, the perchlorate ion has no localized negative charge to which a proton can bind. Consequently, the perchlorate anion has a much lower affinity for a proton than does the hypochlorite ion, and perchloric acid is one of the strongest acids known.

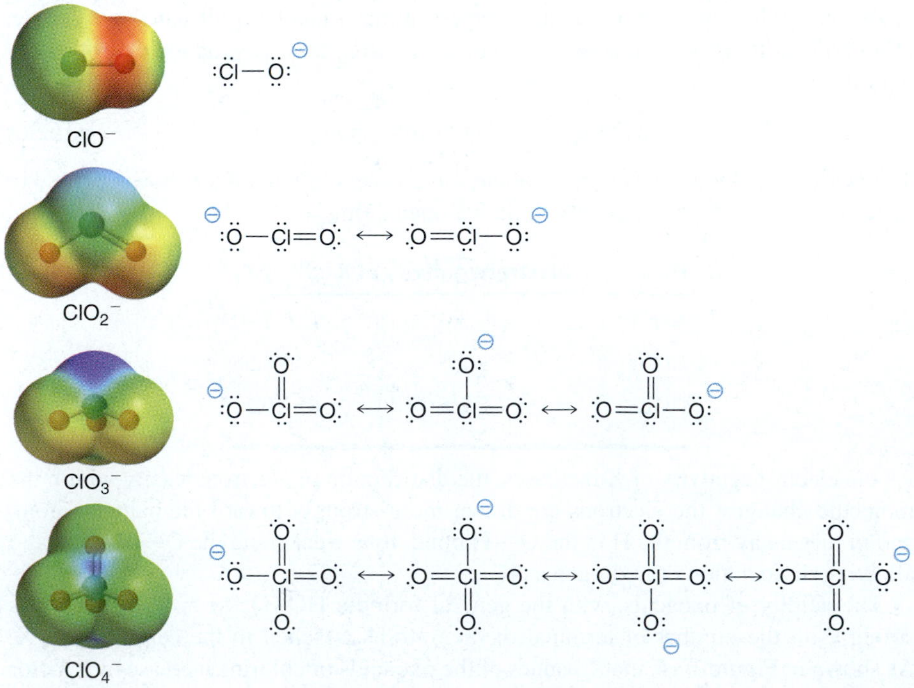

Figure 16.6 The relationship between delocalization of the negative charge in the oxoanions of chlorine and the number of terminal oxygen atoms. As the number of terminal oxygen atoms increases, the number of resonance structures that can be written for the oxoanions of chlorine also increases, and the single negative charge is delocalized over more oxygen atoms. As these electrostatic potential plots demonstrate, the electron density on the terminal oxygen atoms decreases steadily as their number increases. As the electron density on the oxygen atoms decreases, so does their affinity for a proton, making the anion less basic. As a result, the parent oxoacid is more acidic.

Similar inductive effects are also responsible for the trend in the acidities of oxoacids that have the same number of oxygen atoms as we go across a row of the periodic table from left to right. For example, H_3PO_4 is a weak acid, H_2SO_4 is a strong acid, and $HClO_4$ is one of the strongest acids known. The number of terminal oxygen atoms increases steadily across the row, consistent with the observed increase in acidity. In addition, the electronegativity of the central atom increases steadily from P to S to Cl, which causes electrons to be drawn from oxygen to the central atom, weakening the O—H bond and increasing the strength of the oxoacid.

Careful inspection of the data in Table 16.4 shows two apparent anomalies: carbonic acid and phosphorous acid. If carbonic acid (H_2CO_3) were really a discrete molecule with the structure $(HO)_2C{=}O$, it would have a single terminal oxygen atom and should be comparable in acid strength to phosphoric acid (H_3PO_4), for which $pK_{a1} = 2.16$. Instead, the tabulated value of pK_{a1} for carbonic acid is 6.35, making it about 10,000 times weaker than expected. As we shall see in Section 16.6, however, H_2CO_3 is only a minor component of the aqueous solutions of CO_2 that are referred to as carbonic acid. Similarly, if phosphorous acid (H_3PO_3) actually had the structure $(HO)_3P$, it would have no terminal oxygen atoms attached to phosphorous. It would therefore be expected to be about as strong an acid as HOCl ($pK_a = 7.40$). In fact, the value of pK_{a1} for phosphorous acid is 1.30, and the structure of phosphorous acid is $(HO)_2P({=}O)H$ with one H atom directly bonded to P and one P=O bond. Thus, the value of pK_{a1} for phosphorous acid is similar to that of other oxoacids with one terminal oxygen atom, such as H_3PO_4. Fortunately, phosphorous acid is the only common oxoacid in which a hydrogen atom is bonded to the central atom rather than oxygen.

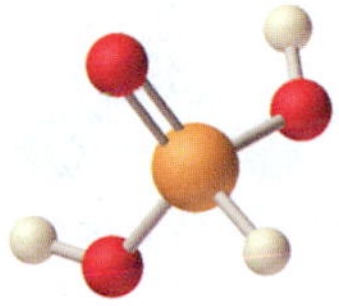

Phosphorous acid, H_3PO_3

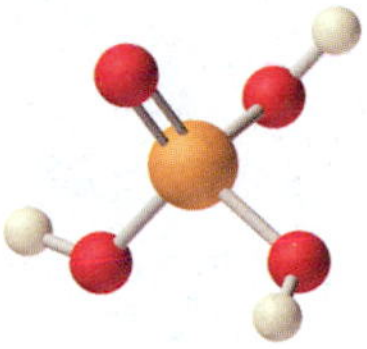

Phosphoric acid, H_3PO_4

Inductive effects are also observed in organic molecules that contain electronegative substituents. The magnitude of the electron-withdrawing effect depends on both

TABLE 16.4 Values of pK_a for selected polyprotic acids and bases

	Formula	pK_{a1}	pK_{a2}	pK_{a3}
Polyprotic Acids				
Carbonic acid[a]	"H_2CO_3"	6.35	10.33	
Citric acid	$HO_2CCH_2C(HOCCO_2H)CH_2CO_2H$	3.13	4.76	6.40
Malonic acid	$HO_2CCH_2CO_2H$	2.85	5.70	
Oxalic acid	HO_2CCO_2H	1.25	3.81	
Phosphoric acid	H_3PO_4	2.16	7.21	12.32
Phosphorous acid	H_3PO_3	1.3	6.70	
Succinic acid	$HO_2CCH_2CH_2CO_2H$	4.21	5.64	
Sulfuric acid	H_2SO_4	−2	1.99	
Sulfurous acid[a]	"H_2SO_3"	1.85	7.21	
Polyprotic Bases		pK_{b1}	pK_{b2}	
Ethylenediamine	$H_2N(CH_2)_2NH_2$	4.08	7.14	
Piperazine	$HN(CH_2)_4NH$	4.27	8.67	
Propylenediamine	$H_2N(CH_2)_3NH_2$	3.45	5.12	

[a] H_2CO_3 and H_2SO_3 are at best minor components of aqueous solutions of $CO_2(g)$ and $SO_2(g)$, respectively, but such solutions are commonly referred to as containing carbonic acid and sulfurous acid, respectively.

the nature and the number of halogen substituents, as shown by the pK_a values for several acetic acid derivatives:

	CH_3CO_2H <	CH_2ClCO_2H <	$CHCl_2CO_2H$ <	CCl_3CO_2H <	CF_3CO_2H
pK_a	4.76	2.87	1.35	0.66	0.52

As you might expect, fluorine, which is more electronegative than chlorine, causes a larger effect than chlorine, and the effect of three halogens is greater than the effect of two or one. Notice from these data that inductive effects can be quite large. For instance, replacing the —CH_3 group of acetic acid by a —CF_3 group results in about a 10,000-fold increase in acidity!

Note the pattern

Any inductive effect that withdraws electron density from an O—H bond increases the acidity of a compound.

EXAMPLE 16.5

Arrange the compounds of each series in order of increasing acid or base strength: (a) sulfuric acid [H_2SO_4, or $(HO)_2SO_2$], fluorosulfonic acid (FSO_3H, or FSO_2OH), sulfurous acid [H_2SO_3, or $(HO)_2SO$]; (b) ammonia (NH_3), trifluoramine (NF_3), hydroxylamine (NH_2OH). The structures are shown in the margin.

Sulfurous acid, H_2SO_3

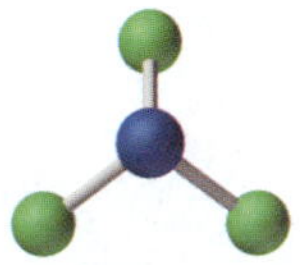

Trifluoramine, NF_3

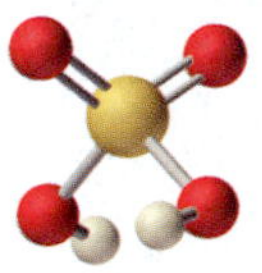

Sulfuric acid, H_2SO_4

Hydroxylamine, NH_2OH

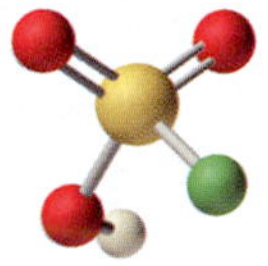

Fluorosulfonic acid, FSO_3H

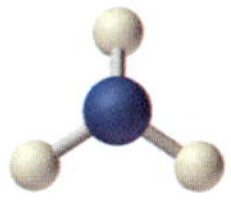

Ammonia, NH_3

Given Series of compounds

Asked for Relative acid or base strengths

Strategy

Use relative bond strengths, stability of the conjugate base, and inductive effects to arrange the compounds in order of increasing tendency to ionize in aqueous solution.

Solution

(a) Although both sulfuric acid and sulfurous acid have two —OH groups, the sulfur atom in sulfuric acid is bonded to two terminal oxygen atoms versus one in sulfurous acid. Because oxygen is highly electronegative, sulfuric acid is the stronger acid because the negative charge on the anion is stabilized by the additional oxygen atom. In comparing sulfuric acid and fluorosulfonic acid, we note that fluorine is more electronegative than oxygen. Thus, replacing an —OH by —F will remove *more* electron density from the central S atom, which will in turn remove electron density from the S—OH bond and the O—H bond. Because its O—H bond is weaker, FSO_3H is a stronger acid than sulfuric acid. The predicted order of acid strengths given here is confirmed by the measured pK_a values for these acids:

	H_2SO_3 <	H_2SO_4 <	FSO_3H
pK_a	1.85	−2	−10

(b) The structures of both trifluoramine and hydroxylamine are similar to that of ammonia. In trifluoramine, all of the hydrogen atoms in NH_3 are replaced by fluorine atoms, whereas in hydroxylamine, one hydrogen atom is replaced by OH. Replacing the three hydrogen atoms by fluorine will withdraw electron density from N, making the lone electron pair on N less available to bond to an H^+ ion. Thus, NF_3 is predicted to be a much weaker base than NH_3. Similarly, since oxygen is more electronegative than hydrogen, replacing one hydrogen atom in NH_3 by OH will make the amine less basic. Because oxygen is less electronegative than fluorine and only one hydrogen atom is replaced, however, the effect will be smaller. The predicted order of increasing base strength shown here is confirmed by the measured pK_b values:

	NF_3 ≪	NH_2OH <	NH_3
pK_b	—	8.06	4.75

Trifluoramine is such a weak base that it does not react with aqueous solutions of strong acids. Hence, its base ionization constant has not been measured.

EXERCISE 16.5

Arrange the following in order of **(a)** decreasing acid strength: H_3PO_4, $CH_3PO_3H_2$, and $HClO_3$; and **(b)** increasing base strength: CH_3S^-, OH^-, and CF_3S^-.

Answer (a) $HClO_3 > CH_3PO_3H_2 > H_3PO_4$; (b) $CF_3S^- < CH_3S^- < OH^-$

16.4 Quantitative Aspects of Acid–Base Equilibria

This section presents a quantitative approach to analyzing acid–base equilibria. You will learn how to determine the values of K_a and K_b, how to use K_a or K_b to calculate the percent ionization and pH of an aqueous solution of an acid or a base, and how to calculate the equilibrium constant for the reaction of an acid with a base from the K_a and K_b of the reactants.

Determining K_a and K_b

MGC Determining Ionic Strength

The ionization constants K_a and K_b are equilibrium constants that are calculated from experimentally measured concentrations, just like the equilibrium constants discussed in Chapter 15. Before proceeding further, it is important to understand exactly what is meant when we talk about the concentration of an aqueous solution of a weak acid or base. Suppose, for example, we have a bottle labeled 1 *M* acetic acid or 1 *M* ammonia. As you learned in Chapter 4, such a solution is usually prepared by dissolving 1 mol of acetic acid or ammonia in water and adding enough water to give a final volume of exactly 1 L. If, however, we were to list the actual concentrations of all the species present in either solution, we would find that *none* of the values is exactly 1 *M* because a weak acid such as acetic acid or a weak base such as ammonia always reacts with water to some extent. The extent of the reaction depends on the value of K_a or K_b, the concentration of the acid or base, and the temperature. Consequently, only the *total concentration* of both the ionized and unionized species is equal to 1 *M*.

The *analytical concentration* (*C*), is defined as the *total* concentration of all forms of an acid or base that are present in solution, regardless of their state of protonation. Thus, a "1 *M*" solution of acetic acid has an analytical concentration of 1 *M*, which is the sum of the *actual* concentrations of unionized acetic acid (CH_3CO_2H) and the ionized form ($CH_3CO_2^-$):

$$C_{CH_3CO_2H} = [CH_3CO_2H] + [CH_3CO_2^-] \tag{16.37}$$

As we shall see shortly, if we know the analytical concentration and the value of K_a, we can calculate the actual values of $[CH_3CO_2H]$ and $[CH_3CO_2^-]$.

The equilibrium equations for the reaction of acetic acid and ammonia with water are

$$K_a = \frac{[H^+][CH_3CO_2^-]}{[CH_3CO_2H]} \tag{16.38}$$

$$K_b = \frac{[NH_4^+][OH^-]}{[NH_3]} \tag{16.39}$$

where K_a and K_b are the ionization constants for acetic acid and ammonia, respectively. In addition to the analytical concentration of the acid (or base), we must have a way to measure the concentration of at least *one* of the species in the equilibrium constant expression in order to determine the value of K_a (or K_b). There are two common ways to obtain the concentrations: by measuring the electrical conductivity of the solution, which is related to the total concentration of ions present, and by measuring the pH of the solution, which gives $[H^+]$ or $[OH^-]$.

The following examples illustrate the procedure for determining K_a for a weak acid and K_b for a weak base. In both cases, we will follow the procedure developed in Chapter 15: the analytical concentration of the acid or base is the *initial* concentration, and the stoichiometry of the reaction with water determines the *change* in concentrations. The *final* concentrations of all species are calculated from the initial concentrations and the changes in the concentrations. Inserting the final concentrations into the equilibrium constant expression enables us to calculate the value of K_a or K_b.

EXAMPLE 16.6

Electrical conductivity measurements indicate that 0.42% of the acetic acid molecules in a 1.000 *M* solution are ionized at 25°C. Calculate the values of K_a and pK_a for acetic acid at this temperature.

Given Analytical concentration and percent ionization

Asked for K_a and pK_a

Strategy

A Write the balanced equilibrium equation for the reaction, and derive the equilibrium constant expression.

B Use the data given and the stoichiometry of the reaction to construct a table showing initial concentrations, changes in concentrations, and final concentrations for all species in the equilibrium constant expression.

C Substitute the final concentrations into the equilibrium constant expression, and calculate the value of K_a. Take the negative logarithm of K_a to obtain pK_a.

Solution

A The balanced equilibrium equation for the dissociation of acetic acid is

$$CH_3CO_2H(aq) \rightleftharpoons H^+(aq) + CH_3CO_2^-(aq)$$

and the equilibrium constant expression is

$$K_a = \frac{[H^+][CH_3CO_2^-]}{[CH_3CO_2H]}$$

B To calculate the value of K_a, we need to know the equilibrium concentrations of CH_3CO_2H, $CH_3CO_2^-$, and H^+. The most direct way to do this is to construct a table that lists the initial concentrations and the changes in concentrations that occur during the reaction to give the final concentrations, using the procedure introduced in Chapter 15. The initial concentration of unionized acetic acid, $[CH_3CO_2H]_i$, is the analytical concentration, 1.000 *M*, and the initial acetate concentration, $[CH_3CO_2^-]_i$, is zero. The initial concentration of H^+ is not zero, however; $[H^+]_i$ is 1.00×10^{-7} *M* due to the autoionization of water. The measured percent ionization tells us that 0.42% of the acetic acid molecules are ionized at equilibrium. Consequently, the change in the concentration of acetic acid is $\Delta[CH_3CO_2H] = -(4.2 \times 10^{-3})(1.000\ M) = -0.0042\ M$. Conversely, the change in the acetate concentration is $\Delta[CH_3CO_2^-] = +0.0042\ M$ because every 1 mol of acetic acid that ionizes gives 1 mol of acetate. Because one proton is produced for each acetate ion formed, $\Delta[H^+] = +0.0042\ M$ as well. These results are summarized in the table.

	$CH_3CO_2H(aq)$	$\rightleftharpoons$ $H^+(aq)$	+ $CH_3CO_2^-(aq)$
	$[CH_3CO_2H]$	**$[H^+]$**	**$[CH_3CO_2^-]$**
Initial	1.000	1.00×10^{-7}	0
Change	−0.0042	+0.0042	+0.0042
Final	0.9958	0.0042	0.0042

The final concentrations of all species are therefore

$$[CH_3CO_2H]_f = [CH_3CO_2H]_i + \Delta[CH_3CO_2H] = 1.000\ M + (-0.0042\ M) = 0.9958\ M$$
$$[CH_3CO_2^-]_f = [CH_3CO_2^-]_i + \Delta[CH_3CO_2^-] = 0\ M + (+0.0042\ M) = 0.0042\ M$$
$$[H^+]_f = [H^+]_i + \Delta[H^+] = 1.00 \times 10^{-7}\ M + (+0.0042\ M) = 0.0042\ M$$

C ✓ We can now calculate K_a by inserting the final concentrations into the equilibrium constant expression:

$$K_a = \frac{[H^+][CH_3CO_2^-]}{[CH_3CO_2H]} = \frac{(0.0042)(0.0042)}{0.9958} = 1.8 \times 10^{-5}$$

The pK_a is the negative logarithm of K_a: $pK_a = -\log K_a = -\log(1.8 \times 10^{-5}) = 4.74$.

EXERCISE 16.6

Picric acid is the common name for 2,4,6-trinitrophenol, a derivative of phenol (C_6H_5OH) in which three H atoms are replaced by nitro (—NO_2) groups. The presence of the nitro groups removes electron density from the phenyl ring, making picric acid a much stronger acid than phenol ($pK_a = 9.99$). The nitro groups also make picric acid potentially explosive, as you might expect based on its chemical similarity to 2,4,6-trinitrotoluene, better known as TNT. A 0.20 *M* solution of picric acid is 73% ionized at 25°C. Calculate K_a and pK_a for picric acid.

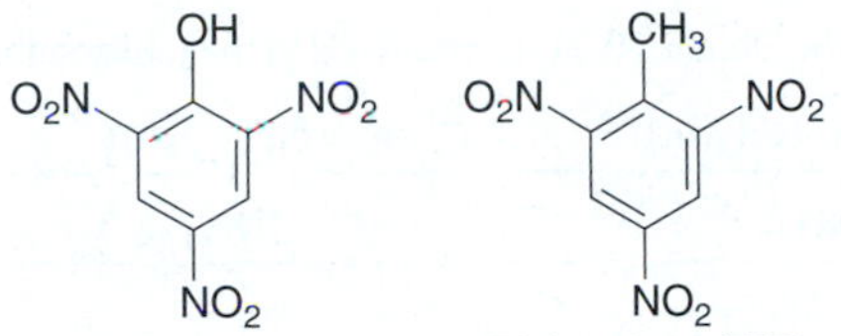

Picric acid   Trinitrotoluene (TNT)

Answer $K_a = 0.39$; $pK_a = 0.41$

EXAMPLE 16.7

A 1.000 *M* aqueous solution of ammonia has a pH of 11.63 at 25°C. Calculate K_b and pK_b for ammonia.

Given Analytical concentration and pH

Asked for K_b and pK_b

Strategy

A Write the balanced equilibrium equation for the reaction, and derive the equilibrium constant expression.

B Use the data given and the stoichiometry of the reaction to construct a table showing initial concentrations, changes in concentrations, and final concentrations for all species in the equilibrium constant expression.

C Substitute the final concentrations into the equilibrium constant expression and calculate the value of K_b. Take the negative logarithm of K_b to obtain pK_b.

Solution

A ✓ The balanced equilibrium equation for the reaction of ammonia with water is

$$NH_3(aq) + H_2O(l) \rightleftharpoons NH_4^+(aq) + OH^-(aq)$$

for which the equilibrium constant expression is

$$K_b = \frac{[NH_4^+][OH^-]}{[NH_3]}$$

(Remember that water does not appear in the equilibrium constant expression for K_b.) **B** ✓ To calculate K_b, we need to know the equilibrium concentrations of NH_3, NH_4^+, and OH^-. The initial concentration of NH_3 is the analytical concentration, 1.000 *M*, and the initial concentrations of NH_4^+ and OH^- are 0 *M* and 1.00×10^{-7} *M*, respectively. In this case, we are given the pH of the solution, which allows us to calculate the final concentration of one species (OH^-) directly, rather than the change in concentration. Recall that $pK_w = pH + pOH = 14.00$

at 25°C. Thus, pOH = 14.00 − pH = 14.00 − 11.63 = 2.37, and $[OH^-]_f = 10^{-2.37} = 4.27 \times 10^{-3}$ *M*. Our data thus far are listed in the table.

$$NH_3(aq) \rightleftharpoons NH_4^+(aq) + OH^-(aq)$$

	$[NH_3]$	$[NH_4^+]$	$[OH^-]$
Initial	1.000	0	1.00×10^{-7}
Change			
Final			4.25×10^{-3}

The final $[OH^-]$ is much greater than the initial $[H^+]$, so the change in $[OH^-]$ is

$$\Delta[OH^-] = (4.25 \times 10^{-3}\,M) - (1.00 \times 10^{-7}\,M) \approx 4.25 \times 10^{-3}\,M$$

The stoichiometry of the reaction tells us that 1 mol of NH_3 is converted to NH_4^+ for each 1 mol of OH^- formed, so

$$\Delta[NH_4^+] = +4.25 \times 10^{-3}\,M \quad \text{and} \quad \Delta[NH_3] = -4.25 \times 10^{-3}\,M$$

Inserting these values for the changes in concentrations in the table enables us to complete the table

$$H_2O(l) + NH_3(aq) \rightleftharpoons NH_4^+(aq) + OH^-(aq)$$

	$[NH_3]$	$[NH_4^+]$	$[OH^-]$
Initial	1.000	0	1.00×10^{-7}
Change	-4.25×10^{-3}	$+4.25 \times 10^{-3}$	$+4.27 \times 10^{-3}$
Final	0.996	4.25×10^{-3}	4.27×10^{-3}

Thus, the final concentrations of all species in the equilibrium constant expression are:

$$[NH_4^+]_f = [NH_4^+]_i + \Delta[NH_4^+] = 0\,M + 4.27 \times 10^{-3}\,M = 4.27 \times 10^{-3}\,M$$
$$[NH_3]_f = [NH_3]_i + \Delta[NH_3] = 1.000\,M + (-4.27 \times 10^{-3}\,M) \approx 0.996\,M$$

Inserting the final concentrations into the equilibrium constant expression gives

$$K_b = \frac{[NH_4^+][OH^-]}{[NH_3]} = \frac{(4.27 \times 10^{-3})^2}{0.996} = 1.83 \times 10^{-5}$$

and $pK_b = -\log K_b = 4.737$.

Note that the values of K_b and pK_b for ammonia are almost exactly the same as the values of K_a and pK_a for acetic acid at 25°C. In other words, ammonia is almost exactly as strong a base as acetic acid is an acid. Consequently, the extent of the ionization reaction in an aqueous solution of ammonia at a given concentration is the same as in an aqueous solution of acetic acid at the same concentration.

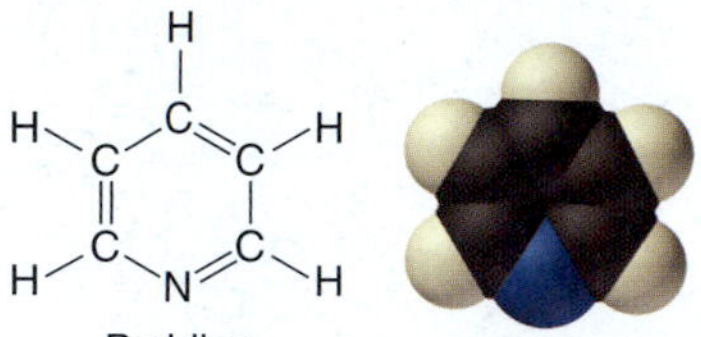

Pyridine

EXERCISE 16.7

The pH of a 0.0500 *M* solution of pyridine (C_6H_5N) is 8.96 at 25°C. Calculate the values of K_b and pK_b for pyridine.

Answer $K_b = 1.7 \times 10^{-9}$; $pK_b = 8.77$

Calculating Percent Ionization from K_a or K_b

MGC pH and Percent Ionization of a Weak Base

When carrying out a laboratory analysis, chemists frequently need to know the concentrations of all species in solution. Because the reactivity of a weak acid or a weak base is usually very different from the reactivity of its conjugate base or acid, we often need to know the percent ionization of a solution of an acid or a base in order to understand a chemical reaction. The percent ionization is defined as

$$\text{percent ionization of acid} = \frac{[H^+]}{C_{HA}} \times 100 \tag{16.40}$$

$$\text{percent ionization of base} = \frac{[OH^-]}{C_B} \times 100 \tag{16.41}$$

One way to determine the concentrations of species in solutions of weak acids and bases is a variation of the tabular method we used above to determine K_a and K_b values. As a demonstration, let's calculate the concentrations of all species and the percent ionization in a 0.150 *M* solution of formic acid at 25°C. The data in Table 16.2 show that formic acid ($K_a = 1.8 \times 10^{-4}$ at 25°C) is a slightly stronger acid than acetic acid. The equilibrium equation for the ionization of formic acid in water is

$$HCO_2H(aq) \rightleftharpoons H^+(aq) + HCO_2^-(aq) \tag{16.42}$$

and the equilibrium constant expression for this reaction is

$$K_a = \frac{[H^+][HCO_2^-]}{[HCO_2H]} \tag{16.43}$$

We set the initial concentration of HCO_2H equal to 0.150 *M*, and that of HCO_2^- is 0 *M*. The initial concentration of H^+ is 1.00×10^{-7} *M* due to the autoionization of water. Because the equilibrium constant for the ionization reaction is small, the equilibrium will lie to the left, favoring the unionized form of the acid. Hence, we can define *x* as the amount of formic acid that dissociates.

If the change in $[HCO_2H]$ is $-x$, then the change in $[H^+]$ and $[HCO_2^-]$ is $+x$. The final concentration of each species is the sum of its initial concentration and the change in concentration, as summarized in the table.

	$HCO_2H(aq)$	$\rightleftharpoons$	$H^+(aq)$	+	$HCO_2^-(aq)$
	$[HCO_2H]$		**$[H^+]$**		**$[HCO_2^-]$**
Initial	0.150		1.00×10^{-7}		0
Change	$-x$		$+x$		$+x$
Final	$(0.150 - x)$		$(1.00 \times 10^{-7} + x)$		x

We can calculate the value of *x* by substituting the final concentrations from the table into the equilibrium constant expression:

$$K_a = \frac{[H^+][HCO_2^-]}{[HCO_2H]} = \frac{(1.00 \times 10^{-7} + x)x}{0.150 - x}$$

Because the ionization constant K_a is small, *x* is likely to be small compared with the initial concentration of formic acid: $(0.150 - x)\,M \approx 0.150\,M$. Moreover, $[H^+]$ due to the autoionization of water (1.00×10^{-7} *M*) is likely to be negligible compared with $[H^+]$ due to the dissociation of formic acid: $(1.00 \times 10^{-7} + x)\,M \approx x\,M$. Inserting these values into the equilibrium constant expression and solving for *x* give

$$K_a = \frac{x^2}{0.150} = 1.8 \times 10^{-4}$$

$$x = 5.2 \times 10^{-3}$$

We can now calculate the concentrations of the species present in a 0.150 *M* formic acid solution by inserting this value of *x* into the expressions in the last line of the table:

$$[HCO_2H] = (0.150 - x)\,M = 0.145\,M$$

$$[HCO_2] = x = 5.2 \times 10^{-3}\,M$$

$$[H^+] = (1.00 \times 10^{-7} + x)\,M = 5.2 \times 10^{-3}\,M$$

Thus, the pH of the solution is $-\log(5.2 \times 10^{-3}) = 2.28$. We can also use these concentrations to calculate the fraction of the original acid that is ionized. In this case, the percent ionization is the ratio of $[H^+]$ (or $[HCO_2^-]$) to the analytical concentration, multiplied by 100 to give a percentage:

$$\text{percent ionization} = \frac{[H^+]}{C_{HA}} \times 100 = \frac{5.2 \times 10^{-3}\ M}{0.150} \times 100 = 3.5\%$$

Always check to make sure that any simplifying assumption was valid. As a general rule of thumb, approximations such as those used above are valid *only* if the quantity being neglected is no more than about 5% of the quantity to which it is being added or from which it is being subtracted. If the quantity that was neglected is much greater than about 5%, then the approximation is probably not valid, and you should go back and solve the problem using the quadratic formula. In the example above, both simplifying assumptions were justified: the percent ionization is only 3.5%, which is well below the approximately 5% limit, and the $1.00 \times 10^{-7}\ M$ $[H^+]$ due to the autoionization of water is much, much less than the 0.0052 M $[H^+]$ due to the ionization of formic acid.

As a general rule, the $[H^+]$ contribution due to the autoionization of water can be ignored as long as the product of the acid (or base) ionization constant and the analytical concentration of the acid (or base) is at least 10 times greater than the $[H^+]$ or $[OH^-]$ from the autoionization of water—that is, if

$$K_aC_{HA} \geq 10(1.00 \times 10^{-7}) = 1.0 \times 10^{-6} \tag{16.44}$$

or

$$K_bC_B \geq 10(1.00 \times 10^{-7}) = 1.0 \times 10^{-6} \tag{16.45}$$

By substituting the appropriate values for the formic acid solution considered above into Equation 16.44, we see that the simplifying assumption is valid in this case:

$$K_aC_{HA} = (1.8 \times 10^{-4})(0.150) = 2.7 \times 10^{-5} > 1.0 \times 10^{-6} \tag{16.46}$$

Doing this simple calculation before solving this type of problem saves time and allows you to write simplified expressions for the final concentrations of species. In practice, it is necessary to include the $[H^+]$ contribution due to the autoionization of water only for extremely dilute solutions of very weak acids or bases.

The next example illustrates how the procedure outlined above can be used to calculate the pH of a solution of a weak base.

EXAMPLE 16.8

Calculate the pH and percent ionization of a 0.225 M solution of ethylamine ($CH_3CH_2NH_2$), which is used in the synthesis of some dyes and medicines. The pK_b of ethylamine is 3.19 at 20°C.

Given Concentration and pK_b

Asked for pH and percent ionization

Strategy

A Write the balanced equilibrium equation for the reaction and the equilibrium constant expression. Calculate the value of K_b from pK_b.

B Use Equation 16.44 to see whether you can ignore $[H^+]$ due to the autoionization of water. Then use a tabular format to write expressions for the final concentrations of all species in solution. Substitute these values into the equilibrium equation, and solve for $[OH^-]$. Use Equation 16.41 to calculate the percent ionization.

C Use the relationship $K_w = [OH^-][H^+]$ to obtain $[H^+]$. Then calculate the pH of the solution.

Solution

A We begin by writing the balanced equilibrium equation for the reaction:

$$CH_3CH_2NH_2(aq) + H_2O(l) \rightleftharpoons CH_3CH_2NH_3^+(aq) + OH^-(aq)$$

The corresponding equilibrium constant expression is

$$K_b = \frac{[CH_3CH_2NH_3^+][OH^-]}{[CH_3CH_2NH_2]}$$

From the value of pK_b, we have $K_b = 10^{-3.19} = 6.46 \times 10^{-4}$. B To calculate the pH, we need to determine the H^+ concentration. Unfortunately, H^+ does not appear in either the chemical equation or the equilibrium constant expression. However, $[H^+]$ and $[OH^-]$ in an aqueous solution are related by the equation $K_w = [H^+][OH^-]$. Hence, if we can determine $[OH^-]$, we can calculate $[H^+]$ and then the pH. The initial concentration of $CH_3CH_2NH_2$ is 0.225 *M*, and the initial $[OH^-]$ is 1.00×10^{-7} *M*. Because ethylamine is a weak base, the extent of the reaction will be small, and it makes sense to let x equal the amount of $CH_3CH_2NH_2$ that reacts with water. The change in $[CH_3CH_2NH_2]$ is therefore $-x$, and the change in both $[CH_3CH_2NH_3^+]$ and $[OH^-]$ is $+x$. To see whether the autoionization of water can safely be ignored, we substitute the values of K_b and C_B into Equation 16.45 to give

$$K_bC_B = (6.46 \times 10^{-4})(0.225) = 1.45 \times 10^{-4} > 1.0 \times 10^{-6}$$

Thus, the simplifying assumption is valid, and we will not include $[OH^-]$ due to the autoionization of water in our calculations.

$$H_2O(l) + CH_3CH_2NH_2(aq) \rightleftharpoons CH_3CH_2NH_3^+(aq) + OH^-(aq)$$

	$[CH_3CH_2NH_2]$	$[CH_3CH_2NH_3^+]$	$[OH^-]$
Initial	0.225	0	1.00×10^{-7}
Change	$-x$	$+x$	$+x$
Final	$(0.225 - x)$	x	x

Substituting the quantities from the last line of the table into the equilibrium constant expression gives

$$K_b = \frac{[CH_3CH_2NH_3^+][OH^-]}{[CH_3CH_2NH_2]} = \frac{(x)(x)}{0.225 - x} = 6.46 \times 10^{-4}$$

As before, we assume the amount of $CH_3CH_2NH_2$ that ionizes is small compared with the initial concentration, so $[CH_3CH_2NH_2]_f = 0.225 - x \approx 0.225$. With this assumption, we can simplify the equilibrium equation and solve for x:

$$K_b = \frac{x^2}{0.225} = 6.46 \times 10^{-4}$$

$$x = 0.0121 = [CH_3CH_2NH_3^+]_f = [OH^-]_f$$

The percent ionization is therefore

$$\text{percent ionization} = \frac{[OH^-]}{C_B} \times 100 = \frac{0.0121\ M}{0.225\ M} \times 100 = 5.4\%$$

which is at the upper limit of the approximately 5% range that can be ignored. The final hydroxide concentration is thus 0.0121 *M*. C We can now determine the value of $[H^+]$ using the expression for K_w:

$$K_w = [OH^-][H^+]$$

$$1.01 \times 10^{-14} = (0.0121\ M)[H^+]$$

$$[H^+] = 8.35 \times 10^{-13}\ M$$

The pH of the solution is $-\log(8.35 \times 10^{-13}) = 12.08$. Alternatively, we could have calculated pOH as $-\log(0.0121) = 1.917$ and determined the pH as follows:

$$pH + pOH = pK_w = 14.00$$

$$pH = 14.00 - 1.917 = 12.08$$

The two methods are equivalent.

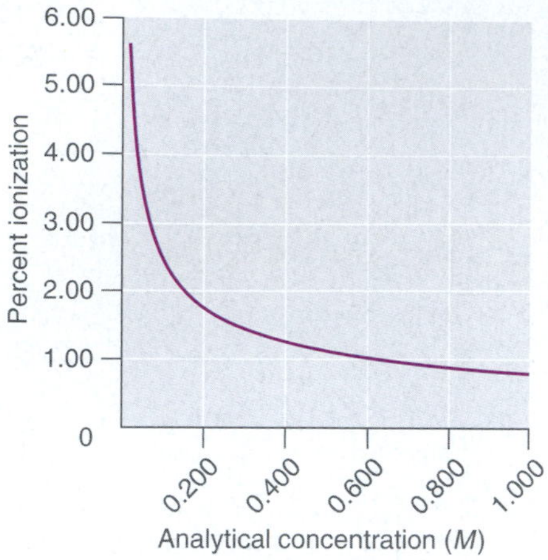

Figure 16.7 The relationship between the analytical concentration of a weak acid and the percent ionization. As shown here for benzoic acid ($C_6H_5CO_2H$), the percent ionization decreases as the analytical concentration of a weak acid increases.

EXERCISE 16.8

Aromatic amines, in which the nitrogen atom is bonded directly to a phenyl ring ($C_6H_5^-$) tend to be much weaker bases than simple alkylamines. For example, aniline ($C_6H_5NH_2$) has a pK_b of 9.13 at 25°C. What is the pH of a 0.050 *M* solution of aniline?

Answer 8.78

The examples worked above illustrate a key difference between solutions of strong acids and bases and solutions of weak acids and bases. Because strong acids and bases ionize essentially completely in water, the percent ionization is always approximately 100%, regardless of the concentration. In contrast, the percent ionization in solutions of weak acids and bases is small and depends on the analytical concentration of the weak acid or base. As illustrated for benzoic acid in Figure 16.7, the percent ionization of a weak acid or a weak base actually *increases* as its analytical concentration decreases. The percent ionization also increases as the magnitude of the ionization constants K_a and K_b increases.

Thus, unlike K_a or K_b, the percent ionization is not a constant for weak acids and bases but depends on *both* the acid or base ionization constant *and* the analytical concentration. Consequently, the procedure outlined above must be used to calculate the percent ionization and pH for solutions of weak acids and bases. The following example and exercise demonstrate that the combination of a dilute solution and a relatively large value of K_a or K_b can give a percent ionization much greater than 5%, making it necessary to use the quadratic equation to determine the concentrations of species in solution.

Note the pattern

The percent ionization in a solution of a weak acid or a weak base increases as the analytical concentration decreases and as the value of K_a or K_b increases.

EXAMPLE 16.9

Benzoic acid ($C_6H_5CO_2H$) is used in the food industry as a preservative and medically as an antifungal agent. Its pK_a at 25°C is 4.20, making it a somewhat stronger acid than acetic acid. Calculate the percentage of benzoic acid molecules that are ionized in **(a)** a 0.0500 *M* solution and **(b)** a 0.00500 *M* solution.

Given Concentrations and pK_a

Asked for Percent ionization

Strategy

Ⓐ Write both the balanced equilibrium equation for the ionization reaction and the equilibrium equation (Equation 16.13). Use Equation 16.21b to calculate the value of K_a from pK_a.

Ⓑ For both the concentrated and dilute solutions, use a tabular format to write expressions for the final concentrations of all species in solution. Substitute these values into the equilibrium equation, and solve for $[C_6H_5CO_2^-]_f$ for each solution.

Ⓒ Use the values of $[C_6H_5CO_2^-]_f$ and Equation 16.40 to calculate the percent ionization.

Solution

Ⓐ If we abbreviate benzoic acid as $PhCO_2H$ where Ph = C_6H_5—, the balanced equilibrium equation for the ionization reaction and the equilibrium equation can be written as

$$PhCO_2H(aq) \rightleftharpoons H^+(aq) + PhCO_2^-(aq)$$

$$K_a = \frac{[H^+][PhCO_2^-]}{[PhCO_2H]}$$

From the value of pK_a, we have $K_a = 10^{-4.20} = 6.31 \times 10^{-5}$.

(a) B For the more concentrated solution, we set up our table of initial concentrations, changes in concentrations, and final concentrations:

	$PhCO_2H(aq)$ ⇌	$H^+(aq)$ +	$PhCO_2^-(aq)$
	$[PhCO_2H]$	**$[H^+]$**	**$[PhCO_2^-]$**
Initial	0.0500	1.00×10^{-7}	0
Change	$-x$	$+x$	$+x$
Final	$(0.0500 - x)$	$(1.00 \times 10^{-7} + x)$	x

Inserting the expressions for the final concentrations into the equilibrium equation and making our usual assumptions, that $[PhCO_2^-]_f$ and $[H^+]$ due to the autoionization of water are negligible, give

$$K_a = \frac{[H^+][PhCO_2^-]}{[PhCO_2H]} = \frac{(x)(x)}{0.0500 - x} = \frac{x^2}{0.0500} = 6.31 \times 10^{-5}$$

$$x = 1.78 \times 10^{-3}$$

This value is less than 5% of 0.0500, so our simplifying assumption is justified, and $[PhCO_2^-]$ at equilibrium is 1.78×10^{-3} M. We reach the same conclusion using C_{HA}: $K_aC_{HA} = (6.31 \times 10^{-5})(0.0500) = 3.16 \times 10^{-6} > 1.0 \times 10^{-6}$. C The percent ionized is the ratio of the concentration of $PhCO_2^-$ to the analytical concentration, multiplied by 100:

$$\text{percent ionized} = \frac{[PhCO_2^-]}{C_{PhCO_2H}} \times 100 = \frac{1.78 \times 10^{-3}}{0.0500} \times 100 = 3.6\%$$

Because only 3.6% of the benzoic acid molecules are ionized in a 0.0500 M solution, our simplifying assumptions are confirmed.

(b) B For the more dilute solution, we proceed in exactly the same manner. Our table of concentrations is therefore

	$PhCO_2H(aq)$ ⇌	$H^+(aq)$ +	$PhCO_2^-(aq)$
	$[PhCO_2H]$	**$[H^+]$**	**$[PhCO_2^-]$**
Initial	0.00500	1.00×10^{-7}	0
Change	$-x$	$+x$	$+x$
Final	$(0.00500 - x)$	$(1.00 \times 10^{-7} + x)$	x

Inserting the expressions for the final concentrations into the equilibrium equation and making our usual simplifying assumptions give

$$K_a = \frac{[H^+][PhCO_2^-]}{[PhCO_2H]} = \frac{(x)(x)}{0.00500 - x} = \frac{x^2}{0.00500} = 6.31 \times 10^{-5}$$

$$x = 5.62 \times 10^{-4}$$

Unfortunately, this number is greater than 10% of 0.00500, so our assumption that the fraction of benzoic acid that is ionized in this solution could be neglected and that $(0.00500 - x) \approx x$ is not valid. Furthermore, we see that $K_a C_{HA} = (6.31 \times 10^{-5})(0.00500) = 3.16 \times 10^{-7} < 1.0 \times 10^{-6}$. Thus, the relevant equation is

$$\frac{x^2}{0.00500 - x} = 6.31 \times 10^{-5}$$

which must be solved using the quadratic formula. Multiplying out the quantities gives

$$x^2 = (6.31 \times 10^{-5})(0.00500 - x) = 3.16 \times 10^{-7} - (6.31 \times 10^{-5})x$$

Rearranging the equation to fit the standard quadratic equation format gives

$$x^2 + (6.31 \times 10^{-5})x - (3.16 \times 10^{-7}) = 0$$

This equation can be solved using the quadratic formula:

$$x = \frac{-b \pm \sqrt{b^2 - 4ac}}{2a}$$

$$= \frac{-(6.31 \times 10^{-5}) \pm \sqrt{(6.31 \times 10^{-5})^2 - 4(1)(-3.16 \times 10^{-7})}}{2(1)}$$

$$= \frac{-(6.31 \times 10^{-5}) \pm (1.12 \times 10^{-3})}{2} = 5.28 \times 10^{-4} \text{ or } -5.92 \times 10^{-4}$$

Because a negative value of x corresponds to a negative $[PhCO_2^-]$, which is not physically meaningful, we use the positive solution: $x = 5.28 \times 10^{-4}$. Thus, $[PhCO_2^-] = 5.28 \times 10^{-4}\,M$. The percent ionized is therefore

$$\text{percent ionized} = \frac{[PhCO_2^-]}{C_{PhCO_2H}} \times 100 = \frac{5.28 \times 10^{-4}}{0.00500} \times 100 = 10.6\%$$

Thus, in the more dilute solution ($C = 0.00500\,M$), 10.6% of the benzoic acid molecules are ionized versus only 3.6% in the more concentrated solution ($C = 0.0500\,M$). Decreasing the analytical concentration by a factor of 10 results in an approximately threefold increase in the percentage of benzoic acid molecules that are ionized.

EXERCISE 16.9

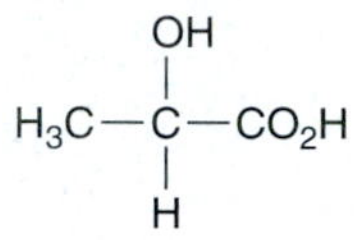

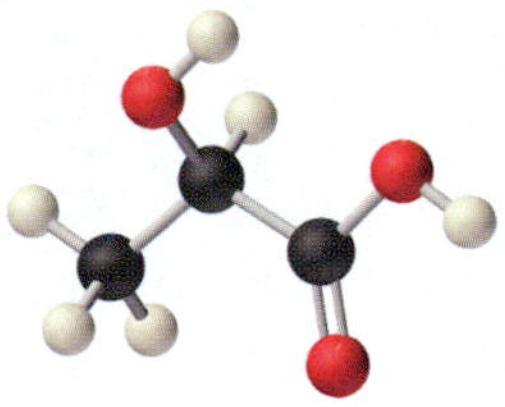

Lactic acid

Lactic acid, $CH_3CH(OH)CO_2H$, is a weak acid with a pK_a of 3.86 at 25°C. What percentage of the lactic acid is ionized in **(a)** a 0.10 M solution and **(b)** a 0.0020 M solution?

Answer **(a)** 3.7%; **(b)** 26%

Note the pattern

Simplifying assumptions in equilibrium calculations for solutions of weak acids and bases are most likely to be valid for concentrated solutions.

Determining K_{eq} from K_a and K_b

In Section 16.2, you learned how to use K_a and K_b values to qualitatively predict whether reactants or products are favored in an acid–base reaction. Tabulated values of K_a (or pK_a) and K_b (or pK_b), plus the value of K_w, enable us to quantitatively determine the direction and extent of reaction for a weak acid and a weak base by calculating the value of K for the reaction. To illustrate how to do this, we begin by writing the dissociation equilibria for a weak acid and a weak base and then summing them:

$$\begin{array}{llll} \textit{Acid} & HA \rightleftharpoons H^+ + A^- & K_a & \\ \textit{Base} & B + H_2O \rightleftharpoons HB^+ + OH^- & K_b & (16.47) \\ \hline \textit{Sum} & HA + B + H_2O \rightleftharpoons H^+ + A^- + HB^+ + OH^- & K_{sum} = K_aK_b & \end{array}$$

Note that the overall reaction has H_2O on the left and H^+ and OH^- on the right, which means it involves the autoionization of water ($H_2O \rightleftharpoons H^+ + OH^-$) in addition to the acid–base equilibrium in which we are interested. We can obtain an equation that includes only the acid–base equilibrium by simply adding the equation for the reverse of the autoionization of water ($H^+ + OH^- \rightleftharpoons H_2O$), for which $K = 1/K_w$, to the overall equilibrium in Equation 16.47 and canceling:

$$\begin{array}{lll} HA + B + \cancel{H_2O} \rightleftharpoons \cancel{H^+} + A^- + HB^+ + \cancel{OH^-} & K_{sum} = K_aK_b & \\ \cancel{H^+} + \cancel{OH^-} \rightleftharpoons \cancel{H_2O} & 1/K_w & (16.48) \\ \hline HA + B \rightleftharpoons A^- + HB^+ & K = (K_aK_b)/K_w & \end{array}$$

Thus, the equilibrium constant for the reaction of a weak acid with a weak base is the product of the ionization constants of the acid and the base divided by K_w. The next example illustrates how to calculate the equilibrium constant for the reaction of a weak acid with a weak base.

EXAMPLE 16.10

Fish tend to spoil rapidly even when refrigerated. The cause of the resulting "fishy" odor is a mixture of amines, particularly methylamine (CH_3NH_2), a volatile weak base ($pK_b = 3.34$). Fish is often served with a wedge of lemon because lemon juice contains citric acid, a triprotic acid with pK_a's of 3.13, 4.76, and 6.40 that can neutralize amines. Calculate the equilibrium constant for the reaction of excess citric acid with methylamine, assuming that only the first dissociation constant of citric acid is important.

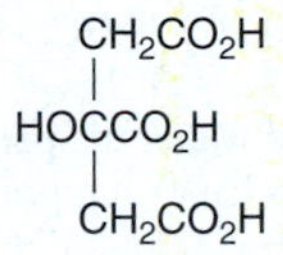

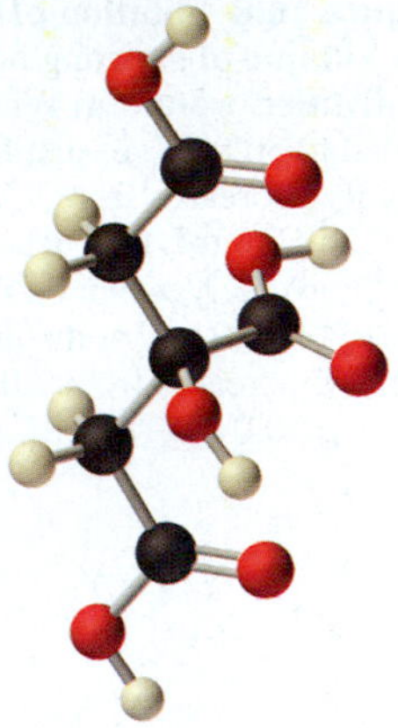

Citric acid

Given pK_b for base and pK_a for acid

Asked for K

Strategy

A Write the balanced equilibrium equation and the equilibrium constant expression for the reaction.
B Convert pK_a and pK_b to K_a and K_b, and then use Equation 16.48 to calculate the value of K.

Solution

A If we abbreviate citric acid as H_3citrate, the equilibrium equation for its reaction with methylamine is

$$CH_3NH_2(aq) + H_3\text{citrate}(aq) \rightleftharpoons CH_3NH_3^+(aq) + H_2\text{citrate}^-(aq)$$

The equilibrium constant expression for this reaction is

$$K = \frac{[CH_3NH_3^+][H_2\text{citrate}^-]}{[CH_3NH_2][H_3\text{citrate}]}$$

B Equation 16.48 is $K = (K_aK_b)/K_w$. Converting pK_a and pK_b to K_a and K_b gives $K_a = 10^{-3.13} = 7.41 \times 10^{-4}$ for citric acid and $K_b = 10^{-3.34} = 4.57 \times 10^{-4}$ for methylamine. Substituting these values into the equilibrium equation gives

$$K = \frac{K_aK_b}{K_w} = \frac{(7.41 \times 10^{-4})(4.57 \times 10^{-4})}{1.01 \times 10^{-14}} = 3.35 \times 10^7 = 3.4 \times 10^7$$

The value of pK can also be calculated directly by taking the negative logarithm of both sides of Equation 16.48, which gives

$$pK = pK_a + pK_b - pK_w = 3.13 + 3.34 - 14.00 = -7.53$$

Thus, $K = 10^{-(-7.53)} = 3.4 \times 10^7$, in agreement with the earlier value. In either case, the K values show that the reaction of citric acid with the volatile, foul-smelling methylamine lies very far to the right, favoring the formation of a much less volatile salt with no odor. This is one reason a little lemon juice helps make less-than-fresh fish more appetizing.

EXERCISE 16.10

Dilute aqueous ammonia solution, often used as a cleaning agent, is also effective as a deodorizing agent. To see why, calculate the equilibrium constant for the reaction of aqueous ammonia with butyric acid ($CH_3CH_2CH_2CO_2H$), a particularly foul-smelling substance associated with the odor of rancid butter and smelly socks. The pK_b of ammonia is 4.75, and the pK_a of butyric acid is 4.83.

Answer 2.6×10^4

16.5 • Acid–Base Titrations

In Chapter 4, you learned that in an acid–base titration, a buret is used to deliver measured volumes of an acid or base solution of known concentration (the *titrant*) to a flask that contains a solution of a base or an acid, respectively, of unknown concentration

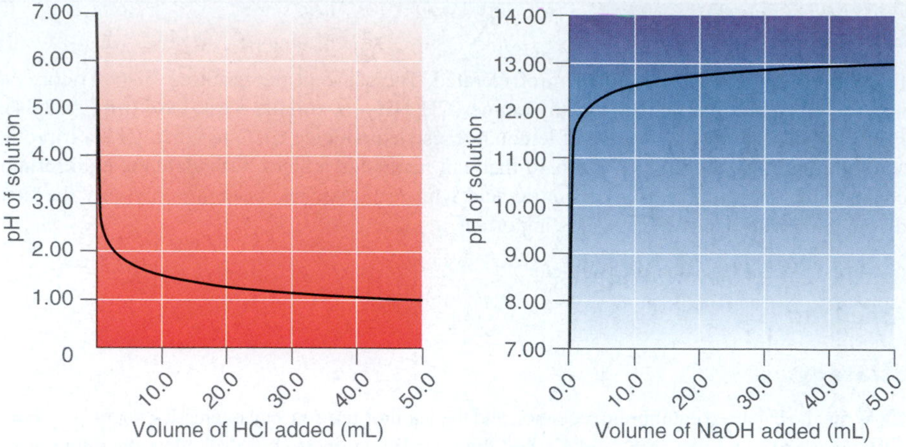

(a) Water with strong acid added **(b) Water with strong base added**

Figure 16.8 Solution pH as a function of the volume of a strong acid or base added to distilled water. (a) When 0.200 *M* HCl is added to 50.0 mL of distilled water, the pH rapidly decreases from 7.0 to 0.70 at a point beyond 50.0 mL (the pH of 0.200 *M* HCl). **(b)** Conversely, when 0.200 *M* NaOH is added to 50.0 mL of distilled water, the pH rapidly increases from 7.0 to 13.30 at a point beyond 50.0 mL (the pH of 0.200 *M* NaOH).

(the *unknown*). If the concentration of the titrant is known, then the concentration of the unknown can be determined. The following discussion focuses on the pH changes that occur during an acid–base titration. Plotting the pH of the solution in the flask against the amount of acid or base added produces a **titration curve**. The shape of the curve provides important information about what is occurring in solution during the titration.

Titrations of Strong Acids and Bases

MGC Titration of Strong Acid with Strong Base

Figure 16.8a shows a plot of the pH as 0.200 *M* HCl is gradually added to 50.00 mL of pure water. The pH of the sample in the flask is initially 7.00 (as expected for pure water), but it drops very rapidly as HCl is added. Eventually the pH becomes constant at a value of 0.70 at a point well beyond its value of 1.00 with the addition of 50.0 mL of HCl, the pH of 0.200 *M* HCl. In contrast, when 0.200 *M* NaOH is added to 50.00 mL of distilled water, the pH (initially 7.00) climbs very rapidly at first, but then more gradually, eventually approaching a limit of 13.30 (the pH of 0.200 *M* NaOH), again well beyond its value of 13.00 with the addition of 50.0 mL of NaOH as shown in Figure 16.8b. As you can see from these plots, the titration curve for the addition of a base is the mirror image of the curve for the addition of an acid.

Suppose that we now add 0.200 *M* NaOH to 50.00 mL of a 0.100 *M* solution of HCl. Because HCl is a strong acid that is completely ionized in water, the initial $[H^+]$ is 0.100 *M*, and the initial pH is 1.00. Adding NaOH decreases the concentration of H^+ because of the neutralization reaction: $OH^- + H^+ \longrightarrow H_2O$ (Figure 16.9a). Thus, the pH of the

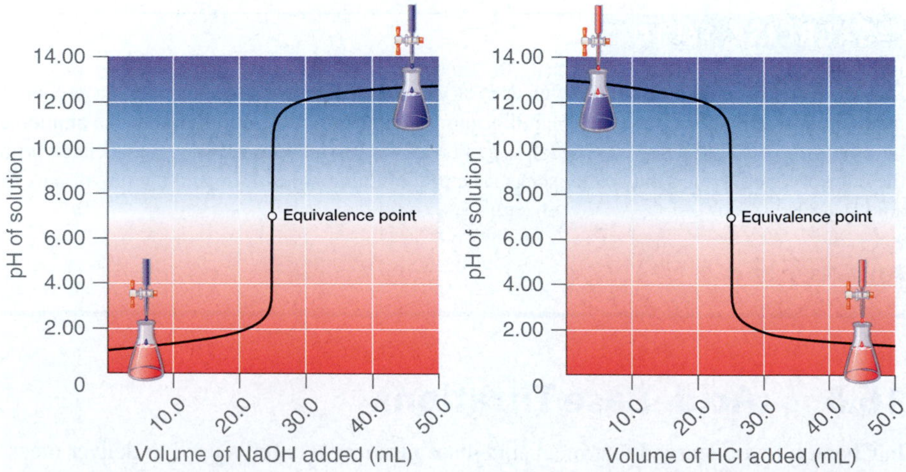

(a) Strong acid titrated with strong base **(b) Strong base titrated with strong acid**

Figure 16.9 The titration of (a) a strong acid with a strong base and (b) a strong base with a strong acid. (a) As 0.200 *M* NaOH is slowly added to 50.0 mL of 0.100 *M* HCl, the pH increases slowly at first, then increases very rapidly as the equivalence point is approached, and finally increases slowly once more. **(b)** Conversely, as 0.200 *M* HCl is slowly added to 50.0 mL of 0.100 *M* NaOH, the pH decreases slowly at first, then decreases very rapidly as the equivalence point is approached, and finally decreases slowly once more.

solution increases gradually. Near the **equivalence point**, however, the point at which the number of moles of base (or acid) added equals the number of moles of acid (or base) originally present in the solution (see Chapter 4) the pH increases much more rapidly because most of the H^+ ions originally present have been consumed. For the titration of a monoprotic strong acid (HCl) with a *monobasic* strong base (NaOH), we can calculate the volume of base needed to reach the equivalence point from the following relationship:

$$\begin{aligned} \text{moles of base} &= \text{moles of acid} \\ (\text{volume})_b(\text{molarity})_b &= (\text{volume})_a(\text{molarity})_a \qquad (16.49) \\ V_b M_b &= V_a M_a \end{aligned}$$

If 0.200 *M* NaOH is added to 50.00 mL of a 0.100 *M* solution of HCl, we solve for V_b to obtain

$$\begin{aligned} V_b(0.200\ \cancel{M}) &= (0.0500\ \text{L})(0.100\ \cancel{M}) \\ V_b &= 0.0250\ \text{L} = 25.0\ \text{mL} \end{aligned}$$

At the equivalence point (when 25.00 mL of NaOH solution has been added), the neutralization is complete: only a salt remains in solution (NaCl), and the pH of the solution is 7.00. Adding more NaOH produces a rapid increase in pH, but eventually the pH levels off at a value of about 13.30, the pH of 0.200 *M* NaOH.

As shown in Figure 16.9b, the titration of 50.0 mL of a 0.100 *M* solution of NaOH with 0.200 *M* HCl produces a titration curve that is nearly the mirror image of the titration curve in Figure 16.9a. The pH is initially 13.00, and it slowly decreases as HCl is added. As the equivalence point is approached, the pH drops rapidly before leveling off at a value of about 0.70, the pH of 0.2 *M* HCl.

Notice that the titration of either a strong acid with a strong base or a strong base with a strong acid produces an S-shaped curve. The curve is somewhat asymmetrical because the steady increase in the volume of the solution during the titration causes the solution to become more dilute. Due to the leveling effect, the shape of the curve for a titration involving a strong acid and a strong base depends *only* on the concentrations of the acid and base, *not* on their identity.

EXAMPLE 16.11

Calculate the pH of the solution after 24.90 mL of 0.200 *M* NaOH has been added to 50.00 mL of 0.100 *M* HCl.

Given Volumes and concentrations of strong base and acid

Asked for pH

Strategy

A Calculate the number of millimoles of H^+ and OH^- to determine which, if either, is in excess after the neutralization reaction has occurred. If one species is in excess, calculate the amount that remains after the neutralization reaction.

B Determine the final volume of the solution. Calculate the concentration of the species in excess, and convert this value to pH.

Solution

A Because 0.100 mol/L is equivalent to 0.100 mmol/mL, the number of millimoles of H^+ in 50.00 mL of 0.100 *M* HCl can be calculated as

$$50.00\ \cancel{\text{mL}}\left(\frac{0.100\ \text{mmol HCl}}{\cancel{\text{mL}}}\right) = 5.00\ \text{mmol HCl} = 5.00\ \text{mmol H}^+$$

The number of millimoles of NaOH added is

$$24.90\ \cancel{\text{mL}}\left(\frac{0.200\ \text{mmol NaOH}}{\cancel{\text{mL}}}\right) = 4.98\ \text{mmol NaOH} = 4.98\ \text{mmol OH}^-$$

Thus, H^+ is in excess. To completely neutralize the acid requires the addition of 5.00 mmol of OH^- to the HCl solution. Because only 4.98 mmol of OH^- has been added, the amount of excess H^+ is 5.00 mmol − 4.98 mmol = 0.02 mmol of H^+. The final volume of the solution is 50.00 mL + 24.90 mL = 74.90 mL, so the final concentration of H^+ is

$$[H^+] = \frac{0.02 \text{ mmol } H^+}{74.90 \text{ mL}} = 3 \times 10^{-4}\,M$$

The pH is $-\log[H^+] = -\log(3 \times 10^{-4}) = 3.5$, which is significantly lower than the pH of 7.00 for a neutral solution.

EXERCISE 16.11

Calculate the pH of a solution prepared by adding 40.00 mL of 0.237 *M* HCl to 75.00 mL of a 0.133 *M* solution of NaOH.

Answer 11.6

Titrations of Weak Acids and Bases

MGC Titration of a Weak Acid with a Strong Base

In contrast to strong acids and bases, the shape of the titration curve for a weak acid or a weak base depends dramatically on the identity of the acid or base and the corresponding value of K_a or K_b. As we shall see, the pH also changes much more gradually around the equivalence point in the titration of a weak acid or a weak base. As you learned in the preceding section, $[H^+]$ of a solution of a weak acid (HA) is *not* equal to the concentration of the acid, but depends on both its pK_a and its concentration. Because only a fraction of a weak acid dissociates, $[H^+]$ is less than [HA]. Thus, the pH of a solution of a weak acid is higher than the pH of a solution of a strong acid of the same concentration. Figure 16.10a shows the titration curve for 50.0 mL of a 0.100 *M* solution of acetic acid with 0.200 *M* NaOH, superimposed on the curve for the titration of 0.100 *M* HCl shown in Figure 16.9a. Below the equivalence point, the two curves are very different. Notice that before any base is added, the pH of the acetic acid solution is higher than the pH of the HCl solution, and the pH changes more rapidly during the first part of the titration. Note also that the pH of the acetic acid solution at the equivalence point is greater than 7.00. That is, at the equivalence point, the solution is basic. In addition, the change in pH around the equivalence point is only about half as large as for the HCl titration; the magnitude of the pH change at the equivalence point depends on the pK_a of the acid being titrated. Above the equivalence point, however, the two curves are identical. Once the acid

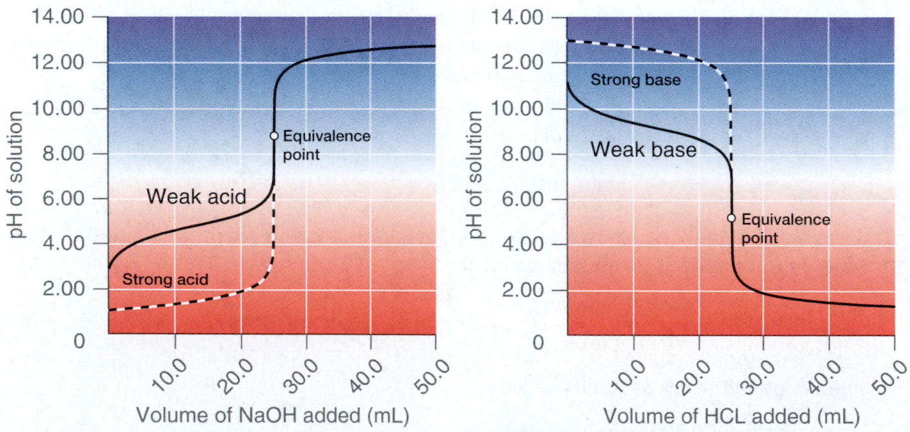

(a) Weak acid titrated with strong base **(b) Weak base titrated with strong acid**

Figure 16.10 The titration of (a) a weak acid with a strong base and (b) a weak base with a strong acid. (a) As 0.200 *M* NaOH is slowly added to 50.0 mL of 0.100 *M* acetic acid, the pH increases slowly at first, then increases rapidly as the equivalence point is approached, and then again increases more slowly. The corresponding curve for the titration of 50.0 mL of 0.100 *M* HCl with 0.200 *M* NaOH is shown in yellow. **(b)** As 0.200 *M* HCl is slowly added to 50.0 mL of 0.100 *M* NH_3, the pH decreases slowly at first, then decreases rapidly as the equivalence point is approached, and then again decreases more slowly. The corresponding curve for the titration of 50.0 mL of 0.100 *M* NaOH with 0.200 *M* HCl is shown in yellow.

has been neutralized, the pH of the solution is controlled only by the amount of excess NaOH present, regardless of whether the acid is weak or strong.

The titration curve in Figure 16.10a was created by calculating the starting pH of the acetic acid solution before any NaOH is added, and then calculating the pH of the solution after the addition of increasing volumes of NaOH. The procedure is illustrated below and in Example 16.12 for three points on the titration curve, using the pK_a of acetic acid (4.76 at 25°C; $K_a = 1.7 \times 10^{-5}$).

Calculating the pH of a Solution of a Weak Acid or Base

As explained in the preceding section, if we know K_a or K_b and the initial concentration of a weak acid or base, we can calculate the pH of a solution of a weak acid or base by setting up a table of initial concentrations, changes in concentrations, and final concentrations. In this example, the initial concentration of acetic acid is 0.100 *M*. If we define x as $[H^+]$ due to the dissociation of the acid, then the table of concentrations for the ionization of 0.100 *M* acetic acid is shown here.

$$CH_3CO_2H(aq) \rightleftharpoons H^+(aq) + CH_3CO_2^-(aq)$$

	$[CH_3CO_2H]$	$[H^+]$	$[CH_3CO_2^-]$
Initial	0.100	1.00×10^{-7}	0
Change	$-x$	$+x$	$+x$
Final	$(0.100 - x)$	x	x

In this and all subsequent examples, we will ignore $[H^+]$ and $[OH^-]$ due to the autoionization of water when calculating the final concentration. However, you should use Equations 16.44 and 16.45 to check that this assumption is justified.

Inserting the expressions for the final concentrations into the equilibrium equation (and using approximations) gives

$$K_a = \frac{[H^+][CH_3CO_2^-]}{[CH_3CO_2H]} = \frac{(x)(x)}{0.100 - x} \approx \frac{x^2}{0.100} = 1.74 \times 10^{-5}$$

Solving this equation gives $x = [H^+] = 1.32 \times 10^{-3}$ *M*. Thus, the pH of a 0.100 *M* solution of acetic acid is

$$\text{pH} = -\log(1.32 \times 10^{-3}) = 2.88$$

Calculating the pH During Titration of a Weak Acid or Base

Now consider what happens when we add 5.00 mL of 0.200 *M* NaOH to 50.00 mL of 0.100 *M* CH_3CO_2H (Figure 16.10a). Because the neutralization reaction proceeds to completion, all of the OH^- ions added will react with the acetic acid to generate acetate ion and water:

$$CH_3CO_2H(aq) + OH^-(aq) \rightarrow CH_3CO_2^-(aq) + H_2O(l) \qquad (16.50)$$

All problems of this type must be solved in two steps: a stoichiometric calculation followed by an equilibrium calculation. In the first step, we use the stoichiometry of the neutralization reaction to calculate the *amounts* of acid and conjugate base present in solution after the neutralization reaction has occurred. In the second step, we use the equilibrium equation (Equation 16.13) to determine $[H^+]$ of the resulting solution.

Step 1 To determine the amount of acid and conjugate base in solution after the neutralization reaction, we calculate the amount of CH_3CO_2H in the original solution and the amount of OH^- in the NaOH solution that was added. The acetic acid solution contained

$$50.00\ \cancel{\text{mL}}\left(\frac{0.100\text{ mmol }CH_3CO_2H}{\cancel{\text{mL}}}\right) = 5.00\text{ mmol }CH_3CO_2H$$

The NaOH solution contained

$$50.00\ \cancel{\text{mL}}\left(\frac{0.200\ \text{mmol NaOH}}{\cancel{\text{mL}}}\right) = 1.00\ \text{mmol NaOH}$$

Comparing the amounts shows that CH_3CO_2H is in excess. Since OH^- reacts with CH_3CO_2H in a 1:1 stoichiometry, the amount of excess CH_3CO_2H is

$$5.00\ \text{mmol}\ CH_3CO_2H - 1.00\ \text{mmol}\ OH^- = 4.00\ \text{mmol}\ CH_3CO_2H$$

Each 1 mmol of OH^- reacts to produce 1 mmol of acetate ion, so the final amount of $CH_3CO_2^-$ is 1.00 mmol.

The stoichiometry of the reaction is summarized in the table, which shows the numbers of moles of the various species, *not* their concentrations.

	$CH_3CO_2H(aq)$	+ $OH^-(aq)$	$\longrightarrow$ $CH_3CO_2^-(aq) + H_2O(l)$
Initial	5.00 mmol	1.00 mmol	0 mmol
Change	− 1.00 mmol	− 1.00 mmol	+ 1.00 mmol
Final	4.00 mmol	0 mmol	1.00 mmol

Note that the table gives the initial amount of acetate and the final amount of OH^- ions as 0. Because an aqueous solution of acetic acid always contains at least a small amount of acetate ion in equilibrium with acetic acid, however, the initial acetate concentration is not actually 0. The value can be ignored in this calculation because the amount of $CH_3CO_2^-$ in equilibrium is insignificant compared to the amount of OH^- added. Moreover, due to the autoionization of water, no aqueous solution can contain 0 mmol of OH^-, but the amount of OH^- due to the autoionization of water is insignificant compared to the amount of OH^- added. We use the initial amounts of the reactants to determine the stoichiometry of the reaction, and defer a consideration of the equilibrium until the second half of the problem.

Step 2 To calculate $[H^+]$ at equilibrium following the addition of NaOH, we must first calculate $[CH_3CO_2H]$ and $[CH_3CO_2^-]$ using the number of millimoles of each and the total volume of the solution at this point in the titration:

$$\text{final volume} = 50.00\ \text{mL} + 5.00\ \text{mL} = 55.00\ \text{mL}$$

$$[CH_3CO_2H] = \frac{4.00\ \text{mmol}\ CH_3CO_2H}{55.00\ \text{mL}} = 7.27 \times 10^{-2}\ M$$

$$[CH_3CO_2^-] = \frac{1.00\ \text{mmol}\ CH_3CO_2^-}{55.00\ \text{mL}} = 1.82 \times 10^{-2}\ M$$

Knowing the concentrations of acetic acid and acetate ion at equilibrium and K_a for acetic acid (1.74×10^{-5}), we can use Equation 16.13 to calculate $[H^+]$ at equilibrium:

$$K_a = \frac{[CH_3CO_2^-][H^+]}{[CH_3CO_2H]}$$

$$[H^+] = \frac{K_a[CH_3CO_2H]}{[CH_3CO_2^-]} = \frac{(1.74 \times 10^{-5})(7.27 \times 10^{-2}\ M)}{1.82 \times 10^{-2}} = 6.95 \times 10^{-5}\ M$$

Calculating $-\log[H^+]$ gives $\text{pH} = -\log(6.95 \times 10^{-5}) = 4.16$.

Comparing the titration curves for HCl and acetic acid in Figure 16.10a, we see that adding the same amount (5.00 mL) of 0.200 *M* NaOH to 50 mL of a 0.10 *M* solution of both acids causes a much smaller pH change for HCl (from 1.00 to 1.14) than for acetic acid (2.88 to 4.16). This is consistent with the qualitative description of the shapes of the titration curves at the beginning of this section. In the next example, we calculate another point for constructing the titration curve of acetic acid.

EXAMPLE 16.12

What is the pH of the solution after 25.00 mL of 0.200 *M* NaOH is added to 50.00 mL of 0.100 *M* acetic acid?

Given Volume and molarity of base and acid

Asked for pH

Strategy

A Write the balanced chemical equation for the reaction. Then calculate the initial numbers of millimoles of OH^- and CH_3CO_2H. Determine which species, if either, is present in excess.
B Tabulate the results showing initial numbers, changes, and final numbers of millimoles.
C If excess acetate is present after the reaction with OH^-, write the equation for the reaction of acetate with water. Use a tabular format to obtain the concentrations of all species present.
D Calculate K_b using the relationship $K_w = K_a K_b$ (Equation 16.20). Calculate $[OH^-]$, and use this to calculate the pH of the solution.

Solution

A Ignoring the spectator ion, Na^+, the equation for this reaction is

$$CH_3CO_2H(aq) + OH^-(aq) \rightarrow CH_3CO_2^-(aq) + H_2O(l)$$

The initial numbers of millimoles of OH^- and CH_3CO_2H are

$$25.00\ \cancel{\text{mL}}\left(\frac{0.200\ \text{mmol OH}^-}{\cancel{\text{mL}}}\right) = 5.00\ \text{mmol OH}^-$$

$$50.00\ \cancel{\text{mL}}\left(\frac{0.100\ \text{mmol CH}_3\text{CO}_2\text{H}}{\cancel{\text{mL}}}\right) = 5.00\ \text{mmol CH}_3\text{CO}_2\text{H}$$

The number of millimoles of OH^- equals the number of millimoles of CH_3CO_2H, so neither species is present in excess.

B Because the number of millimoles of OH^- added corresponds to the number of millimoles of acetic acid in solution, this is the equivalence point. The results of the neutralization reaction can be summarized in tabular form.

	$CH_3CO_2H(aq)$	+	$OH^-(aq)$	$\longrightarrow$	$CH_3CO_2^-(aq) + H_2O(l)$
Initial	5.00 mmol		5.00 mmol		0 mmol
Change	− 5.00 mmol		− 5.00 mmol		+ 5.00 mmol
Final	0 mmol		0 mmol		5.00 mmol

C Because the product of the neutralization reaction is a weak base, we must consider the reaction of the weak base *with water* in order to calculate $[H^+]$ at equilibrium and thus the final pH of the solution. The initial concentration of acetate is obtained from the neutralization reaction:

$$[CH_3CO_2] = \frac{5.00\ \text{mmol CH}_3\text{CO}_2}{(50.00 + 25.00)\ \text{mL}} = 6.67 \times 10^{-2}\ M$$

The equilibrium reaction of acetate with water is

$$CH_3CO_2^-(aq) + H_2O(l) \rightleftharpoons CH_3CO_2H(aq) + OH^-(aq)$$

The equilibrium constant for this reaction is $K_b = K_w/K_a$, where K_a is the acid ionization constant of acetic acid. We therefore define x as $[OH^-]$ produced by the reaction of acetate with water. Here is the completed table of concentrations:

	$H_2O(l)$ + $CH_3CO_2^-(aq)$	$\rightleftharpoons$ $CH_3CO_2H(aq)$ +	$OH^-(aq)$
	$[CH_3CO_2^-]$	**$[CH_3CO_2H]$**	**$[OH^-]$**
Initial	0.0667	0	1.00×10^{-7}
Change	$-x$	$+x$	$+x$
Final	$(0.0667 - x)$	x	x

Substituting the expressions for the final values from this table into Equation 16.16 gives

$$K_b = \frac{[CH_3CO_2H][OH^-]}{[CH_3CO_2^-]} = \frac{(x)(x)}{0.0667 - x} \approx \frac{x^2}{0.0667}$$

We can obtain the value of K_b by rearranging Equation 16.20 and substituting the known values:

$$K_b = \frac{K_w}{K_a} = \frac{1.01 \times 10^{-14}}{1.74 \times 10^{-5}} = 5.80 \times 10^{-10} = \frac{x^2}{0.0667}$$

which we can solve to get $x = 6.22 \times 10^{-6}$. Thus, $[OH^-] = 6.22 \times 10^{-6}$ M, and the pH of the final solution is 8.79 (Figure 16.10a). As expected for the titration of a weak acid, the pH at the equivalence point is greater than 7 because the product of the titration is a base, the acetate ion, which then reacts with water to produce OH^-.

EXERCISE 16.12

Calculate the pH of a solution prepared by adding 45.0 mL of a 0.213 M HCl solution to 125.0 mL of a 0.150 M solution of ammonia. The pK_b of ammonia is 4.75 at 25°C.

Answer 9.23

As shown in Figure 16.10b, the titration curve for NH_3, a weak base, is the reverse of the titration curve for acetic acid. In particular, note that the pH at the equivalence point in the titration of a weak base is *less* than 7.00 because the titration produces an acid.

The identity of the weak acid or base being titrated strongly affects the shape of the titration curve. Figure 16.11 illustrates the shape of titration curves as a function of the pK_a or pK_b. As the acid or base being titrated becomes weaker (its pK_a or pK_b becomes larger), the pH change around the equivalence point decreases significantly.

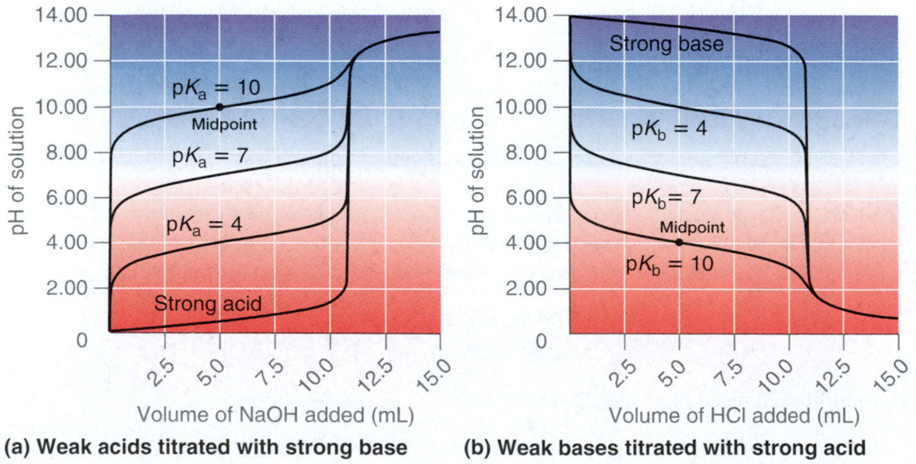

Figure 16.11 Effect of acid or base strength on the shape of titration curves. Unlike strong acids or bases, the shape of the titration curve for a weak acid or base depends on the pK_a or pK_b of the weak acid or base being titrated. **(a)** Solution pH as function of the volume of 1.00 M NaOH added to 10.00 mL of 1.00 M solutions of weak acids with the indicated pK_a values. **(b)** Solution pH as a function of the volume of 1.00 M HCl added to 10.00 mL of 1.00 M solutions of weak bases with the indicated pK_b values. Note that the shapes of the two sets of curves are essentially identical, but one is flipped vertically in relation to the other. Midpoints are indicated for the titration curves corresponding to pK_a = 10 and pK_b = 10. Interactive Graph

With very dilute solutions, the curve becomes so shallow that it can no longer be used to determine the equivalence point.

One point in the titration of a weak acid or a weak base is particularly important: The **midpoint** of a titration is defined as the point at which exactly enough acid (or base) has been added to neutralize one-half of the acid (or base) originally present, and occurs halfway to the equivalence point. The midpoint is indicated in Figure 16.11a and b for the two shallowest curves. By definition, at the midpoint of the titration of an acid, [HA] = [A$^-$]. Recall from Equation 16.13 that the ionization constant for a weak acid is

$$K_a = \frac{[H_3O^+][A^-]}{[HA]}$$

If [HA] = [A$^-$], this reduces to $K_a = [H_3O^+]$. Taking the negative logarithm of both sides gives

$$-\log K_a = -\log[H_3O^+]$$

From the definitions of pK_a and pH, we see that this is identical to

$$pK_a = pH \quad (16.51)$$

Thus, *the pH at the midpoint of the titration of a weak acid is equal to the pK_a of the weak acid*, as indicated in Figure 16.11a for the weakest acid where we see that the midpoint for pK_a = 10 occurs at pH = 10. Titration methods can therefore be used to determine both the concentration *and* the pK_a (or pK_b) of a weak acid (or base).

Note the pattern

The pH at the midpoint of the titration of a weak acid is equal to the pK$_a$ of the weak acid.

Titrations of Polyprotic Acids or Bases

When a strong base is added to a solution of a polyprotic acid, the neutralization reaction occurs in stages. The most acidic group is titrated first, followed by the next most acidic, and so forth. If the pK_a values are separated by at least three pK_a units, then the overall titration curve shows well-resolved "steps" corresponding to the titration of each proton. A titration of H_3PO_4, a triprotic acid, with NaOH is illustrated in Figure 16.12, which shows two well-defined steps: the first midpoint corresponds to pK_{a1}, and the second midpoint to pK_{a2}. Because HPO_4^{2-} is such a weak acid, pK_{a3} has such a high value that the third step cannot be resolved using 0.10 *M* NaOH as the titrant.

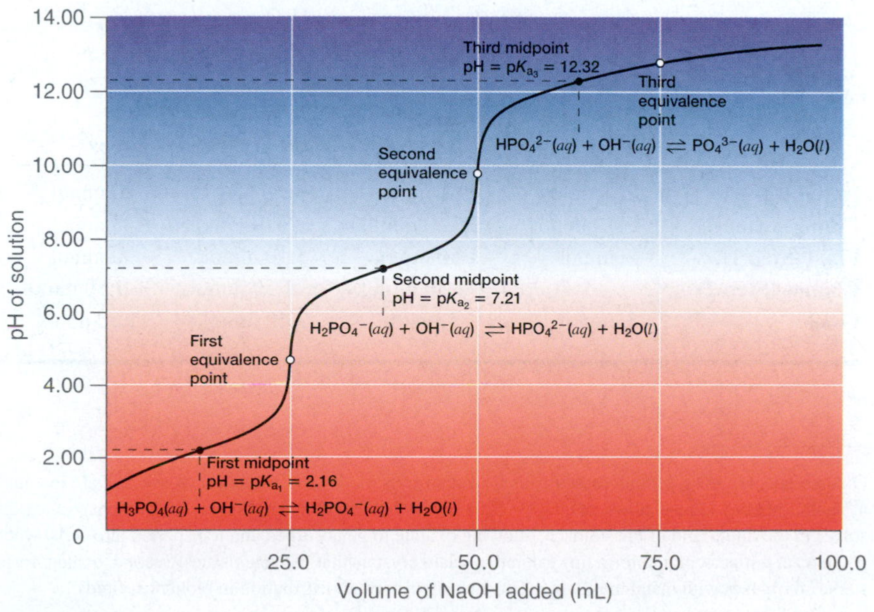

Figure 16.12 Titration curve for phosphoric acid (H_3PO_4), a typical polyprotic acid. Shown here is the curve for the titration of 25.0 mL of a 0.100 *M* H_3PO_4 solution with 0.100 *M* NaOH along with the species in solution at each K_a. Note the two distinct equivalence points corresponding to deprotonation of H_3PO_4 at pH ≈ 4.6 and $H_2PO_4^-$ at pH ≈ 9.8. Because HPO_4^{2-} is a very weak acid, the third equivalence point, at pH ≈13, is not well defined.

The titration curve for the reaction of a polyprotic base with a strong acid is the mirror image of the curve shown in Figure 16.12. The initial pH is high, but as acid is added, the pH decreases in steps if the successive pK_b values are well separated. Table 16.4 lists the ionization constants and pK_a values for some common polyprotic acids and bases.

EXAMPLE 16.13

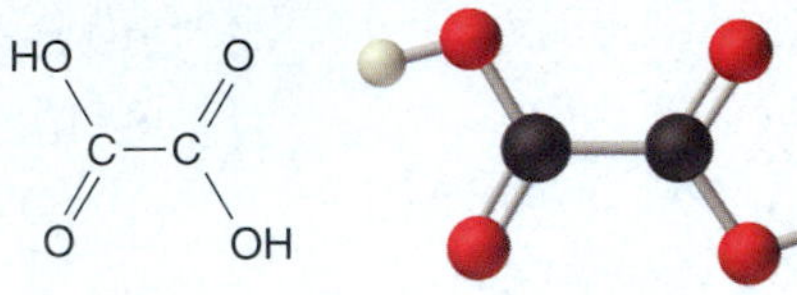

Oxalic acid

Calculate the pH of a solution prepared by adding 55.0 mL of a 0.120 *M* NaOH solution to 100.0 mL of a 0.0510 *M* solution of oxalic acid (HO_2CCO_2H), a diprotic acid (abbreviated as H_2ox). Oxalic acid, the simplest dicarboxylic acid, is found in rhubarb and many other plants. Rhubarb leaves are toxic because they contain the calcium salt of the fully deprotonated form of oxalic acid, the oxalate ion ($^-O_2CCO_2^-$, abbreviated ox^{2-}).*

Given Volume and concentration of acid and base

Asked for pH

Strategy

A Calculate the initial numbers of millimoles of the acid and base. Use a tabular format to determine the amounts of all species in solution.

B Calculate the concentrations of all species in the final solution. Use Equation 16.14 to determine $[H^+]$, and convert this value to pH.

Solution

A Table 16.4 gives the pK_a values of oxalic acid as 1.25 and 3.81. Again we proceed by determining the numbers of millimoles of acid and base initially present:

$$100.0 \cancel{\text{mL}} \left(\frac{0.0510 \text{ mmol } H_2ox}{\cancel{\text{mL}}}\right) = 5.10 \text{ mmol } H_2ox$$

$$55.0 \cancel{\text{mL}} \left(\frac{0.120 \text{ mmol NaOH}}{\cancel{\text{mL}}}\right) = 6.60 \text{ mmol NaOH}$$

The strongest acid (H_2ox) reacts with the base first. This leaves $(6.60 - 5.10) = 1.50$ mmol of OH^- to react with Hox^-, forming ox^{2-} and H_2O. The reactions can be written as

$$\underset{5.10 \text{ mmol}}{H_2ox} + \underset{6.60 \text{ mmol}}{OH^-} \longrightarrow \underset{5.10 \text{ mmol}}{Hox^-} + \underset{5.10 \text{ mmol}}{H_2O}$$

$$\underset{5.10 \text{ mmol}}{Hox^-} + \underset{1.50 \text{ mmol}}{OH^-} \longrightarrow \underset{1.50 \text{ mmol}}{ox^{2-}} + \underset{1.50 \text{ mmol}}{H_2O}$$

In tabular form,

	H_2ox	OH^-	Hox^-	ox^{2-}
Initial	5.10 mmol	6.60 mmol	0 mmol	0 mmol
Change (Step 1)	− 5.10 mmol	− 5.10 mmol	+ 5.10 mmol	0 mmol
Final (Step 1)	0 mmol	1.50 mmol	5.10 mmol	0 mmol
Change (Step 2)	—	− 1.50 mmol	− 1.50 mmol	+ 1.50 mmol
Final	0 mmol	0 mmol	3.60 mmol	1.50 mmol

* Oxalate salts are toxic for two reasons. First, oxalate salts of divalent cations such as Ca^{2+} are insoluble at neutral pH but soluble at low pH, as we shall see in the next chapter. As a result, calcium oxalate dissolves in the dilute acid of the stomach, allowing oxalate to be absorbed and transported into cells, where it can react with calcium to form tiny calcium oxalate crystals that damage tissues. Second, oxalate forms stable complexes with metal ions, which can alter their normal distribution in biological fluids.

B The equilibrium between the weak acid (Hox^-) and its conjugate base (ox^{2-}) in the final solution is determined by the magnitude of the second ionization constant, $K_{a2} = 10^{-3.81} = 1.55 \times 10^{-4}$. To calculate the pH of the solution, we need to know $[H^+]$, which is determined using exactly the same method as in the acetic acid titration in Example 16.12:

$$\text{final volume of solution} = 100.0 \text{ mL} + 55.0 \text{ mL} = 155.0 \text{ mL}$$

Thus, the concentrations of Hox^- and ox^{2-} are

$$[Hox^-] = \frac{3.60 \text{ mmol } Hox^-}{155.0 \text{ mL}} = 2.32 \times 10^{-2}\,M$$

$$[ox^{2-}] = \frac{1.50 \text{ mmol } ox^{2-}}{155.0 \text{ mL}} = 9.68 \times 10^{-3}\,M$$

We can now calculate $[H^+]$ at equilibrium using the equation

$$K_{a2} = \frac{[ox^{2-}][H^+]}{[Hox^-]}$$

Rearranging this equation and substituting the values for the concentrations of Hox^- and ox^{2-} give

$$[H^+] = \frac{K_{a2}\,[Hox^-]}{[ox^{2-}]} = \frac{(1.55 \times 10^{-4})(2.32 \times 10^{-2})}{9.68 \times 10^{-3}} = 3.71 \times 10^{-4}\,M$$

Thus,

$$pH = -\log[H^+] = -\log(3.71 \times 10^{-4}) = 3.43$$

This answer makes chemical sense because the pH is between the first and second pK_a values of oxalic acid, as it must be. We added enough hydroxide ion to completely titrate the first, more acidic proton (which should give us a pH greater than pK_{a1}), but only enough to titrate less than half of the second, less acidic proton, with pK_{a2}. If we had added exactly enough hydroxide to completely titrate the first proton plus half of the second, we would be at the midpoint of the second step in the titration, and the pH would be 3.81, equal to pK_{a2}.

EXERCISE 16.13

Piperazine is a diprotic base used to control intestinal parasites ("worms") in pets and humans. A dog is given 500 mg (5.80 mmol) of piperazine ($pK_{b1} = 4.27$, $pK_{b2} = 8.67$). If the dog's stomach initially contains 100 mL of 0.10 M HCl (pH = 1.00), calculate the pH of the stomach contents after ingestion of the piperazine.

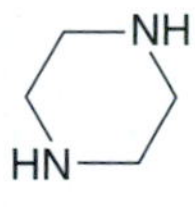

Piperazine

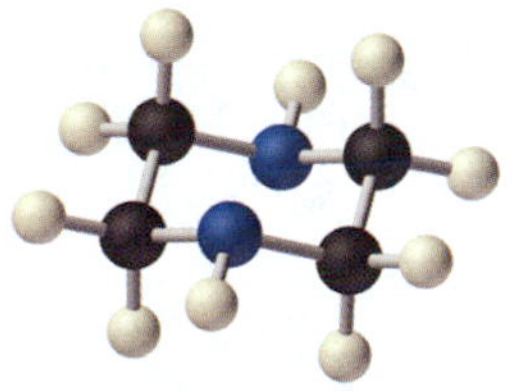

Answer 4.91

Indicators

MGC Indicators

In practice, most acid–base titrations are not monitored by recording the pH as a function of the amount of the strong acid or base solution used as the titrant. Instead, an **acid–base indicator** is often used that, if carefully selected, undergoes a dramatic color change at the pH corresponding to the equivalence point of the titration. Indicators are weak acids or bases that exhibit intense colors that vary with pH. The conjugate acid and conjugate base of a good indicator have very different colors so that they can be distinguished easily. Some indicators are colorless in the conjugate acid form but intensely colored when deprotonated (phenolphthalein, for example), which makes them particularly useful.

We can describe the chemistry of indicators by the general equation

$$HIn(aq) \rightleftharpoons H^+(aq) + In^-(aq)$$

Figure 16.13 Naturally occurring pH indicators in red cabbage juice. Red cabbage juice contains a mixture of substances whose color depends on the pH. Each test tube contains a solution of red cabbage juice in water, but the pH of the solutions vary from pH = 2.0 (far left) to pH = 11.0 (far right). At pH = 7.0, the solution is blue.

where the protonated form is designated by HIn and the conjugate base by In^-. The ionization constant for the deprotonation of indicator HIn is

$$K_{in} = \frac{[H^+][In^-]}{[HIn]} \tag{16.52}$$

The value of pK_{in} (its pK_a) determines the pH at which the indicator changes color.

Many different substances can be used as indicators, depending on the particular reaction to be monitored. For example, red cabbage juice contains a mixture of colored substances that change from deep red at low pH to light blue at intermediate pH's to yellow at high pH (Figure 16.13). In all cases, though, a good indicator must have these properties:

1. The color change must be easily detected.
2. The color change must be rapid.
3. The indicator molecule must not react with the substance being titrated.
4. To minimize errors, the indicator should have a pK_{in} that is within one pH unit of the expected pH at the equivalence point of the titration.

Synthetic indicators have been developed that meet these criteria and cover virtually the entire pH range. Figure 16.14 shows the approximate pH range over which some common indicators change color and their change in color. Some indicators are colorless when protonated but intensely colored when deprotonated, which makes the color change particularly easy to detect. One common example is phenolphthalein, which changes from colorless to pink at a pH of 9.5. In addition, some indicators (such as thymol blue) are polyprotic acids or bases, which change color twice at widely separated pH values.

It is important to be aware that an indicator does not change color abruptly at a particular pH value; instead, it actually undergoes a pH titration just like any other acid or base. As the concentration of HIn decreases and the concentration of In^- increases, the color of the solution slowly changes from the characteristic color of HIn to that of In^-. As we will see in the next section, the $[In^-]/[HIn]$ ratio changes from a value of 0.1 at a pH one unit *below* pK_{in} to a value of 10 at a pH one unit *above* pK_{in}. Thus, most indicators change color over a pH range of about two pH units.

How do you select the right indicator for the titration of a particular acid or base? As we stated above, the indicator should have a pK_{in} value that is close to the expected pH at the equivalence point. For a strong acid–strong base titration, the choice of indicator is not especially critical due to the very large change in pH that occurs around the equivalence point. In contrast, using the wrong indicator for a titration of a weak acid or a weak base can result in relatively large errors, as illustrated in

Indicator	pK_{in}	Color change
Alizarin yellow R	11.0	yellow to red
Thymolphthalein	9.9	colorless to blue
Phenolphthalein	9.5	colorless to pink
Thymol blue	9.2	yellow to blue
Phenol red	7.4	yellow to red
Bromothymol blue	7.3	yellow to blue
Bromocresol purple	6.4	yellow to purple
Methyl red	5.0	red to yellow
Resorcin blue	5.3	red to blue
Ethyl red	5.4	colorless to red
Resazurin	5.1	orange to violet
Bromophenol blue	4.1	yellow to blue
Thymol blue	1.7	red to yellow
Malachite green	1.3	yellow to turquoise
Methyl violet	0.8	yellow to blue

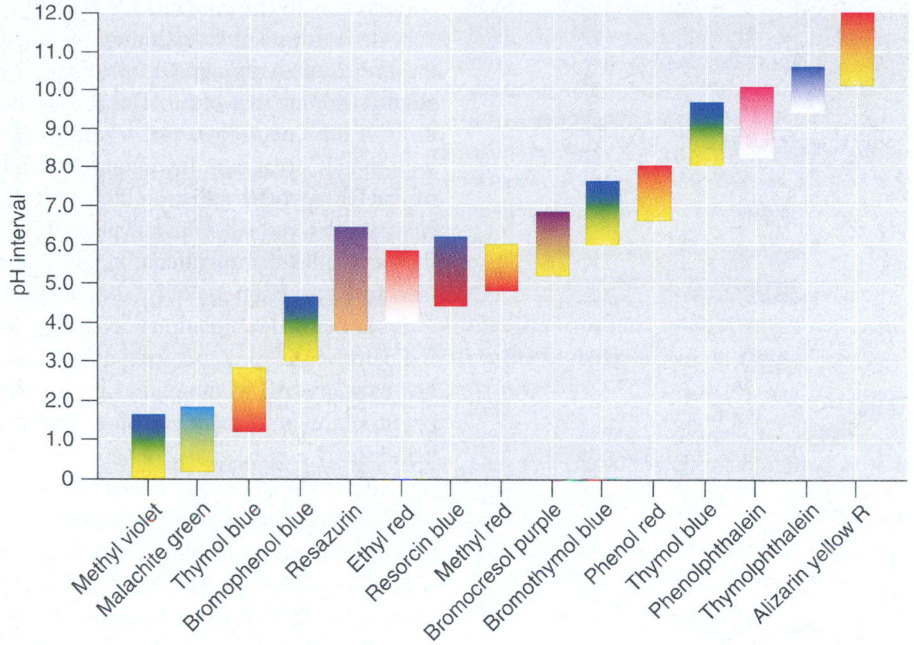

Figure 16.14 Some common acid–base indicators. Approximate colors are shown, along with values of pK_{in} and the pH range over which the color changes.

Figure 16.15. This figure shows plots of pH versus volume of base added for the titration of 50.0 mL of a 0.100 *M* solution of a strong acid (HCl) and a weak acid (acetic acid) with 0.100 *M* NaOH. The pH ranges over which two common indicators (methyl red, pK_{in} = 5.0, and phenolphthalein, pK_{in} = 9.5) change color are also shown. The horizontal bars indicating the pH ranges over which both indicators change color cross the HCl titration curve where it is almost vertical. Hence, both indicators change color when essentially the same volume of NaOH has been added (about 50 mL), which

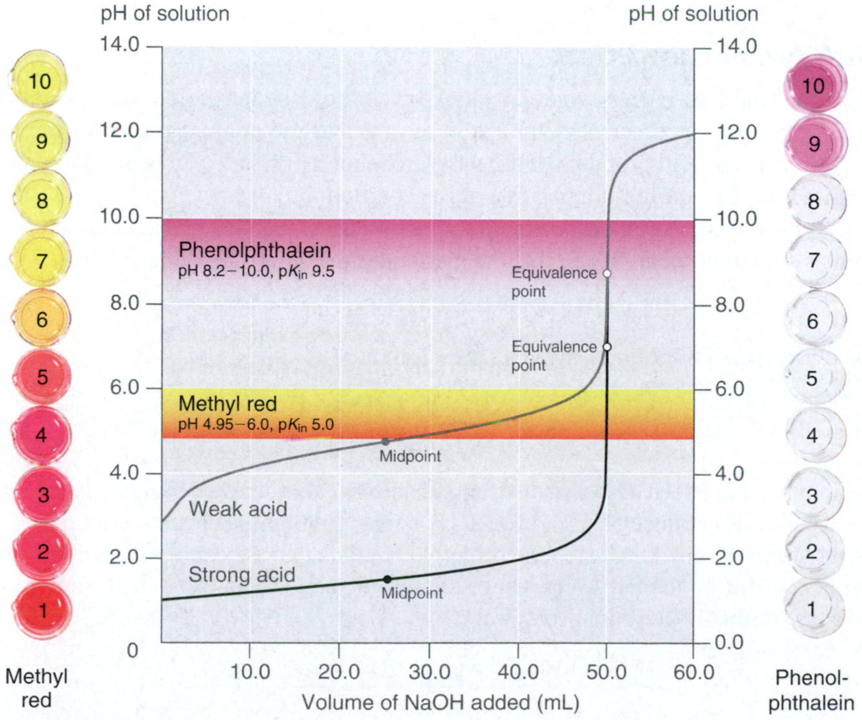

Figure 16.15 Choosing the correct indicator for an acid–base titration. The graph shows the results obtained using two indicators (methyl red and phenolphthalein) for the titration of 0.100 *M* solutions of a strong acid (HCl) and a weak acid (acetic acid) with 0.100 *M* NaOH. Due to the steepness of the titration curve of a strong acid around the equivalence point, either indicator will rapidly change color at the equivalence point for the titration of the strong acid. In contrast, the pK_{in} for methyl red (5.0) is very close to the pK_a of acetic acid (4.76); hence, the midpoint of the color change for methyl red occurs near the midpoint of the titration, rather than at the equivalence point.

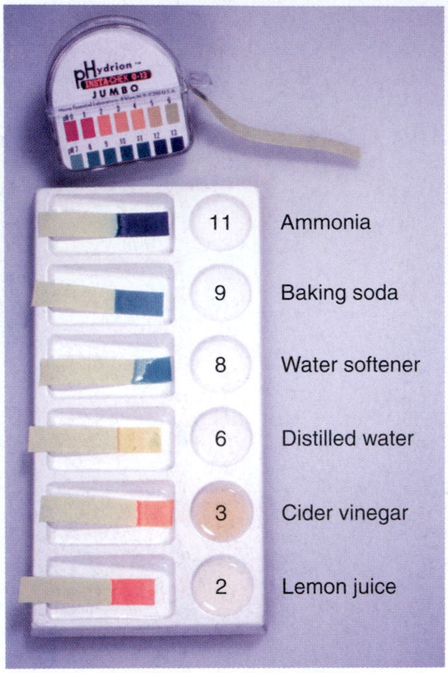

Figure 16.16 **pH paper.** pH paper contains a set of indicators that change color at different pH values. The approximate pH of a solution can be determined by simply dipping a paper strip into the solution and comparing the color to the standards provided.

corresponds to the equivalence point. In contrast, the titration of acetic acid will give very different results depending on whether methyl red or phenolphthalein is used as the indicator. Although the pH range over which phenolphthalein changes color is slightly higher than the pH at the equivalence point of the strong acid titration, the error will be negligible due to the slope of this portion of the titration curve. Just as with the HCl titration, the phenolphthalein indicator will turn pink when about 50 mL of NaOH has been added to the acetic acid solution. In contrast, methyl red begins to change from red to yellow around pH 5, which is near the *midpoint* of the acetic acid titration, not the equivalence point. Addition of only about 25–30 mL of NaOH will therefore cause the methyl red indicator to change color, resulting in a huge error.

In general, for titrations of strong acids with strong bases (and vice versa), any indicator with a pK_{in} between about 4 and 10 will do. For the titration of a weak acid, however, the pH at the equivalence point is greater than 7, and an indicator such as phenolphthalein or thymol blue, with $pK_{in} > 7$, should be used. Conversely, for the titration of a weak base, where the pH at the equivalence point is less than 7, an indicator such as methyl red or bromocresol blue, with $pK_{in} < 7$, should be used.

The existence of many different indicators with different colors and pK_{in} values also provides a convenient way to estimate the pH of a solution without using an expensive electronic pH meter and a fragile pH electrode. Paper or plastic strips that contain combinations of indicators are used as "pH paper," which allows you to estimate the pH of a solution by simply dipping a piece of pH paper into it and comparing the resulting color with the standards printed on the container (Figure 16.16).

16.6 Buffers

Buffers are solutions that maintain a relatively constant pH when an acid or a base is added. They therefore protect, or "buffer," other molecules in solution from the effects of the added acid or base. Buffers contain either a weak acid (HA) and its conjugate base (A^-) or a weak base (B) and its conjugate acid (BH^+), and they are critically important for the proper functioning of biological systems. In fact, every biological fluid is buffered to maintain its physiological pH.

The Common Ion Effect

To understand how buffers work, let's look first at how the ionization equilibrium of a weak acid (HA) is affected by the addition of either the conjugate base of the acid (A^-) or a strong acid (a source of H^+). Le Châtelier's principle can be used to predict the effect on the equilibrium position of the solution.

A typical buffer used in biochemistry laboratories contains acetic acid and a salt such as sodium acetate. Recall that the dissociation reaction of acetic acid is

$$CH_3CO_2H(aq) \rightleftharpoons CH_3CO_2^-(aq) + H^+(aq) \tag{16.53}$$

for which the equilibrium constant expression is

$$K_a = \frac{[H^+][CH_3CO_2^-]}{[CH_3CO_2H]} \tag{16.54}$$

Sodium acetate (CH_3CO_2Na) is a strong electrolyte that ionizes completely in aqueous solution to produce Na^+ and $CH_3CO_2^-$ ions. If sodium acetate is added to a solution of acetic acid, Le Châtelier's principle predicts that the equilibrium in Equation 16.53 will shift to the left, consuming some of the added $CH_3CO_2^-$ and some of the H^+ ions originally present in solution:

$$\underleftarrow{CH_3CO_2H(aq) \rightleftharpoons CH_3CO_2^-(aq) + H^+(aq)}$$
$$+ CH_3CO_2^-$$

(Because Na^+ is a spectator ion, it has no effect on the position of the equilibrium and can be ignored.) The addition of sodium acetate therefore produces a new equilibrium composition, in which the value of $[H^+]$ is lower than the initial value. Since $[H^+]$ has decreased, the pH will be higher. Thus, adding a salt of the conjugate base to a solution of a weak acid increases the pH. This makes sense because sodium acetate is a base, and adding *any* base to a solution of a weak acid should increase the pH.

If we instead add a strong acid such as HCl to the system, $[H^+]$ increases. Once again the equilibrium is temporarily disturbed, but the excess H^+ ions react with the conjugate base ($CH_3CO_2^-$), whether from the parent acid or sodium acetate, to drive the equilibrium to the left. The net result is a new equilibrium composition that has a lower $[CH_3CO_2^-]$ than before. In both cases, *only the equilibrium composition has changed; the ionization constant* K_a *for acetic acid remains the same.* Adding a strong electrolyte that contains one ion in common with a reaction system that is at equilibrium, in this case $CH_3CO_2^-$, will therefore shift the equilibrium in the direction that reduces the concentration of the common ion. The shift in equilibrium is called the **common ion effect**.

EXAMPLE 16.14

In Section 16.4, we calculated that a 0.150 M solution of formic acid at 25°C ($pK_a = 3.75$) has a pH of 2.28 and is 3.5% ionized. **(a)** Is there a change to the pH of the solution if enough solid sodium formate is added to make the final formate concentration 0.100 M (assume that the formic acid concentration does not change)? **(b)** What percentage of the formic acid is ionized if 0.200 M HCl is added to the system?

Given Solution concentration and pH, pK_a, and percent ionization of acid; final concentration of conjugate base or strong acid added

Asked for pH and percent ionization of formic acid

Strategy

A Write a balanced equilibrium equation for the ionization equilibrium of formic acid. Tabulate the initial concentrations, changes, and final concentrations.

B Substitute the expressions for the final concentrations into the expression for K_a. Calculate $[H^+]$ and the pH of the solution.

C Construct a table of concentrations for the dissociation of the formic acid. To determine the percent ionization, determine the anion concentration, divide it by the initial concentration of formic acid, and multiply the result by 100.

Solution

(a) **A** Because sodium formate is a strong electrolyte, it ionizes completely in solution to give formate and sodium ions. The Na^+ ions are spectator ions, so they can be ignored in the equilibrium equation. Because water is both a much weaker acid than formic acid and a much weaker base than formate, the acid–base properties of the solution are determined solely by the formic acid ionization equilibrium:

$$HCO_2H(aq) \rightleftharpoons HCO_2^-(aq) + H^+(aq)$$

The initial concentrations, the changes in concentration that occur as equilibrium is reached, and the final concentrations can be tabulated.

	$HCO_2H(aq)$	$\rightleftharpoons$ $H^+(aq)$	+ $HCO_2^-(aq)$
	$[HCO_2H]$	$[H^+]$	$[HCO_2^-]$
Initial	0.150	1.00×10^{-7}	0.100
Change	$-x$	$+x$	$+x$
Final	$(0.150 - x)$	x	$(0.100 + x)$

B Substituting the expressions for the final concentrations into the equilibrium constant expression and making our usual simplifying assumptions give

$$K_a = \frac{[H^+][HCO_2^-]}{[HCO_2H]} = \frac{(x)(0.100 + x)}{0.150 - x} = \frac{x(0.100)}{0.150} = 10^{-3.75} = 1.8 \times 10^{-4}$$

Rearranging and solving for x give

$$x = (1.8 \times 10^{-4}) \times \frac{0.150\ M}{0.100\ M} = 2.7 \times 10^{-4} = [H^+]$$

The value of x is small compared with 0.150 or 0.100 M, so our assumption about the extent of ionization is justified. Moreover, $K_aC_{HA} = (1.8 \times 10^{-4})(0.150) = 2.7 \times 10^{-5}$, which is greater than 1.0×10^{-6}, so again, our assumption is justified. The final pH is $-\log(2.7 \times 10^{-4}) =$ 3.57, compared with the initial value of 2.29. Thus, adding a salt containing the conjugate base of the acid has increased the pH of the solution, as we expect based on Le Châtelier's principle; the stress on the system has been relieved by the consumption of H^+ ions, driving the equilibrium to the left.

(b) **C** Because HCl is a strong acid, it ionizes completely, and chloride is a spectator ion that can be neglected. Thus, the only relevant acid–base equilibrium is again the dissociation of formic acid, and initially the concentration of formate is zero. We can construct a table of initial concentrations, changes in concentration, and final concentrations.

	$HCO_2H(aq)$	$\rightleftharpoons$	$H^+(aq)$	+	$HCO_2^-(aq)$
	$[HCO_2H]$		$[H^+]$		$[HCO_2^-]$
Initial	0.150		0.200		0
Change	$-x$		$+x$		$+x$
Final	$(0.150 - x)$		$(0.200 + x)$		x

To calculate the percentage of formic acid that is ionized under these conditions, we have to determine the final $[HCO_2^-]$. Substituting final concentrations into the equilibrium constant expression and making the usual simplifying assumptions give

$$K_a = \frac{[H^+][HCO_2^-]}{[HCO_2H]} = \frac{(0.200 + x)(x)}{0.150 - x} = \frac{x(0.200)}{0.150} = 1.80 \times 10^{-4}$$

Rearranging and solving for x give

$$x = (1.80 \times 10^{-4}) \times \frac{0.150\ M}{0.200\ M} = 1.35 \times 10^{-4} = [HCO_2^-]$$

Once again, our simplifying assumptions are justified. The percent ionization of formic acid is

$$\text{percent ionization} = \frac{1.35 \times 10^{-4}\ M}{0.150\ M} \times 100 = 0.090\%$$

Adding the strong acid to the solution, as shown in the table, decreased the percent ionization of formic acid by a factor of approximately 38 (3.45%/0.090%). Again, this is consistent with Le Châtelier's principle: adding H^+ ions drives the dissociation equlibrium to the left.

EXERCISE 16.14

As you learned in Example 16.8, a 0.225 M solution of ethylamine ($CH_3CH_2NH_2$, $pK_b = 3.19$) has a pH of 12.08 and a percent ionization of 5.4% at 20°C. Calculate **(a)** the pH of the solution if enough solid ethylamine hydrochloride ($EtNH_3Cl$) is added to make the solution 0.100 M in $EtNH_3^+$ and **(b)** the percentage of ethylamine that is ionized if enough solid NaOH is added to the original solution to give a final concentration of 0.050 M NaOH.

Answer **(a)** 11.16; **(b)** 1.3%

Now let's suppose we have a buffer solution that contains equimolar concentrations of a weak base (B) and its conjugate acid (BH^+). The general equation for the ionization of a weak base is

$$B(aq) + H_2O(l) \rightleftharpoons BH^+(aq) + OH^-(aq) \qquad (16.55)$$

If the equilibrium constant for the reaction as written in Equation 16.55 is small, for example $K_b = 10^{-5}$, then the equilibrium constant for the reverse reaction is very large: $K = 1/K_b = 10^5$. Adding a strong base such as OH^- to the solution therefore causes the equilibrium in Equation 16.55 to shift to the left, consuming the added OH^-. As a result, the hydroxide ion concentration in solution remains relatively constant, and the pH of the solution changes very little. Le Châtelier's principle predicts the same outcome: when the system is stressed by an increase in the OH^- concentration, the reaction will proceed to the left to counteract the stress.

If the pK_b of the base (B) is 5, the pK_a of its conjugate acid (BH^+) is $pK_a = pK_w - pK_b = 14 - 5 = 9$. Thus, the equilibrium constant for ionization of the conjugate acid (BH^+) is even smaller than that for ionization of the base (B). The ionization reaction for the conjugate acid (BH^+) of a weak base is written

$$BH^+(aq) + H_2O(l) \rightleftharpoons B(aq) + H_3O^+(aq) \qquad (16.56)$$

Again, the equilibrium constant for the reverse of this reaction is very large: $K = 1/K_a = 10^9$. Thus, if a strong acid is added, it is neutralized by reaction with B as the reaction in Equation 16.56 shifts to the left. As a result, the H^+ ion concentration does not increase very much, and the pH changes only slightly. In effect, a buffer solution behaves somewhat like a sponge that can absorb H^+ and OH^- ions, thereby preventing large changes in pH when appreciable amounts of strong acid or base are added to a solution.

Buffers are characterized by the *pH range* over which they can maintain a more or less constant pH, and by their *buffer capacity*, the amount of strong acid or base that can be absorbed before the pH changes significantly. Although the useful pH range of a buffer depends strongly on the chemical properties of the weak acid and weak base used to prepare the buffer (that is, on K), its buffer capacity depends solely on the concentrations of the species in the buffered solution. The more concentrated the buffer solution, the greater its buffer capacity. As illustrated in Figure 16.17, when NaOH is added to solutions that contain different concentrations of an acetic acid/sodium acetate buffer, the observed change in the pH of the buffer is inversely proportional to the concentration of the buffer. If the buffer capacity is ten times larger, then the buffer solution can absorb ten times more strong acid or base before undergoing a significant change in pH.

MGC Henderson–Hasselbalch Equation

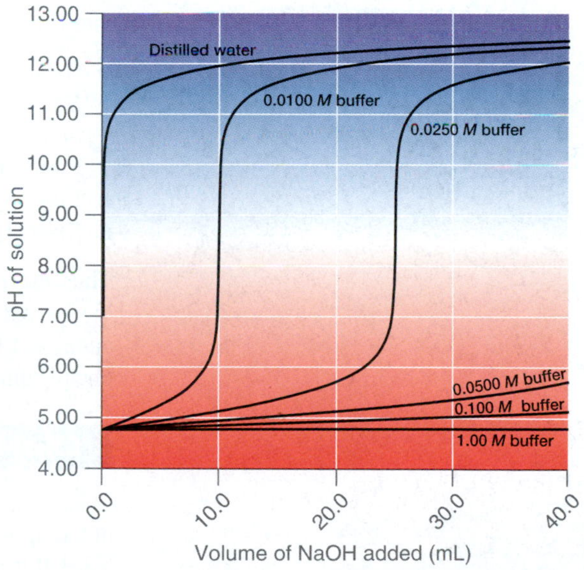

Figure 16.17 Effect of buffer concentration on the capacity of a buffer. A buffer maintains a relatively constant pH when acid or base is added to a solution. The addition of even tiny volumes of 0.10 *M* NaOH to 100.0 mL of distilled water results in a very large change in pH. As the concentration of a 50:50 mixture of sodium acetate/acetic acid buffer in the solution is increased from 0.010 *M* to 1.00 *M*, the change in the pH produced by the addition of the same volume of NaOH solution decreases steadily. For buffer concentrations of at least 0.500 *M*, the addition of even 25 mL of the NaOH solution results in only a relatively small change in pH. Interactive Graph

Calculating the pH of a Buffer

The pH of a buffer can be calculated from the concentrations of the weak acid and the weak base used to prepare it, the concentration of the conjugate base and acid, and the pK_a or pK_b of the weak acid or base. The procedure is analogous to that used in Example 16.14 to calculate the pH of a solution containing known concentrations of formic acid and formate.

An alternative method frequently used to calculate the pH of a buffer solution is based on a rearrangement of the equilibrium equation for the dissociation of a weak acid. The simplified ionization reaction is $HA \rightleftharpoons H^+ + A^-$, for which the equilibrium constant expression is

$$K_a = \frac{[H^+][A^-]}{[HA]} \qquad (16.57)$$

This equation can be rearranged to give

$$[H^+] = \frac{K_a[HA]}{[A^-]} \tag{16.58}$$

Taking the logarithm of both sides and multiplying both sides by −1 give

$$-\log[H^+] = -\log K_a - \log\left(\frac{[HA]}{[A^-]}\right) = -\log K_a + \log\left(\frac{[A^-]}{[HA]}\right) \tag{16.59}$$

Replacing the negative logarithms in Equation 16.59 gives

$$pH = pK_a + \log\left(\frac{[A^-]}{[HA]}\right) \tag{16.60}$$

or, more generally,

$$pH = pK_a + \log\left(\frac{[base]}{[acid]}\right) \tag{16.61}$$

Equations 16.60 and 16.61 are both forms of the **Henderson–Hasselbalch equation**, named after the two early-20th-century chemists who first noticed that this rearranged version of the equilibrium constant expression provides an easy way to calculate the pH of a buffer solution. In general, the validity of the Henderson–Hasselbalch equation may be limited to solutions whose concentrations are at least a hundred times greater than the value of their K_a's.

There are three special cases where the Henderson–Hasselbalch equation is easily interpreted without the need for calculations:

Note the pattern

If [base] = [acid] for a buffer, then pH = pK_a. Changing this ratio by a factor of 10 either way changes the pH by ±1.

- **[base] = [acid].** Under these conditions, [base]/[acid] = 1 in Equation 16.61. Because log 1 = 0, pH = pK_a, *regardless of the actual concentrations of the acid and base*. Recall from Section 16.5 that this corresponds to the midpoint in the titration of a weak acid or base.
- **[base]/[acid] = 10.** In Equation 16.61, because log 10 = 1, pH = pK_a + 1.
- **[base]/[acid] = 100.** In Equation 16.61, because log 100 = 2, pH = pK_a + 2.

Thus, each time we increase the [base]/[acid] ratio by 10, the pH of the solution increases by one pH unit. Conversely, if the [base]/[acid] ratio is 0.1, then pH = pK_a −1. Each additional factor-of-10 decrease in the [base]/[acid] ratio causes the pH to decrease by one pH unit.

EXAMPLE 16.15

What is the pH of a solution that contains **(a)** 0.135 *M* HCO_2H and 0.215 *M* HCO_2Na (the pK_a of formic acid is 3.75); **(b)** 0.0135 *M* HCO_2H and 0.0215 *M* HCO_2Na; **(c)** 0.119 *M* pyridine and 0.234 *M* pyridine hydrochloride (the pK_b of pyridine is 8.77)?

Given Concentration of acid, conjugate base, and pK_a; concentration of base, conjugate acid, and pK_b

Asked for pH

Strategy

Substitute values into either form of the Henderson–Hasselbalch equation (Equation 16.60 or 16.61) to calculate the pH.

Solution

(a) According to the Henderson–Hasselbalch equation, the pH of a solution that contains both a weak acid and its conjugate base is pH = pK_a + log([A^-]/[HA]). Inserting the given values into the equation gives

$$pH = 3.75 + \log\left(\frac{0.215}{0.135}\right) = 3.75 + \log 1.593 = 3.95$$

This result makes sense because the $[A^-]/[HA]$ ratio is between 1 and 10, so the pH of the buffer must be between the pK_a (3.75) and $pK_a + 1$, or 4.75.

(b) This is identical to part (a), except for the concentrations of the acid and the conjugate base, which are ten times lower. Inserting the concentrations into the Henderson–Hasselbalch equation gives

$$pH = 3.75 + \log\left(\frac{0.0215}{0.0135}\right) = 3.75 + \log 1.593 = 3.95$$

This result is identical to the result in part (a), which emphasizes the point that the pH of a buffer depends *only* on the *ratio* of the concentrations of the conjugate base and the acid, *not* on the magnitude of the concentrations. Because the $[A^-]/[HA]$ ratio is the same as in part (a), the pH of the buffer must also be the same (3.95).

(c) In this case, we have a weak base, pyridine (Py), and its conjugate acid, the pyridinium ion (HPy^+). We will therefore use Equation 16.61, the more general form of the Henderson–Hasselbalch equation, in which "base" and "acid" refer to the appropriate members of the conjugate acid–base pair. We are given [base] = [Py] = 0.119 M and [acid] = $[HPy^+]$ = 0.234 M. We also are given $pK_b = 8.77$ for pyridine, but we need pK_a for the pyridinium ion. Recall from Equation 16.23 that the pK_b of a weak base and the pK_a of its conjugate acid are related: $pK_a + pK_b = pK_w$. Thus, pK_a for the pyridinium ion is $pK_w - pK_b = 14.00 - 8.77 = 5.23$. Substituting this pK_a value into the Henderson–Hasselbalch equation gives

$$pH = pK_a + \log\left(\frac{[\text{base}]}{[\text{acid}]}\right) = 5.23 + \log\left(\frac{0.119}{0.234}\right) = 5.23 + \log 0.509 = 4.94$$

Once again, this result makes sense: the $[B]/[BH^+]$ ratio is about 1/2, which is between 1 and 0.1, so the final pH must be between the pK_a (5.23) and $pK_a - 1$, or 4.23.

EXERCISE 16.15

Calculate the pH of a solution that contains **(a)** 0.333 M benzoic acid and 0.252 M sodium benzoate, and **(b)** 0.050 M trimethylamine and 0.066 M trimethylamine hydrochloride. The pK_a of benzoic acid is 4.20, and the pK_b of trimethylamine is also 4.20.

Answer **(a)** 4.08; **(b)** 9.68

The Henderson–Hasselbalch equation can also be used to calculate the pH of a buffer solution after the addition of a given amount of strong acid or base, as demonstrated in the next example.

EXAMPLE 16.16

The buffer solution in Example 16.15 contained 0.135 M HCO_2H and 0.215 M HCO_2Na and had a pH of 3.95. **(a)** What is the final pH if 5.00 mL of 1.00 M HCl is added to 100 mL of this solution? **(b)** What is the final pH if 5.00 mL of 1.00 M NaOH is added?

Given Composition and pH of buffer; concentration and volume of added acid or base

Asked for Final pH

Strategy

A Calculate the amounts of formic acid and formate present in the buffer solution using the procedure from Example 16.14. Then calculate the amount of acid or base added.

B Construct a table showing the amounts of all species after the neutralization reaction. Use the final volume of the solution to calculate the concentrations of all species. Finally, substitute the appropriate values into the Henderson–Hasselbalch equation (Equation 16.61) to obtain the pH.

Solution

The added HCl (a strong acid) or NaOH (a strong base) will react completely with formate (a weak base) or formic acid (a weak acid), respectively, to give formic acid or formate and water. We must therefore calculate the amounts of formic acid and formate present after the neutralization reaction.

(a) A We begin by calculating the numbers of millimoles of formic acid and formate present in 100 mL of the initial pH 3.95 buffer:

$$100\ \cancel{\text{mL}}\left(\frac{0.135\ \text{mmol HCO}_2\text{H}}{\cancel{\text{mL}}}\right) = 13.5\ \text{mmol HCO}_2\text{H}$$

$$100\ \cancel{\text{mL}}\left(\frac{0.215\ \text{mmol HCO}_2^-}{\cancel{\text{mL}}}\right) = 21.5\ \text{mmol HCO}_2^-$$

The number of millimoles of H^+ in 5.00 mL of 1.00 *M* HCl is

$$5.00\ \cancel{\text{mL}}\left(\frac{1.00\ \text{mmol H}^+}{\cancel{\text{mL}}}\right) = 5.00\ \text{mmol H}^+$$

B Next, we construct a table of initial amounts, changes in amounts, and final amounts:

	HCO_2^-	+	H^+	→	HCO_2H
Initial	21.5 mmol		5.00 mmol		13.5 mmol
Change	− 5.00 mmol		− 5.00 mmol		+ 5.00 mmol
Final	16.5 mmol		~0 mmol		18.5 mmol

The final amount of H^+ in solution is given as "~0 mmol." For the purposes of the stoichiometry calculation, this is essentially true, but remember that the point of the problem is to calculate the final $[H^+]$ and thus the pH. We now have all the information we need to calculate the pH. We can use either the lengthy procedure of Example 16.14 or the Henderson–Hasselbach equation. Since we have performed *many* equilibrium calculations in this chapter, we'll take the latter approach. The Henderson–Hasselbalch equation requires the concentrations of HCO_2^- and HCO_2H, which can be calculated using the number of millimoles (n) of each and the total volume (V_T). Substituting these values into the Henderson–Hasselbalch equation gives

$$\text{pH} = \text{p}K_a + \log\left(\frac{[\text{HCO}_2^-]}{[\text{HCO}_2\text{H}]}\right) = \text{p}K_a + \log\left(\frac{n_{\text{HCO}_2^-}/V_f}{n_{\text{HCO}_2\text{H}}/V_f}\right) = \text{p}K_a + \log\left(\frac{n_{\text{HCO}_2^-}}{n_{\text{HCO}_2\text{H}}}\right)$$

Notice that because the total volume appears in both the numerator and denominator, it cancels. We therefore need to use only the ratio of the number of millimoles of the conjugate base to the number of millimoles of the weak acid. Thus,

$$\text{pH} = \text{p}K_a + \log\left(\frac{n_{\text{HCO}_2^-}}{n_{\text{HCO}_2\text{H}}}\right) = 3.75 + \log\left(\frac{16.5\ \text{mmol}}{18.5\ \text{mmol}}\right) = 3.75 + \log(0.892) = 3.70$$

Once again, this result makes sense on two levels. First, the addition of HCl has decreased the pH from 3.95, as expected. Second, the ratio of HCO_2^- to HCO_2H is slightly less than 1, so the pH should be between the pK_a and pK_a − 1.

(b) A The procedure for solving this part of the problem is exactly the same as that used in part (a). We have already calculated the numbers of millimoles of formic acid and formate in 100 mL of the initial pH 3.95 buffer: 13.5 mmol of HCO_2H and 21.5 mmol of HCO_2^-. The number of millimoles of OH^- in 5.00 mL of 1.00 *M* NaOH is

$$5.00\ \cancel{\text{mL}}\left(\frac{1.00\ \text{mmol OH}^-}{\cancel{\text{mL}}}\right) = 5.00\ \text{mmol OH}^-$$

B With this information, we can construct a table of initial amounts, changes in amounts, and final amounts.

Note the pattern

Only the amounts (in moles or millimoles) of the acidic and basic components of the buffer are needed to use the Henderson–Hasselbalch equation, not their concentrations.

	HCO_2H	+	OH^-	⟶	$HCO_2^- + H_2O$
Initial	13.5 mmol		5.00 mmol		21.5 mmol
Change	− 5.00 mmol		− 5.00 mmol		+ 5.00 mmol
Final	8.5 mmol		~0 mmol		26.5 mmol

The final amount of OH^- in solution is not actually zero; this is only approximately true based on the stoichiometric calculation. We can calculate the final pH by inserting the numbers of millimoles of both HCO_2^- and HCO_2H into the simplified Henderson–Hasselbalch expression used in part (a):

$$pH = pK_a + \log\left(\frac{n_{HCO_2^-}}{n_{HCO_2H}}\right) = 3.75 + \log\left(\frac{26.5\text{ mmol}}{8.5\text{ mmol}}\right) = 3.75 + \log(3.12) = 4.24$$

Once again, this result makes chemical sense: the pH has increased, as would be expected upon the addition of a strong base, and the final pH is between the pK_a and $pK_a + 1$, as expected for a solution with a HCO_2^-/HCO_2H ratio between 1 and 10.

MGC Adding Acids or Bases to Buffered Solutions

EXERCISE 16.16

The buffer solution from Example 16.15 contained 0.119 *M* pyridine and 0.234 *M* pyridine hydrochloride and had a pH of 4.94. **(a)** What is the final pH if 12.0 mL of 1.5 *M* NaOH is added to 250 mL of this solution? **(b)** What is the final pH if 12.0 mL of 1.5 *M* HCl is added?

Answer **(a)** 5.30; **(b)** 4.42

Note the pattern

The most effective buffers contain equal concentrations of an acid and its conjugate base.

The results obtained in Example 16.16 and Exercise 16.16 demonstrate how little the pH of a well-chosen buffer solution changes despite the addition of a significant quantity of strong acid or base. Suppose we had added the same amount of HCl or NaOH solution to 100 mL of an *un*buffered solution at pH 3.95 (corresponding to 1.12×10^{-4} *M* HCl). In this case, adding 5.00 mL of 1.00 *M* HCl would lower the final pH to 1.32 instead of 3.70, whereas adding 5.00 mL of 1.00 *M* NaOH would raise the final pH to 12.68 rather than 4.24. (Try verifying these values by doing the calculations yourself.) Thus, the presence of a buffer significantly increases the ability of a solution to maintain an almost constant pH.

A buffer that contains approximately equal amounts of a weak acid and its conjugate base in solution is equally effective at neutralizing either added base or added acid. This is shown in Figure 16.18 for an acetic acid/sodium acetate buffer. Adding a given amount of strong acid shifts the system along the horizontal axis to the left, whereas adding the same amount of strong base shifts the system the same distance to the right. In either case, the change in the ratio of $CH_3CO_2^-$ to CH_3CO_2H from 1:1 reduces the buffer capacity of the solution.

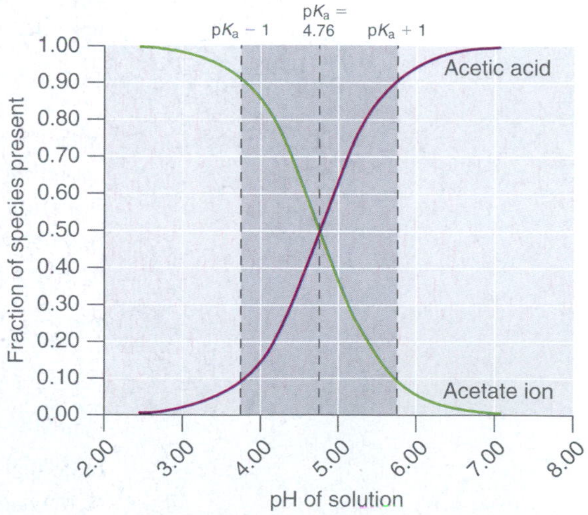

Figure 16.18 Distribution curve showing the fraction of acetic acid molecules and acetate ions as a function of pH in a solution of acetic acid. Note that the pH range over which the acetic acid/sodium acetate system is an effective buffer (the darker shaded region) corresponds to the region in which appreciable concentrations of both species are present (pH 3.76–5.76, corresponding to pH = $pK_a \pm 1$).

The Relationship Between Titrations and Buffers

There is a strong correlation between the effectiveness of a buffer solution and the titration curves discussed in Section 16.5. Consider the schematic titration curve of a weak acid with a strong base shown in Figure 16.19. As indicated by the labels, the region around pK_a corresponds to the midpoint of the titration, when approximately half the weak acid has been neutralized. This portion of the titration curve corresponds to a buffer: it exhibits the smallest change in pH per increment of added strong base, as shown by the nearly horizontal nature of the curve in this region. Note that the nearly flat portion of the curve extends only from approximately a pH value of one unit less than the pK_a to approximately a pH value of one unit greater than the

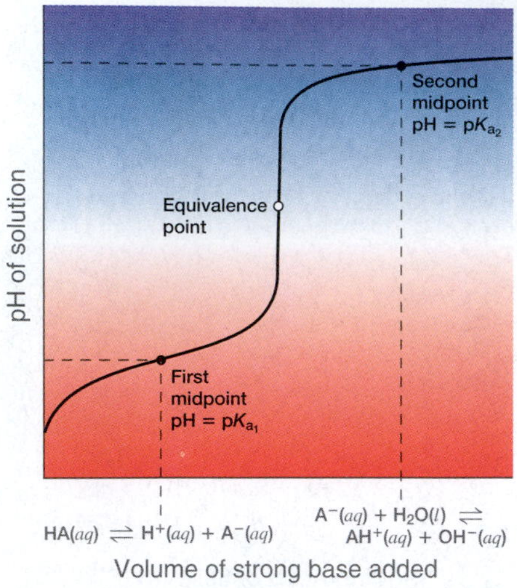

Figure 16.19 The relationship between titration curves and buffers. This schematic plot of pH for the titration of a weak acid with a strong base shows the nearly flat region of the titration curve around the midpoint, which corresponds to the formation of a buffer. At the lower left, the pH of the solution is determined by the equilibrium for dissociation of the weak acid; at the upper right, the pH is determined by the equilibrium for reaction of the conjugate base with water.

pK_a, which is why buffer solutions usually have a pH that is within ±1 pH units of the pK_a of the acid component of the buffer.

In the region of the titration curve at the lower left, before the midpoint, the acid–base properties of the solution are dominated by the equilibrium for dissociation of the weak acid, corresponding to K_a. In the region of the titration curve at the upper right, after the midpoint, the acid–base properties of the solution are dominated by the equilibrium for reaction of the conjugate base of the weak acid with water, corresponding to K_b. Note, however, that we can calculate either K_a or K_b from the other because they are related by K_w.

Blood: A Most Important Buffer

Metabolic processes produce large amounts of acids and bases, yet organisms are able to maintain an almost constant internal pH because their fluids contain buffers. This is not to say that the pH is uniform throughout all cells and tissues of a mammal. The internal pH of a red blood cell is about 7.2, but the pH of most other kinds of cells is lower, around 7.0. Even within a single cell, different compartments can have very different pH values. For example, one intracellular compartment in white blood cells has a pH of around 5.

Because no single buffer system can effectively maintain a constant pH value over the entire physiological range of approximately pH 5 to 7.4, biochemical systems use a set of buffers with overlapping ranges. The most important of these is the CO_2/HCO_3^- system, which dominates the buffering action of blood plasma.

The acid–base equilibrium in the CO_2/HCO_3^- buffer system is usually written as

$$H_2CO_3\,(aq) \rightleftharpoons H^+(aq) + HCO_3^-(aq) \tag{16.62}$$

with $K_a = 4.5 \times 10^{-7}$ and $pK_a = 6.35$ at 25°C. In fact, Equation 16.62 is a grossly oversimplified version of the CO_2/HCO_3^- system because a solution of CO_2 in water contains only rather small amounts of $H_2CO_3(aq)$. Thus, Equation 16.62 does not allow us to understand how blood is actually buffered, particularly at a physiological temperature of 37°C. As shown in Equation 16.63, $CO_2(aq)$ is in equilibrium with $H_2CO_3(aq)$, but the equilibrium lies far to the left, with an $H_2CO_3(aq)/CO_2(aq)$ ratio less than 0.01 under most conditions:

$$CO_2(aq) + H_2O(l) \rightleftharpoons H_2CO_3(aq) \tag{16.63}$$

with $K' = 4.0 \times 10^{-3}$ at 37°C. The true pK_a of carbonic acid at 37°C is 3.70, corresponding to a K_a value of 2.0×10^{-4}, which makes it a much stronger acid than Equation 16.62 suggests. Adding Equations 16.62 and 16.63 and canceling H_2CO_3 from both sides give the following overall equation for the reaction of CO_2 with water to give a proton and the bicarbonate ion:

$$\begin{array}{lll} CO_2(aq) + H_2O(l) \rightleftharpoons \cancel{H_2CO_3(aq)} & K' = 4.0 \times 10^{-3}\,(37°C) & \\ \cancel{H_2CO_3(aq)} \rightleftharpoons H^+(aq) + HCO_3^-(aq) & K_a = 2.0 \times 10^{-4}\,(37°C) & (16.64) \\ \hline CO_2(aq) + H_2O(l) \rightleftharpoons H^+(aq) + HCO_3^-(aq) & K = 8.0 \times 10^{-7}\,(37°C) & \end{array}$$

Note that the equilibrium constant K for the reaction in Equation 16.64 is the product of the *true* ionization constant for carbonic acid K_a and the equilibrium constant K' for the reaction of $CO_2(aq)$ with water to give carbonic acid. The equilibrium equation for the reaction of CO_2 with water to give bicarbonate and a proton is therefore

$$K = \frac{[H^+][HCO_3^-]}{[CO_2]} = 8.0 \times 10^{-7} \tag{16.65}$$

The presence of a gas in the equilibrium constant expression for a buffer is unusual. According to Henry's law (described in Chapter 13), $[CO_2] = kP_{CO_2}$, where k is the Henry's law constant for CO_2, 0.030 mM/mmHg at 37°C. Substituting this expression for $[CO_2]$ in Equation 16.65 gives

$$K = \frac{[H^+][HCO_3^-]}{(3.0 \times 10^{-5}\ M/\text{mmHg})(P_{CO_2})} \tag{16.66}$$

where P_{CO_2} is in mmHg. Taking the negative logarithm of both sides and rearranging give

$$\text{pH} = 6.10 + \log \frac{[HCO_3^-]}{(3.0 \times 10^{-5}\ M/\text{mm Hg})(P_{CO_2})} \tag{16.67}$$

Thus, the pH of the solution depends on both the CO_2 pressure over the solution and $[HCO_3^-]$. Figure 16.20 plots the relationship between pH and $[HCO_3^-]$ under physiological conditions for several different values of P_{CO_2}, with normal pH and $[HCO_3^-]$ values indicated by the dashed lines.

According to Equation 16.64, adding a strong acid to the CO_2/HCO_3^- system causes $[HCO_3^-]$ to decrease as HCO_3^- is converted to CO_2. Excess CO_2 is released in the lungs and exhaled into the atmosphere, however, so there is essentially no change in P_{CO_2}. Because the change in $[HCO_3^-]/P_{CO_2}$ is small, Equation 16.67 predicts that the change in pH will also be rather small. Conversely, if a strong base is added, the OH^- reacts with CO_2 to form HCO_3^-, but CO_2 is replenished by the body, again limiting the change in both $[HCO_3^-]/P_{CO_2}$ and pH. The CO_2/HCO_3^- buffer system is an example of an *open* system, in which the total concentration of the components of the buffer change in order to keep the pH at a nearly constant value.

If a passenger steps out of an airplane in Denver, Colorado, for example, the lower P_{CO_2} at higher elevations (typically 31 mmHg at an elevation of 2000 m versus 40 mmHg at sea level) causes a shift to a new pH and $[HCO_3^-]$. The increase in pH and decrease in $[HCO_3^-]$ in response to the decrease in P_{CO_2} are responsible for the general malaise that many people experience at high altitudes. If their blood pH does not adjust rapidly, the condition can develop into the life-threatening phenomenon known as altitude sickness.

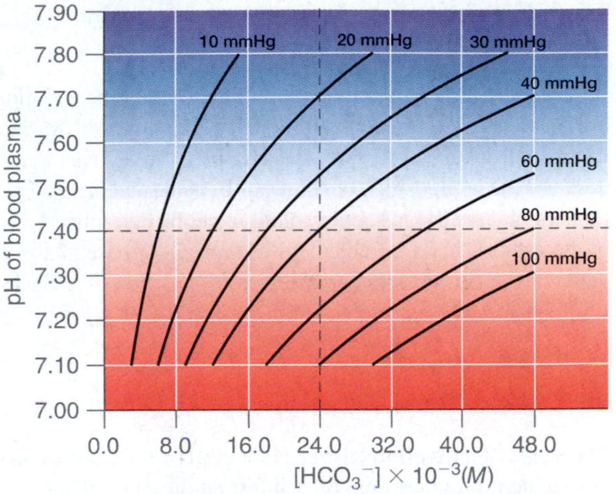

Figure 16.20 Buffering in blood: pH versus $[HCO_3^-]$ curves for buffers with different values of P_{CO_2}. Only those combinations of pH and $[HCO_3^-]$ that lie on a given line are allowed for the particular value of P_{CO_2} indicated. Normal values of blood plasma pH and $[HCO_3^-]$ are indicated by the dashed lines.

SUMMARY AND KEY TERMS

16.1 The Autoionization of Water (p. 722)

Water is **amphiprotic**: it can act as an acid by donating a proton to a base to form the hydroxide ion, or as a base by accepting a proton from an acid to form the **hydronium ion**, H_3O^+. The *autoionization* of liquid water produces OH^- and H_3O^+ ions. The equilibrium constant for this reaction is called the **ion-product constant of liquid water (K_w)** and is defined as $K_w = [H_3O^+][OH^-]$. At 25°C, K_w has a value of 1.01×10^{-14}; hence, pH + pOH = pK_w = 14.00.

16.2 A Qualitative Description of Acid–Base Equilibria (p. 725)

Two species that differ by only a proton constitute a **conjugate acid–base pair**. The magnitude of the equilibrium constant for an ionization reaction can be used to determine the relative strengths of acids and bases. For an aqueous solution of a weak acid, the dissociation constant is called the **acid ionization constant (K_a)**. Similarly, the equilibrium constant for the reaction of a weak base with water is the **base ionization constant (K_b)**. For any conjugate acid–base pair, $K_aK_b = K_w$. Smaller values of pK_a correspond to larger acid ionization constants and hence stronger acids. Conversely, smaller values of pK_b correspond to larger base ionization constants and hence stronger bases. At 25°C, pK_a + pK_b = 14.00. Acid–base reactions always proceed in the direction that produces the weaker acid–base pair. No acid stronger than H_3O^+ and no base stronger than OH^- can exist in aqueous solution, leading to the phenomenon known as the **leveling effect**. Polyprotic acids (and bases) lose (and gain) protons in a stepwise manner, with the fully protonated species being the strongest acid and the fully deprotonated species the strongest base. A salt can dissolve in water to produce a neutral, basic, or acidic solution, depending on whether it contains the conjugate base of a weak acid as the anion (A^-) or the conjugate acid of a weak base as the cation (BH^+), or both. Salts that contain small, highly charged metal ions produce acidic solutions in water. The reaction of a salt with water to produce an acidic or basic solution is called a **hydrolysis reaction**.

16.3 Molecular Structure and Acid–Base Strength (p. 736)

The acid–base strength of a molecule depends strongly on its structure. The weaker the A—H or B—H^+ bond, the more likely it is to dissociate to form an H^+ ion. In addition, any factor that stabilizes the lone pair on the conjugate base favors the dissociation of H^+, making the conjugate acid a stronger acid. Atoms or groups of atoms elsewhere in a molecule can also be important in determining acid or base strength through an *inductive effect*, which can weaken an O—H bond and allow hydrogen to be more easily lost as H^+ ions.

16.4 Quantitative Aspects of Acid–Base Equilibria (p. 741)

If the concentration of one or more of the species in a solution of an acid or a base is determined experimentally, the values of K_a and K_b can be calculated. Values of K_a, pK_a, K_b, and pK_b can be used to quantitatively describe the composition of solutions of acids and bases. The concentrations of all species present in solution can be determined, as can the pH of the solution and the percentage of the acid or base that is ionized. The value of the equilibrium constant for the reaction of a weak acid with a weak base can be calculated from K_a (or pK_a), K_b (or pK_b), and K_w.

16.5 Acid–Base Titrations (p. 751)

The shape of a **titration curve**, a plot of pH versus the amount of acid or base added, provides important information about what is occurring in solution during the titration. The shapes of titration curves for weak acids and bases depend dramatically on the identity of the compound. The **equivalence point** of an acid–base titration is the point at which exactly enough acid or base has been added to react completely with the other component. The equivalence point in the titration of a strong acid or base occurs at pH 7.00. In titrations of weak acids or bases, however, the pH at the equivalence point is greater or less than 7, respectively. The pH tends to change more slowly before the equivalence point is reached in titrations of weak acids and weak bases than in titrations of strong acids and strong bases. The pH at the **midpoint**, the point halfway on the titration curve to the equivalence point, is equal to the pK_a of the weak acid or the pK_b of the weak base. Thus, titration methods can be used to determine both the concentration *and* the pK_a (or pK_b) of a weak acid (or base). **Acid–base indicators** are compounds that change color at a particular pH. They are typically weak acids or bases whose changes in color correspond to deprotonation or protonation of the indicator itself.

16.6 Buffers (p. 764)

Buffers are solutions that resist a change in pH upon the addition of an acid or a base. Buffers contain a weak acid (HA) and its conjugate weak base (A^-). Adding a strong electrolyte that contains one ion in common with a reaction system that is at equilibrium shifts the equilibrium in such a way as to reduce the concentration of the common ion. The shift in equilibrium is called the **common ion effect**. Buffers are characterized by their pH range and their buffer capacity. The useful pH range of a buffer depends strongly on the chemical properties of the conjugate weak acid–base pair used to prepare the buffer (the K_a or K_b), whereas its buffer capacity depends solely on the concentrations of the species in the solution. The pH of a buffer can be calculated using the **Henderson–Hasselbalch equation**, which is valid for solutions whose concentrations are at least a hundred times greater than the value of their K_a's. Because no single buffer system can effectively maintain a constant pH value over the physiological range of approximately 5 to 7.4, biochemical systems use a set of buffers with overlapping ranges. The most important of these is the CO_2/HCO_3^- system, which dominates the buffering action of blood plasma.

KEY EQUATIONS

Ion-product constant of liquid water	$K_w = [H_3O^+][OH^-]$	(16.7)
Definition of pH	$pH = -\log_{10}[H^+]$	(16.9a)
	$[H^+] = 10^{-pH}$	(16.9b)
Definition of pOH	$pOH = -\log_{10}[OH^-]$	(16.10a)
	$[OH^-] = 10^{-pOH}$	(16.10b)
Relationship among pH, pOH, and pK_w	$pK_w = pH + pOH$	(16.11)
Acid ionization constant	$K_a = \frac{[H_3O^+][A^-]}{[HA]}$	(16.14)
Base ionization constant	$K_b = \frac{[BH^+][OH^-]}{[B]}$	(16.16)
Relationship between K_a and K_b of a conjugate acid–base pair	$K_aK_b = K_w$	(16.20)
Definition of pK_a	$pK_a = -\log_{10}K_a$	(16.21a)
	$K_a = 10^{-pK_a}$	(16.21b)
Definition of pK_b	$pK_b = -\log_{10}K_b$	(16.22a)
	$K_b = 10^{-pK_b}$	(16.22b)
Relationship between pK_a and pK_b of a conjugate acid–base pair	$pK_a + pK_b = pK_w$	(16.23a)
	$pK_a + pK_b = 14.00$ (at 25°C)	(16.23b)
Percent ionization of acid	$\frac{[H^+]}{C_{HA}} \times 100$	(16.40)
Percent ionization of base	$\frac{[OH^-]}{C_B} \times 100$	(16.41)
Equilibrium constant for reaction of a weak acid with a weak base	$K = \frac{K_aK_b}{K_w}$	(16.48)
Henderson–Hasselbalch equation	$pH = pK_a + \log\left(\frac{[A^-]}{[HA]}\right)$	(16.60)
	$pH = pK_a + \log\left(\frac{[base]}{[acid]}\right)$	(16.61)

QUESTIONS AND PROBLEMS

Questions and Problems with colored numbers have answers in the Appendix and complete solutions in the Student Solutions Manual.

CONCEPTUAL

16.1 The Autoionization of Water

1. What is the relationship between the value of the equilibrium constant for the autoionization of liquid water and the tabulated value of the ion-product constant of liquid water, K_w?
2. The density of liquid water decreases as the temperature increases from 25°C to 50°C. Will this effect cause the value of K_w to increase or decrease? Why?
3. Show that water is amphiprotic by writing balanced equations for the reactions of water with HNO_3 and NH_3. In which reaction does water act as the acid? In which does it act as the base?
4. Write each equation:
   (a) nitric acid is added to water
   (b) potassium hydroxide is added to water
   (c) calcium hydroxide is added to water
   (d) sulfuric acid is added to water
5. Show that K for the sum of the following reactions is equal to K_w:
   (a) $HMnO_4(aq) \rightleftharpoons H^+(aq) + MnO_4^-(aq)$
   (b) $MnO_4^-(aq) + H_2O(l) \rightleftharpoons HMnO_4(aq) + OH^-(aq)$

16.2 A Qualitative Description of Acid–Base Equilibria

6. Identify the conjugate acid–base pairs in each of these equilibria:
   (a) $HF(aq) + H_2O(l) \rightleftharpoons H_3O^+(aq) + F^-(aq)$
   (b) $CH_3CH_2NH_2(aq) + H_2O(l) \rightleftharpoons CH_3CH_2NH_3^+(aq) + OH^-(aq)$
   (c) $C_3H_7NO_2(aq) + OH^-(aq) \rightleftharpoons C_3H_6NO_2^-(aq) + H_2O(l)$
   (d) $CH_3CO_2H(aq) + 2HF(aq) \rightleftharpoons CH_3C(OH)_2^+(aq) + HF_2^-(aq)$
7. Identify the conjugate acid–base pairs in each equilibrium:
   (a) $HSO_4^-(aq) + H_2O(l) \rightleftharpoons SO_4^{2-}(aq) + H_3O^+(aq)$
   (b) $C_3H_7NO_2(aq) + H_3O^+(aq) \rightleftharpoons C_3H_8NO_2^+(aq) + H_2O(l)$
   (c) $CH_3CO_2H(aq) + NH_3(aq) \rightleftharpoons CH_3CO_2^-(aq) + NH_4^+(aq)$
   (d) $SbF_5(aq) + 2HF(aq) \rightleftharpoons H_2F^+(aq) + SbF_6^-(aq)$
8. Salts such as NaH contain the hydride ion, H^-. When sodium hydride is added to water, it produces hydrogen gas in a highly vigorous reaction. Write a balanced equation for this reaction, and identify the conjugate acid–base pairs.
9. Write the expression for K_a for each reaction:
   (a) $HCO_3^-(aq) + H_2O(l) \rightleftharpoons CO_3^{2-}(aq) + H_3O^+(aq)$
   (b) formic acid$(aq) + H_2O(l) \rightleftharpoons$ formate$(aq) + H_3O^+(aq)$
   (c) $H_3PO_4(aq) + H_2O(l) \rightleftharpoons H_2PO_4^-(aq) + H_3O^+(aq)$
10. Write an expression for the ionization constant K_b for each reaction:
    (a) $OCH_3^-(aq) + H_2O(l) \rightleftharpoons HOCH_3(aq) + OH^-(aq)$
    (b) $NH_2^-(aq) + H_2O(l) \rightleftharpoons NH_3(aq) + OH^-(aq)$
    (c) $S^{2-}(aq) + H_2O(l) \rightleftharpoons HS^-(aq) + OH^-(aq)$
11. Predict whether each equilibrium lies primarily to the left or to the right:
    (a) $HBr(aq) + H_2O(l) \rightleftharpoons H_3O^+(aq) + Br^-(aq)$
    (b) $NaH(soln) + NH_3(l) \rightleftharpoons H_2(soln) + NaNH_2(soln)$
    (c) $OCH_3^-(aq) + NH_3(aq) \rightleftharpoons CH_3OH(aq) + NH_2^-(aq)$
    (d) $NH_3(aq) + HCl(aq) \rightleftharpoons NH_4^+(aq) + Cl^-(aq)$
12. Species that are strong bases in water, such as CH_3^-, NH_2^-, and S^{2-}, are leveled to the strength of OH^-, the conjugate base of H_2O. Because their relative base strengths are indistinguishable in water, suggest a method for identifying which is the strongest base. How would you distinguish between the strong acids HIO_3, H_2SO_4 and $HClO_4$?
13. Is it accurate to say that a 2.0 *M* solution of H_2SO_4, which contains two acidic protons per molecule, is 4.0 *M* in H^+? Explain your answer.
14. The alkalinity of soil is defined by the equation: alkalinity = $[HCO_3^-] + 2[CO_3^{2-}] + [OH^-] - [H^+]$. The source of both HCO_3^- and CO_3^{2-} is H_2CO_3. Explain why the basicity of soil is defined in this way.
15. Why are aqueous solutions of salts such as $CaCl_2$ neutral? Why is an aqueous solution of $NaNH_2$ basic?
16. Predict whether aqueous solutions of the following are acidic, basic, or neutral: (a) Li_3N; (b) NaH; (c) KBr; (d) $C_2H_5NH_3^+Cl^-$.
17. When each of these compounds is added to water, would you expect the pH of the solution to increase, decrease, or remain the same? (a) $LiCH_3$; (b) $MgCl_2$; (c) K_2O; (d) $(CH_3)_2NH_2^+Br^-$
18. Which complex ion would you expect to be more acidic: $Pb(H_2O)_4^{2+}$ or $Sn(H_2O)_4^{2+}$? Why?
19. Would you expect $Sn(H_2O)_4^{2+}$ or $Sn(H_2O)_6^{4+}$ to be more acidic? Why?
20. Is it possible to arrange the hydrides LiH, RbH, KH, CsH, and NaH in order of increasing base strength in aqueous solution? Why or why not?

16.3 Molecular Structure and Acid–Base Strength

21. Section 16.3 presented several factors that affect the relative strengths of acids and bases. For each of the following pairs, identify the most important factor in determining which is the stronger acid or base in aqueous solution:
    (a) $CH_3CCl_2CH_2CO_2H$ vs. $CH_3CH_2CH_2CO_2H$
    (b) CH_3CO_2H vs. CH_3CH_2OH
    (c) HClO vs. HBrO
    (d) $CH_3C(=O)NH_2$ vs. $CH_3CH_2NH_2$
    (e) H_3AsO_4 vs. H_3AsO_3
22. The stability of the conjugate base is an important factor in determining the strength of an acid. Which would you expect to be the stronger acid in aqueous solution: $C_6H_5NH_3^+$ or NH_4^+? Justify your reasoning.
23. Explain why H_2Se is a weaker acid than HBr.
24. Arrange the following in order of decreasing acid strength in aqueous solution: H_3PO_4, $CH_3PO_3H_2$, and $HClO_3$.
25. Arrange the following in order of increasing base strength in aqueous solution: CH_3S^-, OH^-, and CF_3S^-.
26. Arrange the following in order of increasing acid strength in aqueous solution: $HClO_2$, HNO_2, HNO_3.

27. Do you expect H_2SO_3 or H_2SeO_3 to be the stronger acid? Why?
28. Give a plausible explanation for why CF_3OH is a stronger acid than CH_3OH in aqueous solution. Do you expect $CHCl_2CH_2OH$ to be a stronger or weaker acid than CH_3OH? Why?
29. Do you expect Cl_2NH or NH_3 to be the stronger base in aqueous solution? Why?

16.4 Quantitative Aspects of Acid–Base Equilibria

30. Explain why the analytical concentration C of H_2SO_4 is equal to $[H_2SO_4] + [HSO_4^-] + [SO_4^{2-}]$.
31. Write an expression for the analytical concentration C of H_3PO_4 in terms of the concentrations of the species actually present in solution.
32. For a dilute solution of a weak acid such as acetic acid, CH_3CO_2H, the concentration of undissociated acetic acid in solution is often assumed to be the same as the analytical concentration. Explain why this is a valid practice.
33. How does dilution affect the percent ionization of a weak acid or weak base?
34. What is the relationship between the K_a of a weak acid and its percent ionization? Does a compound with a large pK_a value have a higher or lower percent ionization than a compound with a small pK_a value (assuming the same analytical concentration in both cases)? Explain.
35. For a dilute solution of a weak acid, HA, show that the pH of the solution can be approximated using the following equation (where C_{HA} is the analytical concentration of the weak acid):

$$\text{pH} = -\log\sqrt{K_a \cdot C_{HA}}$$

Under what conditions is this approximation valid?

16.5 Acid–Base Titrations

36. Why is the portion of the titration curve that lies below the equivalence point of a solution of a weak acid displaced upward relative to the titration curve of a strong acid? How are the slopes of the curves different at the equivalence point? Why?
37. Predict whether each solution will be neutral, basic, or acidic at the equivalence point of the following titrations: (a) an aqueous solution of NaOH is titrated with 0.100 *M* HCl; (b) an aqueous solution of ethylamine, $CH_3CH_2NH_2$, is titrated with 0.150 *M* HNO_3; (c) an aqueous solution of aniline hydrochloride, $C_6H_5NH_3^+Cl^-$, is titrated with 0.050 *M* KOH.
38. The pK_a values of phenol red, bromophenol blue, and phenolphthalein are 7.4, 4.1, and 9.5, respectively. Which indicator is best suited for each of the following acid–base titrations? (a) titrating a solution of $Ba(OH)_2$ with 0.100 *M* HCl; (b) titrating a solution of trimethylamine, Me_3N, with 0.150 *M* HNO_3; (c) titrating a solution of aniline hydrochloride, $C_6H_5NH_3^+Cl^-$, with 0.050 *M* KOH
39. For the titration of any strong acid with any strong base, the pH at the equivalence point is 7. Why is this not usually the case in titrations of weak acids or bases?
40. Why are the titration curves for a strong acid with a strong base and a weak acid with a strong base identical in shape above the equivalence points but not below?
41. Describe what is occurring on a molecular level during the titration of a weak acid, such as acetic acid, with a strong base, such as NaOH, at the following points along the titration curve: (a) at the beginning of the titration, (b) at the midpoint of the titration, (c) at the equivalence point, and (d) when excess titrant has been added. (e) Which of these points corresponds to the pH = pK_a?
42. On a molecular level, describe what is happening during the titration of a weak base, such as ammonia, with a strong acid, such as HCl, at the following points along the titration curve: (a) at the beginning of the titration, (b) at the midpoint of the titration, (c) at the equivalence point, and (d) when excess titrant has been added. (e) Which of these points corresponds to the pOH = pK_b?
43. For the titration of a weak acid with a strong base, use the K_a expression to show that pH = pK_a at the midpoint of the titration.
44. Chemical indicators can be used to monitor pH rapidly and inexpensively. Nevertheless, electronic methods are generally preferred. Why?
45. Why does adding ammonium chloride to a solution of ammonia in water decrease the pH of the solution?
46. Given the equilibrium system $CH_3NH_2(aq) + H_2O(l) \rightleftharpoons CH_3NH_3^+(aq) + OH^-(aq)$, explain what happens to the position of the equilibrium and the pH in each case: (a) dilute HCl is added; (b) dilute NaOH is added; (c) solid $CH_3NH_3^+Cl^-$ is added.
47. Given the equilibrium system $CH_3CO_2H(aq) \rightleftharpoons CH_3CO_2^-(aq) + H^+(aq)$, explain what happens to the position of the equilibrium and the pH in each case: (a) dilute HCl is added; (b) dilute NaOH is added; (c) solid sodium acetate is added.

16.6 Buffers

48. Explain why buffers are crucial for the proper functioning of biological systems.
49. What is the role of a buffer in chemistry and biology? Is it correct to say that buffers prevent a change in $[H_3O^+]$? Explain your reasoning.
50. Explain why the most effective buffers are those that contain approximately equal amounts of the weak acid and its conjugate base.
51. Which region of the titration curve of a weak acid or base corresponds to the region of the smallest change in pH per amount of added strong acid or base?
52. If you were given a solution of sodium acetate, describe two ways you could convert the solution to a buffer.
53. Why are buffers usually used only within approximately one pH unit of the pK_a or pK_b of the parent weak acid or base?
54. The titration curve for a monoprotic acid can be divided into four regions: the starting point, the region around the midpoint of the titration, the equivalence point, and the region after the equivalence point. For which region would you use each of the following approaches to describe the behavior of the solution? (a) a buffer; (b) a solution of a salt of a weak base; (c) a solution of a weak acid; (d) diluting a strong base
55. Which of the following will produce a buffer solution? Explain your reasoning in each case. (a) mixing 100 mL of 0.1 *M* HCl and 100 mL of 0.1 *M* sodium acetate; (b) mixing 50 mL of a 0.1 *M* solution of HCl and 100 mL of a 0.1 *M* solution of sodium acetate; (c) mixing 100 mL of a 0.1 *M* acetic acid solution and 100 mL of a 0.1 *M* NaOH solution; (d) mixing 100 mL of a 0.1 *M* acetic acid solution and 50 mL of a 0.1 *M* NaOH solution; (e) mixing 100 mL of a 0.1 *M* sodium acetate solution and 50 mL of a 0.1 *M* acetic acid solution
56. Which of the following will produce a buffer solution? Explain your reasoning in each case. (a) mixing 100 mL of 0.1 *M* HCl

and 100 mL of 0.1 *M* sodium fluoride; (b) mixing 50 mL of a 0.1 *M* solution of HCl and 100 mL of a 0.1 *M* solution of sodium fluoride; (c) mixing 100 mL of a 0.1 *M* hydrofluoric acid solution and 100 mL of a 0.1 *M* HCl solution; (d) mixing 100 mL of a 0.1 *M* hydrofluoric acid solution and 50 mL of a 0.1 *M* NaOH solution; (e) mixing 100 mL of a 0.1 *M* sodium fluoride solution and 50 mL of a 0.1 *M* NaOH solution.

57. Use the definition of K_b for a weak base to derive the expression shown here, which is analogous to the Henderson–Hasselbalch equation, but for a weak base, B, rather than a weak acid, HA:

$$\text{pOH} = \text{p}K_b - \log\left(\frac{[\text{base}]}{[\text{acid}]}\right)$$

58. Why do biological systems use overlapping buffer systems to maintain a constant pH?

59. The CO_2/HCO_3^- buffer system of blood has an effective pK_a of approximately 6.1, yet the normal pH of blood is 7.4. Why is CO_2/HCO_3^- an effective buffer when the pK_a is more than one unit below the pH of blood? What happens to the pH of blood when the CO_2 pressure increases? When the O_2 pressure increases?

60. Carbon dioxide produced during respiration is converted to carbonic acid, H_2CO_3. The pK_{a1} of carbonic acid is 6.35, and its pK_{a2} is 10.33. Write the equations corresponding to each pK value, and predict the equilibrium position for each reaction.

NUMERICAL

This section includes paired problems (marked by brackets) that require similar problem-solving skills.

16.1 The Autoionization of Water

61. The autoionization of sulfuric acid can be described by the equation

$$H_2SO_4(l) + H_2SO_4(l) \rightleftharpoons H_3SO_4^+(soln) + HSO_4^-(soln)$$

At 25°C, $K = 3 \times 10^{-4}$. Write an equilibrium constant expression for $K_{H_2SO_4}$ that is analogous to K_w. The density of H_2SO_4 is 1.8 g/cm^3 at 25°C. What is the concentration of $H_3SO_4^+$? What fraction of H_2SO_4 is ionized?

62. An aqueous solution of a substance is found to have $[H_3O]^+ = 2.48 \times 10^{-8}$ *M*. Is the solution acidic, neutral, or basic?

63. The pH of a solution is 5.63. What is the pOH? What is the value of $[OH^-]$? Is the solution acidic or basic?

64. State whether each solution is acidic, neutral, or basic:
(a) solution A: $[H_3O^+] = 8.6 \times 10^{-3}$ *M*
(b) solution B: $[H_3O^+] = 3.7 \times 10^{-9}$ *M*
(c) solution C: $[H_3O^+] = 2.1 \times 10^{-7}$ *M*
(d) solution D: $[H_3O^+] = 1.4 \times 10^{-6}$ *M*

65. Calculate the pH and the pOH of each solution: (a) 0.15 *M* HBr; (b) 0.03 *M* KOH; (c) 2.3×10^{-3} *M* HNO_3; (d) 9.78×10^{-2} *M* NaOH; (e) 0.00017 *M* HCl; (f) 5.78 *M* HI.

66. Calculate the pH and the pOH of each solution: (a) 25.0 mL of 2.3×10^{-2} *M* HCl, diluted to 100 mL; (b) 5.0 mL of 1.87 *M* NaOH, diluted to 125 mL; (c) 5.0 mL of 5.98 *M* HCl added to 100 mL of water; (d) 25.0 mL of 3.7 *M* HNO_3 added to 250 mL of water; (e) 35.0 mL of 0.046 *M* HI added to 500 mL of water; (f) 15.0 mL of 0.0087 *M* KOH added to 250 mL of water.

67. The pH of stomach acid is approximately 1.5. What is the H^+ concentration?

68. Given the pH values in parentheses, what is the H^+ concentration of each solution? (a) household bleach (11.4); (b) milk (6.5); (c) orange juice (3.5); (d) seawater (8.5); (e) tomato juice (4.2)

69. A reaction requires the addition of 250.0 mL of a solution with a pH of 3.50. What mass of HCl(*g*) must be dissolved in 250 mL of water to produce a solution with this pH?

70. If you require 333 mL of a pH 12.50 solution, how would you prepare it using a 0.500 *M* sodium hydroxide stock solution?

16.2 A Qualitative Description of Acid–Base Equilibria

71. Arrange the acids in order of increasing strength:
acid A: pK_a = 1.52
acid B: pK_a = 6.93
acid C: pK_a = 3.86
Given solutions with the same initial concentration of each acid, which would have the highest percent ionization?

72. Arrange the bases in order of increasing strength:
base A: pK_b = 13.10
base B: pK_b = 8.74
base C: pK_b = 11.87
Given solutions with the same initial concentration of each base, which would have the highest percent ionization?

73. Calculate the K_a and pK_a of the conjugate acid of a base with each of these pK_b values: (a) 3.8; (b) 7.9; (c) 13.7; (d) 1.4; (e) −2.5.

74. Determine K_b and pK_b for benzoic acid, a food preservative, which has a pK_a of 4.20.

75. Determine K_a and pK_a of boric acid, $B(OH)_3$, solutions of which are occasionally used as an eyewash, if the pK_b of its conjugate base is 4.8.

16.4 Quantitative Aspects of Acid–Base Equilibria

76. The pK_a of NH_3 is estimated to be 9.25. Its conjugate base, amide ion, NH_2^-, can be isolated as an alkali metal salt, such as sodium amide, $NaNH_2$. Calculate the pH of a solution prepared by adding 0.100 mol of sodium amide to 1.00 L of water. Does the pH differ appreciably from the pH of a NaOH solution of the same concentration? Why or why not?

77. Phenol is a topical anesthetic that has been used in throat lozenges to relieve sore throat pain. Describe in detail how you would prepare a 2.00 *M* solution of phenol, C_6H_5OH, in water; then write equations to show all species present in the solution. What is the equilibrium constant expression for the reaction of phenol with water? Use the information in Table 16.2 to calculate the pH of the phenol solution.

78. Describe in detail how you would prepare a 1.50 *M* solution of methylamine in water; then write equations to show all species present in the solution. What is the equilibrium constant expression for the reaction of methylamine with water? Use the information in Table 16.3 to calculate the pH of the solution.

79. A 0.200 *M* solution of diethylamine, a substance used in insecticides and fungicides, is only 3.9% ionized at 25°C. Write an equation showing the equilibrium reaction, and then calculate the pK_b of diethylamine. What is the pK_a of its conjugate acid, the diethylammonium ion? What is the equilibrium constant expression for the reaction of diethylammonium chloride with water?

80. A 1.00 *M* solution of fluoroacetic acid, FCH_2CO_2H, is 5% dissociated in water. What is the equilibrium constant expression for the dissociation reaction? Calculate the concentration of each species in solution, and then calculate the pK_a of FCH_2CO_2H.

81. The pK_a of 3-chlorobutanoic acid, $CH_3CHClCH_2CO_2H$, is 4.05. What percentage is dissociated in a 1.0 *M* solution? Do you expect the pK_a of butanoic acid to be higher or lower than the pK_a of 3-chlorobutanoic acid? Why?

82. The pK_a of the ethylammonium ion, $C_2H_5NH_3^+$, is 10.64. What percentage of ethylamine is ionized in a 1.00 *M* solution of ethylamine?

83. The pH of a 0.150 *M* solution of aniline hydrochloride, $C_6H_5NH_3^+Cl^-$, is 2.70. What is the pK_b of the conjugate base, aniline, $C_6H_5NH_2$? Do you expect the pK_b of $(CH_3)_2CHNH_2$ to be higher or lower than the pK_b of $C_6H_5NH_2$? Why?

84. The pK_a of Cl_3CCO_2H is 0.64. What is the pH of a 0.580 *M* solution? What percentage of the Cl_3CCO_2H is dissociated?

85. What is the pH of a 0.620 *M* solution of $CH_3NH_3^+Br^-$ if the pK_b of CH_3NH_2 is 10.62?

86. The pK_b of 4-hydroxypyridine is 10.80 at 25°C. What is the pH of a 0.0250 *M* solution?

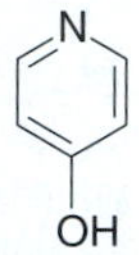

87. The pK_a's of butanoic acid and ammonium ion are 4.82 and 9.24, respectively. Calculate the value of *K* for the reaction

$$CH_3CH_2CH_2CO_2^-(aq) + NH_4^+(aq) \rightleftharpoons CH_3CH_2CH_2CO_2H(aq) + NH_3(aq)$$

88. The pK_a's of formic acid and the methylammonium ion are 3.75 and 10.62, respectively. Calculate the value of *K* for the reaction

$$HCO_2^-(aq) + CH_3NH_3^+(aq) \rightleftharpoons HCO_2H(aq) + CH_3NH_2(aq)$$

89. Use the information in Table 16.2 to calculate the pH of a 0.0968 *M* solution of calcium formate.

90. Calculate the pH of a 0.24 *M* solution of sodium lactate. The pK_a of lactic acid is 3.9.

91. Use the information in Table 16.2 to determine the pH of a solution prepared by dissolving 855 mg of sodium nitrite, $NaNO_2$, in enough water to make 100.0 mL of solution.

92. Use the information in Table 16.3 to determine the pH of a solution prepared by dissolving 750.0 mg of methylammonium chloride, $CH_3NH_3^+Cl^-$, in enough water to make 150.0 mL of solution.

16.5 Acid–Base Titrations

93. Calculate the pH of each solution: (a) 25.0 mL of 6.09 *M* HCl is added to 100.0 mL of distilled water; (b) 5.0 mL of 2.55 *M* NaOH is added to 75.0 mL of distilled water.

94. What is the pH of a solution prepared by mixing 50.0 mL of 0.225 *M* HCl with 100.0 mL of a 0.184 *M* solution of NaOH?

95. What volume of 0.50 *M* HCl is needed to completely neutralize 25.00 mL of 0.86 *M* NaOH?

96. Calculate the final pH when each of the following pairs of solutions is mixed: (a) 100 mL of 0.105 *M* HCl and 100 mL of 0.115 *M* sodium acetate; (b) 50 mL of a 0.10 *M* solution of HCl and 100 mL of a 0.15 *M* solution of sodium acetate; (c) 100 mL of a 0.109 *M* acetic acid solution and 100 mL of a 0.118 *M* NaOH solution; (d) 100 mL of a 0.998 *M* acetic acid solution and 50.0 mL of a 0.110 *M* NaOH solution; (e) 100 mL of a 0.107 *M* sodium acetate solution and 50.0 mL of a 0.987 *M* acetic acid solution.

97. Calculate the final pH when each of the following pairs of solutions is mixed: (a) 100 mL of 0.983 *M* HCl and 100 mL of 0.102 *M* sodium fluoride; (b) 50.0 mL of a 0.115 *M* solution of HCl and 100 mL of a 0.109 *M* solution of sodium fluoride; (c) 100 mL of a 0.106 *M* hydrofluoric acid solution and 50.0 mL of a 0.996 *M* NaOH solution.

98. Calcium carbonate is a major contributor to the "hardness" of water. The amount of $CaCO_3$ in a water sample can be determined by titrating the sample with an acid, such as HCl, which produces water and CO_2. Write a balanced equation for this reaction. Generate a plot of solution pH versus volume of 0.100 *M* HCl added for the titration of a solution of 250 mg of $CaCO_3$ in 200.0 mL of water with 0.100 *M* HCl; assume that the HCl solution is added in 5.00-mL increments. What volume of HCl corresponds to the equivalence point?

99. For a titration of 50.0 mL of 0.288 *M* NaOH, you would like to prepare a 0.200 *M* HCl solution. The only HCl solution available to you, however, is 12.0 *M*.
(a) How would you prepare 500 mL of a 0.200 *M* HCl solution?
(b) Approximately what volume of your 0.200 *M* HCl solution is needed to neutralize the NaOH solution?
(c) After completing the titration, you find that your "0.200 *M*" HCl solution is actually 0.187 *M*. What was the exact volume of titrant used in the neutralization?

100. While titrating 50.0 mL of a 0.582 *M* solution of HCl with a solution labeled "0.500 *M* KOH," you overshoot the endpoint. To correct the problem, you add 10.00 mL of the HCl solution to your flask and then carefully continue the titration. The total volume of titrant needed for neutralization is 71.9 mL.
(a) What is the actual molarity of your KOH solution?
(b) What volume of titrant was needed to neutralize 50.0 mL of the acid?

101. Complete the table and generate a titration curve showing the pH versus volume of added base for the titration of 50.0 mL of 0.288 *M* HCl with 0.321 *M* NaOH. Clearly indicate the equivalence point.

Base Added, mL	10.0	30.0	40.0	45.0	50.0	55.0	65.0	75.0
pH								

102. The following data were obtained while titrating 25.0 mL of 0.156 *M* NaOH with a solution labeled "0.202 *M* HCl." Plot the pH versus volume of titrant added. Then determine the equivalence point from your graph, and calculate the exact concentration of your HCl solution.

Volume of HCl, mL	5	10	15	20	25	30	35
pH	11.46	11.29	10.98	4.40	2.99	2.70	2.52

103. Fill in the data for the titration of 50.0 mL of 0.241 *M* formic acid with 0.0982 *M* KOH. The pK_a of formic acid is 3.75. What is the pH of the solution at the equivalence point?

Volume of Base Added, mL	0	5	10	15	20	25
pH						

104. Glycine hydrochloride, which contains the fully protonated form of the amino acid glycine, has this structure:

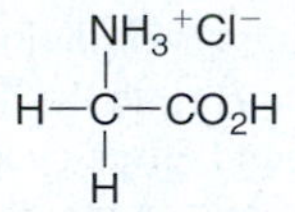

Glycine hydrochloride

It is a strong electrolyte that completely dissociates in water. Titration with base gives two equivalence points: the first corresponds to deprotonation of the carboxylic acid group, and the second to loss of the proton from the ammonium group. The corresponding equilibrium equations are

$$^{+}NH_3{-}CH_2{-}CO_2H(aq) \rightleftharpoons {}^{+}NH_3{-}CH_2{-}CO_2^{-}(aq) + H^{+} \quad pK_{a1} = 2.3$$

$$^{+}NH_3{-}CH_2{-}CO_2^{-}(aq) \rightleftharpoons NH_2{-}CH_2{-}COO^{-}(aq) + H^{+} \quad pK_{a2} = 9.6$$

(a) Given 50.0 mL of solution that is 0.430 *M* in glycine hydrochloride, how many milliliters of 0.150 *M* KOH are needed to fully deprotonate the carboxylic acid group?
(b) How many additional milliliters of KOH are needed to deprotonate the ammonium group?
(c) What is the pH of the solution at each of the equivalence points?
(d) How many milliliters of titrant are needed to obtain a solution in which glycine has no net electrical charge ? The pH at which a molecule such as glycine has no net charge is its *isoelectric point*. What is the isoelectric point of glycine?

105. What is the pH of a solution prepared by adding 38.2 mL of a 0.197 *M* HCl solution to 150.0 mL of a 0.242 *M* solution of pyridine? The pK_b of pyridine is 8.77.

106. Calculate the pH of a solution prepared by adding 40.3 mL of 0.289 *M* NaOH to 150.0 mL of a 0.564 *M* succinic acid, $HO_2CCH_2CH_2CO_2H$, solution ($pK_{a1} = 4.21$, $pK_{a2} = 5.64$).

107. Calculate the pH of a 0.150 *M* solution of malonic acid, $HO_2CCH_2CO_2H$ ($pK_{a1} = 2.85$, $pK_{a2} = 5.70$).

16.6 Buffers

108. Benzenesulfonic acid ($pK_a = 2.55$) is synthesized by treating benzene with concentrated sulfuric acid. Calculate (a) the pH of a 0.286 *M* solution of benzenesulfonic acid, and (b) the pH after adding enough sodium benzenesulfonate to increase the benzenesulfonate concentration to 0.100 *M*.

109. Phenol has a pK_a of 9.99. Calculate (a) the pH of a 0.195 *M* solution, and (b) the percent increase in the concentration of phenol after adding enough solid sodium phenoxide (the sodium salt of the conjugate base) to give a total phenoxide concentration of 0.100 *M*.

110. An intermediate used in the synthesis of perfumes is valeric acid, also called pentanoic acid. The pK_a of pentanoic acid is 4.84 at 25°C.
(a) What is the pH of a 0.259 *M* solution of pentanoic acid?
(b) Sodium pentanoate is added to make a buffered solution. What is the pH of the solution if it is 0.210 *M* in sodium pentanoate?
(c) What is the final pH if 8.00 mL of 0.100 *M* HCl is added to 75.0 mL of the buffered solution?
(d) What is the final pH if 8.00 mL of 0.100 *M* NaOH is added to 75.0 mL of the buffered solution?

111. Salicylic acid is used in the synthesis of acetylsalicylic acid, or aspirin. One gram dissolves in 460 mL of water to create a saturated solution with a pH of 2.40.
(a) What is the K_a of salicylic acid?
(b) What is the final pH of a saturated solution that is also 0.238 *M* in sodium salicylate?
(c) What is the final pH if 10.00 mL of 0.100 *M* HCl is added to 150.0 mL of the buffered solution?
(d) What is the final pH if 10.00 mL of 0.100 *M* NaOH is added to 150.0 mL of the buffered solution?

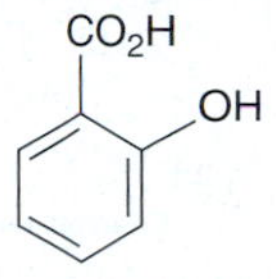

Salicylic acid

APPLICATIONS

112. The analytical concentration of lactic acid in blood is generally less than 1.2×10^{-3} *M*, corresponding to the sum of [lactate] and [lactic acid]. During strenuous exercise, however, oxygen in the muscle tissue is depleted, and overproduction of lactic acid occurs. This leads to a condition known as lactic acidosis, which is characterized by elevated blood lactic acid levels (approximately 5×10^{-3} *M*). The pK_a of lactic acid is 3.9.
(a) What is the actual lactic acid concentration under normal physiological conditions?
(b) What is the actual lactic acid concentration during lactic acidosis?

113. When the internal temperature of a human reaches 105°F, immediate steps must be taken to prevent the person from having convulsions. At this temperature, K_w has a value of approximately 2.94×10^{-14}. (a) Calculate the value of pK_w and the pH and pOH of a neutral solution at 105°F. (b) Is the pH higher or lower than that calculated in Exercise 16.1 for a neutral solution at a normal body temperature of 98.6°F?

114. The compound diphenhydramine is the active ingredient in a number of over-the-counter antihistamine medications used to treat the runny nose and watery eyes associated with hay fever and other allergies. Diphenhydramine is a derivative of trimethylamine (one methyl group is replaced by a more complex organic "arm" containing two phenyl rings) $(C_6H_5)_2\ CHOCH_2CH_2N(CH_3)_2$. The compound is sold as the water-soluble hydrochloride salt, $DPH^{+}Cl^{-}$. A tablet of diphenhydramine hydrochloride contains 25.0 mg of the active ingredient. Calculate the pH of the solution if two tablets are dissolved in 100 mL of water. The pK_b of diphenhydramine is 5.47, and the formula mass of diphenhydramine hydrochloride is 291.81 amu.

115. Fluoroacetic acid is a poison that has been used by ranchers in the western United States. The ranchers place the poison in the carcasses of dead animals to kill coyotes that feed on them; unfortunately, however, eagles and hawks are also killed in the process. How many mL of 0.0953 *M* $Ca(OH)_2$ are needed to completely neutralize 50.0 mL of 0.262 *M* fluoroacetic acid solution ($pK_a = 2.59$)? What is the initial pH of the solution? What is the pH of the solution at the equivalence point?

116. Epinephrine, a secondary amine, is used to counter allergic reactions as well as to bring patients out of anesthesia and cardiac arrest. The pK_b of epinephrine is 4.31. What is the percent ionization in a 0.280 *M* solution? What is the percent ionization after enough solid epinephrine hydrochloride is added to make the final epinephrineH^+ concentration 0.982 *M*? What is the final pH of the solution?

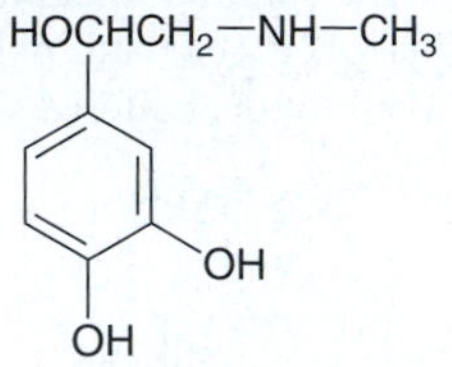

Epinephrine

$$\text{epinephrine}(aq) + H_2O(l) \rightleftharpoons \text{epinephrineH}^+(aq) + OH^-(aq)$$

117. Accidental ingestion of aspirin (acetylsalicylic acid) is probably the most common cause of childhood poisoning. Initially, salicylates stimulate the portion of the brain that controls breathing, resulting in hyperventilation (excessively intense breathing that lowers the P_{CO_2} in the lungs). Subsequently, a potentially serious rebound effect occurs, as the salicylates are converted to a weak acid, salicylic acid, in the body. Starting with the normal values of P_{CO_2} = 40 mmHg, pH = 7.40, and $[HCO_3^-]$ = 24 m*M*, show what happens during (a) the initial phase of respiratory stimulation, and (b) the subsequent phase of acid production. Why is the rebound effect dangerous?

118. Emphysema is a disease that reduces the efficiency of breathing, and as a result less CO_2 is exchanged with the atmosphere. What effect will this have on blood pH, P_{CO_2}, and $[HCO_3^-]$?

17 Solubility and Complexation Equilibria

LEARNING OBJECTIVES

- To be able to calculate the solubility of an ionic compound from its K_{sp}
- To understand the factors that determine the solubility of ionic compounds
- To be able to describe complex-ion formation quantitatively
- To understand why the solubility of many compounds depends on pH
- To know how to separate metal ions by selective precipitation

CHAPTER OUTLINE

Although Chapter 16 focused exclusively on acid–base equilibria in aqueous solutions, equilibrium concepts can also be applied to many other kinds of reactions that occur in aqueous solution. In this chapter, we describe the equilibria involved in the solubility of ionic compounds and the formation of complex ions.

Solubility equilibria involving ionic compounds are important in fields as diverse as medicine, biology, geology, and industrial chemistry. Carefully controlled precipitation reactions of calcium salts, for example, are used by many organisms to produce structural materials, such as bone and the shells that surround mollusks and bird eggs. In contrast, uncontrolled precipitation

Scanning electron micrograph of kettle scale. Hard water is a solution that consists largely of calcium and magnesium carbonate in CO_2-rich water. When the water is heated, CO_2 gas is released and the carbonate salts precipitate from solution producing a solid called *scale*.

reactions of calcium salts are partially or wholly responsible for the formation of scale in coffee makers and boilers, "bathtub rings," and kidney stones, which can be excruciatingly painful. The principles discussed in this chapter will enable you to understand how these apparently diverse phenomena are related. Solubility equilibria are also responsible for the formation of caves and their striking features, such as stalactites and stalagmites, through a long process involving the repeated dissolution and precipitation of calcium carbonate. By the end of this chapter, you will understand why barium sulfate is ideally suited for X-ray imaging of the digestive tract, and why soluble complexes of gadolinium can be used for imaging soft tissue and blood vessels using magnetic resonance imaging (MRI), even though most simple salts of both metals are toxic to humans.

17.1 • Determining the Solubility of Ionic Compounds

We begin our discussion of solubility and *complexation equilibria*, the equilibria associated with the formation of complex ions, by developing quantitative methods for describing dissolution and precipitation reactions of ionic compounds in aqueous solution. Just as with acid–base equilibria, we can describe the concentrations of ions in equilibrium with an ionic solid using an equilibrium constant expression.

The Solubility Product, K_{sp}

MGC The Solubility Product

When a slightly soluble ionic compound is added to water, some of it dissolves to form a solution, establishing an equilibrium between the pure solid and a solution of its ions. For the dissolution of calcium phosphate, one of the two main components of kidney stones, the equilibrium can be written as follows, with the solid salt on the left:*

$$Ca_3(PO_4)_2(s) \rightleftharpoons 3Ca^{2+}(aq) + 2PO_4^{3-}(aq) \tag{17.1}$$

Note the pattern

As with K, the concentration of a pure solid does not appear explicitly in K_{sp}.

The equilibrium constant for the dissolution of a sparingly soluble salt is the **solubility product** of the salt, abbreviated $\boldsymbol{K_{sp}}$. Because the concentration of a pure solid such as $Ca_3(PO_4)_2$ is a constant, it does not appear explicitly in the equilibrium constant expression (see Chapter 15). The equilibrium constant expression for the above reaction is therefore

$$K = \frac{[Ca^{2+}]^3[PO_4^{3-}]^2}{[Ca_3(PO_4)_2]} \tag{17.2}$$

$$[Ca_3(PO_4)_2]K = K_{sp} = [Ca^{2+}]^3[PO_4^{3-}]^2$$

At 25°C and pH 7.00, K_{sp} for calcium phosphate is 2.07×10^{-33}, indicating that the concentrations of $Ca^{2+}(aq)$ and $PO_4^{3-}(aq)$ ions in equilibrium with solid calcium phosphate are very low. The values of K_{sp} for some common salts are listed in Table 17.1; they show that the magnitude of K_{sp} varies dramatically for different compounds. Although in Equation 17.1, the K_{sp} is not a function of pH, changes in pH can affect the solubility of a compound, as you will discover in Section 17.4.

Solubility products are determined experimentally by either directly measuring the concentration of one of the component ions or by measuring the solubility of the compound in a given amount of water. Note, however, that whereas solubility is usually expressed in terms of *mass* of solute per 100 mL of solvent, K_{sp}, like K, is defined in terms of *molar* concentrations of the component ions.

---

* As you will discover in Section 17.4 and in more advanced chemistry courses, basic anions such as S^{2-}, PO_4^{3-}, and CO_3^{2-} react with water to produce OH^- and the corresponding protonated anion. Consequently, their calculated molarities assuming no protonation in aqueous solution are approximate.

TABLE 17.1 Solubility products, K_{sp}, for selected ionic substances at 25°C

Solid	Color	K_{sp}	Solid	Color	K_{sp}
	Acetates			**Iodides**	
$Ca(O_2CCH_3)_2 \cdot 3H_2O$	White	4×10^{-3}	Hg_2I_2[a]	Yellow	5.2×10^{-29}
	Bromides		PbI_2	Yellow	9.8×10^{-9}
AgBr	Off-white	5.35×10^{-13}		**Oxalates**	
Hg_2Br_2[a]	Yellow	6.40×10^{-23}	$Ag_2C_2O_4$	White	5.40×10^{-12}
	Carbonates		$MgC_2O_4 \cdot 2H_2O$	White	4.83×10^{-6}
$CaCO_3$	White	3.36×10^{-9}	PbC_2O_4	White	4.8×10^{-10}
$PbCO_3$	White	7.40×10^{-14}		**Phosphates**	
	Chlorides		Ag_3PO_4	White	8.89×10^{-17}
AgCl	White	1.77×10^{-10}	$Sr_3(PO_4)_2$	White	4.0×10^{-28}
Hg_2Cl_2[a]	White	1.43×10^{-18}	$FePO_4 \cdot 2H_2O$	Pink	9.91×10^{-16}
$PbCl_2$	White	1.70×10^{-5}		**Sulfates**	
	Chromates		Ag_2SO_4	White	1.20×10^{-5}
$CaCrO_4$	Yellow	7.1×10^{-4}	$BaSO_4$	White	1.08×10^{-10}
$PbCrO_4$	Yellow	2.8×10^{-13}	$PbSO_4$	White	2.53×10^{-8}
	Fluorides			**Sulfides**	
BaF_2	White	1.84×10^{-7}	Ag_2S	Black	6.3×10^{-50}
PbF_2	White	3.3×10^{-8}	CdS	Yellow	8.0×10^{-27}
	Hydroxides		PbS	Black	8.0×10^{-28}
$Ca(OH)_2$	White	5.02×10^{-6}	ZnS	White	1.6×10^{-24}
$Cu(OH)_2$	Pale blue	1×10^{-14}			
$Mn(OH)_2$	Light pink	1.9×10^{-13}			
$Cr(OH)_3$	Gray-green	6.3×10^{-31}			
$Fe(OH)_3$	Rust red	2.79×10^{-39}			

[a] These contain the Hg_2^{2+} ion.

EXAMPLE 17.1

Calcium oxalate monohydrate, $Ca(O_2CCO_2) \cdot H_2O$ (also written as $CaC_2O_4 \cdot H_2O$), is a sparingly soluble salt that is the other major component of kidney stones [along with $Ca_3(PO_4)_2$]. Its solubility in water at 25°C is 7.36×10^{-4} g/100 mL. Calculate its K_{sp}.

Given Solubility in g/100 mL

Asked for K_{sp}

Strategy

Ⓐ Write the balanced dissolution equilibrium and the corresponding solubility product expression.
Ⓑ Convert the solubility of the salt to moles per liter. From the balanced dissolution equilibrium, determine the equilibrium concentrations of the dissolved solute ions. Substitute these values into the solubility product expression to calculate K_{sp}.

Solution

Ⓐ We need to write the solubility product expression in terms of the concentrations of the component ions. For calcium oxalate monohydrate, the balanced dissolution equilibrium and the solubility product expression (abbreviating oxalate as ox^{2-}) are

$$Ca(O_2CCO_2) \cdot H_2O(s) \rightleftharpoons Ca^{2+}(aq) + {}^{-}O_2CCO_2^{-}(aq) + H_2O(l) \qquad K_{sp} = [Ca^{2+}][ox^{2-}]$$

Neither solid calcium oxalate monohydrate nor water appears in the solubility product expression because their concentrations are essentially constant. **B** Next, we need to determine $[Ca^{2+}]$ and $[ox^{2-}]$ at equilibrium. We can use the mass of calcium oxalate monohydrate that dissolves in 100 mL of water to calculate the number of moles that dissolve in 100 mL of water. From this we can determine the number of moles that dissolve in 1.00 L of water. For dilute solutions, the density of the solution is nearly the same as that of water, so dissolving the salt in 1.00 L of water gives essentially 1.00 L of solution. Because each 1 mol of dissolved calcium oxalate monohydrate dissociates to produce 1 mol of calcium ions and 1 mol of oxalate ions, we can obtain the equilibrium concentrations that must be inserted into the solubility product expression. The number of moles of calcium oxalate monohydrate that dissolve in 100 mL of water is

$$\frac{7.36 \times 10^{-4}\ \cancel{g}}{146.1\ \cancel{g}/\text{mol}} = 5.04 \times 10^{-6}\ \text{mol}\ Ca(O_2CCO_2) \cdot H_2O$$

The number of moles of calcium oxalate monohydrate that dissolve in 1.00 L of the saturated solution is

$$\left(\frac{5.04 \times 10^{-6}\ \text{mol}\ Ca(O_2CCO_2) \cdot H_2O}{100\ \cancel{\text{mL}}}\right)\left(\frac{1000\ \cancel{\text{mL}}}{1.00\ \text{L}}\right) = 5.04 \times 10^{-5}\ \text{mol/L} = 5.04 \times 10^{-5}\ M$$

Because of the stoichiometry of the reaction, the concentration of Ca^{2+} and ox^{2-} ions are both $5.04 \times 10^{-5}\ M$. Inserting these values into the solubility product expression gives

$$K_{sp} = [Ca^{2+}][ox^{2-}] = (5.04 \times 10^{-5})(5.04 \times 10^{-5}) = 2.54 \times 10^{-9}$$

In our calculation in Example 17.1 we have ignored the reaction of the weakly basic anion with water, which tends to make the actual solubility of salts greater than the calculated value.

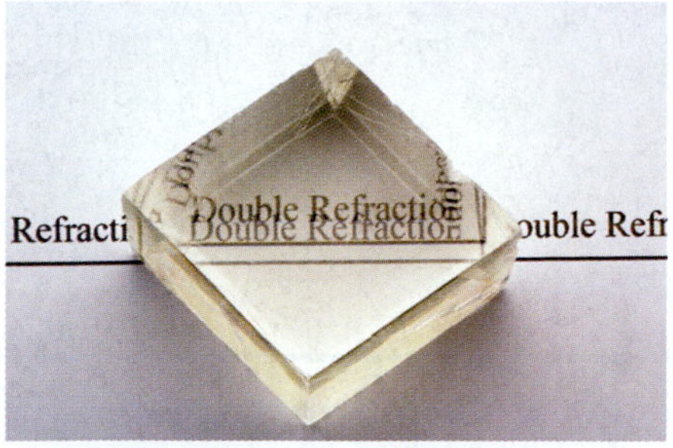

A crystal of calcite ($CaCO_3$), illustrating the phenomenon of double refraction. When a transparent crystal of calcite is placed over a page, we see *two* images of the letters.

EXERCISE 17.1

One crystalline form of calcium carbonate, $CaCO_3$, is the mineral sold as "calcite" in mineral and gem shops. The solubility of calcite in water is 0.67 mg/100 mL. Calculate its K_{sp}.

Answer 4.5×10^{-9}

Tabulated values of K_{sp} can also be used to estimate the solubility of a salt with a procedure that is essentially the reverse of the one used in Example 17.1. In this case, we treat the problem as a typical equilibrium problem and set up a table of initial concentrations, changes in concentration, and final concentrations as we did in Chapter 15, remembering that the concentration of the pure solid is essentially constant.

EXAMPLE 17.2

We saw that the value of K_{sp} for $Ca_3(PO_4)_2$ is 2.07×10^{-33} at 25°C. Calculate the aqueous solubility of $Ca_3(PO_4)_2$ in terms of (a) the molarity of ions produced in solution, and (b) the mass of salt that dissolves in 100 mL of water at 25°C.

Given K_{sp}

Asked for Molar concentration, mass of salt that dissolves in 100 mL of water

Strategy

A Write the balanced equilibrium equation for the dissolution reaction, and construct a table showing the concentrations of the species produced in solution. Insert the appropriate values into the solubility product expression, and calculate the molar solubility at 25°C.

B Calculate the mass of solute in 100 mL of solution from the molar solubility of the salt. Assume that the volume of the solution is the same as the volume of the solvent.

Solution

(a) The dissolution equilibrium for $Ca_3(PO_4)_2$ (Equation 17.1) is shown in the table. Because we are starting with distilled water, the initial concentration of both calcium and phosphate ions is zero. For every 1 mol of $Ca_3(PO_4)_2$ that dissolves, 3 mol of Ca^{2+} and 2 mol of PO_4^{3-} ions are produced in solution. If we let x equal the solubility of $Ca_3(PO_4)_2$ in moles per liter, then the change in $[Ca^{2+}]$ will be $+3x$, and the change in $[PO_4^{3-}]$ will be $+2x$. We can insert these values into the table.

$$Ca_3(PO_4)_2(s) \rightleftharpoons 3Ca^{2+}(aq) + 2PO_4^{3-}(aq)$$

	$Ca_3(PO_4)_2$	$[Ca^{2+}]$	$[PO_4^{3-}]$
Initial	Pure solid	0	0
Change	—	$+ 3x$	$+ 2x$
Final	Pure solid	$3x$	$2x$

Although the *amount* of solid $Ca_3(PO_4)_2$ changes as some of it dissolves, its *molar concentration* does not change. We now insert the expressions for the equilibrium concentrations of the ions into the solubility product expression (Equation 17.2):

$$K_{sp} = [Ca^{2+}]^3[PO_4^{3-}]^2 = (3x)^3(2x)^2$$
$$108x^5 = 2.1 \times 10^{-33}$$
$$x^5 = 1.9 \times 10^{-35}$$
$$x = 1.1 \times 10^{-7}\ M$$

This is the molar solubility of calcium phosphate at 25°C. Notice, however, that the molarity of the ions is $2x$ and $3x$, which means that $[PO_4^{3-}] = 2.2 \times 10^{-7}$ and $[Ca^{2+}] = 3.3 \times 10^{-7}$.

(b) To find the mass of solute in 100 mL of solution, we assume that the density of this dilute solution is the same as the density of water because of the low solubility of the salt, so that 100 mL of water gives 100 mL of solution. We can then determine the amount of salt that dissolves in 100 mL of water:

$$\left(\frac{1.1 \times 10^{-7}\ \text{mol}}{1\ \text{L}}\right) 100\ \text{mL} \left(\frac{1\ \text{L}}{1000\ \text{mL}}\right)\left(\frac{310.18\ \text{g}\ Ca_3(PO_4)_2}{1\ \text{mol}}\right) = 3.4 \times 10^{-6}\ \text{g}\ Ca_3(PO_4)_2 \text{ in 100 mL water}$$

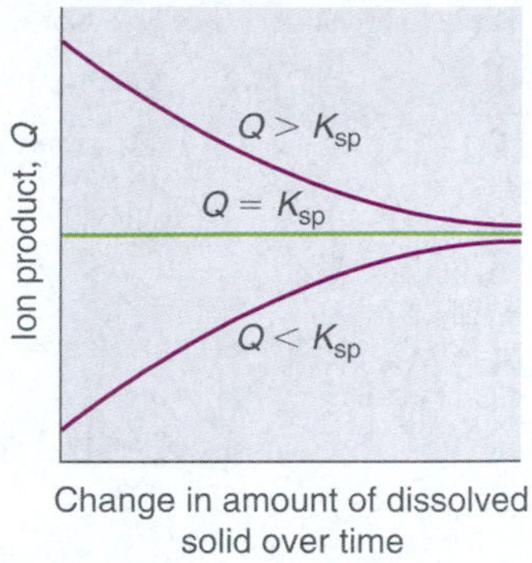

Figure 17.1 The relationship between Q and K_{sp}. If Q is less than K_{sp}, the solution is unsaturated and more solid will dissolve until the system reaches equilibrium ($Q = K_{sp}$). If Q is greater than K_{sp}, the solution is supersaturated and solid will precipitate until $Q = K_{sp}$. If $Q = K_{sp}$, the rate of dissolution is equal to the rate of precipitation; the solution is saturated and no net change in the amount of dissolved solid will occur. Interactive Graph

EXERCISE 17.2

The solubility product of silver carbonate, Ag_2CO_3, is 8.46×10^{-12} at 25°C. Calculate **(a)** the molarity of a saturated solution, and **(b)** the mass of silver carbonate that will dissolve in 100 mL of water at this temperature.

Answer **(a)** $1.28 \times 10^{-4}\ M$; **(b)** 3.54 mg

The Ion Product

The **ion product (Q)** of a salt is the product of the concentrations of the ions in solution raised to the same powers as in the solubility product expression. It is analogous to the reaction quotient (Q) discussed for gaseous equilibria in Chapter 15. Whereas K_{sp} describes equilibrium concentrations, the ion product describes concentrations that are *not* necessarily equilibrium concentrations.

Note the pattern

The ion product Q is analogous to the reaction quotient Q for gaseous equilibria.

As summarized in Figure 17.1, there are three possible conditions for an aqueous solution of an ionic solid:

1. $Q < K_{sp}$: the solution is unsaturated, and more of the ionic solid, if available, will dissolve.

2. $Q = K_{sp}$: the solution is saturated and at equilibrium.
3. $Q > K_{sp}$: the solution is supersaturated, and ionic solid will precipitate.

The process of calculating the value of the ion product and comparing it with the magnitude of the solubility product is a straightforward way to determine whether a solution is unsaturated, saturated, or supersaturated. More important, the ion product tells chemists whether a precipitate will form when solutions of two soluble salts are mixed.

EXAMPLE 17.3

We mentioned that barium sulfate is used in medical imaging of the gastrointestinal tract. Its solubility product is 1.08×10^{-10} at 25°C, so it is ideally suited for this purpose because of its low solubility when a "barium milkshake" is consumed by a patient. The pathway of the sparingly soluble salt can be easily monitored by X-rays. Will barium sulfate precipitate if 10.0 mL of 0.0020 *M* Na_2SO_4 is added to 100 mL of 3.2×10^{-4} *M* $BaCl_2$? Recall that NaCl is highly soluble in water.

Given K_{sp}, and volumes and concentrations of reactants

Asked for Whether precipitate will form

Strategy

Ⓐ Write the balanced equilibrium equation for the precipitation reaction and the expression for K_{sp}.
Ⓑ Determine the concentrations of all ions in solution when the solutions are mixed, and use them to calculate the value of the ion product Q.
Ⓒ Compare the values of Q and K_{sp} to decide whether a precipitate will form.

Solution

Ⓐ The only slightly soluble salt that can be formed when these two solutions are mixed is $BaSO_4$ because NaCl is highly soluble. The equation for the precipitation of $BaSO_4$ is

$$BaSO_4(s) \rightleftharpoons Ba^{2+}(aq) + SO_4{}^{2-}(aq)$$

The solubility product expression is

$$K_{sp} = [Ba^{2+}][SO_4{}^{2-}] = 1.08 \times 10^{-10}$$

Ⓑ To solve this problem we must first calculate the value of the ion product, $Q = [Ba^{2+}][SO_4^{2-}]$, using the concentrations of the ions that are present after the solutions are mixed and before any reaction occurs. The concentration of Ba^{2+} when the solutions are mixed is the total number of moles of Ba^{2+} in the original 100 mL of $BaCl_2$ solution divided by the final volume (100 mL + 10 mL = 110 mL):

$$\text{moles } Ba^{2+} = 100\ \cancel{\text{mL}}\left(\frac{1\ \cancel{\text{L}}}{1000\ \cancel{\text{mL}}}\right)\left(\frac{3.2 \times 10^{-4}\ \text{mol}}{1\ \cancel{\text{L}}}\right) = 3.2 \times 10^{-5}\ \text{mol } Ba^{2+}$$

$$[Ba^{2+}] = \left(\frac{3.2 \times 10^{-5}\ \text{mol } Ba^{2+}}{110\ \cancel{\text{mL}}}\right)\left(\frac{1000\ \cancel{\text{mL}}}{1\ \text{L}}\right) = 2.9 \times 10^{-4}\ M\ Ba^{2+}$$

Similarly, the concentration of $SO_4{}^{2-}$ after mixing is the total number of moles of SO_4^{2-} in the original 10.0 mL of Na_2SO_4 solution divided by the final volume (110 mL):

$$\text{moles } SO_4{}^{2-} = 10.0\ \cancel{\text{mL}}\left(\frac{1\ \cancel{\text{L}}}{1000\ \cancel{\text{mL}}}\right)\left(\frac{0.0020\ \text{mol}}{1\ \cancel{\text{L}}}\right) = 2.0 \times 10^{-5}\ \text{mol } SO_4{}^{2-}$$

$$[SO_4{}^{2}] = \left(\frac{2.0 \times 10^{-5}\ \text{mol } SO_4{}^{2-}}{110\ \cancel{\text{mL}}}\right)\left(\frac{1000\ \cancel{\text{mL}}}{1\ \text{L}}\right) = 1.8 \times 10^{-4}\ M\ SO_4{}^{2-}$$

We can now calculate the ion product Q:

$$Q = [Ba^{2+}][SO_4^{\,2-}] = (2.9 \times 10^{-4})(1.8 \times 10^{-4}) = 5.2 \times 10^{-8}$$

C We now compare Q, the ion product, with the solubility product K_{sp}. If $Q > K_{sp}$, then $BaSO_4$ will precipitate, but if $Q < K_{sp}$, it will not. Since $Q > K_{sp}$, we predict that $BaSO_4$ will precipitate when the two solutions are mixed. In fact, $BaSO_4$ will continue to precipitate until the system reaches equilibrium, which occurs when $[Ba^{2+}][SO_4^{\,2-}] = K_{sp} = 1.08 \times 10^{-10}$.

EXERCISE 17.3

The solubility product of calcium fluoride, CaF_2, is 3.45×10^{-11}. If 2.0 mL of a 0.10 M solution of NaF is added to 128 mL of a 2.0×10^{-5} M solution of $Ca(NO_3)_2$, will CaF_2 precipitate?

Answer yes ($Q = 4.7 \times 10^{-11} > K_{sp}$)

The Common Ion Effect and Solubility

MGC Common Ion Effect on Solubility

The solubility product expression tells us that the equilibrium concentrations of cation and anion are inversely related. That is, as the concentration of the anion increases, the maximum concentration of the cation needed for precipitation to occur decreases, and vice versa, so that K_{sp} is constant. Consequently, the solubility of an ionic compound depends on the concentrations of other salts that contain the same ions. This dependency is another example of the common ion effect discussed in Section 16.6: adding a common cation or anion shifts a solubility equilibrium in the direction predicted by Le Châtelier's principle. As a result, *the solubility of any sparingly soluble salt is almost always decreased by the presence of a soluble salt that contains a common ion.**

Consider, for example, the effect of adding a soluble salt such as $CaCl_2$ to a saturated solution of calcium phosphate. We have seen that the solubility of $Ca_3(PO_4)_2$ in water at 25°C is 1.14×10^{-7} M ($K_{sp} = 2.07 \times 10^{-33}$). Thus, a saturated solution of calcium phosphate in water contains $3 \times 1.14 \times 10^{-7}$ M $= 3.42 \times 10^{-7}$ M Ca^{2+} and $2 \times 1.14 \times 10^{-7}$ M $= 2.28 \times 10^{-7}$ M $PO_4^{\,3-}$ (neglecting hydrolysis to $HPO_4^{\,2-}$ as described in Chapter 16), according to the stoichiometry shown in Equation 17.1. If $CaCl_2$ is added to a saturated solution of $Ca_3(PO_4)_2$, the Ca^{2+} ion concentration will increase such that $[Ca^{2+}] > 3.42 \times 10^{-7}$ M, making $Q > K_{sp}$. The only way the system can return to equilibrium is for the reaction in Equation 17.1 to proceed to the left, resulting in precipitation of calcium phosphate. This will decrease the concentration of both Ca^{2+} and $PO_4^{\,3-}$ until $Q = K_{sp}$.

Note the pattern

The common ion effect usually decreases the solubility of a sparingly soluble salt.

EXAMPLE 17.4

Calculate the solubility of calcium phosphate in 0.20 M $CaCl_2$.

Given Concentration of $CaCl_2$ solution

Asked for Solubility of $Ca_3(PO_4)_2$ in $CaCl_2$ solution

Strategy

A Write the balanced equilibrium equation for the dissolution of $Ca_3(PO_4)_2$. Tabulate the concentrations of all species produced in solution.

B Substitute the appropriate values into the expression for the solubility product, and calculate the solubility of $Ca_3(PO_4)_2$.

* The exceptions generally involve the formation of complex ions, which is discussed in Section 17.3.

Solution

A The balanced equilibrium equation is given in the table. If we let x equal the solubility of $Ca_3(PO_4)_2$ in moles per liter, then the change in $[Ca^{2+}]$ is once again $+3x$, and the change in $[PO_4^{3-}]$ is $+2x$. We can insert these values into the table.

$$Ca_3(PO_4)_2(s) \rightleftharpoons 3Ca^{2+}(aq) + 2PO_4^{3-}(aq)$$

	$Ca_3(PO_4)_2$	$[Ca^{2+}]$	$[PO_4^{3-}]$
Initial	Pure solid	0.20	0
Change	—	$+ 3x$	$+ 2x$
Final	Pure solid	$0.20 + 3x$	$2x$

B The K_{sp} expression is

$$K_{sp} = [Ca^{2+}]^3[PO_4^{3-}]^2 = (0.20 + 3x)^3(2x)^2 = 2.07 \times 10^{-33}$$

Because $Ca_3(PO_4)_2$ is a sparingly soluble salt, we can reasonably expect that $x << 0.20$. Thus, $(0.20 + 3x)$ M is approximately 0.20 M, which simplifies the K_{sp} expression to

$$K_{sp} = (0.20)^3(2x)^2 = 2.07 \times 10^{-33}$$
$$x^2 = 6.5 \times 10^{-32}$$
$$x = 2.5 \times 10^{-16}\ M$$

This value is the solubility of calcium phosphate in 0.20 M $CaCl_2$ at 25°C. It is approximately nine orders of magnitude less than its solubility in pure water, as we would expect based on Le Châtelier's principle. Note that, with one exception, this example was identical to Example 17.2—here the initial $[Ca^{2+}]$ was 0.20 M rather than 0.

EXERCISE 17.4

Calculate the solubility of silver carbonate in a 0.25 M solution of sodium carbonate. The solubility of silver carbonate in pure water is 8.45×10^{-12} at 25°C.

Answer 2.9×10^{-6} M (versus 1.3×10^{-4} M in pure water)

17.2 • Factors That Affect Solubility

The solubility product of an ionic compound describes the concentrations of *ions* in equilibrium with a solid, but what happens if some of the cations become associated with anions rather than being completely surrounded by solvent? Then predictions of the total solubility of the compound based either on the assumption that the solute exists solely as ions or on the measured concentrations of those ions would substantially underestimate the actual solubility. In general, there are four reasons that the solubility of a compound may be other than expected: ion-pair formation, incomplete dissociation of molecular solutes, formation of complex ions, and changes in pH. The first two situations are described in this section, the third is discussed in Section 17.3, and changes in pH are discussed in Section 17.4.

Ion-Pair Formation

An **ion pair** consists of a cation and anion that are in intimate contact in solution, rather than separated by solvent (Figure 17.2). The ions in an ion pair are held together by the same attractive electrostatic forces that we discussed in Chapter 8 for ionic solids. As a result, the ions in an ion pair migrate as a single unit, whose net charge is the sum of the charges on the ions. In many ways, we can view an ion pair as a species intermediate between the ionic solid (in which each ion participates in many cation–anion interactions that hold the ions in a rigid array) and the completely dissociated ions in solution (where each is fully surrounded by water molecules and free to migrate independently).

As illustrated for calcium sulfate in the following equation, a second equilibrium must be included to describe the solubility of salts that form ion pairs:

$$CaSO_4(s) \rightleftharpoons \underset{\text{Ion pair}}{Ca^{2+} \cdot SO_4{}^{2-}(aq)} \rightleftharpoons Ca^{2+}(aq) + SO_4{}^{2-}(aq) \quad (17.3)$$

The ion pair is represented by the symbols of the individual ions separated by a dot, to indicate that they are associated in solution. The formation of an ion pair is a dynamic process, just like any other equilibrium, so a particular ion pair may exist only briefly before dissociating into the free ions, each of which may later associate briefly with other ions.

Note the pattern

The formation of ion pairs increases the solubility of a salt.

Ion-pair formation can have a major effect on the measured solubility of a salt. For example, the measured value of K_{sp} for calcium sulfate is 4.93×10^{-5} at 25°C. The solubility of $CaSO_4$ should be 7.02×10^{-3} *M* if the only equilibrium involved were

$$CaSO_4(s) \rightleftharpoons Ca^{2+}(aq) + SO_4{}^{2-}(aq) \quad (17.4)$$

In fact, the experimentally measured solubility of calcium sulfate at 25°C is 1.6×10^{-2} *M*, almost twice the value predicted from its K_{sp}. The reason for the discrepancy is that the concentration of ion pairs in a saturated $CaSO_4$ solution is almost as high as the concentration of hydrated ions. Recall that the magnitude of attractive electrostatic interactions is greatest for small, highly charged ions. Hence, ion-pair formation is most important for salts that contain M^{2+} and M^{3+} ions, such as Ca^{2+} and La^{3+}, and is relatively unimportant for salts that contain monopositive cations, except for the smallest, Li^+. We therefore expect a saturated solution of $CaSO_4$ to contain a high concentration of ion pairs and its solubility to be greater than predicted from its K_{sp}.

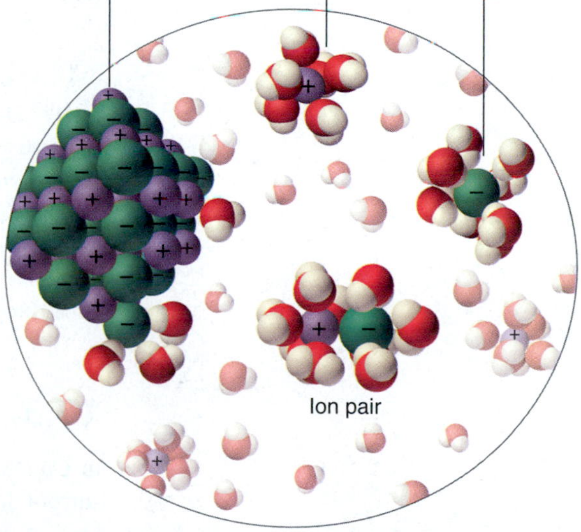

Figure 17.2 Ion-pair formation. In an ion pair, the cation and anion are in intimate contact in solution and migrate as a single unit. They are not completely dissociated and individually surrounded by solvent molecules, as are the hydrated ions, which are free to migrate independently.

Incomplete Dissociation

A molecular solute may also be more soluble than predicted by the measured concentrations of ions in solution due to *incomplete dissociation*. This is particularly common with weak organic acids (see Chapter 16). Although strong acids (HA) dissociate completely into their constituent ions (H^+ and A^-) in water, weak acids such as carboxylic acids do not ($K_a = 1.5 \times 10^{-5}$). However, the molecular (undissociated) form of a weak acid (HA) is often quite soluble in water; for example, acetic acid (CH_3CO_2H) is completely miscible with water. Many carboxylic acids, however, have only limited solubility in water, such as benzoic acid ($C_6H_5CO_2H$), with $K_a = 6.25 \times 10^{-5}$. Just as with calcium sulfate, we need to include an additional equilibrium to describe the solubility of benzoic acid:

$$C_6H_5CO_2H(s) \rightleftharpoons C_6H_5CO_2H(aq) \rightleftharpoons C_6H_5CO_2{}^-(aq) + H^+(aq) \quad (17.5)$$

Note the pattern

Incomplete dissociation of a molecular solute can increase the solubility of the solute.

In a case like this, measuring only the concentration of the ions grossly underestimates the total concentration of the organic acid in solution. In the case of benzoic acid, for example, the pH of a saturated solution at 25°C is 2.85, corresponding to $[H^+] = [C_6H_5CO_2{}^-] = 1.4 \times 10^{-3}$ *M*. The total concentration of benzoic acid in the solution, however, is 2.8×10^{-2} *M*. Thus, approximately

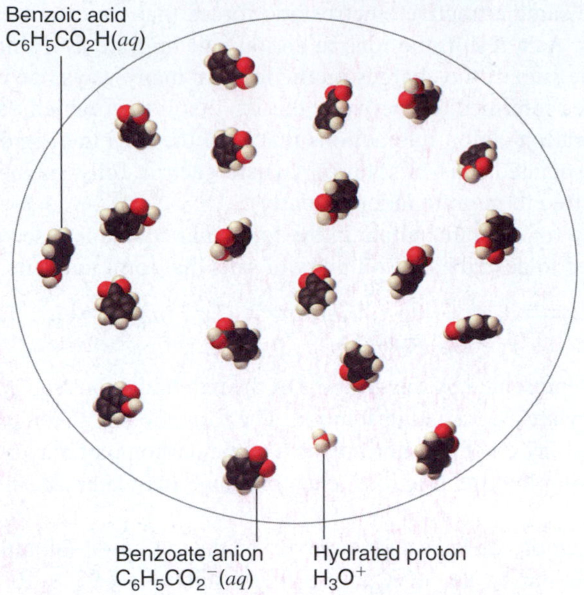

Figure 17.3 Incomplete dissociation of a molecular solute. In a saturated solution of benzoic acid in water at 25°C, only about 5% of the dissolved benzoic acid molecules are dissociated to form benzoate anions and hydrated protons. The remaining 95% exists in solution in the form of hydrated neutral molecules. (H_2O molecules are omitted for clarity.)

95% of the benzoic acid in solution is in the form of hydrated neutral molecules, $C_6H_5CO_2H(aq)$, and only about 5% is present as the dissociated ions (Figure 17.3).

Although ion pairs such as $Ca^{2+} \cdot SO_4^{2-}$, and undissociated electrolytes, such as $C_6H_5CO_2H$, are both electrically neutral, there is a major difference in the forces responsible for their formation. The ion pair is held together by simple electrostatic attractive forces between the cation and the anion, whereas the undissociated acid is held together by a polar covalent O—H bond.

17.3 Complex-Ion Formation

In Chapter 4, you learned that metal ions in aqueous solution are *hydrated*—that is, surrounded by a shell of usually four or six water molecules. A hydrated ion is one kind of **complex ion** (or, simply, *complex*), a species formed between a central metal ion and one or more surrounding **ligands**, molecules or ions that contain at least one lone pair of electrons.

A complex ion forms from a metal ion and a ligand because of a Lewis acid–base interaction. The positively charged metal ion acts as a Lewis acid, and the ligand, with one or more lone pairs of electrons, acts as a Lewis base. Small, highly charged metal ions such as Cu^{2+} or Ru^{3+}, have the greatest tendency to act as Lewis acids, and consequently they have the greatest tendency to form complex ions.

As an example of complex-ion formation, consider the addition of ammonia to an aqueous solution of the hydrated Cu^{2+} ion, $[Cu(H_2O)_6]^{2+}$. Because it is a stronger base than H_2O, ammonia replaces the water molecules in the hydrated ion to form the $[Cu(NH_3)_4(H_2O)_2]^{2+}$ ion. Formation of the $[Cu(NH_3)_4(H_2O)_2]^{2+}$ complex is accompanied by a dramatic color change, as shown in Figure 17.4. The solution changes from the light blue of $[Cu(H_2O)_6]^{2+}$ to the blue-violet characteristic of the $[Cu(NH_3)_4(H_2O)_2]^{2+}$ ion.

Figure 17.4 Complex-ion formation. An aqueous solution of $CuSO_4$ consists of hydrated Cu^{2+} ions in the form of pale blue $[Cu(H_2O)_6]^{2+}$ (*left*). Addition of aqueous ammonia to the solution results in the formation of the intensely blue-violet $[Cu(NH_3)_4(H_2O)_2]^{2+}$, usually written as $[Cu(NH_3)_4{}^{2+}]$ ion (*right*) because ammonia, a stronger base than H_2O, replaces water molecules from the hydrated Cu^{2+} ion.

The Formation Constant

MGC The Formation Constant

The replacement of water molecules from $[Cu(H_2O)_6]^{2+}$ by ammonia occurs in sequential steps. Omitting the water molecules bound to Cu^{2+} for simplicity, we can write the equilibrium reactions as

$$Cu^{2+}(aq) + NH_3(aq) \rightleftharpoons [Cu(NH_3)]^{2+}(aq) \qquad K_1$$
$$[Cu(NH_3)]^{2+}(aq) + NH_3(aq) \rightleftharpoons [Cu(NH_3)_2]^{2+}(aq) \qquad K_2 \qquad (17.6)$$

and so on. The sum of the stepwise reactions is the overall equation for the formation of the complex ion:*

$$Cu^{2+}(aq) + 4NH_3(aq) \rightleftharpoons [Cu(NH_3)_4]^{2+}(aq) \qquad (17.7)$$

The equilibrium constant for the formation of the complex ion from the hydrated ion is called the **formation constant (K_f)**. The equilibrium constant expression for K_f has the same general form as any other equilibrium constant expression. In this case, the expression is

$$K_f = \frac{[[Cu(NH_3)_4]^{2+}]}{[Cu^{2+}][NH_3]^4} = 2.1 \times 10^{13} = K_1K_2K_3K_4 \qquad (17.8)$$

Note the pattern

The formation constant, K_f, has the same general form as any other equilibrium constant expression.

Notice that water a pure liquid, does not appear explicitly in the equilibrium constant expression, and that the hydrated $Cu^{2+}(aq)$ ion is represented as Cu^{2+} for simplicity. As for any equilibrium, the larger the value of the equilibrium constant (in this case, K_f), the more stable the product. With $K_f = 2.1 \times 10^{13}$, the $[Cu(NH_3)_4(H_2O)_2]^{2+}$ complex ion is very stable. The formation constants for some common complex ions are listed in Table 17.2.

EXAMPLE 17.5

If 12.5 g of $Cu(NO_3)_2{\cdot}6H_2O$ is added to 500 mL of 1.00 *M* aqueous ammonia, what is the equilibrium concentration of $Cu^{2+}(aq)$?

Given Mass of Cu^{2+} salt, volume and concentration of ammonia solution

Asked for Equilibrium concentration of $Cu^{2+}(aq)$

Strategy

A Calculate the initial concentration of Cu^{2+}due to the addition of copper(II) nitrate hexahydrate. Use the stoichiometry of the reaction shown in Equation 17.7 to construct a table

* Note that the hydrated Cu^{2+} ion contains six H_2O ligands, but the complex ion that is produced contains only four NH_3 ligands, not six. The reasons for this apparently unusual behavior will be discussed in Chapter 23.

TABLE 17.2 Formation constants, K_f, for selected complex ions in aqueous solution*

	Complex Ion	Equilibrium Equation	K_f
Ammonia Complexes	$[Ag(NH_3)_2]^+$	$Ag^+ + 2NH_3 \rightleftharpoons [Ag(NH_3)_2]^+$	1.1×10^7
	$[Cu(NH_3)_4]^{2+}$	$Cu^{2+} + 4NH_3 \rightleftharpoons [Cu(NH_3)_4]^{2+}$	2.1×10^{13}
	$[Ni(NH_3)_6]^{2+}$	$Ni^{2+} + 6NH_3 \rightleftharpoons [Ni(NH_3)_6]^{2+}$	5.5×10^8
Cyanide Complexes	$[Ag(CN)_2]^-$	$Ag^+ + 2CN^- \rightleftharpoons [Ag(CN)_2]^-$	1.1×10^{18}
	$[Ni(CN)_4]^{2-}$	$Ni^{2+} + 4CN^- \rightleftharpoons [Ni(CN)_4]^{2-}$	2.2×10^{31}
	$[Fe(CN)_6]^{3-}$	$Fe^{3+} + 6CN^- \rightleftharpoons [Fe(CN)_6]^{3-}$	1×10^{42}
Hydroxide Complexes	$[Zn(OH)_4]^{2-}$	$Zn^{2+} + 4OH^- \rightleftharpoons [Zn(OH)_4]^{2-}$	4.6×10^{17}
	$[Cr(OH)_4]^-$	$Cr^{3+} + 4OH^- \rightleftharpoons [Cr(OH)_4]^-$	8.0×10^{29}
Halide Complexes	$[HgCl_4]^{2-}$	$Hg^{2+} + 4Cl^- \rightleftharpoons [HgCl_4]^{2-}$	1.2×10^{15}
	$[CdI_4]^{2-}$	$Cd^{2+} + 4I^- \rightleftharpoons [CdI_4]^{2-}$	2.6×10^5
	$[AlF_6]^{3-}$	$Al^{3+} + 6F^- \rightleftharpoons [AlF_6]^{3-}$	6.9×10^{19}
Other Complexes	$[Ag(S_2O_3)_2]^{3-}$	$Ag^+ + 2S_2O_3^{2-} \rightleftharpoons [Ag(S_2O_3)_2]^{3-}$	2.9×10^{13}
	$[Fe(C_2O_4)_3]^{3-}$	$Fe^{3+} + 3C_2O_4^{2-} \rightleftharpoons [Fe(C_2O_4)_3]^{3-}$	2.0×10^{20}

Source of data: *Lange's Handbook of Chemistry*, 15th Edition (1999).

*Reported values are overall formation constants.

showing the initial concentrations, changes in concentrations, and final concentrations of all species in solution.

Ⓑ Substitute the final concentrations into the expression for the formation constant (Equation 17.8) to calculate the equilibrium concentration of $Cu^{2+}(aq)$.

Solution

Adding an ionic compound that contains Cu^{2+} to an aqueous ammonia solution will result in the formation of $[Cu(NH_3)_4]^{2+}(aq)$, as shown in Equation 17.7. We assume that the volume change caused by adding solid copper(II) nitrate to aqueous ammonia is negligible. Ⓐ✓ The initial concentration of Cu^{2+} from the amount of added copper nitrate prior to any reaction is

$$12.5\ \cancel{g}\ Cu(NO_3)_2 \cdot 6H_2O \left(\frac{1\ mol}{295.65\ \cancel{g}}\right)\left(\frac{1}{500\ \cancel{mL}}\right)\left(\frac{1000\ \cancel{mL}}{1\ L}\right) = 0.0846\ M$$

Because the stoichiometry of the reaction is four NH_3 to one Cu^{2+}, the amount of NH_3 required to react completely with the Cu^{2+} is $4(0.0846) = 0.338\ M$. The concentration of ammonia after complete reaction is $1.00\ M - 0.338\ M = 0.66\ M$. These results are summarized in the first two lines of the table below. Because the equilibrium constant for the reaction is large (2.1×10^{13}) the equilibrium will lie far to the right. Thus, we will assume that the formation of $[Cu(NH_3)_4]^{2+}$ in the first step is complete and then allow some of it to dissociate into Cu^{2+} and NH_3 until equilibrium has been reached. If we define x as the amount of Cu^{2+} produced by the dissociation reaction, then the stoichiometry of the reaction tells us that the change in the concentration of $[Cu(NH_3)_4]^{2+}$ is $-x$, and the change in the concentration of ammonia is $+4x$, as indicated in the table. The final concentrations of all species (in the bottom row of the table) are the sums of the concentrations after complete reaction and the changes in concentrations.

	Cu^{2+} +	$4NH_3$ $\rightleftharpoons$	$[Cu(NH_3)_4]^{2+}$
	$[Cu^{2+}]$	$[NH_3]$	$[[Cu(NH_3)_4]^{2+}]$
Initial	0.0846	1.00	0
After complete reaction	0	0.66	0.0846
Change	$+x$	$+4x$	$-x$
Final	x	$0.66 + 4x$	$0.0846 - x$

B Substituting the final concentrations into the expression for the formation constant (Equation 17.8) and assuming that $x << 0.0846$, which allows us to remove x from the sum and difference, give

$$K_f = \frac{[[Cu(NH_3)_4]^{2+}]}{[Cu^{2+}][NH_3]^4} = \frac{0.0846 - x}{x(0.66 + 4x)^4} \approx \frac{0.0846}{x(0.66)^4} = 2.1 \times 10^{13}$$

$$x = 2.1 \times 10^{-14}$$

The value of x indicates that our assumption was justified. The equilibrium concentration of $Cu^{2+}(aq)$ in a 1.00 M ammonia solution is therefore 2.1×10^{-14} M.

EXERCISE 17.5

The ferrocyanide ion, $[Fe(CN)_6]^{4-}$, is very stable, with a K_f of 1×10^{35}. Calculate the concentration of cyanide ion in equilibrium with a 0.65 M solution of $K_4[Fe(CN)_6]$.

Answer 2×10^{-6} M

The Effect of Complex-Ion Formation on Solubility

What happens to the solubility of a sparingly soluble salt if a ligand that forms a stable complex ion is added to the solution? One such example occurs in black-and-white photography, which was discussed briefly in Chapter 4.

Recall that black-and-white photographic film contains light-sensitive microcrystals of AgBr, or mixtures of AgBr and other silver halides. AgBr is a sparingly soluble salt, with a K_{sp} of 5.35×10^{-13} at 25°C. When the shutter of the camera opens, the light from the object being photographed strikes some of the crystals on the film and initiates a photochemical reaction that converts AgBr to black Ag metal. Well-formed, stable negative images appear in tones of gray, corresponding to the number of grains of AgBr converted, with the areas exposed to the most light being darkest. To fix the image and prevent more AgBr crystals from being converted to Ag metal during processing of the film, the unreacted AgBr on the film is removed using a complexation reaction to dissolve the sparingly soluble salt.

The reaction for the dissolution of silver bromide is

$$AgBr(s) \rightleftharpoons Ag^+(aq) + Br^-(aq) \qquad K_{sp} = 5.35 \times 10^{-13} \text{ at } 25°C \qquad (17.9)$$

The equilibrium lies far to the left, and the equilibrium concentrations of Ag^+ and Br^- ions are very low (7.31×10^{-7} M). As a result, removing unreacted AgBr from even a single roll of film using pure water would require tens of thousands of liters of water and a great deal of time. Le Châtelier's principle tells us, however, that we can drive the reaction to the right by removing one of the products, which will cause more AgBr to dissolve. Bromide ion is difficult to remove chemically, but silver ion forms a variety of stable two-coordinate complexes with neutral ligands such as ammonia or with anionic ligands such as cyanide or thiosulfate ($S_2O_3^{2-}$). In photographic processing, excess AgBr is dissolved using a concentrated solution of sodium thiosulfate.

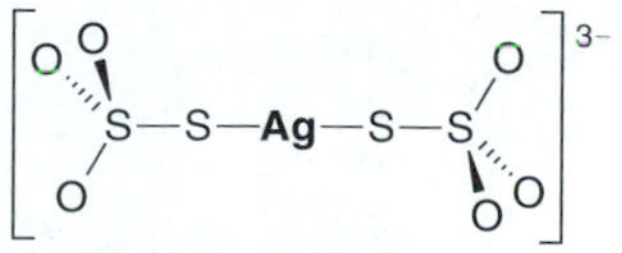

Thiosulfate complex of Ag^+

The reaction of Ag^+ with thiosulfate is

$$Ag^+(aq) + 2S_2O_3^{2-}(aq) \rightleftharpoons [Ag(S_2O_3)_2]^{3-}(aq) \qquad K_f = 2.9 \times 10^{13} \qquad (17.10)$$

The magnitude of the equilibrium constant indicates that almost all Ag^+ ions in solution will be immediately complexed by thiosulfate to form $[Ag(S_2O_3)_2]^{3-}$. We can see the effect of thiosulfate on the solubility of AgBr by writing the appropriate reactions and adding them:

$$\begin{array}{lll} AgBr(s) \rightleftharpoons \cancel{Ag^+(aq)} + Br^-(aq) & K_{sp} = 5.35 \times 10^{-13} & \\ \cancel{Ag^+(aq)} + 2S_2O_3^{2-}(aq) \rightleftharpoons [Ag(S_2O_3)_2]^{3-}(aq) & K_f = 2.9 \times 10^{13} & (17.11) \\ \hline AgBr(s) + 2S_2O_3^{2-}(aq) \rightleftharpoons [Ag(S_2O_3)_2]^{3-}(aq) + Br^-(aq) & K = K_{sp}K_f = 15 & \end{array}$$

Comparing K with K_{sp} shows that the formation of the complex ion increases the solubility of AgBr by approximately 3×10^{13}. The dramatic increase in solubility combined with the low cost and low toxicity explains why sodium thiosulfate is almost universally used for developing black-and-white film. If desired, the silver can be recovered from the thiosulfate solution using any of several methods and recycled.

Note the pattern

If a complex ion has a large K_f, complex-ion formation can dramatically increase the solubility of sparingly soluble salts.

EXAMPLE 17.6

Due to the common ion effect, we might expect a salt such as AgCl to be much less soluble in a concentrated solution of KCl than in water. Such an assumption would be incorrect, however, because it ignores the fact that silver ion tends to form a two-coordinate complex with chloride ions, $AgCl_2^-$. Calculate the solubility of AgCl in (a) pure water; (b) a 1.0 M KCl solution, ignoring complex-ion formation; and (c) the same solution taking complex-ion formation into account, assuming that $AgCl_2^-$ is the only Ag^+ complex that forms in significant concentrations. At 25°C, $K_{sp} = 1.77 \times 10^{-10}$ for AgCl and $K_f = 1.1 \times 10^5$ for $AgCl_2^-$.

Given K_{sp} of AgCl, K_f of $AgCl_2^-$, and KCl concentration

Asked for Solubility of AgCl in water and in KCl solution with and without complex-ion formation

Strategy

A Write the solubility product expression for AgCl, and calculate the concentration of Ag^+ and Cl^- in water.

B Calculate the concentration of Ag^+ in the KCl solution.

C Write balanced equations for the dissolution of AgCl and for the formation of the $AgCl_2^-$ complex. Add the two equations, and calculate the equilibrium constant for the overall equilibrium.

D Write the equilibrium constant expression for the overall reaction. Solve for the concentration of the complex ion.

Solution

(a) **A** If we let x equal the solubility of AgCl, then at equilibrium $[Ag^+] = [Cl^-] = x\ M$. Substituting this value into the solubility product expression gives

$$K_{sp} = [Ag^+][Cl^-] = (x)(x) = x^2 = 1.77 \times 10^{-10}$$
$$x = 1.33 \times 10^{-5}$$

Thus, the solubility of AgCl in pure water at 25°C is $1.33 \times 10^{-5}\ M$.

(b) **B** If x equals the solubility of AgCl in the KCl solution, then at equilibrium $[Ag^+] = x\ M$ and $[Cl^-] = (1.0 + x)\ M$. Substituting these values into the solubility product expression and assuming that $x << 1.0$ give

$$K_{sp} = [Ag^+][Cl^-] = (x)(1.0 + x) \approx x(1.0) = 1.77 \times 10^{-10} = x$$

Thus, if the common ion effect were the only important factor, we would predict that AgCl is approximately five orders of magnitude less soluble in a 1.0 M KCl solution than in water.

(c) **C** To account for the effects of complex-ion formation, we must first write the equation for both the dissolution and complex-ion formation equilibria. Adding the equations corresponding to K_{sp} and K_f gives us an equation that describes the dissolution of AgCl in a KCl solution. The equilibrium constant for the reaction is therefore the product of K_{sp} and K_f:

$$\begin{array}{lr} AgCl(s) \rightleftharpoons \cancel{Ag^+(aq)} + \cancel{Cl^-(aq)} & K_{sp} = 1.77 \times 10^{-10} \\ \underline{\cancel{Ag^+(aq)} + \cancel{2}Cl^- \rightleftharpoons [AgCl_2]^-} & \underline{K_f = 1.1 \times 10^5} \\ AgCl(s) + Cl^- \rightleftharpoons [AgCl_2]^- & K = K_{sp}K_f = 1.9 \times 10^{-5} \end{array}$$

D If we let x equal the solubility of AgCl in the KCl solution, then at equilibrium $[AgCl_2^-] = x$ and $[Cl^-] = 1.0 - x$. Substituting these quantities into the equilibrium constant expression for the net reaction and assuming that $x << 1.0$ give

$$K = \frac{[AgCl_2^-]}{[Cl^-]} = \frac{x}{1.0 - x} \approx \frac{x}{1.0} = 1.9 \times 10^{-5} = x$$

That is, AgCl dissolves in 1.0 M KCl to produce a 1.9×10^{-5} M solution of the $AgCl_2^-$ complex ion. Thus, we predict that AgCl has approximately the same solubility in a 1.0 M KCl solution as it does in pure water, which is 10^5 times greater than that predicted based on the common ion effect. (In fact, the measured solubility of AgCl in 1.0 M KCl is almost a factor of ten *greater* than that in pure water, largely due to the formation of other chloride-containing complexes.)

EXERCISE 17.6

Calculate the solubility of mercury(II) iodide, HgI_2, in (a) pure water, and (b) a 3.0 M solution of NaI, assuming $[HgI_4]^{2-}$ is the only Hg-containing species present in significant amounts. $K_{sp} = 2.9 \times 10^{-29}$ for HgI_2 and $K_f = 6.8 \times 10^{29}$ for $[HgI_4]^{2-}$.

Answer **(a)** 1.9×10^{-10} M; **(b)** 1.4 M

Complexing agents, which are molecules or ions that increase the solubility of metal salts by forming soluble metal complexes, are common components of laundry detergents. Long-chain carboxylic acids, the major components of soaps, form insoluble salts with Ca^{2+} and Mg^{2+}, which are present in high concentrations in "hard" water. Precipitation of these salts produces a bathtub ring and gives a gray tinge to clothing. Adding a complexing agent such as pyrophosphate ($O_3POPO_3^{4-}$, or $P_2O_7^{4-}$) or triphosphate to detergents prevents the magnesium and calcium salts from precipitating because the equilibrium constant for complex-ion formation is large:

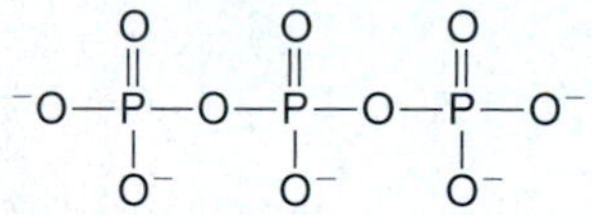

Triphosphate

$$Ca^{2+}(aq) + O_3POPO_3^{4-}(aq) \rightleftharpoons [Ca(O_3POPO_3)]^{2-}(aq)^2 \qquad K_f = 4 \times 10^4 \qquad (17.12)$$

However, phosphates can cause environmental damage by promoting *eutrophication*, the growth of excessive amounts of algae in a body of water, which can eventually lead to large decreases in levels of dissolved oxygen that kill fish and other aquatic organisms. Consequently, many states in the United States have banned the use of phosphate-containing detergents, and France has banned their use beginning in 2007. "Phosphate-free" detergents contain different kinds of complexing agents, such as triacetic acid derivatives or hydroxy polycarboxylic acids. The development of phosphate substitutes is an area of intense research.

Commercial water softeners also use complexing agent to treat hard water by passing the water over ion-exchange resins, which are complex sodium salts. When water

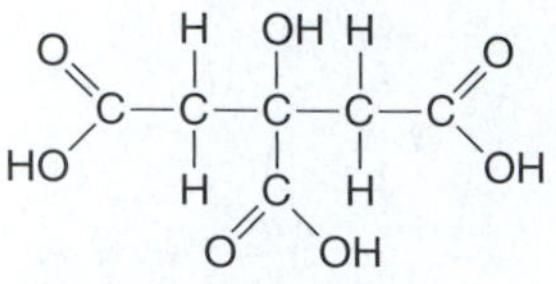

Citric acid
A hydroxy polycarboxylic acid

MGC The Effect of pH on Solubility

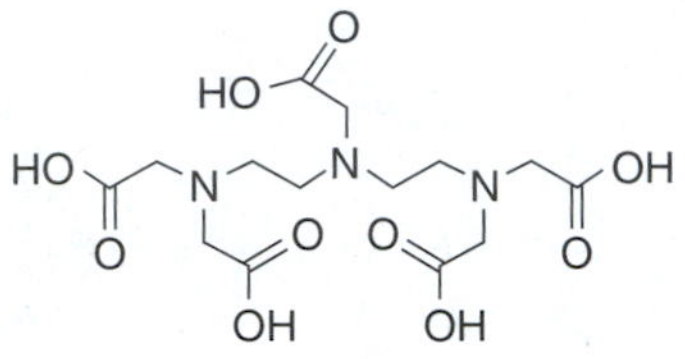

DTPA (diethylenetriaminepentaacetic acid)

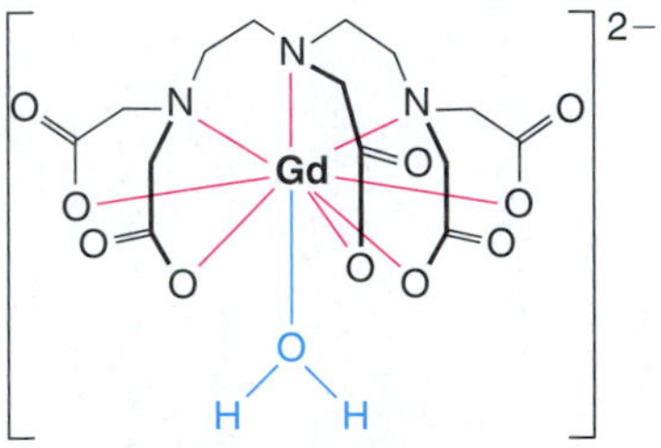

Gadolinium-DTPA complex, $[Gd(DTPA{\cdot}H_2O)]^{2-}$

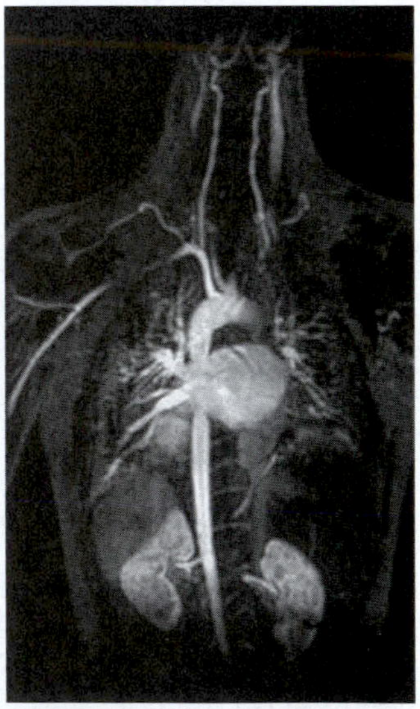

Figure 17.5 An MRI image of heart, arteries, and veins. When a patient is injected with a paramagnetic metal cation in the form of a stable complex known as an MRI contrast agent, the magnetic properties of water in cells are altered. Because the different environments in different types of cells respond differently, a physician can obtain detailed images of soft tissues.

flows over the resin, sodium ion is dissolved and insoluble salts precipitate onto the resin surface. Water treated in this way has a saltier taste due to the presence of Na^+, but it contains fewer dissolved minerals.

Another application of complexing agents is found in medicine. Unlike X rays, magnetic resonance imaging (MRI) can give relatively good images of soft tissues such as internal organs. MRI is based on the magnetic properties of the ^1H nucleus of hydrogen atoms in water, which is a major component of soft tissues. Because the properties of water don't depend very much on whether it is inside a cell or in the blood, it is hard to get detailed images of these tissues that have good contrast. To solve this problem, scientists have developed a class of metal complexes known as "MRI contrast agents." Injecting an MRI contrast agent into a patient selectively affects the magnetic properties of water in cells of normal tissues, in tumors, or in blood vessels, and allows doctors to "see" each of these separately (Figure 17.5). One of the most important metal ions for this application is Gd^{3+}, which with seven unpaired electrons is highly paramagnetic. Because $Gd^{3+}(aq)$ is quite toxic, it must be administered as a very stable complex that does not dissociate in the body and can be excreted intact by the kidneys. The complexing agents used for gadolinium are ligands such as $DTPA^{5-}$.

17.4 Solubility and pH

The solubility of many compounds depends strongly on the pH of the solution. For example, the anion in many sparingly soluble salts is the conjugate base of a weak acid that may become protonated in solution. In addition, the solubility of simple binary compounds such as oxides and sulfides, both strong bases, is often dependent on pH. In this section, we discuss the relationship between the solubility of these classes of compounds and pH.

The Effect of Acid–Base Equilibria on the Solubility of Salts

We begin our discussion by examining the effect of pH on the solubility of a representative salt, M^+A^-, where A^- is the conjugate base of the weak acid HA. When the salt dissolves in water, this reaction occurs:

$$MA(s) \rightleftharpoons M^+(aq) + A^-(aq) \qquad K_{sp} = [M^+][A^-] \tag{17.13}$$

The anion can also react with water in a hydrolysis reaction:

$$A^-(aq) + H_2O(l) \rightleftharpoons OH^-(aq) + HA(aq) \tag{17.14}$$

Because of the reaction described in Equation 17.14, the predicted solubility of a sparingly soluble salt that has a basic anion such as S^{2-}, PO_4^{3-}, or CO_3^{2-} is increased, as described in Section 17.1. If instead a strong acid is added to the solution, the added H^+ will react essentially completely with A^- to form HA. This reaction decreases $[A^-]$, which in turn decreases the magnitude of the ion product, $Q = [M^+][A^-]$. According to Le Châtelier's principle, more MA will dissolve until $Q = K_{sp}$. Hence, *an acidic pH dramatically increases the solubility of virtually all sparingly soluble salts whose anion is the conjugate base of a weak acid.* In contrast, the pH has little to no effect on the solubility of salts whose anion is the conjugate base of a stronger weak acid or a *strong* acid, respectively (chlorides, bromides, iodides, and sulfates, for example).

For example, the hydroxide salt $Mg(OH)_2$ is relatively insoluble in water:

$$Mg(OH)_2(s) \rightleftharpoons Mg^{2+}(aq) + 2OH^-(aq) \qquad K_{sp} = 5.61 \times 10^{-12} \tag{17.15}$$

When acid is added to a saturated solution that contains excess solid $Mg(OH)_2$, the following reaction occurs, removing OH^- from solution:

$$H^+(aq) + OH^-(aq) \longrightarrow H_2O(l) \tag{17.16}$$

The overall equation for the reaction of $Mg(OH)_2$ with acid is thus

$$Mg(OH)_2(s) + 2H^+(aq) \rightleftharpoons Mg^{2+}(aq) + 2H_2O(l) \quad (17.17)$$

As more acid is added to a suspension of $Mg(OH)_2$, the equilibrium shown in Equation 17.17 is driven to the right, and more $Mg(OH)_2$ dissolves.

Such pH-dependent solubility is not restricted to salts that contain anions derived from water. For example, CaF_2 is a sparingly soluble salt:

$$CaF_2(s) \rightleftharpoons Ca^{2+}(aq) + 2F^-(aq) \qquad K_{sp} = 3.45 \times 10^{-11} \quad (17.18)$$

When strong acid is added to a saturated solution of CaF_2, this reaction occurs:

$$H^+(aq) + F^-(aq) \rightleftharpoons HF(aq) \quad (17.19)$$

Because the forward reaction decreases the fluoride ion concentration, more CaF_2 dissolves in order to relieve the stress on the system. The net reaction of CaF_2 with strong acid is thus

$$CaF_2(s) + 2H^+(aq) \longrightarrow Ca^{2+}(aq) + 2HF(aq) \quad (17.20)$$

The next example shows how to calculate the effect of adding a strong acid to a solution of a sparingly soluble salt on its solubility.

Note the pattern

Sparingly soluble salts derived from weak acids tend to be more soluble in an acidic solution.

EXAMPLE 17.7

Lead oxalate (PbC_2O_4), lead iodide (PbI_2), and lead sulfate ($PbSO_4$) are all rather insoluble, with K_{sp} values of 4.8×10^{-10}, 9.8×10^{-9}, and 2.53×10^{-8}, respectively. What effect does adding a strong acid, such as perchloric acid, have on their relative solubilities?

Given K_{sp} values for three compounds

Asked for Relative solubilities in acid solution

Strategy

Write the balanced equation for the dissolution of each salt. Because the strongest conjugate base will be most affected by the addition of strong acid, determine the relative solubilities from the relative basicity of the anions.

Solution

The solubility equilibria for the three salts are

$$PbC_2O_4(s) \rightleftharpoons Pb^{2+}(aq) + C_2O_4^{2-}(aq)$$
$$PbI_2(s) \rightleftharpoons Pb^{2+}(aq) + 2I^-(aq)$$
$$PbSO_4(s) \rightleftharpoons Pb^{2+}(aq) + SO_4^{2-}(aq)$$

The addition of a strong acid will have the greatest effect on the solubility of a salt that contains the conjugate base of a weak acid as the anion. Because HI is a strong acid we predict that adding a strong acid to a saturated solution of PbI_2 will not greatly affect its solubility: the acid will simply dissociate to form $H^+(aq)$ and the corresponding anion. In contrast, oxalate is the fully deprotonated form of oxalic acid, HO_2CCO_2H, which is a weak diprotic acid (pK_a = 1.23 and 4.19). Consequently, the oxalate ion has a significant affinity for one proton, and a lower affinity for a second proton. Adding a strong acid to a saturated solution of lead oxalate will result in these reactions:

$$C_2O_4^{2-}(aq) + H^+(aq) \longrightarrow HO_2CCO_2^-(aq)$$
$$HO_2CCO_2^-(aq) + H^+(aq) \longrightarrow HO_2CCO_2H(aq)$$

These reactions will decrease $[C_2O_4^{2-}]$, causing more lead oxalate to dissolve to relieve the stress on the system.

The pK_a of HSO_4^- (1.99) is similar in magnitude to pK_{a1} of oxalic acid, so adding a strong acid to a saturated solution of $PbSO_4$ will result in this reaction:

$$SO_4^{2-}(aq) + H^+(aq) \rightleftharpoons HSO_4^-(aq)$$

Because HSO_4^- has a pK_a of 1.99, this reaction will lie largely to the left as written. Consequently, we predict that the effect of added strong acid on the solubility of $PbSO_4$ will be significantly less than for PbC_2O_4.

EXERCISE 17.7

Which of the following insoluble salts will be substantially more soluble in 1.0 *M* HNO_3 than in pure water? AgCl, Ag_2CO_3, Ag_3PO_4, AgBr

Answer Ag_2CO_3 and Ag_3PO_4

Caves and their associated pinnacles and spires of stone provide one of the most impressive examples of pH-dependent solubility equilibria (Figure 17.6a). Perhaps the most familiar caves are formed from limestone, such as Carlsbad Caverns in New Mexico, Mammoth Cave in Kentucky, and Luray Caverns in Virginia. The primary reactions that are responsible for the formation of limestone caves are

$$CO_2(aq) + H_2O(l) \rightleftharpoons H^+(aq) + HCO_3^-(aq) \quad (17.21)$$

$$HCO_3^-(aq) \rightleftharpoons H^+(aq) + CO_3^{2-}(aq) \quad (17.22)$$

$$Ca^{2+}(aq) + CO_3^{2-}(aq) \rightleftharpoons CaCO_3(s) \quad (17.23)$$

Limestone deposits that form caves consist primarily of $CaCO_3$ from the remains of living creatures such as clams and corals, which used it for making structures such as

Figure 17.6 The chemistry of cave formation. (a) This cave in Campanet, Mallorca, Spain, and its associated formations are examples of pH-dependent solubility equilibria. (b) A cave forms when groundwater containing atmospheric CO_2, forming an acidic solution, dissolves limestone, $CaCO_3$, in a process that may take tens of thousands of years. As groundwater seeps into a cave, water evaporates from the solution of $CaCO_3$ in CO_2-rich water, producing a supersaturated solution and a shift in equilibrium that causes precipitation of $CaCO_3$. The deposited limestone eventually forms stalactites and stalagmites.

shells. When a saturated solution of $CaCO_3$ in CO_2-rich water rises toward the earth's surface or is otherwise heated, CO_2 gas is released as the water warms. Calcium carbonate then precipitates from the solution according to this equation (Figure 17.6b):

$$Ca^{2+}(aq) + 2HCO_3^-(aq) \rightleftharpoons CaCO_3(s) + CO_2(g) + H_2O(l) \qquad (17.24)$$

The forward direction is the same reaction that produces the solid called *scale* in teapots, coffee makers, water heaters, boilers, and other places where hard water is repeatedly heated.

When groundwater containing atmospheric CO_2 (Equations 17.21 and 17.22) finds its way into microscopic cracks in the limestone deposits, $CaCO_3$ dissolves in the acidic solution in the reverse direction of Equation 17.24. The cracks gradually enlarge from 10–50 μm to 5–10 mm, a process that can take as long as ten thousand years. Eventually, after about another ten thousand years, a cave forms. Groundwater from the surface seeps into the cave and clings to the ceiling, where the water evaporates and causes the equilibrium in Equation 17.24 to shift to the right. A circular layer of solid $CaCO_3$ is deposited, which eventually produces a long, hollow spire of limestone called a *stalactite* that grows down from the ceiling. Below, where the droplets land when they fall from the ceiling, a similar process causes another spire, called a *stalagmite*, to grow up. The same processes that carve out hollows below ground are also at work above ground, in some cases producing fantastically convoluted landscapes like that of Yunnan Province in China (Figure 17.7).

Figure 17.7 Solubility equilibria in the formation of karst landscapes. Landscapes such as the steep limestone pinnacles of the Stone Forest in Yunnan Province, China, are formed from the same process that produces caves and their associated formations.

Acidic, Basic, and Amphoteric Oxides and Hydroxides

One of the earliest classifications of substances was based on their solubility in acidic versus basic solution, which led to the classification of oxides and hydroxides as being either basic or acidic. **Basic oxides** and hydroxides either react with water to produce a basic solution or dissolve readily in aqueous acid. **Acidic oxides** or hydroxides either react with water to produce an acidic solution or are soluble in aqueous base. As shown in Figure 17.8, there is a clear correlation between the acidic or basic character of an oxide and the position of the element combined with oxygen in the periodic table. *Oxides of metallic elements are generally basic oxides, and oxides of nonmetallic elements are acidic oxides.* Compare, for example, the reactions of a typical metal oxide, cesium oxide, and a typical nonmetal oxide, sulfur trioxide, with water:

Note the pattern

Metal oxides generally react with water to produce basic solutions, whereas nonmetal oxides produce acidic solutions.

$$Cs_2O(s) + H_2O(l) \longrightarrow 2Cs^+(aq) + 2OH^-(aq) \qquad (17.25)$$

$$SO_3(g) + H_2O(l) \longrightarrow H_2SO_4(aq) \qquad (17.26)$$

Cesium oxide reacts with water to produce a basic solution of cesium hydroxide, whereas sulfur trioxide reacts with water to produce a solution of sulfuric acid—very different behavior indeed!

The difference in reactivity is due to the difference in bonding in the two kinds of oxides. Because of the low electronegativity of the metals at the far left of the periodic table, their oxides are best viewed as containing discrete M^{n+} cations and O^{2-} anions. At the other end of the spectrum are nonmetal oxides; due to their higher electronegativities, nonmetals form oxides that contain covalent bonds to oxygen. Because of the high electronegativity of oxygen, however, the covalent bond between oxygen and the other atom, E, is usually polarized $E^{\delta+}–O^{\delta-}$. These oxides act as Lewis acids that react with water to produce an oxoacid. Oxides of metals in high oxidation states also tend to be acidic oxides for the same reason: they contain covalent bonds to oxygen. An example of an acidic metal oxide is MoO_3, which is insoluble in both water and acid but dissolves in strong base to give solutions of the molybdate ion, MoO_4^{2-}:

$$MoO_3(s) + 2OH^-(aq) \longrightarrow MoO_4^{2-}(aq) + H_2O(l) \qquad (17.27)$$

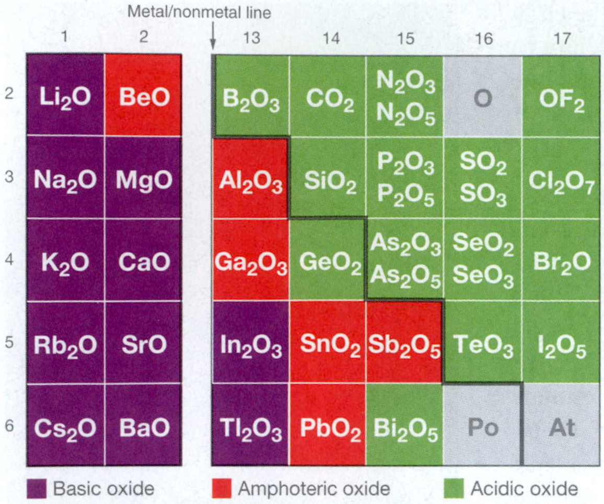

Figure 17.8 Classification of the oxides of the main group elements according to their acidic or basic character. There is a gradual transition from basic oxides to acidic oxides from the lower left to the upper right of the periodic table. Oxides of metallic elements are generally basic oxides, which either react with water to form a basic solution or dissolve in aqueous acid. In contrast, oxides of nonmetallic elements are acidic oxides, which either react with water to form an acidic solution or are soluble in aqueous base. Oxides of intermediate character, called amphoteric oxides, are located along a diagonal line between the two extremes. Amphoteric oxides either dissolve in acid to produce water or dissolve in base to produce a soluble complex ion. (Radioactive elements are not classified.)

As shown in Figure 17.8, there is a gradual transition from basic metal oxides to acidic nonmetal oxides as we go from the lower left to the upper right of the periodic table, with a broad diagonal band of oxides of intermediate character separating the two extremes. Many of the oxides of the elements in this diagonal region of the periodic table are soluble in both acidic *and* basic solutions; consequently, they are called **amphoteric** oxides (from the Greek *ampho*, "both," as in *amphiprotic*, defined in Section 16.1). Amphoteric oxides either dissolve in acid to produce water or dissolve in base to produce a soluble complex. As shown in Figure 17.9, for example, mixing $Cr(OH)_3$ with water gives a muddy, purple-brown suspension. Adding acid causes the $Cr(OH)_3$ to dissolve to give a bright violet solution of $Cr^{3+}(aq)$, which is in the form $[Cr(H_2O)_6]^{3+}$, whereas adding strong base gives a green solution of the $[Cr(OH)_4]^-$ ion. The chemical equations for the reactions are

$$\underset{}{Cr(OH)_3(s)} + 3H^+(aq) \longrightarrow \underset{\text{Violet}}{Cr^{3+}(aq)} + 3H_2O(l) \tag{17.28}$$

$$Cr(OH)_3(s) + OH^-(aq) \longrightarrow \underset{\text{Green}}{[Cr(OH)_4]^-(aq)} \tag{17.29}$$

EXAMPLE 17.8

Aluminum hydroxide, written as either $Al(OH)_3$ or $Al_2O_3 \cdot 3H_2O$, is amphoteric. Write chemical equations to describe the dissolution of aluminum hydroxide in **(a)** acid and **(b)** base.

Given Amphoteric compound

Asked for Dissolution reactions in acid and base

Strategy

Using Equations 17.28 and 17.29 as a guide, write the dissolution reactions in acid and base solutions.

Solution

(a) An acid donates protons to hydroxide to give water and the hydrated metal ion, so aluminum hydroxide, which contains three OH^- ions per Al, needs three H^+ ions:

$$Al(OH)_3(s) + 3H^+(aq) \longrightarrow Al^{3+}(aq) + 3H_2O(l)$$

In aqueous solution, Al^{3+} forms the complex ion $[Al(H_2O)_6]^{3+}$.

(b) In basic solution, OH^- is added to the compound to produce a soluble and stable poly(hydroxo) complex:

$$Al(OH)_3(s) + OH^-(aq) \longrightarrow [Al(OH)_4]^-(aq)$$

Figure 17.9 Chromium(III) hydroxide, $Cr(OH)_3$, is an example of an amphoteric hydroxide. All three beakers originally contained a suspension of brownish purple $Cr(OH)_3(s)$ (*center*). When concentrated acid (6 *M* H_2SO_4) was added to the beaker on the left, $Cr(OH)_3$ dissolved to produce violet $[Cr(H_2O)_6]^{3+}$ ions and water. The addition of concentrated base (6 *M* NaOH) to the beaker on the right caused $Cr(OH)_3$ to dissolve, producing green $[Cr(OH)_4]^-$ ions.

EXERCISE 17.8

Copper(II) hydroxide, written as either $Cu(OH)_2$ or $CuO \cdot H_2O$, is amphoteric. Write chemical equations that describe the dissolution of cupric hydroxide both in an acid and in a base.

Answer

$Cu(OH)_2(s) + 2H^+(aq) \longrightarrow Cu^{2+}(aq) + 2H_2O(l)$

$Cu(OH)_2(s) + 2OH^-(aq) \longrightarrow [Cu(OH)_4]^{2-}(aq)$

Selective Precipitation Using pH

Many dissolved metal ions can be separated by selective precipitation of the cations from solution under specific conditions. In this technique, pH is often used to control the concentration of the anion in solution, which in turn controls which cations precipitate.

Suppose, for example, we have a solution that contains 1.0 mM Zn^{2+} and 1.0 mM Cd^{2+}, and we would like to separate the two metals by selective precipitation as the insoluble sulfide salts, ZnS and CdS. The relevant solubility equilibria can be written as:

$$ZnS(s) \rightleftharpoons Zn^{2+}(aq) + S^{2-}(aq) \qquad K_{sp} = 1.6 \times 10^{-24} \qquad (17.30)$$

$$CdS(s) \rightleftharpoons Cd^{2+}(aq) + S^{2-}(aq) \qquad K_{sp} = 8.0 \times 10^{-27} \qquad (17.31)$$

Because the S^{2-} ion is quite basic and reacts extensively with water to give HS^- and OH^-, the solubility equilibria are more accurately written as $MS(s) \rightleftharpoons M^{2+}(aq) + HS^-(aq) + OH^-$ rather than $MS(s) \rightleftharpoons M^{2+}(aq) + S^{2-}(aq)$. Here we use the simpler form involving S^{2-}, which is justified because we take the reaction of S^{2-} with water into account later in the solution, arriving at the same answer using either equilibrium equation.

The sulfide concentrations needed to cause ZnS and CdS to precipitate are

$$K_{sp} = [Zn^{2+}][S^{2-}]$$
$$1.6 \times 10^{-24} = (0.0010\ M)[S^{2-}]$$
$$[S^{2-}] = 1.6 \times 10^{-21}\ M \qquad (17.32)$$

$$K_{sp} = [Cd^{2+}][S^{2-}]$$
$$8.0 \times 10^{-27} = (0.0010\ M)[S^{2-}] \qquad (17.33)$$
$$[S^{2-}] = 8.0 \times 10^{-24}\ M$$

Thus, sulfide concentrations between $1.6 \times 10^{-21}\ M$ and $8.0 \times 10^{-24}\ M$ will precipitate CdS from solution but not ZnS. How do we obtain such low concentrations of sulfide? A saturated aqueous solution of H_2S contains 0.10 M H_2S at 20°C. The value of pK_{a1} for H_2S is 6.97, and pK_{a2} corresponding to the formation of $[S^{2-}]$ is 12.90. The equations for these reactions are

$$H_2S(aq) \rightleftharpoons H^+(aq) + HS^-(aq) \qquad pK_{a1} = 6.97,\ K_{a1} = 1.1 \times 10^{-7} \qquad (17.34)$$

$$HS^-(aq) \rightleftharpoons H^+(aq) + S^{2-}(aq) \qquad pK_{a2} = 12.90,\ K_{a2} = 1.3 \times 10^{-13}$$

We can show that the concentration of S^{2-} is 1.3×10^{-13} by comparing K_{a1} and K_{a2} and recognizing that the contribution to $[H^+]$ from the dissociation of HS^- is negligible compared with $[H^+]$ from the dissociation of H_2S. Thus, substituting 0.10 M in the equation for K_{a1} for the concentration of H_2S, which is essentially constant regardless of the pH, gives

$$K_{a1} = 1.1 \times 10^{-7} = \frac{[H^+][HS^-]}{[H_2S]} = \frac{x^2}{0.10\ M} \qquad (17.35)$$
$$x = 1.1 \times 10^{-4}\ M = [H^+] = [HS^-]$$

Substituting this value for $[H^+]$ and $[HS^-]$ into the equation for K_{a2} gives

$$K_{a2} = 1.3 \times 10^{-13} = \frac{[H^+][S^{2-}]}{[HS^-]} = \frac{(1.1 \times 10^{-4}\,M)x}{1.1 \times 10^{-4}\,M} = x = [S^{2-}]$$

Although $[S^{2-}]$ in an H_2S solution is very low (1.3×10^{-13} *M*), bubbling H_2S through the solution until it is saturated would precipitate both metal ions because the concentration of S^{2-} would then be much higher than 1.6×10^{-21} *M*. Thus, we must adjust $[S^{2-}]$ to stay within the desired range. The most direct way to do this is to adjust $[H^+]$ by adding acid to the H_2S solution, thereby driving the equilibrium in Equation 17.34 to the left. The overall equation for the dissociation of H_2S is

$$H_2S(aq) \rightleftharpoons 2H^+(aq) + S^{2-}(aq) \tag{17.36}$$

Now we can use the equilibrium constant *K* for the overall reaction, which is the product of K_{a1} and K_{a2}, and the concentration of H_2S in a saturated solution to calculate the H^+ concentration needed to produce $[S^{2-}]$ of 1.6×10^{-21} *M*:

$$K = K_{a1}K_{a2} = (1.1 \times 10^{-7})(1.3 \times 10^{-13}) = 1.4 \times 10^{-20} = \frac{[H^+]^2[S^{2-}]}{[H_2S]} \tag{17.37}$$

$$[H^+]^2 = \frac{K[H_2S]}{[S^{2-}]} = \frac{(1.4 \times 10^{-20})(0.10\,M)}{1.6 \times 10^{-21}\,M} = 0.88 \tag{17.38}$$

$$[H^+] = 0.94\,M$$

Thus, adding a strong acid such as HCl to make the solution 0.94 *M* in H^+ will prevent ZnS from precipitating, while ensuring that the less soluble CdS will precipitate when the solution is saturated with H_2S.

EXAMPLE 17.9

A solution contains 0.010 *M* Ca^{2+} and 0.010 *M* La^{3+}. What concentration of HCl is needed to precipitate $La_2(C_2O_4)_3 \cdot 9H_2O$ but not $Ca(C_2O_4) \cdot H_2O$ if the concentration of oxalic acid is 1.0 *M*? K_{sp} values are 2.32×10^{-9} for $Ca(C_2O_4)$ and 2.5×10^{-27} for $La_2(C_2O_4)_3$; $pK_{a1} = 1.25$ and $pK_{a2} = 3.81$ for oxalic acid.

Given Concentrations of cations, K_{sp} values, concentration and pK_a values for oxalic acid

Asked for Concentration of HCl needed for selective precipitation of $La_2(C_2O_4)_3$

Strategy

A Write each solubility product expression, and calculate the oxalate concentration needed for precipitation to occur. Determine the concentration range needed for selective precipitation of $La_2(C_2O_4)_3 \cdot 9H_2O$.

B Add the equations for the first and second dissociations of oxalic acid to get an overall equation for the dissociation of oxalic acid to oxalate. Substitute the $[ox^{2-}]$ needed to precipitate $La_2(C_2O_4)_3 \cdot 9H_2O$ into the overall equation for the dissociation of oxalic acid to calculate the required $[H^+]$.

Solution

A Because the salts have different stoichiometries, we cannot directly compare the magnitudes of the solubility products. Instead, we must use the equilibrium constant expression for each solubility product to calculate the concentration of oxalate needed for precipitation to occur. Using ox^{2-} for oxalate, we write the solubility product expression for calcium oxalate as

$$K_{sp} = [Ca^{2+}][ox^{2-}] = (0.010)[ox^{2-}] = 2.32 \times 10^{-9}$$

$$[ox^{2-}] = 2.32 \times 10^{-7}\,M$$

The expression for lanthanum oxalate is

$$K_{sp} = [La^{3+}]^2[ox^{2-}]^3 = (0.010)^2[ox^{2-}]^3 = 2.5 \times 10^{-27}$$
$$[ox^{2-}] = 2.9 \times 10^{-8}\ M$$

Thus, lanthanum oxalate is less soluble and will selectively precipitate when the oxalate concentration is between 2.9×10^{-8} M and 2.32×10^{-7} M. B To prevent Ca^{2+} from precipitating as calcium oxalate, we must add enough H^+ to give a maximum oxalate concentration of 2.32×10^{-7} M. We can calculate the required $[H^+]$ by using the overall equation for the dissociation of oxalic acid to oxalate:

$$HO_2CCO_2H(aq) \rightleftharpoons 2H^+(aq) + C_2O_4^{2-}(aq)$$
$$K = K_{a1}K_{a2} = (10^{-1.25})(10^{-3.81}) = 10^{-5.06} = 8.71 \times 10^{-6}$$

Substituting the desired oxalate concentration into the equilibrium constant expression gives

$$8.71 \times 10^{-6} = \frac{[H^+]^2[ox^{2-}]}{[HO_2CCO_2H]} \approx \frac{[H^+]^2(2.32 \times 10^{-7})}{1.0}$$
$$[H^+] = 6.1\ M$$

Thus, adding enough HCl to give $[H^+] = 6.1$ M will cause only $La_2(C_2O_4)_3 \cdot 9H_2O$ to precipitate from the solution.

EXERCISE 17.9

A solution contains 0.015 M Fe^{2+} and 0.015 M Pb^{2+}. What concentration of acid is needed to ensure that Pb^{2+} precipitates as PbS in a saturated solution of H_2S but Fe^{2+} does not precipitate as FeS? K_{sp} values are 6.3×10^{-18} for FeS and 8.0×10^{-28} for PbS.

Answer 0.018 M H^+

17.5 • Qualitative Analysis Using Selective Precipitation

MGC Selective Precipitation

The composition of relatively complex mixtures of metal ions can be determined using **qualitative analysis**, a procedure for discovering the *identity* of metal ions present in the mixture (rather than quantitative information about their amounts).

The procedure used to separate and identify more than 20 common metal cations from a single solution consists of selectively precipitating only a few kinds of metal ions at a time under given sets of conditions. Consecutive precipitation steps become progressively *less* selective until almost all of the metal ions are precipitated, as illustrated in Figure 17.10.

Group 1: Insoluble Chlorides

Most metal chloride salts are soluble in water; only Ag^+, Pb^{2+}, and Hg_2^{2+} form chlorides that precipitate from water. Thus, the first step in a qualitative analysis is to add about 6 M HCl, thereby causing AgCl, $PbCl_2$, and/or Hg_2Cl_2 to precipitate. If no precipitate forms, then these cations are not present in significant amounts. The precipitate can be collected by filtration or centrifugation.

Group 2: Acid-Insoluble Sulfides

Next, the acidic solution is saturated with H_2S gas. Only those metal ions that form very insoluble sulfides, such as As^{3+}, Bi^{3+}, Cd^{2+}, Cu^{2+}, Hg^{2+}, Sb^{3+}, and Sn^{2+}, precipitate as their sulfide salts under these acidic conditions. All others, such as Fe^{2+} and Zn^{2+}, remain in solution. Once again, the precipitates are collected by filtration or centrifugation.

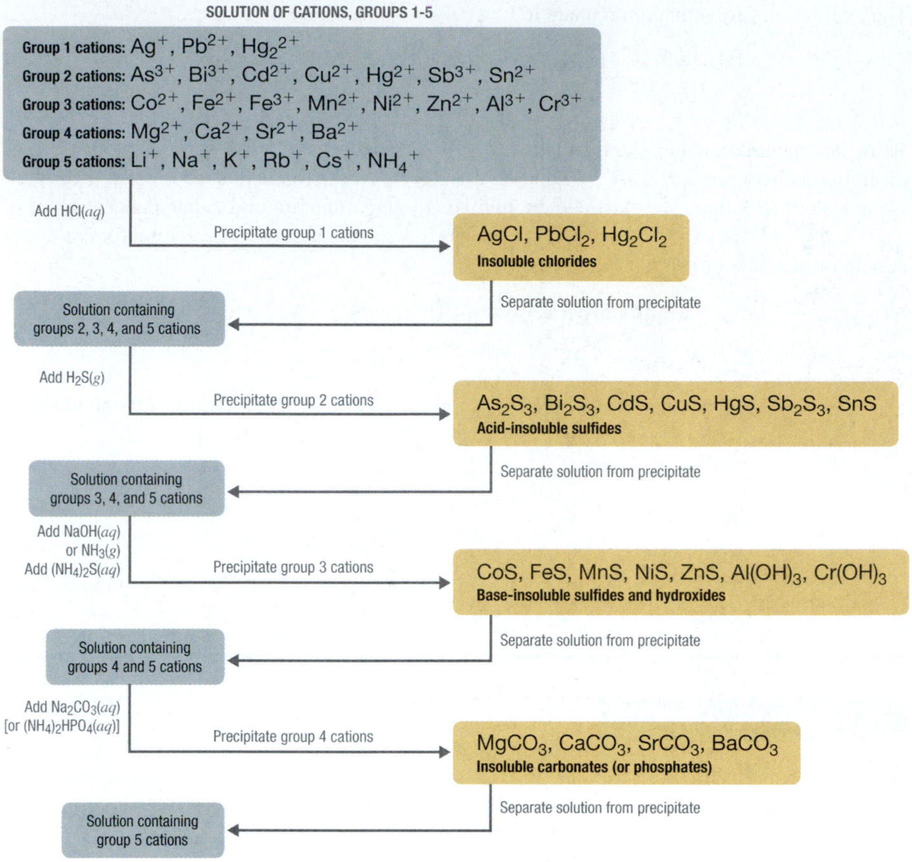

Figure 17.10 **Steps in a typical qualitative analysis scheme for a solution that contains several metal ions.**

Group 3: Base-Insoluble Sulfides (and Hydroxides)

Ammonia or NaOH is now added to the solution until it is basic, and then $(NH_4)_2S$ is added. This treatment removes any remaining cations that form insoluble hydroxides or sulfides. The divalent metal ions Co^{2+}, Fe^{2+}, Mn^{2+}, Ni^{2+}, and Zn^{2+} precipitate as their sulfides, and the trivalent metal ions Al^{3+} and Cr^{3+} precipitate as their hydroxides, $Al(OH)_3$ and $Cr(OH)_3$. If the mixture contains Fe^{3+}, sulfide reduces the cation to Fe^{2+}, which precipitates as FeS.

Group 4: Insoluble Carbonates or Phosphates

The next metal ions to be removed from solution are those that form insoluble carbonates and phosphates. When Na_2CO_3 is added to the basic solution that remains after the precipitated metal ions are removed, insoluble carbonates precipitate and are collected. Alternatively, adding $(NH_4)_2HPO_4$ causes the same metal ions to precipitate as insoluble phosphates.

Group 5: Alkali Metals

At this point, we have removed all the metal ions that form water-insoluble chlorides, sulfides, carbonates, or phosphates. The only common ions that might remain are any alkali metals (Li^+, Na^+, K^+, Rb^+, and Cs^+) and ammonium (NH_4^+). We now take a second sample from the *original* solution and add a small amount of NaOH to neutralize the ammonium ion and produce NH_3. (We can't use the same sample we used for the first four groups because we added ammonium to that sample in earlier steps.) Any ammonia produced can be detected either by its odor or by a litmus paper test. A flame test on another original sample is used to detect sodium, which produces a characteristic bright yellow color. As discussed in Chapter 6, the other alkali metal ions also give characteristic colors in flame tests, which allows them to be identified if only one is present.

Metal ions that precipitate together are separated by various additional techniques, such as forming complex ions, changing the pH of the solution, or increasing the temperature to redissolve some of the solids. For example, the precipitated metal chlorides of group 1 cations, containing Ag^+, Pb^{2+}, and $Hg_2{}^{2+}$, are all quite insoluble in water. Because $PbCl_2$ is much more soluble in hot water than are the other two chloride salts, however, adding water to the precipitate and heating the resulting slurry will dissolve any $PbCl_2$ present. Isolating the solution and adding a small amount of Na_2CrO_4 solution to it will produce a bright yellow precipitate of $PbCrO_4$ if Pb^{2+} was in the original sample (Figure 17.11).

As another example, treating the precipitates from group 1 cations with aqueous ammonia will dissolve any AgCl because Ag^+ forms a stable complex with ammonia, $[Ag(NH_3)_2]^+$. In addition, Hg_2Cl_2 *disproportionates* in ammonia ($2Hg_2{}^{2+} \longrightarrow Hg + Hg^{2+}$) to form a black solid that is a mixture of finely divided metallic mercury and an insoluble mercury(II) compound, which is separated from solution:

$$Hg_2Cl_2(s) + 2NH_3(aq) \longrightarrow Hg(l) + Hg(NH_2)Cl(s) + NH_4{}^+(aq) + Cl^-(aq) \quad (17.39)$$

Any silver ion in the solution is then detected by adding hydrochloric acid, which reverses the reaction and gives a precipitate of white AgCl that slowly darkens when exposed to light:

$$[Ag(NH_3)_2]^+(aq) + 2H^+(aq) + Cl^-(aq) \longrightarrow AgCl(s) + 2NH_4{}^+(aq) \quad (17.40)$$

Similar but slightly more complex reactions are used to separate and identify the individual components of the other groups as well.

Figure 17.11 Separation of metal ions from Group 1 using qualitative analysis. In (a), the cations of group 1 precipitate when $HCl(aq)$ is added to a solution containing a mixture of cations. (b) When a small amount of Na_2CrO_4 solution is added to sample containing Pb^{2+} ions in water, a bright yellow precipitate of $PbCrO_4$ forms. (c) Addition of aqueous ammonia to a second portion of the solid sample produces a black solid that is a mixture of finely divided metallic mercury, an insoluble mercury(II) compound, $Hg(NH_2)Cl$, and a stable $[Ag(NH_3)_2]^+(aq)$ complex. (d) The presence of Ag^+ is detected by decanting the solution from the precipitated mercury and mercury complex and adding hydrochloric acid to the decanted solution, which causes AgCl to precipitate.

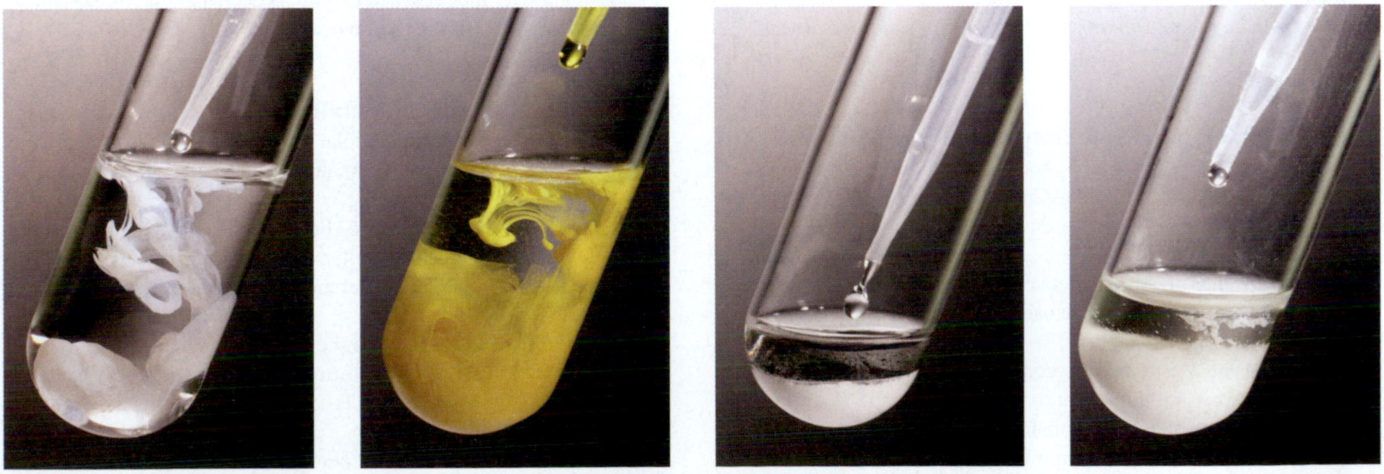

(a) Precipitation of group 1 cations, $AgCl(s) + Hg_2Cl_2(s) + PbCl_2(s)$

(b) Confirmation test for lead: $PbCrO_4(s)$

(c) Confirmation test for mercury: $Hg(l) + Hg(NH_2)Cl(s)$

(d) Confirmation test for silver: $AgCl(s)$

SUMMARY AND KEY TERMS

17.1 Determining the Solubility of Ionic Compounds (p. 782)

The equilibrium constant for a dissolution reaction, called the **solubility product (K_{sp})**, is a measure of the solubility of a compound. Whereas solubility is usually expressed in terms of *mass* of solute per 100 mL of solvent, K_{sp} is defined in terms of *molar* concentrations of the component ions. In contrast, the **ion product (Q)** describes concentrations that are not necessarily equilibrium concentrations. Comparing Q and K_{sp} enables us to determine whether a precipitate will form when solutions of two soluble salts are mixed. Adding a common cation or anion to a solution of a sparingly soluble salt shifts the solubility equilibrium in the direction predicted by Le Châtelier's principle. The solubility of the salt is almost always decreased by the presence of a common ion.

17.2 Factors That Affect Solubility (p. 788)

The solubility of a compound can differ from the solubility indicated by the concentrations of ions for four reasons: (1) **ion-pair** formation, in which an anion and cation are in intimate contact in solution and not separated by solvent, (2) incomplete dissociation of molecular solutes, (3) formation of complex ions, and (4) changes in pH. An ion pair is held together by electrostatic attractive forces between the cation and the anion, whereas incomplete dissociation results from intramolecular forces, such as polar covalent O—H bonds.

17.3 Complex-Ion Formation (p. 790)

A **complex ion** is a species formed between a central metal ion and one or more surrounding **ligands**, molecules or ions that contain at least one lone pair of electrons. Small, highly charged metal ions have the greatest tendency to act as Lewis acids and have the greatest tendency to form complex ions. The equilibrium constant for the formation of the complex ion is the **formation constant (K_f)**. The formation of a complex ion by the addition of a *complexing agent* increases the solubility of a compound.

17.4 Solubility and pH (p. 796)

The anion in many sparingly soluble salts is the conjugate base of a weak acid. At low pH, protonation of the anion can dramatically increase the solubility of the salt. Oxides can be classified as acidic oxides or basic oxides. **Acidic oxides** either react with water to give an acidic solution or dissolve in strong base; most acidic oxides are nonmetal oxides or oxides of metals in high oxidation states. **Basic oxides** either react with water to give a basic solution or dissolve in strong acid; most basic oxides are oxides of metallic elements. Oxides or hydroxides that are soluble in both acidic and basic solutions are called **amphoteric**. Most elements whose oxides exhibit amphoteric behavior are located along the diagonal line separating metals and nonmetals in the periodic table. In solutions that contain mixtures of dissolved metal ions, the pH can be used to control the anion concentration needed to selectively precipitate the desired cation.

17.5 Qualitative Analysis Using Selective Precipitation (p. 803)

In **qualitative analysis**, the identity of metal ions present in a mixture is determined, not their amounts. The technique consists of selectively precipitating only a few kinds of metal ions at a time under given sets of conditions. Consecutive precipitation steps become progressively *less* selective until almost all the metal ions are precipitated. Other additional steps are needed to separate metal ions that precipitate together.

QUESTIONS AND PROBLEMS

MGC For instructor-assigned homework, go to **www.masteringgenchem.com**

Questions and Problems with colored numbers have answers in the Appendix and complete solutions in the Student Solutions Manual.

CONCEPTUAL

17.1 Determining the Solubility of Ionic Compounds

1. Write an expression for K_{sp} for each of these salts: (a) AgI; (b) CaF_2; (c) $PbCl_2$; (d) Ag_2CrO_4.
2. Some species are not represented in a solubility product expression. Why?
3. Describe the differences between Q and K_{sp}.
4. How can an ion product be used to determine whether a solution is saturated?
5. When using K_{sp} to directly compare the solubilities of compounds, why is it important to compare only the K_{sp} values of salts that have the same stoichiometry?
6. Describe the effect of a common ion on the solubility of a salt. Is this effect similar to the common ion effect found in buffers? Explain your answer.
7. Explain why the presence of $MgCl_2$ decreases the molar solubility of the sparingly soluble salt $MgCO_3$.

17.2 Factors That Affect Solubility

8. Do you expect the actual molar solubility of $LaPO_4$ to be higher than, the same as, or lower than the value calculated from its K_{sp}? Explain your reasoning.
9. Do you expect the difference between the calculated molar solubility and the actual molar solubility of $Ca_3(PO_4)_2$ to be greater or less than the difference in the solubilities of $Mg_3(PO_4)_2$? Why?
10. Write equations to describe the interactions in a solution that contains $Mg(OH)_2$, which forms ion pairs, and in one that contains propanoic acid, $CH_3CH_2CO_2H$, which forms a hydrated neutral molecule.
11. Draw circular atom representations of $Ca(IO_3)_2$ in solution (a) as an ionic solid, (b) in the form of ion pairs, and (c) as discrete ions.

17.3 Complex-Ion Formation

12. What is the difference between an equilibrium constant K_{eq} and a formation constant K_f?

13. Which would you expect to have the greater tendency to form a complex ion: Mg^{2+} or Ba^{2+}? Why?

14. How can a ligand be used to affect the concentration of hydrated metal ions in solution? How is K_{sp} affected? Explain your answer.

15. Co(II) forms a complex ion with pyridine, C_5H_5N. Which is the Lewis acid and which is the Lewis base? Use Lewis electron structures to justify your answer.

17.4 Solubility and pH

16. Of the following compounds, which do you expect to be most soluble in 1.0 *M* HNO_3 and which least soluble? (a) $CuCl_2$; (b) $K[Pb(OH)_3]$; (c) $Ba(CH_3CO_2)_2$; (d) $CaCO_3$. Explain your reasoning.

17. Of the compounds $Sn(CH_3CO_2)_2$ and SnS, one is soluble in dilute HCl and the other is soluble only in hot, concentrated HCl. Which is which? Provide a reasonable explanation.

18. Where in the periodic table do you expect to find elements that form basic oxides? Where do you expect to find elements that form acidic oxides?

19. Because water can autoionize, it reacts with oxides either as a base (as OH^-) or as an acid (as H_3O^+). Do you expect oxides of elements in high oxidation states to be more acidic (reacting with OH^-) or more basic (reacting with H_3O^+) than the corresponding oxides in low oxidation states? Why?

20. Given solid samples of CrO, Cr_2O_3, and CrO_3, which would you expect to be the most acidic (reacts most readily with OH^-)? Which would be the most basic (reacts most readily with H_3O^+)? Why?

21. Which of these metals do you expect to form an amphoteric oxide, and why? Be, B, Al, N, Se, In, Tl, Pb

17.5 Qualitative Analysis Using Selective Precipitation

22. Given a solution that contains a mixture of NaCl, $CuCl_2$, and $ZnCl_2$, propose a method for separating the metal ions.

NUMERICAL

17.1 Determining the Solubility of Ionic Compounds

23. Predict the molar solubility of each compound using the K_{sp} values given in Appendix B: (a) $Cd(IO_3)_2$, (b) AgCN, (c) HgI_2.

24. Predict the molar solubility of each compound using the K_{sp} values given: (a) Li_3PO_4, 2.37×10^{-11}; (b) $Ca(IO_3)_2$, 6.47×10^{-6}; (c) $Y(IO_3)_3$, 1.12×10^{-10}.

25. A student prepared 750 mL of a saturated solution of silver sulfate. How many grams of Ag_2SO_4 does the solution contain? $K_{sp} = 1.20 \times 10^{-5}$.

26. Given the K_{sp} values in Table 17.1 and Appendix B, predict the molar concentration of each species in a saturated aqueous solution of the following: (a) silver bromide; (b) lead oxalate; (c) iron(II) carbonate; (d) silver phosphate; (e) copper(I) cyanide.

27. Use the K_{sp} values in Table 17.1 and Appendix B to predict the molar concentration of each species in a saturated aqueous solution of each of the following: (a) copper(I) chloride; (b) lanthanum(III) iodate; (c) magnesium phosphate; (d) silver chromate; (e) strontium sulfate.

28. Silicon dioxide, the most common binary compound of silicon and oxygen, constitutes approximately 60% of the earth's crust. Under certain conditions, this amphoteric compound can react with water to form silicic acid, which is written as either H_4SiO_4 or $Si(OH)_4$. Write balanced chemical equations for the dissolution of SiO_2 under acidic and basic conditions.

29. The K_{sp} of $Mg(OH)_2$ is 5.61×10^{-12}. If you tried to dissolve 24 mg of $Mg(OH)_2$ in 250 mL of water and then filtered the solution and dried the remaining solid, what would you predict to be the mass of the undissolved solid? You discover that only 1.0 mg remains undissolved. Explain the difference between your expected value and the actual value.

30. The K_{sp} of lithium carbonate is 8.15×10^{-4}. If 2.34 g of the salt is stirred with 500 mL of water and any undissolved solid is filtered from the solution and dried, what do you predict to be the mass of the solid? You discover that all of your sample dissolves. Explain the difference between your predicted value and the actual value.

31. You have calculated that 24.6 mg of $BaSO_4$ will dissolve in 1.0 L of water at 25°C. After adding your calculated amount to 1.0 L of water and stirring for several hours, you notice that the solution contains undissolved solid. After carefully filtering the solution and drying the solid, you find that 22.1 mg did not dissolve. According to your measurements, what is the K_{sp} of barium sulfate?

32. In a saturated silver chromate solution, the molar solubility of chromate is 6.54×10^{-5}. What is the K_{sp}?

33. A saturated lead(II) chloride solution contains a chloride concentration of 3.24×10^{-2} mol/L. What is the K_{sp}?

34. From the solubility data given, predict the value of K_{sp} for each compound: (a) AgI, 2.89×10^{-7} g/100 mL; (b) SrF_2, 1.22×10^{-2} g/100 mL; (c) $Pb(OH)_2$, 78 mg/500 mL; (d) $BiAsO_4$, 14.4 mg/2.0 L.

35. From the solubility data given, predict the value of K_{sp} for each compound: (a) $BaCO_3$, 10.0 mg/500 mL; (b) CaF_2, 3.50 mg/200 mL; (c) $Mn(OH)_2$, 6.30×10^{-4} g/300 mL; (d) Ag_2S, 1.60×10^{-13} mg/100 mL.

36. Given the following solubilities, predict the K_{sp} for each compound: (a) $BaCO_3$, 7.00×10^{-5} mol/L; (b) CaF_2, 1.70 mg/100 mL; (c) $Pb(IO_3)_2$, 2.30 mg/100 mL; (d) SrC_2O_4, 1.58×10^{-7}mol/L.

37. From the solubility data given, predict the K_{sp} of each compound? (a) Ag_2SO_4, 4.2×10^{-1} g/100 mL; (b) $SrSO_4$, 1.5×10^{-3} g/100 mL; (c) CdC_2O_4, 6.0×10^{-3} g/100 mL; (d) $Ba(IO_3)_2$, 3.96×10^{-2} g/100 mL

38. The K_{sp} of the phosphate fertilizer $CaHPO_4 \cdot 2H_2O$ is 2.7×10^{-7} at 25°C. What is the molar concentration of a saturated solution? What mass of this compound will dissolve in 3.0 L of water at this temperature?

39. The K_{sp} of zinc carbonate monohydrate is $5.5. \times 10^{-11}$ at 25°C. What is the molar concentration of a saturated solution? What mass of compound will dissolve in 2.0 L of water at this temperature?

40. Silver nitrate eye drops were at one time administered to newborn infants to guard against eye infections contracted during birth. Although silver nitrate is highly water soluble, silver sulfate has a K_{sp} of 1.20×10^{-5} at 25°C. If you add 25.0 mL of 0.015 *M* $AgNO_3$ to 150 mL of 2.8×10^{-3} *M* Na_2SO_4, will you get a precipitate? If so, what will its mass be?

41. Use the data in Appendix B to predict whether precipitation will occur when these pairs of solutions are mixed: (a) 150 mL of 0.142 *M* $Ba(NO_3)_2$ with 200 mL of 0.089 *M* NaF; (b) 250 mL of 0.079 *M* K_2CrO_4 with 175 mL of 0.087 *M* $CaCl_2$; (c) 300 mL of 0.109 *M* $MgCl_2$ with 230 mL of 0.073 *M* $Na_2(C_2O_4)$.

42. Given a 0.048 *M* solution of $Pb(NO_3)_2$, what is the maximum volume that can be added to 250 mL of 0.10 *M* NaSCN before precipitation occurs? $K_{sp} = 2.0 \times 10^{-5}$ for $Pb(SCN)_2$.

43. Given 300 mL of a solution that is 0.056 *M* in lithium nitrate, what volume of 0.53 *M* sodium carbonate can be added before precipitation occurs? $K_{sp} = 8.15 \times 10^{-4}$ for Li_2CO_3.

44. Given the information in the table, calculate the molar solubility of each sparingly soluble salt in 0.95 *M* $MgCl_2$.

Saturated Solution	K_{sp}
$MgCO_3 \cdot 3H_2O$	2.4×10^{-6}
$Mg(OH)_2$	5.6×10^{-12}
$Mg_3(PO_4)_2$	1.04×10^{-24}

45. Ferric phosphate has a molar solubility of 5.44×10^{-16} in 1.82 *M* Na_3PO_4. Predict its K_{sp}. The actual K_{sp} is 9.9×10^{-16}. Explain this discrepancy.

17.3 Complex-Ion Formation

46. Zn(II) forms the complex ion $[Zn(NH_3)_4]^{2+}$ through equilibrium reactions in which ammonia replaces coordinated water molecules in a stepwise manner. If $\log K_1 = 2.37$, $\log K_2 = 2.44$, $\log K_3 = 2.50$, and $\log K_4 = 2.15$, what is the overall K_f? Write the equilibrium equation that corresponds to each stepwise equilibrium constant. Do you expect the $[Zn(NH_3)_4]^{2+}$ complex to be stable? Explain your reasoning.

47. Fe(II) forms the complex ion $[Fe(OH)_4]^{2-}$ through equilibrium reactions in which hydroxide replaces water in a stepwise manner. If $\log K_1 = 5.56$, $\log K_2 = 9.77$, $\log K_3 = 9.67$, and $\log K_4 = 8.58$, what is K_f? Write the equilibrium equation that corresponds to each stepwise equilibrium constant. Do you expect the $[Fe(OH)_4]^{2-}$ complex to be stable? Explain your reasoning.

48. Although thallium has limited commercial applications because it is toxic to humans (10 mg/kg body weight is fatal to children), it is used as a substitute for mercury in industrial switches. The complex ion $[TlBr_6]^{3-}$ is highly stable, with $\log K_f = 31.6$. What is the concentration of Tl(III)(*aq*) in equilibrium with a 1.12 *M* solution of $Na_3[TlBr_6]$?

17.4 Solubility and pH

49. A 1.0-L solution contains 1.98 *M* $Al(NO_3)_3$. What are $[OH^-]$ and $[H^+]$? What pH is required to precipitate the cation as $Al(OH)_3$? $K_{sp} = 1.3 \times 10^{-33}$, $K_a = 1.05 \times 10^{-5}$.

50. A 1.0-L solution contains 2.03 *M* $CoCl_2$. What is $[H^+]$? What pH is required to precipitate the cation as $Co(OH)_2$? $K_{sp} = 5.92 \times 10^{-15}$, $K_a = 1.26 \times 10^{-9}$.

51. Given 100 mL of a solution that contains 0.80 m*M* Ag^+ and 0.80 m*M* Cu^+, can the two metals be separated by selective precipitation as the insoluble bromide salts by adding 10 mL of an 8.0 m*M* solution of KBr? K_{sp} values are 6.27×10^{-9} for CuBr and 5.35×10^{-13} for AgBr. What maximum $[Br^-]$ will separate the ions?

52. Given 100 mL of a solution that is 1.5 m*M* in Tl^+, Zn^{2+}, and Ni^{2+}, which ions can be separated from solution by adding 5.0 mL of a 12.0 m*M* solution of $Na_2C_2O_4$?

Precipitate	K_{sp}
$Tl_2C_2O_4$	2×10^{-4}
$ZnC_2O_4 \cdot 2H_2O$	1.38×10^{-9}
NiC_2O_4	4×10^{-10}

How many milliliters of 12.0 m*M* $Na_2C_2O_4$ should be added to separate Tl^+ and Zn^{2+} from Ni^{2+}?

APPLICATIONS

53. Gypsum ($CaSO_4 \cdot 2H_2O$) is added to soil to enhance plant growth. It dissolves according to the equation: $CaSO_4 \cdot 2H_2O(s) \rightleftharpoons Ca^{2+}(aq) + SO_4{}^{2-}(aq) + 2H_2O(l)$.
(a) The K_{sp} of gypsum is 3.14×10^{-5}. How much gypsum should you add to your 5.0-L watering can to produce a solution that is 0.0050 *M* in Ca^{2+}?
(b) Gibbsite $[Al(OH)_3]$ is a component of clay, found in places such as Molokai, Hawaii. It dissolves according to the equation: $Al(OH)_3(s) \rightleftharpoons Al^{3+}(aq) + 3OH^-(aq)$, with a K_{sp} of 1.3×10^{-33}. You are interested in using gypsum to counteract harmful growth effects of Al^{3+} on plants from dissolved gibbsite, and you have found the pH of your soil to be 8.7. What is the apparent concentration of OH^- in your soil?

54. Egyptian blue is a synthetic pigment developed about 4500 years ago that is difficult to prepare. It was the only blue pigment identified in a study of stocks of dry pigments found in color merchants' shops in Pompeii. The pigment contains 12.9% calcium carbonate (calcite). A major source of $CaCO_3$ is limestone, which also contains $MgCO_3$. Assuming that the masses of $CaCO_3$ and $MgCO_3$ are equal, and that a sample of limestone is dissolved in acidified water to give $[Ca^{2+}] = [Mg^{2+}] = 0.010$ *M* in 5.0 L of solution, would selective precipitation be a viable method for purifying enough $CaCO_3$ from limestone to produce 1.0 g of pigment? Why? K_{sp} values are 3.36×10^{-9} for $CaCO_3$ and 6.8×10^{-6} for $MgCO_3$.

55. One method of mining gold is to extract it through the process of cyanidation. Mined ores are milled and treated with cyanide to produce a gold complex ion that is very stable, $[Au(CN)_2]^-$. Given a sample of AuCl, what is the solubility of AuCl in (a) pure water and (b) a 1.68 *M* solution of NaCN? K_{sp} of AuCl $= 2.0 \times 10^{-13}$; $\log K_f([Au(CN)_2]^-) = 38.3$.

56. Almost all barium carbonate is produced synthetically. The compound is used in manufacturing clay tiles and ceramic products as well as in CRT screens and special optical glasses. $BaCO_3$ is synthesized by allowing barium sulfate to react with coal at 1000–1200°C in a rotary kiln, followed by treatment of a solution of the product with either CO_2 or Na_2CO_3. The reactions are

$$BaSO_4(s) + 4C(s) \longrightarrow 4BaS(s) + 4CO(g)$$

$$(1) \quad BaS(s) + CO_2(g) + H_2O(l) \longrightarrow BaCO_3(aq) + H_2S(g)$$

$$(2) \quad BaS(s) + Na_2CO_3(aq) \rightleftharpoons BaCO_3(aq) + Na_2S(aq)$$

Barium carbonate has a K_{sp} of 2.58×10^{-9}. The pK_a for $H_2S \rightleftharpoons H^+ + HS^-$ is 6.97, and the pK_a for $HS^- \rightleftharpoons H^+ + S^{2-}$ is 12.90. Given this information, answer these questions:
(a) If reaction 1 occurs, what is the pH of the solution if 80.0 g of BaS reacts with CO_2 in 1.00 L of water?

(b) If reaction 2 occurs, how many grams of $BaCO_3$ are produced by the reaction of 80.0 g of BaS with excess Na_2CO_3?

57. A person complaining of chronic indigestion continually consumed antacid tablets containing $Ca(OH)_2$ over a two week period. A blood test at the end of this period showed that the person had become anemic. Explain the reactions that caused this test result.

58. Although the commercial production of radium has virtually ceased since artificial radionuclides were discovered to have similar properties and lower costs, commercial radium is still isolated using essentially the same procedure developed by Marie Curie, as outlined in the figure. Explain what is happening chemically at each step of the purification process. What is precipitate A? What metal ions are present in solution A? What is precipitate B? What metal ions are present in solution B?

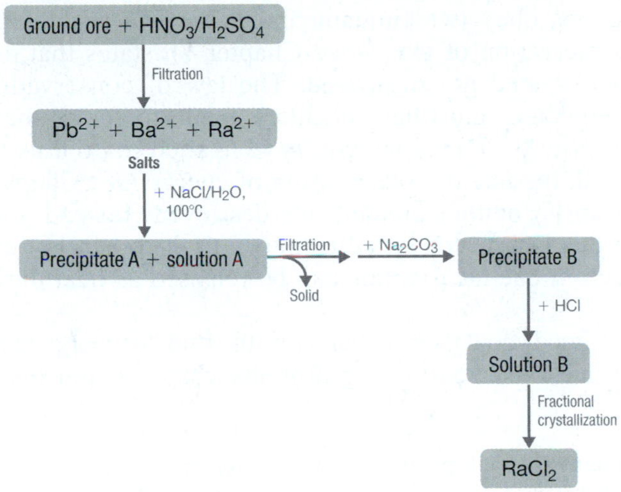

59. In a qualitative analysis laboratory, a student initially treated his sample of metal ions with 6 *M* HNO_3 instead of 6 *M* HCl, and recognized his mistake only after the acid-insoluble sulfides had been precipitated. He decided to simply add 6 *M* HCl to the filtrate from which the sulfides had been removed, but he obtained no precipitate. The student therefore concluded that there were no Ag^+, Hg_2^{2+}, or Pb^{2+} cations in his original sample. Is this conclusion valid?

60. Using qualitative analysis, a student decided to treat her sample with $(NH_4)_2S$ solution directly, skipping the HCl and acidic H_2S treatments because she was running out of time. In a sample that contained Ag^+, Hg_2^{2+}, Cd^{2+}, Sb^{3+}, and Zn^{2+}, which metal ions was she most likely to obtain in the resulting precipitate?

18 Chemical Thermodynamics

LEARNING OBJECTIVES

- To understand the connections among work, heat, and energy
- To be familiar with the concept of *PV* work
- To be able to calculate changes in internal energy
- To understand the relationship between internal energy and entropy
- To be able to use a thermodynamic cycle to calculate changes in entropy
- To understand the relationship between Gibbs free energy and work
- To know the difference between the information that thermodynamics and kinetics provide about a system
- To understand the importance of thermodynamics in biochemical systems

CHAPTER OUTLINE

Chemical reactions obey two fundamental laws. The first of these, the law of conservation of mass (see Chapter 1), states that matter can be neither created nor destroyed. The law of conservation of mass explains why equations must balance and is the basis for all the stoichiometry and equilibrium calculations you have learned to do thus far in chemistry. The second, the law of conservation of energy (see Chapter 5), states that energy can be neither created nor destroyed. Instead, energy takes various forms that can be converted from one to the other. For example, the energy stored in chemical bonds can be released as heat during a chemical reaction.

In Chapter 5, you learned about thermochemistry, the study of energy changes that occur during chemical reactions. Our goal in this chapter is to extend the

The melting of ice is a thermodynamic process. When a cube of ice melts, there is a spontaneous and irreversible transfer of heat from a warm substance, the surrounding air, to a cold substance, the ice cube. The direction of heat flow in this process and the resulting increase in entropy illustrate the second law of thermodynamics.

concepts of thermochemistry to an exploration of **thermodynamics** (from the Greek for "heat" and "power"), the study of the interrelationships among heat, work, and the energy content of a system at equilibrium. Thermodynamics tells chemists *whether* a particular reaction is energetically possible in the direction in which it is written and the composition of the reaction system at equilibrium. It does not, however, say anything about whether an energetically feasible reaction will *actually* occur as written, and it tells us nothing about the rate of the reaction or the pathway by which it will occur. The rate of a reaction and its pathway are described by chemical kinetics (see Chapter 14).

Chemical thermodynamics provides a bridge between the macroscopic properties of a substance and the individual properties of its constituent molecules and atoms. As you will see, thermodynamics explains why graphite can be converted to diamond; how chemical energy stored in molecules can be utilized to perform work; and why certain processes, such as iron rusting and organisms aging and dying, proceed spontaneously in only one direction, requiring no input of energy to occur.

18.1 ● Thermodynamics and Work

We begin our discussion of thermodynamics by reviewing some important terms introduced in Chapter 5. First, we need to distinguish between a system and its surroundings. A *system* is that part of the universe in which we are interested, such as a mixture of gases in a glass bulb or a solution of substances in a flask. The *surroundings* are everything else—the rest of the *universe*. We can therefore state

$$\text{system} + \text{surroundings} = \text{universe} \quad (18.1)$$

A *closed system*, such as the contents of a sealed jar, cannot exchange matter with its surroundings, whereas an *open system* can; in this case, we can convert a closed system (the jar) to an open system by removing the jar's lid.

In Chapter 5, we also introduced the concept of a **state function**, a property of a system that depends only on the present state of the system and not on its history. Thus, a change in a state function depends only on the difference between the initial and final states, not on the pathway used to go from one to the other. To help understand the concept of a state function, imagine a person hiking up a mountain (Figure 18.1). If the person is well trained and fit, he or she may be able to climb almost vertically to the top (path A), whereas another less athletic person may choose a path that winds gradually to the top (path B). If both hikers start from the same point at the base of the mountain and end up at the same point at the top, their *net change in altitude* will be the same regardless of the path chosen. Hence, altitude is a state function. On the other hand, a person may or may not carry a heavy pack, and may climb in hot weather or cold. These conditions would influence changes in the hiker's fatigue level, which depends on the path taken and the conditions experienced. Fatigue, therefore, is not a state function. Thermodynamics is generally concerned with state functions and does not deal with how the change between the initial and final state occurs.

Figure 18.1 Illustration that altitude is a state function. When hiking up a mountain, a person may decide to take path A, which is almost vertical, or path B, which gradually winds up to the top. Regardless of the path taken, the net change in altitude going from the initial state (bottom of the climb) to the final state (top of the climb) is the same. Thus, altitude is a state function.

The Connections Among Work, Heat, and Energy

The **internal energy** (*E*), of a system is the sum of the potential energy and the kinetic energy of all the components; internal energy is a state function. Although a closed system cannot exchange matter with its surroundings, it can exchange energy with its surroundings in two ways: by doing work or by releasing or absorbing heat, the flow of thermal energy. Work and heat are therefore two distinct ways of changing the internal energy of a system. We defined *work* w in Chapter 5 as a force F acting through a distance d:

$$w = Fd \tag{18.2}$$

Because work occurs only when an object, such as a person, or a substance, such as water, moves against an opposing force, work requires that the system and its surroundings be physically connected. In contrast, the flow of *heat*, the transfer of energy due to differences in temperature between two objects, represents a thermal connection between the system and its surroundings. Thus, doing work causes a physical displacement, whereas the flow of heat causes a temperature change. The units of work and heat must be the same because both processes result in the transfer of energy. In the SI system, those units are joules (J), the same unit used for energy. There is no difference between an energy change brought about by doing work on a system and an equal energy change brought about by heating it.

The connections among work, heat, and energy were first described by Benjamin Thompson (1753–1814), an American-born scientist who was also known as Count Rumford. While supervising the manufacture of cannons, Rumford recognized the relationship between the amount of work required to drill out a cannon and the temperature of the water used to cool it during the drilling process (Figure 18.2). At that time, it was generally thought that heat and work were separate and unrelated phenomena. Hence, Rumford's ideas were not widely accepted until many years later, after his findings had been corroborated in other laboratories.

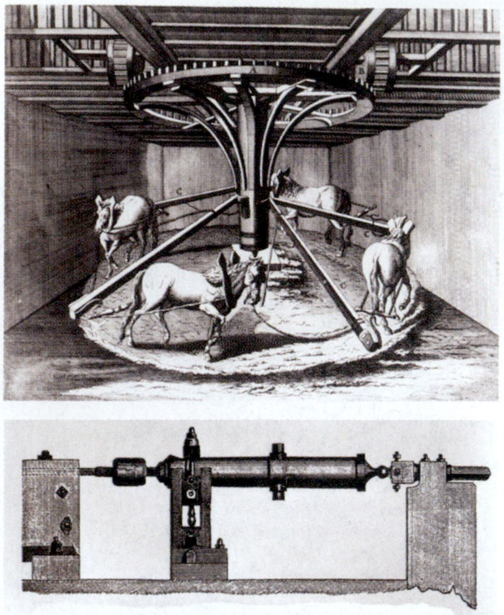

Figure 18.2 The relationship between heat and work. In the 1780s, an American scientist named Benjamin Thompson, also known as Count Rumford, was hired by the Elector of Bavaria to supervise the manufacture of cannons. During the manufacturing process, teams of horses harnessed to a large-toothed wheel supplied the power needed to drill a hole several inches in diameter straight down the center of a solid brass or bronze cylinder, which was cooled by water. Based on his observations, Rumford became convinced that heat and work are equivalent ways of transferring energy.

PV Work

As we saw in Chapter 5, there are many kinds of work, including mechanical work, electrical work, and work against a gravitational or magnetic field. In this section, we will consider only mechanical work, focusing on the work done during changes in the pressure or volume of a gas. To describe this *pressure–volume work* (*PV* work), we will use such imaginary oddities as frictionless pistons, which involve no component of resistance, and ideal gases, which have no attractive or repulsive interactions.

MGC *PV* Work

Imagine, for example, an ideal gas, confined by a frictionless piston, with internal pressure P_{int} and initial volume V_i (Figure 18.3). If $P_{ext} = P_{int}$, the system is at equilibrium; the piston does not move, and no work is done. If the external pressure on the piston, P_{ext}, is less than P_{int}, however, then the ideal gas inside the piston will expand, forcing the piston to perform work on the surroundings; that is, the final volume V_f will be greater than V_i. If $P_{ext} > P_{int}$, then the gas will be compressed and the surroundings will perform work on the system.

If the piston has cross-sectional area A, the external pressure exerted by the piston is by definition the force per unit area: $P_{ext} = F/A$. The volume of any three-dimensional object with parallel sides (such as a cylinder) is the cross-sectional area times the height ($V = Ah$). Rearranging to give $F = P_{ext}A$ and defining the distance the piston moves, d, as Δh, we can calculate the magnitude of the work performed by the piston by substituting into Equation 18.2:

$$w = Fd = P_{ext}A\Delta h \tag{18.3a}$$

The change in the volume of the cylinder, ΔV, as the piston moves a distance d is $\Delta V = A\Delta h$, as shown in Figure 18.4. The work performed is thus

$$w = P_{ext}\Delta V \tag{18.3b}$$

Note that the units of work obtained using this definition are correct: pressure is force per unit area (newton/m^2) and volume has units of cubic meters, so

$$w = \left(\frac{F}{A}\right)_{ext} \Delta V = \frac{\text{newton}}{\text{m}^2} \times \text{m}^3 = \text{newton} \cdot \text{m} = \text{joule}$$

If we use atmospheres for P and liters for V, we obtain units of L·atm for work. These units correspond to units of energy (see Chapter 10), as shown in the different values of the ideal gas constant R:

$$R = \frac{0.08206\ \text{L}\cdot\text{atm}}{\text{mol}\cdot\text{K}} = \frac{8.314\ \text{J}}{\text{mol}\cdot\text{K}}$$

Thus, 0.08206 L·atm = 8.314 J and 1 L·atm = 101.3 J.

Whether work is defined as having a positive or negative sign is a matter of convention. In Chapter 5, we defined heat flow *from* the system *to* the surroundings as negative. Using that same sign convention, we define work done *by* the system *on* the surroundings as having a negative sign because it too results in a transfer of energy from the system to the surroundings.* Because $\Delta V > 0$ for an expansion, Equation 18.3b must be written with a negative sign to describe PV work done *by* the system as negative:

$$w = -P_{ext}\Delta V \tag{18.4}$$

The work done by a gas expanding against an external pressure is therefore negative, corresponding to work done *by* the system *on* the surroundings. Conversely, when a gas is compressed by an external pressure, $\Delta V < 0$ and the work is positive because work is being done *on* the system *by* the surroundings.

Suppose, for example, that the system under study is a mass of steam heated by the combustion of several hundred pounds of coal and enclosed within a cylinder housing a piston attached to the crankshaft of a large steam engine. The gas is not ideal, and the cylinder is not frictionless. Nonetheless, as steam enters the engine chamber and the expanding gas pushes against the piston, the piston moves and useful work is performed. In fact, PV work launched the Industrial Revolution of the 19th century and powers the internal combustion engine on which most of us still rely for transportation.

We can see that work is not a state function by examining Figure 18.5, in which two different, two-step pathways take a gaseous system from an initial state to a final state with corresponding changes in temperature. In pathway A, the volume of a gas is initially increased while its pressure stays constant (step 1), and then its pressure is decreased while the volume remains constant (step 2). In pathway B, the order of the steps is reversed. The temperatures, pressures, and volumes of the initial and final states are identical in both cases, but the amount of work done, indicated by the shaded areas in the figure, is substantially different. As we can see, the amount of work done depends on the pathway taken from (V_1, P_1) to (V_2, P_2), which means that work is not a state function.

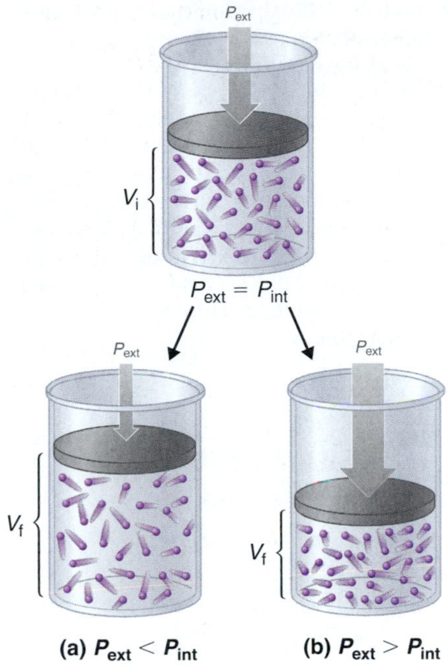

Figure 18.3 ***PV* work.** The frictionless piston is subject to external pressure P_{ext} and encloses an ideal gas with initial volume V_i and internal pressure P_{int}. If $P_{ext} = P_{int}$, the system is at equilibrium, the piston does not move, and no work is done. However, if the external pressure is less than the internal pressure **(a),** the ideal gas inside the piston will expand, forcing the piston to perform work on the surroundings. The final volume V_f will be greater than V_i. Alternatively, if the external pressure is greater than the internal pressure **(b),** the gas will be compressed and the surroundings will perform work on the system.

EXAMPLE 18.1

A small high-performance internal combustion engine has six cylinders with a total nominal displacement (volume) of 2.40 L and a 10:1 compression ratio (meaning that the volume of each cylinder decreases by a factor of 10 when the piston compresses the air–gas mixture inside the cylinder prior to ignition). How much work in joules is done when a gas in one cylinder of the engine expands at constant temperature against an opposing pressure of 40.0 atm during the engine cycle? Assume that the gas is ideal, the piston is frictionless, and no energy is lost as heat.

Given Final volume, compression ratio, external pressure

Asked for Work done

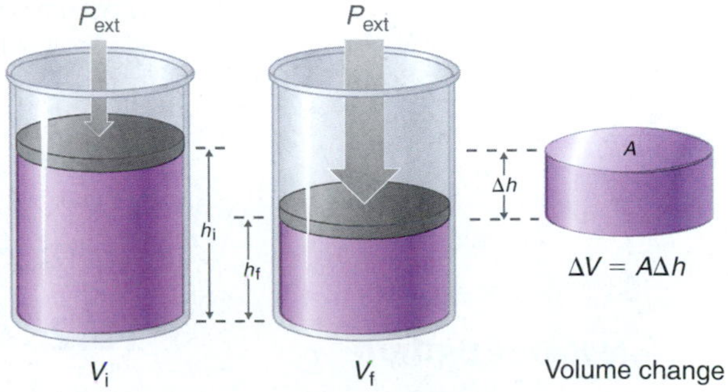

Figure 18.4 Work performed with a change in volume. The volume of a cylinder is equal to its cross-sectional area A times its height h. Thus, the change in the volume, ΔV, of the cylinder housing a piston is $\Delta V = A\ \Delta h$ as the piston moves. The work performed by the surroundings on the system as the piston moves inward is given by $w = P_{ext}\,\Delta V$.

* Note that this is an arbitrary convention and one that is not universally used. Some engineering disciplines are more interested in the work done *on* the surroundings than in the work done *by* the system, and therefore use the opposite convention.

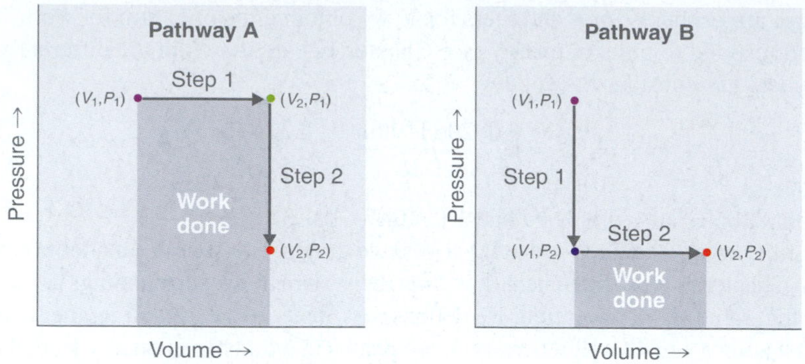

Figure 18.5 Illustration that work is not a state function. Two different pathways, A and B, are taken from an initial state (V_1, P_1) to a final state (V_2, P_2). In pathway A, the volume of a gas is initially increased while its pressure stays constant (step 1). Its pressure is then decreased while the volume remains constant (step 2). The order of the steps is reversed in pathway B. Although V_1, P_1 and V_2, P_2 are identical in both cases, the amount of work done (*shaded area*) depends on the pathway taken. Thus, work is not a state function.

Note the pattern

By convention, both heat flow and work have a negative sign when energy is transferred from the system to the surroundings, and vice versa.

Strategy

Ⓐ Calculate the final volume of gas in a single cylinder. Then compute the initial volume of gas in a single cylinder from the compression ratio.

Ⓑ Use Equation 18.4 to calculate the work done in liter-atmospheres. Convert from liter-atmospheres to joules.

Solution

Ⓐ✓ To calculate the work done, we need to know the initial and final volumes. The final volume is the volume of one of the six cylinders with the piston all the way down: $V_f = 2.40\text{ L}/6 = 0.400\text{ L}$. With a 10:1 compression ratio, the volume of the same cylinder with the piston all the way up is $V_i = 0.400\text{ L}/10 = 0.0400\text{ L}$. Ⓑ✓ To determine the work done, we substitute the values for P_{ext} and ΔV $(V_f - V_i)$ into Equation 18.4:

$$w = -P_{ext}\Delta V = -(40.0\text{ atm})(0.400\text{ L} - 0.0400\text{ L}) = -14.4\text{ L}\cdot\text{atm}$$

Converting from liter-atmospheres to joules gives

$$w = -(14.4\ \cancel{\text{L}\cdot\text{atm}})\,[101.3\text{ J}/(\cancel{\text{L}\cdot\text{atm}})] = -1.46\times10^3\text{ J}$$

In Exercise 18.1, you will see that the concept of work is not confined to engines and pistons. It is found in other applications as well.

EXERCISE 18.1

While you may not be aware of it, breathing requires work. The lung volume of a 70-kg man at rest changed from 2200 mL to 2700 mL when he breathed in, while his lungs maintained a pressure of approximately 1.0 atm. How much work in liter-atmospheres and joules was required to take a single breath? During exercise, his lung volume changed from 2200 mL to 5200 mL on each in-breath. How much additional work in joules did he require to take a breath while exercising?

Answer $-0.500\text{ L}\cdot\text{atm}$ or -50.7 J; -304 J. If he takes a breath every three seconds, this corresponds to 1.4 Calories per minute (1.4 kcal).

Note the pattern

The tendency of all systems, chemical or otherwise, is to move toward the state with the lowest possible energy.

18.2 • The First Law of Thermodynamics

The relationship between the energy change of the system and that of the surroundings is given by the **first law of thermodynamics**, which states that the *energy of the universe is constant.* Using Equation 18.1, we can express this law mathematically as

$$\Delta E_{univ} = \Delta E_{sys} + \Delta E_{surr} = 0 \qquad (18.5)$$
$$\Delta E_{sys} = -\Delta E_{surr}$$

where the subscripts refer to the universe, system, and surroundings. Thus, the change in energy of the system is identical in magnitude but opposite in sign to the change in energy of the surroundings.

One of the most important factors that determine the outcome of a chemical reaction is the tendency of all systems, chemical or otherwise, to move toward the lowest possible energy state. Thus, as a brick dropped from a rooftop falls, its potential energy is converted to kinetic energy, so that when it reaches ground level, it has achieved a state of lower potential energy. Anyone nearby will notice that energy is transferred to the surroundings as the noise of the impact reverberates and the dust rises when the brick hits the ground. Similarly, if a mixture of isooctane and oxygen in an internal combustion engine is ignited by a spark, carbon dioxide and water form spontaneously, while potential energy (in the form of the relative positions of atoms in the molecules) is released to the surroundings as heat and work. The internal energy content of the CO_2/H_2O product mixture is lower than that of the isooctane/O_2 reactant mixture. The two cases differ, however, in the form in which the energy is released to the surroundings. In the case of the falling brick, the energy is transferred as work done on whatever happens to be in the path of the brick; in the case of burning isooctane, the energy can be released as solely heat (if the reaction is carried out in an open container) or as a mixture of heat and work (if the reaction is carried out in the cylinder of an internal combustion engine).

Because heat and work are the only two ways in which energy can be transferred between a system and the surroundings, any change in the internal energy of the system is the sum of the heat transferred, q, and the work done, w:

$$\Delta E_{sys} = q + w \tag{18.6}$$

Although q and w are not state functions on their own, their sum, ΔE_{sys}, is independent of the path taken and is therefore a state function. A major task for the designers of any machine that converts energy to work is to maximize the amount of work obtained and to minimize the amount of energy released to the environment as heat. An example is the combustion of coal to produce electricity. Although the maximum amount of energy available from the process is fixed by the energy content of the reactants and products, the fraction of that energy that can be used to perform useful work is not fixed, as discussed in Section 18.5. Because we focus almost exclusively on the changes in the energy of the system, we will not use the "sys" subscript further unless we need to distinguish explicitly between the system and the surroundings.

Note the pattern

Although q and w are not state functions, their sum, ΔE_{sys}, is independent of the path taken and therefore is a state function.

EXAMPLE 18.2

A sample of an ideal gas in the cylinder of an engine is compressed from 400 mL to 50.0 mL during the compression stroke against a constant pressure of 8.00 atm. At the same time, 140 J of energy is transferred from the gas to the surroundings as heat. What is the total change in the internal energy, ΔE, of the gas in joules?

Given Initial volume, final volume, external pressure, quantity of energy transferred as heat

Asked for Total change in internal energy

Strategy

Ⓐ Determine the sign of q to use in Equation 18.6.
Ⓑ From Equation 18.4, calculate w from the values given. Substitute this value into Equation 18.6 to calculate ΔE.

Solution

Ⓐ From Equation 18.6, we know that $\Delta E = q + w$. We are given the magnitude of q (140 J) and need only determine its sign. Because energy is transferred from the system (the gas) to the surroundings, q is negative by convention. Ⓑ Because the gas is being compressed, we know that work is being done *on* the system, and hence w must be positive. From Equation 18.4,

$$w = -P_{ext}\Delta V = -8.00\ \cancel{\text{atm}}(0.0500\ \cancel{\text{L}} - 0.400\ \cancel{\text{L}})\left(\frac{101.3\ \text{J}}{\cancel{\text{L}}\cdot\cancel{\text{atm}}}\right) = 284\ \text{J}$$

Thus,

$$\Delta E = q + w = -140\ \text{J} + 284\ \text{J} = 144\ \text{J}$$

In this case, although work is done *on* the gas, increasing its internal energy, heat flows *from* the system *to* the surroundings, decreasing its internal energy by 144 J. Notice that the work done and the heat transferred can have opposite signs.

EXERCISE 18.2

A sample of an ideal gas is allowed to expand from an initial volume of 0.200 L to a final volume of 3.50 L against a constant external pressure of 0.995 atm. At the same time, 117 J of heat is transferred from the surroundings to the gas. What is the total change in the internal energy, ΔE, of the gas in joules?

Answer −216 J

Enthalpy

To further understand the relationship between heat flow q and the resulting change in internal energy ΔE, we can look at two sets of limiting conditions: reactions that occur at constant volume and reactions that occur at constant pressure. We will assume that PV work is the only kind of work possible for the system, so that we can substitute its definition from Equation 18.4 into Equation 18.6 to obtain

$$\Delta E = q - P\Delta V \qquad (18.7)$$

where subscripts have been deleted.

If the reaction occurs in a closed vessel, the volume of the system is fixed and ΔV is zero. Under these conditions, the heat flow (often given the symbol q_v to indicate constant volume) must be equal to ΔE:

$$q_v = \Delta E \qquad (18.7a)$$
Constant volume

No PV work can be done, and the change in the internal energy of the system is equal to the amount of heat transferred from the system to the surroundings, or vice versa.

Many chemical reactions are not, however, carried out in sealed containers at constant volume, but in open containers at a more or less constant pressure of about 1 atm. The heat flow under these conditions is given the symbol q_p to indicate constant pressure. Replacing q in Equation 18.7 by q_p and rearranging to solve for q_p give

$$q_p = \Delta E + P\Delta V \qquad (18.7b)$$
Constant pressure

Note the pattern

At constant pressure, the change in the enthalpy of a system is equal to the heat flow: $\Delta H = q_p$.

Thus, at constant pressure, the heat flow for any process is equal to the change in the internal energy of the system plus the PV work done, as we stated in Chapter 5.

Because conditions of constant pressure are so important in chemistry, a new state function called **enthalpy (H)** is defined as $H = E + PV$. At constant pressure, the change in the enthalpy of a system is

$$\Delta H = \Delta E + \Delta(PV) = \Delta E + P\Delta V \qquad (18.8)$$

Comparing the last two equations shows that at constant pressure, the change in the enthalpy of a system is equal to the heat flow: $\Delta H = q_p$. This expression is consistent with our definition of enthalpy in Chapter 5, where we stated that enthalpy is the heat absorbed or produced during any process that occurs at constant pressure.

EXAMPLE 18.3

The molar enthalpy of fusion for ice at 0.0°C and a pressure of 1.00 atm is 6.01 kJ (see Chapter 5), and the molar volumes of ice and water at 0°C are 0.0197 L and 0.0180 L, respectively. Calculate ΔH and ΔE for the melting of ice at 0.0°C.

Given Enthalpy of fusion for ice, pressure, molar volumes of ice and water

Asked for ΔH and ΔE for ice melting at 0.0°C

Strategy

A Determine the sign of q, and set this value equal to ΔH.
B Calculate $\Delta(PV)$ from the information given.
C Determine ΔE by substituting the calculated values into Equation 18.8.

Solution

A Because 6.01 kJ of heat is absorbed from the surroundings when 1 mol of ice melts, $q = +6.01$ kJ. When the process is carried out at constant pressure, $q = q_p = \Delta H = 6.01$ kJ. **B** To find ΔE using Equation 18.8, we need to calculate $\Delta(PV)$. The process is carried out at a constant pressure of 1.00 atm, so

$$\begin{aligned}\Delta(PV) &= P\Delta V = P(V_f - V_i) = (1.00\text{ atm})(0.0180\text{ L} - 0.0197\text{ L})\\ &= (-1.7 \times 10^{-3}\ \cancel{\text{L}\cdot\text{atm}})(101.3\text{ J}/\cancel{\text{L}\cdot\text{atm}}) = -0.0017\text{ J}\end{aligned}$$

C Substituting the calculated values of ΔH and $P\Delta V$ into Equation 18.8 gives

$$\Delta E = \Delta H - P\Delta V = 6010\text{ J} - (-0.0017\text{ J}) = 6010\text{ J} = 6.01\text{ kJ}$$

EXERCISE 18.3

At 298 K and 1 atm, the conversion of graphite to diamond requires the input of 1.850 kJ of heat per mole of carbon. The molar volumes of graphite and diamond are 0.00534 L and 0.00342 L, respectively. Calculate ΔH and ΔE for the conversion of C (graphite) to C (diamond) under these conditions.

Answer $\Delta H = 1.85$ kJ/mol; $\Delta E = 1.85$ kJ/mol

The Relationship Between ΔH and ΔE

If ΔH for a reaction is known, we can use the change in the enthalpy of the system (Equation 18.8) to calculate its change in internal energy. When a reaction involves only solids, liquids, liquid solutions, or any combination of these, the volume does not change appreciably ($\Delta V = 0$). Under these conditions, we can simplify Equation 18.8 to $\Delta H = \Delta E$. If gases are involved, however, ΔH and ΔE can differ significantly. We can calculate ΔE from the measured value of ΔH by using the right side of Equation 18.8 together with the ideal gas law, $PV = nRT$. Recognizing that $\Delta(PV) = \Delta(nRT)$, we can rewrite Equation 18.8 as

$$\Delta H = \Delta E + \Delta(PV) = \Delta E + \Delta(nRT) \qquad (18.9)$$

At constant temperature, $\Delta(nRT) = RT\Delta n$, where Δn is the difference between the final and initial numbers of moles of gas. Thus,

$$\Delta E = \Delta H - RT\Delta n \qquad (18.10)$$

For reactions that result in a net production of gas, $\Delta n > 0$ and therefore $\Delta E < \Delta H$. Conversely, endothermic reactions ($\Delta H > 0$) that result in a net consumption of gas

Note the pattern

For reactions that result in a net production of gas, $\Delta E < \Delta H$. For endothermic reactions that result in a net consumption of gas, $\Delta E > \Delta H$.

have $\Delta n < 0$ and $\Delta E > \Delta H$. The relationship between ΔH and ΔE for systems involving gases is illustrated in the next example.

EXAMPLE 18.4

The combustion of graphite to produce carbon dioxide is described by the equation C (graphite) + $O_2(g) \longrightarrow CO_2(g)$. At 298 K and 1.0 atm, $\Delta H = -393.5$ kJ/mol of graphite for this reaction, and the molar volume of graphite is 0.0053 L. What is the value of ΔE for the reaction?

Given Balanced chemical equation, temperature, pressure, ΔH, molar volume of reactant

Asked for ΔE

Strategy

Ⓐ Use the balanced chemical equation to calculate the change in the number of moles of gas during the reaction.

Ⓑ Substitute this value and the data given into Equation 18.10 to obtain ΔE.

Solution

Ⓐ In this reaction, 1 mol of gas (CO_2) is produced, and 1 mol of gas (O_2) is consumed. Thus, $\Delta n = 1 - 1 = 0$. Ⓑ Substituting this calculated value and the given values into Equation 18.10 gives

$$\Delta E = \Delta H - RT\Delta n = (-393.5 \text{ kJ/mol}) - [8.314 \text{ J/(mol} \cdot \text{K)}](298 \text{ K})(0)$$
$$= (-393.5 \text{ kJ/mol}) - (0 \text{ J/mol}) = -393.5 \text{ kJ/mol}$$

To understand why only the change in the volume of the gases needs to be considered, notice that the molar volume of graphite is only 0.0053 L. A change in the number of moles of gas corresponds to a volume change of 22.4 L/mol of gas at STP, so the volume of gas consumed or produced in this case is (1)(22.4 L) = 22.4 L, which is much, much greater than the volume of 1 mol of a solid such as graphite.

EXERCISE 18.4

Calculate ΔE for the conversion of oxygen gas to ozone at 298 K: $3O_2(g) \longrightarrow 2O_3(g)$. The value of ΔH for the reaction is 285.4 kJ.

Answer 288 kJ

As Exercise 18.4 illustrates, the magnitudes of ΔH and ΔE for reactions that involve gases are generally rather similar, even when there is a net production or consumption of gases.

18.3 • The Second Law of Thermodynamics

The first law of thermodynamics governs changes in the state function we have called internal energy E. As we have just seen, changes in the internal energy, ΔE, are closely related to changes in the enthalpy, ΔH, which is a measure of the heat flow between a system and its surroundings at constant pressure. You also learned in Chapter 5 that the enthalpy change for a chemical reaction can be calculated using tabulated values of enthalpies of formation. This information, however, does not tell us whether a particular process or reaction will occur spontaneously.

Let's consider a familiar example of spontaneous change. If a hot frying pan that has just been removed from the stove is allowed to come into contact with a cooler object, such as cold water in a sink, heat will flow from the hotter object to the cooler one. Eventually both objects will reach the same temperature, at a value between the initial temperatures of the two objects. This transfer of heat from a hot object to a cooler one obeys the first law of thermodynamics: energy is conserved.

Now consider the same process in reverse. Suppose that a hot frying pan in a sink of water were to become hotter while the water became cooler. As long as the same amount of thermal energy was gained by the frying pan and lost by the water, the first law of thermodynamics would be satisfied. Yet we all know that such a process cannot occur: heat always flows from a hot object to a cold one, and never in the reverse direction. That is, *by itself* the magnitude of the heat flow associated with a process does not predict whether the process will occur spontaneously.

For many years, chemists and physicists tried to identify a single measurable quantity that would enable them to predict whether a particular process or reaction would occur spontaneously. Initially, many of them focused on enthalpy changes and hypothesized that an exothermic process would always be spontaneous. But although it is true that many, if not most, spontaneous processes are exothermic, there are also many spontaneous processes that are *not* exothermic. For example, at a pressure of 1 atm, ice melts spontaneously at temperatures higher than 0°C, yet this is an endothermic process: heat is absorbed. Similarly, many salts (such as NH_4NO_3, NaCl, and KBr) dissolve spontaneously in water even though they absorb heat from the surroundings as they dissolve (that is, $\Delta H_{soln} > 0$). Reactions can also be both spontaneous and highly endothermic, like the reaction of barium hydroxide with ammonium thiocyanate shown in Figure 18.6.

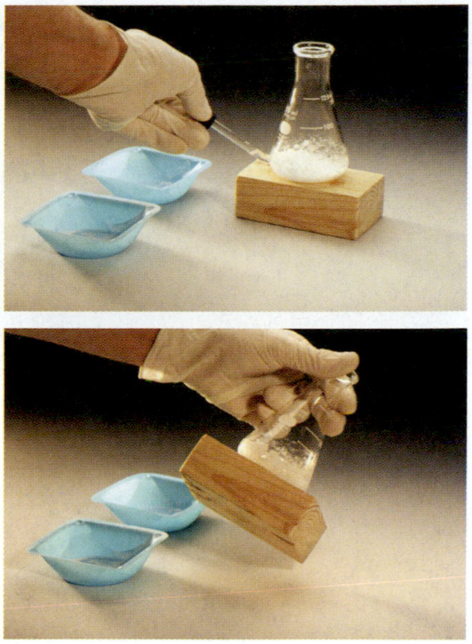

Figure 18.6 An endothermic reaction. The reaction of barium hydroxide with ammonium thiocyanate is spontaneous but highly endothermic, so water, one product of the reaction, quickly freezes into slush. When water is placed on a block of wood under the beaker, the highly endothermic reaction that takes place in the beaker freezes water that has been placed under the beaker, and the beaker becomes frozen to the wood.

Thus, enthalpy is not the only factor that determines whether a process is spontaneous. For example, once a cube of sugar has dissolved in a glass of water so that the sucrose molecules are uniformly dispersed in a dilute solution, they never spontaneously come back together in solution to form a sugar cube. Moreover, the molecules of a gas remain evenly distributed throughout the entire volume of a glass bulb and never spontaneously assemble in only one portion of the available volume. To help explain why these phenomena proceed spontaneously in only one direction requires an additional state function called **entropy (S)**, a thermodynamic property of all substances that is proportional to their degree of disorder. In Chapter 13, we introduced the concept of entropy in relation to solution formation. Here we further explore the nature of this state function and define it mathematically.

Entropy

Chemical and physical changes in a system may be accompanied by either an increase or a decrease in the disorder of the system, corresponding to an increase in entropy ($\Delta S > 0$) or a decrease in entropy ($\Delta S < 0$), respectively. As with any other state function, the change in entropy is defined as the difference between the entropies of the final and initial states: $\Delta S = S_f - S_i$.

When a gas expands into a vacuum, its entropy increases because the increased volume allows for greater atomic or molecular disorder. The greater the number of atoms or molecules in the gas, the greater the disorder. The magnitude of the entropy of a system depends on the number of microscopic states, or *microstates*, associated with it (in this case, the number of atoms or molecules); that is, the greater the number of microstates, the greater the entropy.

We can illustrate the concepts of microstates and entropy using a deck of playing cards, as shown in Figure 18.7. In any new deck, the 52 cards are arranged by suits, with each suit arranged in descending order. If the cards are shuffled, however, there are approximately 10^{68} different ways they might be arranged, which corresponds to 10^{68} different microscopic states. The entropy of an ordered new deck of cards is therefore low, whereas the entropy of a randomly shuffled deck is high. Card games assign a higher value to a hand that has a low degree of disorder. In games like five-card poker, only four of the 2,598,960 different possible hands, or microstates, contain the highly ordered and valued arrangement of cards called a royal flush, almost 1.1 million hands contain one pair, and more than 1.3 million hands are completely disordered and

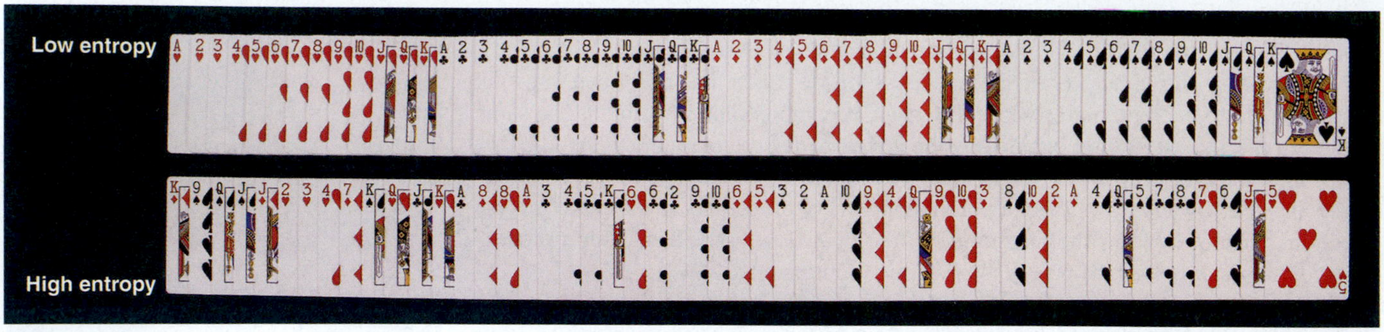

Figure 18.7 Illustrating low- and high-entropy states with a deck of playing cards. A new, unshuffled deck (*top*) contains 52 cards arranged by suits, with each suit arranged in descending order. With only a single arrangement possible, there is only one microstate. The entropy of the deck is therefore low. In contrast, a randomly shuffled deck (*bottom*) can have any one of approximately 10^{68} different arrangements, which correspond to 10^{68} different microstates. The entropy of a randomly shuffled deck is therefore high. One of the possible random arrangements is shown.

therefore have no value. Because the last two arrangements are far more probable than the first, the value of a poker hand is inversely proportional to its entropy.

We can see how to calculate these kinds of probabilities for a chemical system by considering the possible arrangements of a sample of four gas molecules in a two-bulb container (Figure 18.8). There are five possible arrangements: all four molecules in the left bulb (I); three molecules in the left bulb and one in the right bulb (II); two molecules in each bulb (III); one molecule in the left bulb and three molecules in the right bulb (IV); and four molecules in the right bulb (V). If we assign a different color to each molecule in order to keep track of it for this discussion (remember, however, that in reality molecules are indistinguishable from one another), we can see that there are 16 different ways the four molecules can be distributed in the bulbs, each corresponding to a particular microstate. As shown in Figure 18.8, arrangement I is associated with a single microstate, as is arrangement V. Hence each of these arrangements has a probability of $\frac{1}{16}$. Arrangements II and IV each have a probability of $\frac{4}{16}$ because each can exist in four microstates. Similarly, six different microstates can occur as arrangement III, making the probability of this arrangement $\frac{6}{16}$. Thus, the arrangement that we would expect to encounter, with half the gas molecules in each bulb, is the *most probable* arrangement. The others are not impossible, but simply less likely.

Figure 18.8 The possible microstates for a sample of four gas molecules in two bulbs of equal volume. There are 16 different ways to distribute four gas molecules between the bulbs, with each distribution corresponding to a particular microstate. The possible microstates can be grouped into five arrangements. Arrangements I and V each produce a single microstate with a probability of 1/16. This particular arrangement is so improbable that it is likely not observed. Arrangements II and IV each produce four microstates, with a probability of 4/16. Arrangement III, with half the gas molecules in each bulb, has a probability of 6/16. It is the one encompassing the most microstates, so it is the most probable.

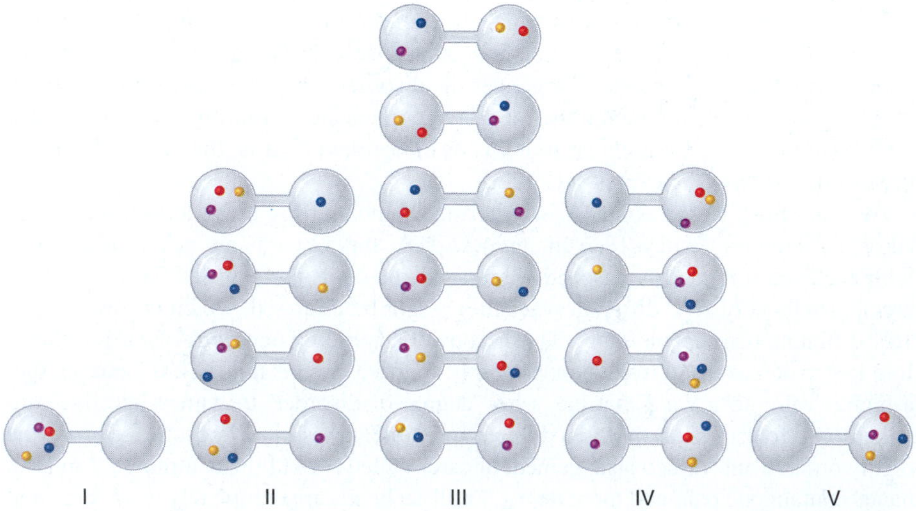

Instead of four molecules of gas, let's now consider 1 L of an ideal gas at STP, which contains 2.69×10^{22} molecules (6.022×10^{23} molecules/22.4 L). If we allow the sample of gas to expand into a second 1-L container, the probability of finding all 2.69×10^{22} molecules in one container and none in the other at any given time is infinitesimally small, approximately $1/(2.69 \times 10^{22})$. The denominator is larger than the total number of atoms in the universe, so the probability of such an occurrence is effectively zero. Thus, although nothing prevents the molecules in the gas sample from occupying only one of the two bulbs, that particular arrangement is so improbable that it is never actually observed. The probability of arrangements with essentially equal numbers of molecules in each bulb is quite high, however, because there are *many* equivalent microstates in which the molecules are distributed equally. Hence, a macroscopic sample of a gas occupies all of the space available to it, simply because this is the most probable arrangement.

A disordered system has a greater number of possible microstates than does an ordered system, so it has a higher entropy. This is most clearly seen in the entropy changes that accompany phase transitions, such as solid to liquid or liquid to gas. As you know from Chapters 11–13, a crystalline solid is composed of an ordered array of molecules that occupy fixed positions in a lattice, whereas the molecules in a liquid are free to move and tumble within the volume of the liquid; molecules in a gas have even more freedom to move than those in a liquid. Each degree of motion increases the number of available microstates, resulting in a higher entropy. Thus, the entropy of a system must increase during melting ($\Delta S_{fus} > 0$). Similarly, when a liquid is converted to a vapor, the greater freedom of motion of the molecules in the gas phase means that $\Delta S_{vap} > 0$. Conversely, the reverse processes (condensing a vapor to form a liquid or freezing a liquid to form a solid) must be accompanied by a decrease in the entropy of the system: $\Delta S < 0$.

Note the pattern

Entropy S is a thermodynamic property of all substances that is proportional to their degree of disorder. The greater the number of possible microstates for a system, the greater the disorder and the higher the entropy.

Experiments show that the magnitude of ΔS_{vap} is 80–90 J/(mol · K) for a wide variety of liquids with different boiling points. However, liquids that have highly ordered structures due to hydrogen bonding or other intermolecular interactions tend to have significantly higher values of ΔS_{vap}. For instance, ΔS_{vap} for water is 102 J/(mol · K).

Another process that is accompanied by entropy changes is the formation of a solution. As illustrated in Figure 18.9, the formation of a liquid solution from a crystalline solid (the solute) and a liquid solvent is expected to result in an increase in the disorder of the system and hence in its entropy. Indeed, dissolving a substance such as NaCl in water disrupts both the ordered crystal lattice of NaCl and the ordered hydrogen-bonded structure of water, leading to an increase in the entropy of the system. At the same time, however, each dissolved Na^+ ion becomes hydrated by an ordered arrangement of at least six water molecules, and the Cl^- ions also cause the water to adopt a particular local structure. Both of these effects *increase* the order of the system, leading to a *decrease* in entropy. The overall entropy change for the formation of a solution therefore depends on the relative magnitudes of these opposing factors. In the case of an NaCl solution, disruption of the crystalline NaCl structure and the hydrogen-bonded interactions in water is quantitatively more important, and thus $\Delta S_{soln} > 0$.

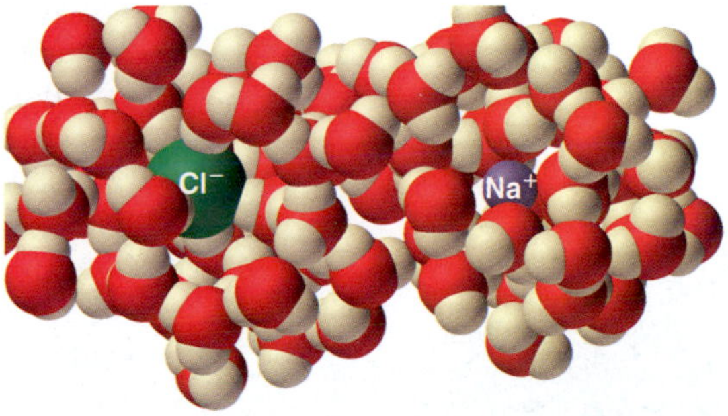

Figure 18.9 The effect of solution formation on entropy. Dissolving NaCl in water disrupts the highly ordered crystalline lattice structure of NaCl and the somewhat ordered hydrogen-bonded structure of liquid water, resulting in an increase in the entropy of the system. Each hydrated Na^+ and Cl^- ion, however, forms an ordered arrangement with at least six water molecules, thereby decreasing the entropy of the system. In the case of NaCl, the magnitude of the factors that increase entropy is greater than the magnitude of those that decrease entropy. Hence, the overall entropy change for the formation of an NaCl solution is positive.

EXAMPLE 18.5

Predict which substance in the following pairs has the higher entropy, and justify your answers: **(a)** 1 mol of $NH_3(g)$ or 1 mol of He(g), both at 25°C; **(b)** 1 mol of Pb(s) at 25°C or 1 mol of Pb(l) at 800°C.

Given Amounts of substances and temperature

Asked for Higher entropy

Strategy

From the number of atoms present and the phase of each substance, predict which has the greater number of available microstates and hence the higher entropy.

Solution

(a) Both substances are gases at 25°C, but one consists of He atoms and the other consists of NH_3 molecules. With four atoms instead of one, the NH_3 molecules have more motions available, leading to a greater number of microstates. Hence, we predict that the NH_3 sample will have the higher entropy.

(b) The nature of the atomic species is the same in both cases, but the phase is different: one sample is a solid and one a liquid. Based on the greater freedom of motion available to atoms in a liquid, we predict that the liquid sample will have the higher entropy.

EXERCISE 18.5

Predict which substance in the following pairs has the higher entropy, and justify your answers: **(a)** 1 mol of He(*g*) at 10 K and 1 atm pressure or 1 mol of He(*g*) at 250°C and 0.2 atm; **(b)** a mixture of 3 mol of $H_2(g)$ and 1 mol of $N_2(g)$ at 25°C and 1 atm or a sample of 2 mol of $NH_3(g)$ at 25°C and 1 atm.

Answer **(a)** 1 mol of He(*g*) at 250°C and 0.2 atm (higher temperature and lower pressure indicate greater volume and more microstates); **(b)** a mixture of 3 mol of $H_2(g)$ and 1 mol of $N_2(g)$ at 25°C and 1 atm (more molecules of gas are present)

Reversible and Irreversible Changes

Changes in entropy, ΔS, together with changes in enthalpy, ΔH, enable us to predict in which direction a chemical or physical change will occur spontaneously. Before discussing how to do so, however, we must understand the difference between a reversible process and an irreversible process. In a **reversible process**, every intermediate state between the extremes is an equilibrium state, regardless of the direction of the change. In contrast, an **irreversible process** is one in which the intermediate states are not equilibrium states and change occurs spontaneously in only one direction. As a result, *a reversible process can change direction at any time, whereas an irreversible process cannot*. When a gas expands reversibly against an external pressure such as a piston, for example, the expansion can be reversed at any time by reversing the motion of the piston; once the gas is compressed, it can be allowed to expand again, and the process can continue indefinitely. In contrast, the expansion of a gas into a vacuum ($P_{ext} = 0$) is irreversible because the external pressure is measurably less than the internal pressure of the gas. No equilibrium states exist, and the gas expands irreversibly. When gas escapes from a microscopic hole in a balloon into a vacuum, for example, the process is irreversible; the direction of air flow cannot change.

Because work done during the expansion of a gas depends on the opposing external pressure ($w = P_{ext}\Delta V$), *work done in a reversible process is always equal to or greater than work done in a corresponding irreversible process*: $w_{rev} \geq w_{irrev}$. Whether a process is reversible or irreversible, $\Delta E = q + w$. Because E is a state function, the magnitude of ΔE does *not* depend on reversibility and is independent of the path taken. Thus,

$$\Delta E = q_{rev} + w_{rev} = q_{irrev} + w_{irrev} \tag{18.11}$$

In other words, ΔE for a process is the same whether that process is carried out in a reversible or an irreversible manner. We now return to our earlier definition of entropy, using the magnitude of the heat flow for a reversible process, q_{rev}, to define entropy quantitatively.

Note the pattern

Work done in a reversible process is always equal to or greater than work done in a corresponding irreversible process: $w_{rev} \geq w_{irrev}$.

The Relationship Between Internal Energy and Entropy

Because the quantity of heat transferred, q_{rev}, is directly proportional to the absolute temperature of an object, T ($q_{rev} \propto T$), the hotter the object, the greater the amount of heat transferred. Moreover, adding heat to a system increases the kinetic energy of the component atoms and molecules and hence their disorder ($\Delta S \propto q_{rev}$). Thus, for any reversible process,

$$q_{rev} = T\Delta S \quad \text{and} \quad \Delta S = \frac{q_{rev}}{T} \tag{18.12}$$

Because the numerator, q_{rev}, is expressed in units of energy (joules), the units of ΔS are joules/kelvin (J/K). Recognizing that the work done in a reversible process at constant pressure is $w_{rev} = -P\Delta V$, we can express Equation 18.11 as

$$\Delta E = q_{rev} + w_{rev} = T\Delta S - P\Delta V \tag{18.13}$$

Thus, the change in the internal energy of the system is related to the change in entropy, the absolute temperature, and the PV work done.

To illustrate the use of Equations 18.12 and 18.13, let's consider two reversible processes before turning to an irreversible process. When a sample of gas is allowed to expand reversibly at constant temperature, heat must be added to the gas during expansion in order to keep its T constant (Figure 18.10). The internal energy of the gas does not change because the temperature of the gas does not change; that is, $\Delta E = 0$ and $q_{rev} = -w_{rev}$. During an expansion, $\Delta V > 0$, so the gas performs work on its surroundings: $w_{rev} = -P\Delta V < 0$. According to Equation 18.13, this means that q_{rev} must increase during the expansion; that is, the gas must absorb heat from the surroundings during the expansion, and the surroundings must give up that same amount of heat. The entropy change of the system is therefore $\Delta S_{sys} = +q_{rev}/T$, and the entropy change of the surroundings is $\Delta S_{surr} = -q_{rev}/T$. The corresponding change in entropy of the universe is

$$\Delta S_{univ} = \Delta S_{sys} + \Delta S_{surr} = \frac{q_{rev}}{T} + \left(-\frac{q_{rev}}{T}\right) = 0 \tag{18.14}$$

Thus, no change in ΔS_{univ} has occurred.

Now consider the reversible melting of a sample of ice at 0°C and 1 atm. The enthalpy of fusion of ice is 6.01 kJ/mol, which means that 6.01 kJ of heat is absorbed reversibly from the surroundings when 1 mol of ice melts at 0°C, as illustrated in Figure 18.11. The surroundings is a sample of low-density carbon foam that is thermally conductive, and the system is the ice cube that has been placed on it. The direction of heat flow along the resulting temperature gradient is indicated with an arrow. From Equation 18.12, we see that the entropy of fusion of ice can be written as

$$\Delta S_{fus} = \frac{q_{rev}}{T} = \frac{\Delta H_{fus}}{T} \tag{18.15}$$

In this case, $\Delta S_{fus} = (6.01 \text{ kJ/mol})/(273 \text{ K}) = 22.0 \text{ J/(mol·K)} = \Delta S_{sys}$. The amount of heat lost by the surroundings is the same as the amount gained by the ice, so $\Delta S_{surr} = q_{rev}/T = -(6.01 \text{ kJ/mol})/(273 \text{ K}) = -22.0 \text{ J/(mol·K)}$. Once again, we see that the entropy of the universe does not change:

$$\Delta S_{univ} = \Delta S_{sys} + \Delta S_{surr} = 22.0 \text{ J/(mol·K)} - 22.0 \text{ J/(mol·K)} = 0$$

In both these examples of reversible processes, the entropy of the universe is unchanged. This is true of all reversible processes and constitutes part of the **second law of thermodynamics**: *The entropy of the universe remains constant in a reversible process, whereas the entropy of the universe increases in an irreversible (spontaneous) process.*

As an example of an irreversible process, consider the entropy changes that accompany the spontaneous and irreversible transfer of heat from a hot object to a

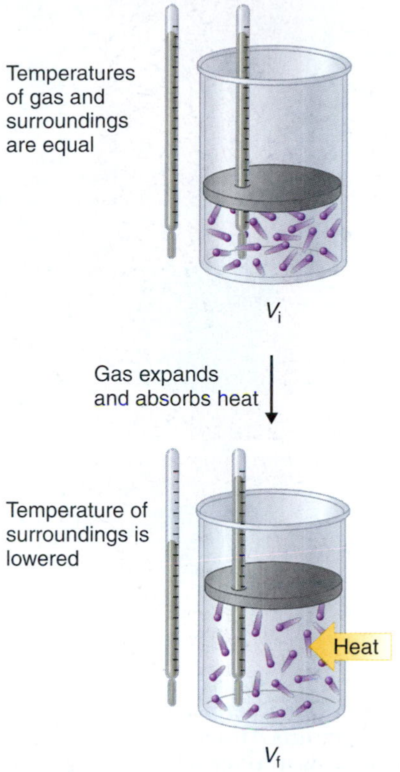

Figure 18.10 Expansion of gas at constant temperature. Internal energy is related to entropy, temperature, and the *PV* work done. In the initial state (*top*), the temperatures of the gas and the surroundings are the same. During the reversible expansion of the gas, heat must be added to the gas to maintain a constant temperature. Because the temperature of the gas does not change, the internal energy of the gas does not change, but work is performed on the surroundings. In the final state (*bottom*), the temperature of the surroundings is lower because the gas has absorbed heat from the surroundings during expansion.

Note the pattern

The entropy of the universe increases during a spontaneous process.

cold one, as occurs when lava spewed from a volcano flows into cold ocean water. The cold substance, the water, gains heat ($q > 0$), so the change in the entropy of the water can be written as $\Delta S_{cold} = q/T_{cold}$. Similarly, the hot substance, the lava, loses heat ($q < 0$), so its entropy change can be written as $\Delta S_{hot} = -q/T_{hot}$. T_{cold} and T_{hot} are the temperatures of the cold and hot substances, respectively. The total entropy change of the universe accompanying this process is therefore

$$\Delta S_{univ} = \Delta S_{cold} + \Delta S_{hot} = \frac{q}{T_{cold}} + \left(-\frac{q}{T_{hot}}\right) \tag{18.16}$$

The numerators on the right side of Equation 18.16 are the same in magnitude but opposite in sign. Whether ΔS_{univ} is positive or negative depends on the relative magnitudes of the denominators. By definition $T_{hot} > T_{cold}$, so $-q/T_{hot}$ must be less than q/T_{cold}, and ΔS_{univ} must be positive. Thus, as predicted by the second law of thermodynamics, the entropy of the universe increases during this irreversible process. Any process for which ΔS_{univ} is positive is by definition a spontaneous one that will occur as written. Conversely, any process for which ΔS_{univ} is negative will not occur spontaneously as written, but will occur spontaneously in the opposite direction. We see, therefore, that heat is spontaneously transferred from a hot substance, the lava, to a cold substance, the ocean water. In fact, if the lava is hot enough (for example, if it is molten), so much heat can be transferred that the water is converted to steam (Figure 18.12).

Figure 18.11 Thermograms showing that heat is absorbed from the surroundings when ice melts reversibly at 0°C. By convention, a thermogram shows cold regions in blue, warm regions in red, and thermally intermediate regions in green. When an ice cube (the system, dark blue) is placed on the corner of a square sample of low-density carbon foam with very high thermal conductivity, the temperature of the foam is lowered (going from red to green). As the ice melts, a temperature gradient appears, ranging from warm to very cold. The direction of heat flow from the surroundings (red and green) to the ice cube is indicated by the arrow. The amount of heat lost by the surroundings is the same as the amount gained by the ice, so the entropy of the universe does not change.

Figure 18.12 Spontaneous transfer of heat from a hot substance to a cold substance. When molten lava flows into cold oceanwater, so much heat is spontaneously transferred to the water that steam is produced.

EXAMPLE 18.6

Tin has two allotropes with different structures. Gray tin (α-tin) has a structure similar to that of diamond, whereas white tin (β-tin) is denser, with a unit cell structure that is based on a rectangular prism. At temperatures higher than 13.2°C, white tin is the more stable phase, but below that temperature, it slowly converts reversibly to the less dense, powdery gray phase. This phenomenon plagued Napoleon's army during his ill-fated invasion of Russia in 1812: the buttons on his soldiers' uniforms were made of tin and disintegrated during the Russian winter, adversely affecting the soldiers' health (and morale). The conversion of white tin to gray tin is exothermic, with $\Delta H = -2.1$ kJ/mol at 13.2°C. **(a)** What is ΔS for this process? **(b)** Which is the more highly ordered form of tin: white or gray?

Given ΔH, temperature

Asked for ΔS, relative degree of order

Strategy

Use Equation 18.12 to calculate the change in entropy for the reversible phase transition. From the calculated value of ΔS, predict which allotrope has the more highly ordered structure.

Solution

(a) We know from Equation 18.12 that the entropy change for any reversible process is the heat transferred (in joules) divided by the temperature at which the process occurs. Because the conversion occurs at constant pressure, and ΔH and ΔE are essentially equal for reactions that involve only solids, we can calculate the change in entropy for the reversible phase transition where $q_{rev} = \Delta H$. Substituting the given values for ΔH and temperature in kelvins (in this case, $T = 13.2°C = 286.4$ K) gives

$$\Delta S = \frac{q_{rev}}{T} = \frac{(-2.1\ \cancel{kJ}/mol)(1000\ J/\cancel{kJ})}{286.4\ K} = -7.3\ J/(mol \cdot K)$$

(b) The fact that $\Delta S < 0$ means that the entropy decreases when white tin is converted to gray tin. Thus, gray tin must be the more highly ordered structure.

EXERCISE 18.6

Elemental sulfur exists in two forms: an orthorhombic form (S_α), which is stable below 95.3°C, and a monoclinic form (S_β), which is stable above 95.3°C. The conversion of orthorhombic sulfur to monoclinic sulfur is endothermic, with $\Delta H = 0.401$ kJ/mol at 1 atm. **(a)** What is ΔS for this process? **(b)** Which is the more highly ordered form of sulfur: S_α or S_β?

Answer **(a)** 1.09 J/(mol·K); **(b)** orthorhombic, S_α

18.4 Entropy Changes and the Third Law of Thermodynamics

The atoms, molecules, or ions that make up a chemical system can undergo several types of molecular motion, including translation, rotation, and vibration (Figure 18.13). The greater the molecular motion of a system, the greater the number of possible microstates and hence the higher the entropy. A perfectly ordered system with only a single microstate available to it would have an entropy of zero. The only system that meets this criterion is a perfect crystal at a temperature of absolute zero (0 K), in which each component atom, molecule, or ion is fixed in place within a crystal lattice and thus exhibits no motion. Such a state of perfect order (or, conversely, zero disorder) corresponds to zero entropy. In practice, absolute zero is an ideal temperature that is unobtainable, and a perfect single crystal is also an ideal that cannot be achieved. Nonetheless, the combination of these two ideals constitutes the basis for the **third law of thermodynamics**: *The entropy of any perfectly ordered, crystalline substance at absolute zero is zero.*

The third law of thermodynamics has two important consequences: it defines the *sign* of the entropy of any substance at temperatures above absolute zero as positive, and it provides a fixed reference point that allows us to measure the *absolute entropy* of any substance at any temperature.* In contrast, other thermodynamic properties, such as internal energy and enthalpy, can be evaluated in only *relative* terms, not absolute terms. In this section, we examine two different ways to calculate ΔS for a reaction or physical change. The first, based on the definition of absolute entropy provided by the

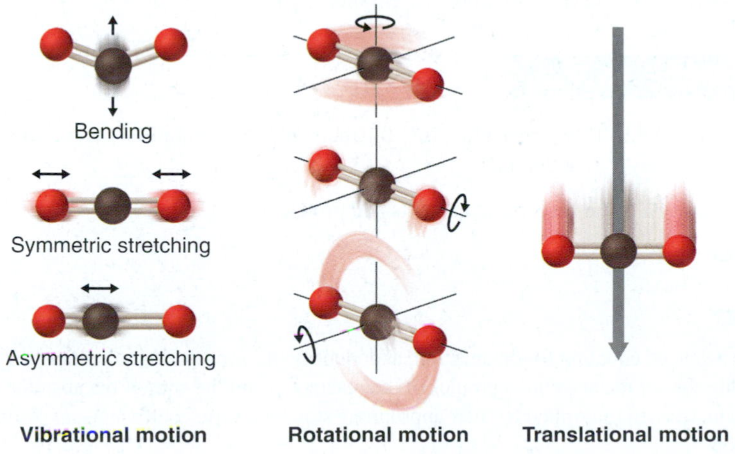

Figure 18.13 Molecular motions. Vibrational, rotational, and translational motions of a carbon dioxide molecule are illustrated here. The more molecular motions the atoms, molecules, or ions of a chemical system undergo, the higher the entropy of the system. Only a perfectly ordered, crystalline substance at absolute zero would exhibit no molecular motion and would thus have zero entropy. In practice, this is an unattainable ideal.

* In practice, chemists determine the value of the absolute entropy of a substance by measuring the molar heat capacity C_p as a function of temperature, and then plotting the quantity (C_p/T) versus T. The *area* under the curve between 0 K and any temperature T is the absolute entropy of the substance at temperature T.

third law, uses tabulated values of absolute entropies of substances. The second, based on the fact that entropy is a state function, uses a thermodynamic cycle similar to those we first encountered in Chapter 5.

TABLE 18.1 Standard molar entropy, $S°$, values of selected substances at 25°C

Substance	$S°$, J/(mol · K)
Gases	
He(g)	126.2
$H_2(g)$	130.7
Ne(g)	146.3
Ar(g)	154.8
Kr(g)	164.1
Xe(g)	169.7
$H_2O(g)$	188.8
$N_2(g)$	191.6
$O_2(g)$	205.2
$CO_2(g)$	213.8
$I_2(g)$	260.7
Liquids	
$H_2O(l)$	70.0
$CH_3OH(l)$	126.8
$Br_2(l)$	152.2
$CH_3CH_2OH(l)$	160.7
$C_6H_6(l)$	173.4
$CH_3COCl(l)$	200.8
$C_6H_{12}(l)$ (cyclohexane)	204.4
$C_8H_{18}(l)$ (isooctane)	329.3
Solids	
C(s) (diamond)	2.4
C(s) (graphite)	5.7
LiF(s)	35.7
$SiO_2(s)$ (quartz)	41.5
Ca(s)	41.6
Na(s)	51.3
$MgF_2(s)$	57.2
K(s)	64.7
NaCl(s)	72.1
KCl(s)	82.6
$I_2(s)$	116.1

MGC Entropy

Calculating ΔS from Standard Molar Entropy Values

One way of calculating ΔS for a reaction is to use tabulated values of the **standard molar entropy ($S°$)**, which is the entropy of 1 mol of a substance at a standard temperature of 298 K; the units of $S°$ are J/(mol · K). Unlike enthalpy or internal energy, it is possible to obtain absolute entropy values by measuring the entropy change that occurs between the reference point of 0 K, corresponding to $S = 0$, and 298 K.

As shown in Table 18.1, for substances with approximately the same molar mass and number of atoms, $S°$ values fall in the order $S°(\text{gas}) > S°(\text{liquid}) > S°(\text{solid})$. For instance, $S°$ for liquid water is 70.0 J/(mol · K), whereas $S°$ for water vapor is 188.8 J/(mol · K). Likewise, $S°$ is 260.7 J/(mol · K) for gaseous I_2 and 116.1 J/(mol · K) for solid I_2. This order makes qualitative sense based on the kinds and extents of motion available to atoms and molecules in the three phases. The correlation between physical state and absolute entropy is illustrated in Figure 18.14, a generalized plot of the entropy of a substance versus temperature.

Closer examination of Table 18.1 also reveals that substances with similar molecular structures tend to have similar $S°$ values. Among crystalline materials, those with the lowest entropies tend to be rigid crystals composed of small atoms linked by strong, highly directional bonds, like diamond [$S° = 2.4$ J/(mol · K)]. In contrast, graphite, the softer, less rigid allotrope of carbon, has a higher $S°$ [5.7 J/(mol · K)] due to more disorder in the crystal. Soft crystalline substances and those that contain larger atoms tend to have higher entropies because of increased molecular motion and disorder. Similarly, the absolute entropy of a substance tends to increase with increasing molecular complexity because the number of available microstates increases with molecular complexity. For example, compare the $S°$ values for $CH_3OH(l)$ [126.8 J/(mol · K)] and $CH_3CH_2OH(l)$ [160.7 J/ (mol · K)]. Finally, substances with strong hydrogen bonds have lower values of $S°$, reflecting a more ordered structure.

To calculate $\Delta S°$ for a chemical reaction from standard molar entropies, we use the familiar "products minus reactants" rule, in which the absolute entropy of each reactant and product is multiplied by its stoichiometric coefficient in the balanced chemical equation. The next example illustrates this procedure for the combustion of the liquid hydrocarbon isooctane (2,2,4-trimethylpentane, C_8H_{18}).

EXAMPLE 18.7

Use the data in Table 18.1 to calculate $\Delta S°$ for the reaction of liquid isooctane with $O_2(g)$ to give $CO_2(g)$ and $H_2O(g)$ at 298 K.

Given Standard molar entropies, reactants and products

Asked for $\Delta S°$

Strategy

Write the balanced equation for the reaction, and identify the appropriate quantities in Table 18.1. Subtract the sum of the absolute entropies of the reactants from the sum of the absolute entropies of the products, each multiplied by their appropriate stoichiometric coefficients, to obtain $\Delta S°$ for the reaction.

Solution

The balanced equation for the complete combustion of isooctane (C_8H_{18}) is

$$C_8H_{18}(l) + \frac{25}{2}O_2(g) \rightarrow 8CO_2(g) + 9H_2O(g)$$

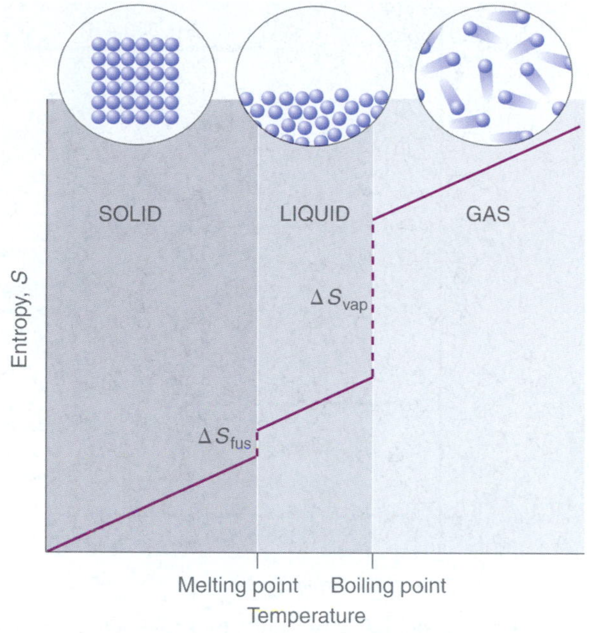

Figure 18.14 A generalized plot of entropy versus temperature for a single substance. According to the third law of thermodynamics, the entropy at 0 K is zero. The absolute entropy increases steadily with increasing temperature until the melting point is reached. At this point, the entropy jumps suddenly as the substance undergoes a phase change from a highly ordered solid to a disordered liquid (ΔS_{fus}). After this, the entropy again increases steadily with increasing temperature until the boiling point is reached and another phase change occurs. The entropy jumps suddenly as the liquid is converted to a highly disordered gas (ΔS_{vap}) and then increases again with increasing temperature. The increase in entropy with increasing temperature is approximately proportional to the heat capacity of the substance. Interactive Graph

We calculate $\Delta S°$ for the reaction using the "products minus reactants" rule, where m and n are the stoichiometric coefficients of each product and each reactant:

$$\Delta S°_{rxn} = \sum mS°(\text{products}) - \sum nS°(\text{reactants})$$
$$= [8S°(CO_2) + 9S°(H_2O)] - \left[S°(C_8H_{18}) + \frac{25}{2}S°(O_2)\right]$$
$$= \{[8 \text{ mol } CO_2 \times 213.8 \text{ J/(mol·K)}] + [9 \text{ mol } H_2O \times 188.8 \text{ J/(mol·K)}]\}$$
$$- \left\{[1 \text{ mol } C_8H_{18} \times 329.3 \text{ J/(mol·K)}] + \left[\frac{25}{2} \text{ mol } O_2 \times 205.2 \text{ J/(mol·K)}\right]\right\}$$
$$= 515.3 \text{ J/K}$$

Note that $\Delta S°$ is positive, as expected for a combustion reaction in which one large hydrocarbon molecule is converted to many molecules of gaseous products.

Note the pattern

Entropy increases with softer, less rigid solids, solids that contain larger atoms, and solids with complex molecular structures.

EXERCISE 18.7

Use the data in Table 18.1 to calculate $\Delta S°$ for the reaction of $H_2(g)$ with liquid benzene (C_6H_6) to give cyclohexane (C_6H_{12}).

Answer −361.1 J/K

Note the pattern

$\Delta S°$ for a reaction can be calculated from absolute entropy values using the same "products minus reactants" rule used to calculate $\Delta H°$.

Calculating ΔS from Thermodynamic Cycles

We can also calculate a change in entropy using a thermodynamic cycle. As you learned in Chapter 5, the molar heat capacity C_p is the amount of heat needed to raise the temperature of 1 mol of a substance by 1°C at constant pressure. Similarly, C_v is the amount of heat needed to raise the temperature of 1 mol of a substance by 1°C at constant volume. The increase in entropy with increasing temperature in Figure 18.14 is approximately proportional to the heat capacity of the substance.

Recall that the entropy change ΔS is related to the heat flow q_{rev} by $\Delta S = q_{rev}/T$. Because $q_{rev} = nC_p\,\Delta T$ at constant pressure or $nC_v\,\Delta T$ at constant volume, where n is

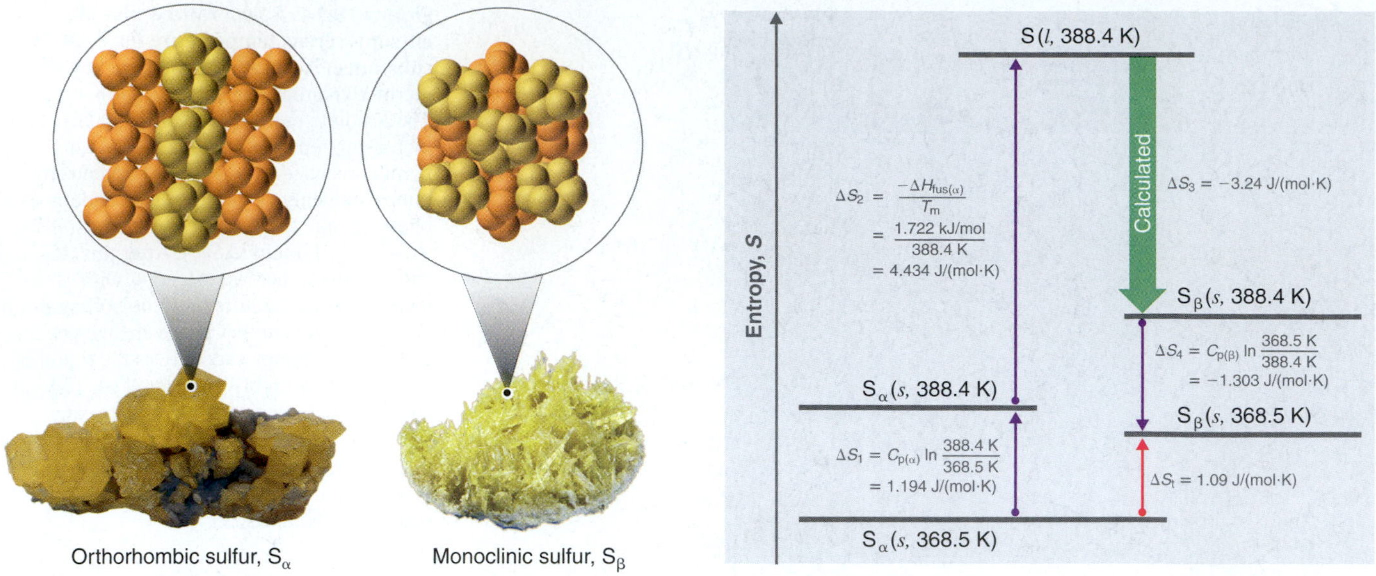

(a) Allotropes of elemental sulfur

(b) Thermodynamic cycle

Figure 18.15 Two forms of elemental sulfur and a thermodynamic cycle showing the transition from one to the other. **(a)** Orthorhombic sulfur, S_α, has a highly ordered structure in which the S_8 rings are stacked in a "crankshaft" arrangement. Monoclinic sulfur, S_β, is also composed of S_8 rings but has a less-ordered structure. **(b)** At 368.5 K, S_α undergoes a phase transition to S_β. Although the change in entropy for the conversion of liquid sulfur to S_β, ΔS_3, cannot be measured directly, it can be calculated using the values shown in this thermodynamic cycle because $\Delta S_t = \Delta S_1 + \Delta S_2 + \Delta S_3 + \Delta S_4$, where the subscripts 1–4 refer to steps in the cycle.

the number of moles of substance present, the change in entropy for a substance whose temperature changes from T_1 to T_2 is

$$\Delta S = \frac{q_{rev}}{T} = nC_p\frac{\Delta T}{T} \quad \text{(constant pressure)}$$

As you will discover in more advanced math courses than is required here, it can be shown that this is equal to*

$$\Delta S = nC_p \ln \frac{T_2}{T_1} \quad \text{(constant pressure)} \qquad (18.17)$$

Similarly,

$$\Delta S = nC_v \ln \frac{T_2}{T_1} \quad \text{(constant volume)} \qquad (18.18)$$

Thus, we can use a combination of heat capacity measurements (Equations 18.17 or 18.18) and experimentally measured values of enthalpies of fusion or vaporization if a phase change is involved (Equation 18.15) to calculate the entropy change corresponding to a change in the temperature of a sample.

We can use a thermodynamic cycle to calculate the entropy change when the phase change for a substance such as sulfur cannot be measured directly. As noted in Exercise 18.6, elemental sulfur exists in two forms (Figure 18.15a): an orthorhombic form with a highly ordered structure (S_α) and a less-ordered monoclinic form (S_β). The orthorhombic (α) form is more stable at room temperature but undergoes a phase transition to the monoclinic (β) form at temperatures higher than 95.3°C (368.5 K). The transition from S_α to S_β can be described by the thermodynamic cycle shown in Figure 18.15b,

* For a review of natural logarithms, see Essential Skills 6.

in which liquid sulfur is an intermediate. The change in entropy that accompanies the conversion of liquid sulfur to S_β ($-\Delta S_{fus(\beta)} = \Delta S_3$ in the cycle) cannot be measured directly. Because entropy is a state function, however, ΔS_3 can be calculated from the overall entropy change, ΔS_t, for the S_α–S_β transition, which equals the sum of the ΔS values for the steps in the thermodynamic cycle, using Equation 18.17 and tabulated thermodynamic parameters (the heat capacities of S_α and S_β, $\Delta H_{fus(\alpha)}$, and the melting point of S_α.

If we know the melting point of S_α (T_m = 115.2°C = 388.4 K) and ΔS_t for the overall phase transition [(calculated to be 1.09 J/(mol·K) in Exercise 18.6], we can calculate ΔS_3 from the values given in Figure 18.15b where $C_{p(\alpha)}$ = 22.70 J/mol·K and $C_{p(\beta)}$ = 24.77 J/mol·K (subscripts on ΔS refer to steps in the cycle):

$$\Delta S_t = \Delta S_1 + \Delta S_2 + \Delta S_3 + \Delta S_4$$

$$1.09 \text{ J/(mol·K)} = C_{p(\alpha)}\ln\left(\frac{T_2}{T_1}\right) + \frac{\Delta H_{fus}}{T_m} + \Delta S_3 + C_{p(\beta)}\ln\left(\frac{T_4}{T_3}\right)$$

$$= 22.70 \text{ J/(mol·K)} \ln\left(\frac{388.4}{368.5}\right) + \left(\frac{1.722 \text{ kJ/mol}}{388.4 \text{ K}}\right) + \Delta S_3 + 24.77 \text{ J/(mol·K)} \ln\left(\frac{368.5}{388.4}\right)$$

$$= [1.194 \text{ J/(mol·K)}] + [4.434 \text{ J/(mol·K)}] + \Delta S_3 + [-1.303 \text{ J/(mol·K)}]$$

Solving for ΔS_3 gives a value of −3.24 J/(mol·K). As expected for the conversion of a less ordered state (a liquid) to a more ordered one (a crystal), ΔS_3 is negative.

MGC Gibbs Free Energy: Spontaneity

18.5 • Free Energy

One of the major goals of chemical thermodynamics is to establish criteria for predicting whether a particular reaction or process will occur spontaneously. We have developed one such criterion, the change in entropy of the universe: if $\Delta S_{univ} > 0$ for a process or reaction, then the process will occur spontaneously as written. Conversely, if $\Delta S_{univ} < 0$, a process cannot occur spontaneously, whereas if $\Delta S_{univ} = 0$, the system is at equilibrium. The sign of ΔS_{univ} is a universally applicable and infallible indicator of the spontaneity of a reaction. Unfortunately, using ΔS_{univ} requires that we calculate ΔS for *both* the system *and* the surroundings. This is not particularly useful for two reasons: we are normally much more interested in the system than in the surroundings, and it is difficult to make quantitative measurements of the surroundings (that is, the rest of the universe). A criterion of spontaneity that is based solely on state functions of the *system* would be much more convenient and is provided by a new state function, the Gibbs free energy.

J. Willard Gibbs (1839–1903). Born in Connecticut, Josiah Willard Gibbs attended Yale, as did his father, a professor of sacred literature at Yale, who was involved in the *Amistad* trial. In 1863 Gibbs was awarded the first engineering doctorate granted in the United States. He was appointed Professor of Mathematical Physics at Yale in 1871, the first such professorship in the United States. His series of papers entitled "On the Equilibrium of Heterogeneous Substances" was the foundation of the field of physical chemistry and is considered one of the great achievements of the 19th century. Gibbs, whose work was translated into French by le Châtelier, lived with his sister and brother-in-law until his death in 1903, shortly before the inauguration of the Nobel Prizes.

Gibbs Free Energy and the Direction of Spontaneous Reactions

The **Gibbs free energy (*G*)**, often called simply *free energy*, was named in honor of J. Willard Gibbs (1838–1903), an American physicist who first developed the concept. It is defined in terms of three other state functions with which you are already familiar: enthalpy, temperature, and entropy:

$$G = H - TS \qquad (18.19)$$

Because it is a combination of state functions, G is also a state function.

The criterion for predicting spontaneity is based on ΔG, the change in G, at constant temperature and pressure. Although very few chemical reactions actually occur under conditions of constant temperature and pressure, most systems can be brought back to the initial temperature and pressure without significantly affecting the value of thermodynamic state functions such as G. At constant temperature and pressure,

$$\Delta G = \Delta H - T\Delta S \qquad (18.20)$$

where all thermodynamic quantities are those of the *system*. Recall that at constant pressure, $\Delta H = q$ whether a process is reversible or irreversible and $T\Delta S = q_{rev}$. Using these expressions, we can reduce Equation 18.20 to $\Delta G = q - q_{rev}$. Thus, ΔG is the difference between the heat released during a process (via a reversible or an irreversible path) and the heat released for the same process occurring in a reversible manner. Under the special condition in which a process occurs reversibly, $q = q_{rev}$ and hence $\Delta G = 0$. As we shall soon see, if ΔG is zero, the system is at equilibrium and there will be no net change.

What about processes for which $\Delta G \neq 0$? To understand how the sign of ΔG for the *system* determines the direction in which change is spontaneous, we can rewrite Equation 18.12 (where $q_p = \Delta H$, Equation 18.8) to give

$$\Delta S_{surr} = -\frac{\Delta H_{sys}}{T} \tag{18.21}$$

Thus, the entropy change of the *surroundings* is related to the enthalpy change of the *system*. We have stated that for a spontaneous reaction, $\Delta S_{univ} > 0$, so substituting we obtain

$$\begin{aligned}\Delta S_{univ} &= \Delta S_{sys} + \Delta S_{surr} > 0 \\ &= \Delta S_{sys} - \frac{\Delta H_{sys}}{T} > 0\end{aligned}$$

Multiplying both sides of the inequality by $-T$ reverses its sign, and rearranging gives

$$\Delta H_{sys} - T\Delta S_{sys} < 0$$

which is equal to ΔG (Equation 18.20). We can therefore see that for a spontaneous process, $\Delta G < 0$.

The relationship between the entropy change of the *surroundings* and the heat gained or lost by the *system* provides the key connection between the thermodynamic properties of the system and the change in entropy of the universe. The relationship shown in Equation 18.20 allows us to predict spontaneity by focusing exclusively on the thermodynamic properties and temperature of the system. We predict that highly exothermic processes ($\Delta H << 0$) that increase the disorder of the *system* ($\Delta S_{sys} >> 0$) would therefore occur spontaneously. An example of such a process is the decomposition of ammonium nitrate fertilizer, which destroyed Texas City, Texas, in 1947; ammonium nitrate was also used to destroy the Murrah Federal Building in Oklahoma City, Oklahoma, in 1995.

For a system at constant temperature and pressure, we can summarize these results:

If $\Delta G < 0$, the process occurs spontaneously.

If $\Delta G = 0$, the system is at equilibrium.

If $\Delta G > 0$, the process is not spontaneous as written but occurs spontaneously in the reverse direction.

To further understand how the various components of ΔG dictate whether a process occurs spontaneously, let's look at a simple and familiar physical change: the conversion of liquid water to water vapor. If this process is carried out at 1 atm and the normal boiling point of 100.00°C (373.15 K), we can calculate ΔG from the experimentally measured value of ΔH_{vap} (40.657 kJ/mol). For vaporizing 1 mol of water, $\Delta H = 40{,}657$ J, so the process is highly endothermic. From the definition of ΔS (Equation 18.12), we know that for 1 mol of water,

$$\Delta S_{vap} = \frac{\Delta H_{vap}}{T_b} = \frac{40{,}657 \text{ J}}{373.15 \text{ K}} = 108.96 \text{ J/K}$$

Hence, there is an increase in the disorder of the system. At the normal boiling point of water,

$$\Delta G_{100°C} = \Delta H_{100°C} - T\Delta S_{100°C} = 40{,}657\ \text{J} - [(373.15\ \cancel{\text{K}})(108.96\ \text{J}/\cancel{\text{K}})] = 0\ \text{J}$$

The energy required for vaporization offsets the increase in disorder of the system. Thus, $\Delta G = 0$ and the liquid and vapor are in equilibrium, as is true of any liquid at its normal boiling point (see Chapter 11).

Now suppose we were to superheat 1 mol of liquid water to a temperature of 110°C. The value of ΔG for the vaporization of 1 mol of water at 110°C, assuming that ΔH and ΔS do not change significantly with temperature, becomes

$$\Delta G_{110°C} = \Delta H - T\Delta S = 40{,}657\ \text{J} - [(383.15\ \cancel{\text{K}})(108.96\ \text{J}/\cancel{\text{K}})] = -1091\ \text{J}$$

Thus, at a temperature of 110°C, $\Delta G < 0$ and vaporization is predicted to occur spontaneously and irreversibly.

We can also calculate ΔG for the vaporization of 1 mol of water at a temperature below its normal boiling point—for example, 90°C—making the same assumptions:

$$\Delta G_{90°C} = \Delta H - T\Delta S = 40{,}657\ \text{J} - [(363.15\ \cancel{\text{K}})(108.96\ \text{J}/\cancel{\text{K}})] = 1088\ \text{J}$$

At a temperature of 90°C, $\Delta G > 0$ and water does not spontaneously convert to water vapor. Notice that when using all digits in the calculator display in carrying out our calculations (see Essential Skills 1), $\Delta G_{110°C} = 1090\ \text{J} = -\Delta G_{90°C}$, as we would predict.

We can also calculate the temperature at which liquid water is in equilibrium with water vapor. Inserting the values of ΔH and ΔS into the definition of ΔG (Equation 18.20), setting $\Delta G = 0$, and solving for T give

$$0\ \text{J} = 40{,}657\ \text{J} - T(108.96\ \text{J/K})$$
$$T = 373.15\ \text{K}$$

Note the pattern

$\Delta G = 0$ only if $\Delta H = T\Delta S$.

Thus, $\Delta G = 0$ at $T = 373.15$ K and 1 atm, indicating that liquid water and water vapor are in equilibrium; this temperature is therefore the *normal boiling point* of water. At temperatures higher than 373.15 K, ΔG is negative, and water evaporates spontaneously and irreversibly. Below 373.15 K, ΔG is positive, and water does not evaporate spontaneously. Instead, water vapor at a temperature lower than 373.15 K and 1 atm will spontaneously and irreversibly condense to liquid water. Figure 18.16 shows how the ΔH and $T\Delta S$ terms vary with temperature for the vaporization of water. Note that $\Delta G = 0$ when the two lines cross, at which point $\Delta H = T\Delta S$.

A similar situation arises in the conversion of liquid egg white to a solid when an egg is boiled. The major component of egg white is a protein called albumin, which is held in a compact, ordered structure by a large number of hydrogen bonds. Breaking them requires an input of energy ($\Delta H > 0$), which converts the albumin to a highly disordered structure in which the molecules aggregate as a disorganized solid ($\Delta S > 0$). At temperatures higher than 373 K, the $T\Delta S$ term dominates and $\Delta G < 0$, so the conversion of a raw egg to a hard-boiled egg is an irreversible and spontaneous process above 373 K.

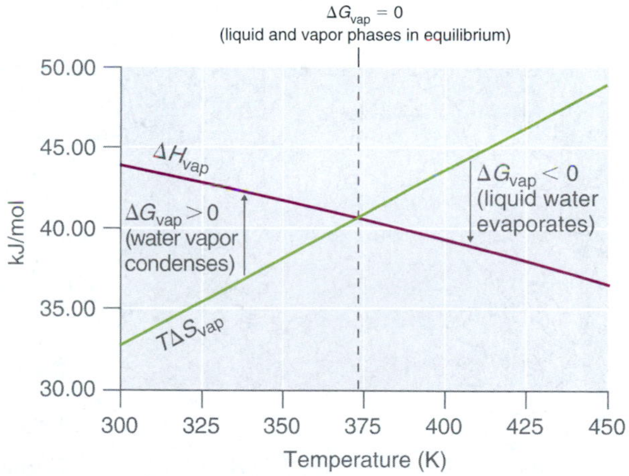

Figure 18.16 Temperature dependence of ΔH and $T\Delta S$ for the vaporization of water. Both ΔH and $T\Delta S$ are temperature dependent, but the lines have opposite slopes and cross at 373.15 K at 1 atm, where $\Delta H = T\Delta S$. Because $\Delta G = \Delta H - T\Delta S$, at this temperature $\Delta G = 0$, indicating that the liquid and vapor phases are in equilibrium. The normal boiling point of water is therefore 373.15 K. Above the normal boiling point, the $T\Delta S$ term is greater than ΔH, making $\Delta G < 0$; hence, liquid water evaporates spontaneously. Below the normal boiling point, the ΔH term is greater than $T\Delta S$, making $\Delta G > 0$. Thus, liquid water does not evaporate spontaneously, but water vapor spontaneously condenses to liquid.

The Relationship Between ΔG and Work

As we've just seen, the value of ΔG allows us to predict the spontaneity of a physical or chemical change. In addition, the magnitude of ΔG for a process provides other important information. The change in free energy, ΔG, is equal to the maximum amount of work that a system can perform on the surroundings while undergoing a spontaneous change (at constant temperature and pressure): $\Delta G = w_{max}$. To see why this is true, let's look again at the relationships among free energy, enthalpy, and entropy expressed in Equation 18.20. We can rearrange this equation to give

$$\Delta H = \Delta G + T\Delta S \tag{18.22}$$

This equation tells us that when energy is released during an exothermic process ($\Delta H < 0$), such as during the combustion of a fuel, for example, some of that energy can be used to do work ($\Delta G < 0$), while some is used to increase the entropy of the universe ($T\Delta S > 0$). Only if the process occurs infinitely slowly in a perfectly reversible manner will the entropy of the universe be unchanged (Section 18.4). Because no real system is perfectly reversible, the entropy of the universe increases during all processes that produce energy. As a result, no process that utilizes stored energy can ever be 100% efficient; that is, ΔH will never equal ΔG because ΔS has a positive value.

One of the major challenges facing engineers is to maximize the efficiency of converting stored energy to useful work or of converting one form of energy to another. As indicated in Table 18.2, the efficiencies of various energy-converting devices vary widely. For example, an internal combustion engine typically uses only 25–30% of the energy stored in the hydrocarbon fuel to perform work; the rest of the stored energy is released in an unusable form as heat. In contrast, gas–electric hybrid engines, now used in several models of automobiles, deliver approximately 50% greater fuel efficiency. A large electrical generator is highly efficient (approximately 99%) in converting mechanical to electrical energy, but a typical incandescent light bulb is one of the least efficient devices known (only approximately 5% of the electrical energy is converted to light). In contrast, a mammalian liver cell is a relatively efficient machine, able to utilize fuels such as glucose with an efficiency of 30–50%.

Standard Free-Energy Change

We have seen that there is no way to measure absolute enthalpies, although we can measure *changes* in enthalpy, ΔH, during a chemical reaction. Because enthalpy is one of the components of Gibbs free energy, we are consequently unable to measure absolute free energies either, only changes in free energy. The **standard free-energy change, $\Delta G°$**, is the change in free energy when one substance or set of substances in their standard states is converted to one or more other substances, also in their standard states. The standard free-energy change can be calculated from the definition of free energy, if the standard enthalpy and entropy changes are known, using

$$\Delta G° = \Delta H° - T\Delta S° \qquad (18.23)$$

TABLE 18.2 Approximate thermodynamic efficiencies of various devices

Device	Energy Conversion	Approximate Efficiency, %
Large electrical generator	Mechanical ⟶ electrical	99
Chemical battery	Chemical ⟶ electrical	90
Home furnace	Chemical ⟶ heat	65
Small electric tool	Electrical ⟶ mechanical	60
Space shuttle engine	Chemical ⟶ mechanical	50
Mammalian liver cell	Chemical ⟶ chemical	30–50
Spinach cell	Light ⟶ chemical	30
Internal combustion engine	Chemical ⟶ mechanical	25–30
Fluorescent light	Electrical ⟶ light	20
Solar cell	Light ⟶ electricity	10
Incandescent light bulb	Electricity ⟶ light	5
Yeast cell	Chemical ⟶ chemical	2–4

Thus, according to Equation 18.23, if $\Delta S°$ and $\Delta H°$ for a reaction have the same sign, then the sign of $\Delta G°$ depends on the relative magnitudes of the $\Delta H°$ and $T\Delta S°$ terms. It is important to recognize that a positive value of $\Delta G°$ for a reaction does *not* mean that no products will form if the reactants in their standard states are mixed, only that at equilibrium the concentrations of the products will be lower than the concentrations of the reactants.

Note the pattern

A positive $\Delta G°$ means that at equilibrium the concentrations of products will be lower than the concentrations of reactants.

EXAMPLE 18.8

Calculate the standard free-energy change, $\Delta G°$, at 25°C for the reaction $H_2(g) + O_2(g) \rightleftharpoons H_2O_2(l)$. At 25°C the standard enthalpy change, $\Delta H°$, is −187.78 kJ/mol, and the absolute entropies of the products and reactants are $S°(H_2O_2)$ = 109.6 J/(mol·K), $S°(O_2)$ = 205.2 J/(mol·K), and $S°(H_2)$ = 130.7 J/(mol·K). Is the reaction spontaneous as written?

Given Balanced chemical equation, $\Delta H°$, $S°$ for reactants and products

Asked for Spontaneity of reaction as written

Strategy

A Calculate $\Delta S°$ from the absolute molar entropy values given.
B Use Equation 18.23, the calculated value of $\Delta S°$, and other data given to calculate $\Delta G°$ for the reaction. Use the value of $\Delta G°$ to determine whether the reaction is spontaneous as written.

Solution

A To calculate $\Delta G°$ for the reaction, we need to know $\Delta H°$, $\Delta S°$, and T. We are given $\Delta H°$, and we know that T = 298.15 K. We can calculate $\Delta S°$ from the absolute molar entropy values provided using the "products minus reactants" rule:

$$\begin{aligned}\Delta S° &= S°(H_2O_2) - [S°(O_2) + S°(H_2)] \\ &= [1\ \cancel{\text{mol}}\ H_2O_2 \times 109.6\ \text{J/}(\cancel{\text{mol}}\cdot\text{K})] - \{[1\ \cancel{\text{mol}}\ H_2 \times 130.7\ \text{J/}(\cancel{\text{mol}}\cdot\text{K})] + [(1\ \cancel{\text{mol}}\ O_2 \times 205.2\ \text{J/}(\cancel{\text{mol}}\cdot\text{K})]\} \\ &= -226.3\ \text{J/K (per mole of } H_2O_2)\end{aligned}$$

Note that $\Delta S°$ is very negative for this reaction, as we might expect for a reaction in which 2 mol of gas is converted to 1 mol of a much more ordered liquid. **B** Substituting the appropriate quantities into Equation 18.23 gives

$$\begin{aligned}\Delta G° = \Delta H° - T\Delta S° &= -187.78\ \text{kJ/mol} - (298.15\ \cancel{\text{K}})[-226.3\ \text{J/}(\text{mol}\cdot\cancel{\text{K}})] \\ &= -187.78\ \text{kJ/mol} + 67.47\ \text{kJ/mol} = -120.31\ \text{kJ/mol}\end{aligned}$$

The negative value of $\Delta G°$ indicates that the reaction is spontaneous as written. Because $\Delta S°$ and $\Delta H°$ for this reaction have the same sign, the sign of $\Delta G°$ depends on the relative magnitudes of the $\Delta H°$ and $T\Delta S°$ terms. Thus, in this particular case, the enthalpy term dominates, indicating that the strength of the bonds formed in the product more than compensates for the unfavorable $\Delta S°$ term and for the energy needed to break bonds in the reactants.

EXERCISE 18.8

Calculate the standard free-energy change, $\Delta G°$, at 25°C for the reaction $2H_2(g) + N_2(g) \rightleftharpoons N_2H_4(l)$. At 25°C the standard enthalpy change, $\Delta H°$, is 50.6 kJ/mol, and the absolute entropies of the products and reactants are $S°(N_2H_4)$ = 121.2 J/(mol·K), $S°(N_2)$ = 191.6 J/(mol·K), and $S°(H_2)$ = 130.7 J/ (mol·K). Is the reaction spontaneous as written?

Answer 149.5 kJ/mol; no

Tabulated values of standard free energies of formation allow chemists to calculate the values of ΔG° for a wide variety of chemical reactions rather than having to measure them in the laboratory. The **standard free energy of formation (ΔG°_f)** of a compound is the change in free energy that occurs when 1 mol of a substance in its standard state is formed from the elements in their standard states. By definition, the standard free energy of formation of an element in its standard state is zero at 298.15 K. One mole of Cl_2 gas at 298.15 K, for example, has $\Delta G^\circ_f = 0$. The standard free energy of formation of a compound can be calculated from the standard enthalpy of formation, ΔH°_f, and the standard entropy of formation, ΔS°_f, using the definition of free energy:

$$\Delta G^\circ_f = \Delta H^\circ_f - T\Delta S^\circ_f \tag{18.24}$$

Using standard free energies of formation to calculate the standard free energy of a reaction is analogous to calculating standard enthalpy changes from standard enthalpies of formation using the familiar "products minus reactants" rule:

Note the pattern

The ΔG° of a reaction can be calculated from tabulated ΔG°_f values using the "products minus reactants" rule.

$$\Delta G^\circ_{rxn} = \Sigma m\Delta G^\circ_f(\text{products}) - \Sigma n\Delta G^\circ_f(\text{reactants}) \tag{18.25}$$

where m and n are the stoichiometric coefficients of each product and reactant in the balanced chemical equation. A very large negative ΔG° indicates a strong tendency for products to form spontaneously from reactants; *it does not, however, necessarily indicate that the reaction will occur rapidly*. To make this determination, we need to evaluate the kinetics of the reaction (see Chapter 14).

EXAMPLE 18.9

Calculate ΔG° for the reaction of isooctane with oxygen gas to give carbon dioxide and water (described in Example 18.7). Use the following data: ΔG°_f (isooctane) = −353.2 kJ/mol, $\Delta G^\circ_f(CO_2)$ = −394.4 kJ/mol, and $\Delta G^\circ_f(H_2O)$ = −237.1 kJ/mol. Is the reaction spontaneous as written?

Given Balanced chemical equation, values of ΔG°_f for isooctane, CO_2, and H_2O

Asked for Spontaneity of reaction as written

Strategy

Use the "products minus reactants" rule to obtain ΔG°_{rxn}, remembering that ΔG°_f for an element in its standard state is zero. From the calculated value, determine whether the reaction is spontaneous as written.

Solution

From Example 18.7, we know that the balanced chemical equation for the reaction is $C_8H_{18}(l) + \frac{25}{2}O_2(g) \rightarrow 8CO_2(g) + 9H_2O(l)$. We are given ΔG°_f values for all the products and reactants except $O_2(g)$. Because oxygen gas is an element in its standard state, $\Delta G^\circ_f(O_2)$ is zero. Using the "products minus reactants" rule, we get

$$\begin{aligned}\Delta G^\circ &= [8\Delta G^\circ_f(CO_2) + 9\Delta G^\circ_f(H_2O)] - \left[1\Delta G^\circ_f(C_8H_{18}) + \frac{25}{2}\Delta G^\circ_f(O_2)\right] \\ &= [(8\ \cancel{\text{mol}})(-394.4\ \text{kJ}/\cancel{\text{mol}}) + (9\ \cancel{\text{mol}})(-237.1\ \text{kJ}/\cancel{\text{mol}})] \\ &\quad - [(1\ \cancel{\text{mol}})(-353.2\ \text{kJ}/\cancel{\text{mol}}) + \left(\frac{25}{2}\ \cancel{\text{mol}}\right)(0\ \text{kJ}/\cancel{\text{mol}})] \\ &= -4935.9\ \text{kJ (per mol of } C_8H_{18})\end{aligned}$$

Because $\Delta G°$ is a large negative number, there is a strong tendency for the spontaneous formation of products from reactants (though not necessarily at a rapid rate). Also notice that the magnitude of $\Delta G°$ is largely determined by the $\Delta G_f°$ of the stable products, water and carbon dioxide.

EXERCISE 18.9

Calculate $\Delta G°$ for the reaction of benzene with hydrogen gas to give cyclohexane using the data $\Delta G_f°$ (benzene) = 124.5 kJ/mol and $\Delta G_f°$ (cyclohexane) = 217.3 kJ/mol. Is the reaction spontaneous as written?

Answer 92.8 kJ; no

Calculated values of $\Delta G°$ are extremely useful in predicting whether a reaction will occur spontaneously if the reactants and products are mixed under standard conditions. We should note, however, that very few reactions are actually carried out under standard conditions, and calculated values of $\Delta G°$ may not tell us whether a given reaction will occur spontaneously under nonstandard conditions. What determines whether a reaction will occur spontaneously is the free-energy change ΔG under the actual experimental conditions, which are usually different from $\Delta G°$. If the ΔH and $T\Delta S$ terms for a reaction have the same sign, for example, then it may be possible to reverse the sign of ΔG by changing the temperature, thereby converting a reaction that is not thermodynamically spontaneous to one that is, or vice versa. Because ΔH and ΔS usually do not vary greatly with temperature in the absence of a phase change, we can use tabulated values of $\Delta H°$ and $\Delta S°$ to calculate $\Delta G°$ at various temperatures, as long as no phase change occurs over the temperature range being considered.

EXAMPLE 18.10

Calculate **(a)** $\Delta G°$ and **(b)** $\Delta G_{300°C}$ for the reaction $N_2(g) + 3H_2(g) \rightleftharpoons 2NH_3(g)$, assuming that ΔH and ΔS do not change between 25°C and 300°C. Use these data: $S°(N_2)$ = 191.6 J/(mol·K), $S°(H_2)$ = 130.7 J/(mol·K), $S°(NH_3)$ = 192.8 J/(mol·K), and $\Delta H_f°$ (NH_3) = −45.9 kJ/mol.

Given Balanced chemical equation, temperatures, $S°$ values, $\Delta H_f°$ for NH_3

Asked for $\Delta G°$ and ΔG at 300°C

Strategy

Ⓐ Convert each temperature to Kelvins. Then calculate $\Delta S°$ for the reaction. Calculate $\Delta H°$ for the reaction, recalling that $\Delta H_f°$ for any element in its standard state is zero.

Ⓑ Substitute the appropriate values into Equation 18.23 to obtain $\Delta G°$ for the reaction.

Ⓒ Assuming that ΔH and ΔS are independent of temperature, substitute values into Equation 18.20 to obtain ΔG for the reaction at 300°C.

Solution

(a) Ⓐ To calculate $\Delta G°$ for the reaction using Equation 18.23, we must know the temperature as well as the values of $\Delta S°$ and $\Delta H°$. At standard conditions, the temperature is 25°C, or 298.15 K. We can calculate $\Delta S°$ for the reaction from the absolute molar entropy values given for the reactants and products using the "products minus reactants" rule:

$$\begin{aligned}\Delta S^\circ_{rxn} &= 2S^\circ(NH_3) - [S^\circ(N_2) + 3S^\circ(H_2)]\\ &= [2\ \cancel{mol}\ NH_3 \times 192.8\ J/(\cancel{mol}\cdot K)]\\ &\quad - \{[1\ \cancel{mol}\ N_2 \times 191.6\ J/(\cancel{mol}\cdot K)] + [3\ \cancel{mol}\ H_2 \times 130.7\ J/(\cancel{mol}\cdot K)]\}\\ &= -198.1\ J/K\ (\text{per mole of } N_2)\end{aligned}$$

We can also calculate ΔH° for the reaction using the "products minus reactants" rule. The value of $\Delta H^\circ_f(NH_3)$ is given and ΔH°_f is zero for both N_2 and H_2:

$$\begin{aligned}\Delta H^\circ_{rxn} &= 2\Delta H^\circ_f(NH_3) - [\Delta H^\circ_f(N_2) + 3\Delta H^\circ_f(H_2)]\\ &= (2 \times -45.9\ kJ/mol) - [(1 \times 0\ kJ/mol) + (3 \times 0\ kJ/mol)]\\ &= -91.8\ kJ\ (\text{per mole of } N_2)\end{aligned}$$

B Inserting the appropriate values into Equation 18.23 gives

$$\Delta G^\circ_{rxn} = \Delta H^\circ - T\Delta S^\circ = (-91.8\ kJ) - (298.15\ \cancel{K})(-198.1\ J/\cancel{K}) = -32.7\ kJ\ (\text{per mole of } N_2)$$

(b) **C** To calculate ΔG for this reaction at 300°C, we assume that ΔH and ΔS are independent of temperature (that is, $\Delta H_{300°C} = \Delta H^\circ$ and $\Delta S_{300°C} = \Delta S^\circ$) and insert the appropriate temperature (573 K) into Equation 18.20:

$$\begin{aligned}\Delta G_{300°C} &= \Delta H_{300°C} - (573\ K)(\Delta S_{300°C}) = \Delta H^\circ - (573\ K)\Delta S^\circ\\ &= (-91.8\ kJ) - (573\ \cancel{K})(-198.1\ J/\cancel{K}) = 21.7\ kJ\ (\text{per mole of } N_2)\end{aligned}$$

In this example, changing the temperature has a major effect on the thermodynamic spontaneity of the reaction. Under standard conditions, the reaction of nitrogen and hydrogen gas to produce ammonia is thermodynamically spontaneous, but in practice, it is too slow to be useful industrially. Increasing the temperature in an attempt to make this reaction occur more rapidly also changes the thermodynamics by causing the $-T\Delta S^\circ$ term to dominate, and the reaction is no longer spontaneous at high temperatures. This is a classic example of the conflict that is encountered in real systems between thermodynamics and kinetics, which is often unavoidable.

EXERCISE 18.10

Calculate **(a)** ΔG° and **(b)** $\Delta G_{750°C}$ for the reaction $2NO(g) + O_2(g) \rightleftharpoons 2NO_2(g)$, which is important in the formation of urban smog. Assume that ΔH and ΔS do not change between 25.0°C and 750°C, and use these data: $S^\circ(NO) = 210.8$ J/(mol · K), $S^\circ(O_2) = 205.2$ J/(mol · K), $S^\circ(NO_2) = 240.1$ J/(mol · K), $\Delta H^\circ_f(NO_2) = 33.2$ kJ/mol, and $\Delta H^\circ_f(NO) = 91.3$ kJ/mol.

Answer **(a)** -70.5 kJ/mol of O_2; **(b)** 35.8 kJ/mol of O_2

The effect of temperature on the spontaneity of a reaction, an important factor in the design of an experiment or an industrial process, depends on the sign and magnitude of both ΔH° and ΔS°. The temperature at which a given reaction is at equilibrium can be calculated by setting $\Delta G^\circ = 0$ in Equation 18.23, as illustrated in the next example.

EXAMPLE 18.11

As you saw in Example 18.10, the reaction of nitrogen and hydrogen gas to produce ammonia is one in which ΔH° and ΔS° are both negative. Such reactions are predicted to be thermodynamically spontaneous at low temperatures, but nonspontaneous at high temperatures. Use the data in Example 18.10 to calculate the temperature at which this reaction changes from spontaneous to nonspontaneous, assuming that ΔH° and ΔS° are independent of temperature.

Given $\Delta H°$, $\Delta S°$

Asked for Temperature at which reaction changes from spontaneous to nonspontaneous

Strategy

Set $\Delta G°$ equal to zero in Equation 18.23 and solve for T, the temperature at which the reaction becomes nonspontaneous.

Solution

In Example 18.10, we calculated that $\Delta H°$ is -91.8 kJ/mol of N_2 and $\Delta S°$ is -198.1 J/K per mole of N_2, corresponding to $\Delta G° = -32.7$ kJ/mol of N_2 at 25°C. Thus, the reaction is indeed spontaneous at low temperatures, as expected based on the signs of $\Delta H°$ and $\Delta S°$. The temperature at which the reaction becomes nonspontaneous is found by setting $\Delta G°$ equal to zero and rearranging Equation 18.23 to solve for T:

$$\Delta G° = \Delta H° - T\Delta S° = 0$$

$$\Delta H° = T\Delta S°$$

$$T = \frac{\Delta H°}{\Delta S°} = \frac{(-91.8\ \cancel{\text{kJ}})(1000\ \cancel{\text{J}}/\cancel{\text{kJ}})}{-198.1\ \cancel{\text{J}}/\text{K}} = 463\ \text{K}$$

This is a case in which a chemical engineer is severely limited by thermodynamics. Any attempt to increase the rate of reaction of nitrogen with hydrogen by increasing the temperature will cause reactants to be favored over products above 463 K.

EXERCISE 18.11

As you found in working Exercise 18.10, $\Delta H°$ and $\Delta S°$ are both negative for the reaction of nitric oxide and oxygen to form nitrogen dioxide. Use the data in Exercise 18.10 to calculate the temperature at which this reaction changes from spontaneous to nonspontaneous.

Answer 792.6 K

18.6 • Spontaneity and Equilibrium

Thus far we have identified three criteria for whether a given reaction will occur spontaneously: $\Delta S_{univ} > 0$, $\Delta G_{sys} < 0$ (both discussed in this chapter), and the relative magnitude of the reaction quotient Q versus the equilibrium constant K (discussed in Chapter 15). Recall that if $Q < K$, then the reaction proceeds spontaneously to the right as written, resulting in the net conversion of reactants to products. Conversely, if $Q > K$, then the reaction proceeds spontaneously to the left as written, resulting in the net conversion of products to reactants. If $Q = K$, then the system is at equilibrium, and no net reaction occurs. Table 18.3 summarizes these criteria and their relative values for spontaneous, nonspontaneous, and equilibrium processes. Because all three

TABLE 18.3 Criteria for the spontaneity of a process

Spontaneous	Equilibrium	Nonspontaneous[a]
$\Delta S_{univ} > 0$	$\Delta S_{univ} = 0$	$\Delta S_{univ} < 0$
$\Delta G_{sys} < 0$	$\Delta G_{sys} = 0$	$\Delta G_{sys} > 0$
$Q < K$	$Q = K$	$Q > K$

[a] Spontaneous in the *reverse* direction.

criteria are assessing the same thing—the spontaneity of the process—it would be most surprising indeed if they were not related. The relationship between ΔS_{univ} and ΔG_{sys} was described in Section 18.5. In this section, we explore the relationship between the standard free energy of reaction $\Delta G°$ and the equilibrium constant K.

Free Energy and the Equilibrium Constant

MGC Gibbs Free Energy: Equilibrium Constant

Because $\Delta H°$ and $\Delta S°$ determine the magnitude of $\Delta G°$ (Equation 18.23), and because the equilibrium constant K is a measure of the ratio of the concentrations of products to the concentrations of reactants, we should be able to express K in terms of $\Delta G°$, and vice versa. As you learned in Section 18.5, ΔG is equal to the maximum amount of work a system can perform on the surroundings while undergoing a spontaneous change. For a reversible process that does not involve external work, we can express the change in free energy in terms of volume, pressure, entropy, and temperature, thereby eliminating ΔH from the equation for ΔG. The general relationship can be shown as

$$\Delta G = V\Delta P - S\Delta T \tag{18.26}$$

If a reaction is carried out at constant temperature ($\Delta T = 0$), then Equation 18.26 simplifies to

$$\Delta G = V\Delta P \tag{18.27}$$

Under normal conditions, the pressure dependence of free energy is not important for solids and liquids because of their small molar volumes. For reactions that involve gases, however, the effect of pressure on free energy is important.

Assuming ideal gas behavior, we can replace the V in Equation 18.27 by nRT/P (where n is the number of moles of gas and R is the ideal gas constant) and express ΔG in terms of the initial and final pressures (P_i and P_f, respectively) as in Equation 18.17:

$$\Delta G = \left(\frac{nRT}{P}\right)\Delta P = nRT\frac{\Delta P}{P} = nRT\ln\left(\frac{P_f}{P_i}\right) \tag{18.28}$$

If the initial state is the standard state with $P_i = 1$ atm, then the change in free energy of a substance upon going from the standard state to any other state with a pressure P can be written as

$$G - G° = nRT\ln P$$

which can be rearranged to

$$G = G° + nRT\ln P \tag{18.29}$$

As you will soon discover, Equation 18.29 allows us to relate $\Delta G°$ and K_p. Any relationship that is true for K_p must also be true for K because K_p and K are simply different ways of expressing the equilibrium constant using different units.

Let's consider the following hypothetical reaction, in which all the reactants and products are ideal gases and the lowercase letters correspond to the stoichiometric coefficients for the various species:

$$a\text{A} + b\text{B} \rightleftharpoons c\text{C} + d\text{D} \tag{18.30}$$

Because the free-energy change for a reaction is the difference between the sum of the free energies of the products and the reactants, we can write this expression for ΔG:

$$\Delta G = \Sigma mG_{products} - \Sigma nG_{reactants} = (cG_C + dG_D) - (aG_A + bG_B) \tag{18.31}$$

Substituting Equation 18.29 for each term into Equation 18.31 gives

$$\Delta G = [(cG°_C + cRT\ln P_C) + (dG°_D + dRT\ln P_D)] - [(aG°_A + aRT\ln P_A) + (bG°_B + bRT\ln P_B)]$$

Combining terms gives the following relationship between ΔG and and the reaction quotient Q:

$$\Delta G = \Delta G^\circ + RT \ln \frac{P_C^c P_D^d}{P_A^a P_B^b} = \Delta G^\circ + RT \ln Q \tag{18.32}$$

where ΔG° indicates that all reactants and products are in their standard states. In Chapter 15, you learned that for gases $Q = K_p$ at equilibrium, and as you've learned in this chapter, $\Delta G = 0$ for a system at equilibrium. Therefore, we can describe the relationship between ΔG° and K_p for gases as

$$0 = \Delta G^\circ + RT \ln K_p \qquad \Delta G^\circ = -RT \ln K_p \tag{18.33}$$

If the products and reactants are in their standard states and $\Delta G^\circ < 0$, then $K_p > 1$ and products are favored over reactants. Conversely, if $\Delta G^\circ > 0$, then $K_p < 1$ and reactants are favored over products. If $\Delta G^\circ = 0$, then $K_p = 1$ and neither reactants nor products are favored: the system is at equilibrium.

Note the pattern

For a spontaneous process under standard conditions, $K > 1$.

EXAMPLE 18.12

In Example 18.10, we calculated that $\Delta G^\circ = -32.7$ kJ/mol of N_2 for the reaction $N_2(g) + 3H_2(g) \rightleftharpoons 2NH_3(g)$. This calculation was for the reaction under standard conditions—that is, with all gases present at a partial pressure of 1 atm and a temperature of 25°C. Calculate ΔG for the same reaction under these nonstandard conditions: $P_{N_2} = 2.00$ atm, $P_{H_2} = 7.00$ atm, $P_{NH_3} = 0.021$ atm, and $T = 100°C$. Does the reaction favor products or reactants?

Given Balanced chemical equation, partial pressure of each species, temperature, ΔG°

Asked for Whether products or reactants are favored

Strategy

A Using the values given and Equation 18.32, calculate Q.
B Substitute the values of ΔG° and Q into Equation 18.32 to obtain ΔG for the reaction under nonstandard conditions.

Solution

A The relationship between ΔG° and ΔG under nonstandard conditions is given in Equation 18.32. Substituting the partial pressures given, we can calculate Q:

$$Q = \frac{P_{NH_3}^2}{P_{N_2} P_{H_2}^3} = \frac{(0.021)^2}{(2.00)(7.00)^3} = 6.4 \times 10^{-7}$$

B Substituting the values of ΔG° and Q into Equation 18.32 gives

$$\begin{aligned}\Delta G = \Delta G^\circ + RT \ln Q &= -32.7 \text{ kJ} + \left[(8.314 \text{ J/K})(373 \text{ K})\left(\frac{1 \text{ kJ}}{1000 \text{ J}}\right) \ln(6.4 \times 10^{-7})\right] \\ &= -32.7 \text{ kJ} + (-44.2 \text{ kJ}) = -76.9 \text{ kJ/mol of } N_2\end{aligned}$$

Because $\Delta G < 0$ and $Q < 1.0$, the reaction is spontaneous to the right as written, so products are favored over reactants.

EXERCISE 18.12

Calculate ΔG for the reaction of nitric oxide with oxygen to give nitrogen dioxide under these conditions: $T = 50°C$, $P_{NO} = 0.0100$ atm, $P_{O_2} = 0.200$ atm, and $P_{NO_2} = 1.00 \times 10^{-4}$ atm. The value of $\Delta G°$ for this reaction is -72.5 kJ/mol of O_2. Are products or reactants favored?

Answer -92.9 kJ/mol of O_2. The reaction is spontaneous to the right as written, so products are favored.

EXAMPLE 18.13

Calculate the equilibrium constant K_p for the reaction of H_2 with N_2 to give NH_3 at 25°C. As calculated in Example 18.10, $\Delta G°$ for this reaction is -32.7 kJ/mol of N_2.

Given Balanced chemical equation from Example 18.10, $\Delta G°$, temperature

Asked for K_p

Strategy

Substitute values for $\Delta G°$ and T (in kelvins) into Equation 18.33 to calculate K_p, the equilibrium constant for the formation of ammonia.

Solution

In Example 18.10, we used tabulated values of $\Delta G_f°$ to calculate $\Delta G°$ for this reaction (-32.7 kJ/mol of N_2). For equilibrium conditions, rearranging Equation 18.33 gives

$$\Delta G° = -RT \ln K_p$$

$$\ln K_p = \frac{-\Delta G°}{RT}$$

Inserting the value of $\Delta G°$ and the temperature (25°C = 298.15 K) into this equation gives

$$\ln K_p = -\frac{(-32.8 \text{ kJ})(1000 \text{ J/kJ})}{(8.314 \text{ J/K})(298.15 \text{ K})} = 13.2$$

$$K_p = 5.6 \times 10^5$$

Thus, the equilibrium constant for the formation of ammonia at room temperature is favorable. As we saw in Chapter 15, however, the *rate* at which the reaction occurs at room temperature is too slow to be useful.

EXERCISE 18.13

Calculate K_p for the reaction of NO with O_2 to give NO_2 at 25°C. As calculated in Exercise 18.10, $\Delta G°$ for this reaction is -72.5 kJ/mol of O_2.

Answer 5.0×10^{12}

Although K_p is defined in terms of the partial pressures of reactants and products, the equilibrium constant K is defined in terms of the concentrations of reactants and products. We described the relationship between the numerical magnitude of K_p and K in Chapter 15 and showed that they are related:

$$K_p = K(RT)^{\Delta n} \quad (18.34)$$

where Δn is the number of moles of gaseous product minus the number of moles of gaseous reactant. For reactions that involve only solutions, liquids, and solids, $\Delta n = 0$ and hence $K_p = K$. Thus, for all reactions that do not involve a change in the number of moles of gas present, the relationship in Equation 18.33 can be written in a more general form:

$$\Delta G° = -RT \ln K \tag{18.35}$$

Only when a reaction results in a net production or consumption of gases is it necessary to correct Equation 18.35 for the difference between K_p and K.*

Combining Equations 18.23 and 18.35 provides insight into how the components of $\Delta G°$ influence the magnitude of the equilibrium constant:

$$\Delta G° = \Delta H° - T\Delta S° = -RT \ln K \tag{18.36}$$

Notice that K becomes larger as $\Delta S°$ becomes more positive, indicating that the magnitude of the equilibrium constant is directly influenced by the tendency of the system to move toward maximum disorder. Moreover, K increases as $\Delta H°$ decreases. Thus, the magnitude of the equilibrium constant is also directly influenced by the tendency of the system to seek the lowest energy state possible.

Note the pattern

The magnitude of the equilibrium constant is directly influenced by the tendency of the system to move toward maximum disorder and to seek the lowest energy state possible.

Temperature Dependence of the Equilibrium Constant

The fact that $\Delta G°$ and K are related provides us with another explanation of why equilibrium constants are temperature dependent. This relationship is shown explicitly in Equation 18.36, which can be rearranged to give

$$\ln K = -\frac{\Delta H°}{RT} + \frac{\Delta S°}{R} \tag{18.37}$$

MGC Gibbs Free Energy: Temperature Dependence

Thus, assuming $\Delta H°$ and $\Delta S°$ are temperature independent, for an exothermic reaction ($\Delta H° < 0$), the magnitude of K decreases with increasing temperature, whereas for an endothermic reaction ($\Delta H° > 0$), the magnitude of K increases with increasing temperature. Note that the quantitative relationship expressed in Equation 18.37 agrees with the qualitative predictions made by applying Le Châtelier's principle, which we discussed in Chapter 15. Because heat is produced in an exothermic reaction, adding heat (by increasing the temperature) will shift the equilibrium to the left, favoring the reactants and decreasing the magnitude of K. Conversely, since heat is consumed in an endothermic reaction, adding heat will shift the equilibrium to the right, favoring the products and increasing the magnitude of K. Equation 18.37 also shows that the *magnitude* of $\Delta H°$ dictates how rapidly K changes as a function of temperature. In contrast, the magnitude and sign of $\Delta S°$ affect the magnitude of K, but not its temperature dependence.

If we know the value of K at a given temperature and the value of $\Delta H°$ for a reaction, we can estimate the value of K at any other temperature even in the absence of information on $\Delta S°$. Suppose, for example, that K_1 and K_2 are the equilibrium constants for a reaction at temperatures T_1 and T_2, respectively. Applying Equation 18.37 gives the following relationship at each temperature:

$$\ln K_1 = \frac{-\Delta H°}{RT_1} + \frac{\Delta S°}{R}$$

$$\ln K_2 = \frac{-\Delta H°}{RT_2} + \frac{\Delta S°}{R}$$

---

* Although we typically use concentrations or pressures in our equilibrium calculations, recall that equilibrium constants are generally expressed as unitless numbers because of the use of activities or fugacities in precise thermodynamic work. Systems that contain gases at high pressures or concentrated solutions that deviate substantially from ideal behavior require the use of fugacities or activities, respectively.

Subtracting $\ln K_1$ from $\ln K_2$ gives

$$\ln K_2 - \ln K_1 = \ln\frac{K_2}{K_1} = \frac{\Delta H^\circ}{R}\left(\frac{1}{T_1} - \frac{1}{T_2}\right) \tag{18.38}$$

Thus, calculating ΔH° from tabulated enthalpies of formation and measuring the equilibrium constant at one temperature, K_1, allow us to calculate the value of the equilibrium constant at any other temperature, K_2, assuming that ΔH° and ΔS° are independent of temperature.

EXAMPLE 18.14

The equilibrium constant for the formation of NH_3 from H_2 and N_2 at 25°C was calculated to be $K_p = 5.6 \times 10^5$ in Example 18.13. What is K_p at 500°C? (Use the data from Example 18.10.)

Given Balanced chemical equation, ΔH°, initial and final T, K_p at 25°C

Asked for K_p at 500°C

Strategy

Convert the initial and final temperatures to kelvins. Then substitute appropriate values into Equation 18.38 to obtain K_2, the equilibrium constant at the final temperature.

Solution

The value of ΔH° for the reaction obtained using Hess's law is −91.8 kJ/mol of N_2. If we set T_1 = 25°C = 298.15 K and T_2 = 500°C = 773 K, then from Equation 18.38 we obtain

$$\ln\frac{K_2}{K_1} = \frac{\Delta H^\circ}{R}\left(\frac{1}{T_1} - \frac{1}{T_2}\right) = \frac{(-91.8\ \cancel{\text{kJ}})(1000\ \cancel{\text{J}}/\cancel{\text{kJ}})}{8.314\ \cancel{\text{J}}/\cancel{\text{K}}}\left(\frac{1}{298.15\ \cancel{\text{K}}} - \frac{1}{773\ \cancel{\text{K}}}\right) = -22.8$$

$$\frac{K_2}{K_1} = 1.3 \times 10^{-10}$$

$$K_2 = (5.6 \times 10^5)(1.3 \times 10^{-10}) = 7.3 \times 10^{-5}$$

Thus, at 500°C, the equilibrium strongly favors the reactants over the products.

EXERCISE 18.14

In Exercise 18.13, you calculated $K_p = 5.0 \times 10^{12}$ for the reaction of NO with O_2 to give NO_2 at 25°C. Use the ΔH°_f values in Exercise 18.10 to calculate K_p for this reaction at 1000°C.

Answer 1.3×10^{-3}

MGC Thermodynamics vs. Kinetics

Note the pattern

Thermodynamics focuses on the energetics of the products and reactants, whereas kinetics focuses on the pathway from reactants to products.

18.7 • Comparing Thermodynamics and Kinetics

Because thermodynamics deals with state functions, it can be used to describe the overall properties, behavior, and equilibrium composition of a system. It is *not* concerned with the particular pathway by which physical or chemical changes occur, however, so it cannot address the *rate* at which a particular process will occur. Thus, although thermodynamics provides a significant constraint on what *can* occur during a reaction process, it does not describe the detailed steps of what *actually* occurs on an atomic or molecular level.

TABLE 18.4 The relationship between K and $\Delta G°$ at 25°C

$\Delta G°$, kJ/mol	K	Physical Significance
500 100	3×10^{-88} 3×10^{-18}	For all practical purposes, the reaction does not proceed in the forward direction: only reactants are present at equilibrium.
10 0 −10	2×10^{-2} 1 60	Both forward and reverse reactions occur: significant amounts of both products and reactants are present at equilibrium.
−100 −500	3×10^{17} 4×10^{87}	For all practical purposes, the forward reaction proceeds to completion: only products are present at equilibrium.

Table 18.4 gives the numerical values of the equilibrium constant K that correspond to various approximate values of $\Delta G°$. Note that $\Delta G° \geq +10$ kJ/mol or $\Delta G° \leq -10$ kJ/mol ensures that an equilibrium lies essentially all the way to the left or to the right, respectively, under standard conditions, corresponding to a reactant-to-product ratio of approxi-mately 10,000:1 (or vice versa). Only if $\Delta G°$ is quite small (± 10 kJ/mol) are significant amounts of both products and reactants present at equilibrium. Most reactions that we encounter have equilibrium constants substantially greater or less than 1, with the equilibrium strongly favoring either products or reactants. In many cases, we will encounter reactions that are strongly favored by thermodynamics but do not occur at a measurable rate. In contrast, we may encounter reactions that are not thermodynamically favored under standard conditions but nonetheless do occur under certain nonstandard conditions.

A typical challenge in industrial processes is a reaction that has a large negative value of $\Delta G°$ and hence a large value of K, but that is too slow to be practically useful. In such cases, mixing the reactants results in only a physical mixture, not a chemical reaction. An example is the reaction of carbon tetrachloride with water to produce carbon dioxide and hydrochloric acid, for which $\Delta G°$ is -232 kJ/mol:

$$CCl_4(l) + 2H_2O(l) \rightleftharpoons CO_2(g) + 2HCl(g) \tag{18.39}$$

The value of K for this reaction is 5×10^{40} at 25°C, yet when CCl_4 and water are shaken vigorously at 25°C, nothing happens: the two immiscible liquids form separate layers, with the denser CCl_4 on the bottom. In comparison, the analogous reaction of $SiCl_4$ with water to give SiO_2 and HCl, which has a similarly large equilibrium constant, occurs almost explosively. Although the two reactions have comparable thermodyna-mics, they have very different kinetics!

There are also many reactions for which $\Delta G° \ll 0$ but that do not occur as written because another possible reaction occurs more rapidly. For example, consider the reaction of lead sulfide with hydrogen peroxide. One possible reaction is

$$PbS(s) + 4H_2O_2(l) \rightleftharpoons PbO_2(s) + SO_2(g) + 4H_2O(l) \tag{18.40}$$

for which $\Delta G°$ is -886 kJ/mol and K is 10^{161}. Yet when lead sulfide is mixed with hydrogen peroxide, the ensuing vigorous reaction does *not* produce PbO_2 and SO_2. Instead, the reaction that actually occurs is

$$PbS(s) + 4H_2O_2(l) \rightleftharpoons PbSO_4(s) + 4H_2O(l) \tag{18.41}$$

This reaction has a $\Delta G°$ value of -1181 kJ/mol, within the same order of magnitude as the previous reaction, but it occurs much more rapidly.

Now consider reactions with $\Delta G^\circ > 0$. Thermodynamically, such reactions do not occur spontaneously under standard conditions. Nonetheless, these reactions can be made to occur under nonstandard conditions. An example is the reduction of chromium(III) chloride by hydrogen gas:

$$CrCl_3(s) + \frac{1}{2}H_2(g) \rightleftharpoons CrCl_2(s) + HCl(g) \tag{18.42}$$

At 25°C, $\Delta G^\circ = 35$ kJ/mol and $K_p = 7 \times 10^{-7}$. However, at 730°C, $\Delta G^\circ = -52$ kJ/mol and $K_p = 5 \times 10^2$; at this elevated temperature, the reaction is a convenient way of preparing chromium(II) chloride in the laboratory. Moreover, removing HCl gas from the system drives the reaction to completion, as predicted by Le Chatelier's principle. Thus, although the reaction is not thermodynamically spontaneous under standard conditions, it becomes spontaneous under nonstandard conditions.

There are also cases in which a compound whose formation appears to be thermodynamically prohibited can be prepared using a different reaction. The reaction for the preparation of chlorine monoxide from the elements, for example, is

$$\frac{1}{2}O_2(g) + Cl_2(g) \rightleftharpoons Cl_2O(g) \tag{18.43}$$

for which ΔG_f° is 97.9 kJ/mol. The large positive value of ΔG_f° for this reaction indicates that mixtures of chlorine and oxygen do not react to any extent to form Cl_2O. Nonetheless, Cl_2O is easily prepared using the reaction

$$HgO(s) + 2Cl_2(g) \rightleftharpoons Cl_2O(g) + HgCl_2(s) \tag{18.44}$$

which has a ΔG° of -22.2 kJ/mol and a K_p of approximately 1×10^4.

Finally, the ΔG° values for some reactions are so positive that the only way to make them proceed in the desired direction is to supply external energy, often in the form of electricity. Consider, for example, the formation of metallic lithium from molten lithium chloride:

$$LiCl(l) \rightleftharpoons Li(l) + \frac{1}{2}Cl_2(g) \tag{18.45}$$

Even at 1000°C, ΔG is very positive (324 kJ/mol), and there is no obvious way to obtain lithium metal using a different reaction. Hence, in the industrial preparation of metallic lithium, electrical energy is used to drive the reaction to the right, as described in Chapter 19.

Often reactions that are not thermodynamically spontaneous under standard conditions can be made to occur spontaneously if coupled, or connected, in some way to another reaction for which $\Delta G^\circ \ll 0$. Because the overall value of ΔG° for a series of reactions is the sum of the ΔG° values for the individual reactions, virtually any unfavorable reaction can be made to occur by chemically coupling it to a sufficiently favorable reaction or reactions. In the preparation of chlorine monoxide from mercuric oxide and chlorine (Equation 18.44), we have already encountered one example of this phenomenon of *coupled reactions*, although we did not describe it as such at the time. We can see how the chemical coupling works if we write Equation 18.44 as the sum of three separate reactions:

$$\begin{array}{lll}
\cancel{\frac{1}{2}O_2(g)} + Cl_2(g) \rightleftharpoons Cl_2O(g) & \Delta G^\circ = 97.9 \text{ kJ/mol} & (1) \\
HgO(s) \rightleftharpoons \cancel{Hg(l)} + \cancel{\frac{1}{2}O_2(g)} & \Delta G^\circ = 58.5 \text{ kJ/mol} & (2) \\
\cancel{Hg(l)} + Cl_2(g) \rightleftharpoons HgCl_2(s) & \Delta G^\circ = -178.6 \text{ kJ/mol} & (3) \\
\hline
HgO(s) + 2Cl_2(g) \rightleftharpoons Cl_2O(g) + HgCl_2(s) & \Delta G^\circ_{rxn} = -22.2 \text{ kJ/mol} &
\end{array}$$

Comparing the $\Delta G°$ values for the three reactions shows that reaction (3) is so energetically favorable that it more than compensates for the other two energetically unfavorable reactions. Hence, the overall reaction is indeed thermodynamically spontaneous as written.

EXAMPLE 18.15

Bronze Age metallurgists were accomplished practical chemists who unknowingly utilized coupled reactions to obtain metals from their ores. Realizing that different ores of the same metal required different treatments, they heated copper oxide ore in the presence of charcoal (carbon) to obtain copper metal, whereas they pumped air into the reaction system if the ore was copper sulfide. Assume that a particular copper ore consists of pure cuprous oxide (Cu_2O). Using the $\Delta G_f°$ values given, calculate **(a)** $\Delta G°$ and K_p for the decomposition of Cu_2O to metallic copper and oxygen gas [$\Delta G_f°(Cu_2O) = -146.0$ kJ/mol] and **(b)** $\Delta G°$ and K_p for the reaction of Cu_2O with carbon to produce metallic copper and carbon monoxide [$\Delta G_f°(CO) = -137.2$ kJ/mol].

Given Reactants and products, $\Delta G_f°$ values for Cu_2O and CO, temperature

Asked for $\Delta G°$ and K_p for the formation of metallic copper from Cu_2O in the absence and presence of carbon

Strategy

A Write the balanced equilibrium equation for each reaction. Using the "products minus reactants" rule, calculate $\Delta G°$ for the reaction.

B Substitute appropriate values into Equation 18.33 to obtain K_p.

Solution

(a) **A** The chemical equation for the decomposition of cuprous oxide is

$$Cu_2O(s) \rightleftharpoons 2Cu(s) + \frac{1}{2}O_2(g)$$

The substances on the right side of this equation are pure elements in their standard states, so their $\Delta G_f°$ values are zero. $\Delta G°$ for the reaction is therefore

$$\Delta G° = \left[2\Delta G_f°(Cu) + \frac{1}{2}\Delta G_f°(O_2)\right] - \Delta G_f°(Cu_2O)$$

$$= \left[(2\ \text{mol})(0\ \text{kJ/mol}) + \left(\frac{1}{2}\ \text{mol}\right)(0\ \text{kJ/mol})\right] - [(1\ \text{mol})(-146.0\ \text{kJ/mol})]$$

$$= 146.0\ \text{kJ}$$

B Rearranging and substituting the appropriate values into Equation 18.33, we obtain

$$\ln K_p = -\frac{\Delta G°}{RT} = -\frac{(146.0\ \text{kJ})(1000\ \text{J/kJ})}{(8.314\ \text{J/K})(298.15\ \text{K})} = -58.90$$

$$K_p = 2.60 \times 10^{-26}$$

This is a very small number, indicating that Cu_2O does not spontaneously decompose to a significant extent at room temperature.

(b) **A** The O_2 produced in the decomposition of Cu_2O can react with carbon to form CO:

$$\frac{1}{2}O_2(g) + C(s) \rightleftharpoons CO(g)$$

Because $\Delta G°$ for this reaction is equal to $\Delta G_f°$ for CO (−137.2 kJ/mol), it is energetically more feasible to produce metallic copper from cuprous oxide by coupling the two reactions:

$$Cu_2O(s) \rightleftharpoons 2Cu(s) + \frac{1}{2}O_2(g) \qquad \Delta G° = 146.0\ \text{kJ/mol}$$

$$\frac{1}{2}O_2(g) + C(s) \rightleftharpoons CO(g) \qquad \Delta G° = -137.2\ \text{kJ/mol}$$

$$Cu_2O(s) + C(s) \rightleftharpoons 2Cu(s) + CO(g) \qquad \Delta G° = 8.8\ \text{kJ/mol}$$

B We can find the corresponding value of K_p:

$$\ln K_p = -\frac{\Delta G^\circ}{RT} = -\frac{(8.8 \cancel{\text{kJ}})(1000 \text{ J}/\cancel{\text{kJ}})}{(8.314 \text{ J}/\cancel{\text{K}})(298.15 \cancel{\text{K}})} = -3.6$$

$$K_p = 0.03$$

Although this value is still less than 1, indicating that reactants are favored over products at room temperature, it is about 24 orders of magnitude greater than K_p for the production of copper metal in the absence of carbon. Because both ΔH° and ΔS° are positive for this reaction, it becomes thermodynamically feasible at slightly elevated temperatures (higher than about 80°C). At temperatures of a few hundred degrees Celsius, the reaction occurs spontaneously, proceeding smoothly and rapidly to the right as written and producing metallic copper and carbon monoxide from cuprous oxide and carbon.

EXERCISE 18.15

Use the ΔG°_f values given to calculate ΔG° and K_p (where appropriate) for **(a)** the decomposition of cuprous sulfide to copper metal and elemental sulfur ΔG°_f for Cu_2S, −86.2 kJ/mol) and **(b)** the reaction of cuprous sulfide with oxygen gas to produce sulfur dioxide and copper metal [ΔG°_f for $SO_2(g)$, −300.1 kJ/mol].

Answer **(a)** $\Delta G^\circ = 86.2$ kJ/mol; **(b)** $\Delta G^\circ = -213.9$ kJ/mol, $K_p = 3 \times 10^{37}$

18.8 • Thermodynamics and Life

MGC Thermodynamics of a Cell

In a thermodynamic sense, a living cell can be viewed as a low-entropy system that is *not* in equilibrium with its surroundings and is capable of replicating itself. A constant input of energy is needed to maintain the cell's highly organized structure, its wide array of precisely folded biomolecules, and its intricate system of thousands of chemical reactions. A cell also needs energy to synthesize complex molecules from simple precursors (for example, to make proteins from amino acids), to create and maintain differences in the concentrations of various substances inside and outside of the cell, and to do mechanical work (such as muscle contraction). In this section, we examine the nature of the energy flow between a cell and its environment as well as some of the chemical strategies cells use to extract energy from their surroundings and to store that energy.

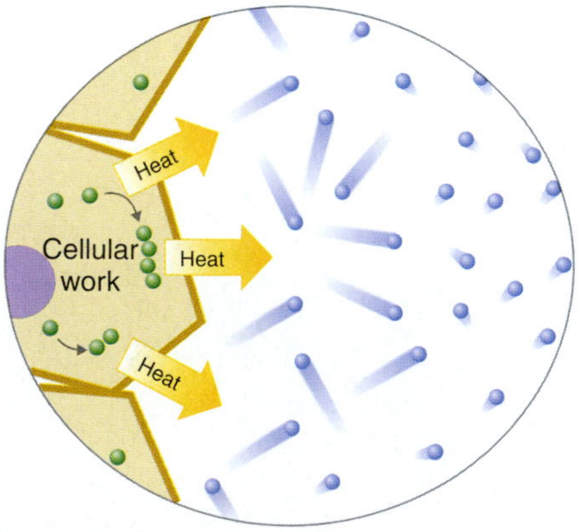

Figure 18.17 Life and entropy. A living cell is in a low-entropy, nonequilibrium state characterized by a high degree of structural organization. To maintain this state, a cell must release some of the energy it obtains from its environment as heat, thereby increasing S_{surr} sufficiently that the second law of thermodynamics is not violated. In this example, the cell combines smaller components into larger, more ordered structures; the accompanying release of heat increases the entropy of the surrounding environment so that $S_{univ} > 0$.

Energy Flow Between the Cell and Its Surroundings

One implication of the first and second laws of thermodynamics is that any closed system must eventually reach equilibrium. With no external input, a clock will run down, a battery will lose its charge, and a mixture of aqueous acid and base will achieve a uniform intermediate pH value. In contrast, a cell is an open system that can exchange matter with its surroundings as well as absorb energy from its environment in the form of heat or light. Cells utilize the energy obtained in these ways to maintain the nonequilibrium state that is essential for life.

Because cells are open systems, they cannot be described using the concepts of classical thermodynamics that we have discussed in this chapter, which have focused on reversible processes occurring in closed chemical systems that can exchange energy, but not matter, with their surroundings. Consequently, a relatively new subdiscipline called *nonequilibrium thermodynamics* has been developed to quantitatively describe open systems such as living cells.

Because a cell cannot violate the second law of thermodynamics, the only way it can maintain a low-entropy, nonequilibrium state characterized by a high degree of structural organization is to increase the entropy of its surroundings. A cell releases some of the energy that it obtains from its environment as heat that is transferred to its surroundings, thereby resulting in an increase in S_{surr} (Figure 18.17). As long as ΔS_{surr} is positive and greater than ΔS_{sys}, the entropy of the universe increases and the second law of thermodynamics is not violated. Releasing heat to the surroundings is necessary but not sufficient for life: the release of energy must be coupled to processes that increase the degree of order within the cell. For example, a wood fire releases heat to its surroundings, but unless energy from the burning wood is also used to do work, there is no increase in order of any portion of the universe.

Note the pattern

Any organism in equilibrium with its environment is dead.

Extracting Energy from the Environment

Although organisms utilize a wide range of specific strategies to obtain the energy they need to live and reproduce, they can generally be divided into two categories. Organisms are either *phototrophs* (from the Greek *photos*, "light," and *trophos*, "feeder"), whose energy source is sunlight, or *chemotrophs*, whose energy source is chemical compounds, usually obtained by consuming or breaking down other organisms. Phototrophs, such as plants, algae, and photosynthetic bacteria, use the radiant energy of the sun directly, converting water and carbon dioxide to energy-rich organic compounds, whereas chemotrophs, such as animals, fungi, and many non-photosynthetic bacteria, obtain energy-rich organic compounds from their environment. Regardless of the nature of their energy and carbon sources, all organisms utilize oxidation–reduction, or redox, reactions to drive the synthesis of complex biomolecules. Organisms that can use only O_2 as the oxidant (a group that includes most animals) are *aerobic organisms* that cannot survive in the absence of O_2. Many organisms that use other oxidants (such as SO_4^{2-}, NO_3^-, or CO_3^{2-}) or oxidized organic compounds can live only in the absence of O_2, which is a deadly poison for them; such species are called *anaerobic organisms*.

The fundamental reaction by which all green plants and algae obtain energy from sunlight is **photosynthesis**, the photochemical reduction of CO_2 to a carbon compound such as glucose. Concurrently, oxygen in water is oxidized to O_2 (recall that hν is energy from light):

$$6CO_2 + 6H_2O \xrightarrow{h\nu} C_6H_{12}O_6 + 6O_2 \qquad (18.46)$$

Photosynthesis

This reaction is not a spontaneous process as written, so energy from sunlight is used to drive the reaction. Photosynthesis is critical to life on Earth—it produces all the oxygen in our atmosphere.

In many ways, chemotrophs are more diverse than phototrophs because the nature of both the reductant (the nutrient) and the oxidant can vary. The most familiar chemotrophic strategy uses compounds such as glucose as the reductant and molecular oxygen as the oxidant in a process called **respiration** (see Chapter 5). The overall reaction of respiration is the reverse of photosynthesis:

$$C_6H_{12}O_6 + 6O_2 \longrightarrow 6CO_2 + 6H_2O \qquad (18.47)$$

Respiration

An alternative strategy utilizes **fermentation** reactions, in which an organic compound is simultaneously oxidized and reduced. Common examples are alcohol fermentation, used in making wine, beer, and bread, and lactic acid fermentation, used in making yogurt:

$$C_6H_{12}O_6 \longrightarrow 2CO_2 + 2CH_3CH_2OH \qquad (18.48)$$

Alcohol fermentation

$$C_6H_{12}O_6 \longrightarrow 2CH_3CH(OH)CO_2H \qquad (18.49)$$

Lactic acid fermentation

In these reactions, some of the carbon atoms of glucose are oxidized, while others are reduced. Recall that a reaction in which a single species is both oxidized and reduced is called a *disproportionation* reaction.

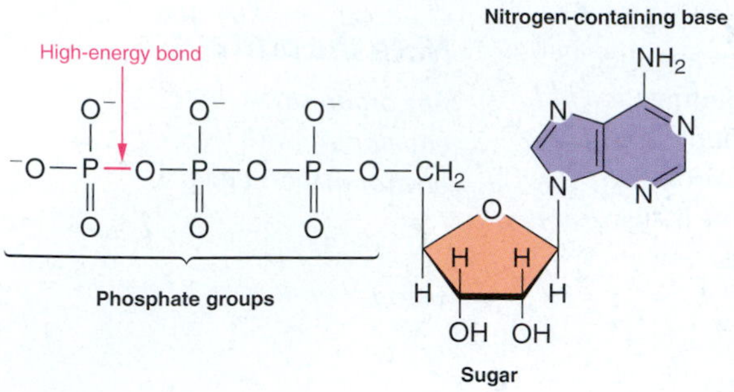

Figure 18.18 **Adenosine triphosphate (ATP), the universal energy currency of all cells.** The ATP molecule contains a nitrogen-containing base, a sugar, and three phosphate groups, as well as a high-energy phosphoric acid anhydride bond.

The Role of NADH and ATP in Metabolism

Regardless of the identity of the substances from which an organism obtains energy, the energy must be released in very small increments if it is to be useful to the cell. Otherwise, the temperature of the cell would rise to lethal levels. Cells store part of the energy that is released as ATP (adenosine triphosphate), a compound that is the universal energy currency of all living organisms (Figure 18.18).

Most organisms use a number of intermediate species to shuttle electrons between the terminal reductant (such as glucose) and the terminal oxidant (such as O_2). In virtually all cases, an intermediate species oxidizes the energy-rich reduced compound, and the now-reduced intermediate then migrates to another site where it is oxidized. The most important of these electron-carrying intermediates is NAD^+ (Figure 18.19), whose reduced form, formally containing H^-, is NADH. The reduction of NAD^+ to NADH can be written as

$$NAD^+ + H^+ + 2e^- \longrightarrow NADH \quad (18.50)$$

Most organisms use NAD^+ to oxidize energy-rich nutrients such as glucose to CO_2 and water; NADH is then oxidized to NAD^+ using an oxidant such as O_2. During the oxidation, a fraction of the energy obtained from the oxidation of the nutrient is stored as ATP. The phosphoric acid anhydride bonds in ATP can then be hydrolyzed by water, releasing energy and forming ADP (adenosine diphosphate). It is through this sequence of reactions that energy from the oxidation of nutrients is made available to cells. Thus, ATP has a central role in metabolism: it is synthesized by the oxidation of nutrients, and its energy is then used by the cell to drive synthetic reactions and to perform work (Figure 18.20).

> **Note the pattern**
>
> *In biochemical reactions, standard state conditions are defined as pH = 7, 1 M, and 1 atm.*

Under standard conditions in aqueous solutions, all reactants are present in concentrations of 1 *M*. For H^+, this corresponds to a pH of zero, and very little biochemistry occurs at pH = 0. For biochemical reactions, chemists have therefore defined a new standard state in which the H^+ concentration is 1×10^{-7} *M* (pH 7) and all other reactants and products are present in their usual standard-state conditions (1 *M* or 1 atm). The free-energy change and corresponding equilibrium constant for a reaction under these new standard conditions are denoted by the addition of a prime sign (′) to the conventional symbol: $\Delta G^{\circ\prime}$ and K'. If protons do not participate in a biological reaction, then $\Delta G^{\circ\prime} = \Delta G^\circ$. Otherwise, the relationship between $\Delta G^{\circ\prime}$ and ΔG° is

$$\Delta G^{\circ'} = \Delta G^\circ + RT\ln(10^{-7})^n \quad (18.51)$$

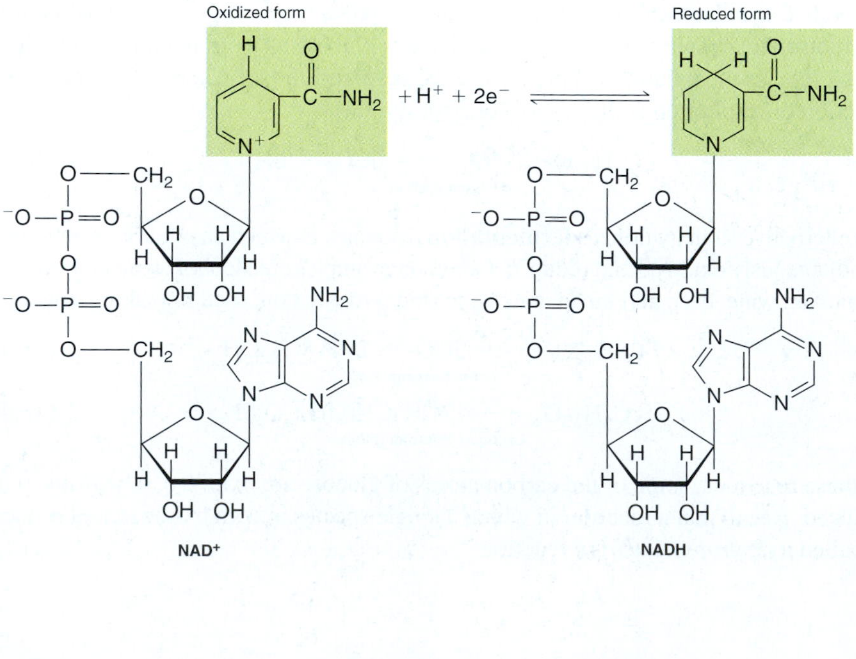

Figure 18.19 **Nicotinamide adenine dinucleotide (NAD^+) and its reduced form (NADH).** This electron carrier is used by biological systems to transfer electrons from one species to another. The oxidized form, NAD^+, is reduced to NADH by energy-rich nutrients such as glucose, and the reduced form, NADH, is in turn oxidized to NAD^+ by O_2 during respiration.

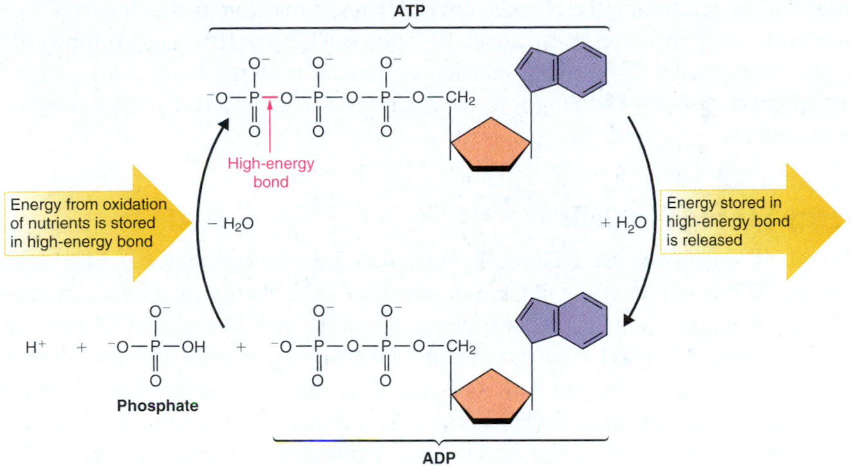

Figure 18.20 The ATP cycle. The high-energy phosphoric acid anhydride bond in ATP stores energy released during the oxidation of nutrients. Hydrolysis of the high-energy bond in ATP releases energy, forming adenosine diphosphate (ADP) and phosphate.

where $\Delta G^{\circ\prime}$ and ΔG° are in kilojoules per mole and n is the number of protons produced in the reaction. At 298 K, this simplifies to

$$\Delta G^{\circ'} = \Delta G^{\circ} - 39.96\, n \tag{18.52}$$

Thus, any reaction that involves the release of protons is thermodynamically more favorable at pH 7 than at pH 0.

The chemical equation that corresponds to the hydrolysis of ATP to ADP and phosphate is

$$ATP^{4-} + H_2O \rightleftharpoons ADP^{3-} + HPO_4^{\ 2-} + H^+ \tag{18.53}$$

This reaction has a $\Delta G^{\circ\prime}$ of -34.54 kJ/mol, but under typical physiological (or biochemical) conditions, the actual value of $\Delta G'$ for the hydrolysis of ATP is about -50 kJ/mol. Organisms use this energy to drive reactions that are energetically uphill, thereby coupling the reactions to the hydrolysis of ATP. One example is found in the biochemical pathway of *glycolysis*, in which the 6-carbon sugar glucose ($C_6H_{12}O_6$) is split into two 3-carbon fragments that are then used as the fuel for the cell. Initially, a phosphate group is added to glucose to form a phosphate ester, glucose-6-phosphate (abbreviated glucose-6-P), in a reaction analogous to that of an alcohol and phosphoric acid:

$$\begin{array}{llll} \text{glucose}(aq) & + \ HPO_4^{\ 2-}(aq) & \rightleftharpoons \text{glucose-6-P}^{2-}(aq) & + \ H_2O(l) \\ \text{RO}\mathbf{H} & + \ \mathbf{HO}PO_3^{\ 2-} & \rightleftharpoons \text{ROPO}_3^{\ 2-} & + \ \mathbf{H_2O} \end{array} \tag{18.54}$$

Due to its electrical charge, the phosphate ester is unable to escape from the cell by diffusing back through the membrane surrounding the cell, ensuring that it remains available for further reactions. For the reaction in Equation 18.54, ΔG° is 17.8 kJ/mol and K is 7.6×10^{-4}, indicating that the equilibrium lies far to the left. To force this reaction to occur as written, it is coupled to a thermodynamically favorable reaction, the hydrolysis of ATP to ADP:

$$\begin{array}{lll} \text{glucose} + \cancel{HPO_4^{\ 2-}} \rightleftharpoons \text{glucose-6-P}^{2-} + \cancel{H_2O} & \Delta G^{\circ\prime} = 17.8 \text{ kJ/mol} & K_1 = 7.6 \times 10^{-4} \\ ATP^{4-} + \cancel{H_2O} \rightleftharpoons ADP^{3-} + \cancel{HPO_4^{\ 2-}} + H^+ & \Delta G^{\circ\prime} = -34.54 \text{ kJ/mol} & K_2 = 2.2 \times 10^{5} \\ \hline \text{glucose} + ATP^{4-} \rightleftharpoons \text{glucose-6-P}^{2-} + ADP^{3-} + H^+ & \Delta G^{\circ\prime} = -16.7 \text{ kJ/mol} & K_3 = 1.7 \times 10^{2} \end{array}$$

Thus, the formation of glucose-6-phosphate is thermodynamically spontaneous if ATP is used as the source of phosphate.

The formation of glucose-6-phosphate is only one of many examples of how cells use ATP to drive an otherwise nonspontaneous biochemical reaction. Under nonstandard physiological conditions, each ATP hydrolyzed actually results in approximately a 10^8 increase in the magnitude of the equilibrium constant, compared with the equilibrium

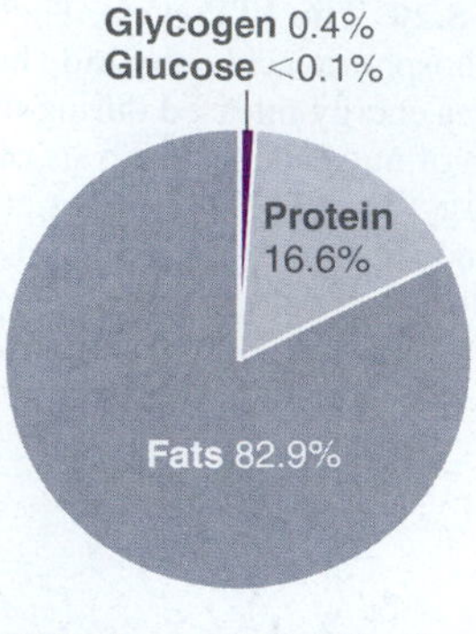

Figure 18.21 Percent distribution of forms of energy storage in adult humans. An average 70-kg adult stores about 6×10^5 kJ of energy in glucose, glycogen, protein, and fats. Fats are by far the most abundant and most efficient form for storing energy.

constant of the reaction in the absence of ATP. Thus, a reaction in which two ATPs are converted to ADP increases K by about 10^{16}, three ATPs by 10^{24}, and so forth. Virtually any energetically unfavorable reaction or sequence of reactions can be made to occur spontaneously by coupling it to the hydrolysis of a sufficiently large number of ATP molecules.

Energy Storage in Cells

Although all organisms use ATP as the *immediate* free-energy source in biochemical reactions, ATP is not an efficient form in which to store energy on a long-term basis. If the caloric intake of an average resting adult human were stored as ATP, two-thirds of the body weight would have to consist of ATP. Instead, a typical 70-kg adult human has a total of only about 50 g of both ATP and ADP, and, far from being used for long-term storage, each molecule of ATP is turned over about 860 times per day. The entire ATP supply would be exhausted in less than 2 minutes if it were not continuously regenerated.

How does the body store energy for the eventual production of ATP? Three primary means are as sugars, proteins, and fats. Combustion of sugars and proteins yields about 17 kJ of energy per gram, whereas combustion of fats yields more than twice as much energy per gram, about 39 kJ/g. Moreover, sugars and proteins are hydrophilic and contain about 2 g of water per gram of fuel even in very concentrated form. In contrast, fats are hydrophobic and can be stored in essentially anhydrous form. As a result, organisms can store about six times more energy per gram as fats than in any other form. A typical 70-kg adult human has about 170 kJ of energy in the form of glucose circulating in the blood, about 2600 kJ of energy stored in the muscles and liver as glycogen (a polymeric form of glucose), about 100,000 kJ stored in the form of protein (primarily muscle tissue), and almost 500,000 kJ in the form of fats (Figure 18.21). Thus, fats constitute by far the greatest energy reserve, while accounting for only about 12 kg of the 70-kg body mass. To store this amount of energy in the form of sugars would require a total body mass of about 144 kg, more than half of which would be sugar.

EXAMPLE 18.16

Glucose is one form in which the body stores energy. **(a)** Calculate $\Delta G^{\circ\prime}$ for the respiration of glucose to CO_2 and H_2O using these values of ΔG°_f: −910.4 kJ/mol for glucose, −394.4 kJ/mol for $CO_2(g)$, and −237.1 kJ/mol for $H_2O(l)$. **(b)** Assuming 50% efficiency in the conversion of the released energy to ATP, how many molecules of ATP can be synthesized by the combustion of one molecule of glucose? At 298.15 K $\Delta G^{\circ\prime}$ for the hydrolysis of ATP is −34.54 kJ/mol.

Given Balanced chemical equation (Equation 18.47), values of ΔG°_f, conversion efficiency, $\Delta G^{\circ\prime}$ for hydrolysis of ATP

Asked for $\Delta G^{\circ\prime}$ for the combustion reaction, molecules of ATP that can be synthesized

Strategy

A Using the "products minus reactants" rule, calculate ΔG°_{rxn} for the respiration reaction.

B Multiply the calculated value of ΔG°_{rxn} by the efficiency to obtain the number of kilojoules available for ATP synthesis. Then divide this value by $\Delta G^{\circ\prime}$ for the hydrolysis of ATP to find the maximum number of ATP molecules that can be synthesized.

Solution

(a) **A** Protons are not released or consumed in the reaction, so $\Delta G^{\circ\prime} = \Delta G^\circ$. We begin by using the balanced equation in Equation 18.47:

$$C_6H_{12}O_6 + 6O_2 \longrightarrow 6CO_2 + 6H_2O$$

From the given values of ΔG°_f (remember that ΔG°_f is zero for an element such as O_2 in its standard state), we can calculate ΔG°_{rxn}:

$$\begin{aligned}\Delta G^\circ_{rxn} &= \Sigma m \Delta G^\circ_f(\text{products}) - \Sigma n \Delta G^\circ_f(\text{reactants}) \\ &= [6(-394.4) + 6(-237.2)] - [(-1333) + 0] \\ &= -2457 \text{ kJ/mol of glucose}\end{aligned}$$

(b) B If we assume that only 50% of the available energy is used, then about 1230 kJ/mol of glucose is available for ATP synthesis. The value of $\Delta G^{\circ\prime}$ for the hydrolysis of ATP under biochemical conditions is −34.54 kJ/mol, so in principle an organism could synthesize

$$\frac{1230 \not{\text{kJ}}\text{/mol glucose}}{34.54 \not{\text{kJ}}\text{/mol ATP}} = 35.6 \approx 36 \text{ ATP/glucose}$$

Most aerobic organisms actually synthesize about 32 molecules of ATP per molecule of glucose, for an efficiency of about 45%.

EXERCISE 18.16

Some bacteria synthesize methane using this redox reaction:

$$CO_2(g) + 4H_2(g) \longrightarrow CH_4(g) + 2H_2O(g)$$

(a) Calculate $\Delta G^{\circ\prime}$ for this reaction using values of ΔG°_f in Appendix A, and **(b)** calculate how many ATP molecules could be synthesized per mol of CO_2 reduced if the efficiency of the process were 100%.

Answer **(a)** −113.3 kJ/mol CO_2; **(b)** 3 ATP/mol CO_2

SUMMARY AND KEY TERMS

18.1 Thermodynamics and Work (p. 811)

Thermodynamics is the study of the interrelationships among heat, work, and the energy content of a system at equilibrium. The sum of the potential and the kinetic energy of all the components of a system is the **internal energy (*E*)** of the system, which is a **state function**. When the pressure or volume of a gas is changed, any mechanical work done is called *PV work*. Work done by the system on the surroundings is given a negative value, whereas work done on the system by the surroundings has a positive value.

18.2 The First Law of Thermodynamics (p. 814)

The **first law of thermodynamics** states that the energy of the universe is constant. The change in the internal energy of a system is the sum of the heat transferred and the work done. At constant pressure, the relationship between the heat flow *q* and the internal energy *E* of a system is its **enthalpy (H)**. The heat flow is equal to the change in the internal energy of the system plus the *PV* work done. When the volume of a system is constant, changes in its internal energy can be calculated by substituting the ideal gas law into the equation for ΔE.

18.3 The Second Law of Thermodynamics (p. 818)

A measure of the disorder of a system is its **entropy (*S*)**, a state function whose value increases with an increase in the number of available microstates. A **reversible process** is one for which all intermediate states between extremes are equilibrium states; it can change direction at any time. In contrast, an **irreversible process** occurs in one direction only. The change in entropy of the system or the surroundings is the quantity of heat transferred divided by the temperature. The **second law of thermodynamics** states that in a reversible process, the entropy of the universe is constant, whereas in an irreversible process, such as the transfer of heat from a hot object to a cold object, the entropy of the universe increases.

18.4 Entropy Changes and the Third Law of Thermodynamics (p. 825)

The **third law of thermodynamics** states that the entropy of any perfectly ordered, crystalline substance at absolute zero is zero. Thus, at temperatures higher than absolute zero, entropy has a positive value, which allows us to measure the *absolute entropy* of a substance. Measurements of the heat capacity of a substance and the enthalpies of fusion or vaporization can be used to calculate the changes in entropy that accompany a physical change. The entropy of 1 mol of a substance at a standard temperature of 298 K is its **standard molar entropy (S°)**. We can use the "products minus reactants" rule to calculate the standard entropy change ΔS° for a reaction using tabulated values of S° for the reactants and products.

18.5 Free Energy (p. 829)

We can predict whether a reaction will occur spontaneously by combining the entropy, enthalpy, and temperature of a system in a new state function called **Gibbs free energy** (G). The change in free energy, ΔG, is the difference between the heat released during a reversible or irreversible process and the heat released for the same process occurring in a reversible manner. If a system is at equilibrium, $\Delta G = 0$. If the process is spontaneous, $\Delta G < 0$. If the process is not spontaneous as written but is spontaneous in the reverse direction, $\Delta G > 0$. At constant temperature and pressure, ΔG is equal to the maximum amount of work a system can perform on the surroundings while undergoing a spontaneous change. The **standard free energy change ($\Delta G°$)** is the change in free energy when one substance or set of substances in their standard states is converted to one or more other substances, also in their standard states. The **standard free energy of formation (ΔG_f°)**, is the change in free energy that occurs when 1 mol of a substance in its standard state is formed from the elements in their standard states. Tabulated values of standard free energies of formation are used to calculate $\Delta G°$ for a reaction.

18.6 Spontaneity and Equilibrium (p. 837)

For a reversible process that does not involve external work, we can express the change in free energy in terms of volume, pressure, entropy, and temperature. If we assume ideal gas behavior, the ideal gas law allows us to express ΔG in terms of the partial pressures of the reactants and products, which gives us a relationship between ΔG and K_p, the equilibrium constant of a reaction involving gases, or K, the equilibrium constant expressed in terms of concentrations. If $\Delta G° < 0$, then K or $K_p > 1$ and products are favored over reactants. If $\Delta G° > 0$, then K or $K_p < 1$ and reactants are favored over products. If $\Delta G° = 0$, then K or $K_p = 1$ and the system is at equilibrium. We can use the measured equilibrium constant K at one temperature and $\Delta H°$ to estimate the equilibrium constant for a reaction at any other temperature.

18.7 Comparing Thermodynamics and Kinetics (p. 842)

Thermodynamics is used to describe the overall properties, behavior, and equilibrium composition of a system; kinetics describes the rate at which a particular process will occur and the pathway by which it will occur. Whereas thermodynamics tells us what can occur during a reaction process, kinetics tells us what actually occurs on an atomic or molecular level. A reaction that is not thermodynamically spontaneous under standard conditions can often be made to occur spontaneously by varying reaction conditions, by using a different reaction to obtain the same product, by supplying external energy, such as electricity, or by coupling the unfavorable reaction to another reaction for which $\Delta G° << 0$.

18.8 Thermodynamics and Life (p. 846)

A living cell is a system that is not in equilibrium with its surroundings; it requires a constant input of energy to maintain its nonequilibrium state. Cells maintain a low-entropy state by increasing the entropy of their surroundings. *Aerobic organisms* cannot survive in the absence of O_2, whereas *anaerobic organisms* can live only in the absence of O_2. Green plants and algae are *phototrophs*, which extract energy from the environment through a process called **photosynthesis**, the photochemical reduction of CO_2 to a reduced carbon compound. Other species, called *chemotrophs*, extract energy from chemical compounds. One of the main processes chemotrophs use to obtain energy is **respiration**, which is the reverse of photosynthesis. Alternatively, some chemotrophs obtain energy by **fermentation**, in which an organic compound is both the oxidant and the reductant. Intermediates used by organisms to shuttle electrons between the reductant and the oxidant include NAD^+ and NADH. Energy from the oxidation of nutrients is made available to cells through the synthesis of ATP, the energy currency of the cell. Its energy is used by the cell to synthesize substances through coupled reactions and to perform work. The body stores energy as sugars, protein, or fats before using it to produce ATP.

KEY EQUATIONS

Definition of *PV* work	$w = -P_{ext}\Delta V$	(18.4)
Internal energy change	$\Delta E_{sys} = q + w$	(18.6)
Enthalpy change	$\Delta H = \Delta E + \Delta(PV)$	(18.8)
Relationship between ΔH and ΔE for an ideal gas	$\Delta E = \Delta H - RT\Delta n$	(18.10)
Entropy change	$\Delta S = \frac{q_{rev}}{T}$	(18.12)
Temperature dependence of entropy at constant pressure	$\Delta S = nC_p \ln\frac{T_2}{T_1}$	(18.17)
Temperature dependence of entropy at constant volume	$\Delta S = nC_v \ln\frac{T_2}{T_1}$	(18.18)
Free-energy change	$\Delta G = \Delta H - T\Delta S$	(18.20)
Standard free-energy change	$\Delta G° = \Delta H° - T\Delta S°$	(18.23)
Relationship between standard free-energy change and equilibrium constant	$\Delta G° = -RT \ln K$	(18.35)
Temperature dependence of equilibrium constant	$\ln K = \frac{-\Delta H°}{RT} + \frac{\Delta S°}{R}$	(18.37)
Calculation of *K* at second temperature	$\ln\frac{K_2}{K_1} = \frac{\Delta H°}{R}\left(\frac{1}{T_1} - \frac{1}{T_2}\right)$	(18.38)

QUESTIONS AND PROBLEMS

Questions and Problems with colored numbers have answers in the Appendix and complete solutions in the Student Solutions Manual.

CONCEPTUAL

18.1 Thermodynamics and Work

1. Thermodynamics focuses on the energetics of the reactants and products, and provides information about the composition of the reaction system at equilibrium. What information on reaction systems is not provided by thermodynamics?
2. Given a system in which a substance can produce either of two possible products, $A \longrightarrow B$ or $A \longrightarrow C$, which of the following can be predicted using chemical thermodynamics?
   (a) At equilibrium, the concentration of product C is greater than the concentration of product B.
   (b) Product C forms more quickly than product B.
   (c) The reaction $A \longrightarrow C$ is exothermic.
   (d) Low-energy intermediates are formed in the reaction $A \longrightarrow B$.
   (e) The reaction $A \longrightarrow C$ is spontaneous.
3. In what two ways can a closed system exchange energy with its surroundings? Are these two processes path dependent or path independent?
4. A microwave oven operates by providing enough energy to rotate water molecules, which in turn produces heat. Can the change in the internal energy of a cup of water heated in a microwave oven be described as a state function? Can the heat produced be described as a state function?

18.2 The First Law of Thermodynamics

5. Describe how a swinging pendulum that slows with time illustrates the first law of thermodynamics.
6. When air is pumped into a bicycle tire, the air is compressed. Assuming that the volume is constant, express the change in internal energy in terms of q and w.
7. What is the relationship between enthalpy and internal energy for a reaction that occurs at constant pressure?
8. An intrepid scientist placed an unknown salt in a small amount of water. All the salt dissolved in the water, and the temperature of the solution dropped several degrees.
   (a) What is the sign of the enthalpy change for this reaction?
   (b) Assuming the heat capacity of the solution is the same as that of pure water, how would the scientist calculate the molar enthalpy change?
   (c) Propose an explanation for the decrease in temperature.
9. For years, chemists and physicists focused on enthalpy changes as a way to measure the spontaneity of a reaction. What arguments would you use to convince them not to use this method?
10. What is the relationship between enthalpy and internal energy for a reaction that occurs at constant volume?
11. The reaction of methyl chloride with water produces methanol and HCl gas at room temperature, despite the fact that $\Delta H^\circ_{rxn} = 7.3$ kcal/mol. Using thermodynamic arguments, why does methanol form?
12. The *enthalpy of combustion* is defined thermodynamically as the enthalpy change for complete oxidation. The complete oxidation of hydrocarbons is represented by this general equation: hydro-carbon $+ O_2(g) \longrightarrow CO_2(g) + H_2O(g)$. Enthalpies of combustion from reactions like this one can be measured experimentally with a high degree of precision. It has been found that the less stable the reactant, the more heat is evolved, so the more negative the value of ΔH_{comb}. In each of the following pairs of hydrocarbons, which member do you expect to have the greater (more negative) heat of combustion? (a) cyclopropane, cyclopentane; (b) butane, 2-methylpropane; (c) hexane, cyclohexane. Justify your answers.
13. Using structural argument, explain why cyclopropane has a positive ΔH°_f (12.7 kJ/mol), whereas cyclopentane has a negative ΔH°_f (−18.4 kJ/mol).
14. Using structural argument, explain why the trans isomer of 2-butene is more stable than the cis isomer. The enthalpies of formation of cis- and trans-2-butene are −7.1 kJ/mol and −11.4 kJ/mol, respectively.

18.3 The Second Law of Thermodynamics

15. A Russian space vehicle developed a leak, which resulted in an internal pressure drop from 1 atm to 0.85 atm. Is this an example of a reversible expansion? Has work been done?
16. Which member of each of pair do you expect to have a higher entropy, and why? (a) solid phenol, liquid phenol; (b) 1-butanol, butane; (c) cyclohexane, cyclohexanol; (d) 1 mol of N_2 mixed with 2 mol of O_2, 2 mol of NO_2; (e) 1 mol of O_2, 1 mol of O_3; (f) 1 mol of propane at 1 atm, 1 mol of propane at 2 atm
17. Which processes are reversible and which are not? (a) ice melting at 0°C; (b) salt crystallizing from a saline solution; (c) evaporation of a liquid in equilibrium with its vapor in a sealed flask; (d) a neutralization reaction
18. Determine whether each process is reversible or irreversible: (a) cooking spaghetti; (b) the reaction between sodium metal and water; (c) oxygen uptake by hemoglobin; (d) evaporation of water at its boiling point.
19. Explain why increasing the temperature of a gas increases its entropy. What effect does this have on the internal energy of the gas?
20. For a series of related compounds, does ΔS_{vap} increase or decrease with an increase in the strength of intermolecular interactions in the liquid state? Why?
21. Is the change in the enthalpy of reaction or the change in entropy of reaction more sensitive to changes in temperature? Explain your reasoning.
22. Solid potassium chloride has a highly ordered lattice structure. Do you expect ΔS_{soln} to be greater or less than zero? Why? What opposing factors must be considered in making your prediction?
23. Aniline ($C_6H_5NH_2$) is an oily liquid at 25°C that darkens on exposure to air and light. It is used in dying fabrics and in staining wood black. One gram of aniline dissolves in 28.6 mL of water, but aniline is completely miscible with ethanol. Do you expect ΔS_{soln} in H_2O to be greater than, less than, or equal to ΔS_{soln} in CH_3CH_2OH? Why?

18.4 Entropy Changes and the Third Law of Thermodynamics

24. Crystalline $MgCl_2$ has $S° = 89.63$ J/(mol·K), whereas aqueous $MgCl_2$ has $S° = -25.1$ J/(mol·K). Is this consistent with the third law of thermodynamics? Explain your answer.

25. Why is it possible to measure absolute entropies but not absolute enthalpies?

26. How many microstates are available to a system at absolute zero? How many are available to a substance in its liquid state?

27. Substance A has a higher heat capacity than substance B. Do you expect the absolute entropy of substance A to be lower than, similar to, or higher than that of substance B? Why? As the two substances are heated, for which substance do you predict the entropy to increase more rapidly?

18.5 Free Energy

28. How does each example illustrate the fact that no process is 100% efficient? (a) burning a log to stay warm; (b) the respiration of glucose to provide energy; (c) burning a candle to provide light

29. Neither the change in enthalpy nor the change in entropy is by itself sufficient to determine whether a reaction will occur spontaneously. Why?

30. If a system is at equilibrium, what must be the relationship between ΔH and ΔS?

31. The equilibrium $2AB \rightleftharpoons A_2B_2$ is exothermic in the forward direction. Which has the higher entropy: the product or the reactant? Why? Which is favored at high temperatures?

32. Is ΔG a state function that describes the system or the surroundings? Do its components, ΔH and ΔS, describe the system or the surroundings?

33. How can you use ΔG to determine the temperature of a phase transition, such as the boiling point of a liquid or the melting point of a solid?

34. Occasionally, an inventor claims to have invented a "perpetual motion" machine, which requires no additional input of energy once the machine has been put into motion. Using your knowledge of thermodynamics, how would you respond to such a claim? Justify your arguments.

35. Must the entropy of the universe increase in a spontaneous process? If not, why is there no process that is 100% efficient?

18.6 Spontaneity and Equilibrium

36. Do you expect products or reactants to dominate at equilibrium in a reaction for which $\Delta G°$ is equal to (a) 1.4 kJ/mol; (b) 105 kJ/mol; (c) −34 kJ/mol?

37. The change in free energy enables us to determine whether a reaction will proceed spontaneously. How is this related to the extent to which a reaction proceeds?

38. What happens to the free energy of the reaction $N_2(g) + 3F_2(g) \longrightarrow 2NF_3(g)$ if the pressure is increased while the temperature remains constant? If the temperature is increased at constant pressure? Why are these effects not so important for reactions that involve liquids and solids?

39. Compare the expressions for the relationship between the free energy of a reaction and its equilibrium constant where the reactants are gases versus liquids. What are the differences between these expressions?

18.7 Comparing Thermodynamics and Kinetics

40. You are in charge of finding conditions to make the reaction $A(l) + B(l) \longrightarrow C(l) + D(g)$ favorable because it is a critical step in the synthesis of your company's key product. You have calculated that $\Delta G°$ for the reaction is negative, yet the ratio of products to reactants is very small. What have you overlooked in your scheme? What can you do to drive the reaction to increase your product yield?

18.8 Thermodynamics and Life

41. The tricarboxylic acid (TCA) cycle in aerobic organisms is one of four pathways responsible for the stepwise oxidation of organic intermediates. The final reaction in the TCA cycle has $\Delta G° = 29.7$ kJ/mol, so it should not occur spontaneously. Suggest an explanation for why this reaction proceeds in the forward direction in living cells.

NUMERICAL

This section includes paired problems (marked by brackets) that require similar problem-solving skills.

18.1 Thermodynamics and Work

42. Calculate the work done in joules in each process:
(a) Compressing 12.8 L of hydrogen gas at an external pressure of 1.00 atm to 8.4 L at a constant temperature.
(b) Expanding 21.9 L of oxygen gas at an external pressure of 0.71 atm to 23.7 L at a constant temperature.

43. How much work in joules per mole is done when oxygen is compressed from a volume of 22.8 L and an external pressure of 1.20 atm to 12.0 L at a constant temperature? Was work done by the system or the surroundings?

44. Champagne is bottled at a CO_2 pressure of about 5 atm. What is the force on the cork if its cross-sectional area is 2.0 cm^2? How much work is done if a 2.0-g cork flies a distance of 8.2 ft straight up when the cork is popped? Was work done by the system or the surroundings?

45. One mole of water is converted to steam at 1.00 atm pressure and 100°C. Assuming ideal behavior, what is the change in volume when the water is converted from a liquid to a gas? If this transformation took place in a cylinder with a piston, how much work could be done by vaporizing the water at 1.00 atm? Is work done by the system or the surroundings?

46. Recall that force can be expressed as mass times acceleration ($F = ma$). Acceleration due to gravity on the earth's surface is 9.8 m/s^2.
(a) What is the gravitational force on a person who weighs 52 kg?
(b) How much work is done if the person leaps from a burning building out of a window 20 m above the ground?
(c) If the person lands on a large rescue cushion fitted with a pressure-release valve that maintains an internal pressure of 1.5 atm, how much air is forced out of the cushion?

47. Acceleration due to gravity on the earth's surface is 9.8 m/s^2. How much work is done by a 175-lb person going over Niagara Falls (approximately 520 ft high) in a barrel that weighs 145 lb?

48. One mole of an ideal gas is allowed to expand reversibly from an initial pressure of 1.0 atm to a final pressure of 0.62 atm at constant temperature against an external pressure of 1.0 atm. How much work has been done?

49. A gas is allowed to expand from a volume of 2.3 L to a volume of 5.8 L at constant temperature. During the process, 138 J of heat is transferred from the surroundings to the gas.

(a) How much work has been done if the gas expands against a vacuum?
(b) How much work has been done if the gas expands against a pressure of 1.3 atm?
(c) What is the change in the internal energy of the system?

18.2 The First Law of Thermodynamics

50. Zinc and HCl react according to the equation

$$Zn(s) + 2HCl(aq) \longrightarrow Zn^{2+}(aq) + 2Cl^-(aq) + H_2(g)$$

When 3.00 g of zinc metal is added to a dilute HCl solution at 1.00 atm and 25°C, and the above reaction is allowed to go to completion at constant pressure, 6.99 kJ of heat must be removed to return the final solution to its original temperature. What are the values of q and w, and what is the change in internal energy?

51. A block of CO_2 weighing 15 g evaporates in a 5.0-L container at 25°C. How much work has been done if the gas is allowed to expand against an external pressure of 0.98 atm under isothermal conditions? The enthalpy of sublimation of CO_2 is 25.1 kJ/mol. What is the change in internal energy (kJ/mol) for the sublimation of CO_2 under these conditions?

52. Acetylene torches, used industrially to cut and weld metals, reach flame temperatures as high as 3000°C. The combustion reaction is

$$2C_2H_2(g) + 5O_2(g) \longrightarrow 4CO_2(g) + 2H_2O(l) \quad \Delta H = -2599 \text{ kJ}$$

Calculate the amount of work done against a pressure of 1.0 atm when 4.0 mol of acetylene is allowed to react with 10 mol of O_2 at 1.0 atm at 20°C. What is the change in internal energy for the reaction?

53. When iron dissolves in 1.00 M aqueous HCl, the products are $FeCl_2(aq)$ and hydrogen gas. Calculate the work done if 30 g of Fe reacts with excess hydrochloric acid in a closed vessel at 20°C. How much work is done if the reaction takes place in an open vessel with an external pressure of 1 atm?

18.3 The Second Law of Thermodynamics

54. Liquid nitrogen, which has a boiling point of −195.79°C, is used as a coolant and as a preservative for biological tissues. Is the entropy of nitrogen higher or lower at −200°C than at −190°C? Explain your answer. Liquid nitrogen freezes to a white solid at −210.00°C, with an enthalpy of fusion of 0.71 kJ/mol. What is its entropy of fusion? Is freezing biological tissue in liquid nitrogen an example of a reversible or irreversible process?

55. Using the second law of thermodynamics, explain why heat flows from a hot body to a cold body but not from a cold body to a hot body.

56. One test of the spontaneity of a reaction is whether the entropy of the universe increases, $\Delta S_{univ} > 0$. Using an entropic argument, show that this reaction is spontaneous at 25°C:

$$4Fe(s) + 3O_2(g) \longrightarrow 2Fe_2O_3(s)$$

Why does the entropy of the universe increase in this reaction even though gaseous molecules, which have a high entropy, are consumed?

57. Calculate the missing data in the table.

Compound	ΔH_{fus}, kJ/mol	ΔS_{fus}, J/(mol · K)	Melting Point, °C
Acetic acid	11.7		16.6
CH_3CN	8.2	35.9	
CH_4	0.94		−182.5
CH_3OH		18.2	−97.7
Formic acid	12.7	45.1	

Based on this table, can you conclude that entropy is related to the nature of functional groups? Explain your reasoning.

58. Calculate the missing data in the table.

Compound	ΔH_{vap}, kJ/mol	ΔS_{vap}, J/(mol · K)	Melting Point, °C
Hexanoic acid	71.1		105.7
Hexane	28.9	85.5	
Formic acid		60.7	100.8
1-Hexanol	44.5		157.5

The text states that the magnitude of ΔS_{vap} tends to be similar for a wide variety of compounds. Based on the values in the table, do you agree?

18.4 Entropy Changes and the Third Law of Thermodynamics

59. Phase transitions must be considered when calculating entropy changes. Why? What is the final temperature of the water when 5.20 g of ice at 0.0°C is added to 250 mL of water in an insulated thermos at 30.0°C? The value of ΔH_{fus} for water is 6.01 kJ/mol, and the heat capacity of liquid water is 75.3 J/(mol·°C). What is the entropy change for this process?

60. Calculate the change in both enthalpy and entropy when a 3.0-g block of ice melts at 0.0°C [$\Delta H_{fus}(H_2O) = 6.01$ kJ/mol]. For the same block of ice, calculate the entropy change for the system when the ice is warmed from 0.0°C to 25°C. The heat capacity of liquid water over this temperature range is 75.3 J/(mol·°C).

61. Use the data in Table 18.1 and Appendix A to calculate $\Delta S°$ for each reaction:
(a) $H_2(g) + \frac{1}{2}O_2(g) \longrightarrow H_2O(l)$
(b) $CH_3OH(l) + HCl(g) \longrightarrow CH_3Cl(g) + H_2O(l)$
(c) $H_2(g) + Br_2(l) \longrightarrow 2HBr(g)$
(d) $Zn(s) + 2HCl(aq) \longrightarrow ZnCl_2(aq) + H_2(g)$

62. Calculate the entropy change (J/K) when 4.35 g of liquid bromine is heated from 30.0°C to 50.0°C if the molar heat capacity of liquid bromine, C_p, is 75.1 kJ/(mol·K).

63. Calculate the molar heat capacity of titanium, C_p[kJ/(mol·K)], if the change in entropy when a 6.00-g sample of $TiCl_4(l)$ is heated from 25.0°C to 40.0°C is 0.153 J/K.

64. When a 1.00-g sample of lead is heated from 298.2 K to its melting temperature of 600.5 K, the change in entropy is 0.130 J/(mol·K). Under standard conditions, the enthalpy of fusion of lead is 4.98 kJ/mol. Determine the molar heat capacity of lead over this temperature range.

65. Phosphorus oxychloride ($POCl_3$) is a chlorinating agent that is frequently used in organic chemistry to replace oxygen with chlorine. Given the following information, does $POCl_3$ spontaneously convert from a liquid to a gas at 110°C? ΔS_{vap} = 93.08 J/(mol · K), ΔH_{vap} = 35.2 kJ/mol. Does it spontaneously crystallize at 0.0°C if ΔH_{fus} = 34.3 kJ/mol and ΔS_{fus} = 125 J/K? Using the information provided, calculate the melting point of $POCl_3$.

66. A useful reagent for the fluorination of alcohols, carboxylic acids, and carbonyl compounds is selenium tetrafluoride (SeF_4). One must be careful when using this compound, however, because it is known to attack glass (such as the glass of a reaction vessel).
(a) Is SeF_4 a liquid or a gas at 100°C given that ΔH_{vap} = 46.9 kJ/mol and ΔS_{vap} = 124 J/K?
(b) Determine the boiling point of SeF_4.
(c) Would you use SeF_4 for a solution reaction at 0°C if ΔH_{fus} = 46 kJ/mol and ΔS_{fus} = 178 J/(mol · K)?

18.5 Free Energy

67. Use the tables in the text to determine whether the following reactions are spontaneous under standard conditions. If the reaction is not spontaneous, write the corresponding spontaneous reaction.
(a) $H_2(g) + \frac{1}{2}O_2(g) \longrightarrow H_2O(l)$
(b) $2H_2(g) + C_2H_2(g) \longrightarrow C_2H_6(g)$
(c) $(CH_3)_2O(g) + H_2O(g) \longrightarrow 2CH_3OH(l)$
(d) $CH_4(g) + H_2O(g) \longrightarrow CO(g) + 3H_2(g)$

68. Use the tables in the text to determine whether the following reactions are spontaneous under standard conditions. If the reaction is not spontaneous, write the corresponding spontaneous reaction.
(a) $K_2O_2(s) \longrightarrow 2K(s) + O_2(g)$
(b) $PbCO_3(s) \longrightarrow PbO(s) + CO_2(g)$
(c) $P_4(s) + 6H_2(g) \longrightarrow 4PH_3(g)$
(d) $2AgCl(s) + H_2S(g) \longrightarrow Ag_2S(s) + 2HCl(g)$

69. Nitrogen fixation is the process by which nitrogen in the atmosphere is reduced to NH_3 for use by organisms. Several reactions are associated with this process; three are listed in the table. Which of these are spontaneous at 25°C? If a reaction is not spontaneous, at what temperature does it become spontaneous?

Reaction	ΔH°_{298}, kcal/mol	ΔS°_{298}, cal/(deg · mol)
$\frac{1}{2}N_2 + O_2 \longrightarrow NO_2$	8.0	−14.4
$\frac{1}{2}N_2 + \frac{1}{2}O_2 \longrightarrow NO$	21.6	2.9
$\frac{1}{2}N_2 + \frac{3}{2}H_2 \longrightarrow NH_3$	−11.0	−23.7

70. A student was asked to propose three reactions for the oxidation of carbon or a carbon compound to CO or CO_2. The reactions are listed in the table. Are any of these reactions spontaneous at 25°C? If a reaction does not occur spontaneously at 25°C, at what temperature does it become spontaneous?

Reaction	ΔH°_{298}, kcal/mol	ΔS°_{298}, cal/(deg · mol)
$C(s) + H_2O(g) \longrightarrow CO(g) + H_2(g)$	42	32
$CO(g) + H_2O(g) \longrightarrow CO_2(g) + H_2(g)$	−9.8	−10.1
$CH_4(g) + H_2O(g) \longrightarrow CO(g) + 3H_2(g)$	49.3	51.3

71. Tungsten trioxide (WO_3) is a dense yellow powder that, because of its bright color, is used as a pigment in oil paints and water colors (although cadmium yellow is more commonly used in artists' paints). Tungsten metal can be isolated by the reaction of WO_3 with H_2 at 1100°C according to the equation $WO_3(s) + 3H_2(g) \longrightarrow W(s) + 3H_2O(g)$. What is the lowest temperature at which the reaction occurs spontaneously? ΔH° = 27.4 kJ/mol and ΔS° = 29.8 J/K.

72. Sulfur trioxide (SO_3) is produced in large quantities in the industrial synthesis of sulfuric acid. Sulfur dioxide is converted to sulfur trioxide by reaction with oxygen gas.
(a) Write a balanced equation for the reaction of SO_2 with $O_2(g)$ and determine its ΔG°.
(b) What is the value of the equilibrium constant at 600°C.
(c) If you had to rely on the equilibrium concentrations alone, would you obtain a higher yield of product at 400°C or at 600°C?

73. Calculate ΔG° for the general reaction $MCO_3(s) \longrightarrow MO(s) + CO_2(g)$ at 25°C, where M is Mg or Ba. At what temperature does each of these reactions become spontaneous?

Compound	ΔH°_f, kJ/mol	S°, J/(mol · K)
MCO_3		
Mg	−1111	65.85
Ba	−1213.0	112.1
MO		
Mg	−601.6	27.0
Ba	−548.0	72.1
CO_2	−393.5	213.8

74. The reaction of aqueous solutions of barium nitrate with sodium iodide is described by the equation

$$Ba(NO_3)_2(aq) + 2NaI(aq) \longrightarrow BaI_2(aq) + 2NaNO_3(aq)$$

You are interested in determining the absolute entropy of BaI_2, but that information is not listed in your tables. However, you have been able to obtain the following information:

	$Ba(NO_3)_2$	NaI	BaI_2	$NaNO_3$
ΔH°_f, kJ/mol	−952.36	−295.31	−605.4	−447.5
S°, J/(mol · K)	302.5	170.3		205.4

You know that ΔG° for the reaction at 25°C is 22.64 kJ/mol. What is ΔH° for this reaction? What is S° for BaI_2?

18.6 Spontaneity and Equilibrium

75. Carbon monoxide, a toxic product from the incomplete combustion of fossil fuels, reacts with water to form CO_2 and H_2, as shown in the equation $CO(g) + H_2O(g) \rightleftharpoons CO_2(g) + H_2(g)$, for which ΔH° = −41.0 KJ/mol and ΔS° = −42.3 J cal/(mol · K) at 25°C and 1 atm.
(a) What is ΔG° for this reaction?
(b) What is ΔG if the gases have the following partial pressures: P_{CO} = 1.3 atm, P_{H_2O} = 0.8 atm, P_{CO_2} = 2.0 atm, and P_{H_2} = 1.3 atm?
(c) What is ΔG if the temperature is increased to 150°C assuming no change in pressure?

76. Methane and water react to form CO and H_2 according to the equation $CH_4(g) + H_2O(g) \rightleftharpoons CO(g) + 3H_2(g)$.
(a) What is the standard free energy change for this reaction?
(b) What is K_p for this reaction under standard conditions?
(c) What is the CO pressure if 1.3 atm of methane reacts with 0.8 atm of water, producing 1.8 atm of hydrogen gas?
(d) What is the hydrogen gas pressure if 2.0 atm of methane is allowed to react with 1.1 atm of water?
(e) At what temperature does the reaction become spontaneous?

77. Calculate the equilibrium constant at 25°C for each equilibrium reaction, and comment on the extent of the reaction:
(a) $CCl_4(g) + 6H_2O(l) \rightleftharpoons CO_2(g) + 4HCl(aq)$; $\Delta G° = -377$ kJ/mol
(b) $Xe(g) + 2F_2(g) \rightleftharpoons XeF_4(s)$; $\Delta H° = -66.3$ kJ/mol, $\Delta S° = -102.3$ J/(mol · K)
(c) $PCl_3(g) + S \rightleftharpoons PSCl_3(l)$; $\Delta G_f°$: $PCl_3 = -272.4$ kJ/mol, $\Delta G_f°$ $PSCl_3 = -363.2$ kJ/mol

78. Calculate the equilibrium constant at 25°C for each equilibrium reaction, and comment on the extent of the reaction:
(a) $2KClO_3(s) \rightleftharpoons 2KCl(s) + 3O_2(g)$; $\Delta G° = -225.8$ kJ/mol
(b) $CoCl_2(s) + 6H_2O(g) \rightleftharpoons CoCl_2{\cdot}6H_2O(s)$; $\Delta H°_{rxn} = -352$ kJ/mol, $\Delta S°_{rxn} = -899$ J/(mol · K)
(c) $2PCl_3(g) + O_2(g) \rightleftharpoons 2POCl_3(g)$; $\Delta G_f°$: $PCl_3 = -272.4$ kJ/mol, $\Delta G_f°$ $POCl_3 = -558.5$ kJ/mol

79. The gas-phase decomposition of N_2O_4 to NO_2 is an equilibrium reaction with $K_p = 4.66 \times 10^{-3}$. Calculate the standard free energy change for the equilibrium reaction between N_2O_4 and NO_2.

80. The standard free energy for the dissolution $K_4Fe(CN)_6 \cdot H_2O(s) \rightleftharpoons 4K^+(aq) + Fe(CN)_6^{4-}(aq) + H_2O(l)$ is 26.1 kJ/mol. What is the equilibrium constant for this process at 25°C?

81. Ammonia reacts with water in liquid ammonia solution (*amm*) according to the equation $NH_3(g) + H_2O(amm) \rightleftharpoons NH_4^+(amm) + OH^-(amm)$. The change in enthalpy for this reaction is 21 kJ/mol, and $\Delta S° = -303$ J/(mol · K). What is the equilibrium constant for the reaction at the boiling point of liquid ammonia (−31°C)?

82. At 25°C, a saturated solution of barium carbonate is found to have a concentration of $[Ba^{2+}] = [CO_3^{2-}] = 5.08 \times 10^{-5}$ *M*. Determine $\Delta G°$ for the dissolution of $BaCO_3$.

83. Lead phosphates are believed to play a major role in controlling the overall solubility of lead in acidic soils. One of the dissolution reactions is $Pb_3(PO_4)_2(s) + 4H^+(aq) \rightleftharpoons 3Pb^{2+}(aq) + 2H_2PO_4^-(aq)$, for which $\log K = -1.80$. What is $\Delta G°$ for this reaction?

84. The conversion of butane to 2-methylpropane is an equilibrium process with $\Delta H° = -2.05$ kcal/mol and $\Delta G° = -0.89$ kcal/mol.
(a) What is the change in entropy for this conversion?
(b) Based on structural arguments, are the sign and magnitude of the entropy change what you would expect? Why?
(c) What is the equilibrium constant for this reaction?

85. The reaction of $CaCO_3(s)$ to produce $CaO(s)$ and $CO_2(g)$ has an equilibrium constant at 25°C of 2×10^{-23}. Values of $\Delta H_f°$ are: $CaCO_3$, −1207.6 kJ/mol; CaO, −634.9 kJ/mol; CO_2, −393.5 kJ/mol.
(a) What is $\Delta G°$ for this reaction?
(b) What is the equilibrium constant at 900°C?
(c) What is the partial pressure of $CO_2(g)$ in equilibrium with CaO and $CaCO_3$ at this temperature?
(d) Are reactants or products favored at the lower temperature? At the higher temperature?

86. In acidic soils, dissolved Al^{3+} undergoes a complex formation reaction with SO_4^{2-} to form $[AlSO_4^+]$. The equilibrium constant at 25°C for the reaction $Al^{3+}(aq) + SO_4^{2-}(aq) \rightleftharpoons AlSO_4^+(aq)$ is 1585.
(a) What is $\Delta G°$ for this reaction?
(b) How does this value compare with $\Delta G°$ for the reaction $Al^{3+}(aq) + F^-(aq) \rightleftharpoons AlF^{2+}(aq)$, for which $K = 10^7$ at 25°C?
(c) Which is the better ligand to use to trap Al^{3+} from the soil?

APPLICATIONS

87. Electric utilities have been exploring thermal energy storage as a potentially attractive energy-storage solution for peak use. Thermal energy is extracted as steam and stored in hot rock, oil, or hot water for later conversion to electricity via heat exchangers. Which steps involve heat transfer? Which involve work done?

88. During World War II, German scientists developed the first rocket-powered airplane to be flown in combat, the *Messerschmitt 163 Komet*. The "Comet" was powered by the reaction of liquid hydrogen peroxide (H_2O_2) and hydrazine (N_2H_4) according to the reaction

$$2H_2O_2(l) + N_2H_4(l) \longrightarrow N_2(g) + 4H_2O(g)$$

(a) Determine the standard molar enthalpy of reaction.
(b) What amount of work can be done by reacting 1.00 kg of hydrazine with a stoichiometric amount of hydrogen peroxide at 25°C at a constant pressure of 1.00 atm?
(c) What is the change in internal energy under these conditions?

89. During the 1950s, pentaborane-9 was tested as a potential rocket fuel. However, the idea was abandoned when it was discovered that B_2O_3, the product of the reaction of pentaborane-9 with O_2, was an abrasive that destroyed rocket nozzles. The reaction is represented by the equation

$$2B_5H_9(l) + 12O_2(g) \longrightarrow 5B_2O_3(s) + 9H_2O(g)$$

(a) Determine the standard molar enthalpy of reaction.
(b) How much work is done against a pressure of 1.0 atm when 50 lb of fuel is allowed to react with sufficient oxygen to cause the reaction to go to completion at 25°C?
(c) What is ΔE for the reaction?

90. Polar explorers must be particularly careful to keep their clothes from becoming damp because the resulting heat loss could be fatal. If a polar explorer's clothes absorbed 1.0 kg of water and the clothes dried from the polar wind, what would be the heat loss? $\Delta H_{vap}(H_2O) = 44$ kJ/mol. How much glucose must be consumed to make up for this heat loss to prevent death? $\Delta H_{comb}(\text{glucose}) = -802$ kJ/mol.

91. Propane gas is generally preferred to kerosene as a fuel for stoves in the boating industry because kerosene stoves require more maintenance. Propane, however, is much more flammable than kerosene and imposes an added risk because it is denser than air and can collect in the bottom of a boat and ignite. The complete combustion of propane produces CO_2 gas

and H_2O vapor and has a value of $\Delta H = -2220$ kJ/mol at 25°C. What is ΔE?

92. The propane in Problem 93 can be produced from the hydrogenation of propene according to the reaction $C_3H_6(g) + H_2(g) \longrightarrow C_3H_8(g)$; $\Delta H = -124$ kJ/mol. Given that the reaction $H_2(g) + \frac{1}{2}O_2(g) \longrightarrow H_2O(g)$ has $\Delta H = -285.8$ kJ/mol, what is (a) the standard enthalpy of combustion of propene, and (b) the value of ΔE for the combustion of propene?

93. The anaerobic conversion of sucrose, a sweetening agent, to lactic acid, which is associated with sour milk, can be described by the equation

$$\underset{\text{Sucrose}}{C_{12}H_{22}O_{11}} + H_2O \longrightarrow \underset{\text{Lactic acid}}{4CH_3CH(OH)CO_2H}$$

The combustion of sucrose, however, occurs as follows:

$$C_{12}H_{22}O_{11}(s) + 12O_2(g) \longrightarrow 12CO_2(g) + 11H_2O(l)$$

(a) Which reaction is thermodynamically more favorable: anaerobic conversion of sucrose to lactic acid, or aerobic oxidation to CO_2 and H_2O? Values of ΔH_f are: lactic acid, −694.1 kJ/mol; sucrose, −222 kJ/mol; CO_2, −393.5 kJ/mol; and H_2O, −285.8 kJ/mol.

(b) What is ΔE for the combustion of 12.0 g of sucrose at normal body temperature (37°C)?

94. Phosphorus exists as several allotropes, the most common being red, black, and white phosphorus. White phosphorus consists of tetrahedral P_4 molecules and melts at 44.15°C; it is converted to red phosphorus by heating at 400°C for several hours. The chemical differences between red and white phosphorus are considerable: white phosphorus burns in air, whereas red phosphorus is stable; white phosphorus is soluble in organic compounds, whereas red phosphorus is not; white phosphorus melts at 44.15°C, whereas red phosphorus melts at 597°C. If the enthalpy of fusion of white phosphorus is 0.659 kJ/mol. What is its ΔS? Black phosphorus is even less reactive than red. Based on this information, which allotrope would you predict to have the highest entropy? The lowest? Why?

95. Ruby and sapphire have a common mineral name, corundum (Al_2O_3). Although they are crystalline versions of the same compound, the nature of the imperfections determines the identity of the gem. Outline a method for measuring and comparing the entropy of a ruby with the entropy of a sapphire. How would you expect the entropies to compare with the entropy of a perfect corundum crystal?

96. Tin has two crystalline forms, α and β, represented in the equilibrium equation

$$\underset{\text{Gray}}{\alpha\text{-tin}} \underset{}{\overset{18^\circ C}{\rightleftharpoons}} \underset{\text{White}}{\beta\text{-tin}} \overset{232^\circ C}{\rightleftharpoons} Sn(l)$$

The earliest known tin artifacts were discovered in Egyptian tombs of the 18th dynasty (1580–1350 B.C.), although archeologists are surprised that so few tin objects exist from earlier eras. It has been suggested that many early tin objects were either oxidized to a mixture of stannous and stannic oxides or transformed to powdery, gray tin. Sketch a thermodynamic cycle similar to Figure 18.15b to show the conversion of liquid tin to gray tin. Then calculate the change in entropy that accompanies the conversion of Sn(l) to α-Sn using the following data: C_p(white) = 26.99, C_p(gray) = 25.77 J/(mol · K), $\Delta H_{fus} = 7.0$ kJ/mol, $\Delta H_{\beta \longrightarrow \alpha} = -2.2$ kJ/mol.

97. The reaction of SO_2 with O_2 to produce SO_3 has great industrial significance because SO_3 is converted to H_2SO_4 by reaction with water. Unfortunately, the reaction is also environmentally important because SO_3 from industrial smokestacks is a primary source of acid rain. ΔH for the reaction of SO_2 with O_2 to form SO_3 is −23.49 kJ/mol, and ΔS is −22.66 J/K. Does this reaction occur spontaneously at 25°C? Does it occur spontaneously at 800°C assuming no change in ΔH and ΔS? Why is this reaction usually carried out at elevated temperatures?

98. Pollutants from industrial societies pose health risks to individuals from exposure to metals such as lead, mercury, and cadmium. The biological effects of a toxic metal may be reduced by removing it from the system using a chelating agent, which binds to the metal and forms a complex that is eliminated from the system without causing more damage. In designing a suitable chelating agent, one must be careful, however, because some chelating agents form metal complexes that are more toxic than the metal itself. Both methylamine (CH_3NH_2) and ethylenediamine ($H_2NCH_2CH_2NH_2$, abbreviated en) are commonly used to treat heavy metal poisoning. In the case of cadmium, the reactions are as follows:

$$Cd^{2+} + 2CH_3NH_2 \longrightarrow Cd(CH_3NH_2)_2^{2+}$$
$$\Delta H = -7.03 \text{ kcal/mol}, \Delta S = -1.58 \text{ cal/(mol} \cdot \text{K)}$$
$$Cd^{2+} + en \longrightarrow Cd(en)^{2+}$$
$$\Delta H = -7.02 \text{ kcal/mol}, \Delta S = 3.15 \text{ cal/(mol} \cdot \text{K)}$$

Based strictly on thermodynamic arguments, which would you choose to administer to a patient suffering from cadmium toxicity, and why? Assume a body temperature of 37°C.

99. Explosive reactions often have a large negative enthalpy change and a large positive entropy change, but the reaction must also be kinetically favorable. For example, the following equation represents the reaction between hydrazine, a rocket propellant, and the oxidizer dinitrogen tetroxide:

$$2N_2H_4(l) + N_2O_4(l) \longrightarrow 4H_2O(g) + 3N_2(g)$$
$$\Delta H^\circ = -249 \text{ kJ/mol}, \Delta S^\circ = 218 \text{ J/(mol} \cdot \text{K)}$$

(a) How much free energy is produced from this reaction at 25°C?

(b) Is the reaction thermodynamically favorable?

(c) What is K?

(d) This reaction requires thermal ignition. Why?

100. Cesium, a silvery-white metal used in the manufacture of vacuum tubes, is produced industrially by the reaction of CsCl with CaC_2:

$$2CsCl(l) + CaC_2(s) \longrightarrow CaCl_2(l) + 2C(s) + 2Cs(g)$$

Compare the free energy produced from this reaction at 25°C and at 1227°C, the temperature at which it is normally run, given these values: $\Delta H^\circ_{298\,K} = 32.0$ kJ/mol, $\Delta S^\circ_{298\,K} = 8.0$ J/(mol·K); $\Delta H^\circ_{1500\,K} = -0.6$ kJ/mol, $\Delta S^\circ_{1500\,K} = 3.6$ J/(mol·K).

(a) If you wanted to minimize energy costs in your production facility by reducing the temperature of the reaction, what is the lowest temperature at which products are favored over reactants (assuming the reaction is kinetically favorable at the lower temperature)? Assume ΔH and ΔS do not vary with temperature.

(b) What is the ratio $K_{1500\ K}/K_{298\ K}$?

101. Dessicants (drying agents) can often be regenerated by heating, although it is generally not economically worthwhile to do so. A dessicant that is commonly regenerated is $CaSO_4 \cdot 2H_2O$:

$$CaSO_4 \cdot 2H_2O(s) \longrightarrow CaSO_4(s) + 2H_2O(g)$$
$$\Delta H^\circ_{298\ K} = 25.1\ \text{kJ/mol},\ \Delta S^\circ_{298\ K} = 69.3\ \text{J/(mol} \cdot \text{K)}$$

Regeneration is carried out at 250°C.

(a) What is ΔG° for this reaction?

(b) What is the equilibrium constant at 25°C?

(c) What is the ratio $K_{250°C}/K_{25°C}$?

(d) What is the equilibrium constant at 250°C?

(e) Is regeneration of $CaSO_4 \cdot 2H_2O$ an enthalpy- or entropy-driven process? Explain your answer.

102. Nitrogen triiodide ($NI_3 \cdot NH_3$) is a simple explosive that, when detonated, produces N_2 and I_2. It can be painted on surfaces when wet, but it is shock sensitive when dry (even touching it with a feather can cause an explosion!). Do you expect ΔG for the explosion reaction to be positive or negative? Why doesn't NI_3 explode spontaneously?

103. Adenosine triphosphate (ATP) contains high-energy phosphate bonds that are used in energy metabolism, coupling energy-yielding and energy-requiring processes. Cleaving a phosphate link by hydrolysis (ATP hydrolysis) can be described by the reaction $ATP + H_2O \rightleftharpoons ADP + P_i + H^+$, where P_i symbolizes phosphate. Glycerol and ATP react to form glycerol-3-phosphate, ADP, and H^+, with an overall $K = 6.61 \times 10^5$ at 37°C. The reaction of glycerol with phosphoric acid to form glycerol-3-phosphate and water has an equilibrium constant of 2.82×10^{-2}. What is the equilibrium constant for ATP hydrolysis? How much free energy is released from the hydrolysis of ATP?

104. Consider the biological reduction of molecular nitrogen, for which the following is the minimal reaction stoichiometry under optimal conditions (P_i = phosphate):

$$8H^+ + 8e^- + N_2 + 16ATP \longrightarrow H_2 + 2NH_3 + 16ADP + 16P_i$$

(a) What is the approximate ratio of K_{eq} for the reaction shown above and for the same reaction in the absence of ATP?

(b) Given the fact that at pH 7 both the reaction of protons and electrons to give H_2 and the reaction of H_2 with N_2 to give ammonia are thermodynamically spontaneous (that is, $K \gg 1$), suggest a reason that nitrogen-fixing bacteria use such a large energy input to drive a reaction that is already spontaneous.

19 Electrochemistry

LEARNING OBJECTIVES

- To distinguish between galvanic and electrolytic cells
- To predict spontaneous reactions using redox potentials
- To balance redox reactions using half-reactions
- To understand the relationship between cell potential and the equilibrium constant
- To be able to measure solution concentrations using cell potentials
- To describe how commercial galvanic cells operate
- To describe the process of corrosion
- To understand electrolysis and to be able to describe it quantitatively

CHAPTER OUTLINE

In oxidation–reduction (redox) reactions, electrons are transferred from one species (the reductant) to another (the oxidant). This transfer of electrons provides a means for converting chemical energy to electrical energy, or vice versa. The study of the relationship between electricity and chemical reactions is called **electrochemistry**, an area of chemistry we introduced in Chapters 4 and 5. In this chapter, we describe electrochemical reactions in more depth and explore some of their applications.

In the first three sections, we review redox reactions; describe how they can be used to generate an electrical potential, or *voltage*; and discuss factors that affect the magnitude of the potential. We then explore the relationships among the electrical potential, the change in free energy, and the equilibrium constant for a redox reaction, which are all measures of the thermodynamic driving force for a reaction. Finally, we examine two kinds of applications of electrochemical

A view from the top of the Statue of Liberty, showing the green patina coating the statue. The patina is formed by corrosion of the copper skin of the statue, which forms a thin layer of an insoluble compound that contains copper(II), sulfate, and hydroxide ions.

principles: those in which a spontaneous reaction is used to provide electricity, and those in which electrical energy is used to drive a thermodynamically nonspontaneous reaction. By the end of this chapter, you will understand why different kinds of batteries are used in cars, flashlights, cameras, and portable computers; how rechargeable batteries operate; and why corrosion occurs and how to slow, if not prevent, it. You will also discover how metal objects can be plated with silver or chromium for protection; how silver polish removes tarnish; and how to calculate the amount of electricity needed to produce aluminum, chlorine, copper, and sodium on an industrial scale.

19.1 Describing Electrochemical Cells

In any electrochemical process, electrons flow from one chemical substance to another, driven by an oxidation–reduction (redox) reaction. As we described in Chapter 3, a redox reaction occurs when electrons are transferred from a substance that is oxidized to one that is being reduced. The **reductant** is the substance that loses electrons and is oxidized in the process; the **oxidant** is the species that gains electrons and is reduced in the process. The associated potential energy is determined by the potential difference between the valence electrons in atoms of different elements (see Section 7.3).

Because it is impossible to have a reduction without an oxidation, and vice versa, a redox reaction can be described as two **half-reactions**, one representing the oxidation process and one the reduction process. For the reaction of zinc with bromine, the half-reactions are

$$\text{Reduction half-reaction:}\quad Br_2(aq) + 2e^- \longrightarrow 2Br^-(aq) \tag{19.1a}$$

$$\text{Oxidation half-reaction:}\quad Zn(s) \longrightarrow Zn^{2+}(aq) + 2e^- \tag{19.1b}$$

Note the pattern

In any redox reaction, the number of electrons lost by the reductant is equal to the number of electrons gained by the oxidant.

Notice that each reaction is written to show what is actually occurring in the system. Adding the two half-reactions gives the overall chemical reaction:

$$\text{Overall reaction:}\quad \underset{\text{Reductant}}{Zn(s)} + \underset{\text{Oxidant}}{Br_2(aq)} \longrightarrow Zn^{2+}(aq) + 2Br^-(aq) \tag{19.1c}$$

A redox reaction is balanced when *the number of electrons lost by the reductant is equal to the number of electrons gained by the oxidant*. Hence, like any balanced chemical equation, the overall process is electrically neutral; that is, the net charge is the same on both sides of the equation.

In most of our discussions of chemical reactions, we have assumed that the reactants are in intimate physical contact with one another. Acid–base reactions, for example, are usually carried out with the acid and the base dispersed in a single phase, such as a liquid solution. With redox reactions, however, it is possible to physically separate the oxidation and reduction half-reactions in space, as long as there is a complete circuit, including an external electrical connection, such as a wire, between the two half-reactions. As the reaction progresses, the electrons flow from reductant to oxidant over this electrical connection, producing an electric current that can be used to do work. An apparatus that is used to generate electricity from a spontaneous redox reaction or, conversely, that uses electricity to drive a nonspontaneous redox reaction is called an **electrochemical cell**.

There are two types of electrochemical cells: galvanic cells and electrolytic cells. A **galvanic cell*** uses the energy released during a spontaneous redox reaction ($\Delta G < 0$) to *generate* electricity. This type of electrochemical cell is often called a **voltaic cell** after its inventor, the Italian physicist Alessandro Volta (1745–1827). In contrast, an **electrolytic cell** *consumes* electrical energy from an external source, using it to cause a

* Named for the Italian physicist and physician Luigi Galvani (1737–1798), who observed that dissected frog leg muscles twitched when a small electric shock was applied, demonstrating the electrical nature of nerve impulses.

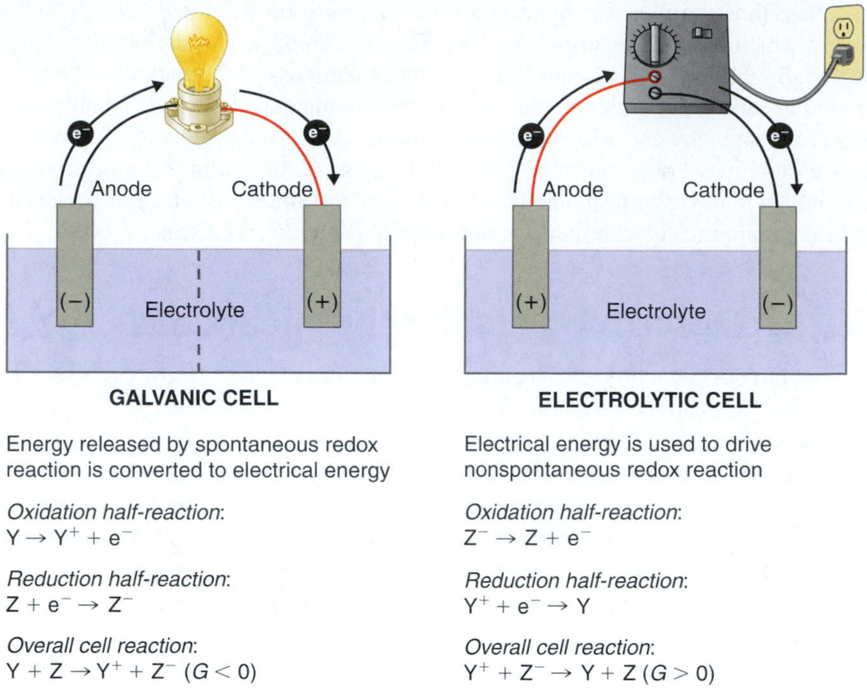

Figure 19.1 Electrochemical cells. A galvanic cell (*left*) transforms the energy released by a spontaneous redox reaction into electrical energy that can be used to perform work. The oxidative and reductive half-reactions usually occur in separate compartments that are connected by an external electrical circuit; in addition, a second connection that allows ions to flow between the compartments (shown here as a vertical dashed line to represent a porous barrier) is necessary to maintain electrical neutrality. The potential difference between the electrodes (voltage) causes electrons to flow from the reductant to the oxidant through the external circuit, generating an electric current. In an electrolytic cell (*right*), an external source of electrical energy is used to generate a potential difference between the electrodes that forces electrons to flow, driving a nonspontaneous redox reaction; only a single compartment is employed in most applications. In both kinds of electrochemical cell, the anode is the electrode at which the oxidation half-reaction occurs, and the cathode is the electrode at which the reduction half-reaction occurs.

nonspontaneous redox reaction to occur ($\Delta G > 0$). Both types contain two **electrodes**, which are solid metals connected to an external circuit that provides an electrical connection between the two parts of the system (Figure 19.1). The oxidation half-reaction occurs at one electrode, the **anode**, and the reduction half-reaction occurs at the other, the **cathode**. When the circuit is closed, electrons flow from the anode to the cathode. The electrodes are also connected by an *electrolyte*, an ionic substance or solution that allows ions to transfer between the electrode compartments, thereby maintaining the system's electrical neutrality. In this section, we focus on reactions that occur in galvanic cells. We discuss electrolytic cells in Section 19.7.

MGC Introduction to Galvanic Cells

$Zn(s) + Cu^{2+}(aq) \rightarrow Zn^{2+}(aq) + Cu(s)$

Figure 19.2 Reaction of metallic zinc with aqueous copper(II) ions in a single compartment. When a zinc rod is inserted into a beaker that contains an aqueous solution of copper(II) sulfate, a spontaneous redox reaction occurs: the zinc electrode dissolves to give $Zn^{2+}(aq)$ ions, while $Cu^{2+}(aq)$ ions are simultaneously reduced to metallic copper. The reaction occurs so rapidly that the copper is deposited as very fine particles that appear black, rather than the usual reddish color of copper.

Galvanic (Voltaic) Cells

To illustrate the basic principles of a galvanic cell, let's consider the reaction of metallic zinc with cupric ion (Cu^{2+}) to give copper metal and Zn^{2+} ion. The balanced chemical equation is

$$Zn(s) + Cu^{2+}(aq) \longrightarrow Zn^{2+}(aq) + Cu(s) \qquad (19.2)$$

We can cause this reaction to occur by inserting a zinc rod into an aqueous solution of copper(II) sulfate. As the reaction proceeds, the zinc rod dissolves and a mass of metallic copper forms (Figure 19.2). These changes occur spontaneously, but all the energy released is in the form of heat rather than in a form that can be used to do work.

This same reaction can be carried out using the galvanic cell illustrated in Figure 19.3a. To assemble the cell, a copper strip is inserted into a beaker that contains a 1 *M* solution of Cu^{2+} ions, and a zinc strip is inserted into a different beaker that contains a 1 *M* solution of Zn^{2+} ions. The two metal strips, which serve as electrodes, are connected by a wire, and the compartments are connected by a **salt bridge**, a U-shaped tube inserted into both solutions that contains a concentrated liquid or gelled electrolyte. The ions in the salt bridge are selected so that they do not interfere with the electrochemical reaction by being oxidized or reduced themselves or by forming a precipitate or complex; commonly used cations and anions are Na^+ or K^+ and NO_3^- or SO_4^{2-}, respectively. (Note that the ions in the salt bridge do *not* have to be the same as those in the redox

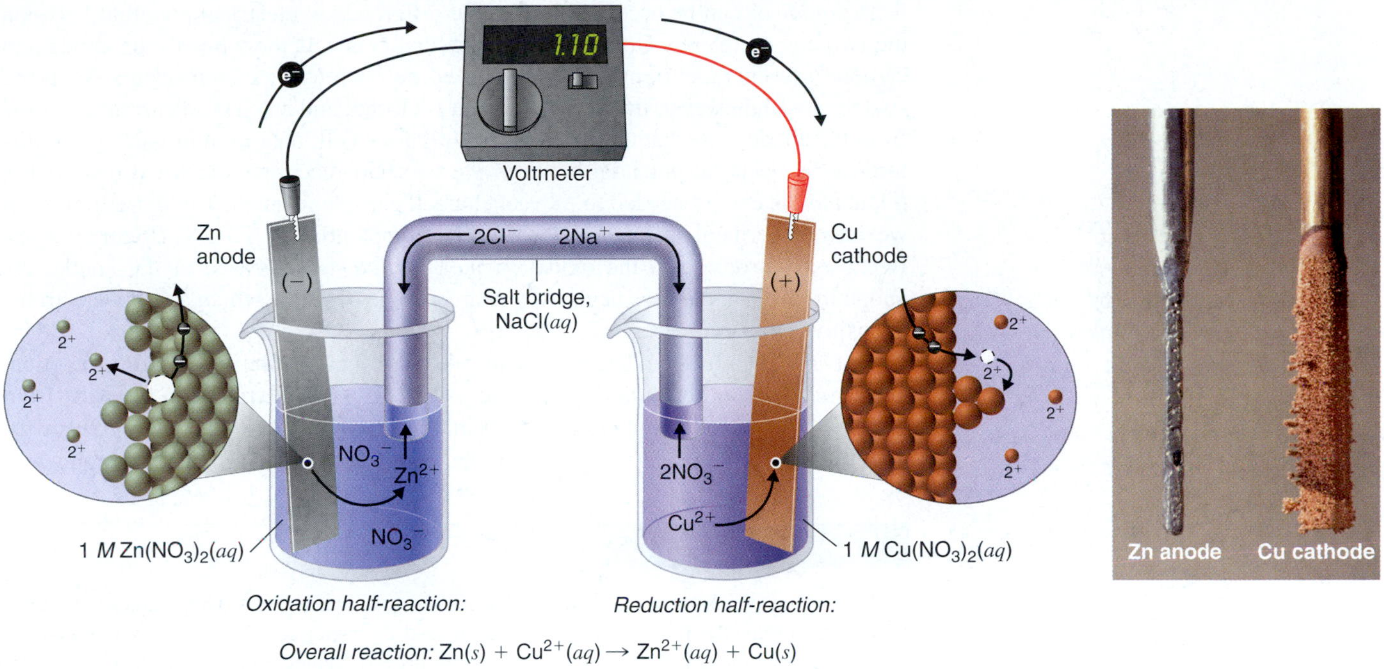

(a) (b)

Figure 19.3 Reaction of metallic zinc with aqueous copper(II) ions in a galvanic cell. (a) A galvanic cell can be constructed by inserting a copper strip into a beaker that contains an aqueous 1 *M* solution of Cu^{2+} ions, and a zinc strip into a different beaker that contains an aqueous 1 *M* solution of Zn^{2+} ions. The two metal strips are connected by a wire that allows electricity to flow, and the beakers are connected by a salt bridge that maintains electrical neutrality. When the switch is closed to complete the circuit, the zinc electrode (the anode) is spontaneously oxidized to Zn^{2+} ions in the left compartment, while Cu^{2+} ions are simultaneously reduced to copper metal at the copper electrode (the cathode). (b) As the reaction progresses, the Zn anode loses mass as it dissolves to give $Zn^{2+}(aq)$ ions, while the Cu cathode gains mass as $Cu^{2+}(aq)$ ions are reduced to copper metal that is deposited on the cathode.

couple in either compartment.) When the circuit is closed, a spontaneous reaction occurs: zinc metal is oxidized to Zn^{2+} ions at the zinc electrode (the anode), and Cu^{2+} ions are reduced to Cu metal at the copper electrode (the cathode). As the reaction progresses, the zinc strip dissolves and the concentration of Zn^{2+} ions in the Zn^{2+} solution increases; simultaneously, the copper strip gains mass and the concentration of Cu^{2+} ions in the Cu^{2+} solution decreases (Figure 19.3b). Thus, we have carried out the same reaction as we did using a single beaker, but this time the oxidative and reductive half-reactions are physically separated from each other. The electrons that are released at the anode flow through the wire, producing an electric current. Galvanic cells therefore transform chemical energy into electrical energy that can then be used to do work.

The electrolyte in the salt bridge serves two purposes: to complete the circuit by carrying electrical charge and to maintain electrical neutrality in both solutions. The identity of the salt in a salt bridge is unimportant, as long as the component ions do not react or undergo a redox reaction under the operating conditions of the cell. The salt bridge maintains electrical neutrality in both compartments by allowing ions to migrate between the two solutions. Without such a connection, the total positive charge in the Zn^{2+} solution would increase as the zinc metal dissolves, and the total positive charge in the Cu^{2+} solution would decrease. The salt bridge allows charges to be neutralized by a flow of anions into the Zn^{2+} solution and a flow of cations into the Cu^{2+} solution. In the absence of a salt bridge or some other similar connection, the reaction would rapidly cease because electrical neutrality could not be maintained.

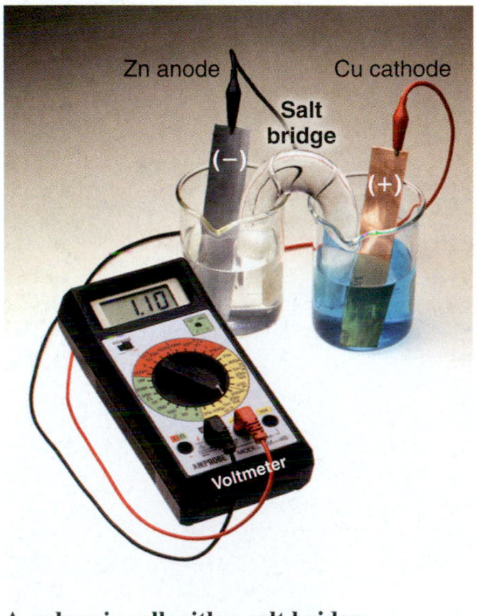

A galvanic cell with a salt bridge.

A *voltmeter* can be used to measure the difference in electrical potential between the two compartments. Opening the switch that connects the wires to the anode and cathode prevents a current from flowing, so no chemical reaction occurs. With the switch closed, however, the external circuit is closed, and an electric current can flow from the anode to the cathode. The **potential** of the cell, measured in volts, is the difference in electrical potential between the two half-reactions; electrical potential is related to the energy needed to move a charged particle in an electric field. In the cell we have described, the voltmeter indicates a potential of 1.10 V (Figure 19.3a). Because electrons from the oxidation half-reaction are released at the anode, the anode in a galvanic cell is negatively charged. The cathode, which attracts electrons, is positively charged.

Note the pattern

A galvanic cell converts the energy released by a spontaneous chemical reaction to electrical energy.

Not all electrodes undergo a chemical tranformation during a redox reaction. To prevent it from reacting during a redox process, the electrode can be made from an inert, highly conducting metal such as platinum, which does not appear in the overall electrochemical reaction. This phenomenon is illustrated in the next example.

EXAMPLE 19.1

A chemist has constructed a galvanic cell consisting of two beakers. One beaker contains a strip of tin immersed in aqueous sulfuric acid, and the other contains a platinum electrode immersed in aqueous nitric acid. The two solutions are connected by a salt bridge, and the electrodes are connected by a wire. Current begins to flow, and bubbles of a gas appear at the platinum electrode. The spontaneous redox reaction that occurs is described by this balanced equation:

$$3Sn(s) + 2NO_3^-(aq) + 8H^+(aq) \longrightarrow 3Sn^{2+}(aq) + 2NO(g) + 4H_2O(l)$$

For this galvanic cell, **(a)** write the half-reaction that occurs at each electrode; **(b)** indicate which electrode is the cathode and which is the anode; and **(c)** indicate which electrode is the positive electrode and which is the negative electrode.

Given Galvanic cell, redox reaction

Asked for Half-reactions, identity of anode and cathode, electrode assignment as positive or negative

Strategy

A Identify the oxidation half-reaction and the reduction half-reaction. Then identify the anode and cathode from the half-reaction that occurs at each electrode.

B From the direction of electron flow, assign each electrode as either positive or negative.

Solution

(a) **A** In the reduction half-reaction, nitrate is reduced to nitric oxide. (The nitric oxide would then react with oxygen in the air to form NO_2, with its characteristic red-brown color.) In the oxidation half-reaction, metallic tin is oxidized. The half-reactions corresponding to the actual reactions that occur in the system are

Reduction: $NO_3^-(aq) + 4H^+(aq) + 3e^- \longrightarrow NO(g) + 2H_2O(l)$

Oxidation: $Sn(s) \longrightarrow Sn^{2+}(aq) + 2e^-$

Thus, nitrate is reduced to NO, while the tin electrode is oxidized to Sn^{2+}.

(b) Because the reduction reaction occurs at the Pt electrode, it is the cathode. Conversely, the oxidation reaction occurs at the tin electrode, so it is the anode.

(c) **B** Electrons flow from the tin electrode through the wire to the platinum electrode, where they transfer to nitrate. The electric circuit is completed by the salt bridge, which permits the diffusion of cations toward the cathode and of anions toward the anode. Because electrons flow *out* of the tin electrode, it must be electrically negative. In contrast, electrons flow *toward* the Pt electrode, so that electrode must be electrically positive.

EXERCISE 19.1

Consider a simple galvanic cell consisting of two beakers connected by a salt bridge. One beaker contains a solution of MnO_4^- in dilute sulfuric acid and has a Pt electrode. The other beaker contains a solution of Sn^{2+} in dilute sulfuric acid, also with a Pt electrode. When the two electrodes are connected by a wire, current flows and a spontaneous reaction occurs that is described by the balanced equation

$$2MnO_4^-(aq) + Sn^{2+}(aq) + 16H^+(aq) \longrightarrow 2Mn^{2+}(aq) + Sn^{4+}(aq) + 8H_2O(l)$$

For this galvanic cell, **(a)** write the half-reaction that occurs at each electrode; **(b)** indicate which electrode is the cathode and which is the anode; and **(c)** indicate which electrode is positive and which is negative.

Answer **(a) (b)** $MnO_4^-(aq) + 8H^+(aq) + 5e^- \longrightarrow Mn^{2+}(aq) + 4H_2O(l)$ occurs at the cathode; $Sn^{2+}(aq) \longrightarrow Sn^{4+}(aq) + 2e^-$ occurs at the anode; **(c)** the cathode (electrode in beaker that contains the permanganate solution) is positive, and the anode (electrode in beaker that contains the tin solution) is negative.

Constructing a Cell Diagram

Because it is somewhat cumbersome to describe any given galvanic cell in words, a more convenient notation has been developed. In this line notation, called a *cell diagram,* the identity of the electrodes and the chemical contents of the compartments are indicated by their chemical formulas, with the anode written on the far left and the cathode on the far right. Phase boundaries are shown by single vertical lines, and the salt bridge, which has two phase boundaries, by a double vertical line. Thus, the cell diagram for the Zn/Cu cell shown in Figure 19.3a is written as

Phase boundary — Two phase boundaries — Phase boundary

$$Zn(s)\,|\,Zn^{2+}(aq, 1\,M)\,||\,Cu^{2+}(aq, 1\,M)\,|\,Cu(s) \qquad (19.3)$$

Anode — Salt bridge — Cathode

Notice that a cell diagram includes solution concentrations when they are provided.

Galvanic cells can have arrangements other than the examples we have seen so far. For example, the voltage produced by a redox reaction can be measured more accurately using two electrodes immersed in a single beaker containing an electrolyte that completes the circuit. This arrangement reduces errors caused by resistance to the flow of charge at a boundary, called the *junction potential.* One example of this type of galvanic cell is

$$Pt(s)\,|\,H_2(g)\,|\,HCl(aq)\,|\,AgCl(s)\,|\,Ag(s) \qquad (19.4)$$

Notice that this cell diagram does not include a double vertical line representing a salt bridge because there is no salt bridge providing a junction between two dissimilar solutions. Moreover, solution concentrations have not been specified, so they are not included in the cell diagram. The half-reactions and the overall reaction for this cell are

Cathode reaction: $AgCl(s) + e^- \longrightarrow Ag(s) + Cl^-(aq)$ (19.5a)

Anode reaction: $\frac{1}{2}H_2(g) \longrightarrow H^+(aq) + e^-$ (19.5b)

Overall: $AgCl(s) + \frac{1}{2}H_2(g) \longrightarrow Ag(s) + Cl^-(aq) + H^+(aq)$ (19.5c)

A single-compartment galvanic cell will initially exhibit the same voltage as a galvanic cell constructed using separate compartments, but it will discharge rapidly because of the direct reaction of the reactant at the anode with the oxidized member of the cathodic redox couple. Consequently, cells of this type are not particularly useful for producing electricity.

EXAMPLE 19.2

Draw a cell diagram for the galvanic cell described in Example 19.1.

Given Galvanic cell, redox reaction

Asked for Cell diagram

Strategy

Using the symbols described, write the cell diagram beginning with the oxidation half-reaction on the left.

Solution

The anode is the tin strip, and the cathode is the Pt electrode. Beginning on the left with the anode, we indicate the phase boundary between the electrode and the tin solution by a vertical bar. The anode compartment is thus $Sn(s)|Sn^{2+}(aq)$. We could include $H_2SO_4(aq)$ with the contents of the anode compartment, but the sulfate ion (as HSO_4^-) does not participate in the overall reaction and does not need to be indicated specifically. The cathode compartment contains aqueous nitric acid, which does participate in the overall reaction, together with the product of the reaction, NO, and the Pt electrode. These are written as $HNO_3(aq)|NO(g)|Pt(s)$, with single vertical bars indicating the phase boundaries. Combining the two compartments and using a double vertical bar to indicate the salt bridge give

$$Sn(s) \mid Sn^{2+}(aq) \parallel HNO_3(aq) \mid NO(g) \mid Pt(s)$$

Notice that solution concentrations were not specified, so they are not included in this cell diagram.

EXERCISE 19.2

Draw a cell diagram for the following reaction, assuming the concentration of Ag^+ and Mg^{2+} are each 1 *M*:

$$Mg(s) + 2Ag^+(aq) \longrightarrow Mg^{2+}(aq) + 2Ag(s)$$

Answer $Mg(s)|Mg^{2+}(aq, 1\ M)\|Ag^+(aq, 1\ M)|Ag(s)$

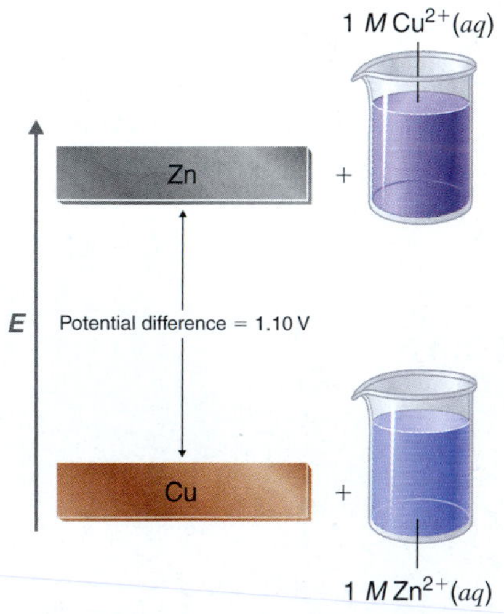

Figure 19.4 Potential energy difference in the Zn/Cu system. The potential energy of a system consisting of metallic zinc and aqueous Cu^{2+} is higher than the potential energy of a system consisting of metallic copper and aqueous Zn^{2+}. Much of this potential energy difference is because the valence electrons of metallic zinc are higher in energy than the valence electrons of metallic copper. Because the $Zn(s) + Cu^{2+}(aq)$ system is higher in energy by 1.10 V than the $Cu(s) + Zn^{2+}(aq)$ system, energy is released when electrons are transferred from zinc to Cu^{2+} to form copper and Zn^{2+}.

19.2 ◦ Standard Potentials

In a galvanic cell, current is produced when electrons flow externally through the circuit from the anode to the cathode because of a difference in potential energy between the two electrodes in the electrochemical cell. In the Zn/Cu system, the valence electrons in zinc have a substantially higher potential energy than the valence electrons in copper because of shielding of the *s* electrons of zinc by the electrons in filled *d* orbitals (see Section 6.5). Hence, electrons flow spontaneously from zinc to copper(II) ions, forming zinc(II) ions and metallic copper (Figure 19.4). Just like water flowing spontaneously downhill, which can be made to do work by forcing a waterwheel, the flow of electrons from a higher potential energy to a lower one can also be harnessed to perform work.

Because the potential energy of valence electrons differs greatly from one substance to another, the voltage of a galvanic cell depends partly on the identity of the reacting substances. If we construct a galvanic cell similar to the one in Figure 19.3a but instead of copper use a strip of cobalt metal and 1 *M* Co^{2+} in the cathode compartment, the measured voltage is not 1.10 V but 0.51 V. Thus, we can conclude that the difference in potential energy between the valence electrons of cobalt and of zinc is less than the difference between the valence electrons of copper and of zinc by 0.59 V.

The measured potential of a cell also depends strongly on the *concentrations* of the reacting species and the *temperature* of the system. Thus, in order to develop a scale of relative potentials that will allow us to predict the direction of an electrochemical reaction and the magnitude of the driving force for the reaction, the potentials for oxidations and reductions of different substances must be measured under comparable conditions. To do this, chemists use the **standard cell potential,** E°_{cell}, defined as the potential of a cell measured under standard conditions—that is, with all species in their standard states (1 *M* for solutions,* 1 atm for gases, pure solids or pure liquids for other substances) and at a fixed temperature, usually 25°C.

Measuring Standard Electrode Potentials

It is physically impossible to measure the potential of a single electrode: only the *difference* between the potentials of two electrodes can be measured. (This is analogous to measuring absolute enthalpies or free energies. Recall from Chapter 18 that only *differences* in enthalpy and free energy can be measured.) We can, however, compare the standard cell potentials for two different galvanic cells that have one kind of electrode in common. This allows us to measure the potential difference between two dissimilar electrodes. For example, the measured standard cell potential, E°, for the Zn/Cu system is 1.10 V, whereas E° for the corresponding Zn/Co system is 0.51 V. This implies that the potential difference between the Co and Cu electrodes is (1.10 V − 0.51 V) = 0.59 V. In fact, that is exactly the potential measured under standard conditions if a cell is constructed with this cell diagram:

$$\mathrm{Co}(s) \mid \mathrm{Co}^{2+}(aq, 1\,M) \parallel \mathrm{Cu}^{2+}(aq, 1\,M) \mid \mathrm{Cu}(s) \qquad E^\circ = 0.59\text{ V} \tag{19.6}$$

This cell diagram corresponds to the oxidation of a cobalt anode and the reduction of Cu^{2+} in solution at the copper cathode.

All tabulated values of *standard electrode potentials* by convention are listed for a reaction written as a reduction, not as an oxidation, in order to be able to compare standard potentials for different substances (see Appendix E). The standard cell potential, E°_{cell}, is therefore the *difference* between the tabulated reduction potentials of the two half-reactions, not their sum:

$$E^\circ_{\text{cell}} = E^\circ_{\text{cathode}} - E^\circ_{\text{anode}} \tag{19.7}$$

Note the pattern

The overall cell reaction is the sum of the two half-reactions, but the cell potential is the difference between the reduction potentials: $E^\circ_{\text{cell}} = E^\circ_{\text{cathode}} - E^\circ_{\text{anode}}$.

In contrast, recall that half-reactions are written to show the reduction and oxidation reactions that actually occur in the cell, so the overall cell reaction is written as the *sum* of the two half-reactions. According to Equation 19.7, when we know the standard potential for any single half-reaction, we can obtain the value of the standard potential of many other half-reactions by measuring the standard potential of the corresponding cell.

Although it is impossible to measure the potential of any electrode directly, we can choose a reference electrode whose potential is defined as 0 V under standard conditions. The **standard hydrogen electrode (SHE)** is universally used for this purpose and is assigned a standard potential of 0 V. It consists of a strip of platinum wire in contact with an aqueous solution containing 1 *M* H^+. The $[H^+]$ in solution is in equilibrium with H_2

* Concentrated solutions of salts (about 1 *M*) generally do not exhibit ideal behavior, and the actual standard state corresponds to an *activity* of 1 rather than a concentration of 1 *M*. Corrections for nonideal behavior are important for precise quantitative work, but not for the more qualitative approach that we are taking here.

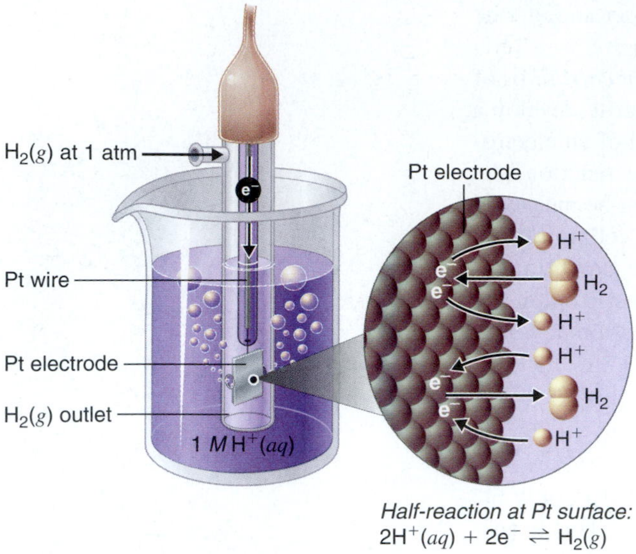

Figure 19.5 The standard hydrogen electrode (SHE). The SHE consists of platinum wire that is connected to a Pt surface in contact with an aqueous solution containing 1 M H^+ in equilibrium with H_2 gas at a pressure of 1 atm. As shown in the molecular view, the Pt surface catalyzes the oxidation of hydrogen molecules to protons or the reduction of protons to hydrogen gas. (Water is omitted for clarity.) The standard potential of the SHE is arbitrarily assigned a value of 0 V.

gas at a pressure of 1 atm at the Pt–solution interface (Figure 19.5). Protons are reduced or hydrogen molecules are oxidized at the Pt surface according to the equation

$$2H^+(aq) + 2e^- \rightleftharpoons H_2(g) \qquad (19.8)$$

One especially attractive feature of the SHE is that the Pt metal electrode is not consumed during the reaction.

Figure 19.6 shows a galvanic cell that consists of a SHE in one beaker and a Zn strip in another beaker containing a solution of Zn^{2+} ions. When the circuit is closed, the voltmeter indicates a potential of 0.76 V. The zinc electrode begins to dissolve to form Zn^{2+}, and H^+ ions are reduced to H_2 in the other compartment. Thus, the hydrogen electrode is the cathode, and the zinc electrode is the anode. The diagram for this galvanic cell is

$$Zn(s) \mid Zn^{2+}(aq) \parallel H^+(aq, 1\ M) \mid H_2(g, 1\ \text{atm}) \mid Pt(s) \qquad (19.9)$$

The half-reactions that actually occur in the cell and their corresponding electrode potentials are

$$\begin{aligned}
&\text{Cathode:} && 2H^+(aq) + 2e^- \longrightarrow H_2(g) && E^\circ_{cathode} = 0\ \text{V} && (19.10a)\\
&\text{Anode:} && Zn(s) \longrightarrow Zn^{2+}(aq) + 2e^- && E^\circ_{anode} = -0.76\ \text{V} && (19.10b)\\
&\text{Overall:} && Zn(s) + 2H^+(aq) \longrightarrow Zn^{2+}(aq) + H_2(g) && && (19.10c)\\
& && E^\circ_{cell} = E^\circ_{cathode} - E^\circ_{anode} = 0.76\ \text{V}
\end{aligned}$$

Notice that although the reaction at the anode is an oxidation, which means that the sum of the two half-reactions gives the overall reaction, by convention its tabulated E° value is reported as a reduction potential. We must therefore subtract E°_{anode} from $E^\circ_{cathode}$ to obtain E°_{cell}.

The potential of a half-reaction measured against the SHE under standard conditions is called the **standard electrode potential** for that half-reaction. The standard electrode potential for $Zn^{2+}(aq) + 2e^- \longrightarrow Zn(s)$ is -0.76 V; thus, the standard electrode potential for the *oxidation* of Zn to Zn^{2+}, often called the Zn/Zn^{2+} *redox*

Figure 19.6 Determining a standard electrode potential using a standard hydrogen electrode (SHE). The voltmeter shows that the standard cell potential of a galvanic cell consisting of a SHE and a Zn/Zn^{2+} couple is $E^\circ_{cell} = 0.76$ V. Because the zinc electrode in this cell dissolves spontaneously to form $Zn^{2+}(aq)$ ions while $H^+(aq)$ ions are reduced to H_2 at the platinum surface, the standard electrode potential of the Zn^{2+}/Zn couple is -0.76 V.

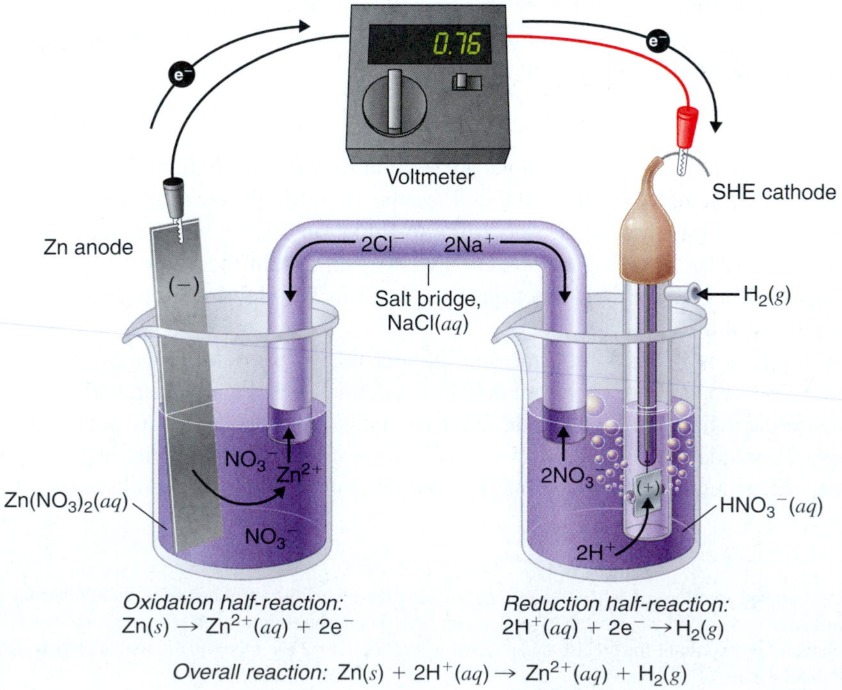

couple, or the Zn/Zn^{2+} *couple*, is $-(-0.76\ \text{V}) = 0.76\ \text{V}$. Because electrical potential is the energy needed to move a charged particle in an electric field, standard electrode potentials for half-reactions are intensive properties and do not depend on the amount of substance involved. Consequently, *$E°$ values are independent of the stoichiometric coefficients for the half-reaction*, and, most important, the coefficients used to produce a balanced overall reaction do not affect the value of the cell potential.

Note the pattern

$E°$ values do not depend on the stoichiometric coefficients for a half-reaction.

To measure the potential of the Cu/Cu^{2+} couple, we can construct a galvanic cell analogous to the one shown in Figure 19.6, but containing a Cu/Cu^{2+} couple in the sample compartment instead of Zn/Zn^{2+}. When we close the circuit this time, the measured potential for the cell is *negative* (-0.34 V) rather than positive. The negative value of $E°_{\text{cell}}$ indicates that the direction of spontaneous electron flow is the *opposite* of that for the Zn/Zn^{2+} couple. Hence, the reactions that occur spontaneously are the reduction of Cu^{2+} to Cu at the copper electrode, which gains mass as the reaction proceeds, and the oxidation of H_2 to H^+ at the platinum electrode. Hence, in this cell the copper strip is the *cathode*, and the hydrogen electrode is the *anode*. The cell diagram therefore is written with the SHE on the left and the Cu^{2+}/Cu couple on the right:

$$\text{Pt}(s) \mid H_2(g, 1\ \text{atm}) \parallel Cu^{2+}(aq, 1\ M) \mid \text{Cu}(s) \qquad (19.11)$$

The half-cell reactions and potentials of the spontaneous reaction are

Cathode:	$Cu^{2+}(aq) + 2e^- \longrightarrow Cu(s)$	$E°_{\text{cathode}} = 0.34\ \text{V}$	(19.12a)
Anode:	$H_2(g) \longrightarrow 2H^+(aq) + 2e^-$	$E°_{\text{anode}} = 0\ \text{V}$	(19.12b)
Overall:	$H_2(g) + Cu^{2+}(aq) \longrightarrow 2H^+(aq) + Cu(s)$		(19.12c)

$$E°_{\text{cell}} = E°_{\text{cathode}} - E°_{\text{anode}} = 0.34\ \text{V}$$

Thus, the standard electrode potential for the Cu^{2+}/Cu couple is 0.34 V.

Balancing Redox Reactions Using the Half-Reaction Method

In Chapter 4, we described a method for balancing redox reactions using oxidation numbers. Oxidation numbers were assigned to each atom in a redox reaction to identify any changes in the oxidation states. Here we present an alternative approach to balancing redox reactions, the half-reaction method, in which the overall redox reaction is divided into an oxidation half-reaction and a reduction half-reaction, each balanced for mass and charge. This method more closely reflects the events that take place in an electrochemical cell, where the two half-reactions may be physically separated from each other.

We can illustrate how to balance a redox reaction using half-reactions with the reaction that occurs when the commercial solid drain cleaner, Drano, is poured into a clogged drain. Drano contains a mixture of sodium hydroxide and powdered aluminum, which in solution reacts to produce hydrogen gas:

$$\text{Al}(s) + OH^-(aq) \longrightarrow Al(OH)_4^-(aq) + H_2(g) \qquad (19.13)$$

In this reaction, Al(s) is oxidized to Al^{3+} and H^+ is reduced to H_2 gas, which bubbles through the solution, agitating it and breaking up the clogs.

The overall redox reaction is composed of a reduction half-reaction and an oxidation half-reaction. From the standard electrode potentials listed in Appendix E, we find the corresponding half-reactions that describe the reduction of H^+ ions in water to H_2 and the oxidation of Al to Al^{3+} in basic solution:

Reduction:	$2H_2O(l) + 2e^- \longrightarrow 2OH^-(aq) + H_2(g)$	(19.14a)
Oxidation:	$Al(s) + 4OH^-(aq) \longrightarrow Al(OH)_4^-(aq) + 3e^-$	(19.14b)

The half-reactions chosen must exactly reflect the reaction conditions, such as the basic conditions shown here. Moreover, the physical states of the reactants and products must be identical to those given in the overall reaction, whether gaseous, liquid, solid, or in solution.

In Equation 19.14a, two H^+ ions gain one electron each in the reduction, and in Equation 19.14b, the aluminum atom loses three electrons in the oxidation. The charges are balanced by multiplying the reduction half-reaction (Equation 19.14a) by 3 and the oxidation half-reaction (Equation 19.14b) by 2 to give the same number of electrons in both half-reactions:

$$\text{Reduction:} \quad 6H_2O(l) + 6e^- \longrightarrow 6OH^-(aq) + 3H_2(g) \quad (19.15a)$$
$$\text{Oxidation:} \quad 2Al(s) + 8OH^-(aq) \longrightarrow 2Al(OH)_4^-(aq) + 6e^- \quad (19.15b)$$

Adding the two half-reactions gives

$$6H_2O(l) + 2Al(s) + 8OH^-(aq) \longrightarrow 2Al(OH)_4^-(aq) + 3H_2(g) + 6OH^-(aq) \quad (19.16)$$

Simplifying by canceling substances that appear on both sides of the equation gives

$$6H_2O(l) + 2Al(s) + 2OH^-(aq) \longrightarrow 2Al(OH)_4^-(aq) + 3H_2(g) \quad (19.17)$$

We have a -2 charge on the left side of the equation and a -2 charge on the right side. Thus, the charges are balanced, but we must also check that atoms are balanced:

$$2Al + 8O + 14H = 2Al + 8O + 14H \quad (19.18)$$

The atoms also balance, so Equation 19.17 is a balanced chemical equation for the redox reaction depicted in Equation 19.13.

We can also balance a redox reaction by first balancing the atoms in each half-reaction and then balancing the charges. With this alternative method, we do not need to use the half-reactions listed in Appendix E, but instead focus on the atoms whose oxidation states change, as illustrated in the following steps:

Step 1 **Write the reduction half-reaction and the oxidation half-reaction.** For the reaction shown in Equation 19.13, hydrogen is reduced from H^+ in OH^- to H_2, and aluminum is oxidized from Al^0 to Al^{3+}:

$$\text{Reduction:} \quad OH^-(aq) \longrightarrow H_2(g) \quad (19.19a)$$
$$\text{Oxidation:} \quad Al(s) \longrightarrow Al(OH)_4^-(aq) \quad (19.19b)$$

Step 2 **Balance the atoms by balancing elements other than O and H. Then balance O atoms by adding H_2O, and finally balance H atoms by adding H^+.** Elements other than O and H in the last two equations are balanced as written, so we proceed with balancing O atoms. We can do this by adding water to the appropriate side of each half-reaction:

$$\text{Reduction:} \quad OH^-(aq) \longrightarrow H_2(g) + H_2O(l) \quad (19.20a)$$
$$\text{Oxidation:} \quad Al(s) + 4H_2O(l) \longrightarrow Al(OH)_4^-(aq) \quad (19.20b)$$

Balancing H atoms by adding H^+, we obtain

$$\text{Reduction:} \quad OH^-(aq) + 3H^+(aq) \longrightarrow H_2(g) + H_2O(l) \quad (19.21a)$$
$$\text{Oxidation:} \quad Al(s) + 4H_2O(l) \longrightarrow Al(OH)_4^-(aq) + 4H^+(aq) \quad (19.21b)$$

We have now balanced the atoms in each half-reaction, but the charges are not balanced.

Step 3 **Balance the charges in each half-reaction by adding electrons.** Two electrons are gained in the reduction of H^+ ions to H_2, and three electrons are lost during the oxidation of Al^0 to Al^{3+}:

$$\text{Reduction:} \quad OH^-(aq) + 3H^+(aq) + 2e^- \longrightarrow H_2(g) + H_2O(l) \quad (19.22a)$$
$$\text{Oxidation:} \quad Al(s) + 4H_2O(l) \longrightarrow Al(OH)_4^-(aq) + 4H^+(aq) + 3e^- \quad (19.22b)$$

Step 4 **Multiply the reductive and oxidative half-reactions by appropriate integers to obtain the same number of electrons in both half-reactions.** In this case, we multiply Equation 19.22a (the reductive half-reaction) by 3 and Equation 19.22b (the oxidative half-reaction) by 2 to obtain the same number of electrons in both half-reactions:

Reduction: $3OH^-(aq) + 9H^+(aq) + 6e^- \longrightarrow 3H_2(g) + 3H_2O(l)$ (19.23a)

Oxidation: $2Al(s) + 8H_2O(l) \longrightarrow 2Al(OH)_4^-(aq) + 8H^+(aq) + 6e^-$ (19.23b)

Step 5 **Add the two half-reactions and cancel substances that appear on both sides of the equation.** Adding and, in this case, canceling $8H^+$, $3H_2O$, and $6e^-$ give

$$2Al(s) + 5H_2O(l) + 3OH^-(aq) + H^+(aq) \longrightarrow 2Al(OH)_4^-(aq) + 3H_2(g) \quad (19.24)$$

Notice that we have three OH^- and one H^+ on the left side. Neutralizing the H^+ gives us a total of $5H_2O + H_2O = 6H_2O$, and leaves $2OH^-$ on the left side:

$$2Al(s) + 6H_2O(l) + 2OH^-(aq) \longrightarrow 2Al(OH)_4^-(aq) + 3H_2(g) \quad (19.25)$$

Step 6 **Check to make sure that all atoms and charges are balanced.** Equation 19.25 is identical to Equation 19.17, obtained using the first method, so the charges and numbers of atoms on each side of the equation balance.

EXAMPLE 19.3

In acidic solution, the redox reaction of dichromate ion ($Cr_2O_7^{2-}$) and iodide (I^-) can be monitored visually. The yellow dichromate solution reacts with the colorless iodide solution to produce a solution that is deep amber due to the presence of a green $Cr^{3+}(aq)$ complex and brown $I_2(aq)$ ions (Figure 19.7):

$$Cr_2O_7^{2-}(aq) + I^-(aq) \longrightarrow Cr^{3+}(aq) + I_2(aq)$$

Balance this equation using half-reactions.

Given Redox reaction, Appendix E

Asked for Balanced equation using half-reactions

Strategy

Follow the steps outlined above to balance the redox reaction using the half-reaction method.

Figure 19.7 The reaction of dichromate with iodide. The reaction of a yellow solution of sodium dichromate with a colorless solution of sodium iodide produces a deep amber solution that contains a green $Cr^{3+}(aq)$ complex and brown $I_2(aq)$ ions.

Solution

From the standard electrode potentials listed in Appendix E, we find the half-reactions corresponding to the overall reaction:

Reduction: $Cr_2O_7^{2-}(aq) + 14H^+(aq) + 6e^- \longrightarrow 2Cr^{3+}(aq) + 7H_2O(l)$

Oxidation: $2I^-(aq) \longrightarrow I_2(aq) + 2e^-$

Balancing the number of electrons by multiplying the oxidation reaction by 3 gives

Oxidation: $6I^-(aq) \longrightarrow 3I_2(aq) + 6e^-$

Adding the two half-reactions and canceling electrons give

$$Cr_2O_7^{2-}(aq) + 14H^+(aq) + 6I^-(aq) \longrightarrow 2Cr^{3+}(aq) + 7H_2O(l) + 3I_2(aq)$$

We must now check to make sure the charges and atoms on each side of the equation balance:

$$(-2) + 14 + (-6) = +6$$

$$+6 = +6$$

$$2Cr + 7O + 14H + 6I = 2Cr + 7O + 14H + 6I$$

The charges and atoms balance, so our equation is balanced.

We can also use the alternative procedure, which does not require the use of the half-reactions listed in Appendix E.

Step 1 Chromium is reduced from Cr^{6+} in $Cr_2O_7^{2-}$ to Cr^{3+}, and I^- ions are oxidized to I_2. Dividing the reaction into two half-reactions gives

Reduction: $Cr_2O_7^{2-}(aq) \longrightarrow Cr^{3+}(aq)$

Oxidation: $I^-(aq) \longrightarrow I_2(aq)$

Step 2 Balancing the atoms other than oxygen and hydrogen gives

Reduction: $Cr_2O_7^{2-}(aq) \longrightarrow 2Cr^{3+}(aq)$

Oxidation: $2I^-(aq) \longrightarrow I_2(aq)$

We now balance the O atoms by adding H_2O—in this case, to the right side of the reduction half-reaction. Because the oxidation half-reaction does not contain oxygen, it can be ignored in this step.

Reduction: $Cr_2O_7^{2-}(aq) \longrightarrow 2Cr^{3+}(aq) + 7H_2O(l)$

Next we balance the H atoms by adding H^+ to the left side of the reduction half-reaction. Again, we can ignore the oxidation half-reaction.

Reduction: $Cr_2O_7^{2-}(aq) + 14H^+(aq) \longrightarrow 2Cr^{3+}(aq) + 7H_2O(l)$

Step 3 We must now add electrons to balance the charges. The reduction half-reaction ($2Cr^{+6}$ to $2Cr^{+3}$) has a $+12$ charge on the left and a $+6$ charge on the right, so six electrons are needed to balance the charge. The oxidation half-reaction ($2I^-$ to I_2) has a -2 charge on the left side and a 0 charge on the right, so it needs two electrons to balance the charge:

Reduction: $Cr_2O_7^{2-}(aq) + 14H^+(aq) + 6e^- \longrightarrow 2Cr^{3+}(aq) + 7H_2O(l)$

Oxidation: $2I^-(aq) \longrightarrow I_2(aq) + 2e^-$

Step 4 To have the same number of electrons in both half-reactions, we must multiply the oxidation half-reaction by 3:

Oxidation: $6I^-(aq) \longrightarrow 3I_2(s) + 6e^-$

Step 5 Adding the two half-reactions and canceling substances that appear in both reactions give

$$Cr_2O_7^{2-}(aq) + 14H^+(aq) + 6I^-(aq) \longrightarrow 2Cr^{3+}(aq) + 7H_2O(l) + 3I_2(aq)$$

Step 6 This is the same equation we obtained using the first method. Thus, the charges and atoms on each side of the equation balance.

EXERCISE 19.3

Copper is commonly found as the mineral *covellite*, CuS. The first step in extracting the copper is to dissolve the mineral in nitric acid (HNO_3), which oxidizes sulfide to sulfate and reduces nitric acid to NO:

$$CuS(s) + HNO_3(aq) \longrightarrow NO(g) + CuSO_4(aq)$$

Balance this equation using the half-reaction method.

Answer $3CuS(s) + 8HNO_3(aq) \longrightarrow 8NO(g) + 3CuSO_4(aq) + 4H_2O(l)$

Calculating Standard Cell Potentials

MGC Cell Potential

The standard cell potential for a redox reaction, E°_{cell}, is a measure of the tendency of the reactants in their standard states to form the products in their standard states; consequently, it is a measure of the driving force for the reaction, which earlier we called *voltage*. We can use the two standard electrode potentials we found earlier to calculate the standard potential for the Zn/Cu cell represented by this cell diagram:

$$Zn(s) \mid Zn^{2+}(aq, 1\ M) \parallel Cu^{2+}(aq, 1\ M) \mid Cu(s) \qquad (19.26)$$

We know the values of E°_{anode} for the reduction of Zn^{2+} and $E^\circ_{cathode}$ for the reduction of Cu^{2+}, so we can calculate E°_{cell}:

Cathode:	$Cu^{2+}(aq) + 2e^- \longrightarrow Cu(s)$	$E^\circ_{cathode} = 0.34$ V	(19.27a)
Anode:	$Zn(s) \longrightarrow Zn^{2+}(aq, 1\ M) + 2e^-$	$E^\circ_{anode} = -0.76$ V	(19.27b)
Overall:	$Zn(s) + Cu^{2+}(aq) \longrightarrow Zn^{2+}(aq) + Cu(s)$	$E^\circ_{cell} = E^\circ_{cathode} - E^\circ_{anode} = 1.10$ V	(19.27c)

This is the same value that is observed experimentally. If the value of E°_{cell} (the standard cell potential) is positive, the reaction will occur spontaneously as written. If the value of E°_{cell} is negative, then the reaction is not spontaneous, and it will not occur as written under standard conditions; it will, however, proceed spontaneously in the opposite direction.* The next example and exercise illustrate how we can use measured cell potentials to calculate standard potentials for redox couples.

Note the pattern

A positive E°_{cell} means that the reaction will occur spontaneously as written. A negative E°_{cell} means that the reaction will proceed spontaneously in the opposite direction.

EXAMPLE 19.4

A galvanic cell with a measured standard cell potential of 0.27 V is constructed using two beakers connected by a salt bridge. One beaker contains a strip of gallium metal immersed in a 1 *M* solution of $GaCl_3$, and the other contains a piece of nickel immersed in a 1 *M* solution of $NiCl_2$. The half-reactions that occur when the compartments are connected are

Cathode: $Ni^{2+}(aq) + 2e^- \longrightarrow Ni(s)$

Anode: $Ga(s) \longrightarrow Ga^{3+}(aq) + 3e^-$

If the potential for the oxidation of Ga to Ga^{3+} is 0.55 V under standard conditions, what is the potential for the oxidation of Ni to Ni^{2+}?

Given Galvanic cell, half-reactions, standard cell potential, potential for the oxidation half-reaction under standard conditions

Asked for Standard electrode potential of reaction occurring at the cathode

* As we shall see in Section 19.7, this does not mean that the reaction cannot be made to occur at all under standard conditions. With a sufficient input of electrical energy, virtually any reaction can be forced to occur.

Strategy

Ⓐ Write the equation for the half-reaction that occurs at the anode along with the value of the standard electrode potential for the half-reaction.

Ⓑ Use Equation 19.7 to calculate the standard electrode potential for the half-reaction that occurs at the cathode. Then reverse the sign to obtain the potential for the corresponding oxidation half-reaction under standard conditions.

Solution

Ⓐ✓ We have been given the potential for the oxidation of Ga to Ga^{3+} under standard conditions, but to report the standard electrode potential, we must reverse the sign. Thus, for the reduction reaction $Ga^{3+}(aq) + 3e^- \longrightarrow Ga(s)$, $E^\circ_{anode} = -0.55$ V. Ⓑ✓ Using the value given for E°_{cell} and the calculated value of E°_{anode}, we can calculate the standard potential for the reduction of Ni^{2+} to Ni from Equation 19.7:

$$E^\circ_{cell} = E^\circ_{cathode} - E^\circ_{anode}$$

$$0.27 \text{ V} = E^\circ_{cathode} - (-0.55 \text{ V})$$

$$E^\circ_{cathode} = -0.28 \text{ V}$$

This is the standard electrode potential for the reaction $Ni^{2+}(aq) + 2e^- \longrightarrow Ni(s)$. Because we are asked for the potential for the oxidation of Ni to Ni^{2+} under standard conditions, we must reverse the sign of $E^\circ_{cathode}$. Thus, $E^\circ = -(-0.28 \text{ V}) = 0.28$ V for the oxidation. With three electrons consumed in the reduction and two produced in the oxidation, the overall reaction is not balanced. Recall, however, that standard potentials are *independent* of stoichiometry.

EXERCISE 19.4

A galvanic cell is constructed with one compartment that contains a mercury electrode immersed in a 1 *M* solution of mercuric acetate [$Hg(NO_3)_2$] and one compartment that contains a strip of magnesium immersed in a 1 *M* solution of $MgCl_2$. When the compartments are connected, a potential of 3.22 V is measured and these half-reactions occur:

Cathode: $Hg^{2+}(aq) + 2e^- \longrightarrow Hg(l)$

Anode: $Mg(s) \longrightarrow Mg^{2+}(aq) + 2e^-$

If the potential for the oxidation of Mg to Mg^{2+} is 2.37 V under standard conditions, what is the standard electrode potential for the reaction that occurs at the anode?

Answer 0.85 V

Reference Electrodes and Measuring Concentrations

When using a galvanic cell to measure the concentration of a substance, we are generally interested in the potential of only one of the electrodes of the cell, the so-called **indicator electrode**, whose potential is related to the concentration of the substance being measured. To ensure that any change in the measured potential of the cell is due to only the substance being analyzed, the potential of the other electrode, the **reference electrode**, must be constant. You are already familiar with one example of a reference electrode, the SHE. The potential of a reference electrode must be unaffected by the properties of the solution and, if possible, it should be physically isolated from the solution of interest. To measure the potential of a solution, we select a reference electrode and an appropriate indicator electrode. Whether reduction or oxidation of the substance being analyzed occurs depends on the potential of the half-reaction for the substance of interest (the sample) and the potential of the reference electrode.

There are many possible choices of reference electrode other than the SHE. The SHE requires a constant flow of highly flammable hydrogen gas, which makes it inconvenient to use. Consequently, two other electrodes are commonly chosen as reference electrodes. One is the **silver–silver chloride electrode**, which consists of a silver wire coated with a very thin layer of AgCl that is dipped into a chloride ion solution with a fixed concentration. The cell diagram and reduction half-reaction are

$$Cl^-(aq) \mid AgCl(s) \mid Ag(s)$$

$$AgCl(s) + e^- \longrightarrow Ag(s) + Cl^-(aq) \qquad (19.28)$$

If a saturated solution of KCl is used as the chloride solution, the potential of the silver–silver chloride electrode is 0.197 V versus the SHE. That is, 0.197 V must be subtracted from the measured value to obtain the standard electrode potential measured against the SHE.

A second common reference electrode is the **saturated calomel electrode (SCE)**, which has the same general form as the silver–silver chloride electrode. The SCE consists of a platinum wire inserted into a moist paste of liquid mercury, Hg_2Cl_2 (called *calomel* in the old chemical literature), and KCl. This interior cell is surrounded by an aqueous KCl solution, which acts as a salt bridge between the interior cell and the exterior solution (Figure 19.8a). Although it sounds and looks complex, this cell is actually easy to prepare and maintain, and its potential is highly reproducible. The SCE cell diagram and corresponding half-reaction are

$$Pt(s) \mid Hg_2Cl_2(s) \mid KCl(aq, sat) \qquad (19.29)$$

$$Hg_2Cl_2(s) + 2e^- \longrightarrow 2Hg(l) + 2Cl^-(aq) \qquad (19.30)$$

At 25°C, the potential of the SCE is 0.2415 V versus the SHE, which means that 0.2415 V must be subtracted from the potential versus an SCE to obtain the standard electrode potential.

One of the most common uses of electrochemistry is to measure the H^+ ion concentration of a solution. A **glass electrode** is generally used for this purpose, in which an internal Ag/AgCl electrode is immersed in a 0.10 *M* HCl solution that is separated from the solution by a very thin glass membrane (Figure 19.8b). The glass membrane

Figure 19.8 Three common types of electrodes. (a) The saturated calomel electrode (SCE) is a reference electrode that consists of a platinum wire inserted into a moist paste of liquid mercury, calomel (Hg_2Cl_2), and KCl. The interior cell is surrounded by an aqueous KCl solution, which acts as a salt bridge between the interior cell and the exterior solution. (b) In a glass electrode, an internal Ag/AgCl electrode is immersed in a 1 *M* HCl solution that is separated from the sample solution by a very thin glass membrane. The potential of the electrode depends on the H^+ ion concentration of the sample. (c) The potential of an ion-selective electrode depends on the concentration of only a single ionic species in solution.

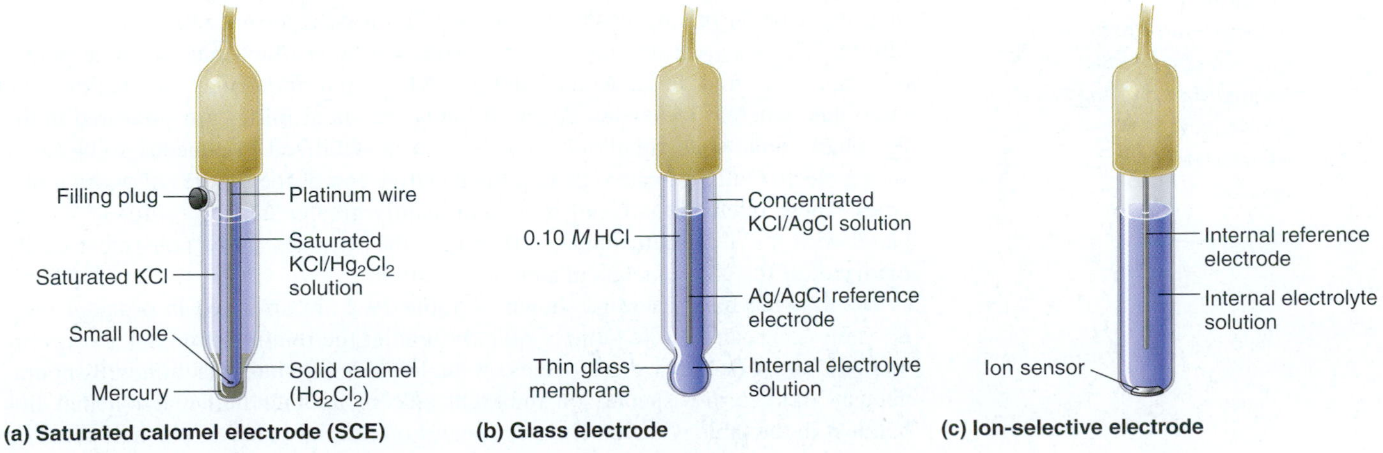

TABLE 19.1 Some species whose aqueous concentrations can be measured using electrochemical methods.

Species	Type of Sample
H^+	Laboratory samples, blood, soil, ground and surface water
NH_3/NH_4^+	Waste water, runoff water
K^+	Blood, wine, soil
CO_2/HCO_3^-	Blood, groundwater
F^-	Groundwater, drinking water, soil
Br^-	Grains, plant extracts
I^-	Milk, pharmaceuticals
NO_3^-	Groundwater, drinking water, soil, fertilizer

absorbs protons, which affects the measured potential. The extent of the adsorption on the inner side is fixed because $[H^+]$ is fixed inside the electrode, but the adsorption of protons on the outer surface depends on the pH of the solution. The potential of the glass electrode depends on $[H^+]$ as follows:

$$E_{glass} = E' + (0.0591\ V \times \log[H^+]) = E' - 0.0591\ V\ (pH) \qquad (19.31)$$

The voltage E' is a constant that depends on the exact construction of the electrode. Although it can be measured, in practice a glass electrode is *calibrated*; that is, it is inserted into a solution of known pH, and the display on the pH meter is adjusted to the known value. Once the electrode is properly calibrated, it can be placed in a solution and used to determine an unknown pH.

Ion-selective electrodes are used to measure the concentration of a particular species in solution; they are designed so that their potential depends on only the concentration of the desired species (Figure 19.8c). These electrodes usually contain an internal reference electrode that is connected by a solution of an electrolyte to a crystalline inorganic material or a membrane, which acts as the sensor. For example, one type of ion-selective electrode uses a single crystal of Eu-doped LaF_3 as the inorganic material. When fluoride ions in solution diffuse to the surface of the solid, the potential of the electrode changes, resulting in a so-called *fluoride electrode*. Similar electrodes are used to measure the concentrations of other species in solution. Some of the species whose concentrations can be determined in aqueous solution using ion-selective electrodes and similar devices are listed in Table 19.1.

19.3 Comparing Strengths of Oxidants and Reductants

MGC Galvanic Cells

We can use the procedure described in Section 19.2 to measure the standard potentials for a wide variety of chemical substances, some of which are listed in Table 19.2. (Appendix E contains a more extensive listing.) These data allow us to compare the oxidative and reductive strengths of a variety of substances. Note that the half-reaction for the standard hydrogen electrode lies more than halfway down the list in Table 19.2. All reactants that lie *above* it in the table are stronger oxidants than H^+, and all those that lie below it are weaker. The strongest oxidant in the table is F_2, with a standard electrode potential of 2.87 V. This high value is consistent with the high electronegativity of fluorine, and tells us that fluorine has a stronger tendency to accept electrons (it is a stronger oxidant) than any other element.

Similarly, all species in Table 19.2 that lie *below* H_2 are stronger reductants than H_2, and those that lie above H_2 are weaker. The strongest reductant in the table is thus metallic lithium, with a standard electrode potential of −3.04 V. This fact might be surprising because cesium is the least electronegative element, not lithium. The apparent anomaly can be explained by the fact that electrode potentials are measured in aqueous solution, where intermolecular interactions are important, whereas ionization potentials and electron affinities are measured in the gas phase. Due to its small size, the Li^+ ion is stabilized in aqueous solution by strong electrostatic interactions with the negative end of the dipoles of water molecules. These interactions result in a significantly greater $\Delta H_{hydration}$ for Li^+ compared with Cs^+. Lithium metal is therefore the strongest reductant (most easily oxidized) of the alkali metals in aqueous solution.

Note the pattern

Species in Table 19.2 that lie below H_2 are stronger reductants (more easily oxidized) than H_2. Species that lie above H_2 are weaker (more easily reduced).

Because the half-reactions shown in Table 19.2 are arranged in order of their $E°$ values, we can use the table to quickly predict the relative strengths of various oxidants and reductants. Any species on the left side of a half-reaction will spontaneously oxidize any species on the right side of another half-reaction that lies *below* it in the table. Conversely, any species on the right side of one half-reaction

TABLE 19.2 Standard potentials for selected reduction half-reactions at 25°C

Half-Reaction	$E°$, V
$F_2(g) + 2e^- \longrightarrow 2F^-(aq)$	2.87
$H_2O_2(aq) + 2H^+(aq) + 2e^- \longrightarrow 2H_2O(l)$	1.78
$Ce^{4+}(aq) + e^- \longrightarrow Ce^{3+}(aq)$	1.72
$PbO_2(s) + HSO_4^-(aq) + 3H^+(aq) + 2e^- \longrightarrow PbSO_4(s) + 2H_2O(l)$	1.69
$Cl_2(g) + 2e^- \longrightarrow 2Cl^-(aq)$	1.36
$Cr_2O_7^-(aq) + 14H^+(aq) + 6e^- \longrightarrow 2Cr^{3+}(aq) + 7H_2O(l)$	1.23
$O_2(g) + 4H^+(aq) + 4e^- \longrightarrow 2H_2O(l)$	1.23
$MnO_2(s) + 4H^+(aq) + 2e^- \longrightarrow Mn^{2+}(aq) + 2H_2O(l)$	1.22
$Br_2(aq) + 2e^- \longrightarrow 2Br^-(aq)$	1.09
$NO_3^-(aq) + 3H^+(aq) + 2e^- \longrightarrow HNO_2(aq) + H_2O(l)$	0.93
$Ag^+(aq) + e^- \longrightarrow Ag(s)$	0.80
$Fe^{3+}(aq) + e^- \longrightarrow Fe^{2+}(aq)$	0.77
$H_2SeO_3(aq) + 4H^+(aq) + 4e^- \longrightarrow Se(s) + 3H_2O(l)$	0.74
$O_2(g) + 2H^+(aq) + 2e^- \longrightarrow H_2O_2(aq)$	0.70
$MnO_4^-(aq) + 2H_2O(l) + 3e^- \longrightarrow MnO_2(s) + 4OH^-(aq)$	0.60
$MnO_4^{2-}(aq) + 2H_2O(l) + 2e^- \longrightarrow MnO_2(s) + 4OH^-(aq)$	0.60
$I_2(s) + 2e^- \longrightarrow 2I^-(aq)$	0.54
$H_2SO_3(aq) + 4H^+(aq) + 4e^- \longrightarrow S(s) + 3H_2O(l)$	0.45
$O_2(g) + 2H_2O(l) + 4e^- \longrightarrow 4OH^-(aq)$	0.40
$Cu^{2+}(aq) + 2e^- \longrightarrow Cu(s)$	0.34
$AgCl(s) + e^- \longrightarrow Ag(s) + Cl^-(aq)$	0.22
$Cu^{2+}(aq) + e^- \longrightarrow Cu^+(aq)$	0.15
$Sn^{4+}(aq) + 2e^- \longrightarrow Sn^{2+}(aq)$	0.15
$2H^+(aq) + 2e^- \longrightarrow H_2(g)$	0.00
$Sn^{2+}(aq) + 2e^- \longrightarrow Sn(s)$	−0.14
$2SO_4^{2-}(aq) + 4H^+(aq) + 2e^- \longrightarrow S_2O_6^{2-}(aq) + H_2O(l)$	−0.22
$Ni^{2+}(aq) + 2e^- \longrightarrow Ni(s)$	−0.26
$PbSO_4(s) + 2e^- \longrightarrow Pb(s) + SO_4^{2-}(aq)$	−0.36
$Cd^{2+}(aq) + 2e^- \longrightarrow Cd(s)$	−0.40
$Fe^{2+}(aq) + 2e^- \longrightarrow Fe(s)$	−0.45
$Ag_2S(s) + 2e^- \longrightarrow 2Ag(s) + S^{2-}(aq)$	−0.69
$Zn^{2+}(aq) + 2e^- \longrightarrow Zn(s)$	−0.76
$Be^{2+}(aq) + 2e^- \longrightarrow Be(s)$	−1.85
$Li^+(aq) + e^- \longrightarrow Li(s)$	−3.04

will spontaneously reduce any species on the left side of another half-reaction that lies *above* it in the table. We can use these generalizations to predict the spontaneity of a wide variety of redox reactions ($E°_{cell} > 0$), as illustrated in the next example.

EXAMPLE 19.5

The black tarnish that forms on silver objects is primarily Ag_2S. The half-reaction for reversing the tarnishing process is

$$Ag_2S(s) + 2e^- \longrightarrow 2Ag(s) + S^{2-}(aq) \qquad E° = -0.69 \text{ V}$$

Referring to Table 19.2, **(a)** predict which of these species can reduce Ag_2S to Ag under standard conditions: $H_2O_2(aq)$, $Zn(s)$, $I^-(aq)$, $Sn^{2+}(aq)$. **(b)** Of these species, identify which is the strongest reducing agent in aqueous solution and thus the best candidate for a commercial product. **(c)** From the data in Table 19.2, suggest an alternative reducing agent that is readily available, inexpensive, and possibly more effective at removing tarnish.

Given Reduction half-reaction and standard electrode potential, list of possible reductants

Asked for Reductants for Ag_2S, strongest reductant, potential reducing agent for removing tarnish

Strategy

A From their positions in Table 19.2, decide which of the species can reduce Ag_2S. Determine which of these is the strongest reductant.
B Use Table 19.2 to identify a reductant for Ag_2S that is a common household product.

Solution

We can solve the problem in one of two ways: by comparing the relative positions of the four possible reductants with that of the Ag_2S/Ag couple in Table 19.2, or by comparing $E°$ for each of the species with $E°$ for the Ag_2S/Ag couple (−0.69 V).

(a) **A✓** The species in Table 19.2 are arranged from top to bottom in order of increasing reducing strength. Of the four species given in the problem, $I^-(aq)$, $Sn^{2+}(aq)$, and $H_2O_2(aq)$ lie above Ag_2S, and one, $Zn(s)$, lies below it. We can therefore conclude that $Zn(s)$ can reduce $Ag_2S(s)$ under standard conditions, whereas $I^-(aq)$, $Sn^{2+}(aq)$, and $H_2O_2(aq)$ cannot. Notice that $Sn^{2+}(aq)$ and $H_2O_2(aq)$ appear *twice* in the table: on the left side (oxidant) in one half-reaction and on the right side (reductant) in another.

(b) The strongest reductant is $Zn(s)$, the species on the right side of the half-reaction that lies closer to the bottom of Table 19.2 than the half-reactions involving $I^-(aq)$, $Sn^{2+}(aq)$, and $H_2O_2(aq)$. (Commercial products that use a piece of zinc are often marketed as a "miracle product" for removing tarnish from silver. All that is required is to add warm water and salt for electrical conductivity.)

(c) **B✓** Of the reductants that lie *below* $Zn(s)$ in Table 19.2, and therefore are stronger reductants, only one is commonly available in household products: $Al(s)$, which is sold as aluminum foil for wrapping foods.

EXERCISE 19.5

Refer to Table 19.2 to predict **(a)** which of these species can oxidize $MnO_2(s)$ to $MnO_4^-(aq)$ under standard conditions, and **(b)** which is the strongest oxidizing agent in aqueous solution: $Sn^{4+}(aq)$, $Cl^-(aq)$, $Ag^+(aq)$, $Cr^{3+}(aq)$, $H_2O_2(aq)$.

Answer **(a)** $Ag^+(aq)$, $H_2O_2(aq)$; **(b)** $H_2O_2(aq)$

EXAMPLE 19.6

Use the data in Table 19.2 to determine whether each of these reactions is likely to occur spontaneously under standard conditions:

(a) $Sn(s) + Be^{2+}(aq) \longrightarrow Sn^{2+}(aq) + Be(s)$
(b) $MnO_2(s) + H_2O_2(aq) + 2H^+(aq) \longrightarrow O_2(g) + Mn^{2+}(aq) + 2H_2O(l)$

Given Redox reaction, list of standard electrode potentials (Table 19.2)

Asked for Reaction spontaneity

Strategy

A Identify the half-reactions in each equation. Using Table 19.2, determine the standard potentials for the half-reactions in the appropriate direction.
B Use Equation 19.7 to calculate the standard cell potential for the overall reaction. From this value, determine whether the overall reaction is spontaneous.

Solution

(a) **A** Metallic tin is oxidized to $Sn^{2+}(aq)$, and $Be^{2+}(aq)$ is reduced to elemental beryllium. We can find the standard electrode potentials for the latter (reduction) half-reaction, −1.85 V, and for the former (oxidation) half-reaction, −0.14 V, directly from Table 19.2. **B** Adding the two half-reactions gives the overall reaction:

Cathode:	$Be^{2+}(aq) + 2e^- \longrightarrow Be(s)$	$E^\circ_{cathode} = -1.85\ V$
Anode:	$Sn(s) \longrightarrow Sn^{2+}(aq) + 2e^-$	$E^\circ_{anode} = -0.14\ V$
Overall:	$Sn(s) + Be^{2+}(aq) \longrightarrow Sn^{2+}(aq) + Be(s)$	$E^\circ_{cell} = E^\circ_{cathode} - E^\circ_{anode} = -1.71\ V$

The standard cell potential is quite negative, so the reaction will *not* occur spontaneously as written. That is, metallic tin cannot be used to reduce Be^{2+} to beryllium metal under standard conditions. Instead, the reverse process, the reduction of stannous ions (Sn^{2+}) by metallic beryllium, which has a positive value of E°_{cell}, will occur spontaneously.

(b) **A** MnO_2 is the oxidant (Mn^{4+} is reduced to Mn^{2+}), while H_2O_2 is the reductant (O^{2-} is oxidized to O_2). We can obtain the standard electrode potentials for the reduction and oxidation half-reactions directly from Table 19.2. **B** The two half-reactions and their corresponding potentials are

Cathode:	$MnO_2(s) + 4H^+(aq) + 2e^- \longrightarrow Mn^{2+}(aq) + 2H_2O(l)$	$E^\circ_{cathode} = 1.22\ V$
Anode:	$H_2O_2(aq) \longrightarrow O_2(g) + 2H^+(aq) + 2e^-$	$E^\circ_{anode} = 0.70\ V$
Overall:	$MnO_2(s) + H_2O_2(aq) + 2H^+(aq) \longrightarrow O_2(g) + Mn^{2+}(aq) + 2H_2O(l)$	$E^\circ_{cell} = E^\circ_{cathode} - E^\circ_{anode} = 0.53\ V$

The standard potential for the reaction is positive, indicating that under standard conditions it will occur spontaneously as written. Hydrogen peroxide will reduce MnO_2, and oxygen gas will evolve from the solution.

EXERCISE 19.6

Use the data in Table 19.2 to determine whether each reaction is likely to occur spontaneously under standard conditions:

(a) $2Ce^{4+}(aq) + 2Cl^-(aq) \longrightarrow 2Ce^{3+}(aq) + Cl_2(g)$
(b) $4MnO_2(s) + 3O_2(g) + 4OH^-(aq) \longrightarrow 4MnO_4^-(aq) + 2H_2O$

Answer **(a)** spontaneous ($E^\circ_{cell} = 0.36\ V$); **(b)** nonspontaneous ($E^\circ_{cell} = -0.20\ V$)

Although the sign of E°_{cell} tells us whether a particular redox reaction will occur spontaneously under standard conditions, it does not tell us to what *extent* the reaction proceeds, nor does it tell us what will happen under nonstandard conditions. To answer these questions requires a more quantitative understanding of the relationship between electrochemical cell potential and chemical thermodynamics, as described in the next section.

Michael Faraday. Faraday (1791–1867) was a British physicist and chemist who was arguably one of the greatest experimental scientists in history. The son of a blacksmith, Faraday was self-educated and became an apprentice bookbinder at age 14 before turning to science. His experiments in electricity and magnetism made electricity a routine tool in science and led to both the electric motor and the electric generator. He discovered the phenomenon of electrolysis and laid the foundations of electrochemistry. In fact, most of the specialized terms introduced in this chapter (*electrode, anode, cathode*, and so on) are due to Faraday. In addition, he discovered benzene and invented the system of oxidation state numbers that we use today. Faraday is probably best known for "The Chemical History of a Candle," a series of public lectures on the chemistry and physics of flames.

MGC Cell Potential and Free Energy

Note the pattern

A spontaneous redox reaction is characterized by a negative value of $\Delta G°$, which corresponds to a positive value of $E°_{cell}$.

19.4 Electrochemical Cells and Thermodynamics

Changes in reaction conditions can have a tremendous effect on the course of a redox reaction. For example, under standard conditions, the reaction of Co(*s*) with Ni^{2+}(*aq*) to form Ni(*s*) and Co^{2+}(*aq*) occurs spontaneously, but if we reduce the concentration of Ni^{2+} by a factor of 100, so that $[Ni^{2+}]$ is 0.01 *M*, then the reverse reaction occurs spontaneously instead. The relationship between voltage and concentration is one of the factors that must be understood in order to predict whether a reaction will be spontaneous.

The Relationship Between Cell Potential and Free Energy

Electrochemical cells convert chemical to electrical energy, and vice versa. The total amount of energy produced by an electrochemical cell, and thus the amount of energy available to do electrical work, depends on both the cell potential and the total number of electrons that are transferred from the reductant to the oxidant during the course of the reaction. The resulting electric current is measured in **coulombs (C)**, an SI unit that measures the number of electrons passing a given point in 1 s. A coulomb relates electrical potential (in volts) to energy (in joules). Electric current is measured in **amperes (A)**; 1 ampere is defined as the flow of 1 coulomb per second past a given point (1 C = 1 A · *s*):

$$\frac{1\ \text{J}}{1\ \text{V}} = 1\ \text{C} = \text{A} \cdot s \tag{19.32}$$

In chemical reactions, however, we need to relate the coulomb to the charge on a *mole* of electrons. Multiplying the charge on the electron by Avogadro's number gives us the charge on 1 mol of electrons, which is called the **faraday (*F*)**, named after the English physicist and chemist Michael Faraday (1791–1867):

$$F = (1.60218 \times 10^{-19}\ \text{C})\left(\frac{6.02214 \times 10^{23}}{1\ \text{mol e}^-}\right) \tag{19.33}$$
$$= 9.64855 \times 10^4\ \text{C/mol e}^- \simeq 96{,}486\ \text{J (V} \cdot \text{mol e}^-)$$

The total charge transferred from the reductant to the oxidant is therefore nF, where n is the number of moles of electrons.

The maximum amount of work that can be produced by an electrochemical cell, w_{max}, is equal to the product of the cell potential, E_{cell}, and the total charge transferred during the reaction, nF:

$$w_{max} = -n F E_{cell} \tag{19.34}$$

Work is expressed as a negative number because work is being done *by* the system (an electrochemical cell with a positive potential) *on* the surroundings.

As you learned in Chapter 18, ΔG is also a measure of the maximum amount of work that can be performed during a chemical process ($\Delta G = w_{max}$). Consequently, there must be a relationship between the potential of an electrochemical cell and the change in free energy, ΔG, the most important thermodynamic quantity discussed in Chapter 18. This relationship is

$$\Delta G = -n F E_{cell} \tag{19.35}$$

A spontaneous redox reaction is therefore characterized by a *negative* value of ΔG and a *positive* value of E_{cell}, consistent with our earlier discussions. When both reactants and products are in their standard states, the relationship between $\Delta G°$ and $E°_{cell}$ is

$$\Delta G° = -n F E°_{cell} \tag{19.36}$$

EXAMPLE 19.7

Suppose you want to prepare elemental bromine from bromide using the dichromate ion as an oxidant. Using data in Table 19.2, calculate the free-energy change, $\Delta G°$, for this redox reaction under standard conditions. Is the reaction spontaneous?

Given Redox reaction

Asked for $\Delta G°$ for the reaction, spontaneity

Strategy

A From the relevant half-reactions and the corresponding values of $E°$, write the overall reaction and calculate $E°_{cell}$ using Equation 19.7.

B Determine the number of electrons transferred in the overall reaction. Then use Equation 19.36 to calculate $\Delta G°$. If $\Delta G°$ is negative, then the reaction is spontaneous.

Solution

A As always, the first step is to write the relevant half-reactions and use them to obtain the overall reaction and the magnitude of $E°$. From Table 19.2, we can find the reduction and oxidation half-reactions and corresponding $E°$ values:

$$\text{Cathode:}\quad Cr_2O_7^{2-}(aq) + 14H^+(aq) + 6e^- \longrightarrow 2Cr^{3+}(aq) + 7H_2O(l) \qquad E°_{cathode} = 1.23\ V$$

$$\text{Anode:}\quad 2Br^-(aq) \longrightarrow Br_2(aq) + 2e^- \qquad E°_{anode} = 1.09\ V$$

To obtain the overall balanced reaction, we must multiply both sides of the oxidation half-reaction by 3 to obtain the same number of electrons as in the reduction half-reaction, remembering that the magnitude of $E°$ is not affected:

$$\text{Cathode:}\quad Cr_2O_7^{2-}(aq) + 14H^+(aq) + 6e^- \longrightarrow 2Cr^{3+}(aq) + 7H_2O(l) \qquad E°_{cathode} = 1.23\ V$$

$$\text{Anode:}\quad 6Br^-(aq) \longrightarrow 3Br_2(aq) + 6e^- \qquad E°_{anode} = 1.09\ V$$

$$\text{Overall:}\quad Cr_2O_7^{2-}(aq) + 6Br^-(aq) + 14H^+(aq) \longrightarrow 2Cr^{3+}(aq) + 3Br_2(aq) + 7H_2O(l) \qquad E°_{cell} = E°_{cathode} - E°_{anode} = 0.14\ V$$

B We can now calculate $\Delta G°$ using Equation 19.36. Because *six* electrons are transferred in the overall reaction, the value of n is 6:

$$\begin{aligned}\Delta G° = -nFE°_{cell} &= -(6\ \cancel{mole})[96{,}486\ J/(\cancel{V}\cdot\cancel{mol})](0.14\ \cancel{V})\\ &= -8.1\times10^4\ J\\ &= -81\ kJ/mol\ Cr_2O_7^{2-}\end{aligned}$$

Thus, $\Delta G°$ is −81 kJ for the reaction as written, and the reaction is spontaneous.

EXERCISE 19.7

Use the data in Table 19.2 to calculate $\Delta G°$ for the reduction of ferric ion by iodide:

$$2Fe^{3+}(aq) + 2I^-(aq) \longrightarrow 2Fe^{2+}(aq) + I_2(s)$$

Is the reaction spontaneous?

Answer −44 kJ/mol I_2; yes

Potentials for the Sums of Half-Reactions

Only a limited number of half-reactions are given in Table 19.2. When the standard potential for a half-reaction is not available, we can use relationships between standard potentials and free energy to obtain the potential of any other half-reaction that can be written as the sum of two or more half-reactions whose standard potentials are available.

For example, the potential for the reduction of $Fe^{3+}(aq)$ to Fe(*s*) is not listed in the table, but two related reductions are given:

$$Fe^{3+}(aq) + e^- \longrightarrow Fe^{2+}(aq) \qquad E° = 0.77\text{ V} \tag{19.37a}$$
$$Fe^{2+}(aq) + 2e^- \longrightarrow Fe(s) \qquad E° = -0.45\text{ V} \tag{19.37b}$$

Although the sum of these two half-reactions gives the desired half-reaction, we cannot simply add the potentials of two reductive half-reactions to obtain the potential of a third reductive half-reaction *because E° is not a state function.* However, because $\Delta G°$ is a state function, the sum of the $\Delta G°$ values for the individual reactions gives us $\Delta G°$ for the overall reaction, which is proportional to *both* the potential *and* the number of electrons (n) transferred. To obtain the value of $E°$ for the overall half-reaction, we first must add the values of $\Delta G°$ ($= -n\,F\,E°$) for each half-reaction (Equations 19.37a and 19.37b) to obtain $\Delta G°$ for the overall half-reaction:

$$\begin{array}{llll} Fe^{3+}(aq) + e^- & \longrightarrow \cancel{Fe^{2+}(aq)} & \Delta G° = -(1)\,F(0.77\text{ V}) & \\ \cancel{Fe^{2+}(aq)} + 2e^- & \longrightarrow Fe(s) & \Delta G° = -(2)\,F(-0.45\text{ V}) & \\ \hline Fe^{3+}(aq) + 3e^- & \longrightarrow Fe(s) & \Delta G° = [-(1)\,F(0.77\text{ V})] + [-(2)\,F(-0.45\text{ V})] & (19.38) \end{array}$$

Solving the last expression for $\Delta G°$ for the overall half-reaction gives

$$\Delta G° = F[(-0.77\text{ V}) + (-2)(-0.45\text{ V})] = F(0.13\text{ V}) \tag{19.39}$$

Note the pattern

Values of E° for half-reactions cannot be added to give E° for the sum of the half-reactions; only values of $\Delta G° = -n\,F\,E°$ for half-reactions can be added.

Three electrons ($n = 3$) are transferred in the overall reaction (Equation 19.38), so substituting into Equation 19.36 and solving for $E°$ give

$$\Delta G° = -n\,F\,E°$$
$$F(0.13\text{ V}) = -3\,F\,E°$$
$$E° = -\frac{0.13\text{ V}}{3} = -0.043\text{ V}$$

Notice that the value of $E°$ is *very* different from the value that is obtained by simply adding the potentials for the two half-reactions (0.32 V), and even has the opposite sign!

The Relationship Between Cell Potential and the Equilibrium Constant

We can use the relationship between $\Delta G°$ and the equilibrium constant K, defined in Chapter 18, to obtain a relationship between $E°_{cell}$ and K. Recall that for a general reaction of the type $a\text{A} + b\text{B} \longrightarrow c\text{C} + d\text{D}$, the standard free-energy change and the equilibrium constant are related by the equation

$$\Delta G° = -RT\ln K \tag{19.40}$$

Given the relationship between the standard free-energy change and the standard cell potential (Equation 19.36), we can write

$$-n\,F\,E°_{cell} = -RT\ln K \tag{19.41}$$

Rearranging this equation gives

$$E°_{cell} = \left(\frac{RT}{n\,F}\right)\ln K \tag{19.42}$$

For $T = 298$ K, Equation 19.42 can be simplified to

$$E°_{cell} = \left(\frac{RT}{n\,F}\right)\ln K = \left[\frac{[8.314\ \cancel{\text{J}}/(\cancel{\text{mol}}\cdot\cancel{\text{K}})](298\ \cancel{\text{K}})}{n[96{,}486\ \cancel{\text{J}}/(\text{V}\cdot\cancel{\text{mol}})]}\right]2.303\log K = \left(\frac{0.0591\text{ V}}{n}\right)\log K \tag{19.43}$$

Thus, the standard cell potential, E°_{cell}, is directly proportional to the logarithm of the equilibrium constant. This means that large equilibrium constants correspond to large positive values of E°_{cell}, and vice versa.

EXAMPLE 19.8

Use the data in Table 19.2 to calculate the equilibrium constant for the reaction of metallic lead with PbO_2 in the presence of sulfate ions to give $PbSO_4$ under standard conditions. (This is the reaction that occurs when a car battery is discharged.) Report your answer to two significant figures.

Given Redox reaction

Asked for Equilibrium constant K

Strategy

A Write the relevant half-reactions and potentials. From these, obtain the overall reaction and E°_{cell}.

B Determine the number of electrons transferred in the overall reaction. Use Equation 19.43 to solve for log K and then K.

Solution

A The relevant half-reactions and potentials from Table 19.2 are

Cathode:	$PbO_2(s) + SO_4^{2-}(aq) + 4H^+(aq) + 2e^- \longrightarrow PbSO_4(s) + 2H_2O(l)$	$E^\circ_{cathode} = 1.69$ V
Anode:	$Pb(s) + SO_4^{2-}(aq) \longrightarrow PbSO_4(s) + 2e^-$	$E^\circ_{anode} = -0.36$ V
Overall:	$Pb(s) + PbO_2(s) + 2SO_4^{2-}(aq) + 4H^+(aq) \longrightarrow 2PbSO_4(s) + 2H_2O(l)$	$E^\circ_{cell} = E^\circ_{cathode} - E^\circ_{anode} = 2.05$ V

B Two electrons are transferred in the overall reaction, so $n = 2$. Solving Equation 19.43 for log K and inserting the values of n and E° give

$$\log K = \frac{nE^\circ}{0.0591\text{ V}} = \frac{2(2.05\text{ V})}{0.0591\text{ V}} = 69.37$$

$$K = 2.3 \times 10^{69}$$

Thus, the equilibrium lies far to the right, favoring a discharged battery (as anyone who has ever tried unsuccessfully to start a car after letting it sit for a long time will know).

EXERCISE 19.8

Use data in Table 19.2 to calculate the equilibrium constant for the reaction of $Sn^{2+}(aq)$ with oxygen to produce $Sn^{4+}(aq)$ and water under standard conditions. Report your answer to two significant figures. The reaction is

$$2Sn^{2+}(aq) + O_2(g) + 4H^+(aq) \rightleftharpoons 2Sn^{4+}(aq) + 2H_2O(l)$$

Answer 1.2×10^{73}

Figure 19.9 summarizes the relationships that we have developed based on properties of the *system*—that is, based on the equilibrium constant, standard free-energy change, and standard cell potential—and the criteria for spontaneity ($\Delta G^\circ < 0$). Unfortunately, these criteria apply only to systems in which all reactants and products are present in their standard states, a situation that is seldom encountered in the real world. A more generally useful relationship between cell potential and reactant and product concentrations, as we are about to see, utilizes the relationship between ΔG and the reaction quotient Q developed in Chapter 18.

Figure 19.9 The relationships among criteria for thermodynamic spontaneity. The three properties of a system that can be used to predict the spontaneity of a redox reaction under standard conditions are the equilibrium constant K, the standard free-energy change $\Delta G°$, and the standard cell potential $E°_{cell}$. If we know the value of one of these quantities, then these relationships enable us to calculate the value of the other two. The signs of $\Delta G°$ and $E°_{cell}$ and the magnitude of K determine the direction of spontaneous reaction under standard conditions.

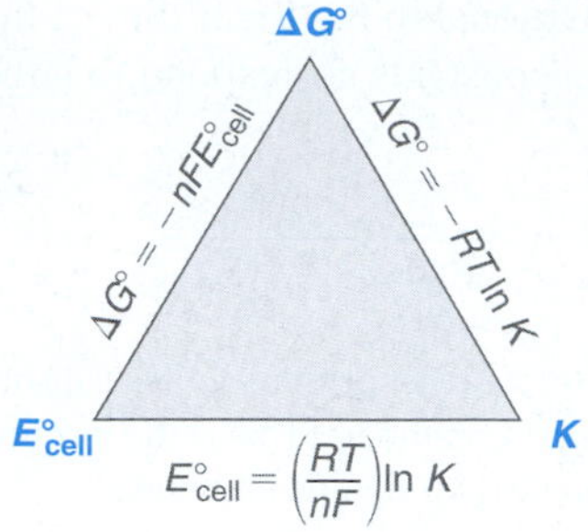

$\Delta G°$	$E°_{cell}$	K	Direction of Reaction
<0	>0	>1	Spontaneous in forward direction
>0	<0	<1	Spontaneous in reverse direction
0	0	1	No net reaction: system at equilibrium

The Effect of Concentration on Cell Potential: The Nernst Equation

MGC The Nernst Equation

Recall from Chapter 18 that the *actual* free-energy change for a reaction under nonstandard conditions, ΔG, is given by

$$\Delta G = \Delta G° + RT \ln Q \tag{19.44}$$

We also know that $\Delta G = -nFE_{cell}$ and $\Delta G° = -nFE°_{cell}$. Substituting these expressions into Equation 19.44, we obtain

$$-nFE_{cell} = -nFE°_{cell} + RT \ln Q \tag{19.45}$$

Dividing both sides of this equation by $-nF$ gives

$$E_{cell} = E°_{cell} - \left(\frac{RT}{nF}\right) \ln Q \tag{19.46}$$

Equation 19.46 is called the **Nernst equation**, after the German physicist and chemist Walter Nernst (1864–1941), who first derived it. The Nernst equation is arguably the most important relationship in electrochemistry. When a redox reaction is at equilibrium ($\Delta G = 0$), Equation 19.46 reduces to Equation 19.42 because $Q = K$ and there is no net transfer of electrons (that is, $E_{cell} = 0$).

Substituting the values of the constants into Equation 19.46 with $T = 298$ K and converting to base-10 logarithms give the relationship of the actual cell potential (E_{cell}), the standard cell potential ($E°_{cell}$), and the reactant and product concentrations at room temperature (contained in Q):

$$E_{cell} = E°_{cell} - \left(\frac{0.0591\text{ V}}{n}\right) \log Q \tag{19.47}$$

Note the pattern

The Nernst equation can be used to determine the value of E_{cell}, and thus the direction of spontaneous reaction, for any redox reaction under any conditions.

Equation 19.47 allows us to calculate the potential associated with any electrochemical cell at 298 K for any combination of reactant and product concentrations under any conditions. We can therefore determine the spontaneous direction of any redox reaction under any conditions, as long as we have tabulated values for the relevant standard electrode potentials. Notice in Equation 19.47 that the cell potential changes by $0.0591/n$ V for each 10-fold change in the value of Q because log 10 = 1.

EXAMPLE 19.9

In Exercise 19.6, you determined that the following reaction proceeds spontaneously under standard conditions because $E°_{cell} > 0$ (which you now know means that $\Delta G° < 0$):

$$2Ce^{4+}(aq) + 2Cl^-(aq) \longrightarrow 2Ce^{3+}(aq) + Cl_2(g) \qquad E°_{cell} = 0.25\text{ V}$$

Calculate E for this reaction under the following nonstandard conditions, and determine whether it will occur spontaneously: $[Ce^{4+}] = 0.013\ M$, $[Ce^{3+}] = 0.60\ M$, $[Cl^-] = 0.0030\ M$, $P_{Cl_2} = 1.0$ atm, and $T = 25°C$.

Given Balanced redox reaction, standard cell potential, nonstandard conditions

Asked for Cell potential

Strategy

Determine the number of electrons transferred during the redox process. Then use the Nernst equation to find the cell potential under the nonstandard conditions.

Solution

We can use the information given and the Nernst equation to calculate E_{cell}. Moreover, because the temperature is 25°C (298 K), we can use Equation 19.47 instead of 19.46. The overall reaction involves the net transfer of two electrons:

$$2Ce^{4+}(aq) + 2e^- \longrightarrow 2Ce^{3+}(aq) \quad \text{and} \quad 2Cl^-(aq) \longrightarrow Cl_2(g) + 2e^-$$

so $n = 2$. Substituting the concentrations given in the problem, the partial pressure of Cl_2, and the value of E°_{cell} into Equation 19.47 gives

$$\begin{aligned} E_{cell} &= E^\circ_{cell} - \left(\frac{0.0591\text{ V}}{n}\right)\log Q \\ &= 0.25\text{ V} - \left(\frac{0.0591\text{ V}}{2}\right)\log\left(\frac{[Ce^{3+}]^2 P_{Cl_2}}{[Ce^{4+}]^2[Cl^-]^2}\right) \\ &= 0.25\text{ V} - [(0.0296\text{ V})(8.37)] = 0.00\text{ V} \end{aligned}$$

Thus, the reaction will *not* occur spontaneously under these conditions (because $E = 0$ V, $\Delta G = 0$). The composition specified is that of an equilibrium mixture.

EXERCISE 19.9

In Exercise 19.6, you determined that molecular oxygen will not oxidize MnO_2 to permanganate via the reaction

$$4MnO_2(s) + 3O_2(g) + 4OH^-(aq) \longrightarrow 4MnO_4^-(aq) + 2H_2O(l) \qquad E^\circ_{cell} = -0.20\text{ V}$$

Calculate E_{cell} for the reaction under the following nonstandard conditions, and decide whether the reaction will occur spontaneously: pH 10, 0.20 atm O_2, 1.0×10^{-4} *M* MnO_4^-, and $T = 25$°C.

Answer $E_{cell} = -0.22$ V; reaction will not occur spontaneously

Applying the Nernst equation to a simple electrochemical cell such as the Zn/Cu cell discussed earlier allows us to see how the cell voltage varies as the reaction progresses and the concentrations of the dissolved ions change. Recall that the overall reaction for this cell is

$$Zn(s) + Cu^{2+}(aq) \longrightarrow Zn^{2+}(aq) + Cu(s) \qquad E^\circ_{cell} = 1.10\text{ V} \qquad (19.48)$$

The reaction quotient is therefore $Q = [Zn^{2+}]/[Cu^{2+}]$. Suppose that the cell initially contains 1.0 *M* Cu^{2+} and 1.0×10^{-6} *M* Zn^{2+}. The initial voltage measured when the cell is connected can be calculated from Equation 19.47:

$$\begin{aligned} E_{cell} &= E^\circ_{cell} - \left(\frac{0.0591\text{ V}}{n}\right)\log\frac{[Zn^{2+}]}{[Cu^{2+}]} \\ &= 1.10\text{ V} - \left(\frac{0.0591\text{ V}}{2}\right)\log\left(\frac{1.0 \times 10^{-6}}{1.0}\right) = 1.28\text{ V} \end{aligned} \qquad (19.49)$$

Thus, the initial voltage is *higher* than E° because $Q < 1$. As the reaction proceeds, $[Zn^{2+}]$ in the anode compartment increases as the zinc electrode dissolves, while $[Cu^{2+}]$ in the cathode compartment decreases as metallic copper is deposited on the electrode. During

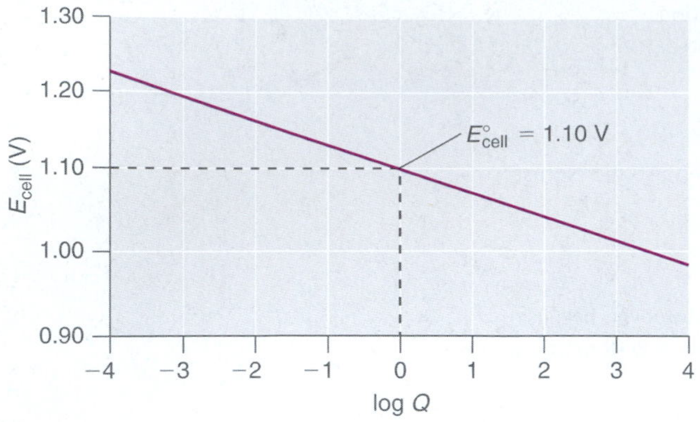

Figure 19.10 The variation of E_{cell} with log Q for a Zn/Cu cell. Initially, log $Q < 0$ and the voltage of the cell is higher than E°_{cell}. As the reaction progresses, log Q increases and E_{cell} decreases. When $[Zn^{2+}] = [Cu^{2+}]$, log $Q = 0$ and $E_{cell} = E^\circ_{cell} = 1.10$ V. As long as the electrical circuit remains intact, the reaction will continue and log Q will increase until $Q = K$ and the cell voltage reaches zero. At this point, the system will have reached equilibrium.

this process, the ratio $Q = [Zn^{2+}]/[Cu^{2+}]$ steadily increases, and the cell voltage therefore steadily decreases. Eventually, $[Zn^{2+}] = [Cu^{2+}]$, so $Q = 1$ and $E_{cell} = E^\circ_{cell}$. Beyond this point, $[Zn^{2+}]$ will continue to increase in the anode compartment and $[Cu^{2+}]$ will continue to decrease in the cathode compartment. Thus, the value of Q will increase further, leading to a further decrease in E_{cell}. When the concentrations in the two compartments are the opposite of the initial concentrations (that is, 1.0 M Zn^{2+} and 1.0×10^{-6} M Cu^{2+}), $Q = 1.0 \times 10^6$ and the cell potential will be reduced to 0.92 V.

The variation of E_{cell} with log Q over this range is linear with a slope of $-0.0591/n$, as illustrated in Figure 19.10. As the reaction proceeds still further, Q continues to increase and E_{cell} continues to decrease. If neither of the electrodes dissolves completely, thereby breaking the electrical circuit, the cell voltage will eventually reach zero. This is the situation that occurs when a battery is "dead." The value of Q when $E_{cell} = 0$ is calculated as follows:

$$E_{cell} = E^\circ_{cell} - \left(\frac{0.0591\text{ V}}{n}\right)\log Q = 0 \qquad (19.50)$$

$$E^\circ = \left(\frac{0.0591\text{ V}}{n}\right)\log Q$$

$$\log Q = \frac{E^\circ n}{0.0591\text{ V}} = \frac{(1.10\text{ V})(2)}{0.0591\text{ V}} = 37.23$$

$$Q = 1.7 \times 10^{37}$$

Recall that at equilibrium, $Q = K$. Thus, the equilibrium constant for the reaction of Zn metal with Cu^{2+} to give Cu metal and Zn^{2+} is 1.7×10^{37} at 25°C.

Concentration Cells

A voltage can also be generated by constructing an electrochemical cell in which each compartment contains the same redox active solution but at different concentrations. The voltage is produced as the concentrations equilibrate. Suppose, for example, we have a cell with 0.01 M $AgNO_3$ in one compartment and 1.0 M $AgNO_3$ in the other. The cell diagram and corresponding half-reactions are

$$Ag(s) \mid Ag^+(aq, 0.010\ M) \parallel Ag^+(aq, 1.0\ M) \mid Ag(s) \qquad (19.51)$$

Cathode: $Ag^+(aq, 1.0\ M) + e^- \longrightarrow Ag(s)$ (19.52a)

Anode: $Ag(s) \longrightarrow Ag^+(aq, 0.01\ M) + e^-$ (19.52b)

Overall: $Ag^+(aq, 1.0\ M) \longrightarrow Ag^+(aq, 0.010\ M)$ (19.52c)

As the reaction progresses, the concentration of Ag^+ will increase in the left (oxidation) compartment as the silver electrode dissolves, while the Ag^+ concentration in the right (reduction) compartment decreases as the electrode in that compartment gains mass. The *total* mass of Ag(s) in the cell will remain constant, however. We can calculate the potential of the cell using the Nernst equation, inserting 0 for E°_{cell} because $E^\circ_{cathode} = -E^\circ_{anode}$:

$$E_{cell} = E^\circ_{cell} - \left(\frac{0.0591\text{ V}}{n}\right)\log Q = 0 - \left(\frac{0.0591\text{ V}}{1}\right)\log\left(\frac{0.010}{1.0}\right) = 0.12\text{ V}$$

An electrochemical cell of this type, in which the anode and cathode compartments are identical except for the concentration of a reactant, is called a **concentration cell**. As the reaction proceeds, the difference between the concentrations of Ag^+ in the two compartments will decrease, and so will E_{cell}. Finally, when the concentration of Ag^+ is the same in both compartments, equilibrium will have been reached, and the measured potential difference between the two compartments will be zero ($E_{cell} = 0$).

EXAMPLE 19.10

Calculate the voltage in a galvanic cell that contains a manganese electrode immersed in a 2.0 *M* solution of $MnCl_2$ as the cathode, and a manganese electrode immersed in a 5.2×10^{-2} *M* solution of $MnSO_4$ as the anode ($T = 25°C$).

Given Galvanic cell, identities of the electrodes, solution concentrations

Asked for Voltage

Strategy

Ⓐ Write the overall reaction that occurs in the cell.

Ⓑ Determine the number of electrons transferred. Substitute this value into the Nernst equation to calculate the voltage.

Solution

Ⓐ This is a concentration cell, in which the electrode compartments contain the same redox active substance but at different concentrations. The anions (Cl^- and SO_4^{2-}) do not participate in the reaction, so their identity is not important. The overall reaction is

$$Mn^{2+}(aq, 2.0\ M) \longrightarrow Mn^{2+}(aq, 5.2 \times 10^{-2}\ M)$$

Ⓑ For the reduction of $Mn^{2+}(aq)$ to $Mn(s)$, $n = 2$. We substitute this value and the given Mn^{2+} concentrations into Equation 19.47:

$$E_{cell} = E^\circ_{cell} - \left(\frac{0.0591\ V}{n}\right)\log Q = 0\ V - \left(\frac{0.0591\ V}{2}\right)\log\left(\frac{5.2 \times 10^{-2}}{2.0}\right) = 0.047\ V$$

Thus, manganese will dissolve from the electrode in the compartment that contains the more dilute solution and will be deposited on the electrode in the compartment that contains the more concentrated solution.

EXERCISE 19.10

Suppose we construct a galvanic cell by placing two identical platinum electrodes in two beakers that are connected by a salt bridge. One beaker contains 1.0 *M* HCl, and the other a 0.010 *M* solution of Na_2SO_4 at pH 7.0. Both cells are in contact with the atmosphere, with $P_{O_2} = 0.20$ atm. If the relevant electrochemical reaction in both compartments is the four-electron reduction of oxygen to water, $O_2(g) + 4H^+(aq) + 4e^- \longrightarrow 2H_2O(l)$, what will be the potential when the circuit is closed?

Answer 0.41 V

Using Cell Potentials to Measure Solubility Products

Because voltages are relatively easy to measure accurately using a voltmeter, electrochemical methods provide a convenient way to determine the concentrations of very dilute solutions and the solubility products, K_{sp}, of sparingly soluble substances. As you learned in Chapter 17, solubility products can be very small, with values of $\leq 10^{-30}$. Equilibrium constants of this magnitude are virtually impossible to measure accurately by direct methods, so we must use alternative methods that are more sensitive, such as electrochemical methods.

To understand how an electrochemical cell is used to measure a solubility product, consider the cell shown in Figure 19.11, which is designed to measure the solubility product of silver chloride, $K_{sp} = [Ag^+][Cl^-]$. In one compartment, the cell contains a silver wire inserted into a 1.0 *M* solution of Ag^+; the other compartment contains a silver wire inserted into a 1.0 *M* Cl^- solution saturated with AgCl. In this system, the Ag^+ ion

Figure 19.11 A galvanic cell for measuring the solubility product of AgCl. One compartment contains a silver wire inserted into a 1.0 *M* solution of Ag^+, and the other compartment contains a silver wire inserted into a 1.0 *M* Cl^- solution saturated with AgCl. The potential due to the difference in $[Ag^+]$ between the two cells can be used to determine K_{sp}.

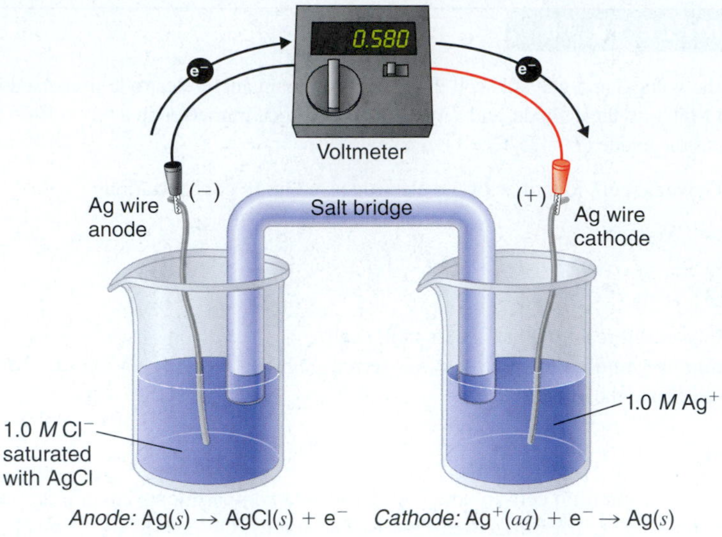

concentration in the first compartment equals K_{sp}. We can see this by dividing both sides of the equation for K_{sp} by $[Cl^-]$ and substituting: $[Ag^+] = K_{sp}/[Cl^-] = K_{sp}/1.0 = K_{sp}$. The overall cell reaction is $Ag^+(aq\text{, concentrated}) \longrightarrow Ag^+(aq\text{, dilute})$. Thus, the voltage of the concentration cell due to the difference in $[Ag^+]$ between the two cells is

$$E_{cell} = 0\text{ V} - \left(\frac{0.0591\text{ V}}{1}\right)\log\left(\frac{[Ag^+]_{dilute}}{[Ag^+]_{concentrated}}\right) = -0.0591\text{ V}\log\left(\frac{K_{sp}}{1.0}\right) = -0.0591\text{ V}\log K_{sp} \quad (19.53)$$

By closing the circuit, we can measure the potential caused by the difference in $[Ag^+]$ in the two cells. In this case, the experimentally measured voltage of the concentration cell at 25°C is 0.580 V. Solving Equation 19.53 for K_{sp} gives

$$\log K_{sp} = \frac{-E_{cell}}{0.0591\text{ V}} = \frac{-0.580\text{ V}}{0.0591\text{ V}} = -9.81 \quad (19.54)$$

$$K_{sp} = 1.5 \times 10^{-10}$$

Thus, a single potential measurement can provide the information we need to determine the value of the solubility product of a sparingly soluble salt.

EXAMPLE 19.11

To measure the solubility product of lead(II) sulfate ($PbSO_4$) at 25°C, you construct a galvanic cell like the one shown in Figure 19.11, which contains a 1.0 *M* solution of a very soluble Pb^{2+} salt [lead(II) acetate trihydrate] in one compartment that is connected by a salt bridge to a 1.0 *M* solution of Na_2SO_4 saturated with $PbSO_4$ in the other. You then insert a Pb electrode into each compartment and close the circuit. Your voltmeter shows a voltage of 230 mV. What is K_{sp} for $PbSO_4$? Report your answer to two significant figures.

Given Galvanic cell, solution concentrations, electrodes, voltage

Asked for K_{sp}

Strategy

A From the information given, write the equation for K_{sp}. Express this equation in terms of the concentration of Pb^{2+}.

B Determine the number of electrons transferred in the electrochemical reaction. Substitute the appropriate values into Equation 19.53 and solve for K_{sp}.

Solution

A You have constructed a concentration cell, with one compartment containing a 1.0 *M* solution of Pb^{2+} and the other containing a dilute solution of Pb^{2+} in 1.0 *M* Na_2SO_4. As for any concentration cell, the voltage between the two compartments can be calculated using the Nernst equation. The first step is to relate the concentration of Pb^{2+} in the dilute solution to K_{sp}:

$$[Pb^{2+}][SO_4^{2-}] = K_{sp}$$

$$[Pb^{2+}] = \frac{K_{sp}}{[SO_4^{2-}]} = \frac{K_{sp}}{1.0\ M} = K_{sp}$$

B The reduction of Pb^{2+} to Pb is a two-electron process, and proceeds according to the reaction Pb^{2+} (*aq*, concentrated) $\longrightarrow$ Pb^{2+} (*aq*, dilute), so

$$E_{cell} = E^{\circ}_{cell} - \left(\frac{0.0591\ V}{n}\right) \log Q$$

$$0.230\ V = 0\ V - \left(\frac{0.0591\ V}{2}\right) \log\left(\frac{[Pb^{2+}]_{dilute}}{[Pb^{2+}]_{concentrated}}\right) = -0.0296\ V \log\left(\frac{K_{sp}}{1.0}\right)$$

$$K_{sp} = 1.7 \times 10^{-8}$$

EXERCISE 19.11

A concentration cell similar to the one described in Example 19.11 contains a 1.0 *M* solution of lanthanum nitrate $[La(NO_3)_3]$ in one compartment and a 1.0 *M* solution of sodium fluoride saturated with LaF_3 in the other. A metallic La strip is inserted into each compartment, and the circuit is closed. The measured potential is 0.32 V. What is the value of K_{sp} for LaF_3? Report your answer to two significant figures.

Answer 5.7×10^{-17}

Using Cell Potentials to Measure Concentrations

Another use for the Nernst equation is to calculate the concentration of a species given a measured potential and the concentrations of all the other species. We saw an example of this in Example 19.11, in which the experimental conditions were defined in such a way that the concentration of the metal ion was equal to K_{sp}. Potential measurements can be used to obtain the concentrations of dissolved species under other conditions as well, which explains the widespread use of electrochemical cells in many analytical devices. Perhaps the most common application is in the determination of $[H^+]$ using a pH meter, as illustrated in the next example.

EXAMPLE 19.12

Suppose a galvanic cell is constructed with a standard Zn/Zn^{2+} couple in one compartment and a modified hydrogen electrode in the second compartment (Figure 19.6). The pressure of hydrogen gas is 1.0 atm, but $[H^+]$ in the second compartment is unknown. The cell diagram is

$$Zn(s)\,|\,Zn^{2+}(aq,\ 1.0\ M)\,\|\,H^+(aq,\ ?\ M)\,|\,H_2(g,\ 1.0\ atm)\,|\,Pt(s)$$

What is the pH of the solution in the second compartment if the measured potential in the cell is 0.26 V at 25°C?

Given Galvanic cell, cell diagram, cell potential

Asked for pH of the solution

Strategy

A Write the overall cell reaction.
B Substitute appropriate values into the Nernst equation and solve for $-\log[H^+]$ to obtain the pH.

Solution

A Under standard conditions, the overall reaction that occurs is the reduction of protons by zinc to give H_2 (note that Zn lies below H_2 in Table 19.2):

$$Zn(s) + 2H^+(aq) \longrightarrow Zn^{2+}(aq) + H_2(g) \qquad E^\circ = 0.76\text{ V}$$

B By substituting the given values into the simplified Nernst equation (Equation 19.47), we can calculate $[H^+]$ under nonstandard conditions:

$$E_{\text{cell}} = E^\circ_{\text{cell}} - \left(\frac{0.0591\text{ V}}{n}\right)\log\left(\frac{[Zn^{2+}]P_{H_2}}{[H^+]^2}\right)$$

$$0.26\text{ V} = 0.76\text{ V} - \left(\frac{0.0591\text{ V}}{2}\right)\log\left(\frac{(1.0)(1.0)}{[H^+]^2}\right)$$

$$16.9 = \log\left(\frac{1}{[H^+]^2}\right) = \log[H^+]^{-2} = (-2)\log[H^+]$$

$$8.46 = -\log[H^+]$$

$$\text{pH} = 8.5$$

Thus, the potential of a galvanic cell can be used to measure the pH of a solution.

EXERCISE 19.12

Suppose you work for an environmental laboratory and you want to use an electrochemical method to measure the concentration of Pb^{2+} in groundwater. You construct a galvanic cell using a standard oxygen electrode in one compartment ($E^\circ_{\text{cathode}} = 1.23$ V). The other compartment contains a strip of lead in a sample of groundwater to which you have added sufficient acetic acid, a weak organic acid, to ensure electrical conductivity. The cell diagram is

$$Pb(s)\,\|\,Pb^{2+}(aq,\ ?\ M)\,\|\,H^+(aq),\ 1.0\ M\,|\,O_2(g,\ 1.0\text{ atm})\,|\,Pt(s)$$

When the circuit is closed, the cell has a measured potential of 1.62 V. Use Table 19.3 and Appendix E to determine the concentration of Pb^{2+} in the groundwater.

Answer $1.18 \times 10^{-9}\ M$

19.5 • Commercial Galvanic Cells

Because galvanic cells can be self-contained and portable, they can be used as batteries and fuel cells. A **battery**, also called a *storage cell*, is a galvanic cell (or a series of galvanic cells) that contains all the reactants needed to produce electricity. In contrast, a **fuel cell** is a galvanic cell that requires a constant external supply of one or more reactants in order to generate electricity. In this section, we describe the chemistry behind some of the more common types of batteries and fuel cells.

Batteries

There are two basic kinds of batteries: *disposable*, or *primary*, batteries, in which the electrode reactions are effectively irreversible and which cannot be recharged; and *rechargeable*, or *secondary*, batteries, which form an insoluble product that adheres to the electrodes. These batteries can be recharged by applying an electrical potential in the reverse direction. The recharging process temporarily converts a rechargeable battery from a galvanic cell to an electrolytic cell.

Batteries are cleverly engineered devices that are based on the same fundamental laws as galvanic cells. The major difference between batteries and the galvanic cells we have described is that commercial batteries use solids or pastes rather than solutions as reactants to maximize the electrical output per unit mass. The use of highly concentrated or solid reactants has another beneficial effect: the concentrations of the reactants and products do not change greatly as the battery is discharged, and consequently the output voltage remains remarkably constant during the discharge process. This behavior is in contrast to that of the Zn/Cu cell, whose output decreases logarithmically as the reaction proceeds (Figure 19.10).

When a battery consists of more than one galvanic cell, the cells are usually connected in series—that is, with the positive (+) terminal of one cell connected to the negative (−) terminal of the next, and so on. The overall voltage of the battery is therefore the sum of the voltages of the individual cells.

Leclanché Dry Cell

The dry cell, by far the most common type of battery, is used in flashlights, Walkmen, and GameBoys. Although the dry cell was patented in 1866 by the French chemist Georges Leclanché and more than 5 *billion* such cells are sold every year, the details of its electrode chemistry are still not completely understood. In spite of its name, the Leclanché dry cell is actually a "wet cell": the electrolyte is an acidic water-based paste containing MnO_2, NH_4Cl, $ZnCl_2$, graphite, and starch (Figure 19.12a). The half-reactions at the anode and cathode can be summarized as follows:

Cathode: $$2MnO_2(s) + 2NH_4^+(aq) + 2e^- \longrightarrow Mn_2O_3(s) + 2NH_3(aq) + H_2O(l) \quad (19.55a)$$

Anode: $$Zn(s) \longrightarrow Zn^{2+}(aq) + 2e^- \quad (19.55b)$$

The Zn^{2+} ions formed by the oxidation of $Zn(s)$ at the anode react with NH_3 formed at the cathode and Cl^- ions present in solution, so the overall cell reaction is

Overall: $$2MnO_2(s) + 2NH_4Cl(aq) + Zn(s) \longrightarrow Mn_2O_3(s) + Zn(NH_3)_2Cl_2(s) + H_2O(l) \quad (19.55c)$$

Figure 19.12 Three kinds of primary (nonrechargeable) batteries. (a) A Leclanché dry cell is actually a "wet cell," in which the electrolyte is an acidic water-based paste containing MnO_2, NH_4Cl, $ZnCl_2$, graphite, and starch. Though inexpensive to manufacture, the cell is not very efficient in producing electrical energy and has a limited shelf life. (b) In a "button" battery, the anode is a zinc–mercury amalgam, and the cathode can be either HgO (shown here) or Ag_2O as the oxidant. Button batteries are reliable and have a high output-to-mass ratio, which allows them to be used in applications such as calculators and watches, where their small size is crucial. (c) A lithium–iodine battery consists of two cells separated by a metallic nickel mesh that collects charge from the anodes. The anode is lithium metal, and the cathode is a solid complex of I_2. The electrolyte is a layer of solid LiI that allows Li^+ ions to diffuse from the cathode to the anode. Although this type of battery produces only a relatively small current, it is highly reliable and long-lived.

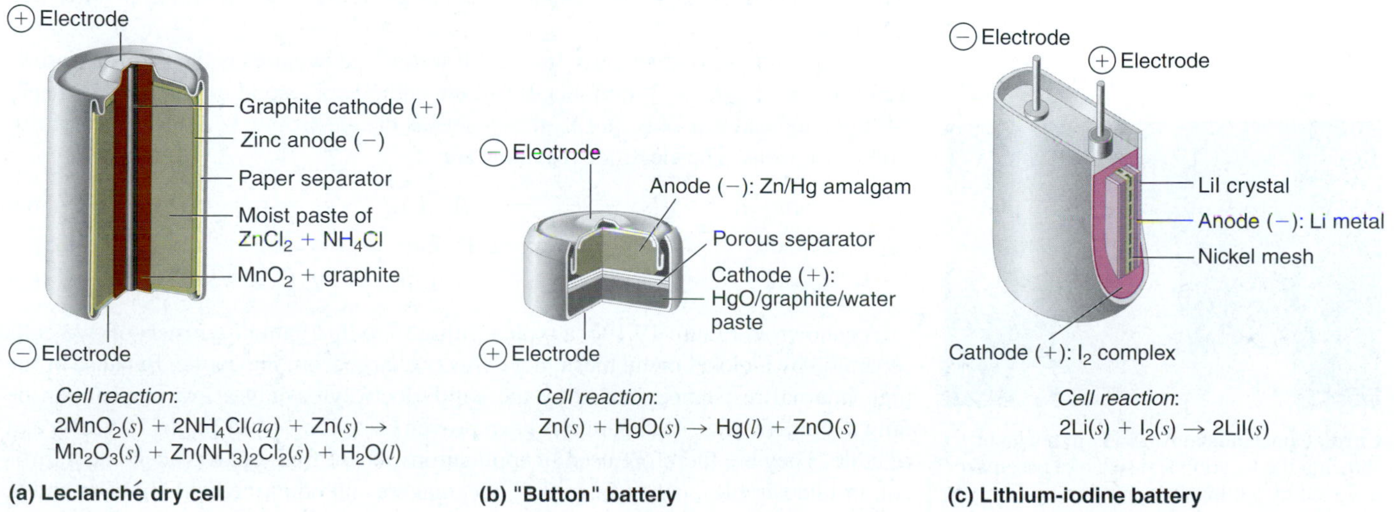

The dry cell produces about 1.55 V and is inexpensive to manufacture. It is not, however, very efficient in producing electrical energy because only the relatively small fraction of the MnO_2 that is near the cathode is actually reduced and only a small fraction of the zinc cathode is actually consumed as the cell discharges. In addition, dry cells have a limited shelf life because the Zn anode reacts spontaneously with NH_4Cl in the electrolyte, causing the case to corrode and allowing the contents to leak out.

The **alkaline battery** is essentially a Leclanché cell adapted to operate under alkaline, or basic, conditions. The half-reactions that occur in an alkaline battery are

$$\text{Cathode:}\quad 2MnO_2(s) + H_2O(l) + 2e^- \longrightarrow Mn_2O_3(s) + 2OH^-(aq) \qquad (19.56a)$$

$$\text{Anode:}\quad Zn(s) + 2OH^-(aq) \longrightarrow ZnO(s) + H_2O(l) + 2e^- \qquad (19.56b)$$

$$\text{Overall:}\quad Zn(s) + 2MnO_2(s) \longrightarrow ZnO(s) + Mn_2O_3(s) \qquad (19.56c)$$

This battery also produces about 1.5 V, but it has a longer shelf life and more constant output voltage as the cell is discharged than the Leclanché dry cell. Although the alkaline battery is more expensive to produce than the Leclanché dry cell, the improved performance makes this battery more cost effective.

"Button" Batteries

Although some of the small *"button" batteries* used to power watches, calculators, and cameras are miniature alkaline cells, most are based on completely different chemistry. In these batteries, the anode is a zinc–mercury amalgam rather than pure zinc, and the cathode uses either HgO or Ag_2O as the oxidant rather than MnO_2 (Figure 19.12b). The cathodic and overall reactions and cell output for these two types of button batteries are

$$\text{Cathode (Hg):}\quad HgO(s) + H_2O(l) + 2e^- \longrightarrow Hg(l) + 2OH^-(aq) \qquad (19.57a)$$

$$\text{Overall (Hg):}\quad Zn(s) + HgO(s) \longrightarrow Hg(l) + ZnO(s) \qquad E_{cell} = 1.35\text{ V} \qquad (19.57b)$$

$$\text{Cathode (Ag):}\quad Ag_2O(s) + H_2O(l) + 2e^- \longrightarrow 2Ag(s) + 2OH^-(aq) \qquad (19.57c)$$

$$\text{Overall (Ag):}\quad Zn(s) + Ag_2O(s) \longrightarrow 2Ag(s) + ZnO(s) \qquad E_{cell} = 1.6\text{ V} \qquad (19.57d)$$

The major advantages of the mercury and silver cells are their reliability and the high output-to-mass ratio. These factors make them ideal for applications where small size is crucial, as in cameras and hearing aids. The disadvantages are the expense and the environmental problems caused by the disposal of heavy metals such as Hg and Ag.

Lithium–Iodine Battery

None of the batteries described above is actually "dry." They all contain small amounts of liquid water, which adds significant mass and causes potential corrosion problems. Consequently, substantial effort has been expended to develop water-free batteries.

One of the few commercially successful water-free batteries is the **lithium–iodine battery**. The anode is lithium metal, and the cathode is a solid complex of I_2. Separating them is a layer of solid LiI, which acts as the electrolyte by allowing the diffusion of Li^+ ions. The electrode reactions are

$$\text{Cathode:}\quad I_2(s) + 2e^- \longrightarrow 2I^-(\text{LiI}) \qquad (19.58a)$$

$$\text{Anode:}\quad Li(s) \longrightarrow Li^+(\text{LiI}) + e^- \qquad (19.58b)$$

$$\text{Overall:}\quad 2Li(s) + I_2(s) \longrightarrow 2LiI(s) \qquad E_{cell} = 3.5\text{ V} \qquad (19.58c)$$

As shown in Figure 19.12c, a typical lithium–iodine battery consists of two cells separated by a nickel metal mesh that collects charge from the anode. Because of the high internal resistance caused by the solid electrolyte, only a low current can be drawn. Nonetheless, such batteries have proven to be long-lived (up to 10 years) and reliable. They are therefore used in applications where frequent replacement is difficult or undesirable, such as in cardiac pacemakers and other medical implants and in

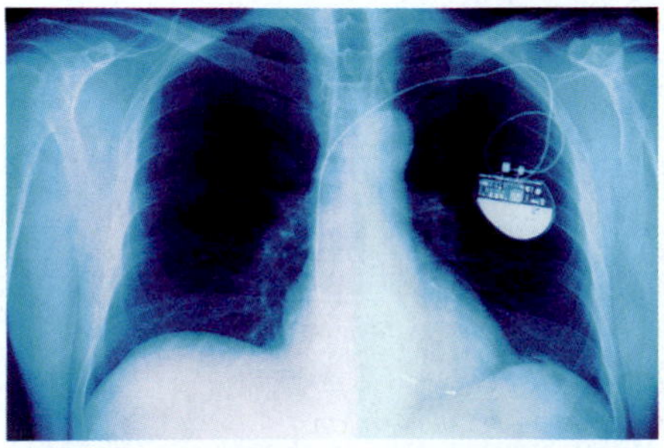

Cardiac pacemaker. An X ray of a patient showing the location and size of a pacemaker powered by a lithium–iodine battery.

computers for memory protection. These batteries are also used in security transmitters and smoke alarms. Other batteries based on lithium anodes and solid electrolytes are under development, using TiS_2, for example, for the cathode.

Dry cells, button batteries, and lithium–iodine batteries are all disposable and cannot be recharged once they are discharged. Rechargeable batteries, in contrast, offer significant economic and environmental advantages because they can be recharged and discharged numerous times. As a result, manufacturing and disposal costs drop dramatically for a given number of hours of battery usage. Two common rechargeable batteries are the nickel–cadmium battery and the lead–acid battery, which we describe next.

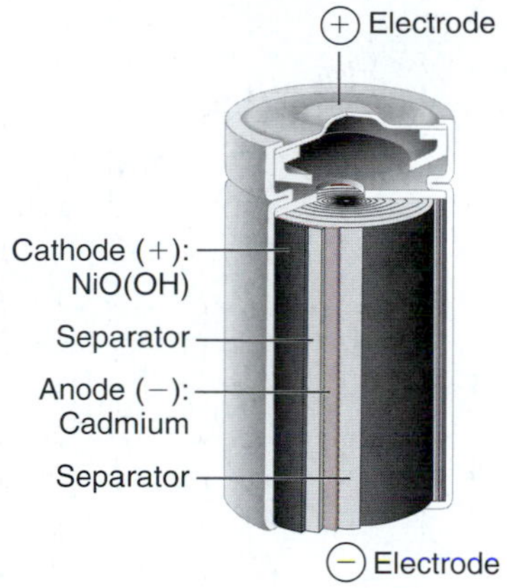

Cell reaction:
$Cd(s) + 2NiO(OH)(s) + 2H_2O(l) \rightarrow Cd(OH)_2(s) + 2Ni(OH_2)(s)$

Figure 19.13 The nickel–cadmium (Nicad) battery, a rechargeable battery. Nicad batteries contain a cadmium anode and a highly oxidized nickel cathode. This design maximizes the surface area of the electrodes and minimizes the distance between them, which gives the battery both a high discharge current and a high capacity.

Nickel–Cadmium (Nicad) Battery

The *nickel–cadmium,* or *Nicad, battery* is used in small electrical appliances and devices like drills, portable vacuum cleaners, and AM and FM digital tuners. It is a water-based cell with a cadmium anode and a highly oxidized nickel cathode that is usually described as the nickel(III) oxo-hydroxide, NiO(OH). As shown in Figure 19.13, the design maximizes the surface area of the electrodes and minimizes the distance between them, which decreases internal resistance and makes a rather high discharge current possible.

The electrode reactions during the discharge of a Nicad battery are

Cathode: $2NiO(OH)(s) + 2H_2O(l) + 2e^- \longrightarrow 2Ni(OH)_2(s) + 2OH^-(aq)$ (19.59a)

Anode: $Cd(s) + 2OH^-(aq) \longrightarrow Cd(OH)_2(s) + 2e^-$ (19.59b)

Overall: $Cd(s) + 2NiO(OH)(s) + 2H_2O(l) \longrightarrow Cd(OH)_2(s) + 2Ni(OH_2)(s)$ $E_{cell} = 1.4$ V (19.59c)

Because the products of the discharge half-reactions are solids that adhere to the electrodes [$Cd(OH)_2$ and $2Ni(OH_2)$], the overall reaction is readily reversed when the cell is recharged. Although Nicad cells are lightweight, rechargeable, and high capacity, they have certain disadvantages. For example, they tend to lose capacity quickly if not allowed to discharge fully before recharging, they do not store well for long periods when fully charged, and they present significant environmental and disposal problems because of the toxicity of cadmium.

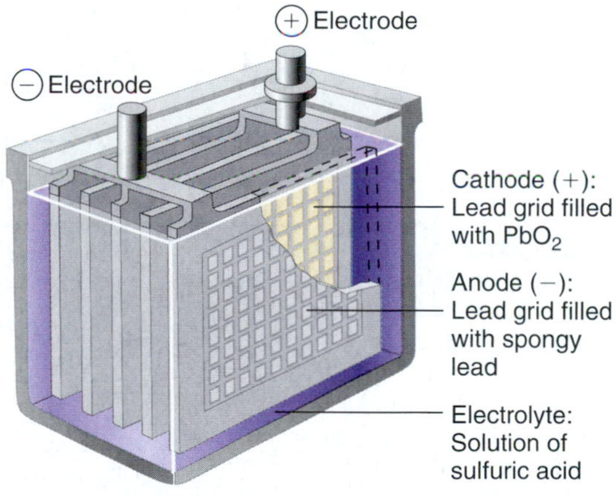

Cell reaction:
$Pb(s) + PbO_2(s) + 2HSO_4^-(aq) + 2H^+(aq) \rightarrow 2PbSO_4(s) + 2H_2O(l)$

Figure 19.14 One cell of a lead–acid battery. The anodes in each cell of a rechargeable battery are plates or grids of lead containing spongy lead metal, while the cathodes are similar grids containing powdered lead dioxide, PbO_2. The electrolyte is an aqueous solution of sulfuric acid. The value of $E°$ for such a cell is about 2 V. Connecting three such cells in series produces a 6-V battery, whereas a typical 12-V car battery contains six. When treated properly, this type of high-capacity battery can be discharged and recharged many times over.

Lead–Acid (Lead Storage) Battery

The lead–acid battery is used to provide the starting power in virtually every automobile and marine engine on the market. Marine and car batteries typically consist of multiple cells connected in series. The total voltage generated by the battery is the potential per cell, $E°_{cell}$, times the number of cells. As shown in Figure 19.14, the anode of each cell in a lead storage battery is a plate or grid of spongy lead metal, and the cathode is a similar grid containing powdered lead dioxide (PbO_2). The electrolyte is usually an approximately 37% solution (by mass) of sulfuric acid in water, with a density of 1.28 g/mL (about 4.5 M H_2SO_4). Because the redox active species are solids, there is no need to separate the electrodes. The electrode reactions in each cell during discharge are

Cathode: $PbO_2(s) + HSO_4^-(aq) + 3H^+(aq) + 2e^- \longrightarrow PbSO_4(s) + 2H_2O(l)$ $E°_{cathode} = 1.685$ V (19.60a)

Anode: $Pb(s) + HSO_4^-(aq) \longrightarrow PbSO_4(s) + H^+(aq) + 2e^-$ $E°_{anode} = -0.356$ V (19.60b)

Overall: $Pb(s) + PbO_2(s) + 2HSO_4^-(aq) + 2H^+(aq) \longrightarrow 2PbSO_4(s) + 2H_2O(l)$ $E°_{cell} = 2.041$ V (19.60c)

As the cell is discharged, a powder of $PbSO_4$ forms on the electrodes. Moreover, sulfuric acid is consumed and water is produced, decreasing the density of the electrolyte and providing a convenient way of monitoring the status of a battery by simply measuring the density of the electrolyte.

When an external voltage in excess of 2.04 V per cell is applied to a lead–acid battery, the electrode reactions reverse, and $PbSO_4$ is converted back to metallic lead and PbO_2. If the battery is recharged too vigorously, however, electrolysis of water can occur (Section 19.7), resulting in the evolution of potentially explosive hydrogen gas. The gas bubbles formed in this way can dislodge some of the $PbSO_4$ or PbO_2 particles from the grids, allowing them to fall to the bottom of the cell, where they can build up and cause an internal short circuit. Thus, the recharging process must be carefully monitored to optimize the life of the battery. With proper care, however, a lead–acid battery can be discharged and recharged thousands of times. In automobiles, the alternator supplies the electric current that causes the discharge reaction to reverse.

Fuel Cells

A fuel cell is a galvanic cell that requires a constant external supply of reactants because the products of the reaction are continuously removed. Unlike a battery, it does not store chemical or electrical energy, but instead allows electrical energy to be extracted directly from a chemical reaction. In principle, this should be a more efficient process than, for example, burning the fuel to drive an internal combustion engine that turns a generator, which is typically less than 40% efficient. Unfortunately, significant cost and reliability problems have hindered the wide-scale adoption of fuel cells. In practice, their use has been restricted to applications in which mass may be a significant cost factor, such as U.S. manned space vehicles.

These space vehicles use a hydrogen/oxygen fuel cell that requires a continuous input of $H_2(g)$ and $O_2(g)$, as illustrated in Figure 19.15. The electrode reactions are

$$\text{Cathode:}\quad O_2(g) + 4H^+ + 4e^- \longrightarrow 2H_2O(g) \qquad (19.61a)$$

$$\text{Anode:}\quad 2H_2(g) \longrightarrow 4H^+ + 4e^- \qquad (19.61b)$$

$$\text{Overall:}\quad 2H_2(g) + O_2(g) \longrightarrow 2H_2O(g) \qquad (19.61c)$$

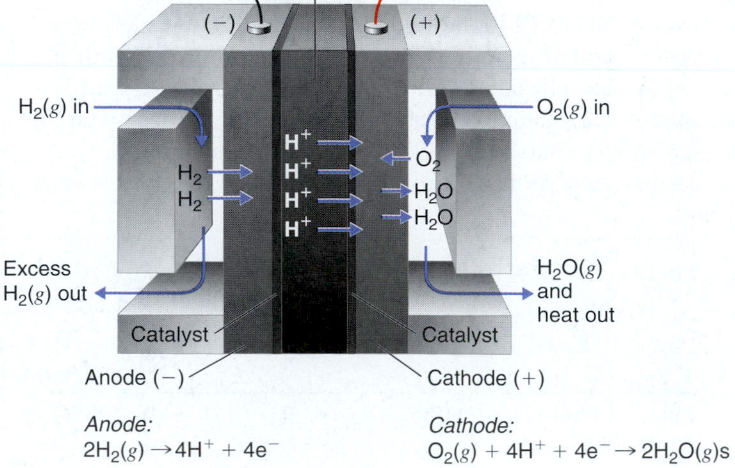

Figure 19.15 A hydrogen fuel cell produces electrical energy directly from a chemical reaction. Hydrogen is oxidized to protons at the anode, and the electrons are transferred through an external circuit to the cathode, where oxygen is reduced and combines with H^+ to form water. A solid electrolyte allows the protons to diffuse from the anode to the cathode. Although fuel cells are an essentially pollution-free means of obtaining electrical energy, their expense and technological complexity have thus far limited their applications.

The overall reaction represents an essentially pollution-free conversion of hydrogen and oxygen to water, which in space vehicles is then collected and used. Although this type of fuel cell should produce 1.23 V under standard conditions, in practice the device achieves only about 0.9 V. One of the major barriers to achieving greater efficiency is the fact that the four-electron reduction of $O_2(g)$ at the cathode is intrinsically rather slow, which limits current that can be achieved. All major automobile manufacturers have major research programs involving fuel cells: one of the most important goals is the development of a better catalyst for the reduction of O_2.

19.6 Corrosion

Note the pattern

Corrosion is a galvanic process.

Corrosion is a galvanic process by which metals deteriorate through oxidation, usually but not always to their oxides. For example, when exposed to air, iron rusts, silver tarnishes, and copper and brass acquire a bluish-green surface called a *patina*. Of the various metals subject to corrosion, iron is by far the most important commercially. An estimated $100 billion per year is spent in the United States alone to replace iron objects destroyed by corrosion. Consequently, the development of methods for protecting metal surfaces from corrosion constitutes a very active area of industrial research. In this section, we describe some of the chemical and electrochemical processes responsible for corrosion. We also examine the chemical basis for some common methods for preventing corrosion and treating corroded metals.

Under ambient conditions, the oxidation of most metals is thermodynamically spontaneous, with the notable exception of gold and platinum. Hence, it is actually somewhat surprising that any metals are useful at all in Earth's moist, oxygen-rich atmosphere. Some metals, however, are resistant to corrosion for kinetic reasons. For example, the aluminum in soft-drink cans and airplanes is protected by a thin coating of metal oxide that forms on the surface of the metal and acts as an impenetrable barrier that prevents further destruction. Aluminium cans also have a thin plastic layer to prevent reaction of the oxide with acid in the soft-drink. Chromium, magnesium, and nickel also form protective oxide films. Stainless steels are remarkably resistant to corrosion because they usually contain a significant proportion of chromium, nickel, or both.

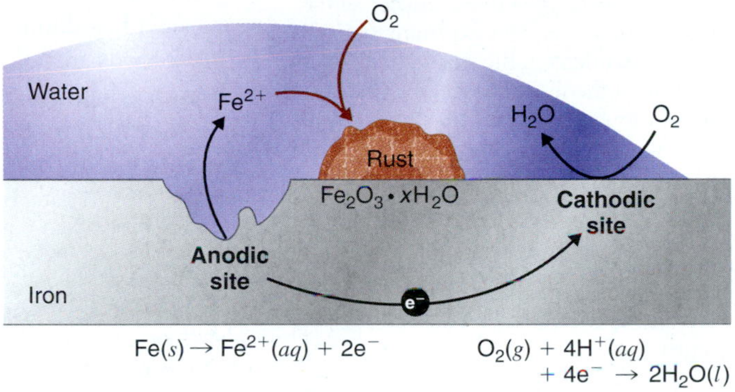

Figure 19.16 Rust, the result of corrosion of metallic iron. Iron is oxidized to $Fe^{2+}(aq)$ at an anodic site on the surface of the iron, which is often an impurity or a lattice defect. Oxygen is reduced to water at a different site on the surface of the iron, which acts as the cathode. Electrons are transferred from the anode to the cathode through the electrically conductive metal. Water provides a salt bridge between the two sites, as well as acting as a solvent for the Fe^{2+} that is produced initially. Rust ($Fe_2O_3 \cdot xH_2O$) is formed by the subsequent nonelectrochemical oxidation of Fe^{2+} by atmospheric oxygen.

In contrast to these metals, when iron corrodes, it forms a red-brown hydrated metal oxide, $Fe_2O_3 \cdot xH_2O$, commonly known as *rust*, that does not provide a tight protective film (Figure 19.16). Instead, the rust continually flakes off to expose a fresh metal surface vulnerable to reaction with oxygen and water. Because both oxygen and water are required for rust to form, an iron nail immersed in deoxygenated water will not rust even over a period of weeks. Similarly, a nail immersed in an organic solvent such as kerosene or mineral oil saturated with oxygen will not rust because of the absence of water.

In the corrosion process, iron metal acts as the anode in a galvanic cell and is oxidized to Fe^{2+}; oxygen is reduced to water at the cathode. The relevant reactions are

$$\text{Cathode:}\quad O_2(g) + 4H^+(aq) + 4e^- \longrightarrow 2H_2O(l) \qquad E^\circ = 1.23\text{ V} \qquad (19.62a)$$

$$\text{Anode:}\quad Fe(s) \longrightarrow Fe^{2+}(aq) + 2e^- \qquad E^\circ = -0.45\text{ V} \qquad (19.62b)$$

$$\text{Overall:}\quad 2Fe(s) + O_2(g) + 4H^+(aq) \longrightarrow 2Fe^{2+}(aq) + 2H_2O(l) \qquad E^\circ = 1.68\text{ V} \qquad (19.62c)$$

The Fe^{2+} ions produced in the initial reaction are then oxidized by atmospheric oxygen in a nonelectrochemical process to produce the insoluble hydrated oxide containing Fe^{3+}, as represented in the equation

$$4Fe^{2+}(aq) + O_2(g) + (2 + 4x)H_2O \longrightarrow 2Fe_2O_3 \cdot xH_2O + 4H^+(aq) \qquad (19.63)$$

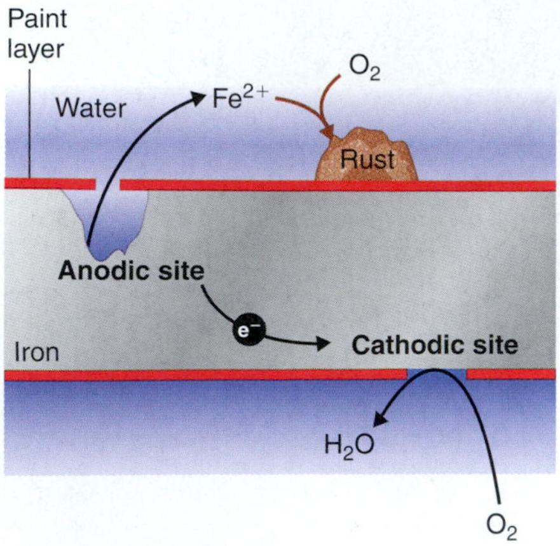

Figure 19.17 Small scratches in a protective paint coating can lead to rapid corrosion of iron. Holes in a protective coating allow oxygen to be reduced at the surface with the greater exposure to air (the cathode), while metallic iron is oxidized to $Fe^{2+}(aq)$ at the less exposed site (the anode). Rust is formed when $Fe^{2+}(aq)$ diffuses to a location where it can react with atmospheric oxygen, which is often remote from the anode. The electrochemical interaction between cathodic and anodic sites can cause a large pit to form *under* a painted surface, eventually resulting in sudden failure with little visible warning that corrosion has occurred.

The sign and magnitude of $E°$ for the corrosion process (Equation 19.62c) indicate that there is a strong driving force for the oxidation of iron by O_2 under standard conditions (1 M H^+). Under neutral conditions, the driving force is somewhat less, but still appreciable (E = 1.25 V at pH 7). Normally, the reaction of atmospheric CO_2 with water to form H^+ and HCO_3^- provides a low enough pH to enhance the rate of the reaction, as does acid rain (see Section 4.7).

Automobile manufacturers spend a great deal of time and money developing paints that adhere tightly to the car's metal surface to prevent oxygenated water, acid, and salt from coming into contact with the underlying metal. Unfortunately, even the best paint is subject to scratching or denting, and the electrochemical nature of the corrosion process means that two scratches relatively remote from each other can operate together as anode and cathode, leading to sudden mechanical failure (Figure 19.17).

One of the most common techniques used to prevent the corrosion of iron is applying a protective coating of another metal that is more difficult to oxidize. Faucets and some external parts of automobiles, for example, are often coated with a thin layer of chromium using an electrolytic process that will be discussed in the next section. With the increased use of polymeric materials in cars, however, the use of chrome-plated steel has diminished in recent years. Similarly, the "tin cans" that hold soups and other foods are actually made of steel coated with a thin layer of tin. Neither chromium nor tin is intrinsically resistant to corrosion, but both form protective oxide coatings.

As with a protective paint, scratching a protective metal coating will allow corrosion to occur. In this case, however, the presence of the second metal can actually increase the rate of corrosion. The values of the standard electrode potentials for Sn^{2+} ($E°$ = −0.14 V) and Fe^{2+} ($E°$ = −0.45 V) in Table 19.2 show that Fe is more easily oxidized than Sn. As a result, the more corrosion-resistant metal (in this case, tin) accelerates the corrosion of iron by acting as the cathode and providing a large surface area for the reduction of oxygen (Figure 19.18). This process is seen in some older homes where copper and iron pipes have been directly connected to each other. The less easily oxidized copper acts as the cathode, causing iron to dissolve rapidly near the connection and occasionally resulting in a catastrophic plumbing failure.

One way to avoid these problems is to use a more easily oxidized metal to protect iron from corrosion. In this approach, called *cathodic protection,* a more reactive metal such as Zn ($E°$ = −0.76 V for $Zn^{2+} + 2e^- \longrightarrow Zn$) becomes the anode and iron becomes the cathode. This prevents oxidation of the iron and protects the iron object from corrosion. The reactions that occur under these conditions are

$$\text{Cathode:}\quad O_2(g) + 4e^- + 4H^+(aq) \longrightarrow 2H_2O(l) \qquad (19.64a)$$

$$\text{Anode:}\quad Zn(s) \longrightarrow Zn^{2+}(aq) + 2e^- \qquad (19.64b)$$

$$\text{Overall:}\quad 2Zn(s) + O_2(g) + 4H^+(aq) \longrightarrow 2Zn^{2+}(aq) + 2H_2O(l) \qquad (19.64c)$$

The more reactive metal reacts with oxygen and will eventually dissolve, "sacrificing" itself to protect the iron object. Cathodic protection is the principle underlying *galvanized steel*, which is steel protected by a thin layer of zinc. Galvanized steel is used in objects ranging from nails to garbage cans. In a similar strategy, *sacrificial electrodes* using magnesium, for example, are used to protect underground tanks or pipes (Figure 19.19). Replacing the sacrificial electrodes is more cost effective than replacing the iron objects they are protecting.

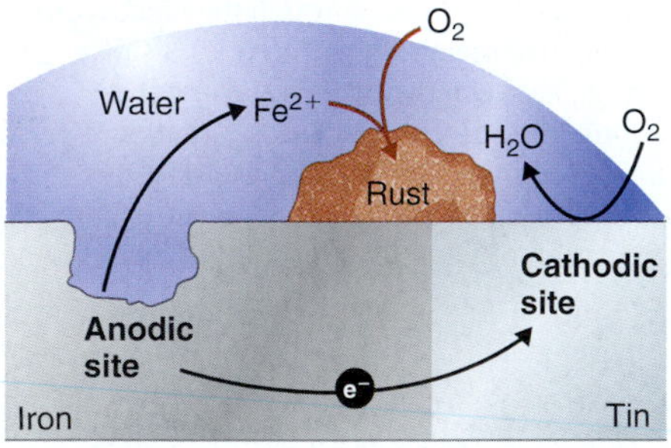

Figure 19.18 Galvanic corrosion. If iron is in contact with a more corrosion-resistant metal such as tin, copper, or lead, the other metal can act as a large cathode that greatly increases the rate of reduction of oxygen. Because reduction of oxygen is coupled to oxidation of iron, this can result in a dramatic increase in the rate at which iron is oxidized at the anode. Galvanic corrosion is likely to occur whenever two dissimilar metals are connected directly, allowing electrons to be transferred from one to the other.

EXAMPLE 19.13

Suppose an old wooden sailboat, held together with iron screws, has a bronze propeller (recall that bronze is an alloy of copper containing about 7–10% tin). (a) If the boat is immersed in seawater, what corrosion reaction will occur? What is $E°_{cell}$? (b) How could you prevent this corrosion from occurring?

Given Identity of metals

Asked for Corrosion reaction, E°_{cell}, preventive measures

Strategy

Ⓐ Write the reactions that occur at the anode and the cathode. From these, write the overall cell reaction and calculate E°_{cell}.

Ⓑ Based on the relative redox activity of various substances, suggest possible preventive measures.

Solution

(a) Ⓐ According to Table 19.2, both copper and tin are less active metals than iron (that is, they have higher positive values of E° than iron). Thus, if tin or copper is brought into electrical contact (by seawater) with iron in the presence of oxygen, corrosion will occur. We therefore anticipate that the bronze propeller will act as the cathode at which O_2 is reduced, and the iron screws will act as anodes at which iron dissolves:

Cathode:	$O_2(g) + 4H^+(aq) + 4e^- \longrightarrow 2H_2O(l)$	$E^\circ_{cathode} = 1.23\ V$
Anode:	$Fe(s) \longrightarrow Fe^{2+}(aq) + 2e^-$	$E^\circ_{anode} = -0.45\ V$
Overall:	$2Fe(s) + O_2(g) + 4H^+(aq) \longrightarrow 2Fe^{2+}(aq) + 2H_2O(l)$	$E^\circ_{cell} = 1.68\ V$

Over time, the iron screws will dissolve and the boat will fall apart.

(b) Ⓑ Possible ways to prevent corrosion, in order of decreasing cost and inconvenience, are: disassembling the boat and rebuilding it with bronze screws; removing the boat from the water and storing it in a dry place; or attaching an inexpensive piece of zinc metal to the propeller shaft to act as a sacrificial electrode and replacing it once or twice a year. Because zinc is a more active metal than iron, it will act as the anode in the electrochemical cell and dissolve (Equation 19.64c).

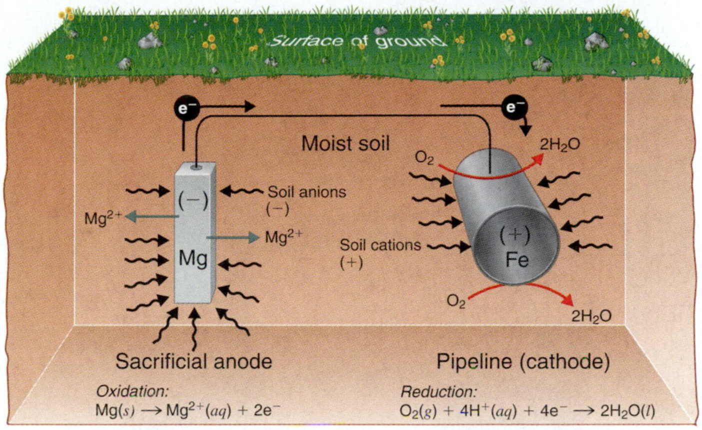

Figure 19.19 The use of a sacrificial electrode to protect against corrosion. Connecting a magnesium rod to an underground steel pipeline protects the pipeline from corrosion. Because magnesium ($E^\circ = -2.37$ V) is much more easily oxidized than iron ($E^\circ = -0.45$ V), the Mg rod acts as the anode in a galvanic cell. The pipeline is therefore forced to act as the cathode at which oxygen is reduced. The soil between the anode and the cathode acts as a salt bridge that completes the electrical circuit and maintains electrical neutrality. As $Mg(s)$ is oxidized to Mg^{2+} at the anode, anions such as nitrate in the soil diffuse toward the anode to neutralize the positive charge. Simultaneously, cations in the soil such as H^+ or NH_4^+ diffuse toward the cathode, where they replenish the protons that are consumed as oxygen is reduced. A similar strategy utilizes many miles of somewhat less reactive zinc wire to protect the Alaska oil pipeline.

EXERCISE 19.13

Suppose the water pipes leading into your house are made of lead, while the rest of the plumbing in the house is iron. To eliminate the possibility of lead poisoning, you call a plumber to replace the lead pipes. He quotes you a very low price if he can use up his existing supply of copper pipe to do the job. **(a)** Do you accept his proposal? **(b)** What else should you have the plumber do while there?

Answer

(a) Not unless you plan to sell the house very soon, because the Cu/Fe pipe joints will lead to rapid corrosion.

(b) Any existing Pb/Fe joints should be examined carefully for corrosion of the iron pipes due to the Pb–Fe junction; the less active Pb will have served as the cathode for the reduction of O_2, promoting oxidation of the more active Fe nearby.

19.7 • Electrolysis

So far we have described various galvanic cells in which a spontaneous chemical reaction is used to generate electrical energy. In an electrolytic cell, however, the opposite process, called **electrolysis**, occurs: an external voltage is applied to drive a nonspontaneous reaction (Figure 19.1). In this section, we look at how electrolytic cells are constructed and explore some of their many commercial applications.

Note the pattern

In an electrolytic cell, an external voltage is applied to drive a nonspontaneous reaction.

Electrolytic Cells

If we construct an electrochemical cell in which one electrode is copper metal immersed in a 1 *M* Cu^{2+} solution and the other electrode is cadmium metal immersed in a 1 *M* Cd^{2+}

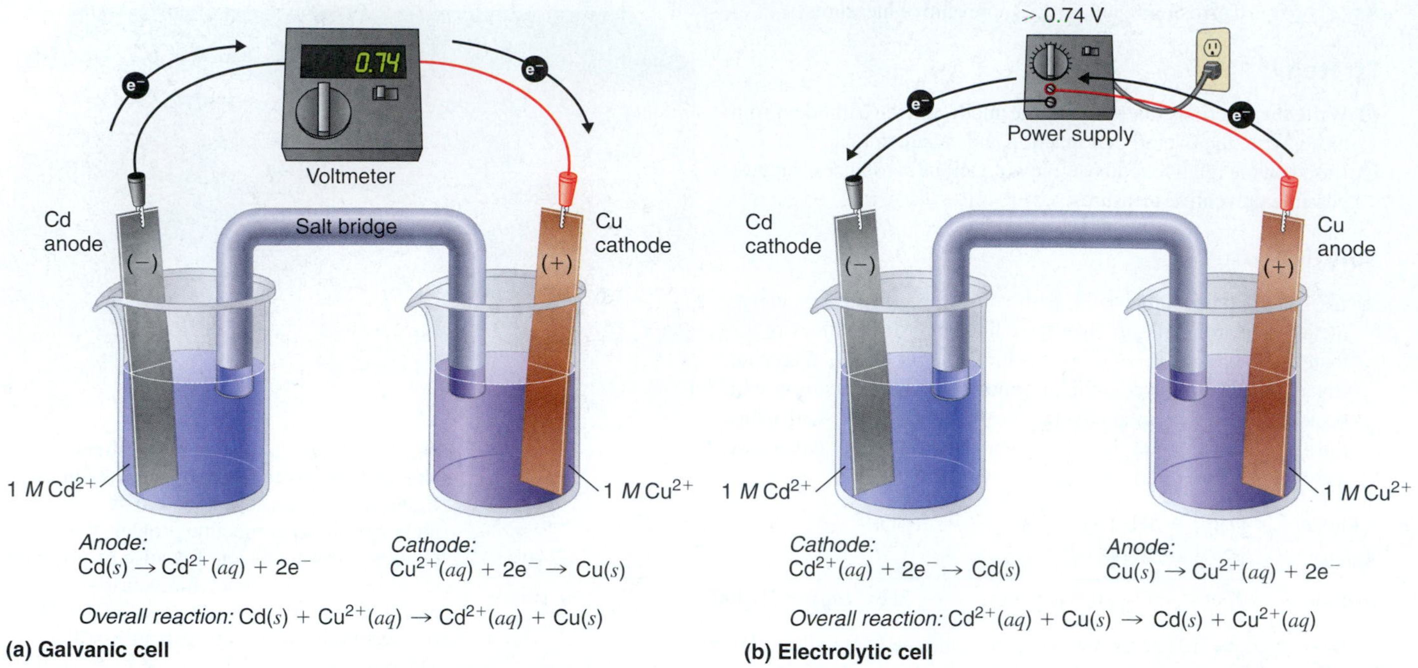

Figure 19.20 An applied voltage can reverse the flow of electrons in a galvanic Cd/Cu cell. (a) When compartments that contain a Cd electrode immersed in 1 *M* $Cd^{2+}(aq)$ and a Cu electrode immersed in 1 *M* $Cu^{2+}(aq)$ are connected to create a galvanic cell, Cd(*s*) is spontaneously oxidized to $Cd^{2+}(aq)$ at the anode, and $Cu^{2+}(aq)$ is spontaneously reduced to Cu(*s*) at the cathode. The potential of the galvanic cell is 0.74 V. (b) Applying an external potential greater than 0.74 V in the reverse direction forces electrons to flow out of the Cu electrode [which is now the anode, at which metallic Cu(*s*) is oxidized to $Cu^{2+}(aq)$] and into the Cd electrode [which is now the cathode, at which $Cd^{2+}(aq)$ is reduced to Cd(*s*)]. Note that the anode in an electrolytic cell is *positive* because electrons are flowing out of it, whereas the cathode is *negative* because electrons are flowing into it.

solution and then close the circuit, the potential difference between the two compartments will be 0.74 V. The cadmium electrode will begin to dissolve (Cd is oxidized to Cd^{2+}) and thus is the anode, while metallic copper will be deposited on the copper electrode (Cu^{2+} is reduced to Cu), which is the cathode (Figure 19.20a). The overall reaction

$$Cd(s) + Cu^{2+}(aq) \longrightarrow Cd^{2+}(aq) + Cu(s) \qquad E^{\circ}_{cell} = 0.74\ V \qquad (19.65)$$

is thermodynamically spontaneous as written ($\Delta G^{\circ} < 0$):

$$\Delta G^{\circ} = -nFE^{\circ} = -(2\ \cancel{mol}\ e^{-})[96{,}486\ J/(\cancel{V} \cdot \cancel{mol})](0.74\ \cancel{V}) = -140\ kJ\ (\text{per mole Cd}) \qquad (19.66)$$

In this direction, the system is acting as a galvanic cell.

The reverse reaction, reduction of Cd^{2+} by Cu, is thermodynamically nonspontaneous and will occur only with an *input* of 140 kJ. We can force the reaction to proceed in the reverse direction by applying an electrical potential greater than 0.74 V from an external power supply. The applied voltage forces electrons through the circuit in the reverse direction, converting a galvanic cell to an electrolytic cell. Thus, the copper electrode is now the anode (Cu is oxidized), and the cadmium electrode is now the cathode (Cd^{2+} is reduced) (Figure 19.20b). Notice that the signs of the cathode and anode have switched to reflect the flow of electrons in the circuit. The half-reactions that occur at the cathode and the anode are

$$\begin{array}{llll}
\text{Cathode:} & Cd^{2+}(aq) + 2e^{-} \longrightarrow Cd(s) & E^{\circ}_{cathode} = -0.40\ V & (19.67a) \\
\text{Anode:} & Cu(s) \longrightarrow Cu^{2+}(aq) + 2e^{-} & E^{\circ}_{anode} = 0.34\ V & (19.67b) \\
\hline
\text{Overall:} & Cd^{2+}(aq) + Cu(s) \longrightarrow Cd(s) + Cu^{2+}(aq) & E^{\circ}_{cell} = -0.74\ V & (19.67c)
\end{array}$$

Because $E^\circ_{cell} < 0$, the overall reaction—the reduction of Cd^{2+} by Cu—clearly cannot occur spontaneously and proceeds only when sufficient electrical energy is applied. The differences between galvanic and electrolytic cells are summarized in Table 19.3.

Electrolytic Reactions

At sufficiently high temperatures, ionic solids melt to form liquids that conduct electricity extremely well due to the high concentrations of ions. If two inert electrodes are inserted into molten NaCl, for example, and an electrical potential is applied, Cl^- is oxidized at the anode and Na^+ is reduced at the cathode. The overall reaction is

$$2NaCl(l) \longrightarrow 2Na(l) + Cl_2(g) \qquad (19.68)$$

This is the reverse of the formation of NaCl from its elements. Note that the product of the reduction reaction is *liquid* sodium because the melting point of sodium metal is 97.8°C, well below that of NaCl (801°C). Approximately 20,000 tons of sodium metal are produced commercially in the United States each year by electrolysis of molten NaCl in a *Downs cell* (Figure 19.21). In this specialized cell, $CaCl_2$ (mp 772°C) is first added to the NaCl to lower the melting point of the mixture to about 600°C, thereby lowering the operating costs.

Similarly, in the *Hall–Heroult process* used to produce aluminum commercially, a molten mixture of about 5% aluminum oxide (Al_2O_3, mp 2054°C) and 95% cryolite (Na_3AlF_6, mp 1012°C) is electrolyzed at about 1000°C, producing molten aluminum at the cathode and CO_2 gas at the carbon anode. The overall reaction is

$$2Al_2O_3(l) + 3C(s) \longrightarrow 4Al(l) + 3CO_2(g) \qquad (19.69)$$

Oxide ions react with oxidized carbon at the anode, producing $CO_2(g)$.

There are two important points to make about these two commercial processes and about the electrolysis of molten salts in general. First, the electrode potentials for molten salts are likely to be very different from the standard cell potentials listed in Table 19.2 and Appendix E, which are compiled for the reduction of the hydrated ions in aqueous solutions under standard conditions. Second, using a mixed salt system means there is a possibility of competition between different electrolytic reactions. When a mixture of NaCl and $CaCl_2$ is electrolyzed, Cl^- is oxidized because it is the

TABLE 19.3 Comparison of galvanic and electrolytic cells

Property	Galvanic Cell	Electrolytic Cell
ΔG	<0	>0
E_{cell}	>0	<0
Electrode process:		
Anode	Oxidation	Oxidation
Cathode	Reduction	Reduction
Sign of electrode:		
Anode	−	+
Cathode	+	−

MGC Electrolysis of Molten Salts
Electrolysis of Aqueous Salts

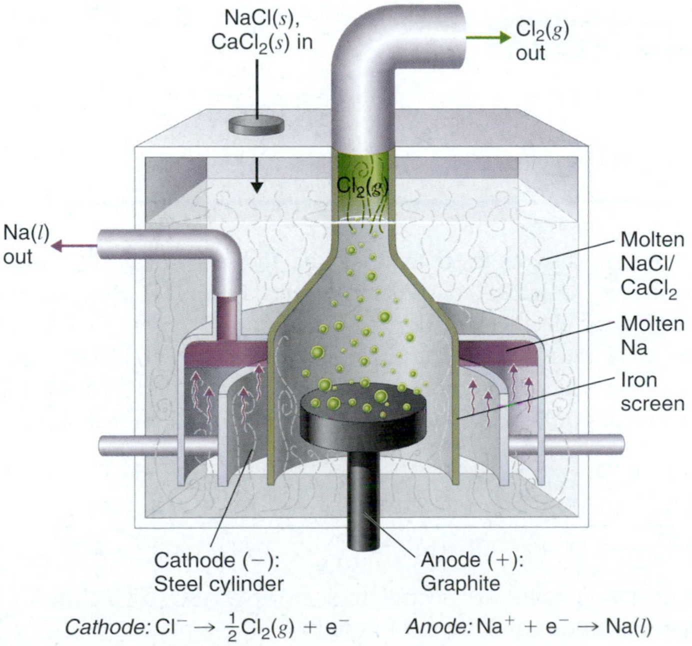

Figure 19.21 A Downs cell for the electrolysis of molten NaCl. Electrolysis of a molten mixture of NaCl and $CaCl_2$ results in the formation of elemental sodium and chlorine gas. Because sodium is a liquid under these conditions and liquid sodium is less dense than molten sodium chloride, the sodium floats to the top of the melt and is collected in concentric capped iron cylinders surrounding the cathode. Gaseous chlorine collects in the inverted cone over the anode. An iron screen separating the cathode and anode compartments ensures that the molten sodium and gaseous chlorine do not come into contact.

only anion present, but either Na^+ or Ca^{2+} can be reduced. Conversely, in the Hall–Heroult process, only one cation is present that can be reduced, Al^{3+}, but there are three species that can be oxidized, C, O^{2-}, and F^-.

In the Hall–Heroult process, C is oxidized instead of O^{2-} or F^- because oxygen and fluorine are more electronegative than carbon, which means that C is a weaker oxidant than either O_2 or F_2. Similarly, in the Downs cell, we might expect electrolysis of a $NaCl/CaCl_2$ mixture to produce calcium rather than sodium because Na is slightly less electronegative than Ca ($\chi = 0.93$ vs. 1.00, respectively), making Na easier to oxidize and, conversely, Na^+ more difficult to reduce. In fact, reduction of Na^+ to Na is the observed reaction. In cases where the electronegativities of two species are similar, other factors, such as the formation of complex ions, become important and may determine the outcome.

EXAMPLE 19.14

If a molten mixture of $MgCl_2$ and KBr is electrolyzed, what products will form at the cathode and the anode, respectively?

Given Identity of salts

Asked for Electrolysis products

Strategy

A List all possible reduction and oxidation products. Based on the electronegativity values shown in Figure 7.15, determine which species will be reduced and which species will be oxidized.

B Identify the products that will form at each electrode.

Solution

A The possible reduction products are Mg and K, and the possible oxidation products are Cl_2 and Br_2. Because Mg is more electronegative than K ($\chi = 1.31$ vs. 0.82), it is likely that Mg will be reduced rather than K. Because Cl is more electronegative than Br (3.16 vs. 2.96), Cl_2 is a stronger oxidant than Br_2.

B Electrolysis will therefore produce Br_2 at the anode and Mg at the cathode.

EXERCISE 19.14

Predict the products if a molten mixture of $AlBr_3$ and LiF is electrolyzed.

Answer Br_2 and Al

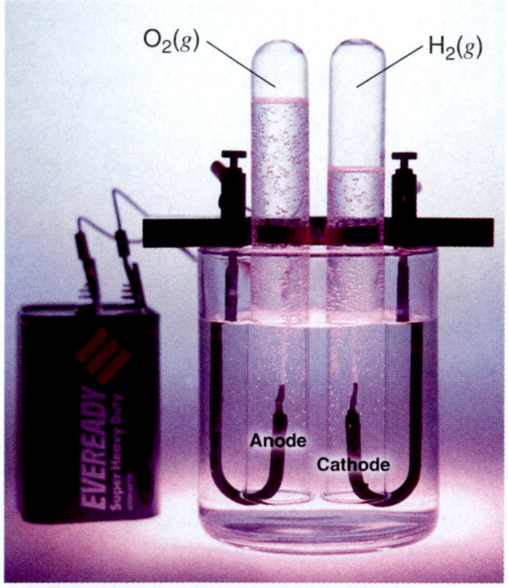

Anode: $2H_2O(l) \rightarrow O_2(g) + 4H^+(aq) + 4e^-$

Cathode: $2H^+(aq) + 2e^- \rightarrow H_2(g)$

Overall: $2H_2O(l) \rightarrow O_2(g) + 2H_2(g)$

Figure 19.22 The electrolysis of water. Applying an external potential of about 1.7–1.9 V to two inert electrodes immersed in an aqueous solution of an electrolyte such as H_2SO_4 or Na_2SO_4 drives the thermodynamically nonspontaneous decomposition of water into H_2 at the cathode and O_2 at the anode.

Electrolysis can also be used to drive the thermodynamically nonspontaneous decomposition of water into its constituent elements, H_2 and O_2. However, because pure water is a very poor electrical conductor, a small amount of an ionic solute (such as H_2SO_4 or Na_2SO_4) must first be added to make it electrically conductive. Inserting inert electrodes into the solution and applying a voltage between them will result in the rapid evolution of bubbles of H_2 and O_2 (Figure 19.22). The reactions that occur are

$$\text{Cathode: } 2H^+(aq) + 2e^- \longrightarrow H_2(g) \qquad E^\circ_{cathode} = 0 \text{ V} \tag{19.70a}$$

$$\text{Anode: } 2H_2O(l) \longrightarrow O_2(g) + 4H^+(aq) + 4e^- \qquad E^\circ_{anode} = 1.23 \text{ V} \tag{19.70b}$$

$$\text{Overall: } 2H_2O(l) \longrightarrow O_2(g) + 2H_2(g) \qquad E^\circ_{cell} = -1.23 \text{ V} \tag{19.70c}$$

For a system that contains an electrolyte such as Na_2SO_4, which has a negligible effect on the ionization equilibrium of liquid water, the pH of the solution will be 7 and $[H^+] = [OH^-] = 1.0 \times 10^{-7}$. Assuming that $P_{O_2} = P_{H_2} = 1$ atm, we can use the standard potentials and Equation 19.47 to calculate E for the overall reaction:

$$E_{cell} = E^\circ_{cell} - \left(\frac{0.0591\ V}{n}\right) \log(P_{O_2} P^2_{H_2})$$

$$E_{cell} = -1.23\ V - \left(\frac{0.0591\ V}{4}\right) \log(1) = -1.23\ V \qquad (19.71)$$

Thus, E_{cell} is −1.23 V, which is the value of E°_{cell} if the reaction is carried out in the presence of 1 M H^+ rather than at pH 7.

In practice, a voltage about 0.4–0.6 V higher than the calculated value is needed to electrolyze water. This added voltage, called an **overvoltage**, represents the additional driving force required to overcome barriers such as the large activation energy for the formation of a gas at a metal surface. Overvoltages are needed in all electrolytic processes, which explains why, for example, approximately 14 V must be applied to recharge the 12-V battery in your car.

In general, any metal that does *not* react readily with water to produce hydrogen can be produced by the electrolytic reduction of an aqueous solution that contains the metal cation. The *p*-block metals and most of the transition metals are in this category, but metals in high oxidation states, which form oxoanions, cannot be reduced to the metal by simple electrolysis. Active metals, such as aluminum and those of Groups 1 and 2, react so readily with water that they can be prepared only by the electrolysis of molten salts. Similarly, any nonmetallic element that does not readily oxidize water to O_2 can be prepared by the electrolytic oxidation of an aqueous solution that contains an appropriate anion. In practice, among the nonmetals, only F_2 cannot be prepared using this method. Oxoanions of nonmetals in their highest oxidation states, such as NO_3^-, SO_4^{2-}, PO_4^{3-}, are usually difficult to reduce electrochemically and usually behave like spectator ions that remain in solution during electrolysis.

Note the pattern

In general, any metal that does not react readily with water to produce hydrogen can be produced by the electrolytic reduction of an aqueous solution that contains the metal cation.

Electroplating

MGC Electroplating

In a process called **electroplating**, a layer of a second metal is deposited on the metal electrode that acts as the cathode during electrolysis. Electroplating is used to enhance the appearance of metal objects and protect them from corrosion. Examples of electroplating include the chromium layer found on many bathroom fixtures or (in earlier days) on the bumpers and hubcaps of cars, as well as the thin layer of precious metal that coats silver-plated dinnerware or jewelry. In all cases, the basic concept is the same. A schematic view of an apparatus for electroplating silverware and a photograph of a commercial electroplating cell are shown in Figure 19.23.

The half-reactions in electroplating a fork, for example, with silver are

Cathode (fork): $Ag^+(aq) + e^- \longrightarrow Ag(s) \qquad E^\circ_{cathode} = 0.80\ V \qquad (19.72a)$

Anode (silver bar): $Ag(s) \longrightarrow Ag^+(aq) + e^- \qquad E^\circ_{anode} = 0.80\ V \qquad (19.72b)$

The overall reaction is the transfer of silver metal from one electrode (a silver bar acting as the anode) to another (a fork acting as the cathode). Because $E^\circ_{cell} = 0$ V, it takes only a small applied voltage to drive the electroplating process. In practice, various other substances may be added to the plating solution to control its electrical conductivity and regulate the concentration of free metal ions, thus ensuring a smooth, even coating.

Quantitative Considerations

If we know the stoichiometry of an electrolysis reaction, the amount of current passed, and the length of time, we can calculate the amount of material consumed or produced in the reaction. Conversely, we can use the stoichiometry to determine the combination of current and time needed to produce a given amount of material.

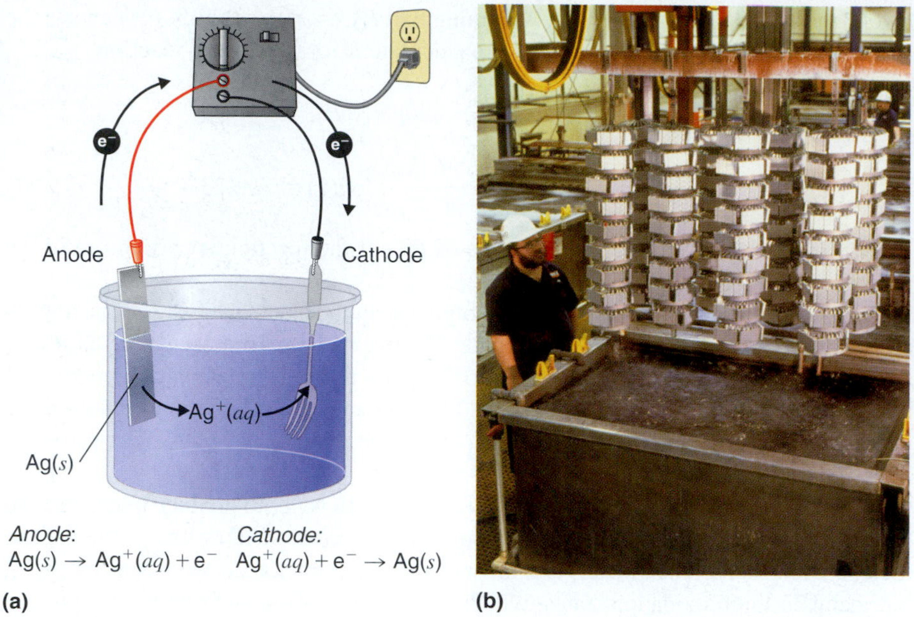

Figure 19.23 Electroplating.
(a) Electroplating uses an electrolytic cell in which the object to be plated, such as a fork, is immersed in a solution of the metal to be deposited. The object being plated acts as the cathode, upon which the desired metal is deposited in a thin layer, while the anode usually consists of the metal that is being deposited (in this case silver) that maintains the solution concentration as it dissolves. (b) In this commercial electroplating apparatus, a large number of objects can be plated simultaneously by lowering the rack into the Ag^+ solution and applying the correct potential.

The quantity of material that is oxidized or reduced at an electrode during an electrochemical reaction is determined by the stoichiometry of the reaction and the amount of charge that is transferred. For example, in the reaction $Ag^+(aq) + e^- \longrightarrow Ag(s)$, 1 mol of electrons reduces 1 mol of Ag^+ to Ag metal. In contrast, in the reaction $Cu^{2+}(aq) + 2e^- \longrightarrow Cu(s)$, 1 mol of electrons reduces only 0.5 mol of Cu^{2+} to Cu metal. Recall that the charge on 1 mol of electrons is 1 faraday (1*F*), which is equal to 96,486 C. We can therefore calculate the number of moles of electrons transferred when a known current is passed through a cell for a given period of time. The total charge transferred is the product of the current and the time:

$$\text{charge (C)} = \text{current (A)} \times \text{time (s)} \tag{19.73}$$

The stoichiometry of the reaction and the total charge transferred enable us to calculate the amount of product formed during an electrolysis reaction or the amount of metal deposited in an electroplating process.

For example, if a current of 0.60 A passes through an aqueous solution of $CuSO_4$ for 6.0 min, the total number of coulombs of charge that passes through the cell is

$$\text{charge} = (0.600\ \text{A})(6.00\ \cancel{\text{min}})(60\ \text{s}/\cancel{\text{min}}) = 216\ \text{A}\cdot\text{s} = 216\ \text{C} \tag{19.74}$$

The number of moles of electrons transferred to Cu^{2+} is therefore

$$\text{moles e}^- = \frac{216\ \cancel{\text{C}}}{96{,}486\ \cancel{\text{C}}/\text{mol}} = 2.24 \times 10^{-3}\ \text{mol e}^- \tag{19.75}$$

Because two electrons are required to reduce a single Cu^{2+} ion, the total number of moles of Cu produced is half the number of moles of electrons transferred, or 1.12×10^{-3} mol. This corresponds to 0.711 g of Cu. In commercial electrorefining processes, much higher currents ($\geq$50,000 A) are used, corresponding to approximately 0.5 *F*/s, and reaction times are on the order of 3 to 4 weeks.

EXAMPLE 19.15

A silver-plated spoon typically contains about 2.00 g of Ag. If 12.0 hours are required to achieve the desired thickness of the Ag coating, what is the average current per spoon that must flow during the electroplating process, assuming an efficiency of 100%?

Given Mass of metal, time, efficiency

Asked for Current required

Strategy

Ⓐ Calculate the number of moles of metal corresponding to the given mass transferred.
Ⓑ Write the reaction, and determine the number of moles of electrons required for the electroplating process.
Ⓒ Use the definition of the faraday to calculate the number of coulombs required. Then convert coulombs to current in amperes.

Solution

Ⓐ We must first determine the number of moles of Ag corresponding to 2.00 g of Ag:

$$\text{moles Ag} = \frac{2.00\ \cancel{\text{g}}}{107.868\ \cancel{\text{g}}/\text{mol}} = 1.85 \times 10^{-2}\ \text{mol Ag}$$

Ⓑ The reduction reaction is $Ag^+(aq) + e^- \longrightarrow Ag(s)$, so 1 mol of electrons produces 1 mol of silver. Ⓒ Using the definition of the faraday gives

$$\text{coulombs} = (1.85 \times 10^{-2}\ \cancel{\text{mol e}^-})(96{,}486\ \text{C}/\cancel{\text{mol e}^-}) = 1.78 \times 10^3\ \text{C}$$

The current in amperes needed to deliver this amount of charge in 12 h is therefore

$$\text{amperes} = \frac{1.78 \times 10^3\ \text{C}}{(12.0\ \cancel{\text{h}})(60\ \cancel{\text{min}}/\cancel{\text{h}})(60\ \text{s}/\cancel{\text{min}})}$$
$$= 4.12 \times 10^{-2}\ \text{C/s} = 4.12 \times 10^{-2}\ \text{A}$$

Because the electroplating process is usually much less than 100% efficient (typical values are closer to 30%), the actual current necessary is higher than 0.1 A.

EXERCISE 19.15

A typical aluminum soft-drink can weighs about 29 g. How much time is needed to produce this amount of $Al(s)$ in the Hall–Herout process, using a current of 15 A to reduce a molten Al_2O_3/Na_3AlF_6 mixture?

Answer 5.8 h

SUMMARY AND KEY TERMS

19.1 Describing Electrochemical Cells (p. 861)

Electrochemistry is the study of the relationship between electricity and chemical reactions. The oxidation–reduction reaction that occurs during an electrochemical process consists of two **half-reactions**, one representing the oxidation process and one the reduction process. The sum of the half-reactions gives the overall chemical reaction. The overall redox reaction is balanced when the number of electrons lost by the **reductant** is equal to the number of electrons gained by the **oxidant**. An electric current is produced from the flow of electrons from reductant to oxidant. An **electrochemical cell** can either generate electricity from a spontaneous redox reaction or consume electricity to drive a nonspontaneous reaction. In a **galvanic (voltaic) cell**, the energy from a spontaneous reaction generates electricity, whereas in an **electrolytic cell**, electrical energy is consumed to drive a nonspontaneous redox reaction. Both types of cells use two **electrodes** that provide an electrical connection between systems that are separated in space. The oxidative half-reaction occurs at the **anode**, and the reductive half-reaction occurs at the **cathode**. A **salt bridge** connects the separated solutions, allowing ions to migrate to either solution to ensure the system's electrical neutrality. A *voltmeter* is a device used to measure the flow of electric current between the two half-reactions. The **potential** of the cell, measured in volts, is the energy needed to move a charged particle in an electric field. An electrochemical cell can be described using line notation called a *cell diagram*, in which vertical lines indicate phase boundaries and the location of the salt bridge. Resistance to the flow of charge at a boundary is called the *junction potential.*

19.2 Standard Potentials (p. 866)

The flow of electrons in an electrochemical cell depends on the identity of the reacting substances, the difference in the potential energy of their valence electrons, and their concentrations. The potential of the cell under standard conditions (1 *M* for solutions, 1 atm for gases, pure solids or liquids for other substances) and at a fixed temperature (25°C) is called the **standard cell potential, E°_{cell}**. Only the *difference* between the potentials of two electrodes can be measured. By convention, all tabulated values of *standard electrode potentials* are listed as standard reduction potentials. The overall cell potential is the reduction potential of the reductive half-reaction minus the reduction potential of the oxidative half-reaction ($E^\circ_{cell} = E^\circ_{cathode} - E^\circ_{anode}$). The potential of the **standard hydrogen electrode (SHE)** is defined as 0 V under standard conditions. The potential of a half-reaction measured against the SHE under standard conditions is called its **standard electrode potential.** The standard cell potential is a measure of the driving force for a given redox reaction. All E° values are independent of the stoichiometric coefficients for the half-reaction. Redox reactions can be balanced using the half-reaction method, in which the overall redox reaction is divided into an oxidation half-reaction and a reduction half-reaction, each one balanced for mass and charge. The half-reactions selected from tabulated lists must exactly reflect reaction conditions. In an alternative method, the atoms in each half-reaction are balanced and then the charges are balanced. Whenever a half-reaction is reversed, the sign of E° corresponding to that reaction must also be reversed. If E°_{cell} is positive, the reaction will occur spontaneously under standard conditions. If E°_{cell} is negative, then the reaction is not spontaneous under standard conditions, although it will proceed spontaneously in the opposite direction. The potential of an **indicator electrode** is related to the concentration of the substance being measured, whereas the potential of the **reference electrode** is held constant. Whether reduction or oxidation occurs depends on the potential of the sample versus the potential of the reference electrode. In addition to the SHE, other reference electrodes are the **silver–silver chloride electrode**, the **saturated calomel electrode (SCE)**, the **glass electrode**, which is commonly used to measure pH, and **ion-selective electrodes**, which depend on the concentration of a single ionic species in solution. Differences in potential between the SHE and other reference electrodes must be included when calculating values for E°.

19.3 Comparing Strengths of Oxidants and Reductants (p. 876)

The oxidative and reductive strengths of a variety of substances can be compared using standard electrode potentials. Apparent anomalies can be explained by the fact that electrode potentials are measured in aqueous solution, which allows for strong intermolecular electrostatic interactions, and are not measured in the gas phase.

19.4 Electrochemical Cells and Thermodynamics (p. 880)

A **coulomb (C)** relates electrical potential, expressed in volts, and energy, expressed in joules. The current generated from a redox reaction is measured in **amperes (A)**, where 1 A is defined as the flow of 1 C/s past a given point. The **faraday (*F*)** is Avogadro's number multiplied by the charge on an electron, and corresponds to the charge on 1 mol of electrons. The product of the cell potential and the total charge is the maximum amount of energy available to do work, which is related to the change in free energy that occurs during the chemical process. Summing ΔG values for half-reactions gives ΔG for the overall reaction, which is proportional to *both* the potential *and* the number of electrons (n) transferred. Spontaneous redox reactions have a negative ΔG and therefore a positive E_{cell}. Because the equilibrium constant K is related to ΔG, E°_{cell} and K are also related. Large equilibrium constants correspond to large positive values of E°. The **Nernst equation** allows us to determine the spontaneous direction of any redox reaction under any reaction conditions from values of the relevant standard electrode potentials. **Concentration cells** consist of anode and cathode compartments that are identical except for the concentrations of the reactant. Because $\Delta G = 0$ at equilibrium, the measured potential of a concentration cell is zero at equilibrium (the concentrations are equal). A galvanic cell can also be used to measure the solubility product of a sparingly soluble substance, and to calculate the concentration of a species given a measured potential and the concentrations of all the other species.

19.5 Commercial Galvanic Cells (p. 890)

A **battery** is a contained unit that produces electricity, whereas a **fuel cell** is a galvanic cell that requires a constant external supply of one or more reactants to generate electricity. One type of battery is the *Leclanché dry cell*, which contains an electrolyte in an acidic water-based paste. This battery is called an **alkaline battery** when adapted to operate under alkaline conditions. *"Button" batteries* have a high output-to-mass ratio; **lithium–iodine batteries** consist of a solid electrolyte; the *nickel–cadmium (Nicad) battery* is rechargeable; and the *lead–acid battery*, which is also rechargeable, does not require the electrodes to be in separate compartments. A fuel cell requires an external supply of reactants as the products of the reaction are continuously removed. In a fuel cell, energy is not stored, but instead electrical energy is provided by a chemical reaction.

19.6 Corrosion (p. 895)

The deterioration of metals through oxidation is a galvanic process called **corrosion**. Protective coatings consist of a second metal that is more difficult to oxidize than the metal being protected. Alternatively, a more easily oxidized metal can be applied to a metal surface, thus providing *cathodic protection* of the surface. *Galvanized steel* is protected by a thin layer of zinc. *Sacrificial electrodes* can also be attached to an object to protect it.

19.7 Electrolysis (p. 897)

In **electrolysis**, an external voltage is applied to drive a nonspontaneous reaction. A *Downs cell* is used to produce sodium metal from a mixture of salts, and the *Hall–Heroult process* is used to produce aluminum commercially. Electrolysis can also be used to produce H_2 and O_2 from water. In practice, an additional voltage, called an **overvoltage**, must be applied to overcome factors such as a large activation energy and a junction potential. **Electroplating** is the process by which a second metal is deposited on a metal surface, thereby enhancing an object's appearance or providing protection from corrosion. The amount of material consumed or produced in a reaction can be calculated from the stoichiometry of an electrolysis reaction, the amount of current passed, and the duration of the electrolytic reaction.

KEY EQUATIONS

Standard cell potential	$E^\circ_{cell} = E^\circ_{cathode} - E^\circ_{anode}$	(19.7)
Charge on a mole of electrons (faraday)	$F \approx 96{,}486\ \text{J/(V}\cdot\text{mol)}$	(19.33)
Maximum work from an electrochemical cell	$w_{max} = -nFE_{cell}$	(19.34)
Relationship between ΔG° and E°	$\Delta G^\circ = -nFE^\circ_{cell}$	(19.36)
Relationship between ΔG° and K for a redox reaction	$\Delta G^\circ = -RT\ln K$	(19.40)
Relationship between E° and K for a redox reaction at 25°C	$E^\circ_{cell} = \left(\frac{RT}{nF}\right)\ln K$	(19.42)
	$E^\circ_{cell} = \left(\frac{0.0591\text{ V}}{n}\right)\log K$	(19.43)
Relationship between ΔG and Q	$\Delta G = \Delta G^\circ + RT\ln Q$	(19.44)
Relationship between E_{cell} and Q at 25°C	$E_{cell} = E^\circ_{cell} - \left(\frac{0.0591\text{ V}}{n}\right)\log Q$	(19.47)
Relationship of charge, current, and time	charge (C) = current (A) × time (s)	(19.73)

QUESTIONS AND PROBLEMS

MGC For instructor-assigned homework, go to **www.masteringgenchem.com**

Questions and Problems with colored numbers have answers in the Appendix and complete solutions in the Student Solutions Manual.

CONCEPTUAL

19.1 Describing Electrochemical Cells

1. Is $2NaOH(aq) + H_2SO_4(aq) \longrightarrow Na_2SO_4(aq) + 2H_2O(l)$ an oxidation–reduction reaction? Why or why not?
2. If the two half-reactions are physically separated, how is it possible for a redox reaction to occur? What is the name of the apparatus in which the two half-reactions are carried out simultaneously?
3. What is the difference between a galvanic cell and an electrolytic cell? Which would you use to generate electricity?
4. What is the purpose of a salt bridge in a galvanic cell? Is it always necessary to use a salt bridge in a galvanic cell?
5. One criterion for a good salt bridge is that it contains ions that have similar rates of diffusion in aqueous solution as K^+ and Cl^- ions do. What would happen if the diffusion rates of the anions and cations differed significantly?
6. It is often more accurate to measure the potential of a redox reaction by immersing two electrodes in a single beaker rather than in two beakers. Why?

19.2 Standard Potentials

7. Is a hydrogen electrode chemically inert? What is the major disadvantage to using a hydrogen electrode?
8. List two factors that affect the measured potential of an electrochemical cell, and explain their impact on the measurements.
9. What is the relationship between electron flow and the potential energy of valence electrons? If the valence electrons of substance A have a higher potential energy than those of substance B, what is the direction of electron flow between them in a galvanic cell?
10. If the components of a galvanic cell include aluminum and bromine, what is the predicted direction of electron flow? Why?
11. Write a cell diagram representing a cell that contains the Ni/Ni^{2+} couple in one compartment and the SHE in the other compartment. What are the values of $E^\circ_{cathode}$, E°_{anode}, and E°_{cell}?
12. Explain why E° values are independent of the stoichiometric coefficients in the corresponding half-reaction.
13. Identify the oxidants and reductants in each redox reaction:
    (a) $Cr(s) + Ni^{2+}(aq) \longrightarrow Cr^{2+}(aq) + Ni(s)$
    (b) $Cl_2(g) + Sn^{2+}(aq) \longrightarrow 2Cl^-(aq) + Sn^{4+}(aq)$
    (c) $H_3AsO_4(aq) + 8H^+(aq) + 4Zn(s) \longrightarrow AsH_3(g) + 4H_2O(l) + 4Zn^{2+}(aq)$
    (d) $2NO_2(g) + 2OH^-(aq) \longrightarrow NO_2^-(aq) + NO_3^-(aq) + H_2O(l)$
14. Identify the oxidants and reductants in each redox reaction:
    (a) $Br_2(l) + 2I^-(aq) \longrightarrow 2Br^-(aq) + I_2(s)$
    (b) $Cu^{2+}(aq) + 2Ag(s) \longrightarrow Cu(s) + 2Ag^+(aq)$
    (c) $H^+(aq) + 2MnO_4^-(aq) + 5H_2SO_3(aq) \longrightarrow 2Mn^{2+}(aq) + 3H_2O(l) + 5HSO_4^-(aq)$
    (d) $IO_3^-(aq) + 5I^-(aq) + 6H^+(aq) \longrightarrow 3I_2(s) + 3H_2O(l)$
15. All reference electrodes must conform to certain requirements. List the requirements and explain their significance.
16. For each application, describe the reference electrode you would use and explain why. In each case, how would the measured potential compare with the corresponding E°?
    (a) measuring the potential of a Cl^-/Cl_2 couple
    (b) measuring the pH of a solution
    (c) measuring the potential of a MnO_4^-/Mn^{2+} couple

19.3 Comparing Strengths of Oxidants and Reductants

17. The order of electrode potentials cannot always be predicted by ionization potentials and electron affinities. Why? Do you expect sodium metal to have a higher or lower electrode potential than predicted from its ionization potential? What is its approximate electrode potential?

18. Without referring to tabulated data, which of the following couples would you expect to have the *least* negative electrode potential and which the *most* negative: Br^-/Br_2, Ca^{2+}/Ca, O_2/OH^-, Al^{3+}/Al? Why?

19. Because of the sulfur-containing amino acids present in egg whites, eating eggs with a silver fork will tarnish the fork. As a chemist, you have all kinds of interesting cleaning products in your cabinet, including a 1 *M* solution of oxalic acid ($H_2C_2O_4$). Would you choose this solution to clean the fork that you have tarnished from eating scrambled eggs?

20. The electrode potential for the reaction $Cu^{2+}(aq) + 2e^- \longrightarrow Cu(s)$ is 0.34 V under standard conditions. Is the potential for the oxidation of 0.5 mol of Cu equal to $-0.34/2$ V? Explain your answer.

19.4 Electrochemical Cells and Thermodynamics

21. State whether you agree or disagree with this reasoning and explain your answer: Standard electrode potentials arise from the number of electrons transferred. The greater the number of electrons transferred, the greater the measured potential difference. If 1 mol of a substance produces 0.76 V when 2 mol of electrons are transferred—as in $Zn(s) \longrightarrow Zn^{2+}(aq) + 2e^-$—then 0.5 mol of the substance will produce 0.76/2 V because only 1 mol of electrons is transferred.

22. What is the relationship between the *measured cell potential* and the *total charge* that passes through the cell? Which of these is dependent on concentration? Which is dependent on the identity of the oxidant or reductant? Which is dependent on the number of electrons transferred?

23. In the equation $w_{max} = -n\,F\,E_{cell}$, which quantities are extensive properties and which are intensive properties?

24. For any spontaneous redox reaction, $E°$ is positive. Use thermodynamic arguments to explain why this is true.

25. State whether you agree or disagree with this statement and explain your answer: Electrochemical methods are especially useful in determining the reversibility or irreversibility of reactions that take place in the cell.

26. Although the sum of two half-reactions gives another half-reaction, the sum of the potentials of the two half-reactions cannot be used to obtain the potential of the net half-reaction. Why? When does the sum of two half-reactions correspond to the overall reaction, and why?

27. Occasionally, you will find high-quality electronic equipment that has its electronic components plated in gold. What is the advantage of this?

28. Blood analyzers, which measure pH, P_{CO_2}, and P_{O_2}, are frequently used in clinical emergencies. For example, blood P_{CO_2} is measured with a pH electrode covered with a plastic membrane that is permeable to CO_2. Based on your knowledge of how electrodes function, explain how such an electrode might work. *Hint*: $CO_2(g) + H_2O(l) \longrightarrow HCO_3^-(aq) + H^+(aq)$.

29. Concentration cells contain the same species in solution in two different compartments. Explain what produces a voltage in a concentration cell. When does V = 0 in such a cell?

30. Describe how an electrochemical cell can be used to measure the solubility of a sparingly soluble salt.

19.5 Commercial Galvanic Cells

31. What advantage is there to using an alkaline battery rather than a Leclanché dry cell?

32. Why does the density of the fluid in lead–acid batteries drop when the battery is discharged?

33. What type of battery would you use for each application and why?
(a) powering an electric motor scooter
(b) a backup battery for a personal digital assistant (PDA)
(c) powering an iPod

34. Why are galvanic cells used as batteries and fuel cells? What is the difference between a battery and a fuel cell? What is the advantage to using highly concentrated or solid reactants in a battery?

19.6 Corrosion

35. Do you expect a bent nail to corrode more or less rapidly than a straight nail? Why?

36. What does it mean when a metal is described as being coated with a sacrificial layer? Is this different from galvanic protection?

37. Why is it important for automobile manufacturers to apply paint to the metal surface of a car? Why is this process particularly important for vehicles in northern climates, where salt is used on icy roads?

19.7 Electrolysis

38. Why might an electrochemical reaction that is thermodynamically favored require an overvoltage in order to occur?

39. How could you use an electrolytic cell to make quantitative comparisons of the strengths of various oxidants and reductants?

40. Why are mixtures of molten salts generally used during electrolysis rather than a pure salt?

41. Two solutions, one containing $Fe(NO_3)_2 \cdot 6H_2O$ and the other the same molar concentration of $Fe(NO_3)_3 \cdot 6H_2O$, were electrolyzed under identical conditions. Which solution produced the most metal, and justify your answer?

NUMERICAL

This section includes paired problems (marked by brackets) that require similar problem-solving skills.

19.1 Describing Electrochemical Cells

42. Copper sulfate forms a bright blue solution in water. If a piece of zinc metal is placed in a beaker of aqueous $CuSO_4$ solution, the blue color fades with time, the zinc strip begins to erode, and a black solid forms around the zinc strip. What is happening? Write half-reactions to show the chemical changes that are occurring. What will happen if a piece of copper metal is placed in a colorless aqueous solution of $ZnCl_2$?

43. Consider this spontaneous redox reaction: $NO_3^-(aq) + SO_3^{2-}(aq) + H^+(aq) \longrightarrow SO_4^{2-}(aq) + HNO_2(aq)$.
(a) Write the two half-reactions for this overall reaction?
(b) If the reaction is carried out in a galvanic cell using an inert electrode in each compartment, which electrode corresponds to which half-reaction?
(c) Which electrode is negatively charged, and which is positively charged?

44. The reaction $Pb(s) + 2VO^{2+}(aq) + 4H^+(aq) \longrightarrow Pb^{2+}(aq) + 2V^{3+}(aq) + 2H_2O(l)$ occurs spontaneously.
(a) Write the two half-reactions for this redox reaction.
(b) If the reaction is carried out in a galvanic cell using an inert electrode in each compartment, which reaction occurs at the cathode and which occurs at the anode?
(c) Which electrode is positively charged, and which is negatively charged?

45. Sulfate is reduced to HS^- in the presence of glucose, which is oxidized to bicarbonate. Write the two half-reactions corresponding to this process. What is the equation for the overall reaction?

46. Phenolphthalein is an indicator that turns pink under basic conditions. When an iron nail is placed in a gel that contains $[Fe(CN)_6]^{3-}$, the gel around the nail begins to turn pink. What is occurring? Write the half-reactions, and then write the overall redox reaction.

47. For each galvanic cell represented by a cell diagram, write the two half-reactions and the overall reaction. Indicate which reaction occurs at the anode and which occurs at the cathode.
(a) $Zn(s)\,|\,Zn^{2+}(aq)\,\|\,H^+(aq)\,|\,H_2(g), Pt(s)$
(b) $Ag(s)\,|\,AgCl(s)\,|\,Cl^-(aq)\,\|\,H^+(aq)\,|\,H_2(g)\,|\,Pt(s)$
(c) $Pt(s)\,|\,H_2(g)\,|\,H^+(aq)\,\|\,Fe^{2+}(aq), Fe^{3+}(aq)\,|\,Pt(s)$

48. Write the half-reactions and the overall reaction for each cell diagram. State which half-reaction occurs at the anode and which occurs at the cathode.
(a) $Pb(s)\,|\,PbSO_4(s)\,|\,SO_4^{2-}(aq)\,\|\,Cu^{2+}(aq)\,|\,Cu(s)$
(b) $Hg(s)\,|\,Hg_2Cl_2(s)\,|\,Cl^-(aq)\,\|\,Cd^{2+}(aq)\,|\,Cd(s)$

49. For each redox reaction, write the half-reactions and draw the cell diagram for a galvanic cell in which the overall reaction occurs as written. Identify each electrode as either positive or negative.
(a) $Ag(s) + Fe^{3+}(aq) \longrightarrow Ag^+(aq) + Fe^{2+}(aq)$
(b) $Fe^{3+}(aq) + \frac{1}{2}H_2(g) \longrightarrow Fe^{2+}(aq) + H^+(aq)$

50. Write the half-reactions for each overall reaction, decide whether the reaction will occur spontaneously, and construct a cell diagram for a galvanic cell in which a spontaneous reaction will occur:
(a) $2Cl^-(aq) + Br_2(l) \longrightarrow Cl_2(g) + 2Br^-(aq)$
(b) $2NO_2(g) + 2OH^-(aq) \longrightarrow NO_2^-(aq) + NO_3^-(aq) + H_2O(l)$
(c) $2H_2O(l) + 2Cl^-(aq) \longrightarrow H_2(g) + Cl_2(g) + 2OH^-(aq)$
(d) $C_3H_8(g) + 5O_2(g) \longrightarrow 3CO_2(g) + 4H_2O(g)$

51. Write the half-reactions for each overall reaction, decide whether the reaction will occur spontaneously, and construct a cell diagram for a galvanic cell in which a spontaneous reaction will occur:
(a) $Co(s) + Fe^{2+}(aq) \longrightarrow Co^{2+}(aq) + Fe(s)$
(b) $O_2(g) + 4H^+(aq) + 4Fe^{2+}(aq) \longrightarrow 2H_2O(l) + 4Fe^{3+}(aq)$
(c) $6Hg^{2+}(aq) + 2NO_3^-(aq) + 8H^+ \longrightarrow 3Hg_2^{2+}(aq) + 2NO(g) + 4H_2O(l)$
(d) $CH_4(g) + 2O_2(g) \longrightarrow CO_2(g) + 2H_2O(g)$

19.2 Standard Potentials

52. Write the cell diagram for a galvanic cell with a SHE and a zinc electrode that carries out this overall reaction: $Zn(s) + 2H^+(aq) \longrightarrow Zn^{2+}(aq) + H_2(g)$.

53. Draw the cell diagram for a galvanic cell utilizing a SHE and a copper electrode that carries out the reaction $H_2(g) + Cu^{2+}(aq) \longrightarrow 2H^+(aq) + Cu(s)$.

54. Balance each reaction and calculate the standard reduction potential for each:
(a) $Cu^+(aq) + Ag^+(aq) \longrightarrow Cu^{2+}(aq) + Ag(s)$
(b) $Sn(s) + Fe^{3+}(aq) \longrightarrow Sn^{2+}(aq) + Fe^{2+}(aq)$
(c) $Mg(s) + Br_2(l) \longrightarrow 2Br^-(aq) + Mg^{2+}(aq)$

55. Balance each reaction and calculate the standard electrode potential for each. Be sure to include the physical state of each product and reactant.
(a) $Cl_2(g) + H_2(g) \longrightarrow 2Cl^- + 2H^+$
(b) $Br_2 + Fe^{2+} \longrightarrow 2Br^- + Fe^{3+}$
(c) $Fe^{3+} + Cd \longrightarrow Fe^{2+} + Cd^{2+}$

56. Write a balanced equation for each redox reaction:
(a) $H_2PO_2^-(aq) + SbO_2^-(aq) \longrightarrow HPO_3^{2-}(aq) + Sb(s)$ in basic solution
(b) $HNO_2(aq) + I^-(aq) \longrightarrow NO(g) + I_2(s)$ in acidic solution
(c) $N_2O(g) + ClO^-(aq) \longrightarrow Cl^-(aq) + NO_2^-(aq)$ in basic solution
(d) $Br_2\,(l) \longrightarrow Br^-(aq) + BrO_3^-(aq)$ in basic solution
(e) $Cl(CH_2)_2OH(aq) + K_2Cr_2O_7(aq) \longrightarrow ClCH_2CO_2H(aq) + Cr^{3+}(aq)$ in acidic solution

57. Balance each redox reaction:
(a) $I^-(aq) + HClO_2(aq) \longrightarrow IO_3^-(aq) + Cl_2(g)$ in acidic solution
(b) $Cr^{2+}(aq) + O_2(g) \longrightarrow Cr^{3+}(aq) + H_2O(l)$ in acidic solution
(c) $CrO_2^-(aq) + ClO^-(aq) \longrightarrow CrO_4^{2-}(aq) + Cl^-(aq)$ in basic solution
(d) $S(s) + HNO_2(aq) \longrightarrow H_2SO_3(aq) + N_2O(g)$ in acidic solution
(e) $F(CH_2)_2OH(aq) + K_2Cr_2O_7(aq) \longrightarrow FCH_2CO_2H(aq) + Cr^{3+}(aq)$ in acidic solution

58. The standard cell potential for the oxidation of Pb to Pb^{2+} with the concomitant reduction of Cu^+ to Cu is 0.39 V. You know that $E°$ for the Pb^{2+}/Pb couple is −0.13 V. What is $E°$ for the Cu^+/Cu couple?

59. You have built a galvanic cell similar to the one in Figure 19.6 using an iron nail, a solution of $FeCl_2$, and a standard hydrogen electrode. When the cell is connected, you notice that the iron nail begins to corrode. What else do you observe? Under standard conditions, what is E_{cell}?

60. Carbon is used to reduce iron ore to metallic iron. The overall reaction is

$$2Fe_2O_3 \cdot xH_2O(s) + 3C(s) \longrightarrow 4Fe(s) + 3CO_2(g) + 2xH_2O(g)$$

Write the two half-reactions for this overall reaction.

61. Will each reaction occur spontaneously under standard conditions?
(a) $Cu(s) + 2H^+(aq) \longrightarrow Cu^{2+}(aq) + H_2(g)$
(b) $Zn^{2+}(aq) + Pb(s) \longrightarrow Zn(s) + Pb^{2+}(aq)$

62. Each reaction takes place in acidic solution. Balance each reaction, and then determine whether it occurs spontaneously as written under standard conditions.
(a) $Se(s) + Br_2(l) \longrightarrow H_2SeO_3(aq) + Br^-(aq)$
(b) $NO_3^-(aq) + S(s) \longrightarrow HNO_2(aq) + H_2SO_3(aq)$
(c) $Fe^{3+}(aq) + Cr^{3+}(aq) \longrightarrow Fe^{2+}(aq) + Cr_2O_7^{2-}(aq)$

63. Calculate $E°_{cell}$ and $\Delta G°$ for the redox reaction represented by the cell diagram $Pt(s)\,|\,Cl_2(g, 1\text{ atm})\,|\,ZnCl_2(aq, 1\,M)\,|\,Zn(s)$. Will this reaction occur spontaneously?

64. If you place Zn-coated (galvanized) tacks in a glass and add an aqueous solution of iodine, the brown color of the iodine solution fades to a pale yellow. What has happened? Write the two half-reactions and the overall balanced equation for this reaction. What is E°_{cell}?

65. Your lab partner wants to recover solid silver from silver chloride by using a 1.0 *M* solution of HCl and 1 atm H_2 under standard conditions. Will this plan work?

19.4 Electrochemical Cells and Thermodynamics

66. The chemical equation for the combustion of butane is

$$C_4H_{10}(g) + \frac{13}{2}O_2(g) \longrightarrow 4CO_2(g) + 5H_2O(g)$$

This reaction has $\Delta H^\circ = -2877$ kJ/mol. Calculate E°_{cell} and then determine ΔG°. Is this a spontaneous process? What is the change in entropy that accompanies this process at 298 K?

67. For the cell represented as $Al(s)\,|\,Al^{3+}(aq)\,||\,Sn^{2+}(aq), Sn^{4+}(aq)\,|\,Pt(s)$, how many electrons are transferred in the redox reaction? What is the standard cell potential? Is this a spontaneous process? What is ΔG°?

68. How many electrons are transferred during the reaction $Pb(s) + Hg_2Cl_2(aq) \longrightarrow PbCl_2(aq) + 2Hg(l)$? What is the standard cell potential? Is the oxidation of Pb by Hg_2Cl_2 spontaneous? Calculate ΔG° for this reaction.

69. Explain why the sum of the potentials for the half-reactions $Sn^{2+}(aq) + 2e^- \longrightarrow Sn(s)$ and $Sn^{4+}(aq) + 2e^- \longrightarrow Sn^{2+}(aq)$ does not equal the potential for the reaction $Sn^{4+}(aq) + 4e^- \longrightarrow Sn(s)$. What is the net cell potential? Compare the values of ΔG° for the sum of the potentials and the actual net cell potential.

70. Based on tables in this text, do you agree that the potential for the half-reaction $Cu^{2+} + 2e^- \longrightarrow Cu(s)$ equals 0.68 V? Why or why not?

71. For each reaction, calculate E°_{cell} and then determine ΔG°. Indicate whether each reaction is spontaneous.

(a) $2Na(s) + 2H_2O(l) \longrightarrow 2NaOH(aq) + H_2(g)$
(b) $K_2S_2O_6(aq) + 2KI(aq) \longrightarrow I_2(s) + 2K_2SO_4(aq)$
(c) $Sn(s) + CuSO_4(aq) \longrightarrow Cu(s) + SnSO_4(aq)$

72. What is the change in free energy for the reaction between Ca^{2+} and sodium metal to give calcium metal and Na^+? Do the sign and magnitude of ΔG agree with what you would expect based on the positions of these elements in the periodic table? Why or why not?

73. In acidic solution, permanganate (MnO_4^-) oxidizes Cl^- to chlorine gas, and MnO_4^- is reduced to $Mn^{2+}(aq)$. (a) Write the balanced equation for this reaction; (b) determine E°_{cell}; and (c) calculate the equilibrium constant.

74. Potentiometric titrations are an efficient method for determining the endpoint of a redox titration. In such a titration, the potential of the solution is monitored as measured volumes of an oxidant or a reductant are added. Data for a typical titration, the potentiometric titration of Fe(II) with a 0.1 *M* solution of Ce(IV), are given in the table. The starting potential has been arbitrarily set equal to zero because it is the *change* in potential with the addition of the oxidant that is important.

Titrant, mL	E, mV
2.00	50
6.00	100
9.00	255
10.00	960
11.00	1325
12.00	1625
14.00	1875

(a) Write the balanced equation for the oxidation of Fe^{2+} by Ce^{4+}.
(b) Plot the data, and then locate the endpoint.
(c) How many millimoles of Fe^{2+} did the solution being titrated originally contain?

75. The standard electrode potential, E°, for the half-reaction $Ni^{2+}(aq) + 2e^- \longrightarrow Ni(s)$ is -0.25 V. What pH is needed for this reaction to take place in the presence of 1.00 atm $H_2(g)$ as the reductant if $[Ni^{2+}]$ is 1.00 *M*?

76. The reduction of Mn(VII) to Mn(*s*) by $H_2(g)$ proceeds in five steps that can be readily followed by changes in the color of the solution. Here is the redox chemistry:

(1) $MnO_4^-(aq) + e^- \longrightarrow MnO_4^{2-}(aq)$
$E^\circ = +0.56$ V (purple $\longrightarrow$ dark green)
(2) $MnO_4^{2-}(aq) + 2e^- + 4H^+(aq) \longrightarrow MnO_2(s)$
$E^\circ = +2.26$ V (dark green $\longrightarrow$ dark brown solid)
(3) $MnO_2(s) + e^- + 4H^+(aq) \longrightarrow Mn^{3+}(aq)$
$E^\circ = +0.95$ V (dark brown solid $\longrightarrow$ red-violet)
(4) $Mn^{3+}(aq) + e^- \longrightarrow Mn^{2+}(aq)$ $E^\circ = +1.51$ V (red-violet $\longrightarrow$ pale pink)
(5) $Mn^{2+}(aq) + 2e^- \longrightarrow Mn(s)$ $E^\circ = -1.18$ V (pale pink $\longrightarrow$ colorless)

(a) Is the reduction of $MnO_4^-(aq)$ to $Mn^{3+}(aq)$ by $H_2(g)$ spontaneous under standard conditions? What is E°_{cell}?
(b) Is the reduction of $Mn^{3+}(aq)$ to Mn(*s*) by $H_2(g)$ spontaneous under standard conditions? What is E°_{cell}?

77. Mn(III) can disproportionate (both oxidize and reduce itself) by means of these half-reactions:

$Mn^{3+}(aq) + e^- \longrightarrow Mn^{2+}(aq)$ $E^\circ = 1.51$ V
$Mn^{3+}(aq) + 2H_2O(l) \longrightarrow MnO_2(s) + 4H^+(aq) + e^-$ $E^\circ = 0.95$ V

(a) What is E° for the disproportionation reaction?
(b) Is disproportionation more or less thermodynamically favored at low pH than at pH 7? Explain your answer.
(c) How could you prevent the disproportionation reaction from occurring?

78. For the reduction of oxygen to water, $E^\circ = 1.23$ V. What is the potential for this half-reaction at pH 7.00? What is the potential in a 0.85 *M* solution of NaOH?

79. Given the following biologically relevant half-reactions, will FAD, a molecule used to transfer electrons, be an effective oxidant for the conversion of acetaldehyde to acetate at pH 4.00?

acetate $+ 2H^+ + 2e^- \longrightarrow$ acetaldehyde $+ H_2O$ $E = -0.58$ V
FAD $+ 2H^+ + 2e^- \longrightarrow FADH_2$ $E = -0.18$ V

80. The biological molecule abbreviated as NADH can be oxidized to NAD^+ via the half-reaction $NADH \longrightarrow NAD^+ + H^+ + 2e^-$,

$E = +0.32$ V. Would NADH be able to reduce (a) acetate to pyruvate or (b) pyruvate to lactate? (c) What potential is needed to convert acetate to lactate?

$$\text{acetate} + CO_2 + 2H^+ + 2e^- \longrightarrow \text{pyruvate} + H_2O \quad E = -0.70\text{ V}$$

$$\text{pyruvate} + 2H^+ + 2e^- \longrightarrow \text{lactate} \quad E = -0.185\text{ V}$$

81. Under acidic conditions, ideally any half-reaction with $E° >$ 1.23 V will oxidize water via the reaction $O_2(g) + 4H^+(aq) + 4e^- \longrightarrow 2H_2O(l)$.
   (a) Will aqueous acidic $KMnO_4$ evolve oxygen with the formation of MnO_2?
   (b) At pH 14.00, what is $E°$ for the oxidation of water by aqueous $KMnO_4$ (1 *M*) with the formation of MnO_2?
   (c) At pH 14.00, will water be oxidized if you are trying to form MnO_2 from MnO_4^{2-} via the reaction $2MnO_4^{2-}(aq) + 2H_2O(l) \longrightarrow 2MnO_2(s) + O_2(g) + 4OH^-(aq)$?
82. Ideally, any half-reaction with $E° > 1.23$ V will oxidize water as a result of the half-reaction $O_2(g) + 4H^+(aq) + 4e^- \longrightarrow 2H_2O(l)$.
   (a) Will FeO_4^{2-} oxidize water if the half-reaction for the reduction of Fe(VI) $\longrightarrow$ Fe(III) is $FeO_4^{2-}(aq) + 8H^+(aq) + 3e^- \longrightarrow Fe^{3+}(aq) + 4H_2O$, $E° = 1.9$ V?
   (b) What pH is needed to cause this reaction to proceed spontaneously if $[Fe^{3+}] = [FeO_4^{2-}] = 1.0\ M$ and $P_{O_2} = 1.0$ atm?
83. Complexing agents can bind to metals and result in net stabilization of the complexed species. What is the net thermodynamic stabilization energy that results from using CN^- as a complexing agent for Mn^{3+}/Mn^{2+}?

$$Mn^{3+}(aq) + e^- \longrightarrow Mn^{2+}(aq) \quad E° = 1.51\text{ V}$$
$$Mn(CN)_6^{3-}(aq) + e^- \longrightarrow Mn(CN)_6^{4-}(aq) \quad E° = -0.24\text{ V}$$

84. You have constructed a cell with zinc and lead amalgam electrodes described by the cell diagram $Zn(Hg)(s)\,|\,ZnSO_4(aq), PbSO_4(aq)\,|\,Pb(Hg)(s)$. If you vary the concentration of $ZnSO_4$ and measure the potential at different concentrations, you obtain these data:

$[ZnSO_4]$, mol/L	E_{cell}, V
0.0005	0.6114
0.002	0.5832
0.01	0.5535

   (a) Write the half-reactions that occur in this cell.
   (b) What is the overall redox reaction?
   (c) $E°$ of the cell is 0.411 V. What is ΔG?
   (d) What is the equilibrium constant for this redox reaction?
85. Hydrogen gas reduces Ni^{2+} according to the reaction $Ni^{2+}(aq) + H_2(g) \longrightarrow Ni(s) + 2\,H^+(aq)$; $E°_{cell} = -0.25$ V, $\Delta H° = 54$ kJ/mol.
   (a) What is *K* for this redox reaction?
   (b) Is this reaction likely to occur?
   (c) What conditions can be changed to increase the likelihood that the reaction will occur as written?
   (d) Is the reaction more likely to occur at higher or lower pH?
86. The silver–silver bromide electrode has a standard potential of 0.073 V. What is K_{sp} of AgBr?

19.5 Commercial Galvanic Cells

87. This reaction is characteristic of a lead storage battery:

$$Pb(s) + PbO_2(s) + 2H_2SO_4(aq) \longrightarrow 2PbSO_4(s) + 2H_2O(l)$$

If you have a battery with an electrolyte that has a density of 1.15 g/cm^3 and contains 30.0% sulfuric acid by mass, is the potential greater or less than that of the standard cell?

19.7 Electrolysis

88. The electrolysis of molten salts is frequently used in industry to obtain pure metals. How many grams of metal are deposited from these salts with each mole of electrons? (a) $AlCl_3$; (b) $MgCl_2$; (c) $FeCl_3$
89. Electrolysis is the most direct way of recovering a metal from its ores. However, the $Na^+(aq)/Na(s)$, $Mg^{2+}(aq)/Mg(s)$, and $Al^{3+}(aq)/Al(s)$ couples all have standard electrode potentials, $E°$, less than zero, indicating that these metals can never by obtained by electrolysis of aqueous solutions of their salts. Why? What reaction would occur instead?
90. Calculate the mass of copper metal that is deposited if a 5.12-A current is passed through a copper nitrate solution for 1.5 h.
91. What volume of chlorine gas at STP is evolved when a solution of $MgCl_2$ is electrolyzed using a current of 12.4 A for 1.0 h?
92. What mass of PbO_2 is reduced when a current of 5.0 A is withdrawn over a period of 2 h from a lead storage battery?
93. Electrolysis of $Cr^{3+}(aq)$ produces $Cr^{2+}(aq)$. If you had 500 mL of a 0.15 *M* solution of $Cr^{3+}(aq)$, how long would it take to reduce the Cr^{3+} to Cr^{2+} using a 0.158-A current?
94. Predict the products when aqueous solutions of the following are electrolyzed: (a) $MgBr_2$; (b) $Hg(CH_3CO_2)_2$; (c) $Al_2(SO_4)_3$.
95. Predict the products obtained at each electrode when aqueous solutions of the following are electrolyzed: (a) $AgNO_3$; (b) RbI.

APPLICATIONS

96. The percent efficiency of a fuel cell is defined as $\Delta G^0/\Delta H^0 \times 100$. If hydrogen gas were distributed for domestic and industrial use from a central electrolysis facility, the gas could be piped to consumers much as methane is piped today. Conventional nuclear power stations have an efficiency of 25–30%. Use tabulated data to calculate the efficiency of a fuel cell in which the reaction $H_2(g) + \frac{1}{2}O_2(g) \longrightarrow H_2O(g)$ occurs under standard conditions.
97. You are about to run an organic reaction and you need a strong oxidant. Although you have BrO_3^- at your disposal, you prefer to use MnO_4^-. You notice you also have MnO_2 in the lab.
   (a) Predict whether you will be able to synthesize MnO_4^- using the materials available to you.
   (b) Write the overall reaction for the synthesis of MnO_4^-.
   (c) What is $\Delta G°$ for this reaction?
   (d) What is the equilibrium constant?
98. It is possible to construct a galvanic cell using amalgams as electrodes, each containing different concentrations of the same metal. One example is the $Pb(Hg)(a_1)\,|\,PbSO_4\,(soln)\,|\,Pb(Hg)(a_2)$ cell, in which a_1 and a_2 represent the concentrations of lead in the amalgams. Notice that no chemical change occurs; rather, the reaction transfers lead from one amalgam to the other, thus altering the Pb concentration in both amalgams. Write an equation for *E* for such a cell.

99. The biological electron transport chain provides for an orderly, stepwise transfer of electrons. Both NADH and $FADH_2$ are energy-rich molecules that liberate a large amount of energy upon oxidation. Free energy released during the transfer of electrons from either of these molecules to oxygen drives the synthesis of ATP formed during respiratory metabolism. The reactions are

$$NADH + H^+ + \tfrac{1}{2}O_2 \longrightarrow NAD^+ + H_2O \qquad \Delta G^{\circ\prime} = -52.6 \text{ kcal/mol}$$

$$FADH_2 + \tfrac{1}{2}O_2 \longrightarrow FAD + H_2O \qquad \Delta G^{\circ\prime} = -43.4 \text{ kcal/mol}$$

The standard potential ($E^{\circ\prime}$) for a biological process is defined at pH = 7.

(a) What is $E^{\circ\prime}_{cell}$ for each reaction?

(b) What is $E^{\circ\prime}_{anode}$ for the reaction that uses NADH?

(c) How does $E^{\circ\prime}_{anode}$ of $FADH_2$ compare with that of NADH?

100. The oldest known metallurgical artifacts are beads made from alloys of copper, produced in Egypt, Mesopotamia, and the Indus Valley in 3000 B.C. To determine the copper content of alloys such as brass, a brass sample is dissolved in nitric acid to obtain $Cu^{2+}(aq)$, and then the pH is adjusted to a value of 4–5. Excess KI is used to reduce the Cu^{2+} to Cu^+ with concomitant oxidation of I^- to I_2. The iodine that is produced is then titrated with thiosulfate solution to determine the amount of Cu^{2+} in the original solution. These reactions are involved in the procedure:

(1) $Cu^{2+}(aq) + I^-(aq) + e^- \longrightarrow CuI(s) \qquad E^\circ = 0.86\text{ V}$

(2) $S_4O_6^{2-}(aq) + 2e^- \longrightarrow 2S_2O_3^{2-}(aq) \qquad E^\circ = 0.08\text{ V}$

(3) $NO_3^-(aq) + 2H^+(aq) + 2e^- \longrightarrow NO_2^-(aq) + H_2O(l) \qquad E^\circ = 0.94\text{ V}$

(4) $I_2(s) + 2e^- \longrightarrow 2I^-(aq) \qquad E^\circ = 0.54\text{ V}$

(a) Write a balanced equation for the reaction between nitric acid and $Cu(s)$.

(b) What is E°_{cell} for this reaction?

(c) What is E°_{cell} for the reaction between thiosulfate and iodine?

(d) When the pH of reaction solution 3 is adjusted to 7.00, what is E for this half-reaction?

(e) Why is it necessary to adjust the pH of reaction solution 3 before proceeding to step 4?

(f) Use tabulated data to explain why rust (Fe_2O_3) is considered a contaminant that renders this method useless in determining Cu concentrations?

101. While working at a nuclear reactor site, you have been put in charge of reprocessing spent nuclear fuel elements. Your specific task is to reduce U(VI) to elemental U without reducing Pu(VI) to elemental Pu. You have the following information at your disposal:

$$E^\circ \quad UO_2^{2+} \xrightarrow{0.05} UO_2^{+} \xrightarrow{0.62} U^{4+} \xrightarrow{-0.61} U^{3+} \xrightarrow{-1.85} U$$

$$E^\circ \quad PuO_2^{2+} \xrightarrow{0.91} PuO_2^{+} \xrightarrow{1.17} Pu^{4+} \xrightarrow{0.98} Pu^{3+} \xrightarrow{-2.03} Pu$$

Use tabulated data to decide what reagent will accomplish your task in a 1 *M* acidic solution.

102. Stainless steels typically contain 11% Cr and are resistant to corrosion because of the formation of an oxide layer that can be approximately described as $FeCr_2O_4$, where the iron is Fe(II). The protective layer forms when Cr(II) is oxidized to Cr(III) and Fe is oxidized to Fe(II). Explain how this film prevents the corrosion of Fe to rust, which has the formula Fe_2O_3.

103. Ion-selective electrodes are a powerful tool for measuring specific concentrations of ions in solution. For example, they are used to measure iodide in milk, copper-ion levels in drinking water, fluoride concentrations in toothpastes, and the silver-ion concentration in photographic emulsions and spent fixing solutions. Describe how ion-selective electrodes work, and then propose a design for an ion-selective electrode that can be used for measuring water hardness (Ca^{2+}, Mg^{2+}) in water-conditioning systems.

104. Enzymes are proteins that catalyze a specific reaction with a high degree of specificity. An example is the hydrolysis of urea by urease:

$$NH_2CONH_2 + 2H_2O + H^+ \xrightarrow{\text{Urease}} 2NH_4^+ + HCO_3^-$$

An enzyme electrode for measuring urea concentrations can be made by coating the surface of a glass electrode with a gel that contains urease.

(a) Explain what occurs when the electrode is placed in contact with a solution that contains urea.

(b) What species diffuses through the gel?

(c) What would be an effective reference electrode?

105. Gas-sensing electrodes can be constructed using a combination electrode that is surrounded by a gas-permeable membrane. For example, to measure CO_2, a pH electrode and a reference electrode are placed in solution on the "inner" side of a CO_2-permeable membrane, and the sample solution is placed on the "external" side. As CO_2 diffuses through the membrane, the pH of the internal solution changes due to the reaction $CO_2(g) + H_2O(l) \longrightarrow HCO_3^-(aq) + H^+(aq)$. Thus, the pH of the internal solution varies directly with the CO_2 vapor pressure in the external sample. Ammonia electrodes operate in the same manner. Describe an electrode that would test for ammonia levels in seawater.

106. U.S. submarines that are not nuclear-powered use a combination of batteries and diesel engines for their power. When submerged, they are battery driven; when on the surface, they are diesel driven. Why aren't batteries used when the submarines are on the surface?

107. List some practical considerations in designing a battery to power an electric car.

108. It is possible to run a digital clock using the power supplied by two potatoes. The clock is connected to two wires, one attached to a copper plate and the other to a zinc plate. Each plate is pushed into a different potato, and when the two potatoes are connected by a wire, the clock begins to run as if it were connected to a battery. (a) Draw a cell diagram of the potato clock. (b) Explain why the clock runs.

109. The silver–zinc battery has the highest energy density of any rechargeable battery available today. Its use is presently limited to military applications, primarily in portable communications, aerospace, and torpedo-propulsion systems. The disadvantages of these cells are their limited life (they typically last no more than about 2 years) and their high cost, which restricts their use to situations in which cost is only a minor factor. The generally accepted equations representing this type of battery are

$$2AgO(s) + Zn(s) + H_2O(l) \longrightarrow Ag_2O(s) + Zn(OH)_2(aq) \quad E° = 1.85\ V$$

$$Ag_2O(s) + Zn(s) + H_2O(l) \longrightarrow 2Ag(s) + Zn(OH)_2(aq) \quad E° = 1.59\ V$$

(a) Write the overall cell reaction, and calculate $E°_{cell}$.
(b) If the cell is 75% efficient, what is the maximum amount of work that can be generated from this type of battery?
(c) Use tabulated data to calculate the maximum work that can be generated by a lead storage cell. If a silver–zinc battery is operating at 100% efficiency, how do the two batteries compare?

110. All metals used in boats and ships are subject to corrosion, particularly when the vessels are operated in salt water, which is a good electrolyte. Based on the data in the table, where potentials are measured using a glass electrode, explain why (a) iron or steel should not be used in bolts in a lead ballast keel; (b) ordinary brass should not be used as a structural fastening, particularly below the waterline; (c) an aluminum hull should not be painted with a copper-based antifouling paint; (d) magnesium sacrificial anodes are preferred over zinc when a vessel is kept in fresh water, and (e) Monel (an alloy that contains mostly nickel and copper) is preferred over stainless steel for freshwater tanks.

Metal	***E* vs. Ag/AgCl, V**
Titanium	0.02
Monel [Ni(Cu)]	−0.06
Ni(Al) bronze	−0.16
Lead	−0.20
Manganese bronze	−0.29
Brass	−0.30
Copper	−0.31
Tin	−0.31
Stainless steel	−0.49
Aluminum	−0.87
Zinc	−1.00
Magnesium	−1.60

111. Parents often coat a baby's first shoes with copper to preserve them for posterity. A conducting powder, such as graphite, is rubbed on the shoe, and then copper is electroplated on the shoe. How much copper is deposited on a shoe if the electrolytic process is run for 60 min at 1.2 A from a 1.0 *M* solution of $CuSO_4$?

112. Before 1886, metallic aluminum was so rare that a bar of it was displayed next to the Crown Jewels at the Paris Exposition of 1855. Today it is obtained commercially from aluminum oxide by the Hall–Heroult process, an electrolytic process that uses molten Al_2O_3 and cryolite (Na_3AlF_6). As the operation proceeds, molten Al sinks to the bottom of the cell. The overall reaction is $2Al_2O_3(l) + 3C(s) \longrightarrow 4Al(l) + 3CO_2(g)$; however, the process is only approximately 90% efficient. (a) Why is cryolite added to the Al_2O_3? (b) How much aluminum is deposited if electrodeposition occurs for 24 h at 1.8 A?

113. Compact audio discs (CDs) are manufactured by electroplating. Information is stored on a CD master in a patter of "pits" (depressions, which correspond to an audio track) and "lands" (the raised areas between depressions). A laser beam cuts the pits into a plastic or glass material. The material is cleaned, sprayed with $[Ag(NH_3)_2]^+$, and then washed with a formaldehyde solution that reduces the complex and leaves a thin silver coating. Nickel is electrodeposited on the disk and then peeled away to produce a master disk, which is used to stamp copies.
(a) Write the half-reactions that correspond to the electrodeposition reaction.
(b) If a CD has a radius of 12 cm and an interior hole with a diameter of 2.5 cm, how long does it take to deposit a 50-μm layer of nickel on one side of the CD using a 1.0 *M* solution of $NiSO_4$ and a current of 0.8 A?

114. Onc of the most important electrolytic processes used in industry is the electrolytic reduction of acrylonitrile ($CH_2{=}CHCN$) to adiponitrile [$NC(CH_2)_4CN$]. The product is then hydrogenated to hexamethylenediamine [$H_2N(CH_2)_6NH_2$], a key component of one form of nylon. Monsanto produces about 200,000 metric tons of adiponitrile annually using this process. The cathode reaction in the electrochemical cell is

$$2CH_2{=}CHCN + 2H^+ + 2e^- \longrightarrow NC(CH_2)_4CN$$

The cost of the electricity consumed is a major portion of the overall cost of the product. Calculate the total number of kilowatt-hours of electricity used by Monsanto each year in this process, assuming a continuous applied potential of 5.0 V and an electrochemical efficiency of 50%.

115. Calculate the total amount of energy consumed in the electrolysis reaction used to make the 16×10^6 metric tons of aluminum produced annually worldwide, assuming a continuous applied potential of 5.0 V and an efficiency of 50%. Express your answer in kilojoules and in kilowatt-hours.

20 Nuclear Chemistry

LEARNING OBJECTIVES

- To understand the factors that affect nuclear stability
- To know the different kinds of radioactive decay
- To be able to balance a nuclear reaction
- To be able to interpret a radioactive decay series
- To know the differences between ionizing and nonionizing radiation and their effects on matter
- To be able to identify natural and artificial sources of radiation
- To be able to calculate a mass–energy balance and a nuclear binding energy
- To understand the differences between nuclear fission and fusion
- To understand how nuclear reactors operate
- To understand how nuclear transmutation reactions led to the formation of the elements in stars and how they can be used to synthesize transuranium elements

CHAPTER OUTLINE

Until now, you have been studying chemical processes in which atoms share or transfer electrons to form new compounds, leaving the atomic nuclei largely unaffected. In this chapter, we examine some properties of the atomic nucleus and the changes that can occur in atomic nuclei.

Nuclear reactions differ from other chemical processes in one critical way: in a nuclear reaction, the identities of the elements change. In addition, nuclear reactions are often accompanied by the release of enormous amounts of energy, as much as a *billion* times more than the energy released by chemical reactions. Moreover, the yields and rates of a nuclear reaction are generally unaffected by changes in temperature, pressure, or the presence of a catalyst.

We begin our discussion by examining the structure of the atomic nucleus and the factors that determine whether a particular nucleus is stable or decays spontaneously to another element. We then discuss the major kinds

The glow caused by intense radiation. The high-energy particles ejected into the surrounding water or air by an intense radioactive source such as this nuclear reactor core produce a ghostly bluish glow.

of nuclear decay reactions as well as the properties and uses of the radiation that is emitted when nuclei decay. In these discussions, you will learn how radioactive emissions can be used to study the mechanisms of chemical reactions and biological processes, and how to calculate the amount of energy released during a nuclear reaction. You will also discover why houses are tested for radon gas, how radiation is used to probe organs such as the brain, and how the energy from nuclear reactions can be harnessed to produce electricity. Last, we explore the nuclear chemistry that takes place in stars, and we describe the role that stars play in producing most of the elements in the universe.

20.1 • The Components of the Nucleus

Although most of the known elements have at least one isotope whose atomic nucleus is stable indefinitely, all elements have isotopes that are unstable and disintegrate, or *decay*, at measurable rates by emitting radiation. Some elements have no stable isotopes and eventually decay to other elements. In contrast to the chemical reactions that were the main focus of earlier chapters and are due to changes in the arrangements of the valence electrons of atoms, the process of nuclear decay is a nuclear reaction that results in changes inside the atomic nucleus. We begin our discussion of nuclear reactions by reviewing the conventions used to describe the components of the nucleus.

The Atomic Nucleus

As you learned in Chapter 1, each element can be represented by the notation ${}^{A}_{Z}X$, where A, the mass number, is the sum of the numbers of protons and neutrons, and Z, the atomic number, is the number of protons. The protons and neutrons that make up the nucleus of an atom are called **nucleons**, and an atom with a particular number of protons and neutrons is called a **nuclide**. Nuclides that have the same number of protons but different numbers of neutrons are called *isotopes*. Isotopes can also be represented by an alternative notation that uses the name of the element followed by the mass number, as we see in *carbon-12*. The stable isotopes of oxygen, for example, can be represented in any of these ways:

${}^{A}_{Z}X$:	${}^{16}_{8}O$	${}^{17}_{8}O$	${}^{18}_{8}O$
${}^{A}X$:	${}^{16}O$	${}^{17}O$	${}^{18}O$
Element-A:	Oxygen-16	Oxygen-17	Oxygen-18

Because the number of neutrons is equal to $A - Z$, we see that the first isotope above has eight neutrons, the second nine, and the third ten. Isotopes of all naturally occurring elements on Earth are present in nearly fixed proportions with each proportion constituting an isotope's *natural abundance*. For example, in a typical terrestrial sample of oxygen, 99.76% of the O atoms are oxygen-16, 0.20% are oxygen-18, and 0.04% are oxygen-17.

Any nucleus that is unstable and decays spontaneously is said to be **radioactive**, emitting subatomic particles and electromagnetic radiation. The emissions are collectively called *radioactivity* and can be measured. Isotopes that emit radiation are called **radioisotopes**. As you learned in Chapter 14, the rate at which radioactive decay occurs is characteristic of the isotope and is generally reported as a *half-life* ($t_{1/2}$), the amount of time required for half the initial number of nuclei present to decay in a first-order reaction (see Section 14.5). An isotope's half-life can range from fractions of a second to billions of years and among other applications, can be used to measure the age of ancient objects. The next example and exercise review the calculations involving radioactive decay rates and half-lives.

EXAMPLE 20.1

Fort Rock Cave in Oregon is the site where archaeologists discovered several Indian sandals, the oldest ever found in Oregon. Analysis of the ^{14}C content of the sagebrush used to make the sandals gave an average decay rate of 5.1 disintegrations per minute (dpm) per gram carbon. The current $^{14}C/^{12}C$ ratio in living organisms is 1.3×10^{-12}, with a decay rate of 15 dpm/g C. How long ago was the sagebrush in the sandals cut? The half-life of ^{14}C is 5730 years.

Given Radioisotope, current $^{14}C/^{12}C$ ratio, initial decay rate, final decay rate, half-life

Asked for Age

Strategy

A Use Equation 14.34 to calculate N_0/N, the ratio of the number of atoms of ^{14}C originally present in the sample to the number of atoms now present.

B Substitute the value for the half-life of ^{14}C into Equation 14.32 to obtain the rate constant for the reaction.

C Substitute the calculated values for N_0/N and the rate constant into Equation 14.36 to obtain the elapsed time t.

Solution

We can use the integrated rate law for a first-order nuclear reaction (Equation 14.36) to calculate the amount of time that has passed since the sagebrush was cut to make the sandals:

$$\ln\frac{N}{N_0} = -kt$$

A From Equation 14.34, we know that $A = kN$. We can therefore use the initial and final activities ($A_0 = 15$ and $A = 5.1$) to calculate N_0/N:

$$\frac{A_0}{A} = \frac{kN_0}{kN} = \frac{N_0}{N} = \frac{15}{5.1}$$

B Now we can calculate the rate constant k from the half-life of the reaction (5730 yr) using Equation 14.32:

$$t_{1/2} = \frac{0.693}{k}$$

Rearranging this equation to solve for k gives us

$$k = \frac{0.693}{t_{1/2}} = \frac{0.693}{5730\ \text{yr}} = 1.21 \times 10^{-4}\ \text{yr}^{-1}$$

C Substituting the calculated values into the equation for t gives

$$t = \frac{\ln(N_0/N)}{k} = \frac{\ln(15/5.1)}{1.21 \times 10^{-4}\ \text{yr}^{-1}} = 8910\ \text{yr}$$

Thus, the sagebrush in the sandals is about 8900 years old.

EXERCISE 20.1

While trying to find a suitable way to protect his own burial chamber, the ancient Egyptian pharoah Sneferu developed the pyramid, a burial structure that protected desert graves from thieves and exposure to wind. Analysis of the ^{14}C content of several items in pyramids built during his reign gave an average decay rate of 8.6 dpm/g C. When were the objects in the chamber created?

Answer About 4600 years ago, 2600 B.C.

Nuclear Stability

As discussed in Chapter 1, the nucleus of an atom occupies a tiny fraction of the volume of the atom and contains the number of protons and neutrons that is characteristic of a given isotope. Electrostatic repulsions would normally cause the positively charged protons to repel each other, but the nucleus does not fly apart because of the **strong nuclear force**, an extremely powerful but very short-range attractive force between nucleons (Figure 20.1). All stable nuclei except the hydrogen-1 nucleus (^{1}H) contain at least one neutron in order to overcome the electrostatic repulsion between protons. As the number of protons in the nucleus increases, the number of neutrons needed for a stable nucleus increases even more rapidly. Too many protons (or too few neutrons) in the nucleus result in an imbalance between forces, which leads to nuclear instability.

The relationship between the numbers of protons and neutrons in stable nuclei, arbitrarily defined as having a half-life longer than ten times the age of the Earth, is shown graphically in Figure 20.2. Note that the stable isotopes form a "peninsula of stability" in a "sea of instability." Only two stable isotopes, ^{1}H and ^{3}He, have a neutron-to-proton ratio less than 1. Several stable isotopes of light atoms have a neutron-to-proton ratio equal to 1 (for example, $^{4}_{2}He$, $^{10}_{5}B$, and $^{40}_{20}Ca$). All other stable nuclei have a higher neutron-to-proton ratio, which increases steadily to about 1.5 for the heaviest nuclei. Regardless of the number of neutrons, however, all elements with $Z > 83$ are unstable and radioactive.

As shown in Figure 20.3, more than half of the stable nuclei (166 out of 279) have *even* numbers of both neutrons and protons; only six of the 279 stable nuclei do not have odd numbers of both. Moreover, certain numbers of neutrons or protons result in especially stable nuclei; these are the so-called *magic numbers* 2, 8, 20, 50, 82, and 126. For example, tin ($Z = 50$) has ten stable isotopes, but the elements on either side of tin in the periodic table, indium ($Z = 49$) and antimony ($Z = 51$), have only two stable isotopes each. Nuclei with magic numbers of *both* protons *and* neutrons are said to be

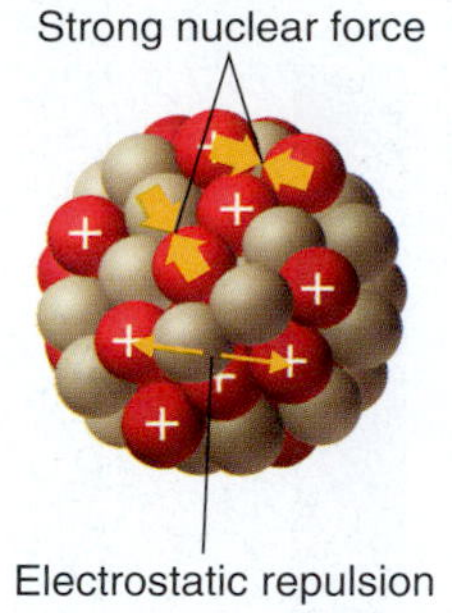

Figure 20.1 Competing interactions within the atomic nucleus. Electrostatic repulsions between positively charged protons would normally cause the nuclei of atoms (except H) to fly apart. In stable atomic nuclei, these repulsions are overcome by the strong nuclear force, a short-range but powerful attractive interaction between nucleons. If the attractive interactions due to the strong nuclear force are weaker than the electrostatic repulsions between protons, the nucleus is unstable, and it will eventually decay.

MGC Predicting Nuclear Stability

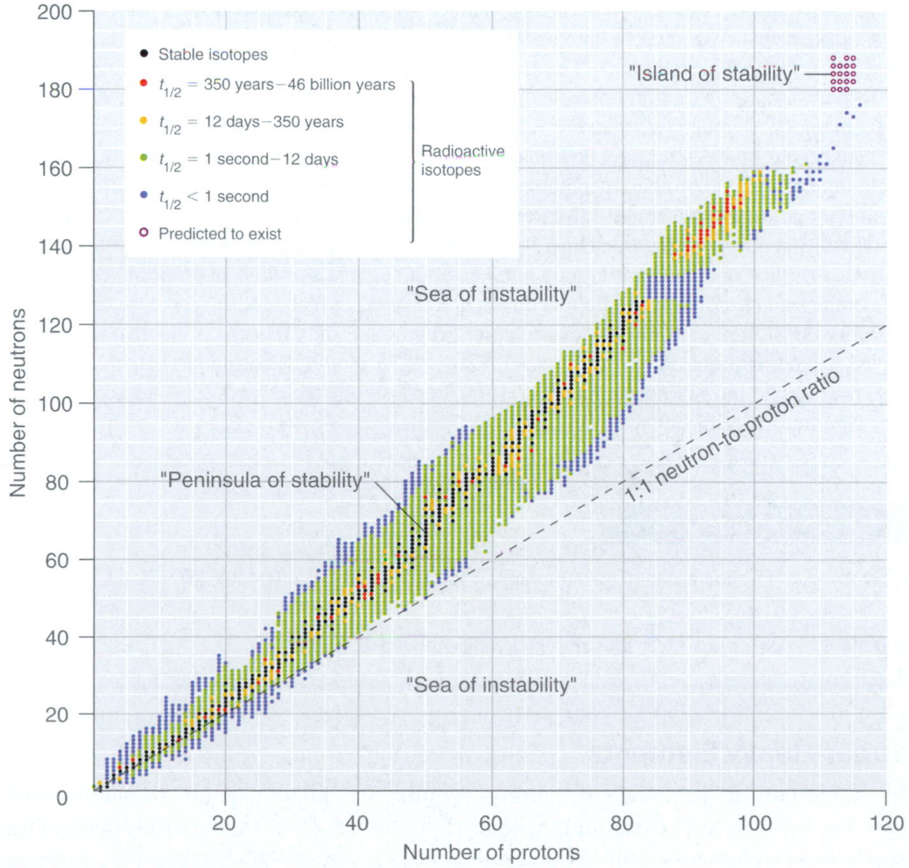

Figure 20.2 The relationship between nuclear stability and the neutron-to-proton ratio. In this plot of the number of neutrons versus the number of protons, each black point corresponds to a stable nucleus. In this classification, a stable nucleus is arbitrarily defined as one with a half-life longer than 46 billion years (ten times the age of Earth). Note that as the number of protons (the atomic number) increases, the number of neutrons required for a stable nucleus increases even more rapidly. Isotopes shown in red, yellow, green, and blue are progressively less stable and more radioactive; the farther an isotope is from the diagonal band of stable isotopes, the shorter its half-life. The purple dots indicate super-heavy nuclei that are predicted to be relatively stable, meaning that they are expected to be radioactive but to have relatively long half-lives. In most cases, these elements have not yet been observed or synthesized. [Data source: National Nuclear Data Center, Brookhaven National Laboratory. Evaluated Nuclear Structure Data File (ENSDF), March 2005.]

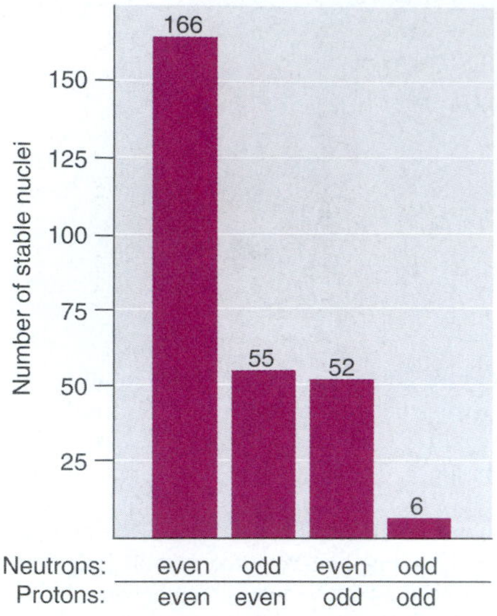

Figure 20.3 The relationship between the numbers of protons and neutrons and nuclear stability. Note that most stable nuclei contain even numbers of *both* neutrons and protons.

"doubly magic" and are even more stable. Examples of elements with doubly magic nuclei are $^{4}_{2}He$, with two protons and two neutrons, and $^{208}_{82}Pb$, with 82 protons and 126 neutrons, which is the heaviest known stable isotope of any element.

The pattern of stability suggested by the magic numbers of nucleons is reminiscent of the stability associated with the closed-shell electron configurations of the noble gases in Group 18, and has led to the hypothesis that the nucleus contains shells of nucleons that are in some ways analogous to the shells occupied by electrons in an atom. As shown in Figure 20.2, the "peninsula" of stable isotopes is surrounded by a "reef" of radioactive isotopes, which are stable enough to exist for varying lengths of time before they eventually decay to produce other nuclei.

EXAMPLE 20.2

Which of these nuclides do you predict should be stable, and which should be radioactive? **(a)** $^{30}_{15}P$; **(b)** $^{98}_{43}Tc$; **(c)** tin-118; **(d)** $^{239}_{94}Pu$

Given Mass number and atomic number

Asked for Predicted nuclear stability

Strategy

Use the number of protons, the neutron-to-proton ratio, and the presence of even or odd numbers of neutrons and protons to predict the stability or radioactivity of each nuclide.

Solution

(a) This isotope of phosphorus, ^{30}P, contains 15 neutrons and 15 protons, giving a neutron-to-proton ratio of 1.0. Although the atomic number, 15, is much less than the value of 83 above which all nuclides are unstable, the neutron-to-proton ratio is lower than expected for stability for an element with this mass. As shown in Figure 20.2, its neutron-to-proton ratio should be greater than 1. Moreover, this isotope has an odd number of both neutrons and protons, which also tends to make a nuclide unstable. Consequently, $^{30}_{15}P$ is predicted to be radioactive, and it is.

(b) This isotope of technetium, ^{98}Tc, contains 55 neutrons and 43 protons, giving a neutron-to-proton ratio of 1.28 and placing $^{98}_{43}Tc$ near the edge of the band of stability. The atomic number, 55, is much less than the value of 83 above which all isotopes are unstable. These facts suggest that $^{98}_{43}Tc$ might be stable. Note, however, that $^{98}_{43}Tc$ contains an odd number of both neutrons and protons, a combination that seldom gives a stable nucleus. Consequently, $^{98}_{43}Tc$ is predicted to be radioactive (which is indeed the case).

(c) Tin-118 contains 68 neutrons and 50 protons, for a neutron-to-proton ratio of 1.36. As in part (b), this value and the atomic number both suggest stability. In addition, the isotope contains an *even* number of both neutrons and protons, which tends to increase nuclear stability. Most important, the nucleus contains 50 protons, and 50 is one of the magic numbers associated with especially stable nuclei. Thus, $^{118}_{50}Sn$ should be particularly stable.

(d) This nuclide has an atomic number of 94. Because all nuclei with $Z > 83$ are unstable, $^{239}_{94}Pu$ must be radioactive.

EXERCISE 20.2

Classify each nuclide as stable or radioactive: **(a)** $^{232}_{90}Th$; **(b)** $^{40}_{20}Ca$; **(c)** $^{15}_{8}O$; **(d)** ^{139}La.

Answer **(a)** radioactive; **(b)** stable; **(c)** radioactive; **(d)** stable

Super-Heavy Elements

In addition to the "peninsula of stability," Figure 20.2 shows a small "island of stability" that is predicted to exist in the upper right corner. This island corresponds to the **super-heavy elements**, with atomic numbers near the magic number 126. Because the next magic number for neutrons should be 184, it has been suggested that an element that contains 114 protons and 184 neutrons might even be stable enough to exist in nature. Although these claims were met with skepticism for many years, since 1999 a few atoms of isotopes with $Z = 114$ and $Z = 116$ have been prepared and found to be surprisingly stable. Thus, a number of relatively long-lived nuclei may well be accessible among the super-heavy elements.

20.2 • Nuclear Reactions

The two general kinds of nuclear reactions are nuclear decay reactions and nuclear transmutation reactions. In a **nuclear decay reaction**, also called *radioactive decay*, an unstable nucleus emits radiation and is transformed into the nucleus of one or more other elements. The resulting *daughter nuclei* have a lower mass and are lower in energy (more stable) than the *parent nucleus* that decayed. In contrast, in a **nuclear transmutation reaction**, a nucleus reacts with a subatomic particle or another nucleus to form a product nucleus that is *more massive* than the starting material. As we shall see, nuclear decay reactions occur spontaneously under all conditions, but nuclear transmutation reactions occur only under very special conditions, such as the collision of a beam of energetic particles with a target nucleus or in the interior of stars. We begin this section by considering the different classes of radioactive nuclei, along with their characteristic nuclear decay reactions and the radiation they emit.

Classes of Radioactive Nuclei

Each of the three general classes of radioactive nuclei is characterized by a different decay process or set of processes:

1. ***Neutron-rich nuclei.*** The nuclei on the upper left side of the band of stable nuclei in Figure 20.2 have a neutron-to-proton ratio that is too high to give a stable nucleus. These nuclei decay by a process that *converts a neutron to a proton*, thereby *decreasing* the neutron-to-proton ratio.
2. ***Neutron-poor nuclei.*** Nuclei on the lower right side of the band of stable nuclei have a neutron-to-proton ratio that is too low to give a stable nucleus. These nuclei decay by processes that have the net effect of *converting a proton to a neutron*, thereby *increasing* the neutron-to-proton ratio.
3. ***Heavy nuclei.*** With very few exceptions, heavy nuclei (those with $A \geq 200$) are intrinsically unstable regardless of the neutron-to-proton ratio, and all nuclei with $Z > 83$ are unstable. This is presumably due to the cumulative effects of electrostatic repulsions between the large number of positively charged protons, which cannot be totally overcome by the strong nuclear force, regardless of the number of neutrons present. Such nuclei tend to decay by emitting an **α particle** (a helium nucleus, ^4_2He), which decreases the number of protons and neutrons in the original nucleus by 2. Since the neutron-to-proton ratio in an α particle is 1, the net result of alpha emission is an increase in the neutron-to-proton ratio.

Note the pattern

Nuclear decay reactions always produce daughter nuclei that have a more favorable neutron-to-proton ratio and hence are more stable than the parent nucleus.

Nuclear Decay Reactions

MGC Modes of Radioactive Decay

Just as we use the number and type of *atoms* present to balance a chemical equation, we can use the number and type of *nucleons* present to write a balanced equation for a nuclear decay reaction. This procedure also allows us to predict the identity of either the parent or the daughter nucleus if the identity of only one is known. Regardless of the mode of decay, *the total number of nucleons is conserved* in all nuclear reactions.

TABLE 20.1 Nuclear decay emissions and their symbols

Identity	Symbol	Charge	Mass, amu
Helium nucleus	${}^{4}_{2}\alpha$	+2	4.001506
Electron	${}^{0}_{-1}\beta$ or β^-	−1	0.000549
Photon	${}^{0}_{0}\gamma$	n/a	n/a
Neutron	${}^{1}_{0}n$	0	1.008665
Proton	${}^{1}_{1}p$	+1	1.007276
Positron	${}^{0}_{+1}\beta$ or β^+	+1	0.000549

To describe nuclear decay reactions, chemists have extended the ${}^{A}_{Z}X$ notation for nuclides to include radioactive emissions. Table 20.1 lists the name and symbol for each type of emitted radiation. We introduced the most common of these, α and β particles and γ rays, in Chapters 1 and 14. The most notable addition is the **positron**, a particle that has the same mass as an electron, but a *positive* charge rather than a negative charge.

Like the notation used to indicate isotopes, the left superscript in the symbol for a particle gives the mass number, which is the total number of protons and neutrons. For a proton or a neutron, $A = 1$. Because neither an electron nor a positron contains protons or neutrons, its mass number is 0. The numbers should not be taken literally, however, as meaning that these particles have zero mass; ejection of a beta particle (an electron) simply has a negligible effect on the mass of a nucleus.

Similarly, the left subscript gives the charge of the particle. Because protons carry a positive charge, $Z = +1$ for a proton. In contrast, a neutron contains no protons and is electrically neutral, so $Z = 0$. In the case of an electron, $Z = -1$, and for a positron, $Z = +1$. Because γ rays are high-energy photons, both A and Z are 0.

In some cases, two different symbols are used for particles that are identical but produced in different ways. For example, the symbol ${}^{0}_{-1}e$, which is usually simplified to e^-, represents a free electron or an electron associated with an *atom*, whereas the symbol ${}^{0}_{-1}\beta$, often simplified to β^-, denotes an electron that originates from within the *nucleus*, which is a β particle. Similarly, ${}^{4}_{2}He^{2+}$ refers to the nucleus of a helium atom, and ${}^{4}_{2}\alpha$ denotes an identical particle that has been ejected from a heavier nucleus.

There are six fundamentally different kinds of nuclear decay reactions, each of which releases a different kind of particle or energy. The essential features of each are shown in Table 20.2. The most common are *alpha* and *beta decay* and *gamma emission*, but the others are essential to an understanding of nuclear decay reactions.

Alpha Decay

Many nuclei with mass numbers greater than 200 undergo **alpha decay**, which results in the emission of a helium-4 nucleus as an α particle, ${}^{4}_{2}\alpha$. The general reaction is

$$\underset{\text{Parent}}{{}^{A}_{Z}X} \longrightarrow \underset{\text{Daughter}}{{}^{A-4}_{Z-2}X'} + \underset{\text{Alpha particle}}{{}^{4}_{2}\alpha} \tag{20.1}$$

The daughter nuclide contains two fewer protons and two fewer neutrons than the parent. Thus, α-particle emission produces a daughter nucleus with a mass number $A - 4$ and a nuclear charge $Z - 2$ compared to the parent nucleus. Radium-226, for example, undergoes alpha decay to form radon-222:

$${}^{226}_{88}Ra \longrightarrow {}^{222}_{86}Rn + {}^{4}_{2}\alpha \tag{20.2}$$

TABLE 20.2 Common modes of nuclear decay

Decay Type	Radiation Emitted	Generic Equation	Model
Alpha decay	${}^{4}_{2}\alpha$	${}^{A}_{Z}X \longrightarrow {}^{A-4}_{Z-2}X' + {}^{4}_{2}\alpha$	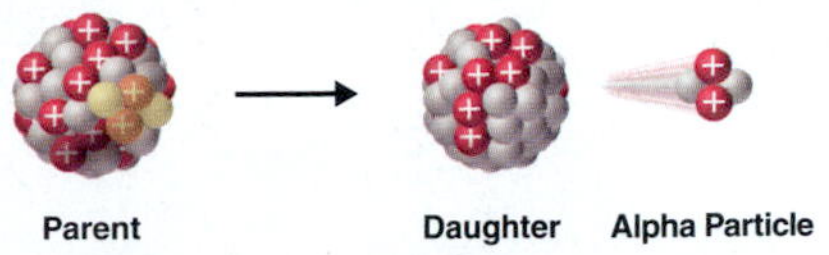
Beta decay	${}^{0}_{-1}\beta$	${}^{A}_{Z}X \longrightarrow {}^{A}_{Z+1}X' + {}^{0}_{-1}\beta$	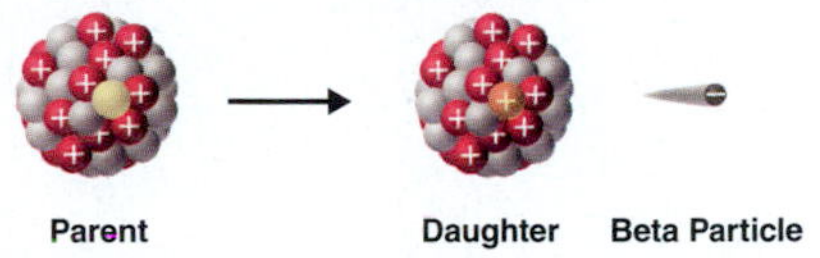
Positron emission	${}^{0}_{+1}\beta$	${}^{A}_{Z}X \longrightarrow {}^{A}_{Z-1}X' + {}^{0}_{+1}\beta$	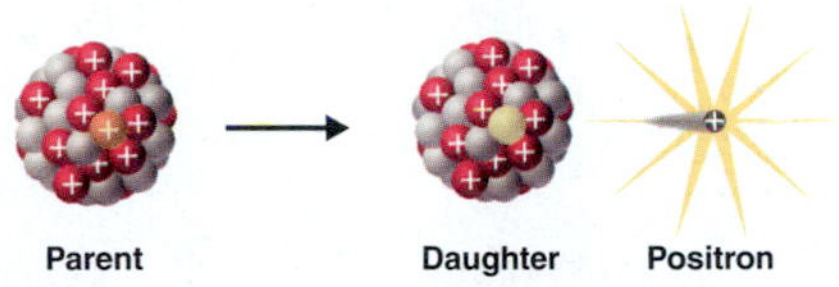
Electron capture	X rays	${}^{A}_{Z}X + {}^{0}_{-1}e \longrightarrow {}^{A}_{Z-1}X' + \text{X ray}$	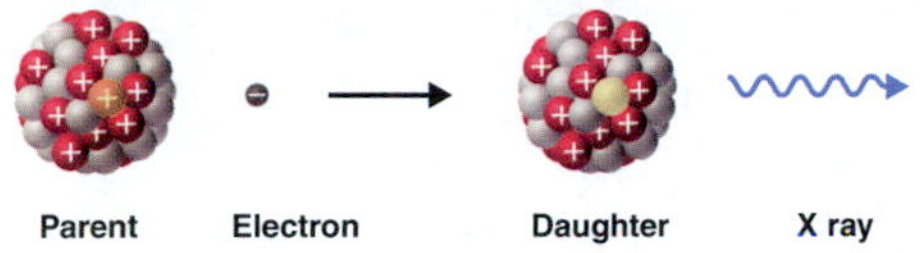
Gamma emission	${}^{0}_{0}\gamma$	${}^{A}_{Z}X^{*} \xrightarrow{\text{Relaxation}} {}^{A}_{Z}X + {}^{0}_{0}\gamma$	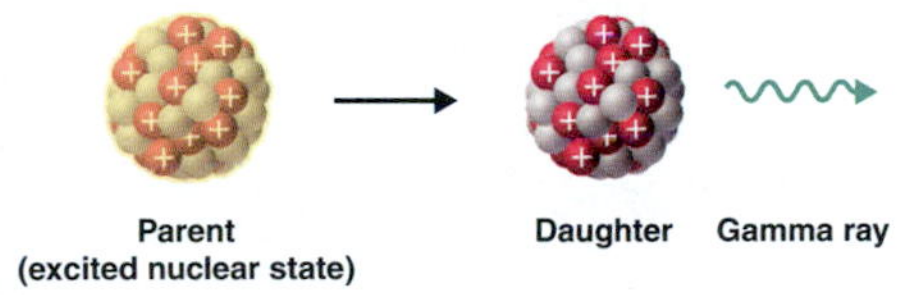
Spontaneous fission	Neutrons	${}^{A+B+C}_{Z+Y}X \longrightarrow {}^{A}_{Z}X' + {}^{B}_{Y}X' + C{}^{1}_{0}n$	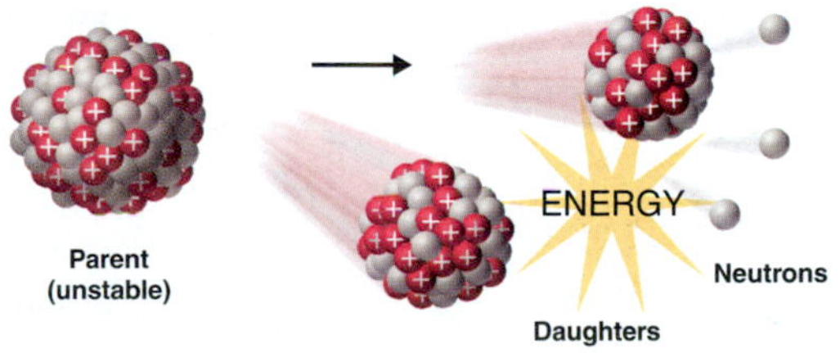

Because nucleons are conserved in this and all other nuclear reactions, the sum of the mass numbers of the products, $222 + 4 = 226$, equals the mass number of the parent. Similarly, the sum of the atomic numbers of the products, $86 + 2 = 88$, equals the atomic number of the parent. Thus, the nuclear equation is balanced.

Beta Decay

Nuclei that contain too many neutrons often undergo **beta decay**, in which a neutron is converted to a proton and a high-energy electron that is ejected from the nucleus as a β particle:

$$\underset{\text{Unstable neutron in nucleus}}{{}^{1}_{0}n} \longrightarrow \underset{\text{Proton retained by nucleus}}{{}^{1}_{1}p} + \underset{\text{Beta particle emitted by nucleus}}{{}^{0}_{-1}\beta} \qquad (20.3)$$

The general reaction for beta decay is therefore

$$\underset{\text{Parent}}{{}^{A}_{Z}X} \longrightarrow \underset{\text{Daughter}}{{}_{Z+1}^{\;A}X'} + \underset{\text{Beta particle}}{{}^{0}_{-1}\beta} \qquad (20.4)$$

Although beta decay does *not* change the mass number of the nucleus, it does result in an increase of +1 in the atomic number because of the addition of a proton in the daughter nucleus. Thus, beta decay decreases the neutron-to-proton ratio, moving the nucleus toward the band of stable nuclei. For example, carbon-14 undergoes beta decay to form nitrogen-14:

$$^{14}_{6}\text{C} \longrightarrow {}^{14}_{7}\text{N} + {}^{0}_{-1}\beta \qquad (20.5)$$

Once again, the number of nucleons is conserved and the charges are balanced. The parent and the daughter nucleus have the same mass number, 14, and the sum of the atomic numbers of the products is 6, the same as the atomic number of the carbon-14 parent.

Positron Emission

Because a positron has the same mass as an electron but opposite charge, **positron emission** is the opposite of beta decay. Thus, positron emission is characteristic of neutron-poor nuclei, which decay by the transformation of a proton to a neutron and a high-energy positron that is emitted:

$$^{1}_{1}\text{p} \longrightarrow {}^{1}_{0}\text{n} + {}^{0}_{+1}\beta \qquad (20.6)$$

The general reaction for positron emission is therefore

$$\underset{\text{Parent}}{^{A}_{Z}\text{X}} \longrightarrow \underset{\text{Daughter}}{_{Z-1}^{\;\;A}\text{X}'} + \underset{\text{Positron}}{^{0}_{+1}\beta} \qquad (20.7)$$

Like beta decay, positron emission does not change the mass number of the nucleus. In this case, however, the atomic number of the daughter nucleus is *lower* by 1 than that of the parent. Thus, the neutron-to-proton ratio has increased, again moving the nucleus closer to the band of stable nuclei. For example, carbon-11 undergoes positron emission to form boron-11:

$$^{11}_{6}\text{C} \longrightarrow {}^{11}_{5}\text{B} + {}^{0}_{+1}\beta \qquad (20.8)$$

Nucleons are conserved, and the charges balance. The mass number, 11, does not change, and the sum of the atomic numbers of the products is 6, the same as the atomic number of the parent carbon-11 nuclide.

Electron Capture

A neutron-poor nucleus can decay by either positron emission or **electron capture** (EC), in which an electron in an inner shell reacts with a proton to produce a neutron:

$$^{1}_{1}\text{p} + {}^{0}_{-1}\text{e} \longrightarrow {}^{1}_{0}\text{n} \qquad (20.9)$$

When a second electron moves from an outer shell to take the place of the lower-energy electron that was absorbed by the nucleus, an X ray is emitted. The overall reaction for electron capture is thus

$$\underset{\text{Parent}}{^{A}_{Z}\text{X}} + \underset{\text{Electron}}{^{0}_{-1}\text{e}} \longrightarrow \underset{\text{Daughter}}{_{Z-1}^{\;\;A}\text{X}'} + \text{X ray} \qquad (20.10)$$

Electron capture does not change the mass number of the nucleus because both the proton that is lost and the neutron that is formed have a mass number of 1. As with positron emission, however, the atomic number of the daughter nucleus is *lower* by 1 than that of the parent. Once again, the neutron-to-proton ratio has increased, moving the nucleus toward the band of stable nuclei. For example, iron-55 decays by electron capture to form manganese-55, which is often written as

$$^{55}_{26}\text{Fe} \xrightarrow{\text{EC}} {}^{55}_{25}\text{Mn} + \text{X ray} \qquad (20.11)$$

Notice that the atomic numbers of the parent and daughter nuclides differ in Equation 20.11, although the mass numbers are the same. To write a balanced reaction for this reaction, we must explicitly include the captured electron in the equation:

$$^{55}_{26}\text{Fe} + {}^{0}_{-1}\text{e} \longrightarrow {}^{55}_{25}\text{Mn} + \text{X ray} \tag{20.12}$$

Both positron emission and electron capture are usually observed for nuclides with low neutron-to-proton ratios, but the decay rates for the two processes can be very different.

Gamma Emission

Many nuclear decay reactions produce daughter nuclei that are in a nuclear excited state, which is similar to an atom in which an electron has been excited to a higher-energy orbital to give an electronic excited state. Just as an electron in an electronic excited state emits energy in the form of a photon when it returns to the ground state (see Chapter 6), a nucleus in an excited state releases energy in the form of a photon when it returns to the ground state. These high-energy photons are γ rays. **Gamma emission** can occur virtually instantaneously, as it does in the alpha decay of uranium-238 to thorium-234:

$$^{238}_{92}\text{U} \longrightarrow \underset{\substack{\text{Excited}\\ \text{nuclear}\\ \text{state}}}{{}^{234}_{90}\text{Th}^{*}} + \underset{\substack{\text{Emitted}\\ \text{alpha}\\ \text{particle}}}{{}^{4}_{2}\alpha} \xrightarrow{\text{Relaxation}} {}^{234}_{90}\text{Th} + {}^{0}_{0}\gamma \tag{20.13}$$

or, if we disregard the decay event that created the excited nucleus:

$$^{234}_{88}\text{Th}^{*} \longrightarrow {}^{234}_{88}\text{Th} + {}^{0}_{0}\gamma \tag{20.14a}$$

or more generally

$$^{A}_{Z}\text{X}^{*} \longrightarrow {}^{A}_{Z}\text{X} + {}^{0}_{0}\gamma \tag{20.14b}$$

Gamma emission can also occur after a significant delay. For example, technetium-99*m* (the *m* is for metastable, as explained in Chapter 14) has a half-life of about 6 hours before emitting a γ ray to form technetium-99.

Because γ rays are energy, their emission does not affect either the mass number or the atomic number of the daughter nuclide. Gamma-ray emission is therefore the only kind of radiation that does not necessarily involve the conversion of one element to another, although it is almost always observed in conjunction with some other nuclear decay reaction.

Spontaneous Fission

Only very massive nuclei with high neutron-to-proton ratios can undergo **spontaneous fission**, in which the nucleus breaks into two pieces that have different atomic numbers and atomic masses. This process is most important for the trans-actinide elements, with $Z \geq 104$. Spontaneous fission is invariably accompanied by the release of large amounts of energy, and it is usually accompanied by the emission of several neutrons as well. An example is the spontaneous fission of $^{254}_{98}\text{Cf}$, which gives a distribution of fission products; one possible set of products is shown in the equation

$$^{254}_{98}\text{Cf} \longrightarrow {}^{118}_{46}\text{Pd} + {}^{132}_{52}\text{Te} + 4{}^{1}_{0}\text{n} \tag{20.15}$$

Once again, the number of nucleons is conserved. Thus, the sum of the mass numbers of the products $(118 + 132 + 4 = 254)$ equals the mass number of the reactant. Similarly, the sum of the atomic numbers of the products $[46 + 52 + (4 \times 0) = 98]$ is the same as the atomic number of the parent nuclide.

Note the pattern

The total number of nucleons is conserved in all nuclear reactions.

EXAMPLE 20.3

Write a balanced equation to describe each nuclear reaction: **(a)** the beta decay of $^{35}_{16}S$; **(b)** the decay of $^{201}_{80}Hg$ by electron capture; **(c)** the decay of $^{30}_{15}P$ by positron emission.

Given Radioactive nuclide and mode of decay

Asked for Balanced equation

Strategy

A Identify the reactants and products from the information given.
B Use the values of A and Z to identify any missing components needed to balance the equation.

Solution

(a) **A✓** We know the identities of the reactant and one of the products (a β particle). We can therefore begin by writing an equation that shows the reactant and one of the products and indicates the unknown product as $^{A}_{Z}X$:

$$^{35}_{16}S \longrightarrow {}^{A}_{Z}X + {}^{\ 0}_{-1}\beta$$

B✓ Because both protons and neutrons must be conserved in a nuclear reaction, the unknown product must have a mass number of $A = 35 - 0 = 35$ and an atomic number of $Z = 16 - (-1) = 17$. The element with $Z = 17$ is chlorine, so the balanced equation for the nuclear reaction is

$$^{35}_{16}S \longrightarrow {}^{35}_{17}Cl + {}^{\ 0}_{-1}\beta$$

(b) **A✓** We know the identities of both reactants: $^{201}_{80}Hg$ and an inner electron, $^{\ 0}_{-1}e$. The reaction is

$$^{201}_{80}Hg + {}^{\ 0}_{-1}e \longrightarrow {}^{A}_{Z}X$$

B✓ Both protons and neutrons are conserved, so the mass number of the product must be $A = 201 + 0 = 201$, and the atomic number of the product must be $Z = 80 + (-1) = 79$, which corresponds to the element gold. The balanced equation for the reaction is thus

$$^{201}_{80}Hg + {}^{\ 0}_{-1}e \longrightarrow {}^{201}_{79}Au$$

(c) **A✓** As in part (a), we are given the identities of the reactant and one of the products—in this case, a positron. The unbalanced equation is therefore

$$^{30}_{15}P \longrightarrow {}^{A}_{Z}X + {}^{\ 0}_{+1}\beta$$

B✓ The mass number of the second product is $A = 30 - 0 = 30$, and its atomic number is $Z = 15 - 1 = 14$, which corresponds to silicon. The balanced equation for the reaction is

$$^{30}_{15}P \longrightarrow {}^{30}_{14}Si + {}^{\ 0}_{+1}\beta$$

EXERCISE 20.3

Write a balanced equation for each nuclear reaction: **(a)** the decay of $^{11}_{6}C$ by positron emission; **(b)** the beta decay of molybdenum-99; **(c)** the emission of an α particle followed by gamma emission from $^{185}_{74}W$.

Answer **(a)** $^{11}_{6}C \longrightarrow {}^{11}_{5}B + {}^{\ 0}_{+1}\beta$; **(b)** $^{99}_{42}Mo \longrightarrow {}^{99m}_{43}Tc + {}^{\ 0}_{-1}\beta$;
(c) $^{185}_{74}W \longrightarrow {}^{181}_{72}Hf + {}^{4}_{2}\alpha + {}^{0}_{0}\gamma$

EXAMPLE 20.4

Predict the kind of nuclear change each unstable nuclide undergoes when it decays: **(a)** $^{45}_{22}Ti$; **(b)** $^{242}_{94}Pu$; **(c)** $^{12}_{5}B$; **(d)** $^{256}_{100}Fm$.

Given Nuclide

Asked for Type of nuclear decay

Strategy

Based on the neutron-to-proton ratio and the value of Z, predict the type of nuclear decay reaction that will produce a more stable nuclide.

Solution

(a) This nuclide has a neutron-to-proton ratio of only 1.05, much lower than the requirement for stability for an element with an atomic number in this range. Nuclei that have low neutron-to-proton ratios decay by converting a proton to a neutron. The two possibilities are positron emission, which converts a proton to a neutron and a positron, and electron capture, which converts a proton and a core electron to a neutron. In this case, both are observed, with positron emission occurring about 86% of the time and electron capture about 14% of the time.

(b) Nuclei with $Z > 83$ are too heavy to be stable and usually undergo alpha decay, which decreases both the mass number and the atomic number. Thus, $^{242}_{94}Pu$ is expected to decay by alpha emission.

(c) This nuclide has a neutron-to-proton ratio of 1.4, which is very high for a light element. Nuclei with high neutron-to-proton ratios decay by converting a neutron to a proton and an electron. The electron is emitted as a β particle, and the proton remains in the nucleus, causing an increase in the atomic number with no change in the mass number. We therefore predict that $^{12}_{5}B$ will undergo beta decay.

(d) This is a massive nuclide, with an atomic number of 100 and a mass number much higher than 200. Nuclides with $A \geq 200$ tend to decay by alpha emission, and even heavier nuclei tend to undergo spontaneous fission. We therefore predict that $^{256}_{100}Fm$ will decay by either or both of these two processes. In fact, it decays by both spontaneous fission and alpha emission, in a 97:3 ratio.

EXERCISE 20.4

Predict the most probable nuclear decay reaction each nuclide undergoes: **(a)** $^{32}_{14}Si$; **(b)** $^{43}_{21}Sc$; **(c)** $^{231}_{91}Pa$.

Answer **(a)** beta decay; **(b)** positron emission or electron capture; **(c)** alpha decay

Radioactive Decay Series

MGC The Uranium Decay Series

The nuclei of all elements with atomic numbers greater than 83 are unstable. Thus, all isotopes of all elements beyond bismuth in the periodic table are radioactive. Because alpha decay decreases Z by only 2, and positron emission or electron capture decreases Z by only 1, it is impossible for any nuclide with $Z > 85$ to decay to a stable daughter nuclide in a single step, except via nuclear fission. Consequently, radioactive isotopes with $Z > 85$ usually decay to a daughter nucleus that is radioactive, which in turn decays to a second radioactive daughter nucleus, and so forth, until a stable nucleus finally results. This series of sequential alpha- and beta-decay reactions is called a **radioactive decay series**. The most common is the uranium-238 decay series, which produces lead-206 in a series of 14 sequential alpha- and beta-decay reactions (Figure 20.4). Although a radioactive decay series can be written for almost any isotope with $Z > 85$, only two others occur naturally: the decay of uranium-235 to lead-207

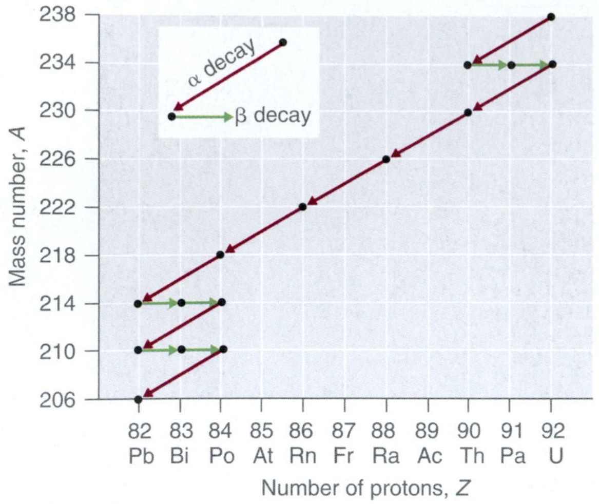

Figure 20.4 A radioactive decay series. The uranium-238 decay series consists of 14 sequential alpha- and beta-decay reactions that eventually produce lead-206. Two other naturally occurring radioactive decay series are known: the decay of uranium-235 to lead-207, which requires 11 steps, and the decay of thorium-232 to lead-208, which takes 10 steps.

(in 11 steps) and thorium-232 to lead-208 (in 10 steps). A fourth series, the decay of neptunium-237 to bismuth-209 in 11 steps, is known to have occurred on the primitive Earth. With a half-life of "only" 2.14 million years, all the neptunium-237 present when the Earth was formed decayed long ago, and today all the neptunium on Earth is synthetic.

Due to these radioactive decay series, small amounts of very unstable isotopes are found in ores that contain uranium or thorium. These rare, unstable isotopes should have decayed long ago to stable nuclei with lower atomic number, and they would no longer be found on Earth. Because they are generated continuously by the decay of uranium or thorium, however, their amounts have reached a steady state, in which their rate of formation is equal to their rate of decay. In some cases, the abundance of the daughter isotopes can be used to date a material or identify its origin, as described in Chapter 14.

Induced Nuclear Reactions

The discovery of radioactivity in the late 19th century showed that some nuclei spontaneously transform into nuclei with a different number of protons, thereby producing a different element. When scientists realized that these naturally occurring radioactive isotopes decayed by emitting subatomic particles, they realized that in principle it should be possible to carry out the reverse reaction, converting a stable nucleus to another more massive nucleus by bombarding it with subatomic particles in a nuclear transmutation reaction.

The first successful nuclear transmutation reaction was carried out in 1919 by Ernest Rutherford, who showed that α particles emitted by radium could react with nitrogen nuclei to form oxygen nuclei. As shown in this equation, a proton is emitted in the process:

$$^{4}_{2}\alpha + ^{14}_{7}\mathrm{N} \longrightarrow ^{17}_{8}\mathrm{O} + ^{1}_{1}\mathrm{p} \qquad (20.16)$$

Rutherford's nuclear transmutation experiments led to the discovery of the neutron. He found that bombarding the nucleus of a light target element with an α particle usually converted the target nucleus to a product that had an atomic number higher by 1 and a mass number higher by 3 than the target nucleus. Such behavior is consistent with the emission of a proton after reaction with the α particle. Very light targets such as Li, Be, and B reacted differently, however, emitting a new kind of highly penetrating radiation rather than a proton. Because these high-energy particles could not be deflected by a magnetic or electrical field, Rutherford concluded that they were electrically neutral (see Chapter 1). Other observations suggested that the mass of the neutral particle was similar to the mass of the proton. In 1932, James Chadwick (Nobel Prize, 1935), who was a student of Rutherford's at the time, named these neutral particles *neutrons* and proposed that they were fundamental building blocks of the atom. The reaction used initially by Chadwick to explain the production of neutrons was

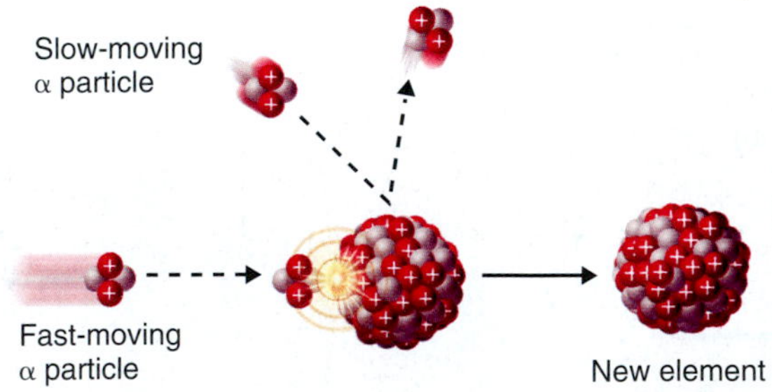

Figure 20.5 A nuclear transmutation reaction. Bombarding a target of one element with high-energy nuclei or subatomic particles can create new elements. Electrostatic repulsions normally prevent a positively charged particle from colliding and reacting with a positively charged nucleus. If the positively charged particle is moving at a very high speed, however, its kinetic energy may be great enough to overcome the electrostatic repulsions and it may collide with the target nucleus. Such collisions can result in a nuclear transmutation reaction.

$$^{4}_{2}\alpha + ^{9}_{4}\mathrm{Be} \longrightarrow ^{12}_{6}\mathrm{C} + ^{1}_{0}\mathrm{n} \qquad (20.17)$$

Because α particles and atomic nuclei are both positively charged, electrostatic forces cause them to repel each other. Only α particles with very high kinetic energy can overcome this repulsion and collide with a nucleus (Figure 20.5). Because neutrons have no electrical charge, however, they are not repelled by the nucleus. Hence, bombardment with neutrons is a much easier way to prepare new isotopes of the lighter elements. In fact, carbon-14 is formed naturally in the atmosphere by the bombardment of nitrogen-14 with neutrons generated by cosmic rays:

$$^{1}_{0}\mathrm{n} + ^{14}_{7}\mathrm{N} \longrightarrow ^{14}_{6}\mathrm{C} + ^{1}_{1}\mathrm{p} \qquad (20.18)$$

EXAMPLE 20.5

In 1933, Frédéric Joliot and Iréne Joliot-Curie (daughter of Marie and Pierre Curie) prepared the first artificial radioactive isotope by bombarding aluminum-27 with α particles. For each ^{27}Al that reacted one neutron was released. Identify the product nuclide, and write a balanced equation for this nuclear transmutation reaction.

Given Reactants in nuclear transmutation reaction

Asked for Product nuclide, balanced equation

Strategy

A Based on the reactants and one product, identify the other product of the reaction. Use conservation of mass and charge to determine the values of Z and A of the product nuclide and thus its identity.

B Write the balanced equation for the nuclear reaction.

Solution

A Bombarding an element with α particles usually produces an element with an atomic number that is 2 greater than the atomic number of the target nucleus. Thus, we expect that aluminum ($Z = 13$) will be converted to phosphorus ($Z = 15$). With one neutron released, conservation of mass requires that the mass number of the other product be 3 greater than the mass number of the target. In this case, the mass number of the target is 27, so the mass number of the product will be 30. The second product is therefore phosphorus-30, $^{30}_{15}P$. **B** The balanced equation for the nuclear reaction is

$$^{27}_{13}Al + ^{4}_{2}\alpha \longrightarrow ^{30}_{15}P + ^{1}_{0}n$$

EXERCISE 20.5

Because all isotopes of technetium are radioactive and have short half-lives, it does not exist in nature. Technetium can, however, be prepared by nuclear transmutation reactions. For example, bombarding a molybdenum-96 target with deuterium nuclei ($^{2}_{1}H$) produces technetium-97. Identify the other product of the reaction, and write a balanced equation for this transmutation reaction.

Answer neutron, $^{1}_{0}n$; $^{96}_{42}Mo + ^{2}_{1}H \longrightarrow ^{97}_{43}Tc + ^{1}_{0}n$

We noted in Section 20.2 that very heavy nuclides, corresponding to $Z \geq 104$, tend to decay by spontaneous fission. Nuclides with slightly lower values of Z, such as the isotopes of uranium ($Z = 92$) and plutonium ($Z = 94$), do not undergo spontaneous fission at any significant rate. Some isotopes of these elements, however, such as $^{235}_{92}U$ and $^{239}_{94}Pu$ undergo *induced nuclear fission* when they are bombarded with relatively low-energy neutrons, as shown in this equation for uranium-235 and in Figure 20.6:

$$^{235}_{92}U + ^{1}_{0}n \longrightarrow ^{236}_{92}U \longrightarrow ^{141}_{56}Ba + ^{92}_{36}Kr + 3^{1}_{0}n \qquad (20.19)$$

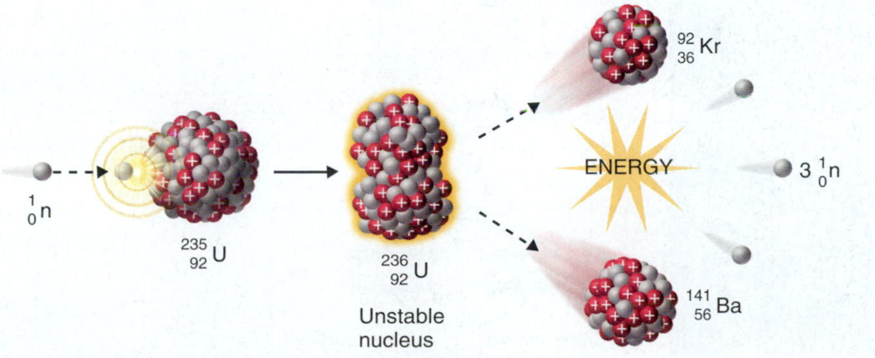

Figure 20.6 Neutron-induced nuclear fission. Collision of a relatively slow-moving neutron with a fissile nucleus can split it into two smaller nuclei with the same or different masses, as shown here. Neutrons are also released in the process, along with a great deal of energy.

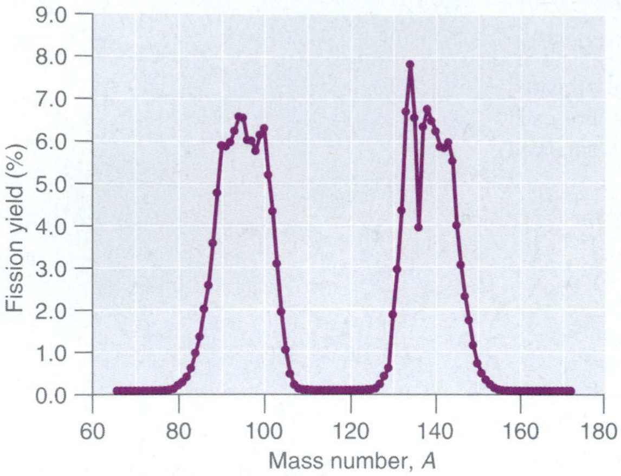

Figure 20.7 Mass distribution of nuclear fission products of ^{235}U. Nuclear fission usually produces a range of products with different masses and yields, although the mass ratio of each pair of fission products from a fission event is approximately 3:2. As shown in this plot, more than 50 different fission products are known for ^{235}U. [Data source: T. R. England and B. F. Rider, Los Alamos National Laboratory, ENDF-349 (1994).]

Any isotope that can undergo a nuclear fission reaction when bombarded with neutrons is called a *fissile isotope*.

During nuclear fission, the nucleus usually divides asymmetrically rather than into two equal parts, as shown in Figure 20.6. Moreover, every fission event of a given nuclide does not give the same products; more than 50 different fission modes have been identified for uranium-235, for example. Consequently, nuclear fission of a fissile nuclide can never be described by a single equation. Instead, as shown in Figure 20.7, a distribution of many pairs of fission products with different yields is obtained, but the mass ratio of each pair of fission products produced by a single fission event is always roughly 3:2.

Synthesis of Transuranium Elements

Uranium ($Z = 92$) is the heaviest naturally occurring element. Consequently, all the elements with $Z > 92$, the **transuranium elements**, are artificial and have been prepared by bombardment of suitable target nuclei with smaller particles. The first of the transuranium elements to be prepared was neptunium ($Z = 93$), which was synthesized in 1940 by bombarding a ^{238}U target with neutrons. As shown in Equation 20.20, this reaction occurs in two steps. Initially, a neutron combines with a ^{238}U nucleus to form ^{239}U, which is unstable and undergoes beta decay to produce ^{239}Np:

$$^{238}_{92}U + ^{1}_{0}n \longrightarrow ^{239}_{92}U \longrightarrow ^{239}_{93}Np + ^{0}_{-1}\beta \qquad (20.20)$$

Subsequent beta decay of ^{239}Np produces the second transuranium element, plutonium ($Z = 94$):

$$^{239}_{93}Np \longrightarrow ^{239}_{94}Pu + ^{0}_{-1}\beta \qquad (20.21)$$

TABLE 20.3 Some reactions used to synthesize transuranium elements

$^{239}_{94}Pu + ^{4}_{2}\alpha \longrightarrow ^{242}_{96}Cm + ^{1}_{0}n$
$^{239}_{94}Pu + ^{4}_{2}\alpha \longrightarrow ^{241}_{95}Am + ^{1}_{1}p + ^{1}_{0}n$
$^{242}_{96}Cm + ^{4}_{2}\alpha \longrightarrow ^{243}_{97}Bk + ^{1}_{1}p + 2^{1}_{0}n$
$^{253}_{99}Es + ^{4}_{2}\alpha \longrightarrow ^{256}_{101}Md + ^{1}_{0}n$
$^{238}_{92}U + ^{12}_{6}C \longrightarrow ^{246}_{98}Cf + 4^{1}_{0}n$
$^{252}_{98}Cf + ^{10}_{5}B \longrightarrow ^{256}_{103}Lr + 6^{1}_{0}n$

Bombarding the target with more massive nuclei creates elements that have atomic numbers significantly greater than that of the target nucleus (Table 20.3). Such techniques have resulted in creation of the super-heavy elements 114 and 116, both of which lie in or near the "island of stability" discussed in Section 20.1.

Accelerating positively charged particles to the speeds needed to overcome the electrostatic repulsions between them and the target nuclei requires a device called a *particle accelerator*, which uses electrical and magnetic fields to accelerate the particles. Operationally, the simplest particle accelerator is the linear accelerator (Figure 20.8), in which

Figure 20.8 A linear particle accelerator. (a) An aerial view of the Stanford Linear Accelerator (SLAC), the longest linear particle accelerator in the world; the overall length of the tunnel is 2 miles. (b) Rapidly reversing the polarity of the electrodes in the tube causes the charged particles to be alternately attracted as they enter one section of the tube and repelled as they leave that section. As a result, the particles are continuously accelerated along the length of the tube.

(a)

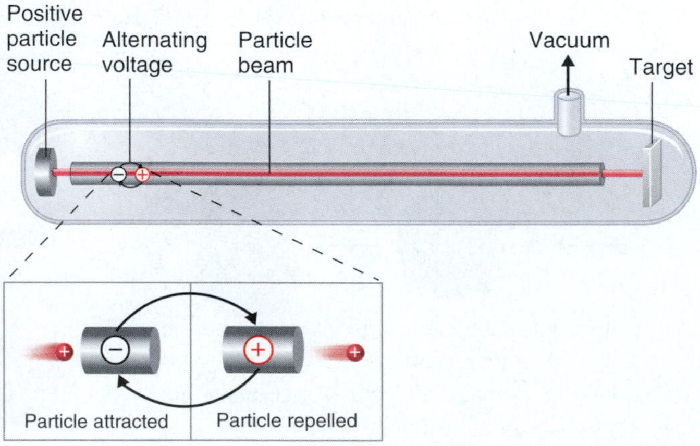

(b)

a beam of particles is injected at one end of a long evacuated tube. Rapid alternation of the polarity of the electrodes along the tube causes the particles to be alternately accelerated toward a region of opposite charge and repelled by a region with the same charge, resulting in a tremendous acceleration as the particle travels down the tube. A modern linear accelerator such as SLAC at Stanford University may be 2 miles long.

To achieve the same outcome in less space, a particle accelerator called a *cyclotron* forces the charged particles to travel in a circular path rather than a linear one. The particles are injected into the center of a ring and accelerated by rapidly alternating the polarity of two large D-shaped electrodes above and below the ring, which accelerates the particles outward along a spiral path toward the target.

Figure 20.9 A synchrotron. An aerial photograph of what is currently the world's most powerful particle accelerator, the Tevatron at the Fermi National Accelerator Laboratory in Illinois. The large tube characteristic of a synchrotron is 4 miles in circumference, contains 1000 superconducting magnets cooled by liquid helium, and can accelerate a beam of protons to almost the speed of light, giving them an energy greater than 1 TeV (teraelectronvolt = 10^3 GeV = 10^{12} eV) for collisions with other particles.

The length of a linear accelerator and the size of the D-shaped electrodes in a cyclotron severely limit the kinetic energy that particles can attain in these device. These limitations can be overcome by using a *synchrotron*, a hybrid of the two designs. A synchrotron contains an evacuated tube similar to that of a linear accelerator, but the tube is circular and can be more than a mile in diameter (Figure 20.9). Charged particles are accelerated around the circle by a series of magnets whose polarities rapidly alternate.

20.3 The Interaction of Nuclear Radiation with Matter

Because nuclear reactions do not typically affect the valence electrons of the atom (although electron capture draws an electron from an orbital of the lowest energy level), they do not cause chemical changes directly. Nonetheless, the particles and photons emitted during nuclear decay are very energetic, and they can indirectly produce chemical changes in the matter surrounding the nucleus that has decayed. For instance, an α particle is an ionized helium nucleus (He^{2+}), which can act as a powerful oxidant. In this section, we describe how radiation interacts with matter and the some of the chemical and biological effects of radiation.

Ionizing Versus Nonionizing Radiation

The effects of radiation on matter are determined primarily by the energy of the radiation, which depends on the nuclear decay reaction that produced it. **Nonionizing radiation** is relatively low in energy, and when it collides with an atom in a molecule or an ion, most or all of its energy can be absorbed without causing a structural or chemical change. Instead, the kinetic energy of the radiation is transferred to the atom or molecule with which it collides, causing it to rotate, vibrate, or move more rapidly. Because this energy can be transferred to adjacent molecules or ions in the form of heat, many radioactive substances are warm to the touch. Highly radioactive elements such as polonium, for example, have been used as heat sources in the U.S. space program. As long as the intensity of the nonionizing radiation is not great enough to cause overheating, it is relatively harmless, and its effects can be neutralized by cooling.

Figure 20.10 Radiation damage. When high-energy particles emitted by radioactive decay interact with matter, they can break bonds or ionize molecules, resulting in changes in physical properties such as ductility or color. The glass electrical insulator on the left has not been exposed to radiation, but the insulator on the right has received intense radiation doses over a long period of time. Radiation damage changed the chemical structure of the glass, causing it to become bright blue.

In contrast, **ionizing radiation** is higher in energy, and some of its energy can be transferred to one or more atoms with which it collides as it passes through matter. If enough energy is transferred, electrons can be excited to very high energy levels, resulting in the formation of positively charged ions:

$$\text{atom} + \text{ionizing radiation} \longrightarrow \text{ion}^+ + e^- \qquad (20.22)$$

Molecules that have been ionized in this way are often highly reactive, and they can decompose or undergo other chemical changes that create a cascade of reactive molecules that can damage biological tissues and other materials (Figure 20.10). Because the energy of ionizing radiation is very high, we often report its energy in units such as *megaelectronvolts (MeV)* per particle: 1 MeV/particle = 96 *billion* J/mol!

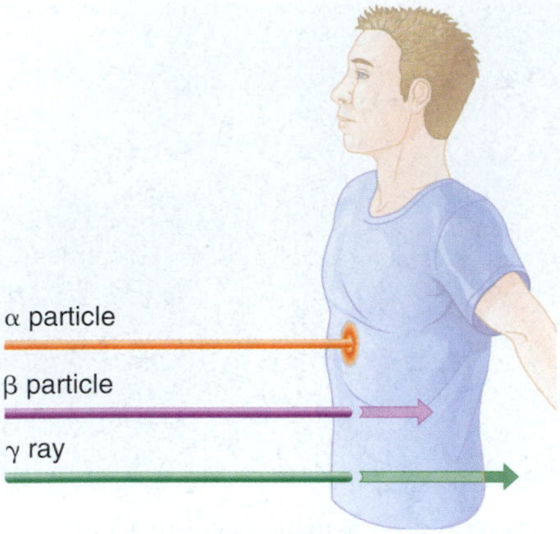

Figure 20.11 Depth of penetration of ionizing radiation. The depth of penetration of alpha, beta, and gamma radiation varies with the particle. Because α particles interact strongly with matter, they do not penetrate deeply into the human body. In contrast, β particles do not interact as strongly with matter and penetrate more deeply. Gamma rays, which have no charge, are stopped by only very dense materials and can pass right through the human body without being absorbed.

The Effects of Ionizing Radiation on Matter

The effects of ionizing radiation depend on four factors:

1. The type of radiation, which dictates how far it can penetrate into matter
2. The energy of the individual particles or photons
3. The number of particles or photons that strike a given area per unit time
4. The chemical nature of the substance exposed to the radiation

The relative abilities of the various forms of ionizing radiation to penetrate biological tissues are illustrated in Figure 20.11. Because of its high charge and mass, α radiation interacts strongly with matter. Consequently, it does not penetrate deeply into an object, and it can be stopped by a piece of paper, clothing, or skin. In contrast, γ rays, with no charge and essentially no mass, do not interact strongly with matter and penetrate deeply into most objects, including the human body. Several inches of lead or more than 12 inches of special concrete is needed to completely stop γ rays. Because β particles are intermediate in mass and charge between α particles and γ rays, their interaction with matter is also intermediate. Beta particles readily penetrate paper or skin, but they can be stopped by a piece of wood or a relatively thin sheet of metal.

Because of their great penetrating ability, γ rays are by far the most dangerous type of radiation when they come from a source *outside* the body. Alpha particles, however, are the most damaging if their source is *inside* the body because all of their energy is absorbed by internal tissues. Recognizing that the danger of radiation depends strongly on its nature and the kind of exposure allows scientists to safely handle many radioactive materials, as long as they take precautions to avoid, for example, inhaling fine particulate dust that contains alpha emitters. Some properties of ionizing radiation are summarized in Table 20.4.

There are many different ways to measure radiation exposure, or the *dose*. The **roentgen (R)**, which is used to measure the amount of energy absorbed by dry air, can be used to describe exposure quantitatively.* Damage to biological tissues, however, is proportional to the amount of energy absorbed by *tissues*, not air. The most common unit used to measure the effects of radiation on biological tissue is the **rad** (radiation absorbed dose); the SI equivalent is the gray (Gy). The *rad* is defined as the amount of radiation that causes 0.01 J of energy to be absorbed by 1 kg of matter, and the *gray* is defined as the amount of radiation that causes 1 J of energy to be absorbed per kilogram:

$$1 \text{ rad} = 0.010 \text{ J/kg} \qquad 1 \text{ Gy} = 1 \text{ J/kg} \tag{20.23}$$

TABLE 20.4 Some properties of ionizing radiation

Type	Energy Range, MeV	Penetration Distance in Water[a]	Penetration Distance in Air[a]
α particles	3–9	<0.05 mm	<10 cm
β particles	≤ 3	<4 mm	1 m
X rays	$<10^{-2}$	<1 cm	<3 m
γ rays	10^{-2}–10^{1}	<20 cm	>3 m

[a]Distance at which half of the radiation has been absorbed.

William Röentgen. Born in the Lower Rhine Province of Germany, Röentgen was the only child of a cloth manufacturer and merchant. His family moved to the Netherlands where he showed no particular aptitude in school, but where he was fond of roaming the countryside. Röentgen was expelled from technical school in Utrecht after being unjustly accused of drawing a caricature of one of the teachers. He began studying mechanical engineering in Zurich, which he could enter without having the credentials of a regular student, and received a Ph.D. at the University of Zurich in 1869. In 1876 he became Professor of Physics at Strasborg University.

* Named after the German physicist Wilhelm Röentgen (1845–1923; Nobel Prize, 1901), who discovered X rays. The roentgen is actually defined as the amount of radiation needed to produce an electrical charge of 2.58×10^{-4} C in 1 kg of dry air.

Thus, a 70-kg human who receives a dose of 1.0 rad over his or her entire body absorbs 0.010 J/70 kg = 1.4×10^{-4} J, or 0.14 mJ. To put this in perspective, 0.14 mJ is the amount of energy transferred to your skin by a 3.8×10^{-5} g droplet of boiling water. Because the energy of the droplet of water is transferred to a relatively large area of tissue, it is harmless. A radioactive particle, however, transfers its energy to a single molecule which makes it the atomic equivalent of a bullet fired from a high-powered rifle.

Because α particles have a much higher mass and charge than β particles or γ rays, the difference in mass between α and β particles is analogous to being hit by a bowling ball instead of a Ping-Pong ball traveling at the same speed. Thus, the amount of tissue damage caused by 1 rad of α particles is much greater than the damage caused by 1 rad of β particles or γ rays. Thus, a unit called the **rem** (roentgen equivalent in man) was devised to describe the *actual* amount of tissue damage caused by a given amount of radiation. The number of rems of radiation is equal to the number of rads multiplied by the *RBE* (relative biological effectiveness) factor, which is 1 for β particles, γ rays, and X rays, and about 20 for α particles. Because actual radiation doses tend to be very small, most measurements are reported in *millirems* (1 mrem = 10^{-3} rem).

Note the pattern

An alpha emitter is the least dangerous type of radiation, unless it is ingested, inhaled, or injected.

Natural Sources of Radiation

We are continuously exposed to measurable background radiation from a variety of natural sources, which on average is equal to about 150–600 mrem/yr (Figure 20.12). One component of background radiation is *cosmic rays*, high-energy particles and γ rays emitted by the sun and other stars that bombard the Earth continuously. Because cosmic rays are partially absorbed by the atmosphere before they reach Earth's surface, the exposure of people living at sea level (about 30 mrem/yr) is significantly lower than the exposure of people living at higher altitudes (about 50 mrem/yr in Denver, Colorado). Every 4 hours spent in an airplane at higher than 30,000 ft adds about 1 mrem to a person's annual radiation exposure.

A second component of background radiation is *cosmogenic radiation*, produced by the interaction of cosmic rays with gases in the upper atmosphere. When high-energy cosmic rays collide with oxygen and nitrogen atoms, neutrons and protons are released. These, in turn, react with other atoms to produce radioactive isotopes such as ^{14}C:

$$^{14}_{7}N + ^{1}_{0}n \longrightarrow ^{14}_{6}C + ^{1}_{1}p \tag{20.24}$$

The carbon atoms react with oxygen atoms to form CO_2, which is eventually washed to Earth's surface in rain and taken up by plants. About 1 atom in 1×10^{12} of the carbon atoms in our bodies is radioactive ^{14}C, which decays by beta emission. About 10,000 ^{14}C nuclei disintegrated in your body during the 30 seconds or so that it took you to read this paragraph. Tritium (^3H) is also produced in the upper atmosphere and falls to Earth in precipitation. The total radiation dose attributable to ^{14}C is estimated to be 1 mrem/yr, while that due to ^3H is about a thousand times less.

The third major component of background radiation is *terrestrial radiation*, which is due to the remnants of radioactive elements that were present on the primordial Earth and their decay products. For example, many of the rocks and minerals in the soil contain small amounts of radioactive isotopes such as ^{232}Th and ^{238}U, as well as radioactive daughter isotopes such as ^{226}Ra. The amount of background radiation from these sources is about the same as that from cosmic rays (~30 mrem/yr). These isotopes are also found in small amounts in building materials derived from rocks and minerals, which significantly increases the radiation exposure for people who live in brick or concrete-block houses (60–160 mrem/yr) instead of houses made of wood (10–20 mrem/yr).

Our tissues also absorb radiation (about 40 mrem/yr) from naturally occurring radioactive elements that are present in our bodies. For example, the average adult contains about 140 g of potassium as the K^+ ion. Naturally occurring potassium

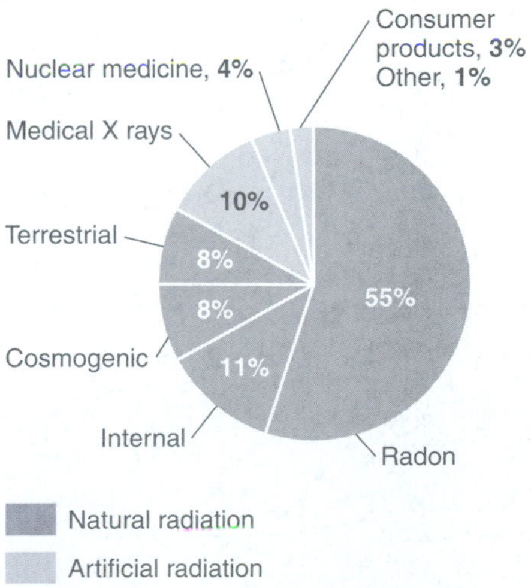

Figure 20.12 The radiation exposure of a typical adult in the United States. The average radiation dose from natural sources for an adult in the United States is about 150–600 mrem/yr. Radon accounts for more than half of an adult's total radiation exposure, whereas background radiation (terrestrial and cosmogenic) and exposure from medical sources account for about 15% each. [Data source: Office of Civilian Radioactive Waste Management.]

contains 0.0117% ^{40}K, which decays by emitting both a β particle and a γ ray. In the last 20 seconds, about the time it took you to read this paragraph, approximately 40,000 ^{40}K nuclei disintegrated in your body.

By far the most important source of background radiation is *radon*, the heaviest of the noble gases (Group 18). Radon-222 is produced during the decay of ^{238}U, and other isotopes of radon are produced by the decay of other heavy elements. Even though radon is chemically inert, all its isotopes are radioactive. For example, ^{222}Rn undergoes two successive alpha-decay events to give ^{214}Pb:

$$^{222}_{86}\text{Rn} \longrightarrow {}^{4}_{2}\alpha + {}^{218}_{84}\text{Po} \longrightarrow {}^{4}_{2}\alpha + {}^{214}_{82}\text{Pb} \qquad (20.25)$$

Because radon is a dense gas, it tends to accumulate in enclosed spaces such as basements, especially in locations where the soil contains greater-than-average amounts of naturally occurring uranium minerals. Under most conditions, radioactive decay of radon poses no problems because of the very short range of the emitted α particle. If an atom of radon happens to be in your lungs when it decays, however, the chemically reactive daughter isotope polonium-218 can become irreversibly bound to molecules in the lung tissue. Subsequent decay of ^{218}Po releases an α particle directly into one of the cells lining the lung, and the resulting damage can eventually cause lung cancer. The ^{218}Po isotope is also readily absorbed by particles in cigarette smoke, which adhere to the surface of the lungs and can hold the radioactive isotope in place. Recent estimates suggest that radon exposure is a contributing factor in about 15% of deaths due to lung cancer. Because of the potential health problem radon poses, many states require houses to be tested for radon before they can be sold. By current estimates, radon accounts for more than half of the radiation exposure of a typical adult in the United States.

Artificial Sources of Radiation

In addition to naturally occurring background radiation, humans are exposed to small amounts of radiation from a variety of artificial sources. The most important of these are the X rays used for diagnostic purposes in medicine and dentistry, which are photons with much lower energy than γ rays. A single chest X ray provides a radiation dose of about 0.07 mrem, and a dental X ray about 0.5 mrem. Other minor sources include television screens and computer monitors with cathode-ray tubes, which also produce X rays. Luminescent paints for watch dials originally used radium, a highly toxic alpha emitter if ingested by those painting the dials. Radium was replaced by tritium (^{3}H) and promethium (^{147}Pr), which emit low-energy β particles that are absorbed by the watch crystal or the glass covering the instrument. Radiation exposure from television screens, monitors, and luminescent dials totals about 2 mrem/yr. Residual fallout from previous atmospheric nuclear-weapons testing is estimated to account for about twice this amount, and the nuclear power industry accounts for less than 1 mrem/yr (about the same as a single 4-hour jet flight).

EXAMPLE 20.6

Calculate the annual radiation dose in rads a typical 70-kg chemistry student receives from the naturally occurring ^{40}K in his or her body, which contains about 140 g of potassium (as the K^+ ion). The natural abundance of ^{40}K is 0.0117%. Each 1.00 mol of ^{40}K undergoes 1.05×10^{7} decays per second, and each decay event is accompanied by the emission of a 1.32-MeV β particle.

Given Mass of student, mass of isotope, natural abundance, rate of decay, energy of particle

Asked for Annual radiation dose in rads

Strategy

A Calculate the number of moles of ^{40}K present using its mass, molar mass, and natural abundance.
B Determine the number of decays per year for this amount of ^{40}K.
C Multiply the number of decays per year by the energy associated with each decay event. To obtain the annual radiation dose, use the mass of the student to convert this value to rads.

Solution

A The number of moles of ^{40}K present in the body is the total number of potassium atoms times the natural abundance of potassium atoms present as ^{40}K divided by the atomic mass of ^{40}K:

$$\text{moles } ^{40}\text{K} = 140\ \cancel{\text{g K}}\ \frac{0.0117\ \text{mol } ^{40}\text{K}}{100\ \cancel{\text{mol K}}} \times \frac{1\ \cancel{\text{mol K}}}{40.0\ \cancel{\text{g K}}} = 4.10 \times 10^{-4}\ \text{mol } ^{40}\text{K}$$

B We are given the number of atoms of ^{40}K that decay per second in 1.00 mol of ^{40}K, so the number of decays per year is

$$\frac{\text{decays}}{\text{year}} = 4.10 \times 10^{-4}\ \cancel{\text{mol } ^{40}\text{K}} \times \frac{1.05 \times 10^{7}\ \text{decays}/\cancel{\text{s}}}{1.00\ \cancel{\text{mol } ^{40}\text{K}}} \times \frac{60\ \cancel{\text{s}}}{1\ \cancel{\text{min}}} \times \frac{60\ \cancel{\text{min}}}{1\ \cancel{\text{h}}} \times \frac{24\ \cancel{\text{h}}}{1\ \cancel{\text{day}}} \times \frac{365\ \cancel{\text{days}}}{1\ \text{yr}}$$
$$= 1.36 \times 10^{11}\ \text{decays/yr}$$

C The total energy the body receives per year from the decay of ^{40}K is equal to the total number of decays per year multiplied by the energy associated with each decay event:

$$\text{total energy per year} = \frac{1.36 \times 10^{11}\ \cancel{\text{decays}}}{\text{yr}} \times \frac{1.32\ \cancel{\text{MeV}}}{\cancel{\text{decays}}} \times \frac{10^{6}\ \cancel{\text{eV}}}{\cancel{\text{MeV}}} \times \frac{1.602 \times 10^{-19}\ \text{J}}{\cancel{\text{eV}}} = 2.87 \times 10^{-2}\ \text{J/yr}$$

We use the definition of the rad (1 rad = 10^{-2} J/kg of tissue) to convert this figure to a radiation dose in rads. If we assume the dose is equally distributed throughout the body, then the radiation dose per year is

$$\text{radiation dose per year} = \frac{2.87 \times 10^{-2}\ \cancel{\text{J}}/\text{yr}}{70.0\ \cancel{\text{kg}}} \times \frac{1\ \text{rad}}{1 \times 10^{-2}\ \cancel{\text{J/kg}}} = 4.10 \times 10^{-2}\ \text{rad/yr} = 41\ \text{mrad/yr}$$

This corresponds to almost half of the normal background radiation most people experience.

EXERCISE 20.6

Because strontium is chemically similar to calcium, small amounts of the Sr^{2+} ion are taken up by the body and deposited in calcium-rich tissues such as bone, using the same mechanism that is responsible for the absorption of Ca^{2+}. Consequently, the radioactive strontium (^{90}Sr) found in fission waste and released by atmospheric nuclear-weapons testing is a major health concern. A normal 70-kg human contains about 280 mg of strontium, and each mole of ^{90}Sr undergoes 4.55×10^{14} decays per second by the emission of a 0.546-MeV β particle. What would be the annual radiation dose in rads for a 70-kg person if 0.10% of the strontium ingested were ^{90}Sr?

Answer 5.7×10^{3} rad/yr (10 times the fatal dose!)

Assessing the Impact of Radiation Exposure

One of the more controversial public policy issues debated today is whether the radiation exposure from artificial sources, when combined with the exposure from natural sources, poses a significant risk to human health. The effects of single radiation doses of different magnitudes on humans are listed in Table 20.5. Because of the many factors involved in radiation exposure (length of exposure, intensity of the source, energy and type of particle), it is difficult to quantify the specific dangers of one

TABLE 20.5 The effects of a single radiation dose on a 70-kg human

Dose, rem	Symptoms/Effects
<5	No observable effect
5–20	Possible chromosomal damage
20–100	Temporary reduction in white bloodcell count
50–100	Temporary sterility in men (up to a year)
100–200	Mild radiation sickness, vomiting, diarrhea, fatigue. Immune system suppressed and bone growth in children retarded
>300	Permanent sterility in women
>500	Fatal to 50% within 30 days. Destruction of bone marrow and intestine
>3000	Fatal within hours

radioisotope versus another. Nonetheless, some general conclusions regarding the effects of radiation exposure are generally accepted as valid.

Radiation doses of 600 rem and higher are invariably fatal, while a dose of 500 rem kills half the exposed subjects within 30 days. Smaller doses (≤50 rem) appear to cause only limited health effects, even though they correspond to tens of years of natural radiation. This does not, however, mean that such doses have no ill effects; they may cause long-term health problems such as cancer or genetic changes that affect offspring. The possible detrimental effects of the much smaller doses attributable to artificial sources (<100 mrem/yr) are more difficult to assess.

Note the pattern

A large dose of radiation spread out over time is less harmful than the same total amount of radiation administered over a short time.

The tissues most affected by large, whole-body exposures are bone marrow, intestinal tissue, hair follicles, and reproductive organs, all of which contain rapidly dividing cells. The susceptibility of rapidly dividing cells to radiation exposure explains why cancers are often treated by radiation: because cancer cells divide faster than normal cells, they are destroyed preferentially by radiation. Long-term radiation-exposure studies on fruit flies show a linear relationship between the number of genetic defects and both the magnitude of the dose and the exposure time. In contrast, similar studies on mice show a much lower number of defects when a given dose of radiation is spread out over a long period of time rather than received all at once. Both patterns are plotted in Figure 20.13, but which of the two is applicable to humans? According to one hypothesis, mice have very low risk from low doses because their bodies have ways of dealing with the damage caused by natural radiation. At much higher doses, however, their natural repair mechanisms are overwhelmed, leading to irreversible damage. Because mice are biochemically much more similar to humans than are fruit flies, many scientists believe that this model also applies to humans. In contrast, the linear model assumes that *all* exposure to radiation is intrinsically damaging, and suggests that stringent regulation of low-level radiation exposure is necessary. Which of these views is more accurate is still in question, and the answer has extremely important consequences for regulating radiation exposure.

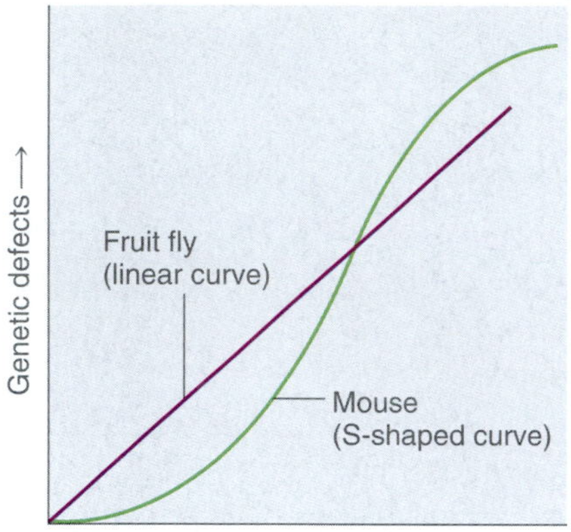

Figure 20.13 Two possible relationships between the number of genetic defects and radiation exposure. Studies on fruit flies show a linear relationship between the number of genetic defects and the magnitude of the radiation dose and exposure time, which is consistent with a cumulative effect of radiation. In contrast, studies on mice show an S-shaped curve, which suggests that the number of defects is lower when radiation exposure occurs over a longer time. Which of these relationships is more applicable to humans is a matter of considerable debate.

20.4 • Thermodynamic Stability of the Atomic Nucleus

Nuclear reactions, like chemical reactions, are accompanied by changes in energy. The energy changes in nuclear reactions, however, are enormous compared with those of even the most energetic chemical reactions. In fact, the energy changes in a typical nuclear reaction are so large that they result in a measurable change of mass. In this section, we describe the relationship between mass and energy in nuclear reactions, and show how the seemingly small changes in mass that accompany nuclear reactions result in the release of enormous amounts of energy.

Mass–Energy Balance

The relationship between mass m and energy E was introduced in Chapter 6 and is expressed in the equation

$$E = mc^2 \tag{20.26}$$

where c is the speed of light (2.998×10^8 m/s), and energy and mass are expressed in units of joules and kilograms, respectively. This relationship was first derived by Albert Einstein in 1905 as part of his special theory of relativity; it states that the mass of a particle is directly proportional to its energy. Thus, according to Equation 20.26, every mass has an associated energy, and similarly, any reaction that involves a change in energy must be accompanied by a change in mass. This implies that all exothermic reactions should be accompanied by a decrease in mass, and all endothermic reactions should be accompanied by an increase in mass. Given the law of conservation of mass (see Chapter 3), how can this be true? The solution to this apparent contradiction is that *chemical reactions are indeed accompanied by changes in mass*, but these changes are simply too small to be detected.*

For example, the chemical equation for the combustion of graphite to produce carbon dioxide is

$$\text{C (graphite)} + \frac{1}{2}\text{O}_2(g) \longrightarrow \text{CO}_2(g) \qquad \Delta H^\circ = -393.5 \text{ kJ/mol} \quad (20.27)$$

Combustion reactions are typically carried out at constant pressure, and under these conditions, the heat released or absorbed is equal to ΔH. As you learned in Chapter 18, when a reaction is carried out at constant volume, the heat released or absorbed is equal to ΔE. For most chemical reactions, however, $\Delta E \approx \Delta H$. If we rewrite Einstein's equation as

$$\Delta E = (\Delta m)c^2 \quad (20.28)$$

we can rearrange the equation to obtain the following relationship between the change in mass and the change in energy:

$$\Delta m = \frac{\Delta E}{c^2} \quad (20.29)$$

Because 1 J = 1 $(\text{kg} \cdot \text{m}^2)/\text{s}^2$, the change in mass is

$$\Delta m = \frac{-393.5 \text{ kJ/mol}}{(2.998 \times 10^8 \text{ m/s})^2} = \frac{-3.935 \times 10^5 \, (\text{kg} \cdot \cancel{\text{m}^2})/(\cancel{\text{s}^2} \cdot \text{mol})}{(2.998 \times 10^8 \, \cancel{\text{m/s}})^2} = -4.38 \times 10^{-12} \text{ kg/mol} \quad (20.30)$$

This is a mass change of about 3.6×10^{-10} g/g carbon that is burned, or about one -hundred-millionths of the mass of an electron per atom of carbon. In practice, this mass change is much too small to be measured experimentally, and it is utterly negligible.

In contrast, a typical nuclear reaction, the radioactive decay of ^{14}C to ^{14}N and an electron (a β particle), produces a much larger change in mass:

$$^{14}\text{C} \longrightarrow {}^{14}\text{N} + {}^{0}_{-1}\beta \quad (20.31)$$

We can use the experimentally measured masses of subatomic particles and common isotopes given in Tables 20.1 and 20.6 to calculate the change in mass directly. The reaction involves the conversion of a neutral ^{14}C *atom* to a positively charged ^{14}N *ion* (with six, not seven, electrons) and a negatively charged β particle (an electron), so the mass of the products is identical to the mass of a neutral ^{14}N atom. The total change in mass during the reaction is therefore the difference between the mass of a neutral ^{14}N atom (14.003074 amu) and the mass of a ^{14}C atom (14.003242 amu):

$$\begin{aligned} \Delta m &= \text{mass}_{\text{products}} - \text{mass}_{\text{reactants}} \\ &= 14.003074 \text{ amu} - 14.003242 \text{ amu} = -0.000168 \text{ amu} \end{aligned} \quad (20.32)$$

The difference in mass, which has been released as energy, corresponds to almost one-third of an electron. The change in mass for the decay of 1 mol of ^{14}C is -0.000168 g $= -1.68 \times 10^{-4}$ g $= -1.68 \times 10^{-7}$ kg. Although a mass change of this magnitude may seem small, it is about a thousand times larger than the mass change for the combustion of graphite. The energy change is

$$\begin{aligned} \Delta E &= (\Delta m)c^2 = (-1.68 \times 10^{-7} \text{ kg})(2.998 \times 10^8 \text{ m/s})^2 \\ &= -1.51 \times 10^{10} \, (\text{kg} \cdot \text{m}^2)/\text{s}^2 = -1.51 \times 10^{10} \text{ J} = -1.51 \times 10^7 \text{ kJ} \end{aligned} \quad (20.33)$$

---

* This situation is similar to the wave–particle duality discussed in Chapter 6. As you may recall, all particles exhibit wavelike behavior, but the wavelength is inversely proportional to the mass of the particle (actually, to its momentum, the product of its mass and velocity). Consequently, wavelike behavior is detectable only for particles with very small masses, such as electrons.

TABLE 20.6 Experimentally measured masses of selected isotopes

Isotope	Mass, amu	Isotope	Mass, amu	Isotope	Mass, amu
^{1}H	1.007825	^{14}N	14.003074	^{208}Po	207.981246
^{2}H	2.014102	^{16}O	15.994915	^{210}Po	209.982874
^{3}H	3.016049	^{52}Cr	51.940508	^{222}Rn	222.017578
^{3}He	3.016029	^{56}Fe	55.934938	^{226}Ra	226.025410
^{4}He	4.002603	^{59}Co	58.933195	^{230}Th	230.033134
^{6}Li	6.015123	^{58}Ni	57.935343	^{234}Th	234.043601
^{7}Li	7.016005	^{60}Ni	59.930786	^{234}Pa	234.043308
^{9}Be	9.012182	^{90}Rb	89.914802	^{233}U	233.039635
^{10}B	10.012937	^{144}Cs	143.932077	^{234}U	234.040952
^{11}B	11.009305	^{206}Pb	205.974465	^{235}U	235.043930
^{12}C	12	^{207}Pb	206.975897	^{238}U	238.050788
^{13}C	13.003355	^{208}Pb	207.976652	^{239}Pu	239.052163
^{14}C	14.003242				

Data source: G. Audi, A. H. Wapstra, and C. Thibault, *The AME2003 atomic mass evaluation.*

The energy released in this nuclear reaction is more than 100,000 times greater than that of a typical chemical reaction, even though the decay of ^{14}C is a relatively low-energy nuclear reaction.

Because the energy changes in nuclear reactions are so large, they are often expressed in kiloelectronvolts (1 keV = 10^3 eV), megaelectronvolts (1 MeV = 10^6 eV), and even gigaelectronvolts (1 GeV = 10^9 eV) *per atom or particle*. The change in energy that accompanies a nuclear reaction can be calculated from the change in mass using the relationship 1 amu = 931 MeV. The energy released by the decay of one atom of ^{14}C is thus

$$(-1.68 \times 10^{-4}\ \cancel{\text{amu}})(931\ \text{MeV}/\cancel{\text{amu}}) = -0.156\ \text{MeV} = -156\ \text{keV} \qquad (20.34)$$

EXAMPLE 20.7

Calculate the changes in mass (in amu) and energy (in J/mol and eV/atom) that accompany the radioactive decay of ^{238}U to ^{234}Th and an α particle. The α particle absorbs two electrons from the surrounding matter to form a helium atom.

Given Nuclear decay reaction

Asked for Changes in mass and energy

Strategy

A Use the mass values in Tables 20.1 and 20.6 to calculate the change in mass for the decay reaction in amu.
B Use Equation 20.29 to calculate the change in energy in J/mol.
C Use the relationship between amu and MeV to calculate the change in energy in eV/atom.

Solution

A Using particle and isotope masses from Tables 20.1 and 20.6, we can calculate the change in mass as follows:

$$\Delta m = \text{mass}_{\text{products}} - \text{mass}_{\text{reactants}} = (\text{mass }^{234}\text{Th} + \text{mass }^{4}_{2}\text{He}) - \text{mass }^{238}\text{U}$$
$$= (234.043601\ \text{amu} + 4.002603\ \text{amu}) - 238.050788\ \text{amu} = -0.004584\ \text{amu}$$

B Thus, the change in mass for 1 mol of ^{238}U is -0.004584 g or -4.584×10^{-6} kg. The change in energy in J/mol is:

$$\Delta E = (\Delta M)c^2 = (-4.584 \times 10^{-6}\ \text{kg})(2.998 \times 10^8\ \text{m/s})^2 = -4.120 \times 10^{11}\ \text{J/mol}$$

C The change in energy in eV/atom is:

$$\Delta E = -4.584 \times 10^{-3}\ \cancel{\text{amu}} \times \frac{931\ \cancel{\text{MeV}}}{\cancel{\text{amu}}} \times \frac{1 \times 10^6\ \text{eV}}{1\ \cancel{\text{MeV}}} = -4.27 \times 10^6\ \text{eV/atom}$$

EXERCISE 20.7

Calculate the changes in mass (in amu) and energy (in kJ/mol and keV/atom) that accompany the radioactive decay of tritium (^{3}H) to ^{3}He and a β particle.

Answer $\Delta m = -2.0 \times 10^{-5}$ amu; $\Delta E = -1.9 \times 10^6$ kJ/mol $= -19$ keV/atom

Nuclear Binding Energies

MGC Nuclear Binding Energy

We have seen that energy changes in both chemical and nuclear reactions are accompanied by changes in mass. Einstein's equation, which allows us to interconvert mass and energy, has another interesting consequence: The mass of an atom is always less than the sum of the masses of its component particles. The only exception to this rule is hydrogen-1 (^{1}H), whose measured mass of 1.007825 amu is identical to the sum of the masses of a proton and an electron. In contrast, the experimentally measured mass of an atom of deuterium (^{2}H) is 2.014102 amu, although its calculated mass is 2.016490 amu:

$$\begin{aligned} m_{^2\text{H}} &= m_{\text{neutron}} + m_{\text{proton}} + m_{\text{electron}} \\ &= 1.008665\ \text{amu} + 1.007276\ \text{amu} + 0.000549\ \text{amu} = 2.016490\ \text{amu} \end{aligned} \tag{20.35}$$

The difference between the sum of the masses of the components and the measured atomic mass is called the **mass defect** of the nucleus. Just as a molecule is more stable than its isolated atoms, a nucleus is more stable (lower in energy) than its isolated components. Consequently, when isolated nucleons assemble into a stable nucleus, energy is released. According to Equation 20.29, this release of energy must be accompanied by a decrease in the mass of the nucleus.

The amount of energy released when a nucleus forms from its component nucleons is the **nuclear binding energy** (Figure 20.14). In the case of deuterium, the mass defect is 0.002388 amu, which corresponds to a nuclear binding energy of 2.22 MeV for the deuterium nucleus. Because the magnitude of the mass defect is proportional to the nuclear binding energy, both values indicate the stability of the nucleus.

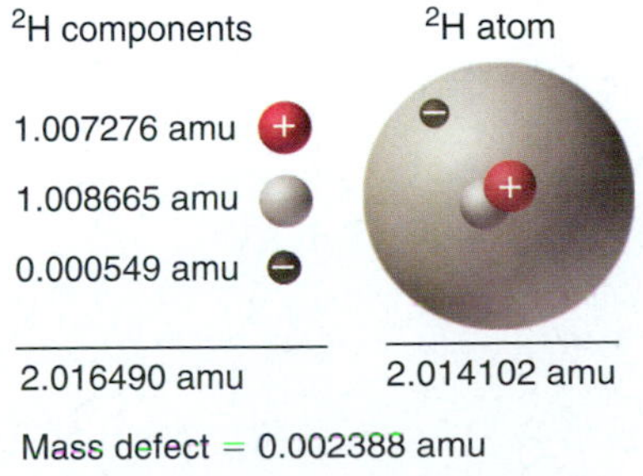

Figure 20.14 Nuclear binding energy in deuterium. The mass of a ^{2}H atom is less than the sum of the masses of a proton, a neutron, and an electron by 0.002388 amu; the difference in mass corresponds to the nuclear binding energy. The larger the value of the mass defect, the greater the nuclear binding energy and the more stable the nucleus.

Not all nuclei are equally stable. Chemists describe the *relative* stability of different nuclei by comparing the binding energy *per nucleon*, which is obtained by dividing the nuclear binding energy by the mass number A of the nucleus. As shown in Figure 20.15, the binding energy per nucleon increases rapidly with increasing atomic number until about $Z = 26$, where it levels off to about 8–9 MeV per nucleon and then decreases slowly. The initial increase in binding energy is not a smooth curve, but exhibits sharp peaks corresponding to the light nuclei that have equal numbers of protons and neutrons (for example, ^{4}He, ^{12}C, ^{16}O). As mentioned earlier, these are particularly stable combinations.

Because the maximum binding energy per nucleon is reached at ^{56}Fe, all other nuclei are thermodynamically unstable with regard to the formation of ^{56}Fe. Consequently, heavier nuclei (toward the right in Figure 20.15) should spontaneously

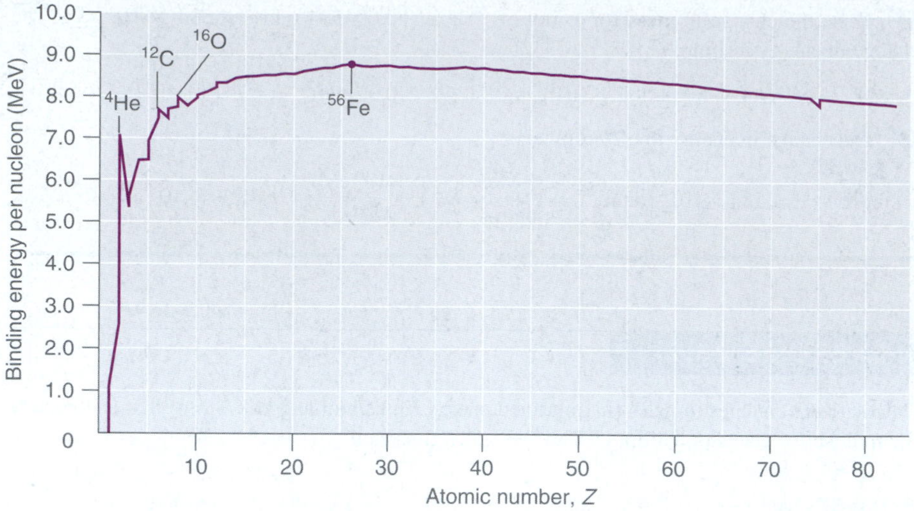

Figure 20.15 The curve of nuclear binding energy. This plot of the average binding energy per nucleon as a function of atomic number shows that the binding energy per nucleon increases with increasing atomic number until about $Z = 26$, levels off, and then decreases. The sharp peaks correspond to light nuclei that have equal numbers of protons and neutrons.

undergo reactions such as alpha decay, which result in a decrease in atomic number. Conversely, lighter elements (on the left in Figure 20.15) should spontaneously undergo reactions that result in an increase in atomic number. This is indeed the observed pattern.

Note the pattern

Heavier nuclei spontaneously undergo nuclear reactions that decrease their atomic number. Lighter nuclei spontaneously undergo nuclear reactions that increase their atomic number.

EXAMPLE 20.8

Calculate the total nuclear binding energy (in MeV) and the binding energy per nucleon for ^{56}Fe. The experimental mass of the nuclide is given in Table 20.6.

Given Nuclide and mass

Asked for Nuclear binding energy, binding energy per nucleon

Strategy

A Sum the masses of the protons, electrons, and neutrons or, alternatively, use the mass of the appropriate number of ^{1}H atoms (since its mass is the same as the mass of one electron and one proton).

B Calculate the mass defect by subtracting the experimental mass from the calculated mass.

C Determine the nuclear binding energy by multiplying the mass defect by the change in energy in electron volts per atom. Divide this value by the number of nucleons to obtain the binding energy per nucleon.

Solution

A An iron-56 atom contains 26 protons, 26 electrons, and 30 neutrons. We could add the masses of these three sets of particles, but noting that 26 protons and 26 electrons are equivalent to 26 ^{1}H atoms, we can calculate the sum of the masses more quickly as follows:

$$\begin{aligned} \text{calculated mass} &= 26(\text{mass } {}^1_1\text{H}) + 30(\text{mass } {}^1_0\text{n}) \\ &= 26(1.007825)\text{ amu} + 30(1.008665)\text{ amu} = 56.463400\text{ amu} \\ \text{experimental mass} &= 55.934938\text{ amu} \end{aligned}$$

B We subtract to find the mass defect:

$$\begin{aligned} \text{mass defect} &= \text{calculated mass} - \text{experimental mass} \\ &= 56.463400\text{ amu} - 55.934938\text{ amu} = 0.528462\text{ amu} \end{aligned}$$

C The nuclear binding energy is thus 0.528462 amu × 931 MeV/amu = 492 MeV. The binding energy per nucleon is 492 MeV/56 nucleons = 8.79 MeV/nucleon.

EXERCISE 20.8

Calculate the total nuclear binding energy (in MeV) and the binding energy per nucleon for ^{238}U.

Answer 1800 MeV/^{238}U; 7.57 MeV/nucleon

Nuclear Fission and Fusion

MGC Fission Reactions

As discussed in Section 20.2, **nuclear fission** is the splitting of a heavy nucleus into two lighter ones. Fission was discovered in 1938 by the German scientists Otto Hahn, Lise Meitner, and Fritz Strassmann, who bombarded a sample of uranium with neutrons in an attempt to produce new elements with $Z > 92$. They observed that lighter elements such as barium ($Z = 56$) were formed during the reaction, and they realized that such products had to originate from the neutron-induced fission of uranium-235:

$$^{235}_{92}U + ^{1}_{0}n \longrightarrow ^{141}_{56}Ba + ^{92}_{36}Kr + 3^{1}_{0}n \quad (20.36)$$

This hypothesis was confirmed by detection of the krypton-92 fission product. As discussed in Section 20.2, the nucleus usually divides asymmetrically rather than into two equal parts, and the fission of a given nuclide does not give the same products every time.

In a typical nuclear fission reaction, more than one neutron is released by each dividing nucleus. When these neutrons collide with and induce fission in other neighboring nuclei, a self-sustaining series of nuclear fission reactions known as a **nuclear chain reaction** can result (Figure 20.16). For example, the fission of ^{235}U releases two to three neutrons per fission event. If absorbed by other ^{235}U nuclei, those neutrons induce additional fission events, and the rate of the fission reaction increases geometrically. Each series of events is called a *generation*. Experimentally, it is found that some minimum mass of a fissile isotope is required to sustain a nuclear chain reaction; if the mass is too low, too many neutrons are able to escape without being captured and inducing a fission reaction. The minimum mass capable of supporting sustained fission is called the **critical mass**. This amount depends on the purity of the material and the shape of the mass,

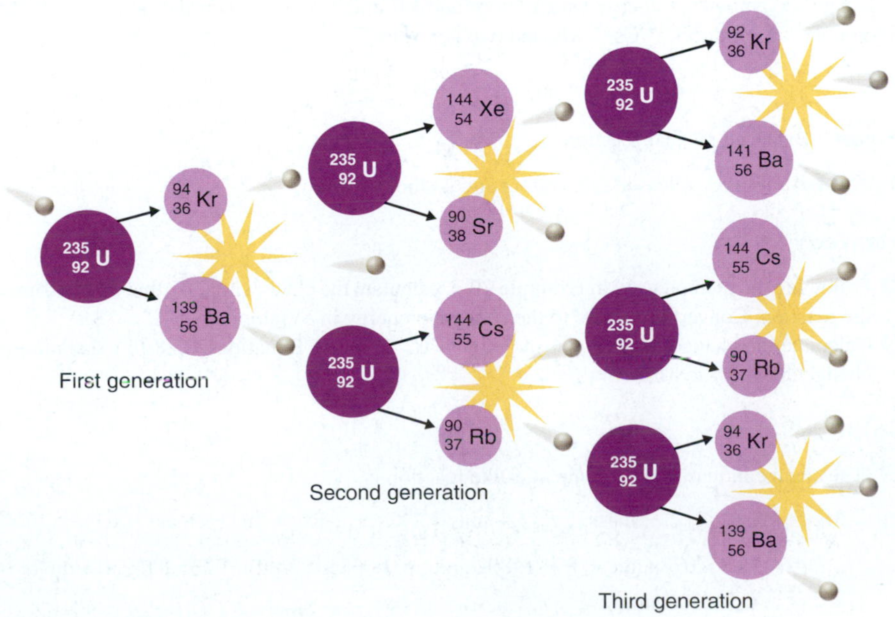

Figure 20.16 A nuclear chain reaction. The process is initiated by the collision of a single neutron with a ^{235}U nucleus, which undergoes fission as shown in Figure 20.6. Because each neutron released can cause the fission of another ^{235}U nucleus, the rate of the fission reaction accelerates geometrically. Each series of events is a generation.

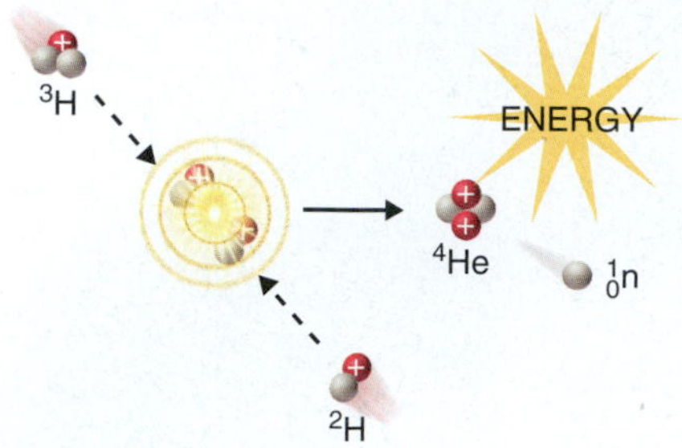

Figure 20.17 Nuclear fusion. In a nuclear fusion reaction, lighter nuclei combine to produce a heavier nucleus. As shown, fusion of ^{3}H and ^{2}H to give ^{4}He and a neutron releases an enormous amount of energy. In principle, nuclear fusion can produce much more energy than fission, but very high temperatures are required to overcome electrostatic repulsions between the positively charged nuclei and initiate the fusion reaction.

which corresponds to the amount of surface area available from which neutrons can escape, and on the identity of the isotope. If the mass of the fissile isotope is greater than the critical mass, then under the right conditions the resulting *supercritical mass* can release energy explosively. The enormous energy released from nuclear chain reactions is responsible for the massive destruction caused by the detonation of nuclear weapons such as fission bombs, but it also forms the basis of the nuclear power industry.

Nuclear fusion, in which two light nuclei combine to produce a heavier, more stable nucleus, is the opposite of a nuclear fission reaction. As in the nuclear transmutation reactions discussed in Section 20.2, the positive charge on both nuclei results in a large electrostatic energy barrier to fusion. This barrier can be overcome if one or both particles have sufficient kinetic energy to overcome the electrostatic repulsions, allowing the two nuclei to approach close enough for a fusion reaction to occur. The principle is similar to adding heat to increase the rate of a chemical reaction (see Chapter 14).

As shown in the plot of nuclear binding energy per nucleon versus atomic number in Figure 20.15, fusion reactions are most exothermic for the lightest element. For example, in a typical fusion reaction, two deuterium atoms combine to produce helium-3:

$$2\,^2_1\text{H} \longrightarrow \,^3_2\text{He} + \,^1_0\text{n} \tag{20.37}$$

In another reaction, a deuterium atom and a tritium atom fuse to produce helium-4 (Figure 20.17):

$$^2_1\text{H} + \,^3_1\text{H} \longrightarrow \,^4_2\text{He} + \,^1_0\text{n} \tag{20.38}$$

Initiating these reactions, however, requires a temperature comparable to that in the interior of the sun (approximately 1.5×10^7 K). Currently, the only method available on Earth to achieve such a temperature is the detonation of a fission bomb. Thus, the so-called hydrogen bomb (or H-bomb) is actually a deuterium-tritium bomb (a D-T bomb), which uses a nuclear fission reaction to create the very high temperatures needed to initiate fusion of solid lithium deuteride (^{6}LiD), which releases neutrons that then react with ^{6}Li, producing tritium. The deuterium-tritium reaction releases energy explosively. The next example and exercise demonstrate the enormous amounts of energy produced by nuclear fission and fusion reactions. In fact, fusion reactions are the power sources for all stars, including our sun.

EXAMPLE 20.9

Calculate the amount of energy (in eV/atom and kJ/mol) released when the neutron-induced fission of ^{235}U produces ^{144}Cs, ^{90}Rb, and two neutrons:

$$^{235}_{92}\text{U} + \,^1_0\text{n} \longrightarrow \,^{144}_{55}\text{Cs} + \,^{90}_{37}\text{Rb} + 2\,^1_0\text{n}$$

Given Balanced nuclear reaction

Asked for Energy released in eV/atom and kJ/mol

Strategy

Ⓐ Following the method used in Example 20.7, calculate the change in mass that accompanies the reaction. Convert this value to the change in energy in eV/atom.

Ⓑ Calculate the change in mass per mole of ^{235}U. Then use Equation 20.28 to calculate the change in energy in kJ/mol.

Solution

Ⓐ The change in mass that accompanies the reaction is

$$\begin{aligned}\Delta m &= \text{mass}_{\text{products}} - \text{mass}_{\text{reactants}} = \text{mass}\left(^{144}_{55}\text{Cs} + \,^{90}_{37}\text{Rb} + \,^1_0\text{n}\right) - \text{mass}\,^{235}_{92}\text{U} \\ &= (143.932077\text{ amu} + 89.914802\text{ amu} + 1.008665\text{ amu}) - 235.043930\text{ amu} \\ &= -0.188386\text{ amu}\end{aligned}$$

The change in energy in eV/atom is

$$\Delta E = (-0.188386 \text{ ~~amu~~})(931 \text{ MeV/~~amu~~}) = -175 \text{ MeV}$$

B The change in mass per mole of $^{235}_{92}U$ is -0.188386 g $= -1.88386 \times 10^{-4}$ kg, so the change in energy in kJ/mol is

$$\Delta E = (\Delta m)c^2 = (-1.88386 \times 10^{-4} \text{ kg})(2.998 \times 10^8 \text{ m/s})^2$$
$$= -1.693 \times 10^{13} \text{ J/mol} = -1.693 \times 10^{10} \text{ kJ/mol}$$

EXERCISE 20.9

Calculate the amount of energy (in eV/atom and kJ/mol) released when deuterium and tritium fuse to give helium-4 and a neutron:

$$^2_1H + ^3_1H \longrightarrow ^4_2He + ^1_0n$$

Answer $\Delta E = -17.6$ MeV/atom $= -1.697 \times 10^9$ kJ/mol

20.5 • Applied Nuclear Chemistry

The ever-increasing energy needs of modern societies have led scientists and engineers to develop ways of harnessing the energy released by nuclear reactions. To date, all practical applications of nuclear power have been based on nuclear fission reactions. Although in principle nuclear fusion offers many advantages, technical difficulties in achieving the high energies required to initiate nuclear fusion reactions have thus far precluded the use of fusion for the controlled release of energy. In this section, we describe the various types of nuclear power plants that are currently used to generate electricity from nuclear reactions, along with some possible ways to harness fusion energy in the future. In addition, we discuss some of the applications of nuclear radiation and radioisotopes, which have innumerable uses in medicine, biology, chemistry, and industry.

Pitchblende, the major uranium ore, consisting mainly of uranium oxide.

Nuclear Reactors

When a critical mass of a fissile isotope has been achieved, the resulting flux of neutrons can lead to a self-sustaining reaction. A variety of techniques can be used to control the flow of neutrons from such a reaction, which allows nuclear fission reactions to be maintained at safe levels. Many levels of control are required, along with a fail-safe design, because otherwise the chain reaction can accelerate so rapidly that it releases enough heat to melt or vaporize the fuel and the container, a situation that can release enough radiation to contaminate the surrounding area. Uncontrolled nuclear fission reactions are relatively rare, but they have occurred at least 18 times in the past.

There is compelling evidence that uncontrolled nuclear chain reactions occurred naturally in the early history of our planet, about 1.7 billion years ago in uranium deposits near Oklo in Gabon, West Africa (Figure 20.18). The natural abundance of ^{235}U 2 billion years ago was about 3% compared with 0.72% today; in contrast, the "fossil nuclear reactor" deposits in Gabon now contain only 0.44% ^{235}U. An unusual combination of geologic phenomena in this region apparently resulted in the formation of deposits of essentially pure uranium oxide containing 3% ^{235}U, which coincidentally happens to be identical to the fuel used in many modern nuclear plants. When rainwater or groundwater saturated one of these deposits, the water acted as a natural *moderator* that decreased the kinetic energy of the neutrons emitted by

Figure 20.18 A "fossil nuclear reactor" in a uranium mine near Oklo in Gabon, West Africa. More than a billion years ago, a number of uranium-rich deposits in West Africa apparently "went critical," initiating uncontrolled nuclear fission reactions that may have continued intermittently for more than 100,000 years, until the concentration of uranium-235 became too low to support a chain reaction. This photo shows a geologist standing in a mine dug to extract the concentrated uranium ore. Commercial interest waned rapidly after it was recognized that the uranium ore was severely depleted in U-235, the isotope of interest.

Figure 20.19 The Chernobyl nuclear power plant. In 1986, mechanical and operational failures during testing at the Chernobyl power plant in the Ukraine caused an uncontrolled nuclear chain reaction. The resulting fire destroyed much of the facility and severely damaged the core of the reactor, resulting in the release of large amounts of radiation that was spread over the surrounding area by the prevailing winds. The effects were devastating to the health of the population in the region and to the Soviet economy.

radioactive decay of ^{235}U, allowing the neutrons to initiate a chain reaction. As a result, the entire deposit "went critical" and became an uncontrolled nuclear chain reaction, which is estimated to have produced about 100 kW of power. It is thought that these natural nuclear reactors operated only intermittently, however, because the heat released would have vaporized the water. Removing the water would have shut down the reactor until the rocks cooled enough to allow water to reenter the deposit, at which point the chain reaction would begin again. This on–off cycle was repeated for more than 100,000 years, until so much ^{235}U was consumed that the deposits could no longer support a chain reaction.

The most recent instance of an uncontrolled nuclear chain reaction occurred on April 25–26, 1986, at the Chernobyl nuclear power plant in the former USSR (now Ukraine), resulting in the worst nuclear power accident in history (Figure 20.19). During testing of the reactor's turbine generator, a series of mechanical and operational failures caused a chain reaction that quickly went out of control, destroying the reactor core and igniting a fire that destroyed much of the facility and released a large amount of radioactivity. Thirty people were killed immediately, and the high levels of radiation in a 20-mile radius forced nearly 350,000 people to be resettled or evacuated. In addition, the accident caused a disruption to the Soviet economy that is estimated to have cost almost $13 billion. It is somewhat surprising, however, that the long-term health effects on the 600,000 people affected by the accident appear to be much less severe than originally anticipated. Initially, it was predicted that the accident would result in tens of thousands of premature deaths, but an exhaustive study almost 20 years after the event suggests that "only" 4,000 people will die prematurely from radiation exposure due to the accident. Although that may seem like a lot, in fact it represents only about a 3% increase in the cancer rate among the 600,000 people most affected, of whom about a quarter would be expected to eventually die of cancer even if the accident had not occurred.

If, on the other hand, the neutron flow in a reactor is carefully regulated so that only enough heat is released to boil water, then the resulting steam can be used to produce electricity. Thus, a nuclear reactor is similar in many respects to the conventional power plants discussed in Chapter 5, which burn coal or natural gas to generate electricity—the only difference is the source of the heat that is used to convert water to steam.

Light-Water Reactors

We begin our description of nuclear power plants with light-water reactors, which are used extensively to produce electricity in countries such as Japan, Israel, South Korea, Taiwan, and France that lack large reserves of fossil fuels. The essential components of a light-water reactor are depicted in Figure 20.20. All existing nuclear power plants have similar components, although different designs utilize different fuels and operating conditions. *Fuel rods* containing a fissile isotope in a structurally stabilized form (such as uranium oxide pellets encased in a corrosion-resistant zirconium alloy) are suspended in a cooling bath that transfers the heat generated by the fission reaction to a secondary cooling system. The heat is used to generate steam for the production of electricity. In addition, *control rods* are utilized to absorb neutrons and thereby control the rate of the nuclear chain reaction. Control rods are made of a substance that efficiently absorbs neutrons, such as boron, cadmium, or, in nuclear submarines, hafnium. Pulling the control rods out increases the neutron flux, allowing the reactor to generate more heat, whereas inserting the rods completely stops the reaction, a process called "scramming the reactor."

Despite this apparent simplicity, many technical hurdles must be overcome for nuclear power to be an efficient and safe source of energy. Uranium contains only 0.72% uranium-235, which is the only naturally occurring fissile isotope of uranium. Because this abundance is not enough to support a chain reaction, the uranium fuel must be at least partially enriched in ^{235}U, to a concentration of about 3%, for it to be able to sustain a chain reaction. At this level of enrichment, a nuclear explosion is

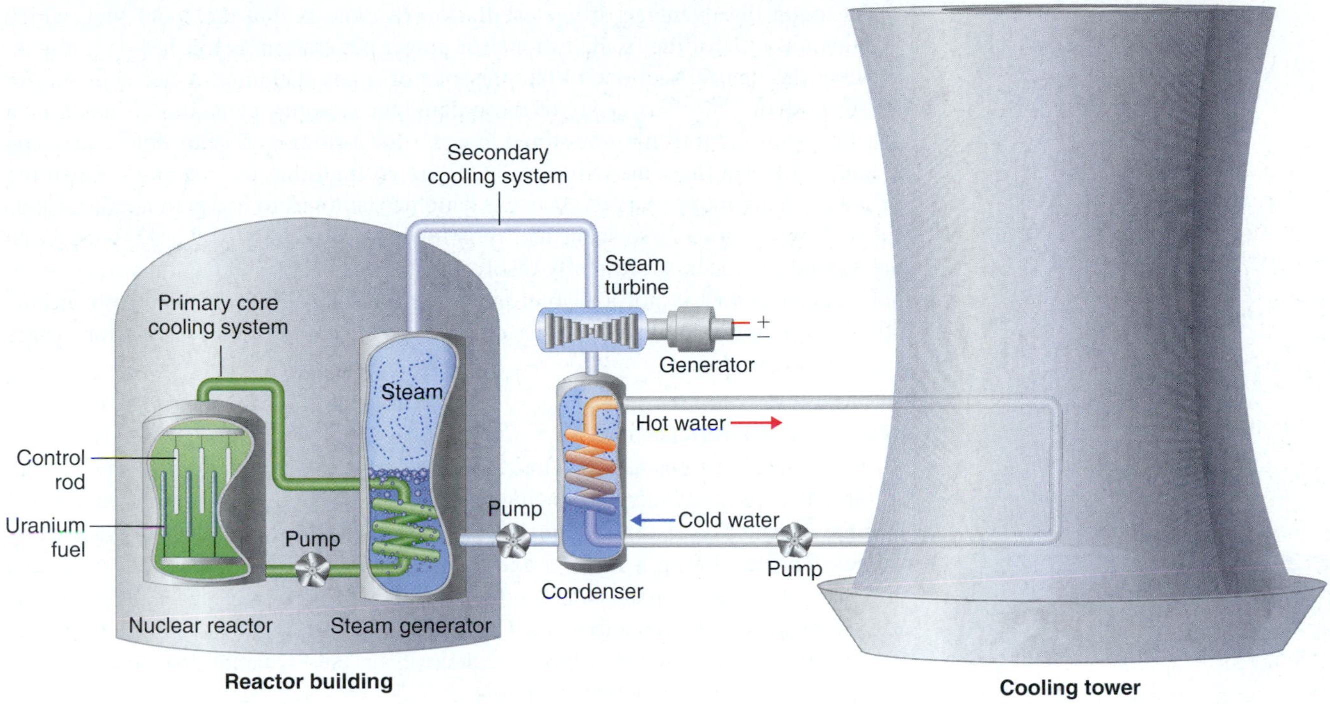

Figure 20.20 A light-water nuclear fission reactor for the production of electric power. The fuel rods are made of a corrosion-resistant alloy that encases the partially enriched uranium fuel; controlled fission of ^{235}U in the fuel produces heat. Water surrounds the fuel rods and moderates the kinetic energy of the neutrons, slowing them to increase the probability that they will induce fission. Control rods that contain elements such as boron, cadmium, or hafnium, which are very effective at absorbing neutrons, are used to control the rate of the fission reaction. A heat exchanger is used to boil water in a secondary cooling system, creating steam to drive the turbine and produce electricity. The large hyperbolic cooling tower, which is the most visible portion of the facility, is used to condense the steam in the secondary cooling circuit; it is often located at some distance from the actual reactor.

impossible; far higher levels of enrichment ($\geq$90%) are required for military applications such as nuclear weapons or the nuclear reactors in submarines. Enrichment is accomplished by converting uranium oxide to UF_6, which is volatile and contains discrete UF_6 molecules. Because $^{235}UF_6$ and $^{238}UF_6$ have different masses, they have different rates of effusion and diffusion, and they can be separated using a gas diffusion process, as described in Chapter 12. Another difficulty is that neutrons produced by nuclear fission are too energetic to be absorbed by neighboring nuclei, and they escape from the material without inducing fission in nearby ^{235}U nuclei. Consequently, a moderator must be used to slow the neutrons enough to allow them to be captured by other ^{235}U nuclei. High-speed neutrons are scattered by substances such as water or graphite, which decreases their kinetic energy and increases the probability that they will react with another ^{235}U nucleus. The moderator in a light-water reactor is the water that is used as the primary coolant. The system is highly pressurized to about 100 atm to keep the water from boiling normal.

All nuclear reactors also require a powerful cooling system to absorb the heat generated in the reactor core and create steam that is used to drive a turbine that generates electricity. In 1979, an accident occurred when the main water pumps used for cooling at the nuclear power plant at Three Mile Island in Pennsylvania stopped running, which prevented the steam generators from removing heat. Eventually, the zirconium casing of the fuel rods ruptured, resulting in a meltdown of about half of the reactor core. Although there was no loss of life and only a small release of radioactivity, the accident produced sweeping changes in nuclear power plant operations. The U.S. Nuclear Regulatory Commission tightened its oversight to improve safety.

The main disadvantage of nuclear fission reactors is that the spent fuel, which contains too little of the fissile isotope for power generation, is much more radioactive than the unused fuel due to the presence of many daughter nuclei with shorter half-lives than ^{235}U. The decay of these daughter isotopes generates so much heat that the spent fuel rods must be stored in water for as long as 5 years before they can be handled. Even then, the radiation levels are so high that the rods must be stored for many, many more years to allow the daughter isotopes to decay to nonhazardous levels. How to store these spent fuel rods for hundreds of years is a pressing issue that has not yet been successfully resolved. As a result, some people are convinced that nuclear power is not a viable option for providing our future energy needs, although a number of other countries continue to rely on nuclear reactors for a large fraction of their energy.

Heavy-Water Reactors

Deuterium (^2H) absorbs neutrons much less effectively than does hydrogen (^1H), but it is about twice as effective at slowing neutrons. Consequently, a nuclear reactor that uses D_2O instead of H_2O as the moderator is so efficient that it can use *unenriched* uranium as fuel. Using a lower grade of uranium reduces the operating costs and eliminates the need for plants that produce enriched uranium. Because of the expense of D_2O, however, only countries like Canada, which has abundant supplies of hydroelectric power for generating D_2O by electrolysis (see Chapter 19), have made a major investment in heavy-water reactors.

(a)

(b)

Figure 20.21 Two possible designs for a nuclear fusion reactor. The extraordinarily high temperatures needed to initiate a nuclear fusion reaction would immediately destroy a container made of any known material. (a) One way to avoid contact with the container walls is to use a high-energy plasma as the fuel. Because a plasma is essentially a gas composed of ionized particles, it can be confined using a strong magnetic field shaped like a torus (a hollow donut). (b) Another approach to nuclear fusion is inertial confinement, which utilizes an icosahedral array of powerful lasers to heat and compress a tiny fuel pellet (a mixture of solid LiD and LiT) to induce fusion.

Breeder Reactors

A *breeder reactor* is a nuclear fission reactor that produces more fissionable fuel than it consumes. This does not violate the first law of thermodynamics because the fuel produced is not the same as the fuel consumed. Under heavy neutron bombardment, the nonfissile ^{238}U isotope is converted to ^{239}Pu, which can undergo fission:

$$^{238}_{92}U + ^{1}_{0}n \longrightarrow ^{239}_{92}U \longrightarrow ^{239}_{93}Np + ^{0}_{-1}\beta \qquad (20.39)$$

$$^{239}_{93}Np \longrightarrow ^{239}_{94}Pu + ^{0}_{-1}\beta$$

The overall reaction is thus the conversion of nonfissile ^{238}U to fissile ^{239}Pu, which can be isolated chemically and used to fuel a new reactor. An analogous series of reactions can be used to convert nonfissile ^{232}Th to ^{233}U, which can also be used as a fuel for a nuclear reactor. Typically, about 8 to 10 years are required for a breeder reactor to produce twice as much fissile material as it consumes, which is enough to fuel a replacement for the original reactor plus a new reactor. The products of fission of ^{239}Pu, however, have substantially longer half-lives than the fission products formed in light-water reactors.

Nuclear Fusion Reactors

Although nuclear fusion reactions such as Equations 20.37 and 20.38 are thermodynamically spontaneous, the positive charge on both nuclei results in a large electrostatic energy barrier to the reaction. (As you learned in Chapter 18, thermodynamic spontaneity has no relationship to reaction rate.) Extraordinarily high temperatures (about 1.0×10^8°C!) are required to overcome the electrostatic repulsions and initiate a fusion reaction. Even the most feasible such reaction, so-called D-T fusion (Equation 20.38), requires a temperature of about 4.0×10^7°C. Achieving these temperatures and controlling the materials to be fused are extraordinarily difficult problems, as is extracting the energy released by the fusion reaction, because a commercial fusion reactor would require such high temperatures to be maintained for long periods of time. Several different technologies are currently being explored, including the use of intense magnetic fields to contain ions in the form of a dense, high-energy plasma at a temperature high enough to sustain fusion (Figure 20.21a).

Another concept utilizes focused laser beams to heat and compress fuel pellets in controlled miniature fusion explosions (Figure 20.21b).

Nuclear reactions such as these are called **thermonuclear reactions** because a great deal of thermal energy must be invested to initiate the reaction. The amount of energy released by the reaction, however, is several orders of magnitude greater than the energy needed to initiate it. Thus, in principle, a nuclear fusion reaction should result in a significant net production of energy. In addition, the Earth's oceans contain an essentially inexhaustible supply of both deuterium and tritium, which suggests that nuclear fusion could provide a limitless supply of energy. Unfortunately, however, the technical requirements for a successful nuclear fusion reaction are so great that net power generation by controlled fusion has yet to be achieved.

Uses of Radioisotopes

Nuclear radiation can damage biological molecules, thereby disrupting normal functions such as cell division. Because radiation is particularly destructive to rapidly dividing cells such as tumor cells and bacteria, it has been used medically to treat cancer since 1904, when radium-226 was first used to treat a cancerous tumor. Many radioisotopes are now available for medical use, and each has specific advantages for certain applications.

In modern *radiation therapy*, radiation is often delivered by a source planted inside the body. For example, tiny capsules containing an isotope such as ^{192}Ir, coated with a thin layer of chemically inert platinum, are inserted into the middle of a tumor that cannot be removed surgically. The capsules are removed when the treatment is over. In some cases, physicians take advantage of the body's own chemistry to deliver a radioisotope to the desired location. For example, the thyroid glands in the lower front of the neck are the only organs in the body that utilize iodine. Consequently, radioactive iodine is taken up almost exclusively by the thyroid (Figure 20.22a). Thus, when radioactive isotopes of iodine (^{125}I or ^{131}I) are injected into the blood of a patient suffering from thyroid cancer, the thyroid glands filter the radioisotope out of the blood and concentrate it in the tissue to be destroyed. In cases where a tumor is surgically inaccessible (for example, when it is located deep in the brain), an external radiation source such as a ^{60}Co "gun" is

Figure 20.22 Medical imaging and treatment with radioisotopes. (a) Radioactive iodine is used both to obtain images of the thyroid and to treat thyroid cancer. Injected iodine-123 or iodine-131 is selectively taken up by the thyroid gland, where it is incorporated into the thyroid hormone, thyroxine. Because iodine-131 emits low-energy β particles that are absorbed by the surrounding tissue, it can be used to destroy malignant tissue in the thyroid. In contrast, iodine-123 emits higher-energy γ rays that penetrate tissues readily, enabling it to image the thyroid gland, as shown in the photo. (b) Some technetium compounds are selectively absorbed by cancerous cells within bones. The yellow spots in this figure show that a primary cancer has metastasized (spread) to the patient's spine (*lower center*) and ribs (*right center*).

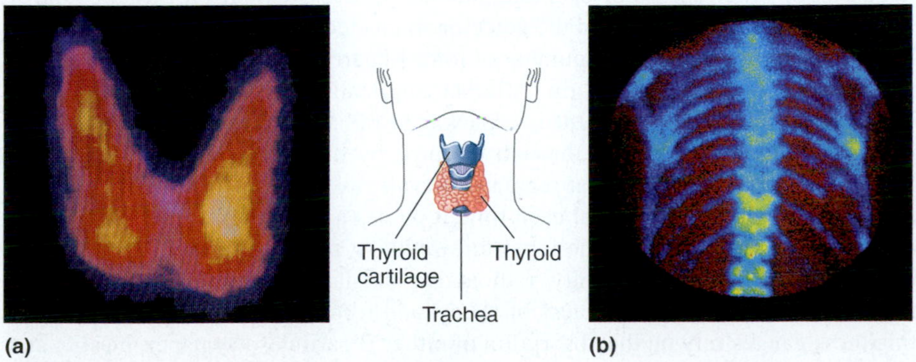

TABLE 20.7 Radioisotopes used in medical imaging and treatment

Isotope	Half-life	Tissue
^{18}F	110 min	Brain
^{24}Na	15 h	Circulatory system
^{32}P	14 days	Eyes, liver, tumors
^{59}Fe	46 days	Blood, spleen
^{60}Co	10.5 mo	External radiotherapy
^{99m}Tc	6 h	Heart, thyroid, liver, kidney, lungs, skeleton
^{125}I	60 days	Thyroid, prostate, brain
^{131}I	8 days	Thyroid
^{201}Tl	3 days	Heart
^{133}Xe	5 days	Lungs

used to aim a tightly focused beam of γ rays at it. Unfortunately, radiation therapy damages healthy tissue in addition to the target tumor and results in severe side effects, such as nausea, hair loss, and a weakened immune system. Although radiation therapy is generally not a pleasant experience, in many cases it is the only choice.

A second major medical use of radioisotopes is *medical imaging*, in which a radioisotope is temporarily localized in a particular tissue or organ, where its emissions provide a "map" of the tissue or organ. Medical imaging uses radioisotopes that cause little or no tissue damage, but are easily detected. One of the most important radioisotopes for medical imaging is ^{99m}Tc. Depending on the particular chemical form in which it is administered, technetium tends to localize in bones and in soft tissues, such as the heart or kidneys, which are almost invisible in conventional X rays (Figure 20.22b). Some properties of other radioisotopes used for medical imaging are listed in Table 20.7.

Because γ rays produced by isotopes such as ^{131}I and ^{99m}Tc are emitted randomly in all directions, it is impossible to achieve high levels of resolution in images that use such isotopes. However, remarkably detailed three-dimensional images can be obtained using an imaging technique called *positron emission tomography (PET)*. The technique uses radioisotopes that decay by positron emission (Section 20.2), and the resulting positron is annihilated when it collides with an electron in the surrounding matter. In the annihilation process, both particles are converted to energy in the form of two γ rays that are emitted simultaneously and at 180° to each other:

$$^{0}_{+1}\beta + ^{0}_{-1}e \longrightarrow 2\,^{0}_{0}\gamma \qquad (20.40)$$

With PET, biological molecules that have been "tagged" with a positron-emitting isotope such as ^{18}F or ^{11}C can be used to probe the functions of organs such as the brain.

Another major health-related use of ionizing radiation is the irradiation of food, an effective way to kill bacteria such as *Salmonella* in chicken and eggs and potentially lethal strains of *E. coli* in beef. Collectively, such organisms cause almost 3 million cases of food poisoning annually in the United States, resulting in hundreds of deaths. Figure 20.23 shows how irradiation dramatically extends the storage life of foods such as strawberries. Although U.S. health authorities have given only limited approval of this technique, the growing number of illnesses caused by antibiotic-resistant bacteria is increasing the pressure to expand the scope of food irradiation.

Figure 20.23 Preservation of strawberries using ionizing radiation. Fruits such as strawberries can be irradiated by high-energy γ rays to kill bacteria and prolong their storage life. The nonirradiated strawberries on the left are completely spoiled after 15 days in storage, but the irradiated strawberries on the right show no visible signs of spoilage under the same conditions.

One of the more unusual effects of radioisotopes is in dentistry. Because dental enamels contain a mineral called feldspar, $KAlSi_3O_8$ (which is also found in granite rocks), teeth contain a small amount of naturally occurring radioactive ^{40}K. The radiation caused by the decay of ^{40}K results in the emission of light (*fluorescence*), which gives the highly desired "pearly white" appearance associated with healthy teeth.

In addition to the medical uses of radioisotopes, radioisotopes have literally hundreds of other uses. For example, smoke detectors contain a tiny amount of ^{241}Am, which ionizes the air in the detector so an electric current can flow through it. Smoke particles reduce the number of ionized particles and decrease the electric current, which triggers an alarm. Another application is the "go-devil" used to detect leaks in long pipelines. It is a packaged radiation detector that is inserted into a pipeline and propelled through the pipe by the flowing liquid. Sources of ^{60}Co are attached to the pipe at regular intervals; as the detector travels along the pipeline, it sends a radio signal each time it passes a source. When a massive leak causes the go-devil to stop, the repair crews know immediately which section of the pipeline is damaged. Finally, radioactive substances are used in gauges that measure and control the thickness of sheets and films. As shown in Figure 20.24, thickness gauges rely on the absorption of either β particles (by paper, plastic, and very thin metal foils) or γ rays (for thicker metal sheets); the amount of radiation absorbed can be measured accurately and is directly proportional to the thickness of the material.

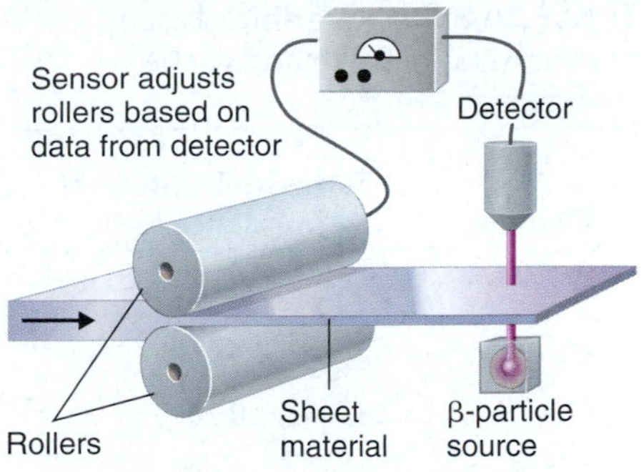

Figure 20.24 Using radiation to control the thickness of a material. Because the amount of radiation absorbed by a material is proportional to its thickness, radiation can be used to control the thickness of plastic film, tin foil, or paper. As shown, a beta emitter is placed on one side of the material being produced and a detector on the other. An increase in the amount of radiation that reaches the detector indicates a decrease in the thickness of the material, and vice versa. The output of the detector can thus be used to control the thickness of the material.

20.6 • The Origin of the Elements

The relative abundances of the elements in the universe vary by more than 12 orders of magnitude. For the most part, these differences in abundance cannot be explained by differences in nuclear stability. Although the ^{56}Fe nucleus is the most stable nucleus known, the most abundant element in the universe is not iron, but hydrogen (^1H), which accounts for about 90% of all atoms. In fact, ^1H is the raw material from which all other elements are formed.

In this section, we explain why ^1H and ^2He together account for at least 99% of all the atoms in the universe. We also describe the nuclear reactions that take place in stars, which transform one nucleus into another and create all the naturally occurring elements.

Relative Abundances of the Elements on Earth and in the Universe

The relative abundances of the elements in the universe and on Earth relative to silicon are shown in Figure 20.25. The data are estimates based on the characteristic emission spectra of the elements in stars, the absorption spectra of matter in clouds of interstellar dust, and the approximate composition of the Earth as measured by geologists. The data in Figure 20.25 illustrate two important points. First, except for hydrogen, the most abundant elements have *even* atomic numbers. Not only is this consistent with the trends in nuclear stability discussed in Section 20.1, but it also suggests that heavier elements are formed by combining helium nuclei ($Z = 2$). Second, the relative

Figure 20.25 The relative abundances of the elements in the universe and on Earth. In this logarithmic plot, the relative abundances of the elements relative to that of silicon (arbitrarily set equal to 1) in the universe (green bars) and on Earth (purple bars) are shown as a function of atomic number. Note that elements with even atomic numbers are generally more abundant in the universe than elements with odd atomic numbers. Note also that the relative abundances of many elements in the universe are *very* different from their relative abundances on Earth.

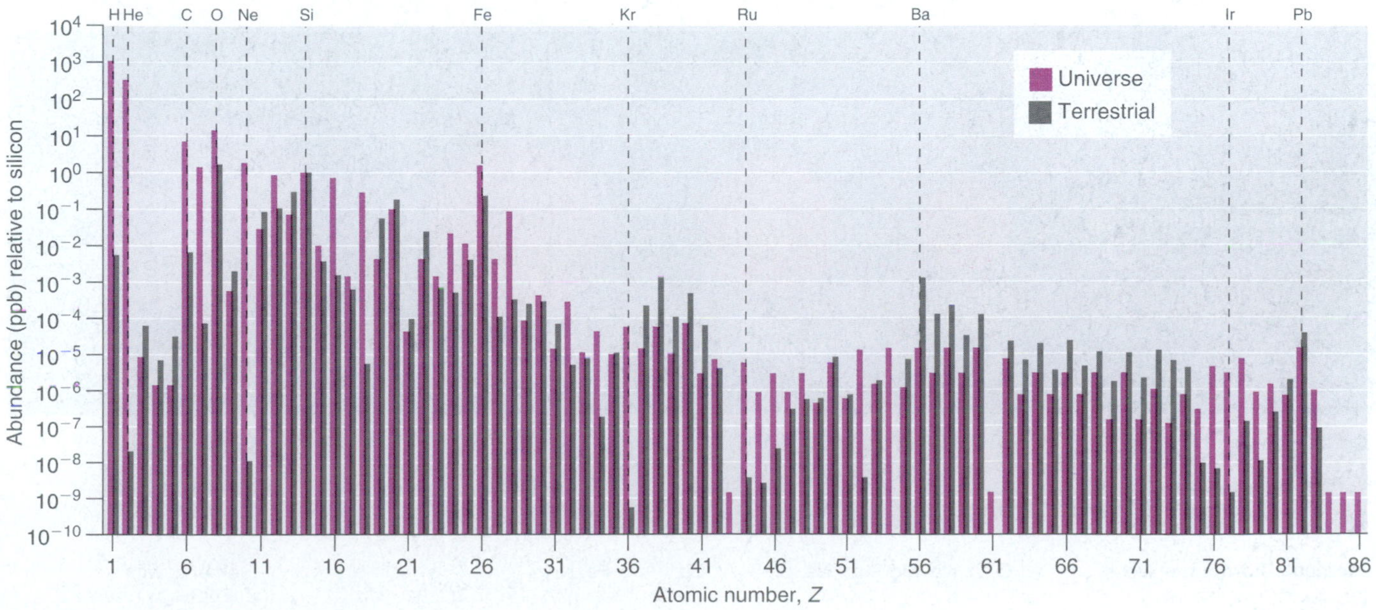

TABLE 20.8 Relative abundances of elements on Earth and in the Universe

Element	Terrestrial/Universal Abundance Ratio
H	0.0020
He	2.4×10^{-8}
C	0.36
N	0.02
O	46
Ne	1.9×10^{-6}
Na	1200
Mg	48
Al	1600
Si	390
S	0.84
K	5000
Ca	710
Ti	2200
Fe	57

abundances of the elements in the universe and on Earth are often very different, as indicated by the data in Table 20.8 for some common elements. Some of these differences are easily explained. For example, nonmetals such as H, He, C, N, O, Ne, and Kr are much less abundant relative to silicon on Earth than they are in the rest of the universe. These elements are either noble gases (He, Ne, Kr) or elements that form volatile hydrides, such as NH_3, CH_4, and H_2O. Because Earth's gravity is not strong enough to hold such light substances in the atmosphere, these elements have been slowly diffusing into outer space ever since our planet was formed. Argon (Ar) is an exception; it is relatively abundant on Earth compared with the other noble gases because it is continuously produced in rocks by the radioactive decay of isotopes such as ^{40}K. In contrast, many metals, such as Al, Na, Fe, Ca, Mg, K, and Ti, are relatively abundant on Earth because they form nonvolatile compounds such as oxides that cannot escape into space. Other metals, however, are much *less* abundant on Earth than in the universe; examples are Ru and Ir. You may recall from Chapter 1 that the anomalously high iridium content of a 66-million-year-old rock layer was a key finding in the development of the asteroid-impact theory for the extinction of the dinosaurs. This section explains some of the reasons for the great differences in abundances of the metallic elements.

All the elements originally present on Earth (and on other planets) were synthesized from hydrogen and helium nuclei in the interiors of stars that have long since exploded and disappeared. Six of the most abundant elements in the universe (carbon, oxygen, neon, magnesium, silicon, and iron) have nuclei that are integral multiples of the helium-4 nucleus, which suggests that helium-4 is the primary building block for heavier nuclei.

Synthesis of the Elements in Stars

Elements are synthesized in discrete stages during the lifetime of a star, and some steps occur only in the most massive stars known (Figure 20.26). Initially, all stars are formed by the aggregation of interstellar "dust," which is mostly hydrogen. As the cloud of dust slowly contracts due to gravitational attraction, its density eventually

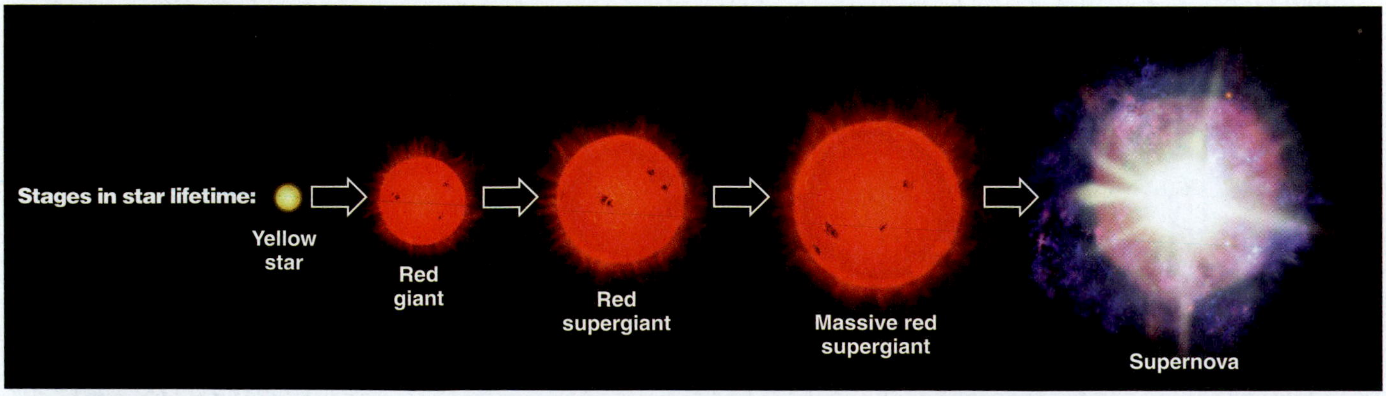

	Yellow star	Red giant	Red supergiant	Massive red supergiant	Supernova
Core Temperature:	1.5×10^7 K	2×10^8 K	7×10^8 K	3×10^9 K	1×10^{11} K
Primary Nuclear Reaction:	^1H fusion	^4He fusion	$^4He + {}^{12}C$ $^{12}C + {}^{12}C$ $^{12}C + {}^{16}O$	Proton–neutron exchange reactions	Multiple neutron captures
Elements Formed:	He	C, O, Ne, Mg	Na, Si, S, Ar, Ca	Fe, Ni	Elements with $Z > 28$

Figure 20.26 Nuclear reactions during the life cycle of a massive star. At each stage in the lifetime of a star, a different fuel is used for nuclear fusion, resulting in the formation of different elements. Fusion of hydrogen to give helium is the primary fusion reaction in young stars. As the star ages, helium accumulates and begins to "burn," undergoing fusion to form heavier elements such as carbon and oxygen. As the adolescent star matures, significant amounts of iron and nickel are formed by fusion of the heavier elements formed previously. The heaviest elements are formed only during the final death throes of the star, formation of a nova or supernova.

reaches about 100 g/cm^3 and the temperature increases to about 1.5×10^7 K, forming a dense plasma of ionized hydrogen nuclei. At this point, self-sustaining nuclear reactions begin, and the star "ignites," creating a *yellow star* like our sun.

In the first stage of its life, the star is powered by a series of nuclear fusion reactions that convert hydrogen to helium:

$$
\begin{aligned}
{}^1_1\text{H} + {}^1_1\text{H} &\longrightarrow {}^2_1\text{H} + {}^{\ 0}_{+1}\beta \qquad (20.41)\\
{}^2_1\text{H} + {}^1_1\text{H} &\longrightarrow {}^3_2\text{He} + {}^0_0\gamma\\
{}^3_2\text{He} + {}^3_2\text{He} &\longrightarrow {}^4_2\text{He} + 2{}^1_1\text{H}
\end{aligned}
$$

The overall reaction is the conversion of four hydrogen nuclei to a helium-4 nucleus, which is accompanied by the release of two positrons, two γ rays, and a great deal of energy:

$$4{}^1_1\text{H} \longrightarrow {}^4_2\text{He} + 2{}^{\ 0}_{+1}\beta + 2{}^0_0\gamma \qquad (20.42)$$

These reactions are responsible for most of the enormous amount of energy that is released as sunlight and solar heat. It takes several billion years, depending on the size of the star, to convert about 10% of the hydrogen to helium.

Once large amounts of helium-4 have been formed, they become concentrated in the core of the star, which slowly becomes denser and hotter. At a temperature of about 2×10^8 K, the helium-4 nuclei begin to fuse, producing beryllium-8:

$$2{}^4_2\text{He} \longrightarrow {}^8_4\text{Be} \qquad (20.43)$$

Although beryllium-8 has both an even mass number and an even atomic number, the low neutron-to-proton ratio makes it *very* unstable, decomposing in only about 10^{-16} s. Nonetheless, this is long enough for it to react with a third helium-4 nucleus to form carbon-12, which is very stable. Sequential reactions of carbon-12 with helium-4 produce the elements with even numbers of protons and neutrons up to magnesium-24:

$${}^8_4\text{Be} \xrightarrow{{}^4_2\text{He}} {}^{12}_{\ 6}\text{C} \xrightarrow{{}^4_2\text{He}} {}^{16}_{\ 8}\text{O} \xrightarrow{{}^4_2\text{He}} {}^{20}_{10}\text{Ne} \xrightarrow{{}^4_2\text{He}} {}^{24}_{12}\text{Mg} \qquad (20.44)$$

So much energy is released by these reactions that it causes the surrounding mass of hydrogen to expand, producing a *red giant* that is about 100 times larger than the original yellow star.

As the star expands, the heavier nuclei accumulate in its core, which contracts further to a density of about 50,000 g/cm^3 and becomes yet hotter. At a temperature of about 7×10^8 K, carbon and oxygen nuclei undergo nuclear fusion reactions to produce sodium and silicon nuclei:

$${}^{12}_{\ 6}\text{C} + {}^{12}_{\ 6}\text{C} \longrightarrow {}^{23}_{11}\text{Na} + {}^1_1\text{H} \qquad (20.45)$$

$${}^{12}_{\ 6}\text{C} + {}^{16}_{\ 8}\text{O} \longrightarrow {}^{28}_{14}\text{Si} + {}^0_0\gamma \qquad (20.46)$$

At these temperatures, carbon-12 reacts with helium-4 to initiate a series of reactions that produce more oxygen-16, neon-20, magnesium-24, and silicon-28, as well as heavier nuclides such as sulfur-32, argon-36, and calcium-40:

$${}^{12}_{\ 6}\text{C} \xrightarrow{{}^4_2\text{He}} {}^{16}_{\ 8}\text{O} \xrightarrow{{}^4_2\text{He}} {}^{20}_{10}\text{Ne} \xrightarrow{{}^4_2\text{He}} {}^{24}_{12}\text{Mg} \xrightarrow{{}^4_2\text{He}} {}^{28}_{14}\text{Si} \xrightarrow{{}^4_2\text{He}} {}^{32}_{16}\text{S} \xrightarrow{{}^4_2\text{He}} {}^{36}_{18}\text{Ar} \xrightarrow{{}^4_2\text{He}} {}^{40}_{20}\text{Ca} \qquad (20.47)$$

The energy released by these reactions causes a further expansion of the star to form a *red supergiant*, and the core temperature increases steadily. At a temperature of about 3×10^9 K, the nuclei that have been formed exchange protons and neutrons freely. This equilibration process forms heavier elements up to iron-56 and nickel-58, which have the most stable nuclei known.

Figure 20.27 A supernova. A view of the remains of Supernova 1987A, showing the circular halo of expanding debris produced by the explosion. Multiple neutron-capture events occur during a supernova explosion, forming both the heaviest elements and many of the less stable nuclides.

Formation of Heavier Elements in Supernovas

None of the processes described so far produces nuclei with $Z > 28$. All naturally occurring elements heavier than nickel are formed in the rare but spectacular cataclysmic explosions called *supernovas* (Figure 20.26). When the fuel in the core of a very massive star has been consumed, its gravity causes it to collapse in about 1 s. As the core is compressed, the iron and nickel nuclei within it disintegrate to protons and neutrons, and many of the protons capture electrons to form neutrons. The resulting *neutron star* is a dark object that is so dense that atoms no longer exist. Simultaneously, the energy released by the collapse of the core causes the supernova to explode in what is arguably the single most violent event in the universe. The force of the explosion blows most of the star's matter into space, creating a gigantic and rapidly expanding dust cloud, a *nebula* (Figure 20.27). During the extraordinarily short duration of this event, the concentration of neutrons is so great that multiple neutron-capture events occur, leading to the production of the heaviest elements and many of the less stable nuclides. Under these conditions, for example, an iron-56 nucleus can absorb as many as 64 neutrons, briefly forming an extraordinarily unstable iron isotope that can then undergo multiple rapid β-decay processes to produce tin-120:

$$^{56}_{26}\text{Fe} + 64\,^{1}_{0}\text{n} \longrightarrow \left[^{120}_{26}\text{Fe}\right] \longrightarrow \,^{120}_{50}\text{Sn} + 24\,^{0}_{-1}\beta \qquad (20.48)$$

MGC Heavy Element Formation

Although a supernova occurs only every few hundred years in a galaxy such as the Milky Way, these rare explosions provide the only conditions under which elements heavier than nickel can be formed. The force of the explosions distributes these elements throughout the galaxy surrounding the supernova, and eventually they are captured in the dust that condenses to form new stars. Based on its elemental composition, our sun is thought to be a second- or third-generation star, which contains a considerable amount of cosmic debris from the explosion of supernovas in the remote past.

EXAMPLE 20.10

The reaction of two carbon-12 nuclei in a carbon-burning star can produce elements other than sodium. Write balanced equations for the formation of **(a)** magnesium-24 and **(b)** neon-20 from two carbon-12 nuclei.

Given Reactant and product nuclides

Asked for Balanced equation

Strategy

Use conservation of mass and charge to determine the type of nuclear reaction that will convert the reactant to the indicated product. Write the balanced equation for the reaction.

Solution

(a) A magnesium-24 nucleus ($Z = 12$, $A = 24$) has the same nucleons as two carbon-12 nuclei ($Z = 6$, $A = 12$). The reaction is therefore a fusion of two carbon-12 nuclei, and no other particles are produced: $^{12}_{6}\text{C} + \,^{12}_{6}\text{C} \longrightarrow \,^{24}_{12}\text{Mg}$.

(b) The neon-20 product has $Z = 10$ and $A = 20$. Conservation of mass requires that the other product have $A = (2 \times 12) - 20 = 4$; because of conservation of charge, it must have $Z = (2 \times 6) - 10 = 2$. These are the characteristics of an α particle. The reaction is therefore $^{12}_{6}\text{C} + \,^{12}_{6}\text{C} \longrightarrow \,^{20}_{10}\text{Ne} + \,^{4}_{2}\alpha$.

EXERCISE 20.10

How many neutrons must an iron-56 nucleus absorb during a supernova explosion to produce an arsenic-75 nucleus? Write a balanced equation for the reaction.

Answer 19 neutrons; $^{56}_{26}\text{Fe} + 19\,^{1}_{0}\text{n} \longrightarrow \left[^{75}_{26}\text{Fe}\right] \longrightarrow {}^{75}_{33}\text{As} + 7\,^{0}_{-1}\beta$

SUMMARY AND KEY TERMS

20.1 The Components of the Nucleus (p. 913)

Subatomic particles of the nucleus (protons and neutrons) are called **nucleons**. A **nuclide** is an atom with a particular number of protons and neutrons. An unstable nucleus that decays spontaneously is **radioactive** whose emissions are collectively called *radioactivity*. Isotopes that emit radiation are called **radioisotopes**. Each nucleon is attracted to other nucleons by the **strong nuclear force**. Stable nuclei generally have even numbers of both protons and neutrons and a neutron-to-proton ratio of at least 1. Nuclei that contain *magic numbers* of protons and neutrons are often especially stable. **Superheavy elements**, with atomic numbers near 126, may even be stable enough to exist in nature.

20.2 Nuclear Reactions (p. 917)

In **nuclear decay reactions**, the *parent nucleus* is converted to a more stable *daughter nucleus*. Nuclei with too many neutrons decay by converting a neutron to a proton, whereas nuclei with too few neutrons decay by converting a proton to a neutron. Very heavy nuclei (with $A \geq 200$ and $Z > 83$) are unstable and tend to decay by emitting an **α particle**. When an unstable nuclide undergoes radioactive decay, the total number of nucleons is conserved, as is the total positive charge. Six different kinds of nuclear decay reactions are known. **Alpha decay** results in the emission of an α particle, $^{4}_{2}\alpha$, and produces a daughter nucleus with a mass number that is lower by 4 and an atomic number that is lower by 2 than the parent nucleus. **Beta decay** converts a neutron to a proton and emits a high-energy electron, producing a daughter nucleus with the same mass number as the parent and an atomic number that is higher by 1. **Positron emission** is the opposite of beta decay, and converts a proton to a neutron plus a positron. Positron emission does not change the mass number of the nucleus, but the atomic number of the daughter nucleus is lower by 1 than the parent. In **electron capture**, an electron in an inner shell reacts with a proton to produce a neutron, with emission of an X ray. The mass number does not change, but the atomic number of the daughter is lower by 1 than the parent. In **gamma emission**, a daughter nucleus in a *nuclear excited state* undergoes a transition to a lower-energy state by emitting a γ ray. Very heavy nuclei with high neutron-to-proton ratios can undergo **spontaneous fission**, in which the nucleus breaks into two pieces that can have different atomic numbers and atomic masses with the release of neutrons. Many very heavy nuclei decay via a **radioactive decay series**—a succession of some combination of alpha- and beta-decay reactions. In **nuclear transmutation reactions**, a target nucleus is bombarded with energetic subatomic particles to give a product nucleus more massive than the original. All the **transuranium elements**—elements with $Z > 92$—are artificial and must be prepared by nuclear transmutation reactions. These reactions are carried out in *particle accelerators* such as *linear accelerators*, *cyclotrons*, and *synchrotrons*.

20.3 The Interaction of Nuclear Radiation with Matter (p. 927)

The effects of radiation on matter depend on the energy of the radiation. **Nonionizing radiation** is relatively low in energy, and the energy is transferred to matter in the form of heat. **Ionizing radiation** is relatively high in energy, and when it collides with an atom, it can completely remove an electron to form a positively charged ion that can damage biological tissues. Alpha particles do not penetrate very far into matter, whereas γ rays penetrate more deeply. Common units of radiation exposure or dose are the **roentgen (R)**, the amount of energy absorbed by dry air, and the **rad**, the amount of radiation that produces 0.01 J of energy in 1 kg of matter. The **rem** measures the actual amount of tissue damage caused by a given amount of radiation. Natural sources of radiation include *cosmic radiation*, consisting of high-energy particles and γ rays emitted by the sun and other stars; *cosmogenic radiation*, which is produced by the interaction of cosmic rays with gases in the upper atmosphere; and *terrestrial radiation*, from radioactive elements present on the primordial Earth and their decay products. The risks of ionizing radiation depend on the intensity of the radiation, the mode of exposure, and the duration of the exposure.

20.4 Thermodynamic Stability of the Atomic Nucleus (p. 932)

Nuclear reactions are accompanied by large changes in energy, which result in detectable changes in mass. The change in mass is related to the change in energy using Einstein's equation, $\Delta E = (\Delta m)c^2$. Large changes in energy are usually reported in units of keV or MeV (thousands or millions of electronvolts). With the exception of ^{1}H, the experimentally determined mass of an atom is always *less* than the sum of the masses of the component particles (protons, neutrons, and electrons) by an amount called the **mass defect** of the nucleus. The energy corresponding to the mass defect is the **nuclear binding energy**, the amount of energy released when a nucleus forms from its component particles. In **nuclear fission**, nuclei split into lighter nuclei with an accompanying release of multiple neutrons and large amounts of energy. The **critical mass** is the minimum mass required to support a self-sustaining **nuclear chain reaction**. **Nuclear fusion** is a process in which two light nuclei combine to produce a heavier nucleus plus a great deal of energy.

20.5 Applied Nuclear Chemistry (p. 939)

In nuclear power plants, nuclear reactions are used to generate electricity. *Light-water reactors* use enriched uranium as a fuel. They include fuel rods, a moderator, control rods, and a powerful cooling system to absorb the heat generated in the reactor core. *Heavy-water reactors* are able to use unenriched uranium as a fuel because they use D_2O as the moderator, which scatters and slows neutrons much more effectively than H_2O. A *breeder reactor* produces more fissionable fuel than it consumes. A *nuclear fusion reactor* requires very high temperatures. Fusion reactions are **thermonuclear reactions** because they require high temperatures for initiation. Radioisotopes are used in both radiation therapy and *medical imaging*.

20.6 The Origin of the Elements (p. 945)

By far the most abundant element in the universe is hydrogen. The fusion of hydrogen nuclei to form helium nuclei is the major process that fuels young stars such as the sun. Elements heavier than helium are formed from hydrogen and helium in the interiors of stars. Successive fusion reactions of helium nuclei at higher temperatures create elements with even numbers of protons and neutrons up to magnesium, then up to calcium. Eventually, the elements up to iron-56 and nickel-58 are formed by exchange processes at even higher temperatures. Heavier elements can only be made by a process that involves multiple neutron-capture events, which can occur only during the explosion of a supernova.

KEY EQUATIONS

Alpha decay	${}^{A}_{Z}X \longrightarrow {}^{(A-4)}_{(Z-2)}X' + {}^{4}_{2}\alpha$	(20.1)
Beta decay	${}^{A}_{Z}X \longrightarrow {}_{(Z+1)}^{\ \ \ A}X' + {}^{\ 0}_{-1}\beta$	(20.4)
Positron emission	${}^{A}_{Z}X \longrightarrow {}_{(Z-1)}^{\ \ \ A}X' + {}^{\ 0}_{+1}\beta$	(20.7)
Electron capture	${}^{A}_{Z}X + {}^{\ 0}_{-1}e \longrightarrow {}_{(Z-1)}^{\ \ \ A}X' + \text{X ray}$	(20.10)
Gamma emission	${}^{A}_{Z}X^* \longrightarrow {}^{A}_{Z}X + {}^{0}_{0}\gamma$	(20.14b)
Rad	$1 \text{ rad} = 0.01 \text{ J/kg}$	(20.23)

QUESTIONS AND PROBLEMS

For instructor-assigned homework, go to **www.masteringgenchem.com**

Questions and Problems with colored number have answers in the Appendix and complete solutions in the Student Solutions Manual.

CONCEPTUAL

20.1 The Components of the Nucleus

1. What distinguishes a nuclear reaction from a chemical reaction? Use an example of each to illustrate the differences.

2. What do chemists mean when they say a substance is *radioactive*?

3. What characterizes an isotope? How is the mass of an isotope of an element related to the atomic mass of the element shown in the periodic table?

4. In a typical nucleus, why doesn't electrostatic repulsion between protons destabilize the nucleus? How does the neutron-to-proton ratio affect the stability of an isotope? Why are all isotopes with $Z > 82$ unstable?

5. What is the significance of a *magic number* of protons or neutrons? What is the relationship between the number of stable isotopes of an element and whether the element has a magic number of protons?

6. Do you expect Bi to have a large number of stable isotopes? Does Ca? Explain your answers.

7. Potassium has three common isotopes, ${}^{39}K$, ${}^{40}K$, and ${}^{41}K$, but only potassium-40 is radioactive (a beta emitter). Suggest two reasons for the instability of ${}^{40}K$.

8. Samarium has 11 relatively stable isotopes, but only four are nonradioactive. One of these four is ${}^{144}Sm$, which has a lower neutron-to-proton ratio than lighter, radioactive isotopes of samarium. Why is ${}^{144}Sm$ more stable?

20.2 Nuclear Reactions

9. Describe the six classifications of nuclear reactions. What is the most common mode of decay for elements that have heavy nuclei? Why?

10. Complete the table for these five nuclear reactions.

	Alpha Decay	Beta Decay	Gamma Emission	Positron Emission	Electron Capture
Identity of Particle or Radiation	Helium-4 nucleus	_____	_____	_____	_____
Mass Number of Parent − Mass Number of Daughter	4	_____	_____	_____	_____
Atomic Number of Parent − Atomic Number of Daughter	_____	−1	_____	_____	_____
Effect on Neutron-to-Proton Ratio	_____	Decreases	_____	_____	_____

11. What is the most common decay process for elements in row 5 of the periodic table that contain too few neutrons for the number of protons present? Why?
12. Explain the difference between the symbols e^- and β^-. What is the difference in meaning between the symbols ${}^{4}_{2}He$ and ${}^{4}_{2}\alpha$?
13. What is a mass number? Which particles have a mass number of zero?
14. What are the key differences between the equations written for chemical reactions and for nuclear reactions? How are they similar?
15. Can all the kinds of nuclear decay reactions discussed be characterized by the general equation: parent $\longrightarrow$ daughter + particle? Explain your answer.
16. Which types of nuclear decay reactions conserve both mass number and atomic number? Which conserve mass number but not atomic number? Which do not convert one element to another?
17. Describe a radioactive decay series. How many series occur naturally? Of these, which one no longer occurs in nature? Why?
18. Only nine naturally occurring elements have an atomic number greater than 83, and all of them are radioactive. Except for some isotopes of uranium that have a very long half-life, the half-lives of the heavy elements are so short that these elements should have been completely converted to lighter, more stable elements long ago. Why are these elements still present in nature?
19. Why are neutrons preferred to protons in the preparation of new isotopes of the lighter elements?
20. Why are particle accelerators and cyclotrons needed to create the transuranium elements?

20.3 The Interaction of Nuclear Radiation with Matter

21. Why are many radioactive substances warm to the touch? Why do many radioactive substances glow?
22. Describe the differences between nonionizing and ionizing radiation in terms of the intensity of energy emitted and the effect each has on an atom or molecule after collision. Which nuclear decay reactions are more likely to produce ionizing radiation? Which are more likely to produce nonionizing radiation?
23. Would you expect nonionizing or ionizing radiation to be more effective at treating cancer? Why?
24. Historically, concrete shelters have been used to protect people from nuclear blasts. Comment on the effectiveness of these shelters.
25. Gamma rays are a very high-energy radiation, yet α particles inflict more damage on biological tissue. Why?
26. List the three primary sources of naturally occurring radiation. Explain the factors that influence the dose that one receives throughout the year. Which is the largest contributor to overall exposure? Which is the most hazardous?
27. Because radon is a noble gas, it is inert and generally unreactive. Despite this, exposure to even low concentrations of radon in air is quite dangerous. Describe the physical consequences of exposure to radon gas. Why are people who smoke more susceptible to these effects?
28. Most medical imaging uses isotopes that have extremely short half-lives. These isotopes usually undergo only one kind of nuclear decay reactions. Which kind of decay reaction is usually utilized, and why? Why would a short half-life be preferred in these cases?
29. Which would you prefer: one exposure of 100 rem, or ten exposures of 10 rem each? Explain your rationale.

20.4 Thermodynamic Stability of the Atomic Nucleus

30. How do chemical reactions compare with nuclear reactions with respect to mass changes? Does either type of reaction violate the law of conservation of mass? Explain your answers.
31. Why is the amount of energy released by a nuclear reaction so much greater than the amount of energy released by a chemical reaction?
32. Explain why the mass of an atom is less than the sum of the masses of its component particles.
33. The stability of a nucleus can be described using two values. What are they, and how do they differ from each other?
34. In the days before true chemistry, ancient scholars (*alchemists*) attempted to find the *philosopher's stone*, a material that would enable them to turn lead into gold. Is the conversion of Pb $\longrightarrow$ Au energetically favorable? Explain why or why not.
35. Describe the energy barrier to nuclear fusion reactions, and explain how it can be overcome.
36. Imagine that the universe is dying, the stars have burned out, and all the elements have undergone fusion or radioactive decay. What would be the most abundant element in this future universe, and why?
37. Numerous elements can undergo fission, but only a few can be utilized as fuels in a reactor. What aspect of nuclear fission allows a nuclear chain reaction to occur?
38. How are transmutation reactions and fusion reactions related? Describe the main impediment to fusion reactions, and suggest one or two ways to surmount this difficulty.

20.5 Applied Nuclear Chemistry

39. In nuclear reactors, two different but interrelated factors must be controlled to prevent a mishap that could cause the release of unwanted radiation. How are these factors controlled?
40. What are the three principal components of a nuclear reactor? What is the function of each component?
41. What is meant by the term *enrichment* with regard to uranium for fission reactors? Why does the fuel in a conventional nuclear reactor have to be "enriched"?
42. The plot in a recent spy/action movie involved the threat of introducing stolen "weapons-grade" uranium, which consists of 93.3% ^{235}U, into a fission reactor that normally uses a fuel containing about 3% ^{235}U. Explain why this could be catastrophic.
43. Compare a heavy-water reactor with a light-water reactor. Why are heavy-water reactors less widely used? How do these two reactor designs compare with a breeder reactor?
44. Conventional light-water fission reactors require enriched fuel. An alternative reactor is the so-called heavy-water reactor. The components of the two different reactors are the same except that instead of using water, H_2O, the moderator in a heavy-water reactor is D_2O, "heavy water." Because D_2O is more efficient than H_2O at slowing neutrons, heavy-water reactors do not require that the fuel be enriched to support fission. Why is D_2O more effective at slowing neutrons, and why does this allow unenriched fuels to be used?
45. Isotopes emit γ rays in random directions. What difficulties does this present for medical imaging, and how are these difficulties overcome?
46. If you needed to measure the thickness of 1.0-mm plastic sheets, what type of radiation would you use? Would the radiation source be the same if you were measuring steel of a similar thickness? What is your rationale? Would you want an isotope with a long or short half-life for this device?

20.6 The Origin of the Elements

47. Why do scientists believe that hydrogen is the building block of all the other elements? Why do scientists believe that helium-4 is the building block of the heavier elements?

48. How does a star produce such enormous amounts of heat and light? How are elements heavier than Ni formed?

49. Propose an explanation for the observation that elements with even atomic numbers are more abundant than elements with odd atomic numbers.

50. During the lifetime of a star, different reactions that form different elements are utilized to power the fusion furnace that keeps the star "lit." Explain the different reactions that dominate in the different stages of a star's life cycle and their effect on the temperature of the star.

51. A line in a popular song from the 1960's by Joni Mitchell stated: "We are stardust. . .". Does this statement have any merit, or is it just poetic? Justify your answer.

52. If the laws of physics were different and the primary element in the universe were boron-11 ($Z = 5$), what would be the next four most abundant elements? Propose nuclear reactions for their formation.

NUMERICAL

This section includes paired problems (marked by brackets) that require similar problem-solving skills.

20.1 The Components of the Nucleus

53. Use $^A_Z X$ notation to write a nuclear symbol for each isotope: (a) chlorine-39; (b) lithium-8; (c) osmium-183; (d) zinc-71.

54. Write the nuclear symbol for each isotope using $^A_Z X$ notation: (a) lead-212; (b) helium-5; (c) oxygen-19; (d) plutonium-242.

55. Give the number of protons, the number of neutrons, and the neutron-to-proton ratio for each isotope: (a) iron-57; (b) ^{185}W; (c) potassium-39; (d) ^{131}Xe.

56. Give the number of protons, the number of neutrons, and the neutron-to-proton ratio for each isotope: (a) technetium-99*m*; (b) ^{140}La; (c) radium-227; (d) ^{208}Bi.

57. Which of these nuclides do you expect to be radioactive? Explain your reasoning. (a) ^{20}Ne; (b) tungsten-184; (c) ^{106}Ti

58. Which of these nuclides do you expect to be radioactive? Explain your reasoning. (a) ^{107}Ag; (b) ^{50}V; (c) lutetium-176

20.2 Nuclear Reactions

59. What type of particle is emitted in each nuclear reaction? (a) $^{218}Po \longrightarrow {}^{214}Po$; (b) $^{32}Si \longrightarrow {}^{32}P$; (c) $^{18}F \longrightarrow {}^{18}O$; (d) $^{206}Tl \longrightarrow {}^{206}Pb$

60. Identify the type of particle emitted in each nuclear reaction: (a) $^{230}Th \longrightarrow {}^{226}Ra$; (b) $^{218}Po \longrightarrow {}^{218}At$; (c) $^{210}Bi \longrightarrow {}^{206}Tl$; (d) $^{36}Cl \longrightarrow {}^{36}S$.

61. Predict the mode of decay and write a balanced nuclear reaction for each of the following: (a) ^{235}U; (b) ^{254}Es; (c) ^{36}S; (d) ^{99}Mo.

62. Predict the mode of decay and write a balanced nuclear reaction for each of the following: (a) ^{13}N; (b) ^{231}Pa; (c) ^{7}Be; (d) ^{77}Ge.

63. Balance each nuclear reaction:

(a) $^{208}Po \longrightarrow \alpha + Pb$

(b) $^{226}Ra \longrightarrow \alpha + Rn$

(c) $^{228}Th \longrightarrow Ra + \alpha + \gamma$

(d) $^{231}Pa \longrightarrow Ac + \alpha + \gamma$

(e) $Ho \longrightarrow {}^{166}Er + \beta^- + \gamma$

(f) $Ac \longrightarrow {}^{223}Th + 3\beta^- + \gamma + \alpha$

64. Complete the nuclear reactions:

(a) $^{210}_{84}Po \longrightarrow {}^{206}Pb$

(b) $^{217}_{85}At \longrightarrow Bi + \alpha$

(c) $Ra \longrightarrow {}^{220}_{86}Rn + \alpha$

(d) $^{208}Tl \longrightarrow {}_{82}Pb + \beta^-$

(e) $Np \longrightarrow {}^{239}Pu + \beta^-$

(f) $Fe \longrightarrow {}^{52}Mn + \beta^+ + \gamma$

65. Write a balanced equation for each nuclear reaction: (a) β^- decay of ^{87}Rb; (b) β^+ decay of ^{20}Mg; (c) α decay of ^{268}Mt.

66. Write a balanced equation for each nuclear reaction: (a) β^- decay of ^{45}K; (b) β^+ decay of ^{41}Sc; (c) α decay of ^{146}Sm.

67. The decay products of several isotopes are listed here. Identify the type of radiation emitted, and write a balanced nuclear equation for each. (a) $^{218}Po \longrightarrow {}^{214}Pb$; (b) $^{32}Si \longrightarrow {}^{32}P$; (c) an excited state of an iron-57 nucleus decaying to its ground state; (d) conversion of thallium-204 to lead-204.

68. The decay products of several isotopes are listed here. Identify the type of radiation emitted, and write a balanced nuclear equation for each: (a) $^{218}Po \longrightarrow {}^{218}At$; (b) $^{230}Th \longrightarrow {}^{226}Ra$; (c) bismuth-211 converted to thallium-207; (d) americium-242 converted to rhodium-107 with the emission of four neutrons.

69. Predict the most likely mode of decay and write a balanced nuclear reaction for each isotope: (a) ^{238}U; (b) ^{208}Po; (c) ^{40}S; (d) molybdenum-93*m*.

70. Predict the most likely mode of decay and write a balanced nuclear reaction for each isotope: (a) ^{16}N; (b) ^{231}Pa; (c) ^{118}In; (d) ^{77}Ge.

71. For each nuclear reaction, identify the type(s) of decay, and write a balanced nuclear equation:

(a) $^{216}Po \longrightarrow ? + At$

(b) $? \longrightarrow \alpha + {}^{231}Pa$

(c) $^{228}Th \longrightarrow ? + \alpha + \gamma$

(d) $^{231}Pa \longrightarrow ? + \beta^- + \gamma$

72. For each nuclear reaction, identify the type(s) of decay, and write a balanced nuclear reaction sequence for each decay:

(a) $^{210}Po \longrightarrow {}^{206}Pb + ?$

(b) $^{192}Ir \longrightarrow Pt + ?$

(c) $^{241}Am \longrightarrow {}^{57}Fe + {}^{184}? + ?$

(d) $Ge \longrightarrow {}^{77}Ge + ?$

73. Identify the parent isotope, and write a balanced nuclear reaction for each of the following: (a) lead-205 is formed via an alpha emission; (b) titanium-46 is formed via beta and gamma emission; (c) krypton-36 is formed via a beta decay and a gamma emission.

74. Identify the parent isotope, and write a balanced nuclear reaction for each of the following: (a) iodine-130 formed by ejection of an electron and a gamma ray from a nucleus; (b) uranium-240 formed by alpha decay; (c) curium-247 formed with the release of a helium dication and a gamma ray.

75. Write a balanced nuclear equation for each:

(a) Bromine undergoes a decay and produces a gas with an atomic mass of 81 amu.

(b) An element decays into two metals while emitting two neutrons, each of which can be extracted and converted to chlorides with the formula MCl_2. The masses of the two salts are 162.9 and 210.9 g/mol, respectively.

76. Write a balanced nuclear equation for each case:
(a) An unknown element emits γ rays plus particles that are readily blocked by paper. The sample also contains a substantial quantity of indium-115.
(b) An unstable element undergoes two different decay reactions: beta decay to produce a material with a mass of 222 amu, and alpha decay to astatine.

77. Bombarding ^{249}Cf with ^{12}C produced a transuranium element with a mass of 257 amu, plus several neutral subatomic particles. Identify the element, and write a nuclear reaction for this transmutation.

78. One transuranium element, ^{253}Es, is prepared by bombarding ^{238}U with 15 neutrons. What is the other product of this reaction? Write a balanced transmutation reaction for this conversion.

79. Complete this radioactive decay series:

$$^{223}_{88}\text{Ra} \xrightarrow{\alpha} \text{Rn} \xrightarrow{\alpha} \text{Po} \xrightarrow{\alpha} \text{Pb} \xrightarrow{\beta^-} \text{Bi} \xrightarrow{\alpha} \text{Tl} \xrightarrow{\beta^-} \text{Pb}$$

80. Complete each nuclear fission reaction:
(a) $^{235}_{92}\text{U} + ^{1}_{0}\text{n} \longrightarrow ^{90}_{36}\text{Kr} + \_\_\_ + 2^{1}_{0}\text{n}$
(b) $\_\_\_ + ^{1}_{0}\text{n} \longrightarrow ^{140}\text{Cs} + ^{96}\text{Y} + 4^{1}_{0}\text{n}$

81. Complete the nuclear fission reactions:
(a) $^{235}_{92}\text{U} + ^{1}_{0}\text{n} \longrightarrow ^{145}_{57}\text{La} + \_\_\_ + 3^{1}_{0}\text{n}$
(b) $\_\_\_ + ^{1}_{0}\text{n} \longrightarrow ^{95}_{42}\text{Mo} + ^{139}_{57}\text{La} + 2^{1}_{0}\text{n} + 7\,^{0}_{-1}\text{e}$

82. A stable nuclide absorbs a neutron, emits an electron, and then splits into two α particles. Identify the nuclide.

83. Using ^{18}O, how would you synthesize an element with atomic number 106 from ^{249}Cf? Write a balanced equation for the reaction.

84. How would you synthesize an element with atomic number 103 using ^{10}B and ^{252}Cf? Write a balanced equation for the reaction.

20.3 The Interaction of Nuclear Radiation with Matter

85. A 2.14-kg sample of rock contains 0.0985 g of uranium. How much energy is emitted over 25 years if 99.27% of the uranium is ^{238}U, which has a half-life of 4.46×10^9 years, if each decay event is accompanied by the release of 4.039 MeV? If a 180-lb individual absorbes all of the emitted radiation, how much radiation has been absorbed in rads?

86. There is a story about a "radioactive boy scout" who attempted to convert thorium-232, which he isolated from about a thousand gas lantern mantles, to uranium-233 by bombarding the thorium with neutrons. The neutrons were generated via bombardment of an aluminum target with α particles from the decay of americium-241, which was isolated from a hundred smoke detectors. Write balanced nuclear reactions for these processes. The "radioactive boy scout" spent approximately 2 hours a day with his experiment for 2 years. Assuming that the alpha emission of americium has an energy of 5.24 MeV/particle and that the americium-241 was undergoing 3.5×10^6 decays per second, what was the exposure of the 60.0-kg scout in rad? The intrepid scientist has shown no ill effects from this exposure. Why?

20.4 Thermodynamic Stability of the Atomic Nucleus

This table is provided as a reference.

Isotope	Mass (amu)	Isotope	Mass (amu)
^{8}B	8.024607	^{209}Fr	208.99592
^{40}K	39.963998	^{210}Po	209.982874
^{52}Cr	51.940508	^{212}At	211.990745
^{58}Ni	57.935343	^{214}Pb	213.999797
^{59}Co	58.933195	^{214}Bi	213.998712
^{60}Co	59.933817	^{216}Fr	216.003198
^{60}Ni	59.930786	^{199}Pb	198.972917
^{90}Sr	89.907738	^{222}Rn	222.017578
^{92}Kr	91.926156	^{226}Ra	226.025410
^{141}Ba	140.914411	^{227}Ra	227.029178
^{143}Xe	142.935110	^{228}Ac	228.031021
^{167}Os	166.971550	^{230}Th	230.033134
^{171}Pt	170.981240	^{233}U	233.039635
^{194}Hg	193.965439	^{234}Th	234.043601
^{194}Tl	193.971200	^{234}Pa	234.043308
^{199}Pb	198.972917	^{233}U	233.039635
^{199}Bi	198.977672	^{234}U	234.040952
^{206}Pb	205.974465	^{235}U	235.043930
^{207}Pb	206.975897	^{238}Pa	238.054500
^{208}Pb	207.976652	^{238}U	238.050788
^{208}Bi	207.979742	^{239}Pu	239.052163
^{208}Po	207.981246	^{245}Pu	245.067747

87. Using information provided in Table 20.6 and the table above, complete each reaction and calculate the amount of energy released from each in kJ:
(a) $^{238}\text{Pa} \longrightarrow \_\_\_ + \beta^-$
(b) $^{209}\text{Fr} \longrightarrow \_\_\_ + \alpha$
(c) $^{199}\text{Bi} \longrightarrow \_\_\_ + \beta^+$

88. Using information provided in Table 20.6 and the table above, complete each reaction and calculate the amount of energy released from each in kJ:
(a) $^{194}\text{Tl} \longrightarrow \_\_\_ + \beta^+$
(b) $^{171}\text{Pt} \longrightarrow \_\_\_ + \alpha$
(c) $^{214}\text{Pb} \longrightarrow \_\_\_ + \beta^-$

89. Using information provided in Table 20.6 and the table above, complete each reaction and calculate the amount of energy released from each in kJ/mol:
(a) $^{234}_{91}\text{Pa} \longrightarrow ? + ^{0}_{-1}\beta$
(b) $^{226}_{88}\text{Ra} \longrightarrow ? + ^{4}_{2}\alpha$

90. Using information provided in Table 20.6 and the table above, complete the reactions and then calculate the amount of energy released from each in kJ/mol:
(a) $^{60}_{27}\text{Co} \longrightarrow ? + ^{0}_{-1}\beta$ (The mass of cobalt-60 is 59.933817 amu.)
(b) technicium-94 (mass = 93.909657) undergoing spontaneous fission to produce chromium-52 and potassium-40

91. Using information provided in Table 20.6 and the table above, predict whether each reaction is favorable and the amount of energy released or required in MeV and kJ/mol:
(a) the beta decay of bismuth-208 (mass = 207.979742 amu)
(b) the formation of lead-206 by alpha decay

92. Using information provided, predict whether each reaction is favorable and the amount of energy released or required in MeV and kJ/mol:
(a) the alpha decay of oxygen-16
(b) alpha decay to produce chromium-52

93. Calculate the total nuclear binding energy (in MeV) and the binding energy per nucleon for ^{87}Sr if the measured mass of ^{87}Sr is 86.908877 amu.

94. Calculate the total nuclear binding energy (in MeV) and the binding energy per nucleon for ^{60}Ni and ^{87}Sr.

95. The experimentally determined mass of ^{53}Mn is 52.941290 amu. What is (a) the calculated mass; (b) the mass defect; (c) the nuclear binding energy; and (d) the nuclear binding energy per nucleon?

96. The experimental mass of ^{29}S is 28.996610 amu. What is (a) the calculated mass; (b) the mass defect; (c) the nuclear binding energy; and (d) the nuclear binding energy per nucleon?

97. Calculate the amount of energy that is released by the neutron-induced fission of ^{235}U to give ^{141}Ba, ^{92}Kr (mass = 91.926156 amu), and three neutrons. Report your answer in eV/atom and kJ/mol.

98. Calculate the amount of energy that is released by the neutron-induced fission of ^{235}U to give ^{90}Sr, ^{143}Xe, and three neutrons. Report your answer in eV/atom and kJ/mol.

99. Calculate the amount of energy that is released or required by the fusion of helium-4 to produce the unstable beryllium-8 (mass = 8.00530510 amu). Report your answer in kJ/mol. Do you expect this to be a spontaneous reaction?

100. Calculate the amount of energy released by the fusion of ^{6}Li and deuterium to give two helium-4 isotopes. Express your answer in eV/atom and kJ/mol.

101. How much energy is released by the fusion of two deuterium nuclei to give one tritium nucleus and one proton? How does this amount compare with the energy released by the fusion of a deuterium nucleus and a tritium nucleus, which is accompanied by ejection of a neutron? Express your answer in MeV and kJ/mol. Pound for pound, which is a better choice for a fusion reactor fuel mixture?

20.5 Applied Nuclear Chemistry

102. Palladium-103, which decays via electron capture and emits X rays with an energy of 3.97×10^{-2} MeV, is often used to treat prostate cancer. Small pellets of the radioactive metal are embedded *in* the prostate gland. This provides a localized source of radiation to a very small area, even though only about 1% of the X rays are absorbed by the tissue. Due to its short half-life, all of the palladium will decay to a stable isotope in less than a year. If a doctor embeds 50 pellets containing 2.50 mg of ^{103}Pd in the prostate gland of a 73.9-kg patient, what is his radiation exposure over the course of a year?

103. A number of medical treatments utilize cobalt-60*m*, which is formed by bombardment of cobalt with neutrons to produce a highly radioactive *gamma* emitter that undergoes 4.23×10^{16} emissions/s per kg of pure cobalt-60. The energy of the gamma emission is 5.86×10^{-2} MeV. Write the balanced nuclear equation for the formation of this isotope. Is this a transmutation reaction? If a 55.3-kg patient received a 0.50-s exposure to a 0.30-kg cobalt-60 source, what would the exposure be in rads? Predict potential side effects of this dose.

20.6 The Origin of the Elements

104. Write a balanced nuclear reaction for the formation of ^{26}Al from two ^{12}C nuclei and of ^{9}Be from two ^{4}He nuclei.

105. At the end of a star's life cycle, it can collapse, resulting in a supernova explosion that leads to the formation of heavy elements by multiple neutron-capture events. Write reactions for the formation of (a) ^{106}Pd from nickel-58 and (b) selenium-79 from iron-56 during such an explosion.

106. When a star reaches middle age, helium-4 is converted to short-lived beryllium-8, which reacts with another helium-4 to produce carbon. How much energy is released in each of these two reactions (in MeV)? How many atoms of helium must be "burned" in this process to produce the same amount of energy obtained from the fusion of 1 mol of hydrogen atoms to give deuterium?

APPLICATIONS

107. Until the 1940s, uranium glazes were popular on ceramic dishware. One brand, Fiestaware, had bright orange glazes that could contain up to 20% uranium by mass. Although this practice is less common today due to the negative association of radiation, it is still possible to buy Depression-era glassware that is quite radioactive. Aqueous solutions in contact with this "hot" glassware can reach uranium concentrations up to 10 ppm by mass. If 1 g of uranium undergoes 1.11×10^{7} decays/s, each to an α particle with an energy of 4.03 MeV, what would be your exposure in rem and rad if you drank 1 L of water that had been sitting for an extended time in a Fiestaware pitcher? Assume that the water and contaminants are excreted only after 18 h and that you weigh 70.0 kg.

108. Neutrography is a technique used to take the picture of an object using a beam of neutrons. How does the penetrating power of a neutron compare with *alpha*, *beta*, and *gamma* radiation? Do you expect similar penetration for protons? How would the biological damage of each particle compare with the other types of radiation? (Recall that a neutron's mass is approximately 2000 times the mass of an electron.)

109. Spent fuel elements in a nuclear reactor contain radioactive fission products in addition to heavy metals. The conversion of nuclear fuel in a reactor is shown in the diagram:

3.2% ^{235}U → 0.76% ^{235}U; 0.44% ^{236}U; 2.0% fission products

96.8% ^{238}U → 1.5% fission products; 94.3% ^{238}U; 0.9% Pu; 0.1% Np, Am, Cm

Neglecting the fission products, write balanced nuclear reactions for the conversion of the original fuel to each product.

110. The first atomic bomb used ^{235}U as a fissile material, but there were immense difficulties in obtaining sufficient quantities of pure ^{235}U. A second fissile element, plutonium, was discovered in 1940, and it rapidly became important as a nuclear fuel. This element was produced by irradiating ^{238}U with neutrons in a nuclear reactor. Complete the series that produced plutonium, all isotopes of which are fissile:

$$^{238}_{92}\text{U} + {}^{1}_{0}\text{n} \longrightarrow \text{U} \longrightarrow \text{Np} \longrightarrow \text{Pu}$$

111. Boron neutron capture therapy is a potential treatment for many diseases. As the name implies, when boron-10, one of the naturally occurring isotopes of boron, is bombarded with neutrons, it absorbs a neutron and emits an α particle. Write a balanced nuclear reaction for this reaction. One advantage of this process is that neutrons cause little damage on their own, but when they are absorbed by boron-10, they can cause localized emission of alpha radiation. Comment on the utility of this treatment and its potential difficulties.

112. An airline pilot typically flies approximately 80 h per month. If 75% of that time is spent at an altitude of about 30,000 ft, how much radiation is that pilot receiving in one month? Over a 30-year career? Is the pilot receiving toxic doses of radiation?

113. At a breeder reactor plant, a 72-kg employee accidentally inhaled 2.8 mg of ^{239}Pu dust. The isotope decays by alpha decay and has a half-life of 24,100 yr. If the energy of the emitted α particles is 5.2 MeV and the dust stays in the employee's body for 18 h, determine (a) the number of plutonium atoms inhaled; (b) the energy absorbed by the body; (c) the physical dose in rads; and (d) the dose in rems. Will the occurrence be fatal?

114. For many years, the standard source for radiation therapy in the treatment of cancer was radioactive ^{60}Co, which undergoes beta decay to ^{60}Ni and emits two γ rays, each with an energy of 1.2 MeV. Show the sequence of reactions. If the half-life of beta decay is 5.27 yr, how many ^{60}Co nuclei are present in a typical source emitting 6,000 disintegrations per second that is used by hospitals? The mass of ^{60}C is 59.93 amu.

115. It is possible to utilize radioactive materials as heat sources to produce electricity. These radioisotope thermoelectric generators (RTGs) have been used in spacecraft and many other applications. Certain Cold War–era Russian-made RTGs used a 5.0-kg strontium-90 source. One mole of strontium-90 releases β particles with an energy of 0.545 MeV and undergoes 2.7×10^{13} decays/s. How many watts of power are available from this RTG? (1 watt = 1 J/s)

116. Potassium consists of three isotopes (potassium-39, potassium-40, and potassium-41). Potassium-40 is the least abundant, and it is radioactive, decaying to argon-40, a stable, nonradioactive isotope, by the emission of a β particle with a half-life of precisely 1.25×10^9 yr. Thus, the ratio of potassium-40 to argon-40 in any potassium-40–containing material can be used to date the sample. In 1952, fragments of an early hominid, *Meganthropus*, were discovered near Modjokerto in Java. The bone fragments were lying on volcanic rock that was believed to be the same age as the bones. Potassium–argon dating on samples of the volcanic material showed that the argon-40-to-potassium-40 molar ratio was 0.00281:1. How old were the rock fragments? Could the bones also be the same age? Could radiocarbon dating have been used to date the fragments?

21 Periodic Trends and the *s*-Block Elements

LEARNING OBJECTIVES

- To be able to describe the physical and chemical properties of hydrogen and to predict its reactivity
- To be able to describe how the alkali and alkaline earth metals are isolated
- To become familiar with the reactions, compounds, and complexes of the alkali and alkaline earth metals
- To know some of the uses of the alkali and alkaline earth metals
- To become familiar with the role of the *s*-block elements in biology

CHAPTER OUTLINE

In this part, "Chemistry of the Elements," we will use the periodic table to guide our discussion of the properties and reactions of the elements and the synthesis and uses of some of their commercially important compounds. We first review periodic trends as an introduction to Chapters 21–23 in this chapter, and then we describe the chemistry of hydrogen and the other *s*-block elements. In Chapter 22, we consider the chemistry of the *p*-block elements, and in Chapter 23, we proceed to the transition metals, in which the *d*-subshell is being filled. The principles of chemical bonding, thermodynamics, and kinetics developed in earlier parts of this text provide a conceptual framework for understanding the chemistry of the elements. In this chapter, for instance, you will learn why potassium chloride is used as a substitute for sodium chloride in

Flame tests. Heating a compound in a very hot flame results in the formation of its component *atoms* in electronically excited states. When an excited atom decays to the ground state, it emits light (Chapter 6). Each element emits light at characteristic frequencies. Flame tests are used to identify many elements based on the color of light emitted in the visible region of the electromagnetic spectrum. As shown here, sodium compounds produce an intense yellow light, whereas calcium compounds produce a crimson color.

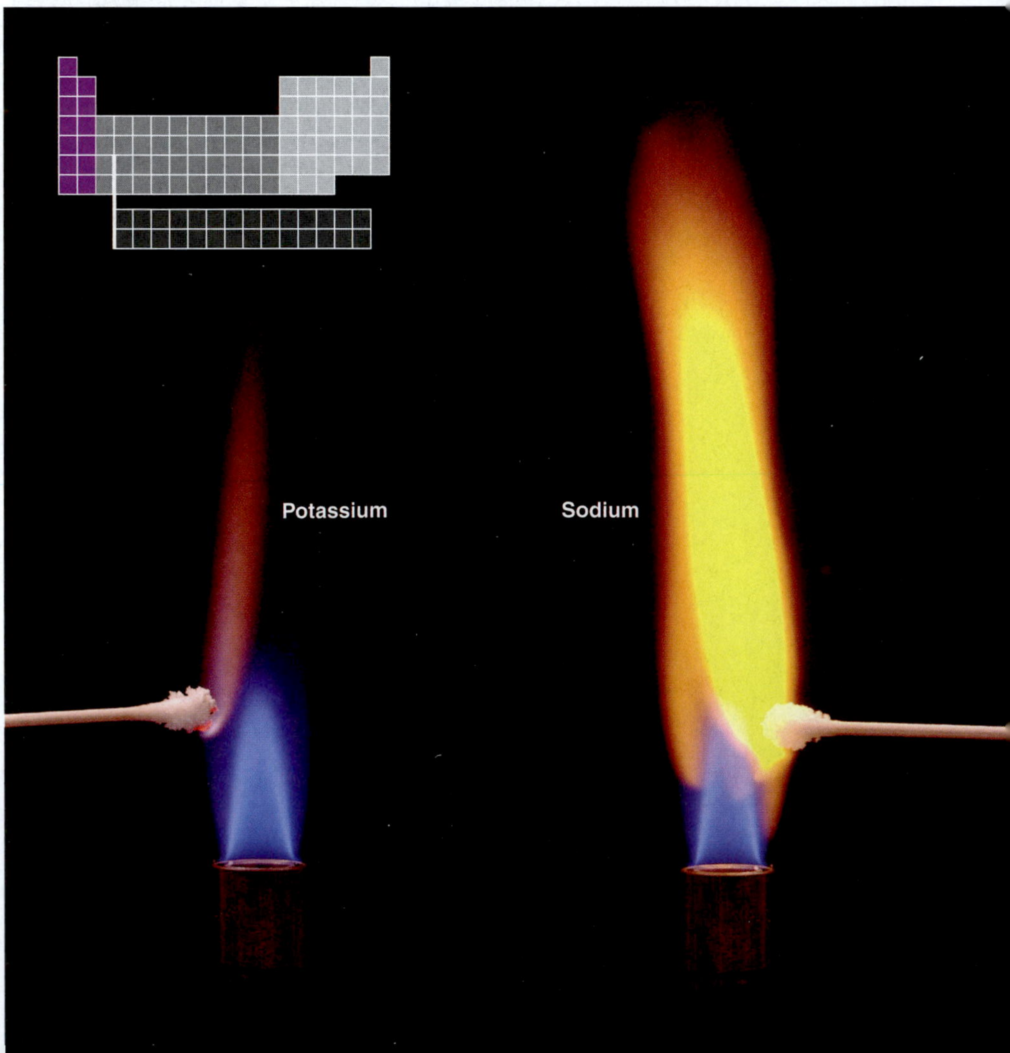

a low-sodium diet, why cesium is used as a photosensor, why the heating elements in electric ranges are coated with magnesium oxide, and why exposure to a radioactive isotope of strontium is more dangerous for children than for adults.

21.1 Overview of Periodic Trends

As you learned in Chapter 7, the single most important unifying principle in understanding the chemistry of the elements is the systematic increase in atomic number, accompanied by the orderly filling of atomic orbitals by electrons, which leads to periodicity in such properties as atomic and ionic size, ionization energy, electronegativity, and electron affinity. The same factors also lead to periodicity in valence-electron configurations, which for each group results in similarities in oxidation states and the formation of compounds with common stoichiometries.

The most important periodic trends in atomic properties are summarized in Figure 21.1. Recall from Chapter 7 that all these trends are based on periodic variations in a single fundamental property, the **effective nuclear charge (Z_{eff})**, which increases from left to right and from top to bottom in the periodic table (Figure 6.29).

The diagonal line in Figure 21.1 separates the metals (to the left of the line) from the nonmetals (to the right). Because metals have relatively low electronegativities, they tend to lose electrons in chemical reactions to elements that have relatively high electronegativities, forming compounds in which they have positive oxidation states. Conversely, nonmetals have high electronegativities, and they therefore tend to gain electrons in chemical reactions to form compounds in which they have negative oxidation states. The semimetals lie along the diagonal line dividing metals and nonmetals. It is not surprising that they tend to exhibit properties and reactivities intermediate between those of metals and nonmetals. Because the elements of Groups 13, 14, and 15 span the diagonal line separating metals and nonmetals, their chemistry is more complex than predicted based solely on their valence-electron configurations.

Unique Chemistry of the Lightest Elements

MGC Chemistry of the Lightest Elements

The chemistry of the second-period element of each group ($n = 2$; Li, Be, B, C, N, O, and F) differs in many important respects from that of the heavier members, or *congeners,* of the group. Consequently, the elements of the third period ($n = 3$; Na,

Figure 21.1 Summary of periodic trends in atomic properties. Ionization energies, the magnitude of electron affinities, and electronegativities generally increase from left to right and from bottom to top. In contrast, atomic size *decreases* from left to right and from bottom to top. Consequently, the elements in the upper right of the periodic table are the smallest and most electronegative; the elements in the bottom left are the largest and least electronegative. The semimetals lie along the diagonal line separating the metals from the nonmetals and exhibit intermediate properties. Interactive Graph

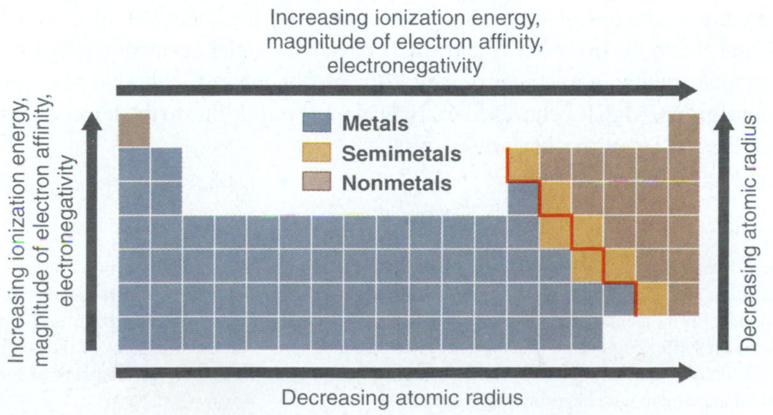

Note the pattern

In contrast to the chemistry of the second-period elements, the chemistry of the third-period elements is more representative of the chemistry of the respective group.

Mg, Al, Si, P, S, and Cl) are generally more representative of the group to which they belong. The anomalous chemistry of second-period elements results from three important characteristics: small radii, energetically unavailable *d* orbitals, and a tendency to form π bonds with other atoms.

Due to their small radii, second-period elements have electron affinities that are less negative than would be predicted from general periodic trends. When an electron is added to such a small atom, increased electron–electron repulsions tend to destabilize the anion. Moreover, the small sizes of these elements prevent them from forming compounds in which they have more than four nearest neighbors. Thus, $A1F_3$ forms the six-coordinate, octahedral $A1F_6^{3-}$ ion, whereas under the same conditions, BF_3 forms only the four-coordinate, tetrahedral BF_4^- ion. Because of the smaller atomic size, simple binary ionic compounds of second-period elements also have more covalent character than the corresponding compounds formed from their heavier congeners. The very small cations derived from second-period elements have a high *charge-to-radius ratio* and can therefore polarize the filled valence shell of an anion. Consequently, the bonding in such compounds has a significant covalent component, giving the compounds properties that can differ significantly from those expected for simple ionic compounds. As an example, LiCl, which is partially covalent in character, is much more soluble than NaCl in solvents with a relatively low dielectric constant, such as ethanol ($\varepsilon = 25.3$ vs. 80.1 for H_2O).

Because *d* orbitals are never occupied for principal quantum numbers less than 3, the valence electrons of second-period elements occupy 2*s* and 2*p* orbitals only. To form an ion such as BF_6^{3-} would require using the lowest-energy *d* orbitals available, the 3*d* orbitals. However, the energy of these orbitals so far exceeds the energy of the 2*s* and 2*p* orbitals that using them in bonding is energetically prohibitive. Consequently, electron configurations with more than four electron pairs around a central, second-period element are simply not observed.*

One of the most dramatic differences between the lightest main group elements and their heavier congeners is the tendency of the second-period elements to form species that contain multiple bonds. For example, N is just above P in Group 15: N_2 contains an N≡N triple bond, but each phosphorus atom in tetrahedral P_4 forms three P—P single bonds. This difference in behavior reflects the fact that within the same group of the periodic table, the relative energies of the pi (π) bond and the sigma (σ) bond differ. A C=C bond, for example, is approximately 80% stronger than a C—C bond. In contrast, an Si=Si bond, with less *p*-orbital overlap between the valence orbitals of the bonded atoms because of the larger atomic size, is only about 40% stronger than an Si—Si bond. Consequently, compounds that contain both multiple and single C to C bonds are common for carbon, but compounds that contain only sigma Si to Si bonds are more energetically favorable for silicon and the other third-period elements.

Another important trend to note in main group chemistry is the chemical similarity between the lightest element of one group and the element immediately below and to the right of it in the next group, a phenomenon known as the *diagonal effect* (Figure 21.2) There are, for example, significant similarities between the chemistry of Li and Mg, Be and Al, and B and Si. Both $BeCl_2$ and $AlCl_3$ have substantial covalent character, so they are somewhat soluble in nonpolar organic solvents. In contrast, although Mg and Be are in the same group, $MgCl_2$ behaves like a typical *ionic halide* due to the lower electronegativity and larger size of magnesium.

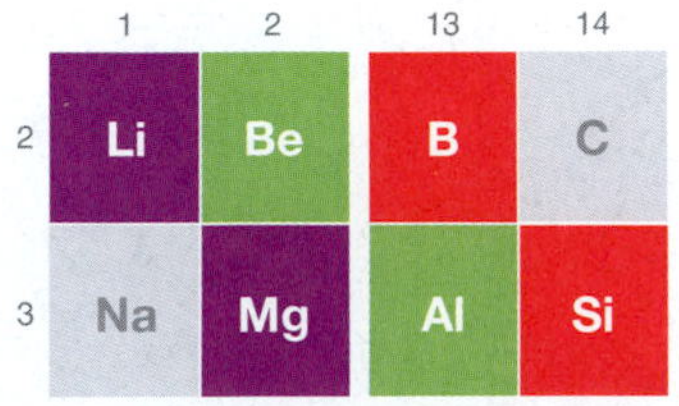

Figure 21.2 The diagonal effect. The properties of the lightest element in a group are often more similar to those of the element below and to the right in the periodic table. For instance, the chemistry of lithium is more similar to that of magnesium in Group 2 than it is to the chemistry of sodium, the next member in Group 1.

* You may recall from Chapter 8 that the role of *d* orbitals in bonding in main group compounds with coordination numbers of 5 or higher remains somewhat controversial. In fact, theoretical descriptions of the bonding in molecules such as SF_6 have been published without mentioning the participation of *d* orbitals on sulfur. Arguments based on *d*-orbital availability and on the small size of the central atom, however, predict that coordination numbers higher than 4 are unusual for the elements of the second period, in agreement with experimental results.

The Inert-Pair Effect

MGC The Inert-Pair Effect

The **inert-pair effect** refers to the empirical observation that the heavier elements of Groups 13–17 often have oxidation states that are lower by 2 than the maximum predicted for their group. For example, although an oxidation state of +3 is common for Group 13 elements, the heaviest element in Group 13, thallium (Tl), is more likely to form compounds in which it has a +1 oxidation state. There appear to be two major reasons for the inert-pair effect: increasing ionization energies and decreasing bond strengths.

The ionization energies increase because filled $(n-1)d$ or $(n-2)f$ subshells are relatively poor at shielding electrons in *ns* orbitals. Thus, the two electrons in the *ns* subshell experience an unusually high effective nuclear charge Z_{eff}, and they are strongly attracted to the nucleus, reducing their participation in bonding. It is therefore substantially more difficult than expected to remove these ns^2 electrons, as shown in Table 21.1 by the difference between the first ionization energies of thallium and aluminum. Because Tl is less likely than Al to lose its two ns^2 electrons, its most common oxidation state is +1 rather than +3.

Going down a group, the atoms generally became larger, and the overlap between the valence orbitals of the bonded atoms decreases. Consequently, bond strengths tend to decrease down a column. As shown by the M—Cl bond energies listed in Table 21.1, the strength of the bond between a Group 13 atom and a chlorine atom decreases by more than 30% from B to Tl. Similar decreases are observed for the atoms of Groups 14 and 15.

The net effect of these two factors—increasing ionization energies and decreasing bond strengths—is that as you go down a group in the *p*-block, the additional energy released by forming two additional bonds eventually is not great enough to compensate for the additional energy required to remove the two ns^2 electrons.

EXAMPLE 21.1

Based on the positions of the Group 13 elements in the periodic table and the general trends outlined in this section, **(a)** classify these elements as metals, semimetals, or nonmetals; **(b)** predict which element forms the most stable compounds in the +1 oxidation state; **(c)** predict which element differs the most from the others in its chemistry; and **(d)** predict which element of another group will exhibit chemistry most similar to that of Al.

Given Positions of elements in periodic table

Asked for Classification, oxidation-state stability, chemical reactivity

TABLE 21.1 Ionization energies (*I*) and average M—Cl bond energies for the Group 13 elements

Element	Electron Configuration	I_1, kJ/mol	$I_1 + I_2 + I_3$, kJ/mol	Average M—Cl Bond Energy, kJ/mol*
B	[He] $2s^2\,2p^1$	801	6828	536
Al	[Ne] $3s^2\,3p^1$	578	5139	494
Ga	[Ar] $3d^{10}\,4s^2\,4p^1$	579	5521	481
In	[Kr] $4d^{10}\,5s^2\,5p^1$	558	5083	439
Tl	[Xe] $4f^{14}\,5d^{10}\,6s^2\,6p^1$	589	5439	373

*Source of data: *Lange's Handbook of Chemistry* (1999)

Strategy

From the position of the diagonal line in the periodic table separating metals and nonmetals, classify the Group 13 elements. Then use the trends discussed in this section to compare their relative stabilities and chemical reactivities.

Solution

(a) Group 13 spans the diagonal line separating the metals from the nonmetals. Although Al and B both lie on the diagonal line, only B is a semimetal; the heavier elements are metals.

(b) All five elements in Group 13 have an ns^2np^1 valence-electron configuration, so they are expected to form ions with a +3 charge from the loss of all valence electrons. The inert-pair effect should be most important for the heaviest element, Tl, so it is most likely to form compounds in an oxidation state that is lower by 2. Thus, the +1 oxidation state is predicted to be most important for thallium.

(c) Among the main group elements, the lightest member of each group exhibits unique chemistry because of its small size resulting in a high concentration of charge, energetically unavailable *d* orbitals, and a tendency to form multiple bonds. In Group 13, we predict that the chemistry of boron will be quite different from that of its heavier congeners.

(d) Within the *s* and *p* blocks, similarities between elements in different groups are most marked between the lightest member of one group and the element of the next group immediately below and to the right. These elements exhibit similar electronegativities and charge-to-radius ratios. Because Al is the second member of Group 13, we predict that its chemistry will be most similar to that of Be, the lightest member of Group 2.

EXERCISE 21.1

Based on the positions of the Group 14 elements C, Si, Ge, Sn, and Pb in the periodic table and the general trends outlined in this section, **(a)** classify these elements as metals, semimetals, or nonmetals; **(b)** predict which element forms the most stable compounds in the +2 oxidation state; **(c)** predict which element differs the most from the others in its chemistry; and **(d)** predict which element of Group 14 will be chemically most similar to a Group 15 element.

Answer **(a)** nonmetal: C; semimetals: Si and Ge; metals: Sn and Pb. **(b)** Pb most stable as M^{2+}. **(c)** C most different. **(d)** C and P most similar in chemistry.

21.2 • The Chemistry of Hydrogen

Note the pattern

Although hydrogen has an ns^1 electron configuration, its chemistry does not resemble that of the metals of Group 1.

We now turn from an overview of periodic trends to a discussion of the *s*-block elements, first by focusing on hydrogen, whose chemistry is sufficiently distinct and important to be discussed in a category of its own. Most versions of the periodic table place hydrogen in the upper left corner immediately above lithium, implying that hydrogen, with a $1s^1$ electron configuration, is a member of Group 1. In fact, the chemistry of hydrogen does *not* greatly resemble that of the metals of Group 1. Indeed, some versions of the periodic table place hydrogen above fluorine in Group 17 because the addition of a single electron to a hydrogen atom completes its valence shell.

Isotopes of Hydrogen

Hydrogen, the most abundant element in the universe, is the ultimate source of all the other elements by the process of nuclear fusion (see Chapter 20). Table 21.2 compares the three isotopes of hydrogen, all of which contain one proton and one electron per atom. The most common isotope is **protium** (^1H or H), followed by **deuterium** (^2H or D), which has an additional neutron. The rarest isotope of hydrogen is **tritium**

TABLE 21.2 The isotopes of hydrogen

	Protium (H)	Deuterium (D)	Tritium (T)
Symbol	^1_1H	^2_1H	^3_1H
Neutrons	0	1	2
Mass, amu	1.00783	2.0140	3.01605
Abundance, %	99.9885	0.0115	$\sim 10^{-17}$
Half-life, years	—	—	12.32
bp of X_2, K	20.28	23.67	25
mp/bp of X_2O, °C	0.0/100.0	3.8/101.4	4.5/?

Harold Urey. Urey (1893–1981) won the Nobel Prize in chemistry for his discovery of deuterium (^2H). Urey was born and educated in rural Indiana. After earning a B.S. in zoology from the University of Montana in 1917, Urey changed career directions. He earned his Ph.D. in chemistry at Berkeley with G. N. Lewis (of Lewis electron structure fame) and subsequently worked with Niels Bohr in Copenhagen. During World War II, Urey was the director of war research for the Atom Bomb Project at Columbia University. In later years, his research focused on the evolution of life. In 1953, he and his graduate student, Stanley Miller, showed that organic compounds, including amino acids, could be formed by passing an electric discharge through a mixture of compounds thought to be present in the atmosphere of primitive Earth.

(^3H or T), which is produced in the upper atmosphere by a nuclear reaction when cosmic rays strike nitrogen and other atoms; it is then washed into the oceans by rainfall. Tritium is radioactive, decaying to ^3He with a half-life of only 12.32 years. Consequently, the atmosphere and ocean contain only a very low, steady-state level of tritium. Normally the term *hydrogen* and the symbol H refer to the naturally occurring mixture of the three isotopes.

The different masses of the three isotopes of hydrogen cause them to have different physical properties. Thus, H_2, D_2, and T_2 differ in their melting points, boiling points, densities, and heats of fusion and vaporization. In 1931, Harold Urey and co-workers discovered deuterium by slowly evaporating several liters of liquid hydrogen until a volume of about 1 mL remained. When that remaining liquid was vaporized and its emission spectrum examined, Urey observed new absorption lines in addition to those previously identified as originating from hydrogen. The natural abundance of tritium, in contrast, is so low that it could not be detected by similar experiments; it was first prepared in 1934 by a nuclear reaction.

Because the normal boiling point of D_2O is 101.4°C (compared to 100.0°C for H_2O), evaporation or fractional distillation can be used to increase the concentration of deuterium in a sample of water by selective removal of the more volatile H_2O. Thus, bodies of water that have no outlet, such as the Great Salt Lake and the Dead Sea, which maintain their level solely by evaporation, have significantly higher concentrations of deuterated water than does lake or seawater with at least one outlet. A more efficient way to obtain water highly enriched in deuterium is by prolonged electrolysis of an aqueous solution. Because a deuteron (D^+) has twice the mass of a proton (H^+), it diffuses more slowly toward the electrode surface. Consequently, the gas evolved at the cathode is enriched in H, the species that diffuses more rapidly, favoring the formation of H_2 over D_2 or HD. Meanwhile, the solution becomes enriched in deuterium. Deuterium-rich water is called *heavy water* because the density of D_2O (1.1044 g/cm^3 at 25°C) is higher than that of H_2O (0.99978 g/cm^3). Heavy water was an important constituent of early nuclear reactors (see Chapter 20).

Because deuterons diffuse so much more slowly, D_2O will not support life and is actually toxic if administered to mammals in large amounts. The rate-limiting step in many important reactions catalyzed by enzymes involves proton transfer. The transfer of D^+ is so slow compared with that of H^+ because bonds to D break more slowly than those to H, so the delicate balance of reactions in the cell is disrupted. Nonetheless, deuterium and tritium are important research tools for biochemists. By incorporating these isotopes into specific positions in selected molecules, where they act as labels, or *tracers*, biochemists can follow the path of a molecule through an organism or cell. Tracers can also be used to provide information about the mechanism of enzymatic reactions.

Bonding in Hydrogen and Hydrogen-Containing Compounds

The $1s^1$ electron configuration of hydrogen indicates a single valence electron. Because the 1*s* orbital has a maximum capacity of two electrons, hydrogen can form compounds with other elements in three ways (Figure 21.3):

- By losing its electron to form a proton (H^+) with an empty 1*s* orbital. The proton is a Lewis acid that can accept a pair of electrons from another atom to form an electron-pair bond. In the acid–base reactions discussed in Chapter 16, for example, the proton always binds to a lone pair of electrons on an atom in another molecule to form a polar covalent bond. If the lone pair of electrons belongs to an oxygen atom of a water molecule, the result is the hydronium ion, H_3O^+.
- By accepting an electron to form a **hydride ion (H^-)**, which has a filled $1s^2$ orbital. Hydrogen reacts with relatively electropositive metals, such as the alkali metals (Group 1) and alkaline earths (Group 2), to form *ionic hydrides*, which contain metal cations and H^- ions.
- By sharing its electron with an electron on another atom to form an electron-pair bond. With a half-filled $1s^1$ orbital, the hydrogen atom can interact with singly occupied orbitals on other atoms to form either a covalent or polar covalent electron-pair bond, depending on the electronegativity of the other atom.

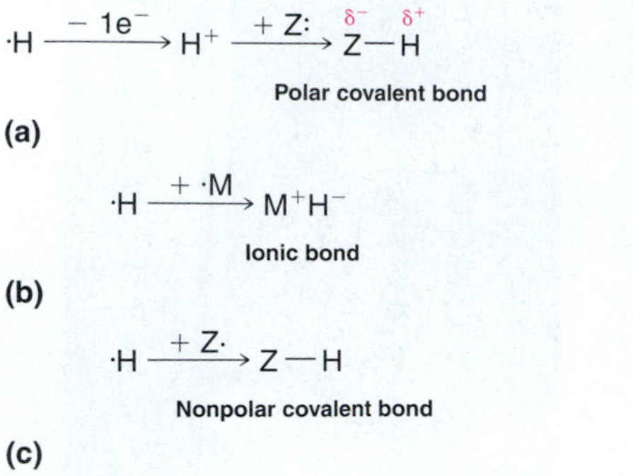

Figure 21.3 Three types of bonding in compounds of hydrogen. Because of its $1s^1$ electron configuration and the fact that the 1*s* orbital can accommodate no more than two electrons, hydrogen can bond to other elements in one of three ways: (a) it can lose an electron to form a proton, which can accept a pair of electrons from a more electronegative atom to form a *polar covalent bond*; (b) hydrogen can gain an electron from an electropositive metal to form a hydride ion, resulting in an *ionic hydride*; or (c) hydrogen can share its half-filled 1*s* orbital with a half-filled orbital on another atom to form a covalent or polar covalent electron-pair bond.

Note the pattern

Hydrogen can lose its electron to form H^+, accept an electron to form H^-, share its electron, hydrogen bond, or form a three-center bond.

Hydrogen can also act as a bridge between two atoms. One familiar example is the **hydrogen bond**, an electrostatic interaction between a hydrogen bonded to an electronegative atom and an atom that has one or more lone pairs of electrons (Figure 21.4). An example of this kind of interaction is the hydrogen-bonded network found in water (see Figure 11.8). Hydrogen can also form a **three-center bond**, in which a hydride bridges two electropositive atoms. Compounds that contain hydrogen bonded to boron and similar elements often have this type of bonding. The B—H—B units found in boron hydrides cannot be described in terms of localized electron-pair bonds. Because the H atom in the middle of such a unit can accommodate a maximum of only two electrons in its 1*s* orbital, the B—H—B unit can be described as containing a hydride that interacts simultaneously with empty sp^3 orbitals on two boron atoms (Figure 21.5). In these bonds, only *two* bonding electrons are used to hold *three* atoms together. Consequently, such bonds are often called *electron-deficient bonds*. You encountered a similar phenomenon in the discussion of π bonding in ozone and the nitrite ion in Section 9.4. Recall that in both these cases, we used the presence of two electrons in a π molecular orbital extending over three atoms to explain the fact that the two O—O bond distances in ozone and the two N—O bond distances in nitrite are the same, which otherwise can be explained only by the use of resonance structures.

MGC Bonding in Hydrogen

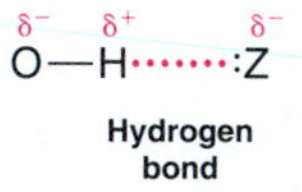

Figure 21.4 The hydrogen bond. The covalent bond between hydrogen and a very electronegative element, such as nitrogen, oxygen, or fluorine, is highly polar. The resulting partial positive charge on H allows it to interact with a lone pair of electrons on another atom to form a hydrogen bond, which is typically a linear arrangement of the three atoms, with the hydrogen atom placed asymmetrically between the two heavier atoms.

Synthesis, Reactions, and Compounds of Hydrogen

The first known preparation of elemental hydrogen was in 1671, when Robert Boyle dissolved iron in dilute acid and obtained a colorless, odorless, gaseous product. Hydrogen was finally identified as an element in 1766, when Henry Cavendish showed that water was the sole product of the reaction of the gas with oxygen. The explosive properties of mixtures of hydrogen with air were not discovered until early in the 18th century; they partially caused the spectacular explosion of the hydrogen-filled dirigible *Hindenburg* in 1937 (Figure 21.6). Due to its extremely low molecular mass, hydrogen gas is difficult to condense to a

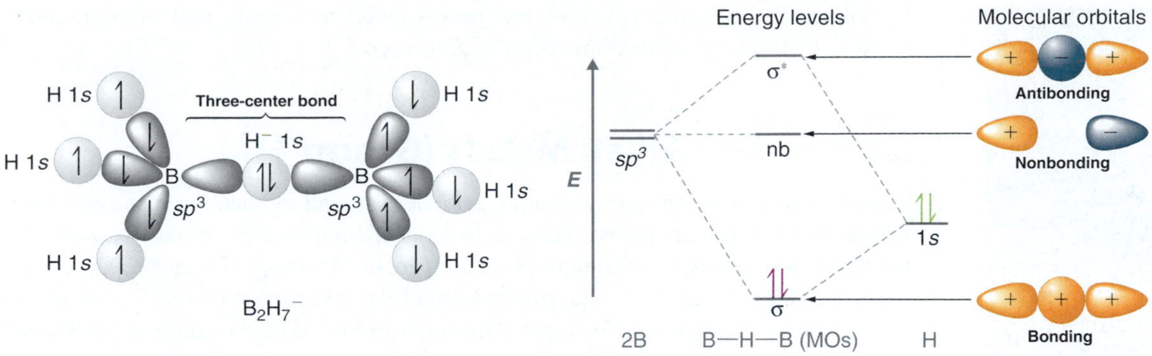

Figure 21.5 A three-center bond uses two electrons to link three atoms. In the B–H–B unit shown, a hydride, with a filled 1*s* orbital, interacts simultaneously with empty sp^3 hybrids on the boron atoms of two BH_3 units to give three molecular orbitals. The two bonding electrons occupy the lowest-energy (σ) bonding orbital, thereby holding all three atoms together.

liquid (bp 20.3 K), and solid hydrogen has one of the lowest melting points known (13.8 K).

The most common way to produce small amounts of highly pure hydrogen gas in the laboratory was discovered by Boyle: the reaction of an active metal (M), such as iron, magnesium, or zinc, with dilute acid:

$$M(s) + 2H^+(aq) \longrightarrow H_2(g) + M^{2+}(aq) \qquad (21.1)$$

Hydrogen gas can also be generated by the reaction of metals such as aluminum or zinc with a strong base:

$$Al(s) + OH^-(aq) + 3H_2O(l) \longrightarrow \frac{3}{2}H_2(g) + [Al(OH)_4]^-(aq) \qquad (21.2)$$

Solid drain cleaners such as Drano use this reaction to generate gas bubbles that help break up clogs in a drainpipe (see Section 19.2). Hydrogen gas is also produced by the reaction of ionic hydrides with water. Because ionic hydrides are expensive, however, this reaction is generally used for only specialized purposes, such as producing HD gas by the reaction of a hydride with D_2O:

$$MH(s) + D_2O(l) \longrightarrow HD(g) + M^+(aq) + OD^-(aq) \qquad (21.3)$$

On an industrial scale, H_2 is produced from methane by means of *catalytic steam reforming*, a method used to convert hydrocarbons to a mixture of CO and H_2 known as *synthesis gas*, or *syngas* (see Chapter 14). The process is carried out at elevated temperatures (800°C) in the presence of a nickel catalyst:

$$CH_4(g) + H_2O(g) \xrightarrow{\text{Ni}} CO(g) + 3H_2(g) \qquad (21.4)$$

Most of the elements in the periodic table form binary compounds with hydrogen, which are collectively referred to as *hydrides*. Binary hydrides in turn can be classified in one of three ways, each with its own characteristic properties. *Covalent hydrides* contain hydrogen bonded to another atom via a covalent or polar covalent bond. Covalent hydrides are usually molecular substances that are relatively volatile and have low melting points. *Ionic hydrides* contain the hydride ion as the anion with cations derived from electropositive metals. Like most ionic compounds, they are typically nonvolatile solids that contain three-dimensional lattices of cations and anions. Unlike most ionic compounds, however, upon heating they often decompose to $H_2(g)$ and the parent metal. *Metallic hydrides* are formed by hydrogen and less electropositive metals such as the transition metals. The properties of metallic hydrides are usually similar to those of the parent metal.

Figure 21.6 The explosive properties of hydrogen. When mixed with air and ignited by a spark, hydrogen gas can explode. The photograph shows the German dirigible *Hindenburg* on fire at Lakehurst, New Jersey, in 1937, after its hydrogen compartment was accidentally ignited as a consequence of an electrical discharge that caused the dirigible's flammable skin to catch fire.

Note the pattern

Covalent hydrides are relatively volatile and have low melting points; ionic hydrides are generally nonvolatile solids in a lattice framework.

Note the pattern

In metallic hydrides, hydrogen atoms are present as interstitial impurities.

Consequently, metallic hydrides are best viewed as metals that contain many hydrogen atoms present as interstitial impurities.

21.3 • The Alkali Metals (Group 1)

The alkali metals are so reactive that they are never found in nature in elemental form. Although some of their ores are abundant, isolating them from their ores is somewhat difficult. For these reasons, the Group 1 elements were unknown until the early 19th century, when Sir Humphry Davy first prepared sodium (Na) and potassium (K) by passing an electric current through molten alkalis. (The ashes produced by the combustion of wood are largely composed of potassium and sodium carbonate.) Lithium (Li) was discovered 10 years later when the Swedish chemist Johan Arfwedson was studying the composition of a new Brazilian mineral. Cesium (Cs) and rubidium (Rb) were not discovered until the 1860s, when Robert Bunsen conducted a systematic search for new elements. Known to chemistry students as the inventor of the Bunsen burner, Bunsen's spectroscopic studies of ores showed sky blue and deep red emission lines that he attributed to two new elements, Cs and Rb, respectively. Francium (Fr) is found in only trace amounts in nature, so our knowledge of its chemistry is limited. All the isotopes of Fr have very short half-lives, in contrast to the other elements in Group 1.

Sir Humphry Davy. Davy (1778–1829) was born in Penzance, Cornwall. He was a bit of a wild man in the laboratory, often smelling and tasting the products of his experiments, which almost certainly shortened his life. He discovered the physiological effects that cause nitrous oxide to be called "laughing gas" (and became addicted to it!), and he almost lost his eyesight in an explosion of nitrogen trichloride (NCl_3), which he was the first to prepare. Davy was one of the first to recognize the utility of Alessandro Volta's "electric piles" (batteries). By connecting several "piles" in series and inserting electrodes into molten salts of the alkali metals and alkaline earths, he was able to isolate six previously unknown elements as pure metals: sodium, potassium, calcium, strontium, barium, and magnesium. He also discovered boron and was the first to prepare phosphine (PH_3) and hydrogen telluride (H_2Te), both of which are highly toxic.

Robert Wilhelm Bunsen. Bunsen (1811–1899) was born and educated in Göttingen, Germany. His early work dealt with organic arsenic compounds, whose highly toxic nature and explosive tendencies almost killed him and did cost him an eye. He designed the "Bunsen burner," a reliable gas burner, and used it and emission spectra to discover the new elements cesium (named for its blue line) and rubidium (named for its red line).

Preparation of Alkali Metals

Because the alkali metals are among the most potent reductants known, obtaining them in pure form requires a considerable input of energy. Pure Li and Na for example, are typically prepared by electrolytic reduction of the molten chlorides:

$$\mathrm{LiCl}(l) \longrightarrow \mathrm{Li}(l) + \frac{1}{2}\mathrm{Cl_2}(g) \qquad (21.5)$$

In practice, $CaCl_2$ is mixed with LiCl to lower the melting point of the lithium salt. The electrolysis is carried out in an argon atmosphere rather than the nitrogen atmosphere typically used for substances that are highly reactive with O_2 and water, because Li reacts with nitrogen gas to form lithium nitride (Li_3N). Metallic sodium is produced by the electrolysis of a molten mixture of NaCl and $CaCl_2$. In contrast, potassium is produced commercially from the reduction of KCl by Na, followed by fractional distillation of $K(g)$. Although rubidium and cesium can also be produced by electrolysis, they are usually obtained by reaction of their hydroxide salts with a reductant such as Mg:

$$2\mathrm{RbOH}(s) + \mathrm{Mg}(s) \longrightarrow 2\mathrm{Rb}(l) + \mathrm{Mg(OH)_2}(s) \qquad (21.6)$$

Massive deposits of essentially pure NaCl and KCl are found in nature and are the major sources of sodium and potassium. The other alkali metals are found in low concentrations in a wide variety of minerals, but ores that contain high concentrations of these elements are relatively rare. No concentrated sources of rubidium are known, for example, even though it is the 16th most abundant element on Earth. Rubidium is obtained commercially by isolating the 2–4% of Rb present as an impurity in micas, minerals composed of sheets of complex hydrated potassium–aluminum silicates.

Alkali metals are recovered from silicate ores in a multistep process that takes advantage of the pH-dependent solubility of selected salts of each metal ion. The steps in this process are *leaching*, which uses sulfuric acid to dissolve the desired alkali metal ion and Al^{3+} from the ore; *basic precipitation* to remove Al^{3+} from the mixture as $Al(OH)_3$; *selective precipitation* of the insoluble alkali metal carbonate; *dissolution* of the salt again in hydrochloric acid; and isolation of the metal by *evaporation* and *electrolysis*. Figure 21.7 illustrates the isolation of liquid lithium from a lithium silicate ore by this process.

General Properties of Alkali Metals

Various properties of the Group 1 elements are summarized in Table 21.3. In keeping with overall periodic trends, the atomic and ionic radii increase smoothly from Li to Cs, and the first ionization energies decrease as the atoms become larger. As a result of their low first ionization energies, the alkali metals have an overwhelming tendency to form ionic compounds where they have a +1 charge. All the alkali metals have relatively high electron affinities because the addition of an electron produces an anion (M^-) with an ns^2 electron configuration. The densities of the elements generally increase from Li to Cs, reflecting another common trend: because the atomic masses of the elements increase more rapidly than the atomic volumes as you go down a group, the densest elements (such as Pb, Au, and U) are near the bottom of the periodic table. An unusual trend in the Group 1 elements is the smooth decrease in the melting and boiling points from Li to Cs. As a result, Cs (mp 28.5°C) is one of only three metals (the others are Ga and Hg) that are liquids at body temperature (37°C).

The standard reduction potentials, $E°$, of the alkali metals do not follow the trend based on ionization energies (see Chapter 19). Unexpectedly, lithium is the strongest reductant, and sodium is the weakest (Table 21.3). Because Li^+ is much smaller than the other alkali metal cations, its hydration energy is the highest. The high hydration energy of Li^+ more than compensates for its higher ionization energy, making lithium metal the strongest reductant in aqueous solution. This apparent anomaly is an example of how the physical or chemical behaviors of the elements in a group are often determined by the subtle interplay of opposing periodic trends.

MGC Introduction to the Alkali Metals

Note the pattern

The alkali metals are among the most potent reductants known.

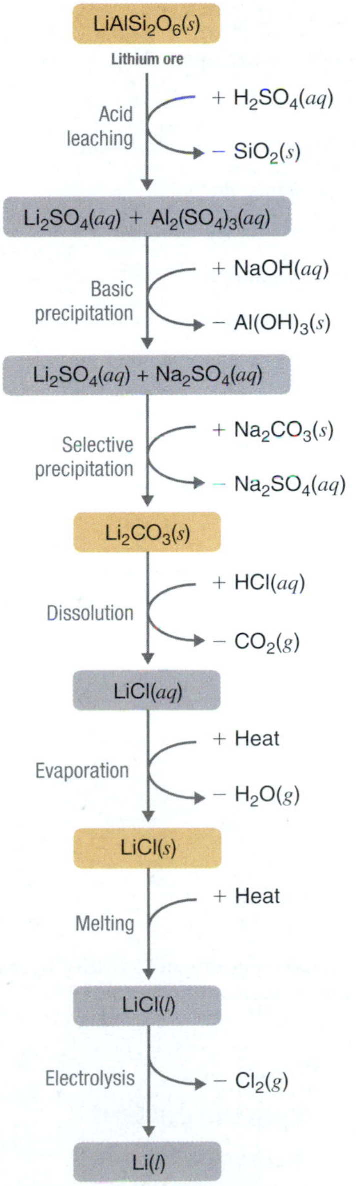

Figure 21.7 Isolating lithium from spodumene, a lithium silicate ore. The key steps are acid leaching, basic precipitation of aluminum hydroxide, selective precipitation of insoluble lithium carbonate, conversion to lithium chloride, evaporation, and electrolysis. The other alkali metals and the alkaline earth metals are recovered from their ores by similar processes.

Reactions and Compounds of Alkali Metals

All alkali metals are electropositive elements with an ns^1 valence-electron configuration, forming the monocation (M^+) by the loss of the single valence electron. Because removing a second electron would require breaking into the $(n - 1)$ closed shell, which is energetically prohibitive, the chemistry of the alkali metals is largely that of ionic compounds that contain the M^+ ions. However, as we discuss later, the lighter Group 1 elements also form a series of organometallic compounds that contain polar covalent M—C bonds.

All the alkali metals react vigorously with the halogens (Group 17) to form the corresponding ionic halides, where X = a halogen:

$$2M(s) + X_2(s, l, g) \longrightarrow 2M^+X^-(s) \quad (21.7)$$

Similarly, the alkali metals react with the heavier chalcogens (sulfur, selenium, and tellurium in Group 16) to produce metal chalcogenides, where Y = S, Se, or Te:

$$2M(s) + Y(s) \longrightarrow M_2Y(s) \quad (21.8)$$

When excess chalcogen is used, however, a variety of products can be obtained that contain chains of chalcogen atoms, such as the sodium polysulfides, Na_2S_n, where $n = 2–6$. For example, Na_2S_3 contains the S_3^{2-} ion, which is V-shaped with an S—S—S angle of about 103°. The one-electron oxidation product of the trisulfide ion, S_3^-, is responsible for the intense blue color of gemstones such as lapis lazuli and blue ultramarine (Figure 21.8).

The reaction of the alkali metals with oxygen, the lightest element in Group 16, is more complex, and the stoichiometry of the product depends on both the metal:oxygen ratio and the size of the metal. For instance, when the alkali metals burn in air, the observed products are Li_2O (white), Na_2O_2 (pale yellow), KO_2 (orange), RbO_2 (brown), and CsO_2 (orange). Only Li_2O has the stoichiometry expected for a substance that contains two M^+ cations and one O^{2-} ion. In contrast, Na_2O_2 contains the O_2^{2-} (peroxide) anion plus two Na^+ cations. The other three salts, with stoichiometry MO_2, contain the M^+ cation and the O_2^- (superoxide) ion. Because O^{2-} is the

TABLE 21.3 Selected properties of the Group 1 elements

	Lithium	Sodium	Potassium	Rubidium	Cesium	Francium
Atomic symbol	Li	Na	K	Rb	Cs	Fr
Atomic number	3	11	19	37	55	87
Atomic mass	6.94	22.99	39.10	85.47	132.91	223
Valence-electron configuration	$2s^1$	$3s^1$	$4s^1$	$5s^1$	$6s^1$	$7s^1$
mp/bp, °C	180.5/1342	97.8/883	63.5/759	39.3/688	28.5/671	27/–
Density, g/cm^3	0.534	0.97	0.89	1.53	1.93	—
Atomic radius, pm	167	190	243	265	298	
First ionization energy, kJ/mol	520	496	419	403	376	393
Most common oxidation state	+1	+1	+1	+1	+1	+1
Ionic radius, pm[a]	76	102	138	152	167	
Electron affinity, kJ/mol	−60	−53	−48	−47	−46	
Electronegativity	1.0	0.9	0.8	0.8	0.8	0.7
Standard electrode potential $E°$, V	−3.04	−2.71	−2.93	−2.98	−3.03	
Product of reaction with O_2	Li_2O	Na_2O_2	KO_2	RbO_2	CsO_2	—
Type of oxide	Basic	Basic	Basic	Basic	Basic	—
Product of reaction with N_2	Li_3N	None	None	None	None	—
Product of reaction with X_2	LiX	NaX	KX	RbX	CsX	—
Product of reaction with H_2	LiH	NaH	KH	RbH	CsH	—

[a]The values cited are for four-coordinate ions except for Rb^+ and Cs^+, for which values for the six-coordinate ion are given.

A crystal of spodumene, $LiAlSi_2O_6$, one of the most important lithium ores.

Note the pattern

The chemistry of the alkali metals is largely that of ionic compounds containing the M^+ ions.

smallest of the three oxygen anions, it forms a stable ionic lattice with the smallest alkali metal cation, Li^+. In contrast, the larger alkali metals, potassium, rubidium, and cesium, react with oxygen in air to give the metal superoxides. Because the Na^+ cation is intermediate in size, sodium reacts with oxygen to form a compound with an intermediate stoichiometry, sodium peroxide. Under specific reaction conditions, however, it is possible to prepare the oxide, peroxide, and superoxide salts of all five alkali metals, except for lithium superoxide, LiO_2.

Figure 21.8 The trisulfide anion, S_3^-, is responsible for the deep blue color of gemstones such as lapis lazuli and ultramarine. (a) The rich blue color of lapis lazuli is due to small amounts of the normally unstable S_3^- anion. (b) The aluminosilicate cages of the minerals (zeolites) that make up the matrix of the gemstone stabilize the reactive anion; excess Na^+ ions in the structure balance the negative charges on the zeolite framework and the S_3^- anion.

(a) Lapis lazuli

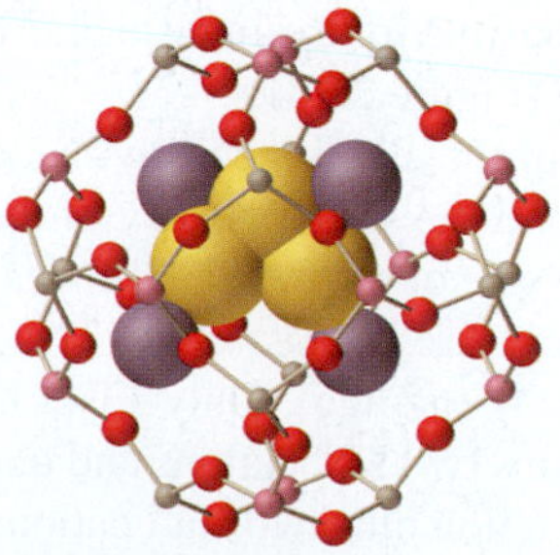

(b) $Na_7[AlSiO_4]_6(S_3)$

The alkali metal peroxides and superoxides are potent oxidants that react, often vigorously, with a wide variety of reducing agents, such as charcoal or aluminum metal. For example, Na_2O_2 is used industrially for bleaching paper, wood pulp, and fabrics such as linen and cotton. In submarines, Na_2O_2 and KO_2 are used to purify and regenerate the air by removing the CO_2 produced by respiration and replacing it with O_2. Both compounds react with CO_2 in a redox reaction in which O_2^{2-} or O_2^- is simultaneously oxidized and reduced, producing the metal carbonate and O_2:

$$2Na_2O_2(s) + 2CO_2(g) \longrightarrow 2Na_2CO_3(s) + O_2(g) \quad (21.9a)$$

$$4KO_2(s) + 2CO_2(g) \longrightarrow 2K_2CO_3(s) + 3O_2(g) \quad (21.9b)$$

The presence of water vapor, the other product of respiration, makes KO_2 even more effective at removing CO_2 because potassium bicarbonate is formed rather than potassium carbonate:

$$4KO_2(s) + 4CO_2(g) + 2H_2O(g) \longrightarrow 4KHCO_3(s) + 3O_2(g) \quad (21.10)$$

Notice that 4 mol of CO_2 are removed in this reaction rather than 2 mol in Equation 21.9b.

Lithium, the lightest alkali metal, is the only one that reacts with atmospheric nitrogen, forming lithium nitride (Li_3N). Lattice energies again explain why the larger alkali metals such as potassium do not form nitrides: packing three large K^+ cations around a single relatively small anion is energetically unfavorable. In contrast, all the alkali metals react with the larger Group 15 elements phosphorus and arsenic to form metal phosphides and arsenides (where Z = P or As):

Note the pattern

Because of lattice energies, only Li forms a stable oxide and nitride.

$$12M(s) + Z_4(s) \longrightarrow 4M_3Z(s) \quad (21.11)$$

The alkali metals react with all the Group 14 elements, but the compositions and properties of the products vary significantly. For example, reaction with the heavier Group 14 elements gives materials that contain polyatomic anions and three-dimensional cage structures, such as K_4Si_4 whose structure is shown in the margin. In contrast, lithium and sodium are oxidized by carbon to produce a compound with the stoichiometry M_2C_2 (where M = Li or Na):

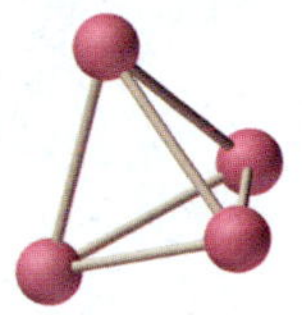

The three-dimensional cage structure of the Si_4^{4-} ion in the ionic compound K_4S_4. The Si_4^{4-} ion is isoelectronic and isostructural with the P_4 molecule.

$$2M(s) + 2C(s) \longrightarrow M_2C_2(s) \quad (21.12)$$

The same compounds can be obtained by the reaction of the metal with acetylene (C_2H_2). In this reaction, the metal is again oxidized and hydrogen is reduced:

$$2M(s) + C_2H_2(g) \longrightarrow M_2C_2(s) + H_2(g) \quad (21.13)$$

The acetylide ion (C_2^{2-}), formally derived from acetylene by the loss of both hydrogens as protons, is a very strong base. Reaction of acetylide salts with water produces acetylene and MOH(*aq*).

The heavier alkali metals (K, Rb, and Cs) also react with carbon in the form of graphite. Instead of disrupting the hexagonal sheets of carbon atoms, however, the metals insert themselves *between* the sheets of carbon atoms to give new substances called **graphite intercalation compounds** (Figure 21.9a). The stoichiometries of these compounds include MC_{60} and MC_{48}, which are black/gray; MC_{36} and MC_{24}, which are blue; and MC_8, which is bronze (Figure 21.9b). The remarkably high electrical conductivity of these compounds (about 200 times higher than graphite) is attributed to a net transfer of the valence electron of the alkali metal to the graphite layers to produce, for example, $K^+C_8^-$.

All the alkali metals react directly with gaseous hydrogen at elevated temperatures to produce ionic hydrides (M^+H^-):

$$2M(s) + H_2(g) \longrightarrow 2MH(s) \quad (21.14)$$

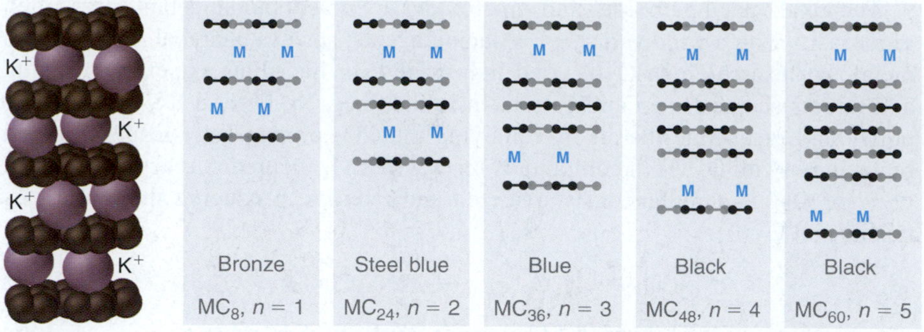

(a) KC_8 **(b) Layer arrangements and stoichiometries**

Figure 21.9 Graphite intercalation compounds. The reaction of graphite with alkali metals such as K, Rb, and Cs results in partial reduction of the graphite and insertion of layers of alkali metal cations between sets of n layers of carbon atoms. **(a)** In KC_8, layers of K^+ ions are inserted between every pair of carbon layers, giving $n = 1$. **(b)** The stoichiometry and color of intercalation compounds depend on the number of layers of carbon atoms (n) between each of the layers of intercalated metal atoms. This schematic diagram illustrates the most common structures that have been observed.

All are also capable of reducing water to produce hydrogen gas:

$$M(s) + H_2O(l) \longrightarrow \frac{1}{2}H_2(g) + MOH(aq) \quad (21.15)$$

Figure 21.10 The reaction of sodium with water. Like most elements in Groups 1 and 2, sodium reacts violently with water. The products are the $Na^+(aq)$ ion and hydrogen gas, which is potentially explosive when mixed with air.

Although lithium reacts rather slowly with water, sodium reacts quite vigorously (Figure 21.10), and the heavier alkali metals (K, Rb, and Cs) react so vigorously that they invariably explode. This trend, which is not consistent with the relative magnitudes of the reduction potentials of the elements, serves as another example of the complex interplay of different forces and phenomena—in this case, kinetics and thermodynamics. Although the driving force for the reaction is greatest for lithium, the heavier metals have lower melting points. The heat liberated by the reaction causes them to melt, and the larger surface area of the liquid metal in contact with water greatly accelerates the reaction rate.

Alkali metal cations are found in a wide variety of ionic compounds. In general, any alkali metal salt can be prepared by reacting the alkali metal hydroxide with an acid and then evaporating the water:

$$2MOH(aq) + H_2SO_4(aq) \longrightarrow M_2SO_4(aq) + 2H_2O(l) \quad (21.16)$$

$$MOH(aq) + HNO_3(aq) \longrightarrow MNO_3(aq) + H_2O(l) \quad (21.17)$$

Hydroxides of alkali metals also can react with organic compounds that contain an acidic hydrogen to produce a salt. An example is the preparation of sodium acetate (CH_3CO_2Na) by the reaction of sodium hydroxide and acetic acid:

$$CH_3CO_2H(aq) + NaOH(s) \longrightarrow CH_3CO_2Na(aq) + H_2O(l) \quad (21.18)$$

Soap is a mixture of the sodium and potassium salts of naturally occurring carboxylic acids, such as palmitic acid [$CH_3(CH_2)_{14}CO_2H$] and stearic acid [$CH_3(CH_2)_{16}CO_2H$]. Lithium salts, such as lithium stearate [$CH_3(CH_2)_{14}CO_2Li$], are used as additives in motor oils and greases.

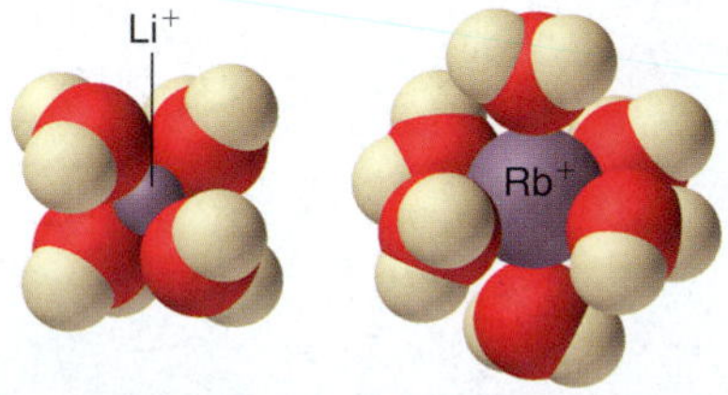

The tetrahedral $[Li(H_2O)_4]^+$ and octahedral $[Rb(H_2O)_6]^+$ complexes. The Li^+ ion is so small that it can accommodate only four water molecules around it, but the larger alkali metal cations tend to bind six water molecules.

Complexes of Alkali Metals

Because of their low positive charge (+1) and relatively large ionic radii, alkali metal cations have only a weak tendency to react with simple Lewis bases to form metal complexes like those discussed in Chapter 17. Complex formation is most significant for the smallest cation, Li^+, and decreases with increasing radius. In aqueous solution, for example, Li^+ forms the tetrahedral $[Li(H_2O)_4]^+$ complex shown in the margin. In contrast, the larger alkali metal cations form octahedral $[M(H_2O)_6]^+$ complexes. Complex formation is primarily due to the electrostatic interaction of the metal cation with the polar water molecules. Because of their high affinity for water, anhydrous salts that contain Li^+ and Na^+ ions (such as Na_2SO_4) are often used as *drying agents*. These compounds absorb trace amounts of water from nonaqueous solutions to form hydrated salts, which are then easily removed from the solution by filtration.

Electrostatic interactions also allow alkali metal ions to form complexes with certain cyclic polyethers and related compounds, such as crown ethers and cryptands. As discussed in Chapter 13, **crown ethers** are cyclic polyethers that contain four or more oxygen atoms separated by two or three carbon atoms. All crown ethers have a central cavity that can accommodate a metal ion coordinated to the ring of oxygen atoms, and crown ethers with rings of different sizes prefer to bind metal ions that fit into the cavity. For example, 14-crown-4, with the smallest cavity that can accommodate a metal ion, has the highest affinity for Li^+, whereas 18-crown-6 forms the strongest complexes with K^+ (see Figure 13.7a).

Cryptands are more nearly spherical analogues of crown ethers and are even more powerful and selective complexing agents. *Cryptands* consist of three chains containing oxygen that are connected by two nitrogen atoms (see Figure 13.7b). They can completely surround (encapsulate) a metal ion of the appropriate size, coordinating to the metal by a lone pair of electrons on each O atom and the two N atoms. Like crown ethers, cryptands with different cavity sizes are highly selective for metal ions of particular sizes. Crown ethers and cryptands are often used to dissolve simple inorganic salts such as $KMnO_4$ in nonpolar organic solvents (see Figure 13.8).

Note the pattern

Because of their low positive charge (+1) and relatively large ionic radii, alkali metal cations have only a weak tendency to form complexes with simple Lewis bases.

Liquid Ammonia Solutions

A remarkable feature of the alkali metals is their ability to dissolve reversibly in liquid ammonia. Just as in their reactions with water, the reaction of alkali metals with liquid ammonia eventually produces hydrogen gas and the metal salt of the conjugate base of the solvent—in this case, the *amide* ion (NH_2^-) rather than hydroxide:

$$M(s) + NH_3(l) \longrightarrow \frac{1}{2}H_2(g) + M^+(am) + NH_2^-(am) \qquad (21.19)$$

where the (*am*) designation refers to an ammonia solution, analogous to (*aq*) used to indicate aqueous solutions. Without a catalyst the reaction in Equation 21.19 tends to be rather slow. In many cases, the alkali metal amide salt (MNH_2) is not very soluble in liquid ammonia and precipitates, but when dissolved, solutions of the alkali metal are produced that can be very concentrated. One mole of Cs metal, for example, will dissolve in as little as 53 mL (40 g) of liquid ammonia. The pure metal is easily recovered when the ammonia evaporates.

Solutions of alkali metals in liquid ammonia are intensely colored and good conductors of electricity due to the presence of *solvated electrons*, $e^-(NH_3)$, which are not attached to single atoms. As shown in the margin, a solvated electron is loosely associated with a cavity in the ammonia solvent that is stabilized by hydrogen bonds. Alkali metal–liquid ammonia solutions of about 3 *M* or less are deep blue (Figure 21.11) and conduct electricity about 10 times better than an aqueous NaCl solution because of the high mobility of the solvated electrons. As the concentration of the metal increases above 3 *M*, the color changes to metallic bronze or gold, and the conductivity increases to a value comparable with that of the pure liquid metals.

In addition to solvated electrons, solutions of alkali metals in liquid ammonia contain the metal cation (M^+), the neutral metal atom (M), metal dimers (M_2), and the metal anion (M^-). The anion is formed by the addition of an electron to the singly occupied *ns* valence orbital of the metal atom. Even in the absence of a catalyst, these solutions are not very stable and eventually decompose to the thermodynamically favored products, $M^+NH_2^-$ and hydrogen gas (Equation 21.19). Nonetheless, the solvated electron is a potent reductant that is often used in synthetic chemistry.

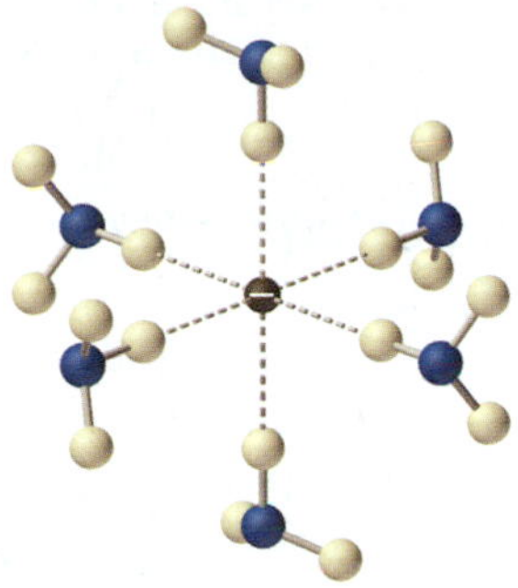

Solvated electrons. The presence of *solvated electrons*, $e^-(NH_3)$ in solutions of alkali metals in liquid ammonia are indicated by the intense color of the solution and its electrical conductivity.

Figure 21.11 Alkali metal–liquid ammonia solutions. Most metals are insoluble in virtually all solvents, but the alkali metals (and the heavier alkaline earth metals) dissolve readily in liquid ammonia to form solvated metal cations and solvated electrons, which give the solution a deep blue color.

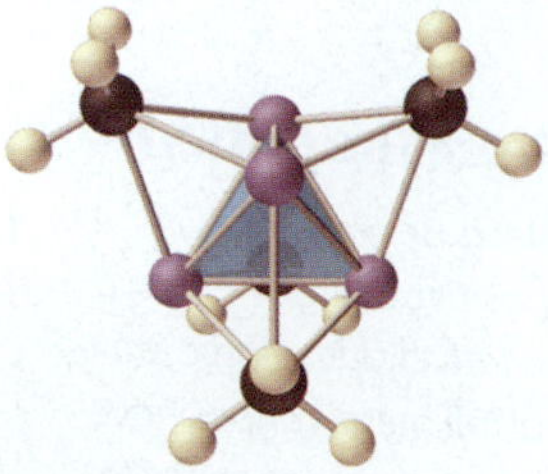

Figure 21.12 The tetrameric structure of methyl lithium. Methyl lithium is not an ionic compound, but exists as tetrameric $(CH_3Li)_4$ molecules. The structure consists of a tetrahedral arrangement of four lithium atoms, with the carbon atom of a methyl group located above the middle of each triangular face of the tetrahedron. The carbon atoms thus bridge three lithium atoms to form four-center, two-electron bonds.

Organometallic Compounds of the Group 1 Elements

Compounds that contain a metal covalently bonded to a carbon atom of an organic species are called **organometallic compounds**. The properties and reactivities of organometallic compounds differ greatly from those of either the metallic or organic components. Because of its small size, lithium, for example, forms an extensive series of covalent organolithium compounds such as methyllithium ($LiCH_3$), which are by far the most stable and best-known Group 1 organometallic compounds. These volatile, low-melting-point solids or liquids can be sublimed or distilled at relatively low temperatures and are soluble in nonpolar solvents. Like organic compounds, the molten solids do not conduct electricity to any significant degree. Organolithium compounds have a tendency to form oligomers with the formula $(RLi)_n$, where R represents the organic component. For example, in both the solid state and solution, methyllithium exists as a tetramer with the structure shown in Figure 21.12, where each triangular face of the Li_4 tetrahedron is bridged by the carbon atom of a methyl group. Effectively, the carbon atom of each CH_3 group is using a single pair of electrons in an sp^3 hybrid lobe to bridge *three* lithium atoms, making this an example of two-electron, *four-center* bonding. Clearly, such a structure, in which each carbon atom is apparently bonded to *six* other atoms, cannot be explained using any of the electron-pair bonding schemes discussed in Chapters 8 and 9. Molecular orbital theory can explain the bonding in methyllithium, but the description is beyond the scope of this text.

Organosodium and organopotassium compounds are more ionic than organolithium compounds. They contain discrete M^+ and R^- ions and are insoluble or only sparingly soluble in nonpolar solvents.

Note the pattern

The properties and reactivities of organometallic compounds differ greatly from those of either the metallic or organic components.

Uses of the Alkali Metals

Because sodium remains liquid over a wide temperature range (97.8–883°C), it is used as a coolant in specialized high-temperature applications, such as nuclear reactors and the exhaust valves in high-performance sports car engines. Because of its low ionization energy, cesium is used in photosensors in automatic doors, toilets, burglar alarms, and other electronic devices. In these devices, cesium is ionized by a beam of visible light, thereby producing a small electric current; blocking the light interrupts the electric current and triggers a response.

Compounds of sodium and potassium are produced on a huge scale in industry. Each year, the top 50 industrial compounds include NaOH, used in a wide variety of industrial processes; Na_2CO_3, used in the manufacture of glass; K_2O, used in porcelain glazes; and Na_4SiO_4, used in detergents.

Several other alkali metal compounds are also important. For example, Li_2CO_3 is one of the most effective treatments available for manic depression or bipolar disorder. It appears to modulate or dampen the effect on the brain of changes in the level of neurotransmitters, which are biochemical substances responsible for the transmission of nerve impulses between neurons. Consequently, patients who take "lithium" do not exhibit the extreme mood swings that characterize this disorder.

EXAMPLE 21.2

For each application, choose the more appropriate substance based on the properties and reactivities of the alkali metals and their compounds. Explain your choice in each case.

(a) For a reaction that requires a strong base in a solution of tetrahydrofuran (THF), would you use LiOH or CsOH?

(b) To extinguish a fire caused by burning lithium metal, would you use water, CO_2, N_2 gas, or sand?

(c) Both $LiNO_3$ and $CsNO_3$ are highly soluble in acetone (2-propanone). Which of these alkali metal salts would you use to precipitate I^- from an acetone solution?

Given Application, selected alkali metals

Asked for Appropriate metal for each application

Strategy

Use the properties and reactivities discussed in this section to determine which alkali metal is most suitable for the indicated application.

Solution

(a) Both LiOH and CsOH are ionic compounds that contain the hydroxide anion. Li^+, however, is much smaller than Cs^+, so the Li^+ cation will be more effectively solvated by the oxygen of THF with its lone pairs of electrons. This difference will have two effects: First, LiOH is likely to be much more soluble than CsOH in the nonpolar solvent, which could be a significant advantage. Second, the solvated Li^+ ions are less likely to form tight ion pairs with the OH^- ions in the relatively nonpolar solution, making the OH^- more basic and thus more reactive. Thus, LiOH is the better choice.

(b) Lithium is a potent reductant that reacts with water to form LiOH and H_2 gas, so adding a source of hydrogen such as water to a lithium fire is likely to produce an explosion! Lithium also reacts with oxygen and nitrogen in the air to form Li_2O and Li_3N, respectively, so we would not expect nitrogen to extinguish a lithium fire. Because CO_2 is a gaseous molecule that contains carbon in its highest accessible oxidation state (+4), adding CO_2 to a strong reductant such as Li should result in a vigorous redox reaction. Thus, water, N_2, and CO_2 are all unsuitable choices for extinguishing a lithium fire. In contrast, sand is primarily SiO_2, which is a network solid that is not readily reduced. Smothering a lithium fire with sand is therefore the best choice.

(c) The salt with the smaller cation has the higher lattice energy, and high lattice energies tend to decrease the solubility of a salt (see Chapter 8). However, the solvation energy of the cation is also important in determining solubility, and small cations tend to have higher solvation energies. Recall from Chapter 13 that high solvation energies tend to increase the solubility of ionic substances. Thus, CsI should be the least soluble of the alkali metal iodides, and LiI the most soluble. Consequently, $CsNO_3$ is the better choice.

EXERCISE 21.2

Indicate which of the alternative alkali metals or their compounds given is more appropriate for each application.

(a) Drying agent for an organic solvent: Li_2SO_4 or Rb_2SO_4?
(b) Removal of trace amounts of N_2 from highly purified Ar gas: Li, K, or Cs?
(c) Reaction with an alkyl halide (formula RX) to prepare an organometallic compound (formula MR): Li or K?

Answer **(a)** Li_2SO_4; **(b)** Li; **(c)** Li.

EXAMPLE 21.3

Predict the products of each reaction, and then balance the equations:

(a) $Na(s) + O_2(g) \longrightarrow$
(b) $Li_2O(s) + H_2O(l) \longrightarrow$
(c) $K(s) + CH_3OH(l) \longrightarrow$
(d) $Li(s) + CH_3Cl(l) \longrightarrow$
(e) $Li_3N(s) + KCl(s) \longrightarrow$

Given Reactants

Asked for Products and balanced chemical equation

Strategy

A Determine whether one of the reactants is an oxidant or a reductant, or a strong acid or base. If so, a redox reaction or an acid–base reaction is likely to occur. Identify the products of the reaction.

B If a reaction is predicted to occur, balance the reaction.

Solution

(a) **A** Sodium is a reductant and oxygen is an oxidant, so a redox reaction is most likely. We expect an electron to be transferred from Na (thus forming Na^+) to O_2. We now need to determine whether the reduced product is a superoxide (O_2^-), peroxide (O_2^{2-}), or oxide (O^{2-}). Under normal reaction conditions, the product of the reaction of an alkali metal with oxygen depends on the identity of the metal. Because of differences in lattice energy, Li produces the oxide (Li_2O), the heavier metals (K, Rb, Cs) produce the superoxide (MO_2), and Na produces the peroxide (Na_2O_2). **B** The balanced equation is $2Na(s) + O_2(g) \longrightarrow Na_2O_2(s)$.

(b) **A** Li_2O is an ionic salt that contains the oxide ion (O^{2-}), which is the completely deprotonated form of water and thus is expected to be a strong base. The other reactant, water, is both a weak acid and a weak base, so we can predict that an acid–base reaction will occur. **B** The balanced equation is $Li_2O(s) + H_2O(l) \longrightarrow 2LiOH(aq)$.

(c) **A** Potassium is a reductant, whereas methanol is both a weak acid and a weak base (similar to water). A weak acid produces H^+, which can act as an oxidant by accepting an electron to form $\frac{1}{2}H_2$. This reaction, therefore, is an acid dissociation that is driven to completion by reduction of the protons as they are released. **B** The balanced equation is $K(s) + CH_3OH(l) \longrightarrow \frac{1}{2}H_2(g) + CH_3OK(soln)$.

(d) **A** One of the reactants is an alkali metal, a potent reductant, and the other is an alkyl halide. Any compound that contains a carbon–halogen bond can, in principle, be reduced, releasing a halide ion and forming an organometallic compound. That outcome seems likely in this case because organolithium compounds are among the most stable organometallic compounds known. **B** Notice that 2 mol of lithium is required to balance the equation: $2Li(s) + CH_3Cl(l) \longrightarrow LiCl(s) + CH_3Li(soln)$.

(e) **A** Lithium nitride and potassium chloride are largely ionic compounds. The nitride ion (N^{3-}) is a very strong base because it is the fully deprotonated form of ammonia, a weak acid. An acid–base reaction requires an acid as well as a base, however, and KCl is not acidic. What about a redox reaction? Both substances contain ions that have closed-shell valence-electron configurations. The nitride ion could act as a reductant by donating electrons to an oxidant and forming N_2. KCl is not an oxidant, however, and a redox reaction requires an oxidant as well as a reductant. **B** We conclude that the two substances will not react with each other.

EXERCISE 21.3

Predict the products of each reaction, and balance the equations:

(a) $K(s) + N_2(g) \longrightarrow$
(b) $Li_3N(s) + H_2O(l) \longrightarrow$
(c) $Na(s) + (CH_3)_2NH(soln) \longrightarrow$
(d) $C_6H_5Li(soln) + D_2O(l) \longrightarrow$

Answer **(a)** no reaction **(b)** $Li_3N(s) + 3H_2O(l) \longrightarrow NH_3(aq) + 3LiOH(aq)$ **(c)** $Na(s) + (CH_3)_2NH(soln) \longrightarrow \frac{1}{2}H_2(g) + Na[(CH_3)_2N](soln)$ **(d)** $C_6H_5Li(soln) + D_2O(l) \longrightarrow C_6H_5D(l) + LiOD(soln)$

21.4 • The Alkaline Earth Metals (Group 2)

Like the alkali metals, the alkaline earth metals are so reactive that they are never found in elemental form in nature. Because they form +2 ions that have very negative reduction potentials, large amounts of energy are needed to isolate them from their ores.

Four of the six Group 2 elements—magnesium (Mg), calcium (Ca), strontium (Sr), and barium (Ba)—were first isolated in the early 19th century by Sir Humphry Davy using a technique similar to the one he used to obtain the first alkali metals. In contrast to the alkali metals, however, compounds of the alkaline earths had been recognized as unique for many centuries. In fact, the name *alkali* comes from the Arabic word *al-qili*, "ashes," which were known to neutralize acids. Medieval alchemists found that a portion of the ashes would melt upon heating, and these substances were later identified as the carbonates of sodium and potassium (M_2CO_3). The ashes that did not melt (but did dissolve in acid), called alkaline earths, were subsequently identified as the alkaline earth oxides (MO). In 1808, Davy was able to obtain pure samples of Mg, Ca, Sr, and Ba by electrolysis of their chlorides or oxides.

Beryllium (Be), the lightest alkaline earth metal, was first obtained in 1828 by Friedrich Wöhler in Germany and simultaneously by Antoine Bussy in France. The method used by both men was reduction of the chloride by the potent "new" reductant, potassium:

$$BeCl_2(s) + 2K(s) \xrightarrow{\Delta} Be(s) + 2KCl(s) \qquad (21.20)$$

Radium was discovered in 1898 by Pierre and Marie Curie, who processed tons of residue from uranium mines to obtain about 120 mg of almost pure $RaCl_2$. For the discovery of radium, Marie Curie was awarded the Nobel Prize in 1911. Because of its low abundance and high radioactivity, radium has few uses and will not be discussed further.

Preparation of Alkaline Earth Metals

The alkaline earths are produced for industrial use by electrolytic reduction of their molten chlorides, as indicated in this equation for calcium:

$$CaCl_2(l) \longrightarrow Ca(l) + Cl_2(g) \qquad (21.21)$$

The Group 2 metal chlorides are obtained from a variety of sources. For example, $BeCl_2$ is produced by the reaction of HCl with beryllia (BeO), which is obtained from the semiprecious stone *beryl* [$Be_3Al_2(SiO_3)_6$].

A crystal of beryl, [$Be_3Al_2(SiO_3)_6$], a gemstone and an important source of beryllium.

Chemical reductants can also be used to obtain the Group 2 elements. For example, magnesium is produced on a large scale by heating a form of limestone called *dolomite* ($CaCO_3 \cdot MgCO_3$) with an inexpensive iron/silicon alloy at a temperature of 1150°C. Initially CO_2 is released, leaving behind a mixture of CaO and MgO; Mg^{2+} is then reduced:

$$2CaO \cdot MgO(s) + Fe/Si(s) \longrightarrow 2Mg(l) + Ca_2SiO_4(s) + Fe(s) \qquad (21.22)$$

An early source of magnesium was an ore called *magnesite* ($MgCO_3$) from the district of northern Greece called Magnesia. Strontium was obtained from *strontianite* ($SrCO_3$) found in a lead mine in the town of Strontian in Scotland. The alkaline earths are somewhat easier than the alkali metals to isolate from their ores because their carbonate, and some sulfate and hydroxide salts are insoluble.

A crystal of strontianite, $SrCO_3$. Both strontianite, one of the most important strontium ores, and strontium are named after the town of Strontian, Scotland, the location of one of the first mines for strontium ores.

General Properties of Alkaline Earth Metals

MGC Introduction to the Alkaline Earth Metals

Several important properties of the alkaline earths are summarized in Table 21.4. Although many of these properties are similar to those of the alkali metals (Table 21.3), certain key differences are attributable to the differences in the valence-electron configurations of the two groups (ns^2 for the alkaline earth metals versus ns^1 for the alkali metals).

As with the alkali metals, the atomic and ionic radii of the alkaline earth metals increase smoothly from Be to Ba, and the ionization energies decrease. As we would expect, the first ionization energy of an alkaline earth, with an ns^2 valence-electron

TABLE 21.4 Selected properties of the Group 2 elements

	Beryllium	Magnesium	Calcium	Strontium	Barium	Radium
Atomic symbol	Be	Mg	Ca	Sr	Ba	Ra
Atomic number	4	12	20	38	56	88
Atomic mass	9.01	24.31	40.08	87.62	137.33	226
Valence-electron configuration	$2s^2$	$3s^2$	$4s^2$	$5s^2$	$6s^2$	$7s^2$
mp/bp, °C	1287/2471	650/1090	842/1434	777/1382	727/1897	700/–
Density, g/cm^3	1.85	1.74	1.54	2.64	3.62	~5
Atomic radius, pm	112	145	194	219	253	
First ionization energy, kJ/mol	900	738	590	549	503	
Most common oxidation state	+2	+2	+2	+2	+2	+2
Ionic radius, pm[a]	45	72	100	118	135	
Electron affinity, kJ/mol	≥0	≥0	−2	−50	−14	
Electronegativity	1.6	1.3	1.0	1.0	0.9	0.9
Standard electrode potential ($E°$) V	−1.85	−2.37	−2.87	−2.90	−2.91	−2.8
Product of reaction with O_2	BeO	MgO	CaO	SrO	BaO_2	—
Type of oxide	Amphoteric	Weakly basic	Basic	Basic	Basic	—
Product of reaction with N_2	None	Mg_3N_2	Ca_3N_2	Sr_3N_2	Ba_3N_2	—
Product of reaction with X_2	BeX_2	MgX_2	CaX_2	SrX_2	BaX_2	—
Product of reaction with H_2	None	MgH_2	CaH_2	SrH_2	BaH_2	—

[a]The values cited are for six-coordinate ions except for Be^{2+}, for which the value for the four-coordinate ion is given.

configuration, is always significantly higher than that of the alkali metal immediately preceding it. The Group 2 elements do exhibit some anomalies, however. For example, the density of Ca is lower than that of Be and Mg, the two lightest members of the group, and Mg has the lowest melting and boiling points. In contrast to the alkali metals, the heaviest alkaline earth, Ba, is the strongest reductant, and the lightest, Be, is the weakest. The standard electrode potentials of Ca and Sr are not very different from that of Ba, indicating that the opposing trends in ionization energies and hydration energies are of roughly equal importance.

One major difference between the Group 1 and Group 2 elements is their electron affinities. With their half-filled ns orbitals, the alkali metals have a significant affinity for an additional electron. In contrast, the alkaline earths generally have little or no tendency to accept an additional electron because their *ns* valence orbitals are already full; an added electron would have to occupy one of the vacant *np* orbitals, which are much higher in energy.

Note the pattern

The Group 2 elements almost exclusively form ionic compounds containing the M^{2+} ions.

Reactions and Compounds of Alkaline Earth Metals

With their low first and second ionization energies, the Group 2 elements almost exclusively form ionic compounds that contain the M^{2+} ions. As expected, however, the lightest element, Be, with its higher ionization energy and small size, forms compounds that are largely covalent, as discussed in Section 21.1. Some compounds of Mg^{2+} also have significant covalent character. Hence, organometallic compounds like those discussed for Li in Group 1 are also important for Be and Mg in Group 2.

All the alkaline earth metals react vigorously with the halogens (Group 17) to form the corresponding halides (MX_2). Except for the beryllium halides, these compounds are all primarily ionic in nature, containing the M^{2+} cation and two X^- anions. The beryllium halides, with properties more typical of covalent compounds, have a polymeric halide-bridged structure in the solid state, as shown in the margin for $BeCl_2$. These compounds are volatile, producing vapors that contain the linear X—Be—X molecules predicted by the valence-shell electron-pair repulsion (VSEPR) model (see Chapter 9). As expected for compounds with only four valence electrons around the central atom, the beryllium halides are potent Lewis acids. They react readily with Lewis bases such as ethers to form tetrahedral adducts in which the central beryllium is surrounded by an octet of electrons:

$$BeCl_2(s) + 2(CH_3CH_2)_2O(l) \longrightarrow BeCl_2[O(CH_2CH_3)_2]_2(soln) \quad (21.23)$$

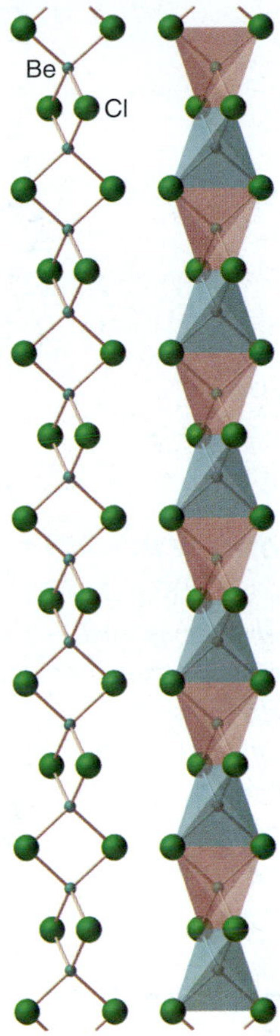

The polymeric, halide-bridged structure of solid beryllium chloride ($BeCl_2$).

The reactions of the alkaline earth metals with oxygen are less complex than those of the alkali metals. All the Group 2 elements except barium react directly with oxygen to form the simple oxide, MO. Barium forms barium peroxide (BaO_2), because the larger O_2^{2-} ion is better able to separate the large Ba^{2+} ions in the crystal lattice. In practice, only BeO is prepared by direct reaction with oxygen, and this reaction requires finely divided Be and high temperatures because Be is relatively inert. The other alkaline earth oxides are usually prepared by thermal decomposition of carbonate salts:

$$MCO_3(s) \xrightarrow{\Delta} MO(s) + CO_2(g) \quad (21.24)$$

The reactions of the alkaline earths with the heavier chalcogens (Y) are similar to those of the alkali metals. When the reactants are present in a 1:1 ratio, the binary chalcogenides (MY) are formed; at higher Y:M ratios, salts containing polychalcogenide ions (Y_n^{2-}) are formed.

In the reverse of Equation 21.24, the oxides of Ca, Sr, and Ba react with CO_2 to regenerate the carbonate. Except for BeO, which has significant covalent character and is therefore amphoteric, all the alkaline earth oxides are basic. Thus, they react with water to form the hydroxides, $M(OH)_2$:

$$MO(s) + H_2O(l) \longrightarrow M^{2+}(aq) + 2OH^-(aq) \quad (21.25)$$

and they dissolve in aqueous acid. The carbonates of the alkaline earths also react with aqueous acid to give CO_2 and H_2O:

$$MCO_3(s) + 2H^+(aq) \longrightarrow M^{2+}(aq) + CO_2(g) + H_2O(l) \quad (21.26)$$

The reaction in Equation 21.26 is the basis of antacids that contain MCO_3, which is used to neutralize excess stomach acid. Hydroxides of the lighter alkaline earths are insoluble in water, but their solubility increases as the atomic number of the metal increases. Because BeO and MgO are much more inert than the other Group 2 oxides, they are used as refractory materials in applications involving high temperatures and mechanical stress. For example, MgO (mp 2825°C) is used to coat the heating elements in electric ranges.

The trend in the reactivities of the alkaline earth metals with nitrogen is the opposite of that observed for the alkali metals. Only the lightest element, Be, does *not* react readily with N_2 to form the nitride (M_3N_2), although finely divided Be will react at high temperatures. The higher lattice energy due to the highly charged M^{2+} and N^{3-} ions is apparently sufficient to overcome the chemical inertness of the N_2 molecule, with its N≡N triple bond. Similarly, all the alkaline earths react with the heavier Group 15 elements to form binary compounds such as phosphides and arsenides with the general formula M_3Z_2.

Note the pattern

Higher lattice energies cause the alkaline earth metals to be more reactive than the alkali metals toward Group 15 elements.

When heated, all the alkaline earths, again except for beryllium, react directly with carbon to form ionic carbides with the general formula MC_2. The most important alkaline earth carbide is calcium carbide (CaC_2), which reacts readily with water to produce acetylene. For many years, this reaction was the primary source of acetylene for welding and lamps on miners' helmets. In contrast, beryllium reacts with elemental carbon to form

Be_2C, which formally contains the C^{4-} ion (although the compound is covalent). Consistent with this formulation, reaction of Be_2C with water or aqueous acid produces methane:

$$Be_2C(s) + 4H_2O(l) \longrightarrow 2Be(OH)_2(s) + CH_4(g) \quad (21.27)$$

Beryllium does not react with hydrogen except at high temperatures (1500°C), although BeH_2 can be prepared at lower temperatures by an indirect route. All the heavier alkaline earths (Mg through Ba) react directly with hydrogen to produce the binary hydrides (MH_2). The hydrides of the heavier alkaline earths are ionic, but both BeH_2 and MgH_2 have polymeric structures that reflect significant covalent character. All alkaline earth hydrides are good reducing agents that react rapidly with water or aqueous acid to produce hydrogen gas:

$$CaH_2(s) + 2H_2O(l) \longrightarrow Ca(OH)_2(s) + 2H_2(g) \quad (21.28)$$

Note the pattern

The solubility of alkaline earth carbonate and sulfates decrease down the group because hydration energies decrease.

Like the alkali metals, the heavier alkaline earths are sufficiently electropositive to dissolve in liquid ammonia. In this case, however, two solvated electrons are formed per metal atom, and no equilibria involving metal dimers or metal anions are known. Also like the alkali metals, the alkaline earths form a wide variety of simple ionic salts with oxoanions such as carbonate, sulfate, and nitrate. The nitrate salts tend to be soluble, but the carbonates and sulfates of the heavier alkaline earths are quite insoluble because of the higher lattice energy due to the doubly charged cation and anion. The solubility of the carbonates and the sulfates decreases rapidly down the group because hydration energies decrease with increasing cation size.

Complexes of Alkaline Earth Metals

Note the pattern

The alkaline earths have a substantially greater tendency to form complexes with Lewis bases than do the alkali metals.

Because of their higher positive charge (+2) and smaller ionic radii, the alkaline earths have a much greater tendency to form complexes with Lewis bases than do the alkali metals. This tendency is most important for the lightest cation, Be^{2+}, and decreases rapidly with increasing radius of the metal ion.

The chemistry of Be^{2+} is dominated by its behavior as a Lewis acid, forming complexes with Lewis bases that produce an octet of electrons around the beryllium. For example, Be^{2+} salts dissolve in water to form acidic solutions that contain the tetrahedral $[Be(H_2O)_4]^{2+}$ ion. Because of its high charge-to-radius ratio, the Be^{2+} ion polarizes the coordinated water molecules, thereby increasing their acidity:

$$[Be(H_2O)_4]^{2+}(aq) \longrightarrow [Be(H_2O)_3(OH)]^{+}(aq) + H^{+}(aq) \quad (21.29)$$

Similarly, in the presence of a strong base, beryllium and its salts form the tetrahedral hydroxo complex, $[Be(OH)_4]^{2-}$. Hence, beryllium oxide is amphoteric. Beryllium also forms a very stable tetrahedral fluoride complex, $[BeF_4]^{2-}$. Recall that beryllium halides behave like Lewis acids by forming adducts with Lewis bases (Equation 21.23).

The heavier alkaline earths also form complexes, but usually with a coordination number of 6 or higher. Again, complex formation is most important for the smaller cations, Mg^{2+} and Ca^{2+}. Thus, aqueous solutions of Mg^{2+} contain the octahedral $[Mg(H_2O)_6]^{2+}$ ion. Like the alkali metals, the alkaline earths form complexes with neutral cyclic ligands like the crown ethers and cryptands discussed in Section 21.3.

Organometallic Compounds Containing Group 2 Elements

Like the alkali metals, the lightest alkaline earths (Be and Mg) form the most covalent-like bonds with carbon, and they form the most stable organometallic compounds. Organometallic compounds of magnesium with the formula RMgX, where R is an alkyl or aryl group and X is a halogen, are universally called *Grignard reagents*, after Victor Grignard (1871–1935), the French chemist who discovered them (see Section 24.5). Grignard reagents can be used to synthesize various organic compounds, such as alcohols, aldehydes and ketones, carboxylic acids and esters, thiols, and amines.

Uses of the Alkaline Earth Metals

Elemental magnesium is the only alkaline earth metal that is produced on a large scale (about 5×10^5 tons per year). Its low density (1.74 g/cm^3 compared with 7.87 g/cm^3 for iron and 2.70 g/cm^3 for aluminum) makes it an important component of the lightweight metal alloys used in aircraft frames and aircraft and automobile engine parts (Figure 21.13). Most commercial aluminum actually contains about 5% magnesium to improve its corrosion resistance and mechanical properties. Elemental magnesium also serves as an inexpensive and powerful reductant for the production of a number of metals, including titanium, zirconium, uranium, and even beryllium, as shown in this equation

Figure 21.13 Magnesium alloys are lightweight and corrosion resistant. Because magnesium is about five times lighter than steel and 50% lighter than aluminum, it was long considered the "material of the future," as shown in this 1950 concept Buick LeSabre sports car made almost entirely of magnesium and aluminum alloys. Modern aluminum alloys usually contain about 5% magnesium to improve their corrosion resistance and mechanical properties.

$$TiCl_4(l) + 2Mg(s) \longrightarrow Ti(s) + 2MgCl_2(s) \qquad (21.30)$$

The only other alkaline earth that is widely used as the metal is beryllium, which is extremely toxic. Ingestion of beryllium or exposure to beryllium-containing dust causes a syndrome called *berylliosis*, characterized by severe inflammation of the respiratory tract or other tissues. A small percentage of beryllium dramatically increases the strength of copper or nickel alloys, which are used in nonmagnetic, nonsparking tools such as wrenches and screwdrivers and in camera springs and electrical contacts. The low atomic number of beryllium gives it a very low tendency to absorb X rays and makes it uniquely suited for applications involving radioactivity. Both elemental Be and BeO, which is a high-temperature ceramic, are used in nuclear reactors, and the windows on all X-ray tubes and sources are made of beryllium foil.

Millions of tons of calcium compounds are used every year. As discussed in earlier chapters, $CaCl_2$ is used as "road salt" to lower the freezing point of water on roads in cold temperatures. In addition, $CaCO_3$ is a major component of cement and an ingredient in many commercial antacids. "Quicklime" (CaO), produced by heating $CaCO_3$ (Equation 21.24), is used in the steel industry to remove oxide impurities, for making many kinds of glass, and to neutralize acidic soil. Other applications of Group 2 compounds described in earlier chapters include the medical use of $BaSO_4$ in "barium milkshakes" for identifying digestive problems by X rays, and the use of various alkaline earth compounds to produce the brilliant colors seen in fireworks.

EXAMPLE 21.4

For each application, choose the most appropriate substance based on the properties and reactivities of the alkaline earths and their compounds. Explain your choice in each case. Use any tables you need in making your decision, such as K_{sp} values (Table 17.1), lattice energies (Table 8.1), and band-gap energies (Section 12.6).

(a) To neutralize excess stomach acid that causes indigestion, would you use $BeCO_3$, $CaCO_3$, or $BaCO_3$?

(b) To remove CO_2 from the atmosphere in a space capsule, would you use MgO, CaO, or BaO?

(c) As a component of the alloy in an automotive spark plug electrode, would you use Be, Ca, or Ba?

Given Application, selected alkaline earths

Asked for Most appropriate substance for each application

Strategy

Based on the discussion in this section and any relevant information elsewhere in the book, determine which substance is most appropriate for the indicated use.

Solution

(a) All the alkaline earth carbonates will neutralize an acidic solution by Equation 21.26. Because beryllium and its salts are toxic, however, $BeCO_3$ cannot be used as an antacid. Of the remaining choices, $CaCO_3$ is somewhat more soluble than $BaCO_3$ (according to K_{sp} values listed in Table 17.1), suggesting that it will act more rapidly. Moreover, the formula mass of $CaCO_3$ is 100.1 amu, whereas that of $BaCO_3$ is almost twice as large. Therefore, neutralizing a given amount of acid would require twice the mass of $BaCO_3$ compared with $CaCO_3$. Furthermore, reaction of $BaCO_3$ with acid produces a solution containing Ba^{2+} ions, which are toxic (Ba^{2+} is a stimulant that can cause ventricular fibrillation of the heart). Finally, $CaCO_3$ is produced on a vast scale, so $CaCO_3$ is likely to be significantly less expensive than any barium compound. Consequently, $CaCO_3$ is the best choice for an antacid.

(b) This application involves the reaction of CO_2 with an alkaline earth oxide to form the carbonate, which is the reverse of the thermal decomposition reaction in which MCO_3 decomposes to CO_2 and the metal oxide, MO (Equation 21.24). Owing to their higher lattice energies (see Chapter 8), the smallest alkaline earths should form the most stable oxides. Hence, their carbonates should decompose at the lowest temperatures, as is observed ($BeCO_3$ decomposes at $<100°C$; $BaCO_3$ at 1360°C). If the carbonate with the smallest alkaline earth decomposes most readily, we would expect the reverse reaction (formation of a carbonate) to occur most readily with the largest metal cation, Ba^{2+}. Hence, BaO is the best choice.

(c) The alloy in a spark plug electrode must release electrons and promote their flow across the gap between the electrodes at high temperatures (see Chapter 12). Of the three metals listed, Ba has the lowest ionization energy and thus releases electrons most readily. Heating a barium-containing alloy to high temperatures will cause some ionization to occur, providing the initial step in forming a spark.

EXERCISE 21.4

Which of the indicated alkaline earth metals or their compounds is most appropriate for each application?

(a) Drying agent for removing water from the atmosphere: $CaCl_2$, $MgSO_4$, or BaF_2?
(b) Removal of scale deposits (largely $CaCO_3$) in water pipes: $HCl(aq)$ or $H_2SO_4(aq)$?
(c) Removal of traces of N_2 from purified argon gas: Be, Ca, or Ba?

Answer **(a)** $MgSO_4$; **(b)** HCl; **(c)** Ba

EXAMPLE 21.5

Predict the products of each reaction, and then balance the equations:

(a) $CaO(s) + HCl(g) \longrightarrow$
(b) $MgO(s) + \text{excess } OH^-(aq) \longrightarrow$
(c) $CaH_2(s) + TiO_2(s) \xrightarrow{\Delta}$

Given Reactants

Asked for Products and balanced chemical equation

Strategy

Follow the procedure given in Example 21.3 to predict the products of each reaction, and then balance the chemical equation.

Solution

(a) **A✓** Gaseous HCl is an acid, and CaO is a basic oxide that contains the O^{2-} ion. This is therefore an acid–base reaction that produces $CaCl_2$ and H_2O. **B✓** The balanced equation is $CaO(s) + 2HCl(g) \longrightarrow CaCl_2(aq) + H_2O(l)$.

(b) A✓ Magnesium oxide is a basic oxide, so it can either react with water to give a basic solution or dissolve in an acidic solution. Hydroxide ion is also a base. Because we have two bases but no acid, an acid–base reaction is impossible. A redox reaction is not likely because MgO is neither a good oxidant nor a good reductant. B✓ We conclude that no reaction occurs.

(c) A✓ Because CaH_2 contains the hydride ion (H^-), it is a good reductant. It is also a strong base because H^- ions can react with H^+ ions to form H_2. Titanium oxide (TiO_2) is a metal oxide that contains the metal in its highest oxidation state (+4 for a Group 4 metal); it can act as an oxidant by accepting electrons. We therefore predict that a redox reaction will occur, in which H^- is oxidized and Ti^{4+} is reduced. The most probable reduction product is metallic titanium, but what is the oxidation product? Oxygen must appear in the products, and both CaO and H_2O are stable compounds. The +1 oxidation state of hydrogen in H_2O is a sign that an oxidation has occurred ($2H^- \longrightarrow 2H^+ + 4e^-$). B✓ The balanced equation is $CaH_2(s) + TiO_2(s) \xrightarrow{\Delta} Ti(s) + CaO(s) + H_2O(l)$. We could also write the products as $Ti(s) + Ca(OH)_2(s)$.

EXERCISE 21.5

Predict the products of each reaction, and then balance the equations:

(a) $BeCl_2(s) + Mg(s) \xrightarrow{\Delta}$
(b) $BaCl_2(aq) + Na_2SO_4(aq) \longrightarrow$
(c) $BeO(s) + OH^-(aq) + H_2O(l) \longrightarrow$

Answer **(a)** $BeCl_2(s) + Mg(s) \longrightarrow Be(s) + MgCl_2(s)$ **(b)** $BaCl_2(aq) + Na_2SO_4(aq) \longrightarrow BaSO_4(s) + 2NaCl(aq)$ **(c)** $BeO(s) + 2OH^-(aq) + H_2O(l) \longrightarrow [Be(OH)_4]^{2-}(aq)$

21.5 • The s-Block Elements in Biology

The s-block elements play important roles in biological systems. Covalent hydrides, for example, are the building blocks of organic compounds, and other compounds and ions containing s-block elements are found in tissues and cellular fluids. In this section, we describe some ways in which biology depends on the properties of the Group 1 and Group 2 elements.

Covalent Hydrides

MGC The s-Block Elements in Biology

There are three major classes of hydrides—covalent, ionic, and metallic—but only covalent hydrides occur in living cells and have any biochemical significance. As you learned in Chapter 7, carbon and hydrogen have similar electronegativities, and the C—H bonds in organic molecules are strong and essentially nonpolar. Little acid–base chemistry is involved in the cleavage or formation of these bonds. In contrast, because hydrogen is less electronegative than oxygen and nitrogen (symbolized by Z), the H—Z bond in the hydrides of these elements is polarized ($H^{\delta+}—Z^{\delta-}$). Consequently, the hydrogen atoms in these H—Z bonds are relatively acidic. Moreover, S—H bonds are relatively weak due to poor s orbital overlap, so they are readily cleaved to give a proton. Hydrides in which H is bonded to O, N, or S atoms are therefore polar, hydrophilic molecules that form hydrogen bonds. They also undergo acid–base reactions by transferring a proton.

Hydrogen bonds are crucial in biochemistry, in part because they help hold proteins in their biologically active folded structures. Hydrogen bonds also connect the two intertwining strands of DNA, the substance that contains the genetic code of all organisms (see Section 24.6). Because hydrogen bonds are easier to break than the covalent bonds that form the individual DNA strands, the two intertwined strands can be separated to give intact single strands, which is essential for the duplication of genetic information.

In addition to the importance of hydrogen bonds in biochemical molecules, the extensive hydrogen-bonding network in water is one of the keys to the existence of life on our planet. Based on its molecular mass, water should be a gas at room temperature (20°C), but the strong intermolecular interactions in liquid water greatly increase its boiling point. Hydrogen bonding also produces the relatively open molecular arrangement found in ice, which causes ice to be less dense than water. Because ice floats on the surface of water, it creates a protective layer that allows aquatic organisms to survive during cold winter months.

Note the pattern

Strong intermolecular hydrogen-bonding interactions are responsible for the high heat capacity of water and its high heat of fusion.

These same strong intermolecular hydrogen-bonding interactions are also responsible for the high heat capacity of water and its high heat of fusion. A great deal of energy must be removed from water for it to freeze. Consequently, as noted in Chapter 5, large bodies of water act as "thermal buffers" that have a stabilizing effect on the climate of adjacent land areas. Perhaps the most striking example of this effect is the fact that humans can live comfortably at very high latitudes. For example, palm trees grow in southern England at the same latitude (51°N) as the southern end of frigid Hudson Bay and northern Newfoundland in North America, areas known more for their moose populations than for their tropical vegetation. Warm water from the Gulf Stream current in the Atlantic Ocean flows clockwise from the tropical climate at the equator past the eastern coast of the United States and then turns toward England, where heat stored in the water is released. The temperate climate of Europe is largely attributable to the thermal properties of water.

Macrominerals

The members of Group 1 and Group 2 that are present in the largest amounts in organisms are sodium, potassium, magnesium, and calcium, all of which form monatomic cations with a charge of +1 (Group 1, M^+) or +2 (Group 2, M^{2+}). Biologically, these elements can be classified as *macrominerals* (see Table 1.4).

For example, calcium is found in the form of relatively insoluble calcium salts that are used as structural materials in many organisms. Hydroxyapatite [$Ca_5(PO_4)_3OH$] is the major component of bones, calcium carbonate ($CaCO_3$) is the major component of the shells of mollusks and the eggs of birds and reptiles, and calcium oxalate (CaO_2CCO_2) is found in many plants. Because calcium and strontium have similar sizes and charge-to-radius ratios, small quantities of strontium are always found in bone and other calcium-containing structural materials. Normally this is not a problem because the Sr^{2+} ions occupy sites that would otherwise be occupied by Ca^{2+} ions. When trace amounts of radioactive ^{90}Sr are released into the atmosphere from nuclear weapons tests or a nuclear accident, however, the radioactive strontium eventually reaches the ground, where it is taken up by plants that are consumed by dairy cattle. The isotope then becomes concentrated in cow's milk along with calcium. Since radioactive strontium co-precipitates with calcium in the hydroxyapatite that surrounds the bone marrow (where white blood cells are produced), children, who typically ingest more of this milk than adults, are at substantially increased risk for leukemia, a type of cancer characterized by the overproduction of white blood cells.

Ion Transport

The Na^+, K^+, Mg^{2+}, and Ca^{2+} ions are important components of intracellular and extracellular fluids. Both Na^+ and Ca^{2+} are found primarily in *extra*cellular fluids, such as blood plasma, whereas K^+ and Mg^{2+} are found primarily in *intra*cellular fluids. Substantial inputs of energy are required to establish and maintain these concentration gradients and prevent the system from reaching equilibrium. Thus, energy is needed to transport each of these ions across the cell membrane toward the side with the higher concentration. The biological machines that are responsible for the selective transport of these metal ions are complex assemblies of proteins called **ion pumps**. Ion pumps recognize and discriminate between metal ions in the same way that crown ethers and cryptands do, with a high affinity for ions of a certain charge and radius.

Defects in the ion pumps or their control mechanisms can result in major health problems. For example, *cystic fibrosis*, the most common inherited disease in the United States, is caused by a defect in the transport system (in this case chloride ions). Similarly, in many cases, *hypertension*, or high blood pressure, is thought to be due to defective Na^+ uptake and/or excretion. If too much Na^+ is absorbed from the diet (or if too little is excreted), water diffuses from tissues into the blood to dilute the solution, thereby decreasing the osmotic pressure in the circulatory system. The increased volume increases the blood pressure, and ruptured arteries called *aneurysms* can result, often in the brain. Because high blood pressure causes other medical problems as well, it is one of the most important biomedical disorders in modern society.

For patients who suffer from hypertension, low-sodium diets that use NaCl substitutes such as KCl are often prescribed. Although KCl and NaCl give similar flavors to foods, the K^+ is not readily taken up by the highly specific Na^+-uptake system. This approach to controlling hypertension is controversial, however, because direct correlations between dietary Na^+ content and blood pressure are difficult to demonstrate in the general population. More important, recent observations indicate that high blood pressure may correlate more closely with inadequate intake of calcium in the diet than with excessive sodium levels. This finding is important because the typical "low-sodium" diet is also low in good sources of calcium, such as dairy products.

Some of the most important biological functions of the Group 1 and Group 2 metals are due to small changes in the cellular concentrations of the metal ion. Transmission of nerve impulses, for example, is accompanied by an increased flux of Na^+ ions into the nerve cell. Similarly, binding of various hormones to specific receptors on the surface of a cell leads to a rapid influx of Ca^{2+} ions; the resulting sudden rise in the intracellular Ca^{2+} concentration triggers other events, such as muscle contraction, release of neurotransmitters, enzyme activation, or secretion of other hormones.

Within cells, K^+ and Mg^{2+} often activate particular enzymes by binding to specific negatively charged sites in the enzyme structure. *Chlorophyll*, the green pigment used by all plants to absorb light and drive the process of *photosynthesis*, contains magnesium. During photosynthesis, CO_2 is reduced to form sugars such as glucose. As shown in the margin, the structure of the central portion of a chlorophyll molecule resembles a crown ether (see Figure 13.7a) with four five-member nitrogen-containing rings linked together to form a large ring that provides a "hole" the proper size to tightly bind Mg^{2+}.

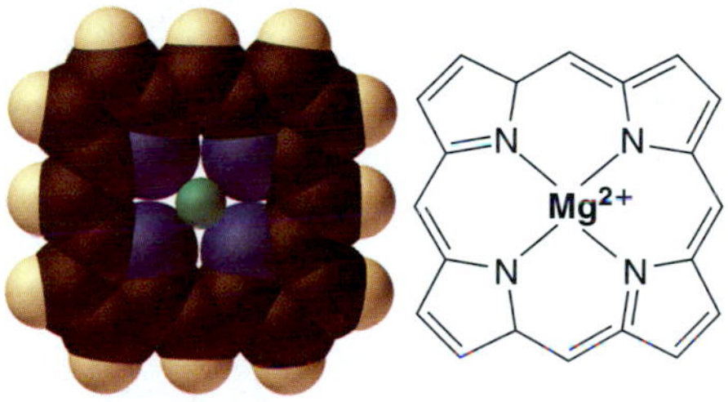

The structure of the central core of chlorophyll, a magnesium complex present in all photosynthetic tissues. Note the resemblance to the crown ether complexes discussed in Chapter 13.

Ionophores

Because the health of cells depends on maintaining the proper levels of cations in intracellular fluids, any change that affects the normal flux of metal ions across cell membranes could well cause an organism to die. Molecules that facilitate the transport of metal ions across membranes are generally called **ionophores** (*ion* plus *phore* from the Greek *phorein*, "to carry"). Many ionophores are potent antibiotics that can kill or inhibit the growth of bacteria. An example is *valinomycin*, a cyclic molecule with a central cavity lined with oxygen atoms (Figure 21.14a) that is similar to the cavity of a crown ether (see Figure 13.7a). Like a crown ether, valinomycin is highly selective: its affinity for K^+ is about a thousand times greater than for Na^+. By increasing the flux of K^+ ions into cells, valinomycin disrupts the normal K^+ gradient across a cell membrane, thereby killing the cell (Figure 21.14b).

EXAMPLE 21.6

A common way to study the function of a metal ion in biology is to replace the naturally occurring metal with one whose reactivity can be traced by spectroscopic methods. The substitute metal ion must bind to the same site as the naturally occurring ion, and it must have a similar (or greater) affinity for that site, as indicated by its charge density. Arrange the following ions in order of increasing effectiveness as a replacement for Ca^{2+}, which has an ionic radius of 100 pm

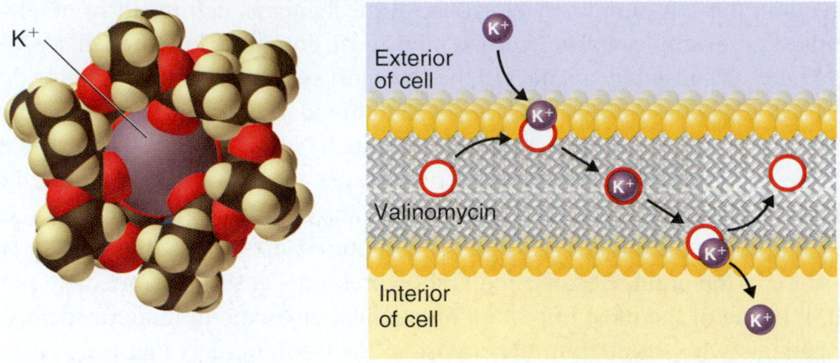

(a) K^+–valinomycin complex (b) Transport of K^+ across a membrane

Figure 21.14 Valinomycin is an antibiotic that functions like an ionophore. (a) This model of the structure of the K^+–valinomycin complex, determined by X-ray diffraction, shows how the valinomycin molecule wraps itself around the K^+ ion, shielding it from the environment, in a manner reminiscent of a crown ether complex (see Chapter 13). (b) Valinomycin kills bacteria by facilitating the transport of K^+ ions across the cell membrane, thereby disrupting the normal distribution of ions in the bacterium. At the surface of the membrane, valinomycin binds a K^+ ion. Because the hydrophobic exterior of the valinomycin molecule forms a "doughnut" that shields the positive charge of the metal ion, the K^+–valinomycin complex is highly soluble in the nonpolar interior of the membrane. After the K^+–valinomycin complex diffuses across the membrane to the interior of the cell, the K^+ ion is released, and the valinomycin is free to diffuse back to the other side of the membrane to bind another K^+ ion. Valinomycin thereby destroys the normal K^+ gradient across the membrane, killing the cell.

(numbers in parentheses are ionic radii): Na^+ (102 pm); Eu^{2+} (117 pm); Sr^{2+} (118 pm); F^- (133 pm); Pb^{2+} (119 pm); La^{3+} (103 pm). Explain your reasoning.

Given Ions and ionic radii

Asked for Suitability as replacement for Ca^{2+}

Strategy

Use periodic trends to arrange the ions from least effective to most effective as a replacement for Ca^{2+}.

Solution

The most important properties in determining the affinity of a biological molecule for a metal ion are the size and charge-to-radius ratio of the metal ion. Of the possible Ca^{2+} replacements listed, the F^- ion has the opposite charge, so it should have no affinity for a Ca^{2+}-binding site. Na^+ is approximately the right size, but with a +1 charge it will bind much more weakly than Ca^{2+}. Although Eu^{2+}, Sr^{2+}, and Pb^{2+} are all a little larger than Ca^{2+}, they are probably similar enough in size and charge to bind. Based on its ionic radius, Eu^{2+} should bind most tightly of the three. La^{3+} is nearly the same size as Ca^{2+} and more highly charged. With a higher charge-to-radius ratio and a similar size, La^{3+} should bind tightly to a Ca^{2+} site and be the most effective replacement for Ca^{2+}. The order is $F^- \ll Na^+ \ll Pb^{2+} \sim Sr^{2+} \sim Eu^{2+} < La^{3+}$.

EXERCISE 21.6

The ionic radius of K^+ is 138 pm. Arrange the following ions in order of increasing affinity for a K^+-binding site in an enzyme (numbers in parentheses are ionic radii): Na^+ (102 pm); Rb^+ (152 pm); Ba^{2+} (135 pm); Cl^- (181 pm); Tl^+ (150 pm).

Answer $Cl^- \ll Na^+ < Tl^+ \sim Rb^+ < Ba^{2+}$

SUMMARY AND KEY TERMS

21.1 Overview of Periodic Trends (p. 957)

The most important unifying principle in describing the chemistry of the elements is that the systematic increase in atomic number and the orderly filling of electron orbitals lead to periodic trends in atomic properties. The most fundamental property leading to periodic variations is the **effective nuclear charge (Z_{eff})**. Because of the position of the diagonal line separating metals and nonmetals in the periodic table, the chemistry of Groups 13, 14, and 15 is relatively complex. The second-period elements ($n = 2$) in each group exhibit unique chemistry compared with their heavier congeners because of their smaller radii, energetically unavailable *d* orbitals, and greater ability to form π bonds with other atoms. Increasing ionization energies and decreasing bond strengths lead to the **inert-pair effect**, which causes the heaviest elements of Groups 13–17 to have a stable oxidation state that is lower by 2 than the maximum predicted for their respective groups.

21.2 The Chemistry of Hydrogen (p. 960)

The three isotopes of hydrogen—**protium** (^1H or H), **deuterium** (^2H or D), and **tritium** (^3H or T)—have different physical properties. Deuterium and tritium can be used as *tracers*, substances that enable biochemists to follow the path of a molecule through an organism or cell. Hydrogen can form compounds that contain a proton (H^+), a **hydride ion (H^-)**, an electron-pair bond to H, a **hydrogen bond**, or a **three-center bond**, also known as an *electron-deficient bond*, in which two electrons are shared between three atoms. Hydrogen gas can be generated by the reaction of an active metal with dilute acid, Al or Zn with strong base, or on an industrial scale by *catalytic steam reforming*, which produces *synthesis gas*, or *syngas*.

21.3 The Alkali Metals (Group 1) (p. 964)

The first alkali metals to be isolated (Na and K) were obtained by passing an electric current through molten potassium and sodium carbonates. The alkali metals are among the most potent reductants known; most can be isolated by electrolysis of their molten salts or, in the case of rubidium and cesium, by reaction of their hydroxide salts with a reductant. They can also be recovered from their silicate ores using a multistep process. Lithium, the strongest reductant, and sodium, the weakest, are examples of the physical and chemical effect of opposing periodic trends. The alkali metals react with halogens (Group 17) to form ionic halides, with the heavier chalcogens (Group 16) to produce metal chalcogenides, and with oxygen to form compounds whose stoichiometry depends on the size of the metal. The peroxides and superoxides are potent oxidants. The only alkali metal to react with atmospheric nitrogen is lithium. Heavier alkali metals react with graphite to form **graphite intercalation compounds**, substances in which metal atoms are inserted between the sheets of carbon atoms. With heavier Group 14 elements, alkali metals react to give polyatomic anions with three-dimensional cage structures. All alkali metals react with hydrogen at high temperature to produce the corresponding hydrides, and all reduce water to produce hydrogen gas. Alkali metal salts are prepared by reaction of the metal hydroxide with an acid, followed by evaporation of water. Both Li and Na salts are used as *drying agents*, compounds used to absorb water. Complexing agents such as **crown ethers** and **cryptands** can accommodate alkali metal ions of the appropriate size. Alkali metals can also react with liquid ammonia to form solutions that slowly decompose to give hydrogen gas and the metal salt of the *amide* ion (NH_2^-). These solutions, which contain unstable solvated electrons loosely associated with a cavity in the solvent, are intensely colored, good conductors of electricity, and excellent reductants. Alkali metals can react with organic compounds that contain an acidic proton to produce salts. They can also form **organometallic compounds**, which have properties that differ from those of their metallic and organic components.

21.4 The Alkaline Earth Metals (Group 2) (p. 972)

Pure samples of most of the alkaline earth metals can be obtained by electrolysis of the chlorides or oxides. Beryllium was first obtained by the reduction of its chloride; radium chloride, which is radioactive, was obtained through a series of reactions and separations. In contrast to the alkali metals, the alkaline earths generally have little or no affinity for an added electron. All alkaline earth metals react with the halogens to produce the corresponding halides, with oxygen to form the oxide (except for barium, which forms the peroxide), and with the heavier chalcogens to form chalcogenides or polychalcogenide ions. All oxides except BeO react with CO_2 to form carbonates, which in turn react with acid to produce CO_2 and H_2O. Except for Be, all the alkaline earth metals react with N_2 to form nitrides, and all react with carbon and hydrogen to form carbides and hydrides. Alkaline earth metals dissolve in liquid ammonia to give solutions that contain two solvated electrons per metal atom. The alkaline earths have a greater tendency than the alkali metals to form complexes with crown ethers, cryptands, and other Lewis bases. The most important alkaline earth organometallic compounds are *Grignard reagents* (RMgX), which are used to synthesize organic compounds.

21.5 The *s*-Block Elements in Biology (p. 979)

Covalent hydrides in which hydrogen is bonded to oxygen, nitrogen, or sulfur are polar, hydrophilic molecules that form hydrogen bonds and undergo acid–base reactions. Hydrogen-bonding interactions are crucial in stabilizing the structure of proteins and DNA, and allow genetic information to be duplicated. The hydrogen-bonding interactions in water and ice also allow life to exist on our planet. The Group 1 and Group 2 metals present in organisms are *macrominerals*, which are important components of intracellular and extracellular fluids. Small changes in the cellular concentration of a metal ion can have a significant impact on biological functions. Metal ions are selectively transported across cell membranes by **ion pumps**, which bind ions based on their charge and radius. **Ionophores**, many of which are potent antibiotics, facilitate the transport of metal ions across membranes.

QUESTIONS AND PROBLEMS

MGC For instructor-assigned homework, go to **www.masteringgenchem.com**

Questions and problems with colored numbers have answers in the Appendix and complete solutions in the Student Solutions Manual.

CONCEPTUAL

21.1 Overview of Periodic Trends

1. List three physical properties that are important in describing the behavior of the main group elements.
2. Arrange the atoms in order of (a) increasing ionization energy, (b) increasing atomic size, and (c) increasing electronegativity: K, Cs, Sr, Ca, Ba, Li.
3. Arrange the atoms in order of (a) decreasing atomic size and (b) decreasing magnitude of electron affinity: Rb, H, Be, Na, Cs, Ca.
4. Which periodic trends are affected by Z_{eff}? Based on the positions of the elements in the periodic table, which element would you expect to have the highest Z_{eff}? The lowest Z_{eff}?
5. Compare the properties of the metals and nonmetals with regard to their electronegativities and preferred oxidation states.
6. Which of these elements have a greater tendency to form positive ions than negative ions? Ca, Br, Li, N, Zr, Ar, Sr, S
7. Arrange these elements in order of decreasing metallic character: As, O, Ca, Sn, Be, Sb.
8. Give three reasons the chemistry of the second-period elements is generally not representative of their groups as a whole.
9. Compare the second-period elements and their heavier congeners with regard to (a) magnitude of electron affinity, (b) coordination number, and (c) solubility of the halides in nonpolar solvents.
10. The heavier main group elements tend to form extended sigma-bonded structures rather than multiple bonds to other atoms. Give a reasonable explanation for this tendency.
11. What is the diagonal effect? How does it explain the similarity in chemistry between, for example, boron and silicon?
12. Although many of the properties of the second- and third-row elements in a group are quite different, one property is similar. Which one?
13. Two elements are effective additives to solid rocket propellant: beryllium and one other element that has similar chemistry. Based on the position of beryllium in the periodic table, identify the second element.
14. Give two reasons for the inert-pair effect. How would this phenomenon explain why Sn^{2+} is a reducing agent but Pb^{2+} is stable?
15. Explain the following trend in electron affinities (in kJ/mol): Al (−41.8); Si (−134.1); P (−72.0); S (−200.4).
16. Using orbital energy arguments, explain why electron configurations with more than four electron pairs around the central atom are not observed for second-period elements.

21.2 The Chemistry of Hydrogen

17. Some periodic tables include hydrogen as a Group 1 element, whereas other periodic tables include it as a Group 17 element. Refer to the properties of hydrogen to propose an explanation for its placement in each group. In each case, give one property of hydrogen that would exclude it from Groups 1 and 17.
18. If there were a planet where the abundances of D_2O and H_2O were reversed and life had evolved to adjust to this difference, what would be the effects of consuming large amounts of H_2O?
19. Describe the bonding in (a) a hydrogen bond and (b) the central B—H bond in $B_2H_7^-$. Why are compounds containing isolated protons unknown?
20. With which elements does hydrogen form (a) ionic hydrides, (b) covalent hydrides, and (c) metallic hydrides? Which of these types of hydrides can behave like acids?
21. Indicate which of these elements are likely to form ionic, covalent, or metallic hydrides, and explain your reasoning: Sr, Si, O, Li, B, Be, Pd, Al.
22. Which has the higher ionization energy: H or H^-? Why?
23. The electronegativities of hydrogen, fluorine, and iodine are 2.20, 3.98, and 2.66, respectively. Why, then, is HI a stronger acid than HF?
24. If H_2O were a linear molecule, would the density of ice be lower or higher than that of liquid water? Explain your answer.
25. In addition to ion–dipole attractions, hydrogen bonding is important in solid crystalline hydrates, such as $Na_4XeO_6{\cdot}8H_2O$. Based on this statement, explain why anhydrous Na_4XeO_6 does not exist.

21.3 The Alkali Metals (Group I)

26. Which of the Group 1 elements reacts least readily with oxygen? Which is most likely to form a hydrated, crystalline salt? Explain your answers.
27. The alkali metals have a significant electron affinity, corresponding to the addition of an electron to give the M^- anion. Why, then, do they commonly lose the ns^1 electron to form the M^+ cation rather than gaining an electron to form M^-?
28. Lithium is a far stronger reductant than sodium; cesium is almost as strong as lithium, which does not agree with the expected periodic trend. What two opposing properties explain this apparent anomaly? Is the same anomaly found among the alkaline earths?
29. Explain why the ionic character of LiCl is less than that of NaCl. Based on periodic trends, would you expect the ionic character of $BeCl_2$ to be greater or less than that of LiCl? Why?
30. Alkali metals and carbon form intercalation compounds with extremely high electrical conductivity. Is this conductivity through the layers or along the layers? Explain your answer.
31. Electrolysis is often used to isolate the lighter alkali metals from their molten halides. Why are halides used rather than the oxides or carbonates, which are easier to isolate? With this in mind, what is the purpose of adding calcium chloride to the alkali metal halide?
32. The only alkali metal that reacts with oxygen to give a compound with the expected stoichiometry is lithium, which gives Li_2O. In contrast, sodium reacts with oxygen to give Na_2O_2, and the heavier alkali metals form superoxides. Explain the difference in the stoichiometries of these products.
33. Classify aqueous solutions of each of the following as acidic, basic, or amphoteric: Li_2O, Na_2O, CsO_2.
34. Although methanol is relatively unreactive, it can be converted to a synthetically more useful form by reaction with LiH. Predict

the products of the reaction of methanol with LiH. Describe the visual changes you would expect to see during this reaction.

35. Lithium reacts with atmospheric nitrogen to form lithium nitride, Li_3N. Why don't the other alkali metals form analogous nitrides? Explain why all the alkali metals react with arsenic to form the corresponding arsenides, M_3As.

21.4 The Alkaline Earth Metals (Group 2)

36. The electronegativities of Li and Sr are nearly identical (0.98 vs 0.95, respectively). Given their positions in the periodic table, how do you account for this?
37. Arrange the atoms in order of increasing Z_{eff}: Na, Ba, Cs, Li.
38. Do you expect the melting point of NaCl to be higher, equal to, or lower than that of $MgCl_2$? Why?
39. Which of the Group 2 elements would you expect to form primarily ionic rather than covalent organometallic compounds? Explain your reasoning.
40. Explain why beryllium forms compounds that are best regarded as covalent in nature, whereas the other elements in Group 2 generally form ionic compounds.
41. Why is the trend in the reactions of the alkaline earths with nitrogen the reverse of the trend seen for the alkali metals?
42. Is the bonding in the alkaline earth hydrides primarily ionic or covalent in nature? Explain your answer. Given the type of bonding, do you expect the lighter or heavier alkaline earths to be better reducing agents?
43. Using arguments based on ionic size, charge, and chemical reactivity, explain why beryllium oxide is amphoteric. What element do you expect to be most similar to beryllium in its reactivity? Why?
44. Explain why the solubility of the carbonates and sulfates of the alkaline earths decreases with increasing cation size.
45. Beryllium oxide is amphoteric, magnesium oxide is weakly basic, and calcium oxide is very basic. Explain how this trend is related to the ionic character of the oxides.
46. Do you expect the ΔH_f° of CaH_2 to be greater than, the same as, or less than that of BaH_2? Why or why not?
47. Which of the *s*-block elements would you select to carry out a chemical reduction on a small scale? Consider cost, reactivity, and stability in making your choice. How would your choice differ if the reduction were carried out on an industrial scale?

21.5 The *s*-Block Elements in Biology

48. Explain the thermochemical properties of water in terms of its intermolecular bonding interactions. How does this affect global climate patterns?
49. Of the three classes of hydrides, which is (are) biochemically significant? How do you account for this?
50. Many proteins are degraded and become nonfunctional when heated higher than a certain temperature, even though the individual protein molecules do not undergo a distinct chemical change. Propose an explanation for this observation.
51. Los Angeles has moderate weather throughout the year, with average temperatures between 57°F and 70°F. In contrast, Palm Springs, which is just 100 miles inland, has average temperatures between 55°F and 95°F. Explain the difference in the average temperature ranges in the two cities.
52. Although all Group 1 ions have the same charge (+1), Na^+ and K^+ ions are selectively transported across cell membranes. What strategy do organisms employ to discriminate between these two cations?

STRUCTURE AND REACTIVITY

This section includes paired problems (marked by brackets) that require similar problem-solving skills.

21.1 Overview of Periodic Trends

53. The table lists the valences, coordination numbers, and ionic radii for a series of cations. Which would you substitute for K^+ in a crystalline lattice? Explain your answer.

Metal	Charge	Coordination Number	Ionic Radius, pm
Li	+1	4	76
Na	+1	6	102
K	+1	6	138
Mg	+2	6	72
Ca	+2	6	100
Sr	+2	6	118

21.2 The Chemistry of Hydrogen

54. One of the largest uses of methane is for the production of syngas to convert hydrogen to ammonia. Write a complete equation to describe this conversion. Calculate ΔG° for this reaction at 298 K, and determine the temperature at which the reaction becomes spontaneous.
55. An alternative method of producing hydrogen is the water–gas shift reaction:

$$CO(g) + H_2O(g) \longrightarrow CO_2(g) + H_2(g)$$

Use Appendix A to calculate ΔG° for this reaction at 298 K, and determine the temperature at which the reaction changes from spontaneous to nonspontaneous (or vice versa).

56. Predict the products of each reaction at 25°C, and then balance the equations:
(a) $CsH(s) + D_2O(l) \longrightarrow$
(b) $CH_3CO_2H(l) + D_2O(l) \longrightarrow$
(c) $H_3PO_4(aq) + D_2O(l) \longrightarrow$
(d) $NH_2CH_2CO_2H(s) + D_2O(l) \longrightarrow$
(e) $NH_4Cl(s) + D_2O(l) \longrightarrow$
57. Using heavy water (D_2O) as the source of deuterium, how could you conveniently prepare (a) D_2SO_4 and (b) LiD?
58. What are the products of the reaction of NaH with D_2O? Do you expect the same products from the reaction of NaD and H_2O? Explain your answer.
59. A 2.50-g sample of zinc metal reacts with 100.0 mL of 0.150 *M* HCl. What volume of H_2 (in liters) is produced at a temperature of 23°C and a pressure of 729 mmHg?
60. A chemical reaction requires 16.8 L of H_2 gas at STP. How many grams of magnesium metal are needed to produce this amount of hydrogen gas?
61. Seawater contains 3.5% dissolved salts by mass and has an average density of 1.026 g/mL. The volume of the ocean is estimated to be 1.35×10^{21}L. Using the data in

Table 21.2, calculate the total mass of deuterium in the ocean.

62. From data in Table 21.2, determine the molarity of DOH in water. Do you expect the molarity of D_2O in water to be similar? Why or why not?

63. Use the data in Table 21.2 to calculate how many liters of water you would have to evaporate to obtain 1.0 mL of TOD. The density of TOD is 1.159 g/mL.

21.3 The Alkali Metals (Group I)

64. Write chemical equations to describe these reactions: (a) the electrolysis of fused (melted) sodium chloride; (b) the thermal decomposition of $KClO_3$; (c) the preparation of hydrogen fluoride from calcium fluoride and sulfuric acid; (d) the oxidation of sodium metal by oxygen.

65. What products are formed at the anode during electrolysis of (a) molten lithium hydride, (b) molten lithium chloride, and (c) aqueous sodium fluoride? Write the corresponding half-reactions for each reaction.

66. Sodium metal is prepared by electrolysis of molten NaCl. If 25.0 g of chlorine gas is produced in the electrolysis of the molten salt using 9.6 amps (C/s) of current, how many hours were required for the reaction? What mass of sodium was produced?

67. Sodium peroxide can remove CO_2 from the air and replace it with oxygen according to the unbalanced equation

$$Na_2O_2(s) + CO_2(g) \longrightarrow Na_2CO_3(s) + O_2(g)$$

(a) Balance the chemical equation.
(b) Identify each oxidation and reduction half-reaction.
(c) Assuming complete reaction, what will be the pressure inside a sealed 1.50-L container after reaction of excess sodium peroxide with carbon dioxide that was initially at 0.133 atm and 37°C?

68. Predict the products of each chemical reaction, and then balance the equations:
(a) $K(s) + CH_3CH_2OH(l) \longrightarrow$
(b) $Na(s) + CH_3CO_2H(l) \longrightarrow$
(c) $NH_4Cl(s) + Li(s) \longrightarrow$
(d) $(CH_3)_2NH(l) + K(s) \longrightarrow$

69. Predict the products of the reaction of (a) an alkyl chloride with lithium metal and (b) rubidium with oxygen.

70. A 655-mg sample of graphite was allowed to react with potassium metal, and 744 mg of product was isolated. What is the stoichiometry of the product?

71. A key step in the isolation of the alkali metals from their ores is selective precipitation. For example, lithium is separated from sodium and potassium by precipitation of Li_2CO_3 ($K_{sp} = 8.15 \times 10^{-4}$). If 500.0 mL of a 0.275 *M* solution of Na_2CO_3 is added to 500.0 mL of a 0.536 *M* lithium hydroxide solution, what mass of Li_2CO_3 will precipitate assuming no further reactions occur? What mass of lithium will remain in solution?

72. Perchloric acid, which is used as a reagent in a number of chemical reactions, is typically neutralized before disposal. When a novice chemist accidentally used K_2CO_3 to neutralize perchloric acid, a large mass of $KClO_4$ ($K_{sp} = 1.05 \times 10^{-2}$) precipitated from solution. What mass of potassium ion is present in 1.00 L of a saturated solution of $KClO_4$?

21.4 The Alkaline Earth Metals (Group 2)

73. Beryllium iodide reacts vigorously with water to produce HI. Write a balanced equation for this reaction, and explain why it is violent.

74. Predict the products of each reaction, and then balance the equations:
(a) $Mg(OH)_2(aq) + (NH_4)_3PO_4(aq) \longrightarrow$
(b) calcium carbonate and sulfuric acid $\longrightarrow$
(c) $CaCl_2(aq) + Na_3PO_4(aq) \longrightarrow$
(d) the decomposition of $SrCO_3$

75. Predict the products of the following reactions, and then balance each equation:
(a) $Sr(s) + O_2(g) \longrightarrow$
(b) thermal decomposition of $CaCO_3(s)$
(c) $CaC_2(s) + H_2O(l) \longrightarrow$
(d) $RbHCO_3(s) + H_2SO_4(aq) \longrightarrow$

76. Indicate whether each pair of substances will react, and, if so, write a balanced chemical equation for the reaction: (a) an alkyl chloride and magnesium metal; (b) strontium metal and nitrogen; (c) magnesium metal and cold water; (d) beryllium and nitrogen.

77. Using a thermodynamic cycle and information presented in Chapters 7 and 8, calculate the lattice energy of magnesium nitride (Mg_3N_2). (ΔH_f° for Mg_3N_2 is -463 kJ/mol, and ΔH° for $N(g) + 3e^- \longrightarrow N^{3-}$ is -1736 kJ.) How does the lattice energy of Mg_3N_2 compare with that of $MgCl_2$ and of MgO (see Appendix A)?

78. The solubility-product constants of the carbonate salts of magnesium, calcium, and strontium are 6.82×10^{-6}, 3.36×10^{-9}, and 5.60×10^{-10}, respectively. How many milligrams of each compound would be present in 200.0 mL of a saturated solution of each? How would the solubility depend on the pH of the solution? Why?

79. The solubility-product constant of $BaSO_4$ is 1.08×10^{-10}, and that of $CaSO_4$ is 4.93×10^{-5}. What accounts for this difference? When 500.0 mL of a solution that contains 1.00 *M* $Ba(NO_3)_2$ and 3.00 *M* $Ca(NO_3)_2$ is mixed with a 2.00 *M* solution of Na_2SO_4, a precipitate forms. What is the identity of the precipitate? How much of it will form before the second salt precipitates?

80. Electrolytic reduction is used to produce magnesium metal from $MgCl_2$. The goal is to produce 200.0 kg of Mg by this method.
(a) How many kilograms of $MgCl_2$ are required?
(b) How many liters of chlorine gas will be released at STP?
(c) How many hours will it take to process the magnesium metal if a total current of 1.00×10^4 amps is utilized?

81. A sample consisting of 20.35 g of finely divided calcium metal is allowed to react completely with nitrogen. What is the mass of the product?

82. What mass of magnesium hydride will react with water to produce 1.51 L of hydrogen gas at STP?

21.4 The *s*-Block Elements in Biology

83. A 0.156-g sample of a chloride salt of an alkaline earth is dissolved in enough water to make 20.5 mL of solution. If this solution has an osmotic pressure of 2.68 atm at 25°C, what is the identity of the alkaline earth metal?

84. The thermal buffering capacity of water is one of the reasons the human body is capable of withstanding a wide range of temperatures. How much heat (in kJ) is required to raise the temperature of a 70.0-kg human from 37.0°C to 38.0°C? Assume that 70% of the mass of the body is solely due to water and that body fluids have the same specific heat as water.

85. During illness, body temperature can increase by more than 2°C. One piece of folklore is that you should "feed a fever." Using the data in Table 5.5, how many pieces of fried chicken would a 70.0-kg person need to eat to generate a 2.0°C change in body temperature? Assume complete conversion of the caloric content of the chicken to thermal energy.

86. Hydrogen bonding is partly responsible for the high enthalpy of vaporization of water (ΔH_{vap} = 43.99 kJ/mol at 25°C), which contributes to cooling the body during exercise. Assume that a 50.0-kg runner produces 20.0 g of perspiration during a race, and all of the perspiration is converted to water vapor at 37.0°C. What amount of heat (in J) is removed from the runner's skin if the perspiration consists of only water?

APPLICATIONS

87. For each application, which of the substances indicated would you select and why? Base your selections on the properties and reactivities of the alkaline earths and their compounds.
(a) Source of CO_2 at low temperature: $BeCO_3$ or $CaCO_3$?
(b) Window material for X-ray tubes: beryllium or strontium?
(c) Source of iodide ions in aqueous solution: BaI_2 or BeI_2?
(d) Formation of a stable organometallic compound by reaction of a metal with an alkyl halide: calcium or magnesium?
(e) Synthesis of refractory materials: magnesium oxide or barium oxide?

88. Ultrahigh-purity tritium needed in the nuclear weapons industry is obtained by allowing a mixture of tritium and its nuclear decay product, helium-3, to diffuse through a thin block of palladium metal. Explain why this is an effective method for separating the two.

89. In one technique for harnessing solar energy, the blue-green algae *Anabaena cylindrica* is used to produce H_2 and O_2 photosynthetically. The resulting gases are collected, passed through palladium metal, and then recombined in a fuel cell to produce electricity. Draw a diagram showing how this process might work.

90. Scientists speculate that sodium atoms react with atmospheric ozone to produce a high-energy species, which is then reduced by atomic oxygen. This process is believed to occur as meteors enter the Earth's atmosphere. Write equations for these reactions. Is sodium regenerated in this process?

91. Propose an effective compound for purifying and regenerating air for breathing in a submarine and justify your choice.

92. Explain why administering cryptands to a person suffering from iron toxicity could be an effective clinical treatment.

93. Calcium magnesium carbonate [$CaMg(CO_3)_2$], also known as dolomite, is a primary constituent of soils. It is formed when water containing magnesium ions comes in contact with calcium carbonate. Do you expect dolomite to be more or less resistant to acid rain than calcium carbonate? Why?

94. Few classes of reagents have proved to be as useful for organic syntheses as the Grignard reagents (RMgX), which are produced by the reaction of an organohalogen compound (RX) with magnesium in an ether solvent. The ease of formation of a Grignard reagent depends on the structure of the organohalogen compound.
(a) For a given R group, arrange the alkyl halides (RX) in order of decreasing reactivity with Mg, and justify your reasoning.
(b) Is it possible to make a Grignard reagent using an organosodium compound and $MgCl_2$? Why or why not?
(c) Is $MgCl_2$ or MgI_2 preferable for the reaction described in part b?

95. The percentage of the population that developed leukemia in the vicinity of the Chernobyl nuclear reactor rose substantially after the nuclear accident in 1986. Why? Based on information in this chapter, what might have been done to reduce the incidence of leukemia in children who lived in the affected region?

96. Explain how ingesting large amounts of NaCl could induce a heart attack.

97. The general mechanism by which valinomycin functions is described in the text. If you were asked to develop a new antibiotic that functions like valinomycin, what type of structural features would you want to incorporate into it?

98. A water sample is known to contain both Mg^{2+} and Ca^{2+} ions. You have been given the task of determining the exact concentration of each ion. To do this, you first add a concentrated KOH solution to the water sample and allow it to stand for several minutes with occasional swirling. You then add an indicator and titrate the solution with a standard solution of EDTA (ethylenediamine tetraacetic acid) until a color change is observed. How will this procedure allow you to determine the concentration of at least one of the ions? Which concentration will you be able to determine first?

99. The primary component of ordinary glass is SiO_2, which is difficult to work with due to its high melting point (1700°C). Adding metal oxides lowers the melting temperature to a workable level (700–900°C) and causes the liquid to flow more easily. Both Na_2O and K_2O are commonly used for this purpose. Which of these oxides produces a glass that hardens more rapidly and at a higher temperature? Why is there a difference?

100. Lead poisoning can be treated by administering EDTA, an organic compound that binds to Pb^{2+}(see question 98), thereby removing lead from the system. If you are thinking of administering EDTA as a clinical treatment for Ni^{2+} poisoning because of its effectiveness in removing Pb^{2+}, what additional factors do you need to consider?

101. Silicon, aluminum, and oxygen, the most abundant constituents of the Earth's crust, are found in many minerals. For example, two feldspars, which are typical components of granites, are orthoclase and anorthite. Orthoclase has the composition $K[AlSi_3O_8]$; in anorthite, Al^{3+} replaces one of the Si^{4+} ions. Suggest a plausible cation in anorthite in place of K^+.

102. When a plant root is immersed in an aqueous environment, its uptake of metal ions is dictated by equilibrium considerations. Metal ions can be captured by reaction with various types of carrier molecules. The resulting complexes can pass through the cell membrane and dissociate, thereby transporting the metal ions into the cell. The Na^+, K^+, Mg^{2+}, and Ca^{2+} ions are all transported using this mechanism. Which of these ions would be captured by (a) oxygen donors with a neutral or low (−1) charge; (b) carboxylates, phosphates, and some neutral oxygen donors; (c) polyphosphates with a total negative charge higher than −2; and (d) nitrogen donors.

103. Sodium carbonate has long been used in the glass industry, for making water softeners, and in the wood pulp and paper industries. Recently, sodium carbonate has proved useful for removing SO_2 and H_2SO_4 from the flue gases of coal- and oil-fired power stations. Write balanced chemical equations showing the reactions that occur when flue gases are passed through solid sodium carbonate.

104. Hard water contains high concentrations of Ca^{2+} and Mg^{2+}. These ions react with soaps (sodium and potassium salts of naturally occurring carboxylic acids) to form an insoluble scum. Explain how adding Na_2CO_3 to hard water can soften the water and improve the ability of soaps to remove dirt and grime.

105. Do you expect $MgCO_3$ or $Mg(OH)_2$ to be more soluble in water? Based on your answer, explain how adding $Ca(OH)_2$ softens water by removing Mg^{2+}. Be sure to show a balanced equation for the reaction.

106. How does a low-sodium diet help lower blood pressure? Do you think a diet high in KCl would also be effective at lowering blood pressure? Explain your answer.

22 The *p*-Block Elements

LEARNING OBJECTIVES

- To understand how valence-electron configurations and periodic trends in atomic properties determine the chemical properties of the *p*-block elements
- To use thermodynamics and kinetics to understand the reactivity of the *p*-block elements
- To understand why the chemistry of the lightest member of each group differs from that of the heavier elements of the group
- To be able to predict the types of reactions the *p*-block elements undergo

CHAPTER OUTLINE

We continue our discussion of the chemistry of the main group elements as we move to the *p* block of the periodic table. We will use the systematic approach developed in Chapter 21, which is based on valence-electron configurations and periodic trends in atomic properties, while applying the unifying principles of chemical bonding, thermodynamics, and kinetics.

The line that divides metals from nonmetals in the periodic table crosses the *p* block diagonally. As a result, the differences between metallic and nonmetallic properties are evident within each group, even though all members of each group have the same valence-electron configuration. The *p* block is the only portion of the periodic table where we encounter the inert-pair effect. Moreover, as with the *s*-block elements, the chemistry of the lightest member of each group in the *p* block differs sharply from that of its heavier congeners, but is

An artist's sketch of tetragonal boron, showing the linked B_{12} icosahedra. The artist, Roger Hayward, served as Linus Pauling's illustrator for several decades, most notably in *The Architecture of Molecules*, published in 1964.

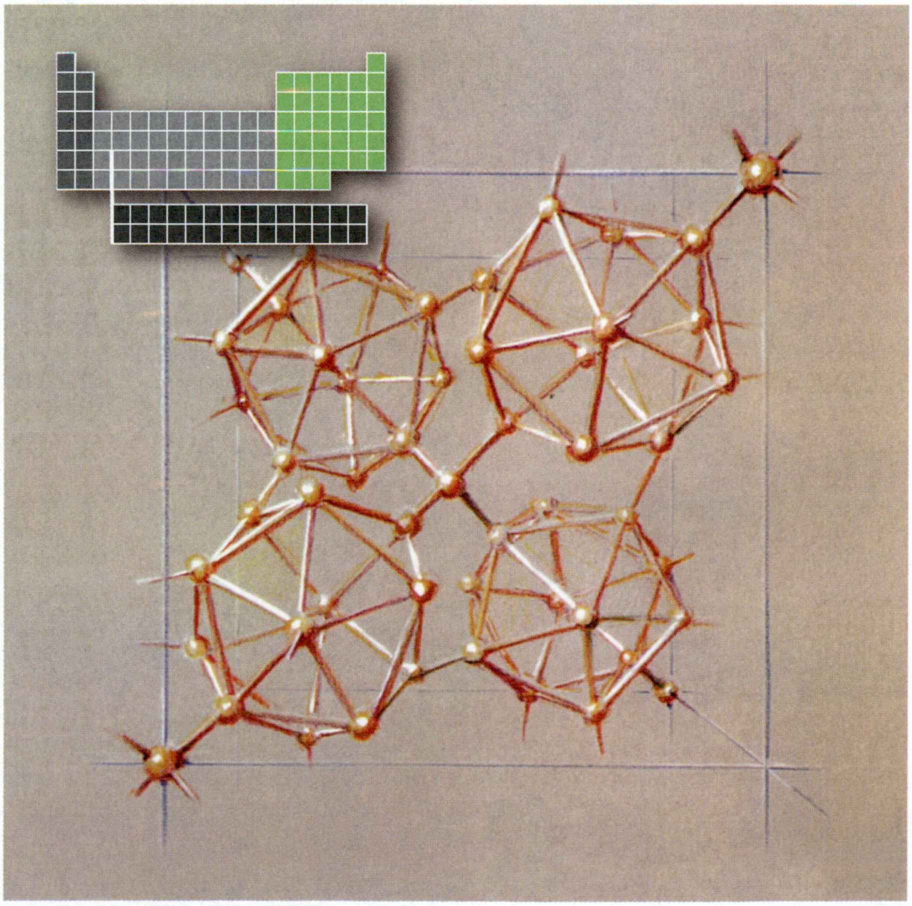

similar to that of the element immediately below and to the right of it in the next group. Thus, diagonal similarities in chemistry are seen across the *p* block.

As you study the periodic trends in properties and reactivity of the elements in Groups 13–18, you will learn how "cobalt blue" glass, rubies, and sapphires are made and why the U.S. military became interested in the use of boron hydrides as rocket fuels, then abandoned its effort. You will also discover the source of diamonds on Earth, why silicon-based life-forms are likely to exist only in science fiction, and why most compounds with N—N bonds are potentially explosive. You will also learn why phosphorus can cause a painful and lethal condition known as "phossy jaw" and why selenium is used in photocopiers.

Note the pattern

Group 13 elements are never found in nature in their free state.

22.1 • The Elements of Group 13

Group 13 is the first group to span the dividing line between metals and nonmetals, and consequently its chemistry is more diverse than that of Groups 1 and 2, which include only metallic elements. Except for the lightest element, boron, the Group 13 elements are all relatively electropositive; that is, they tend to lose electrons in chemical reactions rather than gain them. Although Group 13 includes aluminum, the most abundant metal on Earth, none of these elements was known until the early 19th century because they are never found in nature in their free state. Elemental boron and aluminum, which were first prepared by reducing B_2O_3 and $AlCl_3$, respectively, with potassium, could not be prepared until potassium had been isolated and shown to be a potent reductant. Indium (In) and thallium (Tl) were discovered in the 1860s with the use of spectroscopic techniques, long before methods were available for isolating them. Indium, named for its indigo (deep blue-violet) emission line, was first observed in the spectrum of zinc ores, while thallium (from the Greek *thallos*, a young, green shoot of a plant) was named for its brilliant green emission line. Gallium (Ga) (Mendeleev's *eka*-aluminum) was discovered in 1875 by the French chemist Paul Émile Lecoq de Boisbaudran during a systematic search for Mendeleev's "missing" element in Group 13.

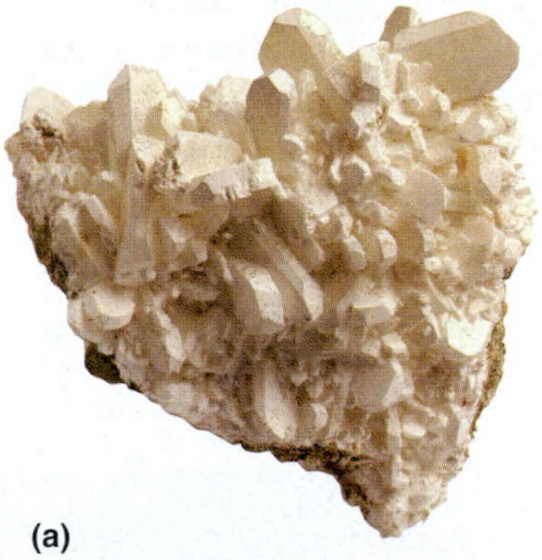

(a)

(b)

Figure 22.1 Borax deposits. (a) Concentrated deposits of crystalline borax ($Na_2B_4O_5(OH)_4{\cdot}8H_2O$) are found in ancient lake beds, such as the Mojave Desert and Death Valley in the western United States. (b) Borax is used in various cleaning products, including 20 Mule Team Borax, a laundry detergent named for the teams of 20 mules that hauled wagons full of borax from desert deposits to railroad terminals in the 1880s.

Preparation and General Properties of Group 13 Elements

The Group 13 elements are not as powerful reductants as are the alkali metals and alkaline earths. Nevertheless, their compounds with oxygen are thermodynamically stable, and large amounts of energy are needed to isolate even the two most accessible elements, boron and aluminum, from their oxide ores.

Although boron is relatively rare (it is about 10,000 times less abundant than aluminum), concentrated deposits of borax ($Na_2B_4O_5(OH)_4{\cdot}8H_2O$) are found in ancient lake beds (Figure 22.1) and were used in ancient times for making glass and glazing pottery. Boron is produced on a large scale by the reaction of borax with acid to produce boric acid [$B(OH)_3$], which is then dehydrated to the oxide (B_2O_3). Reduction of the oxide with magnesium or sodium gives amorphous boron that is only about 95% pure:

$$Na_2B_4O_5(OH)_4\cdot 8H_2O \xrightarrow{\text{Acid}} B(OH)_3(s) \xrightarrow{\Delta} B_2O_3(s) \tag{22.1a}$$

$$B_2O_3(s) + 3Mg(s) \xrightarrow{\Delta} 2B(s) + 3MgO(s) \tag{22.1b}$$

Pure, crystalline boron, however, is extremely difficult to obtain because of its high melting point (2300°C) and the highly corrosive nature of liquid boron. It is usually prepared by reducing pure BCl_3 with hydrogen gas at high temperatures or by thermal decomposition of boron hydrides such as diborane (B_2H_6):

$$BCl_3(g) + \frac{3}{2}H_2(g) \longrightarrow B(s) + 3HCl(g) \tag{22.2}$$

$$B_2H_6(g) \longrightarrow 2B(s) + 3H_2(g) \tag{22.3}$$

The reaction shown in Equation 22.2 is used to prepare boron fibers, which are stiff and light. Hence, they are used as structural reinforcing materials in objects as diverse as the U.S. space shuttle and the frames of lightweight bicycles that are used in races such as the Tour de France. Boron is also an important component of many ceramics and heat-resistant borosilicate glasses such as Pyrex, which is used for ovenware and laboratory glassware.

In contrast to boron, deposits of aluminum ores such as *bauxite*, a hydrated form of Al_2O_3, are abundant. With an electrical conductivity about twice that of copper on a wt. for wt. basis, aluminum is used in more than 90% of the overhead electric power lines in the United States. However, because aluminum–oxygen compounds are stable, obtaining aluminum metal from bauxite is an expensive process. Aluminum is extracted from oxide ores by treatment with a strong base, which produces the soluble hydroxide complex, $[Al(OH)_4]^-$. Neutralization of the resulting solution with gaseous CO_2 results in the precipitation of $Al(OH)_3$:

$$2[Al(OH)_4]^-(aq) + CO_2(g) \longrightarrow 2Al(OH)_3(s) + CO_3^{2-}(aq) + H_2O(l) \quad (22.4)$$

Thermal dehydration of $Al(OH)_3$ produces Al_2O_3, and metallic aluminum is obtained by the electrolytic reduction of Al_2O_3 using the *Hall–Heroult process* described in Chapter 19. Of the Group 13 elements, only aluminum is used on a large scale: for example, each Boeing 777 airplane is about 50% aluminum by mass.

The other members of Group 13 are rather rare: gallium is approximately 5000 times less abundant than aluminum, and indium and thallium are even scarcer. Consequently, these metals are usually obtained as byproducts in the processing of other metals. The extremely low melting point of gallium (29.6°C), however, makes it easy to separate from aluminum. Due to its low melting point and high boiling point, gallium is used as a liquid in thermometers that have a temperature range of almost 2200°C. The heavier Group 13 elements, indium and thallium, are found as trace impurities in sulfide ores of zinc and lead. Indium is used as a crushable seal for high-vacuum cryogenic devices, and its alloys are used as low-melting solders in electronic circuit boards. Thallium, on the other hand, is so toxic that the metal and its compounds have few uses. Both indium and thallium oxides are released in flue dust when sulfide ores are converted to metal oxides and SO_2. Until relatively recently, these and other toxic elements were allowed to disperse in the air, creating large "dead zones" downwind of the smeltery. The flue dusts are now trapped and serve as a relatively rich source of elements such as In and Tl (as well as Ge, Cd, Te, and As).

Table 22.1 summarizes some important properties of the Group 13 elements. Notice the large differences between boron and aluminum in size, ionization energy, electronegativity, and standard reduction potential, which are all consistent with the observation that boron behaves chemically like a nonmetal and aluminum like a metal. All the Group 13 elements have $ns^2\, np^1$ valence-electron configurations, and all tend to lose their three valence electrons to form compounds in the +3 oxidation state. The heavier elements in the group can also form compounds in the +1 oxidation state formed by formal loss of the single *np* valence electron. Because the Group 13 elements generally contain only six valence electrons in their neutral compounds, these compounds are all moderately strong Lewis acids.

In contrast to Groups 1 and 2, the Group 13 elements show no consistent trends in ionization energies, electron affinities, and reduction potentials, whereas electronegativities actually *increase* from aluminum to thallium. Some of these anomalies, especially for the series Ga, In, Tl, can be explained by the increase in the effective nuclear charge (Z_{eff}) that results from poor shielding of the nuclear charge by the filled $(n-1)d^{10}$ and $(n-2)f^{14}$ subshells. Consequently, although the actual nuclear charge increases by 32 as we go from indium to thallium, screening by the filled 5*d* and 4*f* subshells is so poor that Z_{eff} increases significantly from indium to thallium. Thus, the first ionization energy of thallium is actually higher than that of indium.

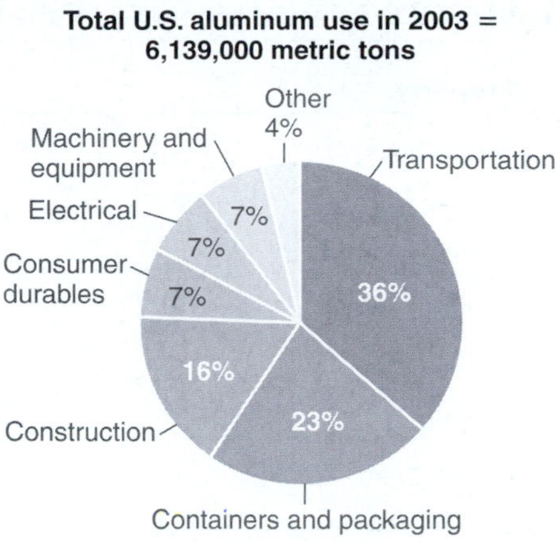

Source of data: U.S. Geological Survey, 2005, Aluminium End-Use Statistics, in Kelly, T.D., and Matos, G.R., comps., Historical statistics for mineral and material commodities in the United States: USGS Data Series 140, available online at http://pubs.usgs.gov/ds/2005/140/.

Note the pattern

Anomalies in periodic trends among Ga, In, and Tl can be explained by the increase in the effective nuclear charge due to poor shielding.

TABLE 22.1 Selected properties of the Group 13 elements

Property	Boron	Aluminum[a]	Gallium	Indium	Thallium
Atomic symbol	B	Al	Ga	In	Tl
Atomic number	5	13	31	49	81
Atomic mass (amu)	10.81	26.98	69.72	114.82	204.38
Valence-electron configuration[b]	$2s^2 2p^1$	$3s^2 3p^1$	$4s^2 4p^1$	$5s^2 5p^1$	$6s^2 6p^1$
mp/bp (°C)	2075/4000	660/2519	29.7/2204	156.6/2072	304/1473
Density (g/cm^3) at 25°C	2.34	2.70	5.91	7.31	11.8
Atomic radius (pm)	87	118	136	156	156
First ionization energy (kJ/mol)	801	578	579	558	589
Most common oxidation state	+3	+3	+3	+3	+1
Ionic radius (pm)[c]	(25)	54	62	80	162
Electron affinity (kJ/mol)	−27	−42	−40	−39	−37
Electronegativity	2.0	1.6	1.8	1.8	1.8
Standard reduction potential, $E°$ (V), of $M^{3+}(aq)$	−0.87	−1.66	−0.55	−0.34	+0.741
Product of reaction with O_2	B_2O_3	Al_2O_3	Ga_2O_3	In_2O_3	Tl_2O
Type of oxide	Acidic	Amphoteric	Amphoteric	Amphoteric	Basic
Product of reaction with N_2	BN	AlN	GaN	InN	None
Product of reaction with X_2[d]	BX_3	Al_2X_6	Ga_2X_6	In_2X_6	TlX
Product of reaction with H_2	None	None	None	None	None

[a]This is the name used in the United States; the rest of the world inserts an extra *i* and calls it *aluminium*.

[b]The configuration shown does not include filled *d* and *f* subshells.

[c]The values cited are for six-coordinate ions in the most common oxidation state, except for Al^{3+}, for which the value for the four-coordinate ion is given. The B^{3+} ion is not a known species; the radius cited is an estimated four-coordinate value.

[d]X = Cl, Br, or I. Reaction with F_2 gives the trifluorides (MF_3) for all Group 13 elements.

Reactions and Compounds of Boron

Note the pattern

Elemental boron forms multicenter bonds, whereas the other Group 13 elements exhibit metallic bonding.

Elemental boron is a semimetal that is remarkably unreactive; in contrast, the other Group 13 elements all exhibit metallic properties and reactivity. We therefore consider the reactions and compounds of boron separately from those of the other elements in the group.

All the Group 13 elements have fewer valence electrons than valence orbitals, which generally results in delocalized, metallic bonding. With its high ionization energy, low electron affinity, low electronegativity, and small size, however, boron does not form a metallic lattice with delocalized valence electrons. Instead, boron forms unique and intricate structures that contain *multicenter bonds*, in which a pair of electrons holds together three or more atoms.

The basic building block of elemental boron is not the individual boron atom, as would be the case in a metal, but rather the B_{12} *icosahedron* (see Section 7.4). Because these icosahedra do not pack together very well, the structure of solid boron contains voids, resulting in its low density (Figure 22.2).

Elemental boron can be induced to react with many nonmetallic elements to give binary compounds that have a variety of applications. For example, plates of boron carbide (B_4C) can stop a 30-caliber armor-piercing bullet, yet they weigh 10–30% less than conventional armor. Other important compounds of boron with nonmetals

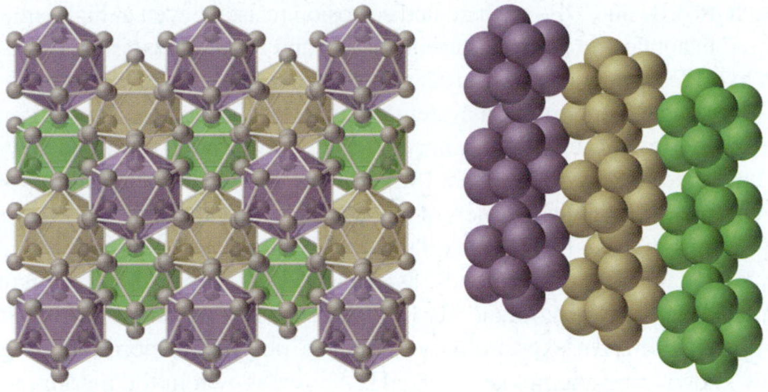

(a) Cubic close-packed (ccp) B_{12} (b) Side view (90° rotation)

Figure 22.2 Solid boron contains B_{12} icosahedra. Unlike metallic solids, elemental boron consists of a regular array of B_{12} icosahedra rather than individual boron atoms. Note that each boron atom in the B_{12} icosahedron is connected to *five* other boron atoms within the B_{12} unit. (a) The allotrope of boron with the simplest structure is α-rhombohedral boron, which consists of B_{12} octahedra in an almost cubic close-packed (ccp) lattice. (b) A side view of the structure shows that icosahedra do not pack as efficiently as spheres, making the density of solid boron lower than expected.

include boron nitride (BN), which is produced by heating boron with excess nitrogen (Equation 22.5); boron oxide (B_2O_3), which is formed when boron is heated with excess oxygen (Equation 22.6); and the boron trihalides (BX_3), which are formed by heating boron with excess halogen (Equation 22.7):

$$2B(s) + N_2(g) \xrightarrow{\Delta} 2BN(s) \tag{22.5}$$

$$4B(s) + 3O_2(g) \xrightarrow{\Delta} 2B_2O_3(s) \tag{22.6}$$

$$2B(s) + 3X_2(g) \xrightarrow{\Delta} 2BX_3(g) \tag{22.7}$$

As is typical of elements lying near the dividing line between metals and nonmetals, many compounds of boron are amphoteric, dissolving in either acid or base.

Boron nitride is similar in many ways to elemental carbon. With eight electrons, the B—N unit is *isoelectronic* with the C—C unit, and B and N have the same average size and electronegativity as C. The most stable form of BN is similar to graphite, containing six-membered B_3N_3 rings arranged in layers. At high temperature and pressure, hexagonal BN converts to a cubic structure similar to diamond, which is one of the hardest substances known. Boron oxide (B_2O_3) contains layers of trigonal planar BO_3 groups (analogous to BX_3) in which the oxygen atoms bridge two boron atoms. It dissolves many metal and nonmetal oxides, including SiO_2, to give a wide range of commercially important borosilicate glasses. A small amount of CoO gives the deep blue color characteristic of "cobalt blue" glass.

At high temperatures, boron also reacts with virtually all metals to give *metal borides* that contain regular three-dimensional networks, or *clusters*, of boron atoms. The structures of two metal borides, ScB_{12} and CaB_6, are shown in Figure 22.3. Because metal-rich

Figure 22.3 The structures of ScB_{12} and CaB_6, two boron-rich metal borides. (a) The structure of ScB_{12} consists of B_{12} clusters and Sc atoms arranged in a faced-centered cubic (fcc) lattice similar to that of NaCl, with B_{12} units occupying the anion positions and scandium atoms the cation positions. The B_{12} units here are not icosahedra but *cubooctahedra*, with alternating square and triangular faces. (b) The structure of CaB_6 consists of octahedral B_6 clusters and calcium atoms arranged in a body-centered cubic (bcc) lattice similar to that of CsCl, with B_6 units occupying the anion positions and calcium atoms the cation positions.

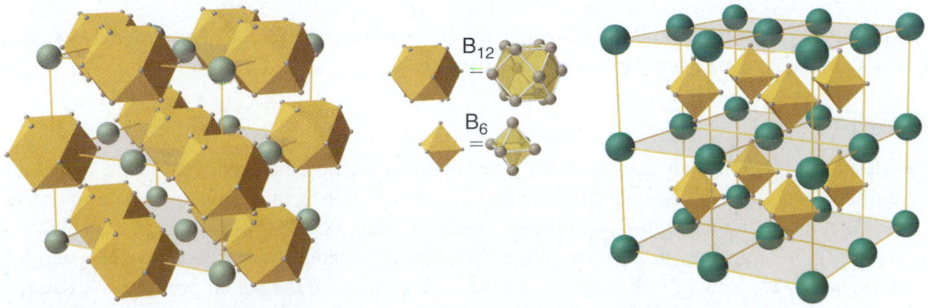

(a) ScB_{12} (b) CaB_6

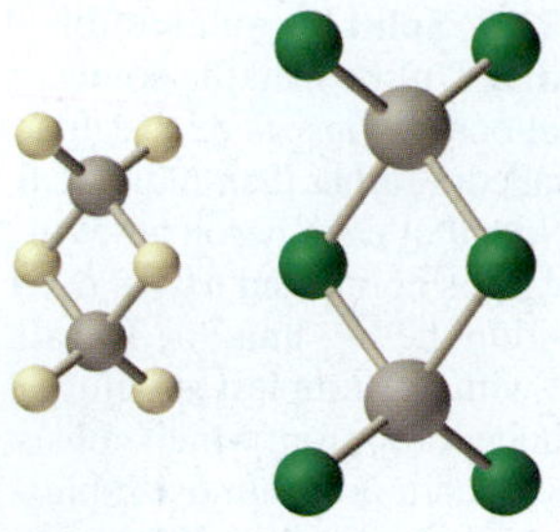

Figure 22.4 The structures of diborane (B_2H_6) and aluminum chloride (Al_2Cl_6). (a) The hydrogen-bridged dimer B_2H_6 contains two three-center, two-electron bonds as described for the $B_2H_7^-$ ion in Figure 21.5. (b) In contrast, the bonding in the halogen-bridged dimer Al_2Cl_6 can be described in terms of electron-pair bonds, in which a chlorine atom bonded to one aluminum atom acts as a Lewis base by donating a lone pair of electrons to another aluminum atom, which acts as a Lewis acid.

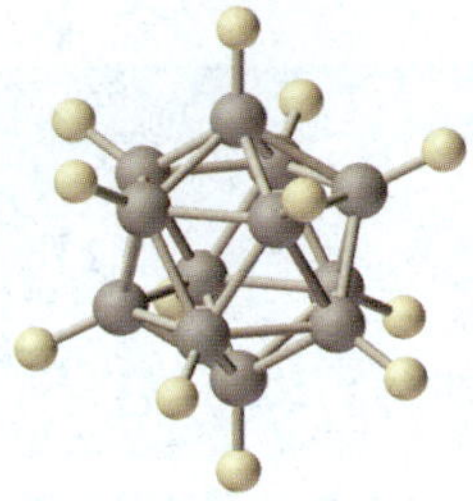

$B_{12}H_{12}^{2-}$

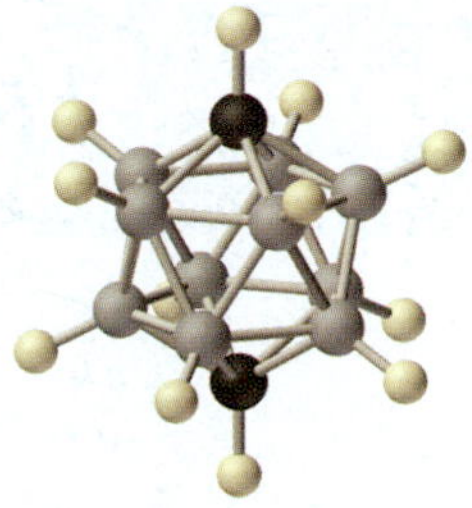

$B_{10}C_2H_{12}$

borides such as ZrB_2 and TiB_2 are hard and corrosion resistant even at high temperatures, they are used in applications such as turbine blades and rocket nozzles.

Boron hydrides were not discovered until the early 20th century, when the German chemist Alfred Stock undertook a systematic investigation of the binary compounds of boron and hydrogen, although binary hydrides of carbon, nitrogen, oxygen, and fluorine have been known since the 18th century. Between 1912 and 1936, Stock oversaw the preparation of a series of boron–hydrogen compounds with unprecedented structures that could not be explained with simple bonding theories. All these compounds contain multicenter bonds, as discussed in Chapter 21 (see Figure 21.5). The simplest example is diborane (B_2H_6), which contains two bridging hydrogen atoms (Figure 22.4a). An extraordinary variety of polyhedral boron–hydrogen clusters is now known; one example is the $B_{12}H_{12}^{2-}$ ion shown in the margin, which has a polyhedral structure similar to the icosahedral B_{12} unit of elemental boron, with a single hydrogen atom bonded to each boron atom.

A related class of polyhedral clusters, the carboranes, contain both CH and BH units; an example is shown in the margin. Replacing the hydrogen atoms bonded to carbon with organic groups produces substances with novel properties, some of which are currently being investigated for their use as liquid crystals and in cancer chemotherapy.

The enthalpy of combustion of diborane (B_2H_6) is −2165 kJ/mol, one of the highest values known:

$$B_2H_6(g) + 3O_2(g) \longrightarrow B_2O_3(s) + 3H_2O(l) \quad \Delta H = -2165 \text{ kJ/mol} \tag{22.8}$$

Consequently, the U.S. military explored using boron hydrides as rocket fuels in the 1950s and 1960s. This effort was eventually abandoned because boron hydrides are unstable, costly, and toxic, and, most important, B_2O_3 proved to be highly abrasive to rocket nozzles. Reactions carried out during this investigation, however, showed that boron hydrides exhibit unusual reactivity.

Because boron and hydrogen have almost identical electronegativities, the reactions of boron hydrides are dictated by the minor differences in the distribution of electron density in a given compound. In general, two distinct types of reaction are observed: electron-rich species such as the BH_4^- ion are reductants, whereas electron-deficient species such as B_2H_6 can act as oxidants.

EXAMPLE 22.1

For each reaction, explain why the given products form:

(a) $B_2H_6(g) + 3O_2(g) \longrightarrow B_2O_3(s) + 3H_2O(l)$
(b) $BCl_3(l) + 3H_2O(l) \longrightarrow B(OH)_3(aq) + 3HCl(aq)$
(c) $2BI_3(s) + 3H_2(g) \xrightarrow{\Delta} \frac{1}{6}B_{12}(s) + 6HI(g)$

Given Balanced chemical equations

Asked for Why the given products form

Strategy

Classify the type of reaction. Using periodic trends in atomic properties, thermodynamics, and kinetics, explain why the reaction products form.

Solution

(a) Molecular oxygen is an oxidant. If the other reactant is a potential reductant, we expect that a redox reaction will occur. Although B_2H_6 contains boron in its highest oxidation state (+3), it also contains hydrogen in the −1 oxidation state (the hydride ion). Because hydride is a strong reductant, a redox reaction will probably occur. We expect that H^- will be oxidized to H^+ and O_2 will be reduced to O^{2-}, but what are the actual products? A reasonable guess is B_2O_3 and H_2O, both stable compounds.

(b) Neither BCl_3 nor water is a powerful oxidant or reductant, so a redox reaction is unlikely; a hydrolysis reaction is more probable. Nonmetal halides are acidic and react with water to form a solution of the hydrohalic acid and a nonmetal oxide or hydroxide. In this case, the most probable boron-containing product is boric acid [$B(OH)_3$].

(c) We normally expect a boron trihalide to behave like a Lewis acid. In this case, however, the other reactant is elemental hydrogen, which usually acts as a reductant. The iodine atoms in BI_3 are in the lowest accessible oxidation state (−1), and boron is in the +3 oxidation state. Consequently, we can write a redox reaction in which hydrogen is oxidized and boron is reduced. Because compounds of boron in lower oxidation states are rare, we expect that boron will be reduced to elemental boron. The other product of the reaction must therefore be HI.

EXERCISE 22.1

Predict the products of the reactions, and write a balanced equation for each:

(a) $B_2H_6(g) + H_2O(l) \xrightarrow{\Delta}$
(b) $BBr_3(l) + O_2(g) \longrightarrow$
(c) $B_2O_3(s) + Ca(s) \xrightarrow{\Delta}$

Answer **(a)** $B_2H_6(g) + 6H_2O(l) \xrightarrow{\Delta} 2B(OH)_3(s) + 6H_2(g)$ **(b)** $BBr_3(l) + O_2(g) \longrightarrow$ no reaction **(c)** $6B_2O_3(s) + 18Ca(s) \xrightarrow{\Delta} B_{12}(s) + 18CaO(s)$

Reactions and Compounds of the Heavier Group 13 Elements

All four of the heavier Group 13 elements (Al, Ga, In, and Tl) react readily with the halogens to form compounds with a 1:3 stoichiometry:

$$2M(s) + 3X_2(s,l,g) \longrightarrow 2MX_3(s) \text{ or } M_2X_6 \quad (22.9)$$

The reaction of Tl with iodine is an exception: although the product has the stoichiometry TlI_3, it is *not* thallium(III) iodide, but rather a thallium(I) compound, the Tl^+ salt of the triiodide ion, I_3^-. This compound forms because iodine is not a powerful enough oxidant to oxidize thallium to the +3 oxidation state.

Of the halides, only the fluorides exhibit behavior typical of an ionic compound: they have high melting points (>950°C) and low solubility in nonpolar solvents. In contrast, the trichorides, tribromides, and triiodides of aluminum, gallium, and indium, as well as $TlCl_3$ and $TlBr_3$, are more covalent in character and form halogen-bridged dimers (Figure 22.4b). Although the structure of these dimers is similar to that of diborane (B_2H_6), the bonding can be described in terms of electron-pair bonds rather than the delocalized electron-deficient bonding found in diborane. Bridging halides are poor electron-pair donors, so the Group 13 trihalides are potent Lewis acids that react readily with Lewis bases such as amines to form a Lewis acid–base adduct:

$$Al_2Cl_6(soln) + 2(CH_3)_3N{:}(soln) \longrightarrow 2(CH_3)_3N{:}AlCl_3(soln) \quad (22.10)$$

In water, the halides of the Group 13 metals hydrolyze to produce the metal hydroxide [$M(OH)_3$]:

$$MX_3(s) + 3H_2O(l) \longrightarrow M(OH)_3(s) + 3HX(aq) \quad (22.11)$$

In a related reaction, $Al_2(SO_4)_3$ is used to clarify drinking water by the precipitation of hydrated $Al(OH)_3$, which traps particulates. The halides of the heavier metals (In and Tl) are less reactive with water because of their lower charge-to-radius ratio. Instead of forming hydroxides, they dissolve to form the hydrated metal ions, $[M(H_2O)_6]^{3+}$.

Note the pattern

Of the Group 13 halides, only the fluorides behave as typical ionic compounds.

Note the pattern

Group 13 trihalides are potent Lewis acids that react with Lewis bases to form a Lewis acid–base adduct.

Note the pattern

All Group 13 oxides dissolve in dilute acid, but Al_2O_3 and Ga_2O_3 are amphoteric.

Like boron (Equation 22.6), all the heavier Group 13 elements react with excess oxygen at elevated temperatures to give the trivalent oxide (M_2O_3), although Tl_2O_3 is unstable:

$$4M(s) + 3O_2(g) \xrightarrow{\Delta} 2M_2O_3(s) \tag{22.12}$$

Aluminum oxide (Al_2O_3), also known as alumina, is a hard, high-melting-point, chemically inert insulator used as a ceramic and as an abrasive in sandpaper and toothpaste. The gemstone known as ruby is formed when a small number of Al^{3+} ions in crystalline alumina are replaced by Cr^{3+} ions, whereas replacing Al^{3+} with a mixture of Fe^{2+}, Fe^{3+}, and Ti^{4+} produces blue sapphires. The gallium oxide compound $MgGa_2O_4$ gives the brilliant green light familiar to anyone who has ever operated a xerographic copy machine. All the oxides dissolve in dilute acid, but Al_2O_3 and Ga_2O_3 are amphoteric, consistent with their location along the diagonal line of the periodic table, also dissolving in concentrated aqueous base to form solutions that contain $M(OH)_4{}^-$ ions.

Aluminum, gallium, and indium also react with the other Group 16 elements (chalcogens) to form chalcogenides with the stoichiometry M_2Y_3. However, because Tl(III) is too strong an oxidant to form a stable compound with electron-rich anions such as S^{2-}, Se^{2-}, and Te^{2-}, thallium forms only the thallium(I) chalcogenides with the stoichiometry Tl_2Y.

Only aluminum, like boron, reacts directly with N_2 (at very high temperatures) to give AlN, which is used in transistors and microwave devices as a nontoxic heat sink because of its thermal stability; GaN and InN can be prepared using other methods. All the metals, again except Tl, also react with the heavier Group 15 elements (pnicogens) to form the so-called III–V compounds, such as GaAs. These are semiconductors, whose electronic properties, such as their band gaps (see Chapter 12), differ from those that can be achieved using either pure or doped Group 14 elements. For example, nitrogen- and phosphorus-doped gallium arsenide ($GaAs_{1-x-y}P_xN_y$) is used in the displays of calculators and digital watches.

Note the pattern

Unlike boron, the heavier Group 13 elements do not react directly with hydrogen. They form a relatively large number of complexes.

Unlike boron, the heavier Group 13 elements do *not* react directly with hydrogen. Only the aluminum and gallium hydrides are known, but they must be prepared indirectly; AlH_3 is an insoluble, polymeric solid that is rapidly decomposed by water, whereas GaH_3 is unstable at room temperature.

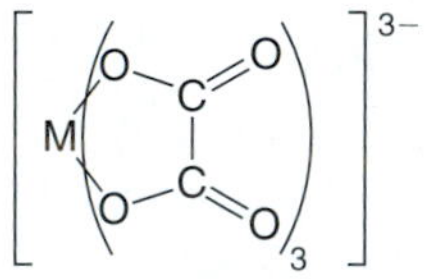

Oxalate ion

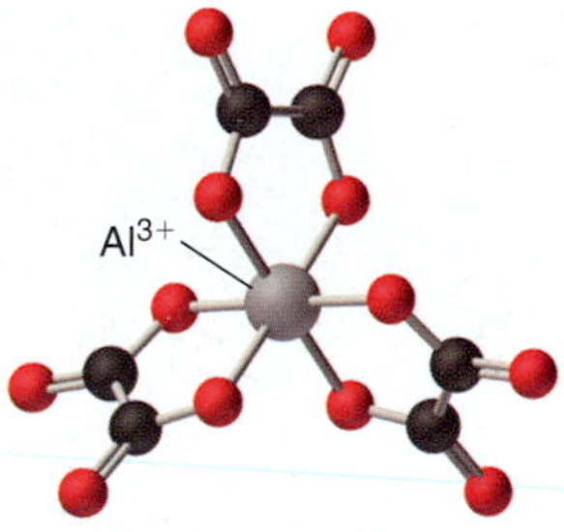

Al^{3+}–oxalate complex

Complexes of Group 13 Elements

Boron has a relatively limited tendency to form complexes, but aluminum, gallium, indium, and, to some extent, thallium form many complexes. Some of the simplest are the hydrated metal ions, $M(H_2O)_6{}^{3+}$, which are relatively strong Brønsted–Lowry acids that can lose a proton to form the $M(H_2O)_5(OH)^{2+}$ ion:

$$M(H_2O)_6{}^{3+}(aq) \longrightarrow M(H_2O)_5(OH)^{2+}(aq) + H^+(aq) \tag{22.13}$$

Group 13 metal ions also form stable complexes with species that contain two or more negatively charged groups, such as the oxalate ion (see structure in margin). The stability of such complexes increases as the number of coordinating groups provided by the ligand increases.

EXAMPLE 22.2

For each reaction, explain why the given products form:

(a) $2Al(s) + Fe_2O_3(s) \xrightarrow{\Delta} 2Fe(l) + Al_2O_3(s)$
(b) $2Ga(s) + 6H_2O(l) + 2OH^-(aq) \longrightarrow 3H_2(g) + 2Ga(OH)_4{}^-(aq)$
(c) $In_2Cl_6(s) \xrightarrow{H_2O(l)} 2In^{3+}(aq) + 6Cl^-(aq)$

Given Balanced chemical equations

Asked for Why the given products form

Strategy

Classify the type of reaction. Using periodic trends in atomic properties, thermodynamics, and kinetics, explain why the reaction products form.

Solution

(a) Aluminum is an active metal and a powerful reductant, and Fe_2O_3 contains Fe(III), a potential oxidant. Hence, a redox reaction is probable, producing metallic Fe and Al_2O_3. Because Al is a main group element that lies above Fe, which is a transition element, it should be a more active metal than Fe. Thus, the reaction should proceed to the right. In fact, this is the thermite reaction, which is so vigorous that it produces molten Fe and can be used for welding (see Section 5.2).

(b) Gallium lies immediately below aluminum in the periodic table and is amphoteric, so it will dissolve in either acid or base to produce hydrogen gas. Because gallium is similar to aluminum in many of its properties, we predict that gallium will dissolve in the strong base.

(c) The metallic character of the Group 13 elements increases with increasing atomic number. Indium trichloride should therefore behave like a typical metal halide, dissolving in water to form the hydrated cation.

EXERCISE 22.2

Predict the products of the reactions, and write a balanced equation for each:

(a) $LiH(s) + Al_2Cl_6(soln) \longrightarrow$

(b) $Al_2O_3(s) + OH^-(aq) \longrightarrow$

(c) $Al(s) + N_2(g) \xrightarrow{\Delta}$

(d) $Ga_2Cl_6(soln) + Cl^-(soln) \longrightarrow$

Answer

(a) $8LiH(s) + Al_2Cl_6(soln) \longrightarrow 2LiAlH_4(soln) + 6LiCl(s)$

(b) $Al_2O_3(s) + 2OH^-(aq) + 3H_2O(l) \longrightarrow 2Al(OH)_4^-(aq)$

(c) $2Al(s) + N_2(g) \xrightarrow{\Delta} 2AlN(s)$

(d) $Ga_2Cl_6(soln) + 2Cl^-(soln) \longrightarrow 2GaCl_4^-(soln)$

22.2 • The Elements of Group 14

The elements of Group 14 show a greater range of chemical behavior than any other family in the periodic table. Three of the five elements—carbon, tin, and lead—have been known since ancient times. For example, some of the oldest known writings are Egyptian hieroglyphics written on papyrus with ink made from *lampblack*, a finely divided carbon soot produced in the incomplete combustion of hydrocarbons (Figure 22.5). *Activated carbon* is an even more finely divided form of carbon that is produced from the thermal decomposition of organic materials such as sawdust. Because it adsorbs many organic and sulfur-containing compounds, activated carbon is used to decolorize foods such as sugar and to purify gases and wastewater.

Tin and lead oxides and sulfides are easily reduced to the metal by heating with charcoal, a discovery that must have occurred by accident when prehistoric humans used rocks containing their ores for a cooking fire. However, because tin and copper ores are often found together in nature, their alloy, which is called *bronze*, was probably discovered before either element, a discovery that led to the Bronze Age. The heaviest element in Group 14, lead, is such a soft and malleable metal that the ancient Romans used thin lead foils as writing tablets, as well as lead cookware and lead pipes for plumbing. (Recall that the atomic symbols for tin and lead come from their Latin names, Sn for *stannum* and Pb for *plumbum*).

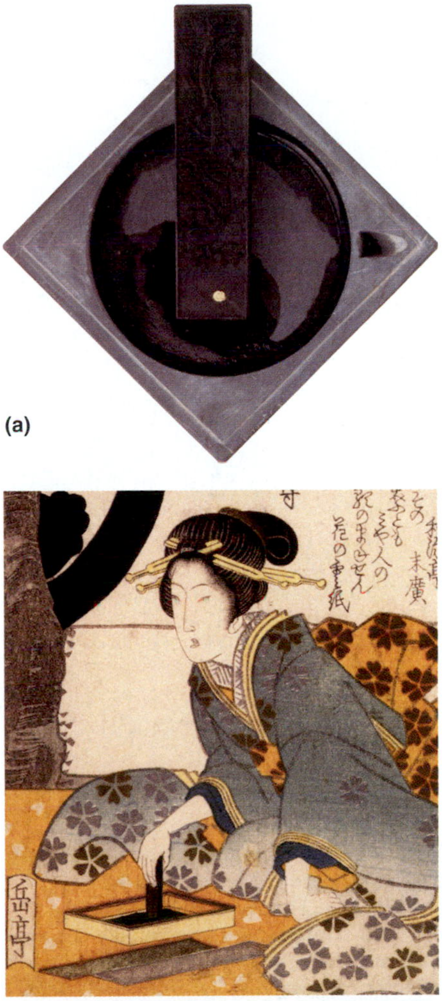

(a)

(b)

Figure 22.5 Very small particles of noncrystalline carbon are used to make black ink. (a) Since ancient times, ink sticks have been the major source of black ink in Asia. Plant oils or resinous woods such as pine are burned, and the resulting soot (lampblack) is collected, mixed with binders such as animal glues and minerals, compressed into a solid stick, and allowed to dry. Liquid ink is made by rubbing the ink stick against the surface of a special stone ink dish with small amounts of water. (b) A 19th-century Japanese painting that illustrates how ink is made from an ink stick.

Although the first glasses were prepared from silica (silicon oxide, SiO_2) around 1500 B.C., elemental silicon was not prepared until 1824 because of its high affinity for oxygen. Jöns Jakob Berzelius was finally able to obtain amorphous silicon by reducing Na_2SiF_6 with molten potassium. The crystalline element, which has a shiny blue-gray luster, was not isolated until 30 years later. The last member of the Group 14 elements to be discovered was germanium, which was found in a newly discovered silver-colored ore by the German chemist Clemens Winkler in 1886, who named the element in honor of his native country.

Preparation and General Properties of Group 14 Elements

The natural abundance of the Group 14 elements varies tremendously. Elemental carbon, for example, ranks only 17th on the list of constituents of the Earth's crust. Pure graphite is obtained by the reaction of coke, an amorphous form of carbon that is used as a reductant in the production of steel, with silica to give silicon carbide (SiC). This is then thermally decomposed at very high temperatures (2700°C) to give graphite:

$$SiO_2(s) + 3C(s) \xrightarrow{\Delta} SiC(s) + 2CO(g) \qquad (22.14)$$

$$SiC(s) \xrightarrow{\Delta} Si(g) + C(graphite) \qquad (22.15)$$

One allotrope of carbon, diamond, is metastable under normal conditions, with a ΔG_f° of 2.9 kJ/mol versus graphite. At pressures higher than 50,000 atm, however, the diamond structure is favored and is the most stable form of carbon. Because the structure of diamond is more compact than that of graphite, its density is significantly higher (3.51 g/cm^3 versus 2.2 g/cm^3). Because of its high thermal conductivity, diamond powder is used to transfer heat in electronic devices.

The most common sources of diamonds on Earth are ancient volcanic pipes that contain a rock called *kimberlite*, a lava that solidified rapidly from deep inside the Earth. Most kimberlite formations, however, are much newer than the diamonds they contain. In fact, the relative amounts of different carbon isotopes in diamond show that diamond is a chemical and geological "fossil" that is older than our solar system, which means that diamonds on Earth predate the existence of our sun. Thus, diamonds were most likely created deep inside the Earth from primordial grains of graphite present when the Earth was formed (Figure 22.6a). Gem-quality diamonds can now be produced synthetically and have chemical, optical, and physical characteristics identical to those of the highest-grade natural diamonds.

(a) (b)

Figure 22.6 Crystalline samples of carbon and silicon, the lightest Group 14 elements. (a) The 78.86-carat Ahmadabad diamond, a historic Indian gem purchased in Gujarat in the 17th century by the French explorer Jean-Baptiste Tavernier and sold in 1995 for $4.3 million, is a rare example of a large single crystal of diamond, the less-stable allotrope of carbon. (b) Large single crystals of highly purified silicon are the basis of the modern electronics industry. They are sliced into very thin wafers that are highly polished and then cut into smaller pieces for use as chips.

While oxygen is the most abundant element on Earth, the next most abundant is silicon, the next member of Group 14. Pure silicon is obtained by the reaction of impure silicon with Cl_2 to give $SiCl_4$, followed by fractional distillation of the impure $SiCl_4$ and reduction with H_2:

$$SiCl_4(l) + 2H_2(g) \xrightarrow{\Delta} Si(s) + 4HCl(g) \tag{22.16}$$

Several million tons of silicon are produced annually using this method. Amorphous silicon that contains residual amounts of hydrogen is used in photovoltaic devices that convert light to electricity, and silicon-based solar cells are used to power pocket calculators, boats, and highway signs, where access to electricity by conventional methods is difficult or expensive. Ultra-pure silicon and germanium form the basis of the modern electronics industry (Figure 22.6b).

In contrast to silicon, the concentrations of germanium and tin in the Earth's crust are only 1–2 ppm. The concentration of lead, which is the end product of the nuclear decay of many radionuclides (see Chapter 20), is 13 ppm, making lead by far the most abundant of the heavy Group 14 elements. No concentrated ores of germanium are known; like indium, germanium is generally recovered from flue dusts obtained by processing the ores of metals such as zinc. Because germanium is essentially transparent to infrared radiation, it is used in optical devices.

Tin and lead are soft metals that are too weak for structural applications, but because tin is flexible, corrosion resistant, and nontoxic, it is used as a coating in food packaging. A "tin can," for example, is actually a steel can whose interior is coated with a thin layer (1–2 μm) of metallic tin. Tin is also used in superconducting magnets and low-melting-point alloys such as solder and pewter. Pure lead is obtained by heating galena (PbS) in air and reducing the oxide (PbO) to the metal with carbon, followed by electrolytic deposition to increase the purity:

$$PbS(s) + \frac{3}{2}O_2(g) \xrightarrow{\Delta} PbO(s) + SO_2(g) \tag{22.17}$$

$$PbO(s) + C(s) \xrightarrow{\Delta} Pb(l) + CO(g) \tag{22.18}$$

or

$$PbO(s) + CO(g) \xrightarrow{\Delta} Pb(l) + CO_2(g) \tag{22.19}$$

By far the single largest use of lead is in lead storage batteries (see Chapter 19).

All the Group 14 elements form compounds in which they formally lose either the two *np* and the two *ns* valence electrons or just the two *np* valence electrons, giving compounds in the +4 or +2 oxidation state, respectively. Because covalent bonds decrease in strength with increasing atomic size and the ionization energies for the heavier elements of the group are higher than expected due to a higher Z_{eff}, the relative stability of the +2 oxidation state increases smoothly from carbon to lead.

Note the pattern

The relative stability of the +2 oxidation state increases and the tendency to form catenated compounds decreases from carbon to lead in Group 14.

Recall that many carbon compounds contain multiple bonds formed by π overlap of singly occupied 2*p* orbitals on adjacent atoms (see Chapter 9). Compounds of silicon, germanium, tin, and lead with the same stoichiometry as those of carbon, however, tend to have different structures and properties. For example, CO_2 is a gas that contains discrete O=C=O molecules, whereas the most common form of SiO_2 is the high-melting solid known as quartz, the major component of sand. Instead of discrete SiO_2 molecules, quartz contains a three-dimensional network of silicon atoms that is similar to the structure of diamond, but with an oxygen atom inserted between each pair of silicon atoms. Thus, each silicon atom is linked to four other silicon atoms by bridging oxygen atoms (see Section 12.1).

The tendency to catenate—to form chains of like atoms—decreases rapidly as we go down Group 14 because bond energies for both the E—E and E—H bonds decrease with increasing atomic number (where E is any Group 14 element). Consequently, inserting a CH_2 group into a linear hydrocarbon such as *n*-hexane is exergonic

TABLE 22.2 Selected properties of the Group 14 elements

Property	Carbon	Silicon	Germanium	Tin	Lead
Atomic symbol	C	Si	Ge	Sn	Pb
Atomic number	6	14	32	50	82
Atomic mass (amu)	12.01	28.09	72.64	118.71	207.20
Valence-electron configuration[a]	$2s^2\,2p^2$	$3s^2\,3p^2$	$4s^2\,4p^2$	$5s^2\,5p^2$	$6s^2\,6p^2$
mp/bp (°C)	4489 (10.3 MPa)/3825	1414/3265	939/2833	232/2602	327/1749
Density (g/cm^3) at 25°C	2.2 (graphite), 3.51 (diamond)	2.33	5.32	7.27(w)	11.30
Atomic radius (pm)	77 (diamond)	111	125	145	154
First ionization energy (kJ/mol)	1087	787	762	709	716
Most common oxidation state	+4	+4	+4	+4	+2
Ionic radius (pm)[b]	(29)	(40)	53	62	119
Electron affinity (kJ/mol)	−122	−134	−119	−107	−35
Electronegativity	2.6	1.9	2.0	2.0	1.8
Standard reduction potential, $E°$ (V) (for $EO_2 \longrightarrow E$ in acidic solution)	0.21	−0.86	−0.18	−0.12	0.79
Product of reaction with O_2	CO_2, CO	SiO_2	GeO_2	SnO_2	PbO
Type of oxide	Acidic (CO_2), neutral (CO)	Acidic	Amphoteric	Amphoteric	Amphoteric
Product of reaction with N_2	None	Si_3N_4	None	Sn_3N_4	None
Product of reaction with X_2[c]	CX_4	SiX_4	GeX_4	SnX_4	PbX_2
Product of reaction with H_2	CH_4	None	None	None	None

[a]The configuration shown does not include filled *d* and *f* subshells.

[b]The values cited are for six-coordinate ions in the most common oxidation state, except for C^{4+} and Si^{4+}, for which values for the four-coordinate ion are estimated.

[c]X = Cl, Br, or I. Reaction with F_2 gives the tetrafluorides (EF_4) for all Group 14 elements.

($\Delta G° = -45$ kJ/mol), whereas inserting an SiH_2 group into the silicon analogue of *n*-hexane (Si_6H_{14}) actually costs energy ($\Delta G° \approx +25$ kJ/mol). As a result of this trend, the thermal stability of catenated compounds decreases rapidly from carbon to lead.

Note the pattern

The Group 14 elements follow the same pattern as the Group 13 elements in their periodic properties.

As you learned in Chapter 7, the Group 14 elements all have $ns^2\,np^2$ valence-electron configurations. In Table 22.2 we see that once again there is a large difference between the lightest element, C, and the others in size, ionization energy, and electronegativity. As in Group 13, the second and third elements (Si and Ge) are similar, and there is a reversal in the trends for some properties, such as ionization energy, between the fourth and fifth elements (Sn and Pb). As for Group 13, these effects can be explained by the presence of filled $(n-1)d$ and $(n-2)f$ subshells, whose electrons are relatively poor at screening the outermost electrons from the higher nuclear charge.

Reactions and Compounds of Carbon

Carbon is the building block of all organic compounds, including biomolecules, fuels, pharmaceuticals, and plastics, whereas inorganic compounds of carbon include metal carbonates, which are found in substances as diverse as fertilizers and antacid tablets, halides, oxides, carbides, and carboranes. Like boron in Group 13, the chemistry of carbon differs sufficiently from that of its heavier congeners to merit a separate discussion.

The structures of the allotropes of carbon—diamond, graphite, fullerenes, and nanotubes—are distinct, but they all contain simple electron-pair bonds (see Figure 7.18). Although it was originally believed that fullerenes were a new form of carbon that could be prepared only in the laboratory, fullerenes have been found in certain types of meteorites. Another possible allotrope of carbon has also been detected in impact fragments of a carbon-rich meteorite; it appears to consist of long chains of carbon atoms linked by alternating single and triple bonds, $(\text{—C}\equiv\text{C—C}\equiv\text{C—})_n$. Carbon nanotubes ("buckytubes") are being studied as potential building blocks for ultra-microscale detectors and molecular computers, and as tethers for space stations; they are currently used in electronic devices such as the electrically conducting tips of miniature electron guns for flat-panel displays in portable computers.

Although all the carbon tetrahalides (CX_4) are known, they are generally not obtained by the direct reaction of carbon with the elemental halogens (X_2) but by indirect methods such as the reaction

CF_4 CI_4

$$CH_4(g) + 2X_2(g) \longrightarrow CX_4(l,s) + 4HX(g) \qquad (X = Cl, Br) \qquad (22.20)$$

The carbon tetrahalides all have the tetrahedral geometry predicted by the valence-shell electron-pair repulsion (VSEPR) model, as shown in the margin for CCl_4 and CI_4. Their stability decreases rapidly as the halogen increases in size because of poor orbital overlap and increased crowding. Because the C—F bond is about 25% stronger than a C—H bond, fluorocarbons are thermally and chemically more stable than the corresponding hydrocarbons, while having a similar hydrophobic character. A polymer of tetrafluoroethylene ($F_2C{=}CF_2$), analogous to polyethylene, is the non-stick Teflon lining found on many cooking pans, and similar compounds are used to make fabrics stain resistant (such as Scotch-Gard) or waterproof but breathable (such as Gore-Tex).

Note the pattern

The stability of the carbon tetrahalides decreases with increasing size of the halogen due to increasingly poor orbital overlap and crowding.

Carbon reacts with oxygen to form either CO or CO_2, depending on the stoichiometry. Carbon monoxide is a colorless, odorless, and poisonous gas that reacts with the iron in hemoglobin to form an Fe—CO unit, which prevents hemoglobin from binding, transporting, and releasing oxygen in the blood (see Figure 23.26 for myoglobin). In the laboratory, carbon monoxide can be prepared on a small scale by dehydrating formic acid with concentrated sulfuric acid:

$$HCO_2H(l) \xrightarrow{H_2SO_4(l)} CO(g) + H_3O^+(aq) + HSO_4^-(aq) \qquad (22.21)$$

Carbon monoxide also reacts with the halogens to form the oxohalides (COX_2). Probably the best known of these is phosgene ($Cl_2C{=}O$), which is highly poisonous and was used as a chemical weapon during World War I:

$$CO(g) + Cl_2(g) \xrightarrow{\Delta} Cl_2C{=}O(g) \qquad (22.22)$$

Despite its toxicity, phosgene is an important industrial chemical that is prepared on a large scale, primarily in the manufacture of polyurethanes.

Carbon dioxide can be prepared on a small scale by the reaction of almost any metal carbonate or bicarbonate salt with strong acid. As is typical of a nonmetal oxide, CO_2 reacts with water to form acidic solutions that contain carbonic acid (H_2CO_3).

In contrast to its reactions with oxygen, the reaction of carbon with sulfur at high temperatures produces only carbon disulfide (CS_2):

$$C(s) + 2S(g) \xrightarrow{\Delta} CS_2(g) \qquad (22.23)$$

The selenium analogue, CSe_2, is also known. Both have the linear structure predicted by the VSEPR model, and both are vile smelling (and in the case of CSe_2, highly toxic), volatile liquids. The sulfur and selenium analogues of carbon monoxide, CS and CSe, are unstable because the C≡Y triple bonds (Y = S, Se) are much weaker than the C≡O triple bond due to poorer π orbital overlap.

Note the pattern

Pi bonds between carbon and the heavier chalcogenides are weak due to poor orbital overlap.

A 19th-century miner's lamp. The lamp uses burning acetylene, produced by the slow reaction of calcium carbide with water, to provide light.

Binary compounds of carbon with less electronegative elements are called *carbides*. The chemical and physical properties of carbides depend strongly on the identity of the second element, resulting in three general classes: ionic carbides, interstitial carbides, and covalent carbides. The reaction of carbon at high temperatures with electropositive metals such as those of Groups 1 and 2 and aluminum produces *ionic carbides*, which contain discrete metal cations and carbon anions. The identity of the anions depends on the size of the second element. For example, smaller elements such as beryllium and aluminum give *methides* such as Be_2C and Al_4C_3, which formally contain the C^{4-} ion derived from methane (CH_4) by the loss of all four H atoms as protons. In contrast, larger metals such as sodium and calcium give carbides with stoichiometries of Na_2C_2 and CaC_2. Because these carbides contain the $C_2^{\ 2-}$ ion, which is derived from acetylene (HC≡CH) by the loss of both H atoms as protons, they are more properly called *acetylides*. As discussed in Section 21.4, the reaction of ionic carbides with dilute aqueous acid results in protonation of the anions to give the parent hydrocarbons, CH_4 or C_2H_2. For many years, miners' lamps used the reaction of calcium carbide with water to produce a steady supply of acetylene, which was ignited to provide a portable lantern.

The reaction of carbon with most transition metals at high temperatures produces *interstitial carbides*. Due to the less electropositive nature of the transition metals, these carbides contain covalent metal–carbon interactions, which result in different properties: most interstitial carbides are good conductors of electricity, have high melting points, and are among the hardest substances known. Interstitial carbides exhibit a variety of nominal compositions, and they are often nonstoichiometric compounds whose carbon content can vary over a wide range. Among the most important are tungsten carbide (WC), which is used industrially in high-speed cutting tools, and cementite (Fe_3C), which is a major component of steel.

Note the pattern

Carbides formed from Group 1 and 2 elements are ionic while transition metals form interstitial carbides that have covalent metal–carbon interactions, and covalent carbides are chemically inert.

Elements with an electronegativity similar to that of carbon form *covalent carbides*, such as silicon carbide (SiC, Equation 22.14) and boron carbide (B_4C). These substances are extremely hard, high melting, and chemically inert. For example, silicon carbide is highly resistant to chemical attack at temperatures as high as 1600°C. Because it also maintains its strength at high temperatures, silicon carbide is used in heating elements for electric furnaces and in variable-temperature resistors.

EXAMPLE 22.3

For each reaction, explain why the given product forms:

(a) $CO(g) + Cl_2(g) \longrightarrow Cl_2CO(g)$
(b) $CO(g) + BF_3(g) \longrightarrow F_3B{:}C{\equiv}O(g)$
(c) $Sr(s) + 2C(s) \xrightarrow{\Delta} SrC_2(s)$

Given Balanced chemical equations

Asked for Why the given products form

Strategy

Classify the type of reaction. Using periodic trends in atomic properties, thermodynamics, and kinetics, explain why the observed reaction products form.

Solution

(a) Because the carbon in CO is in an intermediate oxidation state (+2), CO can be either a reductant or an oxidant; it is also a Lewis base. The other reactant, Cl_2, is an oxidant, so we expect a redox reaction to occur in which the carbon of CO is further oxidized. Because Cl_2 is a two-electron oxidant and the carbon atom of CO can be oxidized by two electrons to the +4 oxidation state, the product is phosgene ($Cl_2C{=}O$).

(b) Unlike Cl_2, BF_3 is not a good oxidant, even though it contains boron in its highest oxidation state, +3. Nor can BF_3 behave like a reductant. Like any other species with only six valence electrons, however, it is certainly a Lewis acid. Hence, an acid–base reaction is the most likely alternative, especially because we know that CO can use the lone pair of electrons on carbon to act as a Lewis base. The most probable reaction is therefore the formation of a Lewis acid–base adduct.

(c) Typically, both reactants behave like reductants. Unless one of them can also behave like an oxidant, no reaction will occur. We know that Sr is an active metal because it lies far to the left in the periodic table, and that it is more electropositive than carbon. Carbon is a nonmetal with a significantly higher electronegativity; it is therefore more likely to accept electrons in a redox reaction. We conclude, therefore, that Sr will be oxidized and C will be reduced. Carbon forms ionic carbides with active metals, so the reaction will produce a species formally containing either C^{4-} or C_2^{2-}. Those that contain C^{4-} usually involve small, highly charged metal ions, so Sr^{2+} will produce the acetylide, SrC_2, instead.

EXERCISE 22.3

Predict the products of the reactions, and write a balanced equation for each:

(a) $C(s) + \text{excess } O_2(g) \xrightarrow{\Delta}$
(b) $C(s) + H_2O(l) \longrightarrow$
(c) $NaHCO_3(s) + H_2SO_4(aq) \longrightarrow$

Answer

(a) $C(s) + \text{excess } O_2(g) \xrightarrow{\Delta} CO_2(g)$
(b) $C(s) + H_2O(l) \longrightarrow$ no reaction
(c) $NaHCO_3(s) + H_2SO_4(aq) \longrightarrow CO_2(g) + NaHSO_4(aq) + H_2O(l)$

Reactions and Compounds of the Heavier Group 14 Elements

Although silicon, germanium, tin, and lead in their +4 oxidation states often form binary compounds with the same stoichiometry as carbon, the structures and properties of these compounds are usually significantly different from those of the carbon analogues. Silicon and germanium are both semiconductors with structures analogous to diamond. Tin has two common allotropes: white (β) tin has a metallic lattice and metallic properties, whereas gray (α) tin has a diamondlike structure and is a semiconductor. The metallic β form is stable above 13.2°C, and the nonmetallic α form is stable below 13.2°C. Lead is the only Group 14 element that is metallic in both structure and properties under all conditions.

Based on its position in the periodic table, we expect silicon to be amphoteric. In fact, it does dissolve in strong aqueous base to produce hydrogen gas and solutions of silicates, but the only aqueous acid that it reacts with is hydrofluoric acid, presumably due to the formation of the stable SiF_6^{2-} ion. Germanium is more metallic in its behavior than silicon. For example, it dissolves in hot oxidizing acids, such as HNO_3 and H_2SO_4, but in the absence of an oxidant, it does not dissolve in aqueous base. Although tin has an even more metallic character than germanium, lead is the only element in the group that behaves purely as a metal. It is not readily attacked by acids, because the solid acquires a thin protective outer layer of a Pb^{2+} salt such as $PbSO_4$.

All the Group 14 dichlorides are known, but their stability increases dramatically as the atomic number of the central atom increases. Thus, CCl_2 is dichlorocarbene, a highly reactive, short-lived intermediate that can be made in solution but cannot be isolated in pure form using standard techniques; $SiCl_2$ can be isolated at very low temperatures, but it decomposes rapidly above −150°C, and $GeCl_2$ is relatively stable at temperatures below 20°C. In contrast, $SnCl_2$ is a polymeric solid that is indefinitely stable at room temperature, whereas $PbCl_2$ is an insoluble crystalline solid with a structure similar to that of $SnCl_2$.

Note the pattern

The stability of the Group 14 dichlorides increases dramatically from carbon to lead.

Note the pattern

Because silicon–oxygen bonds are unusually strong, silicon–oxygen compounds dominate the chemistry of silicon.

Note the pattern

The dioxides of the Group 14 elements become increasingly basic down the group.

Although the first four elements of Group 14 form tetrahalides (MX_4) with all the halogens, only fluorine is able to oxidize lead to the +4 oxidation state, giving PbF_4. The tetrahalides of silicon and germanium react rapidly with water to give amphoteric oxides:

$$EX_4(s,l) + 2H_2O(l) \longrightarrow MO_2(s) + 4HX(aq) \quad (E = Si, Ge) \tag{22.24}$$

In contrast, the tetrahalides of tin and lead react with water to give hydrated metal ions.

Because of the stability of its +2 oxidation state, lead reacts with oxygen or sulfur to form PbO or PbS, respectively, whereas heating the other Group 14 elements with excess O_2 or S_8 gives the corresponding dioxides or disulfides, respectively. The dioxides of the Group 14 elements become increasingly basic as we go down the group.

Because the Si—O bond is even stronger than the C—O bond (~452 kJ/mol versus ~358 kJ/mol), silicon has a strong affinity for oxygen. The relative strengths of the C—O and Si—O bonds contradict the generalization that bond strengths decrease as the bonded atoms become larger. This is because thus far we have assumed that a formal single bond between two atoms can always be described in terms of a single pair of shared electrons. In the case of Si—O bonds, however, the presence of relatively low-energy, empty *d* orbitals on Si and nonbonding electron pairs in the *p* or sp^n hybrid orbitals of O results in a partial π bond (Figure 22.7). Due to its partial π double bond character, the Si—O bond is significantly stronger and shorter than would otherwise be expected. A similar interaction with oxygen is also an important feature of the chemistry of the elements that follow silicon in the third period (P, S, and Cl). Because the Si—O bond is unusually strong, silicon–oxygen compounds dominate the chemistry of silicon.

Compounds with anions that contain only silicon and oxygen are called *silicates*, whose basic building block is the SiO_4^{4-} unit shown below.

SiO_4^{4-}

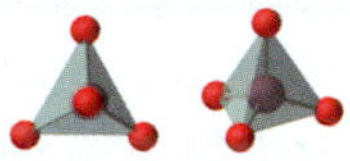

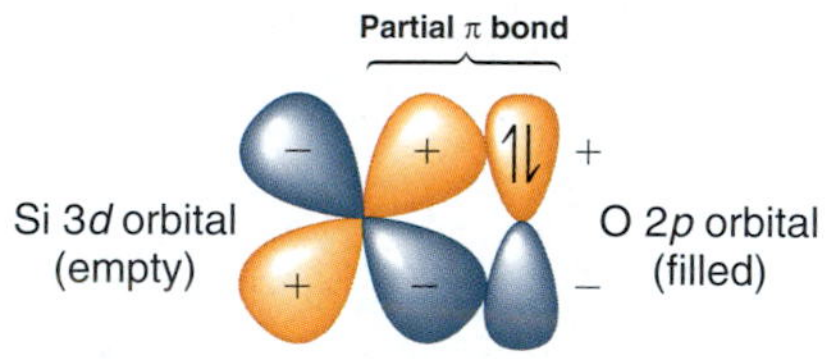

Figure 22.7 Pi bonding between silicon and oxygen. Silicon has relatively low-energy, empty 3*d* orbitals that can interact with filled 2*p* hybrid orbitals on oxygen. This interaction results in a partial π bond in which both electrons are supplied by oxygen, giving the Si—O bond partial double bond character and making it significantly stronger (and shorter) than expected for a single bond.

The number of oxygen atoms shared between silicon atoms and the way in which the units are linked vary considerably in different silicates. Converting one of the oxygen atoms from terminal to bridging generates chains of silicates, while converting two oxygen atoms from terminal to bridging generates double chains. In contrast, converting three or four oxygens to bridging generates a variety of complex layered and three-dimensional structures, respectively.

1 bridging O atom

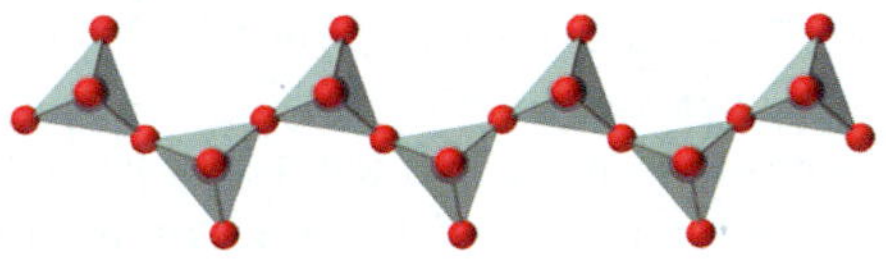

2 bridging O atoms

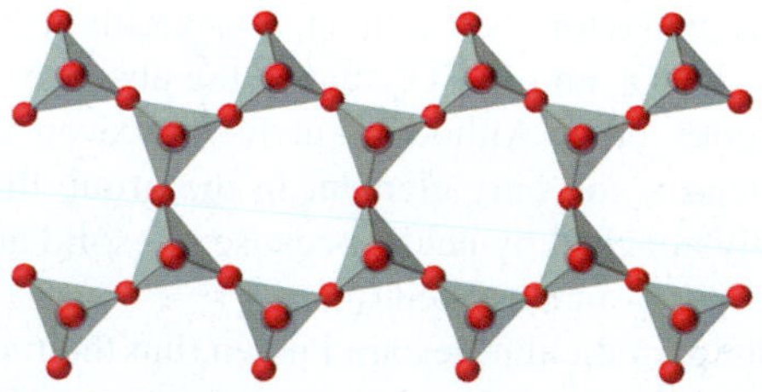

The silicates include many glasses as well as the gemstone known as *opal*, which typically contains 10–15% water. In a large and important class of materials called *aluminosilicates*, some of the Si atoms are replaced by Al atoms to give aluminosilicates such as *zeolites*, whose three-dimensional framework structures have large cavities connected by smaller tunnels (Figure 22.8). Because the cations in zeolites are readily exchanged, zeolites are used in laundry detergents as water-softening agents: the more loosely bound Na^+ ions inside the zeolite cavities are displaced by

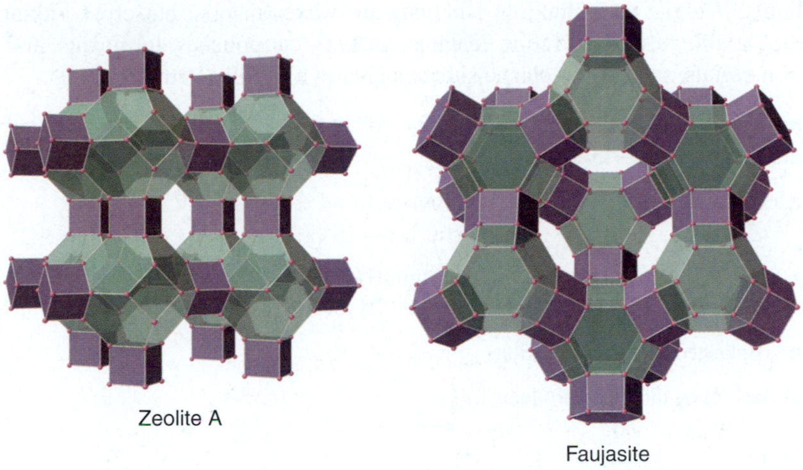

Figure 22.8 Zeolites are aluminosilicates that contain large cavities connected by channels. The cavities normally contain hydrated cations that are loosely bound to the oxygen atoms of the negatively charged framework by electrostatic interactions. The sizes and arrangements of the channels and cavities differ in different types of zeolites. For example, in zeolite A the aluminosilicate cages are arranged in a cubic fashion, and the channels connecting the cavities intersect at right angles. In contrast, the cavities in faujasite are much larger, and the channels intersect at 120° angles. In these idealized models, the oxygen atoms that connect each pair of silicon atoms have been omitted.

the more highly charged Mg^{2+} and Ca^{2+} ions present in hard water, which bind more tightly. Zeolites are also used as catalysts and for water purification.

Silicon and germanium react with nitrogen at high temperature to form *nitrides* (M_3N_4):

$$3Si(l) + 2N_2(g) \longrightarrow Si_3N_4(s) \tag{22.25}$$

Silicon nitride has properties that make it suitable for high-temperature engineering applications: it is strong, very hard, and chemically inert, and it retains these properties to temperatures of about 1000°C.

Because of the diagonal relationship between boron and silicon, metal silicides and metal borides exhibit many similarities. Although metal silicides have structures that are as complex as those of the metal borides and carbides, few silicides are structurally similar to the corresponding borides due to the significantly larger size of Si (atomic radius 117 pm versus 87 pm for B). Silicides of active metals, such as Mg_2Si, are ionic compounds that contain the Si^{4-} ion. They react with aqueous acid to form silicon hydrides such as SiH_4:

$$Mg_2Si(s) + 4H^+(aq) \longrightarrow 2Mg^{2+}(aq) + SiH_4(g) \tag{22.26}$$

Unlike carbon, catenated silicon hydrides become thermodynamically *less* stable as the chain lengthens. Thus, straight-chain and branched *silanes* (analogous to alkanes) are known up to only $n = 10$; the germanium analogues (*germanes*) are known up to $n = 9$. In contrast, the only known hydride of tin is SnH_4, and it slowly decomposes to elemental Sn and H_2 at room temperature. The simplest lead hydride, PbH_4, is so unstable that chemists are not even certain it exists. Because E═E and E≡E bonds become weaker with increasing atomic number, simple silicon, germanium, and tin analogues of alkenes, alkynes, and aromatic hydrocarbones are either unstable (Si═Si and Ge═Ge) or unknown. Silicon-based life-forms are therefore likely to be found only in science fiction.

The only important organic derivatives of lead are compounds such as tetraethyllead [$(CH_3CH_2)_4Pb$]. Because the Pb—C bond is weak, these compounds decompose at relatively low temperatures to produce alkyl radicals (R·), which can be used to control the rate of combustion reactions. For 60 years, hundreds of thousands of tons of lead were burned annually in automobile engines, producing a mist of lead oxide particles along the highways that constituted a potentially serious public health problem (see Example 5.6). The use of catalytic converters reduced the amount of carbon monoxide, nitrogen oxides, and hydrocarbons released into the atmosphere through automobile exhausts, but it did nothing to decrease lead emissions. Because lead poisons catalytic converters, however, its use as a gasoline additive has been banned in most of the world.

Compounds that contain Si—C and Si—O bonds are stable and important. High-molecular-mass polymers called *silicones* contain an $(Si—O—)_n$ backbone with organic groups attached to Si (Figure 22.9). The properties of silicones are determined by the chain length, the type of organic group, and the extent of cross-linking between

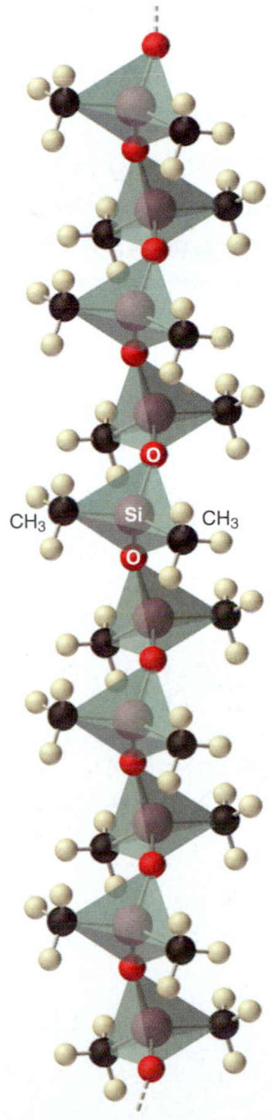

Figure 22.9 Silicones are polymers that contain long chains of alternating silicon and oxygen atoms. The structure of a linear silicone polymer is similar to that of quartz, but two of the oxygen atoms attached to each silicon atom are replaced by the carbon atoms of organic groups, such as the methyl groups (—CH_3) shown here. The terminal silicon atoms are bonded to three methyl groups. Silicones can be oily, waxy, flexible, or elastic, depending on the chain length, the extent of cross-linking between the chains, and the type of organic group.

A child playing with Silly Putty, a silicone polymer with unusual mechanical properties. Gentle pressure causes Silly Putty to flow or stretch, but it cannot be flattened when hit with a hammer.

the chains. Without cross-linking, silicones are waxes or oils, but cross-linking can produce flexible materials used in sealants, gaskets, car polishes, lubricants, and even elastic materials, such as the plastic substance known as Silly Putty.

Note the pattern

The stability of Group 14 hydrides decreases down the group, and E═E and E≡E bonds become weaker.

EXAMPLE 22.4

For each reaction, explain why the given products form:

(a) $Pb(s) + Cl_2(g) \longrightarrow PbCl_2(s)$
(b) $Mg_2Si(s) + 4H_2O(l) \longrightarrow SiH_4(g) + 2Mg(OH)_2(s)$
(c) $GeO_2(s) + 4OH^-(aq) \longrightarrow GeO_4{}^{4-}(aq) + 2H_2O(l)$

Given Balanced chemical equations

Asked for Why the given products form

Strategy

Classify the type of reaction. Using periodic trends in atomic properties, thermodynamics, and kinetics, explain why the observed reaction products form.

Solution

(a) Lead is a metal, and chlorine is a nonmetal that is a strong oxidant. Thus, we can expect a redox reaction to occur in which the metal acts as a reductant. Although lead can form compounds in the +2 and +4 oxidation states, Pb^{4+} is a potent oxidant (the inert-pair effect). Because lead prefers the +2 oxidation state and chlorine is a weaker oxidant than fluorine, we expect $PbCl_2$ to be the product.

(b) This is the reaction of water with a metal silicide, which formally contains the Si^{4-} ion. Water can act as either an acid or a base. Since the other compound is a base, we expect an acid–base reaction to occur in which water acts as an acid. Because Mg_2Si contains Si in its lowest possible oxidation state, however, an oxidation–reduction reaction is also a possibility. But water is a relatively weak oxidant, so an acid–base reaction is more likely. The acid (H_2O) transfers a proton to the base (Si^{4-}), which can accept four protons to form SiH_4. Proton transfer from water produces the OH^- ion, which will combine with Mg^{2+} to give magnesium hydroxide.

(c) We expect germanium dioxide (GeO_2) to be amphoteric because of the position of germanium in the periodic table. It should dissolve in strong aqueous base to give an anionic species analogous to silicate.

EXERCISE 22.4

Predict the products of the reactions, and write a balanced equation for each:

(a) $PbO_2(s) \xrightarrow{\Delta}$
(b) $GeCl_4(s) + H_2O(l) \longrightarrow$
(c) $Sn(s) + HCl(aq) \longrightarrow$

Answer

(a) $PbO_2(s) \xrightarrow{\Delta} PbO(s) + \frac{1}{2}O_2(g)$
(b) $GeCl_4(s) + 2H_2O(l) \longrightarrow GeO_2(s) + 4HCl(aq)$
(c) $Sn(s) + 2HCl(aq) \longrightarrow Sn^{2+}(aq) + H_2(g) + 2Cl^-(aq)$

22.3 The Elements of Group 15 (the Pnicogens)

Like the Group 14 elements, the lightest member of Group 15, nitrogen, is found in nature as the free element, and the heaviest elements have been known for centuries because they are easily isolated from their ores.

(a) (b)

Figure 22.10 The ancient Egyptians used finely ground antimony sulfide for eye makeup. (a) Crystals of the soft black mineral stibnite (Sb_2S_3) on a white mineral matrix. (b) A fragment of an Egyptian painting on limestone from the 16th–13th centuries B.C., showing the use of ground stibnite ("kohl") as black eye shadow. Small vases of ground stibnite have been found among the funeral goods buried with Egyptian pharaohs.

Antimony (Sb) was probably the first of the pnicogens to be obtained in elemental form and to be recognized as an element; its atomic symbol, Sb, comes from its Roman name, *stibium*. It is found in stibnite (Sb_2S_3), a black mineral that has been used as a cosmetic (an early form of mascara) since biblical times, and is easily reduced to the metal in a charcoal fire (Figure 22.10). The Egyptians used antimony to coat copper objects as early as the third millennium B.C., and antimony is still used in alloys to improve the tonal quality of bells.

In the form of its yellow sulfide ore, *orpiment* (As_2S_3), arsenic (As) has been known to physicians and professional assassins since ancient Greece, although elemental arsenic was not isolated until centuries later. The history of bismuth (Bi), in contrast, is more difficult to follow, because early alchemists often confused it with other metals such as lead, tin, antimony, and even silver (due to its slightly pinkish-white luster). Its name comes from the old German *wismut*, "white metal." Bismuth was finally isolated in the 15th century, and it was used to make moveable type for printing shortly after the invention of the Gutenberg printing process in 1440. Bismuth is used in printing because it is one of the few substances known whose solid state is less dense than the liquid. Consequently, its alloys expand as they cool, filling a mold completely and producing crisp, clear letters for typesetting.

Phosphorus was discovered in 1669 by the German alchemist Hennig Brandt, who was looking for the "philosophers' stone," a mythical substance capable of converting base metals to silver or gold. Believing that human urine was the source of the key ingredient, Brandt obtained several dozen buckets full, which he allowed to putrefy. The urine was distilled to dryness at high temperature, and then condensed and the last fumes collected under water, giving a waxy white solid that had unusual properties. For example, it glowed in the dark and burst into flames when removed from the water. (Unfortunately for Brandt, however, it did not turn lead into gold!) The element was given its current name (from the Greek *phos*, "light," and *phoros*, "bringing") in the 17th century. For more than a century, the only way to obtain phosphorus was the distillation of urine, but in 1769 it was discovered that phosphorus could be obtained more easily from bones. During the 19th century, the demand for phosphorus for matches was so great that battlefields and paupers' graveyards were systematically scavenged for bones.

Early matches were pieces of wood coated with elemental phosphorus that were stored in an evacuated glass tube and ignited when the tube was broken (which could cause unfortunate accidents if the matches were kept in a pocket!). Unfortunately, elemental phosphorus is volatile and highly toxic. It is absorbed by the teeth and destroys the bone in the jaw, leading to a painful and fatal condition called "phossy jaw," which for many years was accepted as an occupational hazard of working in the match industry.

Although nitrogen is the most abundant element in the atmosphere, it was the last of the pnicogens to be obtained in pure form. In 1772, Daniel Rutherford, working with Joseph Black (who discovered CO_2), noticed that a gas remained when CO_2 was removed from a combustion reaction. Antoine Lavoisier called the gas *azote*, meaning "no life," because it did not support life. When it was discovered that the same element was also present in nitric acid and nitrate salts such as KNO_3 (nitre), it was named nitrogen. About 90% of the nitrogen produced today is used to provide an inert atmosphere for processes or reactions that are oxygen sensitive, such as the production of steel, petroleum refining, and the packaging of foods and pharmaceuticals.

Preparation and General Properties of Group 15 Elements

Because the atmosphere contains several trillion tons of elemental nitrogen with a purity of about 80%, it is a huge source of nitrogen gas. Distillation of liquefied air yields nitrogen gas that is more than 99.99% pure, but small amounts of very pure nitrogen gas can be obtained from the thermal decomposition of sodium azide:

$$2NaN_3(s) \xrightarrow{\Delta} 2Na(l) + 3N_2(g) \qquad (22.27)$$

In contrast, the Earth's crust is relatively poor in nitrogen. The only important nitrogen ores are large deposits of KNO_3 and $NaNO_3$ in the deserts of Chile and Russia, which were apparently formed by the evaporation of ancient alkaline lakes. Consequently, virtually all nitrogen compounds produced on an industrial scale use atmospheric nitrogen as the starting material. Phosphorus, which constitutes only about 0.1% of the Earth's crust, is much more abundant in ores than nitrogen. Like aluminum and silicon, phosphorus is always found in combination with oxygen, and large inputs of energy are required to isolate it.

The other three pnicogens are much less abundant: arsenic is found in the Earth's crust at a concentration of about 2 ppm, antimony is an order of magnitude less abundant, and bismuth is almost as rare as gold. All three elements have a high affinity for the chalcogens and are usually found as the sulfide ores (M_2S_3), often in combination with sulfides of other heavy elements, such as copper, silver, and lead. Hence, a major source of antimony and bismuth is flue dust obtained by smelting the sulfide ores of the more abundant metals.

In Group 15, as elsewhere in the *p* block, we see large differences between the lightest element, N, and its congeners in size, ionization energy, electron affinity, and electronegativity (Table 22.3). The chemical behavior of the elements can be summarized rather simply: Nitrogen and phosphorus behave chemically like nonmetals, arsenic and antimony like semimetals, and bismuth like a metal. With their $ns^2 np^3$ valence-electron configurations, all form compounds by losing the three *np* valence electrons to form the +3 oxidation state, or by losing the three *np* and the two *ns* valence electrons to give the +5 oxidation state, whose stability decreases smoothly from phosphorus to bismuth. In addition, the relatively large magnitude of the electron affinity of the lighter pnicogens enables them to form compounds in the −3 oxidation state (such as NH_3 PH_3), in which three electrons are formally added to the neutral atom to give a filled *np* subshell. Nitrogen has the unusual ability to form compounds in nine different oxidation states, including −3, +3, and +5. Because neutral covalent compounds of the trivalent pnicogens contain a lone pair of electrons on the central atom, they tend to behave as Lewis bases.

Note the pattern

In Group 15, the stability of the +5 oxidation state decreases from P to Bi.

Reactions and Compounds of Nitrogen

Like carbon, nitrogen has four valence orbitals (one 2*s* and three 2*p*), so it can participate in at most four electron-pair bonds by using sp^3 hybrid orbitals. Unlike carbon, however, nitrogen does not form long chains because of repulsive interactions between

TABLE 22.3 Selected properties of the Group 15 elements

Property	Nitrogen	Phosphorus	Arsenic	Antimony	Bismuth
Atomic symbol	N	P	As	Sb	Bi
Atomic number	7	15	33	51	83
Atomic mass (amu)	14.01	30.97	74.92	121.76	209.98
Valence-electron configuration[a]	$2s^2\,2p^3$	$3s^2\,3p^3$	$4s^2\,4p^3$	$5s^2\,5p^3$	$6s^2\,6p^3$
mp/bp (°C)	−210/−196	44.15/281[c]	817 (3.70 MPa)/ 603 (sublimes)[b]	631/1587	271/1564
Density (g/cm^3 at 25°C)	1.15 (g/L)	1.82[c]	5.75[b]	6.68	9.79
Atomic radius (pm)	56	98	114	133	143
First ionization energy (kJ/mol)	1402	1012	945	831	703
Common oxidation state(s)	−3 to +5	+5, +3, −3	+5, +3	+5, +3	+3
Ionic radius (pm)[d]	146 (−3) 16 (+3)	212 (−3) 44 (+3)	58(+3)	76 (+3) 60 (+5)	103 (+3)
Electron affinity (kJ/mol)	0	−72	−78	−101	−91
Electronegativity	3.0	2.2	2.2	2.1	1.9
Standard reduction potential, E°(V) ($E^{V} \rightarrow E^{III}$ in acidic solution)[e]	+0.93	−0.28	+0.56	+0.65	—
Product of reaction with O_2	NO_2, NO	P_4O_6, P_4O_{10}	As_4O_6	Sb_2O_5	Bi_2O_3
Type of oxide	Acidic (NO_2) Neutral (NO, N_2O)	Acidic	Acidic	Amphoteric	Basic
Product of reaction with N_2	—	None	None	None	None
Product of reaction with X_2	None	PX_3, PX_5	AsF_5, AsX_3	SbF_5, $SbCl_5$, $SbBr_3$, SbI_3	BiF_5, BiX_3
Product of reaction with H_2	None	None	None	None	None

[a]The configuration shown does not include filled *d* and *f* subshells.

[b]For gray arsenic.

[c]For white phosphorus.

[d]The values cited are for six-coordinate ions in the indicated oxidation states. The N^{5+}, P^{5+}, and As^{5+} ions are not known species.

[e]The chemical form of the elements in these oxidation states varies considerably. For N, the reaction is $NO_3^- + 2e^- \longrightarrow HNO_2$; for P and As, it is $H_3EO_4 + 2e^- \rightarrow H_3EO_3$; and for Sb it is $Sb_2O_5 + 4e^- + 10H^+ \rightarrow 2Sb^{3+} + 10H_2O$.

lone pairs of electrons on *adjacent* atoms. These interactions become important at the shorter internuclear distances encountered with the smaller, second-period elements of Groups 15, 16, and 17 (see Chapter 8). Stable compounds with N—N bonds are limited to chains of no more than three N atoms, such as the azide ion (N_3^-).

Nitrogen is the only pnicogen that normally forms multiple bonds with itself and other second-period elements, using π overlap of adjacent *np* orbitals. Thus, the stable form of elemental nitrogen is N_2, whose N≡N triple bond is so strong ($D_{N\equiv N}$ = 942 kJ/mol) compared with the N—N and N=N bonds ($D_{N—N}$ = 167 kJ/mol, $D_{N=N}$ = 418 kJ/mol) that all compounds containing N—N single and N=N double bonds are thermodynamically unstable with respect to the formation of N_2. In fact, the formation of the N≡N triple bond is so thermodynamically favored that virtually all compounds containing N—N bonds are potentially explosive.

Note the pattern

Few binary molecular compounds of nitrogen are formed by direct reaction of the elements.

Again in contrast to carbon, nitrogen undergoes only two important chemical reactions at room temperature: it reacts with metallic lithium to form lithium nitride (see Chapter 21), and it is reduced to ammonia by certain microorganisms. At higher temperatures, however, N_2 reacts with more electropositive elements, such as those in Group 13, to give binary nitrides, which range from covalent to ionic in character. Like the corresponding compounds of carbon, binary compounds of nitrogen with oxygen, hydrogen, or other nonmetals are usually covalent molecular substances.

Few binary molecular compounds of nitrogen are formed by direct reaction of the elements. At elevated temperatures, N_2 reacts with H_2 to form ammonia, with O_2 to form a mixture of NO and NO_2, and with carbon to form *cyanogen* ($N{\equiv}C{-}C{\equiv}N$), but elemental nitrogen does not react with the halogens or the other chalcogens. Nonetheless, all the binary nitrogen halides (NX_3) are known. Except for NF_3, all are toxic, thermodynamically unstable, and potentially explosive, and all are prepared by the reaction of the halogen with NH_3 rather than with N_2.

Both nitrogen monoxide (NO) and nitrogen dioxide (NO_2) are thermodynamically unstable, with positive free energies of formation. Unlike NO, NO_2 reacts readily with excess water, forming a 1:1 mixture of nitrous acid (HNO_2) and nitric acid (HNO_3):

$$2NO_2(g) + H_2O(l) \longrightarrow HNO_2(aq) + HNO_3(aq) \qquad (22.28)$$

NO

N_2O

NO_2

Nitrogen also forms N_2O (dinitrogen monoxide, or nitrous oxide), a linear molecule that is isoelectronic with CO_2 and can be represented as $^-N{=}N^+{=}O$. Like the other two oxides of nitrogen, nitrous oxide is thermodynamically unstable. The structures of the three common oxides of nitrogen are shown in the margin.

At elevated temperatures, nitrogen reacts with highly electropositive metals to form *ionic nitrides*, such as Li_3N and Ca_3N_2. These compounds consist of ionic lattices formed by M^{n+} and N^{3-} ions. Just as boron forms interstitial borides, and carbon forms interstitial carbides, with less electropositive metals nitrogen forms a range of *interstitial nitrides*, in which nitrogen occupies holes in a close-packed metallic structure. Like the interstitial carbides and borides, these substances are typically very hard, high-melting materials that have metallic luster and conductivity.

Nitrogen also reacts with semimetals at very high temperatures to produce *covalent nitrides* such as Si_3N_4 and BN, which are solids with extended covalent network structures similar to those of graphite or diamond. Consequently, they are usually high melting and chemically inert materials.

Ammonia (NH_3) is one of the few thermodynamically stable binary compounds of nitrogen with a nonmetal. It is not flammable in air, but it burns in an O_2 atmosphere:

$$4NH_3(g) + 3O_2(g) \longrightarrow 2N_2(g) + 6H_2O(g) \qquad (22.29)$$

Note the pattern

B, C, and N all react with transition metals to form interstitial compounds that are hard, high-melting materials.

About 10% of the ammonia produced annually is used to make fibers and plastics that contain amide bonds, such as nylons and polyurethanes, while 5% is used in explosives, such as ammonium nitrate, TNT, and nitroglycerine. Large amounts of anhydrous liquid ammonia are used as fertilizer.

Nitrogen forms two other important binary compounds with hydrogen. Hydrazoic acid (HN_3), also called hydrogen azide, is a colorless, highly toxic, and explosive substance. Hydrazine (N_2H_4) is also potentially explosive; it is used as a rocket propellant and to inhibit corrosion in boilers.

EXAMPLE 22.5

For each reaction, explain why the given products form when the reactants are heated:

(a) $Sr(s) + N_2O(g) \xrightarrow{\Delta} SrO(s) + N_2(g)$

(b) $NH_4NO_2(s) \xrightarrow{\Delta} N_2(g) + 2H_2O(g)$

(c) $Pb(NO_3)_2(s) \xrightarrow{\Delta} PbO_2(s) + 2NO_2(g)$

Given Balanced chemical equations

Asked for Why the given products form

Strategy

Classify the type of reaction. Using periodic trends in atomic properties, thermodynamics, and kinetics, explain why the observed reaction products form.

Solution

(a) As an alkali metal, strontium is a strong reductant. If the other reactant can act as an oxidant, then a redox reaction will occur. Nitrous oxide contains nitrogen in a low oxidation state (+1), so we would not normally consider it an oxidant. Nitrous oxide is, however, thermodynamically unstable (ΔH_f° and $\Delta G_f^\circ > 0$), and it can be reduced to N_2, which is a stable species. Consequently, we predict that a redox reaction will occur.

(b) When a substance is heated, a decomposition reaction probably will occur, which often involves the release of stable gases. In this case, ammonium nitrite contains nitrogen in two different oxidation states (−3 and +3), so an internal redox reaction is a possibility. Due to its thermodynamic stability, nitrogen gas (N_2) is the probable nitrogen-containing product, whereas we predict that H and O will combine to form H_2O.

(c) Again, this is probably a thermal decomposition reaction. If one element is in an usually high oxidation state and another in a low oxidation state, a redox reaction will probably occur. Lead nitrate contains the Pb^{2+} cation and the nitrate anion, which contains nitrogen in its highest possible oxidation state, +5. Hence, nitrogen can be reduced, and we know that lead can be oxidized to the +4 oxidation state. Consequently, it is likely that lead(II) nitrate will decompose to lead(IV) oxide and nitrogen dioxide when heated. Even though PbO_2 is a powerful oxidant, the release of a gas such as NO_2 can often drive an otherwise unfavorable reaction to completion (le Chatelier's principle). Note, however, that PbO_2 will probably decompose to PbO and $\frac{1}{2}O_2$ at high temperatures.

EXERCISE 22.5

Predict the products of the reactions, and write a balanced equation for each:

(a) $NO(g) + H_2O(l) \xrightarrow{\Delta}$

(b) $NH_4NO_3(s) \xrightarrow{\Delta}$

(c) $Sr(s) + N_2(g) \longrightarrow$

Answer

(a) $NO(g) + H_2O(l) \xrightarrow{\Delta}$ no reaction

(b) $NH_4NO_3(s) \xrightarrow{\Delta} N_2O(g) + 2H_2O(g)$

(c) $3Sr(s) + N_2(g) \longrightarrow Sr_3N_2(s)$

Reactions and Compounds of the Heavier Pnicogens

Like the heavier elements of Group 14, the heavier pnicogens form catenated compounds that contain only single bonds, whose stability decreases rapidly as we go down the group. For example, phosphorus exists as multiple allotropes, the most common of which is *white phosphorus*, which consists of P_4 tetrahedra and behaves like a typical nonmetal. As is typical of a molecular solid, white phosphorus is volatile, has a low melting point (44.1°C), and is soluble in nonpolar solvents. It is highly strained, with bond angles of only 60°, which partially explains why it is so reactive and so easily converted to more stable allotropes. Heating white phosphorus for several days converts it to *red phosphorus*, a polymer that is air stable, virtually insoluble, denser than white phosphorus, and higher melting, which make it much safer to handle. A third allotrope of phosphorus, *black phosphorus*, is prepared by heating the other allotropes under high pressure; it is even less reactive, denser, and higher melting than

Note the pattern

As in Group 14, the heavier Group 15 elements form catenated compounds that contain only single bonds, whose stability decreases as we go down the group.

red phosphorus. As expected from their structures, white phosphorus is an electrical insulator, and red and black phosphorus are semiconductors. The three heaviest pnicogens—arsenic, antimony, and bismuth—all have a metallic luster, but they are brittle rather than ductile and relatively poor electrical conductors.

Note the pattern

The reactivity of the heavier Group 15 elements decreases as we go down the column.

The reactivity of the heavier pnicogens decreases as we go down the column. Phosphorus is by far the most reactive of the pnicogens, forming binary compounds with every element in the periodic table except antimony, bismuth, and the noble gases. Phosphorus reacts rapidly with O_2, whereas arsenic burns in pure O_2 if ignited, and antimony and bismuth react with O_2 only when heated. None of the pnicogens reacts with nonoxidizing acids such as aqueous HCl, but all dissolve in oxidizing acids such as HNO_3. Only bismuth behaves like a metal, dissolving in HNO_3 to give the hydrated Bi^{3+} cation.

The heavier pnicogens can use energetically accessible 3*d*, 4*d*, or 5*d* orbitals to form dsp^3 or d^2sp^3 hybrid orbitals for bonding. Consequently, these elements often have coordination numbers of 5 or higher. Phosphorus and arsenic form halides (for example, $AsCl_5$) that are generally covalent molecular species and behave like typical nonmetal halides, reacting with water to form the corresponding oxoacids (in this case, H_3AsO_4). All the pentahalides are potent Lewis acids that can expand their coordination to accommodate the lone pair of a Lewis base:

$$AsF_5(soln) + F^-(soln) \longrightarrow AsF_6^-(soln) \tag{22.30}$$

In contrast, bismuth halides have extended lattice structures and dissolve in water to produce hydrated ions, consistent with the stronger metallic character of bismuth.

Except for BiF_3, which is essentially an ionic compound, the *tri*halides are volatile covalent molecules with a lone pair of electrons on the central atom. Like the pentahalides, the trihalides react rapidly with water. In the cases of phosphorus and arsenic, the products are the corresponding acids, H_3PO_3 and H_3AsO_3:

$$EX_3(l) + 3H_2O(l) \longrightarrow H_3EO_3(aq) + 3HX(aq) \qquad (E = P, As) \tag{22.31}$$

Phosphorus halides are also used to produce insecticides, flame retardants, and plasticizers.

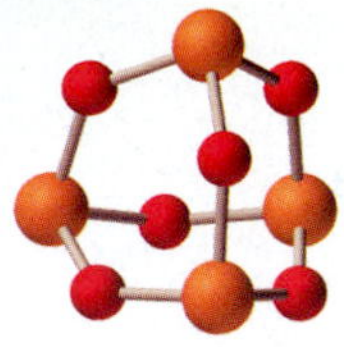

(a) P_4O_6

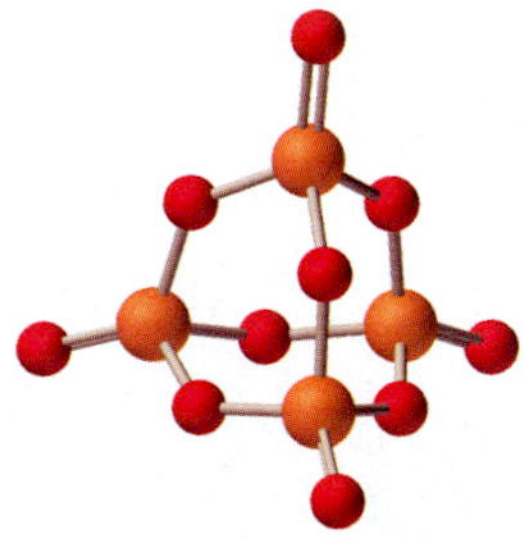

(b) P_4O_{10}

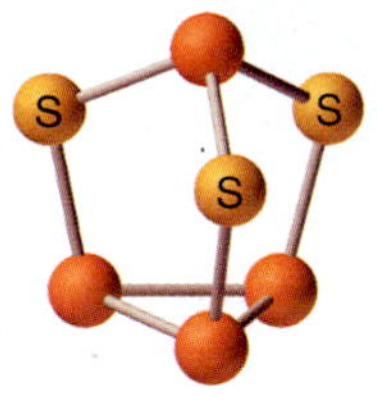

(c) P_4S_3

Figure 22.11 The structures of some cage compounds of phosphorus. (a, b) The structures of P_4O_6 and P_4O_{10} are both derived from the structure of white phosphorus (P_4) by the insertion of an oxygen atom into each of the six edges of the P_4 tetrahedron; P_4O_{10} contains an additional terminal oxygen atom bonded to each phosphorus atom. (c) The structure of P_4S_3 is also derived from the structure of P_4 by the insertion of three sulfur atoms into three adjacent edges of the tetrahedron.

With energetically accessible *d* orbitals, phosphorus and, to a lesser extent, arsenic are able to form π bonds with second-period atoms such as N and O. This effect is even more important for phosphorus than for silicon, resulting in very strong P—O single bonds and even stronger P═O double bonds. The first four elements in Group 15 also react with oxygen to produce the corresponding oxide in the +3 oxidation state. Of these oxides, P_4O_6 and As_4O_6 have cage structures formed by inserting an oxygen atom into each edge of the P_4 or As_4 tetrahedron (Figure 22.11a), and they behave like typical nonmetal oxides. For example, P_4O_6 reacts with water to form phosphorous acid (H_3PO_3). Consistent with its position between the nonmetal and metallic oxides, Sb_4O_6 is amphoteric, dissolving in either acid or base. In contrast, Bi_2O_3 behaves like a basic metallic oxide, dissolving in acid to give solutions that contain the hydrated Bi^{3+} ion. The two least metallic elements of the heavier pnicogens, phosphorus and arsenic, form very stable oxides with the formula E_4O_{10} in the +5 oxidation state (Figure 22.11b). In contrast, Bi_2O_5 is so unstable that there is no absolute proof it exists.

Note the pattern

Phosphorus has the greatest ability to form π bonds with elements such as O, N, and C.

The heavier pnicogens form sulfides that range from molecular species with three-dimensional cage structures, such as P_4S_3 (Figure 22.11c), to layered or ribbon structures, such as Sb_2S_3 and Bi_2S_3, which are semiconductors. The reaction of the heavier pnicogens with metals produces substances whose properties vary with the metal content. Metal-rich *phosphides* (such as M_4P) are hard, high-melting, electrically conductive solids with a metallic luster, whereas phosphorus-rich phosphides (such as MP_{15}) are lower melting and less thermally stable because they contain catenated P_n units. Many organic or organometallic compounds of the heavier pnicogens that contain from one to five alkyl or aryl groups are also known. Because of the decreasing strength of the pnicogen–carbon bond, their thermal stability decreases from phosphorus to bismuth.

EXAMPLE 22.6

For each reaction, explain why the given products form:

(a) $Bi(s) + \frac{3}{2}Br_2(l) \longrightarrow BiBr_3(s)$

(b) $2(CH_3)_3As(l) + O_2(g) \longrightarrow 2(CH_3)_3As{=}O(s)$

(c) $PBr_3(l) + 3H_2O(l) \longrightarrow H_3PO_3(aq) + 3HBr(aq)$

(d) $As(s) + Ga(s) \xrightarrow{\Delta} GaAs(s)$

Given Balanced chemical equations

Asked for Why the given products form

Strategy

Classify the type of reaction. Using periodic trends in atomic properties, thermodynamics, and kinetics, explain why the reaction products form.

Solution

(a) Bromine is an oxidant, and bismuth is a metal that can be oxidized. Hence, a redox reaction is likely to occur. To identify the product, recall that bismuth can form compounds in either the +3 or +5 oxidation state. The heaviest pnicogen, bismuth is rather difficult to oxidize to the +5 oxidation state because of the inert-pair effect. Hence, the product will probably be bismuth(III) bromide.

(b) Trimethylarsine, with a lone pair of electrons on the As atom, can act as either a Lewis base or a reductant. If arsenic is oxidized by two electrons, then oxygen must be reduced, most probably by two electrons to the −2 oxidation state. Since As(V) forms strong bonds to oxygen due to π bonding, the expected product is $(CH_3)_3As{=}O$.

(c) Phosphorus tribromide is a typical nonmetal halide. We expect it to react with water to produce an oxoacid of P(III) and the corresponding hydrohalic acid.*

(d) Gallium is a metal with a strong tendency to act as a reductant and form compounds in the +3 oxidation state. In contrast, arsenic is a semimetal. It can act as a reductant to form compounds in the +3 or +5 oxidation state, or it can act as an oxidant, accepting electrons to form compounds in the −3 oxidation state. If a reaction occurs, then a binary compound will probably form with a 1:1 ratio of the elements. GaAs is an example of a III-V compound, many of which are used in the electronics industry (see Chapter 12).

EXERCISE 22.6

Predict the products of the reactions, and write a balanced equation for each:

(a) $PCl_5(s) + H_2O(l) \longrightarrow$

(b) $Bi_2O_5(s) \xrightarrow{\Delta}$

(c) $Ca_3P_2(s) + H^+(aq) \longrightarrow$

(d) $NaNH_2(s) + PH_3(soln) \longrightarrow$

Answer

(a) $PCl_5(s) + 4H_2O(l) \longrightarrow H_3PO_4(aq) + 5HCl(aq)$

(b) $Bi_2O_5(s) \xrightarrow{\Delta} Bi_2O_3(s) + O_2(g)$

(c) $Ca_3P_2(s) + 6H^+(aq) \longrightarrow 2PH_3(g) + 3Ca^{2+}(aq)$

(d) $NaNH_2(s) + PH_3(soln) \longrightarrow NaPH_2(s) + NH_3(soln)$

Note the pattern

The thermal stability of organic or organometallic compounds of Group 15 decreases down the group due to the decreasing strength of the pnicogen-carbon bond.

* Because of the strength of the P═O double bond, phosphorous acid (H_3PO_3) is actually $HP(O)(OH)_2$, which contains a P═O bond and a P—H bond.

22.4 The Elements of Group 16 (the Chalcogens)

Note the pattern

Group 16 is the first group in the p block that contains no stable metallic elements.

The chalcogens are the first group in the *p* block that contains no stable metallic elements. All isotopes of polonium (Po), the only metal in Group 16, are radioactive, and only one element in the group, tellurium (Te), can even be described as a semimetal. As in Groups 14 and 15, the lightest element of Group 16, oxygen, is found in nature as the free element.

Of the Group 16 elements, only sulfur was known in ancient times; the others were not discovered until the late 18th and 19th centuries. Sulfur is frequently found as yellow crystalline deposits of essentially pure S_8 in areas of intense volcanic activity or around hot springs. As early as the 15th century B.C., sulfur was used as a fumigant in Homeric Greece because, when burned, it produces SO_2 fumes that are toxic to most organisms, including vermin hiding in the walls and under the floors of houses. Hence, references to sulfur are common in ancient literature, frequently in the context of religious purification. In fact, the association of sulfur with the divine was so pervasive that the prefixes *thio-* ("sulfur") and *theo-* ("god") have the same root in ancient Greek. Though used primarily in the production of sulfuric acid, sulfur is also used in the manufacture of gunpowder and as a cross-linking agent for rubber, which enables rubber to hold its shape but retain its flexibility.

Sulfur deposit around a volcanic vent in Kilauea Crater, Hawaii.

Oxygen was not discovered until 1771, when the Swedish pharmacist Carl Wilhelm Scheele found that heating compounds such as KNO_3, Ag_2CO_3, and HgO produced a colorless, odorless gas that supported combustion better than air. The results were not published immediately, however, so Scheele's work remained unknown until 1777. Unfortunately, this was nearly two years *after* a paper by the English chemist Joseph Priestley had been published, describing the isolation of the same gas by using a magnifying glass to focus the sun's rays on a sample of HgO. Oxygen is used primarily in the steel industry during the conversion of crude iron to steel using the Bessemer process (see Chapter 23). Another important industrial use of oxygen is in the production of TiO_2, which is commonly used as a white pigment in paints, paper, and plastics.

Jöns Jakob Berzelius in 1843. Berzelius (1779–1848) was born into a well-educated Swedish family, but both parents died when he was young. He studied medicine at the University of Uppsala, where his experiments with electroshock therapy caused his interests to turn to electrochemistry. Berzelius devised the system of chemical notation that we use today. In addition, he discovered six elements (cerium, thorium, selenium, silicon, titanium, and zirconium).

Tellurium was discovered accidentally in 1782 by the Austrian chemist Franz Joseph Müller von Reichenstein. The silvery-white metal had the same density as antimony but very different properties. Because it was difficult to analyze, Müller called it *metallum problematicum* ("difficult metal"). The name *tellurium* (from the Latin *tellus*, "earth") was coined by another Austrian chemist, Martin Klaproth, who demonstrated in 1798 that Müller's "difficult metal" was actually a new element. Tellurium is used to color glass and ceramics, in the manufacture of blasting caps, and in thermoelectric devices.

Selenium (Se) was first isolated in 1817 by the Swedish chemist Jöns Jakob Berzelius, who also discovered silicon. He had invested money in a sulfuric acid plant and decided to investigate a foul-smelling contaminant that formed a red precipitate. Although he initially thought the contaminant was tellurium, further study showed that it was actually a new element similar to tellurium. To emphasize the similarities, Berzelius named the new element selenium (after the Greek *selene*, "moon"). Selenium is used primarily as a minor ingredient to decolorize glass. Because it is photosensitive, selenium is also used to capture images in the photocopying process (Figure 22.12).

The heaviest chalcogen, polonium, was isolated after an extraordinary effort by Marie Curie (see Section 1.5). Although she was never able to obtain macroscopic quantities of the element, which she named for her native country of Poland, she demonstrated that its chemistry required it to be assigned to Group 16. Marie Curie was awarded her second Nobel Prize in 1911 for the discovery of radium and polonium.

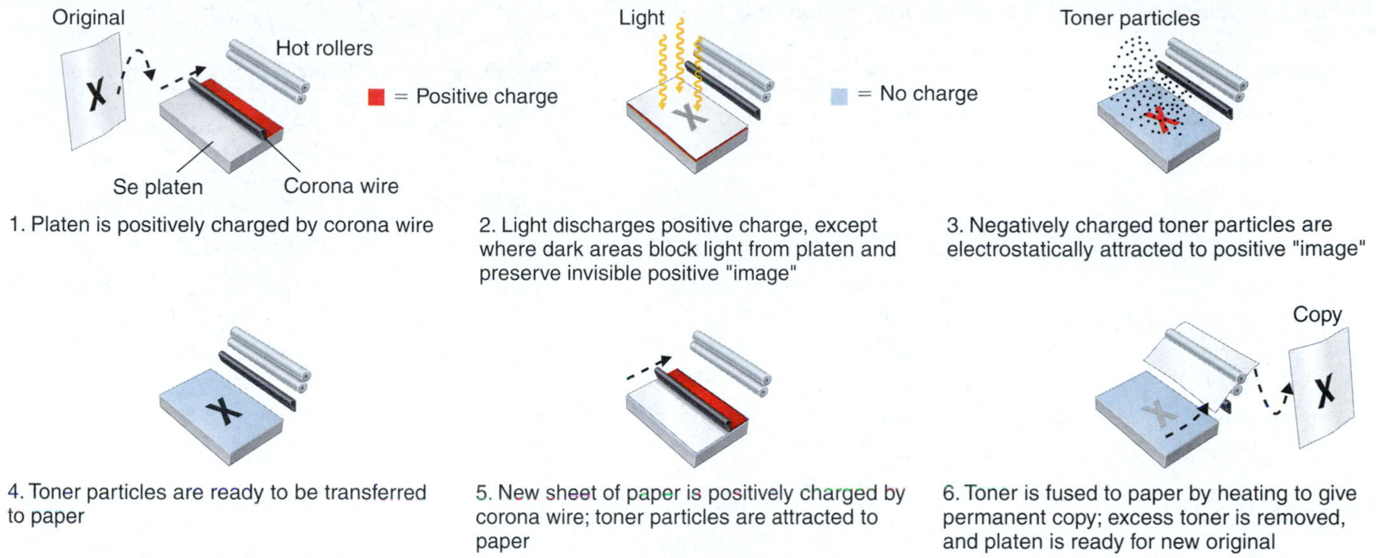

Figure 22.12 The chemistry of photocopying. Because amorphous selenium is a photosensitive semiconductor, exposing an electrostatically charged Se film to light causes the positive charge on the film to be discharged in all areas that are white in the original. Dark areas in the original block the light and generate an invisible, positively charged image. To produce an image on paper, negatively charged toner particles are attracted to the positive image, transferred to a negatively charged sheet of blank paper, and fused with the paper at high temperature to give a permanent image. Before another original can be copied, the platen must be cleaned to remove excess toner and recharged to erase the old image and prepare for the next image.

Preparation and General Properties of Group 16 Elements

Oxygen is by far the most abundant element in the Earth's crust and in the hydrosphere (about 44% and 86% by mass, respectively). The same process that is used to obtain nitrogen from the atmosphere produces pure oxygen. Oxygen can also be obtained by the electrolysis of water (see Chapter 19), the decomposition of alkali metal or alkaline earth peroxides or superoxides (see Chapter 21), or the thermal decomposition of simple inorganic salts, such as potassium chlorate in the presence of a catalytic amount of MnO_2:

$$2KClO_3(s) \xrightarrow[\Delta]{MnO_2(s)} 2KCl(s) + \frac{3}{2}O_2(g) \qquad (22.32)$$

Iron pyrite (FeS_2). Because of its lustrous golden yellow cubic crystals, FeS_2 is sometimes mistaken for gold, giving rise to its common name "fool's gold." Real gold, however, is much denser than FeS_2, and gold is soft and malleable rather than hard and brittle.

Unlike oxygen, sulfur is not very abundant, but it is found as elemental sulfur in rock formations overlying salt domes, which often accompany petroleum deposits (see Figure 2.22). Sulfur is also recovered from H_2S and organosulfur compounds in crude oil and coal, and from metal sulfide ores such as *pyrite* (FeS_2).

Because selenium and tellurium are chemically similar to sulfur, they are usually found as minor contaminants in metal sulfide ores and are typically recovered as byproducts. Even so, they are as abundant in the Earth's crust as silver, palladium, and gold. One of the best sources of selenium and tellurium is the "slime" deposited during the electrolytic purification of copper (see Chapter 19). Both these elements are notorious for the vile odors of many of their compounds. For example, when even trace amounts of tellurium are absorbed by the body, dimethyltellurium [$(CH_3)_2Te$] is produced and slowly released in the breath and perspiration, resulting in an intense garliclike smell that is commonly called "tellurium breath."

With their ns^2np^4 electron configurations, the chalcogens are two electrons short of a filled valence shell. Thus, in reactions with metals they tend to acquire two additional electrons to form compounds in the −2 oxidation state. This tendency is greatest for

TABLE 22.4 Selected properties of the Group 16 elements

Property	Oxygen	Sulfur	Selenium	Tellurium	Polonium
Atomic symbol	O	S	Se	Te	Po
Atomic number	8	16	34	52	84
Atomic mass (amu)	16.00	32.07	78.96	127.60	209
Valence-electron configuration[a]	$2s^2 2p^4$	$3s^2 3p^4$	$4s^2 4p^4$	$5s^2 5p^4$	$6s^2 6p^4$
mp/bp (°C)	−219/−183	115/445	221/685	450/988	254/962
Density (g/cm^3) at 25°C	1.31 (g/L)	2.07	4.81	6.24	9.20
Atomic radius (pm)	48	88	103	123	135
First ionization energy (kJ/mol)	1314	1000	941	869	812
Normal oxidation state(s)	−2	+6, +4, −2	+6, +4, −2	+6, +4, −2	+2 (+4)
Ionic radius (pm)[b]	140 (−2)	184 (−2)	198 (−2)	221 (−2)	(230) (−2)
		29 (+6)	42 (+6)	56 (+6)	97 (+4)
Electron affinity (kJ/mol)	−141	−200	−195	−190	−180
Electronegativity	3.4	2.6	2.6	2.1	2.0
Standard reduction potential, $E°$ (V) ($E^0 \rightarrow H_2E$ in acidic solution)	+1.23	+0.14	−0.40	−0.79	−1.00
Product of reaction with O_2	—	SO_2	SeO_2	SeO_2	PoO_2
Type of oxide	—	Acidic	Acidic	Amphoteric	Basic
Product of reaction with N_2	NO, NO_2	None	None	None	None
Product of reaction with X_2	O_2F_2	SF_6, S_2Cl_2, S_2Br_2	SeF_6, SeX_4	TeF_6, TeX_4	PoF_4, $PoCl_2$, $PoBr_2$
Product of reaction with H_2	H_2O	H_2S	H_2Se	None	None

[a]The configuration shown does not include filled *d* and *f* subshells.

[b]The values cited for the hexacations are for six-coordinate ions and are only estimated values.

oxygen, the chalcogen with the highest electronegativity. The heavier, less electronegative chalcogens can lose either four *np* electrons or four *np* and two *ns* electrons to form compounds in the +4 and +6 oxidation state, respectively, as shown in Table 22.4. Again, the lightest member in the group, in this case oxygen, differs greatly from the others in size, ionization energy, electronegativity, and electron affinity, so its chemistry is unique. As in the other groups, the second and third members (sulfur and selenium) have similar properties because of shielding effects. Only polonium is metallic, forming either the hydrated Po^{2+} or Po^{4+} ion in aqueous solution, depending upon conditions.

Reactions and Compounds of Oxygen

Note the pattern

As in Groups 14 and 15, the lightest element in Group 16 has the greatest tendency to form multiple bonds.

As in Groups 14 and 15, the lightest Group 16 member has the greatest tendency to form multiple bonds. Thus, elemental oxygen is found in nature as a diatomic gas that contains a net double bond, O=O. As with nitrogen, electrostatic repulsion between lone pairs of electrons on adjacent atoms prevents oxygen from forming stable catenated compounds. In fact, except for O_2, all compounds that contain O—O bonds are potentially explosive. Ozone, peroxides, and superoxides are all potentially dangerous in pure form. Ozone (O_3), one of the most powerful oxidants known, is used to purify drinking water because it does not produce the characteristic taste associated with chlorinated water. Hydrogen peroxide (H_2O_2)

TABLE 22.5 Some properties of O_2 and related diatomic species

Species	Bond Order	Number of Unpaired e^-	O—O Distance, pm*
O_2^+	2.5	1	112
O_2	2	2	121
O_2^-	1.5	1	133
O_2^{2-}	1	0	149

* Source of data: L. Vaska, *Acc. Chem. Res.*, 1976, **9**, 175.

is so thermodynamically unstable that it has a tendency to undergo explosive decomposition when impure:

$$2H_2O_2(l) \longrightarrow 2H_2O(l) + O_2(g) \qquad \Delta G^\circ = -119 \text{ kJ/mol} \qquad (22.33)$$

Despite the strength of the O═O bond (D_{O_2} = 494 kJ/mol), O_2 is extremely reactive, reacting directly with nearly all other elements except the noble gases. Some properties of O_2 and related species, such as the peroxide and superoxide ions, are listed in Table 22.5.

With few exceptions, the chemistry of oxygen is restricted to negative oxidation states because of its high electronegativity ($\chi = 3.4$). Unlike the other chalcogens, oxygen does not form compounds in the +4 or +6 oxidation state. Oxygen is second only to fluorine in its ability to stabilize high oxidation states of metals in both ionic and covalent compounds. For example, AgO is a stable solid that contains silver in the unusual Ag(II) state, whereas OsO_4 is a volatile solid that contains Os(VIII). Because oxygen is so electronegative, the O—H bond is highly polar, creating a large bond dipole moment that makes hydrogen bonding much more important for compounds of oxygen than for similar compounds of the other chalcogens.

Note the pattern

Oxides of metals tend to be basic, oxides of nonmetals tend to be acidic, and oxides of elements in or near the diagonal band of semimetals are generally amphoteric.

Metal oxides are usually basic, and nonmetal oxides are acidic, whereas oxides of elements that lie on or near the diagonal band of semimetals are generally amphoteric. A few oxides, such as CO and PbO_2, are neutral and do not react with water, aqueous acid, or aqueous base. Nonmetal oxides are typically covalent compounds in which the bonds between oxygen and the nonmetal are polarized ($E^{\delta+}$—$O^{\delta-}$). Consequently, a lone pair of electrons on a water molecule can attack the partially positively charged E atom to eventually form an oxoacid. An example is the reaction of sulfur trioxide with water to form sulfuric acid:

$$H_2O(l) + SO_3(g) \longrightarrow H_2SO_4(aq) \qquad (22.34)$$

The oxides of the semimetals and of elements such as Al that lie near the metal/nonmetal dividing line are amphoteric, as we expect:

$$Al_2O_3(s) + 6H^+(aq) \longrightarrow 2Al^{3+}(aq) + 3H_2O(l) \qquad (22.35)$$

$$Al_2O_3(s) + 2OH^-(aq) + 3H_2O(l) \longrightarrow 2Al(OH)_4^-(aq) \qquad (22.36)$$

EXAMPLE 22.7

For each reaction, explain why the given products form:

(a) $Ga_2O_3(s) + 2OH^-(aq) + 3H_2O(l) \longrightarrow 2Ga(OH)_4^-(aq)$
(b) $3H_2O_2(aq) + 2MnO_4^-(aq) + 2H^+(aq) \longrightarrow 3O_2(g) + 2MnO_2(s) + 4H_2O(l)$
(c) $2KNO_3(s) \xrightarrow{\Delta} 2KNO_2(s) + O_2(g)$

Given Balanced chemical equations

Asked for Why the given products form

Strategy

Classify the type of reaction. Using periodic trends in atomic properties, thermodynamics, and kinetics, explain why the observed reaction products form.

Solution

(a) Gallium is a metal. We expect the oxides of metallic elements to be basic and therefore not to react with aqueous base. A close look at the periodic table, however, shows that gallium is close to the diagonal line of semimetals. Moreover, aluminum, the element immediately above gallium in Group 13, is amphoteric. Consequently, we predict that gallium will behave like aluminum (Equation 22.36).

(b) Hydrogen peroxide is an oxidant that can accept two electrons per molecule to give two molecules of water. With a strong oxidant, however, H_2O_2 can also act as a reductant, losing two electrons (and two protons) to produce O_2. Because the other reactant is permanganate, which is a potent oxidant, the only possible reaction is a redox reaction in which permanganate is the oxidant and hydrogen peroxide is the reductant. Recall that the reduction of permanganate often gives MnO_2, an insoluble brown solid. The reduction of MnO_4^- to MnO_2 is a three-electron reduction, whereas the oxidation of H_2O_2 to O_2 is a two-electron oxidation.

(c) This is a thermal decomposition reaction. Because KNO_3 contains nitrogen in its highest oxidation state (+5) and oxygen in its lowest oxidation state (−2), a redox reaction is likely. Oxidation of the oxygen in nitrate to atomic oxygen is a two-electron process per oxygen atom. Nitrogen is likely to accept two electrons because oxoanions of nitrogen are known only in the +5 (NO_3^-) and +3 (NO_2^-) oxidation states.

EXERCISE 22.7

Predict the products of the reactions, and write a balanced equation for each:

(a) $SiO_2(s) + H^+(aq) \longrightarrow$
(b) $NO(g) + O_2(g) \longrightarrow$
(c) $SO_3(g) + H_2O(l) \longrightarrow$
(d) $H_2O_2(aq) + I^-(aq) \longrightarrow$

Answer

(a) $SiO_2(s) + H^+(aq) \longrightarrow$ no reaction
(b) $2NO(g) + O_2(g) \longrightarrow 2NO_2(g)$
(c) $SO_3(g) + H_2O(l) \longrightarrow H_2SO_4(aq)$
(d) $H_2O_2(aq) + 2I^-(aq) \longrightarrow I_2(aq) + 2OH^-(aq)$

Reactions and Compounds of the Heavier Chalcogens

Because most of the heavier chalcogens (Group 16) and pnicogens (Group 15) are nonmetals, they often form similar compounds. For example, both third-period elements of these groups (phosphorus and sulfur) form catenated compounds and form multiple allotropes. Consistent with periodic trends, the tendency to catenate decreases as we go down the column.

> **Note the pattern**
> *Just as with the other groups, as we go down the column in Group 16, the tendency to catenate, the strength of single bonds, and the reactivity decrease.*

Sulfur and selenium both form a fairly extensive series of catenated species. For example, elemental sulfur forms S_8 rings packed together in a complex "crankshaft" arrangement (see Figure 18.15), and molten sulfur contains long chains of sulfur atoms connected by S—S single bonds. Moreover, both sulfur and selenium form polysulfides (S_n^{2-}) and polyselenides (Se_n^{2-}), with $n \leq 6$. The only stable allotrope of tellurium is a silvery white substance whose properties and structure are similar to those of one of the selenium allotropes. Polonium, in contrast, shows no tendency to form catenated compounds. The striking decrease in structural complexity from sulfur to polonium is consistent with the decrease in the strength of single bonds and the increase in metallic character as we go down the group.

As in Group 15, the reactivity of elements in Group 16 decreases from lightest to heaviest. For example, selenium and tellurium react with most elements, but not as readily as sulfur does. As expected for nonmetals, sulfur, selenium, and tellurium do not react with water, aqueous acid, or aqueous base, but all dissolve in strongly oxidizing acids such as HNO_3 to form oxoacids such as H_2SO_4. In contrast to the other chalcogens, polonium behaves like a metal, dissolving in dilute HCl to form solutions that contain Po^{2+} ion.

Note the pattern

The stability of the highest oxidation state of the chalcogens decreases as we go down the column.

Fluorine reacts directly with all chalcogens except oxygen to produce the hexafluorides (YF_6), which are extraordinarily stable and unreactive compounds. Four additional stable fluorides of sulfur are known, with sulfur oxidation states ranging from +1 to +5 (Figure 22.13). In contrast, only two fluorides of selenium (SeF_6 and SeF_4) and only three of tellurium (TeF_4, TeF_6, and Te_2F_{10}) are known.

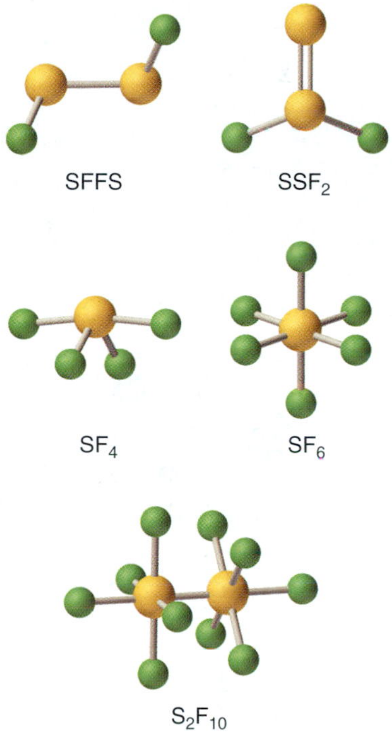

Figure 22.13 The structures of the known fluorides of sulfur. Five sulfur fluorides are known, containing sulfur in oxidation states ranging from +1 to +6. All are volatile molecular compounds that vary tremendously in stability and toxicity. Although both SF_6 and S_2F_{10} are very stable, S_2F_{10} is toxic and SF_6 is not. The other three are highly reactive substances.

Direct reaction of the heavier chalcogens with oxygen at elevated temperatures gives the dioxides (YO_2), which exhibit a dramatic range of structures and properties. As we go down the column, the dioxides become increasingly metallic in character, as expected, and the coordination number of the chalcogen steadily increases. Thus, SO_2 is a gas that contains V-shaped molecules (as predicted by the VSEPR model), SeO_2 is a white solid with an infinite chain structure (each Se is three coordinate), TeO_2 is a light yellow solid with a network structure in which each Te atom is four coordinate, and PoO_2 is a yellow ionic solid in which each Po^{4+} ion is eight coordinate.

The dioxides of sulfur, selenium, and tellurium react with water to produce the weak, diprotic oxoacids (H_2YO_3—sulfurous, selenous, and tellurous acid, respectively). Both sulfuric acid and selenic acid (H_2SeO_4) are strong acids, but telluric acid [$Te(OH)_6$] is quite different. Because tellurium is larger than either sulfur or selenium, it forms weaker π bonds to oxygen. As a result, the most stable structure for telluric acid is $Te(OH)_6$, with six Te—OH single bonds rather than Te=O double bonds. Telluric acid therefore behaves like a weak *triprotic* acid in aqueous solution, successively losing the hydrogen atoms bound to three of the oxygen atoms. As expected for compounds that contain elements in their highest accessible oxidation state (+6 in this case), sulfuric, selenic, and telluric acids are oxidants. Because the stability of the highest oxidation state decreases with increasing atomic number, telluric acid is a stronger oxidant than sulfuric acid.

Sulfur and, to a lesser extent, selenium react with carbon to form an extensive series of compounds that are structurally similar to their oxygen analogues. For example, CS_2 and CSe_2 are both volatile liquids that contain C=S or C=Se double bonds and have the same linear structure as CO_2. Because these double bonds are significantly weaker than the C=O double bond, however, CS_2, CSe_2, and related compounds are less stable and more reactive than their oxygen analogues. The chalcogens also react directly with nearly all metals to form compounds with a wide range of stoichiometries and a variety of structures. Metal chalcogenides can contain either the simple chalcogenide ion (Y^{2-}), as in Na_2S and FeS, or polychalcogenide ions (Y_n^{2-}), as in pyrite (FeS_2) and Na_2S_5.

Note the pattern

The dioxides of the Group 16 elements become increasingly basic and the coordination number of the chalcogen steadily increases down the column.

Ionic chalcogenides like Na_2S react with aqueous acid to produce *binary hydrides* such as hydrogen sulfide (H_2S). Because the strength of the Y—H bond decreases with increasing atomic radius, the stability of the binary hydrides decreases rapidly as we go down the column. It is perhaps surprising that hydrogen sulfide, with its familiar rotten-egg smell, is much more toxic than hydrogen cyanide (HCN), the gas used to execute prisoners in the "gas chamber." Hydrogen sulfide at relatively low concentrations deadens the olfactory receptors in the nose, which allows it to reach toxic levels without detection and makes it especially dangerous.

EXAMPLE 22.8

For each reaction, explain why the given product forms or no reaction occurs:

(a) $SO_2(g) + Cl_2(g) \longrightarrow SO_2Cl_2(l)$
(b) $SF_6(g) + H_2O(l) \longrightarrow$ no reaction
(c) $2Se(s) + Cl_2(g) \longrightarrow Se_2Cl_2(l)$

Given Balanced chemical equations

Asked for Why the given products (or no products) form

Strategy

Classify the type of reaction. Using periodic trends in atomic properties, thermodynamics, and kinetics, explain why the reaction products form or why no reaction occurs.

Solution

(a) One of the reactants, Cl_2, is an oxidant. If the other reactant can be oxidized, then a redox reaction is likely. Sulfur dioxide contains sulfur in the +4 oxidation state, which is 2 less than its maximum oxidation state. Sulfur dioxide is also known to be a mild reducing agent in aqueous solution, producing sulfuric acid as the oxidation product. Hence, a redox reaction is probable. The simplest reaction is the formation of SO_2Cl_2 (sulfuryl chloride), which, as shown in the margin, is a tetrahedral species with two S—Cl and two S═O bonds.

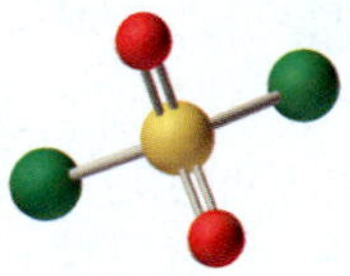

SO_2Cl_2

(b) Sulfur hexafluoride is a nonmetallic halide. Such compounds normally react vigorously with water to produce an oxoacid of the nonmetal and the corresponding hydrohalic acid. In this case, however, we have a highly stable species, presumably because all of sulfur's available orbitals are bonding orbitals. Thus, SF_6 is not likely to react with water.

(c) Here we have the reaction of a chalcogen with a halogen. The halogen is a good oxidant, so we can anticipate that a redox reaction will occur. Only fluorine is capable of oxidizing the chalcogens all the way to the +6 oxidation state, so we have to decide between $SeCl_4$ and Se_2Cl_2 as the product. The stoichiometry of the reaction determines which of the two is obtained, $SeCl_4$ or Se_2Cl_2.

EXERCISE 22.8

Predict the products of the reactions, and write a balanced equation for each:

(a) $Te(s) + Na(s) \xrightarrow{\Delta}$
(b) $SF_4(g) + H_2O(l) \longrightarrow$
(c) $CH_3SeSeCH_3(soln) + K(s) \longrightarrow$
(d) $Li_2Se(s) + H^+(aq) \longrightarrow$

Answer

(a) $Te(s) + 2Na(s) \longrightarrow Na_2Te(s)$
(b) $SF_4(g) + 3H_2O(l) \longrightarrow H_2SO_3(aq) + 4HF(aq)$
(c) $CH_3SeSeCH_3(soln) + 2K(s) \longrightarrow 2KCH_3Se(soln)$
(d) $Li_2Se(s) + 2H^+(aq) \longrightarrow H_2Se(g) + 2Li^+(aq)$

Note the pattern

Because the halogens are highly reactive, none is found in nature as the free element.

22.5 • The Elements of Group 17 (the Halogens)

Because the halogens are highly reactive, none is found in nature as the free element. Hydrochloric acid, which is a component of *aqua regia* (a mixture of HCl and HNO_3 that dissolves gold), and the mineral fluorspar (CaF_2) were well known to alchemists, who used them in their quest for gold. Despite their presence in familiar substances, none of the halogens was even recognized as an element until the 19th century.

Chlorine was the first halogen to be obtained in pure form. In 1774, Carl Wilhelm Scheele (the co-discoverer of oxygen) produced chlorine by the reaction of hydrochloric acid with manganese dioxide. Scheele was convinced, however, that the pale green gas he collected over water was a compound of oxygen and hydrochloric acid. In 1811, Scheele's "compound" was identified as a new element, named from the Greek *chloros*, "yellowish green" (the same stem as in *chlorophyll*, the green pigment in plants). That same year, a French industrial chemist, Bernard Courtois, accidentally added too much sulfuric acid to the residue obtained from burned seaweed. A deep purple vapor was released, which had a biting aroma similar to that of Scheele's "compound." The purple substance was identified as a new element, named iodine from the Greek *iodes*, "violet." Bromine was discovered soon after by a young French chemist, Antoine Jérôme Balard, who isolated a deep red liquid with a strong chlorine-like odor from brine from the salt marshes near Montpellier in southern France. Because many of its properties were intermediate between those of chlorine and iodine, Balard initially thought he had isolated a compound of the two (perhaps ICl). He soon realized, however, that he had discovered a new element, which he named bromine from the Greek *bromos*, "stench." Currently, organic chlorine compounds, such as PVC (polyvinylchloride), consume about 70% of the Cl_2 produced annually, and organobromine compounds are used in much smaller quantities, primarily as fire retardants.

Because of the unique properties of its compounds, fluorine was believed to exist long before it was actually isolated. The mineral fluorspar (now called *fluorite*, CaF_2) had been used since the 16th century as a "flux," a low-melting-point substance that could dissolve other minerals and ores. In 1670, a German glass cutter discovered that heating fluorspar with strong acid produced a solution that could etch glass. The solution was later recognized to contain the acid of a new element, which was named fluorine in 1812. Elemental fluorine proved to be very difficult to isolate, however, because both HF and F_2 are extraordinarily reactive and toxic. After being poisoned three times while trying to isolate the element, the French chemist Henri Moissan succeeded in 1886 in electrolyzing a sample of KF in anhydrous HF to produce a pale green gas (Figure 22.14). For this achievement, among others, Moissan narrowly defeated Mendeleev for the Nobel Prize in chemistry in 1906. Large amounts of fluorine are now consumed in the production

A crystal of the mineral fluorite (CaF_2). The purple color of some fluorite crystals is due to small inclusions of highly oxidizing impurities, which generate detectable amounts of ozone when the crystals are crushed.

Figure 22.14 The French chemist Henri Moissan was the first person to isolate elemental fluorine. (a) Henri Moissan (1852–1907) in his laboratory in Paris, next to the electric furnace that he invented to carry out reactions at high temperature. (b) A reproduction of the U-shaped electrolysis cell with which Moissan first isolated elemental fluorine in 1866, shown with samples of cryolite (left) and fluorspar (right). Fluorspar is the raw material from which anhydrous hydrofluoric acid (HF) is prepared. Cryolite is a rare mineral that contains fluoride ion.

(a)

(b)

of *cryolite* (Na_3AlF_6), a key intermediate in the production of aluminum metal. Fluorine is also found in teeth as *fluoroapatite* [$Ca_5(PO_4)_3F$], which is formed by the reaction of hydroxyapatite [$Ca_5(PO_4)_3OH$] in tooth enamel with fluoride ions in toothpastes, rinses, and drinking water.

The heaviest halogen, astatine (At), is continuously produced by natural radioactive decay. All its isotopes are highly radioactive, and the most stable has a half-life of only about 8 hours. Consequently, astatine is the least abundant naturally occurring element on Earth, with less than 30 g estimated to be present in the Earth's crust at any one time.

Preparation and General Properties of Group 17 Elements

All the halogens except iodine are found in nature as salts of the halide ions (X^-), so the methods used for preparing F_2, Cl_2, and Br_2 all involve oxidizing the halide. The reaction of CaF_2 with concentrated sulfuric acid produces gaseous hydrogen fluoride. Fluorine is produced by the electrolysis of a 1:1 mixture of HF and $K^+HF_2^-$ at 60–300°C in an apparatus made of Monel, a highly corrosion-resistant nickel–copper alloy:

$$CaF_2(s) + H_2SO_4(l) \longrightarrow CaSO_4(s) + 2HF(g) \tag{22.37a}$$

$$KHF_2 \cdot HF(l) \xrightarrow{\text{Electrolysis}} F_2(g) + H_2(g) \tag{22.37b}$$

Fluorine is one of the most powerful oxidants known, and both F_2 and HF are highly corrosive. Consequently, the production, storing, shipping, and handling of these gases pose major technical challenges.

Although chlorine is significantly less abundant than fluorine, elemental chlorine is produced on an enormous scale. Fortunately, large subterranean deposits of rock salt (NaCl) are found around the world (Figure 22.15), and seawater consists of about 2% NaCl by mass, providing an almost inexhaustible reserve. Inland salt lakes such as the Dead Sea and the Great Salt Lake are even richer sources, containing about 23% and 8% NaCl by mass, respectively. Chlorine is prepared industrially by the chlor-alkali process, which uses the reaction

$$2NaCl(aq) + 2H_2O(l) \xrightarrow{\text{Electrolysis}} 2NaOH(aq) + Cl_2(g) + H_2(g) \tag{22.38}$$

Bromine is much less abundant than fluorine or chlorine, but it is easily recovered from seawater, which contains about 65 mg of Br^- per liter. Salt lakes and underground brines are even richer sources; for example, the Dead Sea contains 4 g of Br^-

Figure 22.15 A subterranean salt mine. Subterranean deposits of rock salt are located worldwide, such as this one at Petralia in Sicily.

per liter. Iodine is the least abundant of the nonradioactive halogens, and it is a relatively rare element. Because of its low electronegativity, iodine tends to occur in nature in an oxidized form. Hence, most commercially important deposits of iodine, such as those in the Chilean desert, are iodate salts such as $Ca(IO_3)_2$. The production of iodine from such deposits therefore requires *reduction* rather than oxidation. The process is typically carried out in two steps: reduction of iodate to iodide with sodium hydrogen sulfite, followed by reaction of iodide with additional iodate:

$$2IO_3^-(aq) + 6HSO_3^-(aq) \longrightarrow 2I^-(aq) + 6SO_4^{2-}(aq) + 6H^+(aq) \quad (22.39)$$

$$5I^-(aq) + IO_3^-(aq) + 6H^+(aq) \longrightarrow 3I_2(s) + 3H_2O(l) \quad (22.40)$$

Because the halogens all have $ns^2\,np^5$ electron configurations, their chemistry is dominated by a tendency to accept an additional electron to form the closed-shell ion (X^-). Only the electron affinity and the bond dissociation energy of fluorine differ significantly from the expected periodic trends shown in Table 22.6. Because of the small atomic volume of fluorine, repulsive electron–electron interactions are important in the fluoride ion, making the electron affinity of fluorine lower than that of chlorine. Similarly, repulsions between electron pairs on adjacent atoms are responsible for the unexpectedly low F—F bond dissociation energy. (As discussed earlier, this effect is also responsible for the weakness of O—O, N—N, and N—O single bonds.)

Note the pattern

Electrostatic repulsions between lone pairs of electrons on adjacent atoms cause single bonds between N, O, and F to be weaker than expected.

TABLE 22.6 Selected properties of the Group 17 elements

Property	Fluorine	Chlorine	Bromine	Iodine	Astatine
Atomic symbol	F	Cl	Br	I	At
Atomic number	9	17	35	53	85
Atomic mass (amu)	19.00	35.45	79.90	126.90	210
Valence-electron configuration[a]	$2s^2 2p^5$	$3s^2 3p^5$	$4s^2 4p^5$	$5s^2 5p^5$	$6s^2 6p^5$
mp/bp (°C)	−220/−188	−102/−34.0	−7.2/58.8	114/184	302/—
Density (g/cm^3) at 25°C	1.55 (g/L)	2.90 (g/L)	3.10	4.93	—
Atomic radius (pm)	42	79	94	115	127
First ionization energy (kJ/mol)	1681	1251	1140	1008	926
Normal oxidation state(s)	−1	−1 (+1, +3, +5, +7)	−1 (+1, +3, +5, +7)	−1 (+1, +3, +5, +7)	−1, +1
Ionic radius (pm)[b]	133	181	196	220	—
Electron affinity (kJ/mol)	−328	−349	−325	−295	−270
Electronegativity	4.0	3.2	3.0	2.7	2.2
Standard reduction potential, $E°$ (V) ($X_2 \rightarrow X^-$ in basic solution)	+2.87	+1.36	+1.07	+0.54	+0.30
Dissociation energy of $X_2(g)$ (kJ/mol)	158.8	243.6	192.8	151.1	~80
Product of reaction with O_2	O_2F_2	None	None	None	None
Type of oxide	Acidic	Acidic	Acidic	Acidic	Acidic
Product of reaction with N_2	None	None	None	None	None
Product of reaction with H_2	HF	HCl	HBr	HI	HAt

[a]The configuration shown does not include filled *d* and *f* subshells.

[b]The values cited are for the six-coordinate anion, X^-.

Because it is the most electronegative element in the periodic table, fluorine forms compounds in *only* the −1 oxidation state. Notice, however, that all the halogens except astatine have electronegativities higher than 2.5, making their chemistry exclusively that of nonmetals. The halogens all have relatively high ionization energies, but the energy required to remove electrons decreases substantially as we go down the column. Hence, the heavier halogens also form compounds in positive oxidation states (+1, +3, +5, and +7), derived by the formal loss of *ns* and *np* electrons.

Note the pattern

Because ionization energies decrease as we go down the column, the heavier halogens form compounds in positive oxidation states (+1, +3, +5, and +7).

Reactions and Compounds of the Halogens

Fluorine is the most reactive element in the periodic table, forming compounds with every other element except helium, neon, and argon. The reactions of fluorine with most other elements range from vigorous to explosive; only O_2, N_2, and Kr react slowly. There are three reasons for the high reactivity of fluorine:

1. Because fluorine is so electronegative, it is able to remove or at least share the valence electrons of virtually any other element.
2. Because of its small size, fluorine tends to form very strong bonds to other elements, making its compounds thermodynamically stable.
3. The F—F bond is weak due to repulsion between lone pairs of electrons on adjacent atoms, reducing both the thermodynamic and kinetic barriers to reaction.

With highly electropositive elements, fluorine forms ionic compounds that contain the closed-shell F^- ion. In contrast, with less electropositive elements (or with metals in very high oxidation states), fluorine forms covalent compounds that contain terminal F atoms. Because of its high electronegativity and $2s^2\,2p^5$ valence-electron configuration, fluorine normally participates in only one electron-pair bond. Only a very strong Lewis acid, such as AlF_3, can share a lone pair of electrons with a fluoride ion.

The halogens (X_2) react with metals (M) according to the general equation

$$M(s,l) + nX_2(s,l,g) \longrightarrow MX_n(s,l) \qquad (22.41)$$

For elements that exhibit multiple oxidation states, fluorine tends to produce the highest possible oxidation state and iodine the lowest. For example, vanadium reacts with the halogens to give VF_5, VCl_4, VBr_4, and VI_3.

Metal halides in the +1 or +2 oxidation state are typically ionic halides, which have high melting points and are often soluble in water. As the oxidation state of the metal increases, so does the covalent character of the halide due to polarization of the M—X bond. With its high electronegativity, fluoride is the least polarizable, and iodide, with the lowest electronegativity, is the most polarizable of the halogens. Halides of small trivalent metal ions such as Al^{3+} tend to be relatively covalent. For example, $AlBr_3$ is a volatile solid that contains bromide-bridged Al_2Br_6 molecules. In contrast, the halides of larger trivalent metals such as the lanthanides are essentially ionic. For example, indium tribromide ($InBr_3$) and the lanthanide tribromides ($LnBr_3$) are all high-melting-point solids that are quite soluble in water.

Note the pattern

As the oxidation state of metal increases, the covalent character of the corresponding metal halides also increases due to polarization of the M—X bond.

All the halogens react vigorously with hydrogen to give the hydrogen halides (HX). Because the H—F bond in HF is highly polarized ($H^{\delta+}$—$F^{\delta-}$), liquid HF is extensively hydrogen bonded, giving it an unusually high boiling point and a high dielectric constant. As a result, liquid HF is a polar solvent that is similar in some ways to water and liquid ammonia; after a reaction, the products can be recovered simply by evaporating the HF solvent. (Hydrogen fluoride must be handled with extreme caution, however, because contact of HF with skin causes extraordinarily painful burns that are slow to heal.) Because fluoride has a high affinity for silicon, aqueous hydrofluoric acid is used to etch glass, dissolving SiO_2 to give solutions of the stable SiF_6^{2-} ion.

Except for fluorine, all the halogens react with water in a disproportionation reaction:

$$X_2(g,l,s) + H_2O(l) \longrightarrow H^+(aq) + X^-(aq) + HOX(aq) \quad (X = Cl, Br, I) \quad (22.42)$$

The most stable oxoacids are the *perhalic acids*, which contain the halogens in their highest oxidation state (+7). The acid strengths of the oxoacids of the halogens increase with increasing oxidation state, whereas their stability and acid strength decrease as we go down the group. Thus, perchloric acid ($HOClO_3$, usually written as simply $HClO_4$), is a more potent acid and stronger oxidant than perbromic acid. Although all the oxoacids are strong oxidants, some, such as $HClO_4$, react rather slowly at low temperatures. Consequently, mixtures of the halogen oxoacids or oxoanions with organic compounds are potentially explosive if they are heated or even agitated mechanically to initiate the reaction. Because of the danger of explosions, *oxoacids and oxoanions of the halogens should never be allowed to come into contact with organic compounds*.

Note the pattern

As we go down the column, both the acid strength and the oxidizing power of the halogen oxoacids decrease.

The halogens react with one another to produce *interhalogen compounds* such as ICl_3, BrF_5, and IF_7. In all cases, the heavier halogen, which has the lower electronegativity, is the central atom. The maximum oxidation state and the number of terminal halogens increase smoothly as the ionization energy of the central halogen decreases and the electronegativity of the terminal halogen increases. Thus, depending on conditions, iodine reacts with the other halogens to form IF_n (n = 1–7), ICl or ICl_3, or IBr, whereas bromine reacts with fluorine to form only BrF, BrF_3, and BrF_5, but *not* BrF_7. The interhalogen compounds are among the most powerful Lewis acids known, with a strong tendency to react with halide ions to give complexes with higher coordination numbers, such as the IF_8^- ion:

$$IF_7(l) + KF(s) \longrightarrow KIF_8(s) \quad (22.43)$$

Notice that all of the Group 17 elements except fluorine form compounds in odd oxidation states (−1, +1, +3, +5, +7). The interhalogen compounds are also potent oxidants and strong fluorinating agents; contact with organic materials or water can result in an explosion!

EXAMPLE 22.9

For each reaction, explain why the given products form:

(a) $ClF_3(g) + Cl_2(g) \longrightarrow 3ClF(g)$
(b) $2KI(s) + 3H_2SO_4(aq) \longrightarrow I_2(aq) + SO_2(g) + 2KHSO_4(aq) + 2H_2O(l)$
(c) $Pb(s) + 2BrF_3(l) \longrightarrow PbF_4(s) + 2BrF(g)$

Given Balanced chemical equations

Asked for Why the given products form

Strategy

Classify the type of reaction. Using periodic trends in atomic properties, thermodynamics, and kinetics, explain why the observed reaction products form.

Solution

(a) When the reactants have the same element in two different oxidation states, we expect the product to contain that element in an intermediate oxidation state. We have Cl^{3+} and Cl^0 as reactants, so a possible product would contain Cl in either the +1 or +2 oxidation state. From our discussion, we know that +1 is much more likely. In this case, notice that Cl_2 is behaving like a *reductant* rather than an oxidant.

(b) At first glance, this appears to be a simple acid–base reaction, in which sulfuric acid transfers a proton to I^- to form HI. Recall, however, that I^- can be oxidized to I_2. Sulfuric acid contains sulfur in its highest oxidation state (+6), so it is a good oxidant. In this case, the redox reaction predominates.

Note the pattern

All the Group 17 elements except fluorine form compounds in odd oxidation states (−1, +1, +3, +5, +7), but the importance of the higher oxidation states generally decreases as we go down the column.

(c) This is the reaction of a metallic element with a very strong oxidant. Consequently, a redox reaction will occur. The only question is whether lead will be oxidized to Pb(II) or Pb(IV). Because BrF_3 is a powerful oxidant and fluorine is able to stabilize high oxidation states of other elements, it is likely that PbF_4 will be the product. The two possible reduction products for BrF_3 are BrF and Br_2. The actual product will likely depend on the ratio of the reactants used. With excess BrF_3, we expect the more oxidized product, BrF. With lower ratios of oxidant to lead, we would probably obtain Br_2 as the product.

EXERCISE 22.9

Predict the products of the reactions, and write a balanced equation for each:

(a) $CaCl_2(s) + H_3PO_4(l) \longrightarrow$
(b) $GeO_2(s) + HF(aq) \longrightarrow$
(c) $Fe_2O_3(s) + HCl(g) \xrightarrow{\Delta}$
(d) $NaClO_2(aq) + Cl_2(g) \longrightarrow$

Answer

(a) $CaCl_2(s) + 2H_3PO_4(l) \longrightarrow 2HCl(g) + Ca(HPO_4)_2(soln)$
(b) $GeO_2(s) + 6HF(aq) \longrightarrow GeF_6{}^{2-}(aq) + 2H_2O(l) + 2H^+(aq)$
(c) $Fe_2O_3(s) + 6HCl(g) \xrightarrow{\Delta} 2FeCl_3(s) + 3H_2O(g)$
(d) $2NaClO_2(aq) + Cl_2(g) \longrightarrow 2ClO_2(g) + 2NaCl(aq)$

22.6 The Elements of Group 18 (the Noble Gases)

Sir William Ramsay (1852–1916). Born and educated in Glasgow, Scotland, Ramsay was expected to study for the Calvanist ministry. Instead, he became interested in chemistry while reading about the manufacture of gunpowder. Ramsay earned his Ph.D. in organic chemistry at the University of Tübingen in Germany in 1872. When he returned to England, his interests turned first to physical chemistry and then to inorganic chemistry. He is best known for his work on the oxides of nitrogen and for the discovery of the noble gases with Lord Rayleigh.

The noble gases were all isolated for the first time within a period of only five years at the end of the 19th century. Their very existence was not suspected until the 18th century, when early work on the composition of air suggested that it contained small amounts of gases in addition to oxygen, nitrogen, carbon dioxide, and water vapor. Helium was the first of the noble gases to be identified, when the existence of this previously unknown element on the sun was demonstrated by new spectral lines seen during a solar eclipse in 1868 (see Chapter 6). Samples of helium were not obtained on Earth until almost 30 years later, however. In the 1890s, the English physicist J. W. Strutt (Lord Rayleigh) carefully measured the density of the gas that remained after he had removed all O_2, CO_2, and water vapor from air, and showed that this residual gas was slightly denser than pure N_2 obtained by the thermal decomposition of ammonium nitrite. In 1894, he and the Scottish chemist William Ramsay announced the isolation of a new "substance" (not necessarily a new element) from the residual nitrogen gas. Because they could not force this substance to decompose or react with anything, they named it *argon* (Ar), from the Greek *argos*, "lazy." Because the measured molar mass of argon was 39.9 g/mol, Ramsay speculated that it was a member of a new group of elements located on the right side of the periodic table between the halogens and the alkali metals. He also suggested that these elements should have a preferred valence of 0, intermediate between the +1 of the alkali metals and the −1 of the halogens.

In 1895, Ramsey was able to obtain a terrestrial sample of helium for the first time. Then, in a single year (1898), he discovered the next three noble gases: krypton (Kr), from the Greek *kryptos*, "hidden," was identified by its orange and green emission lines; neon (Ne), from the Greek *neos*, "new," had bright red emission lines; and xenon (Xe), from the Greek *xenos*, "strange," had deep blue emission lines. The last noble gas was discovered in 1900 by the German chemist Friedrich Dorn, who was investigating radioactivity in the air around the newly discovered radioactive elements, radium and polonium. The element was named radon (Rn), and in 1908, Ramsay succeeded in obtaining enough radon to measure its density (and thus its atomic mass). For their discovery of the noble gases, Rayleigh was awarded the Nobel Prize in physics and Ramsay in chemistry in 1904.

Because helium has the lowest boiling point of any substance known (4.2 K), it is used primarily as a cryogenic liquid. Helium and argon are both much less soluble in water (and therefore in blood) than N_2, so scuba divers often use gas mixtures that contain these gases instead of N_2 to minimize the likelihood of the "bends," the painful and potentially fatal formation of bubbles of $N_2(g)$ that can occur when a diver returns to the surface too rapidly.

Preparation and General Properties of Group 18 Elements

Fractional distillation of liquid air is the only source of all the noble gases except helium. Although helium is the second most abundant element in the universe (after hydrogen), the helium originally present in the Earth's atmosphere was lost into space long ago because of its low molecular mass and resulting high mean velocity. Natural gas often contains relatively high concentrations of helium (up to 7%), however, and it is the only practical terrestrial source.

The elements of Group 18 all have closed-shell valence-electron configurations, either $ns^2\ np^6$ for He or $1s^2$. Consistent with periodic trends in atomic properties, these elements have high ionization energies that decrease smoothly as we go down the column. From their electron affinities, the data in Table 22.7 indicate that the noble

TABLE 22.7 Selected properties of the Group 18 elements

Property	Helium	Neon	Argon	Krypton	Xenon	Radon
Atomic symbol	He	Ne	Ar	Kr	Xe	Rn
Atomic number	2	10	18	36	54	86
Atomic mass (amu)	4.00	20.18	39.95	83.80	131.29	222
Valence-electron configuration[a]	$1s^2$	$2s^2\ 2p^6$	$3s^2\ 3p^6$	$4s^2\ 4p^6$	$5s^2\ 5p^6$	$6s^2\ 6p^6$
Triple point/bp (°C)	—/-269[b]	-249 (43 kPa)/-246	-189 (69 kPa)/-189	-157/-153	-112 (81.6 kPa)/-108	-71/-62
Density (g/L) at 25°C	0.16	0.83	1.63	3.43	5.37	9.07
Atomic radius (pm)	31	38	71	88	108	120
First ionization energy (kJ/mol)	2372	2081	1521	1351	1170	1037
Normal oxidation state(s)	0	0	0	0 (+2)	0 (+2, +4, +6, +8)	0 (+2)
Electron affinity (kJ/mol)	≥ 0	≥ 0	≥ 0	≥ 0	≥ 0	≥ 0
Electronegativity	—	—	—	—	2.6	—
Product of reaction with O_2	None	None	None	None	None	None
Type of oxide	—	—	—	—	Acidic	—
Product of reaction with N_2	None	None	None	None	None	None
Product of reaction with X_2	None	None	None	KrF_2	XeF_2, XeF_4, XeF_6	RnF_2
Product of reaction with H_2	None	None	None	None	None	None

[a]The configuration shown does not include filled *d* and *f* subshells.

[b]This is the normal boiling point of He. Solid He does not exist at 1 atm pressure, hence no melting point can be given.

gases are unlikely to form compounds in negative oxidation states. Notice that a potent oxidant is needed to oxidize noble gases and form compounds in positive oxidation states. Like the heavier halogens, xenon and perhaps krypton should form covalent compounds with F, O, and possibly Cl, in which they have *even* formal oxidation states (+2, +4, +6, and possibly +8). These predictions actually summarize the chemistry observed for these elements.

Figure 22.16 The structure of xenon hydrate, an example of a clathrate. Small gaseous atoms or molecules such as Xe or CH_4 can occupy cavities in a lattice of hydrogen-bonded water molecules to produce a stable structure with a fixed stoichiometry (in this case, $Xe{\cdot}5.75H_2O$). (The hydrogen atoms of the water molecules have been omitted for clarity.) Warming the solid hydrate or decreasing the pressure of gas causes it to collapse, with the evolution of gas and formation of liquid water.

Reactions and Compounds of the Noble Gases

For many years, it was thought that the only compounds the noble gases could form were *clathrates*. Clathrates are solid compounds in which a gas, the guest, occupies holes in a lattice formed by a less volatile, chemically dissimilar substance, the host (Figure 22.16). Because clathrate formation does not involve the formation of chemical bonds between the guest (Xe) and the host molecules (H_2O, in the case of xenon hydrate), the guest molecules are immediately released when the clathrate is melted or dissolved. In addition to the noble gases, many other species form stable clathrates. One of the most interesting is methane hydrate, large deposits of which occur naturally at the bottom of the ocean. It is estimated that the amount of methane in such deposits could have a major impact on the world's energy needs later in this century.

"Burning snowballs." Like xenon, methane (CH_4) forms a crystalline clathrate with water, methane hydrate. When the solid is warmed, methane is released and can be ignited to give what appears to be burning snow.

The widely held belief in the intrinsic lack of reactivity of the noble gases was challenged when Neil Bartlett, a British professor of chemistry at the University of British Columbia, showed that PtF_6, a compound used in the Manhattan Project, could oxidize O_2. Because the ionization energy of xenon (1170 kJ/mol) is actually *lower* than that of O_2, Bartlett recognized that PtF_6 should also be able to oxidize xenon. When he mixed colorless xenon gas with deep red PtF_6 vapor, yellow-orange crystals formed immediately (Figure 22.17). Although Bartlett initially postulated that they were $Xe^+PtF_6^-$, it is now generally agreed that the reaction also involves the transfer of a fluorine atom to xenon to give the XeF^+ ion:

$$Xe(g) + PtF_6(g) \longrightarrow [XeF^+][PtF_5^-](s) \qquad (22.44)$$

Subsequent work showed that xenon reacts directly with fluorine under relatively mild conditions to give XeF_2, XeF_4, or XeF_6, depending on conditions; one such reaction is

$$Xe(g) + 2F_2(g) \longrightarrow XeF_4(s) \qquad (22.45)$$

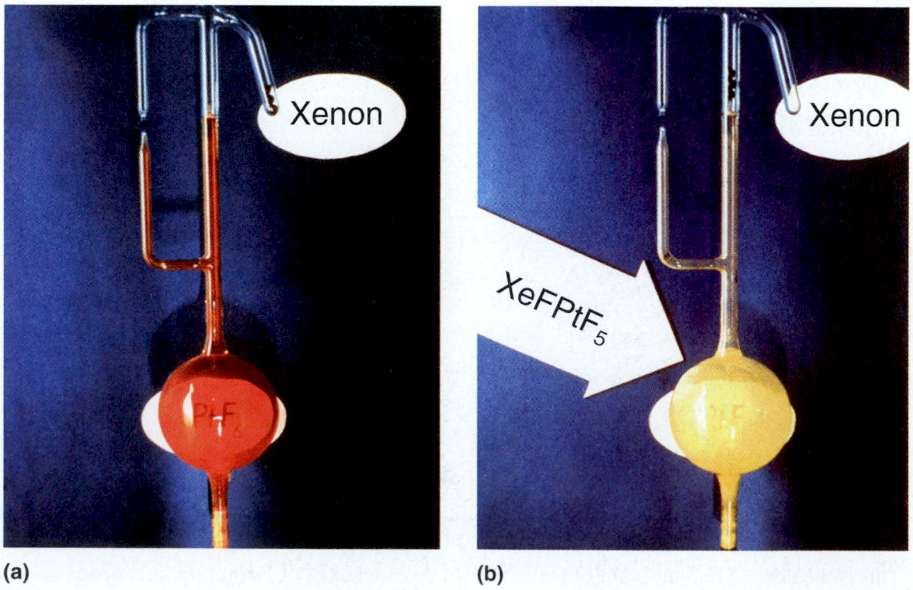

Figure 22.17 The synthesis of the first chemical compound of xenon. (a) An apparatus containing platinum hexafluoride, the red vapor at the bottom left, and xenon, the colorless gas in the small tube at the upper right. (b) When the glass seal separating the two gases is broken and the gases are allowed to mix, a bright yellow solid is formed, which is best described as $XeF^+PtF_5^-$.

The ionization energies of helium, neon, and argon are so high (Table 22.7) that no stable compounds of these elements are known. The ionization energies of krypton and xenon are lower, but still very high, and consequently only highly electronegative elements (F, O, and Cl) can form stable compounds with xenon and krypton without being oxidized themselves. Xenon reacts directly with only two elements, F_2 and Cl_2. Although $XeCl_2$ and KrF_2 can be prepared directly from the elements, they are substantially less stable than the xenon fluorides.

Note the pattern

The ionization energies of helium, neon, and argon are so high that no stable compounds of these elements are known.

Because halides of the noble gases are powerful oxidants and fluorinating agents, they decompose rapidly upon contact with trace amounts of water, and they react violently with organic compounds or other reductants. The xenon fluorides are also Lewis acids; they react with fluoride ion, the only Lewis base that is not oxidized immediately upon contact, to form anionic complexes. For example, the reaction of cesium fluoride with XeF_6 produces $CsXeF_7$, which gives Cs_2XeF_8 when heated:

$$XeF_6(s) + CsF(s) \longrightarrow CsXeF_7(s) \tag{22.46a}$$

$$2CsXeF_7(s) \xrightarrow{\Delta} Cs_2XeF_8(s) + XeF_6(g) \tag{22.46b}$$

The XeF_8^{2-} ion contains eight-coordinate xenon and has the square antiprismatic structure shown in the margin, which is essentially identical to that of the IF_8^- ion. Cs_2XeF_8 is surprisingly stable for a polyatomic ion that contains xenon in the +6 oxidation state, decomposing only at temperatures higher than 300°C. Major factors in the stability of Cs_2XeF_8 are almost certainly the formation of a stable ionic lattice and the high coordination number of xenon, which protects the central atom from attack by other species. (Recall from Section 22.4 that latter effect is responsible for the extreme stability of SF_6.)

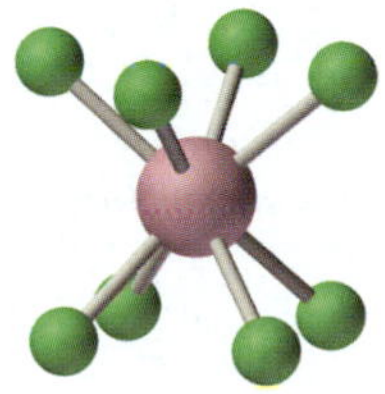

XeF_8^{2-}

Note the pattern

Xenon has a high affinity for both fluorine and oxygen.

For a previously "inert" gas, xenon has a surprisingly high affinity for oxygen, presumably because of π bonding between O and Xe. Consequently, xenon forms an extensive series of oxides and oxoanion salts. For example, hydrolysis of either XeF_4 or XeF_6 produces XeO_3, an explosive white solid:

$$XeF_6(aq) + 3H_2O(l) \longrightarrow XeO_3(aq) + 6HF(aq) \tag{22.47}$$

Treating a solution of XeO_3 with ozone results in further oxidation of xenon to give either XeO_4, a colorless, explosive gas, or the surprisingly stable perxenate ion (XeO_6^{4-}), both of which contain xenon in its highest possible oxidation state, +8. The chemistry of the xenon halides and oxides is best understood by analogy to the corresponding compounds of iodine. For example, XeO_3 is isoelectronic with the iodate ion (IO_3^-), and XeF_8^{2-} is isoelectronic with the IF_8^- ion.

Because the ionization energy of radon is lower than that of xenon, in principle radon should be able to form an even greater variety of chemical compounds than xenon. Unfortunately, however, radon is so radioactive that its chemistry has not been extensively explored.

EXAMPLE 22.10

On a virtual planet similar to Earth, at least one isotope of radon is *not* radioactive. A chemist explored its chemistry and presented her major conclusions in a trailblazing paper on radon compounds, focusing on the kinds of compounds formed and their stoichiometries. Based on periodic trends, how did she summarize the chemistry of radon?

Given Nonradioactive of radon

Asked for Summary of its chemistry

Strategy

Based on the position of radon in the periodic table and periodic trends in atomic properties, thermodynamics, and kinetics, predict the most likely reactions and compounds of radon.

Solution

We expect radon to be significantly easier to oxidize than xenon. Based on its position in the periodic table, however, we also expect its bonds to other atoms to be weaker than those formed by xenon. Radon should be more difficult to oxidize to its highest possible oxidation state (+8) than xenon because of the inert-pair effect. Consequently, radon should form an extensive series of fluorides, including RnF_2, RnF_4, RnF_6, and possibly RnF_8 (due to its large radius). The ion RnF_8^{2-} should also exist. We expect radon to form a series of oxides similar to those of xenon, including RnO_3 and possibly RnO_4. The biggest surprise in radon chemistry is likely to be the existence of stable chlorides, such as $RnCl_2$ and possibly even $RnCl_4$.

EXERCISE 22.10

Predict the stoichiometry of the product formed by the reaction of XeF_6 with a 1:1 stoichiometric amount of KF, and propose a reasonable structure for the anion.

Answer $KXeF_7$. The xenon atom in XeF_7^- has 16 valence electrons, which according to the VSEPR model could give either a square antiprismatic structure with one F atom missing or a pentagonal bipyramid if the $5s^2$ electrons behave like an inert pair that does not participate in bonding.

SUMMARY AND KEY TERMS

22.1 The Elements of Group 13 (p. 990)

Isolation of the Group 13 elements requires a large amount of energy because compounds of the Group 13 elements with oxygen are thermodynamically stable. Boron behaves chemically like a nonmetal, whereas its heavier congeners exhibit metallic behavior. Many of the inconsistencies observed in the properties of the Group 13 elements can be explained by the increase in Z_{eff} that arises from poor shielding of the nuclear charge by the filled $(n-1)d^{10}$ and $(n-2)f^{14}$ subshells. Instead of forming a metallic lattice with delocalized valence electrons, boron forms unique aggregates that contain multicenter bonds, including *metal borides*, in which boron is bonded to other boron atoms to form three-dimensional networks or *clusters* with regular geometric structures. All neutral compounds of the Group 13 elements are electron deficient and behave like Lewis acids. The trivalent halides of the heavier elements form halogen-bridged dimers that contain electron-pair bonds, rather than the delocalized electron-deficient bonds characteristic of diborane. Their oxides dissolve in dilute acid, although the oxides of aluminum and gallium are amphoteric. None of the Group 13 elements reacts directly with hydrogen, and the stability of the hydrides prepared by other routes decreases as we go down the group. In contrast to boron, the heavier Group 13 elements form a large number of complexes in the +3 oxidation state.

22.2 The Elements of Group 14 (p. 997)

The Group 14 elements show the greatest range of chemical behavior of any group in the periodic table. Because the covalent bond strength decreases with increasing atomic size and higher than expected ionization energies due to an increase in Z_{eff}, the stability of the +2 oxidation state increases from carbon to lead. The tendency to form multiple bonds and catenates decreases as the atomic number increases. The stability of the carbon tetrahalides decreases as the halogen increases in size because of poor orbital overlap and steric crowding. With less electronegative elements, carbon forms three kinds of carbides: *ionic carbides*, which contain metal cations and C^{4-} (methide) or C_2^{2-} (acetylide) anions; *interstitial carbides*, which are characterized by covalent metal–carbon interactions and are among the hardest substances known; and *covalent carbides*, which have three-dimensional covalent network structures that make them extremely hard, high melting, and chemically inert. Consistent with periodic trends, metallic behavior increases as we go down the group. Silicon has a tremendous affinity for oxygen because of partial Si—O π bonding. Dioxides of the Group 14 elements become increasingly basic as we go down the group and their metallic character increases. *Silicates* contain anions that consist of only silicon and oxygen. *Aluminosilicates* are formed by replacing some of the Si atoms in silicates by Al atoms; aluminosilicates with three-dimensional framework structures are called *zeolites*. *Nitrides* formed by the reaction of silicon or germanium with nitrogen are strong, hard, and chemically inert. The hydrides become thermodynamically less stable as we go down the group. Moreover, as atomic size increases, multiple bonds between or to the Group 14 elements become weaker. *Silicones*, which contain an Si—O backbone and Si—C bonds, are high-molecular-mass polymers whose properties depend on their compositions.

22.3 The Elements of Group 15 (the Pnicogens) (p. 1006)

In Group 15, nitrogen and phosphorus behave chemically like nonmetals, arsenic and antimony like semimetals, and bismuth like a metal. Nitrogen forms compounds in nine different oxidation states.

The stability of the +5 oxidation state decreases from phosphorus to bismuth because of the inert-pair effect. Due to their higher electronegativity, the lighter pnicogens form compounds in the −3 oxidation state. Because of the presence of a lone pair of electrons on the pnicogen, neutral covalent compounds of the trivalent pnicogens are Lewis bases. Nitrogen does not form stable catenated compounds because of repulsions between lone pairs of electrons on adjacent atoms, but it does form multiple bonds with other second-period atoms. Nitrogen reacts with electropositive elements to produce solids that range from covalent to ionic in character. Reaction with electropositive metals produces *ionic nitrides*, reaction with less electropositive metals produces *interstitial nitrides*, and reaction with semimetals produces *covalent nitrides*. The reactivity of the pnicogens decreases with increasing atomic number. Compounds of the heavier pnicogens often have coordination numbers of 5 or higher and use dsp^3 or d^2sp^3 hybrid orbitals for bonding. Because phosphorus and arsenic have energetically accessible *d* orbitals, these elements form π bonds with second-period atoms such as O and N. The reaction of phosphorus with metals produces *phosphides*. Metal-rich phosphides are hard, high-melting-point, electrically conductive solids with metallic luster, whereas phosphorus-rich phosphides, which contain catenated phosphorus units, are lower melting and less thermally stable.

22.4 The Elements of Group 16 (the Chalcogens) (p. 1014)

Because the electronegativity of the chalcogens decreases as we go down the group, so does their tendency to acquire two electrons to form compounds in the −2 oxidation state. The lightest member, oxygen, has the greatest tendency to form multiple bonds to other elements. It does not form stable catenated compounds, however, due to repulsions between lone pairs of electrons on adjacent atoms. Because of its high electronegativity, the chemistry of oxygen is generally restricted to compounds in which it has a negative oxidation state, and its bonds to other elements tend to be highly polar. Metal oxides are usually basic, and nonmetal oxides are acidic, whereas oxides of elements along the dividing line between metals and nonmetals are amphoteric. As we go down the column, the reactivity, the strength of multiple bonds to oxygen, and the tendency to form catenated compounds all decrease, whereas their maximum coordination numbers increase. Because Te—O double bonds are comparatively weak, the most stable oxoacid of tellurium contains six Te—OH single bonds. The stability of the highest oxidation state (+6) decreases as we go down the group. Double bonds between S or Se and second-row atoms are weaker than the analogous C═O double bonds because of reduced orbital overlap. The stability of the binary hydrides decreases as we go down the column.

22.5 The Elements of Group 17 (the Halogens) (p. 1020)

The halogens are so reactive that none is found in nature as the free element; instead, all but iodine are found as halide salts that contain the X^- ion. Their chemistry is exclusively that of nonmetals. Consistent with periodic trends, ionization energies decrease as we go down the group. Fluorine, the most reactive element in the periodic table, has a low F—F bond dissociation energy due to repulsions between lone pairs of electrons on adjacent atoms. Fluorine forms ionic compounds with electropositive elements and covalent compounds with less electropositive elements and metals in high oxidation states. All the halogens react with hydrogen to produce hydrogen halides. Except for F_2, all react with water to form oxoacids, including the *perhalic acids*, which contain the halogens in their highest oxidation state. Halogens also form *interhalogen compounds*; the heavier halogen, with the lower electronegativity, is the central atom.

22.6 The Elements of Group 18 (the Noble Gases) (p. 1026)

All the noble gases have a closed-shell valence-electron configuration. The ionization energies of the noble gases decrease with increasing atomic number. Only highly electronegative elements can form stable compounds with the noble gases in positive oxidation states without being oxidized themselves. Xenon has a high affinity for both fluorine and oxygen, which form stable compounds that contain xenon in even oxidation states up to +8.

QUESTIONS AND PROBLEMS

MGC For instructor-assigned homework, go to **www.masteringgenchem.com**

Questions and Problems with colored numbers have answers in the Appendix and complete solutions in the Student Solutions Manual.

CONCEPTUAL

22.1 The Elements of Group 13

1. None of the Group 13 elements was isolated until the early 19th century, even though one of these elements is the most abundant metal on Earth. Explain why the discovery of these elements came so late, and describe how they were finally isolated.
2. Boron and aluminum exhibit very different chemistry. Which element forms complexes with the most ionic character? Which element is a metal and which is a nonmetal? What single property best explains the difference in their reactivity?
3. The usual oxidation state of boron and aluminum is +3, whereas the heavier elements in Group 13 have an increasing tendency to form compounds in the +1 oxidation state. Given that all the Group 13 elements have an $ns^2\,np^1$ electron configuration, how do you explain this difference between the lighter and heavier Group 13 elements?
4. Do you expect the Group 13 elements to be highly reactive in air? Why or why not?
5. Which of the Group 13 elements has the least metallic character? Explain why.
6. Boron forms multicenter bonds rather than metallic lattices with delocalized valence electrons. Why does it prefer this type of bonding? Does this explain why boron behaves like a semiconductor rather than a metal? Explain your answer.

7. Because the B—N unit is isoelectronic with the C—C unit, compounds that contain these units tend to have similar chemistry, although they exhibit some important differences. Describe the differences in physical properties, conductivity, and reactivity of these two types of compounds.
8. Boron has a strong tendency to form clusters. Does aluminum have this same tendency? Why or why not?
9. Explain why a B—O bond is much stronger than a B—C bond.
10. The electron affinities of boron and aluminum are −27 and −42 kJ/mol, respectively. Contrary to the usual periodic trends, the electron affinities of the remaining elements in Group 13 are between those of B and Al. How do you explain this apparent anomaly from the expected values?
11. The reduction potentials of B and Al in the +3 oxidation state are −0.87 and −1.66 V, respectively. Do you expect the reduction potentials of the remaining elements of Group 13 to be higher or lower than these values? How do you explain the differences between the expected values and those given in Table 22.1?
12. The compound Al_2Br_6 is a halide-bridged dimer in the vapor phase, similar to diborane (B_2H_6). Draw the structure of Al_2Br_6, and then compare the bonding in this compound with that found in diborane. Explain the differences.
13. The compound AlH_3 is an insoluble, polymeric solid that reacts with strong Lewis bases, such as Me_3N, to form adducts with 10 valence electrons around aluminum. What hybrid orbital set is formed to allow this to occur?

22.2 The Elements of Group 14

14. Why is the preferred oxidation state of lead +2 rather than +4? What do you expect the preferred oxidation state of silicon to be based on its position in the periodic table?
15. Carbon uses $p\pi$–$p\pi$ overlap to form compounds with multiple bonds, but silicon does not. Why? How does this same phenomenon explain why the heavier elements in Group 14 don't form catenated compounds?
16. Diamond is both an electrical insulator and an excellent thermal conductor. Explain this property in terms of its bonding.
17. The lighter chalcogens (Group 16) form π bonds with carbon. Does the strength of these π bonds increase or decrease with increasing atomic number of the chalcogen? Why?
18. The heavier Group 14 elements can form complexes that contain expanded coordination spheres. How does this affect their reactivity compared with the reactivity of carbon? Is this a thermodynamic or a kinetic effect? Explain your answer.
19. Refer to Table 22.2 for the values of the electron affinities of the elements of Group 14. Explain any discrepancies between these actual values and the expected values based on usual periodic trends.
20. Except for carbon, the elements of Group 14 can form five or six electron-pair bonds. What hybrid orbitals are used to allow this expanded coordination? Why doesn't carbon form more than four electron-pair bonds?
21. Which of the Group 14 elements is least stable in the +4 oxidation state? Why?

22.3 The Elements of Group 15 (the Pnicogens)

22. Nitrogen is the first diatomic molecule in the second period of elements. Why is N_2 the most stable form of nitrogen? Draw its Lewis dot structure. What hybrid orbitals are used to describe the bonding in this molecule? Is the molecule polar?
23. The polymer $(SN)_n$ has metallic luster and conductivity. Are the constituent elements in this polymer metals or nonmetals? Why does the polymer have metallic properties?
24. Except for NF_3, all the halides of nitrogen are unstable. Explain why NF_3 is stable.
25. Which of the Group 15 elements forms the most stable compounds in the +3 oxidation state? Explain why.
26. Phosphorus and arsenic react with the alkali metals to produce salts with the composition M_3Z_{11}. Compare these products with those produced in the reaction of P and As with the alkaline earths. What conclusions can you draw about the types of structures favored by the heavier elements in this part of the periodic table?

22.4 The Elements of Group 16 (the Chalcogens)

27. Unlike the other chalcogens, oxygen does not form compounds in the +4 or +6 oxidation state. Why?
28. Classify these oxides as basic, acidic, amphoteric, or neutral: (a) CaO; (b) SO_2; (c) NO; (d) Rb_2O; (e) PbO_2.
29. Classify these oxides as basic, acidic, amphoteric, or neutral: (a) BaO; (b) Br_2O; (c) SnO; (d) B_2O_3; (e) Sb_2O_3.
30. Polarization of an oxide affects its solubility in acids or bases. Based on this, do you expect Ru_2O_3 to be an acidic, basic, or neutral oxide? Is the compound covalent? Justify your answers.
31. Arrange the oxides in order of increasing basicity: CrO_3, Al_2O_3, Sc_2O_3, BaO.
32. As the atomic number of the Group 16 elements increases, the complexity of their allotropes decreases. What factors account for this trend? Which chalcogen do you expect to polymerize the most readily? Why?
33. Arrange the compounds in order of increasing acid strength: H_3BO_3, HIO_4, HNO_2.
34. Of these compounds, select the one that is most ionic: OF_2, SO_2, P_4O_6, SiO_2, Al_2O_3.
35. Indicate which of these compounds you expect to have (a) the most polar covalent bond(s) and (b) the least polar covalent(s): CO_2, NO_2, O_2, SO_2, Cl_2O, H_2O, NH_3, CH_4.
36. Of these oxides, select the one that is most acidic: Na_2O_2, MgO, Al_2O_3, SiO_2.
37. Give an example of (a) a covalent hydride that engages in strong hydrogen bonding, and (b) an amphoteric oxide.
38. The Si—O bond is shorter and stronger than expected. What orbitals are used in this bond? Do you expect Si to interact with Br in the same way? Why or why not?

22.5 The Elements of Group 17 (the Halogens)

39. The lightest elements of Groups 15, 16, and 17 form unusually weak single bonds. Why are their bonds so weak?
40. Fluorine has an anomalously low F—F bond energy. Why? Why does fluorine form compounds only in the −1 oxidation state, whereas the other halogens exist in multiple oxidation states?
41. Compare the following compounds with respect to the type of M—X bond formed, melting point, and solubility in nonpolar solvents: AlI_3, $InCl_3$, GaF_3, $LaBr_3$.
42. Give the formulas of the interhalogen compounds that are most likely to contain the following species in the indicated oxidation states: I (+3), Cl (+3), I (−1), Br (+5).
43. Consider this series of bromides: $AlBr_3$, $SiBr_4$, PBr_5. Does the ionic character of the bond between the Br atoms and the central atom decrease or increase in this series?

44. Chromium forms compounds in the +6, +3, and +2 oxidation states. Which halogen would you use to produce each of these oxidation states? Justify your selections.

45. Of these interhalogen compounds, select the one that is least likely to exist: ClF_7, BrF_5, IF_7, BrF_3, ICl_3, IF_3, IF_5. Justify your selection.

22.6 The Elements of Group 18 (the Noble Gases)

46. The chemistry of the noble gases is largely dictated by a balance between two competing properties. What are these properties, and how do they affect the reactivity of these elements?

47. Of the Group 18 elements, only krypton, xenon, and radon form stable compounds with other atoms, and then only with very electronegative elements. Why?

48. Give the type of hybrid orbitals used by xenon in each of these species: (a) XeF_2; (b) XeF_4; (c) XeO_3; (d) $XeOF_4$; (e) XeO_4; (f) XeO_6^{4-}.

The following questions use concepts from multiple sections of the chapter.

49. Which element is the least metallic? B, Ga, Tl, Pb, Ge

50. Of the following elements, indicate which you expect to form (a) positive ions or (b) negative ions most easily: Br, N, Ar, Bi, Se, S.

51. Which of the following species do you expect to be (a) electron donors; (b) electron acceptors; (c) neither electron donors nor acceptors; or (d) both electron donors and acceptors? BCl_3, BCl_4^-, CH_4, $H_3N{\cdot}BF_3$, PCl_3, PCl_5, H_2O, F^-

52. Which is the strongest acid? HCl, $HClO_4$, HBr, H_2S, HF, PH_3

53. Which is the strongest base? CF_4, NH_3, NF_3, H_2O, OF_2, SiF_4, H_2S, SiH_4

STRUCTURE AND REACTIVITY

This section includes paired problems (marked by brackets) that require similar problem-solving skills.

22.1 The Elements of Group 13

54. Is $B(OH)_3$ a strong or a weak acid? Using bonding arguments, explain why.

55. Using bonding arguments, explain why organoaluminum compounds are expected to be potent Lewis acids that react rapidly with weak Lewis bases.

56. Imagine that you are studying chemistry prior to the discovery of gallium, element 31. Considering its position in the periodic table, predict the following properties of gallium:
   (a) chemical formulas of its most common oxide, most common chloride, and most common hydride
   (b) solubility of its hydride in water, and the acidity or basicity of the resulting solution
   (c) the principal ion formed in aqueous solution

57. The halides of Al, Ga, In, and Tl dissolve in water to form acidic solutions containing the hydrated metal ions, but only the halides of aluminum and gallium dissolve in aqueous base to form soluble metal-hydroxide complexes. Show the formulas of the soluble metal–hydroxide complexes and of the hydrated metal ions. Explain the difference in their reactivities.

58. Complete and balance each chemical equation:
   (a) $BBr_3(g) + H_2(g) \xrightarrow{\Delta}$
   (b) $BF_3(g) + F^-(g) \longrightarrow$
   (c) $LiH(s) + B_2H_6(g) \longrightarrow$
   (d) $B(OH)_3(s) + NaOH(aq) \longrightarrow$

59. Predict the products of these reactions, and balance each equation:
   (a) $BCl_3(g) + H_2(g) \xrightarrow{\Delta}$
   (b) $6C_2H_4(g) + B_2H_6(g) \longrightarrow$
   (c) $B_2H_6(g) + 3Cl_2(g) \longrightarrow$
   (d) $B_2H_6(g) + 2(C_2H_5)_2S(g) \longrightarrow$

60. Write a balanced chemical equation for the reaction that occurs between Al and each of these species: (a) Cl_2; (b) O_2; (c) S; (d) N_2; (e) H_2O; (f) $H^+(aq)$.

61. Write a balanced chemical equation for each reaction:
   (a) the dissolution of Al_2O_3 in dilute acid
   (b) the dissolution of Ga_2O_3 in concentrated aqueous base
   (c) the dissolution of Tl_2O in concentrated aqueous acid
   (d) $Ga(l) + S_8(s)$
   (e) $Tl(s) + H_2S(g)$

62. Write balanced chemical equations that show how you would prepare the following compounds from their elements: (a) BCl_3; (b) $InCl_3$; (c) Tl_2S; (d) $Al(OH)_3$; (e) In_2O_3; (f) AlN.

63. Write a balanced chemical equation that shows how you would prepare each compound from its elements in water: (a) In_2I_6; (b) $B(OH)_3$; (c) Ga_2O_3; (d) $[Tl(H_2O)_6]^{3+}$; (e) $Al(OH)_4^-$; (f) In_4C_3.

64. Diborane is a spontaneously flammable, toxic gas that is prepared by the reaction of $NaBH_4$ with BF_3. Write a balanced equation for this reaction.

65. Draw the Lewis dot structures of each reactant and product in these chemical equations. Then describe the type of bonding found in each of the reactants and products.
   (a) $2B(s) + 3X_2(g) \xrightarrow{\Delta} 2BX_3(g)$
   (b) $4B(s) + 3O_2(g) \xrightarrow{\Delta} 2B_2O_3(s)$
   (c) $2B(s) + N_2(g) \xrightarrow{\Delta} 2BN(s)$

66. Draw the Lewis electron dot structures of each reactant and product in these chemical equations. Then describe the type of bonding found in each of the reactants and products.
   (a) $B_2H_6(g) + 3O_2(g) \longrightarrow B_2O_3(s) + 3H_2O(l)$
   (b) $B_2H_6(g) + 6CH_3CH{=}CH_2(g) \longrightarrow 2[B(CH_2CH_2CH_3)_3](l)$

22.2 The Elements of Group 14

67. Predict the products of the reactions, and balance each equation:
   (a) $CaC_2(s) + HCl(g) \longrightarrow$
   (b) $Pb(s) + Br_2(l) \xrightarrow{\Delta}$
   (c) $(CH_3)_3N(l) + H_2O_2(aq) \longrightarrow$
   (d) $Pb(N_3)_2 \xrightarrow{\Delta}$

68. Write a balanced equation to indicate how you would prepare each species: (a) SiF_6^{2-} from its elements; (b) SiO_2 from $SiCl_4$; (c) GeS_2 from its elements; (d) $Si(CH_3)_4$ from Si.

69. Write a balanced equation to show how you would prepare each compound: (a) CO from CuO; (b) methane from Be_2C; (c) $Si(OH)_4$ from Si; (d) Si_3N_4 from Si.

22.3 The Elements of Group 15 (the Pnicogens)

70. PF_3 reacts with F_2 to produce PF_5, which in turn reacts with F^- to give salts that contain the PF_6^- ion. In contrast, NF_3 does not react with F_2, even under extreme conditions; NF_5 and the NF_6^- ion do not exist. Why?

71. Red phosphorus is safer to handle than white phosphorus, reflecting their dissimilar properties. Given their structures, how do you expect them to compare with regard to reactivity, solubility, density, and melting point?

72. Bismuth oxalate [$Bi(C_2O_4)_3$] is a poison. Draw its structure, and then predict its solubility in (a) H_2O, (b) dilute HCl; and (c) dilute HNO_3. (d) Predict its combustion products. Suggest a method to prepare bismuth oxalate from bismuth.

73. Although pure nitrous acid is unstable, dilute solutions in water are prepared by adding nitrite salts to aqueous acid. Write a balanced equation that represents this type of reaction.

74. Small quantities of NO can be obtained in the laboratory by the reaction of iodide ion with acidic solutions of nitrite. Write a balanced equation for this reaction.

75. Metallic versus nonmetallic behavior becomes apparent in reactions of the elements with an oxidizing acid, such as HNO_3. Write balanced equations for the reaction of each element of Group 15 with nitric acid. Based on the products, predict which of these elements, if any, are metals and which, if any, are nonmetals.

76. Write a balanced equation to show how you would prepare each compound: (a) H_3PO_4 from P; (b) Sb_2O_5 from Sb; (c) SbH_3 from Sb.

77. Predict the products of these reactions, and balance each chemical equation:
(a) $P_4O_{10}(s) + H_2O(l) \longrightarrow$
(b) $AsCl_3(l) + H_2O(l) \longrightarrow$
(c) $Bi_2O_3(s) + H_2O(l) \longrightarrow$
(d) $Sb_4O_6(s) + OH^-(aq) \longrightarrow$
(e) $(C_2H_5)_3Sb(l) + O_2(g) \xrightarrow{\Delta}$
(f) $SbCl_3(s) + LiAlH_4(soln) \longrightarrow$
(g) $Ca(s) + N_2O(g) \xrightarrow{\Delta}$

22.4 The Elements of Group 16 (the Chalcogens)

78. Considering its position in the periodic table, predict the following properties of selenium:
(a) chemical formulas of its most common oxide, most common chloride, and most common hydride
(b) solubility of its hydride in water, and the acidity or basicity of the resulting solution
(c) the principal ion formed in aqueous solution

79. Write balanced chemical equations for the reactions of sulfur with (a) $O_2(g)$; (b) $S^{2-}(aq)$; (c) $F_2(g)$; (d) $HNO_3(aq)$.

80. Using arguments based on electronegativity, explain why ZnO is amphoteric. What product would you expect when ZnO reacts with (a) an aqueous acid and (b) an aqueous base?

22.5 The Elements of Group 17 (the Halogens)

81. SiF_4 reacts easily with NaF to form SiF_6^{2-}. In contrast, CF_4 is totally inert and shows no tendency to form CF_6^{2-} under even extreme conditions. Explain this difference.

82. Predict the products of these reactions, and balance each chemical equation:
(a) $Xe(g) + \text{excess } F_2(g) \longrightarrow$
(b) $Se(s) + Cl_2(g) \longrightarrow$
(c) $SO_2(g) + Br_2(g) \longrightarrow$
(d) $NaBH_4(s) + BF_3(soln) \longrightarrow$

83. Write balanced chemical equations for the reactions of aqueous HF with (a) SiO_2; (b) Na_2CO_3; (c) CaO.

84. Oxyhalides of sulfur, such as the *thionyl halides* (SOX_2, where X = F, Cl, Br), are well known. Because the thionyl halides react vigorously with trace amounts of water, they are used for dehydrating hydrated metal salts. Write a balanced equation to show the products of the reaction of $SOCl_2$ with water.

85. Write a chemical equation describing each reaction:
(a) the burning of sulfur in a chlorine atmosphere
(b) the dissolution of iodine in potassium iodide solution
(c) the hydrolysis of PCl_3
(d) the preparation of HF from calcium fluoride and sulfuric acid
(e) the thermal decomposition of $KClO_3$
(f) the oxidation of sulfide ion by elemental iodine

86. Write the complete electron dot diagram, the type of hybrid used by the central atom, and the number of lone pair electrons present on the central atom for each compound: (a) CF_4; (b) PCl_3; (c) XeF_4.

22.6 The Elements of Group 18 (the Noble Gases)

87. Write a balanced chemical equation showing how you would prepare each compound from its elements: (a) XeF_2; (b) XeF_4; (c) XeF_6; (d) $XeOF_4$; (e) XeO_3.

88. Write balanced chemical equations showing how you would make these compounds: (a) Xe gas from XeF_2; (b) $NaXeF_7$ from its elements; (c) RnO_3 from Rn.

89. In an effort to synthesize XeF_6, a chemist passed fluorine gas through a glass tube containing xenon gas. However, the product was not the one expected. What was the actual product?

The following questions use concepts from multiple sections of the chapter.

90. Write balanced chemical equations to describe the reaction of each of the following with water: (a) B_2H_6; (b) F_2; (c) C^{4+}.

91. Using heavy water (D_2O) as the source of deuterium, how could you prepare these compounds? (a) $LiAlD_4$; (b) D_2SO_4; (c) SiD_4; (d) DF

92. Predict the products of the following reactions, and write a balanced chemical equation for each:
(a) $Al_2O_3(s)$ in $OH^-(aq)$
(b) $Ar(g) + F_2(g)$
(c) $PI_3(s) + H_2O(l)$
(d) $H_3PO_3(l) + OH^-(aq)$
(e) $Bi(s) + \text{excess } Br_2(l)$

APPLICATIONS

93. Borax ($Na_2B_4O_5(OH)_4 \cdot 8H_2O$) is used as a flux during welding operations. As brass is heated during welding, for example, borax cleans the surface of Cu_2O and prevents further oxidation of the fused metal. Explain why borax is effective at cleaning the surface and preventing oxidation of the surface.

94. Extensive research is being conducted into using GaAs as a material for memory chips for computers. It has been found, for example, that chips made from GaAs are up to 10 times faster than those made from silicon. Propose an explanation for this increase in speed.

95. Cement that has a high content of alumina (Al_2O_3) is particularly resistant to corrosion, so it is used for structures that need to be resistant to seawater and acidic conditions. Why is this material so effective under these service conditions?

Failure occurs under prolonged exposure to a hot, wet environment. Why?

96. Aluminum is light and ductile. If you were considering using aluminum rather than steel as a structural material for building a high-speed ferry, what disadvantages would you need to consider in using aluminum for these service conditions?

97. Life on Earth is carbon based. A possible explanation is that no other element in the periodic table forms compounds that are so diverse in their chemistry and physical properties. Discuss the chemistry of carbon with regard to (a) types of bonding; (b) the states of matter of carbon compounds; (c) the properties of elemental C; and (d) the reactivity of elemental C and its compounds. Compare B, Al, Si, N, and P with C in terms of these properties.

98. After a traffic accident in which a tanker truck carrying liquid nitrogen overturned, a reporter at the scene warned of a danger to residents in the vicinity of the accident because nitrogen would react with hydrocarbons in the asphalt to produce ammonia gas. Comment on the credibility of this statement.

99. Nitrogen forms a hydride called *hydrazoic acid* (HN_3), which is a colorless, highly toxic, explosive substance that boils at 37°C. The thermal decomposition of one of its salts, NaN_3, is used to inflate automotive air bags. The N_3^- ion is isoelectronic with CO_2.

(a) Draw the Lewis dot structure of the N_3^- anion.

(b) Write a balanced equation for the thermal decomposition of NaN_3.

(c) Based on your answer to part b, propose an explanation for why many people have suffered skin burns when their air bags exploded.

100. Hydrazine (N_2H_4), a rocket fuel, is a colorless, oily liquid with a melting point of 1.4°C, and it is a powerful reducing agent. The physical properties of hydrazine presumably reflect the presence of multiple hydrogen-bond acceptors and donors within a single molecule. Explain the basis for this statement.

101. Because the N—C single bond is almost twice as strong as the N—H bond, organic analogues of ammonia, hydrazine, and hydroxylamine are stable and numerous. Conceptually at least, they are formed by the successive replacement of H atoms by alkyl or aryl groups. Methylhydrazine and dimethylhydrazine, for example, were used as fuels in the U.S. Apollo space program. They react spontaneously and vigorously with liquid N_2O_4, thus eliminating the need for an ignition source. Write balanced equations for these reactions, and calculate $\Delta G°$ values for the reactions.

102. In an effort to remove a troublesome stain from a sink, one of the cleaning staff of a commercial building first used bleach on the stain and then decided to neutralize the bleach with ammonia. What happened? Why?

103. A slow reaction that occurs on the ocean floor is the conversion of carbonate to bicarbonate, which absorbs CO_2. Write a balanced equation for this reaction. Silicate sediments play an important role in controlling the pH of seawater. Given the reaction, propose a chemical explanation for this.

104. Marketing surveys have shown that customers prefer to buy a bright red steak rather than a dull gray one. It is known that NO combines with hemoglobin to form a bright red NO complex. What would you add to beef during processing to ensure that this reaction occurs and yields the desired appearance?

105. Covalent azides are used as detonators and explosives. Ionic azides, in contrast, are usually much more stable and are used in dyestuffs. Why is there such a difference between these two types of compounds? The N_3^- ion is considered to be a pseudohalide. Why?

106. The heads of modern "strike anywhere" matches contain a mixture of a nonvolatile phosphorus sulfide (P_4S_3) and an oxidizing agent ($KClO_3$), which is ignited by friction when the match is struck against a rough object. Safety matches separate the oxidant and the reductant by putting $KClO_3$ in the head and a paste containing nonvolatile red phosphorus on the match box or cover. Write a balanced equation for the reaction that occurs when a match is rubbed against the abrasive end of a matchbox.

107. Paris green was a common pigment in paints and wallpaper of the Napoleonic era. It is a mixed acetate/arsenite salt of copper with the formula $Cu_2(OAc)_2(AsO_3)$. In damp conditions, certain fungi are able to convert arsenite salts to volatile, toxic organoarsenic compounds. Shortly after his exile in 1815 to the remote island of St. Helena in the southern Atlantic Ocean, Napoleon died. As a forensic scientist investigating the cause of Napoleon's mysterious death, you notice that the walls of his enclosed bedchamber are covered in green wallpaper. What chemical clues would you look for to determine the cause of his death, and how?

108. Selenium, an element essential to humans, appears to function biologically in an enzyme that destroys peroxides. Why is selenium especially suited for this purpose? Would sulfur or tellurium be as effective? Why or why not?

109. One way to distinguish between "fool's gold" (FeS, or iron pyrite) and real gold is to heat the sample over a fire. If your sample were actually fool's gold, what would happen?

110. Calcium hypochlorite is sold as swimming pool bleach. It is formed by the hydrolysis of Cl_2O, which gives only one product, followed by neutralization with lime, $Ca(OH)_2$. Write balanced equations for these reactions.

111. There is much interest in the "super-heavy" elements beyond $Z = 111$ because of their potentially unique properties. Predict the valence electron configurations, preferred oxidation states, and products of the reaction with aqueous acid for elements 113 and 115.

112. Zeolites have become increasingly important in chemical engineering. They can be used as dessicants because the dehydrated zeolite absorbs small molecules such as water. In order to be retained by the zeolite frame, a molecule must satisfy two conditions. What are they? Why can't linear CO_2 and tetrahedral CH_4 be held by a typical zeolite even though they can penetrate it easily?

23 The *d*-Block Elements

LEARNING OBJECTIVES

- To know periodic trends in the *d*-block elements
- To be able to use periodic trends to understand the chemistry of the transition metals
- To understand how metals are extracted from their ores
- To know the most common structures observed for metal complexes
- To be able to predict the relative stabilities of metal complexes
- To understand how crystal field theory can explain the electronic structures and colors of metal complexes
- To become familiar with some of the roles of transition-metal complexes in biological systems

CHAPTER OUTLINE

Chapters 21 and 22 described the chemistry of the *s*- and *p*-block elements. In this chapter, we survey the chemistry of the *d*-block elements, which are also called the *transition metals*. We again use valence-electron configurations and periodic trends, as well as the principles of chemical bonding, thermodynamics, and kinetics, as tools to describe the properties and reactivity of these elements. Because all the *d*-block elements are metals, they do not have the extreme variability in chemistry that we saw among the elements of the *p* block. Instead, these elements exhibit significant horizontal and vertical similarities in chemistry, and all have a common set of characteristic properties due to partially filled *d* subshells.

Alloys and compounds of *d*-block elements are important components of the materials the modern world depends on for its continuing technological

Titanium metal is light and corrosion resistant. The Guggenheim Museum in Bilbao, Spain, is the largest titanium-clad building in the world. The exterior is covered with 344,000 square feet of 0.016-in.-thick titanium pieces, each with a unique shape.

development, while most of the first-row transition metals are essential for life. This chapter introduces some of the key industrial and biological roles of these elements. You will learn, for example, why copper, silver, and gold have been used for coins and jewelry since ancient times, how Cr^{3+} impurities can be responsible for the characteristic colors of both rubies and emeralds, why an iron oxide was used in primitive compasses, why insects have greenish blue blood, and why cobalt is an essential component of vitamin B_{12}.

23.1 • General Trends Among the Transition Metals

MGC General Trends of Transition Metals

The transition metals, Groups 3–12 of the periodic table, are generally characterized by partially filled *d* subshells in the free elements or their cations. (Although the metals of Group 12 do not have partially filled *d* shells, their chemistry is similar in many ways to that of the preceding groups, and we therefore include them in our discussion.) Unlike the *s*- and *p*-block elements, the transition metals exhibit significant horizontal similarities in chemistry in addition to their vertical similarities.

Electronic Structure and Reactivity of the Transition Metals

The valence-electron configurations of the first-row transition metals are given in Table 23.1. As we go across the row from left to right, electrons are added to the 3*d* subshell to neutralize the increase in the positive charge of the nucleus as the atomic number increases. With two important exceptions, the 3*d* subshell is filled as expected based on the aufbau principle and Hund's rule. Unexpectedly, however, chromium has a $4s^1\,3d^5$ electron configuration rather than the $4s^2\,3d^4$ configuration predicted by the aufbau principle, and copper is $4s^1\,3d^{10}$ rather than $4s^2\,3d^9$. In Chapter 7, we attributed these anomalies to the extra stability associated with half-filled subshells. Because the *ns* and $(n-1)d$ subshells in these elements are similar in energy, even relatively small effects are enough to produce apparently anomalous electron configurations.

In the second-row transition metals, electron–electron repulsions within the 4*d* subshell cause additional irregularities in electron configurations that are not easily predicted. For example, Nb and Tc, with atomic numbers 41 and 43, both have a half-filled 5*s* subshell, with $5s^1\,4d^4$ and $5s^1\,4d^6$ valence-electron configurations, respectively. Further complications occur among the third-row transition metals, in which the 4*f*, 5*d*, and 6*s* orbitals are extremely close in energy. Thus, although La has a $6s^2\,5d^1$ valence-electron configuration, the valence-electron configuration of the next element, Ce, is $6s^2\,5d^0\,4f^2$. From this point through element 71, added electrons enter the 4*f* subshell, giving rise to the 14 elements known as the lanthanides. After the 4*f* subshell is filled, the 5*d* subshell is populated, producing the third row of the transition metals. Next comes the seventh period, where the actinides have three subshells (7*s*, 6*d*, and 5*f*) that are so similar in energy that their electron configurations are even more unpredictable.

TABLE 23.1 Valence-electron configurations of the first-row transition metals

Sc	Ti	V	Cr	Mn	Fe	Co	Ni	Cu	Zn
$4s^2\,3d^1$	$4s^2\,3d^2$	$4s^2\,3d^3$	$4s^1\,3d^5$	$4s^2\,3d^5$	$4s^2\,3d^6$	$4s^2\,3d^7$	$4s^2\,3d^8$	$4s^1\,3d^{10}$	$4s^2\,3d^{10}$

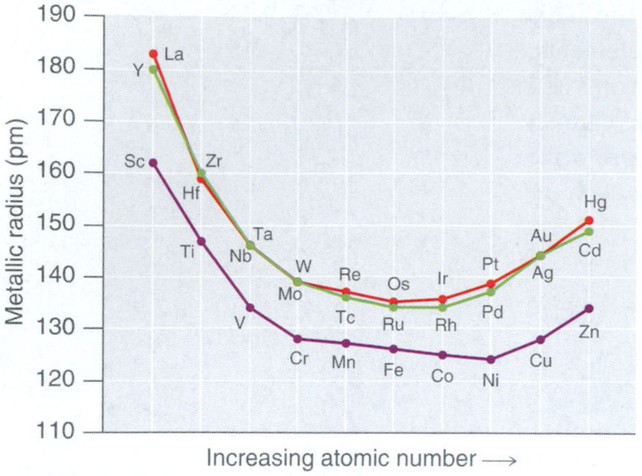

Figure 23.1 The metallic radii of the first-, second-, and third-row transition metals. Because of the lanthanide contraction, the second- and third-row transition metals are very similar in size.

As seen in the *s*- and *p*-block elements, the size of neutral atoms of the *d*-block elements gradually decreases as we go from left to right across a row, due to an increase in the effective nuclear charge (Z_{eff}) with increasing atomic number. In addition, the atomic radius increases as we go down a column, just as it does in the *s* and *p* blocks. Because of the *lanthanide contraction* (see Chapter 7), however, the increase in size between the 3*d* and 4*d* metals is much greater than between the 4*d* and 5*d* metals (Figure 23.1). The effects of the lanthanide contraction are also observed in ionic radii, which explains why, for example, there is only a slight increase in radius from Mo^{3+} to W^{3+}.

As you learned in Chapter 7, electrons in $(n-1)d$ and $(n-2)f$ subshells are only moderately effective at shielding the nuclear charge, and as a result the effective nuclear charge experienced by valence electrons in the *d*- and *f*-block elements does not change greatly as the nuclear charge increases across a row. Consequently, the ionization energies of these elements increase very slowly as we go across a given row (see Figure 7.10). In addition, as we go from the top left to the bottom right corner of the *d* block, electronegativities generally increase, densities and electrical and thermal conductivities increase, and enthalpies of hydration of the metal cations decrease in magnitude, as summarized in Figure 23.2. Consistent with this trend, the transition metals become steadily less reactive and more "noble" in character as we go from left to right *across* a row. The relatively high ionization energies and electronegativities and relatively low enthalpies of hydration are all major factors in the noble character of metals such as Pt and Au.

Trends in Transition Metal Oxidation States

The similarity in ionization energies and the relatively small increase in successive ionization energies lead to the formation of metal ions with the same charge for many of the transition metals. This in turn results in extensive horizontal similarities in chemistry, which are most noticeable for the first-row transition metals and for the lanthanides and actinides. Thus, all the first-row transition metals except Sc form stable compounds that contain the 2+ ion, and, due to the small difference between the second and third ionization energies for these elements, all except Zn also form stable compounds that contain the 3+ ion. The relatively small increase in successive ionization energies causes most of the transition metals to exhibit multiple oxidation states separated by a single electron. Manganese, for example, forms compounds in every oxidation state between −2 and +7. Because of the slow but steady increase in ionization potentials as we go across a row, high oxidation states become progressively less stable for the elements on the right side of the *d* block. The occurrence of multiple oxidation states separated by a single electron causes many, if not most, compounds of the transition metals to be

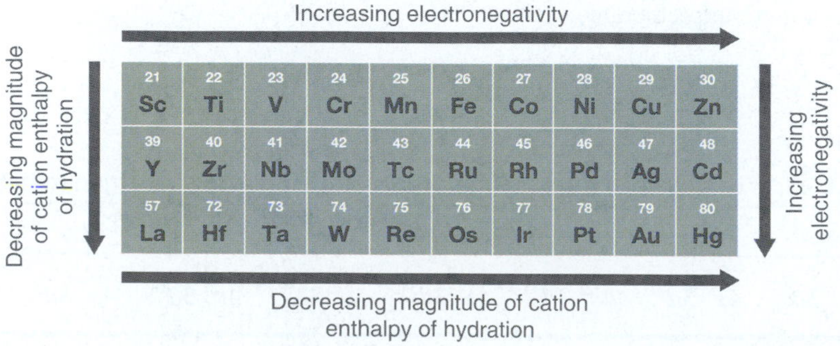

Figure 23.2 Some trends in properties of the transition metals. The electronegativity of the elements increases and the hydration energies of the metal cations decrease in magnitude as we go from left to right and from top to bottom of the *d* block. As a result, the metals in the lower right corner of the *d* block are so unreactive that they are often called the "noble metals."

TABLE 23.2 *d*-Electron configurations of the dications of the first-row transition metals

Ti^{2+}	V^{2+}	Cr^{2+}	Mn^{2+}	Fe^{2+}	Co^{2+}	Ni^{2+}	Cu^{2+}	Zn^{2+}
d^2	d^3	d^4	d^5	d^6	d^7	d^8	d^9	d^{10}

paramagnetic, with one to five unpaired electrons. This behavior is in sharp contrast to that of the *p*-block elements, where the occurrence of two oxidation states separated by two electrons is common, which makes virtually all compounds of the *p*-block elements diamagnetic.

The electronegativities of the first-row transition metals increase smoothly from Sc ($\chi = 1.4$) to Cu ($\chi = 1.9$). Thus, Sc is a rather active metal, whereas Cu is much less reactive. The steady increase in electronegativity is also reflected in the standard reduction potentials: thus, $E°$ for the reaction $M^{2+}(aq) + 2e^- \longrightarrow M^0(s)$ becomes progressively less negative from Ti ($E° = -1.63$ V) to Cu ($E° = +0.34$ V). Exceptions to the overall trends are rather common, however, and in many cases they are attributable to the stability associated with filled and half-filled subshells. For example, the $4s^2\,3d^{10}$ electron configuration of zinc results in its strong tendency to form the stable Zn^{2+} ion, with a $3d^{10}$ electron configuration, whereas Cu^+, which also has a $3d^{10}$ electron configuration, is the only stable monocation formed by a first-row transition metal. Similarly, with a half-filled subshell, Mn^{2+} ($3d^5$) is much more difficult to oxidize than Fe^{2+} ($3d^6$). The chemistry of manganese is therefore primarily that of the Mn^{2+} ion, whereas both the Fe^{2+} and Fe^{3+} ions are important in the chemistry of iron.

The transition metals form cations by the initial loss of the *ns* electrons of the metal, even though the *ns* orbital is *lower* in energy than the $(n-1)d$ subshell in the neutral atoms. This apparent contradiction is due to the small difference in energy between the *ns* and $(n-1)d$ orbitals, together with screening effects. Loss of one or more electrons reverses the relative energies of the *ns* and $(n-1)d$ subshells, making the latter lower in energy. Consequently, *all transition-metal cations possess d^n valence-electron configurations*, as shown in Table 23.2 for the 2+ ions of the first-row transition metals.

The most common oxidation states of the first-row transition metals are shown in Table 23.3. The second- and third-row transition metals behave similarly, but with three important differences:

1. The maximum oxidation states observed for the second- and third-row transition metals in Groups 3–8 increase from +3 for Y and La to +8 for Ru and Os,

Note the pattern

All transition-metal cations have d^n electron configurations; the ns electrons are always lost before the $(n-1)d$ electrons.

Note the pattern

The highest possible oxidation state, corresponding to the formal loss of all valence electrons, becomes increasingly less stable as we go from Group 3 to Group 8, and it is never observed in later groups.

TABLE 23.3 Common oxidation states of the first-row transition metals

Element	Sc	Ti	V	Cr	Mn	Fe	Co	Ni	Cu	Zn
Electronic Structure	$s^2\,d^1$	$s^2\,d^2$	$s^2\,d^3$	$s^1\,d^5$	$s^2\,d^5$	$s^2\,d^6$	$s^2\,d^7$	$s^2\,d^8$	$s^1\,d^{10}$	$s^2\,d^{10}$
Oxidation States				I					I	
	II	II	II	II	II	II	II	II	II	II
	III	III	III	III	III	III	III	III	III	
		IV	IV	IV	IV	IV	IV	IV		
			V	V	V	V	V			
				VI	VI	VI				
					VII					

corresponding to the formal loss of all *ns* and $(n-1)d$ valence electrons. As we go farther to the right, the maximum oxidation state decreases steadily, reaching +2 for the elements of Group 12 (Zn, Cd, Hg), which corresponds to a filled $(n-1)d$ subshell.

2. Within a group, higher oxidation states become more stable as we go down the group. For example, the chromate ion, $[CrO_4]^{2-}$, is a powerful oxidant, whereas the tungstate ion, $[WO_4]^{2-}$, is extremely stable and has essentially no tendency to act as an oxidant. This pattern is in distinct contrast to that observed for the *p*-block elements.

3. Cations of the second- and third-row transition metals in lower oxidation states (+2 and +3) are much more easily oxidized than the corresponding ions of the first-row transition metals. For example, the most stable compounds of chromium are those of Cr(III), but the corresponding Mo(III) and W(III) compounds are highly reactive. In fact, they are often *pyrophoric*, bursting into flames on contact with atmospheric oxygen. As we shall see, the heavier elements in each group form stable compounds in higher oxidation states that have no analogues with the lightest member of the group.

Note the pattern

In the transition metals, the stability of higher oxidation states increases as we go down a column.

Binary transition-metal compounds such as the oxides and sulfides are usually written with idealized stoichiometries, such as FeO or FeS, but these compounds are usually cation deficient and almost never contain a 1:1 cation:anion ratio. Thus, a substance such as ferrous oxide is actually a nonstoichiometric compound that has a range of compositions.

The acid–base character of transition-metal oxides depends strongly on the oxidation state of the metal and its ionic radius. Oxides of metals in lower oxidation states ($\leq +3$) have significant ionic character and tend to be basic. Conversely, oxides of metals in higher oxidation states are more covalent and tend to be acidic, often dissolving in strong base to form oxoanions.

EXAMPLE 23.1

Two of the Group 8 metals (Fe, Ru and Os) form stable oxides in the +8 oxidation state. Identify these metals, predict the stoichiometry of the oxides, describe the general physical and chemical properties, type of bonding, and physical state of the oxides, and decide whether they are acidic or basic oxides.

Given Group 8 metals

Asked for Identity of metals, expected properties of oxides in +8 oxidation state

Strategy

Refer to the trends outlined in Figures 23.1 and 23.2 and Tables 23.1–23.3 to identify the metals. Decide whether their oxides are covalent or ionic in character, and, based on this, predict the general physical and chemical properties of the oxides.

Solution

The +8 oxidation state corresponds to a stoichiometry of MO_4. Since the heavier transition metals tend to be stable in higher oxidation states, we expect Ru and Os to form the most stable tetroxides. Because oxides of metals in high oxidation states are generally covalent compounds, RuO_4 and OsO_4 should be volatile solids or liquids that consist of discrete MO_4 molecules, which the VSEPR model predicts to be tetrahedral. Finally, because oxides of transition metals in high oxidation states are usually acidic, RuO_4 and OsO_4 should dissolve in strong aqueous base to form oxoanions.

EXERCISE 23.1

Predict the identity and stoichiometry of the stable Group 9 bromide in which the metal has the lowest oxidation state, and describe its chemical and physical properties.

Answer Because the lightest element in the group is most likely to form stable compounds in lower oxidation states, the bromide will be $CoBr_2$. We predict that $CoBr_2$ will be an ionic solid with a relatively high melting point and that it will dissolve in water to give the $Co^{2+}(aq)$ ion.

23.2 • A Brief Survey of Transition-Metal Chemistry

We turn now to a brief survey of the chemistry of the transition metals, beginning with Group 3. As we shall see, the two heaviest members of each group usually exhibit substantial similarities in chemical behavior and are quite different from the lightest member.

Groups 3, 4, and 5

MGC Groups 3 through 6

Group 3 (Sc, Y, La, Ac)

As shown in Table 23.4, the observed trends in the properties of the Group 3 elements are similar to those of Groups 1 and 2. Due to their $ns^2\,(n-1)d^1$ valence-electron configurations, the chemistry of all four elements is dominated by the +3 oxidation state formed by the loss of all three valence electrons. Thus, as expected based on periodic trends, these elements are highly electropositive metals and powerful reductants, with La (and Ac) being the most reactive. In keeping with their

Note the pattern

The chemistry of the Group 3 metals is almost exclusively that of the M^{3+} ions.

TABLE 23.4 Some properties of the elements of Groups 3, 4, and 5

Group	Element	Z	Valence-Electron Configuration	Electronegativity	Metallic Radius, pm	mp, °C	Density, g/cm^3
3	Sc	21	$4s^2\,3d^1$	1.36	162	1541	2.99
	Y	39	$5s^2\,4d^1$	1.22	180	1522	4.47
	La	57	$6s^2\,5d^1$	1.10	183	918	6.15
	Ac	89	$7s^2\,6d^1$	1.10	188	1051	10.00
4	Ti	22	$4s^2\,3d^2$	1.54	147	1668	4.51
	Zr	40	$5s^2\,4d^2$	1.33	160	1855	6.52
	Hf	72	$6s^2\,5d^2\,4f^{14}$	1.30	159	2233	13.30
5	V	23	$4s^2\,3d^3$	1.63	134	1910	6.00
	Nb	41	$5s^2\,4d^3$	1.60	146	2477	8.57
	Ta	73	$6s^2\,5d^3\,4f^{14}$	1.50	146	3017	16.40

highly electropositive character, the Group 3 metals react with water to produce the metal hydroxide and hydrogen gas:

$$2M(s) + 6H_2O(l) \longrightarrow 2M(OH)_3(s) + 3H_2(g) \quad (23.1)$$

Moreover, all dissolve readily in aqueous acid to produce hydrogen gas and a solution of the hydrated metal ion, $M^{3+}(aq)$.

The Group 3 metals react with nonmetals to form compounds that are primarily ionic in character. For example, reaction with the halogens produces the corresponding trihalides, MX_3. The trifluorides are insoluble in water because of their high lattice energies, but the other trihalides are very soluble in water and behave like typical ionic metal halides. All the Group 3 elements react with air to form an oxide coating, and all burn in oxygen to form the so-called *sesquioxides*, M_2O_3, which react with water or CO_2 to form the corresponding hydroxides or carbonates, respectively. Commercial uses of the Group 3 metals are limited, but "mischmetal," a mixture of lanthanides containing about 40% La, is used as an additive to improve the properties of steel and to make flints for cigarette lighters.

Group 4 (Ti, Zr, Hf)

Because the elements of Group 4 have a high affinity for oxygen, all three metals occur naturally as oxide ores that contain the metal in the +4 oxidation state resulting from the loss of all four $ns^2\,(n-1)d^2$ valence electrons. They are isolated by initial conversion to the tetrachlorides, as shown for Ti, followed by reduction of the tetrachlorides with an active metal such as Mg:

$$2FeTiO_3(s) + 6C(s) + 7Cl_2(g) \longrightarrow 2TiCl_4(g) + 2FeCl_3(g) + 6CO(g) \quad (23.2)$$

Note the pattern

The chemistry of the Group 4 metals is dominated by the +4 oxidation state. Only Ti has an extensive chemistry in lower oxidation states.

In contrast to the elements of Group 3, the Group 4 elements have important applications. Titanium (mp 1668°C) is often used as a replacement for aluminum (mp 660°C) in applications that require high temperatures or corrosion resistance. For example, friction with the air heats the skin of supersonic aircraft operating above Mach 2.2 to temperatures near the melting point of aluminum; consequently, titanium is used instead of aluminum in many aerospace applications. The corrosion resistance of titanium is increasingly exploited in architectural applications, as shown in the chapter opening photo. Metallic zirconium is used in UO_2-containing fuel rods in nuclear reactors, while hafnium is used in the control rods that modulate the output of high-power nuclear reactors, such as those in nuclear submarines.

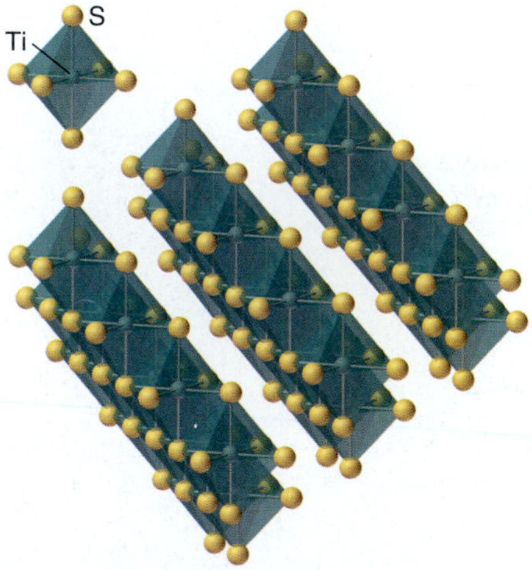

Figure 23.3 The layered structure of TiS_2. Each titanium atom is surrounded by an octahedral arrangement of six sulfur atoms that are shared to form extended layers of atoms. Because the layers are held together by only van der Waals forces between adjacent sulfur atoms, rather than covalent bonds, the layers slide past one another relatively easily when a mechanical stress is applied.

Consistent with the periodic trends shown in Figure 23.2, the Group 4 metals become denser, higher melting, and more electropositive as we go down the column (Table 23.4). Unexpectedly, however, the atomic radius of Hf is slightly *smaller* than that of Zr due to the lanthanide contraction. Because of their $ns^2\,(n-1)d^2$ valence-electron configurations, the +4 oxidation state is by far the most important for all three metals. Only titanium exhibits a significant chemistry in the +2 and +3 oxidation states, although compounds of Ti^{2+} are usually powerful reductants. In fact, the $Ti^{2+}(aq)$ ion is such a strong reductant that it rapidly reduces water to form hydrogen gas.

Reaction of the Group 4 metals with excess halogen forms the corresponding tetrahalides, MX_4. Titanium, the lightest element in the group, also forms dihalides and trihalides ($X \neq F$). The covalent character of the titanium halides increases as the oxidation state of the metal increases because of increasing polarization of the anions by the cation as its charge-to-radius ratio increases. Thus, $TiCl_2$ is an ionic salt, whereas $TiCl_4$ is a volatile liquid that contains tetrahedral molecules. All three metals react with excess oxygen or the heavier chalcogens (Y) to form the corresponding dioxides, MO_2, and dichalcogenides, MY_2. Industrially, TiO_2, which is used as a white pigment in paints, is prepared by the reaction of $TiCl_4$ with oxygen at high temperatures:

$$TiCl_4(g) + O_2(g) \longrightarrow TiO_2(s) + 2Cl_2(g) \quad (23.3)$$

The Group 4 dichalcogenides have unusual layered structures with no M–Y bonds holding adjacent sheets together, which makes them similar in some ways to graphite (Figure 23.3). The Group 4 metals also react with hydrogen, nitrogen, carbon, and boron to form hydrides (such as TiH_2), nitrides (such as TiN), carbides (such as TiC), and borides (such as TiB_2), all of which are hard, high-melting solids. Many of these binary compounds are nonstoichiometric and exhibit metallic conductivity.

Group 5 (V, Nb, Ta)

Like the Group 4 elements, all the Group 5 metals are normally found in nature as oxide ores that contain the metals in their highest oxidation state (+5). Because of the lanthanide contraction, the chemistry of Nb and Ta is so similar that these elements are usually found in the same ores.

Note the pattern

The chemistry of the two heaviest Group 5 metals (Nb and Ta) is dominated by the +5 oxidation state. The chemistry of the lightest element (V) is dominated by lower oxidation states, especially +4.

Three-quarters of the vanadium produced annually is used in the production of steel alloys for springs and high-speed cutting tools. Adding a small amount of vanadium to steel results in the formation of small grains of V_4C_3, which greatly increase the strength and resilience of the metal, especially at high temperatures. The other major use of vanadium is as V_2O_5, an important catalyst for the industrial conversion of SO_2 to SO_3 in the contact process for the production of sulfuric acid (see Chapter 2). In contrast, Nb and Ta have only limited applications, and they are therefore produced in relatively small amounts. Although niobium is used as an additive in certain stainless steels, its primary application is in superconducting wires such as Nb_3Zr and Nb_3Ge, which are used in superconducting magnets for magnetic resonance imaging of soft tissues. Because tantalum is highly resistant to corrosion, it is used as a liner for chemical reactors, in missile parts, and as a biologically compatible material in screws and pins for repairing fractured bones.

As indicated in Table 23.4, the trends in properties of the Group 5 metals are similar to those of Group 4. Only vanadium, the lightest element, has any tendency to form compounds in oxidation states lower than +5. For example, vanadium is the only element in the group that forms stable halides in the lowest oxidation state (+2). All three metals react with excess oxygen, however, to produce the corresponding oxides in the +5 oxidation state, M_2O_5, in which polarization of the oxide ions by the high-oxidation-state metal is so extensive that the compounds are primarily covalent in character. Vanadium–oxygen species provide a classic example of the effect of increasing metal oxidation state on the protonation state of a coordinated water molecule: vanadium(II) in water exists as the violet hydrated ion $V(H_2O)_6^{2+}$, the blue-green $V(H_2O)_6^{3+}$ ion is acidic, dissociating to form small amounts of the $V(H_2O)_5(OH)^{2+}$ ion and a proton; and in water, vanadium(IV) forms the blue vanadyl ion, $(H_2O)_4VO^{2+}$, which contains a formal V=O bond (Figure 23.4). Consistent with its covalent character, V_2O_5 is acidic, dissolving in base to give the vanadate ion, VO_4^{3-}, whereas both Nb_2O_5 and Ta_2O_5 are comparatively inert. Oxides of these metals in lower oxidation states tend to be nonstoichiometric.

Figure 23.4 Aqueous solutions of vanadium ions in oxidation states of +2 to +5. Because vanadium ions with different oxidation states have different numbers of *d* electrons, aqueous solutions of the ions have different colors: in acid V(V) forms the pale yellow VO_2^+ ion; V(IV) is the blue vanadyl ion, VO^{2+}; and V(III) and V(II) exist as the hydrated V^{3+} (blue-green) and V^{2+} (violet) ions, respectively.

Although Group 5 metals react with the heavier chalcogens to form a complex set of binary chalcogenides, the most important are the dichalcogenides, MY_2, whose layered structures are similar to those of the Group 4 dichalcogenides. The elements of Group 5 also form binary nitrides, carbides, borides, and hydrides, whose stoichiometries and properties are similar to those of the corresponding Group 4 compounds. One such compound, tantalum carbide, has the highest melting point of any compound known (3738°C); it is used for the cutting edges of high-speed machine tools.

Groups 6 and 7

Group 6 (Cr, Mo, W)

As an illustration of the trend toward increasing polarizability as we go from left to right across the *d* block, in Group 6 we first encounter a metal (Mo) that occurs naturally as a sulfide ore rather than as an oxide. Molybdenite (MoS_2) is a soft black mineral that can be used for writing, like PbS and graphite. Because of this similarity, people long assumed that these substances were all the same. In fact, the name *molybdenum* is derived from the Greek *molybdos*, "lead." More than 90% of the molybdenum produced annually is used to make steels for cutting tools, which retain their sharp edge even when red hot. In addition, molybdenum is the only second- or third-row transition element that is essential for humans. The major chromium ore is chromite ($FeCr_2O_4$), which is oxidized to the soluble CrO_4^{2-} ion under basic conditions and reduced successively to Cr_2O_3 and Cr with carbon and aluminum, respectively. Pure chromium can be obtained by dissolving Cr_2O_3 in sulfuric acid followed by electrolytic reduction; a similar process is used for electroplating metal objects to give them a bright, shiny, protective surface layer. Pure tungsten is obtained by first converting tungsten ores to WO_3, which is then reduced with hydrogen to give the metal.

Note the pattern

The chemistry of the two heaviest Group 6 metals (Mo and W) is dominated by the +6 oxidation state. The chemistry of the lightest element (Cr) is dominated by lower oxidation states, especially +3.

Consistent with periodic trends, the Group 6 metals are slightly less electropositive than those of the three preceding groups, and the two heaviest metals are essentially the same size because of the lanthanide contraction (Table 23.5). All three elements have a total of six valence electrons, resulting in a maximum oxidation state of +6. Due to extensive polarization of the anions, compounds in the +6 oxidation state are highly covalent. As in Groups 4 and 5, the lightest element exhibits variable oxidation states, ranging from Cr^{2+}, which is a powerful reductant, to CrO_3, a red solid that is a powerful oxidant. For Mo and W, the highest oxidation state (+6) is by far the most important, although compounds in the +4 and +5 oxidation states are known.

As observed in previous groups, the Group 6 halides become more covalent as the oxidation state of the metal increases: their volatility increases, and their melting points decrease. As the electronegativity of the halogens decreases from F to I, they are less able to stabilize high oxidation states, and so the maximum oxidation state of the corresponding metal halides decreases. Thus, all three metals form hexafluorides, but CrF_6 is unstable at temperatures above −100°C, whereas MoF_6 and WF_6 are stable. Consistent with the trend toward increased stability of the highest oxidation state for the second- and third-row elements, the other halogens can oxidize chromium to only the trihalides, CrX_3 (X = Cl, Br, I), while molybdenum forms $MoCl_5$, $MoBr_4$, and MoI_3, and tungsten gives WCl_6, WBr_5, and WI_4.

TABLE 23.5 Some properties of the elements of Groups 6 and 7

Group	Element	Z	Valence-Electron Configuration	Electronegativity	Metallic Radius, pm	mp, °C	Density, g/cm^3
	Cr	24	$4s^1\,3d^5$	1.66	128	1907	7.15
6	Mo	42	$5s^1\,4d^5$	2.16	139	2623	10.20
	W	74	$6s^2\,5d^4\,4f^{14}$	1.70	139	3422	19.30
	Mn	25	$4s^2\,3d^5$	1.55	127	1246	7.30
7	Tc	43	$5s^2\,4d^5$	2.10	136	2157	11.00
	Re	75	$6s^2\,5d^5\,4f^{14}$	1.90	137	3186	20.80

Both Mo and W react with oxygen to form the covalent trioxides, MoO_3 and WO_3, but Cr reacts to form only the so-called sesquioxide, Cr_2O_3. Chromium will form CrO_3, which is a highly toxic compound that can react explosively with organic materials. All the trioxides are acidic, dissolving in base to form the corresponding oxoanions, MO_4^{2-}. Consistent with periodic trends, the sesquioxide of the lightest element in the group (Cr_2O_3) is amphoteric. The aqueous chemistry of molybdate and tungstate is complex, and at low pH they form a series of polymeric anions called *isopolymetallates*, such as the $[Mo_8O_{26}]^{4-}$ ion, whose structure is shown in the margin.

The reaction of molybdenum or tungsten with heavier chalcogens gives binary chalcogenide phases, most of which are nonstoichiometric and electrically conducting. One of the most stable, MoS_2, has a layered structure similar to that of TiS_2 (Figure 23.3), in which the layers are held together by only weak van der Waals forces, which allows them to slide past one another rather easily. Consequently, both MoS_2 and WS_2 are used as lubricants in a variety of applications, including automobile engines. Because tungsten itself has an extraordinarily high melting point (3380°C), lubricants described as containing "liquid tungsten" actually contain a suspension of very small WS_2 particles.

As in Groups 4 and 5, the elements of Group 6 form binary nitrides, carbides, and borides whose stoichiometries and properties are similar to those of the preceding groups. Tungsten carbide (WC), one of the hardest compounds known, is used to make the tips of drill bits.

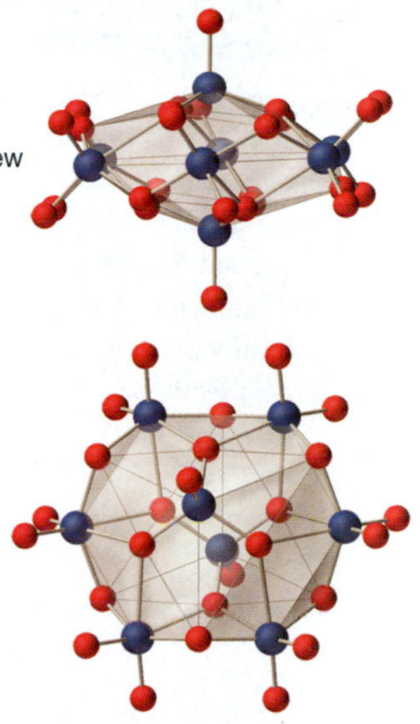

An isopolymolybdate cluster. The $[Mo_8O_{26}]^{4-}$ ion, shown here in both side and top views, is typical of the oxygen-bridged clusters formed by Mo(VI) and W(VI) in aqueous solution.

Group 7 (Mn, Tc, Re)

Continuing across the periodic table, we encounter the Group 7 elements (Table 23.5). One Group 7 metal, Mn, is usually combined with iron in an alloy called *ferromanganese*, which has been used since 1856 to improve the mechanical properties of steel by scavenging sulfur and oxygen impurities to form MnS and MnO. Technetium is named after the Greek *technikos*, "artificial," because all its isotopes are radioactive. One isotope, ^{99m}Tc (*m* for metastable), has become an important biomedical tool for imaging internal organs (see Chapter 20). Because of its scarcity, Re is one of the most expensive elements, and its applications are limited. It is, however, used in a bimetallic Pt/Re catalyst for refining high-octane gasoline.

All three Group 7 elements have seven valence electrons and can form compounds in the +7 oxidation state. Once again, the lightest element exhibits multiple oxidation states. Compounds of Mn in oxidation states ranging from −2 to +7 are known, with the most common being +2 and +4 (Figure 23.5). In contrast, compounds of Tc and Re in the +2 oxidation state are quite rare. Because the electronegativity of Mn is anomalously low, elemental manganese is unusually reactive. In contrast, the chemistry of Tc is similar to that of Re because of their similar size and electronegativity, again a result of the lanthanide contraction. Due to the stability of the half-filled $3d^5$ electron configuration, the aqueous Mn^{3+} ion, with a $3d^4$ valence-electron configuration, is a potent oxidant that is able to oxidize water. It is difficult to generalize about other oxidation states for Tc and Re because their stability depends dramatically on the nature of the compound.

Consistent with higher oxidation states being more stable for the heavier transition metals, the reaction of Mn with F_2 gives only MnF_3, a high-melting red-purple solid, whereas Re reacts with F_2 to give ReF_7, a volatile, low-melting yellow solid. Again, reaction with the less oxidizing, heavier halogens produces halides in lower oxidation states. Thus, reaction with Cl_2, a weaker oxidant than F_2, gives $MnCl_2$ and $ReCl_6$. Reaction of Mn with oxygen forms only Mn_3O_4, a mixed-valent compound that contains two Mn(II) and one Mn(III) per

Figure 23.5 Compounds of manganese in oxidation states of +2 to +7. Like vanadium, compounds of manganese in different oxidation states have different numbers of *d* electrons, which leads to compounds with different colors: the $Mn^{2+}(aq)$ is pale pink; $Mn(OH)_3$, which contains Mn(III), is a dark brown solid; MnO_2, which contains Mn(IV), is a black solid; and aqueous solutions of Mn(VI) and Mn(VII) contain the green manganate ion (MnO_4^{2-}) and the purple permanganate ion (MnO_4^-), respectively.

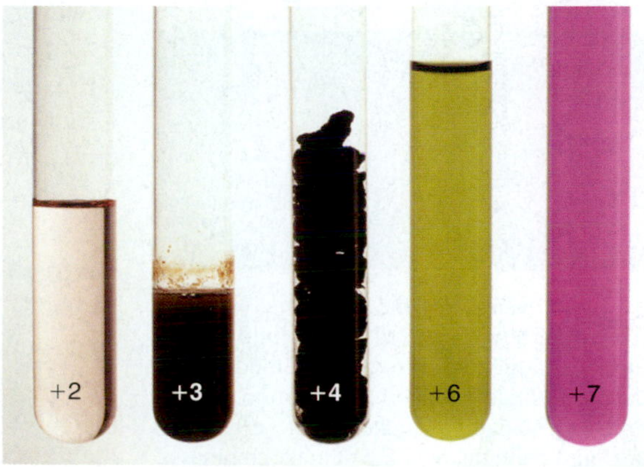

Note the pattern

The chemistry of the Group 7 metals (Mn, Tc, and Re) is dominated by lower oxidation states. Compounds in the maximum possible oxidation state, +7, are readily reduced.

formula unit and is similar in both stoichiometry and structure to magnetite (Fe_3O_4). In contrast, Tc and Re form high-valent oxides, the so-called "heptoxides," M_2O_7, consistent with the increased stability of higher oxidation states for the second and third rows of transition metals. Under forcing conditions, manganese will form Mn_2O_7, an unstable, explosive, green liquid. Also consistent with this trend, the permanganate ion, MnO_4^-, is a potent oxidant, whereas TcO_4^- and ReO_4^- are much more stable. Both Tc and Re form disulfides and diselenides with layered structures analogous to that of MoS_2, as well as more complex "heptasulfides," M_2S_7. As is typical of the transition metals, the Group 7 metals form binary nitrides, carbides, and borides that are generally stable at high temperatures and exhibit metallic properties.

Groups 8, 9, and 10

MGC Groups 7 through 10

In many older versions of the periodic table, Groups 8, 9, and 10 are combined in a single group (Group VIII) because the elements of these three groups exhibit many horizontal similarities in their chemistry, in addition to the similarities within each column. In part, these horizontal similarities are due to the fact that the ionization potentials of the elements, which increase slowly but steadily across the *d* block, have now become so large that the oxidation state corresponding to the formal loss of all valence electrons is encountered only rarely (Group 8) or not at all (Groups 9 and 10). As a result, the chemistry of all three groups is dominated by intermediate oxidation states, especially +2 and +3 for the first-row metals (Fe, Co, and Ni). The heavier elements of these three groups are called "precious metals" because they are rather rare in nature and chemically rather inert.

Group 8 (Fe, Ru, Os)

The chemistry of Group 8 is dominated by iron, whose high abundance in the Earth's crust is due to the extremely high stability of its nucleus. Ruthenium and osmium, on the other hand, are extremely rare elements, with terrestrial abundances of only about 0.1 ppb and 5 ppb, respectively, and they were not discovered until the 19th century. Because of the high melting point of iron (1538°C), early humans could not use it for tools or weapons. The advanced techniques needed to work iron were first developed by the Hittite civilization in Asia Minor sometime before 2000 B.C., and they remained a closely guarded secret that gave the Hittites military supremacy for almost a millennium. With the collapse of the Hittite civilization around 1200 B.C., the technology became widely distributed, however, leading to the Iron Age.

Group 9 (Co, Rh, Ir)

Cobalt is one of the least abundant of the first-row transition metals. Its oxide ores, however, have been used in glass and pottery for thousands of years to produce the brilliant color known as "cobalt blue," and its compounds are consumed in large quantities in the paint and ceramics industries. The heavier elements of Group 9 are also rare, with terrestrial abundances of less than 1 ppb; they are generally found in combination with the heavier elements of Groups 8 and 10 in Ni–Cu–S ores.

Over 2000 years ago, the Bactrian civilization in Western Asia used a 75:25 alloy of copper and nickel for its coins. A modern U.S. nickel has the same composition, but a modern Canadian "nickel" is nickel-plated steel and contains only 2.5% nickel by mass.

Group 10 (Ni, Pd, Pt)

Nickel silicates are easily processed, and consequently nickel has been known and used since antiquity. In fact, a 75:25 Cu:Ni alloy was used for more than 2000 years to mint "silver" coins, and the modern U.S. "nickel" uses the same alloy. In contrast to nickel, palladium and platinum are rare (their terrestrial abundance is about 10–15 ppb), but they are at least an order of magnitude more abundant than the heavier elements of

TABLE 23.6 Some properties of the elements of Groups 8, 9, and 10

Group	Element	Z	Valence-Electron Configuration	Electronegativity	Metallic Radius, pm	mp, °C	Density, g/cm³
	Fe	26	$4s^2\,3d^6$	1.83	126	1538	7.87
8	Ru	44	$5s^1\,4d^7$	2.20	134	2334	12.10
	Os	76	$6s^2\,5d^6\,4f^{14}$	2.20	135	3033	22.60
	Co	27	$4s^2\,3d^7$	1.88	125	1495	8.86
9	Rh	45	$5s^1\,4d^8$	2.28	134	1964	12.40
	Ir	77	$6s^2\,5d^7\,4f^{14}$	2.20	136	2446	22.50
	Ni	28	$4s^2\,3d^8$	1.91	124	1455	8.90
10	Pd	46	$4d^{10}$	2.20	137	1555	12.00
	Pt	78	$6s^2\,5d^8 4f^{14}$	2.20	139	1768	21.50

Groups 8 and 9. Platinum and palladium are used in jewelry, the former as the pure element and the latter as the Pd/Au alloy known as "white gold."

Trends in Group 8, 9, and 10

Some properties of the elements of Groups 8–10 are summarized in Table 23.6. As in earlier groups, similarities in size and electronegativity between the two heaviest members of each group result in similarities in chemistry. We are now at the point in the *d* block where there is no longer a clear correlation between the valence-electron configuration and the preferred oxidation state. For example, all the elements of Group 8 have eight valence electrons, but only Ru and Os have any tendency to form compounds in the +8 oxidation state, and those compounds are powerful oxidants. The predominant oxidation states for all three Group 8 metals are +2 and +3. Although the elements of Group 9 possess a total of nine valence electrons, the +9 oxidation state is unknown for these elements, and the most common oxidation states in the group are +3 and +1. Finally, the elements of Group 10 all have 10 valence electrons, but all three elements are normally found in the +2 oxidation state formed by loss of the ns^2 valence electrons. In addition, Pd and Pt form numerous compounds and complexes in the +4 oxidation state.

Note the pattern

The most important oxidation states are +2 and +3 for Group 8, +3 for Group 9, and +2 and +4 for Group 10.

We stated that higher oxidation states become less stable as we go across the *d*-block elements and more stable as we go down a group. Thus, Fe and Co form trifluorides, but Ni forms only the difluoride, NiF_2. In contrast to Fe, Ru and Os form a series of fluorides up to RuF_6 and OsF_7. The hexafluorides of Rh and Ir are extraordinarily powerful oxidants, and Pt is the only element in Group 10 that forms a hexafluoride. Similar trends are observed among the oxides. For example, Fe forms only FeO, Fe_2O_3, and the mixed-valent Fe_3O_4 (magnetite), all of which are nonstoichiometric. In contrast, Ru and Os form the dioxides, MO_2, and the highly toxic, volatile, yellow tetroxides, which contain formal M=O double bonds. As expected for compounds of metals in such high oxidation states, the latter are potent oxidants. The tendency of the metals to form the higher oxides decreases rapidly as we go farther across the *d* block.

The pattern of reaction with the heavier chalcogens is rather complex. Thus, the oxidation state of Fe, Ru, Os, Co, and Ni in their disulfides is +2 because of the presence of the disulfide ion, S_2^{2-}, but the disulfides of Rh, Ir, Pd, and Pt contain the metal in the +4 oxidation state together with sulfide ions, S^{2-}. This combination of highly charged cations and easily polarized anions results in substances that are not simple ionic compounds and have significant covalent character.

The Groups 8–10 metals form a range of binary nitrides, carbides, and borides. By far the most important of these is cementite (Fe_3C), which is used to strengthen steel. At high temperatures, Fe_3C is soluble in iron, but slow cooling causes the phases to separate and form particles of cementite, which gives a metal that retains much of its strength but is significantly less brittle than pure iron. Palladium is unusual in that it forms a binary hydride with approximate composition $PdH_{0.5}$. Because the H atoms in the metal lattice are highly mobile, thin sheets of Pd are highly permeable to H_2 but essentially impermeable to all other gases, including He. Consequently, diffusion of H_2 through Pd is an effective method for separating hydrogen from other gases.

MGC Groups 11 and 12

Groups 11 and 12

Group 11 (Cu, Ag, Au)

This 1-kg gold nugget was found in Australia; in 2005, it was for sale in Hong Kong at an asking price of more than US$64,000.

The coinage metals—copper, silver, and gold—occur naturally (like the gold nugget shown in the margin), and consequently these were probably the first metals used by ancient humans. For example, decorative gold artifacts dating from the late Stone Age are known, and some gold Egyptian coins are more than 5000 years old. Copper is almost as ancient, with objects dating to about 5000 B.C. Bronze, an alloy of copper and tin that is harder than either of its constituent metals, was used before 3000 B.C., giving rise to the Bronze Age. Deposits of silver are much less common than deposits of gold or copper, yet by 3000 B.C., methods had been developed for recovering silver from its ores, which allowed silver coins to be widely used in ancient times.

Deposits of gold and copper are widespread and numerous, and for many centuries it was relatively easy to obtain large amounts of the pure elements. For example, a single gold nugget discovered in Australia in 1869 weighed more than 150 pounds. Because the demand for these elements has outstripped their availability, methods have been developed to recover them economically from even very low-grade ores (as low as 1% Cu content for copper) by operating on a vast scale, as shown in the photo below of an open-pit copper mine. Copper is used primarily to manufacture electric wires, but large quantities are also used to produce bronze, brass, and alloys for coins. Much of the silver made today is obtained as a byproduct

The Chuquicamata copper mine in northern Chile, the world's largest open-pit copper mine, is 4.3 km long, 3 km wide, and 825 m deep. Each of the gigantic trucks in the foreground (and barely visible in the lower right center) can hold 330 metric tons (330,000 kg) of copper ore.

TABLE 23.7 Some properties of the elements of Groups 11 and 12

Group	Element	Z	Valence-Electron Configuration	Electronegativity	Metallic Radius, pm	mp, °C	Density, g/cm³
	Cu	29	$4s^1\,3d^{10}$	1.90	128	1085	8.96
11	Ag	47	$5s^1\,4d^{10}$	1.93	144	962	10.50
	Au	79	$6s^1\,5d^{10}\,4f^{14}$	2.40	144	1064	19.30
	Zn	30	$4s^2\,3d^{10}$	1.65	134	420	7.14
12	Cd	48	$5s^2\,4d^{10}$	1.69	149	321	8.69
	Hg	80	$6s^2\,5d^{10}\,4f^{14}$	1.90	151	−38.8	13.50

of the manufacture of other metals, especially Cu, Pb, and Zn. In addition to its use in jewelry and silverware, silver is used in Ag/Zn and Ag/Cd "button" batteries (see Chapter 19). Gold is typically found either as tiny particles of the pure metal or as gold telluride ($AuTe_2$). It is used as a currency reserve, in jewelry, in the electronics industry for corrosion-free contacts, and, in very thin layers, as a reflective window coating that minimizes heat transfer.

Some properties of the coinage metals are listed in Table 23.7. The electronegativity of gold (χ = 2.40) is close to that of the nonmetals sulfur and iodine, which suggests that the chemistry of gold should be somewhat unusual for a metal. The coinage metals have the highest electrical and thermal conductivities of all the metals, and they are also the most ductile and malleable. With an $ns^1\,(n-1)d^{10}$ valence-electron configuration, the chemistry of these three elements is dominated by the +1 oxidation state due to loss of the single *ns* electron. Higher oxidation states are also known, however: +2 is common for Cu and, to a lesser extent, Ag, and +3 for Au because of the relatively low values of the second and (for Au) third ionization energies. Due to the half-filled *ns* orbital in the neutral atoms, all three elements have significant electron affinities. As a result, gold reacts with powerful reductants like Cs and solutions of the alkali metals in liquid ammonia to produce the gold *anion*, Au^-, with a $6s^2\,5d^{10}$ valence electron configuration.

Note the pattern

The most important oxidation state for Group 11 is +1.

All the Group 11 elements are relatively unreactive, and their reactivity decreases from Cu to Au. Hence, they are noble metals that are particularly well suited for use in coins and jewelry. Copper reacts with O_2 at high temperatures to produce Cu_2O and with sulfur to form Cu_2S. Neither silver nor gold reacts directly with oxygen, although oxides of these elements can be prepared by other routes. Silver reacts with sulfur compounds to form the black Ag_2S coating known as "tarnish." Gold is the only metal that does not react with sulfur; it also does not react with nitrogen, carbon, or boron. All the coinage metals do, however, react with oxidizing acids. Thus, both Cu and Ag dissolve in HNO_3 and in hot concentrated H_2SO_4, while Au dissolves in the 3:1 $HCl{:}HNO_3$ mixture known as *aqua regia*. Furthermore, all three metals dissolve in basic cyanide solutions in the presence of oxygen to form very stable $[M(CN)_2]^-$ ions, a reaction that is used to separate gold from its ores (see Chapter 4).

All the monohalides except CuF and AuF are known (including AgF). Once again, iodine is unable to stabilize the higher oxidation states (Au^{3+} and Cu^{2+}). Thus, all the copper(II) halides except the iodide are known, but the only dihalide of silver is AgF_2. In contrast, all the gold trihalides, AuX_3, are known, again except the triiodide. No binary nitrides, borides, or carbides are known for the Group 11 elements.

Group 12 (Zn, Cd, Hg)

Continuing across the transition metals, we encounter the Group 12 elements. Because none of the elements of Group 12 has a partially filled $(n - 1)d$ subshell, they are not, strictly speaking, "transition metals." Nonetheless, much of their chemistry is similar to that of the elements that immediately precede them in the *d* block. The Group 12 metals are similar in abundance to those of Group 11, and they are almost always found in combination with sulfur. Because zinc and cadmium are chemically similar, virtually all zinc ores contain significant amounts of cadmium. All three metals are commercially important, although the use of Cd is restricted because of its toxicity. Zinc is used for corrosion protection (see Chapter 19), in batteries, to make brass, and, in the form of ZnO, in the production of rubber and paints. Cadmium is used as the cathode in rechargeable Nicad batteries. Large amounts of mercury are used in the production of chlorine and NaOH by the chlor-alkali process (see Chapter 19), while smaller amounts are consumed in mercury-vapor streetlights and mercury batteries.

As shown in Table 23.7, the Group 12 metals are significantly more electropositive than the elements of Group 11, and they therefore have less "noble" character. They also have much lower melting and boiling points than the preceding transition metals. In contrast to trends in the preceding groups, Zn and Cd are similar to each other, but very different from the heaviest element (Hg). In particular, Zn and Cd are rather active metals, whereas mercury is not. Because mercury, the only metal that is a liquid at room temperature, can "dissolve" many metals by forming amalgams, it was especially valued by medieval alchemists trying to transmute base metals to gold and silver. All three elements in Group 12 have $ns^2(n - 1)d^{10}$ valence-electron configurations, and consequently the +2 oxidation state, corresponding to the loss of the two *ns* electrons, dominates their chemistry. In addition, mercury forms a series of compounds in the +1 oxidation state that contain the diatomic mercurous ion, Hg_2^{2+}.

Note the pattern

The most important oxidation state for Group 12 is +2.

All the possible Group 12 dihalides, MX_2, are known, and they range from ionic (the fluorides) to highly covalent (such as $HgCl_2$). The highly covalent character of many mercuric and mercurous halides is surprising given the large size of the cations, and this has been attributed to the existence of an easily distorted $5d^{10}$ subshell. Zinc and cadmium react with oxygen to form amphoteric MO, whereas mercury forms HgO only within a narrow temperature range (350–400°C). Whereas zinc and cadmium dissolve in mineral acids such as HCl with the evolution of hydrogen, mercury dissolves only in oxidizing acids such as HNO_3 and H_2SO_4. All three metals react with sulfur and the other chalcogens to form the binary chalcogenides, and mercury has an extraordinarily high affinity for sulfur.

EXAMPLE 23.2

For each reaction, explain why the indicated products form:

(a) $TiCl_4(l) + 2H_2O(l) \longrightarrow TiO_2(s) + 4HCl(aq)$

(b) $WO_3(s) + 3C(s) + 3Cl_2(g) \xrightarrow{\Delta} WCl_6(s) + 3CO(g)$

(c) $Sc_2O_3(s) + 2OH^-(aq) + 3H_2O(l) \longrightarrow 2[Sc(OH)_4]^-(aq)$

(d) $2KMnO_4(aq) + 2H_2SO_4(l) \longrightarrow Mn_2O_7(l) + 2KHSO_4(soln) + H_2O(soln)$

(e) $4CrCl_2(aq) + O_2(g) + 4H^+(aq) \longrightarrow 4Cr^{3+}(aq) + 8Cl^-(aq) + 2H_2O(l)$

Given Balanced chemical equation

Asked for Why the indicated products form

Strategy

Refer to the periodic trends in this section, Figures 23.1 and 23.2, Figure 7.11, and Tables 23.1–23.7 to explain why these products form.

Solution

(a) The most stable oxidation state for Ti is +4, and neither reactant is a particularly strong oxidant or reductant; hence, a redox reaction is unlikely. Similarly, neither reactant is a particularly strong acid or base, so an acid–base reaction is unlikely. Because $TiCl_4$ contains Ti in a relatively high oxidation state (+4), however, it is likely to be rather covalent in character, with reactivity similar to that of a semimetal halide such as $SiCl_4$. Covalent halides tend to hydrolyze in water to produce the hydrohalic acid and either the oxide of the other element or a species analogous to an oxoacid.

(b) This reaction involves the oxide of a Group 6 metal in its highest oxidation state (WO_3) and two elements, one of which is a reductant (C) and the other an oxidant (Cl_2). Consequently, some sort of redox reaction will occur. Carbon can certainly react with chlorine to form CCl_4, and WO_3 is a potential source of oxygen atoms that can react with carbon to produce CO, which is very stable. If CO is one of the products, then it seems likely that the other product will contain the metal and chlorine. The most likely combination is WCl_6 (leaving the oxidation state of the metal unchanged).

(c) One of the reactants is a strong base (OH^-), so an acid–base reaction is likely *if* the other reactant is acidic. Because oxides like Sc_2O_3, in which the metal is in an intermediate oxidation state, are often amphoteric, we expect Sc_2O_3 to dissolve in base to form a soluble hydroxide complex.

(d) Concentrated sulfuric acid is both an oxidant and a strong acid that tends to protonate and dehydrate other substances. The permanganate ion already contains manganese in its highest possible oxidation state (+7), so it cannot be oxidized further. A redox reaction is impossible, which leaves an acid–base reaction as the most likely alternative. Sulfuric acid is likely to protonate the terminal oxygen atoms of permanganate, allowing them to be lost as water.

(e) Molecular oxygen is an oxidant, so a redox reaction is likely if the other reactant can be oxidized. Because chromous chloride contains chromium in its lowest accessible oxidation state, a redox reaction will occur in which Cr^{2+} ions are oxidized and O_2 is reduced. In the presence of protons, the reduction product of O_2 is water, so we need to determine only the identity of the oxidation product of Cr^{2+}. Chromium forms compounds in two common higher oxidation states: the Cr^{3+} ion, which is the most stable, and the $Cr_2O_7^{2-}$ ion, which is a more powerful oxidant than O_2. We therefore predict that the reaction will form $Cr^{3+}(aq)$ and water.

EXERCISE 23.2

Predict the products of the reactions and balance each equation:

(a) $Cr^{2+}(aq) + Fe^{3+}(aq) \xrightarrow{H^+}$
(b) $Na_2Cr_2O_7(aq) + H_2SO_4(l) \longrightarrow$
(c) $FeBr_2(aq) + O_2(g) \xrightarrow{H^+}$
(d) $VBr_4(l) + H_2O(l) \longrightarrow$
(e) $V_2O_5(s) + H^+(aq) \longrightarrow$
(f) $ZrO_2(s) + C(s) + Cl_2(g) \xrightarrow{\Delta}$

Answer

(a) $Cr^{2+}(aq) + Fe^{3+}(aq) \xrightarrow{H^+} Cr^{3+}(aq) + Fe^{2+}(aq)$
(b) $Na_2Cr_2O_7(aq) + 2H_2SO_4(l) \longrightarrow 2NaHSO_4(soln) + H_2O(soln) + 2CrO_3(s)$
(c) $4FeBr_2(aq) + O_2(g) + 4H^+(aq) \longrightarrow 4Fe^{3+}(aq) + 2H_2O(l) + 8Br^-(aq)$
(d) $VBr_4(l) + H_2O(l) \longrightarrow VO^{2+}(aq) + 4Br^-(aq) + 2H^+(aq)$
(e) $V_2O_5(s) + 2H^+(aq) \longrightarrow 2VO_2^+(aq) + H_2O(l)$
(f) $ZrO_2(s) + 2C(s) + 2Cl_2(g) \xrightarrow{\Delta} ZrCl_4(s) + 2CO(g)$

23.3 ∘ Metallurgy

MGC Metallurgy

Very few of the transition metals are found in nature as the free metals. Consequently, almost all metallic elements must be isolated from metal oxide or metal sulfide ores. **Metallurgy** is the set of processes by which metals are extracted from their ores and converted to more useful forms.

Metallurgy consists of three general steps: (1) mining the ore, (2) separating and concentrating the metal or the metal-containing compound, and (3) reducing the ore to the metal. Additional processes are sometimes required to improve the mechanical properties of the metal or to increase its purity. Many ores contain relatively low concentrations of the desired metal; for example, copper ores that contain even 1% Cu by mass are considered commercially useful.

Note the pattern

Metallurgy depends on separation of a metal compound from its ore and reduction to the metal at high temperature (pyrometallurgy) or in aqueous solution (hydrometallurgy).

After an ore has been mined, the first step in processing is usually to crush it, because the rate of chemical reactions increases dramatically with an increase in surface area. Next, one of three general strategies is used to separate and concentrate the compound(s) of interest: *settling* and *flotation*, which are based on differences in density between the desired compound and impurities; *pyrometallurgy*, which uses chemical reduction at high temperatures; and *hydrometallurgy*, which employs chemical or electrochemical reduction of an aqueous solution of the metal. Other methods that take advantage of unusual physical or chemical properties of a particular compound may also be used. For example, crystals of magnetite (Fe_3O_4) are tiny but rather powerful magnets; in fact, magnetite (also known as *lodestone*) was used to make the first compasses in China during the first century B.C. If a crushed ore that contains magnetite is passed through a powerful magnet, the Fe_3O_4 particles are attracted to the poles of the magnet, allowing them to be easily separated from other minerals.

Settling and flotation have been used for thousands of years to separate particles of dense metals such as gold, using the technique known as *panning*, in which a sample of gravel or sand is swirled in water in a shallow metal pan. Because the density of gold (19.3 g/cm^3) is so much higher than that of most silicate minerals (about 2.5 g/cm^3), silicate particles settle more slowly and can be poured off with the water, leaving dense gold particles on the bottom of the pan. Conversely, in flotation, the compound of interest is made to float on top of a solution. Blowing air through a suspension of the crude ore in a mixture of water and an organic liquid, such as pine tar, produces a "froth" that contains tiny particles of hydrophobic solids, such as metal sulfides, while more hydrophilic oxide minerals remain suspended in the aqueous phase (Figure 23.6). To make the separation more efficient, small amounts of an anionic sulfur-containing compound, such as $Na^+C_2H_5OCS_2^-$, are added; the additive binds to the sulfur-rich surface of the metal sulfide particles and makes the metal sulfide particles even more hydrophobic. The resulting froth is highly enriched in the desired metal sulfide(s), which can be removed simply by skimming. This method works even for compounds as dense as PbS (7.5 g/cm^3).

Pyrometallurgy

In pyrometallurgy, an ore is heated with a reductant in order to obtain the metal. Theoretically, it should be possible to obtain virtually any metal from its ore by using coke, an inexpensive form of crude carbon, as the reductant. An example of such a reaction is

$$CaO(s) + C(s) \xrightarrow{\Delta} Ca(l) + CO(g) \tag{23.4}$$

Unfortunately, many of the early transition metals, such as Ti, react with carbon to form stable binary carbides. Consequently, more expensive reductants, such as hydrogen, aluminum, magnesium, or calcium, must be used to obtain these metals. Many metals that occur naturally as sulfides can be obtained by heating the sulfide in air, as shown for lead in the equation

$$PbS(s) + O_2(g) \xrightarrow{\Delta} Pb(l) + SO_2(g) \tag{23.5}$$

(a) Froth flotation **(b) Precious-metal–sulfide froth** **(c) Anionic organic sulfur ligands bound to metal-sulfide particle**

Figure 23.6 Froth flotation. (a) When air is blown through a mixture of a finely ground metal sulfide ore and water, the more hydrophobic metal sulfides form a froth that can be easily removed, allowing them to be separated from more hydrophilic metal oxides and silicates. (b) A froth containing precious metal sulfides formed as a byproduct during the production of metallic nickel. (c) An anionic sulfur additive with hydrophobic "tails" can be used to enhance the hydrophobic character of metal sulfide particles, which causes them to be attracted to the air/water interface in the foam.

The reaction is driven to completion by the formation of SO_2, a stable gas.

Pyrometallurgy is also used in the iron and steel industries. The overall reaction for the production of iron in a blast furnace is

$$Fe_2O_3(s) + 3C(s) \xrightarrow{\Delta} 2Fe(l) + 3CO(g) \quad (23.6)$$

The actual reductant is CO, which reduces Fe_2O_3 to give $Fe(l)$ and $CO_2(g)$; the CO_2 is then reduced back to CO by reaction with excess carbon. As the ore, lime, and coke drop into the furnace (Figure 23.7), any silicate minerals in the ore react with the lime

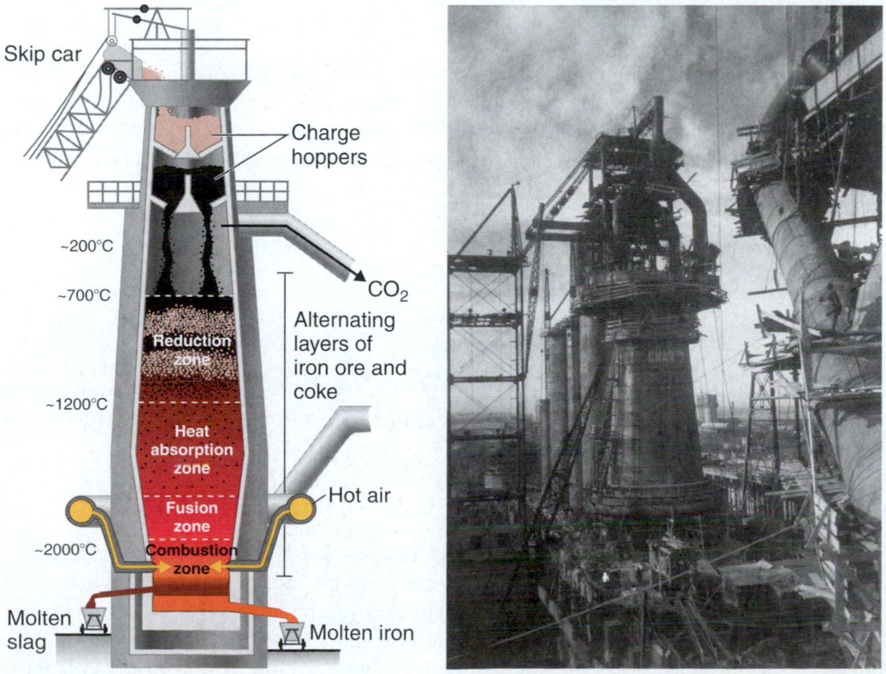

(a) Blast furnace **(b) World's largest blast furnace in 1931**

Figure 23.7 A blast furnace for converting iron oxides to iron metal. (a) The furnace is charged with alternating layers of iron ore (largely Fe_2O_3) and a mixture of coke (C) and limestone ($CaCO_3$). Blasting hot air into the mixture from the bottom causes it to ignite, producing CO and raising the temperature of the lower part of the blast furnace to about 2000°C. As the CO that is formed initially rises, it reduces Fe_2O_3 to form CO_2 and elemental iron, which absorbs heat and melts as it falls into the hottest part of the furnace. Decomposition of $CaCO_3$ at high temperatures produces CaO (lime) and additional CO_2, which reacts with excess coke to form more CO. (b) This blast furnace in Magnitogorsk, Russia, was the largest in the world when it was built in 1931.

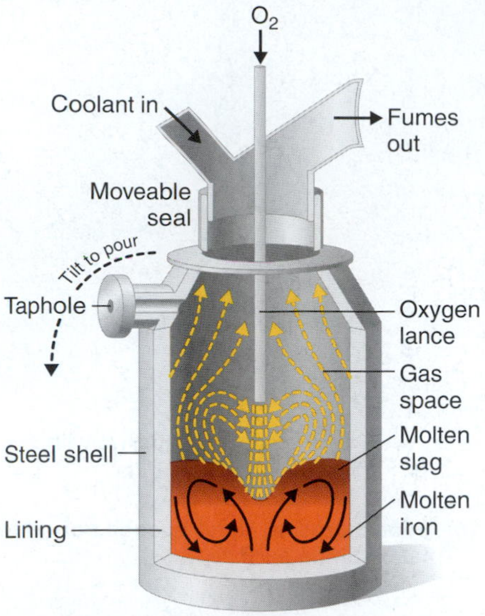

(a) Oxygen furnace

(b) Oxygen furnace in Ohio steel mill

Figure 23.8 A basic oxygen furnace for conversion of crude iron to steel. (a) A blast of oxygen is used to agitate the molten iron and oxidize impurities to products that dissolve in the less dense slag layer. The slag and the molten steel are removed by tilting the entire furnace and pouring the liquids out through the taphole. (b) Filling a basic oxygen furnace with molten iron from a blast furnace.

to produce a low-melting mixture of calcium silicates called *slag*, which floats on top of the molten iron. Molten iron is then allowed to run out the bottom of the furnace, leaving the slag behind. Originally, the iron was collected in pools called *pigs*, which is the origin of the name "pig iron."

Iron that is obtained directly from a blast furnace has an undesirably low melting point (about 1100°C instead of 1539°C) because it contains a large amount of dissolved carbon. It contains other impurities as well (such as Si, S, P, and Mn from contaminants in the iron ore that were also reduced during processing), which must be removed because they make iron brittle and unsuitable for most structural applications. Blowing oxygen through the molten pig iron removes the impurities by selective oxidation because they are more readily oxidized than iron (Figure 23.8). In the final stage of steel production, small amounts of other metals are added at specific temperatures to produce steel with the desired combination of properties.

Hydrometallurgy

The most selective methods for separating metals from their ores are based on the formation of metal complexes. For example, gold is often found as tiny flakes of the metal, usually in association with quartz or pyrite deposits. In those circumstances, gold is typically extracted using *cyanide leaching*, which forms a stable gold–cyanide complex, $Au(CN)_2^-$:

$$4Au(s) + 8NaCN(aq) + O_2(g) + 2H_2O(l) \longrightarrow 4Na[Au(CN)_2](aq) + 4NaOH(aq) \quad (23.7)$$

Virtually pure gold can be obtained by adding powdered zinc to the solution:

$$Zn(s) + 2[Au(CN)_2]^-(aq) \longrightarrow [Zn(CN)_4]^{2-}(aq) + 2Au(s) \quad (23.8)$$

A related method, which is used to separate Co^{3+}, Ni^{2+}, and Cu^+ from Fe, Mn, and Ti, is based on the formation of stable, soluble ammonia complexes of ions of the late transition metals.

EXAMPLE 23.3

Suppose you are working in the chemistry laboratory of a mining company that has discovered a new source of tungsten ore containing about 5% WS_2 in a granite matrix (granite is a complex aluminosilicate mineral). You have been asked to outline an economical procedure for isolating WS_2 from the ore and then converting it to elemental tungsten in as few steps as possible. What would you recommend?

Given Composition of ore

Asked for Procedure to isolate metal sulfide

Strategy

Determine which of the methods described would be most effective for separating the metal sulfide from the ore. Then determine the best method for reducing the metal to the pure element.

Solution

You need to separate and concentrate the WS_2, then convert it to a suitable form so it can be reduced to the metal (if necessary), and then carry out the reduction. Since the new ore is a binary metal sulfide, you could take advantage of the hydrophilic nature of most metal sulfides to separate WS_2 by froth flotation. Then, because most metal sulfides cannot be reduced directly to the metal using carbon, you will probably need to convert WS_2 to an oxide for subsequent reduction. One point to consider is whether the oxide can be reduced using carbon because, as we noted, many transition metals react with carbon to form stable carbides. Here is one possible procedure for producing tungsten from this new ore:

1. Grind the ore and separate WS_2 from the silicate matrix by flotation.
2. Convert the crude WS_2 to an oxide by roasting in air (because W is in Group 6, you anticipate that roasting will yield WO_3, the oxide in the highest possible oxidation state). The reaction will also produce SO_2, which will have to be removed by scrubbing the exhaust gases to minimize environmental pollution.
3. Reduce the oxide with hydrogen gas at high temperature to avoid carbide formation:

$$WO_3(s) + 3H_2(g) \xrightarrow{\Delta} W(s) + 3H_2O(g)$$

EXERCISE 23.3

Propose an economical procedure for converting a mineral deposit of $BaCO_3$ to the pure metal.

Answer (1) Dissolve the barium carbonate in HCl(*aq*) to give $Ba^{2+}(aq)$, which will allow the insoluble silicate minerals to be removed by filtration. (2) Precipitate the barium from solution as $BaCO_3$ by adding solid Na_2CO_3. (3) Dissolve the solid $BaCO_3$ in concentrated HCl and remove the water by evaporation to obtain anhydrous $BaCl_2$. (4) Reduce molten $BaCl_2$ to the metal by electrolysis.

23.4 Coordination Compounds

One of the most important properties of metallic elements is their ability to act as Lewis acids that form complexes with a variety of Lewis bases. A **metal complex** consists of a central metal atom or ion that is bonded to one or more *ligands* (from the Latin *ligare*, "to bind"). **Ligands** are ions or molecules that contain one or more pairs of electrons that can be shared with the metal. Metal complexes can be neutral, such as $[Co(NH_3)_3Cl_3]$; positively charged, such as $[Nd(H_2O)_9]^{3+}$; or negatively charged, such as $UF_8{}^{4-}$. Electrically charged metal complexes are sometimes called **complex ions**. A **coordination compound** contains one or more metal complexes.

Coordination compounds are important for at least three reasons. First, most of the elements in the periodic table are metals, and almost all metals form complexes. Hence, metal complexes are a feature of the chemistry of more than half the elements. Second, many industrial catalysts are metal complexes, and such catalysts are steadily becoming more important as a way to control reactivity. For example, a mixture of a titanium complex and an organometallic compound of aluminum is the catalyst used to produce most of the polyethylene and polypropylene "plastic" items we use every day. Finally, transition-metal complexes are essential in biochemistry. Examples include hemoglobin, an iron complex that transports oxygen in our blood; cytochromes, iron complexes that transfer electrons in our cells; and complexes of Fe, Zn, Cu, and Mo that are crucial components of certain enzymes, the catalysts for all biological reactions. Metal complexes are so important in biology that we consider the topic separately in Section 23.6.

History of Coordination Compounds

MGC Coordination Compounds

Coordination compounds have been known and used since antiquity; probably the oldest is the deep blue pigment called Prussian blue, $KFe_2(CN)_6$. The chemical nature of these substances, however, was unclear for a number of reasons. For example, many compounds called "double salts" were known, such as $AlF_3 \cdot 3KF$, $Fe(CN)_2 \cdot 4KCN$, and $ZnCl_2 \cdot 2CsCl$, which were combinations of simple salts in fixed and apparently arbitrary ratios. Why should $AlF_3 \cdot 3KF$ exist but not $AlF_3 \cdot 4KF$ or $AlF_3 \cdot 2KF$? And why should a 3:1 KF:AlF_3 mixture have different chemical and physical properties than either of its components? Similarly, *adducts* of metal salts with neutral molecules such as ammonia

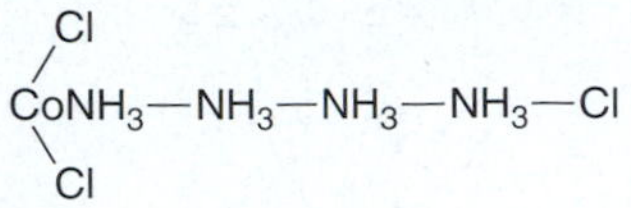

$CoCl_3 \cdot 4NH_3$

were also known—for example, $CoCl_3 \cdot 6NH_3$, which was first prepared sometime before 1798. Like the double salts, the compositions of these adducts exhibited fixed and apparently arbitrary ratios of the components. For example, $CoCl_3 \cdot 6NH_3$, $CoCl_3 \cdot 5NH_3$, $CoCl_3 \cdot 4NH_3$, and $CoCl_3 \cdot 3NH_3$ were all known and had very different properties, but no matter how hard they tried, chemists could not prepare $CoCl_3 \cdot 2NH_3$ or $CoCl_3 \cdot NH_3$.

Thus, the chemical *composition* of such compounds was readily established by existing analytical methods, but their chemical *nature* was puzzling and highly controversial. The major problem was that what we now call *valence* (that is, oxidation state) and *coordination number* were thought to be identical. As a result, highly implausible (to modern eyes at least) structures were proposed for such compounds, including the "Chattanooga choo-choo" model for $CoCl_3 \cdot 4NH_3$ shown in the margin.

The modern theory of coordination chemistry is based largely on the work of Alfred Werner (1866–1919, Nobel Prize 1913). In a series of careful experiments carried out in the late 1880s and early 1890s, he examined the properties of several series of metal halide complexes with ammonia. For example, five different "adducts" of ammonia with $PtCl_4$ were known at the time: $PtCl_4 \cdot nNH_3$ (n = 2–6). Some of Werner's original data on these compounds are shown in Table 23.8. The electrical conductivity of aqueous solutions of these compounds was roughly proportional to the number of ions formed per mole, while the number of chloride ions that could be precipitated as AgCl upon addition of $Ag^+(aq)$ was a measure of the number of "free" chloride ions present. For example, Werner's data on $PtCl_4 \cdot 6NH_3$ in Table 23.8 showed that all the chloride ions were present as free chloride. In contrast, $PtCl_4 \cdot 2NH_3$ was a neutral molecule that contained no free chloride ions.

These data led Werner to postulate that metal ions have two different kinds of valence: a *primary valence* (oxidation state) that corresponds to the positive charge on the metal ion; and a *secondary valence* (coordination number) that is the total number of ligands bound to the metal ion. If Pt had a primary valence of 4 and a secondary valence of 6, Werner could explain the properties of the $PtCl_4$–NH_3 adducts by the following reactions, where the metal complex is enclosed in square brackets:

$$\begin{aligned}
&[Pt(NH_3)_6]Cl_4 \longrightarrow [Pt(NH_3)_6]^{4+}(aq) + 4Cl^-(aq)\\
&[Pt(NH_3)_5Cl]Cl_3 \longrightarrow [Pt(NH_3)_5Cl]^{3+}(aq) + 3Cl^-(aq)\\
&[Pt(NH_3)_4Cl_2]Cl_2 \longrightarrow [Pt(NH_3)_4Cl_2]^{2+}(aq) + 2Cl^-(aq)\\
&[Pt(NH_3)_3Cl_3]Cl \longrightarrow [Pt(NH_3)_3Cl_3]^{+}(aq) + Cl^-(aq)\\
&[Pt(NH_3)_2Cl_4] \longrightarrow [Pt(NH_3)_2Cl_4]^{0}(aq)
\end{aligned} \tag{23.9}$$

Further work showed that the two missing members of the series, $[Pt(NH_3)Cl_5]^-$ and $[PtCl_6]^{2-}$, could be prepared as their mono- and dipotassium salts, respectively. Similar studies established coordination numbers of 6 for Co^{3+} and Cr^{3+} and 4 for Pt^{2+} and Pd^{2+}.

Werner's studies on the analogous Co^{3+} complexes also allowed him to propose a structural model for metal complexes with a coordination number of 6. Thus, he found that $[Co(NH_3)_6]Cl_3$ (yellow) and $[Co(NH_3)_5Cl]Cl_2$ (purple) were 1:3 and 1:2 electrolytes.

Alfred Werner, 1866–1919. Werner, the son of a factory worker, was born in Alsace. He developed an interest in chemistry at an early age, and he did his first independent research experiments at age 18. While doing his military service in southern Germany, he attended a series of chemistry lectures, and he subsequently received his Ph.D. at the University of Zurich in Switzerland, where he was appointed Professor of Chemistry at age 29. He won the Nobel Prize in chemistry in 1913 for his work on coordination compounds, which he performed as a graduate student and first presented at age 26. Apparently, Werner was so obsessed with solving the riddle of the structure of coordination compounds that his brain continued to work on the problem even while he was asleep. In 1891, when he was only 25, he woke up in the middle of the night and, in only a few hours, had laid the foundation for modern coordination chemistry.

TABLE 23.8 Werner's data on complexes of ammonia with $PtCl_4$

Complex	Conductivity, ohm^{-1}	Number of Ions Per Formula Unit	Number of Cl^- Ions Precipitated by Ag^+
$PtCl_4 \cdot 6NH_3$	523	5	4
$PtCl_4 \cdot 5NH_3$	404	4	3
$PtCl_4 \cdot 4NH_3$	299	3	2
$PtCl_4 \cdot 3NH_3$	97	2	1
$PtCl_4 \cdot 2NH_3$	0	0	0

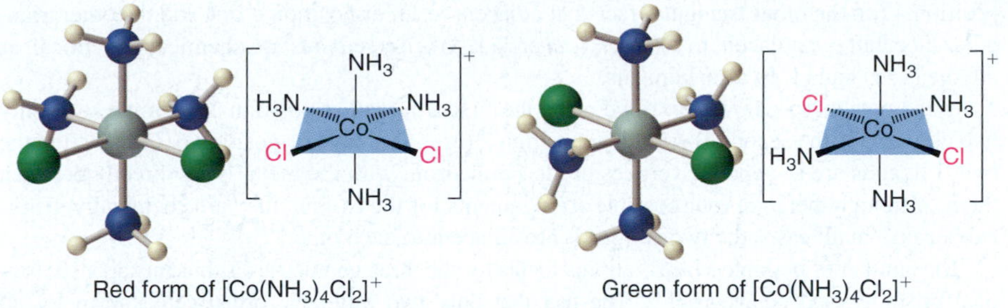

Figure 23.9 Complexes with different arrangements of the same ligands have different properties. The $[Co(NH_3)_4Cl_2]^+$ ion can have two different arrangements of the ligands, which results in different colors: if the two Cl^- ligands are next to each other, the complex is red, but if they are opposite each other, the complex is green.

Unexpectedly, however, *two* different $[Co(NH_3)_4Cl_2]Cl$ compounds were known: one was red, and the other was green (Figure 23.9a). Since both compounds had the same chemical composition and the same number of groups of the same kind attached to the same metal, there had to be something different about the *arrangement* of the ligands around the metal ion. Werner's key insight was that the six ligands in $[Co(NH_3)_4Cl_2]Cl$ had to be arranged at the vertices of an *octahedron*, because that was the only structure consistent with the existence of two, and only two, arrangements of ligands (Figure 23.9b). His conclusion was corroborated by the existence of only two different forms of the next compound in the series, $Co(NH_3)_3Cl_3$.

EXAMPLE 23.4

In Werner's time, many complexes of the general formula MA_4B_2 were known, but no more than two different compounds with the same composition had been prepared for any metal. To confirm Werner's reasoning, calculate the maximum number of different structures that are possible for 6-coordinate MA_4B_2 complexes with each of the three most symmetrical possible structures: a hexagon, a trigonal prism, and an octahedron. What does the fact that no more than two forms of any MA_4B_2 complex were known tell you about the three-dimensional structures of these complexes?

Given Three possible structures, number of different forms known for MA_4B_2 complexes

Asked for Number of different arrangements of ligands for MA_4B_2 complex for each structure

Strategy

Sketch each structure, place a B ligand at one vertex, and see how many different positions are available for the second B ligand.

Solution

The three regular 6-coordinate structures are shown here, with each of the coordination positions numbered so that we can keep track of the different arrangements of ligands. For each structure, all vertices are equivalent. Let's begin with a symmetrical MA_6 complex, and simply replace two of the A ligands in each structure to give an MA_4B_2 complex:

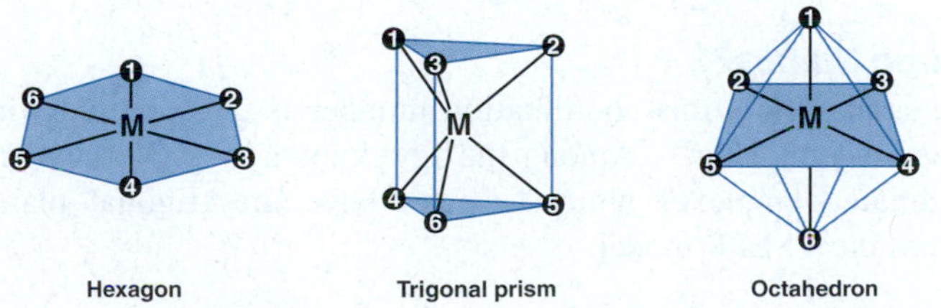

For the hexagon, let's place the first B ligand at position 1. There are now three possible places for the second B ligand: at position 2 (or 6), position 3 (or 5), or position 4. These are the *only* possible arrangements. The (1, 2) and (1, 6) arrangements are chemically identical because the two B ligands are adjacent to each other. The (1, 3) and (1, 5) arrangements are also identical because in both cases the two B ligands are separated by an A ligand.

Turning to the trigonal prism, let's place the first B ligand at position 1. Again, there are three possible choices for the second B ligand: at position 2 or 3 on the same triangular face, position 4 (on the other triangular face, but adjacent to 1), or position 5 or 6 (on the other triangular face, but *not* adjacent to 1). The (1, 2) and (1, 3) arrangements are chemically identical, as are the (1, 5) and (1, 6) arrangements.

In the octahedron, however, if we place the first B ligand at position 1, then we have only two choices for the second B ligand: at position 2 (or 3 or 4 or 5) or position 6. In the latter, the two B ligands are at opposite vertices of the octahedron, with the metal lying directly between them. Although there are four possible arrangements for the former, they are chemically identical because in all cases the two B ligands are adjacent to each other.

The number of possible MA_4B_2 arrangements for the three geometries is thus: hexagon, 3; trigonal prism, 3; and octahedron, 2. The fact that only two different forms were known for all MA_4B_2 complexes that had been prepared *suggested* that the correct structure was the octahedron, but did *not* prove it. For some reason one of the three arrangements possible for the other two structures could have been less stable or harder to prepare, and had simply not yet been synthesized. When combined with analogous results for other types of complexes (e.g., MA_3B_3), however, the data were best explained by an octahedral structure for 6-coordinate metal complexes.

EXERCISE 23.4

Determine the maximum number of structures that are possible for a 4-coordinate MA_2B_2 complex with either of the two most symmetrical structures: square planar and tetrahedral.

Answer square planar, 2; tetrahedral, 1

Structures of Metal Complexes

The coordination numbers of metal ions in metal complexes can range from 2 to at least 9. In general, the differences in energy between different arrangements of ligands are greatest for complexes with low coordination numbers and decrease as the coordination number increases. Thus, usually only one or two structures are possible for complexes with low coordination numbers, whereas several different energetically equivalent structures are possible for complexes with high coordination numbers ($n > 6$). We describe the most commonly encountered structures for coordination numbers 2–9 next. Many of these structures should be familiar to you from our discussion of the VSEPR model in Chapter 9, because they correspond to the lowest-energy arrangements of n electron pairs around a central atom.

Coordination Number 2

Although it is rare for most metals, this coordination number is surprisingly common for d^{10} metal ions, especially Cu^+, Ag^+, Au^+, and Hg^{2+}. An example is the $[Au(CN)_2]^-$ ion, which is used to extract gold from its ores, as described in Section 23.3. As expected based on VSEPR considerations, these complexes have the linear L–M–L structure shown in the margin.

Linear ML_2

Coordination Number 3

Although it is also rare, this coordination number is encountered with d^{10} metal ions such as Cu^+ and Hg^{2+}. Among the few known examples is the HgI_3^- ion. Three-coordinate complexes almost always have the trigonal planar structure expected from the VSEPR model.

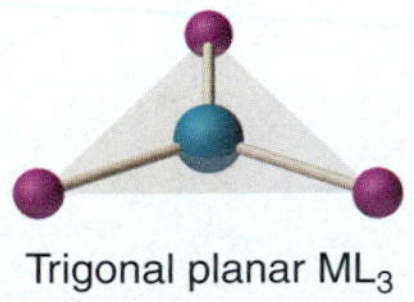

Trigonal planar ML_3

Coordination Number 4

Two common structures are observed for 4-coordinate metal complexes: tetrahedral and square planar. The tetrahedral structure is observed for all 4-coordinate complexes of nontransition metals, such as BeF_4^{2-}, and d^{10} ions, such as $ZnCl_4^{2-}$. It is also found for 4-coordinate complexes of the first-row transition metals, especially those with halide ligands ($FeCl_4^-$, $FeCl_4^{2-}$). In contrast, square planar structures are routinely

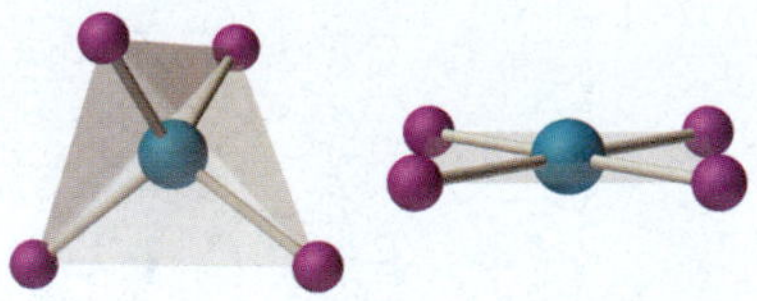

Tetrahedral and square planar ML_4

observed for 4-coordinate complexes of second- and third-row transition metals with d^8 electron configurations, such as Rh^+ and Pd^{2+}, and they are also encountered in some complexes of Ni^{2+} and Cu^{2+}.

Coordination Number 5

This coordination number is less common than 4 and 6, but it is still found frequently in two different structures: trigonal bipyramidal and square pyramidal. Because the energies of these structures are usually rather similar for most ligands, many 5-coordinate complexes have distorted structures that lie somewhere between the two extremes.

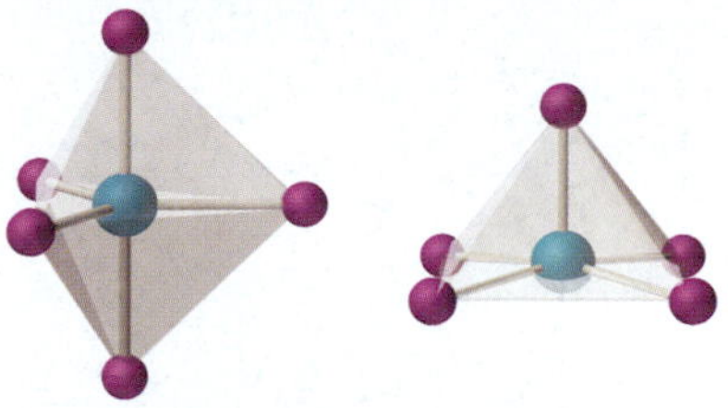

Trigonal bipyramidal and square pyramidal ML_5

Coordination Number 6

This coordination number is by far the most common. The six ligands are almost always at the vertices of an octahedron or a distorted octahedron. The only other 6-coordinate structure is the trigonal prism, which is very uncommon in simple metal complexes.

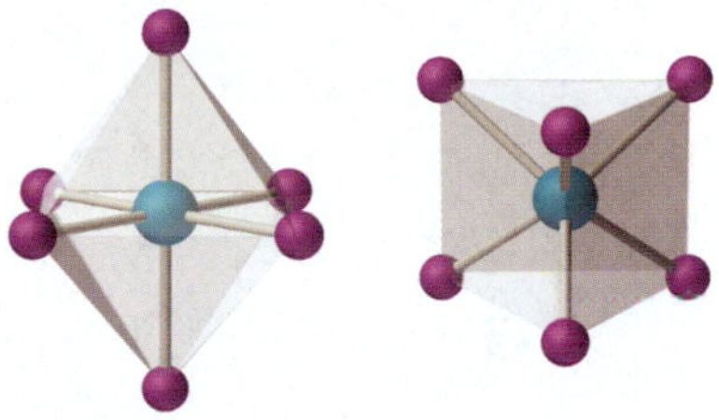

Octahedral and trigonal prismatic ML_6

Coordination Number 7

This relatively uncommon coordination number is generally encountered for only large metals (such as the second- and third-row transition metals, lanthanides, and actinides). At least three different structures are known, two of which are derived from an octahedron or a trigonal prism by adding a ligand to one face of the polyhedron to give a "capped" octahedron or trigonal prism. By far the most common, however, is the pentagonal bipyramid.

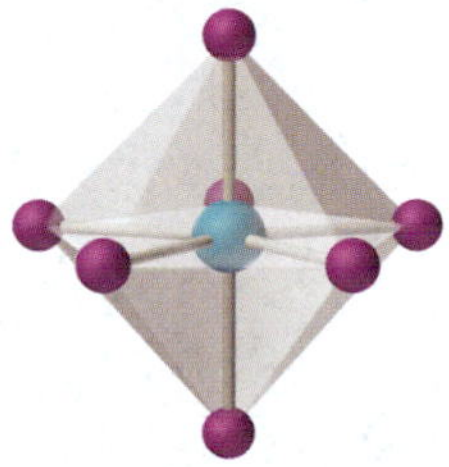

Pentagonal bipyramidal ML_7

Coordination Number 8

This coordination number is relatively common for larger metal ions. The simplest structure is the cube, which is rare because it does not minimize interligand repulsive interactions. Common structures are the square antiprism and the dodecahedron, both of which can be generated from the cube.

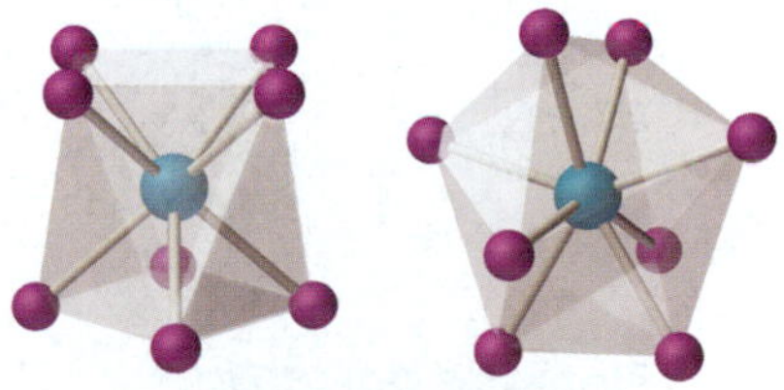

Square antiprismatic and trigonal dodecahedral ML_8

Coordination Number 9

This coordination number is found for larger metal ions, and the most common structure is the tricapped trigonal prism, as in $[Nd(H_2O)_9]^{3+}$.

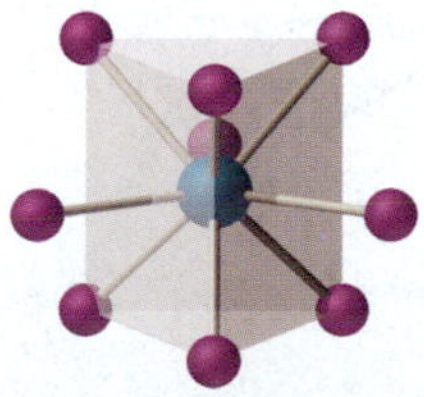

Tricapped trigonal prismatic ML_9

Stability of Metal Complexes

The thermodynamic stability of a metal complex depends greatly on the properties of the ligand and the metal ion and on the type of bonding. Recall that the metal–ligand interaction is an example of a Lewis acid–base interaction. Lewis bases can be divided into two categories: **hard bases**, which contain small, relatively nonpolarizable donor atoms (such as N, O, and F), and **soft bases**, which contain larger, relatively polarizable donor atoms (such as P, S, and Cl). Metal ions with the highest affinities for hard bases are **hard acids**, whereas metal ions with the highest affinity for soft bases are **soft acids**. Some examples of hard and soft acids and bases are given in Table 23.9. Notice that hard acids are usually cations of electropositive metals, and consequently they are relatively nonpolarizable and have higher charge-to-radius ratios. Conversely, soft acids tend to be cations of less electropositive metals, and consequently they have lower charge-to-radius ratios and are more polarizable. Chemists can predict the relative stabilities of complexes formed by the *d*-block metals with a remarkable degree of accuracy by using a simple rule: *Hard acids prefer to bind to hard bases, and soft acids prefer to bind to soft bases.*

Note the pattern

Hard acids prefer hard bases, and soft acids prefer soft bases.

Because the interaction between hard acids and hard bases is primarily electrostatic in nature, the stability of complexes involving hard acids and hard bases increases as the positive charge on the metal ion increases and as its radius decreases. For example, the complex of Al^{3+} (r = 53.5 pm) with four fluoride ligands (AlF_4^-) is about 10^8 times more stable than InF_4^-, the corresponding fluoride complex of In^{3+} (r = 80 pm). In general,

TABLE 23.9 Examples of hard and soft acids and bases

	Acids	Bases
Hard	H^+	NH_3, RNH_2, N_2H_4
	Li^+, Na^+, K^+	H_2O, ROH, R_2O
	Be^{2+}, Mg^{2+}, Ca^{2+}, VO^{2+}	OH^-, F^-, Cl^-, $CH_3CO_2^-$
	Al^{3+}, Sc^{3+}, Cr^{3+}	CO_3^{2-}
	Ti^{4+}	PO_4^{3-}
Soft	BF_3, Al_2Cl_6, CO_2, SO_3	
	Cu^+, Ag^+, Au^+, Tl^+, Hg_2^{2+}	H^-
	Pd^{2+}, Pt^{2+}, Hg^{2+}	CN^-, SCN^-, I^-, RS^-
	$GaCl_3$, $GaBr_3$, GaI_3	CO, R_2S

the stability of complexes of divalent first-row transition metals with a given ligand varies inversely with the radius of the metal ion, as shown in the following series:*

Complex stability	Mn^{2+}	$< Fe^{2+}$	$< Co^{2+}$	$< Ni^{2+}$	$< Cu^{2+}$	$> Zn^{2+}$
Ionic radius, pm	83	78	74.5	69	73	74

Because a hard metal interacts with a base in much the same way as a proton, by binding to a lone pair of electrons on the base, the stability of complexes of hard acids with hard bases increases as the ligand becomes more basic. For example, because ammonia is a stronger base than water, metal ions bind preferentially to ammonia. Consequently, adding ammonia to aqueous solutions of many of the first-row transition-metal cations results in the formation of the corresponding ammonia complexes.

In contrast, the interaction between "soft" metals (such as the second- and third-row transition metals and Cu^+) and "soft" bases is largely covalent in nature. Most soft-metal ions have a filled or nearly filled *d* subshell, which suggests that metal-to-ligand π bonding is important. Complexes of soft metals with soft bases are therefore much more stable than would be predicted based on electrostatic arguments.

Note the pattern

Ligands that are strong bases form the most stable complexes with hard metals.

The hard acid–hard base/soft acid–soft base concept also allows us to understand why metals are found in nature in different kinds of ores. Recall from Section 23.2 that most of the first-row transition metals are isolated from oxide ores, but that copper and zinc tend to occur naturally in sulfide ores. This is consistent with the increase in the "soft" character of the metals as we go across the first row of the transition metals from left to right. Recall also that most of the second- and third-row transition metals occur in nature as sulfide ores, consistent with their greater "soft" character.

Ligands like chloride, water, and ammonia are said to be *monodentate* ("one-toothed," from the Greek *mono*, "one", and the Latin *dent-*, "tooth"), meaning that they are attached to the metal via only a single atom. Ligands can, however, be *bidentate* ("two-toothed"), *tridentate* ("three-toothed"), or, in general, *polydentate* ("many-toothed"), indicating that they are attached to the metal at two, three, or several sites, respectively. Ethylenediamine ($H_2NCH_2CH_2NH_2$, often abbreviated as en) and diethylenetriamine ($H_2NCH_2CH_2NHCH_2CH_2NH_2$, dien) are examples of a bidentate and a tridentate ligand, respectively, because each nitrogen atom has a lone pair that can be shared with a metal ion. As shown in the margin, when a bidentate ligand such as ethylenediamine binds to a metal such as Ni^{2+}, a five-membered ring is formed. A metal-containing ring like that shown is called a *chelate* ring (from the Greek *chele*, "claw"). Correspondingly, a polydentate ligand is a *chelating agent*, and complexes that contain polydentate ligands are called *chelate complexes*.

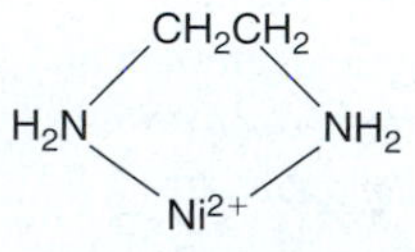

A chelate complex

* The inversion in the order at copper is due to the anomalous structure of copper(II) complexes, which will be discussed shortly.

Experimentally, it is observed that metal complexes of polydentate ligands are significantly more stable than the corresponding complexes of chemically similar monodentate ligands; this increase in stability is called the **chelate effect**. For example, the complex of Ni^{2+} with three ethylenediamine ligands, $[Ni(en)_3]^{2+}$, should be chemically similar to the Ni^{2+} complex with six ammonia ligands, $[Ni(NH_3)_6]^{2+}$. In fact, the equilibrium constant for the formation of $[Ni(en)_3]^{2+}$ is almost 10 orders of magnitude larger than the equilibrium constant for the formation of $[Ni(NH_3)_6]^{2+}$:

$$[Ni(H_2O)_6]^{2+} + 6NH_3 \rightleftharpoons [Ni(NH_3)_6]^{2+} + 6H_2O(l) \qquad K_f = 10^{8.6}$$
$$[Ni(H_2O)_6]^{2+} + 3\text{ en} \rightleftharpoons [Ni(en)_3]^{2+} + 6H_2O(l) \qquad K_f = 10^{18.3} \qquad (23.10)$$

Note the pattern

Chelate complexes are more stable than the analogous complexes with monodentate ligands.

The stability of a chelate complex depends on the *size* of the chelate rings. For ligands with a flexible organic backbone like ethylenediamine, complexes that contain five-membered chelate rings, which have almost no strain, are significantly more stable than complexes with six-membered chelate rings, which are in turn much more stable than complexes with four- or seven-membered rings. For example, the complex of copper(II) with two ethylenediamine ligands is about 1000 times more stable than the corresponding complex with triethylenediamine ($H_2NCH_2CH_2CH_2NH_2$, abbreviated as tn):

$$[Cu(H_2O)_6]^{2+} + 2\text{ en} \rightleftharpoons [Cu(en)_2]^{2+} + 6H_2O(l) \qquad K_f = 10^{20}$$
$$[Cu(H_2O)_6]^{2+} + 2\text{ tn} \rightleftharpoons [Cu(tn)_2]^{2+} + 6H_2O(l) \qquad K_f = 10^{17} \qquad (23.11)$$

EXAMPLE 23.5

Arrange these complexes in order of increasing stability: $[Cr(en)_3]^{3+}$, $[CrCl_6]^{3-}$, $[CrF_6]^{3-}$, $[Cr(NH_3)_6]^{3+}$.

Given Four Cr(III) complexes

Asked for Relative stabilities

Strategy

Ⓐ Determine the relative basicity of the ligands to identify the most stable complexes.
Ⓑ Decide whether any complexes are further stabilized by a chelate effect, and arrange the complexes in order of increasing stability.

Solution

Ⓐ The metal ion is the same in each case: Cr^{3+}. Consequently, we must focus on the properties of the ligands to determine the stabilities of the complexes. Because the stability of a metal complex increases as the basicity of the ligands increases, we need to determine the relative basicity of the four ligands. Our earlier discussion of acid–base properties suggests that ammonia and ethylenediamine, with nitrogen donor atoms, are the most basic ligands. The fluoride ion is a stronger base (it has a higher charge-to-radius ratio) than chloride, so the order of stability expected due to ligand basicity is $[CrCl_6]^{3-} < [CrF_6]^{3-} < [Cr(NH_3)_6]^{3+} \approx [Cr(en)_3]^{3+}$. Ⓑ Because of the chelate effect, we expect ethylenediamine to form a stronger complex with Cr^{3+} than ammonia. Consequently, the likely order of increasing stability is $[CrCl_6]^{3-} < [CrF_6]^{3-} < [Cr(NH_3)_6]^{3+} < [Cr(en)_3]^{3+}$.

EXERCISE 23.5

Arrange the complexes in order of decreasing stability: $[Co(NH_3)_6]^{3+}$, $[CoF_6]^{3-}$, $[Co(en)_3]^{3+}$.

Answer $[Co(en)_3]^{3+} > [Co(NH_3)_6]^{3+} > [CoF_6]^{3-}$

Isomers of Metal Complexes

As we discussed earlier in this section, the existence of coordination compounds with the same formula but different arrangements of the ligands was crucial in the development of coordination chemistry. Two or more compounds that have the same formula but different arrangements of the atoms are called **isomers**. Because isomers usually have different physical and chemical properties, it is important to know which isomer we are dealing with if more than one isomer is possible. Recall from Chapter 2 that in many cases more than one structure is possible for organic compounds that have the same molecular formula; examples discussed previously include *n*-butane versus isobutane, and *cis*-2-butene versus *trans*-2-butene. As we will see, coordination compounds exhibit the same types of isomers as organic compounds, as well as several kinds of isomers that are unique.

Structural Isomers

Isomers that contain the same number of atoms of each kind but differ in which atoms are bonded to one another are called **structural isomers**. Isobutane and *n*-butane are examples of structural isomers. A trivial kind of isomerism consists of two compounds that have the same empirical formula but differ in the number of formula units present in the molecular formula. An example of this kind of isomerism in coordination compounds is two compounds with the empirical formula $Pt(NH_3)_2Cl_2$. One is a simple square planar platinum(II) complex, $[Pt(NH_3)_2Cl_2]$, and the other is an ionic compound that contains the $[Pt(NH_3)_4]^{2+}$ cation and the $[PtCl_4]^{2-}$ anion, $[Pt(NH_3)_4]\,[PtCl_4]$. As you might expect, these compounds have very different properties: the former is yellow, and one arrangement of the Cl^- and NH_3 ligands around the platinum ion gives the anticancer drug Cisplatin, whereas the other has no known biomedical applications.

Geometrical isomers

Metal complexes that differ only in which ligands are adjacent to one another **(cis)** or directly across from one another **(trans)** in the coordination sphere of the metal are called **geometrical isomers**. They are most important for square planar and octahedral complexes.

Because all vertices of a square are equivalent, it does not matter which vertex is occupied by the ligand B in a square planar MA_3B complex; hence, only a single geometrical isomer is possible in this case (and in the analogous MAB_3 case). All four structures shown here are chemically identical because they can be superimposed simply by rotating the complex in space:

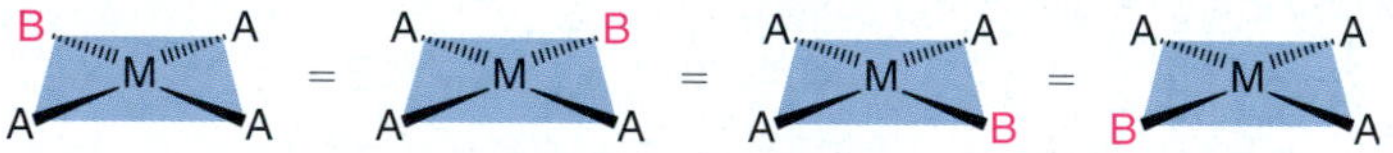

MA_3B square planar complex

For an MA_2B_2 complex, there are two possible isomers: either the A ligands can be adjacent to one another (*cis*), in which case the B ligands must also be *cis*, or the A ligands can be across from one another (*trans*), in which case the B ligands must also be *trans*. Even though it is possible to draw the *cis* isomer in four different ways and the *trans* isomer in two different ways, all members of each set are chemically equivalent:

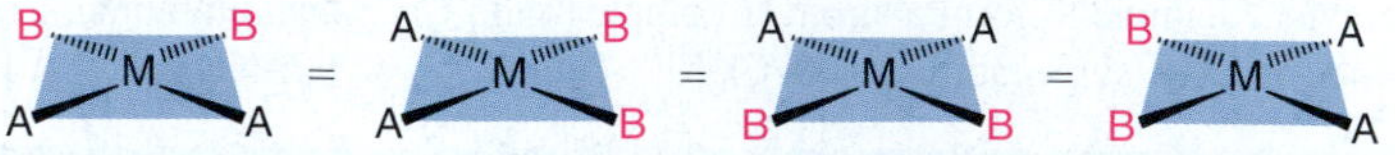

MA_2B_2 square planar complex, *cis* isomer

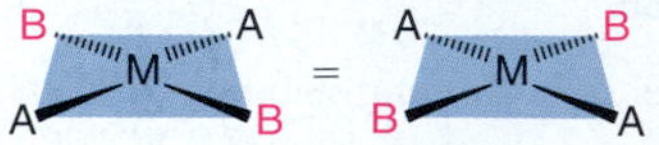

MA_2B_2 square planar complex, *trans* isomer

Because there is no way to convert the *cis* structure to the *trans* by rotating or flipping the molecule in space, they are fundamentally different arrangements of atoms in space. Probably the best-known examples of *cis* and *trans* isomers of an MA_2B_2 square planar complex are *cis*-$[Pt(NH_3)_2Cl_2]$, better known as the anticancer drug Cisplatin (shown in the margin), and *trans*-$[Pt(NH_3)_2Cl_2]$, which is actually toxic rather than therapeutic.

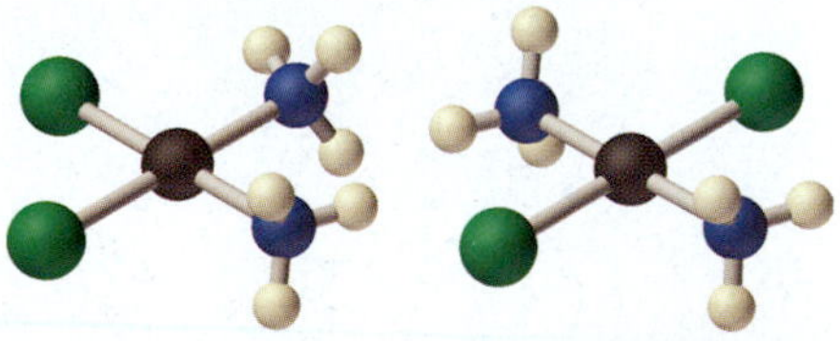

cis-$[Pt(NH_3)_2Cl_2]$ *trans*-$[Pt(NH_3)_2Cl_2]$

The anticancer drug Cisplatin and its inactive *trans* isomer. Cisplatin is especially effective against tumors of the reproductive organs (the testes in males and the ovaries in females), which primarily affect individuals in their 20s and were notoriously difficult to cure. For example, after being diagnosed with metastasized testicular cancer in 1991 and given only a 50% chance of survival, Lance Armstrong was cured by treatment with Cisplatin and went on to win an unprecedented seven Tours de France bicycle races.

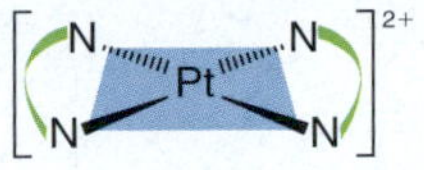

$[Pt(en)_2]^{2+}$

Square planar complexes that contain symmetrical bidentate ligands, such as $[Pt(en)_2]^{2+}$, have only one possible structure, as shown in the margin, in which the ethylenediamine ligands are indicated by the curved lines linking two N atoms.

Let's turn now to octahedral complexes, which can also exhibit *cis* and *trans* isomers. Like square planar complexes, only one structure is possible for octahedral complexes in which only one ligand is different from the other five, MA_5B. Even though we usually draw an octahedron in a way that suggests that the four "in-plane" ligands are different from the two "axial" ligands, in fact all six vertices of an octahedron are equivalent. Consequently, no matter how we draw an MA_5B structure, it can be superimposed on any other representation simply by rotating the molecule in space. Two of the many possible orientations of an MA_5B structure are shown here:

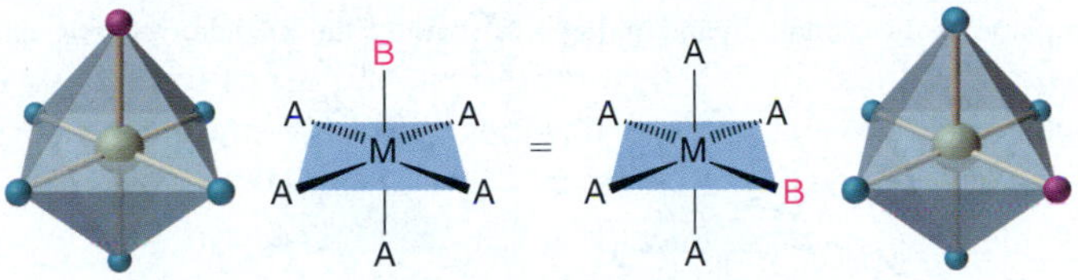

MA_5B octahedral complex

If two ligands in an octahedral complex are different from the other four, giving an MA_4B_2 complex, two isomers are possible. As shown here, the two B ligands can be *cis* or *trans*. *Cis*- and *trans*-$[Co(NH_3)_4Cl_2]Cl$ are examples of this type of system:

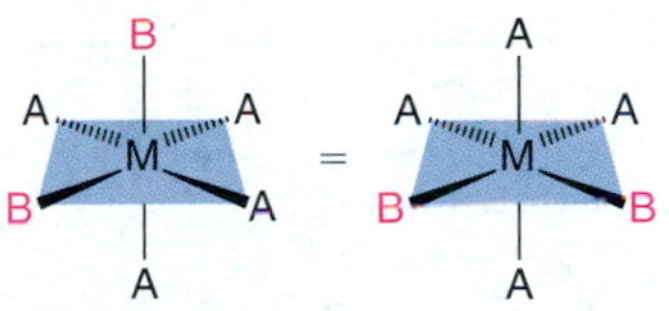

MA_4B_2 octahedral complex, *cis* isomer

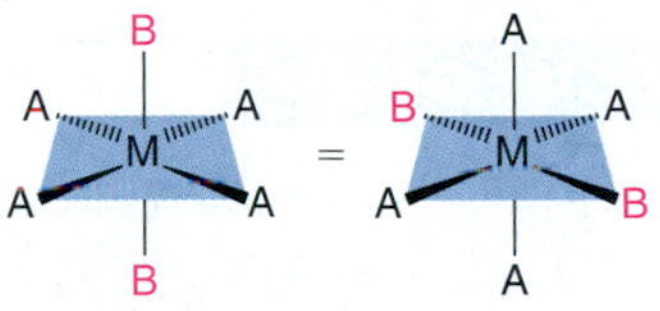

MA_4B_2 octahedral complex, *trans* isomer

Replacing another A ligand by B gives an MA_3B_3 complex for which there are also two possible isomers. In one, the three ligands of each kind occupy opposite triangular faces of the octahedron; this is called the *fac* isomer (for facial). In the other, the three ligands of each kind lie on what would be the meridian if the complex were viewed as a sphere; this is called the *mer* isomer (for meridional):

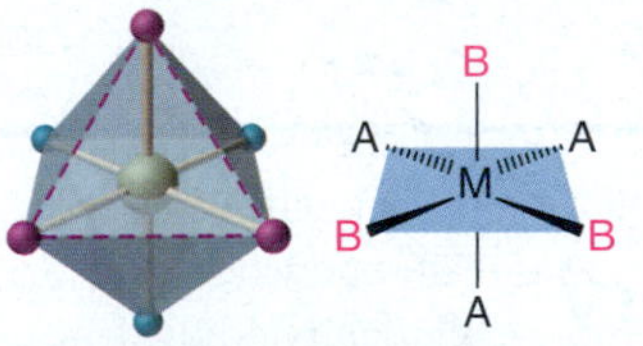

MA_3B_3 octahedral complex, *fac* isomer

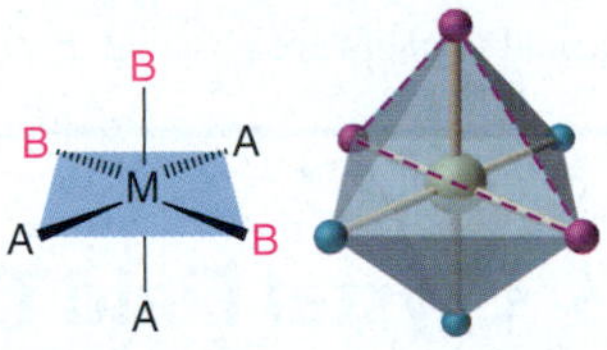

MA_3B_3 octahedral complex, *mer* isomer

EXAMPLE 23.6

Draw all the possible geometrical isomers for the complex $[Co(H_2O)_2(ox)BrCl]^-$, where ox = $^-O_2CCO_2^-$.

Given Formula of complex

Asked for Structures of geometrical isomers

Solution

This complex contains one bidentate ligand (oxalate), which can only occupy adjacent **(cis)** positions, and four monodentate ligands, two of which are identical (H_2O). The easiest way to attack the problem is to go through the various combinations of ligands systematically to determine which ligands can be *trans*. Thus, either the water ligands can be *trans* to one another or the two halide ligands can be *trans* to one another, giving the two geometrical isomers shown here:

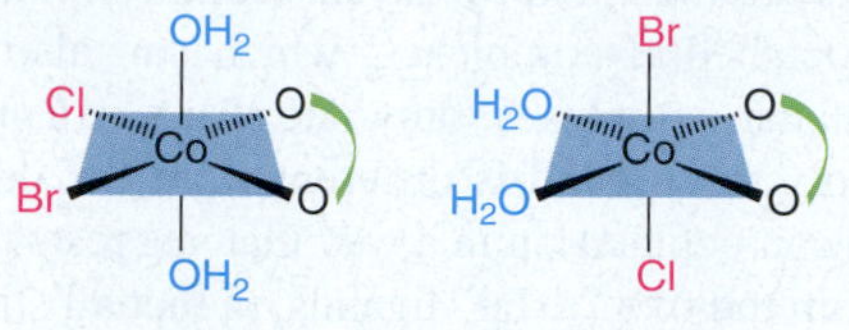

In addition, two structures are possible in which one of the halides is *trans* to a water ligand. In the first, the chloride ligand is in the same plane as the oxalate ligand and *trans* to one of the oxalate oxygens. Exchanging the chloride and bromide ligands gives the other, in which the bromide ligand is in the same plane as the oxalate ligand and *trans* to one of the oxalate oxygens, as shown here:

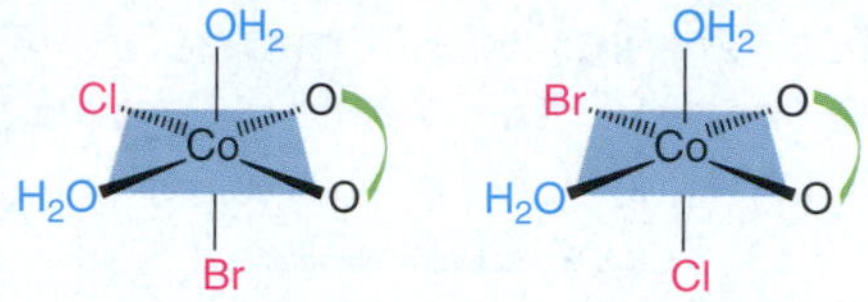

This complex can therefore exist as four different geometrical isomers.

EXERCISE 23.6

Draw all the possible geometrical isomers for the complex $[Cr(en)_2(CN)_2]^+$.

Answer

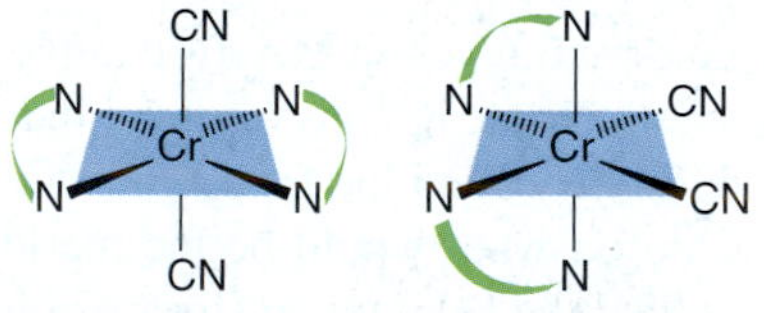

Two geometrical isomers are possible: *trans* and *cis*.

23.5 • Crystal Field Theory

One of the most striking characteristics of transition-metal complexes is the wide range of colors they exhibit (Figures 23.4 and 23.5). In this section, we describe **crystal field theory** (CFT), a bonding model that explains many important properties of transition-metal complexes, including their colors, magnetism, structures, stability, and reactivity. The central assumption of CFT is that metal–ligand interactions are purely electrostatic in nature. Even though this assumption is clearly not valid for many complexes, such as those that contain neutral ligands, like CO, CFT enables chemists to explain many of the properties of transition-metal complexes with a reasonable degree of accuracy.

d-Orbital Splittings

Crystal field theory focuses on the interaction of the five $(n - 1)d$ orbitals with ligands arranged in a regular array around a transition-metal ion. We will focus on the application of CFT to octahedral complexes, which are by far the most common and the easiest to visualize. Other common structures such as square planar complexes can be treated as a distortion of the octahedral model. According to CFT, an octahedral metal complex forms because of the electrostatic interaction of a positively charged metal ion with six negatively charged ligands, or with the negative ends of dipoles associated with the six ligands. In addition, the ligands interact with one other electrostatically. As you learned in our discussion of the VSEPR model in Chapter 9, the lowest-energy arrangement of six identical negative charges is an octahedron, which minimizes repulsive interactions between the ligands.

Let's consider how the energies of the *d* orbitals of a transition-metal ion are affected by an octahedral arrangement of six negative charges. Recall from Chapter 6 that the five *d* orbitals are initially degenerate (have the same energy). If we distribute six negative charges *uniformly* over the surface of a sphere, the *d* orbitals remain degenerate, but their energy will be higher due to repulsive electrostatic interactions between the spherical shell of negative charge and electrons in the *d* orbitals (Figure 23.10a). Placing the six negative charges at the vertices of an octahedron does not change the average energy of the *d* orbitals, but it does remove their degeneracy: the five *d* orbitals split into two groups whose energies depend on their orientations. As shown in Figure 23.10b, the d_{z^2} and $d_{x^2-y^2}$ orbitals point *directly at* the six negative charges located on the *x*, *y*, and *z* axes. Consequently, the energy of an electron in these two orbitals (collectively labeled the "e_g orbitals") will be higher than it will be for a spherical distribution of negative charge because of increased electrostatic repulsions. In contrast, the other three *d* orbitals (d_{xy}, d_{xz}, and d_{yz}, collectively called the "t_{2g} orbitals") are all oriented at a 45° angle to the coordinate axes, and as a result they point *between* the six negative charges. Because the t_{2g} orbitals point at regions of space that contain no negative charge, the energy of an electron in any of the three is lower than the energy for a spherical distribution of negative charge.

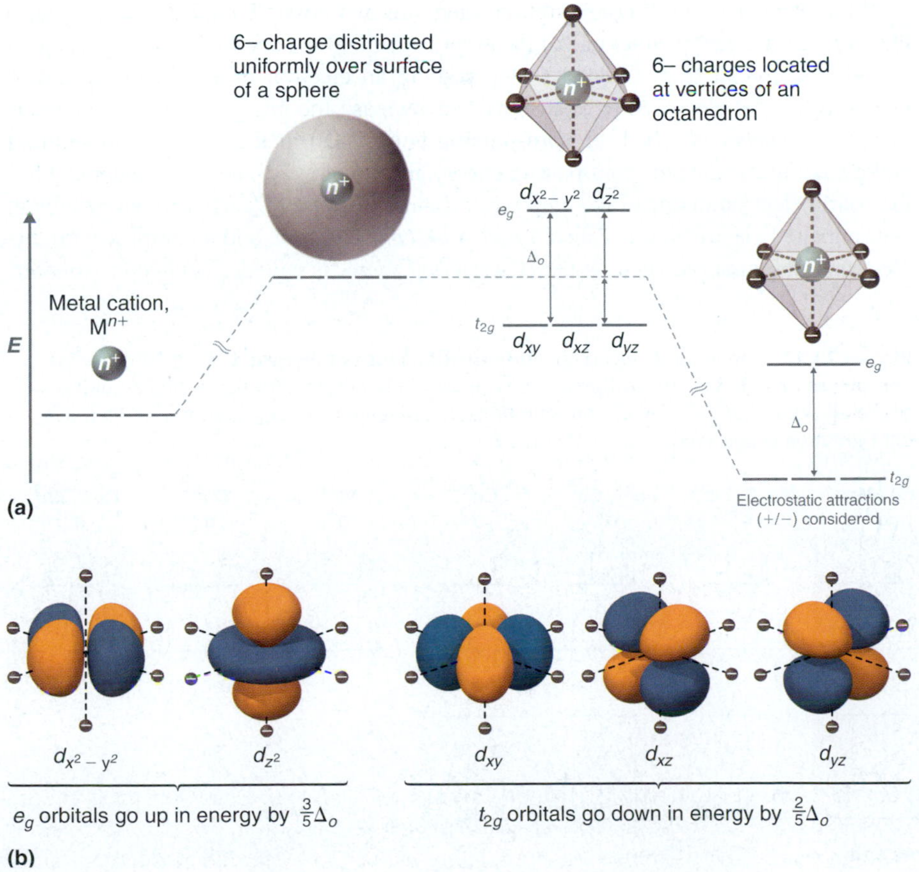

Figure 23.10 An octahedral arrangement of six negative charges around a metal ion causes the five *d* orbitals to split into two sets with different energies. (a) Distributing a charge of −6 *uniformly* over a spherical surface surrounding a metal ion causes the energy of all five *d* orbitals to increase due to electrostatic repulsions, but the five *d* orbitals remain degenerate. Placing a charge of −1 at each of the six vertices of an octahedron instead causes the *d* orbitals to split into two groups with different energies: the $d_{x^2-y^2}$ and d_{z^2} orbitals increase in energy, while the, d_{xy}, d_{xz}, and d_{yz} orbitals decrease in energy. The *average* energy of the five *d* orbitals is the same as for a spherical distribution of a −6 charge, however. Attractive electrostatic interactions between the negatively charged ligands and the positively charged metal ion (*far right*) cause all five *d* orbitals to decrease in energy but does not affect the splittings of the orbitals. (b) The two e_g orbitals (*left*) point *directly at* the six negatively charged ligands, which increases their energy compared with a spherical distribution of negative charge. In contrast, the three t_{2g} orbitals (*right*) point *between* the negatively charged ligands, which decreases their energy compared with a spherical distribution of charge.

Note the pattern
Crystal field splitting does not change the total energy of the d orbitals.

The difference in energy between the two sets of *d* orbitals is called the **crystal field splitting energy**, which is given the symbol Δ_o, where the subscript "o" stands for "octahedral." As we shall see, the magnitude of the splitting depends on the charge on the metal ion, the position of the metal in the periodic table, and the nature of the ligands. It is important to note that the splitting of the *d* orbitals in a crystal field does *not* change the total energy of the five *d* orbitals: the two e_g orbitals *increase* in energy by $0.6\Delta_o$, whereas the three t_{2g} orbitals decrease in energy by $0.4\Delta_o$. Thus, the total change in energy is $2(0.6\Delta_o) + 3(-0.4\Delta_o) = 0$.

Thus far, we have considered only the effect of repulsive electrostatic interactions between electrons in the *d* orbitals and the six negatively charged ligands, which increases the total energy of the system and splits the *d* orbitals. Including the attractive electrostatic interactions between the positively charged metal ion and the ligands results in a net stabilization of the system, which decreases the energy of all five *d* orbitals without affecting their splitting (as shown at the far right in Figure 23.10a).

Electronic Structures of Metal Complexes

We can use the *d*-orbital energy-level diagram in Figure 23.10 to predict the electronic structures and some of the properties of transition-metal complexes. We start with the Ti^{3+} ion, which contains a single *d* electron, and proceed across the first row of the transition metals by adding a single electron at a time. We place additional electrons in the lowest-energy orbital available, while keeping their spins parallel as required by Hund's rule. As shown in Figure 23.11, for d^1–d^3 systems—such as $[Ti(H_2O)_6]^{3+}$, $[V(H_2O)_6]^{3+}$, and $[Cr(H_2O)_6]^{3+}$, respectively—the electrons successively occupy the three degenerate t_{2g} orbitals with their spins parallel, giving one, two, and three unpaired electrons, respectively. We can summarize this for the complex $[Cr(H_2O)_6]^{3+}$, for example, by saying that the chromium ion has a d^3 electron configuration or, more succinctly, Cr^{3+} is a d^3 ion.

When we reach the d^4 configuration, there are two possible choices for the fourth electron: it can occupy either one of the empty e_g orbitals or one of the singly occupied t_{2g} orbitals. Recall from Chapter 6 that placing an electron in an already occupied orbital results in electrostatic repulsions that increase the energy of the system; this increase in energy is called the **spin-pairing energy (P)**. If the crystal field splitting energy, Δ_o, is *less* than the spin-pairing energy, then the lowest-energy arrangement has the fourth electron in one of the empty e_g orbitals. Because this arrangement results in four unpaired electrons, it is called a *high-spin configuration*, and a complex with this electron configuration, such as the $[Cr(H_2O)_6]^{2+}$ ion, is called a *high-spin complex*.

Figure 23.11 The possible electron configurations for octahedral d^n transition-metal complexes (n = 1–10). Two different configurations are possible for octahedral complexes of metals with d^4, d^5, d^6, and d^7 configurations; the magnitude of Δ_o determines which configuration is observed.

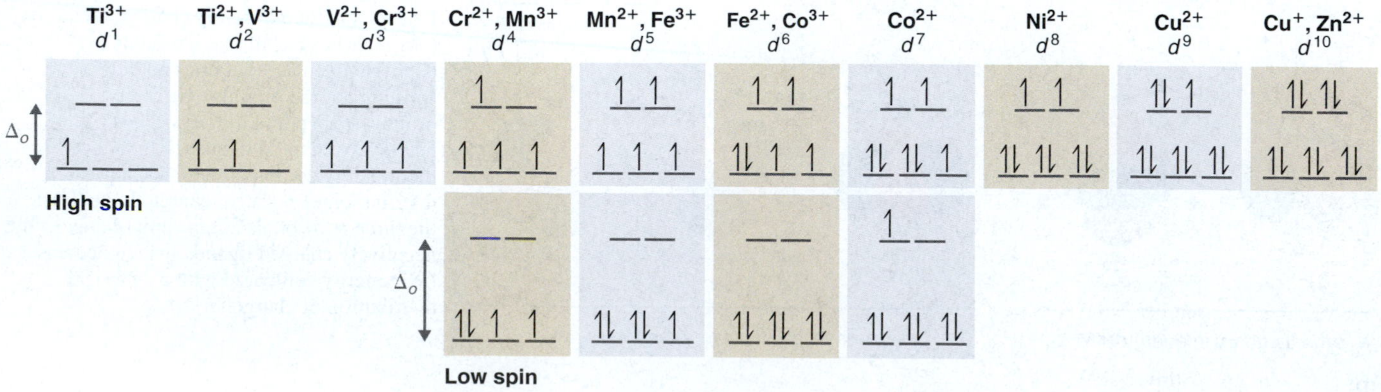

Conversely, if Δ_o is *greater* than the spin-pairing energy, then the lowest-energy arrangement has the fourth electron in one of the occupied t_{2g} orbitals. Because this arrangement results in only two unpaired electrons, it is called a *low-spin configuration*, and a complex with this electron configuration, such as the $[Mn(CN)_6]^{3-}$ ion, is called a *low-spin complex*. Similarly, metal ions with the d^5, d^6, or d^7 electron configurations can be either high spin or low spin, depending on the magnitude of Δ_o.

In contrast, only one arrangement of d electrons is possible for metal ions with d^8–d^{10} electron configurations. For example, the $[Ni(H_2O)_6]^{2+}$ ion is d^8 with two unpaired electrons, the $[Cu(H_2O)_6]^{2+}$ ion is d^9 with one unpaired electron, and the $[Zn(H_2O)_6]^{2+}$ ion is d^{10} with no unpaired electrons.

Factors That Affect the Magnitude of Δ_o

The magnitude of Δ_o dictates whether a complex with four, five, six, or seven d electrons is high spin or low spin, which affects its magnetic properties, structure, and reactivity. Large values of Δ_o ($\Delta_o > P$) yield a low-spin complex, whereas small values of Δ_o ($\Delta_o < P$) produce a high-spin complex. As we noted, the magnitude of Δ_o depends on three factors: the charge on the metal ion, the principal quantum number of the metal (and thus its location in the periodic table), and the nature of the ligand. Values of Δ_o for some representative transition-metal complexes are given in Table 23.10.

Charge on the Metal Ion

Increasing the charge on a metal ion has two effects: the radius of the metal ion decreases, and negatively charged ligands are more strongly attracted to it. Both these

TABLE 23.10 Crystal field splitting energies for some octahedral (Δ_o)[a] and tetrahedral (Δ_t) transition-metal complexes

Octahedral Complexes	Δ_o, cm^{-1}	Octahedral Complexes	Δ_o, cm^{-1}	Tetrahedral Complexes	Δ_t, cm^{-1}
$Ti(H_2O)_6^{3+}$	20,300	$Fe(CN)_6^{4-}$	32,800	VCl_4	9,010
$V(H_2O)_6^{2+}$	12,600	$Fe(CN)_6^{3-}$	35,000	$CoCl_4^{2-}$	3,300
$V(H_2O)_6^{3+}$	18,900	CoF_6^{3-}	13,000	$CoBr_4^{2-}$	2,900
$CrCl_6^{3-}$	13,000	$Co(H_2O)_6^{2+}$	9,300	CoI_4^{2-}	2,700
$Cr(H_2O)_6^{2+}$	13,900	$Co(H_2O)_6^{3+}$	27,000		
$Cr(H_2O)_6^{3+}$	17,400	$Co(NH_3)_6^{3+}$	22,900		
$Cr(NH_3)_6^{3+}$	21,500	$Co(CN)_6^{3-}$	34,800		
$Cr(CN)_6^{3-}$	26,600	$Ni(H_2O)_6^{2+}$	8,500		
$Cr(CO)_6$	34,150	$Ni(NH_3)_6^{2+}$	10,800		
$MnCl_6^{4-}$	7,500	$RhCl_6^{3-}$	20,400		
$Mn(H_2O)_6^{2+}$	8,500	$Rh(H_2O)_6^{3+}$	27,000		
$MnCl_6^{3-}$	20,000	$Rh(NH_3)_6^{3+}$	34,000		
$Mn(H_2O)_6^{3+}$	21,000	$Rh(CN)_6^{3-}$	45,500		
$Fe(H_2O)_6^{2+}$	10,400	$IrCl_6^{3-}$	25,000		
$Fe(H_2O)_6^{3+}$	14,300	$Ir(NH_3)_6^{3+}$	41,000		

[a] Energies obtained by spectroscopic measurements are often given in units of *wave numbers* (cm^{-1}); the wave number is the reciprocal of the wavelength of the corresponding electromagnetic radiation expressed in centimeters: 1 cm^{-1} = 11.96 J/mol.

Source of data: D. F. Shriver, P. W. Atkins, and C. H. Langford, *Inorganic Chemistry*, 2nd ed., 1994.

factors decrease the metal–ligand distance, which in turn causes the negatively charged ligands to interact more strongly with the *d* orbitals. Consequently, the magnitude of Δ_o increases as the charge on the metal ion increases. Typically, Δ_o for a tripositive ion is about 50% higher than for the dipositive ion of the same metal; for example, for $[V(H_2O)_6]^{2+}$, $\Delta_o = 11{,}800\ cm^{-1}$; for $[V(H_2O)_6]^{3+}$, $\Delta_o = 17{,}850\ cm^{-1}$.

Principal Quantum Number of the Metal

For a series of complexes of metals from the same group in the periodic table with the same charge and the same ligands, the magnitude of Δ_o increases with increasing principal quantum number: $\Delta_o\ (3d) << \Delta_o\ (4d) < \Delta_o\ (5d)$. The data for hexaammine complexes of the trivalent Group 9 metals illustrate this point:

$$[Co(NH_3)_6]^{3+} \quad \Delta_o = 22{,}900\ cm^{-1}$$
$$[Rh(NH_3)_6]^{3+} \quad \Delta_o = 34{,}100\ cm^{-1}$$
$$[Ir(NH_3)_6]^{3+} \quad \Delta_o = 40{,}000\ cm^{-1}$$

The increase in Δ_o with increasing principal quantum number is due to the larger radius of valence orbitals as we go down a column. In addition, repulsive ligand—ligand interactions are most important for smaller metal ions. Relatively speaking, this results in shorter M–L distances and stronger *d* orbital–ligand interactions.

Note the pattern

The largest crystal field splitting energies are found in complexes of metal ions from the third row of the transition metals that have charges of at least +3 and ligands that have localized lone pairs of electrons.

The Nature of the Ligands

Experimentally, it is found that the crystal field splitting energy observed for a series of complexes of the same metal ion depends strongly on the nature of the ligands. For a series of chemically similar ligands, the magnitude of Δ_o decreases as the size of the donor atom increases. For example, Δ_o values for halide complexes generally decrease in the order $F^- > Cl^- > Br^- > I^-$ because smaller, more localized charges interact more strongly with the *d* orbitals of the metal ion. In addition, a small neutral ligand with a highly localized lone pair, such as NH_3, results in significantly larger Δ_o values than might be expected. Because the lone pair points directly at the metal ion, the electron density along the M–L axis is higher than for a spherical anion such as F^-. The experimentally observed order of the crystal field splitting energies produced by different ligands is called the *spectrochemical series*, shown here in order of decreasing Δ_o:

$$CO \approx CN^- > NO_2^- > en > NH_3 > SCN^- > H_2O > oxalate^{2-} > OH^- > F^- > acetate^- > Cl^- > Br^- > I^-$$

Strong-field ligands — *Intermediate-field ligands* — *Weak-field ligands*

The values of Δ_o listed in Table 23.10 illustrate the effects of the charge on the metal ion, the principal quantum number of the metal, and the nature of the ligand.

Colors of Transition-Metal Complexes

The striking colors exhibited by transition-metal complexes are caused by the excitation of an electron from a lower-lying *d* orbital to a higher-energy *d* orbital, which is called a *d–d transition* (Figure 23.12). For a photon to effect such a transition, its energy must be equal to the difference in energy between the two *d* orbitals, which depends on the magnitude of Δ_o.

Figure 23.12 A *d–d* transition. In a *d–d* transition, an electron in one of the t_{2g} orbitals of an octahedral complex such as the $[Cr(H_2O)_6]^{3+}$ ion absorbs a photon of light with energy equal to Δ_o, which causes the electron to move to an empty or singly occupied e_g orbital.

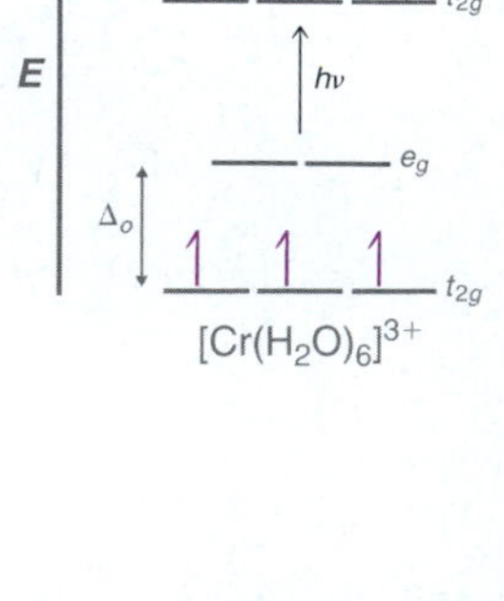

Recall from Chapter 6 that the color we observe when we look at an object or a compound is due to light that is transmitted or reflected, not light that is absorbed, and that reflected or transmitted light is complementary in color to the light that is absorbed. Thus, a green compound absorbs light in the red portion of the visible spectrum, and vice versa, as indicated by the color wheel in Problem 109 of Chapter 6. Because the energy of a photon of light is inversely proportional to its wavelength, the color of a complex depends on the magnitude of Δ_o, which depends on the structure of the complex. For example, the complex $[Cr(NH_3)_6]^{3+}$ has strong-field ligands and a relatively large Δ_o. Consequently, it absorbs relatively high-energy photons, corresponding to blue-violet light, which gives it a yellow color. A related complex with weak-field ligands, the $[Cr(H_2O)_6]^{3+}$ ion, absorbs lower-energy photons corresponding to the yellow-green portion of the visible spectrum, giving it a deep violet color.

We can now understand why emeralds and rubies have such different colors, even though both contain Cr^{3+} in an octahedral environment provided by six oxide ions. Although the chemical identity of the six ligands is the same in both cases, the Cr–O distances are different because the compositions of the host lattices are different (Al_2O_3 in rubies, $Be_3Al_2Si_6O_{18}$ in emeralds). In ruby, the Cr–O distances are relatively short because of the constraints of the host lattice, which increases the *d* orbital–ligand interactions and makes Δ_o relatively large. Consequently, rubies absorb green light and the transmitted or reflected light is red, which gives the gem its characteristic color. In emerald, the Cr–O distances are longer due to relatively large $[Si_6O_{18}]^{12-}$ silicate rings; this results in decreased *d* orbital–ligand interactions and a smaller Δ_o. Consequently, emeralds absorb light of a longer wavelength (red), which gives the gem its characteristic green color. It is clear that the environment of the transition-metal ion, which is determined by the host lattice, dramatically affects the spectroscopic properties of a metal ion.

Gem-quality crystals of ruby and emerald. The colors of both minerals are due to the presence of small amounts of Cr^{3+} impurities in octahedral sites in an otherwise colorless metal oxide lattice.

Crystal Field Stabilization Energies

If the lower-energy set of *d* orbitals (the t_{2g} orbitals) is selectively populated by electrons, then the stability of the complex increases. Recall from Chapter 9 that stable molecules are produced by populating only the lower-energy (bonding) molecular orbitals in an MO diagram. For example, the single *d* electron in a d^1 complex such as $[Ti(H_2O)_6]^{3+}$ is located in one of the t_{2g} orbitals. *Consequently, this complex will be more stable than expected on purely electrostatic grounds by* $0.4\Delta_o$. The additional stabilization of a metal complex by selective population of the lower-energy *d* orbitals is called its *crystal field stabilization energy* (CFSE). The CFSE of a complex can be calculated by multiplying the number of electrons in t_{2g} orbitals by the energy of those orbitals ($-0.4\Delta_o$), multiplying the number of electrons in e_g orbitals by the energy of those orbitals ($+0.6\Delta_o$), and summing the two. Table 23.11 gives CFSE values for octahedral complexes with different *d* electron configurations. Note that the CFSE is highest for low-spin d^6 complexes, which accounts in part for the

TABLE 23.11 CFSEs for octahedral complexes with different electron configurations (in units of Δ_o)

	High Spin	CFSE (Δ_o)	Low Spin	CFSE (Δ_o)
d^0	— — — — —	0	— — — — —	
d^1	↑ — — — —	0.4	— — — — —	
d^2	↑ ↑ — — —	0.8	— — — — —	
d^3	↑ ↑ ↑ — —	1.2	— — — — —	
d^4	↑ ↑ ↑ ↑ —	0.6	↑↓ ↑ ↑ — —	1.6
d^5	↑ ↑ ↑ ↑ ↑	0.0	↑↓ ↑↓ ↑ — —	2.0
d^6	↑↓ ↑ ↑ ↑ ↑	0.4	↑↓ ↑↓ ↑↓ — —	2.4
d^7	↑↓ ↑↓ ↑ ↑ ↑	0.8	↑↓ ↑↓ ↑↓ ↑ —	1.8
d^8	↑↓ ↑↓ ↑↓ ↑ ↑	1.2	— — — — —	
d^9	↑↓ ↑↓ ↑↓ ↑↓ ↑	0.6	— — — — —	
d^{10}	↑↓ ↑↓ ↑↓ ↑↓ ↑↓	0.0	— — — — —	

Note the pattern

Octahedral d^3 and d^8 complexes and low-spin d^6, d^5, d^7, and d^4 complexes exhibit large crystal field stabilization energies.

extraordinarily large number of Co(III) complexes known. The other low-spin configurations also have high CFSEs, as does the d^3 configuration.

CFSEs are important for two reasons. First, the existence of CFSE nicely accounts for the difference between experimentally measured values for bond energies in metal complexes and values calculated based solely on electrostatic interactions. Second, CFSEs represent relatively large amounts of energy (up to several hundred kilojoules per mole), which has important chemical consequences.

Tetragonal and Square Planar Complexes

If two *trans* ligands in an octahedral complex are either chemically different from the other four, as in the *trans*-$[Co(NH_3)_4Cl_2]^+$ ion, or at a different distance from the metal than the other four, the result is a tetragonally distorted octahedral complex. The electronic structures of such complexes are best viewed as the result of distorting an octahedral complex. Consider, for example, an octahedral complex such as $[Co(NH_3)_6]^{3+}$, and suppose that we slowly remove two *trans* NH_3 molecules by moving them away from the metal along the $\pm z$ axes, as shown in the top half of Figure 23.13. As the two axial Co–N distances increase simultaneously, the *d* orbitals that interact most strongly with the two axial ligands will *decrease* in energy due to a decrease in electrostatic repulsions between electrons in these orbitals and the negative ends of the ligand dipoles. The affected *d* orbitals are those that have a component along the $\pm z$ axes—namely, d_{z^2}, d_{xz}, and d_{yz}. They will not be affected equally, however. Because the d_{z^2} orbital points directly at the two ligands that are being removed, its energy will decrease much more rapidly than the energy of the other two, as shown in the bottom half of Figure 23.13. In addition, the positive charge on the metal will increase somewhat as the axial ligands are removed, causing the four remaining in-plane ligands to be more strongly attracted to the metal. This will increase their interactions with the other two *d* orbitals and increase their energy. Again, the two *d* orbitals will not be affected equally. Because the $d_{x^2-y^2}$ orbital points directly at the four in-plane ligands, its energy will increase more rapidly than the energy of the d_{xy} orbital, which points between the in-plane ligands. If we remove the two axial ligands to an infinite distance, we obtain a square planar complex. Note that the energies of the d_{z^2} and d_{xy} orbitals actually cross as the axial ligands are removed, and that the largest orbital spliting in a square planar complex is identical in magnitude to Δ_o.

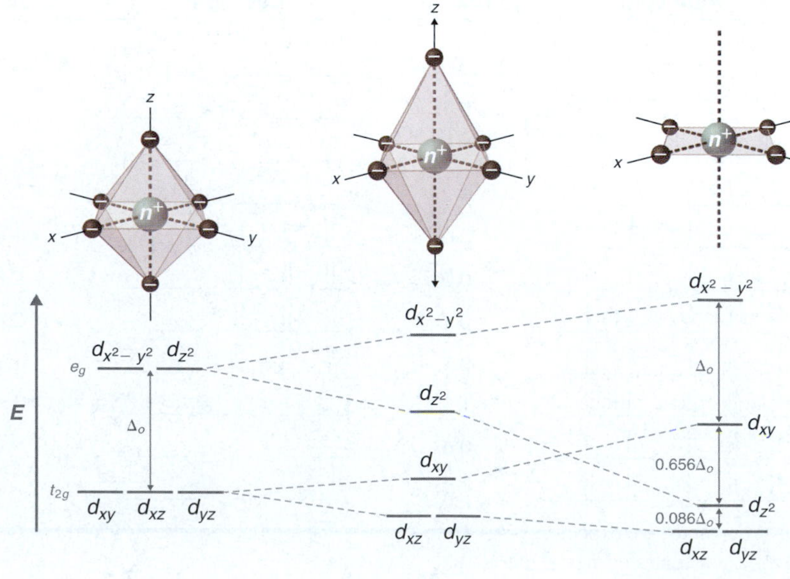

Figure 23.13 ***d*-Orbital splittings for tetragonal and square planar complexes.** Moving the two axial ligands away from the metal ion along the *z* axis initially gives an elongated octahedral complex (*center*) and eventually produces a square planar complex (*right*). As shown below the structures, an axial elongation causes the d_{z^2}, d_{xz}, and d_{yz} orbitals to decrease in energy and the $d_{x^2-y^2}$ and d_{xy} orbitals to increase in energy. As explained in the text, the change in energy is not the same for all five *d* orbitals. Removing the two axial ligands completely causes the energy of the d_{z^2} orbital to decrease so much that the order of the d_{z^2} and d_{xy} orbitals is reversed.

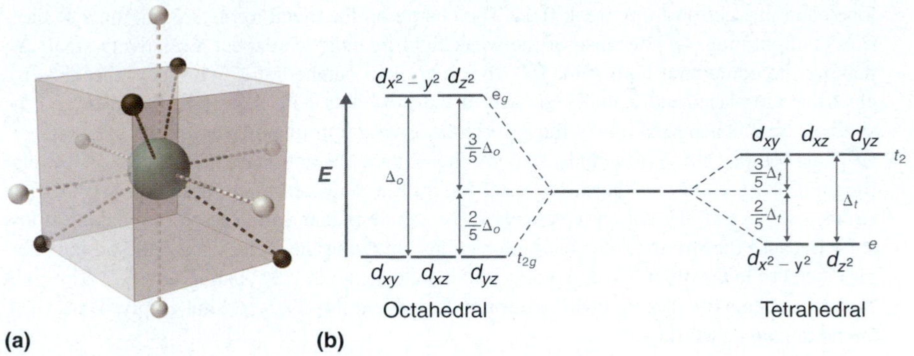

Figure 23.14 ***d*-Orbital splittings for a tetrahedral complex.** (a) In a tetrahedral complex, none of the five *d* orbitals points directly at or between the ligands. (b) Because the d_{xy}, d_{xz}, and d_{yz} orbitals (the "t_2 orbitals") interact more strongly with the ligands than do the $d_{x^2-y^2}$ and d_{z^2} orbitals (the "*e* orbitals"), the order of orbital energies in a tetrahedral complex is the opposite of the order in an octahedral complex.

Tetrahedral Complexes

In a tetrahedral arrangement of four ligands around a metal ion, none of the ligands lies on any of the three coordinate axes (Figure 23.14a), and consequently none of the five *d* orbitals points directly at the ligands. Nonetheless, the d_{xy}, d_{xz}, and d_{yz} orbitals interact more strongly with the ligands than do $d_{x^2-y^2}$ and d_{z_2}, again resulting in a splitting of the five *d* orbitals into two sets. The splitting of the energies of the orbitals in a tetrahedral complex, Δ_t, is much smaller than that in an octahedral complex, Δ_o, however, for two reasons. First, the *d* orbitals interact less strongly with the ligands in a tetrahedral arrangement. Second, there are only four negative charges rather than six, which decreases the electrostatic interactions by one-third if all other factors are equal. It can be shown that, for complexes of the same metal ion with the same charge, the same ligands, and the same M–L distance, $\Delta_t = \frac{4}{9}\Delta_o$. The relationship between the splitting of the five *d* orbitals in octahedral and tetrahedral crystal fields imposed by the same ligands is shown schematically in Figure 23.14b.

Because Δ_o is so large for the second- and third-row transition metals, *all* 4-coordinate complexes of these metals are square planar due to the much higher CFSE for square planar versus tetrahedral structures. The only exception is for d^{10} metal ions such as Cd^{2+}, which have zero CFSE and are therefore tetrahedral as predicted by the VSEPR model. Four-coordinate complexes of the first-row transition metals can be either square planar or tetrahedral. The former is favored by strong-field ligands, whereas the latter is favored by weak-field ligands. For example, the $[Ni(CN)_4]^{2-}$ ion is square planar, while the $[NiCl_4]^{2-}$ ion is tetrahedral.

EXAMPLE 23.7

For each complex, predict its structure, whether it is high spin or low spin, and the number of unpaired electrons present: **(a)** $[CoF_6]^{3-}$; **(b)** $[Rh(CO)_2Cl_2]^-$.

Given Complexes

Asked for Structure, high spin versus low spin, and number of unpaired electrons

Strategy

A From the number of ligands, determine the coordination number of the compound.
B Classify the ligands as either strong field or weak field, and determine the electron configuration of the metal ion.
C Predict the relative magnitude of Δ_o, and decide whether the compound is high spin or low spin.
D Place the appropriate number of electrons in the *d* orbitals, and determine the number of unpaired electrons.

Solution

(a) **A** With six ligands, we expect this complex to be octahedral. **B** The fluoride ion is a small anion with a concentrated negative charge, but compared with ligands that have localized lone pairs of electrons, it is weak field. The charge on the metal ion is +3, giving a d^6 electron configuration. C Because of the weak-field ligands, we expect a relatively small Δ_o, making the compound high spin. D In a high-spin octahedral d^6 complex, the first five electrons are placed individually in each of the *d* orbitals with their spins parallel, and the sixth electron is paired in one of the t_{2g} orbitals, giving four unpaired electrons.

(b) A This complex has four ligands, so it is either square planar or tetrahedral. B C Because rhodium is a second-row transition metal ion with a d^8 electron configuration and CO is a strong-field ligand, the complex is likely to be square planar with a large Δ_o, making it low spin. Because the strongest *d*-orbital interactions are along the *x* and *y* axes, the orbital energies increase in the order d_{z^2}, d_{yz} and d_{xz} (these are degenerate), d_{xy}, and $d_{x^2-y^2}$. D The eight electrons occupy the first four of these orbitals, leaving the $d_{x^2-y^2}$ orbital empty. Thus, there are no unpaired electrons.

EXERCISE 23.7

For each complex, predict its structure, whether it is high spin or low spin, and the number of unpaired electrons: **(a)** $[Mn(H_2O)_6]^{2+}$; **(b)** $[PtCl_4]^{2-}$.

Answer **(a)** octahedral, high spin, five; **(b)** square planar, low spin, no unpaired electrons

Consequences of *d*-Orbital Splitting

The splitting of the *d* orbitals because of their interaction with the ligands in a complex has important consequences for the chemistry of transition-metal complexes; they can be divided into structural effects and thermodynamic effects. Although the two kinds of effects are interrelated, it is convenient to consider them separately.

Structural Effects

Structural effects are of two major kinds: effects on the ionic radius of metal ions with regular octahedral or tetrahedral geometries, and structural distortions that are observed for specific electron configurations.

Figure 23.15 The effect of *d*-orbital splittings on the radii of the divalent ions of the fourth-period metals. Because these radii are based on the structures of octahedral complexes and Cr^{2+} and Cu^{2+} do not form truly octahedral complexes, the points for these ions are shown as open circles. The dashed line represents the behavior predicted based on the effects of screening and variation in effective nuclear charge, Z_{eff}, assuming a spherical distribution of the 3*d* electrons.

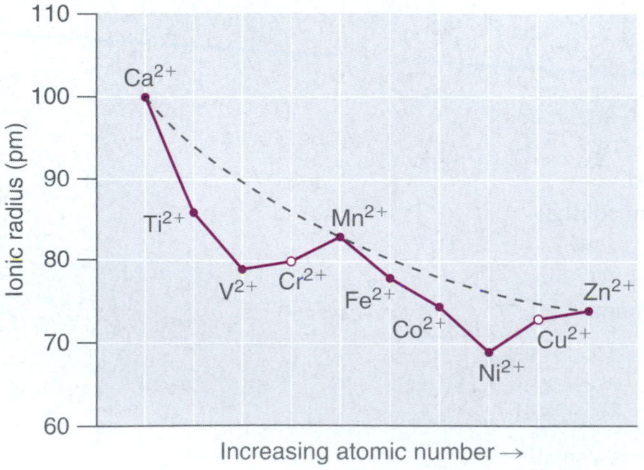

Ionic Radii. Figure 23.15 is a plot of the ionic radii of the divalent fourth-period metal ions versus atomic number. Only Ca^{2+}(d^0), Mn^{2+} (high-spin d^5), and Zn^{2+} (d^{10}) fall on the smooth curve calculated based on the effective nuclear charge, which assumes that the distribution of *d* electrons is spherically symmetrical. This assumption is true for only Ca^{2+}, Mn^{2+}, and Zn^{2+}; *all the other divalent ions fall below this curve because they have asymmetrical distributions of d electrons.* (The points shown for Cr^{2+} and Cu^{2+} are only estimated values; as you will learn shortly, these two ions do not form any truly octahedral complexes.) To see why an asymmetrical distribution of *d* electrons makes a metal ion smaller than expected, consider the Ti^{2+} ion, which has a d^2 configuration with both electrons in the t_{2g} orbitals. Because the t_{2g} orbitals are directed *between the ligands*, the two *d* electrons are unable to shield the ligands from the nuclear charge. Consequently, the ligands experience a higher effective nuclear charge than expected, the metal–ligand distance is unusually short, and the ionic radius is smaller than expected. If instead the two electrons were distributed uniformly over all five *d* orbitals, they would be much more effective at screening the ligands from the nuclear charge, making the metal–ligand distances longer and giving a larger ionic radius.

A similar effect is observed for the V^{2+} ion, which has a d^3 configuration. Because the three electrons in the t_{2g} orbitals provide essentially no shielding of the ligands from the metal, the ligands experience the full increase of +1 in nuclear charge that occurs in going from Ti^{2+} to V^{2+}. Consequently, the observed ionic radius of the V^{2+} ion is significantly smaller than that of the Ti^{2+} ion.

Skipping the Cr^{2+} ion for the moment, let's consider the d^5 Mn^{2+} ion. Because the nuclear charge increases by +2 as we go from V^{2+} to Mn^{2+}, we might expect Mn^{2+} to be smaller than V^{2+}. The two electrons that are also added as we go from V^{2+} to Mn^{2+} occupy the e_g orbitals, however, which point directly at the six ligands. Because these electrons are localized directly between the metal ion and the ligands, they are effective at screening the ligands from the increased nuclear charge. As a result, the ionic radius actually *increases* significantly as we go from V^{2+} to Mn^{2+}, despite the higher nuclear charge of the latter.

Exactly the same effects are seen in the second half of the first-row transition metals. In the Fe^{2+}, Co^{2+}, and Ni^{2+} ions, the extra electrons are added successively to the t_{2g} orbitals, resulting in poor shielding of the ligands from the nuclear charge and abnormally small ionic radii. Skipping over Cu^{2+}, we again see that adding the last two electrons causes a significant increase in the ionic radius of Zn^{2+}, despite its higher nuclear charge.

The Jahn–Teller Effect. Because simple octahedral complexes are not known for the Cr^{2+} and Cu^{2+} ions, only estimated values for their radii are shown in Figure 23.15. We see in Figure 23.11 that both the Cr^{2+} and Cu^{2+} ions have electron configurations with an odd number of electrons in the e_g orbitals. Because the single electron (in the case of Cr^{2+}) or the third electron (in the case of Cu^{2+}) can occupy either one of two degenerate e_g orbitals, they have what is called a *degenerate ground state*. The *Jahn–Teller theorem* states that such systems are not stable and that they will undergo a distortion that makes the complex less symmetrical and splits the degenerate states, which decreases the energy of the system; the distortion and resulting decrease in energy are collectively referred to as the **Jahn–Teller effect**. Neither the nature of the distortion nor its magnitude is specified, and in fact they are difficult to predict. In principle, Jahn–Teller distortions are possible for many transition-metal ions, but in practice they are observed only for systems that have an odd number of electrons in the e_g orbitals, such as the Cr^{2+} and Cu^{2+} ions.

To see how a geometrical distortion can decrease the energy of such a system, consider an octahedral Cu^{2+} complex, the $[Cu(H_2O)_6]^{2+}$ ion, which has been elongated along the z axis. As indicated in Figure 23.16, this kind of distortion splits both the e_g and t_{2g} sets of orbitals. Because the axial ligands interact most strongly with the d_{z^2} orbital, the splitting of the e_g set (δ_1) is significantly larger than the splitting of the t_{2g} set (δ_2), but both δ_1 and δ_2 are much, much smaller than the crystal field splitting energy, Δ_o. This splitting does not change the center of gravity of the energy within each set, so a Jahn–Teller distortion results in no net change in energy for a filled or half-filled set of orbitals. If, however, the e_g set contains one (as in the d^4 ions, Cr^{2+} and Mn^{3+}) or three (as in the d^9 ion, Cu^{2+}) electrons, the distortion decreases the energy of the system. For Cu^{2+}, for example, the change in energy upon distortion is $2(-\delta_1/2) + 1(\delta_1/2) = -\delta_1/2$.

For Cu^{2+} complexes, the observed distortion is *always* an elongation along the z axis by as much as 50 pm; in fact, many Cu^{2+} complexes are so distorted that they are effectively square planar. In contrast, the distortion observed for most Cr^{2+} complexes is a *compression* along the z axis. In both cases, however, the net effect is the same: the distorted system is more stable than the undistorted system.

Note the pattern

Jahn–Teller distortions are most important for d^9 and high-spin d^4 complexes.

Figure 23.16 The Jahn–Teller effect. Increasing the axial metal–ligand distances in an octahedral d^9 complex is an example of a Jahn–Teller distortion, which causes the degenerate pair of e_g orbitals to split in energy by an amount δ_1; δ_1 and δ_2 are much smaller than Δ_o. As a result, the distorted system is more stable (lower in energy) than the undistorted complex by $\delta_1/2$.

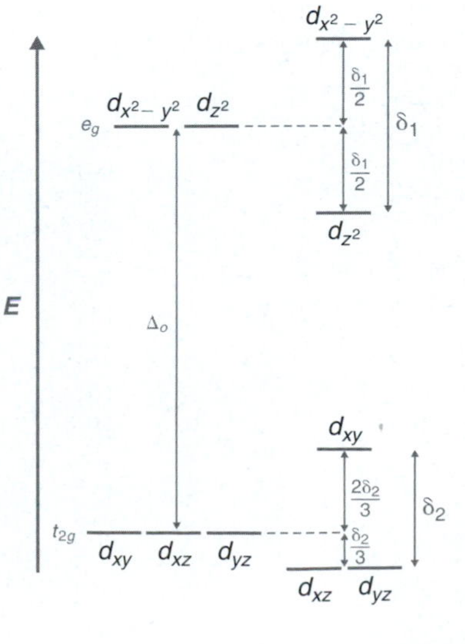

Thermodynamic Effects

As we noted, crystal field stabilization energies can be as large as several hundred kilojoules per mole, which is the same magnitude as the strength of many chemical bonds or the energy change in most chemical reactions. Consequently, CFSEs are important factors in determining the magnitude of hydration energies, lattice energies, and other thermodynamic properties of the transition metals.

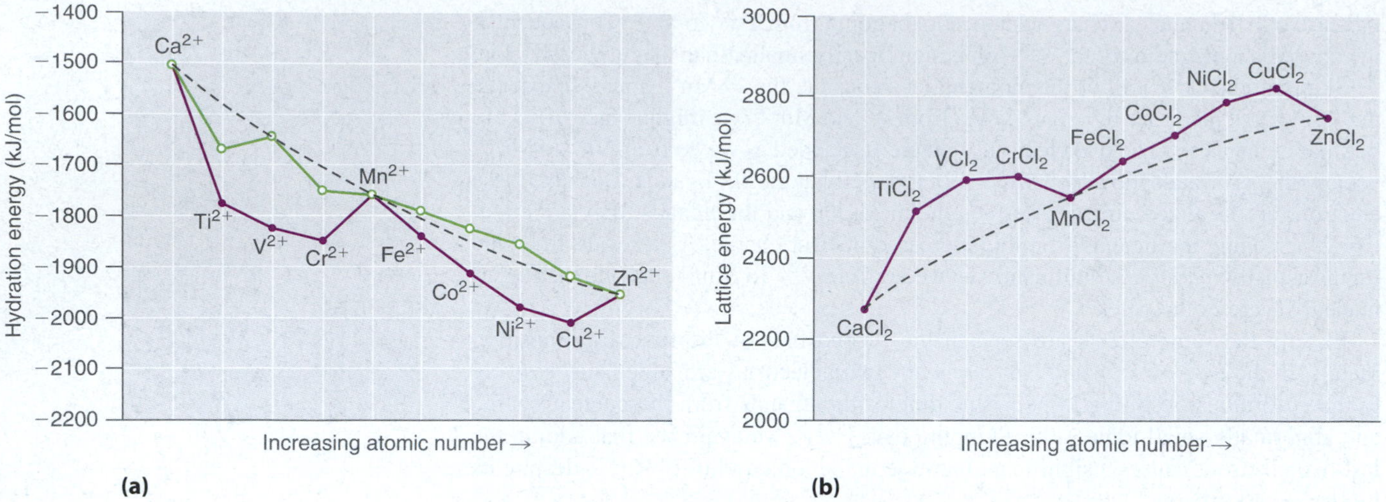

Figure 23.17 Thermochemical effects of *d*-orbital splittings. (a) A plot of the hydration energies of the divalent fourth-period metal ions versus atomic number (solid circles) shows large deviations from the smooth curve calculated assuming a spherical distribution of *d* electrons (dashed line). Correcting for CFSE gives the points shown as open circles, which, except for Ti^{2+} and Cr^{2+}, are close to the calculated values. The apparent deviations for these ions are caused by the fact that solutions of the Ti^{2+} ion in water are not stable, and Cr^{2+} does not form truly octahedral complexes. (b) A plot of the lattice energies for the fourth-period metal dichlorides versus atomic number shows similar deviations from the smooth curve calculated assuming a spherical distribution of *d* electrons (dashed lines), again illustrating the importance of crystal field stabilization energies.

Hydration Energies. The hydration energy of a metal ion is defined as the change in enthalpy for the reaction

$$M^{2+}(g) + H_2O(l) \longrightarrow M^{2+}(aq) \tag{23.12}$$

Although hydration energies cannot be measured directly, they can be calculated from experimentally measured quantities using thermochemical cycles. As shown in Figure 23.17a, a plot of the hydration energies of the fourth-period metal dications versus atomic number gives a curve with two valleys. Note the relationship between the plot in Figure 23.17a and the plot of ionic radii in Figure 23.15: the overall shapes are essentially identical and only the three cations that have spherically symmetrical distributions of *d* electrons (Ca^{2+}, Mn^{2+}, and Zn^{2+}) lie on the dashed lines. In Figure 23.17a, the dashed line corresponds to hydration energies calculated based solely on electrostatic interactions. Subtracting the CFSE values for the $[M(H_2O)_6]^{2+}$ ions from the experimentally determined hydration energies gives the points shown as open circles, which lie very near the calculated curve. Thus, CFSEs are primarily responsible for the differences between the measured and calculated values of hydration energies.

Lattice Energies. Values of the lattice energies for the fourth-period metal dichlorides are plotted versus atomic number in Figure 23.17b. Recall that the lattice energy is defined as the negative of the enthalpy change for the following reaction. Like hydration energies, lattice energies are determined indirectly by using a thermochemical cycle:

$$M^{2+}(g) + 2Cl^{-}(g) \longrightarrow MCl_2(s) \tag{23.13}$$

Note that the shape of the lattice-energy curve is essentially the mirror image of the hydration-energy curve in Figure 23.19a, with only Ca^{2+}, Mn^{2+}, and Zn^{2+} lying on the smooth curve. It is not surprising that the explanation for the deviations from the curve is exactly the same as for the hydration energy data: all the transition-metal dichlorides except $MnCl_2$ and $ZnCl_2$ are more stable than expected due to CFSE.

23.6 Transition Metals in Biology

In Chapter 1, you learned that 19 of the elements in the periodic table are *essential elements* that are necessary for most organisms, including humans, and in Chapter 7 we discussed some of the biological functions of these elements. In this section, we describe several systems that illustrate the roles transition metals play in biological systems. Our goal is for you to understand why the chemical properties of these elements make them essential for life. We begin with a discussion of the strategies organisms use to extract transition metals from their environment. The section continues with a brief discussion of the use of transition metals in reactions that involve the transfer of electrons, reactions of small molecules such as O_2, Lewis-acid catalysis, and the generation of reactive organic radicals.

Uptake and Storage of Transition Metals

MGC Metals in Biological Systems

In Chapter 1, we described the three possible dietary levels for any essential element: deficient, optimal, and toxic, in order of increasing concentration in the diet (see Figure 1.27). If the concentration of an essential element in the diet is too low, an organism must be able to extract the element from the environment and concentrate it. If the concentration of an essential element in the diet is too high, an organism must be able to limit its intake to avoid toxic effects. Moreover, organisms must be able to switch off the uptake process rapidly if dietary levels rise suddenly, and they must be able to store essential elements for future use.

Three distinct steps are involved in transition-metal uptake. First, the metal must be "mobilized" from the environment and brought into contact with the cell in a form that can be absorbed. Second, the metal must be transported across the cell membrane into the cell. Third, the element must be transported to its point of utilization within the cell or to other cells within the organism. In our discussion, we focus on the uptake, transport, and storage of iron, which illustrates the most important points. Because iron deficiency (anemia) is the most widespread nutritional deficiency known in humans, the uptake of iron is especially well understood.

In Chapter 17, you learned that the solubility of metal ions such as Fe^{3+}, which form highly insoluble hydroxides, depends on the pH and the presence of complexing agents. In an oxygen-containing atmosphere, iron exists as Fe(III) because of the positive reduction potential of Fe^{3+} ($Fe^{3+} + e^- \longrightarrow Fe^{2+}$; $E° = +0.77$ V). Because ferric hydroxide, $Fe(OH)_3$, is highly insoluble ($K_{sp} \approx 1 \times 10^{-39}$), the equilibrium concentration of $Fe^{3+}(aq)$ at pH 7 is very low, about 10^{-18} *M*. You would have to drink 2×10^{13} L of iron-saturated water per day (roughly 5 cubic miles!) to consume the recommended daily intake of Fe for humans, which is about 1 mg/day. Animals such as humans can overcome this problem by consuming concentrated sources of iron such as red meat, but microorganisms cannot.

Consequently, most microorganisms synthesize and secrete organic molecules called **siderophores** to increase the total concentration of available iron in the surrounding medium. Siderophores are generally cyclic compounds that use bidentate ligands, such as the *hydroxamate* and *catecholate* groups shown in the margin, to bind Fe^{3+} in an octahedral arrangement. Typical siderophores are *ferrichrome*, a cyclic peptide produced by fungi, and *enterobactin*, a cyclic ester produced by bacteria (Figure 23.18). Attaching the three iron ligands to a cyclic framework greatly increases the stability of the resulting Fe^{3+} complex due to the chelate effect described in Section 23.4. The formation constants for the Fe^{3+} complexes of ferrichrome and enterobactin are about 10^{32} and 10^{40}, respectively, which are high enough to allow them to dissolve almost any Fe(III) compound.

Siderophores simply increase the $[Fe^{3+}]$ in solution, providing the bacterium that synthesized them (as well as any competitors) with a supply of iron. In addition, siderophores neutralize the positive charge on the metal ion and provide a hydrophobic

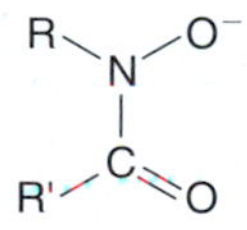

Hydroxamate ligand (1−)

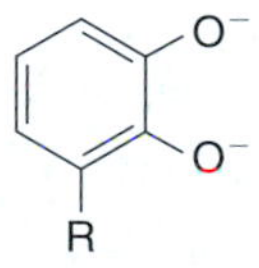

Catecholate ligand (2−)

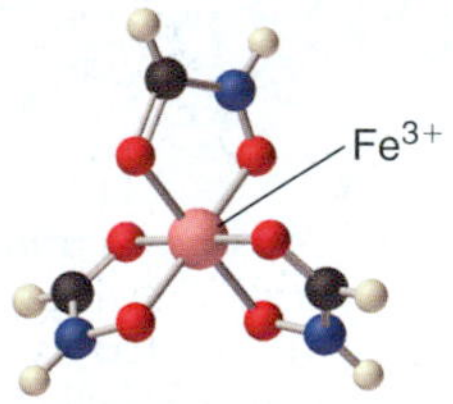

$Fe(hydroxamate)_3$

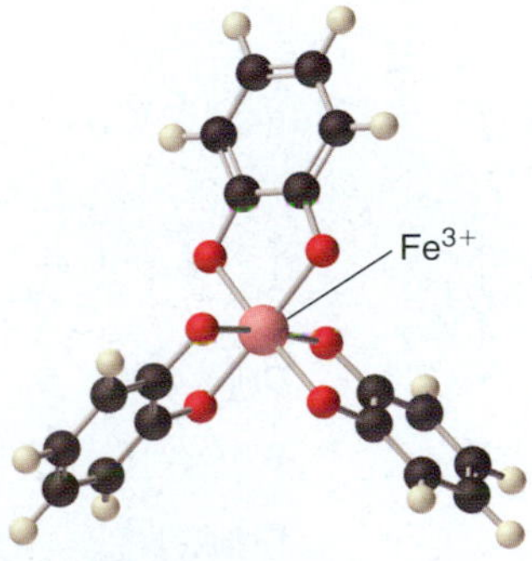

$Fe(catecholate)_3^{3-}$

Iron(III) forms very stable octahedral complexes with hydroxamate and catecholate ligands.

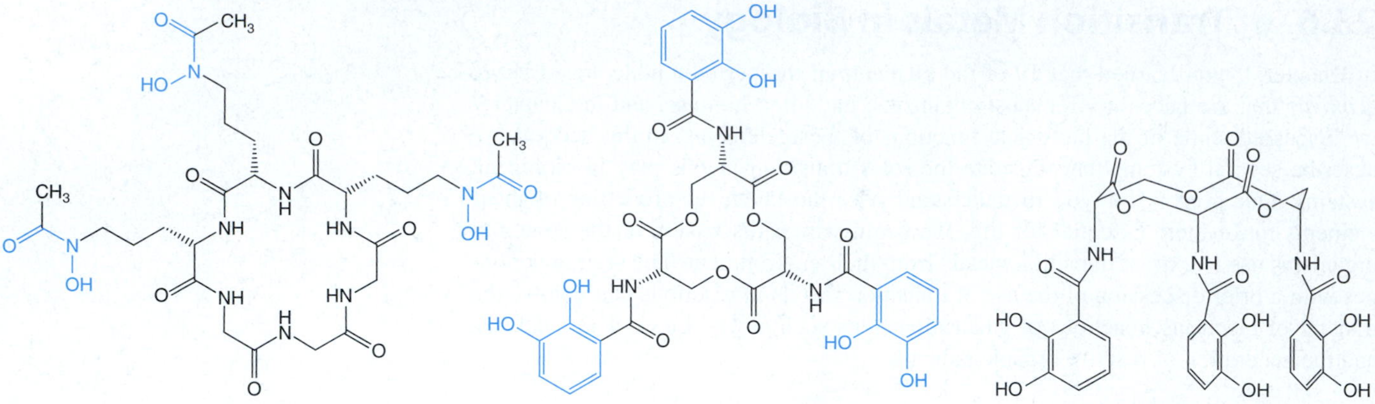

Ferrichrome Enterobactin Enterobactin "wrapping"

Figure 23.18 Siderophores. Ferrichrome (*left*) and enterobactin (*center*) are siderophores that use hydroxamate and catecholate ligands, respectively, to bind Fe^{3+}. The drawing of enterobactin at right shows how the cyclic ester framework places the three catecholate ligands in the correct orientation to bind to a single Fe^{3+} ion, which is an application of the chelate effect. The actual structure of ferrichrome is similar to that of enterobactin, with the three hydroxamate ligands adjacent to one another for optimal binding of Fe^{3+}. *Note:* For clarity, most or all hydrogen atoms have been omitted in this and the following structures.

"wrapping" that enables the Fe^{3+}–siderophore complex to be recognized by a specific protein that transports it into the interior of the cell. Once it is inside the cell, the iron is reduced to Fe^{2+}, which has a much lower affinity for the siderophore and spontaneously dissociates.

In contrast, multicellular organisms can increase the concentration of iron in their diet by simply lowering the pH in the gastrointestinal tract. At pH 1 (the approximate pH of the stomach), most Fe(III) salts dissolve to form $Fe^{3+}(aq)$, which is absorbed by specific proteins in the intestinal wall. A protein called *transferrin* forms a complex with iron(III), allowing it to be transported to other cells. Proteins that bind tightly to Fe(III) can also be used as antibacterial agents because iron is absolutely essential for bacterial growth. For example, milk, tears, and egg white all contain proteins similar to transferrin, and their high affinity for Fe^{3+} allows them to sequester iron, thereby preventing bacteria from growing in these nutrient-rich media.

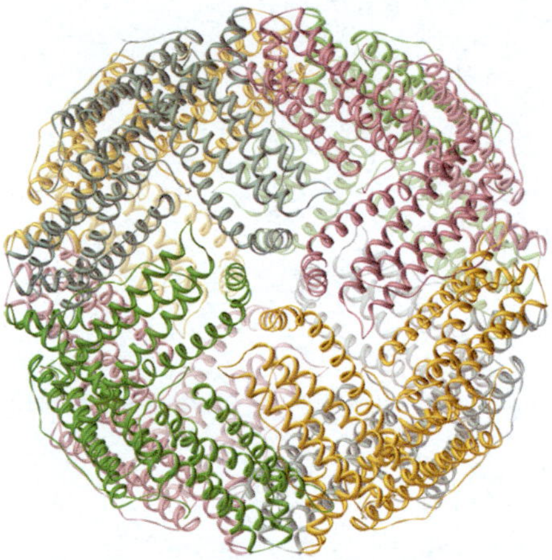

Figure 23.19 Ferritin, an iron storage protein. A schematic drawing of the structure of iron-loaded ferritin, showing the almost spherical protein shell inside which the iron hydroxide/phosphate core is formed.

Iron is released from transferrin by reduction to Fe^{2+}, and then it is either used immediately (for the synthesis of hemoglobin, for example) or stored in a very large protein called *ferritin* for future use (Figure 23.19). Ferritin uses oxygen to oxidize $Fe^{2+}(aq)$ to Fe^{3+}, which at neutral pH precipitates in the central cavity of the protein as a polymeric mixture of $Fe(OH)_3$ and $FePO_4$. Because a fully loaded ferritin molecule can contain as many as 4500 Fe atoms, which corresponds to about 25% Fe by mass, ferritin is an effective way to store iron in a highly concentrated form. When iron is needed by the cell, the Fe^{3+} is reduced to the much more soluble Fe^{2+} by a reductant such as ascorbic acid (vitamin C). The structure of ferritin contains channels at the junctions of the subunits, which provide pathways for iron to enter and leave the interior of the molecule.

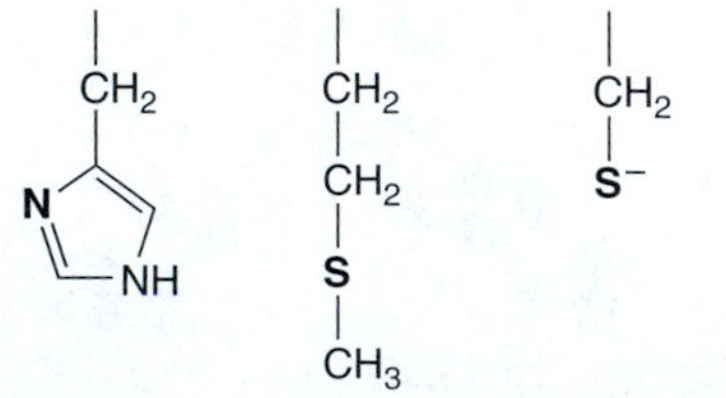

Imidazole (histidine) Thioether (methionine) Thiolate (cysteine)

Some common metal ligands in metalloproteins.

Metalloproteins and Metalloenzymes

A protein that contains one or more metal ions tightly bound to amino acid side chains is called a **metalloprotein**; some of the most common ligands provided by amino acids are shown in the margin. A metalloprotein that catalyzes a chemical reaction is a **metalloenzyme**. Thus, all metalloenzymes are metalloproteins, but the converse is not true. Recent estimates suggest that more than 40% of all known enzymes require at least one metal ion for activity, including almost all the enzymes responsible for the synthesis, duplication, and repair of DNA and RNA.

TABLE 23.12 Some properties of the most common electron-transfer proteins

Protein	Metal Center	M/e$^-$ Transferred	Reduction Potential, V
Iron–sulfur proteins	$[Fe(SR)_4]^{2-}$	1 Fe	−0.1 to +0.1
	$[(RS)_2FeS_2Fe(SR)_2]^{2-}$	2 Fe	−0.2 to −0.4
	$[Fe_3S_4(SR)_3]^{3-}$	3 Fe	−0.1 to −0.2
	$[Fe_4S_4(SR)_4]^{2-}$	4 Fe	−0.3 to −0.5
Cytochromes	Fe-heme (low spin)	1 Fe	~0
Blue copper proteins	$[Cu(Im)_2(SR)(SR_2)]^-$	1 Cu	> +0.20

Electron-Transfer Proteins

Proteins whose function is simply to transfer electrons from one place to another are called *electron-transfer proteins*. Because they do *not* catalyze a chemical reaction, electron-transfer proteins are *not* enzymes, but rather biochemical reductants or oxidants that are consumed in an enzymatic reaction. The general reaction for an electron-transfer protein is

$$M^{n+} + e^- \rightleftharpoons M^{(n-1)+} \tag{23.14}$$

Because many transition metals can exist in more than one oxidation state, electron-transfer proteins usually contain one or more metal ions that can undergo a redox reaction. Incorporating a metal ion into a protein has three important biological consequences:

1. The protein environment can adjust the redox potential, E_0', of the metal ion over a rather large potential range, whereas the redox potential of the simple hydrated metal ion, $M^{n+}(aq)$, is essentially fixed.
2. The protein can adjust the structure of the metal complex to ensure that electron transfer is rapid.
3. The protein environment provides specificity, ensuring that the electron is transferred to only the desired site.

Three important classes of metalloproteins transfer electrons: blue copper proteins, cytochromes, and iron–sulfur proteins, which generally transfer electrons at high (>0.20 V), intermediate (±0 mV), and low (−0.20 to −0.50 V) potentials, respectively (Table 23.12). Although these electron-transfer proteins contain different metals with different structures, they are all designed to ensure rapid electron transfer to and from the metal. Thus, when the protein collides with its physiological oxidant or reductant, electron transfer can occur before the two proteins diffuse apart. For electron transfer to be rapid, the metal sites in the oxidized and reduced forms of the protein must have similar structures.

Blue Copper Proteins

Blue copper proteins were first isolated from bacteria in the 1950s and from plant tissues in the early 1960s. The intense blue color of these proteins is due to a strong absorption band at a wavelength of about 600 nm. Although simple Cu^{2+} complexes such as $[Cu(H_2O)_6]^{2+}$ and $[Cu(NH_3)_4]^{2+}$ are also blue due to an absorption band at 600 nm, the intensity of the absorption band is about a hundred times less than that of a blue copper protein. Moreover, the reduction potential for the Cu^{2+}/Cu^+ couple in a blue copper protein is usually +0.3 to +0.5 V, considerably more positive than that of the aqueous Cu^{2+}/Cu^+ couple (+0.15 V).

The copper center in blue copper proteins has a distorted tetrahedral structure, in which the copper is bound to two histidine imidazole nitrogen atoms, a cysteine thiolate sulfur atom, and a methionine thioether sulfur atom (Figure 23.20). Although the

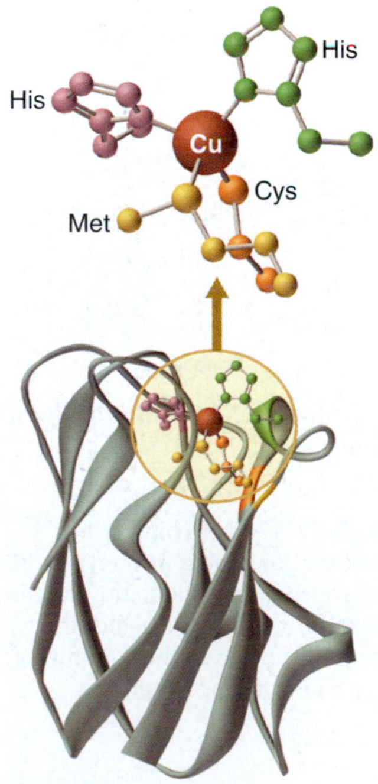

Figure 23.20 A blue copper protein. In both the oxidized and reduced forms of a blue copper protein, the copper is coordinated by four ligands (two histidine imidazole nitrogen atoms, a cysteine thiolate sulfur, and a thioether sulfur of a methionine) in a roughly tetrahedral arrangement.

most common structures for 4-coordinate Cu^{2+} and Cu^{+} complexes are square planar and tetrahedral, respectively, the structures of the oxidized (Cu^{2+}) and reduced (Cu^{+}) forms of the protein are essentially identical. Thus, the protein forces the Cu^{2+} ion to adopt a higher-energy structure that is more suitable for Cu^{+}, which makes the Cu^{2+} form easier to reduce and raises its reduction potential.

Moreover, by forcing the oxidized and reduced forms of the metal complex to have essentially the same structure, the protein ensures that electron transfer to and from the copper site is intrinsically very fast, because only minimal structural reorganization of the metal center is required. Kinetics studies on simple metal complexes have shown that electron-transfer reactions tend to be slow when the structures of the oxidized and reduced forms of a metal complex are very different, and fast when they are similar. You will see that other metal centers used for biological electron-transfer reactions are also set up for minimal structural reorganization upon electron transfer, which ensures that electron transfer is rapid.

Cytochromes

The cytochromes (from the Greek *cytos*, "cell", and *chroma*, "color") were first identified in the 1920s by spectroscopic studies of cell extracts. Based on the wavelength of the maximum absorption in the visible spectrum, they were classified as cytochromes *a* (with the longest wavelength), cytochromes *b* (intermediate wavelength), and cytochromes *c* (shortest wavelength). It quickly became apparent that there was a correlation between their spectroscopic properties and other physical properties. For examples, cytochromes *c* are generally small, soluble proteins with a reduction potential of about +0.25 V, whereas cytochromes *b* are larger, less-soluble proteins with reduction potentials of about 0 V.

All cytochromes contain iron, and the iron atom in all cytochromes is coordinated by a planar array of four nitrogen atoms provided by a cyclic tetradentate ligand called a *porphyrin*. The iron–porphyrin unit is called a *heme* group. The structures of a typical porphyrin (protoporphyrin IX) and its iron complex (protoheme) are shown in the margin. In addition to the four nitrogen atoms of the porphyrin, the iron in a cytochrome is usually bonded to two additional ligands provided by the protein, as shown in Figure 23.21.

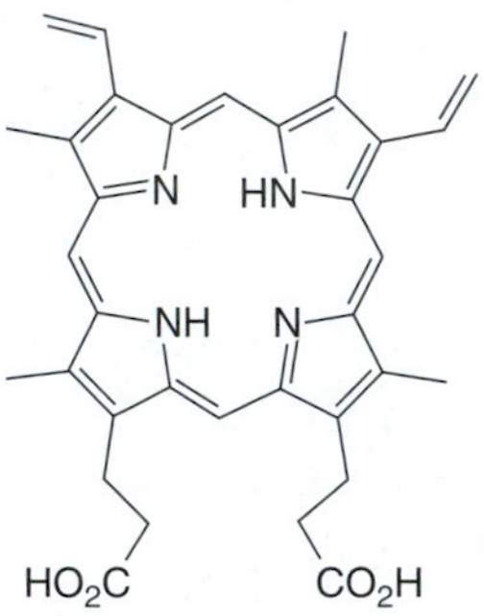

Protoporphyrin IX

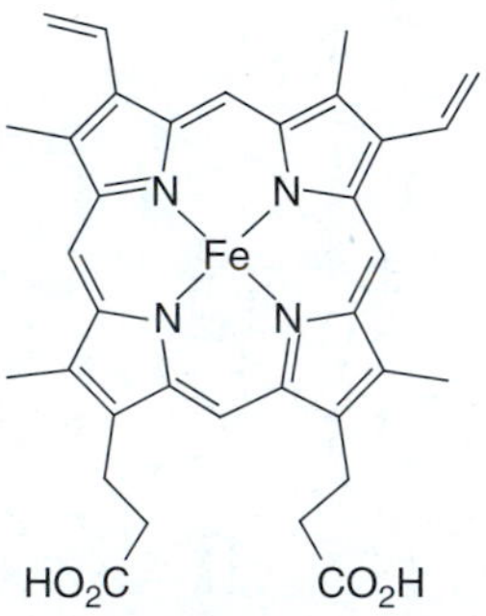

Protoheme

Protoporphyrin IX and its iron complex, protoheme.

In contrast to the blue copper proteins, two electron configurations are possible for both the oxidized and reduced forms of a cytochrome, and this has significant structural consequences. Thus, Fe^{2+} is d^6 and can be either high spin (with four unpaired electrons) or low spin (with no unpaired electrons). Similarly, Fe^{3+} is d^5 and can also be high spin (with five unpaired electrons) or low spin (with one unpaired electron). In low-spin heme complexes, both the Fe^{2+} and the Fe^{3+} ions are small enough to fit into the "hole" in the center of the porphyrin; hence, the iron atom lies almost exactly in the plane of the four porphyrin nitrogen atoms in both cases. Because cytochromes *b* and *c*

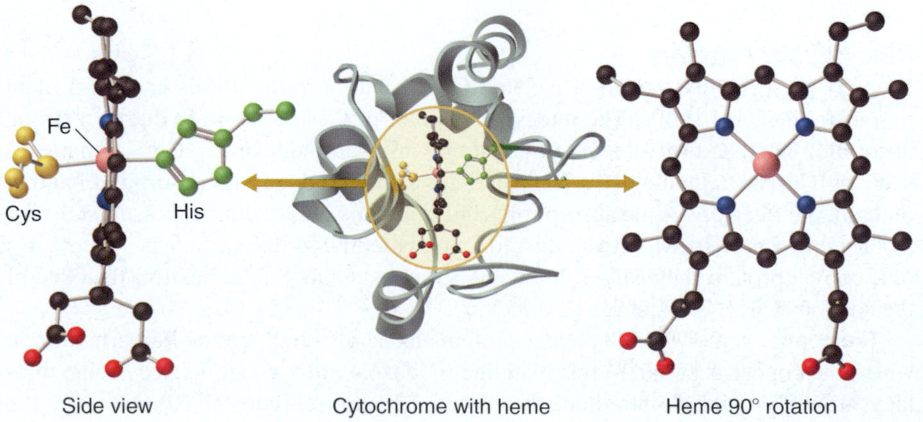

Figure 23.21 Cytochrome *c*. In cytochrome *c*, the heme iron is coordinated to the nitrogen atom of a histidine imidazole and the sulfur atom of a methionine thioether, in addition to the four nitrogen atoms provided by the porphyrin.

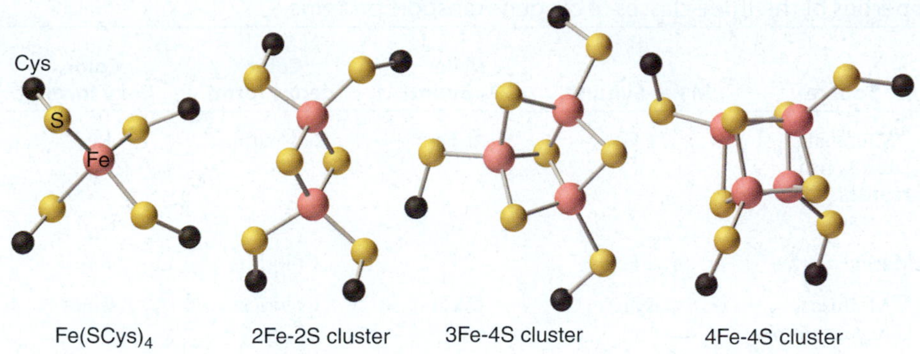

Figure 23.22 Fe–S centers in proteins. Four kinds of iron–sulfur center, containing one, two, three, and four iron atoms, respectively, are known in electron-transfer proteins. Although they differ in the number of sulfur atoms provided by cysteine thiolates versus sulfide, in all cases the iron is coordinated to four sulfur ligands in a roughly tetrahedral environment.

are low spin in both their oxidized and reduced forms, the structures of the oxidized and reduced cytochromes are essentially identical. Hence, minimal structural changes occur upon oxidation or reduction, which makes electron transfer to or from the heme very rapid.

Iron–Sulfur Proteins

Although all known bacteria, plants, and animals use iron–sulfur proteins to transfer electrons, the existence of these proteins was not recognized until the late 1950s. Iron–sulfur proteins transfer electrons over a wide range of reduction potentials, and their iron content can range from 1 to more than 12 Fe atoms per protein molecule. In addition, most iron–sulfur proteins contain stoichiometric amounts of sulfide, S^{2-}.

These properties are due to the presence of four different kinds of iron–sulfur units, which contain one, two, three, or four iron atoms per Fe–S complex (Figure 23.22). In all cases, the Fe^{2+} and Fe^{3+} ions are coordinated to four sulfur ligands in a tetrahedral environment. Due to tetrahedral coordination by weak-field sulfur ligands, the iron is high spin in both the Fe^{3+} and Fe^{2+} oxidation states, which results in similar structures for the oxidized and reduced forms of the Fe–S complexes. Consequently, only small structural changes occur upon oxidation or reduction of the Fe–S center, which results in rapid electron transfer.

Reactions of Small Molecules

Although small molecules such as O_2, N_2, and H_2 do not react with organic compounds under ambient conditions, they do react with many transition-metal complexes. Consequently, virtually all organisms use metalloproteins to bind, transport, and catalyze the reactions of these molecules. Probably the best-known example is hemoglobin, which is used to transport O_2 in many multicellular organisms.

Oxygen Transport

Many microorganisms and most animals obtain energy by *respiration*, the oxidization of organic or inorganic molecules by O_2. At 25°C, however, the concentration of dissolved oxygen in water in contact with air is only about 0.25 m*M*. Because of their high surface area-to-volume ratio, aerobic microorganisms can obtain enough oxygen for respiration by passive diffusion of O_2 through the cell membrane. As the size of an organism increases, however, its volume increases much more rapidly than its surface area, and the need for oxygen depends on its volume. Consequently, as a multicellular organism grows larger, its need for O_2 rapidly outstrips the supply available through diffusion. Unless a transport system is available to provide an adequate supply of oxygen for the interior cells, organisms that contain more than a few cells cannot exist. In addition, O_2 is such a powerful oxidant that the oxidation reactions used to obtain metabolic energy must be carefully controlled in order to avoid releasing so much heat that the water in the cell boils. Consequently, in higher organisms, the respiratory

TABLE 23.13 Some properties of the three classes of oxygen-transport proteins

Protein	Source	M Per Subunit	M Per O_2 Bound	Color (deoxy form)	Color (oxy form)
Hemoglobin (Hb)	Mammals, birds, fish, reptiles, some insects	1 Fe	1 Fe	Red-purple	Red
Hemerythrin (Hr)	Marine worms	2 Fe	2 Fe	Colorless	Red
Hemocyanin (Hc)	Molluscs, crustaceans, spiders	2Cu	2 Cu	Colorless	Blue

apparatus is located in internal compartments called *mitochondria*, which are the power plants of the cell. Oxygen must therefore be transported not just to the cell, but also to the proper compartment within the cell.

Three different chemical solutions to the problem of oxygen transport have developed independently in the course of evolution, as indicated in Table 23.13. Mammals, birds, reptiles, fish, and some insects use a heme protein called *hemoglobin* to transport oxygen from the lungs to the cells, and they use a related protein called *myoglobin* to temporarily store oxygen in the tissues. Several classes of invertebrates, including marine worms, use an iron-containing protein called *hemerythrin* to transport oxygen, whereas other classes of invertebrates (arthropods and molluscs) use a copper-containing protein called *hemocyanin*. Despite the presence of the "hem-" prefix, hemerythrin and hemocyanin do *not* contain a metal–porphyrin complex.

Note the pattern

A vacant coordination site at a metal center in a protein usually indicates that a small molecule will bind to the metal ion, whereas a coordinatively saturated metal center is usually involved in electron transfer.

Myoglobin and Hemoglobin

Myoglobin is a relatively small protein that contains 150 amino acids. The functional unit of myoglobin is protoheme, an iron–porphyrin complex that is embedded in the protein (Figure 23.23). In myoglobin, the heme iron is 5 coordinate, with only a single histidine imidazole ligand from the protein (called the "*proximal histidine*" because it is near the iron) in addition to the four nitrogen atoms of the porphyrin. A second histidine imidazole (the "*distal histidine*" because it is more distant from the iron) is located on the other side of the heme group, too far from the iron to be bonded to it. Consequently, the iron atom has a vacant coordination site, which is where O_2 binds. In the ferrous form (*deoxymyoglobin*), the iron is 5 coordinate and

Figure 23.23 The structure of deoxymyoglobin, showing the heme group. The iron in deoxymyoglobin is 5-coordinate, with one histidine imidazole ligand from the protein. Oxygen binds at the vacant site on iron.

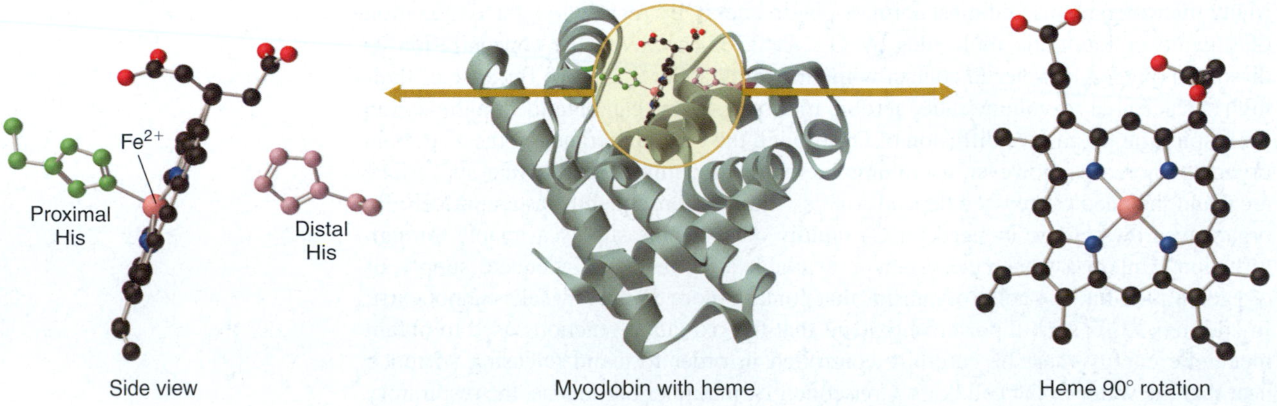

high spin. Because high-spin Fe^{2+} is too large to fit into the "hole" in the center of the porphyrin, it is about 60 pm above the plane of the porphyrin. When O_2 binds to deoxymyoglobin to form *oxymyoglobin*, the iron is converted from 5 coordinate (high spin) to 6 coordinate (low spin) (Figure 23.24). Because low-spin Fe^{2+} and Fe^{3+} are smaller than high-spin Fe^{2+}, the iron atom moves into the plane of the porphyrin ring to form an octahedral complex. The O_2 pressure at which half of the molecules in a solution of myoglobin are bound to O_2 ($P_{1/2}$) is about 1 mm Hg (1.3×10^{-3} atm).

Hemoglobin consists of two subunits of 141 amino acids and two subunits of 146 amino acids, both similar to myoglobin. Because hemoglobin has very different O_2-binding properties, however, it is not simply a "super myoglobin" that can carry four O_2 molecules simultaneously (one per heme group). The shape of the O_2-binding curve of myoglobin (Mb) (Figure 23.25) can be described mathematically by the following equilibrium:

$$MbO_2 \rightleftharpoons Mb + O_2 \qquad K_{diss} = \frac{[Mb][O_2]}{[MbO_2]} \qquad (23.15)$$

In contrast, the O_2-binding curve of hemoglobin is S-shaped (Figure 23.25). At low oxygen pressures, the affinity of deoxyhemoglobin for O_2 is substantially lower than that of myoglobin, whereas at high O_2 pressures the two proteins have comparable O_2 affinities. The physiological consequences of the unusual S-shaped O_2-binding curve of hemoglobin are enormous. In the lungs, where the O_2 pressure is highest, the high oxygen affinity of deoxyhemoglobin allows it to be completely loaded with O_2, giving four O_2 molecules per hemoglobin. In the tissues, however, where the oxygen pressure is much lower, the decreased oxygen affinity of hemoglobin allows it to release O_2, resulting in a net transfer of oxygen to myoglobin.

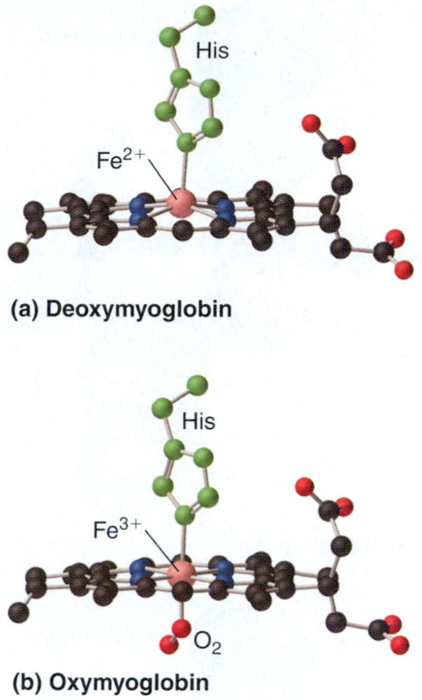

Figure 23.24 Oxygen binding to myoglobin and hemoglobin. (a) The Fe^{2+} ion in deoxymyoglobin is high spin, which makes it too large to fit into the "hole" in the center of the porphyrin. (b) When O_2 binds to deoxymyoglobin, the iron is converted to low-spin Fe^{3+}, which is smaller, allowing the iron to move into the plane of the four nitrogen atoms of the porphyrin to form an octahedral complex.

The S-shaped O_2-binding curve of hemoglobin is due to a phenomenon called *cooperativity*, in which the affinity of one heme for O_2 depends on whether the other hemes are already bound to O_2. Cooperativity in hemoglobin requires an interaction between the four heme groups in the hemoglobin tetramer, even though they are more than 3000 pm apart, and depends on the change in structure of the heme group that occurs upon oxygen binding. The structures of deoxyhemoglobin and oxyhemoglobin are slightly different, and as a result, deoxyhemoglobin has a much lower O_2 affinity than myoglobin, whereas the O_2 affinity of oxyhemoglobin is essentially identical to that of oxymyoglobin. Binding of the first two O_2 molecules to deoxyhemoglobin causes the overall structure of the protein to change to that of oxyhemoglobin, and consequently the last two heme groups have a much higher affinity for O_2 than the first two.

Oxygen is not unique in its ability to bind to a ferrous heme complex; small molecules such as CO and NO bind to deoxymyoglobin even more tightly than does O_2. The interaction of the heme iron with oxygen and other diatomic molecules involves the transfer of electron density from the filled t_{2g} orbitals of the low-spin d^6 Fe^{2+} ion to the empty π^* orbitals of the ligand. In the case of the Fe^{2+}—O_2 interaction, the transfer of electron density is so great that the Fe—O_2 unit can be described as containing low-spin Fe^{3+} (d^5) and O_2^-. We can therefore represent the binding of O_2 to deoxyhemoglobin and its release as a reversible redox reaction:

$$Fe^{2+} + O_2 \rightleftharpoons Fe^{3+} - O_2^- \qquad (23.16)$$

As shown in Figure 23.26, the Fe—O_2 unit is bent, with an Fe—O—O angle of about 130°. Because the π^* orbitals in CO are empty and those in NO are singly occupied, these ligands interact more strongly with Fe^{2+} than does O_2, in which the π^* orbitals of the neutral ligand are doubly occupied.

Although CO has a much higher intrinsic affinity for a ferrous heme than does O_2 (by a factor of about 25,000), the affinity of CO for deoxyhemoglobin is only about 200 times higher than that of O_2, which suggests that something in the protein is decreasing its affinity for CO by a factor of about 100. Both CO and NO bind to ferrous hemes in a

Figure 23.25 The O_2-binding curves of myoglobin and hemoglobin. The curve for myoglobin can be described by a simple equilibrium between deoxy- and oxymyoglobin, but the S-shaped curve for hemoglobin can be described only in terms of a cooperative interaction between the four hemes.

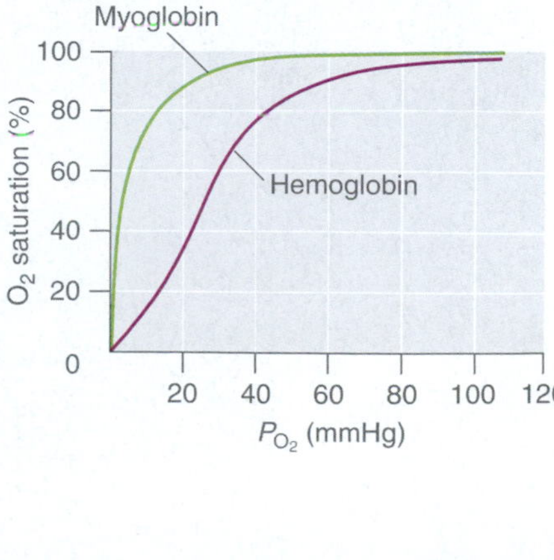

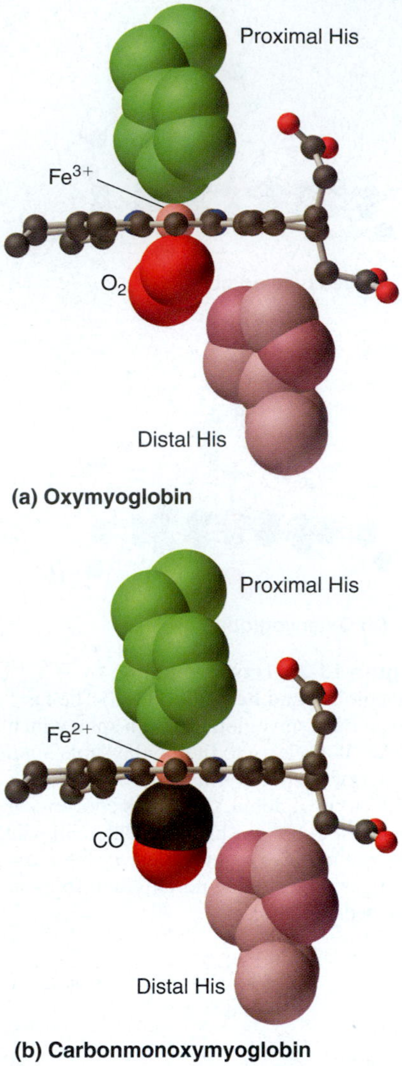

Figure 23.26 Binding of O_2 and CO to the iron of myoglobin. Because the Fe—O—O unit is bent, while the Fe—C—O unit is linear, the imidazole group of the distal histidine in hemoglobin interferes with CO binding and decreases the affinity of hemoglobin for CO.

linear fashion, with an Fe—C(N)—O angle of about 180°, and the difference in the preferred geometry of O_2 and CO provides a plausible explanation for the difference in affinities. As shown in Figure 23.26, the imidazole group of the distal histidine is located precisely where the oxygen atom of bound CO would be if the Fe—C—O unit were linear. Consequently, CO cannot bind to the heme in a linear fashion, and instead it is forced to bind in a bent mode that is similar to the preferred structure for the Fe—O_2 unit. This results in a significant decrease in the affinity of the heme for CO, while leaving the O_2 affinity unchanged, which is important because carbon monoxide is produced continuously in the body by degradation of the porphyrin ligand (even in nonsmokers). Under normal conditions, approximately 1% of the heme sites in hemoglobin and myoglobin are occupied by CO. If the affinity of hemoglobin and myoglobin for CO were 100 times greater (due to the absence of the distal histidine), essentially 100% of the heme sites would be occupied by CO, and no oxygen could be transported to the tissues. Severe carbon-monoxide poisoning, which is frequently fatal, has exactly the same effect. Thus, the primary function of the distal histidine appears to be to decrease the CO affinity of hemoglobin and myoglobin to avoid self-poisoning by CO.

Hemerythrin

Hemerythrin is used to transport O_2 in a variety of marine invertebrates; it is an octamer, with each subunit containing two iron atoms and binding one molecule of O_2. Deoxyhemerythrin contains two Fe^{2+} ions and is colorless, whereas oxyhemerythrin contains two Fe^{3+} ions and is bright reddish violet. These invertebrates also contain a monomeric form of hemerythrin that is located in the tissues, analogous to myoglobin. The binding of oxygen to hemerythrin and its release can be described by the following reaction, where the O_2H^- ligand is the hydroperoxide anion derived by the deprotonation of hydrogen peroxide (H_2O_2):

$$2Fe^{2+} + O_2 + H^+ \rightleftharpoons Fe^{3+}Fe^{3+} - O_2H \qquad (23.17)$$

Thus, O_2 binding is accompanied by the transfer of *two* electrons (one from each Fe^{2+}) and a proton to O_2.

Hemocyanin

Hemocyanin is used for oxygen transport in many arthropods (spiders, crabs, lobsters, and centipedes) and in molluscs (shellfish, octopi, and squid), and it is responsible for the bluish-green color of their blood. The protein is a polymer of subunits that each contain two copper atoms, with an aggregate molecular mass of greater than a million amu. Deoxyhemocyanin contains two Cu^+ ions and is colorless, whereas oxyhemocyanin contains two Cu^{2+} ions and is bright blue. As with hemerythrin, the binding and release of O_2 correspond to a two-electron reaction:

$$2Cu^+ + O_2 \rightleftharpoons Cu^{2+} - {O_2}^{2-} - Cu^{2+} \qquad (23.18)$$

Although hemocyanin and hemoglobin perfom the same basic function as hemoglobin, these proteins are not interchangeable. In fact, hemocyanin is so foreign to humans that it is one of the major factors responsible for the common allergies to shellfish.

Enzymes Involved in Oxygen Activation

Many of the enzymes involved in the biological reactions of oxygen contain metal centers with structures that are similar to those used for O_2 transport. Many of these enzymes also contain metal centers that are used for electron transfer, which have structures similar to those of the electron-transfer proteins discussed earlier. In this section, we briefly describe two of the most important examples: dioxygenases and methane monooxygenase.

Dioxygenases are enzymes that insert both atoms of O_2 into an organic molecule. In humans, dioxygenases are responsible for cross-linking collagen in connective tissue and for the synthesis of complex organic molecules called *prostaglandins*, which trigger inflammation and immune reactions. Iron is by far the most common metal in dioxygenases, and the target of the most commonly used drug in the world, aspirin, is an iron enzyme that synthesizes a specific prostaglandin. Aspirin inhibits this enzyme by binding to the iron atom at the active site, which prevents oxygen from binding.

Dioxygenases also allow some bacteria to grow on organic compounds such as benzene as their sole source of carbon. One of the most intriguing is benzene dioxygenase, which inserts both atoms of O_2 into two adjacent C—H bonds of benzene to form dihydroxybenzene (catechol):

$C_6H_6 + O_2 \longrightarrow$ 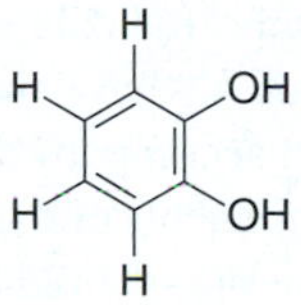 (23.19)

Benzene dioxygenase utilizes Fe^{2+} to bind and activate O_2 for reaction with benzene.

Methane monooxygenase catalyzes the hydroxylation of methane to form methanol. The enzyme is a *mono*oxygenase because only one atom of O_2 is inserted into an organic molecule, while the other is reduced to water:

$$CH_4 + O_2 + 2e^- + 2H^+ \longrightarrow CH_3OH + H_2O \qquad (23.20)$$

Because methane is the major component of natural gas, there is enormous interest in using this reaction to convert methane to a liquid fuel (methanol) that is much more convenient to ship and to store. Because the C—H bond in methane is one of the strongest C—H bonds known, however, an extraordinarily powerful oxidant is needed for this reaction. The active site of methane monooxygenase contains two Fe atoms that bind O_2, but the details of how the bound O_2 is converted to such a potent oxidant remain unclear.

Metal Ions as Lewis Acids

Reactions catalyzed by metal ions that do not change their oxidation states during the reaction are usually *group transfer reactions*, in which a group such as the phosphoryl group, $—PO_3^{2-}$, is transferred. These enzymes usually use metal ions such as Zn^{2+}, Mg^{2+}, and Mn^{2+}, and they range from true metalloenzymes, in which the metal ion is tightly bound, to metal-activated enzymes, which require the addition of metal ions for activity. Because tight binding is usually the result of specific metal–ligand interactions, metalloenzymes tend to be rather specific for a particular metal ion. In contrast, the binding of metal ions to metal-activated enzymes is largely electrostatic in nature, and consequently several different metal ions with similar charges and sizes can often be used to give an active enzyme.

A metal ion that acts as a Lewis acid can catalyze a group transfer reaction in many different ways, but we will focus on only one of these, using a zinc enzyme as an example. *Carbonic anhydrase* is found in red blood cells and catalyzes the reaction of CO_2 with water to give carbonic acid.

$$CO_2(g) + H_2O(l) \rightleftharpoons H^+(aq) + HCO_3^-(aq) \qquad (23.21)$$

Although this reaction occurs spontaneously in the absence of a catalyst, it is too slow to absorb all the CO_2 generated during respiration. Thus, without a catalyst, tissues would explode due to the buildup of excess CO_2 pressure. Carbonic anhydrase contains a single Zn^{2+} ion per molecule, and the Zn^{2+} ion is coordinated by three histidine imidazole ligands and a molecule of water. Because Zn^{2+} is a Lewis acid, the pK_a of the $Zn^{2+}—OH_2$ unit is about 8 versus 14 for pure water. Thus, at pH 7–8, a significant fraction of the enzyme molecules contain a $Zn^{2+}—OH^-$ group, which is much more reactive than bulk water. When carbon dioxide binds in a nonpolar site next to the Zn^{2+}—OH^- unit, it reacts rapidly to give a coordinated bicarbonate ion that dissociates from the enzyme:

$$Zn^{2+}-OH^- + CO_2 \rightleftharpoons Zn^{2+}-OCO_2H^- \rightleftharpoons Zn^{2+} + HCO_3^- \qquad (23.22)$$

Thus, the function of zinc in carbonic anhydrase is to generate the hydroxide ion at pH 7, far lower than the pH required in the absence of the metal ion.

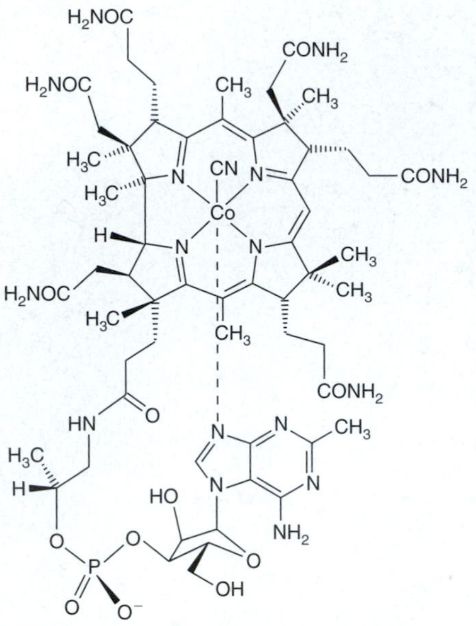

Figure 23.27 The structure of vitamin B_{12}. In the body, the axial cyanide ligand found in the vitamin is replaced by a complex organic unit. Heterolytic cleavage of the Co—C bond in the resulting organometallic complex generates an organic radical for catalysis of rearrangement reactions.

Enzymes That Use Metals to Generate Organic Radicals

An *organic radical* is an organic species that contains one or more unpaired electrons. Chemists often consider organic radicals to be highly reactive species that produce undesirable reactions, and radicals have indeed been implicated in some of the irreversible chemical changes that accompany aging. It is surprising, however, that organic radicals are also essential components of many important enzymes, almost all of which use a metal ion to generate the organic radical within the enzyme. These enzymes are involved in the synthesis of hemoglobin and DNA, among other important biological molecules, and they are the targets of pharmaceuticals for the treatment of diseases such as anemia, sickle-cell anemia, and cancer. In this section, we discuss one class of radical enzymes, those that use *vitamin B_{12}*.

Vitamin B_{12} was discovered in the 1940s as the active agent in the cure of pernicious anemia (an anemia that does not respond to increased iron in the diet) by consumption of large amounts of liver. Humans need only tiny amounts of vitamin B_{12}, and the average blood concentration in a healthy adult is only about 3.5×10^{-8} *M*! The structure of vitamin B_{12}, shown in Figure 23.27, is reminiscent of a heme, but it contains cobalt instead of iron and its structure is much more complex. In fact, vitamin B_{12} has been called the most complex nonpolymeric biological molecule known, and it was the first naturally occurring organometallic compound to be isolated. When vitamin B_{12} (the form present in vitamin tablets) is ingested, the axial cyanide ligand is replaced by a complex organic group.

The cobalt–carbon bond in the enzyme-bound form of B_{12} and related compounds is unusually weak, and it is particularly susceptible to *homolytic cleavage*:

$$CoCH_2R \rightleftharpoons Co^{2+}\cdot + \cdot CH_2R \qquad (23.23)$$

Homolytic cleavage of the Co^{3+}—CH_2R bond produces two species, each of which has an unpaired electron: a d^7 Co^{2+} derivative and an organic radical, $\cdot CH_2R$, which is used by vitamin B_{12}–dependent enzymes to catalyze a wide variety of reactions. Virtually all vitamin B_{12}–catalyzed reactions are rearrangements in which an H atom and an adjacent substituent exchange positions:

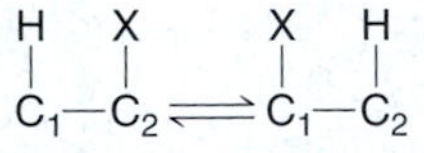

(23.24)

In the conversion of ethylene glycol to acetaldehyde, the initial product is the hydrated form of acetaldehyde, which rapidly loses water:

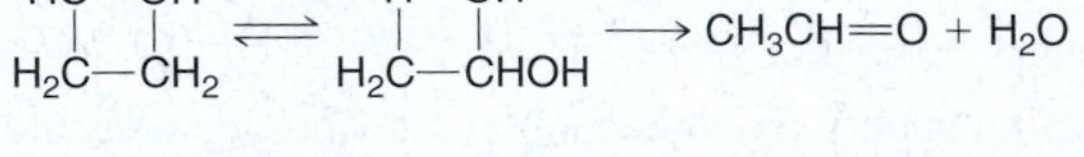

(23.25)

The enzyme uses the $\cdot CH_2R$ radical to temporarily remove a hydrogen atom from the organic substrate, which then rearranges to give a new radical. Transferring the hydrogen atom back to the rearranged radical gives the product and regenerates the $\cdot CH_2R$ radical.

Thus, the metal is not involved in the actual catalytic reaction; it only provides the enzyme with a convenient mechanism for generating an organic radical, which does the actual work. Many examples of similar reactions are now known that utilize metals other than cobalt to generate an enzyme-bound organic radical.

SUMMARY AND KEY TERMS

23.1 General Trends Among the Transition Metals (p. 1037)

The transition metals are characterized by partially filled *d* subshells in the free elements and cations. The *ns* and $(n - 1)d$ subshells have similar energies, so small influences can produce electron configurations that do not conform to the general order in which subshells are filled. In the second- and third-row transition metals, such irregularities can be difficult to predict, particularly for the third row, which has 4*f*, 5*d*, and 6*s* orbitals that are very close in energy. The increase in atomic radius is greater between the 3*d* and 4*d* metals than between the 4*d* and 5*d* metals because of the lanthanide contraction. Ionization energies and electronegativities increase slowly as we go across a row, as do densities and electrical and thermal conductivities, whereas enthalpies of hydration decrease. Anomalies can be explained by the increased stabilization of half-filled and filled subshells. Transition-metal cations are formed by the initial loss of *ns* electrons, and many metals can form cations in several oxidation states. Higher oxidation states become progressively less stable as we go across a row and more stable as we go down a column. Oxides of small, highly charged metal ions tend to be acidic, whereas oxides of metals with a low charge-to-radius ratio are basic.

23.2 A Brief Survey of Transition-Metal Chemistry (p. 1041)

The Group 3 transition metals are highly electropositive metals and powerful reductants. They react with nonmetals to form compounds that are largely ionic, and with oxygen to form *sesquioxides*, M_2O_3. The Group 4 metals also have a high affinity for oxygen. In their reactions with halogens, the covalent character of the halides increases as the oxidation state of the metal increases because the high charge-to-radius ratio causes extensive polarization of the anions. The dichalcogenides have layered structures similar to graphite, and the hydrides, nitrides, carbides, and borides are all hard, high-melting-point solids with metallic conductivity. The Group 5 metals also have a high affinity for oxygen. Consistent with periodic trends, only the lightest, vanadium, has any tendency to form compounds in oxidation states lower than +5. The oxides are sufficiently polarized to make them covalent in character. These elements also form layered chalcogenides, as well as nitrides, carbides, borides, and hydrides that are similar to those of the Group 4 elements. As the metals become more polarizable across the row, their affinity for oxygen decreases. The Group 6 metals are less electropositive and have a maximum oxidation state of +6, making their compounds in high oxidation states largely covalent in character. As the oxidizing strength of the halogen decreases, the maximum oxidation state of the metal also decreases. All three trioxides are acidic, but Cr_2O_3 is amphoteric. The chalcogenides of the Group 6 metals are generally nonstoichiometric and electrically conducting, and these elements also form nitrides, carbides, and borides that are similar to those in the preceding groups. The metals of Group 7 have a maximum oxidation state of +7, but the lightest element, manganese, exhibits an extensive chemistry in lower oxidation states. As with the Group 6 metals, reaction with less oxidizing halogens produces metals in lower oxidation states, and disulfides and diselenides of Tc and Re have layered structures. The Group 7 metals also form nitrides, carbides, and borides that are stable at high temperatures and have metallic properties. In Groups 8, 9, and 10, the ionization potentials of the elements are so high that the oxidation state corresponding to the formal loss of all valence electrons is encountered rarely (Group 8) or not at all (Groups 9 and 10). Compounds of Group 8 metals in their highest oxidation state are powerful oxidants. The reaction of metals in Groups 8, 9, and 10 with the chalcogens is complex, and these elements form a range of binary nitrides, carbides, and borides. The coinage metals (Group 11) have the highest electrical and thermal conductivities and are the most ductile and malleable of the metals. Although they are relatively unreactive, they form halides, but not nitrides, borides, or carbides. The Group 12 elements, whose chemistry is dominated by the +2 oxidation state, are almost always found in nature combined with sulfur. Mercury is the only metal that is a liquid at room temperature, and it "dissolves" many metals to form amalgams. The Group 12 halides range from ionic to covalent. These elements form chalcogenides and have a high affinity for "soft" ligands.

23.3 Metallurgy (p. 1052)

The conversion of metals from their ores to more useful forms is called **metallurgy**, which consists of three general steps: mining, separation and concentration, and reduction. *Settling* and *flotation* are separation methods based on differences in density, whereas *pyrometallurgy* is based on chemical reduction at elevated temperatures and *hydrometallurgy* uses chemical or electrochemical reduction of an aqueous solution. In pyrometallurgy, a reductant must be used that does not form stable compounds with the metal of interest. In hydrometallurgy, metals are separated via the formation of metal complexes.

23.4 Coordination Compounds (p. 1055)

Transition metals form **metal complexes**, polyatomic species in which a metal ion is bound to one or more **ligands**, groups bound to a metal ion. **Complex ions** are electrically charged metal complexes, and a **coordination compound** contains one or more metal complexes. Metal complexes with low coordination numbers generally have only one or two possible structures, whereas those with coordination numbers higher than 6 can have several different structures. Coordination numbers of 2 and 3 are common for d^{10} metal ions. Tetrahedral and square planar complexes have a coordination number of 4, trigonal bipyramidal and square pyramidal complexes 5, and octahedral complexes 6. At least three structures are known for a coordination number of 7, which is generally found for only large metal ions. Coordination numbers of 8 and 9 are also found for larger metal ions. The stability of metal complexes with first-row transition metals in a +2 oxidation state varies inversely with their ionic radius. Lewis bases can be **hard bases**, which have small, relatively nonpolarizable donor atoms, or **soft bases**, with larger, relatively polarizable donor atoms. **Hard acids** have the highest affinity for hard bases, and **soft acids** have the highest affinity for soft bases. Soft metals and soft bases form complexes that are more stable than would be predicted based on electrostatic arguments, which suggests that metal-to-ligand π bonding is important. Ligands that are strong bases form the most stable complexes with metal ions that are hard acids. Exceptionally stable complexes are formed by *chelates*, polyatomic ligands with two or more donor atoms; this enhanced stability is known as the *chelate effect*. Many metal complexes exist as **isomers**, two or more compounds that have the same formula but different arrangements of atoms. **Structural isomers** differ in which atoms are bonded to one another, while **geometrical isomers** differ only in the arrangement of ligands around the metal ion. Ligands adjacent to one another are **cis**, while ligands across from one another are **trans**.

23.5 Crystal Field Theory (p. 1064)

Crystal field theory (CFT) is a bonding model that explains many properties of transition metals that cannot be explained using valence bond theory. In CFT, complex formation is assumed to be due to electrostatic interactions between a central metal ion and a set of negatively charged ligands or ligand dipoles arranged around the metal ion. Depending on the arrangement of the ligands, the *d* orbitals split into sets of orbitals with different energies. The difference between the energy levels in an octahedral complex is called the **crystal field splitting energy** (**Δ_o**), whose magnitude depends on the charge on the metal ion, the position of the metal in the periodic table, and the nature of the ligands. The **spin-pairing energy** is the increase in energy that occurs when an electron is added to an already occupied orbital. A *high-spin* configuration occurs when the crystal field splitting energy is less than the spin-pairing energy, which produces complexes with the maximum number of unpaired electrons possible. Conversely, a *low-spin* configuration occurs when the crystal field splitting energy is greater than the spin-pairing energy, which produces complexes with the minimum number of unpaired electrons possible. *Strong-field ligands* interact strongly with the *d* orbitals of the metal ions and give a large Δ_o, whereas *weak-field ligands* interact more weakly and give a smaller Δ_o. The colors of transition-metal complexes depend on the environment of the metal ion and can be explained by CFT. Distorting an octahedral complex by moving opposite ligands away from the metal produces a tetragonal or square planar arrangement, in which interactions with equatorial ligands become stronger. Because none of the *d* orbitals points directly at the ligands in a tetrahedal complex, these complexes have smaller values of the crystal field splitting energy, Δ_t. The **crystal field stabilization energy (CFSE)** is the additional stabilization of a complex due to placing electrons in the lower-energy set of *d* orbitals. CFSE explains the unusual curves seen in plots of ionic radii, hydration energies, and lattice energies versus atomic number. The **Jahn–Teller effect** refers to the distortions observed for d^4 and d^9 complexes.

23.6 Transition Metals in Biology (p. 1075)

Three separate steps are required for organisms to obtain essential transition metals from their environment: *mobilization* of the metal, *transport* of the metal into the cell, and *transfer* of the metal to where it is needed within the cell or organism. The process of iron uptake is best understood. To overcome the insolubility of $Fe(OH)_3$, many bacteria use organic ligands called **siderophores** that have high affinity for Fe(III) and are secreted into the surrounding medium to increase the total concentration of dissolved iron. The iron–siderophore complex is absorbed by the cell, and the iron is released by reduction for Fe(II). Mammals use the low pH of the stomach to increase the concentration of dissolved iron. Iron is absorbed in the intestine, where it forms an Fe(III) complex with a protein called *transferrin* that is transferred to other cells for immediate use or storage in the form of *ferritin*.

Proteins that contain one or more tightly bound metal ions are called **metalloproteins**, and metalloproteins that catalyze biochemical reactions are called **metalloenzymes**. Proteins that simply transfer electrons from one place to another are called *electron-transfer proteins*. Most electron-transfer proteins are metalloproteins, such as *iron–sulfur proteins*, *cytochromes*, and *blue copper* proteins that accept and donate electrons. The oxidized and reduced centers in all electron-transfer proteins have similar structures to ensure that electron transfer to and from the metal occurs rapidly. Metalloproteins also use the ability of transition metals to bind small molecules, such as O_2, N_2, and H_2, in order to transport or to catalyze the reactions of these small molecules. For example, hemoglobin, hemerythrin, and hemocyanin, which contain heme iron, nonheme iron, and copper, respectively, are used by different kinds of organisms to bind and transfer O_2. Other metalloenzymes use transition-metal ions as Lewis acids to catalyze group transfer reactions. Finally, some metalloenzymes use *homolytic cleavage* of the cobalt–carbon bond in derivatives of *vitamin B_{12}* to generate an organic radical that can abstract a hydrogen atom and thus cause molecular rearrangements to occur.

QUESTIONS AND PROBLEMS

MGC For instructor-assigned homework, go to **www.masteringgenchem.com**

Questions and Problems with colored numbers have answers in the Appendix and complete solutions in the Student Solutions Manual.

CONCEPTUAL

23.1 General Trends Among the Transition Metals

1. The transition metals show significant horizontal similarities in chemistry in addition to their vertical similarities, whereas the same cannot be said of the *s*- and *p*- block elements. Explain why this is so.
2. The energy of the *d* subshell does not change appreciably in a given period. Why? What effect does this have on the ionization potentials of the transition metals? On their electronegativities?
3. Standard reduction potentials vary across the first-row transition metals. What effect does this have on the chemical reactivity of the first-row transition metals? Which two elements in this period are more active than would be expected, and why?
4. Many transition-metal complexes are paramagnetic (have unpaired electrons). How does this affect electrical and thermal conductivities across the rows?
5. What is the lanthanide contraction? What effect does it have on the radii of the transition metals of a given group? What effect does it have on the chemistry of the elements in a group?
6. Why are the atomic volumes of the transition elements low compared with the elements of Groups 1 and 2? Ir has the highest density of any element in the periodic table (22.65 g/cm^3). Why?

7. Of the elements Ti, Mo, Ni, Cu, and Cd, which do you predict has the highest electrical conductivity? Why?
8. The chemistry of As is most similar to the chemistry of which transition metal? Where in the periodic table do you find elements with chemistry similar to that of Ge? Explain your answers.
9. The coinage metals have significant noble character. In fact, they are less reactive than the elements of Group 12. Explain why this is so, referring specifically to their (a) reactivity with mineral acids; (b) electronegativity; and (c) ionization energies. (d) Why are the Group 12 elements more reactive?

23.2 A Brief Survey of Transition-Metal Chemistry

10. The valence-electron configuration of Sc is $4s^2\ 3d^1$, yet it does not lose the d^1 electron to form the 1+ ion. Why?
11. Give the ground-state electron configuration for each: Mn, Mn^{2+}, Au, Au^{3+}, Mo, Mo^{5+}.
12. A great deal of research is being conducted on the use of titanium alloys as materials for transportation applications (airplanes, ships, automobiles, etc.). Why is Ti particularly suited to this purpose? What is the primary disadvantage that needs to be overcome?
13. Both Ti and Ta are used for bioimplants because they are highly resistant to corrosion. Their uses also extend to other applications where corrosion must be avoided. Why are these metals so corrosion resistant?
14. Give two reasons Zr is used to make the casing for UO_2 fuel in water-cooled nuclear reactors.
15. Why is chromium added to steel to form "stainless" steel? What other elements might also be effective additives for this purpose? Why did you select these elements?
16. Tungsten is commonly used as the filament in electric lightbulbs. Why is tungsten particularly suited to this purpose?
17. Palladium metal is used to purify H_2 by removing other gases. Why is Pd so permeable to H_2?
18. Give the valence-electron configuration for each: Sc, Fe, Re, Ag, Zr, Co, V, Pr, Hg, Cr, Ni, Ce, Cu, Tb.
19. The Hg—Hg bond is much stronger than the Cd—Cd bond, reversing the trend found among the other transition-metal groups. Explain this anomaly.
20. Which of the transition metals are most likely to form compounds in the +6 oxidation state? Why?

23.3 Metallurgy

21. The Group 12 elements are never found in their native form but always in combination with one other element. Which element, and why? Which of the Group 12 elements has the highest affinity for the element you selected?
22. Hydrometallurgy is the preferred method for separating late transition metals from their ores. What types of ligands are most effective in this process?
23. Coke is a plentiful and inexpensive reductant that is used to isolate metals from their ores. Which of these elements cannot be isolated using this reductant, and why? Cr, Co, W, Cu, Ni, Os, Fe, Mn, La, Hf

23.4 Coordination Compounds

24. Give two reasons a metal can bind to only a finite number of ligands. Based on this reasoning, what do you predict to be the maximum coordination number of Ti? Of Ac?
25. Can a tetrahedral MA_2B_2 complex exist as *cis* and *trans* isomers? Explain your answer.

23.5 Crystal Field Theory

26. Describe crystal field theory in terms of (a) its assumptions regarding metal–ligand interactions, and (b) its weaknesses and strengths compared with valence bond theory.
27. In CFT, what causes degenerate sets of *d* orbitals to split into different energy levels? What is this splitting called? On what does the magnitude of the splitting depend?
28. Will the value of Δ_o increase or decrease if I^- ligands are replaced by NO_2^- ligands? Why?
29. For an octahedral complex of a metal ion with a d^6 configuration, what factors favor a high-spin versus a low-spin configuration?
30. How can CFT explain the color of a transition-metal complex?

23.6 Transition Metals in Biology

31. What are the advantages of having a metal ion at the active site of an enzyme?
32. Why does the structure of the metal center in a metalloprotein that transfers electrons show so little change in structure upon oxidation or reduction?

NUMERICAL

This section includes paired problems (marked by brackets) that require similar problem-solving skills.

23.1 General Trends Among the Transition Metals

33. Give the valence-electron configurations of the 2+ ions for each of the first-row transition elements. Which two ions do you expect to have the most negative $E°$ value? Why?
34. Arrange the following in order of increasing radius: Ru^{3+}, Cu^{2+}, Zn, Ti^{4+}, Cr^{3+}, Ni^{2+}.
35. Arrange the ions in order of decreasing radius: Pt^{4+}, Hg^{2+}, Fe^{2+}, Zr^{4+}, Fe^{3+}.
36. Which divalent ion has the smallest ionic radius? Ti^{2+}, V^{2+}, Mn^{2+}, Fe^{2+}, Co^{2+}, Ni^{2+}, Zn^{2+}. Explain your reasoning.
37. The ionic radii of V^{2+}, Fe^{2+}, and Zn^{2+} are all roughly the same (approximately 76 pm). Given their positions in the periodic table, explain why their ionic radii are so similar.

23.2 A Brief Survey of Transition-Metal Chemistry

38. Do you expect $TiCl_4$, $TiCl_3$, $TiCl_2$, and Ti to be (a) oxidized; (b) reduced; (c) hydrolyzed by water? Give reasons for your answers.
39. The atomic radii of vanadium, niobium, and tantalum are V (134 pm), Nb (146 pm), and Ta (146 pm). Why does the radius increase from vanadium to niobium, but not from niobium to tantalum?
40. The most stable oxidation state for the metals of Groups 3, 4, and 5 is the highest oxidation state possible. In contrast, for nearly all the metals of Groups 8, 9, and 10, intermediate oxidation states are most stable. Why?
41. Most of the transition metals can form compounds in multiple oxidation states. Ru, for example, can form compounds in the +8, +6, +4, +3, +2, and −2 oxidation states. Give the valence-electron configuration of Ru in each of these oxidation states. Why does Ru exhibit so many oxidation states? Which ones are the most stable? Why?
42. Predict the maximum oxidation state of each element: Cu, Cr, Mo, Rh, Zr, Y, Ir, Hg, Fe.

43. Match each Lewis base with the metal ions with which it is most likely to form a stable complex:

Lewis bases: NH_3, F^-, RS^-, OH^-, Cl^-
Metals: Sc^{3+}, Cu^+, W^{6+}, Mg^{2+}, V^{3+}, Fe^{3+}, Zr^{4+}, Co^{2+}, Ti^{4+}, Au^+, Al^{3+}, Mn^{7+}

44. The +4 oxidation state of the Group 7 metals is the only one in which the three elements form dioxides, MO_2. Which of the three metals do you expect to form the most stable dioxide, and why?
45. Which of the following are not likely to exist, and why? Os^{4+}, Pt^{10+}, Cr^{6+}, Ir^{9+}, Ru^{8+}, Re^{7+}, Ni^{10+}
46. Of these compounds, which do not exist, and why? $[Fe(H_2O)_6]^{2+}$, $OsBr_7$, CoF_4, PtF_6, FeI_3, $[Ni(H_2O)_6]^{2+}$, OsO_4, IrO_4, NiO, RhS_2, PtH
47. The chemistry of gold is somewhat anomalous for a metal. With which elements does it form the Au^- ion? Does it form a stable sulfide?
48. Which of these compounds are not likely to exist? Ag_2S, Cu_2S, AuI_3, CuF, AuF, AgN, AuO
49. There is evidence that the Au^- ion exists. What is its electron configuration? The compound CsAu has been isolated; it does not exhibit a metallic luster, nor does it conduct electricity. Is this compound an alloy? What type of bonding is involved? Explain your answers.
50. Of these compounds, which are not likely to exist? Hg_2Cl_2, ZnO, HgF_2, $[ZnCl_5]^{3-}$, HgNa
51. Mercurous sulfide has never been prepared. What products are formed when H_2S gas is bubbled through an aqueous solution of mercurous nitrate?
52. Mercurous oxide, Hg_2O, and mercurous hydroxide, $Hg_2(OH)_2$, have never been prepared. Why not? What products are formed instead if a solution of aqueous sodium hydroxide is added to an aqueous solution of mercurous nitrate, $Hg_2(NO_3)_2$?
53. Arrange these metal oxides in order of increasing basicity: Sc_2O_3, V_2O_5, CrO_3, Mn_2O_7, MnO_2, VO_2.
54. Arrange these oxides in order of increasing basicity: Sc_2O_3, TiO_2, V_2O_5, CrO_3, Mn_2O_7, OsO_4.
55. Arrange the metal oxides in order of increasing acidity: Sc_2O_3, VO, V_2O_5, Cr_2O_3, Fe_2O_3, Fe_3O_4, ZnO.
56. Predict the products of the reactions, and balance each equation:
    (a) Ti + excess Cl_2, heated
    (b) V_2O_5 in aqueous base
    (c) $K_2Cr_2O_7 + H_2SO_4$
    (d) $RuBr_2 + O_2$, in water
    (e) CrO_4^{2-} in aqueous acid
    (f) Hg^{2+} + Hg, in aqueous acid
57. Predict the products of the reactions, and balance each equation:
    (a) AgBr + $h\nu$
    (b) W + excess Cl_2, heated
    (c) CuO + H_2, heated
    (d) Fe_2O_3 in aqueous acid
    (e) $RhCl_3 + NH_3$, in water
    (f) $Fe^{2+} + MnO_4^-$, in water
58. What do you predict to be the coordination number of each ion? Pt^{2+}, Au^{1+}, Fe^{3+}, Os^{2+}.
59. Which of these metals is most likely to form stable compounds in the +4 oxidation state? La, Sc, Cr, Hf. Why?
60. Give the most common oxidation state for each element: Y, W, Ru, Ag, Hg, Zn, Cr, Nb, Ti.
61. What is the most common oxidation state for each element? Os, Cd, Hf, V, Ac, Ni, Mn, Pt, Fe
62. Give the highest oxidation state observed for each element: Zr, Fe, Re, Hg, Ni, La, Mo.
63. What is the highest oxidation state observed for each element? Ag, Co, Os, Au, W, Mn
64. Arrange these atoms in order of increasing first ionization energy: La, Cs, Y, Pt, Cd, Mo, Fe, Co, Ir.
65. Briefly explain the following trends within the transition metals: (a) transition metals fluorides usually have higher oxidation states than their iodides; (b) for a given metal, the lowest-oxidation-state oxide is basic and the highest-oxidation-state oxide is acidic; (c) halides of the transition metals become more covalent with increasing oxidation state and are more prone to hydrolysis.
66. Propose a method to prepare each compound from its metallic elements: $TiCl_4[(CH_3O)_2]_2$, Na_2TiO_3, V from V_2O_5, $Na_2Cr_2O_7$.
67. Of the Group 5 elements, which (a) has the greatest tendency to form ions in the lower oxidation states; (b) has the greatest tendency to form a polymeric fluoride; (c) does not form an MX_2 species; (d) forms the most basic oxide; (e) has the greatest tendency to form complexes with coordination numbers higher than 6?
68. Which compound are not likely to exist? WF_2, CrF_6, $MoBr_6$, WI_6, CrO_3, MoS_2, W_2S_3, MoH
69. Of these compounds, which are not likely to exist? ReF_2, $ReCl_5$, MnF_6, Mn_2O_7, ReO

23.3 Metallurgy

70. Tantalum and niobium are frequently found together in ores. These elements can be separated by treatment with a solution of HF. Explain why this is an effective separation technique.
71. A commercially important ore of chromium is chromite, $FeCr_2O_4$, an analogue of magnetite, Fe_3O_4. Based on what you know about the oxidation states of iron in magnetite, predict the oxidation states of the metal ions in chromite.
72. Manganese is an important additive in steel because of its reactivity with oxygen and sulfur, both of which contribute to brittleness. Predict the products of the reaction of Mn with these species.
73. Pure vanadium is obtained by reducing VCl_4 with H_2 or Mg, or by reducing V_2O_5 with Ca. Write balanced equations for these reactions. Why isn't carbon used for the reduction?
74. The diagram of a blast furnace in Figure 23.9 illustrates several important features of the reduction of Fe_2O_3 to iron. Write a balanced equation for each step of the process described in the figure, and give the overall equation for the conversion. Oxygen is blown through the final product to remove impurities. Why doesn't this step simply reverse the process and produce iron oxides?
75. Metallic Zr is produced by the Kroll method, which uses Na as the reductant. Write balanced equations for the reactions involved in this process. The product is frequently contaminated with Hf. Propose a feasible method for separating the two elements.
76. The compound Cr_2O_3 is important commercially; among other things, it is used as a pigment in paint and as a catalyst for the manufacture of butadiene. Write balanced equations showing how you would produce this compound from (a) chromium metal; (b) ammonium dichromate; and (c) $CrCl_3$ in a basic solution.

23.4 Coordination Compounds

77. Complexes of metals in the +6 oxidation state usually contain bonds to which two Lewis bases? Why are these bonds best described as covalent rather than ionic? Do Ca, Sr, and Ba also form covalent bonds with these two Lewis bases, or is their bonding best described as ionic?

78. Cr, Mn, Fe, Co, and Ni form highly stable CO complexes. In fact, these complexes are much more stable than similar compounds formed by the early transition metals. Why?

79. The transition metals Cr through Ni form very stable cyanide complexes. Why are these complexes so much more stable than similar compounds formed from the early transition metals?

80. Which species do you expect to be the most stable, and why? $Co(en)_3{}^{3+}$, $CoF_6{}^{3-}$, $Co(NH_3)_6{}^{3+}$, $Co(dien)_2{}^{3+}$ (dien = $NH_2CH_2CH_2NHCH_2CH_2NH_2$)

81. Which of the following divalent metal ions forms the most stable complexes in the +2 oxidation state with ligands such as NH_3, and why? Ca^{2+}, Ti^{2+}, V^{2+}, Mn^{2+}, Fe^{2+}, Co^{2+}, Ni^{2+}, Zn^{2+}

23.5 Crystal Field Theory

82. Do strong-field ligands favor a tetrahedral or square planar structure? Why?

83. For each complex, predict its structure, whether it is high spin or low spin, and the number of unpaired electrons present: (a) $[Cu(NH_3)_4]^{2+}$; (b) $Ni(CN)_4]^{2-}$.

84. For each complex, predict its structure, whether it is high spin or low spin, and the number of unpaired electrons present: (a) $[TiCl_6]^{3-}$; (b) $[CoCl_4]^{2-}$.

23.6 Transition Metals in Biology

85. In enzymes, explain how metal ions are particularly suitable for generating organic radicals.

86. A common method for treating carbon-monoxide poisoning is to have the patient inhale pure oxygen. Explain why this treatment is effective.

APPLICATIONS

87. Tungsten bronzes are semimetallic solids that are inert to strong acids and bases, lustrous, and good conductors of electricity; they are used in the production of bronze or metallic paints. These nonstoichiometric compounds have the general formula M_xWO_3, where M = Na, K, a Group 2 metal, or a lanthanide, and $x < 1$. The properties of tungsten bronzes suggest that at least some of the tungsten atoms are in the +6 oxidation state. Given the high oxidation state, why do these solids conduct electricity?

88. Gout is a painful disorder caused by overproduction of uric acid, which is deposited as sodium urate crystals in joints. The enzyme that catalyzes the production of uric acid contains Fe^{3+} and Mo^{6+}. Molecular oxygen is a substrate in this reaction. Based on the oxidation states of the metals, what do you expect one of the products of the reaction to be?

89. A laboratory technician added aqueous ammonia to an aqueous solution of Mn^{2+}, which produced a pale pink precipitate. She left the solution exposed to air and went home. The next day she returned to the lab and found that her pink precipitate had turned brown-black. Write a balanced equation to show what had happened.

90. Plants use Mn(IV) during photosynthesis for the production of dioxygen from water. Write a balanced equation showing this reaction, and suggest why Mn is well suited for this purpose.

91. Rust stains (Fe_2O_3) can be removed from fabrics by oxalic acid, HO_2CC_2H. Write a balanced equation showing the reaction that occurs, and predict the solubility of the product in water.

92. It has been suggested that complexes that can coordinate N_2 are used by bacteria to fix atmospheric nitrogen. One such complex is $[Ru(NH_3)_5 \cdot N_2]Cl_2$, which was first discovered in 1965. Sigma bonding with N_2 in this complex would be weak because the N_2 molecule is symmetrical and has zero dipole moment.
(a) What is the oxidation state of Ru in this complex?
(b) How does N_2 bond to the metal? Describe the type of bonding involved.
(c) Speculate why this complex might be a suitable vehicle for nitrogen fixation.

93. Monel metal, which contains 68% Ni, 32% Cu, and traces of Fe and Mn, is highly corrosion resistant. It is used, for example, to make items that will be used in marine environments and to hold corrosive fluorine compounds. Based on its composition, why is Monel particularly suitable for these purposes?

94. From 1845 to 1850, a fungus known as "potato blight" caused a potato famine in Ireland. Approximately 25% of the Irish population either died or emigrated as a direct result. A spray called *Bordeaux mixture* is now used to kill the fungus; it is made by the reaction of $CuSO_4$ with $Ca(OH)_2$. What is the formula of Bordeaux mixture? Write a balanced equation for the reaction.

95. Many transition metals and their compounds are used as catalysts. Given MnO_2, $FeCl_3$, Pt, and Ni, which would you select for each purpose and why? (a) removing NO_x from exhaust fumes; (b) producing CCl_4 from CS_2 and Cl_2; (c) producing H_2 from NH_3; (d) decomposing $KClO_3$ to give O_2

96. The Fe^{2+} site in hemoglobin binds oxygen reversibly, and consequently it is suitable for transporting oxygen in blood. Various other small molecules can bind to the iron instead, thus preventing oxygen transport. Based on the type of bonding, why are CO and NO particularly toxic? Would they be as toxic if hemoglobin contained a V^{2+} center instead of Fe^{2+}? Why?

97. In 1951, G. Wilkinson reported a surprising iron–hydrocarbon compound that is now called *ferrocene*. Ferrocene is orange and has a structure in which the metal is sandwiched ("eclipsed") between two planar cyclopentadienyl ($C_5H_5{}^-$) rings. It can be viewed as a compound of Fe^{2+} with two $C_5H_5{}^-$ rings (shown below). Ferrocene does not contain Fe—C σ bonds but another type of bond formed by lateral overlap of orbitals. Which metal orbitals are involved? A similar structure is obtained with Ru^{2+}. One of these metals forms a sandwich complex that has a staggered conformation, and the other a complex that is eclipsed. Which metal produces which conformation? Why?

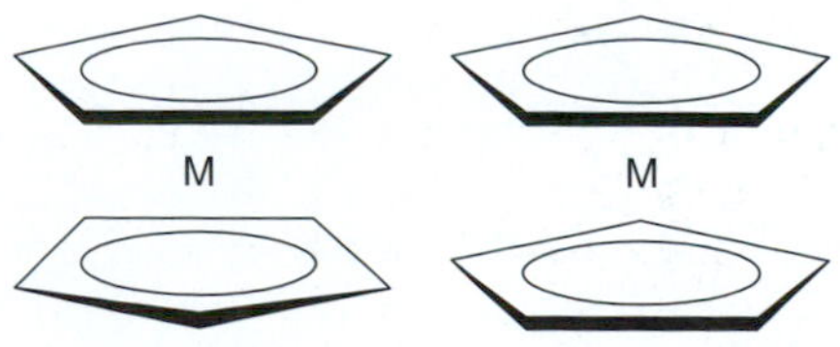

A sandwich complex (staggered) A sandwich complex (eclipsed)

98. The Ziegler–Natta catalyst is used for the polymerization of ethylene to form-high density polyethylene, a widely used lightweight plastic. The active form of the catalyst is believed to be $TiCl_3CH_2CH_3$, and the first step in the polymerization reaction is believed to be binding of the double bond in ethylene to Ti. If you were interested in developing a similar catalyst for this same purpose, would you choose chlorides of Zr, Hf, V, Nb, Ta, Cr, Mo, or W? Why?

99. Cobalt(II) chloride is used as a visual indicator of humidity because it exists as a blue complex when dry and a pink complex when exposed to moisture in the air. The bonding environment around the cobalt in one of these complexes is octahedral, and in the other it is tetrahedral. What reaction occurs to produce the color change? Write a balanced equation for this reaction, indicating the species present.

100. A platinum complex that is widely available commercially is chloroplatinic acid, $H_2[PtCl_6]$, which is used to make platinized asbestos, a catalyst. What is the structure of chloroplatinic acid? Is it distorted from an idealized geometry? Do you expect it to be colored? Justify your answers.

24 Organic Compounds

LEARNING OBJECTIVES

- To learn how organic compounds with the same molecular formula can have different three-dimensional structures
- To understand the relationship between structure and reactivity for a series of related organic compounds
- To become familiar with the common classes of organic reactions
- To understand the general properties of functional groups and the differences in their reactivity
- To be able to identify the common structural units of important biologically relevant molecules

CHAPTER OUTLINE

In Chapter 2, you were introduced to the major classes of *organic compounds*, covalent compounds composed primarily of carbon and hydrogen. Organic substances have been used throughout this text to illustrate the differences between ionic and covalent bonding and to demonstrate the intimate connection between the structures of compounds and their chemical reactivity. You learned, for example, that even though NaOH and alcohols (ROH) both have OH in their formula, NaOH is an ionic compound that dissociates completely in water to produce a basic solution containing Na^+ and OH^- ions, whereas alcohols are covalent compounds that do not dissociate in water and instead form neutral aqueous solutions. You also learned that an amine, with its lone pairs of electrons, is a base, whereas a carboxylic acid, with its dissociable proton, is an acid (see Section 2.5).

The structure of a solid with a hybrid metal-organic framework. Organic and inorganic groups of the proper structure can be used to synthesize solids with very large pores (central sphere) that can accommodate a variety of small molecules. The rigid benzene rings are used as "props" to hold the metal units (carboxylate-bridged copper dimers) apart. Such solids have potential applications in hydrogen storage for use in fuel cells or automobiles.

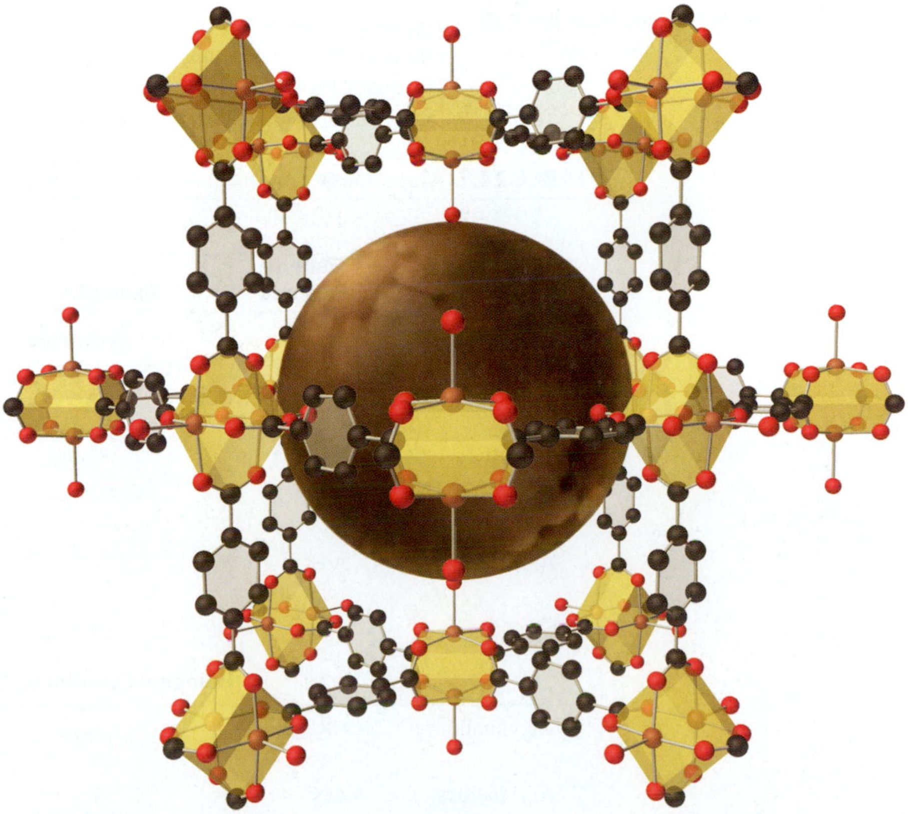

Carbon is unique among the elements in its ability to *catenate*, forming a wide variety of compounds that contain long chains and/or rings of carbon atoms (see Sections 12.8 and 22.2). Some of the most complex chemical structures known are those of the organic molecules found in living organisms (see Section 12.8). In spite of their size and complexity, these biological molecules obey the same chemical principles as simpler organic molecules. Thus, we can use Lewis electron structures to understand the preferred mode of reactivity of a variety of organic compounds (see Section 8.5); relative electronegativities and bond polarities to predict how certain groups of atoms will react (see Section 8.9); and molecular orbital theory to explain why certain organic species that contain multiple bonds are especially stable or undergo particular reactions when they interact with light (see Section 9.4).

In this chapter, we continue our description of organic compounds by focusing on their molecular structures and reactivity; we will also introduce some of the fundamental types of reactions and reaction mechanisms you will encounter in organic and biological chemistry. We discuss why butter is a solid and oils are liquids despite the apparent similarities in their structures, why the widely used anti-inflammatory drug ibuprofen takes longer than half an hour to relieve pain, and the identity of the major carcinogen in grilled meats and cigarette smoke. The chapter concludes with a brief introduction to the molecules of life, which will explain how the consumption of lactose can result in mental retardation and cirrhosis of the liver in some individuals, how hibernating animals survive during the winter, and how certain groups of antibiotics kill bacteria that are harmful to humans.

24.1 Functional Groups and Classes of Organic Compounds

MGC Functional Groups
Aromatic Ring Substitutions
Organic Nomenclature

In Part I, you were introduced to the structural units that chemists use to classify organic compounds and predict their reactivities. These **functional groups**, which determine the chemical reactivity of a molecule under a given set of conditions, can consist of a single atom (such as Cl) or a group of atoms (such as CO_2H). The major families of organic compounds are characterized by their functional groups. Table 24.1

TABLE 24.1 Major classes of organic compounds

Class	General Formula	Example	Common Name (Systematic Name)	Common Suffix/Prefix (Systematic)
		Hydrocarbons		
Alkanes	RH	CH_3CH_3	ethane	-ane
Alkenes	RR′C=CR″R‴	$H_2C{=}CH_2$	ethylene (ethene)	-ene
Alkynes	RC≡CR′	HC≡CH	acetylene (ethyne)	(-yne)
Arenes	ArH[a]		benzene	-ene
		Halogen-Containing Compounds		
Alkyl halides	RX	CH_3CH_2Cl	ethyl chloride (chloroethane)	halide (halo-)
Aryl halides	ArX[a]	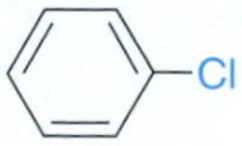	chlorobenzene	halo-

(continued)

TABLE 24.1 Continued

Class	General Formula	Example	Common Name (Systematic Name)	Common Suffix/Prefix (Systematic)
		Oxygen-Containing Compounds		
Alcohols	ROH	CH_3CH_2OH	ethyl alcohol (ethanol)	-ol
Phenols	ArOH[a]	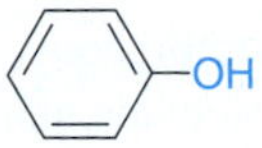	phenol	-ol
Ethers	ROR′	$H_3CH_2COCH_2CH_3$	diethyl ether	
Aldehydes	RCHO	$CH_3\overset{O}{\overset{\|}{C}}H$	acetaldehyde (ethanal)	-dehyde (-al)
Ketones	RR′C=O	$CH_3\overset{O}{\overset{\|}{C}}CH_3$	acetone (2-propanone)	-one
Carboxylic acids	RCO_2H	$CH_3\overset{O}{\overset{\|}{C}}OH$	acetic acid (ethanoic acid)	-ic acid (-oic acid)
		Carboxylic Acid Derivatives		
Esters	RCO_2R'	$CH_3\overset{O}{\overset{\|}{C}}OCH_3$	methyl acetate (methyl ethanoate)	-ate (-oate)
Amides	RCONHR′	$CH_3\overset{O}{\overset{\|}{C}}NHCH_3$	*N*-methylacetamide	-amide
		Nitrogen-Containing Compounds		
Amines	RNH_2, RNHR′, RNR′R″	$CH_3CH_2NH_2$	ethylamine	-amine
Nitriles	RC≡N	$H_3CC{\equiv}N$	acetonitrile	-nitrile
Nitro compounds	$ArNO_2$[a]	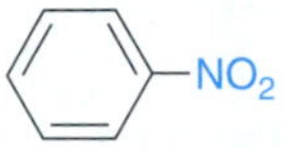	nitrobenzene	nitro-

[a] Ar indicates ***Ar****yl* group.

summarizes five families introduced in earlier chapters, gives examples of compounds that contain each functional group, and lists the suffix or prefix used in the systematic nomenclature of compounds that contain each functional group.

The first family listed in Table 24.1 is the hydrocarbons. These include: alkanes, with the general molecular formula C_nH_{2n+2} where n is an integer; alkenes, represented by C_nH_{2n}; alkynes, represented by C_nH_{2n-2}; and arenes. Halogen-substituted alkanes, alkenes, and arenes form a second major family of organic compounds, which include the alkyl halides and aryl halides. Oxygen-containing organic compounds, a third family, may be divided into two main types: those that contain at least one C—O single bond, which include alcohols, phenols (derivatives of benzene), and ethers, and those that contain a carbonyl group (>C=O), which include aldehydes, ketones, and carboxylic acids. Carboxylic acid derivatives, the fourth family listed, are compounds in which the

OH of the —CO_2H functional group is replaced by either an alkoxy (—OR) group, producing an ester, or by an amido (—NRR′, where R and R′ can be H and/or alkyl groups), forming an amide. Nitrogen-containing organic compounds, the fifth family, include: amines; nitriles, which have a CN triple bond; and nitro compounds, which contain the —NO_2 group.

As you learned in Section 2.4, the systematic nomenclature of organic compounds indicates the positions of substituents using the lowest numbers possible to identify their locations in the carbon chain of the parent compound. If two compounds have the same systematic name, then they are the same compound. Although systematic names are preferred because they are unambiguous, many organic compounds are referred to by common names rather than systematic names. Common nomenclature uses the prefix *form–* for a compound that contains no carbons other than those in the functional group, and *acet–* for those that have one carbon atom in addition [two in the case of acetone, $(CH_3)_2C{=}O$]. Thus, methanal and ethanal, respectively, are the systematic names for formaldehyde and acetaldehyde, shown below.

Formaldehyde (methanal) **Acetaldehyde (ethanal)**

Recall that in the systematic nomenclature of aromatic compounds, the positions of groups attached to the aromatic ring are indicated by numbers, starting with 1 and proceeding around the ring in the direction that produces the lowest possible numbers. For example, the position of the first CH_3 group in dimethyl benzene is indicated with a 1, but the second CH_3 group, which can be placed in any one of three positions, produces 1,2-dimethylbenzene, 1,3-dimethylbenzene, or 1,4-dimethylbenzene (Table 24.2). In common nomenclature, in contrast, the prefixes *ortho-*, *meta-*, and *para-* are used to describe the relative positions of groups attached to an aromatic ring. If the CH_3 groups in dimethylbenzene, whose common name is xylene, are adjacent to each other, the compound is commonly called *ortho*-xylene, abbreviated *o*-xylene. If they are across from each other on the ring, the compound is commonly called *para*-xylene or *p*-xylene. When the arrangement is intermediate between those of *ortho-* and *para-* compounds, the name is *meta*-xylene or *m*-xylene.

TABLE 24.2 Common nomenclature for aromatic ring substitutions

Prefix	Positions of Methyl Groups	Example	Common Name (Systematic Name)
ortho-	1,2	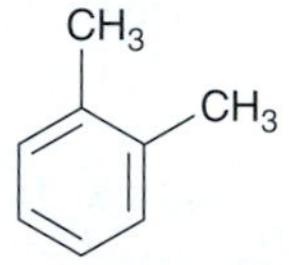	*o*-xylene (1,2-dimethylbenzene)
meta-	1,3	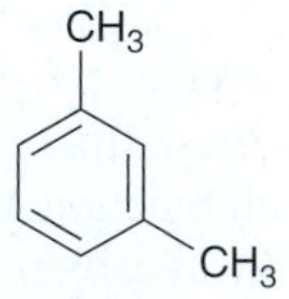	*m*-xylene (1,3-dimethylbenzene)
para-	1,4	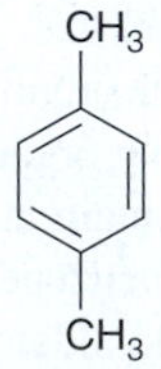	*p*-xylene (1,4-dimethylbenzene)

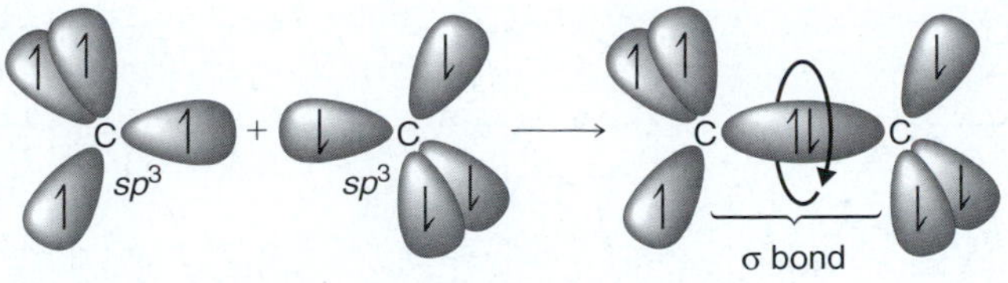

Figure 24.1 **Carbon–carbon bonding in alkanes.** Overlapping sp^3 hybrid orbitals on adjacent carbon atoms form a cylindrically symmetrical σ bond. Because rotation about the bond does not affect the overlap of the bonding orbitals, there is no electronic energy barrier to rotation.

We begin our discussion of the structure and reactivity of organic compounds by exploring structural variations in the simple saturated hydrocarbons known as alkanes. These compounds serve as the scaffolding to which the various functional groups are most often attached.

Note the pattern

Conformational isomers differ in their three-dimensional structure due to rotation about a σ bond.

24.2 ∘ Isomers of Organic Compounds

In earlier discussions of organic compounds, we focused on differences in how the functional groups were connected to the carbon framework. Differences in connectivity resulted in different chemical compounds with different names. You learned, for example, that although 1-propanol (*n*-propanol) and 2-propanol (isopropanol) have the same molecular formula (C_3H_8O), they have different physical and chemical properties. Compounds that have the same molecular formula, but different arrangements of atoms, are called **isomers**. In this section, we describe various types of isomers beginning with those whose three-dimensional structures differ only as the result of rotation about a C—C single bond.

Conformational Isomers

The C—C single bonds in ethane, propane, and other alkanes are formed by the overlap of an sp^3 hybrid orbital on one carbon atom with an sp^3 hybrid orbital on another carbon atom, forming a σ bond (Figure 24.1). Each sp^3 hybrid orbital is cylindrically symmetrical (all cross-sections are circles), resulting in a carbon–carbon single bond that is also cylindrically symmetrical about the C—C axis. Because rotation about the carbon–carbon single bond can occur without changing the overlap of the sp^3 hybrid orbitals, there is no significant electronic energy barrier to rotation. Consequently, many different arrangements of the atoms are possible, each corresponding to different degrees of rotation. Differences in three-dimensional structure resulting from rotation about a σ bond are called differences in *conformation*, and each different arrangement is called a **conformational isomer**, or *conformer*.

The simplest alkane to have a conformational isomer is ethane. Differences between the conformations of ethane are depicted especially clearly in drawings called *Newman projections*, such as those shown in Figure 24.2a. In a Newman projection, the ethane molecule is viewed along the C—C bond axis, with the carbon that is in front shown as a vertex and the carbon that is in back shown as a circle. The three hydrogen atoms nearest the viewer are shown bonded to the front carbon, and the three hydrogen atoms farthest from the viewer are shown bonded to the circle. In one extreme, called the *eclipsed conformation*, the C—H bonds on adjacent carbon atoms lie in the same plane. In the other extreme, called the *staggered conformation*, the hydrogen atoms are positioned as far from one another as possible. Rotation about the C—C bond produces an infinite number of conformations between these two extremes, but the staggered conformation is the most stable because it minimizes electrostatic repulsion between the hydrogen atoms on adjacent carbons.

In a Newman projection, the angles between adjacent C—H bonds on the same carbon are drawn at 120°, although H—C—H angles in alkanes are actually tetrahe-

Figure 24.2 **Eclipsed and staggered conformations of ethane.** (a) In a Newman projection, the ethane molecule is viewed along the C—C bond axis. The carbon in front is represented as a vertex, whereas the carbon that is bonded to it is represented as a circle. The C—H bonds to each carbon are positioned at 120° from each other. In the fully eclipsed conformation, the C—H bonds on adjacent carbon atoms are parallel and lie in the same plane. In the staggered conformation, the hydrogen atoms are positioned as far apart as possible. (b) The eclipsed conformation is 12.6 kJ/mol higher in energy than the staggered conformation because of electrostatic repulsion between hydrogen atoms. An infinite number of conformations of intermediate energy exist between the two extremes.

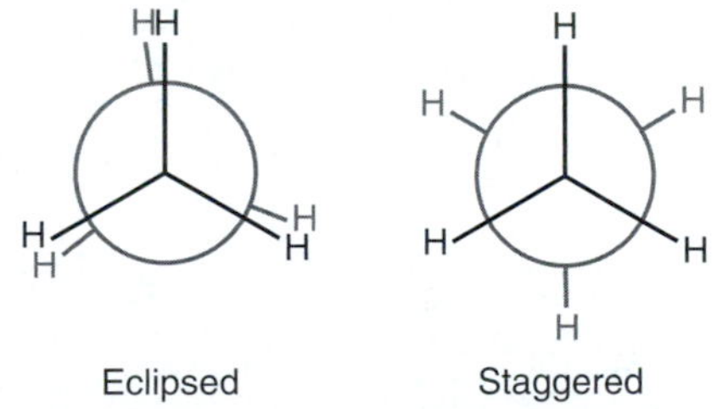

(a) The two extreme conformations of ethane

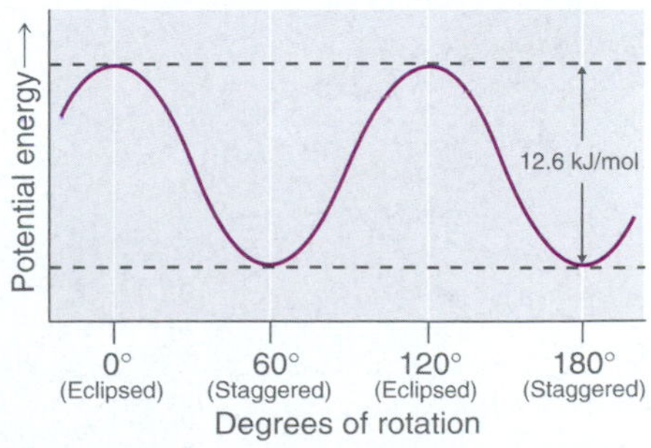

(b) Energy and conformation

Figure 24.3 Potential energy plot and Newman projections of eclipsed and staggered conformations of *n*-butane. In these projections, the molecule is viewed along the C2—C3 bond axis. The least stable structure is the eclipsed conformation in which the two methyl groups (C1 and C4) are adjacent to each other. The most stable structure is the staggered conformation in which the methyl groups are as far apart as possible. Because the substituents on each of the central carbon atoms are not all the same, a 120° rotation about the C2—C3 bond generates energetically nonequivalent eclipsed and staggered conformations.

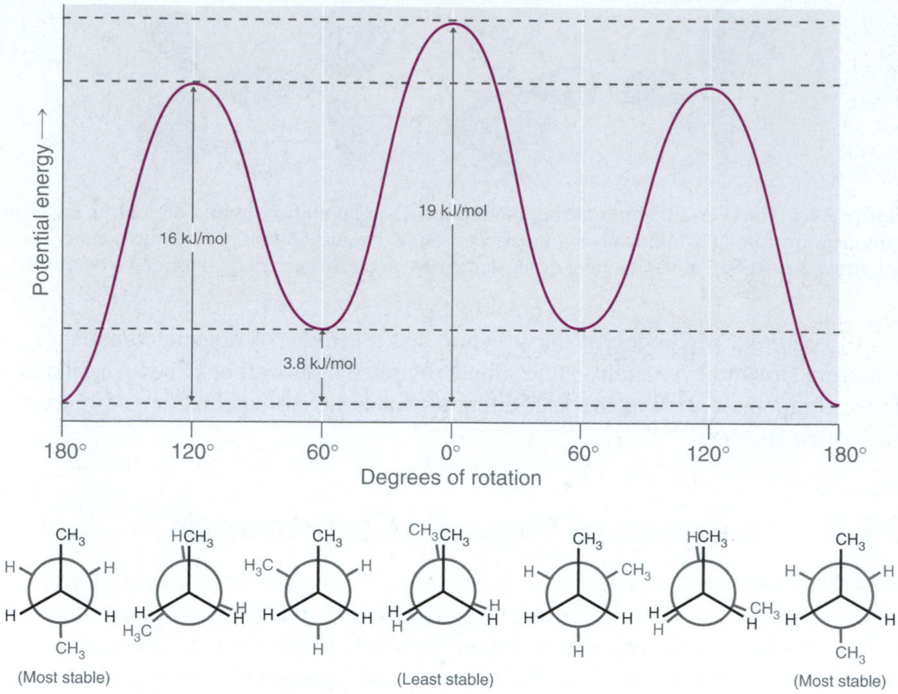

MGC Isomers

dral angles of 109.5°, which for chains of more than three carbon atoms results in a *kinked* structure (see Section 2.4). Despite this three-dimensional inaccuracy, Newman projections are useful for predicting the relative stability of conformational isomers. As shown in Figure 24.2b, the higher energy of the eclipsed conformation represents an energy barrier of 12.6 kJ/mol that must be overcome in order for rotation about the C—C bond to occur. This barrier is so low, however, that rotation about the C—C bond in ethane is very fast at room temperature, occurring several million times per second for each molecule.

Longer-chain alkanes can also be represented by Newman projections. In more complex alkanes and alkane derivatives, rotation can occur about *each* C—C bond in the molecule. Newman projections are therefore useful for revealing *steric barriers* to rotation at a particular C—C bond due to the presence of bulky substituents. Figure 24.3 shows a plot of potential energy versus angle of rotation about the central C—C bond (between carbon atoms 2 and 3) of *n*-butane (C_4H_{10}). The structure that minimizes electrostatic repulsion is the one in which the methyl groups, corresponding to carbon atoms 1 and 4, are as far apart as possible; that is, the staggered conformation. Notice that because the substituents on C2 and C3 in *n*-butane are not all the same, energetically nonequivalent eclipsed and staggered conformations are possible; most molecules interconvert rapidly between these conformations by a series of simple rotations.

EXAMPLE 24.1

Draw Newman projections showing the staggered and eclipsed conformations of 1,1,1-trichloroethane (CCl_3CH_3).

Given Organic molecule

Asked for Staggered and eclipsed conformations

Strategy

A Identify the C—C bond of interest. Then draw the Newman projection by representing one carbon as a vertex and the other as a circle.

B Draw bonds to each carbon at 120° angles from one another, with one arrangement representing the staggered conformation and the other the eclipsed conformation.

C Complete the Newman projections by attaching the appropriate atoms or substituent groups to the central two C atoms in each conformation.

Solution

A There is only one C—C bond: carbon-1 is connected to three Cl atoms and carbon-2 to three H atoms. We draw carbon-1 as a point and carbon-2 as a circle. B Now we draw bonds on each carbon at 120° angles from one another to represent the staggered conformation and the other the eclipsed conformation. C We then attach the H and Cl atoms to the carbon atoms in each conformation as shown in the margin.

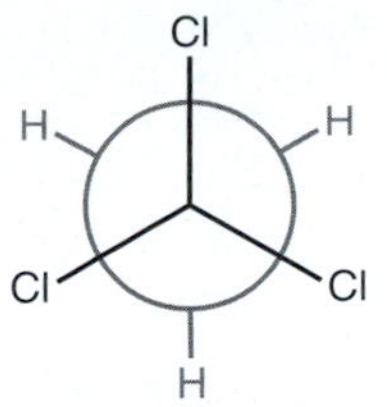

Staggered

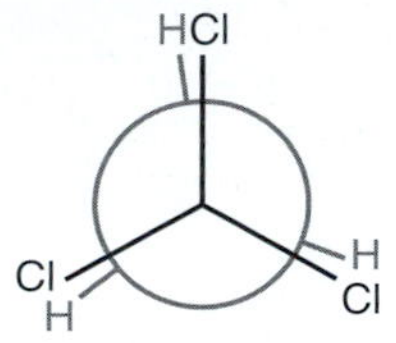

Eclipsed

EXERCISE 24.1

Draw Newman projections to illustrate the staggered and eclipsed conformations of propane (C_3H_8) as viewed along the C1—C2 bond axis.

Answer

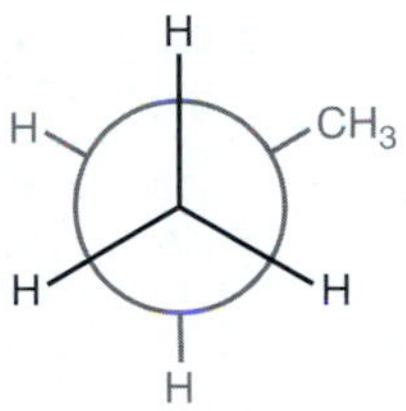

Staggered

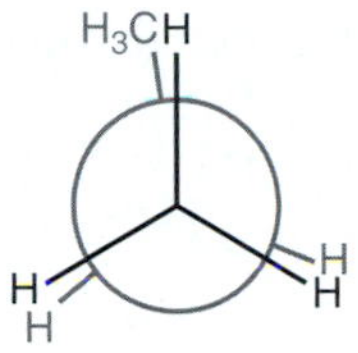

Eclipsed

Structural Isomers

Unlike conformational isomers, which do not differ in connectivity, **structural isomers** differ in connectivity, as illustrated in the margin for 1-propanol and 2-propanol. Although these two alcohols have the same molecular formula (C_3H_8O), the position of the —OH group differs, which leads to differences in their physical and chemical properties.

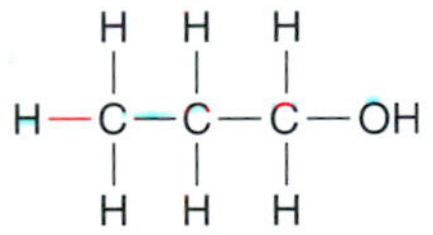

1-Propanol (*n*-propanol)

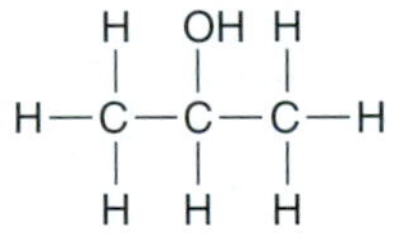

2-Propanol (isopropanol)

In the conversion of one structural isomer to another, at least one bond must be broken and reformed at a different position in the molecule. Consider, for example, the following five structures represented by the formula C_5H_{12}:

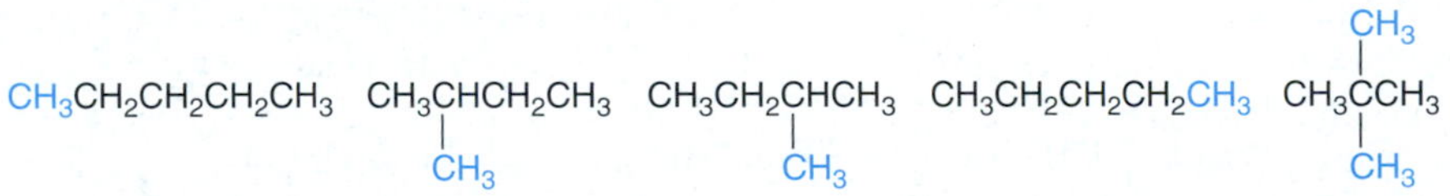

(a) (b) (c) (d) (e)

Of these structures, (a) and (d) represent the same compound, as do (b) and (c), but they are rotated about a 180° vertical axis. Only three—*n*-pentane (a) and (d), 2-methylbutane (b) and (c), and 2,2-dimethylpropane (e)—are structural isomers. Because no bonds are broken in going from (a) to (d) or from (b) to (c), these alternative representations are not structural isomers. The three structural isomers, (a) or (d), (b) or (c), and (e), have distinct physical and chemical properties.

Note the pattern

Structural isomers differ in their connectivity.

EXAMPLE 24.2

Draw all possible structural isomers of C_6H_{14}.

Given Organic molecule

Asked for All structural isomers

Strategy

Ⓐ Draw the simplest structural isomer, which is often the straight-chain alkane.
Ⓑ Obtain branched isomers by substituting one hydrogen along the chain with an appropriate group from the chain.
Ⓒ If possible, substitute more than one hydrogen with appropriate groups to obtain isomers that are more highly branched.

Solution

Ⓐ The simplest structural isomer is the straight-chain alkane *n*-hexane ($CH_3CH_2CH_2CH_2CH_2CH_3$). Ⓑ Removing a methyl group from one end and reattaching it to the adjacent carbons while substituting hydrogen in its place give two other structures:

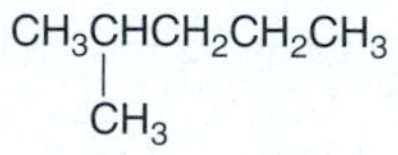
2-Methylpentane

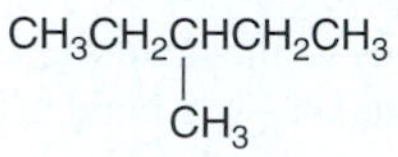
3-Methylpentane

Ⓒ To obtain yet another structural isomer, we move two methyl groups to create a molecule with two branches:

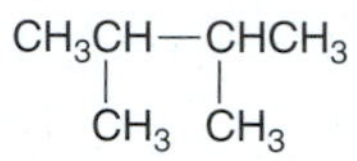
2,3-Dimethylbutane

We create one more structural isomer by attaching two methyl groups to the same carbon atom:

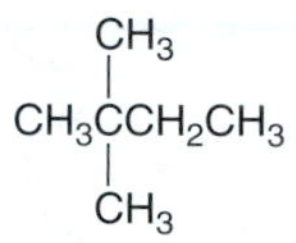
2,2-Dimethylbutane

Thus, there are four structural isomers of C_6H_{14}.

EXERCISE 24.2

Draw all structural isomers of C_4H_9Cl.

Answer

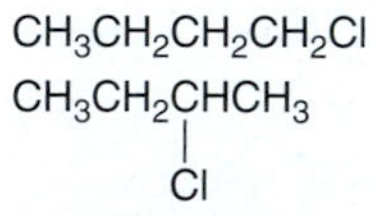
1-Chlorobutane
2-Chlorobutane

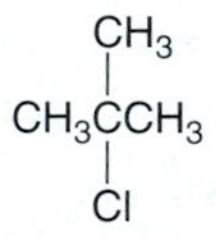
2-Chloro-2-methylpropane

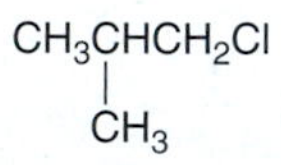
1-Chloro-2-methylpropane

Stereoisomers

Note the pattern
Stereoisomers have the same connectivity, but different arrangements of atoms in space.

Molecules that have the same connectivity but different arrangements of the atoms in space are called **stereoisomers**. There are two types of stereoisomers: geometric and optical. *Geometric isomers* differ in the relative position(s) of substituents in a rigid molecule. Simple rotation about a C—C σ bond in an alkene, for example, cannot occur because of the presence of the π bond. The substituents are therefore rigidly locked into a particular spatial arrangement (see Figure 2.16a). Thus, a carbon–carbon multiple bond, or in some cases a ring, prevents one geometric isomer from being readily converted to the other. The members of an isomeric pair are identified as either *cis* or *trans*, and interconversion between the two forms requires breaking and re-

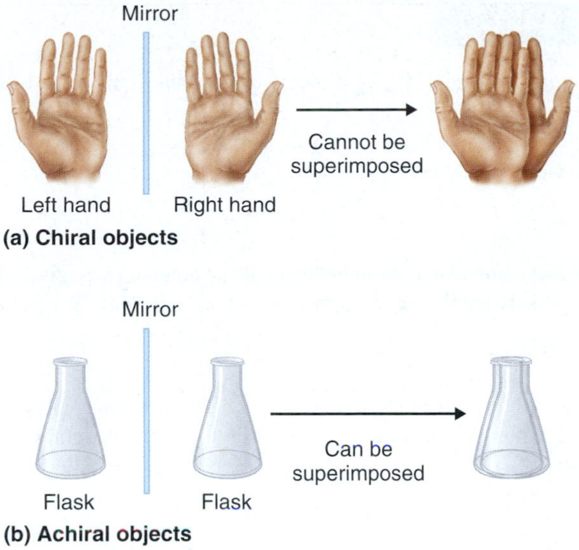

Figure 24.4 Chiral and achiral objects. (a) Objects that are nonsuperimposable mirror images of each other are chiral, such as the left and right hands. (b) The unmarked flask is achiral because it can be superimposed on its mirror image.

MGC Stereoisomers

forming one or more bonds. Because their structural difference causes them to have different physical and chemical properties, *cis* and *trans* isomers are actually two distinct chemical compounds.

Optical isomers are molecules whose structures are mirror images, but cannot be superimposed on one another in any orientation. Optical isomers have *identical physical properties*, although their chemical properties may differ in asymmetric environments. Molecules that are nonsuperimposable mirror images of each other are said to be *chiral* (pronounced "ky-ral," from the Greek *cheir*, "hand"). Examples of some familiar chiral objects are your hands, feet, and ears. As shown in Figure 24.4a, your left and right hands are nonsuperimposable mirror images. (Try putting your right shoe on your left foot—it just doesn't work!) An *achiral* object is one that *can* be superimposed on its mirror image, such as the unmarked flask in Figure 24.4b.

Most organic molecules that are chiral have at least one carbon atom that is bonded to four different groups, as occurs in the bromochlorofluoromethane molecule shown in Figure 24.5a. This carbon, often designated by an asterisk in structural drawings, is called a *chiral center* or *asymmetric carbon atom*. If the bromine atom is replaced by another chlorine (Figure 24.5b), the molecule and its mirror image can now be superimposed by simple rotation. Thus, carbon is no longer a chiral center. Asymmetric carbon atoms are found in many naturally occurring molecules, such as lactic acid, which is present in milk and muscles, and nicotine, a component of tobacco. A molecule and its nonsuperimposable mirror image are called *enantiomers* (from the Greek *enantiou*, "opposite").

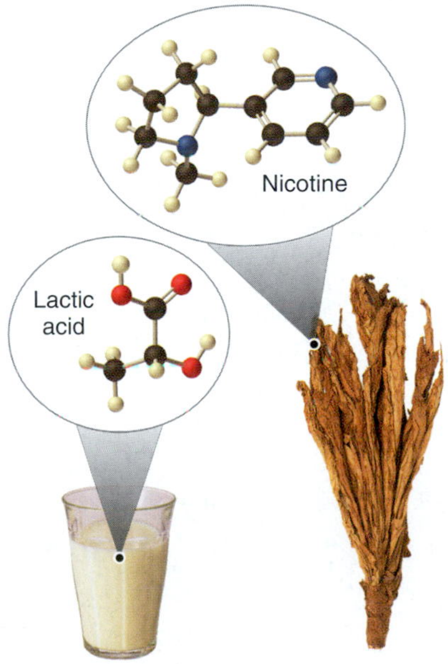

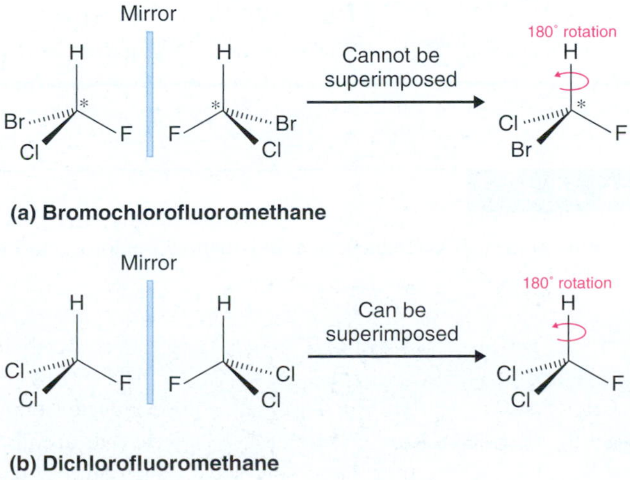

Figure 24.5 Comparison of chiral and achiral molecules. (a) Bromochlorofluoromethane is a chiral molecule whose stereocenter is designated with an asterisk. Rotation of its mirror image does not generate the original structure. To superimpose the mirror images, bonds must be broken and reformed. (b) In contrast, dichlorofluoromethane and its mirror image can be rotated so they are superimposable.

EXAMPLE 24.3

Draw the *cis* and *trans* isomers of **(a)** 1,3-dimethylcyclopentane and **(b)** 3,4-dichloro-3-hexene.

Given Organic compounds

Asked for *Cis* and *trans* isomers

Strategy

Draw the unsubstituted compound corresponding to the systematic name given. Then place substituents on the same side to obtain the *cis* isomer and on opposite sides to obtain the *trans* isomer.

Solution

(a) The name tells us that this compound contains a 5-carbon ring with two methyl groups attached. The 1,3 notation means that the methyl groups are not adjacent in the five-membered ring:

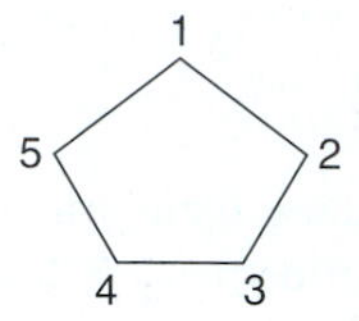

Placing the methyl substituents on the same side of the ring gives the *cis* isomer, whereas placing them on opposite sides of the ring gives the *trans* isomer:

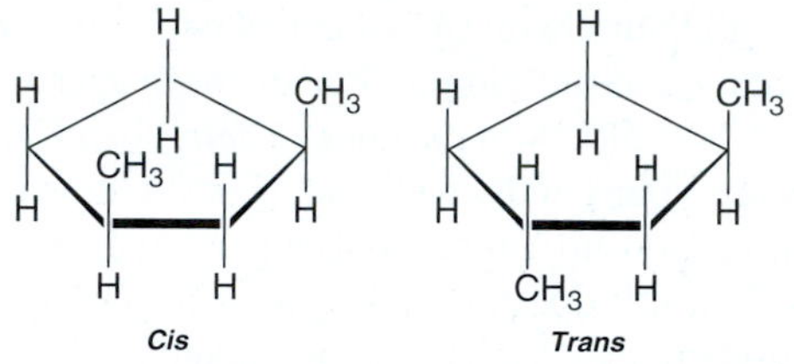

(b) The compound 3-hexene can exist as a *cis* or *trans* isomer:

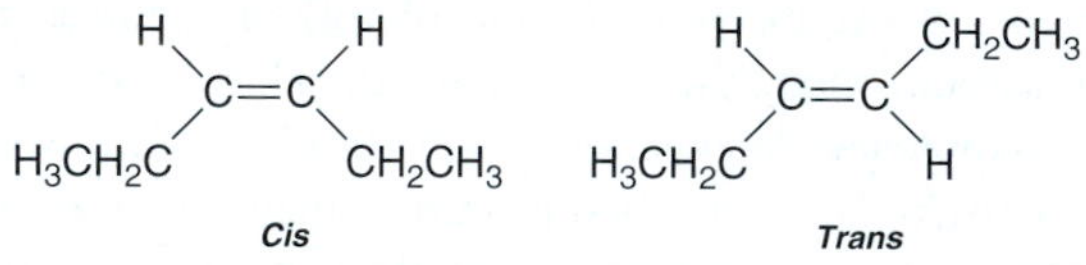

Replacing the hydrogen atoms on the third and fourth carbons by chlorine does not change the overall structures of the isomers:

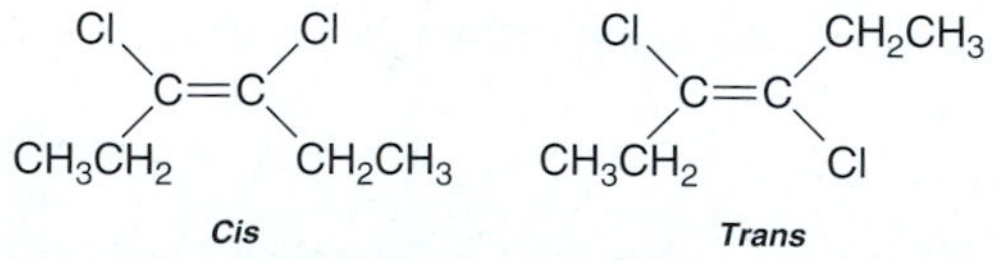

EXERCISE 24.3

Draw the *cis* and *trans* isomers of **(a)** 2-butene and **(b)** 1-methyl-3-chlorocyclopentane.

Answer

(a)

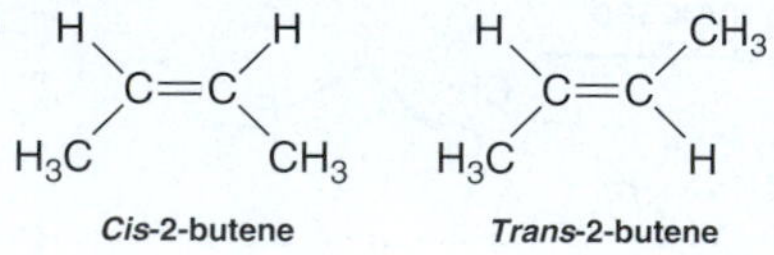

(b)

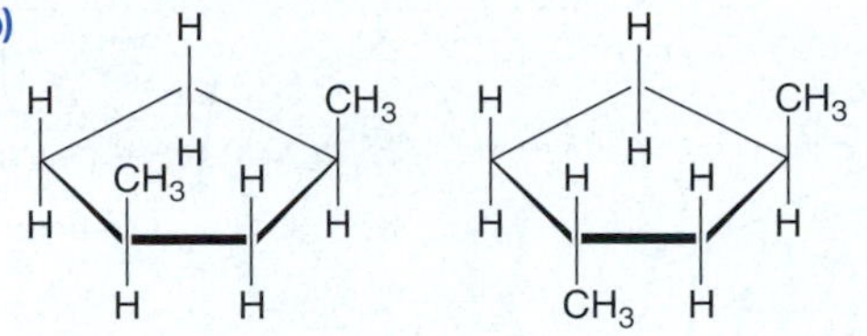

Cis *Trans*

EXAMPLE 24.4

Which of these compounds exist as at least one pair of enantiomers?

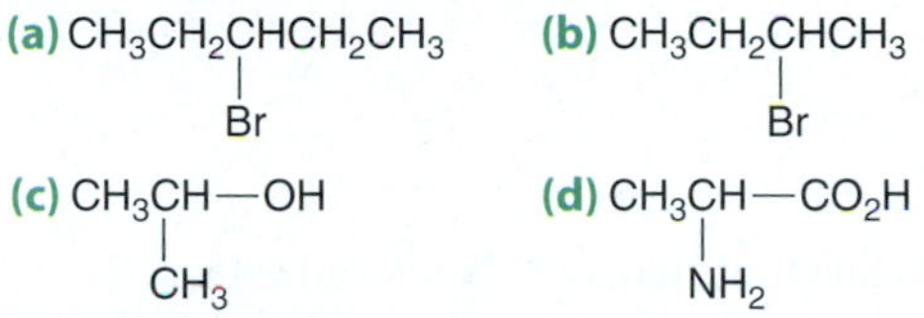

Given Organic compounds

Asked for Existence of enantiomers

Strategy

Determine whether the compound is chiral. In most cases, this means that at least one carbon is bonded to four different groups. If the compound is chiral, it exists as enantiomers.

Solution

(a) The carbons in —CH_3 and —CH_2— are each bound to more than one hydrogen, so they are achiral. The central carbon is bound to two identical ethyl groups, so it is also achiral. Because there are no chiral centers in this compound, it has no enantiomers.

(b) All carbons are bonded to at least two H atoms except for the one bonded to Br, which is bonded to four different groups: —Br, —CH_3, —H, and —CH_2CH_3. The compound therefore has one chiral center and, as a result, exists as enantiomers: the structure shown above and its nonsuperimposable mirror image.

(c) The carbons in each —CH_3 are bonded to more than one hydrogen, so they are achiral. One carbon is bonded to two —CH_3 groups, making it achiral also. This compound has no enantiomers.

(d) One carbon is bonded to three hydrogens, and one to two oxygens, so they are achiral. The central carbon is bonded to four different groups: —CH_3, —H, —NH_2, and —CO_2H, so it is chiral, and the compound has two enantiomers.

EXERCISE 24.4

Which of these compounds have at least one pair of enantiomers?

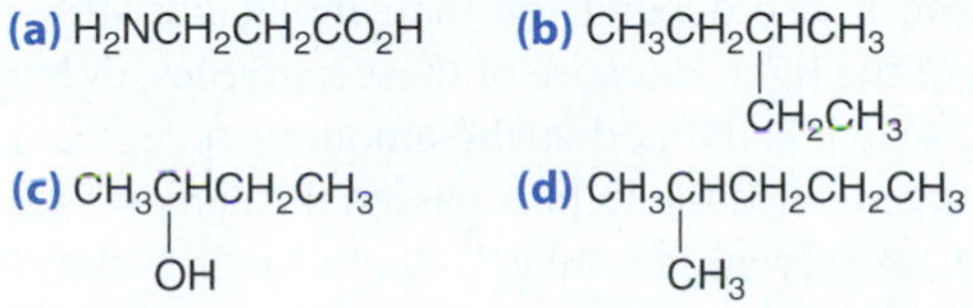

Answer (c)

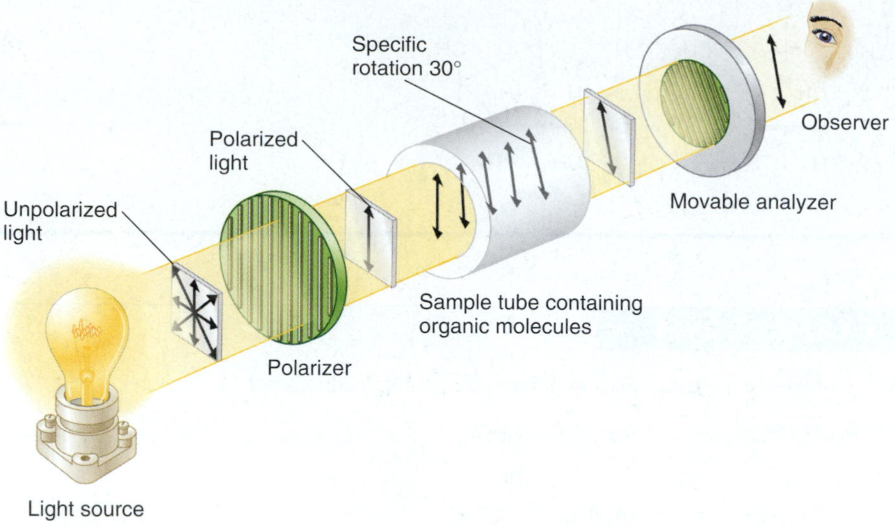

Figure 24.6 Detecting the optical activity of chiral substances. When polarized light is passed through a solution that contains an achiral compound, there is no net rotation of the plane of polarization of the light. In contrast, when polarized light is passed through a solution that contains one enantiomer of a chiral compound, as shown here, the light is rotated either clockwise [*dextrorotatory*, (+) enantiomer] or counterclockwise [*levorotatory*, (−) enantiomer] by an angle that depends on the molecular structure and concentration of the compound, the path length, and the wavelength of the light.

Optical Activity of Enantiomers

Although enantiomers have identical densities, melting and boiling points, colors, and solubility in most solvents, they differ in their interaction with *plane-polarized light*, which consists of electromagnetic waves oscillating in a single plane. In contrast, normal (unpolarized) light consists of electromagnetic waves oscillating in all directions perpendicular to the axis of propagation. When normal light is passed through a substance called a polarizer, only light oscillating in one direction is transmitted. A polarizer selectively filters out light that oscillates in any but the desired plane (Figure 24.6).

When plane-polarized light is passed through a solution, the electromagnetic radiation interacts with the solute and solvent molecules. If the solution contains an achiral compound, the plane-polarized light enters and leaves the solution unchanged because achiral molecules cause it to rotate in random directions. The solute is therefore said to be *optically inactive*. If the solution contains a single enantiomer of a chiral compound, however, the plane-polarized light is rotated in only one direction, and the solute is said to be *optically active*. A clockwise rotation is called *dextrorotatory* (from the Latin for "to the right") and is indicated in the name of the compound by (+), whereas a counterclockwise rotation is called *levorotatory* (from the Latin for "to the left") and is designated (−). As you will soon discover, this designation is important in understanding how chiral molecules interact with one another.

Note the pattern

Chiral molecules are optically active; achiral molecules are not.

The magnitude of the rotation of plane-polarized light is directly proportional to the number of chiral molecules in the solution; it also depends on their molecular structure, the temperature, and the wavelength of the light. Because of these variables, every chiral compound has a **specific rotation,** which is defined as the amount (in degrees) by which the plane of polarized light is rotated when the light is passed through a solution containing 1.0 g of solute per 1.0 mL of solvent in a tube 10.0 cm long. A chiral solution that contains equal concentrations of a pair of enantiomers is called a *racemic mixture*. In such a solution, the optical rotations exactly cancel one another, so there is no net rotation and the solution is optically inactive. The categories of stereoisomers are summarized in Figure 24.7.

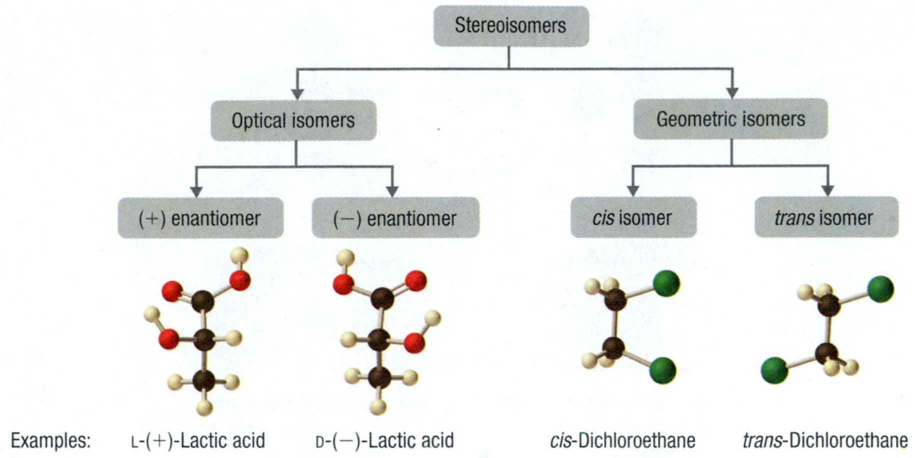

Figure 24.7 Classification of stereoisomers. In both types of stereoisomer—geometric and optical—isomeric molecules have identical connectivity, but the arrangement of atoms in space differs. *Cis* and *trans* isomers exhibit different physical and chemical properties, whereas enantiomers differ only in their interaction with plane-polarized light and reactions in asymmetric environments. Depending on the direction in which they rotate polarized light, enantiomers are identified as (+) or (−).

Interactions of Enantiomers with Other Chiral Molecules

In living organisms, virtually every molecule that contains a chiral center is found as a single enantiomer, not a racemic mixture. Thus, at the molecular level, our bodies are chiral and interact differently with the individual enantiomers of a particular compound. For example, the two enantiomers of carvone, whose structures are shown in the margin, produce very different responses in humans: (−)-carvone is the substance responsible for the smell of spearment oil, and (+)-carvone, the major flavor component of caraway seeds, is responsible for the characteristic aroma of rye bread.

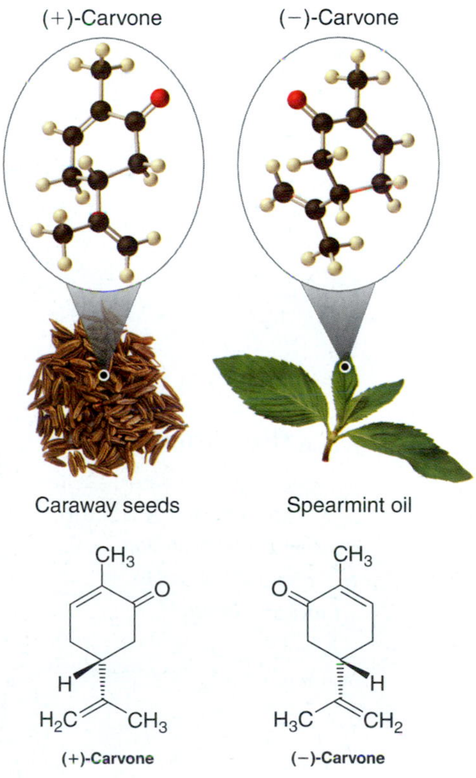

A pharmaceutical example of a chiral compound is ibuprofen, a common analgesic and anti-inflammatory agent that is the active ingredient in pain relievers such as Motrin and Advil (see Figure 24.12). The drug is sold as a racemic mixture, and it takes approximately 38 minutes to achieve its full effect in relieving pain and swelling in an adult human. Because only the (+) enantiomer is active in humans, however, the medication would relieve symptoms in only about 12 minutes if it consisted of only the (+) enantiomer. Unfortunately, isolating only the (+) enantiomer would substantially increase the cost of the drug. Conversion of the (−) to (+) enantiomer in the human body accounts for the delay in feeling the full effects of the drug. A racemic mixture of another drug, the sedative thalidomide, was sold in Europe from 1956 to the early 1960s. It was prescribed to treat nausea during pregnancy, but unfortunately only the (+) enantiomer was safe for that purpose. The (−) enantiomer was discovered to be a relatively potent *teratogen*, a substance that causes birth defects, which caused the children of many women who had taken thalidomide to be born with missing or undeveloped limbs. As a result, thalidomide was quickly banned for this use. It is currently used to treat leprosy, however, and it has also shown promise as a treatment for AIDS.

These examples dramatically illustrate the point that the biological activities of enantiomers may be very different. But how can two molecules that differ only by being nonsuperimposable mirror images cause such different responses? The biological effects of many substances—including molecules such as carvone that have a scent and drugs such as ibuprofen and thalidomide—depend on their interaction with chiral sites on specific receptor proteins. As schematically illustrated in Figure 24.8, only one enantiomer of a chiral substance interacts with a particular receptor, thereby initiating a response. The other enantiomer may not bind at all, or it may bind to *another* receptor, producing a different response.

24.3 • Reactivity of Organic Molecules

Understanding why organic molecules react as they do requires knowing something about the structure and properties of the transient species that are generated during chemical reactions. Identifying transient intermediates enables chemists to elucidate reaction mechanisms, which often allows them to control the products of a reaction. In designing the

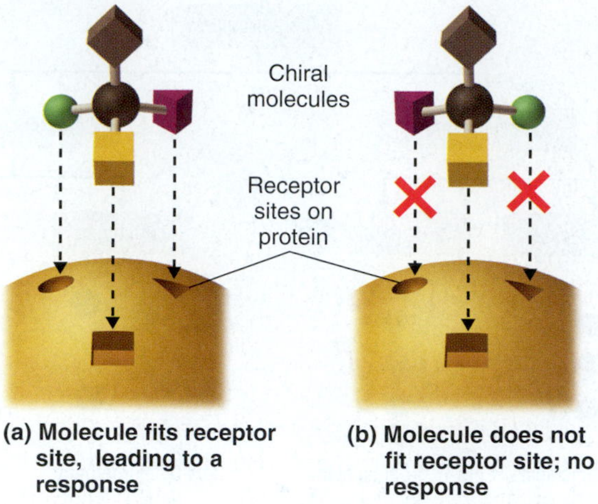

Figure 24.8 Interaction of chiral molecules with biological receptors. Only one enantiomer of a chiral molecule fits into a chiral receptor site, which typically is a small portion of a large protein. Binding of a molecule to its receptor elicits a characteristic response. The other enantiomer cannot fit into the same site and thus elicits no response. It may, however, produce a different response by binding to another site.

synthesis of a molecule such as a new drug, for example, chemists must be able to understand the mechanisms of intermediate reactions in order to maximize the yield of the desired product and minimize the occurrence of unwanted reactions. Moreover, by recognizing the common reaction mechanisms of simple organic molecules, we can understand how more complex systems react, including the much larger molecules encountered in biochemistry.

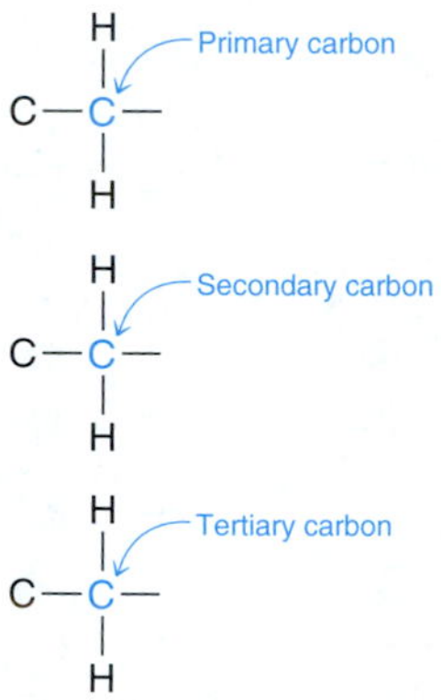

Nearly all chemical reactions, whether organic or inorganic, proceed because atoms or groups of atoms that have a positive charge or partial positive charge interact with atoms or groups of atoms that have a negative charge or partial negative charge. Thus, when a bond in a hydrocarbon is cleaved during a reaction, identifying the transient species formed, some of which are charged, allows chemists to determine the mechanism and predict the products of the reaction. Chemists often find that the reactivity of a molecule is affected by the degree of substitution of a carbon that is bonded to a functional group. These carbons are designated as *primary*, *secondary*, or *tertiary:* a primary carbon is bonded to only one other carbon and a functional group, a secondary carbon to two other carbons and a functional group, and a tertiary carbon to three other carbons and a functional group, as shown in the margin.

Reactive Intermediates

Note the pattern

The reactivity of a molecule is often affected by the degree of substitution of the carbon bonded to a functional group.

Cleaving a C—H bond can generate either C^+— and H^-, C·— and H·, or C^-— and H^+, all of which are unstable and therefore highly reactive. The most common species formed is C^+—, called a **carbocation** (Figure 24.9a), which has only six valence electrons and is therefore electron deficient. A carbocation is an **electrophile** (from "elec-

Figure 24.9 Transient intermediates in organic reactions. (a) The simplest carbocation is the methyl cation (CH_3^+), which has six valence electrons and an electrophile. Its structure is trigonal planar with an sp^2 hybridized carbon and a vacant p orbital. (b) The methyl radical ($\cdot CH_3$) is a radical that, like the carbocation, is trigonal planar and an electrophile. It is also sp^2 hybridized, but there is a single electron in the unhybridized p orbital. (c) The simplest organic carbanion is CH_3^-, which has a trigonal pyramidal structure with an sp^3 hybridized carbon that has a lone pair of electrons. Because it has a strong tendency to share its lone pair with another atom or molecule, a carbanion is a nucleophile.

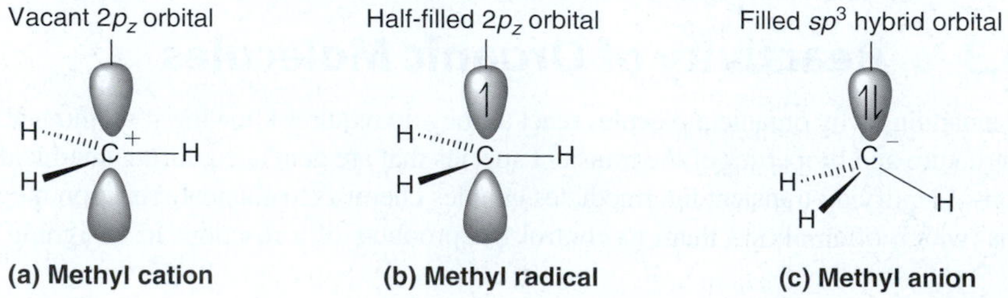

MGC Reactive Intermediates

tron" and the Greek suffix *phile*, "loving"), a species that needs electrons to complete its octet. (Recall that electron-deficient compounds, such as those of the Group 13 elements, act as Lewis acids in inorganic reactions.) In general, when a highly electronegative atom, such as Cl, is bonded to a carbocation, it draws electrons away from the carbon and destabilizes the positive charge. In contrast, alkyl groups and other species are able to stabilize the positive charge by increasing electron density at the carbocation. Thus, a tertiary carbocation (R_3C^+) is more stable than a primary carbocation (RCH_2^+).

Adding one electron to a carbocation produces a neutral species called a **radical** (see Section 14.6). An example is the methyl radical ($\cdot CH_3$), shown in Figure 24.9b. Because the carbon still has less than an octet of electrons, it is electron deficient and it also behaves as an electrophile. Like carbocations, radicals can be stabilized by carbon substituents that can donate some electron density to the electron-deficient carbon center. Thus, like carbocations, a tertiary radical ($R_3C\cdot$) is more stable than a primary radical ($RCH_2\cdot$).

Adding an electron to a radical produces a **carbanion**, which contains a negatively charged carbon with eight valence electrons (Figure 24.9c). The methyl anion (CH_3^-) has a structure that is similar to NH_3 with its lone pair of electrons, but it has a much stronger tendency to share its lone pair with another atom or molecule. A carbanion is a **nucleophile** (from "nucleus" and *phile*), an electron-rich species that has a pair of electrons available to share with another atom. Carbanions are *destabilized* by groups that donate electrons, so the relationship between their structure and reactivity is exactly the opposite of carbocations and radicals. That is, a tertiary carbanion (R_3C^-) is less stable than a primary carbanion (RCH_2^-). Carbanions are most commonly encountered in organometallic compounds such as methyllithium (CH_3Li) or methylmagnesium chloride (CH_3MgCl), where the more electropositive metal ion stabilizes the negative charge on the more electronegative carbon atom.

Electrophiles such as carbocations seek to gain electrons and thus have a strong tendency to react with nucleophiles, which are negatively charged species or substances with lone pairs of electrons. The reaction of electrophiles with nucleophiles is a central theme in organic reactions.

Note the pattern

Electrophiles react with nucleophiles.

EXAMPLE 24.5

Classify each species as an electrophile, a nucleophile, or neither: **(a)** BF_3; **(b)** CH_4; **(c)** $(CH_3)_3C^+$; **(d)** NH_2^-.

Given Molecular formulas

Asked for Mode of reactivity

Strategy

Determine whether the compound is electron deficient, in which case it is an electrophile; electron rich, in which case it is a nucleophile; or neither.

Solution

(a) The BF_3 molecule is a neutral compound that contains a Group 13 element with three bonds to B. The boron atom has only six valence electrons, so it tends to accept an electron pair. The compound is therefore an electrophile.

(b) The CH_4 molecule has four bonds to C, which is typical of a neutral Group 14 compound. The carbon atom has no lone pairs to share and no tendency to gain electrons, and each hydrogen atom forms one bond that contains two valence electrons. Thus, CH_4 is neither an electrophile nor a nucleophile.

(c) The $(CH_3)_3C^+$ cation contains a Group 14 atom (carbon) with only three bonds. It therefore has only six valence electrons and seeks electrons to complete an octet. Hence, $(CH_3)_3C^+$, a carbocation, is an electrophile.

(d) The NH_2^- anion contains a Group 15 element with a lone pair of electrons, two bonds, and a negative charge, giving N a total of eight electrons. With its negative charge, the N atom has two lone pairs of electrons, making it a potent nucleophile.

EXERCISE 24.5

Classify each compound as an electrophile, a nucleophile, or neither: **(a)** C_6H_5OH; **(b)** $AlBr_3$; **(c)** $(CH_3)_4C$.

Answer **(a)** nucleophile; **(b)** electrophile; **(c)** neither

24.4 • Common Classes of Organic Reactions

Certain patterns are encountered repeatedly in organic reactions, many reflecting the interactions of nucleophiles and electrophiles. In this section, we discuss five common types of organic reactions: substitution, elimination, addition, radical reactions, and oxidation–reduction reactions. You have encountered many of these types of reactions, such as the formation of peptides by elimination (see Chapter 12), the oxidation–reduction reactions that generate voltage in batteries (see Chapter 19), and chain reactions that involve organic radicals (see Section 14.6). In this section, we expand our discussion to include some of the mechanisms behind these reactions.

Substitution

In a **substitution reaction**, one atom or group of atoms in a substance is replaced by another atom or group of atoms from another substance. A typical substitution reaction is the reaction of hydroxide ion with methyl chloride:

$$CH_3Cl + OH^- \longrightarrow CH_3OH + Cl^- \qquad (24.1)$$

MGC Organic Reactions

Methyl chloride has a polar C—Cl bond, with the carbon atom having a partial positive charge. Notice in Equation 24.1 that the electronegative Cl atom is replaced by another electronegative species that is a stronger nucleophile, in this case OH^-. Reactions of this sort are called *nucleophilic substitution reactions*. For this type of reaction to occur, the nucleophilic reactant *must* possess a pair of electrons and have a greater affinity for the electropositive carbon atom than the original substituent.

Note the pattern

In nucleophilic substitution reactions, the nucleophile must possess a pair of electrons and have a greater affinity for the electropositive species than the original substituent.

One type of nucleophilic substitution reaction is shown in Equation 24.1. It proceeds by a mechanism in which the lone pair of electrons on the entering nucleophile (OH^-) attacks the partially positively charged carbon atom of the polar C—Cl bond, causing the C—Cl bond to weaken and break:

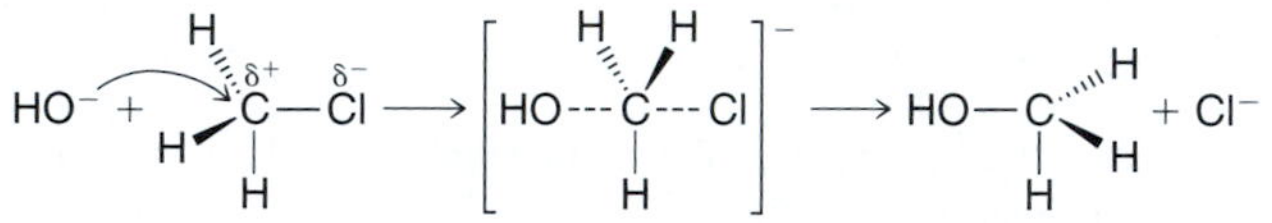

(24.2)

The convention for writing such a mechanism is to draw arrows showing the direction of electron flow—that is, *from* the electron-rich center (the nucleophile) *to* the electron-poor center (the electrophile). The intermediate species, enclosed by square brackets, represents a transient arrangement of atoms that is only postulated to exist. If the atom under attack (in this case, the partially positively charged carbon atom) had $—CH_3$ groups bonded to it rather than H atoms, the bulky methyl groups would interfere with attack by OH^-, making the reaction *sterically hindered*. The reaction would then proceed in two discrete steps in a second type of substitution reaction: the C—Cl bond would break, forming the $(CH_3)_3C^+$ carbocation (the electrophile), which would then react with hydroxide (the nucleophile) in a separate step to give the product, $(CH_3)_3COH$. An example of a nucleophilic substitution reaction involves the chemical warfare agent known as *mustard gas* $(ClCH_2CH_2)_2S$, which caused about 400,000 casualties during World War I. Mustard gas is toxic because it contains a chloride that can be displaced by nucleophilic amino groups in proteins, thereby allowing the molecule to bond to a protein irreversibly. Because the other product of the reaction is HCl, mustard gas causes severe burns to mucous membranes in the respiratory tract. If mustard gas reacts with DNA, cross-linking of DNA strands through sulfur occurs, which results in coding errors, inhibition of replication, and disruption of other DNA functions. If mustard gas reacts with RNA, protein synthesis is altered (see Section 24.6).

Elimination

Reactions in which adjacent atoms are removed, or "eliminated," from a molecule with the formation of a multiple bond and release of a small molecule are called **elimination reactions**. They have the general form

$$\underset{\displaystyle CH_2}{\overset{A}{|}}{-}\underset{\displaystyle CH_2}{\overset{B}{|}} \longrightarrow CH_2{=}CH_2 + A{-}B \tag{24.3}$$

and are similar to cleavage reactions in inorganic compounds (see Section 3.5). A typical example is the conversion of ethyl chloride to ethylene:

$$CH_3CH_2Cl \longrightarrow CH_2{=}CH_2 + HCl \tag{24.4}$$

Much of the approximately 26 million tons of ethylene produced per year in the United States is used for the synthesis of plastics such as polyethylene and polypropylene. In Equation 24.4, the A–B molecule eliminated is HCl, whose components are eliminated as H^+ from the carbon atom on the left and Cl^- from the carbon on the right. When an acid is produced, as occurs here, the reaction is generally carried out in the presence of a base (such as NaOH) to neutralize the acid.

Note the pattern

Elimination reactions are similar to cleavage reactions in inorganic compounds.

Addition

A reaction in which the components of a species A–B are added to adjacent atoms across a carbon–carbon multiple bond is called an **addition reaction**. An example is the reverse of the reaction shown in Equation 24.4, the reaction of HCl with ethylene to give ethyl chloride:

$$HCl + CH_2{=}CH_2 \longrightarrow CH_3CH_2Cl \tag{24.5}$$

Note the pattern

An addition reaction is the reverse of an elimination reaction.

Although a multiple bond is stronger than a single bond, the π bonds of the multiple bond are weaker than the σ bond. The high electron density located between multiply bonded carbon atoms causes alkenes and alkynes to behave like nucleophiles, where nucleophilic attack occurs from the more weakly bound π electrons. Hence, alkenes and alkynes are regarded as functional groups. Nucleophilic attack occurs on the $H^{\delta+}$ atom of the polar HCl bond, initially producing a species with a carbon that has only three bonds, a carbocation. In a second nucleophilic attack, Cl^-, the electrophile in Equation 24.5, attacks the carbocation:

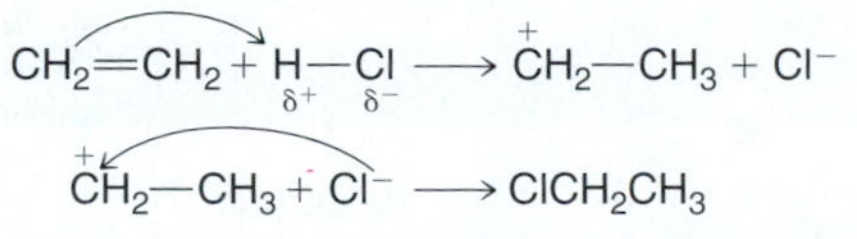

(24.6)

Alcohols, an important class of organic compounds, are often produced by addition reactions. Initial attack by the π bond of an alkene on a $H^{\delta+}$ of H_3O^+ produces a carbocation. The carbocation then undergoes nucleophilic attack by a lone pair of electrons from H_2O followed by elimination of H^+ to form the alcohol.

Radical Reactions

Many important organic reactions involve radicals, such as the combustion of fuels. Probably the best known is the reaction of a saturated hydrocarbon, such as ethane, with a halogen, such as Br_2. The overall reaction is

$$CH_3CH_3 + Br_2 \xrightarrow[\text{or } h\nu]{400°C} CH_3CH_2Br + HBr \tag{24.7}$$

Radical chain reactions occur in three stages: *initiation, propagation*, and *termination* (see Section 14.6). At high temperature or in the presence of light, the relatively weak Br—Br bond is broken in an initiation step that produces an appreciable number of Br atoms (Br·). During propagation, a bromine atom attacks ethane, producing a radical, which then reacts with another bromine molecule to produce ethyl bromide:

$$\begin{array}{l} \cancel{Br\cdot} + CH_3CH_3 \longrightarrow \cancel{CH_3CH_2\cdot} + HBr \\ \cancel{CH_3CH_2\cdot} + Br_2 \longrightarrow CH_3CH_2Br + \cancel{Br\cdot} \\ \hline Br_2 + CH_3CH_3 \longrightarrow CH_3CH_2Br + HBr \end{array} \tag{24.8}$$

Notice that the sum of the two propagation steps corresponds to the balanced equation for the overall reaction. There are three possible termination steps: the combination of two bromine atoms, of two ethyl radicals, or of an ethyl and a bromine radical:

$$\begin{array}{l} Br\cdot + Br\cdot \longrightarrow Br_2 \\ CH_3CH_2\cdot + \cdot CH_2CH_3 \longrightarrow CH_3CH_2CH_2CH_3 \\ CH_3CH_2\cdot + Br\cdot \longrightarrow CH_3CH_2Br \end{array} \tag{24.9}$$

Note the pattern

Radical reactions are not very selective.

Because radicals are powerful nucleophiles and hence highly reactive, such reactions are not very selective. For example, chlorination of *n*-butane gives a roughly 70:30 mixture of 2-chlorobutane, formed from the more stable radical, and 1-chlorobutane.

Oxidation–Reduction Reactions

Oxidation–reduction reactions, which are common in organic chemistry, can often be identified by changes in the number of oxygen atoms at a particular position in the hydrocarbon skeleton or in the number of bonds between carbon and oxygen at that position. An increase in either corresponds to an oxidation, whereas a decrease corresponds to a reduction. Conversely, an increase in the number of hydrogen atoms in a hydrocarbon is often an indication of a reduction. We can illustrate these points by considering how the oxidation state of the carbon atom changes in the series of compounds shown in Figure 24.10a. (For a review of oxidation states, see Sections 3.5 and 4.8.) Notice that the number of oxygen atoms or the number of bonds to oxygen changes throughout the series. Hence, the conversion of methane to formic acid is an oxidation, whereas the conversion of carbon dioxide to methanol is a reduction. Also notice that the number of hydrogen atoms increases in going from the most oxidized to least oxidized

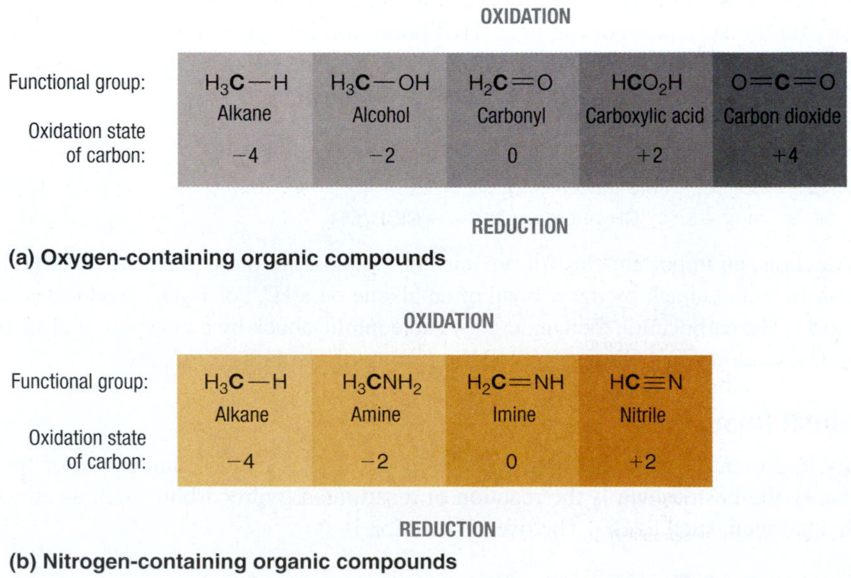

Figure 24.10 Oxidation state of carbon in oxygen- and nitrogen-containing functional groups. (a) In a hydrocarbon, oxidation is indicated by an increase in the number of oxygen atoms or carbon–oxygen bonds, or a decrease in the number of hydrogen atoms. (b) In nitrogen-containing compounds, the number of carbon–nitrogen bonds changes with the oxidation state of carbon.

compound. As expected, as the oxidation state of carbon increases, the carbon becomes a more potent electrophile. Thus, the carbon of CO_2 is a stronger electrophile (that is, more susceptible to nucleophilic attack) than the carbon of an alkane such as methane.

Similarly, in compounds with a carbon–nitrogen bond, the number of bonds between the C and N atoms increases as the oxidation state of the carbon increases (Figure 24.10b). Notice that in a nitrile, which contains the —C≡N group, the carbon has the same oxidation state (+2) as in a carboxylic acid, characterized by the $—CO_2H$ group. We therefore expect the carbon of a nitrile to be a rather strong electrophile.

EXAMPLE 24.6

Write an equation to describe **(a)** the nucleophilic substitution reaction of potassium cyanide with 1-chloropropane to give $CH_3CH_2CH_2CN$ (butyronitrile) and **(b)** the electrophilic addition reaction of HBr with *cis*-2-butene. Identify the electrophile and the nucleophile in each reaction.

Given Reactants, products, and reaction mechanism

Asked for Equation and identification of electrophile and nucleophile

Strategy

Use the mechanisms described to show how the indicated products are formed from the reactants.

Solution

(a) The CN^- ion of KCN is a potent nucleophile that can displace the chlorine atom of 1-chloropropane, releasing a chloride ion. Substitution results in the formation of a new C—C bond:

$$CN^- + \underset{\text{1-Chloropropane}}{CH_3CH_2CH_2Cl} \longrightarrow \underset{\text{Butyronitrile}}{CH_3CH_2CH_2CN} + Cl^-$$

The carbon bonded to chlorine is an electrophile because of the highly polar C—Cl bond.

(b) In the electrophilic addition of a hydrogen halide to an alkene, the reaction is

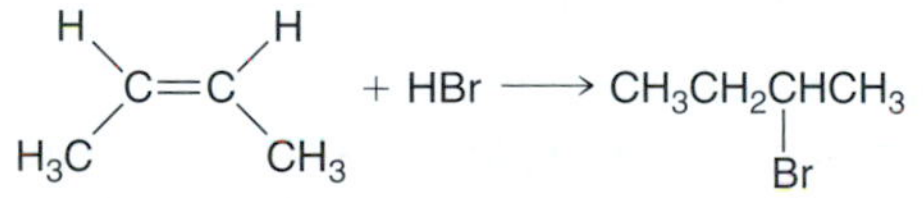

The first step is nucleophilic attack of the π electrons of the double bond on the electrophilic hydrogen of the polar H—Br bond to generate the transient carbocation shown in the margin, followed by nucleophilic attack by the halide to give the product. Thus, the alkene is the nucleophile, and the proton of the acid is the electrophile.

$$CH_3CH_2\overset{+}{C}HCH_3$$

Carbocation intermediate

EXERCISE 24.6

Write an equation to describe **(a)** the nucleophilic substitution reaction of sodium methoxide ($NaOCH_3$) with benzyl bromide ($C_6H_5CH_2Br$), and **(b)** the acid-catalyzed electrophilic addition reaction of water with cyclopentene. In each reaction, identify the electrophile and nucleophile.

Answer

(a) $\underset{\text{Methoxide ion (nucleophile)}}{OCH_3^-} + \underset{\text{Benzyl bromide (electrophile)}}{C_6H_5CH_2Br} \longrightarrow \underset{\text{Benzylmethyl ether}}{C_6H_5CH_2OCH_3} + Br^-$

(b)

Cyclopentene (nucleophile) | Hydronium ion (electrophile) | Cyclopentanol

24.5 • Common Classes of Organic Compounds

The functional groups characteristic of each class of organic compounds (Table 24.1) largely determine the general properties and reactivity of that class. In this section, we describe the relationships between structure, physical properties, and reactivity for the major classes of organic compounds. We also show you how to apply these relationships to understand some common reactions that chemists use to synthesize organic compounds.

Alkanes, Alkenes, and Alkynes

MGC Alkane, Alkene, Alkyne

The boiling points of alkanes increase smoothly with increasing molecular mass. They are similar to those of the corresponding alkenes and alkynes because of similarities in molecular mass between analogous structures (Table 24.3). In contrast, the melting points of alkanes, alkenes, and alkynes with similar molecular masses show a much wider variation, because the melting point depends strongly on how the molecules stack in the solid state. It is therefore sensitive to relatively small differences in structure.

Because alkanes contain only C—C and C—H bonds, which are strong and not very polar (the electronegativities of C and H are similar—see Figure 7.15), they are not easily attacked by nucleophiles or electrophiles. Consequently, their reactivity is limited, and often their reactions occur only under extreme conditions. For example, *catalytic cracking* can be used to convert straight-chain alkanes to highly branched alkanes, which are better fuels for internal combustion engines. Catalytic cracking is one example of a **pyrolysis reaction** (from the Greek *pyros*, "fire," and *lysis*, "loosening"), in which alkanes are heated to a sufficiently high temperature to induce cleavage of the weakest bonds, the C—C single bonds. The result is a mixture of radicals derived from essentially random cleavage of the various C—C bonds in the chain. Pyrolysis of *n*-pentane, for example, is nonspecific and can produce four different radicals:

$$2CH_3CH_2CH_2CH_2CH_3 \xrightarrow{\Delta} CH_3\cdot + CH_3CH_2CH_2\cdot + CH_3CH_2\cdot + CH_3CH_2CH_2\cdot \qquad (24.10)$$

Recombination of these radicals (a termination step) can produce ethane, propane, butane, *n*-pentane, *n*-hexane, *n*-heptane, and *n*-octane. Radicals that are formed in the middle of the chain tend to produce branched hydrocarbons. Lighter alkanes are removed from the mixture by distillation.

Radicals are also produced during the combustion of alkanes, with CO_2 and H_2O as the final products. As discussed in Section 24.3, radicals are stabilized by the presence of multiple carbon substituents that can donate electron density to the electron-deficient carbon. The chemical explanation of octane ratings, described in Section 2.6, rests partly on the stability of radicals produced from the different hydrocarbon fuels. Recall that *n*-heptane, which does not burn smoothly, has an octane rating of 0, and 2,2,4-trimethylpentane ("isooctane"), which burns quite smoothly, has a rating of 100 (see Table 2.13). Isooctane has a branched structure (see structure in margin) and is capable of forming tertiary radicals that are comparatively stable. In contrast, the radicals formed during the

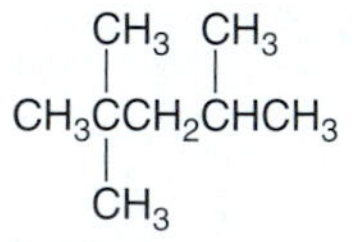
2,2,4-Trimethylpentane ("isooctane")

TABLE 24.3 Boiling points (in °C) of alkanes, alkenes, and alkynes of comparable molecular mass

	Length of Carbon Chain		
Class	**Two C Atoms**	**Three C Atoms**	**Four C Atoms**
Alkane	−88.7	−42.1	−0.5
Alkene	−103.8	−47.8	−6.3
Alkyne	−83.8	−23.2	8.0

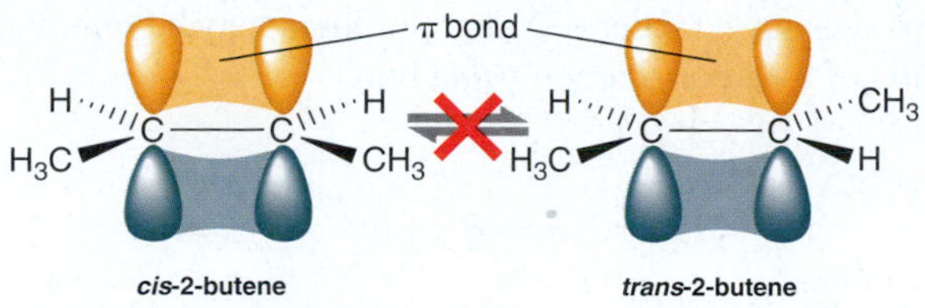

Figure 24.11 Carbon–carbon bonding in alkenes and interconversion of *cis* and *trans* isomers. In butane, there is only a small energy barrier to rotation about the C2–C3 σ bond. In the formation of *cis*- or *trans*-2-butene from butane, the *p* orbitals on C2 and C3 overlap to form a π bond. To convert *cis*-2-butene to *trans*-2-butene, or vice versa, through rotation about the double bond, the π bond must be broken. Because this interconversion is energetically unfavorable, *cis* and *trans* isomers are distinct compounds that generally have different physical and chemical properties.

combustion of *n*-heptane, a straight-chain molecule, are less stable and hence more reactive, which partly explains why burning *n*-heptane causes premature firing and engine knocking.

In Section 24.2, we explained that rotation about the carbon–carbon multiple bonds of alkenes and alkynes cannot occur without breaking a π bond, which therefore constitutes a large energy barrier to rotation (Figure 24.11). Consequently, the *cis* and *trans* isomers of alkenes generally behave as distinct compounds with different chemical and physical properties. A four-carbon alkene has four possible isomeric forms: three structural isomers, which differ in their connectivity; in addition, one of these structural isomers, 2-butene, has a pair of geometric isomers, *cis*-2-butene and *trans*-2-butene:

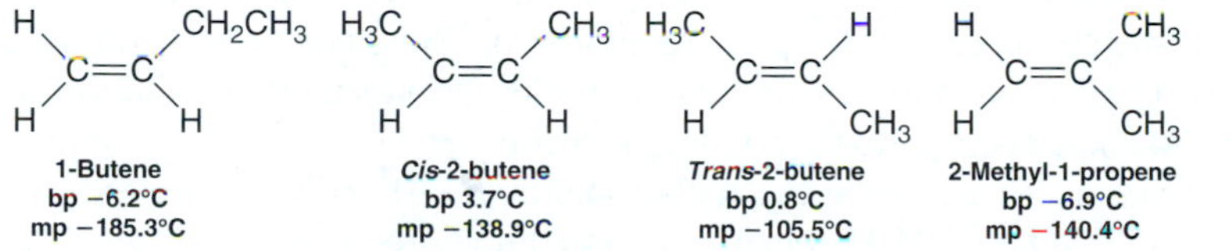

(24.11)

Notice that the four isomers have significantly different physical properties.

Alkynes in which the triple bond is located at one end of a carbon chain are called *terminal alkynes*, and contain a hydrogen atom attached directly to a triply bonded carbon: R—C≡C—H. Terminal alkynes are unusual in that the hydrogen atom can be removed relatively easily as H^+, forming an *acetylide ion* (R—C≡C^-). Acetylide ions are potent nucleophiles that are especially useful reactants for making longer carbon chains by a nucleophilic substitution reaction. As in earlier examples of such reactions, the nucleophile attacks the partially positively charged atom in a polar bond, which in this reaction is the carbon of the Br—C bond:

$$RC{\equiv}C^- + Br{-}CH_2R' \longrightarrow RC{\equiv}C{-}CH_2R' + Br^- \tag{24.12}$$

Alkenes and alkynes are most often prepared by elimination reactions (Equation 24.3). A typical example is the preparation of 2-methyl-1-propene, whose derivative, 3-chloro-2-methyl-1-propene, is used as a fumigant and insecticide. The parent compound can be prepared from either 2-hydroxy-2-methylpropane or 2-bromo-2-methylpropane:

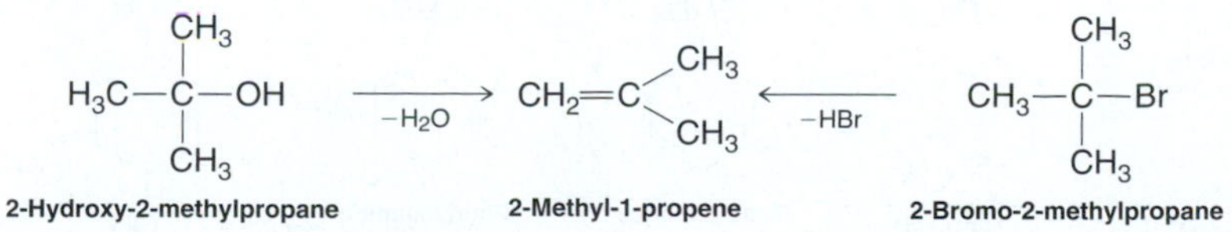

(24.13)

The reaction on the left proceeds by elimination of the elements of water (H^+ plus OH^-), so it is a *dehydration* reaction. If an alkane contains *two* properly located

Note the pattern
Alkenes and alkynes are most often prepared by elimination reactions.

functional groups such as —OH or —X, both of them may be removed as H_2O or HX with the formation of a carbon–carbon *triple* bond:

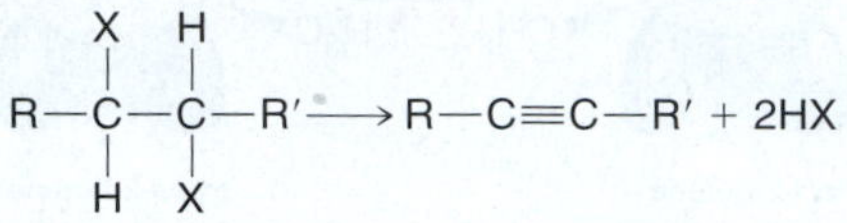

(24.14)

Arenes

Most arenes that contain a single six-membered ring are volatile liquids, such as benzene and the xylenes, although some arenes with substituents on the ring are solids at room temperature. In the gas phase, the dipole moment of benzene is zero, but the presence of electronegative or electropositive substituents can result in a net dipole moment that increases intermolecular attractive forces and raises the melting and boiling points. For example, 1,4-dichlorobenzene (see Section 11.5), a compound used as an alternative to naphthalene in the production of mothballs, has a melting point of 52.7°C, considerably higher than the melting point of benzene (5.5°C).

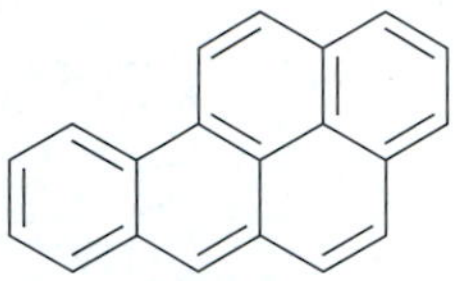

Benz[a]pyrene

Certain aromatic hydrocarbons, such as benzene and benz[a]pyrene (see structure in margin), are potent liver toxins and carcinogens. In 1775, a British physician, Percival Pott, described the high incidence of cancer of the scrotum among small boys used as chimney sweeps and attributed it to their exposure to soot. His conclusions were correct: benz[a]pyrene, a component of chimney soot, charcoal-grilled meats, and cigarette smoke, was the first chemical carcinogen to be identified.

Although arenes are usually drawn with three C=C double bonds, benzene is about 150 kJ/mol more stable than would be expected if it contained three double bonds. This increased stability is due to the delocalization of the π electron density over all the atoms of the ring (see Section 9.3). Thus, compared with alkenes, arenes are poor nucleophiles. Consequently, they do not undergo addition reactions like alkenes, but instead they undergo a variety of *electrophilic aromatic substitution* reactions that involve the replacement of —H on the arene by a group —E, such as $—NO_2$, $—SO_3H$, a halogen, or an alkyl group, in a two-step process. The first step involves addition of the electrophile (E) to the π system of benzene, forming a carbocation. In the second step, a proton is lost from the adjacent carbon on the ring:

Note the pattern
Arenes undergo substitution reactions rather than elimination because of increased stability arising from delocalization of their π electron density.

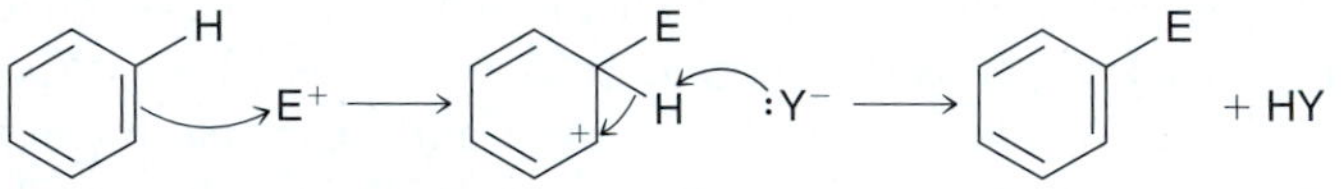

(24.15)

The carbocation formed in the first step is stabilized by resonance.

Many substituted arenes have potent biological activity. Examples include common drugs and antibiotics such as aspirin and ibuprofen, illicit drugs such as amphetamines and peyote, the amino acid phenylalanine, and hormones such as adrenaline (Figure 24.12).

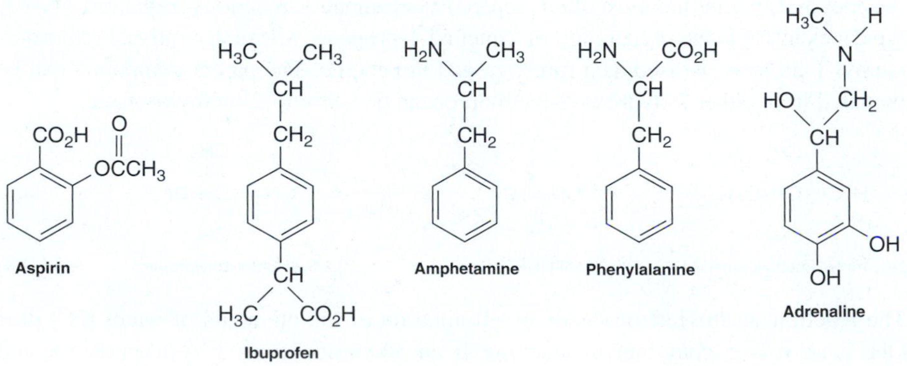

Figure 24.12 Biologically active substituted arenes. Aspirin (anti-fever activity), ibuprofen (anti-fever and anti-inflammatory activity), and amphetamine (stimulant) have pharmocological effects. Phenylalanine is an amino acid. Adrenaline is a hormone that elicits the "fight or flight" response to stress.

TABLE 24.4 Boiling points of alkanes, ethers, and alcohols of comparable molecular mass

	Name	Formula	Molecular Mass (amu)	bp (°C)
Alkane	Propane	C_3H_8	44	−42.1
	n-Pentane	C_5H_{12}	72	36.1
	n-Heptane	C_7H_{16}	100	98.4
Ether	Dimethylether	$(CH_3)_2O$	46	−24.9
	Diethylether	$(CH_3CH_2)_2O$	74	34.6
	Di-*n*-propylether	$(CH_3CH_2CH_2)_2O$	102	89.6
Alcohol	Ethanol	CH_3CH_2OH	46	78.3
	n-Butanol	$CH_3(CH_2)_3OH$	74	117.7
	n-Hexanol	$CH_3(CH_2)_5OH$	102	157.5

Alcohols and Ethers

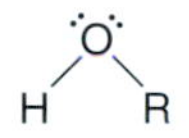
Alcohol

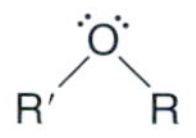
Ether

Both alcohols and ethers can be thought of as derivatives of water in which at least one hydrogen atom has been replaced by an organic group, as shown in the margin. Because of the electronegative oxygen atom, the individual O—H bond dipoles in alcohols cannot cancel one another, resulting in a substantial dipole moment allows alcohols to form hydrogen bonds. Alcohols therefore have significantly higher boiling points than alkanes or alkenes of comparable molecular mass, whereas ethers, without a polar O—H bond, have intermediate boiling points due to the presence of a small dipole moment. (Table 24.4). The larger the alkyl group in the molecule, however, the more "alkane-like" the alcohol is in its properties. Because of their polar nature, alcohols and ethers tend to be good solvents for a wide range of organic compounds.

Alcohols are usually prepared by the addition of water across a carbon–carbon double bond or by a nucleophilic substitution reaction of an alkyl halide using hydroxide, a potent nucleophile (Equation 24.2). As you will see in the next section, alcohols can also be prepared by the reduction of compounds that contain the carbonyl functional group (>C═O, Figure 24.10a). Alcohols are classified as primary, secondary, or tertiary, depending on whether the —OH group is bonded to a primary, secondary, or tertiary carbon. For example, the compound 5-methyl-3-hexanol is a secondary alcohol.

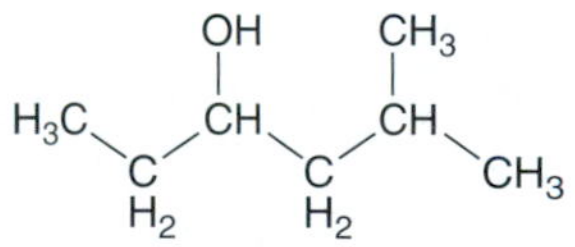
5-methyl-3-hexanol

Ethers, especially those with two different alkyl groups (ROR′), can be prepared by a substitution reaction in which a nucleophilic alkoxide ion (RO^-) attacks the partially positively charged carbon atom of the polar C—X bond of an alkyl halide (R′X):

$$RO^- + \overset{\delta+}{R'}\text{—}\overset{\delta-}{X} \longrightarrow ROR' + X^- \qquad (24.16)$$

Although both alcohols and phenols have an —OH functional group, phenols are 10^6–10^8 more acidic than alcohols. This is largely because simple alcohols have the —OH unit attached to an sp^3 hybridized carbon, whereas phenols have an sp^2 hybridized carbon atom bonded to the oxygen atom. The negative charge of the phenoxide ion can therefore interact with the π electrons in the ring, thereby delocalizing and stabilizing the negative charge through resonance (see Section 8.5). In contrast, the negative charge on an alkoxide ion cannot be stabilized by these types of interactions.

Alcohols undergo two major types of reactions: those involving cleavage of the O—H bond and those involving cleavage of the C—O bond. Cleavage of an O—H bond is a reaction characteristic of an acid, but alcohols are even weaker acids than

Note the pattern

Alcohols undergo two major types of reactions: cleavage of the O—H bond and cleavage of the C—O bond.

water. The acid strength of phenols, however, is about a million times greater than that of ethanol, making the pK_a of phenol comparable to that of the NH_4^+ ion (9.89 versus 9.25, respectively):

$$C_6H_5OH + H_2O \rightleftharpoons H_3O^+ + C_6H_5O^- \qquad (24.17)$$

Cleavage of the C—O bond in alcohols occurs under acidic conditions. The —OH is first protonated, and then nucleophilic substitution follows:

$$-\overset{|}{\underset{|}{C}}-OH_2^+ + X^- \longrightarrow -\overset{|}{\underset{|}{C}}-X + H_2O \qquad (24.18)$$

In the absence of a nucleophile, however, elimination can occur, producing an alkene (Equation 24.3).

Ethers lack the —OH unit that is central to the reactivity of alcohols, so they are comparatively unreactive. Their low reactivity makes them highly suitable as solvents in which to carry out organic reactions.

Aldehydes and Ketones

MGC Aldehydes, Ketones, Carboxylic Acids

Aromatic aldehydes, which have intense and characteristic flavors and aromas, are the major components of such well-known flavorings as vanilla and cinnamon (Table 24.5). Many ketones, such as camphor and jasmine, also have intense aromas. Ketones are found in many of the hormones responsible for sex differentiation in humans, such as progesterone and testosterone (see Section 4.1).

Aldehydes and ketones contain the carbonyl functional group, which has an appreciable dipole moment because of the polar C═O bond. The presence of the carbonyl group results in strong intermolecular interactions that cause aldehydes and ketones to have higher boiling points than alkanes or alkenes of comparable molecular mass (Table 24.6). As the mass of the molecule increases, the carbonyl group becomes less important to the overall properties of the compound, and the boiling points approach those of the corresponding alkanes

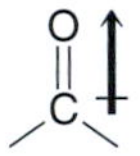

Carbonyl group

Aldehydes and ketones are typically prepared by the oxidation of alcohols (Figure 24.10a). In their reactions, the partially positively charged carbon atom of the carbonyl group is an electrophile that is subject to nucleophilic attack. Conversely, the lone pairs of electrons on the oxygen atom of the carbonyl group allow electrophilic

TABLE 24.5 Some familiar aldehydes and their uses

Name	Formula	Use
Formaldehyde	CH_2O	Biological preservative
Cinnamaldehyde	$C_6H_5CH{=}CHCHO$	Cinnamon
Citral	$(CH_2)_2C{=}CHCH_2CH_2\underset{\displaystyle CH_3}{\underset{|}{C}}{=}CHCHO$	Antiseptic, perfumes
Benzaldehyde	C_6H_5CHO	Manufacture of dyes, almond flavoring
Vanillin	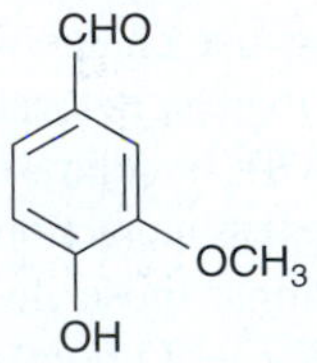	Vanilla flavoring

TABLE 26.6 Boiling points of alkanes, aldehydes, and ketones of comparable molecular mass

	Name	Formula	Molecular Mass (amu)	bp (°C)
Alkane	*n*-Butane	C_4H_{10}	58	−0.5
	n-Pentane	C_5H_{12}	72	36.1
Aldehyde	Propionaldehyde (propanal)	C_3H_6O	58	49.5
	Butyraldehyde (butanal)	C_4H_8O	72	75.0
Ketone	Acetone (2-propanone)	C_3H_6O	58	56.1
	Methyl ethyl ketone (2-butanone)	C_4H_8O	72	79.6

attack to occur. Aldehydes and ketones can therefore undergo both nucleophilic attack (at the carbon atom) and electrophilic attack (at the oxygen atom).

Aldehydes and ketones react with many organometallic compounds that contain stabilized carbanions. One of the most important classes of such compounds are the **Grignard reagents**, organomagnesium compounds with the formula RMgX (X = Cl, Br, I) that are so strongly polarized that they can be viewed as containing R^- and MgX^+. These reagents are named for the French chemist Victor Grignard (1871–1935), who won a Nobel Prize in 1912 for their development. In a Grignard reaction, the carbonyl functional group is converted to an alcohol and the carbon chain of the carbonyl compound is lengthened by the addition of the R group from the reagent. One example is the reaction of cyclohexylmagnesium chloride, a Grignard reagent, with formaldehyde:

Note the pattern

Nucleophilic attack occurs at the partially positively charged carbon of a carbonyl functional group. Electrophilic attack occurs at the lone pairs of electrons on the oxygen atom.

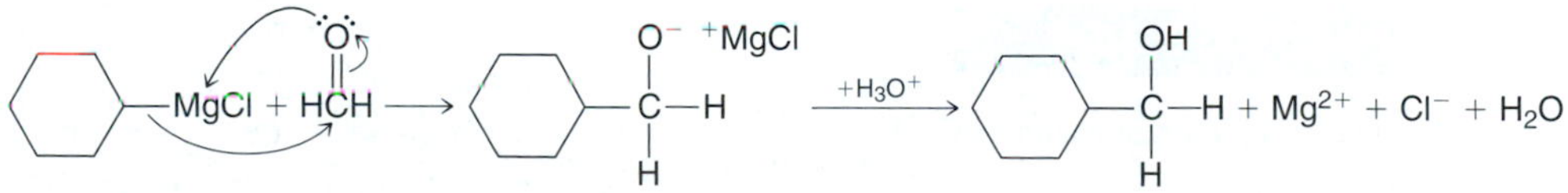

(24.19)

The nucleophilic carbanion of the cyclohexyl ring attacks the electrophilic carbon atom of the carbonyl group. Acidifying the solution results in protonation of the intermediate to give the alcohol. Aldehydes and ketones can also be prepared by reduction of a carboxylic acid group (—CO_2H) (Figure 24.10a) or by reaction of a carboxylic acid derivative with a Grignard reagent, respectively. The former reaction requires a powerful reducing agent, such as a metal hydride.

EXAMPLE 24.7

Explain how each reaction proceeds to form the indicated product:

(a)

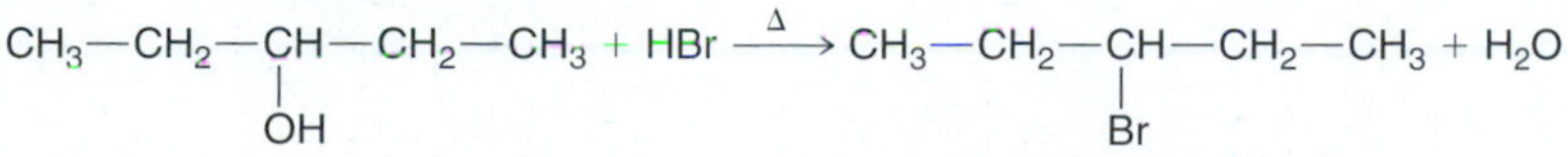

(b)

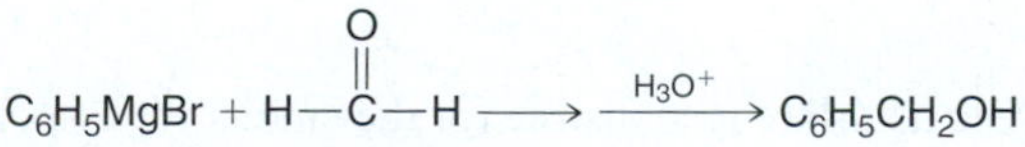

Given Chemical reaction

Asked for How products are formed

Strategy

Ⓐ Identify the functional group, and classify the reaction.
Ⓑ Use the mechanisms described to propose the initial steps in the reaction.

Solution

(a) Ⓐ One reactant is an alcohol that undergoes a substitution reaction. Ⓑ In the product, a bromide group is substituted for a hydroxyl group. The first step in this reaction must therefore be protonation of the —OH group of the alcohol by H^+ of HBr, followed by elimination of water to give the carbocation:

$$CH_3{-}CH_2{-}\underset{\displaystyle OH}{CH}{-}CH_2{-}CH_3 + H^+ \longrightarrow CH_3{-}CH_2{-}\underset{\displaystyle {}^+OH_2}{CH}{-}CH_2{-}CH_3 \xrightarrow{-H_2O} CH_3{-}CH_2{-}\overset{+}{C}H{-}CH_2{-}CH$$

The bromide ion is a good nucleophile that can react with the carbocation to give an alkyl bromide:

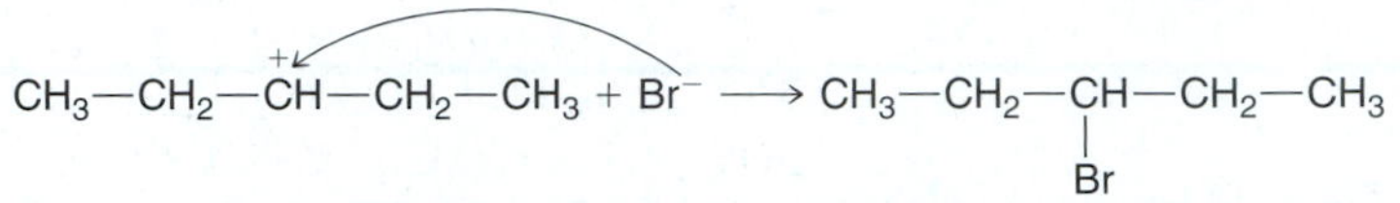

(b) Ⓐ One reactant is a Grignard reagent, and the other contains a carbonyl functional group. Carbonyl compounds act as electrophiles, undergoing nucleophilic attack at the carbonyl carbon. Ⓑ The nucleophile is the phenyl carbanion of the Grignard reagent:

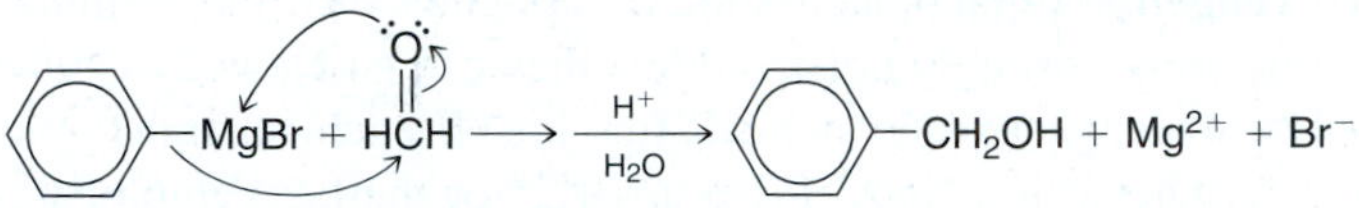

Benzyl alcohol

The product is benzyl alcohol.

EXERCISE 24.7

Predict the product of each reaction:

(a) 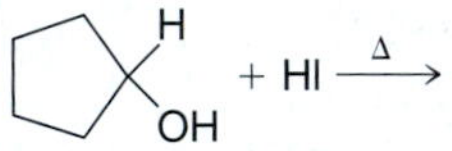

(b) CH_3MgCl + 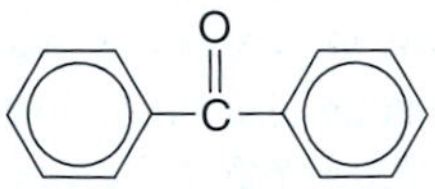

(benzophenone), followed by reaction with aqueous acid

Answer

(a) 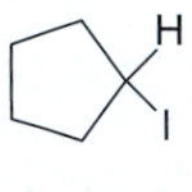

(b) 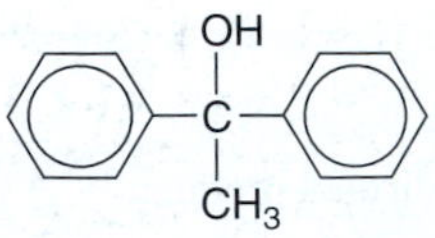

Carboxylic Acids

The pungent odors of many carboxylic acids are responsible for the smells we associate with sources as diverse as Swiss cheese, rancid butter, manure, goats, and sour milk. The boiling points of carboxylic acids tend to be somewhat higher than would be expected from their molecular masses because of strong hydrogen-bonding interactions between molecules (see structure in margin). In fact, most simple carboxylic acids form dimers in the liquid and even in the vapor phase (see Section 11.2). The four lightest carboxylic acids are completely miscible with water, but as the alkyl chain lengthens, they become more "alkane-like," and their solubility in water decreases.

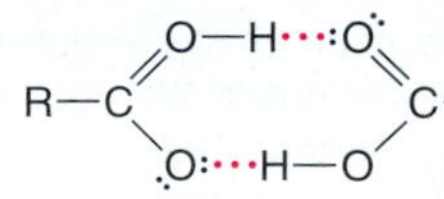

Hydrogen-bonded carboxylic acid molecules

Compounds that contain the carboxyl functional group are acidic, because carboxylic acids can easily lose a proton: the negative charge in the carboxylate ion (RCO_2^-) is stabilized by delocalization of the π electrons:

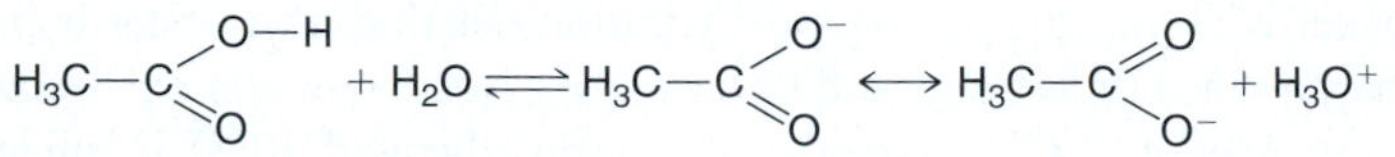

(24.20)

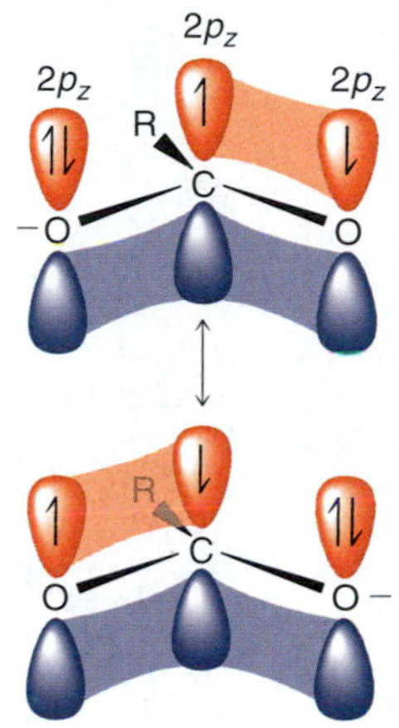

As a result, carboxylic acids are about 10^{10} times more acidic than the corresponding simple alcohols whose anions (RO^-) are not stabilized through resonance.

Carboxylic acids are typically prepared by oxidation of the corresponding alcohols and aldehydes (Figure 24.10a). They can also be prepared by the reaction of a Grignard reagent with CO_2, followed by acidification:

$$CO_2 + RMgCl \longrightarrow \xrightarrow{H_3O^+} RCO_2H + Mg^{2+} + Cl^- + H_2O \qquad (24.21)$$

The initial step in the reaction is nucleophilic attack by nucleophilic R^- group of the Grignard reagent on the electrophilic carbon of CO_2:

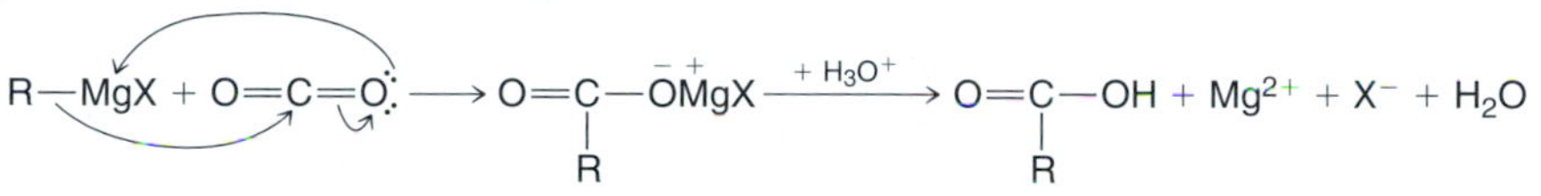

(24.22)

Delocalization of π bonding over *three* atoms (O—C—O) makes carboxylic acids and their derivatives less susceptible to nucleophilic attack than aldehydes and ketones with their single π bond. The reactions of carboxylic acids are dominated by two factors: their polar $—CO_2H$ group and their acidity. Reaction with strong bases, for example, produce carboxylate salts, such as sodium stearate (see Section 13.6):

$$RCO_2H + NaOH \longrightarrow RCO_2^-Na^+ + H_2O \qquad (24.23)$$

where $R = CH_3(CH_2)_{16}$. As you learned in Chapter 13, long-chain carboxylate salts are used as soaps.

Note the pattern

Delocalization of π bonding over three atoms makes carboxylic acids and their derivatives less susceptible to nucleophilic attack than aldehydes and ketones.

Carboxylic Acid Derivatives

Replacing the —OH of a carboxylic acid with groups that have different tendencies to participate in resonance with the C=O functional group produces derivatives with rather different properties. Resonance structures have significant effects on the reactivity of carboxylic acid derivatives, but their influence varies substantially, being least important for halides and most important for the nitrogen of amides. In this section, we take a brief look at the chemistry of two of the most familiar and important carboxylic acid derivatives: esters and amides.

Esters

Esters have the general formula RCO_2R', where R and R′ can be virtually any alkyl or aryl group. Esters are often prepared by the reaction of an alcohol (R′OH) with a carboxylic acid (RCO_2H) in the presence of a catalytic amount of strong acid (see Section 3.4). The purpose of the acid (an electrophile) is to protonate the doubly bonded oxygen atom of the carboxylic acid (a nucleophile) to give a species that is *more* electrophilic than the parent carboxylic acid, as shown in the margin.

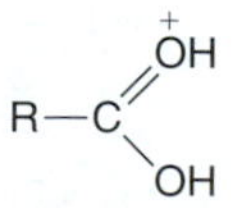

Protonated carboxylic acid

Amyl acetate (pentyl acetate) (banana scent)

Ethyl cinnamate (a pheromone)

The nucleophilic oxygen atom of the alcohol attacks the electrophilic carbon atom of the protonated carboxylic acid to form a new C—O bond. The overall reaction can be written as

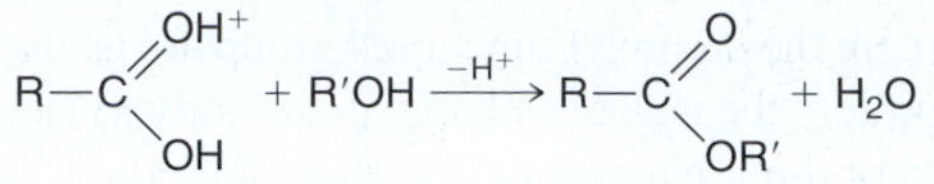
(24.24)

Because water is eliminated, this is a dehydration reaction. If an ester is heated with water in the presence of a strong acid or base, the reverse reaction will occur, producing the parent alcohol, R′OH, and either the carboxylic acid, RCO_2H (under strongly acidic conditions), or the carboxylate anion, RCO_2^- (under basic conditions).

As stated earlier, esters are familiar to most of us as fragrances, such as banana (shown in the margin) and pineapple. Other esters with intense aromas function as sex attractants, or *pheromones*, such as the pheromone from the oriental fruit fly (also shown in the margin). Research on the use of synthetic insect pheromones as a safer alternative to insecticides for controlling insect populations, such as cockroaches, is a rapidly growing field in organic chemistry.

Amides

In the general structure of an amide,

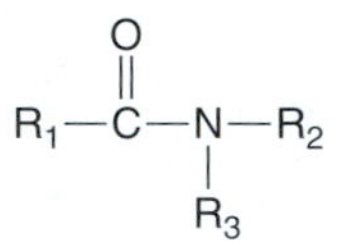

An amide

the two substituents on the amide nitrogen can be hydrogen atoms, alkyl groups, aryl groups, or any combination of those species. Although amides appear to be derived from an acid and an amine, in practice they usually cannot be prepared by this synthetic route. In principle, nucleophilic attack by the lone electron pair of the amine on the carbon of the carboxylic acid could occur, but because carboxylic acids are weak acids and amines are weak bases, an acid–base reaction generally occurs instead:

$$RCO_2H + R'NH_2 \longrightarrow RCO_2^- + R'NH_3^+ \qquad (24.25)$$

Amides are therefore usually prepared by the nucleophilic reaction of amines with other, more electrophilic carboxylic acid derivatives, such as esters.

Note the pattern

Amides are the least reactive of the carboxylic acid derivatives because amides participate in π bonding with the carbonyl group.

The lone pair of electrons on the nitrogen atom of an amide can participate in π bonding with the carbonyl group, thus reducing the reactivity of the amide (Figure 24.13) and inhibiting free rotation about the C—N bond. Amides are therefore the least reactive of the carboxylic acid derivatives. The stability of the amide bond is crucially important in biology because amide bonds form the backbones of peptides and proteins (see Section 12.8). The amide bond is also found in many other biologically active and commercially important molecules, including penicillin (see Section 3.2); urea, used as fertilizer; saccharin, a sugar substitute; and valium, a potent tranquilizer.

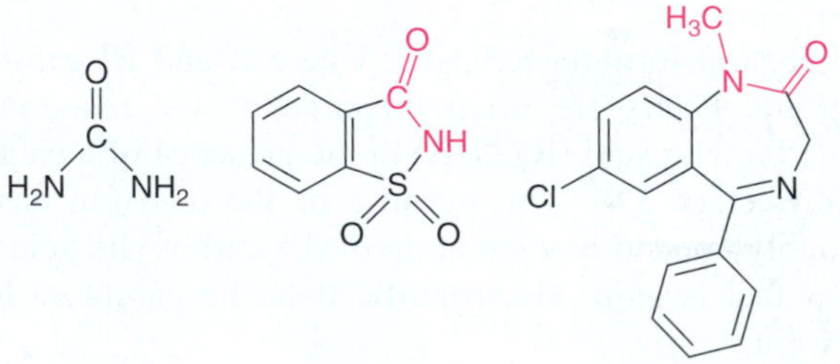

Urea **Saccharin** **Valium**

Amines

Amines are derivatives of ammonia in which one or more hydrogen atoms have been replaced by alkyl or aryl groups. They are therefore analogous to alcohols and ethers. Like alcohols, amines are classified as primary, secondary, or tertiary, but in this case the designation refers to the number of alkyl groups bonded to the *nitrogen* atom, *not* to the adjacent carbon atoms. In primary amines, the nitrogen is bonded to two hydrogen atoms and one alkyl group; in secondary amines, the nitrogen is bonded to one hydrogen and two alkyl groups; and in tertiary amines, the nitrogen is bonded to three alkyl groups. With one lone pair of electrons and C—N bonds that are less polar than C—O bonds, ammonia and simple amines have much lower boiling points than water or alcohols with similar molecular masses. Primary amines tend to have boiling points intermediate between those of the corresponding alcohol and alkane. Moreover, secondary and tertiary amines have lower boiling points than primary amines of comparable molecular mass.

Tertiary amines form cations analogous to the ammonium ion (NH_4^+), in which all four H atoms are replaced by alkyl groups. Notice that such substances, called **quaternary ammonium salts**, can be chiral if all four substituents are different. (Amines with three different substituents are also chiral because the lone pair of electrons represents a fourth substituent.)

Alkylamines can be prepared by nucleophilic substitution reactions of alkyl halides with ammonia or other amines:

$$RCl + NH_3 \longrightarrow RNH_2 + HCl \qquad (24.26a)$$

$$RCl + R'NH_2 \longrightarrow RR'NH + HCl \qquad (24.26b)$$

$$RCl + R'R''NH \longrightarrow RR'R''N + HCl \qquad (24.26c)$$

The primary amine formed in the first reaction (Equation 24.25a) can react with more alkyl halide to generate a secondary amine (Equation 24.25b), which in turn can react to form a tertiary amine (Equation 24.25c). Consequently, the actual reaction mixture contains primary, secondary, and tertiary amines, and even quaternary ammonium salts.

Reactions of amines are dominated by two properties: their ability to act as weak bases and their tendency to act as nucleophiles, both of which are due to the presence of the lone pair of electrons on the nitrogen atom. Amines typically behave as bases by accepting a proton from an acid to form an ammonium salt, as in the reaction of triethylamine with aqueous HCl:

$$Et_3N{:}(l) + HCl(aq) \longrightarrow Et_3NH^+Cl^-(aq) \qquad (24.27)$$

to give triethylammonium chloride. Amines can react with virtually any electrophile, including the carbonyl carbon of an aldehyde, ketone, or ester. Aryl amines such as aniline ($C_6H_5NH_2$) are much weaker bases than alkylamines because the lone pair of electrons on nitrogen interacts with the π bonds of the aromatic ring, delocalizing the lone pair through resonance (Figure 24.14).

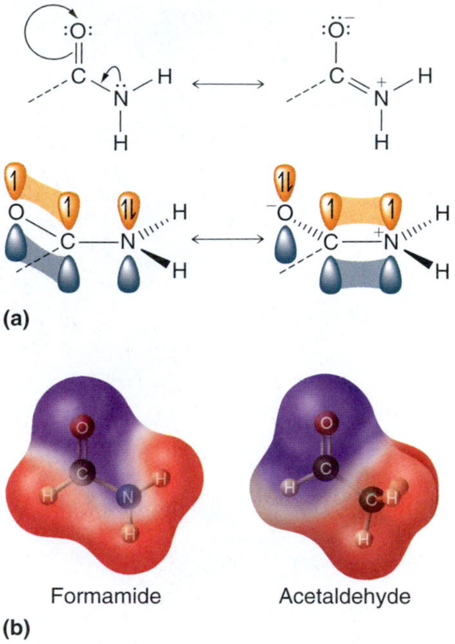

Figure 24.13 The electronic structure of an amide. (a) An unhybridized $2p_z$ orbital on nitrogen, containing a lone electron pair of electrons, can interact with the π orbital of the carbonyl group to give a three-center, four-electron bond. This interaction reduces the reactivity of the amide, making amides the least reactive of the carboxylic acid derivatives. (b) Comparison of the electrostatic potential maps of acetaldehyde and formamide shows that the negative charge (indicated in blue) is more localized on the oxygen atom of acetaldehyde than it is in formamide. Formamide is therefore less reactive.

Note the pattern

Reactions of amines are dominated by their ability to act as weak bases and their tendency to act as nucleophiles.

EXAMPLE 24.8

Predict the products formed in the reactions, and for **(b)** show the initial site of attack and the final products:

(a) $C_6H_5CH_2CO_2H + KOH \rightarrow$

(b)

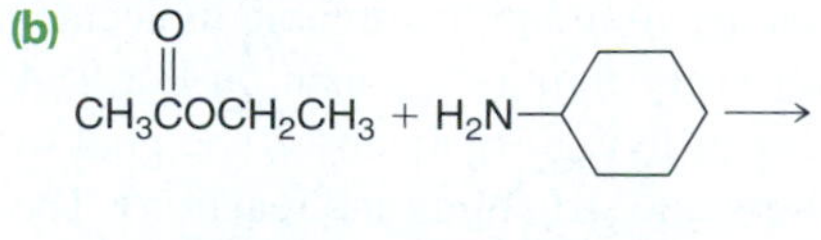

Ethyl acetate — Cyclohexylamine

Given Reactants

Asked for Products and mechanism of reaction

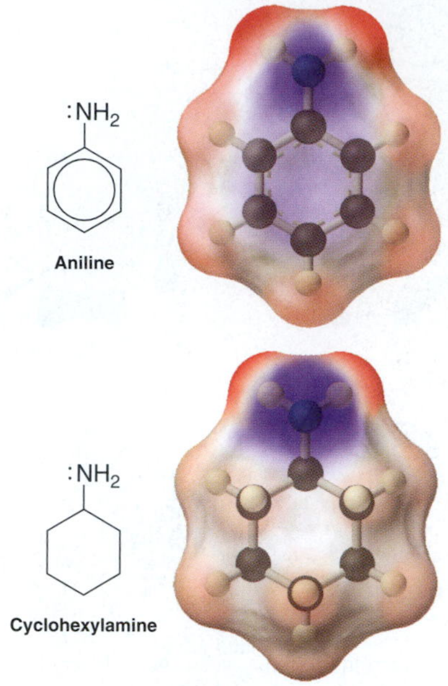

Figure 24.14 Structures and basicity of cyclohexylamine and aniline. Delocalization of the lone electron pair on N over the benzene ring reduces the basicity of aryl amines, such as aniline, compared with that of alkylamines, such as cyclohexylamine. These electrostatic potential maps show that the electron density on the N of cyclohexylamine is more localized than it is in aniline, which makes cyclohexylamine a stronger base.

Strategy

Use the strategy outlined in Example 24.7.

Solution

(a) The proton on the carboxylic acid functional group is acidic. Thus, the reaction of a carboxylic acid with a strong base is an acid–base reaction, whose products are a salt—in this case, $C_6H_5CH_2CO_2^-K^+$—and water.

(b) The nitrogen of cyclohexylamine contains a lone pair of electrons, making it an excellent nucleophile, whereas the carbonyl carbon of ethyl acetate is a good electrophile. We therefore expect a reaction in which nucleophilic attack on the carbonyl carbon of the ester produces an amide and ethanol. The initial site of attack and the reaction products are

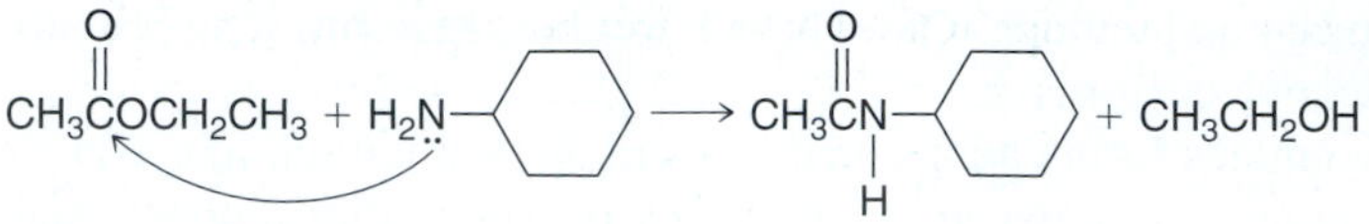

EXERCISE 24.8

Predict the products of each reaction: **(a)** acetic acid with 1-propanol, and **(b)** aniline ($C_6H_5NH_2$) with propyl acetate [$CH_3C({=}O)OCH_2CH_2CH_3$].

Answer

(a) 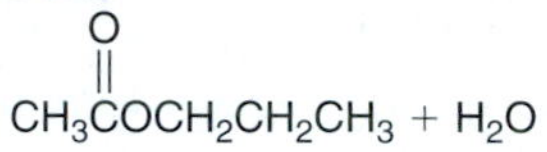

(b) 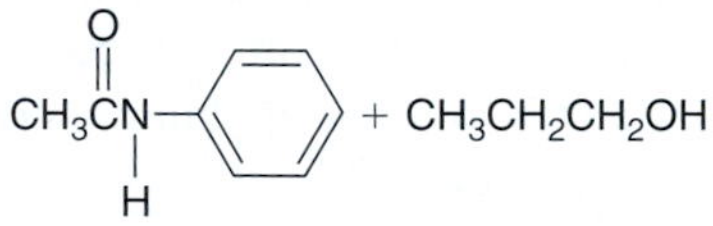

Reactions such as those we have discussed in this and the preceding section are used to synthesize a wide range of organic compounds. When chemists plan the synthesis of an organic molecule, however, they must take into consideration various factors, such as the availability and cost of reactants, the need to minimize the formation of undesired products, and the proper sequencing of reactions to maximize the yield of the target molecule and minimize the formation of undesired products. Because the synthesis of many organic molecules requires multiple steps, in designing a synthetic scheme for such molecules, chemists must often work backward from the desired product in a process called *retrosynthesis.* Using this process, they can identify the reaction steps needed to synthesize the desired product from the available reactants.

24.6 • The Molecules of Life

All the functional groups described in this chapter are found in the organic molecules that are constantly synthesized and destroyed by every living organism on Earth. A detailed understanding of the reactions that occur in living organisms is the goal of biochemistry, which deals with a wide variety of organic structures and reactions. The most abundant substances found in living systems belong to four major classes: proteins, carbohydrates, lipids, and nucleic acids. Here we briefly describe the structure and some functions of these biological molecules.

Proteins

In Section 12.8, we described **proteins** as biologically active polymers formed from amino acids linked together by amide bonds. In addition to an amine group and a carboxylic acid group, each amino acid contains a characteristic R group (see Figure 5.16). In the simplest amino acid, glycine, the R group is hydrogen (—H), but in other naturally occurring amino acids, the R group may be an alkyl or a substituted alkyl group, carboxylic group, or aryl group. The nature of the R group determines the particular chemical properties of each amino acid. Notice in Figure 5.16 that all the amino acids found in proteins except glycine are chiral compounds, which suggests that their interactions with other chiral compounds are selective. Some proteins, called enzymes, catalyze biological reactions, whereas many others have structural, contractile, or signaling functions. Because we have described proteins previously, we will not discuss them further.

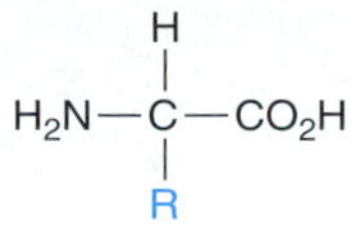

General structure of an amino acid, which is chiral except when R is an H atom

MGC The Molecules of Life

Carbohydrates

Carbohydrates are the most abundant of the organic compounds found in nature. They constitute a substantial portion of the food we consume to provide us with the energy needed to support life. Table sugar, milk, honey, and fruits all contain low-molecular-mass carbohydrates that are easily assimilated into the human body. In contrast, the walls of plant cells and wood contain high-molecular-mass carbohydrates that we cannot digest.

Once thought to be hydrates of carbon with the general formula $C_n(H_2O)_m$, **carbohydrates** are actually polyhydroxy aldehydes or polyhydroxy ketones (that is, aldehydes or ketones with several —OH groups attached to the parent hydrocarbon). The simplest carbohydrates consist of unbranched chains of three to eight carbon atoms: one carbon atom is part of a carbonyl group, and some or all of the others are bonded to hydroxyl groups. The structure of a carbohydrate can be drawn either as a hydrocarbon chain, using a *Fischer projection*, or as a ring, using a *Haworth projection* (Figure 24.15). The Haworth projection is named after the British chemist Sir Walter Norman Haworth, who was awarded a Nobel Prize in 1937 for his discovery that sugars exist mainly in their cyclic forms (and for his collaboration on the synthesis of vitamin C). The cyclic form is the product of nucleophilic attack by the oxygen of a hydroxyl group on the electrophilic carbon of the carbonyl group within the same molecule, producing a stable ring structure composed of five or six carbons that minimizes bond strain (Figure 24.15). Notice that the substituents on the right side of the carbon chain in a Fischer projection are in the "down" position in the corresponding Haworth projection. Also notice that attack by the hydroxyl group on either side of the carbonyl group leads to the formation of two cyclic forms, called *anomers:* an α form, with the —OH in the "up" position, and a β form, with the —OH in the "down" position.

At age 14, Walter Norman Haworth left school to join his father to learn linoleum design and manufacturing, but he became interested in chemistry through his use of dyes. Private tutoring enabled him to pass the entrance exam of the University of Manchester, where he received his doctorate in 1911. During World War I, Haworth organized the laboratories at St. Andrews for the production of chemicals and drugs, returning to the investigation of carbohydrates after the war.

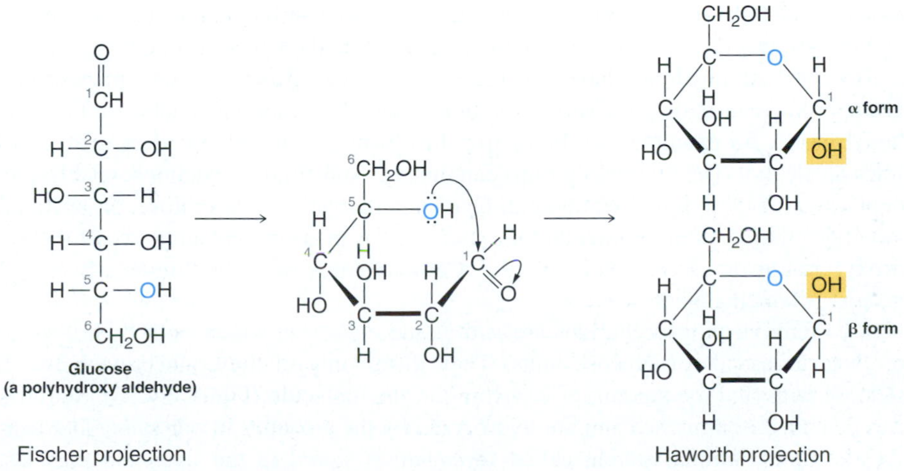

Figure 24.15 Fischer projection and Haworth projection of glucose. In solution, simple sugars exist predominantly in the ring form, the product of nucleophilic attack by the oxygen of a hydroxyl group on the electrophilic carbon of the carbonyl group. The α and β forms, called anomers, differ in the configuration at C1.

Carbohydrates are classified according to the number of single *saccharide,* or sugar, units they contain (the term is from the Latin *saccharum*, "sugar"). The simplest are *monosaccharides*; a *disaccharide* consists of two linked monosaccharide units; a *trisaccharide* contains three linked monosaccharide units; and so forth. Glucose is a monosaccharide, and sucrose (common table sugar) is a disaccharide. Hydrolysis of sucrose produces glucose and another monosaccharide, fructose, in a reaction catalyzed by an enzyme (shown here) or by acid:

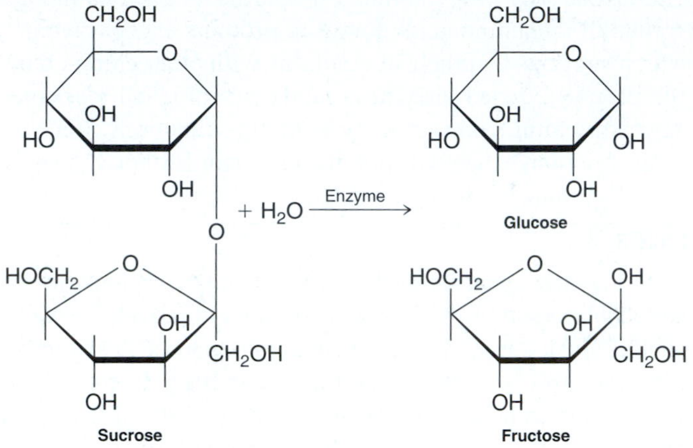

(24.28)

Polysaccharides hydrolyze to produce more than 10 monosaccharide units.

The common monosaccharides contain several chiral carbons and exist in several isomeric forms. One isomer of glucose, for example, is galactose, which differs from glucose in the position of the —OH bond at carbon-4:

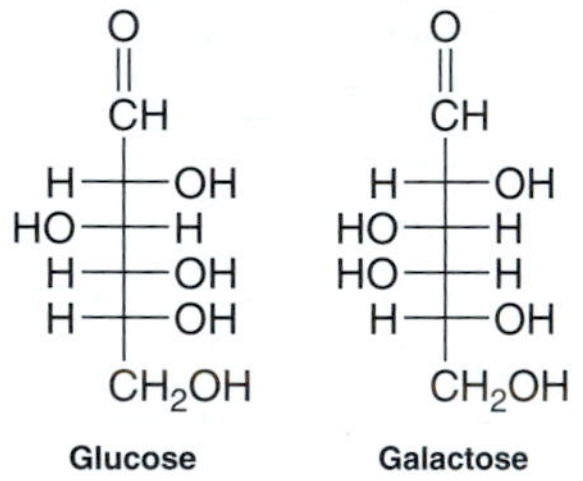

Because carbons -2, -3, -4, and -5 of glucose are chiral, changing the position of the —OH on carbon-4 does not produce an enantiomer of glucose, but rather a different compound, galactose, with distinct physical and chemical properties. Galactose is a hydrolysis product of lactose, a disaccharide found in milk. People who suffer from *galactosemia* lack the enzyme needed to convert galactose to glucose, which is then metabolized to CO_2 and H_2O, releasing energy. Galactose accumulates in their blood and tissues, leading to mental retardation, cataracts, and cirrhosis of the liver.

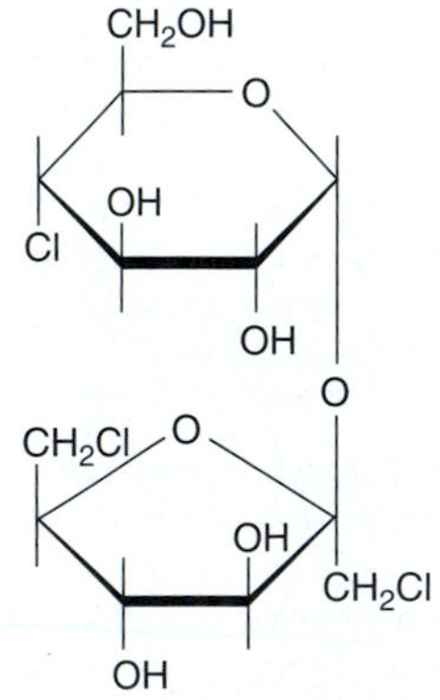

Sucralose, an artificial sweetner

Note the pattern

Because carbohydrates have a carbonyl functional group and several hydroxyl groups, they can undergo a variety of reactions.

Because carbohydrates have a carbonyl functional group and several hydroxyl groups, they can undergo a variety of biochemically important reactions. The carbonyl group, for example, can be oxidized to form a carboxylic acid or reduced to form an alcohol. The hydroxyl groups can undergo substitution reactions, resulting in derivatives of the original compound. One such derivative is Sucralose, an artificial sweetener that is six times sweeter than sucrose; it is made by replacing two of the hydroxyl groups on sucrose with chlorine. Carbohydrates can also eliminate hydroxyl groups, producing alkenes.

Two familiar polysaccharides are *starch* and *cellulose*, which both hydrolyze to produce thousands of glucose units. They differ only in the connection between glucose units and the amount of branching in the molecule (Figure 24.16). Starches can be coiled or branched and are hydrolyzed by the enzymes in our saliva and pancreatic juices. Animal starch, called *glycogen*, is stored in the liver and muscles.

It consists of branched glucose units linked by bonds that produce a coiled structure. The glucose units in cellulose, in contrast, are linked to give long, unbranched chains. The chains in cellulose stack in parallel rows held together by hydrogen bonds between hydroxyl groups. This arrangement produces a rigid structure that is insoluble in water.

Cellulose is the primary structural material of plants and one of the most abundant organic substances on Earth. Because our enzymes are not able to hydrolyze the bonds between the glucose units in cellulose, we are unable to digest it. A recently marketed product containing a high percentage of cellulose was sold as a dietetic substance for rapid weight loss, but those who consumed it experienced severe intestinal discomfort because the cellulose could not be digested. The product was quickly removed from the market.

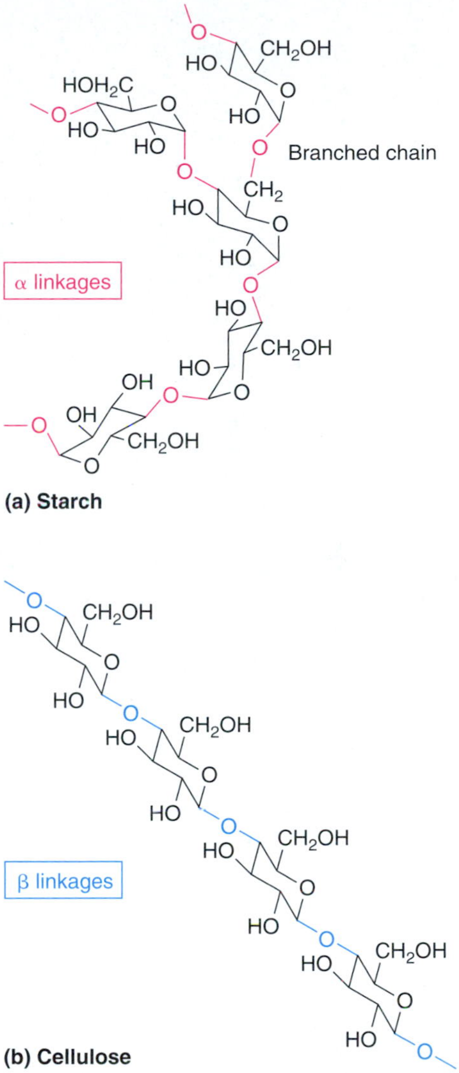

Figure 24.16 The polysaccharides starch and cellulose. Starches and cellulose differ in the connection between glucose units and the amount of branching in the molecule. Starches can be coiled or branched, whereas cellulose, the primary structural material of plants, has long, unbranched chains held together by hydrogen bonds.

EXAMPLE 24.9

The Fischer projection of xylose, found in many varieties of apples, is shown. Draw the ring form (Haworth projection) of xylose.

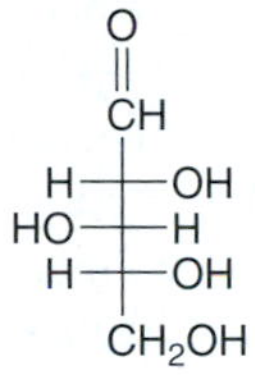

Given Fischer projection of a sugar

Asked for Cyclic structure

Strategy

A Identify the nucleophile and the electrophile. Indicate the point of attack, remembering that cyclic structures are most stable when they contain at least five atoms in the ring to prevent bond strain from bond angles that are too small.

B Draw the cyclic form of the structure.

Solution

A The carbonyl carbon (C1) is a good electrophile, and each oxygen is a good nucleophile. Nucleophilic attack occurs from the —OH group on C4, producing a stable five-membered ring. **B** Because of rotation about the bond between C1 and C2, ring formation gives both α and β anomers, with the following structures (H atoms have been omitted for clarity):

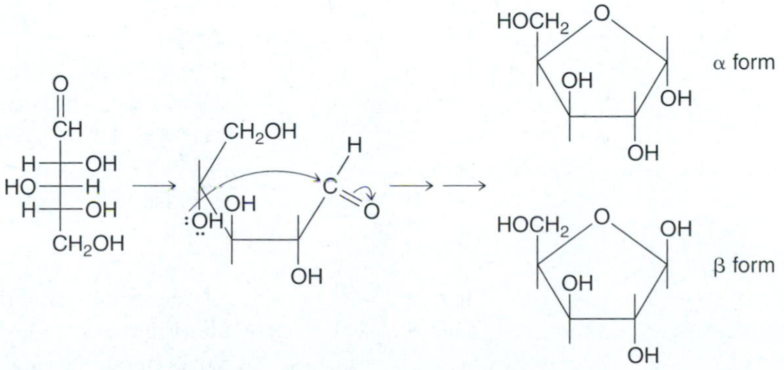

EXERCISE 24.9

Draw the cyclic form(s) of galactose, whose Fischer projection is shown on page 1122.

Answer

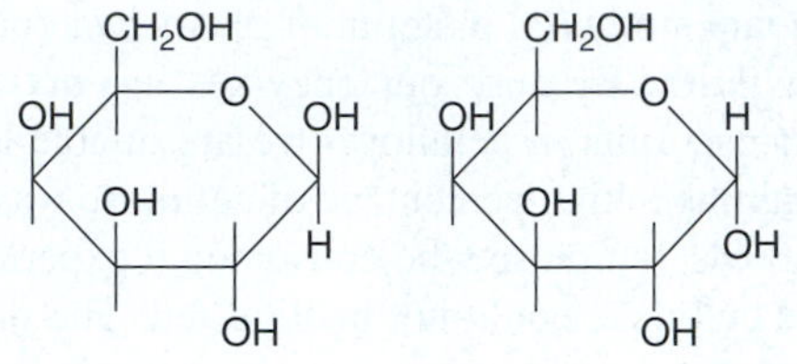

Lipids

Lipids (from the Greek *lipos*, "fat" or "lard") are characterized by their insolubility in water. They form a family of compounds that includes fats, waxes, some vitamins, and steroids. *Fatty acids*, the simplest lipids, have a long hydrocarbon chain that ends with a carboxylic acid functional group. In saturated fatty acids, the hydrocarbon chains contain only C—C single bonds, so they can stack in a regular array (Figure 24.17a). In contrast, unsaturated fatty acids have a single double bond in the hydrocarbon chain (monounsaturated) or more than one double bond (polyunsaturated). These double bonds give fatty acid chains a kinked structure, which prevents the molecules from packing tightly (Figure 24.17b). As a result of reduced van der Waals interactions (see Section 11.2), the melting point of an unsaturated fatty acid is lower than that of a saturated fatty acid of comparable molecule mass, thus producing an oil rather than a solid. Fish oils and vegetable oils, for example, have a higher concentration of unsaturated fatty acids than does butter.

The double bonds of unsaturated fatty acids can by hydrogenated in an addition reaction that produces a saturated fatty acid (see Section 14.8):

$$\underset{\text{Unsaturated fatty acid}}{\text{—CH=CH—}} + H_2 \xrightarrow{\text{Ni}} \underset{\text{Saturated fatty acid}}{\text{—}CH_2CH_2\text{—}} \qquad (24.29)$$

They can also be oxidized to produce an aldehyde or carboxylic acid.

Unsaturated fatty acids are the starting compounds for the biosynthesis of *prostaglandins*. These hormone-like substances are involved in the regulation of blood pressure, inflammation of tissues, and contraction and relaxation of smooth muscles. Drugs such as aspirin and ibuprofen inhibit the production of prostaglandins, thereby reducing inflammation.

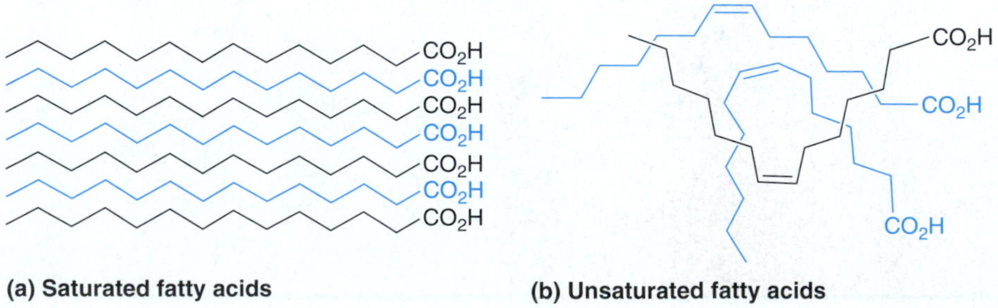

Figure 24.17 Fatty acids, the simplest class of lipids. Fatty acids are composed of a long chain that terminates in a carboxylic acid functional group. (a) Molecules of saturated fatty acids, which contain no carbon–carbon double bonds, can stack in a regular array. (b) Molecules of unsaturated fatty acids, which contain one or more cis carbon–carbon double bonds, have kinked structures that cannot pack closely.

Waxes are esters produced by nucleophilic attack of an alcohol on the carbonyl carbon of a long-chain carboxylic acid (Equation 24.24). For example, the wax used in shoe polish and wax paper, which is derived from beeswax, is formed from a straight-chain alcohol with 15 carbon atoms and a fatty acid with 31 carbon atoms. *Triacylglycerols* are a particularly important type of ester in living systems; they are used by the body to store fats and oils. These compounds are formed from one molecule of glycerol (1,2,3-trihydroxypropane) and three fatty acid molecules. During warmer months of the year, animals that hibernate consume large quantities of plants, seeds, and nuts that have a high fat and oil content. They convert the fatty acids to triacylglycerols and store them. Hydrolysis of stored triacylglycerols during hibernation (the reverse of Equation 24.24) releases alcohols and carboxylic acids that the animal uses to generate energy for maintaining cellular activity, respiration, and heart rate. Derivatives of triacylglycerols that contain a phosphate group are major components of all cell membranes.

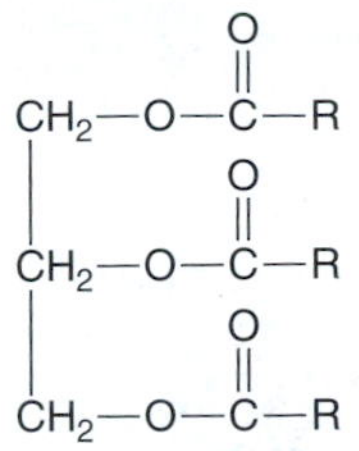

General structure of a triacylglycerol

Steroids are lipids whose structure is made up of three cyclohexane rings and one cyclopentane ring fused together. The presence of various substituents, including double bonds, on the basic steroid ring structure produces a large family of steroid compounds with different biological activities. For example, *cholesterol*, a steroid found in cellular membranes, contains a double bond in one ring and four substituents: a hydroxyl group, two methyl groups, and a hydrocarbon chain (see structure below). Cholesterol is the starting point for the biosynthesis of steroid hormones, including testosterone, the primary male sex hormone, and progesterone, which helps maintain pregnancy. These cholesterol derivatives lack the long hydrocarbon side chain, and most contain one or more ketone groups.

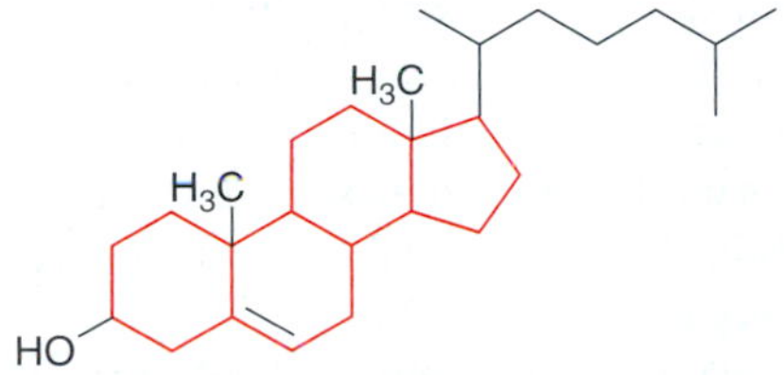

Cholesterol

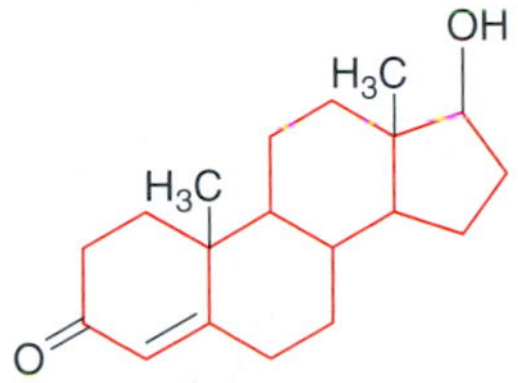

Testosterone
The structure in red is derived from cholesterol.

Cholesterol is synthesized in the human body in a multistep pathway that begins with a derivative of acetic acid. We also consume cholesterol in our diets: eggs, meats, fish, and diary products all contain cholesterol, but vegetables and other plant-derived foods do not. Excess cholesterol in the human body can cause gallstones, which are composed of nearly 100% cholesterol, or lipid deposits called plaque in arteries. A buildup of plaque can block a coronary artery and result in a heart attack (Figure 24.18).

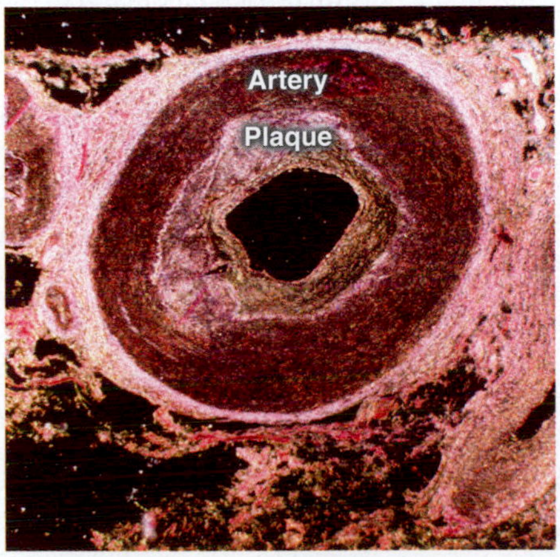

Figure 24.18 Plaque in an artery. Plaque, a lipid deposit, forms from excess cholesterol in the body. This artery is nearly blocked by a thick deposit of plaque, which greatly increases the risk of a heart attack due to reduced blood flow to the heart.

Nucleic Acids

Nucleic acids are the basic structural components of DNA and RNA, the biochemical substances found in the nuclei of all cells that transmit the information needed to direct cellular growth and reproduction. Their structures are derived from cyclic nitrogen-containing compounds called *pyrimidines* and *purines*, which can hydrogen bond through the lone electron pair on nitrogen (in pyrimidine and purine) or through the hydrogen of the amine (in purine):

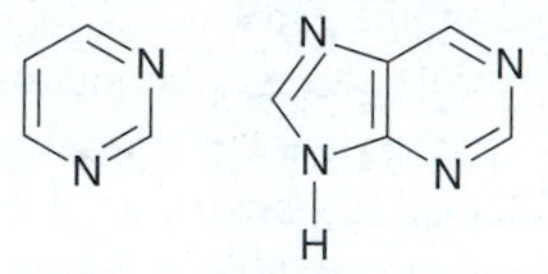

Pyrimidine Purine

The same cyclic structures are found in substances such as caffeine (see Section 3.2), a purine that is a stimulant, and the antifungal agent flucytosine, a primidine.

When a pyrimidine or purine is linked to a sugar by a bond called a *glycosidic bond*, a *nucleoside* is formed. Addition of a phosphoric acid group to the sugar then produces a *nucleotide* (Figure 24.19a). Linkage of nucleotides forms a polymeric chain that consists of alternating sugar and phosphate groups, which is the backbone of DNA and RNA (Figure 24.19b).

While the function of DNA is to preserve genetic information, RNA translates the genetic information in DNA and carries that information to cellular sites where proteins are synthesized. Many antibiotics function by interfering with the synthesis of proteins in one or more kinds of bacteria. Chloramphenicol, for example, is used against infections of the eye or outer ear canal; it inhibits the formation of peptide bonds between amino acids in a protein chain. Puromycin, used against herpes simplex type I, interrupts extension of a peptide chain, causing the release of an incomplete protein and subsequent death of the organism.

Mutations in the DNA of an organism may lead to the synthesis of defective proteins. *Phenylketonuria (PKU)*, for example, is a condition caused by a defective enzyme. Left untreated, it produces severe brain damage and mental retardation. *Albinism* is caused by a defective enzyme that is unable to produce melanin, the pigment responsible for the color of skin and hair. *Cystic fibrosis*, the most common inherited disease in the United States, blocks pancreatic function and causes thick mucus secretions that make breathing difficult. An area of intense research in combating cancer involves the synthesis of drugs that stop uncontrolled cell growth by interfering with DNA replication.

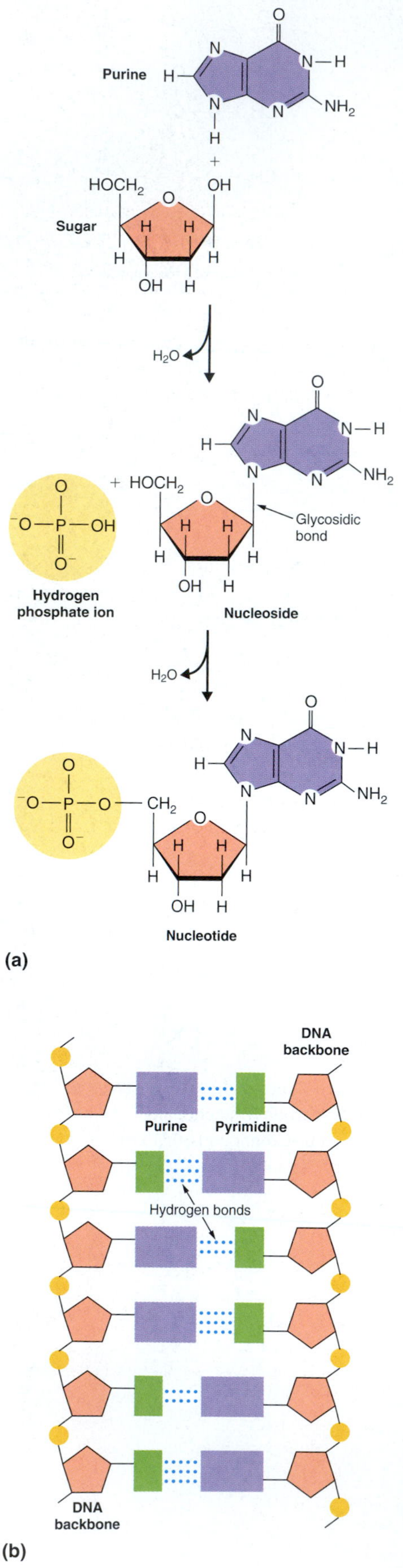

Figure 24.19 Formation of nucleic acids. (a) When pyrimidine or purine and a sugar react to form a glycosidic bond, a nucleoside is produced. Addition of a phosphoric acid group to the sugar of a nucleoside produces a nucleotide. (b) Nucleotides link together to form long polymeric chains. A DNA molecule consists of two such chains held together by hydrogen bonding between the purine and pyrimidine components on different chains.

SUMMARY AND KEY TERMS

24.1 Functional Groups and Classes of Organic Compounds (p. 1092)

Functional groups are structural units that determine the chemical reactivity of a molecule under a given set of conditions. Organic compounds are classified into several major categories based on the functional groups they contain. In the systematic names of organic compounds, numbers indicate the positions of functional groups in the basic hydrocarbon framework. Many organic compounds also have common names, which use the prefix *form–* for a compound that contains no carbons other than those in the functional group, and *acet–* for those that have one additional carbon atom.

24.2 Isomers of Organic Compounds (p. 1094)

Isomers are different compounds that have the same molecular formula. For an organic compound, rotation about a σ bond can produce different three-dimensional structures called **conformational isomers**. In a *Newman projection*, which represents the view along a C—C bond axis, the *eclipsed conformation* has the C—H bonds on adjacent carbon atoms parallel to each other and in the same plane, representing one conformational extreme. In the *staggered conformation*, the opposite extreme, the hydrogen atoms are as far from one another as possible. Electrostatic repulsions are minimized in the staggered conformation. **Structural isomers** differ in the connectivity of the atoms. Structures that have the same connectivity but whose components differ in their orientations in space are called **stereoisomers**. Stereoisomers can be *geometric isomers*, which differ in the placement of substituents in a rigid molecule, or *optical isomers*, nonsuperimposable mirror images. Molecules that are nonsuperimposable mirror images are *chiral*. A molecule and its nonsuperimposable mirror image are called *enantiomers*. These differ in their interaction with *plane-polarized light*, light that oscillates in only one direction. A compound is *optically active* if its solution rotates plane-polarized light in only one direction, and *optically inactive* if its rotations cancel to produce no net rotation. A clockwise rotation is called *dextrorotatory* and is indicated in the compound's name by (+), whereas a counterclockwise rotation is called *levorotatory*, designated by (−). The **specific rotation** is the amount (in degrees) by which the plane of polarized light is rotated when the light is passed through a solution containing 1.0 g of solute per 1.0 mL of solvent in a tube 10.0 cm long. A solution that contains equal concentrations of each enantiomer in a pair is a *racemic mixture*; such solutions are optically inactive.

24.3 Reactivity of Organic Molecules (p. 1103)

The reactivity of a molecule is often affected by the degree of substitution of the carbon bonded to a functional group; the carbon is designated as *primary, secondary*, or *tertiary*. Identifying the transient species formed in a chemical reaction, some of which are charged, enables chemists to predict the mechanism and products of the reaction. One common transient species is a **carbocation**, a carbon with six valence electrons that is an **electrophile**; that is, it needs electrons to complete its octet. A **radical** is a transient species that is neutral but electron deficient and thus acts as an electrophile. In contrast, a **carbanion** contains eight valence electrons and is negatively charged. It is an electron-rich species that is a **nucleophile** because it can share a pair of electrons. In chemical reactions, electrophiles react with nucleophiles.

24.4 Common Classes of Organic Reactions (p. 1106)

There are common patterns to how organic reactions occur. In a **substitution reaction**, one atom or group of atoms in a substance is replaced by another atom or group of atoms from another substance. Bulky groups that prevent attack cause the reaction to be *sterically hindered*. In an **elimination reaction**, adjacent atoms are removed with subsequent formation of a multiple bond and a small molecule. An **addition reaction** is the reverse of an elimination reaction. *Radical reactions* are not very selective and occur in three stages: initiation, propagation, and termination. Oxidation–reduction reactions in organic chemistry are identified by the change in the number of oxygens in the hydrocarbon skeleton or the number of bonds between carbon and oxygen or carbon and nitrogen.

24.5 Common Classes of Organic Compounds (p. 1110)

There are strong connections among the structure, physical properties, and reactivity for the compounds that contain the major functional groups. Hydrocarbons that are alkanes undergo *catalytic cracking*, which can convert straight-chain alkanes to highly branched alkanes. Catalytic cracking is one example of a **pyrolysis reaction**, in which the weakest bond is cleaved at high temperature, producing a mixture of radicals. The multiple bond of an alkene produces geometric isomers (*cis* and *trans*). *Terminal alkynes* contain a hydrogen atom directly attached to a triply bonded carbon. Removal of the hydrogen forms an *acetylide ion*, a potent nucleophile used to make longer carbon chains. Arenes undergo substitution rather than elimination because of enhanced stability from delocalization of their π electron density. An alcohol is prepared by the addition of the elements of water across a double bond or by a substitution reaction. Alcohols undergo two major types of reactions: those in which the O—H bond is cleaved and those involving cleavage of the C—O bond. Phenols are acidic because of π interactions between the oxygen atom and the ring. Ethers are comparatively unreactive. Aldehydes and ketones are generally prepared from the oxidation of alcohols. Their chemistry is characterized by nucleophilic attack at the carbon atom of the carbonyl functional group and electrophilic attack at the oxygen atom. **Grignard reagents** (RMgX, where X = Cl, Br, I) convert the carbonyl functional group to an alcohol and lengthen the carbon chain. Compounds that contain the carboxyl functional group are weakly acidic because of delocalization of the π electrons, which causes them to easily lose a proton and form the carboxylate anion. Carboxylic acids are generally prepared from the oxidation of alcohols and aldehydes or through the reaction of a Grignard reagent with CO_2. Carboxylic acid derivatives include esters, prepared by the reaction of a carboxylic acid and an alcohol, and amides, prepared by the nucleophilic reaction of amines with other, more electrophilic carboxylic acid derivatives, such as esters. Amides are relatively unreactive because of π bonding interactions between the lone pair on nitrogen and the carbonyl group. Amines can also be primary, secondary, or tertiary, depending on the number of alkyl groups bonded to the amine. **Quaternary ammonium salts** have four substituents attached to nitrogen and can be chiral. Amines are often prepared by a nucleophilic substitution reaction between a polar alkyl halide and ammonia or other amines. They are nucleophiles, but their base strength depends on their substituents.

24.6 The Molecules of Life (p. 1120)

Proteins are biologically active polymers formed from amino acids linked together by amide bonds. All the amino acids in proteins are chiral compounds except glycine. The most common organic compounds found in nature are the **carbohydrates**, polyhydroxy aldehydes or polyhydroxy ketones in unbranched chains of three to eight carbons. They are classified according to their number of sugar, or *saccharide*, units, and they can be drawn as a chain in a *Fischer projection* or in a cyclic form called a *Haworth projection*. The two cyclic forms in a Haworth projection are called *anomers*. Many sugars contain at least one chiral center. With their carbonyl and hydroxyl functional groups, carbohydrates can undergo a variety of biochemically relevant reactions. *Starch* and *cellulose* differ only in the connectivity between glucose units. Starches can be branched or unbranched, but cellulose, the structural material of plants, is unbranched, and cannot be digested by humans. **Lipids** are insoluble in water. The simplest lipids, *fatty acids*, have a long hydrocarbon chain ending in a carboxylic acid functional group. Their physical properties depend on the number of double bonds in the chain. *Prostoglandins*, hormone-like substances, are formed from unsaturated fatty acids, and *waxes* are long-chain esters of saturated fatty acids. *Triacylglycerols*, which the body uses to store fats and oils, consist of glycerol esterified to three fatty acid molecules. *Steroids*, which include cholesterol and the steroid hormones, are characterized by three cyclohexane rings and one cyclopentane ring fused together. The basic structural units of DNA and RNA are the **nucleic acids**, whose structures are derived from nitrogen-containing cyclic compounds called *pyrimidines* and *purines*. These structures are linked to a sugar through a *glycosidic bond*, forming a *nucleoside*. Addition of a phosphoric acid group produces a *nucleotide*. Nucleotides link to form a polymeric chain that is the backbone of DNA and RNA.

QUESTIONS AND PROBLEMS

MGC For instructor-assigned homework, go to **www.masteringgenchem.com**

Questions and Problems with colored numbers have answers in the Appendix and complete solutions in the Student Solutions Manual.

CONCEPTUAL

24.2 Isomers of Organic Compounds

1. What hybrid orbitals are used to form C—C single bonds in saturated hydrocarbons? Describe the bond.
2. How are conformational isomers related? Sketch two conformational isomers of propane, looking along the C1—C2 bond axis.
3. Why do alkanes with more than two carbons have a kinked structure? Explain why a kinked structure is so stable.
4. Are *n*-pentane and 3-methylbutane conformational isomers or structural isomers? How would you separate these compounds from a mixture of the two?
5. How are structural isomers different from stereoisomers? Do stereoisomers have free rotation about all carbon–carbon bonds? Explain your answers.
6. Which of these objects is chiral? (a) a shoe; (b) a laced football; (c) an automobile; (d) a fork
7. Which of these objects is chiral? (a) a rollerblade; (b) an unmarked baseball bat; (c) a bicycle; (d) your arms; (e) a spoon
8. Are all stereoisomers also enantiomers? Are all enantiomers stereoisomers? Explain your answers.

24.3 Reactivity of Organic Molecules

9. Order these cations by increasing stability: CH_2F^+, $CHCl_2^+$, CH_3^+, CHF_2^+. Explain your reasoning.
10. Place the cations in order of decreasing stability: $CH_3CH_2^+$, $CHBr_2^+$, CH_3^+, $CHBrCl^+$. Explain your reasoning.
11. For each pair, identify the electrophile and the nucleophile: (a) CH_3^-; Li^+; (b) formaldehyde, CH_3ONa; (c) H^+, propene; (d) benzene, Cl^-.
12. Identify the electrophile and the nucleophile in each pair: (a) CH_3^+, Br^-; (b) HC≡CNa, pentanal; (c) acetone, CN^-; (d) $(CH_3)_2S$, CH_3I.

24.4 Common Classes of Organic Reactions

13. Identify the nucleophile and the electrophile in the nucleophilic substitution reaction of 2-bromobutane with KCN.
14. Identify the nucleophile and the electrophile in the nucleophilic substitution reaction of 1-chloro-4-methyl butane with sodium methoxide.
15. Do you expect an elimination reaction to be favored by a strong or a weak nucleophile? Why?
16. Why do molecules with π bonds behave as nucleophiles when mixed with strong electrophiles?

24.5 Common Classes of Organic Compounds

17. Why do branched-chain alkanes have lower melting points than straight-chain alkanes of comparable molecular mass?
18. Describe alkanes in terms of their orbital hybridization, polarity, and reactivity. What is the geometry about each carbon of a straight-chain alkane?
19. Why do alkenes form *cis* and *trans* isomers whereas alkanes do not? Do alkynes form *cis* and *trans* isomers? Why or why not?
20. Which of these componds can exist as *cis* and *trans* isomers? (a) 2,3-dimethyl-1-butene; (b) 3-methyl-1-butene; (c) 2-methyl-2-pentene; (d) 2-pentene
21. Which of these compounds can exist as *cis* and *trans* isomers? (a) 3-ethyl-3-hexene; (b) 1,1-dichloro-1-propene; (c) 1-chloro-2-pentene; (d) 3-octene
22. Identify which of these compounds have a net dipole moment: (a) *o*-nitrotoluene; (b) *p*-bromonitrobenzene; (c) *p*-dibromobenzene.
23. Why is the boiling point of alcohols so much higher than that of alkanes of comparable molecular mass? Why are low-molecular-mass alcohols reasonably good solvents for some ionic compounds whereas alkanes are not?
24. Is an alcohol a nucleophile or an electrophile? What determines the mode of reactivity of an alcohol? How does the reactivity of an alcohol differ from that of an ionic compound containing OH, such as KOH?

25. How does the reactivity of ethers compare with that of alcohols? Why? Ethers can be cleaved under strongly acidic conditions. Explain how this can occur.

26. What functional group is common to aldehydes, ketones, carboxylic acids, and esters? This functional group can react with both nucleophiles and electrophiles. Where does nucleophilic attack on this functional group occur? Where does electrophilic attack occur?

27. What key feature of a Grignard reagent allows it to engage in a nucleophilic attack on a carbonyl carbon?

28. Do you expect carboxylic acids to be more or less water soluble than ketones of comparable molecular mass? Why?

29. Since amides are formally derived from an acid plus an amine, why can't they actually be prepared from the reaction of an acid with an amine? How are they generally prepared?

30. Is an amide susceptible to nucleophilic attack, electrophilic attack, or both? Specify where attack occurs.

31. What factors determine the reactivity of amines?

STRUCTURE AND REACTIVITY

This section includes paired problems (marked by brackets) that require similar problem-solving skills.

24.2 Isomers of Organic Compounds

32. Single bonds between carbon atoms are free to rotate 360°.
(a) Explain what happens to the potential energy of an *n*-hexane molecule as rotation occurs around the C2–C3 bond.
(b) Draw Newman projections of the *n*-hexane conformations corresponding to the energy minima and maxima in the diagram, which shows potential energy versus degrees of rotation about the C3–C4 axis.

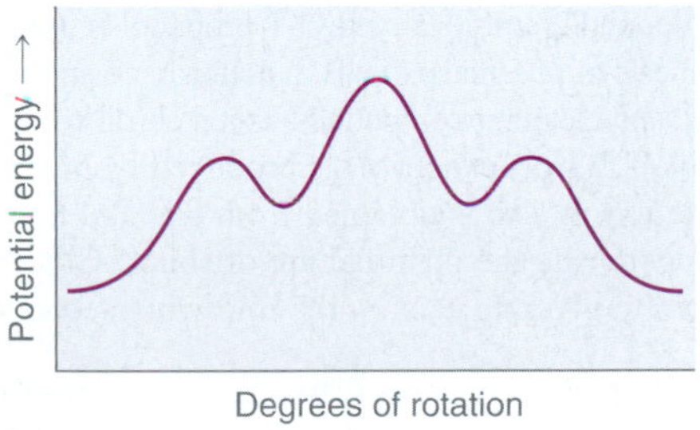

33. Draw all the possible structural isomers of (a) $C_5H_{11}Br$ and (b) $C_5H_{10}Cl_2$.

34. Sketch all the structural isomers of (a) $C_6H_{13}Br$ and (b) $C_6H_{12}Cl_2$.

35. Sketch all structural isomers of (a) dichlorohexene and (b) bromochlorohexene. Identify the *cis*-dichlorohexenes and *trans*-dichlorohexenes.

36. Indicate which of these molecules are chiral: (a) 1,1-dichloropropane; (b) 1,2-dichloropropane; (c) 1,3-dichloropropane; (d) 1,1,2-trichloropropane. On the structural formulas of the chiral molecules, identify any chiral centers with an asterisk.

37. Indicate which molecules are chiral: (a) 2,2-dimethylpentane; (b) 2,3-dimethylpentane; (c) 2,4-dimethylpentane; (d) 1-chloro-2-methylpentane. On the structural formulas of the chiral molecules, identify any chiral centers with an asterisk.

38. Draw the structures of the enantiomers of (a) $CH_2CHCHClBr$; (b) 4-chlor-2-hexene; and (c) methylethylpropylamine.

39. Draw the structures of the enantiomers of (a) 4-methyl-2-hexene; (b) methylethylhexylamine; and (c) $CH_3CH_2CH(CH_3)Cl$.

40. Draw the structures of the enantiomers of these compounds:

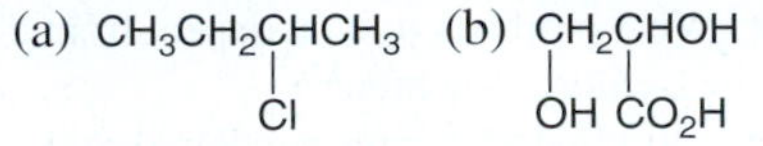

41. Draw the structures of the enantiomers of these compounds:

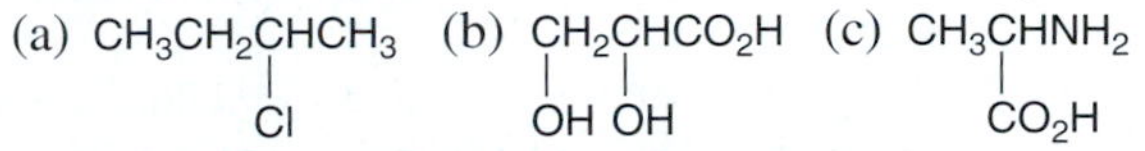

24.3 Reactivity of Organic Molecules

42. Draw Lewis dot structures of the products of carbon–hydrogen cleavage reactions. What is the charge on each of these species?

43. Identify the electrophile and the nucleophile in each reaction; then complete the equations:
(a) $CH_3^+ + Cl\cdot \longrightarrow$
(b) $CH_3CH{=}CH_2 + HBr \longrightarrow$
(c) $(CH_3)_3N + BCl_3 \longrightarrow$

24.4 Common Classes of Organic Reactions

44. Show a detailed mechanism for the nucleophilic substitution reaction of potassium cyanide with iodoethane.

45. Sketch the detailed mechanism of the reaction between NaSH and 1-bromopropane.

46. Show a mechanism for the elimination reaction of cyclohexylchloride with potassium ethoxide. Identify the electrophile and the nucleophile in this reaction.

47. What is the product of the elimination reaction of 1-bromo-2-methylpropane with sodium ethoxide?

48. Write the structure of the product expected from the electrophilic addition of HBr to *cis*-2-butene.

49. What is the addition product of the reaction of 1-methylcyclopentene with HBr? Identify the electrophile and the nucleophile, and then write a mechanism for this reaction.

50. Show how you would synthesize ethylene from ethane. After synthesizing ethylene, how would you make iodoethane?

51. Write a synthetic scheme for making propene from propane. How would you convert propene to 2-bromopropane?

52. For the photochemical reaction of Cl_2 with 2,4-dimethylpentane, (a) how many different monochlorinated isomers would you expect to be produced, and (b) which one is produced from the most stable precursor radical?

53. How many monobrominated isomers will be produced from the high-temperature reaction of Br_2 with 3-methylpentane? Which is favored from the most stable radical?

54. How many monobrominated isomers would you expect from the photochemical reaction of Br_2 with (a) isobutane and (b) 2,2,3-trimethylpentane?

55. In the photochemical reaction of Cl_2 with 3,3,4-trimethylhexane, how many different radicals can be formed?

56. Arrange the compounds in order of increasing oxidation state of carbon: acetone, ethane, carbon dioxide, acetaldehyde, ethanol.

57. What product(s) do you expect from the reduction of formaldehyde? From the oxidation of ethanol?

58. What product(s) do you expect from the oxidation of a ketone? From the reduction of an aldehyde?

24.5 Common Classes of Organic Compounds

59. What is the product of the reaction of 2-butyne with excess HBr?

60. What is the product of the reaction of 3-hexyne with excess HCl?

61. What elements are eliminated during the dehydrohalogenation of an alkylhalide? What products do you expect from the dehydrohalogenation of 2-chloro-1-pentene?

62. What elements are eliminated during the dehydration of an alcohol? What products do you expect from the dehydration of ethanol?

63. Predict the products of these reactions: (a) sodium phenoxide with ethylchloride; (b) 1-chloropropane with NaOH.

64. Show the mechanism and predict the organic product of each reaction: (a) 2-propanol + HCl; (b) cyclohexanol + H_2SO_4.

65. A Grignard reagent can be used to generate a carboxylic acid. Show the mechanism for the first step in this reaction using CH_3CH_2MgBr as the Grignard reagent. What is the geometry about the carbon of the —CH_2 of the intermediate species formed in this first step?

66. Draw a molecular orbital picture showing the bonding in an amide. What orbital is used for the lone pair of electrons on nitrogen?

67. What is the product of the reaction of (a) acetic acid with ammonia, and (b) methyl acetate with ethylamine, followed by heat.

68. Develop a synthetic scheme to generate (a) 1,1-dichloroethane from 1,1-dibromoethane and (b) 2-bromo-1-heptene from 1-bromopentane.

24.6 The Molecules of Life

69. Propose a method for synthesizing the dipeptide alanylglycine (Ala-Gly) starting with the amino acids (see Figure 5.16).

70. Are all the naturally occurring amino acids chiral compounds? Do you expect proteins to contain both enantiomers of alanine and other amino acids? Explain your answer.

APPLICATIONS

71. After exercise, the concentration of lactic acid increases in both muscle tissue and blood. In fact, it is responsible for muscle cramps that may develop after vigorous exercise. The structure of lactic acid is shown here. Draw the conformational isomers of lactic acid as viewed along the C1–C2 bond axis, where C1 is the carbon of the methyl group. Which would you predict to be the most stable conformation? Why?

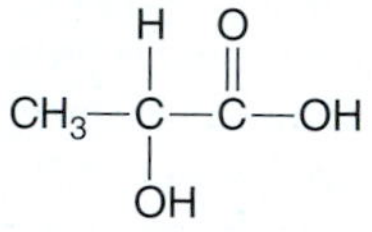

Lactic acid

72. Cyclohexanecarboxylic acid is a liquid used in insecticide formulations. A derivative of this compound, 2-methylcyclohexanecarboxylic acid is shown here. Sketch the *cis* and *trans* isomers of this derivative. Arrange the conformations of its geometric isomers in order of increasing energy.

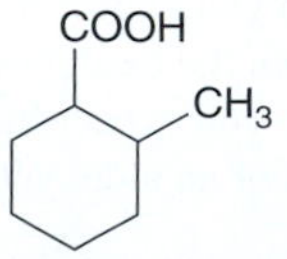

73. Coniine is a naturally occurring compound that has insect-paralyzing properties. Ingestion of this compound causes weakness, vomiting, labored respiration, and eventual death. Does coniine have a chiral carbon? If so, indicate the carbon with an asterisk.

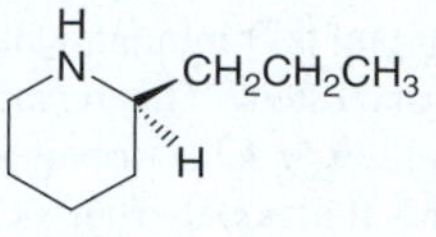

Coniine

74. A compound that has been found to be an effective hypertensive is Captopril, sold commercially as Hypertil and Tensoprel, among other names. Captopril has a chiral center at C2. Draw the enantiomers that result from this chiral center. Indicate any other chiral centers in Captopril with an asterisk.

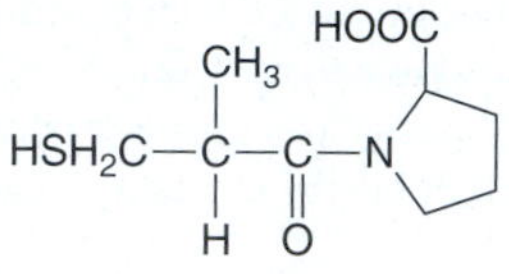

Captopril

75. The structure of phenobarbital, a compound used to treat epilepsy, is shown. Identify the functional groups.

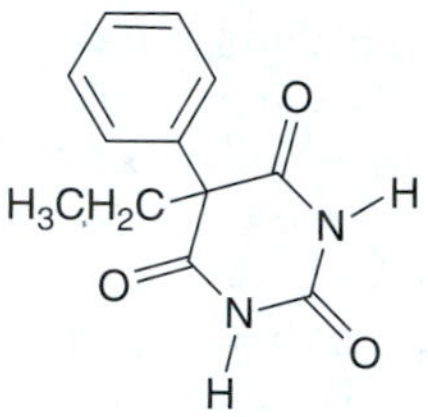

Phenobarbital

76. The compound 2-amino-2-methyl-1-propanol is used not only in the synthesis of pharmaceuticals, but also in cosmetic creams, polishes, and cleaning compounds. Is it a chiral compound?

77. Glycerol (1,2,3-propanetriol) is produced from sugars by fermentation. It also is obtained from oils and fats as a byproduct during the manufacture of soaps. Glycerol can be converted to glyceric acid by the following sequence:

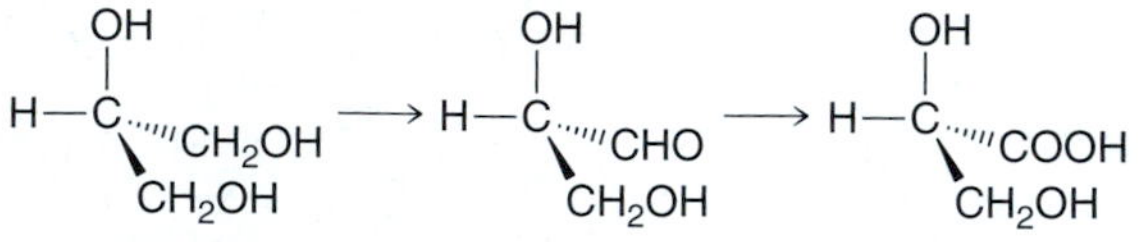

Glycerol Glyceraldehyde Glyceric acid

Both glyceraldehyde and glyceric acid are derivatives of biochemical intermediates in sugar metabolism.

(a) Is the conversion of glycerol to glyceric acid an oxidative or a reductive process?

(b) How many of these compounds are chiral? Indicate any chiral centers with an asterisk.

78. When exposed to light or heat, peroxides (ROOR) can undergo radical reactions. The first step is the dissociation of the peroxide into two alkoxy radicals. Each alkoxy radical then reacts with a species such as HBr to form an alcohol.

(a) Write equations for these reactions.

(b) One of the products of the second step can react with an alkene to generate another radical. Show an equation for this reaction, using ethylene as the alkene.

(c) The product of part b can then react with HBr to form two products. What are those products?

(d) Identify any propagation reactions in the above descriptions. Show all possible termination reactions.

79. The structure of isophytol is shown. This compound is used in the preparation of vitamins E and K. Does isophytol form *cis* and *trans* isomers? If so, draw these isomers.

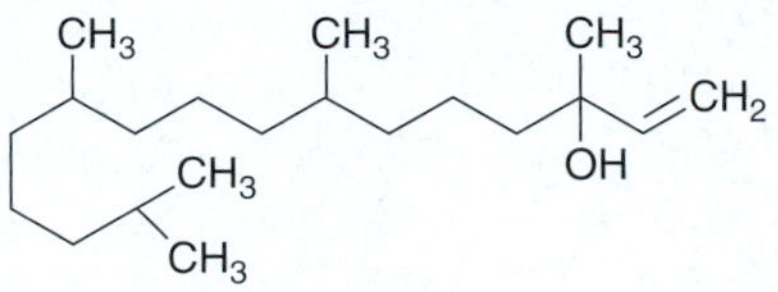

Isophytol

80. Isopentyl acetate, also known as "oil of banana," has the structure shown here. Show the first step in the mechanism for the reaction of isopentyl acetate with C_6H_5MgBr, which gives the first ionic adduct.

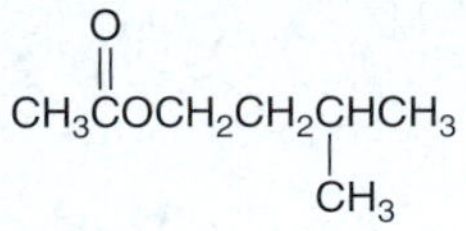

Isopentyl acetate

81. An enkephalin is a pentapeptide that controls pain in humans. Enkephalins function by binding to the same specific sites in brain cells that are known to bind morphine and heroin. Methionine enkephalin has the structure Tyr-Gly-Gly-Phe-Met.

(a) Use Figure 5.16 to draw the structure of methionine enkephalin.

(b) Rotation about the amide bond that connects residues 3 and 4 is restricted because of electron delocalization. Draw the *cis* and *trans* isomers.

APPENDIX A

Standard Thermodynamic Quantities for Chemical Substances at 25°C

Substance	ΔH_f° (kJ/mol)	ΔG_f° (kJ/mol)	S° (J/mol·K)
Aluminum:			
$Al(s)$	0.0	0.0	28.3
$Al(g)$	330.0	289.4	164.6
$AlCl_3(s)$	−704.2	−628.8	109.3
$Al_2O_3(s)$	−1675.7	−1582.3	50.9
Barium:			
$Ba(s)$	0.0	0.0	62.5
$Ba(g)$	180.0	146.0	170.2
$BaO(s)$	−548.0	−520.3	72.1
$BaCO_3(s)$	−1213.0	−1134.4	112.1
$BaSO_4(s)$	−1473.2	−1362.2	132.2
Beryllium:			
$Be(s)$	0.0	0.0	9.5
$Be(g)$	324.0	286.6	136.3
$Be(OH)_2(s)$	−902.5	−815.0	45.5
$BeO(s)$	−609.4	−580.1	13.8
Bismuth:			
$Bi(s)$	0.0	0.0	56.7
$Bi(g)$	207.1	168.2	187.0
Bromine:			
$Br(g)$	111.9	82.4	175.0
$Br_2(l)$	0.0	0.0	152.2
$Br^-(aq)$	−121.6	−104.0	82.4
$Br_2(g)$	30.9	3.1	245.5
$HBr(g)$	−36.3	−53.4	198.7
$HBr(aq)$	−121.6	−104.0	82.4
Cadmium:			
$Cd(s)$	0.0	0.0	51.8
$Cd(g)$	111.8	—	167.7
$CdCl_2(s)$	−391.5	−343.9	115.3
$CdS(s)$	−161.9	−156.5	64.9
Calcium:			
$Ca(s)$	0.0	0.0	41.6
$Ca(g)$	177.8	144.0	154.9
$CaCl_2(s)$	−795.4	−748.8	108.4
$CaF_2(s)$	−1228.0	−1175.6	68.5
$Ca(OH)_2(s)$	−985.2	−897.5	83.4
$CaO(s)$	−634.9	−603.3	38.1
$CaSO_4(s)$	−1434.5	−1322.0	106.5
$CaCO_3(s$, calcite)	−1207.6	−1129.1	91.7
$CaCO_3(s$, aragonite)	−1207.8	−1128.2	88.0

Substance	ΔH°_f (kJ/mol)	ΔG°_f (kJ/mol)	S° (J/mol·K)
Carbon:			
C(*s*, graphite)	0.0	0.0	5.7
C(*s*, diamond)	1.9	2.9	2.4
C(*s*, fullerene—C_{60})	2327.0	2302.0	426.0
C(*s*, fullerene—C_{70})	2555.0	2537.0	464.0
C(*g*)	716.7	671.3	158.1
C(*g*, fullerene—C_{60})	2502.0	2442.0	544.0
C(*g*, fullerene—C_{70})	2755.0	2692.0	614.0
$CBr_4(s)$	29.4	47.7	212.5
$CBr_4(g)$	83.9	67.0	358.1
$CCl_2F_2(g)$	−477.4	−439.4	300.8
$CCl_2O(g)$	−219.1	−204.9	283.5
$CCl_4(l)$	−128.2	−62.6	216.2
$CCl_4(g)$	−95.7	−53.6	309.9
$CF_4(g)$	−933.6	−888.3	261.6
$CHCl_3(l)$	−134.1	−73.7	201.7
$CHCl_3(g)$	−102.7	6.0	295.7
$CH_2Cl_2(l)$	−124.2	—	177.8
$CH_2Cl_2(g)$	−95.4	−68.9	270.2
$CH_3Cl(g)$	−81.9	−58.5	234.6
$CH_4(g)$	−74.6	−50.5	186.3
$CH_3CO_2H(l)$	−484.3	−389.9	159.8
$CH_3OH(l)$	−239.2	−166.6	126.8
$CH_3OH(g)$	−201.0	−162.3	239.9
$CH_3NH_2(l)$	−47.3	35.7	150.2
$CH_3NH_2(g)$	−22.5	32.7	242.9
$CH_3CN(l)$	40.6	86.5	149.6
$CH_3CN(g)$	74.0	91.9	243.4
$CO(g)$	−110.5	−137.2	197.7
$CO_2(g)$	−393.5	−394.4	213.8
$CS_2(l)$	89.0	64.6	151.3
$CS_2(g)$	116.7	67.1	237.8
$C_2H_2(g)$	227.4	209.9	200.9
$C_2H_4(g)$	52.4	68.4	219.3
$C_2H_6(g)$	−84.0	−32.0	229.2
$C_3H_8(g)$	−103.8	−23.4	270.3
$C_3H_6O_3(l)$ (lactic acid)	−674.5	−518.2	192.1
$C_6H_6(l)$	49.1	124.5	173.4
$C_6H_6(g)$	82.9	129.7	269.2
$C_6H_{12}O_6(s)$ (glucose)	−1273.3	−910.4	212.1
$C_2H_5OH(l)$	−277.6	−174.8	160.7
$C_2H_5OH(g)$	−234.8	−167.9	281.6
$(CH_3)_2O(l)$	−203.3	—	—
$(CH_3)_2O(g)$	−184.1	−112.6	266.4
$CH_3CO_2^-(aq)$	−486.0	−369.3	86.6
$C_8H_{18}(l)$ (isooctane)	−259.2	—	—
n-$C_{12}H_{26}(l)$ (dodecane)	−350.9	28.1	490.6
Cesium:			
$Cs(s)$	0.0	0.0	85.2
$Cs(g)$	76.5	49.6	175.6
$CsCl(s)$	−443.0	−414.5	101.2
Chlorine:			
$Cl(g)$	121.3	105.3	165.2
$Cl_2(g)$	0.0	0.0	223.1
$Cl^-(aq)$	−167.2	−131.2	56.5
$HCl(g)$	−92.3	−95.3	186.9
$HCl(aq)$	−167.2	−131.2	56.5
$ClF_3(g)$	−163.2	−123.0	281.6

Substance	ΔH_f° (kJ/mol)	ΔG_f° (kJ/mol)	S° (J/mol·K)
Chromium:			
Cr(*s*)	0.0	0.0	23.8
Cr(*g*)	396.6	351.8	174.5
$CrCl_3$(*s*)	−556.5	−486.1	123.0
CrO_3(*g*)	−292.9	—	266.2
Cr_2O_3(*s*)	−1139.7	−1058.1	81.2
Cobalt:			
Co(*s*)	0.0	0.0	30.0
Co(*g*)	424.7	380.3	179.5
$CoCl_2$(*s*)	−312.5	−269.8	109.2
Copper:			
Cu(*s*)	0.0	0.0	33.2
Cu(*g*)	337.4	297.7	166.4
CuCl(*s*)	−137.2	−119.9	86.2
$CuCl_2$(*s*)	−220.1	−175.7	108.1
CuO(*s*)	−157.3	−129.7	42.6
Cu_2O(*s*)	−168.6	−146.0	93.1
CuS(*s*)	−53.1	−53.6	66.5
Cu_2S(*s*)	−79.5	−86.2	120.9
CuCN(*s*)	96.2	111.3	84.5
Fluorine:			
F(*g*)	79.4	62.3	158.8
F^-(*aq*)	−332.6	−278.8	−13.8
F_2(*g*)	0.0	0.0	202.8
HF(*g*)	−273.3	−275.4	173.8
HF(*aq*)	−332.6	−278.8	−13.8
Hydrogen:			
H(*g*)	218.0	203.3	114.7
H_2(*g*)	0.0	0.0	130.7
H^+(*aq*)	0.0	0.0	0.0
Iodine:			
I(*g*)	106.8	70.2	180.8
I^-(*aq*)	−55.2	−51.6	111.3
I_2(*s*)	0.0	0.0	116.1
I_2(*g*)	62.4	19.3	260.7
HI(*g*)	26.5	1.7	206.6
HI(*aq*)	−55.2	−51.6	111.3
Iron:			
Fe(*s*)	0.0	0.0	27.3
Fe(*g*)	416.3	370.7	180.5
Fe^{2+}(*aq*)	−89.1	−78.9	−137.7
Fe^{3+}(*aq*)	−48.5	−4.7	−315.9
$FeCl_2$(*s*)	−341.8	−302.3	118.0
$FeCl_3$(*s*)	−399.5	−334.0	142.3
FeO(*s*)	−272.0	−251.4	60.7
Fe_2O_3(*s*)	−824.2	−742.2	87.4
Fe_3O_4(*s*)	−1118.4	−1015.4	146.4
FeS_2(*s*)	−178.2	−166.9	52.9
$FeCO_3$(*s*)	−740.6	−666.7	92.9
Lead:			
Pb(*s*)	0.0	0.0	64.8
Pb(*g*)	195.2	162.2	175.4
PbO(*s*, red or litharge)	−219.0	−188.9	66.5

Substance	ΔH_f° (kJ/mol)	ΔG_f° (kJ/mol)	S° (J/mol·K)
Lead (continued):			
PbO(*s*, yellow or massicot)	−217.3	−187.9	68.7
$PbO_2(s)$	−277.4	−217.3	68.6
$PbCl_2(s)$	−359.4	−314.1	136.0
PbS(*s*)	−100.4	−98.7	91.2
$PbSO_4(s)$	−920.0	−813.0	148.5
$PbCO_3(s)$	−699.1	−625.5	131.0
$Pb(NO_3)_2(s)$	−451.9	—	—
$Pb(NO_3)_2(aq)$	−416.3	−246.9	303.3
Lithium:			
Li(*s*)	0.0	0.0	29.1
Li(*g*)	159.3	126.6	138.8
$Li^+(aq)$	−278.5	−293.3	13.4
LiCl(*s*)	−408.6	−384.4	59.3
$Li_2O(s)$	−597.9	−561.2	37.6
Magnesium:			
Mg(*s*)	0.0	0.0	32.7
Mg(*g*)	147.1	112.5	148.6
$MgCl_2(s)$	−641.3	−591.8	89.6
MgH_2	−75.3	−35.9	31.1
MgO(*s*)	−601.6	−569.3	27.0
$Mg(OH)_2(s)$	−924.5	−833.5	63.2
$MgSO_4(s)$	−1284.9	−1170.6	91.6
MgS(*s*)	−346.0	−341.8	50.3
Manganese:			
Mn(*s*)	0.0	0.0	32.0
Mn(*g*)	280.7	238.5	173.7
$MnCl_2(s)$	−481.3	−440.5	118.2
MnO(*s*)	−385.2	−362.9	59.7
$MnO_2(s)$	−520.0	−465.1	53.1
$KMnO_4(s)$	−837.2	−737.6	171.7
$MnO_4^-(aq)$	−541.4	−447.2	191.2
Mercury:			
Hg(*l*)	0.0	0.0	75.9
Hg(*g*)	61.4	31.8	175.0
$HgCl_2(s)$	−224.3	−178.6	146.0
$Hg_2Cl_2(s)$	−265.4	−210.7	191.6
HgO(*s*)	−90.8	−58.5	70.3
HgS(*s*, red)	−58.2	−50.6	82.4
$Hg_2(g)$	108.8	68.2	288.1
Molybdenum:			
Mo(*s*)	0.0	0.0	28.7
Mo(*g*)	658.1	612.5	182.0
$MoO_2(s)$	−588.9	−533.0	46.3
$MoO_3(s)$	−745.1	−668.0	77.7
Nickel:			
Ni(*s*)	0.0	0.0	29.9
Ni(*g*)	429.7	384.5	182.2
$NiCl_2(s)$	−305.3	−259.0	97.7
$Ni(OH)_2(s)$	−529.7	−447.2	88.0
Nitrogen:			
N(*g*)	472.7	455.5	153.3
$N_2(g)$	0.0	0.0	191.6

Substance	ΔH_f° (kJ/mol)	ΔG_f° (kJ/mol)	S° (J/mol·K)
Nitrogen (continued):			
$NH_3(g)$	−45.9	−16.4	192.8
$NH_4^+(aq)$	−132.5	−79.3	113.4
$N_2H_4(l)$	50.6	149.3	121.2
$N_2H_4(g)$	95.4	159.4	238.5
$NH_4Cl(s)$	−314.4	−202.9	94.6
$NH_4OH(l)$	−361.2	−254.0	165.6
$NH_4NO_3(s)$	−365.6	−183.9	151.1
$(NH_4)_2SO_4(s)$	−1180.9	−901.7	220.1
$NO(g)$	91.3	87.6	210.8
$NO_2(g)$	33.2	51.3	240.1
$N_2O(g)$	81.6	103.7	220.0
$N_2O_4(l)$	−19.5	97.5	209.2
$N_2O_4(g)$	11.1	99.8	304.4
$N_2O_5(g)$	13.3	117.1	355.7
$HNO_2(g)$	−79.5	−46.0	254.1
$HNO_3(l)$	−174.1	−80.7	155.6
$HNO_3(g)$	−133.9	−73.5	266.9
$HNO_3(aq)$	−207.4	−111.3	146.4
$NF_3(g)$	−132.1	−90.6	260.8
$HCN(l)$	108.9	125.0	112.8
$HCN(g)$	135.1	124.7	201.8
Osmium:			
$Os(s)$	0.0	0.0	32.6
$Os(g)$	791.0	745.0	192.6
$OsO_4(s)$	−394.1	−304.9	143.9
$OsO_4(g)$	−337.2	−292.8	293.8
Oxygen:			
$O(g)$	249.2	231.7	161.1
$O_2(g)$	0.0	0.0	205.2
$O_3(g)$	142.7	163.2	238.9
$OH^-(aq)$	−230.0	−157.2	−10.8
$H_2O(l)$	−285.8	−237.1	70.0
$H_2O(g)$	−241.8	−228.6	188.8
$H_2O_2(l)$	−187.8	−120.4	109.6
$H_2O_2(g)$	−136.3	−105.6	232.7
Phosphorus:			
P(*s*, white)	0.0	0.0	41.1
P(*s*, red)	−17.6	−12.5	22.8
P(*s*, black)	−39.3	—	—
$P(g)$	316.5	280.1	163.2
$P_2(g)$	144.0	103.5	218.1
$P_4(g)$	58.9	24.4	280.0
$PCl_3(l)$	−319.7	−272.3	217.1
$PCl_3(g)$	−287.0	−267.8	311.8
$POCl_3(l)$	−597.1	−520.8	222.5
$POCl_3(g)$	−558.5	−512.9	325.5
$PCl_5(g)$	−374.9	−305.0	364.6
$PH_3(g)$	5.4	13.5	210.2
$H_3PO_4(s)$	−1284.4	−1124.3	110.5
$H_3PO_4(l)$	−1271.7	−1123.6	150.8
$H_3PO_4(aq)$	−1277.4	−1018.8	222.0
Potassium:			
$K(s)$	0.0	0.0	64.7
$K(g)$	89.0	60.5	160.3
$KBr(s)$	−393.8	−380.7	95.9
$KCl(s)$	−436.5	−408.5	82.6
$KClO_3(s)$	−397.7	−296.3	143.1

Substance	ΔH_f° (kJ/mol)	ΔG_f° (kJ/mol)	S° (J/mol·K)
Potassium (continued):			
$K_2O(s)$	−361.5	−322.1	94.1
$K_2O_2(s)$	−494.1	−425.1	102.1
$KNO_2(s)$	−369.8	−306.6	152.1
$KNO_3(s)$	−494.6	−394.9	133.1
$KSCN(s)$	−200.2	−178.3	124.3
$K_2CO_3(s)$	−1151.0	−1063.5	155.5
$K_2SO_4(s)$	−1437.8	−1321.4	175.6
Rubidium:			
$Rb(s)$	0.0	0.0	76.8
$Rb(g)$	80.9	53.1	170.1
$RbCl(s)$	−435.4	−407.8	95.9
Selenium:			
$Se(s)$	0.0	0.0	42.4
$Se(g$, gray)	227.1	187.0	176.7
$H_2Se(g)$	29.7	15.9	219.0
Silicon:			
$Si(s)$	0.0	0.0	18.8
$Si(g)$	450.0	405.5	168.0
$SiCl_4(l)$	−687.0	−619.8	239.7
$SiCl_4(g)$	−657.0	−617.0	330.7
$SiH_4(g)$	34.3	56.9	204.6
$SiC(s$, cubic)	−65.3	−62.8	16.6
$SiC(s$, hexagonal)	−62.8	−60.2	16.5
Silver:			
$Ag(s)$	0.0	0.0	42.6
$Ag(g)$	284.9	246.0	173.0
$Ag^+(aq)$	105.6	77.1	72.7
$AgBr(s)$	−100.4	−96.9	107.1
$AgCl(s)$	−127.0	−109.8	96.3
$AgNO_3(s)$	−124.4	−33.4	140.9
$Ag_2O(s)$	−31.1	−11.2	121.3
$Ag_2S(s)$	−32.6	−40.7	144.0
Sodium:			
$Na(s)$	0.0	0.0	51.3
$Na(g)$	107.5	77.0	153.7
$Na^+(aq)$	−240.1	−261.9	59.0
$NaF(s)$	−576.6	−546.3	51.1
$NaF(aq)$	−572.8	−540.7	45.2
$NaCl(s)$	−411.2	−384.1	72.1
$NaCl(aq)$	−407.3	−393.1	115.5
$NaBr(s)$	−361.1	−349.0	86.8
$NaBr(g)$	−143.1	−177.1	241.2
$NaBr(aq)$	−361.7	−365.8	141.4
$NaO_2(s)$	−260.2	−218.4	115.9
$Na_2O(s)$	−414.2	−375.5	75.1
$Na_2O_2(s)$	−510.9	−447.7	95.0
$NaCN(s)$	−87.5	−76.4	115.6
$NaNO_3(aq)$	−447.5	−373.2	205.4
$NaNO_3(s)$	−467.9	−367.0	116.5
$NaN_3(s)$	21.7	93.8	96.9
$Na_2SO_3(s)$	−1100.8	−1012.5	145.9
$Na_2CO_3(s)$	−1130.7	−1044.4	135.0
$Na_2SO_3(s)$	−1100.8	−1012.5	145.9
$Na_2SO_4(s)$	−1387.1	−1270.2	149.6

Substance	ΔH_f° (kJ/mol)	ΔG_f° (kJ/mol)	S° (J/mol·K)
Sulfur:			
$S(s$, rhombic$)$	0.0	0.0	32.1
$S(g)$	277.2	236.7	167.8
$SO_2(g)$	−296.8	−300.1	248.2
$SO_3(g)$	−395.7	−371.1	256.8
$SO_4^{2-}(aq)$	−909.3	−744.5	20.1
$SOCl_2(g)$	−212.5	−198.3	309.8
$H_2S(g)$	−20.6	−33.4	205.8
$H_2SO_4(aq)$	−909.3	−744.5	20.1
$H_2SO_4(l)$	−814	−690.0	156.9
Tin:			
$Sn(s$, white$)$	0.0	0.0	51.2
$Sn(s$, gray$)$	−2.1	0.1	44.1
$Sn(g)$	301.2	266.2	168.5
$SnCl_4(l)$	−511.3	−440.1	258.6
$SnCl_4(g)$	−471.5	−432.2	365.8
$SnO_2(s)$	−577.6	−515.8	49.0
Titanium:			
$Ti(s)$	0.0	0.0	30.7
$Ti(g)$	473.0	428.4	180.3
$TiCl_2(s)$	−513.8	−464.4	87.4
$TiCl_3(s)$	−720.9	−653.5	139.7
$TiCl_4(l)$	−804.2	−737.2	252.3
$TiCl_4(g)$	−763.2	−726.3	353.2
$TiO_2(s)$	−944.0	−888.8	50.6
Uranium:			
$U(s)$	0.0	0.0	50.2
$U(g)$	533.0	488.4	199.8
$UO_2(s)$	−1085.0	−1031.8	77.0
$UO_2(g)$	−465.7	−471.5	274.6
$UF_4(s)$	−1914.2	−1823.3	151.7
$UF_4(g)$	−1598.7	−1572.7	368.0
$UF_6(s)$	−2197.0	−2068.5	227.6
$UF_6(g)$	−2147.4	−2063.7	377.9
Vanadium:			
$V(s)$	0.0	0.0	28.9
$V(g)$	514.2	754.4	182.3
$VCl_3(s)$	−580.7	−511.2	131.0
$VCl_4(l)$	−569.4	−503.7	255.0
$VCl_4(g)$	−525.5	−492.0	362.4
$V_2O_5(s)$	−1550.6	−1419.5	131.0
Zinc:			
$Zn(s)$	0.0	0.0	41.6
$Zn(g)$	130.4	94.8	161.0
$ZnCl_2(s)$	−415.1	−369.4	111.5
$Zn(NO_3)_2(s)$	−483.7	—	—
$ZnS(s$, sphalerite$)$	−206.0	−201.3	57.7
$ZnSO_4(s)$	−982.8	−871.5	110.5
Zirconium:			
$Zr(s)$	0.0	0.0	39.0
$Zr(g)$	608.8	566.5	181.4
$ZrCl_2(s)$	−502.0	−386	110
$ZrCl_4(s)$	−980.5	−889.9	181.6

Source of data: *CRC Handbook of Chemistry and Physics,* 84th Edition (2004).

APPENDIX B

Solubility-Product Constants (K_{sp}) for Compounds at 25°C

Compound Name	Compound Formula	K_{sp}
Aluminum phosphate	$AlPO_4$	9.84×10^{-21}
Barium bromate	$Ba(BrO_3)_2$	2.43×10^{-4}
Barium carbonate	$BaCO_3$	2.58×10^{-9}
Barium chromate	$BaCrO_4$	1.17×10^{-10}
Barium fluoride	BaF_2	1.84×10^{-7}
Barium iodate	$Ba(IO_3)_2$	4.01×10^{-9}
Barium nitrate	$Ba(NO_3)_2$	4.64×10^{-3}
Barium sulfate	$BaSO_4$	1.08×10^{-10}
Barium sulfite	$BaSO_3$	5.0×10^{-10}
Beryllium hydroxide	$Be(OH)_2$	6.92×10^{-22}
Bismuth arsenate	$BiAsO_4$	4.43×10^{-10}
Bismuth iodide	BiI_3	7.71×10^{-19}
Cadmium carbonate	$CdCO_3$	1.0×10^{-12}
Cadmium fluoride	CdF_2	6.44×10^{-3}
Cadmium hydroxide	$Cd(OH)_2$	7.2×10^{-15}
Cadmium iodate	$Cd(IO_3)_2$	2.5×10^{-8}
Cadmium phosphate	$Cd_3(PO_4)_2$	2.53×10^{-33}
Cadmium sulfide	CdS	8.0×10^{-27}
Calcium carbonate	$CaCO_3$	3.36×10^{-9}
Calcium fluoride	CaF_2	3.45×10^{-11}
Calcium hydroxide	$Ca(OH)_2$	5.02×10^{-6}
Calcium iodate	$Ca(IO_3)_2$	6.47×10^{-6}
Calcium phosphate	$Ca_3(PO_4)_2$	2.07×10^{-33}
Calcium sulfate	$CaSO_4$	4.93×10^{-5}
Cesium perchlorate	$CsClO_4$	3.95×10^{-3}
Cesium periodate	$CsIO_4$	5.16×10^{-6}
Cobalt(II) arsenate	$Co_3(AsO_4)_2$	6.80×10^{-29}
Cobalt(II) hydroxide	$Co(OH)_2$	5.92×10^{-15}
Cobalt(II) phosphate	$Co_3(PO_4)_2$	2.05×10^{-35}
Copper(I) bromide	$CuBr$	6.27×10^{-9}
Copper(I) chloride	$CuCl$	1.72×10^{-7}
Copper(I) cyanide	$CuCN$	3.47×10^{-20}
Copper(I) iodide	CuI	1.27×10^{-12}
Copper(I) thiocyanate	$CuSCN$	1.77×10^{-13}
Copper(II) arsenate	$Cu_3(AsO_4)_2$	7.95×10^{-36}
Copper(II) oxalate	CuC_2O_4	4.43×10^{-10}
Copper(II) phosphate	$Cu_3(PO_4)_2$	1.40×10^{-37}
Copper(II) sulfide	CuS	6.3×10^{-36}
Europium(III) hydroxide	$Eu(OH)_3$	9.38×10^{-27}
Gallium(III) hydroxide	$Ga(OH)_3$	7.28×10^{-36}
Iron(II) carbonate	$FeCO_3$	3.13×10^{-11}
Iron(II) fluoride	FeF_2	2.36×10^{-6}
Iron(II) hydroxide	$Fe(OH)_2$	4.87×10^{-17}
Iron(III) hydroxide	$Fe(OH)_3$	2.79×10^{-39}
Iron(III) sulfide	FeS	6.3×10^{-18}
Lanthanum iodate	$La(IO_3)_3$	7.50×10^{-12}
Lead(II) bromide	$PbBr_2$	6.60×10^{-6}
Lead(II) carbonate	$PbCO_3$	7.40×10^{-14}
Lead(II) chloride	$PbCl_2$	1.70×10^{-5}
Lead(II) fluoride	PbF_2	3.3×10^{-8}
Lead(II) hydroxide	$Pb(OH)_2$	1.43×10^{-20}
Lead(II) iodate	$Pb(IO_3)_2$	3.69×10^{-13}
Lead(II) iodide	PbI_2	9.8×10^{-9}

Compound Name	Compound Formula	K_{sp}
Lead(II) selenate	$PbSeO_4$	1.37×10^{-7}
Lead(II) sulfate	$PbSO_4$	2.53×10^{-8}
Lead(II) sulfide	PbS	8.0×10^{-28}
Lithium carbonate	Li_2CO_3	8.15×10^{-4}
Lithium fluoride	LiF	1.84×10^{-3}
Lithium phosphate	Li_3PO_4	2.37×10^{-11}
Magnesium carbonate	$MgCO_3$	6.82×10^{-6}
Magnesium fluoride	MgF_2	5.16×10^{-11}
Magnesium hydroxide	$Mg(OH)_2$	5.61×10^{-12}
Magnesium phosphate	$Mg_3(PO_4)_2$	1.04×10^{-24}
Manganese(II) carbonate	$MnCO_3$	2.24×10^{-11}
Manganese(II) iodate	$Mn(IO_3)_2$	4.37×10^{-7}
Mercury(I) bromide	Hg_2Br_2	6.40×10^{-23}
Mercury(I) carbonate	Hg_2CO_3	3.6×10^{-17}
Mercury(I) chloride	Hg_2Cl_2	1.43×10^{-18}
Mercury(I) fluoride	Hg_2F_2	3.10×10^{-6}
Mercury(I) iodide	Hg_2I_2	5.2×10^{-29}
Mercury(I) oxalate	$Hg_2C_2O_4$	1.75×10^{-13}
Mercury(I) sulfate	Hg_2SO_4	6.5×10^{-7}
Mercury(I) thiocyanate	$Hg_2(SCN)_2$	3.2×10^{-20}
Mercury(II) bromide	$HgBr_2$	6.2×10^{-20}
Mercury (II) iodide	HgI_2	2.9×10^{-29}
Mercury(II) sulfide (red)	HgS	4×10^{-53}
Mercury(II) sulfide (black)	HgS	1.6×10^{-52}
Neodymium carbonate	$Nd_2(CO_3)_3$	1.08×10^{-33}
Nickel(II) carbonate	$NiCO_3$	1.42×10^{-7}
Nickel(II) hydroxide	$Ni(OH)_2$	5.48×10^{-16}
Nickel(II) iodate	$Ni(IO_3)_2$	4.71×10^{-5}
Nickel(II) phosphate	$Ni_3(PO_4)_2$	4.74×10^{-32}
Palladium(II) thiocyanate	$Pd(SCN)_2$	4.39×10^{-23}
Potassium hexachloroplatinate	K_2PtCl_6	7.48×10^{-6}
Potassium perchlorate	$KClO_4$	1.05×10^{-2}
Potassium periodate	KIO_4	3.71×10^{-4}
Praseodymium hydroxide	$Pr(OH)_3$	3.39×10^{-24}
Rubidium perchlorate	$RbClO_4$	3.00×10^{-3}
Scandium fluoride	ScF_3	5.81×10^{-24}
Scandium hydroxide	$Sc(OH)_3$	2.22×10^{-31}
Silver(I) acetate	$AgCH_3CO_2$	1.94×10^{-3}
Silver(I) arsenate	Ag_3AsO_4	1.03×10^{-22}
Silver(I) bromate	$AgBrO_3$	5.38×10^{-5}
Silver(I) bromide	$AgBr$	5.35×10^{-13}
Silver(I) carbonate	Ag_2CO_3	8.46×10^{-12}
Silver(I) chloride	$AgCl$	1.77×10^{-10}
Silver(I) chromate	Ag_2CrO_4	1.12×10^{-12}
Silver(I) cyanide	$AgCN$	5.97×10^{-17}
Silver(I) iodate	$AgIO_3$	3.17×10^{-8}
Silver(I) iodide	AgI	8.52×10^{-17}
Silver(I) oxalate	$Ag_2C_2O_4$	5.40×10^{-12}
Silver(I) phosphate	Ag_3PO_4	8.89×10^{-17}
Silver(I) sulfate	Ag_2SO_4	1.20×10^{-5}
Silver(I) sulfide	Ag_2S	6.3×10^{-50}
Silver(I) sulfite	Ag_2SO_3	1.50×10^{-14}
Silver(I) thiocyanate	$AgSCN$	1.03×10^{-12}
Strontium arsenate	$Sr_3(AsO_4)_2$	4.29×10^{-19}
Strontium carbonate	$SrCO_3$	5.60×10^{-10}
Strontium fluoride	SrF_2	4.33×10^{-9}
Strontium iodate	$Sr(IO_3)_2$	1.14×10^{-7}
Strontium sulfate	$SrSO_4$	3.44×10^{-7}
Thallium(I) bromate	$TlBrO_3$	1.10×10^{-4}
Thallium(I) bromide	$TlBr$	3.71×10^{-6}
Thallium(I) chloride	$TlCl$	1.86×10^{-4}
Thallium(I) chromate	Tl_2CrO_4	8.67×10^{-13}
Thallium(I) iodate	$TlIO_3$	3.12×10^{-6}
Thallium(I) iodide	TlI	5.54×10^{-8}

Compound Name	Compound Formula	K_{sp}
Thallium(I) thiocyanate	$TlSCN$	1.57×10^{-4}
Thallium(III) hydroxide	$Tl(OH)_3$	1.68×10^{-44}
Tin(II) hydroxide	$Sn(OH)_2$	5.45×10^{-27}
Tin(II) sulfide	SnS	1.0×10^{-25}
Yttrium carbonate	$Y_2(CO_3)_3$	1.03×10^{-31}
Yttrium fluoride	YF_3	8.62×10^{-21}
Yttrium hydroxide	$Y(OH)_3$	1.00×10^{-22}
Yttrium iodate	$Y(IO_3)_3$	1.12×10^{-10}
Zinc arsenate	$Zn_3(AsO_4)_2$	2.8×10^{-28}
Zinc carbonate	$ZnCO_3$	1.46×10^{-10}
Zinc fluoride	ZnF_2	3.04×10^{-2}
Zinc hydroxide	$Zn(OH)_2$	3×10^{-17}
Zinc selenide	$ZnSe$	3.6×10^{-26}
Zinc sulfide (wurtzite)	ZnS	1.6×10^{-24}
Zinc sulfide (sphalerite)	ZnS	2.5×10^{-22}

Source of data: *CRC Handbook of Chemistry and Physics,* 84th Edition (2004); sulfide data from *Lange's Handbook of Chemistry,* 15th Edition (1999).

APPENDIX C

Ionization Constants and pK_a Values for Acids at 25°C

Name	Formula	K_{a1}	pK_{a1}	K_{a2}	pK_{a2}	K_{a3}	pK_{a3}	K_{a4}	pK_{a4}
Acetic acid	CH_3CO_2H	1.75×10^{-5}	4.756						
Arsenic acid	H_3AsO_4	5.5×10^{-3}	2.26	1.7×10^{-7}	6.76	5.1×10^{-12}	11.29		
Benzoic acid	$C_6H_5CO_2H$	6.25×10^{-5}	4.204						
Boric acid	H_3BO_3	5.4×10^{-10}*	9.27*	$>10^{-14}$*	>14*				
Bromoacetic acid	CH_2BrCO_2H	1.3×10^{-3}	2.90						
Carbonic acid	H_2CO_3	4.5×10^{-7}	6.35	4.7×10^{-11}	10.33				
Chloric acid	$HClO_3$	1.00×10^{2}	−2.00						
Chloroacetic acid	CH_2ClCO_2H	1.3×10^{-3}	2.87						
Chlorous acid	$HClO_2$	1.15×10^{-2}	1.94						
Chromic acid	H_2CrO_4	1.8×10^{-1}	0.74	3.2×10^{-7}	6.49				
Citric acid	$C_6H_8O_7$	7.4×10^{-4}	3.13	1.7×10^{-5}	4.76	4.0×10^{-7}	6.40		
Cyanic acid	HCNO	3.5×10^{-4}	3.46						
Dichloroacetic acid	$CHCl_2CO_2H$	4.5×10^{-2}	1.35						
Fluoroacetic acid	CH_2FCO_2H	2.6×10^{-3}	2.59						
Formic acid	CH_2O_2	1.8×10^{-4}	3.75						
Hydrazoic acid	HN_3	3×10^{-5}	4.6						
Hydrocyanic acid	HCN	6.2×10^{-10}	9.21						
Hydrofluoric acid	HF	6.3×10^{-4}	3.20						
Hydroselenic acid	H_2Se	1.3×10^{-4}	3.89	1×10^{-11}	11.0				
Hydrosulfuric acid	H_2S	1.1×10^{-7}	6.97	10^{-13}	12.90				
Hydrotelluric acid	H_2Te	3×10^{-3}‡	2.6‡	10^{-11}	11				
Hypobromous acid	HBrO	2.8×10^{-9}	8.55						
Hypochlorous acid	HClO	2.88×10^{-8}	7.54						
Hypoiodous acid	HIO	3×10^{-11}	10.5						
Iodic acid	HIO_3	1.7×10^{-1}	0.78						
Iodoacetic acid	CH_2ICO_2H	6.6×10^{-4}	3.18						
Nitrous acid	HNO_2	5.6×10^{-4}	3.25						
Oxalic acid	$C_2H_2O_4$	5.6×10^{-2}	1.25	1.5×10^{-4}	3.81				
Perchloric acid	$HClO_4$	2.00×10^{7}	−7.3						
Periodic acid	HIO_4	2.3×10^{-2}	1.64						
Phenol	C_6H_5OH	1.0×10^{-10}	9.99						
Phosphoric acid	H_3PO_4	6.9×10^{-3}	2.16	6.2×10^{-8}	7.21	4.8×10^{-13}	12.32		
Phosphorous acid	H_3PO_3	5×10^{-2}*	1.3*	2.0×10^{-7}*	6.70*				
Pyrophosphoric acid	$H_4P_2O_7$	1.2×10^{-1}	0.91	7.9×10^{-3}	2.10	2.0×10^{-7}	6.70	4.8×10^{-10}	9.32
Resorcinol	$C_6H_4(OH)_2$	4.8×10^{-10}	9.32	8×10^{-12}	11.1				
Selenic acid	H_2SeO_4	Strong	Strong	2×10^{-2}	1.7				
Selenious acid	H_2SeO_3	2.4×10^{-3}	2.62	4.8×10^{-9}	8.32				
Sulfuric acid	H_2SO_4	Strong	Strong	1.0×10^{-2}	1.99				
Sulfurous acid	H_2SO_3	1.4×10^{-2}	1.85	6×10^{-8}	7.2				
meso-Tartaric acid	$C_4H_6O_6$	6.8×10^{-4}	3.17	1.2×10^{-5}	4.91				
Telluric acid	H_2TeO_4	2.1×10^{-8}‡	7.68‡	1×10^{-11}‡	11.0‡				
Tellurous acid	H_2TeO_3	5.4×10^{-7}	6.27	3.7×10^{-9}	8.43				
Trichloroacetic acid	CCl_3CO_2H	2.2×10^{-1}	0.66						
Trifluoroacetic acid	CF_3CO_2H	3.0×10^{-1}	0.52						

Source of data: *CRC Handbook of Chemistry and Physics,* 84th Edition (2004), except those for chloric acid, chlorous acid, hypochlorous acid, and perchloric acid, which came from J.R. Bowser, *Inorganic Chemistry*, 1993; sulfide data came from *Lange's Handbook of Chemistry*, 15th Edition (1999).

* Measured at 20°C, not 25°C.

‡ Measured at 18°C, not 25°C.

APPENDIX D

Ionization Constants and pK_b Values for Bases at 25°C

Name	Formula	K_b	pK_b
Ammonia	NH_3	1.8×10^{-5}	4.75
Aniline	$C_6H_5NH_2$	7.4×10^{-10}	9.13
n-Butylamine	$C_4H_9NH_2$	4.0×10^{-4}	3.40
sec-Butylamine	$(CH_3)_2CHCH_2NH_2$	3.6×10^{-4}	3.44
tert-Butylamine	$(CH_3)_3CNH_2$	4.8×10^{-4}	3.32
Dimethylamine	$(CH_3)_2NH$	5.4×10^{-4}	3.27
Ethylamine	$C_2H_5NH_2$	4.5×10^{-4}	3.35
Hydrazine	N_2H_4	1.3×10^{-6}	5.9
Hydroxylamine	NH_2OH	8.7×10^{-9}	8.06
Methylamine	CH_3NH_2	4.6×10^{-4}	3.34
Propylamine	$C_3H_7NH_2$	3.5×10^{-4}	3.46
Pyridine	C_5H_5N	1.7×10^{-9}	8.77
Trimethylamine	$(CH_3)_3N$	6.3×10^{-5}	4.20

Source of data: *CRC Handbook of Chemistry and Physics,* 84th Edition (2004).

APPENDIX E

Standard Reduction Potentials at 25°C

Half-Reaction	$E°$ (V)
$Ac^{3+} + 3e^- \longrightarrow Ac$	−2.20
$Ag^+ + e^- \longrightarrow Ag$	0.7996
$AgBr + e^- \longrightarrow Ag + Br^-$	0.07133
$AgCl + e^- \longrightarrow Ag + Cl^-$	0.22233
$Ag_2CrO_4 + 2e^- \longrightarrow 2Ag + CrO_4^{2-}$	0.4470
$AgI + e^- \longrightarrow Ag + I^-$	−0.15224
$Ag_2S + 2e^- \longrightarrow 2Ag + S^{2-}$	−0.691
$Ag_2S + 2H^+ + 2e^- \longrightarrow 2Ag + H_2S$	−0.0366
$AgSCN + e^- \longrightarrow Ag + SCN^-$	0.08951
$Al^{3+} + 3e^- \longrightarrow Al$	−1.662
$Al(OH)_4^- + 3e^- \longrightarrow Al + 4OH^-$	−2.328
$Am^{3+} + 3e^- \longrightarrow Am$	−2.048
$As + 3H^+ + 3e^- \longrightarrow AsH_3$	−0.608
$H_3AsO_4 + 2H^+ + 2e^- \longrightarrow HAsO_2 + 2H_2O$	0.560
$Au^+ + e^- \longrightarrow Au$	1.692
$Au^{3+} + 3e^- \longrightarrow Au$	1.498
$H_3BO_3 + 3H^+ + 3e^- \longrightarrow B + 3H_2O$	−0.8698
$Ba^{2+} + 2e^- \longrightarrow Ba$	−2.912
$Be^{2+} + 2e^- \longrightarrow Be$	−1.847
$Bi^{3+} + 3e^- \longrightarrow Bi$	0.308
$BiO^+ + 2H^+ + 3e^- \longrightarrow Bi + H_2O$	0.320
$Br_2(aq) + 2e^- \longrightarrow 2Br^-$	1.0873
$Br_2(l) + 2e^- \longrightarrow 2Br^-$	1.066
$BrO_3^- + 6H^+ + 5e^- \longrightarrow \frac{1}{2}Br_2 + 3H_2O$	1.482
$BrO_3^- + 6H^+ + 6e^- \longrightarrow Br^- + 3H_2O$	1.423
$CO_2 + 2H^+ + 2e^- \longrightarrow HCO_2H$	−0.199
$Ca^{2+} + 2e^- \longrightarrow Ca$	−2.868
$Ca(OH)_2 + 2e^- \longrightarrow Ca + 2OH^-$	−3.02
$Cd^{2+} + 2e^- \longrightarrow Cd$	−0.4030
$CdSO_4 + 2e^- \longrightarrow Cd + SO_4^{2-}$	−0.246
$Cd(OH)_4^{2-} + 2e^- \longrightarrow Cd + 4OH^-$	−0.658
$Ce^{3+} + 3e^- \longrightarrow Ce$	−2.336
$Ce^{4+} + e^- \longrightarrow Ce^{3+}$	1.72
$Cl_2(g) + 2e^- \longrightarrow 2Cl^-$	1.35827
$HClO + H^+ + e^- \longrightarrow \frac{1}{2}Cl_2 + H_2O$	1.611
$HClO + H^+ + 2e^- \longrightarrow Cl^- + H_2O$	1.482
$ClO^- + H_2O + 2e^- \longrightarrow Cl^- + 2OH^-$	0.81
$ClO_3^- + 6H^+ + 5e^- \longrightarrow \frac{1}{2}Cl_2 + 3H_2O$	1.47
$ClO_3^- + 6H^+ + 6e^- \longrightarrow Cl^- + 3H_2O$	1.451
$ClO_4^- + 8H^+ + 7e^- \longrightarrow \frac{1}{2}Cl_2 + 4H_2O$	1.39
$ClO_4^- + 8H^+ + 8e^- \longrightarrow Cl^- + 4H_2O$	1.389
$Co^{2+} + 2e^- \longrightarrow Co$	−0.28
$Co^{3+} + e^- \longrightarrow Co^{2+}$	1.92
$Cr^{2+} + 2e^- \longrightarrow Cr$	−0.913
$Cr^{3+} + e^- \longrightarrow Cr^{2+}$	−0.407
$Cr^{3+} + 3e^- \longrightarrow Cr$	−0.744
$Cr_2O_7^- + 14H^+ + 6e^- \longrightarrow 2Cr^{3+} + 7H_2O$	1.232
$CrO_4^{2-} + 4H_2O + 3e^- \longrightarrow Cr(OH)_3 + 5OH^-$	−0.13
$Cs^+ + e^- \longrightarrow Cs$	−3.026
$Cu^+ + e^- \longrightarrow Cu$	0.521
$Cu^{2+} + e^- \longrightarrow Cu^+$	0.153
$Cu^{2+} + 2e^- \longrightarrow Cu$	0.3419
$CuI_2^- + e^- \longrightarrow Cu + 2I^-$	0.00

Half-Reaction	$E°$ (V)
$Cu_2O + H_2O + 2e^- \longrightarrow 2Cu + 2OH^-$	−0.360
$Dy^{3+} + 3e^- \longrightarrow Dy$	−2.295
$Er^{3+} + 3e^- \longrightarrow Er$	−2.331
$Es^{3+} + 3e^- \longrightarrow Es$	−1.91
$Eu^{2+} + 2e^- \longrightarrow Eu$	−2.812
$Eu^{3+} + 3e^- \longrightarrow Eu$	−1.991
$F_2 + 2e^- \longrightarrow 2F^-$	2.866
$Fe^{2+} + 2e^- \longrightarrow Fe$	−0.447
$Fe^{3+} + 3e^- \longrightarrow Fe$	−0.037
$Fe^{3+} + e^- \longrightarrow Fe^{2+}$	0.771
$[Fe(CN)_6]^{3-} + e^- \longrightarrow [Fe(CN)_6]^{4-}$	0.358
$Fe(OH)_3 + e^- \longrightarrow Fe(OH)_2 + OH^-$	−0.56
$Fm^{3+} + 3e^- \longrightarrow Fm$	−1.89
$Fm^{2+} + 2e^- \longrightarrow Fm$	−2.30
$Ga^{3+} + 3e^- \longrightarrow Ga$	−0.549
$Gd^{3+} + 3e^- \longrightarrow Gd$	−2.279
$Ge^{2+} + 2e^- \longrightarrow Ge$	0.24
$Ge^{4+} + 4e^- \longrightarrow Ge$	0.124
$2H^+ + 2e^- \longrightarrow H_2$	0.00000
$H_2 + 2e^- \longrightarrow 2H^-$	−2.23
$2H_2O + 2e^- \longrightarrow H_2 + 2OH^-$	−0.8277
$H_2O_2 + 2H^+ + 2e^- \longrightarrow 2H_2O$	1.776
$Hf^{4+} + 4e^- \longrightarrow Hf$	−1.55
$Hg^{2+} + 2e^- \longrightarrow Hg$	0.851
$2Hg^{2+} + 2e^- \longrightarrow Hg_2{}^{2+}$	0.920
$Hg_2Cl_2 + 2e^- \longrightarrow 2Hg + 2Cl^-$	0.26808
$Ho^{2+} + 2e^- \longrightarrow Ho$	−2.1
$Ho^{3+} + 3e^- \longrightarrow Ho$	−2.33
$H_2SeO_3 + 4H^+ + 4e^- \longrightarrow Se + 3H_2O$	0.74
$H_2SO_3 + 4H^+ + 4e^- \longrightarrow S + 3H_2O$	0.449
$I_2 + 2e^- \longrightarrow 2I^-$	0.5355
$I_3{}^- + 2e^- \longrightarrow 3I^-$	0.536
$2IO_3{}^- + 12H^+ + 10e^- \longrightarrow I_2 + 6H_2O$	1.195
$IO_3{}^- + 6H^+ + 6e^- \longrightarrow I^- + 3H_2O$	1.085
$In^+ + e^- \longrightarrow In$	−0.14
$In^{3+} + 2e^- \longrightarrow In^+$	−0.443
$In^{3+} + 3e^- \longrightarrow In$	−0.3382
$Ir^{3+} + 3e^- \longrightarrow Ir$	1.156
$K^+ + e^- \longrightarrow K$	−2.931
$La^{3+} + 3e^- \longrightarrow La$	−2.379
$Li^+ + e^- \longrightarrow Li$	−3.0401
$Lr^{3+} + 3e^- \longrightarrow Lr$	−1.96
$Lu^{3+} + 3e^- \longrightarrow Lu$	−2.28
$Md^{3+} + 3e^- \longrightarrow Md$	−1.65
$Md^{2+} + 2e^- \longrightarrow Md$	−2.40
$Mg^{2+} + 2e^- \longrightarrow Mg$	−2.372
$Mn^{2+} + 2e^- \longrightarrow Mn$	−1.185
$MnO_2 + 4H^+ + 2e^- \longrightarrow Mn^{2+} + 2H_2O$	1.224
$MnO_4{}^- + 8H^+ + 5e^- \longrightarrow Mn^{2+} + 4H_2O$	1.507
$MnO_4{}^- + 2H_2O + 3e^- \longrightarrow MnO_2 + 4OH^-$	0.595
$Mo^{3+} + 3e^- \longrightarrow Mo$	−0.200
$N_2 + 2H_2O + 6H^+ + 6e^- \longrightarrow 2NH_4OH$	0.092
$HNO_2 + H^+ + e^- \longrightarrow NO + H_2O$	0.983
$NO_3{}^- + 3H^+ + 2e^- \longrightarrow HNO_2 + H_2O$	0.934
$NO_3{}^- + 4H^+ + 3e^- \longrightarrow NO + 2H_2O$	0.957
$Na^+ + e^- \longrightarrow Na$	−2.71
$Nb^{3+} + 3e^- \longrightarrow Nb$	−1.099
$Nd^{3+} + 3e^- \longrightarrow Nd$	−2.323
$Ni^{2+} + 2e^- \longrightarrow Ni$	−0.257
$No^{3+} + 3e^- \longrightarrow No$	−1.20
$No^{2+} + 2e^- \longrightarrow No$	−2.50
$Np^{3+} + 3e^- \longrightarrow Np$	−1.856
$O_2 + 2H^+ + 2e^- \longrightarrow H_2O_2$	0.695
$O_2 + 4H^+ + 4e^- \longrightarrow 2H_2O$	1.229

Half-Reaction	***E*° (V)**
$O_2 + 2H_2O + 2e^- \longrightarrow H_2O_2 + 2OH^-$	−0.146
$O_3 + 2H^+ + 2e^- \longrightarrow O_2 + H_2O$	2.076
$OsO_4 + 8H^+ + 8e^- \longrightarrow Os + 4H_2O$	0.838
$P + 3H_2O + 3e^- \longrightarrow PH_3(g) + 3OH^-$	−0.87
$PO_4^{3-} + 2H_2O + 2e^- \longrightarrow HPO_3^{2-} + 3OH^-$	−1.05
$Pa^{3+} + 3e^- \longrightarrow Pa$	−1.34
$Pa^{4+} + 4e^- \longrightarrow Pa$	−1.49
$Pb^{2+} + 2e^- \longrightarrow Pb$	−0.1262
$PbO + H_2O + 2e^- \longrightarrow Pb + 2OH^-$	−0.580
$PbO_2 + SO_4^{2-} + 4H^+ + 2e^- \longrightarrow PbSO_4 + 2H_2O$	1.6913
$PbSO_4 + 2e^- \longrightarrow Pb + SO_4^{2-}$	−0.3588
$Pd^{2+} + 2e^- \longrightarrow Pd$	0.951
$Pm^{3+} + 3e^- \longrightarrow Pm$	−2.30
$Po^{4+} + 4e^- \longrightarrow Po$	0.76
$Pr^{3+} + 3e^- \longrightarrow Pr$	−2.353
$Pt^{2+} + 2e^- \longrightarrow Pt$	1.18
$[PtCl_4]^{2-} + 2e^- \longrightarrow Pt + 4Cl^-$	0.755
$Pu^{3+} + 3e^- \longrightarrow Pu$	−2.031
$Ra^{2+} + 2e^- \longrightarrow Ra$	−2.8
$Rb^+ + e^- \longrightarrow Rb$	−2.98
$Re^{3+} + 3e^- \longrightarrow Re$	0.300
$Rh^{3+} + 3e^- \longrightarrow Rh$	0.758
$Ru^{3+} + e^- \longrightarrow Ru^{2+}$	0.2487
$S + 2e^- \longrightarrow S^{2-}$	−0.47627
$S + 2H^+ + 2e^- \longrightarrow H_2S(aq)$	0.142
$2S + 2e^- \longrightarrow S_2^{2-}$	−0.42836
$H_2SO_3 + 4H^+ + 4e^- \longrightarrow S + 3H_2O$	0.449
$SO_4^{2-} + H_2O + 2e^- \longrightarrow SO_3^{2-} + 2OH^-$	−0.93
$Sb + 3H^+ + 3e^- \longrightarrow SbH_3$	−0.510
$Sc^{3+} + 3e^- \longrightarrow Sc$	−2.077
$Se + 2e^- \longrightarrow Se^{2-}$	−0.924
$Se + 2H^+ + 2e^- \longrightarrow H_2Se$	−0.082
$SiF_6^{2-} + 4e^- \longrightarrow Si + 6F^-$	−1.24
$Sm^{3+} + 3e^- \longrightarrow Sm$	−2.304
$Sn^{2+} + 2e^- \longrightarrow Sn$	−0.1375
$Sn^{4+} + 2e^- \longrightarrow Sn^{2+}$	−0.151
$2SO_4^{2-} + 4H^+ + 2e^- \longrightarrow S_2O_6^{2-} + H_2O$	−0.22
$Sr^{2+} + 2e^- \longrightarrow Sr$	−2.899
$Ta^{3+} + 3e^- \longrightarrow Ta$	−0.6
$TcO_4^- + 4H^+ + 3e^- \longrightarrow TcO_2 + 2H_2O$	0.782
$TcO_4^- + 8H^+ + 7e^- \longrightarrow Tc + 4H_2O$	0.472
$Tb^{3+} + 3e^- \longrightarrow Tb$	−2.28
$Te + 2e^- \longrightarrow Te^{2-}$	−1.143
$Te^{4+} + 4e^- \longrightarrow Te$	0.568
$Th^{4+} + 4e^- \longrightarrow Th$	−1.899
$Ti^{2+} + 2e^- \longrightarrow Ti$	−1.630
$Tl^+ + e^- \longrightarrow Tl$	−0.336
$Tl^{3+} + 2e^- \longrightarrow Tl^+$	1.252
$Tl^{3+} + 3e^- \longrightarrow Tl$	0.741
$Tm^{3+} + 3e^- \longrightarrow Tm$	−2.319
$U^{3+} + 3e^- \longrightarrow U$	−1.798
$VO_2^+ + 2H^+ + e^- \longrightarrow VO^{2+} + H_2O$	0.991
$V_2O_5 + 6H^+ + 2e^- \longrightarrow 2VO^{2+} + 3H_2O$	0.957
$W_2O_5 + 2H^+ + 2e^- \longrightarrow 2WO_2 + H_2O$	−0.031
$XeO_3 + 6H^+ + 6e^- \longrightarrow Xe + 3H_2O$	2.10
$Y^{3+} + 3e^- \longrightarrow Y$	−2.372
$Yb^{3+} + 3e^- \longrightarrow Yb$	−2.19
$Zn^{2+} + 2e^- \longrightarrow Zn$	−0.7618
$Zn(OH)_4^{2-} + 2e^- \longrightarrow Zn + 4OH^-$	−1.199
$Zn(OH)_2 + 2e^- \longrightarrow Zn + 2OH^-$	−1.249
$ZrO_2 + 4H^+ + 4e^- \longrightarrow Zr + 2H_2O$	−1.553
$Zr^{4+} + 4e^- \longrightarrow Zr$	−1.45

Source of data: *CRC Handbook of Chemistry and Physics,* 84th Edition (2004).

APPENDIX F

Properties of Water

Density:	0.99984 g/cm^3 at 0°C 0.99970 g/cm^3 at 10°C 0.99821 g/cm^3 at 20°C 0.98803 g/cm^3 at 50°C 0.95840 g/cm^3 at 100°C
Enthalpy (heat) of vaporization:	45.054 kJ/mol at 0°C 43.990 kJ/mol at 25°C 42.482 kJ/mol at 60°C 40.657 kJ/mol at 100°C
Surface tension:	74.23 J/m^2 at 10°C 71.99 J/m^2 at 25°C 67.94 J/m^2 at 50°C 58.91 J/m^2 at 100°C
Viscosity:	1.793 mPa·s at 0°C 0.890 mPa·s at 25°C 0.547 mPa·s at 50°C 0.282 mPa·s at 100°C
Ion-product constant, K_w:	1.15×10^{-15} at 0°C 1.01×10^{-14} at 25°C 5.18×10^{-14} at 50°C 4.99×10^{-13} at 100°C
Specific heat (C_s):	4.2176 J/(g-°C) at 0°C 4.1818 J/(g-°C) at 20°C 4.1806 J/(g-°C) at 50°C 4.2159 J/(g-°C) at 100°C

Vapor pressure of water (kPa)

T(°C)	*P*(kPa)	*T*(°C)	*P*(kPa)	*T*(°C)	*P*(kPa)	*T*(°C)	*P*(kPa)
0	0.61129	30	4.2455	60	19.932	90	70.117
5	0.87260	35	5.6267	65	25.022	95	84.529
10	1.2281	40	7.3814	70	31.176	100	101.32
15	1.7056	45	9.5898	75	38.563	105	120.79
20	2.3388	50	12.344	80	47.373	110	143.24
25	3.1690	55	15.752	85	57.815	115	169.02

Vapor pressure of water (mmHg)

T(°C)	*P*(mmHg)	*T*(°C)	*P*(mmHg)	*T*(°C)	*P*(mmHg)	*T*(°C)	*P*(mmHg)
0	4.585	30	31.844	60	149.50	90	525.92
5	6.545	35	42.204	65	187.68	95	634.02
10	9.211	40	55.364	70	233.84	100	759.96
15	12.793	45	71.929	75	289.25	105	906.00
20	17.542	50	92.587	80	355.33	110	1074.39
25	23.769	55	118.15	85	433.65	115	1267.75

Source of data: *CRC Handbook of Chemistry and Physics*, 84th Edition (2004).

APPENDIX G

Glossary

A

A: See "Activity (*A*)" (*Section 14.5*), "Frequency factor (*A*)" (*Section 14.7*), or "Mass number (*A*)" (*Section 1.6*).

Absolute zero (0 K): The lowest possible temperature that could theoretically be achieved. It corresponds to −273.15°C. (*Section 10.3*)

Absorption spectrum: A spectrum produced by the absorption of light by ground-state atoms (c.f. emission spectrum). Each element has a characteristic absorption spectrum, so scientists can use these spectra to analyze the composition of matter. (*Section 6.3*)

Accuracy: The degree to which a measured value is the same as the true value of the quantity. (*Essential Skills 1*)

Accurate: When a measured value is the same as the true value of the quantity. (*Essential Skills 1*)

Acid (Arrhenius definition): A substance with at least one hydrogen atom that can dissociate to form an anion and an H^+ ion (a proton) in aqueous solution, thereby forming an acidic solution (c.f. the Brønsted–Lowry and Lewis definitions of an acid). (*Sections 2.5 and 4.6*)

Acid (Brønsted–Lowry definition): Any substance that can donate a proton (c.f. the Arrhenius and Lewis definitions of an acid). This is essentially the same as the Arrhenius definition, but it is more general because it is not restricted to aqueous solutions. (*Section 4.6*)

Acid (Lewis definition): Any species that can *accept* a pair of electrons (c.f. the Arrhenius and Brønsted–Lowry definitions of an acid). The Lewis definition expands the Brønsted–Lowry definition to include substances other than the H^+ ion. (*Section 8.7*)

Acid–base indicator: A compound added in small amounts to an acid–base titration to signal the equivalence point by changing color. Indicators are weak acids or weak bases that exhibit intense colors that vary with pH. The conjugate acid and conjugate base of a good indicator have different colors so that they can be distinguished easily. The point in the titration at which the indicator changes color is called the endpoint. (*Sections 4.9 and 16.5*)

Acid–base reaction: A reaction of the general form acid + base $\longrightarrow$ salt (see Table 3.1 for an example). (*Section 3.5*)

Acidic oxides: Oxides, such as SO_3, that react with water to produce an acidic solution or are soluble in aqueous base (c.f. basic oxides and amphoteric oxides). Oxides of nonmetallic elements are generally acidic oxides. (*Section 17.4*)

Acid ionization constant (K_a): An equilibrium constant for the ionization (dissociation) of a weak acid (HA) in water, $HA(aq) + H_2O(l) \rightleftharpoons H_3O^+(aq) + A^-(aq)$, in which the concentration of water is treated as a constant. Thus, $K_a = [H_3O^+][A^-]/[HA]$. (*Section 16.2*)

Acid rain: Rain and snow that are dramatically more acidic because of human activities during the last 150 years. Pollutants expelled from industrial plants, such as nitrogen and sulfur oxides, combine with water in the atmosphere to form acids such as HNO_3 and H_2SO_4 that get washed back into surface water supplies via precipitation. (*Section 4.7*)

Actinide element: Any of the 14 elements between $Z = 90$ (thorium) and $Z = 103$ (lawrencium). The 5*f* orbitals are filling for the actinides and their chemistry is dominated by M^{3+} ions. The actinides and lanthanides are usually grouped together in two rows beneath the main body of the periodic table. (*Sections 1.7 and 7.4*)

Activated complex: Also called the transition state of the reaction, the activated complex is the arrangement of atoms that first forms when molecules are able to overcome the activation energy and react. The activated complex is not a reaction intermediate; it does not last long enough to be detected readily. (*Section 14.7*)

Activation energy (E_a): The energy barrier or threshold that corresponds to the amount of energy the particles in a reaction (atoms, molecules, or ions) must have in order to react when they collide. The activation energy is the minimum amount of energy needed for a reaction to occur. Reacting molecules must have enough energy to overcome electrostatic repulsion, and a minimum amount of energy is required to break chemical bonds so that new ones may be formed. (*Section 14.7*)

Active metals: The metals at the top of the activity series (the alkali metals, alkaline earths, and Al), which have the greatest tendency to lose electrons (to be oxidized) (c.f. inert metals). (*Section 4.8*)

Activity (*A*): Also called the rate of decay, the activity of a radioactive isotope is the decrease in the number of the radioisotope's nuclei per unit time (Equation 14.33): $A = -\Delta N/\Delta t$, where N is the number of atoms of the radioactive isotope. Activity is usually measured in units of disintegrations per second (dps) or disintegrations per minute (dpm). (*Section 14.5*)

Activity series: A list of metals and hydrogen in order of their relative tendency to be oxidized (see Table 4.4). The metals at the top of the series (the alkali metals, alkaline earths, and Al) have the greatest tendency to lose electrons (to be oxidized), whereas those at the bottom of the series (Pt, Au, Ag, Cu, and Hg) have the least tendency to be oxidized. (*Section 4.8*)

Actual yield: The measured mass of products obtained from a reaction. It is almost always less than the theoretical yield (often much less). (*Section 3.4*)

Addition reactions: Chemical reactions in which the components of a species A—B are added to adjacent atoms across a carbon–carbon multiple bond, addition reactions result in the reduction of carbon–carbon multiple bonds. Addition reactions have the general form $CH_2{=}CH_2 + A{-}B \longrightarrow CH_2A{-}CH_2B$ and are the reverse of elimination reactions. (*Section 24.4*)

Adduct: The product of a reaction between a Lewis acid and a Lewis base. The resulting acid–base adduct contains a coordinate covalent bond because both electrons are provided by only one of the atoms. (*Section 8.7*)

Adhesive forces: The attractive intermolecular forces between a liquid and the substance comprising the surface of a capillary (see capillary action; c.f. cohesive forces). (*Section 11.3*)

Adsorption: A physical process in which a reactant interacts with the solid surface of a heterogeneous catalyst (c.f. desorption). Adsorption causes a chemical bond in the reactant to become weak and then break. (*Section 14.8*)

Aerobic organisms: Organisms, such as humans, that cannot survive in the absence of O_2 (c.f. anaerobic organisms). (*Section 18.8*)

Aerosol: A dispersion of solid or liquid particles in a gas. An aerosol is one of three kinds of colloids (the other two are sols and emulsions). (*Section 13.7*)

Alcohol: A class of organic compounds obtained by replacing one or more of the hydrogen atoms of a hydrocarbon with an $-OH$ group. The simplest alcohol is methanol (CH_3OH). (*Section 2.4*)

Aldehyde: A class of organic compounds that has the general form RCHO, in which the carbon atom of the carbonyl group is bonded to a hydrogen atom and an R group. The R group may be either another hydrogen atom or an alkyl group (c.f. ketone). (*Section 4.1*)

Aliphatic hydrocarbons: Alkanes, alkenes, alkynes, and cyclic hydrocarbons (hydrocarbons that are not aromatic). (*Section 2.4*)

Alkali metal: Any of the elements in Group 1 of the periodic table (Li, Na, K, Rb, Cs, and Fr). All of the Group 1 elements react readily with nonmetals to give ions with a +1 charge, such as Li^+ and Na^+. (*Section 1.7*)

Alkaline battery: A battery that consists of a Leclanché cell adapted to operate under alkaline (basic) conditions. (*Section 19.5*)

Alkaline earths: The elements in Group 2 of the periodic table (Be, Mg, Ca, Sr, Ba, and Ra). All are metals that react readily with non-metals to give ions with a +2 charge, such as Mg^{2+} and Ca^{2+}. (*Section 1.7*)

Alkanes: One of the four major classes of hydrocarbons, alkanes contain only carbon–hydrogen and carbon–carbon single bonds. Alkanes are saturated hydrocarbons. The other three classes of hydrocarbons are the alkenes, alkynes, and aromatics (all of which are unsaturated hydrocarbons). (*Section 2.4*)

Alkenes: One of the four major classes of hydrocarbons, alkenes contain at least one carbon–carbon double bond. Alkenes are unsaturated hydrocarbons. The other three classes of hydrocarbons are the alkanes, alkynes, and aromatics. (*Section 2.4*)

Alkynes: One of the four major classes of hydrocarbons, alkynes contain at least one carbon–carbon triple bond. Alkynes are unsaturated hydrocarbons. The other three classes of hydrocarbons are the alkanes, alkenes, and aromatic compounds. (*Section 2.4*)

Alloy: A solid solution of two or more metals whose properties differ from those of the constituent elements. (*Sections 1.3 and 12.5*)

Alpha (α) decay: A nuclear decay reaction that results in the emission of a helium-4 nucleus as an α particle. The daughter nuclide contains two fewer protons and two fewer neutrons than the parent nuclide. Thus, α decay produces a daughter nucleus with a mass number (A) that is lower by 4 and a nuclear charge (Z) that is lower by 2 than the parent nucleus. (*Section 20.2*)

Alpha (α) particle: A helium nucleus, ${}^4_2He$. (*Section 20.2*)

Amalgams: Solutions (usually solid solutions) of metals in liquid mercury. (*Section 13.3*)

Amide (peptide) bonds: The covalent bond that links one amino acid to another in peptides and proteins. The carbonyl carbon atom of one amino acid residue bonds to the amino nitrogen atom of the other residue, eliminating a molecule of water in the process (see Equation 12.4). Thus, peptide bonds form in a condensation reaction between a carboxylic acid and an amine. (*Sections 3.5 and 12.8*)

Amine: An organic compound that has the general formula RNH_2, where R is an alkyl group. Amines can be thought of as being obtained by replacing one (or more) of the hydrogen atoms of NH_3 (ammonia) with an alkyl group. Amines, like ammonia, are bases. (*Section 2.5*)

Amorphous solid: The type of solid that forms when the atoms, molecules, or ions of a compound aggregate with no particular order (c.f. crystalline solid). Amorphous solids have irregular or curved surfaces, do not give well-resolved X-ray diffraction patterns, and melt over a wide range of temperatures. (*Section 12.1*)

Ampere (A): The fundamental SI unit of electric current, 1 ampere is defined as the flow of 1 coulomb (C) per second past a given point (i.e., $1\ A = 1\ C/s$). (*Section 19.4*)

Amphiprotic: Substances that can behave as either an acid or a base in a chemical reaction, depending on the nature of the other reactant(s). Water is amphiprotic. (*Section 16.1*)

Amphoteric oxides: Oxides that can dissolve in acid to produce water and dissolve in base to produce a soluble complex (c.f. acidic oxides and basic oxides). Most elements whose oxides exhibit amphoteric behavior are located along the diagonal line that separates the metals from the nonmetals in the periodic table. (*Section 17.4*)

Amplification mechanism: A process by which elements that are present in trace amounts can exert large effects on the health of an organism. For example, a molecule containing a trace element may be an essential part of a larger molecule that acts in turn to regulate the concentrations of other molecules. The amplification mechanism enables small variations in the concentration of the trace element to have large biological effects. (*Section 7.5*)

Amplitude: The vertical height of a wave. The amplitude is defined as half the peak-to-trough height. As the energy of a wave with a given frequency increases, so does its amplitude. (*Section 6.1*)

Anaerobic organisms: Organisms that can live only in the absence of O_2 (c.f. aerobic organisms). (*Section 18.8*)

Anion: An ion that contains more electrons than protons, resulting in a net negative charge. Anions may be monatomic (e.g., Cl^-) or polyatomic (e.g., SO_4^{2-}). (*Section 2.1*)

Anisotropic: An arrangement of molecules in which their properties depend on the direction they are measured (c.f. isotropic). Liquid crystals are anisotropic because they are not as disordered as a liquid (their molecules have some degree of alignment). (*Section 11.8*)

Anode: One of two electrodes in an electrochemical cell (galvanic/voltaic or electrolytic). The oxidation half-reaction occurs at the anode (c.f. cathode, the other electrode). (*Section 19.1*)

Antibonding molecular orbital: A molecular orbital that forms when atomic orbitals or orbital lobes of opposite sign interact to give decreased electron probability between the nuclei due to destructive reinforcement of the wave functions. Antibonding molecular orbitals are always higher in energy than the parent atomic orbitals. They are one of three types of molecular orbitals that can form when atomic orbitals interact (the other two are bonding and nonbonding molecular orbitals). (*Section 9.3*)

Aqueous solution: A solution in which water is the solvent (c.f. nonaqueous solution). (*Section 3.3 and Chapter 4 introduction*)

Arenes: See aromatic hydrocarbons. (*Section 2.4*)

Aromatic hydrocarbons: One of the four major classes of hydrocarbons, aromatics usually contain rings of six carbon atoms that can be drawn with alternating single and double bonds. Aromatics are unsaturated hydrocarbons and are sometimes called arenes. The other three classes of hydrocarbons are the alkanes, alkenes, and alkynes. (*Section 2.4*)

Arrhenius equation: An equation (Equation 14.43), $k = Ae^{-E_a/RT}$, that summarizes the collision model of chemical kinetics, where k is the rate constant, A is the frequency factor, E_a is the activation energy, R is the ideal gas constant [8.314 J/(K · mol)], and T is the absolute temperature (in k). (*Section 14.7*)

Atmosphere (atm): Also referred to as standard atmospheric pressure, it is the atmospheric pressure required to support a column of mercury exactly 760 mm tall. The atmosphere is related to other pressure units as follows: 1 atm = 760 torr = 760 millimeters of mercury (mmHg) = 101325 pascals (Pa) = 101.325 kPa. (*Section 10.2*)

Atom: The fundamental, indivisible particles of which matter is composed. (*Section 1.4*)

Atomic mass unit (amu): One-twelfth of the mass of one atom of ^{12}C; 1 amu = 1.66×10^{-24} g. (*Section 1.6*)

Atomic number (Z): The number of protons in the nucleus of an atom of an element. The atomic number is different for each element. (*Section 1.6*)

Atomic orbital: A wave function with an allowed combination of n, l, and m_l quantum numbers; a particular spatial distribution for an electron. Thus, for a given set of quantum numbers, each principal shell contains a fixed number of subshells, and each subshell contains a fixed number of orbitals. (*Section 6.5*)

Aufbau principle: The process used to build up the periodic table by adding protons one by one to the nucleus and by adding the corresponding electrons to the lowest-energy orbital available without violating the Pauli exclusion principle. (*Section 6.6*)

Average reaction rate: The reaction rate calculated for a given time interval from the concentrations of either the reactant or one of the products at the beginning of the interval (time = t_0) and at the end of the interval (t_1) (c.f. instantaneous reaction rate). (*Section 14.2*)

Avogadro's hypothesis: Equal volumes of different gases contain equal numbers of gas particles when measured at the same temperature and pressure (c.f. Avogadro's law). (*Sections 1.4 and 10.3*)

Avogadro's law: A corollary to Avogadro's hypothesis that describes the relationship between the volume and amount of a gas. According to Avogadro's law, the volume (V) of a sample of gas at constant temperature (T) and pressure (P) is directly proportional to the number of moles of gas in the sample (n) (see Equation 10.8): $V \propto n$ (at constant T and P). (*Section 10.3*)

Avogadro's number: The number of units (e.g., atoms, molecules, or formula units) in one mole: 6.022×10^{23}. (*Section 3.1*)

Azimuthal quantum number (l): One of three quantum numbers (the other two are n and m_l) used to specify any wave function, the azimuthal quantum number describes the shape of the region of space occupied by the electron. The allowed values of l depend on the value of n and can range from 0 to $n - 1$. (*Section 6.5*)

B

Band gap: In band theory, the difference in energy between the highest level of one energy band and the lowest level of the band above it. The band gap represents a set of forbidden energies that do not correspond to any allowed combinations of atomic orbitals. (*Section 12.6*)

Band theory: A theory used to describe the bonding in metals and semiconductors. Band theory assumes that the valence orbitals of the atoms in the solid interact with one another to generate a set of molecular orbitals that extend throughout the solid. (*Section 12.6*)

Bandwidth: The difference in energy between the highest and lowest energy levels in the energy band (see band theory). The energy band is proportional to the strength of the interaction between orbitals on adjacent atoms: the stronger the interaction, the larger the band width. (*Section 12.6*)

Barometer: A device used to measure atmospheric pressure (c.f. manometer). A barometer may be constructed from a long glass tube that is closed at one end. It is then filled with mercury and placed upside down in a dish of mercury without allowing any air to enter the tube (see Figure 10.4). The height of the mercury column is proportional to the atmospheric pressure. (*Section 10.2*)

Base (Arrhenius definition): A substance that produces one or more hydroxide ions (OH^-) and a cation when dissolved in aqueous solution, thereby forming a basic solution (c.f. the Brønsted–Lowry and Lewis definitions of a base). (*Sections 2.5 and 4.6*)

Base (Brønsted–Lowry definition): Any substance that can accept a proton (c.f. the Arrhenius and Lewis definitions of a base). This definition is far more general than the Arrhenius definition because the hydroxide ion (OH^-) is just one of many substances that can accept a proton. (*Section 4.6*)

Base (Lewis definition): Any species that can *donate* a pair of electrons (c.f. the Arrhenius and Brønsted–Lowry definitions of a base). All Brønsted–Lowry bases (proton acceptors) are also electron-pair donors, so the Lewis definition of a base does not contradict the Brønsted–Lowry definition. (*Section 8.7*)

Base ionization constant (K_b): An equilibrium constant for the reaction of a weak base (B) with water, $B(aq) + H_2O(l) \rightleftharpoons BH^+(aq) + OH^-(aq)$, in which the concentration of water is treated as a constant. Thus, $K_b = [BH^+][OH^-]/[B]$. (*Section 16.2*)

Basic oxides: Oxides, such as Cs_2O, that react with water to produce a basic solution or dissolve readily in aqueous acid (c.f. acidic oxides and amphoteric oxides). Oxides of metallic elements are generally basic oxides. (*Section 17.4*)

Battery: Also called a storage cell, a battery is a galvanic cell (or series of galvanic cells) that contains all the reactants needed to produce electricity (c.f. fuel cell). Batteries can be disposable (primary) or rechargeable (secondary). (*Section 19.5*)

Bent: One of two possible molecular geometries for an AB_2 molecular species (the other is linear; see Table 9.1). In the bent geometry, the B—A—B bond angle is less than 180°. (*Section 9.1*)

Beta (β) decay: A nuclear decay reaction in which a neutron is converted to a proton and a high-energy electron that is ejected from the nucleus as a β particle (c.f. positron emission). Beta decay is common in nuclei that contain too many neutrons. Although β decay does not change the mass number of the nucleus, it does increase the atomic number by 1. Thus, β decay decreases the neutron-to-proton ratio, moving the nucleus toward the band of stable nuclei. (*Section 20.2*)

Beta (β) particle: A high-energy electron is that is ejected from a nucleus, ${}_{-1}^{0}\beta$. (*Section 20.2*)

Bilayer: A two-dimensional sheet consisting of a double layer of phospholipid molecules arranged tail to tail. As a result, the hydrophobic tails of the phospholipids are located in the center of the bilayer, where they are *not* in contact with water, and their hydrophilic heads are on the two surfaces, in contact with the surrounding aqueous solution. (*Section 13.7*)

Blackbody radiation: The energy emitted by an object when it is heated. Blackbody radiation is electromagnetic radiation whose wavelength (and color) depends on the temperature of the object (see Figure 6.5). (*Section 6.2*)

Body-centered cubic (bcc) unit cell: A cubic unit cell that contains eight component atoms, molecules, or ions located at the corners of a cube, as well as an identical component in the center of the cube (see Figure 12.5b). (*Section 12.2*)

Boiling-point elevation (ΔT_b): The difference between the boiling point of a solution and the boiling point of the pure solvent (see Equation 13.20): $\Delta T_b = T_b - T_b^\circ$, where T_b is the boiling point of the solution and T_b° is the boiling point of the pure solvent. ΔT_b is proportional to the molality of the solution (see Equation 13.21). (*Section 13.6*)

Boltzmann distribution: A curve that shows the distribution of molecular speeds at a given temperature (see Figure 10.14). The actual values of speed and kinetic energy are not the same for all particles of a gas but are given by a Boltzmann distribution, in which some molecules have higher or lower speeds (and kinetic energies) than average. (*Section 10.7*)

Bomb calorimeter: A device used to measure enthalpy changes in chemical processes at constant volume (calorimetry). A bomb calorimeter is one kind of constant-volume calorimeter. The heat released by a reaction carried out at constant volume is identical to the change in internal energy (ΔE) rather than the enthalpy change (ΔH), although the difference is usually quite small (on the order of a few percent). (*Section 5.3*)

Bond angles: The angles between bonds. (*Section 9.1*)

Bond distance (r_0): The optimal internuclear distance between two bonded atoms in a molecule. (*Section 8.1*)

Bond energy: The enthalpy change that occurs when a given bond in a gaseous molecule is broken (c.f. lattice energy). (*Section 8.1*)

Bonding molecular orbital: A molecular orbital that forms when atomic orbitals or orbital lobes with the same sign interact to give increased electron probability between the nuclei due to constructive reinforcement of the wave functions. Bonding molecular orbitals are always lower in energy than the parent atomic orbitals. They are one of three types of molecular orbitals that can form when atomic orbitals interact (the other two are antibonding and nonbonding molecular orbitals). (*Section 9.3*)

Bonding pair: A pair of electrons in a Lewis structure that is shared by two atoms, thus forming a covalent bond (c.f. lone pair). (*Section 8.5*)

Bond order (Lewis bonding model): The number of electron pairs that hold two atoms together. For a single bond, the bond order is 1; for a double bond, the bond order is 2; and for a triple bond, the bond order is 3. (*Section 8.8*)

Bond order (molecular orbital theory): One-half the net number of bonding electrons in a molecule. When calculating the bond order, electrons in antibonding molecular orbitals cancel electrons in bonding molecular orbitals, while electrons in nonbonding molecular orbitals have no effect and are not counted. (*Section 9.3*)

Born–Haber cycle: A thermochemical cycle developed by Max Born and Fritz Haber in 1919, the Born–Haber cycle describes a process in which an ionic solid is conceptually formed from its component elements in a stepwise manner (see Figure 8.4). The Born–Haber cycle, combined with Hess's law and experimentally determined enthalpy changes for other chemical processes, can be used to determine enthalpy changes for processes that otherwise could *not* be measured experimentally. (*Section 8.3*)

Bragg equation: The equation that describes the relationship between two X-ray beams diffracted from different planes of atoms (see Equation 12.1): $2d \sin \theta = n\lambda$, where d is the distance separating the two planes of atoms, θ is the angle of incidence of the two X-ray beams (i.e., the angle between the X-ray beams and the planes of the crystal), n is an integer (to distinguish one layer from another), and λ is the wavelength of the X rays. Two X rays that are in phase reinforce each other, whereas two X rays that are out of phase interfere destructively, effectively canceling each other. (*Section 12.3*)

Breeder reactor: A nuclear fission reactor that produces more fissionable fuel than it consumes. This process does not violate the first law of thermodynamics because the fuel produced is different from the fuel consumed. (*Section 20.5*)

BSC theory: A theory, first formulated by J. Bardeen, L. Cooper, and J. R. Schrieffer, that attempts to explain the phenomenon of superconductivity. According to BSC theory, electrons can travel through a superconducting material without resistance because they couple to one another to form pairs of electrons (called Cooper pairs). (*Section 12.7*)

Buffers: Solutions that maintain a relatively constant pH when an acid or a base is added. Buffers protect other molecules in solution from the effects of the added acid or base. Buffers contain either a weak acid (HA) and its conjugate base (A^-) or a weak base (B) and its conjugate acid (BH^+). (*Section 16.6*)

Buffer capacity: The amount of strong acid or strong base that a buffer solution can absorb before the pH changes dramatically. The buffer capacity depends solely on the concentrations of the species in the buffered solution. The more concentrated the buffer solution, the greater its buffer capacity. (*Section 16.6*)

C

***c*:** See "Speed of light (c)" (*Section 6.1*).

Calorie (cal): A non-SI unit of energy, 1 calorie = 4.184 joules exactly. As a result, 1 J = 0.2390 cal. (*Section 5.1*)

Calorie (Cal): The nutritional Calorie used to indicate the caloric content of food. It is equal to 1 kilocalorie (kcal). (*Section 5.4*)

Calorimetry: The set of techniques (experimental procedures) used to measure enthalpy changes in chemical processes with devices called calorimeters. (*Section 5.3*)

Capillary action: The tendency of a polar liquid to rise against gravity into a small-diameter glass tube (a capillary). Capillary action is the net result of two opposing forces: cohesive forces (which hinder capillary action) and adhesive forces (which encourage capillary action). (*Section 11.3*)

Carbanion: One of three unstable, highly reactive species that can form when a C—H bond is cleaved (the other two are carbocations and radicals), carbanions form when both bonding electrons remain with the carbon, generating a C^- species that has eight valence electrons (and is thus electron rich). Carbanions are nucleophiles. Carbanions are stabilized by groups that withdraw electrons (they are destabilized by groups that donate electrons), so primary carbanions ($^-CH_2R$) are more stable than secondary carbanions ($^-CHR_2$), and secondary carbanions are more stable than tertiary carbanions ($^-CR_3$). (*Section 24.3*)

Carbocation: One of three unstable, highly reactive species that can form when a C—H bond is cleaved (the other two are carbanions and radicals), carbocations form when both bonding electrons remain with the hydrogen, generating a C^+ species that has only six valence electrons (and is thus electron deficient). Carbocations are electrophiles. Carbocations are stabilized by groups that donate electrons, so tertiary carbocations ($^+CR_3$) are more stable than secondary carbocations ($^+CHR_2$), and secondary carbocations are more stable than primary carbocations ($^+CH_2R$). (*Section 24.3*)

Carbohydrates: Polyhydroxy aldehydes or polyhydroxy ketones (i.e., aldehydes or ketones with multiple —OH groups attached to the parent hydrocarbon), carbohydrates have the general formula $C_n(H_2O)_m$. (*Section 24.6*)

Carbon cycle: The distribution and flow of carbon throughout the planet (see Figure 5.21). (*Section 5.5*)

Carbonyl group: A carbon atom double-bonded to an oxygen atom. It is a characteristic feature of many organic compounds, including aldehydes, ketones, and carboxylic acids. (*Section 2.5*)

Carboxylic acid: An organic compound that contains an $-OH$ group covalently bonded to the carbon atom of a carbonyl group. The general formula of carboxylic acids is RCO_2H. Carboxylic acids are covalent compounds, but they dissociate to produce H^+ and RCO_2^- ions when they dissolve in water. (*Section 2.5*)

Catalysis: The acceleration of a chemical reaction by a catalyst. (*Section 3.5*)

Catalyst: A substance that participates in a reaction and causes it to occur more rapidly but that can be recovered unchanged at the end of the reaction and reused. Catalysts, which may be homogeneous or heterogeneous, may also control which products are formed in a reaction. Catalysts are not involved in the overall stoichiometry of the reaction, but they must appear in at least one of the elementary steps in the mechanism for the catalyzed reaction. The catalyzed pathway has a lower activation energy (E_a), but the net change in energy that results from the reaction (i.e., the difference between the energy of the reactants and the energy of the products) is unaffected by the presence of a catalyst. (*Sections 3.5 and 14.8*)

Cathode: One of two electrodes in an electrochemical cell (galvanic/voltaic or electrolytic). The reduction half-reaction occurs at the cathode (c.f. anode, the other electrode). (*Section 19.1*)

Cation: An ion that contains fewer electrons than protons, resulting in a net positive charge. Cations may be monatomic (e.g., Na^+) or polyatomic (e.g., NH_4^+). (*Section 2.1*)

Cell: A collection of molecules, capable of reproducing itself, that is surrounded by a phospholipid bilayer. (*Section 13.7*)

Cell membrane: A mixture of phospholipids that form a phospholipid bilayer around the cell. (*Section 13.7*)

Cell potential: See "Potential (E_{cell})" (*Section 19.1*).

Ceramic: Any nonmetallic inorganic solid that is strong enough to be used in structural applications. Ceramics are typically strong and have high melting points, but they are brittle. Ceramics can be classified as ceramic oxides or nonoxide ceramics. (*Section 12.9*)

Ceramic-matrix composites: A composite consisting of reinforcing fibers embedded in a ceramic matrix. The fibers are generally ceramic, too. (*Section 12.9*)

Cesium chloride structure: The unit cell for many ionic compounds that contain relatively large cations and a 1:1 cation:anion ratio (including CsCl). The cesium chloride structure consists of a simple cubic lattice of Cl^- anions with a Cs^+ cation in the center of the cubic cell (i.e., in the cubic hole) (see Figure 12.9). (*Section 12.3*)

Chain reactions: Reaction mechanisms in which one or more elementary reactions that contain a highly reactive species repeat again and again during the reaction process. Chain reactions have three stages: initiation, propagation, and termination. (*Section 14.6*)

Chalcogens: The elements in Group 16 of the periodic table (O, S, Se, Te, and Po). (*Section 7.4*)

Change in enthalpy (ΔH): At constant pressure, the change in enthalpy of a system is identical to the heat transferred from the surroundings to the system (or vice versa): $\Delta H = q_p$. Because enthalpy is a state function, the magnitude of ΔH depends only on the initial and final states of the system, not on the path taken. (*Section 5.2*)

Chemical bond: The attractive interactions between atoms that hold them together in compounds. Chemical bonds are generally divided into two fundamentally different kinds: ionic and covalent. (*Section 2.1*)

Chemical energy: One of the five forms of energy, chemical energy is stored within a chemical compound because of a particular arrangement of atoms. The other four forms of energy are radiant, thermal, nuclear, and electrical. (*Section 5.1*)

Chemical equation: An expression that gives the identities and quantities of the substances in a chemical reaction. Chemical formulas and other symbols are used to indicate the reactants (on the left) and the products (on the right). An arrow points from the reactants to the products. (*Section 3.3*)

Chemical equilibrium: The point at which the rates of the forward and reverse reactions become the same, so that the net composition of the system no longer changes with time. (*Chapter 15 introduction*)

Chemical kinetics: The study of reaction rates. Factors that affect reaction rates include reactant concentrations, temperature, physical states and surface areas of reactants, as well as solvent and catalyst properties if either is present. By studying the kinetics of a reaction, chemists gain insights into how to control reaction conditions to achieve a desired outcome. (*Chapter 14 introduction*)

Chemical nomenclature: The systematic language of chemistry, which makes it possible to recognize and name the most common kinds of compounds. Nomenclature can be used to name a compound from its structure or draw its structure from its name. (*Chapter 2 introduction*)

Chemical property: The characteristic ability of a substance to react to form new substances (e.g., flammability and susceptibility to corrosion). (*Section 1.3*)

Chemical reaction: A process in which a substance is converted to one or more other substances that have different compositions and properties. A chemical reaction changes only the distribution of atoms, *not* the number of atoms. In most chemical reactions, bonds are broken in the reactants and new bonds are formed to create the products. (*Chapter 3 introduction*)

Chemistry: The study of matter and the changes that material substances undergo. (*Section 1.1*)

Chemotrophs: Organisms, such as animals and fungi, whose source of energy is chemical compounds from their environment, usually obtained by consuming or breaking down other organisms (c.f. phototrophs). (*Section 18.8*)

Cholesteric phase: One of three different ways that most liquid crystals can orient themselves (the other two are the nematic and smectic phases). In the cholesteric phase, the molecules are arranged in planes (similar to the smectic phase), but each layer is rotated by a certain amount with respect to those above and below it, giving it a helical structure. (*Section 11.8*)

cis: One of two types of geometrical isomers (the other type is trans), the cis isomer has ligands or substituents adjacent to one another in a rigid molecule or metal complex. (*Section 23.4*)

Clausius–Clapeyron equation: A linear relationship that expresses the nonlinear relationship between the vapor pressure of a liquid and temperature (see Equation 11.1): $\ln P = -\Delta H_{vap}/RT + C$, where P is pressure, ΔH_{vap} is the heat of vaporization, R is the universal gas constant, T is the absolute temperature, and C is a constant. The Clausius–Clapeyron equation can be used to calculate the heat of vaporization of a liquid from its measured vapor pressure at two or more temperatures. (*Section 11.4*)

Cleavage reaction: A chemical reaction that has the general form $AB \longrightarrow A + B$ (see Table 3.1 for examples). Cleavage reactions are the reverse of condensation reactions. (*Section 3.5*)

Closed system: One of three kinds of system, a closed system can exchange energy but not matter with its surroundings. The other kinds of system are open and isolated. (*Sections 5.2 and 18.1*)

Coal: A complex solid material derived primarily from plants that died and were buried hundreds of millions of years ago and were subsequently subjected to high temperatures and pressures. It is used as a fuel. (*Section 5.5*)

Coefficient: The number greater than 1 that precedes a formula in a balanced chemical equation. When no coefficient is written in front of a species, the coefficient is assumed to be 1. The coefficient indicates the number of atoms, molecules, or formula units of a reactant or product in a balanced chemical equation. (*Section 3.3*)

Cohesive forces: The intermolecular forces that hold a liquid together (c.f. adhesive forces). Cohesive forces hinder capillary action. (*Section 11.3*)

Colligative properties: Properties of solutions (e.g., changes in freezing point or changes in boiling point) that depend primarily on the *number* of solute particles rather than the *kind* of solute particles. (*Section 13.6*)

Colloid: A heterogeneous mixture of particles with diameters of about 2–1000 nm that are distributed throughout a second phase. The dispersed particles do not separate from the dispersing phase upon standing (c.f. suspension). Colloids can be classified as sols, aerosols, or emulsions. (*Section 13.7*)

Combustion: The burning of a material in an oxygen atmosphere. (*Section 1.4*)

Combustion reaction: An oxidation–reduction reaction in which the oxidant is O_2. (*Section 3.5*)

Common ion effect: The shift in equilibrium that results when a strong electrolyte is added that contains one ion in common with a reaction system that is at equilibrium. The equilibrium shifts in the direction that reduces the concentration of the common ion (in accordance with Le Châtelier's principle). (*Section 16.6*)

Complete ionic equation: A chemical equation that shows which ions and molecules are hydrated and which are present in other forms and phases. The complete ionic equation shows what is actually going on in solution (c.f. overall equation and net ionic equation). (*Section 4.4*)

Complex ion: An ionic species formed between a central metal ion and one or more surrounding ligands. A complex ion forms from a metal ion and a ligand because of a Lewis acid–base interaction. The positively charged metal ion acts as the Lewis acid and the ligand acts as the Lewis base. Small, highly charged metal ions have the greatest tendency to act as Lewis acids, so they have the greatest tendency to form complex ions, too. Metal complexes that are either positively charged, such as $[Nd(H_2O)_9]^{3+}$, or negatively charged, such as UF_8^{4-}. (*Sections 17.3 and 23.4*)

Composite materials: Materials that consist of at least two distinct phases, the matrix (which constitutes the bulk of the material) and fibers or granules that are embedded within the matrix and that limit the growth of cracks by pinning defects in the bulk material. Composite materials are stronger, tougher, stiffer, and more resistant to corrosion than either component alone. (*Section 12.9*)

Compound: A pure substance that contains two or more elements and has chemical and physical properties that are usually different from those of the elements of which it is composed. With only a few exceptions, a particular compound has the same elemental composition (the same elements in the same proportions) regardless of its source or history. (*Section 1.3*)

Concentration: The quantity of solute that is dissolved in a particular quantity of solvent or solution. (*Section 4.2*)

Concentration cell: An electrochemical cell in which the anode and cathode compartments are identical except for the concentration of a reactant. As the electrochemical reaction proceeds, the difference between the concentrations of the reactant in the two cells decreases, and so does E_{cell}. Equilibrium is reached when the concentration of the reactant is the same in both cells, in which case the measured potential difference between the two compartments is zero ($E_{cell} = 0$). (*Section 19.4*)

Condensation: The physical process by which atoms or molecules in the vapor phase enter the liquid phase. To condense, vapor molecules must first collide with the surface of the liquid. Condensation is the opposite of evaporation or vaporization. (*Section 11.4*)

Condensation reaction: A chemical reaction that has the general form $A + B \longrightarrow AB$ (see Table 3.1 for examples). Condensation reactions are the reverse of cleavage reactions. Some, but not all, condensation reactions are also oxidation–reduction reactions. (*Section 3.5*)

Conduction band: The band of empty molecular orbitals in a semiconductor. Exciting electrons from the filled valence band to the empty conduction band increases the electrical conductivity of a material. (*Section 12.6*)

Conformational isomers: Isomers whose three-dimensional structures differ because of rotation about a sigma bond (c.f. structural isomer, and stereoisomer). (*Section 24.2*)

Conjugate acid: The substance formed when a Brønsted–Lowry base accepts a proton. (*Section 4.6*)

Conjugate acid–base pair: An acid and a base that differ by only one hydrogen ion. All acid–base reactions involve two conjugate acid–base pairs, the Brønsted–Lowry acid and the base it forms after donating its proton, and the Brønsted–Lowry base and the acid it forms after accepting a proton. (*Sections 4.6 and 16.2*)

Conjugate base: The substance formed when a Brønsted–Lowry acid donates a proton. (*Section 4.6*)

Constant-pressure calorimeter: A device used to measure enthalpy changes in chemical processes at constant pressure (calorimetry). Because ΔH is defined as the heat flow at constant pressure, measurements made using a constant-pressure calorimeter give ΔH values directly. (*Section 5.3*)

Cooling curve: A plot of the temperature of a substance versus the heat removed or versus the cooling time at a constant rate of cooling. Cooling curves relate temperature changes to phase transitions (c.f. heating curve). A cooling curve is not necessarily the opposite of a heating curve because supercooled liquids may form. (*Section 11.5*)

Cooper pairs: Pairs of electrons that migrate through a superconducting material as a unit. Cooper pairs may be the reason electrons are able to travel through a superconducting solid without resistance. (*Section 12.7*)

Coordinate covalent bond: A covalent bond in which both electrons come from the same atom (see adduct). Coordinate covalent bonds are particularly common in compounds that contain atoms with fewer than an octet of electrons. Once formed, a coordinate covalent bond behaves like any other covalent single bond. (*Section 8.5*)

Coordination compound: A chemical compound that contains one or more metal complexes. (*Section 23.4*)

Coordination number: The number of nearest neighbors in a solid structure such as the different cubic and close-packed structures. The hexagonal close-packed and cubic close-packed arrangements result in a coordination number of 12 for each atom in the lattice, whereas the simple cubic and body-centered cubic lattices have coordination numbers of 6 and 8, respectively. (*Section 12.2*)

Corrosion: A galvanic process by which metals deteriorate through oxidation, usually but not always to their oxides. (*Section 19.6*)

Coulomb (C): The SI unit of measure for the number of electrons that pass a given point in 1 second, the coulomb is defined as 6.25×10^{18} e^-/s. The coulomb relates electrical potential (in volts) to energy (in joules): 1 J/1 V = 1 coulomb = 6.25×10^{18} e^-/s (Equation 19.32). (*Section 19.4*)

Covalent atomic radius: Half the distance between the nuclei of two like atoms joined by a covalent bond in the same molecule (i.e., Cl_2 or N_2; see Figure 7.5a). (*Section 7.2*)

Covalent bond: The electrostatic attraction between the positively charged nuclei of the bonded atoms and the negatively charged electrons they share. (*Section 2.1*) A chemical bond in which the electrons are shared equally between the two bonding atoms in a molecule or polyatomic ion (c.f. polar covalent bond). (*Section 8.1*)

Covalent compound: A compound that consists of discrete molecules (c.f. ionic compound). (*Section 2.1*)

Covalent solid: A solid that consists of two- or three-dimensional networks of atoms held together by covalent bonds (c.f. ionic solid, molecular solid, and metallic solid). Covalent solids tend to be very hard and have high melting points. (*Section 12.5*)

C_p: See "Molar heat capacity (C_p)" (*Section 5.3*).

Cracking: A process in petroleum refining in which the larger and heavier hydrocarbons in the kerosene and higher-boiling-point fractions are heated to temperatures as high as 900°C to convert them to lighter molecules similar to those in the gasoline fraction. The high temperature causes C—C bonds to break ("crack"), thus converting less volatile, lower-value fractions to more volatile, higher-value mixtures that have carefully controlled formulas. (*Section 2.6*)

Critical mass: The minimum mass of a fissile isotope capable of supporting sustained fission (a nuclear chain reaction). The critical mass depends on the purity of the fissile material, the shape of the mass, and the identity of the isotope. (*Section 20.4*)

Critical point: The combination of the critical temperature and the critical pressure of a substance. (*Section 11.6*)

Critical pressure: The minimum pressure needed to liquefy a substance at its critical temperature. (*Section 11.6*)

Critical temperature (T_c): The highest temperature at which a substance can exist as a liquid, regardless of the applied pressure. Above the critical temperature, molecules have too much kinetic energy for their intermolecular attractive forces to be able to hold them together in a separate liquid phase. Instead, the substance forms a single phase that completely occupies the volume of the container. (*Section 11.6*)

Crown ethers: Cyclic polyethers that contain four or more oxygen atoms separated by two or three carbon atoms, all crown ethers have a central cavity that can accommodate a metal ion coordinated to the ring of oxygen atoms (c.f. cryptand). Crown ethers with rings of different sizes prefer to bind metal ions that fit into the cavity. Thus, 14-crown-4 favors Li^+, whereas 18-crown-6 favors the larger K^+. Crown ethers (and cryptands) are often used to dissolve simple inorganic salts such as $KMnO_4$ in nonpolar organic solvents. (*Sections 13.3 and 21.3*)

Cryogenic liquids: The ultracold liquids formed from the liquefaction of gases. Cryogenic liquids have applications as refrigerants in both industry and biology. (*Section 10.8*)

Cryptands: More nearly spherical analogues of crown ethers, cryptands consist of three $(-OCH_2CH_2O-)_n$ chains connected by two nitrogen atoms. Cryptands can completely surround (encapsulate) a metal ion of the appropriate size, coordinating to the metal by a lone pair of electrons on each O atom and the two N atoms (see Figure 13.7b). Like crown ethers, cryptands with different cavity sizes are highly selective for metal ions of particular sizes. Cryptands (and crown ethers) are often used to dissolve simple inorganic salts such as $KMnO_4$ in nonpolar organic solvents. (*Sections 13.3 and 21.3*)

Crystal field splitting energy (Δ_o): The difference in energy between the e_g set of d orbitals (d_{z^2} and $d_{x^2-y^2}$) and the t_{2g} set of d orbitals (d_{xy}, d_{xz}, and d_{yz}) that results when the five d orbitals are placed in an octahedral crystal field. The magnitude of the splitting depends on the charge on the metal ion, the position of the metal in the periodic table, and the nature of the ligands. The splitting of the d orbitals in a crystal field does not change the total energy of the five d orbitals. (*Section 23.5*)

Crystal field stabilization energy (CFSE): The additional stabilization of a metal complex by selective population of the lower-energy d orbitals (the t_{2g} orbitals), the crystal field stabilization energy of a complex can be calculated by multiplying the number of electrons in t_{2g} orbitals by the energy of those orbitals ($-0.4\Delta_o$), multiplying the number of electrons in e_g orbitals by the energy of those orbitals ($+0.6\Delta_o$), and summing the two. Table 23.11 lists CFSE values for octahedral complexes with different d-electron configurations. (*Section 23.5*)

Crystal field theory (CFT): A bonding model that explains many important properties of transition-metal complexes, including their colors, magnetism, structures, stability, and reactivity. The central assumption of CFT is that metal–ligand interactions are purely electrostatic in nature. (*Section 23.5*)

Crystal lattice: The regular, repeating three-dimensional structure that a crystalline solid forms. A crystal lattice is one of two general ways in which the atoms, molecules, or ions of a solid can be arranged (the other is as an amorphous solid). (*Section 12.1*)

Crystalline solid: The type of solid that forms when the atoms, molecules, or ions of a compound produce a crystal lattice (c.f. amorphous solid). Crystalline solids have well-defined edges and faces, diffract X rays, and tend to have sharp melting points. (*Section 12.1*)

Crystallization: A physical process used to separate homogeneous mixtures (solutions) into their component substances. Crystallization separates mixtures based on differences in their solubilities. (*Section 1.3*)

C_s: See "Specific heat (C_s)" (*Section 5.3*).

Cubic close-packed (ccp) structure: One of two variants of the close-packed arrangement, the most efficient way to pack spheres in a lattice [the other is the hexagonal close-packed (hcp) arrangement]. The cubic close-packed structure results from arranging the atoms of the solid so that their positions alternate from layer to layer in the pattern ABCABC . . . (see Figure 12.7b). (*Section 12.2*)

Cubic hole: The hole located at the center of the simple cubic lattice. The hole is equidistant from all eight atoms or ions at the corners of the unit cell. An atom or ion in a cubic hole has a coordination number of 8. (*Section 12.3*)

Cyclic hydrocarbon: A hydrocarbon in which the ends of the carbon chain are connected to form a ring of covalently bonded carbon atoms. The simplest cyclic hydrocarbon is cyclopropane (C_3H_6). (*Section 2.4*)

D

***d*:** See "Density (*d*)" (*Section 1.3*).

Dalton's law of partial pressures: The total pressure exerted by a mixture of gases is the sum of the partial pressures of the component gases. (*Section 10.5*)

***d* block:** The 10 columns of the periodic table that fall between the *s* block and the *p* block. The elements in the *d* block are filling their $(n - 1)d$ orbitals (see Figure 6.34; c.f. *s* block, *p* block, and *f* block). (*Section 6.6*)

***d*-block elements:** The elements of the 10 columns of the periodic table that fall between the *s* block and the *p* block. The elements in the *d* block are filling their $(n - 1)d$ orbitals (see Figure 6.34; c.f. *s*-block elements, *p*-block elements, and *f*-block elements). (*Section 6.6*)

Defects: Errors in an idealized crystal lattice. Real crystals contain large numbers of defects (typically 10^4 per milligram), ranging from variable amounts of impurities to places where atoms or ions are missing or misplaced. Defects can affect a single point in the lattice (point defects), a row of lattice points (line defects), or a plane of points (plane defects). (*Section 12.4*)

Degenerate: Having the same energy. All orbitals with the same value of *n* (e.g., the three 2*p* orbitals) are degenerate because they all have the same energy. (*Section 6.5*)

Density (*d*): An intensive property of matter, density is the mass per unit volume (and is usually expressed in grams per cubic centimeter, g/cm^3). At a given temperature and pressure, the density of a pure substance is a constant. (*Section 1.3*)

Desorption: The process by which products of heterogeneous catalysis leave the surface of the solid catalyst. Desorption is the opposite of adsorption. (*Section 14.8*)

Deuterium: One of the three isotopes of hydrogen, deuterium (^2H or D) consists of one proton, one neutron, and one electron. The other two isotopes of hydrogen are protium and tritium. (*Section 21.2*)

Dialysis: A process that uses a semipermeable membrane with pores large enough to allow small solute molecules and solvent molecules to pass through, but not large solute molecules (e.g., proteins). (*Section 13.6*)

Dielectric constant (ε): A constant that expresses the ability of a bulk substance (e.g., a particular solvent) to decrease the electrostatic forces between two charged particles. The dielectric constant indicates the tendency of a solvent to dissolve ionic compounds. A solvent that has a high dielectric constant causes the charged particles of an ionic substance to behave as if they have been moved farther apart, so their electrostatic attraction is reduced. (*Section 13.3*)

Differential rate law: A rate law that expresses the rate of a reaction in terms of changes in the concentration of one or more reactants, $\Delta[R]$, over a specific time interval, Δt (c.f. integrated rate law). (*Section 14.2*)

Diffusion: The gradual mixing of gases due to the motion of their component particles even in the absence of mechanical agitation such as stirring (c.f. effusion). The result is a gas mixture with a uniform composition. The rate of diffusion of a gaseous substance is inversely proportional to the square root of its molar mass: rate $\propto 1/(M)^{1/2}$ (Graham's law). (*Section 10.7*)

Dipole–dipole interactions: A kind of intermolecular interaction (force) that results between molecules that have net dipole moments. These molecules tend to align themselves so that the positive end of one dipole is near the negative end of another, and vice versa (see Figure 11.3a and b). These attractive interactions are more stable than arrangements in which the positive and negative ends of two dipoles are adjacent. Dipole–dipole interactions and London dispersion forces are often called van der Waals forces. (*Section 11.2*)

Dipole moment (μ): Produced by the asymmetrical charge distribution in a polar substance, the dipole moment is the product of the partial charge *Q* on the bonded atoms and the distance *r* between the partial charges: $\mu = Qr$, where *Q* is measured in coulombs (C) and *r* in meters (m). The unit for dipole moments is the debye (D): $1\ D = 3.3356 \times 10^{-30}$ C.m. Any diatomic molecule that contains a polar covalent bond has a dipole moment, but in polyatomic molecules the presence or absence of a net dipole moment depends on the molecular geometry. For some symmetrical AB_n structures, the individual bond dipole moments cancel one another, giving a net dipole moment of zero. (*Sections 8.9 and 9.1*)

Diprotic acid: A compound that can donate two hydrogen ions per molecule in separate steps (e.g., H_2SO_4). Diprotic acids are one kind of polyprotic acids (see also triprotic acid). (*Section 4.6*)

Disposable batteries: Also called primary batteries, the electrode reactions in disposable batteries are effectively irreversible and cannot be recharged (c.f. rechargeable batteries). (*Section 19.5*)

Distillation: A physical process used to separate homogeneous mixtures (solutions) into their component substances. Distillation makes use of differences in the volatilities of the component substances. (*Section 1.3*)

Doping: The process of deliberately introducing small amounts of impurities into commercial semiconductors in order to tune their electrical properties for specific applications (see *n*-type semiconductor and *p*-type semiconductor). (*Section 12.6*)

Double bond: A chemical bond formed when two atoms share two pairs of electrons. (*Section 2.1*)

Ductile: The ability to be pulled into wires. Metals are ductile, whereas nonmetals are usually brittle instead. (*Section 1.7*)

Dynamic equilibrium: A state in which two opposing processes (e.g., evaporation and condensation) occur at the same rate, thus producing no *net* change in the system. (*Section 11.4*)

E

***E*:** See "Internal energy (*E*)" (*Section 5.2*).

***EA*:** See "Electron affinity (*EA*)" (*Section 7.3*).

E_a: See "Activation energy (E_a)" (*Section 14.7*).

E_{cell}: See "Potential (E_{cell})" (*Section 19.1*)

E°_{cell}: See "Standard cell potential (E°_{cell})" (*Section 19.2*).

Edge dislocation: A crystal defect that results from the insertion of an extra plane of atoms into part of the crystal lattice. The edge dislocation causes the planes of atoms in the lattice to become kinked where the extra plane of atoms begins (see Figure 12.16). (*Section 12.4*)

Effective nuclear charge (Z_{eff}): The nuclear charge an electron actually experiences because of shielding from other electrons closer to the nucleus. Z_{eff} is less than the actual nuclear charge (*Z*) because the intervening electrons neutralize a portion of the positive charge of the nucleus and thereby decrease the attractive interaction between it and the electron farther away. (*Sections 6.5 and 21.1*)

Effusion: The escape of a gas through a small (usually microscopic) opening into an evacuated space (c.f. diffusion). The rate of effusion of a gaseous substance is inversely proportional to the square root of its molar mass: rate $\propto 1/(M)^{1/2}$ (Graham's law). Heavy molecules effuse through a porous material more slowly than light molecules (c.f. Figure 10.17). (*Section 10.7*)

Electrical energy: One of the five forms of energy, electrical energy results from the flow of electrically charged particles. The other four forms of energy are radiant, thermal, chemical, and nuclear. (*Section 5.1*)

Electrical insulators: Materials that conduct electricity poorly because their valence bands are full. The energy gap between the highest filled levels and the lowest empty levels is so large in an electrical insulator that the empty levels are inaccessible. Thermal energy cannot excite an electron from a filled level to an empty one. (*Section 12.6*)

Electrochemical cell: An apparatus that is used to generate electricity from a spontaneous oxidation–reduction (redox) reaction or, conversely, that uses electricity to drive a nonspontaneous redox reaction. (*Section 19.1*)

Electrochemistry: The study of the relationship between electricity and chemical reactions. (*Chapter 19 introduction*)

Electrodes: Solid metals connected to an external circuit that provides an electrical connection between systems in an electrochemical cell (galvanic or electrolytic). The oxidation half-reaction occurs at one electrode (the anode) and the reduction half-reaction occurs at the other (the cathode). The electrodes in an electrochemical cell are also connected by an electrolyte. (*Section 19.1*)

Electrolysis: An electrochemical process in which an external voltage is applied in an electrolytic cell to drive a nonspontaneous reaction. Electrolysis is the opposite process to that which occurs in galvanic cells. (*Section 19.7*)

Electrolyte: An ionic substance or solution that allows ions to transfer between the two electrodes of an electrochemical cell (galvanic/voltaic or electrolytic), thereby maintaining electrical neutrality (see salt bridge). (*Sections 4.1 and 19.1*)

Electrolytic cell: One of two types of electrochemical cells, an electrolytic cell consumes electrical energy from an external source, using it to cause a nonspontaneous oxidation–reduction (redox) reaction ($\Delta G > 0$) to occur (c.f. galvanic cell or voltaic cell, the other type of electrochemical cell). (*Section 19.1*)

Electromagnetic radiation: Energy that is transmitted, or radiated, through space in the form of periodic oscillations of an electric and a magnetic field (see Figure 6.3). All forms of electromagnetic radiation (e.g., microwaves, visible light, and gamma rays) consist of perpendicular oscillating electric and magnetic fields. (*Section 6.1*)

Electron: A subatomic particle with a mass of 9.109×10^{-28} g (0.0005486 amu) and an electrical charge of -1.602×10^{-19} C. The charge on the electron is equal in magnitude but opposite in sign to the charge on a proton. (*Section 1.5*)

Electron affinity (*EA*): The energy change that occurs when an electron is added to a gaseous atom (c.f. ionization energy): $E(g) + e^- \longrightarrow E^-(g)$. Electron affinities can be negative (energy is released when an electron is added), positive (energy must be added to produce an anion), or zero (the process is energetically neutral). (*Section 7.3*)

Electron capture: A nuclear decay reaction in which an electron in an inner shell reacts with a proton to produce a neutron. When a second electron moves from an outer shell to take the place of the lower-energy electron that was absorbed by the nucleus, an X ray is emitted. Like positron emission, electron capture occurs in neutron-poor nuclei. (*Section 20.2*)

Electron configuration: The arrangement of an element's electrons in its atomic orbitals. (*Section 6.6*)

Electron-deficient molecules: Compounds that contain fewer than an octet of electrons around one atom. Electron-deficient molecules (e.g., BCl_3) have a strong tendency to gain an additional pair of electrons by reacting with substances that possess a lone pair of electrons. (*Section 8.7*)

Electronegativity (χ): The relative ability of an atom to attract electrons to itself in a chemical compound. Elements with high electronegativities tend to acquire electrons and are found in the upper right corner of the periodic table, whereas elements with low electronegativities tend to lose electrons and are found in the lower left corner of the periodic table. (*Section 7.3*)

Electron-pair geometry: The three-dimensional arrangement of electron pairs around the central atom of a molecule or polyatomic ion (c.f. molecular geometry). (*Section 9.1*)

Electron sea: Valence electrons that are delocalized throughout a metallic solid. A simple model used to describe the bonding in metals is to view the solid as consisting of positively charged nuclei embedded in an electron sea. (*Section 12.5*)

Electron shielding: The effect by which electrons closer to the nucleus neutralize a portion of the positive charge of the nucleus and thereby decrease the attractive interaction between the nucleus and an electron farther away. Because of electron shielding, the electron experiences an effective nuclear charge (Z_{eff}) that is less than the actual nuclear charge (Z). (*Section 6.5*)

Electron spin: The magnetic moment that results when an electron (an electrically charged particle) spins. In an external magnetic field, the electron has two possible orientations, which are described by a fourth quantum number, m_s [the two possible values of m_s are $+\frac{1}{2}$ (up) and $-\frac{1}{2}$ (down)]. (*Section 6.6*)

Electrophile: A species that needs electrons to complete its octet, an electrophile (from *electron* plus the Greek suffix *phile*, "loving") is electron deficient (c.f. nucleophile). Carbocations and free radicals, which are electron deficient, are electrophiles in organic reactions, whereas the Group-13 elements, which are also electron deficient, are electrophiles in inorganic reactions. (*Section 24.3*)

Electroplating: A process in which a layer of a second metal is deposited on the metal electrode that acts as the cathode during electrolysis. Electroplating is used to enhance the appearance of metal objects and to protect them from corrosion. (*Section 19.7*)

Electrostatic attraction: An electrostatic interaction between oppositely charged species (positive and negative) that results in a force that causes them to move toward each other. (*Section 2.1*)

Electrostatic interaction: An interaction between electrically charged particles such as protons and electrons. (*Section 2.1*)

Electrostatic repulsion: An electrostatic interaction between two species that have the same charge (both positive or both negative) that results in a force that causes them to repel each other. (*Section 2.1*)

Element: A pure substance that cannot be broken down into simpler ones by chemical changes. (*Section 1.3*)

Elementary reaction: Each of the complex series of reactions that take place in a stepwise fashion to convert reactants to products. Each elementary reaction involves one, two, or (rarely) three atoms, molecules, or ions. The overall sequence of elementary reactions is the mechanism of the reaction. The sum of the elementary reactions in a mechanism must give the balanced chemical equation for the overall reaction. (*Section 14.6*)

Elimination reactions: Chemical reactions in which adjacent atoms are removed, or "eliminated," from a molecule, elimination reactions result in the formation of a multiple bond and a small molecule. Elimination reactions have the general form $CH_2A—CH_2B \longrightarrow CH_2{=}CH_2 + A—B$ and are the reverse of addition reactions. (*Section 24.4*)

Emission spectrum: A spectrum produced by the emission of light by atoms in excited states (c.f. absorption spectrum). Each element has a characteristic emission spectrum, so scientists can use these spectra to analyze the composition of matter. (*Section 6.3*)

Empirical formula: A formula for a compound that consists of the atomic symbol for each component element accompanied by a subscript indicating the *relative* number of atoms of that element in the compound, reduced to the smallest whole numbers. An empirical formula is based on experimental measurements of the numbers of atoms in a sample of a compound, so it indicates only the ratios of the numbers of the elements present (c.f. molecular formula). (*Section 2.2*)

Empirical formula mass: Another name for formula mass. (*Section 3.1*)

Emulsion: A dispersion of one liquid phase in another liquid with which it is immiscible. An emulsion is one of three kinds of colloids (the other two are sols and aerosols). (*Section 13.7*)

Endothermic: A process in which heat (q) is transferred to the system from the surroundings (c.f. exothermic). By convention, $q > 0$ for an endothermic reaction. (*Section 5.2*)

Endpoint: The point in a titration at which an indicator changes color. Ideally the endpoint matches the equivalence point of the titration. (*Section 4.9*)

Energy: The capacity to do work. The five forms of energy are radiant, thermal, chemical, nuclear, and electrical. (*Section 5.1*)

Energy band: The continuous set of allowed energy levels generated in band theory when the valence orbitals of the atoms in the solid interact with one another, thus creating a set of molecular orbitals that extend throughout the solid. (*Section 12.6*)

Energy-level diagram: A schematic drawing that compares the energies of the molecular orbitals (bonding, antibonding, and nonbonding) with the energies of the parent atomic orbitals. The molecular orbitals are arranged in order of energy, with the lowest-energy orbitals at the bottom and the highest-energy orbitals at the top. The bonding in a molecule can be described by inserting the total number of valence electrons into the energy-level diagram, filling the orbitals according to the Pauli principle and Hund's rule (i.e., each molecular orbital can accommodate a maximum of two electrons of opposite spins and the orbitals are filled in order of increasing energy). (*Section 9.3*)

Enthalpy (*H*): The sum of a system's internal energy E and the product of its pressure P and volume V: $H = E + PV$. Because internal energy, pressure, and volume are all state functions, enthalpy is a state function, too. Enthalpy is also an extensive property (like mass), so the magnitude of the enthalpy change for a reaction is proportional to the amounts of the substances that react. (*Sections 5.2 and 18.2*)

Enthalpy of combustion (ΔH_{comb}): The change in enthalpy that occurs during a combustion reaction. (*Section 5.2*)

Enthalpy of formation (ΔH_f): The enthalpy change for the formation of 1 mol of a compound from its component elements, such as the formation of CO_2 from C and O_2. (*Section 5.2*)

Enthalpy of fusion (ΔH_{fus}): The enthalpy change that accompanies the melting (fusion) of 1 mol of a substance. (*Section 5.2*)

Enthalpy of reaction (ΔH_{rxn}): The change in enthalpy that occurs during a chemical reaction. If heat flows from the system to the surroundings, the enthalpy of the system decreases and ΔH_{rxn} is negative. If heat flows from the surroundings to the system, the enthalpy of the system increases and ΔH_{rxn} is positive. Thus, $\Delta H_{rxn} < 0$ for an exothermic reaction, and $\Delta H_{rxn} > 0$ for an endothermic reaction. (*Section 5.2*)

Enthalpy of solution (ΔH_{soln}): The change in enthalpy that occurs when a specified amount of solute dissolves in a given quantity of solvent. (*Section 5.2*)

Enthalpy of sublimation (ΔH_{sub}): The enthalpy change that accompanies the conversion of a solid directly to a gas. ΔH_{sub} is always positive because energy is always required to evaporate a solid. The enthalpy of sublimation of a substance equals the enthalpy of fusion plus the enthalpy of vaporization of the same substance. (*Sections 8.3 and 11.5*)

Enthalpy of vaporization (ΔH_{vap}): The enthalpy change that accompanies the vaporization of 1 mol of a substance. (*Section 5.2*)

Entropy (*S*): The degree of disorder in a thermodynamic system. The greater the number of possible microstates for a system, the greater the disorder and the higher the entropy. (*Sections 13.2 and 18.3*)

Enzyme: Catalysts that occur naturally in living organisms and that catalyze biological reactions. Most biological reactions do not occur without a biological catalyst (i.e., an enzyme). Almost all enzymes are protein molecules with molecular masses of 20,000–100,000 amu. Some are homogeneous catalysts, while others are heterogeneous catalysts. (*Sections 3.5, 12.8, and 14.8*)

Enzyme inhibitors: Substances that decrease the rate of an enzyme-catalyzed reaction by binding to a specific portion of the enzyme, thus slowing or preventing a reaction from occurring. Irreversible inhibitors are the equivalent of poisons in heterogeneous catalysis. (*Section 14.8*)

Equilibrium constant expression: The right side of the equilibrium equation (Equation 15.5)—namely, $[C]^c[D]^d/[A]^a[B]^b$, where A and B are reactants, C and D are products, and a, b, c, and d are their stoichiometric coefficients in the balanced chemical equation for the reaction. (*Section 15.2*)

Equilibrium constant (*K*): The ratio of the rate constants for the forward reaction (k_f) and the reverse reaction (k_r); that is, $K = k_f/k_r$. It is also the equilibrium constant calculated from solution concentrations instead of partial pressures [c.f. equilibrium constant (K_p)]. $K = (C)^c(D)^d/(A)^a(B)^b$ for the general reaction $aA + bB \rightleftharpoons cC + dD$, in which all of the components are dissolved in solution. The higher the value of the equilibrium constant, the more stable the products of the reaction. (*Section 15.2*)

Equilibrium constant (K_p): The equilibrium constant calculated from partial pressures instead of solution concentrations [c.f. equilibrium constant (K)]. $K_p = (P_C)^c(P_D)^d/(P_A)^a(P_B)^b$ for the general reaction $aA + bB \rightleftharpoons cC + dD$, in which all of the components are gases. (*Section 15.2*)

Equilibrium equation: The equation that expresses the equilibrium constant as the ratio of the product of the equilibrium concentrations of the products (raised to their stoichiometric coefficients in the balanced chemical equation) to the product of the equilibrium concentrations of the reactants (also raised to their stoichiometric coefficients in the balanced chemical equation). The equilibrium equation is $K = [C]^c[D]^d/[A]^a[B]^b$ (Equation 15.5), where K is the equilibrium constant for the reaction, A and B are reactants, C and D are products, and a, b, c, and d are their stoichiometric coefficients in the balanced chemical equation for the reaction. (*Section 15.2*)

Equilibrium vapor pressure: The pressure exerted by a vapor in dynamic equilibrium with its liquid. (*Section 11.4*)

Equivalence point: The point in a titration where a stoichiometric amount of the titrant has been added (i.e., the amount required to react completely with the unknown, the substance whose concentration is to be determined). In an acid–base titration, the equivalence point is reached when the number of moles of base (or acid) added equals the number of moles of acid (or base) originally present in the solution. (*Sections 4.9 and 16.5*)

Essential element: Any of the 19 elements (H, C, N, O, Na, Mg, P, S, Cl, K, Ca, Mn, Fe, Co, Cu, Zn, Se, Mo, and I) that are absolutely required in the diets of humans in order for them to survive (see Figure 1.26). Seven more elements (F, Si, V, Cr, Ni, As, and Sn) are thought to be essential for humans. (*Section 1.8*) Approximately 28 elements are known to be essential for the growth of at least one biological species. (*Section 7.5*)

Evaporation: The physical process by which atoms or molecules in the liquid phase enter the gas or vapor phase. Evaporation, also called vaporization, occurs only for those atoms or molecules that have enough kinetic energy to overcome the intermolecular attractive forces holding the liquid together. To escape the liquid, though, the atoms or molecules must be at the surface of the liquid. The atoms or molecules that undergo evaporation create the vapor pressure of the liquid. (*Section 11.4*)

Exact number: An integer obtained either by counting objects or from definitions (e.g., 1 in. = 2.54 cm). Exact numbers have infinitely many significant figures. (*Essential Skills 1*)

Excess reactant: The reactant that remains after a reaction has gone to completion (c.f. limiting reactant). (*Section 3.4*)

Exchange reaction: A chemical reaction that has the general form $AB + C \longrightarrow AC + B$ or $AB + CD \longrightarrow AD + CB$ (see Table 3.1 for examples). (*Section 3.5*)

Excited state: Any arrangement of electrons that is higher in energy than the ground state. When an atom in an excited state undergoes a transition to the ground state, it loses energy by emitting a photon whose energy corresponds to the difference in energy between the two states (c.f. ground state). (*Section 6.3*)

Exothermic: A process in which heat (q) is transferred from the system to the surroundings (c.f. endothermic). By convention, $q < 0$ for an exothermic reaction. (*Section 5.2*)

Expanded-valence molecules: Compounds that have more than an octet of electrons around an atom (c.f. octet rule). The expanded valence shell can be achieved only by elements in Period 3 or higher because they have empty d orbitals that can accommodate the additional electrons. (*Section 8.6*)

Experiments: Systematic observations or measurements, preferably made under controlled conditions—that is, conditions in which the variable of interest is clearly distinguished from any others. (*Section 1.2*)

Extensive property: A physical property that varies with the amount of the substance (e.g., mass, weight, and volume; c.f. intensive property). (*Section 1.3*)

F

Face-centered cubic (fcc) unit cell: A cubic unit cell that contains eight component atoms, molecules, or ions located at the corners of a cube, as well as an identical component in the center of each face of the cube (see Figure 12.5c). (*Section 12.2*)

Faraday (F): The charge on 1 mole of electrons, the faraday is obtained by multiplying the charge on the electron by Avogadro's number (see Equation 19.33). (*Section 19.4*)

***f* block:** The 14 columns usually placed below the main body of the periodic table, the f block consists of elements in which the $(n - 2)f$ orbitals are being filled (see Figure 6.34; c.f. s block, p block, and d block). (*Section 6.6*)

***f*-block elements:** The elements in the 14 columns usually placed below the main body of the periodic table. The elements in the f block are filling their $(n - 2)f$ orbitals (see Figure 6.34; c.f. s-block elements, p-block elements, and d-block elements). (*Section 6.6*)

Fermentation: A process used by some chemotrophs to obtain energy from their environment, fermentation is a chemical reaction in which both the oxidant and the reductant are organic compounds. Common examples include alcohol fermentation (Equation 18.48) and lactic acid fermentation (Equation 18.49). (*Section 18.8*)

Fiber: A particle of a synthetic polymer that is more than 100 times longer than it is wide. Fibers can be formed by pyrolysis. (*Section 12.8*)

First law of thermodynamics: The energy of the universe is constant. Mathematically, $\Delta E_{\text{universe}} = \Delta E_{\text{system}} + \Delta E_{\text{surroundings}} = 0$ (Equation 18.5), so $\Delta E_{\text{system}} = -\Delta E_{\text{surroundings}}$. The first law of thermodynamics relates the energy change of the system to that of the surroundings. (*Section 18.2*)

First-order reaction: A reaction whose rate is directly proportional to the concentration of one of the reactants (c.f. zeroth-order reaction and second-order reaction; see Table 14.6). First-order reactions have the general form $A \longrightarrow$ products. The differential rate law for a first-order reaction is rate = k[A], where k is the rate constant for the reaction. (*Section 14.3*)

Fissile isotope: Any isotope that can undergo a nuclear fission reaction upon bombardment with neutrons. (*Section 20.4*)

Formal charge: The difference between the number of valence electrons in a free atom and the number of electrons assigned to it in a particular Lewis electron structure (see Equation 8.15). The formal charge is a way of computing the charge distribution within a Lewis structure; the sum of the formal charges on the atoms within a molecule or ion must equal the overall charge on the molecule or ion. A formal charge does not represent a true charge on an atom but is simply used to predict the most likely structure when a compound has more than one Lewis structure. (*Section 8.5*)

Formation constant (K_f): The equilibrium constant for the formation of a complex ion from a hydrated metal ion. The equilibrium constant expression for K_f has the same general form as any other equilibrium constant expression; that is, $K_f = [C]^c[D]^d/[A]^a[B]^b$, where A and B are reactants, C and D are products, and a, b, c, and d are their stoichiometric coefficients in the balanced chemical equation for the reaction. Water, a pure liquid, does not appear explicitly in the equilibrium constant expression. (*Section 17.3*)

Formula mass: Also called the empirical formula mass, the formula mass is the sum of the atomic masses of all the elements in the empirical formula, each multiplied by its subscript (written or implied). The formula mass is particularly convenient for ionic compounds, which do not have a readily identifiable molecular unit. Thus, the formula mass is directly analogous to the molecular mass of a covalent compound and is expressed in units of amu. (*Section 3.1*)

Formula unit: The absolute grouping of atoms or ions represented by the empirical formula of a compound, either ionic or covalent. Butane, for example, has the empirical formula C_2H_5, but it contains two C_2H_5 formula units, giving it a molecular formula of C_4H_{10}. (*Section 2.2*)

Fractional crystallization: The separation of compounds based on their relative solubilities in a given solvent. (*Section 13.5*)

Free energy (*G*): See "Gibbs free energy (*G*)" (*Section 18.5*).

Free radicals: One of three unstable, highly reactive species that can form when a C—H bond is cleaved (the other two are carbanions and carbocations), free radicals form when the two bonding electrons are split evenly between the carbon (generating a C· species) and the hydrogen (generating an H · species). The C· species (e.g., $\cdot CH_3$) has fewer than eight valence electrons, so it is electron deficient (i.e., it is an electrophile). Free radicals are stabilized by groups that donate electrons, so tertiary free radicals ($\cdot CR_3$) are more stable than secondary free radicals ($\cdot CHR_2$), and secondary free radicals are more stable than primary free radicals ($\cdot CH_2R$). (*Section 24.3*)

Freezing-point depression (ΔT_f): The difference between the freezing point of a pure solvent and the freezing point of the solution (see Equation 13.22): $\Delta T_f = T_f^\circ - T_f$, where T_f° is the melting point of the pure solvent and T_f is the melting point of the solution. ΔT_f is proportional to the molality of the solution (see Equation 13.23). (*Section 13.6*)

Frenkel defect: A defect in an ionic lattice that occurs when one of the ions is in the wrong position. A cation, for example, may occupy a tetrahedral hole instead of an octahedral hole in the lattice. To preserve electrical neutrality, one of the normal cation sites (usually octahedral) must be vacant. (*Section 12.4*)

Frequency (*V*): The number of oscillations (i.e., of a wave) that pass a particular point in a given period of time. The usual units of frequency are oscillations per second or $1/s = s^{-1}$, which in the SI system is called the hertz (Hz). (*Section 6.1*)

Frequency factor (*A*): A constant in the Arrhenius equation (Equation 14.43), the frequency factor converts concentrations to collisions per second. The frequency factor represents the product of the collision frequency and the steric factor (the fraction of collisions with the proper orientation for reaction). (*Section 14.7*)

Fuel cell: A galvanic cell that requires a constant external supply of one or more reactants in order to generate electricity (c.f. battery). (*Section 19.5*)

Fullerenes: One of at least four allotropes of carbon (the other three are graphite, diamond, and nanotubes). The fullerenes (e.g., buckminsterfullerene, C_{60}) are a group of related caged structures (see Figure 7.18). (*Section 7.4*)

Functional groups: The structural units that chemists use to classify organic compounds and predict their reactivities, functional groups determine the chemical reactivity of a molecule under a given set of conditions. Functional groups can consist of a single atom (such as H) or a group of atoms (such as CO_2H). The major families of organic compounds are characterized by their functional groups. (*Section 24.1*)

Fundamental vibration: The lowest-energy standing wave. (*Section 6.4*)

Fusion: The conversion of a solid to a liquid. Fusion is also called melting. (*Section 11.5*)

G

***G*:** See "Gibbs free energy (*G*)" (*Section 18.5*).

Galvanic cell: One of two types of electrochemical cells, a galvanic cell (or voltaic cell) uses the energy released during a spontaneous oxidation–reduction (redox) reaction ($\Delta G < 0$) to generate electricity (c.f. electrolytic cell, the other type of electrochemical cell). (*Section 19.1*)

Gamma (γ) emission: A nuclear decay reaction that results when a nucleus in an excited state releases energy in the form of a high-energy photon (a γ ray) when it returns to the ground state. Gamma rays are energy, so emitting them leaves the mass number and atomic number of the daughter nuclide unaffected. (*Section 20.2*)

Gas: One of three distinct states of matter under normal conditions, gases have neither fixed shapes nor fixed volumes and expand to fill their containers completely. The volume of a gas depends on its temperature and pressure. The other two states of matter are solid and liquid. (*Section 1.3*)

Gas constant (*R*): A proportionality constant in the ideal gas law (among other equations); $R = 0.08206$ (L.atm)/(K.mol) $= 8.31447$ J/(K.mol) $= 1.9872$ cal/(K.mol). (*Section 10.4*)

Gel: A semisolid sol in which all of the liquid phase has been absorbed by the solid particles. (*Section 13.7*)

Geometrical isomers: One of two types of stereoisomers (the other is optical isomers), geometrical isomers differ in the relative placement of substituents in a rigid molecule. Geometrical isomers may be either cis or trans, and these structural differences cause them to have different physical and chemical properties. In metal complexes, geometrical isomers differ only in which ligands are adjacent to one another (cis) or directly across from one another (trans) in the coordination sphere of the metal. (*Sections 23.4 and 24.2*)

Gibbs free energy (*G*): A state function that is defined in terms of three other state functions—namely, enthalpy (*H*), entropy (*S*), and temperature (*T*): $G = H - TS$ (Equation 18.19). (*Section 18.5*)

Glass: An amorphous, translucent solid. A glass is a solid that has been cooled too quickly to form ordered crystals. (*Section 12.1*)

Glass electrode: An electrode used to measure the H^+ ion concentration of a solution, the glass electrode consists of an internal Ag/AgCl electrode immersed in a 1 *M* HCl solution that is separated from the solution by a very thin glass membrane. The glass membrane absorbs protons, which affects the measured potential. The extent of adsorption on the inner side is fixed because $[H^+]$ is fixed inside the electrode, but the adsorption of protons on the outer surface depends on the pH of the solution. (*Section 19.2*)

Grain boundary: The place where two grains in a solid intersect. Most materials consist of many microscopic grains that are randomly oriented with respect to one another, and each grain boundary can be viewed as a two-dimensional dislocation. Defect motion tends to stop at grain boundaries, so controlling the size of the grains in a material controls its mechanical properties. (*Section 12.4*)

Graphite intercalation compounds: Compounds that form when the heavier alkali metals (K, Rb, and Cs) react with carbon in the form of graphite. The metals insert themselves between the sheets of carbon atoms, instead of disrupting the hexagonal sheets of carbon atoms, thus yielding substances with remarkably high electrical conductivities. (*Section 21.3*)

Gray (Gy): The SI unit used to measure the effects of radiation on biological tissue, the gray is the amount of radiation that causes 1 joule of energy to be absorbed per kilogram of matter (c.f. rad and roentgen). (*Section 20.3*)

Greenhouse effect: The phenomenon in which substances (e.g., CO_2, water vapor, methane, and chlorofluorocarbons) absorb thermal energy radiated by the earth, thus trapping thermal energy in the earth's atmosphere (analogous to the glass in a greenhouse). (*Section 5.5*)

Greenhouse gases: The substances (e.g., CO_2, water vapor, methane, and chlorofluorocarbons) that absorb thermal energy radiated by the earth, thus trapping thermal energy in the earth's atmosphere (analogous to the glass in a greenhouse). (*Section 5.5*)

Grignard reagents: A class of organometallic compounds that contain stabilized carbanions, Grignard reagents are organomagnesium compounds with the formula RMgX (X = Cl, Br, or I). These reagents are named for the French chemist Victor Grignard (1871–1935). (*Section 24.5*)

Ground state: The most stable arrangement of electrons for an element or compound (c.f. excited state). (*Section 6.3*)

Group: Any of the vertical columns of elements in the periodic table. There are 18 groups in the periodic table, numbered from 1 to 18. The rows of elements in the periodic table are arranged such that elements with similar chemical properties reside in the same group (column). (*Section 1.7*)

Group transfer reaction: A reaction in which a recognizable functional group, such as a phosphoryl unit ($-PO_3^-$), is transferred from one molecule to another. (*Section 7.5*)

H

***H*:** See "Enthalpy (H)" (*Section 5.2*).

Half-life ($t_{1/2}$): The period of time it takes for the concentration of a reactant to decrease to one-half its initial value. Thus, the half-life of a reaction is the time required for the reactant concentration to decrease from $[A]_0$ to $[A]_0/2$. If two reactions have the same order, the faster reaction will have a shorter half-life and the slower reaction will have a longer half-life. Unlike zeroth-order or second-order reactions, the half-life of a first-order reaction under a given set of reaction conditions is a constant because it is independent of the concentrations of the reactants. (*Section 14.5*)

Half-reactions: Reactions that represent either the oxidation half or the reduction half of an oxidation–reduction (redox) reaction. Redox reactions can be described as two half-reactions (one for oxidation and one for reduction) because it is impossible to have an oxidation without a reduction, and vice versa. (*Section 19.1*)

Halogen: Any of the elements in Group 17 of the periodic table (F, Cl, Br, I, and At). All of the halogens react readily with metals to give ions with a -1 charge, such as Cl^- and Br^-. (*Section 1.7*)

Hard acids: The metal ions (Lewis acids) with the highest affinity for hard bases, hard acids are usually cations of electropositive metals (see Table 23.9), so they are relatively nonpolarizable and have relatively high charge-to-radius ratios (c.f. soft acids). (*Section 23.4*)

Hard bases: One type of Lewis base (the other is soft bases), hard bases contain small, relatively nonpolarizable donor atoms (such as N, O, and F; see the examples in Table 23.9). (*Section 23.4*)

Hardness (of ionic materials): The resistance of ionic materials to scratching or abrasion. Hardness depends on the lattice energy of the compound because hardness is directly related to how tightly the ions are held together electrostatically. (*Section 8.3*)

Heat (q): Thermal energy that can be transformed from an object at one temperature to an object at another temperature. The net transfer of thermal energy stops when the two objects reach the same temperature. (*Sections 5.1 and 18.1*)

Heat capacity (C): The amount of energy needed to raise the temperature of an object 1°C. The units of heat capacity are joules per degree Celsius (J/°C). (*Section 5.3*)

Heating curve: A plot of the temperature of a substance versus the heat added or versus the heating time at a constant rate of heating. Heating curves relate temperature changes to phase transitions (c.f. cooling curve). (*Section 11.5*)

Heisenberg uncertainty principle: A principle stating that the uncertainty in the position of a particle (Δx) multiplied by the uncertainty in its momentum $[\Delta(mv)]$ is greater than or equal to Planck's constant (h) divided by 4π: $(\Delta x)[\Delta(mv)] \geq h/4\pi$. In short, the more accurately we know the exact position of a particle (as $\Delta x \longrightarrow 0$), the less accurately we know the speed and hence the kinetic energy of the particle ($\frac{1}{2}mv^2$) because $\Delta(mv) \longrightarrow \infty$, and vice versa. Because Planck's constant is a very small number, the Heisenberg uncertainty principle is important for only very small particles such as electrons. (*Section 6.4*)

Henderson–Hasselbalch equation: A rearranged version of the equilibrium constant expression that provides a direct way to calculate the pH of a buffer solution. The equation is $pH = pK_a + \log([A^-]/[HA])$ or, more generally, $pH = pK_a + \log([\text{base}]/[\text{acid}])$. (*Section 16.6*)

Henry's law: An equation that quantifies the relationship between the pressure and the solubility of a gas: $C = kP$, where C is the concentration of dissolved gas at equilibrium, P is the partial pressure of the gas, and k is the Henry's law constant, which must be determined experimentally for each combination of gas, solvent, and temperature. (*Section 13.5*)

Hess's law: The enthalpy change (ΔH) for an overall reaction is the sum of the ΔH values for the individual reactions. Hess's law makes it possible to calculate ΔH values for reactions that are difficult to carry out directly by summing known ΔH values for individual steps that give the overall reaction when added, even though the overall reaction may not actually occur via those steps. (*Section 5.2*)

Heterogeneous catalysis: Catalysis in which the catalyst is in a different phase from the reactants (c.f. homogeneous catalysis). At least one of the reactants in heterogeneous catalysis interacts with the solid surface of the catalyst via adsorption, thus leading first to a weakening of a chemical bond and then to a breaking of the bond. (*Section 14.8*)

Heterogeneous catalyst: A catalyst that is in a different physical state than the reactants (c.f. homogeneous catalyst). (*Section 3.5*)

Heterogeneous equilibrium: An equilibrium in which the reactants of an equilibrium reaction, the products, or both are in more than one phase (c.f. homogeneous equilibrium). The reaction of a gas with a solid or liquid would form a heterogeneous equilibrium. (*Section 15.2*)

Heterogeneous mixture: A mixture in which a material is not completely uniform (e.g., chocolate chip cookie dough, blue cheese, and dirt). (*Section 1.3*)

Heteronuclear diatomic molecule: A molecule that consists of two atoms of different elements (e.g., CO or NO). (*Section 9.3*)

Hexagonal close-packed (hcp) structure: One of two variants of the close-packed arrangement, the most efficient way to pack spheres in a lattice [the other is the cubic close-packed (ccp) arrangement]. The hexagonal close-packed structure results from arranging the atoms of the solid so that their positions alternate from layer to layer in the pattern ABABAB . . . (see Figure 12.7a). (*Section 12.2*)

High-temperature superconductors: Materials that become superconductors at temperatures higher than 30 K (i.e., superconductors with superconducting transition temperatures, T_c, higher than 30 K). (*Section 12.7*)

Homogeneous catalysis: Catalysis in which the catalyst is in the same phase as the reactant(s) (c.f. heterogeneous catalysis). The number of collisions between reactants and catalyst is maximized because the catalyst is uniformly dispersed throughout the reaction mixture. (*Section 14.8*)

Homogeneous catalyst: A catalyst that is uniformly dispersed throughout the reactant mixture to form a solution (c.f. heterogeneous catalyst). (*Section 3.5*)

Homogeneous equilibrium: An equilibrium in which the reactants and products of an equilibrium reaction form a single phase, whether gas or liquid (c.f. heterogeneous equilibrium). When the system is a homogeneous equilibrium, the concentrations of the reactants and products can vary over a wide range. (*Section 15.2*)

Homogeneous mixture: A mixture in which all portions of the material are in the same state, have no visible boundaries, and are uniform throughout (e.g., air and tap water). Homogeneous mixtures are also called solutions. (*Section 1.3*)

Homonuclear diatomic molecule: A molecule that consists of two atoms of the same element (e.g., H_2 or Cl_2). (*Section 9.3*)

Hund's rule: Named for F. H. Hund, the rule states that the lowest-energy electron configuration for an atom is the one that has the maximum number of electrons with parallel spins in degenerate orbitals. (*Section 6.6*)

Hybrid atomic orbitals: The new atomic orbitals formed from the process of hybridization. Hybrid atomic orbitals are equivalent in energy and oriented properly for forming bonds. (*Section 9.2*)

Hybridization: A process in which two or more atomic orbitals that are similar in energy but not equivalent are combined mathematically to produce sets of equivalent orbitals that are properly oriented to form bonds. These new combinations are called hybrid atomic orbitals because they are produced by combining (hybridizing) two or more atomic orbitals from the same atom. (*Section 9.2*)

Hydrate: A compound that contains specific ratios of loosely bound water molecules, called waters of hydration. These loosely bound water molecules can often be removed by simply heating the compound. (*Section 2.2*)

Hydrated ions: Individual cations and anions that are each surrounded by their own shell of water molecules. These shells form when an ionic solid dissolves in water because water molecules are polar. The partially negatively charged oxygen atoms of the H_2O molecules surround the cations (e.g., Na^+ in the case of NaCl), whereas the partially positively charged hydrogen atoms surround the anions (e.g., Cl^-). (*Section 4.1*)

Hydration: The process of surrounding solute particles (atoms, molecules, or ions) with water molecules. Solvation when the solvent is water. (*Section 13.2*)

Hydride ion (H^-): The anion formed when a hydrogen atom accepts an electron, the hydride ion has an electron configuration of $1s^2$ (making it isoelectronic with He). (*Section 21.2*)

Hydrocarbons: The simplest class of organic molecules, hydrocarbons consist entirely of carbon and hydrogen. The four major classes of hydrocarbons are the alkanes, alkenes, alkynes, and aromatics. Hydrocarbons may be saturated (alkanes) or unsaturated (alkenes, alkynes, and aromatics). Structurally they may be chains (linear or branched) or cyclic. They may also be aliphatic (alkanes, alkenes, alkynes, and cyclic hydrocarbons) or aromatic. (*Section 2.4*)

Hydrogenation: A chemical reaction in which hydrogen atoms are added to the double bond of an alkene, such as ethylene, to give a product that contains C—C single bonds, in this case ethane. (*Section 14.8*)

Hydrogen bonds: Unusually strong dipole–dipole interactions (intermolecular forces) that result when hydrogen is bonded to very electronegative elements such as O, N, and F. A hydrogen bond consists of a hydrogen bond donor (the hydrogen attached to O, N, or F) and a hydrogen bond acceptor (an atom, usually O, N, or F, that has a lone pair of electrons). (*Sections 11.2 and 21.2*)

Hydrolysis reactions: Chemical reactions in which a salt reacts with water to yield an acidic or basic solution. Hydrolysis reactions are just acid–base reactions in which the acid is a cation or the base is an anion; they obey the same principles and rules as all other acid–base reactions. (*Section 16.2*)

Hydronium ion: The H_3O^+ ion. H_3O^+ is a more accurate representation of $H^+(aq)$. (*Sections 4.6 and 16.1*)

Hydrophilic: Substances are classified as hydrophilic if they are attracted to water (c.f. hydrophobic). Hydrophilic substances are polar and often contain O—H or N—H groups that can form hydrogen bonds to water. (*Section 13.3*)

Hydrophobic: Substances are classified as hydrophobic if they are repelled by water (c.f. hydrophilic). Hydrophobic substances are non-polar and usually contain C—H bonds that do not interact favorably with water. (*Section 13.3*)

Hypothesis: A tentative explanation for scientific observations. A hypothesis may not be correct, but it puts the scientist's understanding of the system being studied into a form that can be tested. (*Section 1.2*)

I

***I*:** See "Ionization energy (*I*)" (*Section 7.3*).

Ideal gas: A hypothetical gaseous substance whose behavior is independent of attractive and repulsive forces and can be completely described by the ideal gas law. In reality, there is no such thing as an ideal gas, but it is a useful conceptual model that makes it easier to understand how real gases respond to changing conditions. (*Section 10.4*)

Ideal gas law: $PV = nRT$, where P = pressure, V = volume, n = amount of a gas, R = the gas constant, and T = temperature. The ideal gas law can be obtained by combining the empirical relationships among the volume, temperature, pressure, and amount of a gas (i.e., Boyle's law, Charles's law, and Avogadro's law). (*Section 10.4*).

Ideal solution: A solution that obeys Raoult's law (see Equation 13.13). Like an ideal gas, an ideal solution is a hypothetical system whose properties can be described in terms of a simple model. (*Section 13.6*)

Indicator electrode: The electrode of a galvanic (voltaic) cell whose potential is related to the concentration of the substance being measured. To ensure that any change in the measured potential of the cell is due to only the substance being analyzed, the potential of the other electrode (the reference electrode) must be constant. (*Section 19.2*)

Indicators: Intensely colored organic molecules whose colors change dramatically depending on the pH of the solution. They can be used to determine the approximate pH of a solution. (*Section 4.6*)

Induced dipole: The short-lived dipole moment that is created in atoms and nonpolar molecules adjacent to atoms or molecules that have an instantaneous dipole moment (resulting from the shortlived and ever-fluctuating asymmetrical distribution of its electrons). These instantaneous and induced dipole moments form the basis of London dispersion forces. (*Section 11.2*)

Inert metals: The metals at the bottom of the activity series (Pt, Au, Ag, Cu, and Hg), which have the least tendency to be oxidized (c.f. active metals). (*Section 4.8*)

Inert-pair effect: The empirical observation that the heavier elements of Groups 13–17 often have oxidation states that are lower by 2 than the maximum predicted for their group. For example, an oxidation state of +3 is common for the Group-13 elements, but the heaviest element in Group 13, thallium (Tl), is more likely to form compounds in which it has a +1 oxidation state. The two major reasons for the inert-pair effect seem to be increasing ionization energies and decreasing bond strengths. (*Section 21.1*)

Initiation: The first stage in a chain reaction, initiation produces one or more reactive intermediates (see radicals). The other stages in a chain reaction are propagation and termination. (*Section 14.6*)

Inorganic compound: An ionic or covalent compound that consists primarily of elements other than carbon and hydrogen (c.f. organic compound). (*Section 2.1*)

Instantaneous dipole moment: The short-lived dipole moment in atoms and nonpolar molecules caused by the constant motion of their electrons, which results in an asymmetrical distribution of charge at any given instant. The instantaneous dipole moment on one atom can interact with the electrons of an adjacent atom, inducing a temporary dipole moment in the second atom. These instantaneous and induced dipole moments form the basis of London dispersion forces. (*Section 11.2*)

Instantaneous reaction rate: The rate of a chemical reaction at any given point in time (c.f. average reaction rate). As the period of time used to calculate an average rate of a reaction becomes shorter and shorter, the average rate approaches the instantaneous rate. (*Section 14.2*)

Integrated rate law: A rate law that expresses the rate of a reaction in terms of the initial concentration, $[R]_0$, and the measured concentration of one or more reactants, [R], after a given amount of time, t (c.f. differential rate law). The integrated rate law can be found by using calculus to integrate the differential rate law. (*Section 14.2*)

Intensive property: A physical property that does not depend on the amount of the substance (e.g., color, melting point, boiling point, electrical conductivity, and physical state at a given temperature; c.f. extensive property). (*Section 1.3*)

Intermediate: A species in a reaction mechanism that does not appear in the balanced chemical equation for the overall reaction. Intermediates are formed as products in one step (elementary reaction) of the mechanism and consumed again as reactants in a later step in the mechanism. (*Section 14.6*)

Intermetallic compound: An alloy that consists of certain metals that combine in only specific proportions. The structures and physical properties of intermetallic compounds are frequently quite different from those of their constituent elements. (*Section 12.5*)

Internal energy (E): The sum of the kinetic and potential energies of all of a system's components. Additionally, $\Delta E = q + w$, where q is the heat produced by the system and w is the work performed by the system. Internal energy is a state function. (*Sections 5.2 and 18.1*)

Interstitial alloy: An alloy formed by inserting smaller atoms into holes in the metal lattice (c.f. substitutional alloy and intermetallic compound). (*Section 12.5*)

Interstitial impurity: A point defect that results when an impurity atom occupies an octahedral or tetrahedral hole in the lattice between atoms (c.f. vacancy and substitutional impurity). (*Section 12.4*)

Ion: A charged particle produced when one or more electrons is removed from or added to an atom or molecule. Ions may be cations (positively charged) or anions (negatively charged). (*Section 1.6*)

Ionic compound: A compound consisting of positively charged ions (cations) and negatively charged ions (anions) held together by strong electrostatic forces (c.f. covalent compound). The ratio of cations to anions in an ionic compound is such that there is no net electrical charge. In an ionic compound, the cations and anions are arranged in space to form an extended three-dimensional array that maximizes the number of attractive electrostatic interactions and minimizes the number of repulsive electrostatic interactions. (*Section 2.1*)

Ionic liquids: Ionic substances that are liquids at room temperature. Ionic liquids consist of small, symmetrical anions (e.g., PF_6^- and BF_4^-), combined with larger, symmetrical organic cations that prevent the formation of a highly organized structure, resulting in a low melting point. (*Section 11.6*)

Ionic radius: The radius of a cation or anion. The internuclear distance between a cation and an adjacent anion in an ionic compound is measured experimentally and then divided proportionally between the smaller cation and larger anion (see Figure 7.8). The different methods used to proportionally divide the experimentally measured internuclear distance give slightly different values for the ionic radii, but they do give sets of ionic radii that are internally consistent from one ionic compound to another. (*Section 7.2*)

Ionic solid: A solid that consists of positively and negatively charged ions held together by electrostatic forces. Ionic solids tend to have high melting points and to be rather hard and brittle (c.f. molecular solid, covalent solid, and metallic solid). (*Section 12.5*)

Ionization energy (I): The minimum amount of energy needed to remove an electron from the gaseous atom E in its ground state. I is therefore the energy change for the reaction, $E(g) + I \longrightarrow E^+(g) + e^-$. Because an input of energy is needed, the ionization energy is always positive ($I > 0$) for the reaction as written here (c.f. electron affinity). Higher values of I mean that the electron is more tightly bound to the atom and harder to remove. (*Section 7.3*)

Ionizing radiation: Higher in energy than nonionizing radiation, some of the energy of ionizing radiation can be transferred to one or more atoms with which it collides as it passes through matter. If enough energy is transferred, electrons can be excited to very high energy levels, resulting in the formation of positively charged ions. (*Section 20.3*)

Ionophores: Molecules that facilitate the transport of metal ions across membranes, many ionophores (*ion* plus *phore* from the Greek *phorein*, "to carry") are potent antibiotics (e.g., valinomycin) that can kill or inhibit the growth of bacteria. (*Section 21.5*)

Ion pair: A cation and anion that are in intimate contact in solution rather than separated by solvent (see Figure 17.2). The ions in an ion pair are held together by the same attractive forces as in ionic solids. The ions in an ion pair migrate in solution as a single unit (whose net charge is the sum of the charges on the ions). An ion pair can be viewed as a species intermediate between the ionic solid (in which each ion participates in many of the cation–anion interactions that hold the ions in a rigid array) and the completely dissociated ions in solution (where each is fully surrounded by solvent molecules and free to migrate independently). (*Sections 13.6 and 17.2*)

Ion-product constant of liquid water (K_w): An equilibrium constant for the autoionization of water, $2H_2O(l) \rightleftharpoons H_3O^+(aq) + OH^-(aq)$, in which the concentration of water is treated as a constant. Thus, $K_w = [H_3O^+][OH^-] = 1.006 \times 10^{-14}$ at 25°C. Like any other equilibrium constant, the value of K_w varies with temperature, ranging from 1.15×10^{-15} at 0°C to 4.99×10^{-13} at 100°C. (*Section 16.1*)

Ion product (Q): A quantity that has precisely the same form as the solubility product (K_{sp}) for the dissolution of a sparingly soluble salt, except that the concentrations used for Q are not necessarily equilibrium concentrations. If $Q < K_{sp}$, the solution is unsaturated and more of the ionic solid will dissolve. If $Q = K_{sp}$, the solution is saturated and at equilibrium. If $Q > K_{sp}$, the solution is supersaturated and some of the ionic solid will precipitate. (*Section 17.1*)

Ion pumps: Complex assemblies of proteins that selectively transport ions (e.g., Na^+, Ca^{2+}, Cl^-, K^+, Mg^{2+}, and phosphate) across cell membranes toward the side with the higher concentration, ion pumps recognize and discriminate among metal ions (e.g., Na^+, K^+, Mg^{2+}, and Ca^{2+}) in the same way that crown ethers and cryptands do, with a high affinity for ions of a certain charge and radius. (*Sections 7.5 and 21.5*)

Ion-selective electrodes: Electrodes used to measure the concentration of a particular species in solution, ion-selective electrodes are designed so that their potential depends on only the concentration of the desired species. (*Section 19.2*)

Irreversible process: A process in which the intermediate states between the extremes are not equilibrium states and change occurs spontaneously in only one direction (c.f. reversible process). A reversible process can change direction at any time, whereas an irreversible process cannot. (*Section 18.3*)

Isoelectronic series: A group of ions or atoms and ions that have the same number of electrons (and thus the same ground-state electron configurations). For example, the isoelectronic series of ions with the neon closed-shell electron configuration ($1s^22s^22p^6$) is N^{3-}, O^{2-}, F^-, Na^+, Mg^{2+}, and Al^{3+}. (*Section 7.2*).

Isolated system: One of three kinds of system, an isolated system can exchange neither energy nor matter with its surroundings. A truly isolated system does not exist, however, because energy is always exchanged between a system and its surroundings. The other kinds of systems are open and closed. (*Section 5.2*)

Isomers: Two or more compounds that have the same molecular formula but different arrangements of their atoms, isomers can be one of several types (see conformational isomers, structural isomers, and stereoisomers). Isomers usually have different physical and chemical properties. (*Sections 23.4 and 24.2*)

Isotopes: Atoms that have the same number of protons, and hence the same atomic number (Z), but different numbers of neutrons. All isotopes of an element have the same number of protons and electrons, so they exhibit the same chemistry. (*Section 1.6*)

Isotropic: The arrangement of molecules that is equally disordered in all directions (c.f. anisotropic). Normal liquids are isotropic because their molecules possess enough thermal energy to overcome their intermolecular attractive forces and tumble freely. (*Section 11.8*)

J

Jahn–Teller theorem: A theory that states that octahedral complexes with an odd number of electrons in the e_g orbitals (a so-called degenerate ground state) are unstable, and that they will undergo a distortion that makes the complex less symmetric and thus split the degenerate states, which will decrease the energy of the system. Jahn–Teller distortions are possible for many transition-metal ions, but in practice they are observed only for systems that have an odd number of electrons in the e_g orbitals, such as the Cr^{2+} and Cu^{2+} ions. (*Section 23.5*)

Joule (J): The SI unit of energy; 1 joule = 1 (kilogram · meter^2)/second^2. (*Section 5.1*)

K

k: See "Rate constant (k)" (*Section 14.2*).

K: See "Equilibrium constant (K)" (*Section 15.2*).

K_a: See "Acid ionization constant (K_a)" (*Section 16.2*).

K_b: See "Base ionization constant (K_b)" (*Section 16.2*).

KE: See "Kinetic energy (KE)" (*Section 5.1*).

Ketone: A class of organic compounds with the general form RC(O)R′, in which the carbon atom of the carbonyl group is bonded to two alkyl groups (c.f. aldehyde). The alkyl groups may be the same or different. (*Section 4.1*)

K_f: See "Formation constant (K_f)" (*Section 17.3*).

Kinetic control: The altering of reaction conditions to control reaction rates, thereby obtaining a single desired product or set of products (c.f. thermodynamic control). (*Section 15.6*)

Kinetic energy (KE): Energy due to the motion of an object. $KE = \frac{1}{2}mv^2$, where m is the mass of the object and v is its velocity. (*Section 5.1*)

Kinetic molecular theory of gases: A theory that describes (on the molecular level) why ideal gases behave the way they do. The basic postulates of kinetic molecular theory are that gas particles are in constant random motion, they are so far apart that their individual volumes are negligible, their intermolecular interactions are negligible, all of their collisions are elastic, and the average kinetic energy of the molecules of any gas depends only on the temperature, and at a given temperature, the molecules of all gases have the same average kinetic energy. (*Section 10.7*)

K_p: See "Equilibrium constant (K_p)" (*Section 15.2*).

K_{sp}: See "Solubility product (K_{sp})" (*Section 17.1*).

K_w: See "Ion-product constant of liquid water (K_w)" (*Section 16.1*).

L

l: See "Azimuthal quantum number (l)" (*Section 6.5*).

Lanthanide element: Any of the 14 elements between $Z = 58$ (cerium) and $Z = 71$ (lutetium). The $4f$ orbitals are filling for the lanthanides and their chemistry is dominated by M^{3+} ions. The lanthanides and actinides are usually grouped together in two rows beneath the main body of the periodic table. (*Sections 1.7 and 7.4*)

Lattice energy: The enthalpy change that occurs when a solid ionic compound (whose ions form a three-dimensional array called a lattice) is transformed into gaseous ions (c.f. bond energy). Lattice energies are highest for substances that contain small, highly charged ions. (*Section 8.1*)

Law: A verbal or mathematical description of a phenomenon that allows for general predictions. A law says *what* happens, not *why* it happens. (*Section 1.2*)

Law of conservation of energy: The total amount of energy in the universe remains constant. Energy cannot be created or destroyed. Energy can be converted from one form to another, however. (*Section 5.1*)

Law of conservation of mass: In any chemical reaction, the mass of the substances that react equals the mass of the products that are formed. (*Section 1.4*)

Law of definite proportions: Formulated by the French scientist Joseph Proust (1754–1826), the law states that a chemical substance always contains the same proportions of elements by mass. (*Section 1.2*)

Law of mass action: The ratio of the product of the equilibrium concentrations of the products (raised to their stoichiometric coefficients in the balanced chemical equation) to the product of the equilibrium concentrations of the reactants (also raised to their stoichiometric coefficients in the balanced chemical equation). For any reversible reaction of the general form $aA + bB \rightleftharpoons cC + dD$, where A and B are reactants, C and D are products, and *a*, *b*, *c*, and *d* are the stoichiometric coefficients in the balanced chemical equation for the reaction, the law of mass action can be stated as $K = [C]^c[D]^d/[A]^a[B]^b$, where *K* is the equilibrium constant for the reaction. (*Section 15.2*)

Law of multiple proportions: When two elements form a series of compounds, the ratios of the masses of the second element that are present per gram of the first element can almost always be expressed as the ratios of integers. (The same law holds for the mass ratios of compounds forming a series that contains more than two elements.) (*Section 1.4*)

LCAOs: See "Linear combination of atomic orbitals (LCAOs)" (*Section 9.3*).

Le Châtelier's principle: Formulated by the French chemist Henri Louis Le Châtelier (1850–1936), the principle states that if a stress is applied to a system at equilibrium, the composition of the system will change to relieve the applied stress. The types of stresses that can change the composition of an equilibrium mixture are (1) a change in the concentrations (or partial pressures) of the components by the addition or removal of reactants or products, (2) a change in the total pressure or volume, and (3) a change in the temperature of the system. (*Section 15.5*)

Leclanché dry cell: Actually a "wet cell" battery, the electrolyte in the Leclanché cell is an acidic water-based paste that contains MnO_2, NH_4Cl, $ZnCl_2$, graphite, and starch (see Figure 19.12a). (*Section 19.5*)

Leveling effect: The phenomenon that makes H_3O^+ the strongest acid that can exist in water. As a result, any species (e.g., HI or HNO_3) that is a stronger acid than the conjugate acid of water (namely, H_3O^+) is leveled to the strength of H_3O^+ in aqueous solution. The leveling effect makes it impossible to distinguish in aqueous solution between the strengths of acids that are stronger than H_3O^+. (*Section 16.2*)

Lewis acid: Any species that can *accept* a pair of electrons (c.f. the Arrhenius and Brønsted–Lowry definitions of an acid). The Lewis definition expands the Brønsted–Lowry definition to include substances other than the H^+ ion. (*Section 8.7*)

Lewis base: Any species that can *donate* a pair of electrons (c.f. the Arrhenius and Brønsted–Lowry definitions of a base). All Brønsted–Lowry bases (proton acceptors) are also electron-pair donors, so the Lewis definition of a base does not contradict the Brønsted–Lowry definition. (*Section 8.7*)

Lewis dot symbols: A system of symbols developed by G. N. Lewis in the early 20th century that can be used to predict the number of bonds formed by most elements in their compounds. Each Lewis dot symbol consists of the chemical symbol for an element surrounded by dots that represent its valence electrons. (*Section 8.4*)

Ligands: Ions or molecules that contain one or more pairs of electrons that can be shared with the central metal in a metal complex. The word *ligand* comes from the Latin *ligare*, "to bind." (*Sections 17.3 and 23.4*)

Limiting reactant: The reactant that restricts the amount of product obtained in a chemical reaction (c.f. excess reactant). (*Section 3.4*)

Linear: The lowest-energy arrangement for compounds that have two electron pairs around the central atom. In the linear arrangement, the electron pairs are on opposite sides of the central atom with a 180° angle between them (see Table 9.1). Linear is also one of two possible molecular geometries for an AB_2 molecular species (the other is bent). (*Section 9.1*)

Linear combination of atomic orbitals (LCAOs): Molecular orbitals created from the sum and the difference of two wave functions (atomic orbitals). According to the LCAO method, a molecule has as many molecular orbitals as there are atomic orbitals. (*Section 9.3*)

Line defect: A defect in a crystal that affects a row of points in the lattice (c.f. point defect and plane defect). (*Section 12.4*)

Line spectrum: A spectrum in which light of only certain wavelengths is emitted or absorbed, rather than a continuous range of wavelengths (colors). The line spectrum of each element is characteristic of that element. (*Section 6.3*)

Liquefaction: The condensation of gases into a liquid form. Gases invariably condense to form liquids because they no longer possess enough kinetic energy to overcome the intermolecular attractive forces. (*Section 10.8*)

Lipids: A family of compounds that includes fats, waxes, some vitamins, and steroids, lipids (from the Greek *lipos*, meaning "fat" or "lard") are characterized by their insolubility in water. Fatty acids, the simplest lipids, have a long hydrocarbon chain that ends with a carboxylic acid functional group. (*Section 24.6*)

Liquid: One of three distinct states of matter under normal conditions, liquids have fixed volumes but flow to assume the shape of their containers. The volume of a liquid is virtually independent of temperature and pressure. The other two states of matter are solid and gas. (*Section 1.3*)

Liquid crystals: Substances that exhibit phases that have properties intermediate between those of a crystalline solid and those of a normal liquid. Liquid crystals possess long-range molecular order (like a solid) but still flow (like a liquid). Liquid crystals typically consist of long, rigid molecules that can interact strongly with one another. (*Section 11.8*)

Lithium–iodine battery: A battery that consists of an anode of lithium metal and a cathode containing a solid complex of I_2. Separating the anode and the cathode is a layer of solid LiI, which acts as the electrolyte by allowing the diffusion of Li^+ ions. A typical lithium–iodine battery is shown in Figure 19.12c. (*Section 19.5*)

London dispersion forces: A kind of intermolecular interaction (force) that results from temporary fluctuations in the electron distribution within atoms and nonpolar molecules. These fluctuations produce instantaneous dipole moments that induce similar dipole moments in adjacent molecules, resulting in attractive forces between otherwise nonpolar substances. These attractive interactions are weak and fall off rapidly with increasing distance (as $1/r^6$, where *r* is the distance between dipoles). London dispersion forces and dipole–dipole interactions are often called van der Waals forces. (*Section 11.2*)

Lone pair: A pair of electrons in a Lewis structure that is not involved in covalent bonding (c.f. bonding pair). Lone pairs are valence electrons, but they are not used to bond with other atoms. (*Section 8.5*)

Luster: The ability to reflect light. Metals, for instance, have a shiny surface that reflects light (metals are lustrous), whereas nonmetals do not. (*Section 12.5*)

Lustrous: Having a shiny appearance. Metals are lustrous, whereas nonmetals are not. (*Section 1.7*)

M

Macromineral: Any of the six essential elements—sodium, magnesium, potassium, calcium, chlorine, and phosphorus—that provide essential ions in body fluids and form the major structural components of the body. They are found in large amounts in biological tissues and are present as inorganic compounds, either dissolved or precipitated. (*Sections 1.8 and 7.5*)

Magnetic quantum number (m_l): One of three quantum numbers (the other two are n and l) used to specify any wave function, the magnetic quantum number describes the orientation of the region of space occupied by the electron with respect to an applied magnetic field. The allowed values of m_l depend on the value of l and can range from $-l$ to l in integral steps. (*Section 6.5*)

Main group element: Any of the elements in Groups 1, 2, and 13–18 in the periodic table. These groups contain metals, nonmetals, and semimetals. (*Section 1.7*)

Malleable: The ability to be hammered or pressed into thin sheets or foils. Metals are malleable, whereas nonmetals are usually brittle instead. (*Section 1.7*)

Manometer: A device used to measure the pressures of samples of gases contained in an apparatus (c.f. barometer). The key feature of a manometer is a U-shaped tube that contains mercury (or some other nonvolatile liquid). A closed-end manometer is shown schematically in Figure 10.5a. (*Section 10.2*)

Mass: The quantity of matter an object contains. Mass is a fundamental property of an object that does not depend on its location (c.f. weight). (*Section 1.3*)

Mass defect: The difference between the sum of the masses of the components of an atom (protons, neutrons, and electrons) and the measured atomic mass. The mass defect occurs because a nucleus is more stable (lower in energy) than its isolated components. When isolated nucleons assemble into a stable nucleus, energy is released, which must be accompanied by a decrease in mass (because $E = mc^2$, where E is energy, m is mass, and c is the speed of light). (*Section 20.4*)

Mass number (A): The number of protons and neutrons in the nucleus of an atom of an element. The isotopes of an element differ only in their atomic mass, which is given by the mass number (A). (*Section 1.6*)

Mass percentage: A common unit of concentration, the mass percentage of a solution is the ratio of the mass of the solute to the total mass of the solution (see Equation 13.8). (*Section 13.4*)

Matter: Anything that occupies space and possesses mass. (*Section 1.3*)

Mechanical work: The energy required to move an object a distance d when opposed by a force F, such as gravity: work (w) = force (F) × distance (d). (*Section 5.1*)

Meissner effect: The phenomenon, first described by W. Meissner in 1933, in which a superconductor completely expels a magnetic field from its interior. (*Section 12.7*)

Melting point: The temperature at which the individual ions in a lattice or the individual molecules in a covalent compound have enough kinetic energy to overcome the attractive forces that hold them in place in the solid. At the melting point, the ions or molecules can move freely and the substance becomes a liquid. Thus, the solid and the liquid coexist in equilibrium at the melting point. (*Section 8.3*)

Meniscus: The upper surface of a liquid in a tube. The shape of the meniscus depends on the relative strengths of the cohesive and adhesive forces. Liquids (e.g., water) for which the adhesive forces are stronger than the cohesive forces have a concave meniscus, whereas liquids (e.g., mercury) for which the cohesive forces are stronger than the adhesive forces have a convex meniscus. (*Section 11.3*)

Metal: Any of the elements to the left of the zigzag line in the periodic table that runs from boron (B) down to astatine (At). Metals are good conductors of electricity and heat, they can be pulled into wires (ductility), they can be hammered into thin sheets or foils (malleability), and most are shiny (lustrous). In chemical reactions, metals tend to lose electrons to form positively charged ions (cations). All metals except mercury (Hg) are solids at room temperature and pressure. (*Section 1.7*)

Metal complex: A chemical compound composed of a central metal atom or ion that is bonded to one or more ligands. Metal complexes can be neutral, such as $[Co(NH_3)_3Cl_3]$; positively charged, such as $[Nd(H_2O)_9]^{3+}$; or negatively charged, such as UF_8^{4-}. (*Section 23.4*)

Metallic atomic radius: Half the distance between the nuclei of two adjacent metal atoms (see Figure 7.5b). (*Section 7.2*)

Metallic solid: A solid that consists of metal atoms held together by metallic bonds. Metallic solids have high thermal and electrical conductivities, they are malleable and ductile, and they have luster (c.f. ionic solid, molecular solid, and covalent solid). (*Section 12.5*)

Metallurgy: The set of processes by which metals are extracted from their ores and converted to more useful forms. Metallurgy consists of three general steps: (1) mining the ore, (2) separating and concentrating the metal or the metal-containing compound, and (3) reducing to the metal. (*Section 23.3*)

Metal-matrix composites: A composite that consists of reinforcing fibers embedded in a metal or metal alloy matrix. The fibers tend to consist of boron, graphite, or ceramic. (*Section 12.9*)

Micelle: A spherical or cylindrical aggregate of detergents or soaps in water that minimizes contact between the hydrophobic tails of the detergents or soaps and water. Only the hydrophilic heads of the detergents or soaps are in direct contact with water, while the hydrophobic tails are located in the interior of the micelle aggregate. Micelles form spontaneously above a certain concentration. (*Section 13.7*)

Midpoint: The point in an acid–base titration at which exactly enough acid (or base) has been added to neutralize one-half of the base (or acid) originally present. The midpoint occurs halfway to the equivalence point. By definition, $[HA] = [A^-]$ at the midpoint of the titration of an acid (HA). (*Section 16.5*)

Millimeters of mercury (mmHg): A unit of pressure, often called the torr (after Evangelista Torricelli, the inventor of the mercury barometer); 760 mmHg = 760 torr = 1 atmosphere (atm) = 101325 pascals (Pa) = 101.325 kPa. (*Section 10.2*)

Miscible: Capable of forming a single homogeneous phase, regardless of the proportions with which the substances are mixed. Ethanol and water are miscible because they form a solution when mixed in all proportions. (*Section 13.1*)

Mixture: A combination of two or more pure substances in variable proportions in which the individual substances retain their identities. (*Section 1.3*)

m_l: See "Magnetic quantum number (m_l)" (*Section 6.5*).

MO: See "Molecular orbital (MO)" (*Section 9.3*).

Molality (m): A common unit of concentration, the molality of a solution is the number of moles of solute present in exactly one kilogram of solvent (c.f. molarity). (*Section 13.4*)

Molar heat capacity (C_p): The amount of energy needed to increase the temperature of 1 mol of a substance by 1° C. The units of C_p are J/(mol·°C). (*Section 5.3*)

Molarity (M): A common unit of concentration, the molarity of a solution is the number of moles of solute present in exactly 1 L of solution (c.f. molality). The molarity is also the number of millimoles of solute present in exactly 1 mL of solution. The units of molarity are moles per liter of solution (mol/L), abbreviated as *M*. (*Section 4.2*)

Molar mass: The mass in grams of one mole of a substance. The molar mass of any substance is numerically equivalent to its atomic mass, molecular mass, or formula mass in grams per mole. (*Section 3.1*)

Molar volume: The molar mass (g/mol) of an element divided by its density (g/cm^3). Molar volume, therefore, has units of cm^3/mol. (*Section 7.1*)

Mole (mol): The quantity of a substance that contains the same number of units (e.g., atoms or molecules) as the number of carbon atoms in exactly 12 g of isotopically pure carbon-12. According to the most recent experimental measurements, 12 g of carbon-12 contains 6.022142×10^{23} atoms. (*Section 3.1*)

Molecular formula: A representation of a covalent compound that consists of the atomic symbol for each component element (in a prescribed order) accompanied by a subscript indicating the number of atoms of that element in the molecule. The subscript is written only if the number is greater than 1. Molecular formulas give only the elemental composition of molecules (see Figure 2.4a; c.f. structural formula and empirical formula). (*Section 2.1*)

Molecular geometry: The arrangement of the bonded atoms in a molecule or polyatomic ion in space (c.f. electron-pair geometry). Knowing the molecular geometry of a compound is crucial to understanding its chemistry. (*Section 9.1*)

Molecular mass: The sum of the average masses of the atoms in one molecule of the substance. The molecular mass is calculated by summing the atomic masses of the elements in the substance, each multiplied by its subscript (written or implied) in the molecular formula, and is expressed in units of amu. It is analogous to the formula mass of an ionic compound. (*Section 3.1*)

Molecularity: The number of molecules that collide during any step (elementary reaction) in a reaction mechanism. An elementary reaction is unimolecular if only a single reactant molecule is involved; it is bimolecular if two reactant molecules are involved; and it is termolecular (a relatively rare situation) if three reactant molecules are involved (see Table 14.8). The reaction order of an elementary reaction is the same as its molecularity. (*Section 14.6*)

Molecular orbital (MO): A particular spatial distribution of electrons in a molecule that is associated with a particular orbital energy. Unlike an atomic orbital, which is localized on a single atom, a molecular orbital extends (is delocalized) over all of the atoms in a molecule or ion. (*Section 9.3*)

Molecular orbital theory: A delocalized bonding model (c.f. valence bond theory). In molecular orbital theory, the linear combination of atomic orbitals creates molecular orbitals that can be used to explain the bonding in molecules and polyatomic ions as well as their concomitant molecular geometries. (*Section 9.3*)

Molecular solid: A solid that consists of molecules held together by relatively weak forces, such as dipole–dipole interactions, hydrogen bonds, and London dispersion forces (c.f. ionic solid, covalent solid, and metallic solid). (*Section 12.5*)

Molecule: A group of atoms in which one or more pairs of electrons are shared between bonded atoms. Covalent compounds consist of molecules. (*Section 2.1*)

Mole fraction (*X*): The ratio of the number of moles of any component of a mixture to the total number of moles of all species present in the mixture (n_t): mole fraction of component A = X_A = moles of A/total moles = n_A/n_t (see Equation 10.29). The mole fraction, a dimensionless quantity between 0 and 1, can be used to describe the composition of a gas mixture. In a mixture of gases, the partial pressure of each of the component gases is the product of the total pressure and the mole fraction of that gas. (*Section 10.5*)

Mole ratio: The ratio of the number of moles of one substance to the number of moles of another, as depicted by a balanced chemical equation. (*Section 3.3*)

Molten salt: A salt that has been heated to its melting point. Molten salts conduct electricity, have high heat capacities, attain very high temperatures as liquids, and are useful as solvents because of their relatively low toxicity. (*Section 11.6*)

Monatomic ion: An ion that contains only a single atom (e.g., Na^+, Al^{3+}, Cl^-, or S^{2-}). (*Section 2.1*)

Monomers: The basic structural units of polymers. Many monomer molecules are connected in polymers, forming chains or networks via covalent bonds. (*Section 12.8*)

Monoprotic acid: A compound that is capable of donating a single proton per molecule (e.g., HF and HNO_3; c.f. polyprotic acid). (*Section 4.6*)

N

***n*:** See "Principal quantum number (*n*)" (*Section 6.5*).

Nanotubes: One of at least four allotropes of carbon (the other three are graphite, diamond, and the fullerenes). Nanotubes are cylinders of carbon atoms (see Figure 7.18) and are intermediate in structure between graphite and the fullerenes. Carbon nanotubes can be described as sheets of graphite that have been rolled up into a cylinder, or as fullerene cages that have been stretched in one direction. (*Section 7.4*)

Natural logarithm: The power *x* to which *e* (an irrational number whose value is approximately 2.7183) must be raised to obtain a particular number. (*Essential Skills 6*)

Nematic phase: One of three different ways that most liquid crystals can orient themselves (the other two are the smectic and cholesteric phases). In the nematic phase, only the long axes of the molecules are aligned, so they are free to rotate or to slide past one another. (*Section 11.8*)

Nernst equation: An equation for calculating cell potentials (E_{cell}) under nonstandard-state conditions, the Nernst equation can be used to determine the direction of spontaneous reaction for any redox reaction under any conditions. $E_{cell} = E^\circ_{cell} - (RT/nF) \ln Q$, where E_{cell} is the cell potential, E°_{cell} is the standard cell potential, *R* is the ideal gas constant, *T* is the absolute temperature, *n* is the number of moles of electrons, *F* is the faraday constant, and *Q* is the reaction quotient (Equation 19.46). (*Section 19.4*)

Net ionic equation: A chemical equation that shows only those species that participate in the chemical reaction. Canceling the spectator ions from a complete ionic equation gives the net ionic equation. (*Section 4.4*)

Neutralization reaction: A chemical reaction in which an acid and a base react in stoichiometric amounts to produce water and a salt. (*Section 4.6*)

Neutral solution: A solution in which the total positive charge from all of the cations is matched by an identical total negative charge from all of the anions. Pure water is a neutral solution in which $[H^+] = [OH^-] = 1.0 \times 10^{-7}$ *M* at 25°C, and which has a pH of 7.0. (*Section 4.6*)

Neutron: A subatomic particle that resides in the nucleus of almost all atoms. A neutron has a mass of 1.675×10^{-24} g (1.008665 amu), which is only slightly greater than the mass of a proton, but no charge (neutrons are electrically neutral). Neutrons and protons constitute by far the bulk of the mass of atoms. (*Section 1.5*)

Nickel–cadmium (nicad) battery: A water-based electrochemical cell, the nicad battery has a cadmium anode and a highly oxidized nickel cathode that is usually described as the nickel(III) oxo-hydroxide, NiO(OH). (*Section 19.5*)

Noble gas: Any of the elements in Group 18 of the periodic table (He, Ne, Ar, Kr, Xe, and Rn). All are unreactive monatomic gases at room temperature and pressure. (*Section 1.7*)

Nodes: Points where the amplitude of a wave is zero. For a standing wave, such as a plucked guitar string, the string does not move at the nodes. All overtones of the fundamental standing wave have one or more nodes. (*Section 6.4*)

Nonaqueous solution: A solution in which any substance other than water is the solvent (c.f. aqueous solution). (*Chapter 4 introduction*)

Nonbonding molecular orbital: A molecular orbital that forms when atomic orbitals or orbital lobes interact only very weakly, creating essentially no change in the electron probability density between the nuclei. Nonbonding molecular orbitals have approximately the same energy as the parent atomic orbitals. They are one of three types of molecular orbitals that can form when atomic orbitals interact (the other two are bonding and antibonding molecular orbitals). (*Section 9.3*)

Nonelectrolyte: A substance (e.g., ethanol or glucose) that dissolves in water to form neutral molecules. Nonelectrolytes have essentially no effect on the conductivity of water (c.f. electrolytes). (*Section 4.1*)

Nonionizing radiation: Radiation that is relatively low in energy, so that when it collides with an atom in a molecule or ion, most or all of its energy can be absorbed without causing a structural or chemical change (c.f. ionizing radiation). Instead, the kinetic energy of the radiation is transferred to the atom or molecule with which it collides, causing the atom or molecule to rotate, vibrate, or move more rapidly. (*Section 20.3*)

Nonmetal: Any of the elements to the right of the zigzag line in the periodic table that runs from boron (B) down to astatine (At). Nonmetals are generally poor conductors of heat and electricity, they are not shiny (lustrous), and solid nonmetals tend to be brittle (so they cannot be pulled into wires or hammered into sheets or foils). Nonmetals tend to gain electrons in reactions with metals to form negatively charged ions (anions) or to share electrons in reactions with other nonmetals. Nonmetals may be solids, liquids, or gases at room temperature and pressure. (*Section 1.7*)

Nonstoichiometric compounds: Solids that have intrinsically variable stoichiometries. Nonstoichiometric compounds contain large numbers of defects, usually vacancies, which give rise to stoichiometries that can depart significantly from simple integral ratios without affecting the fundamental structure of the crystal. (*Section 12.4*)

Nonvolatile liquids: Liquids that have relatively low vapor pressures. Nonvolatile liquids tend to evaporate more slowly from an open container than volatile liquids do. (*Section 11.4*)

Normal boiling point: The temperature at which a substance boils when the external pressure is one atmosphere. For water, the normal boiling point is exactly 100°C. (*Section 11.4*)

***n*-Type semiconductor:** A semiconductor that has been doped with an impurity that contains more valence electrons than the atoms of the host lattice (c.f. *p*-type semiconductor). The increased number of valence electrons helps populate the conduction band, thus increasing the electrical conductivity of the semiconductor. (*Section 12.6*)

Nuclear binding energy: The amount of energy released when a nucleus forms from its component nucleons. The nuclear binding energy is the amount of energy that corresponds to the mass defect of the nucleus because energy and mass are related by $E = mc^2$, where E is energy, m is mass, and c is the speed of light. (*Section 20.4*)

Nuclear chain reaction: A self-sustaining series of nuclear fission reactions. A minimum mass of a fissile isotope is required to sustain a nuclear chain reaction (see critical mass). (*Section 20.4*)

Nuclear decay reaction: Also called radioactive decay, a nuclear decay reaction occurs when an unstable nucleus emits radiation and is transformed into the nucleus of one or more other elements. The resulting daughter nuclei have a lower mass and are lower in energy (more stable) than the parent nucleus that decayed (c.f. nuclear transmutation reaction). (*Section 20.2*)

Nuclear energy: One of the five forms of energy, nuclear energy is stored in the nucleus of an atom. The other four forms of energy are radiant, thermal, chemical, and electrical. (*Section 5.1*)

Nuclear fission: The splitting of a heavy nucleus into two lighter ones (c.f. nuclear fusion). (*Section 20.4*)

Nuclear fusion: The combining of two light nuclei to produce a heavier, more stable nucleus (c.f. nuclear fission). (*Section 20.4*)

Nuclear transmutation reaction: A nuclear reaction in which a nucleus reacts with a subatomic particle or another nucleus to give a product nucleus that is more massive than the starting material (c.f. nuclear decay reaction). An artificially induced reaction that converts an otherwise stable nuclide to another, more massive nuclide. (*Section 20.2*)

Nucleic acids: Linear polymers of nucleotides, nucleic acids are the basic structural components of DNA and RNA, the biochemical substances found in the nuclei of cells that transmit the information needed to direct cellular growth and reproduction. The backbone of DNA and RNA is the alternating sugar and phosphate groups of the nucleic acids. The unique structures of nucleic acids are due to nitrogen-containing cyclic compounds called pyrimidines and purines, which are attached to the sugar portions of the backbone. (*Section 24.6*)

Nucleons: The protons and neutrons that make up the nucleus of an atom. (*Section 20.1*)

Nucleophile: A species that has a pair of electrons available to be shared with another atom, a nucleophile (from *nucleus* plus the Greek suffix *phile*, "loving") is electron rich (c.f. electrophile). Carbanions, which are electron rich, are nucleophiles in organic reactions. (*Section 24.3*)

Nucleophilic substitution reaction: A substitution reaction in which the nucleophilic reactant possesses a pair of electrons and has a greater affinity for the electropositive species than the original substituent does. (*Section 24.4*)

Nucleoside: The compound formed when a pyrimidine or purine base is linked to a sugar by a glycosidic bond (see Figure 24.19a). (*Section 24.6*)

Nucleotide: The compound formed when a phosphoric acid group is attached to the sugar portion of a nucleoside (see Figure 24.19b). Nucleotides are the monomers used to make the nucleic acids DNA and RNA. (*Section 24.6*)

Nucleus: The central core of an atom, where the protons and any neutrons reside. The nucleus comprises most of the mass of an atom but very little of the volume. (*Section 1.5*)

Nuclide: An atom with a particular number of nucleons (protons and neutrons). Nuclides that have the same number of protons, but different numbers of neutrons, are called isotopes. (*Section 20.1*)

O

Octahedral: The lowest-energy arrangement for compounds that have six electron pairs around the central atom. In the octahedral arrangement, each electron pair is positioned at 90° angles to four adjacent electron pairs and at a 180° angle to the fifth electron pair (that is, they are positioned at the vertices of an octahedron) (see Table 9.1). Octahedral is also the most common molecular geometry for an AB_6 molecular species. (*Section 9.1*)

Octahedral hole: One of two kinds of holes in a face-centered cubic array of atoms or ions (the other is a tetrahedral hole). One octahedral hole is located in the center of the face-centered cubic unit cell, and there is a shared one in the middle of each edge. An atom or ion in an octahedral hole has a coordination number of 6. (*Section 12.3*)

Octane rating: A measure of a fuel's ability to burn in a combustion engine without knocking or pinging (indications of premature combustion). The higher the octane rating, the higher quality the fuel. *n*-Heptane, for example, which causes a great deal of knocking upon combustion, has an octane rating of 0, whereas isooctane, a very smooth-burning fuel, has an octane rating of 100. (*Section 2.6*)

Octaves: Groups of seven elements, corresponding to the horizontal rows in the main group elements (not counting the noble gases, which were unknown at the time). John Newlands arranged the elements that were known in the mid-19th century in order of increasing atomic mass and discovered that every seventh element had similar properties. He thus proposed the "law of octaves" to explain this pattern, but it turned out that the law did not seem to work for elements heavier than calcium. (*Section 7.1*)

Octet rule: The tendency for atoms to lose, gain, or share electrons to reach a total of eight valence electrons (c.f. expanded-valence molecules). The octet rule explains the stoichiometry of most compounds in the *s* and *p* blocks of the periodic table. (*Section 8.4*)

Open system: One of three kinds of systems, an open system can exchange both matter and energy with its surroundings. The other kinds of system are closed and isolated. (*Sections 5.2 and 18.1*)

Optical isomers: One of two types of stereoisomers (the other is geometrical isomers), optical isomers are mirror images of each other but cannot be superimposed on one another in any orientation. Optical isomers have identical physical properties, although their chemical properties may differ. (*Section 24.2*)

Organic compound: A covalent compound that contains predominantly carbon and hydrogen (c.f. inorganic compound). (*Section 2.1 and Chapter 24 introduction*)

Organometallic compounds: Compounds that contain a metal covalently bonded to a carbon atom of an organic species, organometallic compounds have properties and reactivities that differ greatly from those of either their metallic or organic components. (*Section 21.3*)

Osmosis: The net flow of solvent through a semipermeable membrane. The direction of solvent flow is always from the side with the lower concentration of solute to the side with the higher concentration. (*Section 13.6*)

Osmotic pressure (Π): The pressure difference between the two sides of a semipermeable membrane that separates a pure solvent from a solution prepared from the same solvent. The osmostic pressure develops because the flow of solvent through the membrane in opposing directions is unequal. Osmotic pressure is a colligative property of a solution, so it depends on the concentration of dissolved solute particles (see Equation 13.24): $\Pi = MRT$, where M is the molarity of the solution, R is the ideal gas constant, and T is the absolute temperature. (*Section 13.6*)

Overall equation: A chemical equation that shows all of the reactants and products as undissociated, electrically neutral compounds. Although an overall equation gives the identity of the reactants and products, it does not show the identities of the actual species in solution, especially if ionic substances that are strong electrolytes are involved (c.f. complete ionic equation and net ionic equation). (*Section 4.4*)

Overall reaction order: For an experimentally determined rate law of the form rate $= k[A]^m[B]^n$, the overall reaction order is the sum of the exponents of the rate law, $m + n$ (c.f. reaction order). (*Section 14.2*)

Overlapping bands: Molecular orbitals derived from two or more different kinds of valence electrons that have similar energies. Overlapping bands result when the width of adjacent bands is relatively large compared with the energy gap between them. (*Section 12.6*)

Overtones: Vibrations of a standing wave that are higher in energy than the fundamental vibration (the lowest-energy vibration). All overtones have one or more nodes. (*Section 6.4*)

Overvoltage: The voltage that must be applied in electrolysis in addition to the calculated (theoretical) value to overcome factors such as a high activation energy and the formation of bubbles on a surface. Overvoltages are needed in all electrolytic processes. (*Section 19.7*)

Oxidant: A compound that is capable of accepting electrons (c.f. reductant). Oxidants (also called oxidizing agents) can oxidize other compounds. In the process of accepting electrons, an oxidant is reduced. (*Sections 3.5 and 19.1*)

Oxidation: The loss of one or more electrons in a chemical reaction (c.f. reduction). The substance that loses electrons is said to be oxidized. (*Section 3.5*)

Oxidation–reduction (redox) reaction: A chemical reaction in which there is a net transfer of one or more electrons from one reactant to another. This transfer of electrons provides a means for converting chemical energy to electrical energy, or vice versa. The total number of electrons lost by some reactants must equal the total number of electrons gained by other reactants. Thus, oxidation must be accompanied by reduction, and vice versa. (*Section 3.5 and Chapter 19 introduction*)

Oxidation state: The charge that each atom in a compound would have if all of its bonding electrons were transferred to the atom with the greater attraction for electrons. (*Section 3.5*)

Oxidation state method: A procedure for balancing oxidation–reduction (redox) reactions in which the overall reaction is conceptually separated into two parts: an oxidation and a reduction (see Table 4.3 for a step-by-step summary of this method). (*Section 4.8*)

Oxidizing agent: Another name for an oxidant (c.f. reducing agent). (*Section 3.5*)

Oxoacid: An acid in which the dissociable H^+ ion is attached to an oxygen atom of a polyatomic anion (e.g., H_2SO_4 or HNO_3). Oxoacids are occasionally called oxyacids. (*Section 2.5*)

Oxoanion: A polyatomic anion that contains a single metal or nonmetal atom plus one or more oxygen atoms. Sometimes oxoanions are called oxyanions. (*Section 2.3*)

Ozone: An unstable form of oxygen that consists of three oxygen atoms bonded together (O_3). A layer of ozone in the stratosphere helps protect the plants and animals on earth from harmful ultraviolet radiation. Ozone is responsible for the pungent smell we associate with lightning discharges and electric motors. It is also toxic. (*Section 3.6*)

Ozone layer: A concentration of ozone in the stratosphere (about 10^{15} ozone molecules per liter) that acts as a protective screen, absorbing ultraviolet light that would otherwise reach the surface of the earth, where it would harm plants and animals. (*Section 3.6*)

P

***p*:** See "Steric factor (*p*)" (*Section 14.7*).

***P*:** See "Pressure (*P*)" (*Sections 1.3 and 10.2*) or "Spin-pairing energy (*P*)" (*Section 23.5*).

Partial pressure: The pressure a gas in a mixture would exert if it were the only one present (at the same temperature and volume). The pressure exerted by each gas in a mixture (its partial pressure) is independent of the pressure exerted by all other gases present. As a result, the total pressure exerted by a mixture of gases is the sum of the partial pressures of the components (Dalton's law of partial pressures). Additionally, the partial pressure of each of the component gases in a mixture is the product of the total pressure and the mole fraction of that gas. (*Section 10.5*)

Parts per billion (ppb): A common unit of concentration, parts per billion is micrograms of solute per kilogram of solvent (see Equation 13.10; c.f. parts per million). (*Section 13.4*)

Parts per million (ppm): A common unit of concentration, parts per million is milligrams of solute per kilogram of solvent (see Equation 13.9; c.f. parts per billion). (*Section 13.4*)

Pascal (Pa): The SI unit for pressure. Derived from the SI units for force (newtons) and area (square meters), the pascal is newtons per square meter, N/m^2: 1 pascal (Pa) = 1 newton/meter^2 (N/m^2). The pascal is named after the French mathematician Blaise Pascal (1623–1662). (*Section 10.2*)

Pauli exclusion principle: Developed by Wolfgang Pauli, this principle states that no two electrons in an atom can have the same value of all four quantum numbers (n, l, m_l, and m_s). This principle arises from Pauli's determination that no two electrons can have the same four quantum numbers. (*Section 6.6*)

***p* block:** The six columns on the right side of the periodic table, consisting of the elements in which the *np* orbitals are being filled (see Figure 6.34; c.f. *s* block, *d* block, and *f* block). (*Section 6.6*)

***p*-block elements:** The elements in the six columns on the right side of the periodic table. The elements of the *p* block are filling their *np* orbitals (see Figure 6.34; c.f. *s*-block elements, *d*-block elements, and *f*-block elements). (*Section 6.6*)

***PE*:** See "Potential energy (*PE*)" (*Section 5.1*).

Peptide (amide) bonds: The covalent bond that links one amino acid to another in peptides and proteins. The carbonyl carbon atom of one amino acid residue bonds to the amino nitrogen atom of the other residue, eliminating a molecule of water in the process (see Equation 12.4). Thus, peptide bonds form in a condensation reaction between a carboxylic acid and an amine. (*Sections 3.5 and 12.8*)

Peptides: Biological polymers that contain fewer than about 50 amino acid residues (c.f. proteins). (*Section 12.8*)

Percent composition: The percentage of each element present in a pure substance. With few exceptions, the percent composition of a chemical compound is constant (see law of definite proportions). (*Section 3.2*)

Percent yield: The ratio of the actual yield of a reaction to the theoretical yield, multiplied by 100% to give a percentage. Percent yields can range from 0% to 100%. (*Section 3.4*)

Period: The rows of elements in the periodic table. At present the periodic table consists of seven periods, numbered from 1 to 7. (*Section 1.7*)

Periodic: Phenomena, such as waves, that repeat regularly in both space and time. (*Section 6.1*)

Periodic table: A chart of the chemical elements arranged in rows of increasing atomic number (Z), so that the elements in each column (group) have similar chemical properties. Each element in the periodic table is assigned a unique one-, two-, or three-letter symbol. (*Section 1.7*)

Perovskite structure: A structure that consists of a body-centered array of two metal ions, with one set (M) located at the corners of the cube, and the other set (M′) in the centers of the cube. The anions are in the centers of the square faces (see Figure 12.12). (*Section 12.3*)

pH: The negative base-10 logarithm of the hydrogen ion concentration: $pH = -\log[H^+]$. Because it is a negative logarithm, pH decreases with increasing $[H^+]$. (*Section 4.6*)

Phase changes: The changes of state that occur when any of the three forms of matter (solids, liquids, and gases) is converted to either of the other two (see Figure 11.17). All phase changes are accompanied by changes in the energy of the system. Changes from a more ordered to a less ordered state (e.g., from liquid to gas) are endothermic, whereas changes from a less ordered to a more ordered state (e.g., from liquid to solid) are always exothermic. Phase changes are also called phase transitions. (*Section 11.5*)

Phase diagram: A graphic summary of the physical state of a substance as a function of temperature and pressure in a closed system (see Figure 11.22). A typical phase diagram consists of regions that represent the different phases possible for the substance (solid, liquid, and gas). The solid and liquid regions are separated by the melting curve of the substance, while the liquid and gas phases are separated by the vapor pressure curve (which ends at the critical point). Only a single phase is stable within a given region, but two phases are in equilibrium at the given temperatures and pressures along the lines that separate the regions. (*Section 11.7*)

Phase transitions: Another name for phase changes. (*Section 11.5*)

Phospholipids: A large class of biological, detergent-like molecules that contain a hydrophilic head and two hydrophobic tails (detergents and soaps have just one tail). The additional tail results in a cylindrical shape that prevents phospholipids from forming spherical micelles. Instead, phospholipids form bilayers. (*Section 13.7*)

Photoelectric effect: A phenomenon in which electrons are ejected from the surface of a metal that has been exposed to light. Each metal has a characteristic threshold frequency of light (ν_0). Below the threshold frequency, no electrons are emitted regardless of the light's intensity. Above the threshold frequency, the number of electrons emitted is proportional to the intensity of the light, and the kinetic energy of the ejected electrons is proportional to the frequency of the light. (*Section 6.2*)

Photons: "Particles" (quantums) of light (radiant energy), each of which possesses a particular energy E given by $E = h\nu$. (*Section 6.2*)

Photosynthesis: The fundamental reaction by which all green plants and algae obtain energy from sunlight, photosynthesis is the photochemical reduction of CO_2 to a carbon compound such as glucose (c.f. respiration). Oxygen in water is concurrently oxidized to O_2: $6CO_2 + 6H_2O \xrightarrow{h\nu} C_6H_{12}O_6 + 6O_2$ (Equation 18.46). (*Section 18.8*)

Phototrophs: Organisms, such as green plants and algae, that extract energy from the environment through photosynthesis (c.f. chemotrophs). Sunlight, therefore, is the ultimate energy source for phototrophs. (*Section 18.8*)

pH scale: A logarithmic scale used to express the hydrogen ion (H^+) concentration of a solution, making it possible to describe acidity or basicity quantitatively. The pH scale is convenient because very large and very small concentrations can be expressed as relatively simple numbers. (*Section 4.6*)

Physical property: A characteristic that scientists can measure without changing the composition of the sample under study (e.g., mass, color, and volume). Physical properties can be extensive or intensive. (*Section 1.3*)

Pinning: A process that introduces multiple defects into a material so that the presence of one defect prevents the motion of another. By preventing the motion of the defects, pinning increases the mechanical strength of the material. (*Section 12.4*)

Pi (π) orbital: A bonding molecular orbital formed from the sum of the side-to-side interactions of two or more parallel *np* atomic orbitals. The sum of the interactions increases electron probability in the region above and below a line connecting the nuclei [c.f. pi star (π^*) orbital]. A π molecular orbital, like a π^* molecular orbital, possesses a nodal plane that contains the nuclei. (*Section 9.3*)

Pi star (π^*) orbital: An antibonding molecular orbital formed from the difference of the side-to-side interactions of two or more parallel *np* atomic orbitals. The difference of the interactions decreases electron probability in the region above and below a line connecting the nuclei [c.f. pi (π) orbital]. A π^* molecular orbital, like a π molecular orbital, possesses a nodal plane that contains the nuclei. (*Section 9.3*)

Plane defect: A defect in a crystal that affects a plane of points in the lattice (c.f. point defect and line defect). (*Section 12.4*)

Plastic: The property of a material that allows it to be molded into almost any shape. Although many plastics are polymers, not all polymers are plastics. (*Section 12.8*)

Pnicogens: The elements in Group 15 of the periodic table (N, P, As, Sb, and Bi). (*Section 7.4*)

Point defect: A defect in a crystal that affects a single point in the lattice c.f. line defect and plane defect). Point defects can consist of an atom missing from a site in the crystal (a vacancy) or an impurity atom that occupies either a normal lattice site (a substitutional impurity) or a hole in the lattice between atoms (an interstitial impurity). (*Section 12.4*)

Poisons: Substances that bind irreversibly to catalysts, thus preventing reactants from adsorbing to the catalyst, which reduces or destroys the catalyst's efficiency (see enzyme inhibitors, too). (*Section 14.8*)

Polar bond: A chemical bond in which there is an unequal distribution of charge between the bonding atoms. One atom is electron rich (so it has a partial negative charge, δ^-), whereas the other atom is electron poor (so it has a partial positive charge, δ^+). (*Section 4.1*)

Polar covalent bond: A covalent bond in which the electrons are shared unequally between the bonded atoms (c.f. covalent bond). (*Section 8.9*)

Polyatomic ions: Groups of two or more atoms that have a net electrical charge, although the atoms that make up a polyatomic ion are held together by the same covalent bonds that hold atoms together in molecules. Just as there are many more kinds of molecules than simple elements, there are many more kinds of polyatomic ions than monatomic ions (see Table 2.4). (*Section 2.1*)

Polyatomic molecules: Molecules that contain more than two atoms. (*Section 2.1*)

Polymer-matrix composites: Composites that consist of reinforcing fibers embedded in a polymer matrix. Fiberglass, for instance, is a polymer-fiber matrix consisting of glass fibers embedded in a polymer matrix. (*Section 12.9*)

Polymers: Giant molecules that consist of many basic structural units (monomers) connected in chains or networks by covalent bonds. (*Section 12.8*)

Polyprotic acid: A compound that can donate more than one hydrogen ion per molecule (e.g., H_2SO_4 or H_3PO_4; c.f. monoprotic acid). (*Section 4.6*)

Positron: A particle that has the same mass as an electron, but has a positive charge instead of a negative charge. (*Section 20.2*)

Positron emission: The opposite of β decay, positron emission is a nuclear decay reaction in which a proton is transformed into a neutron and a high-energy positron is emitted. Positron emission is characteristic of neutron-poor nuclei. (*Section 20.2*)

Potential (E_{cell}): Measured in volts, the potential of an electrochemical cell is the difference in electrical potential between the two half-reactions (oxidation at the anode and reduction at the cathode). Electrical potential is related to the energy needed to move a charged particle in an electric field. (*Section 19.1*)

Potential energy (*PE*): Energy stored in an object because of the relative positions or orientations of its components. Electrical, nuclear, and chemical energy are different forms of potential energy. (*Section 5.1*)

Precipitate: The insoluble product that forms in a precipitation reaction. (*Section 4.5*)

Precipitation reaction: A chemical reaction that yields an insoluble product (a precipitate) when two solutions are mixed. Precipitation reactions are a subclass of exchange reactions that occur between ionic compounds when one of the products is insoluble. Because both components of each compound change partners, such reactions are sometimes called double-displacement reactions. (*Section 4.5*)

Precise: When multiple measurements give nearly identical values for a quantity. (*Essential Skills 1*)

Precision: The degree to which multiple measurements give nearly identical values for a quantity. (*Essential Skills 1*)

Pressure (*P*): The amount of force (*F*) exerted on a given area (*A*) of surface: $P = \frac{F}{A}$ (*Sections 1.3 and 10.2*)

Principal quantum number (*n*): One of three quantum numbers (the other two are *l* and m_l) used to specify any wave function, the principal quantum number tells the average relative distance of the electron from the nucleus. The allowed values of *n* are positive integers ($n = 1, 2, 3, 4, \ldots$). As *n* increases for a given atom, so does the average distance of the electron from the nucleus. (*Section 6.5*)

Principal shell: All wave functions that have the same value of *n* constitute a principal shell because those electrons have similar average distances from the nucleus. (*Section 6.5*)

Product: The final compound or compounds produced in a chemical reaction. The products are usually written to the right of the arrow in the chemical equation. (*Section 3.3*)

Promotion: The excitation of an electron from a filled ns^2 atomic orbital to an empty *np* or $(n-1)d$ valence orbital. The formation of hybrid atomic orbitals can be viewed as occurring via promotion followed by hybridization. (*Section 9.2*)

Propagation: The second stage in a chain reaction, propagation takes the reactive intermediates formed in initiation and continuously consumes and regenerates them in reactions that also form products. The other stages of a chain reaction are initiation and termination. (*Section 14.6*)

Proteins: Biological polymers that contain more than 50 amino acid residues linked together by amide bonds (c.f. peptides). In addition to an amine group and a carboxylic acid group, each amino acid in the protein contains a characteristic R group, the nature of which determines the physical and chemical properties of the protein. (*Sections 12.8 and 24.6*)

Protium: One of the three isotopes of hydrogen, protium (^1H or H) is the most common, consisting of one proton and one electron. The other two isotopes of hydrogen are deuterium and tritium. (*Section 21.2*)

Proton: A subatomic particle that resides in the nucleus of all atoms. A proton has a mass of 1.673×10^{-24} g (1.007276 amu) and an electrical charge of -1.602×10^{-19} C. The charge on the proton is equal in magnitude but opposite in sign to the charge on an electron. Protons and neutrons constitute by far the bulk of the mass of atoms. (*Section 1.5*)

Pseudo inert gas configuration: The $(n-1)d^{10}$ and similar electron configurations that are particularly stable and are often encountered in the heavier *p*-block elements. The electron configuration of Ga^{3+}, for example, is [Ar] $3d^{10}$, not [Ar] $4s^2\ 3d^8$. (*Section 7.3*)

***p*-Type semiconductor:** A semiconductor that has been doped with an impurity that contains fewer valence electrons than the atoms of the host lattice (c.f. *n*-type semiconductor). The reduced number of valence electrons creates holes in the valence band, thus increasing the electrical conductivity of the semiconductor. (*Section 12.6*)

Pyrolysis reaction: A high-temperature decomposition reaction that can be used to form fibers of synthetic polymers. Catalytic cracking is one type of pyrolysis reaction (from the Greek *pyros*, "fire," plus *lysis*, "loosening"), in which alkanes are heated to a sufficiently high temperature to induce cleavage of the weakest bonds, the C—C single bonds. The result is a mixture of free radicals derived from essentially random cleavage of the various C—C bonds in the alkane chain. (*Sections 12.8 and 24.5*)

Q

***q*:** See "Heat (*q*)" (*Section 5.1*).

***Q*:** See "Reaction quotient (*Q*)" (*Section 15.4*) or "Ion product (*Q*)" (*Section 17.1*).

Q_p: See "Reaction quotient (Q_p)" (*Section 15.4*).

Qualitative analysis: A procedure for determining the identity of metal ions present in a mixture (rather than quantitative information about their amounts). (*Section 17.5*)

Quantitative analysis: A methodology that combines chemical reactions and stoichiometric calculations to determine the amounts or concentrations of substances present in a sample. (*Section 4.9*)

Quantum: The smallest possible unit of energy. The energy of electromagnetic waves is quantized rather than continuous, so energy can be gained or lost only in integral multiples of some smallest unit of energy, a quantum. (*Section 6.2*)

Quantum mechanics: A theory developed by Erwin Schrödinger to describe the energies and spatial distributions of electrons in atoms and molecules. (*Section 6.5*)

Quantum numbers: A unique set of numbers that specifies a wave function (a solution to the Schrödinger equation). Quantum numbers provide important information about the energy and spatial distribution of an electron. (*Section 6.5*)

Quaternary ammonium salts: Salts that consist of an anion (e.g., Cl^-) and a cation in which all four H atoms of the ammonium ion (NH_4^+) are replaced by alkyl groups. Quaternary ammonium salts can be chiral if all four alkyl substituents are different. (*Section 24.5*)

R

R: The abbreviation used for alkyl groups and aryl groups in general formulas and structures. (*Section 2.4*)

***R*:** See "Gas constant (*R*)" (*Section 10.4*).

r_0: See "Bond distance (r_0)" (*Section 8.1*).

Rad (radiation absorbed dose): A unit used to measure the effects of radiation on biological tissues, the rad is the amount of radiation that causes 0.01 joule of energy to be absorbed by 1 kilogram of matter (c.f. gray and roentgen). (*Section 20.3*)

Radiant energy: One of the five forms of energy, radiant energy is carried by light, microwaves, and radio waves (the other forms of energy are thermal, chemical, nuclear, and electrical). Objects left in bright sunshine or exposed to microwaves become warm because much of the radiant energy they absorb is converted to thermal energy. (*Section 5.1*)

Radiation dose: The amount of radiation an object has been exposed to (see roentgen, rad, gray, and rem). (*Section 20.3*)

Radicals: Species that have one or more unpaired valence electrons. Radicals are often formed as reaction intermediates in chain reactions. (*Section 14.6*)

Radioactive: Any nucleus that is unstable and decays spontaneously, emitting subatomic particles and electromagnetic radiation. (*Section 20.1*)

Radioactive decay series: A series of sequential alpha- and beta-decay reactions that occur until a stable nucleus is finally obtained. The uranium-238 decay series, for example, produces lead-206 in a series of 14 sequential α- and β-decay reactions (Figure 20.4). (*Section 20.2*)

Radioactivity: The spontaneous emission of energy rays (radiation) by matter. (*Section 1.5*)

Radioisotopes: Isotopes that emit radiation. (*Section 20.1*)

Raoult's law: The equation that quantifies the relationship between solution composition and vapor pressure (see Equation 13.13): $P_A = X_A P_A^\circ$, where P_A is the vapor pressure of component A of the solution, X_A is the mole fraction of component A present in solution, and P_A° is the vapor pressure of pure A. (*Section 13.6*)

Rate constant (*k*): A proportionality constant, the value of which is characteristic of the reaction and reaction conditions. A given reaction has a particular value of the rate constant under a given set of conditions, such as temperature, pressure, and solvent. Varying the temperature or the solvent usually changes the value of the rate constant. The numerical value of *k*, however, does not change as the reaction progresses under a given set of conditions. (*Section 14.2*)

Rate-determining step: The slowest step (elementary reaction) in a reaction mechanism. The rate law for an overall reaction is the same as the rate law for the rate-determining step of the reaction. (*Section 14.6*)

Rate laws: Mathematical expressions that describe the relationships between reactant rates and reactant concentrations in a chemical reaction (see differential rate law and integrated rate law). Rate laws are mathematical descriptions of experimentally verifiable data. (*Section 14.2*)

Rate of decay: Also called the activity (A), the rate of decay of a radioactive isotope is the decrease in the number of the radioisotope's nuclei per unit time (Equation 14.33): $A = -\Delta N/\Delta t$, where N is the number of atoms of the radioactive isotope. Rate of decay is usually measured in units of disintegrations per second (dps) or disintegrations per minute (dpm). (*Section 14.5*)

Reactant: The starting material or materials in a chemical reaction. The reactants are usually written to the left of the arrow in the chemical equation. (*Section 3.3*)

Reaction mechanism: The sequence of events (the stepwise changes) that occur at the molecular level during a reaction, the reaction mechanism describes how individual atoms, ions, or molecules interact to form particular products. (*Section 14.6*)

Reaction order: Numbers that indicate the degree to which the rate of the reaction depends on the concentration of each reactant. For an experimentally determined rate law of the form rate $= k[\mathrm{A}]^m[\mathrm{B}]^n$, the values of m and n are derived from experimental measurements of the changes in reactant concentrations over time and indicate the reaction order (c.f. overall reaction order). The values of m and n need not be integers, and they are unrelated to the stoichiometric coefficients of A and B in the balanced chemical equation. (*Section 14.2*)

Reaction quotient (Q): A quantity that has precisely the same form as the equilibrium constant expression, except that the reaction quotient may be derived from a set of values measured at *any* time during the reaction of *any* mixture of the reactants and products, regardless of whether the system is at equilibrium. Thus, for the general reaction $a\mathrm{A} + b\mathrm{B} \rightleftharpoons c\mathrm{C} + d\mathrm{D}$, where A and B are reactants, C and D are products, and a, b, c, and d are the stoichiometric coefficients in the balanced chemical equation for the reaction, $Q = [\mathrm{C}]^c[\mathrm{D}]^d/[\mathrm{A}]^a[\mathrm{B}]^b$ (Equation 15.25). (*Section 15.4*)

Reaction quotient (Q_p): A quantity that has precisely the same form as K_p, the equilibrium constant calculated from partial pressures instead of solution concentrations, except that Q_p may be derived from a set of values measured at *any* time during the reaction of *any* mixture of the reactants and products, regardless of whether the system is at equilibrium. $Q_\mathrm{p} = (P_\mathrm{C})^c(P_\mathrm{D})^d/(P_\mathrm{A})^a(P_\mathrm{B})^b$ for the general reaction $a\mathrm{A} + b\mathrm{B} \rightleftharpoons c\mathrm{C} + d\mathrm{D}$, in which all the components are gases. (*Section 15.4*)

Reaction rates: The changes in the concentrations of reactants and products with time. The factors that affect reaction rates include reactant concentrations, temperature, physical states and surface areas of reactants, as well as solvent and catalyst properties if either is present. (*Chapter 14 introduction*)

Rechargeable batteries: Also called secondary batteries, rechargeable batteries can be recharged by applying an electrical potential in the reverse direction (c.f. disposable batteries). The recharging process temporarily converts a rechargeable battery from a galvanic cell to an electrolytic cell. (*Section 19.5*)

Redox reaction: Another name for an oxidation–reduction reaction. (*Section 3.5*)

Reducing agent: Another name for a reductant (c.f. oxidizing agent). (*Section 3.5*)

Reductant: A compound that is capable of donating electrons (c.f. oxidant). Reductants (also called reducing agents) can reduce other compounds. In the process of donating electrons, a reductant is oxidized. (*Sections 3.5 and 19.1*)

Reduction: The gain of one or more electrons in a chemical reaction (c.f. oxidation). The substance that gains electrons is said to be reduced. (*Section 3.5*)

Reference electrode: An electrode in a galvanic (voltaic) cell whose potential is unaffected by the properties of the solution (c.f. indicator electrode). Ideally, the reference electrode should be physically isolated from the solution whose concentration is being measured. (*Section 19.2*)

Reforming: The chemical conversion of straight-chain alkanes to either branched-chain alkanes or mixtures of aromatic hydrocarbons. Reforming is the second process used in petroleum refining to increase the amounts of more volatile, higher-value products (cracking is the first process). The necessary chemical reactions in reforming are brought about by the use of metal catalysts such as platinum. (*Section 2.6*)

Rem (roentgen equivalent in man): A unit that describes the actual amount of tissue damage caused by a given amount of radiation, the number of rems of radiation equals the number of rads multiplied by a relative biological effectiveness (RBE) factor, which is 1 for β particles, γ rays, and X rays, and about 20 for α particles. (*Section 20.3*)

Resonance structures: Lewis dot structures that have different arrangements of electrons. The positions of the atoms are the same in the various resonance structures of a compound, but the positions of the electrons are different. The different resonance structures of a compound are linked by double-headed arrows. (*Section 8.5*)

Respiration: A process by which chemotrophs obtain energy from their environment, the overall reaction of respiration is the reverse of photosynthesis: $C_6H_{12}O_6 + 6O_2 \longrightarrow 6CO_2 + 6H_2O$ (Equation 18.47). (*Section 18.8*)

Reverse osmosis: A process that uses the application of an external pressure greater than the osmotic pressure of a solution to reverse the flow of solvent through the semipermeable membrane. Reverse osmosis can be used to produce pure water from seawater. (*Section 13.6*)

Reversible process: A process in which every intermediate state between the extremes is an equilibrium state, regardless of the direction of the change (c.f. irreversible process). A reversible process can change direction at any time, whereas an irreversible process cannot. (*Section 18.3*)

Roentgen (R): A unit that describes the amount of energy absorbed by dry air, the roentgen can be used to measure radiation exposure or dose (c.f. rad and gray). (*Section 20.3*)

Root mean square (rms) speed (v_rms): The square root of $\overline{v^2}$, where $\overline{v^2}$ is the average of the squares of the speeds of the particles of a gas at a given temperature (see Equation 10.33). The root mean square speed is the speed of a gas particle that has average kinetic energy. As a result, the root mean square speed and the average speed are different (although the difference is typically less than 10%). (*Section 10.7*)

S

S: See "Entropy (S)" (*Sections 13.2 and 18.3*).

Salt: The general term for any ionic substance that does not have OH^- as the anion or H^+ as the cation. (*Section 4.6*)

Salt bridge: A U-shaped tube inserted into both solutions of a galvanic (voltaic) cell that contains a concentrated liquid or gelled electrolyte. The ions in the salt bridge are selected so they do not interfere with the electrochemical reaction by being oxidized or reduced themselves. The salt bridge completes the circuit between the anode and cathode. (*Section 19.1*)

Saturated calomel electrode (SCE): A reference electrode that consists of a platinum wire inserted into a moist paste of liquid mercury, Hg_2Cl_2, and KCl. These substances, which make up the interior cell, are surrounded by an aqueous KCl solution, which acts as a salt bridge between the interior cell and the exterior solution. The half-reaction for the saturated calomel electrode is $Hg_2Cl_2(s) + 2e^- \longrightarrow 2Hg(l) + 2Cl^-(aq)$ (Equation 19.30). (*Section 19.2*)

Saturated hydrocarbons: Alkanes are also called saturated hydrocarbons because they contain only carbon–carbon and carbon–hydrogen single bonds. Each carbon atom has four single bonds, the maximum number possible (c.f. unsaturated hydrocarbons). (*Section 2.4*)

Saturated solution: A solution that contains the maximum possible amount of a solute under a given set of conditions (e.g., temperature and pressure). (*Section 13.3*)

***s* block:** The two columns on the left side of the periodic table, consisting of the elements in which the *ns* orbitals are being filled (see Figure 6.34; c.f. *p* block, *d* block, and *f* block). (*Section 6.6*)

***s*-block elements:** The elements of the two columns on the left side of the periodic table. The elements of the *s* block are filling their *ns* orbitals (see Figure 6.34; c.f. *p*-block elements, *d*-block elements, and *f*-block elements). (*Section 6.6*)

SCE: See "Saturated calomel electrode (SCE)" (*Section 19.2*).

Schottky defects: A coupled pair of vacancies, one cation and one anion, that maintains the electrical neutrality of an ionic solid, (*Section 12.4*)

Scientific method: The procedure scientists use to search for answers to questions and solutions to problems. The procedure consists of making observations, formulating hypotheses, and designing experiments, which lead in turn to additional observations, hypotheses, and experiments in repeated cycles. (*Section 1.2*)

Scientific notation: A system that expresses numbers in the form $N \times 10^n$, where N is greater than or equal to 1 and less than 10 ($1 \leq N \leq 10$) and n is an integer that can be either positive or negative ($10^0 = 1$). The purpose of scientific notation is to simplify the manipulation of numbers with large or small magnitudes. (*Essential Skills 1*)

Second law of thermodynamics: The entropy of the universe remains constant in a reversible process, whereas the entropy of the universe increases in an irreversible (spontaneous) process. (*Section 18.3*)

Second-order reaction: A reaction whose rate is proportional to the square of the concentration of the reactant (for a reaction of the general form $2A \longrightarrow$ products) or is proportional to the product of the concentrations of two reactants (for a reaction of the general form $A + B \longrightarrow$ products) (c.f. zeroth-order reaction and first-order reaction; see Table 14.6). The differential rate law for the second-order reaction $2A \longrightarrow$ products is rate $= k[A]^2$, where k is the rate constant of the reaction. (*Section 14.3*)

Seed crystal: A solid sample of a substance that can be added to a supercooled liquid or a supersaturated solution to help induce crystallization. (*Sections 11.5 and 13.3*)

Semiconductor: Substances such as Si and Ge that have conductivities between those of metals and insulators. (*Section 12.6*)

Semimetal: Any of the seven elements (B, Si, Ge, As, Sb, Te, and At) that lie adjacent to the zigzag line in the periodic table that runs from boron (B) down to astatine (At). Semimetals (also called metalloids in some texts) exhibit properties intermediate between those of metals and nonmetals. (*Section 1.7*)

Semipermeable membrane: A barrier with pores small enough to allow solvent molecules to pass through, but not solute molecules or ions. (*Section 13.6*)

SHE: See "Standard hydrogen electrode (SHE)" (*Section 19.2*).

Sigma (σ) orbital: A bonding molecular orbital in which the electron density along the internuclear axis and between the nuclei has cylindrical symmetry [c.f. sigma star (σ*) orbital]. All cross-sections perpendicular to the internuclear axis are circles. (*Section 9.3*)

Sigma star (σ*) orbital: An antibonding molecular orbital in which the electron density along the internuclear axis and between the nuclei has cylindrical symmetry [c.f. sigma (σ) orbital]. All cross-sections perpendicular to the internuclear axis are circles. (*Section 9.3*)

Significant figures: The numbers that describe a value without exaggerating the degree to which it is known to be accurate. An additional figure is often reported to indicate the degree of uncertainty. (*Essential Skills 1*)

Silver–silver chloride electrode: A reference electrode that consists of a silver wire coated with a very thin layer of AgCl that is dipped into a chloride ion solution with a fixed concentration. The half-reaction for the silver–silver chloride electrode is $AgCl(s) + e^- \longrightarrow Ag(s) + Cl^-(aq)$ (Equation 19.28). (*Section 19.2*)

Simple cubic unit cell: A cubic unit cell that consists of eight component atoms, molecules, or ions located at the corners of a cube (see Figure 12.5a). (*Section 12.2*)

Single bond: A chemical bond formed when two atoms share a single pair of electrons. (*Section 2.1*)

Single-displacement reaction: A chemical reaction in which an ion in solution is displaced through oxidation of a metal. Single-displacement reactions include the oxidation of certain metals by aqueous acid and the oxidation of certain metals by aqueous solutions of various metal salts. (*Section 4.8*)

Sintering: A process that fuses the grains of a ceramic into a dense, strong material. Sintering is used to produce high-strength ceramics. (*Section 12.9*)

Slip plane: The plane along which the motion of a deformation of a solid occurs. To shape a solid without shattering it, planes of closepacked atoms must move past one another to a new position that is energetically equivalent to the old one. (*Section 12.4*)

Smectic phase: One of three different ways that most liquid crystals can orient themselves (the other two are the nematic and cholesteric phases). In the smectic phase, the long axes of the molecules are aligned (similar to the nematic phase), but the molecules are arranged in planes, too. (*Section 11.8*)

Sodium chloride structure: The solid structure that results when the octahedral holes of a face-centered cubic lattice of anions are filled with cations. The sodium chloride structure has a 1:1 cation:anion ratio, and each ion has a coordination number of 6. (*Section 12.3*)

Soft acids: The metal ions (Lewis acids) with the highest affinity for soft bases, soft acids tend to be cations of the less electropositive metals (see Table 23.9), so they have low charge-to-radius ratios and are more polarizable than hard acids. (*Section 23.4*)

Soft bases: One type of Lewis base (the other is hard bases), soft bases contain large, relatively polarizable donor atoms (such as P, S, and Cl; see the examples in Table 23.9). (*Section 23.4*)

Sol: A dispersion of solid particles in a liquid or solid. A sol is one of three kinds of colloids (the other two are aerosols and emulsions). (*Section 13.7*)

Sol–gel process: A process used to manufacture ceramics. The sol–gel process produces the fine powders of ceramic oxides with uniformly sized particles that are necessary to manufacture high-quality ceramics. (*Section 12.9*)

Solid: One of three distinct states of matter under normal conditions, solids are relatively rigid and have fixed shapes and volumes. The volume of a solid is virtually independent of temperature and pressure. The other two states of matter are liquid and gas. (*Section 1.3*)

Solid electrolytes: Solid materials with very high electrical conductivities. Cations in compounds with Frenkel defects are often able to move rapidly from one site in the crystal to another, resulting in the high electrical conductivities. (*Section 12.4*)

Solubility: A measure of how much of a solid substance remains dissolved in a given amount of a specified liquid at a specified temperature and pressure. Most substances are more soluble at higher temperatures and pressures. (*Sections 1.3 and 13.3*)

Solubility product (K_{sp}): The equilibrium constant expression for the dissolution of a sparingly soluble salt. The concentration of a pure solid is a constant, so it does not appear explicitly in the equilibrium constant expression. (*Section 17.1*)

Solute: The substance or substances present in lesser amounts in a solution (c.f. solvent). (*Chapter 4 introduction*)

Solution: A homogeneous mixture of two or more substances in which the substances present in lesser amounts (called solutes) are dispersed uniformly throughout the substance present in greater amount (the solvent). (*Chapters 4 and 13 introductions*)

Solvation: The process of surrounding each solute particle (atom, molecule, or ion) with particles of solvent (see hydration). (*Section 13.2*)

Solvent: The substance present in the greater amount in a solution (c.f. solute). (*Chapter 4 introduction*)

***sp* hybrid orbitals:** The two equivalent hybrid orbitals that result when one *ns* orbital and one *np* orbital are combined (hybridized). The two *sp* hybrid orbitals are oriented at 180° from each other. They are equivalent in energy, and their energy is between the energy values associated with pure *s* and pure *p* orbitals. (*Section 9.2*)

sp^2 hybrid orbitals: The three equivalent hybrid orbitals that result when one *ns* orbital and two *np* orbitals are combined (hybridized). The three sp^2 hybrid orbitals are oriented in a plane at 120° from each other. They are equivalent in energy, and their energy is between the energy values associated with pure *s* and pure *p* orbitals. (*Section 9.2*)

sp^3 hybrid orbitals: The four equivalent hybrid orbitals that result when one *ns* orbital and three *np* orbitals are combined (hybridized). The four sp^3 hybrid orbitals point at the vertices of a tetrahedron, so they are oriented at 109.5° from each other. They are equivalent in energy, and their energy is between the energy values associated with pure *s* and pure *p* orbitals. (*Section 9.2*)

sp^3d hybrid orbitals: The five hybrid orbitals that result when one *ns*, three *np*, and one $(n - 1)d$ orbitals are combined (hybridized). The five sp^3d hybrid orbitals point at the vertices of a trigonal bipyramid. As a result, the five hybrid orbitals are not all equivalent: three form a triangular array oriented at 120° angles, while the other two are oriented at 90° to the first three and at 180° to each other. These kinds of hybrid orbitals, along with sp^3d^2 hybrid orbitals, are invoked to explain the bonding in molecules that contain more than an octet of electrons around the central atom. (*Section 9.2*)

sp^3d^2 hybrid orbitals: The six equivalent hybrid orbitals that result when one *ns*, three *np*, and two $(n - 1)d$ orbitals are combined (hybridized). The six sp^3d^2 hybrid orbitals point at the vertices of an octahedron, so each is oriented at 90° from the four adjacent orbitals and at 180° from the fifth orbital. These kinds of hybrid orbitals, along with sp^3d hybrid orbitals, are invoked to explain the bonding in molecules that contain more than an octet of electrons around the central atom. (*Section 9.2*)

Specific heat (C_s): The amount of energy needed to increase the temperature of 1 g of a substance by 1°C. The units of C_s are J/(g·°C). (*Section 5.3*)

Specific rotation: The amount (in degrees) by which the plane of polarized light is rotated when the light is passed through a solution that contains 1.0 g of a solute per 1.0 mL of solvent in a tube 10.0 cm long. Every chiral compound has a specific rotation, where one enantiomer rotates the light clockwise and the other enantiomer rotates the light an equal amount counterclockwise. (*Section 24.2*)

Spectator ions: Ions that do not participate in the actual reaction. Spectator ions appear on both sides of a complete ionic equation and their coefficients are the same on both sides. Canceling the spectator ions from a complete ionic equation gives a net ionic equation. (*Section 4.4*)

Spectrochemical series: The experimentally observed order of the crystal field splitting energy (Δ_o) produced by different ligands, the spectrochemical series (from strong-field ligands to weak-field ligands) is as follows: $CO \approx CN^- > NO^{2-} > en > NH_3 > SCN^- > H_2O > oxalate^{2-} > OH^- > F^- > acetate^- > Cl^- > Br^- > I^-$. (*Section 23.5*)

Speed (ν) (of a wave): The distance traveled by a wave per unit time. The speed of a wave equals the product of its wavelength and frequency ($v = \lambda\nu$) and is typically measured in meters per second (m/s). (*Section 6.1*)

Speed of light (c): The speed with which *all* forms of electromagnetic radiation (e.g., microwaves, visible light, and gamma rays) travel in a vacuum. The speed of light is a fundamental constant with a value of 2.99792458×10^8 m/s. Because the various kinds of electromagnetic radiation all have the same speed (c), they differ only in wavelength (λ) and frequency (ν); $c = \lambda\nu$. (*Section 6.1*)

Spin-pairing energy (P): The energy that must be overcome in order to place an electron in an orbital that already has one electron. That is, placing an electron in an already occupied orbital results in electrostatic repulsions that increase the energy of the system by an amount that is called the spin-pairing energy. The relative magnitudes of P and Δ_o (the crystal field splitting energy) determine whether octahedral metal complexes are low spin ($P < \Delta_o$) or high spin ($P > \Delta_o$). (*Section 23.5*)

Spontaneous fission: A nuclear decay reaction in which the nucleus breaks into two pieces with different atomic numbers and atomic masses. Spontaneous fission is most important for the transactinide elements ($Z \geq 104$)—that is, very massive nuclei with high neutron-to-proton ratios. It is accompanied by the release of large amounts of energy and usually the emission of several neutrons. (*Section 20.2*)

Square planar: One of two possible molecular geometries for an AB_4 molecular species (the other is tetrahedral; see Table 9.1). In the square planar geometry, all five atoms lie in the same plane and all adjacent B—A—B bond angles are 90°. (*Section 9.1*)

Standard atmospheric pressure: The atmospheric pressure required to support a column of mercury exactly 760 mm tall. This pressure is also referred to as 1 atmosphere (atm) and is related to other pressure units as follows: 1 atm = 760 torr = 760 millimeters of mercury (mmHg) = 101325 pascals (Pa) = 101.325 kPa. (*Section 10.2*)

Standard cell potential (E°_{cell}): The potential of an electrochemical cell measured under standard conditions—that is, with all species in their standard states (1 *M* for solutions, 1 atm for gases, and pure solids or pure liquids for other substances) and at a fixed temperature (usually 298 K). (*Section 19.2*)

Standard conditions: The conditions under which most thermochemical data are tabulated. The standard conditions are a pressure of 1 atmosphere (atm) for all gases and a concentration of 1.0 *M* for all species in solution. In addition, each pure substance must be in its standard state. (*Section 5.2*)

Standard electrode potential: The potential of a half-reaction measured against the standard hydrogen electrode (SHE) under standard conditions. (*Section 19.2*)

Standard enthalpy of formation (ΔH°_f): The enthalpy change for the formation of 1 mol of a compound from it component elements, when the component elements are each in their standard states (most stable forms). The standard enthalpy of formation of any element in its most stable form is zero by definition. (*Section 5.2*)

Standard enthalpy of reaction (ΔH°_{rxn}): The enthalpy change that occurs when a reaction is carried out with all reactants and products in their standard states. The magnitude of ΔH°_{rxn} is the sum of the standard enthalpies of formation of the products, each multiplied by its appropriate stoichiometric coefficient, minus the sum of the standard enthalpies of formation of the reactants, also multiplied by their coefficients. (*Section 5.2*)

Standard free-energy change (ΔG°): The change in free energy when one substance or set of substances in their standard states is converted to one or more other substances, also in their standard states. The standard free-energy change can be calculated from the definition of free energy, if the standard enthalpy and entropy changes are known, using Equation 18.23: $\Delta G^\circ = \Delta H^\circ - T\Delta S^\circ$. (*Section 18.5*)

Standard free energy of formation (ΔG°_f): The change in free energy that occurs when 1 mole of a substance in its standard state is formed from the elements in their standard states. By definition, the standard free energy of formation of an element in its standard state is taken to be zero at 298 K. $\Delta G^\circ_f = \Delta H^\circ_f - T\Delta S^\circ_f$ (Equation 18.24). (*Section 18.5*)

Standard hydrogen electrode (SHE): The electrode chosen as a reference for all others, the standard hydrogen electrode has been assigned a standard potential of 0 V. It consists of a strip of platinum wire in contact with an aqueous solution that contains 1 *M* H^+. The $[H^+]$ in solution is in equilibrium with H_2 gas at a pressure of 1 atm at the Pt–solution interface. Protons are reduced or hydrogen molecules are oxidized at the Pt surface according to the equation $2H^+(aq) + 2e^- \rightleftharpoons H_2(g)$ (Equation 19.8). (*Section 19.2*)

Standard molar entropy (S°): The entropy of 1 mole of a substance at a standard temperature of 298 K. The units of S° are J/(mol•K). Unlike enthalpy or internal energy, it is possible to obtain absolute entropy values by measuring the entropy change that occurs between the reference point of 0 K, corresponding to $S = 0$, and 298 K. (*Section 18.4*)

Standard molar volume: The volume of 1 mol of an ideal gas at STP (0°C and 1 atm pressure). The standard molar volume corresponds to 22.41 L (see Table 10.3 for the molar volumes of several real gases). (*Section 10.4*)

Standard reduction potentials: The standard cell potentials (E°_{cell}) for reactions written as reductions, not oxidations. All tabulated values of standard cell potentials by convention are listed as standard reduction potentials in order to be able to compare standard potentials for different substances. (*Section 19.2*)

Standard solution: A solution whose concentration is known precisely. Only pure crystalline compounds that do not react with water or CO_2 (such as potassium hydrogen phthalate, KHP) are suitable for use in preparing a standard solution. (*Section 4.9*)

Standard state: The most stable form of a pure substance at a pressure of 1 atm at a specified temperature [e.g., 25°C (298 K)]. (*Section 5.2*)

Standard temperature and pressure (STP): A particular set of reference conditions—namely, 0°C (273.15 K) and 1 atm pressure. (*Section 10.4*)

Standing wave: A wave that does not travel in space. An example of a standing wave is the motion of a string of a violin or guitar. When the string is plucked, it vibrates at certain fixed frequencies because it is fastened at both ends. (*Section 6.4*)

State (of a system): A complete description of the system at a given time, including its temperature and pressure, the amount of matter it contains, its chemical composition, and the physical state of the matter. (*Section 5.2*)

State function: A property of a system whose magnitude depends on only the present state of the system and not its previous history. Temperature, pressure, volume, and potential energy are all state functions, whereas heat and work are not (they are path dependent). (*Sections 5.2 and 18.1*)

Stereoisomers: Molecules that have the same connectivity but whose component atoms have different orientations in space, stereoisomers come in two types: geometrical isomers and optical isomers. (*Section 24.2*)

Steric factor (*p*): The fraction of orientations of particles (atoms, molecules, or ions) that result in a chemical reaction. In general, the value of the steric factor can range from 0 (no orientations of molecules result in reaction) to 1 (all orientations result in reaction). (*Section 14.7*)

Stock solution: A commercially prepared solution of known concentration. A solution of desired concentration is often prepared by diluting a small volume of a more concentrated stock solution with additional solvent. (*Section 4.2*)

Stoichiometric quantity: The amount of product or reactant specified by the coefficients in a balanced chemical equation. (*Section 3.4*)

Stoichiometry: A collective term for the quantitative relationships among the masses, numbers of moles, and numbers of particles (atoms, molecules, and ions) of the reactants and products in a balanced reaction. (*Section 3.4*)

STP: See "Standard temperature and pressure (STP)" (*Section 10.4*).

Stratosphere: The layer of the atmosphere above the troposphere, the stratosphere extends from an altitude of 13 km (8 miles) to about 44 km (27 miles). Of all of the layers in earth's atmosphere, the stratosphere contains the highest concentration of ozone. (*Section 3.6*)

Strong acid: An acid that reacts essentially completely with water to give H^+ and the corresponding anion (c.f. weak acid). Strong acids are strong electrolytes. (*Section 4.6*)

Strong base: A base that dissociates essentially completely in water to give OH^- and the corresponding cation (c.f. weak base). Strong bases are strong electrolytes, (*Section 4.6*)

Strong electrolyte: An electrolyte that dissociates completely into ions when dissolved in water, thus producing an aqueous solution that conducts electricity very well (e.g., $BaCl_2$ or NaOH). (*Section 4.1*)

Strong nuclear force: An extremely powerful but very short-range attractive force between nucleons that keeps the nucleus of an atom from flying apart (due to electrostatic repulsions between the positively charged protons). (*Section 20.1*)

Structural formula: A representation of a molecule that shows which atoms are bonded to one another and, in some cases, the approximate arrangement of the atoms in space (see Figure 2.4b; c.f. molecular formula). (*Section 2.1*)

Structural isomers: Structural isomers have the same molecular formulas but differ in which atoms are bonded to one another (c.f. conformational isomers and stereoisomers). For example, 1-propanol and 2-propanol have the same molecular formula (C_3H_8O), but the position of the —OH group differs between the two, which causes differences in their physical and chemical properties. (*Sections 23.4 and 24.2*)

Sublimation: The conversion of a solid directly to a gas (without an intervening liquid phase). (*Sections 8.3 and 11.5*)

Subshell: A group of wave functions that have the same values of both n (the principal quantum number) and l (the azimuthal quantum number). The regions of space occupied by electrons in the same subshell usually have the same shape, but they are oriented differently in space. (*Section 6.5*)

Substitutional alloy: An alloy formed by the substitution of one metal atom for another of similar size in the lattice (c.f. interstitial alloy and intermetallic compound). (*Section 12.5*)

Substitutional impurity: A point defect that results when an impurity atom occupies a normal lattice site (c.f. vacancy and interstitial impurity). (*Section 12.4.*)

Substitution reaction: A chemical reaction in which one atom or group of atoms in a substance is replaced by another atom or group of atoms from another substance. A typical substitution reaction is the reaction of hydroxide ion with methyl chloride: $CH_3Cl + OH^- \longrightarrow CH_3OH + Cl^-$. (*Section 24.4*)

Substrate: The reactant in an enzyme-catalyzed reaction. (*Section 14.8*)

Superalloys: High-strength alloys, often of complex composition, that are used in applications (e.g., aerospace) that require mechanical strength, high surface stability (minimal flaking or pitting), and resistance to high temperatures. Superalloys are new metal phases based on cobalt, nickel, and iron. (*Section 12.9*)

Superconducting transition temperature (T_c): The temperature at which a material becomes superconducting (i.e., the temperature at which the electrical resistance of a substance drops to zero). (*Section 12.7*)

Superconductivity: The phenomenon in which a solid at low temperatures exhibits zero resistance to the flow of electrical current. (*Section 12.7*)

Superconductors: Solids that at low temperatures exhibit zero resistance to the flow of electrical current. (*Section 12.7*)

Supercooled liquid: A metastable liquid phase that exists below the normal melting point of the substance (c.f. superheated liquid). Supercooled liquids usually crystallize upon standing or when a seed crystal is added. (*Section 11.5*)

Supercritical fluid: The single, dense fluid phase that exists above the critical temperature of a substance. A supercritical fluid resembles a gas (because it completely fills its container), but it has a density comparable to a liquid. (*Section 11.6*)

Supercritical mass: A mass of a fissile isotope that exceeds the critical mass for that isotope. Under the right conditions, a supercritical mass can release energy explosively. (*Section 20.4*)

Superheated liquid: A metastable liquid phase that exists at a temperature and pressure at which the substance should be a gas (i.e., above its normal boiling point; c.f. supercooled liquid). Superheated liquids eventually boil, sometimes violently. (*Section 11.5*)

Super-heavy elements: Elements with atomic numbers near the magic number of 126. The super-heavy elements form a small "island of stability" in the upper right corner of the plot of number of neutrons versus number of protons (Figure 20.2). (*Section 20.1*)

Supersaturated solution: An unstable solution that contains more dissolved solute than it would normally contain under the given set of conditions (e.g., temperature and pressure). (*Section 13.3*)

Surface tension: The energy required to increase the surface area of a liquid by a certain amount. Surface tension is measured in units of energy per area (e.g., J/m^2). The stronger the intermolecular interactions between molecules of the liquid, the higher the surface tension. (*Section 11.3*)

Surfactants: Substances (surface-active agents), such as soaps and detergents, that disrupt the attractive intermolecular interactions between molecules of a polar liquid (e.g., water), thereby reducing the surface tension of the liquid. (*Section 11.3*)

Surroundings: All of the universe that is not the system; that is, system + surroundings = universe (c.f. system). (*Sections 5.2 and 18.1*)

Suspension: A heterogeneous mixture of particles with diameters of about 1 μm (1000 nm) that are distributed throughout a second phase. The dispersed particles separate from the dispersing phase upon standing (c.f. colloid). (*Section 13.7*)

System: The small, well-defined part of the universe in which we are interested (e.g., a chemical reaction; c.f. surroundings). A system can be open, closed, or isolated. (*Sections 5.2 and 18.1*)

Système Internationale d'Unités (SI): The International System of Units is based on metric units and requires measurements to be expressed in decimal form. There are seven base units in the SI system; all other SI units of measurement are derived from these seven. (*Essential Skills 1*)

T

T_c: See "Critical temperature (T_c)" (*Section 11.6*).

$t_{1/2}$: See "Half-life ($t_{1/2}$)" (*Section 14.5*).

Temperature: A measure of an object's thermal energy content. (*Section 5.1*)

Termination: The third stage in a chain reaction, termination consumes the reactive intermediates (see radicals) produced in the initiation and propagation stages, usually by forming stable products. (*Section 14.6*)

Tetrahedral: The lowest-energy arrangement for compounds that have four electron pairs around the central atom. In the tetrahedral arrangement, the electron pairs point at the vertices of a tetrahedron with 109.5° angles between adjacent electron pairs (see Table 9.1). Tetrahedral is also one of two possible molecular geometries for an AB_4 molecular species (the other is square planar). (*Section 9.1*)

Tetrahedral hole: One of two kinds of holes in a face-centered cubic array of atoms or ions (the other is an octahedral hole). Tetrahedral holes are located between an atom at a corner and the three atoms at the centers of the adjacent faces of the face-centered cubic unit cell. An atom or ion in a tetrahedral hole has a coordination number of 4. (*Section 12.3*)

Theoretical yield: The maximum amount of product that can be formed from the reactants in a chemical reaction. The theoretical yield is the amount of product that would be obtained if the reaction occurred perfectly and the method of purifying the product were 100% efficient (c.f. actual yield). (*Section 3.4*)

Theory: A statement that attempts to explain *why* nature behaves the way it does. Theories tend to be incomplete and imperfect, evolving with time to explain new facts as they are discovered. (*Section 1.2*)

Thermal energy: One of five forms of energy, thermal energy results from atomic and molecular motion; the faster the motion, the higher the thermal energy. The other four kinds of energy are radiant, chemical, nuclear, and electrical. (*Section 5.1*)

Thermochemistry: The branch of chemistry that describes the energy changes that occur during chemical reactions. (*Chapter 5 introduction*)

Thermodynamic control: The altering of reaction conditions so that a single desired product or set of products is present in significant quantities at equilibrium (c.f. kinetic control). (*Section 15.6*)

Thermodynamics: The study of the interrelationships among heat, work, and the energy content of a system at equilibrium. Thermodynamics can be used to determine whether a particular reaction is energetically possible in the direction in which it is written and the composition of the reaction system at equilibrium. It cannot be used to determine whether an energetically feasible reaction will actually occur as written, and it reveals nothing about the rate of the reaction or pathway by which it will occur. (*Chapter 18 introduction*)

Thermonuclear reactions: Nuclear reactions, such as fusion, for which a great deal of thermal energy must be invested in order to initiate the reaction. The amount of energy released by the reaction, however, is several orders of magnitude greater than the energy needed to initiate it. (*Section 20.5*)

Third law of thermodynamics: The entropy of any perfectly ordered, crystalline substance at absolute zero is zero. In practice, absolute zero is an ideal temperature that is unobtainable, and a perfect single crystal is also an ideal that cannot be achieved. (*Section 18.4*)

Three-center bond: A bond in which a hydride ion bridges two electropositive ions, three-center bonds are often found in compounds that contain hydrogen bonded to boron and similar elements. Only two bonding electrons are used in these bonds to hold three atoms together, so three-center bonds are often called electron-deficient bonds. (*Section 21.2*)

Titrant: The solution of known concentration that is reacted with a compound in a solution of unknown concentration in a titration. For a successful titration, the chemical reaction must be fast, complete, and specific (i.e., only the compound of interest must react with the titrant). (*Section 4.9*)

Titration: An experimental procedure used to determine the concentration of a compound of interest. In a titration, a carefully measured volume of a solution of known concentration (called the titrant) is added to a measured volume of a solution containing a compound whose concentration is to be determined (the unknown). The reaction used in a titration can be an acid–base, precipitation, or oxidation–reduction reaction, as long as it is fast, complete, and specific (i.e., only the compound of interest reacts with the titrant). (*Section 4.9*)

Titration curve: A plot of the pH of the solution being titrated versus the amount of acid or base (of known concentration) added. The shape of the curve is indicative of the species involved in the acid–base reaction (see Figure 16.9 for the titration curves for strong acid–strong base titrations and Figure 16.10 for weak acid–strong base and weak base–strong acid titrations). (*Section 16.5*)

Torr: A unit of pressure named for Evangelista Torricelli, the inventor of the mercury barometer. One torr is the same as one millimeter of mercury (mmHg); 760 torr = 760 mmHg = 1 atmosphere (atm) = 101325 pascals (Pa) = 101.325 kPa. (*Section 10.2*)

trans: One of two types of geometrical isomers (the other type is cis), the trans isomer has ligands or substituents opposite or directly across from one another in a rigid molecule or metal complex. (*Section 23.4*)

Transition element: Any of the elements in Groups 3–12 in the periodic table. All of the transition elements are metals. (*Section 1.7*)

Transition state: Also called the activated complex, the transition state of the reaction is the arrangement of atoms that first forms when molecules are able to overcome the activation energy and react. The transition state is not a reaction intermediate; it does not last long enough to be detected readily. (*Section 14.7*)

Transmutation: The process of converting one element to another. (*Section 1.4*)

Transuranium elements: All the elements with $Z > 92$, the transuranium elements are artificial and have been prepared by bombarding suitable target nuclei with smaller particles. (*Section 20.2*)

Triads: Sets of three elements that have similar properties. Two examples are chlorine, bromine, and iodine, and copper, silver, and gold. (*Section 7.1*)

Trigonal bipyramidal: The lowest-energy arrangement for compounds that have five electron pairs around the central atom. In the trigonal bipyramidal arrangement, three of the electron pairs are in the same plane with 120° angles between them and the other two are above and below the plane, positioned at 90° to the plane and 180° to each other (see Table 9.1). Trigonal bipyramidal is also the most common molecular geometry for an AB_5 molecular species. (*Section 9.1*)

Trigonal planar: The lowest-energy arrangement for compounds that have three electron pairs around the central atom. In the trigonal planar arrangement, the electron pairs are in the same plane with 120° angles between them (see Table 9.1). Trigonal planar is also one of three possible molecular geometries for an AB_3 molecular species (the other two are trigonal pyramidal and T-shaped). (*Section 9.1*)

Trigonal pyramidal: One of three possible molecular geometries for an AB_3 molecular species (the other two are trigonal planar and T-shaped). In the trigonal pyramidal geometry, the central atom lies above the plane of the three atoms bonded to it. As a result, the B—A—B bond angles are less than 120°. (*Section 9.1*)

Triple bond: A chemical bond formed when two atoms share three pairs of electrons. (*Section 2.1*)

Triple point: The point in a phase diagram where the solid/liquid, liquid/gas, and solid/gas lines intersect. The triple point is the only combination of temperature and pressure at which all three phases (solid, liquid, and gas) are in equilibrium and can therefore exist simultaneously. (*Section 11.7*)

Triprotic acid: A compound that can donate three hydrogen ions per molecule in separate steps (e.g., H_3PO_4). Triprotic acids are one kind of polyprotic acids (see also diprotic acid). (*Section 4.6*)

Tritium: One of the three isotopes of hydrogen, tritium (^3H or T) is the rarest, consisting of one proton, two neutrons, and one electron. The other two isotopes of hydrogen are protium and deuterium. Unlike protium or deuterium, tritium is radioactive, decaying to ^3He with a half-life of only 12.3 years. (*Section 21.2*)

Troposphere: The lowest layer of the atmosphere, the troposphere extends from earth's surface to an altitude of about 11–13 km (7–8 miles). The temperature of the troposphere decreases steadily with increasing altitude. (*Section 3.6*)

T-shaped: One of three possible molecular geometries for an AB_3 molecular species (the other two are trigonal planar and trigonal pyramidal). The T-shaped molecular geometry is achieved when the central atom forms three bonds and has two lone pairs of electrons. The T-shaped geometry is based on the trigonal bipyramidal geometry, with the two lone pairs occupying equatorial positions (thus, the three bonds occupy the two axial positions and the third equatorial position). (*Section 9.1*)

Tyndall effect: The phenomenon of scattering a beam of visible light. The particles of a colloid exhibit the Tyndall effect, but the particles of a solution do not. (*Section 13.7*)

U

Ultraviolet light: Higher-energy radiation than visible light, ultraviolet (uv) light cannot be detected by the human eye but can cause a wide variety of chemical reactions that are harmful to organisms (e.g., sunburn). (*Section 3.6*)

Uncertainty: The estimated degree of error in a measurement. The degree of uncertainty in a measurement can be indicated by reporting all significant figures plus one. (*Essential Skills 1*)

Unit cell: The smallest repeating unit of a crystal lattice. (*Section 12.2*)

Unsaturated hydrocarbons: Hydrocarbons that contain at least one carbon–carbon multiple bond—that is, alkenes, alkynes, and aromatics (c.f. saturated hydrocarbons). (*Section 2.4*)

V

Vacancy: A point defect that consists of a single atom missing from a site in the crystal (c.f. substitutional impurity and interstitial impurity). (*Section 12.4*)

Valence bond theory: A localized bonding model (c.f. molecular orbital theory) that assumes that the strength of a covalent bond is proportional to the amount of overlap between atomic orbitals and that an atom can use different combinations of atomic orbitals (hybrids) to maximize the overlap between bonded atoms. (*Section 9.2*)

Valence electrons: Electrons in the outermost shell of an atom. The chemistry of an atom depends mostly on its valence electrons. (*Section 6.6*)

Valence-shell electron-pair repulsion (VSEPR) model: A model used to predict the shapes of many molecules and polyatomic ions, based on the idea that the lowest-energy arrangement for a compound is the one in which its electron pairs (bonding and non-bonding) are as far apart as possible. The VSEPR model provides no information about bond lengths or the presence of multiple bonds, nor does it attempt to explain any observations about molecular structure. (*Section 9.1*)

van der Waals atomic radius: Half the internuclear distance between two nonbonded atoms in the solid (see Figure 7.5c). The van der Waals atomic radius is particularly useful for elements such as the noble gases, most of which form no stable compounds. (*Section 7.2*)

van der Waals equation: A modification of the ideal gas law designed to describe the behavior of real gases by explicitly including the effects of molecular size and intermolecular forces: $(P + an^2/V^2)(V - nb) = nRT$, where P = pressure, V = volume, n = amount of a gas, R = the gas constant, T = temperature, and a and b are empirical constants that are different for each gas. The pressure term corrects for intermolecular attractive forces that tend to reduce the pressure from that predicted by the ideal gas law. The volume term corrects for the volume occupied by the gaseous molecules. (*Section 10.8*)

van der Waals forces: The intermolecular forces known as dipole–dipole interactions and London dispersion forces. (*Section 11.2*)

van't Hoff factor (i): The ratio of the apparent number of particles in solution to the number predicted by the stoichiometry of the salt (see Equation 13.25): i = (moles of particles in solution)/(moles of solute dissolved). (*Section 13.6*)

Vaporization: The physical process by which atoms or molecules in the liquid phase enter the gas or vapor phase. Vaporization, also called evaporation, occurs only for those atoms or molecules that have enough kinetic energy to overcome the intermolecular attractive forces holding the liquid together. To escape the liquid, though, the atoms or molecules must be at the surface of the liquid. The atoms or molecules that undergo vaporization create the vapor pressure of the liquid. (*Section 11.4*)

Vapor pressure: The pressure created over a liquid by the molecules of the liquid substance that have enough kinetic energy to escape to the vapor phase. (*Section 11.4*)

Viscosity (η): The resistance of a liquid to flow. Liquids that flow readily (e.g., water or gasoline) have a low viscosity, whereas liquids that flow very slowly (e.g., motor oil or molasses) have a high viscosity. Viscosity is expressed in units of the poise (mPa·s). (*Section 11.3*)

Visible light: Radiation that the human eye can detect. (*Section 3.6*)

Volatile liquids: Liquids with relatively high vapor pressures. Volatile liquids tend to evaporate readily from an open container (c.f. nonvolatile liquids). (*Section 11.4*)

Volume: The amount of space occupied by a sample of matter. (*Section 1.3*)

Voltaic cell: One of two types of electrochemical cells, a voltaic cell (or galvanic cell) uses the energy released during a spontaneous oxidation–reduction (redox) reaction ($\Delta G < 0$) to generate electricity (c.f. electrolytic cell, the other type of electrochemical cell). (*Section 19.1*)

W

Waters of hydration: The loosely bound water molecules in hydrate compounds. These waters of hydration can often be removed by simply heating the compound. (*Section 2.2*)

Wave: A periodic oscillation that transmits energy through space. (*Section 6.1*)

Wave function: A mathematical function that relates the location of an electron at a given point in space (identified by x, y, z coordinates) to the amplitude of its wave, which corresponds to its energy. As a result, each wave function Ψ is associated with a particular energy E. That is, wave functions are mathematical equations that describe atomic orbitals. (*Sections 6.5 and 9.3*)

Wavelength (λ): The distance between two corresponding points in a wave—between the midpoints of two peaks, for example, or two troughs. Wavelengths are described by any appropriate unit of distance, such as meters. (*Section 6.1*)

Wave–particle duality: A principle that matter and energy (e.g., light) have properties typical of both waves and particles. (*Section 6.4*)

Weak acid: An acid in which only a fraction of the molecules react with water to produce H^+ and the corresponding anion (c.f. strong acid). Weak acids are weak electrolytes. (*Section 4.6*)

Weak base: A base in which only a fraction of the molecules react with water to produce OH^- and the corresponding cation (c.f. strong base). Weak bases are weak electrolytes. (*Section 4.6*)

Weak electrolyte: A compound (e.g., CH_3CO_2H) that produces relatively few ions when dissolved in water, thus producing an aqueous solution that conducts electricity, but not as well as solutions of strong electrolytes (c.f. strong electrolyte). (*Section 4.1*)

Weight: A force caused by the gravitational attraction that operates on an object. The weight of an object depends on its location (c.f. mass). (*Section 1.3*)

Work (*w*): The product of a force F acting through a distance d: $w = Fd$ (Equation 18.2). Because work occurs only when an object or a substance moves against an opposing force, work requires that the system and its surroundings be physically connected. (*Section 18.1*)

Work hardening: The practice of introducing a dense network of dislocations throughout a solid, making it very tough and hard. (*Section 12.4*)

X

***X*:** See "Mole fraction (X)" (*Section 10.5*).

X-ray diffraction: An experimental technique used to obtain information about the structures of crystalline substances. A beam of X rays (whose wavelengths are approximately the same magnitude as the distances between atoms in molecules or ions) is aimed at a sample of a crystalline material, and the X rays are diffracted by the layers of atoms in the crystalline lattice. When the beams strike photographic film, an X-ray diffraction pattern is produced, which consists of dark spots on a light background. Interatomic distances in crystals can be obtained mathematically from these diffraction data. (*Section 12.3*)

Z

Z: See "Atomic number (Z)" (*Section 1.6*).

Z_{eff}: See "Effective nuclear charge (Z_{eff})" (*Section 6.5*).

Zeroth-order reaction: A reaction whose rate is independent of concentration (c.f. first-order reaction and second-order reaction; see Table 14.6). The differential rate law of a zeroth-order reaction is rate $= k$, where k is the rate constant of the reaction. (*Section 14.3*)

Zinc blende structure: The solid structure that results when half of the tetrahedral holes in a face-centered cubic lattice of anions are filled with cations. The zinc blende structure has a 1:1 cation:anion ratio and each ion has a coordination number of 4. (*Section 12.3*)

APPENDIX H

Answers to Selected Problems

Chapter 1

3. (a) law (b) theory (c) law (d) theory

5. (a) qualitative (b) qualitative (c) quantitative (d) quantitative

11. (a) homogenous (b) homogeneous (c) homogeneous (d) homogeneous (e) heterogeneous (f) heterogeneous

15. (a) compound (b) element (c) element (d) compound (e) compound

19. (a) intensive (b) mass, extensive; color, intensive (c) intensive (d) intensive (e) intensive

25. Yes—Avogadro's hypothesis states that equal volumes of different gases contain equal numbers of gas particles.

37.

Fluorine	${}^{19}_{9}F$
Helium	${}^{4}_{2}He$
Terbium	${}^{159}_{65}Tb$
Iodine	${}^{127}_{53}I$
Aluminum	${}^{27}_{13}Al$
Scandium	${}^{45}_{21}Sc$
Sodium	${}^{23}_{11}Na$
Niobium	${}^{93}_{41}Nb$
Manganese	${}^{55}_{25}Mn$

41.

Symbol	Type
Fe	Metal: transition metal
Ta	Metal: transition metal
S	Nonmetal
Si	Semimetal
Cl	Nonmetal (halogen)
Ni	Metal: transition metal
K	Metal: alkali metal
Rn	Nonmetal (noble gas)
Zr	Metal: transition metal

47. Cr, Rb, Cu, Al, Bi, and Nd

49. 8.0×10^1 kg

51.

(a) Cu	1.12 cm^3
(b) Ca	6.49 cm^3
(c) Ti	2.22 cm^3
(d) Ir	0.4376 cm^3

Volume decreases: Ca > Ti > Cu > Ir

53. 629 g

57. 1.74 g/cm^3

61. Yes. Multiplying all of the subscripts by ten gives a formula of $Al_2Si_2O_5$.

65. (a) electrons: 0.02734% (b) neutrons: 50.20%

67. 24 electrons

69. ${}^{131}I$:53 protons, 78 neutrons, 53 electrons; ${}^{60}Co$: 27 protons, 33 neutrons, 27 electrons

71. technetium-97: 43 protons, 54 neutrons, 43 electrons; americium-240: 95 protons, 145 neutrons, 95 electrons

73. 63.55 amu

75. No—these two isotopes have different atomic numbers, $Z = 28$ and 29, respectively.

77.

Isotope	${}^{238}X$ (Am)	${}^{238}U$	${}^{187}Re$
Number of protons	95	92	**75**
Number of neutrons	143	146	**112**
Number of electrons	**95**	92	75

79. Average mass: 32.06 amu; reported value: 32.06. The scientist's data are accurate.

81. 5.52004. The mass of iodine is greater.

83. (a) 29 protons and 27 electrons (b) 42 protons and 38 electrons (c) 53 protons and 54 electrons (d) 31 protons and 28 electrons (e) 70 protons and 67 electrons (f) 21 protons and 18 electrons

91. 78.91 amu

Chapter 2

7. Empirical formula: smallest whole number ratio of elements in a compound. Structural formula: connectivity and arrangement of atoms in space. $(CH_3)_2S$.

11. Intermolecular interactions that hold the molecules together in a molecular solid are weaker than the electrostatic attractions that hold oppositely charged ions together in an ionic solid.

21. (a) sulfate (b) cyanide (c) dichromate (d) nitride (e) hydroxide (f) iodide (g) peroxide

25. (a) rubidium bromide (b) manganese(III) sulfate (c) sodium hypochlorite (d) ammonium sulfate (e) sodium bromide (f) potassium iodate (g) sodium chromate

35. (a) ROH (b) RNH_2 (as well as R_2NH or R_3N for amines with more than one R group).

47. Sulfuric acid converts insoluble salts containing PO_4^{3-} ions to the soluble $H_2PO_4^-$ and HPO_4^{2-} ions, which plants can readily absorb via their roots.

49. (a) 18 (b) 36 (c) 46 (d) 36 (e) 30 (f) 10 (g) 80

51. (a) +1 (b) −2 (c) +2 (d) +1 (e) −3 (f) +3

53. (a) $MgSO_4$ (b) $C_2H_6O_2$ (c) $C_2H_4O_2$ (d) $KClO_3$ (e) $NaOCl \cdot 5\,H_2O$

55.

Ion	K^+	Fe^{3+}	NH_4^+	Ba^{2+}
Cl^-	KCl	$FeCl_3$	NH_4Cl	$BaCl_2$
SO_4^{2-}	K_2SO_4	$Fe_2(SO_4)_3$	$(NH_4)_2SO_4$	$BaSO_4$
PO_4^{3-}	K_3PO_4	$FePO_4$	$(NH_4)_3PO_4$	$Ba_3(PO_4)_2$
NO_3^-	KNO_3	$Fe(NO_3)_3$	NH_4NO_3	$Ba(NO_3)_2$
OH^-	KOH	$Fe(OH)_3$	NH_4OH	$Ba(OH)_2$

57. (a) Li_3N (b) CsCl (c) GeO_2 (d) Rb_2S (e) Na_3As
59. (a) $AlCl_3$ (b) $K_2Cr_2O_7$ (c) CH_2 (d) CH_5N_3 (e) CH_2O
61. (a) The cation is Be^{2+} (the beryllium ion); the anion is O^{2-} (the oxide ion). (b) The cation is Pb^{2+} (the lead(II) ion); the anion is OH^- (the hydroxide ion). (c) The cation is Ba^{2+} (the barium ion); the anion is S^{2-} (the sulfide ion). (d) The cation is Na^+ (the sodium ion); the anion is $Cr_2O_7^{-2}$ (the dichromate ion). (e) The cation is Zn^{2+} (the zinc ion); the anion is SO_4^{2-} (the sulfate ion). (f) The cation is K^+ (the potassium ion); the anion is the ClO^- (the hypochlorite ion). (g) The cation is Na^+ (the sodium ion); the anion is $H_2PO_4^-$ (the dihydrogen phosphate ion).
63. (a) $MgCO_3$ (b) $Al_2(SO_4)_3$ (c) K_3PO_4 (d) PbO_2 (e) Si_3N_4 (f) NaClO (g) $TiCl_4$ (h) $Na_2NH_4PO_4$
65. (a) $Zn(CN)_2$ (b) Ag_2CrO_4 (c) PbI_2 (d) C_6H_6 (e) $Cu(ClO_4)_2$
67. (a) NaOH (b) $Ca(CN)_2$ (c) $Mg_3(PO_4)_2$ (d) Na_2SO_4 (e) $NiBr_2$ (f) $CaClO_2$ (g) $TiBr_4$
69. (a) N_2O (b) SiF_4 (c) BCl_3 (d) NF_3 (e) PBr_3
71. (a) Tl_2Se (b) NpO_2 (c) FeS (d) CuCN (e) NCl_3
73. (a) niobium(IV) oxide
(b) molybdenum(IV) sulfide
(c) tetraphosphorus decasulfide
(d) copper(I) oxide
(e) rhenium(V) fluoride

75. (a) 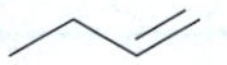 (b) 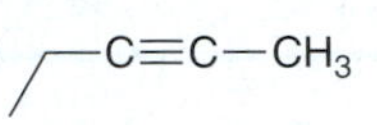

(c) 

(d) 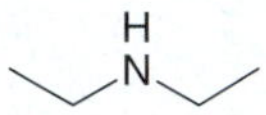

(e) 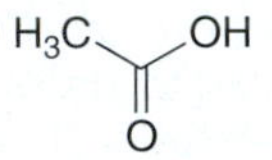

(f) 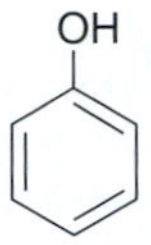

77. (a) HClO (b) $HBrO_4$ (c) HBr (d) H_2SO_3 (e) NaN_3
79. (a) hydrobromic acid (b) sulfurous acid (c) hydrocyanic acid (d) perchloric acid (e) sodium hydrogen sulfate (or sodium bisulfate)

81. 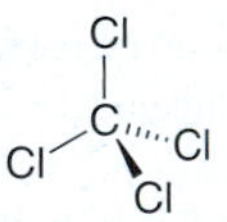

83. MgO, magnesium oxide
85. Carbonic acid is H_2CO_3; lithium carbonate is Li_2CO_3.
87. (a) Sodamide is $NaNH_2$, and sodium cyanide is NaCN. (b) Sodium hydroxide (NaOH) and ammonia (NH_3).
89. (a) 68 (b) 52 g of MTBE must be added to 48 g of the crude distillate.

Chapter 3

5. g nitroglycerin $\xrightarrow{\times A}$ gN $\xrightarrow{\times B}$ mol N
A = %N by mass, expressed as a decimal

$$B = \frac{1}{\text{molar mass of nitrogen in g}}$$

9. S_8 refers to a molecule of elemental sulfur, in which 8 S atoms are connected to form an 8-membered ring. In contrast, 8 S refers to 8 sulfur atoms. Changing the subscript in S_8 to some other number would change the meaning, and indicate a different form of sulfur with a different number of S atoms in the ring.
15. Sulfur, because there is essentially an infinite amount of oxygen in the atmosphere.
19. One or more of the atoms in the reactants gives up at least one electron, and one or more of the atoms in the reactants gains at least one electron. The total number of electrons gained must equal the total number of electrons lost.

27.

	Homogeneous	Heterogeneous
Number of phases	Single phase	At least two phases
Ease of separation from product	Difficult	Easy
Ease of recovery of catalyst	Difficult	Easy

33. To three decimal places, the answers are: (a) 165.880 amu (b) 116.161 amu (c) 260.432 amu (d) 60.052 amu (e) 311.797 amu (f) 105.988 amu (g) 60.096 amu
35. (a) 8.91 mol CaO = 5.37×10^{24} formula units of CaO
(b) 4.99 mol $CaCO_3$ = 3.00×10^{24} formula units of $CaCO_3$
(c) 1.46 mol $C_{12}H_{22}O_{11}$ = 8.79×10^{23} molecules of $C_{12}H_{22}O_{11}$
(d) 6.72 mol NaOCl = 4.04×10^{24} formula units of NaOCl
(e) 11.4 mol CO_2 = 6.84×10^{24} molecules of CO_2
37. (a) 7.83×10^{21} molecules (b) 6.20×10^{23} molecules (c) 1.60×10^{23} molecules
39. 5.22×10^{23} I atoms
41. (a) 2.60×10^{23} atoms (b) 1.006×10^{25} atoms (c) 1.09×10^{23} atoms (d) 2.414×10^{24} atoms
43. (a) False. The number of molecules in 0.5 moles of a substance does not depend upon the nature of the substance. (b) False. Each H_2 molecule contains two H atoms, a mole of H_2 contains 1.204×10^{24} H atoms. (c) True. (d) False for two reasons: (1) benzene is a molecular substance, and formula units apply only to ionic compounds; and (2) the units of formula mass are amu, just as for molecular mass—molar mass is expressed in g/mol.
45. To three decimal places (where allowed), the answers are: (a) 273.23 amu (b) 69.619 amu (c) 330.035 amu (d) 426.984 amu
47. To two decimal places, the percentages are: (a) 5.97% (b) 37.12% (c) 43.22%
49. % oxygen: $KMnO_4$, 40.50%; $K_2Cr_2O_7$, 38.07%; Fe_2O_3, 30.06%
51. To two decimal places, the percentages are: (a) 66.32% Br (b) 22.79% As (c) 25.40% P (d) 73.43% C
53. No. CrO_3 contains 48.00% O and 52.00% Cr by mass.
55. To two decimal places, the percentages are: (a) 29.82% (b) 51.16% (c) 25.40%
57. $NiSO_4 \cdot 6H_2O$ and $CoCl_2 \cdot 6H_2O$
59. Both $NaHSO_4$ and K_2SO_4 contain one S atom per formula unit, but the atomic mass of K_2SO_4 (174.23 amu) is significantly greater than that of $NaHSO_4$ (120.02 amu). Hence, $NaHSO_4$ contains more sulfur per gram of material.
61. (a) 72.71% (b) 69.55% (c) 65.99% (d) 0%
63. $C_4H_8O_2$
65. (a) 27.6 mg C and 1.98 mg H (b) 5.2 mg O (c) 15% (d) C_7H_6O (e) C_7H_6O
67. hydrocyanic acid, HCN
69. To two decimal places, the values are: (a) 273.23 amu (b) 69.62 amu (c) 157.01 amu (d) 42.08 amu (e) 82.15 amu (f) 97.99 amu (g) 45.09 amu
71. Urea (46.6% N) contains much more N per gram than $(NH_4)_2SO_4$ (21.2% N).
73. (a) $2KI(aq) + Br_2(l) \longrightarrow 2KBr(aq) + I_2(s)$
(b) $MnO_2(s) + 4HCl(aq) \longrightarrow MnCl_2(aq) + Cl_2(g) + 2H_2O(l)$
(c) $Na_2O(s) + H_2O(l) \longrightarrow 2NaOH(aq)$
(d) $Cu(s) + 2AgNO_3(aq) \longrightarrow Cu(NO_3)_2(aq) + 2Ag(s)$
(e) $SO_2(g) + H_2O(l) \longrightarrow H_2SO_3(aq)$
(f) $6S_2Cl_2(l) + 16NH_3(l) \longrightarrow S_4N_4(s) + S_8(s) + 12NH_4Cl(s)$

75. (a) $2N_2O_5(g) \longrightarrow 4NO_2(g) + O_2(g)$
(b) $2NaNO_3(s) \longrightarrow 2NaNO_2(s) + O_2(g)$
(c) $2Al(s) + 3NH_4NO_3(s) \longrightarrow 3N_2(g) + 6H_2O(l) + Al_2O_3(s)$
(d) $4C_3H_5N_3O_9(l) \longrightarrow 12CO_2(g) + 6N_2(g) + 10H_2O(g) + O_2(g)$
(e) $2C_4H_{10}(l) + 13O_2(g) \longrightarrow 8CO_2(g) + 10H_2O(g)$
(f) $3IO_2F(s) + 4BrF_3(l) \longrightarrow 3IF_5(l) + 2Br_2(l) + 3O_2(g)$

77. (a) $2Mg(s) + O_2(g) \longrightarrow 2MgO(s)$
(b) $CO_2(g) + Na_2O(s) \longrightarrow Na_2CO_3(s)$
(c) $2Al(s) + 6HCl(aq) \longrightarrow 2AlCl_3(aq) + 3H_2(g)$
(d) $AgNO_3(aq) + KCl(aq) \longrightarrow AgCl(aq) + KNO_3(aq)$
(e) $CH_4(g) + 2O_2(g) \longrightarrow CO_2(g) + 2H_2O(g)$
(f) $2NaNO_3(aq) + H_2SO_4(aq) \longrightarrow Na_2SO_4(aq) + 2HNO_3(aq)$

79. (a) 53.49 amu (b) 49.01 amu (c) 58.32 amu (d) 310.17 amu (e) 73.89 amu

81. (a) 0.383 mol Si (b) 4.2×10^{-2} mol Pb (c) 0.102 mol Mg (d) 6.8×10^{-3} mol La (e) 3.78×10^{-2} mol Cl_2 (f) 1.2×10^{-3} mol As

83. (a) 9.80×10^{-3} mol or 9.80 mmol $Ba(OH)_2$ (b) 8.08×10^{-3} mol or 8.08 mmol H_3PO_4 (c) 2.91×10^{-2} mol or 29.1 mmol K_2S (d) 4.634×10^{-3} mol or 4.634 mmol $Cu(NO_3)_2$ (e) 1.769×10^{-2} mol or 17.69 mmol $Ba_3(PO_4)_2$ (f) 4.38×10^{-2} mol or 43.8 mmol $(NH_4)_2SO_4$ (g) 4.06×10^{-3} mol or 4.06 mmol $Pb(C_2H_3O_2)_2$ (h) 1.96×10^{-2} mol or 19.6 mmol $CaCl_2 \cdot 6H_2O$

85. (a) 613 g or 6.13×10^5 mg Ag (b) 296 g or 2.96×10^5 mg Sn (c) 16.6 g or 1.66×10^4 mg Os (d) 48.9 g or 4.89×10^4 mg Si (e) 0.764 g or 764 mg H_2 (f) 92.05 g or 9.205×10^4 mg Zr

87. (a) 1.81 kg P_4O_{10} (b) 0.387 kg $Ba(OH)_2$ (c) 0.923 kg K_3PO_4 (d) 0.458 kg $Ni(ClO_3)_2$ (e) 0.118 kg $(NH_4)NO_3$ (f) 0.109 kg $Co(NO_3)_3$

89. (a) 1.91×10^4 mg Pt (b) 965.6 mg Hg (c) 2841 mg Cl

91. The balanced chemical equation for this reaction is

$$2\,NH_3(g) + 2\,O_2(g) \longrightarrow N_2O(g) + 3\,H_2O(l)$$

(a) NH_3 (b) NH_3 (c) O_2 (d) NH_3

93. (a) 150 g NaI and 35 g Cl_2 (b) 29 g NaCl and 25 g H_2SO_4 (c) 140 g NO_2 and 27 g H_2O

95. (a) 80% (b) 30% (c) 35.74%

97. 45%

99. (a) $CO(g) + 2H_2(g) \longrightarrow CH_3OH(l)$ (b) 58.28%

101. (a) 2.24 g Cl_2 (b) 4.95 g (c) 2.13 g $CH_3CH_2CH_2Cl$ plus 2.82 g $CH_3CHClCH_3$

103. (a) chlorobenzene (b) ammonia (c) 8.74 g ammonium chloride (d) 54.7% (e)

Theoretical yield (NH_4Cl) =

$$\frac{\text{mass of chlorobenzene (g)} \times 0.92 \times 53.49\text{ g/mol}}{112.55\text{ g/mol}}$$

105. (a) redox reaction (b) exchange (c) acid–base (d) condensation

107. (a) S, −2; N, −3; H, +1 (b) P, +5; O, −2 (c) F, −1; Al, +3 (d) S, −2; Cu, +2 (e) H, +1; O, −2; C, +4 (f) H, +1; N, −3 (g) H, +1; O, −2; S, +6 (h) H, +1, O, −2; C, +2

(i)

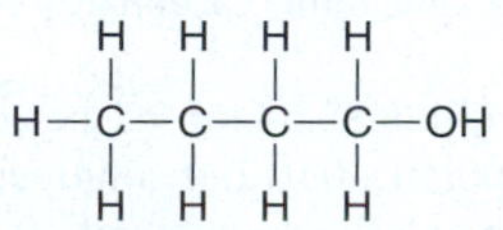

butanol: O, −2; H, +1
From left to right: C, −3 −2 −2 −1

109. $2NaHCO_3(aq) + H_2SO_4(aq)\ Na_2SO_4(aq) + 2CO_2(g) + 2H_2O(l)$
acid–base reaction

111. (a) Ca, +2; O, −2; C, +4
(b) Na, +1; Cl, −1
(c) O, −2; C, +4
(d) K, +1; O, −2; Cr, +7
(e) K, +1; O, −2; Mn, +7
(f) O, −2; Fe, +3
(g) O, −2; H, +1; Cu, +2
(h) O, −2; S, +6
(i)

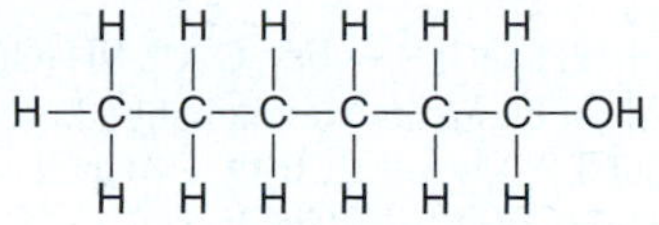

hexanol O, −2; H, +1
From left to right: C, −3 −2 −2 −2 −2 −1

113. (a) Na is the reductant and is oxidized. Cl_2 is the oxidant and is reduced. (b) Mg is the reductant and is oxidized. Si is the oxidant and is reduced. (c) H_2O_2 is both the oxidant and reductant. One molecule is oxidized, and one molecule is reduced.

115. (a) $H_2O(g) + C(s) \longrightarrow H_2(g) + CO(g)$
C is the reductant and is oxidized. H_2O is the oxidant and is reduced.
(b) $8Mn(s) + S_8(s) + 8CaO(s) \longrightarrow 8CaS(s) + 8MnO(s)$
Mn is the reductant and is oxidized. The S_8 is the oxidant and is reduced.
(c) $2C_2H_4(g) + O_2(g) \longrightarrow 2C_2H_4O(g)$
Ethylene is the reductant and is oxidized. O_2 is the oxidant and is reduced.
(d) $8ZnS(s) + 8H_2SO_4(aq) + 4O_2(g) \longrightarrow 8ZnSO_4(aq) + S_8(s) + 8H_2O(l)$
Sulfide in ZnS is the reductant and is oxidized. O_2 is the oxidant and is reduced.

117. (a) $Na_2SO_4 + 2C + 4NaOH \cdot 2Na_2CO_3 + Na_2S + 2H_2O$ (b) The sulfate ion is the oxidant, and the reductant is carbon. (c) 473 g (d) 3300 g (e) carbon

119. (a) 22.1 g (b) 9.9×10^5 g (c) 1.0×10^7 g (d) heterogeneous

121. 4.32×10^{20} molecules, 7.15×10^{-4}

123. $PbCrO_4$

125. To two decimal places, the percentages are: H: 0.54%; O: 51.39%; Al: 19.50%; Si: 24.81%; Ca: 3.75%

127. $C_{16}H_{19}O_5N_3S$

133. (a) $Hg(NO_3)_2$ (b) 1.3×10^{22} molecules (c) 86.96% mercury by mass

135. (a) $Ni(O_2CCH_3)_2$; $Pb_3(PO_4)_2$; $Zn(NO_3)_2$; BeO (b) To four significant figures, the values are: $Ni(O_2CCH_3)_2$, 176.8 amu; $Pb_3(PO_4)_2$, 811.5 amu; $Zn(NO_3)_2$, 189.4 amu; BeO, 25.01 amu. (c) Because Cd lies directly below Zn in Group 12, its chemistry is likely to be similar to that of Zn. Consequently, one might expect that compounds containing the Cd^{2+} ion would also be carcinogenic.

137. C, 40.98%; O, 23.39%; S, 15.63%

139. 140.22 amu

141. $Al(s) + 3NH_4ClO_4(s) \longrightarrow Al_2O_3(s) + AlCl_3(g) + 3NO(g) + 6H_2O(g)$

	3Al	$3NH_4ClO_4$	Al_2O_3
(a)	3 atoms	30 atoms, 6 ions	5 atoms
(b)	3 mol	3 mol	1 mol
(c)	81 g	352 g	102 g
(d)	6×10^{23}	6×10^{23}	2×10^{23}

	$AlCl_3$	3NO	$6H_2O$
(a)	4 atoms, 1 molecule	6 atoms, 3 molecules	18 atoms, 6 molecules
(b)	1 mol	3 mol	6 mol
(c)	130 g	90 g	108 g
(d)	2×10^{23}	6×10^{23}	1.2×10^{24}

143. (a) $5Fe_2O_3(s) + 9H_2O(l) \longrightarrow Fe_{10}O_{15} \cdot 9H_2O(s)$ (b) No. The oxidation numbers of all elements are unchanged. (c) 1090 kg
145. Equation 1: $8CuCl + O_2 \longrightarrow 2Cu_2O + 4CuCl_2$;
Equation 2: $CuCl_2(s) + Cu(s) \longrightarrow 2CuCl(s)$;
Equation 3: $12CuCl(s) + 3O_2(g) + 8H_2O(s) \longrightarrow 2[CuCl_2 \cdot 3Cu(OH)_2 \cdot H_2O(s)] + 4CuCl_2(s)$
(a) Equation 1: Oxygen is the oxidant, and CuCl is the reductant. Equation 2: Copper is the reductant, and copper(II) chloride is the oxidant. Equation 3: Copper(I) chloride is the reductant, and oxygen is the oxidant. (b) 46 pounds (c) temperature, humidity, and wind (to bring more O_2 into contact with the statue)
147. $2PbS(s) + 3O_2(g) \longrightarrow 2PbO(s) + 2SO_2(g)$ Sulfur in PbS has been oxidized, and oxygen has been reduced. 1.1×10^3 g SO_2 is produced. Lead is a toxic metal. Sulfur dioxide reacts with water to make acid rain.
149. 10.8 g benzene; 7.9 g benzene
151. (a) $SiO_2(s) + 6HF(g) \longrightarrow SiF_6^{2-}(aq) + 2H^+(aq) + 2H_2O(l)$ (b) 2.8 g (c) 6400 g HF
153. (a) C_8H_3 (b) C_8H_3 (c) C_4H_2; 6.0 g (d) Complex molecules are essential for life. Also reactions that help block UV may have implications regarding life on other planets.
155. The disaster occurred because organic compounds are highly flammable in a pure oxygen atmosphere. Using a mixture of 20% O_2 and an inert gas such as N_2 or He would have prevented the disaster.

Chapter 4

5. The favorable electrostatic interactions between water and the ions are stronger than the electrostatic interactions between ions in the solid.
9. (a) Toluene lacks polar groups, so it is unlikely to form a homogenous solution with water. (b) Acetic acid contains a polar carboxylic acid group attached to a small alkyl group (a methyl group). Acetic acid will form a homogenous solution with water. (c) Because most sodium salts are soluble, sodium acetate should form a homogenous solution with water. (d) The nonpolar character of the alkyl group is likely to be as important as the polar character of the –OH, decreasing the likelihood that butanol will form a homogeneous solution with water. (e) The nonpolar character of the alkyl group is likely to be as important as the polar character of the carboxylic acid group, making it likely that pentanoic acid will form a heterogeneous mixture with water.
17. A strong electrolyte dissociates completely to give the constituent ions. A weak electrolyte produces relatively few ions in solution.
27. The solute is dispersed in a much larger mass of solvent, which allows chemists to measure the quantity of the substance more accurately.
37. (a) weakly acidic (b) strongly acidic (c) weakly basic (d) strongly basic (e) weakly acidic (f) weakly basic (g) weakly acidic (h) strongly basic
53. The unknown is the limiting reactant, and the titrant is the reactant that is present in excess. At the equivalence point, exactly enough titrant has been added to react with the unknown. Addition of more titrant often causes a color change, either directly if the titrant is colored or indirectly if an indicator is used.
55. (a) 39.13 g $NaBrO_3$ (b) 161.0 g KNO_3 (c) 93.62 g acetic acid (d) 202 g KIO_3

57.

Compound	Mass, g	Moles	Concentration, M
Calcium sulfate, $CaSO_4$	**4.86**	0.0357	0.0714
Acetic acid, CH_3CO_2H	217	**3.62**	7.24
Hydrogen iodide dihydrate, $HI \cdot 2\,H_2O$	104.3	0.6365	**1.273**
Barium bromide, $BaBr_2$	**3.92**	0.0132	0.0264
Glucose, $C_6H_{12}O_6$	88.5	0.492	**0.983**
Sodium acetate, CH_3CO_2Na	199	**2.42**	4.84

59. (a) 0.815 *M* sulfate and 0.815 *M* nickel (b) 4.18 *M* bromide and 2.09 *M* magnesium (c) 0.183 *M* glucose (d) 2.05 *M* chloride and 0.684 *M* cerium(III)
61. (a) 0.17 *M* $CaBr_2$ (b) 0.30 *M* Li_2SO_4 (c) 4.83×10^{-2} *M* sucrose (d) 0.135 *M* $Fe(NO_3)_3 \cdot 6H_2O$
63. Dilute 44.9 mL of the stock solution, which is 1.47 *M*, to a final volume of 200.0 mL.
65. 0.24 *M* $Ca(ClO)_2$
67. 1.74×10^{-3} *M* caffeine
69. 18.1 g sodium sulfate
71. 20.4 g NH_4Br; FeS; 9.14 g FeS
73. 0.181 *M* silver nitrate
75. 3.75 g Ag_2CrO_4; 5.02×10^{-2} *M* nitrate
77. $[H_3O^+] = [HA]$ *M*
79. (a) H_2SO_4 and $Ba(OH)_2$; $2H^+(aq) + SO_4^{2-}(aq) + Ba^{2+}(aq) + 2OH^-(aq) \longrightarrow 2H_2O(l) + BaSO_4(s)$
(b) HNO_3 and LiOH; $H^+(aq) + NO_3^-(aq) + Li^+(aq) + OH^-(aq) \longrightarrow H_2O(l) + Li^+(aq) + NO_3^-(aq)$
(c) HBr and NaOH; $H^+(aq) + Br^-(aq) + Na^+(aq) + OH^-(aq) \longrightarrow H_2O(l) + Na^+(aq) + Br^-(aq)$
(d) $HClO_4$ and $Ca(OH)_2$; $2H^+(aq) + 2ClO_4^-(aq) + Ca^{2+}(aq) + 2OH^-(aq) \longrightarrow 2H_2O(l) + Ca^{2+}(aq) + 2ClO_4^-(aq)$
81. (a) 7.9×10^{-6} *M* H^+ (b) 5.01×10^{-7} to 2.51×10^{-8} *M* H^+ (c) 1.00×10^{-4} to 3.98×10^{-5} *M* H^+
83. pH = 1.402
85. 25 mL
87. 0.13 *M* HCl; magnesium carbonate, $MgCO_3$, or aluminum hydroxide, $Al(OH)_3$
89. 1.00 *M* solution: dilute 82.58 mL of the concentrated solution to a final volume of 1000 mL. 0.012 *M* solution: dilute 12.0 mL of the 1.00 *M* stock solution to a final volume of 1000 mL.
91. (a) 4.65×10^{-2} mol NaOH (b) 4.49×10^{-3} mol HCl (c) 1.42×10^{-3} mol HBr
93. (a) $HClO_4(aq) + KOH(aq) \longrightarrow KClO_4(aq) + H_2O(l)$
(b) $2HNO_3(aq) + Ca(OH)_2(aq) \longrightarrow Ca(NO_3)_2(aq) + 2H_2O(l)$

95. The acid is nitric acid, and the base is calcium hydroxide. The other product is water.

$$2HNO_3(aq) + Ca(OH)_2(aq) \longrightarrow Ca(NO_3)_2(aq) + 2H_2O(l)$$

The acid is hydroiodic acid, and the base is cesium hydroxide. The other product is water.

$$HI(aq) + CsOH(aq) \longrightarrow CsI(aq) + H_2O(l)$$

97. (a) $3CuS(s) + 8NO_3^-(aq) + 8H^+(aq) \longrightarrow 3Cu^{2+}(aq) + 3SO_4^{2-}(aq) + 8NO(g) + 4H_2O(l)$
(b) $6Ag(s) + 3HS^-(aq) + 2CrO_4^{2-}(aq) + 5H_2O(l) \longrightarrow 3Ag_2S(s) + 2Cr(OH)_3(s) + 7OH^-(aq)$
(c) $Zn(s) + 2H^+(aq) \longrightarrow Zn^{2+}(aq) + H_2(g)$
(d) $3O_2(g) + 2Sb(s) + 2OH^-(aq) + 2H_2O(l) \longrightarrow 3H_2O_2(aq) + 2SbO_2^-(aq)$
(e) $3UO_2^{2+}(aq) + Te(s) + 4H^+(aq) \longrightarrow 3U^{4+}(aq) + TeO_4^{2-}(aq) + 2H_2O(l)$

99. (a) no reaction (b) manganese metal will reduce $Fe^{2+}(aq)$

$Fe^{2+}(aq) + Mn(s) \longrightarrow Mn^{2+}(aq) + Fe(s)$
$Fe^{2+}(aq) + 2Cl^-(aq) + Mn(s) \longrightarrow Mn^{2+}(aq) + 2Cl^-(aq) + Fe(s)$

(c) no reaction (d) no reaction

101. (a) Copper has been oxidized, and nitrate has been reduced.
(b) $3Cu^{2+}(s) + 8HNO_3(aq) \longrightarrow 3Cu^{+2}(aq) + 6NO_3^-(aq) + 2NO(g) + 4H_2O(l)$

103. (a) redox reaction: $Pt^{+2}(aq) + 2Ag(s) \longrightarrow Pt(s) + 2Ag^+(aq)$
(b) acid–base reaction: $HCN(g) + NaOH(aq) \longrightarrow NaCN(aq) + H_2O(l)$
(c) precipitation (exchange) reaction: $Fe(NO_3)_3(aq) + 3NaOH(aq) \longrightarrow Fe(OH)_3(s) + 3NaNO_3(aq)$
(d) redox reaction: $CH_4(g) + 2O_2(g) \longrightarrow CO_2(g) + 2H_2O(l)$

105. 14.9 ml NaOH, 1.46×10^{-2} mol NaOH

107. 0.106 *M* acetaminophen; acetaminophen is an organic compound that is much more soluble in ethanol than water, so using an ethanol/water mixture as the solvent allows a higher concentration of the drug to be used.

109. (a) 0.34 *M* $Ca(HCO_3)_2$ (b) 7.00 g H_2CO_3

111. 2.646 g citric acid, $Ca_3(C_6H_5O_7)_2$

113. (a) $5CaO(s) + 3HPO_4^{2-}(aq) + 2H_2O(l) \longrightarrow Ca_5(PO_4)_3OH(s) + 6OH^-(aq)$ (b) This is an acid–base reaction, in which the acid is the HPO_4^{2-} ion and the base is CaO. Transfering a proton from the acid to the base produces the PO_4^{3-} ion and the hydroxide ion. (c) 2.2 lbs (1 kg) of lime (d) 3.6×10^{-4} *M* HPO_4^{-2}

115. (a) Chloride is oxidized, and protons are reduced.
(b) Oxidation: $2Cl \longrightarrow Cl_2 + 2e^-$
Reduction: $2H_2O + 2e^- \longrightarrow H_2 + 2OH^-$
(c) $2Cl^- + 2H_2O \longrightarrow Cl_2 + H_2 + 2OH^-$
(d) Any alkali metal or alkaline earth chloride could be used; the products would be the same, chlorine gas, hydrogen gas, and a solution of the metal hydroxide.

117. (a) Step 1: $2Fe(s) + 6HCl(aq) \longrightarrow 2FeCl_3(aq) + 3H_2(g)$
Step 2: $2FeCl_3(aq) + Zn(s) \longrightarrow 2FeCl_2(aq) + ZnCl_2(aq)$
Step 3: $10FeCl_2(aq) + 2MnO_4^-(aq) + 7H_2O(l) \longrightarrow 5Fe_2O_3(s) + 2MnCl_2(aq) + 14HCl(aq) + 2Cl^-(aq)$
(b) Step 1: $2Fe(s) + 6H^+(aq) \longrightarrow 2Fe^{3+}(aq) + 3H_2(g)$
Step 2: $2Fe^{3+}(aq) + Zn(s) \longrightarrow 2Fe^{+2}(aq) + Zn^{2+}(aq)$
Step 3: $10Fe^{2+}(aq) + 2MnO_4^-(aq) + 7H_2O(l) \longrightarrow 2Mn^{2+}(aq) + 5Fe_2O_3(s) + 14H^+(aq)$
(c) 7.965 g Fe
(d) 15.83% iron

119. (a) Chlorine will oxidize bromide to bromine. $Cl_2(g) + 2Br^-(aq) \longrightarrow Br_2(aq) + 2Cl^-(aq)$ (b) No reaction. Bubbling chlorine gas through a brine solution will remove any bromide present by oxidizing it to bromine.

121. 1.5×10^7 kg

Chapter 5

3. No, because the object does not move.

11. The kinetic energy of the hammer is transferred to the metal, causing it to deform under stress and increasing the magnitude of atomic vibrations in the metal, which increases its thermal energy.

23. (a) enthalpy of fusion, ΔH_{fus} (b) enthalpy of vaporization, ΔH_{vap} (c) enthalpy of fusion, ΔH_{fus} (d) enthalpy of solution, ΔH_{soln}

31. potassium hydroxide > sodium acetate > lithium nitrate > potassium permanganate

41. A greenhouse gas is any gas that traps thermal energy radiated by the Earth. Carbon dioxide, methane, and water vapor are three important greenhouse gases.

43. (a) directly proportional; (b) directly proportional

45. (a) 1.3 kJ stored (b) 0.26 kJ released (c) 0.251 kJ released

47. 250 kJ released

49. (a) -1203 kJ/mol O_2 (b) 179.2 kJ (c) -59.3 kJ

53. -174.1 kJ/mol

55. 20.3 kJ

57. -34.3 kJ/mol Cl_2; exothermic

59. $\Delta H = -2.86$ kJ; $CaCl_2$: -4.6 kJ; K_2CO_3, -0.65 kJ; mixing, 0.28 kJ

61. (a) methanol: ΔH/g $= -22.6$ kJ
C_9H_{12}: ΔH/g $= -43.3$ kJ
octane: ΔH/g $= -47.9$ kJ

Octane provides the largest amount of heat per gram upon combustion.

(b) $\Delta H_f(C_9H_{12}) = -46.1$ kJ/mol

63. 79 g crude oil

65. $C_p = C_s \times$ (molar mass)

67. For Cu: $q = 58$ J; For Al: $q = 130$ J; Al; Cu; $C_s = 0.385$ and 0.897 J/(g·K) for Cu and Al, respectively

69. 4.7 kJ

71. $\Delta H_{soln} = -0.56$ kJ/g; based on reaction with $AgNO_3$, salt contains halide; dividing ΔH_{soln} values in Table 5.2 by molar mass of salts gives lithium bromide as best match, with -0.56 kJ/g.

73. $T_{final} = 43.1°C$; the combustion reaction is $4C_7H_5N_3O_6(s) + 21O_2(g) \longrightarrow 28CO_2(g) + 10H_2O(l) + 6N_2(g)$; $\Delta H_f°$ (TNT) $= -65.5$ kJ/mol

75. -410 kJ or -98 Cal

77. 9.8 kJ/g = 2.4 Cal/g

79. 130 kg CO_2, 56°C

81. (a) -75.7 kJ/mol (b) This is significantly less than the energy obtained by combustion of glucose (-2803 kJ/mol). (c) Soreness may be due to a buildup of lactic acid during periods of low oxygen.

83. $q = 1.3 \times 10^{-6}$ J, 8.5×10^{-11}g glucose

85. $\Delta H_{rxn} = 184.1$ kJ/mol SnO_2, 43 kJ

87. (a) $\Delta H_{rxn} = -75.7$ kJ/mol (b) $\Delta H_{rxn} = -68.9$ kJ/mol (c) $\Delta H_{rxn} = 10.2$ kJ/mol; pathway a releases the most energy per mole of glucose.

89. (a) endothermic (b) 682 g glucose (c) 0.45 kg CO_2 (d) 306 g glucose $\times$ -2803 kJ/mol = 4770 kJ (e) The two reactions release the same amount of energy per glucose
91. six apples or one hamburger
93. 16.5 Cal is required to warm the milk in the cereal, but only 2.5 Cal is required for the oatmeal. It costs energy for an infant to warm a synthetic milk formula to body temperature.

Chapter 6

5. They are directly proportional.
9. A harp, tire sizes, waves from a speedboat, and the Bohr model of the atom are all quantized.
13. Differences in energy between one set of electronic excited states and the ground state of a neon atom correspond to photons of light in the red portion of the visible spectrum. They contain other gases, such as Ar, Kr, and Xe, or mixtures of gases.
19. From left to right by rows: Plank; Balmer (and others), Bohr, Einstein; Heisenberg, Schrödinger, de Broglie.
25. The energy of an orbital increases as its average radius increase; an emission spectrum is produced because a photon of light is released.
33. (a) $3s$ (b) $4d$ (c) $2p$ (d) $4s$
41. (a) 5.2×10^{14} Hz (b) 8.1×10^{17} Hz (c) 3.5 Hz (d) 4.8×10^{10} Hz (e) 8.1×10^{16} Hz (f) 1.3×10^{18} Hz
43. (a) microwave, 12 J (b) infrared, 13 kJ (c) X-ray, 1.7×10^5 kJ
47. 1209 kHz
49. (a) 4.59×10^{-31} J/photon, radio (b) 3.1×10^{-15} J/photon, gamma ray (c) 2.53×10^{-16} J/photon, X ray
51. 532 nm
55. For a transition from $n = 3$ to $n = 2$, we replace n_1 (final orbit) in Equation 6.10 by 2 and n_2 (initial orbit) by 3. Because $n_2 > n_1$, the energy change is negative, indicating that energy is released; thus, this corresponds to an emission line. For a transition from $n = 2$ to $n = 3$, we reverse the identity of n_1 and n_2, which gives a positive energy change, corresponding to an absorption line. The energy change associated with this transition is 3.03×10^{-19} J. Emission: $\Delta E = -3.03 \times 10^{-19}$ J; absorption: $\Delta E = 3.03 \times 10^{-19}$ J. An energy of 3.03×10^{-19} J/photon corresponds to a wavelength of 656 nm, which is red light in the visible spectrum.
57. $n = 2$, blue light
59. 97.2 nm, 2.04×10^{-18} J/photon, ultraviolet light, absorption spectrum is a single dark line at a wavelength of 97.2 nm
61. Violet: 390 nm, 509×10^{-19} J/photon; Blue-purple: 440 nm, 4.51×10^{-19} J/photon; Blue-green: 500 nm, 3.97×10^{-19} J/photon; Orange: 580 nm, 3.42×10^{-19} J/photon; Red: 650 nm, 3.06×10^{-19} J/photon
63. 2.18×10^{-18} J/atom, $\lambda \leq 91.1$ nm
65. $E = 112$ kJ, $\Delta T = 0.390$°C, over ten times more light is needed for a 4.0°C increase in temperature
67. 1.7×10^{-35} m, uncertainty in position is $\geq 1.4 \times 10^{-34}$ m
69. 9.1×10^{-39} kg, uncertainty in position ≥ 2.6 m
71. Three subshells, with $l = 0$ (s), $l = 1$ (p), and $l = 2$ (d).
73. A d subshell has $l = 2$ and contains 5 orbitals.
75. (a) 2 electrons (b) 6 electrons (c) 10 electrons (d) 14 electrons
77. 36 orbitals and 6 subshells. The energies of the orbitals with $n = 6$ are higher than those of the corresponding orbitals with the same value of l for $n = 4$. For l values ≥ 3, however, the energies of the $n = 4$ orbitals will be higher than those of the lowest energy $n = 6$ orbitals.
79. A $4p$ subshell has three nodes, and a $5s$ orbital has four nodes.
81. Three, corresponding to three $4p$ orbitals. Five, corresponding to five $3d$ orbitals.
83. $n = 3, l = 2, m_l = 2, m_s = +\frac{1}{2}$; $n = 3, l = 2, m_l = 2, m_s = -\frac{1}{2}$
85. (a) Not allowed: m_l cannot be larger than l. (b) Not allowed: if $l = 0$, m_l can only be 0. (c) Not allowed: if $n = 2$, the only allowed values of l are 0 and 1. (d) allowed
87. (a) Be: [He] $2s^2$; $[n, l, m_l, m_s] = [2, 0, 0, +\frac{1}{2}], [2, 0, 0, -\frac{1}{2}]$
(b) Xe: [Kr] $5s^2 4d^{10} 5p^6$; $5s^2$ has $[n, l, m_l, m_s] = [5, 0, 0, +\frac{1}{2}], [5, 0, 0, -\frac{1}{2}]$; $5p^6$ has $[n, l, m_l, m_s] = [5, 1, 1, +\frac{1}{2}], [5, 1, 1, -\frac{1}{2}], [5, 1, 0, +\frac{1}{2}], [5, 1, 0, -\frac{1}{2}], [5, 1, -1, +\frac{1}{2}], [5, 1, -1, -\frac{1}{2}]$
(c) Li: [He] $2s^1$; $[n, l, m_l, m_s] = [2, 0, 0, +\frac{1}{2}]$
(d) F: [He] $2s^2 2p^5$; $2s^2$ has $[n, l, m_l, m_s] = [2, 0, 0, +\frac{1}{2}], [2, 0, 0, -\frac{1}{2}]$; $2p^5$ has $[n, l, m_l, m_s] = [2, 1, 1, +\frac{1}{2}], [2, 1, 1, -\frac{1}{2}]$; $[2, 1, 0, +\frac{1}{2}], [2, 1, 0, -\frac{1}{2}]$; $[2, 0, -1, +\frac{1}{2}]$
(e) S: [He] $2s^2 2p^4$; $2s^2$ has $[n, l, m_l, m_s] = [2, 0, 0, +\frac{1}{2}], [2, 0, 0, -\frac{1}{2}]$; $2p^4$ has $[n, l, m_l, m_s] = [2, 1, 1, +\frac{1}{2}], [2, 1, 1, -\frac{1}{2}]$; $[2, 1, 0, +\frac{1}{2}], [2, 0, -1, +\frac{1}{2}]$

89.

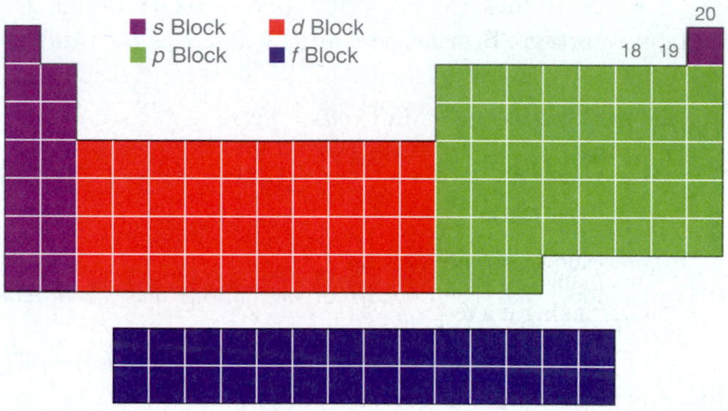

91. (a) Na, $Z = 11$: $1s^3 2s^3 2p^5$
(b) Ta, $Z = 22$: $1s^3 2s^3 2p^6 3s^3 3p^6 4s^1$
(c) F, $Z = 9$: $1s^3 2s^3 2p^3$
(d) Ca, $Z = 20$: $1s^3 2s^3 2p^6 3s^3 3p^5$

93. (a) B: $[He]2s^2 2p^1$
(b) Rb: $[Kr]5s^1$
(c) Br: $[Ar]4s^2 3d^{10} 4p^5$
(d) Ge: $[Ar]4s^2 3d^{10} 4p^2$
(e) V: $[Ar]4s^2 3d^3$
(f) Pd: $[Kr]4d^{10}$
(g) Bi: $[Xe]6s^2 4f^{14} 5d^{10} 6p^3$
(h) Eu: $[Xe]6s^2 4f^7$

95. Sn: $1s^2 2s^2 2p^6 3s^2 3p^6 4s^2 3d^{10} 4p^6 5s^2 4d^{10} 5p^2$
Cu: $1s^2 2s^2 2p^6 3s^2 3p^6 4s^1 3d^{10}$
F: $1s^2 2s^2 2p^5$
Hg: $1s^2 2s^2 2p^6 3s^2 3p^6 4s^2 3d^{10} 4p^6 5s^2 4d^{10} 5p^6 6s^2 4f^{14} 5d^{10}$
Th: $1s^2 2s^2 2p^6 3s^2 3p^6 4s^2 3d^{10} 4p^6 5s^2 4d^{10} 5p^6 6s^2 4f^{14} 5d^{10} 6p^6 7s^2 6d^2$
Y: $1s^2 2s^2 2p^6 3s^2 3p^6 4s^2 3d^{10} 4p^6 5s^2 4d^2$

97. (a) Sm: $6s^2 4f^6$ (b) Pr: $6s^2 4f^3$ (c) B: $2s^2 2p^1$ (d) Co: $4s^2 3d^7$

99.

Ba: $6s$ ↑↓

Nd: $6s$ ↑↓ $4f$ ↑ ↑ ↑ ↑ _ _ _

I: $5s$ ↑↓ $4d$ ↑↓ ↑↓ ↑↓ ↑↓ ↑↓ $5p$ ↑↓ ↑↓ ↑

101. (a) 2 (b) 1 (c) 1 (d) 2 (e) 2
103. (a) Carbon: Aufbau principle, 2*s* filled before 2*p*; Pauli principle, only two electrons per orbital; Hund's rule, two electrons in different 2*p* orbitals with spins parallel.

C: 2*s* ↑↓ 2*p* ↑ ↑ __

(b) Sulfur: Aufbau principle, 3*s* filled before 3*p*; Pauli principle, only two electrons per orbital; Hund's rule, three electrons in different 3*p* orbitals with spins parallel and last electron paired in one of the 3*p* orbitals

S: 3*s* ↑↓ 3*p* ↑↓ ↑ ↑

105. (a) 2.83×10^{13} Hz, 11.3 kJ/mol (b) Monochromatic light is easier to focus on small area of tissue. (c) The light emitted by a CO_2 laser is in the infrared region of the electromagnetic spectrum, so it is easily absorbed by water, which causes local heating.
107. $v = 2 \times 10^8$ m/s, gamma rays in electromagnetic spectrum
109. (a) A, orange; B, red; C, yellowish-green. (b) Compound A will most likely fade most rapidly, because it absorbs the highest energy photons. (c) Its absorption spectrum would show a broad peak centered around 440 nm. A violet contaminant would absorb yellow-green light, resulting in a smaller absorption peak around 570 nm and giving the pigment an orange or brownish-orange tint.
111. (a) 1.4 μm (b) 200 nm (c) 2.0 nm
115. The paint absorbs blue light (450 nm) plus some green light (530 nm), making it a reddish orange.

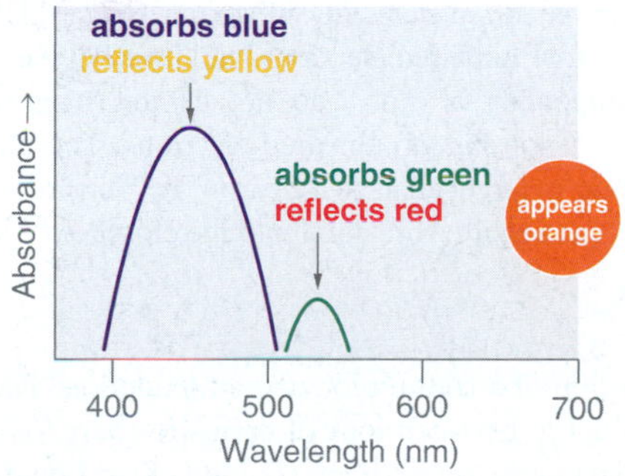

(a)

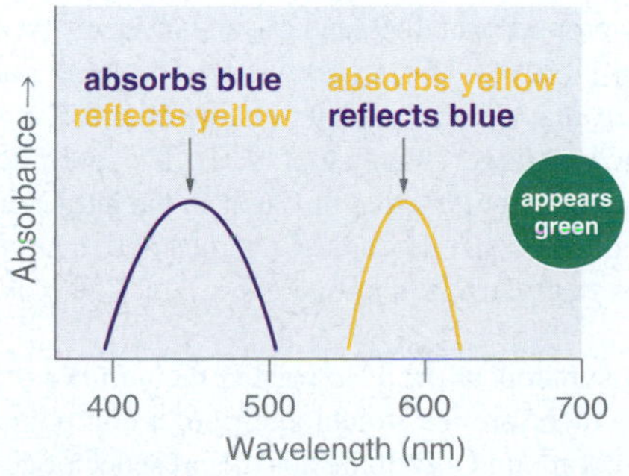

(b)

Mixing red and yellow paints produces a paint that absorbs all light except yellow and red, which the eye sees as a single intermediate color, orange.

117. $[n, l, m_l] = [3, 1, -1]\ [3, 1, 0]\ [3, 1, +1]$. On Earth, each orbital can contain two electrons, and each *p* subshell can contain six electrons.
119.

$4s^2\ 3d^3$

$4s^2$: $[n, l, m_l, m_s] = [4, 1, 0, +\frac{1}{2}]$ and $[4, 1, 0, -\frac{1}{2}]$.

For $3d^3$, many combinations of quantum numbers are possible, including $[n, l, m_l, m_s] = [3, 2, +2, +\frac{1}{2}]$, $[3, 2, +1, +\frac{1}{2}]$, $[3, 2, 0, +\frac{1}{2}]$. Vanadium has three unpaired electrons.

Chapter 7

5. The 1*s* shell is closer to the nucleus and therefore experiences a greater electrostatic attraction. In addition, the electrons in the 2*s* subshell are shielded by the filled $1s^2$ shell, which further decreases the electrostatic attraction to the nucleus.
7. Ba > K > Pb > I > B > F
11. The sodium cation is significantly smaller than a neutral sodium atom (102 versus 154 pm), whereas the chloride ion is much larger than a neutral chlorine atom (181 versus 99). Thus, transferring an electron from sodium to chlorine decreases the radius of sodium by about 50%, but causes the radius of chlorine to almost double.
23. Both Al and Nd. Aluminum is in Group 13, and loss of all three valence electrons will produce the Al^{3+} ion with a noble gas configuration. Neodymium is a lanthanide, and all of the lanthanides tend to form +3 ions because the ionization potentials do not vary greatly across the row, and a +3 charge can be achieved with many oxidants.
29. K < Mo ≈ Sb < P ≈ H < O
41. Nitrogen forms compounds with each of these elements. N_2O_5 or N_2O_3, AlN, NH_3, and NCl_3.

49.

Species	Molar Mass, g/mol	Density, g/cm^3	Molar Volume, cm^3/mol
A	**40.078**	1.550	**25.85**
B	**39.09**	**0.856**	45.7
C	**32.065**	1.961	**16.35**
D	30.95	**1.823**	**16.98**
E	**26.98**	2.700	**9.992**
F	**22.98**	**0.968**	23.7

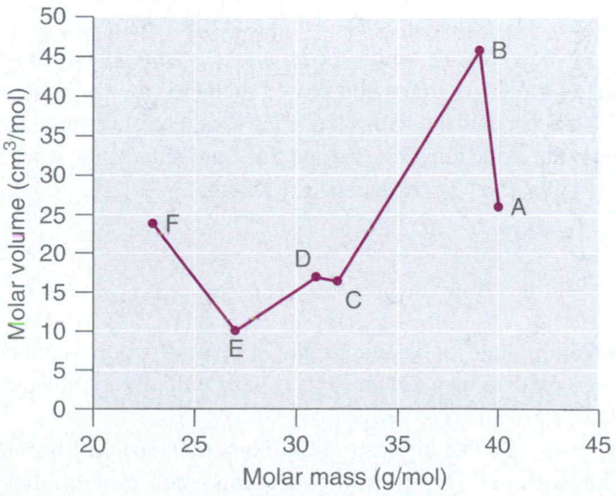

A, B, F are metals, whereas C and D are most likely to be nonmetals.

51. (a) F^-, 133 pm; Cl^-, 181 pm; Br^-, 196 pm; I^-, 220 pm (b) Na^+, 102 pm

53.

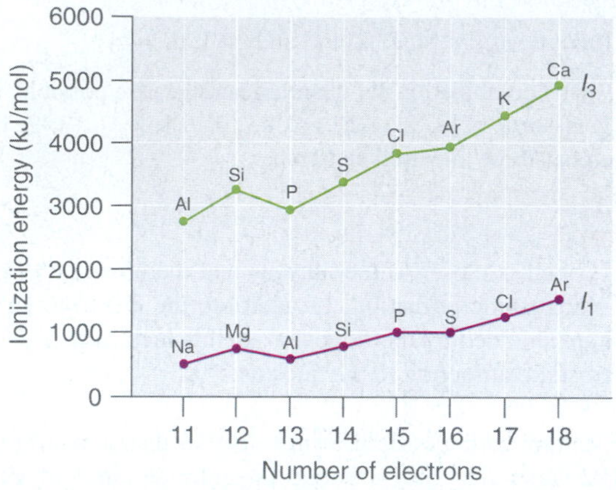

I_3 values correspond to removing an electron from an ion with a +2 charge rather than a neutral atom. The increased charge increases the effect of the steady increase in effective nuclear charge across the row.

55. Si has the highest first ionization energy. The third ionization energy of Mg is the highest.

59. (a) Bi > As > N (b) F > O >> Ar (c) Rb > Cs > Ba

61. $Hg^{2+} > H^+ > O^- > N^{2-}$; Hg^{2+} has the highest positive charge plus a relatively low energy vacant set of orbitals (the 6*p* subshell) to accommodate an added electron, giving it the highest electron affinity; N^{2-} has a greater negative charge than O^-, so electron–electron repulsions will cause its electron affinity to be even lower (more positive) than that of O^-.

63. insulators: S, O, C (diamond), P, B; conductors: Be, Ba, Fe, Al, C (graphite), Sc, W, Na, Rb; B and Te are semimetals and exhibit electrical conductivities between those of metals and insulators.

65. Mg, Al, and H_2; magnesium and aluminum are electropositive metals that tend to donate their valence electrons to nonmetals to form cations; hydrogen has a higher electronegativity and exists as the covalent H_2 molecule, but it has a strong tendency to form covalent compounds in the +1 oxidations state, in which it has formally lost an electron.

67. oxidants: O_2; reductants: Ba, Al, Ni; neither: Mo, Xe

69. I_2 is the best oxidant, with a moderately strong tendency to accept an electron to form the I^- ion, with a closed shell electron configuration. O^- would also be an oxidant, with a tendency to add an electron to form salts containing the oxide ion, O^{2-}. Zn and Sn^{2+} are reductants, while K^+ has no tendency to act as an oxidant or a reductant.

71.

$$2Xe(g) + 3O_2(g) \longrightarrow 2XeO_3(s)$$

The oxidation state of xenon in XeO_3 is +6. Argon is much more difficult to oxidize than xenon, so it is very unlikely to undergo such a reaction to form an argon oxide.

73. +6, +4, −2. The highest oxidation state (+6) will be stabilized by bonds to highly electronegative atoms such as F (SeF_6) and O (SeO_3), while the lowest oxidation state will be stabilized in covalent compounds by bonds to less electronegative atoms such as H (H_2Se) or C [$(CH_3)_2Se$], or in ionic compounds with cations of electropositive metals (Na_2Se).

75. +5, +3, −3. Bonds to highly electronegative atoms such as F and O will stabilize the higher oxidation states, while bonds to less electronegative atoms such as H and C will stabilize the lowest oxidation state, as will formation of an ionic compound with the cations of electropositive metals. +5

77. Uuq = *eka*-lead

79. Transition metal; atomic mass of *Z* is 48 amu, which identifies it as titanium.

81. The element forms two oxides, most likely ZO_2 and ZO_3, suggesting that it is a chalcogen. The atomic mass of *Z* is 32 amu, which identifies it as sulfur.

83.

	Mg	**C**	**Ne**	**Fe**	**Br**
Valence electron configuration	$3s^2$	$2s^22p^2$	$2s^22p^6$	$4s^23d^6$	$4s^24p^5$
Common oxidation states	+2	−4, +4	0	+2, +3	−1, +1, +3, +5, +7
Oxidizing strength of elemental form	None	Weak	None	None	Strong

85. $4Cs(s) + O_2(g) \longrightarrow 2Cs_2O(s)$
$2F_2(g) + O_2(g) \longrightarrow 2OF_2(g)$
$4Al(s) + 3O_2(g) \longrightarrow 2Al_2O_3(s)$
$He(g) + O_2(g) \longrightarrow$ no reaction

87. (a) potassium (b) xenon (c) bromine

89. +6, +4, and −2. Examples of each are: −2 oxidation state, the sulfide anion, S^{2-} or hydrogen sulfide, H_2S; +4 oxidation state, the sulfite ion, SO_3^{2-}; +6 oxidation state, the sulfate ion, SO_4^{2-}. The sulfate ion.

91. Iron(II) has four unpaired electrons, and iron(III) has five unpaired electrons. Compounds of Zn^{2+} do not exhibit magnetic behavior, because the Zn^{2+} ion has no unpaired electrons. The third ionization potential of zinc is larger than that of iron, because removing a third electron from zinc requires breaking into the closed $3d^{10}$ subshell.

Chapter 8

5. The interaction of a sodium ion and an oxide ion. The electrostatic attraction energy between ions of opposite charge is directly proportional to the charge on each ion (Q_1 and Q_2 in Equation 8.1). The larger value of $Q_1 \times Q_2$ for the sodium ion–oxide ion interaction means it will release more energy.

9. *E* is directly proportional to Q_1 and Q_2, and inversely proportional to *r*.

17. In (a), Mg^{2+}; in (b), Li^+; in (c), Cu^{2+}. The ionic radius of Mg^{2+} is smaller than that of Na^+ and its charge is greater, so the lattice energy of $MgCl_2$ is larger than that of NaCl. The internuclear distance in LiCl is much smaller than that of CsCl, so the lattice energy of LiCl is larger than that of CsCl. Because Cu^{2+} has both a larger charge and smaller radius than Cu^+, the lattice energy of $CuCl_2$ is greater than that of CuCl.

23. Lewis dot symbols allow us to predict the number of bonds atoms will form, and therefore the stoichiometry of a compound. The Lewis structure of a compound also indicates the presence or absence of lone pairs of electrons, which provides information on its chemical reactivity and physical properties.

47. Endothermic. Because Σ(bond energies of bonds broken) − Σ(bond energies of bonds formed) is a positive number, ΔH_{rxn} will be positive.

51. *E* will be three times larger for the +3/−1 ions. *E* will be nine times larger than for +1/−1 ions.

53.

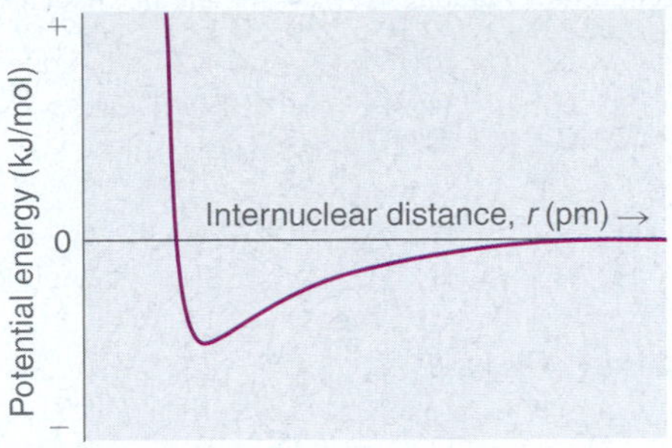

At $r < r_0$, the energy of the system increases due to electron–electron repulsions between the overlapping electron distributions on *adjacent* ions. At very short internuclear distances, electrostatic repulsions between adjacent *nuclei* also become important.

55. 624 kJ

57. $PrI_3 > SrO > PbS$

59. 2522.2 kJ/mol

61. CaO

63. Se: $[Ar]4s^2 3d^{10} 4p^4$ ·S̈e: Two
Kr: $[Ar]4s^2 3d^{10} 4p^6$:K̈r: Zero
Li: $1s^2 2s^1$ L̇i One
Sr: $[Kr]5s^2$ ·Ṡr Zero
H: $1s^1$ Ḣ One

65. Be^{2+}, F^-, B^{3+}, C^{4+}, Cs^+

67. Ag, 11; Pt^{2+}, 8; H_2S, 8; OH^-, 8; I_2, 14; CII_4, 8; SO_4^{2-}, 32; NH_4^+, 8

69.

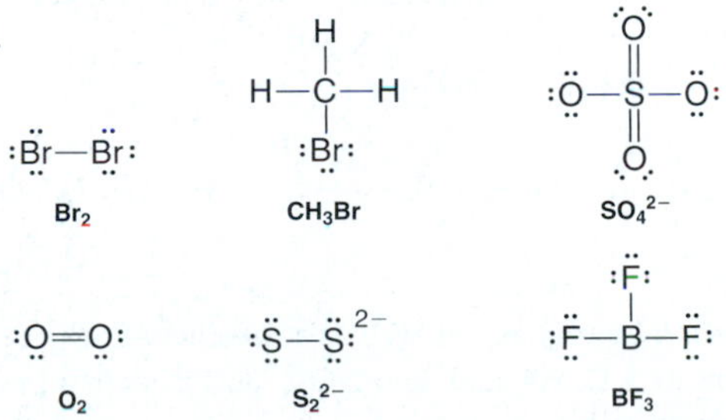

71.

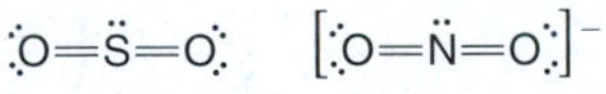

Several resonance structures can be drawn for both SO_2 and NO_2^-, but these two show that both species have similar electronic structures.

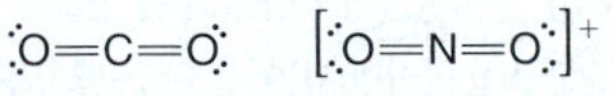

Again, several resonance structures can be drawn for both CO_2 and NO_2^+, but these two show that both species have similar electronic structures.

73. (a) ·K, K is the reductant; :Ṡ· S is the oxidant. K_2S.
(b) ·Sr·, Sr is the reductant; :B̈r: Br is the oxidant. $SrBr_2$.
(c) ·Al·, Al is the reductant; :Ö· O is the oxidant. Al_2O_3.
(d) ·Mg, Mg is the reductant; :C̈l: Cl is the oxidant. $MgCl_2$.

75.

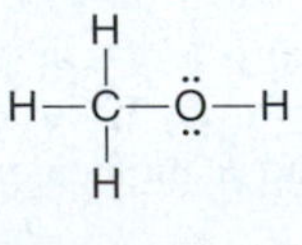

77. The student's proposed structure has three flaws: the hydrogen atom with the double bond has four valence electrons, the carbon on the left has five bonds, and the carbon bound to oxygen only has six valence electrons.

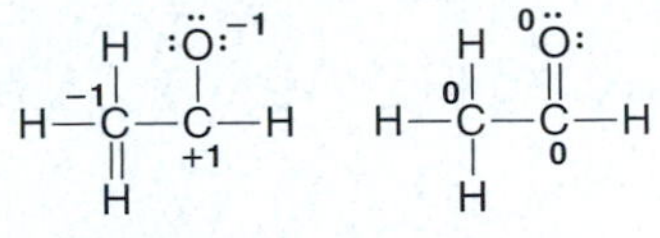

79.

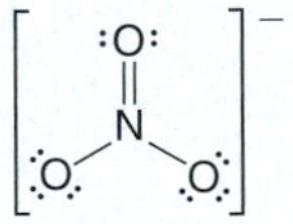

There are three equivalent resonance structures for nitrate.

This Lewis structure is *not* plausible:

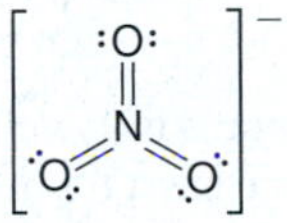

81. With four S—O single bonds, each oxygen in SO_4^{2-} has a formal charge of −1, and the central sulfur has a formal charge of +2. With two S═O double bonds, only two oxygens have a formal charge of –1, and sulfur has a formal charge of zero. Lewis structures that minimize formal charges tend to be lowest in energy.

83. At least five Lewis structures can be drawn for the azide ion:

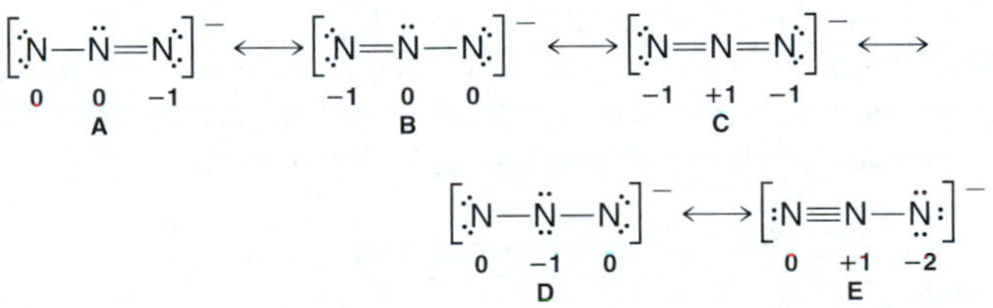

Structures A and B are equivalent resonance structures. In both cases, only one N atom has a formal charge of –1, but another N atom has only six electrons. Structure C, with two N═N double bonds, gives each atom an octet, but all atoms have non-zero formal charges. Structure D, with no N—N multiple bonds, also has a non-zero formal charge for only one atom, but both terminal N atoms have only six electrons. Structure E has an N≡N triple bond and an N—N single bond, but results in high formal charges. Structures A and B, with one N═N double, minimize formal charges.

85. Yes. The formal charge on all atoms is zero, and each atom has an octet of electrons.

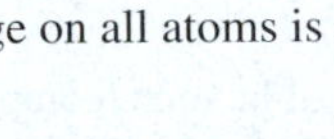

87. (a)

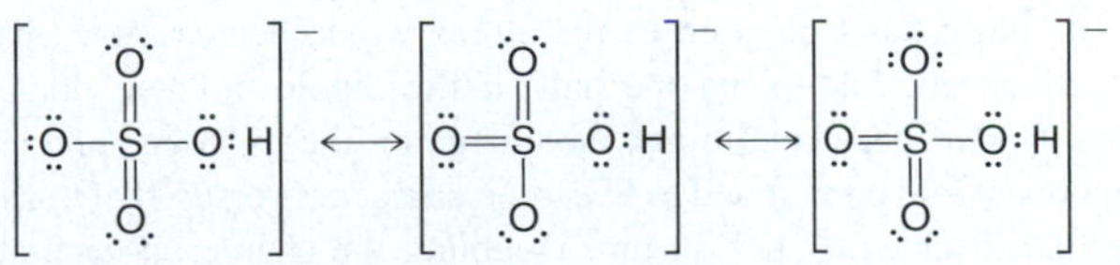

(b)

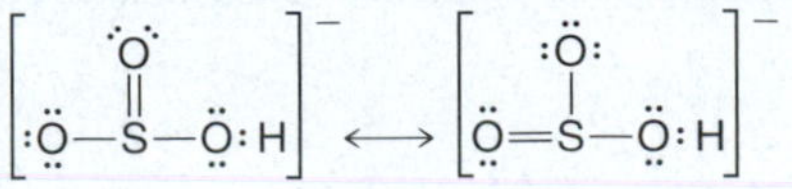

91. ClO_4^- (one of four equivalent resonance structures)

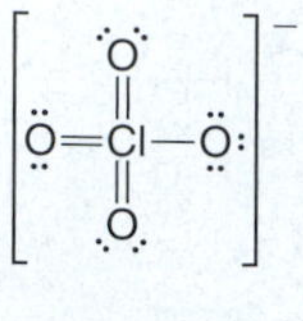

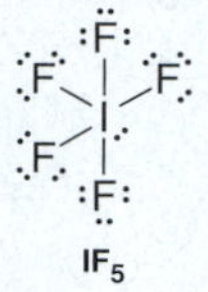

IF₅

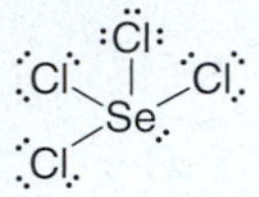

SeCl₄

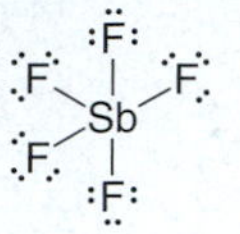

SbF₅

93.

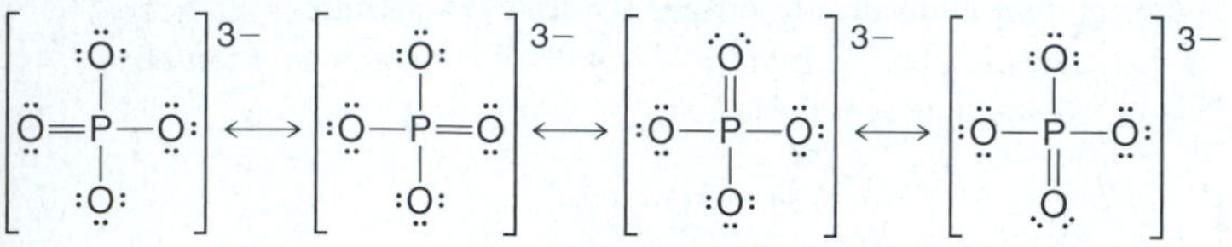

The formal charge on phosphorus is 0, while three oxygen atoms have a formal charge of −1 and one has a formal charge of zero.

95.

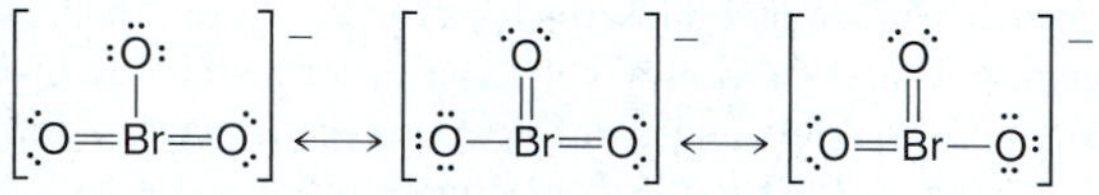

97. (a) $AlCl_3$ is the Lewis acid, and the ether is the Lewis base; $AlCl_3 + (CH_3)_2O \longrightarrow AlCl_3{\cdot}O(CH_3)_2$. (b) $SnCl_4$ is the Lewis acid, and chloride is the Lewis base; $SnCl_4 + 2Cl^- \longrightarrow SnCl_6^{2-}$.

99. N_2H_4, bond order 1; N_2H_2, bond order 2; N_2, bond order 3; N. . .N bond distance: $N_2 < N_2H_2 < N_2H_4$; Largest bond energy: N_2; Highest bond order correlates with strongest and shortest bond.

101.

(a) **S—S;** greater overlap between smaller atoms makes for stronger bonds.

(b) **C—H;** greater overlap between C and H versus larger Ge and H makes for a stronger bond.

(c) **Si—Si;** lone pair–lone pair repulsions between adjacent bonded atoms causes the P—P bond strength to be somewhat less than the Si–Si bond strength.

(d) **Cl—Cl;** lone pair–lone pair repulsions between adjacent bonded atoms decrease the F—F bond strength.

(e) **Al—H;** better overlap between Al and H versus larger Ga and H makes for a stronger bond.

103. Ethanol: 3218 kJ/mol; formaldehyde, 1571 kJ/mol; hydrazine: 1711 kJ/mol

105. RbCl, $TiCl_2$, and LiI will be ionic compounds, Br_2 and S_8 will be purely covalent, while $SbCl_3$ will be polar covalent.

107. NaBr: ionic; OF_2, BCl_3, the C—Cl bond in CH_2Cl_2, and the O—H bond in CH_3OH: polar covalent; bond polarities: $^{\delta+}OF^{\delta-}$, $^{\delta+}BCl^{\delta-}$, $^{\delta+}CCl^{\delta-}$, $^{\delta+}HO^{\delta-}$. The S—S bond in $CH_3SSCH_2CH_3$: covalent.

109. Doubling the separation of the atoms would decrease the charge localized on each atom by one half, if the dipole moment does not change. Under these conditions, the charge on the hydrogen and chlorine atoms would correspond to 9% ionic character, versus 18% for HCl at an HCl distance of 127.46 pm. Doubling the charge on each atom would double the dipole moment. This would double the ionic character of the bond.

111. The percent ionic character is ~2%, indicating that CO is not very polar.

113.

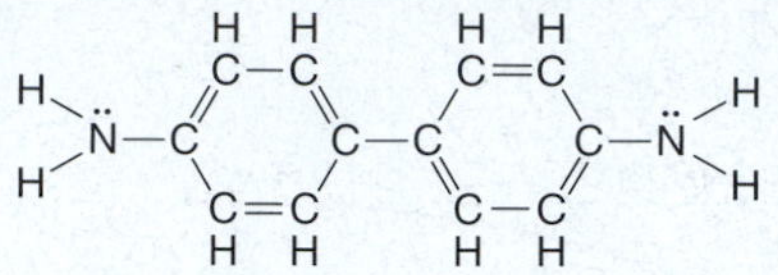

This molecule is a Lewis base.

115.

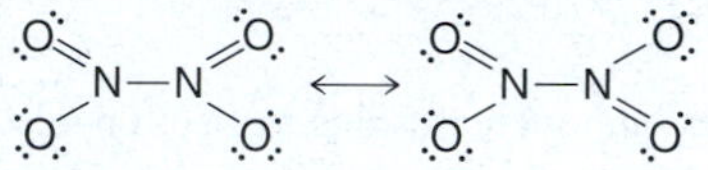

117. The balanced chemical reaction is:

$N_2H_4(g) + O_2(g) \longrightarrow N_2(g) + 2H_2O(g)$

Hydrazine:

$\Delta H_{comb} = -573$ kJ/mol or -17.9 kJ/ml

Methanol (CH_3OH):

$\Delta H_{comb} = -643$ kJ/mol or -15.9 kJ/ml

Hydrazine is both extremely toxic and potentially explosive.

119. (a) $ClO + NO_2 \longrightarrow ClONO_2$

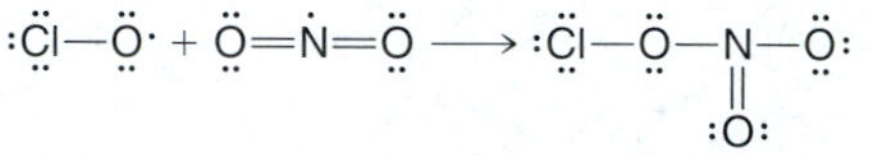

(b) Both reactants have one unpaired electron, which makes them more reactive than might otherwise be expected.

(c) $ClONO_2 + H_2O \longrightarrow HClO + HONO_2$

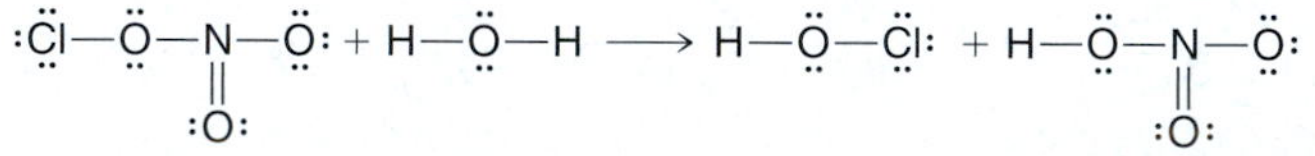

(d) Water is acting as a Lewis base, as well as a Brønsted–Lowry acid. Chlorine nitrate acts as a Lewis acid, and NO_2^+ acts as both a Lewis and a Brønsted–Lowry acid.

Chapter 9

3. VSEPR treats multiple bonds like single bonds. Only when considering fine points of molecular structure does VSEPR recognize that multiple bonds occupy more space around the central atom than single bonds.

11. In general, molecules that have substantial dipole moments are likely to exhibit greater intermolecular interactions, resulting in higher melting points and boiling points.

13. The term "polar" is generally used to mean that a molecule has an asymmetrical structure and contains polar bonds. The resulting dipole moment causes the substance to have a higher boiling or melting point than a nonpolar substance.

29. An atomic orbital is a region of space around an atom that has a non-zero probability for an electron with a particular energy. Analogously, a molecular orbital is a region of space in a molecule that has a non-zero probability for an electron with a particular energy. Both an atomic orbital and a molecular orbital can contain two electrons.

43. No. Because an np_x orbital on one atom is perpendicular to an np_y orbital on an adjacent atom, the net overlap between the two is zero. This is also true for np_y and np_z orbitals on adjacent atoms.

65. BF_3: trigonal planar (all electron pairs are bonding)
PCl_3: tetrahedral (one lone pair on P)
XeF_2: trigonal bipyramidal (three lone pairs on Xe)
$AlCl_4^-$: tetrahedral (all electron pairs on Al are bonding)
CH_2Cl_2: tetrahedral (all electron pairs on C are bonding)

67. HCl: tetrahedral electron-pair geometry, linear molecular geometry
NF_3: tetrahedral electron-pair geometry, trigonal pyramidal molecular geometry

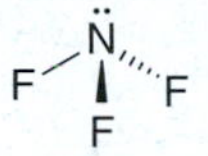

ICl_2^+: tetrahedral electron-pair geometry, bent molecular geometry

N_3^-: linear electron-pair and molecular geometry
H_3O^+: tetrahedral electron-pair geometry, trigonal pyramidal molecular geometry

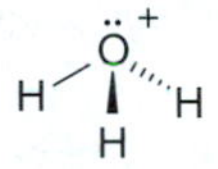

69.

The idealized geometry is T-shaped, but the two lone pairs of electrons on Cl will distort the structure, making the F—Cl—F angle *less* than 180°.

73. $Cl_2C{=}CCl_2$: Although the C—Cl bonds are rather polar, the individual bond dipoles cancel one another in this symmetrical structure, and $Cl_2C{=}CCl_2$ does not have a net dipole moment.

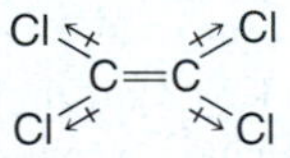

IF_3: The individual I—F bond dipoles cannot cancel one another, giving IF_3 a net dipole moment.

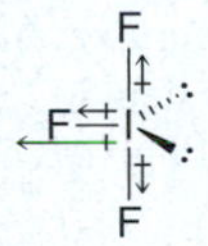

SF_6: The S—F bonds are quite polar, but the individual bond dipoles cancel one another in an octahedral structure. Thus, SF_6 has no net dipole moment.

75.

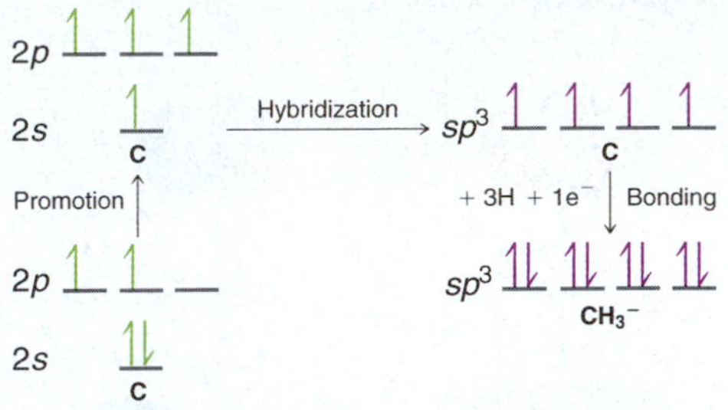

The C atom uses the four singly occupied sp^3 hybrid orbitals to form electron-pair bonds with only *three* H atoms, and an electron is added to the fourth hybrid orbital to give a charge of 1–. The electron-pair geometry is tetrahedral, but the molecular geometry is pyramidal, as in NH_3.

77. BBr_3:

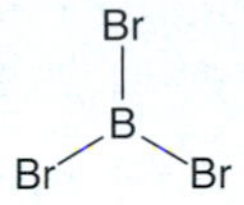

, sp^2, trigonal planar

PCl_3:

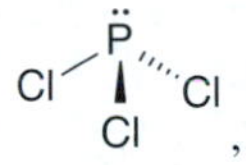

, sp^3, trigonal pyramidal

NO_3^-:

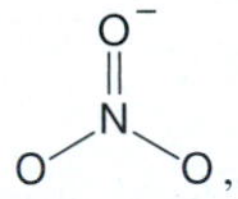

, sp^2, trigonal planar

79. They are all sp^3 hybridized.

81. The phosphorus atom in the PF_6^- ion is d^2sp^3 hybridized, and the ion is octahedral. The PF_6^- ion is isoelectronic with SF_6 and has essentially the same structure. It should therefore be a stable species.

83.

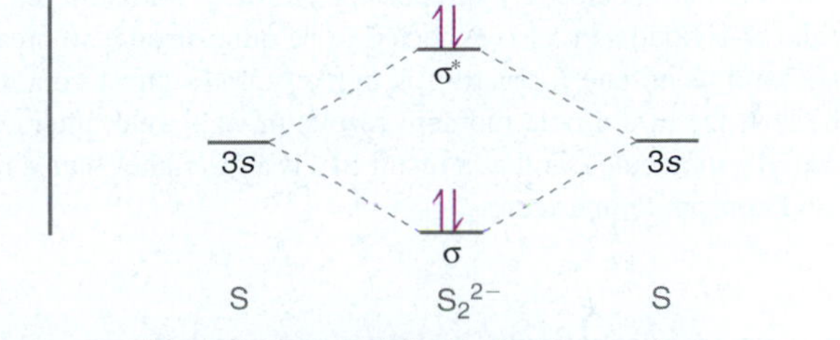

Each S atom contributes 6 valence electrons
+ 2 e⁻ for charge = 14 e⁻

The bond order is 1, and the ion has no unpaired electrons.

85. Three; four or five, respectively.

87.

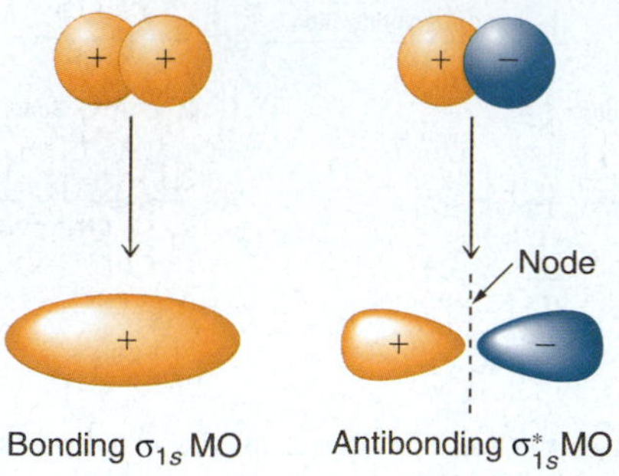

Combining two atomic *s* orbitals gives two molecular orbitals, a σ (bonding) orbital with no nodal planes, and a σ* (antibonding) orbital with one nodal plane.

91. (a) Decreases the bond order, thereby increasing the bond distance. (b) Increases the bond order, thereby decreasing the bond distance.

93.

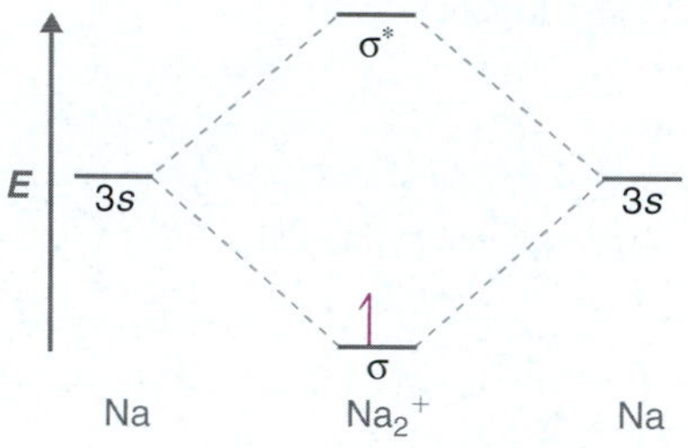

The bond order is 0.5. Consequently, Na_2^+ should be a relatively stable species. Reduction to Na_2 would produce a more stable species than oxidation to Na_2^{2+}. Na_2 is the most stable of the three species.

99. (a) NO^+, 3. (b) NO^-, 2.

101. BN and C_2, and N_2 and CO

105. (a)

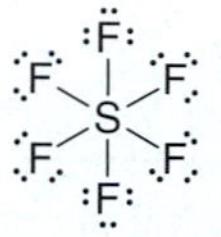

(a) Both the electron-pair geometry and the molecular geometry are octahedral, and the hybridization of S is d^2sp^3. (b) With six fluorine atoms packed around the central sulfur atom, there is no room for another species to approach the sulfur to initiate a reaction. The polar S–F bonds are also expected to be quite strong, so breaking an S–F bond to initiate a reaction is unlikely under most conditions. (c) The absence of a dipole moment results in very weak interactions between SF_6 molecules, and as a result SF_6 is a gas rather than a liquid or a solid at room temperature.

Chapter 10

5. The molecular properties of a substance control its state of matter under a given set of conditions, *not* the other way around.

7. H, He, N, O, F, Ne, Cl, Ar, Kr, Xe, and Rn. Mercury and bromine (although gallium, with a melting point of 29.76°C = 85.57°F, can be a liquid on a hot summer day!). With the exception of the noble gases, the upper right portion of the periodic table also includes most of the elements whose binary hydrides are gases. In addition, the binary hydrides of the elements of Groups 14–16 are gases.

13. It increases. A heavy person requires larger snowshoes than a lighter person. Spreading the force exerted on the heavier person by gravity (that is, their weight) over a larger area decreases the pressure exerted per unit of area, such as a square inch, and makes them less likely to sink into the snow.

19.

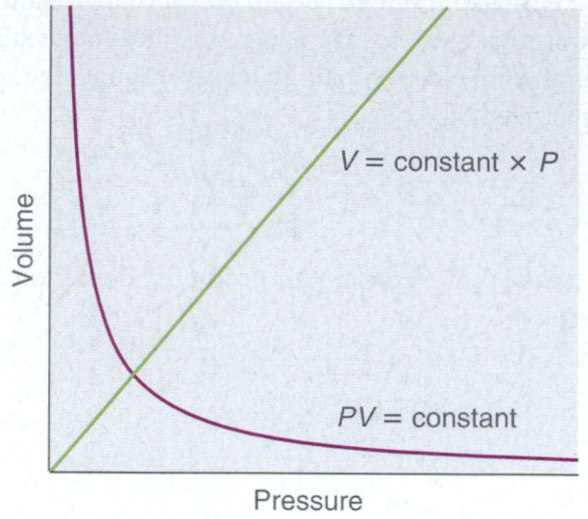

31. (a) P/T = constant (b) V/T = constant (Charles' law) (c) P/n = constant (d) PV = constant (Boyle's law) (e) V/n = constant (Avogadro's law)

61. 5.4 kPa or 5.3×10^{-2} atm; 11 kPa, or 1.1×10^{-3} atm; the same force acting on a smaller area results in a greater pressure.

63.

Atm	kPa	mmHg	torr
1.40	142	1060	1060
0.951	96.4	**723**	723
0.426	**43.2**	324	324

65.

	Liquid Density, g/cm³	Column Height, m
Isopropanol	0.785	18.3
Coconut oil	0.924	15.5
Glycerine	1.259	11.4
Mercury	13.5	1.06

Due to its high density, mercury permits the use of a much shorter barometer tube than any other liquid.

67. (a) 1.99×10^5 mL (b) 6.19×10^4 mmHg (c) 6640°C (d) 24.5 L (e) 51.4 K

69. (a) 7.05×10^{-2} mol (b) 3.78×10^{-2} mol (c) 0.214 mol (d) 3.36×10^{-2} mol

71. (a) 0.21 g HI; (b) 837 g H_2S; (c) 0.986 g CH_4

73. (a) 0.449 L Kr (b) 1510 L C_3H_8 (c) 3.13×10^{-4} L $(CH_3)_2O$

75. (a) 1.48 L (b) 0.0049 L or 4.9 mL (c) 0.654 L

77. 281 mmHg

79. 20.7 kg Al, 5.7×10^4 L HCl

81. (a) 1.0 g/L (b) 1.1 g/L (c) 4.6 g/L

83. 2175 L

85. 30.2 lb/in.², 28.0 lb/in.²

87. 1.78 atm
89. (a) P_{CH_4} = 1.54 atm, P_{CO_2} = 0.851 atm, P_T = 2.39 atm (b) P_{CO} = 0.0919 atm, P_{NO_2} = 0.0264 atm, P_T = 0.1183 atm (c) P_{CH_3Cl} = 34.5 atm, P_{SO_2} = 14.0 atm, P_T = 48.5 atm
91. 52.6 kPa, 66.2 kPa
93. (a)

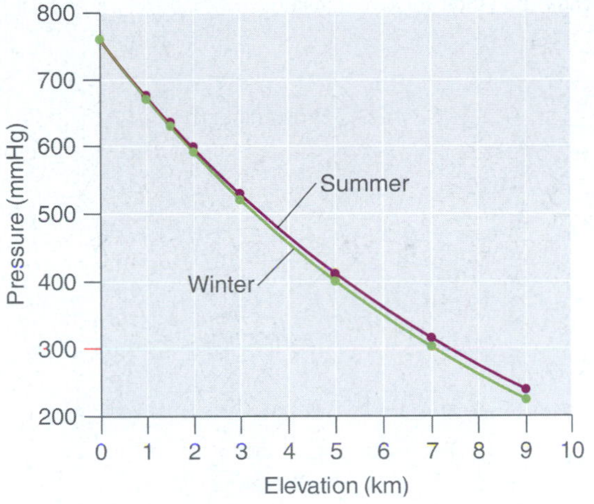

(b) Mt. Elbrus: 376 mmHg (0.494 atm, 50.1 kPa) in the summer, and 366 mmHg (0.482 atm, 48.8 kPa) in the winter; Mt. Aconcagua: 319 mmHg (0.420 atm, 42.5 kPa) in the summer, and 308 mmHg (0.405 atm, 41.1 kPa) in the winter (c) Mt. Elbrus: 0.103 atm O_2, Mt. Aconcagua: 0.088 atm O_2
95. (a) 2.20 g $KClO_3$ (b) 0.862 g O_2 (c) 604 mL O_2
97. Percent composition: 58.20% C, 4.87% H, 23.28% O, and 13.67% N; empirical formula: $C_{10}H_{10}O_3N_2$
99. At any temperature, the rms speed of hydrogen is 4.47 times that of argon.
103. 26 passes
105. Ideal behavior: P = 1.03 atm; real behavior: P = 1.01 atm
107. 5.75 g oxalic acid, 7.6 L
109. 278 L

Chapter 11

11. Electrostatic attractions favor the condensed phase. Each water molecule can simultaneously form hydrogen bonds to *four* other water molecules, resulting in an extensive hydrogen-bonding network similar to that seen in the cage-like structure of ice.
15. London dispersion forces are strong enough to favor the condensed states for bromine and iodine under normal conditions of temperature and pressure.
17. SO_2: dipole–dipole interactions will be most important. HF: hydrogen bonding will be most important. CO_2: London dispersion forces are most important. CCl_4: London dispersion forces will dominate. CH_2Cl_2: dipole–dipole interactions and London dispersion forces.
21. Methanol can only form *two* hydrogen bonds per molecule on average, versus four for water. Hydrogen bonding therefore has a much greater effect on the boiling point of water.
27. No difference in boiling point; vigorous boiling does not affect the temperature of the liquid.
35. Adding a soap or a surfactant to water decreases the surface tension. Because water is a polar molecule, a soap or a surfactant would also disrupt the attractive interactions responsible for adhesion of water to the surface of a glass capillary. This would decrease the height of the water column inside the capillary, as well as making the meniscus less concave.

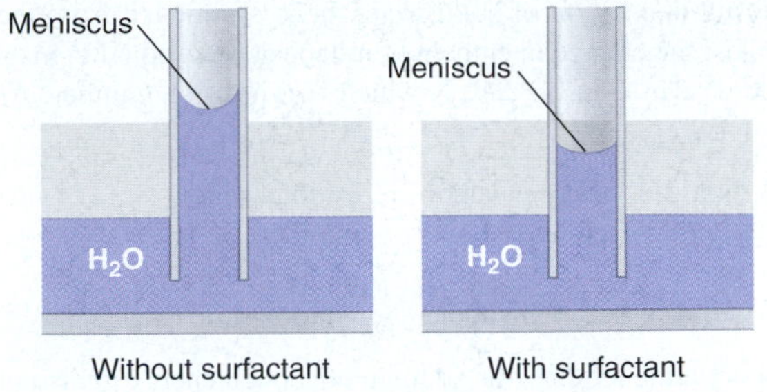

37. Cyclohexanol is a polar substance that can engage in hydrogen bonding. It has a higher surface tension due to stronger intermolecular interactions.
39. Cohesive and adhesive forces. If the adhesive forces are stronger than the cohesive forces, the liquid is pulled up into the capillary and the meniscus is concave. Conversely, if the cohesive forces are stronger than the adhesive forces, the level of the liquid inside the capillary will be lower than the level outside the capillary, and the meniscus will be convex.
41. Viscous substances often consist of molecules that are much longer than they are wide and whose structures are often rather flexible. As a result, the molecules tend to become tangled with one another, which decreases the rate at which they can move through the liquid.
43. Volatile substances have high vapor pressures, which allow them to evaporate readily from an open container. In contrast, nonvolatile liquids have lower vapor pressures and do not evaporate readily. If two substances have similar molecular masses, the less volatile substance will be more polar, resulting in strong intermolecular interactions due to dipole–dipole interactions or hydrogen bonding that stabilize the condensed phase.
45. As water evaporates from grass, the water absorbs heat from the surrounding air due to the large ΔH_{vap} of water. In contrast, sand does not release water, and simply absorbs sunlight and emits the energy as heat.
47. (a) Vapor pressure increases exponentially with temperature. (b) There is no relationship between vapor pressure and surface area. (c) The pressure of other gases above the liquid has no effect on the vapor pressure. (d) There is no direct relationship between viscosity and vapor pressure *per se.* Experimentally, however, it is found that most viscous substances consist of large molecules, and the high molecular masses tend to result in high boiling points and low vapor pressures.
49. It sublimes. Altitude (and changes in atmospheric pressure) will not affect the rate of sublimation directly. The relative humidity at high elevations in winter tends to be very low, however, which will indirectly increase the rate of sublimation.
51.

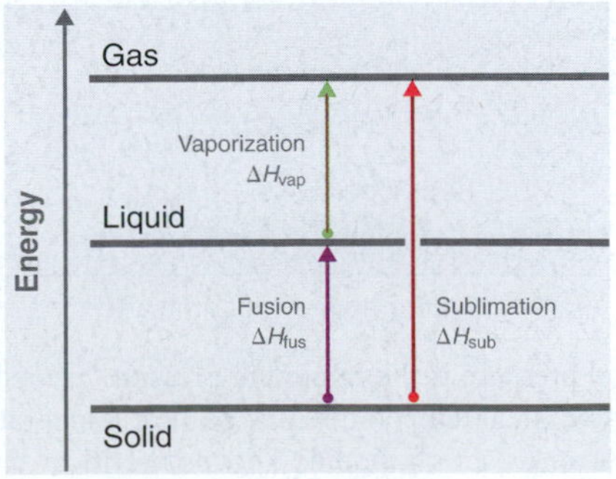

53. During deposition of solid water, heat is removed from the water molecules; the change in enthalpy for deposition is equal in magnitude, but opposite in sign, to ΔH_{sub}, which is a positive number: $\Delta H_{sub} = \Delta H_{fus} + \Delta H_{vap}$.

55. (a) liquid + heat ⟶ vapor endothermic
(b) liquid ⟶ solid + heat exothermic
(c) gas ⟶ liquid + heat exothermic
(d) solid + heat ⟶ vapor endothermic

57. Vaporization requires the addition of enough energy to disrupt all intermolecular interactions and create a gas in which the molecules move essentially independently. Fusion requires only enough energy to overcome the intermolecular interactions that lock molecules in place in a lattice, thereby allowing them to move more freely.

59.

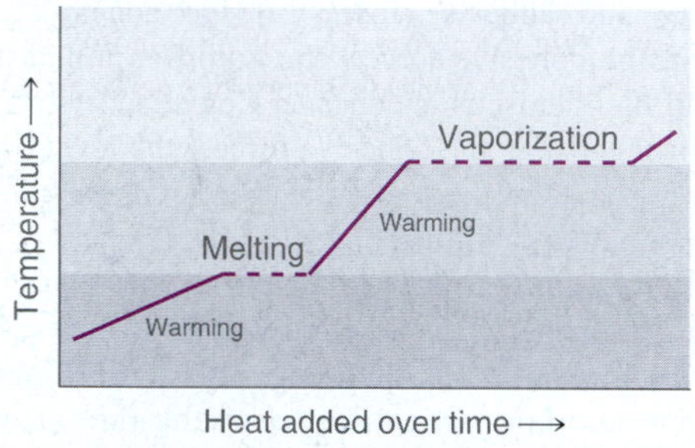

The portions of the curve with a positive slope correspond to heating a single phase, while the horizontal portions of the curve correspond to phase changes. During a phase change, the added heat is melting the solid at its melting point or evaporating the liquid at its boiling point.

61.

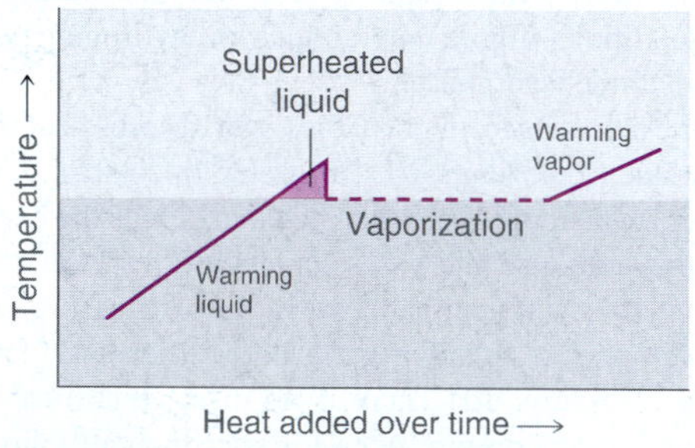

A superheated liquid exists temporarily as liquid with a temperature above the normal boiling point of the liquid.

Conversely, a supercooled liquid exists temporarily as a liquid with a temperature lower than the normal melting point of the solid. When it crystallizes, the temperature increases as the liquid is converted to a solid.

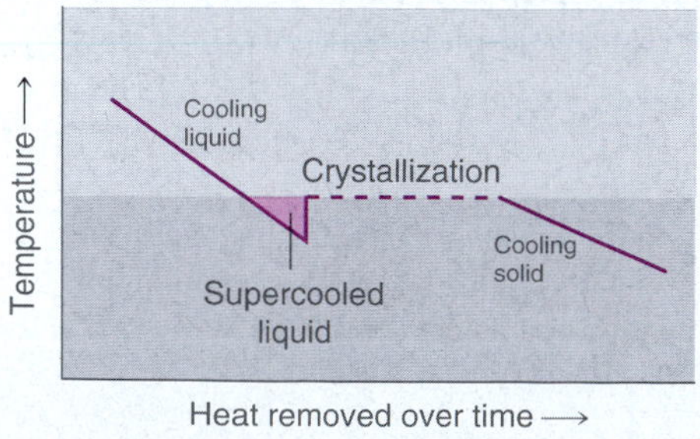

63. The critical pressure is the minimum pressure needed to liquefy a substance at its critical temperature, the highest temperature at which the substance can exist as a liquid. Above its critical temperature, a substance cannot be liquefied regardless of the pressure, and it forms a supercritical fluid.

67. The lines in a phase diagram represent boundaries between different phases; at any combination of temperature and pressure that lies on a line, two phases are in equilibrium. In principle there can be more than one triple point in a phase diagram. The slope of the line separating two phases depends upon their relative densities.

73.

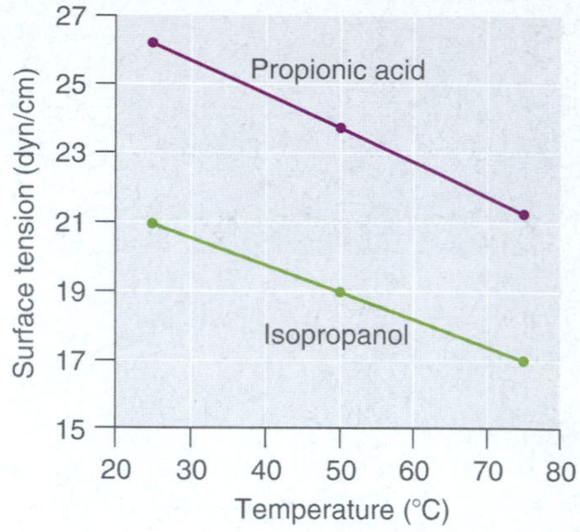

At all temperatures the surface tension of propionic acid is about 30% greater than for isopropanol. These data suggest that propionic acid is more polar than isopropanol.

75.

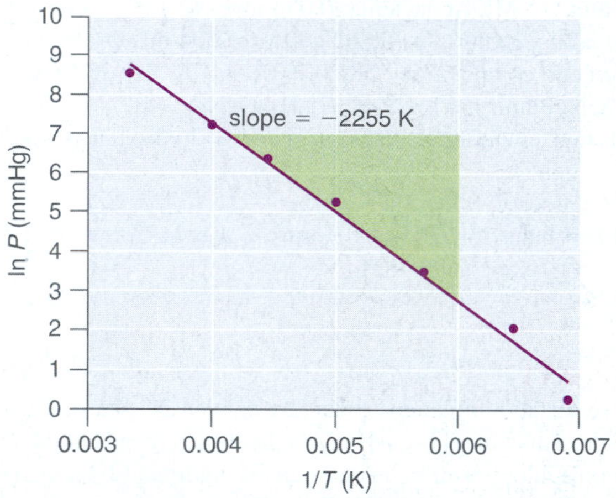

vapor pressure at 293 K is 4040 mmHg; ΔH_{vap} = 18.7 kJ/mol, 1.44 kJ

77. 12.6°C

79. ΔH_{vap} = 28.9 kJ/mol, *n*-hexane

81. ΔH_{vap} = 7.81 kJ/mol, 36°C

85.

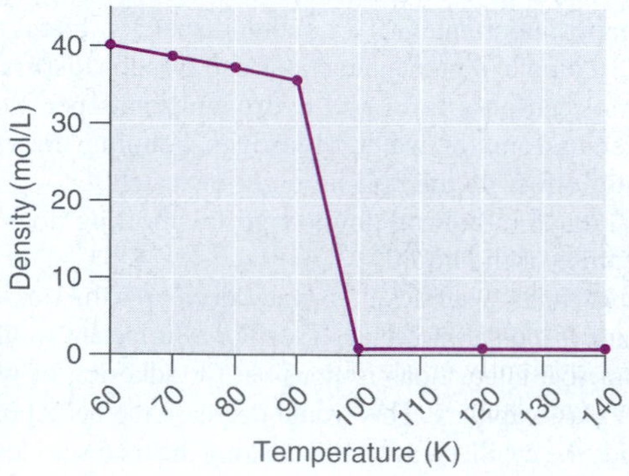

The boiling point of liquid oxygen is between 90 and 100 K (actually 90.2 K).

87.

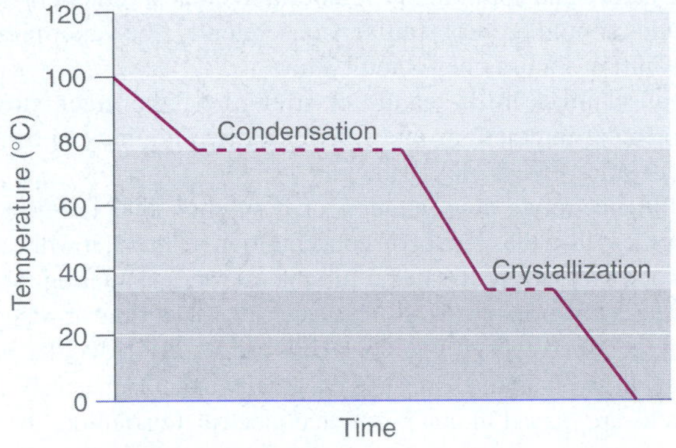

89. 45.0 kJ/mol
91. 490 kJ
93. 32.6 kJ
95. 57 g
97.

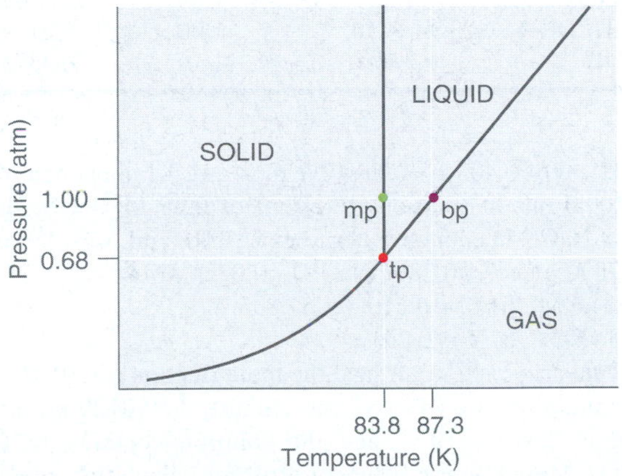

99. (a) 81% (b) Relative humidity is the *ratio* of the actual partial pressure of water vapor to its vapor pressure at a given temperature, expressed as a percent. Thus, at lower temperatures a given value of the relative humidity corresponds to less water vapor in the air, making the air feel drier.
101. (a) If the pressure is lower than the vapor pressure of water at a given temperature, water will sublime to form water vapor, which can be pumped off. (b) sublimation (c) At a temperature of 263 K, the vapor pressure of water is still appreciable (2.15 mmHg), which allows water to be removed from a frozen object by using very low pressures.

Chapter 12

3. Solids with the same or very similar compositions can have very different arrangements of the atoms or molecules, which often results in very different properties. Unlike liquids, the arrangement of atoms or molecules in a solid usually does not change with time.
7. The initial product was probably amorphous. After melting the first time, it apparently crystallized when it solidified upon cooling.
9. Yes. Cleavage planes do *not* have to be parallel to the faces of the unit cell.
15. Packing efficiency: simple cubic (~52%) < body-centered cubic (68%) < face-centered cubic (cubic close-packed) (74%) = hexagonal close-packed (74%). The hexagonal close-packed and cubic close-packed structures have the same packing efficiency.
21. In cesium chloride, both the cesium and chloride ions are arranged in simple cubic arrays that are offset in space, resulting in a cesium ion occupying the center of the cubic array of chlorides and a chloride ion occupying the center of the cubic array of cesiums. Metallic iron has a body-centered cubic structure.
25. $d = \dfrac{n\gamma}{2 \sin \theta}$ The diffraction angle would be smaller upon switching from a copper to a molybdenum X-ray source.
31. Impurity atoms of similar size and with similar chemical properties would be most likely to maintain the ductility of the metal, because they are unlikely to have a large effect on the ease with which one layer of atoms can move past another under mechanical stress. Interstitial atoms that form polar covalent bonds with the metal atoms act as a "glue" that holds layers of metal atoms together, which greatly decreases the ductility.
33. The oxide ion would introduce an extra negative charge. There are at least three ways to maintain electrical neutrality: (i) a fluoride ion vacancy elsewhere in the crystal; (ii) introduction of an additional cation as an impurity; and (iii) simultaneous replacement of a monocation in the lattice by a dication of approximately the same size.
35. By gaining electrons to compensate for the absence of anions, by the presence of additional metal ions, or by losing electrons to compensate for the presence of additional anions or the absence of metal ions.
37. A: NaCl, ionic solid
B: quartz, covalent solid
C: zinc, metal
D: sucrose, molecular solid
41. In a substitutional alloy, the impurity atoms are similar in size and chemical properties to the atoms of the host lattice. In an interstitial alloy, the impurity atoms are generally much smaller and have very different chemical properties. Comparing the mechanical properties of an alloy with those of the parent metal could be used to decide if the alloy were a substitutional or interstitial alloy.
45. The low electrical conductivity of diamond implies a very large band gap, corresponding to the energy of a photon of ultraviolet light rather than visible light. Consequently, diamond should be colorless. Pink or green diamonds contain small amounts of highly colored impurities that are responsible for their color.
47. The band gap will also increase due to a decrease in orbital overlap.
51. The interactions that lead to formation of Cooper pairs of electrons are so weak that they should be disrupted by thermal vibrations of lattice atoms above about 30 K.
59. four
61. fcc
63. molybdenum
65. sodium, unit cell edge = 429 pm, r = 186 pm
67. $d = 0.5337\ \text{g/cm}^3$, $r = 152$ pm
69. (a) $418 \times 10^6\ \text{pm}^3$ (b) 200 pm (c) $800 \times 10^6\ \text{pm}^3$ (d) 52.3%
73. Unit cell volume: $163.0 \times 10^6\ \text{pm}^3$; unit cell mass: 5.1585×10^{-22} g;

$$\text{Avogadro's number} = \frac{(4\ \text{formula units/unit cell}) \times (78.074\ \text{g/mol})}{(5.1585 \times 10^{-22}\ \text{g/unit cell})}$$

$$= 6.022 \times 10^{23}\ \text{formula units/mol}$$

75. Both have same stoichiometry, $CaTiO_3$.
77. Stoichiometry is MX_2; coordination number of cations is 8.
79. No, the structure shown has an empirical formula of Cd_3Cl_8.
81. Mg: 17.3°, Zn: 18.2°, Ni: 20.9°
83. No. The potassium is much larger than the sodium ion.
85. (a) 7.87 g/cm^3 (b) 7.86 g/cm^3 (c) Without defects, the mass is 0.15% greater.
87. The lower charge of K^+ makes it the best candidate for inducing an oxide vacancy. Substituting two K^+ ions for two Ca^{2+} ions will decrease the total positive charge by two, and an oxide vacancy will maintain electrical neutrality.
89. Os versus Hf: osmium has a higher melting point, due to more valence electrons for metallic bonding. SnO_2 versus ZrO_2: zirconium oxide has a higher melting point, because it has more ionic character. Al_2O_3 versus SiO_2: aluminum oxide has a higher melting point, again because it has more ionic character.
93. phosphorus; an *n*-type semiconductor has more electrons than the parent Group 14 semimetal, and a Group 15 element will provide an extra valence electron without greatly perturbing the local structure.
97. Given the mismatch in size between Au and Al, this is likely to be an interstitial alloy. Gold (174 pm) has a greater atomic radius than aluminum (118 pm), so gold forms the fcc lattice and aluminum occupies the tetrahedral holes. In a fcc unit cell, there are four gold atoms and eight tetrahedral holes, so the empirical formula of the alloy is $AuAl_2$.

Chapter 13

1. Homogeneous mixtures: aqueous ammonia, liquid decongestant, vinegar, and gasoline. Heterogeneous mixtures: seawater and fog. The homogeneous mixtures have the same composition regardless of location or magnification, while the composition of the heterogeneous mixtures varies depending upon location and magnification.
7. All are chemical reactions except dissolving iodine crystals in CCl_4.
17. London dispersion forces increase with increasing atomic mass, thus it requires more energy to separate I_2 molecules than Br_2 molecules.
21. (a) A third solvent with intermediate polarity and/or dielectric constant can effectively dissolve both of the immiscible solvents, creating a single liquid phase. (b) *n*-butanol—it is intermediate in polarity between methanol and *n*-hexane, while water is more polar than either and cyclohexane is comparable to *n*-hexane.
25. The mercury atoms are locked in the solid, and cannot readily diffuse to the surface where they could undergo chemical reactions.
31. Dissolve the mixture of A and B in a solvent in which they are both soluble when hot and relatively insoluble when cold, filter off any undissolved B, and cool slowly. Pure A should crystallize, while B stays in solution. If B were less soluble, it would be impossible to obtain pure A by this method in a single step, because some of the less soluble compound (B) will always be present in the solid that crystallizes from solution.
33. Only if the mass of 1000 mL of solution is exactly the same as the sum of the mass of solute and 1000 g solvent will molarity be equal to molality. The molality and molarity of an aqueous solution will be most similar at low concentrations. At higher concentrations, the solute contributes appreciably to the total volume of the solution, and 1000 mL of solution contains significantly less than 1000 g of water, making the molarity significantly greater than the molality.
39. When water is boiled, all of the dissolved oxygen and nitrogen are removed. If fish are added to deaerated water, they will suffocate. Allowing the water to stand overnight allows oxygen in the air to dissolve.
41. Evacuating the flask to remove gases decreases the partial pressure of oxygen above the solution. Consequently, dissolved oxygen escapes from solution into the gas phase, where it is removed by the vacuum pump. Filling the flask with nitrogen gas and repeating this process several times effectively removes almost all of the dissolved oxygen. The temperature of the solvent decreases because of evaporative cooling.
45. The nature and strength of the intermolecular interactions in the two liquids should be very similar. Consequently, the vapor pressure of the solution should obey Raoult's law.
47. Small changes in the nature or strength of the much stronger intermolecular interactions in a solution of two liquids can result in major deviations from ideal behavior.
53. If the electrolyte concentration of a solution that is injected is greater or less than the electrolyte concentration in blood, it will change the osmotic pressure of the blood plasma, as well as altering the balance of electrolytes in the blood. This can cause a net flow of water into or out of red blood cells. If red blood cells are placed in distilled water, the cells will swell dramatically and eventually burst. Conversely, if red blood cells are placed in a concentrated electrolyte solution, the cells decrease in volume and shrivel up until they are not functional.

59.

Compound	Molarity, *M*	Solution Density, g/mL	Mole Fraction, *X*
H_2SO_4	18.0	1.84	**0.81**
CH_3COOH	$\mathbf{3.94 \times 10^{-4}}$	1.00	7.21×10^{-3}
KOH	3.60	1.16	$\mathbf{6.33 \times 10^{-2}}$

61. 100.0 ml of 0.40 M KI: dissolve 6.64 g of KI in enough water to make 100.0 mL of solution; 100.0 ml of 0.65 M NaCN: dissolve 3.18 g of NaCN in enough water to make 100.0 mL of solution.
63. 0.473 *M* glucose, 0.28 *M* Na_3PO_4, 0.0221 *M* I_2
65. (a) 14.6 M (b) $X = 0.510$ (c) 57.8 m
67. (a) 14.8 *M* (b) $X = 0.291$
69. The molarity is 0.743 *M*, and the mole fraction is 0.0134.
71. The molarity is 0.0129 *M*, the molality is 0.0129 *m*, the mole fraction is 2.32×10^{-4}, and the solution contains 1830 ppm Na_2HPO_4. Mole fraction is most useful for calculating vapor pressure, because using Raoult's law, the mole fraction of the solvent is just one minus the mole fraction of solute.
73. 6.65×10^{-5} mol sodium
75. 1.63×10^{-3} g
77. 2.22×10^{4} mL or 22.2 L
79. 4.67 *M* HCl
81. 0.777 g $BaSO_4$ 0.0134 *M* Na_2SO_4
83. 0.634 L CO_2; $k_{0°C} = 7.61 \times 10^{-2}$ *M*/atm $k_{20°C} = 3.84 \times 10^{-2}$ *M*/atm
85. 6.7×10^{4} amu
87. 9.24 atm
89. The $CaCl_2$ solution will have a lower vapor pressure, because it contains three times as many particles as the glucose solution.
91. 0.363 *m* NaCl, 2.57 g NaCl
93. 59.2 g NaBr
95. 688 g NaCl
97. $MgCl_2$ produces three particles in solution versus two for NaCl, so the same molal concentration of $MgCl_2$ will produce a 50% greater freezing point depression than for NaCl. Nonetheless, the molar mass of $MgCl_2$ is 95.3 g/mol versus 48.45 g/mol for NaCl. Consequently, a solution containing 1 g NaCl per 1000 g H_2O will produce a freezing point depression of 0.064°C versus 0.059°C for a solution containing 1 g $MgCl_2$ per 1000 g H_2O. Thus, given

equal cost per gram, NaCl is more effective. Yes, $MgCl_2$ would be effective at −8°C; a 1.43 *m* solution (136 g per 1000 g H_2O) would be required.

99.

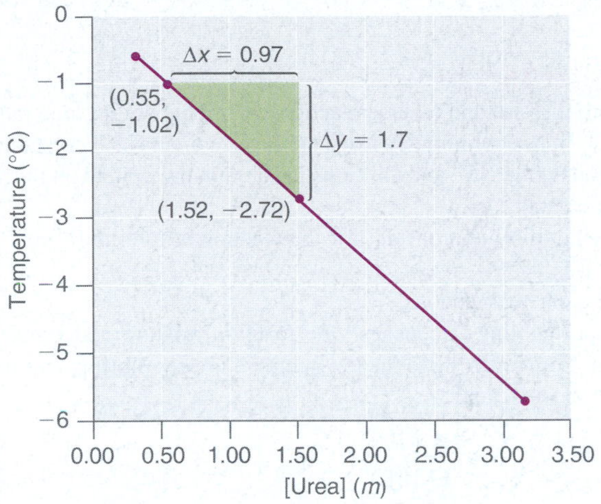

$k_f = -1.81$(°C · kg)/mol, molecular mass of urea = 60.0 g/mol

101. (a) glucose, 1.5 osmole; NaCl, 3.0 osmole; $CaCl_2$, 4.5 osmole (b) The concentration of water decreases as the osmolarity increases. (c) Water will flow from the NaCl solution into the $CaCl_2$ solution.

105. 1 atm: 2.69×10^{-4} *M* O_2 and 5.14×10^{-4} *M* N_2
4 atm: 1.08×10^{-3} *M* O_2 and 2.05×10^{-3} *M* N_2

107. 2.6×10^{-6} mg/L, 550 ppm

109. 1.4×10^{-5} mol, 1.7×10^{-6} *m*

111. To obtain the same concentration of dissolved oxygen in their "blood" at a lower partial pressure of oxygen, the value of the Henry's law constant would have to be higher.

113. The large, coarse particles would precipitate, the medium particles would form a suspension, and the fine ones would form a colloid. A colloid consists of solid particles in a liquid medium, so it is not an emulsion, which consists of small particles of one liquid suspended in another liquid. The finest particles would be used for painting.

Chapter 14

1. Kinetics gives information on the rate and mechanism of a reaction; the balanced chemical equation tells only the stoichiometry of the reaction.

3. Reaction rates generally increase with increasing reactant concentration, temperature, and addition of a catalyst. Physical properties such as high solubility also increase reaction rates. Solvent polarity can either increase or decrease the rate of a reaction, but increasing solvent viscosity generally decreases reaction rates.

5. Increasing the temperature increases the average kinetic energy of molecules and ions, causing them to collide more frequently and with greater energy, which increases the rate of reaction. First dissolve sugar in hot tea, and then add ice.

9. Reactant concentrations are highest at the beginning of a reaction. The plot of [C] vs. *t* is a curve with a slope that becomes steadily less positive.

11. Faster. The H_3O^+/OH^- reaction is faster due to the decreased relative size of reactants and the higher electrostatic attraction between reactants.

15. Only reactions (b) and (d). Zeroth-order reactions are usually those for which the rate is limited by the number of available reaction sites or the amount of available surface.

17. (a) The reaction is first order in that reactant; (b) the overall order is ≥ 1.

19. (a) For a given reaction under particular conditions, the magnitude of the first-order rate constant does not depend on whether a differential or integrated rate law is used. (b) The differential rate law requires multiple experiments to determine reactant order, the integrated rate law needs only one experiment. (c) Using the differential rate law, a graph of concentration versus time is a curve with a slope that becomes less negative with time, whereas for the integrated rate law, a graph of ln[reactant] versus time gives a straight line with slope = $-k$. The integrated rate law allows you to calculate the concentration of a reactant at any time during the reaction; the differential rate law does not.

21. The reaction rate increases as the rate constant increases. We cannot directly compare reaction rates and rate constants for reactions of different orders because they are not mathematically equivalent.

23. The substance with more disintegrations per second has the shorter half-life.

25. The order of a reaction depends upon the molecularity of the rate-determining step; there is no relationship between the order of a reaction and its stoichiometry.

31. (a) An increase in electrostatic repulsions between reactants will increase the activation energy; (b) bond formation in the activated complex tends to lower the activation energy; (c) an increase in the energy of the activated complex increases the activation energy of the reaction.

35. The activation energy is not related to either the temperature or the frequency factor *A*, but the frequency factor is usually proportional to the square root of the temperature. An increase in the frequency factor increases the rate.

37. A catalyst lowers the activation energy of a reaction. Some catalysts can also orient the reactants and thereby increase the frequency factor. Catalysts have no effect on the change in potential energy for a reaction.

39. In the process of adsorption, a reactant binds tightly to a surface. Because intermolecular interactions between the surface and the reactant weaken or break bonds in the reactant, its reactivity is increased, and the activation energy for a reaction is often decreased.

41. (a) Heterogeneous catalysts are easier to recover; (b) collision frequency is greater for homogeneous catalysts; (c) homogeneous catalysts are often more sensitive to temperature; and (d) homogeneous catalysts are often more expensive.

43. The Mn^{2+} ion donates an electron to Ce^{4+} and then accepts an electron from Tl^+. It functions as an electron conduit.

47. Average rate: 41 mi/h; instantaneous rate was 0 at t = 13 min and 38 mi/h at 28 min.

49. (a) rate will double; (b) rate will double; (c) rate will quadruple; (d) no change in rate.

51. 298 s, 1270 s

53. First order in Fe^{3+}; second order in I^-; third order overall; rate = $k[Fe^{3+}][I^-]^2$.

55. 1.29×10^{-4} *M*/s; 3.22×10^{-5} *M*/s

59. No. The reaction is second order: the half-life decreases with increasing reactant concentration according to $t_{1/2} = 1/k[A_0]$.

61. 1.92×10^3 s or 1920 s

63. The k_2 step is likely to be rate limiting; the reaction cannot proceed any faster than the second step.

65. $$\text{rate} = k_2 \frac{k_1 [O_2NNH_2]}{k_{-1} [H^+]} = k \frac{[O_2NNH_2]}{[H^+]}$$

67. Rate will approximately double: 20°C to 30°C, rate increases by about $2^1 = 2$; 20°C to 70°C, rate increases by about $2^5 = 32$-fold. Plot of rate versus temperature will give an exponential increase: rate $\propto 2^{\Delta T/10}$.
69. (a) 1.0×10^{-5} *M*/s; (b) 6.6×10^{-5} *M*/s; (c) 3.5×10^{-4} *M*/s
71. 96 kJ/mol

73. $$\frac{\Delta[ES]}{\Delta t} = -(k_2 + k_{-1})[ES] + k_1[E][S] + k_{-2}[E][P] \approx 0;$$
first order in substrate.
75. In both cases, pathway A's product. All of the *Z* produced in the catalyzed reversible pathway B ($X \rightleftharpoons Z$) will eventually be converted to *X* as *X* is converted irreversibly to *Y* by pathway A:

$$Z \underset{\text{B}}{\rightleftharpoons} X \xrightarrow{\text{A}} Y.$$

77. rate $= k_f [SO_2][CO] - k_r[SO][CO_2]$
rate $= k_f [SO][CO] - k_r[S][CO_2]$
rate $= k_f [SO][SO_2] - k_r[S][SO_3]$
79. Reaction is second order, first order in O and first order in O_3. Yes, ozone is being produced faster than it is being destroyed. If ozone concentrations are not increasing, then either some other reaction must be consuming some of the ozone produced in this reaction or the ozone-producing reaction does not operate at this rate continuously.
81. Yes, the object is about 2300 years old.
83. (a) $k = 0.0565$ yr^{-1}; (b) 0.482 g of ^3H
87. (a) Second order, first order in O and first order in O_3; (b) 17 kJ/mol; (c) 0.44 kJ/mol;
(d)

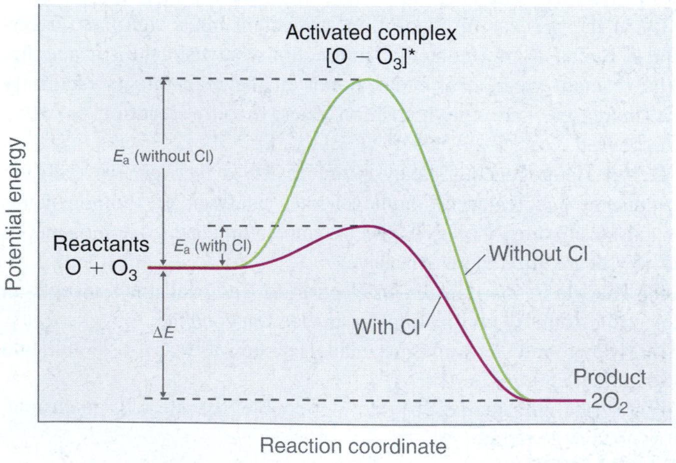

(e) Cl is a potent catalyst for ozone destruction.

89.

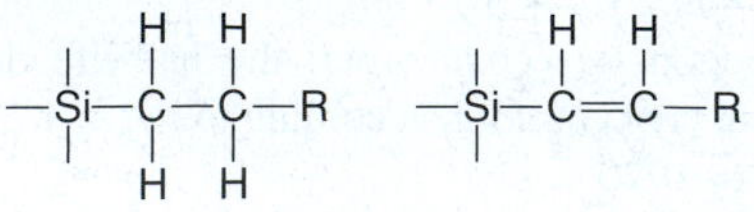

91. $Ni_{1-x}O$ is a nonstoichiometric oxide that contains a fraction of Ni(I) sites. These can react with oxygen to form a Ni(III)-oxide site, which is reduced by CO to give Ni(I) and CO_2.
93. 0.35 g ^{32}P
95. 4.1×10^{15}

Chapter 15

3. Both forward and reverse reactions occur, but at the same rate. Na^+ and Cl^- ions continuously leave the surface of an NaCl crystal to enter solution, while Na^+ and Cl^- ions in solution precipitate on the surface of the crystal.
5. The equilibrium constant for the reaction written in reverse (K') is $K' = 1/K$.
7. All are heterogeneous.
9. Rapid cooling "quenches" the reaction mixture and prevents the system from reverting to the low-temperature equilibrium composition that favors reactants.

11. $K = k_f/k_r$; $K = \dfrac{[C]^c[D]^d}{[A]^a[B]^b}$

13. (a) $K = \dfrac{[NO_2]^2}{[NO]^2[O_2]}$; $K_p = \dfrac{(P_{NO_2})^2}{(P_{NO})^2\,(P_{O_2})}$

(b) $K = \dfrac{[HI]}{[H_2]^{1/2}[I_2]^{1/2}}$; $K_p = \dfrac{P_{HI}}{(P_{H_2})^{1/2}\,(P_{I_2})^{1/2}}$

(c) $K = \dfrac{[\textit{trans}\text{-stilbene}]}{[\textit{cis}\text{-stilbene}]}$; $K_p = \dfrac{P_{\textit{trans}\text{-stilbene}}}{P_{\textit{cis}\text{-stilbene}}}$

15. (a) $K = \dfrac{[SO_3]^2}{[O_2]^3}$; $K_p = \dfrac{(P_{SO_3})^2}{(P_{O_2})^3}$

(b) $K = \dfrac{[CO]^2}{[CO_2]}$; $K_p = \dfrac{(P_{CO})^2}{P_{CO_2}}$

(c) $K = \dfrac{[SO_2]^2}{[O_2]^3}$; $K_p = \dfrac{(P_{SO_2})^2}{(P_{O_2})^3}$

17. At equilibrium, $[A] = \sqrt{[B]}$; $\Delta n = -1$, so $K_p = K(RT)^{\Delta n} = \dfrac{K}{RT}$; difference increases as *T* increases.

23. (a) $K = \dfrac{[CH_4][H_2S]^2}{[CS_2][H_2]^4}$; doubling $[CS_2]$ would require multiplying $[H_2]$ by $(1/2)^4 = 1/16$.

(b) $K = \dfrac{[PCl_3][Cl_2]}{[PCl_5]}$; if $[Cl_2]$ is halved, $[PCl_5]$ must also be halved.

(c) $K = \dfrac{[NO]^4[H_2O]^6}{[NH_3]^4[O_2]^5}$; if [NO] is doubled, $[H_2O]$ is multiplied by $\sqrt[6]{1/16} = 1/2^{2/3} = 1/0.587$.

25. $K = \dfrac{[NO]^4[H_2O]^6}{[NH_3]^4[O_2]^5}$; increasing *P* will favor reactants. Increasing *P* favors the side with fewer moles of gas.
27. The ratio k_f/k_r will decrease. Increasing *P* favors products, while increasing *T* favors reactants, so the effect of increasing both *T* and *P* depends on the magnitude of ΔH.
29. (a) No effect; (b) P_{H_2} will decrease; (c) P_{Cl_5} will decrease
33. If $Q < K$, reactants will be converted to products; if $Q > K$, products will be converted to reactants.

35. $K = 0.892$: concentrations of products and reactants are approximately equal at equilibrium; $K = 3.25 \times 10^8$: concentration of products to reactants at equilibrium is very large; $K = 5.26 \times 10^{-11}$: concentration of products to reactants at equilibrium is very small.

37. (a) $K' = \dfrac{[NH_3]}{[N_2]^{1/2}[H_2]^{3/2}}$; (b) $K'' = \dfrac{[NH_3]^{2/3}}{[N_2]^{1/3}[H_2]}$; $K = \dfrac{[NH_3]^2}{[N_2][H_2]^3}$ for $N_2(g) + 3H_2(g) \rightleftharpoons 2NH_3(g)$; $K' = \sqrt{K}$, $K'' = \sqrt[n]{K}$

39. (a) $K = \dfrac{[NO_2]^2}{[NO]^2[O_2]}$; (b) $K = \dfrac{[HI]}{[H_2]^{1/2}[I_2]^{1/2}}$;

(c) $K = \dfrac{[Ca^{2+}][OCl^-]^2 P_{CO_2}}{[HOCl]^2}$

41. (a) $K = 1.25 \times 10^{-5}$, $K_p = 2.40 \times 10^{-4}$; (b) $K = 9.43$, $K_p = 5.60 \times 10^{-3}$

43. (a) $K = \dfrac{[Cl_2][NO]^2}{[NOCl]^2} = 4.66 \times 10^{-4}$, $K_p = 1.91 \times 10^{-2}$;

(b) $K = \dfrac{[PCl_5]}{[PCl_3][Cl_2]} = 28.2$, $K_p = 0.658$

45. $NH_3 + 5/4O_2 \rightleftharpoons NO + 3/2H_2O$, also written as: $4NH_3(g) + 5O_2(g) \rightleftharpoons 4NO(g) + 6H_2O(g)$

47. (a) 3.29×10^3; (b) 1.17×10^{14}; (c) 3.04×10^{-4}

49. No; since $\Delta H > 0$, K_p (and P_{O_2}) should increase with increasing T.

51. $K = 54 = K_p$

53. 0.135

55. (a) $K_{375} = 7.8 \times 10^{-2}$, $K_{303} = 1.2 \times 10^{-3}$; (b) 0.26 M; (c) 11 atm

59. $K = 2.50$; 0.71 mol isobutane, 0.29 mol n-butane

61. $P_{COCl_2} = 0.0934$ atm, $P_{Cl_2} = P_{CO} = 4.5 \times 10^{-6}$ atm; because K_p is very small, assume that the decrease in P_{COCl_2} is negligible compared with its initial pressure.

63. [IBr] = 0.049 M; $P_{IBr} = 1.5$ atm, $P_{I_2} = 0.12$ atm, $P_{Br_2} = 0.43$ atm

65. Not at equilibrium: in both cases, the sum of the equilibrium partial pressures is *less* than the total pressure, so the reaction will proceed to the right to decrease the pressure.

67. (a) $K = \dfrac{[H_2O]}{[H_2]}$, $K = \dfrac{[CO_2]}{[CO]}$; (b) $P_{H_2O} = 21.1$atm, $P_{H_2} = 0.27$ atm, $P_{CO_2} = 21.3$ atm, $P_{CO} = 0.05$ atm; (c) $K_p = 0.14$; (d) the amount of CoO has no effect on the shape of a graph of products versus reactants *as long as some solid CoO is present.*

69. (a) 1: Q and K do not change; 2: Q does not change, K decreases; 3: Q and K do not change; 4: Q decreases, K does not change.
(b) 1: Q decreases, K does not change; 2: Q does not change, K increases; 3: Q increases, K does not change; 4: Q decreases, K does not change.
(c) 1: Q and K do not change; 2: Q does not change, K increases; 3: Q and K do not change; 4: Q decreases, K does not change.

71. None of the changes would affect K. (a) Q doubles; (b) Q is halved; Q decreases.

73. K would not change; it does not depend on volume.

75. Endothermic; K increases with increasing T.

77. [CO] = [H_2O] = 0.630 M, [CO_2] = [H_2] = 0.703 M; no effect on K

79. (a) At 800°C, [CO] = 0.678 M, [CO_2] = 0.231 M; at 1000°C, [CO] = 0.649 M, [CO_2] = 0.260 M; (b) at 800°C, $P_{CO} = 59.7$ atm, $P_{CO_2} = 20.3$ atm; at 1000°C, $P_{CO} = 67.8$ atm, $P_{CO_2} = 27.2$ atm;
(c) removing CO would cause the reaction to shift to the right, causing P_{CO_2} to decrease.

81.

	CH_4	O_2	CO_2	H_2O	Q	K
Initial (moles)	0.45	0.90	0	0	**0**	1.29
At equilibrium	0.215	0.43	0.235	0.47	**1.29**	1.29
Add 0.50 mol of methane	0.715	0.43	0.235	0.47	**0.39**	1.29
New equilibrium	0.665	0.33	0.285	0.57	1.29	**1.29**
Remove water	0.665	0.33	0.285	0	**0**	1.29
New equilibrium	0.57	0.14	0.38	0.19	**1.29**	1.29

83. Use low temperature and high pressure (small volume).

87. (a) reactants; (b) [Cl_2] = [PCl_3] = 0.082 M, [PCl_5] = 0.168 M; (c) increasing pressure favors reactant (PCl_5); (d) 1.60×10^3 kg, 52.6%.

89. Products are favored; $K = \dfrac{[\text{chloral hydrate}]}{[Cl_3CHO][H_2O]}$; high concentrations of water will favor chloral hydrate formation.

93. $K = \dfrac{[CO_2]^4[S_2]}{[CO]^4[SO_2]^2}$; (a) $K' = \dfrac{[CO_2]^2[S_2]^{1/2}}{[CO]^2[SO_2]} = K^{1/2}$;

(b) $K'' = \dfrac{[CO_2][S_2]^{1/4}}{[CO][SO_2]^{1/2}} = K^{1/4}$;

(c) $K_p = 1.6$, $K = 93$, $[S_2] = 5.1 \times 10^{-10}$ M.

95. $P_{CO} = P_{Cl_2} = 0.420$ atm, $P_{COCl_2} = 0.261$ atm; reactants are slightly favored.

99. Both reactions are favored by increasing [HCl] and decreasing volume.

101. Isocitrate is immediately consumed by the next reaction in the cycle, which favors formation of products and drives the conversion of citrate to isocitrate to completion. The reaction is initially driven toward the formation of isocitrate, and then a new equilibrium state is reached in which the isocitrate concentration has increased proportionally.

Chapter 16

1. $K_{auto} = [H_3O^+]\,[OH^-]/[H_2O]^2$
$K_w = [H_3O^+]\,[OH^-] = K_{auto}\,[H_2O]^2$

3. $H_2O(l) + HNO_3(g) \longrightarrow H_3O^+(aq) + NO_3^-(aq)$; water is the base
$H_2O(l) + NH_3(g) \longrightarrow OH^-(aq) + NH_4^-(aq)$; water is the acid

7. (a) $HSO_4^-(aq)$ [acid(1)] $+ H_2O(l)$ [base(2)] $\rightleftharpoons SO_4^{2-}(aq)$ [base(1)] $+ H_3O^+(aq)$ [acid(2)]

(b) $C_3H_7NO_2(aq)$ [base(2)] $+ H_3O^+(aq)$ [acid(1)] $\rightleftharpoons C_3H_8NO_2^+(aq)$ [acid(2)] $+ H_2O(l)$ [base(1)]

(c) $HOAc(aq)$ [acid(1)] $+ NH_3(aq)$ [base(2)] $\rightleftharpoons CH_3CO_2^-(aq)$ [base(1)] $+ NH_4^+(aq)$ [acid(2)]

(d) $SbF_5(aq)$ [acid(1)] $+ 2HF(aq)$ [base(2)] $\rightleftharpoons H_2F^+(aq)$ [acid(2)] $+ SbF_6^-(aq)$ [base(1)]

9. (a) $K_a = \dfrac{[CO_3^{2-}][H^+]}{[HCO_3^-]}$

(b) $K_a = \dfrac{[HCO_2^-][H^+]}{[HCO_2H]}$

(c) $K_a = \dfrac{[H_2PO_4^-][H^+]}{[H_3PO_4]}$

11. (a) right; (b) right; (c) left; (d) right
13. No, only the first dissociation reaction goes to completion; $2.0\ M < [H^+] < 4.0\ M$
17. (a) increase; (b) stay approximately the same; (c) increase; (d) decrease
19. $Sn(H_2O)_6^{4+}$; its higher charge to radius ratio will cause greater polarization of bound water molecules.
25. $CF_3S^- < CH_3S^- < OH^-$ (strongest base)
29. NH_3; Cl atoms withdraw electron density from N in Cl_2NH.
31. $C_{H_3PO_4} = [H_3PO_4] + [H_2PO_4^-] + [HPO_4^{2-}] + [PO_4^{3-}]$
33. Percent ionization increases upon dilution.
37. (a) neutral; (b) acidic; (c) basic
39. At the equivalence point of the titration of a weak acid, the solution contains the conjugate base, which reacts with water to form hydroxide ions, making the solution basic. At the equivalence point of the titration of a weak base, the solution contains the conjugate acid, which reacts with water to form hydronium ions, making the solution acidic.
45. Adding NH_4^+ shifts the ionization equilibrium of ammonia ($H_2O(l) + NH_3(aq) \rightleftharpoons OH^-(aq) + NH_4^+(aq)$) to the left, decreasing $[OH^-]$ and decreasing the pH.
47. (a) Shifts to left, pH decreases; (b) shifts to right, pH increases; (c) shifts to left, pH increases
51. The midpoint
55. (a) Not a buffer; the HCl completely neutralizes the sodium acetate to give acetic acid and NaCl(*aq*).
(b) Buffer; the HCl neutralizes only half of the sodium acetate to give a solution containing equal amounts of acetic acid and sodium acetate.
(c) Not a buffer; the NaOH completely neutralizes the acetic acid to give sodium acetate.
(d) Buffer; the NaOH neutralizes only half of the acetic acid to give a solution containing equal amounts of acetic acid and sodium acetate.
(e) Buffer; the solution will contain a 2:1 ratio of sodium acetate and acetic acid.
59. The CO_2/HCO_3^- buffer is an open system, whose pH and buffering capacity depend on the CO_2 pressure as well as $[HCO_3^-]$. Increasing the CO_2 pressure decreases the pH of blood; increasing the O_2 pressure has no direct effect on the pH of blood.
61. $K_{H_2SO_4} = [H_3SO_4^+][HSO_4^-] = K[H_2SO_4]^2$; $[H_3SO_4^+] = 0.3\ M$; fraction ionized is 0.02.
63. pOH = 8.37, $[OH^-] = 4.3 \times 10^{-9}\ M$; acidic
65. (a) pH = 0.82, pOH = 13.18; (b) pH = 12.5, pOH = 1.5; (c) pH = 2.64, pOH = 11.36; (d) pH = 12.99, pOH = 1.01; (e) pH = 3.77, pOH = 10.23; (f) pH = −0.76, pOH = 14.76
69. 2.88 mg of HCl
71. acid B < acid C < acid A (strongest)
73. (a) $K_a = 6.3 \times 10^{-11}$, $pK_a = 10.2$; (b) $K_a = 7.9 \times 10^{-7}$, $pK_a = 6.1$; (c) $K_a = 0.50$, $pK_a = 0.30$; (d) $K_a = 2.5 \times 10^{-13}$, $pK_a = 12.6$; (e) $K_a = 3.2 \times 10^{-17}$, $pK_a = 16.5$
75. $K_a = 6.3 \times 10^{-11}$; $pK_a = 9.2$
81. 0.94% dissociated; butanoic acid will have the higher pK_a value; the inductive effect due to the chlorine substituent will make 3-chlorobutanoic acid a somewhat stronger acid.
83. $pK_b = 9.43$; $(CH_3)_2CHNH_2$ will be a stronger base and have a lower pK_b; aniline is a weaker base because the lone pair on the nitrogen atom can be delocalized onto the aromatic ring.
85. 1.79
87. 3.8×10^{-5}
91. 8.12
95. 0.86 mL
99. (a) dilute 8.33 mL of 12.0 *M* HCl to 500.0 mL; (b) about 72 mL; (c) 77.0 mL
103. pH at equivalence point = 8.46

Volume of Base Added, mL	0	5	10	15	20	25
pH	2.21	3.23	3.67	4.11	4.87	12.36

107. 1.82
111. (a) 1.35×10^{-3}; (b) 4.05; (c) 3.88; (d) 4.30
113. (a) $pK_w = 13.53$, pH = pOH = 6.77; (b) lower
115. 68.4 mL; 1.59; 7.82

Chapter 17

1. (a) $K_{sp} = [Ag^+][I^-]$
(b) $K_{sp} = [Ca^{2+}][F^-]^2$
(c) $K_{sp} = [Pb^{2+}][Cl^-]^2$
(d) $K_{sp} = [Ag^+]^2[CrO_4^{2-}]$
5. For a 1:1 salt, the molar solubility is simply $\sqrt{K_{sp}}$, for a 2:1 salt, the molar solubility is $\sqrt[3]{K_{sp}/4}$. Consequently, the magnitudes of K_{sp} can be correlated with molar solubility *only* if the salts have the same stoichiometry.
7. Because of the common ion effect. Adding a soluble Mg^{2+} salt increases $[Mg^{2+}]$ in solution, and le Châtelier's principle predicts that this will shift the solubility equilibrium of $MgCO_3$ to the left, decreasing its solubility.
9. Because Mg^{2+} is smaller than Ca^{2+}, ion-pair formation will be more important for $Mg_3(PO_4)_2$. Consequently, the difference for $Ca_3(PO_4)_2$ will be less.
13. Mg^{2+} has a higher charge-to-radius ratio than Ba^{2+}. Consequently, Mg^{2+} will be a significantly stronger Lewis acid and will have a greater tendency to form complex ions.
15. Co^{2+} is the Lewis acid, and pyridine, with a lone pair of electrons on the nitrogen atom, is the Lewis base.
17. The S^{2-} ion is a stronger base than the acetate ion. Consequently, SnS will dissolve more readily in dilute HCl.
21. Elements that form amphoteric oxides generally lie on or near the diagonal line dividing metals (which tend to form basic oxides) from nonmetals (which form acidic oxides). Thus, Be, B, and Al are most likely to form amphoteric oxides.
23. (a) $2.05 \times 10^{-4}\ M$; (b) $7.73 \times 10^{-9}\ M$; (c) $3.1 \times 10^{-10}\ M$
25. 3.38 g
27. (a) $4.15 \times 10^{-4}\ M$; (b) $9.3 \times 10^{-6}\ M$; (c) $6.3 \times 10^{-6}\ M$; (d) $6.5 \times 10^{-5}\ M$; (e) $1.03 \times 10^{-3}\ M$
29. 22.3 mg. A secondary reaction occurs, where OH^- from the dissociation of the salt reacts with H^+ from the dissociation of water. This reaction causes further dissociation of the salt (le Châtelier's principle).
31. 1.2×10^{-10}
33. 1.70×10^{-5}
37. (a) 3.86×10^{-71}; (b) 6.7×10^{-9}; (c) 9.0×10^{-8}; (d) 2.15×10^{-9}
39. $7.4 \times 10^{-6}\ M$, 2.1 mg
41. Precipitation will occur in all cases.
43. 290 mL
45. 1.59×10^{-11}. The common ion effect decreases the solubility of the salt.
47. $[Fe(H_2O)_6]^{2+}(aq) + OH^-(aq) \rightleftharpoons [Fe(H_2O)_5(OH)]^+(aq) \quad \log K_1 = 5.56$
$[Fe(H_2O)_5(OH)]^+(aq) + OH^-(aq) \rightleftharpoons [Fe(H_2O)_4(OH)_2](aq) \quad \log K_2 = 9.77$

$[Fe(H_2O)_4(OH)_2](aq) + OH^-(aq) \rightleftharpoons$
$[Fe(H_2O)_3(OH)_3]^-(aq)$ $\log K_3 = 9.67$
$[Fe(H_2O)_3(OH)_3]^-(aq) + OH^-(aq) \rightleftharpoons$
$[Fe(OH)_4]^{2-}(aq) + 2H_2O(l)$ $\log K_4 = 8.58$
$[Fe(H_2O)_6]^{2+}(aq) + 4OH^-(aq) \rightleftharpoons$
$[Fe(OH)_4]^{2-}(aq) + 6H_2O(l)$ $\log K_f$
$= \log K_1 + \log K_2 + \log K_2 + \log K_4$
$= 33.58$
$[Fe(OH)_4]^{2-}$ has a large value of K_f, so it should be stable in the presence of excess OH^-.
49. 2.94
51. No, both metal ions will precipitate; AgBr will precipitate as Br^- is added, and CuBr will begin to precipitate at $[Br^-] = 8.6 \times 10^{-6}\,M$.
53. (a) 4.3 g; (b) $4.9 \times 10^{-3}\,M$
55. (a) $4.5 \times 10^{-7}\,M$; (b) 0.84 *M*
59. No; these cations would precipitate as sulfides.

Chapter 18

1. Thermodynamics tells us nothing about the rate at which reactants are converted to products.
3. Heat and work; path dependent
7. At constant pressure, $\Delta H = \Delta E + P\Delta V$.
11. For a spontaneous process, $\Delta G = \Delta H - T\Delta S < 0$. Since ΔH is positive, $T\Delta S$ must also be positive and greater in magnitude than ΔH.
13. With bond angles of 60°, cyclopropane is highly strained, causing it to be less stable than an unstrained cyclic hydrocarbon like cyclopentane.
15. Irreversible expansion, no work is done because the external pressure is effectively zero.
17. Reversible: (a) and (c); irreversible: (b) and (d)
23. ΔS_{soln} in ethanol must be more positive than ΔS_{soln} in water to make ΔG_{soln} more negative for ethanol.
25. The third law of thermodynamics gives an absolute reference point for entropies, but no such absolute reference point is available for enthalpy (or any other function involving energy).
27. At any temperature, A will have a higher value of *S* than B, and the entropy of A will increase faster with increasing temperature.
29. Only a combination of the change in enthalpy and the change in entropy of the system can be used to predict how changes in the system will affect the entropy of the universe.
31. Because there are more molecules of reactants than products, ΔS is negative for the reaction as written. The $-T\Delta S$ term is therefore positive for the reaction and increases in magnitude as *T* increases, making ΔG more positive and favoring reactants.
35. Yes. All real processes involve some loss of heat (due to friction, etc.), which is used to increase the entropy of the universe.
37. Because $\Delta G° = -RT \ln K$, the equilibrium constant becomes larger and the extent of the reaction increases as $\Delta G°$ becomes more negative.
39. $\Delta G° = -RT \ln K$ in both cases. For gases, the equilibrium constant is K_p. *K* and K_p are related by: $K_p = K(RT)^{\Delta n}$, where Δn is the number of moles of gaseous product minus the number of moles of gaseous reactant.
41. It is coupled to another reaction that is spontaneous, which drives this reaction forward (le Châtelier's principle).
43. 13.1 kJ work done on the surroundings
45. $\Delta V = 30.6$ L, $w = -3.1$ kJ, work done on the surroundings
47. 220 kJ
49. (a) 0; (bq) −455 J; (c) −317 J
51. −350 J; 8.2 kJ
55. $\Delta S_{univ} = \Delta S_{cold} + \Delta S_{hot} = \dfrac{q}{T_{cold}} + \left(-\dfrac{q}{T_{hot}}\right)$; only if $T_{cold} < T_{hot}$ can ΔS_{univ} be > 0.

59. Phase transitions have large ΔS values; 13.1°C; 4.3 J.
61. (a) −164.3 J/K; (b) −9.1 J/K; (c) 114.5 J/K; (d) −173.2 J/K
63. 25 J/ (mol · K)
65. yes; 274 K
67. (a) −237.1 kJ/mol, spontaneous as written; (b) −100.4 kJ/mol, spontaneous as written; (c) 16.6 kJ/mol, spontaneous in reverse direction; (d) 141.9 kJ/mol, spontaneous in reverse direction.
69. (a) $\frac{1}{2}N_2(g) + O_2(g) \longrightarrow NO_2(g)$; not spontaneous at any *T*; (b) $\frac{1}{2}N_2(g) + \frac{1}{2}O_2(g) \longrightarrow NO(g)$; not spontaneous at 25°C, spontaneous above 7500 K; (c) $\frac{1}{2}N_2(g) + \frac{3}{2}H_2 \longrightarrow NH_3(g)$; spontaneous at 25°C
71. 919 K, 646°C
73. $MgCO_3$: $\Delta G° = 63$ kJ/mol, spontaneous above 658 K; $BaCO_3$: $\Delta G° = 220$ kJ/mol, spontaneous above 1560 K
75. (a) $\Delta G° = -28.5$ kJ/mol; (b) −26.4 kJ/mol; (c) −23.0 kJ/mol
77. (a) 1.21×10^{66}; (b) 1.89×10^6; (c) 8.25×10^{15}
79. −13.3 kJ/mol
81. 5.1×10^{-21}
83. 10.3 kJ/mol
85. (a) 129.5 kJ/mol; (b) 6.0; (c) 6.0 atm; (d) products are favored at high *T*, reactants are favored at low *T*.
89. (a) −4520 kJ/mol B_5H_9; (b) 1300 kJ; (c) -1.63×10^6 kJ/50 lb of B_5H_9
91. −2220 kJ/mol
93. (a) aerobic conversion; (b) −268 kJ
97. Yes, reaction is spontaneous at 25°C; not spontaneous at 800°C ($\Delta G = 0.82$ kJ/mol); reaction rate is much faster at 800°C
99. (a) −314 kJ/mol; (b) yes; (c) 1.1×10^{55}; (d) ignition is required to overcome high activation energy to reaction.
101. (a) 4.4 kJ/mol; (b) 0.17; (c) 78; (d) 13; (e) entropy driven
103. 2.3×10^7; −43.7 kJ/mol

Chapter 19

5. A large difference in cation/anion diffusion rates would increase resistance in the salt bridge and limit electron flow through the circuit.

11. $Ni(s)\,|\,Ni^{2+}(aq)\,||\,H^+(aq)\,|\,H_2(g)\,|\,Pt(s)$
$E°_{anode}$ $Ni^{2+} + 2e^- \longrightarrow Ni$; − 0.257 V
$E°_{cathode}$ $2H^+ + 2e^- \longrightarrow H_2$; 0.000 V
$E°_{cell}$ $2H^+ + Ni \longrightarrow H_2 + Ni^{2+}$; 0.257 V

13. (a) oxidant: $Ni^{2+}(aq)$; reductant: Cr(*s*)
(b) oxidant: $Cl_2(g)$; reductant: $Sn^{2+}(aq)$
(c) oxidant: $H_3AsO_4(aq)$; reductant: Zn(*s*)
(d) oxidant: $NO_2(g)$; reductant: $NO_2(g)$

19. No. $E° = -0.691$ V for $Ag_2S(s) + 2e^- \rightarrow Ag(s) + S^{2-}(aq)$, which is too negative for Ag_2S to be spontaneously reduced by oxalic acid [$E° = 0.49$ V for $2CO_2(g) + 2H^+(aq) + 2e^- \rightarrow H_2C_2O_4(aq)$].
23. extensive: w_{max} and *n*; intensive: E_{cell}
27. Gold is highly resistant to corrosion because of its very positive reduction potential.
33. (a) lead storage battery; (b) lithium–iodine battery; (c) Nicad or lithium ion battery (rechargeable)
37. Paint keeps oxygen and water from coming into direct contact with the metal, which prevents corrosion. Paint is more necessary because salt is an electrolyte that increases the conductivity of water and facilitates the flow of electric current between anodic and cathodic sites.
41. For the same electrolysis time and current, the $Fe(NO_3)_2 \cdot 6H_2O$ solution will produce a greater mass of metallic iron; reduction of

1 mol of Fe^{2+} ions requires only 2 mol of electrons versus 3 mol of electrons for 1 mol of Fe^{3+} ions.

43. (a) Reductive half-reaction: $NO_3^-(aq) + 3H^+(aq) + 2e^- \longrightarrow HNO_2(g) + H_2O(l)$

Oxidative half-reaction: $SO_3^{2-}(g) + 2OH^-(aq) \longrightarrow SO_4^{2-}(aq) + H_2O(l) + 2e^-$

(b) Reduction of nitrate occurs at the cathode; oxidation of sulfur dioxide occurs at the cathode.

(c) Cathode is positive, anode is negative.

45.

Reduction: $SO_4^{2-}(aq) + 9H^+(aq) + 8e^- \longrightarrow HS^-(aq) + 4H_2O(l)$

Oxidation: $C_6H_{12}O_6(aq) + 12H_2O(l) \longrightarrow 6HCO_3^-(g) + 30H^+(aq) + 24e^-$

Overall: $C_6H_{12}O_6(aq) + 3SO_4^{2-}(aq) \longrightarrow 6HCO_3^-(g) + 3H^+(aq) + 3HS^-(aq)$

47. (a) Reduction: $2H^+(aq) + 2e^- \longrightarrow H_2(g)$; cathode
Oxidation: $Zn(s) \longrightarrow Zn^{2+}(aq) + 2e^-$; anode
Overall: $Zn(s) + 2H^+(aq) \longrightarrow Zn^{2+}(aq) + H_2(g)$

(b) Reduction: $AgCl(s) + e^- \longrightarrow Ag(s) + Cl^-(aq)$; cathode
Oxidation: $H_2(g) \longrightarrow 2H^+(aq) + 2e^-$; anode
Overall: $2AgCl(s) + H_2(g) \longrightarrow 2H^+(aq) + Ag(s) + Cl^-(aq)$

(c) Reduction: $Fe^{3+}(aq) + e^- \longrightarrow Fe^{2+}(aq)$; cathode
Oxidation: $H_2(g) \longrightarrow 2H^+(aq) + 2e^-$; anode
Overall: $2Fe^{3+}(aq) + H_2(g) \longrightarrow 2H^+(aq) + 2Fe^{2+}(aq)$

53. $Pt(s) \mid H_2(g) \mid H^+(aq) \parallel Cu^{2+}(aq) \mid Cu(s)$

55. (a) $Cl_2(g) + H_2(g) \longrightarrow 2Cl^-(aq) + 2H^+(aq)$; $E° = 1.358$ V
(b) $Br_2(l) + 2Fe^{2+}(aq) \longrightarrow 2Br^-(aq) + 2Fe^{3+}(aq)$; $E° = 0.316$ V
(c) $2Fe^{3+}(aq) + Cd(s) \longrightarrow 2Fe^{2+}(aq) + Cd^{2+}(aq)$; $E° = 1.174$ V

59. (a) $I^-(aq) + 2HClO_2(aq) \longrightarrow IO_3^-(aq) + Cl_2(g) + H_2O(l)$
(b) $4Cr^{2+}(aq) + O_2(g) + 4H^+(aq) \longrightarrow 4Cr^{3+}(aq) + 2H_2O(l)$
(c) $2CrO_2^-(aq) + 3ClO^-(aq) + 2OH^-(aq) \longrightarrow 2CrO_4^{2-}(aq) + 3Cl^-(aq) + H_2O(l)$
(d) $S(s) + 2HNO_2(aq) \longrightarrow H_2SO_3(aq) + N_2O(g)$
(e) $3F(CH_2)_2OH(aq) + 2K_2Cr_2O_7(aq) + 16H^+(aq) \longrightarrow 4K^+(aq) + 3FCH_2CO_2H(aq) + 4Cr^{3+}(aq) + 11H_2O(l)$

61. (a) no; (b) no

65. Yes, $H_2(g)$ will reduce $AgCl(s)$ under standard conditions.

67. $6e^-$; $E°_{cell} = 1.813$ V; reaction is spontaneous; $\Delta G° = -525$ kJ/mol of Al.

73. (a) $10Cl^-(aq) + 2MnO_4^-(aq) + 16H^+(aq) \longrightarrow 5Cl_2(g) + 2Mn^{2+}(aq) + 8H_2O(l)$
(b) $E°_{cell} = 0.149$ V
(c) $K = 1.6 \times 10^{25}$

79. yes, $E° = 0.40$ V

81. (a) yes, $E° = 0.47$ V; (b) 0.194 V; (c) yes, $E° = 0.20$ V

87. $[H_2SO_4] = 3.52\ M$; $E > E°$

91. 5.2 L

95. (a) cathode: $Ag(s)$; anode: $O_2(g)$; (b) cathode: $H_2(g)$; anode: $I_2(s)$

99. (a) $E°'_{cell} = 1.14$ V for NADH, 0.94 V for $FADH_2$; (b) -0.09 V; (c) -0.29 V

107. High power density, light weight, able to be recharged many times, able to hold charge for a long time without damage, long lifetime before replacement, operational safety, lack of toxic components for environmentally friendly disposal of discarded batteries

109. (a) $AgO(s) + Zn(s) + H_2O(l) \longrightarrow Ag(s) + Zn(OH)_2(aq)$, $E° = 3.44$ V; (b) 498 kJ/mol of Zn; (c) 396 kJ/mol Pb (lead storage battery) versus 665 kJ/mol of Zn (silver–zinc battery)

111. 1.42 g Cu

113. (a) $Ni^{2+}(aq) + 2e^- \longrightarrow Ni(s)$; $Ni(s) \longrightarrow Ni^{2+}(aq) + 2e^-$

Chapter 20

5. Isotopes with "magic numbers" of protons and/or neutrons tend to be especially stable. Elements with magic numbers of protons tend to have more stable isotopes than elements that do not.

7. Potassium-40: 19 protons and 21 neutrons. Nuclei with odd numbers of both protons and neutrons tend to be unstable. In addition, the neutron-to-proton ratio is very low for an element with this mass, which decreases nuclear stability.

11. Both positron decay and electron capture increase the neutron-to-proton ratio; electron capture is more common for heavier elements such those of row 5.

13. The mass number is the sum of the numbers of protons and neutrons present. Particles with a mass number of zero include β particles (electrons) and positrons; gamma rays and X rays also have a mass number of zero.

19. Unlike protons, neutrons have no charge, which minimizes the electrostatic barrier to colliding and reacting with a positively charged nucleus.

23. Ionizing radiation causes greater tissue damage, so it is more likely to destroy cancerous cells.

29. Ten exposures of 10 rem are less likely to cause major damage.

33. The mass defect is proportional to the total nuclear binding energy of the nucleus, but the binding energy *per nucleon* is a better measure of the relative stability of a nucleus.

39. The neutron flow is regulated by the use of control rods that absorb neutrons, whereas the speed of the neutrons produced by fission is controlled by the use of a moderator that slows the neutrons enough to allow them to react with nearby fissile nuclei.

45. It is difficult to pinpoint the exact location of the nucleus that decayed. In contrast, the collision of a positron with an electron causes both particles to be annihilated, and in the process two gamma rays are emitted in opposite directions, which makes it possible to locate precisely where a positron emitter is located and to create detailed images of tissues.

49. The raw material for all elements with $Z > 2$ is helium ($Z = 2$), and fusion of helium nuclei will always produce nuclei with an even number of protons.

53. (a) ${}^{39}_{17}Cl$; (b) ${}^{8}_{3}Li$; (c) ${}^{183}_{76}Os$; (d) ${}^{71}_{30}Zn$

55. (a) 26 protons, 31 neutrons, 1.19; (b) 74 protons, 111 neutrons, 1.50; (c) 19 protons, 20 neutrons, 1.05; (d) 54 protons, 77 neutrons, 1.43

61. (a) α decay; ${}^{235}_{92}U \longrightarrow {}^{4}_{2}\alpha + {}^{231}_{90}Th$
(b) α decay; ${}^{254}_{99}Es \longrightarrow {}^{4}_{2}\alpha + {}^{250}_{97}Bk$
(c) β decay; ${}^{36}_{16}S \longrightarrow {}^{0}_{-1}\beta + {}^{36}_{17}Cl$
(d) β decay; ${}^{99}_{42}Mo \longrightarrow {}^{0}_{-1}\beta + {}^{99m}_{43}Tc$

63. (a) ${}^{208}_{84}Po \longrightarrow {}^{4}_{2}\alpha + {}^{204}_{82}Pb$
(b) ${}^{226}_{88}Ra \longrightarrow {}^{4}_{2}\alpha + {}^{222}_{86}Rn$
(c) ${}^{228}_{90}Th \longrightarrow {}^{224}_{88}Ra + {}^{4}_{2}\alpha + \gamma$
(d) ${}^{231}_{91}Pa \longrightarrow {}^{227}_{89}Ac + {}^{4}_{2}\alpha + \gamma$
(e) ${}^{166}_{67}Ho \longrightarrow {}^{166}_{68}Er + {}^{0}_{-1}\beta + \gamma$
(f) ${}^{223}_{89}Ac \longrightarrow {}^{223}_{90}Th + {}^{0}_{-1}\beta + \gamma$

65. (a) ${}^{87}_{37}Rb \longrightarrow {}^{87}_{38}Sr + {}^{0}_{-1}\beta$
(b) ${}^{20}_{12}Mg \longrightarrow {}^{20}_{11}Na + {}^{0}_{+1}\beta$
(c) ${}^{268}_{109}Mt \longrightarrow {}^{4}_{2}\alpha + {}^{264}_{107}Bh$

67. (a) α particle; $^{218}_{84}Po \longrightarrow ^{4}_{2}\alpha + ^{214}_{82}Pb$
(b) β particle; $^{32}_{14}Si \longrightarrow ^{0}_{-1}\beta + ^{32}_{15}P$
(c) γ ray; $^{57m}_{26}Fe \longrightarrow ^{57}_{26}Fe + \gamma$
(d) β particle; $^{204}_{81}Tl \longrightarrow ^{0}_{-1}\beta + ^{204}_{82}Pb$

69. (a) α emission; $^{238}_{92}U \longrightarrow ^{4}_{2}\alpha + ^{234}_{90}Th$
(b) EC; $^{208}_{84}Po \longrightarrow ^{208}_{83}Bi + \gamma$
(c) β emission; $^{40}_{16}S \longrightarrow ^{0}_{-1}\beta + ^{40}_{17}Cl$
(d) γ emission; $^{93m}_{42}Mo \longrightarrow ^{93}_{42}Mo + \gamma$

71. (a) β decay; $^{216}_{84}Po \longrightarrow ^{0}_{-1}\beta + ^{216}_{85}At$
(b) α decay; $^{235}_{93}Np \longrightarrow ^{4}_{2}\alpha + ^{231}_{91}Pa$
(c) α decay; γ emission; $^{228}_{90}Th \longrightarrow ^{224}_{88}Ra + ^{4}_{2}\alpha + \gamma$
(d) β decay, γ emission; $^{231}_{91}Pa \longrightarrow ^{231}_{92}U + ^{0}_{-1}\beta + \gamma$

75. (a) $^{81}_{35}Br \longrightarrow ^{81}_{36}Kr + ^{0}_{-1}\beta$
(b) $^{234}_{94}Pu \longrightarrow ^{140}_{56}Ba + ^{92}_{38}Sr + 2^{1}_{0}n$

77. $^{249}_{98}Cf + ^{12}_{6}C \longrightarrow ^{257}_{104}Rf + 4^{1}_{0}n$

81. (a) $^{235}_{92}U + ^{1}_{0}n \longrightarrow ^{145}_{57}La + ^{88}_{35}Br + 3^{1}_{0}n$
(b) $^{235}_{92}U + ^{1}_{0}n \longrightarrow ^{95}_{42}Mo + ^{139}_{57}La + 2^{1}_{0}n + 7^{0}_{-1}\beta$

87. (a) $^{238}_{91}Pa \longrightarrow ^{238}_{92}U + ^{0}_{-1}\beta$; 4.721×10^{-16} kJ
(b) $^{216}_{87}Fr \longrightarrow ^{212}_{85}At + ^{4}_{2}\alpha$; 1.470×10^{-15} kJ
(c) $^{199}_{83}Bi \longrightarrow ^{199}_{82}Pb + ^{0}_{+1}\beta$; 6.277×10^{-16} kJ

89. (a) $^{234}_{91}Pa \longrightarrow ^{234}_{92}U + ^{0}_{-1}\beta$; 1.624×10^{8} kJ/mol
(b) $^{226}_{88}Ra \longrightarrow ^{222}_{86}Rn + ^{4}_{2}\alpha$; 4.700×10^{8} kJ/mol

91. (a) Beta decay of bismuth-208 to polonium is endothermic (ΔE = 1.415 MeV/atom) (b) Formation of lead-208 by alpha decay of polonium-210 is exothermic ($\Delta E = -5.408$ MeV/atom).
93. 757 MeV/atom, 8.70 MeV/nucleon
95. (a) calculated mass = 53.438245 amu; (b) mass defect = 0.241235 amu; (c) 463 MeV/atom; (d) 8.73 MeV/nucleon
97. 173 MeV/atom, 1.67×10^{10} kJ/mol
99. $\Delta E = +9.0 \times 10^{6}$ kJ/mol beryllium-8
101. D–D fusion: $\Delta E = -4.03$ MeV/tritium nucleus formed = -3.89×10^{8} kJ/mol tritium; D–T fusion: $\Delta E = -17.6$ MeV/tritium nucleus = -1.70×10^{9} kJ/mol; D–T fusion releases about 3 1/2 times as much energy per unit mass as D–D fusion.
103. $^{59}_{27}Co + ^{1}_{0}n \longrightarrow ^{60}_{27}Co + \gamma$; not a transmutation reaction; 110 rad; nausea, diarrhea, hair loss, decreased white blood cell count, fatigue

105. (a) $^{58}_{28}Ni + 48^{1}_{0}n \longrightarrow ^{106}_{28}Ni \longrightarrow ^{106}_{46}Pd + 18^{0}_{-1}\beta$
(b) $^{56}_{26}Fe + 23^{1}_{0}n \longrightarrow ^{79}_{26}Fe \longrightarrow ^{79}_{34}Se + 8^{0}_{-1}\beta$

107. 6.6×10^{-3} rad, 0.13 rem

109. $^{235}_{92}U + ^{1}_{0}n \longrightarrow ^{236}_{92}U + \gamma$
$^{238}_{92}U + ^{1}_{0}n \longrightarrow ^{239}_{92}U + \gamma$
$^{239}_{92}U \longrightarrow ^{239}_{93}Np + ^{0}_{-1}\beta$
$^{239}_{93}Np \longrightarrow ^{239}_{94}Pu + ^{0}_{-1}\beta$
$^{239}_{94}Pu \longrightarrow ^{239}_{95}Am + ^{0}_{-1}\beta$
$^{239}_{95}Am \longrightarrow ^{239}_{96}Cm + ^{0}_{-1}\beta$

113. (a) 7.1×10^{18} atoms of Pu; (b) 0.35 J; (c) 0.48 rad; (d) 9.6 rem—this dose is unlikely to be fatal.
115. 130 W

Chapter 21

3. (a) Cs > Rb > Ca > Na > Be > H; (b) H > Na > Rb > Cs >> Ca > Be
7. Ca > Be > Sn > Sb > As > O
13. Al
15. The magnitude of the electron affinity increases from left to right in a row due to the increase in Z_{eff}; P has a lower electron affinity than expected due to its half-filled 3*p* shell, which requires the added electron to enter an already occupied 3*p* orbital.
17. H has one electron in an *s* orbital, like the Group 1 metals, but it is also one electron short of a filled principal shell, like the Group 17 elements. Unlike the alkali metals, hydrogen is not a metal. Unlike the halogens, elemental hydrogen is not a potent oxidant.
21. Ionic: Li, Sr; covalent: Be, B, Al, O, Si; metallic: Pd
25. Hydrogen bonding with waters of hydration will partially neutralize the negative charge on the terminal oxygen atoms on the XeO_6^{4-} ion, which stabilizes the solid.
27. The alkali metals have less of an affinity for an electron than most other elements, while their ionization energies are lower than those of most other elements. Consequently, most other elements preferentially remove an electron from the alkali metals rather than donating an electron to them.
29. The Li^+ ion is smaller than the Na^+ ion, so Li^+ causes greater polarization of the Cl^- anion, resulting in polar covalent character. Be^{2+} is smaller and more highly charged than Li^+, so it should cause greater polarization, giving $BeCl_2$ less ionic character.
31. Due to their lower lattice energies, halide salts have lower melting points than oxides, and most metal carbonates decompose to metal oxides and carbon dioxide upon heating. $CaCl_2$ is added to further decrease the melting point.
33. They are all basic.
35. Only the Li^+ ion is small enough to form a stable ionic lattice containing three M^+ ions around the relatively small N^{3-} ion. The As^{3-} ion is much larger than N^{3-} and can accommodate three metal cations.
37. Li < Na < Cs < Ba
41. The smallest alkaline earth metal, Be, has the highest electronegativity of the *s*-block metals, so it is least reactive with most elements, including N_2. All of the alkaline earth metal ions, M^{2+}, are able to form a stable ionic lattice with the N^{3-} ion, giving ionic nitrides with the stoichiometry M_3N_2.
43. Because the Be^{2+} ion is small and highly charged, coordinated water molecules are acidic. Thus, Be and BeO dissolve in either acid or base, to form the $Be(H_2O)_4^{2+}$ and $Be(OH)_4^{2-}$ ions, respectively. Diagonal relationships suggest that Al should exhibit similar chemistry.
47. Sodium is the cheapest, but the greater reactivity of potassium might make it better for a small-scale reaction, despite its greater cost. On an industrial scale, the lower cost of sodium makes it the only choice for an alkali metal.
49. Only covalent hydrides are important in biochemistry. Ionic hydrides are powerful reductants that react with water to form H_2, while metallic hydrides are generally insoluble and unreactive.
51. Because of the high heat capacity of liquid water, the temperature of large bodies of water increases more slowly than that of a land mass in the spring; conversely, a large body of water cools more slowly than a land mass in the fall and winter. Thus, cities near an ocean tend to have milder climates with smaller temperature ranges.
53. Sr^{2+}
55. $\Delta G° = -28.6$ kJ/mol; nonspontaneous above 981 K.
57. (a) $D_2O(l) + SO_3(g) \longrightarrow D_2SO_4(l)$; (b) $2D_2O(l) + 2Li(s) \longrightarrow 2LiOD(aq) + D_2(g)$

59. 0.190 L
61. 4.48×10^{13} kg
63. 5×10^{15} L
65. (a) anode: $2H^- \longrightarrow H_2(g) + 2e^-$; cathode: $Li^+ + e^- \longrightarrow Li(l)$; (b) anode: $2Cl^- \longrightarrow Cl_2(g) + 2e^-$; cathode: $Li^+ + e^- \longrightarrow Li(l)$; (c) anode: $2H_2O(l) \longrightarrow O_2(g) + 4e^-$; cathode: $2H_2O(l) + 2e^- \longrightarrow H_2(g) + 2OH^-(aq)$
67. (a) $2Na_2O_2(s) + 2CO_2(g) \longrightarrow 2Na_2CO_3(s) + O_2(g)$; (b) oxidation: $O_2^{2-} \longrightarrow O_2 + 2e^-$; reduction: $O_2^{2-} + 2e^- \longrightarrow 2\ O^{2-}$; (c) 0.0665 atm
69. (a) $RX(soln) + 2Li(s) \longrightarrow RLi(soln) + LiX(s)$; (b) $Rb(s) + O_2(g) \longrightarrow RbO_2(s)$
71. 5.54 g Li_2CO_3; 0.82 g Li^+
75. (a) $2Sr(s) + O_2(g) \longrightarrow 2SrO(s)$
(b) $CaCO_3(s) \xrightarrow{\Delta} CaO(s) + CO_2(g)$
(c) $CaC_2(s) + 2H_2O(l) \longrightarrow C_2H_2(g) + Ca(OH)_2(aq)$
(d) $2RbHCO_3(s) + H_2SO_4(aq) \longrightarrow Rb_2SO_4(aq) + 2CO_2(g) + 2H_2O(l)$
79. The Ba^{2+} ion is larger and has a lower hydration energy than the Ca^{2+} ion. The precipitate is $BaSO_4$. 117 g $BaSO_4$.
81. 25.09 g Ca_3N_2
83. Cs
87. (a) $BeCO_3$; less stable to decomposition due to small size of Be^{2+}; (b) Be; low Z results in little absorption of X-rays; (c) BaI_2 is ionic; (d) Mg; smaller size results in stronger polar covalent bonds to carbon; (e) MgO has higher lattice energy and mp.
91. Yes. $4KO_2(s) + 4CO_2(g) + 2H_2O(g) \longrightarrow 4KHCO_3(s) + 3O_2(g)$
93. Dolomite will be less resistant to acid rain, because $MgCO_3$ is more soluble than $CaCO_3$.
95. Restrict milk consumption and supplement diet with calcium to minimize uptake of strontium.
99. The glass containing sodium will have a higher softening or melting temperature, because the smaller sodium ions will link the silicate chains more strongly.
101. Ca^{2+} or Sr^{2+}
103. $Na_2CO_3(s) + SO_2(g) \longrightarrow Na_2SO_3(s) + CO_2(g)$; $Na_2CO_3(s) + H_2SO_4(g) \longrightarrow Na_2SO_4(s) + CO_2(g) + H_2O(g)$
105. $MgCO_3$ is more soluble. $Ca(OH)_2$ is a source of OH^- ions, which causes Mg^{2+} to precipitate as $Mg(OH)_2$.

Chapter 22

1. The high stability of compounds of the Group 13 elements with oxygen required powerful reductants such as metallic potassium to be isolated. Al and B were initially prepared by reduction of molten $AlCl_3$ and B_2O_3 with potassium, respectively.
5. Boron is least metallic; due to its high electronegativity and small size, it is an unreactive semimetal.
7. The B—N bond is significantly more polar than the C—C bond, which makes B—N compounds more reactive and generally less stable than the corresponding carbon compounds. Increased polarity results in less delocalization and makes the planar form of BN less conductive than graphite.
9. Partial pi bonding between O and B increases the B—O bond strength.
11. Periodic trends predict that the cations of the heavier elements should be easier to reduce, so the elements should have less negative reduction potentials. In fact, the reverse is observed because the heavier elements have anomalously high Z_{eff} values due to poor shielding by filled $(n–1)d$ and $(n–2)f$ subshells.
13. dsp^3
17. C—Y bond strength decreases as Z of chalcogen increases, due to decreased orbital overlap.
19. Electron affinities decrease in magnitude with increasing Z as expected, *except* that the electron affinity of C is less than that of Si, presumably due to greater electron-electron repulsions upon adding an electron to a small $2p$ orbital.
21. Lead is least stable in the +4 oxidation state due to the inert pair effect.
23. Both S and N are nonmetals, but metallic character is due to the presence of a partially filled band derived from overlapping atomic orbitals (as discussed in Chapter 12).
25. All of the Group 15 elements form stable compounds in the +3 oxidation state, but the relative importance increases as you go down the column due to the inert pair effect. As a result, the chemistry of Bi is almost exclusively that of the +3 oxidation state.
27. Oxygen has the second highest electronegativity of any element; consequently, it prefers to share or accept electrons from other elements, and it only forms compounds in positive oxidation states with fluorine.
29. Basic: BaO; acidic: Br_2O, B_2O_3; amphoteric, SnO, Sb_2O_3
31. $CrO_3 << Al_2O_3 < Sc_2O_3 < BaO$
33. $H_3BO_3 < HNO_2 < HIO_4$
35. Most polar: H_2O; least polar: O_2 (nonpolar) and Cl_2O and CH_4 (slightly polar)
37. (a) H_2O, HF, or NH_3; (b) BeO, Al_2O_3
39. Electrostatic repulsions between lone pairs on *adjacent atoms* decrease bond strength.
43. Ionic character decreases as $\Delta\chi$ decreases from Al to P.
45. ClF_7
47. These three elements have the lowest ionization energies of the noble gases, but their compounds are still potent oxidants. Only highly electronegative elements can form bonds to these elements without being oxidized.
49. Boron
51. Electron donors: F^-, H_2O, BCl_4^-; electron acceptors: BCl_3, PCl_5; both: PCl_3; neither: CH_4, $H_3N{\cdot}BF_3$
53. NH_3
55. Neutral organoaluminum compounds, R_3Al, have only six valence electrons around the aluminum atom, which makes them strong Lewis acids.
57. $M(H_2O)_6^{3+}$ (M = Al, Ga, In, Tl), $M(OH)_4^-$ (M = Al, Ga); smaller size of Al^{3+} and Ga^{3+} makes these metals amphoteric.
59. (a) $12BCl_3(g) + 18H_2(g) \xrightarrow{\Delta} B_{12}(s) + 36HCl(g)$
(b) $6C_2H_4(g) + B_2H_6(g) \longrightarrow 2B(C_2H_5)_3(l)$
(c) $6B_2H_6(g) + 18Cl_2(g) \longrightarrow B_{12}(s) + 36HCl(g)$
(d) $B_2H_6(g) + 2(C_2H_5)_2S(g) \longrightarrow 2H_3B{\cdot}S(C_2H_5)_2(l)$
61. (a) $Al_2O_3(s) + 6H^+(aq) + 6H_2O(l) \longrightarrow 2Al(H_2O)_6^{3+}(aq)$
(b) $Ga_2O_3(s) + 2OH^-(aq) + 3H_2O(l) \longrightarrow 2Ga(H_2O)_4^-(aq)$
(c) $Tl_2O(s) + 2H^+(aq) + 11H_2O(l) \longrightarrow 2Tl(H_2O)_6^+(aq)$
(d) $16Ga(l) + 3S_8(s) \longrightarrow 8Ga_2S_3(s)$
(e) $2Tl(s) + H_2S(g) \longrightarrow Tl_2S(s) + H_2(g)$
63. (a) $2In(s) + 3I_2(s) \xrightarrow{\Delta} In_2I_6(s)$
(b) $B_{12}(s) + 18Cl_2(g) \longrightarrow 12BCl_3(l)$
$12BCl_3(l) + 36H_2O(l) \longrightarrow 12B(OH)_3(s) + 36HCl(g)$
(c) $4Ga(s) + 3O_2(s) \xrightarrow{\Delta} 2Ga_2O_3(s)$
(d) $2Tl(s) + 3Cl_2(g) \longrightarrow 2TlCl_3(s)$
$2TlCl_3(s) + 12H_2O(l) \longrightarrow 2Tl(H_2O_6^{3+}(aq) + 6Cl^-(aq)$

(e) $2Na(s) + 2H_2O(l) \longrightarrow 2NaOH(aq) + H_2(g)$
$2Al(s) + 2NaOH(aq) + 6H_2O(l) \longrightarrow$
$2Al(OH)_4^- + 3H_2(g) + 2Na^+(aq)$

(f) $4In(s) + 3C(s) \xrightarrow{\Delta} In_4C_3(s)$

67. (a) $CaC_2(s) + 2HCl(g) \longrightarrow CaCl_2(s) + C_2H_2(g)$
(b) $Pb(s) + Br_2(l) \longrightarrow PbBr_2(s)$
(c) $(CH_3)_3N(l) + H_2O_2(aq) \longrightarrow (CH_3)_3N{-}O(l) + H_2O(l)$
(d) $Pb(N_3)_2 \xrightarrow{\Delta} Pb(s) + 3N_2(g)$

69. (a) $CuO(s) + C(s) \xrightarrow{\Delta} Cu(s) + CO(g)$
(b) $Be_2C(s) + 4H_2O(l) \longrightarrow 2Be(OH)_2(s) + CH_4(g)$
(c) $Si(s) + 2Cl_2(g) \longrightarrow SiCl_4(l)$
$SiCl_4(l) + 4H_2O(l) \longrightarrow Si(OH)_4(s)$
(d) $3Si(s) + 2N_2(g) \xrightarrow{\Delta} Si_3N_4(s)$

71. White phosphorus is a molecular solid that contains tetrahedral P_4 molecules, while red phosphorus has a layered network structure. The highly strained 3-membered rings make white phosphorus much more reactive, while its molecular nature makes it more soluble, less dense, and lower melting.

73. $NaNO_2(s) + HCl(aq) \longrightarrow HNO_2(aq) + NaCl(aq)$

77. (a) $2P_4O_{10}(s) + 12H_2O(l) \longrightarrow 8H_3PO_4(aq)$
(b) $AsCl_5(l) + 4H_2O(l) \longrightarrow H_3AsO_4(aq) + 5HCl(aq)$
(c) $Bi_2O_3(s) + 3H_2O(l) \longrightarrow 2Bi(OH)_3(s)$
(d) $AsCl_3(l) + 3H_2O(l) \longrightarrow H_3AsO_3(aq) + 3HCl(aq)$
(e) $2(C_2H_5)_3Sb(l) + 21O_2(g) \longrightarrow$
$12CO_2(g) + 15H_2O(g) + Sb_2O_5(s)$
(f) $SbCl_3(s) + 3LiAlH_4(soln) \longrightarrow$
$SbH_3(soln) + 3LiCl(soln) + 3AlH_3(s)$
(g) $Ca(s) + N_2O(g) \xrightarrow{\Delta} CaO(s) + N_2(g)$

79. (a) $S_8(s) + 8O_2(g) \xrightarrow{\Delta} 8SO_2(g)$
(b) $S_8(s) + 8S^{2-}(aq) \longrightarrow 8S_2^{2-}(aq)$
(c) $S_8(s) + 24F_2(g) \longrightarrow 8SF_6(g)$
(d) $5S_8(s) + 48HNO_3(aq) + 16H_2O(l) \longrightarrow$
$40H_2SO_4(aq) + 24N_2(g)$

81. Carbon has no low energy *d* orbitals to make a set of d^2sp^3 hybrid orbitals, and it is so small that it is impossible for six fluorine atoms to fit around it at a distance that would allow for formation of strong C—F bonds.

83. (a) $SiO_2(s) + 6HF(aq) \longrightarrow SiF_6^{2-}(aq) + 2H^+(aq) + 2H_2O(l)$
(b) $Na_2CO_3(s) + 2HF(aq) \longrightarrow CO_2(g) + 2NaF(aq) + H_2O(l)$
(c) $CaO(s) + 2HF(aq) \longrightarrow CaF_2(s) + H_2O(l)$

85. (a) $S_8(s) + 4Cl_2(g) \longrightarrow 4S_2Cl_2(l)$
(b) $I_8(s) + KI(aq) \longrightarrow I_3^-(aq) + K^+(aq)$
(c) $PCl_3(l) + 3H_2O(l) \longrightarrow H_3PO_3(aq) + 3HCl(aq)$
(d) $CaF_2(s) + H_2SO_4(aq) \longrightarrow 2HF(aq) + CaSO_4(s)$
(e) $2KClO_3(s) \xrightarrow{\Delta} 2KCl(s) + 3O_2(g)$
(f) $8S^{2-}(aq) + 8I_2(aq) \longrightarrow S_8(s) + 16I^-(aq)$

87. (a) $Xe(g) + 1F_2(g) \longrightarrow XeF_2(s)$
(b) $Xe(g) + 2F_2(g) \longrightarrow XeF_4(s)$
(c) $Xe(g) + 3F_2(g) \longrightarrow XeF_6(s)$
(d) $2XeF_6(s) + SiO_2(s) \longrightarrow 2XeOF_4(l) + SiF_4(l)$
(e) $XeF_6(s) + 3H_2O(l) \longrightarrow XeO_3(s) + 6HF(aq)$

89. SiF_4; $SiO_2(s) + 2F_2(g) \longrightarrow SiF_4(l) + 2O_2(g)$

91. (a) $2Na(s) + 2D_2O(l) \longrightarrow D_2(g) + 2NaOD(aq)$
$2Li(s) + D_2(g) \longrightarrow 2LiD(s)$
$8LiD(s) + 2AlCl_3(soln) \longrightarrow 2LiAlD_4(s) + 6LiCl(soln)$
(b) $D_2O(l) + SO_3(g) \longrightarrow D_2SO_4(l)$
(c) $SiCl_4(l) + LiAlD_4(s)$ [from (a)] $\longrightarrow$
$SiH_4(g) + LiCl(s) + AlCl_3(s)$
(d) $CaF_2(s) + D_2SO_4(l)$ [from (b)] $\longrightarrow 2DF(g) + CaSO_4(s)$

99. (a) $[\ddot{\underset{..}{N}}{=}N{=}\ddot{\underset{..}{N}}]$
(b) $2NaN_3(s) \xrightarrow{\Delta} 2Na(l) + 3N_2(g)$
(c) One of the products of decomposition of sodium azide is elemental sodium, which is highly reactive and can ignite in the presence of water and air.

107. Examine samples of Napoleon's hair and/or fingernails form museums or collections to determine arsenic concentrations.

109. Gold will melt upon heating, but pyrite will react with oxygen to form $SO_2(g)$, which has a pungent smell.

111. Element 113: $5f^{14}6d^{10}7s^27p^1$, +1, $E^+(aq)$; element 115: $5f^{14}6d^{10}7s^27p^3$, +3, $E^{3+}(aq)$

Chapter 23

1. Electrons in *d* subshells are rather effective at screening one another from the nuclear charge. Consequently, as you go across the row from left to right, the effective nuclear charge experienced by the valence electrons does not increase rapidly, resulting in similar atomic sizes and ionization energies, which lead to similarities in chemistry.

5. The lanthanide contraction refers to the steady decrease in atomic size as you go from left to right across the lanthanides; it is due to relatively poor screening of the increasing nuclear charge by the electrons that are added to the 4*f* subshell, which results in a steady increase in Z_{eff}. The lanthanide contraction causes the second- and third-row transition metals to be very similar in size and causes the two heaviest members of a group to have rather similar chemistry.

9. Due to interpenetration of the n*s* and (n-1)*d* orbitals, the valence electrons of the Group 11 metals experience relatively high values of Z_{eff}, which results in unusually high values of the ionization energies and electronegativities. Consequently, these metals do not react readily with mineral acids. In contrast, by the time we reach Group 12, the filled *d* subshell has dropped in energy and is closer to the nucleus, resulting in a decrease in Z_{eff} and lower ionization energies and electronegativities. Consequently, the Group 12 metals are much more reactive.

11. Mn: $3d^54s^2$; Mn^{2+}: $3d^5$
Au: $4f^{14}5d^{10}6s^1$; Au^{3+}: $4f^{14}5d^8$
Mo: $4d^45s^2$; Mo^{5+}: $4d^1$

13. Both Ti and Ti form protective oxide coatings that adhere tightly to the surface of the metal and prevent corrosion.

17. Palladium reacts with hydrogen gas to form a non-stoichiometric hydride, PdH_x ($x < 1$). Because the H atoms in this hydride are highly mobile, they can migrate from one side of a palladium sheet to the other, where they recombine to form H_2.

21. The Group 12 metals are rather "soft" and prefer to bind to a "soft" anion such as sulfur rather than to a hard anion like oxide; hence, they are usually found in nature as sulfide ores. Because it is the "softest" of these metals, mercury has the highest affinity for sulfide.

23. Coke cannot be used as a reductant for metals that form stable carbides, such as the early transition metals (La, Hf, and W).

25. No, all four vertices of a tetrahedron are equivalent, and each vertex is adjacent to the other three.

27. Because the five *d* orbitals are not identical in shape and orientation, some of them will always interact more strongly with a set of negatively charged ligands than the others, which causes them to split into sets with different energies. The energy difference is called the

crystal field splitting, the magnitude of which depends on the nature and number of the ligands, the charge on the metal ion, and the location of the metal in the periodic table.
29. Anything that favors a large value of Δ_o will favor a low-spin configuration, including strong-field ligands, high charge on the metal ion, and second- or third-row metal.
31. Enzymes can use metal ions as Lewis acids whose acidity can be fine tuned or to accept or donate electrons.
33. Ti^{2+}: $3d^2$; V^{2+}: $3d^3$; Cr^{2+}: $3d^4$; Mn^{2+}: $3d^5$; Fe^{2+}: $3d^6$; Co^{2+}: $3d^7$; Ni^{2+}: $3d^8$; Cu^{2+}: $3d^9$; Zn^{2+}: $3d^{10}$; because Z_{eff} increases from left to right, Ti^{2+} and V^{2+} will have the most negative reduction potentials (be most difficult to reduce).
35. $Hg^{2+} > Fe^{2+} > Zr^{4+} > Fe^{3+} > Pt^{4+}$
37. Because electrons in *d* subshells are rather effective at screening one another from the nuclear charge, the effective nuclear charge experienced by the valence electrons does not increase rapidly as you go across the row from left to right. This results in similar sizes for ions of the same charge.
39. The increase in radius from vanadium to niobium is expected based on periodic trends; niobium and tantalum have the same radius because of the lanthanide contraction, which is due to the increase in nuclear charge as the 4*f* subshell is filled.
43. Hard acids (Mn^{7+}, W^{6+}, Ti^{4+}, Zr^{4+}, Fe^{3+}, Sc^{3+}, Al^{3+}, Mg^{2+}) prefer hard bases (F^-, OH^-). Intermediate acids (V^{3+}, Co^{2+}) prefer intermediate bases (Cl^-, NH_3). Soft acids (Cu^+, Au^+) prefer soft bases (RS^-)
45. Pt^{10+}, Ir^{9+}, and Ni^{10+}; because ionization energies increase from left to right across the *d*-block, by the time you reach Group 9 it is impossible to form compounds in the oxidation state corresponding to loss of all valence electrons.
49. $4f^{14}5d^{10}6s^2$; lack of luster and metallic conductivity indicate it is not an alloy, but an ionic compound, Cs^+Au^-. Cs is a powerful reductant, and Au can accept an electron to give a filled 6*s* subshell.
51. $Hg_2^{2+}(aq) + H_2S(g) \longrightarrow Hg(l) + HgS(s) + 2H^+(aq)$
53. $Mn_2O_7 < CrO_3 < V_2O_5 < MnO_2 \approx VO_2 < Sc_2O_3$
55. $VO < ZnO < Sc_2O_3 < Cr_2O_3 < Fe_2O_3 < Fe_3O_4 < V_2O_5$
57. (a) $2AgBr(s) \xrightarrow{\text{light}} 2Ag(s) + Br_2(l)$
(b) $W(s) + \text{excess } Cl_2(g) \xrightarrow{\Delta} WCl_6(s)$
(c) $CuO(s) + H_2(g) \xrightarrow{\Delta} Cu(s) + H_2O(g)$
(d) $Fe_2O_3(s) + 6H^+(aq) \longrightarrow 2Fe^{3+}(aq) + 3H_2O(l)$
(e) $RhCl_3(s) + 6NH_3(aq) \longrightarrow Rh(NH_3)_6^{3+}(aq) + 3Cl^-(aq)$
(f) $3Fe^{2+}(aq) + MnO_4^-(aq) + 4H^+(aq) \longrightarrow Fe^{3+}(aq) + MnO_2(s) + 2H_2O(l)$
59. Hf is in Group 4 and has four valence electrons.
61. W: +6; Cd: +2; Zr: +4; V: +5; Ac: +3; Ni: +2; Mn: +2; Pt: +2; Fe: +3
63. Ag: +2; Co: +5; Os: +8; Au: +3; W: +6; Mn: +7
67. (a) V; (b)V; (c) Nb, Ta; (d) V; (e) Nb, Ta
69. Re^{2+} is a *very* soft cation, and F^- and O^{2-} are very hard bases, so ReO and ReF_2 are unlikely to exist. MnF_6 is also unlikely to exist: although fluoride should stabilize high oxidation states, but in this case Mn^{6+} is probably too small to accommodate six F^- ions.
71. Two Cr^{3+} and one Fe^{2+}
73. $VCl_4(l) + 2\,H_2(g) \xrightarrow{\Delta} V(s) + 4\,HCl(g)$

$VCl_4(l) + 2\,Mg(s) \xrightarrow{\Delta} V(s) + 2\,MgCl_2(g)$

$V_2O_5(s) + 5\,Ca(s) \xrightarrow{\Delta} 2V(s) + 5\,CaO(s)$

Carbon cannot be used as a reductant, because vanadium forms stable carbides such as VC and VC_2.
77. Metals in the +6 oxidation state are stabilized by oxide (O^{2-}) and fluoride (F^-). The M—F and M—O bonds are polar covalent due to extreme polarization of the anions by the highly charged metal. Ca, Sr, and Ba can only be oxidized to the dications, M^{2+}, which form ionic oxides and fluorides.
79. Cyanide is a relatively soft base, and the early transition metal cations are harder acids than the cations of the later transition metals. In addition, the later transition metals have more electrons in their partially filled *d* subshell, which result in an additional contribution to the bonding due to metal to ligand pi bonding.
81. Formation of complexes between NH_3 and a divalent cation is largely due to electrostatic interactions between the negative end of the ammonia dipole moment and the positively charged cation. Thus, the smallest divalent cations (Ni^{2+} and Cu^{2+}) will form the most stable complexes with ammonia.
83. Strong field ligands favor a square planar structure for metal ions that have nine or fewer *d* electrons due to the greater CFSE arising from the much larger orbital splitting in a square planar structure.
85. (a) $[Cu(NH_3)_4]^{2+}$ is d^9, square planar, with one unpaired electron, neither high nor low spin; (b) $Ni(CN)_4]^{2-}$ is d^9, square planar, low-spin, with no unpaired electrons.
87. Transition metals can accept or donate one electron at a time, allowing them to form radicals with an odd number of electrons. In addition, some metals can form covalent bonds to organic groups, and the resulting M—C bonds can be cleaved homolytically to give an organic radical.
89. The tungsten bronzes can be viewed as the products of partial reduction of WO_3 by an active metal to give a mixture of W(VI) and W(V). In the solid, the *d* orbitals of the metal overlap to forma partially filled band, which is responsible for the intense color and the electrical conductivity.
93. $Fe_2O_3(s) + 6HO_2CCO_2H(aq) \longrightarrow 2\,Fe(ox)_3^{3-}(aq) + 3\,H_2O(l)$, where $ox^{2-} = {}^-O_2CCO_2^-$; the highly charged iron(III) complex is water soluble.
97. (a) Pt; (b) $FeCl_3$; (c) Ni; (d) MnO_2
99. The metals use a degenerate pair of *d* orbitals, probably the d_{xz} and d_{yz}. The iron complex is most likely to be staggered to minimize nonbonded interactions between the rings, while the larger ruthenium atom keeps the rings far enough apart for an eclipsed structure to be more stable.

Chapter 24

1. sp^3. It is spherically symmetrical (all cross sections are circles).
3. The sp^3 hybridized orbitals form a tetrahedron. It minimizes electrostatic repulsions.
7. Only (a), (c), and (d).
9. CHF_2^+, $CHCl_2^+$, CH_2F^+, CH_3^+. Electronegative substituents destablize the positive charge. The greater the number of electronegative substituents and the higher their electronegativity the more unstable the carbocation.
11. (a) CH_3^- nucleophile, Li^+ electrophile; (b) CH_3O^- nucleophile, formaldehyde electrophile; (c) H^+ electrophile, propene nucleophile; (d) benzene electrophile; Cl^- nucleophile
13. CN^- is the nucleophile, $C_2H_5C^{\delta+}HBrCH_3$ is the electrophile
21. Only (c) and (d).
27. The presence of a nucleophilic $C^{\delta-}$ resulting from a highly polar interaction with an electropositive Mg.
31. Their ability to act as weak bases and their tendency to act as nucleophiles.

33. Bromopentane: 2-Methylbutane:

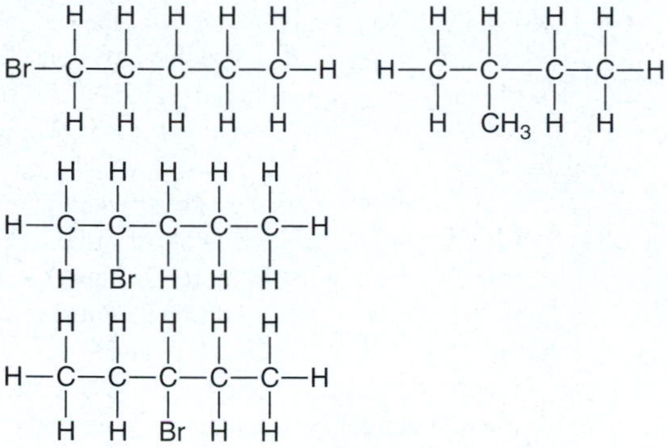

37. Only (b) and (d):

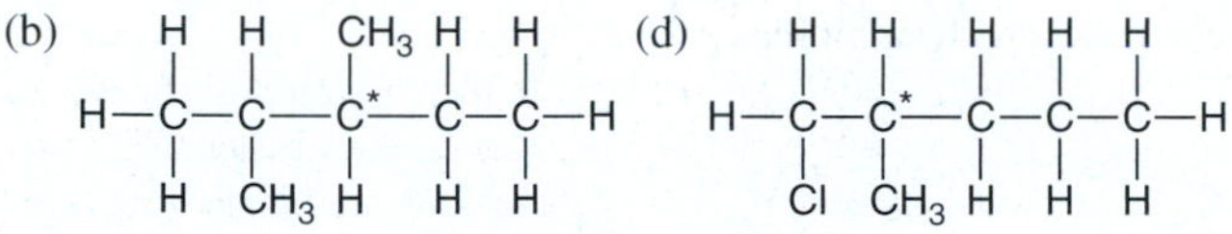

41.

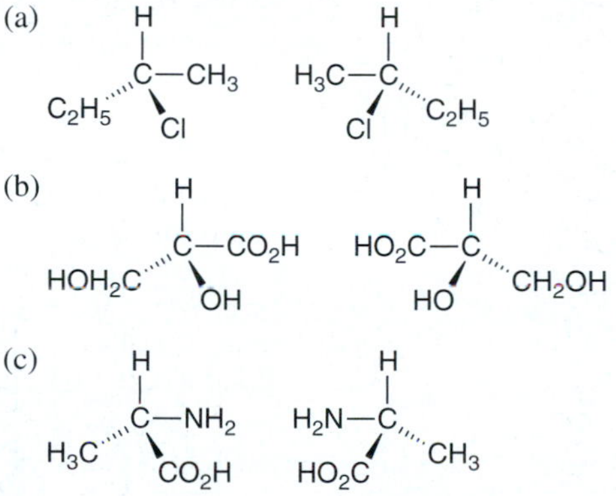

43. (a) Cl^- nucleophile, CH_3^+ electrophile: $CH_3^+ + Cl^- \longrightarrow CH_3Cl$. (b) $CH_3CH{=}CH_2$ nucleophile, H^+ electrophile, followed by $CH_3C^+HCH_3$ as the electrophile and Br^- as the nucleophile: $CH_3CH = CH_2 + HBr \longrightarrow CH_3CHBrCH_3$. (c) BCl_3 nucleophile, $(CH_3)_3N$ nucleophile: $(CH_3)_3N + BCl_3 \longrightarrow (CH_3)_3NBCl_3$.

47. 2-Methyl-1-propene

51.

$$CH_3CH_2CH_3 + 1/2\ Cl_2 \xrightarrow{hv} CH_3CH_2Cl \xrightarrow{+NaOH} CH_3CH{=}CH_2 + NaCl + H_2O$$

Use an addition reaction: $CH_3CH = CH_2 + HBr \longrightarrow CH_3CH_2BrCH_3$

53. Four; 3-bromo-3-methylpentane

55. Seven

57. Methanol. Acetaldehyde, followed by acetic acid and finally $2CO_2$

59. 2,3-dibromobutane

63. (a) $C_6H_5OC_2H_5 + NaCl$; (b) propene + NaCl + H_2O

67. (a) $CH_3CO_2NH_4^+$ (an acid-base reaction); (b) $CH_3CONHC_2H_5 + CH_3OH$

69. Use a condensation reaction:

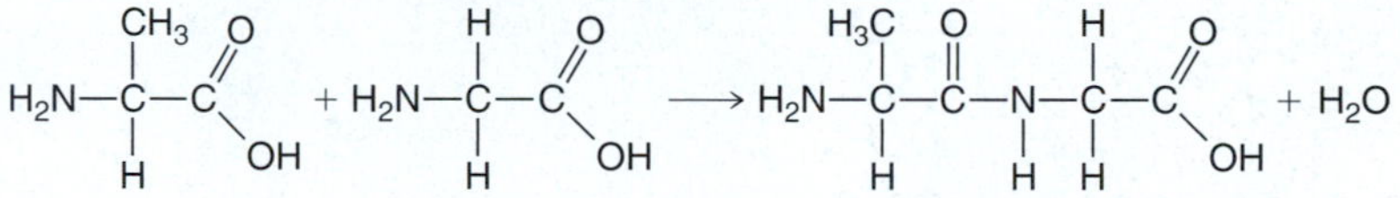

71. The first staggered conformation minimizes electrostatic repulsions between adjacent atoms:

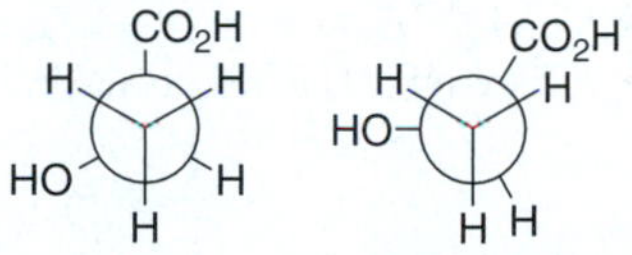

Staggered Eclipsed

73. Yes.

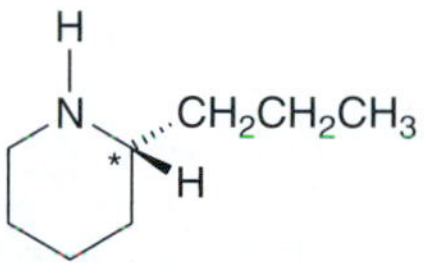

75. 3 amides and a phenyl group

81. (a)

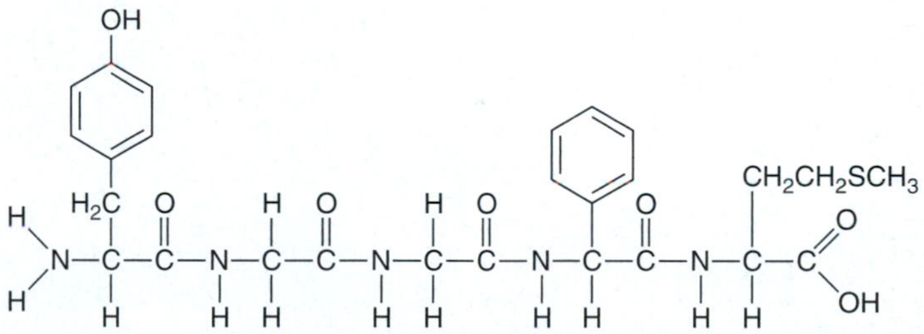

(b) The bond is indicated in red in (a). Using R to represent the left side of the bond and R′ the right:

(c)

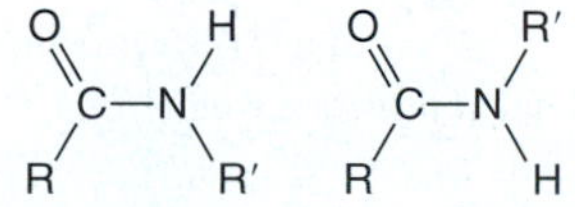

Art and Photo Credits

Molecular Models

We wish to thank the Cambridge Crystallographic Data Centre (CCDC) and the Fachinformationszentrum Karlsruhe (FIZ Karlsruhe) for allowing Imagineering Media Services (IMS) to access their databases of atomic coordinates for experimentally determined three-dimensional structures. CCDC's **Cambridge Structural Database (CSD)** is the world repository of small molecule crystal structures (distributed as part of the CSD System), and in FIZ Karlsruhe's **Inorganic Crystal Structure Database (ICSD)** is the world's largest inorganic crystal structure database. The coordinates of organic and organometallic compounds in CSD and inorganic and intermetallic compounds in ICSD were invaluable in ensuring the accuracy of the molecular models produced by IMS for this textbook. The authors, the publisher, and IMS gratefully acknowledge the assistance of both organizations. Any errors in the molecular models in this text are entirely the responsibility of the authors, the publisher, and IMS.

The CSD System: The Cambridge Structural Database: a quarter of a million crystal structures and rising. Allen, F.H., *Acta Cryst.* (2002), **B58,** 380–388. *ConQuest:* New Software for searching the Cambridge Structural Database and visualizing crystal structures. Bruno, I.J., Cole, J.C., Edgington, P.R., Kessler, M., Macrae, C.F., McCabe, P., Pearson, J., Taylor, R., *Acta Cryst.* (2002), **B58,** 389–397. *IsoStar:* IsoStar: A Library of Information about Nonbonded Interactions. Bruno, I.J., Cole, J.C., Lommerse, J.P.M., Rowland, R.S., Taylor, R., Verdonk, M., *Journal of Computer-Aided Molecular Design* (1997), **11-6,** 525–537.

The Inorganic Crystal Structure Database (ICSD) is produced and owned by Fachinformationszentrum Karlsruhe (FIZ Karlsruhe) and National Institute of Standards and Technology, an agency of the U.S. Commerce Department's Technology Administration (NIST).

Photo Credits

Chapter 1: Opening photo IBM Almaden Research Center Visualization Laboratory; **1.1** Kristin Piljay, Benjamin Cummings Publishers, Pearson Education; **1.2a** Lawrence Berkeley National Laboratory; **1.2b left and right** Lawrence Berkeley National Laboratory; **1.3b** NASA; **1.6 center** Richard Megna/Fundamental Photographs; **1.6 left and right** Dorling Kindersley; **1.7 left** Dorling Kindersley; **1.7 right** Michael Dalton/Fundamental Photographs; **1.8** Richard Megna/Fundamental Photographs; **1.9** Richard Megna/Fundamental Photographs; **1.10** Charles D. Winters/Photo Researchers; **1.12 left and right** Dorling Kindersley; **1.13** The Alchemist's Workshop, 1570, Jan van der Straet (Joannes Stradanus), Palazzo Vecchio, Florence, Italy; Bridgeman Art Library; **1.16** Richard Megna/Fundamental Photographs; **p 19** Laboratoire Curie, Institut du Radium, Paris

Chapter 2: Opening photo Courtesy of ConocoPhillips; **2.7** Jeremy Burgess/Photo Researchers, Inc.; **2.9 top and bottom** Richard Megna/Fundamental Photographs; **p 62** Dave G. Houser/CORBIS; **p 63 top and bottom** Dorling Kindersley; **p 77** RDF/Visuals Unlimited; **p 79** The Canadian National Railway Historic Photograph Collection

Chapter 3: Opening photo Chip Clark; **3.1** Chip Clark; **3.3a** Christine Chase; **3.3b** David Scharf/Peter Arnold, Inc.; **3.7 left and right** Chip Clark; **p 105** Associated Press; **3.9** Richard Megna/Fundamental Photographs; **3.10** Carey B. Van Loon; **p 108 top** Mason Morfit/Taxi/Getty Images; **bottom** Stephen J. Kron, University of Chicago; **p 110** NASA; **p 113** Michael Freeman/CORBIS; **p 117 top** Dorling Kindersley; **p 117 bottom** Chip Clark; **p 123** Chip Clark; **p 125** Johnson Matthey PLC. Science Photo Library/Photo Researchers; **3.14** NASA

Chapter 4: Opening photo Richard Megna/Fundamental Photographs; **4.4 a–c** Richard Megna/Fundamental Photographs; **4.9** Dorling Kindersley; **p 159** Richard Megna/Fundamental Photographs; **4.11** Richard Megna/Fundamental Photographs; **p 166 all photos** Richard Megna/Fundamental Photographs; **4.13** PhotoDisc; **p 169 top and bottom** Dorling Kindersley; **4.14** Richard Megna/Fundamental Photographs; **p 175** Digital Vision; **4.16** Richard Megna/Fundamental Photographs; **4.18** Spencer Platt/Getty Images; **4.19** Oliver Strewe/Stone/Getty Images; **4.20** Ferrell McCollough/Visuals Unlimited; **4.21** Peticolas/Megna/Fundamental Photographs; **p 188** Ed Degginger/Color-Pic; **4.22** Richard Megna/Fundamental Photographs; **4.23 left and right** Richard Megna/Fundamental Photographs

Chapter 5: Opening photo Richard Megna/Fundamental Photographs; **5.1a** NASA; **5.1b** Joanna B. Pinneo/Aurora & Quanta Productions Inc.; **5.1c** Herrmann/Starke/CORBIS; **5.1d** Los Alamos National Laboratory; **5.1e** Robert Llewellyn/CORBIS; **5.2** David W. Hamilton/Image Bank/Getty Images; **5.3** Bettmann/CORBIS; **5.10** General Electric Corporate Research & Development Center; **5.12** Richard Megna/Fundamental Photographs; **5.20** Brian Lightfoot/AGEfotostock; **p 239** Reuters/CORBIS

Chapter 6: Opening photo Richard Megna/Fundamental Photographs; **6.1a** Alex Howe/Image State; **6.1b** David Pu'u/CORBIS; **6.4** Andrew Davidhazy; **6.5 left** PhotoDisc Red/Getty Images; **6.5 right** Dorling Kindersley; **p 261 top** AIP, Emilio Segre Archives; **p 261 bottom** AIP, Emilio Segre Archives, W.F. Meggers Gallery of Nobel Laureates; **6.8** AGEfotostock; **6.9a** Charles Winters/Photo Researchers; **6.9b top** Richard Megna/Fundamental Photographs; **p 264** AIP Emilio Segre Visual Archives, W. F. Meggers Gallery of Nobel Laureates; **p 265** Princeton U./AIP Neils Bohr Library; **6.13a–c** "Simultaneous Display of Spectral Images and Graphs Using a Web Camera and Fiber Optic Spectrometer" by Brian Niece. Journal of Chemical Education. **p 268 all photos** the International Dark-Sky Association, www.darksky.org; **6.15a** Jeff Hunter/The Image Bank/Getty Images; **p 270** Laboratory for

Microscopy and Micro-analysis, University of Pretoria, South Africa; **6.17 left and right** Chris Hollis; **p 275** A. K. Kleinschmidt/American Institute of Physics; **p 297** "Simultaneous Display of Spectral Images and Graphs Using a Web Camera and Fiber Optic Spectrometer" by Brian Niece. Journal of Chemical Education.

Chapter 7: Opening photo Science & Society Picture Library/Science Museum, London; **p 302 top** University of Pennsylvania; **p 302 bottom** Novosti/Science Photo Library/Photo Researchers; **p 305** Bettman/CORBIS; **p 322** J.R. Eyerman/Time Life Pictures/Getty Images; **p 329** NOAA, Geodesy Collection; **p 330** Richard Megna/Fundamental Photographs; **p 334** Alison Wright/CORBIS

Chapter 8: Opening photo Richard Megna/Fundamental Photographs; **8.6** University Archives, the Bancroft Library, University of California, Berkeley; **8.9** Justin Urgitis/www.chemicalforums.com

Chapter 9: Opening photo Jian-Min Zuo, Miyoung Kim, Michael O'Keefe and John Spence, Arizona State University; **9.24** Richard Megna/Fundamental Photographs

Chapter 10: Opening photo CORBIS; **p 435** Time Life Pictures/Mansell/Getty Images; **p 439 left** Science & Society Picture Library/Science Museum, London; **p 439 right** National Institutes of Health; **p 441 left** Library of Congress; **p 441 center** CORBIS; **p 441 right** Science & Society Picture Library/Science Museum, London; **10.15** Richard Megna/Fundamental Photographs; **10.16** Richard Megna/Fundamental Photographs; **10.18** U.S. Department of Energy/ Photo Researchers, Inc.; **10.25** Network Photographers/Alamy

Chapter 11: Opening photo Barbara and Zafer Baran/Alamy **11.2** Kristen Brochmann/Fundamental Photographs; **11.10a** Chip Clark; **11.10b** Herman Eisenbeiss/Photo Researchers, Inc.; **11.11** Richard Megna/Fundamental Photographs; **11.12b** Richard Megna/Fundamental Photographs; **11.18** Richard Megna/Fundamental Photographs; **11.21a–d** Division of Chemical Education, Inc., American Chemical Society; **11.26** Richard Megna/Fundamental Photographs; **11.29** Liquid Crystal Resources

Chapter 12: Opening photo M. C. Escher's "Symmetry Drawing E128" © 2005 The M. C. Escher Company—Holland. All rights reserved. www.mcescher.com; **p 526 all photos** Dorling Kindersley; **12.16 left** Dorling Kindersley; **p 540** Pascal Goetgheluck/Photo Researchers, Inc.; **p 541 all photos** Dorling Kindersley; **12.29b** J.H. Rector courtesy of R. Griessen, Vrije Universiteit, Amsterdam, The Netherlands; **12.33** Suminar Pratapa and Brian O'Connor (Curtin University of Technology) and Brett Hunter (ANSTO), Bragg Institute, Australian Nuclear Science and Technology Organisation

Chapter 13: Opening photo TPL Distribution/Photolibrary; **13.2** Dorling Kindersley; **13.5** Richard Megna/Fundamental Photographs; **13.8** Richard Megna/Fundamental Photographs; **13.19a–c** Sam Singer/ArsNatura; **13.22a–c** Richard Megna/Fundamental Photographs; **13.23** Oliver Meckes & Nicole Ottawa/Photo Researchers, Inc.; **13.24** John F. Kennedy Space Center/NASA

Chapter 14: Opening photo Fritz Goro; **14.1** Chip Clark; **14.2** Chip Clark; **14.3** Chip Clark; **14.4** Chip Clark; **14.23** Thomas Eisner

Chapter 15: Opening photo James Whitlow Delano/Redux; **15.1** Richard Megna/Fundamental Photographs; **p 690** Deutsches Museum, Munich; **15.12** Richard Megna/Fundamental Photographs; **15.13** Richard Megna/Fundamental Photographs

Chapter 16: Opening photo Richard Megna/Fundamental Photographs; **16.13** Richard Megna/Fundamental Photographs; **16.15** Richard Megna/Fundamental Photographs; **16.16** Richard Megna/Fundamental Photographs

Chapter 17: Opening photo Andrew Syred/Photo Researchers; **p 784** Chip Clark; **17.4** Richard Megna/Fundamental Photographs; **17.5** Wesley Vick and Taylor Chung, Baylor College of Medicine, Houston; **17.6a** Martin Siepmann/AGEfotostock; **17.6b** Chase Studio/Photo Researchers; **17.7** Carl & Ann Purcell/CORBIS; **17.9** Richard Megna/Fundamental Photographs; **17.11** Richard Megna/Fundamental Photographs

Chapter 18: Opening photo Robert Llewellyn/Image State; **18.1** Robert Harding Picture Library Ltd./Photolibrary; **18.2 top** Bettmann/CORBIS; **18.2 bottom** Bettmann/CORBIS; **18.6** Richard Megna/Fundamental Photographs; **18.7** Richard Megna/Fundamental Photographs; **18.11** James Klett, Oak Ridge National Laboratory; **18.12** Olivier Grunewald/Photolibrary; **18.15 left** Dorling Kindersley; **18.15 right** Andrew Lambert Photography/Photo Researchers; **p 829** Library of Congress

Chapter 19: Opening photo Paul Chesley/National Geographic/ Getty Images; **19.2** Richard Megna/Fundamental Photographs; **19.3b** Stephen Frisch/Stock Boston; **p 863** Richard Megna/Fundamental Photographs; **19.7** Richard Megna/Fundamental Photographs; **p 880** Hulton Archive/Getty Images; **p 892** Charles O'Rear/CORBIS; **19.22** Charles D. Winters/Photo Researchers; **19.23** Sam Ogden/Photo Researchers

Chapter 20: Opening photo US Dept. of Energy/SPL/Photo Researchers; **20.8a** Michael Collier; **20.9** Fermilab Visual Media Services; **20.10** Dwayne Anthony and the National Insulator Association; **p 928** Bettmann/CORBIS; **p 939** Thomas Seilnacht; **20.18** Robert D. Loss, WAISRC; **20.19** NOVOSTI/SIPA; **20.21a** Plasma Physics Laboratory, Princeton University; **20.21b** Lawrence Livermore National Laboratory; **20.22a** Chris Priest/SPL/Photo Researchers; **20.22b** Simon Fraser/SPL/Photo Researchers; **20.23** International Atomic Energy Agency; **20.27** Space Telescope Science Institute

Chapter 21: Opening photo Journal of Chemical Education; **p 961** George Karger/Pix Inc./Time Life Pictures/Getty Images; **21.6** Bettmann/CORBIS; **p 964 top** Library of Congress; **p 964 bottom** Hulton-Deutsch Collection/CORBIS; **p 966 margin** Dorling Kindersley; **21.8** Dorling Kindersley; **21.10** Richard Megna/Fundamental Photographs; **21.11** Richard Megna/Fundamental Photographs; **p 973** Dorling Kindersley; **21.13** Hulton Archive/Getty Images

Chapter 22: Opening photo Roger Hayward; **22.1a** Dorling Kindersley; **22.1b** The Dial Corporation; **p 993** Indus Global Superabrasives; **22.5a** Dorling Kindersley; **22.5b** Brooklyn Museum of Art/CORBIS; **22.6a** AP/Wide World Photos; **22.6b** Texas Instruments Incorporated; **p 1002** Inner Mountain Outfitters; **p 1006** Roger Ressmeyer/CORBIS; **22.10a** Dorling Kindersley; **22.10b** Erich Lessing/Art Resource, NY; **p 1014 top** David Cavagnaro/Visuals Unlimited; **p 1014 bottom** Private Collection, Archives

Charmet/Bridgeman Art Library; **p 1015** Photolibrary; **p 1021** Paul Silverman/Fundamental Photographs; **22.14a** Bettmann/CORBIS; **22.14b** Science & Society Picture Library/Science Museum, London; **22.15** Ferdinando Scianna/Magnum Photos; **p 1026** The Dibner Library of the History of Science and Technology, Smithsonian Institution; **p 1028** Kazuhiro Nogi/AFP/Getty Image; **22.17** Walter Gruber, University of British Columbia. Collection of Neil Bartlett

Chapter 23: Opening photo Txomin Sáez/AGEfotostock; **23.4** Richard Megna/Fundamental Photographs; **23.5** Richard Megna/Fundamental Photographs; **p 1046** Doug Smith; **p 1048 margin** Ted Aljibe/AFP/Getty Images; **p 1048 bottom** Martin Bernetti/AFP/Getty Images; **23.6** Johnson Matthey; **23.7** Margaret Bouke-White/Time Life Pictures/Getty Images; **23.8** Alex Webb/Magnum Photos; **p 1056** SPL/Photo Researchers; **p 1069** Dorling Kindersley

Chapter 24: Opening photo After Eddaoudi, M.; Kim, J.; O'Keeffe, M.; Yaghi, O. M. J. Am. Chem. Soc. 2002, 124, 376, Figure 1. Crystallographic data (ja017154e_s2.cif) available at http://pubs.acs.org; **24.18** Eye of Science/Photo Researchers; **p 1099** Dorling Kindersley; **p 1103 left** Dorling Kindersley; **p 1103 right** James Baigrie/Foodpix/Jupiter Images; **p 1118** Dorling Kindersley; **p 1121** Nobelstiftelsen/The Nobel Foundation

Index

Page references followed by italicized *f*, *n*, and *t* refer to figures, footnotes, and tables, respectively.

B

F

K

L

N

S

U

V